WORLD DIRECTORY OF CRYSTALLOGRAPHERS

AND OF OTHER SCIENTISTS

EMPLOYING CRYSTALLOGRAPHIC METHODS

NINTH EDITION

1995

GENERAL EDITOR

Y. EPELBOIN

REGIONAL EDITORS

G. FILIPPINI, H. HASHIZUME, I. L. TORRIANI, W. L. DUAX

SPRINGER-SCIENCE+BUSINESS MEDIA, B.V.

ISBN 978-0-7923-3180-3   ISBN 978-94-017-3699-2 (eBook)
DOI 10.1007/978-94-017-3699-2

TABLE OF CONTENTS

PREFACE TO THE NINTH EDITION

The eighth edition of the *World Directory of Crystallographers and of Other Scientists using Crystallographic Methods* was published at the end of 1990. The 9th edition, *WDC9*, appears four years later. It may look rather similar. However drastic changes have been made; not only have the contents been updated, but the methods used to acquire the information, to disseminate it and to produce this directory have involved the latest advances in technology.

When Professor André Authier, as President of the International Union of Crystallography, offered me the position of General Editor of *WDC9* in 1991 I was at first reluctant. I had been working in 1978 at IBM Research at the time when Dr Allan L. Bednowitz was preparing the first computerised edition, *i.e.* the 6th edition. Many National Editors were able to provide only punched cards and you may imagine what was left to the General Editor. The Editors for the eighth edition, Professor E. N. Maslen and Dr A. L. Bednowitz again, were in a better position, since diskettes and microcomputers were now available. However, it was tedious to check all the data, to correct them and to introduce all the control words necessary for final printing: a useful job but a tiresome one!

After my first reaction against accepting this proposal, I realized that it was a very exciting challenge to build a database of crystallographers. Advances in technology now allow us to search very quickly through a large database by means of keywords, and thus to extract very precise information. International networks between computers make it feasible to access a central database which contains the latest information. The production of a directory could be made automatically from the data stored in the database. This was a new and interesting challenge. I have not been disappointed.

I could not begin with the previous edition. Data were indeed written on a magnetic medium, but resembling a long text file without any structure except for control words to produce the required fonts and format for the directory. Worse than anything else the keywords describing the scientific interests, as written by the scientists themselves, were of no use. Everyone had used his or her own wording, and only a few names would be obtained when searching on a given keyword. It was obvious that many scientists would be missed. Thus one had to start again, with a fresh set. The published specifications for the database contents [*Acta Cryst.* (1993). A**49**, 222–225] are reproduced as Annex 1.

It was essential to build something that would endure for a long time: the structure of the database and its software would certainly change in the future. This is not too important so long as the data are structured and the database makes provision for all possible information. Thus it was necessary to include all information of interest, not only for today but also for tomorrow. It took about one year to discuss the contents of a future database by postal mail with all National Committees, Chairpersons of Commissions and other people in positions of responsibility in the IUCr. It is quite surprising how few modifications have been made. Most people were satisfied with the existing contents of the directory. The few changes are the following:

- Provision has been made to include the address of the institution and the postal address, since abbreviations and box numbers are increasingly used for postal deliveries. If you read my own address as '75252 Paris Cedex 05', I am very doubtful that you will be able to arrive in my office! So both the 'real' and the postal address may now be included. As you may see, this distinction has not been very consistently applied, but I believe this will improve in future updates.

- E-mail and fax addresses are now routinely included. Telex numbers, though still included, are rapidly becoming obsolete.

- There are one or two lists of interest keywords for each entry. This has been the most controversial point. The first list is mandatory and should contain a maximum of four keywords. It is printed in a bold typeface. It is taken from a list published in *Acta Cryst.* (1992). A**48**, 949–954. It was established with the help of Mike Dacombe, at that time the Technical Editor of the IUCr Journals. It is a compilation of keywords from the previous edition and from the Journals, where I have removed all equivalences, misspellings and sentences which were not really keywords. Astonishingly I received only one comment when it was published, but I know that it has been heavily criticised in some places. However, I am convinced that everybody has been able to describe his or her interests – the list contains more than one thousand words and the combination of keywords makes it much larger – and a mandatory list is the only means for an efficient search in a database. The full list is included as Annex 2. The second list of keywords is not restricted. It allows each scientist to choose his or her own wording to describe professional and scientific interests. I shall use this list later to add new keywords for future editions. I believe that it is not very efficient to search for individuals using this field, but it may give a better insight into the interests of a given person.

At the time when the structure of the projected database was under discussion, I was lucky enough to meet Professor Syd Hall of the University of Western Australia, who was the developer of the STAR format used by the Union in its new archive file format, the Crystallographic Information File (CIF). He was so enthusiastic and so ready to help that I was very quickly convinced that we must use this format. One good reason was that the editorial staff in the Chester office already had substantial expertise in handling such files, and would be very helpful in establishing the database. This was my best decision: Brian McMahon, Research and Development Officer at the Chester office, undertook the technical editing. Without Brian and Syd little would have been achieved, and I am very grateful to them both for the wonderful time we have had working together.

It took about six months of intense discussion by electronic mail to establish the format of the database and to write some software to help the National Editors to input their data in the correct format. This was a very exciting period: the code was passing from Australia to Europe and *vice versa*, improving day and night due to the time difference between Australia and Europe (I suspect that Syd was not only working during the Australian hours of daylight!), and all this was ready by mid-92.

In the mean time the search for National Editors had started. An important innovation has been the nomination of Regional Editors: for Australasia Professor H. Hashizume, for Europe Dr G. Filippini, for South America Dr I. Torriani, for North America Dr W. Duax. Their role has been very important, since they were in charge of all relations with the National Editors after their nomination. I am very grateful to them, since they have had the difficult job of assembling and checking the data, reminding National Editors if they were overdue, and sometimes establishing contact in countries not belonging to the Union. Among them Giuseppe's duty was more a quest than a job. Everything was ready by the end of 1992, so that the collection of data could start. Instructions and software were shipped to the National Editors either by means of diskettes or through the networks. The deadline for receipt of entries was established as the end of 1993. As usual with deadlines, it had to be shifted later to mid-94, which explains why the directory is published only now.

In the mean time other work had not stopped. Brian wrote the software to produce the printed edition automatically from the data files. Syd, with the help of Mark Favas, wrote an interface to search the database either by electronic mail or by interactive sessions. Establishing the specifications for this software provided us with a reason to continue our lengthy discussions over the networks! Brian decided to explore new tools such as *gopher* and *World-Wide Web* to provide other access mechanisms. If you are lucky enough to have effective access to the Internet, you will discover (if you have not already done so) how successful he has been. He has installed a wonderful interface which allows very easy consultation of the information. The first demonstration of this approach was presented at the IUCr XVIth General Assembly in August 1993, and it has been continuously improved. Full explanations of all methods of accessing the database are given in Annex 3.

At a meeting in Chester in March 1994 the Committee on the Electronic Dissemination of Information gave strong support to the project and asked the Executive Committee to upgrade the computer in the Union office in Chester to enhance access to the database. It also recommended that such access should be free. These proposals were accepted by the Executive Committee and I am grateful to them for this. Since June 1994 the database has been operating experimentally and its use is increasing very rapidly. This is really impressive since the user's guide has not yet been published, in October 1994, at the time of writing this introduction!

What about the contents?

The National and Regional Editors have done a wonderful job. It must be emphasised that the Union has not provided any funds for their work. Thus they have had to find the means of sending questionnaires to all scientists in their country and to do any necessary typing by themselves. It has been their own policy how to advertise for the directory and how to collect the data.

The table showing the number of entries per country cannot be compared directly with previous editions due to the political turmoil in Europe at the beginning of the 90s. Giuseppe has had the greatest difficulties to reach representatives in the new countries born from the former Soviet Union. Postal links have been difficult, fax almost impossible to send, and very few had e-mail access. In other European countries, too, political changes were reflected in the time it took to nominate a National Editor. However, he has been extremely successful and I believe we are only lacking a few people. The directory reflects the political and economical situation in the world today. Africa is largely missing, except for South Africa. The few African nationals appearing in previous editions never responded to my efforts to contact them. I was not successful either in North Africa. There are few names from the Middle East, except from the few countries which are traditionally very active in our Community. The scientists around the Pacific Ocean are growing in number. The total number of entries is about 8000 which is less than in previous editions.

I hope this will improve in the near future and that ever more scientists and ever more countries will be listed in the database. We intend, Brian McMahon and I, to update the database very regularly if the National Editors are ready and able to provide updates. The information will be immediately accessible through the networks. It will be technically easy to publish a new version or an addendum to the directory. Thus I hope that we will now be able to avoid long delays between succeeding editions.

Paris, October 1994

Y. Epelboin, General Editor
Laboratoire de Minéralogie-Cristallographie,
Université P.M. Curie, case 115,
75252 Paris Cedex 05, France

Table 1. *Number of Listings by Country*

Country	9th edition 1995	8th edition 1990	7th edition 1986	6th edition 1981	5th edition 1977	Country	9th edition 1995	8th edition 1990	7th edition 1986	6th edition 1981	5th edition 1977
Albania	8	-	-	-	-	(FYR) Macedonia	7	-	-	-	-
Algeria	-	-	-	2	2	Malaysia	17	30	30	20	15
Argentina	57	49	44	28	29	Mexico	53	45	42	39	19
Armenia	40	-	-	-	-	Moldova	44	-	-	-	-
Australia	173	192	197	200	170	Myanmar (Burma)	9	6	6	6	4
Austria	79	97	86	90	73	Netherlands	148	163	163	166	174
Azerbaijan	14	-	-	-	-	New Zealand	21	46	42	49	65
Bangladesh	-	22	22	25	23	Niger	1	-	-	-	-
Belgium	80	68	86	87	68	Nigeria	-	9	9	9	3
Bolivia	-	13	11	12	10	Norway	57	60	68	69	75
Botswana	1	-	-	-	-	Pakistan	91	64	60	31	29
Brazil	90	86	81	108	109	Panama	-	-	-	-	1
Bulgaria	82	83	83	74	68	Peru	-	11	11	11	4
Canada	186	170	203	165	153	Philippines	-	14	13	8	4
Chile	29	43	44	43	42	Poland	122	175	150	153	136
China	113	245	223	10	-	Portugal	41	24	20	20	20
Colombia	1	30	30	28	22	Romania	23	31	31	32	30
Croatia	69	-	-	-	–	Russia	559	-	-	-	-
Cuba	28	19	12	-	-	Saudi Arabia	-	16	14	10	9
Czech Republic	73	-	-	-	-	Serbia	34	-	-	-	-
Czechoslovakia	-	116	103	91	82	Singapore	10	12	10	6	4
Denmark	69	80	71	62	62	Slovakia	31	-	-	-	-
Egypt	101	71	71	77	77	Slovenia	47	-	-	-	-
Estonia	10	-	-	-	-	South Africa	66	74	98	80	72
Ethiopia	-	-	-	-	2	Spain	227	167	113	113	87
Finland	88	108	109	118	119	Sri Lanka	-	2	3	5	4
France	191	180	312	348	320	Sudan	-	2	2	2	1
Georgia	2	-	-	-	-	Sweden	112	180	206	206	207
Germany	588	-	-	-	-	Switzerland	97	116	121	133	132
German Dem. Rep.	-	347	374	376	377	Syria	1	2	2	-	1
German Fed. Rep.	-	650	589	455	428	Tadjikistan	5	-	-	-	-
Ghana	-	2	2	8	4	Taiwan	40	42	30	25	16
Greece	75	88	88	82	72	Tanzania	-	-	1	-	-
Hong Kong	8	6	7	6	4	Thailand	43	44	36	28	20
Hungary	40	94	74	71	63	Tunisia	-	37	32	18	4
Iceland	5	5	5	5	5	Turkey	40	79	84	60	27
India	318	350	370	340	318	Turkmenistan	5	-	-	-	-
Indonesia	5	1	13	13	4	Uganda	-	-	-	-	2
Iran	-	40	20	25	23	Ukraine	119	-	-	-	-
Iraq	-	6	6	6	2	USSR	-	907	697	657	595
Ireland	8	9	8	5	-	UK	420	636	732	651	598
Israel	52	66	79	71	56	USA	1609	2044	1621	1622	1642
Italy	323	373	344	276	248	Uruguay	4	6	3	3	4
Ivory Coast	-	8	10	8	6	Uzbekistan	8	-	-	-	-
Japan	605	595	565	516	446	Venezuela	51	18	13	10	7
Kazakhstan	19	-	-	-	-	Vietnam	46	21	-	-	-
Kenya	-	-	-	4	2	West Indies	-	-	-	-	2
Korea	42	26	17	15	11	Yugoslavia	-	154	138	128	118
Latvia	27	-	-	-	-	Zambia	-	3	-	-	-
Libya	-	10	4	-	-	Zimbabwe	-	-	-	2	5

Totals: 5th edition (1977): 71 countries, 7638 entries
6th edition (1981): 68 countries, 8174 entries
7th edition (1986): 69 countries, 8968 entries
8th edition (1990): 70 countries, 9589 entries
9th edition (1995): 72 countries, 7907 entries

HOW IT WAS DONE

National Editors were allowed discretion in the methods they employed to collect their entries. A few relied on submission of details by e-mail from individuals using the template provided by the Chester office (see Annex 1); others devised their own version of such a form; most collected the data in traditional hard copy, and input the information using the software supplied by Chester or the General Editor, or devised their own input programs. Whatever the chosen method, the entries for each country were sent as a STAR file to the Chester office by e-mail.

A preliminary set of proofs was generated for inspection by the National Editors. Proofs were obtained by using the IUCr program *ciftex* to translate the data files into files containing formatting commands for use by the TEX typesetting system. Output from the typesetting program was mailed to the National Editor, or in some cases transmitted by e-mail as a PostScript file. Preliminary proofs were often dispatched within minutes of receipt of the data files.

Corrections to the data files already received were made during the prolonged period of data collection. When all National Editors had submitted their files, the program *ciftex* was again used to generate individual data sheets, one for each person represented in the directory. These sheets contained the password needed to access the online database implementation of the directory, and instructions for so doing; but they also summarised all the details held for each entry. Every individual was therefore given the opportunity to proofread his or her own entry. This exercise resulted in more than 2500 late revisions, greatly improving the accuracy of the published information. Notable among these revisions was the large number of *additions* of fax and e-mail addresses, reflecting the spread of modern methods of rapid communication.

When all revisions had been incorporated, the national data files were concatenated into a single large (almost 3 Mb) file, which was once more processed using *ciftex* to produce the final TEX file. After some manual editing of this file to handle the relatively few formatting problems, the typeset output was sent as PostScript files to a commercial typesetter for printing at high resolution on photographic bromide paper. Plates were made from the bromides and the directory was printed and bound in the conventional way.

From the stage when the majority of entries had been received, the data files were processed daily to load the information into the *qi* database format developed at the University of Illinois. Qi (or 'phonebook') databases are client/server databases that can be queried over a TCP/IP network (*i.e.* using standard Internet protocols). They are also capable of interrogation by most client programs that obey *gopher* or *WorldWide Web* protocols. This allowed users of several Internet information services to access the data as they were received.

The final data files, in STAR format, may also be queried by e-mail or online telnet sessions, using the powerful *StarBase* (or *sb*) query language of Spadaccini & Hall [*J. Chem. Inf. Comput. Sci.* (1994). **34**, 509–516]. The *StarBase* interface allows users to annotate their entries with brief messages – indicating, for example, a temporary change of address. Such annotations are immediately available to other users of the *StarBase* system; and are propagated in the daily database rebuild to users of the other Internet services.

EXPLANATORY NOTES

Although the data included in this directory are stored in a machine-readable and structured file format, the typographic layout has been designed to resemble that of earlier editions, both to provide continuity of style and for compactness.

a. *Alphabetical order*

The sections of this edition are arranged in alphabetical order by countries, and by individuals within each country. Prefixes were handled differently depending on whether they are capitalized in spelling. For example, in **De Camp** with a capital 'D', the name is placed in the alphabetical 'D' group, whereas in **van der Meer** with uncapitalized prefixes the name will be found in the 'M' group. Names that contain diacritical marks are generally handled as if there were no such marks, unless a Sub-Editor indicated a different practice for that country.

b. *Contents of the biographical entries*

Study of the categories of information collected (Annex 1) will reveal more than in previous editions. Some typographical cues have been employed to identify the different information within a single compact listing. Thus a complete individual entry may contain:

 (i) family name (in bold type), followed by title and given names
 (ii) year of birth, in parentheses
 (iii) current position
 (iv) positions of responsibility within the IUCr, national crystallographic organisations, or, occasionally, other relevant bodies
 (v) full institution address
 (vi) full postal address within square brackets
 (vii) highest degree, with awarding institution and date of degree in parentheses, followed by field in which the degree was granted
 (viii) interests from the list of approved keywords in bold italic type
 (ix) other interests, in italic type
 (x) e-mail, telephone, fax and telex addresses or codes

Sub-Editors have interpreted differently the requirements for the institution and postal address, especially in cases where there is rarely any distinction; and so either or both of items (v) and (vi) may be present. Where both occur, the preferred address for postal correspondence is that in square brackets.

The use of the underbar to link words in the interests fields reflects the manner in which compound terms are stored within the database, although it does in some cases reduce the readability of the entry.

A typical (hypothetical) example is appended below to illustrate the layout of individual entries.

> **Dolittle**, Dr Alphonsus (1923). Research chemist. Member, IUCr Commission on Crystal Studies. Dept of Chemistry, U. Rutland, 1234 Little Road, Anytown, Rutland, England. [Dept of Chemistry, U. Rutland, PO Box 1234, Rutland, RU1 1UR, England.] PhD (U. Rutland, 1946) chemistry. ***Organic_chemistry*** *phenoxide_complexes*
> E-mail nemo@rut.ac.uk Tel. 44(32)2341234 Fax 44(32)4321432

The name of each crystallographer listed is contained in the Name Index, together with the country in which the name is listed. Cross references are hence unnecessary, and have not been used.

Explanatory *Notes* precede some country listings, as in previous editions, but an effort has been made to make each entry as far as possible self-contained, so that these *Notes* are of less importance than before.

Telephone and fax numbers include the international telephone country code, followed by the area code (where it exists) in parentheses. Extension numbers follow the main number and are indicated by an 'x'.

The language used throughout the bulk of this edition is English. Attempts have been made to give equivalents for local addresses, degrees and positions that will be universally understood, but in some cases the *Notes* should be consulted for further explanation.

ANNEX 1

International Union of Crystallography
Electronic Data Entry Form for *World Directory of Crystallographers*, Ninth Edition

Abstract

A data entry form allowing biographical data to be entered in the *World Directory of Crystallographers* is described. The form is a STAR file allowing automated typesetting of the Directory and the establishment of an electronic database of crystallographers' names, addresses and interests.

Introduction

The International Union of Crystallography has since 1957 been responsible for producing eight editions of an international directory of crystallographers. The projected ninth edition of this work will be established as an electronic database, allowing constant updating of changing information. The database will be established from electronic files supplied by the national Subeditors who are responsible for their own country's entries. The master input files will be in the STAR format described by Hall [(1991). *J. Chem. Inf. Comput. Sci.* **31**, 326–333]. This file format is already used by the IUCr as the basis of the Crystallographic Information File (CIF) used by authors to transmit structural papers to the journal *Acta Crystallographica*. Software similar to that used to prepare journal articles from CIF input will be used to publish a typeset edition of the *World Directory*.

Individuals may transmit their own entries to their national Subeditor for inclusion in the *World Directory* using the template electronic file described in this article. Copies of the file may be obtained by network email from the Technical Editor's office by sending a message to the address **sendcif@iucr.ac.uk** (or, within the UK, **sendcif@uk.ac.iucr**) containing the text send form.wdc. (Please note that all mail sent to this address is handled electronically by an information server program, and so must conform to the syntax indicated.)

A list of national Subeditors may also be retrieved by sending the request send subedit.lst to the **sendcif** address.

Completion of data form

The electronic form distributed from the Technical Editor's Office is shown in the Appendix. Any text to the right of a hash character is treated by STAR file processing software as a comment and ignored. This allows the file to include comments and an example describing how it should be completed. Apart from the comments, the file includes a data block declaration (data_wdc9: the 'wdc9' is an arbitrary choice), a list of datanames to follow in a 'loop_' structure, and the data itself. The query marks in the latter part of the file serve two purposes – they indicate where the contributor's information should be placed, but they also act as valid (but information-free) data items, thus maintaining the integrity of the file.

Descriptions of the fields are given below in a Dictionary of allowed data names.

General guidelines

The structure of data stored in a STAR file is determined by the presence of data names indicating the nature of the data to follow (the 'S' in the acronym means 'self-defining'). In the data entry form file, a set of twelve data fields is declared within a loop structure. It is therefore essential for the proper parsing of this file that each entry contains twelve data fields. In the blank template, twelve dummy fields are supplied as query marks (fields of type *text* are indicated with the appropriate delimiting semicolons). These query marks should be replaced with the contributor's personal details. If a particular category is not appropriate, the matching query mark should be left intact, and not deleted.

A field of type *char* is a continuous string of no more than 80 printable ASCII characters. It must completely fit on one line of text. If it represents a phrase, embedded blanks may be replaced by underscore characters, or matching single or double quote marks may bracket the phrase. For example, the two component family name 'Van Dyk' may be entered in the _name_family field as Van_Dyk, 'Van Dyk' or "Van Dyk". If the expression to be entered contains quote marks (as apostrophes or components of a code for an accented letter), quote marks of a different type should be used, or the conversion of blank spaces to underscores may also be performed. Thus, enter the name O'Neill as "O'Neill" and not 'O'Neill'.

Longer text, extending over more than one line, may be entered in fields of type *text* which are bounded by a semicolon as the leftmost character of the first and trailing lines. The first line may contain the start of the text, or may be blank, according to taste, but the last line contains only the closing semicolon. Thus, the address

```
; Case 115
  75252 Paris Cedex 05
  France

;
```

is entered correctly, as is the example

```
;
Case 115
75252 Paris Cedex 05
France
;
```

but a final line saying France ; would be wrong, as would a last line with a space or tab character before the semicolon.

In the examples of text fields given above, the alignment of lines of text is for purely aesthetic reasons – lines need not be offset from the left margin, and can be given in free form, with the proviso that certain individual fields must be formatted according to the rules given in the relevant Dictionary entry.

Accented or other special characters may be indicated in any field using the coding scheme given in the Dictionary under the _name_family entry.

General Editor

The General Editor for the *World Directory* project will be pleased to answer any questions and give any assistance required. Crystallographers who do not have access to a national Subeditor may also send their data forms direct to him. The address of the General editor is:

Y. Epelboin
LMCP, Universités P. M. Curie et Paris VII,
URA 009, CNRS,
Case 115, 75252 Paris CEDEX 05,
France
[e-mail: epelboin@lmcp.jussieu.fr]

Dictionary of data names for the *World Directory*
Version: STAR Dictionary (WDC 1992)

_name_family *(char)*

Family name. The name used for primary identification. Multiple components should be hyphenated or capitalised according to local custom. If quote delimiters are used, they should not conflict with embedded diacritical marks. Accents and special alphabetic characters should be indicated as follows:

\'	acute	\"	umlaut	\=	overbar
\`	grave	\~	tilde	\.	overdot
\^	circumflex	\;	hook (ogonek)	\<	hacek ("vee")
\,	cedilla	\>	long umlaut	\(	breve
\%a	a-ring	\?i	dotless i	\&s	German "ss"
\/o	o-slash	\/l	barred l	\/d	barred d

Dynastic components ('Jr', 'III') may also be given.
Example(s): Epelboin, 'Van Dyk', von_Graben, 'O'Neill"

_name_other *(char)*

Given names or initials. Title may precede the name, and should be shortened with a trailing point if it is an abbreviation (*e.g.* 'Prof.' for 'Professor') and without a trailing point if it is a contraction ('Dr' for 'Doctor').

Example(s): 'Prof. Yves', 'Dr Richard J.', 'Mr Zhong-he'

_address_institution *(text)*

Institution where resident. This will usually include department or section. Addresses spanning several lines should not have trailing commas or full points for punctuation. This field is intended to characterize the institution to which you belong.
Example(s):
; Lab. Min\'eralogie--Cristallographie
Universit\'es P. M. Curie et Paris VII
U. A. 009 CNRS
;

_address_postal *(text)*

Full postal address for correspondence. May differ from _address_institution and may contain duplication of information from _address_institution, if appropriate. Addresses spanning several lines should not have trailing commas or full points for punctuation. This field must contain all information necessary for postal delivery. Country must be given as the only information on the last line.
Example(s):
; Universit\'e P.M. Curie
Case 115
75252 Paris Cedex 05
France
;

_phone *(char)*

Telephone number. Should be given in the form
$<$*country code*$>$($<$*area code*$>$)$<$*local number*$>$[x$<$*ext.*$>$]
where parentheses surround the area or regional code, and an extension number, if given, is prefixed by an 'x'.
Example(s): 33(1)44275211, 44(244)342878x27

_fax *(char)*

Fax number. Should be given in the form
$<$*country code*$>$($<$*area code*$>$)$<$*local number*$>$
where parentheses surround the area or regional code.
Example(s): 33(1)44273785

_telex *(char)*

Telex number.
Example(s): '669755 OFFICE G attn UNICRYSTAL'

_email *(char)*

Email address. Should be given all in lower case unless the relevant network addressing software is known to be case-sensitive. Domain-type addresses should be given as
$<$*user*$>$@$<$*machine*$>$.$<$*site*$>$.$<$*net*$>$.$<$*country*$>$
(*i.e.* in little-endian order), but sites which have addresses on different networks may use alternative ordering where there is danger of ambiguity. Bitnet addresses should contain the routing suffix '.bitnet'. Addresses on networks other than Internet or Bitnet should give routing information *via* known Internet/Bitnet gateways, where possible.
Example(s): epelboin@lmcp.jussieu.fr,
teeter@bcchem.bitnet,
'bm@iucr.ac.uk (Internet) bm@uk.ac.iucr (JANET)',
xray%neosoft.ineos.free@suearn2.earn,
...!uucp1!uucp2!uucp3!addressee

_degree_highest *(char)*

Highest degree. Given in abbreviated form without periods. Avoid titles which are not well known outside your own country such as DiplMet, StudLic, IngDoct, CSc. Try to find an equivalent internationally known.
Example(s): DSc, PhD, DPhil, MS, MSc, Cand

_degree_institution *(char)*

Institution awarding degree. Use the abbreviations U. for University, Coll. for College, Inst. for Institute or Institution.
Example(s): 'Paris VI U.', "Tech. U. \/L\'od\'z",
Imperial Coll. London'

_degree_date *(char)*

Year in which highest degree was conferred.

_degree_field *(char)*

Degree field.
Example(s): Physics, 'Chemical crystallography'

_birth_date *(char)*

Year of birth.

_current_position *(char)*

Current position. Avoid titles peculiar to your institution or country, such as Akad. Oberrat, Head, Senior registrar, Chercheur. Describe your position in English and do not use abbreviations.
Example(s): Lecturer, Associate_professor, 'Post-Doctoral Research Assistant'

_interests_key_words *(text)*

Current interests by keywords. The entry should contain no more than four terms from the list of allowed keywords published in *Acta Cryst.* (1992), A**48**, 949–954. Compound terms may be generated by affixing keywords from the 'attributes' list to keywords from the 'methods' or 'compounds' lists. Words in a compound term should be separated by underscores; space or tab characters should separate terms.
Example(s):
; computing X-ray_topography silicon_compounds
;

_interests_other *(text)*

Interests not precisely covered by keywords in previous field. There is no limit on the number or nature of phrases that may be given. Words in a compound term should be separated by underscores; space or tab characters should separate terms. *Warning*: no database searches will be performed on this field.

Example(s):

```
; real-time_online_scattering_experiments
computerised_typesetting
;
```

_iucr_responsibilities *(text)*

Position(s) of responsibility within the International Union of Crystallography. Responsibility in Commissions and in National Unions may be indicated here.

Example(s):

```
; Chairman, Commission on Crystallographic Computing
;
```

_permission_to_distribute *(char)*

Authorization for the International Union of Crystallography to distribute information in this entry.

Permitted responses are yes no. Where no value is given, the assumed value is 'yes'.

APPENDIX: Listing of the data entry form

```
♦♦♦♦♦♦♦♦♦♦♦♦♦♦♦♦♦♦♦♦♦♦♦♦♦♦♦♦♦♦♦♦♦♦♦♦♦♦♦♦♦♦♦♦♦♦♦♦♦♦♦♦♦♦♦♦♦♦♦♦
♦♦♦ Data Entry Form for World Directory of Crystallographers ♦♦♦
♦♦♦                   (Ninth Edition)                       ♦♦♦
♦♦♦♦♦♦♦♦♦♦♦♦♦♦♦♦♦♦♦♦♦♦♦♦♦♦♦♦♦♦♦♦♦♦♦♦♦♦♦♦♦♦♦♦♦♦♦♦♦♦♦♦♦♦♦♦♦♦♦♦
♦
♦ This is an electronic "form" for entries into the 9th Edition
♦ of the IUCr World Directory of Crystallographers. This file
♦ will be used for automatic typesetting of a printed version
♦ of the World Directory, and as a basis for electronic access
♦ to this information.
♦
♦ Please fill in the details as indicated below and send to the
♦ appropriate Sub-editor. This data is entered in the format of
♦ a STAR File. If you require further details on this type of
♦ file structure refer to the papers (Hall (1991) J Chem Inf
♦ Comp Sci 31,326-333) or (Hall, Allen & Brown (1991) Acta Cryst
♦ A47,655-685.)
♦
♦----------------------------------------------------------------
♦ The following items are requested for each entry. These are
♦ given here as a STAR data 'name' and a brief description. Note
♦ that a 'char' data item is a string of less than 80 characters
♦ on the same line -- if this string contains imbedded blanks
♦ then it must be bounded by single or double quotes. If the data
♦ item is of type 'text' it must be bounded by semicolons as the
♦ first character of a line. An example entry is provided. Note
♦ that missing data items are flagged with a single question mark.
♦----------------------------------------------------------------

data_wdc9  ♦ this is a STAR data block identifier
loop_      ♦ this is STAR command to repeat the following data items
_name_family            ♦char Name used for primary identification
_name_other             ♦char Other name(s). Title is optional.
_address_institution    ♦text Primary institution address.
_address_postal         ♦text Postal address
_phone                  ♦char <country>(<city>)<local>x<extension>
_fax                    ♦char Same format as phone number.
_telex                  ♦char Complete international telex number.
_email                  ♦char <namecode>@<host>.<net>.<country>
_degree_highest         ♦char Highest degree received. Initials.
_degree_institution     ♦char Institution where degree received.
_degree_date            ♦char Year that the degree was awarded.
_degree_field           ♦char Field of study for the degree.
_birth_date             ♦char Year of birth.
_current_position       ♦char Position in the above institution.
_interests_key_words    ♦text Key words* describing interests.
_interests_other        ♦text Interests other than key words*.
_iucr_responsibilities  ♦text Current positions held in the IUCr.
_permission_to_distribute ♦char 'yes' or 'no' to electronic distr.

♦               * See (Acta Cryst.(1992) A48,949-954).
♦
♦----------------------- EXAMPLE ENTRY -----------------------
♦ 'Epelboin'
♦ 'Prof Yves'
♦ ;Lab. Min\'eralogie-Cristallographie
♦  Universit\'es P. M. Curie et Paris VII
♦  U.A. 009 CNRS
♦ ;
♦ ;Case 115
♦  75252 Paris Cedex 05
♦  France
♦ ;
```

```
♦  33(1)44275211
♦  33(1)44273785
♦ ?
♦ 'epelboin@lmcp.jussieu.fr'
♦ DSc
♦ 'Paris VI U'
♦ 1974
♦ 'Physics'
♦ 1944
♦ 'Group leader'
♦ ;computing  X-ray_topography  silicon_compounds
♦ ;
♦ ;Scientific_visualization
♦ ;
♦ ;Editor of ninth edition of World Directory of Crystallographers
♦ ;
♦  yes
♦
♦----- PLEASE FILL IN THE FIELDS BELOW FOR YOUR OWN ENTRY ------
♦ family name
♦   ?
♦ other name(s)
♦   ?
♦ institution
♦ ; ?
♦   ;
♦ postal address
♦ ; ?
♦   ;
♦ phone
♦   ?
♦ fax
♦   ?
♦ telex
♦   ?
♦ email
♦   ?
♦ highest degree
♦   ?
♦ degree institution
♦   ?
♦ degree date
♦   ?
♦ degree field
♦   ?
♦ year of birth
♦   ?
♦ current position
♦   ?
♦ interests (by keyword)
♦ ; ?
♦   ;
♦ other interests
♦ ; ?
♦   ;
♦ IUCr responsibilities
♦ ; ?
♦   ;
♦ permission to distribute
♦   ?
♦----------------------------------------------------------------
```

This is an example of the data entry form (slightly modified for typographical representation). It is a complete STAR file, defined by a datablock name identifier, loop declaration and list of data items. All data items currently have dummy values (indicated by query marks), *but their presence as placeholders is nevertheless essential to ensure the logical integrity of the file structure.* The author should *replace* each query mark with the appropriate information. For text fields, where information may extend over more than one line, it is *essential* that the delimiting semicolons are retained at the leftmost positions of the initial and concluding lines. All text on a line to the right of a hash character (#) is treated as a comment, and ignored by file processing software. This allows an explanatory text and example to be included in the file.

ANNEX 2

International Union of Crystallography
Keywords for the Database of Crystallographers and the *World Directory*

Y. Epelboin, General Editor, LMCP, Universités P. M. Curie et Paris VII, URA 009,
CNRS, Case 115, 75252 Paris CEDEX 05, France [e-mail: epelboin@lmcp.jussieu.fr]

The International Union of Crystallography will set up a world database of crystallographers. The next issue of the *World Directory* will be a by-product of this database.

The aim is to allow any scientist to retrieve useful information on other scientists: addresses, interests . . . . The database will be accessed by e-mail and later *via* telnet sessions. Security will be enforced to ensure that the data are not used for non-scientific purposes.

One of the main uses is to find specialists on given topics. This means that it will be possible to search the database by keywords and these must be defined in advance. The present list has been established on the basis of the keywords used in the eighth edition of the *World Directory*. Some additional keywords corresponding to new fields have been added. Some, too specific or misspelled, have been suppressed. Altogether there are about 1500 keywords.

Scientists will be able to use their own keywords for a better definition of their fields of interest but electronic searching of the database will be based on this printed list.

The collection of data will start at the beginning of 1993 and instructions will be distributed by the national Sub-Editors.

The list is divided into three parts: – *Methods, Properties and Applications* – *Compounds* – *Attributes*. The *Attributes* list must be used in conjunction with the other two lists and defines additional keywords for a better description of entries in those lists.

I hope that everybody will find appropriate definitions in the present lists. For maximum efficiency of the search process it is necessary to bear in mind that a too strict definition will be useless. This is one of the key points for the success of this database.

Methods, Properties and Applications

This list contains the keywords for methods of study, properties (physical, chemical, biological, . . .) and applications. It may be used with words defined in the *Attributes* list.

Aberration	Anvil cell	Biophysics	Chemotherapy	Conformation
Absolute configuration	Aperiodic material	Biosynthesis	Chirality	Conformational change
Absolute structure	Apparatus	Birefringence	Chromatography	Contaminant clean-up
Absorption correction	Archeology	Bloch structure	Circular dichroism	Contractile system
Absorption edge	Archeometallurgy	Bloch wall	Classification	Contrast
Absorption spectroscopy	Archeometry	Bond length	Clinker	Control
Accuracy	Area detector	Bond method	Close packing	Convective heat
Accurate intensity	Art conservation	Bond order	Cloud physics	Convergent-beam diffrac-
Acoustics	Arthropatic disease	Bonding	Clustering	tion
Acoustooptics	Artificial intelligence	Born approximation	Coagulation	Cooperative interaction
Activity	ARUPS	Bormann absorption	Coalification	Cooperative phenomena
Adhesion	ASAXS	Boundaries	Coarsening	Coordination
Adrenergics	Association theory	Bragg intensity	Coating	Corrosion
Adsorbate	Astronomy	Bravais lattice	Codification	Cosmochemistry
Adsorption	Astrophysics	Bridgman Stockbarger	Cohesion	Crack
AEM	Asymmetric synthesis	technique	Cohesive energy	Cracking
Aerodynamics	Asymmetry	Brillouin spectroscopy	Colour center	Creep
Aerosol	Athletic medicine	Burial diagenesis	Colour symmetry	Critical phenomena
Aerospace	Atomic weight	Calcification	Combinatorial theory	Cross section
AES	Attenuation coefficient	Calibration	Combustion	Cryogenics
Affinity	Auger spectroscopy	Calorimetry	Complexation	Crystal field
Ageing process	Automation	Camera	Compliance sampling	Crystal force
Agriculture	Autometasomatism	Carboxylation	Compression	Crystal form
Algorithm		Carcinogenesis	Compton scattering	Crystal growth
ALISUVAX	Back-reflection	Catalysis	Computer	Crystallinity
Allostery	Ballistic	Centrosymmetry	Computer-aided education	Crystallite
Alteration	Band calculation	Chandler wobble	Computer architecture	Crystallization
Amorphization	Basicity relationship	Channelling	Computer-assisted design	Crystallogeny
Amorphous phase	Battery	Characterization	Computer automation	Crystallography
Analgesics	Bijvoet absorption edge	Charge density	Computer graphics	CVD
Anharmonicity	Biochemistry	Charge-density wave	Computer management	Cycloaddition
Anisotropy	Biocoordination	Charge localization	Computer modelling	Czochralski technique
Annealing	Biocrystallography	Charge transfer	Computer sciences	
Anomalous dispersion	Bioelectret	Chelation	Computer technology	Damage
Anomeric effect	Bioenergetics	Chemisorption	Computing	Data collection
Antiferroelectricity	Biology	Chemistry	Condensed matter	Data processing
Antiferromagnetism	Biomaterial	Chemometrics	Conductivity	Database
Antiphases	Biomechanics	Chemotaxis	Conductor	Debye Scherrer
	Biomolecule			Debye temperature

Misorientation
Mobility
MO calculation
Modelling
Modulated structures
Moire
Molecular beam
Molecular crystal
Molecular mechanics
Molecular rectifier
Molecular replacement
Molecular vibration
Momentum density
Momentum distribution
Monitoring
Monochromator
Monocrystal
Monolayer
Monte Carlo
Morphology
MOS
Mosaicity
Mossbauer
Motion
Multibeam
Multicrystal
Multidomain
Multilayer
Multiphase
Multiple-crystal diffractom-
etry
Multiple scattering
Multislice method
Mutagenesis
Mycology

Nanoanalysis
Neurochemistry
Non-destructive analysis
Non equilibrium
Non-linear property
Nonstoichiometry
NPR
NQR
Nuclear filter
Nuclear fusion
Nuclear magnetic reso-
nance
Nuclear reactor
Nucleation
Number theory
Nutrition

Occupancy
OD
Oncology
One dimension
Ontogeny
Optical activity
Optical property
Optical transform
Optics
Optimization
Optoelectrical property
Optoelectronics
Orbital calculation
Order
Order-disorder
Ordered structure
Ordering
Orientation
Orogenic belt
Oscillation camera
Ostwald ripening

Overcrowding
Oxidation

Packing
Paleomagnetism
Paracrystal
Paragenesis
Paramagnetic resonance
Paramagnetics
Parameter
Patent
Pattern recognition
Patterson method
Perfect crystal
Perfection
Performance
Pericyclic reaction
Permittivity
Petrography
Petrology
Pharmacology
Phase determination
Phase diagram
Phase equilibrium
Phase formation
Phase kinetics
Phase refinement
Phase separation
Phase transition
Philosophy
Philosophy of science
Phonon resonance
Phonon softening
Photochemistry
Photochromism
Photoconductivity
Photodimerization
Photoelasticity
Photoelectron
Photoemission
Photogeology
Photography
Photon effect
Photoreaction centre
Photorearrangement
Photorefraction
Photostimulated process
Photosynthesis
Phylogeny
Physical property
Physics
Physiology
Pi electron
Piezoelectricity
Pigment
Pitch
Planar defect
Planetology
Planning
Plasmon
Plastic flow
Plasticity
Plastics
Platelet
Point defect
Point group
Poisoning
Polarity
Polarization
Polarization microscopy
Polarized neutron
Pole figure
Pollution
Polycrystal

Polymerization
Polymorphism
Polytypism
Porosity
Positron annihilation
Potential energy
Powder
Precession
Precipitation
Precise measurement
Prediction
Preparation
Pressure
Processing
Profile analysis
Proportional counter
Prosthesis
Pseudomorphism
Pseudosymmetry
Publishing
Pulsed neutron
Purification
PVD
Pyroelectricity

QSAR
Quadrupole resonance
Qualitative analysis
Quantum mechanics
Quasicrystal

Radiation
Radiation protection
Radioactivity
Radiochemistry
Radiotracer
Raman
Random phasing method
Random system
Random walk
Rayleigh scattering
Reactivity
Real crystal
Real structure
Real-time control
Real-time imaging
Rearrangement
Receptor
Recognition
Recombination
Reconstruction
Recrystallization
Refinement method
Reflectance
Reflected light microscopy
Reflectivity
Refractive index
Regulation
Relaxation
Reliability
REM
Remote control
Repair
Replacement
Replication
Representation theory
Research
Residual electron density
Residual stress
Resistivity
Resonance
Resonance spectrometry
Resonant scattering
Restrained least squares

Reversible reaction
RHEED
Rietveld method
Rigid-body analysis
Risk assessment
Rocking curves
Rotatory dispersion

Safety
Satellite reflection
SAXS
Scale factor
Scale mechanism
Scanning electron
microscopy
Scanning tunnel mi-
croscopy
Scattering
Scattering factor
Sciences
Search and match
Secondary bonding
Secondary electron emis-
sion
Sedimentation
Seismology
Selectivity
Semiconductor
Semi-empirical calculation
Sensor
Sequencing
Service
Shape
Shape memory
Shock metamorphism
Shock wave
Short hydrogen bond
Short-range order
SIMS
Simulation
Simultaneous diffraction
Single crystal
Sintering
Size distribution
Size effect
Slow neutron
Small-angle scattering
Soft mode
Software
Soft X-ray
Solar cell
Solar collector
Solar energy
Solidification
Solid phase
Solid solution
Solid state
Solubility
Solution
Sorption
Sound propagation
Space
Space group
Space processing
Specific heat
Spectrography
Spectrometry
Spectrophotometry
Spectroscopy
Spectrum analysis
Spin
Spin density
Spinel
Spin resonance

Spin wave
Sport
Sputtering
Stability
Stacking
Stacking fault
Standing wave
Statistical mechanics
Statistical method
Statistical model
Statistical thermodynamics
Statistics
Stepanov method
Stereochemistry
Stereoselectivity
Stoichiometry
Strain
Strain deformation
Strain determination
Strain hardening
Streaks
Strength
Stress
Structural change
Structural disorder
Structure
Structure determination
Structure factor
Structure-activity
relationship
Subconductor
Sublimation
Substitution
Substructure
Supercomputer
Superconductivity
Superconductor
Superfluid
Superlattice
Supermagnetism
Superstructure
Surface
Survey
Symbolism
Symmetry
Symmetry breaking
Symmetry group
Synchrotron radiation
Syncrystallization
Synthesis
Systematics
System dynamics
System integration

Tautomerism
Technique
Technology
Television
Temperature
Tensor
Tensometry
Tensor property
Termination effect
Tertiary structure
Testing
Texture
TGA
Theory
Thermal expansion
Thermal motion
Thermal property
Thermal stress
Thermal vibration
Thermistor

Thermoanalysis
Thermodynamics
Thermogravimetry
Thermoluminescence
Thermostability
Thick film
Thin film
Thin layer
Three-dimensional reconstruction
Time-resolved effect
Time-of-flight diffraction
Topochemistry
Topography
Topology
Topotacticity
Topotaxy
Toxicity
Toxicology
Trace
Trace analysis
Track detector
Transcription
Transducer
Transduction
Transformation
Transmission electron microscopy
Tribology
Triplet
Tube
Tunnelling
Twin
Twinning
Typomorphism
Ultra high pressure
Ultra high vacuum
Ultra pure compound
Ultrasonics
Ultraviolet
Unit cell
Unusual bonding
UPS
Vacancy
Vacuum
Valence charge density
van der Waals radius
Vector search
Vibration
Vitreous state
Volatility
Volcanology
VVPES
Wavelength
WAXS
WDS
Weak-beam electron diffraction
Weak interaction
Weathering
Welding
Whisker
White-beam radiation
Wide-angle scattering
Wigner crystal
XANES
XPS
X-ray fluorescence
X-ray fluorescence spectroscopy
Yeast expression system

Compounds

This list contains classes of compounds and more general names to define classes of materials, such as *Magnets*. It may be used with words defined in the *Attributes* list.

Acetylene
Acids
Actin
Actinides
Adenoviruses
Adrenergic compounds
Aggregates
AIDS
Air
Albumin
Alkaline
Alkalis
Alkaloids
Alkanes
Alkoxides
Allergenics
Alloys
Alumina
Aluminate
Aluminium compounds
Aluminosilicates
Amino acids
Analgesics
Anorthosite
Antiallergenics
Antiamoebic compounds
Antiangional compounds
Antiarrhythmic compounds
Antiarthritic compounds
Antiasthmatic compounds
Antibacterial compounds
Antibiotics
Antibodies
Anticancer compounds
Anticholinergic compounds
Anticoagulants
Anticonvulsants
Antidepressants
Antiemetics
Antiestrogen compounds
Antifolates
Antigelling compounds
Antigens
Antihistaminic compounds
Antihypertensive compounds
Anti-inflammatory compounds
Anti-influenza compounds
Antileprosy compounds
Antileukemia compounds
Antimalarial compounds
Antimicrobial compounds
Antimitotic compounds
Antimony compounds
Antimuscarinic compounds
Antioxidants
Antiparasitic compounds
Anti-Parkinsonian compounds
Antipsychotic compounds
Antipyretics
Antirheumatic compounds
Antischistosomal compounds
Antischizophrenia compounds
Antisickling compounds
Antispasmodics
Antithrombotic compounds
Antitumour compounds
Antiulcer compounds
Antiviral compounds
Anxiolitic compounds
Apatite
Archeological materials
Arsenic compounds
Asbestos
Austenite
Bacterial compounds
Bacterial toxin
Barbiturates
Barium compounds
Basaltic rock
Bases
Bauxite
Beryl
Beryllium compounds
Bile pigments
Bimetallic compounds
Binary alloys
Bioceramics
Biopolymers
Bismuth compounds
Blende
Blood
Boehmite
Bone
Boron compounds
Borophosphates
Borosilicates
Bromium compounds
Bronzes
Buffers
Bushveld complex
Cadmium compounds
Cage molecules
Calcium compounds
Cancer
Carbanions
Carbides
Carbohydrates
Carbonates
Carbonyls
Carbon compounds
Carboranes
Carboxylates
Carboxylic acids
Carboxypeptidases
Carcinogens
Carcinostats
Cardenolides
Cardiac compounds
Cascade proteins
Catalysts
Celluloses
Cements
Ceramics
Chalcogenides
Chalcogens
Chalcopyrites
Chelates
Chlorine compounds
Chlorites
Chromatin
Chromite
Chromium compounds
Chrysotile
Clathrates
Clays
Clusters
Coal
Cobalt compounds
Coke
Colloids
Conglomerates
Copper compounds
Cordierite
Crown compounds
Crust
Cryptates
Cubanes
Cyanide
Cyanins
Cyclic polyethers
Cyclodextrins
Cyclophosphazenes
Cytochrome
Cytoplasm
Cytotoxins
Dehydrogenases
Dental material
Detergents
Diamond
Diaspores
Dichalcogenides
Dielectrics
Dihydrofolate
Dipeptide
Disease
Dismutases
Diterpenes
Diuretics
DNA
Drug
Dust
Dyes
Elastomers
Electroceramics
Electrolytes
Energetic compounds
Enkephalins
Enzyme inhibitors
Enzymes
Estrogens
Expectorants
Explosives
Fab fragments
Fats
Feldspars
Ferrites
Fertilizers
Fibres
Fire-resistant compounds
Flavonoids
Fluids
Fluoride
Fluorine compounds
Fluorometallates
Fluoroorganics
Fossils
Free radicals
Fuel
Fulgides
Fungicides
Fused rings
Gallium compounds
Gallstones
Gases
Gelatins
Gels
Gemstones
Genes
Germanates
Germanium compounds
Glasses
Glycogens
Glycoproteins
Glycosaminoglycans
Glycosides
Gold compounds
Grains
Granites
Graphites
Halides
Halogens
Hemes
Hemoglobins
Hemoproteins
Herbicides
Heterocycles
Heteropolyacids
Heusler alloys
Histamine agonists
Hormones
HSLA steels
Humic compounds
Hydrates
Hydrides
Hydrogen compounds
Hydroxides
Hypnotics
Ice
II-VI compounds
III-V compounds
Immunoglobulins
Immuno modulators
Immunosuppressants
Indium compounds
Inhibitors
Insecticides
Insulin
Intercalates
Interstitial compounds
Invar
Iodine compounds

Ionic conductors
Ionophores
Iridium compounds
Iron compounds
Isomers
Isopolymetallates
Isotopes
IV-VI compounds

Jahn Teller compounds

Lamellar compounds
Lamprophyres
Lamp materials
Lanthanides
Layered compounds
Lead compounds
Ligands
Lipases
Lipids
Lipoproteins
Liquid crystals
Liquids
Lithium compounds
Living systems
Lubricants
Luminescent compounds
Lymphocytes
Lymphokines

Macrocycles
Macromolecules
Magnesium compounds
Magnets
Main-group compounds
Manganese compounds
Mantle
Martensites
Materials
Melts
Membranes
Mercury compounds
Metallacarboranes
Metalloenzymes
Metallophthalocyanines
Metalloporphyrins
Metalloproteins
Metals
Meteorites
Micas
Micelles

Microcrystallite compounds
Minerals
Mixed-layer compounds
Mixed-valence compounds
Modulated structures
Molecular complexes
Molecules
Molybdates
Molybdenum compounds
Moon rocks
Multilayers
Muscarinic compounds
Muscles
Mutagenic compounds

Narcotics
Natural products
Nematogenic compounds
Nervous system
Neuroleptics
Neuropeptides
Neurotoxins
Nickel compounds
Niobium compounds
Nitrates
Nitrides
Nitrogenases
Nitrogen compounds
Noble gases
Noble metals
Nuclear materials
Nucleic acids
Nucleoproteins
Nucleosides
Nucleotides

Oils
Oligomers
Oligonucleotides
Oligopeptides
Oligosaccharides
Oncogenes
Opiates
Ores
Osmium compounds
Oxides
Oxygenases
Oxygen compounds
Oxyhydrides

Palladium compounds

Paper
Parasites
Particles
Penicillins
Peptaibols
Peptides
Perovskites
Pesticides
Phosphatases
Phosphates
Phosphorus compounds
Phosphorylases
Photochromic compounds
Photoconductors
Phyllosilicates
Pigments
Plagioclases
Plants
Plasmas
Platinum compounds
Plutonium compounds
Polar compounds
Polyamides
Polyanions
Polydentates
Polyelectrolytes
Polyesters
Polyimidazoles
Polyiodides
Polymerases
Polymers
Polyolefins
Polyoxoanions
Polypeptides
Polyphosphides
Polyproteins
Polysaccharides
Polythionates
Porous materials
Porphyrins
Potassium compounds
Powders
Precipitates
Propellants
Prostaglandins
Proteases
Proteins
Proteinases
Protein kinases

Prothrombins
Psychoactive compounds
Radicals
Radical salts
Radiopharmaceutical
    compounds
Radium compounds
Rare-earth compounds
Reductases
Refractory compounds
Renins
Rhenium compounds
Rhodium compounds
Ribosomes
Ring molecules
RNA
Rock
Rubber
Ruthenium compounds
Saccharides
Salts
Sandwich compounds
Sapidants
Sediments
Selenium compounds
Semiconductors
Semicrystalline compounds
Serums
Sesquiterpenes
Siderophores
Silicates
Silicon compounds
Silver compounds
Small molecules
Soaps
Sodium compounds
Soils
Solids
Sols
Solvents
Steels
Steroids
Sterols
Strontium compounds
Sulfates
Sulfides
Sulfur compounds
Superalloys
Superconductors

Superoxides
Surfactants
Sweeteners
Tantalum compounds
Technetium compounds
Tellurides
Tellurium compounds
Terpenes
Textiles
Therapeutic compounds
Thermoelectric materials
Thorium compounds
Thyrotoxic compounds
Tin compounds
Tissues
Titanates
Titanium compounds
Tooth compounds
Toxins
Tranquillizers
Transcriptases
Transition elements
Trypanosomes
Trypsins
Tumours
Tungstates
Tungsten compounds
Unidirectional compounds
Unsaturated compounds
Uranides
Uranium compounds
Uricosuric compounds
Vanadium compounds
Vasodilators
Venums
Viruses
Vitamins
Waste
Water
Waxes
Ylides
Ytterbium compounds
Zeolites
Zinc compounds
Zirconium compounds
Zymogen

Attributes

This list contains additional keywords which may be used together with those defined in the *Compounds* and *Methods, Properties and Applications* lists.

Absolute
Absorbing
Accurate
Acid
Acoustic
Activation
Active site
Active surface
Acyclic
Adduct
Agrochemical
Amorphous
Amphibole
Amphiphilic
Analysis
Analytical
Anharmonic

Anhydrous
Anion
Anisotropic
Anomalous
Anorthic
Antiferroelastic
Antiferroelectric
Antiferromagnetic
Application
Applied
Aqueous
Asymmetric
Asymptotic
Atmospheric
Atom
Atomic

Behaviour

Binary
Binding
Bioactive
Biochemical
Biogenic
Bioinorganic
Biological
Biomedical
Bioorganic
Bond
Boundary
Bragg
Bridged
Building
Bulk

Catalytic
Cation

Chain
Channel
Charge
Chemical
Chiral
Chiroptical
Chromatic
Clinical
Close-packed
Coherent
Colour
Commensurate
Comparison
Complex
Composite
Composition
Compound

Condensed
Conducting
Conformational
Constituent
Cosmic
Crystal
Crystalline
Cubic
Cyclic
Density
Dependence
Deposit
Dielectric
Difference
Diffuse
Disordered
Displacive

Domain
Donor
Doped
Double
Drug
Dynamic
Dynamical

Efflorescent
Elastic
Electrical
Electromagnetic
Electron
Electronic
Electrooptic
Electrostatic
Elongation
Emission
Energy
Energy-dispersive
Environmental
Enzymatic
Epitaxic
Equilibrium
Evolution
Exchange
Excitation
Experimental
Exploration
Extended

Ferroelastic
Ferroelectric
Ferroic
Ferromagnetic
Fibrillous
Fibrous
Five-dimensional
Focusing
Forensic
Formation
Four-dimensional

Gamma-ray
Genetic
Geochemical
Geometric
Geothermal

Globular
Glycolytic

Halophilic
Heavy
Helical
Heterocyclic
Heterogeneous
Hexagonal
High
High-precision
Holographic
Homogeneous
Hydrothermal
Hydrous
Hygroscopic

Icosahedral
Ideal
Incoherent
Inelastic
Infrared
Inorganic
Interaction
Interatomic
Intercrystalline
Interfacial
Intermetallic
Intermolecular
Internal
Interstitial
Intracrystalline
Intramolecular
Intrazeolitic
Inverse
Ion
Ionic
Irradiated
Isometric

Laminated
Large-angle
Laue
Layered
Light
Linear
Liquid
Local

Long-period
Low
Low-dimensional

Macrocyclic
Macromolecular
Magmatic
Magnetic
Marine
Martensitic
Mass
Material
Mathematical
Mechanical
Medical
Medicinal
Medium-size
Mesogenic
Metallic
Metalloorganic
Metallurgical
Metamorphic
Metastable
Method
Mineralized
Mixed
Model
Modulated
Molecular
Monochromatic
Monoclinic
Monoclonal
Mosaic
Multiple

Non-ideal
Nematic
Neurological
Neutron
Non-
Non-bonded
Non-crystalline
Non-crystallographic
Non-linear
Nuclear-one-dimensional

Optical
Organic

Organometallic
Orthorhombic

Pathological
Perfect
Pharmaceutical
Phase
Phonon
Phosphoorganic
Photochromic
Photon
Photovoltaic
Physical
Piezoelectric
Plastic
Polar
Polychromatic
Polycyclic
Polymeric
Polymorphic
Polytypic
Porous
Process
Property
Pulsed
Pyroelectric

Qualitative
Quantitative
Quantum
Quasielastic
Quaternary

Rapid
Reaction
Refinement
Reflection
Relationship
Relative
Residue
Resolution
Respiratory
Restrained
Rhombohedral
Ring
Rolled

Secondary

Separation
Sequence
Short-period
Site
Size
Slag
Small
Small-angle
Smectic
Solid
Soluble
Spectral
Stainless
Static
Stereographic
Strained
Structural
Substituent
Superionic
Synthetic

Tensile
Ternary
Tetragonal
Theoretical
Thermal
Thermophile
Toxic
Transfer
Transport
Treatment
Two-dimensional

Unidirectional
Unsaturated

Vacancy
Vibrating
Viral
Volatile
Volcanic

Wet

X-ray

Zone

ANNEX 3

International Union of Crystallography
Accessing the WDC9 Database

Y. Epelboin, WDC9 General Editor, LMCP, Université P. M. Curie, case 115, 75252 Paris CEDEX 05, France
[epelboin@lmcp.jussieu.fr]
B. Mc Mahon, Research & Development Officer, IUCr, 5 Abbey Square, Chester CH1 2HU, England [bm@iucr.ac.uk]
S. R. Hall, Crystallography Centre, University of Western Australia, Nedlands 6009, Australia [syd@crystal.uwa.edu.au]

Abstract

A database of scientists with crystallographic interests has been established and located at the IUCr Office in Chester (UK). Methods for remotely accessing this database are described.

Introduction

A database of scientists with crystallographic interests has been constructed (Epelboin, 1992a). National sub-editors have been assigned by the IUCr to collect information on addresses and research interests of scientists in each country. These data were collated by Regional Editors for America, Asia and Europe to form the WDC9 database. The format of the database complies with that of the STAR File (Hall, 1991) which is the accepted IUCr standard for exchanging and archiving data. A hardcopy version of the WDC database – the *World Directory of Crystallographers* – will be printed periodically.

The contents of the WDC9 database were described in *Acta Cryst.* (1993), A49, 222–225. Appendix A lists the data items stored in the database. The research interests of scientists are listed in terms of standard keywords (Epelboin, 1992b). This permits searches on particular research topics.

Access to the WDC database requires a registered security code, and a connection to an international computer network. The several different access approaches which are available for searching the WDC database are described below. The appropriateness of these approaches depends on the application and the network connection. Those with minimum facilities can use e-mail access, while those with an Internet connection may use several modes of interactive access.

E-mail and telnet access modes

E-mail mode

E-mail queries to the WDC database are sent to
wdc-search@iucr.ac.uk
The query should form the body of the message. It *must* begin with a 'user' command to identify yourself to the database software. The rest of the query is given using the WDC query language discussed below, and in Appendix C.

Telnet mode

Interactive telnet access to the WDC database is initiated with
telnet wdc.iucr.ac.uk
The telnet session proceeds as follows:

(i) Enter the login sequence **wdc** when prompted.
(ii) In response to the prompts, enter your registered user name and access code.
(iii) You are given an opportunity to add a temporary message to your entry, or to modify or delete one already supplied by you in an earlier session. (Delete the old message by entering a new one with no contents.)

(iv) Enter your query as a sequence of commands. Close the query sequence with a control-D. A detailed description of the WDC query language is given in Appendix C.

The query input by the user will be displayed on the screen for checking. Confirm if it is correct. The search results of the query will also appear on the screen.

WDC query language

E-mail and telnet searches use a common query language. This was developed for the WDC access software by Mark Favas and Syd Hall, and is a subset of the general *Star_Base* query language (Spadaccini & Hall, 1994).

A quick introduction to the WDC query language will be provided *via* several annotated examples. These examples show how a hypothetical 'Dr Maxwell' would retrieve information from the WDC database using e-mail or telnet.

Example 1

Dr Maxwell needs information about a person with the family name 'van Bloggs'. Here are the commands used to do this search.

(i) The sequence 'user', followed by the security code and name of the user must be used to open a search session. The combination of the code and name must match with an entry in the WDC database. The security code is supplied to registered users by the IUCr office. In a telnet session, these are supplied by the user as part of the login sequence. For an e-mail query, this command *must* constitute the first line of the message.

    user ukfd0876 Maxwell

Queries are insensitive to case and embedded blanks. Each security code has a pre-assigned access limit for the number of successful matches (*i.e.* 'hits') per month. For most users this limit is 150.

(ii) The following sequence is used to request information on all persons in the world named 'van Bloggs'. Note that some knowledge of the WDC data names will be needed to request a data item (such as _name_family). These are listed in Appendix A.

    if _name_family == vanbloggs

Again, the search string is case insensitive, and all blanks and special accents are ignored. The 'double equals' operator specifies an 'exact match' with 'vanbloggs'. If one were uncertain of the exact spelling, the use of '~= vanblog' would search for partial matches. All search operators are listed in Appendix C.

(iii) The search is started with the specific command

    search

This will result in all information about persons with the family name 'vanbloggs' to be returned as e-mail output or to the screen. Multiple searches are possible within the same e-mail or telnet session.

Example 2

In this example Dr Maxwell searches for scientists with similar research interests in a specific country.

(i) The allocated number of search matches per month is fixed by the database administrator for each user. To prevent excessive hits in a search, the hit limit is set here to 10.

```
user ukfd0876 Maxwell
limit 10
```

(ii) Because Dr Maxwell only wants to know about scientists working in France, the search is constrained by entering

```
location fr
```

The abbreviated code for each country is listed in Appendix B. Country names may also be used. Regional codes ('north_america', "south_america', asia' and 'europe') may also be used.

(iii) The research interests of each scientist are stored as the two data items _interests_key_words and _interests_other. In this example the specific topic of interest is 'incommensurate structures'. Because keyword combinations exist in the WDC, the best form of a search query is:

```
if _interest* ~= incommensur
```

The wild card character '*' is used to signal that both 'interests' data items should be searched for the string 'incommensur'.

(iv) Dr Maxwell only wants the names, institution and interests of people satisfying the above. This is requested with the following command.

```
want _name* _address_institution _interest*
search
```

The 'want' command is only used when specific items are needed. The default request is for all data (i.e. _*), as was seen in the first example. This command serves to restrict the amount of data returned.

Example 3

The WDC database also permits registered users to store or replace short messages as part of their accessible information. This message facility is only available to the e-mail or telnet access methods. Messages are automatically output when any data item associated with the user's name is accessed. Messages cannot be searched for.

In the telnet mode of access, the user is prompted to enter such a message. In an e-mail transaction, the message is placed between 'message' and 'message_end' commands. Here Dr Maxwell deposits a message. Lucky fellow!

```
user   ukxxxfd0876 Maxwell
message
I am on sabbatical leave from April to
December 1994 at the Dept of Polynesian
Studies, University of Tahiti. Please do
not contact me unless you are desperate.
message_end
```

Note that only Maxwell is able to insert or replace a message for his entry.

qi/ph access

The STAR file format of WDC9 supports complex queries using the query language described above. However, to support casual inquiries from Internet sites, an alternative database representation is also employed. This is the *qi* database format developed at the University of Illinois at Urbana-Champaign. Many sites already have standalone query software (usually known as *ph*) able to interrogate such a database. If you have such software, you may supply the address

gopher.iucr.ac.uk

to extract information from WDC9 – your local documentation or system administrator should be able to explain how to do this.

However, a more convenient approach is to use one of the gopher or WWW client programs that have built-in gateways

to qi databases (also known to these programs as CSO). These techniques are described below.

Gopher access mode

Gopher is a search tool available on TCP/IP networks linked to the Internet. It is available for computers ranging from Unix workstations to Apple microcomputers. Check the availability and usage of gopher with your system administrator. Note that your gopher program *must* be able to support searches on qi (CSO) databases for this approach to work properly.

A gopher session on the WDC data base will use the address

gopher.iucr.ac.uk

and standard port number of 70.

On entry, a menu will offer different services. Choose 'World Database of Crystallographers'. A search sub-menu is provided. Fill in the data items to be searched for. The wild card character '*' may be used at the end of a search string only. (Note that this differs from the usage in the e-mail or telnet access methods.) The range of search options available with gopher will depend on the characteristics of your local gopher program.

You may find a selection of query fields in which you can pose specific questions. Alternatively, there may be only a single field. Entering a string in this field will usually initiate a search on family_name. Sometimes additional search fields may be explicitly specified in this field. For example it may be possible to search on all French crystallographers with interests in the field of 'incommensurate structures' by entering

```
key_interests=incommensur*   country=france
```

and separately

```
other_interests=incommensur* country=france
```

Note that queries entered in the same field will be searched for jointly – *i.e.* using Boolean *and* logic. Note also that full country names need to be specified in this approach – the short location codes of Appendix B may not be used.

Appendix A lists the names of the searchable gopher data items.

WWW and mosaic access modes

World-Wide Web is a general approach to accessing global databases which also provides facilities for gopher, telnet, or ftp access. A WWW user is automatically linked to the computer where the database exists. Several programs exist to connect to WWW sites.

Mosaic, one of the most popular of such programs, uses a graphical interface which establishes links to data by clicking on highlighted words or phrases. Check the availability and usage of WWW with your system administrator.

To use mosaic to connect to the WWW server in Chester, open the URL (Uniform Resource Locator)

http://www.iucr.ac.uk/welcome.html

An introductory page will appear. The section on 'World Database of Crystallographers' offers options to access the database through the gopher service described above, or by using *forms*, a device for requesting specific information from a WWW site that not all client programs support. If you select the general option, a new window will open containing a writable field for entering a query similar to that described for gopher above. For example,

```
family_name=dupon*   country=france
```

will return all names starting with 'dupon' in France. The number of hits per query is limited to 25 – no information at all will be returned if the number of hits would exceed this figure.

If your WWW program supports forms, and you choose the advanced option, a page of text appears with several writable fields in which search terms may be placed. There are also options to search on additional fields, or to alter the information returned by a successful search. The form is intended to be self-explanatory.

Appendix A

Searchable data items

e-mail, telnet access	ph, gopher access	
_name_family	family_name	family name
_name_other	forenames	first names
_address_institution	institution_address	institution address
_address_postal	postal_address	preferred postal address
_phone	phone	phone number
_fax	fax	fax number
_telex	telex	telex number
_email	email	electronic mail address
_degree_highest	highest_degree	highest degree
_degree_institution	degree_institution	institution conferring degree
_degree_date	degree_date	year when degree was conferred
_degree_field	degree_field	degree field
_birth_date	year_of_birth	year of birth
_current_position	position	current position
_interests_key_words	key_interests	combination of keywords (Epelboin 1992b)
_interests_other	other_interests	unrestricted list of interests
_iucr_responsibilities	iucr_responsibilities	IUCr or Regional Associate responsibilities
	country	country name (for e-mail or telnet access, use location command)

Appendix B

Location codes

National codes

al	Albania	ee	Estonia	mk	Macedonia	si	Slovenia
ar	Argentina	fi	Finland	my	Malaysia	za	South Africa
am	Armenia	fr	France	mx	Mexico	es	Spain
au	Australia	ge	Georgia	md	Moldova	se	Sweden
at	Austria	de	Germany	mm	Myanmar (Burma)	ch	Switzerland
az	Azerbaijan	gr	Greece	nl	Netherlands	sy	Syria
be	Belgium	hk	Hong	nz	New	tj	Tadjikistan
bw	Botswana		Kong		Zealand	tw	Taiwan
br	Brazil	hu	Hungary	ne	Niger	th	Thailand
bg	Bulgaria	is	Iceland	no	Norway	tr	Turkey
ca	Canada	in	India	pk	Pakistan	tm	Turkmenistan
cl	Chile	id	Indonesia	pl	Poland	ua	Ukraine
cn	China	ie	Ireland	pt	Portugal	gb or	United
co	Colombia	il	Israel			uk	Kingdom
hr	Croatia	it	Italy	ro	Romania	us	USA
cu	Cuba	jp	Japan	ru	Russia	uy	Uruguay
cz	Czech Rep.	kz	Kazakhstan	xx	Serbia	uz	Uzbekistan
dk	Denmark	kr	Korea	sg	Singapore	ve	Venezuela
eg	Egypt	lv	Latvia	sk	Slovakia	vn	Vietnam

Regional codes

asia	europe	n_america	s_america

Appendix C

Query language summary

user <security code> <family name>   [mandatory]
This sequence is the first entry by a user. The registered security code and user family name form a unique combination.

access <access level>   [*general, journal, national, regional*]
The access code specifies the current search and hit permissions. The codes a user may enter depend on the access class assigned to the user. The default code is 'general'.

limit <upper hit limit>   [optional]
This sets the maximum number of successful matches (i.e. 'hits') for the current search. This number may not exceed the assigned limits associated with the access class, or the accumulated number of hits for the current calendar month.

location <location code>   [optional]
This command restricts searches according to geographical location. The country or regional codes are listed in Appendix B. The default is 'world'.

message   [optional]
This command starts a text message to be inserted into the user information. If any information for this user is accessed this message will be listed in the search output. The message must be closed by message_end.

if <conditional search string>   [optional]
This command is used to restrict the information output according to specific matching of text entries. The construction of the conditional search string is

          <data item name> <operator> <value>

The permitted search operators are

==	'exact match'
~=	'containing string'
!=	'NOT exact match'
!~	'NOT containing string'

Sequences of the conditional search string may be concatenated with the Boolean operators:

&&	'and'
||	'or'

The wild card character '*' may be used in data names only; not in the search values. The default is 'all entries'.

want <data items to be returned>   [optional]
This specifies which data items are to be returned. The default is all items.

ignore <data items NOT to be returned>   [optional]
This specifies those data items which should not be returned. The default is null. The 'want' and 'ignore' commands should not be used in the same search.

search   [optional]
This command starts a search which is based on the preceding query parameters. Multiple searches may be performed in a single session. Unless search conditions are re-specified, those used in preceding searches will apply.

References

Epelboin, Y. (1992a). *IUCr Newsletter* Vol. 1, No. 2, 12-13.
Epelboin, Y. (1992b). *Acta Cryst.* A48, 949-954.
Hall, S.R. (1991). *J. Chem. Inf. Comput. Sci.* 31, 326-333.
Spadaccini, N. & Hall, S.R. (1994). *J. Chem. Inf. Comput. Sci.* 34, 509–516.

ALBANIA

Sub-Editor: I. Mandia

Andoni, Dr Eduard (1949). Lecturer. Inorganic Chemistry Chair, Faculty of Natural Sciences, University of Tirane, Tirane, Albania. DSc (U. Tirane, 1987) amorphous state - powder method. *Structure_of_organometallic_compounds chemical_education_in_high_school*
Tel. 355(42)27669 Fax 355(42)27669 Telex 2211 UNISIT AB

Buri, Dr Stavri (1932). Lecturer, crystallography. Chair of Mineralogy, Petrography and Geochemistry, Polytechnic University, Tirane, Albania. DSc (U. Tirane, 1972) nickel-silicate lateritic clay mineralogy and crystallography. *Geometric_crystallography clay_minerals*
Tel. 355(42)22592 Fax 355(42)22592

Çina, Dr Aleksander (1934). President, Albanian Geologists Association. Geological Research and Development Institute, Blloku "Vasil Shanto", Tirane, Albania. DSc (U. Tirane, 1978) association of hydrothermal cu minerals. *Mineralogy_and_crystallography_of_platinum_ores education_of_mineralogists_in_high_school*
Tel. 355(42)26597 Fax 355(42)22592

Guda, Dr Bardhyl (1955). Lecturer. Crystal Physics Chair, Faculty of Natural Sciences, University of Tirane, Tirane, Albania. DSc (U. Tirane, 1988) amorphous state - powder method. *Structure_of_ceramics atomic_physics*
Tel. 355(42)27669 Fax 355(42)27669 Telex 2211 UNISIT AB

Klosi, Dr Fatos (1946). Head, Crystal Physics Chair. Crystal Physics Chair, Faculty of Natural Sciences, University of Tirane, Tirane, Albania. DSc (U. Tirane, 1983) silicium refinement. *Crystallography_of_thin_layers properties_of_microdispersed_ores*
Tel. 355(42)27669 Fax 355(42)27669 Telex 2211 UNISIT AB

Mandia, Ilir (1950). Crystallographer. Sub-Editor, World Directory of Crystallographers 9. Central Geology Laboratory, Blloku "Vasil Shanto", Tirane, Albania. U. Diploma (U. Tirane, 1975) colour metrology. *Minerals X-ray_diffractometry Debye_Scherrer_method X-ray_fluorescence_spectrometry*
Tel. 355(42)26597 Fax 355(42)42704

Sinojmeri, Dr Agim (1957). Lecturer, crystallography and mineralogy. Chair of Mineralogy, Faculty of Geology and Mining, Polytechnic University, Tirane, Albania. PhD (U. Orleans, France, 1990) sulphide mineralogy of volcanic mineralization. *Crystallography_of_sulphides_and_oxides_mineral mineralogy_of_zeolites*
Tel. 355(42)22592 Fax 355(42)22592

Tashko, Dr Artan (1940). Head, Chair of Mineralogy, Petrography and Geochemistry. Chair of Mineralogy, Petrography and Geochemistry, Polytechnic University, Tirane, Albania. DSc (U. Tirane, 1986) petrology of ultramafic rocks. *Crystal_chemistry_of_minerals crystallographic_education_of_geochemists*
Tel. 355(42)22592 Fax 355(42)22592

ARGENTINA

Sub-Editor: G. Punte

Notes

1. Degrees conferred by Argentine Universities are Doctor (Dr) (equivalent to PhD), Licenciado (Lic) (equivalent to MSc), Ingeniero (Ing) (between PhD and MSc).

2. Abbreviations: UNBA - Universidad Nacional de Buenos Aires
   UNC - Universidad Nacional de Córdoba
   UNCu - Universidad Nacional de Cuyo
   UNLP - Universidad Nacional de La Plata
   UNL - Universidad Nacional del Litoral
   UNR - Universidad Nacional de Rosario
   UNS - Universidad Nacional del Sur
   UNSL - Universidad Nacional de San Luis
   CAB - Centro atómico Bariloche
   CICPBA - Comisión de Investigaciones Científicas de la Provincia de Buenos Aires
   CIG - Centro de Investigaciones Geológicas
   CITEFA - Centro de Investigaciones Téctinas de las Fuerzas Armadas
   CNEA - Comisión Nacional de Energía Atómica
   CONICET - Consejo Nacional de Investigaciones Científicas y Técnicas
   LEMIT - Laboratorio de Ensayo de Materiales de Interés Tecnológico
   PRINSO - Program of Research in Solid State Physics
   FI - Facultad de Ingeniería

Alconchel, Dr Silvia Alejandra (1963). Research Assistant. [Dept. de Físicoquímica, Facultad de Ingeniería Química, UNL, Santiago del Estero 2829, 3000 Santa Fe, Argentina.] Dr (UNL, 1993) material science. *Crystal_structure mixed_oxides*
E-mail nfisico@arcride.edu.ar Tel. 54(42)36861 Fax 54(42)553727

Alvarez, Dr Alberto Guillermo Alvarez (1931). Group leader; Principal Researcher. [Lab. de Cristalografía y Microscopía Electrónica, CINDECA, Calle 47 No 257 - CC 59, 1900 La Plata, Argentina.] Dr (UN Cuyo, 1965) physics. *Electron_microscopy image_processing X-ray_diffraction*
Tel. 54(21)210711 Fax 54(21)254277

Andrade Gamboa, Mr Julio (1958). Fellowship holder. [Centro Atómico Bariloche, Av. Bustillo, Km 9,500 (8400) S. C. de Bariloche, Río Negro, Argentina.] Lic. (UNLP, 1985) physical chemistry. *X-ray_powder_diffraction quantitative_analysis refractory_compounds*
E-mail fcoqca@cab.edu.ar Tel. 54(944)61009 Fax 54(944)61006 Telex 80723cabar

Augsburger, Ms Marta Susana (1954). Researcher. [Química Inorgánica, Area de Qca Gral e Inorgánica, UNSL, Chacabuco y Pedernera, 5700 San Luis, Argentina.] Lic. (UNSL, 1978) chemistry. *Crystal_structure silicates arsenic_compounds phosphates*
E-mail estrada@imasl.edu.ar Tel. 54(652)23789x07 Fax 54(652)30224

Baggio, Dr Ricardo Fortunato (1946). Group Leader. [Comisión Nacional de Energía Atómica, Dpto. de Física - Div. Física de Sólidos, Av. del Libertador 8250, 1429 Buenos Aires, Argentina.] Dr (UNBA, 1975) physics. *Inorganic_structure organic_structure*
E-mail baggio@cnea.edu.ar Tel. 54(1)7520381 Fax 54(1)7558710 Telex 23458CNEASCAR

Baggio, Dr Sergio (1940). Professor of Chemistry. Departamento de Química, Universidad Nacional de la Patagonia, PO Box 164, 9120 Puerto Madryn, Chubut, Argentina. Dr (UNBA, 1970) chemistry. *Crystal_structure_determination powder_diffraction*
E-mail UNBAGGIO@CENPAT.EDU.AR Tel. 54(965)51024 Fax 54(965)51375 Telex 54(965)51543

Bengochea, Dr Amado Leandro (1945). Associate Professor. [Departamento de Geologia, Universidad Nacional del Sur, Av. Alem 1253, 8000 Bahía Blanca, Buenos Aires, Argentina.] Dr (Universidad Nacional del Sur, 1976) ore deposits. *Geology powder_diffraction mineralogy computing*
Tel. 54(91)25196x254 Fax 54(91)551447

de Benyacar, Prof. María Angélica Rodriguez (1928). Group leader. [Comisión Nacional de Energía Atómica, Div. Física de Sólidos, Dto. de Física de Sólidos, Av. del Libertador 8250, 1429 Buenos Aires, Argentina.] Lic. (UNBA, 1952) chemistry. *Irradiated_crystal crystalline_structure*
E-mail solidos@cnea.edu.ar Tel. 54(1)7520381 Fax 54(1)7558710 Telex 23458CNEASCAR

Boneto, Ms Rita (1951). Researcher. [Lab. de Cristalografía y Microscopía Electrónica, CINDECA, Calle 47 No. 257, CC 59, 1900 La Plata, Argentina.] Lic. (UNC, 1976) physics. *Electron_microscopy image_processing*
E-mail cindeca1@cespivm2.unlp.edu.ar Tel. 54(21)210711 Fax 54(21)254533

Cabanillas, Mr Edgardo Domingo (1947). Researcher. [CONICET Dpto. de Ciéncias de Materiales, CAC CNEA, Av. del Libertador 8250, 1429 Buenos Aires, Argentina.] Lic. (UNLP, 1978) physics. *Materials electron_microscopy*
E-mail cmag@atgcnea2.bitnet Tel. 54(1)7542644 Telex 26110CACAR

Caneiro, Dr Alberto (1950). Group leader. [Centro Atómico Bariloche, Av. Bustillo, Km 9,500 (8400) S. C. de Bariloche, Río Negro, Argentina.] Dr (UNCu, 1983) physics. *X-ray_powder_diffraction Rietveld_method superconductors*
E-mail fcoqca@cab.edu.ar Tel. 54(944)61009 Fax 54(944)61006 Telex 80723cabar

Canepa, Dr Horacio Ricardo (1944). Researcher. [PRINSO, CITEFA-CONICET, Zufri-ategui No 4380, 1603 Villa Martelli, Buenos Aires, Argentina.] Dr (U. Rennes I, France, 1983) solid state physics. *Semiconductors IV-VI_compounds crystal_growth detector*
E-mail rtprinso@arcriba.edu.ar Tel. 54(1)7610031x158 Fax 54(1)7603210

Carbonio, Dr Raul Ernesto (1957). Associate Professor. [INFIQC Depto. de Fisico Química, Facultad de Ciencias Químicas UNC, CC 61, 5016 Cordoba, Argentina.] Dr (UNC, 1982) physical chemistry. *Inorganic_solids powder_diffraction superconductors catalysts*
E-mail carbonio@uncbde.edu.ar Tel. 54(51)606373 Fax 54(51)694724

Casanova, Mr Jorge Ramón (1944). Researcher. [PRINSO, CITEFA-CONICET, Zufri-ategui No 4380, 1603 Villa Martelli, Buenos Aires, Argentina.] Lic. (UNBA, 1979) physics. *X-ray_diffraction intercalation_compound X-ray_topography crystal_growth*
E-mail rtprinso@arcriba.edu.ar Tel. 54(1)7610031x158 Fax 54(1)7603210

Conconi, María Susana (1962). Technical assistant. Centro de Tecnología de Recursos Minerales y Cerámica, Cno. Centenario y 506, M.B.Gonnet, Pcia. de Buenos Aires, Argentina. [CICPBA, CETMIC, CC 49, (1897) M.B.Gonnet, Buenos Aires, Argentina.] Lic. (U. Nacional de La Plata (UNLP), 1986) chemistry. *Clay_minerals ceramics structure_refinement Rietveld_method*
E-mail Postmaster@Cetmic.Edu.Ar Tel. 54(21)840247 Fax 54(21)710075

Cortelezzi, Dr Cesar Rafael (1926). Group leader. [LEMIT-CICPBA, Calle 52 entre 121 y 122, 1900 La Plata, Argentina.] Dr (UNLP, 1952) geology. *Mineralogy petrography*
Tel. 54(21)31141 Fax 54(21)250471

Diodati, Dr Francisco Piero (1940). Senior Researcher. [Departamento de Física, FI UNBA, Paseo Colón 850, 1063 - Buenos Aires, Argentina.] Dr (UNLP, 1970) physics. *Electron_diffraction gases solid_laser*
Tel. 54(1)346441

Dristas, Dr Jorge Anastasio (1944). Professor. [Departamento de Geología, UNS, Av. Alem 1253, 8000 Bahía Blanca, Buenos Aires, Argentina.] Dr (UNS, 1972) economic geology-mineralogy. *Powder_diffraction electron_diffraction*
Tel. 54(91)25196x254 Fax 54(91)551447

Echeverría, Mr Gustavo Alberto (1965). Research fellow. [Dept de Física - UNLP, CC 67, 1900 La Plata, Buenos Aires, Argentina.] Lic. (UNLP, 1991) physics. *Organic_crystalline_structure inorganic_crystalline_structure X-ray_diffraction artificial_heterolayers*
E-mail rivero@ayelen.unlp.edu.ar Tel. 54(21)39061 Fax 54(21)252006

Ellena, Mr Javier Alcides (1968). Research fellow. [Dept. de Física, Facultad de Ciencias Exactas, UNLP, CC 67, 1900 La Plata, Argentina.] Lic. (UNLP, 1993) physics. *Crystal_structure non-linear_optics conjugated_organic_compounds*
E-mail rivero@ayelen.unlp.edu.ar Tel. 54(21)40640 Fax 54(21)252006 Telex 311151BU-LAPAR

Fernández, Ing. Juan Carlos (1951). Asst. Professor. [Departamento de Física, FI UNBA, Paseo Colón 850, 1063 - Buenos Aires, Argentina.] Eng (UNLP, 1974) electrical engineering. *LEED surface_crystallography Auger_spectroscopy epitaxy surface_science adsorbates*
Tel. 54(1)346441

Fernández de Rapp, Ms María Emilia (1948). Researcher. [PRINSO, CITEFA-CONICET, Zufriategui No 4380, 1603 Villa Martelli, Buenos Aires, Argentina.] Eng (UNS, 1972) engineering. *X-ray_diffraction*
E-mail rtprinso@arcriba.edu.ar Tel. 54(1)7610031x158 Fax 54(1)7603210

Gay, Hebe Dina (1922). Group leader. [Museo de Mineralogía y Geología, Facultad de Ciencias Exactas Físicas y Nat., Universidad Nacional de Córdoba, Av. Vélez Sarsfield, (5000) Córdoba, Argentina.] Dr (UNC, 1952) mineralogy. *Minerals_structure*
Tel. 54(51)222284 Fax 54(51)244092

Gilabert, Mr Ulises (1963). Research assistant. [PRINSO, CITEFA-CONICET, Zufri-ategui No 4380, 1603 Villa Martelli, Buenos Aires, Argentina.] Lic. (UNBA, 1988) chemistry. *Semiconductors*
E-mail rtprinso@arcriba.edu.ar Tel. 54(1)7610031x158 Fax 54(1)7603210

Goeta, Mr Andrés Eduardo (1965). Research Fellow. [Dept. de Física, Facultad de Ciencias Exactas, UNLP, CC 67, 1900 - La Plata, Argentina.] Lic. (UNLP, 1990) physics. *Charge_density synchrotron_radiation powder_diffraction Rietveld_method superconductivity*
E-mail goeta@fisilp.edu.ar Tel. 54(21)40640 Fax 54(21)252006 Telex 31151BULA-PAR

Guerin, Dr Diego Marcelo (1955). Associate Professor. [Laboratorio de Biología Molecular, INIBIBB-CONICET, CRIBABB - Camino La Carrindanga, Km 7, 8000 Bahía Blanca, Buenos Aires, Argentina.] Dr (Universidad Nacional de La Plata, 1985) physics. *Proteins*
Tel. 54(91)25196x210 Fax 54(91)551447

Guimpel, Julio J. (1957). Researcher. Centro Atómico Bariloche, 8400 - Bariloche, Río Negro, Argentina. Dr (UNCu, 1987) physics. *Computing superlattice_structure thin_film*
E-mail jguimpel@arib51.bitnet Tel. 54(944)61014 Fax 54(944)61006

Heredia, Mr Eduardo Armando (1959). Research assistant. [PRINSO, CITEFA-CONICET, Zufriategui No 4380, 1603 Villa Martelli, Buenos Aires, Argentina.] Lic. (UNBA, 1989) physics. *Semiconductors crystal_growth detector*
E-mail rtprinso@arcriba.edu.ar Tel. 54(1)7610031x158 Fax 54(1)7603210

Hermida, Mr Jorge Daniel (1946). Research scientist. [Dept. Materiales, CNEA, Av. del Libertador 8250, 1429 Buenos Aires, Argentina.] Lic. (UNBA, 1971) physics. *X-ray_diffraction defect_structure*
Tel. 54(1)7550181x268 Fax 54(1)7542644

Iñiguez Rodríguez, Dr Adrian Mario (1937). Head CIG. [CONICET Centro de Investigaciones Geologicas, Calle 1 No. 644, 1900 La Plata, Argentina.] Dr (UNLP, 1967) geology. *Clays mineralogy zeolites*
Tel. 54(21)215677 Fax 54(21)258696

Ipohorski Lenkiewicz, Dr Miguel (1939). Research scientist. [Dept. Metalurgia, CNEA, Av. del Libertador 8250, 1429 Buenos Aires, Argentina.] Dr (UNCu, 1967) physics. *Electron_microscopy metals*
Tel. 54(1)7550181 Fax 54(1)7542644

König de Perazzo, Ms Patricia Veronica (1941). Research scientist. [Div. Física de Sólidos, Dto. de Física, CNEA, Av. del Libertador 8250, 1429 Buenos Aires, Argentina.] Lic. (UNBA, 1965) physics. *Inorganic_crystal_structure powder_diffraction Rietveld_method*
E-mail solidos@cnea.edu.ar Tel. 54(1)7520381 Fax 54(1)7558710 Telex 23458CNEASCAR

Lamas, Mr Diego Germán (1968). Research Assistant. [PRINSO, CITEFA-CONICET, Zufriategui No 4380, 1603 Villa Martelli Buenos Aires, Argentina.] Lic. (UNBA, 1992) physics. *X-ray_diffraction solid_electrolytes*
E-mail rtprinso@arcriba.edu.ar Tel. 54(1)7610031x158 Fax 54(1)7603210

Lombardo, Prof. Eduardo (1939). Professor; Chairman of the Department of Physical Chemistry. [Dept de Fisicoquímica, Facultad de Ingeniería Química, UNL, Santiago del Estero 2829, 3000 Santa Fé, Argentina.] Eng. (U. Nacional del Litoral UNL, 1964) chemical engineering. *Structure zeolites mixed_oxides heterogeneous_catalysis*
E-mail nfisico@arcride.edu.ar Tel. 54(42)36861 Fax 54(42)553727

Lovey, Dr Francisco Carlos (1966). Research Fellow. [Centro Atómico Bariloche, S. C. de Bariloche, 8400 Río Negro, Argentina.] MSc (UNCu, 1990) physics. *X-ray_powder_diffraction Rietveld_method superconductors*
E-mail fcoqca@cab.edu.ar Tel. 54(944)61009 Fax 54(944)61006 Telex 80723cabar

Maiza, Dr Pedro Jose (1943). Prof. [Departamento de Geología, UNS, Av. Alem 1253, 8000 Bahía Blanca, Buenos Aires, Argentina.] Dr (UNS, 1972) ore deposits. *Geochemistry ores_geology powder_diffraction mineralogy*
Tel. 54(91)25196x254 Fax 54(91)551447

Malachevsky, Maria Teresa (1961). Fellow. [Division Desarrollo, Centro Atómico Bariloche, CNEA, 8400 - S.C. de Bariloche, Río Negro, Argentina.] PhD (IB (Instituto Balseiro - Universidad Nacional de Cuyo), 1994) physics. *Ceramics X-ray_diffraction Rietveld_refinement*
E-mail malache@cab.cnea.edu.ar Tel. 54(944)61002/005 - int. 5282 Fax 54(944)61006 Telex 80723CABAR

Manghi, Estela Margarita (1940). Research scientist. [Div. Física de Sólidos, Dto. de Física, CNEA, Av. del Libertador 8250, 1429 Buenos Aires, Argentina.] Lic. (UNBA, 1964) physics. *Structural_defect crystal_growth*
E-mail solidos@cnea.edu.ar Tel. 54(1)7520381 Fax 54(1)7558710 Telex 23458CNEASCAR

Marbec, Miss Emma Rosa (1939). Researcher. [Dept. Química, INTI, CC 157 San Martin, 1650 Buenos Aires, Argentina.] Lic. (UNBA, 1966) chemistry. *X-ray_fluorescence metals X-ray_diffraction iron_oxides uranyl_compounds*
Tel. 54(1)7556161x76

Mas, Dra Graciela Raquel (1948). Associate Professor. [Departamento de Geología, UNS, Av. Alem 1253, 8000 Bahía Blanca, Buenos Aires, Argentina.] Dr (UNS, 1976) mineralogy. *Zeolites clays mineralogy powder_diffraction computing fluid_inclusions*
Tel. 54(91)25196x254 Fax 54(91)551447

Narda, Dra Griselda Edith (1960). Researcher. [Química Inorgánica, Area de Qca. Gral. e Inorgánica, UNSL, Chacabuco y Pedernera, 5700 San Luis, Argentina.] Dr (UNSL, 1990) chemistry. *Crystalline_structure oxometallates*
E-mail estrada@imasl.edu.ar Tel. 54(652)23789x07 Fax 54(652)30224

Parisi, Francisco Eduardo Alberto (1963). Research assistant. [Div. Física de Sólidos, Dto. de Física, Comisión Nacional de Energía Atómica, Av. del Libertador 8250, 1429 Buenos Aires, Argentina.] Lic. (UNBA, 1988) physics. *Modulated_structures*
E-mail solidos@cnea.edu.ar Tel. 54(1)7520381 Fax 54(1)7558710 Telex 23458CNEASCAR

Pedregosa, Dr Jose Carmelo (1945). Group leader. [Química Inorgánica, Area de Qca Gral e Inorgánica, UNSL, Chacabuco y Pedernera, 5700 San Luis, Argentina.] Dr (UNSL, 1975) chemistry. *Crystalline_structure metallic_complex*
E-mail estrada@imasl.edu.ar Tel. 54(652)23789x07 Fax 54(652)30224

Piro, Dr Oscar Enrique (1944). Professor. Dept. Física, Facultad de Ciencias Exactas, UNLP, CC 67, 1900 La Plata, Argentina. Dr (UNLP, 1977) physics. *Crystalline_structure optical_property magnetism electronic_magnetic_properties_of_coordination_solids*
E-mail oscar@fisilp.edu.ar Tel. 54(21)39061 Fax 54(21)252006 Telex 31151BULA-PAR

Pochettino, Dr Alberto Antonio (1947). Senior scientist. [Dept. Metalurgia, CNEA, Av. del Libertador 8250, 1429 Buenos Aires, Argentina.] Dr (UNLP, 1978) physics. *Electron_microscopy metals*
Tel. 54(1)7550181x282 Fax 54(1)7542644

Prado, Fernando (1966). Research Fellow. [Centro Atómico Bariloche., Av. Bustillo, Km 9,500 (8400) S. C. de Bariloche, Río Negro, Argentina.] MSc (UNCu, 1990) physics. *X-ray_powder_diffraction Rietveld_method superconductors*
E-mail fcoqca@cab.edu.ar Tel. 54(944)61009 Fax 54(944)61006 Telex 80723cabar

Punte, Prof. Graciela (1944). Professor. Argentina Sub-Editor of ninth edition of WDC. [Dept. Física, Facultad de Ciencias Exactas, UNLP, CC 67, 1900 La Plata, Argentina.] Dr (UNLP, 1972) physics. *Conformation QSAR perovskites Rietveld_method charge_density non-linear_optics*
E-mail punte@fisilp.edu.ar Tel. 54(21)39061 Fax 54(21)252006

Rigotti, Dr Graciela Ester (1948). Researcher. [Dept. Física, Facultad de Ciencias Exactas, UNLP, CC 67, 1900 La Plata, Argentina.] Dr (UNLP, 1979) chemistry. *Organic_structure inorganic_structure solid_state_physical_chemistry*
E-mail rigotti@fisilp.edu.ar Tel. 54(21)39061 Fax 54(21)252006

Rivero, Prof. Blas Eduardo (1942). Group leader. [Dept. Física, Facultad de Ciencias Exactas, UNLP, CC 67, 1900 La Plata, Argentina.] Dr (UNLP, 1976) physics. *Organic_structure inorganic_structure crystallography* computing
E-mail rivero@ayelen.edu.ar Tel. 54(21)39061 Fax 54(21)252006

Trigubo, Ms Alicia Beatriz (1944). Researcher. [PRINSO, CITEFA-CONICET, Zufriategui No 4380, 1603 Villa Martelli, Buenos Aires, Argentina.] MSc (USC, 1981) material science. *Crystal_growth X-ray_diffraction semiconductors electron_microscopy defect_structure*
E-mail rtprinso@arcriba.edu.ar Tel. 54(1)7610031x158 Fax 54(1)7603210

Vega, Mr Daniel Roberto (1961). Research Assistant. [Div. Física de Sólidos, Dto. de Física, CNEA, Av. del Libertador 8250, (1429) Buenos Aires, Argentina.] Lic. (UNBA, 1987) physics. *Crystal_structures phase_transition electron_density*
E-mail vega@cnea.edu.ar Tel. 54(1)7520381 Fax 54(1)7558710

Versaci, Dr Raúl Antonio (1945). Research Scientist. [Dept. Materiales, CNEA, Av. del Libertador 8250, 1429 Buenos Aires, Argentina.] Dr (UNLP, 1979) physics. *Metals alloys crystalline_structure*
Tel. 54(1)7550181x268 Fax 54(1)7542644

Vidal, Miss Haydeé Marta (1938). Researcher. [Dept. Química, INTI, CC 157, San Martin, 1650 Buenos Aires, Argentina.] Lic. (UNBA, 1970) chemistry. *Powder_diffraction fibrous_polymers*
Tel. 54(1)7556161x76

Viña, Mr Raúl Oscar (1947). General Supervisor. [Depto. Catálisis YPF S.A. and UNLP, Dept. Física, Lab. de Cristalografía, CC 67, 1900 La Plata, Argentina.] BSc (UNLP, 1984) chemistry. *Powder_diffraction electron_microscopy, surface_structure*
E-mail rivero@ayelen.edu.ar Tel. 54(21)39061 Fax 54(21)252006

Walsöe de Reca, Dr Noemí Elisabeth (1939). Head PRINSO; post-degree Professor. [PRINSO, CITEFA-CONICET, Zufristegui No 4380, 1603 Villa Martelli, Buenos Aires, Argentina.] Dr (UNBA, 1965) chemistry. *Ionic_materials intercalation_materials semiconductors sensors infrared_detectors*
E-mail postmaster@cinso.edu.ar Tel. 54(1)7610031 54(1)7610191 54(1)7610081x158 Fax 54(1)7603210 54(1)7606365 Telex 26057 AR

Zalba, Dr Patricia Eugenia (1944). Senior Researcher; Head of the Geology and Mineralogy Department (CETMIC). Cno. Centenario y 506, CC 49 (1897) M. B. Gonnet, Buenos Aires, Argentina. [CICPBA, Centro de Tecnología de Recursos Minerales y Cerámica, CC 49, 1897 M. B. Gonnet, Buenos Aires, Argentina.] Dr (UNLP, 1978) geology. *Clays mineralogy electron_microscopy zeolites*
E-mail Postmaster@cetmic.edu.ar Tel. 54(21)840247 Fax 54(21)710075

Zimmerman, Miss Rosa (1928). Senior Researcher. [Departamento de Física, FI, UNBA, Paseo Colón 850, 1063 - Buenos Aires, Argentina.] Lic. (UNBA, 1953) physics. *Thin_film crystalline_structure*
Tel. 54(1)346441

ARMENIA

Sub-Editor: K. G. Truni

Abovian, Dr Eduard S. (1945). Associate professor. [Yerevan State University, Faculty of Physics, Department of Solid State Physics, Alek Manoukian-1, Yerevan 375049, Armenia.] PhD (Institute of Applied Physics, Academy of Sciences of Moldova, Kishinev, 1979) crystallography and crystallophysics. *Structural_analysis_of_molecular_crystals metallic_compounds synchrotron_radiation*
Tel. 7(8852)553187

Adamian, Dr Victor E. (1936). Head Researcher. [Yerevan State University, Alek Manoukian-1, Yerevan 375049, Armenia.] DSc (Institute of Applied Physics Problems, Armenian National Academy Sciences, 1992) solid state physics. *Electric_magnetic_properties*
Tel. 7(8852)553187 Fax 7(8852)553187

Aivazian, Dr Ashot P. (1954). Senior Researcher. [Yerevan State University, Faculty of Physics, Department of Solid State Physics, Alek Manoukian-1, Yerevan 375049, Armenia.] PhD (Institute for Physical Research, Armenian National Academy, 1987) crystallography and crystallophysics. *X-ray_diffraction real_crystals*
Tel. 7(8852)553187

Alexanian, Dr Petros L. (1953). Researcher. [Yerevan State University, Faculty of Physics, Department of Solid State Physics, Alek Manoukian-1, Yerevan 375049, Armenia.] MS solid state physics. *Electron_diffraction_crystal*
Tel. 7(8852)553187

Balian, Dr Minas K. (1958). Researcher. [Yerevan State University, Faculty of Physics, Department of Solid State Physics, Alek Manoukian-1, Yerevan 375049, Armenia.] MS solid state physics. *X-ray_dynamical_diffraction optics electron_diffraction*
Tel. 7(8852)553187

Bezirganian, Dr Siranoush E. (1954). Senior Researcher. [Yerevan State University, Faculty of Physics, Department of Solid State Physics, Alek Manoukian-1, Yerevan 375049, Armenia.] PhD (Institute of Applied Physics Problems, Armenian National Academy, Yerevan, 1987) solid state physics. *X-ray_optics electron_optics ultracold_neutron*
Tel. 7(8852)553187

Bezirganyan, Dr Hakob P. (1954). Leading Researcher. [Yerevan State University, Faculty of Physics, Department of Solid State Physics, Alek Manoukian-1, Yerevan 375049, Armenia.] PhD (Yerevan State University, 1990) solid state physics. *X-ray_optics electrooptics ultracold_neutron_optics*
Tel. 7(8852)553187

Bezirganyan, Prof. Petros H. (1916). Professor. [Yerevan State University, Faculty of Physics, Department of Solid State Physics, Alek Manoukian-1, Yerevan 375049, Armenia.] DSc (Yerevan State University, 1969) solid state physics. *X-ray_optics*
Tel. 7(8852)553187

Egikian, Dr David (1951). Head of Group. [Yerevan Physics Institute, Alikhanian Brothers St., 375036, Yerevan, Armenia.] PhD (Institute for Physical Research, Armenian National Academy of Sciences, 1986) solid state physics. *Synchrotron_radiation crystal_perfection synchrotron_radiation_application*
E-mail ERPI@ADONIS.IASNET.COM Tel. 7(8852)351192 Fax 7(8852)350030

Elramjian, Dr Ferdinand H. (1937). Leading Researcher. [Yerevan State University, Faculty of Physics, Department of Solid State Physics, Alek Manoukian-1, Yerevan 375049, Armenia.] PhD (Yerevan State University, 1975) solid state physics. *X-ray_topography microdefects*
Tel. 7(8852)553187

Elramjian, Dr Tigran H. (1946). Researcher. [Yerevan State University, Faculty of Physics, Department of Solid State Physics, Alek Manoukian-1, Yerevan 375049, Armenia.] MS solid state physics. *X-ray_optics resonator monochromator collimator*
Tel. 7(8852)553187

Gabrielian, Dr Karen T. (1951). Leading Researcher. [Yerevan State University, Faculty of Physics, Department of Solid State Physics, Alek Manoukian-1, Yerevan 375049, Armenia.] PhD (Institute of Radiophysics and Electronics, Armenian National Academy, 1981) solid state physics. *X-ray_dynamical_diffraction X-ray_optics electron_diffraction synchrotron_radiation superlattices heterostructures*
E-mail armtsb@adonis.iasnet.com Tel. 7(8852)553246 Fax 7(8852)567511

Gasparian, Dr Laura G. (1938). Senior Researcher. [Yerevan State University, Alek Manoukian-1, Yerevan 375049, Armenia.] PhD (Institute of Radiophysics and Electronics, Armenian National Academy Sciences, 1978) solid state physics. *X-ray_diffraction interferometry*
Tel. 7(8852)553187 Fax 7(8852)553187

Grigorian, Dr Arshak H. (1947). Leader lector. [Yerevan State University, Alek Manoukian-1, Yerevan 375049, Armenia.] MS solid state physics. *X-ray_optics synchrotron_diffraction*
Tel. 7(8852)553187 Fax 7(8852)553187

Haroutyunian, Dr Levon A. (1958). Senior Researcher. [Yerevan State University, Faculty of Physics, Department of Solid State Physics, Alek Manoukian-1, Yerevan 375049, Armenia.] PhD (Institute for Physical Research, Armenian National Academy of Sciences, 1987) crystallography and crystallophysics. *X-ray_interferometry dynamical_diffraction optics topography*
Tel. 7(8852)553187

Haroutyunian, Dr Valery S. (1955). Senior Researcher. [Yerevan State University, Faculty of Physics, Department of Solid State Physics, Alek Manoukian-1, Yerevan 375049, Armenia.] PhD (Yerevan State University, 1988) solid state physics. *X-ray_kinematical_dynamical_diffraction*
Tel. 7(8852)553187

Karakhanian, Dr Robert K. (1945). Associate Professor. [Yerevan State University, Faculty of Physics, Department of Solid State Physics, Alek Manoukian-1, Yerevan 375049, Armenia.] PhD (Yerevan State University, 1975) solid state physics. *Electron_diffraction*
Tel. 7(8852)553187

Karapetian, Dr Harutyun A. (1954). Senior Researcher, Head of Research Group. [Institute of Fine Organic Chemistry, Armenian National Academy of Sciences, Azatutian Ave. 26, Yerevan 375022, Armenia.] PhD (Institute of Applied Physics, Academy of Sciences of Moldova, 1984) crystallography and crystallophysics. *Structural_activity small_molecules*

Tel. 7(8852)281764 Fax 7(8852)288332

Karapetian, Dr Sveta V. (1943). Senior Researcher. [Yerevan State University, Faculty of Physics, Department of Solid State Physics, Alek Manoukian-1, Yerevan 375049, Armenia.] PhD (Yerevan State University, 1989) solid state physics. *X-ray_diffraction dislocations*

Tel. 7(8852)553187

Kazarian, Dr Levon (1946). Senior Researcher, Head of research group. [Institute for Physical Research, Armenian National Academy of Sciences, Ashtarak-2, 378410, Armenia.] PhD (Institute for Physical Research, Armenian National Academy of Sciences, 1981) crystallography and crystallophysics. *Crystal_growth*

E-mail karas@adonis.iasnet.com Tel. 7(8852)288150 Fax 7(8852)523640

Kokanian, Dr Edward P. (1954). Senior Researcher, Head of Research Group. [Institute for Physical Research, Armenian National Academy of Sciences, Ashtarak-2, 378410, Armenia.] PhD (Institute for Physical Research, Armenian National Academy of Sciences, 1987) crystallography and crystallophysics. *Crystal_growth superconductor spectroscopic_properties oxygen_crystal*

E-mail karas@adonis.iasnet.com Tel. 7(8852)21047 Fax 7(8852)523640

Levonian, Dr Levon V. (1950). Leading Researcher. [Yerevan State University, Faculty of Physics, Department of Solid State Physics, Alek Manoukian-1, Yerevan 375049, Armenia.] PhD (Institute of Applied Physics Problems, Armenian National Academy, Yerevan, 1984) solid state physics. *X-ray_dynamical_diffraction optics*

Tel. 7(8852)553187

Manoukian, Dr Hasmik M. (1951). Senior researcher. [Yerevan State University, Faculty of Physics, Department of Solid State Physics, Alek Manoukian-1, Yerevan 375049, Armenia.] PhD (Institute of Applied Physics Problems, Armenian National Academy, 1985) solid state physics. *X-ray_dynamical_diffraction superlattices*

Tel. 7(8852)553187

Martirosian, Dr Aida H. (1941). Associate Professor. [Yerevan State University, Alek Manoukian-1, Yerevan 375049, Armenia.] PhD (Yerevan State University, 1975) solid state physics. *Organic_crystal_X-ray*

Tel. 7(8852)553187 Fax 7(8852)553187

Melkonian, Dr Haroutyun S. (1964). Researcher. [Yerevan State University, Faculty of Physics, Department of Solid State Physics, Alek Manoukian-1, Yerevan 375049, Armenia.] PhD (Yerevan State University, 1992) solid state physics. *X-ray_diffraction dislocation*

Tel. 7(8852)553187

Mesropian, Dr Mesrop H. (1952). Leading Researcher. [Yerevan State University, Faculty of Physics, Department of Solid State Physics, Alek Manoukian-1, Yerevan 375049, Armenia.] PhD (Yerevan State University, 1992) solid state physics. *Dynamical_diffraction monochromators collimators*

Tel. 7(8852)553187

Oganesian, Dr Levone A. (1947). Senior Researcher. [Institute for Physical Research, Armenian National Academy of Sciences, Ashtarak-2, 378410, Armenia.] PhD (Institute for Physical Research, Armenian National Academy of Sciences, 1981) crystallography and crystallophysics. *Melt-growth oxides crystal_chemistry defects physical_properties*

E-mail karas@adonis.iasnet.com Tel. 7(8852)288150 Fax 7(8852)523640

Ovanesian, Dr Karine L. (1947). Senior Researcher. [Laboratory for Crystal Growth of High Melting Oxide-Based Laser Materials, Institute for Physical Research, Armenian National Academy of Sciences, Ashtarak-2, 378410, Armenia.] PhD (Institute for Physical Research, Armenian National Academy of Sciences, 1988) crystallography and crystallophysics. *Oxides growth crystal_chemistry defects physical_properties*

E-mail karas@adonis.iasnet.com Tel. 7(8852)288150 Fax 7(8852)523640

Petrosian, Dr Ashot G. (1943). Head of Laboratory. [Laboratory for Crystal Growth of High-Melting Oxide-based Laser Materials, Institute for Physical Research, Armenian National Academy of Sciences, Ashtarak-2, 378410, Armenia.] DSc (Institute for Physical Research, Armenian National Academy of Sciences, 1992) laser physics. *Crystal_growth oxides defects physical_properties*

E-mail karas@adonis.iasnet.com Tel. 7(8852)288150 Fax 7(8852)523640

Rostomian, Dr Armand (1936). Leading Researcher. [Yerevan State University, Faculty of Physics, Department of Solid State Physics, Alek Manoukian-1, Yerevan 375049, Armenia.] PhD (Institute for Physical Research, Armenian National Academy, 1980) optics. *X-ray_optics resonators monochromators giroscopy*

Tel. 7(8852)553187

Rostomian, Dr Armen M. (1958). Leading Researcher. [Yerevan State University, Faculty of Physics, Department of Solid State Physics, Alek Manoukian-1, Yerevan 375049, Armenia.] PhD (Yerevan State University, 1992) solid state physics. *X-ray_optics resonator giroscopy*

Tel. 7(8852)553187

Rusian, Dr Peter R. (1952). Senior Researcher. [Laboratory for High Temperature Superconductors, Institute for Physical Research, Armenian National Academy of Sciences, Ashtarak-2, 378410, Armenia.] PhD (Institute for Physical Research, Armenian National Academy of Sciences, 1985) crystallography and crystallophysics. *Crystal_growth superconductor*

E-mail karas@adonis.iasnet.com Tel. 7(8852)288150 Fax 7(8852)523640

Sacanian, Dr Martin S. (1945). Senior Researcher. [Yerevan State University, Faculty of Physics, Department of Solid State Physics, Alek Manoukian-1, Yerevan 375049, Armenia.] PhD (Yerevan State University, 1989) solid state physics. *X-ray_dislocation*

Tel. 7(8852)553187

Shirinyan, Dr Grikor (1943). Senior Researcher. [Institute for Physical Research, Armenian National Academy of Sciences, Ashtarak-2, 378410, Armenia.] PhD (Institute for Physical Research of Armenian National Academy, 1985) crystallography and crystallophysics. *X-ray_melt-growth_oxide_compounds_crystal_chemistry*

E-mail karas@adonis.iasnet.com Tel. 7(8852)288150 Fax 7(8852)523640

Terzyan, Dr Simon S. (1954). Senior Researcher. [Institute of Fine Organic Chemistry, Armenian National Academy of Sciences, Azatutian Ave. 26, Yerevan 375022, Armenia.] PhD (Institute of Crystallography of USSR Academy of Sciences, 1984) crystallography and crystallophysics. *X-ray_crystallography proteins superconductor*

E-mail karas@adonis.iasnet.com Tel. 7(8852)281764

Truni, Dr Karapet G. (1946). Professor. Sub-Editor, WDC9. [Yerevan State University, Faculty of Physics, Department of Solid State Physics, Alek Manoukian-1, Yerevan 375049, Armenia.] DSc (Institute of Applied Physics Problems, Armenian National Academy of Sciences, 1989) solid state physics. *X-ray_optics diffraction_theory*

Tel. 7(8852)553187

Truni, Dr Lusine K. (1970). Researcher. [Yerevan State University, Alek Manoukian-1, Yerevan 375049, Armenia.] MS solid state physics. *X-ray_dynamical_diffraction topography*

Tel. 7(8852)553187 Fax 7(8852)553187

Vardanian, Dr David M. (1950). Researcher. [Yerevan State University, Alek Manoukian-1, Yerevan 375049, Armenia.] PhD (Yerevan State University, Armenian National Academy Sciences, 1981) solid state physics. *X-ray_superlattices*

Tel. 7(8852)553187 Fax 7(8852)553187

Vermishian, Dr Garnik A. (1953). Researcher. [Yerevan State University, Faculty of Physics, Department of Solid State Physics, Alek Manoukian-1, Yerevan 375049, Armenia.] MS solid state physics. *Electron_microscopy*

Tel. 7(8852)553187

Yeritsian, Dr Grant (1937). Head of Laboratory. [Radiation Physics Laboratory, Yerevan Physics Institute, Alikhanian Bros. 2, 375036 Yerevan, Armenia.] PhD (Physics Institute of Ukrainian Academy of Sciences, 1968) solid state physics. *Solid_state radiation_physics synchrotron_radiation_application*

E-mail ERPI@ADONIS.IASNET.COM Tel. 7(8852)341065 7(8852)344324 Fax 7(8852)350030

AUSTRALIA

Sub-Editor: A. W. Stevenson

1. The abbreviation CSIRO is used for the Commonwealth Scientific and Industrial Research Organisation.

Amjad, Mrs Nuzhat I. Patail (1963). Lecturer. Department of Geology, University of Karachi, Karachi, Pakistan. [Department of Earth Sciences, James Cook University, Townsville 4811, Queensland, Australia.] MSc (Karachi U., 1988) mineralogy, sedimentology. *Sedimentation mineralogy clays SEM XRD*

E-mail Nuzhat.Amjad@jcu.edu.au Tel. 61(77)815138 Fax 61(77)251501

Anstis, Dr Geoffrey Richard (1949). Senior Lecturer. [Department of Applied Physics, University of Technology, Sydney, PO Box 123, Broadway, NSW 2007, Australia.] PhD (Adelaide, 1975) mathematical physics. *Electron_diffraction electron_microscopy computing*

E-mail granstis@phys.uts.edu.au Tel. 61(2)3302193 Fax 61(2)3302219

Avey, Hugh Philip (1934). Dean of Sciences. [University of Southern Queensland,

Toowoomba, Qld. 4350, Australia.] PhD (London University, England, 1968) crystallography. *Biological_macromolecules instrumentation computing crystallography astronomical_instrumentation*

E-mail avey@zeus.usq.edu.au Tel. 61(76)312254 Fax 61(76)312721

Baker, Anthony Thomas (1954). Senior Lecturer. [Department of Chemistry, UTS, PO Box 123, Broadway, NSW 2007, Australia.] PhD (UNSW, 1985) inorganic chemistry/crystallography. *Coordination*

E-mail t.baker@uts.edu.au Tel. 61(2)3301881 Fax 61(2)3301755

Bakshi, Eduard Nicolaevitch (1952). Senior Lecturer. [Dept. of Physics, Swinburne University of Technology, PO Box 218, Hawthorn, Vic. 3122, Australia.] PhD (Monash University, 1983) neutron diffraction. *Neutron_diffraction X-ray_diffraction time-*

correlated_diffraction magnetism magnetic_materials
Tel. 61(3)8198892 Fax 61(3)8190856

Balaic, David Xavier (1965). PhD student. [School of Physics, University of Melbourne, Parkville, VIC 3052, Australia.] BSc(Hons) (University of Melbourne, 1991) physics. *Diffraction_theory extinction crystallography computer_modelling Laue_diffraction optics microbeam_analysis synchrotron_radiation capillary_X-ray_optics*
E-mail dxb@tauon.ph.unimelb.edu.au Tel. 61(3)3445465 Fax 61(3)3474783 Telex AA35185

Barnea, Zwi (1932). Reader in Physics. [School of Physics, University of Melbourne, Parkville, VIC 3052, Australia.] PhD (University of Melbourne, 1974) physics. *Extinction crystallography anharmonicity instrumentation*
E-mail barnea@tauon.ph.unimelb.edu.au Tel. 61(3)3447074 Fax 61(3)3474783 Telex AA35185

Barry, John C. (1954). Senior Research Fellow. [Centre for Microscopy and Microanalysis, University of Queensland, Brisbane, Queensland 4072, Australia.] PhD physics. *High-resolution_electron_microscopy carbon_compounds superconductors tooth_compounds*
E-mail barry@uqimage1.emc.uq.oz.au Tel. 61(7)3654210 Fax 61(7)3654422

Bayliss, Peter (1936). Professor Emeritus. [Australian Museum, Mineralogy, PO Box A285, Sydney, NSW 2000, Australia.] PhD (Uni. NSW, 1967) mineralogy. *Mineralogy*
E-mail rossp@amsg.austmus.oz.au Tel. 61(43)434294 Fax 61(2)3604350

Beretka, Julius (1930). Principal Research Scientist. CSIRO Div. Building Constr. and Engineering, Graham Rd., Highett, Vic. 3190, Australia. [PO Box 56, Highett, Vic. 3190, Australia.] MSc (University of Adelaide, SA, Australia, 1962) physical chemistry. *X-ray_diffraction powder cements waste_materials*
Tel. 61(3)2526000 Fax 61(3)2526244 Telex AA33766

Boehm, Dr James M. (1950). Chief Physicist. [Cancer Care Centre, St George Hospital, Kogarah, NSW 2217, Australia.] PhD (University of Melbourne, 1978) physics. *X-ray_diffraction radiation_protection medical_physics*
Tel. 61(2)3503904 Fax 61(2)3503958

Bott, Raymond Clinton (1966). PhD student. [Griffith University, Faculty of Science and Technology, Nathan, Qld. 4111, Australia.] MAppSci (Queensland University of Technology, 1993) chemistry. *Structural_chemistry metals group_15_metals*
E-mail r.bott@sct.gu.edu.au Tel. 61(7)8757242 Fax 61(7)8757656

du Boulay, Douglas John (1967). Postgraduate Student (PhD). Department of Physics, The University of Western Australia, Nedlands 6009, Western Australia. [Crystallography Centre, The University of Western Australia, Nedlands 6009, Western Australia.] BSc(hons) (U. Western Australia, 1990) physics. *Accurate_intensity extinction high-precision_structure perovskites rare-earth_compounds residual_electron_density*
E-mail ddb@crystal.uwa.edu.au Tel. 61(9)3803482 Fax 61(9)3801118

Bryant, Dr Peter John (1954). Lab. Manager. [GEC Marconi Systems Pty Ltd, Faraday Park, Railway Rd., Meadowbank, NSW 2114, Australia.] PhD (University of Sydney, 1981) physical chemistry. *Electronic_ceramics battery_and_charging_systems*
Tel. 61(2)8099762 Fax 61(2)8099777

Bursill, Leslie A. (1941). Reader-in-Physics. [School of Physics, University of Melbourne, Parkville, Vic. 3052, Australia.] DSc (University of Melbourne, 1981) physics. *Aperiodic_material high-resolution_electron_microscopy transmission_electron_microscopy quasicrystallography spiral-nanostructures*
E-mail bursill@tauon.ph.unimelb.edu.au Tel. 61(3)3445431 Fax 61(3)3474783

Butler, Dr Brent Dennis (1962). Postdoctoral Fellow. [Research School of Chemistry, Australian National University, Canberra City, 0200, Australia.] PhD (Northwestern University, 1991) materials science and engineering. *Diffuse_scattering disordered_materials*
E-mail butler@rschp2.anu.edu.au Tel. 61(6)2493579 Fax 61(6)2490750

Byriel, Karl Alwyn (1959). Research Officer. [Small Molecule X-ray Diffraction Laboratory, Department of Chemistry, The University of Queensland, QLD 4072, Australia.] BAppSc (Queensland Institute of Technology, 1983) applied chemistry. *Computing X-ray_crystallography computer_graphics fullerenes*
E-mail K.Byriel@mailbox.uq.oz.au Tel. 61(7)3654300 Fax 61(7)3654299 Telex UNIVQLD AA40315

Carr, Paul David (1959). Research Fellow. [Research School of Chemistry, Australian National University, GPO Box 4, Canberra, ACT 2601, Australia.] PhD (University of Keele, UK, 1984) biophysics. *Macromolecular_crystallography time-resolved_effect Laue_diffraction synchrotron_radiation fibre_diffraction*
E-mail pdc@rsc.anu.edu.au Tel. 61(6)2493766 Fax 61(6)2490750

Cashion, John Dixon (1942). Associate Professor. [Physics Department, Monash University, Clayton, Victoria 3168, Australia.] PhD (Oxford U., 1969) physics. *Mossbauer coal iron_compounds gold_compounds magnets polyesters_active_sites*
E-mail JohnC@physics1.physics.monash.edu.au Tel. 61(3)9053680Fax 61(3)9053637 Telex AA32691

Chatterjee, Ajoy (1952). Part-time Tutor. [Department of Physics, Murdoch University, Murdoch, WA 6150, Australia.] PhD (Department of Physics, U. Western Australia, 1984) electron density distributions in crystalline rare earth complexes. *Crystal_structure charge_density chemical_bonding crystallographic_computing electron_microscopy*
Tel. 61(9)3602870 Fax 61(9)3101711

Chung, Liping (1959). PhD student. [School of Physics, University of New South Wales, PO Box 1, Kensington, NSW 2033, Australia.] MSc (University of New South Wales, 1993) physics. *Crystallography peptides synthesis myosin_light_chain protein_purification folding structure*
E-mail lcg@newt.phys.unsw.edu.au Tel. 61(2)6974730 Fax 61(2)6633420

Clapp, Rodney Alexander (1942). General Manager. [Sietronics Manufacturing Pty. Ltd, PO Box 3066, Belconnen, ACT 2617, Australia.] MAppSc Physics (West Australian Institute of Technology, 1981) applied nuclear physics. *Instrumentation software data_collection powder_diffraction Rietveld_methods*
E-mail rod@sietron.com.au Tel. 61(6)2516611 Fax 61(6)2516659

Cockayne, Prof. David (1942). Professor of Physics, Director. Deputy Chair, Academic Board; Chair, Research Policy Committee; General Secretary, International Federation of Societies of Electron Microscopy. University of Sydney, NSW 2006, Australia. [Electron Microscope Unit, University of Sydney, NSW 2006, Australia.] DPhil (U. Oxford, 1969) materials science. *Electron_microscopy electron_diffraction crystalline_defect*
E-mail djhc@emu.su.oz.au Tel. 61(2)3512351 Fax 61(2)5521967

Colman, Peter Malcolm (1944). Chief of Division. [CSIRO Division of Biomolecular Engineering, 343 Royal Parade, Parkville, Victoria 3052, Australia.] PhD (University of Adelaide, 1970) physics. *Structural_biology*
Tel. 61(3)3424211 Fax 61(3)3424221

Colmanet, Silvano Francesco (1955). Senior Research Scientist. Australian Radiation Laboratory, Yallambie, Victoria. [Australian Radiation Laboratory, Lower Plenty Road, Yallambie, Victoria 3085, Australia.] PhD (La Trobe U., 1989) chemistry. *Single_crystal X-ray_diffraction technetium_compounds transition_elements coordination_chemistry radiopharmaceutical_compounds*
E-mail xrc@arl.oz.au Tel. 61(3)4332211 Fax 61(3)4321835

Corbett, Madeline (1941). [37 O'Briens Lane, Templestowe, Victoria 3106, Australia.] PhD (University of Melbourne, 1971) chemistry. *Computing inorganic_structure*

Coyle, Richard Alan (1922). Hon. Research Associate. [Physics Department, Monash University, Clayton, Victoria 3168, Australia.] MAppSc (Royal Melbourne Institute of Technology, 1993) applied physics. *Powder instrumentation residual_stress size_effect ICDD_powder_diffraction_file*
E-mail dick.coyle@physics.monash.edu.au Tel. 61(3)5653601 Fax 61(3)5653637 Telex monash AA32691

Craig, Donald Chadwick (1936). Professional Officer. School of Chemistry, The University of New South Wales, Sydney, Australia 2052. [School of Chemistry, The University of New South Wales, PO Box 1, Kensington, Sydney, Australia 2033.] MSc (UNSW, 1964) geology. *Small_molecules diffractometry computing*
E-mail Crystallography@gmq.chem.unsw.edu.au Tel. 61(2)3854595Fax 61(2)6622835

Cranswick, Lachlan Michael David (1969). Experimental Scientist. CSIRO Division of Mineral Products, 339 Williamstown Road, Port Melbourne, Victoria, Australia 3207. [CSIRO Division of Mineral Products, PO Box 124, Port Melbourne, Victoria, Australia 3207.] BSc (Monash U., 1988) chemistry. *Computing X-ray_diffraction Rietveld_method*
E-mail lachlan@dmp.csiro.au Tel. 61(3)6470367 Fax 61(3)6463223 Telex AA34349

Creagh, Dudley Cecil (1935). Associate Professor. Consultant, Commission on Crystallographic Apparatus; Australian National Beamline Facility: Member, Management Committee, Chairman, Technical Committee, Member, Programme Committee; Immediate Past President, Society of Crystallographers in Australia; Councillor, International Radiation Physics Society; Councillor, Asian Crystallographic Society; Programme Chairman, AXAA96. [Physics Department, University College, U. New South Wales, ADFA, Northcott Dr, Canberra, ACT 2600, Australia..] PhD (UNSW, 1975) physics. *Anomalous_dispersion material_sciences synchrotron_radiation apparatus_design museology music*
E-mail dudley@phadfa.ph.adfa.oz.au Tel. 61(6)2688766 Fax 61(6)2473320

Cuff, Christopher (1944). Managing Director, National Key Centre, J. C. U. [James Cook University, PO Douglas, Qld. 4811, Australia.] PhD (Imperial, 1969) crystallography/geochemistry. *Minerals structure clays clay/water_interchanges_in_the_environment*
Tel. 61(77)814486 61(77)815185 61(77)815186 Fax 61(77)252371 61(77)815522

Curmi, Paul (1957). Senior Lecturer. [University of New South Wales, School of Physics, PO Box 1, Kensington, NSW 2033, Australia.] PhD (University of Sydney, 1983) biophysics. *Proteins structure*
E-mail pmgc@newt.phys.unsw.edu.au Tel. 61(2)6974552 Fax 61(2)6633420

Davis, Paul Christopher (1948). Head, Computer Sciences. [Biomolecular Research Institute, 343 Royal Pde, Parkville, Victoria 3052, Australia.] Assoc diploma of Electronic Eng (RMIT, 1971) electronics. *Computer detector*
E-mail paul@mel.dbe.csiro.au Tel. 61(3)3424300 Fax 61(3)3424301

Davis, R. Lindsay (1940). Principal research scientist. [Australian Nuclear Science and Technology Organisation, Private Mail Bag 1, Menai, New South Wales 2234, Australia.] PhD (Monash University, 1970) solid state physics. *Neutron_diffraction neutron_scattering magnetic_structure chemical_structure*
E-mail rld@photon.ansto.gov.au Tel. 61(2)7173607 Fax 61(2)7179265

Davis, Timothy John (1958). Senior Research Scientist. CSIRO Materials Science & Technology, Normanby Road, Clayton, Victoria, Australia. [CSIRO Materials Science & Technology, Private Mail Bag 33, Rosebank MDC, Clayton, Victoria, 3169, Australia.] PhD (University of Melbourne, 1987) physics. *Dynamical_diffraction imperfection semiconductors physics*
E-mail davis@rivett.mst.csiro.au Tel. 61(3)5422915 Fax 61(3)5441128

Delaney, William Timothy (1944). Consultant. [Central Metropolitan College, 25 Aberdeen St., Perth 6000, Australia.] PhD (U. Western Australia, 1972) physics. *Electron_density_distribution*
Tel. 61(9)4272287 Fax 61(9)2275943

Drennan, Dr John (1951). Principal Research Scientist. [CSIRO Div. of Materials Science and Technology, Private Bag 33, Rosebank MDC, Clayton 3169, Australia.] PhD (Flinders U., 1979) chemistry. *Electron_microscopy diffraction solid_state chemistry*
E-mail drennan@rivett.mst.csiro.au Tel. 61(3)5422777 Fax 61(3)5441128

Edwards, Alison J. (1962). Lecturer. [School of Chemistry, University of Melbourne, Parkville, Victoria 3052, Australia.] PhD (U. Melbourne, 1990) inorganic chemistry. *Chemical_crystallography absolute_configuration polymorphism phase_transition dynamic_processes_in_crystals solid-state_NMR*
Tel. 61(3)3444000 Fax 61(3)3475180

Edwards, Karen Jennifer (1964). Postdoctoral research fellow. [Research School of Chemistry, Australian National University, Canberra, ACT 0200, Australia.] PhD (University of London, UK, 1992) crystallography. *Macromolecular_crystallography molecular_modelling*
E-mail karen@rsc3.anu.edu.au Tel. 61(6)2493766 Fax 61(6)2490750

Eggleton, Richard A. (1937). Reader. [Geology Department, Australian National University, Canberra, ACT 0200, Australia.] PhD (Wisconsin, 1965) geology. *Clay_mineralogy electron_microscopy*
E-mail Tony.Eggleton@anu.edu.au Tel. 61(6)2492060 Fax 61(6)2495544

Elcombe, Dr Margaret Marion (1942). Principal Research Scientist. Australian Nuclear Science and Technology Organisation, Lucas Heights, New South Wales, Australia. [Australian Nuclear Science and Technology Organisation, Private Mail Bag 1, Menai, NSW 2234, Australia.] PhD (Cambridge U., 1966) lattice dynamics. *Lattice_dynamics neutron_scattering*
E-mail mme@atom.ansto.gov.au Tel. 61(2)7173611 Fax 61(2)7179265 Telex AA24562

Fallon, Dr Gary D. (1948). Professional Officer. [Chemistry Department, Monash University, Clayton, Victoria, Australia 3168.] PhD (Monash University, 1976) chemistry. *Structure_determination computing crystallography*
E-mail gdfallon@ccs1.cc.monash.edu.au Tel. 61(3)5654514 Fax 61(3)5654597

Field, Dr Donald William (1947). Assistant Dean. [Faculty of Science, Queensland University of Technology, GPO Box 2434, Brisbane 4001, Australia.] PhD (University of Adelaide, 1973) physics. *X-ray_powder metals strain line_broadening particle_size dynamically_compacted_powders*
E-mail d.field@qut.edu.au Tel. 61(7)8642860 Fax 61(7)8641508

Figgis, Brian N. (1930). Professor. [University of Western Australia, Nedlands, WA 6009, Australia.] DSc (U. Western Australia, 1966) chemistry. *Charge_density polarized_neutron neutron_diffraction chemical_bonding transition_metal_complexes*
E-mail bnf@crystal.uwa.edu.au Tel. 61(9)3803157 Fax 61(9)3801005

Finlayson, Trevor Roy (1943). Associate Professor. [Department of Physics, Monash University, Clayton, Victoria, Australia 3168.] PhD (Monash University, 1969) physics. *Cryogenics magnetic_susceptibility phonon_softening residual_stress Rietveld_method superconductivity thermal_expansion martensites alloy_chemistry Raman_and_neutron_scattering*
E-mail trevorf@physics1.physics.monash.edu.au Tel. 61(3)9053683 Fax 61(3)9053637 Telex Monash AA32691

Fletcher, Neville Horner (1930). Chief research scientist. CSIRO. [RSPhysSE, Australian National University, Canberra 0200, Australia.] DSc (Sydney University, 1973) physics. *Crystal_growth surface interface defect acoustics biophysics environmental physics*
E-mail neville.fletcher@anu.edu.au Tel. 61(6)2494406 Fax 61(6)2490511

Forwood, Christopher Thomas (1940). SPRS (Project Leader). [Division of Materials Science & Technology, CSIRO, Private Bag 33, Rosebank MDC, Clayton, Victoria 3169, Australia.] PhD (University of Bristol, UK, 1966) physics. *Interface electron_microscopy defect intermetallic_materials mechanical_properties*
Tel. 61(3)5422865 Fax 61(3)5441128

Freeman, Hans Charles (1929). Professor of Inorganic Chemistry. School of Chemistry, University of Sydney, NSW 2006, Australia. [Department of Inorganic Chemistry, University of Sydney, NSW 2006, Australia.] PhD (University of Sydney, 1957) chemical crystallography. *Structure_determination EXAFS proteins metalloproteins metalloenzymes coordination_compounds biological_electron_transfer*
E-mail freeman_h@summer.chem.su.oz.au Tel. 61(2)3512757 Fax 61(2)3513329

Gable, Dr Robert William (1954). Professional Officer. [School of Chemistry, University of Melbourne, Parkville, Victoria 3052, Australia.] PhD (University of Melbourne, 1988) inorganic chemistry. *Structure_determination computing coordination*
E-mail U6144414@ucsvc.ucs.unimelb.edu.au Tel. 61(3)3447145 Fax 61(3)3475180 Telex AA35185UNIMELB

Gan, Miss Bee Kwan (1968). PhD student. [Department of Applied Physics, Curtin University of Technology, GPO Box U1987, Perth, WA 6001, Australia.] BSc (Applied Physics) (Curtin University, 1990) crystallography. *Powder Rietveld_method alumina crystallography structure_determination_using_X-ray_and_neutron_diffraction*
E-mail tganbk@cc.curtin.edu.au Tel. 61(9)3517192 Fax 61(9)3512377

Gao, Dachao (1950). Experimental scientist. CSIRO Div. of Materials Science and Technology, Normanby Road, Clayton, Vic. 3169, Australia. [CSIRO Div. of Materials Science and Technology, Private Bag 33, Rosebank MDC, Clayton, Vic. 3169, Australia.] PhD (Royal Melbourne Institute of Technology, 1993) applied physics. *X-ray_topography X-ray_diffraction semiconductors X-ray_monochromator*
E-mail gao@rivett.mst.csiro.au Tel. 61(3)5422929 Fax 61(3)5441128 Telex AA32945

Garrett, Richard Federick (1955). Manager, Australian National Beamline Facility. Australian National Beamline Facility, Australian Nuclear Science and Technology Organisation. [Ansto Building 22, Private Mail Bag 1, Menai, NSW, 2234, Australia.] PhD (University of Newcastle, 1984) physics. *Synchrotron_radiation instrumentation medical_imaging EXAFS X-ray_focusing X-ray_microprobe*
E-mail garrett@photon.ansto.gov.au Tel. 61(2)7173657 Fax 61(2)7173145

Garrett, Dr Thomas P. J. (1959). Senior Research Scientist. [Biomolecular Research Institute, 343 Royal Parade, Parkville, VIC 3052, Australia.] PhD (U. Sydney, 1988) chemistry. *Proteins macromolecular_interaction immunology*
E-mail tomg@rox.mel.dbe.csiro.au Tel. 61(3)3424300 Fax 61(3)3424301

Gatehouse, Dr Bryan Michael Kenneth Cummings (1932). Reader in Chemistry. [Chemistry Department, Monash University, Clayton Vic 3168, Australia.] DSc (University of London, 1977) chemistry spectroscopy. *Structure_determination metallic_oxides teaching_of_crystallography*
E-mail Bryan.Gatehouse@sci.Monash.edu.au Tel. 61(3)9054561 Fax 61(3)9054597

Gerson, Andrea Ruth (1963). Research Fellow. [School of Chemical Technology, The Levels, University of South Australia, Pooraka 5095, South Australia, Australia.] PhD (U. Strathclyde, 1991) crystallography/crystallisation kinetics. *Rietveld_method molecular_modelling crystal_interaction alkanes*
E-mail ctarg@levels.unisa.edu.au Tel. 61(8)3023044 Fax 61(8)3495805

Glaisher, Robert William (1953). Lecturer. La Trobe University, Bendigo, Victoria 3550, Australia. [PO Box 199, Bendigo, Victoria 3550, Australia.] PhD (Melbourne University, 1986) physics. *Electron_microscopy semiconductors*
Tel. 61(54)447950 Fax 61(54)447476

Goodman, Peter (1928). Senior Research Associate. [School of Physics, University of Melbourne, Parkville 3052, Australia.] DSc (Melbourne University, 1978) physics. *Electron_diffraction scattering_physics materials_investigation*
E-mail pag@muon.ph.unimelb.edu.au Tel. 61(3)3445442 Fax 61(3)3474783

Grey, Dr Ian Edward (1944). Chief research scientist. [CSIRO Division of Mineral Products, PO Box 124, Port Melbourne, Victoria 3207, Australia.] PhD (University of Tasmania, 1969) inorganic chemistry. *Structure nonstoichiometry powder minerals mixed_metal_oxide_structures*
E-mail iang@dmp.csiro.au Tel. 61(3)6470211 Fax 61(3)6463223 Telex AA34349

Guss, Dr J. Mitchell (1946). Senior Research Fellow. University of Sydney, NSW 2006, Australia. [Department of Biochemistry - G08, University of Sydney, NSW 2006, Australia.] PhD (U. Sydney, 1970) inorganic chemistry. *Proteins refinement_method accuracy visualization multiple_wavelength_phasing*
E-mail guss_m@summer.chem.su.oz.au Tel. 61(2)3514302 Fax 61(2)3514726

Hall, Prof. Eric O. (1925). Retired (Emeritus Prof., U. Newcastle, NSW). [770 Sandy Bay Road, Tasmania 7005, Australia.] PhD (Cambridge, UK, 1952) metal physics. *Intermetallic_structure crystal_defects*
Tel. 61(2)253408

Hall, Sydney Reading (1939). Associate Professor. Editor, Acta Crystallographica Section C; Member Journals Commission; Member Data Commission. [Crystallography Centre, University of Western Australia, Nedlands 6009, Australia.] PhD (U. Western Australia, 1964) physics. *Software chemical_database symmetry structure_determination*
E-mail syd@crystal.uwa.edu.au Tel. 61(9)3802725 Fax 61(9)3801118

Hambley, Trevor William (1955). Senior Lecturer. [Department of Inorganic Chemistry, University of Sydney, NSW 2006, Australia.] PhD (University of Adelaide, 1984) chemistry. *Drug_design molecular_mechanics crystallography small_molecules metal_complexes DNA_crystallography*
E-mail hambley_t@summer.chem.su.oz.au Tel. 61(2)6922830 Fax 61(2)6923329 Telex UNISYD 26169

Hammond, Mr Lloyd C. (1961). Experimental Scientist. [Defence Science and Technology Organisation, Materials Research Laboratory, PO Box 50, Ascot Vale, Victoria 3032, Australia.] MSc (Curtin U. Tech., Bentley, Western Australia, 1991) physics. *Rietveld_method structure_determination instrumentation computing powder_diffraction alloy_structure_determination mineralogy residual_stress*
E-mail hammond@scm.mrl.cis.dsto.gov.au Tel. 61(3)2468754 Fax 61(3)2468751 Telex AA35230

Hartley, Richard H. (1939). Head, Sensor Development. [Land, Space & Optoelectronics Div. of the Surveillance Research Lab., PO Box 1650, Salisbury, South Australia 5108, Australia.] PhD (University of Western Australia, 1971) thin film physics, electrical engineering. *Solid_state physics thin_film MBE II-VI_semiconductor_compounds*
Tel. 61(8)2596377 Fax 61(8)2595796

Hay, Dr David Gilbert (1948). Senior Research Scientist. [CSIRO Div. of Materials Science and Technology, Private Bag 33, Rosebank MDC, Clayton 3169, Australia.] PhD (La Trobe U., 1982) chemistry. *Powder X-ray_diffraction ceramics*
E-mail hay@rivett.mst.csiro.au Tel. 61(3)5422777 Fax 61(3)5441128

Hester, James R. (1967). Postgraduate Research Student. [Crystallography Centre, University of Western Australia, Nedlands 6009, Australia.] BSc(Hons) (U. Western Australia, 1989) physics. *Charge_density extinction synchrotron_radiation single_crystal*
E-mail jrh@crystal.uwa.edu.au Tel. 61(9)3803482 Fax 61(9)3801118

Hill, Dr Roderick Jeffrey (1949). Program Manager/Assistant Chief. Chairman of Commission on Powder Diffraction. CSIRO Division of Mineral Products, 339 Williamstown Road, Port Melbourne, Victoria 3207, Australia. [CSIRO Division of Mineral Products, PO Box 124, Port Melbourne, Victoria 3207, Australia.] DSc (U. Adelaide, 1991) crystal chemistry. *Mineralogy X-ray_powder Rietveld_method crystallography solid_state*
E-mail rodh@dmp.csiro.au Tel. 61(3)6470211 Fax 61(3)6463223 Telex AA34349

Hockless, Dr David (1965). Research Officer. [Research School of Chemistry, Australian National University, Canberra, ACT 0200, Australia.] PhD (U. Newcastle-Upon-Tyne, UK, 1992) chemistry. *Coordination organic_structure computing*
E-mail david@rscir2.anu.edu.au Tel. 61(6)2494109 Fax 61(6)2490750

Hoskins, Dr Bernard Foster (1935). Honorary Research Associate. [School of Chemistry, University of Melbourne, Parkville, Victoria 3052, Australia.] DSc (University of Melbourne, 1976) structural inorganic chemistry. *Structure_determination framework_structure coordination*
E-mail U6101512@ucsvc.ucs.unimelb.edu.au Tel. 61(3)3448141 Fax 61(3)3475180 Telex AA35185UNIMELB

Howard, Dr Christopher J. (1942). Manager, Neutron Scattering. Co-Editor, Journal of Applied Crystallography. Australian Nuclear Science and Technology Organisation, Lucas Heights, New South Wales, Australia. [Australian Nuclear Science and Technology Organisation, Private Mail Bag 1, Menai, NSW 2234, Australia.] PhD (U. Nottingham, 1970) physics. *Neutron_scattering powder accurate_structure_determination phase_transition ceramics*
E-mail cjh@atom.ansto.gov.au Tel. 61(2)7173609 Fax 61(2)7173606 Telex AA24562

Hsu, Rebeka Min-fang (1956). Postgraduate Student (PD). Department of Physics, The University of Western Australia, Nedlands 6009, Western Australia. [Crystallography Centre, The University of Western Australia, Nedlands, WA 6009, Australia.] MS (National Cheng Kung University, 1981) physics. *Absorption_correction accurate_intensity high-precision_structure perovskites residual_electron_density superconductivity ferroelectricity Jahn_Teller_effect stoichiometry synchrotron_radiation*
E-mail rhsu@earwax.pd.uwa.edu.au Tel. 61(9)3803482 Fax 61(9)3801118 Telex AA92992

Hu, Dr Shu-Hong (1963). Research Officer. [Centre for Drug Design and Development, University of Queensland, Brisbane, QLD 4072, Australia.] DPhil (Oxford University, 1991) protein crystallography. *Proteins structure function modelling drug_design*
E-mail s.hu@mailbox.uq.oz.au Tel. 61(7)3654941 Fax 61(7)3651990

Huq, Fazlul (1945). Senior Lecturer. [University of Sydney, Department of Biological Sciences, Faculty of Health Sciences, East St., Lidcombe, NSW 2141, Australia.] PhD (U. London (Imperial College), 1971) chemical crystallography. *Anticancer_compounds drug platinum_compounds palladium_compounds teratogenic_effects proteins exercise writing_books education*
E-mail f.huq@cchs.su.edu.au Tel. 61(2)6466522 Fax 61(2)6466520

Hyde, Bruce (1925). Retired. [9 Ogilvie Place, Garran, ACT 2605, Australia.] DSc (Bristol, 1972) solid state chemistry. *Solid_state crystal_chemistry crystal_structures*
Tel. 61(6)2810945

Izard, Tina (1967). Graduate Student. [CSIRO, 343 Royal Parade, Parkville 3052, Australia.] MSc (Biocenter, Univ. of Basel, Switzerland, 1990) protein crystallography. *Proteins enzyme_mechanisms*
E-mail tina@tauon.pb.unimelb.edu.au Tel. 61(3)3424251 Fax 61(3)3475481

Jagadish, Chennupati (1957). Head, Semiconductor Epitaxy Group. Chairman, IEEE Australian Chapter of LEOS and EDS. Australian National University. [Electronic Materials Eng. Dept., RSPhysSE, Canberra, ACT 0200, Australia.] PhD (U. Delhi, 1986) physics. *Crystal_growth epitaxy III–V_compounds ion_implantation electron_transport DLTS X-ray_diffraction PL defects*
E-mail cxj109@phyvs1.anu.edu.au Tel. 61(6)2490020 61(6)2490363 Fax 61(3)2490511

James, Veronica Jean. Associate Professor. [School of Physics, University of New South Wales, PO Box 1, Kensington, New South Wales 2033, Australia.] PhD (U. New South Wales, 1970) physics. *Crystal_structure tissues disease collagen_structure keratin*
E-mail vjs@newt.phys.unsw.edu.au Tel. 61(2)6974631 Fax 61(2)6633420 Telex AA26054

Johnson, Dr Andrew William Syme (1936). Director. [Centre for Microscopy and Microanalysis, University of Western Australia, Nedlands 6009, Western Australia, Australia.] PhD (U. Western Australia, 1965) physics. *Convergent-beam_diffraction electron_diffraction*
E-mail andy@earwax.pd.uwa.edu.au Tel. 61(9)3802764 Fax 61(9)3801087

Kalceff, Dr Walter (1950). Senior lecturer in Applied Physics. Department of Applied Physics, University of Technology, Sydney, Tower Building, 15-73 Broadway, Sydney, NSW, Australia. [Department of Applied Physics, University of Technology, Sydney, PO Box 123, Broadway, NSW 2007, Australia.] PhD (U. New South Wales, 1983) physics. *Magnetism superconductivity diffraction computing pulsed_field_magnetometry*
E-mail wkalceff@phys.uts.edu.au Tel. 61(2)3302191 Fax 61(2)3302219

Kelly, Dr Pat (1935). Associate Professor. [Department of Mining & Metallurgical Engineering, University of Queensland, Brisbane, Queensland 4072, Australia.] ScD (Cambridge U., 1978) materials science. *Crystallography phase_transformation electron_diffraction*
E-mail pat@minmet.uq.oz.au Tel. 61(7)3653728 Fax 61(7)3653888 Telex UNIVQLD AA40315

Kennard, Colin Harold Leslie (1935). Associate Professor. IUCr Computing Commission (member); IUCr Teaching Commission (consultant). Department of Chemistry, The University of Queensland, Brisbane, Q 4072, Australia. [The University of Queensland, Brisbane, Q 4072, Australia.] PhD (New South Wales U. Technology, 1961) crystallography. *Single_crystal grazing_incidence Rietveld_method computing teaching*
E-mail C.Kennard@mailbox.uq.oz.au Tel. 61(7)3653662 Fax 61(7)3654299 Telex UNIVQLD AA40315

Kharisun, (1961). Student. School of Physical Sciences, The Flinders University of South Australia, Bedford Park S.A., Australia. [School of Physical Sciences, The Flinders University of South Australia, GPO Box 2100, Adelaide 5001, Australia.] BSc (Honours) (UNSOED (Indonesia), 1985) agriculture. *Minerals inorganic_chemistry computing Pb-transitional_AsO4_system powder_diffraction*
E-mail chkh@cc.flinders.edu.au Tel. 61(8)2012467 Fax 61(8)2013035

Killen, Peter David (1953). Lecturer. [School of Physics, Queensland University of Technology, GPO Box 2434, Brisbane 4001, Australia.] PhD (U. Queensland, 1984) mechanical engineering. *X-ray_powder crystal_size_effect line_broadening strain dynamic_compaction_of_powders*
E-mail p.killen@qut.edu.au Tel. 61(7)8642326 Fax 61(7)8641521

Kisi, Dr Erich (1960). Lecturer in Materials Science. [Dept. Mechanical Engineering, University of Newcastle, NSW 2308, Australia.] PhD (Newcastle U., 1988) materials science. *Neutron_diffraction powder zirconium_compounds diffraction_applied_to_materials_science_problems*
E-mail meehk@cc.newcastle.edu.au Tel. 61(49)216213 Fax 61(49)216946

Kucharski, Dr Edward Stanislaw (1953). Research Scientist. CSIRO, Division of Exploration and Mining, 39 Fairway, Nedlands, WA 6009, Australia. [PO Box 437, Nedlands, WA 6009, Australia.] PhD (U. Western Australia, 1984) physical and inorganic chemistry. *Mineralization crystallization neutron_diffraction magnetization_density charge_density magnetochemistry*
E-mail e.kucharski@dem.csiro.au Tel. 61(9)3898421 Fax 61(9)3891906

Lambert-Smith, John Ernle Warwick (1931). Chemist - Water Management. [Callide "B" Power Station, Queensland Electricity Commission, PO Box 392, Biloela 4715, Australia.] BSc (U. Sydney, 1957) coordination chemistry and X-ray crystal structure analysis. *Computing crystal_structure copper_compounds X-ray_diffraction teaching*
Tel. 61(79)929389 Fax 61(79)929328

Lawrence, Michael Colin (1955). Principal Research Scientist. [Biomolecular Research Institute, 343 Royal Parade, Parkville, Victoria 3052, Australia.] PhD (U. Cape Town, 1980) theoretical physics. *Proteins crystallography macromolecular_crystallography electron_crystallography drug_design*
E-mail mike@mel.dbe.csiro.au Tel. 61(3)3424200 Fax 61(3)3424301

Leverett, Prof. Peter (1944). Head, Department of Chemistry. Faculty of Science & Technology, UWS Nepean. [PO Box 10, Kingswood, NSW 2747, Australia.] PhD (Monash U., 1970) chemistry. *X-ray_structure_determination chirality metals molybdenum_(VI)_oxy_compounds mineral_structures*
E-mail p.leverett@st.nepean.uws.edu.au Tel. 61(47)360808 Fax 61(47)360713

Lincoln, Frank John (1935). Senior Lecturer. [Research Centre for Advanced Mineral & Materials Processing, Dept. of Chemistry, Univ. of Western Australia, Nedlands, WA 6009, Australia.] PhD (U. Western Australia, 1967) chemistry/materials science. *Structure crystallography solid_state chemistry materials_science electron_microscopy/diffraction X-ray_diffraction inorganic_systems minerals alloys disordered_materials*
E-mail fj1@chem.uwa.edu.au Tel. 61(9)3803142 Fax 61(9)3801116

Lloyd, Douglas James (1946). Head of School, School of Education. [La Trobe University, Bendigo, PO Box 199, Bendigo 3550, Australia.] PhD (Monash U., 1972) solid state chemistry, inorganic structural studies. *Structure-property_measurement relationship_structure_and_properties education*
E-mail lloyd@major.ucnv.edu.au Tel. 61(54)447298 Fax 61(54)447800

Lucas, Dr Brian William. Senior Lecturer. [Physics Department, The University of Queensland, QLD 4072, Australia.] PhD (London U., 1965) physics. *X-ray_diffraction neutron_scattering computing Rietveld_method*
E-mail lucas@physics.uq.oz.au Tel. 61(7)3653421 Fax 61(7)3651242 Telex UNIVQLD AA40315

Lukaszewski, George Michael (1935). Principal Research Scientist. [CSIRO Division of Mineral Products, PO Box 124, Port Melbourne, Victoria, Australia 3207.] PhD (U. London, UK, 1960) inorganic chemistry. *Minerals chemistry oxidation thermoanalysis magnesite_processing*
Tel. 61(3)6470211 Fax 61(3)6463223 Telex AA34349

Lynch, Denis Francis (1941). Deputy Chief. [CSIRO Division of Materials Science & Technology, Private Bag 33, Rosebank MDC, Clayton, Vic. 3169, Australia.] PhD (U. New South Wales, 1965) physics. *Electron_diffraction dynamical_scattering RHEED*
E-mail lynch@rivett.mst.csiro.au Tel. 61(3)5422777 Fax 61(3)5441128

Mackay, Dr Maureen Florence (1927). Emeritus Scholar. Commission Member on Small Molecules. [La Trobe University, Department of Chemistry, Bundoora, Victoria 3083, Australia.] PhD (U. Melbourne, 1968) crystallography. *Crystal_structure small_molecules natural_products biological_molecules*
E-mail xraymm2@lure.latrobe.edu.au Tel. 61(3)4792520 Fax 61(3)4791300

Mackenzie, Dr James Kenneth (1920). Retired. [15 Ronald St., Box Hill North, Victoria 3129, Australia.] PhD (U. Bristol, UK, 1949) physics. *Crystal_physics crystallography phase_transformation*
Tel. 61(3)8907458

Madsen, Ian Charles (1951). Senior Experimental Scientist. [CSIRO Division of Mineral Products, PO Box 124, Port Melbourne, Victoria 3207, Australia.] BAppSc (South Australian Institute of Technology, 1977) applied physics. *X-ray_diffraction Rietveld_method structure_determination minerals crystallographic_computing*
E-mail ianm@sol.dmp.csiro.au Tel. 61(3)6470211 Fax 61(3)6463223 Telex AA34349

Maheswaran, Saravanamuthu (1959). Associate Lecturer. [Dept. of Applied Physics, University of Technology, Sydney, PO Box 123, Broadway, NSW 2007, Australia.] PhD (Simon Fraser U., Canada, 1989) physics. *Dynamical_theory extinction strained_materials*
E-mail mahesh@acacia.itd.uts.edu.au Tel. 61(2)3302191 Fax 61(2)3302219

Majumdar, Dr Amit (1958). Research Scientist. Aeronautical & Maritime Research Laboratory, Defence Science & Technology Organisation. [Aeronautical & Maritime Research Laboratory, Defence Science & Technology Organisation, GPO Box 4331, Melbourne, Australia 3001.] PhD (Monash U., Clayton, Australia, 1989) materials engineering. *Electron_microscopy electron_diffraction structure-property_correlation alloy_development* aerospace_alloys rapid_solidification_processing
E-mail amit.majumdar@dsto.defence.gov.au Tel. 61(3)6267279 Fax 61(3)6267087

Malby, Ms Robyn L. (1966). Postgraduate student. [Biomolecular Research Institute, 343 Royal Parade, Parkville, VIC. 3052, Australia.] BSc(Hons) (U. Sydney, 1988) chemistry. *Molecular_replacement immunoglobulins proteins data_collection* crystallization protein-protein_interactions
E-mail robyn@rox.mel.dbe.csiro.au robyn@dpca.dn.mu.oz.au Tel. 61(3)3424300 Fax 61(3)3424301

Martin, Dr Jennifer L. (1961). Queen Elizabeth II Fellow. [Centre for Drug Design and Development, University of Queensland, St Lucia, QLD 4072, Australia.] DPhil (U. Oxford, 1989) molecular biophysics. *Proteins enzymes structure_determination crystallography protein_folding inhibitor_design*
E-mail J.Martin@mailbox.uq.oz.au Tel. 61(7)3654942 Fax 61(7)3651990

Maslen, Edward Norman (Ted) (1935). Director. Chairman, Committee on Electronic Publishing, Dissemination and Storage of Information. [Crystallography Centre, University of Western Australia, Nedlands, WA 6009, Australia.] DPhil (Oxford, 1960) chemical crystallography. *Structure charge_density information_science*
E-mail enm@crystal.uwa.edu.au Tel. 61(9)3802727 Fax 61(9)3801118

Mathieson, Alexander McLeod (1920). Honorary Professor. [Chemistry Dept., La Trobe University, Bundoora, Victoria 3083, Australia.] DSc (Melbourne University, 1956) structural crystallography. *Single_crystal profile_analysis instrumentation extinction delta_omega, delta_2_theta_space*
Tel. 61(3)4792506 Fax 61(3)4791399 Telex AA33143

Mayo, Sheridan Clare (1968). Postdoctoral Fellow. [Research School of Chemistry, Australian National University, GPO Box 4, Canberra, ACT 2601, Australia.] DPhil (Oxford U., 1993) physics. *Diffuse_scattering disorder phase_transition optical_materials high-temperature_diffraction_techniques*
E-mail scmayo@rschp2.anu.edu.au Tel. 61(6)2493579 Fax 61(6)2490750

McKenzie, David Robert (1946). Reader (Physics). [University of Sydney, NSW 2006, Australia.] PhD (U. New South Wales, 1972) solid state physics. *Amorphous_materials electron_diffraction X-ray_diffraction fullerenes electronic_properties hard_materials*
E-mail mckenzie@physics.su.oz.au Tel. 61(2)6923180 Fax 61(2)6600703

McLaren, Alexander Clark (1928). Professor. [Research School of Earth Sciences, Australian National University, GPO Box 4, Canberra, ACT, Australia.] DSc (Cambridge U., UK, 1981) physics. *Minerals defect transformation electron_microscopy*
Tel. 61(6)2492494 Fax 61(6)2490738

McLaughlin, George (1947). Head MIS Division. [The Australian National University, Canberra, ACT, 0200, Australia.] Grad RIC, FRSC (Brighton Polytechnic, 1971) chemistry. *Information_system innovative_uses_of_information_technology_for_information_dissemination*
E-mail George.McLaughlin@anu.edu.au Tel. 61(6)2494472 Fax 61(6)2490449

Moodie, Alexander Forbes (1923). Adjunct Professor. [Royal Melbourne Institute of Technology, 124 Latrobe St., Melbourne, Vic. 3000, Australia.] DSc (St Andrews University, 1949) physics. *Dynamical_diffraction convergent-beam_diffraction electron_microscopy*
Tel. 61(3)6602434 61(3)6602205 Fax 61(3)6603837

Moon, Prof. Tony (1945). Dean of Science & Professor of Physics. Department of Applied Physics, Faculty of Science, University of Technology, Sydney, Australia. [PO Box 123, Broadway, Sydney, NSW 2007, Australia.] PhD (U. Melbourne, Australia, 1970) physics. *Electron_diffraction computing alumina spectroscopy*
E-mail tonym@phys.uts.edu.au Tel. 61(2)3302210 Fax 61(2)33301656

Morton, Dr Allan James (1939). Manager, Alloys Research and Development Program. Member, National Committee for Electron Microscopy. CSIRO Division of Materials Science and Technology, Gate 4 Normanby Road, Clayton, Vic. 3168, Australia. [CSIRO Division of Materials Science & Technology, Private Bag 33, Rosebank MDC, Clayton, Vic. 3169, Australia.] PhD (U. New South Wales, 1964) applied science (metallurgy). *Intermetallic_materials superalloys martensitic_transformation electron_microscopy aluminium magnesium alloy_development*
E-mail morton@rivett.mst.csiro.au Tel. 61(3)5422777 (direct 2860) Fax 61(3)5441128

Muddle, Dr Barry C. (1948). Reader in Materials Engineering. [Dept. of Materials Engineering, Monash University, Clayton, Victoria 3168, Australia.] PhD (U. New South Wales, 1975) metallurgy. *Solid_state phase_transformation electron_microscopy electron_diffraction*
E-mail muddle@eng2.eng.monash.edu.au Tel. 61(3)5654908 Fax 61(3)5654940 Telex AA32691

Nikulin, Dr Andrei Yurievich (1962). Research Fellow. School of Physics, The University of Melbourne, Parkville, Vic 3052, Australia. [3/60 O'Shanassy Street, North Melbourne, Vic 3051, Australia.] PhD (Institute of Solid State Physics, Chernogolovka, Russia, 1990) X-ray optics and its applications. *X-ray_diffractometry optics computing two-dimensional_deformation surface_and_interfaces thin_films heterostructures X-ray_tomography*
E-mail ayn@muon.ph.unimelb.edu.au or ayn@tauon.ph.unimelb.edu.au Tel. 61(3)3445458 Fax 61(3)3474783

Nimmo, John Kenneth. Lecturer. [Department of Physics, University of Queensland, Queensland 4072, Australia.] PhD (Queensland, 1976) structural studies of phase transitions in solids. *Phase_transition neutron_diffraction Rietveld_method*
E-mail nimmo@physics.uq.oz.au Tel. 61(7)3652364 Fax 61(7)3651242

O'Connor, Brian (1941). Professor of Applied Physics (Personal Chair). Curtin University of Technology. [GPO Box U1987, Perth, WA 6001, Australia.] PhD (U. Western Australia, 1965) *Powder Rietveld_method minerals ceramics materials_characterization_using_X-rays,_neutrons_and_synch._radiation*
E-mail oconnorb@curtin.cc.edu.au Tel. 61(9)3517192 Fax 61(9)3512377

Ollis, David (1952). Senior Fellow. [Research School of Chemistry, Australian National Uni., Canberra, ACT 0200, Australia.] PhD (U. Sydney, 1981) protein crystallography. *Proteins structure engineering protein–DNA_interactions*
E-mail ollis@rsc3.anu.edu.au Tel. 61(6)2494377 Fax 61(6)2490750

Olsson, Christina L. (1965). Professional Officer. Defence Research and Science Technology, Aeronautical Research Laboratory, Melbourne, Australia 3207. [Aeronautical Research Laboratory, Airframes and Engines Division, 506 Lorimer St, Fishermens Bend 3207, Australia.] BSc(Hons) (Monash U., Melbourne, Australia, 1990) physics. *X-ray_diffraction residual_stress XPS AES alloys*
E-mail olssonc@fencer.cis.dsto.gov.au Tel. 61(3)6477279 Fax 61(3)6466769

Parker, Dr Michael William (1959). Senior research fellow. [St. Vincent's Institute of Medical Research, 41 Victoria Parade, Fitzroy 3065, Victoria, Australia.] DPhil (Oxford U., 1985) protein crystallography. *Proteins channel_proteins therapeutic_compounds toxins*
E-mail u5605504@ucsvc.ucs.unimelb.edu.au Tel. 61(3)2882480 Fax 61(3)4162676

Poppleton, Bruce James [CSIRO, Division of Materials Science & Technology, Private Bag 33, Rosebank MDC, Clayton, Vic. 3169, Australia.] PhD (Monash, 1972) chemical crystallography. *Membranes porous_materials crystal_packing astronomy spectrum_analysis*
Tel. 61(3)5422650 Fax 61(3)5441128

Prager, Dr Peter Robert (1944). [PO Box 17, Black Rock, Vic. 3193, Australia.] PhD (University of Melbourne, 1971) physics. *X-ray_diffraction*
Tel. 61(3)5894059 Fax 61(3)5891241

Pring, Allan (1956). Senior Curator of Minerals & Meteorites. [South Australian Museum, North Terrace, Adelaide, South Australia 5000, Australia.] PhD (U. Cambridge, 1983) mineralogy/solid state chemistry. *Structural_mineralogy crystal_chemistry defect high-resolution_electron_microscopy perovskite_structures*
Tel. 61(8)2077449 Fax 61(8)2077222

Rae, Professor Alan David (1941). Professor. [Research School of Chemistry, The Australian National University, Canberra 0200, Australia.] PhD (Auckland, 1964) chemistry. *Structure_determination refinement_method modulated_structures twinning disorder pseudosymmetry*
E-mail Rae@RSC3.anu.edu.au Tel. 61(6)2493895 Fax 61(6)2490750

Raftery, Tony (1956). Technologist. [School of Physics, Queensland University of Technology, GPO Box 2434, Brisbane 4001, Australia.] BSc (Griffith U., Brisbane, 1979) physics. *X-ray_powder crystal_size_effect orientation line_profile line_broadening quantitative_XRPD compaction_of_powders*
E-mail a.raftery@qut.edu.au Tel. 61(7)8642271 Fax 61(7)8641521

Raston, Prof. Colin Llewellyn (1950). Professor of Chemistry. [Faculty of Science & Technology, Griffith University, Nathan, Brisbane, QLD 4111, Australia.] DSc (Griffith U., 1993) chemistry. *Main-group_compounds fullerenes*
E-mail c.raston@sct.gu.edu.au Tel. 61(7)8757564 Fax 61(7)8755369

Reynolds, Philip Andrew (1947). ARC Fellow. [Department of Chemistry, University of Western Australia, Nedlands, W.A. 6009, Australia.] DPhil (Oxford U., 1973) chemistry. *Polarized_neutron neutron_diffraction X-ray_diffraction magnetism transition_metal_complexes*
E-mail pr@crystal.uwa.edu.au Tel. 61(9)3803515 Fax 61(9)3801005

Rossouw, Chris (1951). Research Scientist. [CSIRO Division of Materials Science and Technology, Private Bag 33, Rosebank MDC, Clayton, Victoria, Australia 3169.] DPhil (Oxford U., 1977) physics. *Dynamical_diffraction electron_diffraction inelastic_scattering HOLZ_analysis keen_squash_player*
E-mail rossouw@rivett.mst.csiro.au Tel. 61(3)5422777 Fax 61(3)5441128

Sabine, Terence Murray (1930). Professorial Fellow. [ANSTO, Private Mail Bag 1, Menai, NSW 2234, Australia.] DSc (U. Melbourne, 1971) physics. *Diffraction neutron_scattering material_sciences synchrotron_radiation*
E-mail tms@atom.oz.au Tel. 61(2)7173472 61(2)6920832 Fax 61(2)5437179

Schmid, Dr Siegbert (1959). Postdoctoral Fellow. [Research School of Chemistry, Australian National University, Canberra, ACT 0200, Australia.] PhD (U. Tübingen, 1990) inorganic chemistry. *Solid_state chemistry modulated_structures structure_determination*
E-mail schmid@rsc.anu.edu.au Tel. 61(6)2495408 Fax 61(6)2490750

Scudder, Marcia Lorraine (1948). Research Fellow. [School of Chemistry, The University of New South Wales, 2052 Sydney, Australia.] PhD (U. Sydney, 1973) chemistry. *Inclusion computer_graphics metal_thiolates*
E-mail M.Scudder@unsw.edu.au Tel. 61(2)3854750 Fax 61(2)6622835

Self, Dr Peter Geoffrey (1952). Senior Research Scientist. [CSIRO, Division of Soils, PMB#2, Glen Osmond, SA 5064, Australia.] PhD (U. Melbourne, 1979) physics. *Electron_microscopy clays high-resolution_electron_microscopy clay intercalates soil minerals*
E-mail self@adl.soils.csiro.au Tel. 61(8)3038490 Fax 61(8)3038550 Telex 82406

Sellar, Jeffrey Ronald (1946). Senior Lecturer. [Dept. of Materials Engineering, Monash University, Clayton, Victoria, Australia.] PhD (Arizona State U., 1976) physics. *Electron_microscopy diffraction ceramics zirconium_compounds*
Tel. 61(3)5654910 Fax 61(3)5654940

Shen, Wei (1957). Postdoctoral Research Associate. [Department of Chemistry, La Trobe University, Bundoora, Victoria 3083, Australia.] PhD (La Trobe U., 1991) surface science. *Crystal_growth surface electrochemistry*
E-mail gauws@lure.latrobe.edu.au Tel. 61(3)4792519 Fax 61(3)4791399

Shi, Dashuang (1964). PhD Student. [Department of Inorganic Chemistry, University of Sydney, NSW 2006, Australia.] MS (Xiamen U., 1987) inorganic chemistry. *Proteins crystallization refinement_method computer_modelling*
E-mail shi_d@summer.chem.su.oz.au Tel. 61(2)6922830 Fax 61(2)6923329

Skelton, Brian Warwick (1948). Research Officer. [Chemistry Department, University of Western Australia, Nedlands, Western Australia 6009, Australia.] PhD (Auckland U., Auckland, New Zealand, 1974) chemistry. *X-ray_structure_determination small_molecules coordination chemistry*
E-mail bws@crystal.uwa.edu.au Tel. 61(9)3803481 Fax 61(9)3801118

Skipworth, Dr Margaret Helene (1969). Lecturer Level A. School of Physical Sciences, The Flinders University of South Australia, Bedford Park S.A., Australia. [School of Physical Sciences, The Flinders University of South Australia, GPO Box 2100, Adelaide 5001, Australia.] Hons (Flinders U. South Australia, 1989) chemistry. *Charge_density bioinorganic_chemistry crystallographic_teaching*
E-mail chmhs@pippin.cc.flinders.edu.au Tel. 61(8)2013071 Fax 61(8)2013035

Slade, Phillip Garland (1941). Principal Research Scientist. [CSIRO Division of Soils, Private Bag No. 2, Glen Osmond, S. Aust. 5064, Australia.] PhD (U. Adelaide, 1967) mineralogy. *Clays mineralogy intercalation structure_determination*
Tel. 61(8)3038499 Fax 61(8)3038550

Smith, Dr Graham (1941). Senior Lecturer. [Queensland University of Technology, School of Chemistry, GPO Box 2434, Brisbane 4001, Australia.] PhD (U. Queensland, 1978) chemistry. *Structural_chemistry carboxylic_acids silver_compounds chemistry*
E-mail smithg@qut.edu.au Tel. 61(7)8642293 Fax 61(7)8641804

Smith, Katherine Leah (1955). Group Leader, Senior Research Scientist. [Advanced Materials Prog., Aust. Nuc. Sci. & Tech. Org. (ANSTO), Private Mail Bag 1, Menai, NSW 2234, Australia.] PhD (Monash U., 1982) physics. *Electron_microscopy mineralogy ceramics yoga*
E-mail kls@nucleus.ansto.gov.au Tel. 61(2)7173448 Fax 61(2)5437179 Telex AA24562

Smith, Prof. T. Fred (1939). Deputy Vice Chancellor (Research). [La Trobe University, Bundoora, Victoria 3083, Australia.] PhD (U. Sheffield, UK, 1963) physics. *Lattice_stability instability neutron_Kikuchi_effect thermal_expansion magnetism superconductivity*
E-mail vcomr@lure.latrobe.edu.au Tel. 61(3)4791185 Fax 61(3)4791499

Spackman, Mark A. (1954). Senior Lecturer. [Department of Chemistry, University of New England, Armidale, NSW 2351, Australia.] PhD (U. Western Australia, 1980) chemistry. *Electron_density_distribution theoretical_chemistry chemistry computing*
E-mail mspackma@metz.une.edu.au Tel. 61(67)732722 Fax 61(67)711563

Spink, John Arthur (1925). Honorary Fellowship. [CSIRO Division of Materials Science & Technology, Private Bag 33, Rosebank MDC, Clayton, Vic. 3169, Australia.] MSc (U. Melbourne, 1955) surface chemistry and electron diffraction. *High-energy_electron_diffraction Auger_spectroscopy history_of_electron_diffraction history_of_physical_science_research_laboratories*
Tel. 61(3)5422777 Fax 61(3)5441128

Stevenson, Andrew Wesley (1957). Senior Research Scientist. Sub-Editor of Ninth Edition of World Directory of Crystallographers. CSIRO Division of Materials Science and Technology, Gate 5, Normanby Rd, Clayton, Victoria 3168, Australia. [CSIRO Division of Materials Science and Technology, Private Bag 33, Rosebank MDC, Clayton, Victoria 3169, Australia.] PhD (U. Melbourne, 1984) physics. *Diffractometry microstructure semiconductors X-ray_optics thermal_vibration accurate_intensity twinning polarity dynamical_diffraction synchrotron_radiation software multiple-crystal_diffractometry pole_figure thin-layer epitaxy extinction Laue_diffraction least-squares_refinement optoelectronics rocking_curves crystallinity two-dimensional_Bragg_intensity*
E-mail stevens@rivett.mst.csiro.au Tel. 61(3)5422917 Fax 61(3)5441128

Stewart, Glen Alan (1949). Senior Lecturer. [Dept of Physics, University College, ADFA, The University of New South Wales, Canberra, ACT 2600, Australia.] PhD (Monash U., 1976) physics. *Crystal_field rare-earth_compounds Mossbauer low_temperature_physics NMRON*
E-mail glen@phadfa.ph.adfa.oz.au Tel. 61(6)2688770 Fax 61(6)2688786

Streltsov, Dr Victor A. (1959). Research Fellow. Crystallography Centre, University of Western Australia, Western Australia, and Institute of Crystallography, Russian Academy of Sciences, Moscow Russia. [Crystallography Centre, University of Western Australia, Nedlands, Perth 6009, Western Australia.] PhD (Institute of Applied Physics, Kishinev, 1986) physics and mathematics. *Charge_density computing X-ray_diffraction inorganic_materials synchrotron_radiation*
E-mail strel@crystal.uwa.edu.au Tel. 61(9)3802768 Fax 61(9)3801118

Streltsova, Dr Natalie (1961). Research Associate. [Crystallography Centre, University of Western Australia, Nedlands, Perth, Western Australia 6009.] PhD (Karpov Institute of Physical Chemistry, Moscow, Russia, 1989) chemistry. *Chemistry charge_density crystal_growth rare-earth_compounds coordination_chemistry crown_compounds*
E-mail nat@crystal.uwa.edu.au Tel. 61(9)3802768 Fax 61(9)3801118

Stuart, Dr Sue-Anne (1965). Postdoctoral Research Fellow. CSIRO Division of Forest Products, Bayview Ave, Clayton, Victoria 3168, Australia. [CSIRO Division of Forest Products, Private Bag 10, Rosebank MDC, Clayton, Victoria 3169, Australia.] DAppSc (Royal Melbourne Institute of Technology U., 1993) applied physics. *X-ray_microtomography X-ray_diffraction celluloses CVD diamond convergent_beam_electron_diffraction faults_in_diamond*
E-mail S.Stuart@forprod.csiro.au Tel. 61(3)5422208 Fax 61(3)5436613

Sutherland, Dr Elizabeth Eldred (1955). Research Fellow. [School of Chemistry, University of Melbourne, Parkville 3052, Victoria, Australia.] PhD (U. Cape Town, 1986) chemistry. *Framework_structure X-ray_diffraction osmium_compounds carbonyls*
Tel. 61(3)3447145 Fax 61(3)3475180

Taylor, Dr John Charles (1935). Senior Principal Research Scientist. [CSIRO Division of Coal and Energy Technology, Private Bag 7, Menai, NSW 2234, Australia.] DSc (U. New South Wales, 1980) chemistry. *Rietveld_method X-ray_diffraction quantitative_measurement neutron_diffraction chemical_crystallography*
Tel. 61(2)7106777 Fax 61(2)7106800

Taylor, Max Ronald (1936). Senior Lecturer. School of Physical Sciences, The Flinders University of South Australia, Bedford Park S.A., Australia. [School of Physical Sciences, The Flinders University of South Australia, GPO Box 2100, Adelaide 5001, Australia.] PhD (U. Sydney, 1964) chemistry. *Charge_density bioinorganic_chemistry computing crystallographic_teaching*
E-mail chmrt@cc.flinders.edu.au Tel. 61(8)2012467 Fax 61(8)2013035

Thompson, Dr John Gerard (1954). Fellow. [Research School of Chemistry, Australian National University, Canberra, ACT 0200, Australia.] PhD (James Cook U., Australia, 1986) geology. *Crystallography clays modulated_structures intercalates synthesis crystal_growth powder_diffraction non-molecular_solids kaolinite_derivatives*
E-mail thompson@rsc3.anu.edu.au Tel. 61(6)2494190 Fax 61(6)2490750

Tibballs, Dr John E. (1947). Principal Consultant, Fabricated Products. [Comalco Research & Technology, PO Box 314, Thomastown, Victoria 3074, Australia.] PhD (U. Melbourne, 1974) physics. *Intermetallic_materials ordering defect thermochemistry phase_transitions crystal_growth physical_metallurgy*
E-mail tibballj@crc.cra.com.au Tel. 61(3)4690725 Fax 61(3)4622700

Tiekink, Edward Richard Tom (1960). University Lecturer. [Department of Chemistry, University of Adelaide, South Australia 5005, Australia.] PhD (U. Melbourne, 1985) inorganic chemistry. *Structure_determination pharmacology*
E-mail etiekink@chemistry.adelaide.edu.au Tel. 61(8)3035943 Fax 61(8)2236771

Todd, Dr Donald Douglas (1933). Director. [X-ray Centre, Aviation Building, The University of Newcastle, Callaghan, NSW 2308, Australia.] PhD (U. Newcastle, 1973) metallurgy/solid state transformations. *Computing X-ray_diffraction X-ray_fluorescence food wine*
Tel. 61(49)215763 Fax 61(49)674946

Trefry, Dr Michael (1961). Computer Systems Administrator. Division of Water Resources, CSIRO, Western Australia. [CSIRO, Private Bag, PO Wembley, Western Australia 6014, Australia.] PhD (U. Western Australia, 1989) molecular physics. *Computing simulation historical_exploration*
E-mail M.Trefry@per.dwr.csiro.au Tel. 61(9)3870200 Fax 61(9)3878211

Usher, Dr Brian Francis (1949). Principal Scientist–Photonics. [Telecom Research Laboratory, 770 Blackburn Road, Clayton, VICTORIA 3168, Australia.] PhD (U. Western Australia, 1981) physics. *MBE III–V_compounds rocking_curves dislocation lattice_distortion synchrotron_radiation topography*
E-mail b.usher@trl.oz.au Tel. 61(3)2536681 Fax 61(3)2536665

Vagg, Robert Sylvester (1945). Associate Professor in Chemistry. [School of Chemistry, Macquarie University, NSW 2109, Australia.] PhD (Macquarie U., 1971) chemistry. *Transition_elements metals*
Tel. 61(2)8058269 Fax 61(2)8058313 Telex MACUNI AA122377

Varghese, Dr Joseph Noozhumurry (1949). Senior Principal Research Scientist. [CSIRO Division of Biomolecular Engineering, 343 Royal Parade, Parkville, Victoria 3052, Australia.] PhD (U. Western Australia, 1974) physics. *Proteins structure function viruses anti_virals enzyme_inhibitors computing*
E-mail jose@venus.mel.dbe.csiro.au Tel. 61(3)3424277 Fax 61(3)3475481

Wagenfeld, Heinrich Karsten (1928). Professor, Institute Fellow. [Royal Melbourne Institute of Technology (R.M.I.T.), GPO Box 2476V, Melbourne 3001, Australia.] Habil, Dr rer. nat. (Technical U. Berlin (Habil), Free U. Berlin (Dr rer. nat.), 1958 (Dr rer. nat.)) theoretical physics (Habil), experimental physics (Dr rer. nat.). *Dynamical_diffraction SAXS inelastic_scattering solid_state_physics*
Tel. 61(3)6603405 Fax 61(3)6603837 Telex AA36406

Warmiński, Tadeusz Piotr (1940). Principal Scientist. [Telecom Australia Research Laboratories, 770 Blackburn Road, Clayton, Vic. 3168, Australia.] DSc (Polish Academy of Sciences, 1974) physics. *Imperfection III–VI_compounds optical_glasses optical_fibres physics_phenomena_(electrons,_X-ray)*
E-mail t.warminski@trl.oz.au Tel. 61(3)2536688 Fax 61(3)2536665 Telex AA152026

Warner, Dr Joanne Kathleen (1961). P/T lecturer. [School of Physical Sciences, University of Technology, Sydney, PO Box 123, Broadway NSW 2007, Australia.] DPhil (U. Oxford, 1992) chemical crystallography. *X-ray_diffraction neutron_diffraction synchrotron_radiation powder XANES solid-state_chemistry anomalous_scattering resonant_diffraction*
E-mail jwarner@phys.uts.edu.au Tel. 61(2)3302206 Fax 61(2)3302219

Watts, John Andrew (1932). Senior Experimental Scientist. Secretary, Commission of Classification of Minerals. [CSIRO Division of Mineral Products, 339 Williamstown Road, Port Melbourne, Vic. 3207, Australia.] BSc (Ballarat School of Mines, SMB, 1958) chemistry. *Powder diffraction Rietveld_method minerals cycling*
E-mail johnw@dmp.csiro.au Tel. 61(3)6470272 Fax 61(3)6463223

Welberry, (Thomas) Richard (1945). Senior Fellow. [Research School of Chemistry, Australian National University, GPO Box 4, Canberra City, ACT 0200, Australia.] PhD (London, 1970) chemical crystallography. *X-ray_diffuse_scattering disorder computer_modelling*
E-mail welberry@rsc3.anu.edu.au Tel. 61(6)2494122 Fax 61(6)2490750

White, Allan Henry (1938). Associate Professor. [Crystallography Centre, University of Western Australia, Nedlands, Western Australia 6009, Australia.] DSc (U. Melbourne, 1981) inorganic chemistry. *Structural_chemistry inorganic_chemistry coordination chemical_crystallography*
Tel. 61(9)3803144 Fax 61(9)3801005

White, John William (1937). Professor of Physical and Theoretical Chemistry. Chairman, Neutron Scattering Commission, IUCr; Pro Vice Chancellor, A. N. U.; Chairman Board Institute of Advanced Studies, A. N. U.; Chairman, Australian National Committee on Crystallography. [Research School of Chemistry, The Australian National University, Canberra, ACT 0200, Australia.] DPhil (Oxford U., 1962) physical chemistry. *Adsorption structure polymers interface neutron_inelastic_scattering molecular_dynamics*
E-mail jww@rsc.anu.edu.au Tel. 61(6)2493767 Fax 61(6)2494903

White, Timothy John (1958). Professor. [Ian Wark Research Institute, The University of South Australia, The Levels, S.A. 5095, Australia.] PhD (Australian National U., 1982) chemistry. *Electron_microscopy crystal_chemistry ceramics environmental_science*
Tel. 61(8)3023694 Fax 61(8)3023683

Whitfield, Harold John (1931). Adjunct Professor. [Department of Applied Physics, RMIT, 124 Latrobe St., Melbourne, Australia.] PhD (Victoria U., Wellington, NZ, 1967) chemistry. *Convergent-beam_diffraction nuclear_magnetic_resonance sols gels*
E-mail harry@bunyip.ph.rmit.edu.au Tel. 61(3)6602205 Fax 61(3)6603837

Wielunski, Leszek (1943). Principal Research Scientist. Division of Applied Physics CSIRO, Sydney, Australia. [PO Box 218, Lindfield, NSW, Australia 2070.] PhD (Institute of Nuclear Research, Warsaw, Poland, 1972) applied physics. *Semiconductors material_characterization ion_beam channeling epitaxial_structure compound_semiconductors polarity superlattices impurity_detection nuclear_reaction_analysis oxygen_detection defects dechanneling_on_dislocations direct_ion_scattering_on_dislocations*
E-mail leszek@dap.csiro.au Tel. 61(2)4137162 Fax 61(2)4137238 Telex AA26296

Wilkins, Dr Stephen William (1946). Senior Principal Research Scientist. Member IUCr Commission on Synchrotron Radiation. [CSIRO Division of Materials Science & Technology, Private Bag 33, Rosebank MDC, Clayton, Vic 3169, Australia.] PhD (University of Melbourne, 1973) physics. *X-ray_instrumentation diffraction_theory high-precision_diffractometry maximum-entropy_method crystal_growth crystallography diffuse_scattering direct_method dynamical_diffraction epitaxy EXAFS extinction Fankuchen_effect image_processing infrared_detector Kikuchi_effect Kossel_diffraction Laue_diffraction microbeam_analysis micro-diffraction microlithography microstrain microstructure monochromator multiple-crystal_diffractometry optoelectronics order-disorder perfect_crystal perfection*

phase_transition phonon_softening physics pole_figure powder residual_stress SAXS shape_memory order-disorder sensor statistical_mechanics structure_determination X-ray_fluorescence
E-mail wilkins@rivett.mst.csiro.au Tel. 61(3)5422918 Fax 61(3)5441128

Williams, Mr Donald Allan (1937). Consultant; Director ASC & Sietronics. Analytical Science Consultants PL and Sietronics PL. [178 Buckley Street, Essendon, Vic 3040, Australia.] MSc (La Trobe University, 1974) chemistry. *Rietveld_method instrumentation XRF electron_microscopy*
Tel. 61(3)3377211 Fax 61(3)8175358

Williams, Geoffrey Allan (1950). Senior Research Scientist. [Australian Radiation Laboratory, Lower Plenty Road, Yallambie, Victoria 3085, Australia.] PhD (U. Melbourne, 1976) chemistry. *Single_crystal X-ray_diffraction technetium_compounds transition_elements coordination_chemistry EXAFS topotaxy*
E-mail geofw@arl.oz.au Tel. 61(3)4332211 Fax 61(3)4321835

Willis, Dr Anthony C. (1951). Research officer. [Research School of Chemistry, Australian National University, Canberra, ACT 0200, Australia.] PhD (U. Western Australia, 1977) physical and inorganic chemistry. *Organometallic_chemistry small_molecules*
E-mail willis@rsc3.anu.edu.au Tel. 61(6)2494109 Fax 61(6)2490750

Wilson, Alan Richard (1953). Senior Research Scientist; Task Manager. Secretary, Australian Society for Electron Microscopy. Ship Structures and Materials Division, Aeronautical and Maritime Research Laboratory, DSTO, 506 Lorimer Street, Fishermens' Bend, Victoria 3207, Australia. [GPO Box 4331, Melbourne, Victoria 3001, Australia.] PhD (U. Melbourne, 1982) physics. *Electron_microscopy surface microanalysis computing*
E-mail wilsona@blackjack.cis.dsto.gov.au Tel. 61(3)6267508 Fax 61(3)6267087

Withers, Dr Ray (1954). Group Leader - Solid State Chemistry. [Research School of Chemistry, Australian National University, Canberra, ACT 0200, Australia.] PhD (Melbourne U., 1982) physics. *Incommensurate modulated_structures electron_microscopy phase_transition*
E-mail withers@rsc3.anu.edu.au Tel. 61(6)2493714 Fax 61(6)2490750

Zou, Dr Jin (1959). Postdoctoral Fellow. [Electron Microscope Unit, The University of Sydney, NSW 2006, Australia.] PhD (U. Sydney, 1993) physics. *Electron_diffraction elasticity defect crystal_structure_determination semiconductor_heterostructures*
E-mail zou@emu.su.oz.au Tel. 61(2)3512351 Fax 61(2)5521967

Zubak, Vilma Maria (1966). Research Assistant. [Department of Inorganic Chemistry, The University of Sydney, Sydney, NSW 2006, Australia.] MSc (U. Melbourne, Australia, 1993) biochemistry. *Proteins crystallography structure*
E-mail bubbles@summer.chem.su.oz.au Tel. 61(2)6924268 Fax 61(2)6923329

AUSTRIA

Sub-Editor: A. Preisinger

Amthauer, Prof. Dr Georg (1942). Full professor. [Institut für Mineralogie, Universität Salzburg, Hellbrunnerstraße 34, A-5020 Salzburg, Austria.] PhD (Universität Saarbrücken/FRG, 1974) mineralogy and crystallography. *Mineralogy crystallography Mossbauer solid_state absorption_spectroscopy charge_transfer magnetism order-disorder phase_transition soft_X-ray materials*
E-mail amthauer@edvz.sbg.ac.at Tel. 43(662)80445402 Fax 43(662)80445485

Badurek, Dr Gerald (1948). Assistant professor. [Institut für Kernphysik, Technische Universität Wien, Schüttelstraße 115, A-1020 Wien, Austria.] DSc (Technische Universität Wien, 1977) experimental neutron physics. *Polarized_neutron pulsed_neutron magnetic_domain high_field quantum_mechanics neutron_interferometry*
E-mail badurek@eati2.una.ac.at Tel. 43(1)72701x229 Fax 43(1)7289220

Bauer, Prof. Dr Günther (1942). Professor of physics. [Institut für Halbleiterphysik, Johannes Kepler Universität Linz, Altenbergerstraße 69, A-4040 Linz, Austria.] PhD (Universität Wien, 1966) physics. *Molecular_beam epitaxy semiconductor high_resolution_X-ray_diffractometry reciprocal_space_mapping scanning_tunneling_microscopy low_dimensional_semiconductor_systems*
E-mail k352470@edvz.linz.ac.at Tel. 43(732)24689601 Fax 43(732)246810

Baumgartner, Dr Oswald (1948). Assistant professor. [Institut für Mineralogie, Kristallographie und Strukturchemie, Technische Universität Wien, Getreidemarkt 9/171, A-1060 Wien, Austria.] DSc (Technische Universität Wien, 1977) chemistry. *Crystal_structure neutron_diffraction*
E-mail obaumgar@fbch.tuwien.ac.at Tel. 43(1)58801x4743

Beran, Prof. Dr Anton (1944). Assistant professor, lecturer in mineralogy. [Institut für Mineralogie und Kristallographie, Universität Wien, Dr. Karl Lueger-Ring 1, A-1010 Wien, Austria.] PhD (Universität Wien, 1970) mineralogy. *Spectroscopy infrared minerals microscopy trace_hydrogen_in_minerals reflected_light_microscopy mineralization optical_property gemology*
E-mail anton.beran@min.univie.ac.at Tel. 43(1)40103x2320 Fax 43(1)4037622

Blaschko, Prof. Dr Oskar (1948). Associate professor. [Institut für Experimentalphysik, Universität Wien, Strudlhofgasse 4, A-1090 Wien, Austria.] PhD (Universität Wien, 1974) physics. *Phase_transition phase_separation neutron_scattering*
Tel. 43(1)31367x3037 Fax 43(1)3102683 Telex 116222

Boller, Prof. Dr Herbert (1937). Full professor for general and inorganic chemistry. [Institut für Chemie, Universität Linz, Altenbergerstraße 69, A-4040 Linz, Austria.] PhD (Universität Wien, 1963) physical chemistry. *Chalcogenides metals synthesis magnetochemistry inorganic_solid_state_chemistry materials_science soft_chemistry*
Tel. 43(732)2468801 Fax 43(732)246810

Brandstätter, Dr Franz (1953). Curator. [Naturhistorisches Museum Wien, PO Box 417, A-1014 Wien, Austria.] PhD (Universität Wien, 1979) mineralogy and petrography. *Electron_probe_micro-analysis energy-dispersive_analysis meteorites*
Tel. 43(1)52177x270 Fax 43(1)935254

Breiter, Prof. Dr Manfred W. (1925). Full professor. [Institut für Technische Elektrochemie, Technische Universität Wien, Getreidemarkt 9/158, A-1060 Wien, Austria.] DSc (Technische Universität München, 1957) physical chemistry and electrochemistry. *Electrochemistry ionic_conductivity thermoanalysis electrode_kinetics batteries fuel_cells sensors*
Tel. 43(1)58801x4762 Fax 43(1)5874835

Dirl, Prof. Dr Rainer (1941). Assistant professor. [Institut für Theoretische Physik, Technische Universität Wien, Wiedner Hauptstraße 8-10/136, A-1040 Wien, Austria.] PhD (Technische Universität Wien, 1977) mathematical physics. *Group_theory space_group_symmetry representation_theory solution*
E-mail rdirl@email.tuwien.ac.at Tel. 43(1)58801x5684 Fax 43(1)567760

Eder, Prof. Dr Otto J. (1936). Division head. Vice President, Austrian Society of Nondestructive Testing; Board Member, Austrian Space Agency. [Österreichisches Forschungszentrum Seibersdorf, A-2440 Seibersdorf, Austria.] DSc (Technische Universität Wien, 1967) physics. *Liquid_state dynamics structure_factor small-angle_scattering research_and_development_management science_and_society international_science*
E-mail eder@zdfzs.arcs.ac.at Tel. 43(2254)780x3100 Fax 43(2254)74060 Telex 14-353 fzs

Effenberger, Dr Herta (1954). Lecturer. [Institut für Mineralogie und Kristallographie, Universität Wien, Dr. Karl Lueger-Ring 1, A-1010 Wien, Austria.] PhD (Universität Wien, 1978) crystal chemistry. *Crystal_chemistry synthesis minerals inorganic_compounds*
Tel. 43(1)40103x2687 Fax 43(1)4037622

Ettmayer, Prof. Dr Peter (1934). Associate professor. [Institut für Chemische Technologie anorganischer Stoffe, Technische Universität Wien, Getreidemarkt 9/161, A-1060 Wien, Austria.] DSc (Technische Universität Wien, 1964) chemistry. *Phase_diagram phase_transition structure_determination sintering high-temperature_X-ray_diffraction*
Tel. 43(1)58801x4799 Fax 43(1)5868271

Gieren, Prof. Dr Alfred (1939). Full professor. [Institut für Analytische Chemie und Radiochemie, Leopold-Franzens-Universität Innsbruck, Innrain 52a, A-6020 Innsbruck, Austria.] DSc (Technische Universität München, 1974) chemistry. *Structure_determination stereochemistry heterocycles biomolecules S–N_multiple_bonding_systems charge_densities*
Tel. 43(512)5075148x72 Fax 43(512)580519

Giester, Mag. Dr Gerald (1960). Postdoctoral research assistant. [Institut für Mineralogie und Kristallographie, Universität Wien, Geozentrum, Althanstr. 14, A-1090 Wien, Austria.] PhD (Universität Wien, 1990) mineralogy and crystallography. *Inorganic_structure mineralogy inorganic_synthesis inorganic_stereochemistry sulfates selenates iron_compounds copper_compounds hydrogen_bonding powder_diffraction*
E-mail gerald.giester@min.univie.ac.at Tel. 43(1)31336 Fax 43(1)4037622

Glatter, Prof. Dr Otto (1945). Associate professor; Head of Institute. [Institut für Physikalische Chemie, Universität Graz, Heinrichstraße 28, A-8010 Graz, Austria.] DSc (Technische Universität Graz, 1972) theoretical physics. *SAXS inverse_problem computing light_scattering*
E-mail glatter@bkfug.kfunigraz.ac.at Tel. 43(316)380x5433 Fax 43(316)322248

Götzinger, Dr Michael Alois (1949). Assistant professor. [Institut für Mineralogie und Kristallographie, Universität Wien, Dr. Karl Lueger-Ring 1, A-1010 Wien, Austria.] PhD (Universität Wien, 1976) mineralogy and petrography. *Mineralogy electron_microscopy EDX infrared_spectroscopy inclusions_in_minerals genesis_of_industrial_minerals*
E-mail Michael.Goetzinger@min.univie.ac.at Tel. 43(1)40103x2688 Fax 43(1)4037622

Gruber, Mr Karl (1968). PhD student. [Institut für Physikalische Chemie, Universität Graz, Heinrichstraße 28, A-8010 Graz, Austria.] MSc (Universität Graz, 1991) physical chemistry. *Biocrystallography low_temperature disorder computing*
E-mail gruber@bkfug.kfunigraz.ac.at Tel. 43(316)380x5469 Fax 43(316)32248

Haditsch, Prof. Dr Johann Georg (1934). Full professor. [Institut für Geowissenschaften, Montan-Universität Leoben, Franz-Josef-Straße 18, A-8700 Leoben, Austria.] PhD (Universität Graz, 1958) geology, mineralogy, petrography. *Environment_protection archeometry ores sediments industrial_minerals_and_rocks*
Tel. 43(3842)402x452 Fax 43(3842)47016

Halwax, Dr Erich Johann (1951). Assistant professor. [Institut für Mineralogie, Kristallographie und Strukturchemie, Technische Universität Wien, Getreidemarkt 9/171, A-1060 Wien, Austria.] DSc (Technische Universität Wien, 1985) chemistry. *Inorganic_crystal_structure germanates diffractometry computing powder_diffraction quantitative_phase_analysis*
E-mail ehalwax@fbch.tuwien.ac.at Tel. 43(1)58801x4742 Fax 43(1)568136

Hiebl, Dr Kurt (1948). Assistant professor. [Institut für Physikalische Chemie, Universität Wien, Währingerstraße 42, A-1090 Wien, Austria.] PhD (Universität Wien, 1973) physical chemistry. *Condensed_matter magnetism superconductivity fermion*
Tel. 43(1)31367x2504 Fax 43(1)3104597

Higatsberger, Prof. Dr Dr h. c. Michael J. (1924). Full professor. [Institut für Experimentalphysik, Universität Wien, Boltzmanngasse 5, A-1090 Wien, Austria.] PhD (Universität Wien, 1949) physics. *Computer-aided_education electron_probe_microanalysis tunnelling vacuum surface_physics nuclear_and_reactor_physics*
E-mail higat@apap1.pap.univie.ac.at Tel. 43(1)31367x3007 Fax 43(1)3102683 Telex 116222

Jeglitsch, Prof. Dr Dr h. c. Franz (1934). Full professor. [Institut für Metallkunde und Werkstoffprüfung, Montanuniversität Leoben, Franz-Josef-Straße 18, A-8700 Leoben, Austria.] DSc (Montanuniversität Leoben, 1963) mining sciences. *Material_sciences microstructure solidification failure_analysis mechanical_properties crystallization phase_diagram PVD laser steels*
Tel. 43(3842)402x304 Fax 43(3842)402737

Jogl, Mr Gerwald (1968). Student. [Institut für Physikalische Chemie, Universität Graz, Heinrichstraße 28, A-8010 Graz, Austria.] *Computing biomolecule low_temperature biocrystallography area_detectors proteins*
E-mail jogl@bkfug.kfunigraz.ac.at Tel. 43(316)380x5468 Fax 43(316)32248

Kabelka, Dr Heinz (1949). Assistant professor. [Institut für Experimentalphysik, Universität Wien, Strudlhofgasse 4, A-1090 Wien, Austria.] PhD (Universität Wien, 1976) physics. *Computing phase_transition*
E-mail kab@pap.univie.ac.at Tel. 43(1)31367x3044 Fax 43(1)3102683 Telex 116222

Kahlert, Dr Hartmut (1940). Full professor. [Institut für Festkörperphysik, Technische Universität Graz, Petersgasse 16, A-8010 Graz, Austria.] PhD (U. Wien, 1965) physics. *Conducting_polymers superconductors crystallinity semiconductors*
Tel. 43(316)8738460 Fax 43(316)8738466

Kirchner, Prof. Dr Elisabeth Ch. (1935). Associate professor. [Institut für Mineralogie, Universität Salzburg, Hellbrunnerstraße 34, A-5020 Salzburg, Austria.] PhD (Universität Wien, 1965) mineralogy and petrography. *Mineralogy diffractometry volcanology Rietveld_method optical_property chemistry main/trace_elements petrography*
E-mail kirchner@dsb835.edvz.sbg.ac.at Tel. 43(662)80445403 Fax 43(662)80445485

Klintschar, Mr Gerd (1967). Research student. [Institut für Physikalische Chemie, Universität Graz, Heinrichstraße 28, A-8010 Graz, Austria.] *Biocrystallography low_temperature solvent_structure*
E-mail klintschar@bkfug.kfunigraz.ac.at Tel. 43(316)380x5469 Fax 43(316)32248

Kohlbeck, Dr Franz (1943). Assistant professor. [Institut für Theoretische Geodäsie und Geophysik, Technische Universität Wien, Gußhausstraße 27-29/128, A-1040 Wien, Austria.] DSc (Technische Universität Wien, 1974) physics. *Computing minerals crystallography inverse_problem geophysics*
E-mail fkohlbec@email.tuwien.ac.at Tel. 43(1)58801x3803 Fax 43(1)5044232

Komarek, Prof. Dr Kurt L. (1926). Full professor. [Institut für Anorganische Chemie, Universität Wien, Währingerstraße 42, A-1090 Wien, Austria.] PhD (Universität Wien, 1950) chemistry. *Thermodynamics phase_diagram nonstoichiometry chalcogenides science_policy enthalpies_of_mixing vapor_pressure_measurements advanced_materials liquid_alloys*
Tel. 43(1)31367x2001 Fax 43(1)3104597

Kratky, Dr Christoph (1946). Associate professor. [Institut für Physikalische Chemie, Universität Graz, Heinrichstraße 28, A-8010 Graz, Austria.] PhD (ETH Zürich, 1977) chemistry. *Low_temperature_crystallography diffractometry proteins vitamins solid_state_reactions*
E-mail kratky@bkfug.kfunigraz.ac.at Tel. 43(316)380x5417 Fax 43(316)32248

Kratky, Prof. Dr Dr h. c. mult. Otto (1902). Professor emeritus. Institut für Physikalische Chemie, Universität Graz, retired 1972, Institut für Röntgenfeinstrukturforschung der Forschungsgesellschaft Joanneum, und der Österreichischen Akademie der Wissenschaften, retired 1982. [Drosselweg 15, A-8010 Graz, Austria.] DSc (Technische Universität Wien, 1929) chemistry. *High_polymers small-angle_scattering physical_chemistry_of_high_polymers X-ray-behaviour_of_high_polymers*
Tel. 43(316)472073

Krischner, Prof. Dr Harald (1930). Professor retired. Institut für Physikalische und Theoretische Chemie, Technische Universität Graz, Rechbauerstraße 12, A-8010 Graz, Austria. [Stenggstraße 63, A-8043 Graz, Austria.] DSc (Technische Universität Graz, 1959) structural research. *X-ray diffraction structure_determination azides chemistry*
Tel. 43(316)348895

Kuzmany, Prof. Dr Hans (1940). Associate professor. [Institut für Festkörperphysik, Universität Wien, Strudlhofgasse 4, A-1090 Wien, Austria.] PhD (Universität Wien, 1966) experimental physics. *High_temperature_superconductivity polymers structure vibrational_spectroscopy fullerenes*
E-mail kuzman@awirap Tel. 43(1)31367x3206 Fax 43(1)3103888

Laggner, Prof. Dr Peter (1944). Research director. [Institut für Biophysik und Röntgenstrukturforschung, Österreichische Akademie der Wissenschaften, Steyrergasse 17, A-8010 Graz, Austria.] PhD (Karl-Franzens-Universität Graz, 1971) chemistry/physics. *Small-angle_scattering synchrotron_radiation lipoproteins membranes phase_transition porous_materials fractal_materials crystallization*
E-mail fscm1.tu-graz.ac.at Tel. 43(316)812003 Fax 43(316)812367

Lengauer, Dr Christian Leopold (1959). Assistant professor. [Institut für Mineralogie und Kristallographie, Universität Wien, Dr. Karl Lueger-Ring 1, A-1010 Wien, Austria.] PhD (Universität Salzburg, 1989) geology and mineralogy. *Rietveld_method zeolites diffraction_technique optical_property ore mineralogy*
E-mail Christian.Lengauer@min.univie.ac.at Tel. 43(1)40103x2333 Fax 43(1)4037622

Lengauer, Dr Walter Oskar Franz (1958). Assistant professor. [Institut für Chemische Technologie anorganischer Stoffe, Technische Universität Wien, Getreidemarkt 9/161, A-1060 Wien, Austria.] DSc (Technische Universität Wien, 1987) metallurgy. *Diffusion high_temperature phase_equilibrium solid_state carbides nitrides solid_state_property technology*
E-mail lengauer@email.tuwien.ac.at Tel. 43(1)58801x4812 Fax 43(1)5868271

Lottermoser, Dr Werner (1955). Research assistant. [Institut für Mineralogie, Universität Salzburg, Hellbrunnerstraße 34/III, A-5020 Salzburg, Austria.] PhD (Universität Frankfurt, FRG, 1986) solid state physics. *Mossbauer spectroscopy physics minerals neutron_diffraction cryophysics computing single_crystals silicates magnetism*
Tel. 43(662)80445422 Fax 43(662)80445485

Lux, Prof. Dr Benno (1930). Full professor. [Institut für Chemische Technologie anorganischer Stoffe, Technische Universität Wien, Getreidemarkt 9/161, A-1060 Wien, Austria.] DSc (Technische Universität Graz, 1956) inorganic and physical chemistry. *Tungsten_compounds solidification low_pressure_diamond_synthesis tungsten CVD_thin_layer CVD_deposit cast_iron*
E-mail blux@email.tuwien.ac.at Tel. 43(1)58801x4787 Fax 43(1)5868271

Mantler, Dr Michael K. (1945). Group leader. [Institut für Angewandte und Technische Physik, Technische Universität Wien, Wiedner Hauptstraße 8-10/1374, A-1040 Wien, Austria.] DSc (Technische Universität Wien, 1973) X-ray fluorescence. *X-ray_fluorescence instrumentation computer_automation computer_modelling crystallography*
E-mail mmantler@atpibm6000.tuwien.ac.at Tel. 43(1)58801x5668 Fax 43(1)5878876

Mautner, Dr Franz Andreas (1955). Assistant professor. [Institut für Physikalische und Theoretische Chemie, Technische Universität Graz, Rechbauerstraße 12, A-8010 Graz, Austria.] DSc (Technische Universität Graz, 1988) chemistry. *X-ray structure_determination single_crystal powder azides chemistry*
E-mail mautner@ftug.denet.tu-graz.ac.at Tel. 43(316)8738221 Fax 43(316)8736427

Mayer, Dr Helmut (1939). Staff scientist, lecturer. [Institut für Mineralogie, Kristallographie und Strukturchemie, Technische Universität Wien, Getreidemarkt 9/171, A-1060 Wien, Austria.] DSc (Technische Universität Wien, 1971) chemistry. *Crystallography mineralogy thermoanalysis archeometry*
Tel. 43(1)58801x4742

Mayr, Dr Michael (1945). Manager of laboratory. [VOEST-Alpine Stahl Linz GmbH, Dept. SFP 3, Postbox 3, A-4031 Linz, Austria.] DSc (Technische Universität Wien, 1975) applied physics. *Characterization diffractometry phase_determination texture refractories slags steel metallic_coatings_on_steel*
Tel. 43(732)5852031 Fax 43(732)59809572 Telex 2207461 va a

Mereiter, Prof. Dr Kurt (1945). Assistant professor. [Institut für Mineralogie, Kristallographie und Strukturchemie, Technische Universität Wien, Getreidemarkt 9/171, A-1060 Wien, Austria.] PhD (Universität Wien, 1975) mineralogy and crystallography. *Crystallography structure_determination mineralogy mineral_crystal_structures organometallic_crystal_structures organic_crystal_structures*
E-mail kmereite@fbch.tuwien.ac.at Tel. 43(1)58801x4747 Fax 43(1)568136

Mikenda, Dr Werner (1946). Assistant professor. [Institut für Organische Chemie, Universität Wien, Währingerstraße 38, A-1090 Wien, Austria.] PhD (Universität Wien, 1976) chemistry. *Spectroscopy structure hydrogen_bonding structure-spectroscopy_relationship*
Tel. 43(1)31367x2209 Fax 43(1)3196312

Mirwald, Prof. Dr Peter W. (1937). Professor; Director of the Institute. Institut für Mineralogie und Petrographie der Universität Innsbruck, Austria. [Institut für Mineralogie und Petrographie der Universität Innsbruck, Innrain 52, A-6020 Innsbruck, Austria.] Dr rer.nat. (U. München, 1971) mineralogy and crystallography. *Experimental_mineralogy silicates carbonates metamorphism physical_and_chemical_properties_of_minerals*
E-mail mineralogie-e714@uibk.ac.at Tel. 43(512)5075501 Fax 43(512)5072926

Neckel, Prof. Dr Adolf (1926). Full professor. [Institut für Physikalische Chemie, Universität Wien, Währingerstraße 42, A-1090 Wien, Austria.] PhD (Universität Wien, 1954) chemistry. *Electrochemistry solid_state band_calculation thermodynamics*
Tel. 43(1)31367x2501 Fax 43(1)3104597

Niedermayr, Dr Gerhard (1941). Curator, mineral collection; head of the Staatliches Edelstein-Institut. [Naturhistorisches Museum Wien, Mineralogisch-Petrographische Abteilung, PO Box 417, A-1014 Wien, Austria.] PhD (Universität Wien, 1965) sedimentary petrology. *Gemology hydrothermal_mineralization sediments petrology history_of_mineralogy Austrian_topographic_mineralogy*
Tel. 43(1)52177274 Fax 43(1)935254 43(1)52177264 Telex 134441 natur a

Pertlik, Prof. Dr Franz (1943). Associate professor. [Institut für Mineralogie und Kristallographie, Universität Wien, Geozentrum, Althanstr. 14, A-1090 Wien, Austria.] PhD (Universität Wien, 1969) mineralogy and petrography. *Crystal_structure_determination crystal_chemistry crystal_physics minerals_synthesis X-ray_techniques synchrotron_radiation inorganic_crystal_chemistry hydrothermal_synthesis*
Tel. 43(1)40103x2328 Fax 43(1)4037622

Preisinger, Prof. Dr Anton (1925). Full professor and head. Austrian representative; Sub-Editor WDC9. [Institut für Mineralogie, Kristallographie und Strukturchemie, Technische Universität Wien, Getreidemarkt 9/171, A-1060 Wien, Austria.] PhD (Universität Wien, 1953) mineralogy and crystal chemistry. *Crystallography mineralogy crystal_growth structure_determination phase_determination phase_transition structure_classification biomolecule*
E-mail e1711dab@awiuni11.edvz.univie.ac.at Tel. 43(1)58801x4749 Fax 43(1)568136

Rauch, Prof. Dr Helmut (1939). Full professor. Member of IUPAP Commission C-10. [Atominstitut der Österreichischen Universitäten, Schüttelstraße 115, A-1020 Wien, Austria.] DSc (Technische Universität Wien, 1965) physics. *Neutron interferometry quantum_mechanics perfect_crystal reactor_physics neutron_scattering*
E-mail rauch@ati.ac.at Tel. 43(1)72701x266 Fax 43(1)7289220

Reissner, Dr Michael (1956). Postdoctoral research assistant. [Institut für Angewandte und Technische Physik, Technische Universität Wien, Wiedner Hauptstraße 8-10/137, A-1040 Wien, Austria.] DSc (Technische Universität Wien, 1990) solid state physics. *Magnetism superconductivity Mossbauer physical_property*
E-mail reissner@email.tuwien.ac.at Tel. 43(1)58801x5635 Fax 43(1)5868814

Rogl, Dr Peter Franz (1945). Assistant professor. [Institut für Physikalische Chemie, Universität Wien, Währingerstraße 42, A-1090 Wien, Austria.] PhD (Universität Wien, 1971) chemistry. *Crystallography structure thermodynamics phase_diagram high-melting_systems solid_state_chemistry alloys refractories borides carbides magnetism*
Tel. 43(1)31367x2537

Schattschneider, Dr Peter (1950). Assistant professor. [Institut für Angewandte und Technische Physik, Technische Universität Wien, Wiedner Hauptstraße 8-10/137, A-1040 Wien, Austria.] DSc (Technische Universität Wien, 1976) technical physics. *EELS plasmon solid_state microbeam_analysis science_fiction*
E-mail schatt@email.tuwien.ac.at Tel. 43(1)58801x5626 Fax 43(1)5868814

Schranz, Dr Wilfried (1960). Assistant. [Institut für Experimentalphysik, Universität Wien, Strudlhofgasse 4, A-1090 Wien, Austria.] PhD (Universität Wien, 1988) solid state physics. *Phase_transition ultrasonics domain_structure structural_disorder group_theory birefringence symmetry statistical_thermodynamics defects*
E-mail schranz@pap.univie.ac.at Tel. 43(1)3191366x3071 Fax 43(1)3102683 Telex 116222

Schubert, Prof. Dr Ulrich (1946). Professor of Inorganic Chemistry. Institut für Anorganische Chemie der Technischen Universität Wien, Austria. [Institut für Anorganische Chemie der Technischen Universität Wien, Getreidemarkt 9, A-1060 Wien, Austria.] Dr (TU München, 1975) chemistry. *Chemistry molecular_crystal main-group_compounds materials*
Tel. 43(1)588014633 Fax 43(1)5874835

Schuster, Dr Julius (1952). Assistant professor. [Institut für Physikalische Chemie, Universität Wien, Währingerstraße 42, A-1090 Wien, Austria.] PhD (Universität Wien, 1977) physics/chemistry. *Intermetallic_compounds ceramics metal_ceramic_interfaces*
E-mail jcs.phc@univie.ac.at Tel. 43(1)31367x2518 Fax 43(1)3104597

Schwarz, Prof. Dr Karlheinz (1941). Associate professor. Commission on Charge, Spin and Momentum Densities. Wien, Gumpendorferstr. 1a, A-1060 Vienna, Austria. [Institut für Technische Elektrochemie, Technische Universität Wien, Getreidemarkt 9/158, A-1060 Wien, Austria.] PhD (Universität Wien, 1968) physical chemistry. *Band_calculation condensed_matter quantum_mechanics computing cohesive_energy magnetism spin_densities density_distribution electron_densities transition_metal_compounds oxides carbides nitrides molecular_dynamics phase_transitions*
E-mail kschwarz@email.tuwien.ac.at Tel. 43(1)58801x5188 Fax 43(1)5868937

Seidl, Dr Erwin (1939). Group leader. [Atominstitut der Österreichischen Universitäten, Schüttelstraße 115, A-1020 Wien, Austria.] PhD (Universität Wien, 1966) physics. *Crystal_growth neutron_interferometry neutron_optics superconductivity Bridgman_Stockbarger_technique Czochralski_technique education etching float_zone_growth intermetallic_alloys luminescent_compounds orientation perfect_crystals ultrahigh_vacuum_preparation_of_perfect_silicon_crystals_for_neutron_interferometry_and_neutron_optics*
Tel. 43(1)72701x259 Fax 43(1)7289220 Telex 3222467=tuw

Sitter, Prof. Dr Helmut (1951). Associate professor. [Institut für Experimentalphysik, Universität Linz, Altenbergerstraße 69, A-4040 Linz, Austria.] PhD (Universität Linz, 1982) solid state physics. *Epitaxy semiconductor AES luminescence II-VI_compounds optoelectronics*
E-mail H.Sitter@jk.uni-linz.ac.at Tel. 43(732)24689623 Fax 43(732)2468650 Telex 022323 uni li a

Skalicky, Prof. Dr Peter (1941). Full professor. [Institut für Angewandte und Technische Physik, Technische Universität Wien, Wiedner Hauptstraße 8-10/137, A-1040 Wien, Austria.] DSc (Technische Universität Wien, 1965) physics. *Crystal_defects electron_diffraction electron_microscopy X-ray_diffraction*
E-mail skalicky@email.tuwien.ac.at Tel. 43(1)58801x3004 Fax 43(1)5878905

Spindler, Dr Peter (1959). Research assistant. [BVFA Arsenal / Geotechnisches Institut, Faradaygasse 3, A-1030 Wien, Austria.] PhD (Universität Wien, 1991) mineralogy-geochemistry. *Mineralogy analytical_geochemistry*
Tel. 43(1)79747457 Fax 43(1)79747520

Steiner, Prof. Dr Walter (1942). Associate professor. [Institut für Angewandte und Technische Physik, Technische Universität Wien, Wiedner Hauptstraße 8-10/137, A-1040 Wien, Austria.] DSc (Technische Universität Wien, 1970) low temperature physics. *Lattice_dynamics magnetism Mossbauer neutron_diffraction Rayleigh_scattering superconductivity*
E-mail steiner@email.tuwien.ac.at Tel. 43(1)58801x5636 Fax 43(1)5868814

Stumpfl, Prof. Dr Eugen F. (1931). Full professor of mineralogy and petrology. [Institut für Geowissenschaften, Montan-Universität Leoben, Franz-Josef-Straße 18, A-8700 Leoben, Austria.] PhD (Universität Heidelberg, 1956) mineralogy. *Geochemistry reflected_light_microscopy petrology microbeam_analysis economic_geology ore_mineralogy*
Tel. 43(3842)402x451 Fax 43(3842)47016

Tillmanns, Prof. Dr Ekkehart (1941). Full professor. [Institut für Mineralogie und Kristallographie, Universität Wien, Geozentrum, Althanstr. 14, A-1090 Wien, Austria.] PhD (Universität Bochum, FRG, 1968) mineralogy/crystallography. *Crystallography mineralogy diffraction*
E-mail tillmanns@min.univie.ac.at Tel. 43(1)40103x2333 Fax 43(1)4037622

Vana, Prof. Dr Norbert (1940). Professor; Head of Division for dosimetry and radiation protection. [Atominstitut der Österreichischen Universitäten, Schüttelstraße 115, A-1020 Wien, Austria.] DSc (Technische Universität Wien, 1967) solid state physics. *Dosimetry radiation_protection thermoluminescence spectroscopy archeometry*
E-mail vana@ati.ac.at Tel. 43(1)72701x277 Fax 43(1)7289220 Telex 3222467=tuw

Vogl, Prof. Dr Gero (1941). Full professor. [Institut für Festkörperphysik, Universität Wien, Strudlhofgasse 4, A-1090 Wien, Austria.] DSc (Technische Universität München, 1974) experimental physics. *Diffusion structural_disorder neutron_scattering gamma-ray_resonance_spectroscopy*
Tel. 43(1)31367x3204 Fax 43(1)3100183 Telex 116222

Völlenkle, Dr Horst (1938). Assistant professor. [Institut für Mineralogie, Kristallographie und Strukturchemie, Technische Universität Wien, Getreidemarkt 9/171, A-1060 Wien, Austria.] PhD (Universität Wien, 1965) chemistry. *Diffractometry structure_determination silicates germanates software*
E-mail hvoellen@fbch.tuwien.ac.at Tel. 43(1)58801x4742

Wagner, Dr Ulrike Gabriella (1960). Research assistant. [Institut für Physikalische Chemie, Karl-Franzens-Universität Graz, Heinrichstraße 28, A-8010 Graz, Austria.] PhD (Universität Graz, 1987) chemistry. *Proteins low_temperature_crystallography crystallization*
E-mail wagneru@bkfug.kfunigraz.ac.at Tel. 43(316)380x5431 Fax 43(316)32248

Walltzl, Prof. Dr Eva Maria (1930). Associate professor, group leader. [Institut für Mineralogie-Kristallographie und Petrologie, Karl-Franzens-Universität Graz, Universitätsplatz 2, A-8010 Graz, Austria.] PhD (Karl-Franzens-Universität Graz, 1958) mineralogy-geology. *Mineralogy crystallography silicates crystal_chemistry minerals_of_pegmatites*
E-mail ettinger@edvz.uni-graz.ada.at Tel. 43(316)380x5541 Fax 43(316)386093

Walter, Dr Franz (1952). Assistant professor. [Institut für Mineralogie-Kristallographie und Petrologie, Karl-Franzens-Universität Graz, Universitätsplatz 2, A-8010 Graz, Austria.] PhD (Karl-Franzens-Universität Graz, 1980) mineralogy-geology. *Mineralogy crystallography silicates phosphates minerals_of_pegmatites beryllophosphates*
E-mail ettinger@edvz.uni-graz.ac.at Tel. 43(316)380x5547 Fax 43(316)386093

Warhanek, Prof. Dr Hans (1926). Full professor. [Institut für Experimentalphysik, Universität Wien, Strudlhofgasse 4, A-1090 Wien, Austria.] PhD (Universität Wien, 1951) physics. *Phase_transition ferroelectricity crystal_growth orientational_glasses* E-mail warhanek@pap.univie.ac.at Tel. 43(1)31367x3005 Fax 43(1)3102683 Telex 116222

Wildner, Mag. Dr Manfred (1962). Postdoctoral research assistant. [Institut für Mineralogie und Kristallographie, Universität Wien, Geozentrum, Althanstr. 14, A-1090 Wien, Austria.] PhD (Universität Wien, 1991) mineralogy and crystallography. *Inorganic_structure stereochemistry optical_absorption_spectroscopy inorganic_cobalt_compounds mineralogy inorganic_synthesis crystal_field_theory Jahn_Teller_effect hydrogen_bonding selenites sulfates* E-mail manfred.wildner@min.univie.ac.at Tel. 43(1)31336 Fax 43(1)4037622

Winter, Prof. Dr Hannspeter (1941). Full professor; Institute Director. [Institut für Allgemeine Physik, Technische Universität Wien, Wiedner Hauptstraße 8-10/134, A-1040 Wien, Austria.] DSc (Technische Universität Wien, 1970) physics. *Surface_physics plasmas atomic_collisions* E-mail winter@eapv38.tuwien.ac.at Tel. 43(1)58801x5710 Fax 43(1)5864203

Wobrauschek, Dr Peter (1939). Associate professor. [Atominstitut der Österreichischen Universitäten, Schüttelstraße 115, A-1020 Wien, Austria.] DSc (Technische Universität Wien, 1975) physics. *X-ray_fluorescence-spectroscopy grazing_incidence X-ray_polarization diffraction_technique instrumentation_and_development_of_ EDXRF_spectrometer total_reflection_X-ray_fluorescence*

Zeilinger, Prof. Dr Anton (1945). Full professor; Institute Director. [Institut für Experimentalphysik, Universität Innsbruck, Technikerstraße 25/4, A-6020 Innsbruck, Austria.] PhD (Universität Wien, 1971) physics. *Dynamical_diffraction perfect_crystal slow_neutron quantum_mechanics philosophy_of_science* E-mail anton.zeilinger@uibk.ac.at Tel. 43(512)2186300 Fax 43(512)2182921

Zemann, Emer. Prof. Dr Josef (1923). Professor emeritus. [Institut für Mineralogie und Kristallographie, Universität Wien, Geozentrum, Althanstr. 14, A-1090 Wien, Austria.] PhD (Universität Wien, 1946) mineralogy. *Inorganic_stereochemistry minerals optics* Tel. 43(1)31336 Fax 43(1)4037622

Zipper, Prof. Dr Peter (1941). Associate professor. [Institut für Physikalische Chemie, Universität Graz, Heinrichstraße 28, A-8010 Graz, Austria.] PhD (Universität Graz, 1970) chemistry. *Small-angle_scattering wide-angle_scattering polymers biopolymers* E-mail zipper@bkfug.kfunigraz.ac.at Tel. 43(316)380x5415 Fax 43(316)32248

Zobetz, Dr Erich (1950). Assistant professor. [Institut für Mineralogie, Kristallographie und Strukturchemie, Technische Universität Wien, Getreidemarkt 9/171, A-1060 Wien, Austria.] PhD (Universität Wien, 1983) mineralogy and petrology. *Crystallography structure_determination quasicrystals homometry diffraction_enhancement_of_symmetry* Tel. 43(1)58801x5053

AZERBAIJAN

Sub-Editor: N. Mustafayev

Agaiev, Dr Gahraman Agaiev (1938). Associate Professor. [Baku State University, Z. Halilov St. 23, 370145 Baku, Azerbaijan.] PhD (Baku U., 1968) crystallography and crystallophysics. *Electronography chalcogenides* Tel. 7(8922)2390523

Agayev, Dr Agamehti Agayev (1935). Lecturer. [Azerbaijan Oil Academy, Azadlyg Prospekti 20, 370003 Baku, Azerbaijan.] PhD (Inst.of Inorganic and Physical Chemistry, Acad. Sci. Azerbaijan, 1969) chemistry, chemistry and crystallography. *Mineralogy crystallography boron_compounds history_of_science* Tel. 7(8922)934461

Amiraslanov, Dr Imameddin Amiraslanov (1948). Head of Laboratory. [Lab. of Structural Chemistry, Institute of Inorganic and Physical Chemistry, Academy of Sciences Azerbaijan, Husein Javid Prospekti 29, 370143 Baku, Azerbaijan.] PhD (Inst.of Applied Physics, Academy of Sciences Moldavia, 1979) physics, crystallography and crystallophysics. *X-ray_study inorganic_layered_compounds crystallographic_ornaments* Tel. 7(8922)394163

Amirov, Prof. Amirov (1939). Lecturer. [Azerbaijan Technical University, Husein Javid Prospekti 25, 370073 Baku, Azerbaijan.] PhD (Inst.of Chemistry of Academy of Sciences Uzbekistan, 1991) chemistry, inorganic chemistry. *Crystal_structure_analysis silicates crystal_chemistry_silicates* Tel. 7(8922)389677 7(8922)381303

Asadov, Prof. Ysif Gazanfar oglu (1934). Head of Laboratory. [Lab. of Structural Transitions, Institute of Physics, Academy of Sciences Azerbaijan, Husein Javid Prospekti 33, 370143 Baku, Azerbaijan.] PhD (Inst. Physics of Academy of Sciences Azerbaijan, 1980) physics, semiconductors and dielectrics. *Polymorphism phase_transitions organic_crystal inorganic_crystal crystal growth* Tel. 7(8922)392135

Chiragov, Prof. Mamed Isa oglu (1938). Lecturer. Chairman, National Committee for Crystallography, Azerbaijan. [Baku State University, Z. Halilov St. 23, 370145 Baku, Azerbaijan.] DSc (Leningrad U., 1989) crystallography and crystallophysics. *X-ray_study inorganic_modulated_structures mathematics statistics* Tel. 7(8922)2390981 Fax 7(8922)2390521

Gasymov, Dr Vagif Akber ogly (1947). Group leader. [Lab. of Chemistry of Silicates, Institute Inorganic and Physical Chemistry, Academy of Sciences Azerbaijan, Husein Javid Prospekti 29, 370143 Baku, Azerbaijan.] PhD (Inst.of Inorganic and Physical Chemistry of Academy Sciences, 1990) chemistry, inorganic chemistry. *X-ray_study_inorganic_semiconductors crystal_synthesis_and_phase_transitions* Tel. 7(8922)394163 7(8922)770787

Guseinov, Dr Gahraman Gusein oglu (1937). Group leader. [Lab. of Structural Transitions, Institute of Physics, Academy of Sciences Azerbaijan, Husein Javid Prospekti 33, 370143 Baku, Azerbaijan.] PhD (Inst. of Inorganic and Physical Chemistry of Academy of Sciences, 1968) chemistry, inorganic chemistry. *X-ray_study_inorganic_semiconductors crystal_synthesis_and_phase_transition* Tel. 7(8922)392135 7(8922)948672

Hamidov, Prof. Hamidov (1927). Lecturer. [Azerbaijan Technical University, Husein Javid Prospekti 25, 370073 Baku, Azerbaijan.] PhD (Azerbaijan U., 1973) chemistry, inorganic chemistry. *History_of_sciences* Tel. 7(8922)678757 7(8922)391425

Jafarov, Dr Jafarov (1939). Research Fellow. [Lab. of Phase Analysis, Institute of Inorganic and Physical Chemistry, Academy of Sciences Azerbaijan, Husein Javid Prospekti 29, 370143 Baku, Azerbaijan.] PhD (Inst. of Inorganic and Physical Chemistry of Academy of Sciences, 1989) chemistry, inorganic chemistry. *Structure boron_compounds crystal_chemistry_of_boron_compounds* Tel. 7(8922)394163

Mustafayev, Dr Nariman (1929). Group leader. Sub-Editor of ninth edition of World Directory of Crystallographers. [Lab. of Crystallochemistry, Institute Inorganic and Physical Chemistry, Academy of Sciences Azerbaijan, Husein Javid Prospekti 29, 370143 Baku, Azerbaijan.] PhD (Inst. of Crystallography of Academy of Sciences USSR, 1966) geology, crystallography and crystallophysics. *X-ray_study silicates phosphates molecular_sieves* Tel. 7(8922)625180 7(8922)389654 Fax 7(8922)676128 Telex 142232 Bilik

Osmanzadeh Shener, Dr Osmanzadeh (1935). Head of Laboratory. [Lab.of Phase Analysis, Institute of Inorganic and Physical Chemistry, Academy of Sciences Azerbaijan, Husein Javid Prospekti 29, 370143 Baku, Azerbaijan.] PhD (Inst. of Physical Chemistry of Academy of Sciences USSR, 1967) chemistry, structural chemistry. *Structure_chemistry zeolites structural_aspects_of_hydratation_of_silicates* Tel. 7(8922)690294

Ragimov, Dr Kerim Ragimov (1952). Researcher. [Department of Chemistry, Baku State University, Z. Khalilov Street 23, 370145 Baku, Azerbaijan.] PhD (Baku U., 1994) geology. *X-ray_study silicates philosophy* Tel. 7(8922)387293

Shnulin, Dr Anatoly Shnulin (1936). Leader of Laboratory. [Institute for Theoretical Problems of Chemical Technology, Academy of Sciences Azerbaijan, Husein Javid Prospekti 29, 370143 Baku, Azerbaijan.] PhD (Inst. of Applied Physics, Academy of Sciences of Moldavia, 1978) physics, chemical crystallography. *X-ray_conformational_analysis compounds philosophy_of_christianity* Tel. 7(8922)985501

BELGIUM

Sub-Editor: G. S. D. King

Aertgeerts, Ms Kathleen (1969). Teaching assistant. Laboratorium voor Analytische Chemie en Medicinale Fysicochemie, Faculteit Farmaceutische Wetenschappen, Katholieke U. Leuven, Van Evenstraat 4, B-3000 Leuven, Belgium. Pharmacist (U. Leuven, 1991) pharmacy. *Structural_chemistry macromolecules proteins computer_modelling serpins*
E-mail kathleen%ana%far@cc3.kuleuven.ac.be Tel. 32(16)323420 Fax 32(16)283414

Amelinckx, Prof. Severin (1922). Professor (retired). Member, Committee on Electron Diffraction. University of Antwerpen (RUCA), Dept. Natuurkunde, EMAT, Groenenborgerlaan 171, B-2020 Antwerpen, Belgium. PhD (U Gent, 1952) physics. *Defect_photography electron_microscopy electron_diffraction*
Tel. 32(3)2180245 Fax 32(3)2180257

Bender, Dr Hugo (1957). Senior researcher. IMEC, Materials and Packaging Division, Analysis and Reliability, Kapeldreef 75, B-3001 Leuven, Belgium. PhD (U Antwerpen, 1984) physics. *Electron_microscopy semiconductors silicon_compounds*
E-mail bender@imec.be Tel. 32(16)321304 Fax 32(16)281501 Telex 26152

Blaton, Prof. Norbert Maurice (1945). Senior lecturer. Laboratorium voor Analytische Chemie en Medicinale Fysicochemie, Faculteit Farmaceutische Wetenschappen, Katholieke U. Leuven, Van Evenstraat 4, B-3000 Leuven, Belgium. PhD (U. Leuven, 1974) chemistry. *Structure_determination powder_diffraction structural_chemistry*
E-mail norbert%ana%far@cc3.kuleuven.ac.be Tel. 32(16)323419 Fax 32(16)283414

Boesman, Prof. Etienne Roland (1932). Professor. Vakgroep Vaste-Stofwetenschappen, Universiteit Gent, Krijgslaan 281 S1, B-9000 Gent, Belgium. PhD (U Gent, 1962) paramagnetic resonance. *EPR optics defect_structures ENDOR*
E-mail etienne.boesman@rug.ac.be Tel. 32(9)2644341 Fax 32(9)2644996

Bouckaert, Ms Julie (1967). Research student. Instituut Moleculaire Biologie, Vrije U Brussel, Paardenstraat 65, B-1640 Sint Genesius Rode, Belgium. MSc (VUBrussel, 1990) biotechnology. *Proteins carbohydrates metalloproteins enzymology crystallization metalloporphyrins peptide_ligands*
E-mail bouckaej@vub.ac.be Tel. 32(2)3590209 Fax 32(2)3590390

Brasseur, Prof. Henri Alphonse Lambert (1905). Professor (retired). Laboratoire de Cristallographie, Université de Liège au Sart Tilman, Institut de Physique, bât. B5, B-4000 Liège, Belgium. Hab. Dipl. (U Liège, 1934) physics. *Crystallography*
Tel. 32(41)515652 Fax 32(41)562355 Telex 41397 UNIVLG B

Callens, Dr Freddy Johan (1956). Research Associate, National Foundation for Scientific Research. Vakgroep Vaste-Stofwetenschappen, Universiteit Gent, Krijgslaan 281 S1, B-9000 Gent, Belgium. DSc (U Gent, 1991) solid state physics. *EPR defect_structures ENDOR*
Tel. 32(9)2644352 Fax 32(9)2644996

Clauws, Prof. Paul Cyriel (1944). Senior lecturer. Vakgroep Vaste-Stofwetenschappen, Universiteit Gent, Krijgslaan 281 S1, B-9000 Gent, Belgium. DSc (U. Gent, 1991) semiconductor physics. *Semiconductors optics infrared_spectroscopy defect_structures*
E-mail Paul.Clauws@rug.ac.be Tel. 32(9)2644365 Fax 32(9)2644996

Culot, Ms Christine M. S. (1968). Research assistant. Laboratoire de Chimie Moléculaire Structurale, Facultés Universitaires Notre-Dame de la Paix, Rue de Bruxelles 61, B-5000 Namur, Belgium. PhD (FUNDP, Namur, 1994) chemistry. *Polymorphism quantum_mechanics powder_diffraction lipids fats NMR_&_modelization_of_lipids lipid_crystallization lipoproteins*
E-mail culot@scf.fundp.ac.be Tel. 32(81)724556 Fax 32(81)724530 Telex 59222 FACNAMB

De Bondt, Dr Hendrik Leon Augusta Jozef (1964). Postdoctoral fellow. Laboratorium voor Analytische Chemie en Medicinale Fysicochemie, Faculteit Farmaceutische Wetenschappen, Katholieke U. Leuven, Van Evenstraat 4, B-3000 Leuven, Belgium. PhD (U. Leuven, 1991) pharmacy. *Electron_density structural_chemistry proteins*
E-mail hendrik%ana%far@cc3.kuleuven.ac.be Tel. 32(16)323419 Fax 32(16)283414

Declercq, Prof. Jean-Paul (1948). Professor. Unité de chimie physique moléculaire et de cristallographie, Université Catholique de Louvain, Place L. Pasteur 1, B-1348-Louvain-la-Neuve, Belgium. PhD (U Louvain, 1972) chemistry. *Computing phase_determination proteins*
E-mail declercq@cpmc.ucl.ac.be Tel. 32(10)472924 Fax 32(10)472836 Telex 59037 UCL B

De Gryse, Prof. Roger Marc (1941). Lecturer. Vakgroep Vaste-Stofwetenschappen, Universiteit Gent, Krijgslaan 281 S1, B-9000 Gent, Belgium. PhD (U Gent, 1973) applied science. *Crystal_physics surface_science vacuum_technology*
Tel. 32(9)2644363 Fax 32(9)2644996

Deliens, Dr Michel (1939). Section head. Section Minéralogie et Pétrographie, Institut royal des Sciences naturelles de Belgique, Rue Vautier 29, B-1040 Bruxelles, Belgium. PhD (U Louvain, 1972) mineralogy. *Minerals structural_chemistry X-ray_diffraction*
Tel. 32(2)6274323 Fax 32(2)6464433

De Ranter, Prof. Camiel Joseph (1937). Professor. Correspondent to the CSM. Laboratorium voor Analytische Chemie en Medicinale Fysicochemie, Faculteit Farmaceutische Wetenschappen, Katholieke Universiteit Leuven, Van Evenstraat 4, B-3000 Leuven, Belgium. PhD (KU Leuven, 1964) chemistry. *Structural_chemistry structure-activity_relationship polymorphism*
E-mail camiel%ana%far@cc3.kuleuven.ac.be Tel. 32(16)323416 Fax 32(16)283414

Desmedt, Dr Amelia N. M. G. (1964). Research assistant. Laboratoire de Chimie Moléculaire Structurale, Facultés Universitaires Notre-Dame de la Paix, Rue de Bruxelles 61, B-5000 Namur, Belgium. PhD (FUNDP, Namur, 1993) chemistry. *Fats polymorphism surfactants powder_diffraction calorimetry fats_synthesis*
E-mail desmedt@scf.fundp.ac.be Tel. 32(81)724555 Fax 32(81)724530 Telex 59222 FACNAMB

De Winter, Dr Hans Louis Jos (1963). Postdoctoral fellow. Laboratorium voor Analytische Chemie en Medicinale Fysicochemie, Faculteit Farmaceutische Wetenschappen, Katholieke Universiteit Leuven, Van Evenstraat 4, B-3000 Leuven, Belgium. PhD (U. Leuven, 1992) pharmacy. *Structural_chemistry structure-activity_relationship*
E-mail hans%ana%far@cc3.kuleuven.ac.be Tel. 32(16)323421 Fax 32(16)283414

Dosière, Dr Marcel (1945). Professor; Head of Department, Physical Chemistry of Polymers. Université de Mons-Hainault, Place du Parc 20, B-7000 Mons, Belgium. PhD (U. Mons-Hainault, 1975) physical chemistry of polymers. *Polymers solid_state FTIR_spectroscopy Raman_spectroscopy SAXS WAXS X-ray_diffraction neutron_diffraction electron_diffraction*
E-mail s.dosiere@vm1.umh.ac.be Tel. 32(65)373351 Fax 32(65)373054 Telex 57764 UEMONS B

Dupont, Dr Léon (1941). Lecturer. President of the Belgian National Committee of Crystallography; Belgian correspondent of the IUCr newsletter. Laboratoire de Cristallographie, Université de Liège au Sart Tilman, Institut de Physique, bât. B5, B-4000 Liège, Belgium. PhD (U Liège, 1969) physics. *Phase_determination structural_biology molecular_modelling structure-activity_relationship*
E-mail u210406@vm1.ulg.ac.be Tel. 32(41)663762 Fax 32(41)662355 Telex 41397 UNIVLG B

Durant, Prof. François (1939). Group leader. Laboratoire de Chimie Moléculaire Structurale, Facultés Universitaires Notre-Dame de la Paix, Rue de Bruxelles 61, B-5000 Namur, Belgium. PhD (U Louvain, 1965) chemistry. *QSAR graphics bioactive_structure structure_crystallography bioorganic_compounds ligand-receptor_interactions*
E-mail durant@scf.fundp.ac.be Tel. 32(81)724549 Fax 32(81)724530 Telex 59222 FACNAMB

Elisabettini, Ms Paola (1970). Research student. Laboratoire de Chimie Moléculaire Structurale, Facultés Universitaires Notre-Dame de la Paix, Rue de Bruxelles 61, B-5000 Namur, Belgium. MSc (FUNDP, Namur, 1991) chemistry. *Fats polymorphism surfactants powder_diffraction DSC calorimetry graphics*
E-mail elisabet@scf.fundp.ac.be Tel. 32(81)724556 Fax 32(81)724530 Telex 59222 FACNAMB

Everaert, Mr Dirk Herman Maria (1965). Teaching assistant. Laboratorium voor Analytische Chemie en Medicinale Fysicochemie, Katholieke Universiteit Leuven, Van Evenstraat 4, B-3000 Leuven, Belgium. Pharmacist (U. Leuven, 1988) pharmacy. *Structural_chemistry structure-activity_relationship*
E-mail dirk%ana%far@cc3.kuleuven.ac.be Tel. 32(16)323421 Fax 32(16)283414

Evrard, Miss Christine (1970). Research student. Unité de chimie physique moléculaire et de cristallographie, Université Catholique de Louvain, Place L. Pasteur 1, B-1348-Louvain-la-Neuve, Belgium. MSc (Facultés Universitaires Notre Dame de la Paix, Namur, 1992) chemistry. *Structural_crystallography proteins biopolymers*
E-mail evrard@cpmc.ucl.ac.be Tel. 32(10)472924 Fax 32(10)472836 Telex 59037 ucl b

Evrard, Prof. Guy Henri (1943). Professor. Laboratoire de Chimie Moléculaire Structurale, Facultés Universitaires Notre-Dame de la Paix, Rue de Bruxelles 61, B-5000 Namur, Belgium. PhD (U Louvain, 1969) chemistry. *Structure_crystallography bioactive_structure organometallic_structure*
E-mail Guy.Evrard@fundp.ac.be Tel. 32(81)724548 Fax 32(81)724530

Feneau-Dupont, Ms Janine (1932). Assistant. Unité de chimie physique moléculaire et de cristallographie, Université Catholique de Louvain, Place L. Pasteur 1, B-1348-Louvain-la-Neuve, Belgium. MSc (U Louvain, 1953) chemistry. *Crystallography structure_determination*
E-mail feneau@cpmc.ucl.ac.be Tel. 32(10)472921 Fax 32(10)472836

Flermans, Prof. Lucien Victor August (1937). Senior lecturer. Vakgroep Vaste-Stofwetenschappen, Universiteit Gent, Krijgslaan 281 S1, B-9000 Gent, Belgium. DSc (U Gent, 1974) crystallography and solid state physics. *Crystal_physics surface_physics teaching aids in crystallography defect_structures*
Tel. 32(9)2644370 Fax 32(9)2644996

Fontaine, Dr Frédéric (1932). Lecturer (retired). Laboratoire de Cristallographie, Université de Liège au Sart Tilman, Institut de Physique, bât. B5, B-4000 Liège, Belgium. PhD (U Liège, 1968) chemistry. *Small-angle_scattering polymers morphology*
E-mail u217304@vm1.ulg.ac.be Tel. 32(41)663631 or 32(41)584920 Fax 32(41)663715 Telex 41397 UNIVLG B

Fransolet, Dr André-Mathieu (1947). Senior lecturer. Université de Liège au Sart Tilman, Institut de Minéralogie, bât. B18, B 4000 Liège, Belgium. PhD (U Liège, 1975) Geology and Mineralogy. *Minerals crystal_chemistry*
Tel. 32(41)562206 Fax 32(41)562202 Telex 41397 UNIVLG B

Gaspard, Prof. Jean-Pierre (1945). Professor; Chairman of the council of studies in Physics. Condensed Matter Physics, Institut de Physique, B5, University of Liège, B 4000 Sart-Tilman, Belgium. PhD (U Paris-Sud, 1975) physics. *Neutron_diffraction liquids EXAFS SAXS structural_chemistry applied_mathematics*
E-mail gaspard@gw.unipc.ulg.ac.be Tel. 32(41)663745 Fax 32(41)662990

Geise, Prof. Herman J. V. H. (1937). Group leader. Structural Chemistry Laboratory, Chemistry Department, Universitaire Instelling Antwerpen, Universiteitsplein 1, B-2610 Wilrijk, Belgium. PhD (U. Leiden, Netherlands, 1964) chemistry. *Electron_diffraction structural_chemistry functional_dyes_and_polymers*
E-mail alsenoy@uia.ac.be Tel. 32(3)8202349 Fax 32(3)8202276 Telex 33646

Germain, Prof. Gabriel (1933). Professor. Unité de chimie physique moléculaire et de cristallographie, Université Catholique de Louvain, Place L. Pasteur 1, B-1348-Louvain-la-Neuve, Belgium. PhD (U Louvain, 1958) chemistry. *Direct_method computing*
E-mail germain@cpmc.ucl.ac.be Tel. 32(10)472833 Fax 32(10)472707

Hoogewijs, Prof. Robert (1950). Senior lecturer. Vakgroep Vaste-Stofwetenschappen, Universiteit Gent, Krijgslaan 281 S1, B-9000 Gent, Belgium. [Sint-Pietersnieuwstraat 25, B-9000 Gent, Belgium.] DSc (U Gent, 1980) solid state physics. *Crystal_physics surface_physics academic_management*
E-mail Robert.Hoogewijs@rug.ac.be Tel. 32(9)2643110 Fax 32(9)2644297

Kartheuser, Dr Edward Peter (1938). Senior lecturer. Département de physique théorique, Université de Liège au Sart Tilman, Institut de Physique, bât. B5, B 4000 Liège, Belgium. PhD (U Liège, 1968) physics. *Crystal_physics semiconductor defect_structures*
E-mail karth@gw.unipc.ulg.ac.be Tel. 32(41)563639 Fax 32(41)562355 Telex 41397 UNIVLG B

King, Prof. Geoffrey S. D. (1924). Professor (retired). WDC Sub-Editor, Belgium. Fysico-chemische Geologie, Katholieke U. Leuven, Celestijnenlaan 200C, B-3001 Heverlee, Belgium. MSc (U London UK, 1950) crystallography. *Molecular_biology chemistry mineralogy computing*
E-mail fgeea01@cc1.kuleuven.ac.be Tel. 32(16)327582 Fax 32(16)201368 Telex 23674 kuleuv b

Lamotte-Brasseur, Dr Josette (1943). Lecturer. Laboratoire de Cristallographie, Université de Liège au Sart Tilman, Institut de Physique, bât. B5, B 4000 Liège, Belgium. PhD (U Liège, 1973) physics. *Proteins modelling*
E-mail u210405@vm1.ulg.ac.be Tel. 32(41)563499 Fax 32(41)563364 Telex 41397 UNIVLG B

Lenstra, Dr Albert T. H. (1942). Group leader. Structural Chemistry Laboratory, Universitaire Instelling Antwerpen, Universiteitsplein 1, B-2610 Wilrijk, Belgium. DSc (U. Antwerpen, 1992) chemical physics. *Accurate_measurement modelling statistics quantum_chemistry solid_state_modelling*
E-mail lenstra@sch2.uia.ac.be Tel. 32(3)8202365 Fax 32(3)8202276 Telex 33646

Loris, Mr Remy (1966). Research assistant. Instituut Moleculaire Biologie, Vrije Universiteit Brussel, Paardenstraat 65, B-1640 Sint Genesius Rode, Belgium. MSc (U Brussels (VUB), 1989) biotechnology. *Proteins carbohydrates nucleic_acids enzymology molecular_biology*
E-mail reloris@vub.ac.be Tel. 32(2)3590209 Fax 32(2)3590390

Maene, Mr Norbert (1935). Research Physicist. VITO, Afdeling Materialen, Boeretang 200, B-2400 Mol, Belgium. MSc(Eng) (U Gent, 1958) *Diffraction soft_X-ray*
Tel. 32(14)333183 Fax 32(14)321186

Maenhout - Van Der Vorst, Dr Wenefride Marguerite Romain (1930). Lecturer (retired). Vakgroep Vaste-Stofwetenschappen, Universiteit Gent, Krijgslaan 281 S1, B-9000 Gent, Belgium. DSc (U Gent, 1957) physics. *Crystal_physics surface_structure*
Tel. 32(9)2644342 Fax 32(9)2644996

Maes, Dr Dominique (1961). Research associate. Instituut Moleculaire Biologie, Vrije Universiteit Brussel, Paardenstraat 65, B-1640 Sint Genesius Rode, Belgium. PhD (U Brussels (VUB), 1989) physics. *Crystallography proteins nuclear_magnetic_resonance molecular_biology*
E-mail dommaes@vub.ac.be Tel. 32(2)3590262 Fax 32(2)3590390

Maes, Mr Stefan (1970). Research student. Structural chemistry laboratory, Chemistry department, Universitaire Instelling Antwerpen, Universiteitsplein 1, B-2610 Wilrijk, Belgium. MSc (U. Antwerpen, 1992) chemistry. *Accurate_measurement deformation_density*
E-mail maes@sch2.uia.ac.be Tel. 32(3)8202365 Fax 32(3)8202276 Telex 33646

Matthys, Dr Paul Frederik André Edmond (1949). Senior lecturer. Vakgroep Vaste-Stofwetenschappen, Universiteit Gent, Krijgslaan S1, B-9000 Gent, Belgium. DSc (U Gent, 1990) physics. *Physics paramagnetic_resonance general_physics*
Tel. 32(9)2644357 Fax 32(9)2644996

Meunier-Piret, Dr Jacqueline (1934). Lecturer. Unité de chimie physique moléculaire et de cristallographie, Université Catholique de Louvain, Place L. Pasteur 1, B-1348-Louvain-la-Neuve, Belgium. PhD (U Louvain, 1961) chemistry. *Organometallic_structure crystal_structure*
E-mail meunier@cpmc.ucl.ac.be Tel. 32(10)472922 Fax 32(10)472774

Michel, Prof. Karl H. (1939). Professor of physics. Physics Department, Universitaire Instelling Antwerpen, Universiteitsplein 1, B-2610 Wilrijk, Belgium. PhD (U. Nijmegen, Netherlands, 1965) physics. *Statistical_mechanics disorder molecular_crystals*
Tel. 32(3)8202458 Fax 32(3)8202245

Moreau, Prof. Jules (1931). Professor. Laboratoire de Minéralogie, Université Catholique de Louvain, Bâtiment Mercator, Place L. Pasteur 3, B-1348 Louvain-la-Neuve, Belgium. Engineer (U Louvain, 1954) mining. *Mineralogy Crystallography economic_geology*
Tel. 32(10)472855 Fax 32(10)472429

Moureau, Dr Florence G. C. G. (1965). Teaching assistant. Laboratoire de Chimie Moléculaire Structurale, Facultés Universitaires Notre-Dame de la Paix, Rue de Bruxelles 61, B-5000 Namur, Belgium. PhD (FUNDP, Namur, 1992) chemistry. *Computer-assisted_design LCAO_method drug_design QSAR monoamine_oxidase-inhibitors muscarinic_compounds*
E-mail moureau@scf.fundp.ac.be Tel. 32(81)724556 Fax 32(81)724530 Telex 59222 FACNAMB

Naud, Dr Jean (1942). Lecturer; manager of XRD and XRF research centre. Laboratoire de Minéralogie, Université Catholique de Louvain, Bâtiment Mercator, Place L. Pasteur 3, B-1348 Louvain-la-Neuve, Belgium. PhD (U Louvain, 1968) chemistry. *Mineralogy crystallography inorganic_chemistry X-ray_diffraction X-ray_spectrometry*
E-mail naud@gem.ucl.ac.be Tel. 32(10)472851 Fax 32(10)472429

Norberg, Ms Bernadette (1951). Staff technician. Laboratoire de Chimie Moléculaire Structurale, Facultés Universitaires Notre-Dame de la Paix, Rue de Bruxelles 61, B-5000 Namur, Belgium. BSc (IPL, Liège, 1972) chemistry. *Single_crystal powder molecular_crystal*
E-mail norberg@scf.fundp.ac.be Tel. 32(81)724556 Fax 32(81)724530 Telex 59222 FACNAMB

Peeters, Ms Anik (1967). Research student. Structural Chemistry Laboratory, Universitaire Instelling Antwerpen, Universiteitsplein 1, B-2610 Wilrijk, Belgium. MSc (U. Antwerpen, 1989) chemistry. *Quantum_mechanics computer_modelling structural_chemistry*
E-mail anik@uia.ac.be Tel. 32(3)8202366 Fax 32(3)8202376 Telex 33646

Peeters, Prof. Oswald Maurice (1945). Lecturer. Laboratorium voor Analytische Chemie en Medicinale Fysicochemie, Faculteit Farmaceutische Wetenschappen, Katholieke Universiteit Leuven, Van Evenstraat 4, B-3000 Leuven, Belgium. PhD (U. Leuven, 1977) chemistry. *Structure_determination structural_chemistry structure-activity_relationship*
E-mail maurice%ana%far@cc3.kuleuven.ac.be Tel. 32(16)323418 Fax 32(16)283414

Pirard, Mr Bernard G. G. (1968). Research student. Laboratoire de Chimie Moléculaire Structurale, Facultés Universitaires Notre-Dame de la Paix, Rue de Bruxelles 61, B-5000 Namur, Belgium. MSc (FUNDP, Namur, 1990) chemistry. *Biochemistry computer-assisted_design LCAO_method single_crystal pharmacophore_identification_for_GABAb_receptor QSAR*
E-mail pirard@scf.fundp.ac.be Tel. 32(81)724556 Fax 32(81)724530 Telex 59222 FACNAMB

Piret, Prof. Paul (1932). Professor. Unité de chimie physique moléculaire et de cristallographie, Université Catholique de Louvain, Place L. Pasteur 1, B-1348-Louvain-la-Neuve, Belgium. DSc (U Louvain, 1956) chemistry. *Crystal_structure mineralogy*
E-mail piret@cpmc.ucl.ac.be Tel. 32(10)472769 Fax 32(10)472774

Pletinckx, Mr Jurgen (1968). Research student. Instituut Moleculaire Biologie, Vrije Universiteit Brussel, Paardenstraat 65, B-1640 Sint Genesius Rode, Belgium. MSc (U Brussels, 1990) biotechnology. *Macromolecules carbohydrates structural_hydration enzymatic structure-activity relation*
E-mail jgpletin@vub.ac.be Tel. 32(2)3590209 Fax 32(2)3590390

Poortmans, Dr Freddy (1938). Group Leader. VITO, Afdeling Materialen, Boeretang 200, B-2400 Mol, Belgium. PhD (U Gent, 1966) physics. *Diffraction condensed_matter small-angle_scattering biological_macromolecules*
E-mail poortmaf@vito.be Tel. 32(14)333164(34) Fax 32(14)321186

Quere, Mr Luc (1968). Research student. Laboratoire de Chimie Moléculaire Structurale, Facultés Universitaires Notre-Dame de la Paix, Rue de Bruxelles 61, B-5000 Namur, Belgium. MSc (FUNDP, Namur, 1990) chemistry. *Biochemistry molecular_crystal QSAR philosophy peptides proteins*
E-mail quere@scf.fundp.ac.be Tel. 32(81)724548 Fax 32(81)724530 Telex 59222 FACNAMB

Raty, Jean-Yves (1971). Assistant Professor. Condensed Matter Physics, Institut de Physique, B5, University of Liège, B 4000 Sart-Tilman, Belgium. Graduate in Physics (U. Liège, 1993) physics. *Neutron_diffraction liquids EXAFS high_pressure SAXS structural_chemistry applied_mathematics*
E-mail U222501@vm1.ulg.ac.be Tel. 32(41)663747 Fax 32(41)662990

Reynaers, Prof. Harry (1938). Professor. Lab. Macromoleculaire Structuurchemie, Dept. Scheikunde, Katholieke Univerzsiteit Leuven, Celestijnenlaan 200F, B-3001 Leuven (Heverlee), Belgium. PhD (U. Leuven, 1964) chemistry. *Polymers small-angle_scattering*
E-mail fgcab01@cc1.kuleuven.ac.be Tel. 32(16)327446 Fax 32(16)201215 Telex 23674

Schryvers, Dr Dominique (1959). Lecturer. University of Antwerpen (RUCA), Dept. Natuurkunde, EMAT, Groenenborgerlaan 171, B-2020 Antwerpen, Belgium. DSc (U Brussels (VUB), 1991) physics. *Electron_microscopy alloys martensites small_particles*
E-mail schryver@ruca.ua.ac.be Tel. 32(3)2180247 Fax 32(3)2180257

Schuerman, Ms Geertrui Simone Maria (1967). Teaching assistant. Laboratorium voor Analytische Chemie en Medicinale Fysicochemie, Faculteit Farmaceutische Wetenschappen, Katholieke Universiteit Leuven, Van Evenstraat 4, B-3000 Leuven, Belgium. Pharmacist (U. Leuven, 1990) pharmacy. *Structure_determination QSAR Oligonucleotides*
E-mail geertrui%ana%far@cc3.kuleuven.ac.be Tel. 32(16)323419 Fax 32(16)283414

Sobry, Dr Roger (1946). Lecturer. Physique expérimentale, Université de Liège au Sart Tilman, Institut de Physique, bât. B5, B 4000 Liège, Belgium. PhD (U Liège, 1972) physics. *Small-angle_scattering polymers group_theory*
E-mail u217302@vm1.ulg.ac.be Tel. 32(41)563715 Fax 32(41)563715 Telex 41397 UNIVLG B

Spirlet, Dr Marie-Rose (1946). Lecturer. Physique expérimentale, Université de Liège au Sart Tilman, Institut de Physique, bât. B5, B-4000 Liège, Belgium. PhD (U Liège, 1976) chemistry. *X-ray_diffraction neutron_diffraction organometallic_structure lanthanides actinides*
E-mail u217303@vm1.ulg.ac.be Tel. 32(41)663758 Fax 32(41)662813 Telex 41397 UNIVLG B

Tinant, Dr Bernard Guy André Francois (1951). Lecturer. Unité de chimie physique moléculaire et de cristallographie, Université Catholique de Louvain, Place L. Pasteur 1, B-1348-Louvain-la-Neuve, Belgium. PhD (U Louvain, 1978) chemistry. *Structure_determination proteins calcium_compounds crystal_growth*
E-mail tinant@cpmc.ucl.ac.be Tel. 32(10)472923 Fax 32(10)472836

Toussaint, Prof. Jean (1916). Professor (retired). Laboratoire de Cristallographie, Université de Liège au Sart Tilman, Institut de Physique, bât. B5, B-4000 Liège, Belgium. DSc (U Liège, 1945) physics. *Crystallography*
Tel. 32(41)334579 Fax 32(41)562355 Telex 41397 UNIVLG B

Van Acker, Mr Karel (1968). Research assistant. Dept. Metaalkunde en Toegepaste Materiaalkunde, Katholieke U. Leuven, de Croylaan 2, B-3001 Heverlee, Belgium. MSc (U. Leuven, 1991) metallurgy. *Materials residual_stresses*
E-mail Karel=Van=Acker%MEC%MTM@cc3.kuleuven.ac.be Tel. 32(16)321111 Fax 32(16)321990

Van Alsenoy, Dr Christian (1948). Group leader. Structural Chemistry Laboratory, Universitaire Instelling Antwerpen, Universiteitsplein 1, B-2610 Wilrijk, Belgium. DSc (U Antwerpen, 1990) chemistry. *Quantum_mechanics computer_modelling structural_chemistry spectroscopy*
E-mail alsenoy@sch2.uia.ac.be Tel. 32(3)8202366 Fax 32(3)8202276 Telex 33646

Van den Bossche, Dr Guy Ghislain Remy (1941). Lecturer. Member of the Belgian National Committee for Crystallography. Laboratoire de Cristallographie, Université de Liège au Sart Tilman, Institut de Physique, bât. B5, B 4000 Liège, Belgium. PhD (U Liège, 1973) physics. *Small-angle_scattering morphology_of_polymers crystal_physics*
E-mail u210408@vm1.ulg.ac.be Tel. 32(41)663763 Fax 32(41)662355 Telex 41397 UNIVLG B

Van Dyck, Dr Dirk Ernest Maria (1948). Senior Lecturer; Professor/Head of Department. University of Antwerpen (RUCA), Dept. Natuurkunde, EMAT, Groenenborgerlaan 171, B-2020 Antwerpen, Belgium. DSc, habilitation (U Antwerpen (UIA), 1987) physics. *Electron_microscopy solid_state Image_processing artificial_intelligence*
E-mail dvd@ruca.ua.ac.be Tel. 32(3)2180258 Fax 32(3)2180257

Vanhellemont, Dr Jan H. (1953). Senior researcher. IMEC, Advanced Semiconductor Processing Division, Ultra Clean Processing, Kapeldreef 75, B-3001 Leuven, Belgium. PhD (U Antwerpen, 1990) physics. *Defects_in_semiconductors characterisation_techniques ultra_clean_processing*
E-mail vanhellemont@imec.be Tel. 32(16)321307 Fax 32(16)281214 Telex 26152

Van Houtte, Prof. Paul (1948). Professor. Dept. Metaalkunde en Toegepaste Materiaalkunde, Katholieke U. Leuven, de Croylaan 2, B-3001 Heverlee, Belgium. PhD (U. Leuven, 1975) metallurgy. *Materials texture*
E-mail paul.vanhoutte@mtm.kuleuven.ac.be Tel. 32(16)321304 Fax 32(16)321990

Van Landuyt, Prof. Jef Florent (1938). Professor. Chairman, National Committee for Crystallography. University of Antwerpen (RUCA), Dept. Natuurkunde, EMAT, Groenenborgerlaan 171, B-2020 Antwerpen, Belgium. PhD (U Gent, 1965) physics. *Electron_microscopy crystallography solid_state materials* phase transition electron diffraction
E-mail jovalan@ruca.ua.ac.be Tel. 32(3)2180259 Fax 32(3)2180257

Van Meervelt, Dr Luc (1958). Research associate. Lab. Macromoleculaire Structuurchemie, Dept. Scheikunde, Katholieke Universiteit Leuven, Celestijnenlaan 200F, B-3001 Heverlee, Belgium. PhD (U. Leuven, 1986) chemistry. *Structural_biology structural_chemistry Cambridge structural database*
E-mail fgcae01@cc1.kuleuven.ac.be Tel. 32(16)327609 Fax 32(16)201215 Telex 23674

Van Tendeloo, Prof. Gustaaf (1950). Senior Lecturer. University of Antwerpen (RUCA), Dept. Natuurkunde, EMAT, Groenenborgerlaan 171, B-2020 Antwerpen, Belgium. DSc (U Brussels (VUB), 1981) physics. *Superconductors phase_transition electron_microscopy fullerenes*
E-mail gvt@ruca.ua.ac.be Tel. 32(3)2180262 Fax 32(3)2180257

Vennik, Prof. Joost (1927). Professor (retired). Vakgroep Vaste-Stofwetenschappen, Universiteit Gent, Krijgslaan 281 S1, B-9000 Gent, Belgium. MSc (U Gent, 1952) electronic engineering. *Physics Materials Semiconductors*
Tel. 32(9)2644343 Fax 32(9)2644996

Verbist, Prof. Jacques Jozef (1943). Professor Director, Laboratoire Interdisciplinaire de Spectroscopie Electronique. Laboratoire Interdépartemental de Spectroscopie Electronique, Facultés Universitaires Notre-Dame de la Paix, Rue de Bruxelles 61, B-5000 Namur, Belgium. PhD (U Louvain, 1969) chemistry. *Film interface geometry electronic_structure electron_spectroscopies*
E-mail lise@scf.fundp.ac.be Tel. 32(81)724608 Fax 32(81)724595 Telex 59222 FACNAMB

Verboven, Ms Christel Clara Remigius (1967). Teaching assistant. Laboratorium voor Analytische Chemie en Medicinale Fysicochemie, Katholieke Universiteit Leuven, Van Evenstraat 4, B-3000 Leuven, Belgium. Pharmacist (U. Leuven, 1991) pharmacy. *Structure_determination crystallization proteins cadmium_compounds*
E-mail christel%ana%far@cc3.kuleuven.ac.be Tel. 32(16)323420 Fax 32(16)283414

Verhulst, Mr Koen (1966). Research student. Structural Chemistry Laboratory, Universitaire Instelling Antwerpen, Universiteitsplein 1, B-2610 Wilrijk, Belgium. MSc (U. Antwerpen, 1990) physical chemistry. *Crystal_field modelling structural chemistry solid_state_model*
E-mail verhulst@sch2.uia.ac.be Tel. 32(3)8202365 Fax 32(3)8202276 Telex 33646

Wegener, Dr Wolter (1941). Research Physicist. VITO, Afdeling Materialen, Boeretang 200, B-2400 Mol, Belgium. PhD (T. H. Aachen, Germany, 1970) physics. *Diffraction condensed_matter small-angle_scattering biological_macromolecules molecular_crystals*
Tel. 32(14)333183 Fax 32(14)321186

Wouters, Mr Johan F. A. M. (1969). Research assistant. Laboratoire de Chimie Moléculaire Structurale, Facultés Universitaires Notre-Dame de la Paix, Rue de Bruxelles 61, B-5000 Namur, Belgium. MSc (FUNDP, Namur, 1991) chemistry. *Antidepressants computer-assisted_design membranes proteins biochemistry monoamine_oxidase_inhibitors*
E-mail wouters@scf.fundp.ac.be Tel. 32(81)724556 Fax 32(81)724530 Telex 59222 FACNAMB

Wyns, Prof. Lode (1946). Group leader. Instituut Moleculaire Biologie, Vrije Universiteit Brussel, Instituut Moleculaire Biologie VUB, Paardenstraat 65, B-1640 Sint Genesius Rode, Belgium. PhD (U Brussels (VUB), 1978) chemistry. *Proteins carbohydrates nucleic_acids enzymology molecular_biology*
E-mail ljwyns@vub.ac.be Tel. 32(2)3590262 Fax 32(2)3590390

Zegers, Ms Ingrid (1965). Research assistant. Instituut Moleculaire Biologie, Vrije Universiteit Brussel, Instituut Moleculaire Biologie VUB, Paardenstraat 65, B-1640 Sint Genesius Rode, Belgium. MSc (U Brussels (VUB), 1989) applied chemistry. *Proteins crystallography nucleic_acids enzymology molecular_biology*
E-mail igzegers@vub.ac.be Tel. 32(2)3590209 Fax 32(2)3590390

BOTSWANA

Leckebusch, Dr Rudolf (1937). Project Manager. Botswana Ministry of Commerce and Industry, Department of Commerce and Consumer Affairs, Botswana. [Botswana Ministry of Commerce and Industry, Department of Commerce and Consumer Affairs, c/o GTZ-PAS, private bag X12, Gaborone, Botswana.] Dr rer.nat.habil. (U. Bonn, 1977) mineralogy/crystallography. *Absorption_spectroscopy colour_center luminescence thermoluminescence*

Tel. 267( )356361 Fax 267( )374474 Telex 2957gtzga

BRAZIL

Sub-Editor: I. L. Torriani

Abrão Pereira, Mr Romeo (1947). Researcher. Centro Brasileiro de Pesquisas Físicas, Rua Xavier Sigaud 150, Urca, Rio de Janeiro, 22290-180 RJ, Brazil. MSc (Centro Brasileiro de Pesquisas Físicas, 1988) physics. *Crystal_physics polymers instrumentation* E-mail laudo@brlncc.bitnet Tel. 55(21)5410337x140 Fax 55(21)5412047

Adusumilli, Prof. Maria P. S. (1929). Full Professor. Instituto de Geociências, U. de Brasilia (UnB), CP 04465, 70910-900 Brasilia DF, Brazil. PhD (U. Federal de Minas Gerais, 1976) mineralogy. *X-ray_diffractometry crystalline_form lantalum_compounds* Tel. 55(61)2734735 Fax 55(61)2744286 Telex (61)2730UNBSBR

Alves, Dr Oswaldo Luiz (1947). Group Leader. Solid State Chem. Lab., Inst. de Química, U. Estadual de Campinas, CP 6154, 13081-970 Campinas SP, Brazil. DrSc (U. Estadual de Campinas, 1977) chemistry. *SAXS Rietveld_method layered_compounds ceramics NLO glasses nanocrystallites inclusion_compounds* E-mail oalves@iqm.unicamp.ansp.br Tel. 55(192)397201 Fax 55(192)393805 Telex (19)1150

Arguello, Dr Zoraide Primerano (1938). Assoc. Prof. Instituto de Física, U. Est. de Campinas, CP 6165, 13083-970 Campinas SP, Brazil. PhD (U. Estadual de Campinas, 1972) physics. *Crystal_growth* E-mail arguello@ccvax.unicamp.br Tel. 55(192)397240 Fax 55(192)393137 Telex (19)1150

Arni, Dr Raghuvir Krishnaswamy (1953). Assistant Professor. UNESP/IBILCE, Department of Physics, C. Postal 136, 15054 São José do Rio Preto SP, Brazil. Dr (Technical U. - Free U. - Berlin, 1987) protein crystallography. *Proteins_crystallography proteins_crystallization enzymes_structure minerals_structure* E-mail arni@minerva.ibilce.unesp.br Tel. 55(172)244966 Fax 55(172)248692 Telex (11)19020

Baran, Dr Zbigniew (1930). Professor. Inst. de Física, U. Federal da Bahia, Rua Caetano Moura 123, 40210-340 Salvador BA, Brazil. PhD (U. Warsaw - Poland, 1970) physics. *Structural_defect crystalline_solid* E-mail dioni@brufba.bitnet Tel. 55(71)2472714 Fax 55(71)2355592

Barelli, Dr Nilso (1920). Assistant Professor. Dep. de Tecnologia e Química de Aplicação, Instituto de Química, UNESP Araraquara, CP 355, 14800-970 Araraquara SP, Brazil. PhD (Fac. de Ciências e Letras de Araraquara, 1944) mineralogy. *Mineralogy crystal_growth morphology epitaxy* E-mail ueara@brfapesp.bitnet Tel. 55(162)322022 Fax 55(162)227932 Telex (11)19006UJMFBR

Beltran Abrego, Dr José Ramón (1952). Assistant Professor. Inst. de Biociências Letras e Cs. Exatas, U. Estadual Paulista (UNESP), Dep. de Física, Rua Cristovão Colombo 2265, 15054 São José do Rio Preto SP, Brazil. PhD (U. São Paulo, 1987) physics. *SAXS macromolecules* E-mail ramon@minerva.ibilce.unesp.br Tel. 55(172)244966 Fax 55(172)248692 Telex (11)19020

Bittencourt, Dr Diomar da Rocha (1959). Assistant Professor. Instituto de Física, Dpto. de Física Aplicada, U. de São Paulo, CP 20516, 01498-970 São Paulo SP, Brazil. Dr (U. São Paulo, 1987) physics. *SAXS polymers liquid_crystals* E-mail diomar@uspif.if.usp.br Tel. 55(11)8187110 Fax 55(11)8140503

Bristoti, Dr Anildo (1936). Associate Professor. Departamento de Física, U. Federal do Rio Grande do Sul, 90000 Porto Alegre RS, Brazil. PhD (UCLA - USA, 1970) engineer. *Metallurgy powder_diffractometry crystallography* Tel. 55(51)3369822x6541

Campelo Farias, Prof. Carlinda (1940). Associate Professor. Dep. de Eng. de Minas, U. Federal de Pernambuco, Cidade Universitária, 50000 Recife PE, Brazil. DSc (U. Federal de Pernambuco, 1977) mineralogy. *Minerals powder_diffractometry crystallography* Tel. 55(81)2718245 Fax 55(81)2718242

Campos, Dr Cícero (1948). Assistant Professor. Dep. de Estado Sólido, Inst. de Física CP 6165, U. Estadual de Campinas, 13083-970 Campinas SP, Brazil. PhD (U. Estadual de Campinas, 1983) physics. *Multiple_scattering dynamical_theory* E-mail cicero@ifi.unicamp.br Tel. 55(192)397291 Fax 55(192)393137 Telex (19)1150

Cardoso de Lima, Dr João (1950). Assistant Professor. Depto. de Física, U. Federal de Santa Catarina, Campus Universitário, Trindade CP 476, 88040 Florianopolis SC, Brazil. Dr (U. Paris-Sud, 1989) physics. *Anomalous_scattering EXAFS* E-mail fsc1jcd@brufsc.bitnet Tel. 55(482)319434 Fax 55(482)319688 Telex (56)(482)240

Carvalho, Dr Carlos A. M. (1962). Assistant Professor. Instituto de Física, Dpto. de Física Aplicada, U. de São Paulo - CP 20516, 01498-970 São Paulo SP, Brazil. PhD (U. Paris - France, 1990) physics. *Dynamical_theory defect_structure computing* E-mail cadena@uspif.if.usp.br Tel. 55(11)8187110 Fax 55(11)8140503 Telex (11)80923IFUSPBR

Cassedane, Dr Jeannine (1927). Professor. Instituto de Geociências, U. Federal do Rio de Janeiro, Ilha do Fundão, 20000 Rio de Janeiro RJ, Brazil. PhD (U. Strasbourg - France, 1969) physics. *Crystallography minerals* Tel. 55(21)5983292 Fax 55(21)5983280

Castellano, Dr Eduardo Ernesto (1941). Full Professor. Dep. de Física e Ciência dos Materiais, Inst. de Física e Química de São Carlos, U. de São Paulo, CP 369, 13560-970 São Carlos SP, Brazil. PhD (U. Nacional de La Plata - Argentina, 1968) physics. *Direct_method macromolecules* E-mail pino@ifqsc.ansp.br Tel. 55(162)749188 Fax 55(162)713616

Caticha Ellis, Prof. Stephenson (1927). Professor. Fac. de Eng. Mecânica, U. Estadual de Campinas, 13083-970 Campinas SP, Brazil. Engineer (U. Montevideo - Uruguay, 1954) engineering. *Crystal_physics imperfection instrumentation* E-mail caticha@fem.unicamp.br Tel. 55(192)391301x591

Correia Neves, Prof. José Marques (1929). Full Professor. Instituto de Geociências, Dep. de Geologia, U. Federal de Minas Gerais, 31270-901 Belo Horizonte MG, Brazil. DSc (U. Coimbra - Portugal, 1963) mineralogy-geochemistry. *Inorganic_structure mineralogy* Tel. 55(31)4481322 Fax 55(31)4485410

Costa Gouveia, Prof. Albany H. (1941). Associate Professor. Dep. de Engenharia de Minas, U. Federal de Pernambuco, 50000 Recife PE, Brazil. DSc (U. Federal de Pernambuco, 1977) mineralogy. *Optical_crystallography mineralogy* Tel. 55(81)2718245 Fax 55(81)2718242

Costa Viana, Prof. Carlos Sérgio da (1942). Associate Professor; Head of materials science graduate programme. Instituto Militar de Engenharia (IME), Praça Gal. Tiburcio 80 - Urca, 22230-270 Rio de Janeiro RJ, Brazil. PhD (Cambridge University - England, 1978) philosophy. *Texture anisotropy mechanical_property plasticity foaming phase_transformations* E-mail SAVIANA@IMERJ.BITNET Tel. 55(21)5422049 Fax 55(21)2759047

Craievich, Dr Aldo Félix (1939). Full Scientist. Laboratório Nacional de Luz Síncrotron, CP 6192, 13087-410 Campinas SP, Brazil. PhD (U. Nacional de Cuyo - Argentina, 1969) physics. *Synchrotron_radiation SAXS glasses phase_separation instrumentation* E-mail aldo@lnls.ansp.br Tel. 55(192)542624 Fax 55(192)512458 Telex (19)7517

Cusatis, Dr César (1939). Full Professor. Dep. de Física, U. Federal do Paraná, CP 81531-970 Curitiba PR, Brazil. PhD (U. São Paulo, 1973) physics. *X-ray_optics X-ray_interferometry* E-mail $gorxi@lnls.ansp.br Tel. 55(41)2669271 Fax 55(41)2674236

Denicoló, Dr Ireno (1954). Assistant Professor. Depto. de Física, U. Federal do Paraná, CP 19091, 81531 Curitiba PR, Brazil. Dr (U. São Paulo, 1984) applied physics. *Single_crystal X-ray_diffraction materials characterization* E-mail $gorxi@lnls.ansp.br Tel. 55(41)2669271 Fax 55(41)2674236

Dias Rodrigues, Mrs Ana Maria Gonçalves (1954). Assistant Professor. Dep. de Química e Física Molecular, Inst. de Química de São Carlos, 13560-970 São Carlos, SP, Brazil. DSc (U. São Paulo, 1986) chemistry. *Structure_determination* E-mail nana@ifqsc.ansp.br Tel. 55(162)749171 Fax 55(162)749163

Fantini, Dr Marcia C. A. (1956). Assistant Professor. Instituto de Física, Física Aplicada, U. São Paulo, CP 20516, 01498-970 São Paulo SP, Brazil. PhD (U. Estadual de Campinas, 1985) physics. *Crystalline_materials thin_film X-ray_diffraction electrochromic_materials* E-mail mfantini@uspif.if.usp.br Tel. 55(11)8186882 Fax 55(11)8140503 Telex (11)80923IFUSPBR

Fernandes, Dr Nelson Gonçalves (1950). Lecturer. Dep. de Química, U. Federal de Minas Gerais, CP 702, Av. Antônio Carlos 6627, 31270-901 Belo Horizonte MG, Brazil. PhD (Uppsala University - Sweden, 1989) inorganic chemistry. *Accuracy structure_determination hydrogen_bonding neutron_diffraction* E-mail tgquimic@brufmg.bitnet Tel. 55(31)4412718 Fax 55(31)4433986

Ferreira de Souza, Prof. Milton (1932). Professor. Dep. de Física e Ciência dos Materiais, Inst. de Física e Química de São Carlos, CP 369, Av. Dr. Carlos Botelho 1465, 13560-970 São Carlos SP, Brazil. PhD (U. São Paulo, 1969) physics. *Crystal_growth crystalline_defect* E-mail pino@ifqsc.ansp.br Tel. 55(162)711016 Fax 55(162)721125

Figueiredo Neto, Prof. Antônio Martins (1953). Group Leader. Lab. de Optica de Cristais Líquidos, Instituto de Física, U. de São Paulo, CP 20516, 01498-990 São Paulo SP, Brazil. PhD (U. São Paulo, 1981) physics. *Liquid_crystals ferrofluids* E-mail afigueiredo@if.usp.ansp.br Tel. 55(11)8186830 Fax 55(11)8140503 Telex 1180923IFUSPBR

Formoso, Dr Milton Luiz (1927). Professor. Instituto de Geociências, U. Federal do Rio Grande do Sul, 90000 Porto Alegre RS, Brazil. PhD (U. São Paulo, 1973) geology. *Geochemistry mineralogy geology* Tel. 55(51)281633 Fax 55(51)336501

Francisco, Dr Regina Helena Porto (1952). Assistant Professor. Dep. de Química e Física Molecular, Inst. de Física e Química de São Carlos, Universidade de São Paulo, CP 369, 13560-970 São Carlos SP, Brazil. DSc (U. São Paulo, 1983) chemistry. *Structure_determination* E-mail porto@ifqsc.ansp.br Tel. 55(162)749171 Fax 55(162)749163 Telex (16)5122FQSC

Freire D'Aguiar, Dr Manoel Marcos (1947). Assistant Professor. Inst. de Física, U. Federal da Bahia, Rua Caetano Moura 123, 40210-340 Salvador BA, Brazil. PhD (U. Göttingen - Germany, 1985) physics. *Crystal_defect* Tel. 55(71)2472714 Fax 55(71)2355592

Fulfaro, Dr Roberto (1938). Chief Scientist. Inst. de Pesquisas Energéticas e Nucleares, CP 11049, 05508-900 São Paulo SP, Brazil. PhD (U. Estadual de Campinas, 1970) physics. *Neutron_scattering lattice_dynamics*
Tel. 55(11)2127904 Fax 55(11)8146909

Gambardella, Miss Maria Teresa do Prado (1954). Assistant Professor. Departamento de Química e Física Molecular, Inst. de Química de São Carlos, U. de São Paulo, CP 369, 13560-250 Sao Carlos, SP, Brazil. DSc (U. São Paulo, 1985) chemistry. *Structure_determination*
E-mail teca@ifqsc.sc.usp.br Tel. 55(162)749171 Fax 55(162)749163

Herdade, Dr Sílvio B. (1926). Associate Professor. Instituto de Física, U. São Paulo, CP 20516, 01498-970 São Paulo SP, Brazil. PhD (U. Estadual de Campinas, 1969) physics. *Geophysics nuclear_physics*
E-mail sherdade@uspif.if.usp.br Tel. 55(11)8155599 Fax 55(11)8140503 Telex (11)80923IFSPBR

Imakuma, Dr Kengo (1943). Section Head. Inst. de Pesquisas Energéticas e Nucleares, CP 11049, 05508-900 São Paulo SP, Brazil. PhD (U. São Paulo, 1973) physics. *Phase_transition radiation_damage ceramics alloys*
E-mail ioy000@brusp.bitnet Tel. 55(11)2116011x242 Fax 55(11)2123546

Itri, Dr Rosangela (1961). Assistant Professor. Instituto de Física, U. São Paulo, CP 20516, 01498-970 São Paulo SP, Brazil. PhD (U. São Paulo, 1991) physics. *Liquid_crystals micelles biophysics crystallography*
E-mail itri@uspif.if.br Tel. 55(11)8187110 Fax 55(11)8180503 Telex (11)80923IF-SPBR

Kunrath, Mr Josè Irineu (1931). Assistant Professor. Inst. de Física, U. Federal do Rio Grande do Sul, 90000 Porto Alegre RS, Brazil. MSc (U. Federal do Rio Grande do Sul, 1960) physics. *Crystallography solid_state*
Tel. 55(51)3369822x6541 Fax 55(51)3361762

Labaki, Dr Lucila (1943). Assistant Professor. Faculdade de Engenharia Civil, U. Estadual de Campinas, CP 6021, 13083-970 Campinas SP, Brazil. PhD (U. Estadual de Campinas, 1990) physics. *X-ray diffractometry polymers*
Tel. 55(192)394823

Lariucci, Dr Carlito (1958). Assistant Professor. Depto. de Física, U. Federal de Goiás, Campus II - CP 131, 74000 Goiania GO, Brazil. Dr (U. São Paulo, 1988) physics. *X-ray_diffraction materials characterization*
Tel. 55(62)2051000x168 Telex (62)2206UFGO

Lechat, Dr Johannes Rudiger (1943). Assistant Professor. Departamento de Química e Física Molecular, Inst. de Física e Química de São Carlos, U. de São Paulo, CP 369, 13560-970 São Carlos SP, Brazil. PhD (U. São Paulo, 1972) chemistry. *Crystal_structure organic_compound*
E-mail lechat@ifqsc.ansp.br Tel. 55(162)749171 Fax 55(162)749163

Leite, Dr Cirano Rocha (1941). Professor. Dep. de Química Tecnológica e Aplicação, U. Estadual Paulista (UNESP), Instituto de Química, CP 355, 14800-970 Araraquara SP, Brazil. PhD (U. São Paulo, 1969) mineralogy. *Mineralogy crystal_growth morphology epitaxy*
E-mail uearq@brfapesp.bitnet Tel. 55(162)322022 Fax 55(162)227932 Telex (11)19006UJMFBR

Madureira Filho, Prof. José Barbosa de (1940). Assistant Professor. Dep. de Mineralogia e Petrografia, Inst. de Geociências, U. de São Paulo, CP 20899 Butantã, 05508 São Paulo SP, Brazil. PhD (U. São Paulo, 1983) mineralogy and petrology. *Solid_solution*
Tel. 55(11)2122011 Fax 55(11)2104958

Martinez, Mr Luiz Gallego (1957). Researcher. Inst. de Pesquisas Energéticas e Nucleares, IPEN CP 11049, 05508-900 São Paulo SP, Brazil. MSc (Inst. de Pesquisas Energéticas e Nucleares-USP, 1988) physics. *Powder_diffractometry metals ceramics defect*
E-mail ioy000@brusp.bitnet Tel. 55(11)2116011x242 Fax 55(11)2123546

Mascarenhas, Dr Sérgio (1928). Chief Researcher. Empresa Brasileira de Pesquisas Agropequárias, EMBRAPA - Unidade de Apoio à Pesquisa, CP 741, 13560 São Carlos SP, Brazil. PhD (U. São Paulo, 1958) physics. *Macromolecules molecular_structure biophysics*
E-mail sergimasc@ifqsc.ans.br Tel. 55(162)725741 Fax 55(162)715381

Mascarenhas, Prof. Yvonne Primerano (1931). Full Professor. Member of IUCr teaching commission. Dep. de Física e Ciência dos Materiais, Inst. de Física e Química de São Carlos, U. de São Paulo, CP 369, 13560-970 São Carlos SP, Brazil. PhD (U. São Paulo, 1963) physics. *Crystal_structure natural_products macromolecules proteins_crystallography*
E-mail yvonne@ifqsc.ansp.br Tel. 55(162)749188 Fax 55(162)713616

Mastelaro, Dr Valmor Roberto (1960). Researcher. Dep. de Ciência dos Materiais, U. Federal de São Carlos, Rod. Washington Luiz Km 235, 13565-905 São Carlos SP, Brazil. PhD (U. Paris XI - France, 1992) solid state physics. *X-ray_diffraction absorption_spectroscopy EXAFS*
E-mail dvrm@power.ufscar.br Tel. 55(162)717691

Mazzaro, Dr Irineu (1953). Assistant Professor. Dep. de Física, U. Federal do Paraná, CP 19091, 81531-970 Curitiba PR, Brazil. PhD (U. São Paulo, 1989) physics. *X-ray_optics multiple_diffraction crystal_defect*
E-mail $gorxi@lnls.ansp.br Tel. 55(41)2669271 Fax 55(41)2674236

Mazzochi, Dr Vera (1955). Researcher. Inst. de Pesquisas Energéticas e Nucleares, IPEN CP 05508-900 São Paulo SP, Brazil. PhD (IPEN/USP, 1990) physics. *Neutron_diffraction multiple_diffraction texture*
E-mail vlmazzo@iho.ipen.br Tel. 55(11)2116011x1365 Fax 55(11)2123546

Medeiros Rodrigues, Dr Maria Mabel (1930). Professor. Dep. de Química e Física Molecular, Inst. de Física e Química de São Carlos, U. de São Paulo, CP 369, 13560-970 São Carlos SP, Brazil. PhD (U. São Paulo, 1968) chemistry. *Structure_determination organic_compound*
E-mail mabelrdgues@ifqsc.ansp.br Tel. 55(162)749171 Fax 55(162)749163

Mestnik, Dr José (1949). Researcher. Inst. de Pesquisas Energéticas e Nucleares, IPEN - CP 05508-900 São Paulo SP, Brazil. PhD (IPEN/USP, 1987) neutron physics. *Neutron_scattering neutron_diffraction lattice_dynamics metal_hydrides*
E-mail ioy000@brusp.bitnet Tel. 55(11)2116011x141 Fax 55(11)2123546

Mosca, Dr Dante Homero (1960). Professor. Dep. de Física, U. Federal do Paraná, Centro Politécnico, 81531-970 Curitiba PR, Brazil. PhD (U. Federal do Rio Grande do Sul, 1992) physics. *Metallic_superlattice heterostructure surface_physics*
E-mail mosca@brufpr.bitnet Tel. 55(41)3662323x102

Murta, Prof. Clécio (1929). Head Instrumental Analysis Lab. Companhia Brasileira de Tecnologia Nuclear, U. Federal de Minas Gerais, Belo Horizonte, 30000 Minas Gerais MG, Brazil. MSc (U. Federal de Minas Gerais, 1971) nuclear science. *Diffractometry X-ray_fluorescence electron_microanalysis thermoanalysis_mineralogy*
Tel. 55(31)425422

Oliva, Dr Glaucius (1959). Associate Professor. Instituto de Física de São Carlos, U. São Paulo, CP 369, 13560-970 São Carlos, SP, Brazil. PhD (Birkbeck College - UK, 1988) physics. *Proteins_crystallography small_molecules*
E-mail oliva@ifqsc.sc.usp.br Tel. 55(162)749188 Fax 55(162)713616

Oliveira Lopes, Prof. César (1949). Assistant Professor. Dep. de Física, Inst. de Ciências Exatas, U. Federal Rural do Rio de Janeiro, Estrada Rio-S.Paulo Km 47, 23851 Itaguai RJ, Brazil. MSc (U. Estadual de Campinas, 1975) physics. *Thin_film ferromagnetic_alloys*
Tel. 55(21)7821220x227

Olivieri, Dr Johnny Rizzieri (1950). Assistant Professor. Inst. de Biociências Letras e Cs. Exatas, U. Estadual Paulista - UNESP, Dep. de Física, Rua Cristovão Colombo 2265, 15054 São José do Rio Preto SP, Brazil. PhD (U. São Paulo, 1992) physics. *SAXS amorphous_materials*
E-mail johnny@minerva.ibilce.unesp.br Tel. 55(172)324966x59 Fax 55(172)248692 Telex (11)19020

Paiva Santos, Dr Carlos de Oliveira (1953). Assist. Professor. Dep. de Fisico Química, Instituto de Química, U. Estadual Paulista - UNESP, CP 355, 14800-900 Araraquara (SP), Brazil. DSc (U. São Paulo, 1990) physics. *Rietveld_method structure_determination ceramics polymers*
E-mail COPSANTOS@IFQSC.IQ.USP.BR Tel. 55(162)322022 Fax 55(162)227932

Pavie Cardoso, Dr Lisandro (1950). Assistant Professor. Inst. de Física CP 6165, U. Estadual de Campinas, 13083-970 Campinas SP, Brazil. PhD (U. Estadual de Campinas, 1983) physics. *Crystal_defect multiple_diffraction*
E-mail iisandro@ifi.unicamp.br Tel. 55(192)397291 Fax 55(192)393137 Telex (19)1150

Pulcinelli, Dr Sandra Helena (1957). Assistant Professor. Instituto de Química, U. Estadual Paulista - UNESP, CP 355, 14800-970 Araraquara SP, Brazil. PhD (U. São Paulo, 1987) applied physics. *Powder_diffraction single_crystal*
E-mail sandra@arq000.uesp.ansp.br Tel. 55(162)322022x147 Fax 55(162)227932 Telex (11)1900UJMFBR

Queiroz do Amaral, Dr Lia (1941). Full Professor. Inst. de Física, U. de São Paulo, CP 20516, 01498-970 São Paulo SP, Brazil. PhD (U. São Paulo, 1972) physics. *Liquid_crystals SAXS phase_transition membranes*
E-mail amaral@uspif.if.usp.br Tel. 55(11)8187110 Fax 55(11)8140503

Ramos Parente, Dr Carlos Benedicto (1937). Researcher. Inst. de Pesquisas Energéticas e Nucleares, CP 11049, 05508-900 São Paulo SP, Brazil. PhD (U. São Paulo, 1973) neutron physics. *Neutron_diffraction texture*
E-mail cparente@iho.ipen.br Tel. 55(11)2116011x1365 Fax 55(11)2123546

Regueira Teodósio, Prof. Joel (1943). Assistant Professor. U. Federal do Rio de Janeiro, Dep. de Metalurgia, CP 68529 Cidade Univ., 21945-970 Rio de Janeiro RJ, Brazil. MSc (U. Federal do Rio de Janeiro, 1973) metallurgy. *X-ray_crystallography texture*
Tel. 55(21)5900441 Fax 55(21)2601092

Riella, Mr Humberto Gracher (1953). Researcher. Metallurgical Engineering Div., Comissão Nacional de Energia Nuclear, CNEN - Travessa R 400, 05508 São Paulo SP, Brazil. Engineer (U. Federal do Parana, 1975) metallurgy. *Metals diffusion_metals*
Tel. 55(11)8177480 Fax 55(11)8144695

Rodrigues, Dr Antonio Ricardo Droher (1951). Project Director LNLS. Laboratório Nacional de Luz Síncrotron, LNLS CP 6192, 13087-410 Campinas SP, Brazil. PhD (Kings College - London U. - England, 1979) X-ray optics. *X-ray_optics instrumentation*
E-mail ricardo@lnls.ansp.br Tel. 55(192)542624 Fax 55(192)512458 Telex (19)7517

Rodrigues, Dr Edson (1928). Professor. Dep. de Química e Física Molecular, Inst. de Química de São Carlos, University of São Paulo, CP 369, 13560-970 São Carlos, SP, Brazil. PhD (U. São Paulo, 1964) physics. *Crystal_physics magnetic_resonance solid_state_chemistry*
E-mail edsonro@uspqsc.iqsc.sc.usp.br edsonro@ifqsc.sc.usp.br Tel. 55(162)749174 55(162)749175 Fax 55(162)749163

Rodrigues da Silva, Dr Rilson (1932). Full Professor. Dep. de Engenharia de Minas, Centro de Tecnologia, U. Federal de Pernambuco, 50000 Recife Pernambuco, Brazil. [Av. Beira Mar 4050 Apt.604, 54420-000 Jaboatão PE, Brazil.] DrSci (U. Louis Pasteur - France, 1969) crystallography-mineralogy. *Crystal_growth geometric_crystallography phosphates_minerals X-ray_crystallography*
Tel. 55(81)3612789 Fax 55(81)27118242 Telex (81)1267

Santos, Dr Pérsio de Souza (1928). Professor. Dep. de Engenharia Química, Escola Politécnica - U. de São Paulo, CP 61548, 05424-970 São Paulo SP, Brazil. PhD (U. São Paulo, 1960) physical chemistry. *Minerals clays zeolites*
Tel. 55(11)8185627 Fax 55(11)2113020

Santos, Dr Regina Helena de Almeida (1947). Assistant Professor. Dep. de Química e Física Molecular, Inst. de Física e Química de São Carlos, U. de São Paulo, CP 369, 13560-970 São Carlos SP, Brazil. Dr (U. São Paulo, 1979) physical chemistry. *Structural_crystallography*
E-mail reginas@ifqsc.ansp.br Tel. 55(162)749171 Fax 55(162)749163

Silva Campos, Dr Helio (1959). Assistant Professor. Instituto de Física, U. Federal da Bahia, Rua Caetano Moura 123, 40210-340 Salvador BA, Brazil. Dr (Centro Brasileiro de Pesquisas Físicas, 1988) physics. *Diffraction_theory perfect_crystal computing computer_simulation*
E-mail campos@brufba.bitnet Tel. 55(71)2472714 Fax 55(71)2355592

Simone, Dr Carlos Alberto de (1951). Assistant Professor. Dep. de Química, U. Federal de Alagoas, Campus Universitário, Tabuleiro do Martins, 57000 Maceió AL, Brazil. Dr (U. São Paulo, 1989) applied physics. *Crystal_structure natural_products refinement_method*
Tel. 55(82)3222318 Fax 55(82)3222399

Soares de Vasconcelos, Dr Dionicarlos (1958). Assistant Professor. Instituto de Física, U. Federal da Bahia, Rua Caetano Moura 123, 40210-340 Salvador BA, Brazil. Dr (CBPF (Centro Bras. de Pesquisas Físicas), 1991) physics. *Diffraction_theory perfect_crystal diffraction_simulation computing*
E-mail dioni@brufba.bitnet Tel. 55(71)2472714 Fax 55(71)2355592

Soledade Jr, Prof. Teomar (1948). Research Associate. Inst. de Física, U. Federal da Bahia, Rua Caetano Moura 123, 40210-340 Salvador BA, Brazil. MSc (U. Estadual de Campinas, 1976) physics. *Crystal_defect divergent_beam_methods*
Tel. 55(71)2472714 Fax 55(71)2355592

de Sousa, Dr José Carlos (1948). Assistant Professor. Dep. de Física, Fundação U. Est. de Maringá, Av. Colombo 3690, 87100 Maringá, Brazil. PhD (U. Estadual de Campinas, 1991) physics. *Crystal_microstructure*
E-mail jcsousa@brfuem.bitnet Tel. 55(442)262727x330 Fax 55(442)222754 Telex (442)198

de Sousa, Dr Christina Franco (1959). Researcher. Telebrás - CPqD, Rodovia Campinas - Mogi Mirim Km 118, CP 1579, 13010-142 Campinas SP, Brazil. PhD (U. Estadual de Campinas, 1990) physics. *Heterostructure crystal_growth multilayers*
E-mail rxcssouza@venus.cpqd.ansp.br Tel. 55(192)396442

Souza, Ms Dulce Helena Ferreira (1962). Assistant Professor. U. Estadual Paulista - UNESP, Campus de Ilha Solteira, Av. Brasil Centro 56, 15378 Ilha Solteira SP, Brazil. MSc (U. Federal de São Carlos, 1989) chemistry. *Ruthenium_compounds proteins_structure*
E-mail unesp%ipmet1@brfapesp.bitnet Tel. 55(187)623113 Fax 55(187)622735

Souza, Prof. Irineu Marques (1940). Professor. Inst. de Geociências, Dep. de Geologia Econômica, U. de São Paulo, Cidade Universitária, CP 05508 São Paulo SP, Brazil. BSc (U. São Paulo, 1964) geology. *Geology crystal_optics*
Tel. 55(11)8184079x4144 Fax 55(11)2104958

Speziali, Dr Nivaldo Lúcio (1953). Lecturer. Dep. de Física, U. Federal de Minas Gerais, Av. Antonio Carlos 6627 CP 702, 31270-901 Belo Horizonte MG, Brazil. PhD (U. Lausanne - Switzerland, 1989) crystallography. *Structural_crystallography modulated_structure phase_transition*
E-mail nspezial@brufmg.bitnet Tel. 55(31)4419466 Fax 55(31)4481372

Stojanoff, Dr Vivian (1955). Assistant Professor. Inst. de Física, U. de São Paulo, CP 20516, 01498-970 São Paulo SP, Brazil. PhD (U. São Paulo, 1984) physics. *Semiconductor_structure defect*
E-mail stojanoff@uspif.if.usp.br Tel. 55(11)8187110 Fax 55(11)8140503 Telex (11)80923IFUSPBR

Suzuki, Dr Carlos Kenichi (1945). Associate Professor. Fac. de Engenharia Mecânica, Depto. de Engenharia de Materiais, CP 6122, U. Estadual de Campinas, 13083-970 Campinas SP, Brazil. PhD (U. Tokyo - Japan, 1980) applied physics engineering. *Materials_engineering X-ray_diffraction synchrotron_radiation*

Santos, Dr Regina Helena de Almeida (1947). [...]

Svisero, Dr Darcy Pedro (1940). Full Professor; Head, Department of Mineralogy and Petrology. Dep. of Mineralogy and Petrology, Institute of Geosciences, University of São Paulo, CP 20899, 05508 São Paulo SP, Brazil. PhD (U. São Paulo, 1971) geology. *Mineralogy high-pressure_minerals diamond*
Tel. 55(11)8184079 Fax 55(11)2104958

Távora, Prof. Elysiário (1920). Professor. Comissão Nacional de Energia Nuclear, 22294 Rio de Janeiro RJ, Brazil. DSc (Univ. do Brasil, 1946) natural sciences. *X-ray_crystallography mineralogy*
Tel. 55(21)5462405

Tolentino, Dr Hélio Cesar Nogueira (1963). Researcher. Laboratório Nacional de Luz Síncrotron, LNLS CP 6192, 13087-410 Campinas SP, Brazil. PhD (U. Paris Sud, 1990) physics. *X-ray_optics EXAFS instrumentation*
E-mail helio@lnls.ansp.br Tel. 55(192)542624 Fax 55(192)512458

Tomita, Dr Koychi (1943). Professor. Inst. de Química, UNESP, CP 355, 14800-900 - Araraquara - SP, Brazil. PhD (Fac. de Filos. Ciências e Letras de Araraquara, 1967) chemistry. *Structural_crystallography*
E-mail uearq@brfapesp.bitnet Tel. 55(162)322022x147 Fax 55(162)227932 Telex 1119006 UJMF BR

Torriani, Dr Iris Linares (1934). Associate Professor. WDC9 Regional Ed. for South America & Sub-Editor for Brazil. Inst. de Física, U. Estadual de Campinas, CP 6165, 13083-970 Campinas SP, Brazil. PhD (U. Nacional de La Plata - Argentina, 1975) physics. *SAXS synchrotron_radiation biological_application multilayer_structure nanostructured_materials*
E-mail torriani@bruc.bitnet Tel. 55(192)398473 Fax 55(192)393137 Telex (19)1150

Udron, Dr Dominique (1962). Visiting Researcher. Dep. de Física, U. Federal do Paraná, CP 19091, 81531 Curitiba PR, Brazil. PhD (U. Paris VII, 1990) physics. *X-ray_optics instrumentation synchrotron_radiation absorption_spectroscopy anomalous_scattering*
E-mail domi@inf.ufpr.br Tel. 55(41)2669271 Fax 55(41)2674236

Varela, Dr José Arana (1944). Professor. Instituto de Química, U. Estadual Paulista - UNESP, CP 355, 14800-970 Araraquara SP, Brazil. PhD (U. of Washington - USA, 1981) materials science. *Ceramics sintering characterization biomaterials*
E-mail uearq@brfapesp.bitnet Tel. 55(162)220015 Fax 55(162)227932

Vencato, Dr Ivo (1944). Assistant Professor. Dep. de Física, U. Federal de Santa Catarina, 88040 Florianópolis SC, Brazil. PhD (U. São Paulo, 1984) physics. *Liquid_crystals micelles vanadium_compounds metalloproteins computer_aided_education*
E-mail fsc1ivo@brufsc.bitnet Tel. 55(482)319544 Fax 55(482)340059

Villaroel, Prof. Hugo S. (1931). Professor. Dep. de Engenharia de Minas, U. Federal de Pernambuco, Cidade Universitária, 50000 Recife PE, Brazil. MSc (U. Austral de Chile, 1970) crystallography. *X-ray_diffraction X-ray_spectroscopy minerals*
Tel. 55(81)2718245 Fax 55(81)2718242

Vinhas, Dr Laércio A. (1943). Head of Research Div. Comm. Nacional de Energia Nuclear (CNEN), Rua Gal. Severiano, 90 - Botafogo, 20000 Rio de Janeiro RJ, Brazil. PhD (U. Estadual de Campinas, 1970) physics. *Neutron_scattering neutron_diffraction molecular_crystals*
Tel. 55(21)2950645 Fax 55(21)5462379

Zaghete, Dr Maria Aparecida (1954). Assistant Professor. Inst. de Química, U. Estadual Paulista - UNESP, CP 355, 14800-900 Araraquara (SP), Brazil. PhD (U. Federal de São Carlos, 1993) materials science. *Ceramics sintering characterization*
E-mail UEARQ@BRFAPESP.BITNET Tel. 55(162)322022 Fax 55(162)227932

Zukerman Schpector, Dr Julio (1948). Associate Professor. Depto. de Química, U. Federal de São Carlos, CP 676, 13565-970 São Carlos SP, Brazil. PhD (U. São Paulo, 1984) chemistry. *Crystal_structure polymers*
E-mail julio@ifqsc.ansp.br Tel. 55(162)727344 Fax 55(162)723616

BULGARIA

Sub-Editor: I. Bonev

Alexandrova, Dr Daniela (1958). Assistant. Faculty of Physics, Sofia University, 1126 Sofia, Bulgaria. PhD (Sofia U., 1993) physics. *Colour_symmetry group_theory phase_transition hydrides*
Tel. 359(2)62561x469 Fax 359(2)689085

Angelova, Dr Olyana (1963). Research associate. Central Laboratory of Mineralogy and Crystallography, Bulgarian Academy of Sciences, Rakovska 92, 1000 Sofia, Bulgaria. PhD (Sofia U., 1991) chemistry. *Structure_determination crystal_chemistry molecular_crystals protein_crystallography*
E-mail jmacicek@bgcict.bitnet Tel. 359(2)872450 Fax 359(2)884979

Apostolov, Dr Anton (1951). Research associate. Faculty of Chemistry, Sofia University, 1126 Sofia, Bulgaria. PhD (Sofia U., 1993) chemistry. *X-ray_diffraction polymers SAXS*
Tel. 359(2)62561x418 Fax 359(2)682808

Apostolov, Prof. Andrei (1935). Professor. Chairman, National Committee for Crystallography. Faculty of Physics, Sofia University, 1126 Sofia, Bulgaria. DSc (Sofia U., 1974) physics. *Solid-state_magnetism crystallography ecology*
E-mail phymvax@bgcict.bitnet Tel. 359(2)62561x472 Fax 359(2)682808

Aroyo, Dr Mois (1952). Senior scientist. Faculty of Physics, Sofia University, 1126 Sofia, Bulgaria. PhD (Sofia U., 1983) physics. *Symmetry mathematical_crystallography phase_transition representation_theory*
Tel. 359(2)62561x8377 Fax 359(2)689085

Aslanian, Dr Selma (1937). Senior scientist. Geological Institute, Bulgarian Academy of Sciences, 1113 Sofia, Bulgaria. PhD (Leipzig U., 1969) mineralogy. *Crystal_growth crystal_chemistry X-ray_diffractometry*
Tel. 359(2)7133470 Fax 359(2)724638

Atanassov, Dr Vassil (1933). Associate professor. Faculty of Geology, University of Mining and Geology, 1113 Sofia, Bulgaria. PhD (Inst. of Geology and Geophysics, Novosibirsk, 1973) mineralogy. *Mineral_morphology*
Tel. 359(2)62581x385 Fax 359(2)621042

Avramov, Dr Isak (1946). Senior scientist. Institute of Physical Chemistry, Bulgarian Academy of Sciences, 1113 Sofia, Bulgaria. DSc (Inst. of Physical Chemistry, 1993) chemistry. *Crystal_growth glasses phase_transition thin_films*
E-mail banchem@bgearn.bitnet Tel. 359(2)7132566 Fax 359(2)720038

Bliznakov, Prof. Georgi (1920). Professor. Institute of General and Inorganic Chemistry, Bulgarian Academy of Sciences, 1113 Sofia, Bulgaria. Academician (Bulgarian Acad. Sci., 1979) chemistry. *Crystal_growth adsorption inorganic_synthesis*
E-mail banchem@bgearn.bitnet Tel. 359(2)700478 Fax 359(2)705024

Bonev, Dr Ivan (1936). Senior scientist. Sub-Editor of ninth edition of World Directory of Crystallographers; Member, National Committee for Crystallography. Geological Institute, Bulgarian Academy of Sciences, 1113 Sofia, Bulgaria. PhD (Geological Inst., 1972) mineralogy. *Crystal_growth whiskers topotaxy minerals ore-forming_processes*
E-mail geology@bgcict.bitnet Tel. 359(2)7132236 Fax 359(2)724638

Bostanov, Dr Vesselin (1933). Senior scientist. Central Laboratory for Electrochemical Power Sources, Bulgarian Academy of Sciences, 1040 Sofia, Bulgaria. DSc (Inst. of Physical Chemistry, 1993) chemistry. *Electrocrystallization*
E-mail banchem@bgearn.bitnet Tel. 359(2)722146 Fax 359(2)722544

Bozukov, Latchezar Nikolov. Researcher. Member, National Committee for Crystallography. Faculty of Physics, Sofia University, 1126 Sofia, Bulgaria. MS (Sofia U., 1984) solid state physics. *X-ray_powder_diffraction Rietveld_method hydride_batteries*
E-mail phymvax@bgcict.bitnet Tel. 359(2)62561x472 Fax 359(2)682808

Budevski, Prof. Evgeni (1922). Professor. Central Laboratory for Electrochemical Power Sources, Bulgarian Academy of Sciences, 1040 Sofia, Bulgaria. Corresponding member (Bulgarian Acad. Sci., 1984) chemistry. *Electrocrystallization*
E-mail banchem@bgearn.bitnet Tel. 359(2)723454 Fax 359(2)722544

Budurov, Prof. Stoyan (1930). Professor. Faculty of Chemistry, University of Sofia, 1126 Sofia, Bulgaria. DSc (Sofia U., 1976) chemistry. *Crystallization metals phase_transition metallic_glasses*
Tel. 359(2)62561x337 Fax 359(2)682808

Delineshev, Dr Svetoslav (1940). Senior scientist. Institute of General and Inorganic Chemistry, Bulgarian Academy of Sciences, 1113 Sofia, Bulgaria. PhD (Inst. of General and Inorganic Chemistry, 1978) chemistry. *Crystal_growth*
E-mail banchem@bgearn.bitnet Tel. 359(2)7132542 Fax 359(2)705024

Djarova, Dr Maria (1945). Associate professor. Faculty of Chemistry, Sofia University, 1126 Sofia, Bulgaria. PhD (Sofia U., 1979) chemistry. *Mass_crystallization nucleation*
Tel. 359(2)62561x215 Fax 359(2)682808

Dobrev, Dr Dobri (1935). Senior scientist. Institute of Physical Chemistry, Bulgarian Academy of Sciences, 1113 Sofia, Bulgaria. PhD (Inst. of Physical Chemistry, 1976) chemistry. *Thin_film*
E-mail banchem@bgearn.bitnet Tel. 359(2)7132598 Fax 359(2)720038

Filizova, Dr Lyudmila (1937). Senior scientist. Central Laboratory of Mineralogy and Crystallography, Bulgarian Academy of Sciences, Rakovska 92, 1000 Sofia, Bulgaria. PhD (Inst. of Geology and Geophysics, Novosibirsk, 1974) mineralogy. *Minerals synthesis zeolites*
E-mail mincryst@bgcict.bitnet Tel. 359(2)872450 Fax 359(2)884979

Gospodinov, Prof. Marin (1944). Professor. Institute of Solid State Physics, Bulgarian Academy of Sciences, Tzarigradsko Chaussee 72, 1784 Sofia, Bulgaria. PhD (Inst. of Solid State Physics, 1980) physics. *Crystal_growth characterization crystallography crystal_synthesis Czochralski_method*
Tel. 359(2)74311x240 Fax 359(2)757032

Grozdanov, Mr Lyudmil (1934). Research associate. Geological Institute, Bulgarian Academy of Sciences, 1113 Sofia, Bulgaria. MSc (Sofia U., 1957) geochemistry. *Mineralogy silicates crystal_chemistry amphiboles*
Tel. 359(2)7132260 Fax 359(2)724638

Gutzov, Prof. Ivan (1933). Professor. Member, National Committee for Crystallography. Institute of Physical Chemistry, Bulgarian Academy of Sciences, 1113 Sofia, Bulgaria. DSc (Inst. of Physical Chemistry, 1972) chemistry. *Nucleation crystal_growth glasses*
E-mail banchem@bgearn.bitnet Tel. 359(2)719305 Fax 359(2)720038

Iwanov, Dr Dantsho (1939). Senior scientist. Institute of Physical Chemistry, Bulgarian Academy of Sciences, 1113 Sofia, Bulgaria. PhD (Inst. of Physical Chemistry, 1985) chemistry. *Crystal_growth thin_film morphology surfaces*
Tel. 359(2)7132507 Fax 359(2)722808

Kaischew, Prof. Rostislaw (1908). Professor. Institute of Physical Chemistry, Bulgarian Academy of Sciences, 1113 Sofia, Bulgaria. Academician (Bulgarian Acad. Sci., 1961) chemistry. *Physical_chemistry crystal_growth nucleation*
E-mail banchem@bgearn.bitnet Tel. 359(2)715865 Fax 359(2)720038

Kashchiev, Prof. Dimcho (1942). Professor. Institute of Physical Chemistry, Bulgarian Academy of Sciences, 1113 Sofia, Bulgaria. DSc (Inst. of Physical Chemistry, 1987) chemistry. *Crystal_growth nucleation electrocrystallization*
E-mail banchem@bgearn.bitnet Tel. 359(2)7132557 Fax 359(2)720038

Kerestedjian, Dr Thomas (1957). Research associate. Geological Institute, Bulgarian Academy of Sciences, 1113 Sofia, Bulgaria. PhD (Geological Inst., 1989) mineralogy. *Sulfides information_systems information_storage crystal_anatomy*
Tel. 359(2)7132244 Fax 359(2)724638

Kirkova, Prof. Elena (1923). Professor. Faculty of Chemistry, Sofia University, 1126 Sofia, Bulgaria. DSc (Sofia U., 1983) chemistry. *Crystal_growth adsorption*
Tel. 359(2)62561x215 Fax 359(2)682808

Kirov, Mr Georgi Kirilov (1932). Senior scientist. Central Laboratory of Mineralogy and Crystallography, Bulgarian Academy of Sciences, Rakovska 92, 1000 Sofia, Bulgaria. MSc (Sofia U., 1955) geochemistry. *Minerals synthesis crystal_growth morphology quartz*
E-mail mincryst@bgcict.bitnet Tel. 359(2)885115x671 Fax 359(2)884979

Kirov, Prof. Georgi Nikolov (1930). Professor. Faculty of Geology and Geography, Sofia University, 1000 Sofia, Bulgaria. DSc (Sofia U., 1988) mineralogy. *Minerals zeolites X-ray_diffractometry*
Tel. 359(2)8581x256 Fax 359(2)446487

Konstantinov, Dr Ivan (1943). Director. Central Laboratory of Photoprocesses, Bulgarian Academy of Sciences, 1040 Sofia, Bulgaria. PhD (Inst. of Physical Chemistry, 1975) chemistry. *Crystal_growth crystal_chemistry photochemistry*
E-mail banchem@bgearn.bitnet Tel. 359(2)720073 Fax 359(2)722465

Kostov, Prof. Ivan (1913). Professor. National Natural History Museum, Bulgarian Academy of Sciences, Tzar Osvoboditel 1, 1000 Sofia, Bulgaria. Academician (Bulgarian Acad. Sci., 1966) mineralogy. *Mineralogy crystal_growth morphology crystal_structures*
E-mail mincryst@bgcict.bitnet Tel. 359(2)882554 Fax 359(2)884979

Kostov, Dr Ruslan I. (1956). Geological Institute, Bulgarian Academy of Sciences, 1113 Sofia, Bulgaria. PhD (IGEM, Moscow, 1984) mineralogy. *Mineral_physics symmetry electron_spin_resonance gemology*
Tel. 359(2)7132273 Fax 359(2)724638

Kotzev, Dr Joseph (1942). Associate professor. Faculty of Physics, Sofia University, 1126 Sofia, Bulgaria. PhD (Moscow U., 1975) physics. *Symmetry colour_symmetry group_theory phase_transition*
Tel. 359(2)62561x8346 Fax 359(2)689085

Kovachev, Dr Peter (1943). Senior scientist. Institute of Metal Science, Bulgarian Academy of Sciences, Shipchenski Prochod 67, 1574 Sofia, Bulgaria. PhD (Inst. of Physical Chemistry, 1975) chemistry. *Thermodynamics metals alloys diffusion*
Tel. 359(2)7142327 Fax 359(2)703207

Krestev, Mr Venelin (1940). Research associate. Faculty of Physics, Sofia University, 1126 Sofia, Bulgaria. MSc (Polytechnical Inst., Leningrad, 1966) physics. *X-ray_diffraction diffractometry polymers*
Tel. 359(2)62561x8301 Fax 359(2)689085

Kresteva, Dr Manya (1943). Research associate. Faculty of Physics, Sofia University, 1126 Sofia, Bulgaria. PhD (Sofia U., 1982) physics. *Polymers X-ray_diffractometry*
Tel. 359(2)62561x669 Fax 359(2)689085

Maciček, Dr Josef (1953). Senior scientist. Secretary, National Committee for Crystallography. Central Laboratory of Mineralogy and Crystallography, Bulgarian Academy of Sciences, Rakovska 92, 1000 Sofia, Bulgaria. PhD (Moscow U., 1981) chemistry. *Structure_determination geometry_analysis crystal_chemistry indexing adducts_of_inorganic_salts*
E-mail jmacicek@bgcict.bitnet Tel. 359(2)872450 Fax 359(2)884979

Maleev, Dr Michael (1941). Director. Member, National Committee for Crystallography. "Earth and Man" National Museum, Blvd Cherni Vrach 4, 1421 Sofia, Bulgaria. PhD (Moscow U., 1968) mineralogy. *Minerals electron_microscopy whiskers*
Tel. 359(2)656639 Fax 359(2)661455

Malinowski, Prof. Yordan (1923). President. Bulgarian Academy of Sciences, 15th November Street, 1, 1000 Sofia, Bulgaria. Academician (Bulgarian Acad. Sci., 1989) chemistry. *Crystal_physics crystal_growth photochemistry*
Tel. 359(2)874086 Fax 359(2)803023

Marinov, Dr Miko (1939). Senior scientist. Institute of Physical Chemistry, Bulgarian Academy of Sciences, 1113 Sofia, Bulgaria. PhD (Inst. of Physical Chemistry, 1981) chemistry. *Crystal_growth electron_microscopy thin_films environment*
E-mail banchem@bgearn.bitnet Tel. 359(2)7132533 Fax 359(2)720038

Markov, Prof. Ivan (1941). Professor. Institute of Physical Chemistry, Bulgarian Academy of Sciences, 1113 Sofia, Bulgaria. DSc (Inst. of Physical Chemistry, 1988) chemistry. *Crystal_growth nucleation epitaxy thin_film*
E-mail banchem@bgearn.bitnet Tel. 359(2)7132557 Fax 359(2)720038

Michailov, Dr Evgeni (1940). Research associate. Institute of Physical Chemistry, Bulgarian Academy of Sciences, 1113 Sofia, Bulgaria. PhD (Inst. of Physical Chemistry, 1990) chemistry. *Crystal_growth thin_film adsorption*
E-mail banchem@bgearn.bitnet Tel. 359(2)7132565 Fax 359(2)720038

Michailov, Mr Michail (1954). Research associate. Institute of Physical Chemistry, Bulgarian Academy of Sciences, 1113 Sofia, Bulgaria. MSc (Sofia U., 1979) physics. *Crystal_growth critical_phenomena phase_transition LEED*
E-mail banchem@bgearn.bitnet Tel. 359(2)7132529 Fax 359(2)720038

Mikhov, Dr Michael (1944). Associate professor. Faculty of Physics, Sofia University, 1126 Sofia, Bulgaria. PhD (Sofia U., 1981) physics. *Magnetism magnetic_recording rare-earth_compounds particles teaching_in_material_science*
Tel. 359(2)62561x8377 Fax 359(2)689085

Milchev, Dr Alexander (1943). Senior scientist. Institute of Physical Chemistry, Bulgarian Academy of Sciences, 1113 Sofia, Bulgaria. PhD (Inst. of Physical Chemistry, 1982) chemistry. *Crystal_growth electrocrystallization*
E-mail banchem@bgearn.bitnet Tel. 359(2)7132558 Fax 359(2)720038

Milchev, Dr Andrei (1946). Senior scientist. Institute of Physical Chemistry, Bulgarian Academy of Sciences, 1113 Sofia, Bulgaria. DSc (Inst. of Physical Chemistry, 1993) chemistry. *Phase_transition glasses*
E-mail banchem@bgearn.bitnet Tel. 359(2)7132566 Fax 359(2)720038

Miloshev, Prof. Georgi (1933). Professor. Institute of Geophysics, Bulgarian Academy of Sciences, 1113 Sofia, Bulgaria. DSc (Inst. of Geophysics, 1974) physics. *Crystal_growth*
E-mail geophys@bgearn.bitnet Tel. 359(2)7133334 Fax 359(2)700226

Minčeva-Stefanova, Prof. Yordanka (1923). Professor. Geological Institute, Bulgarian Academy of Sciences, 1113 Sofia, Bulgaria. MSc (Sofia U., 1946) mineralogy. *Minerals crystal_growth morphology*
Tel. 359(2)7132282 Fax 359(2)724638

Nanev, Prof. Christo (1938). Director. Institute of Physical Chemistry, Bulgarian Academy of Sciences, 1113 Sofia, Bulgaria. DSc (Inst. of Physical Chemistry, 1990) chemistry. *Crystal_growth electrocrystallization thin_film*
E-mail banchem@bgearn.bitnet Tel. 359(2)727550 Fax 359(2)720038

Nenow, Prof. Dimiter (1931). Professor. Institute of Physical Chemistry, Bulgarian Academy of Sciences, 1113 Sofia, Bulgaria. DSc (Inst. of Physical Chemistry, 1989) chemistry. *Nucleation crystal_growth crystal_surfaces surface_roughening*
E-mail banchem@bgearn.bitnet Tel. 359(2)7132557 Fax 359(2)720038

Nihtianova, Dr Diana (1954). Research associate. Central Laboratory of Mineralogy and Crystallography, Bulgarian Academy of Sciences, 1040 Sofia, Bulgaria. PhD (Sofia U., 1987) chemistry. *Transmission_electron_microscopy thin_film convergent-beam_diffraction molybdates*
E-mail mincryst@bgcict.bi Tel. 359(2)872450 Fax 359(2)884979

Obretenov, Dr Willy (1958). Senior scientist. Central Laboratory for Electrochemical Power Sources, Bulgarian Academy of Sciences, 1040 Sofia, Bulgaria. PhD (Inst. of Physical Chemistry, 1987) chemistry. *Electrocrystallization*
Tel. 359(2)7132750 Fax 359(2)722544

Pashov, Prof. Nikolai (1929). Professor. Institute of Solid State Physics, Bulgarian Academy of Sciences, Tzarigradsko Chaussee 72, 1784 Sofia, Bulgaria. DSc (Sofia U., 1952) physics. *Crystal_imperfection image_processing electron_microscopy electron_diffraction*
Tel. 359(2)7341x641 Fax 359(2)755019

Paunov, Dr Michael (1939). Senior scientist. Institute of Physical Chemistry, Bulgarian Academy of Sciences, 1113 Sofia, Bulgaria. PhD (Inst. of Physical Chemistry, 1969) chemistry. *Crystal_growth nucleation surfaces*
E-mail banchem@bgearn.bitnet Tel. 359(2)7132529 Fax 359(2)720038

Peneva, Dr Stefka (1937). Senior scientist. Member, National Committee for Crystallography. Faculty of Chemistry, Sofia University, 1126 Sofia, Bulgaria. PhD (U. of Delhi, 1970) physics. *Electron_microscopy thin_film tin_compounds imperfections*
Tel. 359(2)62561x281 Fax 359(2)682808

Peshev, Prof. Pavel (1933). Director. Institute of General and Inorganic Chemistry, Bulgarian Academy of Sciences, 1113 Sofia, Bulgaria. DSc (Inst. of General and Inorganic Chemistry, 1981) chemistry. *Inorganic_synthesis crystal_growth*
E-mail banchem@bgearn.bitnet Tel. 359(2)7132573 Fax 359(2)705024

Petkov, Dr Valeri (1958). Lecturer. Faculty of Physics, Sofia University, 1126 Sofia, Bulgaria. PhD (Sofia U., 1990) physics. *X-ray_diffraction amorphous_phase computing Rietveld_analysis magnetism*
E-mail Petkov%v@bgcict.bitnet Tel. 359(2)623015 Fax 359(2)463589

Petrov, Prof. Kostadin (1940). Professor. Member, National Committee for Crystallography. Institute of General and Inorganic Chemistry, Bulgarian Academy of Sciences, 1113 Sofia, Bulgaria. DSc (Inst. of General and Inorganic Chemistry, 1990) chemistry. *X-ray_diffraction topotaxy condensed_matter crystal_chemistry*
E-mail banchem@bgearn.bitnet Tel. 359(2)7132582 Fax 359(2)705024

Petrov, Dr Ognyan (1952). Senior scientist. Central Laboratory of Mineralogy and Crystallography, Bulgarian Academy of Sciences, Rakovska 92, 1000 Sofia, Bulgaria. PhD (Sofia U., 1988) mineralogy. *X-ray_diffraction crystal_chemistry minerals*
E-mail mincryst@bgcict.bitnet Tel. 359(2)872450 Fax 359(2)884979

Philipov, Mr Alexander (1944). Assistant. Faculty of Geology and Geography, Sofia University, 1000 Sofia, Bulgaria. MSc (Sofia U., 1972) geochemistry. *Crystal_morphology minerals speleology*
Tel. 359(2)8581x256

Platikanova, Dr Vesselina (1939). Senior scientist. Central Laboratory of Photoprocesses, Bulgarian Academy of Sciences, 1040 Sofia, Bulgaria. PhD (Inst. of Physical Chemistry, 1971) chemistry. *Crystal_growth crystal_physics*
Tel. 359(2)723713 Fax 359(2)722465

Popov, Dr Alexander (1942). Senior scientist; Vice-Director. Central Laboratory for Electrochemical Power Sources, Bulgarian Academy of Sciences, 1113 Sofia, Bulgaria. PhD (Inst. of Physical Chemistry, 1980) chemistry. *Electrocrystallization electrochemistry*
Tel. 359(2)7132758 Fax 359(2)722544

Rainov, Dr Nikola (1939). Laboratory head. Geochemical Laboratories, Geological Survey, G. M. Dimitrov Str. 16, 1113 Sofia, Bulgaria. PhD (Sofia U., 1975) mineralogy. *X-ray_diffractometry infrared_spectroscopy minerals feldspars*
Tel. 359(2)723651x232

Rashkov, Prof. Stefan (1935). Professor. Institute of Physical Chemistry, Bulgarian Academy of Sciences, 1113 Sofia, Bulgaria. DSc (Inst. of Physics, 1984) chemistry. *Thin_film corrosion*
E-mail banchem@bgearn.bitnet Tel. 359(2)7133537 Fax 359(2)720038

Russev, Dr Krassimir (1946). Senior scientist. Institute of Metal Science, Bulgarian Academy of Sciences, Shipchenski Prochod 67, 1574 Sofia, Bulgaria. PhD (Inst. of Physical Chemistry, 1975) chemistry. *Solid_state crystallization*
Tel. 359(2)7142241 Fax 359(2)703207

Shumov, Dr Dimiter (1954). Research associate. Central Laboratory of Mineralogy and Crystallography, Bulgarian Academy of Sciences, Rakovska 92, 1000 Sofia, Bulgaria. PhD (Moscow U., 1986) physics. *Crystal_growth*
Tel. 359(2)872450 Fax 359(2)884979

Simov, Dr Stefan (1934). Senior scientist. Institute of Solid State Physics, Bulgarian Academy of Sciences, Tzarigradsko Chaussee 72, 1784 Sofia, Bulgaria. PhD (Inst. of Solid State Physics, 1989) physics. *Crystal_growth thin_film electron_microscopy whisker*
Tel. 359(2)7431x340 Fax 359(2)757032

Spassov, Dr Tony (1961). Assistant. Faculty of Chemistry, Sofia University, 1126 Sofia, Bulgaria. PhD (Sofia U., 1988) chemistry. *Phase_transitions metallic_glasses alloys*
Tel. 359(2)62561x236 Fax 359(2)682808

Staykov, Dr Georgi (1943). Senior scientist. Central Laboratory for Electrochemical Power Sources, Bulgarian Academy of Sciences, 1040 Sofia, Bulgaria. PhD (Inst. of Physical Chemistry, 1981) chemistry. *Electrocrystallization*
Tel. 359(2)7132773 Fax 359(2)722544

Stefanov, Mr Dechko (1931). Senior scientist. Geological Institute, Bulgarian Academy of Sciences, 1113 Sofia, Bulgaria. MSc (Sofia U., 1954) physics. *X-ray_diffractometry clays*
Tel. 359(2)7132282 Fax 359(2)724638

Stoilova, Prof. Margarita (1930). Professor. Faculty of Geology, University of Mining and Geology, 1156 Sofia, Bulgaria. DSc (Sofia U., 1989) mineralogy. *Mineralogy morphology*
Tel. 359(2)62851x384 Fax 359(2)621042

Stojanova, Dr Liljana (1948). Research associate. Institute of Metal Science, Bulgarian Academy of Sciences, Shipchenski Prochod 67, 1574 Sofia, Bulgaria. PhD (Inst. of Physical Chemistry, 1987) chemistry. *Metallic_glasses alloys*
Tel. 359(2)7142327 Fax 359(2)703207

Stoyanov, Dr Stoyan (1941). Senior scientist. Institute of Physical Chemistry, Bulgarian Academy of Sciences, 1113 Sofia, Bulgaria. PhD (Inst. of Physical Chemistry, 1977) chemistry. *Crystal_growth epitaxy thin_film*
E-mail banchem@bgearn.bitnet Tel. 359(2)7132557 Fax 359(2)720038

Stoyanova, Dr Valeriya (1949). Research associate. Institute of Physical Chemistry, Bulgarian Academy of Sciences, 1113 Sofia, Bulgaria. PhD (Inst. of Physical Chemistry, 1989) chemistry. *Crystal_growth surfaces morphology ice surface_roughening*
Tel. 359(2)7132507 Fax 359(2)720038

Stoychev, Dr Nikola (1939). Senior scientist. Institute of Metal Science, Bulgarian Academy of Sciences, Shipchenski Prochod 67, 1574 Sofia, Bulgaria. PhD (Inst. of Metallophysics, Kiev, 1977) physics. *Crystallization metals alloys*
Tel. 359(2)7142337 Fax 359(2)703207

Tomov, Dr Ivan (1939). Senior scientist. Institute of Physical Chemistry, Bulgarian Academy of Sciences, 1113 Sofia, Bulgaria. PhD (Inst. of Physical Chemistry, 1981) chemistry. *X-ray_diffractometry textures thin_layer*
E-mail banchem@bgearn.bitnet Tel. 359(2)7132534 Fax 359(2)720038

Topalova-Kalitzova, Dr Maria (1940). Research associate. Institute of Solid State Physics, Bulgarian Academy of Sciences, Tzarigradsko Chaussee 72, 1184 Sofia, Bulgaria. PhD (Inst. of Solid State Physics, 1990) physics. *Electron_microscopy implantation imperfection*
Tel. 359(2)7341340 Fax 359(2)757032

Vassilev, Dr Ivan (1932). Senior scientist. Institute of Solid State Physics, Bulgarian Academy of Sciences, Tzarigradsko Chaussee 72, 1784 Sofia, Bulgaria. PhD (Inst. of Solid State Physics, 1976) physics. *X-ray_diffraction X-ray_topography imperfections*
Tel. 359(2)878517 Fax 359(2)755019

Vesselinov, Mr Iliya (1942). Senior scientist. Geological Institute, Bulgarian Academy of Sciences, 1113 Sofia, Bulgaria. MSc (Sofia U., 1967) geochemistry. *Crystal_growth crystal_morphology internal_morphology*
E-mail geology@bgcict.bitnet Tel. 359(2)7132236 Fax 359(2)724638

Vitanov, Prof. Todor (1926). Professor. Central Laboratory for Electrochemical Power Sources, Bulgarian Academy of Sciences, 1040 Sofia, Bulgaria. DSc (Inst. of Physical Chemistry, 1986) chemistry. *Electrocrystallization electrochemistry catalysis*
E-mail banchem@bgearn.bitnet Tel. 359(2)718751 Fax 359(2)722544

Yaneva, Dr Svetlana (1942). Senior scientist. Institute of Metal Science, Bulgarian Academy of Sciences, Shipchenski Prochod 67, 1574 Sofia, Bulgaria. PhD (Sofia U., 1970) chemistry. *Crystallization eutectic_crystallization*
Tel. 359(2)7142337 Fax 359(2)703207

Zidarova, Dr Bogdana (1943). Senior scientist. Central Laboratory of Mineralogy and Crystallography, Bulgarian Academy of Sciences, Rakovska 92, 1000 Sofia, Bulgaria. DSc (Inst. of Applied Mineralogy, 1991) mineralogy. *Morphology synthesis minerals fluorite*
E-mail mincryst@bgcict.bitnet Tel. 359(2)872450 Fax 359(2)884979

Zotov, Dr Nikolay (1957). Research associate. Central Laboratory of Mineralogy and Crystallography, Bulgarian Academy of Sciences, Rakovska 92, 1000 Sofia, Bulgaria. PhD (Inst. of General and Inorganic Chemistry, 1991) chemistry. *X-ray_diffraction computing structural_disorder glasses*
E-mail mincryst@bgcict.bitnet Tel. 359(2)872450 Fax 359(2)884979

CANADA

Sub-Editor: Y. Le Page

Notes

1. Abbreviations used for the Canadian provinces are as follows: Alta, Alberta; B. C., British Columbia; Man., Manitoba; N. B., New Brunswick; Nfld, Newfoundland; N. S., Nova Scotia; Ont., Ontario; P. E. I., Prince Edward Island; Qué., Québec; Sask., Saskatchewan.

Ahmed, Dr Farid Ramadan (1924). Retired. Institute for Biological Sciences, National Research Council of Canada. [Inst. for Biological Sci., M-54, NRC, Ottawa, Ont. K1A 0R6, Canada.] PhD (U. Leeds, UK, 1953) physical chemistry. *Proteins computer_graphics area_detector*
E-mail cph@nrcbsa.bio.nrc.ca Tel. 1(613)9900850 Fax 1(613)9414475

Bagchi, Prof. Subodh Nath (1915). Retired. [5550 Bellerive, Brossard, Qué. J4Z 3C8, Canada.] DSc (U. Calcutta, India, 1946) chemistry. *Amorphous_structure diffraction_theory small-angle_scattering*
Tel. 1(514)4623626

Bakshi, Mr Pradip Kumar (1959). PhD student. Dept of Chemistry, Dalhousie University. [Dept. of Chemistry, Dalhousie University, Halifax, N.S. B3H 4J3, Canada.] MSc (Dhaka Univ., Bangladesh, 1985) inorganic chemistry. *X-ray_structure valence_charge_density*
E-mail bakshi@ac.dal.ca Tel. 1(902)4943759 Fax 1(902)4941310

Barbier, Dr Jacques (1952). Associate professor. Dept of Chemistry, McMaster University. [Dept of Chemistry, McMaster University, Hamilton, Ont. L8S 4M1, Canada.] PhD (Australian National University, 1985) crystal chemistry of oxide compounds. *Inorganic_oxides minerals crystal_chemistry phase_transformation X-ray/neutron/electron_diffraction*
E-mail barbier@mcmaster.ca Tel. 1(905)5259140x23477 Fax 1(905)5222509

Barrington-Leigh, Dr John (1943). Associate director; biomedical technology transfer management; senior licensing manager. U. of Alberta. [Industry Liaison Office, 1-3 University Hall, U. of Alberta, Edmonton, Alta, T6G 2J9, Canada.] PhD (Cambridge U., UK, 1970) physics. *Synchrotron_radiation X-ray_optics biological_structure energy_transduction diffraction_by_fibres structure_prediction immune_regulation cancer_diagnostic*
E-mail jbarring@vm.ucs.ualberta.ca Tel. 1(403)4920230 Fax 1(403)4926446

Barton, Prof. Richard J. (1928). Associate professor. University of Regina, Saskatchewan. [Dept of Physics, U. of Regina, Regina, Sask. S4S 0A2, Canada.] PhD (Iowa State U., USA, 1956) chemistry. *Metalloorganic_coordination small_molecules valence_charge_density*
E-mail barton@meena.cc.uregina.ca Tel. 1(306)5854653 Fax 1(306)5854894

Bassignana, Dr Isabella C. (1955). Staff scientist. Vice-Chair, CNC/IUCr. Bell-Northern Research Ltd [Advanced Technology Lab., Bell-Northern Research Ltd, Box 3511, Stn C., Ottawa, Ont. K1Y 4H7 Canada.] PhD (U. California, Los Angeles, USA, 1982) chemistry. *Topography X-ray_multiple-crystal_diffractometry optoelectronics X-ray_powder high-resolution_diffractometry*
E-mail isa@bnr.ca Tel. 1(613)7633550 Fax 1(613)7632626 or 1(613)7633404 Telex 0533175

Batchelor, Dr Raymond John (1952). Research fellow. Simon Fraser U. [Dept of Chemistry, Simon Fraser Univ., Burnaby, B.C. V5A 1S6, Canada.] PhD (McMaster U., 1983) chemistry. *Inorganic_chemistry crystal_structure_determination analysis_bond_length*
E-mail batchelo@sfu.ca Tel. 1(604)2914878 Fax 1(604)2913765

Beauchamp, Prof. André (1940). Professor. Université de Montréal. [Dépt. de Chimie, U. de Montréal, CP 6128, Succ. A, Montréal, Qué. H3C 3J7, Canada.] PhD (U. de Montréal, 1967) chemistry. *Bioinorganic_materials metalloorganic_coordination*
E-mail beauchmp@ere.umontreal.ca Tel. 1(514)3436446 Fax 1(514)3437586

Bélanger, Ms Suzanne (1969). PhD student. Dépt. de Chimie, Université de Montréal. [Dépt de Chimie, U. de Montréal, CP 6128, Succ. A, Montréal, Qué. H3C 3J7, Canada.] MSc inorganic chemistry. *Rhenium_compounds technetium-compounds metalloorganic_coordination complexation charge_transfer electrochemistry*
E-mail belanges@ere.umontreal.ca Tel. 1(514)3436111x3939 Fax 1(514)3437586

Bélanger-Gariépy, Ms Francine (1955). Research assistant. Dépt. de Chimie, Université de Montréal. [Dépt de Chimie, U. de Montréal, CP 6128, Succ. A, Montréal, Qué. H3C 3J7, Canada.] MSc (U. de Montréal, 1981) chemistry. *Synthesis characterization crystal_structure_determination*
E-mail belangef@tornade.ere.umontreal.ca Tel. 1(514)3436111x3937 Fax 1(514)3437586

Bensimon, Dr Corinne (1962). Research associate. Chemistry Dept, Ottawa University. [Chemistry Dept, Ottawa Univ., Ottawa, Ont. K1N 6N5, Canada.] PhD (U. de Montréal, 1990) chemistry. *Small_molecules computing inorganic_chemistry natural_products*
E-mail corinne@sg1.chem.nrc.ca Tel. 1(613)5644228 Fax 1(613)5646793

Bergmann, Dr Ernst Michael (1962). Postdoctoral fellow. Biochemistry Department, University of Alberta. [Biochemistry Dept, U. of Alberta, Edmonton, Alta T6G 2H7, Canada.] PhD (U. California, USA, 1993) biochemistry. *Proteins enzymatic_structure-activity_relationship*
E-mail berg@medievil.biochem.ualberta.ca Tel. 1(403)4922422 Fax 1(403)4920886

Bird, Prof. Peter Hans (1942). Professor. Dept of Chemistry and Biochemistry, Concordia University, Montréal. [Dept of Chemistry and Biochemistry, Concordia University, 1455 Blvd de Maisonneuve, Montréal, Qué. H3G 1M8, Canada.] PhD (U. Sheffield, UK, 1966) inorganic chemistry. *Metalloorganic_complexation transition_elements small_molecules*
E-mail BIRP@vax2.concordia.ca Tel. 1(514)8482089 Fax 1(514)8482868

Birnbaum, Dr Karin Bjämer. Head, development support & data administration. National Research Council of Canada. [M60, NRC, Ottawa, Ont. K1A 0R6, Canada.] PhD (U. Glasgow, UK, 1968) crystallography. *Antiviral_compounds nucleosides nucleotides computer_modelling databases*
E-mail karin@informatics.lan.nrc.ca Tel. 1(613)9934753 Fax 1(613)9520778

Blom, Dr Nick S. (1956). Research officer. Dept of Biochemistry, Université de Montréal. [Dépt de Biochimie, U. de Montréal, CP 6128 Succ Centre Ville, Montréal, Qué. H3C 3J7, Canada.] PhD (Delft U. of Technology, The Netherlands, 1987) convergent-beam electron diffraction. *Macromolecular_structure computer_modelling substrate_docking*
E-mail nick@bch.umontreal.ca Tel. 1(514)3436111x5352 Fax 1(514)3432210

Booth, Dr Andrew Donald (1918). Chairman of the board. Autonetics Research Associates Inc. [Autonetics Res. Assoc. Inc., Box 518, Sooke, BC V0S 1N0, Canada.] DSc (U. London, UK, 1951) crystallography. *Computing sound_propagation numerical_analysis sound_propagation_in_the_ocean*
E-mail abooth@ios.bc.ca Tel. 1(604)6425352 Fax 1(604)6425352

Bottomley, Prof. Frank (1941). Professor. Dept of Chemistry, U. of New Brunswick. [Dept of Chemistry, U. of New Brunswick, PO Box 4400, Fredericton, N.B. E3B 5A3 Canada.] PhD (U. Toronto, 1968) chemistry. *Inorganic_chemistry structural_chemistry*
E-mail bottomly@unb.ca Tel. 1(506)4534774 Fax 1(506)4534599

Brandon, Dr James Kenneth (1940). Associate professor. Physics Dept, U. of Waterloo. [Physics Dept., Univ. of Waterloo, Waterloo, Ont. N2L 3G1, Canada.] PhD (McMaster U., 1967) physics. *Inorganic_structure metals alloys*
Tel. 1(519)8851211x3494 Fax 1(519)7468115

Brayer, Prof. Gary David (1953). Professor. Dept of Biochemistry, U. of British Columbia. [Dept of Biochemistry, U. of British Columbia, 2146 Health Sciences Mall, Vancouver, B.C. V6T 1Z3, Canada.] PhD (U. Alberta, 1979) biochemistry. *Macromolecular_crystallography structure-activity_relationship enzymes proteins proteins-nucleic_acids_interaction complexation electron_transfer_mechanism*
E-mail usergbbc@mtsg.ubc.ca Tel. 1(604)8225216 Fax 1(604)8225227

Bridson, Dr John N. (1945). Associate professor. Chemistry Department, Memorial University. [Chemistry Dept, Memorial University, St-Johns, Nfld A1B 3X7, Canada.] PhD (Oxford U., UK, 1970) organic chemistry. *Conformation ring_molecules macrocyclic_ligands*
E-mail jbridson@kean.ucs.mun.ca Tel. 1(709)7372014

Brisse, Dr François (1935). Professor. Member, CNC/IUCr. Département de Chimie, Université de Montréal. [Département de Chimie, U. de Montréal, CP 6128, Succ. A, Montréal, Qué. H3C 3J7, Canada.] PhD (Dalhousie U., 1967) chemistry. *Crystal_structure_determination polymers polyesters polysaccharides clathrates electron_microscopy electron_diffraction model_molecules_related_to_polymers*
E-mail brisse@ere.umontreal.ca Tel. 1(514)3437604 Fax 1(514)3437586

Britten, Dr James Francis (1955). X-ray manager. Dept of Chemistry, McMaster University. [Dept of Chemistry, McMaster University, Hamilton, Ont. L8S 4M1, Canada.] PhD (McMaster Univ., 1984) chemistry. *Inorganic_chemistry bioinorganic_chemistry solid_nuclear_magnetic_resonance computer_graphics*
E-mail xman@xraysg.chemistry.mcmaster.ca Tel. 1(905)5259140x23481 Fax 1(905)5222509

Brown, Dr Ian David (1932). Professor of physics. Member, Teaching Commission; Data Commission Consultant; Co-Editor Acta Crystallographica; COMCIFS; ACA Council member; ex-officio member, CNC/IUCr. Institute for Materials Research, McMaster University. [Inst. for Materials Research, McMaster Univ., Hamilton, Ont. L8S 4M1, Canada.] PhD (U. London, UK, 1959) crystallography. *Inorganic_bonding inorganic_solid_state inorganic_database*
E-mail idbrown@mcmaster.ca Tel. 1(905)5259140x24710 Fax 1(905)5212773 Telex 0618347

Bushnell, Prof. Gordon William (1936). Professor. Dept of Chemistry, Univ. of Victoria. [Dept of Chemistry, U. of Victoria, PO Box 3055, Victoria, B.C. V8W 3P6, Canada.] PhD (U. West Indies, Kingston, Jamaica, 1966) inorganic chemistry. *Inorganic_chemistry metalloorganic_coordination proteins nucleic_acids*
E-mail bushnell@uvvm.uvic.ca Tel. 1(604)7217163 Fax 1(604)7217147

Buyers, FRSC, Dr William James Leslie (1937). Scientist. President, Canadian Institute for Neutron Scattering; Secretary, IUPAP Commission for Magnetism. AECL. [Neutron and Condensed Matter Sci., AECL, Chalk River, Ont. K0J 1J0, Canada.] PhD (U. Aberdeen, UK, 1963) physics. *Neutron_scattering phase_transitions magnetism quantum_magnetism_in_low_dimensions*

E-mail buyersw@crl.aecl.ca Tel. 1(613)5843311x4532 Fax 1(613)5844040 Telex 0533455

Camerman, Dr Norman (1939). Professor. Dept of Biochemistry, U. of Toronto. [Dept of Biochemistry, U. of Toronto, Toronto, Ont. M5S 1A8, Canada.] PhD (U. British Columbia, 1964) chemistry. *Biological_structure biological_structure-activity_relationship molecular_design*
E-mail camerman@medac.med.utoronto.ca Tel. 1(416)9787027 Fax 1(416)9788548

Cameron, Prof. Theodore Stanley (1942). Professor. Member, CNC/IUCr. Dept of Chemistry, Dalhousie U. [Dept. of Chemistry, Dalhousie University, Halifax, N.S. B3H 4J3, Canada.] DPhil (Oxford U., UK, 1969) chemistry. *Hydrogen_bonding molecular_packing inorganic_structure organometallic_structure sulphur_compounds and_phosphorus_compounds residual_electron_density graphical_investigation*
E-mail cameron@ac.dal.ca Tel. 1(902)4943305 Fax 1(902)4941310 Telex 01921863

Campbell, Dr Robert Laurence (1960). Assistant research officer. Biotechnology Research Institute, National Research Council of Canada. [Biotechnology Res. Inst., M-54, NRC, Ottawa, Ont. K1A 0R6, Canada.] PhD (Massachusetts Inst. of Technology, USA, 1988) biological chemistry. *Macromolecular_structure-activity_relationship enzymatic_activity computer_graphics macromolecular_crystallography*
E-mail rlc@nrcbsa.bio.nrc.ca Tel. 1(613)9900857 Fax 1(613)9414475

Černý, Dr Petr (1934). Professor. Dept of Geological Sci., U. of Manitoba. [Dept of Geological Sciences, U. of Manitoba, Winnipeg, Man. R3T 2N2, Canada.] PhD (Czechoslovak Acad of Sci., Prague, 1966) mineralogy. *Morphology crystal_chemistry inorganic_oxides silicates igneous_petrology geochemistry*
Tel. 1(204)4748765 Fax 1(204)2617581

Charland, Dr Jean-Pierre (1957). Research scientist. Treasurer, CNC/IUCr. Department of Natural Resources Canada, CANMET. [DNRC, CANMET, BCC Bldg 3, Energy Research Labs, 555 Booth St., Ottawa, Ont., K1A 0G1, Canada.] PhD (Univ. de Montréal, 1984) chemistry. *Inorganic_chemistry bioorganic_molecules bioinorganic_chemistry surface*
E-mail charland@emr.ca Tel. 1(613)9955751 Fax 1(613)9959584

Chen, Dr Jie (1958). Postdoctoral fellow. Dept of Biochemistry, University of British Columbia. [Dept of Biochemistry, University of British Columbia, 2146 Health Science Mall, Vancouver, B.C. V6T 1Z3, Canada.] PhD (Osaka Univ., Japan, 1991) physical chemistry. *Macromolecular_crystallography macromolecular_structure-activity_relationship metalloorganic_structure*
E-mail userjcbc@mtsg.ubc.ca Tel. 1(604)8225007 Fax 1(604)8225227

Cheng, Dr Pei-Tak (1940). Staff scientist. Mount-Sinai Hospital. [Dept of Pathology, Mount Sinai Hospital, 600 University Avenue, Toronto, Ont. M5G 1X5, Canada.] PhD (Univ. of Toronto, 1972) crystallography. *Bone bioinorganic_minerals deposit_disease phosphates calcium_phosphate osteoporosis tissue_crystallography ultrastructure*
Tel. 1(416)5864468 Fax 1(416)5868589

Chenite, Dr Abdellatif (1955). Postdoctoral assistant. Inst. for Environmental Research and Technology, National Research Council of Canada. [IERT, NRC, Ottawa, Ont. K1A 0R6, Canada.] PhD (U. de Montréal, 1992) chemistry. *Polymers intercalation electron_diffraction surface*
E-mail gjccarp@sg1.chem.nrc.ca Tel. 1(613)9932527 Fax 1(613)9932451

Chieh, Prof. Chung (1939). Professor. Dept of Chemistry, U. of Waterloo. [Dept of Chemistry, Univ. of Waterloo, Waterloo, Ont. N2L 3G1, Canada.] PhD (U. British Columbia, 1969) chemistry. *Crystal_systematics structural_theory theory_of_crystal_structure*
E-mail cchieh@chemistry.watstar.uwaterloo.ca Tel. 1(519)8884567x5816 Fax 1(519)7460435

Clark, Dr Malcolm John Roy (1944). Environmental chemist. Ministry of Environment. [Environmental Protection Div., Ministry of Environment, 777 Broughton St., Victoria, B.C. V8V 1X5 Canada.] PhD (U. New Brunswick, 1971) chemistry. *Environmental_chemistry computing statistics quality_assurance*
E-mail mclark@epdiv1.env.gov.bc.ca Tel. 1(604)3879947 Fax 1(604)3566337

Codding, Dr Penelope W. (1946). Professor and Head. Ordinary Member, IUCr Executive Committee; ex-officio member, CNC/IUCr; Chair, IUCr Calendar Subcommittee; IUCr representative to ACA. U. of Calgary. [Dept of Chemistry, U. of Calgary, Calgary, Alta T2N 1N4, Canada.] PhD (Michigan State U., USA, 1971) chemistry. *Structure-activity_relationship drug neurotoxins peptides DNA_and_proteins_crystallography*
E-mail pcodding@acs.ucalgary.ca Tel. 1(403)2205340 Fax 1(403)2841372 Telex 03821545

Collins, Dr M. F. (1935). Professor. Dept of Physics, McMaster University. [Dept of Physics, McMaster University, Hamilton, Ont. L8S 4M1, Canada.] PhD (Cambridge Univ., UK, 1962) physics. *Neutron_scattering magnetism phase_transition*
E-mail mcollins@mcmail.cis.mcmaster.ca Tel. 1(905)5259140x24172 Fax 1(905)5461252

Coulombe, Mr René (1967). Technical officer. Biotechnology Research Institute, National Research Council of Canada. [Biotechnology Research Institute, NRC, 6100 Royalmount Ave., Montréal, Qué. H4P 2R2, Canada.] PhD (U. de Sherbrooke, 1993) biochemistry. *Macromolecular_purification crystallization ribonucleotide_reductase HSV aldolase*
E-mail Rene.Coulombe@bri.nrc.ca Tel. 1(514)4966341 Fax 1(514)4965143

Cowie, Prof. Martin (1947). Professor. Dept of Chemistry, University of Alberta. [Dept of Chemistry, U. of Alberta, Edmonton, Alta T6G 2G2, Canada.] PhD (U. Alberta, 1974) X-ray crystallography. *Hydrides metalloorganic_chemistry rhodium_compounds iridium_compounds bimetallic_compounds small_molecule_activation metal-metal_cooperativity binuclear_complexes mixed-metal_complexes mixed-valence_compounds*

E-mail martin_cowie@dept.chem.ualberta.ca Tel. 1(403)4925581 Fax 1(403)4928231 Telex 0372979

Curzon, Prof. Albert Edward (1934). Professor. Simon Fraser U [Physics Dept., Simon Fraser University, Burnaby, B.C. V5A 1S6, Canada.] PhD (Imperial College, UK, 1959) physics. *Electron_microscopy electron_diffraction thin_film crystal_structure*
Tel. 1(604)2914181 Fax 1(604)2913592

Cygler, Dr Mirek (1947). Senior research officer. Biotechnology Research Institute, National Research Council of Canada. [Biotechnology Research Institute, NRC, 6100 Royalmount Ave., Montréal, Qué. H4P 2R2, Canada.] PhD (U. of Łódź, Poland, 1976) crystallography. *Macromolecular_structure-activity_relationship proteins nucleic_acids molecular_replacement*
E-mail mirek@bri.nrc.ca Tel. 1(514)4966321 Fax 1(514)4965143

Dabkowska, Dr Hanna A. (1951). Research scientist. Institute for Materials Research, McMaster University. [Inst. for Materials Research, McMaster Univ., Hamilton, Ont. L8S 4M1, Canada.] PhD (Inst. of Physics, Polish Acad. of Sci., Warsaw, 1983) solid-state physics. *Crystal_growth superconductors thermoanalysis optical_property high-melting_oxide_materials*
E-mail dabko@mcmail.cis.mcmaster.ca Tel. 1(905)5297070x27092 Fax 1(905)5212773

Dabkowski, Dr Antoni (1948). Research scientist. Institute for Materials Research, McMaster University. [Inst. for Materials Res. ABB 433, McMaster Univ., Hamilton, Ont. L8S 4M1, Canada.] PhD (Inst. of Physics, Polish Acad. of Sci., Warsaw, 1987) solid-state physics. *Crystal_growth Czochralski_technique liquid_epitaxy magnetic_oxides superconductors*
E-mail delbaere@mcmail.cis.mcmaster.ca Tel. 1(905)5297070x27092 Fax 1(905)5212773

Delbaere, Dr Louis Theophil Joseph (1943). Professor. Regional Director, Medical Research Council of Canada. Dept of Biochemistry, U. of Saskatchewan. [Dept of Biochemistry, U. of Saskatchewan, Saskatoon, Sask. S7N 0W0, Canada.] PhD (U. Manitoba, 1970) chemistry. *Crystal_structure proteins molecular_recognition structure-activity_relationship biologically_important_molecules*
E-mail delbaere@sask.usask.ca Tel. 1(306)9664373 Fax 1(306)9664390 Telex 0742659

Derewenda, Dr Zygmunt (1953). Associate professor of biochemistry. Biochemistry Dept, U. of Alberta. [Biochemistry Dept, U. of Alberta, Edmonton, Alta T6G 2H7, Canada.] PhD (U. of Łódź, Poland, 1981) chemistry. *Macromolecular_crystallography proteins lipids macromolecular_structure-activity_relationship protein-lipid_interactions*
E-mail zygmunt@hal.biochem.ualberta.ca Tel. 1(403)4922136 Fax 1(403)4920886

Dichman, Dr Klaus (1942). Professor. Vanier College. [Chemistry Dept, Vanier College, 821 Blvd Ste-Croix, St-Laurent, Qué. H4L 3X9, Canada.] PhD (U. Toronto, 1972) X-ray crystallography. *Crystal_structure_determination inorganic_chemistry*
Tel. 1(514)4882341x4052

Dolling, Dr Gerald (1935). Vice-president, physical sciences. AECL Research, Chalk River Laboratories. [Physical Sciences, AECL Research, Chalk River, Ont. K0J 1J0, Canada.] PhD (Cambridge U., UK, 1961) physics. *Neutron_diffraction phonon_phase_transition magnon neutron_inelastic_scattering*
E-mail dollingg@crl.aecl.ca Tel. 1(613)5843311x3108 Fax 1(613)5844660

Drake, Prof. John E. (1936). Professor. U. of Windsor. [Chemistry Dept, Univ. of Windsor, Windsor, Ont. N9B 3P4, Canada.] DSc (U. Southampton, UK, 1978) inorganic chemistry. *Spectroscopy structure organometallic_coordination*
E-mail ak4@uwindsor.ca Tel. 1(519)2534232x3551 Fax 1(519)9737098

Drouin, Mr Marc (1965). X-ray crystallography services. Structural Chemistry Laboratory, Université de Sherbrooke. [Structural Chemistry Lab., U. de Sherbrooke, Sherbrooke, Qué. J1K 2R1, Canada.] MSc (McGill Univ., 1989) physical chemistry. *Amino_acids polypeptides steroids cobalt_compounds palladium_compounds platinum_compounds dipeptides*
E-mail mdrouin@structure.chimie.usherb.ca Tel. 1(819)8217813 Fax 1(819)8218017

Duke, Dr Norma (1959). Postdoctoral fellow. Department of Medical Microbiology and Infectious Diseases, University of Alberta. [Medic. Microb. & Inf. Diseases, Univ. of Alberta, Edmonton, Alta T6G 2H7, Canada.] PhD (Univ. of Calgary, 1989) physical chemistry. *Proteins receptors DNA/RNA membranes*
E-mail norma@mycroft.mmid.ualberta.ca Tel. 1(403)4924696 Fax 1(403)4927521

Einstein, Prof. Frederick W. B. (1940). Professor. Dept of Chemistry, Simon Fraser University. [Dept of Chemistry, Simon Fraser University, Burnaby, B.C. V5A 1S6, Canada.] PhD (U. Canterbury, NZ, 1965) inorganic structural chemistry. *Inorganic_chemistry low_temperature molecular_modelling macromolecular_structure*
E-mail einstein@sfu.ca Tel. 1(604)2914878 Fax 1(604)2913765

Enright, Dr Gary (1946). Associate research officer. Steacie Inst. of Molecular Science, National Research Council of Canada. [Steacie Inst. of Molecular Sci., NRC, Ottawa, Ont. K1A 0R6, Canada.] PhD (Univ. of Toronto, 1975) molecular physics. *Small_molecules X-ray_diffractometry guest-host_structures*
E-mail enright@ned1.sims.nrc.ca Tel. 1(613)9937393 Fax 1(613)9545242

Ercit, Dr Timothy Scott (1957). Research scientist. Canadian Museum of Nature. [Research Div., Can Mus. Nature, PO Box 3443, Stn D, Ottawa, Ont. K1P 6P4, Canada.] PhD (U. Manitoba, 1986) mineralogy. *Oxides phosphates inorganic_structure paragenesis pegmatite_mineralogy*
Tel. 1(613)9523516 Fax 1(613)9523510

Evans, Dr Stephen (1959). Assistant professor. Department of Biochemistry, University of Ottawa. [Department of Biochemistry, University of Ottawa, Ottawa, Ont. K1H 8M5, Canada.] PhD (Univ. of British Columbia, 1986) chemistry. *Immunobiology molecular_replacement enzymes DNA glycoproteins*
E-mail stephen.evans@nrc.ca Tel. 1(613)7876576 Fax 1(613)7876732

Falk, Dr Michael (1931). Senior research officer. Institute for Marine Biosciences, National Research Council of Canada. [Inst. for Marine Biosciences, NRC, 1411 Oxford St., Halifax, N.S. B3H 3Z1, Canada.] DSc (U. Laval, 1958) physical chemistry. *Infrared_spectroscopy crystalline_hydrates NMR_spectroscopy molecular_modelling configuration_and_comformation_of_biomolecules*
E-mail falkm@imb.lan.nrc.ca Tel. 1(902)4268265 Fax 1(902)4269413

Farrar, Dr David H. (1951). Associate professor. Dept of Chemistry, University of Toronto. [Dept of Chemistry, U. of Toronto, Toronto, Ont. M5S 1A1, Canada.] PhD (U. Western Ontario, 1979) inorganic chemistry. *Transition_elements carbonyls*
E-mail dfarrar@alchemy.chem.utoronto.ca Tel. 1(416)9783568 Fax 1(416)9788775

Ferguson, Dr George (1936). Prof. of chemistry. Coeditor, Acta Crystallographica. Chemistry Dept., University of Guelph. [Chemistry Dept., U. of Guelph, Guelph, Ont. N1G 2W1, Canada.] DSc (U. Glasgow, UK, 1969) crystal structure analysis. *Inorganic_structure organic_structure*
E-mail george@xray.chembio.uoguelph.ca or ferguson@chembio.uoguelph.ca Tel. 1(519)8244120x3548 or x3800 Fax 1(519)7661499

Ferguson, Prof. Robert Bury (1920). Prof. emeritus; retired. Dept of Geological Sciences, U. of Manitoba. [Dept of Geological Sci., U. of Manitoba, Winnipeg, Man. R3T 2N2, Canada.] PhD (U. Toronto, 1948) mineralogy. *Structural_crystallography feldspars minerals aluminosilicates*
E-mail rbferguson@bldgwall.lan1.umanitoba.ca Tel. 1(204)4747406Fax 1(204)2617581

Fethiere, Dr James (1963). Postdoctoral fellow. Biotechnology Research Institute, National Research Council of Canada. [Biotechnology Research Institute, NRC, 6100 Royalmount Ave., Montréal, Qué. H4P 2R2, Canada.] PhD (Univ. de Montréal, 1992) pharmacology. *Macromolecular_crystal_structure proteins drug_design structure-assisted_drug_design*
E-mail james.fethiere@bri.nrc.ca Tel. off: 1(514)4966173 lab: 1(514)4966376 Fax 1(514)4965143

Fleet, Dr Michael Edward (1938). Professor. Dept of Earth Sciences, U. of Western Ontario. [Dept of Earth Sciences, U. Western Ontario, London, Ont. N6A 5B7, Canada.] PhD (U. Manchester, UK, 1963) geology. *Inorganic_structure structural_physical_property crystal_chemistry*
Tel. 1(519)6613184 Fax 1(519)6613198

Fortier, Dr Suzanne (1949). Professor. Chair, CNC/IUCr. Dept of Chemistry, Queen's University. [Dept of Chemistry, Queen's Univ., Kingston, Ont. K7L 3N6, Canada.] PhD (McGill U., 1976) crystallography. *Direct_method artificial_intelligence structure_determination*
E-mail fortiers@qucdn.queensu.ca Tel. 1(613)5452654 Fax 1(613)5456669

Fraser, Dr Marie (1962). Research associate. Biochemistry Dept., University of Alberta. [Biochemistry Dept., University of Alberta, Edmonton, Alta T6G 2H7, Canada.] PhD (Queen's U., 1987) chemistry. *Macromolecules*
E-mail mari@zeus.biochem.ualberta.ca Tel. 1(403)4922422 Fax 1(403)4920886

Fujinaga, Dr Masao (1958). Research associate. Biochemistry Department, University of Alberta. [Biochemistry Dept, U. of Alberta, Edmonton, Alta T6G 2H7, Canada.] PhD (U. Alberta, 1986) biochemistry. *Macromolecular_crystallography enzymatic_structure-activity_relationship computing*
E-mail fujinaga@manitou.biochem.ualberta.ca Tel. 1(403)4928249 Fax 1(403)4920886

Gabe, Dr Eric James (1933). Retired. Inst. for Environmental Research and Technology, National Research Council of Canada. [104 Rothwell Dr., Ottawa, Ont. K1J 8L9, Canada.] PhD (U. Wales, UK, 1960) crystallography. *Automation small_computer structure*
E-mail gabe@sg1.chem.nrc.ca Tel. 1(613)7466795

Galt, Dr Robert Irwin (1938). Curator of mineralogy. Royal Ontario Museum. [Dept of Mineralogy, ROM, 100 Queen's Park, Toronto, Ont. M5S 2C6, Canada.] PhD (U. Manitoba, 1967) mineralogy. *Mineralogy morphology X-ray_crystallography descriptive_mineralogy*
E-mail robertg@rom.on.ca Tel. 1(416)5865818 Fax 1(416)5865814

Gibbons, Dr Cyril Stephen (1945). Manager, patents and contract administration. ORTECH Corporation. [ORTECH, Sheridan Park, Mississauga, Ont. L5K 1B3, Canada.] PhD (U. British Columbia, 1971) structure analysis. *Crystal_structure*
Tel. 1(905)8224111x244 Fax 1(905)8221092

Graham, Dr A. Ronald (1917). Retired. [515 St George St. E, Fergus, Ont. N1M 1L1, Canada.] PhD (U. Toronto, 1950) mineralogy. *Geochemical_economy ores economic_mineralogy*
Tel. 1(519)8431150

Grattan-Bellew, Dr Patrick Edward (1934). Research officer. Institute for Research in Construction, National Research Council of Canada. [Inst. for Res. in Construction, NRC, Ottawa, Ont. K1A 0R6, Canada.] PhD (Cambridge U., UK, 1969) experimental mineralogy. *Composition_cements reaction_cements mineralogy*
Tel. 1(613)9930096 Fax 1(613)9545984

Greedan, Dr John E. (1942). Professor of chemistry. Dept of Chemistry, McMaster University. [Dept of Chemistry, McMaster University, Hamilton, Ont. L8S 4M1, Canada.] PhD (Tufts University, USA, 1969) inorganic solid state chemistry. *Magnetic_structure neutron_diffraction superconductor intercalation frustrated_magnets lithium_intercalation_compounds*
Tel. 1(905)5259140x24725 Fax 1(905)5212773

Grenier, Ms Lucie. Quality assurance specialist. Novacor Chemicals (Canada) Ltd [Div. Plastiques, Novacor Chemicals, 11625 Sherbrooke Ouest, Montréal, Qué H1B 5L9, Canada.] MSc (Univ. de Montréal, 1986) chemistry.
Tel. 1(514)6406682 Fax 1(514)6454126

Grice, Dr Joel Denison (1946). Curator of minerals. Canadian Museum of Nature. [Mineral Sci. Div., Can. Mus. Nature, PO Box 3443, Station D, Ottawa, Ont. K1P 6P4, Canada.] PhD (U. Manitoba, 1973) mineralogy. *Crystal_systematics crystal_growth minerals*
Tel. 1(613)9523513 Fax 1(613)9523510

Groat, Dr Lee Andrew (1959). Assistant professor. Dept of Geological Sciences, U. of British Columbia. [Dept of Geological Sci., U. of British Columbia, Vancouver, B.C. V6T 1Z4, Canada.] PhD (U. Manitoba, 1988) geology. *Minerals optical_crystallography electron_probe_microanalysis structure_determination*
E-mail groat@unixg.ubc.ca Tel. 1(604)8228238 Fax 1(604)8226088

Grochulski, Dr Pawel (1955). Visiting fellow. Biotechnology Research Institute, National Research Council of Canada. [Biotechn. Res. Inst., NRC, 6100 Royalmount Avenue, Montréal, Qué H4P 2R2, Canada.] PhD (Technical Univ. of Łódź, Poland, 1988) physical chemistry. *Proteins enzymes lipases ionophores steroids_structure-activity_relationship*
E-mail pawel@bri.nrc.ca Tel. 1(514)4962619 Fax 1(514)4965143

Grundy, Dr Harry Douglas (1941). Professor. Dept. of Geology, McMaster University. [Dept of Geology, McMaster Univ., Hamilton, Ont. L8S 4M1, Canada.] PhD (U. Manchester, UK, 1966) mineralogy. *Silicates minerals characterization inorganic_solids*
E-mail grundy@mcmaster.ca Tel. 1(416)5259140x4516 Fax 1(416)5223141

Hawthorne, Dr Frank Christopher (1946). Professor. Member, CNC/IUCr. Dept of Geological Sciences, University of Manitoba. [Dept of Geological Sciences, U. of Manitoba, Winnipeg, Man. R3T 2N2, Canada.] PhD (McMaster U., 1973) geology. *Minerals graph_theory electronic_structure Rietveld_method spectroscopy topological_aspects_of_structure*
E-mail FCHAWTHORN@bldgwall.lan1.umanitoba.ca Tel. 1(204)4748861 Fax 1(204)2617581

Hazes, Dr Bart (1966). Research Associate. Department of Medical Microbiology and Infectious Diseases, University of Alberta. [Medic. Microb. & Inf. Diseases, Univ. of Alberta, Edmonton, Alta T6G 2H7, Canada.] PhD (U. Groningen, The Netherlands, 1993) physical chemistry. *X-ray_crystallography macromolecular_structure-activity_relationship computing*
E-mail bart@mycroft.mmid.ualberta.ca Tel. 1(403)4924696 Fax 1(403)4927521

Hempel, Dr Andrew (1944). Research associate. Dept of Biochemistry, University of Toronto. [Dept of Biochemistry, U. of Toronto, Toronto, Ont. M5S 1A8, Canada.] PhD (Techn. U. of Gdansk, Poland, 1975) organic chemistry. *Biological_structure structure-activity_relationship*
E-mail hempel@medac.med.utoronto.ca Tel. 1(416)9783726 Fax 1(416)9788548

Henderson, Dr Grant S. (1956). Associate Professor. Geology Dept., University of Toronto. [Geology Dept., U. of Toronto, 22 Russell St., Toronto, Ont. M5S 3B1, Canada.] PhD (U. Western Ontario, 1983) geology. *Vitreous_state inorganic_surface EXAFS XANES AFM/STM_studies_of_minerals_and_glasses surface_structure_and_relaxation_amorphization_etc.*
E-mail henders@afm1.geology.utoronto.ca Tel. 1(416)9786041 or 1(416)9780668 Fax 1(416)9783938

Héroux, Ms Annie. PhD student. Dépt. de Chimie, Université de Montréal. [Dépt de Chimie, U. de Montréal, CP 6128, Succ. A, Montréal, Qué. H3C 3J7, Canada.] MSc (U. de Montréal, 1992) chemistry. *Polymers structure_determination conformation polyamides*
Tel. 1(514)3436111x3937 Fax 1(514)3437586

Heyding, Dr Robert Donald (1925). Professor emeritus. Dept of Chemistry, Queen's University. [Dept of Chemistry, Queen's U., Kingston, Ont. K7L 3N6, Canada.] PhD (McGill U., 1951) physical chemistry. *X-ray_powder phase_transition inorganic_structure polymers_structure*
Tel. 1(613)5452607 Fax 1(613)5456669

Howell, Dr P. Lynne (1960). Research scientist; assistant professor. Hospital for Sick Children, Toronto. [Biochemistry Research, Hospital for Sick Children, Toronto, Ont. M5G 1X8, Canada.] PhD (U. London, UK, 1986) crystallography. *Macromolecular_crystallography Laue_diffraction enzymatic_catalysis structure-activity_relationship*
E-mail howell@aragorn.psf.sickkids.on.ca Tel. 1(416)8135917 Fax 1(416)8135022

Huber, Dr Carol P. (1937). Senior research officer. Secretary, CNC/IUCr. Biotechnology Research Institute, National Research Council of Canada. [Biotechnology Res. Inst., M-54, NRC, Ottawa, Ont. K1A 0R6, Canada.] DPhil (Oxford U., UK, 1963) chemical crystallography. *Biological_crystallography enzymatic_structure-activity_relationship*
E-mail cph@nrcbsa.bio.nrc.ca Tel. 1(613)9900856 Fax 1(613)9414475

Hutcheon, Dr Wendy Lou (Brooks) (1946). [4437 W 13th Avenue, Vancouver, B.C. V6R 2V2, Canada.] PhD (U. Alberta, 1971) crystallography. *Macromolecular_crystallography enzymatic_structure-activity_relationship proteins nucleic_acids*
Tel. 1(604)2247058

Hynes, Dr Rosemary Catherine (1963). Postdoctoral fellow. Department of Biochemistry, University of Toronto. [Dept of Biochemistry, U. of Toronto, Toronto, Ont. M5S 1A8, Canada.] PhD (U. Western Ontario, 1989) chemistry. *Proteins hemoglobin computer*
E-mail rosi@hera.med.utoronto.ca Tel. 1(416)9780560 Fax 1(416)9788548

James, Dr Margaret Ann (1939). Associate professor. Dept of Chemistry, Mt St Vincent University. [Department of Chemistry, Mt St Vincent University, Halifax, N.S. B3M 2J6, Canada.] PhD (Dalhousie Univ., 1986) inorganic chemistry. *Solid_hydrogen_bonding inorganic_structure*
E-mail mjames@linden.msvu.ca Tel. 1(902)4576141 Fax 1(902)4570579

James, Prof. Michael N. G. (1940). University professor. Biochemistry Dept, U. of Alberta. [Biochemistry Dept, U. of Alberta, Edmonton, Alta T6G 2H7, Canada.] DPhil (Oxford U., UK, 1966) chemical crystallography. *Macromolecular_structure proteins enzymatic_structure-activity_relationship biological_molecules*
E-mail miik@manitou.biochem.ualberta.ca Tel. 1(403)4924550 Fax 1(403)4920886

Jennings, Dr M. C. (1961). Research associate. Surface Science Western, Univ. of Western Ontario. [Surface Science Western, U. Western Ontario, London, Ont. N6A 5B7, Canada.] PhD (U. Western Ontario, 1989) organometallic chemistry. *Organometallic_chemistry*
E-mail jennings@surf.ssw.uwo.ca Tel. 1(519)6792111x6734

Jones, Dr Stephen John (1942). Senior research officer. Institute for Marine Dynamics, National Research Council of Canada. [Inst. for Marine Dynamics, NRC, Box 12093, Stn A, St Johns, Nfld A1B 3T5, Canada.] PhD (U. Birmingham, UK, 1967) physics. *Marine_ice physical_property sea_ice_properties*
E-mail sjones@minnie.imd.nrc.ca Tel. 1(709)7725403 Fax 1(709)7722462

Kayden, Dr Catherine Sheila (1960). Postdoctoral fellow. Dept of Molecular and Medical Genetics, U. of Toronto. [Dept of Molecular and Medical Genetics, U. of Toronto, Toronto, Ont. M5S 1A8, Canada.] PhD (U. Victoria, 1989) chemistry. *Macromolecular_crystallography*
E-mail kayden@lec.med.utoronto.ca Tel. 1(416)9780742 Fax 1(416)9786885

King, Prof. Hubert Wylam (1930). Professor. Dept of Materials Engineering, Univ. of Western Ontario. [Dept of Materials Engineering, Univ. of Western Ontario, London, Ont. N6A 5B9, Canada.] PhD (U. Birmingham, UK, 1956) physical metallurgy. *Phase_transition ceramics thermal_expansion electromagnetic_physical_property high_and_low_temperature_powder_diffraction*
Tel. 1(519)6792111x8437 Fax 1(519)6613808

Klug, Dr Dennis D. (1942). Senior research officer. Steacie Inst. of Molecular Science, National Research Council of Canada. [Steacie Inst. of Molecular Sci., NRC, Ottawa, Ont. K1A 0R6, Canada.] PhD (Univ. of Wisconsin, Madison, USA, 1968) chemical physics. *High-pressure_spectroscopy phase_transition molecular_dynamics*
E-mail klug@jtsg.sims.nrc.ca Tel. 1(613)9911238 Fax 1(613)9545242

Knop, Prof. Osvald (1922). Harry Shireff professor of chemical research. Dept of Chemistry, Dalhousie University. [Dept of Chemistry, Dalhousie U., Halifax, N.S. B3H 4J3, Canada.] DSc (U. Laval, 1957) physical chemistry. *Structural_chemistry inorganic_chemistry*
Tel. 1(902)4943317 or 4943305 Fax 1(902)4941310

Kocman, Dr Vladimir (1937). Senior research scientist; analytical group. DOMTAR Inc. Research Centre. [DOMTAR Inc. Res. Centre, Box 300, Senneville, Qué. H9X 3L7, Canada.] PhD (Masaryk Univ., Brno, Czech., 1968) inorganic chemistry. *X-ray_powder X-ray_fluorescence thermogravimetry experimental_instrumentation analytical_chemistry*
Tel. 1(514)4578284 Fax 1(514)4574527

Kodama, Dr Hideomi (1931). Senior research scientist. Agriculture Canada. [Centre for Land & Biol. Resource Res., CEF, Agriculture Canada, Carling Ave., Ottawa, Ont. K1A 0C6, Canada.] PhD (Tokyo U. of Education, Japan, 1961) mineralogy. *Clays layered_compounds silicates structural_disorder mineral_characterization_as_soil_component*
Tel. 1(613)9955011x7888 Fax 1(613)9951823 Telex 0533283

Lea, Prof. Sydney George (1933). Professor. Ryerson Polytechnical University. [Chemical and Biological Sci., Ryerson Polytech. Univ., 350 Victoria St., Toronto, Ont., M5B 2K3, Canada.] PhD (U. London, UK, 1968) crystallography. *Spectroscopy inorganic_structure crystal_physics*
Tel. 1(416)9795000x6349 Fax 1(416)9795044

Lebuis, Dr Anne-Marie (1962). Manager, Crystallographic Service Lab. Department of Chemistry, McGill University. [Dept of Chemistry, McGill Univ., Montréal, Qué H3A 2K6, Canada.] PhD (Université de Montréal, 1993) chemistry. *Rhenium_compounds radiochemistry crystal_data_collection solid_state*
E-mail lebuis@omc.lan.mcgill.ca Tel. 1(514)3986728 Fax 1(514)3983797

Leong, Mr Weng Kee (1960). PhD student. Dept of Chemistry, Simon Fraser University. [Dept of Chemistry, Simon Fraser University, Burnaby, B.C. V5A 1S6, Canada.] MSc (National University of Singapore, 1992) organometallic clusters. *Metallic_clusters transition_elements metallic_main-group_compounds metal_clusters_containing_transition_and_main-group_metals*
E-mail wengl@sfu.ca Tel. 1(604)2914408 Fax 1(604)2913765

Le Page, Dr Yvon (1943). Senior research officer. Sub-Editor of World Directory for Canada. Inst. for Environmental Research and Technology, National Research Council of Canada. [IERT, NRC, Ottawa, Ont. K1A 0R6, Canada.] PhD (U. de Montréal, 1974) physics engineering. *Environmental_chemistry electron_diffraction toxic_microcrystal computing*
E-mail yvon@sg1.chem.nrc.ca Tel. 1(613)9932527 Fax 1(613)9932451

Litster, Dr Stephen (1966). Postdoctoral fellow. U. of Calgary. [Dept of Chemistry, U. of Calgary, Calgary, Alta, T2N 1N4, Canada.] PhD (University of Salford, UK, 1992) chemical crystallography. *Macromolecular_crystallography*
E-mail litster@acs.ucalgary.ca Tel. 1(403)2205069 Fax 1(403)2899488 Telex 03821545

Lobsanov, Dr Yuri D. (1955). Postdoctoral fellow. Dept of Molecular and Medical Genetics, U. of Toronto. [Dept of Molecular and Medical Genetics, U. of Toronto, Toronto, Ont. M5S 1A8, Canada.] PhD (Institute of Crystallography, Moscow, 1992) physics and mathematics. *Macromolecular_crystallography crystal_structure proteins carbohydrates protein-carbohydrate_interaction*
E-mail yura@lec.med.utoronto.ca Tel. 1(416)9780742 Fax 1(416)9786885

Lock, Prof. Colin James Lyne (1933). Professor of chemistry and pathology. Labs for Inorganic Medicine, McMaster University. [Labs for Inorganic Medicine, McMaster U., ABB-266A, Hamilton, Ont. L8S 4M1, Canada.] DSc (London U., UK, 1986) bioinorganic chemistry/crystallography. *X-ray_structure metalloorganic_complexation medicinal_activity*
E-mail lock@mcmaster.ca Tel. 1(905)5259140x24760 Fax 1(905)5222509 Telex 0618347

Lough, Dr Alan John (1960). X-ray service manager. Dept of Chemistry, University of Toronto. [Dept of Chemistry, U. of Toronto, Toronto, Ont. M5S 1A1, Canada.] PhD (Napier Univ., Edinburgh, UK, 1988) crystal structure analysis. *Inorganic_structure organic_structure*
E-mail alough@alchemy.chem.utoronto.ca Tel. 1(416)9786275 Fax 1(416)9788775

Luo, Dr Yao Guang (1956). Research associate. Dept. of Biochemistry, U. of British Columbia. [Dept. of Biochemistry, U. of British Columbia, 2146 Health Sciences Mall, Vancouver, B.C. V6T 1Z3, Canada.] PhD (U. Regina, 1989) chemistry. *Macromolecular_crystallography structure-activity_relationship computing*
E-mail ylbc@unixg.ubc.ca Tel. 1(604)8225007 Fax 1(604)8225227

Marchessault, Dr Robert Henry (1928). E. B. Eddy professor. Dept of Chemistry, McGill University. [Dept of Chemistry, McGill U., 3420 University St, Montréal, Qué. H3A 2A7, Canada.] PhD (McGill U., 1954) physical chemistry. *Polymers crystal_structure morphology biodegradable_polymers*
E-mail ch21@musica.mcgill.ca Tel. 1(514)3986276 Fax 1(514)3987249 Telex 05268510

Masut, Prof. Remo A. (1948). Professor. Ecole Polytechnique, Montréal. [Ecole Polytechnique, CP 6079, Stn A, Montréal, Qué, H3C 3A7, Canada.] PhD (U. Massachussetts, USA, 1982) physics. *Epitaxic_growth III-V_compounds quantum_heterostructure high-resolution_diffractometry photovoltaics transport_and_optical_properties semiconductor_devices*
E-mail bk00@polytec1.polymtl.ca Tel. 1(514)3404310 Fax 1(514)3403218

Matte, Mr Allan (1963). PhD student. Dept of Biochemistry, U. of Saskatchewan. [Dept of Biochemistry, U. of Saskatchewan, Saskatoon, Sask. S7N 0W0, Canada.] MSc (Univ. of Guelph, 1991) microbiology. *Multiple_isomorphous_replacement macromolecular_structure_determination proteins enzymatic_activity*
E-mail matte@sask.usask.ca Tel. 1(306)9664366 Fax 1(306)9668718

Matthews, Prof. Frederick White (1915). Retired, but teaching. Dalhousie University. [1168 Studley Ave., Halifax, N.S. B3H 3R7, Canada.] PhD (McGill U., 1941) physical chemistry. *X-ray_powder database information_system*
Tel. 1(902)4232155

Maurus, Mr Robert (1964). Graduate student. Dept of Biochemistry, U. of British Columbia. [Dept of Biochemistry, U. of British Columbia, 2146 Health Sciences Mall, Vancouver, B.C. V6T 1Z3, Canada.] BSc (Univ. of Guelph, 1991) biochemistry. *Enzymatic_structure-activity_relationship macromolecular_crystallography*
E-mail maurus@laue.biochem.ubc.ca Tel. 1(604)8225007 Fax 1(604)8225227

Maxwell, Prof. George (1946). Associate professor. Mathematics Dept., University of British Columbia. [Mathematics Dept., UBC, #121-1984 Mathematics Rd, Univ. Campus, Vancouver, B.C. V6T 1Y4, Canada.] PhD (Queen's U., 1970) mathematics. *Group_theory geometry mathematical_crystallography*
Tel. 1(604)8226402

McDonald, Dr Robert (1962). Director, crystallographic service lab. Chemistry Dept, U. of Alberta. [Chemistry Dept, U. of Alberta, Edmonton, Alta T6G 2G2, Canada.] PhD (U. Alberta, 1991) inorganic chemistry. *Structural_service small_molecules*
E-mail Bob_McDonald@dept.chem.ualberta.ca Tel. 1(403)4922485 Fax 1(403)4928231

McPhalen, Dr Catherine A. (1956). Research associate. MRC Group in Protein Structure and Function, Dept of Biochemistry, University of Alberta. [Dept of Biochemistry, University of Alberta, Edmonton, Alta T6G 2H7, Canada.] PhD (University of Alberta, 1986) biochemistry. *Macromolecular_structure_determination structure-activity_relationship proteins*
E-mail ceej@meditate.biochem.ualberta.ca Tel. 1(403)4922422 Fax 1(403)4920886

de Médicis, Dr Rinaldo M. (1934). Research assistant. Faculté de Médecine, Université de Sherbrooke. [Faculté de Médecine, U. de Sherbrooke, Sherbrooke, Qué. J1H 5N4, Canada.] PhD (U. de Louvain, Belgium, 1967) crystal chemistry. *Crystal_chemistry X-ray_diffraction electron_diffraction microcrystal-induced_synovitides*
E-mail rmedicis@vm1.si.usherb.ca Tel. 1(819)5645252 Fax 1(819)5645265

Mihichuk, Prof. Lynn Michael (1947). Associate professor. Chemistry Dept., University of Regina. [Chemistry Dept., U. of Regina, Regina, Sask. S4S 0A2, Canada.] PhD (U. British Columbia, 1975) inorganic chemistry. *Bioinorganic_chemistry organometallic_chemistry structure_determination*
E-mail mihichuk@meena.cc.uregina.ca Tel. 1(306)5854793 Fax 1(306)5854894

Mitchell, Dr Crighton Maurice (1917). Retired. [8 Kingsford Crescent, Kanata, Ont. K2K 1T3, Canada.] PhD (U. Toronto, 1952) physics. *Deformation residual_stress anisotropic_orientation crystal_physics*
Tel. 1(613)5921186

Mitchell, Dr Keith A. R. (1938). Professor. Dept of Chemistry, U. British Columbia. [Dept of Chemistry, U. British Columbia, Vancouver, B.C. V6T 1Z1, Canada.] PhD (U. London, UK, 1963) chemistry. *Crystal_surface electron_diffraction LEED*
Tel. 1(604)8225831 Fax 1(604)8222847

Montgomery, Prof. Henry (1916). Emeritus professor. [502-2910 Cook St., Victoria, B.C. V8T 3S7, Canada.] PhD (U. Washington, USA, 1961) inorganic chemistry. *X-ray_structure inorganic_chemistry metalloorganic_chemistry* Tel. 1(604)3838423

Nuffield, Prof. Edward Wilfrid (1914). Professor emeritus. University of Toronto. [1835 Morton Ave, Apt 1603, Vancouver, B.C. V6G 1V3, Canada.] PhD (U. Toronto, 1944) mineralogy. *Crystal_chemistry ores X-ray_diffractometry* Tel. 1(604)6886062

Owen, Mr Charles Gordon (1957). Manager. Dalhousie University. [Technology Transfer Office, Dalhousie University, Halifax, N.S. B3H 3J5, Canada.] MSc (Dalhousie U., 1982) inorganic chemistry. *Transfer_technology technology_transfer (medical_dental_physical_sci.)* E-mail Gordon.Owen@dal.ca Tel. 1(902)4941648 Fax 1(902)4945189 Telex 01921863

Pai, Dr Emil F. (1950). Professor. Departments of Biochemistry and Molecular and Medical Genetics, University of Toronto. [Dept of Biochemistry, U. of Toronto, Toronto, Ont. M5S 1A8, Canada.] PhD (University of Heidelberg, Germany, 1978) chemistry. *Proteins enzymatic_mechanism flavoproteins signal_transduction* E-mail pai@hera.med.utoronto.ca Tel. 1(416)9787015 Fax 1(416)9788548

Pang, Dr Li (1958). Postdoctoral fellow. Dépt. de Chimie, Université de Montréal. [Dépt de Chimie, U. de Montréal, CP 6128, Succ. A, Montréal, Qué. H3C 3J7, Canada.] PhD (U. de Genève, Switzerland, 1989) inclusion compounds. *Molecular_inclusion macromolecules fullerenes* E-mail pangl@tornade.ere.umontreal.ca Tel. 1(514)3436111

Parvez, Dr Masood (1947). Senior professional associate. [Dept of Chemistry, U. of Calgary, 2500 University Dr, N.W., Calgary, Alta, T2N 1N4, Canada.] PhD (Queen's U. Belfast, N. Ireland, 1977) chemistry. *Valence_charge_density X-ray_crystallography_of_organic_inorganic_and_organometallic_compounds anti-allergic_drugs* E-mail parvez@acs.ucalgary.ca Tel. 1(403)2205348 Fax 1(403)2899488 Telex 03821545

Payne, Dr Nicholas Charles (1942). Professor. Chemistry Dept, Univ. of Western Ontario. [Chemistry Dept., U. Western Ontario, London, Ont. N6A 5B7, Canada.] PhD (U. Sheffield, UK, 1967) inorganic chemistry. *Synthesis transition_elements catalysis chirality single_crystal asymmetric_synthesis absolute_configuration Bijvoet_absorption_edge* E-mail noggin@uwo.ca Tel. 1(519)6613793 or -2167 Fax 1(519)6613022

Pazdernik, Prof. LeRoy J. (1942). Professor. Univ. du Québec à Trois-Rivières. [Dept de Chimie-Biologie, UQTR, CP 500, Trois-Rivières, Qué. G9A 5H7, Canada.] PhD (U. Iowa, USA, 1970) inorganic chemistry. *Metallic_trace_analysis organometallic_complexation environmental_system_dynamics metallic_coordination* E-mail pazdernk@neptune.uqtr.uquebec.ca Tel. 1(819)3765052 Fax 1(819)3765084

Pearson, Ms Céline (1969). PhD student. Dépt. de Chimie, Université de Montréal. [Dépt de Chimie, U. de Montréal, CP 6128, Succ. A, Montréal, Qué. H3C 3J7, Canada.] MSc inorganic chemistry. *Complexation coordination rhenium_compounds technetium_compounds NMR* E-mail pearsonc@ere.umontreal.ca Tel. 1(514)3436111x3939 Fax 1(514)3437586

Pearson, Prof. William Burton (1921). Professor Emeritus. University of Waterloo. Physics Dept, U. of Waterloo, Waterloo, Ont. N2L 3G1, Canada. [RRZ ARISS, Ont. N0B 1B0, Canada.] DSc metallurgy, chemistry. *Intermetallic_structure physical_properties_of_matter food_production* E-mail epe@uoguelph.ca Tel. 1(519)8241208 Fax 1(519)8241208

Peterson, Dr Ronald Charles (1953). Associate professor. Dept of Geology, Queen's University. [Dept of Geology, Queen's Univ., Kingston, Ont. K7L 3N6, Canada.] PhD (Virginia Polytechnic Inst., USA, 1981) geology. *Silicates mineralogy crystal_chemistry* E-mail peterson@qucdn.queensu.ca Tel. 1(613)5452597x6180 Fax 1(613)5456592

Phipps, Dr Barry (1957). Research associate. Biotechnology Research Institute, National Research Council of Canada. [Biotechnology Res. Inst., M-54, NRC, Ottawa, Ont. K1A 0R6, Canada.] PhD (U. Victoria, 1988) biochemistry. *Proteinases biocrystallography three-dimensional_reconstruction macromolecular_structure-activity_relationship electron_microscopy membrane_proteins chaperones* E-mail phi@nrcbsa.bio.nrc.ca Tel. 1(613)9900889 Fax 1(613)9414475

Post, Dr Michael Leonard (1945). Senior research officer. Inst. for Environmental Research and Technology, National Research Council of Canada. [IERT, NRC, Ottawa, Ont. K1A 0R6, Canada.] PhD. (U. Surrey, UK, 1971) chemistry. *Ceramics sensor solid_state thin_film* E-mail mike.post@nrc.ca Tel. 1(613)9932101 Fax 1(613)9932451

Poulin, Mrs Suzie (1954). Research associate; Surface Laboratory manager. École Polytechnique de Montréal. [Département de Physique, École Polytechnique, CP 6079, Succ. Centre-ville, Montréal, Qué. H3C 3A7, Canada.] MSc (UQAM, 1978) chemistry. *Surface materials atomic-force_microscopy surfaces_analysis thin-film_devices* E-mail suziepoulin@phys.polymtl.ca Tel. 1(514)3404308 Fax 1(514)3403218 Telex 0524146

Powell, Dr Brian Mathieson (1938). Branch manager. Member, Neutron Diffraction Commission. AECL Research. [Neutron & Condensed Matter Sci., AECL Research, Chalk River Labs, Chalk River, Ont. K0J 1J0, Canada.] PhD (U. London, UK, 1964) physics. *Neutron_diffraction profile_analysis neutron_inelastic_scattering* E-mail bmpowell@crl2.crl.aecl.ca Tel. 1(613)5843311x3974 Fax 1(613)5844040

Prasad, Dr Lata (1945). Professional research associate. Biochemistry Dept., University of Saskatchewan. [Biochemistry Dept, U. of Saskatchewan, Saskatoon, Sask. S7N 0W0, Canada.] PhD (Flinders U. of South Australia, 1972) crystallography. *Macromolecular_structure* E-mail prasad@sask.usask.ca Tel. 1(306)9664366 Fax 1(306)9668718

Pugazhenthi, Ms Umarani (1956). Research technical assistant. Dept of Biochemistry, Univ. of Saskatchewan. [Dept of Biochemistry, Univ. of Saskatchewan, Saskatoon, Sask. S7N 0W0, Canada.] MSc (U. Saskatchewan, 1993) chemistry. *Small_molecules structure-activity_relationship antiviral_compounds microgravity_crystallization* E-mail pugazhenthi@sask.usask.ca Tel. 1(306)9664366 Fax 1(306)9668718

Quail, Dr J. Wilson (1936). Professor. Dept of Chemistry, U. of Saskatchewan. [Dept of Chemistry, U. of Saskatchewan, Saskatoon, Sask. S7N 0W0, Canada.] PhD (McMaster U., 1963) chemistry. *Macromolecular_structure-activity_relationship drug_structure-activity_relationship microgravity crystallization* E-mail quail@sask.usask.ca Tel. 1(306)9664663 Fax 1(306)9664730

Ranger, Dr Georges Joseph (1955). Senior research chemist. Sterling Pulp Chemicals Ltd [Sterling Pulp Chemicals Ltd, R&D Div., 2 Gibbs Rd., Islington, Ont. M9B 1R1, Canada.] PhD (Wayne State U., USA, 1984) inorganic chemistry. *Inorganic_chemistry crystal_structure_determination heavy_metals complexation* Tel. 1(416)2397111x301 Fax 1(416)2322146

Raudsepp, Dr Mati (1947). Research scientist, hon. professor. Dept of Geological Sciences, U. of British Columbia. [Dept. of Geological Sciences, University of British Columbia, Vancouver, B.C. V6T 1Z4, Canada.] PhD (U. Manitoba, 1984) mineralogy. *Mineralogy crystallography* E-mail raudsepp@unixg.ubc.ca Tel. 1(604)8225065x6396 Fax 1(604)8226088

Read, Dr Randy John (1957). Associate professor. Member, CNC/IUCr. U. of Alberta. [Medical Microbiology and Infectious Diseases, U. of Alberta, Edmonton, Alta T6G 2H7, Canada.] PhD (U. of Alberta, 1986) biochemistry. *Macromolecular_crystallography drug_design macromolecular_derivative_structure structure-factor_probabilities* E-mail rndy@mycroft.mmid.ualberta.ca Tel. 1(403)4924305 Fax 1(403)4927521

Rehse, Dr Peter (1961). Research associate. Biotechnology Research Institute, National Research Council of Canada. [Biotechnology Research Institute, NRC, 6100 Royalmount Ave., Montréal, Qué. H4P 2R2, Canada.] PhD (Univ. of London (Birkbeck), UK, 1989) crystallography. *Macromolecular_crystal_structure proteins* E-mail peter.rehse@bri.nrc.ca Tel. off: 1(514)4962557 lab: 1(514)4966376 Fax 1(514)4965143

Restivo, Dr Roderic John (1943). Professor. Heritage College. [Heritage College, Science Dept., 325 boul. Cité des Jeunes, Hull, Qué. J8Y 6T3, Canada.] PhD (U. Waterloo, 1969) chemistry. *Bioinorganic_chemistry organometallic_chemistry* Tel. 1(819)7782270x2141 Fax 1(819)7787364

Rettig, Dr Steven John (1948). Manager, structural chemistry lab. U. of British Columbia. [Chemistry Dept, U.B.C., 2036 Main Mall, U. Campus, Vancouver, B.C. V6T 1Z1, Canada.] PhD (U. British Columbia, 1974) chemistry. *X-ray_crystallography organic_chemistry organometallic_chemistry* E-mail xtal@xray1.chem.ubc.ca Tel. 1(604)8224865 Fax 1(604)8222847

Richardson, Dr Mary Frances (1941). Professor. Brock University. [Brock University, Chemistry Dept., St Catherines, Ont. L2S 3A1, Canada.] PhD (U. Kentucky, USA, 1967) inorganic. *X-ray_structure polytypism molecular_modelling solid_nuclear_magnetic_resonance* E-mail mrichard@abacus.ac.brocku.ca Tel. 1(905)6885550x3400 Fax 1(905)6829020

Rini, Dr James (1958). Assistant professor. Dept of Molecular and Medical Genetics, U. of Toronto. [Dept of Molecular and Medical Genetics, U. of Toronto, Toronto, Ont. M5S 1A8, Canada.] PhD (Univ. of Toronto, 1986) medical biophysics. *Macromolecular_crystallography proteins carbohydrates structure-activity_relationship* E-mail rini@gene4d.med.utoronto.ca Tel. 1(416)9780557 Fax 1(416)9786885

Roberts, Mr Andrew Clifford (1950). X-ray diffraction mineralogist. Geological Survey of Canada. [Geological Survey of Canada, EMR, 762-601 Booth St., Ottawa, Ont. K1A 0E8, Canada.] MSc (Queen's U., 1976) mineralogy. *Single_crystal X-ray_powder diffraction* Tel. 1(613)9922802 Fax 1(613)9431286

Robertson, Prof. Beverly Ellis (1939). Professor. National executive, Canadian Association of Physicists. University of Regina. [University of Regina, Dept. of Physics, Regina, Sask. S4S 0A2, Canada.] PhD (McMaster U., 1967) physics. *Small_molecules charge_density health_physics* E-mail brob@max.cc.uregina.ca Tel. 1(306)5854264 Fax 1(306)5854894

Robertson, Ms Katherine Nancy (1959). PhD student. Dept of Chemistry, Dalhousie U. [Dept. of Chemistry, Dalhousie University, Halifax, N.S. B3H 4J3, Canada.] MSc (Dalhousie U., 1991) chemistry. *Crystallography* E-mail kathy%dalx@ac.dal.ca Tel. 1(902)4943759 Fax 1(902)4941310

Rochon, Dr Fernande D. Professor. U. du Québec à Montréal. [Dépt. de Chimie, UQAM, CP 8888, Succ. A, Montréal, Qué. H3C 3P8, Canada.] PhD (U. de Montréal, 1971) inorganic chemistry. *Inorganic_chemistry platinum_compounds technetium_compounds metalloorganic_complexation* E-mail rochon@mips1.info.uqam.ca Tel. 1(514)9874896 or 9874119 Fax 1(514)9874054

Rodgers, Dr John R. (1944). Senior research officer. Chair, Commission on Crystallographic Data; Member, Commission on Crystallographic Computing; Representative of IUCr to CODATA. Canada Institute for Scientific and Technical Information, National Research Council of Canada. [CISTI, NRC, Ottawa, Ont. K1A 0S2, Canada.] PhD (Univ. of London, UK, 1979) crystallography. *Inorganic_database intermetallic_database crystal_chemistry material_classification* E-mail rodgers@snd.cisti.nrc.ca Tel. 1(613)9933294 Fax 1(613)9528246

Root, Dr J. H. (1957). Research scientist. AECL Research. [Neutron & Condensed Matter Sci., AECL Research, Chalk River Labs, Chalk River, Ont. K0J 1J0, Canada.] PhD (Univ. of Guelph, 1986) physics. *Materials neutron_diffraction* E-mail rootj@crl.aecl.ca Tel. 1(613)5843311

Rose, Dr David Richard (1955). Senior scientist. Ontario Cancer Institute, University of Toronto. [Ontario Cancer Institute, University of Toronto Toronto Ont. M4X 1K9, Canada.] PhD (Oxford U., UK, 1981) molecular biophysics. *Proteins antibodies membranes molecular_recognition homology_prediction*
E-mail drose@oci.utoronto.ca Tel. 1(416)9240671x5446 Fax 1(416)9266529

Roszak, Dr Aleksander W. (1954). Asst. prof. Department of Chemistry, Queen's University. [Department of Chemistry, Queen's University, Kingston, Ont. K7L 3N6, Canada.] PhD (A. Mickiewicz Univ., Poznan, Poland, 1986) crystallography. *Structure–activity_relationship bioactive_molecules organometallic_structure macromolecular_crystallography organic_structure*
E-mail roszaka@qucdn.queensu.ca roszaka@qucdn.bitnet Tel. 1(613)5452655 Fax 1(613)5456669

Schrag, Dr Joseph D. (1955). Research associate. Biotechnology Res. Inst. [Biotechnology Res. Inst., 6100 Royalmount Ave., Montréal, Qué. H4P 2R2, Canada.] PhD (U. Illinois, USA, 1984) physiology. *Macromolecular_structure structure_determination structure–activity_relationship macromolecules*
E-mail joe@bobino.bri.nrc.ca Tel. 1(514)4962557 Fax 1(514)4966232

Secco, Prof. Anthony Silvio (1956). Associate Professor and Associate Head. Secretary of the Canadian Division of the ACA. U. of Manitoba. [Dept. of Chemistry, U. of Manitoba, Winnipeg, Man. R3T 2N2, Canada.] PhD (U. British Columbia, 1982) chemistry. *Biochemical_crystallography nucleic_acids proteins*
E-mail secco@iris.chem.umanitoba.ca Tel. 1(204)4748379 Fax 1(204)2750905 Telex 07587721

Sielecki, Dr Anita R. (1940). Research associate. Biochemistry Dept., U. of Alberta. [Biochemistry Department, University of Alberta, Edmonton, Alta T6G 2H7, Canada.] PhD (Hebrew U., Israel, 1969) hydrodynamics. *Macromolecular_crystallography*
E-mail siel@manitou.biochem.ualberta.ca Tel. 1(403)4922422 Fax 1(403)4920886

Simard, Dr Michel (1958). Manager X-ray lab. Université de Montréal. [Dépt. de Chimie, U. de Montréal, CP 6128, Succ. A, Montréal, Qué. H3C 3J7, Canada.] PhD (U. de Montréal, 1986) chemistry. *Inorganic_chemistry*
E-mail simardmi@ere.umontreal.ca Tel. 1(514)3436111x3937 Fax 1(514)3437586

Skowron, Dr Aniceta (1955). Inst. for Materials Research, McMaster University. [Inst. for Materials Res., McMaster University, Hamilton, Ont. L8S 4M1, Canada.] PhD (McMaster University, 1991) materials science. *X-ray_crystallography high-resolution_electron_microscopy chalcogenides*
E-mail skowron@mcmaster.ca Tel. 1(905)5259140x27092 Fax 1(905)5212773 Telex 0618347

Smith Jr, Prof. Vedene H. (1935). Professor. Queen's University. [Chemistry Department, Queen's University, Kingston, Ont. K7L 3N6, Canada.] PhD (U. Uppsala, Sweden, 1967) quantum chemistry. *Electron_density_distribution*
E-mail smithvh@qucdn.queensu.ca Tel. 1(613)5452650 Fax 1(613)5456669

Spinney, Mr Richard (1959). PhD student. U. of Calgary. [Dept of Chemistry, U. of Calgary, Calgary, Alta, T2N 1N4, Canada.] BSc (U. Calgary, 1990) chemistry. *Molecular_modelling structure–activity_relationship drug_design ligand-receptor_interactions rational_drug-design_methods*
E-mail rspinney@acs.ucalgary.ca Tel. 1(403)2205069 Fax 1(403)2899488 Telex 03821545

Strynadka, Dr Natalie (1963). Postdoctoral fellow. Biochemistry Department, University of Alberta. [Biochemistry Dept, U. of Alberta, Edmonton, Alta T6G 2H7, Canada.] PhD (U. Alberta, 1991) biochemistry. *Macromolecular_crystallography enzymatic_structure–activity_relationship*
E-mail tali@manitou.biochem.ualberta.ca Tel. 1(403)4922422 Fax 1(403)4920886

Sui, Ms Xiaoling (1963). PhD student. U. of Calgary. [Dept. of Chemistry, U. of Calgary, Calgary, Alta T2N 1N4, Canada.] PhD (U. Calgary, 1994) biophysical chemistry. *Crystallography computer_modelling medicinal_chemistry structure–activity_relationship*
E-mail xsui@acs.ucalgary.ca Tel. 1(403)2205069 Fax 1(403)2899488 Telex 03821545

Sundararajan, Dr Pudupadi R. (1943). Principal scientist. Xerox Research Center of Canada. [Xerox Research Center of Canada, 2600 Speakman Dr., Mississauga, Ont. L5K 2L1, Canada.] DSc (Madras U., India, 1982) physics. *Polymers conformation crystallography*
E-mail Sundar.XRCC@xerox.com Tel. 1(416)8237091x219 Fax 1(416)8227022

Sunder, Dr Sham (1942). Research officer. AECL, Whiteshell Labs. [Res. Chemistry Branch, AECL, Whiteshell Laboratories, Pinawa, Man. R0E 1L0, Canada.] PhD (U. Alberta, 1972) chemistry. *Surface ESCA X-ray_diffraction phase_transition XPS vibrational_spectra nuclear_fuel corrosion*
E-mail sunders@wl.aecl.ca Tel. 1(204)7532311x2749 Fax 1(204)7532455

Sutton, Dr Mark (1951). Associate professor. Physics Dept, McGill University. [Physics Dept, McGill University, Montréal, Qué. H3A 2T8, Canada.] PhD (Univ. of Toronto, 1981) physics. *Phase_transition dynamics*
E-mail mark@physics.mcgill.ca Tel. 1(514)3986523 Fax 1(514)3986526

Swainson, Dr Ian Peter (1968). Research associate. AECL Research. [Neutron & Condensed Matter Sci., AECL Research, Chalk River Labs, Chalk River, Ont. K0J 1J0, Canada.] PhD (Cambridge, UK, 1993) mineral physics. *Neutron_scattering silicates property_framework_structure structural_phase_transition disorder_in_solids*
E-mail swainsoni@crl.aecl.ca Tel. 1(613)5843311x3995 Fax 1(613)5844040

Sygusch, Prof. Jurgen (1945). Professor. Département de Biochimie, Université de Montréal. [Dépt. de Biochimie, U. de Montréal, CP 6128, Succ. A, Montréal, Qué. H3C 3J7, Canada.] PhD (U. de Montréal, 1975) chemical crystallography. *Macromolecular_crystallography enzymatic_catalysis crystallization proteins_engineering*
E-mail syguschj@bch.umontreal.ca Tel. 1(514)3432389 Fax 1(514)3432210

Szkaradzinska, Ms Maria (1951). Research assistant. U. of Calgary. [Dept of Chemistry, U. of Calgary, Calgary, Alta, T2N 1N4, Canada.] MSc (Jagellonian U., Poland, 1974) physical chemistry. *Structure–activity_relationship conformation_analysis*
E-mail mbszkara@acs.ucalgary.ca Tel. 1(403)2205069 Fax 1(403)2899488 Telex 03821545

Szymański, Dr Jan Tomasz (1938). Research scientist. Department of Natural Resources Canada, CANMET. [DNRC, CANMET, 555 Booth St., Ottawa, Ont. K1A 0G1, Canada.] PhD (King's College, London U., UK, 1963) inorganic chemistry. *Minerals X-ray_powder inorganic_complexation*
E-mail jan_szymanski@cc2smtp.emr.ca Tel. 1(613)9954077 Fax 1(613)9969673

Taylor, Dr Peter (1949). Research chemist. AECL, Whiteshell Labs. [Res. Chemistry Branch, AECL, Whiteshell Labs, Pinawa, Man. R0E 1L0, Canada.] PhD (U. Birmingham, UK, 1972) inorganic chemistry. *Structural_solubility structural_stability inorganic_oxides salts*
Tel. 1(204)7532311x3054 Fax 1(204)7532455

Tong, Dr Hua Harry (1963). Postdoctoral fellow. Dept of Biochemistry, U. of British Columbia. [Dept of Biochemistry, U. of British Columbia, 2146 Health Sciences Mall, Vancouver, B.C. V6T 1Z3, Canada.] PhD (Univ. of Sydney, Australia, 1992) biochemistry. *Macromolecular_crystallography enzymatic_structure–activity_relationship transfer_mechanism computing*
E-mail htong@unixg.ubc.ca Tel. 1(604)8225007 Fax 1(613)8225227

Trotter, Prof. James (1933). Professor. University of British Columbia. [Department of Chemistry, University of British Columbia, Vancouver, B.C. V6T 1Z1, Canada.] DSc (U. Glasgow, UK, 1963) chemistry. *Organic_structure inorganic_structure solid_state photochemistry*
E-mail james.trotter@mtsg.ubc.ca Tel. 1(604)8224865 Fax 1(604)8222847

Tse, Dr John S. (1953). Senior research officer. Steacie Inst. of Molecular Science, National Research Council of Canada. [Steacie Inst. of Molecular Sci., NRC, Ottawa, Ont. K1A 0R6, Canada.] PhD (Univ. of Western Ontario, 1980) chemistry. *Solid_phase_transition electronic_structure solid_state*
E-mail tse@jtsg.sims.nrc.ca Tel. 1(613)9911237 Fax 1(613)9545242

Tun, Dr Zin (1957). Research physicist. AECL Research, Chalk River. [AECL Research, Chalk River Labs, Chalk River, Ont. K0J 1J0, Canada.] PhD (McMaster U., 1985) physics. *Order-disorder inorganic_structure phase_transition neutron_diffraction magnetic_structures_and_excitations thin_films_and_multilayers*
E-mail tunz@crl.aecl.ca Tel. 1(613)5843311x3994 Fax 1(613)5844040

Turner, Dr Mary (1960). Postdoctoral fellow. Hospital for Sick Children, Toronto. [Biochemistry Research, Hospital for Sick Children, Toronto, Ont. M5G 1X8, Canada.] PhD (McMaster U., 1989) chemistry. *Proteins Laue_diffraction macromolecular_structure time-resolved_Laue_diffraction*
E-mail turner@gandalf.psf.sickkids.on.ca Tel. 1(416)8135359 Fax 1(416)8135022

Van der Heijden, Dr Simon Petrus Nicolaas (1943). Chief chemist. SaskPower. [Techn. Serv. & Res., SaskPower, 2025 Victoria Ave., Regina, Sask. S4P 0S1, Canada.] PhD (U. Saskatchewan, 1974) crystallography. *Structural_crystallography coal gasification database coal_gasification_kinetics*
Tel. 1(306)5663073 Fax 1(306)5663348

Vandonselaar, Mrs Margaret (1943). Technician. Dept of Biochemistry, U. of Saskatchewan. [Dept of Biochemistry, U. of Saskatchewan, Saskatoon, Sask. S7N 0W0, Canada.] MSc (University of Saskatchewan, 1968) organic chemistry. *Biochemistry crystallization data_collection structure_determination*
E-mail vandonselaar@sask.usask.ca Tel. 1(306)9664385 Fax 1(306)9668718

Vittal, Dr Jagadese (1956). Manager, X-ray structure facility. Chemistry Dept, Univ. of Western Ontario. [Chemistry Dept., U. Western Ontario, London, Ont. N6A 5B7, Canada.] PhD (Indian Institute of Science, Bangalore, India, 1982) inorganic chemistry. *Small_molecules structural_service inorganic_synthesis organometallic_reactivity*
E-mail jvittal@uwovax.uwo.ca Tel. 1(519)6612167 Fax 1(519)6613022

Vrielink, Dr Alice (1959). Assistant Professor of Biochemistry. [Biochemistry Department, McIntyre Medical Sciences Building, McGill University, 3655 Drummond St., Montreal, Quebec H3G 1Y6, Canada.] PhD (Imperial Coll. London, 1989) physics. *Macromolecular_crystallography enzyme_active_site drug_design domain_structure*
E-mail alice@bri.nrc.ca Tel. 1(514)4966129 or 1(514)3981918 Fax 1(514)3987384

Wang, Mr Weibin (1960). PhD student. Dept of Chemistry, Simon Fraser University. [Dept of Chemistry, Simon Fraser University, Burnaby, B.C. V5A 1S6, Canada.] MSc (Xiamen Univ., China, 1985) photochemistry. *Metallic_clusters proteins macromolecular_structure*
E-mail wwang@sfu.ca Tel. 1(604)2914878 or 4408 Fax 1(604)2913765

White, Mr Andre (1966). PhD student. Dept of Medical Biophysics, Ontario Cancer Institute. [Ontario Cancer Institute, 500 Sherbourne St., 7th floor, Toronto, Ont. M4X 1K9, Canada.] MSc (Universitde Sherbrooke, 1991) biochemistry. *Macromolecular_crystallography computer_modelling*
E-mail awhite@oci.utoronto.ca Tel. 1(416)9240671x5053 Fax 1(416)9266529

Whitlow, Dr Simon Hugh (1943). Chief, informatics and systems. Atmospheric Environment Service. [Atmospheric Environment Service, 700-1200w 73 Ave., Vancouver, B.C. V6P 6H9, Canada.] PhD (U. British Columbia, 1969) chemistry. *Electronic_data_collection computing database telecommunications*
Tel. 1(604)6649166 Fax 1(604)6649195

Wicks, Dr Frederick John (1937). Curator. Royal Ontario Museum. [Dept. of Mineralogy, ROM, 100 Queen's Park, Toronto, Ont. M5S 2C6, Canada.] PhD (Oxford U., UK, 1969) mineralogy. *Minerals atomic_force differential_thermal_analysis X-ray_diffraction*
Tel. 1(416)5865820 Fax 1(416)5865814

Wood, Dr Gordon H. (1940). CAN/SND manager. Canada Institute for Scientific and Technical Information, National Research Council of Canada. [CISTI, NRC, Ottawa, Ont. K1A 0S2, Canada.] PhD (U. British Columbia, 1969) physics. *Crystal_database*
E-mail gordon.wood@nrc.ca Tel. 1(613)9933294 Fax 1(613)9528426 Telex 0533115

Xia, Mr Qi. Graduate Student. Department of Physics, University of Regina. [Department of Physics, U. of Regina, Regina, Sask. S4S 0A2, Canada.] BSc (Yangzhou University, China, 1982) physics. *Crystallography*
E-mail xiaqi@meena.cc.uregina.ca Tel. 1(306)5854308 Fax 1(306)5854894

Yan, Dr Xiaoqian (1956). Postdoctoral fellow. Simon Fraser University. [Chemistry Department, Simon Fraser University, Burnaby, B.C. V5A 1S6, Canada.] PhD (Simon Fraser U., 1993) inorganic chemistry. *Structure inorganic_molecules*
Tel. 1(604)2914406 Fax 1(604)2913765

Yang, Prof. Daniel Shun-Chung (1953). Assistant professor. McMaster University. [Biochemistry Department, McMaster University, 1200 Main St West, Hamilton, Ont. L8N 3Z5, Canada.] PhD (University of Pittsburgh, USA, 1983) crystallography. *Macromolecular_structure Fourier_transform*
E-mail yang@xtliris.csu.mcmaster.ca Tel. 1(905)5259140x2455 Fax 1(905)5229033

Yang, Mr Jian (1968). Graduate student. Dept of Chemistry, U. of Saskatchewan. [Dept of Chemistry, U. of Saskatchewan, Saskatoon, Sask. S7N 0W0, Canada.] BSc (Peking University, Beijing, China, 1989) biochemistry (tumor biology). *Bioactive_molecules drug*
E-mail yangj@sask.usask.ca Tel. 1(306)9664366 Fax 1(306)9668718

Zaworotko, Dr Michael John. Associate Professor and Chairperson. Dept of Chemistry, Saint Mary's University. [Dept of Chemistry, St Mary's Univ., Halifax, N.S. B3H 3C3, Canada.] PhD (U. Alabama, USA, 1982) inorganic chemistry. *Crystal_engineering bonding noncovalent_bonding supramolecular_chemistry structure/function_relationship_in_solids*
E-mail mzaworot@science.stmarys.ca Tel. 1(902)4205661 Fax 1(902)4205261

CHILE

Sub-Editor: O. Wittke

Aguilar, Mrs Adela (1934). Head. Lab. de Petrografía y Mineralogía, Langerfeldt y Aguilar Ltda., Presidente Errázuriz 4361, Departamento 20, Casilla 4003, Santiago, Chile. BSc *Minerals optical_microscopy*
Tel. 56(2)2060315

Almendras, Mrs Eliana (1940). Researcher. Departamento de Minas, Facultad de Ciencias Físicas y Matemáticas, Universidad de Chile, Casilla 2777, Santiago, Chile. MSc (U. Chile, 1962) mining engineering. *Minerals optical_microscopy*
Tel. 56(2)6982071x502 Fax 56(2)6712799

Atria, Dr Ana María (1952). Assistant Professor. Departamento de Química Inorgánica y Analítica, Facultad de Ciencias Químicas y Farmacéuticas, Universidad de Chile, Casilla 233, Santiago 1, Chile. Dr (U. Chile, 1991) chemistry. *Inorganic_chemistry magnetochemistry X-ray_diffraction copper_compounds*
Tel. 56(2)7778853 Fax 56(2)7378920

Barbagelata, Mr Franco (1942). G. M. A. Ltda., Asesorías y Servicios de Geología y Mineralogía Aplicada, Los Jazmines 881, ūōa, Casilla 6-60, Santiago, Chile. MSc (U. Chile, 1967) chemistry. *Minerals optical_microscopy ore_dressing_products*
Tel. 56(2)2395599 Fax 56(2)2395599

Besoain, Dr Eduardo (1929). Lab. de Fisicoquímica y Mineralogía, Instituto de Investigaciones Agropecuarias, Estación Experimental La Platina, Santa Rosa 11610 Paradero 33, Casilla 4393, Santiago, Chile. Dr (U. Bonn, 1969) soil mineralogy. *Soils X-ray_diffraction X-ray_spectrography infrared_spectrophotometry*
Tel. 56(2)5417223x149

Boys, Dr Daphne (1941). Associate professor. Secretary, National Committee. Departamento de Física, Facultad de Ciencias Físicas y Matemáticas, Universidad de Chile, Casilla 487-3, Santiago, Chile. PhD (Wales U., 1972) X-ray diffraction. *X-ray_diffraction coordination_compound*
E-mail dboys@uchcecvm.cec.uchile.cl Tel. 56(2)6960148x520 Fax 56(2)6712799

Brito, Dr Ivan (1958). Associate professor. Departamento de Química, Facultad de Ciencias Básicas, Universidad de Antofagasta, Casilla 170, Antofagasta, Chile. Dr (U. La Laguna, Spain, 1992) chemistry. *Organic_crystal_structure natural_products*
Tel. 56(55)242160x216 Fax 56(55)247835

Chornik, Dr Boris (1941). Professor. Departamento de Física, Facultad de Ciencias Físicas y Matemáticas, Universidad de Chile, Casilla 487-3, Santiago, Chile. PhD (U. California, 1970) physics. *Surface_structure XPS tunnelling_microscopy thin_film materials_science magnetism*
E-mail bchornik@tamarugo.cec.uchile.cl Tel. 56(2)6784333 Fax 56(2)6967359

Cid, Dr Hilda (1933). Professor. Dept. Biológia Molecular, Facultad de Ciencias Biológicas, Universidad de Concepción, Casilla 4077, Correo 3, Concepción, Chile. PhD (MIT, 1964) crystallography. *Proteins macromolecules structural_biological_function*
Tel. 56(41)234985 Fax 56(41)240280

Costamagna, Dr Juan Alberto (1940). Departamento de Química, Facultad de Ciencia, Universidad de Santiago de Chile, Departamento de Química, Ave. B. O'Higgins 3363, Santiago, Chile. Dr (U. Buenos Aires, 1970) chemistry. *Coordination_compound*
Tel. 56(2)6811100x2425

Fonseca, Dr Eugenia. Servicio Nacional de Geología y Minería, Til-Til 1993, ūōa, Santiago, Chile. Dr (U. Carolina of Praga, 1984) mineralogy. *X-ray_diffractometry clays_mineralogy geochemistry petrology*
Tel. 56(2)2385332

Garín, Dr Jorge Leonidas (1943). Professor. Departamento de Metalurgia, Facultad de Ingeniería, Universidad de Santiago de Chile, Casilla 10233, Santiago, Chile. PhD (U. Pennsylvania, 1972) metallurgy and materials science. *Metals alloys inorganic_crystal_chemistry X-ray_diffraction industrial_X-ray_diffraction*
E-mail jgarin@usachvm1.bitnet Tel. 56(2)6811545 Fax 56(2)6811545

Garland, Mrs María Teresa (1944). Associate professor. Departamento de Física, Facultad de Ciencias Físicas y Matemáticas, Universidad de Chile, Casilla 487-3, Santiago, Chile. DrSc (U. Rennes 1, 1986) crystallography. *Coordination_compound organometallic_compound orbital_calculation*
E-mail mtgarlan@uchcecvm.cec.uchile.cl Tel. 56(2)6982071x517 Fax 56(2)6712799

Greene, Mr Fernando (1943). G. M. A. Ltda., Asesorías y Servicios de Geología y Mineralogía Aplicada, Los Jazmines 881, ūōa, Casilla 6-60, Santiago, Chile. MSc (U. Chile, 1967) chemistry. *Minerals optical_microscopy ore_dressing_products*
Tel. 56(2)2395599 Fax 56(2)2395599

Henríquez, Mr Fernando (1942). Departamento de Minas, Facultad de Ingeniería, Universidad de Santiago de Chile, Casilla 10233, Santiago, Chile. MSc (McGill U., 1972) geology. *Minerals*
Tel. 56(2)2393136

Hervé, Dr Francisco (1942). Dept. de Geología y Geofísica, Facultad de Ciencias Físicas y Matemáticas, Universidad de Chile, Casilla 13518, Correo 21, Santiago, Chile. DSc (Hokkaido U., 1974) metamorphic petrology. *Minerals petrology*
E-mail alahsen@uchcecvm.cec.uchile.cl Tel. 56(2)6982071x514 Fax 56(2)6712799

Llanos, Dr Jaime (1952). Departamento de Química, Facultad de Ciencias, Universidad del Norte, Casilla 1260, Antofagasta, Chile. Dr (U. Stuttgart, 1984) *Inorganic_crystal*
E-mail jllanos@socompa.cecun.ucn.cl Tel. 56(55)241148

Manríquez, Dr Víctor (1953). Associate professor. Departamento de Química, Facultad de Ciencias, Universidad de Chile, Casilla 653, Santiago, Chile. Dr (U. Stuttgart, 1983) *Inorganic_crystal*
Tel. 56(2)2712865x259

Martínez, Dr Vicente de Paul (1955). Departamento de Metalurgia, Facultad de Ingeniería, Universidad de Santiago de Chile, Casilla 10233, Santiago, Chile. Dr (U. Navarra, Spain, 1990) metallurgy. *Material_sciences*
Tel. 56(2)6811545 Fax 56(2)681155

Mujica, Dr Carlos (1954). Professor. Departamento de Química, Facultad de Ciencias, Universidad del Norte, Casilla 1280, Antofagasta, Chile. Dr (U. Stuttgart, 1984) *Inorganic_crystal*
E-mail cmujica@socompa.cecun.ucn.cl Tel. 56(55)241148

Roeschmann, Mr Carlos. Servicio Nacional de Geología y Minería, Til-Til 1993, ūōa, Santiago, Chile. MSc (U. Chile, 1963) geology. *X-ray_diffractometry clays_mineralogy geochemistry petrology geology*
Tel. 56(2)2385292 Fax 56(2)2385332

Silva, Dr Elisa (1927). Associate professor. Departamento de Física, Facultad de Ciencias Físicas y Matemáticas, Universidad de Chile, Casilla 487-3, Santiago, Chile. Dr (U. Paris, 1963) sciences. *Electron_microscopy*
E-mail esilva@uchcecvm.uchile.cec.cl Tel. 56(2)6982071x521 Fax 56(2)6712799

Spodine, Dr Evgenia (1942). Professor. Departamento de Química Inorgánica y Analítica, Facultad de Ciencias Químicas y Farmacéuticas, Universidad de Chile, Casilla 233, Santiago, Chile. Dr (U. Chile, 1987) chemistry. *Inorganic_compound X-ray_diffraction magnetic_copper_compounds bioinorganic*
Tel. 56(2)7778853 Fax 56(2)7378920

Suwalsky, Dr Mario (1936). Professor; Director of Research and Graduate Studies, Faculty of Chemical Sciences. Departamento de Polimeros, Facultad de Ciencias Químicas, Universidad de Concepción, Casilla 3-C, Concepción, Chile. PhD (Weizmann Inst., 1969) crystallography. *Polymers structural_biology biomembranes*
E-mail msuwalsk@buho.dpi.udec.cl Tel. 56(41)234985x2171 Fax 56(41)240280

Valero, Mr Ricardo (1958). Laboratorio de Productos Metalúrgicos, Centro de Investigación Minera y Metalúrgica, Casilla 170, Correo 10, Santiago, Chile. MSc (U. Tecnica del Estado, 1979) chemistry. *X-ray_fluorescence X-ray_diffraction*
Tel. 56(2)2184311 Fax 56(2)2426278

Varschavsky, Ari (1940). Sección Metales, Departamento de Ciencias de los Materiales, Facultad de Ciencias Físicas y Matemáticas, Universidad de Chile, Casilla 1420, Santiago, Chile. MSc (U. Chile, 1965) electrical engineering. *Metals material_sciences*
Tel. 56(2)6982071x150 Fax 56(2)6712799

Wittke, Prof. Oscar (1929). Professor. Chairman, National Committee - Sub-Editor for 9th Ed. of WDC. Departamento de Física, Facultad de Ciencias Físicas y Matemáticas, Universidad de Chile, Casilla 487-3, Santiago, Chile. MSc (U. Chile, 1961) mathematics and physics. *X-ray_diffraction structure_determination crystal_optics symmetry*
E-mail owittke@uchcecvm.cec.uchile.cl Tel. 56(2)6982071x517 Fax 56(2)6712799

Zelada, Mr Gabriel (1955). Laboratorio de Física, Comisión Chilena de Energía Nuclear, Casilla 188-D, Santiago, Chile. MSc (U. Chile, 1982) physics. *X-ray_diffraction neutron_diffraction*
Tel. 56(2)2731827x841

Zlosilo, Mr Mario (1943). Laboratorio Espectroscopía Rayos X, Centro de Investigación Minera y Metalúrgica, Casilla 170, Correo 10, Santiago, Chile. MSc (U. Chile, 1967) chemistry. *X-ray_fluorescence X-ray_diffraction*
Tel. 56(2)2184311 Fax 56(2)2426278

CHINA

Sub-Editor: Han Yuzhen

Notes

1. There was no degree system in China before 1981.

Bai, Prof. Chunli (1953). Deputy Director of the Institute. Inst. of Chem., Chinese Academy of Sciences, China. [Inst. of Chem., Chinese Academy of Sciences, Beijing 100080, China.] PhD (Institute of Chemistry, Chinese Academy of Sciences, 1985) chemistry. *Scanning_tunnel_microscopy structure chemistry*
E-mail clbai@bepc2.ihep.ac.cn Tel. 86(1)2568158 Fax 86(1)2557908

Bi, Prof. Ruchang (1940). Group Leader. Institute of Biophysics, Academia Sinica, China. [Institute of Biophysics, 15 Datun Road, Chaoyang District, Beijing 100101, China.] BS (Leningrad U., 1965) molecular physics. *Crystallography proteins crystal_growth protein_engineering*
E-mail birc@bepc2.ihep.ac.cn Tel. 86(1)2020077x517 Fax 86(1)2027837

Cai, Dr Guanliang (1948). Associate Professor. [Central Laboratory, Fuzhou University, Fuzhou, Fujian 350002, China.] PhD (Fuzhou University, 1992) physical chemistry. *Materials complex*
Fax 86(591)713886

Cao, Dr Rong (1966). State Key Lab. on Structure Chemistry, Fujian Institute of Research on the Structure of Matter, China. [PO Box 143, Fuzhou, Fujian 350002, China.] PhD (Fujian Inst. Research on the Structure of Matter, 1993) physical chemistry. *X-ray_structure transition_elements*
Tel. 86(591)714578 Fax 86(591)714946 Telex 92219FIRSMCN

Chang, Prof. Wenrui (1940). Institute of Biophysics, Academia Sinica, Beijing 100101, China. (Nankai U., 1964) chemistry. *Proteins crystallography*
Tel. 86(1)2020077x458 Fax 86(1)2027837

Chen, Prof. Benming (1938). Leader of the Institute. Institute of Chemistry, Academia Sinica, China. [Institute of Chemistry, Academia Sinica, Beijing 100080, China.] BS (U. Sciences and Technology of China, 1964) chemistry. *Structure chemistry*
Tel. 86(1)2554448 Fax 86(1)2569564

Chen, Prof. Liquan (1940). Group Head. Institute of Physics, Academia Sinica, China. [PO Box 603, Beijing 100080, China.] (U. Sciences and Technology of China, 1963) physics. *Solid_state ionic superconductivity*
Tel. 86(1)2559131x523 Fax 86(1)2562605

Chen, Prof. Minqin (1939). Research Center of Analysis and Measurement, Fudan University, Shanghai 200433, China. (Fudan U., 1962) chemistry. *Transition_element drug proteins structure-properties_relationship*
Tel. 86(21)5492222x3016 Fax 86(21)5491875

Chen, Prof. Shizhi (1935). Institute of Biophysics, Academia Sinica, China. [Institute of Biophysics, 15 Datun Road, Chaoyang District, Beijing 100101, China.] BS (Peking U., 1958) physical chemistry. *Proteins crystallography*
Tel. 86(1)2020077x338 Fax 86(1)2027837

Cui, Prof. Wenyuan (1934). Professor, Director of Department. Department of Geology, Peking University, Beijing 100871, China. (Changchun Inst. Geology, 1958) geology. *Mineralogy*
Tel. 86(1)2501146 Fax 86(1)2564095

Cui, Prof. Xiushan (1940). Department of Chemical Engineering, Beijing Institute of Technology, Beijing 100081, China. MS (Peking U., 1981) structural chemistry. *Crystallography chemistry*
Tel. 86(1)8416688x2657 Fax 86(1)8412889

Deng, Dr Shuiquan (1964). Group Leader. State Key Laboratory, Institute of Research on the Structure of Matter, Chinese Academy of Sciences, Fuzhou, Fujian 350002, China. PhD (Fujian Inst. Research on the structure of Matter, 1992) chemistry. *Solid_state electronic_structure physical_properties*
Tel. 86(591)713074 Fax 86(591)714946

Dong, Prof. Yicheng (1939). Institute of Biophysics, Academia Sinica, Beijing 100101, China. MS (Peking U., 1966) structural chemistry. *Proteins crystallography chemistry*
Tel. 86(1)2020077 Fax 86(1)2027837

Fan, Prof. Chenggao (1946). Professor. Structure Research Laboratory, University of Science and Technology of China, Hefei, Anhui 230026, China. BS (U. Science and Technology of China, 1969) physics. *Electron_microscopy electron_diffraction*
Tel. 86(551)302544 Fax 86(551)331760 Telex 90028USTCCN

Fan, Prof. Haifu (1933). Group Leader, Member of Chinese Academy of Sciences. Member of National Committee of Crystallography of China. Institute of Physics, Chinese Academy of Sciences, China. [Institute of Physics, Beijing 100080, China.] BS (Peking U., 1956) physical chemistry. *Direct_method image_processing computing structural_modulation high_Tc_superconductors*
E-mail fan@bepc2.ihep.ac.cn or fan@aphy01.iphy.ac.cn Tel. 86(1)2559131x383 Fax 86(1)2562605 Telex 210208 SVCIP CN

Fan, Prof. Yuguo (1940). Institute Director, Professor. Institute of Theoretical Chemistry, Jilin University, Changchun 130024, China. MS (Jilin University, 1981) quantum chemistry. *Crystallography structure_determination*
Tel. 86(431)823189 Fax 86(431)823907

Fu, Prof. Heng (1929). Institute of Chemistry, Academia Sinica, Beijing 100080, China. (Peking U., 1953) chemistry. *Crystallography chemistry structure*
Tel. 86(1)2554245 Fax 86(1)2569564

Fu, Prof. Pingqiu (1933). Director of Laboratory of Mineralogy. Institute of Geochemistry, Chinese Academy of Sciences, 73 Guan Shui Road, Guiyang, Guizhou 550002, China. (Peking U., 1935) chemistry. *Crystallography mineralogy*
Tel. 86(851)25502x484 Fax 86(851)522982

Fu, Prof. Zhengmin (1938). Group Leader. Institute of Physics, Chinese Academy of Sciences, PO Box 603, Beijing 100080, China. BS (Wuhan University, 1961) physics. *Phase_transition phase_diagram structure crystal_growth*
E-mail heqy@ihepvx.slac.stanford.edu Tel. 86(1)2559131x524 Fax 86(1)2562605 Telex 210208SVCIPCN

Fu, Prof. Zhuji (1949). Fujian Institute of Research on the Structure of Matter, Chinese Academy of Sciences, Fuzhou, Fujian 350002, China. MS (Fujian Inst. Research on the Structure of Matter, 1981) physical chemistry. *Proteins crystallography structural_chemistry*
Tel. 86(591)715544 Fax 86(591)714946 Telex 92219FIRSMCN

Ge, Prof. Zhongjiu (1941). Group Leader. Analysis Research Department, Changchun Institute of Physics, Academia Sinica, China. [1 Yanan Road, Changchun 130021, China.] (Jinlin U., 1965) physics. *Semiconductors thin_layer*
Tel. 86(431)552215x115 Fax 86(431)55317

Gu, Prof. Xiaocheng (1930). Professor; Chairman, Academic Committee, National Lab. Protein Engineering and Plant Genetic Engineering, Peking U.. [College of Life Sciences, Peking University, Beijing 100871, China.] (Peking U., 1952) biology. *Protein_crystallography protein_engineering_and_biotechnology*
E-mail xcg@iris.lsc.pku.edu.cn Tel. 86(1)2501843 Fax 86(1)2501843 Telex '22239 PKUNI CN (please specify Xiaocheng Gu, Life Science College)

Gu, Prof. Yuanxin (1938). Institute of Physics, Chinese Academy of Sciences, PO Box 603, Beijing 100080, China. BS (Lanzhou U., 1964) physics. *Crystallography direct_method computing*
Tel. 86(1)2559131x316 Fax 86(1)2562605 Telex 210208SVCIPCN

Guo, Prof. Changlin (1937). Group Leader. X-ray Diffraction Laboratory, Shanghai Institute of Ceramics, Chinese Academy of Sciences, 1295 Dingxi Road, Shanghai 200050, China. (Shandong U., 1959) physics. *Diffraction polytypism topography*
Tel. 86(21)2512990 Fax 86(21)2513903 Telex 33309ASSICCN

Han, Prof. Yuzhen (1940). Sub-Editor, WDC9. Department of Chemistry, Peking University, Beijing 100871, China. (Peking U., 1965) computing mathematics. *Crystallography computing molecular_design quantum_mechanics*
Tel. 86(1)2501490 Fax 86(1)2501725

Hong, Prof. Maochun (1953). Division head; Professor, Assistant director. Fujian Institute of Research on the Structure of Matter, PO Box 143, Fuzhou, Fujian 350002, China. MS (Chinese Academy of Sciences, 1981) physical chemistry. *Crystallography structural_chemistry*
Tel. 86(591)714578 Fax 86(591)714946 Telex 92219FIRSMCN

Hu, Prof. Shengzhi (1932). Group Leader. Department of Chemistry, Xiamen University, Xiamen 361005, China. (Wuhan U., 1954) chemistry. *Structure_determination coordination_compounds natural_products*
Tel. 86(592)2087267 Fax 86(592)2088054

Hu, Dr Zhengwei (1963). Associate Professor. National Laboratory of Solid State Microstructures, Nanjing University, Nanjing 210008, China. PhD (Shanghai Institute of Metallurgy, Academia Sinica, 1990) material physics. *Topography diffraction phase thin_film_materials transmission_electron_microscopy*
E-mail naiben%bepc2@slacvx.bitnet Tel. 86(25)3300535 Fax 86(25)3300535

Hua, Prof. Ziqian (1933). Professor. Department of Biology, Peking University, Beijing 100871, China. (Peking U., 1960) biology. *Crystallography proteins biochemistry*
Tel. 86(1)2501864 Fax 86(1)2564095 Telex 211215BEUNCN

Huang, Prof. Jinling (1932). Professor. Central Laboratory, Fuzhou University, Fuzhou, Fujian 350002, China. (Fuzhou U., 1954) chemistry. *Complex clusters materials*
Tel. 86(591)713218 Fax 86(591)713866

Huang, Prof. Jinshun (1938). Director of Fujian Institute. Fujian Institute of Research on the Structure of Matter, Chinese Academy of Sciences, PO Box 143, Fuzhou, Fujian 350002, China. MS (Fujian Inst. Research on the Structure of Matter, 1965) physical chemistry. *Crystallography*
Tel. 86(591)711174 Fax 86(591)714946 Telex 92219FIRSMCN

Huang, Prof. Qichen (1945). Associate Professor. Institute of Physical Chemistry, Peking University, China. [Department of Chemistry, Peking University, Beijing 100871, China.] PhD (Shanghai Institute of Organic Chemistry, CAS, 1986) organic chemistry. *X-ray_crystallography proteins biochemistry*
Tel. 86(1)2501702 Fax 86(1)2564095 Telex 211215BEUNCN

Huang, Prof. Taishan (1935). Associate Professor. Department of Chemistry, Xiamen University, Xiamen 361005, China. (Xiamen U., 1957) chemistry. *Powder_diffraction anticancer_compounds*
Tel. 86(592)280665

Jiang, Prof. Shusheng (1940). Deputy Director. National Lab. of Microstructures, Nanjing University, Nanjing 210008, China. BS (Nanjing U., 1963) physics. *Diffraction crystallography topography TEM superlattices*
E-mail naiben%bepc2@scs.bitnet Tel. 86(25)3300535 Fax 86(25)3300535 Telex 34151PRCNUCN

Jiang, Prof. Xiaolong (1940). Shanghai Institute of Metallurgy, Chinese Academy of Sciences, 865 Changning Road, Shanghai 200050, China. MS (Shanghai Inst. Metallurgy, Chinese Academy of Sciences, 1966) metal physics. *Crystallography topography powder_diffraction materials_sciences*
Tel. 86(21)2511070x227 Fax 86(21)2513510

Jiang, Dr Xiaoming (1963). Deputy Director. Synchrotron Radiation Laboratory, Institute of High Energy Physics, Chinese Academy of Sciences, China. [PO Box 918, Beijing 100039, China.] PhD (U. Science and Technology of China, 1988) physics. *Multilayer diffuse_scattering*
E-mail jiangxm%bepc2@scs.slac.stanford.edu Tel. 86(1)8213344x574 Fax 86(1)8213374 Telex 22082IHEPCN

Jiang, Prof. Yadao (1936). Group Leader. Institute of Physics, Chinese Academy of Sciences, China. [PO Box 603, Beijing 100080, China.] BS (Nanjing U., 1958) physics. *Crystal_growth materials*
Tel. 86(1)2559131x501 Fax 86(1)2562605 Telex 210208SVCIPCN

Jin, Prof. Xianglin (1940). Department of Chemistry, Peking University, Beijing 100871, China. MS (Peking U., 1967) physical chemistry. *Crystallography chemistry*
Tel. 86(1)2501490 Fax 86(1)2501725

Kuang, Prof. Bao (1943). Associate Professor. Department of Biology, Peking University, Beijing 100871, China. (Peking U., 1966) biology. *Crystallography proteins computing computer_application*
Tel. 86(1)2501864 Fax 86(1)2501844 Telex 211215BEUNCN

Kuo, Prof. Ke-hsin (1923). Professor, Academia Sinica Member. Beijing Electron Microscopy Lab., Academia Sinica, PO Box 2724, Beijing 100080, China. (Zhejiang U., 1947) chemical engineering. *Crystallography electron_microscopy*
Tel. 86(1)2568304 Fax 86(1)2561422

Lai, Prof. Luhua (1963). Professor. Department of Chemistry, Peking University, Beijing 100871, China. PhD (Peking U., 1989) chemistry. *Proteins drug_design*
E-mail xuxj%bepc2@scs.slac.stanford.edu Tel. 86(1)2501490 Fax 86(1)2501725

Li, Prof. Fanghua (1932). Professor; Member of Chinese Academy of Sciences (elected 1993). Institute of Physics, Chinese Academy of Sciences, China. [Institute of Physics, Beijing 100080, China.] BS (Leningrad U., 1956) physics. *Electron_microscopy diffraction*
E-mail lifh@aphy01.iphy.ac.cn Tel. 86(1)2559131x270 Fax 86(1)2562605 Telex 210208SVCIPCN

Li, Prof. Genpei (1940). Department of Chemistry, Peking University, Beijing 100871, China. (Peking U., 1965) inorganic chemistry. *Crystallography proteins*
Tel. 86(1)2501490 Fax 86(1)2501725

Li, Ms Mei (1954). Analysis Research Department, Changchun Institute of Physics, Academia Sinica, China. [1 Yanan Road, Changchun 130021, China.] (Jilin U., 1980) physics. *Semiconductors thin_layer*
Tel. 86(431)552515x115 Fax 86(431)55317

Li, Prof. Runshen (1942). Deputy Director, Department of Material Sciences. Shanghai Institute of Metallurgy, Chinese Academy of Sciences, 865 Changning Road, Shanghai 200050, China. MS (Shanghai Inst. Metallurgy, Chinese Academy of Sciences, 1981) physics. *Crystallography diffraction topography scientific_instruments*
Tel. 86(21)2511070x227 Fax 86(21)2513510

Li, Prof. Wanmao (1936). Department of Geology, Lanzhou University, Lanzhou 730000, China. (Lanzhou U., 1959) geology. *Structure mineralogy*
Tel. 86(931)412904x611

Liang, Prof. Dongcai (1932). Professor, Academia Sinica Member. Institute of Biophysics, Academia Sinica, Beijing 100101, China. Cand (Acad. Sci., USSR, 1960) chemistry. *Proteins crystallography*
Tel. 86(1)2020077x458 Fax 86(1)2027837

Liang, Prof. Jingkui (1931). Professor. Institute of Physics, Chinese Academy of Sciences, China. [8, South 3rd Street, Zhong Guan Cun, PO Box 603, Beijing 100080, China.] PhD (Institute of Metallurgy, Academy of Sciences, USSR, 1960) metallic physics. *Phase_diagram phase_transition structure powder*
E-mail jkliang@aphy01.iphy.ac.cn Tel. 86(1)2559131x284 Fax 86(1)2562605 Telex 210208SVCIPCN

Lin, Prof. Chichang (1933). Professor. Central Laboratory, Fuzhou University, Fuzhou, Fujian 350002, China. (Fuzhou U., 1955) chemistry. *Complex clusters*
Tel. 86(591)715144 Fax 86(591)713866

Lin, Prof. Shaofan (1935). Professor, Chairman of Central Laboratory. Member of the IUCr Commission on Powder Diffraction. Central Laboratory, Nankai University, Tianjin 300071, China. MS (Nankai U., 1966) organic chemistry. *Structural_chemistry database computational_chemistry*
E-mail fengcb@bepc2.ihep.ac.cn Tel. 86(22)3371380 86(22)3371183 Fax 86(22)3371555 Telex 23133 NAANKICN

Lin, Prof. Xisheng (1943). Group Leader. Lab. Center, Scientific Research Institute of Petroleum Exploration and Development, Beijing 100083, China. (Peking U., 1966) physics. *Clay_mineralogy physics computer application_software*
Tel. 86(1)2097266 Fax 86(1)2097414 Telex 22007CCLBJCN

Lin, Prof. Yonghua (1937). Group Leader. Director, Changchun Physics Society, China. Changchun Institute of Applied Chemistry, Changchun 130022, China. [Changchun Stalin Street 109, Changchun 130022, China.] BS (Peking U., 1960) geochemistry. *Rare-earth_compounds crystallography crystallography_of_biological_small_molecules*
Tel. 86(431)682801 Fax 86(431)685653 Telex 83063CHIACCN

Lin, Prof. Yujuan (1939). Fujian Institute of Research on the Structure of Matter, Academia Sinica, Fuzhou, Fujian 350002, China. BS (Fuzhou U., 1964) physical chemistry. *Proteins crystallography tetraaza_macrocyclic transition_elements*
Tel. 86(591)715544 Fax 86(591)714946 Telex 92219FIRSMCN

Lin, Prof. Zhengjiong (1935). Group Leader. Institute of Biophysics, Academia Sinica, China. [15 Datun Road, Chaoyang District, Beijing 100101, China.] BS (Peking U., 1958) physical chemistry. *Crystallography proteins*
Tel. 86(1)2020077x338 Fax 86(1)2027837

Liu, Prof. Hanqin (1938). Fujian Institute of Research on the Structure of Matter, Chinese Academy of Sciences, China. [PO Box 143, Fuzhou, Fujian 350002, China.] PhD (Chicago U., 1967) chemical physics. *Crystallography*
Tel. 86(591)714578 Fax 86(591)714946 Telex 92219FIRSMCN

Liu, Prof. Shixiong (1943). Professor. Central Laboratory, Fuzhou University, Fuzhou, Fujian 350002, China. (Fuzhou U., 1965) chemistry. *Complex clusters materials powder_crystal*
Tel. 86(591)713786 Fax 86(591)713866

Liu, Prof. Xiaolan (1943). Institute of Crystallography, Tianjin Normal University, Tianjin, China. MS (Tianjin U., 1980) structural chemistry. *Crystallography structure macromolecule structure*
Tel. 86(22)3345026x552 Fax 86(22)3358489

Lu, Prof. Guangying (1937). Professor. Department of Biology, Peking University, Beijing 100871, China. (Peking U., 1960) biology. *X-ray_crystallography electron_crystallography biochemistry*
Tel. 86(1)2501864 Fax 86(1)2564095 Telex 211215BEUNCN

Lu, Prof. Jiaxi (1915). Honorary Chairman, Presidium of Academic Divisions, CAS. Chinese Academy of Sciences, Sanlihe Road, Beijing 100864, China. PhD (U. London, 1939) physical chemistry. *Crystallography structural_chemistry transition_elements*
Tel. 86(1)8597283 Fax 86(1)8511095

Lu, Prof. Kunquan (1939). Professor, Group Leader. Institute of Physics, Chinese Academy of Sciences, PO Box 603, Beijing 100080, China. (Academia Sinica, 1967) crystallography. *EXAFS liquid_state crystal_growth*
Tel. 86(1)2559131x235 Fax 86(1)2562605 Telex 210208SVCIPCN

Lu, Prof. Shaofang (1941). Professor. Fujian Institute of Research on the Structure of Matter, Chinese Academy of Sciences, China. [PO Box 143, Fuzhou, Fujian 350002, China.] BS (Xiamen U., 1962) physics. *Structure_determination clustering structure-property_relationship*
Tel. 86(591)3711368x2108 Fax 86(591)3714946 Telex 92219FIRSMCN

Lu, Ms Yang (1959). Associate Professor; Group Leader. Institute of Materia Medica, Academy of Medical Sciences, China. [No. 1 Xian Nong Tan Street, Beijing 100050, China.] BS (Branch School of Qinghua U., 1983) electronics. *Crystallography_of_natural_organic_molecules crystallography_of_biological_macromolecules*
E-mail ZhouHF@bepc2.ihep.ac.cn Tel. 86(1)3013366x271 Fax 86(1)3017757 Telex 8602(Beijing)

Luo, Prof. Jingchu (1947). Associate Professor. Department of Biology, Peking University, Beijing 100871, China. (Peking U., 1970) biology. *Molecular_modelling computing protein_conformation*
Tel. 86(1)2501843 Fax 86(1)2501844 Telex 211215BEUNCN

Ma, Prof. Lidun (1935). Group Leader. Research Centre for Analysis and Measurement, Fudan University, 220 Handan Road, Shanghai 200433, China. (Fudan U., 1957) chemistry. *EXAFS catalysis materials*
Tel. 86(21)5492222x3010 Fax 86(21)5490653 Telex 33317HUAFUCN

Ma, Prof. Xingqi (1936). Institute of Biophysics, Academia Sinica, Beijing 100101, China. (Peking U., 1962) chemistry. *Proteins crystallography*
Tel. 86(1)2020077 Fax 86(1)2027837

Ma, Prof. Zhesheng (1937). Director of Research Section. Research Section of Crystal Chemistry, Department of Material Sciences, China University of Geosciences, China. [X-ray Lab., China University of Geosciences, Xue Yuan Road 29, Beijing 100083, China.] (China U. Geosciences, 1961) geology. *Structural_determination mineralogy surface_structure new_mineral*
Tel. 86(1)2022244x2243

Mai, Prof. Zhenhong (1942). Assistant Director, Group Leader. Institute of Physics, Chinese Academy of Sciences, China. [Institute of Physics, Chinese Academy of Sciences, Beijing 100080, China.] MS (Institute of Physics, Chinese Academy of Sciences, 1969) physics. *Crystal_defect topography multilayer_structures liquid_structure*
E-mail user404@aphy01.iphy.ac.cn Tel. 86(1)2559131x513 Fax 86(1)2562605 Telex 210208SVCIPCN

Mao, Mr Zhihua (1948). Lab. Structural Chemistry, Sichuan University, China. [Box 35, Department of Chemistry, Sichuan University, Chengdu, Sichuan 610064, China.] MS (Sichuan U., 1984) physical chemistry. *Computing crystallography coordination_compound*
Tel. 86(28)5583875 Fax 86(28)5582844

Miao, Prof. Fangming (1935). Leader of Institute of Crystallography. Institute of Crystallography, Tianjin Normal University, Tianjin, China. MS (Peking U., 1959) structural chemistry. *Structure molecular_design macromolecule_structure*
Tel. 86(22)3345026x552 Fax 86(22)3358489

Pan, Prof. Kezhen (1933). Group Leader. Fujian Institute of Research on the Structure of Matter, Chinese Academy of Sciences, Fuzhou, Fujian 350002, China. [PO Box 143, Fuzhou, Fujian 350002, China.] MS (Xiamen U., 1958) chemistry. *Proteins crystallography structural_chemistry*
Tel. 86(591)715544 Fax 86(591)714946 Telex 92219FIRSMCN

Pei, Prof. Guangwen (1939). Group Leader. Central Laboratory, Nankai University, 94 Weijin Road, Tianjin 300071, China. BS (Nankai U., 1963) chemistry. *Polycrystal structure_determination SAXS*
Tel. 86(22)3371380 Fax 86(22)3344853

Shao, Prof. Meicheng (1931). Professor. Department of Chemistry, Peking University, Beijing 100871, China. MS (Peking U., 1959) structural chemistry. *Crystallography chemistry*
Tel. 86(1)2501490 Fax 86(1)2501725

Shen, Prof. Fuling (1941). Institute of Biophysics, Academia Sinica, China. [Institute of Biophysics, 15 Datun Road, Chaoyang District, Beijing 100101, China.] BS (Nankai U., 1965) chemistry. *Crystallography proteins crystal_growth*
Tel. 86(1)2020077x517 Fax 86(1)2027837

Shen, Prof. Jinchuan (1936). Professor. Test Center of Rocks and Mineral, China University of Geosciences, Wuhan 430074, China. (China U. Geosciences, 1960) geology. *Mineralogy rare-earth_compounds crystallography*
Tel. 86(27)702136

Shi, Mr Lei (1959). Director of X-ray Laboratory, Associate Research Fellow. Structure Research Laboratory, University of Science and Technology of China, Hefei, Anhui 230026, China. PhD (U. Science and Technology of China, 1994) solid state physics. *Superconductor X-ray_crystallography crystal_structure_analysis*
Tel. 86(551)3602810 Fax 86(551)3631760 Telex 90028USTCCN

Shi, Prof. Nicheng (1937). Group Leader. Research Section of Crystal Chemistry, Department of Material Sciences, China University of Geosciences, China. [X-ray Lab., China University of Geosciences, Xue yuan Road 29, Beijing 100083, China.] (China U. Geosciences, 1961) geology. *Structure_determination mineralogy quasicrystallography surface_structure computing*
Tel. 86(1)2022244x2243

Song, Prof. Shiying (1938). Group Leader. Institute of Biophysics, Academia Sinica, China. [15 Datun Road, Chaoyang District, Beijing 100101, China.] BS (Peking U., 1964) physical chemistry. *Crystallography proteins*
Tel. 86(1)2020077x338 Fax 86(1)2027837

Tang, Prof. Youqi (1920). Institute Director, Academia Sinica Member. Department of Chemistry, Peking University, Beijing 100871, China. PhD (Caltech, 1950) chemistry. *Crystallography symmetry proteins*
Tel. 86(1)2502510 Fax 86(1)2501725

Wan, Ms Jiayi (1945). Chemistry Department, Sichuan University, Chengdu 610064, China. BS (Sichuan U., 1968) chemistry. *Crystallography complex_structure X-ray_spectrometry*
Tel. 86(28)5581554x6841

Wang, Prof. Dacheng (1940). Professor. Institute of Biophysics, Chinese Academy of Sciences, 15 Datun Road, Chaoyang District, Beijing 100101, China. (U. Science and Technology of China, 1963) biophysics. *Structure proteins antiviral_protein trichosanthin RGH*
E-mail chnrs%bepc2@slacvx.bitnet Tel. 86(1)2020077x523 Fax 86(1)2027837

Wang, Prof. Jinling (1943). Institute of Crystallography, Tianjin Normal University, Tianjin, China. MS (Nankai U., 1982) structural chemistry. *Structure crystallography macromolecule_structure*
Tel. 86(22)3345026x552 Fax 86(22)3358489

Wang, Prof. Qiguang (1945). Instrumental Analysis and Research Center, Lanzhou University, Lanzhou 730000, China. MS (Lanzhou U., 1982) physics. *Direct_method*
Tel. 86(931)8843000x3102 86(931)8843000x8334 Fax 86(931)8885076

Wang, Prof. Ruji (1945). Associate Professor. Central Laboratory, Nankai University, Tianjin 300071, China. MS (Nankai U., 1982) chemistry. *Structure metalloorganic inorganic_compounds*
Tel. 86(22)3371380 Fax 86(22)3344853

Wang, Prof. Yuming (1932). Head of Dept. Lab. Thin Film Physics, Dept. Materials Sci., Jilin University, China. [Jilin University, Jifang Road 79, Changchun 130023, China.] PhD (Moscow Institute of Steels, 1962) physics. *Powder_diffraction thin_film interface software_design*
E-mail gangchen@jilin.ihep.ac.cn Tel. 86(431)8922331x2830 Fax 86(431)8923907 Telex 83040JUNCN

Wang, Prof. Zutao (1926). President, Tianjin Society of X-ray Analysis. Central Laboratory, Nankai University 730000, China. 94 Weijin Road, Tianjin 300071, China. BS (Nankai U., 1951) physical chemistry. *Medicine metalloorganic powder_diffraction*
Tel. 86(22)3344200x2483 Fax 86(22)3344853

Wei, Prof. Xincheng (1937). Associate Professor. Department of Biology, Peking University, Beijing 100871, China. (Peking U., 1964) biology. *Crystallography proteins biochemistry*
Tel. 86(1)2501864 Fax 86(1)2564095 Telex 211215BEUNCN

Wu, Prof. Bomu (1937). Associate Professor. Institute of Biophysics, Academia Sinica, Beijing 100101, China. (Peking U., 1964) crystallography. *Crystallography computing viruses_structure*
Tel. 86(1)2020077x338 Fax 86(1)2027837

Wu, Mr Wanguo (1945). Engineer. Central Laboratory, Fuzhou University, Fuzhou, Fujian 350002, China. (Fuzhou U., 1968) chemistry. *Powder_crystal*
Tel. 86(591)715935 Fax 86(591)713866

Wu, Prof. Xintao (1939). Group Leader. State Key Laboratory of Structural Chemistry, China. [Fujian Institute of Research on the Structure of Matter, Fuzhou 350002, China.] MS (Fuzhou U., 1966) physical chemistry. *Transition_elements cluster_compounds*
Tel. 86(591)3714517 Fax 86(591)3714946 Telex 92219FIRSMCN

Xia, Prof. Zongxiang (1942). Professor. Shanghai Institute of Organic Chemistry, Chinese Academy of Sciences, 354 Fenglin Road, Shanghai 200032, China. (Fudan U., 1964) chemistry. *Structure_determination structure-function_relationship proteins*
Tel. 86(21)4313300x1119 Fax 86(21)4335712 Telex 33354SIOCCN

Xian, Prof. Dingchang (1935). Group Leader. Synchrotron Radiation Laboratory, Institute of High Energy Physics, Chinese Academy of Sciences, China. [PO Box 918, Beijing 100039, China.] BS (Peking U., 1956) physics. *Theory nanoanalysis multilayer*
E-mail xian%bepc2@scs.slac.stanford.edu Tel. 86(1)8213344x365 Fax 86(1)8213374 Telex 22082IHEPCN

Xu, Prof. Jingyang (1935). Group Leader. Shanghai Institute of Metallurgy, Chinese Academy of Sciences, 865 Changning Road, Shanghai 200050, China. MS (Shanghai Inst. Metallurgy, Chinese Academy of Sciences, 1965) X-ray diffraction. *Materials defects_in_crystal*
Tel. 86(21)2511070x227 Fax 86(21)2513510

Xu, Prof. Xiaojie (1937). Chairman of Department. Department of Chemistry, Peking University, Beijing 100871, China. (Peking U., 1960) physical chemistry. *Crystallography proteins modelling drug_design*
Tel. 86(1)2501490 Fax 86(1)2501725

Yang, Prof. Guangdi (1942). Institute of Theoretical Chemistry, Jilin University, Changchun 130023, China. (Jilin U., 1964) chemistry. *Crystallography structure_determination*
Tel. 86(431)823189 Fax 86(431)823907

Yang, Prof. Qingchuan (1940). Department of Chemistry, Peking University, Beijing 100871, China. (Peking U., 1964) physical chemistry. *Structure chemistry QSAR drug_design accurate_structure_determination*
Tel. 86(1)2501490 Fax 86(1)2501725

Yao, Prof. Xinkan (1939). Deputy Director of Central Laboratory. Central Laboratory, Nankai University, Tianjin 300071, China. BS (Nankai U., 1962) chemistry. *Structure_determination clustering metalloorganic synchrotron_radiation*
Tel. 86(22)3371380 Fax 86(22)3344853

Ye, Prof. Chuntang (1934). Head of Lab. Thermal Neutron Scattering Lab., China Institute of Atomic Energy, China. [PO Box 275-30, Beijing 102413, China.] (Peking U., 1956) physics. *Neutron_scattering structure lattice_dynamics*
Tel. 86(1)9357727 Fax 86(1)9357008 Telex 222373IAECN

Yu, Prof. Ruihuang (1906). Professor, Member Mathematics & Physics Comm. Chinese Acad. Science. Department of Physics and Institute of Materials Science, Jilin University, China. [Institute of Materials Science, Jilin University, Changchun 130023, China.] PhD (Manchester U., 1937) X-ray crystallography. *Crystallography X-ray_analysis crystal_structure*
Tel. 86(431)823189x3352 Telex 83040JUNCN

Yu, Prof. Xiufen (1932). Professor. Department of Chemistry, Fuzhou University, Fuzhou, Fujian 350002, China. (Fuzhou U., 1954) chemistry. *Structure complex clusters*
Tel. 86(591)711536 Fax 86(591)713866

Zhang, Prof. Hanqing (1936). Vice-Chairman X-ray Cryst. of Mineral Branch GSC. Institute of Mineral Deposits CAGS, China. [26 Baiwanzhuang Road, Beijing 100037, China.] BS (CGU (Beijing), 1960) cryst. of mineral. *Crystallography structure*
Tel. 86(1)8311133x217 Fax 86(1)8310894 Telex 222721CAGSCN

Zhang, Prof. Ruilin (1934). Professor. Institute of Materials Science, Jilin University, Changchun 130023, China. BS (Jilin U., 1961) condensed matter physics. *Electron_theory structure_determination*
Tel. 86(431)823189x3132 Telex 83040JUNCN

Zhang, Prof. Shaohui (1938). Professor. Department of Chemistry, Wuhan University, Wuhan 430072, China. MS (Peking U., 1966) physical chemistry. *Direct_method metalloorganic_compounds*
Tel. 86(27)722712

Zhang, Prof. Shiwei (1945). Department of Chemistry, Peking University, Beijing 100871, China. PhD (Peking U., 1990) structural chemistry. *Crystallography chemistry*
Tel. 86(1)2501490 Fax 86(1)2501725

Zhang, Prof. Ze (1953). Prof.; Director of Beijing Lab. of Electron Microscopy. Secretary of Asian Crystallographic Association. Beijing Lab. of Electron Microscopy, CAS, China. [PO Box 2724, Beijing 100080, China.] PhD (Institute of Metal Res. CAS, 1987) condensed physics. *Quasicrystal diamond thin_film electro_diffraction imaging*
Tel. 86(1)2568304 Fax 86(1)2561422

Zhang, Prof. Zhishun (1936). Group Leader. Analysis Research Department, Changchun Institute of Physics, Academia Sinica, China. [1 Yanan Road, Changchun 130021, China.] (Jilin U., 1960) physics. *Electron_microscopy thin_layer*
Tel. 86(431)552215x115 Fax 86(431)55317

Zheng, Prof. Peiju (1933). Group Leader. Research Center for Analysis and Measurement, Fudan University, 220 Handan Road, Shanghai 200433, China. (Fudan U., 1955) chemistry. *Structure_determination crystallography chemistry thermodynamics*
Tel. 86(21)5492222x3016 Fax 86(21)5490323 Telex 33317HUAFUCN

Zheng, Prof. Qi-Tai (1938). Prof. of organic crystallography; Group Leader. Institute of Materia Medica, Academy of Medical Sciences, China. [No.1 Xian Nong Tan Street, Beijing 100050, China.] BS (Nankai U., 1961) physics. *Crystallography_of_natural_organic_molecules crystallography_of_biological_macromolecules*
E-mail ZhouHF@bepc2.ihep.ac.cn Tel. 86(1)3013366x271 Fax 86(1)3017757 Telex 8602(Beijing)

Zheng, Prof. Weitao (1963). Associate Professor. Institute of Materials Science, Jilin University, Changchun 130023, China. PhD (Jilin U., 1990) condensed matter physics. *Solid_phase phase_diagram*
Tel. 86(431)823189x4229 Telex 83040JUNCN

Zhou, Prof. Kangjing (1939). Fujian Institute of Research on the Structure of Matter, Chinese Academy of Sciences, Fuzhou, Fujian 350002, China. MS (Peking U., 1963) chemistry. *Proteins crystallography*
Tel. 86(591)715544 Fax 86(591)714946 Telex 92219FIRSMCN

Zhou, Prof. Zhongyuan (1938). Group Leader. Analysis and Testing Center of Chengdu Branch, Academia Sinica, China. [9 Renminnanlu, Block 4, Chengdu, Sichuan 610041, China.] (Sichuan U., 1962) chemistry. *Structure_determination tunneling_and_disorder crystal_structure electron_density_distribution_in_bonds*
Tel. 86(28)5581260x270 Fax 86(28)5582846 Telex 600321SICDCN

Zhou, Prof. Zonghua (1935). Lab. Structural Chemistry, Sichuan University, China. [Box 35, Department of Chemistry, Sichuan University, Chengdu, Sichuan 610064, China.] BS (Sichuan U., 1957) physical chemistry. *X-ray_crystallography molecular_structure molecular_design*
Tel. 86(28)5583875 Fax 86(28)5582844

Zhu, Dr Wenjie (1965). Researcher. National Laboratory for Superconductivity, Institute of Physics, Academia Sinica, PO Box 603, Beijing 100080, China. PhD (Peking U., 1990) solid state chemistry. *Structure_determination materials superconductivity*
Tel. 86(1)2559131x506 Fax 86(1)2562605

Zhu, Prof. Ying (1954). Group Leader of X-ray Crystallography; PhD student. Instrumental Analysis and Research Center, Lanzhou University, Lanzhou 730000, China. BS (Lanzhou U., 1976) chemistry. *Anticancer_compounds metalloorganic clusters*
Tel. 86(931)8843000x3102 86(931)8419652 Fax 86(931)8885076

COLOMBIA

Moreno Fuquen, Rodolfo (1951). Professor. Departamento de Quimica, Fac. de Ciencias, Universidad del Valle, Apartado 25360, Cali, Colombia. PhD (U. São Paulo, Brazil, 1991) physical chemistry. *Structure_determination non-linear_optical_properties*
E-mail romoreno@hypatia.univalle.edu.co Tel. 57(923)393248 Fax 55(923)392440

CROATIA

Sub-Editor: D. Matković-Čalogović

Alujević-Stipanov, Dr Višnja (1948). Research assistant. Farmaceutsko-biokemijski fakultet, University of Zagreb, Ul. Ante Kovačića 1, 41000 Zagreb, Croatia. PhD (Zagreb U., 1981) chemistry. *X-ray_diffraction polycrystal*
Tel. 385(41)445311x36

Antolić, Ms Snježana (1970). Research assistant. Ruđer Bošković Institute, Bijenička cesta 54, PO Box 1016, 41001 Zagreb, Croatia. BSc (Zagreb U., 1992) chemistry. *Biocrystallography X-ray_diffraction crystallization*
Tel. 385(41)461111x1264 Fax 385(41)425497 Telex 21383RHIRBZG

Balen, Mr Dražen (1967). Research assistant. Dept. of Mineralogy and Petrography, Faculty of Science, Demetrova 1, 41000 Zagreb, Croatia. BSc (Zagreb U., 1991) mineralogy. *Structural_mineralogy structural_properties_of_minerals*
Tel. 385(41)422136 Fax 385(41)432526

Balen, Ms Milka (1951). Research assistant. Željezara Sisak, IRI, 44103 Sisak, Croatia. BSc (Zagreb U., 1990) chemistry. *Qualitative_analysis quantitative_analysis diffraction_technique extraction_spectroscopy*
Tel. 385(44)35807 Fax 385(44)30261

Balzar, Dr Davor (1957). Scientific associate. Ruđer Bošković Institute, PO Box 1016, 41001 Zagreb, Croatia. PhD (Zagreb U., 1993) physics. *Alloys ceramics powder_diffraction*
E-mail dbalzar@dominis.phy.hr Tel. 385(41)461120 Fax 385(41)425497

Banić, Ms Zdravka (1966). Research assistant. Pliva Pharmaceutical, Chemical, Food and Cosmetic Industry, Prilaz baruna Filipovića 89, 41000 Zagreb, Croatia. BSc (Zagreb U., 1990) chemistry. *Biocrystallography computer_modelling X-ray_structure_determination structure-activity_relationship*
Tel. 385(41)181562 Fax 385(41)174719

Bermanec, Dr Vladimir (1955). Assistant professor. Dept. of Mineralogy and Petrography, Faculty of Science, Demetrova 1, 41000 Zagreb, Croatia. PhD (Zagreb U., 1992) geology. *Structural_mineralogy*
E-mail bermanec@x400.srce.hr Tel. 385(41)422136 Fax 385(41)432526

Bezjak, Prof. Dr Aleksandar (1928). Professor. Farmaceutsko-biokemijski fakultet, University of Zagreb, Ul. Ante Kovačića 1, 41000 Zagreb, Croatia. PhD (Zagreb U., 1964) chemistry. *X-ray_diffraction polycrystal*
Tel. 385(41)446061

Blažina, Dr Želimir (1946). Senior scientific associate. Ruđer Bošković Institute, Bijenička cesta 54, PO Box 1016, 41001 Zagreb, Croatia. PhD (Zagreb U., 1979) chemistry. *X-ray_diffraction intermetallic_compounds*
Tel. 385(41)461111 Fax 385(41)425497

Bonefačić, Dr Antun M. (1925). Professor. Dept. of Physics, Faculty of Science, Bijenička cesta 32, PO Box 162, 41001 Zagreb, Croatia. PhD (Zagreb U., 1963) physics. *Physical_crystallography crystal_imperfection metal_physics*
E-mail antunb@phy.hr Tel. 385(41)432480 Fax 385(41)432526

Bruvo, Mr M. (1934). Research assistant. Lab. of General and Inorganic Chemistry, Faculty of Science, Ul. kralja Zvonimira 8, 41000 Zagreb, Croatia. MSc (Zagreb U., 1972) chemistry. *Inorganic_carboxylates*
Tel. 385(41)442823 Fax 385(41)432526

Drašner, Dr Antun (1955). Research assistant. Ruđer Bošković Institute, Bijenička cesta 54, PO Box 1016, 41001 Zagreb, Croatia. PhD (Zagreb U., 1991) chemistry. *Metallic_hydrides intermetallic_structure*
Tel. 385(41)461111 Fax 385(41)425497

Duževic, Prof. Dr Davor (1936). Associate professor. Faculty of Textile Technology, Pierottijeva 6, 41000 Zagreb, Croatia. PhD (Zagreb U., 1979) physics. *Phase_kinetics metals hard_metals allotropy*
Tel. 385(41)571403 Fax 385(41)571649

Gladic, Mr Jadranko (1959). Research assistant. Institute of Physics, Bijenička cesta 46, PO Box 304, 41000 Zagreb, Croatia. MSc (Zagreb U., 1991) physics. *Symmetry incommensurate_structure equilibrium_crystal_shape*
E-mail gladic@olimp.irb.hr Tel. 385(41)271211 Fax 385(41)421156 Telex 22203RHIFSZG

Grdenić, Prof. Dr Drago (1919). Professor retired. Lab. of General and Inorganic Chemistry, Faculty of Science, Ul. kralja Zvonimira 8, 41000 Zagreb, Croatia. PhD (Zagreb U., 1951) chemistry. *Crystal_chemistry mercury_compounds history_of_chemistry*
Tel. 385(41)442823 Fax 385(41)432526

Gržeta, Dr Biserka (1949). Senior scientific associate. Ruđer Bošković Institute, Bijenička cesta 54, PO Box 1016, 41001 Zagreb, Croatia. PhD (Zagreb U., 1980) physics. *Powder_diffraction phase_transition material_sciences*
E-mail grzeta@olimp.irb.hr Tel. 385(41)461120 Fax 385(41)425497 Telex 21383 IRB ZG RH

Herceg, Dr Marija (1938). Senior scientific associate. Co-editor of Croatica Chemica Acta. Ruđer Bošković Institute, PO Box 1016, 41001 Zagreb, Croatia. PhD (Zagreb U., 1970) chemistry. *Coordination_chemistry crown_compounds biologically_interesting_synthetic_compounds*
E-mail herceg@olimp.irb.hr Tel. 385(41)461111x1335 Fax 385(41)425497 Telex 21383RHIRBZG

Hergold-Brundić, Dr Antonija (1942). Assistant professor. Lab. of General and Inorganic Chemistry, Faculty of Science, Ul. kralja Zvonimira 8, 41000 Zagreb, Croatia. PhD (Zagreb U., 1980) chemistry. *Structural_crystallography organic_structure*
Tel. 385(41)442823 Fax 385(41)432526

Horvatić, Mr Davor (1957). Research assistant. Ruđer Bošković Institute, Bijenička cesta 54, PO Box 1016, 41001 Zagreb, Croatia. MSc (Zagreb U., 1984) mathematics. *Computing*
Tel. 385(41)461111 Fax 385(41)425497 Telex 21383RHIRBZG

Ilakovac, Ms Vita (1965). Research assistant. Ruđer Bošković Institute, Bijenička 54, PO Box 1016, 41001 Zagreb, Croatia. BSc (Zagreb U., 1989) physics. *Phase_diagram diffuse_scattering*
Tel. 385(41)461111x1320 Fax 385(41)425497 Telex 21383RHIRBZG

Ivanković, Mr Hrvoje (1958). Research assistant. Faculty of Chemical Engineering and Technology, Marulićev trg 20, 41000 Zagreb, Croatia. MSc (Zagreb U., 1990) chemistry. *Ceramics*
Tel. 385(41)452477

Jurković, Prof. Dr Ivan B. (1917). Professor retired. Rudarsko-geološko-naftni fakultet, Pierottijeva 6, 41000 Zagreb, Croatia. PhD (Zagreb U., 1956) geology. *Crystal_chemistry optical_property*
Tel. 385(41)441839 Fax 385(41)440008

Kaitner, Dr Branko (1942). Associate professor. Lab. of General and Inorganic Chemistry, Faculty of Science, Ul. kralja Zvonimira 8, 41000 Zagreb, Croatia. PhD (Zagreb U., 1979) chemical crystallography. *Coordination_chemistry transition_elements X-ray_structure*
E-mail Branko.Kaitner@public.srce.hr Tel. 385(41)442823 Fax 385(41)432526

Kamenar, Prof. Dr Boris (1929). Professor. Chairman of the Croatian Committee of Crystallography. Lab. of General and Inorganic Chemistry, Faculty of Science, Ul. kralja Zvonimira 8, 41000 Zagreb, Croatia. PhD (Zagreb U., 1960) chemistry. *Crystal_chemistry inorganic_crystal_structure organic_crystal_structure structures_of_pharmaceutically_interesting_compounds*
E-mail Boris.Kamenar@public.srce.hr Tel. 385(41)442823 Fax 385(41)432526

Kiralj, Mr Rudolf (1966). Research assistant. Ruđer Bošković Institute, Bijenička cesta 54, PO Box 1016, 41001 Zagreb, Croatia. BSc (Zagreb U., 1990) chemistry. *X-ray_diffraction small_molecules molecular_mechanics*
Tel. 385(41)461111x1335 Fax 385(41)425497 Telex 21383RHIRBZG

Kojić-Prodić, Dr Biserka (1938). Senior scientist. Ruđer Bošković Institute, Bijenička cesta 54, PO Box 1016, 41001 Zagreb, Croatia. PhD (Zagreb U., 1968) chemistry. *Biocrystallography structure–activity_relationship hydrogen_bonding computer_modelling*
E-mail kojic@olimp.irb.hr Tel. 385(41)425386 Fax 385(41)425497 Telex 21383RHIRBZG

Košutić Hulita, Mrs Nada (1966). Research assistant. Lab. of General and Inorganic Chemistry, Faculty of Science, Ul. kralja Zvonimira 8, 41000 Zagreb, Croatia. MSc (Zagreb U., 1994) chemistry. *Organic_crystal_structure*
Tel. 385(41)442823 Fax 385(41)432526

Kunstelj, Dr Drago (1941). Scientific associate. Dept. of Physics, Faculty of Science, Bijenička cesta 32, PO Box 162, 41001 Zagreb, Croatia. PhD (Zagreb U., 1979) physics. *High-resolution_electron_microscopy structure_analysis_by_X-rays_(phase_transformations)*
Tel. 385(41)432480 Fax 385(41)432525

Luić, Dr Marija (1953). Scientific associate. Ruđer Bošković Institute, Bijenička cesta 54, PO Box 1016, 41001 Zagreb, Croatia. PhD (Zagreb U., 1985) geology. *Biomolecule_structure_determination*
Tel. 385(41)461111x1319 Fax 385(41)425497 Telex 21383RHIRBZG

Maksić, Prof. Dr Zvonimir (1938). Senior scientist. Ruđer Bošković Institute, Bijenička cesta 54, PO Box 1016, 41001 Zagreb, Croatia. PhD (Zagreb U., 1968) chemistry. *Molecular_geometry electron_density_distribution*
Tel. 385(41)434461 Fax 385(41)425497 Telex 21383RHIRBZG

Marjanović, Ms Tihana (1960). Research assistant. Željezara Sisak, IRI, 44105 Sisak, Croatia. BSc (Sisak Faculty of Metallurgy, 1987) metallurgy. *Quantitative_analysis qualitative_analysis Debye_Scherrer polycrystal_phase_determination metallurgy X-ray_fluorescence_spectroscopy*
Tel. 385(44)35395 Fax 385(44)30261

Marković, Mr Berislav (1957). Research assistant. Ruđer Bošković Institute, Bijenička cesta 54, PO Box 1016, 41001 Zagreb, Croatia. MSc (Zagreb U., 1985) chemistry. *Powder_diffraction phase_determination*
Tel. 385(41)461111 Fax 385(41)425497 Telex 213HRIRBZG

Matak, Ms Dijana (1967). Research assistant. Lab. of General and Inorganic Chemistry, Faculty of Science, Ul. kralja Zvonimira 8, 41000 Zagreb, Croatia. MSc (Zagreb U., 1994) chemistry. *Calcification complexation thermoanalysis thermogravimetry calcium_compounds carboxylates*
E-mail sikirica@olimp.irb.hr Tel. 385(41)442823 Fax 385(41)432526

Matijašić, Dr Ivanka (1944). Assistant professor. Lab. of Organic Chemistry and Biochemistry, Faculty of Science, Strossmayerov trg 14, 41000 Zagreb, Croatia. PhD (Zagreb U., 1984) chemistry. *Organic_structure organometallic_crystallography*
Tel. 385(41)432581 Fax 385(41)432526

Matković, Dr Boris (1927). Senior scientist. Ruđer Bošković Institute, Bijenička cesta 54, PO Box 1016, 41001 Zagreb, Croatia. PhD (Zagreb U., 1961) chemistry. *Cements X-ray_powder_diffraction*
E-mail matkovic@olimp.irb.hr Tel. 385(41)461111x1335 Fax 385(41)425497 Telex 21383RHIRBZG

Matković, Dr Prošper (1945). Assistant professor. Metalurški fakultet, Aleja nar. heroja 1, 44000 Sisak, Croatia. PhD (Stuttgart U., BRD, 1977) chemistry. *Intermetallic structure*
Tel. 385(44)32756 Fax 385(44)32961

Matković, Dr Tanja (1948). Assistant professor. Metalurški fakultet, Aleja nar. heroja 3, 44000 Sisak, Croatia. PhD (Stuttgart U., BRD, 1977) chemistry. *Intermetallic structure*
Tel. 385(44)32756 Fax 385(44)32961

Matković-Čalogović, Dr Dubravka (1957). Research assistant. Sub-Editor for Croatia of ninth edition of World Directory of Crystallographers. Lab. of General and Inorganic Chemistry, Faculty of Science, Ul. kralja Zvonimira 8, 41000 Zagreb, Croatia. PhD (Zagreb U., 1994) chemistry. *Mercury_compounds organometallic_structure crown_compounds transition_elements*
E-mail Dubravka.Matkovic.Calogovic@public.srce.hr Tel. 385(41)442823 Fax 385(41)432526

Medimorec, Mr Stanislav S. (1939). Research assistant. Dept. of Mineralogy and Petrography, Faculty of Science, Demetrova 1, 41000 Zagreb, Croatia. MSc (Zagreb U., 1977) mineralogy. *Minerals optical_property*
Tel. 385(41)422136 Fax 385(41)432526

Meštrović, Mr Ernest (1967). Research assistant. Lab. of General and Inorganic Chemistry, Faculty of Science, Ul. kralja Zvonimira 8, 41000 Zagreb, Croatia. BSc (Zagreb U., 1992) chemistry. *Coordination_chemistry X-ray_structure transition_elements coordination_polymers superconductivity*
E-mail sikirica@olimp.irb.hr Tel. 385(41)442823 Fax 385(41)432526

Milat, Dr Ognjen (1949). Scientific associate. Institute of Physics, Bijenička cesta 46, PO Box 304, 41001 Zagreb, Croatia. PhD (Zagreb U., 1990) physics. *Phase_transition incommensurate_structure*
E-mail ognjen.milat@olimp.irb.hr Tel. 385(41)271211 Fax 385(41)421156 Telex 22203RHIFSZG

Milinković, Mr Vjekoslav (1962). Research assistant. Ruđer Bošković Institute, Bijenička cesta 54, PO Box 1016, 41001 Zagreb, Croatia. BSc (Zagreb U., 1989) physics. *X-ray_diffraction biocrystallography computer_modelling*
E-mail milinkov@olimp.irb.hr Tel. 385(41)461111x1264 Fax 385(41)425497 Telex 21383RHIRBZG

Moguš-Milanković, Dr Andrea (1953). Scientific associate. Ruđer Bošković Institute, Bijenička cesta 54, PO Box 1016, 41001 Zagreb, Croatia. PhD (Zagreb U., 1989) chemistry. *Electrical_property electric_properties_of_crystals*
Tel. 385(41)461111 Fax 385(41)425497 Telex 21383RHIRBZG

Mrvoš-Sermek, Dr Draginja (1961). Research assistant. Lab. of General and Inorganic Chemistry, Faculty of Science, Ul. kralja Zvonimira 8, 41000 Zagreb, Croatia. PhD (Zagreb U., 1994) chemistry. *Organic_crystal_structure*
Tel. 385(41)442823 Fax 385(41)432526

Nagl, Dr Ante (1942). Associate professor. Faculty of Textile Technology, Pierottijeva 6, 41000 Zagreb, Croatia. PhD (Bern U. Switzerland, 1973) chemistry. *Organic_structure transition_elements biocrystallography biologically_active_compounds*
Tel. 385(41)442823 Fax 385(41)432526

Novosel Radović, Dr Vjera (1937). Senior scientific associate. Željezara Sisak, IRI, 44105 Sisak, Croatia. PhD (Zagreb U., 1983) chemistry. *Powder_material inclusion thin_film X-ray_fluorescence_spectroscopy X-ray_detectability ores steels environment graphites catalysts*
Tel. 385(44)35395 Fax 385(44)30261

Paljević, Dr Matija (1943). Senior scientific associate. Ruđer Bošković Institute, Bijenička cesta 54, PO Box 1016, 41001 Zagreb, Croatia. PhD (Zagreb U., 1978) chemistry. *High_temperature oxidation gas-solid_reactions*
Tel. 385(41)461111 Fax 385(41)425497 Telex 21383RHIRBZG

Pavlović, Ms Gordana (1965). Research assistant. Lab. of General and Inorganic Chemistry, Faculty of Science, Ul. kralja Zvonimira 8, 41000 Zagreb, Croatia. MSc (Zagreb U., 1993) chemistry. *Molybdenum_compounds structures_of_metal_complexes*
Tel. 385(41)442823 Fax 385(41)432526

Penavić, Dr Maja (1941). Assistant professor. Lab. of General and Inorganic Chemistry, Faculty of Science, Ul. kralja Zvonimira 8, 41000 Zagreb, Croatia. PhD (Zagreb U., 1977) chemistry. *X-ray_crystallography molybdenum_compounds*
Tel. 385(41)442823 Fax 385(41)432526

Popović, Prof. Dr Stanko (1938). Senior scientist, Professor. Secretary of the Croatian Committee of Crystallography. Ruđer Bošković Institute, Bijenička cesta 54, PO Box 1016, 41001 Zagreb, Croatia. PhD (Zagreb U., 1968) physics. *Powder_diffraction phase_transition line_profile_analysis metallic_alloys*
Tel. 385(41)461111x1320 Fax 385(41)425497 Telex 21383RHIRBZG

Puntarec, Mr Vitomir (1961). Research assistant. Ruđer Bošković Institute, Bijenička cesta 54, PO Box 1016, 41001 Zagreb, Croatia. BSc (Zagreb U., 1989) physics. *X-ray_diffraction computer_graphics computer_modelling*
E-mail puntarec@olimp.irb.hr Tel. 385(41)425386 Fax 385(41)425497 Telex 21383RHIRBZG

Radović, Ms Nikol (1963). Research assistant. Geodetski fakultet, Kačićeva 26, 41000 Zagreb, Croatia. BSc (Zagreb U., 1990) mathematics. *Computing powder_diffraction mathematics*
Tel. 385(41)442600x231

Sikirica, Prof. Dr Milan (1934). Professor. Lab. of General and Inorganic Chemistry, Faculty of Science, Ul. kralja Zvonimira 8, 41000 Zagreb, Croatia. PhD (Zagreb U., 1963) chemistry. *Chemistry_education structural_chemistry mercury_compounds*
E-mail Milan.Sikirica@public.srce.hr Tel. 385(41)442823 Fax 385(41)432526

Slovenec, Prof. Dr Dragutin (1941). Professor. Rudarsko-geološko-naftni fakultet, Pierottijeva 6, 41000 Zagreb, Croatia. PhD (Zagreb U., 1980) geology. *Phyllosilicates powder_diffraction*
Tel. 385(41)441839 Fax 385(41)440008

Stefanović, Mr Aleksandar (1957). Research assistant. Lab. of General and Inorganic Chemistry, Faculty of Science, Ul. kralja Zvonimira 8, 41000 Zagreb, Croatia. MSc (Zagreb U., 1985) chemistry. *Coordination_chemistry transition_elements X-ray_structure transition_metal_complexes*
E-mail Aleksandar.Stefanovic@public.srce.hr Tel. 385(41)442823 Fax 385(41)432526

Strukan, Mr Neven (1961). Research assistant. Lab. of General and Inorganic Chemistry, Faculty of Science, Ul. kralja Zvonimira 8, 41000 Zagreb, Croatia. MSc (Zagreb U., 1990) chemistry. *Structural_crystallography molybdenum_compounds*
E-mail Neven.Strukan@public.srce.hr Tel. 385(41)442823 Fax 385(41)432526

Stubičar, Dr Mirko (1940). Assistant professor. Dept. of Physics, Faculty of Science, Bijenička cesta 32, PO Box 162, 41001 Zagreb, Croatia. PhD (Zagreb U., 1986) physics. *Microstructure_hardness_relationship*
Tel. 385(41)432480 Fax 385(41)432525

Šarc-Lahodny, Prof. Dr Olga (1928). Professor. Rudarsko-geološko-naftni fakultet, Pierottijeva 6, 41000 Zagreb, Croatia. PhD (Zagreb U., 1962) physical chemistry. *Clays structure_of_clay_minerals*
Tel. 385(41)442409 Fax 385(41)440008

Šćavničar, Prof. Dr Stjepan (1923). Professor. Vice-chairman of the Croatian Committee of Crystallography. Dept of Mineralogy and Petrography, Faculty of Science, Demetrova 1, 41000 Zagreb, Croatia. PhD (Zagreb U., 1956) chemistry. *Inorganic_crystal_structure*
Tel. 385(41)428610 Fax 385(41)432526

Šmit, Dr Ivan (1948). Scientific associate. Ruđer Bošković Institute, Bijenička cesta 54, PO Box 1016, 41001 Zagreb, Croatia. PhD (Zagreb U., 1979) chemistry. *Polymers polymer_structure multicomponent_polymer_systems*
Tel. 385(41)461111 Fax 385(41)425497 Telex 21383RHIRBZG

Tibljaš, Mr Darko (1957). Research assistant. Dept. of Mineralogy and Petrography, Faculty of Science, Demetrova 1, 41000 Zagreb, Croatia. MSc (Zagreb U., 1987) geology. *Structural_mineralogy*
Tel. 385(41)422136 Fax 385(41)432526

Tkalčec, Dr Emilija (1931). Associate professor. Faculty of Technology, Marulićev trg 20, 41000 Zagreb, Croatia. PhD (Zagreb U., 1975) chemistry. *X-ray_diffraction polycrystal silicates*
Tel. 385(41)452477

Tomić, Dr Sanja (1958). Research assistant. Ruđer Bošković Institute, Bijenička cesta 54, PO Box 1016, 41001 Zagreb, Croatia. PhD (Zagreb U., 1993) chemistry. *Molecular_mechanics modelling biocrystallography*
E-mail tomic@olimp.irb.hr Tel. 385(41)461111x1264 Fax 385(41)425497 Telex 21383RHIRBZG

Tonejc, Dr Anđelka (1942). Assistant professor. Dept. of Physics, Faculty of Science, Bijenička cesta 32, PO Box 162, 41001 Zagreb, Croatia. PhD (Zagreb U., 1980) physics. *High-resolution_electron_microscopy X-ray_diffraction line_profile_analysis crystal_defect*
Tel. 385(41)432480 Fax 385(41)432525

Tonejc, Prof. Dr Anton (1942). Associate professor. Dept. of Physics, Faculty of Science, Bijenička cesta 32, PO Box 162, 41001 Zagreb, Croatia. PhD (Zagreb U., 1972) physics. *Physical_crystallography phase_transition*
Tel. 385(41)432480 Fax 385(41)432525

Trojko, Mr Rudolf (1942). Research assistant. Ruđer Bošković Institute, Bijenička cesta 54, PO Box 1016, 41001 Zagreb, Croatia. MSc (Zagreb U., 1974) chemistry. *Intermetallic intermetallic_compounds*
Tel. 385(41)461111x1319 Fax 385(41)425497 Telex 21383RHIRBZG

Vicković, Dr Ivan (1945). Scientific associate. Lab. of General and Inorganic Chemistry, Faculty of Science, Ul. kralja Zvonimira 8, 41000 Zagreb, Croatia. PhD (Zagreb U., 1977) physics. *Crystal_structure computing*
E-mail vickovic@srce.hr Tel. 385(41)442823 Fax 385(41)432526

Vinković, Dr Mladen (1966). Research assistant. PLIVA Research Institute, PLIVA - Pharmaceutical, Chemical, Food and Cosmetic Industry, Prilaz baruna Filipovića 89, 41000 Zagreb, Croatia. PhD (Zagreb U., 1994) chemistry. *Computer-assisted_design computer_modelling antimicrobial_compounds*
E-mail vinkovic@olimp.irb.hr Tel. 385(41)181737 Fax 385(41)174719

Vučić, Dr Zlatko (1947). Scientific associate. Institute of Physics, Bijenička cesta 46, PO Box 304, 41001 Zagreb, Croatia. PhD (Zagreb U., 1989) physics. *Phase_transition incommensurate_structure*
E-mail vucic@olimp.irb.hr Tel. 385(41)271211 Fax 385(41)421156 Telex 22203RHIF-SZG

CUBA

Sub-Editor: R. Pomes Hernandez

Calzadilla, Prof. Octavio (1949). Assistant Professor. Dept. of General Physics, Fac. of Physics, U. of Havana, San Lázaro y L. Vedado, Havana, Cuba. MSc (Havana U., 1976) crystallography. Semiconductor_structure
E-mail imre@ceniai.cu Tel. 53(7)78956

Canizares, Mr Hian (1970). Researcher. National Center for Scientific Research, X-ray Laboratory, PO Box 6990, Havana, Cuba. BSc (Nuclear Inst. of Havana, 1993) physics. Powder_diffractometry Mossbauer_spectroscopy automatic_control
E-mail infrared@ceniai.cu Tel. 53(7)218066 Fax 53(7)219446

Capo, Dr Luis (1952). Professor. Pedagogical Institute Frank Pais, Autopista Nacional s.n., Santiago de Cuba, Cuba. Dr (Leningrad Pedagogical Inst., Russia, 1987) solid state physics. Semiconductor_structure
E-mail labsem@ceniai.cu Tel. 53(7)22642019

Cerpa, Mr Arisbel (1959). Assistant Professor. Politechnical Institute Julio A. Mella, Ave. Las Américas s.n., Santiago, Cuba. Engineer (Polytechnical Institute Julio A. Mella, 1982) chemical analysis. Powder_diffraction minerals surface_chemistry_and_rheology
E-mail inguim%ispjam@ceniai.cu Tel. 53(7)22642019 Fax 53(7)338212

Cruz, Dr Carlos (1950). Professor. Centre of Appl. Studies to Nuclear Dev., PO Box 6122, Havana, Cuba. Dr (Dresden U., Germany, 1981) crystallography. Powder_diffraction ceramics
E-mail ceaden@ceniai.cu

Cruz, Dr Francisco (1951). Professor. Structural Analysis Lab., Inst. of Materials and Reagents for Electronics, Univ. of Havana, San Lázaro y L. Vedado, Havana, Cuba. Dr (Havana U., 1992) crystallography. Texture powder_diffraction electron_microscopy
E-mail imre@ceniai.cu Tel. 53(7)78956 Telex (51)12210

Curbelo, Mr Ciro (1960). Assistant Professor. Structural Analysis Lab., Inst. of Materials and Reagents for Electronics, Univ. of Havana, San Lázaro y Vedado, Havana, Cuba. BSc (U. Havana, 1983) chemistry. Powder_diffraction
E-mail imre@ceniai.cu Tel. 53(7)78956 Telex (51)12210

Dago, Dr Angel (1949). Professor. National Center for Scient. Research, X-ray Laboratory, PO Box 6990, Havana, Cuba. Dr (Moscow U., Russia, 1986) crystallography. Structure_determination powder_diffraction
E-mail infrared@ceniai.cu Tel. 53(7)210135 Fax 53(7)219446

Duque, Mr Julio (1958). Assistant Professor. National Center for Scientific Research, X-ray Laboratory, PO Box 6990, Havana, Cuba. BSc (Havana U., 1988) chemistry. Structure_determination crystal_chemistry
E-mail infrared@ceniai.cu Tel. 53(7)218066 Fax 53(7)219446

Durruthy, Mr Obel (1945). Assistant Professor. Polytechnical Inst. Julio A. Mella, Ave. Las Américas s.n., Santiago de Cuba, Cuba. BSc (Havana U., 1970) solid state physics. Structure_determination crystallography
E-mail frisica%ispjam@ceniai.cu Tel. 53(7)22642019 Fax 53(7)338212

Fajardo, Dr Fabio (1948). Professor. Inst. of Meteorology, PO Box 17032, Havana, Cuba. Dr (U. Santiago de Cuba, 1985) crystallography. Structure_determination crystallography
E-mail fabio@ceniai.cu Tel. 53(7)603411

Fuentes, Dr Juan (1940). Professor. Fac. of Physics, Havana University, San Lázaro y L. Vedado, Havana, Cuba. Dr (Moscow U., Russia, 1977) solid state physics. Semiconductors piezoelectric_materials
E-mail imre@ceniai.cu Tel. 53(7)78956 Telex (51)1277

Fuentes, Dr Luis (1945). Professor. Inst. of Mathematics and Physics, Acad. of Sciences, 15 e/ C y D, Havana, Cuba. Dr (Havana U., 1982) crystallography. Neutron_diffraction X-ray_diffraction texture
E-mail icimaf@ceniai.cu

Gomes, Mr Ariel (1969). Researcher. National Center for Scientific Research, X-ray Laboratory, PO Box 6990, Havana, Cuba. BSc (Nuclear Institute of Havana, 1992) physics. Powder_diffraction
E-mail infrared@ceniai.cu Tel. 53(7)218066 Fax 53(7)219446

Guardiola, Dr Rene (1955). Assistant Professor. Inst. of Mining and Metallurgy of Moa, Las Coloradas, Moa, Holguin, Cuba. Dr (Natl. Center for Scient. Research - Havana, 1990) crystallography. Structure_determination crystallography
E-mail ismm@ceniai.cu

Herrera, Ms Victoria (1953). Assistant Professor. Center of Appl. Studies of Nuclear Dev., PO Box 6122, Havana, Cuba. MSc (Moscow U., Russia, 1977) solid state physics. Phase_analysis corrosion
E-mail ceaden@ceniai.cu

Infante, Mr Guillermo (1950). Assistant Professor. Center of Research for Mining and Metallurgy, Carr. de Varona Km 3, Havana, Cuba. BSc (Havana U., 1975) chemistry. Powder_diffraction
E-mail cipimm@ceniai.cu

Lopez, Dr Silio (1944). Professor. Centre of Chemical Research, Washington 169 esq. Churruca, Havana, Cuba. Dr (Havana U., 1982) chemistry. Powder_diffraction
E-mail ciq@ceniai.cu

Melian, Ms Maria Antonia (1957). Assistant Professor. Polytechnical Institute Julio A. Mella, Ave. Las Americas s.n., Santiago de Cuba, Cuba. BSc (U. Oriente, 1981) chemistry. Powder_diffraction minerals zeolites
E-mail fquim%ispjam@ceniai.cu Tel. 53(7)22642019 Fax 53(7)338212

Novoa, Mr Hector (1968). Researcher. Chemical Pharmaceutical Center, PO Box 6990, Havana, Cuba. BSc (Havana U., 1991) physics. Crystallography structure_determination
E-mail cpf@ceniai.cu Tel. 53(7)218066 Fax 53(7)219446

Paneque, Mr Armando (1968). Researcher. Chemical Pharmaceutical Center, PO Box 6990, Havana, Cuba. BSc (Havana U., 1991) solid state chemistry. Crystallography structure_determination
E-mail cqf@ceniai.cu Tel. 53(7)218066 Fax 53(7)219446

Perez, Ms Hiram (1958). Assistant Professor. Fac. of Chemistry, University of Havana, San Lázaro y L. Vedado, Havana, Cuba. BSc (Havana U., 1987) solid state chemistry. Powder_diffraction
E-mail imre@ceniai.cu Tel. 53(7)78956

Pomes Hernandez, Dr Ramon (1947). Full Professor. Cuban Sub-Editor of the World Directory of Crystallographers. National Center for Scient. Research, X-ray Laboratory, PO Box 6880, Havana, Cuba. Dr. Sc. (Humboldt U., Germany, 1982) crystallography. Structure_determination powder_diffraction
E-mail xray@ceniai.cu Tel. 53(7)210135 Fax 53(7)336321 Telex (51)1581

Quinones, Dr Jose (1944). Professor. Fac. of Physics, Havana University, Havana, Cuba. Dr (Havana U., 1983) crystallography. Electron_microscopy powder_diffraction
E-mail imre@ceniai.cu Tel. 53(7)78956 Telex (51)2210

Rodriguez, Dr Carlos (1947). Professor. Faculty of Physics, Havana University, San Lázaro y L., Havana, Cuba. Dr (Moscow U., 1980) solid state physics. Solid_state theoretical_physics
E-mail imre@ceniai.cu Tel. 53(7)78956 Telex (51)2210

Serra, Mr Alberto (1943). Assistant Professor. Fac. of Physics, Univ. of Havana, San Lázaro y L., Havana, Cuba. MSc (Havana U., 1980) semiconductor devices. Powder_diffraction texture
E-mail imre@ceniai.cu Tel. 53(7)78956 Telex (51)2210

Tanus, Ms Mercedes (1955). Assistant Professor. Fac. of Physics, Havana University, San Lázaro y L., Havana, Cuba. BSc (Havana U., 1988) solid state physics. Texture powder_diffraction
E-mail imre@ceniai.cu Tel. 53(7)78956 Telex (51)2210

Tintorero, Mr Oscar (1948). Assistant Professor. University of Matanzas, Km 3, Carr a Varadero, Matanzas, Cuba. BSc (Havana U., 1972) solid state physics. Electron_microscopy powder_diffraction
E-mail isam@ceniai.cu

CZECH REPUBLIC

Sub-Editor: M. Čeřňanský

Notes

1. The first standard degree awarded by universities and technical universities (after 5 years study) corresponds to MSc in the British system of degrees. The contraction 'Dr' is used for the title 'Doctor' which is conferred by universities. The abbreviation 'Ing.' indicates the title 'Ingenieur' of persons graduated from technical universities. The higher degree PhD is indicated inside the country by CSc (candidatus scientiarum) - at least 3 years post-graduate study. The highest degree DSc, locally indicated by DrSc (doctor scientiarum) - for outstanding contributions to the development of the scientific field. The abbreviation 'Prof.' is indicated in addition to a title for the highest current position of a university teacher - (full) professor. The abbreviation 'Doc.' is used for the 'Docent' - associate professor.

2. Additional abbreviations used:
Acad. Sci. Cz. Rep. - Academy of Sciences of the Czech Republic
Cz. Tech. U. - Czech Technical University
Tech. U. - Technical University
P. A. - private address

Balcárek, Ing. Jan (1952). Group leader. [Precheza a.s. Central Laboratory, Nábřeží Dr. E. Beneše 24, 750 62 Přerov, Czech Republic.] MSc (Inst. of Chemical Technology, Pardubice, 1977) chemistry. *X-ray_diffractometry X-ray_spectroscopy* Tel. 42(641)252344 Fax 42(641)202028 Telex 066254

Baldrian, Dr Josef (1938). Scientist. [Lab. of X-ray Polymer Structure Analysis, Inst. of Macromolecular Chemistry, Acad. Sci. Cz. Rep., Heyrovského nám. 2, 162 06 Praha 6, Czech Republic.] PhD (Acad. Sci. Cz. Rep., Praha, 1965) physical chemistry. *SAXS wide-angle_scattering polymers structural_change* Tel. 42(2)360341x207 Fax 42(2)367981 Telex 122019

Bořecký, Ing. Karel (1960). Research worker. [Research Institute for Brown Coal, Budovatelu 2830, 434 37 Most, Czech Republic.] MSc (Inst. of Chemical Technology, Praha, 1984) chemistry. *Coal clays qualitative_phase_determination quantitative_phase_determination calibration coalification combustion crystallinity geochemistry mineralogy petrography* Tel. 42(35)0312x4129 Fax 42(35)4580

Brádler, Ing. Jaroslav (1938). Scientist. [Dept. of Metal Physics, Inst. of Physics, Acad. Sci. Cz. Rep., Na Slovance 2, 180 40 Praha 8, Czech Republic.] MSc (Tech. U., Plzeň, 1961) material sciences. *Instrumentation X-ray_topography real_structure single_crystal X-ray_diffractometry* E-mail bradlen@fzu.cz Tel. 42(2)66052608 Fax 42(2)821227·Telex 122018

Březina, Ing. Bohuslav (1928). Scientist. [Dept. of Dielectrics, Inst. of Physics, Acad. Sci. Cz. Rep., Na Slovance 2, 180 40 Praha 8, Czech Republic.] PhD (Inst. of Chemical Technology, Praha, 1956) chemistry. *Crystal_growth ferroelectricity phase_transition molecular_crystal* Tel. 42(2)66052695 Fax 42(2)821227 Telex 122018

Buchal, Dr Antonín (1946). Scientist. [Technical University of Brno, Technická 2, 616 69 Brno, Czech Republic.] PhD (Military Academy, Brno, 1983) physics. *Powder_diffraction real_structure stress strain phase_determination profile_analysis diffractometry software* E-mail buchal@kinf.fme.vutbr.cz Tel. 42(5)41143109 Fax 42(5)41143198

Čapková, Doc. Dr Pavla (1945). Associate professor. [Dept. of Physics of Semiconductors, Faculty of Mathematics and Physics, Charles U., Ke Karlovu 5, 121 16 Praha 2, Czech Republic.] PhD (Charles U., Praha, 1975) physics. *X-ray_diffraction layered_structure intercalates texture thermal_vibration chemical_bonding education* E-mail capkova@ns.karlov.mff.cuni.cz 42(2)24915014x389 Fax 42(2)24911061 Telex 121673

Čejka, Ing. Jiří (1929). Senior research scientist. [Res. Chem. Lab., National Museum, Václavské nám. 68, 115 79 Praha 1, Czech Republic.] PhD (Inst. of Chemical Technology, Praha, 1970) chemistry. *Uranium_compounds minerals thermoanalysis infrared_absorption_spectroscopy* E-mail muzeum@earn.cvut.cz Tel. 42(2)24222550 Fax 42(2)24226488

Čejková, Dr Iva (1963). Research worker. [Res. Chem. Lab., National Museum, Václavské nám. 68, 115 79 Praha 1, Czech Republic.] MSc (Charles U., Praha, 1989) chemistry. *Electron_microscopy bone trace_analysis archeological_materials* E-mail muzeum@earn.cvut.cz Tel. 42(2)24230485x242 Fax 42(2)24226488

Čepera, Dr Milan (1963). Scientist. [Dept. of Material and Technology Development, Military Technical Institute, PO Box 547, Rybkova 2a, 602 00 Brno, Czech Republic.] MSc (Masaryk U., Brno, 1988) physics. *Diffractometry ceramics texture thin_layer X-ray phase_transition ferroelasticity line_broadening lattice_distortion zirconium_compounds nitrides* Tel. 42(5)41183119 Fax 42(5)41211850

Čeřňanský, Ing. Marian (1946). Scientist. Sub-Editor of ninth edition of World Directory of Crystallographers. [Dept. of Metal Physics, Inst. of Physics, Acad. Sci. Cz. Rep., Na Slovance 2, 180 40 Praha 8, Czech Republic.] PhD (Acad. Sci. Cz. Rep., Praha, 1981) physics. *Quantitative_phase_determination line_profile_analysis deconvolution surface instrumentation non-crystalline_materials X-ray_diffraction* E-mail cernan@fzu.cz Tel. 42(2)66052898 Fax 42(2)821227 Telex 122018

Černý, Dr Radovan (1957). Scientist. [Dept. of Physics of Semiconductors, Faculty of Mathematics and Physics, Charles U., Ke Karlovu 5, 121 16 Praha 2, Czech Republic.] PhD (Charles U., Praha, 1990) physics. *Rietveld_method thin_film real_structure crystal_structure powder_diffraction* E-mail cerny@ns.karlov.mff.cuni.cz Tel. 42(2)24915014x455 Fax 42(2)24911061 Telex 121673

Červinka, Dr Ladislav (1935). Group leader. [Dept. of Structures and Bonding, Inst. of Physics, Acad. Sci. Cz. Rep., Cukrovarnická 10, 162 00 Praha 6, Czech Republic.] DSc (Charles U., Praha, 1991) physics. *X-ray_diffraction neutron_diffraction non-crystalline_materials modelling chalcogenide_and_oxide_glasses polymers liquids* E-mail cervinka@fzu.cz 42(2)357714 Fax 42(2)3123184 Telex 122018

Čevelík, Dr Radomír (1961). Group leader. [Central Laboratories - Dept. of Physical Chemistry, Inst. of Rubber Technology and Testing, 764 22 Zlin - Louky, Czech Republic.] MSc (Masaryk U., Brno, 1985) chemistry. *Macromolecules liquid_crystal light_scattering X-ray_diffraction* Tel. 42(67)601x772 Fax 42(67)61754 Telex 067311

Císařová, Dr Ivana (1955). Senior lecturer. [Dept. of Inorganic Chemistry, Faculty of Sciences, Charles U., Hlavova 8/2030, 128 40 Praha 2, Czech Republic.] PhD (Charles U., Praha, 1986) chemistry. *Aperiodic_material modulated_structures structure_determination* E-mail cisarova@prfdec.natur.cuni.cz Tel. 42(2)24915472x2343 Fax 42(2)291958

Dlouhá, Ing. Maja (1938). Scientist. [Dept. of Solid State Engineering, Faculty of Nuclear Sci. and Physical Eng., Cz. Tech. U., Břehová 7, 115 19 Praha l, Czech Republic.] PhD (Cz. Tech. U., Praha, 1984) physics. *Neutron_structure_determination perovskites superconductors zeolites* E-mail dlouha@troja.fjfi.cvut.cz Tel. 42(2)85762410 Fax 42(2)2320861 Telex 121254

Dubský, Ing. Jiří (1939). Scientist. [Dept. of Plasma Physics, Acad. Sci. Cz. Rep., Za Slovankou 3, PO Box 17, 182 11 Praha 8, Czech Republic.] PhD (Cz. Tech. U., Praha, 1981) physical metallurgy. *X-ray_diffraction metallography material_deposition phase_transition surface_coatings ceramics metals plasma_spraying* E-mail dubsky@tokamak.ipp.cas.cz Tel. 42(2)8586340 Fax 42(2)8586389 Telex 122018

Eichler, Dr František (1956). Research worker. [P.A.: Jáchymovská 282, 460 10 Liberec 10, Czech Republic.] MSc (Charles U., Praha, 1980) chemistry. *Silicates high_temperatures stress strain factor_analysis*

Fábry, Dr Jan (1957). Scientist. [Dept. of Structures and Bonding, Inst. of Physics, Acad. Sci. Cz. Rep., Cukrovarnická 10, 162 00 Praha 6, Czech Republic.] PhD (Inst. of Chemical Technology, Pardubice, 1988) chemistry. *Single_crystal X-ray_structure_determination* E-mail fabry@fzu.cz Tel. 42(2)24311137x596 Fax 42(2)3123184

Fiala, Doc. Dr Jaroslav (1940). Associate professor, Scientist. Member of IUCr Commission on Powder Diffraction; Member of JCPDS - International Centre for Diffraction Data; Member of the Regional Committee of Czech and Slovak Crystallographers, IUCr. [Dept. of Metallurgy, Central Res. Inst. Škoda, Tylova 46, 316 00 Plzeň, Czech Republic.] PhD (Slovak Tech. U., Bratislava, 1990) physics. *Quantitative_phase_determination real_structure epitaxy* Tel. 42(19)2154335 Fax 42(19)220762 Telex 154247

Ganev, Ing. Nikolaj (1953). Scientist. [Dept. of Solid State Engineering, Faculty of Nuclear Sci. and Physical Eng., Cz. Tech. U., Břehová 7, 115 19 Praha l, Czech Republic.] PhD (Cz. Tech. U., Praha, 1986) physics. *X-ray_diffraction residual stress polycrystal X-ray_strain_determination* E-mail ganev@troja.fjfi.cvut.cz Tel. 42(2)85762413 Fax 42(2)66445095 Telex 121254

Gosmanová, Dr Galina (1936). Senior lecturer. [Dept. of Solid State Engineering, Faculty of Nuclear Sci. and Physical Eng., Cz. Tech. U., Břehová 7, 115 19 Praha l, Czech Republic.] MSc (Charles U., Praha, 1976) physics. *X-ray_diffraction residual_stress polycrystal X-ray_strain_determination* E-mail gosmanov@troja.fjfi.cvut.cz Tel. 42(2)85762414 Fax 42(2)66445095 Telex 121254

Gruber, Doc. Dr Boris (1921). Associate professor. Dept. of Applied Mathematics, Faculty of Mathematics and Physics, Charles U., Malostranské nám. 25, 118 00 Praha 1, Czech Republic. [P.A.: Sochařská 14, 170 00 Praha 7, Czech Republic.] PhD (Charles U., Praha, 1965) physics. *Bravais_lattice lattice_parameter unit_cell mathematical_crystallography algorithm superlattice* Tel. 42(2)381886

Had, Ing. Jiří (1942). Scientist. [Central Laboratories - Lab. of X-ray Diffractometry, Inst. of Chemical Technology, Technická 5, 166 28 Praha 6, Czech Republic.] PhD (Inst. of Chemical Technology, Pardubice, 1974) chemistry. *X-ray_diffractometry X-ray_qualitative_analysis lattice_parameter corrosion X-ray_quantitative_analysis pigments zeolites stomatological_amalgams*
E-mail hadj@vscht.cz Tel. 42(2)24354141 Fax 42(2)3114769 Telex 122744

Hašek, Dr Jindřich (1945). Group leader. Member of IUCr Commission on Crystallographic Computing, Chairman of the regional Crystallographic Association. [Lab. of X-ray Polymer Structure Analysis, Inst. of Macromolecular Chemistry, Acad. Sci. Cz. Rep., Heyrovského nám. 2, 162 06 Praha 6, Czech Republic.] DSc (Charles U., Praha, 1990) physics. *Structure_determination molecular_recognition computing polymers structure–activity_relation drug cyclic_polyethers photochromism*
E-mail hasekj@imc.cas.cz Tel. 42(2)360341x390 Fax 42(2)367981 Telex 122019

Holý, Doc. Dr Václav (1953). Associate professor. Member of the Regional Committee of Czech and Slovak Crystallographers, IUCr. [Dept. of Solid State Physics, Faculty of Sciences, Masaryk U., Kotlářská 2, 611 37 Brno, Czech Republic.] PhD (Masaryk U., Brno, 1982) physics. *Diffraction_theory diffuse_scattering dynamical_diffraction thin_film X-ray_reflection*
E-mail holy@elanor.sci.muni.cz Tel. 42(5)41129441 Fax 42(5)41211214

Hrdý, Dr Jaromír (1938). Group leader. Member of IUCr Commission on Sychrotron Radiation. [Dept. of Applied Optics, Inst. of Physics, Acad. Sci. Cz. Rep., Na Slovance 2, 180 40 Praha 8, Czech Republic.] DSc (Acad. Sci. Cz. Rep., Praha, 1988) physics. *Monochromator synchrotron_radiation*
E-mail hrdy@fzu.cz Tel. 42(2)66052148 Fax 42(2)8581448 Telex 122018 ATOM C

Huml, Dr Karel (1934). Group leader. Chairman of the European Crystallographic Committee. [Lab. of X-ray Polymer Structure Analysis, Inst. of Macromolecular Chemistry, Acad. Sci. Cz. Rep., Heyrovského nám. 2, 162 06 Praha 6, Czech Republic.] DSc (Cz. Tech. U., Praha, 1984) applied physics. *Drug structure_determination computing modelling*
E-mail huml@imc.cas.cz Tel. 42(2)360341x390 Fax 42(2)367981 Telex 122019

Hušák, Ing. Michal (1967). Research assistant. [Dept. of Solid State Chemistry, Inst. of Chemical Technology, Technická 5, 166 28 Praha 6, Czech Republic.] MSc (Inst. of Chemical Technology, Praha, 1991) chemical technology. *Structure drug software zeolites X-ray_diffraction crystal_structure*
E-mail husakm@vscht.cz Tel. 42(2)3323692 Fax 42(2)3119919 Telex 122744

Hybler, Dr Jiří (1949). Scientist. Secretary of the regional Crystallographic Association. [Dept. of Structures and Bonding, Inst. of Physics, Acad. Sc. Cz. Rep., Cukrovarnická 10, 162 00 Praha 6, Czech Republic.] MSc (Charles U., 1973) mineralogy. *Structure_determination orientation minerals halogens diffraction_data mineralogy phosphates superconductors superstructure twinning sulfides micas*
E-mail hybler@fzu.cz Tel. 42(2)24311137x596 Fax 42(2)3123184 Telex 122018

Janovec, Prof. Dr Václav (1930). Senior scientist. [Inst. of Physics, Acad. Sci. Cz. Rep., Na Slovance 2, 180 40 Praha 8, Czech Republic.] Prof. (Technical U., Liberec, 1994) physics. *Domain_structure twinning phase_transition symmetry_breaking ferroelasticity ferroelectricity modulated_structures*
E-mail janovec@fzu.cz Tel. 42(2)66052144 Fax 42(2)821227 Telex 122018

Kacerovský, Ing. Pavel (1946). Department head. [Lab. of X-ray Diagnostic, Inst. of Radioengineering and Electronics, Acad. Sci. Cz. Rep., Chaberská 57, 182 51 Praha 8, Czech Republic.] PhD (Inst. of Chemical Technology, Praha, 1980) chemistry. *X-ray_diffraction X-ray_characterization single_crystal multilayers material_sciences defect*
Tel. 42(2)66411804 Fax 42(2)66410222

Kameníček, Dr Jiří (1948). Senior lecturer. [Dept. of Inorganic and Physical Chemistry, Faculty of Sciences, Palacký U., Křížkovského 10, 771 47 Olomouc, Czech Republic.] PhD (Inst. of Chemical Technology, Pardubice, 1988) chemistry. *Crystallography diffraction_technique single_crystal nickel_compounds*
E-mail kamen@risc.upol.cz Tel. 42(68)5508256

Karmazin, Dr Lubomír (1938). Scientist. [Dept. of Phase Transformation, Inst. of Physics of Materials, Acad. Sci. Cz. Rep., Žižkova 22, 616 62 Brno, Czech Republic.] PhD (Acad. Sci. Cz. Rep., Praha, 1966) physics. *Alloys phase_transition electron_microscopy powder_diffraction phase_determination material_sciences*
Tel. 42(5)746555x387 Fax 42(5)746378 Telex 62345

Knížek, Dr Karel (1966). Research assistant. [Institute of Physics, Acad. Sci. Cz. Rep., Laboratory of Oxidic Materials, Cukrovarnická 10, 162 00 Praha 6, Czech Republic.] MSc (Charles U., Praha, 1989) chemistry. *Superconductors Rietveld_method phase_determination Jahn_Teller_compounds*
E-mail knizek@fzu.cz Tel. 42(2)24311137x562 Fax 42(2)3123184 Telex 122018

Kolega, Ing. Michal (1964). Research worker. [X-ray Laboratory, Inst. of Technology and Reliability, West Bohemia U., Veleslavínova 11, 301 04 Plzeň, Czech Republic.] MSc (Tech. U., Plzeň, 1987) electroengineering. *Amorphization coating diffraction lattice_distortion line_broadening paracrystal profile_analysis PVD strain thin_film X-ray_fluorescence_spectroscopy nitrides*
Tel. 42(19)36415x6 Fax 42(19)220787

Kopský, Dr Vojtěch (1936). Senior research scientist. Member of IUCr Commission on Crystallographic Nomenclature; Editor of International Tables for Crystallography Vol. E: Subperiodic Groups. [Dept. of Theoretical Physics, Inst. of Physics, Acad. Sci. Cz. Rep., Na Slovance 2, 180 40 Praha 8, Czech Republic.] PhD (Acad. Sci. Cz. Rep., Praha, 1970) theoretical physics. *Group_theory mathematical_crystallography phase_transition structural_change*
E-mail kopsky@fzu.cz Tel. 42(2)66052912 Fax 42(2)821227 Telex 122018

Králová, Dr Rudolfa (1941). Assistant professor. [Dept. of Physics, Faculty of Mechanical Engineering, Cz.Tech.U., Technická 4, 166 07 Praha 6, Czech Republic.] PhD (Cz. Tech. U., Praha, 1984) physics. *Diffraction_technique residual_stress laser_and_other_techniques_of_surface_treatments_of_metallic_materials*
E-mail kralova@fsid.cvut.cz Tel. 42(2)24352423 Fax 42(2)24310292 Telex 121254

Kraus, Prof. Dr Ivo (1936). Professor. Member of the Board of Czech Crystallographic Society. [Dept. of Solid State Engineering, Faculty of Nuclear Sci. and Physical Eng., Cz. Tech. U., Břehová 7, 115 19 Praha l, Czech Republic.] DSc (Cz. Tech. U., Praha, 1989) physics. *Condensed_matter diffraction_technique residual_stress ceramics history_of_physics*
E-mail kraus@troja.fjfi.cvut.cz Tel. 42(2)85762416 Fax 42(2)85762407 Telex 121254

Krausová, Dr Dagmar (1946). Research worker. [Dept. of Inorganic and Physical Chemistry, Faculty of Sciences, Palacký U., Křížkovského 10, 771 47 Olomouc, Czech Republic.] MSc (Palacký U., Olomouc, 1981) chemistry. *X-ray_diffraction phase_determination real_structure diffractometry*
E-mail kraus@risc.upol.cz Tel. 42(68)5508257

Kuběna, Doc. Dr Josef (1935). Associate professor. [Dept. of Solid State Physics, Faculty of Sciences, Masaryk U., Kotlářská 2, 611 37 Brno, Czech Republic.] PhD (Masaryk U., Brno, 1969) physics. *X-ray_topography X-ray_reflectivity*
E-mail kubena@elanor.sci.muni.cz Tel. 42(5)41129441 Fax 42(5)41211214

Kužel, Dr Radomír Jr (1955). Scientist. Secretary of the regional Crystallographic Association. [Dept. of Physics of Semiconductors, Faculty of Mathematics and Physics, Charles U., Ke Karlovu 5, 121 16 Praha 2, Czech Republic.] PhD (Charles U., Praha, 1989) physics. *X-ray_diffraction thin_film real_structure education powder_diffraction software texture material_sciences stress-strain*
E-mail kuzel@karlov.mff.cuni.cz Tel. 42(2)24915014x394 Fax 42(2)24911061 Telex 121673

Loub, Prof. Dr Josef (1929). Professor. [Dept. of Inorganic Chemistry, Faculty of Sciences, Charles U., Hlavova 8/2030, 128 40 Praha 2, Czech Republic.] PhD (Charles U., Praha, 1967) chemistry. *Inorganic_structural_chemistry inorganic_crystal crystal_structure X-ray_diffraction*
E-mail loub@prfdec.natur.cuni.cz Tel. 42(2)24915472x2343 Fax 42(2)291958

Maixner, Dr Jaroslav (1961). Research worker; Head of Laboratory of X-ray Diffractometry. [Central Laboratories - Lab. of X-ray Diffractometry, Inst. of Chemical Technology, Technická 5, 166 28 Praha, Czech Republic.] PhD (Charles U., Praha, 1993) physics. *Rietveld_method single_crystal nucleotides zeolites crystallography metallography minerals powders polymorphism pharmacology*
E-mail maixnerj@vscht.cz Tel. 42(2)24354201 Fax 42(2)3114769 Telex 122744 vsch/c

Mejstříková, Ing. Lubomíra (1957). Research worker. [Research Institute for the Brown Coal, Budovatelu 2830, 434 37 Most, Czech Republic.] MSc (Inst. of Chemical Technology, Praha, 1981) chemistry. *Coal geochemistry phase_determination beryllium_compounds mineralogy*
Tel. 42(35)0312x4129 Fax 42(35)4580

Melka, Dr Karel (1930). Scientist. [X-ray Laboratory, Czech Geological Survey, Klárov 3, 118 21 Praha 1, Czech Republic.] PhD (Charles U., Praha, 1964) mineralogy. *Silicates X-ray_diffraction phase_determination minerals sheet_silicates powder_diffraction*
Tel. 42(2)7980724 Fax 42(2)7980965

Mikula, Dr Pavol (1947). Department head. [Dept. of Neutron Physics, Nuclear Physics Institute, Acad. Sci. Cz. Rep., 250 68 Řež near Prague, Czech Republic.] PhD (Charles U., Praha, 1976) physics. *Neutron_small-angle_scattering neutron_interferometry diffraction_technique residual_stress neutron_optics*
E-mail mikula@ujf.cas.cz Tel. 42(2)66412171x3553 Fax 42(2)6857003 Telex 122626

Nehasil, Dr Miroslav (1921). Senior scientist. [P.A.: Novorosijska 14, 100 00 Praha 10, Czech Republic.] PhD (Tech. U., Praha, 1981) physics. *Diffraction_technique strain strain_deformation thin_layer instrumentation phase_determination metals stress diffractometry material_sciences*
Tel. 42(2)7376203

Ondráček, Ing. Jan (1955). Senior lecturer. [Dept. of Solid State Chemistry, Inst. of Chemical Technology, Technická 5, 166 28 Praha 6, Czech Republic.] PhD (Inst. of Chemical Technology, Praha, 1988) chemistry. *Small_molecules symmetry structure_determination proteins*
E-mail ondracej@vscht.cz Tel. 42(2)24353692 Fax 42(2)3119919 Telex 122744

Ondruš, Ing. Petr (1960). Scientist. [Dept. of Mineralogy - XRD Laboratory, Czech Geological Survey, Geologická 6, 150 00 Praha 5, Czech Republic.] Ing. (Engineer) (Inst. of Chemical Technology, Pardubice, 1986) physical and analytical chemistry. *Powder_diffraction computer software minerals programming*
Tel. 42(2)5816741 Fax 42(2)5818748

Osiková, Dr Jana (1941). Research worker. [Dept. of Mineralogy and Petrography, Faculty of Sciences, Masaryk U., Kotlářská 2, 611 37 Brno, Czech Republic.] MSc (Masaryk U., Brno, 1968) mineralogy. *Minerals zeolites inorganic_crystals silicates*
Tel. 42(5)41129244 Fax 42(5)41211214

Petříček, Dr Václav (1948). Group leader. [Dept. of Structures and Bonding, Inst. of Physics, Acad. Sci. Cz. Rep., Cukrovarnická 10, 162 00 Praha 6, Czech Republic.] PhD (Acad. Sci. Cz. Rep., Praha, 1981) physics. *Crystal_structure_determination computing modulated_structures aperiodic_material*
E-mail petricek@fzu.cz Tel. 42(2)24311137x598 Fax 42(2)3123184 Telex 122018

Pleštil, Ing. Josef (1946). Group leader. [Lab. of X-ray Polymer Structure Analysis, Inst. of Macromolecular Chemistry, Acad. Sci. Cz. Rep., Heyrovského nám. 2, 162 06 Praha 6, Czech Republic.] PhD (Charles U., Praha, 1974) physics. *Neutron_small-angle_scattering SAXS polymers micelles*
E-mail plestil@imc.cas.cz Tel. 42(2)360341x388 Fax 42(2)367981 Telex 122019

Podlahová, Doc. Dr Jana (1937). Associate professor. [Dept. of Inorganic Chemistry, Faculty of Sciences, Charles U., Hlavova 8/2030, 128 40 Praha 2, Czech Republic.] (Charles U., Praha, 1964) chemistry. *Structural_chemistry X-ray_diffraction coordination_compounds crystal_structure education database molecular_structure chemical_bonding preparation_chemistry*
Tel. 42(2)24915472x2366 Fax 42(2)291958

Polcarová, Dr Milena (1931). Group leader. [Dept. of Metal Physics, Inst. of Physics, Acad. Sci. Cz. Rep., Na Slovance 2, 180 40 Praha 8, Czech Republic.] PhD (Acad. Sci. Cz. Rep., Praha, 1961) physics. *X-ray_topography X-ray_diffractometry real_structure single_crystal*
E-mail polcar@fzu.cz Tel. 42(2)66052608 Fax 42(2)821227 Telex 122018

Pučálka, Dr Vladimír (1929). Senior scientist. [P.A.: Korunní 14, 709 00 Ostrava 1, Czech Republic.] MSc (Masaryk U., Brno, 1952) physics. *Material_sciences thin_films X-ray_diffraction crystal_structure vacuum_coating electrical_conductivity*

Renner, Ing. Oldřich (1944). Group leader. [Dept. of Gas Laser Physics, Inst. of Physics, Acad. Sci. Cz. Rep., Cukrovarnická 10, 162 00 Praha 6, Czech Republic.] PhD (Acad. Sci. Cz. Rep., Praha, 1972) physics. *Diffuse_scattering thin_film surface_structure instrumentation X-ray_diffraction scattering_theory synchrotron_radiation*
E-mail renner@fzu.cz Tel. 42(2)24311137x429 Fax 42(2)3123184 Telex 122018

Rieder, Prof. Dr Milan (1940). Professor. [Inst. of Geochemistry, Mineralogy and Mineral Resources, Charles U., Albertov 6, 128 43 Praha 2, Czech Republic.] PhD (Johns Hopkins U., Baltimore, MD, USA, 1968) geology. *Diffraction_technique phase_equilibrium epitaxy minerals*
E-mail rieden@prfdec.natur.cuni.cz Tel. 42(2)24915472x2409 Fax 42(2)296084

Schneider, Bohdan (1957). Scientist. [Heyrovsky Institute of Physical Chemistry, Acad. Sci. Cz. Rep., Dolejškova 3, 182 23 Praha, Czech Republic.] PhD (Acad. Sci. Cz. Rep, Praha, 1989) chemistry. *Biomolecule computer_modelling intermolecular_recognition*
E-mail schneider@jh-inst.cas.cz Tel. 42(2)66053776 Fax 42(2)8582307

Síchová, Dr Hana (1931). Senior lecturer. [Dept. of Physics of Semiconductors, Faculty of Mathematics and Physics, Charles U., Ke Karlovu 5, 121 16 Praha 2, Czech Republic.] PhD (Charles U., Praha, 1965) physics. *Education superconductors real_structure X-ray_diffraction powder_diffraction diffractometry texture phase_determination*
E-mail sichova@ns.karlov.mff.cuni.cz Tel. 42(2)24915014x394 Fax 42(2)24911061 Telex 121673

Skála, Dr Roman (1967). Research worker. [Dept. of Mineralogy and Petrography, National Museum, Václavské nám. 68, 115 79 Praha 1, Czech Republic.] MSc (Charles U., Praha, 1990) mineralogy. *Geosciences shock_metamorphism meteorites minerals*
E-mail muzeum@earn.cvut.cz Tel. 42(2)24230485x243 Fax 42(2)24226488

Sopková, Dr Jana (1967). Research assistant. [Dept. of Chemical Physics, Faculty of Mathematics and Physics, Charles U., Ke Karlovu 3, 121 16 Praha 2, Czech Republic.] MSc (Charles U., Praha, 1990) physics. *Proteins crystallography structure modelling structural_biology biopolymers*
Tel. 42(2)24915014x267 Fax 42(2)299272 Telex 121673

Šourek, Dr Zbyněk (1948). Group leader. [Dept. of Structures and Bonding, Inst. of Physics, Acad. Sci. Cz. Rep., Cukrovarnická 10, 162 00 Praha 6, Czech Republic.] PhD (Charles U., Praha, 1980) physics. *X-ray_diffraction X-ray_topography real_structure high-resolution_diffractometry*
E-mail sourek@fzu.cz Tel. 42(2)24311137x587 Fax 42(2)3123184 Telex 122018

Steinhart, Dr Miloš (1955). Senior lecturer. [Dept. of Physics, Inst. of Chemical Technology, Nám. Čs. legií 565, 532 10 Pardubice, Czech Republic.] PhD (Charles U., Praha, 1989) physics. *Biomolecule pressure SAXS synchrotron_radiation*
E-mail stein@nw1.upce.cz Tel. 42(40)47461x543 Fax 42(40)48400

Studnička, Dr Václav (1951). Research worker. [Dept. of Metal Physics, Inst. of Physics, Acad. Sci. Cz. Rep., Na Slovance 2, 180 40 Praha 8, Czech Republic.] MSc (Charles U., Praha, 1979) physics. *X-ray_diffractometry X-ray_diffraction X-ray_diffraction_technique X-ray_diffraction_data*
E-mail studnic@fzu.cz Tel. 42(2)66052612 Fax 42(2)821227 Telex 122018

Šubrtová, Ing. Věra (1936). Scientist. [Dept. of Structures and Bonding, Inst. of Physics, Acad. Sci. Cz. Rep., Cukrovarnická 10, 162 00 Praha 6, Czech Republic.] MSc (Inst. of Mining and Metallurgy, Moscow, Russia, 1959) geology. *X-ray_diffraction chemical_bonding structural_chemistry molecular_crystal*
Tel. 42(2)24311137x596 Fax 42(2)3123184 Telex 122018

Valvoda, Prof. Dr Václav (1937). Group leader, Vice-dean. Member of the Regional Committee of Czech and Slovak Crystallographers, IUCr. [Dept. of Physics of Semiconductors, Faculty of Mathematics and Physics, Charles U., Ke Karlovu 5, 121 16 Praha 2, Czech Republic.] PhD (Charles U., Praha, 1968) physics. *Thick_film microstructure ion_implantation texture*
E-mail valvoda@karlov.mff.cuni.cz Tel. 42(2)24915014x395 Fax 42(2)24911061 Telex 121673

Vojtěchovský, Dr Jaroslav (1964). Scientist. [Lab. of X-ray Polymer Structure Analysis, Inst. of Macromolecular Chemistry, Acad. Sci. Cz. Rep., Heyrovského nám. 2, 162 06 Praha 6, Czech Republic.] MSc (Charles U., Praha, 1987) physics. *Material_sciences X-ray_diffraction molecular_structure molecular_crystal diffractometry database polymers*
E-mail vojtech@imc.cas.cz Tel. 42(2)360341x390 Fax 42(2)367981 Telex 122019

Vratislav, Ass. Prof. Ing. Stanislav (1939). Department head. [Dept. of Solid State Engineering, Faculty of Nuclear Sci. and Physical Eng., Cz. Tech. U., Břehová 7, 115 19 Praha 1, Czech Republic.] PhD (Cz. Tech. U., Praha, 1978) physics. *Neutron_structure_determination perovskites superconductors zeolites diffraction_methods texture_analysis material_research*
E-mail vratislav@troja.fjfi.cvut.cz Tel. 42(2)85762380 Fax 42(2)2320861 42(2)85762407 Telex 121254 fjfi

Weiss, Dr Zdeněk (1942). Associate Professor, Director of Central Analytical Laboratory. [Central Analytical Laboratory, Technical University Ostrava, 17. listopadu 1, 708 33 Ostrava - Poruba, Czech Republic.] DSc (Masaryk U., Brno, 1990) mineralogy and crystallography. *Minerals polytypism disorder phase_determination X-ray_diffraction silicates crystal_structure software*
E-mail zdenek.weiss@vsb.cz Tel. 42(69)441079 Fax 42(69)6918647

Žák, Doc. Dr Zdirad (1941). Associate professor. Vicechairman, Regional Committee of Czech and Slovak Crystallographers, IUCr. [Dept. of Inorganic Chemistry, Faculty of Sciences, Masaryk U., Kotlářská 2, 611 37 Brno, Czech Republic.] PhD (Masaryk U., Brno, 1982) chemistry. *Chemical_bonding IV–VI_compounds low_temperature rare-earth_compounds nitrogen_compounds phosphorus_compounds sulfur_compounds main-group_compounds diffractometry crystallography*
E-mail zak@csbrmu11.bitnet Tel. 42(5)41129323 Fax 42(5)41211214

Zeman, Ing. Jiří (1940). Group leader. [Dept. of Material and Technology Development, Military Technical Institute, PO Box 547, Rybkova 2a, 602 00 Brno, Czech Republic.] DSc (Tech. U., Brno, 1985) material sciences. *Texture material_sciences ceramics thin_film strain stress high_strength_steel residual_stress*
Tel. 42(5)41183122 Fax 42(5)41211850

Zikmund, Dr Zdeněk (1939). Scientist. [Dept. of Dielectrics, Inst. of Physics, Acad. Sci. Cz. Rep., Na Slovance 2, 180 40 Praha 8, Czech Republic.] PhD (Acad. Sci. Cz. Rep., Praha, 1977) physics. *Domain_structure group_theory symmetry symmetry_breaking phase_transitions ferroelectrics ferroelastics crystal_structures*
E-mail zikmund@fzu.cz Tel. 42(2)66052677 Fax 42(2)821227 Telex 122018

DENMARK

Sub-Editor: R. K. Feidenhans'l

Andersen, Gregers Rom (1967). PhD student. [Dept. of Chemistry, Aarhus University, DK-8000 Aarhus C., Denmark.] MSc (Aarhus U., 1992) protein crystallography. *Proteins crystallization structure_determination*
E-mail gregers@kaktus.kemi.aau.dk Tel. 45()89423333x3864 Fax 45()86196199

Andersen, Niels Hessel (1945). Senior Scientist. [Department of Solid State Physics, Risø National Laboratory, PO Box 49, DK-4000 Roskilde, Denmark.] PhD (U. Copenhagen, 1976) solid state physics. *Superconductors*
E-mail fys-nian@risoe.dk Tel. 45()46774711 Fax 45()42370115 Telex 43116

Andersen, Dr Peter (1938). [Department of Inorganic Chemistry, Chemical Institute, University of Copenhagen, Universitetsparken 5, DK-2100 Copenhagen Ø, Denmark.] *Co-ordination_chemistry*
Tel. 45()35320113 Fax 45()35320133

Balić Žunić, Tonči (1952). Researcher. [Haldor Topsøe A/S, Nymøllevej 55, 2800 Lyngby, Denmark.] DSc (Zagreb U., 1984) geology. *Crystal_structure powder catalysis cultural_history arts philosophy*
E-mail tbz@htas.dk Tel. 45()45272115 Fax 45()45272999 Telex 37 444 htas dk

Berg, Rolf W. (1945). Ass. Professor. [Chemistry Department A., Technical University of Denmark, DK-2800 Lyngby, Denmark.] PhD (Technical U. Denmark, 1972) spectrochemistry. *Solid_state inorganic_chemistry phase_transitions spectroscopy*
E-mail kearolf@urm.uni-c.dk klarwb@k-databar.dth.dk Tel. 45()45931222x2316 Fax 45()42883136

Bohr, Jakob (1957). Research Professor. [Department of Solid State Physics, Risø National Laboratory, DK-4000 Roskilde, Denmark.] Dr. Scient. (U. Copenhagen, 1991) science. *Critical_phenomena diffraction fractal magnetism melting molecular_beam Patterson_method synchrotron_radiation*
E-mail bohr@risoe.dk Tel. 45()46774745 Fax 45()42370115

Brehm, Lotte (1940). Assoc. Prof. [Department of Medicinal Chemistry, Royal Danish School of Pharmacy, 2 Universitetsparken, DK-2100 Copenhagen, Denmark.] PhD (Department of Medicinal Chemistry, Royal Danish School of Pharmacy, 1969) organic chemistry, crystallography. *Organic_molecules biological_molecules medicinal_chemistry*
Tel. 45()35370850 Fax 45()35372209

Buchwald, Vagn Fabritius (1929). Associate Professor. [Department of Metallurgy, Technical University of Denmark, Building 204, DK-2800 Lyngby, Denmark.] Dr.

Scient (Copenhagen U., 1977) (iron meteorites) metallurgy. *Archaeometallurgy bronze_iron_and_slags_from_ancient_production steels*
Tel. 45( )42884022 Fax 45( )936213

Christensen, Dr Axel Nørlund (1934). Associate professor. [Department of Chemistry, Aarhus University, DK-8000 Aarhus C, Denmark.] DPhil (Aarhus U., 1967) inorganic chemistry. *Crystal_growth crystal_structures powder_diffraction synchrotron_radiation*
E-mail anc@kemi.aau.dk Tel. 45( )89423894 Fax 45( )86196199

Christensen, Finn Erland (1954). Physicist. [Danish Space Research Institute, Gl. Lundtoftevej 7, DK-2800 Lyngby, Denmark.] PhD (Technical U., 1982) solid state physics, liquid crystals. *High-resolution_diffractometry synchrotron_radiation astrophysics diffuse_scattering surfaces multilayers diffraction_techniques reflectivity STM Langmuir–Blodgett_film liquid_crystals*
E-mail finn@dsri.dk Tel. 45( )874077x127 Fax 45( )45930283 Telex 37198 dahnru dk

Clausen, Kurt Nørgaard (1952). Section Leader. [Risø National Laboratory, DK-4000 Roskilde, Denmark.] PhD (DTH, 1981) physics. *Condensed_matter cooperative_phenomena magnetism diffraction inelastic_scattering*
E-mail clausen@risoe.dk Tel. 45( )46774709 Fax 45( )42370115

Danielsen, Jacob (1937). Associate professor. Member of Commission on Crystallographic Data. [Department of Inorganic Chemistry, Aarhus University, DK-8000 Aarhus C, Denmark.] DSc (Aarhus U., 1961) inorganic chemistry. *Computing crystal_growth*
E-mail jda@kemi.aau.dk Tel. 45( )89423884 Fax 45( )86196199 Telex 64767 aausci dk

Feidenhans'l, Robert Krarup (1958). Senior scientist. Member of the Danish National Committee for Crystallography. [Department of Solid State Physics, Risø National Laboratory, PO Box 49, DK 4000, Roskilde, Denmark.] lic.scient. (U. Aarhus, 1986) physics. *X-ray_scattering surface thin_film structure synchrotron_radiation superlattice*
E-mail fys-rofe@risoe.dk Tel. 45( )46774708 Fax 45( )42370115 Telex 43116

Flensburg, Claus (1966). Student. Department of Chemistry, Chemical Laboratory IV, University of Copenhagen, Copenhagen, Denmark. [H.C. Ørsted Institute, Universitetsparken 5, DK-2100, Copenhagen Ø, Denmark.] BSc (U. Copenhagen, 1991) chemistry. *Small_molecules charge_density computing short_hydrogen_bonds synchrotron_radiation*
E-mail claus@laue.ki.ku.dk Tel. 45( )35320294 Fax 45( )35320299 Telex Fonotelex

Frydenvang, Karla (1958). Postdoctoral Research Assistant. [Department of Medicinal Chemistry, Royal Danish School of Pharmacy, 2, Universitetsparken, DK-2100 Copenhagen, Denmark.] PhD (Department of Medicinal Chemistry, Royal Danish School of Pharmacy, 1989) organic chemistry, crystallography and molecular modelling. *Molecular_modelling molecular_interaction biological_molecules proteins medicinal_chemistry*
E-mail karla@medchem.dfh.dk Tel. 45( )35370850 Fax 45( )35372209

Gajhede, Michael (1954). Associate Professor. [Chemical Institute, Copenhagen University, Universitetsparken 5, DK-2100 Copenhagen, Denmark.] PhD (Copenhagen U., 1987) chemistry. *Macromolecular_crystallography*
E-mail michael@bragg.ki.ku.dk Tel. 45( )35320280 Fax 45( )35320299

Gerstenberg, Michael Christian (1967). PhD Student. [Risø National Laboratory, Department of Solid State Physics, DK-4000 Roskilde, Denmark.] Master of Science in Engineering (Technical U. Denmark, 1993) physics. *Magnetism superlattice polymers neutron_scattering X-ray_scattering reflectivity rare_earth_superlattice*
E-mail fys-mig@risoe.dk Tel. 45( )46774741 Fax 45( )42370115

Gerward, Dr Leif (1939). Associate professor. Member of the Danish National Committee for Crystallography. Physics Department, Technical University of Denmark, Lyngby, Denmark. [Physics Department, Building 307, Technical University of Denmark, DK-2800 Lyngby, Denmark.] DSc (Chalmers Tech. U., Gothenburg, Sweden, 1974) physics. *High_pressure X-ray_diffraction X-ray_attenuation*
E-mail gerward@fysik.dtu.dk Tel. 45( )42882488x3146 Fax 45( )45932399

Gråbæk, Lars Friis (1958). Research Assistant. [Haldor Topsøe, Research Laboratories, Nymøllevej 55, DK-2800 Lyngby, Denmark.] PhD (U. Denmark, 1990) physics. *X-ray_powder_diffraction phase_identification catalyst_structure size_distribution in_situ_X-ray_powder_diffraction*
E-mail TOPEOT@vm.uni-c.dk Tel. 45( )45272115 Fax 45( )45272999 Telex 37444 htas dk

Grundvig, Sidsel (1941). Associate Professor. [Department of Geology, Aarhus University, C.F. Møllers alle, DK-8000 Aarhus C, Denmark.] PhD (Royal Danish School of Pharmacy, 1967) crystallography. *Microbeam_analysis X-ray_fluorescence*
E-mail sidsel@geo.aau.dk Tel. 45( )89422510 Fax 45( )86139248

Hazell, Alan Charles (1935). Lektor. [Department of Inorganic Chemistry, Aarhus University, DK-8000 Aarhus C, Denmark.] PhD (U. Leeds, 1961) structural and inorganic chemistry. *Structural_determination bioinorganic_chemistry*
E-mail ach@kemi.aau.dk Tel. 45( )89423891 Fax 45( )86196199 Telex 64767 aausci dk

Hjorth, Michael (1962). Assistant Professor. [Chemistry Department B, Technical University of Denmark, DTU 207, DK-2800 Lyngby, Denmark.] PhD (Tech. U. Denmark, 1990) chemical crystallography. *Crystallography charge_density high-resolution_electron_microscopy computing*
E-mail kelbnt@vms2.uni-c.dk Tel. 45( )42882222 until February 1995; then 45( )45252525 45( )42882239 until February 1995; then 45( )45881639 Telex 37529 DTHDIA DK

Honoré, Tage (1951). Vice President. [Novo Nordisk A/S, Pharmaceutical Division, Drug Discovery, Novo Nordisk Park, DK-2760 Måløv, Denmark.] DSc (Royal Danish School of Pharmacy, 1993) neurochemistry.
Tel. 45( )44668888 Fax 45( )44663939

Jensen, Birthe (1938). Lecturer. [Department of Medicinal Chemistry, Royal Danish School of Pharmacy, 2, Universitetsparken, DK-2100 Copenhagen, Denmark.] DSc (Department of Medicinal Chemistry, Royal Danish School of Pharmacy, 1984) organic chemistry. *Organic_structures biological_structures interatomic_forces molecular_modelling medicinal_chemistry*
Tel. 45( )35370850 Fax 45( )35372209

Jensen, Dorte Juul (1957). Senior Scientist. [Materials Department, Risø National Laboratory, DK-4000 Roskilde, Denmark.] PhD (Technical U. Denmark, 1983) materials science. *Recrystallisation deformation texture neutron_diffraction*
E-mail dorte.juul.jensen@risoe.dk Tel. 45( )42371212x5728 Fax 45( )42351173

Jensen, Stig Jorgo (1932). Ext. Lecturer in Dental Materials Science. [Forsbakken 15, 8520 Lystrup, Denmark.] Dr. Phil (U. Aarhus, 1969) chemistry. *Dental_materials*
Tel. 45( )86221252

Johnsen, Ole (1940). Ass. Professor. [Geological Museum, University of Copenhagen, Øster Voldgade 5-7, DK-1350 Copenhagen, Denmark.] Cand. Scient (U. Copenhagen, 1971) mineralogy. *Silicates*
Tel. 45( )35322337 Fax 45( )35322325

Jørgensen, Jens (1955). Lecturer. [Department of Inorganic Chemistry, Aarhus Universitet, DK-8000 Aarhus C, Denmark.] PhD (Aarhus U., 1984) chemistry. *Solid_state_chemistry X-ray_diffraction neutron_diffraction*
E-mail jenserik@kemi.aau.dk Tel. 45( )89423888 Fax 45( )86196199 Telex 64767 aausci dk

Kadziola, Anders (1961). Postdoctoral Research Assistant. [Chemical Laboratory IV, H.C. Ørsted Institute, Universitetsparken 5, DK-2100 Copenhagen, Denmark.] PhD (U. Copenhagen, 1993) macromolecular crystallography. *Biocrystallography structure_determination*
E-mail anders@xray.ki.ku.dk Tel. 45( )35320282 Fax 45( )35320199

Kalsbeek, Nicoline (1961). Lecturer. [The Royal Danish Academy of Fine Arts, School of Conservation, Esplanaden 34, DK-1263 Copenhagen K., Denmark.] PhD. Cand. Scient (1993) chemistry, geology. *Short_hydrogen_bonds mineral_structures NMR_spectroscopy*
E-mail kosknico@inet.uni-c.dk Tel. 45( )33126860 Fax 45( )33320801

Kastrup, Jette Sandholm Jensen (1961). Post doc. [Department of Medicinal Chemistry, Royal Danish School of Pharmacy, 2, Universitetsparken, DK-2100 Copenhagen, Denmark.] PhD (Department of Medicinal Chemistry, Royal Danish School of Pharmacy, 1988) molecular modelling and qsar. *X-ray_crystallography proteins*
E-mail kastrup@medchem.dfh.dk Tel. 45( )35370850 Fax 45( )35372209

Kjær, Kristian (1955). Senior Scientist. [Physics Department, Risø National Laboratory, DK-4000 Roskilde, Denmark.] PhD (Danmarks Tekniske Hoejskole (Technical U. Denmark), 1984) physics. *Surface Langmuir_monolayer X-ray_reflectivity neutron_reflectivity X-ray_scattering*
E-mail kkjaer@risoe.dk Tel. 45( )46774709 Fax 45( )42370115 Telex 43116

Kjeldgaard, Morten (1953). Associate Professor. [Dept. of Chemistry, Aarhus University, DK-8000 Aarhus C., Denmark.] PhD (Aarhus University, 1990) protein crystallography. *Proteins nucleic_acids graphics*
E-mail morten@oase.kemi.aau.dk Tel. 45( )89423333x3877 Fax 45( )86196199

Larsen, Finn Krebs (1941). Associate professor. [Department of Inorganic Chemistry, Aarhus University, DK-8000 Aarhus C, Denmark.] Cand. Scient. (MSc) (Aarhus U., 1966) X-ray and neutron crystallography. *Structure diffraction low_temperature_crystallography charge_density modulated_structures synchrotron_radiation*
E-mail kre@kemi.aau.dk Tel. 45( )89423897 Fax 45( )86196199 Telex 64767 aausci dk

Larsen, Ingrid Kjoeller (1935). Assoc. Prof. [Department of Organic Chemistry, Royal Danish School of Pharmacy, 2, Universitetsparken, DK-2100 Copenhagen, Denmark.] PhD (Department of Organic Chemistry, Royal Danish School of Pharmacy, 1965) organic chemistry, crystallography. *Proteins molecular_interaction medicinal_chemistry*
E-mail KJOLLER@dfhvax.nbi.dk Tel. 45( )35370850 Fax 45( )35375744

Larsen, Sine (1943). Associate Professor. Member of the Danish National Committee for Crystallography. [Center for Crystallographic Studies, Department of Chemistry, University of Copenhagen, Universitetsparken 5, DK-2100 Copenhagen, Denmark.] Cand. Scient. (U. Copenhagen, 1968) physical chemistry. *Relationships_structure_properties chiral_compounds proteins*
E-mail sine@xray.ki.ku.dk Tel. 45( )35320282 Fax 45( )35320299

Lebech, Bente (1937). Senior scientist. Member of the Neutron Diffraction Commission. [Department of Solid State Physics, Risø National Laboratory, PO Box 49, DK 4000, Roskilde, Denmark.] Cand. Polyt. (Technical U. Denmark, 1962) physics. *Neutron_scattering magnetic_structures phase_transitions quasi_crystals modulated_structures high_Tc_superconductors*
E-mail lebech@risoe.dk Tel. 45( )46774705 Fax 45( )42370115 Telex 43116

Leonardsen, Erik (1934). Lecturer. [Dep. of Mineralogy, University of Copenhagen, Østervoldgade 10, DK - 1350, Denmark.] Cand. real (U. Oslo, 1961) inorganic chemistry. *Minerals structure computing powder_methods*
E-mail eleon@ujarak.geomus.ku.dk Tel. 45( )35322429 Fax 45( )35322499

Lindegaard-Andersen, Asger (1925). Professor. Physics Department, Technical University of Denmark, Lyngby, Denmark. [Physics Department, Building 307, Technical University of Denmark, DK-2800 Lyngby, Denmark, 1967) crystal growth, crystal defects. *X-ray_topography crystal_growth scale_mechanism*
E-mail ala@fysik.dth.dk Tel. 45( )42882488x3148 Fax 45( )45932399 Telex 37529 dthdia dk

Lindgreen, Holger (1946). Senior Research assistant. Geological Survey Denmark, Toravej 8, DK-2400, Copenhagen NV, Denmark. [Esrum Park 8, Esrum, DK-3230 Graested, Denmark.] DSc (U. Copenhagen, 1991) clay mineralogy. *Clays burial_diagenesis structure_of_layered_silicates STM AFM X-ray_diffraction Mössbauer_spectroscopy DTA EGA*
Tel. 45()42290724 Fax 45()31196868

Lopez de Diego, Heidi (1964). PhD Student. [Department of Chemistry Laboratory 4, University of Copenhagen, H.C. Ørsted Institute, Universitetsparken 5, DK-2100 Copenhagen Ø, Denmark.] Cand. Scient (U. Copenhagen, 1991) chemistry, crystallography. *Chirality small_molecules hydrogen_bonding thermoanalysis isomers calorimetry*
E-mail heidi@laue.ki.ku.dk Tel. 45()35320294 Fax 45()35320299 Telex Fonotelex

Lorentzen, Torben (1959). Senior Scientist. [Materials Department, Risø National Laboratory, PO Box 49, DK-4000 Roskilde, Denmark.] PhD (U. Aalborg, 1990) materials science. *Neutron_diffraction residual_stress metal_matrix_composites*
E-mail torben.lorentzen@risoe.dk Tel. 45()46775805 Fax 45()42351173 Telex 43116

Makovicky, Emil (1940). Assoc. Prof. [University of Copenhagen, Dept of Mineralogy, Østervoldgade 10, DK - 1350, Denmark.] PhD (Mc Gill U., 1970) mineralogy. *Crystal_chemistry sulphides sulphosalts minerals symmetry*
Tel. 45()35322432 Fax 45()35322499

Marthi, Katalin (1968). PhD Student. [Centre for Crystallographic Studies, University of Copenhagen, Universitetsparken 5, 2100 Copenhagen, Denmark.] MSc (Technical U. Budapest, 1992) organic and biological industrial chemistry. *Chirality phase_diagram drug thermoanalysis resolution_of_racemic_compounds enantiomeric_and_diastereomeric_mixtures*
E-mail kati@xray.ki.ku.dk Tel. 45()35320279 Fax 45()35320299

McMorrow, Desmond Francis (1961). Senior Scientist. [Department of Solid State Physics, Risø National Laboratory, DK-4000 Roskilde, Denmark.] PhD (U. Manchester, 1987) NMR spectroscopy of rare-earth compounds. *X-ray_scattering neutron_scattering phase_transition*
E-mail momorrow@fys-hp-1.risoe.dk Tel. 45()46774723 Fax 45()42370115

Mortensen, Kell (1952). Senior Research. [Dept. Solid State Physics, Risø National Laboratory, DK-4000 Roskilde, Denmark.] PhD (Techn. U. Denmark, 1981) physics. *Polymers small-angle_scattering superconductors*
E-mail mortensen@risoe.dk Tel. 45()46774710 Fax 45()42370115

Nielsen, Anders (1919). Director of Research and Development. [Haldor Topsøe A/S, Nymøllevej 55, DK-2800 Lyngby, Denmark.] Dr. Techn. (Danish Technical University, 1951) ammonia catalysis. *Catalysis reactivity*
Tel. 45()45272000 Fax 45()45272999 Telex 37444 HTAS DK

Nielsen, Kurt (1943). Associate Professor. [Chemistry Department B, Technical University of Denmark, DTU 207, DK-2800 Lyngby, Denmark.] MSc (U. Aarhus, 1969) crystallography. *Structural_chemistry powder_diffraction structure-property_relations computing*
Tel. 45()42882222 Fax 45()42883136

Nissen, Poul (1967). PhD student. [Dept. of Chemistry, Aarhus University, DK-8000 Aarhus C., Denmark.] Cand. Scient. (Aarhus U., 1993) protein crystallography. *Proteins nucleic_acids crystallization*
E-mail nissen@kemi.aau.dk Tel. 45()89423333x3871 Fax 45()86196199

Nørskov-Lauritsen, Dr Leif (1953). Research scientist. [Biostructure Dept., Novo Nordisk A/S, Novo Allé, DK-2880 Bagsværd, Denmark.] PhD (U. Copenhagen, 1984) physical organic chemistry. *Structure_design computer_modelling*
E-mail lnl@novo.dk Tel. 45()44448888x2834 Fax 45()44490555 Telex 37173

Nyborg, Jens (1942). Associate Professor. [Dept. of Chemistry, Aarhus University, DK-8000 Aarhus C., Denmark.] Fil.dr (Gothenburg U., 1971) protein crystallography. *Proteins nucleic_acids direct_methods*
E-mail jnb@kemi.aau.dk Tel. 45()89423333x3866 Fax 45()86196199

Pedersen, Jan Skov (1959). Senior Scientist. Co-Editor of Journal of Applied Crystallography. [Department of Solid State Physics, Risø National Laboratory, DK-4000 Roskilde, Denmark.] PhD (U. Copenhagen, 1988) solid state physics/surface crystallography. *Small-angle_scattering structural_biology biophysics polymers materials_science colloids reflectivity*
E-mail skov@risoe.dk Tel. 45()46774718 Fax 45()42370115 Telex 43116

Petersen, Jens Frølund Winthe'r (1966). PhD Student. [Center for Crystallographic Studies, Department of Chemistry, University of Copenhagen, Universitetsparken 5, DK 2100 Kbh. Ø., Denmark.] Cand. Scient (U. Copenhagen, 1991) protein chemistry. *Biocrystallography proteins crystallization cryogenics molecular_replacement*
E-mail Jens@xray.ki.kv.dk Tel. 45()35320283 Fax 45()35320290

Petersen, Ole V (1939). Curator. [Geologisk Museum, Øster Voldgade 5-7, DK-1350 Koebenhavn K., Denmark.] Lic. Scient (Koebenhavns U., 1970) mineralogy. *Mineralogy*
Tel. 45()35322338 Fax 45()35322325

Polekhina, Galina (1967). PhD student. [Dept. of Chemistry, Aarhus University, DK-8000 Aarhus C., Denmark.] MSc (Moscow U., 1989) applied mathematics. *Proteins crystallization structure_determination*
E-mail galina@kemi.aau.dk Tel. 45()89423333x3871 Fax 45()86196199

Ranløv, Jens (1964). Doctorate candidate. [Materials Department, Risø National Laboratory, Frederiksborgvej 399, DK-4000 Roskilde, Denmark.] MSc (U. Copenhagen, 1991) mineralogy. *Structural_chemistry perovskites ionic_conductors*
E-mail afm-jran@risoe.dk Tel. 45()46774677 Fax 45()42882239 Telex 43116 RISØ DK

Rasmussen, Hanne (1961). Postdoc. [Department of Medicinal Chemistry, Royal Danish School of Pharmacy, 2, Universitetsparken, DK-2100 Copenhagen, Denmark.] PhD (Department of Chemistry, U. Aarhus, 1988) protein crystallography. *Protein_crystallography*
E-mail hanne@medchem.dfh.dk Tel. 45()35370850 Fax 45()35372209

Rasmussen, Svend Erik (1925). Professor of inorganic chemistry. [Department of Inorganic Chemistry, Aarhus University, DK-8000 Aarhus C, Denmark.] Dr.phil. (U. Copenhagen, 1960) crystal structure determination. *Instrumentation powder_diffraction oxides*
E-mail ser@kemi.aau.dk Tel. 45()89423882 Fax 45()86196199 Telex 64767 aausci dk

Schmidt Nielsen, Søren (1942). Teacher (Gymnasium). Counsellor. [Sct. Knuds Gymnasium, Læssøegade 154, DK-5000 Odense, Denmark.] Cand. Scient. (1971) small-angle scattering, crystallography.
Tel. 45()66123120

Simonsen, Ole (1937). Assoc. prof. [Dept. of Chemistry, Univ of. Odense, Campusvej 55, DK-5230 Odense M, Denmark.] Cand. Scient. (Copenhagen U., 1966) organic chemistry. *Structure_determination nitrogen_compounds sulfur_compounds*
E-mail osi@oukemi.ou.dk Tel. 45()66158696x2533 Fax 45()66158780

Sommer-Larsen, Peter (1958). Lektor. [The Engineering Academy of Denmark, Dept. of Chemistry (DIA-K), Bygning 376, DK-2800 Lyngby, Denmark.] PhD (U. Copenhagen, 1990) physical chemistry. *Band_calculation modulated_structure physical_property*
E-mail peter@cismirx.symbion.ki.ku.dk Tel. 45()42885600 Fax 45()42881770

Sørensen, Mr Ole (1935). Senior Scientist. [Haldor Topsøe, Research Laboratories, Nymøllevej 55, DK-2800 Lyngby, Denmark.] MSc (Tech. U. Denmark, 1960) physics. *Electron_microscopy porous_materials electron_probe_microanalysis scanning_electron_microscopy catalysis*
E-mail TOPEOT@vm.uni-c.dk Tel. 45()45272484 Fax 45()45272999 Telex 37444 htas dk

Søtofte, Inger (1940). Associate Professor. [Chemistry Department B, Technical University of Denmark, DTU 207, DK-2800 Lyngby, Denmark.] MSc (U. Copenhagen, 1965) chemistry. *Crystal_structures intermolecular_interactions*
E-mail is@tkemi.klb.dtu.dk Tel. 45()42882222 Fax 45()42883136

Thirup, Søren (1956). Assistant professor. [Dept. of Chemistry, Aarhus University, DK-8000 Aarhus C., Denmark.] PhD (Aarhus U., 1992) protein crystallography. *Proteins nucleic_acids structure_determination*
E-mail soren@kemi.aau.dk Tel. 45()89423333x3878 Fax 45()86196199

Thoft, Nina Bjørn (1966). PhD. Student. [Department of Solid State Physics, Risø National Laboratory, PO Box 49, DK-4000 Roskilde, Denmark.] Master of Science (Department of Physics and Astronomy, U. Aarhus, 1991) atomic physics. *X-ray_diffraction melting magnetic_superlattice*
E-mail thoft@risoe.dk Tel. 45()46774719 Fax 45()42370115

Thorup, Niels (1939). Associate Professor. [Chemistry Department B, Technical University of Denmark, DTU 207, DK-2800 Lyngby, Denmark.] MSc (Tech. U. Denmark, 1963) chemistry. *Structural_chemistry radical_salts superconductors databases*
E-mail klbnt@kbar.dtu.dk Tel. 45()45252525 Fax 45()45251639

Villadsen, Jørgen (1930). Senior scientist. [Haldor Topsøe A/S, Research Laboratories, Nymøllevej 55, DK-2800 Lyngby, Denmark.] MSc (Tech. U. Denmark, 1953) chemical engineering. *X-ray_powder_diffraction phase_identification crystal_chemistry catalyst_structure*
E-mail TOPEOT@vm.uni-c.dk Tel. 45()45272483 Fax 45()45272999 Telex 37444 htas dk

Zachau-Christiansen, Birgit (1951). Senior Scientist. [Department of Physical Chemistry, Technical University of Denmark, DTH-206, DK-2800 Lyngby, Denmark.] PhD (1986) solid state electrochemistry. *Solid_state_electrochemistry solid_oxide_fuel_cells intercalation_chemistry mixed_conducting_oxides*
Tel. 45()45932380x2438 Fax 45()45934808 Telex 37529 DTHDIA DK

Zimmermann, Dr Hans Dieter (1940). Associate Professor. Geologisk Institut, Aarhus Universitet, Denmark. [Geologisk Institut, Aarhus Universitet, C. F. Møllers Alle 120, DK-8000 Aarhus C, Denmark.] Dr rer.nat. (U. Göttingen, 1969) mineralogy. *Geochemistry thermodynamics polarizing_microscopy phase_equilibrium*
E-mail geolhans@aau.dk Tel. 45()89422509 Fax 45()86139248

EGYPT

Sub-Editor: K. Elsayed

Abbase, Prof. Yehia (1944). Professor. Faculty of Science, Suez Canal University, Ismailia, Egypt. PhD (Grenoble U., France, 1976) physics. *X-ray_diffraction neutron_diffraction*

Abdallah, Dr Atef (1938). Lecturer. Physics Dept., Faculty of Science, Assiut University, Assiut, Egypt. PhD (Assiut U., 1976) physics. *X-ray_diffraction Mossbauer*
Tel. 20(88)322564 Fax 20(88)322564

Abdelaal, Mr Fawzi Amer. Researcher. Geological Survey and Mineral Res., Abbassia, Egypt. BSc (Cairo U., 1945) geology. *Geology economics*
Tel. 20(2)830782

Abdelatif, Dr A. Y. (1955). Lecturer. Physics Dept., Faculty of Science, Assiut University, Assiut, Egypt. PhD physics. *Metals crystallography*
Tel. 20(88)326293 Fax 20(88)322564

Abdelaziz, Dr A. Abuelfadl (1954). Lecturer. Physics Dept., Faculty of Science, Assiut University, Assiut, Egypt. PhD (Assiut U., 1985) physics. *Crystal_growth ferroelectrics*
Tel. 20(88)323000 Fax 20(88)322564

Abdelhady, Prof. Siham (1934). Vice-Dean, Faculty of science. Physics Dept., Faculty of Science, Helwan University, Helwan, Egypt. PhD (Cairo U., 1965) physics. *Crystal_diffraction material_sciences*
Tel. 20(2)784244 Fax 20(2)784244

Abdelkader, Mrs Zinab Mohamed. Ass. Lecturer. Faculty of Science, Cairo University, University Street, Giza, Egypt. MSc (Cairo U., 1969) geology. *Geology economics*
Tel. 20(2)841722

Abdelkader, Dr Adelaziz. Lecturer. X-ray Unit, National Research Center, Dokkiy, Cairo, Egypt. PhD (London U., UK, 1964) geology. *Inorganic_petrology*
Tel. 20(2)802129 Fax 20(2)700931

Abdelmohsen, Prof. Hussein. Professor. Nuclear Research Center, Atomic Authority, Anshas, Egypt. DSc (Nancy U., France, 1959) geology. *Radioactivity mineralogy*
Tel. 20(2)862013

Abdelmoniem, Ms Eman Mohamed (1968). Ass. Lecturer. Physics Dept., Faculty of Science, Minia University, Minia, Egypt. BSc (Minia U., 1990) physics. *Physical_properties structure*
Tel. 20(86)326243 Fax 20(86)326243

Abdelrahman, Dr Abdelrahman Mohamed (1947). Lecturer. Physics Dept., Faculty of Science, Ain Shams University, Cairo, Egypt. PhD (Bonn U., Germany, 1982) physics. *Crystal_structure*
Tel. 20(2)822284 Fax 20(2)822284

Abdu, Prof. Fayez Madi. Professor. Faculty of Agriculture, Shoubra el kheima, Ain Shams University, Abbassia, Egypt. PhD (Ain Shams U., 1957) soils. *Soils mineralogy*
Tel. 20(2)948716

Abohilal, Dr Hassan (1938). Ass. Professor. Nuclear Research Center, Atomic Authority, Anshas, Egypt. PhD (1976) physics. *X-ray_diffraction neutron_diffraction*
Tel. 20(2)3540982 Fax 20(2)3540982

Aboushama, Dr Ali (1956). Lecturer. Physics Dept., Faculty of Science, Ain Shams University, Cairo, Egypt. PhD (Ain Shams U., 1994) physics. *Structure amorphous_material*
E-mail Shama@frcu.eun.Eg Tel. 20(2)822284 Fax 20(2)822284

Adam, Dr Alia Elmonein Ebrahim (1955). Professor. Physics Dept., Faculty of Science, Azhar University Girls Branch, Naser City, Cairo, Egypt. PhD (Hiroshima U., Japan, 1992) physics. *X-ray_diffraction neutron_diffraction*

Afify, Prof. Nasser (1946). Professor. Physics Dept., Faculty of Science, Assiut University, Assiut, Egypt. PhD (Assiut U., 1979) physics. *Semiconductors structure*
Tel. 20(88)322564 Fax 20(88)322564

Ahmed, Prof. Mohamed Saleh. Professor. Physics Dept., Faculty of Science, Alexandria University, Alexandria, Egypt. PhD (London U., UK, 1950) physics. *Diffuse_scattering crystal_analysis*

Ahmed, Prof. Naima Abdelkader (1937). Head of Solid State Physics Lab. Physics Dept., National Research Center, Dokkiy, Cairo, Egypt. PhD (Cairo U., 1965) physics. *X-ray_diffraction crystal_structure*
Tel. 20(2)700931 Fax 20(2)700931

Akkad, Mr Salah eldin. Researcher. Geological Survey and Mineral Res., Abbassia, Egypt. BSc (Cairo U., 1942) geology. *Petrology*
Tel. 20(2)830782

Akl, Mr Alaa Ahmed (1960). Specialist of Science. Central Lab. Minia University, Minia, Egypt. MSc (Minia U., 1988) physics. *X-ray_diffraction*
Tel. 20(86)326243 Fax 20(86)326243

Anwar, Prof. Yehia. Professor. Faculty of Science, Alexandria University, Alexandria, Egypt. PhD (Durham U., UK, 1950) geology. *Petrology*

Arafa, Prof. Salah. Professor. Science Dept., American University, Cairo, Egypt. PhD (Cairo U., 1982) physics. *Powder_diffraction*
Tel. 20(2)2968169

Attie, Prof. Abdelkader (1958). Professor. CMRDI, Tebbine, Helwan, Egypt. PhD (Columbia U., USA, 1964) physics. *Crystallography mineralogy*
Tel. 20(2)790898 Fax 20(2)790898

Azer, Dr Nazima. Ass. Professor. Faculty of Science, Cairo University, University Street, Giza, Egypt. PhD (Wien U., Austria, 1956) geology. *Geology economics*
Tel. 20(2)841722

Bader, Dr Yehia Abdelhamid. Lecturer. Physics Dept., Faculty of Science, Cairo University, University Street, Giza, Egypt. PhD (Leningrad State U., USSR, 1974) physics. *Phase_transition ionic_crystal*
Tel. 20(2)841722

Bahgat, Prof. Alaa Eldien (1949). Professor. Physics Dept., Faculty of Science, Azhar University, Naser City, Cairo, Egypt. PhD (Azhar U., 1975) physics. *X-ray_spectroscopy structure glasses ceramics*
Tel. 20(2)2611404 Fax 20(2)2611404

Barakat, Prof. Nayel. Professor. Physics Dept., Faculty of Science, Ain Shams University, Abbassia, Egypt. PhD (London U., UK, 1952) physics. *Spectrography*
Tel. 20(2)821096

Basha, Dr Ahmed Fouad (1941). Lecturer. Physics Dept., Faculty of Science, Cairo University, University Street, Giza, Egypt. PhD (Moscow State U., USSR, 1974) physics. *Dielectrics molecular_solid*
Tel. 20(2)841722

Bassiouny, Prof. Mohamed Khafagi. Professor. Geology Dept., Faculty of Science, Ain Shams University, Abbassia, Egypt. PhD (Wien U., Austria, 1969) geology. *Petrology*
Tel. 20(2)821096

Ebied, Dr Mohamed (1949). Lecturer. Physics Dept., Faculty of Science, Minia University, Minia, Egypt. PhD (Minia U., 1991) physics. *X-ray_diffraction amorphous_materials*
Tel. 20(86)326243 Fax 20(86)326243

Elbadri, Dr Ahmed Fouad. Lecturer. H. Dept. of Mining, Faculty of Eng., Cairo University, University Street, Giza, Egypt. PhD (London U., UK, 1951) geology. *Crystal_optics*
Tel. 20(2)896926

Eldosoky, Dr Shokry (1960). Lecturer. Physics Dept., Faculty of Science, Azhar University, Naser City, Cairo, Egypt. PhD (Azhar U., 1989) physics. *Amorphous_material*
Tel. 20(2)2611404 Fax 20(2)2611404

Elhagry, Mr Magdy A. (1966). Ass. Lecturer. Physics Dept., Faculty of Science, Helwan University, Helwan, Egypt. BSc (Helwan U., 1993) physics. *Thin_film*
Tel. 20(2)784244 Fax 20(2)784244

Elhinnaur, Prof. Essam E. (1936). Professor. Physics Dept., National Research Center, Dokkiy, Cairo, Egypt. PhD (1961) physics. *Mineralogy crystallography*
Tel. 20(2)700931 Fax 20(2)700931

Elkardy, Dr Mhamed Abdelmoneim (1944). Lecturer. Physics Dept., National Research Center, Dokkiy, Cairo, Egypt. PhD (Cairo U., 1985) physics. *X-ray_diffraction crystal_structure*
Tel. 20(2)700931 Fax 20(2)700931

Elkoiy, Dr Esmat M. (1940). Lecturer. Physics Dept., National Research Center, Dokkiy, Cairo, Egypt. PhD (Suez Canal U., 1993) physics. *X-ray_diffraction crystal_structure*
Tel. 20(2)700931 Fax 20(2)700931

Elmaghraby, Dr E. Mohamed (1964). Ass. lecturer. Physics Dept., Faculty of Science, Assiut University, Assiut, Egypt. MSc (Assiut U., 1992) physics. *Crystal_growth electrical_conductivity*
Tel. 20(88)333555 Fax 20(88)322564

Elsayed, Prof. Karimat (1933). Head of Physics Dept. Consultant for the Teaching Commission for North Africa and Middle East; Sub-Editor, WDC9. Physics Dept., Faculty of Science, Ain Shams University, Cairo, Egypt. [PO Box 8014, Masaken Nassr, Cairo 11371, Egypt.] PhD (London U., UK, 1965) crystallography. *Small_molecules powder_diffraction_analysis material_sciences*
E-mail Karima@frcu.eun.Eg Tel. 20(2)2601742 Fax 20(2)822284

Elshabiny, Prof. Aida M. (1939). Professor. Physics Dept., National Research Center, Dokkiy, Cairo, Egypt. PhD (TU Dresden, 1979) physics. *X-ray_diffraction material_sciences*
Tel. 20(2)700931 Fax 20(2)700931

Elshanshury, Dr Ismail. Ass. Professor. Nuclear Research Center, Atomic Authority, Anshas, Egypt. PhD (Rensselaer Polytechnic Inst., USA, 1963) physics. *Metals*
Tel. 20(2)862013

Elsharkawi, Dr Mohamed Abdelhamid. Lecturer. Faculty of Science, Cairo University, University Street, Giza, Egypt. PhD (London U., UK, 1964) geology. *Ores_deposit*
Tel. 20(2)841722

Elshazli, Prof. Elshazli Mohamed. Professor. Nuclear Research Center, Atomic Authority, Anshas, Egypt. PhD (London U., UK, 1950) geology. *Ores_deposit*
Tel. 20(2)862013

Elshora, Dr A. Ebrahim (1947). Ass. Professor. Physics Dept., Faculty of Science, Tanta University, Tanta, Egypt. PhD (London U., 1981) physics. *Crystal_structure amyrin_benzoate*

Farag, Prof. Ibrahim S. A. (1938). Professor. Physics Dept., National Research Center, Dokkiy, Cairo, Egypt. PhD (Moscow U., 1971) physics. *X-ray_diffraction crystal_structure*
Tel. 20(2)700931 Fax 20(2)700931

Fayed, Prof. Leila. Professor. Faculty of Science, Cairo University, University Street, Giza, Egypt. PhD (Sheffield U., UK, 1966) geology. *Rock_mechanics*
Tel. 20(2)841722

Fayek, Prof. Mohamed Khorshed (1941). Professor. Nuclear Research Center, Atomic Authority, Anshas, Egypt. PhD (Cairo U., 1971) physics. *X-ray_diffraction neutron_diffraction*
Tel. 20(2)3540982 Fax 20(2)3540982

Gabalaah, Dr G. Abdelstar (1949). Ass. Professor. Physics Dept., Faculty of Science, Tanta University, Tanta, Egypt. PhD (London U., 1983) physics. *Crystal_structure amyrin_iode*

Gaber, Dr Abdelfattah (1950). Associate Professor. Physics Dept., Faculty of Science, Assiut University, Assiut, Egypt. PhD (Assiut U., 1984) physics. *Metals_physics materials_structure*
Tel. 20(88)322564 Fax 20(88)322564

Gad, Prof. Gamal Mohamed. Professor. Refractories & Ceramic Lab., National Research Center, Dokkiy, Cairo, Egypt. PhD (London U., UK, 1950) geochemistry. *Clays mineralogy*
Tel. 20(2)802129 Fax 20(2)700931

Gaffar, Prof. M. A. (1941). Professor. Physics Dept., Faculty of Science, Assiut University, Assiut, Egypt. PhD (Moscow U., 1975) physics. *Crystal_growth ferroelectricity*
Tel. 20(88)333555 Fax 20(88)322564

Guindi, Prof. Amin Riad. Professor. Faculty of Science, Alexandria University, Alexandria, Egypt. PhD (London U., UK, 1950) geology. *Petrology*

Gweifel, Dr Ismail. Lecturer. Faculty of Agriculture, Alexandria University, Alexandria, Egypt. PhD (Alexandria U., 1967) soils morphology. *Soils*

Hafiz, Prof. Mohamed M. (1943). Professor. Physics Dept., Faculty of Science, Assiut University, Assiut, Egypt. PhD,(Assiut U., 1973) physics. *Semiconductors thin_film*
Tel. 20(88)322564 Fax 20(88)322564

Hamdi, Prof. Hassan Mahmoud. Professor. Faculty of Agriculture, Ain Shams University, Abbassia, Egypt. DSc (ETH Zurich, Switzerland, 1942) clay minerals. *Clays mineralogy structure*
Tel. 20(2)948716

Hammad, Prof. Elsayed Mohamed (1940). Professor. Physics Dept., National Research Center, Dokkiy, Cairo, Egypt. PhD (Cairo U., 1983) physics. *X-ray_diffraction clays mineralogy*
Tel. 20(2)700931 Fax 20(2)700931

Hasan, Dr Zinab Abdelkhlek (1956). Professor. Physics Dept., Faculty of Science, Azhar University Girls Branch, Naser City, Cairo, Egypt. PhD (Azhar U., 1992) physics. *X-ray_diffraction neutron_diffraction*

Hassan, Prof. Mohamed Youssef. Professor. Geology Dept., Faculty of Science, Ain Shams University, Abbassia, Egypt. PhD (Bristol U., UK, 1951) geology. *Crystal_structure paleontology*
Tel. 20(2821069

Heba, Dr Zein Elabdeen Kamel (1954). Lecturer. Physics Dept., Faculty of Science, Ain Shams University, Cairo, Egypt. PhD (1987) physics. *Computing X-ray_diffraction*
Tel. 20(2)822284 Fax 20(2)822284

Hewaidy, Prof. I. F. (1932). Professor. CMRDI, Tebbine, Helwan, Egypt. PhD (Cairo U., 1954) physics. *X-ray_diffraction inorganic*
Tel. 20(2)790898 Fax 20(2)790898

Hilmi, Prof. Mohamed Ezzeldin. Professor. Geology Dept., Faculty of Science, Ain Shams University, Egypt. PhD (Michigan U., USA, 1951) geology. *Mineralogy crystal_structure*
Tel. 20(2)821096

Hilmy, Prof. Mohamed Ezzeldin (1924). Professor. Geology Dept., Faculty of Science, Ain Shams University, Cairo 11566, Egypt. PhD (Michigan U., USA, 1952) mineralogy. *Mineralogy crystallography*
Tel. 20(2)822284 Fax 20(2)822284

Kabish, Prof. Lotfi. Professor. National Research Center, Dokkiy, Cairo, Egypt. PhD (Cairo U., Egypt, 1948) geology. *Petrology igneous_materials*
Tel. 20(2)802129 Fax 20(2)700931

Kaid, Ms Mayssa Fathi (1963). Ass. Lecturer. Physics Dept., Faculty of Science, Minia University, Minia, Egypt. MSc (Minia U., 1991) physics. *Physical_property_structure_materials*
Tel. 20(86)326243 Fax 20(86)326243

Kamel, Prof. Omar A. (1935). Professor. Geology Dept., Faculty of Science, Minia University, Minia, Egypt. PhD (1963) physics. *Mineralogy*
Tel. 20(86)326243 Fax 20(86)326243

Kamel, Prof. Raafat Wasef. Professor. Physics Dept., Faculty of Science, Cairo University, University Street, Giza, Egypt. DSc (Cairo U., 1968) physics. *Defect_structure crystalline_solids*
Tel. 20(2)894095

Khadr, Prof. Moustafa. Lecturer. Faculty of Agriculture, Alexandria University, Alexandria, Egypt. PhD (Alexandria U., 1956) soils. *Soils*

Khallfa, Prof. Berlant. Professor. Physics Dept., Faculty of Science, Ain Shams University, Abbassia, Egypt. PhD (London U., UK, 1967) geology. *Electron_diffraction*
Tel. 20(2)862013

Khalifa, Dr B. Abdelmeguid. Researcher. Nuclear Research Center, Atomic Authority, Anshas, Egypt. PhD (London U., UK, 1968) chemistry. *Inorganic_structure*
Tel. 20(2)862013

Khidr, Prof. Fatma Abdelhakim. Professor. X-ray Crystallography Unit, National Research Center, Dokkiy, Cairo, Egypt. PhD (Cairo U., 1967) physics. *X-ray_metallography*
Tel. 20(2)802129 Fax 20(2)700931

Kholif, Dr Mahmoud. Lecturer. National Research Center, Dokkiy, Cairo, Egypt. PhD (Moscow U., USSR, 1970) geology. *Ores_deposit*
Tel. 20(2)802129 Fax 20(2)700931

Kishk, Dr Fawzi Mohamed. Ass. Prof. Faculty of Agriculture, Alexandria University, Alexandria, Egypt. PhD (California U., Berkeley, USA, 1967) soils morphology. *Soils*

Labib, Dr Fawkia. Lecturer. X-ray Crystallography Unit, National Research Center, Dokkiy, Cairo, Egypt. PhD (Ain Shams U., Egypt, 1970) geology. *Soils_chemistry soils_mineralogy*
Tel. 20(2)802129 Fax 20(2)700931

Lotfy, Prof. Mohamed. Professor. Faculty of Science, Cairo University, University Street, Giza, Egypt. PhD (Cairo U., 1957) geology. *Petrology geology*
Tel. 20(2)841722

Mady, Prof. Khairy A. (1938). Professor. Physics Dept., National Research Center, Dokkiy, Cairo, Egypt. PhD (Cairo U., 1975) physics. *Crystals thin_film semiconductors*
Tel. 20(2)700931 Fax 20(2)700931

Mobed, Dr A. Mohamed (1966). Ass. lecturer. Physics Dept., Faculty of Science, Assiut University, Assiut, Egypt. MSc (Assiut U., 1992) physics. *Crystal_growth ferroelectric*
Tel. 20(88)333555 Fax 20(88)322564

Mohamed, Mr Ahmed Aly (1966). Ass. Lecturer. Physics Dept., Faculty of Science, Ain Shams University, Cairo, Egypt. MSc (1994) physics. *Computing X-ray_diffraction semiconductors*
E-mail rady@frcu.eun.Eg Tel. 20(2)822284 Fax 20(2)822284

Mohamed, Mr Almoktar Mohamed (1966). Ass. Lecturer. Physics Dept., Faculty of Science, Assiut University, Assiut, Egypt. MSc (Assiut U., 1993) physics. *Potassium_compounds ferrocyanide_compounds*
Tel. 20(88)326293 Fax 20(88)322564

Mohamed, Dr Esmat A. (1948). Ass. Professor. Physics Dept., Faculty of Science, Azhar University Girls Branch, Naser City, Cairo, Egypt. PhD (Azhar U., 1980) physics. *X-ray_diffraction thermal_properties*
Tel. 20(2)4010019 Fax 20(2)4010019

Mohamed, Dr Galal A. (1952). Lecturer. Physics Dept., Faculty of Science, Assiut University, Assiut, Egypt. PhD (Boland U., 1986) physics. *Crystal_growth ferroelectric*
Tel. 20(88)333555 Fax 20(88)322564

Mostafa, Mr Said H. (1965). Ass. Lecturer. Physics Dept., Faculty of Science, Helwan University, Helwan, Egypt. BSc (Helwan U., 1993) physics. *Thin_film*
Tel. 20(2)784244 Fax 20(2)784244

Naga, Prof. Mohamed Abdelhamid. Lecturer. Faculty of Agriculture, Cairo University, University Street, Giza, Egypt. PhD (Cairo U., 1954) soils. *Soils_mineralogy*
Tel. 20(2)896586

Nakhla, Dr Fakhry (1924). Lecturer; Professor of Applied Mineralogy and Mining Geology. Faculty of Engineering, Cairo University, University Street, Giza, Egypt. [3 Mourad Street, Giza Square, Giza, Egypt.] PhD (London U., UK, 1950) mining geology. *Minerals ore_mineralogy crystal_chemistry mineral_deposits coal industrial_rocks exploration_and_mining_geology*
Tel. 20(2)5720250 Fax 20(2)723486

Philipp, Dr George. Ass. Prof. Faculty of Science, Cairo University, University Street, Giza, Egypt. PhD (Cairo U., 1956) geology. *Sediments petrology*
Tel. 20(2)841722

Rabie, Dr Farida Hamed. Ass. Professor. Faculty of Agriculture, Ain Shams University, Abbassia, Egypt. PhD (California U., Davis, USA, 1967) soils. *Soils*
Tel. 20(2)948716

Radwan, Prof. Mostafa Mohsen. Professor. Science Dept., American University, Cairo, Egypt. PhD (Dundee U. UK, 1982) physics. *Biological_structure viruses_structure*
Tel. 20(2)853983

Ragab, Prof. Dr Abdelghani (1939). Professor. Geology Dept., Faculty of Sciences, Ain Shams University, Abbassia, Cairo, Egypt. PhD (Ain Shams U., 1972) geology. *Petrology plate_tectonics*
Tel. 20(2)821096 Fax 20(2)822284

Ramadan, Prof. Ahmed Ahmed (1941). Head of Physics Dept. Physics Dept., Faculty of Science, Helwan University, Helwan, Egypt. PhD (TU Dresden, 1979) physics. *Physical_properties material_structure*
Tel. 20(2)784244 Fax 20(2)784244

Sadek, Prof. Gamil. Professor. National Research Center, Dokkiy, Cairo, Egypt. DSc (Lyon U., France, 1964) geology. *Ores_deposit*
Tel. 20(2)802129 Fax 20(2)700931

Said, Dr Fikria (1944). Ass. Professor. Nuclear Research Center, Atomic Authority, Anshas, Egypt. PhD (1978) physics. *X-ray_diffraction neutron_diffraction*
Tel. 20(2)3540982 Fax 20(2)3540982

Salem, Dr Safia Mahmoud. Ass. Prof. Faculty of Science, Cairo University, University Street, Giza, Egypt. PhD (Cairo U., 1961) X-ray crystallography. *X-ray_crystallography*
Tel. 20(2)802129

Shamah, Prof. Abdelmoniem Mohamed (1942). Professor. Faculty of Petroleum and Mining Eng., Suez Canal University, Suez, Egypt. PhD (Sheffield U., UK, 1979) physics. *X-ray_diffraction neutron_diffraction*

Shoukri, Prof. Nasri. Professor. Faculty of Science, Cairo University, University Street, Giza, Egypt. PhD (London U., UK, 1940) geology. *Petrology*
Tel. 20(2)841722

Soliman, Mr F. Abdelaal. Researcher. Geological Survey and Mineral Res., Abbassia, Egypt. MSc (Cairo U., 1944) geology. *Petrology*
Tel. 20(2)830782

Soliman, Dr Mohamed Soliman. Ass. Professor. Geology Dept., Faculty of Science, Ain Shams University, Abbassia, Egypt. PhD (Stanford U., USA, 1958) geology. *Sedimentation*
Tel. 20(2)821096

Tahoun, Prof. Salah (1937). Professor. Faculty of Agriculture, Zagazig University, Zagazig, Egypt. [PO Box 2893, Heliopolis El-Horria, Cairo 11361, Egypt.] PhD (Michigan State U., USA, 1965) clay mineralogy. *Structure minerals environmental_affairs*
Tel. 20(2)2601742 Fax 20(2)3610764

Takla, Dr Maher Azmi (1942). Professor of Mineralogy, Petrology and Economic Geology. Faculty of Science, Cairo University, University Street, Giza, Egypt. PhD (Cairo U., 1971) geology. *Crystallography mineralogy petrology ore_deposits*
Tel. 20(2)5850757

Thabet, Dr Atef. Researcher. Geological Survey and Mineral Res., Abbassia, Egypt. PhD

(Cairo U., 1963) geology. *Petrology*
Tel. 20(2)830782

Tosson, Dr Salama (1928). Professor of Mineralogy. Member of the National Committee of Mineralogy. Faculty of Science, Alexandria University, Alexandria, Egypt. [49 Ahmed Kamha Street, Camp Caesar, Alexandria, Egypt.] DSc (Nancy U., France, 1957) geology. *Mineral_structure ore_deposits*
Tel. 20(3)5978739 Fax 20(3)4836618

Youssef, Prof. I. Mourad. Professor. Geology Dept., Faculty of Science, Ain Shams University, Abbassia, Egypt. PhD (Alexandria U., 1950) geology. *Structure_geology*
Tel. 20(2)821096

Zaghloul, Prof. Mohamed Zaki. Professor. Faculty of Science, Mansoura University, Egypt. PhD (Bristol U., UK, 1950) geology. *Structural_geology*

Zaghloul, Prof. Zaki M. (1925). Professor. Geology Dept., Faculty of Science, Mansora University, Mansora, Egypt. PhD(1958) physics. *Mineralogy crystallography*

ESTONIA

Sub-Editor: H. Mändar

Aabloo, Mr Alvo (1965). Scientist. Institute of Experimental Physics and Technology, University of Tartu, 4 Tähe Street, EE2400 Tartu, Estonia. PhD (U. Tartu, 1994) solid state physics. *Computer crystal_energy X-ray_diffraction polysaccharides molecular_mechanics computer_modelling force_field carbohydrates celluloses*
E-mail alvo@physic.ut.ee Tel. 372(7)435356 Fax 372(7)441102 Telex 173243 TAUN SU

Aarik, Dr Jaan (1951). Researcher; Group leader. Institute of Experimental Physics and Technology, University of Tartu, Ülikooli 18, EE2400 Tartu, Estonia. MSc (U. Tartu, 1994) solid state physics. *Epitaxy oxides adsorption_kinetics*
E-mail yeftijaa@raud.ut.ee Tel. 372(7)433276 Fax 372(7)435440 Telex 173243 TAUN SU

Haav, Dr Aksel (1928). Lecturer. Institute of Experimental Physics and Technology, University of Tartu, 4 Tähe Street, EE2400 Tartu, Estonia. PhD (U. Tartu, 1967) solid state physics. *X-ray_powder_diffraction mixed_crystal lattice_parameter*
E-mail hugo@physic.ut.ee Tel. 372(7)435356 Fax 372(7)441102 Telex 173243 TAUN SU

Kiiranen, Dr Kalle (1953). Scientist. Technique Center of Institute of General and Molecular Pathology, University of Tartu, 4 Narva Street, EE2400 Tartu, Estonia. PhD (U. Tartu, 1989) field theory. *Atomic_molecular_modelling ionic_crystal_modelling field_theory elementary_particles*
Tel. 372(7)435496 Fax 372(7)441102 Telex 173243 TAUN SU

Kirs, Dr Juho (1946). Lecturer. Institute of Geology, Department of Geology and Mineralogy, University of Tartu, 46 Vanemuise Street, EE2400 Tartu, Estonia. *Geology mineralogy diffractometry polarization_microscopy philosophy_of_science*
E-mail geol@geogr.ut.ee Tel. 372(7)430607 Fax 372(7)441102 Telex 173243 TAUN SU

Mändar, Dr Hugo (1961). Docent. Head, Estonian Crystallographic Society; Sub-Editor, World Directory of Crystallographers 9. Institute of Experimental Physics

and Technology, Department of Physics, University of Tartu, 4 Tähe Street, EE2400 Tartu, Estonia. PhD (U. Tartu, 1990) solid state physics. *X-ray_powder_diffraction line_profile_analysis crystallite_size zeolites powder_diffraction_programming diffractometer_automation sodalites*
E-mail hugo@physic.ut.ee Tel. 372(7)435356 Fax 372(7)441102 Telex 173243 TAUN SU

Mikelsaar, Dr Raik-Hiio (1939). Group leader. Institute of General and Molecular Pathology, University of Tartu, 34 Veski Street, EE2400 Tartu, Estonia. PhD (Moscow U., 1989) bioorganic chemistry. *Molecular_modelling celluloses DNA biopolymers conformation hydrogen_bonding stereochemistry amino_acids RNA proteins carbohydrates lipids polysaccharides*
E-mail raik@igpm.ut.ee Tel. 372(7)435169 Fax 372(7)430365 Telex 173243 TAUN SU

Rimm, Mr Karel (1966). Engineer. Laboratory of Electron Microscopy and X-ray Analysis, Centre for Materials Research, Tallinn Technical University, 5 Ehitajate Road, EE0026 Tallinn, Estonia. *X-ray_powder_diffraction phase_determination apatite_compound programming powder_diffraction_pattern_analysis*
E-mail karla@ttu.ee Tel. 372(2)532014 Fax 372(2)532446

Tamm, Prof. Jüri (1937). Dean. Institute of Physical Chemistry, University of Tartu, 2 Jakobi Street, EE2400 Tartu, Estonia. PhD (U. Tartu, 1969) physical chemistry. *Electrochemistry conducting_polymers structure synthetic_methods*
Tel. 372(7)435159 Fax 372(7)441102 Telex 173243 TAUN SU

Uustare, Dr Teet (1952). Senior Researcher. Institute of Experimental Physics and Technology, University of Tartu, Ülikooli 18, EE2400 Tartu, Estonia. PhD (U. Tartu, 1984) solid state physics. *AES RHEED oxides surfaces CVD*
E-mail yeftitu@raud.ut.ee Tel. 372(7)433276 Fax 372(7)435440 Telex 173243 TAUN SU

FINLAND

Sub-Editor: A. M. Vahvaselkä

Åberg, Prof. Teijo (1937). Associate professor. [Laboratory of Physics, Helsinki University of Technology, FIN-02150 Espoo, Finland.] PhD (U. Helsinki, 1969) physics. *Inelastic_scattering absorption_spectroscopy fluorescence anomalous_dispersion ion-atom_collisions multiphoton_ionization*
E-mail teijo.aberg@hut.fi Tel. 358(0)4513103 Fax 358(0)4513116 Telex 125161 htkk sf

Ahlgrén, Markku Jouko (1944). Associate professor. [Department of Chemistry, University of Joensuu, PO Box 111, FIN-80101 Joensuu, Finland.] PhD (U. Helsinki, 1979) chemistry. *Coordination clusters molecular_complexes*
E-mail ahlgren@joyl.joensuu.fi Tel. 358(73)1513368 Fax 358(73)1513344 Telex 462 23 joy sf

Antson, Olli Kalervo (1955). Senior research scientist. [VTT Chemical Technology, PO Box 1404 (Otakaari 3 A), FIN-02044 VTT, Finland.] DTech (Helsinki U. Technology, 1991) nuclear engineering. *Neutron_diffraction neutron_time-of-flight instrument_design*
E-mail olli.antson@vtt.fi Tel. 358(0)4566354 Fax 358(0)4566390 Telex 122972 vttha sf

Blomberg, Merja Kristiina (1957). Assistant. University of Helsinki, Department of Physics, Siltavuorenpenger 20 D, Helsinki, Finland. [Department of Physics, PO BOx 9, FIN-00014 University of Helsinki, Finland.] PhD (U. Helsinki, 1994) physics. *X-ray_crystallography*
E-mail merja.blomberg@helsinki.fi Tel. 358(0)1918329 Fax 358(0)1918680 Telex 100 2125 attn Physics Department

Glumoff, Dr Tuomo (1961). Research scientist. Turku Centre for Biotechnology, University of Turku, Finland. [PO Box 123, FIN-20521 Turku, Finland.] PhD (Swiss Federal Inst. of Technology, Zürich, 1992) biochemistry. *Biological_crystallography proteins structure_determination structure-activity_relationship lignin_degradation protein_purification*
E-mail tglumoff@ra.abo.fi Tel. 358(21)6338031 Fax 358(21)6338000

Goldman, Adrian (1958). Senior research scientist. Turku Centre for Biotechnology, University of Turku, PO Box 123, Tykistökatu 6, FIN-210521 Turku, Finland. PhD (Yale U., Connecticut USA, 1985) macromolecular X-ray crystallography. *Biocrystallography catalysis_macromolecules computer_modelling biochemistry macromolecular_structure-function_relationships membrane_proteins*
E-mail goldman@ala.btk.utu.fi agoldman@finabo.bitnet Tel. 358(21)6338029 Fax 358(21)6338000

Haapala, Ilmari J. (1939). Professor. [Department of Geology, PO Box 11 (Snellmaninkatu 3), FIN-00014 University of Helsinki, Finland.] PhD (U. Helsinki, 1966) geology and mineralogy. *Crystallography mineralogy petrology*
Tel. 358(0)1913426 Fax 358(0)1913466

Hakanen, Arvi Tapani (1968). Researcher. [Department of Physics, University of Turku, FIN-20500 Turku, Finland.] MSc (U. Turku, 1992) physics. *Chirality SAXS polysaccharides X-ray_diffraction*
E-mail ahakanen@sara.vtu.fi Tel. 358(21)6335674 Fax 358(21)6335070

Hämäläinen, Keijo Johannes (1963). Docent. [Department of Physics, PO Box 9, FIN-00014 University of Helsinki, Finland.] PhD (U. Helsinki, 1990) physics. *Inelastic_X-ray_scattering absorption_spectroscopy synchrotron_radiation*
E-mail keijo.hamalainen@helsinki.fi Tel. 358(0)1918329 Fax 358(0)1918680

Hämäläinen, Reijo Pertti (1940). Docent. [Department of Chemistry, PO Box 6, FIN-00014 University of Helsinki, Finland.] PhD (U. Helsinki, 1972) chemical crystallography. *Crystallography transition_elements chelates*
E-mail hamaloff@fltu.helsinki.fi Tel. 358(0)1913624 Fax 358(0)657288

Harte, Säde Pirjo Anneli (1962). Geologist. [Geological Survey of Finland, Betonimiehenkuja 4, FIN-02150 Espoo, Finland.] MSc (U. Helsinki, 1987) geology and mineralogy. *X-ray_diffraction mineralogy*
Tel. 358(0)46931 Fax 358(0)462205 Telex 123185

Heikinheimo, Pirkko (1962). Research scientist. [University of Turku, Department of Biochemistry/Biocity, FIN-20520 Turku, Finland.] MSc (U. Turku, 1990) biochemistry. *Structure-activity_relationship*
E-mail pheikinh@ala.btk.utu.fi Tel. 358(21)6338056 Fax 358(21)6338050

Helin, Sari Katariina (1963). PhD student. [Turku Centre for Biotechnology, PO Box 123, FIN-20521 Turku, Finland.] MSc (U. Turku, 1990) biochemistry. *Macromolecular_crystallography proteins structure*
E-mail shelin@ala.btk.utu.fi Tel. 358(21)6338024 Fax 358(21)6338000

Hiismäki, Pekka Eljas (1939). Research professor. VTT Chemical Technology, Otakaari 3A, Espoo, Finland. [VTT Chemical Technology, PO Box 1404, FIN-02044 Espoo, Finland.] Dr (Technical U. Helsinki, 1970) physics. *Neutron_diffraction time-of-flight_diffraction powder_diffraction neutron_physics Rietveld_refinement*
E-mail pekka.hiismaki@vtt.fi Tel. 358(0)4566320 Fax 358(0)4566390

Hiltunen, Lassi Ilmari (1943). Laboratory manager. [Helsinki University of Technology, Chemistry Department, Kemistintie 1, FIN-02150 Espoo, Finland.] MSc (Helsinki U. Technology, 1969) chemistry. *Direct_method spectroscopy*
E-mail lhiltune@leka.hut.fi Tel. 358(0)4512591 Fax 358(0)462373

Hölsä, Jorma Pertti Kalervo (1952). Professor, Head, Laboratory of Inorganic Chemistry. [Department of Chemistry, University of Turku, FIN-20500 Turku, Finland.] DTech (Helsinki U. Technology, 1983) chemical engineering. *X-ray_powder_diffraction neutron_powder_diffraction Rietveld_method rare-earth_compounds spectroscopy quantum_mechanics solid_state rare-earth_elements*
E-mail jholsa@utu.fi Tel. 358(21)6336730 and 358(21)6336737 Fax 358(21)6336730 Telex 62683 tyf sf

Hongisto, Ossi Valtteri (1969). Research assistant. [Department of Physics, University of Turku, FIN-20500 Turku, Finland.] MSc (U. Turku, 1993) physics. *X-ray_diffraction acoustics lipids micelles*
E-mail vahongis@sara.cc.utu.fi Tel. 358(21)6335674 Fax 358(21)6335070

Honkimäki, Veijo (1962). Assistant. [Department of Physics, PO Box 9, FIN-00014 University of Helsinki, Finland.] PhD (U. Helsinki, 1991) physics. *Powder_diffraction anomalous_dispersion EXAFS*
E-mail honkimaki@phcu.helsinki.fi Tel. 358(0)1918329 Fax 358(0)1918680

Jääskeläinen, Sirpa (1961). Assistant. [Department of Chemistry, University of Joensuu, PO Box 111, FIN-80101 Joensuu, Finland.] PhD (U. Joensuu, 1992) physical chemistry. *Structure_determination clusters organometallic_compounds*
Tel. 358(73)1513335 Fax 358(73)1513344

Järvinen, Dr Matti J. (1934). Head manager. [Lappeenranta University of Technology, Department of Physics, PO Box 20, FIN-53851 Lappeenranta, Finland.] PhD (U. Helsinki, 1969) physics. *X-ray_diffraction materials_science*
E-mail matti.jarvinen@lut.fi Tel. 358(53)6212807 Fax 358(53)6212899 Telex 58290 lttk sf

Johanson, Bo Stefan (1954). Head of Electron Microprobe Laboratory, Geologist. [Geological Survey of Finland, Betonimiehenkuja 4, FIN-02150 Espoo, Finland.] MSc (U. Helsinki, 1984) geology and mineralogy. *Ore_mineralogy tellurides gold archean electron_microprobe analytical_mineralogy*
E-mail bo.johanson@gsf.fi Tel. 358(0)46932321 Fax 358(0)462205

Kallio, Mr Pekka Yrjö Juhani (1937). Geologist. [Geological Survey of Finland, PO Box 1237, FIN-70701 Kuopio, Finland.] MSc (U. Helsinki, 1964) geology and mineralogy. *Mineralogy X-ray_diffraction*
E-mail pekka.kallio@kuovax.gsf.fi Tel. 358(71)205750 Fax 358(71)205215

Kansikas, Dr Jarno Juhani (1947). Docent, Amanuensis. [Department of Chemistry, PO Box 6, FIN-00014 University of Helsinki, Finland.] PhD (U. Helsinki, 1985) inorganic chemistry. *X-ray_structure_determination coordination powder synchrotron_radiation*
E-mail kansikas@cc.helsinki.fi Tel. 358(0)1913607 Fax 358(0)657288

Karvinen, Saila Marjatta (1959). Project manager. [R & D, Kemira Oy, FIN-28840 Pori, Finland.] LicTech (Helsinki U. Technology, 1986) inorganic chemistry. *Crystal_structure titanium_sulphates*
Tel. 358(39)341000 Fax 358(39)341852 Telex 66141 kepor sf

Ketolainen, Prof. Pertti Pekka Juhani (1937). Associate professor. [Department of Physics, University of Joensuu, PO Box 111, FIN-80101 Joensuu, Finland.] PhD (U. Turku, 1969) physics. *Point_defect colour_centre*
E-mail ketolainen@finujo.bitnet Tel. 358(73)1513190 Fax 358(73)1513290

Kivekäs, Raikko Terjo Ilari (1944). Instructor. [Department of Chemistry, PO Box 6, FIN-00014 University of Helsinki, Finland.] PhD (U. Helsinki, 1977) inorganic and analytical chemistry. *Structure_determination copper_compounds boron_compounds*
E-mail kivekas@phcu.helsinki.fi Tel. 358(0)1913649 Fax 358(0)657288

Kivikoski, Dr Jussi Heikki (1964). Research assistant. [Department of Chemistry, University of Jyväskylä, PO Box 35, FIN-40351 Jyväskylä, Finland.] PhD (U. Jyväskylä, 1993) chemistry. *Absolute_configuration structure_determination small_molecules*
E-mail kivikoski@jykem.jyu.fi Tel. 358(41)601211 Fax 358(41)602501

Klinga, Martti (1942). Researcher. [Department of Chemistry, PO Box 55, FIN-00014 University of Helsinki, Finland.] PhD (U. Helsinki, 1987) inorganic chemistry. *Crystallography catalysis nickel_compounds copper_compounds*
E-mail martti.klinga@helsinki.fi Tel. 358(0)1911 Fax 358(0)657288

Korvenranta, Jorma Artturi (1938). Lecturer. [Department of Inorganic Chemistry, PO Box 6, FIN-00014 University of Helsinki, Finland.] PhD (U. Helsinki, 1974) inorganic chemistry. *Structure_determination copper_compounds nickel_compounds*
E-mail korvenranta@finuha.bitnet Tel. 358(0)1913635 Fax 358(0)657288

Kotila, Sirpa Kristiina (1962). Senior assistant. [University of Jyväskylä, Department of Chemistry, Survontie 9, FIN-40500 Jyväskylä, Finland.] PhD (U. Jyväskylä, 1994) chemical crystallography. *Synthesis structure_determination thermogravimetry copper_compounds*
E-mail kotila@jykem.jyu.fi Tel. 358(41)602612 Fax 358(41)602501

Kurki-Suonio, Prof. Dr Kaarle V. J. (1933). Professor. [Department of Physics, PO Box 9, FIN-00014 University of Helsinki, Finland.] PhD (U. Helsinki, 1959) physics. *Charge_density physics education*
E-mail kurkisuonio@hyflt.helsinki.fi Tel. 358(0)1918328 Fax 358(0)1918680

Kyröläinen, Antero Johannes (1951). Research engineer. [Outokumpu Polarit Oy, FIN-95400 Tornio, Finland.] MSc (U. Oulu, 1976) physical metallurgy. *Electron_microscopy stainless_steels*
Tel. 358(698)452580 Fax 358(698)452350 Telex 3518 okto sf

Lähdeniemi, Matti J. I. (1949). Docent, Senior lecturer. [Satakunta Polytechnic, PO Box 30, FIN-28601 Pori, Finland.] PhD (U. Turku, 1982) physics. *Electronic_structure semiconductors molecular image_processing automation sensors*
E-mail mattil@pori.tut.fi Tel. 358(39)802540 Fax 358(39)802600

Lahti, Seppo I. (1947). Geologist. [Geological Survey of Finland, Betonimiehenkuja 4, FIN-02150 Espoo, Finland.] PhD (U. Helsinki, 1981) geology and mineralogy. *Miner-als powder_diffraction mineral_crystal_structures single_crystal_X-ray_methods*
Tel. 358(0)46932248 Fax 358(0)462205 Telex 123 185 geolo sf

Laine, Ensio Sulo Uolevi (1940). Head of the Department of Physics. [Department of Physics, University of Turku, FIN-20500 Turku, Finland.] PhD (U. Turku, 1973) physics. *X-ray_diffraction pharmaceutical_physics phase_transition poly-morphism*
E-mail enls@sara.utu.fi Tel. 358(21)6335673 Fax 358(21)6335070

Lehtinen, Martti Kalevi (1941). Head of the Geological Museum. [Geological Mu-seum, PO Box 11 (Snellmaninkatu 3), FIN-00014 University of Helsinki, Finland.] PhD (U. Helsinki, 1976) geology and mineralogy. *Mineralogy petrology meteoritics shock_metamorphism meteorite_craters*
Tel. 358(0)1913424 Fax 358(0)1913466

Lehto, Vesa-Pekka (1967). Researcher. [Department of Physics, University of Turku, FIN-20500 Turku, Finland.] MSc (U. Turku, 1993) physics. *X-ray_diffraction pharmaceu-tical_physics phase_transition line_profile_analysis*
E-mail vlehto@sara.utu.fi Tel. 358(21)6335675 Fax 358(21)6335070

Leiro, Jarkko Albert (1949). Acting lecturer, Docent. [Department of Applied Physics, University of Turku, Itäinen pitkäkatu 1, FIN-20520 Turku, Finland.] PhD (U. Turku, 1982) physics. *Electronic_structure X-ray_spectroscopy photoemis-sion_spectroscopy surface_physics*
E-mail jleiro@finabo.bitnet Tel. 358(21)6338698 Fax 358(21)6338692

Leppä-aho, Jaakko Antero (1958). Assistant. [Department of Chemistry, University of Jyväskylä, PO Box 35, FIN-40351 Jyväskylä, Finland.] LicPhil (U. Jyväskylä, 1989) inorganic chemistry. *Inorganic_crystal_structure*
E-mail leppa_aho@jylk.jyu.fi Tel. 358(41)602614 Fax 358(41)602501

Leskelä, Markku Antero (1950). Professor. [Department of Chemistry, PO Box 6, FIN-00014 University of Helsinki, Finland.] DTech (Helsinki U. Technology, 1980) inor-ganic chemistry. *Thin_film surface rare-earth_compounds catalysis thermoanalysis*
E-mail markku.leskela@helsinki.fi Tel. 358(0)1913644 Fax 358(0)657288

Lindqvist, Kristian Vilhelm (1949). Geologist. [Geological Survey of Finland, Petrologi-cal Department, Betonimiehenkuja 4, FIN-02150 Espoo, Finland.] MSc (U. Helsinki, 1981) geology and mineralogy. *X-ray_diffraction*
Tel. 358(0)46931 Fax 358(0)462205 Telex 123185 geolo sf

Lindroos, Veikko Kalervo (1938). Professor. [Helsinki University of Technology, Depart-ment of Materials Science and Engineering, Vuorimiehentie 2 A, FIN-02150 Espoo, Finland.] DSc (Helsinki U. Technology, 1968) physical metallurgy. *Metallography electronic_materials magnetic_materials composite_materials nanophase_materials sustainable_development*
Tel. 358(0)4512673 Fax 358(0)4512677 Telex 125161

Lumme, Paavo Olavi (1923). Professor emeritus. [Department of Chemistry, PO Box 55, FIN-00014 University of Helsinki, Finland.] PhD (U. Helsinki, 1956) physical chem-istry. *Bioinorganic complex organometallic structural_chemistry magnetochemistry spectroscopy thermal_chemistry*
E-mail paavo.lumme@cc.helsinki.fi lumme@convex.csc.fi Tel. 358(0)1911 Fax 358(0)40918

Mäki, Jouko Kalervo (1943). Laboratory manager. [University of Turku, Laboratory of Electron Microscopy, Kiinamyllynkatu 10, FIN-20520 Turku, Finland.] PhD (U. Turku, 1993) physics. *Electron_microscopy long-period_order structure macro-molecules atomic_force_microscopy*
E-mail jouko.maki@utu.fi Tel. 358(21)6337318 Fax 358(21)6337380

Manninen, Seppo Olavi (1944). Senior lecturer. Consultant to Commission on Charge, Spin and Momentum Densities of IUCr. [Department of Physics, PO Box 9, FIN-00014 University of Helsinki, Finland.] PhD (U. Helsinki, 1972) solid state physics. *Inelas-tic_scattering resonant_scattering synchrotron_radiation XANES*
E-mail seppo.manninen@helsinki.fi Tel. 358(0)1918336 Fax 358(0)1918680

Mansikka, Kauko (1932). Professor. [Department of Physics, University of Turku, FIN-20500 Turku, Finland.] PhD (U. Turku, 1961) physics. *Theory metals binary_alloys surface foundations_of_quantum_mechanics_and_cosmology*
E-mail mansikka@sara.cc.utu.fi Tel. 358(21)6335680 Fax 358(21)6335070

Merisalo, Matti Juhani (1937). Laboratory manager in chief, Docent. [Depart-ment of Physics, PO Box 9 (Siltavuorenpenger 20 D), FIN-00014 University of Helsinki, Finland.] PhD (U. Helsinki, 1967) physics. *X-ray_crystallography neu-tron_crystallography charge_density materials_science*
E-mail matti.merisalo@helsinki.fi Tel. 358(0)1918330 Fax 358(0)1918680 Telex 100 2125 attn Physics Department

Muhonen, Dr Heikki Juhani (1946). Instructor. [Department of Chemistry, PO Box 6, FIN-00014 University of Helsinki, Finland.] PhD (U. Helsinki, 1987) inorganic chem-istry. *Crystal_structure_determination magnetochemistry coordination_compounds organic_small_molecules*
E-mail muhonen@csc.fi Tel. 358(0)1913623 Fax 358(0)657288 Telex 124690 unih sf

Mutikainen, Dr Ilpo Pellervo (1947). Docent. [Department of Chemistry, PO Box 6, FIN-00014 University of Helsinki, Finland.] PhD (U. Helsinki, 1988) inorganic chemistry. *Chemical_crystallography*
E-mail mutikainen@finuha.bitnet Tel. 358(0)1913608 Fax 358(0)657288

Näsäkkälä, Matti Eerik (1944). Instructor, Docent. [Inorganic Chemistry Laboratory, PO Box 6, FIN-00014 University of Helsinki, Finland.] PhD (U. Helsinki, 1977) inor-ganic chemistry. *Crystallography structure_determination copper_compounds sul-fur_compounds*
E-mail nasakkala@finuha.bitnet Tel. 358(0)1913634 Fax 358(0)657288

Nenonen, Pertti Olavi (1943). Chief research scientist. VTT Manufacturing Technology, Metallimiehenkuja 4, Espoo, Finland. [VTT Manufacturing Technology, PO Box 1703, FIN-02044 VTT, Finland.] LicTech (Helsinki U. Technology, 1974) physical metal-lurgy. *Electron_microscopy materials_science*
E-mail pertti.nenonen@vtt.fi Tel. 358(0)4561 Fax 358(0)463118 Telex 122972 vttha sf

Nieminen, Kari Veikko Juhani (1946). Senior research chemist. [Research Centre of the Defence Forces, PO Box 5, FIN-34111 Lakiala, Finland.] PhD (U. Helsinki, 1983) inorganic chemistry. *Crystal_structure aerosol_research NBC-protection_material*
Tel. 358(31)2843336 Fax 358(31)2843333

Orama, Olli Antero (1944). Docent. [Department of Chemistry, PO Box 6, FIN-00014 University of Helsinki, Finland.] PhD (U. Helsinki, 1976) inorganic chemistry. *Organometallic*
Tel. 358(0)1913609 Fax 358(0)657288

Paakkari, Prof. Timo L. P. (1937). Associate professor. [Department of Physics, PO Box 9, FIN-00014 University of Helsinki, Finland.] PhD (U. Helsinki, 1968) physics. *Pow-der_diffraction disordered_materials cellulases Compton_scattering*
E-mail timo.paakkari@helsinki.fi Tel. 358(0)1918354 Fax 358(0)1918680

Pajunen, Aarne Veikko (1939). Associate professor. [Department of Chemistry, PO Box 6, FIN-00014 University of Helsinki, Finland.] PhD (U. Helsinki, 1967) inorganic chem-istry. *Structure_determination copper_compounds*
E-mail aarne.pajunen@cc.helsinki.fi Tel. 358(0)1913626 Fax 358(0)657288

Pakkanen, Tapani Antti (1949). Professor. [Department of Chemistry, University of Joen-suu, PO Box 111, FIN-80101 Joensuu, Finland.] PhD (SUNY at Stony Brook, NY, USA, 1977) chemistry. *Surface_chemistry catalysis cluster_chemistry*
E-mail tap@joyl.joensuu.fi Tel. 358(73)1513345 Fax 358(73)1513344

Pakkanen, Dr Tuula T. (1949). Associate professor. Department of Chemistry, University of Joensuu, Finland. [Department of Chemistry, University of Joensuu, PO Box 111, FIN-80101 Joensuu, Finland.] PhD (SUNY at Stony Brook, USA, 1978) chemistry. *Organometallic_clustering catalysis material_nuclear_magnetic_resonance*
E-mail tuulapakkane@joyl.joensuu.fi Tel. 358(73)1513340 Fax 358(73)1513344 Telex 462 23 joy sf

Papunen, Prof. Heikki Tapani (1936). Professor. [Department of Geology, University of Turku, FIN-20500 Turku, Finland.] PhD (U. Turku, 1971) economic geology. *Miner-alogy*
E-mail papunen@utu.fi Tel. 358(21)6335480 Fax 358(21)6336580

Pessa, Markus (1941). Professor, Research director. [Tampere University of Technol-ogy, PO Box 692, FIN-33101 Tampere, Finland.] PhD (U. Turku, 1971) elec-tron spectroscopy. *X-ray_crystallography compound_semiconductor_technology electron_spectroscopy catalysis epitaxial_semiconductor_layer_growth surface_physics*
E-mail pessa@ee.tut.fi Tel. 358(31)3162548 Fax 358(31)3162600 Telex 22313

Pirttimäki, Jukka Olavi (1965). Researcher. [University of Turku, Department of Physics, FIN-20500 Turku, Finland.] MSc (U. Turku, 1992) physics. *X-ray_diffraction phar-maceutical_physics phase_transition small-angle_scattering*
E-mail jpirttimaki@sara.utu.fi Tel. 358(21)6335675 Fax 358(21)6335070

Pitkänen, Ilkka Pellervo (1937). Lecturer. [Department of Chemistry, University of Jyväskylä, PO Box 35, FIN-40351 Jyväskylä, Finland.] PhD (U. Jyväskylä, 1980) in-organic chemistry. *Structural_chemistry co-crystals*
Tel. 358(41)602604 Fax 358(41)602501

Punkkinen, Matti (1939). Professor. [Department of Physics, University of Turku, FIN-20500 Turku, Finland.] PhD (U. Turku, 1967) physics. *Nuclear_magnetic_reso-nance NMR in solids*
Tel. 358(21)6335947 Fax 358(21)319836

Pyykkö, Prof. Pekka (1941). Professor. [Department of Chemistry, PO Box 19, FIN-00014 University of Helsinki, Finland.] PhD (U. Turku, 1967) physics. *Magnetic_resonance_spectroscopy relativistic_quantum_chemistry heavy-element_structures*
E-mail pyykko@cc.helsinki.fi Tel. 358(0)1917242 Fax 358(0)406159 Telex 12 11 99 seism

Rajala, Riitta Helena (1960). Research scientist, PhD student. [University of Turku, Department of Physics, FIN-20500 Turku, Finland.] Phil. Lic. (U. Turku, 1984) applied physics. *Organic_hydrates pharmaceutical_physics crys-tal_structure_determination polymorphic_structure*
E-mail riitta.rajala@utu.fi Tel. 358(21)6336562 Fax 358(21)6335070

Rautioaho, Risto Heikki (1945). Associate professor. [Materials Engineering Laboratory, University of Oulu, Linnanmaa, FIN-90570 Oulu, Finland.] PhD (U. Oulu, 1981) ma-terials science. *Structure ceramics*
Tel. 358(81)5532141 Fax 358(81)5533973 Telex 32375 oylin sf

Rissanen, Kari (1959). Associate professor. [Department of Chemistry, University of Joensuu, PO Box 111, FIN-80101 Joensuu, Finland.] PhD (U. Jyväskylä, 1990) chemistry. *Structure_determination synthesis absolute_configuration area_detector supramolecular_structures organic_compounds*
E-mail rissanen@joyl.joensuu.fi Tel. 358(73)1513367 Fax 358(73)1513390

Rouvinen, Juha (1960). Assistant. [Department of Chemistry, University of Joensuu, PO Box 111, FIN-80101 Joensuu, Finland.] PhD (U. Joensuu, 1991) chemistry. *Biocrys-tallography modelling*
E-mail jpro@joyl.joensuu.fi Tel. 358(73)1513318 Fax 358(73)1513390

Salminen, Tiina (1965). PhD student. [University of Turku, Department of Biochemistry/Biocity, Tykistökatu 6, FIN-20520 Turku, Finland.] MSc (U. Turku, 1992) biochemistry. *Structure-activity_relationship macromolecular_crystallography proteins*
E-mail tsalmine@ala.btk.utu.fi Tel. 358(21)6338055 Fax 358(21)6338050

Salo, Päivi Tuulikki (1963). Physics teacher, Research assistant. [University of Turku, Department of Physics, FIN-20500 Turku, Finland.] MSc (U. Turku, 1993) applied physics. *Organic_hydrates pharmaceutical_physics crystal_structure_determination phase_determination*
E-mail patusa@utu.fi Tel. 358(21)6335675 Fax 358(21)6335070

Serimaa, Ritva Elina (1957). Assistant. [Department of Physics, PO Box 9, FIN-00014 University of Helsinki, Finland.] PhD (U. Helsinki, 1990) physics. *Computer_modelling amorphous_phase scattering ASAXS AWAXS*
E-mail rserimaa@convex.csc.fi Tel. 358(0)1918349 Fax 358(0)1918680

Silvola, Prof. Jaakko Uolevi (1938). Professor. [Department of Geology, PO Box 11 (Snellmaninkatu 3), FIN-00014 University of Helsinki, Finland.] PhD (U. Helsinki, 1971) geology and mineralogy. *Crystallography mineralogy geology Antarctic_geology*
Tel. 358(0)1913457 Fax 358(0)1913466 Telex 124690 unih sf

Sivonen, Seppo Juhani (1942). Director. [Institute of Electron Optics, University of Oulu, PO Box 400, FIN-90571 Oulu, Finland.] MSc (U. Oulu, 1967) technical physics. *X-ray_microanalysis X-ray_fluorescence_analysis*
Tel. 358(81)5533140 Fax 358(81)5561278 Telex 32375 oylin sf

Smolander, Dr Kimmo Juhani Nils-Eric (1944). Lecturer. [Department of Chemistry, University of Joensuu, PO Box 111, FIN-80101 Joensuu, Finland.] PhD (U. Helsinki, 1983) inorganic chemistry. *Structure_determination transition_element_complexes*
Tel. 358(73)1513364 Fax 358(73)1513390 Telex 46223 joy sf

Sundius, Tom Robert (1942). Docent. [Department of Physics, PO Box 9, FIN-00014 University of Helsinki, Finland.] PhD (U. Helsinki, 1981) physics. *Molecular_force_field molecular_mechanics spectrum_analysis ab_initio_calculations*
E-mail tom.sundius@helsinki.fi Tel. 358(0)1918361 Fax 358(0)1918680 Telex 100 2125 attn Physics Department

Suoninen, Prof. Eero Juhani (1929). Professsor. [University of Turku, Department of Applied Physics, Materials Science, Itäinen Pitkäkatu 1, FIN-20520 Turku, Finland.] PhD (Massachusetts Inst. of Technology USA, 1957) physics. *Electron_spectroscopy materials_research surface_science basic_studies_of_flotation*
E-mail suoninen@polaris.utu.fi Tel. 358(21)6338694 Fax 358(21)6338692

Suortti, Prof. Pekka (1938). Associate professor. [Department of Physics, PO Box 9, FIN-00014 University of Helsinki, Finland.] PhD (U. Helsinki, 1967) physics. *Anomalous_dispersion synchrotron_radiation*
E-mail pekka.suortti@helsinki.fi Tel. 358(0)1918336 Fax 358(0)1918680

Taikina-aho, Olavi Seppo Allan (1954). Assistant. [Institute of Electron Optics, University of Oulu, PO Box 400, FIN-90571 Oulu, Finland.] MSc (U. Oulu, 1983) geology and mineralogy. *Analytical_electron_microscopy X-ray_diffractometry*
Tel. 358(81)5533142 Fax 358(81)5561278 Telex 32375 oylin sf

Tarna, Toivo (1940). Lecturer. [Turku Institute of Technology, Sepänkatu 1, FIN-20700 Turku, Finland.] LicPhil (U. Turku, 1976) physics. *X-ray_powder_diffraction energy-dispersive_diffraction_analysis data_processing teaching_(physics)*
Tel. 358(21)6370691 Fax 358(21)6370791

Tiitta, Antero Tapani (1946). Senior research scientist. VTT Chemical Technology, Physics Bldg., Otakaari 3A, Espoo, Finland. [VTT Chemical Technology, PO Box 1404, FIN-02044 VTT, Finland.] LicTech (Helsinki U. Technology, 1980) nuclear engineering. *Neutron_diffraction neutron_time-of-flight instrument_design*
E-mail antero.tiitta@vtt.fi Tel. 358(0)4566350 Fax 358(0)4566390 Telex 122972 vttha sf

Toivonen, Jukka Tapio (1951). Lecturer. [Espoo-Vantaa Institute of Technology, Leiritie 1, FIN-01600 Vantaa, Finland.] LicTech (Helsinki U. Technology, 1983) inorganic chemistry. *Diffraction_technique crystal_structure uranium_compounds*
E-mail jukkatt@evitech.fi Tel. 358(0)5119738 Fax 358(0)5119977

Torkkeli, Hannu Mika (1966). Research assistant. [Department of Physics, PO Box 9, FIN-00014 University of Helsinki, Finland.] MSc (U. Helsinki, 1992) physics. *SAXS disorder*
E-mail mika.torkkeli@helsinki.fi Tel. 358(0)1918329 Fax 358(0)1918680

Tuomi, Turkka Olavi (1939). Professor. [Helsinki University of Technology, FIN-02150 Espoo, Finland.] DSc (Helsinki U. Technology, 1968) technical physics. *Semiconductors synchrotron_X-ray_topography optical_properties_of_solids*
Tel. 358(0)4513120 Fax 358(0)465077

Turpeinen, Urho Taneli (1944). Instructor. [Department of Chemistry, PO Box 6, FIN-00014 University of Helsinki, Finland.] PhD (U. Helsinki, 1977) chemical crystallography. *Crystallography carboxylates transition_elements*
E-mail turpeinen@flta.helsinki.fi Tel. 358(0)1913610 Fax 358(0)657288

Turunen, Dr Markus (1947). Research scientist, Docent. [Vaisala Oy, PO Box 26, FIN-00421 Helsinki, Finland.] DSc (Helsinki U. Technology, 1975) physical metallurgy. *Sensor semiconductors new_materials signal_processing neural_networks*
E-mail mturunen@finsun.csc.fi Tel. 358(0)8949549 Fax 358(0)8949225

Vahvaselkä, Dr Aino Margit (1942). Amanuensis. Secretary of the Finnish National Committee for Crystallography, Finnish Sub-Editor of the WDC9. [Department of Physics, PO Box 9, FIN-00014 University of Helsinki, Finland.] PhD (U. Helsinki, 1978) physics. *Charge_density*
E-mail aino.vahvaselka@helsinki.fi Tel. 358(0)1918334 Fax 358(0)1918680

Vahvaselkä, Dr Sakari (1942). Assistant. [Department of Physics, PO Box 9, FIN-00014 University of Helsinki, Finland.] PhD (U. Helsinki, 1977) physics. *Diffraction_technique polymers liquids liquid_crystals*
E-mail sakari.vahvaselka@helsinki.fi Tel. 358(0)1918350 Fax 358(0)1918680

Valkonen, Jussi Uolevi (1947). Professor. University of Jyväskylä, Department of Chemistry, Survontie 9, FIN-40500 Jyväskylä, Finland. [University of Jyväskylä, Department of Chemistry, PO Box 35, FIN-40351 Jyväskylä, Finland.] PhD (Helsinki U. Technology, 1979) inorganic chemistry. *Inorganic_chemistry structure_determination*
E-mail valkonen@jykem.jyu.fi Tel. 358(41)601211 Fax 358(41)602501

Ylinen, Dr Eero E. (1944). Research physicist, Docent. [Wihuri Physical Laboratory, University of Turku, FIN-20500 Turku, Finland.] PhD (U. Turku, 1978) physics. *Nuclear_magnetic_resonance NMR_in_solids*
Tel. 358(21)6335944 Fax 358(21)319836

FRANCE

Sub-Editor: B. Capelle

Aberdam, Dr Daniel (1937). Director of Research, CNRS. Laboratoire de Cristallographie, 25 avenue des martyrs, Grenoble, France. [BP 166X, 38042 Grenoble Cedex, France.] DSc (U. Grenoble, 1971) physics. *Crystallography surface electrochemistry LEED*
E-mail aberdam@labs.polycnrs-gre.fr Tel. 33()76887941 Fax 33()76881038

Alzari, Dr Pedro M. (1956). Chef de Laboratoire Institut Pasteur. Unité d'Immunologie Structurale, Institut Pasteur, 25 rue du Dr Roux, Paris Cedex 15, France. PhD (La Plata U., Argentina, 1985) physics. *X-ray_crystallography biological_macromolecules*
E-mail alzari@pasteur.fr Tel. 33(1)45688607 Fax 33(1)45688639

Andonov-Dandigna, Dr Paulette (1932). Structural studies group leader. Lab. de Magnétisme et Matériaux magnetiques, CNRS, 1 place A. Briand, 92195 Meudon Cedex, France. DSc (U. Paris-Sud, 1979) physics. *Local_order liquid_state clustering local_order_X-ray_and_neutron_diffraction*
Tel. 33(1)45075076 Fax 33(1)45075899 Telex LABOBEL634002F

Andreazza, Dr Pascal (1961). Centre de Recherche sur la Matiere Divisée, Université d'Orleans, URM 0131, BP 6759, rue de Chartres, 45067 Orleans Cedex 02, France. DU (U. Paris VI, 1990) physics. *Grazing_incidence crystal_growth non_linear_optics phyllosilicates*
E-mail andreazz@avion.univ-orleans.fr Tel. 33()38417005 Fax 33()38417080

Arnoux, Dr Bernadette (1952). Laboratoire de Biologie Structurale, Bât. 34 CNRS, 91198 Gif sur Yvette, France. DSc (U. Orsay, 1985) physics. *Macromolecular_crystallography collagenase membrane_protein*
E-mail arnoux@cygne.lbs.cnrs-gif.fr Tel. 33(1)69823491 Fax 33(1)69823129

Authier, Prof. André (1932). Professor. Past President IUCr; Editor, Section A, Acta Crystallographica. Lab. Minéralogie-Cristallographie, Université P.M. Curie, 4 place Jussieu, 75252 Paris Cedex 05, France. DSc (Paris University, 1961) physics. *Topography dynamical_diffraction crystal_growth characterization*
E-mail authier@lmcp.jussieu.fr Tel. 33(1)44273784 Fax 33(1)44273785

Averbuch-Pouchot, Dr Marie-Thérèse (1940). Lab. Cristallographie, CNRS, 25 avenue des Martyrs, 38042 Grenoble Cedex, France. DSc (U. Grenoble, 1974) structural chemistry. *Phosphates structure_determination chemistry*
E-mail mt_averbuch@labs.polycnrs-gre.fr Tel. 33()76887801 Fax 33()76881038 Telex 320254CNRSALPGRENO

Bachet, Mr Bernard (1933). Research ingenior. Laboratoire de Minéralogie-Cristallographie, URA009 CNRS, Universités P.M. Curie et Paris 7, 4 place Jussieu. Paris 5, France. [Université P.M. Curie, case 115, 75252 Paris Cedex 05, France.] MSc (U. Paris Sorbonne, 1961) mineralogy. *Structure small_molecules macromolecules X-ray_diffraction*
E-mail bachet@lmcp.jussieu.fr Tel. 33(1)44274586 Fax 33(1)44273785

Baruchel, Dr José (1947). Topography group leader. Experiment Division ESRF, BP 220, 38043 Grenoble Cedex, France. DSc (U. Grenoble, 1980) physics. *Topography magnetism phase_transition domain*
E-mail baruchel@ill.fr Tel. 33()76882101 Fax 33()76882542 Telex 308352F

Baudet-Maze, Dr Mona (1938). Laboratoire OCM/MPA, CNET, route de Tregastel, 22301 Lannion, France. DSc (U. Paris, 1970) physics. *Epitaxy semiconductors superlattice computer_simulation strain_determination*

Tel. 33()96053353 Fax 33()96053239 Telex 740970F

Bavoux, Dr Claude (1943). Laboratoire de Cristallographie, Université Claude Bernard, ER 60, 43 boulevard du 11 Novembre 1918, 69622 Villeurbanne Cedex, France. DSc (U. Lyon, 1986) physics. *Macrocycles structure_determination molecular_complexes molecular_recognition*
E-mail bavoux@cdlyon.univ-lyon1.fr Tel. 33()72448000 Fax 33()72431160

Becker, Dr Paul (1941). Crystal growth leader. Centre Lorrain Opt. Electron. Solide, Université de Metz, France. [2 rue Edouard Belin, 57078 Metz, France.] DSc (University of Metz, 1978) metallurgical chemistry. *Crystal_growth organic inorganic optical_property*
Tel. 33()87749938 Fax 33()87751724

Bentley, Dr Graham (1946). Head of Unit. Unité d'Immunologie Structurale, Institut Pasteur, 25 rue du Dr Roux, 75724 Paris, France. PhD (U. Auckland, NZ, 1971) chemistry. *Biocrystallography immunology*
E-mail bentley@pasteur.fr Tel. 33(1)45688610 Fax 33(1)45688639

Berar, Dr Jean-François (1949). CRG beamline D2 (anomalous scattering) at ESRF. Laboratoire de Cristallographie, CNRS, 38000 Grenoble, France. [Lab. Cristallographie, CNRS, BP 166, 38042 Grenoble Cedex 9, France.] DSc (U. Paris 6, 1980) physical chemistry. *Rietveld_method synchrotron_radiation powder real-time_imaging*
E-mail berar@rx-ctg1.polycnrs-gre.fr Tel. 33()76887414 Fax 33()76881038

de Bergevin, Dr François (1934). Research scientist. Lab. Cristallographie, CNRS, 25 avenue des Martyrs, Grenoble, France. [BP 166, 38042 Grenoble Cedex 09, France.] DSc (U. Grenoble, 1968) physics. *X-ray_diffraction*
Tel. 33()76881144 Fax 33()76881038

Bertaut, Dr Erwin Felix (1913). Directeur Honoraire, Membre de l'Institut de France. Lab. Cristallographie, CNRS, 25 avenue des Martyrs, 38042 Grenoble Cedex, France. [BP 166, 38042 Grenoble Cedex, France.] DSc (U. Grenoble, 1949) solid state physics. *Crystallography magnetism phase_transition*
E-mail bertaut@labs.polycnrs-gre.fr Tel. 33()76881000x1484 Fax 33()76881038

Berthet-Colominas, Dr Carmen (1940). European Molecular Biology Laboratory, Grenoble Outstation, France. [EMBL c/o ILL, BP 158X, 38042 Grenoble Cedex, France.] DSc (U. Grenoble, 1967) physics. *Crystallography diffraction crystal_growth enzymes*
E-mail berthet@embl.embl-grenoble.fr Tel. 33()76207276 Fax 33()76207199

Bessiere, Dr Michel (1950). Scientific and technical coordinator of DCI (X-ray ring); in charge of beamline D23. LURE, Bât. 209d, Université Paris Sud, 91405 Orsay Cedex, France. DSc (U. Paris 6, 1984) physics. *Diffraction diffuse_scattering order-disorder quasicrystal*
E-mail bessiere@lure.u-psud.fr Tel. 33(1)64468125 Fax 33(1)64468148

Besson, Dr Jean-Michel (1936). Director of laboratory. Physique des Milieux Condensés, URA 782, Université Paris 6, Paris 5, France. [case 77, Université Paris 6, 75252 Paris Cedex 05, France.] DSc (U. Paris 6, 1967) physics. *Diffraction high_pressure neutron_diffraction*
Tel. 33(1)44274461 Fax 33(1)44274469

Bodot, Prof. Hubert (1933). Director Doctoral School Spectrometry. Université de Provence, URA CNRS 773, Centre Saint Jerome, Case 542, 13397 Marseille Cedex 20, France. DSc (U. Marseille, 1957) chemistry. *Stereochemistry tautomerism photochromic_compounds FTIR photochemistry*
Tel. 33()91636510 Fax 33()91020550 Telex AMIUP402014

Bonnet, Dr Michel (1944). Centre d'Etudes Nucleaires de Grenoble, DRFMC/MDN, 85X, 38041 Grenoble Cedex, France. DSc (U. Grenoble, 1976) physics. *Neutron_diffraction structure magnetism*
E-mail bonnet@drfmc.ceng.cea.fr Tel. 33()76883956 Fax 33()76885109

Bonpunt, Dr Louis (1944). Lab. Cristallographie, URA CNRS 144, Université Bordeaux 1, 33405 Talence, France. DSc (U. Bordeaux, 1981) physics. *Defect molecular_crystal*
Tel. 33()56846155 Fax 33()56846686

Bouillot, Prof. Jacques (1943). Group leader. Faculté Annecienne de Sciences et Techniques, 41 avenue de la Plaine, Annecy, France. [BP 806, 74016 Annecy Cedex, France.] DSc (U. Dijon, 1977) condensed matter physics. *Ferroelectricity optical_property electrooptics non-linear property*
Tel. 33()50675611 Fax 33()50574885

Bourret, Dr Alain (1938). Director department. Former president Société Française de Microscopie. Dept de Recherche Fondamentale/SP2M, CENG, 17 rue des Martyrs, 38054 Grenoble Cedex 9, France. DSc (U. Grenoble, 1970) physics. *Interface high-resolution_electron_diffraction microstructure dislocation imaging diffraction structures_and_interfaces*
Tel. 33()76883458 Fax 33()76885197

Brunie, Dr Simone (1936). Group leader. Laboratoire de Biologie Cellulaire et Moléculaire, INRA, 78352 Jouy en Josas Cedex, France. DSc (U. Lyon, 1970) physics. *X-ray_crystallography proteins molecular_modelling*
E-mail brunie@sgi1.jouy.inra.fr Tel. 33(1)34652565 Fax 33(1)34652273

Burggraf, Prof. Charles (1933). Group leader. Laboratoire de Cristallographie, 1 rue Blessig, 67000 Strasbourg, France. [3 Rue Saint-Paul, 67300 Schiltigheim, France.] DSc (U. Strasbourg, 1969) physics. *Crystallography surface thin_layer alloys*
E-mail charles@iutlp1.u-strasbg.fr Tel. 33()88815506 Fax 33()88811600

Burggraf, Dr Christiane (1933). Group leader. Laboratoire de Cristallographie, 1 rue Blessig, 67000 Strasbourg, France. [3 Rue Saint-Paul, 67300 Schiltigheim, France.] DSc (U. Metz, 1974) physics. *Crystallography surface thin_layer alloys*
E-mail christiane@iutlp1.u-strasbg.fr Tel. 33()88815506 Fax 33()88811600

Busetta, Dr Bernard (1945). Laboratoire de Cristallographie, URA 144, Université de Bordeaux I, cours de la Liberation, 33405 Talence, France. DSc (U. Bordeaux, 1973) physics. *Crystallography*
Tel. 33()56846164 Fax 33()56846686

Callebaut, Dr Isabelle (1966). CR2 CNRS. Laboratoire de Minéralogie-Cristallographie, Universités Paris 6/7, Paris 5, France. [t.16, case 115, 4 place Jussieu, F75252 Paris Cedex 05, France.] DSc (Faculté des Sciences Agronomiques, Gembloux, Belgium, 1993) agronomy. *Proteins structure macromolecules virology protein_sequence_analysis*
E-mail callebau@lmcp.jussieu.fr Tel. 33(1)44277247 Fax 33(1)44273785

Capelle, Dr Bernard (1949). Group leader. Secretary French Crystallographic Committee; Sub-Editor WDC9. Laboratoire de Minéralogie-Cristallographie, URA 009 CNRS, Case 115, 4 place Jussieu, 75252 Paris Cedex 05, France. DSc (U. Paris VI, 1982) physics. *Interface piezoelectric_materials X-ray_topography standing_wave thin_film synchrotron_radiation*
E-mail capelle@lmcp.jussieu.fr Tel. 33(1)44275217 Fax 33(1)44273785

Caranoni, Prof. Claude (1941). Group leader. Laboratoire des Matériaux: organisation et propriétés, associé au CNRS, case 151, Faculté des Sciences et Techniques, 13397 Marseille Cedex 20, France. DSc (U. Aix-Marseille 3, 1979) physics. *X-ray_diffraction electron_microscopy DTA disordered_ferroelectric_oxides nanostructure microdomains phase_transition physical_properties*
E-mail nmenguy@matop.univ-mrs.fr Tel. 33()91288120 Fax 33()91288775 Telex 402876 F

Cavarelli, Dr Jean (1957). Associate Professor. IGBMC, Parc d'Innovation, BP163, 67404 Illkirch Cedex, France. PhD (U. Strasbourg, 1987) biophysics. *Data_collection refinement_method proteins DNA aminoacyl-tRNA_synthetases*
E-mail cava@brouilly.u-strasbg.fr Tel. 33()88653301 Fax 33()88653201

Chattopadhyay, Dr Tapan (1941). Institut Laue-Langevin, BP 156X, 38042 Grenoble Cedex, France. PhD (Indian Institute of Technology Kharagpur, India, 1972) physics. *Modulated_structures magnetism superconductors X-ray and neutron diffraction*
Fax 33()76207066 Telex 320621F

Chevrier, Dr Bernard (1941). Research Associate. IGBMC, BP 163, 67404 Illkirch Cedex, France. DSc (U. Louis Pasteur, Strasbourg, 1971) chemistry. *Computing X-ray structure proteins*
E-mail chevrier@igbmc.u-strasbg.fr Tel. 33()88653313 Fax 33()88653201

Chomilier, Dr Jacques (1955). Laboratoire de Minéralogie-Cristallographie, URA 009 CNRS, Universités P.M. Curie et Paris VII, 4 place Jussieu, Paris 5, France. [Université P.M. Curie, case 115, 75252 Paris Cedex 05, France.] DSc (U. Paris VII, 1987) physics. *Proteins structure computer_modelling*
E-mail chomilie@lmcp.jussieu.fr Tel. 33(1)44275079 Fax 33(1)44273785

Cohen-Addad, Dr Claudine (1939). Institut de Biologie Structurale, 41 av. des Martyrs, 38047 Grenoble Cedex, France. DSc (Grenoble U., 1969) physics. *Proteins structure_crystallography crystallization*
E-mail addad@ibs.fr Tel. 33()76889588 Fax 33()76885494

Colloc'h, Dr Nathalie (1962). Laboratoire de Minéralogie-Cristallographie, CNRS, URA 009, Université P.M. Curie, France. [case 115, université P.M. Curie, 75252 Paris Cedex 05, France.] DU (U. Paris 6, 1988) physics. *Proteins structure_determination computer_modelling isomorphous replacement secondary_structure proteins_prediction*
E-mail colloch@lmcp.jussieu.fr Tel. 33(1)44274585 Fax 33(1)44273785

Comes, Dr Robert (1937). Director. LURE, Bât. 209D, Centre Universitaire, 91405 Orsay Cedex, France. DSc (U. Orsay, 1969) physics. *Crystallography phase_transition X-ray_scattering neutron_scattering ferroelectrics 1D_conductors disorder_in_solids phonons*
E-mail luremail@lure.ups.circe.fr Tel. 33(1)64468001 Fax 33(1)64464102 Telex LURELAB603340F

Convert, Dr Pierre (1941). Scientist-ingenior. Contributor Vol. C Int. Tables for Crystallography. Institut Laue-Langevin, BP 156, 38042 Grenoble Cedex 9, France. DU (U. Grenoble, 1975) physics. *Neutron_diffractometry neutron_linear_detector kinetics strain_determination neutron_powder_diffractometry*
E-mail convert@ill.fr Tel. 33()76207295 Fax 33()76483906 Telex 320621F

Curien, Prof. Hubert (1924). President of CERN. Lab. Minéralogie-Cristallographie, Université P.M. Curie, 4 place Jussieu, 75252 Paris Cedex 05, France. [Tour 16, Case 115, Université P.M. Curie, 75252 Paris Cedex 05, France.] DSc (U. Paris, 1951) physics. *Defect diffuse_scattering*
E-mail curien@lmcp.jussieu.fr Tel. 33(1)44273783 Fax 33(1)44273785

Dahan, Dr Françoise (1941). Technical direction. Laboratoire de Chimie de Coordination CNRS, 205 route de Narbonne, 31077 Toulouse Cedex, France. DSc (U. Paris 6, 1980) physics. *X-ray_structure computing*
E-mail dahan@lcctou.lcc-toulouse.fr Tel. 33()61333194 Fax 33()61553003

Daran, Dr Jean-Claude (1941). Directeur de Recherche CNRS; group leader. Laboratoire de Chimie de Coordination, CNRS, UPR8241, 205 Route de Narbonne, 31077 Toulouse Cedex, France. DSc (U. Toulouse, 1973) chemistry. *Coordination_chemistry organometallic_compound X-ray_structure_analyses chiral_compunds non-linear_optical_material*
E-mail daran@lcctoul.lcc-toulouse.fr Tel. 33()61333174

Dartyge, Prof. Elisabeth (1941). LURE, Bât. 209D, Université Paris Sud, 91405 Orsay Cedex, France. DSc (U. Orsay, 1979) physics. *X-ray_spectroscopy circular_dichroism magnetism rare-earth_compounds synchrotron_radiation*
E-mail dartyge@lure.ups.circe.fr Tel. 33(1)64468020 Fax 33(1)64464148 Telex LURELAB603340F

Dautant, Dr Alain (1957). Laboratoire de Cristallographie, URA 144 CNRS, Université de Bordeaux I, 351 Cours de la Libération, 33405 Talence, France. DSc (U. Bordeaux I, 1988) physics. *X-ray macromolecular_structure*
E-mail dautant@lotus.cristal.u-bordeaux.fr Tel. 33()56846163 Fax 33()56846686

Delettré, Dr Jean (1943). Laboratoire de Minéralogie-Cristallographie, URA 009 CNRS, Universités P.M. Curie et Paris VII, France. [Université P.M. Curie, Case 115, 75252 Paris Cedex 05, France.] DSc (U. Paris VI, 1978) physics. *Biological_structure X-ray_diffraction macromolecules*
E-mail delettre@lmcp.jussieu.fr Tel. 33(1)44275241 Fax 33(1)44273785

Denoyer, Dr Françoise (1945). Lab. Physique des Solides, Bât. 510, Université d'Orsay, 91405 Orsay, France. DSc (U. Paris XI, 1977) physics. *Quasicrystal incommensurate phase_transition*
Tel. 33(1)69416060 Fax 33(1)69416086 Telex 602166FFACORS

Despujols, Prof. Jacques (1925). Emeritus Professor. L.A.S.S.I., Université de Reims-Champagne-Ardenne, Faculté des Sciences, BP 347, 51052 Reims Cedex, France. [30 rue de la Bienfaisance, 75008 Paris, France.] DSc (U. Paris, 1956) physics. *X-ray_spectrometry X-ray_optics thin_film defect*
Tel. 33(1)45222876

Dideberg, Dr Otto (1942). Group leader. Institut de Biologie Structurale, LCM, 41 avenue des Martyrs, 38027 Grenoble Cedex 01, France. DSc (University of Liège, 1969) physics. *Structural_biology macromolecular synchrotron_radiation drug*
E-mail otto@ibs.fr Tel. 33(1)76885609 Fax 33(1)76885609

Dominguez Perez, Mr Roberto (1963). Student. Institut Pasteur, 25 rue du Dr Roux, 75724 Paris Cedex 15, France. MSc (U. Odessa, Ukraine, 1987) physics. *Proteins crystallography structure function*
E-mail dominigue@pasteur.fr Tel. 33(1)45688608 Fax 33(1)45688639

Dubourg, Dr Antoine (1945). Groupe de Biochimie Structurale, Faculté de Pharmacie, avenue Charles Flahaut, 34060 Montpellier, France. DSc (U. Perpignan, 1990) physics. *X-ray_diffraction phosphorus_compound molecular_modelling*
E-mail dubourg@cnusc.fr Tel. 33( )67522301 Fax 33( )67042140

Ducruix, Dr Arnaud (1947). Group leader. Ex. Com. Groupe Fr. Croiss. Cristall., Soc. Fr. Biophysique. Laboratoire de Biologie Structurale, Bât. 34, CNRS, 91198 Gif sur Yvette, France. DSc (U. Paris XI, 1976) physics. *Crystal_growth proteins structure_determination crystallogenesis*
E-mail ducruix@cygne.lbs.cnrs-gif.fr Tel. 33(1)69823474 Fax 33(1)69823129

Durif, Dr André (1929). Coeditor Acta Crystallographica. Lab. Cristallographie, CNRS, 25 av. des martyrs, Grenoble, France. [BP 166, 38042 Grenoble Cedex 09, France.] DSc (U. Grenoble, 1958) structural chemistry. *Phosphates structure_determination chemistry*
E-mail. mt_averbuch@labs.polycnrs-gre.fr Tel. 33( )76881045 Fax 33( )76881038 Telex 320254CNRSALPGRENO

Ehrlich, Grant M. (1967). Graduate Student - PhD. ISITEM, La Chantrerie, Rue Christian Pauc, CP 3023, F 44087 Nantes Cedex 03, France. PhD (Cornell U.; 1995) inorganic chemistry. *Synthesis ceramics refractory_compounds conducting_materials solid_state*
E-mail donald@isirec4.isitem.univ-nantes.fr Tel. 33( )40683127 Fax 33( )40683199

Epelboin, Prof. Yves (1944). Group leader. General Editor World Directory of Crystallographers. Lab. Minéralogie-Cristallographie, URA009 CNRS, Tour 16, 4 place Jussieu, Paris 5, France. [Case 115, Université Paris VI, 75252 Paris Cedex 05, France.] DSc (U. Paris VI, 1974) physics. *X-ray_topography dynamical_diffraction computing image_processing computer_graphics communication*
E-mail epelboin@lmcp.jussieu.fr Tel. 33(1)44275211 Fax 33(1)44273785

Estienne, Dr Jacques (1946). Project leader. Laboratoire de Spectrométrie Moléculaire, Université de Provence, Centre de Saint Jérôme, France. [Case 542, 13 397 Escadrille Normandie-Niémen, 13 397 Marseille Cedex 20, France.] DSc (U. Aix-Marseille I, 1986) chemistry. *Single_crystal structure_determination X-ray_powder_diffraction structural_modelling infrared_spectroscopy*
Tel. 33(1)91288581 Fax 33(1)91020550 Telex AMIUP402014

Fanchon, Dr Eric (1959). Institut de Biologie Structurale, 41 avenue des Martyrs, 38027 Grenoble Cedex, France. DU (U. Grenoble, 1987) physics. *Biomolecule synchrotron_radiation anomalous_diffusion software*
E-mail fanchon@ibs.fr Tel. 33(1)76889591 Fax 33( )76885494

Faure, Prof. René (1946). Lab. Chimie Analytique 2, Université Claude Bernard Lyon I, Bât. 305, France. [69622 Villeurbanne Cedex, France.] DSc (U. Lyon 1, 1973) chemistry. *Crystallography*
Tel. 33( )72431153 Fax 33( )72431020

Filhol, Dr Alain (1944). Technical staff. Institut Laue-Langevin, BP 156, 38042 Grenoble Cedex, France. DSc (U. Bordeaux, 1985) physics. *Graphics modelling statistical_method organic_conductor superconductors*
E-mail filhol@ill.fr Tel. 33( )76207156 Fax 33( )76483906 Telex 320621F

Fitch, Dr Andrew (1956). Beamline scientist. ESRF, BP 220, 38043 Grenoble Cedex, France. PhD (Oxford U., UK, 1982) chemistry. *Neutron_diffraction synchrotron_radiation chemistry Rietveld_method*
E-mail fitch@ill.fr Tel. 33( )76882532 Fax 33( )76882542

Fontaine, Dr Alain (1944). Scientific coordinator at DCI. Vice-chairman EXAFS society. LURE, Bât. 209D, 91405 Orsay, France. DSc (U. Orsay-Paris Sud, 1975) physics. *EXAFS multilayer superconductivity magnetism circular_magnetism X-ray_dichroism X-ray_optics*
E-mail fontaine@lure.ups.circe.fr Tel. 33(1)64468055 Fax 33(1)64464148

Fourme, Prof. Roger (1942). Group leader. Chairman French Crystallographic Association. LURE, Bât. 209d, Université Paris Sud, 91405 Orsay Cedex, France. DSc (U. Paris, 1970) physics. *Macromolecular_structures synchrotron_radiation anomalous_diffraction area_detector*
E-mail fourme@lure.u-psud.fr Tel. 33(1)64468126 Fax 33(1)64464148

Frey, Dr Michel (1938). Directeur de Recherches CNRS. Lab. Cristallogénèse et de Cristallographie des Proteines, Institut de Biologie Structurale - J. P. Ebel, 38027 Grenoble Cedex 01, France. [IBS/LCCP, 41 avenue des Martyrs, 38027 Grenoble Cedex 01, France.] DSc (U. Caen, 1970) physics. *Biocrystallography bioenergetics crystal_growth hydrogenase redox_proteins*
E-mail frey@lccp.ibs.fr Tel. 33( )76885924 Fax 33( )76885122

Gallois, Dr Bernard (1953). Laboratoire de Cristallographie, Université de Bordeaux I, 351 Cours de la Libération, 33405 Talence Cedex, France. DSc (U. Bordeaux I, 1987) physics. *Macromolecular_structure*
E-mail precigou@lotus.cristal.u-bordeaux.fr Tel. 33( )56846163 Fax 33( )56846686

Galy, Dr Jean (1938). Director of CEMES/CNRS. Centre d'Elaboration des Matériaux et d'Etudes Structurales, CNRS, 29 rue J. Marvig, 31400 Toulouse, France. [CEMES-LOE/CNRS, BP 4347, 31055 Toulouse Cedex, France.] DSc (Université de Bordeaux I, 1966) physics. *Crystallography solid_state oxides electron_microscopy crystal_chemistry*
Tel. 33( )62257800 Fax 33( )62257999

Gandais, Dr Madeleine (1934). Group leader. Treasurer Société Française de Physique. Lab. Minéralogie-Cristallographie, Université Paris 6, 4 place Jussieu, 75252 Paris Cedex 05, France. DSc (U. Orsay, 1969) physics. *Semiconductor_microcrystal crystal_growth high-resolution_electron_microscopy*
E-mail gandais@lmcp.jussieu.fr Tel. 33(1)44275212 Fax 33(1)44273785

Garnier, Dr Emmanuel (1947). Assistant Professor. Laboratoire de Chimie Theorique, Université de Poitiers, UA D0350 CNRS, 40 avenue du Recteur Pineau, 86022 Poitiers Cedex, France. DSc (U. Poitiers, 1985) chemistry. *Catalytic_structure_determination crystal_growth ab_initio_and_Hückel_calculations*
E-mail eg@ss3.univ-poitiers.fr Tel. 33( )49454035 Fax 33( )49453600

Gauthier, Prof. Jean-Pierre (1943). Laboratoire de Pétrologie et Tectonique, 43 bld du 11 Novembre 1918, 69622 Villeurbanne Cedex, France. DSc (U. Lyon, 1978) crystallography. *Polytypism gemology electron_diffraction mineralogy microscopy*
Tel. 33( )72448490 Fax 33( )72448382

George, Dr Amand (1946). Director. Lab. Metallurgie Physique - Sciences des Materiaux, Ecole des Mines de Nancy, URA 155 CNRS, Parc de Saurupt, 54042 Nancy, France. DSc (INPL, 1977) materials science. *Defect dislocation plastic_deformation X-ray_topography*
Tel. 33( )83574156 Fax 33( )83579794

Gérard, Dr François (1941). Computer center head. Lab. Acquisition et Traitement des Données, CUPF, Université du Pacifique, Tahiti, Polynèsie Française, France. [CUPF, BP 6570, Faaa, 98735 Iles du Vent, France.] DSc (U. Poitiers, 1976) organic chemistry. *Energetical_materials molecular_crystal natural_products computational_assistance_to_researchers*
Fax 33( )689803804

Ghelis, Dr Marianne (1938). Laboratoire de Minéralogie-Météorites, Museum d'Histoire Naturelle, 61 rue Buffon, 75005 Paris, France. DU (U. Paris VI, 1980) mineralogy. *Meteorites separation phosphates*
Tel. 33( )40793533 Fax 33(1)40793524

Ghermani, Dr Nour-Eddine (1957). Assistant Professor in the Faculty of Pharmacy. Laboratoire de Minéralogie-Cristallographie et Physique Infrarouge, URA CNRS 809, Faculté des Sciences, BP 239, Université Henri Poincare, Nancy 1, 54506 Vandoeuvre-les-Nancy Cedex, France. DU (U. Henri Poincare, Nancy 1, 1983) physics, crystallography. *Crystallography electrostatic_properties electron_density zeolites synchrotron computing_in_crystallography*
E-mail ghermani@lmcpi.u-nancy.fr Tel. 33( )83912569 Fax 33( )83406492

Ginderow, Dr Daria (1933). Laboratoire de Minéralogie-Cristallographie, URA 009 CNRS, 4 place Jussieu, 75005 Paris, France. [Université Paris 6, case 115, 75252 Paris Cedex 05, France.] DSc (U. Orleans, 1969) physics. *Crystallography monocrystal antiviral_compounds*
E-mail ginderow@lmcp.jussieu.fr Tel. 33(1)44277205 Fax 33(1)44273785

Giorgi, Mr (1965). Technician. Lab. de Bioinorganique Structurale, C12, Université de Saint Jérôme, av. Escadrille Normandie-Niemen, 13397 Marseille Cedex 20, France. MSc (U. Marseille-Saint Jérôme, 1993) chemistry. *Computer-modelling monocrystal_structure molecular_modelling organic_inorganic_materials*
E-mail bigpdp@frmrs11.univ-mrs.fr Tel. 33( )91288424 Fax 33( )91288030

Graafsma, Dr Heinz (1964). Beamline scientist. ESRF, BP 220, 38043 Grenoble Cedex, France. PhD (U. Twente, Netherlands, 1992) physics. *X-ray_diffraction crystal_physics electron_density crystals_in_electric_fields*
E-mail graafsma@esrf.fr Tel. 33( )76882475 Fax 33( )76882542

Gravereau, Prof. Pierre (1944). Group leader "X-ray services". Laboratoire de Chimie du Solide du CNRS, 351 Cours de la Libération, 33405 Talence Cedex, France. DSc (U. Poitiers, 1975) material sciences. *Diffraction intermetallic_compound fluoride structure nasicon-type_phosphates*
E-mail graver@cribx1.u-bordeaux.fr Tel. 33( )56846326 Fax 33( )56846634

Guillin, Mr Jacques (1958). Ing. Siemens S.A., 39/47 boulevard Ornano, 93527 Saint Denis Cedex 2, France. DU (U. Paris VI, 1985) physics. *X-ray_diffraction instrumentation*
Tel. 33( )49223931 Fax 33( )49223062

Guinier, Prof. André (1911). Membre de l'Académie des Sciences. Lab. de Physique des Solides, Université Paris Sud, France. [87 avenue Denfert-Rochereau, 75014 Paris, France.] DSc (U. Paris, 1939) physics. *Crystallography*
Tel. 33(1)46333805

Hagelstein, Dr Michael (1957). Physicist. European Synchrotron Radiation Facility, France. [European Synchrotron Radiation Facility, BP 220, F-38043 Grenoble Cedex, France.] Dr (U. Kiel, 1991) mineralogy. *EXAFS heterogeneous_catalysis diffusion zeolites*
E-mail hagelstein@ill.fr Tel. 33()76882147 Fax 33()76882160

Hansen, Dr Niels (1950). Laboratoire de Minéralogie-Cristallographie et Physique In-frarouge, URA CNRS 809, Université de Nancy I, France. [Faculté des Sciences, BP 239, 54506 Vandoeuvre-les-Nancy Cedex, France.] PhD (U. Aarhus, Denmark, 1978) chemistry. *Non-linear_property computing valence_charge_density quantum_mechanics*
E-mail hansen@lmcpi.u-nancy.fr Tel. 33()3383912265 Fax 33()3383406492

Hardy, Prof. Antoine (1929). Manager. Lab. Cristallochimie Minérale, Université de Poitiers, 40 av. du Recteur Pineau, 86022 Poitiers Cedex, France. DSc (Université de Bordeaux, 1962) chemistry. *X-ray crystal_structure explosives*
Tel. 33()49453637 Fax 33()49453600

Hårtwig, Dr Jürgen (1947). ESRF, BP 220, 38043 Grenoble Cedex, France. DSc (U. Jena, Germany, 1989) experimental physics. *X-ray_topography dynamical_diffraction X-ray_diffractometry crystal_defect*
E-mail haertwig@esrf.fr Tel. 33()76882500 Fax 33()76882542

Haser, Dr Richard (1942). Group leader, director GDR "Biomolecules, structures and functions". Laboratoire de Cristallographie et Cristallisation des Macromolécules Biologique, s, URA 1296 et GDR 1000 CNRS, France. [Faculté de Medecine Nord, boulevard Pierre Dramard, 13916 Marseille Cedex 20, France.] DSc (U. Aix-Marseille, 1972) physics. *X-ray_crystallography structural_biology enzymes redox proteins*
Tel. 33()91164057 Fax 33()91717896

Hewat, Dr Alan W. (1942). Head, Diffraction Group. Commission on Powder Diffraction founding member. Institut Laue-Langevin, Avenue des Martyrs, Grenoble, France. [BP 156X Cedex, Grenoble 38042, France.] PhD (U. Melbourne, 1971) physics. *Neutron_powder_diffraction high_Tc_superconductors phase_transition computing*
E-mail hewat@ill.fr Tel. 33()76207213 Fax 33()76483906 Telex ILL320-621F

Hewat, Dr Elizabeth Ann (1947). Staff scientist. Member of the Council Société Française de Microscopie Electronique. Institut de Biologie Structurale, 41 avenue des martyrs, 38027 Grenoble Cedex 1, France. PhD (U. Oxford, UK, 1975) physics. *Transmission_electron_microscopy viruses image_processing structure_determination cryo-electron_microscopy biological_molecules*
E-mail hewat@bsiris.ibs.fr Tel. 33()76884568 Fax 33()76885494

Hodeau, Dr Jean Louis (1952). Directeur de Recherche CNRS; group leader. Laboratoire de Cristallographie, CNRS, BP 166X, 38042 Grenoble Cedex, France. PhD (U. J. Fourier, Grenoble, and Institut National Polytechnique de Grenoble, 1984) physical chemistry. *Powder_diffraction single-crystal_diffraction diffraction_anomalous_fine_structure synchrotron_radiation scientific_teaching (HERCULES)*
E-mail hodeau@labs.polycnrs-gre.fr Tel. 33()76881142 Fax 33()76881038

Hospital, Dr Michel (1935). Director of research. Lab. de Cristallographie, Université Bordeaux I, 33405 Talence, France. DSc (U. Bordeaux, 1968) physics. *Peptides polypeptides proteins nucleic_acids synthesis*
E-mail precigou@lotus.u-bordeaux.fr Tel. 33()56846153 Fax 33()56846686

Isnard, Mr Olivier (1965). Institut Laue-Langevin, BP 156, 38042 Grenoble Cedex, France. DU (U. Grenoble, 1993) physics. *Diffractometry hydrides_compound intermetallic_alloys magnetism iron_compound*
Tel. 33()76207091 Fax 33()76483906 Telex 320621F

James-Surcouf, Dr Evelyne (1949). Senior Research Advisor. Member of the Board, Groupe Graphique Moléculaire; Member of the Board, Société Française de Biophysique. Rhône-Poulenc Rorer, Centre de Recherche de Vitry-Alfortville, 13 quai Jules Guesde, BP14, 94403 Vitry sur Seine Cedex, France. DSc (U. Paris VI, 1982) physics. *Molecular_modelling computer-assisted_design organic_compound macromolecular_modelling crystallography database*
Tel. 33()45738086 Fax 33()45738014 Telex 265864F

Janin, Prof. Joël (1943). Director. Member EMBO. Lab. de Biologie Structurale, Bât. 43 CNRS, 91198 Gif sur Yvette, France. DSc (U. Orsay, 1969) biology. *Proteins enzymes modelling proteins_interaction proteins_structure_and_function*
E-mail janin@cygne.lbs.cnrs-gif.fr Tel. 33()69823477 Fax 33()69823129

Janot, Prof. Christian (1936). Member Commission on Aperiodic Crystals of IUCr. Institut Laue-Langevin, 156X, Grenoble Cedex 9, France. DSc (U. Nancy, 1963) physics. *Non-crystalline_growth scattering vibration aperiodic_materials*
E-mail janot@ill.fr Tel. 33()76207327 Fax 33()76483906 Telex 320621F

Jardin, Dr Christian (1947). Laboratoire de Minéralogie-Cristallographie, Université Claude Bernard Lyon I, 43 boulevard du 11 Novembre 1918, 69622 Ville-urbanne Cedex, France. DSc (U. Lyon, 1981) physics. *Surface interface electron_spectroscopy ceramics cathodoluminescence*
Tel. 33()72448151 Fax 33()72441160

Jensen, Anette Frost (1965). Postdoctoral Fellow. [European Synchrotron Radiation Facility, Experimental Division, BP 220, F-38043 Grenoble Cedex, France.] PhD (Aarhus U., 1994) chemical crystallography. *Anomalous_dispersion charge_density modulated_structures synchrotron_radiation*
E-mail frost@esrf.fr Tel. 33()76882447 Fax 33()76882542

Kappenstein, Prof. Charles (1946). Group leader. URA CNRS 350 Catalyse, Groupe Chimie Minérale, Faculté des Sciences, 40 av. du Recteur Pineau, 86022 Poitiers, France. DSc (U. Reims, 1977) inorganic chemistry. *EXAFS powder_diffraction XPS solid_catalysts preparation characterization solid_catalysts*
Tel. 33()49453860 Fax 33()49453499

Kern, Prof. Raymond (1928). Group leader. Centre de Recherche Mécanismes Croissance Cristalline, Luminy, case 901, 13288 Marseille, France. DSc (U. Strasbourg, 1953) physics. *Crystal_growth surface*
Tel. 33()91172837 Fax 33()91418916

Knossow, Dr Marcel (1951). Group leader. Lab. Biologie Structurale, CNRS, 91 198 Gif sur Yvette Cedex, France. DSc (U. Paris VI, 1980) physics. *Crystallography proteins enzymes viruses*
E-mail knossow@cygne.lbs.cnrs-gif.fr Tel. 33()69823462 Fax 33()69823129

Kulda, Dr Jiří (1953). Staff scientist. Member IUCr Commission Neutron Diffraction. Institut Laue Langevin, Grenoble, France. [BP 156, 38042 Grenoble Cedex 9, France.] PhD (Czech. Tech. U. Praha (Czechoslovakia), 1985) physics. *Dynamical_diffraction neutron_inelastic_scattering extinction instrumentation neutron_interferometry topography diffraction*
E-mail kulda@ill.fr Tel. 33()76207256 Fax 33()76483906 Telex 320621F

Kvick, Prof. Åke (1942). Head diffraction group. Chairman IUCr Commission on Synchrotron Radiation. European Synchrotron Radiation Facility, 38043 Grenoble Cedex, France. [European Synchrotron Radiation Facility, BP 220, 38043 Grenoble Cedex, France.] DSc (U. Uppsala, Sweden, 1974) chemistry. *Synchrotron_radiation instrumentation structure_determination zeolites accurate_structure_determination rapid_data_collection*
E-mail kvick@esrf.fr Tel. 33()882116 Fax 33()76882542 Telex 308352F

Labbé, Prof. Philippe (1939). Group leader. Lab. CRISMAT, ISMRA, boulevard du Maréchal Juin, 14050 Caen Cedex, France. DSc (U. Caen, 1978) physics. *X-ray structure_determination oxides modulated_structures*
Tel. 33()31452612 Fax 33()31951600

Lajzerowicz, Prof. Janine (1932). Director. Lab. de Spectrométrie Physique (LA 08), BP 87, 38042 Saint Martin d'Hères Cedex, France. DSc (U. Grenoble, 1964) physics. *Molecular_crystal order_disorder chirality*
Tel. 33()76514358 Fax 33()76514544

Lambert, Prof. Marianne (1932). Lab. de Physique des Solides, Bât. 510, Université Paris Sud, 91405 Orsay, France. DSc (U. Paris, 1958) physics. *Phase_transition disorder quasicrystal*
Tel. 33(1)69416057 Fax 33(1)69416086 Telex 692166FFACORS

Lascombe, Dr Marie-Bernard (1955). Laboratoire d'Immunologie Structurale, Institut Pasteur, 25 rue du Dr Roux, 75274 Paris Cedex 15, France. DSc (U. Paris VI, 1989) biophysics. *Crystallography biological_structure macromolecular_interaction*
Tel. 33(1)45688605 Fax 33(1)45688639

Leadbetter, Dr Alan (1934). Directeur Adjoint. Institut Laue-Langevin, Avenue des Martyrs, BP 156, 38042 Grenoble Cedex 9, France. PhD (Liverpool U., 1957) physical chemistry. *Neutron_scattering synchrotron_radiation amorphous_scattering liquid_crystals*
Tel. 33()76207100 Fax 33(76)961195

Lecomte, Prof. Claude E. P. (1948). Group leader. Member Sagamore Committee (1993–97). Laboratoire de Minéralogie-Cristallographie et Physique Infra Rouge, URA CNRS 809, Faculté des Sciences, Université de Nancy I, BP 239, 54506 Vandoeuvre-les-Nancy, France. DSc (U. Nancy, 1979) material science. *Charge_density structure_determination accuracy porphyrins electrostatic properties modelisation X-ray_diffraction*
E-mail lecomte@lmcpi.u-nancy.fr Tel. 33()83912267 Fax 33()83406492

Lefebvre, Dr Simone (1937). Scientist; co-responsible for beamline D23 (diffraction in material science). LURE, Bât. 209D, 91405 Orsay Cedex, France. DSc (U. Paris, 1975) physics. *Powder_diffraction diffuse_scattering order-disorder quasicrystal multilayers alloys*
E-mail lefebvre@lure.u-psud.fr Tel. 33(1)64468125 Fax 33(1)64464148

Legros, Prof. Jean-Pierre (1943). Coleader X-ray and computing facilities. Laboratoire de Chimie de Coordination, CNRS, UP 8241, 205 route de Narbonne, 31077 Toulouse Cedex, France. DSc (U. Toulouse, 1976) chemistry. *Molecular_conductor superconductors thin_film OMCVD*
E-mail legros@lcctoul.lcc-toulouse.fr Tel. 33()61333121 Fax 33()61553003

Lehmann, Dr Mogens Steen (1942). Staff Scientist. Institut Laue-Langevin, Av. des Martyrs, F38042 Grenoble, France. PhD (Århus University Denmark, 1973) chemistry. *Biocrystallography structural_change neutron_diffraction synchrotron_radiation*
E-mail lehmann@ill.fr Tel. 33()76207382 Fax 33()76483906 Telex 320621

Lewit-Bentley, Dr Anita (1948). Instrument responsible. LURE, Bât. 209D, Université Paris-Sud, 91405 Orsay Cedex, France. DU (U. Prague (Czechoslovakia), 1972) physical chemistry. *Structure-activity relationship biological_molecules crystallization*
E-mail anita@lure.u-psud.fr Tel. 33(1)64468050 Fax 33(1)64464148

Longueville, Dr Willy (1937). Laboratoire de Dynamique et des Structures des Matériaux Moléculaires, URA 801, Villeneuve d'Ascq, France. [UFR de Physique Fondamentale, Université des Sciences et Technologies de Lille, 59655 Villeneuve d'Ascq Cedex, France.] DSc (U. Lille, 1987) physics. *Condensed_matter X-ray_topography spectroscopy nuclear_magnetic_resonance*
Tel. 33()20434924 Fax 33()20472688 Telex 136339F

Louër, Dr Daniel (1942). Group leader. Secretary Commission on Powder Diffraction. CSIM, Laboratoire de Cristallochimie, Université de Rennes I, avenue du Général Leclerc, 35042 Rennes Cedex, France. DSc (U. Rennes, 1969) chemistry. *Powder_diffraction solid_state_chemistry*
E-mail louer@univ-rennes1.fr Tel. 33()99286248 Fax 33()99383487

Loupias, Prof. Geneviève (1940). Group leader. Consultant Commission on Charge, Spin and Momentum Densities. Laboratoire de Minéralogie-Cristallographie, URA009 CNRS, Universités P.M. Curie et Paris VII, 4 place Jussieu Paris 5, France. [Université P.M. Curie, case 115, 75252 Paris Cedex 05, France.] DSc (U. Paris VI, 1978) physics. *Compton_scattering inelastic_scattering electronic_density*
E-mail loupias@lmcp.jussieu.fr Tel. 33(1)44275230 Fax 33(1)44273785

Luzzati, Dr Vittorio (1923). Emeritus. Centre de Génétique Moléculaire, CNRS, 91198 Gif sur Yvette, France. DSc (U. Paris, 1951) physics. *Biophysics order-disorder pattern_recognition SAXS lipid_polymorphism myelin*
Tel. 33(1)69823185 Fax 33(1)69823150

Magerl, Dr Andreas (1949). Group leader. Institut Laue-Langevin, BP 156, 38042 Grenoble Cedex, France. Dr Hab (U. Bochum, Germany, 1993) experimental physics. *Diffractometry crystal_growth excitation disordered_structure neutron X-ray*
E-mail magerl@ill.fr Tel. 33()76207383 Fax 33()76483906

Malgrange, Prof. Cécile (1939). Group leader. Laboratoire de Minéralogie-Cristallographie, Tour 16, 4 place Jussieu, Paris 5, France. [case 115, 75252 Paris Cedex 05, France.] DSc (U. Paris, 1967) physics. *X-ray_optics dynamical_theory standing_wave_technique X-ray_topography*
E-mail malgrang@lmcp.jussieu.fr Tel. 33(1)44275221 Fax 33(1)44273785 Telex 200145F

Marezio, Dr Massimo (1930). Director of Research. Lab. de Cristallographie, CNRS, BP 166, 38042 Grenoble Cedex, France. Libera Docenza (U. Rome, 1964) physics. *Superconductivity phase_transition oxides structure_determination_at_the_local_level*
E-mail marezio@labs.polycnrs-gre.fr Tel. 33()76881040 Fax 33()76881038

Marsau, Prof. Pierre (1937). Group leader. Lab. de Cristallographie et Physique Cristalline, Université de Bordeaux I, 351 Cours de la Libération, 33405 Talence, France. DSc (Bordeaux I U., 1972) physics. *Structure_determination macro-cyclic_complex_materials*
E-mail marsau@zita.cristal.u-bordeaux.fr Tel. 33()56846152 Fax 33()56846686

Mason, Dr Sax Anton (1946). Staff Scientist. Member, Commission Neutron Scattering. Institut Laue Langevin, Grenoble, France. [ILL, BP 156, 38042 Grenoble Cedex 9, France.] PhD (U. Melbourne, Australia, 1971) chemistry. *Neutron X-ray_diffraction macromolecules two-dimensional detector cyclodextrins H/D_exchange charge_density_studies synchrotron_radiation*
E-mail mason@ill.fr Tel. 33()76207067 Fax 33()76483906 Telex 320621F

Massaux, Dr Michel Louis (1934). Group leader. President section Auvergne of Société Française de Physique. Université Blaise Pascal, UFR Sciences, Clermont-Ferrand, France. [Physics Dept, UFR Sciences, Les Cezeaux, 63177 Aubière Cedex, France.] DU (Blaise Pascal U., 1957) physics. *Coordination_compounds structure dielectric_materials_characterization microwave_materials materials_structure_and_characterization*
Tel. 33()73407329 Fax 33()73407650

Maveyraud, Mr Laurent (1969). Student. Laboratoire de Cristallographie Biologique, LPTF, CEMES-CNRS, BP 4347, 31055 Toulouse, France. MSc (U. Strasbourg, 1992) biological crystallography. *Crystallography*
E-mail maveyrau@cemes.fr Tel. 33()62257963 Fax 33()62257960

May, Dr Roland Peter (1948). Staff scientist. Institut Max von Laue - Paul Langevin, Avenue des Martyrs, F-38042 Grenoble Cedex 9, France. [Institut Max von Laue - Paul Langevin, BP 156, F-38042 Grenoble Cedex 9, France.] Dr. rer. nat. (Technische U. München, 1978) biophysics. *Macromolecular_structures neutron_small-angle_scattering ribosomes polymerases chaperones computing modelling*
E-mail may@ill.fr Tel. 33()76207047 Fax 33()76483906

McIntyre, Dr Garry James (1951). Staff scientist. Member IUCr Commission on Crystallographic Apparatus. Institut Laue Langevin, Grenoble, France. [BP 156, 38042 Grenoble Cedex, France.] PhD (U. Melbourne, Australia, 1978) physics. *X-ray_neutron_diffractometry charge_density thermal_vibration two-dimensional_detector magnetism extinction incommensurates noncentrosymmetry*
E-mail mcintyre@ill.fr Tel. 33()76207090 Fax 33()76483906 Telex 320621F

Meinnel, Prof. Jean (1926). Group leader. Groupe Matière Condensée et Matériaux, Université Rennes I, UA 804 CNRS, av. du Général Leclerc, 35042 Rennes Cedex, France. DSc (U. Paris Sorbonne, 1958) physics. *Tunnelling lattice_dynamics intramolecular_force molecular_crystal ring_molecule*
Tel. 33()99286058 Fax 33()99286717

Michel, Prof. André Gustave (1944). Professor; Director of Research Centre. Institut de recherches Servier, 11 rue des Moulineaux, Suresnes, 92150, France. [14, rue du Petit Moutesson, Le Vesinet, 78110 France.] PhD (U. Namur, Belgium, 1976) physical chemistry. *Biomedical_molecules drug_design computer-assisted_design*
Tel. 33()141182296 Fax 33()141182640

Moras, Dr Dino (1944). Research Director. IGBMC, BP 163, 67404 Illkirch Cedex, France. DSc (U. Louis Pasteur, Strasbourg, 1971) chemistry. *Crystallization proteins nucleic_acids structure_determination*
E-mail moras@igbmc.u-strasbg.fr Tel. 33()88653351 Fax 33()88653201

Moreau, Dr Jean-Michel (1944). Head dept. Faculté Annecienne des Sciences et Techniques, 74942 Annecy-le-Vieux Cedex, France. [IUT, 9 rue de l'Arc-en-Ciel, 74942 Annecy-le-Vieux Cedex, France.] DSc (Grenoble U., 1976) physics. *Crystallography*
Tel. 33()50092383 Fax 33()50276535

Morniroll, Prof. Jean-Paul (1945). Group leader. Laboratoire de Métallurgie Physique, URA CNRS 234, Université de Lille I, Bât. C6, 59655 Villeneuve d'Ascq Cedex, France. DSc (U. Nancy, 1974) physics. *Electron_microscopy electronic_microdiffraction convergent-beam_diffraction defect*
Tel. 33()20436937 Fax 33()20434040 Telex 136339F

Mornon, Dr Jean-Paul (1942). Group leader. Scientific Director of the Research program Organibio. Laboratoire de Minéralogie-Cristallographie, URA 009 CNRS, Universités P.M. Curie et D. Diderot, 4 place Jussieu, Paris 5, France. [case 115, 75252 Paris Cedex 05, France.] DSc (U. Paris VI, 1969) crystallography. *Protein_structure protein_growth sequences macromolecules folding*
E-mail mornon@lmcp.jussieu.fr Tel. 33(1)44274587 Fax 33(1)44273785

Mosset, Prof. Alain (1946). Group leader. Centre d'Elaboration des Matériaux et d'Etudes Structurales, CEMES, UPR 8011, CNRS, 29 rue J. Marvig, Toulouse, France. [CEMES, BP 4345, 31055 Toulouse Cedex, France.] DSc (U. Toulouse III, 1981) chemistry. *Lanthanides sulfides glasses wide-angle_scattering*
E-mail mosset@cemes.fr Tel. 33()62257849 Fax 33()62257999

Naudon, Dr André (1939). Group leader. Laboratoire de Métallurgie Physique, URA 131, CNRS, Université de Poitiers, 40 avenue du Recteur Pineau, 86022 Poitiers Cedex, France. DSc (U. Poitiers, 1971) physics. *Small-angle_scattering anomalous_dispersion grazing_incidence porous_silicon_compounds*
E-mail naudon@zeus.univ-poitiers.fr Tel. 33()49453682 Fax 33()49453759

Nguyen-Ba, Dr Chanh (1934). Directeur de Recherches au CNRS. Lab. de Cristallographie et Physique Cristalline, Université de Bordeaux I, 33405 Talence, France. DSc (Bordeaux U., 1965) physical chemistry. *Structural_disorder phase_transition phase_diagram thermodynamics low_dimensional_molecular_composites*
E-mail chanh@zita.cristal.u-bordeaux.fr Tel. 33()56846154 Fax 33()56846686

Nierlich, Dr Martine (1944). Group leader. CEA.CE Saclay, SCM, Bât. 125, 91191 Gif sur Yvette Cedex, France. DSc (1975) physics. *Structure uranium_compounds*
Tel. 33(1)69083222 Fax 33(1)69086640

Nouet, Prof. Jean (1938). Group leader. Equipe de Physique de l'Etat Condensé, URA CNRS 807, Université du Maine, avenue Olivier Messiaen, 72017 Le Mans Cedex, France. DSc (U. Paris VI, 1973) physics. *Crystal growth phase_transition lattice_dynamics ultrasonics fluoride_compounds Raman ferroelasticity diffractometry*
Tel. 33()43833264 Fax 33()43833518

Olivier-Deyris, Mrs Laurence (1966). Institut de Biologie Structurale, 41 avenue des Martyrs, 38027 Grenoble Cedex, France. Ing (U. Clermont-Ferrand, 1990) biology. *Biocrystallography proteins molecular_replacement biological_refinement_method biological_structure_determination*
E-mail olivier@bsiris.ibs.fr Tel. 33()76889592 Fax 33()76885494

Ouladdiaf, Dr Bachir (1958). Staff scientist. Institut Laue Langevin, Grenoble, France. [BP 156, 38042 Grenoble Cedex, France.] DSc (U. Joseph Fourier - Grenoble I, 1986) physics. *Diffractometry magnetism group_theory manganese_compounds magnetic_structures magnetic_frustration intermetallic_compounds*
E-mail ouladdiaf@ill.fr Tel. 33()76207089 Fax 33()76483906 Telex 320621F

Pannetier, Dr Jean (1947). Staff scientist. Institut Laue Langevin, Grenoble, France. [BP 156, 38042 Grenoble Cedex, France.] DSc (U. Rennes, 1973) chemistry. *Powder_diffractometry Rietveld_method time-resolved_effect solid_state_chemistry neutron_X-ray_structural_simulation_of_solids crystal_defect*
E-mail pannetier@ill.fr Tel. 33()76207091 Fax 33()76483906 Telex 320621F

Pascard, Dr Claudine (1929). Director. Coeditor Acta Crystallographica. Lab. Cristallochimie, Institut de Chimie des Substances Naturelles, CNRS, 91198 Gif sur Yvette, France. DSc (Paris U., 1959) physics. *Cryptates macrocycles non-bonded_interaction*
E-mail pascard@icsn01.icsn.cnrs-gif.fr Tel. 33(1)69823050 Fax 33(1)69077247

Pepe, Dr Gérard (1945). Group leader. CRMC2, CNRS, Campus de Luminy, Case 913, 13288 Marseille Cedex 9, France. DSc (Marseille U., 1976) physics. *Molecular_mechanics computer_graphics modelling electrostatic_potential molecular_and_crystal_modelling*
E-mail pepe@mccir3.univ-mrs.fr Tel. 33()91172855 Fax 33()91418916

Perez, Dr Serge (1947). Group leader. President French Mol. Graph. Soc. and French Carbohydrate Soc. Ingénierie Moléculaire, INRA, BP 527, 44026 Nantes Cedex 03, France. DSc (Grenoble U., 1978) physics. *X-ray_fibres_diffraction electron_diffraction computing carbohydrates protein_carbohydrate_interaction polysaccharide_engineering*
E-mail perez@nantes.inra.fr Tel. 33()40675043 Fax 33()40675092

Perrin, Prof. Monique (1936). Group leader. Laboratoire de cristallographie, Université Claude Bernard Lyon I, ER060, 43 boulevard du 11 Novembre 1918, 69622 Villeurbanne Cedex, France. DSc (U. Lyon, 1974) physics. *Structure_determination macrocycles molecular_complexes molecular_recognition*
E-mail perrin@cdlyon.univ-lyon1.fr Tel. 33()72448220 Fax 33()72431160

Petroff, Prof. Jean-François (1935). Laboratoire de Minéralogie-Cristallographie, Universités Paris 6 et Paris 7, 4 place Jussieu, case 115, 75252 paris Cedex 05, France. DSc (U. Paris, 1971) physics. *Synchrotron_radiation X-ray_standing_wave surface epitaxy*
E-mail petroff@lmcp.jussieu.fr Tel. 33(1)44275219 Fax 33(1)44273785

Pichon-Pesme, Dr Virginie (1960). Laboratoire de Minéralogie-Cristallographie, URA 809 CNRS, Faculté des Sciences, Université de Nancy I, 54506 Vandoeuvre-les-Nancy, France. DU (U. Nancy I, 1986) physics. *Crystallography valence_charge_density structure_determination peptides*
E-mail pichon@lmcpi.u-nancy.fr Tel. 33()83912000x2767 Fax 33()83406492

Pierrot, Dr Marcel (1938). Group leader. Laboratoire de Bioorganique Structurale, Centre scientifique Saint Jerome C12, 13397 Marseille Cedex 20, France. DSc (U. Marseille, 1968) physics. *Bioinorganic_structure biological_structure crystal_structure*
Tel. 33()91983208 Fax 33()91288030

Podjarny, Dr Alberto (1950). Group leader. Coordinator, Group Macromolecules, French Cryst. Ass. IGBMC, BP 163, 67404 Illkirch Cedex, France. PhD (Weizmann Inst. of Science, 1977) chemistry. *Macromolecular_phase_determination phase_refinement_method structure_determination molecular_replacement*
E-mail podjarny@igbmc.u-strasbg.fr Tel. 33()88653311 Fax 33()88653201

Precigoux, Dr Gilles (1946). Group leader. Member, Board French Cryst. Ass. Laboratoire de Cristallographie, Université de Bordeaux I, 351 Cours de la Libération, 33405 Talence, France. DSc (U. Bordeaux I, 1978) physics. *X-ray_macromolecular_structure*
E-mail precigou@zebre.cristal.u-bordeaux.fr Tel. 33()56846163 Fax 33()56846686

Pyka, Dr Niels Michael (1958). Research Associate. Laboratoire Léon Brillouin, France. [Laboratoire Léon Brillouin, CEN-Saclay, F-91191 Gif sur Yvette Cedex, France.] Dr rer.nat. (TU Berlin, 1988) physics. *Lattice_dynamics superconductivity magnetism heavy_fermion residual_stress Jahn-Teller-effect*
Tel. 33(1)69086039 Fax 33(1)69088261 Telex energ690641flbs+

Rabu, Dr Pierre (1964). Searcher at the CNRS (grade CR1). IPCMS - UMR 046, Groupe des Matériaux Inorganiques, 23 rue du Loess, BP20/CR, 67037 Strasbourg, France. DU (U. Nantes, 1990) chemistry. *Low-dimensional_magnetism structure_determination solid_state chemistry numerical_methods_and_simulation_techniques organic/inorganic_materials*
E-mail rabu@teutates.u-strasbg.fr Tel. 33()88107135 Fax 33()88107247

Rambaud, Dr Jöelle (1942). Laboratoire de Chimie Générale et Minérale, Faculté de Pharmacie, France. [Université de Montpellier I, Faculté de Pharmacie, 15 av. Ch. Flahaut, 34060 Montpellier, France.] DSc (Faculté de Pharmacie Montpellier I, 1982) physical chemistry. *X-ray_diffraction crystallization proteins drug_structure*
E-mail rambaud@cnusc.fr Tel. 33()67522301 Fax 33()67042140

Raoux, Dr Denis (1942). Director of research at CNRS; director of the crystallography laboratory. Lab. de Cristallographie, CNRS, 25 avenue des Martyrs, 38042 Grenoble, France. [CNRS, BP 166, 38042 Grenoble Cedex 09, France.] DSc (Orsay U., 1974) solid state physics. *Synchrotron-radiation anomalous_diffraction EXAFS*
Tel. 33()76881044 Fax 33()76881038

Rees, Dr Bernard (1940). Research Director. IGBMC, BP 163, 67404 Illkirch Cedex, France. DSc (Strasbourg U., 1969) chemistry. *Macromolecular_crystallography RNA toxins phase_refinement aminoacyl_tRNA_synthetases*
E-mail rees@igbmc.u-strasbg.fr Tel. 33()88653312 Fax 33()88653201

Renaud, Dr Jean-Paul (1960). Research Associate. IGBMC, BP 163, 67404 Illkirch Cedex, France. DU (U. Paris VI, 1986) chemistry. *Biocrystallography receptor steroids DNA_interaction retinoic_acid retinoids heterodimers DNA_complex*
E-mail renaud@igbmc.u-strasbg.fr Tel. 33()88653348 Fax 33()88653201

Ritter, Dr Clemens (1957). Staff scientist. Institut Laue-Langevin, Grenoble, France. [BP 156, 38042 Grenoble Cedex, France.] DSc (Westfälische Wilhelms-U. Münster, Germany, 1985) chemistry. *Powder_diffraction rare-earth_compounds intercalates texture*
E-mail ritter@ill.fr Tel. 33()76207460 Fax 33()76483906 Telex 320621F

Rodríguez-Carvajal, Dr Juan (1953). Instrument responsible. Laboratoire Léon Brillouin, CEA-CNRS, Centre d'Etudes de Saclay, 91191 Gif sur Yvette, France. DSc (U. Barcelona (Spain), 1984) physics. *Diffractometry Rietveld_method magnetic_structure_determination computer_modelling group_theory magnetism transition_metal-rare_earth_oxides_and_intermetallics*
E-mail juan@bali.saclay.cea.fr rodriguez2@ill.fr Tel. 33(1)69083343 Fax 33(1)69088261 Telex Energ 690 641 F LBS +

Rondeau, Dr Jean-Michel (1959). Senior associate scientist. Marion Merrell Dow Research Institute, Strasbourg Center, 16 rue d'Ankara, 67080 Strasbourg Cedex, France. DU (U. Louis Pasteur Strasbourg, 1988) enzymology/crystallography. *Crystallization biological_crystallography enzymes_mechanism drug_design*
E-mail jeanmichelrondeau@mmd.com Tel. 33()88414616 Fax 33()88459070 Telex 890252F

Roquet, Ms Françoise Jeanne Valentine (1963). Postdoctoral associate. UMR 9921 - Faculté de Pharmacie, 15 avenue Ch. Flahault, 34060 Montpellier Cedex 01, France. PhD (U. Louvain-la-Neuve, Belgium, 1993) biochemistry. *Biocrystallography biochemistry immunology proteins*
E-mail roquet@ljcrf.edu Tel. 33()67040414 Fax 33()67040341

Sauvage, Dr Michèle (1941). Group leader. Coeditor, Journal of Synchrotron Radiation. Laboratoire de Minéralogie-Cristallographie, Université P.M. Curie, URA CNRS 009, & LURE, Université Paris Sud, France. [LURE, Bât. 209D, Campus Orsay, 91405 Orsay Cedex, France.] DSc (U. Paris, 1968) physics. *Surface surface_structure interface epitaxy synchrotron_radiation*
E-mail sauvage@lure.ups.fr Tel. 33(1)64468018 Fax 33(1)64464148

Schiltz, Mr Marc (1969). Student. LURE, Centre Universitaire Paris Sud, Bât. 209d, 91405 Orsay Cedex, France. MSc (U. Paris-Sud, 1993) crystallography. *Biocrystallography phase_determination isomorphous_replacement data_collection*
E-mail schiltz@lure.u-psud.fr Tel. 33(1)64468027 Fax 33(1)64464148 Telex 603340F

Schlenker, Prof. Michel (1940). Professeur, Institut National Polytechnique de Grenoble. Former editor J. Appl. Cryst. (1984-90). Laboratoire Louis Néel du CNRS, BP 166, 38042 Grenoble Cedex, France. DSc (U. Grenoble, 1970) physics. *X-ray_topography neutron_topography neutron_imaging X-ray_imaging optical_imaging domain_structure defects magnetism magnetic_materials*
E-mail schlenk@labs.polycnrs-gre.fr Tel. 33()76881092 Fax 33()76881191

Schreuder, Dr Herman A. (1958). Senior associate scientist. Marion Merrell Dow Research Institute, 16 Rue d'Ankara, 67080 Strasbourg Cedex, France. PhD (U. Groningen, Netherlands, 1988) chemistry. *Macromolecular_crystallography enzymes_mechanism drug_design medicine*
E-mail hermanschreuder@mmd.com Tel. 33()88414616 Fax 33()88459070 Telex 890252F

Schuster, Dr Isabelle (1961). X-ray general service manager. CENG, Service de Physique des Matériaux et Microstructures, 17 avenue des Martyrs, 38054 Grenoble Cedex 09, France. DU (U. Technologie de Compiègne, 1986) material sciences. *Characterization crystallography diffractometry condensed_matter*
Tel. 33()76884877 Fax 33()76885097

Schwegle, Dr Wolfgang (1962). Research Associate. European Synchrotron Radiation Facility, BP 220, F-38043 Grenoble Cedex, France. Dr (U. Karlsruhe, 1993) physics. *Multiple_scattering dynamical_theory_of_diffraction synchrotron_radiation diffractometer*
E-mail schwegle@esrf.fr

Semertzidis, Mr Michel (1966). Lab. Minéralogie-Cristallographie, UA 009 CNRS, Université P.M. Curie, 4 place Jussieu, 75005 Paris, France. [70 Rue Monge, F-75005 Paris, France.] PhD (Paris 7 U., 1991) biomathematics-biocomputing. *Computing biocrystallography proteins folding*
E-mail semerz@lmcp.jussieu.fr Tel. 33(1)44276247 Fax 33(1)44273785

Silvestre, Dr Jean-Paul (1942). Lab. Chimie & physico-chimie moléculaires, Ecole Centrale Paris, ERS 070 CNRS, France. [Grande Voie des Vignes, 92295 Chatenay-Malabry Cedex, France.] DSc (Paris VI U., 1978) chemistry. *Organometallic_crystal_structure inorganic_crystal_structure lanthanides phosphorus_compounds polyacid alcohol salts*
E-mail silves@pcm.ecp.fr Tel. 33(1)41131297 Fax 33(1)41131437

Tasset, Dr Francis (1944). Staff scientist. User representative in "Comité de la Spectroscopie et de la Diffusion Neutronique". Institut Laue Langevin, Grenoble, France. [BP 156, 38042 Grenoble Cedex, France.] DSc (U. Grenoble, 1975) physics. *Neutron_polarization_analysis magnetization_density magnetic_neutron_scattering transition_elements superconductivity magnetism neutron_spin_filtering neutron_polarimetry*
E-mail tasset@ill.fr Fax 33()76483906 Telex 320621F

Theobald, Prof. François (1942). Director. Chimie du Solide Cristallin, Université Paris Sud, rue Clémenceau, 91405 Orsay, France. [Bât. 490, 91405 Orsay Cedex, France.] DSc (U. Besançon, 1975) chemistry. *Structure_determination oxides ionic_conductivity vanadium_compounds*
Tel. 33(1)69417211

Thozet, Dr Alain (1941). Director Centre de Diffractométrie. Laboratoire de Cristallographie, Université Claude Bernard Lyon I, 43 boulevard du 11 Novembre 1918, 69622 Villeurbanne, France. DSc (U. Lyon, 1981) physics. *Computing macrocycles diffraction_technique molecular_conformation*
E-mail thozet@cdlyon.univ-lyon1.fr Tel. 33()72448219 Fax 33()72431160

Timmins, Dr Peter (1946). Group manager. Institut Laue-Langevin, 156X, 38042 Grenoble Cedex, France. PhD (London U., 1972) crystallography. *Neutron_crystallography X-ray_crystallography small-angle_scattering biological_macromolecules*
E-mail timmins@ill.fr Tel. 33()76207263 Fax 33()76483906

Toudic, Dr Bertrand (1956). Coleader of group. Groupe Matière Condensée et Matériaux, URA 040804 CNRS, Campus de Beaulieu, 35042 Rennes Cedex, France. DSc (U. Rennes, 1986) physics. *Phase_transition neutron_diffraction molecular_compound*
Tel. 33()99886719 Fax 33()99286717

Tougard, Dr Pierre (1942). Researcher. Enzymologie Physicochimique, Bât. 433, Université Paris Sud, 91405 Orsay Cedex, France. DSc (U. Paris VI, 1974) physics. *Proteins crystallography*

Toupet, Dr Loïc (1949). Technical responsible centre of diffractometry. Groupe Matière Condensée et Matériaux, URA 804 CNRS, Rennes, France. [Faculté des Sciences, Campus de Beaulieu, 35042 Rennes Cedex, France.] DU (U. Rennes, 1976) chemistry. *Computing X-ray_structure phase_transition cryogenics*
E-mail toupet@univ-rennes.fr Tel. 33()99286065 Fax 33()99286717

Turco, Prof. Guy (1927). Institute leader. Institut de Geologie, Faculté des Sciences, laboratoire de Petrologie-Minéralogie, Parc Valrose, 06108 Nice Cedex 2, France. DSc (U. Paris, 1962) physics. *X-ray_crystallography synthesis silicon_compounds minerals_inclusion*
Tel. 33()93529939 Fax 33()93529919 Telex 970281F

Vaney, Dr Marie-Christine (1959). UMR09, laboratoire de Minéralogie-Cristallographie, 4 place Jussieu, 75252 Paris Cedex 05, France. [Laboratoire de Biologie Structurale, CNRS, Université Paris Sud, UMR 9920, Bât. 34, 1 avenue de la Terrasse, 91 Gif sur Yvette, France.] DU (U. Paris VI, 1986) material sciences. *Crystallography proteins macromolecules X-ray*
E-mail vaney@cygne.lbs.cnrs-gif.fr Tel. 33(1)69823476 Fax 33(1)68823129

Vellieux, Dr Fred (1958). IBS/LCCP, 41 avenue des Martyrs, 38047 Grenoble Cedex 1, France. DSc (Rijksuniversiteit Groningen Netherlands, 1990) chemistry. *Biocrystallography macromolecules computing*
E-mail vellieux@lccp.ibs.fr Tel. 33()76889605 Fax 33()76885122

Vettier, Dr Christian (1946). Group leader. ESRF, BP 220, 38043 Grenoble Cedex, France. DSc (U. Grenoble, 1975) physics. *Magnetism resonant_scattering superconductivity circular_dichroism magnetic_X-ray_scattering*
E-mail vettier@ill.fr Tel. 33()76882251

Veysseyre, Dr Renée (1934). Laboratoire de Chimie Physique du Solide, Ecole Centrale de Paris, UA 453 CNRS, France. [Grande voie des Vignes, 92295 Chatenay-Malabry Cedex, France.] DSc (U. Paris VI, 1987) physics. *Geometric_symmetry n-dimensional quasicrystal*
E-mail lamoureux@cti.ecp.fr Tel. 33(1)41131283 Fax 33(1)41131261 Telex 634991F

Viani, Prof. Robert (1942). Computing group leader. Lab. Biophysique, Université de Nice-Sophia Antipolis, Parc Valrose, 06108 Nice Cedex 2, France. DSc (University of Marseille, 1978) crystallography. *Biological_materials organic_materials drug design*
E-mail viani@naxos.unice.fr Tel. 33()93529850 Fax 33()93529851 Telex UNINICEF970281

Vicat, Prof. Jean (1941). Coeditor, Journal of Applied Crystallography. Institut de Biologie Structurale, 41 avenue des martyrs, 38027 Grenoble Cedex 1, France. DSc (U. Joseph Fourier Grenoble, 1977) physics. *Crystallization X-ray_structure_determination proteins*
E-mail vicat@ibs.fr Tel. 33()76889593 Fax 33()76885494

Vidal, Prof. Geneviève (1944). Professor. Commission on Charge, Spin and Momentum Densities. Departement Matière Condensée, GDPC, URA 233, case courrier 26, Université Montpellier II, 34095 Montpellier Cedex 5, France. DSc (U. Montpellier, 1975) physics. *Density_distribution multipole_expansion_in_3D*
E-mail vidal@frmop22.bitnet Tel. 33()67591069 Fax 33()67143031

Vidal, Dr Jean-Pierre (1940). Maître de Conférences. Commission on Charge, Spin and Momentum Densities. Departement Matière Condensée, GDPC, URA 233, case courrier 26, Université Montpellier II, 34095 Montpellier Cedex 5, France. DSc (U. Montpellier, 1974) physics. *Density_distribution multipole_expansion_in_3D*
E-mail vidal@frmop22.bitnet Tel. 33()67591069 Fax 33()67143031

Villeret, Mr Vincent (1967). Institut de Biologie Structurale, 41 avenue des Martyrs, 38027 Grenoble Cedex 1, France. MSc (U. Liège, Belgium, 1990) biochemistry. *Biocrystallography proteins biochemistry structure_determination_and_analysis*
E-mail villeret@ibs.fr Tel. 33()76889592 Fax 33()76885494

Vincent, Prof. Henri (1941). Group leader. Laboratoire de Matériaux et du Génie Physique, ENS de Physique de Grenoble, URA 1109 CNRS, France. [BP 46, 38042 Saint Martin d'Hères, France.] DSc (U. Grenoble, 1975) physics. *Physical_property_structure_relationship doped_ferrites semiconductors_alloys organometallic_small_molecules*
Fax 33()76826394

Vuilhorgne, Dr Marc (1950). Group leader. Rhône-Poulenc Rorer Research and Development, France. [Centre de Recherche de Vitry-Alfortville, 13 quai Jules Guesde, 94403 Vitry Cedex, France.] DU (Orsay U., 1977) chemistry. *Structural_analysis conformation natural_products nuclear_magnetic_resonance*
Tel. 33(1)45738062 Fax 33(1)45738058

Wakatsuki, Dr Soichi (1959). Responsible for Beamline 20 (protein crystallography) at ESRF. ESRF, BP 220, F-38043 Grenoble Cedex, France. PhD (Stanford U., 1990) chemistry. *Protein_crystallography synchrotron_radiation Laue_diffraction instrumentation*
E-mail wakatsuki@esrf.fr Tel. 33()76882362 Fax 33()76882542

Westhof, Prof. Eric (1948). Group leader. Treasurer group Molecular Graphics. UPR 9002, Structure des Macromolécules biologiques et Mécanismes de Reconnaissance, France. [IBMC-CNRS, 15 rue Descartes, 67084 Strasbourg, France.] DSc (U. Liège Belgium, 1974) physics. *RNA DNA modelling simulation modelling_and_simulation_of_nucleic_acids*
E-mail westhof@ibmc.u-strasbg.fr Tel. 33()88417046 Fax 33()88602218

Wöhler, Dr Annick (1946). Laboratoire de Minéralogie-Cristallographie, UA009 CNRS, Universités P.M. Curie et Paris VII, 4 place Jussieu, 75252 Paris Cedex 05, France. DSc (U. Paris VI, 1978) physics. *Crystallography X-ray_structure_determination pharmaceutical_compound*
Tel. 33(1)44275082 Fax 33(1)44273785

Wilkinson, Dr Clive (1941). Group leader. European Molecular Biology Laboratory, Grenoble, France. [BP 156, 38042 Grenoble Cedex, France.] PhD (U. Cambridge (UK), 1965) physics. *Proteins Laue_diffraction area_detector image_processing synchrotron_instrumentation image_plates neutron_X-ray*
E-mail wilkinson@ill.fr Tel. 33()76207448 Fax 33()76207199 Telex 320621F

Willaime, Prof. Christian (1940). Group leader. Lab. Minéralogie Physique, Géosciences Rennes, Université Rennes I, 35042 Rennes Cedex, France. DSc (Paris VI U., 1973) physics. *Defect electron_microscopy minerals phase_transformation*
E-mail willaime@univ-rennes.fr Tel. 33()99286378 Fax 33()99286780

Wintenberger, Dr Micheline (1929). Retired. [79 rue du Théatre, 75015 Paris, France.] DSc (U. Paris, 1962) physics. *Transition_elements neutron_diffraction magnetic_order representation_theory*

Witz, Dr Jean (1935). Immunochimie des Peptides et des Virus, UPR 9021, IBMC, CNRS, 15 rue Descartes, 67000 Strasbourg, France. DSc (U. Strasbourg, 1964) physics. *Structure small-angle_scattering calorimetry viruses assembly_decapsidation_of_viruses*
Tel. 33()88417018 Fax 33()88610680

Zaccai, Dr Joseph (1947). Institut Laue-Langevin, avenue des Martyrs, Grenoble, France. [ILL, BP 156, 38042 Grenoble Cedex, France.] PhD (U. Edinburgh, UK, 1972) physics. *Neutron_diffraction scattering biophysics structure_dynamics molecular_biology*
E-mail zaccai@ill.fr Tel. 33()76207046 Fax 33()76483906

Zarka, Dr Albert (1942). Lab. Minéralogie-Cristallographie, Université P.M. Curie, 4 place Jussieu, Paris 5, France. [case 115, Université Paris VI, 75252 Paris Cedex 05, France.] DSc (U. Paris, 1973) physics. *Topography piezoelectricity X-ray stroboscopic_study_of_piezoelectric_crystals*
Tel. 33(1)44275225 Fax 33(1)44273785

Zelwer, Dr Charles (1940). Group leader. Centre de Biophysique Moléculaire, CNRS, rue Charles Sadron, 45071 Orléans, Cedex 2, France. DSc (U. Paris, 1967) physics. *Biological_crystallography phase_refinement_method proteins nucleic_acids*
E-mail zelwer@rubi.cnrs-orleans.fr Tel. 33()38517803 Fax 33()38631517

Zeyen, Dr Claude Mathias Emile (1947). Physicist; head of development branch. Institut Laue-Langevin, Grenoble, France. [BP 156, 38042 Grenoble Cedex 09, France.] PhD (TU Munich, Germany, 1975) physics. *Neutron_spectrometry anharmonic_condensed_matter spin_precession_spectrometry dynamical_diffraction fundamental_physics measurement_science technology*
E-mail zeyen@ill.fr Tel. 33()76207148 Fax 33()76483906 Telex 320621F

GEORGIA

Sub-Editor: G. Tsintsadze

Shvelashvili, Prof. Arsen (1935). Laboratory and Department Leader. Member, National Committee for Crystallography. Medical Institute, Institute of Physical and Organic Chemistry, Georgian Academy of Sciences, Tbilisi, Republic of Georgia. [Bakhtrioni 9, case 106, 380004 Tbilisi, Republic of Georgia.] DSc (Inst. General and Inorganic Chemistry, Russian Academy Sciences, Moscow, 1974) chemistry. *Prospective_bio-coordinative_compounds bio-active_compounds*
Tel. 7(8832)377762 Fax 7(8832)998823

Tsintsadze, Prof. Givi (1933). Laboratory and Department Leader. Chairman, National Committee for Crystallography; Sub-Editor, WDC9. Georgian Technical University, Institute of Inorganic and Electrochemistry, Georgian Academy of Sciences, Tbilisi, Republic of Georgia. [Tsagareli str. 29/7, 380060 Tbilisi, Republic of Georgia.] DSc (Inst. General and Inorganic Chemistry, Russian Academy Sciences, Moscow, 1971) chemistry. *Crystallochemistry_of_coordination_compounds bioinorganic chemistry*
Tel. 7(8832)384555 Fax 7(8832)998823

GERMANY

Sub-Editor: H. W. Zimmermann

Notes

1. The degrees conferred by the universities are the *Doctor habilitatus* (Dr habil.) (comparable to DSc), *Doctor scientiae naturalis* (Dr sc.nat.) (comparable to DSc), *Doctor rerum naturalium* (Dr rer.nat.) (equivalent to PhD), *Diplom-Chemiker* (Dipl. Chem.), *Diplom-Metallurge* (Dipl. Met.), *Diplom-Mineraloge* (Dipl. Min.), *Diplom-Physiker* (Dipl. Phys.), *Diplom-Kristallograph* (Dipl. Krist.), *Diplom-Mathematiker* (Dipl. Math.), *Diplom-Geologe* (Dipl. Geol.) (these seven equivalent to MSc). At the *Technische Universität* (TU) and *Technische Hochschulen* (TH) the degrees *Doctor Ingenieur* (Dr Ing.), *Diplom Ingenieur* (Dipl. Ing.) and *Physik Ingenieur* (Phys. Ing.) can be obtained.

Abram, Dr Ulrich (1957). Lecturer. Institut für Anorganische Chemie der Universität Tübingen, Germany. [Institut für Anorganische Chemie der Universität Tübingen, Auf der Morgenstelle 18, D-72076 Tübingen, Germany.] Dr habil. (TU Dresden, 1990) inorganic chemistry. *Coordination spectroscopy rhenium_compounds technetium_compounds*

E-mail caaar01@mailserv.zdv.uni-tuebingen.de   Tel.   49(7071)296230   Fax 49(7071)292436

Abriel, Dr Walter (1949). Manager. AWHchemconsult München, Germany. [AWHchemconsult, Weilheimer Str. 15, D-81373 München, Germany.] Dr rer.nat.habil. (U. Marburg, 1984) inorganic chemistry. *Solid_state chemistry reactivity phase_transition environment_protection*

Tel. 49(89)7148956 Fax 49(89)7147342

Abs-Wurmbach, Prof. Dr Irmgard (1938). Professor. Institut für Mineralogie und Kristallographie der Technischen Universität Berlin, Germany. [Institut für Mineralogie und Kristallographie der Technischen Universität Berlin, Ernst-Reuter-Platz 1, D-10587 Berlin, Germany.] Dr rer.nat. (U. Bonn, 1973) mineralogy. *Physical_properties_of_silicates_and_oxides Mossbauer_spectroscopy experimental_mineralogy_and petrology applied_mineralogy*

Tel. 49(30)31425639 Fax 49(30)31421124 Telex 184262tubln-d

Alex, Dr Volker Ernst (1939). Research Associate. Institut für Kristallzüchtung im Forschungsverbund Berlin e.V., Germany. [Institut für Kristallzüchtung im Forschungsverbund Berlin e.V., Geb. 18.46, Rudower Chaussee 6, D-12484 Berlin, Germany.] Dr rer.nat. (U. Halle-Wittenberg, 1970) physics. *Dynamical_diffraction crystal_growth semiconductors superconductors diffraction_theory topography*

Tel. 49(30)63923098 Fax 49(30)63923003

Amstutz, Prof. Dr G. Christian (1922). Professor emeritus. Mineralogisches Institut der Universität Heidelberg, Germany. [Mineralogisches Institut der Universität Heidelberg, Postfach 104040, D-69030 Heidelberg, Germany.] Dr sc.nat. (ETH Zürich, 1952) mineralogy-petrology. *Mineralogy petrography oregenesis crystal_growth ore_deposits history_and_philosophy_of_science*

Tel. 49(6221)564812 Fax 49(6221)564805

André, Dr Christoph (1962). Postdoctoral Research Assistant. Institut für Kristallographie der Freien Universität Berlin, Germany. [Institut für Kristallographie der Freien Universität Berlin, Takustr. 6, D-14195 Berlin, Germany.] Dr (FU Berlin, 1993) biochemistry. *Systematics_of_crystal_packings_of_organic_compounds molecular_crystals conformational_and_packing_behaviour_of_acyclic_carbohydrate_derivatives*

E-mail andre@chemie.fu-berlin.de Tel. 49(30)8384270 Fax 49(30)8383464

Anselment, Dr Bernhard (1955). Head of a producing firm. BASF Aktiengesellschaft, RCK/K Business Unit Catalysts, Germany. [BASF Aktiengesellschaft, Business Unit Catalysts, RCK/K-A 520, D-67056 Ludwigshafen, Germany.] Dr rer.nat. (U. Karlsruhe, 1985) inorganic chemistry. *Diffraction single_crystal phase_transition catalysts*

Tel. 49(621)6021072 Fax 49(621)6022538

Arnold, Prof. Dr Heinrich Günther Alfred (1930). Professor. Institut für Kristallographie der RWTH Aachen, Germany. [Institut für Kristallographie der RWTH Aachen, Jägerstr. 17-19, D-52056 Aachen, Germany.] Dr rer.nat. (U. Würzburg, 1964) mineralogy. *Powder_diffraction synchrotron_radiation kinetics symmetry*

E-mail fg050kr@dacth11 (EARN) fg050kr@vm1.rz.rwth-aachen.de (Internet) Tel. 49(241)806901 Fax 49(241)8888184 Telex 0832704thacd

Auffermann, Dr Gudrun (1957). Postdoctoral Research Assistant. Institut für Anorganische Chemie der RWTH Aachen, Germany. [Institut für Anorganische Chemie der RWTH Aachen, Prof.-Pirlet Str. 1, D-52056 Aachen, Germany.] Dr (RWTH Aachen, 1987) chemistry. *Neutron_diffraction X-ray_diffraction hydrides high_pressure powder_diffraction solid_state_chemistry*

E-mail fb010au@dacth11.bitnet Tel. 49(241)804669 Fax 49(241)8888288

Baars, Dr Jan W. (1931). Head of surface physics division. Fraunhofer Institut für Angewandte Festkörperphysik (IAF), Germany. [Fraunhofer Institut für Angewandte Festkörperphysik (IAF), Tullastr. 72, D-79108 Freiburg i.Br., Germany.] Dr-Ing. (TU Berlin, 1967) physics. *XPS surface chalcogenides LEED electrical_and_optical_characterization*

E-mail baars@iaf.fhg.de Tel. 49(761)51590 Fax 49(761)5159400 Telex 772510

Babel, Prof. Dr Dietrich (1930). Professor. Fachbereich Chemie der Universität Marburg, Germany. [Fachbereich Chemie der Universität Marburg, Hans-Meerwein-Straße, D-35043 Marburg, Germany.] Dr (U. Tübingen, 1961) chemistry. *Magnetochemistry structure cyanide fluorometallates transition_metal_compounds*

Tel. 49(6421)285625 Fax 49(6421)288917

Backhaus, Dr Karl-Otto (1936). [Baumschulenstr. 94, D-12437 Berlin, Germany.] Dr sc.nat. (Akad. d. Wiss. d. DDR Berlin, 1988) crystallography. *OD_structure polytypism symmetry minerals*

Tel. 49(30)6329617

Bärnighausen, Prof. Dr Hartmut (1933). Professor. Institut für Anorganische Chemie der Universität Karlsruhe, Germany. [Institut für Anorganische Chemie der Universität Karlsruhe, Engesserstr., Gebäude-Nr 30.45, D-76128 Karlsruhe, Germany.] Dr rer.nat. (U. Freiburg/Br., 1959) chemistry. *Inorganic_rare_earth_compound phase_transition twinning symmetry crystal chemistry*

E-mail bhn@achibm1.chemie.uni-karlsruhe.de   Tel.   49(721)6083484   Fax 49(721)6084290 Telex 17-721166UNIKar

Barbier, Mr Bruno (1952). Assistant. Mineralogisches Institut der Universität Bonn, Germany. [Mineralogisches Institut der Universität Bonn, Poppelsdorfer Schloß, D-53115 Bonn, Germany.] Dipl. Ing. (Rouen/France, 1974) physical metrology. *Powder lattice parameter medicine human_stones*

Tel. 49(228)733557 Fax 49(228)732770

Bartels, Mrs Heike (1964). Research Assistant. Arbeitsgruppe für Ribosomenstruktur der Max-Planck-Gesellschaft, Germany. [Arbeitsgruppe für Ribosomenstruktur der Max-Planck-Gesellschaft, c/o DESY, Notkestr. 85, D-22603 Hamburg, Germany.] Dipl. Min. mineralogy. *Macromolecular_crystallography ribosomes crystal_growth*

E-mail bartels@mpgars.desy.de Tel. 49(40)894696 Fax 49(40)891314

Bartels, Dr Matthias (1955). Head of Laboratory. E. MERCK, Germany. [E. MERCK, ZD-A/F+E6, D-64271 Darmstadt, Germany.] Dr rer.nat. (U. Göttingen, 1986) chemistry. *Diffractometer quantitative_phase_determination phase_transition calibration polymorphism amorphous_phase*

Tel. 49(6151)722061 Fax 49(6151)714494

Bartl, Prof. Dr Hans (1933). Professor. Institut für Kristallographie und Mineralogie der Universität Frankfurt, Germany. [Institut für Kristallographie und Mineralogie der Universität Frankfurt, Senckenberganlage 30, D-60054 Frankfurt/M., Germany.] Prof. (U. Frankfurt/M., 1972) crystallography and mineralogy. *Crystallography hydrogen_bonding diffractometry minerals instrumentation teaching classic_cars skiing*

Tel. 49(69)7982105 Fax 49(69)7982101

Bats, Dr Jan Willem (1949). Research Assistant. Institut für Organische Chemie, Universität Frankfurt, Germany. [Institut für Organische Chemie, Universität Frankfurt, Mertonviertel, Marie-Curie-Str. 11, D-60439 Frankfurt/M., Germany.] Dr (U. Twente Netherlands, 1976) chemistry. *Structure_determination data_collection accuracy molecular_crystal small molecules*

Tel. 49(69)58009124 Fax 49(69)58009250

Baum, Dr Elke (1960). [Michaelstr. 8, D-76137 Karlsruhe, Germany.] Dr rer.nat. (U. Marburg, 1989) chemistry. *Database X-ray_structure_determination chemistry magnetism_of_minerals*

Tel. 49(721)814251

Bautsch, Prof. Dr Hans-Joachim (1929). Museum für Naturkunde der Humboldt-Universität zu Berlin, Germany. [Museum für Naturkunde der Humboldt-Universität zu Berlin, Invalidenstr. 43, D-10115 Berlin, Germany.] Dr rer.nat.habil. (Humboldt-U. Berlin, 1965) mineralogy. *Crystallography mineralogy petrology gemology*

Tel. 49(30)28972560 Fax 49(30)28972561

Becherer, Prof. Dr Gerhard (1915). Professor. Chairman. Fachbereich Physik der Universität Rostock, Germany. [Fachbereich Physik der Universität Rostock, Parkstr. 18, D-18095 Rostock, Germany.] Dr rer.nat.habil. (U. Halle, 1953) physics. *Amorphous_phase crystallography diffraction physics camera charge_density clustering density_distribution desmearing liquids glasses*

Tel. 49(381)23587

Beck, Prof. Dr Horst P. (1941). Full Professor. Fachrichtung Anorganische und Analytische Chemie und Radiochemie der Universität Saarbrücken, Germany. [Fachrichtung Anorganische und Analytische Chemie und Radiochemie der Universität Saarbrücken, Im Stadtwald, Postfach 1150, D-66041 Saarbrücken, Germany.] Dr (U. Karlsruhe, 1972) chemistry. *Structure_determination high_pressure high_temperature diffraction_technique phase_transformation symmetry*

Tel. 49(681)3022481 Fax 49(681)3024233

Beck, Prof. Dr Johannes (1956). Full Professor. Institut für Anorganische und Analytische Chemie I der Universität Gießen, Germany. [Institut für Anorganische und Analytische Chemie I der Universität Gießen, Heinrich-Buff-Ring 58, D-35392 Gießen, Germany.] Prof. (U. Gießen, 1992) chemistry. *Transition_elements solid_state magnetism area_detector metal_chalcogenides polycations_of_main_group_elements complexes_with_polyazenido_ligands*

Tel. 49(641)7025660 Fax 49(641)7025669

Becker, Prof. Dr Gerd (1940). Professor. Institut für Anorganische Chemie der Universität Stuttgart, Germany. [Institut für Anorganische Chemie der Universität Stuttgart, Pfaffenwaldring 55, D-70550 Stuttgart, Germany.] Dr habil. (U. Karlsruhe, 1976) chemistry. *Structure_of_phosphorus_compounds silicon_compounds lithium_compounds*
Tel. 49(711)6854172 Fax 49(711)6854241

Behlke, Prof. Dr Joachim (1934). Head of Research Group. Humboldt-Universität zu Berlin im Max-Delbrück-Centrum für Molekulare Medizin, Germany. [Max-Delbrück-Centrum für Molekulare Medizin, Robert-Rössle-Str. 10, D-13122 Berlin, Germany.] Dr habil. (U. Greifswald, 1970) biophysics. *Association_theory crystallization hydrodynamics light_scattering biocrystallography biophysics modelling_of_proteins*
E-mail behlke@orion.rz.mdc-berlin.de Tel. 49(30)94062205 Fax 49(30)94062802

Behm, Dr Helmut (1947). Technical Leader. TTC Microelectronic GmbH, Germany. [TTC Microelectronic GmbH, Nordstr. 22, D-31653 Stadthagen, Germany.] Dr (U. Freiburg, 1976) chemistry. *Boron_compounds electrochemistry iodine_compounds Patterson_method*
Tel. 49(5721)970952 Fax 49(5721)970985

Behrens, Dr Heinrich (1937). Head of Department III. FIZ Karlsruhe, Germany. [FIZ Karlsruhe, Postfach 2465, D-76012 Karlsruhe, Germany.] Dr rer.nat. (TH Karlsruhe, 1966) physics. *Data_collection data_processing database*
E-mail DG@fiz-karlsruhe.de Tel. 49(7247)808250 Fax 49(7247)808666 Telex 17724710+

Behrens, Prof. Dr Peter (1957). Professor. Institut für Anorganische Chemie, Ludwig-Maximilians-Universität, Germany [Institut für Anorganische Chemie, Ludwig-Maximilians-Universität, Meiserstrasse 1, D-80333 München, Germany] Prof. (U. München, 1994) chemistry. *EXAFS XANES Rietveld_method porous_material*
E-mail pbe@anorg.chemie.uni-muenchen.de Tel. 49(89)5902356 Fax 49(89)590257

Behrens, Prof. Dr Ulrich (1946). Lecturer. [Institut für Anorganische Chemie der Universität, Martin-Luther-King-Platz 6, D-20146 Hamburg, Germany.] Dr habil. (U. Hamburg, 1975) inorganic chemistry. *Single_crystal X-ray organometallic copper_compounds*
Tel. 49(40)41232894 Fax 49(40)41232893

Benedict, Dr Ulrich (1930). Unit Head. European Commission, Joint Research Centre, Institute for Transuranium Elements, Germany. [European Commission, Joint Research Centre, Institute for Transuranium Elements, Postfach 2340, D-76125 Karlsruhe, Germany.] Dr rer.nat. (Saarbrücken U., 1959) metal physics. *High_pressure phase_transition actinides lanthanides*
Tel. 49(7247)951377 Fax 49(7247)951590

Bennett, Dr William (1949). Research Scientist. Max-Planck-Gesellschaft, Arbeitsgruppe für Ribosomenstruktur, Germany. [Max-Planck-Gesellschaft, Arbeitsgruppe für Ribosomenstruktur, c/o DESY, Notkestr. 85, D-22603 Hamburg, Germany.] PhD (Yale U./USA, 1978) molecular biophysics and biochemistry. *Macromolecular_crystallography computing_in_crystallography enzyme_mechanism DNA–protein_interaction*
E-mail bennett@mpgars.desy.de Tel. 49(40)89982833 Fax 49(40)891314 Telex 215124desyd

Bensch, Dr Wolfgang (1953). Lecturer. Institut für Anorganische Chemie der Universität Frankfurt, Germany. [Institut für Anorganische Chemie der Universität Frankfurt, Marie-Curie-Str. 11, D-60439 Frankfurt/M., Germany.] Dr habil. (U. Frankfurt, 1993) inorganic chemistry. *Magnetism_of_chalcogenides XPS_of_chalcogenides electronic_structure_of_chalcogenides low-dimensional_metals electronic_band_structure_calculations relations_between_crystal_structure_and_physical_properties*
E-mail bensch@chemie.uni-frankfurt.d400.de Tel. 49(69)58009151 Fax 49(69)58009260

Benz, Prof. Dr Klaus-Werner (1938). Director. Commission on Crystal Growth and Characterization of Materials. Kristallographisches Institut der Universität Freiburg, Germany. [Kristallographisches Institut der Universität Freiburg, Hebelstr. 25, D-79104 Freiburg/Br., Germany.] Dr Ing.habil. (U. Stuttgart, 1986) electrical engineering. *Characterization crystallography crystallization semiconductors growth_kinetics numerical_modelling*
E-mail benz@sun1.ruf.uni-freiburg.de Tel. 49(761)2036448 Fax 49(761)2036434

Berg, Dr Liselotte (1933). Chief Editor. Gmelin-Institut der Max-Planck-Gesellschaft, Germany. [Gmelin-Institut der Max-Planck-Gesellschaft, Varrentrappstr. 40/42, D-60486 Frankfurt/M., Germany.] Dr rer.nat. (TU Braunschweig, 1964) chemistry and physics. *Chemistry physics crystallography microstructure*
Tel. 49(69)7917385

Berger, Dr Hans (1940). Lecturer. Institut für Physik der Humboldt-Universität zu Berlin, Invalidenstraße 110, D-10115 Berlin, Germany. [Institut für Physik der Humboldt-Universität zu Berlin, Sitz: Invalidenstraße 110, Unter den Linden 6, D-10099 Berlin, Germany.] Dr sc.nat. (Humboldt-U. Berlin, 1985) crystallography. *Multiple_crystal_diffractometry precise_measurement of_lattice_parameters nonstoichiometry_of_II–VI_compounds solid_solutions orientation_relationships_and_lattice_distortion superlattices X-ray_topography*
Tel. 49(30)2803367 Fax 49(30)2803304 Telex 304156huphy

Bergerhoff, Prof. Dr Günter (1926). Professor. Institut für Anorganische Chemie der Universität Bonn, Germany. [Institut für Anorganische Chemie der Universität Bonn, Gerhard-Domagk-Str. 1, D-53121 Bonn, Germany.] Dr (U. Bonn, 1954) chemistry. *Structural_database structural_classification*
E-mail unc412@ibm.rhrz.uni-bonn.de Tel. 49(228)732657 Fax 49(228)735660 Telex 886657unibod

Bergunde, Dr Thomas (1959). Postdoctoral Research Assistant. Ferdinand-Braun-Institut für Höchstfrequenztechnik, Germany. [Ferdinand-Braun-Institut für Höchstfrequenztechnik, Rudower Chaussee 5, D-12489 Berlin, Germany.] Dr rer.nat. (Humboldt-U. Berlin, 1990) crystallography. *CVD layer luminescence semiconductor*
Tel. 49(30)63922671 Fax 49(30)63922642

Bernhardt, Prof. Dr Wolfgang (1934). Vice Director. Institut für Elektrotechnik, Werkstoff- und Verfahrenstechnik, Humboldt-Universität zu Berlin, Germany. [Institut für Elektrotechnik, Werkstoff- und Verfahrenstechnik, Invalidenstr. 110, D-10099 Berlin, Germany.] Dr rer.nat.habil. (Bergakademie Freiberg/Sa., 1970) metallography. *Metallography real_structure semiconductors microelectronics electrical_properties_of_matter*
Tel. 49(30)2803508 Fax 49(30)2803394

Berthold, Mr Thomas (1955). Scientist. SIEMENS AG, ZFE BT MR 22, Germany. [SIEMENS AG, ZFE BT MR 22, Otto-Hahn-Ring 6, D-81739 München, Germany.] *Crystal_growth luminescence structural_disorder*
Tel. 49(89)6362684 Fax 49(89)63648131 Telex 898250 siemcp

Betzel, Dr Christian (1956). Scientific Staff Member. European Molecular Biology Laboratory, Germany. [European Molecular Biology Laboratory, Notkestr. 85, D-22603 Hamburg, Germany.] Dr (FU. Berlin, 1986) chemistry. *Synchrotron_radiation proteins crystal_growth high_resolution_diffractometry*
E-mail christian@embl-hamburg.de Tel. 49(40)89902136 Fax 49(40)89902141

Bieniok, Dr Anna (1959). Research Assistant. Institut für Kristallographie und Mineralogie der Universität Frankfurt, Germany. [Institut für Kristallographie und Mineralogie der Universität Frankfurt, Senckenberganlage 30, D-60054 Frankfurt/M., Germany.] Dr (U. Frankfurt, 1992) crystallography. *Zeolites powder_and_single_crystal_diffraction physical_property structure_correlation*
E-mail bieniok@kristall.uni-frankfurt.d400.de Tel. 49(69)7983104 Fax 49(69)7982101

Binas, Dr Horst (1934). Scientific Assistant. Bundesanstalt für Materialforschung und -prüfung (BAM), Germany. [Bundesanstalt für Materialforschung und -prüfung (BAM), Unter den Eichen 87, D-12203 Berlin, Germany.] Dr rer.nat. (Humboldt-U. Berlin, 1963) crystallography. *Phase_determination residual_stress_of_metals_and_ceramics phase_transition*
Tel. 49(30)81043109 Fax 49(30)8112029

Bissert, Dr Gertrud (1933). Research Assistant. Mineralogisches Institut der Universität Kiel, Germany. [Mineralogisches Institut der Universität Kiel, Olshausenstr. 40, D-24098 Kiel, Germany.] Dr (U. Kiel, 1969) crystallography. *Structure_determination silicates_classification*
E-mail nmp13@rz.uni-kiel.d400.de Tel. 49(431)8802893 Fax 49(431)8804457

Blanke, Mrs Frauke (1967). Post-Graduate Student. Institut für Mineralogie der Universität Münster, Germany. [Institut für Mineralogie der Universität Münster, Corrensstr. 24, D-48149 Münster, Germany.] Dipl. Chem. (U. Münster, 1992) chemistry. *Zeolites hydrothermal_synthesis thermoanalysis spectroscopy Rietveld_method aluminophosphate_molecular_sieves*
Tel. 49(251)833405

Blödner, Mr Ralph-Uwe (1965). Post-Graduate Student. Institut für Kristallographie und Materialforschung, Fachbereich Physik, Humboldt-Universität zu Berlin, Germany. [Institut für Kristallographie und Materialforschung, Fachbereich Physik, Humboldt-Universität zu Berlin, Unter den Linden 6, D-10099 Berlin, Germany.] Dipl. Kristallogr. (Humboldt-U. Berlin, 1991) crystallography. *Crystal_growth_from_solution hydrodynamics alloys ACRT*
Tel. 49(30)2803294 Fax 49(30)2803360

Blüthgen, Mr Waldemar (1940). Patent Manager. Leica Industrieverwaltung GmbH, Germany. [Leica Industrieverwaltung GmbH, Ernst-Leitz-Straße, D-35578 Wetzlar, Germany.] Dipl. Min. (U. Bonn, 1968) mineralogy. *Patent optical_microscopy polarization_microscopy reflected_light_microscopy biological optics fluorescence*
Tel. 49(6441)292466 Fax 49(6441)292559 Telex 483849leizd

Bluhm, Dr Karsten (1961). Postdoctoral Research Assistant. Institut für Anorganische Chemie der Universität Kiel, Germany. [Institut für Anorganische Chemie der Universität Kiel, Otto-Hahn-Platz 6/7, D-24118 Kiel, Germany.] Dr (U. Kiel, 1990) chemistry. *Chemistry diffraction_data oxides boron_compounds*
E-mail nac76@rz.uni-kiel.d400.de Tel. 49(431)8802096 Fax 49(431)8801520

Bode, Dr Wolfram (1942). Scientific Assistant. Max-Planck-Institut für Biochemie, Germany. [Max-Planck-Institut für Biochemie, D-82152 Martinsried, Germany.] Dr habil. (U. München, 1983) physics and chemistry. *Crystallography proteinases drug_design anticoagulants inhibitors protein_crystallography*
E-mail bode@vms.biochem.mpg.de Tel. 49(89)85782676 Fax 49(89)85783516

Bögge, Dr Hartmut (1953). Research Associate. Fakultät für Chemie (ACI) der Universität Bielefeld, Germany. [Fakultät für Chemie (ACI) der Universität Bielefeld, Universitätsstraße, Postfach 100131, D-33501 Bielefeld, Germany.] Dr (U. Bielefeld, 1980) chemistry. *X-ray_structure_determination transition_elements inorganic_clusters bond_order*
E-mail hboegge@post.uni-bielefeld.de Tel. 49(521)1066140 Fax 49(521)1066146

von Böhlen, Dr Klaus J. (1959). Research Associate. Arbeitsgruppe Ribosomenstruktur der Max-Planck-Gesellschaft, Germany. [Arbeitsgruppe Ribosomenstruktur der Max-Planck-Gesellschaft, c/o DESY, Notkestr. 85, D-22603 Hamburg, Germany.] Dr (U. Hamburg, 1990) biology. *Ribosomes cryogenics oscillation_camera synchrotron_radiation biological_macromolecules area_detector cryostats focussing_mirrors*
E-mail boehlen@mpgars.desy.de Tel. 49(40)89982834 Fax 49(40)891314 Telex 215124desyd

Böhm, Prof. Dr Horst (1937). Professor. Institut für Geowissenschaften der Universität Mainz, Germany. [Institut für Geowissenschaften der Universität Mainz, D-55099 Mainz, Germany.] Dr (ETH Zürich, 1968) crystallography. *Crystallography modulated_structure ionic_conductivity phase_transition*
E-mail boehm@vgemia.geo.uni-mainz.de Tel. 49(6131)392848 Fax 49(6131)393070

Boehnke, Dr Undine-Constanze (1957). Research Assistant. Fachbereich Physik der Universität Leipzig, Germany. [Fachbereich Physik der Universität Leipzig, Linnéstr. 5, D-04103 Leipzig, Germany.] Dr rer.nat. (U. Leipzig, 1985) crystallography. *Phase_diagram superconductor semiconductor liquid_crystal*
Tel. 49(341)6858215 Fax 49(341)6858221

Boese, Dr Roland (1945). Professor. Institut für Anorganische Chemie der Universität Essen, Germany. [Institut für Anorganische Chemie der Universität Essen, Universitätsstr. 5-7, D-45117 Essen, Germany.] Professor (U. Essen, 1991) inorganic chemistry. *Structure_determination_of_boron_compounds low_temperature_diffractometry_and_crystal_growth electron_density_distribution strained_hydrocarbons*
E-mail boese@structchem.uni-essen.de Tel. 49(201)1832416 Fax 49(201)1832535

Bohatý, Prof. Dr Ladislav (1948). Professor. Institut für Kristallographie der Universität zu Köln, Germany. [Institut für Kristallographie der Universität zu Köln, Zülpicher Str. 49 b, D-50674 Köln, Germany.] Dr rer.nat. (U. Köln, 1975) crystallography. *Single_crystal physical_property crystal_growth acentric_crystals*
E-mail bohaty@kri.uni-koeln.de Tel. 49(221)4703154 Fax 49(221)4704963

Bohm, Prof. Dr Joachim (1935). Honorary Professor. Inst. Mineralogie, TU Bergakademie, D-09596 Freiberg, Germany. [Trützschlerstr. 14, D-12487 Berlin, Germany.] Dr habil. (Humboldt-U. Berlin, 1969) crystallography. *Crystal_growth real_structure symmetry oxides history crystallography education mineralogy*
Tel. 49(30)6363284 49(30)63923019 Fax 49(30)63923003

Bohnen, Mr Frank Michael (1966). Research Associate. Max-Planck-Institut für Kohlenforschung, Germany. [Max-Planck-Institut für Kohlenforschung, Kaiser-Wilhelm-Platz 1, D-45470 Mühlheim/Ruhr, Germany.] Dipl. Chem. (U. Göttingen, 1993) chemistry. *Heterogeneous_catalysis*
E-mail bohnen@mpi-muelheim.mpg.dbp.de Tel. 49(208)3062449

Bollmann, Dr Ulrich (1952). Group Leader. VAW aluminium AG, Forschung/Entwicklung, Germany. [VAW aluminium AG, Forschung/Entwicklung, Georg-von-Boeselager-Str. 25, Postfach 2468, D-53117 Bonn, Germany.] Dr. habil. (Bergakademie Freiberg, 1993) chemistry. *Metallurgy technology carbon_compounds coke pitch graphites preparation_and_characterization_of_catalysts*
Tel. 49(228)5522322 Fax 49(228)5522268 Telex 8869607

Bolte, Dr Michael (1958). Senior Assistant. Institut für Organische Chemie der Universität Frankfurt, Germany. [Institut für Organische Chemie der Universität Frankfurt, Marie-Curie-Str. 11, D-60439 Frankfurt/M., Germany.] Dr rer.nat. (U. Göttingen, 1988) chemistry. *Molecular_modelling force_field_method structure-activity_relationship*
E-mail bolte@chemie.uni-frankfurt.d400.de Tel. 49(69)58009136 Fax 49(69)58009128

Bonse, Prof. Dr Dr h. c. Ulrich (1928). Full Professor. Experimentelle Physik I, Fachbereich Physik der Universität Dortmund, Germany. [Experimentelle Physik I, Fachbereich Physik der Universität Dortmund, D-44221 Dortmund, Germany.] Dr rer.nat.habil (U. Münster, 1963) physics. *X-ray_diffraction neutron_diffraction X-ray_neutron-interferometry microtomography SAXS*
E-mail bonse@fkp.physik.uni-dortmund.de Tel. 49(231)7553504 Fax 49(231)7553569 Telex 822445unido

Borchardt-Ott, Dr Walter (1933). Academic Director. Institut für Mineralogie der Universität Münster, Germany. [Institut für Mineralogie der Universität Münster, Corrensstr. 24, D-48149 Münster, Germany.] Dr rer.nat. (U. Münster, 1964) mineralogy. *Symmetry morphology mineralogy teaching_of_crystallography_and_mineralogy*
Tel. 49(251)833453 Fax 49(251)838397 Telex 892529unimsd

Boysen, Dr Hans H. (1944). Research Assistant. Institut für Kristallographie der Universität München, Germany. [Institut für Kristallographie der Universität München, Theresienstr. 41, D-80333 München, Germany.] Dr (U. München, 1978) physics. *Neutron_scattering diffuse_scattering disorder phase_transition powder_diffraction*
E-mail boysen@neutronenbeugung.geologie.uni-muenchen.d400.de Tel. 49(89)32094040 Fax 49(89)32094015 Telex 529815

Bram, Mr Andreas (1966). Research Assistant. Institut für Angewandte Physik, Lehrstuhl für Kristallographie der Universität Erlangen-Nürnberg, Germany. [Institut für Angewandte Physik, Lehrstuhl für Kristallographie der Universität Erlangen-Nürnberg, Bismarckstr. 10, D-91054 Erlangen, Germany.] Dipl. Phys. (U. Erlangen-Nürnberg, 1991) physics. *Electron_density_distribution superconductor diffractometry structure_determination*
E-mail mpkr00@cd4680fs.rrze.uni-erlangen.de Tel. 49(9131)852119 Fax 49(9131)852733

Brand, Prof. Dr Paul (1931). Professor. Institut für Anorganische Chemie, Fachbereich Chemie, Technische Universität Bergakademie Freiberg, Germany. [Institut für Anorganische Chemie, Fachbereich Chemie, Technische Universität Bergakademie Freiberg, D-09596 Freiberg, Germany.] Dr rer.nat.habil. (U. Halle-Wittenberg, 1967) chemistry. *Decomposition structure-activity_relationship diffraction_technique aluminium_compounds gels sol-gel_transitions*
Tel. 49(3731)512050 Fax 49(3731)514386

Brandmüller, Prof. Dr Josef (1921). Professor emeritus. Fakultät für Physik der Universität München, Germany. [Hubertusstr. 61, D-82131 Gauting, Germany.] Dr rer.nat. (LMU München, 1945) experimental physics. *Raman group_and_representation_theory tensor_property symmetry symmetry_in_science_and_art*
E-mail heinz.schroetter@lehrstuhl-haensch.physik.uni-muenchen.dbp.de Tel. 49(89)8504197 Fax 49(89)8932026

Brendel, Mr Uwe (1961). Institut für Kristallographie und Mineralogie der Universität Frankfurt, Germany. [Institut für Kristallographie und Mineralogie der Universität Frankfurt, Senckenberganlage 30, D-60054 Frankfurt/M., Germany.] Dipl. Ing. (Fachhochschule Frankfurt/M., 1987) technical informatics. *Crystal_growth physical_property diffraction*
E-mail ubrendel@kristall.uni-frankfurt.d400.de Tel. 49(69)7982100 Fax 49(69)7982101

Brokmeier, Dr Heinz-Günther (1952). Research Assistant. Institut für Metallkunde und Metallphysik, Außenstelle am GKSS-Forschungszentrum, Germany. [GKSS Forschungszentrum (W-TUC), Postfach 1160, D-21494 Geesthacht, Germany.] Dr rer.nat. (TU Clausthal, 1983) mineralogy and crystallography. *Texture neutron_scattering multiphase_materials metals deformation_behaviour anisotropic_properties polycrystal_diffraction*
E-mail brokmeier@gkss.de Tel. 49(4152)871207 Fax 49(4152)871338 Telex 218712 gkss d

Bronger, Prof. Dr Welf (1932). Full Professor. Institut für Anorganische Chemie der RWTH Aachen, Germany. [Institut für Anorganische Chemie der RWTH Aachen, Prof.-Pirlet-Str. 1, D-52056 Aachen, Germany.] Dr rer.nat. (U. Münster, 1961) chemistry. *Solid_state_chemistry*
Tel. 49(241)804643 Fax 49(241)8888288

Brüderl, Mr Georg (1963). Research Associate. Institut für Angewandte Physik, Lehrstuhl für Kristallographie der Universität Erlangen-Nürnberg, Germany. [Institut für Angewandte Physik, Lehrstuhl für Kristallographie der Universität Erlangen-Nürnberg, Bismarckstr. 10, D-91054 Erlangen, Germany.] Dipl. Phys. (U. Regensburg, 1991) physics. *Powder structure_determination Rietveld_method deconvolution*
E-mail mpkr00@cd4680fs.rrze.uni-erlangen.de Tel. 49(9131)852118 Fax 49(9131)852733

Buck, Prof. Dr Peter (1939). Professor. Institut für Mineralogie, Petrologie und Kristallographie der Universität Marburg, Germany. [Institut für Mineralogie, Petrologie und Kristallographie der Universität Marburg, Hans-Meerwein-Str., D-35032 Marburg, Germany.] Dr rer.nat. (U. Freiburg, 1967) crystallography. *Growth physical_property piezoelectricity*
Tel. 49(6421)285500 Fax 49(6421)285831

Büttner, Dr Wolfgang (1948). Sales and Marketing. SIEMENS AG, Germany. [SIEMENS AG, AUT V 371, Postfach 301166, D-50781 Köln, Germany.] Dr rer.nat. (U. Köln, 1982) mineralogy. *Diffractometer instrumentation X-ray_fluorescence*
Tel. 49(221)5762574 Fax 49(221)5763895

Bunge, Prof. Dr Dr h. c. Hans-Joachim (1929). Full Professor. Dept Physical Metallurgy der Technischen Universität Clausthal, Grosser Bruch 23, D-38678 Clausthal-Zellerfeld, Germany. [Ahornweg 15, D-38678 Clausthal-Zellerfeld, Germany.] Dr rer.nat.habil. (Humboldt-U. Berlin, 1964) physics. *Metallurgy anisotropic_physical_property powder_diffraction texture*
E-mail meed@ibm.rz.tu-clausthal.de Tel. 49(5323)722244 Fax 49(5323)722340

Burgäzy, Dr Frank (1960). Product Specialist X-ray Diffraction. SIEMENS AG, Germany. [SIEMENS AG, AUT V 371 F2, Siemensallee 84, D-76187 Karlsruhe, Germany.] Dr (U. Stuttgart, 1990) physics. *High-precision_diffractometry X-ray_diffractometer_instrumentation microdiffraction III-V_semiconductors thin_film_analysis*
Tel. 49(721)5954265 Fax 49(721)5954506 Telex 72139-0=siekhe

Burkel, Prof. Dr Eberhard (1952). Professor. Institut für Angewandte Physik, Lehrstuhl für Kristallographie der Universität Erlangen-Nürnberg, Germany. [Institut für Angewandte Physik, Lehrstuhl für Kristallographie der Universität Erlangen-Nürnberg, Bismarckstr. 10, D-91054 Erlangen, Germany.] Dr rer.nat.habil. (U. München, 1991) experimental physics. *X-ray_spectroscopy thin_film dynamics high_pressure order-disorder synchrotron_radiation neutrons*
E-mail mpkr00@cd4680fs.rrze.uni-erlangen.de Tel. 49(9131)852711 Fax 49(9131)852733

Burschka, Dr Christian (1946). Education and Technical Management. Institut für Anorganische Chemie der Universität Würzburg, Germany. [Institut für Anorganische Chemie der Universität Würzburg, Am Hubland, D-97074 Würzburg, Germany.] Dr (RWTH Aachen, 1975) chemistry. *X-ray_crystallography inorganic_chemistry semi-empirical_calculation safety education*
Tel. 49(931)8885286 Fax 49(931)8884605

Burzlaff, Dr Hans (1932). Full Professor. Lehrstuhl für Kristallographie, Institut für Angewandte Physik der Universität Erlangen-Nürnberg, Germany. [Lehrstuhl für Kristallographie, Institut für Angewandte Physik der Universität Erlangen-Nürnberg, Bismarckstr. 10, D-91054 Erlangen, Germany.] Dr rer.nat.habil (U. Marburg, 1968) crystallography. *Crystallography symmetry_theory systematics_of_crystal_structures*
E-mail mpkr00@cd4680fs.rrze.uni-erlangen.dbp.de Tel. 49(9131)852700Fax 49(9131)852733

Buschmann, Dr Jürgen F. (1939). Scientist. Institut für Kristallographie der Freien Universität Berlin, Germany. [Institut für Kristallographie der Freien Universität Berlin, Takustr. 6, D-14195 Berlin, Germany.] Dr (FU. Berlin, 1980) crystallography. *Single-crystal structure_determination low_temperature_crystallization valence_charge_density*
E-mail buschmann@kristall.chemie.fu-berlin.de Tel. 49(30)8383408 Fax 49(30)8383464

Cammenga, Dr Heiko Karl (1938). Professor. Institut für Physikalische und Theoretische Chemie der Universität Braunschweig, Germany. [Institut für Physikalische und Theoretische Chemie der Universität Braunschweig, Hans-Sommer-Str. 10, D-38106 Braunschweig, Germany.] Dr rer.nat. (TH Braunschweig, 1967) chemistry. *DTA DSC calorimetry topochemistry kinetics time/temperature_resolved_diffraction*
Tel. 49(531)3915333 Fax 49(531)3915832 Telex 952526tubs^w

Claus, Mrs Cornelia (1967). Student. Mineralogisch-Petrographisches Institut der Universität Hamburg, Grindelallee 48, D-20146 Hamburg, Germany. [Veringstr. 48, D-21107 Hamburg, Germany.] *Germanates hydrothermal synthesis structure_determination crystal_growth gemology minerals refinement_methods rare-earth_compound*
Tel. 49(40)7532331

Claus, Mr Karl Heinz (1940). Technician for X-ray structure analysis. Max-Planck-Institut für Kohlenforschung, Germany. [Max-Planck-Institut für Kohlenforschung, Postfach 101353, D-45466 Mülheim/Ruhr, Germany.] *Diffraction_technique low_temperature preparation organometallic*
Tel. 49(208)3062180 Fax 49(208)3062989

Czank, Priv. Doz. Dr Michael (1941). Senior Research Scientist. Mineralogisches Institut der Universität Kiel, Germany. [Mineralogisches Institut der Universität Kiel, Olshausenstr. 40, D-24098 Kiel, Germany.] Dr rer.nat. (ETH Zürich/Switzerland, 1973) crystallography. *Electron_microscopy real_structure minerals silicates crystal_chemistry structure_determination modulated_structures classification*
Tel. 49(431)8802903 Fax 49(431)8804457

Däweritz, Dr Lutz (1943). Group Leader; Head of Research. Paul-Drude-Institut für Festkörperelektronik, Germany. [Paul-Drude-Institut für Festkörperelektronik, Hausvogteiplatz 5-7, D-10117 Berlin, Germany.] Dr rer.nat.habil. (U. Halle-Wittenberg, 1993) experimental physics. *MBE RHEED semiconductors surface_reconstruction real-time_control growth_kinetics III-V_compounds*
E-mail daeweritz@pdi1.iaas-berlin.de Tel. 49(30)20377359 Fax 49(30)20377201

Dahlems, Mr Thomas (1963). Post-Graduate Student. Institut für Anorganische Chemie und Strukturchemie, Lehrstuhl II der Universität Düsseldorf, Germany. [Institut für Anorganische Chemie und Strukturchemie, Lehrstuhl II der Universität Düsseldorf, Universitätsstr. 1, D-40225 Düsseldorf, Germany.] Dipl. Chem. (U. Düsseldorf, 1992) chemistry. *Acids_hydrogen_bonding DTA low_temperature aqueous_acids crystal_structure_of_aqueous_acids_at_low_temperature*
E-mail Dahlems@mail.rz.uni-duesseldorf.de Tel. 49(211)3113857 Fax 49(211)3113085

Daniels, Dr Peter (1959). Postdoctoral Research Assistant. Institut für Mineralogie der Universität Bochum, Germany. [Institut für Mineralogie der Universität Bochum, D-44780 Bochum, Germany.] Dr rer.nat. (U. Bochum, 1990) mineralogy. *Crystal_chemistry X-ray_diffraction disorder X-ray_optics*
Tel. 49(234)7007545 Fax 49(234)7094179

Dauter, Dr Zbigniew (1948). Staff Scientist. European Molecular Biology Laboratory, Außenstelle Hamburg, Germany. [European Molecular Biology Laboratory, Außenstelle Hamburg, c/o DESY, Notkestr. 85, D-22603 Hamburg, Germany.] Dr (U. Gdansk/Poland, 1975) organic chemistry. *X-ray crystallography structural macromolecular*
E-mail dauter@embl-hamburg.de Tel. 49(40)899020 Fax 49(40)89902149

Decanniere, Dr Klaas (1969). Institut für Kristallographie der Freien Universität Berlin, Germany. [Institut für Kristallographie der Freien Universität Berlin, Takustr. 6, D-14195 Berlin, Germany.] Grote onderscheiding (Vrije U. Brussel/Belgium, 1992) ing. in scheikunde en landbouwindustrien. *Proteins nucleic_acids single_crystal diffraction*
E-mail dnapro@chemie.fu-berlin.de Tel. 49(30)8386325 Fax 49(30)8386702

Deiseroth, Prof. Dr Hans-Jörg (1945). Professor. Institut für Anorganische Chemie der Universität Stuttgart, Germany. [Institut für Anorganische Chemie der Universität Stuttgart, Pfaffenwaldring 55, D-70550 Stuttgart, Germany.] Dr rer.nat.habil. (U. Stuttgart, 1983) chemistry. *Chemistry solid_state alloys ceramics inorganic_solid_state_chemistry*
Tel. 49(711)6854239 Fax 49(711)6854241

Depmeier, Prof. Dr Wulf (1944). Professor. Institut für Mineralogie der Universität Kiel, Germany. [Institut für Mineralogie der Universität Kiel, Olshausenstr. 40, D-24098 Kiel, Germany.] Dr rer.nat.habil. (U. Genève/Switzerland, 1983) solid state chemistry. *Phase_transition modulated_structure zeolites pseudosymmetry*
E-mail depmeier@iris.min.uni-kiel.d400.de Tel. 49(431)8802839 Fax 49(431)8804457

Diehl, Dr Roland (1944). Deputy Director. Fraunhofer-Institut für Angewandte Festkörperphysik, Germany. [Fraunhofer-Institut für Angewandte Festkörperphysik, Tullastr. 72, D-79108 Freiburg, Germany.] Dr rer.nat. (U. Freiburg/Br., 1972) crystallography. *Semiconductor III-V_compounds epitaxy integrated_circuit gemstones*
E-mail diehl@iaf.fhg.de Tel. 49(761)5159416 Fax 49(761)5159400

Dietrich, Prof. Dr Burkhart (1939). Scientist. Institut für Halbleiterphysik GmbH, Frankfurt (Oder), Germany. [Institut für Halbleiterphysik GmbH, Postfach 409, D-15204 Frankfurt (Oder), Germany.] Dr rer.nat.habil. (U. Jena, 1979) physics. *Diffractometry heterostructure Raman semiconductors bond_method diffraction_theory lattice_dynamics superlattice germanium silicon*
E-mail dietrich@ihp.d400.de Tel. 49(335)373125 Fax 49(335)326195

Dörfel, Dr Ilona (1953). Senior Scientist. Bundesanstalt für Materialforschung und -prüfung, Germany. [Bundesanstalt für Materialforschung und -prüfung, Außenstelle Adlershof, Rudower Chaussee 5, Geb. 8.15, D-12484 Berlin, Germany.] Dr rer.nat. (Humboldt-U. Berlin, 1988) crystallography. *Characterization_of_microstructure_of_ceramics AEM transmission_electron_microscopy diffraction*
Tel. 49(30)63925833 Fax 49(30)63925787

Dräger, Prof. Dr Martin (1940). Lecturer. Inst. für Anorgan. Chemie und Analyt. Chemie der Universität Mainz, Germany. [Inst. für Anorgan. Chemie und Analyt. Chemie der Universität Mainz, D-55099 Mainz, Germany.] Dr rer.nat. (U. Mainz, 1970) chemistry. *Organometallic_chemistry stereochemistry antimony_compounds tin_compounds*
E-mail draeger@mzdmza.zdv.uni-mainz.de Tel. 49(6131)395757 Fax 49(6131)395380

Dressler, Dr Ludwig (1944). Research Associate. Institut für Optik und Quantenelektronik der Universität Jena, Germany. [Institut für Optik und Quantenelektronik der Universität Jena, Max-Wien-Platz 1, D-07743 Jena, Germany.] Dr sc.nat. (U. Jena, 1981) physics. *X-ray_characterization silicon_compounds crystal_growth surface calcium_compounds optical_materials*
Tel. 49(3641)8225625 Fax 49(3641)23843

Driesel, Dr Wolfgang (1946). Postdoctoral Research Assistant. Max-Planck-Institut für Mikrostrukturphysik, Germany. [Max-Planck-Institut für Mikrostrukturphysik, Weinberg 2, D-06120 Halle/Saale, Germany.] Dr rer.nat. (U. Leipzig, 1974) physics. *Focused_ion_beam microelectronics cross_section diagnostic liquid_metal_ion_source custom_FIB_micromachining*
E-mail drie@secundus.mpi-msp-halle.mpg.de Tel. 49(345)558250 Fax 49(345)5511223

Dünkel, Dr Lothar (1942). Project Leader. KAI e.V. WIP-Arbeitsgruppe, Dr. Dünkel, Germany. [KAI e.V. WIP-Arbeitsgruppe, Dr. Dünkel, Rudower Chaussee 5-6, Geb. 10.2, D-12489 Berlin, Germany.] Dr sc.nat. (Akad. Wiss. der DDR Berlin, 1989) physical chemistry. *Photography photosynthesis holography electron_microscopy*
Tel. 49(30)63923187

Durchschlag, Dr Helmut (1944). Research Chemist. Institut für Biophysik und Physikalische Biochemie der Universität Regensburg, Germany. [Institut für Biophysik und Physikalische Biochemie der Universität Regensburg, Universitätsstr. 31, D-93040 Regensburg, Germany.] Dr phil. (U. Graz/Austria, 1971) physical chemistry. *Small-angle_scattering biophysics proteins structural_change biopolymers damage antioxidants repair*
E-mail durchschlag@alf1.ngate.uni-regensburg.de Tel. 49(941)9433041 Fax 49(941)9432813 Telex 65658unired

Eberhard, Prof. Dr Emil (1928). Professor. Institut für Mineralogie der Universität Hannover, Germany. [Institut für Mineralogie der Universität Hannover, Postfach 6009, D-30060 Hannover, Germany.] Dr rer.nat. (U. Fribourg/Switzerland, 1954) mineralogy. *Mineralogy silicates diffraction*
Tel. 49(511)7622443 Fax 49(511)7623456 Telex 923868unihnd

Egert, Prof. Dr Ernst (1949). Professor. Institut für Organische Chemie, Universität Frankfurt, Germany. [Institut für Organische Chemie, Universität Frankfurt, Marie-Curie-Str. 11, D-60439 Frankfurt/M., Germany.] Dr rer.nat.habil. (U. Göttingen, 1988) structural chemistry. *Molecular_replacement molecular_modelling force_field method structure-activity_relationship*
E-mail egert@chemie.uni-frankfurt.d400.de Tel. 49(69)58009230 Fax 49(69)58009128

Egner, Dr Ursula (1958). Scientist. SCHERING AG, Germany. [SCHERING AG, PCH-TCH, D-13342 Berlin, Germany.] Dr rer.nat. (U. Heidelberg, 1986) biology-biophysics. *Macromolecules computer_modelling drug peptides protein–ligand_interaction molecular_dynamics*
Tel. 49(30)4681522 Fax 49(30)46916741

Ehses, Dr Karl-Heinz (1943). Research Associate. Fachrichtung Kristallographie, Fachbereich Physik, Universität Saarbrücken, Germany. [10.4 Kristallographie, Universität Saarbrücken, Postfach 151150, D-66041 Saarbrücken, Germany.] Apl. Prof. (U. Saarbrücken, 1987) crystallography. *Phase_transition critical_phenomena diffuse_scattering ferroelectricity pseudosymmetry incommensurate*
Tel. 49(681)3023460 Fax 49(681)3024439

Eichhorn, Dr Gerd (1939). Assistant Professor. Technische Universität Ilmenau, Germany. [Technische Universität Ilmenau, Bergrat-Voigt-Str. 24, D-98693 Ilmenau, Germany.] Dr Ing.habil. (TU Ilmenau, 1984) solid-state technology. *Semiconductors dielectrics thin_film technology chemical_vapor_deposition rapid_thermal_processing*
Tel. 49(3677)692979 Fax 49(3677)693132

Eichhorn, Dr Klaus D. (1949). Research Associate. Institut für Kristallographie der Universität Karlsruhe, Germany. [Institut für Kristallographie der Universität Karlsruhe, c/o DESY-F41, HASYLAB, Notkestr. 85, D-22603 Hamburg, Germany.] Dr (U. Saarbrücken, 1982) crystallography. *Synchrotron_radiation anomalous_dispersion accurate_structure_analysis diffractometry*
E-mail eichhorn@vxdesy.desy.de Tel. 49(40)89982695 Fax 49(40)99982787 Telex 215124desyd

Eisenschmidt, Dr Christian (1953). Research Associate. Universität Halle-Wittenberg, Germany. [Universität Halle-Wittenberg, Herderstr. 12, D-06114 Halle/Saale, Germany.] Dr rer.nat. (U. Halle-Wittenberg, 1984) physics. *X-ray_polarization X-ray_diffraction dynamical_diffraction X-ray_anisotropy*
Tel. 49(345)25788

Elf, Dr Frank (1944). Mineralogisches Institut der Universität Bonn, Germany. [Mineralogisches Institut der Universität Bonn, Poppelsdorfer Schloß, D-53115 Bonn, Germany.] Dr rer.nat. (U. Bonn, 1989) crystallography. *EDS high_pressure high_temperature synchrotron_radiation phase_kinetics phase_transition profile_analysis*
Tel. 49(228)732771 Fax 49(228)732770

Ellner, Dr Martin Oliver (1938). Senior Research Scientist. MPI für Metallforschung, Institut für Werkstoffwissenschaft, Germany. [MPI für Metallforschung, Institut für Werkstoffwissenschaft, Seestr. 75, D-70174 Stuttgart, Germany.] Dr rer.nat. (U. Stuttgart, 1971) chemistry. *Crystal_chemistry intermetallic_compound phase_transition instrumentation metastable_crystalline_and_amorphous_phases structural_disorder twinning*
E-mail ellner@vaxww1.mpi-stuttgart.mpg.de Tel. 49(711)2095244 Fax 49(711)2265722 Telex 723742mpimd

Endriss, Mr Axel (1964). Research Associate. Institut für Kristallographie der Universität Tübingen, Germany. [Institut für Kristallographie der Universität Tübingen, Charlottenstr. 33, D-72070 Tübingen, Germany.] Dipl. Min. (U. Tübingen, 1991) mineralogy. *Ferroelectricity powder Rietveld_method phase_transition fullerenes*
Tel. 49(7071)296389 Fax 49(7071)296058

Engel, Dr Walter (1935). Senior Scientist. Fraunhofer-Institut für Chemische Technologie, J.-v.-Fraunhofer-Str. 7, D-76327 Pfinztal-Wöschbach, Germany. [Fruehlingsstr. 21, D-76327 Pfinztal-Wöschbach, Germany.] Dr phil.nat. (U. Gießen, 1962) inorganic chemistry. *Temperature_resolution_in_powder_diffraction phase_transition solid_state reaction*
Tel. 49(721)4640144 Fax 49(721)4640111

Engelen, Dr Bernward (1944). Lecturer. FB8/Anorganische Chemie, Universität-Gesamthochschule Siegen, Germany. [FB8/Anorganische Chemie, Universität-Gesamthochschule Siegen, D-57068 Siegen, Germany.] Dr (U. Köln, 1974) chemistry. *Solid_state chemistry hydrates structure_determination chalcogenites*
E-mail engelen@chemie.uni-siegen.d400.de Tel. 49(271)7404220 Fax 49(271)7402330
Telex 271383

Engelhardt, Dr Günter (1936). Lecturer. Institut für Technische Chemie I der Universität Stuttgart, Germany. [Institut für Technische Chemie I der Universität Stuttgart, D-70550 Stuttgart, Germany.] Dr rer.nat.habil. (Humboldt-U. Berlin, 1969) chemistry. *Nuclear_magnetic_resonance solid_state zeolites catalysis*
Tel. 49(711)6854309 Fax 49(711)6854065 Telex 7255445univd

Engels, Prof. Dr Siegfried (1932). Professor. Institut für Anorganische Chemie der Universität Halle-Wittenberg, Germany. [Institut für Anorganische Chemie der Universität Halle-Wittenberg, Geusaer Straße, D-06217 Merseburg, Germany.] Dr habil. (Humboldt-U. Berlin, 1968) chemistry. *Heterogeneous_catalysis inorganic_solid_state chemistry phase_determination thermoanalysis transition_metals crystalline_and_disordered_alumina*
Tel. 49(3461)462115 Fax 49(3461)462370 Telex 471320

Englert, Dr Ulli (1957). Research Assistant. Institut für Anorganische Chemie der RWTH Aachen, Germany. [Institut für Anorganische Chemie der RWTH Aachen, Prof.-Pirlet-Str. 1, D-52074 Aachen, Germany.] Dr (U. Tübingen, 1987) chemistry. *Packing_disorder topotaxy_in_molecular_crystal*
E-mail ie010en@dacth11.(EARN) Tel. 49(241)804666 Fax 49(241)871984

Ermer, Prof. Dr Otto (1940). Professor. Institut für Organische Chemie der Universität zu Köln, Germany. [Institut für Organische Chemie der Universität zu Köln, Greinstr. 4, D-50939 Köln, Germany.] Dr sc.techn. (ETH Zürich/Switzerland, 1970) chemistry. *Organic_crystal_chemistry molecular_complexes hydrogen_bonding strained_molecules*
Tel. 49(221)4704104 Fax 49(221)4705151

Ermrich, Dr Martin (1955). Staff Scientist. STOE & Cie GmbH, Hilpertstr. 10, D-64295 Darmstadt, Germany. [STOE & Cie GmbH, Postfach 101302, D-64213 Darmstadt, Germany.] Dr rer.nat. (Akad. d. Wiss. d. DDR Berlin, 1987) physics. *X-ray_method high_temperature powder_diffractometry detector micro_absorption imaging_plate position_sensitive_detector surface_characterization*
Tel. 49(6151)891225 Fax 49(6151)891293

Euler, Dr Harald (1958). Senior Research Scientist. Mineralogisches Institut der Universität Bonn, Germany. [Mineralogisches Institut der Universität Bonn, Poppelsdorfer Schloß, D-53115 Bonn, Germany.] Dr (U. München, 1989) mineralogy. *High_pressure diffraction_technique crystallography silicates kinetics twinning superstructure*
Tel. 49(228)732768 Fax 49(228)732770 Telex 886657unibod

Euler, Dr Robert (1925). Lecturer. Kanalstr. 13, D-63512 Hainburg, Germany. Dr phil. (U. Marburg, 1958) mineralogy. *Refractory_compounds metallurgical_slag non-metallic_inclusion_in_steels*
Tel. 49(6182)69231

Eysel, Prof. Dr Walter (1935). Professor. Mineralogisch-Petrographisches Institut der Universität Heidelberg, Germany. [Mineralogisch-Petrographisches Institut der Universität Heidelberg, Im Neuenheimer Feld 236, D-69120 Heidelberg, Germany.] Dr (RWTH Aachen, 1968) mineralogy/crystallography. *Crystallography mineralogy inorganic_oxygen_compounds thermoanalysis crystal_structure_relations_in_oxide_minerals crystal_structure_determination*
E-mail ey0@vm.urz.uni-heidelberg.de Tel. 49(6221)568217 Fax 49(6221)564805
Telex 461745unikld

Fanter, Dr Detlef (1945). Postdoctoral Research Assistent. Fraunhofer-Gesellschaft e.V., Institut für Angewandte Materialforschung, Aussenstelle Teltow (Polymerverbunde), Germany. [Fraunhofer-Gesellschaft e.V., Institut für Angewandte Materialforschung, Aussenstelle Teltow (Polymerverbunde), Kantstr. 55, D-14513 Teltow, Germany.] Diplom (Humboldt-U. Berlin, 1969) crystallography. *Polymers thin_layer spectroscopy computer_modelling ESCA ellipsometry FTIR adhesion surface software*
E-mail fanter@epv.ifam.fhg.de Tel. 49(3328)46288 Fax 49(3328)46282

Finster, Prof. Dr Joachim (1936). Retired. Fachbereich Physik der Universität Rostock, Germany. [Fachbereich Physik der Universität Rostock, D-18051 Rostock, Germany.] Dr habil. (U. Leipzig, 1981) physical chemistry. *XPS EXAFS amorphous_dielectrics structural_modelling deposition_of_thin_layers silicon_oxynitrides*
Tel. 49(381)4981668 Fax 49(381)4981667 Telex 31140unird

Fischer, Prof. Dr Karl (1925). Professor Emeritus. Institut für Kristallographie, Physik Department, Universität Saarbrücken, Germany. [Institut für Kristallographie, Physik Department, Universität Saarbrücken, Postfach 1150, D-66041 Saarbrücken, Germany.] Dr habil. (U. Frankfurt/M., 1962) crystallography/mineralogy. *Anomalous_dispersion instrumentation pseudosymmetry synchrotron_radiation energy_dispersive_Laue_diffraction phase_problem tensorial_scattering_factors*
Tel. 49(681)3023410 Fax 49(681)3024439

Fischer, Prof. Dr Werner (1931). Professor. Institut für Mineralogie, Petrologie und Kristallographie, Universität Marburg, Germany. [Institut für Mineralogie, Petrologie und Kristallographie, Universität Marburg, Hans-Meerwein-Straße, D-35032 Marburg, Germany.] Dr habil. (U. Marburg, 1970) crystallography. *Mathematical_crystallography*

E-mail fischerw@mailer.uni-marburg.de Tel. 49(6421)285704 Fax 49(6421)288919
Telex 482372

Flörke, Dr Ulrich (1952). Senior Scientist. Anorganische und Analytische Chemie der Universität-GH Paderborn, Germany. [Anorganische und Analytische Chemie der Universität-GH Paderborn, Warburgerstr. 100, D-33098 Paderborn, Germany.] Dr rer.nat. (U. Münster, 1980) chemistry. *Structure_determination organometallic_structure rhenium_compounds*
E-mail floe@mvaxac.uni-paderborn.de Tel. 49(5251)602496 Fax 49(5251)603423

Follner, Prof. Dr Heinz (1938). Professor. Institut für Mineralogie und Mineralische Rohstoffe, Fachabteilung Kristallographie der Technischen Universität Clausthal, Germany. [Institut für Mineralogie und Mineralische Rohstoffe, Fachabteilung Kristallographie der Technischen Universität Clausthal, Postfach 1253, D-38670 Clausthal-Zellerfeld, Germany.] Dr rer.nat. (TU Clausthal, 1971) crystallography/mineralogy. *Crystal_growth structure physical_property*
Tel. 49(5323)722394 Fax 49(5323)723500

Frahm, Dr Ronald R. (1954). Scientific Staff Member. HASYLAB am DESY, Germany. [HASYLAB am DESY, Notkestr. 85, D-22603 Hamburg, Germany.] Dr (U. Kiel, 1983) experimental physics. *Absorption_spectroscopy solid_state anomalous_dispersion synchrotron_radiation*
E-mail frahm@vxdesy.desy.de Tel. 49(40)89983705 Fax 49(40)89982787

Francke, Mr Rudolf (1937). Forschungsinstitut-FEE, Germany. [Forschungsinstitut-FEE, Prof.-Schloßmacher-Str. 1, D-55743 Idar-Oberstein, Germany.] Dipl. Phys. (1965) physics. *Crystal_growth Czochralski_technique Fourier_transform temperature remote_control convection measurement_and_computing_of_temperature*
Tel. 49(6781)41003

Frank, Prof. Dr Walter (1957). Professor. Fachbereich Chemie der Universität Kaiserslautern, Germany. [Fachbereich Chemie der Universität Kaiserslautern, Postfach 3049, D-67653 Kaiserslautern, Germany.] Dr rer.nat.habil. (U. Saarbrücken, 1990) inorganic chemistry. *Main-group_compounds structure_determination one_dimension_disorder pseudosymmetry*
Tel. 49(631)2052986 Fax 49(631)2052187 Telex 45627unikld

Frey, Prof. Dr Friedrich (1942). Professor. Institut für Kristallographie und Mineralogie der Universität München, Germany. [Institut für Kristallographie und Mineralogie der Universität München, Theresienstr. 41, D-80333 München, Germany.] Dr habil. (U. München, 1980) crystallography. *Disorder diffraction diffuse_scattering quasicrystal neutron_diffraction zirconium_compounds*
E-mail frey@kri.physik.uni-muenchen.de Tel. 49(89)23944332 Fax 49(89)23944334
Telex 529815

Friese, Mrs Karen (1964). Research Associate. Mineralogisch-Petrographisches Institut der Universität Hamburg, Germany. [Mineralogisch-Petrographisches Institut der Universität Hamburg, Grindelallee 48, D-20146 Hamburg, Germany.] Dipl. Min. (U. Hamburg, 1990) mineralogy. *Crystal_structure_determination modulated_structures disorder*
E-mail Mi2a015@Miaix1.mineralogie.uni-hamburg.de Tel. 49(40)41232073

Fritzsch, Dr Günter (1940). Postdoctoral Research Associate; Group leader; representative for X-ray security. Max-Planck-Institut für Biophysik, Germany. [Max-Planck-Institut für Biophysik, Heinrich-Hoffmann-Str. 7, D-60528 Frankfurt/M., Germany.] Dr (U. Leipzig, 1970) physics. *Absorption_spectroscopy bioenergetics photosynthesis proteins kinetics membrane_biophysics*
E-mail fritzsch@mpibp-frankfurt.mpg.de Tel. 49(69)96769417 Fax 49(69)96769423

Fröhlich, Dr Armin (1958). Head of Laboratory. Wülfrather Zement GmbH, Germany. [Wülfrather Zement GmbH, Am Mühlenberg 5, D-53925 Kall-Sötenich, Germany.] Dr rer.nat. (U. Hannover, 1990) mineralogy. *X-ray cements clinker phase_determination fractals*
Tel. 49(2441)5992

Frye, Mr Thomas (1962). Post-Graduate Student. Institut für Mineralogie der Universität Münster, Germany. [Institut für Mineralogie der Universität Münster, Corrensstr. 24, D-48149 Münster, Germany.] Dipl. Chem. (U. Münster, 1992) chemistry. *Zeolites ion_exchange silicates sodalites*
Tel. 49(251)833493

Fuess, Prof. Dr-Ing. Hartmut (1941). Professor. Chairman, European Crystallographic Committee. Fachbereich Materialwissenschaft-Strukturforschung, Technischen Hochschule Darmstadt, Germany. [Fachbereich Materialwissenschaft-Strukturforschung, Technischen Hochschule Darmstadt, Petersenstr. 20, D-64287 Darmstadt, Germany.] Dr-Ing. (TH Darmstadt, 1968) chemistry. *Material_sciences neutron_scattering powder_diffraction zeolites*
E-mail dd9n@rs11.hrz.th-darmstadt.de Tel. 49(6151)162298 Fax 49(6151)166023

Fütterer, Mr Klaus (1963). Institut für Mineralogie und Kristallographie der Technischen Universität Berlin, Germany. [Institut für Mineralogie und Kristallographie der Technischen Universität Berlin, Ernst-Reuter-Platz 1, Sekr. BH1, D-10587 Berlin, Germany.] Dipl. Phys. (U. Karlsruhe, 1989) physics. *High_pressure phase_transition neutron_diffraction inorganic_materials*
E-mail futt1637@mailszrz.zrz.tu-berlin.d400.de Tel. 49(30)31422986 Fax 49(30)31421124

Fuhrmann, Mr Peter (1965). Research Assistant. Institut für Kristallographie der Freien Universität Berlin, Germany. [Institut für Kristallographie der Freien Universität Berlin, Takustr. 6, D-14195 Berlin, Germany.] Dipl. Kristallogr. (U. Leipzig, 1991) crystallography. *Superconductors charge_density computer low_temperature_single_crystal_diffractometry*
E-mail fuhrmann@chemie.fu-berlin.de Tel. 49(30)8383408 Fax 49(30)8383464

Gaebel, Mr Rainer (1965). Post-Graduate Student. Fraunhofer-Institut für Werkstoffmechanik, Germany. [Fraunhofer-Institut für Werkstoffmechanik, Wöhlerstr. 11, D-79108 Freiburg/Br., Germany.] Dipl. Min. (U. Göttingen, 1992) mineralogy. *Phase_transition sintering crystallization coarsening* computer_modelling
Tel. 49(761)5142180 Fax 49(761)5142110

Garsche, Mr Markus (1963). Research Assistant. Institut für Mineralogie und Kristallographie der Technischen Universität Berlin, Germany. [Institut für Mineralogie und Kristallographie der Technischen Universität Berlin, Ernst-Reuter-Platz 1, D-10587 Berlin, Germany.] Dipl. Min. (U. Würzburg, 1990) mineralogy. *X-ray_diffraction_technique optical_spectroscopy minerals*
Tel. 49(30)31423225 Fax 49(30)31421124

Gebert, Dr Walter (1938). Postdoctoral Research Assistant. Institut für Mineralogie der Universität Bochum, Germany. [Institut für Mineralogie der Universität Bochum, Universitätsstr. 150, D-44780 Bochum, Germany.] Dr (U. Bochum, 1970) mineralogy. *Structure_determination symmetry minerals*
Tel. 49(234)7004380 Fax 49(234)7094179

Geiger, Dr Charles A. (1956). Assistant Professor. Mineralogisch-Petrographisches Institut der Universität Kiel, Germany. [Mineralogisch-Petrographisches Institut der Universität Kiel, Olshausenstr. 40, D-24098 Kiel, Germany.] PhD (U. Chicago/USA, 1986) geology. *Crystal_growth ultra_high_pressure mineralogy phase_equilibrium silicate_crystal_chemistry geosciences*
E-mail nmp46@rz.uni-kiel.d400.de Tel. 49(431)8802895 Fax 49(431)8804457

Geiger, Dr Martin (1960). Research Associate. Fachrichtung Kristallographie der Universität Saarbrücken, Germany. [Fachrichtung Kristallographie der Universität Saarbrücken, Im Stadtwald, D-66041 Saarbrücken, Germany.] Dr rer.nat. (U. Saarbrücken, 1994) mineralogy. *Energy-dispersive_diffraction texture_polycrystal crystallography_and_computing* capillary_beamline synchrotron_radiation
Tel. 49(681)3024507 Fax 49(681)3024439

Geist, Dr Volker (1942). Lecturer. Institut für Kristallographie, Mineralogie und Materialwissenschaft, FB Chemie der Universität Leipzig, Germany. [Institut für Kristallographie, Mineralogie und Materialwissenschaft, FB Chemie der Universität Leipzig, Scharnhorststr. 20, D-04275 Leipzig, Germany.] Dr rer.nat.habil. (U. Leipzig, 1986) physics. *Kossel_diffraction channelling semiconductor quasicrystal crystal_physics*
E-mail geist@rz.uni-leipzig.dbp.de Tel. 49(341)310502 Fax 49(341)310502

Gernat, Mrs Christine (1952). Research Assistant. Max-Delbrück-Zentrum für Molekulare Medizin, Germany. [Max-Delbrück-Zentrum für Molekulare Medizin, Robert-Rössle-Str. 10, D-13122 Berlin, Germany.] Dipl. Kristallogr. (Humboldt-U. Berlin, 1974) crystallography. *SAXS WAXS biopolymers biocrystallography* crystal_growth polysaccharides gels
Tel. 49(30)94063259 Fax 49(30)9494161

Gertel-Kloos, Dr Heike (1963). Research Scientist. [GKSS-Forschungszentrum Geesthacht GmbH, Max-Planck-Straße, D-21502 Geesthacht, Germany.] Dr-Ing. (TU Clausthal, 1992) materials science. *Material_sciences neutron_diffraction texture metals*
E-mail heike.gertel@gkss.de Tel. 49(4152)871661 Fax 49(4152)871618

Gesemann, Prof. Dr-Ing. Dr Renate (1936). Professor. Hochschule für Technik und Wirtschaft Mittweida, Germany. [Hochschule für Technik und Wirtschaft Mittweida, Technikumsplatz, D-09648 Mittweida, Germany.] Dr-Ing.habil. (TU Ilmenau, 1981) electrotechnical materials. *Electrodeposition film materials multilayers ferroelectricity*
Tel. 49(3727)581228 Fax 49(3727)581379

Gies, Prof. Dr Hermann (1952). Professor. Institut für Mineralogie, Ruhr-Universität Bochum, Universitätsstr. 150, Germany. [Institut für Mineralogie, Ruhr-Universität Bochum, Postfach 102148, D-44780 Bochum, Germany.] Dr habil. (U. Kiel, 1987) crystallography. *Zeolites structure synthesis Rietveld_method NMR*
E-mail hermann.gies@ruba.rz.ruhr-uni-bochum.de Tel. 49(234)7003521 Fax 49(234)7094179

Gille, Dr Peter (1954). Research Assistant. Institut für Werkstoffwissenschaften VI der Universität Erlangen-Nürnberg, Germany. [Institut für Werkstoffwissenschaften VI der Universität Erlangen-Nürnberg, Martensstr. 7, D-91058 Erlangen, Germany.] Dr rer.nat. (Humboldt-U. Berlin, 1984) crystallography. *Crystal_growth semiconductor solution alloys*
Tel. 49(9131)857757 Fax 49(9131)858495 Telex 629755tferld

Gille, Dr Wilfried (1953). Research Assistant. FB Physik/Metallphysik der Universität Halle-Wittenberg, Germany. [FB Physik/Metallphysik der Universität Halle-Wittenberg, PF8, Kröllwitzer Str. 44/Hoher Weg 7, D-06099 Halle/S., Germany.] Dr rer.nat. (PH. Halle/S., 1983) physics. *Small-angle_scattering desmearing Fourier_transform theory geometry distribution_functions random_systems*
Tel. 49(345)38201 Fax 49(345)38211

Gjudjenov, Mrs Karin (1966). Museum Educationist. Optisches Museum Jena, Germany. [Optisches Museum Jena, Musäusring 49, D-07747 Jena, Germany.] Dipl. Kristallogr. (U. Leipzig, 1989) crystallography. *Densitometry dislocation microscopy*
Tel. 49(3641)372573

Glaremin, Mr Peter (1959). Consultant for Hardware. SNI AG, Germany. [SNI AG, Reitwiesstr. 9, D-85560 Ebersberg, Germany.] Dipl. Phys. (U. Dortmund, 1985) physics. *Computer_technology statistics*
Tel. 49(8092)21297

Glinnemann, Dr Jürgen (1951). Senior Scientist. Institut für Kristallographie der RWTH Aachen, Germany. [Institut für Kristallographie der RWTH Aachen, D-52056 Aachen, Germany.] Dr rer.nat. (RWTH Aachen, 1987) mineralogy. *Inorganic_crystal_chemistry structural_change crystal_structure_determination high_pressure phase_transitions symmetry teaching_crystallography*
E-mail fg050gl@dacth11. Tel. 49(241)806988 Fax 49(241)806916

Goddard, Dr Richard John (1952). Research Assistant. MPI für Kohlenforschung, Germany. [MPI für Kohlenforschung, Kaiser-Wilhelm-Platz 1, D-45470 Mülheim/Ruhr, Germany.] PhD (U. Bristol/UK, 1977) chemistry. *Structural_chemistry catalysis charge_density theory*
E-mail goddard@mpi-muelheim.mpg.d400.de Tel. 49(208)3062172 Fax 49(208)3062989

Göbel, Dr Herbert E. (1940). Senior Scientist. SIEMENS AG, ZFE T MR 3, Germany. [SIEMENS AG, ZFE T MR 3, Postfach, D-81730 München, Germany.] Dr rer.nat. (TU München, 1969) physics. *X-ray_powder_diffraction methodology_of_diffraction_analysis thin_film material_characterization* reflectivity double/triple_crystal_diffraction
E-mail herb.goebel@zfe.siemens.de Tel. 49(89)6363274 Fax 49(89)63642256 Telex 52109-0snid

Görls, Dr Helmar (1955). Research Associate. Max-Planck-Gesellschaft AG "CO2-Chemie" an der Universität Jena, Germany. [Max-Planck-Gesellschaft AG "CO2-Chemie" an der Universität Jena, Am Steiger 3/Hs 3, D-07743 Jena, Germany.] Dr rer.nat. (U. Jena, 1986) chemistry. *Metalloorganic organic_and_inorganic_structure_determination molecular_modelling*
E-mail goerls@xa.nlwl.uni-jena.de Tel. 49(3641)636336 Fax 49(3641)635360

Göttlicher, Mr Jörg (1964). Post-Graduate Student. Institut für Mineralogie der Universität Münster, Corrensstr. 24, D-48149 Münster/W., Germany. [Friedensstr. 18, D-35580 Wetzlar, Germany.] Dipl. Min. (U. Münster, 1989) mineralogy. *Transmission_electron_microscopy lattice_parameter_refinement synthetic_feldspars glasses_phase_separation* hydrothermal_synthesis REM replica_technique
E-mail gottlic@dmswwu1a.uni-muenster.de Tel. 49(6441)25399 Fax 49(251)838397

Götz, Dr Wolfgang (1928). Research Scientist. Otto-Schott-Institut der Universität Jena, Germany. [Institut für Glaschemie, Fraunhoferstr. 6, D-07743 Jena, Germany.] Dr rer.nat.habil. (U. Jena, 1969) chemistry. *Glasses_ceramics diffractometry crystal_growth crystal_chemistry*
Tel. 49(3461)8225460 Fax 49(3461)8225461 Telex 331506unidd

Gomm, Dr Martin (1943). Research Scientist. Institut für Angewandte Physik, Lehrstuhl für Kristallographie der Universität Erlangen-Nürnberg, Germany. [Institut für Angewandte Physik, Lehrstuhl für Kristallographie der Universität Erlangen-Nürnberg, Bismarckstr. 10, D-91054 Erlangen, Germany.] Dr rer.nat. (U. Erlangen, 1977) physics. *Diffractometry electronics computer data_collection least-squares_refinement*
E-mail mpkr00@cd4680fs.rrze.uni-erlangen.dbp.de Tel. 49(9131)852712 Fax 49(9131)852733

Gonschorek, Dr Walter (1937). Assistant. Materialwissenschaft/Strukturforschung der Technischen Hochschule Darmstadt, Petersenstr. 20, D-64287 Darmstadt, Germany. [Georgstr. 14, D-52078 Aachen, Germany.] Dr rer.nat. (RWTH Aachen, 1971) physics. *Density_distribution band_calculation disorder statistics X-ray_and_neutron_diffraction*
E-mail iff182@djukfa11. Tel. 49(2461)613339 Fax 49(2461)612610

Graf, Dr Hans Anton (1945). Staff Scientist. Hahn-Meitner-Institut, Germany. [Hahn-Meitner-Institut, Glienicker Str. 100, D-14109 Berlin, Germany.] Dr (U. München, 1975) chemistry. *Diffraction_theory disorder neutron_diffraction*
E-mail graf@vax.hmi.d400.de Tel. 49(30)80622778 Fax 49(30)80622999 Telex 01-85763

Granzin, Dr Joachim (1957). Research Assistant. Institut für Biologische Informationsverarbeitung, Structural Biology, Forschungszentrum Jülich GmbH, Germany. [Institut für Biologische Informationsverarbeitung, Forschungszentrum Jülich GmbH, D-52425 Jülich, Germany.] Dr (U. Hamburg, 1987) mineralogy/crystallography. *Area_detector biocrystallography synchrotron_radiation Patterson_method* isomorphous_replacement structural_change
E-mail joachimG@ibi024.ibi.kfa-juelich.de Tel. 49(2461)612024 Fax 49(2461)612020

Greis, Dr Ortwin (1941). Head Electron Microscopy. Zentralbereich Elektronenmikroskopie der Technischen Universität Hamburg-Harburg, Germany. [Zentralbereich Elektronenmikroskopie der Technischen Universität Hamburg-Harburg, Eißendorfer Str. 42, 21071 Hamburg, Germany.] Dr rer.nat.habil. (U. Heidelberg, 1980) crystallography. *Electron_microscopy energy-dispersive_analysis structure_determination material_sciences inorganic_compounds minerals order-disorder superstructures*
Tel. 49(40)77183543 or 49(40)77183544 Fax 49(40)77182684

Griewatsch, Mr Carsten (1966). Student. Mineralogisch-Petrographisches Institut der Universität Kiel, Germany. [Mineralogisch-Petrographisches Institut der Universität Kiel, Waisenhofstr. 39, D-24103 Kiel, Germany.] *Crystallography*
Tel. 49(431)95666

Grin, Dr Yuri N. (1955). Research. Max-Planck-Institut für Festkörperforschung, Heisenbergstr.1, D-70569 Stuttgart, Germany. [Max-Planck-Institut für Festkörperforschung, Postfach 800665, D-70506 Stuttgart, Germany.] DSc (U. Lviv, 1980) inorganic chemistry. *Solid_state chemistry structure_and_bonding structural_relationship gallium_compounds computing*
E-mail grin@vsibm1.mpi-stuttgart.mpg.de Tel. 49(711)6891473 Fax 49(711)6891010

Grosse, Mr Mirco (1964). Research Associate. Forschungszentrum Rossendorf eV., Germany. [Forschungszentrum Rossendorf eV., PF 510 119, D-01314 Dresden, Germany.] Dipl. Ing. (TU-Bergakademie Freiberg, 1990) metallography. *Nuclear_reactor microstructure Rietveld_method liquid_state*
Tel. 49(351)5913155 Fax 49(351)4605813

Grosse, Prof. Dr Peter (1932). Director. I. Physikalisches Institut der RWTH Aachen, Germany. [I. Physikalisches Institut der RWTH Aachen, Templergraben 55, D-52056 Aachen, Germany.] Dr rer.nat. (U. Köln, 1965) physics. *Czochralski_technique_of_semiconductors epitaxy_of_semiconductors thin_film_of_transition_elements_and_their_oxides infrared_spectroscopy dielectric_and_optical_properties_of_inhomogeneous_materials*
E-mail grosse@acphyz.hep.rwth-aachen.de Tel. 49(241)807155 Fax 49(241)8888331 Telex 832704

Gruehn, Prof. Dr Reginald (1929). Professor. Institut für Anorganische und Analytische Chemie der Universität Gießen, Germany. [Institut für Anorganische und Analytische Chemie der Universität Gießen, Heinrich-Buff-Ring 58, D-35392 Gießen, Germany.] Dr rer.nat.habil. (U. Münster, 1969) inorganic and analytical chemistry. *Transmission_electron_microscopy microstructure nonstoichiometry superconductivity inorganic_crystal_structures chemical_transport materials_science*
E-mail Gruehn@sp10ac.anorg.Chemie.uni-giessen.de Tel. 49(641)7025670 Fax 49(641)7025673

Günter, Dr Christina (1961). Postdoctoral Research Assistant. Geo Forschungszentrum, Germany. [Geo Forschungszentrum, Telegrafenberg A50, D-14473 Potsdam, Germany.] Dr (TH Darmstadt, 1992) materials science. *Phase_determination diffractometry phase_kinetics X-ray fluorescence spectroscopy high_pressure high_temperature powder_diffraction*
E-mail guente@gfz.potsdam.de Tel. 49(331)2881468 Fax 49(331)2881474

Günther, Prof. Dr H. Fritz (1912). Professor Emeritus. Technische Universität, Bergakademie Freiberg, Germany. [Technische Universität, Bergakademie Freiberg, Anton-Günther-Str. 42, D-09599 Freiberg/Sa., Germany.] Dr rer.nat.habil. (Bergakademie Freiberg, 1956) metallography. *Diffraction_technique real_structure line_profile_analysis*
Tel. 49(3731)48825

Haake, Mrs Annegret (1933). Retired. Institut für Kristallographie der Universität Frankfurt, Germany. [Jaminstr. 11B, D-61476 Kronberg, Germany.] *Symmetry two-dimensional_symmetry design textiles ethnography*
Tel. 49(6173)5306 Fax 49(6173)5306 49(69)7982101

Haase, Prof. Dr Wolfgang (1936). Professor. Institut für Physikalische Chemie der Technischen Hochschule Darmstadt, Germany. [Institut für Physikalische Chemie der Technischen Hochschule Darmstadt, Petersenstr. 20, D-64287 Darmstadt, Germany.] Dr rer.nat. (U. Jena, 1964) chemistry. *Magnetism dielectric_relaxation liquid_crystals cooperative_phenomena nonlinear_optics*
E-mail D58L@rs11.hrz.th-darmstadt.de Tel. 49(6151)163398 Fax 49(6151)164298 Telex 419579thd

Hadan, Dr Marianne (1939). Research Associate. uve-Institut für Technische Chemie und Umweltschutz GmbH, Germany. [uve-Institut für Technische Chemie und Umweltschutz GmbH, Rudower Chaussee 5, D-12489 Berlin, Germany.] Dr rer.nat. (Humboldt-U. Berlin, 1971) crystallography. *Powder_diffraction polarization_microscopy zeolites environment_protection*
Tel. 49(30)63924722 Fax 49(30)6774006

Hädicke, Dr Erich (1940). BASF AG, Germany. [BASF AG, ZKM-B1, 67056 Ludwigshafen, Germany.] Dr (TU München, 1969) chemistry. *Molecular_modelling molecular_mechanics polymeric_packing catalytic_polymerization MO-calculation reactivity*
E-mail erich.haedicke@zk.basf-ag.de Tel. 49(621)6043786 Fax 49(621)6020313 Telex 464769basfd

Hähnert, Dr Irmela (1941). Research Assistant. Humboldt-Universität zu Berlin, Institut für Physik/Kristallographie, Germany. [Humboldt-Universität zu Berlin, Math.-nat. Fakultät 1, Institut für Physik/Kristallographie, Invalidenstr. 110, D-10115 Berlin, Germany.] Dr rer.nat. (Humboldt-U. Berlin, 1971) crystallography. *Transmission_electron_microscopy crystal_defect chemical_etching II–VI_compounds EDX twinning dislocation high-Tc_superconductors*
Tel. 49(30)2803361 Fax 49(30)2803360

Hahn, Mr Michael (1965). Research Associate. Institut für Kristallographie der Freien Universität Berlin, Germany. [Institut für Kristallographie der Freien Universität Berlin, Takustr. 6, D-14153 Berlin, Germany.] Dipl. Biochem. (FU. Berlin, 1991) biochemistry. *Proteins nucleic_acids single_crystal diffraction*
E-mail mhahn@chemie.fu-berlin.de Tel. 49(30)8384937 Fax 49(30)8386702

Hahn, Prof. Dr Theo (1928). Professor. Former President IUCr (1984–87); Chairman Commission on International Tables; Editor Volume A of International Tables. Institut für Kristallographie der RWTH Aachen, D-52056 Aachen, Germany.] Dr rer.nat. (U. Frankfurt/M., 1952) mineralogy. *Structural_crystallography symmetry physical_property oxides crystal_chemistry crystal_physics crystal_structures crystal_growth theoretical_crystallography*
E-mail fg050kr@dacth11 (EARN) fg050kr@vm1.rz.rwth-aachen.de (INTERNET) Tel. 49(241)806900 or 49(241)806909 Fax 49(241)8888184 Telex 0832704thacd

Haibach, Mr Torsten (1966). Research Assistant. Institut für Mineralogie der Universität Hannover, Germany. [Institut für Mineralogie der Universität Hannover, Welfengarten 1, D-30167 Hannover, Germany.] Dipl. Min. (U. München, 1992) mineralogy. *Quasicrystal maximum-entropy_method synchrotron_radiation alloys*
E-mail torsten.haibach@cdc2.mineralogie.erdwissenschaften.uni-hannover.d400.de Tel. 49(511)7629086 Fax 49(511)7623456

Hanglelter, Dr Thomas (1948). Assistant Professor. Fachbereich Physik der Universität Paderborn, Germany. [Fachbereich Physik der Universität Paderborn, Warburger Str. 100, D-33098 Paderborn, Germany.] Dr (U. Paderborn, 1979) physics. *Crystal_growth Czochralski_technique Bridgman_Stockbarger_technique halides optical_spectroscopy photoemission photostimulated_process*

E-mail sp_ha@physik.uni-paderborn.de Tel. 49(5251)602761 Fax 49(5251)603422

Hanke, Dr Kurt (1935). Research Scientist. Mineralogisch-Kristallographisches Institut, Universität Göttingen, Germany. [Mineralogisch-Kristallographisches Institut, Universität Göttingen, Goldschmidtstr. 1, D-37077 Göttingen, Germany.] Dr (U. Göttingen, 1968) mineralogy. *Powder_diffraction_technique*
Tel. 49(551)393937 Fax 49(551)399521

Hansen, Mr Harly Andreas Smedemark (1952). Scientist. Abteilung für Ribosomenstruktur der Max-Planck-Gesellschaft, Germany. [Abteilung für Ribosomenstruktur der Max-Planck-Gesellschaft, Notkestr. 85 Geb. 25B, D-22603 Hamburg, Germany.] Cand.scient. (U. Aarhus/Denmark, 1985) chemistry/physics. *Crystallography ribosomes synchrotron_radiation data_processing*
E-mail harly@mpgars.desy.de Tel. 49(40)894686 Fax 49(40)891314

Harbrecht, Prof. Dr Bernd (1950). Professor. Institut für Anorganische Chemie der Universität Bonn, Germany. [Institut für Anorganische Chemie der Universität Bonn, Gerhard-Domagk-Str. 1, D-5321 Bonn, Germany.] Dr habil. (U. Dortmund, 1989) inorganic chemistry. *Solid_state chemistry synthesis chalcogenides tantalum_compounds*
Tel. 49(228)735353 Fax 49(228)735660

Harms, Mr Joerg M. (1963). PhD Student. Arbeitsgruppe für Ribosomenstruktur der Max-Planck-Gesellschaft, Germany. [MPG-AGRS c/o DESY, Notkestr. 85, D-22603 Hamburg, Germany.] Physik-Diplom (U. Hamburg, 1994) physics. *Macromolecular_crystallography ribosomes cutting_of_macromolecular_crystals isomorphism_of_macromolecular_crystals heavy_atom_soaks*
E-mail joerg@mpgars.desy.de Tel. 49(40)89982809 Fax 49(40)891314

Hartl, Prof. Dr Hans (1940). Professor. Institut für Anorganische und Analytische Chemie der Freien Universität Berlin, Germany. [Institut für Anorganische und Analytische Chemie der Freien Universität Berlin, Fabeckstr. 34/36, D-14195 Berlin, Germany.] Dr rer.nat. (TU München, 1969) inorganic chemistry. *Structure_determination inorganic_halides inorganic_oxides inorganic_synthesis thermoanalysis IR_and_Raman_spectroscopy*
Tel. 49(30)8384003 Fax 49(30)8382424

Hartmann, Mr Frank (1966). Post-Graduate Student. Institut für Anorganische Chemie und Strukturchemie, Lehrstuhl II der Universität Düsseldorf, Germany. [Institut für Anorganische Chemie und Strukturchemie, Lehrstuhl II der Universität Düsseldorf, Universitätsstr. 1, D-40225 Düsseldorf, Germany.] Dipl. Chem. (U. Düsseldorf, 1993) chemistry. *Aqueous_bases hydrogen_bonding low_temperature DSC crystal_structures_of_aqueous_bases_at_low_temperature*
E-mail fhartman@mail.rz.uni-duesseldorf.de Tel. 49(211)3113857 Fax 49(211)3113085

Hartung, Prof. Dr Helmut (1935). Professor. Institut für Physikalische Chemie, Universität Halle-Wittenberg, Germany. [Institut für Physikalische Chemie, Universität Halle-Wittenberg, Mühlpforte 1, D-06108 Halle/S., Germany.] Dr rer.nat.habil. (U. Halle, 1978) chemistry. *Structural_chemistry X-ray_analysis molecular_packing liquid_crystal coordination_chemistry*
E-mail hartung@chemie.uni-halle.de Tel. 49(345)2025081 Fax 49(345)2025083 Telex 4353

Hausen, Dr Hans-Dieter (1937). Research Scientist. Institut für Anorganische Chemie der Universität Stuttgart, Germany. [Institut für Anorganische Chemie der Universität Stuttgart, Pfaffenwaldring 55, D-70511 Stuttgart, Germany.] Dr (U. Stuttgart, 1966) chemistry. *Crystallography structure_determination main-group_compounds molecular_complexes organometallic_compounds H-bridges*
Tel. 49(711)6854220

Heide, Prof. Dr Klaus (1938). Professor; Chairman of the Mineralogy Department. Institut für Geowissenschaften, Lehrstuhl für Mineralogie der Universität Jena, Germany. [Institut für Geowissenschaften, Lehrstuhl für Mineralogie der Universität Jena, Burgweg 11, D-07749 Jena, Germany.] Dr habil. (U. Jena, 1968) mineralogy. *Kinetics microscopy glasses crystallization thermal_analysis melting*
E-mail ckh@RZ.UNI-Jena.De Tel. 49(3641)27041 Fax 49(3641)24647 Telex 331553 GEO D

Helme, Prof. Dr Klaus (1935). Director. Institut für Halbleitertechnik, Lehrstuhl I der RWTH Aachen, Germany. [Institut für Halbleitertechnik, Lehrstuhl I der RWTH Aachen, Templergraben 55, D-52056 Aachen, Germany.] Dr rer.nat. (TH Darmstadt, 1968) physics. *Metalloorganic_epitaxy II–VI_compounds III–V_compounds high_temperature_superconductors SiGe*
E-mail mailbox@enterprise.iht.rwth-aachen.de Tel. 49(241)807745 Fax 49(241)8888199

Heinemann, Dr Frank Wilhelm (1964). Postdoctoral Research Assistant. Martin-Luther-Universität Halle-Wittenberg, Germany. [Martin-Luther-Universität Halle-Wittenberg, Institut für Anorganische Chemie, Weinbergweg 16, D-06120 Halle (Saale), Germany.] Dr rer.nat. (U. Halle-Wittenberg, 1993) physical chemistry. *Electronic_ligand_influence organometallic_compounds organosulfur_compounds structure–activity_relationship NMR_spectroscopic_investigations*
E-mail heinemann@chemie.uni-halle.d400.de Tel. 49(345)622210 Fax 49(345)5511182

Heinemann, Dr Udo (1953). Head of Research Group. [FG Kristallographie, Max-Delbrück-Centrum für Molekulare Medizin, Robert-Rössle-Straße 10, 13122 Berlin, Germany.] Dr rer.nat. (U. Göttingen, 1982) chemistry. *Single_crystal_diffraction proteins nucleic_acids*
E-mail uh@orion.rz.mdc-berlin.de Tel. 49(30)94063420/2270 Fax 49(30)94062548

Helgesson, Dr Göran (1953). General Manager. Molecular Structure Corporation, Germany. [Molecular Structure Corporation, Grossenbaumer Weg 6, D-40472 Düsseldorf, Germany.] PhD (U. Göteborg/Sweden, 1990) chemistry. *Crystallography diffractometry area_detector*
Tel. 49(211)424177 Fax 49(211)4791919

Hellner, Prof. Dr Erwin E. (1920). Professor Emeritus. Mineralogisches Institut der Universität Marburg, Germany. [Mineralogisches Institut der Universität Marburg, Hans-Meerwein-Straße, D-35032 Marburg, Germany.] Dr habil. (U. Marburg, 1953) mineralogy. *Classification_of_crystal_structure high_pressure transformation charge_density magnetism electron_theory_of_intermetallic_compounds*
Tel. 49(6421)282045 Fax 49(6421)285831

Helmreich, Dr Dieter (1939). Research and Development. WACKER-CHEMITRONIC GmbH, Germany. [WACKER-CHEMITRONIC GmbH, D-84479 Burghausen, Germany.] Dr rer.nat. (TU München, 1967) physics. *Semiconductor crystal_growth solar_energy characterization*
Tel. 49(8677)832763 Fax 49(8677)835824

Henke, Dr Henning (1943). Research Associate. Institut für Anorganische Chemie II der Universität Karlsruhe, Germany. [Institut für Anorganische Chemie II der Universität Karlsruhe, Kaiserstr. 12, Postfach 6980, D-76128 Karlsruhe, Germany.] Dr rer.nat. (U. Karlsruhe, 1971) chemistry. *Hydrogen_bonding phase_transition symmetry_group twinning*
Tel. 49(721)6082977

Henkel, Prof. Dr Gerald (1948). Professor. Institut für Anorganische Chemie der Universität Duisburg, Germany. [Institut für Anorganische Chemie der Universität Duisburg, Lotharstr. 1, D-47048 Duisburg, Germany.] Dr rer.nat. (U. Bielefeld, 1976) chemistry. *Transition_elements chalcogens bioinorganic_chemistry EXAFS_and_XANES inorganic_and_organic_compounds*
Tel. 49(203)3793186 Fax 49(203)3793333

Herbst-Irmer, Dr Regine (1962). Lecturer. Institut für Anorganische Chemie der Universität Göttingen, Germany. [Institut für Anorganische Chemie der Universität Göttingen, Tammannstr. 4, D-37077 Göttingen, Germany.] Dr (U. Göttingen, 1990) chemistry. *Structure_determination database education*
E-mail rherbst@shelx.uni-ac.gwdg.de Tel. 49(551)393007 Fax 49(551)392582

Herdtweck, Dr Eberhardt F. Ch. (1948). Research Associate. Institut für Anorganische Chemie I der Technischen Universität München, Germany. [Institut für Anorganische Chemie I der Technischen Universität München, Lichtenbergstr. 4, D-85748 Garching, Germany.] Dr rer.nat. (U. Tübingen, 1978) inorganic chemistry. *Small_molecular_complexes fluorometallates oxides hydrates magnetochemistry bismuth_compounds_in_medicine*
E-mail · drhe@zaphod.anorg.chemie.tu-muenchen.de Tel. 49(89)32093143 Fax 49(89)32093473

Herres, Dr Nikolaus (1954). Staff Scientist. Fraunhofer-Institut für Angewandte Festkörperphysik, Germany. [Fraunhofer-Institut für Angewandte Festkörperphysik, Tullastr. 72, D-79108 Freiburg, Germany.] Dr rer.nat. (RWTH Aachen, 1991) mineralogy. *X-ray_diffraction instrumentation semiconductors diamond laboratory_automation microscopy epitaxy X-ray_topography texture_analysis*
E-mail herres@iaf.fhg.de Tel. 49(761)5159330 Fax 49(761)5159400

Herting-Agthe, Dr Susanne (1954). Curator. Institut für Mineralogie und Kristallographie der Technischen Universität Berlin, Germany. [Institut für Mineralogie und Kristallographie der Technischen Universität Berlin, Sekr. BH1, Ernst-Reuter-Platz 1, D-10587 Berlin, Germany.] Dr (TU Berlin, 1984) mineralogy. *Minerals systematics geosciences X-ray_fluorescence mineralogical_collections*
E-mail hert1637@mailszrz.zrz.tu-berlin.de Tel. 49(30)31422254 Fax 49(30)31421124

Herzberg, Dr Armin (1948). Vice Managing Director. UCW Umwelt Consulting Wolfen GmbH, Puschkinplatz 1, Geb. 122, D-06766 Wolfen, Germany. [c/o UCW, Hessenweg 11, D-34233 Fuldatal, Germany.] Dr rer.nat. (TU Clausthal, 1977) mineralogy. *Granites micas environment_protection environmental_technology*
Tel. 49(561)811309

Hesse, Dr Karl-Friedrich (1939). Scientist. Mineralogisch-Petrographisches Institut der Universität Kiel, Germany. [Mineralogisch-Petrographisches Institut der Universität Kiel, Olshausenstr. 40, D-24098 Kiel, Germany.] Dr (U. Kiel, 1973) mineralogy. *Diffractometer mineralogy structure_determination silicates*
Tel. 49(431)8802901 Fax 49(431)8804457

Heydenreich, Prof. Dr Johannes (1930). Professor. Max-Planck-Institut für Mikrostrukturphysik, Germany. [Max-Planck-Institut für Mikrostrukturphysik, Weinberg 2, D-06120 Halle/S., Germany.] Dr (Akad. d. Wiss. d. DDR Berlin, 1973) electron microscopy. *Electron_microscopy diffraction_technique semiconductors*
Tel. 49(345)601512 Fax 49(345)27155

Hildebrandt, Prof. Dr Gerhard (1922). Professor. Fritz-Haber-Institut der Max-Planck-Gesellschaft, Germany. [Fritz-Haber-Institut der Max-Planck-Gesellschaft, Katzwanger Steig 2, D-14089 Berlin, Germany.] Dr habil. (TU Berlin, 1975) physics. *Solid_state_physics X-ray_diffraction dynamical_theory crystal_defect X-ray_diffraction_apparatus*
Tel. 49(30)3655132

Hilgenfeld, Dr Rolf (1954). Group Leader. Central Research G865A, HOECHST AG, Germany. [Central Research G865A, HOECHST AG, D-65926 Frankfurt/M., Germany.] Dr rer.nat. (FU Berlin, 1987) chemistry. *Proteins_crystallography molecular_modelling proteins_engineering insulin aids drug_design*
E-mail r-hilgenfeld@zf1.hoechst-ag.dbp.de Tel. 49(69)30516207 Fax 49(69)30580169

Hiller, Prof. Dr Wolfgang Paul (1953). Professor. Anorganisch-Chemisches Institut der Technischen Universität München, Germany. [Anorganisch-Chemisches Institut der Technischen Universität München, Lichtenbergstr. 4, D-85748 Garching, Germany.] Dr rer.nat.habil. (U. Tübingen, 1991) chemistry. *Chemistry crystallography modelling*
E-mail hiller@hi1d25.anorg.chemie.tu-muenchen.de Tel. 49(89)32093133 Fax 49(89)32093133

Hinrichs, Dr Winfried (1950). Assistant Professor. Institut für Kristallographie der Freien Universität Berlin, Germany. [Institut für Kristallographie der Freien Universität Berlin, Takustr. 6, D-14195 Berlin, Germany.] Dr rer.nat. (U. Hamburg, 1983) chemistry. *Biocrystallography structure_activity_relationship molecular_biology anti-_biotics cobalamins*
E-mail winfried@chemie.fu-berlin.de 49(30)8386326 Fax 49(30)8386702

Hinrichsen, Prof. Dr Georg (1941). Director. Institut für Nichtmetallische Werkstoffe der Technischen Universität Berlin, Germany. [Institut für Nichtmetallische Werkstoffe der Technischen Universität Berlin, Englische Str. 20, D-10587 Berlin, Germany.] Dr rer.nat. (U. Mainz, 1970) physics. *Orientation plastics SAXS WAXS*
Tel. 49(30)31424464 Fax 49(30)31421100

Hinz, Dr Dietrich (1945). Research Assistant. Institut für Festkörper- und Werkstofforschung e.V., Germany. [Institut für Festkörper- und Werkstofforschung e.V., Postfach, D-01171 Dresden, Germany.] Dr rer.nat. (TU Dresden, 1974) physics. *Metallography microstructure texture_of_magnetic_materials magnets hard_magnetic_materials domain_structure RE_compounds properties_of_permanent_magnets*
Tel. 49(351)4659237 Fax 49(351)4659541

Hippler, Mr Bernd (1960). Research Associate. Staatliches Materialprüfungsamt NRW, Germany. [Staatliches Materialprüfungsamt NRW, Marsbruchstr. 186, D-44285 Dortmund, Germany.] Dipl. Min. (U. Münster, 1986) mineralogy. *Scanning_electron_microscopy powder_diffraction asbestos silicates*
Tel. 49(231)4502392 Fax 49(231)458549

Höbler, Dr Hans-Joachim (1951). Mineral Curator. Institut für Mineralogie, Kristallographie und Materialwissenschaften, Fakultät für Chemie und Mineralogie, Universität Leipzig, Germany. [Institut für Mineralogie, Kristallographie und Materialwissenschaften, Fakultät für Chemie und Mineralogie, Universität Leipzig, Scharnhorststr. 20, D-04275 Leipzig, Germany.] Dr rer.nat. (U. Leipzig, 1980) crystallography. *Mineralogy minerals*
Tel. 49(341)310502 Fax 49(341)310502 Telex 311432

Höche, Prof. Dr Hans-Reiner (1942). Dean and Professor. Fachbereich Physik der Universität Halle-Wittenberg, Germany. [Fachbereich Physik der Universität Halle-Wittenberg, Friedemann-Bach-Platz 6, D-06099 Halle/S., Germany.] Dr rer.nat.habil. (U. Halle/S., 1980) physics. *Dynamical_diffraction X-ray_polarization defect X-ray_optics nearly_perfect_crystals topography high_resolution_diffractometry*
Tel. 49(345)37761 Fax 49(345)29515 Telex 4353uni/hall/d

Höfer, Dr Harry (1962). Assistant. Fachbereich Physik der Universität Halle-Wittenberg, Germany. [Fachbereich Physik der Universität Halle-Wittenberg, Ernst-Barlach-Ring 35, D-06124 Halle-Neustadt, Germany.] Dr rer.nat. (U. Halle-Wittenberg, 1991) physics. *X-ray_diffraction polarization semiconductor multilayer diffractometry dynamical_theory Si GaAs BaTiO3 BixSbxTe3*
Tel. 49(345)2002742 Fax 49(345)21259

Hoeffken, Dr Hans Wolfgang (1953). Research Scientist. BASF AG, Germany. [BASF AG, ZHB/W-A30, D-67056 Ludwigshafen, Germany.] Dr rer.nat. (TU München, 1987) physics. *Protein crystallography drug_design*
E-mail hoeffken@brh.dnet.basf-ag.de Tel. 49(621)6049418 Fax 49(621)6020440

Höhling, Prof. Dr Hans Jürgen (1930). Professor. Institut für Medizinische Physik und Biophysik der Universität Münster, Germany. [Institut für Medizinische Physik und Biophysik der Universität Münster, Robert-Koch-Str. 31, D-48149 Münster, Germany.] Dr rer.nat.habil. (U. Münster, 1964) mineralogy. *Biocrystallography calcification electron_microscopy mineralization normal_and_pathological_hard_tissue_formation biomineralization*
Tel. 49(251)835191 Fax 49(251)835144

Höhne, Dr Wolfgang (1948). Head. Institut für Biochemie der Humboldt-Universität zu Berlin, Germany. [Institut für Biochemie der Humboldt-Universität zu Berlin, Hessische Str. 3-4, D-10115 Berlin, Germany.] Dr sc.nat. (Humboldt-U. Berlin, 1978) biochemistry. *Crystallography_and_engineering_of_proteins_in_medicine structure_modelling protein_evolution*
E-mail bioch@comcom.rz.charite.hu-berlin.de Tel. 49(30)28026119 Fax 49(30)28026252

Hölzel, Mr Alexander R. (1945). Systematik in der Mineralogie, Germany. [Systematik in der Mineralogie, Ulmenring 11, D-55270 Oberolm, Germany.] Dipl. Ing. (FH) (FH. Wiesbaden, 1970) chemistry. *Database minerals identification information_system software_for_mineral_identification*
Tel. 49(6136)89688 Fax 49(6136)889770

Hönle, Dr Wolfgang Johannes (1947). Research Associate. Max-Planck-Institut für Festkörperforschung, Heisenbergstr. 1, D-70569 Stuttgart, Germany. [Max-Planck-Institut für Festkörperforschung, Postfach 800665, D-70506 Stuttgart, Germany.] Dr rer.nat. (U. Münster, 1975) inorganic chemistry. *Solid_state_chemistry structure_and_bonding polycations_and_polyanions structural_relationship crystallographic_journal edition history_of_crystallography*
Tel. 49(711)6891464 Fax 49(711)6891010 Telex 7255555mpifd

Hofeldt, Mr Jochen (1965). Student. Institut für Kristallographie der RWTH Aachen, Germany. [Institut für Kristallographie der RWTH Aachen, Jägerstr. 17-19, D-52056 Aachen, Germany.] *Crystal_growth crystallography physical_property powder phase_determination semiconductor*

Hoffmann, Dr Klaus (1939). Marketing. Freiberger Elektronikwerkstoffe GmbH, Berthelsdorfer Str. 113, Germany. [Freiberger Elektronikwerkstoffe GmbH, Postfach 211, D-09584 Freiberg/Sa., Germany.] Dr rer.nat. (Bergakademie Freiberg, 1962) inorganic chemistry. *Crystal_growth_of_semiconductors defect_characterization preparation geometric_measurement wafer_production*
Tel. 49(3731)278292 Fax 49(3731)278233

Hoffmann, Prof. Dr Wolfgang (1935). Professor and Director. Member of the Commission: History and Teaching. Institut für Mineralogie der Universität Münster, Germany. [Institut für Mineralogie der Universität Münster, Corrensstr. 24, D-48149 Münster, Germany.] Dr rer.nat. (U. Hamburg, 1961) mineralogy/crystallography. *Mineralogy crystallography crystal_structure_research order-disorder_phenomena crystal_chemistry_and_reactions_in_solids*
Tel. 49(251)833461 Fax 49(251)838397 Telex 892529unimsd

Hoffmann, Mr Rolf R. (1944). International Sales Manager. SIEMENS AG, Analytical X-ray-systems (AUT V37), Germany. [SIEMENS AG, Analytical X-ray-systems (AUT V37), Postfach 211262, D-76181 Karlsruhe, Germany.] Dipl. Phys. (U. Erlangen, 1971) physics/materials science. *Diffractometer X-ray_fluorescence area_detector*
Tel. 49(721)5956213 Fax 49(721)5954506 Telex 78255-69sid

Hofmeister, Dr Wolfgang (1952). Assistant Professor and Head of Institute. Institut für Edelsteinforschung, Fachbereich Geowissenschaften der Universität Mainz, Germany. [Institut für Edelsteinforschung, Fachbereich Geowissenschaften der Universität Mainz, D-55099 Mainz, Germany.] Dr rer.nat.habil. (U. Mainz, 1990) mineralogy/crystallography. *Crystallography spectroscopy gemstones minerals pseudosymmetry twinning*
Tel. 49(6131)394365 Fax 49(6131)393070

Hoier, Dr Helga (1956). Postdoctoral Research Assistant. Laboratory of Biophysical Chemistry, University of Groningen, Nijenborg 4, 9747 AG Groningen, The Netherlands. [Zweibrückener Str. 13, D-70499 Stuttgart, Germany.] Dr rer.nat. (U. Stuttgart, 1988) chemistry. *Biocrystallography computer_modelling enzymes molecular_mechanics protein_engineering*
E-mail hoier@rugch2.chem.rug.nl Tel. 31(50)636478 Fax 31(50)634800

Holinski, Dr Rüdiger (1939). Research Scientist. MOLYKOTE, Dow Corning GmbH, Germany. [MOLYKOTE, Dow Corning GmbH, Pelkovenstr. 152, D-80992 München, Germany.] Dr rer.nat. (TU Clausthal, 1967) crystallography. *Solid_lubricants molybdenum_compounds brake_linings friction*
Tel. 49(89)14971265 Fax 49(89)14971254 Telex 5215654

Holzapfel, Prof. Dr Wilfried B. (1938). Professor. Fachbereich Physik der Universität-GH Paderborn, Warburger Str. 100, Germany. [Fachbereich Physik der Universität-GH Paderborn, D-33095 Paderborn, Germany.] Dr.rer.nat. habil. (U. Stuttgart, 1977) physics. *High_pressure enery_dispersive X-ray_diffraction structure_determination phase_diagram optical_properties solid_state_physics*
E-mail holz_we@physik.uni-paderborn.de Tel. 49(5251)602673 Fax 49(5251)603216

Hooft, Dr Rob W. W. (1967). Post-Doc. European Molecular Biology Laboratory, Heidelberg, Germany. [EMBL, Meyerhofstr. 1, D-69117 Heidelberg, Germany.] PhD (U. Utrecht, The Netherlands, 1993) chemistry. *Proteins*
E-mail hooft@embl-heidelberg.de Tel. 49(6221)387534 Fax 49(6221)387517

Hoppe, Prof. Dr Dr h. c. mult. Rudolf (1922). Professor Emeritus. Institut für Anorganische und Analytische Chemie I der Universität Gießen, Germany. [Institut für Anorganische und Analytische Chemie I der Universität Gießen, Heinrich-Buff-Ring 58, D-35392 Gießen, Germany.] Dr rer.nat. (U. Münster, 1954) chemistry. *Fluoride_and_oxides crystal_structure lattice_energy thermochemistry bond_length/bond_strength*
E-mail hoppe@dgihrz01.eam Tel. 49(641)7025700 Fax 49(641)7025712

Hosemann, Prof. Dr Dr h. c. Rolf (1912). Professor Emeritus. Institut für Kristallographie der Freien Universität Berlin, Takustr. 6, D-14195 Berlin, Germany. [Königsallee 55c, D-14193 Berlin, Germany.] Dr rer.nat.habil. (U. Freiburg/Br., 1939) physics. *Metallography polymerization proteins macromolecules paracrystals*
Tel. 49(30)8263503

Hovestreydt, Dr Eric Robert (1957). Product Specialist. SIEMENS AG, Dept. AUT V371, Germany. [SIEMENS AG, Dept. AUT V371, Postfach 211262, D-76181 Karlsruhe, Germany.] Dr (U. Genève/Switzerland, 1986) crystallography. *Single_crystal area_detector microdiffraction software data_collection data_processing structure_determination SAXS*
Tel. 49(721)5956573 Fax 49(721)5954506 Telex 78255-69sid

Hoyer, Prof. Dr Walter (1944). Full Professor. Fachbereich Physik der Technischen Universität Chemnitz-Zwickau, Germany. [Fachbereich Physik der Technischen Universität Chemnitz-Zwickau, PSF 964, D-09009 Chemnitz, Germany.] Dr rer.nat.habil. (TU Chemnitz, 1986) experimental physics. *Liquid_state phase_transition WAXS neutron_scattering small-angle_scattering alloy_melts real_structure phase_formation*
E-mail hoyer@physik.tu-chemnitz.de Tel. 49(371)852413 Fax 49(371)852491

Hu, Dr Xiaorui (1962). Postdoctoral Fellow. Hahn-Meitner-Institut, Glienicker Str. 100, D-14109 Berlin, Germany. [Hahn-Meitner-Institut, Postfach 390128, D-14091 Berlin, Germany.] Dr (TU Berlin, 1992) mineralogy/crystallography. *Neutron_and_X-ray_scattering phase_transition structure_determination symmetry computing condensed_matter data_collection diffraction_technique group_theory electron_microscopy Rietveld_method maximum_entropy_method*
E-mail hu@vax.hmi.dbp.de Tel. 49(30)80093074 Fax 49(30)80092999

Hübener, Dr Rainer (1962). Research Associate. Institut für Anorganische Chemie der Universität Tübingen, Germany. [Institut für Anorganische Chemie der Universität Tübingen, Auf der Morgenstelle 18, D-72076 Tübingen, Germany.] Dr rer.nat. (U. Tübingen, 1994) inorganic chemistry. *X-ray_diffraction crystal_structure_determination rhenium_compounds*
E-mail huebener@mailserv.zdv.uni-tuebingen.de Tel. 49(7071)296233 Fax 49(7071)292436

Hümmer, Prof. Dr Kurt (1939). Full Professor. Institut für Kristallographie der Universität Karlsruhe, Kaiserstr. 12, Germany. [Institut für Kristallographie der Universität Karlsruhe, Postfach 6980, D-76128 Karlsruhe, Germany.] Dr rer.nat.habil. (U. Erlangen-Nürnberg, 1978) physics. *Absolute_structure dynamical_diffraction multiple_scattering phase_determination*
E-mail bj05@dkauni2.bitnet Tel. 49(721)6083320 Fax 49(721)697720 Telex 17-721166unikar

Hummel, Dr Hans-Ulrich (1954). Associate Professor. Institut für Anorganische Chemie der Universität Erlangen-Nürnberg, Germany. [Institut für Anorganische Chemie der Universität Erlangen-Nürnberg, Egerlandstr. 1, D-91054 Erlangen, Germany.] Dr rer.nat.habil. (U. Erlangen-Nürnberg, 1989) inorganic chemistry. *X-ray crystal bonding layered phase_diagrams thermal_methods*
Tel. 49(9131)857351 Fax 49(9131)857387

Iberl, Angelika R. (1966). PhD student. Siemens AG, ZFE BT MR 32, Germany. [Siemens AG, ZFE BT MR 32, Otto-Hahn-Ring 6, D-81739 München, Germany.] Dipl. Phys. (U. Regensburg, 1991) physics. *High-resolution_diffractometry rocking_curves semiconductors topography epitaxy powder_diffraction high-temperature_X-ray_diffraction reflectometry*
E-mail jmr@cube.net Tel. 49(89)6363274 Fax 49(89)63642256 Telex 52109-11sin d

Ihringer, Dr Jörg (1943). Research Associate. Institut für Kristallographie der Universität Tübingen, Germany. [Institut für Kristallographie der Universität Tübingen, Charlottenstr. 33, D-72070 Tübingen, Germany.] Dr rer.nat.habil. (U. Tübingen, 1988) physics. *Profile_analysis powder real_structure pseudosymmetry apparatus statistics*
Tel. 49(7071)295242 Fax 49(7071)296058

Imhof, Dr Wolfgang (1962). Research Assistant. Max-Planck-Arbeitsgruppe "CO2-Chemie" an der Universität Jena, Germany. [Max-Planck-Arbeitsgruppe "CO2-Chemie" an der Universität Jena, August-Bebel-Str. 2, D-07743 Jena, Germany.] Dr rer.nat. (U. Heidelberg, 1992) chemistry. *Iron_cluster_compounds ruthenium_cluster_compounds osmium_cluster compounds carbon_dioxide CO2_activation catalytic_reactions_or_processes_involving_CO2*
E-mail cwi@rz.uni-jena.de Tel. 49(3641)635409 Fax 49(3641)635538

Irngartinger, Prof. Dr Hermann (1938). Professor. Organisch-Chemisches Institut der Universität Heidelberg, Germany. [Organisch-Chemisches Institut der Universität Heidelberg, Im Neuenheimer Feld 270, D-69120 Heidelberg, Germany.] Dr habil. (U. Heidelberg, 1972) chemistry. *High-precision_structures molecular_crystal topochemistry X-ray_crystallography organic_photochemistry solid_state_reaction strained_cage_molecules*
E-mail e56@ix.urz.uni-heidelberg.de Tel. 49(6221)562422 Fax 49(6221)564205

Jacobi, Dr Hans (1934). Lecturer. Fachabteilung Kristallographie des Instituts für Mineralogie und mineralische Rohstoffe der Technischen Universität Clausthal, Germany. [Fachabteilung Kristallographie des Instituts für Mineralogie und mineralische Rohstoffe der Technischen Universität Clausthal, Postfach 1253, D-38670 Clausthal-Zellerfeld, Germany.] Dr phil. (U. Marburg, 1963) mineralogy. *Modulated_structures satellite_reflection corrosion pseudosymmetry*
Tel. 49(5323)722392 Fax 49(5323)723500

Jacobs, Prof. Dr Herbert (1936). Professor. Fachbereich Chemie der Universität Dortmund, Germany. [Fachbereich Chemie der Universität Dortmund, Otto-Hahn-Str. 6, D-44221 Dortmund, Germany.] Dr rer.nat. (U. Kiel, 1966) chemistry. *Chemistry solid_state inorganic*
Tel. 49(231)7553802 Fax 49(231)7553771 Telex 822445

Jagodzinski, Prof. Dr Dr h. c. Heinz E. (1916). Retired Full Professor. Institut für Kristallographie und Mineralogie der Universität München, Germany. [Institut für Kristallographie und Mineralogie der Universität München, Theresienstr. 41, D-80333 München, Germany.] Dr rer.nat. (U. Göttingen, 1941) physics. *Disorder diffraction_theory diffuse_scattering modulated_structures surface_structure quasicrystal phase_transitions*
Tel. 49(89)23944357 Fax 49(89)23944334

Jahn, Dr Irmin Rudolf (1939). Lecturer. Institut für Kristallographie der Universität Tübingen, Germany. [Institut für Kristallographie der Universität Tübingen, Charlottenstr. 33, D-72070 Tübingen, Germany.] Dr rer.nat. (U. Tübingen, 1971) physics. *Crystal_optics phase_transition phase_diagram perovskites_layered_compounds*
Tel. 49(7071)296389 Fax 49(7071)296058

Jansen, Prof. Dr Martin (1944). Professor and Director. Institut für Anorganische Chemie der Universität Bonn, Germany. [Institut für Anorganische Chemie der Universität Bonn, Gerhard-Domagk-Str. 1, D-53121 Bonn, Germany.] Dr rer.nat. (U. Gießen, 1973) chemistry. *Chemistry crystal_growth crystallography solid_state chemistry binary_and_multinary_oxides*
E-mail unc404@ibmrhrz.uni-bonn.de Tel. 49(228)733114 Fax 49(228)735660 Telex 886657unibod

Jarchow, Prof. Dr Otto (1931). Professor. Mineralogisch-Petrographisches Institut der Universität Hamburg, Germany. [Mineralogisch-Petrographisches Institut der Universität Hamburg, Grindelallee 48, D-20146 Hamburg, Germany.] Dr rer.nat. (U. Saarbrücken, 1961) mineralogy/crystallography. *Crystallography structural_disorder structure_determination crystal_chemistry subgroup_relations_in_domain_structures*
Tel. 49(40)41232056 Fax 49(40)41232422

Jauch, Dr Wolfgang (1947). Staff Scientist; 4-circle gamma-ray diffractometer, 4-circle neutron diffractometer. Hahn-Meitner-Institut, Glienicker Str. 100, Germany. [Hahn-Meitner-Institut, Postfach 390128, D-14109 Berlin, Germany.] Dr rer.nat.habil. (TU Berlin, 1992) crystallography. *Gamma-ray_diffraction single_crystal neutron_diffraction high-precision_structure statistical_methods charge_density phase_transitions*

E-mail jauch@vax.hmi.d400.de Tel. 49(30)80092767 Fax 49(30)80092999

Jeitschko, Prof. Dr Wolfgang (1936). Director. Anorganisch-Chemisches Institut der Universität Münster, Germany. [Anorganisch-Chemisches Institut der Universität Münster, Wilhelm-Klemm-Str. 8, D-48149 Münster, Germany.] Dr phil. (U. Wien/Austria, 1964) chemistry. *Inorganic_and_intermetallic_solids structure_and_property* Tel. 49(251)833121 Fax 49(251)278532

Jex, Prof. Dr Hartmut (1940). Professor. Abteilung Festkörperphysik der Universität Ulm, Germany. [Abteilung Festkörperphysik der Universität Ulm, Albert-Einstein-Allee 11, D-89081 Ulm, Germany.] Dr (U. Frankfurt/M., 1972) physics. *Lattice_dynamics inelastic_scattering interferometry time-resolved_effect X-quantum_beats neutron_scattering* Tel. 49(731)5022972 Fax 49(731)5022987

Jockel, Mr Dietmar (1956). Research Associate. Zentrum für Funktionswerkstoffe gem. GmbH, Göttingen/Clausthal, Germany. [Zentrum für Funktionswerkstoffe gem. GmbH, Göttingen/Clausthal, Windausweg 2, D-37073 Göttingen, Germany.] Dipl. Kristallogr. (Humboldt-U. Berlin, 1983) crystallography. *Crystal_growth crystallography III–V_compounds semiconductors* E-mail jockel@umpsun1.gwdg.de Tel. 49(551)399733 Fax 49(551)5071750

Joswig, Dr Werner (1940). Institut für Kristallographie der Universität Frankfurt, Senckenberganlage 30, Germany. [Institut für Kristallographie der Universität Frankfurt, Postfach 111932, D-60054 Frankfurt/M., Germany.] Dr (U. Frankfurt/M., 1972) mineralogy. *Inorganic_crystal_structure zeolites clays* Tel. 49(69)7982104 Fax 49(69)7982101 Telex 413932

Kabsch, Dr Wolfgang (1941). Research Associate. Max-Planck-Institut für Medizinische Forschung, Germany. [Max-Planck-Institut für Medizinische Forschung, Jahnstr. 29, D-69028 Heidelberg, Germany.] Dr (U. Heidelberg, 1972) physics. *Biocrystallography computing data_processing* E-mail kabsch@mpimf-heidelberg.mpg.de Tel. 49(6221)486276 Fax 49(6221)486437 Telex 461505

Kaiser, Dr Volker (1961). Manager of Central Department for Crystal Structure Determination. Institut für Anorganische und Analytische Chemie der Universität Clausthal, Germany. [Institut für Anorganische und Analytische Chemie der Universität Clausthal, Paul-Ernst-Str. 4, D-38670 Clausthal-Zellerfeld, Germany.] Dr rer.nat. (U. Marburg, 1992) chemistry. *Structure_determination single_crystal powder magnetism neutron_scattering magnetic_structures spin_glasses* E-mail cavk@idefix.rz.tu-clausthal.de Tel. 49(5323)723012 Fax 49(5323)722995

Kalbe, Mrs Ute (1959). Research Associate. Institut für Ökologie, FG Bodenkunde der Technischen Universität Berlin, Am Salzufer 11-12, D-10587 Berlin, Germany. [Schönefelder Chaussee 231, D-12524 Berlin, Germany.] Dipl. Kristallogr. (Humboldt-U. Berlin, 1983) crystallography. *Environmental_pollution soils X-ray_fluorescence_spectroscopy diffractometry polarization_microscopy cross_section asbestos coals* Tel. 49(30)31423307

Kammann, Mrs Elke (1966). Research Associate. Institut für Mineralogie der Universität Hannover, Germany. [Institut für Mineralogie der Universität Hannover, Welfengarten 1, D-30167 Hannover, Germany.] Dipl. Min. mineralogy. *Crystallography surface kinetics mineralogy* Tel. 49(511)7623583

Kandler, Mr Andreas (1950). Sales Engineer. SIEMENS AG, Germany. [SIEMENS AG, Chemnitzer Str. 12, D-09599 Freiberg/Sa., Germany.] Dipl. Phys. (TU Dresden, 1974) physics. *Area_detector crystal_structure_determination diffractometer X-ray_fluorescence high_precision_diffractometry microdiffraction SAXS* Tel. 49(3731)48616 Fax 49(3731)48616

Karl, Prof. Dr Norbert (1939). Professor. 3. Physikalisches Institut der Universität Stuttgart, Germany. [3. Physikalisches Institut der Universität Stuttgart, Pfaffenwaldring 57, D-70550 Stuttgart, Germany.] Dr habil. (U. Stuttgart, 1975) physics. *Molecular_crystal thin_film LEED high_purity molecular_electronics charge_carrier_transport* Tel. 49(711)6855195 Fax 49(711)6855281

Kastner, Dr Berthold (1952). Research Scientist. Institut für Molekularbiologie und Tumorforschung der Universität Marburg, Germany. [Institut für Molekularbiologie und Tumorforschung der Universität Marburg, Emil-Mannkopff-Str. 2, D-35037 Marburg, Germany.] Dr rer.nat. (TU Berlin, 1982) chemistry. *Biochemistry nucleoproteins crystallization electron_microscopy molecular_biology RNA_splicing* Tel. 49(6421)285064 Fax 49(6421)287008

Katzke, Mrs Hannelore (1965). Research Associate. Mineralogisch-Petrographisches Institut der Universität Kiel, Germany. [Mineralogisch-Petrographisches Institut der Universität Kiel, Olshausenstr. 40, D-24118 Kiel, Germany.] Dipl. Chem. (TU Berlin, 1989) chemistry. *Electrochemistry intercalation_compound high_resolution_crystallography phase_transition* Tel. 49(431)8802906 Fax 49(431)8804457

Keitel, Mr Thomas (1962). Postdoctoral Fellow. Institut für Biochemie der Charité, Humboldt-Universität zu Berlin, Germany. [Institut für Biochemie der Charité, Humboldt-Universität zu Berlin, Hessische Str. 3-4, D-10115 Berlin, Germany.] Dipl. Chem. (FU. Berlin, 1989) chemistry. *Macromolecular X-ray Fab_fragments* Tel. 49(30)28468688 Fax 49(30)28468600

Kek, Dr Stefan (1961). Research Associate. Fachrichtung Kristallographie der Universität Saarbrücken, Postfach 1150, D-66041 Saarbrücken, Germany. [Fachrichtung Kristallographie der Universität Saarbrücken, c/o HASYLAB/DESY, D-22603 Hamburg, Germany.] Dr rer.nat. (U. Stuttgart, 1991) metallurgy. *Crystallography anomalous_dispersion quasicrystal pseudosymmetry* E-mail kek@vxdesy.desy.de Tel. 49(40)89982601 Fax 49(40)89982787

Keller, Dr Egbert (1950). Research Assistant. Kristallographisches Institut der Universität Freiburg, Germany. [Kristallographisches Institut der Universität Freiburg, Hebelstr. 25, D-79104 Freiburg/Br., Germany.] Dr rer.nat. (U. Freiburg/Br., 1978) chemistry. *Computer_graphics structure_determination superstructure* E-mail kell@sun1.ruf.uni-freiburg.de Tel. 49(761)2036438 Fax 49(761)2036434

Keller, Prof. Dr Paul (1940). Professor. Institut für Mineralogie und Kristallchemie der Universität Stuttgart, Germany. [Institut für Mineralogie und Kristallchemie der Universität Stuttgart, Pfaffenwaldring 55, D-70550 Stuttgart, Germany.] Prof. (U. Stuttgart, 1980) mineralogy. *Structure modelling mineralization oxides crystal_chemistry* Tel. 49(711)6854112 Fax 49(711)6853500

Kemmler-Sack, Prof. Dr Sibylle (1934). Professor. Institut für Anorganische Chemie der Universität Tübingen, Germany. [Institut für Anorganische Chemie der Universität Tübingen, Auf der Morgenstelle 18, D-72076 Tübingen, Germany.] Prof. (U. Tübingen, 1973) inorganic chemistry. *Catalysis condensed_matter magnetism superconductivity crystallography luminescence structure* E-mail kemmler-sack@mailserv.zdv.uni-tuebingen.de Tel. 49(7071)296231 Fax 49(7071)296918

Kern, Mr Arnt (1965). Doctorand. Mineralogisch-Petrographisches Institut der Universität Heidelberg, Germany. [Mineralogisch-Petrographisches Institut der Universität Heidelberg, INF 236, D-69120 Heidelberg, Germany.] Dipl. Min. (U. Heidelberg, 1992) mineralogy. *Rietveld_method high_temperature phase_transition accuracy thermal_expansion* E-mail bm4@vm.urz.uni-heidelberg.de Tel. 49(6221)564814 Fax 49(6221)564805

Kies, Dr Jörg (1942). Patent Assessor. WITEGA e.V., Landsberger Allee 253, D-13055 Berlin, Germany. Dr rer.nat. (Akad.d. Wiss. Berlin, 1971) physics. *Biomaterial structural_change X-ray_spectroscopy patent semiconductor III–V-compounds biosensor* Tel. 49(30)9754682

Kießling, Dr Frank-Michael (1960). Postdoctoral Research Assistant. Fachbereich Physik der Humboldt Universität zu Berlin, Invalidenstr. 110, Germany. [Fachbereich Physik der Humboldt Universität zu Berlin, Unter den Linden 6, D-10099 Berlin, Germany.] Dr rer.nat. (Humboldt-U. Berlin, 1991) crystallography. *Crystal_growth II–VI_compounds positron_annihilation point_defect* Tel. 49(30)2803250 Fax 49(30)2803477

Kirchhoff, Dr Andreas (1960). Research Assistant. Gmelin-Institut, Germany. [Gmelin-Institut, Varrentrappstr. 40-42, D-60486 Frankfurt/M., Germany.] Dr rer.nat. (U. Göttingen, 1988) mineralogy. *Database data_collection zeolites silicates* E-mail kirchhoff@mpigm-frankfurt.mpg.dbp.de Tel. 49(69)7917578 Fax 49(69)7917338

Kirfel, Prof. Dr Armin Harald (1943). Professor. Mineralogisches Institut der Universität Würzburg, Germany. [Mineralogisches Institut der Universität Würzburg, Am Hubland, D-97074 Würzburg, Germany.] Dr rer.nat.habil. (U. Bonn, 1983) crystallography. *Charge_density X-ray_fluorescence anomalous_dispersion synchrotron_radiation cation_distribution* E-mail kirfel@vax.rz.uni-wuerzburg.dbp.de Tel. 49(931)8885430 Fax 49(931)8884620

Klapdor, Mr Martin Frank (1964). Doctoral Candidate. Institut für Anorganische Chemie und Strukturchemie, Lehrstuhl II der Universität Düsseldorf, Germany. [Institut für Anorganische Chemie und Strukturchemie, Lehrstuhl II der Universität Düsseldorf, Universitätsstr. 1 (Geb.26.42), D-40225 Düsseldorf, Germany.] Dipl. Chem. (U. Düsseldorf, 1992) chemistry. *Low_temperature DTA crystal_growth fluoride hydrogen_bonding* E-mail klapdor@mail.rz.uni-duesseldorf.de Tel. 49(211)3114143 Fax 49(211)3113085

Klapper, Prof. Dr Helmut (1937). Professor. Chairman of the Commission of Crystal Growth and Characterization. Mineralogisch-Petrologisches Institut der Universität Bonn, Germany. [Mineralogisch-Petrologisches Institut der Universität Bonn, Poppelsdorfer Schloß, D-53115 Bonn, Germany.] Dr rer.nat.habil. (RWTH Aachen, 1975) crystallography and crystal physics. *Crystal_growth crystal_physics real_structure topography twinning phase_transitions* E-mail une110@ibm.rhrz.uni-bonn.de Tel. 49(228)732769 Fax 49(228)732770 Telex 886657 unibo d

Klee, Prof. Dr Wilfrid Edgar (1935). Professor. Institut für Kristallographie der Universität Karlsruhe, D-76128 Karlsruhe, Germany. [Am Kirchberg 77, D-76229 Karlsruhe, Germany.] Dr rer.nat. (U. Freiburg, 1960) physical chemistry. *Topology infrared_spectroscopy apatite* Tel. 49(721)6082136 Fax 49(721)697720

Klein, Mr Volker (1961). Doctoral Fellow. Institut für Anorganische Chemie der Universität Tübingen, Germany. [Institut für Anorganische Chemie der Universität Tübingen, Auf der Morgenstelle 18, D-72076 Tübingen, Germany.] Dipl. Chem. (U. Tübingen, 1991) inorganic chemistry. *Structure magnetism spectroscopy topotaxy intercalation/deintercalation spin_glasses solids mineralogy nickel_compounds Mössbauer* Tel. 49(7071)296231 Fax 49(7071)296918

Kleint, Prof. Dr Christian (1926). Professor Emeritus. Fachbereich Physik Leipzig der Universität, Germany. [Fachbereich Physik Leipzig der Universität, Linnéstr. 5, D-04103 Leipzig, Germany.] Dr habil. (U. Leipzig, 1968) experimental physics. *Adsorption semiconductor metals EELS high-Tc_superconductors field_emission* Tel. 49(341)6858222 Fax 49(341)6858221

Klement, Dr Ulrich (1930). Assistant Professor. Institut für Anorganische Chemie der Universität, D-93040 Regensburg, Germany. Dr (U. München, 1959) chemistry. *Structure_determination* Tel. 49(941)9434542

Klimakow, Dr Alexander (1949). Fachbereich Physik, Lehrstuhl für Physikalische Grundlagen der Photonik der Humboldt-Universität zu Berlin, Invalidenstr. 110, D-10115 Berlin, Germany. [Fachbereich Physik, Lehrstuhl für Physikalische Grundlagen der Photonik der Humboldt-Universität zu Berlin, Unter den Linden 6, D-10099 Berlin, Germany.] Dr rer.nat. (MGU. Moscow/Russia, 1975) inorganic chemistry. *Crystal_growth diffusion II–VI_compounds heterostructure IV–VI_semiconductor*
Tel. 49(30)2803288 Fax 49(30)2803486 Telex 304156huphy

Klimm, Dr Detlef (1957). Postdoctoral Research Assistant. Institut für Kristallographie, Mineralogie und Materialwissenschaft der Universität Leipzig, Germany. [Institut für Kristallographie, Mineralogie und Materialwissenschaft der Universität Leipzig, Scharnhorststr. 20, D-04275 Leipzig, Germany.] Dr rer.nat. (U. Leipzig, 1986) crystallography. *Dislocation plastic_deformation internal_friction III–V_compounds fatigue*
Tel. 49(341)6858391

Klinkhammer, Dr Karl Wilhelm (1958). Assistant professor. Institut für Anorganische Chemie der Universität Stuttgart, Germany. [Institut für Anorganische Chemie der Universität Stuttgart, Pfaffenwaldring 55, D-70550 Stuttgart, Germany.] Dr rer.nat. (U. Stuttgart, 1992) chemistry. *Molecular_crystal organometallic_alkalis silicon_compounds heteronuclear_main_group_metal_clusters disordering twinning*
E-mail karl.klinkhammer@rus.uni-stuttgart.de   Tel.   49(711)6854207   Fax 49(711)6854241

Kniep, Prof. Dr Rüdiger (1945). Professor. Eduard-Zintl-Institut der Technischen Hochschule Darmstadt, Germany. [Eduard-Zintl-Institut der Technischen Hochschule Darmstadt, Hochschulstr. 10, D-64289 Darmstadt, Germany.] Dr rer.nat. (TU Braunschweig, 1973) chemistry. *Crystallization phase_diagram nitrides solids*
E-mail d133@rs11.hrz.th-darmstadt.de Tel. 49(6151)162192 Fax 49(6151)164073

Knoch, Dr Falk (1953). Research Assistant. Institut für Anorganische Chemie II der Universität Erlangen-Nürnberg, Germany. [Institut für Anorganische Chemie II der Universität Erlangen-Nürnberg, Egerlandstr. 1, D-91058 Erlangen, Germany.] Dr (U. Bonn, 1984) chemistry. *Structure_determination computer_technology*
E-mail knoch@anorganik.chemie.uni-erlangen.de   Tel.   49(9131)857373   Fax 49(9131)857367

Koch, Prof. Dr Elke (1943). Wiss. Ang. Institut für Mineralogie, Petrologie und Kristallographie der Universität Marburg, Germany. [Institut für Mineralogie, Petrologie und Kristallographie der Universität Marburg, Hans-Meerwein-Straße, D-35032 Marburg, Germany.] Dr rer.nat.habil. (U. Marburg, 1987) crystallography. *Mathematical_crystallography symmetry_group packing International_Tables_for_Crystallography minimal_surface systematics_of_inorganic_crystal_structures*
E-mail kochelke@mailer.uni-marburg.de Tel. 49(6421)285610 Fax 49(6421)288919 Telex 482372

Ködderitzsch, Dr Horst (1931). Formerly Chemie-Kombinat Bitterfeld, Germany. [Ignaz-Stroof-Str. 15, D-06749 Bitterfeld, Germany.] Dr rer.nat. (U. Halle-Wittenberg, 1979) chemistry. *Electron_microscopy scanning_electron_microscopy polarization_microscopy*
Tel. 49(3493)69214

Köhler, Dr Rolf (1942). Leader. Max-Planck-Arbeitsgruppe Röntgenbeugung, Germany. [Max-Planck-Arbeitsgruppe Röntgenbeugung, Hausvogteiplatz 5-7, D-10117 Berlin, Germany.] Dr habil. (Humboldt-U. Berlin, 1993) physics. *Topography high_resolution_diffractometry diffraction_theory real_structure strain_in_nearly_perfect_crystals X-ray_contrast_simulation*
E-mail koehler@x-ray0.roen.ipp-garching.mpg.de   Tel.   49(30)20366234   Fax 49(30)2004536

Koellner, Dr Gertraud (1960). Postdoctoral Research Assistant. Institut für Kristallographie der Freien Universität Berlin, Germany. [Institut für Kristallographie der Freien Universität Berlin, Takustr. 6, D-14195 Berlin, Germany.] Dr (FU. Berlin, 1993) chemistry. *Biocrystallography proteins cyclodextrins packing*
E-mail koellner@chemie.fu-berlin.de Tel. 49(30)8386325 Fax 49(30)8386702

Köpke, Dr Jürgen (1956). Staff Scientist. Max-Planck-Institut für Biophysik, Abteilung für Molekulare Membranbiologie, Germany. [Max-Planck-Institut für Biophysik, Abteilung für Molekulare Membranbiologie, Heinrich-Hoffmann-Str. 7, D-60528 Frankfurt/M., Germany.] Dr (U. Hamburg, 1985) mineralogy. *Macromolecular_structure_determination area_detector computer_modelling phase_refinement membrane_proteins*
E-mail koepke@mpibp-frankfurt.mpg.de Tel. 49(69)96769411 Fax 49(69)96769423

Körber, Dr Wolfgang (1956). Development Engineer. ALCATEL-SEL, Germany. [ALCATEL-SEL, Lorenzstr.10, D-70435 Stuttgart, Germany.] Dr rer.nat. (U. Stuttgart, 1988) physics. *Epitaxy semiconductors compound*
Tel. 49(711)8214898 Fax 49(711)8216355 Telex 72526-0

Kohl, Prof. Dr Helmut (1955). Professor. Physikalisches Institut der Universität Münster, Germany. [Physikalisches Institut der Universität Münster, Wilhelm-Klemm-Str. 10, D-48149 Münster, Germany.] Dr habil. (TH Darmstadt, 1989) physics. *High_resolution_electron_microscopy EELS channelling nanoanalysis*
E-mail kohl@nwz00.uni-muenster.de Tel. 49(251)833640 Fax 49(251)833602

Kopf, Dr Jürgen (1942). Research Scientist. Institut für Anorganische und Angewandte Chemie der Universität Hamburg, Germany. [Institut für Anorganische und Angewandte Chemie der Universität Hamburg, Martin-Luther-King-Platz 6, D-20146 Hamburg, Germany.] Dr rer.nat.habil. (U. Hamburg, 1987) chemistry. *Diffractometry computer_graphics carbohydrates organometallic_compound*
E-mail kopf@rz.informatik.uni-hamburg.d400.de   Tel.   49(40)41232897   Fax 49(40)41236348

Kosten, Dr Klaus (1949). Postdoctoral Research Assistant. Institut für Kristallographie der RWTH Aachen, Jägerstr. 17-19, D-52066 Aachen, Germany. Dr rer.nat. (RWTH Aachen, 1989) mineralogy. *Computer-aided_education computer_graphics education powder_method*
E-mail fg050kr@dacth11.(EARN) fg050kr@vm1.rz.rwth-aachen.de (INTERNET) Tel. 49(241)806902 Fax 49(241)806916 Telex 0832704thacd

Krämer, Prof. Dr Volker (1940). Professor. Kristallographisches Institut der Universität Freiburg, Germany. [Kristallographisches Institut der Universität Freiburg, Hebelstr. 25, D-79104 Freiburg/Br., Germany.] Dr rer.nat. (TU München, 1968) crystallography. *Crystallography crystal_growth structure_determination phase_diagram*
Tel. 49(761)2036436 Fax 49(761)2036434

Kräußlich, Dr Jürgen (1945). Research Assistant. Institut für Optik und Quantenelektronik der Universität Jena, Germany. [Institut für Optik und Quantenelektronik der Universität Jena, Max-Wien-Platz 1, D-07743 Jena, Germany.] Dr rer.nat. (U. Jena, 1972) physics. *Thin_film multilayer X-ray_diffraction X-ray_topography silicon_compounds*
Tel. 49(3641)8225625 Fax 49(3641)23843

Krane, Dr Hans-Georg (1958). Postdoctoral Research Assistant; responsible for Beamline D3 at Hasylab (Hamburg). Fachrichtung Kristallographie, Universität Karlsruhe, Germany. [Fachrichtung Kristallographie, Universität Karlsruhe, c/o HASYLAB am DESY, D-22603 Hamburg, Germany.] Dr (U. Saarbrücken, 1991) mineralogy. *Single-crystal_diffraction anomalous_dispersion white-beam_radiation Laue_diffraction anomalous_dispersion energy-dispersive_analysis*
E-mail krane@vxdesy.desy.de Tel. 49(40)89982695 Fax 49(40)89982787

Kranold, Prof. Dr Rainer (1943). Professor. Fachbereich Physik der Universität Rostock, Germany. [Fachbereich Physik der Universität Rostock, Universitätsplatz 3, D-18051 Rostock, Germany.] Dr rer.nat.habil. (U. Rostock, 1991) experimental physics. *Small-angle_scattering non-crystalline_materials vitreous_state phase_separation SAXS ASAXS SANS WAXS EXAFS intermediate-range_order_of_glasses glasses glass_ceramics porous_materials fractals*
E-mail baade@saphir.physik1.uni-rostock.de Tel. 49(381)23212 Fax 49(381)4981626 Telex 398460unird

Krause, Dr Christian (1955). Research Associate. Bundesamt für Strahlenschutz, Germany. [Bundesamt für Strahlenschutz, Postfach 100149, D-38201 Salzgitter, Germany.] Dr (U. Münster, 1986) mineralogy. *Plutonium_compounds uranium_compounds radioactivity radiochemistry nuclear_fuel_cycle*
E-mail krause%mindendorf%bfs@bfs.de Tel. 49(5341)225162 Fax 49(5341)225225

Krauß, Dr Norbert (1960). Postdoctoral Research Assistant. Institut für Kristallographie der Freien Universität Berlin, Germany. [Institut für Kristallographie der Freien Universität Berlin, Takustr. 6, D-14195 Berlin, Germany.] Dr rer.nat. (U. Köln, 1988) *Biocrystallography photosynthesis inorganic_crystal_structure*
E-mail phosys@chemie.fu-berlin.de Tel. 49(30)8386326 Fax 49(30)8386702

Krauße, Dr Joachim (1933). Postdoctoral Research Assistant. Institut für Physikalische Chemie der Universität Jena, Germany. [Institut für Physikalische Chemie der Universität Jena, Lessingstr. 10, D-07743 Jena, Germany.] Dr rer.nat. (U. Jena, 1965) chemistry. *Least-squares_refinement line_profile_analysis molecular_complexes RHEED infrared*
Tel. 49(3641)8226044

Krebs, Prof. Dr Bernt (1938). Full Professor. Member of the German National Committee. Anorganisch-Chemisches Institut der Universität Münster, Germany. [Anorganisch-Chemisches Institut der Universität Münster, Wilhelm-Klemm-Str. 8, D-48149 Münster, Germany.] Dr rer.nat. (U. Göttingen, 1965) chemistry. *Structural_chemistry bioinorganic_chemistry molecular_modelling chalcogenides. transition_metal_complexes metalloprotein_structures ionic_conductors hydrogen_bonding*
E-mail krebs@vnwz01.uni-muenster.de Tel. 49(251)833131 Fax 49(251)838366 Telex 892529unimsd

Kretschmar, Mr Martin. Institut für Anorganische Chemie der Universität Tübingen, Germany. [Institut für Anorganische Chemie der Universität Tübingen, Auf der Morgenstelle 18, A9, D-72076 Tübingen, Germany.] *Crystallography*
E-mail   caakr01@mailserv.zdv.uni-tuebingen.de   Tel.   49(7071)295245Fax 49(7071)292436

Kretzschmar, Dr Ulrike (1945). Research Assistant. Institut für Angewandte Chemie i.Gr. e.V., Rudower Chaussee 5, D-12489 Berlin, Germany. [Steinbinderweg 31, D-12527 Berlin, Germany.] Dr (Akad. d. Wiss. d. DDR Berlin, 1980) chemistry/physics. *Wide-angle_scattering small-angle_scattering phase_formation density_distribution in_situ-reactions_solid/gas catalysts*
Tel. 49(30)63924315 Fax 49(30)63924350

Kreutz, Dr Ernst Wolfgang (1940). Vice-Director. Lehrstuhl für LASERtechnik der RWTH Aachen, Germany. [Lehrstuhl für LASERtechnik der RWTH Aachen, Steinbachstr. 15, D-52074 Aachen, Germany.] Dr rer.nat. (TH Darmstadt, 1969) physics. *Computer_modelling PVD hydrodynamics ceramics Auger_spectroscopy epitaxy physics*
E-mail postmaster@llt.rwth-aachen.de Tel. 49(241)8906146 Fax 49(241)8906121 Telex 832267fhiltd

Kriegel, Mr Steffen (1965). Research Associate. Fachbereich Physik, Abteilung Halbleiterphysik der Universität Leipzig, Germany. [Fachbereich Physik, Abteilung Halbleiterphysik der Universität Leipzig, Linnéstr. 5, D-04103 Leipzig, Germany.] Dipl. Phys. (U. Leipzig, 1991) physics. *Rocking_curves simulation misorientation grazing_incidence lattice_parameters ordering III–V_compounds*
Tel. 49(341)6858558 Fax 49(341)6858281

Kroll, Prof. Dr Herbert (1940). Professor. Institut für Mineralogie der Universität Münster, Germany. [Institut für Mineralogie der Universität Münster, Corrensstr. 24, D-48149 Münster, Germany.] Dr (U. Münster, 1971) mineralogy. *Kinetics solid_solution phase_transition modelling*
Tel. 49(251)833455 Fax 49(251)838397

Krüger, Dr Carl (1933). Professor and Director. Member of the Commission on Small Molecules. Max-Planck-Institut für Kohlenforschung, Röntgenlabor, Germany. [Max-Planck-Institut für Kohlenforschung, Röntgenlabor, Postfach 101353, D-45466 Mühlheim/Ruhr, Germany.] Dr rer.nat. (RWTH Aachen, 1961) inorganic chemistry. *Structural_chemistry molecular_crystal deformation_density_distribution molecular_modelling solid_state_chemistry chemical_bonding*
E-mail krueger@mpi-muelheim.mpg.d400.de Tel. 49(208)3062174 Fax 49(208)3062980

Krug, Prof. Dr Detlef (1936). Professor. Institut für Anorganische Chemie der Universität Tübingen, Germany. [Institut für Anorganische Chemie der Universität Tübingen, Auf der Morgenstelle 18, D-72076 Tübingen, Germany.] Dr (U. Tübingen, 1972) inorganic chemistry. *DTA TGA thermoanalysis oxides analytical_chemistry*
Tel. 49(7071)296227

Kryschi, Dr Carola (1957). Assistant Professor. Universität Düsseldorf, Germany. [Im Riedbusch 20, D-41564 Kaarst, Germany.] Dr habil. (U. Düsseldorf, 1993) physics. *Crystallization molecular_crystal photochemistry photoreaction_centre laser phonon_resonance*
Tel. 49(2131)601239 Fax 49(2131)667287

Kuban, Dr Ralf-Jürgen (1953). Research Associate. Universitätsklinikum Charité, Institut für Biochemie der Humboldt-Universität zu Berlin, Germany. [Universitätsklinikum Charité, Institut für Biochemie der Humboldt-Universität zu Berlin, Hessische Str. 3-4, D-10115 Berlin, Germany.] Dr sc.rer.nat. (Akad. d. Wiss. d. DDR Berlin, 1989) physical chemistry. *Structure_determination biomolecule diffraction proteins*
E-mail kuban@rz.charite.hu-berlin.de Tel. 49(30)28468688 Fax 49(30)28468600

Kühlbrandt, Dr Werner (1951). Group Leader. European Molecular Biology Laboratory, Germany. [European Molecular Biology Laboratory, Meyerhofstr. 1, D-69117 Heidelberg, Germany.] Dr (U. Cambridge/UK., 1981) molecular biology. *Electron_crystallography membranes_proteins structure_and_function_of_proteins electron_microscopy_technique*
E-mail kuehlbrandt@embl-heidelberg.de Tel. 49(6221)387245 Fax 49(6221)387306

Kühn, Prof. Dr Robert (1911). Lecturer in München. Universität Heidelberg and Technische Universität München, Germany. [Richard-Wagner-Str. 31, D-69259 Wilhelmsfeld, Germany.] Dr phil. (U. Kiel, 1938) mineralogy. *Mineralogy petrography geochemistry_of_salts teaching*
Tel. 49(6220)8924

Küppers, Prof. Dr Horst (1933). Professor. Mineralogisch-Petrographisches Institut der Universität Kiel, Germany. [Mineralogisch-Petrographisches Institut der Universität Kiel, Olshausenstr. 40, D-24098 Kiel, Germany.] Dr rer.nat. (U. Freiburg/Br., 1966) physics. *Elasticity thermal_expansion structure_determination hydrogen_bonding*
E-mail nmp26@rz.uni-kiel.d400.de Tel. 49(431)8802897 Fax 49(431)8804457

Kuhs, Prof. Dr Werner F. (1952). Full Professor. Mineralogisch-Kristallographisches Institut, Universität Göttingen, Germany. [Mineralogisch-Kristallographisches Institut, Universität Göttingen, Goldschmidtstr. 1, D-37077 Göttingen, Germany.] Dr rer.nat.habil. (U. Karlsruhe, 1991) physics/crystallography. *High_pressure synchrotron_radiation nuclear_reactor hydrogen_bonding ice clathrates ferroelectricity melting*
E-mail kuhs@silly.uni-mki.gwdg.de Tel. 49(551)393891 Fax 49(551)399521

Lacmann, Prof. Dr-Ing. Rolf (1927). Full Professor. Institut für Physikalische und Theoretische Chemie der Technischen Universität, Braunschweig, Hans-Sommer-Str. 10, D-38106 Braunschweig, Germany. [Institut für Physikalische und Theoretische Chemie der Technischen Universität, Braunschweig, Postfach 3329, D-38023 Braunschweig, Germany.] Dr-Ing. (TU Berlin, 1958) chemistry. *Crystal_growth nucleation electro-crystallization kinetics*
Tel. 49(531)3915326 Fax 49(531)3915832 Telex 952526tubsw

Lambert, Dr Ulrich (1955). Scientist (materials characterization). WACKER-CHEMITRONIC GmbH, Germany. [WACKER-CHEMITRONIC GmbH, Postfach 1140, D-84479 Burghausen, Germany.] Dr rer.nat. (U. Heidelberg, 1988) mineralogy/crystallography. *Semiconductors silicon_compounds characterization defect bond_length diffractometry copper_compounds statistics oxides*
Tel. 49(8677)835003 Fax 49(8677)62171 Telex 56923-15

Lange, Dr Joachim Reinhard (1961). Postdoctoral Research Assistant. Institut für Angewandte Physik, Lehrstuhl für Kristallographie der Universität Erlangen-Nürnberg, Germany. [Institut für Angewandte Physik, Lehrstuhl für Kristallographie der Universität Erlangen-Nürnberg, Bismarckstr. 10, D-91054 Erlangen, Germany.] Dr rer.nat. (U. Erlangen-Nürnberg, 1990) physics/crystallography. *Laue_spectrum_analysis Laue_diffraction absolute_configuration phase_determination detector_properties X-ray_properties multichannel_analyzer*
E-mail mpkr00@cd6480fs.rrze.uni-erlangen.dbp.de Tel. 49(9131)852705 Fax 49(9131)852733

Langer, Prof. Dr Klaus (1936). Professor. Institut für Mineralogie und Kristallographie der Technischen Universität Berlin, Germany. [Institut für Mineralogie und Kristallographie der Technischen Universität Berlin, Ernst-Reuter-Platz 1, D-10587 Berlin, Germany.] Dr rer.nat. (U. Kiel, 1965) inorganic chemistry. *Synthesis resonance_spectrometry high_pressure kinetics silicates stability solid_solution FTIR electronic_spectrum point_defect*
Tel. 49(30)31425325 Fax 49(30)31421124

Laube, Dr Gert (1957). Group Leader. ALCATEL-SEL AG, Research Center, Germany. [ALCATEL-SEL AG, Research Center, Lorenzstr. 10, D-70435 Stuttgart, Germany.] Dr rer.nat. (U. Stuttgart, 1988) chemistry. *Epitaxy semiconductor optoelectronics III-V_compounds MOVPE quantum_well_structures III-V_semiconductor_technology*
Tel. 49(711)8213827 Fax 49(711)8216355 Telex 72526-0

Lehmann, Mrs Alice (1964). Postgraduate. Institut für Kristallographie der Freien Universität Berlin, Germany. [Institut für Kristallographie der Freien Universität Berlin, Takustr. 6, D-14195 Berlin, Germany.] Dipl. Chem. (FU. Berlin, 1991) chemistry. *Small_molecules powder_diffraction*
E-mail lehmann@chemie.fu-berlin.de Tel. 49(30)8383460 Fax 49(30)8383464

Leipner, Dr Hartmut S. (1958). Postdoctoral Research Assistant. Fachbereich Physik der Martin-Luther-Universität Halle-Wittenberg, Germany. [Fachbereich Physik der Martin-Luther-Universität Halle-Wittenberg, Friedemann-Bach-Platz 6, D-06108 Halle, Germany.] Dr rer.nat. (U. Halle/S., 1988) physics. *Electron_microscopy positron_annihilation semiconductor defect*
E-mail leipner@mluep4s.physik.uni-halle.de Tel. 49(345)21902741 Fax 49(345)2021259 Telex 04353unihald

Lenz, Mrs Annett (1969). Postgraduate Student. Technikum Analytikum, Zimmer 327, Linnéstr. 5, D-04103 Leipzig, Germany. [Ella-Kay-Str. 38, D-10405 Berlin, Germany.] *Crystallography crystal_growth ternary_chalcopyrites thin_film analysis biocrystallography solar_energy*
Tel. 49(341)6858390

Lerf, Dr Anton (1948). Postdoctoral Research Assistant. Walther-Meissner-Institut, Germany. [Walther-Meissner-Institut, Walther-Meissner-Str. 8, D-85748 Garching, Germany.] Dr habil. (TU München, 1991) chemistry. *Dichalcogenides intercalation superconductors clays graphite intercalation_compounds nanochemistry colloids*
Tel. 49(89)32094225 Fax 49(89)32094206

Leuschner, Dr Dieter (1940). Disablement Pensionist. [Anton-Graff-Str. 14, D-01309 Dresden, Germany.] Dr rer.nat. physics. *Biophysics symmetry symmetry_breaking thermodynamics biology free_energy structural_disorder structure systematics*

Leute, Prof. Dr Volkmar (1938). Professor. Institut für Physikalische Chemie der Universität Münster, Germany. [Institut für Physikalische Chemie der Universität Münster, Schloßplatz 4, D-48149 Münster, Germany.] Dr rer.nat. (U. München, 1969) physical chemistry. *Diffusion electron_probe microanalysis order-disorder phase_diagram chalcogenides point_defect*
Tel. 49(251)833431 Fax 49(251)839138

Liebau, Prof. Dr Friedrich (1926). Emeritus. Mineralogisches Institut der Universität Kiel, Germany. [Mineralogisches Institut der Universität Kiel, Olshausenstr. 40, D-24098 Kiel, Germany.] Dr habil. (U. Würzburg, 1963) crystal-structure science. *Inorganic_crystal_chemistry silicates porous_materials antimony_compounds zeolites clathrates*
Tel. 49(431)8802888 Fax 49(431)8804457

Liehr, Dr Günter (1939). Scientist for Structure Analysis. Institut für Anorganische Chemie I der Universität Erlangen-Nürnberg, Germany. [Institut für Anorganische Chemie I der Universität Erlangen-Nürnberg, Egerlandstr. 1, D-91058 Erlangen, Germany.] Dr (U. Erlangen-Nürnberg, 1972) chemistry. *Chemistry structure_determination classification symmetry modulated_structures superstructure satellite_reflection diffuse_scattering crystal_growth*
E-mail liehr@anorganik.chemie.uni-erlangen.de Tel. 49(9131)857355 Fax 49(9131)857387

Lindemann, Prof. Dr Willi (1921). Emeritus. Lehrstuhl für Kristallstrukturlehre und Mathematische Kristallographie der Universität Würzburg, Germany. [Lehrstuhl für Kristallstrukturlehre und Mathematische Kristallographie der Universität Würzburg, Am Hubland, D-97074 Würzburg, Germany.] Dr phil.nat. (U. Erlangen, 1951) crystallography. *Mathematical_crystallography theoretical_crystallography mathematics group_theory ring_theory theoretical_physics theoretical_chemistry structure_analysis*
Tel. 49(931)8885429

Lindner, Prof. Dr Hans Jörg (1939). Professor. Institut für Organische Chemie der Technischen Hochschule Darmstadt, Germany. [Institut für Organische Chemie der Technischen Hochschule Darmstadt, Petersenstr. 22, D-64287 Darmstadt, Germany.] Dr-Ing. (TH Darmstadt, 1967) chemistry. *Organic_molecular_crystal conformation molecular_mechanics computer_modelling*
E-mail lindner@oc1.oc.chemie.th-darmstadt.de Tel. 49(6151)163576 Fax 49(6151)165591

Linke, Mr Dieter (1956). Research Associate; leader of working group. Gesellschaft zur Förderung angewandter Optik, Optoelektronik, Quantenelektronik, und Spektroskopie e.V. (GOS), Rudower Chaussee 5, D-12489 Berlin, Germany. [Alt-Schmöckwitz 5, D-12527 Berlin, Germany.] Dipl.-Kristallogr. (Humboldt-U. Berlin, 1983) crystallography. *III-V_compounds scanning_electron_microscopy crystal_growth_of_III-V_compounds scribing_and_clearing behaviour_of_III-V_compounds*
Tel. 49(30)63922636 Fax 49(30)63922612

Lippmann, Prof. Dr Friedrich (1928). Retired (1993). [Dreifürstensteinstr. 22, D-72116 Mössingen, Germany.] Apl. Prof. (U. Tübingen, 1978) mineralogy. *Crystal_chemistry structure_of_carbonates mixed_crystal decomposition clay_minerals iridescence_of_feldpars*
Tel. 49(7473)21291

Lippmann, Mr Thomas (1964). Research Assistant. Fachrichtung Kristallographie der Universität Saarbrücken, Germany. [Fachrichtung Kristallographie der Universität Saarbrücken, c/o HASYLAB/DESY, Postfach, D-22603 Hamburg, Germany.] Dipl. Phys. (U. Münster, 1988) physics. *Anisotropic_anomalous_dispersion birefringence*
E-mail lippmann@vxdesy.desy.de Tel. 49(40)89982601 Fax 49(40)89982787

Löchner, Dr Ulrich (1948). Research Associate. c/o Materialwissenschaft, FB 21, Technische Hochschule Darmstadt, Germany. [Alsenstr. 55, D-44789 Bochum, Germany.] Dr (U. Karlsruhe, 1980) chemistry. *Powder synchrotron_radiation EXAFS rare-earth_compounds crystallography hydrates crystallization disordered systems catalysis nonstoichiometry*
E-mail loechner@vxdesy.desy.de Tel. 49(234)312680 Fax 49(234)300686

Löns, Dr Friedrich Jürgen (1939). Lecturer. Institut für Mineralogie der Universität Münster, Germany. [Institut für Mineralogie der Universität Münster, Corrensstr. 24, D-48149 Münster, Germany.] Dr (U. Hamburg, 1969) mineralogy. *Crystallography crystal_growth single_crystal_applications*
E-mail lons@vnwz01.uni-muenster.de Tel. 49(251)833456 Fax 49(251)838397

Lois, Mr Christos (1966). Postgraduate Student. Institut für Mineralogie der Universität Münster, Germany. [Institut für Mineralogie der Universität Münster, Corrensstr. 24, D-48149 Münster, Germany.] Dipl. Phys. (U. Iraklio/Kreta/Greece, 1988) physics. *Crystallography least-squares_refinement molecular_modelling zeolites catalysis computer_graphics dynamics force_field lattice_dynamics energy_minimization simulation salts LCAO_method X-ray*
E-mail lois@dmswwu1a.uni-muenster.de Tel. 49(251)833463

Lueken, Prof. Dr Heiko (1942). Professor. Institut für Anorganische Chemie der RWTH Aachen, Germany. [Institut für Anorganische Chemie der RWTH Aachen, Prof.-Pirlet-Str. 1, D-52056 Aachen, Germany.] Dr habil. (RWTH Aachen, 1979) chemistry. *Solid_state magnetochemistry lanthanides metals intermetallic_compounds mixed-valence_compounds*
E-mail ie010fe@dacth11.de Tel. 49(241)804648 Fax 49(241)8888288

Luger, Prof. Dr Peter (1943). Professor. Institut für Kristallographie, Fachbereich Chemie der Freien Universität Berlin, Germany. [Institut für Kristallographie, Fachbereich Chemie der Freien Universität Berlin, Takustr. 6, D-14195 Berlin, Germany.] Dr habil. (FU. Berlin, 1974) crystallography. *Single_crystal analysis low_temperature structure–activity_relationship charge_density*
E-mail luger@chemie.fu-berlin.de Tel. 49(30)8383411 Fax 49(30)8383464

Lutz, Prof. Dr Heinz Dieter (1934). Professor. Anorganische Chemie I der Universität Siegen, D-57068 Siegen, Germany. Dr habil. (U. Köln, 1967) chemistry. *Solid_state hydrogen_bonding ionic_conductivity lattice_dynamics hydrates halides Raman infrared*
E-mail lutz@hrz.uni-siegen.dbp.de Tel. 49(271)7404217 Fax 49(271)7402330 Telex 271383

Maas, Prof. Dr Gerhard (1949). Professor. Universität Ulm, Germany. [Abteilung Organische Chemie I der Universität Ulm, Albert-Einstein-Allee 11, D-89069 Ulm, Germany.] Dr (U. Saarbrücken, 1974) chemistry. *Organic_molecules metallo-organic_compound molecular_crystal silicon_compounds*
Tel. 49(731)5022790 Fax 49(731)5022803 Telex 49(731)502712567

Madden, Dean R. (1963). Postdoctoral researcher. Max-Planck-Institut fuer med. Forschung, Abt. Biophysik, Jahnstr. 29, 69120 Heidelberg, Germany. [Max-Planck-Institut fuer med. Forschung, Abt. Biophysik, Postfach 10 38 20, 69028 Heidelberg, Germany.] PhD (Harvard U., 1992) biophysics. *Protein_structure ion_channels electron_diffraction membrane_proteins immunology major_histocompatibility_complex peptide_ligands*
E-mail madden@mpimfhd.mpimf-heidelberg.mpg.de Tel. 49(6221)486286 Fax 49(6221)486437 Telex 461505

Maier, Dr Horst (1939). Leader. AEG Aktiengesellschaft, Bereich Infrarot- und Nachtsichtkomponenten, Germany. [AEG Aktiengesellschaft, Bereich Infrarot- und Nachtsichtkomponenten, Theresienstr. 2, D-74072 Heilbronn, Germany.] Dr habil. (TU Berlin, 1974) physics. *II–VI_compounds infrared_detector semiconductor_epitaxy optoelectronics*
Tel. 49(7131)621290 Fax 49(7131)621233

Malcherek, Mr Thomas (1966). Institut für Mineralogie der Universität Münster, Germany. [Institut für Mineralogie der Universität Münster, Corrensstr. 24, D-48149 Münster, Germany.] (1993) mineralogy/crystallography. *Phase_transition feldspars microstructure solid_state physics*
E-mail malcher@nwz.uni-muenster.de Tel. 49(251)833477 Fax 49(251)838397

Mannherz, Prof. Dr Hans Georg (1943). Professor. [Institut für Anatomie und Embryologie, Ruhr-Universität Bochum, Universitätsstr. 150, 44780 Bochum, Germany.] Prof. (U. Marburg, 1980) cytobiology. *Actin actin_binding_proteins*
Tel. 49(234)7004553 Fax 49(234)7094339

Maresch, Prof. Walter (1944). Professor. Institut für Mineralogie der Universität Münster, Germany. [Institut für Mineralogie der Universität Münster, Corrensstr. 24, D-48149 Münster, Germany.] PhD (U. Princeton/USA, 1972) geology. *Synthesis petrology phase_equilibrium solid_solution*
E-mail maresch@vnwz00.uni-muenster.de Tel. 49(251)833047 Fax 49(251)838397 Telex 892529unimsd

Mariolacos, Dr Konstantin (1936). Research Assistant. Mineralogisch-Kristallographisches Institut der Universität Göttingen, Germany. [Mineralogisch-Kristallographisches Institut der Universität Göttingen, Goldschmidtstr. 1, D-37077 Göttingen, Germany.] Dr (U. Wien/Austria, 1972) crystallography. *Crystal_growth phase_diagram structure_determination thermodynamics*
Tel. 49(551)393931 Fax 49(551)399521

Martin-Vosshage, Dr Dagmar (1963). Post-Doc (Fedor-Lynen) Fellow. Department of Physics, National University of Singapore, Lower Kent Ridge Road BLK S12, Faculty of Science, Singapore 0511, Singapore. [Lessingstr. 8, D-30159 Hannover, Germany.] Dr rer.nat. (U. Hannover, 1990) mineralogy. *Crystallography XPS polymers electrolytes structure structural_disorder thin_films*
E-mail phydm@nusvm.bitnet. Tel. 49(511)320756 Fax 49(2324)30944 Telex unispors33943

Massa, Prof. Dr Werner (1944). Head of X-ray Laboratory. Fachbereich Chemie der Universität Marburg, Germany. [Fachbereich Chemie der Universität Marburg, Hans-Meerwein-Str., D-35043 Marburg, Germany.] Dr habil. (U. Marburg, 1982) chemistry. *Structure_determination fluorometallates low-dimensional_magnetism Jahn_Teller_effect*
E-mail massa@ps1515.chemie.uni-marburg.de Tel. 49(6421)285525 Fax 49(6421)288917

Mattes, Prof. Dr Rainer (1937). Professor. Anorganisch-Chemisches Institut der Universität Münster, Germany. [Anorganisch-Chemisches Institut der Universität Münster, Wilhelm-Klemm-Str. 8, D-48149 Münster, Germany.] Dr habil. (U. Münster, 1971) chemistry. *Inorganic_chemistry magnetism structure_determination bonding*
Tel. 49(251)833117 Fax 49(251)833169

Maue, Mrs Erika (1962). Post-Graduate Student. Institut für Angewandte Physik, Lehrstuhl für Kristallographie der Universität Erlangen-Nürnberg, Bismarckstr. 10, D-91054 Erlangen, Germany. [Schumannstr. 3, D-93128 Regenstauf, Germany.] Dipl. Phys. (TH Karlsruhe, 1990) physics. *Mathematical crystallography symmetry crystallography_in_higher_dimensions*
Tel. 49(9402)7368

Mayer, Mrs Bärbel (1964). Research Fellow. Fachbereich Physik der Universität Osnabrück, Germany. [FB Physik, Barbarastrasse 7, D-49069 Osnabrück, Germany.] Dipl. Min. (U. Heidelberg, 1990) mineralogy. *Oxides crystallography xps surface*
E-mail BMAYER@fb4-301.physik.uni-osnabrueck.de or BMAYER@titan.rz.uni-osnabrueck.de Tel. 49(541)9692682 Fax 49(541)9692670

Meissner, Mrs Elke (1964). Research Associate. [Bayerisches Forschungsinstitut für Experimentelle Geochemie und Geophysik, Univ. Bayreuth, D-95440 Bayreuth, Germany] Dipl. Min. (U. Erlangen-Nürnberg, 1989) mineralogy. *TEM STEM EDS diffusion electron_diffraction crystal_growth devitrification high_pressure ceramics*
E-mail elke.meissner@uni-bayreuth.de Tel. 49(921)553765 Fax 49(921)553769

Melzer, Dr Rolf (1960). Postdoctoral Research Assistant. Hahn-Meitner-Institut/N2, Germany. [Hahn-Meitner-Institut/N2, Glienicker Str. 100, D-14109 Berlin, Germany.] Dr rer.nat. (TU Berlin, 1992) mineralogy. *Structure_determination phase_transition order-disorder modulated_structures*
E-mail melzer@vax.hmi.dbp.de Tel. 49(30)80093081 Fax 49(30)80092523

Mertin, Dr Wilhelm (1938). Lecturer. Institut für Anorganische und Analytische Chemie der Universität Gießen, Germany. [Institut für Anorganische und Analytische Chemie der Universität Gießen, Heinrich-Buff-Ring 58, D-35392 Gießen, Germany.] Dr rer.nat. (U. Münster, 1967) inorganic chemistry. *High-resolution_electron_microscopy solid_state chemistry oxides structural_disorder chemical_transport*
Tel. 49(641)7025676 Fax 49(641)7025669

Messerschmidt, Dr Albrecht Matthias Wilhelm (1945). Senior Scientist. Max-Planck-Institut für Biochemie, Germany. [Max-Planck-Institut für Biochemie, Am Klopferspitz 18A, D-82152 Martinsried, Germany.] Dr (Humboldt-U. Berlin, 1972) crystallography. *Biocrystallography bioinorganic_chemistry proteins X-ray structure_determination*
E-mail messerschmid@vms.biochem.mpg.de Tel. 49(89)85782662 Fax 49(89)85783516

Meyer, Prof. Dr Gerd (1949). Professor. Institut für Anorganische Chemie der Universität Hannover, Germany. [Institut für Anorganische Chemie der Universität Hannover, Callinstr. 9, D-30167 Hannover, Germany.] Dr rer.nat. (U. Gießen, 1976) chemistry. *Synthesis crystal_growth structure_determination magnetism ionic_conductivity*
Tel. 49(511)7623696 Fax 49(511)7623006

Meyer-Ehmsen, Prof. Dr Gerhard (1932). Professor. Fachbereich Physik der Universität Osnabrück, Germany. [Fachbereich Physik der Universität Osnabrück, Barbarastr. 7, D-49069 Osnabrück, Germany.] Dr rer.nat. (U. Hamburg, 1961) physics. *RHEED structure_determination structural_disorder diffraction_theory*
Tel. 49(541)969628 Fax 49(541)9692670

Meyerheim, Dr Holger L. (1961). Postdoctoral Research Assistant. Institut für Kristallographie und Mineralogie der Universität München, Germany. [Institut für Kristallographie und Mineralogie der Universität München, Theresienstr. 41, D-80333 München, Germany.] Dr rer.nat. (FU. Berlin, 1990) physics. *GIXS structure_determination superstructure surface*
Tel. 49(89)23944329 Fax 49(89)23944334 Telex 529815univmd

Michel, Prof. Dr Bernd (1949). Director. Centrum für Mikromechanik Chemnitz, Germany. [Centrum für Mikromechanik Chemnitz, Horststr. 8, D-09119 Chemnitz, Germany.] Dr rer.nat.habil. (U. Halle/S., 1979) physics. *Mechanics fracture microstrain reliability micromechanics*
Tel. 49(371)5614623 Fax 49(371)5614625

Michel, Prof. Dr Hartmut (1948). Professor and Director. Max-Planck-Institut für Biophysik, Germany. [Max-Planck-Institut für Biophysik, Heinrich-Hoffmann-Str. 7, D-60528 Frankfurt/M., Germany.] Dr rer.nat. (U. Würzburg, 1977) biochemistry. *Biocrystallography proteins photosynthesis photoreaction_centre membrane_proteins*
Tel. 49(69)96769401 Fax 49(69)96769423

Milius, Dr Wolfgang (1958). Postdoctoral Research Assistant. Laboratorium für Anorganische Chemie I der Universität Bayreuth, Germany. [Laboratorium für Anorganische Chemie I der Universität Bayreuth, Universitätsstr. 30, D-95440 Bayreuth, Germany.] Dr rer.nat. (U. Erlangen-Nürnberg, 1987) chemistry. *Structure_determination of molecular_compounds symmetry structural_disorder theory crystallization_of_compounds_on_diffractometers*
E-mail wolfgang.milius@uni-bayreuth.de Tel. 49(921)552547 Fax 49(921)552535

Modlich, Mr Jörn. (1967). Student. Mineralogisch-Petrographisches Institut der Universität, Germany. [Grindelallee 48, D-20146 Hamburg, Germany.] *Orthorhombic iron_compound perovskites single_crystal X-ray_diffraction chemical_reactivity thermoanalysis* Mössbauer_spectroscopy nonstoichiometric_oxides Czochralski_technique

Möck, Dr Peter (1957). Postdoctoral Research Assistant. Institut für Kristallographie und Materialforschung der Humboldt-Universität zu Berlin, Germany. [Institut für Kristallographie und Materialforschung der Humboldt-Universität zu Berlin, Invalidenstr. 110, D-10115 Berlin, Germany.] Dr rer.nat. (Humboldt-U. Berlin, 1992) crystallography. *High-resolution_electron_microscopy epitaxic_misorientation X-ray_diffractometry geometry_analysis electron_microscopy transmission_electron_microscopy high-resolution_diffractometry*
Tel. 49(30)2803294 Fax 49(30)2803360

Möhling, Dr Werner (1934). Max-Planck-Arbeitsgruppe "Röntgenbeugung", Germany. [Max-Planck-Arbeitsgruppe "Röntgenbeugung", Hausvogteiplatz 5-7, D-10117 Berlin, Germany.] Dr rer.nat. (Humboldt-U. Berlin, 1967) physics. *Crystal_defect topography multiple_crystal_diffraction semiconductors*
Tel. 49(30)20366216 Fax 49(30)2004536

Mootz, Prof. Dr Dietrich (1933). Professor. Institut für Anorganische Chemie und Strukturchemie der Universität Düsseldorf, Germany. [Institut für Anorganische Chemie und Strukturchemie der Universität Düsseldorf, Universitätsstr. 1, D-40225 Düsseldorf, Germany.] Dr rer.nat. (TU Berlin, 1959) chemistry. *Binary_phase_diagram hydrogen_bonding hydrates fluorine_compounds crystal_growth_of_low_melting_materials hydrates_and_clathrate hydrates_of_acids_and_bases adducts_of_hydrogen_halides*
Tel. 49(211)3113135 Fax 49(211)3113085 Telex 08587348 uni d

Morgenroth, Dr Wolfgang (1962). Postdoctoral Research Associate; Energy Dispersive Laue (EDL) Diffractometer at beamline F1, Hasylab. Fachrichtung Kristallographie, Universität des Saarlandes, Postfach 151150, D-66041 Saarbrücken, Germany. [c/o HASYLAB/DESY, D-22603 Hamburg, Germany.] Dr (U. des Saarlandes, 1993) mineralogy. *Single_crystal diffractometry synchrotron_radiation anomalous_dispersion*
E-mail morgenroth@vxdesy.desy.de Tel. 49(40)89982601 Fax 49(40)89982787

Mühlberg, Prof. Dr Manfred (1949). Professor. Institut für Kristallographie der Universität zu Köln, Germany. [Institut für Kristallographie der Universität zu Köln, Zülpicher Str. 49b, D-50674 Köln, Germany.] Dr sc.nat. (Humboldt-U. Berlin, 1990) crystallography. *Crystal_growth phase_transition real_structure thermodynamics inclusions thermoanalysis tensor_properties*
E-mail muehlberg@kri.uni-koeln.de Tel. 49(221)4704420 Fax 49(221)4704963

Müller, Prof. Dr Eberhard (1942). Professor. Institut für Keramische Werkstoffe der Technischen Universität - Bergakademie, Freiberg, Germany. [Institut für Keramische Werkstoffe der Technischen Universität - Bergakademie, Freiberg, Gustav-Zeuner-Str. 3, D-09596 Freiberg/Sa., Germany.] Dr rer.nat.habil. (U. Jena, 1986) physics. *Composite_ceramics electron_diffraction amorphous_phase crystallization*
Tel. 49(3731)392014 Fax 49(3731)393662

Müller, Prof. Dr Gerd (1942). Director. Fraunhofer-Institut für Silicatforschung, Germany. [Fraunhofer-Institut für Silicatforschung, Neunerplatz 2, D-97082 Würzburg, Germany.] Dr rer.nat. (U. Karlsruhe; 1969) mineralogy. *Chemistry mineralogy microstructure sintering*
E-mail leisner@isc.fhg.de Tel. 49(931)419090 Fax 49(931)4190980

Müller, Prof. Dr Gerhard (1953). Professor. Fakultät für Chemie der Universität Konstanz, Germany. [Fakultät für Chemie der Universität Konstanz, Postfach 5560-M723, D-78434 Konstanz, Germany.] Dr rer.nat.habil. (TU München, 1989) inorganic and analytical chemistry. *Structural_chemistry organometallic_chemistry coordination_chemistry weak_interaction*
E-mail xanorg@vg10.chemie.uni-konstanz.de Tel. 49(7531)883735 Fax 49(7531)883140

Müller, Prof. Dr Horst (1929). Professor (retired). Chemisches Laboratorium, Abteilung Radiochemie der Universität Freiburg, Albertstr. 21, D-79104 Freiburg/Br., Germany. [Mauracherstr. 21, D-79211 Denzlingen, Germany.] Dr rer.nat. (U. Freiburg/Br., 1958) chemistry. *Cosmochemistry damage radiochemistry solid_solution philosophy_of_science history_of_science*
Tel. 49(7666)3369

Müller, Dr Jürgen Joachim (1943). Research Associate. Max-Delbrück-Zentrum für Molekulare Medizin, Germany. [Max-Delbrück-Zentrum für Molekulare Medizin, Robert-Rössle-Str. 10, D-13122 Berlin, Germany.] Dr habil. (U. Rostock, 1991) applied physics. *SAXS biophysics computer_modelling X-ray_crystallography*
E-mail jjm@orion.rz.mdc-berlin.de Tel. 49(30)94063421 Fax 49(30)94062548

Müller, Dr Paul (1951). Postdoctoral Research Assistant. Institut für Anorganische Chemie der RWTH Aachen, Germany. [Institut für Anorganische Chemie der RWTH Aachen, Prof. Pirlet-Str. 1, D-52056 Aachen, Germany.] Dr (RWTH Aachen, 1980) chemistry. *Neutron_diffraction X-ray_diffraction magnetism Rietveld_method solid_state_chemistry*
E-mail fbmull@dacth11.bitnet Tel. 49(241)804469 Fax 49(241)8888288

Müller, Mrs Sabine (1962). Student. IFBE GmbH, Standort Leipzig, Spinnereistr. 7, D-04179 Leipzig, Germany. [Reiskestr. 14, D-04317 Leipzig, Germany.] Dipl. Kristallogr. crystallography. *Environment_protection solar_energy physical_property_of_III-V compounds plasticity*
Tel. 49(341)66905

Müller, Prof. Dr Ulrich (1940). Professor of Inorganic Chemistry. Fachbereich 19 Biologie/Chemie der Universität Kassel, D-34109 Kassel, Germany. Dr rer.nat. (TH Stuttgart, 1966) chemistry. *Inorganic_chemistry structure_determination group_theory diffuse_scattering*
Tel. 49(561)8044425 Fax 49(561)8044466

Müller-Fahrnow, Dr Anke (1960). Scientist. SCHERING AG, Germany. [SCHERING AG, Müllerstr. 170-178, D-13342 Berlin, Germany.] Dr rer.nat. (FU. Berlin, 1990) crystallography. *Crystallization macromolecules molecular_crystal steroids*
Tel. 49(30)4687699 Fax 49(30)46916741

Münninghoff, Dr Günter (1953). Product Manager. SIEMENS AG, Abt. AUTV371, Germany. [SIEMENS AG, Abt. AUTV371, Siemensallee 84, D-76181 Karlsruhe, Germany.] Dr (U. Marburg, 1980) mineralogy. *Instrumentation structure crystallography diffractometer*
Tel. 49(721)5956574 Fax 49(721)5954506 Telex 78255-69sid

Mundt, Dr Otto (1950). Research Assistant. Institut für Anorganische Chemie der Universität Stuttgart, Germany. [Institut für Anorganische Chemie der Universität Stuttgart, Pfaffenwaldring 55, D-70550 Stuttgart, Germany.] Dr rer.nat. (U. Karlsruhe, 1979) chemistry. *Inorganic_chemistry organometallic_chemistry crystal_packing intermolecular_interaction*
Tel. 49(711)6854221 Fax 49(711)6854241

Murad, Dr Enver (1941). Department Head. Bayerisches Geologisches Landesamt, Germany. [Bayerisches Geologisches Landesamt, Postfach 120141, D-96033 Bamberg, Germany.] Dr rer.nat.habil. (TU München, 1986) mineralogy. *Magnetism mineralogy Mossbauer paramagnetics_supermagnetism environment*
Tel. 49(951)57280 Fax 49(951)51251

Nägele, Dr Erhard (1933). Publisher. E.Schweizerbart'sche Verlagsbuchhandlung, (Nägele & Obermiller), Germany. [E.Schweizerbart'sche Verlagsbuchhandlung, (Nägele & Obermiller), Johannesstr. 3A, D-70176 Stuttgart, Germany.] Dr rer.nat. (U. Tübingen, 1959) geology. *Crystallography*
Tel. 49(711)625001 Fax 49(711)625005 Telex 723363schbd

Nagel, Dr Wolfgang (1948). Manager. SYSTEC Microprocessor System-Technologie GmbH, Nottulner Landweg 104, D-48161 Münster-Roxel, Germany. [Postfach 410220, D-48046 Münster-Roxel, Germany.] Dr (U. Münster, 1986) mineralogy. *Computer_technology automation monitoring*
Tel. 49(2534)8090 Fax 49(2534)80941

Nar, Dr Herbert (1963). Postdoctoral Research Assistant. Max-Planck-Institut für Biochemie, Abteilung Strukturforschung, Germany. [Max-Planck-Institut für Biochemie, Abteilung Strukturforschung, Am Klopferspitz, D-82152 Martinsried, Germany.] Dr (TU München, 1992) chemistry. *Proteins_crystallography enzymes bioinorganic_chemistry*
E-mail nar@nmrvex.biochem.mpg.de Tel. 49(89)85782662 Fax 49(89)85783516

Nauer-Gerhardt, Carola U. (1955). Product Specialist. SIEMENS AG, AUT 37 PM, Germany. [SIEMENS AG, AUT 37 PM, Östliche Rheinbrückenstr. 50, D-76181 Karlsruhe, Germany.] Dr-Ing. (TU Clausthal, 1987) metal physics. *Texture X-ray_diffraction*
Tel. 49(721)5956572 Fax 49(721)5954506

Neder, Dr Reinhard B. (1959). Postdoctoral Research Assistant. Institut für Kristallographie und Mineralogie der Universität München, Germany. [Institut für Kristallographie und Mineralogie der Universität München, Theresienstr. 41, D-80333 München, Germany.] Dr rer.nat. (U. München, 1989) mineralogy. *Diffuse_scattering structural_disorder neutron_diffraction structural_simulation*
E-mail neder@yoda.kri.physik.uni-muenchen.de Tel. 49(89)23944314 Fax 49(89)23944134

Neff, Prof. Dr Hans (1920). Retired. [Grünbergerstr. 17A, D-76139 Karlsruhe, Germany.] Dr rer.nat. (TH Karlsruhe, 1951) physics. *X-ray_powder_diffraction X-ray_fluorescence_spectroscopy history*
Tel. 49(721)681970

Neubert, Mr Michael (1960). Assistant. Institut für Kristallzüchtung, Rudower Chaussee 6 / Geb. 18.46, 12489 Berlin, Germany. [Grunowstraße 21, 13187 Berlin, Germany.] Dipl. Kristallogr. (Humboldt-U. Berlin, 1987) crystallography. *Atomic_diffusion compound_semiconductors annealing growth computing imperfect_crystals defect_analysis II-VI_compounds*
E-mail ur@ikz.FTA-berlin.de, subject: ATTN: IKZ, M. Neubert Tel. 49(30)63923092 Fax 49(30)63923003

Neumann, Dr Wolfgang (1944). Max-Planck-Institut für Mikrostrukturphysik, Germany. [Max-Planck-Institut für Mikrostrukturphysik, Weinberg 2, D-06120 Halle/S., Germany.] Dr rer.nat.habil. (U. Halle/S., 1993) applied physics. *High_resolution electron_microscopy electron_diffraction image_processing nanoanalysis quasicrystals disordered_structure*
Tel. 49(345)601512 Fax 49(345)27155

Nickel, Prof. Dr Klaus G. (1953). Professor. Mineralogisches Institut der Universität Tübingen, Germany. [Eberhard-Karls-Universität Tübingen, Mineralogisches Institut, Wilhelmstr. 56, D-72074 Tübingen, Germany.] PhD (U. of Tasmania, 1983) geosciences. *Ceramics corrosion nitrides carbides*
E-mail amprof@amprof.minpetgeochem.geowissenschaften.uni-tuebingen.de Tel. 49(7071)296802 Fax 49(7071)293060

Nieger, Dr Martin (1959). Postdoctoral Research Assistant. Institut für Anorganische Chemie der Universität Bonn, Germany. [Institut für Anorganische Chemie der Universität Bonn, Gerhard-Domagk-Str. 1, D-53121 Bonn, Germany.] Dr (U. Bonn, 1989) inorganic chemistry. *Phosphorus_compounds silicon_compounds organic_carbon_compounds organometallic_compound elementorganic_compounds low_coordinated_phosphorus_and_silicon_compounds supramolecular_compounds host-guest_compounds*

Tel. 49(228)733307 Fax 49(228)735327

Nippus, Dr Michael (1945). Manager. HUBER Diffraktionstechnik GmbH, Germany. [HUBER Diffraktionstechnik GmbH, Sommerstr. 4, D-83253 Rimsting/Chiemsee, Germany.] Dr rer.nat. (U. München, 1977) physics. *Diffractometry computer_technology texture residual_stress Raman_spectroscopy high-power_laser*
Tel. 49(8051)4472 Fax 49(8051)61680

Nitsche, Mr Robert (1964). Research Associate. Fachbereich Materialwissenschaft, Fachgruppe Dünne Schichten der Technischen Hochschule Darmstadt, Germany. [Fachbereich Materialwissenschaft, Fachgruppe Dünne Schichten der Technischen Hochschule Darmstadt, Hilpertstr. 31, D-64295 Darmstadt, Germany.] Mag.rer.nat. (U. Salzburg/Austria, 1991) mineralogy. *Nanoanalysis EXAFS Rietveld_method diffusion materials_science*
Tel. 49(6151)813266 Fax 49(6151)813270

Nitschmann, Dr Günter Max Alfred (1914). [Am Entenspiel 1, D-35578 Wetzlar, Germany.] Dr rer.nat. (U. Breslau, 1940) mineralogy/crystallography. *Crystal_growth germination crystallization real_crystal*
Tel. 49(6441)23748

Nowack, Mrs Ellen Carla (1956). Post-Graduate Student. Institut für Kristallographie der RWTH Aachen, D-52056 Aachen, Germany. [Institut für Kristallographie der RWTH Aachen, Jägerstr. 17-19, D-52066 Aachen, Germany.] Dipl. Min. (RWTH Aachen, 1983) mineralogy. *Charge_density dichalcogenides crystal_chemistry computing_in_crystallography*
E-mail fg050no@dacth11.bitnet (EARN) fg050no@vm1.rz.rwth-aachen.de (INTERNET) Tel. 49(241)806900 Fax 49(241)8888184 Telex 0832704thacd

Oesten, Dr Rüdiger (1961). Scientist. Zentrum für Sonnenenergie- und Wasserstofforschung, GB3: Energiespeicher und Energiewandlung, Germany. [Zentrum für Sonnenenergie- und Wasserstofforschung, GB3: Energiespeicher und Energiewandlung, Helmholtzstr. 8, D-89091 Ulm/Donau, Germany.] Dr rer.nat. (U. Mainz, 1992) mineralogy. *Crystallography diffraction electrochemistry energy_conversion nickel_compounds perovskite hydroxides*
Tel. 49(731)9530611 Fax 49(731)9530666

Okrusch, Prof. Dr Martin (1934). Head of Institute. Mineralogisches Institut der Universität Würzburg, Germany. [Mineralogisches Institut der Universität Würzburg, Am Hubland, D-97074 Würzburg, Germany.] Dr rer.nat.habil. (U. Würzburg, 1968) mineralogy. *Mineralogy petrology geochemistry*
Tel. 49(931)8885420 Fax 49(931)8884620

R. Oldenbourg Verlag, Zeitschrift für Kristallographie, Rosenheimer Str. 145, D-81671 München, Germany. [R. Oldenbourg Verlag, Zeitschrift für Kristallographie, Postfach 801360, D-81613 München, Germany.] *Publishing*
E-mail zkrist@fiz-karlsruhe.de Tel. 49(89)45051324 Fax 49(89)45051204

Otten, Mr Peter (1938). Deputed. Universität Bremen, Germany. [Hüttenstr. 25, D-24790 Schacht-Audorf, Germany.] Dipl. Min. (U. Bonn, 1967) crystallography. *Diffraction_technique zinc_compounds phase_transition*
Tel. 49(4331)951207 Fax 49(4331)951205 Telex 29431

Otto, Prof. Dr Hans Hermann (1938). Professor. Institut für Mineralogie und Mineralische Rohstoffe, Fachgebiet Materialwissenschaftliche Kristallographie der Technischen Universität Clausthal, Germany. [Institut für Mineralogie und Mineralische Rohstoffe, Fachgebiet Materialwissenschaftliche Kristallographie der Technischen Universität Clausthal, Adolph-Roemer-Str. 2A, D-38670 Clausthal-Zellerfeld, Germany.] Dr rer.nat.habil. (TU Berlin, 1983) crystallography and mineralogy. *Ferroic synthesis structure_determination superconductors infrared_detectors imaging_plate_system synchrotron_radiation*
Tel. 49(5323)723860 Fax 49(5323)722779

Pabst, Dr Ingeborg (1956). Research Assistant. Fachbereich Materialwissenschaften, Strukturforschung der Technischen Hochschule Darmstadt, Germany. [Fachbereich Materialwissenschaften, Strukturforschung der Technischen Hochschule Darmstadt, Petersenstr. 20, D-64287 Darmstadt, Germany.] Dr phil.nat. (U. Frankfurt/M., 1990) geosciences. *Phase_transition spectroscopy single_crystal*
E-mail de0u@mvs.hrz.th-darmstadt.de Tel. 49(6151)164363 Fax 49(6151)165551

Parak, Prof. Dr Fritz (1940). Full Professor. Fakultät für Physik, E17, Germany. [Fakultät für Physik, E17, James-Franck-Straße, 85748 Garching, Germany.] Dr rer.nat.habil. (TU München, 1979) experimental physics. *Biophysics proteins_structure_determination proteins_dynamics Mossbauer_spectroscopy*
Tel. 49(89)32092551 Fax 49(89)32092548

Paufler, Prof. Dr Peter (1940). Full Professor. Consultant, Commission on Crystallographic Teaching; Chairman of the National Committee. Institut für Kristallographie und Festkörperphysik der Technischen Universität Dresden, Germany. [Institut für Kristallographie und Festkörperphysik der Technischen Universität Dresden, Mommsenstr. 13, D-01069 Dresden, Germany.] Dr rer.nat.habil. (TU Dresden, 1971) physics. *Diffraction defect alloys crystal_chemistry data_bases*
E-mail paufler@pbtrs1.phy.tu-dresden.de Tel. 49(351)4634670 Fax 49(351)4637048 Telex 2279 tuni d

Paulitsch, Dr Peter (1922). Emeritus. Institut für Mineralogie der Technischen Hochschule Darmstadt, Germany. [Institut für Mineralogie der Technischen Hochschule Darmstadt, Landskrone 79, D-64285 Darmstadt, Germany.] Dr rer.nat. (U. Graz, 1944) mineralogy. *Twinning deformation orientation texture*

Paulmann, Mr Carsten (1963). Research Associate. Institut für Mineralogie der Universität Hannover, Germany. [Institut für Mineralogie der Universität Hannover, Welfengarten 1, D-30167 Hannover, Germany.] Dipl. Min. (U. Hannover, 1991) mineralogy. *High_resolution_electron_microscopy diffuse_scattering real_structure of minerals modulated_structures image_reconstruction computer_modelling_and_simulation_of_real_structures multislice_method*
Tel. 49(511)7622222 Fax 49(511)7623045

Paulus, Dr Erich Friedrich (1937). Head of Laboratory. Allg. Pharma Forschung, G864, HOECHST AG, D-65926 Frankfurt/M., Germany. [Allg. Pharma Forschung, G864, HOECHST AG, D-65926 Frankfurt/M., Germany.] Dr (U. München, 1965) chemistry. *Single_crystal powder pharmaceutical inorganic_chemistry Rietveld_method*
E-mail paulus@aph.hoechst-ag.dbp.de Tel. 49(69)3056360 Fax 49(69)307941

Pavlov, Konstantin (1966). Guest scientist. Fraunhofer-Institut für Angewandte Festkörperphysik, Germany. [Fraunhofer-Institut für Angewandte Festkörperphysik, Tullastr. 72, D-79108 Freiburg, Germany.] Diploma (Syktyvkar State U., Russia, 1990) physics. *X-ray_diffraction semiconductors computer_modelling*
E-mail pavlow@iaf.fhg.de Tel. 49(761)5159326 Fax 49(761)5159400

Pennartz, Dr Paul Ulrich (1961). Product Manager. ENRAF-NONIUS, Germany. [ENRAF-NONIUS, Postfach 101023, D-42648 Solingen, Germany.] Dr (TH Darmstadt, 1991) materials science. *Powder time-resolved_effect X-ray_optics*
Tel. 49(212)58750 Fax 49(212)587549

Pense, Prof. Dr Jürgen K. E. (1931). Professor. Institut für Geowissenschaften der Universität Mainz, D-55099 Mainz, Germany. Dr (U. Mainz, 1965) mineralogy. *Spectrography optics diffractometry gemology electron_microscopy polarization_microscopy structure_determination electron_probe_microanalysis charge_transfer colour_center crystal_growth*
Tel. 49(6131)381756 Fax 49(6131)393070

von Philipsborn, Prof. Dr Henning (1934). Professor. Universität Regensburg, D-93040 Regensburg, Germany. Dr phil. (U. Zürich/Switzerland, 1964) crystallography. *Crystallography symmetry radioactivity semiconductors electronic_materials*
Tel. 49(941)9432481 Fax 49(941)9433316

Pickardt, Prof. Dr.-Ing. Joachim (1939). Associate Professor. Institut für Anorganische und Analytische Chemie der Technischen Universität Berlin, Germany. [Institut für Anorganische und Analytische Chemie der Technischen Universität Berlin, Straße des 17. Juni 135, D-10623 Berlin, Germany.] Dr (TU Berlin, 1971) chemistry. *Inorganic chemistry transition_elements macrocycles supramolecular_chemistry*
E-mail pickardt@wap0201.chem.tu-berlin.de Tel. 49(30)31422469

Pierer, Dr Gerhard (1934). Research Assistant. Institut für Kristallographie und Mineralogie der Universität Frankfurt, Germany. [Institut für Kristallographie und Mineralogie der Universität Frankfurt, Senckenberganlage 30, D-60054 Frankfurt/M., Germany.] Dr phil.nat. (U. Frankfurt/M., 1967) mineralogy. *Powder Rietveld_method*
Tel. 49(69)7982104 Fax 49(69)79821

Pietsch, Mr Hans-Henning (1965). Postgraduate Student. Institut für Mineralogie der Universität Münster, Germany. [Institut für Mineralogie der Universität Münster, Corrensstr. 24, D-48749 Münster, Germany.] Dipl. Chem. (U. Münster, 1993) chemistry. *Zeolites hydrothermal_synthesis spectroscopy thermoanalysis sodalites cation-exchange catalysis*
Tel. 49(251)833499

Pietsch, Dr Ullrich (1952). Professor. Structuranalyse. Institut für Festkörperphysik, Universität Potsdam, Am Neuen Palais 10, Germany. [Universität Potsdam, Postfach 601553, D-14415 Potsdam, Germany.] Dr habil. (U. Leipzig, 1988) experimental physics. *Synchrotron_radiation valence_charge_density diffractometry multilayer Langmuir-Blodgett_film III-V_compounds reflectometry*
E-mail upietsch@hp.rz.uni-potsdam.de Tel. 49(331)9771286 Fax 49(331)9771083

Pietzuch, Mr Walter (1963). Research Associate. Fachbereich Chemie, Arbeitskreis Prof. Reinen der Universität Marburg, Germany. [Fachbereich Chemie, Arbeitskreis Prof. Reinen der Universität Marburg, Hans-Meerwein-Straße, D-35032 Marburg, Germany.] Dipl. Chem. (U. Marburg, 1989) chemistry. *Crystallography disorder band_calculation bismuth_compounds colour_due_to_clustered_lattice_defects*
E-mail pietz_wa@ps1515.chemie.uni-marburg.de Tel. 49(6421)285668 Fax 49(6421)285547

Pilz, Mr Edgar (1962). Research Assistant. Fachrichtung Kristallographie der Universität des Saarlandes, Germany. [Fachrichtung Kristallographie der Universität des Saarlandes, c/o HASYLAB/DESY, Postfach, D-22603 Hamburg, Germany.] Dipl. Kristallogr. (U. Leipzig, 1988) crystallography. *Energy-dispersive_analysis white-beam_radiation anomalous_dispersion Laue_diffraction*
E-mail pilz@vxdesy.desy.de Tel. 49(40)89983176 Fax 49(40)89982787

Pilz, Mrs Katrin (1964). Research Assistant. Fachrichtung Kristallographie der Universität des Saarlandes, Germany. [Fachrichtung Kristallographie der Universität des Saarlandes, c/o HASYLAB/DESY, Postfach, D-22603 Hamburg, Germany.] Dipl. Kristallogr. (U. Leipzig, 1989) crystallography. *Anomalous_dispersion pseudosymmetry direct_method*
E-mail pilz@vxdesy.desy.de Tel. 49(40)89982601 Fax 49(40)89982787

Platzbecker, Mr Rolf (1957). [Erich-Böger-Str. 29, D-53127 Bonn, Germany.] Dipl. Min. (U. Bonn, 1990) mineralogy. *Structural_disorder structure_determination*

Plesken, Prof. Dr Wilhelm (1950). Professor. Lehrstuhl B für Mathematik der RWTH Aachen, Germany. [Lehrstuhl B für Mathematik der RWTH Aachen, Templergraben 64, D-52056 Aachen, Germany.] Dr rer.nat. (RWTH Aachen, 1974) mathematics. *Superlattice superstructure modulated_structure mathematics group_theory algorithmic_algebra*
E-mail plesken@willi.math.rwth-aachen.de Tel. 49(241)804535

Pöllmann, Dr Herbert (1956). Lecturer. Mineralogisches Institut der Universität Erlangen-Nürnberg, Germany. [Mineralogisches Institut der Universität Erlangen-Nürnberg, Schloßgarten 5, D-91054 Erlangen, Germany.] Dr rer.nat.habil. (U. Erlangen-Nürnberg, 1990) mineralogy. *Mineralogy crystallization ceramics environment*
Tel. 49(9131)853986 Fax 49(9131)852131

Pöttgen, Dr Rainer (1966). Anorganisch Chemisches Institut der Universität Münster, Germany. [Anorganisch Chemisches Institut der Universität Münster, Wilhelm-Klemm-Str. 8, D-48149 Münster, Germany.] Dr rer.nat. (U. Münster, 1993) chemistry. *Structure_determination intermetallic_compound magnetism*
E-mail pottgen@vnwz00.uni-muenster.de Tel. 49(251)833121 Fax 49(251)833169

Pohl, Prof. Dr Dieter (1940). Professor. Mineralogisch-Petrographisches Institut der Universität Hamburg, Germany. [Mineralogisch-Petrographisches Institut der Universität Hamburg, Grindelallee 48, D-20146 Hamburg, Germany.] Dr (U. Hamburg, 1969) physics. *Optical_property phase_transition structure minerals*
Tel. 49(40)41232482 Fax 49(40)41232422

Pohl, Mr Ehmke (1967). Post-Graduate Student. Institut für Anorganische Chemie der Universität Göttingen, Germany. [Institut für Anorganische Chemie der Universität Göttingen, Tammannstr. 4, D-37077 Göttingen, Germany.] Dipl. Chem. (U. Göttingen, 1991) chemistry. *Structure_determination proteins immunoglobulins synchrotron_radiation*
E-mail epohl@shelx.uni-ac.gwdg.de Tel. 49(551)393075 Fax 49(551)392582

Pohl, Prof. Dr Siegfried (1943). Full Professor. Fachbereich Chemie der Universität Oldenburg, Germany. [Fachbereich Chemie der Universität Oldenburg, Postfach 2503, D-26111 Oldenburg, Germany.] Dr habil. (U. Bielefeld, 1978) chemistry. *Iron_compounds synthesis nitrogenases*
Tel. 49(441)7983656 Fax 49(441)7983329

Polborn, Dr Kurt Volkmar (1946). Research Associate. Institut für Organische Chemie der Universität München, Germany. [Institut für Organische Chemie der Universität München, Karlstr. 23, D-80333 München, Germany.] Dr rer.nat. (U. München, 1975) chemistry. *Structure_determination molecular_crystal*
Tel. 49(89)5902423 Fax 49(89)5902483

Poll, Dr Wolfgang (1954). Lecturer. Institut für Anorganische Chemie und Strukturchemie, Heinrich-Heine-Universität Düsseldorf, Germany. [Institut für Anorganische Chemie und Strukturchemie, Heinrich-Heine-Universität Düsseldorf, Universitätsstr. 1, D-40225 Düsseldorf, Germany.] Dr rer.nat. (U. Düsseldorf, 1983) chemistry. *Computing crystal_growth hydrogen_bonding low_temperature X-ray_instrumentation thermoanalysis fluorine_compounds small_molecules*
E-mail poll@clio.rz.uni-duesseldorf.de Tel. 49(211)3113146 Fax 49(211)3113085 Telex 08587348unid

Postma, Dr Johannes Petrus Maria (1951). Staff Scientist. EMBL, Germany. [EMBL, Postfach 102209, D-69012 Heidelberg, Germany.] PhD (U Groningen/Netherlands, 1985) physical chemistry. *Biocrystallography database computer_automation computer_graphics image_reconstruction artificial_intelligence*
E-mail postma@embl-heidelberg.de Tel. 49(6221)387469 Fax 49(6221)387306

Prandl, Prof. Dr Wolfram (1935). Professor and Director. Member of Neutron Diffraction Commission (1984–1993). Institut für Kristallographie, Fakultät für Physik der Universität Tübingen, Germany. [Institut für Kristallographie, Fakultät für Physik der Universität Tübingen, Charlottenstr. 33, D-72070 Tübingen, Germany.] Dr habil. (U. München, 1973) crystallography/mineralogy. *Disorder magnetism statistical_thermodynamics molecular_crystal spin_glasses molecular_dynamics neutron_scattering*
Tel. 49(7071)296058 Fax 49(7071)296058

Preuß, Dr Heinz Hans Walter (1934). Retired. Technische Hochschule Zittau, Germany. [Technische Hochschule Zittau, Schillerstr. 38, D-02763 Zittau, Germany.] Dr rer.nat.habil. (U. Leipzig, 1977) crystal physics. *Electron_microscopy goniometry plasticity surface fatigue statistical_method polycrystal*
Tel. 49(3583)703383

Preut, Dr Hans (1940). Research Associate. Fachbereich Chemie, Universität Dortmund, Germany. [Fachbereich Chemie, Universität Dortmund, Otto-Hahn-Str. 6, D-44227 Dortmund, Germany.] Dr (U. Dortmund, 1972) chemistry. *Data_collection structure_determination service education lattice_energy_calculation history_of_X-ray_structure_analysis*
E-mail uch002@ux9.hrz.uni-dortmund.de Tel. 49(231)7553813 Fax 49(231)7553771

Queisser, Prof. Dr Hans Joachim (1931). Director. Max-Planck-Institut für Festkörperforschung, Germany. [Max-Planck-Institut für Festkörperforschung, Postfach 70506, D-70569 Stuttgart, Germany.] Dr rer.nat. (U. Göttingen, 1958) physics. *Epitaxy crystal_growth dislocation semiconductor*
E-mail queisser@quasix.mpi-stuttgart.mpg.de Tel. 49(711)6891600 Fax 49(711)6891602 Telex 7255555

Raabe, Dr Gerhard P. (1950). Research Associate. Institut für Organische Chemie der RWTH Aachen, Germany. [Institut für Organische Chemie der RWTH Aachen, Prof.-Pirlet-Str. 1, D-52056 Aachen, Germany.] Dr rer.nat. (RWTH Aachen, 1984) chemistry. *Molecular_crystal quantum_chemistry MO_calculation lattice_energy*
E-mail gk010ra@vm1.rz.rwth-aachen.de gk010ra@dacth11.bitnet Tel. 49(241)804709 Fax 49(241)8888127 Telex 48(241)832704 thac d

Ramm, Dr Matthias (1955). Research Assistant. Zentrum für Selektive Organische Synthese, Germany. [Zentrum für Selektive Organische Synthese, Geb. 9.9, D-12484 Berlin, Germany.] Dr rer.nat. (FU. Berlin, 1992) crystallography. *X-ray_structure_determination absolute_configuration crystallography MO_calculation*
Tel. 49(30)63924179 Fax 49(30)63924103

Rehfeldt, Dr Angeline (1959). Siemens AG, Germany. [Siemens AG, Siemensallee 84, D-76187 Karlsruhe, Germany.] Dr (U. Münster, 1986) mineralogy. *Computer_graphics diffractometry electron_microscopy Rietveld_method*
Tel. 49(721)5954503 Fax 49(721)5954506 Telex 72139362

Reimers, Dr Walter (1954). Research Associate. Hahn-Meitner-Institut GmbH, Germany. [Hahn-Meitner-Institut GmbH, Glienicker Str. 100, D-14109 Berlin, Germany.] Dr habil. (U. Dortmund, 1989) mechanical engineering. *Residual_stress microstrain strain_determination non-destructive_analysis*
Tel. 49(30)80092773 Fax 49(30)80093059

Reinemer, Dr Peter (1963). Postdoctoral Research Assistant. Max-Planck-Institut für Biochemie, Abt. Strukturforschung, Germany. [Max-Planck-Institut für Biochemie, Abt. Strukturforschung, Postfach, D-82143 Martinsried, Germany.] Dr rer.nat. (TU München, 1993) chemistry. *Biochemistry biocrystallography proteins nucleic_acids*
Tel. 49(89)85782703 Fax 49(89)85783516

Reinen, Prof. Dr Dirk (1930). Professor. Fachbereich Chemie der Universität Marburg, Hans-Meerwein-Str., Germany. [Fachbereich Chemie der Universität Marburg, Postfach 1929, D-35032 Marburg, Germany.] Dr (U. Bonn, 1960) inorganic chemistry. *Inorganic_solids spectroscopy colour phase_transition mixed_valence_chemistry*
Tel. 49(6421)285668 Fax 49(6421)285547 Telex 482372

Reinke, Dr Helmut (1946). Lecturer. Fachbereich Chemie der Universität Rostock, Dr.-Lorenz-Weg 1, D-18059 Rostock, Germany. [Fachbereich Chemie der Universität Rostock, Buchbinderstr. 9, D-18051 Rostock, Germany.] Dr rer.nat.habil. (Güstrow, 1990) chemistry. *X-ray_structure_determination polymorphism hydrogen_bonding infrared_spectroscopy DSC carbohydrates macrocycles*
E-mail helmut.reinke@chemie.uni-rostock.de (SMTP) c=de; a=d400; p=uni-rostock; ou=chemie; s=reinke; gi=helmut (X.400) Tel. 49(381)4923667

Reuber-Kürbs, Dr-Ing. Ellen (1923). [Quantzstr. 16, D-14129 Berlin, Germany.] Dr-Ing. (TU Berlin, 1953) physics. *Densitometry image_processing electron_microscopy kidney_stones_analysis*
E-mail reuber@kristall.chemie.fu-berlin.d400.de Tel. 49(30)8035265

Richter, Dr Rainer (1946). Lecturer. Fachbereich Chemie der Universität Leipzig, Germany. [Fachbereich Chemie der Universität Leipzig, Linnéstr. 3, D-04103 Leipzig, Germany.] Dr habil. (U. Leipzig, 1991) chemistry. *X-ray_structure_determination single_crystal chelates organic_sulfur_compounds education*
E-mail richter@serverl.rz.uni-leipzig.de Tel. 49(341)6858545 Fax 49(341)6858387

Richter, Dr Waltraut (1944). Consultant and Group Leader. Sächsische Landesanstalt für Landwirtschaft, Fachbereich Landwirtschaftliche Untersuchungen, Germany. [Sächsische Landesanstalt für Landwirtschaft, Fachbereich Landwirtschaftliche Untersuchungen, Gustav-Kühn-Str. 8, D-04159 Leipzig, Germany.] Dr rer.nat. (Sächsische Landesanstalt Landwirtschaft, Leipzig, 1975) soil science and polarization microscopy. *Soils sorption polarization_microscopy plants soil_physics phosphates*
Tel. 49(341)59390 Fax 49(341)5939211

Riedel, Mr Gernot (1962). Research Associate. Institut für Kristallographie der Universität Tübingen, Germany. [Institut für Kristallographie der Universität Tübingen, Charlotenstr. 33, D-72072 Tübingen, Germany.] Dipl. Min. (Tübingen, 1991) mineralogy. *Amorphization amorphous_phase spin crystallography garnet spessartine recrystallisation low_temperature computer*
E-mail pkiri01@mailserv.zdv.uni-tuebingen.de Tel. 49(7071)296389 Fax 49(7071)296058

Ries, Mr Ronald (1946). Unemployed. [Simon-Bolivar-Str. 8, D-13055 Berlin, Germany.] Dipl. Krist. (Humboldt-U. Berlin, 1971) crystallography. *Electron_microscopy cross_section optoelectronics III–V_compounds II–VI_compounds EDS REM heterostructures failure_analysis semiconductor*
Tel. 49(30)9765029

Robarick, Dr Eckhard (1946). Institut für Mineralogie und Kristallographie der Technischen Universität Berlin, Germany. [Institut für Mineralogie und Kristallographie der Technischen Universität Berlin, Mittenwalder Str. 30, D-12053 Berlin, Germany.] Dr rer.nat. (TU Berlin, 1988) mineralogy/petrology. *Order-disorder polymerization crystal_growth_of_silicates steric_relationships_and_structures_of_solid_phases*
Tel. 49(30)6925693

Rossmanith, Prof. Dr Elisabeth (1943). Professor. Mineralogisch-Petrographisches Institut der Universität Hamburg, Germany. [Mineralogisch-Petrographisches Institut der Universität Hamburg, Grindelallee 48, D-20146 Hamburg, Germany.] Dr phil. habil. (U. Hamburg, 1981) crystallography. *Multiple_diffraction line_profile_analysis forbidden_reflection*
E-mail mi2a000@miaixl.mineralogie.uni-hamburg.de Tel. 49(40)41232485 Fax 49(40)41232422

Roth, Dr Georg (1954). Staff Scientist. Institut für Nukleare Festkörperphysik, Kernforschungszentrum Karlsruhe, Germany. [Institut für Nukleare Festkörperphysik, Kernforschungszentrum Karlsruhe, Postfach 3640, D-76021 Karlsruhe, Germany.] Dr (U. Münster, 1985) mineralogy/crystallography. *X-ray_diffractometry superconductivity neutron_diffractometry superionic_conductor fullerenes*
Tel. 49(7247)823987 Fax 49(7247)824624 Telex 17724716+

Rothammel, Dr Walter (1960). [Unteres Weiler 7, D-74594 Kreßberg/Bergbronn, Germany.] Dr rer.nat. (U. Erlangen/Nürnberg, 1991) physics/crystallography. *Structure_determination Patterson_method symmetry computer_graphics structural_relations X-ray_detectors*
Tel. 49(7957)8103

Rothbauer, Dr Richard (1938). [Weidenstr. 11, D-65795 Hattersheim, Germany.] Dr phil.nat. (U. Frankfurt/M., 1971) crystallography. *Method_of_structure_determination direct_method diffractometry elastic_neutron_diffractometry least_squares_methods*
Tel. 49(6190)2643

Ruck, Dr Michael (1963). Postdoctoral Research Assistant. Institut für Anorganische Chemie der Universität Karlsruhe, Germany. [Institut für Anorganische Chemie der Universität Karlsruhe, Engesserstr., D-76128 Karlsruhe, Germany.] Dr (U. Stuttgart, 1992) chemistry. *Solid_state twinning structure_determination ternary_bismuth_compounds* Tel. 49(721)6082975

Rudert, Dr Rainer (1956). Research Assistant. Bundesanstalt für Materialforschung und -prüfung, Germany. [Bundesanstalt für Materialforschung und -prüfung, Rudower Chaussee 5, D12489 Berlin, Germany.] Dr rer.nat. (FU Berlin, 1990) physics. *Amphiphilic_molecules crystallography structure_determination charge_density* E-mail nwfrieden@gn.apc.org Tel. 49(30)63925858

Rudolph, Prof. Dr Peter (1945). Professor. Institut für Kristallzüchtung im Forschungsverbund Berlin e.V., Germany. [Institut für Kristallzüchtung im Forschungsverbund Berlin e.V., Rudower Chaussee 6, D-12489 Berlin, Germany.] Dr sc.nat. (habil) (Humboldt-U. Berlin, 1979) crystallography. *Crystal_growth point_defect real_structure II–VI,IV–VI,III–V_compounds stoichiometry transport_phenomena theory growth_kinetics distribution precipitation optoelectronics* E-mail u@ikz.wtza-berlin.de Tel. 49(30)63923034 Fax 49(30)63923003

Rypniewski, Dr Wojciech R. (1958). Staff Scientist. EMBL c/o DESY, Notkestrasse 85, D-22603 Hamburg, Germany. PhD (Cambridge U., 1987) protein crystallography. *Biochemistry proteins catalysis computing data_validation evolution* E-mail wojtek@embl-hamburg.de Tel. 49(40)89902142 Fax 49(40)89902149 Telex 215 124 (desy d)

Saalfeld, Prof. Dr Horst (1920). Emeritus. Mineralogisch-Petrographisches Institut der Universität, Grindelallee 48, D-20146 Hamburg, Germany. [Am Hochsitz 1, D-22850 Norderstedt, Germany.] Prof. (U. Saarbrücken, 1961) mineralogy. *Mineralogy crystallography disorder clays silicates oxides_alumina_spinels_ZrO2* Tel. 49(40)5252835

Sabrowsky, Prof. Dr Horst (1934). Lecturer. Anorganische Chemie I, Germany. [Universitätsstr. 150, D-44780 Bochum, Germany.] Dr chemistry. *Solid_phase DTA magnetism chalcogenides* Tel. 49(234)7004151

Sachsenröder, Mr Steffen (1952). EHU Thale AG, UB BAA, Abt. BATE, Germany. [EHU Thale AG, UB BAA, Abt. BATE, Steinbachstraße, D-06502 Thale, Germany.] Dipl. Kristallogr. crystallography. *Coating corrosion glasses enamel* Tel. 49(3947)72160 Fax 49(3947)2552 Telex 351834ehwd

Saenger, Prof. Dr-Ing. Wolfram H. E. (1939). Full Professor. Co-Editor of Acta Crystallographica. Institut für Kristallographie, Freie Universität Berlin, Germany. [Institut für Kristallographie, Freie Universität Berlin, Takustr. 6, D-14195 Berlin, Germany.] Prof. (TH Darmstadt, 1965) chemistry. *Biocrystallography crystallization hydrogen_bonding inclusion_complexes proteins nucleic_acids cyclodextrins* E-mail saenger@kristall.chemie.fu-berlin.de Tel. 49(30)8383412 Fax 49(30)8386702

Sahm, Prof. Dr-Ing. Peter R. (1934). Director. Gießerei-Institut der RWTH Aachen, Germany. [Gießerei-Institut der RWTH Aachen, Intzestr. 5, D-52072 Aachen, Germany.] Dr-Ing. (RWTH Aachen, 1961) foundry engineering. *Single_crystal alloys solidification computer_simulation_of_solidification single_crystal_superalloys oriented_solidification computer_simulation_of_casting and_solidification* Tel. 49(241)805880 Fax 49(241)407217

Schaefer, Mr Klaus (1921). Manager. KLAUS SCHAEFER, Gesellschaft für Verfahrenstechnik mbH, Germany. [KLAUS SCHAEFER, Gesellschaft für Verfahrenstechnik mbH, Postfach 1660, D-63206 Langen, Germany.] Dipl. Chem/Dipl. Ing (TH Berlin/Darmstadt, 1941/1964) chemistry/electrical engineering. *Instrumentation X-ray_diffraction crystal_growth furnace vacuum cryogenics magnetism* Tel. 49(6103)79085 Fax 49(6103)71799

Schäfer, Dr Wolfgang (1942). Research Associate; head of neutron diffraction group. Mineralogisches Institut der Universität Bonn, Forschungszentrum Jülich, Arbeitsgruppe MIN/ZFR, Germany. [Mineralogisches Institut der Universität Bonn, Forschungszentrum Jülich, Arbeitsgruppe MIN/ZFR, D-52425 Jülich, Germany.] Dr (U. Bonn, 1971) crystallography. *Crystal_and_magnetic_structure texture neutron_diffractometry neutron_detector time-of-flight_techniques spallation_source_applications* E-mail iff080@djukfa11.bitnet Tel. 49(2461)616024 Fax 49(2461)613841 Telex 833556 kfa d

Schaetzle, Dr Peter (1962). Post-Doc. Institut für Festkörper- und Werkstoffforschung e.V., Germany. [Institut für Festkörper- und Werkstoffforschung e.V., Winterbergstr. 28, D-01277 Dresden, Germany.] Dr rer.nat. (U. Freiburg/Br., 1992) mineralogy. *Crystallization measurement metallography semiconductor high-temperature_superconductor silicon high_temperature nitrides etching* E-mail schaetzle@ifw-dresden.de400.de Tel. 49(351)2322-225/272 Fax 49(351)2322-314

Schenk, Prof. Dr Manfred (1934). Professor. Institut für Physik/Kristallographie, Humboldt-Universität zu Berlin, Invalidenstr. 110, Germany. [Institut für Physik/Kristallographie, Humboldt-Universität zu Berlin, Unter den Linden 6, D-10099 Berlin, Germany.] Dr rer.nat.habil. (Humboldt-U. Berlin, 1970) crystallography. *Crystallography real_structure semiconductor material_sciences development_of_mankind* Tel. 49(30)2803361 Fax 49(30)2803360 Telex 304156huphy

Schetelich, Mr Christoph (1964). Research Assistant. Institut für Kristallographie, Mineralogie und Materialwissenschaft der Universität Leipzig, Germany. [Institut für Kristallographie, Mineralogie und Materialwissenschaft der Universität Leipzig, Linnéstr. 3-5, D-04103 Leipzig, Germany.] Dipl. Kristallogr. (U. Leipzig, 1991) crystallography. *Kossel_diffraction quasicrystal scanning_electron_microscopy dynamical_diffraction dislocation´computer_modelling TEM* E-mail scheteli@rz.uni-leipzig.de Tel. 49(341)6858383

Schilling, Dr Frank Rüdiger (1963). Postdoctoral Research Assistant. Institut für Mineralogie der Freien Universität Berlin, Germany. [Institut für Mineralogie der Freien Universität Berlin, Takustr. 6, D-14195 Berlin, Germany.] Dr rer.nat. (U. Tübingen, 1991) mineralogy. *Thermal_property X-ray_fluorescence_spectroscopy high_pressure physical_property petrophysics* Tel. 49(30)8383434 Fax 49(30)8383469

Schinzer, Mr Carsten (1965). Doctoral Fellow. Institut für Anorganische Chemie der Universität Tübingen, Germany. [Institut für Anorganische Chemie der Universität Tübingen, Auf der Morgenstelle 18, D-72076 Tübingen, Germany.] Dipl. Chem. (U. Tübingen, 1993) inorganic chemistry. *Catalysis condensed_matter magnetism spectroscopy neutron_diffraction_(elastic_&_inelastic) crystallography refinement_methods* E-mail schinzer@student.uni-tuebingen.de Tel. 49(7071)296231 Fax 49(7071)296918

Schloemer, Prof. Dr Hermann Johannes (1923). Emeritus. Department für Physik, Technische Mineralogie der Universität Saarbrücken, Germany. [Department für Physik, Technische Mineralogie der Universität Saarbrücken, Postfach 1150, D-66041 Saarbrücken, Germany.] Prof. (U. Saarbrücken, 1968) mineralogy. *Porous_ceramics crystal_growth hydrothermal_synthesis epitaxy membrane_technology macro/micro_separation nano_crystallinity* Tel. 49(681)3022912 Fax 49(681)3022272 Telex 17-7533=unis

Schlünzen, Mr Frank (1963). Doctoral Fellow. Max-Planck-Gesellschaft, AG Ribosomenstruktur, Germany. [Max-Planck-Gesellschaft, AG Ribosomenstruktur, c/o DESY, Notkestr. 85, D-22603 Hamburg, Germany.] *Macromolecular_crystallography computing_in_crystallography ribosomes anomalous_dispersion contrast maximum_entropy_method* E-mail schluenzen@mpgars.desy.de Tel. 49(40)89982834 Fax 49(40)891314

Schmahl, Dr Wolfgang W. (1958). Research Assistant. Fachgebiet Strukturforschung der Technischen Hochschule Darmstadt, Germany. [Fachgebiet Strukturforschung der Technischen Hochschule Darmstadt, Petersenstr. 20, D-64287 Darmstadt, Germany.] Dr (U. Kiel, 1986) mineralogy. *Phase_transition thermodynamics Rietveld_method crystallography* E-mail schmahl@eddy.materie.th-darmstadt.de Tel. 49(6151)165464 Fax 49(6151)165551

Schmidbauer, Dr Martin (1962). Max-Planck-Arbeitsgruppe "Röntgenbeugung", Germany. [Max-Planck-Arbeitsgruppe "Röntgenbeugung", Hausvogteiplatz 5-7, D-10117 Berlin, Germany.] Dr rer.nat. (TU Berlin, 1992) physics. *Grazing_incidence diffraction semiconductor_superlattice surface_sciences synchrotron_radiation photoionization_of_free_molecules* E-mail martin%x-ray0.roen.ipp-garching.mpg.de@sat.ipp-garching.mpg.de (international) Tel. 49(30)20366219 Fax 49(30)2004536

Schmidt, Prof. Dr Günther (1921). Emeritus. Fachbereich Physik der Universität Halle-Wittenberg, Germany. [Fachbereich Physik der Universität Halle-Wittenberg, Friedemann-Bach-Platz 6, D-06108 Halle/S., Germany.] Dr rer.nat.habil. (U. Halle/Wittenberg, 1960) experimental physics. *Phase_transition ferroelectricity electromechanics perovskites* Tel. 49(345)832580 Fax 49(345)29515 Telex 318282unihad

Schmitz, Dr Werner (1943). Lecturer. Institut für Kristallographie, Mineralogie und Materialwissenschaft der Universität Leipzig, Germany. [Institut für Kristallographie, Mineralogie und Materialwissenschaft der Universität Leipzig, Scharnhorststr. 20, D-04275 Leipzig, Germany.] Dr rer.nat.habil. (U. Leipzig, 1986) mineralogy. *X-ray_diffraction powder identification zeolites quantitative_X-ray_analysis* Tel. 49(341)310502 Fax 49(341)310502

Schneider, Prof. Dr Jochen R. (1941). Director of HASYLAB. HASYLAB at DESY, Germany. [HASYLAB at DESY, Notkestr. 85, D-22603 Hamburg, Germany.] Dr rer.nat. (U. Hamburg, 1973) physics. *Charge_density phase_transition scattering* Tel. 49(40)89983815 Fax 49(40)89983815

Schneider, Dr Julius (1942). Research Associate. Institut für Kristallographie und Mineralogie der Universität München, Germany. [Institut für Kristallographie und Mineralogie der Universität München, Theresienstr. 41, D-80333 München, Germany.] Dr rer.nat. (TU München, 1975) physics. *Powder Rietveld_method high_temperature* E-mail schneider@gekrmu.neutronenbeugung.geologie.uni-muenchen.de Tel. 49(89)23944354 Fax 49(89)23944334

Schneider, Mr Thomas R. (1966). Doctoral Fellow. European Molecular Biology Laboratory, Außenstelle Hamburg, Germany. [EMBL c/o DESY, Notkestr. 85, D-22603 Hamburg, Germany.] Dipl. Phys. (TU München, 1991) physics. *Macromolecular_crystallography proteins dynamics hydration* E-mail thomas@embl-hamburg.de Tel. 49(40)89902116 Fax 49(40)89902149

Schneider, Dr Walter (1936). Group Leader. Kernforschungszentrum Karlsruhe, IMF 1, Germany. [Kernforschungszentrum Karlsruhe, IMF 1, Postfach 3640, D-76021 Karlsruhe, Germany.] Dr rer.nat. (U. Göttingen, 1960) mineralogy. *Electron_microscopy high-resolution_electron_microscopy X-ray_diffraction metals_physics* Tel. 49(7247)824093 Fax 49(7247)824567

von Schnering, Prof. Dr Dr h. c. Hans Georg (1931). Director. Max-Planck-Institut für Festkörperforschung, Germany. [Max-Planck-Institut für Festkörperforschung, Heisenbergstr. 1, D-70569 Stuttgart, Germany.] Dr rer.nat.habil. (U. Münster, 1964) inorganic chemistry. *Solid_state chemistry structural_chemistry clusters_compound structural_systematics structural_relations* Tel. 49(711)6891560 Fax 49(711)6891562

Schnick, Prof. Dr Wolfgang (1957). Professor. Laboratorium für Anorganische Chemie der Universität Bayreuth, Germany. [Laboratorium für Anorganische Chemie der Universität Bayreuth, NW1, Universitätsstr. 30, D-95440 Bayreuth, Germany.] Dr rer.nat. (U. Hannover, 1986) chemistry. *Solid_state chemistry ionic_conductivity materials* E-mail wolfgang.schnick@uni-bayreuth.de Tel. 49(921)552530 Fax 49(921)552788

Schnieders, Mr Felix (1961). Doctoral Fellow. Institut für Anorganische Chemie und Strukturchemie, Lehrstuhl II, Geb. 26.42 der Universität Düsseldorf, Germany. [Institut für Anorganische Chemie und Strukturchemie, Lehrstuhl II, Geb. 26.42 der Universität Düsseldorf, Universitätsstr. 1, D-40225 Düsseldorf, Germany.] Dipl. Chem. (U. Düsseldorf, 1991) chemistry. *Chalcogenides silver_compounds ionic_conductors crystal_growth*
Tel. 49(211)3114257

Schöllhorn, Prof. Dr Robert (1935). Professor. Institut für Anorganische und Analytische Chemie der Technischen Universität Berlin, Germany. [Institut für Anorganische und Analytische Chemie der Technischen Universität Berlin, Straße des 17. Juni 135, D-10623 Berlin, Germany.] Dr (U. Heidelberg, 1963) inorganic chemistry. *Reactivity_of_solids inclusion_compound*
E-mail schoe@wap0209.chem.tu-berlin.de Tel. 49(30)31422740 Fax 49(30)31421106

Schollmeyer, Dr Dieter (1961). Research Assistant. Institut für Organische Chemie der Universität Mainz, Germany. [Institut für Organische Chemie der Universität Mainz, J.-J.-Becherweg 18-20, D-55099 Mainz, Germany.] Dr rer.nat. (U. Halle/S., 1989) chemistry. *Absolute_structure molecular_crystal*
E-mail scholli@uacdr0.chemie.uni-mainz.de Tel. 49(6131)395320 Fax 49(6131)394778

Schramm, Dr Volker (1939). Lecturer. FR 10.4 - Kristallographie der Universität Saarbrücken, D-66041 Saarbrücken, Germany. Dr (U. Saarbrücken, 1972) mineralogy. *Computer-aided_education mineralogy structure_determination symmetry*
Tel. 49(681)3023470 Fax 49(681)3024439

Schreuer, Mr Jürgen (1963). Research Assistant. Institut für Kristallographie der Universität zu Köln, Germany. [Institut für Kristallographie der Universität zu Köln, Zülpicher Str. 49b, D-50674 Köln, Germany.] Dipl. (U. Köln, 1989) mineralogy. *Non-linear_physical_property thermal_expansion structure_determination phase_transition quasicrystals*
E-mail schreuer@kri.uni-koeln.de Tel. 49(221)4703194 Fax 49(221)4704963

Schröder, Dr-Ing. Jens (1960). Postdoctoral Research Assistant. GKSS Forschungszentrum GmbH, Germany. [GKSS Forschungszentrum GmbH, Postfach 1160, D-21496 Geesthacht, Germany.] Dr-Ing. (TU Hamburg, 1993) materials science. *Neutron high-resolution diffractometry residual_stress analysis texture_analysis material_sciences*
E-mail jens.schroeder@dvnetz.gkss.dbp.de Tel. 49(4152)871249 Fax 49(4152)871338

Schröder, Dr Winfried (1937). Director. Chairman, German Crystal Growth Association. Institut für Kristallzüchtung, Germany. [Institut für Kristallzüchtung, Rudower Chaussee 6, D-12484 Berlin, Germany.] Dr habil. (TU/BA Freiberg/Sax., 1993) crystallography. *Crystal_growth diagnostic hiking climbing*
E-mail ur@ikz.wtza-berlin.de Tel. 49(30)63923000 Fax 49(30)63923003

Schubert, Prof. Dr Ulrich (1946). Professor of Inorganic Chemistry. Institut für Anorganische Chemie der Technischen Universität Wien, Austria. [Institut für Anorganische Chemie der Technischen Universität Wien, Getreidemarkt 9, A-1060 Wien, Austria.] Dr (TU München, 1975) chemistry. *Chemistry molecular_crystal main-group_compounds materials*
Tel. 43(1)5880014633 Fax 43(1)5874835

Schüller, Prof. Dr Karl-Heinz (1928). Professor. Fachbereich Werkstofftechnik der Fachhochschule Nürnberg, Germany. [Fachbereich Werkstofftechnik der Fachhochschule Nürnberg, Wassertorstr. 8, D-90489 Nürnberg, Germany.] Prof. (U. Erlangen-Nürnberg, 1972) technical mineralogy. *Applied mineralogy microstructure_of_ceramics clay mineralogy thermoanalysis*
Tel. 49(911)5880172 Fax 49(911)5880177

Schürmann, Dr Kay Uwe (1939). Lecturer and Head of Museum. Institut für Mineralogie, Fachbereich Geowissenschaften der Universität Marburg, D-35032 Marburg, Germany. Dr rer.nat. (U. Marburg, 1966) mineralogy. *Experimental_mineralogy special_mineralogy*
Tel. 49(6421)282228 Fax 49(6421)288919 Telex 482372umrd

Schulz, Prof. Dr Heinz Hermann (1935). Chair for Crystallography; Head of institute. Institut für Kristallographie und Mineralogie, Universität München, Germany. [Institut für Kristallographie und Mineralogie, Universität München, Theresienstr. 41, D-80333 München, Germany.] Dr rer.nat. (U. Saarbrücken, 1964) physics. *Anharmonicity microcrystal high_pressure surface ionic_conductor diamond_anvil_cells*
E-mail uk40304@sunmail.lrz-muenchen.de Tel. 49(89)23944311 Fax 49(89)23944334

Schumann, Dr Bernd (1947). Assistant. Institut für Kristallographie der Universität Leipzig, Germany. [Institut für Kristallographie der Universität Leipzig, Scharnhorststr. 20, D-04275 Leipzig, Germany.] Dr rer.nat. (U. Leipzig, 1974) chemistry/crystallography. *Crystallography epitaxy defect symmetry semiconductors RHEED solar_cells*
Tel. 49(341)3310502 Fax 49(341)310502

Schuster, Prof. Dr Hans-Uwe (1930). Director. Institut für Anorganische Chemie der Universität zu Köln, Greinstr. 6, D-50939 Köln, Germany. [Nikolaus-Ehlen-Str. 45, D-50374 Erftstadt, Germany.] Prof. (U. Köln, 1971) inorganic chemistry. *Solid_state chemistry intermetallic_compound physical_property structure_determination magnetism conductivity optical_investigations differential_thermoanalysis*
Tel. 49(221)4703262 Fax 49(221)4705196

Schuster, Dr Manfred R. (1953). Research Manager. SIEMENS AG, ZFE BT MR 32, [SIEMENS AG, ZFE BT MR 32, Otto-Hahn-Ring 6, D-81739 München, Germany.] Dr (U. München, 1986) physics. *High-resolution diffractometry rocking_curves semiconductors X-ray_fluorescence_spectroscopy epitaxy grazing_incidence ion_channeling Auger_spectroscopy*
E-mail manfred.schuster@zfe.siemens.de Tel. 49(89)63644165 Fax 49(89)63642256 Telex 52109-11sin d

Schwahn, Dr Dietmar (1943). Senior Scientist. Institut für Festkörperforschung, Forschungszentrum Jülich GmbH, Germany. [Institut für Festkörperforschung, Forschungszentrum Jülich GmbH, Postfach 1913, D-52425 Jülich, Germany.] Dr (U. Bochum, 1976) physics. *Conformation critical_phenomena phase_separation macromolecular SAXS*
E-mail schwahn@ella02.iff.kfa-juelich.de Tel. 49(2461)616661 Fax 49(2461)612610

Schwarz, Dr Ulrich (1961). Research Associate. Max-Planck-Institut für Festkörperforschung, Germany. [Max-Planck-Institut für Festkörperforschung, Heisenbergstr. 1, D-70569 Stuttgart, Germany.] Dr rer.nat. (TH Darmstadt, 1992) chemistry. *Anvil_cell spectroscopy phase_transition structure_determination high_pressure Raman reflectivity Rietveld_method absorption_edge anisotropy*
E-mail schwarz@servix.mpi-stuttgart.mpg.de Tel. 49(711)6891469 Fax 49(711)6891010 Telex 7255555

Schweda, Dr Eberhard (1950). Lecturer. Institut für Anorganische Chemie der Universität Tübingen, Germany. [Institut für Anorganische Chemie der Universität Tübingen, Auf der Morgenstelle 18, D-72076 Tübingen, Germany.] Dr rer.nat.habil. (U. Tübingen, 1991) inorganic chemistry. *Ferroic_phase_transition solid_state chemistry defect_structure electron_microscopy X-ray_single_crystal powder_diffraction*
E-mail caase01@mailserv.zdv.uni-tuebingen.de Tel. 49(7071)296217 Fax 49(7071)292436

Seifert, Prof. Dr Karl-Friedrich (1927). [Norbertstr. 12, D-53925 Kall-Steinfeld, Germany.] Dr rer.nat. (U. Münster, 1957) mineralogy. *History_of_crystallography_and_mineralogy archeometry biocrystallography*
Tel. 49(2441)5779

Serafin, Dr Michael (1949). Research Assistant. Institut für Anorganische und Analytische Chemie der Universität Gießen, Germany. [Institut für Anorganische und Analytische Chemie der Universität Gießen, Heinrich-Buff-Ring 58, D-35392 Gießen, Germany.] Dr rer.nat. (U. Gießen, 1980) chemistry. *Chemistry solid_state structure_determination computer_management*
E-mail serafin@anorg.chemie.uni-giessen.d400.de Tel. 49(641)7025666 Fax 49(641)7025669

Sheldrick, Prof. Dr George M. (1942). Professor. Institut für Anorganische Chemie der Universität Göttingen, Germany. [Institut für Anorganische Chemie der Universität Göttingen, Tammannstr. 4, D-37077 Göttingen, Germany.] PhD (U. Cambridge/UK, 1966) chemistry. *Structure_determination computing stereochemistry education*
E-mail gsheldr@shelx.uni-ac.gwdg.de Tel. 49(551)393021 Fax 49(551)392582

Sheldrick, Prof. Dr William Stephen (1945). Chairman. Lehrstuhl für Analytische Chemie, Universität Bochum, Germany. [Lehrstuhl für Analytische Chemie, Universität Bochum, Universitätsstr. 150, D-44780 Bochum, Germany.] PhD (Cambridge/UK, 1969) inorganic chemistry. *Structural_inorganic_chemistry polyanions amino_acids nucleotides arsenic_antimony_and_tin_compounds*
E-mail heeb@analyt.chemie.uni-bochum.d400.de Tel. 49(234)7004192 Fax 49(234)7094420 Telex 17-234356

Sieber, Dr Norbert H. W. (1960). Manager. Firma SIEBER, Heddenheimer Landstr. 22, D-60439 Frankfurt/M., Germany. Dr (U. Würzburg, 1989) crystallography. *Minerals zeolites phase_transition Jahn_Teller_compounds*
Tel. 49(69)573853

Sieger, Dr Peter (1963). Postdoctoral Research Assistant. AG Felsche, Fakultät für Chemie der Universität Konstanz, Germany. [AG Felsche, Fakultät für Chemie der Universität Konstanz, Postfach 5560-M730, D-78434 Konstanz, Germany.] Dr rer.nat. (U. Konstanz, 1992) chemistry. *Zeolites synthesis structure_determination crystallography thermoanalysis DSC TGA alumino-silicates solid_state_magnetic_resonance*
E-mail sieger@vg3.chemie.uni-konstanz.de Tel. 49(7531)882014 Fax 49(7531)883142

Sieler, Prof. Dr Joachim (1939). Lecturer. Institut für Anorganische Chemie der Universität Leipzig, Germany. [Institut für Anorganische Chemie der Universität Leipzig, Linnéstr. 3, D-04103 Leipzig, Germany.] Prof. (U. Leipzig, 1992) chemistry. *X-ray_diffractometry reactivity bonding alkoxides*
E-mail sieler@serverl.rz.uni-leipzig.de Tel. 49(341)6858384 Fax 49(341)6858387

Soa, Prof. Dr Ernst-Adolf (1930). TU Ilmenau, Germany. [Fritz-Krieger-Str. 4, D-07743 Jena, Germany.] Prof. (TU Ilmenau, 1992) applied physics. *Infrared_detector surface thin_film monolayer philosophy*
Tel. 49(3641)25754

Sommer, Dr Olaf (1957). Manager. Institut für Festkörperanalytik GmbH, Sommer, Langner & Partner, Germany. [Institut für Festkörperanalytik GmbH, Sommer, Langner & Partner, Yorckstr. 36, D-76185 Karlsruhe, Germany.] Dr rer.nat. mineralogy. *Pharmaceutical_structure_determination order_disorder scanning_electron_microscopy thin_film fractals incommensurate_modulated_structures image_processing failure_analysis*
Tel. 49(721)853038 Fax 49(721)853039

Sondermann, Dr Ulrich (1936). Postdoctoral Research Assistant. Institut für Mineralogie der Universität Marburg, Germany. [Institut für Mineralogie der Universität Marburg, Hans-Meerwein-Str., D-35032 Marburg, Germany.] Dr (U. Marburg, 1970) physics. *X-ray_diffraction magnetic_property phase_stability hydrates_stability*
Tel. 49(6421)282226 Fax 49(6421)285831 Telex 482372

Sowa, Dr Heidrun (1954). Research Scientist. Institut für Mineralogie, Petrologie und Kristallographie der Universität Marburg, Germany. [Institut für Mineralogie, Petrologie und Kristallographie der Universität Marburg, Hans-Meerwein-Str., D-35032 Marburg, Germany.] Dr (U. Marburg, 1983) mineralogy. *High_pressure_crystallography crystal_chemistry anion_packing*
E-mail sowa@mailer.uni-marburg.de Tel. 49(6421)285675 Fax 49(6421)288919 Telex 482372

Spengler, Mr Roland Helmuth Michael (1966). Research Assistant. Institut für Angewandte Physik, Lehrstuhl für Kristallographie der Universität Erlangen-Nürnberg, Germany. [Institut für Angewandte Physik, Lehrstuhl für Kristallographie der Universität Erlangen-Nürnberg, Bismarckstr. 10, D-91054 Erlangen, Germany.] Dipl. Phys. (U. Erlangen-Nürnberg, 1993) physics. *Powder single_crystal structure_determination structure_refinement space-group_determination_from_powders*
E-mail mpkr00@cd4680fs.rrze.uni-erlangen.de Tel. 49(9131)852119 Fax 49(9131)852733

Stalke, Dr Dietmar (1958). Lecturer. Institut für Anorganische Chemie der Universität Göttingen, Germany. [Institut für Anorganische Chemie der Universität Göttingen, Tammannstr. 4, D-37077 Göttingen, Germany.] Dr (U. Göttingen, 1987) chemistry. *Structure_determination nuclear_magnetic_resonance education low_temperature chemistry_of_alkali_and_alkaline_earth_metals copper_compounds reactive_intermediates*
E-mail dstalke@shelx.uni-sc.gwdg.de Tel. 49(551)393000 Fax 49(551)392582

Steeb, Prof. Dr Siegfried (1931). Head of Department. Max-Planck-Institut für Metallforschung, Institut für Werkstofforschung, Germany. [Max-Planck-Institut für Metallforschung, Institut für Werkstofforschung, Seestr. 92, D-70174 Stuttgart, Germany.] Prof. (U. Stuttgart, 1969) metal physics. *Amorphous_phase diffraction grazing_incidence electron_probe microanalysis small_angle_scattering*
Tel. 49(711)2095384 Fax 49(711)2265722 Telex 723742mpimd

Steffes-Tun, Mr Wolfgang (1960). Manager. EnTeCo - Gesellschaft für Umweltschutztechnik, und Auftragsforschung GmbH, Germany. [EnTeCo - Gesellschaft für Umweltschutztechnik, und Auftragsforschung GmbH, Technologiehof, Mendelstr. 11, D-48149 Münster, Germany.] Dipl. Min. (U. Münster, 1987) mineralogy. *Environment_protection electron_microscopy mineralogy combustion consulting immobilisation_in_reservoir_minerals waste_treatment*
Tel. 49(251)9802173 Fax 49(251)9802176

Steiner, Prof. Dr Michael (1943). Head of Department. Hahn-Meitner-Institut, Germany. [Hahn-Meitner-Institut, Glienicker Str. 110, D-14109 Berlin, Germany.] Dr rer.nat.habil. (TU Berlin, 1979) experimental physics. *Magnetism magnon phase_transition neutron_scattering critical_phenomena nonlinear_excitation*
E-mail steiner@vax.hmi.d400.de Tel. 49(30)80622741 Fax 49(30)80622999 Telex 185763

Steiner, Dr Thomas (1961). Postdoctoral Research Assistant. Institut für Kristallographie der Freien Universität Berlin, Germany. [Institut für Kristallographie der Freien Universität Berlin, Takustr. 6, D-14159 Berlin, Germany.] Dr (FU. Berlin, 1990) physics. *Neutron_crystallography cyclodextrins hydrogen_bonding van_der_Waals_contacts*
E-mail steiner@chemie.fu-berlin.de Tel. 49(30)8386758 Fax 49(30)8386702

Steinke, Prof. Dr Ursula (1935). Group Leader; Head of scientific group. Institut für Angewandte Chemie i.Gr. e.V., (former Zentrum f. heterogene Katalyse), Rudower Chaussee 5, D-12489 Berlin, Germany. [Waldstr. 8, D-12527 Berlin, Germany.] Dr rer.nat.habil. (Akad. d. Wiss. d. DDR Berlin, 1976) crystallography. *Wide-angle_scattering phase_formation catalysts ceramics in_situ_reactions gas/solid_interactions*
E-mail steinike@aca.fta-berlin.de Tel. 49(30)63924314 Fax 49(30)63924350

Steins, Dr Manfred (1960). Research Associate. Institut für Mineralogie und Kristallographie der Technischen Universität Berlin, Germany. [Institut für Mineralogie und Kristallographie der Technischen Universität Berlin, Ernst-Reuter-Platz 1, D-10587 Berlin, Germany.] Dr rer.nat. (U. Göttingen, 1991) mineralogy. *Superconductors sulfides structure texture computing_methods_in_crystallography multilayer thin_films*
E-mail stei1637@mailszrz.zrz.tu-berlin.de Tel. 49(30)31424245 Fax 49(30)31421124

Strähle, Prof. Dr Joachim (1937). Full Professor. Institut für Anorganische Chemie der Universität Tübingen, Germany. [Institut für Anorganische Chemie der Universität Tübingen, Auf der Morgenstelle 18, D-72076 Tübingen, Germany.] Dr habil. (TH Karlsruhe, 1973) inorganic chemistry. *Structure_determination structural_disorder clusters coordination_compound nitrides gold_compounds*
E-mail jstraehle@mailserv.zdv.uni-tuebingen.de Tel. 49(7071)296102 Fax 49(7071)292436

Strasdeit, Dr Henry (1957). Lecturer. Fachbereich Chemie der Universität Oldenburg, D-26111 Oldenburg, Germany. Dr rer.nat.habil. (U. Oldenburg, 1993) inorganic chemistry. *Coordination_chemistry bioinorganic_chemistry cadmium_compounds structure_determination EXAFS metalloproteins irradation clusters secondary_bonding stereochemistry philosophy_of_science*
Tel. 49(441)7983677 Fax 49(441)7983329

Strumpel, Dr Marianna (1939). Research Associate. Institut für Kristallographie der Freien Universität Berlin, Germany. [Institut für Kristallographie der Freien Universität Berlin, Takustr. 6, D-14195 Berlin, Germany.] Dr (FU. Berlin, 1983) crystallography. *Structure_determination conformation valence_charge_density vector_search*
E-mail struempl@kristall.chemie.fu-berlin.de Tel. 49(30)8383408 Fax 49(30)8383464

Stuckenschmidt, Dr Elli (1941). Research Associate. Institut für Kristallographie und Mineralogie der Universität Frankfurt. [Institut für Kristallographie und Mineralogie der Universität Frankfurt, Senckenberganlage 30, D-60054 Frankfurt/M., Germany.] Dr rer.nat. (U. Frankfurt/M., 1986) physics. *Zeolites chemical_bonding X-ray_diffraction electron_density_distribution*
Tel. 49(69)7983104 Fax 49(69)7982101

Stuhrmann, Prof. Dr Heinrich B. (1938). Head of Department. GKSS Forschungszentrum, Postfach 1160, D-21502 Geesthacht, Germany. [Beckedorfer Str. 154, D-21218 Seevetal, Germany.] Prof. (U. Mainz, 1972) physical chemistry. *Biological_macromolecules anomalous_dispersion polarized_neutron scattering contrast biophysics condensed_matter synchrotron_radiation neutrons*

E-mail heinrich.stuhrmann@dvnetz.gkss.dbp.de Tel. 49(4152)871290 Fax 49(4152)871338 Telex 0218712gkssg

Suck, Dr Dietrich (1944). Group leader. European Molecular Biology Laboratory, Germany. [European Molecular Biology Laboratory, Meyerhofstr.1, Postfach 102209, D-69012 Heidelberg, Germany.] Dr (U. Braunschweig, 1971) chemistry. *Macromolecular_crystallography nucleic_acids structural_biology proteins–DNA_interaction*
E-mail suck@embl-heidelberg.de Tel. 49(6221)387307 Fax 49(6221)387306 Telex 49(6221)461613(embl d)

Susse, Prof. Dr Peter (1939). Director. Mineralogische Sammlungen, Universität Göttingen, Germany. [Mineralogische Sammlungen, Universität Göttingen, Goldschmidtstr. 1, D-37077 Göttingen, Germany.] Dr phil. (U. Marburg, 1963) mineralogy and crystallography. *Mineralogy crystallography database computer-aided education*
Tel. 49(551)393936 Fax 49(551)399521

Tebbe, Prof. Dr Karl-Friedrich (1941). Associate Professor. Institut für Anorganische Chemie der Universität zu Köln, Germany. [Institut für Anorganische Chemie der Universität zu Köln, Greinstraße 6, D-50939 Köln, Germany.] Dr (U. Münster, 1971) chemistry. *Structure_determination computing phase_transition polyiodides solid_state_chemistry halogens inorganic_molecules crystal_structures*
E-mail tebbe@rr2.uni-koeln.de Tel. 49(221)4703285 Fax 49(221)4705196

Tempelhoff, Dr Klaus (1938). Technische Universität Berlin, Germany. [Technische Universität Berlin, Heinrich-Heine-Str. 2, D-10179 Berlin, Germany.] Dr sc.nat. (Akad. d. Wiss. d. DDR Berlin, 1977) physical chemistry. *Monolayer epitaxy III–V_compounds growth_of_semiconductor_material*
Tel. 49(30)2752410 Fax 49(30)20366111

Teske, Dr Christoph Ludwig (1942). Research Assistant. Institut für Anorganische Chemie der Universität Kiel, Germany. [Institut für Anorganische Chemie der Universität Kiel, Otto-Hahn-Platz 6/7, D-24098 Kiel, Germany.] Dr (U. Gießen, 1970) chemistry. *Oxides sulfides structure_determination preparation oxocuprates superconductor DTA/TG*
Tel. 49(431)8802408 Fax 49(431)8801520

Thurn, Dr Herbert (1937). Research Assistant. Institut für Anorganische Chemie der Universität Stuttgart, Germany. [Institut für Anorganische Chemie der Universität Stuttgart, Pfaffenwaldring 55, D-70569 Stuttgart, Germany.] Dr rer.nat. (U. Stuttgart, 1967) chemistry. *Inorganic_solid_state diffraction crystal_structure_determination liquid_state diffraction symmetry*
Tel. 49(711)6854237 Fax 49(711)6854241

Thygesen, Mr Jesper (1965). Scientific Scholar. Max-Planck-Gesellschaft, Arbeitsgruppe Ribosomenstruktur, Germany. [Max-Planck-Gesellschaft, Arbeitsgruppe Ribosomenstruktur, c/o DESY, Notkestr. 85, D-22603 Hamburg, Germany.] Cand.scient. (U. Aarhus/Denmark, 1992) chemistry. *Crystallography ribosomes data_collection data_processing area_detector isomorphous_replacement crystallization contrast*
E-mail jesper@mpgars.desy.de Tel. 49(40)89982826 Fax 49(40)891314 Telex 215124desyd

Tigges, Mr Hartmut (1945). Marketing. RIGAKU EUROPE GmbH, Germany. [RIGAKU EUROPE GmbH, Monschauer Str. 7, D-40549 Düsseldorf, Germany.] Dipl. Chem. (U. Münster, 1972) chemistry. *X-ray_diffractometry X-ray_spectroscopy*
Tel. 49(211)502186 Fax 49(211)504496

Tippelt, Mrs Birgit (1966). Research Assistant. Institut für Kristallographie, Mineralogie und Materialwissenschaft der Universität Leipzig, Germany. [Institut für Kristallographie, Mineralogie und Materialwissenschaft der Universität Leipzig, Scharnhorststr. 20, D-04275 Leipzig, Germany.] Dipl. Kristallogr. (U. Leipzig, 1990) crystallography. *Dislocation plastic_deformation transmission_electron_microscopy*
Tel. 49(341)6858382

Töpel-Schadt, Dr Jutta (1952). [Am Rottfeld 38, D-50374 Erftstadt, Germany.] Dr phil.nat. (U. Frankfurt/M., 1980) mineralogy. *Transmission_electron_microscopy real_structure phase_transition mineralogy editing*
Tel. 49(2235)72046

Trah, Dr Hans-Peter (1958). Research and Development Scientist. ROBERT BOSCH GmbH, K8/ESE, Germany. [ROBERT BOSCH GmbH, K8/ESE, Tübinger Str. 123, D-72703 Reutlingen, Germany.] Dr rer.nat. (U. Freiburg/Br., 1988) crystallography. *Structure crystal_growth epitaxy crystallography micromachining silicon smart_sensors_and_actuators micro_system_technology*
E-mail trah@ese.k8.rt.bosch.de Tel. 49(7121)352578 Fax 49(7121)351493 Telex 2627-7121148rbrt

Trampert, Dr Achim Peter (1961). Postdoctoral Research Assistant. Max-Planck-Institut für Festkörperforschung, Germany. [Max-Planck-Institut für Festkörperforschung, Postfach 800665, D-70506 Stuttgart, Germany.] Dr (U. Stuttgart, 1993) metallography/chemistry. *High-resolution_electron_microscopy epitaxy heterostructure defect strained_layer_heteroepitaxy relaxation electron_diffraction*
E-mail trampert@wselix.mpi-stuttgart.mpg.de Tel. 49(711)6891841 Fax 49(711)6891010 Telex 7255555mpifd

Trauth, Dr Jürgen (1959). Scientist. Fraunhofer-Institut für Kurzzeitdynamik, Germany. [Fraunhofer-Institut für Kurzzeitdynamik, Eckerstr. 4, D-79104 Freiburg/Br., Germany.] Dipl. Min. (U. Freiburg, 1987) mineralogy. *Crystal_growth phase_determination diffractometry residual_stress mechanical_alloying*
Tel. 49(761)2714369 Fax 49(761)2714316

Trömel, Prof. Dr Martin (1934). Professor. Institut für Anorganische Chemie der Universität Frankfurt. [Institut für Anorganische Chemie der Universität Frankfurt, Marie-Curie-Str. 11, D-60439 Frankfurt/M., Germany.] Dr (U. Frankfurt/M., 1963) inorganic chemistry. *Bond_length coordination disorder oxides crystal_chemistry aperiodic_material Madelung_factor*

E-mail troemel@chemie.uni-frankfurt.d400.de   Tel.   49(69)58009159   Fax 49(69)58009235

Tschulena, Dr Guido R. (1945). Manager. SGT Sensor Gruppe Dr. Tschulena, Germany. [SGT Sensor Gruppe Dr. Tschulena, Reichenberger Str. 5, D-61273 Wehrheim, Germany.] Dr phil. (U. Wien/Austria, 1971) physics. *Sensor ceramics semiconductor micro_systems gas_sensor screen_printing micromechanics*
Tel. 49(6081)56168 Fax 49(6081)57222

Uecker, Mrs Doris-Christiane (1953). Research Associate. Institut für Angewandte Chemie i.Gr. e.V., Rudower Chaussee 5, D-12489 Berlin, Germany. [Wotanstr. 27a, D-10365 Berlin, Germany.] Dipl. Kristallogr. (Humboldt-U. Berlin, 1975) crystallography. *X-ray_powder_diffraction phase_formation_of_inorganic_materials ceramics_filter purification_of_coal_gases in_situ_reactions gas/solid_reactions*
Tel. 49(30)63924320 Fax 49(30)63924350

Uecker, Mr Reinhard (1951). Research Associate. Institut für Kristallzüchtung, im Forschungsverbund Berlin e.V., Germany. [Institut für Kristallzüchtung, im Forschungsverbund Berlin e.V., Rudower Chaussee 6, Geb. 18.46, 12489 Berlin, Germany.] Dipl. Kristallogr. (Humboldt-U. Berlin, 1975) crystallography. *Crystal_growth Czochralski_technique oxides laser_crystal*
Tel. 49(30)63923020 Fax 49(30)63923003

Uhlig, Dr Stefan Helmut (1954). Manager. SIEMENS Röntgenanalytik, Germany. [SIEMENS Röntgenanalytik, Siemensallee 84, D-76181 Karlsruhe, Germany.] Dr rer.nat. (U. Gießen, 1987) geology. *X-ray_fluorescence automation management instrumentation X-ray_diffraction oregenesis trace_analysis powder*
Tel. 49(721)5956571 Fax 49(721)5954506

Ulbricht, Prof. Dr Heinz (1931). Retired. Fachbereich Physik der Universität Rostock, Germany. [Fachbereich Physik der Universität Rostock, Thomas-Müntzer-Platz 11, D-18057 Rostock, Germany.] Dr rer.nat.habil. (U. Rostock, 1971) physics. *Phase_transition nucleation statistical_thermodynamics fluids electrolytes history_of_science*
Tel. 49(381)25420

Ullrich, Prof. Dr Hans-Jürgen (1938). Professor. Institut für Werkstoffwissenschaft der Technischen Universität Dresden, Germany. [Institut für Werkstoffwissenschaft der Technischen Universität Dresden, Helmholtzstr. 7, D-01062 Dresden, Germany.] Dr rer.nat.habil. (TU Dresden, 1975) solid state physics. *Kossel_diffraction microtexture electron_probe microanalysis diffractometry Kikuchi_effect lattice_parameters residual_stress synchrotron_radiation diffusion*
Tel. 49(351)4632366 Fax 49(351)4637125

Untersteller, Mr Eugen (1953). Post-Graduate Student. Institut für Mineralogie der Universität Marburg, Germany. [Institut für Mineralogie der Universität Marburg, Hans-Meerwein Str., D-35032 Marburg, Germany.] Dipl. Min. (U. Marburg, 1985) mineralogy. *Software magnetism neutron_diffraction silicates*
E-mail unterstell@mailer.uni-marburg.de Tel. 49(6421)283466 Fax 49(6421)288919

Urland, Prof. Dr Werner (1944). Professor. Institut für Anorganische und Analytische Chemie der Universität Hannover, Germany. [Institut für Anorganische und Analytische Chemie der Universität Hannover, Callinstr. 9, D-30167 Hannover, Germany.] Dr rer.nat. (U. Gießen, 1971) inorganic chemistry. *Preparation solid_state rare-earth_compounds physical_property band_calculation magnetism ionic_conductors superconductors*
Tel. 49(511)7622567 Fax 49(511)7623006

Van Smaalen, Dr Sander (1958). Res. fellow. Mineralogisch-Petrographisches Institut, University, Olshausenstraße 40, D-24098 Kiel, Germany. Dr (U. Groningen, 1985) chemistry. *Aperiodic_material physical_properties_of_solids*
Tel. 49(431)8802892 Fax 49(431)8804457

Veith, Prof. Dr Michael (1944). Chemistry. Institut für Anorganische Chemie der Universität Saarbrücken, Germany. [Institut für Anorganische Chemie der Universität Saarbrücken, Am Stadtwald 6, Postfach 15 11 50, D-66041 Saarbrücken, Germany.] Dr rer.nat. (U. München, 1971) chemistry. *Chemistry low_temperature structure_determination molecular_crystal metal_clusters nanocomposites CVD molecular_motions*
E-mail veith@sbusol.rz.uni-sb.de Tel. 49(681)3023415 Fax 49(681)3023995

Vielhaber, Dr Edmund Antonius (1931). Research and Teaching. Institut für Anorganische Chemie I der Universität Gießen, Germany. [Hagener Str. 55, D-58642 Iserlohn-Lethmathe, Germany.] Dr rer.nat. (U. Münster, 1964) chemistry. *Inorganic_crystal_structure crystal_growth astronomy cosmochemistry*
Tel. 49(641)7025693

Volkmann, Mr Niels Heinz Paul (1962). Post-Graduate Student. Max-Planck-Arbeitsgruppe, für Ribosomenstruktur, Germany. [Max-Planck-Arbeitsgruppe, für Ribosomenstruktur, c/o DESY Geb. 25b, Notkestr. 85, D-22603 Hamburg, Germany.] Dipl. Phys. (U. Hamburg, 1991) physics. *Maximum-entropy_method phase_determination ribosomes isomorphous_replacement phase_extension molecular_replacement direct_methods density_modifications*
E-mail niels@mpgars.desy.de Tel. 49(40)89982809 Fax 49(40)891314

Vorbach, Dr Angelika (1953). Student of Pharmacy. Universität Freiburg, Germany. [Wiesenweg 1, D-79241 Ihringen, Germany.] Dr rer.nat. (U. Karlsruhe, 1983) mineralogy. *Petrology pharmacology mineralogy*
Tel. 49(7668)1650

Vorderwisch, Dr Peter (1940). Research Associate. Hahn-Meitner-Institut Berlin GmbH, Germany. [Hahn-Meitner-Institut Berlin GmbH, Postfach 390128, D-14091 Berlin, Germany.] Dr rer.nat. (FU. Berlin, 1974) physics. *Inelastic_neutron_scattering lattice_dynamics crystal_field interstitial_compounds*
E-mail vorderwisch@vax.hmi.dbp.de Tel. 49(30)80092171 Fax 49(30)80092523 Telex 185763

Wacker, Dr Klaus (1954). Lecturer. Wissenschaftliches Zentrum für Materialwissenschaften der Universität Marburg, Germany. [Wissenschaftliches Zentrum für Materialwissenschaften der Universität Marburg, Hans-Meerwein-Str., D-35032 Marburg, Germany.] Dr rer.nat.habil. (U. Marburg, 1983) crystallography. *Chalcogenides structural_thermal_and_physical_property impedance_spectroscopy*
E-mail wacker@papin.hrz.uni-marburg.de Tel. 49(6421)285443 Fax 49(6421)285831 Telex 482372

Wagner, Dr Gerald (1950). Research Associate. Institut für Kristallographie, Mineralogie und Materialwissenschaft der Universität Leipzig, Germany. [Institut für Kristallographie, Mineralogie und Materialwissenschaft der Universität Leipzig, Linnéstr. 3-5, D-04103 Leipzig, Germany.] Dr rer.nat. (U. Leipzig, 1980) crystallography. *III-V_compound semiconductors defect transmission_electron_microscopy twinning crystal_growth dislocations ordering interfaces*
Tel. 49(341)6858382 Fax 49(341)6858387

Wagner, Mrs Trixie (1966). Research Assistant. Institut für Anorganische Chemie der RWTH Aachen, Germany. [Institut für Anorganische Chemie der RWTH Aachen, Prof.-Pirlet-Str. 1, D-52074 Aachen, Germany.] Dipl. Chem. (RWTH Aachen, 1992) chemistry. *Topotaxy structure_determination crystallinity solid_state reaction*
Tel. 49(241)804655

Walker, Dr Nigel P. C. (1955). Research Scientist. BASF AG, Germany. [BASF AG, ZHB/W-A30, D-67056 Ludwigshafen, Germany.] PhD (U. Bristol/UK, 1983) protein crystallography. *Biocrystallography computer_modelling structure_determination drug_design*
E-mail walker@brh.dnet.basf-ag.de Tel. 49(621)6042638 Fax 49(621)6020440

Wallrafen, Dr Franz (1938). Research Associate. Mineralogisches Institut der Universität Bonn, Germany. [Mineralogisches Institut der Universität Bonn, Poppelsdorfer Schloß, D-53115 Bonn, Germany.] Dr rer.nat. (U. Bonn, 1972) mineralogy. *Bridgman_Stockbarger_technique crystal_growth Czochralski_technique flux aluminium_compounds binary_alloys fluorine_compounds halides oxides polar_compounds*
E-mail crystal.growth@uni-bonn.de Tel. 49(228)732961 Fax 49(228)732770 Telex 886657unibod

Wang, Dr Naiding (1935). Research Associate. Mineralogisch-Petrographisches Institut der Universität Heidelberg, Germany. [Mineralogisch-Petrographisches Institut der Universität Heidelberg, Im Neuenheimer Feld 236, D-69120 Heidelberg, Germany.] Dr rer.nat. (U. Heidelberg, 1968) crystallography. *Sulfides synthesis phase_equilibrium*
Tel. 49(6221)564804 Fax 49(6221)562805

Wang, Dr Xiqu (1963). Research Associate. Mineralogisches Institut der Universität Kiel, Germany. [Mineralogisches Institut der Universität Kiel, Olshausenstr. 40, D-24098 Kiel, Germany.] Dr rer.nat. (U. Kiel, 1993) mineralogy. *Crystal_chemistry zeolites antimony_compounds*
Tel. 49(431)8802912 Fax 49(431)8804457

Wartchow, Dr Rudolf (1940). Research Assistant. Institut für Anorganische Chemie der Universität Hannover, Germany. [Institut für Anorganische Chemie der Universität Hannover, Callinstr. 9, D-30167 Hannover, Germany.] Dr rer.nat. (U. Hannover, 1975) chemistry. *Computing data_collection structure_determination symmetry*
E-mail wartchow@mbox.aca.uni-hannover.de Tel. 49(511)7622216 Fax 49(511)7623006

Weber, Dr Jürgen (1959). Development Engineer. ALCATEL-SEL, Germany. [ALCATEL-SEL, Lorenzstr. 10, D-70435 Stuttgart, Germany.] Dr rer.nat. (U. Stuttgart, 1991) physics. *Epitaxy III-V_compounds semiconductors optoelectronics*
Tel. 49(711)8215150 Fax 49(711)8216355 Telex 725526-0

Weckert, Dr Edgar F. (1960). Postdoctoral Research Assistant. Institut für Kristallographie der Universität Karlsruhe, Germany. [Institut für Kristallographie der Universität Karlsruhe, Kaiserstr. 12, D-76128 Karlsruhe, Germany.] Dr rer.nat. (U. Erlangen-Nürnberg, 1988) crystallography. *Synchrotron_radiation multiple_scattering absolute_structure dynamical_diffraction three-beam_diffraction*
E-mail edgar.weckert@physik.uni-karlsruhe.de Tel. 49(721)6086045 Fax 49(721)697720

Weise, Mrs Sylvia (1969). Post-Graduate Student. Fachbereich Physik/Festkörperphysik der Universität Leipzig, Linnéstr. 3-5, D-04103 Leipzig, Germany. [Breisgaustr. 33/1002, D-04209 Leipzig, Germany.] Dipl. Kristallogr. (U. Leipzig, 1992) crystallography. *Semiconductors solar_cell crystal_growth thin_film selenides lithium_compounds laser_ablation thermoanalysis X-ray electrical_characterization*
Tel. 49(341)6858390

Weitzel, Dr Hans (1941). Academic Director. Fachgebiet Strukturforschung im Fachbereich Materialwissenschaft der Technischen Hochschule Darmstadt, Germany. [Fachgebiet Strukturforschung im Fachbereich Materialwissenschaft der Technischen Hochschule Darmstadt, Petersenstr. 20, D-64287 Darmstadt, Germany.] Dr rer.nat. (U. Tübingen, 1969) physics. *Neutron_diffraction crystal_structure magnetic_structure phase_transition*
Tel. 49(6151)162698 Fax 49(6151)165551

Wenda, Prof. Dr Richard (1955). Professor. Fachbereich Werkstofftechnik der Fachhochschule Nürnberg, Germany. [Fachbereich Werkstofftechnik der Fachhochschule Nürnberg, Postfach 210320, D-90121 Nürnberg, Germany.] Dr rer.nat. (U. Erlangen-Nürnberg, 1984) mineralogy. *Chemistry_and_mineralogy_of_cements*
Tel. 49(911)5880247 Fax 49(911)5880177

Wendschuh-Josties, Dr Michael Thomas (1955). Postdoctoral Research Assistant. Mineralogisch-Kristallographisches Institut der Universität Göttingen, Germany. [Mineralogisch-Kristallographisches Institut der Universität Göttingen, Goldschmidtstr. 1, D-37077 Göttingen, Germany.] Dr rer.nat. (U. Göttingen, 1990) mineralogy. *Synchrotron_radiation intracrystalline_ordering*

E-mail mwendsc@gwdgv1.gwdg.de Tel. 49(551)393894 Fax 49(551)399521

Wenig, Prof. Dr Werner (1943). Professor. Laboratorium für Angewandte Physik der Universität Duisburg, Germany. [Laboratorium für Angewandte Physik der Universität Duisburg, Lotharstr. 1, D-47048 Duisburg, Germany.] Dr rer.nat. (U. Ulm, 1974) physics. *SAXS WAXS morphology polycrystal polymers*
E-mail hm359we@unidui.uni-duisburg.de Tel. 49(203)3792386 Fax 49(203)3793163

Werthmann, Mrs Ulrike (1966). Student. [Stephanstr. 4, D-52064 Aachen, Germany.] (1990) *Crystal_growth crystallography International_Tables_for_Crystallography diffractometry*
Tel. 49(241)37983

Wetzel, Mrs Andrea (1963). Research Assistant. Mineralogisch-Petrographisches Institut der Universität Heidelberg, Germany. [Mineralogisch-Petrographisches Institut der Universität Heidelberg, Im Neuenheimer Feld 236, D-69120 Heidelberg, Germany.] Dipl. Min. (U. Heidelberg, 1990) mineralogy. *Powder_diffraction_data phosphates phase_diagram Rietveld_method structure_determination*
E-mail T50@VM.URZ.UNI-HEIDELBERG.DE Tel. 49(6221)562533 Fax 49(6221)564805

Wetzig, Prof. Dr Klaus (1940). Director. Institut für Festkörperanalytik und Strukturforschung, im IFW Dresden e.V., Germany. [Institut für Festkörperanalytik und Strukturforschung, im IFW Dresden e.V., Postfach, D-01171 Dresden, Germany.] Dr rer.nat.habil. (TU Dresden, 1973) physics. *Characterization electron_microscopy layer surface*
Tel. 49(351)4659217 Fax 49(351)4659452

Weyrich, Prof. Dr-Ing. Dr phil. h. c. Wolf (1941). Full Professor. Lehrstuhl für Physikalische Chemie I, Fakultät für Chemie der Universität Konstanz, Universitätstr. 10, D-78464 Konstanz, Germany. [Lehrstuhl für Physikalische Chemie I, Fakultät für Chemie der Universität Konstanz, Postfach 5560 M721, D-78434 Konstanz, Germany.] Habilitation (TH Darmstadt, 1978) physical chemistry. *Chemical_bonding Compton_scattering momentum_density quantum_mechanics electronic_structure_of_matter density_matrices phase_space*
E-mail weyrich@chclu.chemie.uni-konstanz.de Tel. 49(7531)882002 Fax 49(7531)883139 Telex 0733359univd

Wiedemann, Mr Peter (1960). Development Engineer. ALCATEL-SEL, Germany. [ALCATEL-SEL, Lorenzstr. 10, D-70435 Stuttgart, Germany.] Dipl. Phys. (U. Stuttgart, 1986) physics. *Epitaxy semiconductors X-ray*
Tel. 49(711)8213782 Fax 49(711)8216355 Telex 72526-0

Wieder, Dr Thomas (1961). Head of Laboratory. Institut für Innovation und Transfer der Fachhochschule Furtwangen, Germany. [Institut für Innovation und Transfer der Fachhochschule Furtwangen, Jakob-Kienzle-Str. 17, D-78027 Villingen-Schwenningen, Germany.] Dr rer.nat. (U. Kassel; 1988) physics. *Powder_diffraction grazing_incidence profile_analysis residual_stress*
E-mail wieder@physik.uni-kassel.de Tel. 49(7720)307297 Fax 49(7720)307210

Wiehl, Dr Leonore (1953). Lecturer. Mineralogisch-Petrologisches Institut und Museum der Universität Bonn, Germany. [Mineralogisch-Petrologisches Institut und Museum der Universität Bonn, Poppelsdorfer Schloß, D-53115 Bonn, Germany.] Dr rer.nat.habil. (U. Bonn, 1992) crystallography. *Crystal_structure_determination diffraction_theory physical_property_of_crystal thermal_motion phase_transitions*
E-mail une104@ibm.rhrz.uni-bonn.de Tel. 49(228)732771 Fax 49(228)732770

Wieser, Prof. Dr Egbert (1938). Research Associate. Institut für Ionenstrahlphysik und Materialforschung, Forschungszentrum Rossendorf e.V., Germany. [Institut für Ionenstrahlphysik und Materialforschung, Forschungszentrum Rossendorf e.V., PSF 510119, D-01314 Dresden, Germany.] Dr rer.nat.habil. (TU Dresden, 1980) physics. *Ion_implantation thin_layer ceramics metals*
E-mail wieser@fz-rossendorf.de Tel. 49(351)5912345 Fax 49(351)5913285 Telex 328683 fzr d

Wilke, Prof. Dr Wolfgang (1934). Professor. Abteilung Experimentalphysik der Universität Ulm, Germany. [Abteilung Experimentalphysik der Universität Ulm, Albert-Einstein-Allee 11, D-89069 Ulm, Germany.] apl. Prof. (U. Ulm, 1974) physics. *Real_crystal paracrystal small-angle_scattering synchrotron_radiation polymers*
E-mail wilke@dulruu51. Tel. 49(731)5023010 Fax 49(731)502)3036

Wilken, Dr Gerdt (1940). Research and Development. MOBIL Erdgas Erdöl GmbH, Germany. [MOBIL Erdgas Erdöl GmbH, Postfach 3107, D-29231 Celle, Germany.] Dr rer.nat. (U. Saarbrücken, 1974) physics. *Corrosion sulfides sulfates crystallization*
Tel. 49(5141)15420 Fax 49(5141)28198

Will, Dr Dr Georg Professor. Co-Editor of Journal of Applied Crystallography. Mineralogisches Institut, Universität Bonn, Germany. [Mineralogisches Institut, Universität Bonn, Poppelsdorfer Schloß, D-53115 Bonn, Germany.] Dr rer.nat. (TH München, 1958) crystallography. *Neutron_diffraction synchrotron_radiation high_pressure_research powder_diffraction*
Tel. 49(228)732761 Fax 49(228)732770 Telex 886657unibod

Wilmanns, Dr Matthias (1959). European Molecular Biology Laboratory, c/o Saraste Group, Postfach 10.2209, Meyerhofstrasse 1, 69012 Heidelberg, Germany. [Department of Chemistry, Boston College, Chestnut Hill, MA 02167, USA.] PhD (U. Basel, Switzerland, 1989) biochemistry. *Beta/alpha_barrels signal_transduction cytochrome_oxidase inverse_protein_folding*
E-mail wilmanns@embl-heidelberg.de Tel. 49(6221)387386

Winkler, Dr Björn (1962). Research Assistant. Mineralogisch-Petrographisches Institut und Museum der Universität Kiel, Germany. [Mineralogisch-Petrographisches Institut und Museum der Universität Kiel, Olshausenstr. 40, D-24098 Kiel, Germany.] PhD (U. Cambridge/UK, 1991) mineral physics. *Computer_modelling neutron_scattering mineralogy*
E-mail bjoern@anatas.min.uni-kiel.de Tel. 49(431)8802909 Fax 49(431)8804457

Winnacker, Prof. Dr Albrecht (1942). Director. Institut für Werkstoffwissenschaften der Universität Erlangen-Nürnberg, Germany. [Institut für Werkstoffwissenschaften der Universität Erlangen-Nürnberg, Martensstr. 7, D-91058 Erlangen, Germany.] Dr rer.nat. (1970) physics. *Semiconductors photochromic_compounds X-ray_fluorescence defect*
Tel. 49(9131)857633 Fax 49(9131)858495

Winter, Dr Gabriele (1961). SCHERING AG, Allgemeine Physikochemie, D-13342 Berlin, Germany. Dr rer.nat. (U. Halle/S., 1987) chemistry. *Pharmacology polymorphism phase_transition X-ray_powder*
Tel. 49(30)4687518 Fax 49(30)46918018

Winter, Prof. Dr Werner (1943). Head of Chemistry Research. GRÜNENTHAL GmbH, Forschungszentrum, Germany. [GRÜNENTHAL GmbH, Forschungszentrum, Zieglerstr. 6, D-52078 Aachen, Germany.] Prof. (U. Tübingen, 1982) organic chemistry. *Medicinal_chemistry drug_design computer_modelling asymmetric_synthesis*
Tel. 49(241)5692326 Fax 49(241)5692614

Wiskemann, Dr René (1960). Postdoctoral Research Assistant. Institut für Anorganische Chemie und Strukturchemie II, Universität Düsseldorf, Universitätsstr. 1, D-40225 Düsseldorf, Germany. [Bilker Allee 176, D-40217 Düsseldorf, Germany.] Dr (U. Düsseldorf, 1993) chemistry. *Symmetry coordination superstructure order-disorder cluster databases*
E-mail wiskeman@mail.rz.uni-duesseldorf.de Tel. 49(211)318713 Fax 49(211)318713

Witte, Prof. Dr Helmut (1909). Emeritus. Institut für Physikalische Chemie der Technischen Hochschule Darmstadt, Germany. [Institut für Physikalische Chemie der Technischen Hochschule Darmstadt, Petersenstr. 20, D-64287 Darmstadt, Germany.] Dr phil. (U. Göttingen, 1934) physics. *Electron_density_distribution_in_crystallite solubility_of_hydrogen_in_crystals*
Tel. 49(6151)165107

Wittmüss, Mrs Dörte (1965). [Jungmannstr. 36, D-24105 Kiel, Germany.] Dipl. Min. (U. Kiel, 1991) mineralogy. *Brillouin_spectroscopy computer_modelling software zeolites*
Tel. 49(431)563847

Wögerbauer, Dr Rupert (1941). Sworn Expert. [Am Schlehenbusch 15, D-97320 Mainstockheim, Germany.] Dr rer.nat. (U. Würzburg, 1974) crystal-structure science. *Corrosion crystal_growth water gemstones phase_equilibrium phase_determination*
Tel. 49(9321)5628 Fax 49(9321)21502

Wölfel, Prof. Dr Erich R. (1922). Emeritus. Former Co-Editor of Journal of Applied Crystallography. [Mozartweg 1b, D-64287 Darmstadt, Germany.] Dr rer.nat. (TH Darmstadt, 1952) chemistry. *Apparatus imaging_detector*
Tel. 49(6151)74596

Woitok, Dr Joachim (1955). Postdoctoral Research Assistant. I. Physikalisches Institut der RWTH Aachen, Germany. [I. Physikalisches Institut der RWTH Aachen, Sommerfeldstr. 14, Turm 2B, D-52056 Aachen, Germany.] Dr (RWTH Aachen, 1989) physics. *Epitaxy semiconductors high-resolution diffractometry grazing_incidence*
E-mail woitok@acds10.physik.rwth-aachen.de Tel. 49(241)807208 Fax 49(241)807196

Wolf, Dr Eberhard Günter (1932). Research Associate. Institut für Kristallzüchtung, im Forschungsverbund Berlin e.V., Germany. [Institut für Kristallzüchtung, im Forschungsverbund Berlin e.V., Rudower Chaussee 6, Gebäude 18.46, D-12489 Berlin, Germany.] Dr rer.nat. (Akad. d. Wiss. d. DDR Berlin, 1985) physics. *Semiconductor_crystal real_structure analysis defect_behaviour characterization_method optical_microscopy X-ray_topography infrared_transmission_polarization_microscopy crystallographic_orientation multicrystalline_silicon scanning_electron_microscopy*
E-mail ur@ikz.wtza-berlin.de Tel. 49(30)63923092 Fax 49(30)63923003

Wolmershäuser, Dr Gotthelf (1945). Research Associate. Fachbereich Chemie der Universität Kaiserslautern, Germany. [Fachbereich Chemie der Universität Kaiserslautern, Postfach 3049, D-67653 Kaiserslautern, Germany.] Dr (U. Kaiserslautern, 1976) chemistry. *Charge_transfer conductor selenium_compounds sulfur_compounds*
E-mail wolmersh@chemie.uni-kl.de Tel. 49(631)2052468 Fax 49(631)2052187 Telex 45627unikld

Wondratschek, Prof. Dr Hans (1925). Emeritus. Institut für Kristallographie der Universität Karlsruhe, Germany. [Institut für Kristallographie der Universität Karlsruhe, Postfach 6980, D-76128 Karlsruhe, Germany.] Dr rer.nat. (U. Bonn, 1953) mineralogy. *Apatite feldspars group_theory symmetry representation_theory*
E-mail bj01@dkauni2.bitnet or bjo1@ibm3090.rz.uni-karlsruhe.de Tel. 49(721)6083858 Fax 49(721)697720 Telex 17-721166unikar

Worzala, Prof. Dr Horst (1938). Project Leader. Zentrum für Anorganische Polymere, Germany. [Zentrum für Anorganische Polymere, Rudower Chaussee 5, D-12484 Berlin, Germany.] Dr sc.nat. (Akad.d. Wiss.d. DDR Berlin, 1985) chemistry. *Solid_state chemistry isotopaxy phosphorus_compounds powder_diffraction*
Tel. 49(30)63924406

Wulff, Dr Harm (1944). Lecturer. Institut für Physikalische Chemie der Universität Greifswald, Germany. [Institut für Physikalische Chemie der Universität Greifswald, Soldtmannstr. 16, D-17489 Greifswald, Germany.] Dr habil. (U. Greifswald, 1988) chemistry. *Diffractometry luminescence line_profile_analysis thin_film*
Tel. 49(3834)75470 Fax 49(3834)75495

Wunderlich, Dr Hartmut (1938). Research Scientist. Institut für Anorganische Chemie und Strukturchemie der Universität Düsseldorf, Germany. [Institut für Anorganische Chemie und Strukturchemie der Universität Düsseldorf, Universitätsstr. 1, D-40225 Düsseldorf, Germany.] Dr rer.nat. (TU Braunschweig, 1969) physics. *Structure_determination_of_coordination_and_pharmaceutical_compounds instrumentation teaching service_crystallography*
E-mail wunderl@ze8.rz.uni-duesseldorf.de Tel. 49(211)3113144 Fax 49(211)3113085 Telex 08587348unid

Wutzler, Mrs Ronny (1963). Post-Graduate Student. Zentrum für Heterogene Katalyse in der KAI e.V., Germany. [Zentrum für Heterogene Katalyse in der KAI e.V., Rudower Chaussee 5, D-12484 Berlin, Germany.] Dipl. Kristallogr. (Humboldt-U. Berlin, 1985) crystallography. *Phase_determination morphology zinc_compounds ferrites hot_gas_desulfurization_of_coal_gases*
Tel. 49(30)63924317

Yonath, Prof. Dr Ada (1939). Max-Planck-Laboratory for Ribosomal Structure, Germany. [Max-Planck-Laboratory for Ribosomal Structure, c/o DESY, Notkestr. 85, D-22603 Hamburg, Germany.] Dr (Weizmann Inst. Rehovot/Israel, 1969) chemistry. *Proteins_crystallography ribosomes cryogenics_crystallography electron_microscopy image_reconstruction*
E-mail yonath@mpgars.desy.de Tel. 49(40)89982827 Fax 49(40)891314

Ysker, Dr Jan Stinus (1941). Bundesmarine (Navy), Germany. [Mozartstr. 41, D-26382 Wilhelmshaven, Germany.] Dr (U. Münster, 1972) chemistry. *Geophysics geosciences fractal*
Tel. 49(4421)12525

Zaumsell, Dr Peter (1953). Research Assistant. Institut für Halbleiterphysik GmbH, Postfach 409, D-15204 Frankfurt/Oder, Germany. Dr sc.nat. (U. Jena, 1989) physics. *Diffractometry real_structure semiconductor heterostructure*
Tel. 49(335)373131 Fax 49(335)326195

Zedler, Dr Achim (1934). Research Associate. Institut für Kristallographie und Mineralogie der Universität Frankfurt, Germany. [Institut für Kristallographie und Mineralogie der Universität Frankfurt, Senckenberganlage 30, D-60054 Frankfurt/M., Germany.] Dr rer.nat. (Akad. d. Wiss. d. DDR Berlin, 1973) physics. *OD_theory diffraction_theory analysis_of_disordered_structure structural_pseudosymmetry studies_of_building_units stacking_faults_in_inorganic_structures theoretical_ structure_modelling*
Tel. 49(69)7982532 Fax 49(69)7982101

Zheng, Dr Yueqing (1963). Chemistry Department, Ningbo Normal University, Ningbo City, Zhejiang, People's Republic of China. [Institut für Anorganische Chemie der Universität Köln, Greinstr. 6, D-50939 Köln, Germany.] Dr (U. Köln, 1993) chemistry. *Diffraction molecular_crystal catalysts carbonates crystal_growth spectrometry electrochemistry physical_property*

Ziel, Mr Rainer (1966). Additional Scientific Worker. Zentralbereich Elektronenmikroskopie der Technischen Universität Hamburg-Harburg, Germany. [Zentralbereich Elektronenmikroskopie der Technischen Universität Hamburg-Harburg, Eißendorfer Str. 42, D-21071 Hamburg, Germany.] *Structure_determination transmission_electron_microscopy energy-dispersive_analysis computer least-squares_ refinement Rietveld_method absorption_correction computer_modelling image_*

processing high-energy_electron_diffraction order-disorder inorganic_ compounds profile_analysis
Tel. 49(40)77183543 49(40)77183544 Fax 49(40)77182684

Ziemer, Dr Burkhard (1945). Research Assistant. Fachbereich Chemie der Humboldt-Universität zu Berlin, Germany. [Fachbereich Chemie der Humboldt-Universität zu Berlin, Hessische Str. 1-2, D-10115 Berlin, Germany.] Dr rer.nat. (Akad. d. Wiss. d. DDR Berlin, 1974) chemistry. *Powder_diffraction_of_tungsten_compounds structure_determination of dyes high_temperature_X-ray_powder_diffraction*
Tel. 49(30)28468284 Fax 49(30)28468343

Zimmermann, Dr Helmuth W. (1945). Associate Professor. Sub-Editor, WDC9. Institut für Angewandte Physik, Lehrstuhl für Kristallographie, Universität Erlangen-Nürnberg, Germany. [Institut für Angewandte Physik, Lehrstuhl für Kristallographie, Universität Erlangen-Nürnberg, Bismarckstr. 10, D-91054 Erlangen, Germany.] Dr rer.nat.habil. (U. Erlangen-Nürnberg, 1982) crystallography. *Mathematical_crystallography symmetry Laue_technique n-dimensional_crystallography*
E-mail mpkr00@cd4680fs.rrze.uni-erlangen.de Tel. 49(9131)852705 Fax 49(9131)852733

Zink, Mr Ullrich N. (1964). Research Scientist. Institut für Metallkunde und Metallphysik der Technischen Universität Clausthal, Außenstelle am GKSS Forschungszentrum, Geesthacht, Germany. [IMMTUC c/o GKSS-W-TUC, Postfach 1160, D-21494 Geesthacht, Germany.] Dipl. Ing. (TU Clausthal, 1992) materials science. *Physical_metallurgy magnetism neutron texture texture_analysis*
E-mail ulrich.zink@gkss.de Tel. 49(4152)871327 Fax 49(4152)871338

Zobel, Dr Dieter (1941). Research Associate. Institut für Kristallographie der Freien Universität Berlin, Germany. [Institut für Kristallographie der Freien Universität Berlin, Takustr. 6, D-14195 Berlin, Germany.] Dr rer.nat. (FU. Berlin, 1968) chemistry. *Low_temperature charge_density molecular_crystal diffractometer*
E-mail dieter@kristall.chemie.fu-berlin.d400.de Tel. 49(30)8383453 Fax 49(30)8383464

Zorn, Dr Gerhard M. (1956). Research Manager. SIEMENS AG, Corporate Research and Development, Germany. [SIEMENS AG, Corporate Research and Development, Otto-Hahn-Ring 6, D-81730 München, Germany.] Dr rer.nat. (TU München, 1984) mineralogy. *Diffraction_technique high_temperature thin_film interferometry total_reflection*
E-mail wsmchp0348@laocoon.zfe.siemens.de Tel. 49(89)63641411 Fax 49(89)63642256

Zotov, Dr Nikolay Stamenov (1957). A.v. Humboldt Fellow. Institut für Kristallographie der Universität München, Germany. [Institut für Kristallographie der Universität München, Theresienstr. 41, D-80333 München, Germany.] Dr (U. Sofia/Bulgaria, 1991) crystallography. *Crystallography diffraction disorder glasses diffuse_scattering solid_solutions computer_modelling distribution_functions*
E-mail zotov@kri.physik.uni-muenchen.de Tel. 49(89)32094041 Fax 49(89)32094015

Zschach, Dr Siegfried (1938). CARL ZEISS, Geschäftsbereich Mikroskopie, Germany. [CARL ZEISS, Geschäftsbereich Mikroskopie, Postfach 1380, D-73446 Oberkochen, Germany.] Dr (U. Leipzig, 1972) mineralogy. *Mineralogy geology microscopy polarization_microscopy cosmomineralogy cosmochemistry phase_diagram materials*
Tel. 49(7364)202748 Fax 49(7364)204561 Telex 713751-57

GREECE

Sub-Editor: P. I. Rentzeperis

Alexandropoulos, Prof. Nikos G. (1934). Professor of Physics. President, Hellenic Society for Science and Technology of Condensed Matter. Dept. of Physics, University of Ioannina, PO Box 1186, 45110 Ioannina, Greece. PhD (U. Athens, 1964) X-ray physics. *X-ray_physics electron_distribution instrumentation X-ray_diffraction*
E-mail nga@grioanun.bitnet Tel. 30(651)45396 Fax 30(651)45396 Telex 30(651)322160

Alexopoulos, Prof. Cesar (1909). Professor Emeritus, U. Athens. Academy of Athens, 11 Platonos St., Athens 15151, Greece. PhD (ETH Zurich, Switzerland, 1935) crystal physics. *Crystal_physics structure_defects*
Tel. 30(1)6715697 Fax 30(1)223815

Antonopoulos, Prof. Ioannis (1939). Professor. Dept. of Physics, Aristotle University of Thessaloniki, 54006 Thessaloniki, Greece. PhD (U. Thessaloniki, 1972) electron microscopy. *Dislocation structure_properties_of_materials*
Tel. 30(31)998104 Fax 30(31)998019

Argyrakis, Prof. Panos (1950). Associate Professor. Dept. of Physics, Aristotle University of Thessaloniki, 54006 Thessaloniki, Greece. PhD (U. Michigan, USA, 1979) chemical physics. *Disordered_materials complex_systems fractals lattice_properties diffusion*
E-mail cacz11@grtheun1 Tel. 30(31)998043 Fax 30(31)206138

Bozopoulos, Prof. Anastasios (1944). Assistant Professor. Lab of Applied Physics, Aristotle University of Thessaloniki, 54006 Thessaloniki, Greece. DSc (U. Thessaloniki, 1985) crystal structure analysis. *Inorganic_and_organic_crystal_structures*
E-mail bozopoulos@olymp.ccf.auth.gr Tel. 30(31)998194 Fax 30(31)206138

Calamiotou, Prof. Maria S. (1949). Associate Professor. Dept. of Physics, University of Athens, Panepistimioupolis, 15784 Athens, Greece. PhD (U. Athens, 1978) solid state physics. *Dynamical_theory_of_X-ray_diffraction high-resolution_X-ray_diffractometry powder_diffractometry_and_crystal_physics*
E-mail mcalam@atlas.uoa.ariadne-t.gr Tel. 30(1)7284384 Fax 30(1)7234100

Christidis, Prof. Panayiotis Chrysostomos (1942). Associate Professor. Lab. of Applied Physics, Aristotle University of Thessaloniki, 54006 Thessaloniki, Greece. [6 Komninon St., 55236 Panorama, Thessaloniki, Greece.] DSc (U. Thessaloniki, 1975) crystal structure analysis. *Crystal_structure_analysis electron_density_distribution powder_diffraction*
E-mail christidis@olymp.ccf.auth.gr Tel. 30(31)998193 Fax 30(31)206138

Christofides, Prof. George (1945). Professor. Dept. of Mineralogy-Petrology-Economic Geology, Aristotle University of Thessaloniki, 54006 Thessaloniki, Greece. PhD (U. Thessaloniki, 1977) mineralogy-petrology. *Crystallography mineralogy petrology geochemistry*
Tel. 30(31)998541 Fax 30(31)998568

Dimitriadis, Prof. Charalambos (1950). Associate Professor. Dept. of Physics, Aristotle University of Thessaloniki, 54006 Thessaloniki, Greece. PhD (U. Manchester, UK, 1979) solid state electronics. *Semiconductor_devices*
Tel. 30(31)998094 Fax 30(31)206138

Dionysiou-Kouimtzi, Prof. Semiramis (1943). Associate Professor. Dept. of Physics, Aristotle University of Thessaloniki, 54006 Thessaloniki, Greece. PhD (U. Reading, UK, 1977) radiation damage. *Defects_in_semiconductors*
Tel. 30(31)998073 Fax 30(31)332997

Doni, Prof. Efthimia G. (1956). Assistant Professor. Dept. of Physics, Aristotle University of Thessaloniki, 54006 Thessaloniki, Greece. [14 Cornarou St, 54655 Thessaloniki, Greece.] PhD (U. Thessaloniki, 1978) solid state physics. *Lattice-symmetries structure_and_energy of crystal_interfaces triple_junctions*
Tel. 30(31)998155 Fax 30(31)998019

Economou, Prof. Nicolas Alkiviadis (1926). Professor Emeritus. Dept. of Physics, Aristotle University of Thessaloniki, 54006 Thessaloniki, Greece. PhD (Wayne State Univ. Detroit, Mich. USA, 1960) chemical physics. *Semiconductor_physics*
Tel. 30(31)998121 Fax 30(31)214276

Eleftheriadis, Prof. George (1944). Associate Professor. Dept. of Mineralogy-Petrology-Economic Geology, Aristotle University of Thessaloniki, 54006 Thessaloniki, Greece. PhD (U. Thessaloniki, 1977) mineralogy petrology. *Geochemistry volcanology*
Tel. 30(31)998455 Fax 30(31)998568

Eliopoulos, Prof. Elias E. Assistant Professor. Agricultural University of Athens, 75 Iera Odos, 11855 Athens, Greece. PhD (Leeds U., UK, 1986) mineralogy-petrology. *Protein_crystallography_modelling molecular_graphics*
E-mail eliop@isosun.ariadne-t.gr Tel. 30(1)3422888 Fax 30(1)3426011

Euthymiou, Prof. Paraskevi (1923). Professor Emeritus. Physics Lab., University of Athens, 104 Solonos St., Athens, Greece. PhD (U. Athens, 1952) solid state physics. *Semiconductor-physics transport_phenomena_in_crystals*
Tel. 30(1)3611927 Telex 30(1)223815

Filippakis, Dr Sophocles (1933). Senior Scientist. Physics Dept., NCSR "Demokritos", PO Box 60228, 15310 Agia Paraskevi, Athens, Greece. PhD (Weizmann Inst. of Science, Rehovot, Israel, 1966) crystal structure analysis. *Crystal_structures crystal_physics*
Tel. 30(1)6513112/9 Fax 30(1)6511767 Telex 30(1)216199ATOM

Ftikos, Prof. Christos (1949). Assistant professor. Chemistry Lab., Technical University of Athens, Athens, Greece. PhD (Technical U. Athens, 1972) X-ray diffraction. *X-ray_diffraction electron_microscopy cement_chemistry_and_technology*
Tel. 30(1)3691385 Fax 30(1)7728126 Telex 30(1)221682

Gountsidou, Vassiliki (1955). Assistant. Dept. of Physics, Aristotle University of Thessaloniki, 54006 Thessaloniki, Greece. MSc (Aristotle U. Thessaloniki, 1984) electronic physics. *Electronic_physics*
Tel. 30(31)998038

Gregoriadis, Prof. Panagiotis (1943). Assistant Professor. Dept. of Physics, Aristotle University of Thessaloniki, 54006 Thessaloniki, Greece. PhD (Aristotle U. Thessaloniki, 1967) solid state physics. *Electron_microscopy simulation*
E-mail pgrigoriadis@olymp.ccf.auth.gr Tel. 30(31)998064

Hamodrakas, Dr Stavros (1947). Res. asst. Physics Dept., NCSR "Demokritos", PO Box 60228, 15310 Agia Paraskevi, Athens Greece. PhD (U. Leeds, UK, 1980) biophysics. *Molecular_biology proteins pharmacology*
Tel. 30(1)6513112/9 Fax 30(1)6511767 Telex 30(1)216199ATOM

Hatzikraniotis, Evripidis (1957). Lecturer. Dept. of Physics, Aristotle University of Thessaloniki, 54006 Thessaloniki, Greece. PhD (Aristotle U. Thessaloniki, 1991) solid state physics. *Electron_properties_of_semiconductor_devices*
E-mail hatzikranio@olymp.ccf.auth.gr Tel. 30(31)998216 Fax 30(31)998015

Hatzisymeon, Constantinos (1969). Postgraduate student. Dept. of Physics, Aristotle University of Thessaloniki, 54006 Thessaloniki, Greece. BSc (Aristotle U. Thessaloniki, 1991) solid state physics. *Crystal_structures powder_diffraction crystal_growth*
E-mail hatzisymeon@olymp.ccf.auth.gr Tel. 30(31)998126 Fax 30(31)207319

Hountas, Prof. Athanasios S. (1945). Assistant Professor. Physics Lab., Agricultural University of Athens, 75 Iera Odos, 11855 Athens, Greece. PhD (Aristotle U. Thessaloniki, 1978) crystallography. *Dynamical_diffraction crystal_structures direct_methods*
E-mail hountas@aua.ariadne-t.gr Tel. 30(1)5294217 Fax 30(1)5294233

Kagarakis, Prof. Constantine. Professor. Dept. Electrical Institution, Technical University of Athens, Greece. DSc (Technical U. Athens, 1955) chemical engineering. *Semiconductors*
Tel. 30(1)3623770 Fax 30(1)7728126 Telex 30(1)221682

Kambas, Prof. Costas (1945). Assoc. Professor. Dept. of Physics, Aristotle University of Thessaloniki, 54006 Thessaloniki, Greece. PhD (Aristotle U. Thessaloniki, 1981) solid state physics. *Optical_properties_of_solids*
Tel. 30(31)998093 Fax 30(31)998019

Kanellis, Prof. George (1942). Assoc. Professor. Dept. of Physics, Aristotle University of Thessaloniki, 54006 Thessaloniki, Greece. DSc (Aristotle U. Thessaloniki, 1977) solid state physics. *Lattice_dynamics spectroscopy band_structure superlattices*
Tel. 30(31)998103 Fax 30(31)246484

Karakostas, Prof. Theodoros Professor. Dept. of Physics, Aristotle University of Thessaloniki, 54006 Thessaloniki, Greece. PhD (U. Thessaloniki, 1976) semiconductor physics. *Condensed_matter material_science*
E-mail karakostas@olymp.ccf.auth.gr Tel. 30(31)998061 Fax 30(31)998019 Telex 30(31)2181

Katagas, Prof. Christos (1944). Lecturer. Geology Lab., University of Patras, Patras, Greece. PhD (U. Manchester, UK, 1975) semiconductor physics. *Crystal_structures minerals*
Tel. 30(61)991972 Telex 30(61)312239

Kavounis, Prof. Constantin A. (1945). Assistant Professor. Lab of Applied Physics, Aristotle University of Thessaloniki, 54006 Thessaloniki, Greece. DSc (U. Thessaloniki, 1985) crystallography. *Crystal_structures protein_crystallography*
E-mail cavounis@olymp.ccf.auth.gr Tel. 30(31)998134 Fax 30(31)206138

Keramidas, Constantinos (1960). Postdoctoral fellow. Lab of Applied Physics, Aristotle University of Thessaloniki, 54006 Thessaloniki, Greece. PhD (U. Thessaloniki, 1990) crystallography. *Crystal_structures powder_diffractometry archaeometry*
E-mail keramidask@olymp.ccf.auth.gr Tel. 30(31)998168 Fax 30(31)248602

Kokkou, Prof. Sokrates (1940). Associate Professor. Lab of Applied Physics, Aristotle University of Thessaloniki, 54006 Thessaloniki, Greece. DSc (U. Thessaloniki, 1975) crystallography. *Crystal_structures direct_methods powder_diffractometry*
E-mail kokkou@olymp.ccf.auth.gr Tel. 30(31)998134 Fax 30(31)206138

Komninou, Prof. Philomela Assistant Professor. Dept. of Physics, Aristotle University of Thessaloniki, 54006 Thessaloniki, Greece. PhD (U. Thessaloniki, 1987) solid state physics. *Condensed_matter electron_microscopy*
E-mail komninou@olymp.ccf.auth.gr Tel. 30(31)998195 Fax 30(31)998019

Konguetsof, Dr Helen (1938). Lecturer. Dept. of Electr. Engineering, Demokritos University of Thrace, Xanthi, Greece. PhD (U. Sussex, 1970) solid state physics. *Crystal_structures magnetic_structures rare_earth_compounds*
Tel. 30(541)26475 Telex 30(531)462205

Kosmopoulos, Dr John (1945). Assistant Professor. Department of Physics, University of Patras, Patras 26500, Greece. PhD (U. Patras, 1975) physics. *Liquid_crystals*
E-mail kosmop@physics.upatras.gr Tel. 30(61)997450 Fax 30(61)991980

Kotsanidis, Dr Panagiotis (1938). Lecturer. Physics Lab., Dept. of Electr. Engineering, Demokritos University of Thrace, Xanthi, Greece. DSc (Demokritos U. Thrace, 1975) magnetic structures. *Crystal_structures magnetic_structures rare_earth_compounds*
Tel. 30(541)26475 Telex 30(531)462205

Kotsis, Dr Konstadinos T. (1959). Assistant Professor. Dept. of Physics, University of Ioannina, PO Box 1186, 45110 Ioannina, Greece. PhD (U. Ioannina, 1987) X-ray diffraction. *Kinematical_and_dynamical_X-ray_diffraction X-ray_spectroscopy gamma-ray_spectroscopy*
E-mail kkotsis@cc.uoi.gr Tel. 30(651)98564 Fax 30(651)45631 Telex 30(651)322160

Koumelis, Dr Christos (1931). Assistant Professor. University of Athens, Section of Mineralogy and Petrology, Panepistimiopolis, Ano Ilissia, Athens 157 84, Greece. DSc, Reader (U. Athens, 1963, 1973) solid state physics. *Crystal_structures solid_state X-ray_spectroscopy*
Tel. 30(1)7284405 30(1)7243254 Fax 30(1)7249567

Kyriakos, Prof. Dimitris S. (1941). Assoc. Professor. Dept. of Physics, Aristotle University of Thessaloniki, 54006 Thessaloniki, Greece. DSc (U. Thessaloniki, 1979) solid state physics. *Condensed_matter electrical_magnetic_properties conductivity_phenomena* transport properties
Tel. 30(31)998123 Fax 30(31)998019

Leventouri, Dr Dora (1943). Teaching assistant. Physics Lab., University of Athens, 104 Solonos St., Athens, Greece. DSc (U. Athens, 1972) solid state physics. *Solid_state X-ray_spectroscopy*
Tel. 30(1)3633413 Telex 30(1)223815

Loizos, Mr Zafiris (1950). Assistant professor. Chemistry Lab., Technical University of Athens, Athens, Greece. BSc (Technical U. Athens, 1973) chemical engineering. *Inorganic_crystal_structures*
Tel. 30(1)3691244 Telex 30(1)221682

Lontos, Dr Charalambos (1947). Assistant Professor. Physics Lab., University of Athens, 104 Solonos St., Athens, Greece. DSc (U. Athens, 1979) solid state physics. *Solid_state inelastic_scattering*
Tel. 30(1)3633413 Telex 30(1)223815

Manolikas, Prof. Constantinos (1944). Professor. Dept. of Physics, Aristotle University of Thessaloniki, 54006 Thessaloniki, Greece. PhD (U. Thessaloniki, 1976) solid state physics. *Ferroelasticity ferroelectricity superstructures phase_transformations*
Tel. 30(31)998081 Fax 30(31)998019

Mavridis, Prof. Aristides A. (1944). Full Professor; Head of the Physical Chemistry Laboratory. Dept. of Chemistry, University of Athens, Panepistimioupolis 15771, Athens, Greece. PhD (Michigan State U., USA, 1975) physical chemistry. *Electron_distribution_of_small_molecules theoretical_chemistry* history_of_science
E-mail amav@arnold.phchem.uoa.ariadne-t.gr Tel. 30(1)7233219 Fax 30(1)7233219

Mentzafos, Prof. Dimitrios E. Associate Professor. Physics Lab., Agricultural University of Athens, 75 Iera Odos, 11855 Athens, Greece. DSc (U. Grenoble, France, 1971) crystallography. *Inclusion_compounds*
E-mail mentzafos@aua.ariadne-t.gr Tel. 30(1)5294215 Fax 30(1)5294233

Michaelides, Dr Adonis (1952). Group leader. Dept. of Chemistry, University of Ioannina, 45110 Ioannina, Greece. DSc (U. Lyon, France, 1982) chemistry. *X-ray_structure_determination crystal_engineering coordination_compounds crystal_chemistry*
E-mail amihail@cc.uoi.gr Fax 30(651)44112 Telex 30(651)322160

Moustakali-Mavridis, Dr Irene E. (1946). Senior Researcher. Institute Phys. Chemistry, NCSR "Demokritos", PO Box 60228, 15310 Agia Paraskevi, Athens, Greece. PhD (Michigan State U., USA, 1975) physical chemistry. *Structural_properties macrocycle_chemistry macromolecular_crystallography*
E-mail mavridi@cyclades.nrcps.ariadne-t.gr Tel. 30(1)6513112/634 Fax 30(1)6511767

Nastopoulos, Dr Vassilios (1957). Assistant Professor. Dept. of Chemistry, University of Patras, 265 00 Patras, Greece. DSc (U. Patras, 1987) chemistry. *Crystal_structure_determination protein_crystallography*
E-mail nastopoulo@upatras.gr Tel. 30(61)997133 Fax 30(61)997118

Panagos, Prof. Athanasios (1926). Professor Emeritus. Geosciences Lab., Technical University of Athens, 15780 Athens, Greece. PhD (ETH Zurich, Switzerland, 1960) crystallography. *Inorganic_structures mineralogy*
Tel. 30(1)7783525 Fax 30(1)3618094

Papadimitriou, Prof. Leonidas (1944). Associate Professor. Dept. of Physics, Aristotle University of Thessaloniki, 54006 Thessaloniki, Greece. PhD (U. Thessaloniki, 1981) solid state physics. *Semiconductors semiconductor_devices*
Tel. 30(31)998214 Fax 30(31)998019

Papadopoulos, Dr Dimitrios (1945). Assistant Professor. Dept. of Physics, Aristotle University of Thessaloniki, 54006 Thessaloniki, Greece. BSc (U. Thessaloniki, 1969) electron microscopy. *Electron_microscopy solid_state_physics*
Tel. 30(31)998086 Fax 30(31)998019

Papaioannou, Prof. John (1947). Assistant Professor. Dept. of Chemistry, University of Athens, Panepistimioupolis, 15771 Athens, Greece. PhD (U. Athens, 1984) physics. *Electrical_properties_of_solids crystallography_of_small_molecules*
E-mail jpapaio@arnold.phchem.uoa.ariadne-t.gr Tel. 30(1)7284517 Fax 30(1)7233219

Paraskevopoulos, Prof. Konstantinos (1950). Associate Professor. Dept. of Physics, Aristotle University of Thessaloniki, 54006 Thessaloniki, Greece. PhD (U. Thessaloniki, 1969) solid state physics. *Optical_properties semiconductors ceramic_materials*
E-mail kpar@olymp.ccf.auth.gr Tel. 30(31)998015 Fax 30(31)998015

Parissakis, Prof. George (1929). Professor; Director, Laboratory of Inorganic and Analytic Chemistry. Chemistry Lab., Technical University of Athens, Athens, Greece. PhD (ETH Zurich, Switzerland, 1955) chemistry. *X-ray_diffraction electron_microscopy cement_chemistry*
Tel. 30(1)7758759 Fax 30(1)7758759 30(1)7700989

Perdikatsis, Dr Basilios (1942). Group leader. Mineralogy Lab., Inst. Geol. and Mineral. Res., 70 Mesogion St., Athens 11527, Greece. PhD (U. Erlangen, Germany, 1972) crystallography. *Powder_diffraction microanalysis mineralogy*
Tel. 30(1)7798412 30(1)7784508 Fax 30(1)7752211 Telex 30(1)223915

Petratos, Dr Kyriacos (1955). Research Assistant Professor. Institude of Molecular Biology and Biotechnology, Foundation for Research and Technology, PO Box 1527, 711 10 Heraklion, Crete, Greece. PhD (Wayne State U., Detroit, USA, 1984) biochemistry. *Metalloproteins structure_determination enzymatic_catalysis environment electron_transfer structure_comparison*
E-mail petratos@venus.imbb.forth.gr petratos@nefeli.cc.uch.gr Tel. 30(81)210091 Fax 30(81)230469 Telex 262896 IMBB GR

Polychroniadis, Prof. Evstathios K. (1946). Associate Professor. Dept. of Physics, Aristotle University of Thessaloniki, 54006 Thessaloniki, Greece. PhD (U. Thessaloniki, 1981) solid state physics. *Electron_microscopy thin_films*
Tel. 30(31)998163 Telex 30(31)206138

Poulios, Prof. Ioannis (1954). Assistant Professor. Dept. of Chemistry, Aristotle University of Thessaloniki, 54006 Thessaloniki, Greece. PhD (Technical U. Graz, Austria, 1982) physical chemistry. *Semiconductors photoelectrochemistry solar_cells conductive_polymers heterogeneous_photocatalysis*
Tel. 30(31)997785 Fax 30(31)997709 30(31)206138

Priftis, Prof. George (1937). Professor. Physics Lab., University of Patras, Patras, Greece. DSc (U. Athens, 1969) physics. *Diffuse_scattering inorganic_structures*
Tel. 30(61)997453 Telex 30(61)991996

Profi, Ms Stella (1949). Res. Asst. Physics Dept., NCSR "Demokritos", PO Box 60228, 15310 Agia Paraskevi, Athens, Greece. BSc (Athens U., 1972) physics. *Inorganic_and_mineral_structures mineral_chemistry*
Tel. 30(1)6513112 Fax 30(1)6511766 Telex 30(1)216184

Rentzeperis, Prof. Panagiotis I. (1928). Professor. WDC9 Sub-Editor for Greece. Lab of Applied Physics, Aristotle University of Thessaloniki, 54006 Thessaloniki, Greece. [13 M. Alexandrou Ave, 54640 Thessaloniki, Greece.] DSc (Aristotle U. Thessaloniki, 1956) crystallography, mineralogy and petrography. *Crystal_structures crystal_physics archaeology music stamps photography*
E-mail rentzeperis@olymp.ccf.auth.gr Tel. 30(31)998151 Fax 30(31)206138

Rontoyianni, Dr Aliki (1961). Research Associate. Institute Phys. Chemistry, NCSR "Demokritos", PO Box 60228, 15310 Agia Paraskevi, Athens, Greece. PhD (U. Athens, 1994) crystallography. *Structural_chemistry supramolecular_crystallography*
E-mail aront@cyclades.nrcps.ariadne-t.gr Tel. 30(1)6513112/634 Fax 30(1)6511766 Telex 30(1)216184

Routsi, Prof. Christine (1950). Assistant Professor. Physics Lab., Dept. of Electr. Engineering, Demokritos University of Thrace, 67100 Xanthi, Greece. PhD (Demokritos U. Thrace, 1985) magnetic structures. *Magnetic_materials hyperfine_interactions*
Tel. 30(541)29310 Fax 30(541)20373

Sahalos, John (1943). Professor. Dept. of Physics, Aristotle University of Thessaloniki, 54006 Thessaloniki, Greece. PhD (U. Thessaloniki, 1974) electromagnetics. *Microwaves magnetic_materials*
E-mail sahalos@olymp.ccf.auth.gr Tel. 30(31)998161 Fax 30(31)248639

Skoulika, Dr Stavroula (1953). Group leader. Dept. of Chemistry, University of Ioannina, 45110 Greece. DU (Aix–Marseille I, 1982) chemistry. *X-ray_structure_determination crystal_growth inhibitors crystal_chemistry organic_organometallic compounds*
E-mail vskoull@cc.uoi.gr Tel. 30(651)98446 Telex 30(651)44112

Soldatos, Prof. Triantafyllos (1957). Assistant Professor. Dept. of Mineralogy-Petrology-Economic Geology, Aristotle University of Thessaloniki, 54006 Thessaloniki, Greece. PhD (U. Thessaloniki, 1985) mineralogy-petrology-geochemistry. *Igneous_rocks mineralogy petrology geochemistry*
Tel. 30(31)998497 Fax 30(31)998568

Stergiou, Prof. Anagnostis (1941). Assistant Professor. Lab of Applied Physics, Aristotle University of Thessaloniki, 54006 Thessaloniki, Greece. PhD (U. Thessaloniki, 1984) crystal structure analysis. *Crystal_growth crystal_structures magnetic properties*
E-mail anag@olymp.ccf.auth.gr Tel. 30(31)998075 Fax 30(31)248602

Stergioudis, Prof. George (1946). Assistant Professor. Lab of Applied Physics, Aristotle University of Thessaloniki, 54006 Thessaloniki, Greece. DSc (U. Thessaloniki, 1984) crystal structure analysis. *Amorphous_materials local_structure*
E-mail stergioudis@olymp.ccf.auth.gr Tel. 30(31)998085 Fax 30(31)206138

Stoemenos, Prof. Ioannis (1934). Professor. Dept. of Physics, Aristotle University of Thessaloniki, 54006 Thessaloniki, Greece. PhD (U. Thessaloniki, 1969) solid state physics. *Solid_state semiconductors electron_microscopy*
Tel. 30(31)998191 Fax 30(31)998191

Terzis, Dr Aristides (1941). Senior Researcher. Physics Dept., NCSR "Demokritos", PO Box 60228, 15310 Agia Paraskevi, Athens, Greece. PhD (Princeton U., USA, 1970) crystallography. *Metal_complexes unidirectional_compounds low_temperatures*
Tel. 30(1)6513112/123 Fax 30(1)6511767 Telex 30(1)216199

Triantafyllou, Spyros (1968). Postgraduate student. Dept. of Physics, Aristotle University of Thessaloniki, 54006 Thessaloniki, Greece. BSc (U. Thessaloniki, 1991) physics. *Crystal_growth crystal_structures powder_diffraction*
E-mail triantafilou@olymp.ccf.auth.gr Tel. 30(31)998186 Fax 30(31)207319

Trontsios, Prof. George (1945). Lecturer. Dept. of Mineralogy, Aristotle University of Thessaloniki, 54006 Thessaloniki, Greece. PhD (U. Thessaloniki, 1991) sedimentary rocks. *Industrial_minerals_and_rocks*
Tel. 30(31)998518 Fax 30(31)998568

Tsirambides, Prof. Ananias (1949). Associate Professor. Dept. of Mineralogy, Aristotle University of Thessaloniki, 54006 Thessaloniki, Greece. PhD (U. Thessaloniki, 1983) sedimentary rocks. *Industrial_minerals_and_rocks*
Tel. 30(31)998504 Fax 30(31)998568

Vavelidis, Prof. Michael (1952). Associate Professor. Dept. of Mineralogy-Petrology-Economic Geology, Aristotle University of Thessaloniki, 54006 Thessaloniki, Greece. PhD (U. Heidelberg, Germany, 1984) ore deposits. *Ore_deposits archeometry* Tel. 30(31)998474 Fax 30(31)998568

Voutsas, Prof. George (1945). Assistant Professor. Lab of Applied Physics, Aristotle University of Thessaloniki, 54006 Thessaloniki, Greece. PhD (U. Thessaloniki, 1983) crystal structure analysis. *Crystal_growth crystal_structures semiconductors*

archeometry powder_diffraction E-mail voutsas@olymp.ccf.auth.gr Tel. 30(31)998054 Fax 30(31)248602

Yakinthos, Prof. Ioannis (1937). Professor; Director of the Phyical and Applied Mathematics Section. Physics Lab., Dept. of Electr. Engineering, Demokritos University of Thrace, 67100 Xanthi, Greece. DSc (U. Grenoble, 1971) solid state physics. *Crystal_properties magnetic_structures hyperfine_interactions physical_properties* E-mail yakinthos@xanthi.cc.duth.gr Tel. 30(541)26475 Telex 30(531)462205

HONG KONG

Sub-Editor: T. C. W. **Mak**

Cheng, Mr Graham Cheng-Hsun (1936). Director. [Taching Petroleum Company Limited, 19/F., Fung House, 20 Connaught Road Central, Hong Kong.] MSc (Manchester U., 1962) geology-mineralogy. *Crystal_structures high-temperature_materials* Tel. 852()23366169 Fax 852()23382553 Telex 39843 TAOIL HX

Cheung, Dr Kung Kai (1935). Senior Lecturer. [Department of Chemistry, The University of Hong Kong, Pokfulam Road, Hong Kong.] PhD (Glasgow U., 1965) chemistry. *Crystal_structure determination structural_chemistry* E-mail kkcheung@hkucc.bitnet Tel. 852()28592167 Fax 852()28571587 Telex 71919 CEREB HX

Hon, Dr Ping-Kay (1936). Lecturer. [Department of Chemistry, The Chinese University of Hong Kong, Shatin, New Territories, Hong Kong.] PhD (U. Illinois, 1964) chemistry. *Analytical_chemistry instrumentation X-ray_fluorescence* Tel. 852()26096371 Fax 852()26035057 Telex 50301 CUHK HX

Lai, Dr Ting Fong (1930). Research Associate. [Department of Chemistry, The University of Hong Kong, Pokfulam Road, Hong Kong.] PhD (U. Oxford, 1960) chemistry. *Crystal_structures coordination_compounds* Tel. 852()28592159 Fax 852()28571586

Mak, Prof. Thomas Chung Wai (1936). Professor of Chemistry; Chairman. WDC9 Sub-Editor for Hong Kong. [Department of Chemistry, The Chinese Univer-

sity of Hong Kong, Shatin, New Territories, Hong Kong.] PhD (U. British Columbia, 1963) chemistry. *Structural_chemistry inclusion_compounds hydrogen-bonded_molecular_adducts teaching* E-mail tcwmak@cuhk.hk or b005783@cucsc.bitnet Tel. 852()26096343 Fax 852()26035057 Telex 50301 CUHK HX

Williams, Dr Ian Duncan (1959). Lecturer. [Department of Chemistry, Hong Kong University of Science and Technology, Clearwater Bay, Hong Kong.] PhD (U. Bristol, 1985) inorganic chemistry. *Crystal_engineering ferroics nonlinear_optics materials_chemistry catalysis* E-mail chwill@usthk.bitnet Tel. 852()23587384 Fax 852()23581594

Wong, Dr Wing Tak (1963). Lecturer. [Department of Chemistry, The University of Hong Kong, Pokfulam Road, Hong Kong.] PhD (U. Cambridge, 1991) chemistry. *Metal_clusters organometallic_compounds inorganic_polymers computation* E-mail wtwong@hkucc.bitnet Tel. 852()28592157 Fax 852()28571586

Wong, Dr Yau-Shing (1946). Senior Chemist. [Hong Kong Government Laboratory, 7/F Homantin Government Offices, 88 Chung Hau Street, Kowloon, Hong Kong.] PhD (U. Waterloo, 1976) chemistry. *Heavy_metal amino_acid complexes forensic sciences* Tel. 852()27623819 Fax 852()27144299

HUNGARY

Sub-Editor: A. **Borbély**

Arató, Dr Péter (1941). Department leader. [Research Institute for Technical Physics of the Hungarian Academy of Sciences, H-1325, PO Box 76, Budapest, Hungary.] PhD (Eötvös U. Budapest, 1974) natural sci. *Powder_diffraction computing ceramics* E-mail h843bar@ella.uucp Tel. 36(1)1692100 Fax 36(1)1698037

Argay, Mr Gyula (1939). Res. scient. [Dept. X-ray Diffraction, Central Res. Inst. Chemistry, Hung. Acad. Sci., H-1525 Budapest, PO Box 17, Hungary.] MSc (ELTE, 1962) physics. *Computing X-ray_crystallography* E-mail H1250KAL@ella.hu Tel. 36(1)2124790 Fax 36(1)2125020

Böcskei, Dr Zsolt (1962). Research Associate. [Chinoin Pharmaceuticals, Department of Chemical Research, Budapest, PO Box 110, H-1325 Hungary.] PhD (Eötvös U. Budapest, 1985) chemistry. *Proteins_crystallography small_molecules* E-mail H2959BOC@huella.Bitnet Tel. 36(1)1692500x2234 Fax 36(1)1690293 Telex 224236ChinH

Bombicz, Dr Petra (1966). Res. scient. [Dept. X-ray Diffraction, Central Res. Inst. Chemistry, Hungarian Academy Sciences, H-1525 Budapest, PO Box 17, Hungary.] PhD (Technical U. Budapest, 1993) chemistry. *Molecular_recognition computer_modelling clathrates crystal_growth crystallization* E-mail H1475CZU@ELLA.HU or H1475CZU@HUELLA.BITNET Tel. 36(1)2124790 Fax 36(1)2125020

Borbély, Mr András (1960). Lecturer. Hungarian Sub-Editor of Ninth Edition of WDC. Inst. General Physics, Eötvös University, 1088 Budapest, Muzeum Krt. 6-8, Hungary. [1445 Budapest 8, Muzeum Krt. 6-8, Pf. 323, Hungary.] MSc (U. Babes-Bolyai Kolozsvár, 1984) physics. *Computing X-ray_diffraction real_structure metals alloys* E-mail borbely@ludens.elte.hu Tel. 36(1)2669833 Fax 36(1)2660206

Bortel, Mr Gábor (1969). Postgraduate student. [Research Institute for Solid State Physics of the Hungarian Academy of Sciences, 1525 Budapest, PO Box 49, Hungary.] M. S. (Roland Eötvös U., 1993) physics. *Computer_modelling crystallography diffraction phase_transition computing structural_change synchrotron_radiation fullerenes* E-mail gb@power.szfki.kfki.hu Tel. 36(1)1699499x2371 Fax 36(1)1551193

Czugler, Dr Mátyás (1948). Sen. res. fellow. [Dept. X-ray Diffraction, Central Res. Inst. Chemistry, Hung. Acad. Sci., H-1525 Budapest, PO Box 17, Hungary.] PhD (Hungarian Academy of Sciences, 1989) chemistry. *Molecular_recognition*

computer_modelling nucleosides organic sulfur_compounds clathrates coordination_compounds E-mail H1475CZU@ELLA.HU or H1475CZU@HUELLA.BITNET Tel. 36(1)2124790 Fax 36(1)2125020

Faigel, Dr Gyula (1954). Group leader. [Research Institute for Solid State Physics of the Hungarian Academy of Sciences, 1525 Budapest, PO Box 49, Hungary.] PhD (Roland Eötvös U., 1982) physics. *Amorphous_phase charge-density_wave cooperative_phenomena holography crystallography diffraction EXAFS interferometry lattice_dynamics magnetism Mossbauer multilayer phase_transition resonant_scattering structure synchrotron_radiation fullerenes* E-mail gf@power.szfki.kfki.hu Tel. 36(1)1699499x2371 Fax 36(1)1551193

Farkas-Jahnke, Dr Mária (1932). Senior research fellow; Head of Laboratory. Member of Commission of Aperiodic Crystals. [Research Institute for Technical Physics, Hungarian Academy of Sciences, H-1325 Budapest, PO Box 76, Hungary.] PhD (Hungarian Academy of Sciences, 1974) polytypes. *Phase_transition oxides polytypes ZnS* Tel. 36(1)1692-100 Fax 36(1)1698-037

Griger, Dr Ágnes (1945). Lab. leader. Laboratory of Powder Metallurgy and Composite Materials, ALUTERV-FKI Engineering and Development Centre, HUNGALU, H-1116 Budapest, Fehérvári út 144, Hungary. [1502 Budapest, PO Box 308, Hungary.] PhD (Hungarian Academy of Sciences, 1993) physics. *Powder_diffraction computing_crystallography intermetallic_phase_in_aluminium_compounds mechanical_alloying composite_materials* Tel. 36(1)1812979 Fax 36(1)1810181 Telex 226029fkibbh

Gurbán, Mr Sándor (1958). Research fellow. [Research Institute for Technical Physics of the Hungarian Academy of Sciences, H-1325 Budapest, PO Box 76, Hungary.] *Topography single_crystal diffraction structural_disorder* Tel. 36(1)1692100 Fax 36(1)1698037

Hajdu, Dr Ferenc (1926). Retired, Senior Research Fellow. Central Research Inst. for Chemistry, Hungarian Academy of Sciences, H-1025 Budapest, Rákóczi u. 12, Hungary. [Central Research Inst. for Chemistry, Hungarian Academy of Sciences, H-1525 Budapest, PO Box 17, Hungary.] PhD (Hungarian Academy of Sciences, 1975) physical chemistry. *X-ray_short-range_order liquids amorphous_metals diffraction_theory* Tel. 36(1)1422698

Hange, Mr Ferenc (1954). Head of Lab. [Röntgen Lab., Tungsram Co. Ltd, H-1340 Budapest, Váci út 77, Hungary.] MSc (ELTE, 1979) physics. *Powder_diffractometry spectrometry*
Tel. 36(1)692800x1663

Hargittai, Prof. István (1941). Head. Structural Chemistry Research Group of the Hungarian Academy of Sciences, Eötvös University, Puskin utca 11-13, H-1088 Budapest,. Hungary. [Institute of General and Analytical Chemistry, Technical University of Budapest, Szt. Gellért tér 4, H-1521 Budapest, Budapest Hungary.] Full Member Hungarian Academy of Sciences (Hungarian Academy of Sciences, 1993) chemistry. *Structure chemistry electron_microscopy symmetry*
E-mail hargittai@ch.bme.hu Tel. 36(1)1669035 Fax 36(1)1853493

Hargittai, Prof. Magdolna (1945). Scientific advisor. Structural Chemistry Research Group of the Hungarian Academy of Sciences, Eötvös University, Puskin utca 11-13, H-1088 Budapest, Hungary. [Structural Chemistry Research Group of the Hungarian Academy of Sciences, Eötvös University, Pf. 117, H-1431 Budapest, Hungary.] DSc (Hungarian Academy of Sciences, 1991) chemistry. *Inorganic_molecular_structure electron_microscopy symmetry photography*
E-mail hargittaim@ludens.elte.hu Tel. 36(1)2663942 Fax 36(1)2663899

Hartmann, Prof. Ervin (1935). Scientific councillor. [Res. Lab. Crystal Physics, Hungarian Academy of Sciences, H-1112 Budapest, Budaörsi ut 45, Hungary.] DSc (Hungarian Academy of Sciences, 1988) physics. *Crystal_growth crystal_physics*
E-mail h4359har@ella.hu Tel. 36(1)1850777 Fax 36(1)227030ecnath

Horváth, Dr Zsolt E. (1964). Scientist. KFKI Res. Inst. for Materials Science, Budapest XII, Konkoly Thege ut 29-33, Hungary. [KFKI RIMS, PO Box 49, 1525 Budapest, Hungary.] PhD (Eötvös U. Budapest, 1993) physics. *High-resolution_diffractometry phase_determination rocking_curves transmission_electron_microscopy*
E-mail xray@r1.atki.kfki.hu Tel. 36(1)1699499 Fax 36(1)1550694

Kálmán, Prof. Alajos (1935). Head. Coeditor of Acta Crystallographica. [Dept. X-ray Diffraction, Central Res. Inst. Chemistry, Hung. Acad. Sci., H-1525 Budapest, PO Box 17, Hungary.] DSc (Hungarian Academy Sciences, 1975) chemistry. *Cardiac_compounds steroids nucleosides organic_sulfur_compounds superstructure polymorphism isomorphism*
E-mail H1250KAL@ELLA.HU or h1250kal.huella.bitnet Tel. 36(1)2124790 Fax 36(1)2125020

Koritsánszky, Dr Tibor (1954). Research associate. Central Institute for Chemistry of the Hungarian Academy of Sciences, H-1525 Budapest, PO Box 17, Hungary. [Free University Berlin, Institute for Crystallography, Takustr. 6, 14195 Berlin, Germany.] PhD (FU–Berlin, 1992) chemistry. *Charge_density theoretical_crystallography computing*
E-mail tibor@kristall.chemie.fu-berlin.de Tel. 49(30)8386779 Fax 49(30)8383464

Kőszegi, Mr László (1945). Group member. Research Institute for Solid State Physics of the Hungarian Academy of Sciences, Budapest XII, Konkoly-Thege str. 29-33, Hungary. [Research Institute for Solid State Physics, H-1525 Budapest, PO Box 49, Hungary.] MSc (ELTE–TTK, 1968) physics. *Neutron_scattering metals internal_stress magnetism*
E-mail H6398KOS@HUELLA.BITNET Tel. 36(1)1699499x1469 Fax 36(1)1698566

Lovas, Dr György Antal (1947). Sr. asst. [Dept. Mineralogy, ELTE, Muzeum Krt. 4/A, 1088 Budapest, Hungary.] PhD (ELTE, 1985) physics. *Powder_diffraction crystal_and_molecular_structure_determination minerals*
E-mail lovas@ludens.elte.hu Tel. 36(1)2669833 Fax 36(1)2660206

Malickó, Ass. Prof. Dr László (1934). Senior research fellow. Member of the Hungarian National Committee of IUCr. [Research Laboratory for Crystal Physics of the Hungarian Academy of Sciences, H-1502 Budapest 112, PO Box 132, Hungary.] Candidate of physical sciences (Hungarian Academy of Sciences, 1968) physics. *Crystal_growth real_structure_of_crystal optical_and_electron_microscopy*
Tel. 36(1)1850777/363 Fax 36(1)1851158 Telex TX227039ECNAT

Oszlányi, Dr Gábor (1963). Scientific worker. [Research Institute for Solid State Physics of the Hungarian Academy of Sciences, 1525 Budapest, PO Box 49, Hungary.] PhD (Roland Eötvös U., 1990) physics. *Computing crystallography data_collection diffraction high_pressure phase_transition structural_change structure synchrotron_radiation fullerenes*
E-mail go@power.szfki.kfki.hu Tel. 36(1)1699499x2371 Fax 36(1)1551499

Párkányi, Dr László (1940). Group leader. [Dept. X-ray Diffraction, Central Res. Inst. Chemistry of the Hungarian Academy of Sciences, H-1525 Budapest, PO Box 17, Hungary.] DSc (Hungarian Academy Sciences, 1990) chemistry. *Computing silicon_compounds transition_elements organic_sulfur_compounds coordination_compounds nucleosides*
E-mail H1072PAR@ELLA.HU or h1072par@huella.bitnet Tel. 36(1)2124790 Fax 36(1)2125020

Petrás, Mr László (1957). Research Assistant. [Structure Research Department, Research Institute for Technical Physics of the Hungarian Academy of Science, PO Box 76, H-1325 Budapest, Hungary.] MSc (Eötvös U., Budapest, 1981) physics. *High_temperature_powder_X-ray_diffractometry phase_transition*
E-mail H1935PET@ELLA.HU Tel. 36(1)1692100 Fax 36(1)1698037

Radnai, Dr Tamás (1949). Senior research fellow. Central Research Institute for Chemistry, Hungarian Academy of Sciences Budapest, Pusztaszeri ut 59-67, H-1025 Hungary. [Central Research Institute for Chemistry, Hungarian Academy of Sciences Budapest, PO Box 17, H-1525 Hungary.] PhD (Hungarian Academy of Sciences, 1991)

chemistry. *Liquid_X-ray_diffraction structure_chemistry complex_chemistry computer_simulations solution_chemistry*
E-mail H1204RAD@ELLA.HU Tel. 36(1)1351540 Fax 36(1)1352148 Telex 226686

Radnóczy, Dr György (1946). Head of division. [Research Institute for Technical Physics of the Hungarian Academy of Sciences, H-1325 Budapest, PO Box 76, Hungary.] PhD (Hungarian Academy of Sciences, 1982) electron microscopy of grain boundaries. *Electron_microscopy thin_film multilayers*
E-mail hrad2994@huella.bitnet Tel. 36(1)169-0315 Fax 36(1)169-8037

Redler, Dr László (1954). Res. scient. [Dept. Silicate Chemistry, Central Research and Design Inst. Silicate Industry, H-1300 Budapest, PO Box 112, Hungary.] DPhil (Chemical U. Veszprém, 1991) physics silicate chemistry. *Powder_diffraction clays phase_determination cement*
Tel. 36(1)2501311x273 Fax 36(1)1687626 Telex 226827

Rozsondai, Dr Béla (1934). Senior research scientist. Structural Chemistry Research Group of the Hungarian Academy of Sciences, Eötvös University, Puskin utca 11-13, H-1088 Budapest, Hungary. [PO Box 117, H-1431 Budapest, Hungary.] PhD (Hungarian Academy of Sciences, 1977) chemistry. *Molecular_structure electron_diffraction*
E-mail h2378roz@ella.hu Tel. 36(1)2663942 Fax 36(1)2663899

Sasvári, Dr Kálmán (1912). Res. Scient. emeritus (organic crystal structures, computer programming). Dept. X-ray Diffraction, Central Res. Inst. Chemistry, Hung. Acad. Sci., H-1525 Budapest, PO Box 17, Hungary. [1118 Budapest, Ugron G.u. 80, Hungary.] PhD (Hungarian Academy of Sciences, 1938) organic structures. *Geometry_of_amino_acids_in_complex_molecules*
Tel. 36(1)186-5130

Schultz, Dr György (1938). Senior research scientist. Structural Chemistry Research Group of the Hungarian Academy of Sciences, Eötvös University, Puskin utca 11-13, H-1088 Budapest, Hungary. [PO Box 117, H-1431 Budapest, Hungary.] PhD (Hungarian Academy of Sciences, 1982) chemistry. *Molecular_structure electron_diffraction*
E-mail H2365SCH@ELLA.HU Tel. 36(1)2663942 Fax 36(1)2663899

Simon, Dr Kálmán (1946). Head of the Analytical Unit. Member of the Data Commission. [Chinoin Research and Development, Chemical Research Department, Budapest, PO Box 110, H-1325 Hungary.] PhD (Hungarian Academy of Sciences, 1981) chemistry. *Computer-assisted_design chirality polymorphism*
E-mail 112959BOC@Huella.Bitnet Tel. 36(1)1692500x2234 Fax 36(1)1690293 Telex 224236ChinH

Szent-Királyi, Mrs Zsuzsanna (1950). Developing engineer. EGIS Pharmaceuticals, H-1106 Budapest, Keresztturi ut 30-38, Hungary. [1475 Budapest 10, PO Box 100, Hungary.] Speciality physicist (Eötvös-Lóránd U., 1994) material sciences. *Polymorphism pharmaceutical_compound X-ray_powder_diffraction minerals*
Tel. 36(1)2527222x292 Fax 36(1)1834554 Telex 22-4101 egish

Tarján, Prof. Imre (1912). Prof. emeritus, scientific adviser. Honorary President of the Society of Hungarian Crystallographers. [Inst. of Biophysics, Semmelweis University of Medicine, H-1088 Budapest, Puskin u. 9, Hungary.] Member of the Hungarian Academy of Sciences (Hungarian Academy of Sciences, 1970) physics. *Crystal_growth crystal_defect molecular_biophysics*
E-mail h6295ron@ella.hu Tel. 36(1)2666656 Fax 36(1)2666656

Tegze, Dr Miklós (1954). Senior scientist. [Research Institute for Solid State Physics of the Hungarian Academy of Sciences, 1525 Budapest, PO Box 49, Hungary.] PhD (Roland Eötvös U., 1982) physics. *Amorphous_phase band_calculation crystallography diffraction holography Mossbauer resonant_scattering structural_disorder superconductors synchrotron_radiation fullerenes*
E-mail mt@power.szfki.kfki.hu Tel. 36(1)1699499x2371 Fax 36(1)1551193

Tóth, Dr Lajos (1952). Senior research fellow. [Research Institute for Technical Physics of the Hungarian Academy of Sciences, H-1325 Budapest, PO Box 76, Hungary.] PhD (Hung. Academy of Sciences, 1990) physics. *Transmission_electron_microscopy thin_film phase_transition*
E-mail h3650tot@ella.hu Tel. 36(1)1692100 Fax 36(1)1693541

Ungár, Prof. Tamás (1943). Full professor. Inst. General Physics, Eötvös University, 1088 Budapest, Muzeum Krt. 6-8, Hungary. [1445 Budapest 8, Muzeum Krt. 6-8, Pf. 323, Hungary.] DSc (Hungarian Academy of Sciences, 1989) physics. *Line_profile_analysis residual_stress real_structure metals alloys*
E-mail ungar@ludens.elte.hu Tel. 36(1)2669833 Fax 36(1)2660206

Viczián, Dr István (1940). Scientific adviser. [Hungarian Geological Survey, H-1143 Budapest, Stefánia ut 14, Hungary.] DSc (Hungarian Academy of Sciences, 1989) earth sciences. *X-ray_diffraction minerals sedimentary_rocks*
E-mail h6543tit@ella.hu Tel. 36(1)2510999 Fax 36(1)2510703 Telex 36(1)225220mafih

Zsoldos, Mrs Éva (1934). Scientist. KFKI Res. Inst. for Materials Science, Budapest XII, Konkoly Thege ut 29-33, Hungary. [KFKI RIMS, PO Box 49, 1525 Budapest, Hungary.] MSc (Eötvös U., Budapest, 1957) physics. *Phase_determination thin_layer lattice_parameter topography*
E-mail xray@r1.atki.kfki.hu Tel. 36(1)1699499 Fax 36(1)1550694

Zsoldos, Dr Lehel (1931). Senior scientist. Chairman of the Hungarian National Committee of Crystallography. Res. Inst. for Technical Physics of the Hungarian Academy of Sciences, Budapest, Hungary. [Bibó I. u. 16/b, 2092 Budakeszi, Hungary.] PhD (Hungarian Academy of Sciences, 1965) solid state physics. *X-ray_topography dynamical_diffraction multiple-crystal_diffractometry radiation_protection*
E-mail h2835zso@huella.bitnet Tel. 36(1)1386241 Fax 36(1)1698037

ICELAND

Sub-Editor: S. Steinthórsson

Eiríksson, Dr Vésteinn Rúni (1944). Lecturer. Hamrahlíd College, Reykjavík, Iceland. [Birkigrund 24, IS-200 Kópavogur, Iceland.] PhD (U. Edinburgh, UK, 1974) physics. *Phase_transitions_and_structure ferroelectrics* Tel. 354(1)44710

Gíslason, Dr Haflidi Pétur (1952). Professor. Science Institute, University of Iceland, Reykjavík, Iceland. [Dunhagi 3, IS-107 Reykjavík, Iceland.] Tech. Dr (U. Lund, Sweden, 1982) semiconductor physics. *Defects_in_semiconductors optical_detection_of_magnetic_resonance transport_in_semiconductors* E-mail haflidi@raunvis.hi.is Tel. 354(1)694680 Fax 354(1)28911

Sigvaldason, Dr Gudmundur (1932). Director. The Nordic Volcanologic Institute, Reykjavík, Iceland. [Geology Building, IS-101 Reykjavík, Iceland.] Dr. Rer. Nat. (U. Göttingen, 1959) geochemistry mineralogy. *Crystal_optics silicate_crystal_structure*

clay_minerals E-mail ges@norvol.hi.is Tel. 354(1)694491 Fax 354(1)29450

Steinthórsson, Dr Sigurdur (1940). Professor. Sub-Editor, World Directory of Crystallographers 9th Edition. Department of Geology, Faculty of Sciences, University of Iceland, Reykjavík, Iceland. [Geology Building, IS-101 Reykjavík, Iceland.] PhD (Princeton U., USA, 1974) geochemistry petrology. *Crystal_optics structure_of_silicates_and_oxides igneous_petrology geochemistry* E-mail sigst@raunvis.hi.is Tel. 354(1)694476 Fax 354(1)694499

Tómasson, Cand. Real. Jens (1925). Section Leader. National Energy Authority, Geothermal Division, Grensásvegur 9, IS-108 Reykjavík, Iceland. Cand. Real. (U. Oslo, Norway, 1962) mineralogy petrology. *Low_temperature_metamorphic_minerals* E-mail jt@os.is Tel. 354(1)696000 Fax 354(1)688896

INDIA

Sub-Editor: K. K. Kannan

Agarwal, Dr Bhagwati Prasad (1943). Reader. Physics Dept., University School of Sciences, Gujrat University, Ahmedabad 380009, India. PhD (Sardar Patel U. (Gujrat), 1972) physics. *Crystal_growth single_crystal* Tel. 91()423041

Aggarwal, Dr Prem Sarup (1934). Scientist. Central Glass and Ceramic Research Institute, Calcutta 700032, India. PhD (Poona U., 1958) chemistry. *X-ray diffraction ceramics materials electron_diffraction thermal_analysis*

Agrawal, Mr Jawahar Lal (1950). Res. Scholar. Physics Dept., Ranchi University, Ranchi 834008, India. PhD (Ranchi U., 1976) physics. *Crystallography*

Ananthamurthy, Dr Rayasa V. (1948). Scientist. Crystal Growth and Charecterisation Section, National Physical Laboratory, Dr K. S. Krishnan Road, New Delhi 110012, India. PhD (IIT Madras, 1976) X-ray crystallography. *Structure analysis crystal_growth*

Aravindakshan, Dr Cheetambadi Aravindakshan (1934). Addl. Superintendent. Physical Research Wing, Projects and Dev. India Ltd, Sindri, Dhanbad 828122, India. PhD (Madras U., 1974) X-ray crystallography. *Metals physics alloys corrosion fertilizer materials crystal_structure*

Arora, Dr Narinder Kumar (1952). Scient. C. Materials Division, National Physical Laboratory, Dr K. S. Krishnan Road, New Delhi 110012, India. PhD (Delhi U., 1983) semiconductors. *Crystal_growth*

Bahadur, Mr S. Asath (1963). Res. Scholar. School of Physics, Madurai Kamaraj University, Madurai 625021, India. MPhil (Madurai Kamaraj U., 1987) physics. *Crystal_structure_analysis*

Balasingh, Dr C. (1941). Scient. E1. Materials Science Division, National Aeronautical Laboratory, Bangalore 560017, India. PhD (IISc, 1981) *Powder_diffractometry stress_measurement instrumentation*

Balasubramanian, Mr T. (1949). Senior Lecturer. Nehru Memorial College, Puthanampatti, India. MPhil (Bharathidasan U., 1985) physics. *X-ray_crystallography*

Bamzai, Mr Krishen Kumar (1964). Research Fellow. P. G. Dept. of Physics, University of Jammu, F-167, Durgay Niwas, Old Janipura, Jammu, India. MPhil (U. Jammu, 1991) crystal characterization and growth. *Crystal_growth characterization*

Banerjee, Dr Srikumar (1946). Head, Metallurgy Div., BARC. Metallurgy Div., BARC, Bombay 400094, India. PhD (IIT Khargpur, 1974) metallurgy. *Physical_metallurgy electron_microscopy radiation_damage_studies* E-mail sbanerji@magnum.barctl.ernet.in Tel. 91()5519949 Fax 91()5560750 Telex 1171017BARCIN

Bansal, Dr Manju (1950). Associate Prof. Molecular Biophysics Unit, Indian Institute of Science, Bangalore, India. PhD (Indian Institute of Science, Bangalore, 1977) molecular biophysics. *Computer_modelling nucleic_acids sequence_and_structure_analysis biomolecular_structures* E-mail mb@mbu.iisc.ernet.in Tel. 91()344411 Fax 91()3341683

Betal, Mr Badal Kumar (1943). Lecturer. Physics Dept., Presidency C., College St., Calcutta 700073, India. MSc (Calcutta U., 1963) physics. *Sterols_structure*

Bhagwat, Dr Vasant (1947). Lecturer. Dept. of Chemistry, Vikram University, Ujjain, 456001, India. PhD (Vikram U., 1970) chemistry. *X-ray_crystallography*

Bhaktapriya, Dr S. R. Y. (1935). Prof. Physics Dept., Ranchi University, Ranchi 834008, India. PhD (Ranchi U., 1981) X-ray crystallography. *Crystal_structure*

Bharati Rao, Ms T. (1965). Res. Scholar. Dept. of Organic Chemistry, IISc, Bangalore 560012, India. MSc (IIT Madras, 1987) chemistry. *Structure*

Bhat, Ms Sushma (1964). Research Scholar. University Girls Hostel, Jammu University, Jammu, India. MSc (Jammu U., 1990) crystal growth and materials research. *Crystal_growth characterization materials_research*

Bhatia, Dr Subhash Chander (1950). Lecturer. Chemistry Dept., Kurukshetra University, Kurukshetra 132119, India. PhD (Kurukshetra U., 1977) chemistry. *Organometallic_compound*

Bhattacharya, Ms Archana (1951). Res. Scholar. Dept. of Physics, Jadavpur University, Calcutta 7000329, India. MSc (Jadavpur U., 1974) physics. *X-ray_crystallography*

Bhattacharya, Dr Ramendranarayan (1934). Lecturer. Dept. of Magnetism, Indian Association for the Cultivation of Science, Calcutta 700032, India. PhD (Calcutta U., 1975) physics. *Crystal_physics*

Bhattacherjee, Dr Satyananda (1935). Prof. Physics Dept., IIT Kharagpur, Kharagpur 721302, India. PhD (IIT Kharagpur, 1970) X-ray structure of matter. *X-ray_diffraction_and_structure structure_property_relationship_in_solids*

Bist, Dr B. M. S. (1950). Postdoctoral Fellow. Physics Dept., Banaras Hindu University, Varanasi 221005, India. PhD (Banaras Hindu U., 1974) physics. *Electron_microscopy*

Biswas, Mr Gautam (1960). Res. Scholar. Dept. of Biophysics, Bose Inst., P-1/12, C. I. T. Scheme, Kankurgachi, Calcutta, India. MSc (Calcutta U., 1982) physics. *Biological_macromolecules*

Biswas, Dr Subhash Chandra (1939). Lecturer. Brahmananda Keshab Chandra College, 111/2 B T road, Calcutta 700035, India. [Dept. of Solid State Physics, I.A.C.S, Jadavpur, Calcutta 700032, India.] PhD (Calcutta U., 1987) physics. *Organic_structure* E-mail sspmm@iacs.ernet.in Tel. 91(33)4733073 Fax 91(33)4732805 Telex 0215501 IACS IN

Bora, Mr Mohendra Nath (1942). Professor. Dept. of Physics, Gauhati University, Guwahati 781014, India. PhD (Gauhati U., 1974) *Crystallography textiles fibres* Tel. 91()570531 Fax 91(361)570133

Bose, Mr Shyamal Kumar (1936). Scient. C. X-ray Crystallography Group, National Metallurgical Lab., Jamshedpur 831007, India. MSc (Allahabad U., 1957) physics. *Metals alloys structure_transformation*

Chacko, Dr K. K. (1942). Reader. Dept. of Crystallography and Biophysics, University of Madras, Guindy Campus, Madras 600025, India. PhD (Madras U., 1971) physics. *Anomalous_dispersion*

Chadha, Dr Gopal Krishan (1942). Professor. Dept. of Physics and Astrophysics, Delhi University, Delhi 110007, India. PhD (Delhi U., 1968) physics. *Crystal_growth X-ray_diffraction polytypism* E-mail esec@doe.ernet.in Tel. 91()7257793 Fax 91(11)7257336

Chakrabarti (Chatterjee), Dr Mrs Chandana (1951). Reader. Crystallography and Molecular Biology Division, Saha Inst. of Nuclear Physics, 1/AF Bidhan Nagar, Calcutta 700 064, India. PhD (Calcutta U., 1984) physics. *Biomolecules protein_crystallography* E-mail cc@saha.ernet.in Tel. 91()370659 Fax 91(33)374637

Chakrabarti, Dr Chandra (1951). Reader. Crystallography and Molecular Biology Division, S.I.N.P., 1/AF Bidhan Nagar, Calcutta 700064, India. PhD (Calcutta U., 1984) physics. *Biocrystallography proteins_crystallography* Tel. 91(33)370659

Chakrabarti, Dr Pinak (1955). Scientist. Division of Physical Chemistry, National Chemical Laboratory, Pune 411008, India. PhD (Indian Institute of Science, Bangalore, 1982) X-ray crystallography. *Proteins structure function folding intermolecular_interactions_and_packing_in_small_molecule_crystals* E-mail pinak@ncl.ernet.in Tel. 91()346973 Fax 91()330233 Telex 0145-266

Chakrabarty, Mr Dipak Kumar (1953). Lecturer. Dept. of Physics, X-ray Laboratory, Jadabpur University, Calcutta 700032, India. MSc (Calcutta U., 1976) physics. *Lattice_dynamics crystal_structure*

Chakrabarty, Mr Subhasis (1959). Res. scholar. Dept. of Biophysics, Bose Inst., P-1/12, C. I. T. scheme, Kankurgach, Calcutta 700054, India. MSc (Calcutta U., 1982) physics. *Molecular_biophysics*

Chakrabarty, Mr Sugoto (1957). Scient. officer. Solid State Physics Div., BARC, Trombay, Bombay 400085, India. MSc (Delhi U., 1979) physics. *Proteins crystallography* Tel. 91(22)5563060

Chandrasekhar, Prof. Sivaramakrishna (1930). Prof. and Director. Centre for Liquid Crystal Research, PO Box 1329, Bangalore 560013, India. ScD (U. Cambridge, UK, 1987) liquid crystals. *Liquid_crystals condensed_matter* Tel. 91(80)3315924 Fax 91(80)3346044

Chandrasekharaiah, Dr M. N. (1945). Res. assoc. Dept of Metallurgical Engineering, Institute of Technology, Benaras Hindu University, Varanasi 221005, India. PhD (Benaras Hindu U., 1972) crystal growth. *Crystal_growth crystallography*

Chatterjee, Dr Amitava (1948). Pool officer. Dept. of General Physics, X-ray, I.A.C.S, Calcutta 700032, India. PhD (Calcutta, 1978) macromolecular structure. *Macromolecular_structure conformational_analysis*

Chatterjee, Dr Sanat Kumar (1945). Asst Prof. Physics Dept, Regional Eng. C., Durgapur 731209, India. PhD (Calcutta U., 1976) lattice imperfections. *Phase_transformation amorphous_metals*

Chattopadhyay, Dr Debasish (1957). Res. fellow. Crystallography & Molecular Biology Division, Saha Institute of Nuclear Physics, 1/AF Bidhannagar, Calcutta 700064, India. MSc (Calcutta U., 1979) biochemistry. *X-ray_crystallography coordination_chemistry proteins_crystallography*

Chaudhuri, Dr Ahindra Kumar (1941). Scient. Solid State Laboratory, Central Glass & Ceramic Res Institute, Calcutta 700032, India. PhD (IIT Kharagpur, 1974) ceramic research. *Nucleation crystallization order_disorder*

Chelliah, Dr Mahadevan (1958). Asst Prof. Dept of Physics, S.T. Hindu College, Nagarcoil, Tamilnadu, 629002, India. PhD (IIT Madras, 1984) X-ray crystallography. *X-ray_crystallography crystal_growth characterization* Tel. 91()3127

Chetal, Prof. Amritlal (1938). Prof. Dept of Applied Science, Indian School of Mines, Dhanbad 826004, India. PhD (Poona U., 1966) crystal structure. *Crystal_structure* Tel. 91()3127

Chidambaram, Dr Rajagopala (1936). Chairman, Atomic Energy Commission; Scientist, Bhabha Atomic Research Centre. Department of Atomic Energy, Anushakti Bhaven, CSM Marg, Bombay 400 039, India. DSc (IISc, Bangalore, 1962) materials science, crystallography. *High_pressure_physics crystallography neutron_diffraction nuclear_technology* E-mail CHMN@TIFRVAX.TIFR.RES.IN or CHMN@TIFRVAX.BITNET Tel. 91(22)2022543 Fax 91(22)2048476 Telex 011-82355 ATOM IN

Chinnakali, Mr K. (1939). Res. Scholar. Dept. of Physics, Anna University, Madras 600025, India. MSc (Madras U., 1960) physics. *X-ray_crystallography*

Dalel, Mrs Snehlata (1939). Lecturer. Physics Dept., Ranchi University, Ranchi 834008, India. MSc (Ranchi U., 1962) crystallography. *Crystallography*

Das, Mr Amit Kumar (1964). Research Fellow. Crystallography and Molecular Biology Div., Saha Institute of Nuclear Physics, Sector-1, Block AF, Bidhannagar, Calcutta 700064, India. MSc (Calcutta U., 1987) physical chemistry. *X-ray_crystallography protein_crystallography structure-based_drug_design_crystallography* E-mail adas@saha.ernet.in Tel. 91(37)5345-49 (5 lines) Fax 91(33)374637 Telex 21-4103 SINP IN

Das, Dr Birendra Nath (1941). Lecturer. Physics Dept., Vivekananda C., D. H. Rd, Calcutta 700063, India. PhD (Calcutta U., 1981) biologically interesting organic compounds. *Structure_and_function* Tel. 91()7773434

Das, Dr Indu Mohan (1935). Lecturer. USIC, Gauhati University, Guwahati 781014, India. PhD (Guahati, 1967) X-ray crystallography. *X-ray_fluorescence crystallography* Tel. 91()88531

Das, Dr Kalyan (1962). Postdoctoral Fellow. MBU, Indian Institute of Science, Bangalore 560012, India. PhD (Indian Institute of Science, 1991) X-ray crystallography. *Proteins_crystallography computing small_molecule_crystallography* E-mail kd@mbu.iisc.ernet.in Tel. 91()3344411 Fax 91()3341683 Telex 845-8349IIScIn

Das, Dr Sabita (1934). Reader. Victoria Institution, 78 B. A. P. C. Road, Calcutta 700009, India. PhD (Calcutta U., 1969) solid state physics. *Phase_determination crystallography powder_diffraction macromolecules maximum_entropy* Tel. 91()3501959

Datta, Mr Manuja (1959). Research Associate. Crystallography and Molecular Biology Division, Saha Institute of Nuclear Physics, 1/AF Bidhan Nagar, Calcutta - 700 064, India. PhD (Calcutta U., 1992) X-ray crystallography. *Structure_function proteins_crystallography* E-mail jiban@saha.ernet.in Tel. 91(33)370659 Fax 91(33)374637 Telex (21)4103SIN-PIN

Dattagupta, Prof. Jiban Karati (1945). Professor. Crystallography & Molecular Biology Div., Saha Institute of Nuclear Physics, 1/AF Bidhannagar, Calcutta 700 064, India. PhD (Calcutta U., 1972) physics. *Structure_conformation proteins_crystallography medicinal_compounds* E-mail jiban@saha.ernet.in Tel. 91(33)370659 Fax 91(33)374637 Telex 21-4103SINPIN

Dayal, Dr Radha Raman (1941). Scient. E. Glass Tech. Div., Def Sci. Center, Metcalfe House, Delhi 110054, India. PhD (Aberdeen U., UK, 1965) phase equilibria. *Crystal_growth crystal_chemistry glass_science_and_technology*

De, Dr Amitabha (1958). Res. Scholar. Physics Dept., Calcutta University, 92 A. P. C. Road, Calcutta 700009, India. PhD (Calcutta U., 1988) physics. *Biocrystallography*

De, Dr Madhusudhan (1941). Reader. Dept. of Materials Science, Indian Association for the Cultivation of Science, Jadavpur, Calcutta 700032, India. PhD (Calcutta U., 1970) physics. *Metals_physics ceramics powder_crystallography composites electron_microscopy* E-mail msmd@iacs.ernet.in Tel. 91(473)4971 Fax 91(473)2805 Telex 021-5501IACSIN

Deopura, Dr B. L. (1946). Asst. Prof. Textile Technology Dept., IIT Delhi, New Delhi 110016, India. PhD (IIT Kanpur, 1970) physics. *Small-angle_scattering polymers X-ray_diffraction*

Desiraju, Prof. Gautam R. (1952). Prof. School of Chemistry, University of Hyderabad, Hyderabad 500134, India. PhD (U. Illinois, 1976) chemistry. *Structural_chemistry molecular_chemistry* E-mail GRD-CH@UOHYD.ERNET.IN Tel. 91(40)289567 Fax 91(40)253145

Dhanaraj, Dr V. (1957). Res. Assoc. Molecular Biophysics Unit, IISc, Bangalore 560012, India. PhD (IISc, 1986) X-ray crystallography. *Biological_crystallography molecular_biology* Tel. 91()344411x2389

Dhaneshwar, Dr Narayandatta Nagesh (1936). Scientist E-1. National Chemical Laboratory, Pune 411008, India. PhD (U. Poona, 1973) chemistry (crystallography). *Structure_analysis_of_small_molecules* E-mail nnd@ncl.ernet.in Tel. 91()336451 Fax 91()330233 Telex 0145-266

Dutta, Dr BishnuPada (1942). University Professor. Physics Dept., Science College, Patna 800 005, India. PhD (Ranchi U., 1975) X-ray crystallography. *X-ray_crystallography* Tel. 91()350839

Dwivedi, Dr Ganpat Lal (1942). Sr. Mineralogist; Head, Mineral Physics Division, GSI(WR), Jaipur. Mineral Physics Division, Geological Survey of India (W.R.), Jhalana Institutional Area, Jaipur 302004, India. [11A Shaheed Colony 'A', Malaviya Nagar, Jaipur 302017, India.] PhD (IIT Kanpur, 1971) X-ray crystallography. *Minerals crystal_structure molecular_structure powder_diffraction_for_unknown_samples* E-mail cmplruj@sirnetd.ernet.in Tel. 91(141)510753 Fax 91(141)511582 Telex 9190365-2233

Elango, Mr N. (1957). Res Scholar. Dept of Crystallography & Biophysics, Guindy Campus, Madras University, Madras 600025, India. MSc (Madras U., 1980) physics. *Theoretical_crystallography crystal_structure_analysis*

Eswara Prasad, Dr Gummuluri (1940). Scientific Officer. Physical Metallurgy Division, Bhabha Atomic Research Center, Trombay 400085, India. PhD (Bombay U., 1980) physical chemistry. *Ceramics failure_analysis metals alloys rapid_solidification_oxides*

Fernandis, Mr Jacob Richard (1951). Res Fellow. Raman Research Institute, Bangalore 560006, India. MSc. (IIT Kharagpur, 1973) physics. *Liquid_crystal* Tel. 91()30124

Gadgil, Mr Ajit K. (1969). Research Fellow. Solid State Physics Divn., BARC, Bombay 400085, India. MSc (Bombay U., 1992) physics. *Macromolecular_crystallography* E-mail kannan@magnum.barct1.ernet.in Tel. 91(22)5563060x3614 Fax 91(22)5560750 Telex 011-71017BARCIN

Ganesan, Mr V. (1959). Res. fellow. IIT Madras, Madras, India. MSc (IIT, 1980) physics. *Lattice_dynamics*

Garg, Dr Ajay Kumar (1952). Lecturer. Dept of Physics, N. A. S. College, Shanti Sadan, Sharda Road, Delhi Gate, Meerut, UP 25002, India. PhD (IIT Delhi, 1975) solid state physics. *Crystal_growth X-ray_diffraction powder_diffraction microscopy inorganic_crystals* Tel. 91()24528

Gautham, Mr Namasivayam (1955). Senior Lecturer. Dept. of Crystallography & Biophysics, University of Madras, Guindy Campus, Madras 600 025, India. PhD (Indian Institute of Science, Bangalore, 1983) X-ray crystallography of DNA fragments. *Structure_macromolecules modelling bioinformatics optimization_algorithms* Tel. 91(44)2351367 Telex 416376UNOMIN

Ghosh, Ms Manuja (1959). Res. Fellow. Crystallography & Microbiology Division, Saha Inst of Nuclear Physics, 1/AF Biddhanagar, Calcutta 700064, India. MSc (Calcutta, 1981) physics. *Biological*

Ghosh, Dr Mrs Minakshi (1948). Lecturer. Dept. of Physics, Presidency C, Calcutta 700073, India. PhD (Calcutta U., 1981) physics. *Phase_transition disorder*

Ghosh, Miss Soma (1967). S. R. F. I. A. C. S., Jadavpur, Calcutta 700032, India. MSc (Calcutta U., 1991) physics. *Crystal_structure liquid_crystal* E-mail sspmm@iacs.ernet.in Tel. 91(33)4733073 Fax 91(33)4732805

Ghosh, Dr Sutapa (1959). Scient. Crystallography & Molecular Biol Division, Saha Inst of Nuclear Physics, Sector 1 Block-AF, Biddhanagar, Calcutta 700064, India. MSc (Calcutta U., 1981) physics. *Molecules*

Girl, Mr Anil K. (1960). Res. Scholar. C. S. S. Dept., Indian Association for the Cultivation of Science, Jadavpur, Calcutta 700032, India. PhD (Jadavpur U., 1988) science. *Structure_property_relationship minerals_characterization amorphous_materials*

Girl, Mr Siba Narayan (1936). Reader. Dept. of Physics, Bidhan Chandra Krishi Viswavidyalaya, Mohanpur, Nadia, India. MSc (Allhabad U., 1960) physics. *Crystal_structure*

Girirajan, Dr K. S. (1946). Lecturer. Dept of Biophysics & Crystallography, University of Madras, Guindy Campus, Madras 600025, India. PhD (IIT Madras, 1974) physics. *Crystallography*

Godavarthi, Mr Bhagavannarayana (1955). Scient. C. G. C. Section, National Physical Laboratory, New Delhi 110012, India. MSc (Andhra U., 1979) physics. *High_resolution_X-ray_diffraction*

Gomes, Mr Albert Cardinal (1947). Res. Scholar. Dept of Biophysics, Bose Institute, P-1/12 C. I. T. Scheme 7 M, Kankurgachi, Calcutta 700054, India. MSc (Calcutta, 1974) physics. *Organic_crystal_structure biological_macromolecules*

Goswami, Prof. K. N. (1934). Professor; former Head, Physics Dept., Jammu U. Physics Dept., Jammu University, Jammu - 180001, India. PhD (Sardar Patel U., V. V. Nagar (Gujarat), India, 1963) solid state physics. *X-ray_crystallography structure_analysis phase_transition solid_state_physics protein_crystallography* Fax 91()48021

Goswami, Dr S. Niranjana N. (1949). Scientist 'C'. Material Characterization Division, National Physics Laboratory, New Delhi 110012, India. PhD (Jiwaji U., 1975) physics. *Perfection X-ray_diffraction*
Tel. 91(11)5741733 Fax 91(11)5752678 Telex 031-77099 NPL IN

Guha, Dr Romel (1941). Sr. Mineralogist. Mineral Physics Division, Geological Survey of India, 29 J. N. Rd, Calcutta 700016, India. PhD (Calcutta U., 1969) physics. *Powder_diffractometry*
Tel. 91()297645

Guha, Dr Sankarananda (1943). Proprietor. Gayatri Chemicals, 37 Garia Park, Calcutta 700084, India. PhD (Calcutta U., 1968) physics. *Structure ferroelectric biomolecule*
Tel. 91()722035

Gupta, Mr Manoj Kumar (1955). Res Fellow. Physics Dept., Lucknow University, Lucknow 226007, India. MSc (Kanpur U., 1973) physics. *Conformation structure biomolecule*

Gupta, Dr Satish Chander (1951). Scientific Officer; Group Leader. High-Pressure Physics Division, Bhabha Atomic Research Center, Trombay, Bombay 400085, India. PhD (Bombay U., 1980) physics. *High_pressure_crystallography shock_wave_in_material phase_transitions electronic_structure*
E-mail satish@magnum.barct1.ernet.in Tel. 91(22)5513848 Fax 91(22)5560750

Gupta, Miss Sunita (1952). Res. Fellow. Physics Dept., Lucknow University, Lucknow 226007, India. MSc (Lucknow U., 1972) physics. *Lattice_dynamics*
Tel. 91()25140

Gupta, Dr Vijai Prakash (1950). Sr. Lecturer. Physics Dept., Lucknow University, Lucknow 226007, India. PhD (Lucknow U., 1982) physics. *Conformation vibrational_optical_activity*
E-mail lkuniv@sernet.ernet.in Tel. 91(522)75947 Fax 91(522)381583

Gupta, Prof. Vishwambhar Dayal (1934). Prof. Physics Dept., Lucknow University, Lucknow 226007, India. PhD (Allahabad U., 1958) X-ray structure of cellulose. *Biomolecule small-angle_X-ray_scattering lattice_dynamics*
Tel. 91()25140

Gururow, Dr Tayur N. (1951). Scient. Physical Chemistry Division, National Chemistry Laboratory, Pune 411008, India. PhD (IISc Bangalore, 1976) X-ray crystallography. *Crystal_structure electron_density*

Halder, Dr Sujit Kumar (1948). Scient. Crystal Growth and Characterization Section, National Physical Laboratory, New Delhi 110012, India. PhD (Calcutta U., 1976) physics. *X-ray_diffraction microstructure alloys crystal_growth characterization*

Hanan, Ms Salima Shaukat (1942). Lecturer. Dept. of Physics, Assam Engineering College, Guwahati -781013, India. MSc (Gauhati U., 1964) X-ray crystallography. *X-ray_diffraction ferroelectricity materials*
Tel. 91()570531 Fax 91(361)70133

Hariharan, Ms Meena (1957). Res. Scholar. Dept. of Crystallography and Biophysics, University of Madras, Madras 600025, India. MSc (U. Madras, 1957) physics. *Biocrystallography*

Hiremath, Mr Chaitanya N. (1962). Res. student. Molecular Biophysics Unit, IISc Bangalore, Bangalore 560012, India. MSc (Karnataka U., 1985) physics. *Viruses_crystallography molecular_biophysics*

Hosur, Dr Madhusudhan V. (1950). Scientist. Solid State Physics Division, BARC, Bombay 400085, India. PhD (Indian Institute of Science, 1976) crystallography. *Macromolecular_crystallography*
E-mail hosur@magnum.barct1.ernet.in Tel. 91(22)5563060x3614 Fax 91(22)5567050 Telex 011-71017BARCIN

Inbanathan, Mr S. Stephen R. (1916). Research Scholar. Dept. of Physics, Banaras Hindu University, Varanasi 221 005, India. MSc (Banares Hindu U., 1987) solid state physics. *Superconductivity crystal_growth electron_microscopy polytypes modulated_structures*

Iyengar, Dr Leela (1943). Lecturer. Physics Dept., C. of Sci., Osmania University, Hyderabad 500007, India. PhD (Osmania U., 1970) X-ray crystallography. *Crystal_structure*

Jagadeesh Kumar, Dr N. (1966). Student (PhD Scholar). Dept. of Physics, Sri Venkateshwara University, Tirupati 517502, India. MPhil (Sri Venkateshwara U., Tirupati, 1992) X-ray crystallography. *Small_molecules_crystallography macro-molecules_structures*
Tel. 91()24166 Fax 91()21211

Jagannathan, Mr N. R. (1954). Lecturer. Dept. of Crystallography & Biophysics, University of Madras, Guindy Campus, Madras 600 025, India. PhD (U. Madras, 1983) structural studies by NMR & X-ray crystallography. *Structure crystallography NMR imaging*
Tel. 91(44)2351367 Telex 416376UNOMIN

Jain, Dr Arun Kumar (1954). Scientist S-2. Physics Division, Jute Technical Research Laboratory, Indian Council of Agricultural Research, 12 Regent Park, Calcutta 700040, India. PhD (IIT Delhi, 1988) fibre science. *Fibres structure*

Jakkal, Mr Vasant Shankar (1936). Scient. officer. Water Chemistry Division, BARC, Trombay, Bombay 400085, India. MSc (Bombay U., 1961) physics. *Organic_crystal_structure*

Jayadevan, Dr Naduviledath Chennuvittil (1936). Head, Fuel Development Chemistry Section. Fuel Chemistry Division, BARC, Trombay, Bombay 400085, India. PhD (McMaster U., Canada, 1968) inorganic chemistry. *Actinides_complex mixed_oxides mixed_carbides*

Jayanty, Dr Ashok (1951). Lecturer. Physics Dept., Ranchi University, Ranchi 8340089, India. PhD (Ranchi U., 1980) X-ray crystallography. *Crystal_structure*

Jayashree, Ms A. N. (1963). Res. Student. Molecular Biophysics Unit, IISc Bangalore, Bangalore 560012, India. MSc (Karnatak U., 1984) physics. *Biological_crystallography*

Jeyabalan Moses, Mr Mohandas Prabhu (1969). Research in the field of X-ray crystallography. Molecular Biophysics Unit, Indian Institute of Science, Bangalore 560012, India. MSc (The American College, Madurai, India, 1992) physics. *Proteins_crystallography software_development_in_related_field*
E-mail prabhu@mbu.iisc.ernet.in Tel. 91()344411 Fax 91()3341683

Joshi, Dr Shri Krishna (1935). Prof. and Head, Director Gen. (C. S. I. R). Physics Dept., Roorkee University, Roorkee 247667, India. DPhil (Allahabad U., 1962) crystal physics. *Solid_state_theory magnetism*

Kabaleeswaran, Mr V (1966). Research Scholar. Dept. of Crystallography and Biophysics, University of Madras, Guindy Campus, Madras 600 025, India. MPhil (U. Madras, 1991) X-ray crystallography. *X-ray_crystallography crystallography_study_of_inclusion_complexes*
E-mail nicrys@iitm.ernet.in Tel. 91(44)2351367 Telex 416376UNOMIN

Kalyanaraman, Dr A. R (1937). Deputy Director. The South India Textile Research Association, PB No. 3205, Aerodrome PO, Avanashi Road, Coimbatore 641 014, India. PhD (Madras U., 1967) X-ray crystallography. *Fibres diffraction crystallography instrumentation machinery_design_and_energy*

Kanagaraj, Dr Sekar (1961). Postdoctoral Fellow. Molecular Biophysics Unit, Indian Institute of Science, Bangalore 560012, India. PhD (U. Madras, 1992) X-ray crystallography. *Theoretical_and_protein_crystallography crystallographic_computing*
E-mail sekar@mbu.iisc.ernet.in Tel. 91()3344411 Fax 91()3341683

Kannan, Dr K. K. (1939). Scientific Officer (G). Chairman Indian National Committee for IUCr; Sub-Editor WDC9 for India. Solid State Physics Division, Bhabha Atomic Research Centre, Bombay - 400085, India. PhD (Indian Institute of Science, Bangalore, India, 1966) X-ray crystallography. *Macromolecular_crystallography*
E-mail kannan@magnum.barct1.ernet.in Tel. 91(22)5563060x2152 Fax 91(22)5560750 Telex (011)710117&(011)72322

Kannan, Dr S. (1963). Scientific Officer SD. Fuel Chemistry Division, BARC, Trombay, Bombay 400 085, India. PhD (Bombay U., 1993) inorganic and organic chemistry. *Crystallography NMR XRD thermal_studies*
Tel. 91()5563060

Kant, Dr Rajni (1962). Lecturer. Dept. of Physics, University of Jammu, Jammu Tawi 180 001, India. PhD (U. Jammu, 1988) x ray crystallography. *Crystallography small_molecules*
Tel. 91()48021

Kar, Dr (Mrs) Tanusree (1954). Lecturer. Dept. of Materials Science, IACS, Jadavpur, Calcutta 700032, India. PhD (Calcutta U., 1984) physics. *X-ray_crystallography crystal_growth non-linear_materials X-ray_topography SEM*
E-mail mstk@iacs.ernet.in Tel. 91(33)4734971 Fax 91(33)4732805 Telex Cable-INDASSON,Jadavpur

Karthe, Mr P. (1967). Research Scholar. Department of Crystallography & Biophysics, University of Madras, Guindy Campus, Madras 600 025, India. MPhil (Madras U., 1993) X-ray crystallography. *Crystallography biophysics DNA*
Tel. 91(22)2351367

Kashyap, Dr Ram Prasad (1947). Lecturer. Chemistry Dept., Guru Nanak Dev University, Amritsar 143005, India. PhD (Kurukshetra U., 1974) crystallography. *X-ray_crystallography*

Kodandapani, Mr R. (1961). Res. student. Molecular Biophysics Unit, IISc, Bangalore 560012, India. MSc (Karnataka U., 1983) physics. *Proteins_crystallography biophysics*

Kohli, Dr Vijay Kumar (1948). Scient. Crystal Growth and Characterization Section, National Physical Laboratory, New Delhi 110012, India. PhD (Delhi U., 1984) physics. *Crystal_growth topography instrumentation*

Kotru, Dr P. N. (1941). Professor. Member, Secretary National Committee for IUCr. Dept. of Physics, University of Jammu, Jammu 180 001, India. PhD (Sardar Patel U., Vallabh Vidyanagar (Gujarat), 1969) solid state crystallography. *Crystal_growth materials_research characterization scientific_planning*
Tel. 91()48021

Krishna, Prof. Padmanabhan (1938). Rector. Fellow of the Indian National Science Academy. Rajghat Education Center, Krishnamurti Foundation India, Rajghat Fort, Varanasi 221001, India. PhD (Banaras Hindu U., 1962) physics. *Crystal_growth_and_perfection phase_transformation polytypism education*
Tel. 91(542)330218/330718 Fax 91(542)330218

Krishnaiah, Dr Musali (1950). Lecturer. Dept. of Physics, Sri Venkateshwara University, Tirupati 517502, India. PhD (Sri Venkateshwara University, Tirupati, 1985) X-ray crystallography. *Small_molecules crystallography phosphazenes sulphones organophosphorus heterocyclic_compounds*
Tel. 91()24166 Fax 91()21211

Krishnamachari, Dr Ramaswamy (1935). Prof. Physics Dept., Annamalai University, Annamalainagar 608002, India. PhD (Annamalai U., 1961) molecular spectroscopy. *Spectroscopy nuclear_magnetic_resonance archaeomagnetism petrophysics*
Tel. 91()04422106

Krishnaswamy, Dr S. (1959). Scientist. School of Biotechnology, Madurai Kamaraj University, Madurai 625 021, India. PhD (U. Madras, 1986) X-ray crystallography. *Macromolecular_crystallography modelling bioinformatics*
E-mail krishna%bic-mku@dbt.ernet.in Tel. 91(452)85741 Fax 91(452)85205 Telex 445 337

Kumar, Mr V. Amarendra (1962). Res. Scholar. Dept. of Organic Chemistry, Indian Inst. of Science, Bangalore 560012, India. MSc (Sri Sathya Sai Inst. of Higher Learning, 1986) chemistry. *Crystallography*

Kumar, Dr Vinay (1960). Scientific Officer (SE). Solid State Physics Division, Bhabha Atomic Research Centre, Bombay 400 085, India. PhD (U. Bombay, 1992) protein crystallography. *Proteins_crystallography X-ray_diffraction biomolecular_structure-function*
E-mail Kannan@magnum.barct1.ernet.in Tel. 91()5563060 Fax 91()5560750 Telex 011-71017BARCIN

Kumereson, Dr P. (1948). Lecturer (S. G). Vivekananda College, Agasteeswaram, Kanyakumari Dist., Tamil Nadu, India. [No.40, State Bank Colony View, Ramavarmapuram, Nagercoil 629 001, Tamil Nadu, India.] PhD (U. Kerala, Trivandrum, 1991) crystal growth. *Crystal_characterization electronics biological_crystallization*

Kundra, Dr Krishan Dev (1934). Scient. Materials Characterization (X-rays) Division, National Physical Laboratory, Hillside Road, New Delhi 110012, India. PhD (Delhi U., 1977) X-ray crystallography. *Characterization_of_materials X-ray_diffraction fluorescence*
Tel. 91()584179

Kundu, Dr Mohanlal (1942). Deputy Superintendent. Physical Research Wing, Projects and Dev. India Ltd, Sindri 828122, Dhanbad, India. PhD (Burdwan U., 1979) physics. *Powder_diffraction catalysts minerals*

Kunte, Mr Vivek P (1950). Product manager. AIMIL Instrumentation, 44 Millers Road, Bangalore 560052, India. MSc (Inst. of Science, 1973) physics. *Instrumentation music birdwatching cooking gardening*
Tel. 91(80)2281180 Fax 91(80)2267437 Telex 0845-8269AMILIN

Lahiri, Dr Barendra Nath (1938). Lecturer. Physics Dept., Burdwan University, Burdawan, India. PhD (Calcutta U., 1973) X-ray crystallography. *X-ray_crystallography silicates minerals*
Tel. 91()341121

Lakshminarayanan, Mr Muthuvijayan (1960). Res. scholar. Inorganic and Physical Chemistry Dept., IISc Bangalore, Bangalore 560012, India. MSc (Madurai Kamaraj U., 1985) chemistry. *Crystallography*

Lal, Dr Krishan (1941). Director Grade Scientist. National Physical Laboratory, Hillside Road, New Delhi 110012, India. PhD (Delhi U., 1969) physics. *Crystal_growth high_resolution_X-ray_diffraction_technique topography multicrystal_X-ray_diffractometry diffuse_X-ray_scattering*
E-mail klal%npl@simetd.ernet.in Tel. 91(11)5741733 (office); 91(11)5754891 (residence) Fax 91(11)5752678 Telex 031-77099 NPL IN and 031-77384 RSD IN

Lekshmanasarma, Mr R (1946). Lecturer (Selection Grade) in College. Vivekananda College, Agasteeswaram 629 701, India. PhD (Bangalore U., Bangalore, 1982) electro crystallization. *Electrocrystallization mechanistic_aspects_of_electrocrystallisation metal_solution_interface*
Tel. 91()71245

Lokanatha, Mr S. (1958). Lecturer. Physics Dept., IIT Kharagpur, Kharagpur 7213023, India. MSc (Bangalore U., 1981) physics. *Minerals structure-property_relationships*

Madhusudan, Mr (1962). Res. student. Molecular Biophysics Unit, IISc Bangalore, Bangalore 560012, India. MSc (Delhi U., 1985) physics. *Proteins_crystallography biophysics*
Tel. 91()344411x2389

Madhusudan, Mr (1962). Res Fellow. Molecular Biophysics Unit, IISc, Bangalore 560012, India. Msc (Delhi U., 1985) physics. *Biophysics proteins_crystallography*

Madhusudhana, Dr N. V. (1944). Scient.; Head, Liquid Crystal Laboratory. Liquid Crystals Laboratory, Raman Res. Institute, Bangalore 560 080, India. PhD (Mysore U., 1971) liquid crystals. *Liquid_crystal*
E-mail nvmadhu@rri.ernet.in Tel. 91(80)30124 91(80)3311015 Fax 91(80)3340492 Telex 845 2671 RRI IN

Mahalingam, Mr Bhuvaneswari (1967). Research Scholar. Molecular Biophysics Unit, Indian Institute of Science, Bangalore 560012, India. MSc (Indian Institute of Technology, Delhi, 1990) chemistry. *X-ray_crystallography proteins_folding viruses_crystallography*
E-mail bhuvaneswari@mbu.iisc.ernet.in Tel. 91()3344411 Fax 91()3341683 Telex 0845-8439IIScIn

Mahata, Dr Akhil (1939). Res. Scholar. Physics Dept, Ranchi University, Ranchi 834008, India. PhD (Ranchi U., 1978) X-ray crystallogsathy. *X-ray_crystallography*

Maiti, Dr Gobinda Chandra (1944). Physical Research Wing, Projects & Div. India Ltd, Sindri 828122, India. PhD (Calcutta U., 1977) chemistry. *Catalysis*

Mandal, Dr Pradip Kumar (1958). Lecturer. North Eastern Hill University, Bijni Complex, Bhagyakul, India. PhD (North Bengal U., 1986) physics. *Liquid_crystal structure_determination*

Mande, Mr Sekhar Chintamani (1962). Res. student. Molecular Biophysics Unit, IISc, Bangalore 560012, India. MSc (Nagpur U., 1984) physics. *Proteins_crystallography molecular_biophysics*

Mani, Mr A. (1950). Scient. Electrochemical Research Institute, Karaikudi 623006, India. MSc (Anna U., 1981) materials science. *X-ray_crystallography method apparatus phase_transition organic_structures thin_film_deposits*

Manickkavachgam, Dr Ramanathan (1955). Lecturer. Christian C., Madras, India. PhD (Madurai Kamaraj U., 1988) physics. *Crystal_structure phase_transition*

Mathews, Mr Irimpan I. (1962). Res. scholar. Inorganic and Physical Chemistry Dept., IISc, Bangalore 560012, India. MSc (Calicut U., 1984) chemistry. *X-ray_crystallography bio-coordination_compounds*

Mathur, Dr Balbir Kumar (1948). Sr. Scient. Physics Dept., IIT Kharagpur, Kharagpur 721302, India. PhD (IIT Kharagpur, 1978) physics. *Crystal_defect computer_programming optical_transforms*

Misra, Prof. Nirmal Kumar (1939). Prof. Dept. of Physics and Meteorology, IIT, Kharagpur 721302, India. PhD (IIT Kharagpur, 1967) physics. *Defect_crystal_structure crystal_growth disorder_materials thin_films metallic_glass*

Misra, Prof. Somnath (1936). Principal. Regional Eng. C., Rourkela 769008, India. DSc (MIT, USA, 1963) metallurgy. *Intermetallic_compound order-disorder_transitions*

Mitra, Prof. Girija Bhushan (1923). Em. Scient. C.S.S. Dept., Indian Assoc. for the Cultivation of Science, Calcutta 700032, India. DSc (Calcutta U., 1967) X-ray studies of lattice defects. *Crystal_defect lattice_vibration dynamical_diffraction_theory material_characterization structure instrumentation small-angle_X-ray_scattering*

Mohanlal, Dr Sembu Krishnaiyer (1940). Head & Coordinator, School of Physics, MKU. School of Physics, Madurai Kamaraj University, Madurai 625021, Tamilnadu, India. PhD (Madurai U., 1971) X-ray crystallography. *X-ray_diffraction solid_solution crystal_defect instrumentation photography drawing*
E-mail prabu%bic-mku@dbt.ernet.in Tel. 91(452)85252 91(452)520514 Fax 91(452)85239 Telex 445-308 MKU IN

Mohanty, Mr Arun K. (1968). Research Fellow. Solid State Physics Divn., BARC, Bombay 400085, India. MSc (Utkal U., 1990) microbiology. *Molecular_biology macro-molecular_crystallography*
E-mail kannan@magnum.barct1.ernet.in Tel. 91(22)5563060x2693 Fax 91(22)5560750 Telex 011-71017BARCIN

Mostafa, Mr Golam (1962). S. R. F. IACS, Jadavpur, Calcutta 700032, India. MSc (Jadavpur U., 1989) physics. *Disorder software_development*
E-mail sspm@iacs.ernt.in Tel. 91(33)4733073 Fax 91(33)4732805 Telex)215501IAC-SIN

Mudher, Dr Khush Dev Singh (1946). Scientific officer. Fuel Chemistry Division, BARC, Bombay 400094, India. PhD (IIT Bombay, 1972) X-ray crystallography. *X-ray_crystallography powder_diffraction X-ray_fluorescence thermogravimetry actinide_chemistry*
Tel. 91()5563060

Mukherjee, Dr Alok Kumar (1950). Reader. Dept. of Physics, Jadavpur University, Calcutta 700032, India. PhD (Visva-Bharati, 1978) physics. *Disordered_structure direct_methods*
E-mail sspmm@iacs.ernet.in Tel. 91(33)730399 Fax 91(33)732443 Telex 0214160 VCJUIN

Mukherjee, Dr Amal Bikash (1935). Prof. Dept. of Geology and Geophysics, IIT, Kharagpur 721302, India. PhD (IIT Kharagpur, 1961) geology. *Crystal_structure minerals structure*

Mukherjee, Dr Biswanath (1936). Asst. Dir. X-ray Crystallography Dept., Central Glass and Ceramic Res. Inst., Calcutta 700032, India. PhD (Calcutta U., 1965) physics. *Crystal_stability structure material_science*

Mukherjee (Mondal), Dr (Mrs) Monika (1947). Reader. Dept. of Solid State Physics, IACS, Jadavpur, Calcutta 700032, India. PhD (Calcutta U., 1977) physics. *Direct_method macromolecular_crystallography liquid_crystals*
E-mail sspmm@iacs.ernet.in Tel. 91(33)4733073 Fax 91(33)4732805 Telex 0215501 IACS IN

Mukherjee, Dr Partha Sarathi (1952). Scient. Materials Division, Regional Research Laboratory, Trivandrum 695019, India. PhD (IIT Khargpur, 1982) physics. *Semicrystalline_compounds minerals polymers characterization high_temperature_superconductors*

Mukhopadhyay, Dr (Mrs) Anuradha (1955). Lecturer. Dept. of Physics, Jadavpur University, Calcutta 700032, India. PhD (Calcutta U., 1987) physics. *Liquid_crystal crystallographic_statistics*
E-mail sspmm@iacs.ernet.in Tel. 91(33)730399 Fax 91(33)732443 Telex 0214160 VCJUIN

Mukhopadhyay, Mr Bishnu Prasad (1954). Res. fellow. Crystallography and Molecular Biology Division, SINP, 1/AF, Bidhannagar, Calcutta 700064, India. MSc (Burdwan U., 1977) chemistry. *Biomolecule_structure adrenergics biomolecule_structure_and_conformation*

Mukhopadhyay, Dr Pradip (1943). Scient. officer. Physical Metallurgy Division, BARC, Bombay 4000856, India. PhD (Bombay U., 1980) physics. *Phase_transformation electron_microscopy_and_diffraction electron_spectroscopy defect_structure_properties_correlation*

Munirathinam, Mr Nethaji (1958). Res. scholar. Dept. of Crystallography and Biophysics, Madras University, Madras 60025, India. MSc (Madras U., 1981) physics. *Biomolecule crystal_structure*

Munshi, Mr Sanjeev Kumar (1962). Res. student. Molecular Biophysics Unit, IISc, Bangalore 560012, India. MSc (Kashmir U., 1984) biochemistry. *Molecular_biophysics viruses_crystallography*

Muralidharan, Mr K. V. (1937). Scient. officer. Chemistry Division, BARC, Bombay 400085, India. MSc (Bombay U., 1971) physics. *Biomolecule structure crystallography*
Tel. 91()523321x288

Murthy, Mr Mathur R N (1950). Associate Prof. Molecular Biophysics Unit, Indian Institute of Science, Bangalore 560012, India. PhD (Indian Institute of Science, 1977) X-ray crystallography. *Macromolecular_crystallography proteins_structure_and_function molecular_evolution*
E-mail mrn@mbu.iisc.ernet.in Tel. 91()344411 Fax 91()3341683 Telex 0845-8349IIScIn

Nag, Dr Dilip Kumar (1945). Mineralogist. Minerals Physics Division, Geological Survey of India, 29 Chowringhee Road, Calcutta 700016, India. PhD (Calcutta U., 1974) X-ray crystallography. *Instrumentation crystal_physics*
Tel. 91()238321x8

Nag, Mrs Jhumjhumi (1955). Res. Scholar. Dept. of General Physics and X-rays, Indian Association for the Cultivation of Science, Calcutta 700032, India. MSc (Calcutta U., 1976) physics. *Structural characterization amorphous materials thin film crystallography X-ray diffraction*

Nagbhushana Rao, Mr Chemboli (1941). Scient. officer. Metallurgy Division, BARC, Bombay 400085, India. MTech (IIT Kanpur, 1973) metallurgy. *X-ray metallography texture studies*

Nagendra, Mr (1966). Research Student. Molecular Biophysics Unit, Indian Institute of Science, Bangalore 560012, India. MSc (Bangalore University, 1990) physics. *Proteins crystallography*
E-mail mbu!hgn@vigyan.ernet.in Tel. 91()3344411 Fax 91()3341683

Nair, Ms Bindu (1970). Scientist. Solid State Physics Division, BARC, Bombay 400085, India. MSc (IIT Delhi, 1992) chemistry. *Macromolecular crystallography*
E-mail hosur@magnum.barct1.ernet.in Tel. 91(22)5563060 Fax 91(22)5560750 Telex 011-71017BARCIN

Nandi, Mr Asok Kumar (1944). Scient. X-ray Section, Central Glass and Ceramic Research Institute, Calcutta 700032, India. MSc (Calcutta U., 1965) *Crystal structure amorphous materials organic compounds*

Nandi, Dr Ranjan Kumar (1950). Principal Res. Eng. 2E Neelamber, 28B Shakespeare Sarani, R & D Center, Steel Authority of India Ltd, Calcutta 700017, India. PhD (Calcutta U., 1978) physics. *Crystal imperfection EXAFS electrical steels thin films texture supported metal catalysts*

Narasimhamurthy, Mr Narasappa (1960). Res. scholar. Inorganic and Physical Chemistry Dept., IISc, Bangalore 560012, India. MSc (Mysore U., 1983) chemistry. *Organometallic compound*

Narasimhan, Dr P. (1944). Prof. Physics Dept., A. M. Jain C., Madras 600114, India. PhD (Madras U., 1984) physics. *Crystal structure organic compound small molecules dipeptides*

Natarajan, Dr S. (1941). Prof. and Head. Physics Dept., Anna University, Madras 600025, India. PhD (IIT Madras, 1969) physics. *Crystal structure high pressure X-ray diffraction*

Natarajan, Dr Subramanian (1949). Reader. School of Physics, Madurai Kamaraj University, Madurai 625021, India. PhD (Madurai Kamaraj U., 1979) physics. *Crystal growth aminoacid complexes crystal structure*
E-mail rvkk%bic-mku@dbt.ernet.in Tel. 91()85252 Fax 91(425)85205 Telex 445 337 MKU IN

Natesan, Mr Elango (1957). Res. Scholar. Dept. of Crystallography and Biophysics, Madras University, Madras 600025, India. MSc (Madras U., 1980) physics. *Theoretical X-ray crystallography structure*

Nath, Mr Kashi (1951). Res. fellow. Physics Dept., Lucknow University, Lucknow 226007, India. MSc (Lucknow U., 1973) physics. *Polypeptides biomolecules conformation conformational transition*

Nigam, Dr Gur Dayal (1939). Professor. Physics Dept., IIT, Kharagpur 721302, India. PhD (IIT Madras, 1969) physics. *Crystallography statistics crystallography symmetry direct methods*
E-mail gdnigam@hijli.iitkgp.ernet.in Tel. 91(3222)2221 to 2224, ext. off. 4921, res. 7921 Fax 91(3222)2303 Telex 06401-201 ITKG IN

Nigli, Ms Selina (1949). Res. Scholar. Dept. of Physics and Astrophysics, University of Delhi, Delhi 110007, India. MPhil (Delhi U., 1982) physics. *Crystal growth X-ray diffraction*

Nirmala, Ms Kusuma Ananthanamiah (1947). Reader in Physics. Physics Dept., Bangalore University, India. [Amruth Kunj, Do No. 16, I-C-floor, IV Cross, IV Main Road, Gandhinagar, Bangalore 560009, India.] PhD (Bangalore U., 1981) X-ray crystallography. *Small molecules pharmacologically important molecules*
Tel. 91()2266518

Noor, Mrs Sahina Begum (1949). Res. Scholar. Inorganic and Physical Chemistry Dept., IISc, Bangalore 560012, India. MSc (Bangalore U., 1982) physics. *X-ray crystallography metal nucleotide complexes*

Pal, Mr Hiranmay (1965). S. R. F. Dept. of Material Science, IACS, Jadavpur, Calcutta 700032, India. MSc (U. Kalyan, 1989) physics. *Material characterization*
E-mail Msmd@iacs.ernet.in Tel. 91()4733073 Fax 91(473)2805

Pan, Dr Nitya Ranjan (1935). Lecturer. Physics Dept., Calcutta University, India. [92 A. P. C. Road, Calcutta 700009, India.] PhD (Calcutta U., 1975) physics. *Orientation electret orientation and related properties*
Tel. 91()359186

Pandey, Dr Dhananjai (1952). Reader. School of Material Science and Technology, Banaras Hindu University, Varanasi 221005, India. PhD (Banaras Hindu U., 1976) physics. *Structural imperfection solid state transformation*

Pandya, Prof. Janardhan Rameshchandra (1934). Prof. Dept. of Physics, MSU, Baroda 390002, India. PhD (MSU, 1961) physics. *Crystal growth dissolution (etch phenomenon) hardness electrical conductivity*

Papavinasam, Dr E (1943). Prof. Dept. of Physics, Thiagarajar C., Madurai 625009, India. PhD (Madurai Kamaraj U., 1986) X-ray crystallography. *Crystal growth structure analysis*

Parthasarathy, Dr S. (1940). Prof. Dept. of Crystallography and Biophysics, Guindy Campus, Madras University, Madras 600025, India. PhD (Madras U., 1967) crystallography. *Crystal structure analysis theoretical crystallography*

Parthasarathy, Dr Tiruvallur Eachambadi (1949). Senior Lecturer. Dept. of Physics, A. M. Jain College, 114/229A, 28th Street, Nanganallur, Madras 600061, India. PhD (Madras U., 1992) biophysics. *Crystallography biological materials biophysics material science Indology Sanskrit carnatic music*
Tel. 91()2341705

Patel, Dr Prabhudas Revandas (1938). Silical Lab., Sandhporepardi, Valsad 396001, India. PhD (MSU, Baroda, 1967) liquid crystals. *Liquid crystal*
Tel. 91()2281

Patel, Dr Rajan Prafulbhai (1953). Scientific Officer. Chemistry Division, Bhabha Atomic Research Centre, Bombay 400 085, India. PhD (Bombay U., 1988) inorganic chemistry. *Synthesis spectroscopy structure transition metal organic compounds*

Patel, Dr Tankadhar (1943). Asst. Prof. Physics Dept., Regional Engineering College, Rourkela, 769008, Orissa, India. PhD (IIT Bombay, 1982) physics. *Crystal structure organic and inorganic substances*

Pathinettam, Dr Pandian (1956). Lecturer. D. V. H. N. S. N. College, Virudhunagar 626002, India. PhD (Madurai Kamraj U., 1988) physics. *X-ray diffraction bonding semiconductors anharmonicity*

Pattabhi, Dr Mrs Vasantha (1944). Professor. Dept. of Crystallography & Biophysics, University of Madras, India. [Dept of Crystallography, University of Madras, Guindy Campus, Madras 600025, India.] PhD (Madras U., 1972) X-ray crystallography. *Crystallography biomolecules conformational analysis*
E-mail nicrys@iitm.ernet.in Tel. 91(44)2351367 Telex 41 6376 UNOM IN

Podder, Dr Alok (1937). Associate Professor. Saha Institute of Nuclear Physics, 1/AF Bidhannagar, Calcutta 700 064, India. PhD (Calcutta U., 1976) physics. *Proteins crystallography structure small molecules*
E-mail aloka@saha.ernet.in Tel. 91(33)370659 Fax 91(33)374637 Telex 21-4103 SIN-PIN

Poojary, Dr M. Damodara (1955). Scient. Asst. Inorganic and Physical Chemistry Dept., Indian Institute of Science, Bangalore, Bangalore 560012, India. PhD (IISc, 1984) physics. *X-ray crystallography metal-nucleotide complexes*

Pradhan, Dr Dukhabandhu (1948). Reader. Dept. of Physics, Deogarh College, Sambalpur, India. PhD (IIT Kharagpur, 1987) physics. *Crystal structure organic compounds intensity statistics*

Pradhan, Dr Swapan Kumar (1957). Research associate. Dept. of Materials Science, Indian Association for the Cultivation of Science, Jadavpur, Calcutta 700032, India. PhD (J. U., 1990) physics. *Powder crystallography*
E-mail msmd@iacs.ernet.in Tel. 91(473)4971 Fax 91(473)2805 Telex 021-5501-IACSIN

Prakash, Mr Balaji (1968). Research Scholar. Molecular Biophysics Unit, Indian Institute of Science, Bangalore 560012, India. MSc (U. Hyderabad, 1990) physics. *X-ray crystallography proteins folding*
E-mail balaji@mbu.iisc.ernet.in Tel. 91()3344411 Fax 91()3341683 Telex 0845-8439IIScIn

Prasad, Dr Narayan (1944). Lect. Physics Dept, Ranchi University, Ranchi 834008, India. PhD (Ranchi U., 1978) X-ray crystallography. *X-ray crystallography organometallic compounds*

Prasad, Dr Ravindra (1943). Sr. Mineralogist. Geological Survey of India, B-192 Niralanagar, Lucknow, India. PhD (Banaras Hindu U., 1971) defects. *Crystal growth structure phase transformations stacking faults*

Prasad, Dr Satya Murti (1943). Prof. Physics Dept., Ranchi University, Ranchi 834008, India. PhD (Ranchi U., 1970) X-ray crystallography. *X-ray crystallography organic and organometallic compounds proteins*

Prasad, Dr Y. R. Ananth (1942). Fellow NAL. Materials Division, National Aeronautical Laboratory, PO Box No 1779, Bangalore 560017, India. PhD (IIT Delhi, 1970) semiconductors. *Single crystal growth solar energy ion-implantation and microscopy*
Tel. 91()573351

Pulya, Mr Umamaheswara Sastry (1964). Scientific officer. Solid State Physics Division, Bhabha Atomic Research Centre, Bombay 400 085, India. MSc (U. Hyderabad, 1986) physics. *Crystallography*
E-mail sspd@magnum.barct1.ernet.in Tel. 91()5563060 Fax 91()5560750 Telex 011-71017BARCIN

Puranik, Mrs Vedavati Gururaj (1953). Scientist. Physical Chemistry Division, National Chemical Laboratory, Pashan, Pune 411008, India. PhD (Bangalore U., 1983) X-ray crystallography. *X-ray structure analysis of small molecules organic inorganic organometallic bioorganic and natual products structure activity correlations intermolecular interactions*
E-mail pinak@ncl.ernet.in Tel. 91()336451 Fax 91()330233 Telex 0145-266

Raghavacharyalu, Dr Iyyunni Venkata Veera (1934). Scientific Officer. Nuclear Physics Divn, Bhabha Atomic Research Centre, Bombay 400085, India. DSc (Andhra U., 1958) group theory & representation of space groups. *Mathematical crystallography solid state physics*

Raghunatha, Mr Chary (1955). Res. Student. Dept. of Physics, Indian Institute of Science, Bangalore 560012, India. MPhil (Hyderabad U., 1979) physics. *Crystal growth crystal structure*

Rajagopal, Mr Hariharasubramonia Iyer (1941). Scient. Officer. Solid State Physics Division, Bhabha Atomic Research Centre, Bombay 400 085, India. BSc (Kerala U., 1962) physics. *Neutron diffraction computer programming software development*

Rajan, Mr R. D. (1945). Lecturer. Dept. of Physics, Anna University, Madras 600 025, India. MPhil (Anna U., 1984) physics. *Crystal structure organic compounds*

Rajan, Dr S. S. (1949). Reader. Crystallography & Biophysics Dept, Madras University, Madras 600025, India. PhD (Madras U., 1977) physics. *Crystal structure analysis organic compounds proteins*

Rajaram, Dr Ramaswamy Karunandam (1951). Reader. School of Physics, Madurai Kamaraj University, Madurai 625021, India. PhD (Madurai, Kamraj U., 1978) physics. *X-ray crystallography hydrogen bonding phase transitions*

Rajashekharan, Dr T. (1950). Scientist. Defence Metallurgical Res. Lab., Hyderabad 500258, India. PhD (IIT Madras, 1978) physics. *Structure intermetallic phases*

Raju, Dr I. V. K. Bhagavan (1946). Reader. Dept. of Physics, Kakatiya University, Warangal 506009, India. PhD (Osmania U., Hyderabad, 1973) physics. *Crystal_growth_defect plastic_materials*

Raju, Dr K. S. (1937). Reader. Dept. of Crystallography & Biophysics, Guindy Campus, Madras 600025, India. PhD (Sardar Patel U., 1969) physics. *Crystal_growth_defect_and_characterization*

Ram, Dr Purushottam (1948). Lect. Physics Dept., Ranchi University, Ranchi 834008, India. PhD (Ranchi U., 1979) physics. *X-ray_crystallography organometallic_compounds*

Ram Kishore, Dr (1948). Scientist. Materials Divn, National Physical Laboratory, Krishnan Road, New Delhi 110012, India. PhD (Agra U., 1983) physics. *Single_crystal_growth multicrystalline_silicon_ingot_technology*
Tel. 91( )5726058

Ramachandran, Prof. Gopalasamudram N. (1922). INSA Albert Einstein Prof. Mathematical Philosophy Group, Indian Institute of Science, Bangalore, Bangalore 560012, India. DSc (Madras U., 1949) crystallography. *Theoretical_crystallography*

Ramakrishnan, Prof. Chandrasekharan (1939). Prof. Molecular Biophysics Unit, Indian Institute of Science, Bangalore 560012, India. PhD (Madras U., 1966) biophysics. *Biomolecule_conformation molecular_modelling proteins_structure statistics_analysis curricular_X-ray_crystallography algorithmic_methods*
E-mail ramki@mbu.iisc.ernet.in Tel. 91(80)3344411 Fax 91(80)3341683 Telex 0845-8349 IISC IN

Ramanadham, Dr Muthyala (1945). Scientific Officer(SG). Member Crystallographic Computing Commission IUCr. Solid State Physics Division, Bhabha Atomic Research Centre, Bombay - 400 085, India. PhD (Bombay U., 1975) physics. *Macromolecular_structure_determination X-ray_and_neutron_diffraction hydrogen_bonding parallel_processors graphics*
E-mail ramu@magnum.barct1.ernet.in Tel. 91(22)5563060 Fax 91(22)5560750 Telex 01171017BARCIN

Ramaswamy, Dr Krishnamachari (1935). Head, Physics Dept. Physics Dept., Annamalai University, Annamalai Nagar 608002, India. PhD (Annamalai U., 1961) molecular spectroscopy. *Raman_spectroscopy infrared_spectroscopy nuclear_magnetic_resonance X-ray_techniques*

Ramaswamy, Mr S. (1964). Res. Stu. Molecular Biophysics Unit, Indian Institute of Science, Bangalore, Bangalore 560012, India. MSc (Bharatidasan U., 1987) physics. *Biological_crystallography*
Tel. 91( )344411

Ranganath, Dr G. S. (1944). Scientist. Liquid Crystals Lab., Raman Research Institute, Bangalore 560006, India. PhD (Bangalore U., 1974) crystal optics. *Crystal_optics liquid_crystals*
Tel. 91( )30124

Rao, Dr Keshavamurthy Narayana Swamy (1934). Sr. Res. Asst. Physics Dept., Indian Institute of Technology, Kanpur, Kanpur 208016, India. PhD (Kanpur U., 1974) crystal physics. *Crystal_growth characterization intercrystalline_boundaries crystal_defects*
Tel. 91( )40066

Ratna, Miss B. R. (1949). Res. fellow. Liquid Crystals Laboratory, Raman Research Institute, Bangalore 560006, India. MTech (IISc, 1972) physical eng. *Liquid_crystal*
Tel. 91( )30124

Ravichandran, Dr Veena (1959). Senior scientific officer. Committee on Science & Technology in Developing Countries (COSTED), 24, Gandhi Mandapam Road, Madras 600025, India. PhD (U. Madras, 1988) crystallography, biomolecular structure. *Crystallography structural protein_conformation folding crown_ethers*
E-mail costed@sirnetm.ernet.in Tel. 91( )419466 Fax 91( )4914543

Ravishankar, Mr Ramachandran (1970). Research Scholar. Molecular Biophysics Unit, Indian Institute of Science, Bangalore 560012, India. MSc (Sri Sathya Sai Institute of Higher Learning, 1992) physics. *Biocrystallography*
Tel. 91( )3344411 Fax 91( )3341683

Ray, Dr Pankaj Narayan (1934). Reader. Crystallography & Molecular Biology Division, Saha Institute of Nuclear Physics, Calcutta 700040, India. PhD (Calcutta U., 1974) physics. *Amino_acids biocompounds*

Ray, Dr Pradip Kumar (1938). Sr. Scientist. Physics Division, Jute Technical Research Laboratory, Indian Council of Agricultural Research, 12 Regent Park, Calcutta 700040, India. PhD (Calcutta U., 1968) fibre structure. *Celluloses fibres structure*
Tel. 91( )723192

Ray, Dr Mrs (1935). Managing Dir. Radon House (P) Ltd, 7 Sirdar Sankar Road, Calcutta 700026, India. PhD (Calcutta U., 1966) physics. *Crystal_physics instrumentation*

Roychowdhury, Dr Priyobroto (1940). Reader. Physics Dept., Calcutta University, Calcutta 700009, India. PhD (Calcutta U., 1974) physics. *X-ray_crystallography biomolecule_proteins*

Roychowdhury, Dr Mrs S. (1919). Lect. X-ray Laboratory, Dept. of Physics, Presidency College, Calcutta, India. PhD (Calcutta U., 1919) physics. *Small_molecules_crystallography*

Saha, Mr Bishwa Nath (1936). Res. scholar. Physics Dept., Ranchi University, Ranchi 834008, India. MSc (Bihar U., 1960) physics. *X-ray_crystallography organic_and_organometallic_compound*

Sahaymary, Mrs J. James (1958). Res. Scholar. Dept. of Biophysics and crystallography, ACC, Madras University, Madras 600025, India. MSc (Madras U., 1981) physics. *Molecular_biophysics*

Sahu, Dr Bhola Nath (1943). Lecturer. Physics Dept., Ranchi University, Ranchi 834008, India. PhD (Ranchi U., 1970) X-ray crystallography. *X-ray_crystallography organic_structures*

Sahu, Dr Ram Gopal (1934). Reader. Dept. of Physics, GALC, Daltonganj 822102, India. PhD (Ranchi U., 1968) physics. *Crystal_growth crystal_structure organic_crystals*

Samanta, Ms Chitra (1949). Res. scholar. Dept. of Physics, Jadavpur University, Calcutta 700032, India. MSc (Jadavpur U., 1971) applied mathematics. *X-ray_crystallography molecular_lattice_dynamics computer_simulation*

Samantaray, Dr Biswas Kumar (1947). Res. Asst. Dept. of Physics, IIT Kharagpur, Kharagpur 721302, India. PhD (IIT Kharagpur, 1977) physics. *Applied_crystallography mineralogy thin_films*

Sanjeeviraja, Dr C. (1954). Lecturer. Alagappa University, Karaikudi 623003, India. PhD (Madurai Kamaraj U., 1985) physics. *X-ray_diffraction instrumentation crystalline_solid_solutions*

Sankaran, Ms Hema (1962). Scient. Officer. Neutron Physics Division, Bhabha Atomic Research Center, Trombay, Bombay 400085, India. MSc (IIT Bombay, 1983) physics. *Crystal_structure solid_state_physics high_pressure*
Tel. 91( )5513848

Sankaranarayanan, Mr Rajan (1968). Research Fellow. Molecular Biophysics Unit, Indian Institute of Science, Bangalore 560012, India. MSc (Madurai Kamaraj U., 1990) physics. *Proteins_crystallography*
E-mail sn@mbu.iisc.ernet.in Tel. 91( )3344411 Fax 91( )3341683 Telex 845-8349IISCIN

Sanker, Mr B. N. (1951). Lecturer. Dept. of Physics, College of Engineering, Anna University, Madras 60025, India. MPhil (Anna U., 1983) physics. *Crystal_structure organic_compound*

Sarkar, Ms Chitra (1951). Res. Fellow. Dept. of Physics, Jadavpur University, Calcutta 700032, India. MSc (Jadavpur U., 1975) physics. *X-ray_crystallography*

Sarkar, Dr Satyabrata (1944). Tech. Superintendent. Dept of Physics, Jadavpur University, Calcutta 700032, India. PhD (Jadavpur U., 1983) physics. *X-ray_crystallography diffuse_scattering lattice_dynamics structure_solution computer_simulation*

Sarma, Dr J. A. R. P. (1957). Scientist, Inorganic & Physical Chemistry Division. Indian Institute of Chemical Technology, Tarnaka, Hyderabad 500 007, India. [I&PC Division, Indian Institute of Chemical Technology, Tarnaka, Hyderabad 500 007, India.] PhD (U. Hyderabad, 1986) chemistry. *Chemical_crystallography structural_organic_chemistry photochemistry spectroscopy molecular_modelling organic_synthesis*
E-mail jarps%bic-ccmb@dbt.ernet.in Tel. 91(40)673874x214 Fax 91(40)673387 or 91(40)673757 Telex 91-0425-7061 iict in

Sastry, Dr G. V. S. (1952). Reader. Dept of Metallurgy Eng., Banaras Hindu University, Varanasi 221005, India. PhD (Banaras Hindu U., 1982) metallurgical eng. *Electron_diffraction alloys_metastable_phase*

Sathyamurthy, Ms Padma (1942). Scientist. Solid State Physics Division, BARC, Bombay 400085, India. BSc (Bombay U., 1960) physics. *Macromolecular_crystallography*
E-mail hosur@magnum.barct1.ernet.in Tel. 91(22)5563060x3614 Fax 91(22)5560750 Telex 011-71017BARCIN

Savithramma, Miss K. L. (1952). Res. Fellow. Liquid Crystal Laboratory, Raman Research Institute, Bangalore 560006, India. MSc (Mysore U., 1974) theoretical physics. *Liquid_crystal*
Tel. 91( )30124

Seal, Dr Alpana (1955). Res. Scholar. Magnetism Dept., Indian Association for the Cultivation of Science, Calcutta 700032, India. PhD (Calcutta U., 1985) physics. *Disorder liquid_crystal neutron_proteins_crystallography*

Seal, Prof. Arun Kumar (1955). Principal. Bengal Engineering College, Howrah 711103, India. PhD (Sheffield U., 1956) physical metallurgy. *Crystal_structure metals*

Seetharaman, Mr Venkataramakrishanan (1950). Scientific Officer. Metallurgy Division, BARC, Trombay, Bombay 400085, India. BTech (IIT Madras, 1971) metallurgy. *Electron_microscopy microstrain radiation_damage*
Tel. 91( )523321

Selladurai, Mr S. (1961). Res. Fellow. Dept of Physics, Anna University, Madras 600025, India. MSc (Anna U., 1984) physics. *X-ray_crystallography*

Sen Gupta, Dr Amitava (1944). Deputy Superintendent. Physical Research Wing, Projects & Div India Ltd, Sindri 828122, India. PhD (Jadavpur U., 1981) physics. *Powder_diffraction metals alloys*

Sen Gupta, Prof. Siba Prasad (1941). Prof. & Head. Editor-in-chief, Indian Journal of Physics; Past Chairman, National Committee of Crystallography. Dept. of Materials Science, IACS, Jadavpur, Calcutta 700032, India. DPhil (Calcutta U., 1968) physics. *Crystal_growth crystal_structure topography non-linear_materials microstructures metals alloys intermetallics*
E-mail msspsg@iacs.ernet.in Tel. 91(33)4734971 Fax 91(33)4732805

Sen, Dr Deb Kumar (1942). Mineralogist(Sr). Mineral Physics Division, Geological Survey of India, 29, J. N. Rd, Calcutta 700016, India. PhD (Calcutta U., 1972) X-ray crystallography. *X-ray_crystallography minerals electron_microscopy*
Tel. 91( )297645

Sen, Miss Mina (1947). Lecturer in Vidyasagar College. Physics Dept., Presidency C., Calcutta 700073, India. MSc (Calcutta U., 1970) physics. *X-ray_crystallography*

Sen, Dr Ranjit Kumar (1919). Em. Scient. Dept of General Physics & X-rays, Indian Association for the Cultivation of Science, Calcutta 700032, India. DSc (Allahabad U., 1956) physics. *Structure_of_metals*

Sen, Dr Suchitra (1950). Scient. Central Glass & Ceramic Research Institute, Calcutta 700032, India. PhD (Calcutta U., 1977) physics. *Ceramics_materials X-ray_diffraction electron_microscopy*

Sen, Mr Udayaditya (1966). Senior Research Fellow. Crystallography & Molecular Biology Division, Saha Institute of Nuclear Physics, Calcutta 700 064, India. MSc (Burdwan U., 1989) chemistry. *Crystallography modelling*
E-mail jiban@Saha.ernet.in Tel. 91(33)370659 Fax 91(33)374637 Telex 21-4103SINPIN

Sengupta, Ms Suparna (1962). S. R. F. IACS, Jadavpur, Calcutta 700032, India. MPhil (Rani Durgawati Vishwavidyalaya (Jabalpur), 1984) physics. *Crystal_growth X-ray_topography etching microhardness*
E-mail msspsg@iacs.ernet.in Tel. 91(33)4734971 Fax 91(33)4732805 Telex cable-INDASSON,Jadavpur

Sequeira, Dr Anisbert S. (1915). Head. Crystallography & Characterisation Section. Member, NDC (1987–1993). Solid State Physics Division, Bhabha Atomic Research Centre, Trombay, Bombay 400085, India. PhD (U. Bombay, 1970) physics. *Neutron_diffraction instrumentation phase_transition diffractometry synchrotron_instrumentation*
E-mail sequeira@magnum.barct1.ernet.in Tel. 91()5517273 Fax 91()5560750 Telex 011-71017-BARCIN

Seshasayee, Ms Maha (1943). Professor. Indian Institute of Technology, Dept. of Physics, IIT, Madras 600036, India. PhD (Rensselaer Polytechnic Institute, Troy, NY, USA, 1968) solid state physics. *Crystallography small_molecules EXAFS*
Tel. 91(44)2351365 Fax 91(44)2350509 Telex 41-8926IITIN

Sharma, Mr Braj Bhushan (1941). Sr. Scient. Solid State Physics Laboratory, Ministry of Defence, Lucknow Road, Delhi 110007, India. MSc (Agra U., 1961) physics. *Semiconductors X-ray_topography*

Sharma, Mr Girish B. (1970). Research Scholar. Molecular Biophysics Unit, Indian Institute of Science, Bangalore 560012, India. MSc (Bangalore U., 1993) physics. *X-ray_crystallography proteins folding*
E-mail balaji@mbu.iisc.ernet.in Tel. 91()3344411 Fax 91()3341683 Telex 0845-8439-IISc-In

Sharma, Mr K. K. (1965). Research Scholar. P. G. Dept. of Physics, University of Jammu, Canal Road, Jammu 180 001, India. MPhil (U. Jammu, 1991) crystal growth and characterization. *Crystal_growth characterization*

Sharma, Dr Surinder Dutt (1947). Scient. Crystal Growth Characterization Section, National Physical Laboratory, New Delhi 110012, India. PhD (Allahabad U., 1975) physics. *Crystal_growth X-ray_topography*

Shashidhar, Dr R. (1946). Scient. Liquid Crystals Laboratory, Raman Research Institute, Bangalore 560006, India. PhD (Mysore U., 1972) liquid crystals. *Liquid_crystal*

Shivaprakash, Dr N. C. (1955). Scient. Instrumentation and Service Unit, IISc, Bangalore 560012, India. PhD (Mysore U., 1982) physics. *Liquid_crystal_and_structure*
E-mail ncshiva@isu.iisc.ernet.in Tel. 91()3092242 (office) 91()3092544 (residence) Fax 91(80)3341683

Sikka, Dr Satinder Kumar (1942). Scient. H; Head, High-Pressure Physics Division. High-Pressure Physics Division, BARC, Bombay 400085, India. PhD (Bombay U., 1970) physics. *High_pressure_crystallography neutron_diffraction phase_problem*
E-mail sksikka@magnum.barct1.ernet.in Tel. 91(22)5550979 Fax 91(22)5565773

Singh, Dr Bhanu Pratap (1947). Scient. High Pressure Technical Division, National Physical Laboratory, New Delhi 110012, India. PhD (Delhi U., 1980) physics. *Defect_characterization diffuse_X-ray_scattering high_pressure_phase_transformation*

Singh, Dr Govind (1940). Prof. Physics Dept., Banaras Hindu University, Varanasi 221005, India. PhD (Banaras Hindu U., 1967) physics. *X-ray_crystallography electron_microscopy crystal_growth_and_imperfections*

Singh, Dr S. N. (1947). Scient. C. Materials Division, National Physical Laboratory, Dr K. S. Krishnan Road, New Delhi 110012, India. PhD (Agra U., 1975) physics. *Single_crystal_growth*

Singh, Mr Surendra Prakash (1954). Res. scholar. Physics Dept., Ranchi University, Ranchi 834008, India. MSc (Ranchi U., 1976) physics. *X-ray_crystallography*

Sinha, Dr Umesh Chandra (1936). Asst. Prof. Physics Dept., IIT Bombay, Bombay 40066, India. DPhil (Allahabad U., 1961) X-ray crystallography. *X-ray_diffraction crystallography instrumentation*

Sirdeshmukh, Dr Dinker (1935). Prof. Physics Dept., Kakatia University, Warangal 506009, India. PhD (Osmania U., 1964) X-ray crystallography. *X-ray_crystallography crystal_growth*
Tel. 91()7701

Sivakumar, Dr K. (1960). Visiting lecturer. Dept. of Physics, Anna University, Madras 600025, India. PhD (Anna U., 1988) physics. *Crystal_structure*

Soman, Ms Jayashree (1963). Res. Student. Molecular Biophysics Unit, IISc, Bangalore 560012, India. MSc (Karnataka U., 1984) physics. *Biological_crystallography*
Tel. 91()344411

Sridhar Prasad, Mr G. (1963). Res. Fellow. Molecular Biophysics Unit, IISc, Bangalore 560012, India. MSc (Bangalore U., 1985) physics. *Molecular_biophysics biocrystallography*
Tel. 91()344411

Srinivasa, Dr V. K. (1934). Deputy Superintendent. Physical Research Wing, Projects & Dev. India Ltd, PO Sindri, Dhanbad 828122, India. PhD (Indian School of Mines, 1978) physics. *Phase_transformation stress_analysis double_salts*
Tel. 91()2613 Telex 0629-216FPDIL

Srinivasan, Prof. R. (1940). Senior Prof. Dept. of Crystallography & Biophysics, University of Madras, Guindy Campus, Madras 600 025, India. PhD (U. Madras, 1970) X-ray crystallography. *Crystallography*
E-mail nicry@iitm.ernet.in Tel. 91(44)2351367 Telex 416376UNOMIN

Srinivasan, Dr Sampat (1943). Scientific officer. Radiochemistry Division, BARC, Trombay, 400085 Bombay, India. PhD (IIT Madras, 1971) inorganic chemistry. *Actinides crystal_chemistry oxides carbides*
Tel. 91()523321

Srivastava, Dr Ramesh Chandra (1936). Asst. Prof. Physics Dept., IIT Kanpur, Kanpur 208016, India. DPhil (Allahabad U., 1960) physics. *Thermal_diffuse_scattering crystal_structure neutron_diffraction*
E-mail rcs@iitk.ernet.in Tel. 91(512)258688 Fax 91(512)250260

Stephen, Mr Suresh (1965). Research Scholar. Indian Institute of Science, Bangalore, India. MSc (Indian Institute of Science, 1989) physics. *Macromolecules structure biomolecular_interactions*
E-mail mbu!ss@vigyan.ernet.in Tel. 91()3344411 Fax 91()3341683 Telex 0845-8349-IIScIn

Subhadra, Dr K. G. (1947). Reader. Physics Dept., Kakatiya University, Warangal 506009, India. PhD (Osmania U., 1976) physics. *Crystal_growth inorganic_crystal_structure chemical_crystallography*

Subramanian, Prof. E. (1938). Prof. Dept of Physics, Crystallography & Biophysics, University of Madras, Madras, Tamilnadu 600025, India. PhD (Madras U., 1965) physics. *X-ray_diffraction proteins_crystallography structure_peptides structure-based_drug_design molecular_graphics_modelling_studies*
E-mail nicrys@iitm.ernet.in Tel. 91(44)2351367 or 91(44)4918512 Fax 91(44)415856 Telex 41 6376 UNOM IN

Subramanian, Dr K. (1947). Lecturer. Physics Dept., College of Engineering, Anna University, Madras 600025, India. PhD (Madras U., 1983) physics. *Crystallography*

Subramanya, Mr H. S. (1963). Res. Fellow. Molecular Biophysics Unit, IISc, Bangalore 560012, India. MSc (Mysore U., 1986) physics. *Biological_crystallography*
Tel. 91()344411

Sudaramoorthy, Mr M. (1960). Res. student. Molecular Biophysics unit, IISc, Bangalore, Bangalore 560012, India. MSc (Anna U., 1983) material science. *Molecular_biophysics fibres_diffraction*
Tel. 91()344411x2534

Sudarsanakumar, Dr Chellappan Pillai (1963). Research Associate. Molecular Biophysics Unit, Indian Institute of Science, Bangalore, India. PhD (Indian Institute of Technology, Madras, 1992) X-ray crystallography. *Macromolecular_crystallography*
E-mail sudarsan@mbu.iisc.ernet.in Tel. 91()3344411 Fax 91()3341683

Suguna, Dr K. (1956). Senior Scientific Officer. Molecular Biophysics Unit, Indian Institute of Science, Bangalore 560 012, India. PhD (Indian Institute of Science, 1982) physics. *Proteins_structure_and_function*
E-mail suguna@mbu.iisc.ernet.in Tel. 91(80)3344411 Fax 91(80)3341683

Sundaramoorthy, Mr M. (1960). Res. Fellow. Molecular Biophysics Unit, IISc, Bangalore 560012, India. MSc (Anna U., 1983) material science. *Molecular_biophysics fibres_diffraction*
Tel. 91()344411

Suresh, Dr C. G. (1955). Scientist. Division of Biochemical Sciences, National Chemical Laboratory, Pune 411008, India. PhD (Indian Institute of Science, Bangalore, 1985) biomolecular crystallography. *Biomolecule_proteins*
E-mail suresh@ncl.ernet.in Tel. 91()338234 Fax 91()330233 Telex 0145-266

Suresh, Mr K. A. (1948). Res. Fellow. Liquid Crystals Laboratory, Raman Research Institute, Bangalore 560006, India. MSc (Mysore U., 1969) solid state physics. *Liquid_crystal*
Tel. 91()30124

Suri, Mr D. K. (1953). Scient. X-ray Section, Material Characterization Division, National Physical Laboratory, New Delhi 110012, India. MSc (Meerut U., 1972) physics. *Crystal_structure phase_transformation inorganic_materials*

Suri, Mr D. K. (1953). Scient. X-ray Section, Material Characterization Division, National Physical Laboratory, New Delhi 110012, India. MSc (Meerut U., 1972) physics. *Inorganic_materials structure_of_crystal phase_transformation*

Suryanarayana, Dr Challapalli (1945). Reader. Dept. of Metallurgical Engineering, Banaras Hindu University, Varanasi 221005, India. PhD (Banaras Hindu U., 1970) physical metallurgy. *Crystallography electron_microscopy defects*
Tel. 91()54290x250

Suryanarayana, Dr Shambhuni V. (1943). Lecturer. Physics Dept., C. of Sci., Osmania University, Hyderabad 5600007, India. PhD (Osmania U., 1971) X-ray crystallography. *X-ray_crystallography alloy_phases thermal_expansion Debye_temperatures*

Suta, Ms Elizabeth (1959). Scient. asst. Dept. of Physics, IISc, Bangalore, Bangalore 560012, India. MSc (Kerala U., 1983) physics. *Crystal_growth*

Talukdar, Dr Amarendra Nath (1934). Reader. Dept. of Physics, Gauhati University, Guwahati 781014, India. PhD (Gauhati U., 1974) physics. *Crystal_structure_analysis X-ray_fluorescence_spectrometry*
Tel. 91()570531 Fax 91(361)70133

Tavale, Dr Sudam Shankar (1934). Scientist E-II. Physical Chemistry Division, National Chemistry Laboratory, Pune 411008, India. PhD (Poona U., 1964) X-ray crystallography. *Single_crystal_X-ray_diffraction*
E-mail nnd@ncl.ernet.in Tel. 91()336451 Fax 91()330233 Telex 0145-266

Tewari, Dr Raghavendra (1945). Lecturer. Computer Center, Aligarh Muslim University, Aligarh 202001, India. PhD (IIT Kanpur, 1973) X-ray crystallography. *X-ray_crystallography*

Thomas, Mr P. Muthiah (1955). Senior Lecturer. Chemistry Dept, Bharatidasan University, Tiruchi, Tamilnadu, India. PhD (Calcutta U., 1985) chemistry. *Crystallography bioinorganic_chemistry*
Tel. 91()896353 Telex 455-253

Uma Devi, Ms B (1967). Bharathidasan University, Tiruchi 620024, India. MPhil (Bharathidasan U., 1990) chemistry. *Crystallography_macromolecules*

Usha, Ms R. (1953). Lecturer. Centre for Plant Molecular Biology, Madurai Kamraj University, Madurai 625021, India. PhD (Indian Institute of Science, Bangalore, 1981) crystallography. *Crystallography_macromolecules molecular_biology*
E-mail USHA%BIC-MKU@imtech.ernet.in Tel. 91(452)85214 Fax 91(452)85205

Varadarajan, Dr Raghavan (1960). Asst. Prof. Molecular Biophysics Unit, Indian Institute of Science, Bangalore 560012, India. PhD (Stanford U., 1988) chemistry. *Proteins_structure_and_folding*
E-mail mbu!varadar@vigyan.ernet.in Tel. 91(80)3092612 Fax 91(80)3341683

Vasanth, Dr K. L. (1940). Prof. and Head. Chemistry Dept., PSG College of Technology, Coimbatore 641004, India. PhD (MSU, Baroda, 1968) chemistry. *Liquid_crystal metal_&_alloy_electrodeposit_structures*

Velankar, Mr Sameer Sudhakar (1970). Research Scholar. Molecular Biophysics Unit, Indian Institute of Science, Bangalore 560012, India. MSc (U. Poona, 1992) physics. *X-ray_crystallography proteins_folding*
E-mail Sameen@mbu.iisc.ernet.in Tel. 91()3344411 Fax 91()3341683 Telex 0845-8439IIScIn

Venkataraman, Dr Ravichandran (1954). Senior Lecturer. Dept. of Nuclear Physics, Guindy Campus, University of Madras, Madras 600 025, India. PhD (U. Madras, 1985) biocrystallography: molecular biophysics and X-ray crystallography. *Structure_conformation_macromolecules peptides nucleic_acids PIXE lattice_studies_at_high_temperatures crystallite_size diffuse_scattering*
E-mail madphy@iitm.ernet.in Tel. 91()2351269 Fax 91(44)2352870 Telex 416376UN-OMIN

Venkatesan, Dr Kailash (1932). INSA–Senior Scientist. Dept. of Organic Chemistry, Indian Institute of Science, Bangalore 560012, India. PhD (Madras U., 1959) crystallography. *Chemical_crystallography music*
E-mail ockven@orgchem.iisc.ernet.edu Tel. 91()3344411 Fax 91()3341683

Venudhar, Dr Y. C. (1950). Reader. Physics Dept., Osmania Univ., Hyderabad 500007, India. PhD (Osmania U., 1981) X-ray crystallography. *X-ray_crystallography solid_state_physics materials_science*

Venugopalan, Mr P. (1963). Res. Sch. Organic Chemistry Dept., Indian Inst. of Science, Bangalore 560012, India. MSc (Mysore U., 1986) chemistry. *Structure_reactivity_correlations*

Verma, Dr Ajit Ram (1921). Em. Scient. National Physical Laboratory, Hillside Road, New Delhi 110012, India. DSc (London U., 1969) physics. *Crystal_growth crystal_defect*

Vijayan, Dr Mrs Kalyani (1942). Scient. Material Science Division, National Aeronautical Laboratory, Bangalore 560017, India. PhD (IISc, 1969) X-ray crystallography. *Liquid_crystal polymers_diffraction crystallography_method organic-bio_structures*
Tel. 91()570098

Vijayan, Prof. Mamannamana (1941). Prof. Chairman, IUCr Commission on Biological Macromolecules; Co-Editor, Acta Crystallographica. Molecular Biophysics Unit, Indian Institute of Science, Bangalore 560012, India. PhD (Indian Institute of Science, Bangalore, 1967) X-ray crystallography. *Biological_crystallography molecular_biophysics chemical_evolution_and_origin_of_life*
E-mail mv@mbu.iisc.ernet.in Tel. 91(80)3346765 Fax 91(80)3341683 Telex 0845-8349 IISC IN

Vimla, Ms T. M. (1941). Asst Prof. Dept. of Physics, Indian Institute of Technology, Madras 600 036, India. PhD (Indian Institute of Technology, Madras 600 036, 1987) X-ray crystallography. *Structure X-ray_diffraction_method*
Tel. 91(44)2351365 Fax 91(44)2350509 Telex 41-8926IITIN

Viswamitra, Prof. Mysore Ananthamurthy (1932). Prof. Physics Dept, Indian Institute of Science, Bangalore 560012, India. PhD (IISc, 1960) *Biological_crystallography high_temperature_crystallography*

Vora, Dr Rasiklal Amulakhbhai (1938). Prof. and Head. Applied Chemistry Dept., Faculty of Technology & Engineering, MSU, Baroda 390001, India. PhD (MSU, Baroda, 1975) liquid crystals. *Liquid_crystal*

Vyas, Mr K (1958). Res. Sch. Physical Chemistry Dept, Indian Institute of Science, Bangalore 560012, India. MSc (Mysore U., 1980) chemistry. *Solid_state_reaction crystallography small_molecules topochemistry*

Wadhawan, Dr Vinod Kumar (1944). Head, Crystal Growth Group, Scientific Officer (SG). Regional Editor (India), 'Phase Transitions'. Crystal Growth Group, Laser Programme, Centre for Advanced Technology, Indore - 452013, India. PhD (U. Bombay, 1975) physics. *Crystal_growth crystal_morphology phase_transition ferroelastic_materials composites smart_structures*
E-mail wadhawan@cat.ernet.in Tel. 91(731)481078 or 91(731)488356 Fax 91(731)481426 Telex 735-275 CAT IN

Yadav, Dr Asheshwar (1945). University Professor of Physics. Dept. of Physics, Patna University, Patna 800 005, India. PhD (Patna U., Patna, 1983) solid state physics. *Crystallography X-ray_diffraction*
Tel. 91()660012

Yadav, Dr Tapaswi (1939). Reader. Physics Dept., G. J. College, Rambagh, Bihta (Patna) 801103, India. PhD (Ranchi U., 1984) X-ray crystallography. *Crystal_structure X-ray_diffraction_techniques*

Yadava, Dr Bishwanath (1935). Res. Sch. Physics Dept., Ranchi University, Ranchi 834008, India. PhD (Ranchi U., 1981) X-ray crystallography. *Crystallography*

Yadava, Dr Vijay Singh (1942). Scientific Officer. Solid State Physics Division, Bhabha Atomic Research Centre, Bombay - 400 085, India. PhD (Bombay U., 1974) crystallography. *Macromolecular_structure_determination X-ray_and_neutron_diffraction biocrystallography*
E-mail kannan@magnum.barct1.ernet.in Tel. 91(22)5563060 Fax 91(22)5560750 Telex 01171017BARCIN

INDONESIA

Sub-Editor: Dr Effendy

Arryanto, Dr Yateman (1951). Lecturer. Department of Chemistry, Faculty of Mathematics and Science, Gajah Mada University, Sekip Utara, Jogyakarta, Indonesia. PhD (Salford U., England, 1990) inorganic chemistry. *Inorganic_material*
Tel. 62(274)902370

Effendy, Dr (1956). Lecturer. National Sub-Editor, WDC9. Department of Chemical Education, Faculty of Mathematics and Science Education, Institute of Teacher Training and Education, FPMIPA IKIP Malang, Malang, East Java 65145, Indonesia. PhD (U. Western Australia, 1994) physical inorganic chemistry. *Silver_complex synthesis_and_structural_characterization_of_silver(I)_complexes*
Tel. 62(341)51312 Fax 62(341)51921

Loeksmanto, Prof. Dr Waloejo (1933). Lecturer. Department of Physics, Faculty of Mathematics and Science, Bandung Institute of Technology, Bandung 40132, Indonesia. [Dept of Physics, Faculty of Mathematics and Science, Bandung Institute of Technology, Jln. Ganesha 10, Bandung, Indonesia.] PhD (U. Sci. et Techn., Languedoc, France, 1979) structural chemistry.

Setiaji, Dr Bambang (1949). Lecturer. Department of Chemistry, Faculty of Mathematics and Science, Gajah Mada University, Sekip Utara, Jogyakarta, Indonesia. PhD (UMIST, England, 1986) physical chemistry. *Materials_science catalysts*
Tel. 62(274)902370

Syarif, Dr Shamid Lecturer. Department of Soil Science, Bogor Institute of Agriculture, Jln. Raya Pajajaran, Bogor, Indonesia. PhD (U. Western Australia, 1992) soil science. *Mineralogy*

IRELAND

Sub-Editor: P. Mcardle

Cardin, Dr Christine (1947). Senior lecturer in Chemistry. Department of Chemistry, Trinity College, Dublin, Ireland. PhD (U. Sussex, 1973) organometallic chemistry. *Organometallic_chemistry bioinorganic anticancer_compounds oligonucleotides fullerenes non-linear_optics DNA_binding*
E-mail ccardin@vax1.tcd.ie Tel. 353(1)7022026 Fax 353(1)6712826

Cunningham, Prof. Des (1942). Professor. Department of Chemistry, University College, Galway, Ireland. DSc (National U. Ireland, 1993) chemistry. *Mossbauer_spectroscopy tin_chemistry*
E-mail checunningha@bodkin.ucg.ie Tel. 353(91)24411x2483 Fax 353(91)25700

Higgins, Dr Tim (1955). Lecturer. Department of Chemistry, University College, Galway, Ireland. PhD (National U. Ireland, 1980) organometallic chemistry. *Macromolecular_crystallography bioinorganic_chemistry*
E-mail xrahiggins@bodkin.ucg.ie Tel. 353(91)24411x2464 Fax 353(91)25700

Kelly, Dr Thomas (1954). Director, Ceramics Research Centre. [Ceramics Research Centre, Forbairt, Glasnevin, Dublin, Ireland.] PhD (National U. Ireland, 1980) chemistry. *Powder_diffraction crystallite_size_strain ceramics_materials thin_films*
E-mail kellyt@venus.eolas.ie Tel. 353(1)8370101 Fax 353(1)8370173 Telex 32501

Mcardle, Prof. Patrick (1945). Professor. National Sub-Editor, WDC9. Department of Chemistry, University College, Galway, Ireland. DSc (National U. Ireland, 1992) chemistry. *Organometallic_chemistry computing*
E-mail xramcardle@bodkin.ucg.ie Tel. 353(91)24411x2487 Fax 353(91)25700

Mernagh, Ms Victoria (1968). Scientific Officer. EOLAS, Ireland. [Ceramics Research Eolas, The Irish Science and Technology Agency, Glasnevin, Dublin 9, Ireland.] MSc (Trinity College Dublin, 1993) ceramic thin films. *Ceramics powders line_broadening coatings*
E-mail mernagh@venus.eolas.ie Tel. 353(1)370101 Fax 353(1)370173

O Brien, Mr Terence P. (1965). Research Scientist. Ceramics Dept, EOLAS, Ballymun Rd, Dublin 9, Ireland. BEng (National U. Ireland, 1987) electronic engineering. *Superconductors diamond ceramics*
E-mail TPOBRIEN@venus.eolas.ie Tel. 353(1)370101x2311 Fax 353(1)370173

Walsh, Dr Martin (1967). Post-doc. Department of Chemistry, University College, Galway, Ireland. PhD (National U. Ireland, 1994) chemistry. *Macromolecular_crystallography biological_electron_transfer*
E-mail xrawalsh@bodkin.ucg.ie Tel. 353(91)24411x3112 Fax 353(91)25700

ISRAEL

Sub-Editor: M. Harel

Agmon, Ilana (1949). Associate staff scientist. Department of Structural Biology, Weizmann Institute of Science, Rehovot 76100, Israel. PhD (Technion–Israel Inst. of Technology, 1989) crystallography. *Structure_ribosomes phase_transition*
E-mail Csagmon@weizmann.weizmann.ac.il chr03il@technion.technion.ac.il Tel. 972(8)342541 Fax 972(8)344154

Benghiat, Dr Victor (1932). Dept. of Biological Services, The EM Center, The Weizmann Institute of Science, Rehovot 76100, Israel. PhD (The Weizmann Institute of Science, 1970) X-ray crystallography. *Electron_microscopy*
E-mail LSVICTOR@WEIZMANN.WEIZMANN.AC.IL Tel. 972(8)343223

Berkovitch-Yellin, Dr Ziva (1946). Associate Professor. Department of Structural Biology, Weizmann Institute of Science, Rehovot 76100, Israel. PhD (Weizmann Institute of Science, 1976) solid state chemistry and crystallography. *Structure_ribosomes protein crystal_growth*
E-mail csberkov@weizmann.weizmann.ac.il Tel. 972(8)342541 Fax 972(8)344154 Telex 381300 WIX IL

Bernstein, Prof. Joel (1941). Professor. Chairman, Israeli Crystallographic Society. Department of Chemistry, Ben-Gurion University, PO Box 653, Beer Sheva 84105, Israel. PhD (Yale U., USA, 1967) physical chemistry. *Polymorphism organic_solid_state_chemistry hydrogen_bond structure_property_relationships crystal_structure_analysis conformational_polymorphism organic_conductors*
E-mail Yoel@BGUVMS.bitnet Tel. 972(7)461187 Fax 972(7)237787 Telex 5253 UNASI-IL

Bino, Prof. Avi (1949). Professor. Department of Inorganic and Analytical Chemistry, The Hebrew University of Jerusalem, Jerusalem 91904, Israel. PhD (1978) inorganic chemistry. *Structure_inorganic bioinorganic*

E-mail shmuel@vms.huli.ac.il Tel. 972(2)585722 Fax 972(2)585319

Botoshansky, Dr Mark (1947). Scientist. Department of Chemistry, Technion, Israel Inst. of Technology, Haifa, 32000, Israel. PhD (Acad. Sci. Moldavian SSR, 1977) crystallography. *Structure high_temperature low_temperature small_molecules phase_transition*
E-mail CHR03BM@TECHNION.AC.IL Tel. 972(4)293716 Fax 972(4)293735

Cohen, Dr Shmuel (1944). Crystallographer. Department of Inorganic and Analytical Chemistry, The Hebrew University, Jerusalem 91904, Israel. PhD (Hebrew U., 1976) inorganic chemistry. *Solid_state_chemistry crystallography data_collection structure_determination powder_diffraction*
E-mail shmuel@vms.huji.ac.il Tel. 972(2)585482 Fax 972(2)585319

Deutsch, Prof. M. (1946). Professor and Department Chairman. Physics Department, Bar-Ilan University, Ramat-Gan, 52900, Israel. PhD (Bar-Ilan U., 1979) physics. *Surface X-ray_optics mathematical_methods crystal_bonding synchrotron_radiation liquids monolayers critical_phenomena X-ray_spectroscopy*
E-mail F67234@BARILAN.BITNET moshe@solids.phy.bnl.gov Tel. 972(3)5318476 Fax 972(3)5353298

Dym, Orly (1961). PhD student. Department of Structural Biology, Weizmann Institute of Science, Rehovot 76100, Israel. MSc (Weizmann Institute of Science, 1987) structural biology. *Proteins*
E-mail csdym@weizmann.weizmann.AC.IL Tel. 972(8)342647 Fax 972(8)344159

Eisenberg, Dr Henryk (1921). Professor. Structural Biology Department, Weizmann Institute of Science, Rehovot 76100, Israel. PhD (Hebrew U., 1952) chemistry. *Chromatin halophilic_enzymes thermodynamics SAXS sedimentation neutron_scattering light_scattering nucleic_acids*

E-mail bpeisenb@weizmann.weizmann.ac.il Tel. 972(8)343252 Fax 972(8)344136 Telex 381300 WIX IL

Eisenstein, Dr Miriam (1951). Associate staff scientist. Department of Structural Biology, Weizmann Institute of Science, Rehovot 76100, Israel. PhD (Weizmann Institute of Science, 1981) crystallography and theoretical chemistry. *Surface_recognition molecular_replacement homology_modelling energy_minimization computer_ modelling force_field hydration modelling_service*

E-mail cseisens@weizmann.weizmann.ac.il Tel. 972(8)343031 Fax 972(8)344136 Telex 381300 WIX IL

Ellern, Dr Arkady M. (1955). Senior Research Fellow. Department of Chemistry, Ben-Gurion University of the Negev, PO Box 653 Beer-Sheva, 84105, Israel. PhD (Institute of General and Inorganic Chemistry Ac. Sc. USSR, 1978) inorganic chemistry. *Solid_state_chemistry computer_modelling low_temperature_structures polymorphism*

E-mail ellern@bgumail.bgu.ac.il Tel. 972(7)461187 Fax 972(7)281340

Erez, Prof. Gidon (1929). Professor. Physics Department, Nuclear Research Center - Negev, PO Box 9001, Beer Sheva 84190, Israel. [Physics Department, Ben Gurion University of the Negev, Beer Sheva 84105, Israel.] DSc (Technion - Israel Inst. of Technology, 1960) physics.

Tel. 972(7)567222 Fax 972(7)554848

Felsteiner, Joshua (1938). Professor. Department of Physics, Technion, Haifa 32000, Israel. PhD (U. Toronto, 1967) physics. *Compton_scattering momentum_density Monte_Carlo multiple_scattering magnetic_ordering*

E-mail phr01ya@technion.bitnet Tel. 972(4)293869 Fax 972(4)221514

Frolow, Dr Felix (1947). Senior Research Associate. Department of Chemical Services, The Weizmann Institute of Science, Rehovot 76100, Israel. PhD (The Weizmann Inst. of Science, 1980) crystallography. *Proteins measurement structure_solution_methods*

E-mail cvfrolow@weizmann.weizmann.ac.il Tel. 972(8)343580 Fax 972(8)344136

Gilboa, Dr A. Joseph (1937). Senior Scientist. Dept. of Structural Biology, The Weizmann Institute of Science, Rehovot 76100, Israel. PhD (U. California, 1963) chemistry. *Protein_structure protein_saccharide_interaction EXAFS metalloproteins hemoproteins cytochrome iron_compounds*

E-mail BFGILBOA@WEIZMANN.WEIZMANN.AC.IL Tel. 972(8)342218 Telex 381300 WIX IL

Goldberg, Prof. Israel (1945). Professor of Chemistry; Head of the School of Chemistry, 1993-. School of Chemistry, Tel-Aviv University, 69978 Ramat-Aviv, Tel-Aviv, Israel. PhD (Tel-Aviv U., 1974) crystallography physical chemistry. *Inclusion_chemistry crystal_engineering chemical_crystallography crystallographic_computing*

E-mail goldberg@chemsg2.tau.ac.il goldberg@chemsg7.tau.ac.il Tel. 972(3)6408258 972(3)6409965 Fax 972(3)6409293

Goldgur, Mr Yehuda (1959). PhD student. Department of Structural Biology, Weizmann Institute of Science, Rehovot 76100, Israel. MSc (Moscow State U., 1981) physical chemistry. *Protein_crystallography molecular_modelling enzymes*

E-mail CSGOLD@WEIZMANN.WEIZMANN.AC.IL Tel. 972(8)342591 Fax 972(8)344136

Greenblatt, Harry (Tzvi) (1962). PhD student. Dept. of Inorganic and Analytical Chemistry, Hebrew University, Jerusalem 91904, Israel. MSc (U. Alberta, 1988) crystallography. *Enzyme_mechanisms*

E-mail harry@vms.huji.ac.il Tel. 972(2)585610

Guzikevich-Guerstein, Gali (1964). PhD student. Department of Structural Biology, Weizmann Institute of Science, Rehovot, 76100, Israel. MSc (Weizmann Institute of Science, 1989) crystallography. *Macromolecules*

E-mail csguzi@weizmann.weizmann.ac.il Tel. 972(8)342479 Telex 381300 WIX IL

Harel, Michal (1940). Researcher. Israel Sub-Editor of World Directory of Crystallographers; Secretary of Israeli Crystallographic Society. Department of Structural Biology, Weizmann Institute of Science, Rehovot 76100, Israel. PhD (Weizmann Institute of Science, 1976) structural chemistry. *Proteins computer_modelling*

E-mail csharel@weizmann.weizmann.ac.il Tel. 972(8)342647 Fax 972(8)344159 Telex 381300 WIX IL

Heller-Kallai, Lisa (1926). Professor. Institute of Earth Science, The Hebrew University, Jerusalem 91904, Israel. PhD (London U., England, 1951) crystallography. *Clay_minerals clay_diagenesis structural_change catalysis*

E-mail Liza@vms.huji.ac.il Tel. 972(2)584883 Fax 972(2)662581 Telex 25391HUIL

Herbstein, Prof. Frank H. (1926). Prof. of Chemistry. Chairman, IUCr Commission on Small Molecules; Co-Editor, Acta Crystallographica. Department of Chemistry, Technion, Israel Institute of Technology, Haifa, 32000, Israel. DSc (U. Cape Town, 1968) chemical and physical crystallography. *Polyiodides thermal_decomposition small_molecules phase-transitions*

E-mail CHR03FH@TECHNION.AC.IL Tel. 972(4)293715 Fax 972(4)233735

Kaftory, Prof. Menahem (1943). Professor. Department of Chemistry, Technion, Israel Institute of Technology, Haifa 32000, Israel. DSc (Technion, 1973) chemistry. *Chemical_crystallography solid_state_chemistry reaction_mechanism*

E-mail chr13mk@technion.technion.ac.il Tel. 972(4)293761 Fax 972(4)233735

Kapon, Dr Moshe (1941). Departmental Crystallographer. Department of Chemistry, Technion, Israel Inst. of Technology, Haifa, 32000, Israel. PhD (Technion, I. I. T, 1974) crystallography of polyiodides. *Organic_polyiodides thermal_decomposition small_molecules crystal_structures*

E-mail CHR03KP@TECHNION.AC.IL Tel. 972(4)293716 Fax 972(4)233735

Kimmel, Dr Giora (1939). Associate Professor. Division of Research and Development, Nuclear Research Center Negev, PO Box 9001, Beer Sheva 84190, Israel. DSc (Technion Israel Institute of Technology, Haifa, 1973) materials science. *Metals alloys physical_metallurgy applied_crystallography alloy_phases powder_diffraction thin_film computing*

E-mail giokim@bgumail.bgu.ac.il giokim@bguvms.bgu.ac.il Tel. 972(7)567746 Fax 972(7)554848

Korkhin, Yacov M. (1962). PhD Student. Department of Structural Biology, The Weizmann Institute of Science, Rehovot 76100, Israel. MSc (Leningrad State U., 1985) atmospheric physics. *DNA_protein_complex transcription_control thermophile_enzymes allostery phasing_methods*

E-mail cskorkhi@wicc.weizmann.ac.il cskorkhi@weizmann.weizmann.ac.il Tel. 972(8)342479 Fax 972(8)344136 Telex 381300 WIX IL

Leiserowitz, Prof. Leslie (1934). Professor. Department of Materials and Interfaces, Weizmann Institute of Science, Rehovot 76100, Israel. PhD (Hebrew U., 1965) crystallography. *Molecular_recognition solid_state surface_crystallography crystallization*

E-mail CSLES@WEIZMANN.weizmann.ac.il Tel. 972(8)343727 Fax 972(8)344138 Telex 381300 WIX IL

Livnah, Oded (1958). Post-Doctorate fellow. Dept. of Structural Biology, The Weizmann Inst. of Science, Rehovot 76100, Israel. PhD (The Weizmann Inst. of Science, 1993) macromolecular crystallography. *Biological_macromolecules*

E-mail csoded@weizmann.weizmann.ac.il Tel. 972(8)342647

Mayer, Dr Itzchak (1927). Professor. Department of Inorganic Chemistry, The Hebrew University of Jerusalem, Jerusalem 91904, Israel. PhD (Hebrew U., 1960) inorganic chemistry. *Inorganic_crystal_structures synthetic_apatites biological_apatites*

E-mail isaacm@vms.huji.ac.il Tel. 972(2)585214 Fax 972(2)585319

Minkoff, Prof. I. (1922). Professor. Department of Materials Eng., Technion, Israel Inst. of Tech., Haifa 32000, Israel. DSc (M. I. T., 1957) metallurgy. *Crystallization_defect growth_metallurgy structure_twinning crystal_form laser_technology*

Tel. 972(4)294591/2 Fax 972(4)321978

Mosyak, Ms Lidia G. (1961). PhD student. Department of Structural Biology, Weizmann Institute of Science, Rehovot 76100, Israel. MSc (Kishinev State U., 1985) solid state physics. *Proteins molecular_modelling small_molecules*

E-mail CSLIDIA@WEIZMANN.WEIZMANN.AC.IL Tel. 972(8)342591 Fax 972(8)344136 Telex 381300 WIX IL

Rabinovich, Prof. Dov (1929). Professor. Department of Structural Biology, Weizmann Institute of Science, Rehovot 76100, Israel. PhD (Hebrew U., 1964) X-ray crystallography. *Structure_solution_methods molecular_replacement direct_methods DNA_structures*

E-mail csrabin1@weizmann.weizmann.ac.il Tel. 972(8)343029 Fax 972(8)344136 Telex 381300 WIX IL

Raves, Drs Mia L. (1969). PhD student. Department of Structural Biology, Weizmann Institute of Science, Rehovot 76100, Israel. Drs (Utrecht U., Netherlands, 1993) crystal and structural chemistry. *Proteins*

E-mail mia@sgjs2.weizmann.ac.il Tel. 972(8)343759 Fax 972(8)344159

Rozenberg, Haim (1964). PhD student. Department of Structural Biology, The Weizmann Institute of Science, Rehovot, 76100, Israel. D. E. A (U. Paris 7, 1990) molecular biophysics. *DNA transcription_control DNA_binding_protein*

E-mail csrozen@weizmann.weizmann.ac.il Tel. 972(8)342479 Fax 972(8)344136

Safro, Dr Mark G. (1946). Associate Professor. Department of Structural Biology, Weizmann Institute of Science, Rehovot 76100, Israel. PhD (Institute of Crystallography, Moscow, 1979) protein crystallography. *Proteins computer_modelling proteases computing*

E-mail CSSAFRO@WEIZMANN.WEIZMANN.AC.IL Tel. 972(8)343320 Fax 972(8)344136 Telex 381300 WIX IL

Sariel, Joseph (1947). Researcher. Metallurgy Department, Nuclear Research Center, Negev, PO Box 9001, Beer-Sheva, 84190, Israel. MSc (Ben-Gurion U., Beer-Sheva, 1981) materials science. *Alloy_phases powder_diffraction crystal_structure multilayers uranium_alloys*

E-mail sariel@bgumail.bgu.ac.il Tel. 972(7)568786 Fax 972(7)568786

Shaanan, Dr Boaz (1946). Associate Professor. Dept. of Biological Chemistry, The Inst. of Life Sciences, The Hebrew Univ. of Jerusalem, Givat-Ram, Jerusalem 91904, Israel. PhD (Tel-Aviv U., 1979) physical chemistry. *Proteins_structure_function protein_crystallography structural_biology protein_dynamics*

E-mail boazsh@vms.huji.ac.il boazsh@hujivms.bitnet Tel. 972(2)585241,585411 Fax 972(2)666804

Shach, Romem (1964). PhD student. School of Chemistry, Faculty of Exact Sciences, Tel Aviv University, Tel Aviv 69978, Israel. MSc (Tel Aviv U., 1990) crystallography. *Small_molecules diffraction structure_determination*

E-mail romem@chemsg2.tau.ac.il romem1@chemsg2.tau.ac.il

Shaked, Prof. Hagai (1931). Professor. Department of Physics, Ben Gurion University of the Negev, Beer Sheva 84105, Israel. PhD (U. California at Berkeley, 1963) solid state physics. *Magnetic_crystal_structures high_temperature_superconductors hydrides magnetic_structural_phase_transitions neutron_scattering*

E-mail hshaked@bguvms.bgu.ac.il Tel. 972(7)461567 Fax 972(7)281340 Telex UNASI5253

Shakked, Prof. Zippora (1943). Associate Professor. Department of Structural Biology, Weizmann Institute of Science, Rehovot 76100, Israel. PhD (Weizmann Institute of Science, 1975) structural chemistry. *Biological_molecules DNA hydration protein-DNA_interactions*

E-mail cshaked2@weizmann.weizmann.ac.il Tel. 972(8)342672 Fax 972(8)344136 Telex 381300 WIX IL

Shatzky-Schwartz, Michal (1963). PhD student. Department of Structural Biology, The Weizmann Institute of Science, Rehovot 76100, Israel. MSc (The Weizmann Institute of Science, 1991) organic chemistry. *DNA_structure*
E-mail CSSHATZ@WEIZMANN.WEIZMANN.AC.IL Tel. 972(8)342479 Fax 972(8)344136 Telex 381300 WIX IL

Shimon, Dr Linda J. W. (1958). Dept of Chemical Services, Weizmann Institute of Science, Rehovot 76100, Israel. PhD (Weizmann Institute of Science, 1989) structural chemistry. *Protein_nucleic_acid_interactions molecular_recognition*
E-mail cvlinda@weizmann.weizmann.ac.il Tel. 972(8)342479 Fax 972(8)344114

Shmueli, Prof. Uri (1928). Professor. Editor of Volume B of International Tables for Crystallography. School of Chemistry, Tel Aviv University, Tel Aviv 69978, Israel. PhD (Weizmann Institute of Science, 1966) crystallography. *Intensity_statistics direct_methods crystallographic_computing symmetry molecular_crystals Fourier_methods*
E-mail shmueli@chemsg2.tau.ac.il uri@chemsg2.tau.ac.il Tel. 972(3)6408258 Fax 972(3)6409293 Telex 342171 VERSY IL

Shoham, Gil (1952). Senior Lecturer. Department of Inorganic Chemistry, and The Laboratory for Structural Chemistry and Biology, The Hebrew University of Jerusalem, Jerusalem 91904, Israel. PhD (Harvard U., 1984) physical chemistry. *Macromolecules proteinases macrocycles computer_modelling metalloproteins enzyme_inhibitors protein_crystallization*
E-mail gil@vms.huji.ac.il Tel. 972(2)585611 Fax 972(2)585319

Stein, Zafra (1953). Research Associate. School of Chemistry, Tel-Aviv University, Tel-Aviv 69978, Israel. MSc (Technion - Israel Institute of Technology, 1980) chemical engineering. *Crystallographic_computing crystal_structure_analysis*
E-mail zafra@ccsg.tau.ac.il Tel. 972(3)6409965 Fax 972(3)6409293

Steinberger, Prof. Itzhak (1928). Professor. Racah Institute of Physics, Hebrew University, Jerusalem 91904, Israel. PhD (Hebrew U., 1957) physics. *Inorganic_compounds crystal_physics fluid_electronic_properties fluid_optical_properties*
E-mail steinber@vms.huji.ac.il Tel. 972(2)584648

Sussman, Prof. Joel Leonard (1943). Professor. Department of Structural Biology, Weizmann Institute of Science, Rehovot 76100, Israel. PhD (MIT, 1972) biophysics. *Proteins nucleic_acids refinement halophilic_proteins protein_structure homology model_building*
E-mail csjoel@weizmann.weizmann.ac.il csjoel@weizmann.bitnet Tel. 972(8)342638 Fax 972(8)344159

Traub, Prof. Wolfie (1927). Professor. Department of Structural Biology, Weizmann Institute of Science, Rehovot 76100, Israel. PhD (U. London, 1955) X-ray crystallography. *Proteins bone apatite*
E-mail cstraub@weizmann.weizmann.ac.il Tel. 972(8)468133 Fax 972(8)344136 Telex 381300 WIX IL

Yonath, Prof. Ada (1939). Professor; in parallel, Head of Max-Planck Research Unit for Ribosomal Structure in Hamburg at DESY. Department of Structural Biology, Weizmann Institute of Science, Rehovot 76100, Israel. PhD (Weizmann Institute of Science, 1969) crystallography. *Macromolecular_crystallography ribosome_structure three-dimensional_image_reconstruction*
E-mail csyonath@weizmann.weizmann.ac.il yonath@mpgars.desy.de Tel. 972(8)343028 Fax 972(8)344154 Telex 381300 WIX IL

Zamir, Sharona (1962). PhD student. Department of Chemistry, Ben-Gurion University of the Negev, PO Box 653, Beer Sheva 84105, Israel. MSc (Ben-Gurion U., Beer-Sheva, 1991) organic solid-state chemistry. *Polymorphism solid_state_chemistry crystal_forces*
E-mail cryst@bgumail.bgu.ac.il Tel. 972(7)461187 Fax 972(7)237787 Telex 5253 UNASI-IL

Zaytzev-Bashan, Anat (1964). PhD student. Department of Stuctural Biology, The Weizmann Institute of Science, Rehovot 76100, Israel. MSc (Weizmann Institute of Science, 1989) X-ray crystallography. *Proteins ribosomes macromolecules direct_methods maximum-entropy_method contrast isomorphous_replacement*
E-mail cszaytz@weizmann.weizmann.ac.il Tel. 972(8)342541 Fax 972(8)344154 Telex 381300 WIX IL

ITALY

Sub-Editor: G. Zanotti

Adovasio, Dr Victor (1935). Technical coordinator. Dip. di Chimica Generale ed Inorganica, Chimica Analitica, Chimica Fisica, Università di Parma, Viale delle Scienze 78, 43100 Parma, Italy. Dr (Parma U., 1972) geology. *X-ray_structure_determination organic_compounds inorganic_compounds*
E-mail chimic3@ipruniv.cce.unipr.it Tel. 39(521)905424 Fax 39(521)905557

Ajò, Dr David (1946). Research leader. Istituto di Chimica e Tecnologie Inorganiche e dei Materiali Avanzati, C.N.R., Corso Stati Uniti 4, I-35020 Padova, Italy. Dr (Roma U., 1970) chemistry. *Molecular_structure electronic_structure magnetochemistry spectroscopy inorganic_molecules inorganic_materials*
Tel. 39(49)8295939 Fax 39(49)8295649 Telex 430302 CNRPDI

Albano, Prof. Vincenzo Giulio (1937). Full professor. Dipartimento di Chimica "G. Ciamician", Università di Bologna, Via F. Selmi 2, 40126 Bologna, Italy. Dr (Bari U., 1960) chemistry. *Organometallic_crystal_chemistry clusters synthesis_reactivity metal_complexes*
E-mail vgxray@ciam01.cineca.it or vn3bodg4@icineca2.it Tel. 39(51)259538 Fax 39(51)259456

Alberti, Prof. Alberto (1938). Full professor. Istituto di Mineralogia, Università di Ferrara, Corso Ercole I D'Este 32, I-44100 Ferrara, Italy. Dr (Modena U., 1962) physics. *Mineralogy crystallography X-ray_diffraction zeolites lattice_energy phase_transition*
E-mail m83femaz@icineca.cineca.it Tel. 39(532)202987 Fax 39(532)202304

Albinati, Prof. Alberto (1945). Associate professor. Member of the IUCr Neutron Diffraction Commission. Ist. di Chimica Farmaceutica, Università di Milano, 42 Viale Abruzzi, I 20131 Milano, Italy. Dr (Milano U., 1970) chemistry. *Coordination_chemistry neutron_diffraction powder_diffraction synchrotron_radiation solid_state_NMR*
E-mail albi@albi.farma.unimi.it Tel. 39(2)29502254 Fax 39(2)29514197

Allegra, Prof. Giuseppe (1933). Full professor. Director of "Consorzio Interuniversitario Macromolecole" (Italian Interuniversity Consortium of Macromolecular Science). Dipartimento di Chimica, Politecnico di Milano, via Mancinelli 7, 20131 Milano, Italy. Dr (Politecnico, Milano, 1958) chemical engineering. *Polymers_crystallography polymers_physics X-ray_line_profiles direct_methods*
E-mail allegra@dept.polimi.it Tel. 39(2)23993023 Fax 39(2)23993080

Altomare, Dr Angela (1963). Researcher. CNR IRMEC - c/o Dipartimento Geomineralogico, Via Orabona 4, I-70125 Bari, Italy. Dr (Bari U., 1988) physics. *Computing direct_method powder mineralogy*
E-mail cryst@arba.ba.cnr.it Tel. 39(80)5442624 Fax 39(80)5442591

Amorese, Mr Alvaro (1947). Senior technician. Istituto di Strutturistica Chimica "G. Giacomello", Area della Ricerca di Roma - CNR, CP no. 10, 00016 Monterotondo Stazione (Roma), Italy. Mr (Technical High School, 1973) *X-ray_diffractometry*
E-mail struta@irmias.bitnet Tel. 39(6)90625142 Fax 39(6)90672630 Telex 624809 CN-RMLI

Andreetti, Prof. Giovanni Dario (1939). Full professor. Dip. di Chimica Generale ed Inorganica, Chimica Analitica, Chimica Fisica, Viale delle Scienze, 43100 Parma, Italy. Dr (Parma U., 1963) chemistry. *Structure_determination computing*
E-mail andre@risc.cce.unipr.it Tel. 39(521)905552 Fax 39(521)905556

Angela, Prof. Marinella (1936). Full professor. Dipartimento di Scienze Mineralogiche e Petrologiche, Università di Torino, Via Valperga Caluso 37, I-10125 Torino, Italy. Dr (Torino U., 1959) natural sciences. *Crystal_growth soil_minerals*
Tel. 39(11)6707129 Fax 39(11)6502657

Aquilano, Prof. Dino (1940). Associate professor. Dipartimento di Scienze Mineralogiche e Petrologiche, Università di Torino, Via Valperga Caluso 37, I-10125 Torino, Italy. Dr (Torino U., 1963) physics. *Crystal_growth defect_structures*
Tel. 39(11)6707125 Fax 39(11)6707128

Artioli, Prof. Gilberto (1957). Associate professor. Dip. Scienze della Terra, Sezione di Mineralogia, Università degli Studi di Milano, Via Botticelli 23, 20133 Milano, Italy. PhD (Chicago U., 1985) geophysical sciences. *X-ray_diffraction neutron_diffraction synchrotron_radiation mineralogy*
E-mail artioli@iummi4.terra.unimi.it Tel. 39(2)23698320 Fax 39(2)70638681

Auriemma, Dr Finizia (1961). Researcher. Dipartimento di Chimica, Università di Napoli Federico II, via Mezzocannone 4, 80134 Napoli, Italy. PhD (Napoli U., 1984) chemistry. *Macromolecular_crystallography computer_modelling statistical_thermodynamics defects*
E-mail auriemma@chemna.dichi.unina.it Tel. 39(81)5476536 Fax 39(81)5527771

Bacchi, Dr Alessia (1966). Postdoctoral fellow. Dip. di Chimica Generale ed Inorganica, Chimica Analitica, Chimica Fisica, Università di Parma, Viale delle Scienze 78, 43100 Parma, Italy. Dr (Parma U., 1991) chemistry. *Organometallic coordination structural_chemistry drug_molecules*
E-mail chimic6@ipruniv.cce.unipr.it Tel. 39(521)905421 Fax 39(521)905557

Bachechi, Dr Fiorella (1939). Senior researcher. Istituto di Strutturistica Chimica "G. Giacomello", Area della Ricerca di Roma - CNR, CP no. 10, 00016 Monterotondo Stazione (Roma), Italy. Dr (Toronto U., 1970) crystallography. *Coordination organometallic*
E-mail struta@irmias.bitnet Tel. 39(6)9005142 Fax 39(6)90672630 Telex 624809 CN-RMLI

Balzarotti, Prof. Adalberto (1939). Full professor. Dipartimento di Fisica - Università di Roma "Tor Vergata", Via della Ricerca Scientifica 1, 00133 Roma, Italy. Professor (Pavia U., 1961) physics. *X-ray_spectroscopy electron_spectroscopy superconductivity ternary_alloys synchrotron_radiation*
E-mail balza@roma2.infn.it Tel. 39(6)72594565 Fax 39(6)2023507 Telex 626382 FI-UNIV I

Bandoli, Prof. Giuliano (1941). Associate professor. Dip. di Scienze Farmaceutiche, Università di Padova, Via Marzolo 5, 35131 Padova, Italy. Dr (Padova, Ferrara U., 1966,

1975) chemistry, pharmacy. *Structure_determination psychoactive_compounds con-formation computer-assisted_design technetium_radiopharmaceuticals*
E-mail Giulio@pdfarm.dsfarm.unipd.it Tel. 39(49)831638 Fax 39(49)831639

Barbieri, Prof. Renato (1930). Full professor. Dipartimento di Chimica Inorganica, Università di Palermo, Via Archirafi, 90123 Palermo, Italy. Dr (Padova U., 1956) chemistry. *Proteins structural_biology structural_chemistry*
Tel. 39(91)6161474

Baricco, Dr Marcello (1958). Researcher. Dipartimento di Chimica Inorganica, Chimica-Fisica e Chimica dei Materiali, Università di Torino, Via P. Giuria 9, I-10125 Torino, Italy. PhD (Torino U., 1986) chemistry. *Powder_diffraction EXAFS metals_alloys*
E-mail baricco@silver.ch.unito.it Tel. 39(11)6707567 Fax 39(11)6707855

Basso, Prof. Riccardo (1947). Full professor. Dipartimento di Scienze della Terra, Sezione di Mineralogia, Università di Genova, Corso Europa 26, 16132 Genova, Italy. Dr (Genova U., 1971) mathematics. *Minerals structure_refinement crystal_chemistry deriva-tive_structure*
E-mail mineral@hp433.dister.unige.it Tel. 39(10)3538311 Fax 39(10)352169

Battaglia, Prof. Luigi Pietro (1943). Associate professor. Dip. di Chimica Generale ed Inorganica, Chimica Analitica, Chimica Fisica, Università di Parma, Viale delle Scienze 78, 43100 Parma, Italy. Dr (Parma U., 1970) chemistry. *Coordination struc-tural_chemistry organometallic_compounds*
E-mail chimic9@ipruniv.cce.unipr.it Tel. 39(521)905416 Fax 39(521)905557

Battezzati, Prof. Livio (1950). Associate professor. Dipartimento di Chimica Inorganica, Chimica-Fisica e Chimica dei Materiali, Università di Torino, Via P. Giuria 9, I-10125 Torino, Italy. Dr (Torino U., 1974) chemistry. *Powder_diffraction rapid_solidification metals_alloys glasses*
E-mail battezzati@silver.ch.unito.it Tel. 39(11)6707567 Fax 39(11)6690957

Bedarida, Prof. Federico (1924). Full professor; Director of Research. Dip. di Scienze della Terra, U. di Genova, Corso Europa 26, I-16132 Genova, Italy. Dr (Genova U., 1958) physics, mathematics. *Crystal_growth microgravity*
Tel. 39(10)3538301 Fax 39(10)352169

Bellon, Prof. Pier Luigi (1931). Full professor. Dip. di Chimica Strutturale e Stereochimica Inorganica, Università di Milano, Via G. Venezian 21, 20133 Milano, Italy. Dr (Padova U., 1955) industrial chemistry. *Electron_tomography X-ray tomography im-age_processing living_systems signal_processing*
E-mail bellon@stinch9.csmtbo.mi.cnr.it Tel. 39(2)2361410 Fax 39(2)70635288

Bellotto, Dr Maurizio (1958). Research scientist. CISE Tecnologie Innovative, Dept of Materials and Technology, PO Box 12081, I-20134 Milan, Italy. Dr (Polytechnic of Milan, 1982) chemical engineering. *Disorder microstructure phase_kinetics Rietveld_method applied_catalysis sintering phase_transition*
E-mail 00990bell@s1.cise.it Tel. 39(2)21672242 Fax 39(2)21672620 Telex 311643 CISE I

Belluso, Dr Elena (1960). Researcher. Dipartimento di Scienze Mineralogiche e Petro-logiche, Università di Torino, Via Valperga Caluso 37, I-10125 Torino, Italy. PhD (Torino U., 1989) mineralogy, crystallography. *Crystal_chemistry fibrous_materials opaque_minerals*
Tel. 39(11)6707135 Fax 39(11)6502657

Benedetti, Prof. Ettore (1940). Full professor. Dip. di Chimica, Università di Napoli "Fed-erico II", Via Mezzocannone 4, 80134 Napoli, Italy. Dr (Napoli U., 1965) chemistry. *Biocrystallography peptides neuropeptides nuclear_magnetic_resonance molecu-lar_mechanics biomolecules*
E-mail benedetti@chemna.dichi.unina.it Tel. 39(81)5476553 Fax 39(81)5527771

Benetollo, Dr Franco (1947). Researcher. Istituto di Chimica e Tecnologie Inorganiche e dei Materiali Avanzati, C.N.R., Corso Stati Uniti 4, I-35020 Padova, Italy. Dr (Padova U., 1981) pharmacy. *Lanthanides actinides inorganic_compounds*
E-mail Benetollo@ictr02.pd.cnr.it Tel. 39(49)8295930 Fax 39(49)8295649

Benna, Dr Piera (1954). Researcher. Dipartimento di Scienze Mineralogiche e Petro-logiche, Università di Torino, Via Valperga Caluso 37, I-10125 Torino, Italy. Dr (Torino U., 1978) geological sciences. *Silicates crystal_structures*
Tel. 39(11)6707131 Fax 39(11)6707128

Berti, Dr Giovanni (1949). Researcher. Dipartimento di Scienze della Terra, Uni-versità di Pisa, Via S. Maria 53, 56126 Pisa, Italy. Dr (Pisa U., 1977) physics. *X-ray_powder_diffractometry magnetism computing signal_analysis microcrys-tallinity*
E-mail berti@vsg.dst.unipi.it Tel. 39(50)568254 Fax 39(50)500932

Bertolasi, Prof. Valerio (1948). Associate professor. Dipartimento di Chimica, Università di Ferrara, Via L. Borsari 46, I-44100 Ferrara, Italy. Dr (Ferrara U., 1973) chemistry. *Chemical_bonding hydrogen_bonding molecular_crystals reaction_pathways*
E-mail m38a@icineca.cineca.it Tel. 39(532)291131 Fax 39(532)240709 Telex 510.850 UNIVFE-I

Biagini-Cingi, Prof. Marina (1925). Full professor. Dip. di Chimica Generale ed Inorgan-ica, Chimica Analitica, Chimica Fisica, Università di Parma, Viale delle Scienze 78, 43100 Parma, Italy. Dr (Parma U., 1949) chemistry. *Structural_chemistry coordina-tion bioinorganic_chemistry*
Tel. 39(521)905419 Fax 39(521)905557

Bianchi, Dr Riccardo (1942). Senior researcher. Centro di Studio per le Relazioni tra Strut-tura e Reattività Chimica, C.N.R., via Golgi 19, 20133 Milano, Italy. Dr (Milano U., 1972) mathematics. *Computing quantum_Monte_Carlo electrostatic_properties*
E-mail bian@rs6.csrsrc.mi.cnr.it Tel. 39(2)26603272 Fax 39(2)70638129

Bianchi Orlandini, Dr Annabella (1944). Senior researcher. Istituto per lo Studio della Stereochimica ed Energetica dei Composti, di Coordinazione (I.S.S.E.C.C. - C.N.R.), Via J. Nardi 39, 50132 Firenze, Italy. Dr (Firenze U., 1968) chemistry. *Inor-ganic_chemistry structure_determination structure-magnetism_relationships*
E-mail orlandin@cacao.issecc.fi.cnr.it Tel. 39(55)243990 Fax 39(55)2478366

Bigi, Prof. Adriana (1951). Associate professor. Dipartimento di Chimica "G. Ciami-cian", Università di Bologna, Via F. Selmi 2, 40126 Bologna Italy. Dr (Bologna U., 1975) chemistry. *Calcification SAXS synchrotron_radiation bone biopolymers bio-logical_phosphates biological_carbonates biomaterial*
E-mail bigi@ciam01.cineca.it Tel. 39(51)259551 Fax 39(51)259456

Bigoli, Prof. Francesco (1936). Associate professor. Dip. di Chimica Generale ed Inorgan-ica, Chimica Analitica, Chimica Fisica, Università di Parma, Viale delle Scienze 78, 43100 Parma, Italy. Dr (Parma U., 1964) chemistry. *Coordination X-ray_diffraction structure_determination organic_compounds*
E-mail chimic10@ipruniv.cce.unipr.it Tel. 39(521)905428 Fax 39(521)905557

Blasi, Prof. Achille (1940). Full professor. Dipartimento di Scienze della Terra, Università degli Studi, Via Botticelli 23, 20133 Milano, Italy. Dr (Milano U., 1968) geological sciences. *Mineralogy crystal_chemistry polymorphism order-disorder feldspars*
E-mail umimin@imicilea.cilea.it Tel. 39(2)23698319 Fax 39(2)70638681

Bocchi, Dr Claudio (1948). Researcher. Istituto MASPEC - C.N.R., Via Chiavari 18/a, 43100 Parma, Italy. Dr (Parma U., 1986) physics. *Defect materials semiconductors diffuse_scattering*
Tel. 39(521)269221 Fax 39(521)269206

Bocelli, Dr Gabriele (1942). Senior researcher. Centro di Studio per la Strutturistica Diffrat-tometrica CNR, Viale delle Scienze, 43100 Parma, Italy. Dr (Parma U., 1969) chem-istry. *X-ray_diffraction magnets alloys*
E-mail bocelli@ipruniv.cce.unipr.it Tel. 39(521)905448 Fax 39(521)905556

Bolognesi, Prof. Martino (1951). Full professor. Member, IUCr Commission on Biologi-cal Macromolecules. Dip.Genetica e Microbiologia, Università di Pavia, Via Abbiate-grasso 207, I-27100 Pavia, Italy. Dr (Pavia U., 1974) chemistry. *Biocrystallography proteins protein_crystal_growth*
E-mail bolog@ipvgen.unipv.it Tel. 39(382)505557 Fax 39(382)528496

Bombieri, Prof. Gabriella (1936). Full professor. Istituto di Chimica Farmaceutica, Università, Viale Abruzzi 42, I-20131 Milano, Italy. Dr (Padova U., 1962) chem-istry. *Small_molecules lanthanides structure–activity_relationship drug_modelling drugs*
E-mail bombieri@pamela.farma.unimi.it Tel. 39(2)29502224 Fax 39(2)29514197

Bonamartini-Corradi, Prof. Anna (1942). Associate professor. Dipartimento di Chimica, Via Campi 183, 41100 Modena, Italy. Dr (Parma U., 1967) chemistry. *Coordination inorganic_molecules X-ray_structure_determination*
Tel. 39(59)378427 Fax 39(59)373543

Bonamico, Dr Mario (1930). Research leader. Istituto di Chimica dei Materiali, Area della Ricerca di Roma - CNR, CP no. 10, 00016 Monterotondo Stazione, (Roma), Italy. Dr (Roma U., 1956) chemistry. *Crystallography X-ray_crystal_structure_determination*
Tel. 39(6)90672316 Fax 39(6)90005849 Telex 624809 CNRMLI

Bonazzi, Dr Paola (1960). Researcher. Dipartimento di Scienze della Terra, Università di Firenze, Via La Pira 4, 50121 Firenze, Italy. PhD (Firenze U., 1985) mineralogy, petrology. *Structural_mineralogy*
Tel. 39(55)2757488 Fax 39(55)284571

Bovio, Prof. Bruna (1928). Associate professor. Dipartimento di Chimica Generale, Uni-versità di Pavia, Viale Taramelli 12, I-27100 Pavia, Italy. Dr (Pavia U., 1960) chemistry. *Inorganic_molecules organic_molecules*
Tel. 39(382)507350 Fax 39(382)507575

Braga, Prof. Dario (1953). Associate professor. Dipartimento di Chimica "G. Ciami-cian", Università di Bologna, Via F. Selmi 2, 40126 Bologna, Italy. Dr (Bologna U., 1977) chemistry. *Organometallic_crystal_chemistry clusters solid_state com-puter_chemistry computer_graphics dynamics molecular_crystals*
E-mail dbxray@ciam01.cineca.it, vn3bodg2@icineca1.it Tel. 39(51)259555 Fax 39(51)259456

Bresciani Pahor, Prof. Nevenka (1949). Associate professor. Dipartimento di Scienze Chimiche, Università di Trieste, Via Giorgieri 1, I-34127 Trieste, Italy. Dr (Trieste U., 1973) chemistry. *Crystallography*
Tel. 39(40)5603111 Fax 39(40)6763903

Brigatti, Prof. Maria Franca (1946). Dipartimento di Scienze della Terra, Via S. Eufemia 19, I-41100 Modena, Italy. Dr (Modena U., 1971) geological sciences. *Mineralogy crystal_chemistry*
E-mail brigatti@imovx2.unimo.it Tel. 39(59)218062 Fax 39(59)223605

Brückner, Prof. Sergio (1943). Full professor. Dipartimento di Scienze e Tecnolo-gie Chimiche, Facoltà di Ingegneria, Università di Udine, Via del Cotonificio 108, I-33100 Udine, Italy. Dr (Trieste U., 1967) chemistry. *Polymer_crystallography Rietveld_method conformational_analysis*
E-mail bruckner@udiniv.cilea.it Tel. 39(432)558836 Fax 39(432)558803

Bruno, Prof. Emiliano (1938). Full professor. Dipartimento di Scienze Mineralogiche e Petrologiche, Università di Torino, Via Valperga Caluso 37, I-10125 Torino, Italy. Dr (Torino U., 1962) geological sciences. *Crystal_chemistry minerals experimen-tal_mineralogy*
Tel. 39(11)6707124 Fax 39(11)6502657

Bruno, Prof. Giuseppe (1952). Associate professor. Dipartimento di Chimica Inorganica, Analitica e Struttura Molecolare, Via Salita Sperone 31, I-98166 Villaggio S. Agata (Messina), Italy. Dr (Messina U., 1976) chemistry. *Structure_determination inor-ganic_compounds organic_compounds*
E-mail cendiff@imeuniv.net. 39(90)6763129 Fax 39(90)393725

Bruzzone, Prof. Giacomo (1931). Full professor. Ist. di Chimica Fisica, U. di Genova, Corso Europa 26, I-16132 Genova, Italy. Dr (Genova U., 1959) industrial chemistry. *Crystal_chemistry intermetallic_molecules intermetallic_hydrides*
E-mail metalli@pchem2.chimica.unige.it Tel. 39(10)3538221 Fax 39(10)515076

Burla, Dr Maria Cristina (1953). Researcher. Dip. di Scienze della Terra, Università di Perugia, Piazza Università, I-06100 Perugia, Italy. Dr (Perugia U., 1976) mathematics. *Direct_method computing*
E-mail sir2@ipguniv Tel. 39(75)5853255 Fax 39(75)5853203 Telex 662078 UNIPG I

Busetti, Prof. Vilma (1934). Associate professor. Dipartimento di Chimica Organica, Università di Padova, Via Marzolo 1, 35131 Padova, Italy. Dr (Padova U., 1960) chemistry. *Organic_molecules inorganic_molecules X-ray_crystallography*
E-mail bustop@chor00.unipd.it Tel. 39(49)831273 Fax 39(49)831222

Caglioti, Prof. Giuseppe Full professor. Dip. di Ingegneria Nucleare, Politecnico di Milano, via Ponzio 34/3, 20133 Milano, Italy. Dr (Milano U., 1953) physics. *Metallic_materials alloys acoustic and thermal emission aluminium_alloys*
E-mail caglioti@ipmce1.polimi.it Tel. 39(2)23996310 Fax 39(2)23996309

Cairati, Dr Paolo A. (1965). PhD student. Dip. Chimica Strutturale e Stereochimica Inorganica, Università di Milano, Via Venezian 21, 20133 Milano, Italy. Dr (Milano U., 1991) chemistry. *Powder_diffraction organometallic_molecules materials solid_state*
E-mail paolo@stinch0.csmtbo.mi.cnr.it Tel. 39(2)70635120 Fax 39(2)70635288

Calestani, Prof. Gianluca (1952). Associate professor. Dipartimento di Chimica Fisica ed Inorganica, Via Risorgimento 4, 40136 Bologna, Italy. PhD (U. des Saarlandes, Saarbruecken, 1980) chemistry. *Magnets structure_determination electron_microscopy*
E-mail struttu@ipruniv.cce.unipr.it Tel. 39(51)432544

Callegari, Dr Athos (1958). Museum curator. Università di Pavia, Via Abbiategrasso 209, I-27100 Pavia, Italy. Dr (Pavia U., 1984) petrography. *X-ray_spectrometry*
E-mail callegari@crystal.unipv.it Tel. 39(382)505873 Fax 39(382)505890

Calleri, Prof. Mariano (1934). Full professor. Dipartimento di Scienze Mineralogiche e Petrologiche, Università di Torino, Via Valperga Caluso 37, I-10125 Torino, Italy. Dr (Torino U., 1958) chemistry. *Structure_morphology material_sciences oxysalts quasicrystals*
Tel. 39(11)6707134 Fax 39(11)6502657

Calligaris, Prof. Mario (1939). Full professor. Dipartimento di Scienze Chimiche, Università di Trieste, Via Giorgieri 1, I-34127 Trieste, Italy. Dr (Trieste U., 1964) chemistry. *Coordination molecular_mechanics crystallography*
E-mail mario.calligaris@univ.trieste.it Tel. 39(40)6763933 Fax 39(40)6763903

Camalli, Dr Mercedes (1947). Researcher. Istituto di Strutturistica Chimica "G. Giacomello", Area della Ricerca di Roma - CNR, CP no. 10, 00016 Monterotondo Stazione (Roma), Italy. Dr (Buenos Aires U., Napoli U., 1974, 1978) chemistry. *Computing direct_method synchrotron_radiation crystallography*
E-mail struta@irmias.bitnet Tel. 39(6)90625142 Fax 39(6)90672630 Telex 624809 CN-RMLI

Campanelli, Dr Anna Rita (1953). Researcher. Dipartimento di Chimica, Universita di Roma "La Sapienza", Piazzale Aldo Moro 5, I-00185 Roma, Italy. Dr (Roma U., 1978) chemistry. *Conformational_analysis micellar_aggregates*
E-mail Campanelli@sci.uniroma1.it Tel. 39(6)49913655 Fax 39(6)490631

Candeloro De Sanctis, Prof. Sofia (1942). Associate professor. Delegate to ECC. Dip. di Chimica, Università di Roma "La Sapienza", Piazzale A. Moro 5, 00185 Roma, Italy. PhD (Oxford U., 1970) crystallography. *Micelles structural_biology potential_energy_calculations*
E-mail candeloro@sci.uniroma1.it Tel. 39(6)49913322 Fax 39(6)490631

Cannas, Prof. Mario (1926). Full professor. Università di Cagliari, Dipartimento di Scienze Chimiche, Via Ospedale 72, 09124 Cagliari, Italy. Dr (Cagliari U., 1951) chemistry. *Coordination_compound structural_chemistry*
Tel. 39(70)668047 Fax 39(70)669272

Cannillo, Dr Elio (1938). CNR chief researcher; Director of the Centre. Centro di Studio per la Cristallochimica e la Cristallografia - C.N.R., Via Abbiategrasso 209, I-27100 Pavia, Italy. Dr (Pavia U., 1962) chemistry. *Crystal_chemistry computing rock_forming_minerals*
E-mail cannillo@crystal.unipv.it Tel. 39(382)505875 Fax 39(382)505887

Capasso, Prof. Sante (1944). Associate professor. Dip. di Chimica, Università di Napoli "Federico II", 80134 Napoli, Italy. Dr (Napoli U., 1969) chemistry. *Proteins peptides crystallography crystallization kinetics*
E-mail zagari@chemna.dichi.unina.it Tel. 39(81)5476512 Fax 39(81)5527771

Cappuccio, Dr Giorgio (1939). Senior Researcher. INFNLNF, C.N.R., CP no. 13, 00044 Frascati (Roma), Italy. Dr (Roma U., 1970) physics. *Methodology polycrystal powder_diffraction synchrotron_radiation thin_films*
E-mail struta@irmias.bitnet Tel. 39(6)90625142 Fax 39(6)9403597 Telex 624809 CN-RMLI

Carbonin, Prof. Susanna (1948). Associate professor. Dipartimento di Geologia e Petrologia, Corso Garibaldi 37, 35100 Padova, Italy. Dr (Padova U., 1972) natural sciences. *Minerals refinement_method*
Tel. 39(49)663122 Fax 39(49)8753813

Carnasciali, Dr Maria M. (1955). Researcher. Ist. di Chimica Fisica, U. di Genova, Corso Europa 26, I-16132 Genova, Italy. Dr (Genova U., 1980) chemistry. *Structure_determination superconductivity intermetallic_compounds Rietveld_method*
E-mail pani@chimica.unige.it Tel. 39(10)3538221 Fax 39(10)515076

Caruso, Dr Francesco (1947). Researcher. Istituto di Strutturistica Chimica "G. Giacomello", Area della Ricerca di Roma - CNR, CP no. 10, 00016 Monterotondo Stazione (Roma), Italy. Dr (Buenos Aires U., Napoli U., 1974, 1977) chemistry. *Coordination crystal_growth Mossbauer anticancer_compounds bioinorganic organometallic*
E-mail struta@irmias.bitnet Tel. 39(6)90625142 Fax 39(6)90672630 Telex 624809 CN-RMLI

Casalone, Dr Gianluigi (1939). Research leader. Centro di Studio per le Relazioni tra Struttura e Reattività Chimica - CNR, via Golgi 19, 20133 Milano, Italy. Dr (Milano U., 1965) industrial chemistry. *Surface crystallography organic_crystal_structures*
E-mail casa@rs6.csrsrc.mi.cnr.it Tel. 39(2)70635452 Fax 39(2)70638129

Cascarano, Dr Giovanni Luca (1953). Researcher. CNR IRMEC - c/o Dipartimento Geomineralogico, Via Orabona 4, I-70125 Bari, Italy. PhD (Bari U., 1990) earth sciences. *Computing direct_method powder mineralogy*
E-mail cryst@arba.ba.cnr.it Tel. 39(80)5442624 Fax 39(80)5442591

Castellari, Prof. Carlo (1938). Associate professor. Dipartimento di Chimica "G. Ciamician", Università di Bologna, Via F. Selmi 2, 40126 Bologna, Italy. Dr (Bologna U., 1969) chemistry. *Organometallic_crystal_chemistry clusters transition_metal_complexes*
E-mail vn3bodg4@icineca2.it Tel. 39(51)391755 39(51)354327 Fax 39(51)259456

Catti, Prof. Michele (1945). Full professor. Dipartimento di Chimica Fisica ed Elettrochimica, Università di Milano, via Golgi 19, 20133 Milano, Italy. Dr (Torino U., 1969) industrial chemistry. *Disorder materials minerals phase_transition*
E-mail catti@ri55.csrsrc.mi.cnr.it Tel. 39(2)26603258 Fax 39(2)70638129

Caucia, Dr Franca (1957). Researcher. Dipartimento di Scienze della Terra, Università di Pavia, Via Abbiategrasso 209, I-27100 Pavia, Italy. Dr (Pavia U., 1980) geological sciences. *Minerals_crystal_chemistry*
E-mail caucia@crystal.unipv.it Tel. 39(382)505872 Fax 39(382)505890

Cellai, Dr Luciano (1944). Senior scientist. Istituto di Strutturistica Chimica "G. Giacomello", Area della Ricerca di Roma - CNR, CP no. 10, 00016 Monterotondo Stazione (Roma), Italy. Dr (Roma U., 1969) chemistry. *Molecular_pharmacology enzyme_inhibitors antibiotics antiviral_compounds structure–activity_relationship AIDS glycosaminoglycanes chemotherapy*
E-mail struta@irmias.bitnet Tel. 39(6)90625142 Fax 39(6)90672630 Telex 624809 CN-RMLI

Celotti, Dr Giancarlo (1943). Researcher. Istituto LAMEL, C.N.R., Via dei Castagnoli 1, 40126 Bologna, Italy. Dr (Bologna U., 1966) physics. *Semiconductors organic_compounds inorganic_compounds defect_structures*
Tel. 39(51)519593

Cerrini, Dr Silvio (1937). Research leader. Istituto di Strutturistica Chimica "G. Giacomello", Area della Ricerca di Roma - CNR, CP no. 10, 00016 Monterotondo Stazione (Roma), Italy. Dr (Roma U., 1963) physics. *Peptides X-ray_diffraction antibiotics computing toxins crystal_structure natural_compounds*
E-mail struta@irmias.bitnet Tel. 39(6)90625142 Fax 39(6)90672630 Telex 624809 CN-RMLI

Chiari, Prof. Giacomo (1943). Associate professor. Dipartimento di Scienze Mineralogiche e Petrologiche, Università di Torino, Via Valperga Caluso 37, I-10125 Torino, Italy. Dr (Torino U., 1967) chemistry. *Synchrotron_radiation powder synthetic_zeolites single_crystal conservation_of_monuments pigments minerals computer_programming organic_compounds crystal_structure_determination*
Tel. 39(11)6707122 Fax 39(11)6502657

Chiesi-Villa, Prof. Angiola (1944). Full professor. Dip. di Chimica Generale ed Inorganica, Chimica Analitica, Chimica Fisica, Viale delle Scienze, 43100 Parma, Italy. Dr (Parma U., 1964) chemistry. *Structure_determination molecular_recognition organometallic_compounds*
E-mail struchi@ipruniv.cce.unipr.it Tel. 39(521)905449 Fax 39(521)905556

Ciani, Prof. Gianfranco (1944). Full professor. Dip. Chimica Strutturale e Stereochimica Inorganica, Università di Milano, Via Venezian 21, 20133 Milano, Italy. Dr (Milano U., 1968) chemistry. *X-ray_diffraction structural_chemistry chemical_bonding organometallic_compounds*
E-mail guest@stinch0.csmtbo.mi.cnr.it Tel. 39(2)70635120 Fax 39(2)70635288

Cini, Dr Renzo (1949). Associate professor. Università di Siena, Dipartimento di Chimica, Pian dei Mantellini 44, Siena, Italy. Dr (Pisa U., 1974) chemistry. *Coordination_chemistry organometallic bioinorganic*
E-mail cini@sivax.cineca.it Tel. 39(577)298041 Fax 39(577)280405

Cirafici, Prof. Salvino (1944). Associate professor. Ist. di Chimica Fisica, Università di Genova, Corso Europa 26, I-16132 Genova, Italy. Dr (Genova U., 1971) industrial chemistry. *Intermetallic_compounds*
E-mail metalli@chimica.unige.it Tel. 39(10)3538221 Fax 39(10)515076

Cirilli, Dr Maurizio (1957). Senior technician. Istituto di Strutturistica Chimica "G. Giacomello", Area della Ricerca di Roma - CNR, CP no. 10, 00016 Monterotondo Stazione (Roma), Italy. Dr (Roma U., 1984) biological sciences. *Biocrystallography biomolecule crystallization biochemistry*
E-mail struta@irmias.bitnet Tel. 39(6)9005142 Fax 39(6)90672630 Telex 624809 CN-RMLI

Clemente, Prof. Dore Augusto (1940). Full professor. Dipartimento di Scienza dei Materiali, Università di Lecce, Via Monteroni, 73100 Lecce, Italy. Dr (Padova U., 1966) chemistry. *Computing electron_density*
Tel. 39(832)621529

Coda, Prof. Alessandro (1936). Full professor. Dipto di Genetica e Microbiologia, Via Abbiategrasso 207, Pavia, Italy. Dr (Pavia U., 1959) chemistry. *Biocrystallography molecular_biology*
E-mail cauda@unipv85.unipv.it Tel. 39(382)505555 Fax 39(382)528496

Coiro, Dr Vincenza Maria (1937). Senior researcher. Istituto di Strutturistica Chimica "G. Giacomello", Area della Ricerca di Roma - CNR, CP no. 10, 00016 Monterotondo Stazione, (Roma), Italy. Dr (Napoli U., 1961) chemistry. *X-ray_crystal_structure carbohydrates micelles structure_determination_amphiphiles*
E-mail struta@irmias.bitnet Tel. 39(6)90625142 Fax 39(6)90672630 Telex 624809 CN-RMLI

Cojazzi, Dr Gianna (1935). Senior researcher. Centro Studio Fisica Macromolecole, C.N.R., Via F. Selmi 2, 40126 Bologna, Italy. Dr (Padova U., 1961) chemistry. *Biopolymers biomaterial DSC TGA WAXS*
Tel. 39(51)259552 Fax 39(51)259456

Colapietro, Prof. Marcello (1937). Associate professor. Member, IUCr Commission on Synchrotron Radiation. Dipartimento di Chimica, Università di Roma "La Sapienza", Piazzale Aldo Moro 5, 00185 Roma, Italy. Dr (Roma U., 1962) physics. *Synchrotron_radiation small_molecules_crystallography instrumentation methodology*
E-mail colapietro@sci.uniroma1.it Tel. 39(6)49913715 Fax 39(6)490631

Colombo, Dr Arturo (1935). Senior researcher. Istituto di Chimica delle Macromolecole - C.N.R., via Bassini 15, 20133 Milano, Italy. Dr (Pavia U., 1960) chemistry. *Polymers_crystallography Rietveld_method*
Tel. 39(2)70636333 Fax 39(2)2362946

Comodi, Dr Paola (1960). Researcher. Dip. di Scienze della Terra, Università di Perugia, Piazza Università, I-06100 Perugia, Italy. PhD (Firenze U., 1990) mineralogy and petrology. *Mineralogy high_pressure high_temperature X-ray_diffraction crystal_chemistry*
E-mail crystal@ipguniv Tel. 39(75)5853217 Fax 39(75)5853203 Telex 662078 UNIPG I

Corradini, Prof. Paolo (1930). Full professor. Dipartimento di Chimica, Università di Napoli Federico II, via Mezzocannone 4, 80134 Napoli, Italy. Dr (Roma U., 1951) chemistry. *Macromolecular_crystallography stereochemistry reaction_mechanisms defects*
Tel. 39(81)5476540 Fax 39(81)5476539

Costa, Prof. Giorgio Andrea (1948). Associate professor. Ist. di Chimica Fisica, U. di Genova, Corso Europa 26, I-16132 Genova, Italy. Dr (Genova U., 1972) physics. *Calorimetry superconductivity crystal_growth metallurgy magnetism electrochemistry transduction cryogenics dielectrics astronomy*
E-mail costa@chimica.unige.it Tel. 39(10)3538221 Fax 39(10)515076

Dal Negro, Prof. Alberto (1941). Full professor. Dipartimento di Mineralogia e Petrologia, Corso Garibaldi 37, 35100 Padova, Italy. Dr (Pavia U., 1964) chemistry. *Refinement_method minerals*
E-mail a45700@pdadr1.pd.cnr.it Tel. 39(49)663122 Fax 39(49)8753813

Dapporto, Prof. Paolo (1942). Full professor. Dipartimento di Energetica, Università di Firenze, Via S. Marta 3, 50139 Firenze, Italy. Dr (Firenze U., 1966) chemistry. *Computing pharmacological_compounds*
E-mail def@ingfi1.ing.unifi.it Tel. 39(55)4796209 Fax 39(55)4796342

De Angelis, Prof. Giuseppe (1944). Associate professor. Dip. di Scienze della Terra, Università di Roma "La Sapienza", Piazzale A. Moro 5, 00185 Roma, Italy. Dr (Roma U., 1970) geological sciences. *Computing Mossbauer diffractometry*
Tel. 39(6)49914895 Fax 39(6)4454729

Della Giusta, Prof. Antonio (1941). Full professor. Dipartimento di Geologia e Petrologia, Corso Garibaldi 37, 35100 Padova, Italy. Dr (Genova U., 1964) geological sciences. *Refinement_method computing minerals*
Tel. 39(49)663122 Fax 39(49)8753813

Del Piero, Dr Gastone (1943). Researcher. Eniricerche S.p.A., via F. Maritano 26, I-20097 San Donato Milanese (Milano), Italy. Dr (Padova U., 1968) chemistry. *X-ray_crystallography catalysis*
Tel. 39(2)52024904 Fax 39(2)5204422 Telex 310246 ENI I

Del Prà, Prof. Antonio (1932). Full professor. Istituto di Chimica Farmaceutica, Viale Abruzzi 42, I-20131 Milano, Italy. Dr (Padova U., 1962) chemistry. *Coordination diffuse_scattering structural_biology*
Tel. 39(2)29514197 Fax 39(2)29514197

Demartin, Prof. Francesco (1953). Associate professor. Dip. Chimica Strutturale e Stereochimica Inorganica, Università di Milano, Via Venezian 21, 20133 Milano, Italy. Dr (Milano U., 1978) chemistry. *Structural_chemistry EDS coordination single_crystal_X-ray_diffraction coordination_compounds*
E-mail guest@stinch0.csmtbo.mi.cnr.it Tel. 39(2)70635120 Fax 39(2)70635288

De Montis, Prof. Pierfranco (1954). Associate professor. Dipartimento di Chimica, Università di Sassari, via Vienna 2, 07100 Sassari, Italy. Dr (Sassari U., 1979) chemistry. *Molecular_dynamics disordered_and_amorphous_solids zeolites*
E-mail demontis@mvchss.cineca.it Tel. 39(79)229551 Fax 39(79)229559

De Munno, Prof. Giovanni (1951). Associate professor. Dipartimento di Chimica, Università della Calabria, Arcavacata di Rende, 87030 Cosenza, Italy. Dr (Roma U., 1975) chemistry. *Coordination liquid_crystals magnetism*
Tel. 39(98)4839321

Depero, Dr Laura Eleonora (1960). Researcher. Dipartimento di Chimica Fisica dei Materiali, Università di Brescia, Via Branze 38, I-25060 Brescia, Italy. Dr (Milano U., 1984) physics. *Materials accurate_intensity_measurement*
E-mail mzocchi@icil64.cilea.it Tel. 39(30)3715574 Fax 3702448

De Pol Blasi, Prof. Carla (1936). Associate professor. Dipartimento di Scienze della Terra, Università degli Studi, Via Botticelli 23, 20133 Milano, Italy. Dr (Milano U., 1960) geological sciences. *Mineralogy crystal_chemistry polymorphism order_disorder_feldspars*
E-mail umimin@imicilea.bitnet Tel. 39(2)23698322 Fax 39(2)70638681

De Rosa, Prof. Claudio (1960). Associate professor. Dipartimento di Chimica, Università di Napoli "Federico II", via Mezzocannone 4, 80134 Napoli, Italy. PhD (Napoli U., 1983) chemistry. *Macromolecular_crystallography computer_modelling defects*
E-mail derosa@chemna.dichi.unina.it Tel. 39(81)5476535 Fax 39(81)5527771

De Santis, Prof. Pasquale (1959). Full professor. Dip. di Chimica, Università di Roma "La Sapienza", Piazzale A. Moro 5, 00185 Roma, Italy. Dr (Bari U., 1935) chemistry. *Polymers proteins small-angle_scattering macromolecules*
Tel. 39(6)49913730 Fax 39(6)490631

Destro, Prof. Riccardo (1940). Professor. Dipartimento di Chimica Fisica ed Elettrochimica dell'Università, via Golgi 19, 20133 Milano, Italy. Dr (Milano U., 1964) industrial chemistry. *Charge_density low_temperature electrostatic_properties*
E-mail dest@rs6.csrsrc.mi.cnr.it Tel. 39(2)26603285 Fax 39(2)70638129

Di Blasio, Prof. Benedetto (1945). Associate professor. Dipartimento di Chimica, Università di Napoli "Federico II", via Mezzocannone 4, 80134 Napoli, Italy. Dr (Napoli U., 1970) chemistry. *Peptides conformation crystal_structures*
Tel. 39(81)5476549 Fax 39(81)5476550

Di Vaira, Prof. Massimo (1940). Full professor. Dipartimento di Chimica, Università di Firenze, Via Maragliano 77, I-50144 Firenze, Italy. Dr (Firenze U., 1963) chemistry. *Molecular_structure organometallic coordination_compounds*
E-mail divaira@ifichim Tel. 39(55)354841 Fax 39(55)354845 Telex 570123 CHIMFI

Djinovic Carugo, Dr Kristina (1963). Postdoctoral research assistant. Università di Pavia, Dipartimento di Genetica e Microbiologia, Sezione di Biologia Molecolare e Biofisica, Via Abbiategrasso 207, 27100 Pavia, Italy. PhD (Ljubljana U., 1992) chemistry. *Biocrystallography proteins hydrogen_bonding computing*
E-mail djinovic@ipvgen.unipv.it Tel. 39(382)505560 Fax 39(382)528496

Dolmella, Dr Alessandro (1962). Researcher. Dip. di Scienze Farmaceutiche, Università di Padova, Via Marzolo 5, 35131 Padova, Italy. Dr (Padova U., 1985) pharmacy. *Computing computer_modelling psychoactive_compounds structure_determination bioinorganic*
E-mail Dolly@pdfarm.dsfarm.unipd.it Tel. 39(49)831638 Fax 39(49)831639

Domeneghetti, Dr Maria Chiara (1954). Researcher. Centro di Studio per la Cristallochimica e la Cristallografia, C.N.R., Via Abbiategrasso 209, I-27100 Pavia, Italy. Dr (Pavia U., 1978) natural sciences. *Minerals crystal_chemistry*
E-mail domeneghetti@crystal.unipv.it Tel. 39(382)505871 Fax 39(382)505887

Domenicano, Prof. Aldo (1938). Full professor. Dipartimento di Chimica, Ingegneria Chimica e dei Materiali, Università dell'Aquila, I-67100 L'Aquila, Italy. Dr (Roma U., 1965) chemistry. *Small_molecules_crystallography electron_diffraction*
Tel. 39(862)432595 Fax 39(862)432573

Domiano, Prof. Paolo (1936). Associate professor. Dip. di Chimica Generale ed Inorganica, Chimica Analitica, Chimica Fisica, Università di Parma, Viale delle Scienze 78, I-43100 Parma, Italy. Dr (Parma U., 1960) chemistry. *Computing*
E-mail dp@mmlab.cce.unipr.it Tel. 39(521)905446 Fax 39(521)905556

Dovesi, Prof. Roberto (1947). Associate professor. Dipartimento di Chimica Inorganica, Chimica-Fisica e Chimica, dei Materiali, Università di Torino, Via P. Giuria 5, I-10125 Torino, Italy. Dr (Torino U., 1971) chemistry. *Computing solid_state charge_density molecular_crystal*
E-mail U107@ITOCSIVM.CSI.IT Tel. 39(11)6707561 Fax 39(11)6690957

Fagherazzi, Prof. Giuliano (1938). Full professor. Dipartimento di Chimica Fisica, Università di Venezia, Calle Larga S. Marta 2137, 30123 Venezia, Italy. Dr (Padova U., 1961) physics. *Powder_diffraction SAXS inorganic_materials*
E-mail fagheraz@unive.it Tel. 39(41)5298545 Fax 39(81)5298594 Telex 470638 UNIVVE-I

Fanfani, Prof. Luca (1941). Full professor. Dipartimento di Scienze della Terra, Università di Cagliari, Via Trentino, I-09127 Cagliari, Italy. Dr (Firenze U., 1964) chemistry. *Mineralogy*
Tel. 39(70)2006219 Fax 39(70)282236

Fares, Dr Vincenzo (1942). Senior researcher. Istituto TSECS Composti di Coordinazione, Area della Ricerca di Roma - CNR, Via Salaria Km 29.500, CP no. 10, 00016 Monterotondo Stazione (Roma), Italy. Dr (Roma U., 1966) chemistry. *Transition_elements X-ray_crystal_structure_analysis*
Tel. 39(6)90672285 Fax 39(6)9005849 Telex 624809 CNRMLI

Fedeli, Prof. Walter (1933). Full professor. Dipartimento di Chimica, Ingegneria Chimica e dei Materiali, Università dell'Aquila, I-67100 L'Aquila, Italy. Dr (Roma U., 1961) chemistry. *Biologically_important_substances*
Tel. 39(862)25387

Ferracini, Prof. Elena (1938). Associate professor. Dipartimento di Chimica "G. Ciamician", Università di Bologna, Via F. Selmi 2, 40126 Bologna, Italy. Dr (Bologna U., 1962) industrial chemistry. *SAXS macromolecules WAXS morphology crystallinity*
E-mail macmol@ciam01.cineca.it Tel. 39(51)259556 Fax 39(51)259456

Ferrari, Dr Claudio (1958). Researcher. Istituto MASPEC, C.N.R., Via Chiavari 18/a, 43100 Parma, Italy. Dr physics. *Materials semiconductors electron_microscopy defect_structures*
Tel. 39(521)269221 Fax 39(521)269206

Ferrari-Belicchi, Prof. Marisa (1942). Associate professor. Dip. di Chimica Generale ed Inorganica, Chimica Analitica, Chimica Fisica, Università di Parma, Viale delle Scienze 78, 43100 Parma, Italy. Dr (Parma U., 1966) chemistry. *X-ray_structure_determination coordination_chemistry bioinorganic_chemistry organic_compounds*
E-mail chimic8@ipruniv.cce.unipr.it Tel. 39(521)905420 Fax 39(521)905557

Ferraris, Prof. Giovanni (1937). Full professor; Head of Department. Dipartimento di Scienze Mineralogiche e Petrologiche, Università di Torino, Via Valperga Caluso 37, I-10125 Torino, Italy. Dr (Torino U., 1960) physics. *Crystal_chemistry_of_minerals hydrogen_bonding neutron_diffraction electron_microscopy*
E-mail ferraris@itocsivm.csi.it Tel. 39(11)6707121 Fax 39(11)6707128

Ferrero, Prof. Adele (1925). Associate professor. Dipartimento di Chimica "G. Ciamician", Università di Bologna, Via F. Selmi 2, 40126 Bologna, Italy. Dr (Bologna U., 1950) chemistry. *SAXS macromolecules WAXS morphology crystallinity*
E-mail macmol@ciam01.cineca.it Tel. 39(51)259450 Fax 39(51)259456

Ferretti, Dr Maurizio (1955). Researcher. Istituto di Chimica Fisica, Università di Genova, Corso Europa 26, I-16132 Genova, Italy. Dr (Genova U., 1981) physics. *Thermoanalysis TGA_DTA phase_diagram superconductor_ceramics crystal_growth internal_friction lattice_distortion*
E-mail ferretti@chimica.unige.it Tel. 39(10)3538221 Fax 39(10)515076

Ferretti, Dr Valeria (1958). Researcher. Dipartimento di Chimica, Università di Ferrara, Via L. Borsari 46, I-44100 Ferrara, Italy. Dr (Ferrara U., 1981) chemistry. *Chemical_bonding hydrogen_bonding organometallic_compounds molecular_crystals_reaction_pathways*
E-mail vg3a@icineca Tel. 39(532)291132 Fax 39(532)40709

Fichera, Dr Annamaria (1939). Senior researcher. Centro Studio Fisica Macromolecole, C.N.R., Via F. Selmi 2, 40126 Bologna, Italy. Dr (Bologna U., 1965) chemistry. *WAXS polymers*
E-mail xray@ciam01.cineca.it Tel. 39(51)259552 Fax 39(51)259456

Filippini, Dr Giuseppe (1940). Senior researcher. Regional Editor of 9th edition of World Directory of Crystallographers; Scientific Secretary, Italian Crystallographic Committee. Centro di Studio per le Relazioni tra Struttura e Reattività Chimica del CNR, via Golgi 19, 20133 Milano, Italy. Dr (Milano U., 1968) industrial chemistry. *Lattice_dynamics force_field intermolecular_potential_energy intermolecular_potentials computing*
E-mail fili@rs6.csrsrc.mi.cnr.it Tel. 39(2)26603279 Fax 39(2)70638129

Foresti, Prof. Elisabetta (1943). Associate professor. Dipartimento di Chimica "G. Ciamician", Università di Bologna, Via F. Selmi 2, 40126 Bologna, Italy. Dr (Bologna U., 1968) chemistry. *Powder_diffraction structural_chemistry crystallographic_data inorganic_crystals organic_compounds*
Tel. 39(51)259553 Fax 39(51)259456

Fornasini, Prof. Maria L. (1941). Associate professor. Ist. di Chimica Fisica, U. di Genova, Corso Europa 26, I-16132 Genova, Italy. Dr (Genova U., 1965) chemistry. *Intermetallic_compounds*
E-mail cfmet@chimica.unige.it Tel. 39(10)3538221 Fax 39(10)515076

Franceschi, Prof. Enrico (1942). Associate professor. Ist. di Chimica Fisica, U. di Genova, Corso Europa 26, I-16132 Genova, Italy. Dr (Genova U., 1965) chemistry. *Archeometallurgy intermetallic_compounds*
E-mail metalli@chimica.unige.it Tel. 39(10)3538221 Fax 39(10)515076

Franzosi, Dr Paolo (1948). Researcher. Istituto MASPEC, C.N.R., Via Chiavari 18/a, 43100 Parma, Italy. Dr (Parma U., 1972) physics. *Defect materials semiconductors electron_microscopy*
Tel. 39(521)269222 Fax 39(521)269206

Frigeri, Dr Cesare (1950). Researcher. Istituto MASPEC, C.N.R., Via Chiavari 18/a, 43100 Parma, Italy. Dr physics. *Defect materials semiconductors electron_diffraction electron_microscopy*
Tel. 39(521)96841 Fax 39(521)96315

Galli, Prof. Ermanno (1937). Full professor. Dipartimento di Scienze della Terra, Via S. Eufemia 19, I-41100 Modena, Italy. Dr (Modena U., 1963) geological sciences. *Zeolites silicates inorganic_crystal_structures*
E-mail mineralogia@imovx2.infnet Tel. 39(59)218062 Fax 39(59)223605

Ganazzoli, Dr Fabio (1955). Associate professor. Dip. di Chimica, Politecnico di Milano, Via Mancinelli 7, 20133 Milano, Italy. Dr (Milano U., 1978) chemistry. *Polymer_physics coordination*
E-mail fabio@iris.chem.polimi.it Tel. 39(2)23993024 Fax 39(2)23993080

Ganis, Prof. Paolo (1933). Full professor. Dipartimento di Chimica, U. di Napoli, Via Mezzocannone, Napoli, Italy. Dr (Padova U., 1957) chemistry. *Structure_determination organometallic physicochemical_properties*

Gasparri-Fava, Prof. Giovanna (1930). Full professor. Dip. di Chimica Generale ed Inorganica, Chimica Analitica, Chimica Fisica, Viale delle Scienze 78, 43100 Parma, Italy. Dr (Parma U., 1955) chemistry. *X-ray_structure_determination coordination_chemistry inorganic_chemistry bioinorganic_chemistry organic_compounds*
E-mail chimic8@ipruniv.cce.unipr.it Tel. 39(521)905420 Fax 39(521)905557

Gavezzotti, Prof. Angelo (1944). Full professor. Dip. Chimica Fisica ed Elettrochimica, Università, via Golgi 19, 20133 Milano, Italy. Dr (Milano U., 1968) chemistry. *Organic_crystal_chemistry intermolecular_potentials*
E-mail gave@rs6.csrsrc.mi.cnr.it Tel. 39(2)70633120 Fax 39(2)70638129

Gavuzzo, Dr Enrico (1943). Senior researcher. Istituto di Strutturistica Chimica "G. Giacomello", Area della Ricerca di Roma - CNR, CP no. 10, 00016 Monterotondo Stazione (Roma), Italy. Dr (Roma U., 1972) chemistry. *X-ray_crystallography computing peptides semi-empirical_calculation natural synthetic_products potential_energy*
E-mail struta@irmias.bitnet Tel. 39(6)90625142 Fax 39(6)90672630 Telex 624809 CN-RMLI

Gazzano, Dr Massimo (1958). Researcher. Centro Studio Fisica Macromolecole, C.N.R., Via F. Selmi 2, 40126 Bologna, Italy. PhD (Bologna U., 1987) chemistry. *Phosphates polymers catalysts powder_diffraction crystal_growth*
E-mail xray@ciam01.cineca.it Tel. 39(51)259549 Fax 39(51)259456

Gazzoni, Prof. Giuseppe (1937). Senior researcher. Centro di Studi sulla Geodinamica delle Catene Collisionali, Sezione Mineropetrologica, Via Valperga Caluso 37, I-10125 Torino, Italy. Dr (Torino U., 1963) physics. *Crystal_chemistry experimental_mineralogy nesosilicates tectosilicates*
Tel. 39(11)6707123 Fax 39(11)6502657

Gervasio, Prof. Giuliana (1943). Associate professor. Dipartimento di Chimica Inorganica Chimica-Fisica e Chimica dei Materiali, Università di Torino, Via P. Giuria 7, I-10125 Torino, Italy. Dr (Torino U., 1967) chemistry. *Crystal_structure coordination organometallic_compounds*
E-mail gervasio@silver.ch.unito.it Tel. 39(11)6707504 Fax 39(11)6690957

Ghilardi, Dr Carlo Alfredo (1943). Senior researcher. Istituto per lo Studio della Stereochimica ed Energetica dei Composti, di Coordinazione (I.S.S.E.C.C. - C.N.R.), Via J. Nardi 39, 50132 Firenze, Italy. Dr (Firenze U., 1968) chemistry. *Inorganic_chemistry structure_determination relations_structure_magnetism*
E-mail ghilardi@cacao.issecc.fi.cnr.it Tel. 39(55)243990 Fax 39(55)2478366

Giacovazzo, Prof. Carmelo (1940). Full professor; Director of CNR IRMEC Bari. Dipartimento Geomineralogico, Università, Via Orabona 4, I-70125 Bari, Italy. Dr (Bari U., 1966) physics. *Computing direct_method powder mineralogy methods_for_protein_structure_solution*
E-mail Giacovazzo@arba.ba.cnr.it Tel. 39(80)5442590 Fax 39(80)5442591

Gianfagna, Dr Antonio (1952). Researcher. Dip. di Scienze della Terra, Università di Roma "La Sapienza", Piazzale A. Moro 5, 00185 Roma, Italy. Dr (Roma U., 1977) geological sciences. *Minerals electron_microscopy*
Tel. 39(6)49914921 Fax 39(6)4454729

Giglio, Prof. Edoardo (1931). Full professor. Dip. di Chimica, Università di Roma "La Sapienza", Piazzale A. Moro 5, 00185 Roma, Italy. Dr (Bari U., 1955) chemistry. *Micelles biocrystallography bile_salt_micellar_aggregates bile_salt_interaction_complexes*
E-mail giglio@sci.uniroma1.it Tel. 39(6)4452993 Fax 39(6)490631

Gilli, Prof. Gastone (1937). Full professor. Dipartimento di Chimica, Università di Ferrara, Via L. Borsari 46, I-44100 Ferrara, Italy. Dr (Ferrara U., 1962) chemistry. *Chemical_bonding hydrogen_bonding organic structural_biology teaching X-ray_diffraction organometallic_compounds molecular_crystals_reaction_pathways*
E-mail M38A@ICINECA.CINECA.IT Tel. 39(532)210370-291141 Fax 39(532)240709

Gilli, Dr Paola (1964). Researcher. Dipartimento di Chimica, Università di Ferrara, Via L. Borsari 46, I-44100 Ferrara, Italy. Dr (Ferrara U., 1990) chemistry. *Chemical_bonding hydrogen_bonding organometallic_compounds molecular_crystals_reaction_pathways*
E-mail M38A@ICINECA.CINECA.IT Tel. 39(532)291141 Fax 39(532)240709

Giordano, Prof. Federico (1939). Associate professor. Dip. di Chimica, Università di Napoli "Federico II", 80134 Napoli, Italy. Dr (Napoli U., 1964) chemistry. *Natural_products molecular_complexes synthetic_molecules computing chemical_bonding*
E-mail giordano@chemna.dichi.unina.it Tel. 39(81)5476512 Fax 39(81)5527771

Giunchi, Dr Giovanni (1946). Researcher. SIR Industriale, via Bellini 35, 20050 Macherio MI, Italy. [via don Gallotti 15, 28100 Novara, Italy.] Dr (Bologna U., 1970) physics. *Polymers ceramics resins composite_materials*
Tel. 39(321)451044

Gramaccioli, Prof. Carlo Maria (1935). Full professor. Chairman, IUCr Commission on Crystallographic Teaching. Dipartimento di Scienze della Terra, Università di Milano, via Botticelli 23, 20133 Milano, Italy. Dr (Milano U., 1959) industrial chemistry. *Minerals molecular_crystal thermal_vibration force_fields*
E-mail pila@sg1.csrsrc.mi.cnr.it Tel. 39(2)23698314 Fax 39(2)70638681

Graziani, Prof. Giorgio (1933). Full professor. Dip. di Scienze della Terra, Università di Roma "La Sapienza", Piazzale A. Moro 5, 00185 Roma, Italy. Dr (Roma U., 1958) geological sciences. *Minerals crystallography gemology archeometry*
Tel. 39(6)49914686 Fax 39(6)4454729

Grepioni, Dr Fabrizia (1960). Research assistant. Dipartimento di Chimica "G. Ciamician", Università di Bologna, Via F. Selmi 2, 40126 Bologna, Italy. PhD (Bologna U., 1989) chemistry. *Organometallic_crystal_chemistry clusters molecular_crystal computer_chemistry computer_graphics metal-complexes dynamics solid_state_properties*
E-mail fab@ciam01.cineca.it Tel. 39(51)259536 Fax 39(51)259456

Guagliardi, Dr Antonietta (1962). Postdoctoral fellow. CNR IRMEC - c/o Dipartimento Geomineralogico, Via Orabona 4, I-70125 Bari, Italy. Dr (Bari U., 1987) geological sciences. *Computing direct_method powder mineralogy*
E-mail cryst@arba.ba.cnr.it Tel. 39(80)5442624 Fax 39(80)5442591

Guerra, Prof. Gaetano (1953). Associate professor. Dipartimento di Chimica, Università di Napoli "Federico II", via Mezzocannone 4, 80134 Napoli, Italy. Dr (Napoli U., 1976) chemistry. *Macromolecules defect reaction_mechanisms*
Tel. 39(81)5476535 Fax 39(81)5476539

Iandelli, Prof. Aldo (1912). Professor emeritus. Ist. di Chimica Fisica, U. di Genova, Corso Europa 26, I-16132 Genova, Italy. Dr (Firenze U., 1933) chemistry. *Intermetallic_compounds*
E-mail metalli@chimica.unige.it Tel. 39(10)3538221 Fax 39(10)515076

Ianelli, Dr Sandra (1951). Researcher. Dip. di Chimica Generale ed Inorganica, Chimica Analitica, Chimica Fisica, Università di Parma, Viale delle Scienze 78, 43100 Parma, Italy. Dr (Parma U., 1976) chemistry. *Coordination X-ray_diffraction structure_determination organic_compounds*
E-mail chimic3@ipruniv.cce.unipr.it Tel. 39(521)905425 Fax 39(521)905557

Iannelli, Dr Pio (1958). Researcher. Dip. di Fisica, Università di Salerno, Via Baronissi, 84081 Baronissi (Salerno), Italy. Dr (Napoli U., 1981) chemistry. *Polymers liquid_crystals macromolecules computing*
Tel. 39(89)822366 Fax 39(89)953804

Immirzi, Prof. Attilio (1938). Full professor. Dip. di Fisica, Università di Salerno, Via Baronissi, 84100 Salerno, Italy. Dr (Napoli U., 1961) industrial chemistry. *Polymers Rietveld_method fibers*
Tel. 39(89)822231 Fax 39(89)958291

Imperatori, Dr Patrizia (1957). Researcher. Istituto di Chimica dei Materiali, Area della Ricerca di Roma - CNR, via Salaria KM 29,500, 00016 Monterotondo Stazione (Roma), Italy. PhD (Roma U., 1981) chemistry. *X-ray_diffraction semiconductors grazing incidence*
E-mail cryst1@icmat.mlib.cnr.it Tel. 39(6)90672346 Fax 39(6)90625849

Ivaldi, Prof. Gabriella (1950). Associate professor. Dipartimento di Scienze Mineralogiche e Petrologiche, Università di Torino, Via Valperga Caluso 37, I-10125 Torino, Italy. Dr (Torino U., 1974) natural sciences. *Crystal_chemistry mineralogy X-ray_diffractometry*
Tel. 39(11)6707133 Fax 39(11)6502657

Kovàcs, Prof. Alessandro Leone (1936). Full professor; Group Coordinator, Theoretical Physics of Condensed Matter. Città Universitaria, Piazzale A. Moro, 00185 Roma, Italy. [Chair of Physics, Dept. Biochemical Sciences, Città Universitaria, Piazzale A. Moro, 00185 Roma, Italy.] PhD (Bologna U., 1961) physics. *System_dynamics mathematical_physics statistical_physics*
Tel. 39(6)4463971 Fax 39(6)4453881

Krajewski, Dr Adriano (1947). Senior Researcher. Dipartimento di Ceramiche Classiche, Ist. di Ricerche Tecnologiche per la Ceramica del C.N.R., Via Granarolo 64, 48018 Faenza (Ravenna), Italy. Dr (Bologna U., 1971) chemistry. *Crystal_growth bioceramics biological_glasses ceramics refractories electronic_ceramics*
E-mail kraxi@irtec1.irtec.bo.cnr.it Tel. 39(546)46147 Fax 39(546)46381

Lagomarsino, Dr Stefano (1948). Senior researcher. Istituto Elettronica Stato Solido - CNR, V. Cineto Romano 42, 00156 Roma, Italy. Dr (Roma U., 1972) physics. *X-ray_diffraction thin_film crystal_growth*
E-mail stefano@iess.rm.cnr.it Tel. 39(396)415221 Fax 39(396)41522220 Telex 61076

Lamba, Dr Doriano (1955). Researcher. Istituto di Strutturistica Chimica "G. Giacomello", Area della Ricerca di Roma - CNR, CP no. 10, 00016 Monterotondo Stazione (Roma), Italy. Dr (Roma U., 1980) chemistry. *Direct_method nuclear_magnetic_resonance modelling structure_determination carbohydrates antibiotics natural_compounds*
E-mail struta@irmias.bitnet Tel. 39(6)9005142 Fax 39(6)90672630 Telex 624809 CN-RMLI

Lanfranchi, Prof. Maurizio (1942). Associate professor. Dip. di Chimica Generale ed Inorganica, Chimica Analitica, Chimica Fisica, Università di Parma, Viale delle Scienze 78, 43100 Parma, Italy. Dr (Parma U., 1985) chemistry. *X-ray_structure_determination clusters organometallic_chemistry coordination*
E-mail chimic4@ipruniv.cce.unipr.it Tel. 39(521)905424 Fax 39(521)905557

Lanzavecchia, Dr Salvatore (1962). Research assistant. Dip. Chimica Strutturale e Stereochimica Inorganica, Università di Milano, Via Venezian 21, 20133 Milano, Italy. Dr (Milano U., 1985) physics. *Electron_microscopy image_reconstruction living_systems signal_processing image_processing*
E-mail guest@stinch0.csmtbo.mi.cnr.it Tel. 39(2)70635120 Fax 39(2)70635288

Leccabue, Dr Fabrizio (1947). Researcher. Istituto MASPEC, C.N.R., Via Chiavari 18/a, 43100 Parma, Italy. Dr (Parma U., 1972) chemistry. *Crystal_growth magnetic_materials ferroelectrics nanocrystals inorganic_crystals rapid_cooling pulsed_laser_ablation*
E-mail leccabue@pr.masp.pr.cnr.it Tel. 39(521)269207 Fax 39(521)269206

Leoni, Prof. Leonardo (1944). Full professor. Dipartimento di Scienze della Terra, Università di Pisa, Via S. Maria 53, 56126 Pisa, Italy. Dr (Pisa U., 1967) geological sciences. *Microscopy minerals powder_diffraction*
Tel. 39(50)501457

Leporati, Prof. Enrico (1934). Associate professor. Dip. di Chimica Generale ed Inorganica, Chimica Analitica, Chimica Fisica, Università di Parma, Viale delle Scienze 78, 43100 Parma, Italy. Dr (Bologna U., 1964) chemistry. *Spectrophotometry siderophores inorganic_compounds*
E-mail chimic4@ipruniv.cce.unipr.it Tel. 39(521)905426 Fax 39(521)905557

Licci, Dr Francesca (1945). Researcher. Member, IUCr Commission on Crystal Growth and Characterization of Materials. Istituto MASPEC, C.N.R., Via Chiavari 18/a, 43100 Parma, Italy. Dr (Bologna U., 1969) chemistry. *Superconductors high_Tc_superconductors*
Tel. 39(521)269204 Fax 39(521)269206

Licheri, Prof. Giovanni (1939). Full professor. Università di Cagliari, Dipartimento di Scienze Chimiche, Via Ospedale 72, 09124 Cagliari, Italy. Dr (Cagliari U., 1962) physics. *Disorder materials powder_diffraction*
Tel. 39(70)668047 Fax 39(70)669272

Liquori, Prof. Alfonso Maria (1926). Full professor. Dip. di Scienze Chimiche, Università di Roma "Tor Vergata", Via della Ricerca Scientifica, 00193 Roma, Italy. Dr (Roma U., 1948) chemistry. *Macromolecules conformation proteins conformational_transition*

Locchi, Dr Stelio (1929). Associate Professor. Dipartimento di Chimica Generale, Università di Pavia, Viale Taramelli 12, I-27100 Pavia, Italy. Dr (Pavia U., 1952) chemistry. *Inorganic_crystal_structures organic_crystal_structures*
Tel. 39(382)507350 Fax 39(382)507575

Loreto, Prof. Lucio (1937). Associate professor. Dip. di Scienze della Terra, Università di Roma "La Sapienza", Piazzale A. Moro 5, 00185 Roma, Italy. Dr (Roma U., 1965) geological sciences. *Computer_graphics quasicrystal symmetry minerals teaching*
E-mail loreto@sci.uniroma1.it Tel. 39(6)49914890 Fax 39(6)49914891

Lucchesi, Dr Sergio (1958). Researcher. Dip. di Scienze della Terra, Università di Roma "La Sapienza", Piazzale A. Moro 5, 00185 Roma, Italy. Dr (Roma U., 1981) geological sciences. *Minerals crystallography*
Tel. 39(6)4463731 Fax 39(6)4454729

Lucchetti, Prof. Gabriella (1947). Associate professor. Dip. di Scienze della Terra, U. di Genova, Corso Europa 26, I-16132 Genova, Italy. Dr (Genova U., 1976) natural sciences. *Electron_microscopy minerals powder_diffraction crystallographic_data*
Tel. 39(10)3538304 Fax 39(10)352169

Malpezzi, Dr Luciana (1946). Researcher. Dip. di Chimica, Politecnico di Milano, via Mancinelli 7, 20131 Milano, Italy. Dr (Bologna U., 1970) chemistry. *Crystallography conformational_analysis*
E-mail 2chim11@ipmch2.polimi.it Tel. 39(2)23993022 Fax 39(2)23993080

Malta, Dr Viscardo (1938). Researcher. Centro Studio Fisica Macromolecole, C.N.R., Via F. Selmi 2, 40126 Bologna, Italy. Dr (Bologna U., 1968) chemistry. *Polymers macromolecules powder_diffraction*
E-mail vs3bodi3@icineca Tel. 39(51)259552 Fax 39(51)259456

Mammi, Prof. Mario (1932). Full professor. Chairman, Italian Crystallographic Committee. Dipartimento di Chimica Organica, e Centro di Studio sui Biopolimeri, Università di Padova, Via Marzolo 1, 35131 Padova, Italy. Dr (Padova U., 1956) chemistry. *Biocrystallography*
E-mail chim09@ipdunivx.unipd.it Tel. 39(49)831234 Fax 39(49)831222

Manassero, Prof. Mario (1944). Associate professor. Dip. Chimica Strutturale e Stereochimica Inorganica, Università di Milano, Via Venezian 21, 20133 Milano, Italy. Dr (Milano U., 1968) chemistry. *Coordination structural_chemistry chemical_bonding*
E-mail guest@stinch0.csmtbo.mi.cnr.it Tel. 39(2)70635120 Fax 39(2)70635288

Mancini, Prof. Annamaria (1944). Associate professor. Dipartimento di Scienza dei Materiali, Università di Lecce, Via Arnesano, I-73100 Lecce, Italy. Dr (Bari U., 1970) chemistry. *Crystal_growth structural_characterization*
Tel. 39(832)627247

Mangani, Dr Stefano (1951). Assistant professor. Dipartimento di Chimica, Università di Siena, Pian dei Mantellini 44, 53100 Siena, Italy. Dr (Firenze U., 1977) chemistry. *Biocrystallography EXAFS metalloenzymes bioinorganic_chemistry*
E-mail crystl@ifichim.bitnet Tel. 39(577)46264 Fax 39(577)280405

Manotti-Lanfredi, Prof. Anna Maria (1933). Full professor. Dip. di Chimica Generale ed Inorganica, Chimica Analitica, Chimica Fisica, Università di Parma, Viale delle Scienze 78, 43100 Parma, Italy. Dr (Parma U., 1957) chemistry. *Structural_chemistry organometallic_chemistry coordination*
E-mail chimic2@ipruniv.cce.unipr.it Tel. 39(521)905417 Fax 39(521)905557

Mantovani, Prof. Giorgio (1925). Full professor; Director of the Chemistry Department. Dipartimento di Chimica, Università di Ferrara, Via L. Borsari 46, I-44100 Ferrara, Italy. Dr (Ferrara U., 1948) chemistry; pharmacy. *Crystal_growth sucrose*
Tel. 39(532)291168 Fax 39(532)240709 Telex 510850 UNIV FE

Marchetti, Dr Fabio (1950). Researcher. Dipartimento di Chimica e Chimica Industriale, Università di Pisa, Via Risorgimento 35, 56126 Pisa, Italy. Dr (Pisa U., 1974) chemistry. *Chemical_bonding coordination structural_chemistry molecular_crystals*
E-mail fama@vm.cnuce.cnr.it Tel. 39(50)918233 Fax 39(50)20237

Marciante, Mrs Clara (1956). Senior technician. Istituto di Strutturistica Chimica "G. Giacomello", Area della Ricerca di Roma - CNR, CP no. 10, 00016 Monterotondo Stazione (Roma), Italy. *Synchrotron_radiation crystallography small_molecules coordination_compounds*
E-mail struta@irmias.bitnet Tel. 39(6)90625142 Fax 39(6)90672630 Telex 624809 CN-RMLI

Marigo, Prof. Antonio (1950). Associate professor. Dipartimento di Chimica Inorganica Metallorganica ed Analitica, Università di Padova, via Loredan 4, 35131 Padova, Italy. Dr (Padova U., 1974) chemistry. *Polymers catalysts powder_diffraction small-angle_scattering*
E-mail marigo@chim02.unipd.it Tel. 39(49)831210 Fax 39(49)831249

Marongiu, Prof. Giaime (1939). Full professor. Università di Cagliari, Dipartimento di Scienze Chimiche, Via Ospedale 72, 09124 Cagliari, Italy. Dr (Cagliari U., 1963) chemistry. *Amorphous_materials nanostructured_materials*
Tel. 39(70)668047 Fax 39(70)669272

Martinelli, Prof. Giuliano (1938). Associate professor. Dipartimento di Fisica, Università di Ferrara, Via Paradiso 12, 44100 Ferrara, Italy. Dr (Ferrara U., 1969) physics. *Crystal_growth electron_microscopy semiconductors*
Tel. 39(532)781853 Fax 39(532)781810

Martorana, Prof. Antonio (1952). Associate professor. Dipartimento di Chimica Inorganica, U. di Palermo, Via Archirafi 26-28, I-90123 Palermo, Italy. Dr (Padova U., 1979) physics. *WAXS_characterization stacking_fault size_distribution powder_intensity_simulation SAXS*
E-mail cric1@ipacuc.cuc.unipa.IT Tel. 39(91)6165571 Fax 39(91)6166281

Masciocchi, Dr Norberto (1959). Researcher. Dip. Chimica Strutturale e Stereochimica Inorganica, Università di Milano, Via Venezian 21, 20133 Milano, Italy. PhD (Milano U., 1989) chemistry. *Single_crystal X-ray_diffraction powder_diffraction organometallic_compounds synchrotron_radiation*
E-mail norbert@stinch0.csmtbo.mi.cnr.it Tel. 39(2)70635120 Fax 39(2)70635288

Masi, Mr Dante (1948). Technician. Istituto per lo Studio della Stereochimica ed Energetica dei Composti, di Coordinazione (I.S.S.E.C.C. - C.N.R.), Via J. Nardi 39, 50132 Firenze, Italy. Mr (High School, 1966) chemistry. *Structure_determination*
E-mail masi@cacao.issecc.fi.cnr.it Tel. 39(55)243990 Fax 39(55)2478366

Massarotti, Prof. Vincenzo (1944). Associate professor. Dip. Chimica Fisica, Università di Pavia, Viale Taramelli 16, 27100 Pavia, Italy. Dr (Pavia U., 1968) chemistry. *Phase_transition powder_diffraction solid_state_reactions defects*
E-mail chimfi@ipvccn.it Tel. 39(382)507203 Fax 39(382)507575

Mattevi, Dr Andrea (1965). Researcher. Università di Pavia, Dipartimento di Genetica e Microbiologia, Sezione di Biologia Molecolare e Biofisica, Via Abbiategrasso 207, 27100 Pavia, Italy. PhD (Groningen U., 1992) biocrystallography. *Biocrystallography*
E-mail mattevi@ipvgen.unipv.it Tel. 39(382)505557 Fax 39(382)528496

Mattia, Prof. Carlo Andrea (1946). Associate professor. Dip. di Chimica, Università di Napoli "Federico II", Via Mezzocannone 4, 80134 Napoli, Italy. Dr (Napoli U., 1970) chemistry. *Natural_products peptides computing data_collection*
E-mail mattia@chemna.dichi.unina.it Tel. 39(81)5476514 Fax 39(81)5527771

Mazza, Prof. Fernando (1938). Associate professor. Dipartimento di Chimica, Ingegneria Chimica e dei Materiali, Universita dell'Aquila, I-67100 L'Aquila, Italy. Dr (Roma U., 1964) chemistry. *Peptides_conformation energy_calculation*
Tel. 39(862)432572 Fax 39(862)432573

Mazzarella, Prof. Lelio (1938). Full professor. Dip. di Chimica, Università di Napoli "Federico II", Via Mezzocannone 4, 80134 Napoli, Italy. Dr (Napoli U., 1961) chemistry. *Proteins natural_products computing data_collection computer_graphics*
E-mail mazzarella@chemna.dichi.unina.it Tel. 39(81)5476506 Fax 39(81)5527771

Mazzi, Prof. Fiorenzo (1924). Full professor. Dipartimento di Scienze della Terra, Università di Pavia, Via Abbiategrasso 209, I-27100 Pavia, Italy. Dr (Firenze U., 1947) chemistry. *Minerals crystal_chemistry crystallography*
E-mail mazzi@crystal.unipv.it Tel. 39(382)505878 Fax 39(382)505890

Mealli, Dr Carlo (1946). Research leader. Member of the CNR Committee for Crystallography (CIC). Istituto per lo Studio della Stereochimica ed Energetica dei Composti, di Coordinazione (I.S.S.E.C.C. - C.N.R.), Via J. Nardi 39, 50132 Firenze, Italy. Dr (Firenze U., 1969) chemistry. *Inorganic_chemistry structure_determination structure-activity_relationship MO_calculation*
E-mail mealli@cacao.issecc.fi.cnr.it Tel. 39(55)2346653 Fax 39(55)2478366

Meille, Dr Stefano Valdo (1952). Researcher. Dip. di Chimica, Politecnico di Milano, via Mancinelli 7, 20131 Milano, Italy. Dr (Milano U., 1976) chemistry. *Polymers_crystallization polymers_morphology polymers_crystallography conformational_analysis small_organic_molecules*
E-mail 2chimi12@ipmcc3.polimi.it Tel. 39(2)23993021 Fax 39(2)23993080

Mellini, Prof. Marcello (1949). Full professor. Dip. Scienze Terra, Università, Via d. Cerchia 3, 53100 Siena, Italy. Dr (Pisa U., 1974) chemistry. *Silicates_structure electron_microscopy defects*
E-mail mellini@sivax.cineca.it Tel. 39(577)298813 Fax 39(577)298815

Menchetti, prof. Silvio (1937). Full professor. Dipartimento di Scienze della Terra, Università di Firenze, Via La Pira 4, 50121 Firenze, Italy. Dr (Firenze U., 1961) geological sciences. *Structural_mineralogy*
Tel. 39(55)2757488 Fax 39(55)284571

Menzinger, Prof. Filippo (1937). Full professor. Dip. di Fisica, Università di Roma II, Via O. Raimondo, 00173 Roma, Italy. Dr (Roma U., 1961) physics. *Neutron_diffraction spin_density thermal_vibration*
Tel. 39(6)24990441 Fax 39(6)6133074

Merati, Dr Felicita (1955). Researcher. Dipartimento di Chimica Fisica ed Elettrochimica dell' Università, via Golgi 19, 20133 Milano, Italy. Dr (Milano U., 1980) chemistry. *Charge_density electrostatic_properties*
E-mail feli@rs6.csrsrc.mi.cnr.it Tel. 39(2)26603286 Fax 39(2)70638129

Meriani, Prof. Sergio (1941). Full professor; Chair of Department. Materials Engineering Department, Università di Trieste, Via Valerio 2, I-34127 Trieste, Italy. Dr (Trieste U., 1966) chemistry. *Materials phase_determination phase_transition materials_engineering_teaching_programme*
E-mail tk7t01@chimmap.bitnet Tel. 39(40)6763705 Fax 39(40)572044

Merlino, Prof. Stefano (1938). Full professor. Dipartimento di Scienze della Terra, Università di Pisa, Via S. Maria 53, 56126 Pisa, Italy. Dr (Pisa U., 1962) chemistry. *Mineralogy structural_disorder OD systematics*
E-mail merlino@vm.cnuce.cnr.it Tel. 39(50)568203 Fax 39(50)40976

Merlo, Prof, Franco (1940). Full professor. Ist. di Chimica Fisica, U. di Genova, Corso Europa 26, I-16132 Genova, Italy. Dr (Genova U., 1963) industrial chemistry. *Intermetallic_compounds*
E-mail cfmet@chimica.unige.it Tel. 39(10)3538221 Fax 39(10)515076

Millini, Dr Roberto (1959). Senior researcher. Eniricerche S.p.A./SOLS, via F. Maritano 26, I-20097 San Donato Milanese (Milano), Italy. Dr (Pavia U., 1983) chemistry. *X-ray_diffraction Rietveld_analysis molecular_modelling catalysis*
Tel. 39(2)5207543 Fax 39(2)52036347

Molin, Gianmario (1948). Associate professor. Dipartimento di Geologia e Petrologia, Corso Garibaldi 37, 35100 Padova, Italy. Dr (Padova U., 1975) geological sciences. *Silicates X-ray_refinement ion_exchange_reactions_in_terrestrial_and_extra-terrestrial_minerals*
E-mail a45700@pdadr1.pd.cnr.it Tel. 39(49)663122 Fax 39(49)8753813

Moliterni, Dr Anna Grazia (1967). Postdoctoral fellow. CNR IRMEC - c/o Dipartimento Geomineralogico, via Orabona 4, I-70125 Bari, Italy. Dr (Bari U., 1991) physics. *Computing direct_method powder mineralogy*
E-mail CRYST@ARBA.BA.CNR.IT Tel. 39(80)5442624 Fax 39(80)5442591

Monaco, Prof. Hugo L. (1947). Associate professor. Department of Genetics, University of Pavia, via Abbiategrasso 207, 27100 Pavia, Italy. PhD (Harvard U., 1978) chemistry. *Biocrystallography ligand-binding_proteins protein_purification protein_crystallization*
E-mail Monaco@ipvgen.unipv.it Tel. 39(382)505559 Fax 39(382)528496 Telex 312841 UNIPAVI

Monari, Dr Magda (1955). Researcher. Dipartimento di Chimica "G. Ciamician", Università di Bologna, via Selmi F. Selmi 2, 40126 Bologna, Italy. PhD (Bologna U., 1987) chemistry. *Organometallic_crystal_chemistry clusters synthesis_reactivity metal-complexes*
E-mail vgxray@ciam01.cineca.it or vn3bodg4@icineca2.it Tel. 39(51)259559 Fax 39(51)259456

Monlorgi, Prof. Romano (1940). Associate professor. Dipartimento di Scienze Mineralogiche, Università di Bologna, Piazza di Porta S. Donato 1, 40126 Bologna, Italy. Dr (Bologna U., 1969) geological sciences. *Crystal_growth electron_microscopy biominerals*
Tel. 39(51)243556 Fax 39(51)243336

Montenero, Prof. Angelo (1944). Associate professor; President of CoRIVe (Consorzio Ricerca Innovazione Vetro); Vice-President ATTV (Associazione Tecnici Italiani del Vetro). Dip. di Chimica Generale ed Inorganica, Chimica Analitica, Chimica Fisica, Viale delle Scienze, 43100 Parma, Italy. Dr (Parma U., 1968) chemistry. *Glasses ceramics structure_properties*
E-mail monte@ipruniv.cce.unipr.it Tel. 39(521)905553 Fax 39(521)905556

Morandi, Prof. Noris (1938). Associate professor. Dipartimento di Scienze Mineralogiche, Università di Bologna, Piazza di Porta S. Donato 1, 40126 Bologna, Italy. Dr (Bologna U., 1961) geology. *Minerals phase_determination powder_diffraction teaching*
Tel. 39(51)243556 Fax 39(51)243336

Moret, Dr Massimo (1961). Researcher. Dip. Chimica Strutturale e Stereochimica Inorganica, Università di Milano, Via Venezian 21, 20133 Milano, Italy. PhD (Milano U., 1991) chemistry. *Single_crystal organometallic_crystal_chemistry molecular_mechanics scanning_tunnel_microscopy atomic_force_microscopy*
E-mail max@stinch0.csmtbo.mi.cnr.it Tel. 39(2)70635120 Fax 39(2)70635288

Mottana, Prof. Annibale (1940). Full professor; Board of Governors, III Univ. Roma. Dip. di Scienze Geologiche, III Università di Roma, Via Ostiense 169, 00154 Roma, Italy. Dr (Milano U., 1963) geological sciences. *Powder crystal_growth EXAFS XANES petrology*
E-mail mottana@lnf.infn.it Tel. 39(6)5756334 Fax 39(6)5756320

Moze, Prof. Oscar (1954). Associate professor. Member of the Commission for computation and instrumentation of the Italian Crystallographic Association. Dipartimento di Fisica, Università, Viale delle Scienze, 43100 Parma, Italy. PhD (Monash U., 1981) physics. *Materials neutron_diffraction diffuse_scattering disorder magnetism superconductivity*
E-mail moze@vaxpr.cineca.it Tel. 39(521)905268 Fax 39(521)905223 Telex 530327 univpr I

Mugnoli, Prof. Angelo (1933). Full professor. Ist. di Chimica Fisica, U. di Genova, Corso Europa 26, I-16132 Genova, Italy. Dr (Milano U., 1958) industrial chemistry. *Molecular_structure chemical_reactivity computing molecular_pharmacology structure-activity_relationships*
E-mail mugnoli@chimica.unige.it Tel. 39(10)3538221 Fax 39(10)515076

Mura, Dr Pasquale (1944). Senior researcher. Istituto di Strutturistica Chimica "G. Giacomello", Area della Ricerca di Roma - CNR, CP no. 10, 00016 Monterotondo Stazione (Roma), Italy. Dr (Roma U., 1975) chemistry. *Platinum_compounds nuclear_magnetic_resonance infrared_spectroscopy synthesis crystal_structure*
E-mail struta@irmias.bitnet Tel. 39(6)90625142 Fax 39(6)90672630 Telex 624809 CN-RMLI

Musatti, Prof. Amos (1935). Associate professor. Dip. di Chimica Generale ed Inorganica, Chimica Analitica, Chimica Fisica, Viale delle Scienze, 43100 Parma, Italy. Dr (Parma U., 1962) chemistry. *Crystallography computer*
E-mail struttur@ipruniv.cce.unipr.it Tel. 39(521)905450 Fax 39(521)905556

Napolitano, Prof. Roberto (1949). Associate professor. Dipartimento di Chimica, Università della Basilicata, via N. Sauro 85, 85100 Potenza, Italy. Dr (Napoli U., 1974) chemistry. *Computing macromolecules powder_diffraction crystallographic_data*
Tel. 39(971)334246

Nardelli, Prof. Mario (1922). Full professor. IUCr Past President; Member of IUCr Executive; Co-editor of Acta Crystallographica. Dip. di Chimica Generale ed Inorganica, Chimica Analitica, Chimica Fisica, Università di Parma, Viale delle Scienze 78, I-43100 Parma, Italy. Dr (Parma U., 1946) chemistry. *Structure_determination inorganic_chemistry organic_crystal_chemistry computing*
E-mail nardelli@ipruniv.cce.unipr.it Tel. 39(521)905433 Fax 39(521)905557 Telex 530327 UNIVPR I

Nardin, Prof. Giorgio (1940). Associate professor. Dipartimento di Scienze Chimiche, Università di Trieste, Via Giorgieri 1, I-34127 Trieste, Italy. Dr (Trieste U., 1964) chemistry. *Inorganic_crystal_chemistry minerals ceramics superconductors*
E-mail struttur@univ.trieste.it Tel. 39(40)5064111 Fax 39(40)6763903

Navarra, Dr Gabriele (1954). Researcher. Università di Cagliari, Dipartimento di Scienze Chimiche, Via Ospedale 72, 09124 Cagliari, Italy. Dr (Cagliari U., 1980) chemistry. *Computing materials powder_diffraction*
Tel. 39(70)668047 Fax 39(70)669272

Nicolò, Dr Francesco (1959). Researcher. Dip. di Chimica Inorganica, Analitica e Struttura Molecolare, Via Salita Sperone 31, I-98166 Villaggio S. Agata (Messina), Italy. Dr (Messina U., 1983) chemistry. *Structure_determination inorganic_crystal_chemistry computing modelling*
E-mail cendif@imeuniv.bitnet Tel. 39(90)6763129 Fax 39(90)393756

Nunzi, Prof. Antonio (1943). Associate professor. Dip. di Scienze della Terra, Università di Perugia, Piazza Università, I-06100 Perugia, Italy. Dr (Perugia U., 1966) chemistry. *Computing direct_method*
E-mail dipsctn@ipguniv Tel. 39(75)5853215 Fax 39(75)5853203 Telex 662078 UNIPG I

Oberti, Dr Roberta (1952). Researcher. Centro di Studio per la Cristallochimica e la Cristallografia, Via Abbiategrasso 209, I-27100 Pavia, Italy. Dr (Pavia U., 1976) chemistry. *Crystal_chemistry structure_modelling rock-forming_minerals organic_crystal_structures*
E-mail oberti@crystal.unipv.it Tel. 39(382)505885 Fax 39(382)505887

Olcese, Prof. Giorgio L. (1933). Full professor. Ist. di Chimica Fisica, Università di Genova, Corso Europa 26, I-16132 Genova, Italy. Dr (Genova U., 1957) chemistry. *Solid_state magnetochemistry structure rare-earth_compounds ceramics superconductors*
E-mail metalli@chimica.unige.it Tel. 39(10)3538221 Fax 39(10)515076

Orioli, Prof. Pierluigi (1933). Full professor. Dip. Chimica, Università Firenze, via G. Capponi 7, 50121 Firenze, Italy. Dr (Firenze U., 1956) chemistry. *Biocrystallography EXAFS metalloenzymes nucleotides*
E-mail cryst1@ifichim.bitnet Tel. 39(55)2757554 Fax 39(55)2757555

Palenzona, Prof. Andrea (1935). Full professor. Ist. di Chimica Fisica, Università' di Genova, Corso Europa 26, I-16132 Genova, Italy. Dr (Genova U., 1961) chemistry. *Phase_diagram DTA structure_determination rare-earth_compounds mineralogy geochemistry intermetallic_compounds*
E-mail metalli@chimica.unige.it Tel. 39(10)3538221 Fax 39(10)515076

Pani, Dr Marcella (1959). Researcher. Ist. di Chimica Fisica, Università di Genova, Corso Europa 26, I-16132 Genova, Italy. Dr (Genova U., 1984) industrial chemistry. *Crystal_structure_determination Rietveld_method intermetallic_compounds*
E-mail pani@chimica.unige.it Tel. 39(10)3538221 Fax 39(10)515076

Paorici, Prof. Carlo (1936). Full professor. Dipartimento di Fisica, Università di Parma, Viale delle Scienze, I-43100 Parma, Italy. Dr (Roma U., 1962) chemistry. *Crystal_growth semiconductors non-linear_optics*
Tel. 39(521)905271 Fax 39(521)905223

Parisini, Dr Emilio (1964). Postdoctoral Fellow. Dipartimento di Chimica "G. Ciamician", Università di Bologna, Via F. Selmi 2, 40126 Bologna, Italy. Dr (Bologna U., 1990) chemistry. *Organometallic_crystal_chemistry clusters molecular_crystal_chemistry structural_correlation solid_state_properties dynamics*
E-mail emilio@ciam01.cineca.it Tel. 39(51)259555 Fax 39(51)259456

Pasero, Dr Marco (1958). Researcher. Dipartimento di Scienze della Terra, Università di Pisa, Via S. Maria 53, I-56126 Pisa, Italy. PhD (Pisa U., 1988) mineralogy. *Inorganic_crystal_chemistry OD transmission_electron_microscopy*
E-mail zuppa@vm.cnuce.cnr.it Tel. 39(50)568215 Fax 39(50)40976

Passaglia, Prof. Elio (1941). Full professor. Dipartimento di Scienze della Terra, Università di Modena, Via S. Eufemia 19, 41100 Modena, Italy. Dr (Modena U., 1965) geological sciences. *Minerals powder_diffraction*
Tel. 39(59)218062 Fax 39(59)223605

Pasti, Mr Fabio (1937). Maker of scientific instruments. Ital Structures, Zona Industriale Baltera, I-38066 Riva del Garda (TN), Italy. Mr (High School, 1957) electrotechnique. *Diffractometry instrumentation powder_diffraction texture X-ray_fluorescence X-ray_diffraction*
Tel. 39(464)553426 Fax 39(464)555270

Pavel, Prof. Nicolae Viorel (1949). Full professor. Dip. di Chimica, Università di Roma "La Sapienza", Piazzale A. Moro 5, 00185 Roma, Italy. Dr (Roma U., 1973) chemistry. *EXAFS SAXS micelles liquids xas_methodologies*
E-mail pavel@sci.uniroma1.it Tel. 39(6)49913652 Fax 39(6)490631

Pavese, Dr Alessandro (1964). Researcher. Dipartimento di Scienze della Terra, Università degli Studi, Via Botticelli 23, 20133 Milano, Italy. Dr (Torino U., 1987) physics. *Mineralogy crystal_chemistry computer_modelling neutron_radiation X-ray_diffraction*
E-mail umimin@imicilea.cilea.it Tel. 39(2)23698321 Fax 39(2)70638681

Pedone, Prof. Carlo (1938). Full professor. Dipartimento di Chimica, Università di Napoli "Federico II", via Mezzocannone 4, 80134 Napoli, Italy. Dr (Napoli U., 1961) chemistry. *Macromolecules structural_biology bio_molecules*
Tel. 39(81)5476549 Fax 39(81)5476550

Pelizzi, Prof. Giancarlo (1939). Full professor. Dip. di Chimica Generale ed Inorganica, Chimica Analitica, Chimica Fisica, Università di Parma, Viale delle Scienze 78, 43100 Parma, Italy. Dr (Parma U., 1964) chemistry. *Organometallic_chemistry coordination structural_chemistry drug_molecules*
E-mail chimic6@ipruniv.cce.unipr.it Tel. 39(521)905421 Fax 39(521)905557

Pellinghelli, Prof. Maria Angela (1943). Full professor. Dip. di Chimica Generale ed Inorganica, Chimica Analitica, Chimica Fisica, Università di Parma, Viale delle Scienze 78, 43100 Parma, Italy. Dr (Parma U., 1966) chemistry. *Structural_chemistry coordination organometallic_chemistry*
E-mail dipchige@ipruniv.cce.unipr.it Tel. 39(521)905428 Fax 39(521)905557

Pelosi, Dr Giorgio (1962). Researcher. Dip. di Chimica Generale ed Inorganica, Chimica Analitica, Chimica Fisica, Università di Parma, Viale delle Scienze 78, 43100 Parma, Italy. PhD (Parma U., 1991) chemistry. *X-ray_structure_determination transition_elements molecular_complexes protein_crystallography*
E-mail giorgio@ipruniv.cce.unipr.it Tel. 39(521)905477 Fax 39(521)905557

Perego, Dr Giovanni (1938). Manager, Physical Chemistry Dept. Eniricerche S.p.A., via F. Maritano 26, I-20097 San Donato Milanese (Milano), Italy. Doctorat (U. Strasbourg, France, 1971) sciences. *Structure morphology catalysts polymers*
Tel. 39(2)5205961 Fax 39(2)52036347 Telex 310246 ENI I

Petraccone, Prof. Vittorio (1943). Associate professor. Dipartimento di Chimica, Università di Napoli "Federico II", via Mezzocannone 4, 80134 Napoli, Italy. Dr (Napoli U., 1966) chemistry. *Macromolecules computer_modelling defects*
Tel. 39(81)5476538 Fax 39(81)5476539

Pifferi, Dr Augusto (1955). Researcher. Istituto di Strutturistica Chimica "G. Giacomello", Area della Ricerca di Roma - CNR, CP no. 10, 00016 Monterotondo Stazione (Roma), Italy. Dr (Roma U., 1979) physics. *Computer_automation methodology synchrotron_radiation diffractometer*
E-mail struta@irmias.bitnet Tel. 39(6)90625142 Fax 39(6)90672630 Telex 624809 CN-RMLI

Pilati, Dr Tullio (1946). Senior researcher. Centro di Studio per le Relazioni tra Struttura e Reattività Chimica, C.N.R., via Golgi 19, 20133 Milano, Italy. Dr (Milano U., 1971) chemistry. *Lattice_dynamics computing structure_and_crystal_chemistry_of_minerals structure_of_organic_compounds*
E-mail pila@sg1.csrsrc.mi.cnr.it Tel. 39(2)26603285 Fax 39(2)70638129

Pochetti, Dr Giorgio (1957). Researcher. Istituto di Strutturistica Chimica "G. Giacomello", Area della Ricerca di Roma - CNR, CP no. 10, 00016 Monterotondo Stazione (Roma), Italy. Dr (Roma U., 1981) chemistry. *Computer_graphics diffraction_technique structure_determination*
E-mail struta@irmias.bitnet Tel. 39(6)90625142 Fax 39(6)90672630 Telex 624809 CN-RMLI

Polidori, Dr Giampiero (1950). Researcher. Dip. di Scienze della Terra, Università di Perugia, Piazza Università, I-06100 Perugia, Italy. Dr (Perugia U., 1973) mathematics. *Computing direct_method computer_graphics*
E-mail sir1@ipguniv Tel. 39(75)5853213 Fax 39(75)5853203 Telex 662078 UNIPG I

Poppi, Prof. Luciano (1940). Associate professor. Istituto di Mineralogia e Petrologia, Università di Modena, Via S. Eufemia 19, 41100 Modena, Italy. Dr (Modena U., 1968) geology. *Minerals powder_diffraction crystal_chemistry*
Tel. 39(59)218062 Fax 39(59)223605

Porta, Prof. Piero (1933). Full professor. Dip. di Chimica, Università di Roma "La Sapienza", Piazzale A. Moro 5, 00185 Roma, Italy. Dr (Bari U., 1958) chemistry. *Inorganic_crystal_chemistry solid_state*
Tel. 39(6)49913378 Fax 39(6)490631

Portalone, Dr Gustavo (1952). Researcher. Dip. di Chimica, Università di Roma "La Sapienza", Piazzale A. Moro 5, 00185 Roma, Italy. Dr (Roma U., 1977) pharmaceutical chemistry. *Synchrotron_radiation electron_diffraction organic_crystal_chemistry*
E-mail portalone@sci.uniroma1.it Tel. 39(6)49913106 Fax 39(6)4958251

Porzio, Dr William (1951). Researcher. Istituto di Chimica delle Macromolecole, C.N.R., via Bassini 15, 20133 Milano, Italy. Dr (Milano U., 1976) industrial chemistry. *Polymers_crystallography polymers_physics powder_diffraction materials*
E-mail rayx@imicilea.cilea.it Tel. 39(2)70636333 Fax 39(2)2362946

Previde Massara, Dr Elisabetta (1962). Researcher. Eniricerche S.p.A., via F. Maritano 26, I-20097 San Donato Milanese (Milano), Italy. PhD (Pavia U., 1990) crystallography and mineralogy. *X-ray_powder_diffraction Rietveld_method clay_minerals zeolites*
Tel. 39(2)5207543 Fax 39(2)52036347 Telex 310246 ENI I

Proserpio, Dr Davide M. (1962). Researcher. Dip. Chimica Strutturale e Stereochimica Inorganica, Università di Milano, Via Venezian 21, 20133 Milano, Italy. Dr (Pavia U., 1986) chemistry. *Single_crystal X-ray_diffraction organometallic_crystal_chemistry computer_modelling structure-activity_relationship*
E-mail davide@stinch0.csmtbo.mi.cnr.it Tel. 39(2)70635120 Fax 39(2)70635288

Puliti, Dr Raffaella (1936). Research leader. Ist. per la Chimica di Molecole di Interesse Biologico, C.N.R., via Toiano 6, 80072 Arco Felice (Napoli), Italy. [Dipartimento di Chimica, Via Mezzocannone, 4, 80134 Napoli, Italy.] Dr (Bari U., 1961) chemistry. *Natural_products peptides biomaterial computing*
E-mail puliti@chemna.dichi.unina.it Tel. 39(81)5476514 Fax 39(81)5527771

Quagliata, Prof. Claudio (1945). Associate professor. Dip. di Chimica, Università di Roma "La Sapienza", Piazzale A. Moro 5, 00185 Roma, Italy. Dr (Roma U., 1970) chemistry. *Micelles conformational_analysis inclusion_compounds*
Tel. 39(6)49913687 Fax 39(6)490631

Quartieri, Dr Simona (1955). Researcher. Dipartimento di Scienze della Terra, Via S. Eufemia 19, I-41100 Modena, Italy. PhD (Modena U., 1987) mineralogy. *Zeolites phase_transition EXAFS XANES Mossbauer Debye–Waller_factor silicate_crystal_structures*
E-mail quartieri@imomn1.unimo.it Tel. 39(59)218062 Fax 39(59)223605

Randaccio, Prof. Lucio (1940). Full professor. Dipartimento di Scienze Chimiche, Università di Trieste, Via Giorgieri 1, I-34127 Trieste, Italy. Dr (Napoli U., 1963) chemistry. *Structural_chemistry inorganic_crystal_chemistry bioinorganic metalloorganic_chemistry*
E-mail struttur@univ.trieste.it Tel. 39(40)6763935 Fax 39(40)6763903

Rigault de la Longrais, Prof. Germain (1930). Full professor. Dipartimento di Scienze Mineralogiche e Petrologiche, Università di Torino, Via Valperga Caluso 37, I-10125 Torino, Italy. Dr (Torino U., 1953) chemistry. *Morphology crystal_optics symmetry*
Tel. 39(11)6707132 Fax 39(11)6502657

Rinaldi, Prof. Romano (1944). Full-professor. Dip. di Scienze della Terra, Università di Perugia, Piazza Università, I-06100 Perugia, Italy. Dr (Modena U., 1969) geological sciences. *Minerals electron_microscopy crystal_chemistry*
E-mail crystal@ipguniv.unipg.it Tel. 39(75)5853214 Fax 39(75)5853203 Telex 662078 UNIPG I

Rinaudo, Prof. Caterina (1951). Associate professor. Dipartimento di Scienze Mineralogiche, Università di Torino, Via Valperga Caluso 37, I-10125 Torino, Italy. Dr (Torino U., 1974) natural sciences. *Crystal_growth X-ray_topography*
Tel. 39(11)6707130 Fax 39(11)6707128

Ripamonti, Prof. Alberto (1930). Full professor. Dipartimento di Chimica "G. Ciamician", Università di Bologna, Via F. Selmi 2, 40126 Bologna, Italy. Dr (Roma U., 1953) chemistry. *Biomaterial calcification biopolymers phosphates bioinorganic_crystal_growth*
E-mail xray@ciam01.cineca.it Tel. 39(51)259550 Fax 39(51)259456

Riva di Sanseverino, Prof. Lodovico (1939). Full professor. Dipartimento di Scienze Mineralogiche, Università di Bologna, Piazza di Porta S. Donato 1, 40126 Bologna, Italy. Dr (Firenze U., 1962) chemistry. *Structure-activity_relationship crystal_growth teaching*
E-mail t54bom12@icineca.cineca.it Tel. 39(51)243556 Fax 39(51)243316

Riva, Dr Fernando (1933). Researcher. Istituto Polimeri, C.N.R., Via Mezzocannone 4, 80134 Napoli, Italy. Dr (Pavia U., 1958) chemistry. *Macromolecules polymers powder_diffraction small-angle_scattering*
Tel. 39(81)205730

Rizzi, Dr Menico (1965). PhD student. Dip. Genetica e Microbiologia, Sez. Cristallografia, Università di Pavia, Via Abbiategrasso 207, 27100 Pavia, Italy. Dr (Pavia U., 1990) chemistry. *Biocrystallography*
E-mail Rizzi@ipvgen.unipv.it Tel. 39(382)505557 Fax 39(382)528496

Rizzoli, Dr Corrado (1957). Researcher. Dip. di Chimica Generale ed Inorganica, Chimica Analitica, Chimica Fisica, Viale delle Scienze, 43100 Parma, Italy. Dr (Parma U., 1981) chemistry. *Structure_determination molecular_recognition organometallic_crystal_chemistry*
E-mail struchi@ipruniv.cce.unipr.it Tel. 39(521)905669 Fax 39(521)905556

Roetti, Prof. Carla (1943). Associate professor. Dipartimento di Chimica Inorganica, Chimica-Fisica e Chimica dei Materiali, Università di Torino, Via P. Giuria 5, I-10125 Torino, Italy. Dr (Torino U., 1967) chemistry. *Solid_state charge_density molecular_crystals*
E-mail roetti@itocsivm.csi.it Tel. 39(11)6707564 Fax 39(11)6690957

Rosa, Dr Rodolfo (1944). Senior researcher. Istituto LAMEL, C.N.R., Via dei Castagnoli 1, 40126 Bologna, Italy. Dr (Bologna U., 1968) physics. *Electron_microscopy computing*
Tel. 39(51)287912 Fax 39(51)229702

Rossi, Dr Marco (1961). Researcher. Dip. di Energetica, Universita di Roma "La Sapienza", Via Scarpa 14, 00161 Roma, Italy. PhD (Roma U., 1991) material sciences. *Electron_diffraction electron_microscopy semiconductors thin_films*
E-mail redone@itcaspur.caspur.it Tel. 39(6)49766543 Fax 39(6)44240183

Roveri, Prof. Norberto (1947). Full professor. Dipartimento di Chimica "G. Ciamician", Università di Bologna, Via F. Selmi 2, 40126 Bologna, Italy. Dr (Bologna U., 1972) chemistry. *Calcification SAXS synchrotron_radiation biopolymers biological_carbonates biomaterial bone biological_phosphates biomechanics*
E-mail xray@ciam01.cineca.it Tel. 39(51)259551 Fax 39(51)259456

Rubbo, Prof. Marco (1946). Associate professor. Dipartimento di Scienze Mineralogiche e Petrologiche, Università di Torino, Via Valperga Caluso 37, I-10125 Torino, Italy. Dr (Torino U., 1971) chemistry. *Crystal_growth materials mineralogy petrology*
Tel. 39(11)6707127 Fax 39(11)6707128

Sabatino, Dr Piera (1950). Researcher. Dipartimento di Chimica "G. Ciamician", Università di Bologna, Via F. Selmi 2, 40126 Bologna, Italy. Dr (Bologna U., 1974) pharmaceutical sciences. *Organometallic_crystal_chemistry clusters solid_state organic_crystal_chemistry molecular_crystals*
E-mail crsytal@ciam01.cineca.it or vn3bodg2@icineca2.it Tel. 39(51)259537 Fax 39(51)259456

Sabbioni, Dr Cristina (1954). Researcher. Istituto FISBAT - C.N.R., Via Gobetti 101, 40129 Bologna, Italy. Dr (Bologna U., 1978) physics. *Electron_microscopy diffractometry material_damage*
Tel. 39(51)6399572 Fax 39(51)6399649

Sabelli, Dr Cesare (1934). Research leader. Centro di Studio per la Mineralogia e la Geochimica dei Sedimenti, Via La Pira 4, 50121 Firenze, Italy. Dr (Firenze U., 1959) geological sciences. *Minerals crystallographic_data*
Tel. 39(55)2757308 Fax 39(55)284571

Sacerdoti, Prof. Michele (1935). Full professor. Istituto di Mineralogia, Università di Ferrara, Corso Ercole I D'Este 32, 44100 Ferrara, Italy. Dr (Padova U., 1959) geological sciences. *Minerals inorganic_crystal_chemistry*
E-mail m83fem12@iceneca Tel. 39(532)202987 Fax 39(532)202304

Saiviati, Dr Giancarlo (1950). Researcher. Istituto MASPEC - C.N.R., Via Chiavari 18/a, 43100 Parma, Italy. Dr physics. *Materials semiconductors electron_diffraction defect_structures electron_microscopy*
Tel. 39(521)2691 Fax 39(521)269206

Sansoni, Prof. Mirella (1939). Associate professor. Dip. Chimica Strutturale e Stereochimica Inorganica, Università di Milano, Via Venezian 21, 20133 Milano, Italy. Dr (Milano U., 1964) physics. *Coordination_crystal_chemistry structural_chemistry chemical_bonding*
E-mail guest@stinch0.csmtbo.mi.cnr.it Tel. 39(2)70635120 Fax 39(2)70635288

Santini, Dr Antonello (1958). Researcher. Dip. di Chimica, Università di Napoli "Federico II", Via Mezzocannone 4, 80134 Napoli, Italy. Dr (Napoli U., 1986) industrial chemistry. *Biocrystallography peptides molecular_mechanics biomolecules*
E-mail santini@chemna.dichi.unina.it Tel. 39(81)5476553 Fax 39(81)5727771

Scandale, Prof. Eugenio (1943). Full professor. Dipartimento Geomineralogico, Università di Bari, Via Orabona 4, I-70125 Bari, Italy. Dr (Bari U., 1971) physics. *Crystal_growth X-ray_topography minerals*
Tel. 39(80)242585 Fax 39(80)242591

Scardi, Prof. Paolo (1960). Associate professor. Dip. di Ingegneria dei Materiali, Università di Trento, 38050 Mesiano (TN), Italy. Dr (Napoli U., 1985) physics. *Powder_diffraction profile_analysis stress_texture Rietveld_method materials_characterization thin_films coatings*
E-mail scardi@itnux2.ing.unitn.it Tel. 39(461)882417 Fax 39(461)881977

Scatturin, Prof. Vladinmiro (1922). Full professor. Dip. Chimica Strutturale e Stereochimica Inorganica, Università di Milano, Via Venezian 21, 20133 Milano, Italy. Dr (Padova U., 1946) chemistry. *Structural_chemistry chemical_education*
E-mail guest@stinch0.csmtbo.mi.cnr.it Tel. 39(2)70635120 Fax 39(2)70635288

Schiavinato, Prof. Giuseppe (1915). Professor emeritus. Dipartimento di Scienze della Terra, Università degli Studi, Via Botticelli 23, 20133 Milano, Italy. Dr (Padova U., 1939) geological sciences. *Inorganic_crystal_chemistry instrumentation crystal_optics clay_minerals*
E-mail umimin@imicilea.cilea.it Tel. 39(2)23698325 Fax 39(2)70638681

Scordari, Prof. Fernando (1944). Full professor. Dipartimento Geomineralogico, Università di Bari, Via E. Orabona 4, I-70125 Bari, Italy. Dr (Bari U., 1968) geological sciences. *Inorganic_crystal_chemistry crystal_chemistry*
Tel. 39(80)242587 Fax 39(80)242591

Servidori, Dr Marco (1943). Researcher; Department Head of Structural Characterization of LAMEL. Istituto LAMEL-CNR, Via Gobetti 101, I-40129 Bologna, Italy. Dr (Bologna U., 1967) industrial chemistry. *X-ray_topography electron_microscopy electronics_materials multiple-crystal_diffractometry*
E-mail marco@iolanda.lamel.bo.cnr.it Tel. 39(51)6399156 39(51)6399184 Fax 39(51)6399216

Sgarabotto, Prof. Paolo (1940). Full professor. Dip. di Chimica Generale ed Inorganica, Chimica Analitica, Chimica Fisica, Viale delle Scienze, 43100 Parma, Italy. Dr (Parma U., 1967) chemistry. *Crystallography diffractometry structure_determination*
E-mail struttur@ipruniv.cce.unipr.it Tel. 39(521)905447 Fax 39(521)905556

Sguaidino, Dr Giulio (1947). Researcher. Dipartimento di Chimica, Università di Ferrara, Via L. Borsari 46, I-44100 Ferrara, Italy. Dr (Ferrara U., 1974) chemistry. *Crystal_growth*
Tel. 39(532)291111 Fax 39(532)40709

Sica, Dr Filomena (1959). Researcher. Dip. di Chimica, Università di Napoli "Federico II", Via Mezzocannone 4, 80134 Napoli, Italy. Dr (Napoli U., 1984) chemistry. *Proteins peptides crystallography crystallization microgravity*
E-mail zagari@chemna.dichi.unina.it Tel. 39(81)5476513 Fax 39(81)5527771

Sironi, Prof. Angelo (1948). Associate professor. Dip. Chimica Strutturale e Stereochimica Inorganica, Università di Milano, Via Venezian 21, 20133 Milano, Italy. Dr (Milano U., 1972) chemistry. *Single_crystal X-ray_diffraction powder_diffraction computer_modelling scanning_probe_microscopy*
E-mail angelo@stinch0.csmtbo.mi.cnr.it Tel. 39(2)70635120 Fax 39(2)70635288

Spadon, Prof. Paola (1950). Associate professor. Dipartimento di Chimica Organica, e Centro di Studio sui Biopolimeri, Università di Padova, Via Marzolo 1, 35131 Padova, Italy. Dr (Padova U., 1974) chemistry. *Biocrystallography*
E-mail paola@chor00.unipd.it Tel. 39(49)831327 Fax 39(49)831222

Spagna, Dr Riccardo (1941). Senior researcher. Istituto di Strutturistica Chimica "G. Giacomello", Area della Ricerca di Roma - CNR, CP no. 10, 00016 Monterotondo Stazione (Roma), Italy. Dr (Roma U., 1967) chemistry. *Crystallography computing direct_method synchrotron_radiation*
E-mail struta@irmias.bitnet Tel. 39(6)90625142 Fax 39(6)90672630 Telex 624809 CN-RMLI

Stasi, Prof. Francesca (1939). Associate professor. Dipartimento Geomineralogico, U. di Bari, Campus Universitario, I-70121 Bari, Italy. Dr (Bari U., 1971) physics. *Inorganic_crystal_chemistry organic_crystal_chemistry X-ray_topography*
Tel. 39(80)242615 Fax 39(80)242591

Suffritti, Prof. Giuseppe Baldovino (1947). Associate professor. Dipartimento di Chimica, Università di Sassari, Via Vienna 2, I-07100 Sassari, Italy. Dr (Milano U., 1972) physics. *Molecular_dynamics disordered_and_amorphous_solids zeolites*
E-mail suffritti@mvchss.cineca.it Tel. 39(79)229552 Fax 39(79)229559

Tadini, Prof. Carla (1924). Associate professor. Dipartimento di Scienze della Terra, Università di Pavia, Via Abbiategrasso 209, I-27100 Pavia, Italy. Dr (Pavia U., 1956) chemistry. *Minerals crystal_chemistry crystallography*
E-mail tadini@crystal.unipv.it Tel. 39(382)505877 Fax 39(382)505890

Tarricone, Dr Aldo (1967). PhD student. Università di Pavia, Dipartimento di Genetica e Microbiologia, Sezione di Biologia Molecolare e Biofisica, Via Abbiategrasso 207, 27100 Pavia, Italy. Dr (Bari U., 1992) biology. *Biocrystallography*
E-mail tarricone@ipvgen.unipv.it Tel. 39(382)505557 Fax 39(382)528496

Tazzoli, Prof. Vittorio (1938). Full professor. Dipartimento di Scienze della Terra, Università di Pavia, Via Abbiategrasso 209, I-27100 Pavia, Italy. Dr (Pavia U., 1963) chemistry. *Minerals crystal_chemistry*
E-mail tazzoli@crystal.unipv.it Tel. 39(382)505870 Fax 39(382)505890

Tieghi, Prof. Giuseppe (1943). Associate professor. Dip. di Chimica Industriale ed Ingegneria Chimica, Politecnico di Milano, Piazza L. da Vinci 32, 20133 Milano, Italy. Dr (Politecnico, Milano, 1968) chemical engineering. *Polymers macromolecules materials powder_diffraction*
Tel. 39(2)23993218 Fax 39(2)70638173

Tiripicchio, Prof. Antonio (1936). Full professor. Dip. di Chimica Generale ed Inorganica, Chimica Analitica, Chimica Fisica, Università di Parma, Viale delle Scienze 78, 43100 Parma, Italy. Dr (Parma U., 1959) chemistry. *Structural_chemistry organometallic_chemistry clusters coordination*
E-mail dipchige@ipruniv.cce.unipr.it Tel. 39(521)905418 Fax 39(521)905557

Tiripicchio-Camellini, Prof. Marisa (1938). Associate professor. Dip. di Chimica Generale ed Inorganica, Chimica Analitica, Chimica Fisica, Università di Parma, Viale delle Scienze 78, 43100 Parma, Italy. Dr (Parma U., 1962) natural sciences. *X-ray_structure_determination organometallic_chemistry clusters coordination*
E-mail chimcll@ipruniv.cce.unipr.it Tel. 39(521)905417 Fax 39(521)905557

Tosi, Prof. Giorgio (1942). Associate professor. Dipartimento di Scienza dei Materiali e della Terra, Facoltà di Ingegneria, Via Brecce Bianche, I-60100 Ancona, Italy. Dr (Bologna U., 1966) industrial chemistry. *Clusters chemical_physical_relationships*
Tel. 39(71)5893723 Fax 39(71)5893714

Tribaudino, Dr Mario (1960). Researcher. Dipartimento di Scienze Mineralogiche e Petrologiche, Università di Torino, Via Valperga Caluso 37, I-10125 Torino, Italy. PhD (Torino U., 1989) mineralogy and crystallography. *Minerals_crystal_chemistry experimental_mineralogy*
Tel. 39(11)6707131 Fax 39(11)6707128

Trosti-Ferroni, Prof. Renza (1947). Associate professor. Dipartimento di Scienze della Terra, Università di Firenze, Via La Pira 4, 50121 Firenze, Italy. Dr (Firenze U., 1969) natural sciences. *Inorganic_crystal_chemistry structural_chemistry minerals*
Tel. 39(55)287140

Ughetto, Dr Giovanni (1934). Senior researcher. Istituto di Strutturistica Chimica "G. Giacomello", Area della Ricerca di Roma - CNR, CP no. 10, 00016 Monterotondo Stazione (Roma), Italy. Dr (Roma U., 1969) chemistry. *DNA anticancer_compounds drug*
E-mail struta@irmias.bitnet Tel. 39(6)90625142 Fax 39(6)90672630 Telex 624809 CN-RMLI

Ugliengo, Prof. Piero (1957). Associate professor. Dipartimento di Chimica Inorganica, Chimica-Fisica e Chimica dei Materiali, Università di Torino, Via P. Giuria 7, I-10125 Torino, Italy. Dr (Torino U., 1981) chemistry. *Surface molecular_graphics hydrogen_bonding theoretical_calculations*
E-mail ugliengo@silver.ch.unito.it Tel. 39(11)6707515 Fax 39(11)6707855

Ugozzoli, Dr Franco (1947). Researcher. Dip. di Chimica Generale ed Inorganica, Chimica Analitica, Chimica Fisica, Università di Parma, Viale delle Scienze 78, 43100 Parma, Italy. Dr (Parma U., 1983) physics. *Computing X-ray_diffraction structural_chemistry structure–activity_relationship*
E-mail chimic2@ipruniv.cce.unipr.it Tel. 39(521)905430 Fax 39(521)905557

Ungaretti, Prof. Luciano (1942). Full professor. Centro di Studio per la Cristallochimica e la Cristallografia, Università di Pavia, Via Abbiategrasso 209, I-27100 Pavia, Italy. Dr (Pavia U., 1965) chemistry. *Crystallography crystal_chemistry mineralogy*
E-mail ungaretti@crystal.unipv.it Tel. 39(382)505892 Fax 39(382)505890

Vaccari, Prof. Giuseppe (1948). Associate professor. Dipartimento di Chimica, Università di Ferrara, Via L. Borsari 46, I-44100 Ferrara, Italy. Dr (Ferrara U., 1972) chemistry. *Crystal_growth industrial_crystallization sucrose*
Tel. 39(532)291169 Fax 39(532)240709 Telex 510850 UNIV FE

Valdrè, Prof. Ugo (1926). Full professor. Dipartimento di Fisica, Università di Bologna, Via Imerio 46, 40126 Bologna, Italy. Dr (Bologna U., 1954) physics. *Electron_microscopy semiconductors superconductors instrumentation*
Tel. 39(51)241134 Fax 39(51)247244

Valle, Dr Giovanni Carlo (1930). Research leader. Centro di Studio sui Biopolimeri, Via Marzolo 1, 35131 Padova, Italy. Dr (Padova U., 1961) industrial chemistry. *Small_molecules_crystallography peptides*
E-mail valle@chor00.unipd.it Tel. 39(49)831229 Fax 39(49)831222

Venzo, Dr Alfonso (1950). Senior researcher. Centro di Studio sugli Stati Molecolari Radicalici ed Eccitati, Via Loredan 2, 35131 Padova, Italy. Dr (Padova U., 1974) chemistry. *Organometallic_crystal_chemistry nuclear_magnetic_resonance reactivity*
E-mail venzo@pdchfi.unipd.it Tel. 39(49)831357 Fax 39(49)831328

Vezzalini, Prof. Maria Giovanna (1951). Associate professor. Dipartimento di Scienze della Terra, Via S. Eufemia 19, I-41100 Modena, Italy. Dr (Modena U., 1974) geology. *Mineralogy zeolites phase_transition microanalysis material_science X-ray_diffraction*
E-mail vezzalini@imomn1.unimo.it Tel. 39(59)417238 39(59)417111 Fax 39(59)417399

Vitali, Prof. Gianfranco (1934). Full professor. Dip. di Energetica, Università di Roma "La Sapienza", Via A. Scarpa 14-16, 00161 Roma, Italy. Dr (Roma U., 1963) physics. *Electron_microscopy semiconductors radiation_damage*
E-mail redone@itcaspur.caspur.it Tel. 39(6)49766545 Fax 39(6)44240183

Viterbo, Prof. Davide (1939). Full professor. Chairman, IUCr Commission on Crystallographic Computing. Dipartimento di Chimica Inorganica, Chimica-Fisica e Chimica dei Materiali, Università di Torino, Via P. Giuria 7, I-10125 Torino, Italy. Dr (Torino U., 1962) chemistry. *Structure_determination molecular_graphics small_biological_molecules*
E-mail viterbo@silver.ch.unito.it Tel. 39(11)6707515 Fax 39(11)6707855

Zagari, Prof. Adriana (1946). Associate professor. Dip. di Chimica, Università di Napoli "Federico II", Via Mezzocannone 4, 80134 Napoli, Italy. Dr (Napoli U., 1969) chemistry. *Proteins polypeptides peptides computing crystallization computer_graphics microgravity molecular_mechanics*
E-mail zagari@chemna.dichi.unina.it Tel. 39(81)5476513 Fax 39(81)5527771

Zanazzi, Prof. Pierfrancesco (1939). Full professor. Dip. di Scienze della Terra, Università di Perugia, Piazza Università, 06100 Perugia, Italy. Dr (Firenze U., 1962) chemistry. *Crystal_chemistry mineralogy high_pressure high_temperature*
E-mail crystal@ipguniv Tel. 39(75)5853212 Fax 39(75)5853203 Telex 662068 UNIPG I

Zannetti, Prof. Roberto (1929). Full professor. Dipartimento di Chimica Inorganica Metallorganica ed Analitica, Università di Padova, via Loredan 4, 35131 Padova, Italy. Dr (Padova U., 1953) chemistry. *Polymers catalysts powder_diffraction small-angle_scattering*
E-mail marigo@chim02.unipd.it Tel. 39(49)831257 Fax 39(49)831249

Zanotti, Prof. Giuseppe (1950). Associate professor. Sub-Editor, World Directory of Crystallographers 9. Dipartimento di Chimica Organica e Centro di Studio sui Biopolimeri, Università di Padova, Via Marzolo 1, 35131 Padova, Italy. Dr (Padova U., 1974) chemistry. *Biocrystallography*
E-mail zanotti@chor00.unipd.it Tel. 39(49)831229 Fax 39(49)831222

Zanotti, Dr Lucio (1944). Research leader. Istituto MASPEC - C.N.R., Via Chiavari 18/a, 43100 Parma, Italy. Dr (Bologna U., 1969) chemistry. *Crystal_growth materials semiconductors defect_structures*
E-mail d11200@pr.maspr.pr.cnr.it Tel. 39(521)2691 Fax 39(521)269206

Zefiro, Dr Livio (1948). Researcher. Dipartimento di Scienze della Terra, Sezione di Mineralogia, Università di Genova, Corso Europa 26, 16132 Genova, Italy. Dr (Genova U., 1972) physics. *Minerals structure_refinement crystal_chemistry crystal_growth holographic_interferometry inverse_problem*
Tel. 39(10)3538311 Fax 39(10)352169

Zerbi, Prof. Giuseppe (1933). Full professor of material science; Director, Advanced School of Polymer Science, Politecnico di Milano. Dipartimento di Chimica Industriale, Politecnico di Milano, Piazza L. da Vinci 32, I-20133 Milano, Italy. Dr (Pavia U., 1956) chemistry. *Liquid_crystals materials polymers semiconductors*
Tel. 39(2)23993235 Fax 39(2)23993231

Zocchi, Prof. Marcello (1929). Full professor. Dipartimento di Chimica Fisica dei Materiali, Università di Brescia, Via Branze 38, I-25060 Brescia, Italy. Dr (Roma U., 1956) chemistry. *E-mail mzocchi@icil64.cilea.it Tel. 39(30)3715573 Fax 39(30)3702448

Zosi, Prof. Gianfranco Luigi (1940). Associate professor. Istituto Fisica Generale "A. Avogadro", Università di Torino, via P. Giuria 1, I-10125 Torino, Italy. Dr (Torino U., 1962) physics. *X-ray_diffraction Avogadro_constant X-ray_interferometry*
E-mail zosi@toxd37.to.infn.it Tel. 39(11)6707426 Fax 39(11)6699579 Telex INFN 211041

JAPAN

Sub-Editors: H. Hashizume and K. Ogawa

Abe, Dr Toshiya (1959). Research associate. Dept. of Mineralogical Sciences and Geology, Yamaguchi University, 1677-1 Yoshida, Yamaguchi 753, Japan. PhD (Tohoku U., 1988) science. *Crystal_growth mineralogy*
Tel. 81(839)226111 Fax 81(839)322041

Achiwa, Prof. Norio (1940). Professor. Department of Physics, Kyushu University, Fukuoka 812, Japan. DSc (Kyoto U., 1969) science. *Four-dimensional_crystallography X-ray_diffraction neutron_scattering incommensurate magnetism*
Tel. 81(92)6411101x4176 Fax 81(92)6334525

Ajimura, Mr Shoji (1956). Fine Materials Section, Advanced Technology R&D Center, Fujikura Ltd, 1-5-1 Kiba, Koto-ku, Tokyo 135, Japan. MEng (Hokkaido U., 1982) *Crystal_growth non-linear_property barium_compounds*
Tel. 81(3)56061068 Fax 81(3)56061516

Akai, Dr Shin-ichi (1938). Director. Semiconductor Division, Itami Research Laboratories, Sumitomo Electric Industries, Ltd, 1-1-1 Koya-kita, Itami, Hyogo 664, Japan. DEng (Kyoto U., 1976) crystal growth. *Growth semiconductors*
Tel. 81(727)721404 Fax 81(727)706727

Akamatsu, Dr Tadashi (1960). Associate professor. Faculty of Education, Kochi University, 2-5-1 Akebono-cho, Kochi 780, Japan. DSc (Nagoya U., 1989) earth sciences. *Computer physical_property solid_solution ultra_high_pressure molecular_dynamics cation_distribution phase_diagram thermodynamics*
E-mail akamatsu@cc.kochi-u.ac.jp Tel. 81(888)440111x8419 Fax 81(888)448453

Akao, Dr Masaru (1945). Associate professor. Institute for Medical and Dental Engineering, Tokyo Medical and Dental University, 2-3-10 Surugadai, Kanda, Chiyoda-ku, Tokyo 101, Japan. DEng (Tokyo Inst. of Techn., 1975) chemical eng.. *Biomaterial bioactive_ceramics bone_mineralization calcium_compounds structure structural_change*
Tel. 81(3)52808018 Fax 81(3)52808005

Akasaki, Prof. Isamu (1929). Professor. Dept. of Electrical and Electronic Engineering, Faculty of Science and Technology, Meijo Univ., I-501 Shiogamaguchi, Tempaku-ku, Nagoya 468, Japan. DEng (Nagoya U., 1964) semiconductor electronics. *Crystal_growth semiconductors electronic_materials*
Tel. 81(52)8321151 Fax 81(52)8321244

Akimoto, Dr Toshio (1941). Manager. Fuji Gotemba Laboratories, Chugai Pharmaceuticals Co. Ltd, Komakado, 1-135 Gontenba-shi, Shizuoka 412, Japan. PhD (U. Tokyo, 1970) pharmaceutical science. *Pharmaceutical single_crystal computer-assisted_design computer_modelling crystallization*

Tel. 81(550)873411x228 Fax 81(550)871960

Akiya, Dr Takaji (1947). Senior researcher. Dept. of Chemical Systems, National Institute of Materials and Chemical Research, 1-1 Higashi, Tsukuba, Ibaraki 305, Japan. DEng (Waseda U., 1988) crystallization. *Crystal_growth Nucleation*

Tel. 81(298)544661 Fax 81(298)544660

Amano, Hiroki (1968). Student. Department of Chemistry, Faculty of Science, Tokyo Institute of Technology, Ookayama, Meguro-ku, Tokyo 152, Japan. MSc (Tokyo Institute of Technology, 1994) chemistry. *Chirality reactivity time-resolved_effect cobalt_compounds crystalline_state_reactions time-resolved_structure_analysis*

E-mail hamano@chem.titech.ac.jp Tel. 81(3)57342608 Fax 81(3)37206206

Amemiya, Dr Yoshiyuki (1952). Associate professor. Member of the Commission on Synchrotron Radiation; Editorial Board of Journal of Synchrotron Radiation. Photon Factory, National Laboratory for High Energy Physics, 1-1 Oho, Tsukuba, Ibaraki 305, Japan. PhD (The University of Tokyo, 1979) physics. *Detector synchrotron_radiation diffraction small-angle_scattering polarization*

E-mail amemiya@kekvax.kek.go Tel. 81(298)645642 Fax 81(298)642801

Ando, Prof. Yoshinori (1942). Professor. Dept. of Physics, Meijo Univ., 1-501 Shiogamaguchi, Tenpaku-ku, Nagoya 468, Japan. DEng (Nagoya U., 1970) X-ray diffraction. *Crystal_growth electron_microscopy carbides sintering carbon_nanotubes fullerenes ultra_fine_particles*

Tel. 81(52)8321151x5280 Fax 81(52)8321170

Aoki, Prof. Katsuyuki (1945). Professor. Department of Materials Science, Toyohashi Univ. of Technology, Tempaku-cho, Toyohashi 441, Japan. Dr of Pharmacy (U. Tokyo, 1978) chemistry. *Biological_structure-activity_relationship biological_molecular_complexes bioinorganic_chemistry molecular_recognition antibiotics drug nucleotides vitamins peptides amino_acids*

Tel. 81(532)470111x442 Fax 81(532)485833 Telex 4322201 JPNTUT

Aoki, Prof. Dr Yoshikazu (1939). Professor. Department of Earth and Planetary Sciences, Faculty of Science, Kyushu Univ., Hakozaki, Higashi-ku, Fukuoka 812, Japan. DSc (Kyushu U., 1977) crystal growth. *Mineralogy crystal_growth*

Tel. 81(92)6411101x4311 Fax 81(92)6322736

Aoki, Dr Yoshihira (1940). Corporate R&D Center, Mitsui Mining & Smelting Co., Ltd, 1333-2 Haraichi, Ageo 362, Japan. DSc (Tohoku U., 1968) magnetism. *Crystal_growth magnetism metallurgy solidification nucleation_and_growth high_magnetic_field*

Tel. 81(48)7753211 Fax 81(48)7753210

Arima, Dr Yoshiyasu (1960). Research associate. Dept. of Physics, Gakushuin Univ., 1-5-1 Mejiro, Toshima-ku, Tokyo 171, Japan. DSc (Gakushuin U., 1988) crystal growth. *Crystal_growth*

E-mail arima@gakushuin.ac.jp Tel. 81(3)39860221x6480 Fax 81(3)59921029

Arimitsu, Dr Yutaka (1955). Associate professor. Dept. of Mechanical Eng., Ehime Univ., 3 Bunkyocho, Matsuyama 790, Japan. DEng (Osaka U., 1989) mechanical eng.. *Crystal_growth*

Tel. 81(899)247111 Fax 81(899)230672

Asada, Dr Eiichi (1924). Lecturer. [6-13 Miyamae, 1-chome, Suginami-ku, Tokyo 168, Japan.] DEng (U. Tokyo, 1979) industrial analytical chemistry. *Qualitative_identification quantitative_identification X-ray_reflectivity X-ray_fluorescence chemical_shift_of_X-ray_spectra*

Tel. 81(3)33345489 Fax 81(3)33355360

Asami, Harumi (1965). Research scientist. Analytical Sciences Lab., Mitsubishi Chemical Corp., 1000 Kamoshida, Aobaku, Yokohama 227, Japan. MEng. (Nagoya U., 1990) physics. *Powder materials Rietveld_method thin_film interface*

E-mail maruha2@atlas.rc.m-kasei.co.jp Tel. 81(45)9633158 Fax 81(45)9634206

Asaoka, Mr Hidehito (1965). Research scientist. Material Innovation Laboratory, Japan Atomic Energy Research Institute, Tokai-mura, Naka-gun, Ibaraki 319-11, Japan. MSc (U. Tokyo, 1991) mineralogy. *Crystal_growth single_crystal*

Tel. 81(292)826081 Fax 81(292)825460

Ashida, Mr Atsushi (1961). Research associate. Dept. of Applied Materials Science, College of Engineering, Univ. of Osaka Prefecture, 1-1 Gakuen-cho, Sakai, Osaka 593, Japan. MEng (U. Osaka Prefecture, 1987) engineering (electricity). *Chalcopyrites semiconductors solar_cell crystal_growth photoluminescence*

Tel. 81(722)521161x2346 Fax 81(722)593340

Ashida, Dr Sakichi (1935). Director Senior Engineer. Nihon-Dempakogyo Co., Ltd, 1275-2 Kamihirose, Sayama, Saitama, Japan. DSc (Tohoku U., 1965) crystal phase transition. *Crystal_growth Czochralski_technique phase_transition ferroic twin niobium_compounds oxides*

Tel. 81(429)527211x1403 Fax 81(429)543968

Ashida, Prof. Tamaichi (1933). Professor. Co-Editor of Acta Crystallographica. Department of Biotechnology, School of Engineering, Nagoya University, Nagoya 464-01, Japan. DSc (Osaka U., 1964) chemistry. *Computing structure_determination proteins peptides*

E-mail a40206a@nucc.cc.nagoya-u.ac.jp Tel. 81(52)7815111 Fax 81(52)7831574 Telex 4477355 ENUNAG J

Chikaura, Dr Yoshinori (1946). Professor. Fac. of Eng., Dept. of Phys., Kyushu Institute of Technology, Sensui-cho, Tobata-ku, Kitakyushu 804, Japan. DEng (Tokyo Institute of Technology, 1973) applied crystallography. *X-ray_topography scattering_topography synchrotron_radiation crystal_growth X-ray_scattering_topography*

E-mail chikaura@hakobera.isct.kyuteh.ac.jp Tel. 81(93)8711931 Fax 81(93)8832231

Doi, Mitsunobu (1957). Assistant professor. Lab. Physical Chemistry, Osaka University of Pharmaceutical Sciences, 2-10-65 Kawai, Matsubara, Osaka 580, Japan. PhD (Osaka U., 1988) X-ray analysis. *Peptides proteins DNA ionophores antibiotics peptide_synthesis*

E-mail a61020g@center.osaka-u.ac.jp Tel. 81(723)321015 Fax 81(723)329929

Endoh, Dr Hisamitsu (1944). Associate Professor. Dept. of Electronics and Information Science, Kyoto Institute of Technology, Matsugasaki, Sakyo-ku, Kyoto 606, Japan. DEng (Osaka U., 1985) applied physics. *Computing metals electron_microscopy diffraction_theory image_processing electron_holography*

E-mail endoh@dj.kit.ac.jp Tel. 81(75)7247441 Fax 81(75)7247400

Esaka, Dr Hisao (1953). Senior researcher. Steelmaking Process, Process Technology Research Laboratories, 20-1 Shintomi, Futtsu, Chiba 299-12, Japan. PhD (Ecole Polytechnique Federal de Lausanne, France, 1986) materials science. *Crystal_growth*

Tel. 81(439)802160 Fax 81(439)802742

Eto, Mr Tetsujiro (1968). Master course student. Department of Physics, Kyushu University, Fukuoka 812, Japan. BEn (Kyushu U., 1992) physics. *X-ray_high-resolution_diffractometry crystal_characterization X-ray_anomalous_dispersion phase_determination structure_determination*

Tel. 81(92)6411101x4184 Fax 81(92)6334525

Fujii, Mr Mitsuhiro (1943). Associate professor. General Education, Nagasaki Institute of Applied Science, 536 Abamachi, Nagasaki 851-01, Japan. B (Nagasaki U., 1965) *Crystal_growth*

Tel. 81(958)393111 Fax 81(958)390584

Fujii, Dr Satoshi (1946). Associate professor. Faculty of Pharmaceutical Sciences, Osaka University, 1-6 Yamada Oka, Suita, Osaka 565, Japan. PhD (Osaka U., 1978) pharmacy. *Biomolecule nucleic_acids proteins structure-activity_relationship drug_design conformational_change*

E-mail fujii@pxews1.protein.osaka-u.ac.jp Tel. 81(6)8775111x6212 Fax 81(6)8774489

Fujii, Mr Takashi (1955). Senior research engineer. Development Group II, R&D Division, Murata Mfg. Co., Ltd, 2-26-10 Tenjin, Nagaokakyo, Kyoto 617, Japan. MEng (Nagoya U., 1982) *Crystal_growth*

Tel. 81(75)9519111 Fax 81(75)9566259

Fujii, Mr Tomomi (1967). Research associate. Institute for Chemical Research, Kyoto University, Uji, Kyoto-fu 611, Japan. MSc (Tokyo Inst. of Technology, 1992) life science. *Biocrystallography proteins*

E-mail fujii@icrccc.kuicr.kyoto-u.ac.jp Tel. 81(744)323111x2160 Fax 81(744)331247

Fujimoto, Mr Hiroshi (1943). Lecturer. Dept. of Applied Electronics, Daido Institute of Technology, 2-21 Daido-cho, Minami-ku, Nagoya 457, Japan. MEng (Nagoya Institute of Technology, 1972) *Crystal_growth*

Tel. 81(52)6110513 Fax 81(52)6125653

Fujimura, Dr Norifumi (1960). Assistant professor. Dept. of Applied Materials Science, College of Engineering, Univ. of Osaka Prefecture, 1-1 Gakuen-cho, Sakai, Osaka 593, Japan. DEng (U. Osaka Prefecture, 1993) thin film. *Ferroelectricity ordering stress thin_film*

Tel. 81(722)521161x2346 Fax 81(722)593340

Fujishita, Mr Hideshi (1951). Associate professor. College of Liberal Arts, Kanazawa Univ., Kakuma-machi, Kanazawa-shi 920-11, Japan. DSc (Hokkaido U., 1980) physics. *X-ray_diffraction neutron_scattering dielectrics superconductors conductors*

E-mail fujishit@icews1.ipc.kanazawa-u.ac.jp Tel. 81(762)645796 Fax 81(762)645987

Fujita, Mr Keiichiro (1947). General manager. Semiconductor Materials R&D Dept., Itami Laboratories, Sumitomo Electric Industries Ltd, 1-1-1 Keya-kita, Itami, Hyogo 664, Japan. MEng (Tohoku U., 1971) material science. *Crystal_growth*

Tel. 81(727)724581 Fax 81(727)722440

Fujita, Prof. Shigeo (1941). Professor. Dept. of Electrical Engineering, Faculty of Engineering, Kyoto Univ., Yoshida-Honmachi, Sakyo-ku, Kyoto 606-01, Japan. DEng (Kyoto U., 1975) optoelectronic materials. *CVD MBE optoelectronics semiconductors low-dimensional quantum organic*

Tel. 81(75)7535363 Fax 81(75)7535363

Fujiwara, Prof. Takaji (1936). Professor. Faculty of Science, Shimane University, 1060 Nishi-kawatsu, Matsue 690, Japan. DSc (Osaka U., 1971) chemistry. *Structure_determination inclusion hydrogen_bonding structural_disorder computer_sciences clathrates cyclodextrins drug dyes computer_graphics information_science*

E-mail fuji@cis.shimane-u.ac.jp Tel. 81(852)326475 Fax 81(852)326489

Fujiwara, Miss Yohko (1964). Research assistant. Faculty of Science, Gakushuin Univ., 1-5-1 Mejiro, Toshima-ku, Tokyo 171, Japan. MSc (Gakushuin U., 1990) chemistry. *Area_detector grazing_incidence surface liquids fractal SAXS carbon_compounds clusters porous_materials water Fourier_transform*

E-mail yohko.fujiwara@gakushuin.ac.jp Tel. 81(3)39860221x6421 Fax 81(3)59921029

Fukamachi, Prof. Tomoe (1943). Professor. Dept. of Electronic Enginnering, Saitama Institute of Technology, 1690 Fusaiji, Okabe, Saitama 369-02, Japan. DSc (U. Tokyo, 1976) physics. *Dynamical_diffraction anomalous_dispersion Compton_scattering instrumentation energy-dispersive_diffraction synchrotron_radiation X-ray_optics*

Tel. 81(485)852521x2411 Fax 81(485)857030

Fukami, Dr Takeshi (1943). Associate professor. Dept. of Physics, Kyushu Univ., 6-10-1 Hakozaki, Higashi-ku, Fukuoka 812, Japan. PhD (Kyushu U., 1971) solid state physics. *Superconductivity thin_film ultrasonics semimetals*

Tel. 81(92)6411101x4192 Fax 81(92)6334525

Fuke, Dr Shunro (1943). Professor. Dept. of Electronics, Faculty of Engineering, Shizuoka Univ., 3-5-1 Johoku, Hamamatsu 432, Japan. DEng (U. Tokyo, 1971) crystal growth. *CVD II-VI_compounds III-V_compounds*

Tel. 81(53)4711171 Fax 81(53)4748845

Fukuda, Prof. Akeharu (1937). Professor. Dept. of Materials Science and Engineering, Muroran Institute of Technology, 27-1 Mizumoto-cho, Muroran, Hokkaido 050, Japan. DEng (Hokkaido U., 1975) mechanical properties. *Dislocation growth ice*
Tel. 81(143)444181 Fax 81(143)473371

Fukuhara, Dr Akira (1933). Senior Consulting Scientist. Advanced Research Laboratory, Hitachi, Ltd, Hatoyama, Saitama 350-03, Japan. PhD (U. Tokyo, 1961) physics. *Dynamical_diffraction multiple_diffraction strain_determination EXAFS*
Tel. 81(492)966111 Fax 81(492)965999

Fukui, Prof. Takashi (1950). Professor. Research Center for Interface Quantum Electronics, Hokkaido Univ., North 13, West 8, Kita-ku, Sapporo 060, Japan. DEng (Hokkaido U., 1983) crystal growth. *Epitaxy semiconductor*
Tel. 81(11)7066870 Fax 81(11)7166004

Fukuyama, Dr Keiichi (1949). Associate professor. Department of Biology, Faculty of Science, Osaka University, 1-1 Machikaneyama-cho, Toyonaka, Osaka 560, Japan. DSc (Osaka U., 1979) macromolecular science. *Proteins electron_transfer_proteins metalloenzymes membrane_protein_complexes*
E-mail fukuyama@pxews1.protein.osaka-u.ac.jp Tel. 81(6)8505424 Fax 81(6)8505425

Furuhata, Mr Yoshio (1929). Manager. No. 4 Research Laboratory, Space Technology Corporation, Japan. [Hitachi Central Research Laboratory, 1-280 Higashi-koigakubo, Kokubunji, Tokyo, Japan.] MS (Waseda U., 1955) *Ferroelectricity optoelectronics crystal_growth dielectrics space_material_processing*
Tel. 81(423)231111

Furukawa, Dr Yoshinori (1951). Associate professor. Institute of Low Temperature Science, Hokkaido Univ., North 19, West 8, Kita-ku, Sapporo 060, Japan. DSc (Hokkaido U., 1981) geophysics. *Ice microgravity surface morphology*
E-mail furukawa@lt.hines.hokudai.ac.jp Tel. 81(11)7065467 Fax 81(11)7067142

Fusegawa, Mr Izumi (1959). Isobe R&D Center, Shin-Etsu Handotai Co., Ltd, 2-13-1 Isobe, Annaka-shi, Gunma 379-01, Japan. (Gunma U., 1982) *Crystal_growth*
Tel. 81(273)852511 Fax 81(273)852774

Fushino, Mr Tetsuo (1949). Research Dept., Taki Chemical Co, Ltd, 2 Midorimachi, Befu-cho, Kakogawa-shi, Hyogo 675-01, Japan. MS (Chuo U., 1974) inorganic chemistry. *Crystal_growth*
Tel. 81(794)372111 Fax 81(794)379138

Gonda, Prof. Dr Takehiko (1937). Professor. Division of General Education, Aichi-Gakuin Univ., Iwasaki, Nisshin-cho, Aichi 470-01, Japan. DSc (Nagoya U., 1977) crystal growth. *Crystal_growth cloud_physics ice micromorphology ice_physics*
Tel. 81(5617)31111 Fax 81(5617)31860

Goto, Ms Midori (1934). Senior Researcher. Department of Analytical Chemistry, National Institute for Materials and Chemical Research, Tsukuba, Ibaraki 305, Japan. Cand (Tokyo U., 1958) chemistry. *Phase_transition structure_determination fats*
E-mail ck649@nimc.go.jp Tel. 81(298)544627 Fax 81(298)551397

Goto, Dr Yoshiaki (1940). Associate professor. Dept. of Materials Chemistry, Ryukoku Univ., Seta, Otsu 520-21, Japan. DSc (Hokkaido U., 1976) mineral chemistry. *Dynamical_diffraction surface thin_film single_crystal_characterization*
Tel. 81(775)437469 Fax 81(775)437483

Gotoh, Mr Hirohito (1949). General Manager. Development Dev., Matels Inc., 1147-13 Nagakuni, Tsutiura, Ibaraki 300, Japan. (Science Univ. of Tokyo, 1971) *Crystal_growth*
Tel. 81(298)236705 Fax 81(298)236705

Habuka, Hitoshi (1957). Senior researcher. Isobe R&D Center, Shin-Etsu Handotai Co., Ltd, 2-13-1 Isobe, Annaka-shi, Gunma 379-01, Japan. MSc (Kyoto U., 1981) polymer science. *Semiconductor epitaxy CVD simulation*
Tel. 81(273)852575 Fax 81(273)852774

Haga, Dr Yumiko (1966). Research associate. Dept. Metall. Eng., Tokyo Inst. Tech., Oh-okayama, Meguro, Tokyo 152, Japan. DEng (Tokyo Inst. Tech., 1994) metallurgy. *Multilayer characterizarion thin_film polycrystal*
Tel. 81(3)37261111x3145 Fax 81(3)37263419

Hagiya, Dr Kenji (1963). Research Associate. Department of Life Science, Faculty of Science, Himeji Institute of Technology, Kanaji 1479-1, Kamigori, Akogun, Hyogo 678-12, Japan. DEng (U. of Tsukuba, 1990) materials science. *X-ray_diffraction inorganic_materials modulated_structures phase_transition*
E-mail hagiya@sci.himeji-tech.ac.jp Tel. 81(7915)80216 Fax 81(7915)80216

Hamada, Dr Kensaku (1951). Associate professor. Faculty of Science, Shimane University, Nishikawatsu, Matsue 690, Japan. DPharm (Osaka U., 1983) pharmaceutical science. *Biocrystallography proteins biomolecule computer-assisted_design enzymes computer_sciences synchrotron_radiation*
E-mail hamadak@cis.shimane-u.ac.jp Tel. 81(852)326480 Fax 81(852)326489

Harada, Prof. Jimpei (1931). Professor. IUCr Executive Committee Member; Co-Editor of Acta Crystallographica; Member of National Committee. Department of Applied Physics, Nagoya University, Furo-cho, Chikusa-ku, Nagoya 464-01, Japan. DSc (Tokyo Institute of Technology, 1964) physics. *Phase_transition lattice_dynamics order-disorder surface interface thin_film crystal_growth fine_particles diffuse_scattering grazing_incidence X-ray_reflectivity instrumentation powder_diffraction synchrotron_radiation*
E-mail a40361a@nucc.cc.nagoya-u.ac.jp Tel. 81(52)7815111x4464 Fax 81(52)7822129 Telex 4477355ENUNAGJ

Harada, Dr Shigeharu (1954). Assistant professor. Department of Applied Chemistry, Faculty of Engineering, Osaka University, Suita, Osaka 565, Japan. DSc (Osaka U., 1982) physical chemistry. *Biocrystallography protein_crystallography_with_synchrotron_radiation*
E-mail a63486a@center.osaka-u.ac.jp Tel. 81(6)8775111x4322 Fax 81(6)8776994

Hasebe, Dr Masami (1961). Senior researcher. Semiconductor Materials Laboratory, Electronics Research Laboratories, Nippon Steel Corp., 3434 Shimata, Hikari, Yamaguchi 743, Japan. PhD (State Univ. of New York, 1992) crystal characterization. *Czochralski_technique transmission_electron_microscopy defect*
E-mail hasebe@sml.erl.nsc.co.jp Tel. 81(833)725855 Fax 81(833)723098

Hasebe, Prof. (1942). Professor. Department of Physics, Faculty of Liberal Arts, Yamaguchi Univ., Yoshida, Yamaguchi 753, Japan. DSc (Kyoto U., 1982) physics. *Diffraction solid_state phase_transition dielectrics*
Tel. 81(839)226111x584 Fax 81(839)282710

Hasegawa, Dr Masashi (1961). Assistant professor. Materials Development Division, Institute for Solid State Physics, Univ. of Tokyo, 7-22-1 Roppongi, Minato-ku, Tokyo 106, Japan. DEng (Nagoya U., 1989) materials science. *Crystal_growth*
Tel. 81(3)34786811x5722 Fax 81(3)34015169

Hashimoto, Prof. Dr Mituru (1939). Professor. Dept. of Applied Physics and Chemistry, Univ. of Electro-Communications, 1-5-1 Chofugaoka, Chofu, Tokyo 182, Japan. PhD (The Tokyo Metropolitan U., 1967) physics of ferromagnetism. *Thin_film crystal_growth epitaxy deposition plasma_sputtering magnetic_properties*
E-mail mituru@diamond.crc.uec.ac.jp Tel. 81(424)832161x3841 Fax 81(424)844518

Hashizume, Daisuke (1968). Graduate student. Department of Chemistry, Faculty of Science, Tokyo Institute of Technology, Ookayama, Meguro-ku, Tokyo 152, Japan. MSc (Tokyo Institute of Technology, 1994) chemistry. *Chirality reactivity time-resolved_effect cobalt_compounds crystalline_state_reactions time-resolved_structure_analysis*
E-mail hashi@chem.titech.ac.jp Tel. 81(3)57342608 Fax 81(3)37206206

Hashizume, Prof. Hiroo (1940). Professor. Chairman of Apparatus Commission; Co-editor of Journal of Applied Crystallography; Regional and National Sub-Editor of Ninth Edition of World Directory of Crystallographers. Res. Lab. of Engineering Materials, Tokyo Institute of Technology, Nagatsuta, Midori, Yokohama 226, Japan. DSc (U. Tokyo, 1970) applied physics. *Surface dynamical_diffraction X-ray_optics instrumentation interface thin_film diffraction_technique standing_wave grazing_incidence X-ray_reflectivity X-ray_diffractometry single_crystal_characterization X-ray_topography detector synchrotron_radiation powder structure_determination maximum-entropy_method*
E-mail hhashizu@nc.titech.ac.jp Tel. 81(45)9245333 Fax 81(45)9225169 Telex 3823553 TITNAGJ

Hata, Dr Tadashi (1944). Chief researcher. Analytical and Metabolic Res. Labs., Sankyo Co. Ltd, Hiromachi, Shinagawa, Tokyo 140, Japan. PhD (Tokyo Institute of Technology, 1986) chemistry. *Bioactive_molecules drug_topography structure-activity_relationship synchrotron_radiation structure_determination database computer-assisted_drug_design*
Tel. 81(3)34923131x4121 Fax 81(3)34923543

Hata, Dr Yasuo (1951). Associate prof. Institute for Chemical Research, Kyoto University, Uji, Kyoto 611, Japan. DSc (Osaka U., 1979) macromolecular science. *Biocrystallography proteins structure_function_relationship reaction_mechanism*
E-mail yasuo@icrccc.kuicr.kyoto-u.ac.jp Tel. 81(744)323111x2159 Fax 81(744)331247

Hayakawa, Dr Yasuhiro (1953). Associate professor. Research Institute of Electronics, Shizuoka Univ., 3-5-1 Johoku, Hamamatsu 432, Japan. DEng (U. Tokyo, 1988) crystal growth. *Crystal_growth electronics semiconductor antimony_compounds*
E-mail hayakawa@rie.shizuoka.ac.jp Tel. 81(53)4711171 Fax 81(53)4740630

Hayashi, Dr Koya (1947). Professor. Lab. for Solid State Chemistry, Okayama University of Science, 1-1 Ridai-cho, Okayama 700, Japan. PhD (Tokyo Institute of Technology, 1975) solid state chemistry. *Chalcogenides phase_equilibrium charge-density_wave intercalation exfoliation nano-composites*
Tel. 81(86)2523161x4267 Fax 81(86)2542891

Hayashi, Dr Shigeyuki (1935). Associate professor. Crystal Physics, Institute for Materials Research, 2-1-1 Katahira, Aoba-ku, Sendai 980, Japan. PhD (Hokkaido U., 1974) crystal growth. *Crystal_growth*
Tel. 81(22)2152012 Fax 81(22)2152011

Hibiya, Taketoshi (1945). Research fellow. NEC Corporation, Research and Development Group, 34 Miyukigaoka, Tsukuba 305, Japan (Keio U., 1978) applied chemistry. *Crystal_growth semiconductors magnetic_oxides microgravity epitaxy thermophysical_properties heat_and_mass_transfer*
E-mail hibiya@sci.cl.nec.co.jp Tel. 81(298)501147 Fax 81(298)566136

Higashi, Dr Akira (1922). DSc (Hokkaido U., 1951) physics. *Ice morphology crystal_growth melting international_melt_figure*

Higuchi, Prof. Taiichi (1929). Professor. Fac. of Eng., Kansai University, 3-3-35 Yamate-cho, Suita, Osaka 564, Japan. DSc (Osaka City U., 1967) chemistry. *Enzymes clathrates_cyclodextrins bile_pigments organometallic_compound*
Tel. 81(6)3881121

Higuchi, Yoshiki (1956). Associate Professor. Dept Life Science, Faculty of Science, Himeji Inst. of Techn., Kanaji, 1479-1 Kamigori, Akou-gun, Hyogo 678-12, Japan. PhD (Osaka U., 1984) macromolecular science. *Structure_determination proteins enzymes computing gardening tennis*
E-mail higuchi@sci.himeji-tech.ac.jp Tel. 81(7915)80178 Fax 81(7915)80177

Hikosaka, Prof. Masamichi (1944). Associate professor. Faculty of Engineering, Yamagata Univ., 4-3-16 Jonan, Yonezawa 992, Japan. DSc (Tokyo Metropolitan U., 1987) polymer crystallization. *Polymers crystal_growth kinetics diffusion*
Tel. 81(238)225181x362 Fax 81(238)225362

Hirabayashi, Prof. Makoto (1925). President. Kitami Institute of Technology, Koen-cho 165, Kitami 090, Japan. DSc (Tohoku U., 1956) materials science. *Electron_microscopy incommensurate quasicrystal interstitial_compounds antiphase ordered_structures alloys superconductor hydrogen_compounds*
Tel. 81(157)241010x201 Fax 81(157)227198

Hiraga, Prof. Kenji (1939). Professor. Institute for Materials Research, Tohoku Univ., Katahira 2-1-1, Sendai 980, Japan. PhD (Tohoku U., 1976) metallurgy. *High-resolution_electron_microscopy quasicrystal inorganic_materials*
Tel. 81(22)2152125 Fax 81(22)2152126 Telex 852238 KINKENJ

Hiramatsu, Dr Kazumasa (1952). Associate professor. Dept. of Electronics, School of Engineering, Nagoya Univ., Furo-cho, Chikusa-ku, Nagoya 464-01, Japan. DEng (Nagoya U., 1986) semiconductor. *Semiconductor epitaxy CVD optoelectronics wide_band_gap_semiconductor*
E-mail hiramatu@itakura.nuee.nagoya-u.ac.jp Tel. 81(52)7815111x4431 Fax 81(52)7813359

Hirano, Prof. Dr Shin-ichi (1942). Professor. Dept. of Applied Chem., School of Eng., Nagoya Univ., 1 Furo-cho, Chikusa-ku, Nagoya 464-01, Japan. DEng (Nagoya U., 1970) eng., applied chem. *Crystal_growth chemistry hydrothermal sol-gel film*
Tel. 81(52)7893343 Fax 81(52)7893182

Hirasawa, Prof. Dr Izumi (1954). Associate professor. Applied Chem., Faculty Science & Engineering, Waseda Univ., 3-4-1 Ohkubo, Shinjuku, Tokyo 169, Japan. DEng (Waseda U., 1989) ind. crystallization. *Crystal_surface optical_resolution functional_fine_crystal resources_recovery*
Tel. 81(3)32034141x733326 Fax 81(3)32086896

Hirata, Dr Hiroshi (1951). Senior research engineer. Telecommunication Networks Laboratories, NTT Corp., 1-2356 Take, Yokosuka, Kanagawa 238-03, Japan. DEng (U. Tokyo, 1991) engineering. *Crystal_growth*
E-mail hirata@nttmhs.ntt.jp Tel. 81(468)593526 Fax 81(468)592546

Hirayama, Prof. Noriaki (1948). Professor. Dept. Biological Science and Technology, Tokai Univ., 317 Nishino, Numazu, Shizuoka 410-03, Japan. DSc (Tokyo Inst Technology, 1982) chemistry. *Structure–activity_relationship pharmacology toxicology biochemistry proteins small_molecules macromolecules*
E-mail hirayama@ashitaka.ncc.u-tokai.ac.jp Tel. 81(559)681111x4502 Fax 81(559)681156

Hirotsu, Prof. Ken (1942). Professor. Department of Chemistry, Faculty of Science, Osaka City University, Sugimoto, Sumiyoshi-ku, Osaka 558, Japan. DSc (Osaka City U., 1974) chemistry. *Small_molecules clathrates proteins enzymes*
Tel. 81(6)6052557 Fax 81(6)6052522

Homma, Prof. Teiichi (1931). Professor. Dept. of Precision Engineering, Chiba Institute of Technology, Tsudanuma, Narashino, Chiba 275, Japan. DEng (U. Tokyo, 1964) applied physics. *Thin_film X-ray topography surface_characterization*
Tel. 81(474)780505 Fax 81(474)780529

Honda, Dr Kazumasa (1956). Senior Researcher. Department of Analytical Chemistry, National Institute for Materials and Chemical Research, Tsukuba, Ibaraki 305, Japan. PhD (Tokyo U., 1985) pharmaceutical sciences. *Microcrystallography powder_structure_determination EXAFS XANES*
E-mail honda@nimc.go.jp Tel. 81(298)544625 Fax 81(298)544487

Honma, Mr Shigeru (1936). Chief scientist. 13 Group, National Institute Research in Inorganic Materials, 1-1 Namiki, Tsukuba 305, Japan. (Hokkaido U., 1962) *Crystal_growth X-ray X-ray_topography*
Tel. 81(298)513351 Fax 81(298)527449

Hori, Dr Kayako (1944). Associate Professor. Dept. of Chem., Ochanomizu University, Otsuka, Bunkyo-ku, Tokyo 112, Japan. DSc (Osaka U., 1981) physical chemistry. *Liquid_crystals thermoanalysis crystallography polymorphism crystallization ferroelectricity vibrational_spectroscopy*
E-mail a0053@jpnocha1.bitnet Tel. 81(3)39433151x552 Fax 81(3)39442120

Horiuchi, Dr Hiroyuki (1940). Associate professor. Mineralogical Institute, Faculty of Science, University of Tokyo, Hongo 7-3-1, Bunkyo, Tokyo 113, Japan. DSc (U. Tokyo, 1969) mineralogy. *Structure phase_transition texture instrumentation software_development*
Tel. 81(3)38122111x4542 Fax 81(3)38165714

Horn, Dr Ernst (1952). Senior Researcher. Inorganic Analysis Laboratory, Department of Analytical Chemistry, National Institute of Materials and Chemical Research, Higashi 1-1, Tsukuba, Ibaraki 305, Japan. PhD (U. Adelaide (Australia), 1984) transition metal carbonyl complexes. *Crystal_and_powder_X-ray_diffraction_structure analysis_of_metals_complex organic_compound minerals_and_peptides computer_software_development*
Tel. 81(298)544626 Fax 81(298)544487

Hoshikawa, Prof. Dr Keigo (1942). Professor. Faculty of Education, Shinshu Univ., 6-12 Nishinagano, Nagano 380, Japan. DEng (Tokyo Institute of Technology, 1987) electronics. *Crystal_growth semiconductor*
Tel. 81(262)376128 Fax 81(262)376128

Hoshino, Prof. Sadao (1926). Emeritus professor. [41-5 Kamaya-cho, Hodogaya-ku, Yokohama 240, Japan.] DSc (Osaka U., 1958) physics. *Neutron scattering X-ray diffraction phase_transition ferroelectricity instrumentation superionic conductor*
Tel. 81(45)3314752 Fax 81(45)3314752

Hosomi, Dr Satoru (1932). Managing director. Tomei Diamond Co., Ltd, 4-5-1 Joto, Oyama 323, Japan. DEng (Osaka U., 1991) physical property. *Crystal_growth*
Tel. 81(285)225821 Fax 81(285)225827

Hozawa, Prof. Dr Mitsunori (1940). Professor. Institute for Chemical Reaction Science, Tohoku Univ., 2-1-1 Katahira, Aoba-ku, Sendai 980, Japan. DEng (Tohoku U., 1974) chemical engineering. *Extraction surface melting mass_transfer*
Tel. 81(22)2276200 Fax 81(22)2238956

Ibata, Dr Koichi (1947). Research manager. Basic Research Lab., Graphtec Corp., 503-10 Shinanocho, Totsuka-ku, Yokohama 244, Japan. DSc (Tokyo Institute of Technology, 1975) chemistry. *Electron_microscopy infrared chromatography molecular_mechanics scanning_tunnel_microscopy*
Tel. 81(45)8256282 Fax 81(45)8256393

Ichimiya, Prof. Ayahiko (1940). Professor. Department of Applied Physics, School of Engineering, Nagoya University, Furo-cho, Chikusa-ku, Nagoya 464-01, Japan. DSc (Nagoya U., 1969) physics. *Surface_physics RHEED dynamical_diffraction crystal_growth positron_diffraction multiply_charged_ions*
E-mail a40278a@nucc.cc.nagoya-u.ac.jp Tel. 81(52)7894459 Fax 81(52)7893724 Telex 4477355 ENUNAG-J

Ichinose, Prof. Noboru (1935). Professor. School of Sci. & Eng., Waseda Univ., 3-4-1 Ohkubo, Shinjuku-ku, Tokyo 169, Japan. PhD (Waseda U., 1967) physics. *Magnetism ferroelectricity superconductivity piezoelectricity*
Tel. 81(3)32034141 Fax 81(3)32002567

Ido, Prof. Toshiyuki (1943). Professor. Dept. of Electrical Eng., College of Eng., Chubu Univ., 1200 Matsumoto-cho, Kasugai, Aichi 487, Japan. DEng (Nagoya U., 1973) electronic eng. *Crystal_growth*
Tel. 81(568)511111 Fax 81(568)520134

Iijima, Dr Kenji (1953). Senior researcher. Central Research Laboratories, Matsushita Electric Industrial Co., Ltd, Seikacho, Kyoto 619-02, Japan. DSc (Tokyo Institute of Technology, 1982) X-ray crystallography. *Thin_film ferroelectricity diffractometry*
Tel. 81(7749)82515 Fax 81(7749)82583

Iijima, Dr Kinya (1941). Associate professor. Department of Chemistry, Shizuoka Univ., 836 Oya, Shizuoka 422, Japan. DSc (Hokkaido U., 1976) chemistry. *Crystal_structure X-ray_diffraction amino_acids molecular_structure molecular_internal_rotation gaseous_electron_diffraction*
E-mail sckiiji@sci.shizuoka.ac.jp Tel. 81(54)2371111x5602 Fax 81(54)2373384

Iijima, Dr Sumio (1939). Research fellow. R & D group, NEC Corp., 34 Miyukigaoka, Tsukuba 305, Japan. PhD (Tohoku U., 1969) physics. *Electron_microscopy high-resolution_electron_microscopy microcrystallography crystal_growth surface microscopy semiconductor*
E-mail iijima@tgn.cl.nec.co.jp Tel. 81(298)501117 Fax 81(298)566136

Iijima, Prof. Takao (1934). Professor. Department of Chemistry, Gakushuin University, Toshimaku, Tokyo 171, Japan. DSc (The University of Tokyo, 1962) structural chemistry. *Electron_diffraction X-ray_diffraction gases liquids*
Tel. 81(3)39860221x6424 Fax 81(3)59921029

Iishi, Prof. Dr Kazuaki (1942). Professor. Dept. of Mineralogical Science and Geology, Faculty of Science, Yamaguchi Univ., 1677-1 Yoshida, Yamaguchi 753, Japan. DSc (Hiroshima U., 1973) mineralogy. *Float_zone_growth flux electron_microscopy*
Tel. 81(839)2261111 Fax 81(839)322041

Iitaka, Prof. Yoichi (1927). Professor. Dept. of Biological Sciences, Fac. of Science and Engineering, Nishi-Tokyo Univ., Uenohara, Kitatsuru, Yamanashi 409-01, Japan. DSc (U. Tokyo, 1959) crystallography. *Biocrystallography structure–activity_relationship synchrotron_radiation computer instrumentation detector interatomic_interaction molecular_mechanics molecular_dynamics*
Tel. 81(554)634411x2524 Fax 81(554)634431

Ikemiya, Mr Norihito (1965). Research associate. Department of Materials Science and Processing, Osaka University, 2-1 Yamada-oka, Suita 505, Japan. MEng (Osaka U., 1989) *Crystal_growth atomic_force_microscopy*
Tel. 81(6)8775111 Fax 81(6)8764729

Imashimizu, Dr Yuji (1942). Lecturer. Department of Materials Engineering, Mining College, Akita Univ., 1-1 Tegata Gakuen-cho, Akita 010, Japan. DEng (Nagoya U., 1993) crystal growth. *Crystal_growth imperfection*
Tel. 81(188)335261 Fax 81(188)370403

Inaoka, Dr Kimio (1940). Faculty of Electro-Mechanical Engineering, Yuge Mercantile Marine College, Yuge-cho, Ochi-gun, Ehime 794-25, Japan. DEng (Hiroshima U., 1985) thin film. *Crystal_growth*
Tel. 81(897)773000 Fax 81(897)772308

Ino, Prof. Shozo (1936). Dept. Phys., Faculty of Science, Univ. of Tokyo, 7-3-1 Hongo, Bunkyo-ku, Tokyo 113, Japan. DSc (Tohoku U., 1968) physics. *Crystal_growth*
Tel. 81(3)38122111x4208 Fax 81(3)38149717

Inoue, Dr Morio (1937). Councillor. Kyoto Res. Lab. of Matsushita Electronics Corp., 19 Nishikujo-Kasugacho, Minami-ku, Kyoto 601, Japan. DSc (Kyoto U., 1947) chemistry of crystals. *Semiconductor_crystal crystal_growth thin_film crystal_characterization ULSI_process_technology semiconductor_chemistry defect_engineering*
Tel. 81(75)6813181 Fax 81(75)6810705

Inoue, Dr Tsuyoshi (1966). Dept. of Applied Chemistry, Faculty of Engineering, Osaka Univ., Yamadaoka 2-1, Suita, Osaka 565, Japan. PhD (Osaka U., 1994) engineering. *Biological_copper_compounds X-ray_crystallography bioinorganic_chemistry metalloproteins biophysics electron_transportation blue_copper_proteins biocrystallography*
Tel. 81(6)8775111x4322 Fax 81(6)8776994

Ioku, Dr Koji (1960). Associate Professor. Faculty of Engineering, Yamaguchi University, 2557 Tokiwadai, Ube-shi, Yamaguchi 755, Japan. DEng (Tokyo Institute of Technology, 1989) material science. *Biomaterial synthesis sintering apatite hydrothermal*
Tel. 81(836)359963 Fax 81(836)359965

Ishibashi, Dr Hiroyuki (1958). Researcher. Tsukuba Research Laboratory, Hitachi Chemical Co., Ltd, 48 Wadai, Tsukuba, Ibaraki 300-42, Japan. DSc (Rikkyo U., 1990) crystal growth. *Oxides scintillator*
E-mail tsu00141@jpnjrdc.bitnet Tel. 81(298)644000 Fax 81(298)644008

Ishibashi, Mr Hisato (1957). Development Section Saga Plant, Saga Electronics Co., Ltd, 950 Tateno, Mitagawa-machi, Kanzaki-gun, Saga 842, Japan. BSc (Kumamoto U., 1971) *Crystal_growth*
Tel. 81(952)523181 Fax 81(952)523185

Ishibashi, Mr Tsutomu (1954). Komukai Works, Toshiba Corp., 1 Komukai Toshiba-cho, Saiwai-ku, Kawasaki-city 210, Japan. MEng (Science University of Tokyo, 1978) *Crystal_growth*
Tel. 81(44)5485020 Fax 81(44)5411125

Ishibashi, Prof. Yoshihiro (1935). Professor. Synthetic Crystal Res. Lab., School of Engineering, Nagoya Univ., Chikusa-ku, Nagoya 464-01, Japan. DSc (U. Tokyo, 1963) physics. *Ferroelectricity ferroelasticity light_scattering lattice_dynamics*
E-mail a40102a@nucc.cc.nagoya-u.ac.jp Tel. 81(52)7815111x3597 Fax 81(52)7829209

Ishida, Prof. Ishida (1946). Professor. Lab. Physical Chemistry, Osaka University of Pharmaceutical Sciences, 2-10-65 Kawai, Matsubara, Osaka 580, Japan. PhD (Osaka U., 1979) pharmacy. *Peptides proteins DNA biomolecule structure_determination recognition molecular_recognition*
E-mail a61020g@center.osaka-u.ac.jp. Tel. 81(723)321015 Fax 81(723)329929

Ishihara, Mr Makoto (1964). Ashigara Research Lab., Fuji Photo Film Co. Ltd, 210 Nakanuma, Minami-Ashigara, Kanagawa 250-01, Japan. MS (Keio U., 1988) chemistry. *Electron_density_distribution metalloorganic_complex photochemistry*
Tel. 81(465)737080 Fax 81(465)737923

Ishii, Prof. Makoto (1937). Professor. Kanazawa Institute of Technology, 7-1 Ohgigaoka, Nonoichi, PO Kanazawa-South, Ishikawa 921, Japan. DEng (Osaka U., 1980) crystal growth. *Crystal_growth*
Tel. 81(762)946706 Fax 81(762)946707

Ishii, Prof. Mitsuru (1930). Professor. Dept. of Materials Science and Ceramic Technology, Shonan Institute of Technology, 1-1-25 Tsujido-Nishikaigan, Fujisawa 251, Japan. DEng (Tokyo Institute of Technology, 1971) metal working. *Bridgman_Stockbarger_technique oxides fluorine_compounds radiation_detectors scintillation_materials*
Tel. 81(466)344111 Fax 81(466)361594

Ishikawa, Dr Tetsuya (1954). Associate professor. Dept. of Applied Physics, Univ. of Tokyo, 7-3-1 Hongo, Bunkyo-ku, Tokyo 113, Japan. DEng (U. Tokyo, 1982) applied physics. *Crystal_growth*
E-mail pxolab1@tansei.cc.u-tokyo.ac.jp Tel. 81(3)38122111x6826 Fax 81(3)56898258

Ishitani, Dr Akihiko (1947). Senior manager. ULSI Device Labs., NEC Corp., 1120 Shimokuzawa, Sagamihara 229, Japan. DEng (U. Tokyo, 1989) electronic engineering. *Crystal_growth*

Ishizaki, Mr Junya (1969). Research Center for Interface Quantum Electronics, Hokkaido Univ., North 13, West 8, Kita-ku, Sapporo 060, Japan. (Hokkaido U., 1992) crystal growth. *Crystal_growth*
Tel. 81(11)7162111 Fax 81(11)7079750

Ishizawa, Dr Nobuo (1949). Associate Professor. Research Laboratory of Engineering Materials, Tokyo Institute of Technology, Nagatsuta, Midori, Yokohama 226, Japan. DSc (Tokyo Institute of Technology, 1979) material science. *Structure_determination crystallography EXAFS charge_density amorphization anomalous_dispersion anvil_cell bonding crystal_field crystallization diffraction_data diffuse_scattering high_precision_structures high_pressure high_temperature Jahn_Teller_effect lattice_vibration lattice_dynamics LCAO_method microcrystal structural_disorder superstructure synchrotron_radiation chalcogenides oxides*
E-mail nishizaw@nc.titech.ac.jp Tel. 81(45)9221111x2312 Fax 81(45)9221015

Ishizuka, Dr Kazuo (1947). Group leader. Tonomura Electron Wavefront Project, JRDC, c/o Toyo University, Kawagoe 350, Japan. DSc (Kyoto U., 1978) electron microscopy. *High-resolution_electron_microscopy image_processing electron_holography multislice_method dynamical_electron_diffraction EELS electron_crystallography*
Tel. 81(492)342691 Fax 81(492)342697

Isoda, Prof. Dr Seiji (1947). Associate professor. Institute for Chemical Research, Kyoto University, Gokasho, Uji, Kyoto 611, Japan. DSc (Kyoto U., 1983) polymer crystal. *Crystal_growth*
Tel. 81(774)335725 Fax 81(774)337096

Isogami, Dr Mineo (1948). Department manager. R&D Center for Devices, Dept. Functional Materials, Kyocera Corp., 11-17 Koga Hon-machi, Fushimi-ku, Kyoto 612, Japan. DSc (Tohoku U., 1976) mineralogy. *Crystal_growth optoelectrical_property optoelectronics*
Tel. 81(75)9335121 Fax 81(75)9337593

Ito, Prof. Taichiro (1940). Professor. Department of Applied Materials Science, College of Engineering, Univ. of Osaka Prefecture, 1-1 Gakuen-cho, Sakai, Osaka 593, Japan. DEng (U. Osaka Prefecture, 1973) metal physics. *Crystal_growth thin_film interface surface*
Tel. 81(722)521161 Fax 81(722)593340

Ito, Prof. Tetsuzo (1936). Prof. Dept. of Chemical Technology, Kanagawa Inst. of Techn., 1030 Shimoogino Atsugi, Kanagawa 243-02, Japan. DSc (Tokyo U., 1965) chemistry. *Charge_density molecular_crystal*
Tel. 81(462)411211 Fax 81(462)423737

Itoh, Dr Hideaki (1944). Associate professor. School of Engineering, Nagoya Univ., Furocho, Chikusa-ku, Nagoya 464-01, Japan. DEng (Nagoya U., 1972) crystal growth. *CVD high_pressure_synthesis*
Tel. 81(52)7815111x2751 Fax 81(52)7829209

Itoh, Dr Hiroyuki (1961). Research associate. Institute for Advanced Materials Processing, Tohoku Univ., 2-1-1 Katahira, Aoba-ku, Sendai 980, Japan. DEng (Waseda U., 1989) *Crystal_growth*
Tel. 81(22)2276200x2820 Fax 81(22)2666097

Itoh, Dr Nobuo (1940). Lecturer. College of Engineering, Univ. of Osaka Prefecture, 1-1 Gakuen-cho, Sakai, Osaka 593, Japan. DEng (U. Osaka Prefecture, 1983) material science. *Characterization diffractometry semiconductor*
Tel. 81(722)521161 Fax 81(722)593340

Iwai, Dr Kunimoto (1943). Associate professor. Faculty of Education, Shinshu Univ., 6 Nishi-Nagano, Nagano 380, Japan. DSc (Nagoya U., 1986) meteorology. *Ice*
Tel. 81(262)328106 Fax 81(262)325144

Iwaki, Prof. Dr Toshihiro (1942). Professor. Dept. of Mechanical Systems Eng., Faculty of Eng., Toyama Univ., 3190 Gofuku, Toyama 930, Japan. DEng (U. Tokyo, 1989) mechanical eng. *Thermal_stress*
Tel. 81(764)411271 Fax 81(764)418432

Iwami, Prof. Dr Motohiro (1939). Professor, Dean of the faculty of science. Research Laboratory for Surface Science, Faculty of Science, Okayama Univ., 3-1-1 Tsushimanaka, Okayama 700, Japan. DEng (Kyoto U., 1971) semiconductor. *Photoelectron non-destructive_analysis soft_X-ray interface silicide electron_spectroscopy semiconductor_device*
Tel. 81(86)2517897 Fax 81(86)2548467

Iwanaga, Prof. Dr Hiroshi (1938). Professor. Faculty of Liberal Arts, Nagasaki Univ., 1-14 Bunkyo-cho, Nagasaki 852, Japan. DSc (Tokyo Institute of Technology, 1978) crystal growth. *Dislocation tensile_strength*
Tel. 81(958)471111 Fax 81(958)431379

Iwasaki, Prof. Fujiko (1937). Professor. Dept. of Applied Physics and Chemistry, The Univ. of Electro-Communications, 1-5-1 Chofugaoka, Chofu-shi, Tokyo 182, Japan. DSc (U. Tokyo, 1966) chemistry. *Molecular_crystal charge_density phase_transition organic_sulfur_compounds rapid_X-ray_measurement_system unstable_crystal heteroatom_compounds hypervalent_compounds*
E-mail fuji@struct.pc.uec.ac.jp Tel. 81(424)832161x3821 Fax 81(424)844518

Iwasaki, Prof. Hiroshi (1933). Professor. Member of the National Committee for Crystallography Japan. Department of Physics, Ritsumeikan University, Kusatsu, Shiga 525, Japan. DSc (Tohoku U., 1966) physics. *Synchrotron_radiation high_pressure phase_transition*
Tel. 81(775)612719 Fax 81(775)612657

Iwasaki, Dr Hitoshi (1935). Chief Scientist. Inst. for Physical and Chemical Research, Wako-shi, Saitama 351-01, Japan. DSc (U. Tokyo, 1965) chemistry. *X-ray_crystallography symmetry organic_structure synchrotron_radiation nonperiodic_structure*
Tel. 81(48)4621111x3461 Fax 81(48)4624645

Iwata, Dr Yutaka (1937). Associate professor. Research Reactor Institute, Kyoto Univ., Kumatori-cho, Sen-nan gun, Osaka 590-04, Japan. DSc (Kyushu U., 1977) physics. *Neutron_diffraction ferroelectric_material*
Tel. 81(724)520901 Fax 81(724)535810

Izui, Prof. Kazuhiko (1929). Professor. Japan Atomic Energy Research Institute, Migawa, 5-1251-12 Mito, Ibaraki 310, Japan. DSc (Hiroshima U., 1965) physics. *Electron_microscopy electron_diffraction radiation_damage defect structural_change_by_irradiation ion_irradiation*
Tel. 81(292)826360 Fax 81(292)825922

Izumi, Dr Kunihide (1937). Research associate. Department of Physics, Faculty of Science, Kyoto Univ., Kitashirakawa, Sakyo-ku, Kyoto, Japan. DSc (Kyoto U., 1969) science (physics). *Crystal_growth dislocation electron_microscopy topography lattice_defect organic_crystal*
Tel. 81(75)7533754 Fax 81(75)7533791

Izumi, Prof. Yoshinobu (1944). Professor. Macromolecular Res. Lab., Fac. of Engineering, Yamagata University, Jo-nan 4-3-16, Yonezawa 992, Japan. DSc (Hokkaido U., 1972) polymer physics. *Macromolecules phase_transition diffuse_scattering small-angle_scattering synchrotron_radiation neutron_scattering*
Tel. 81(238)225181x246 Fax 81(238)225362

Jang, Dr Wen-Jye (1960). Visiting scientist. Division III, Superconductivity Research Laboratory, International Superconductivity Technology Center, 1-10-13 Shinonome, Koto-ku, Tokyo 135, Japan. PhD (U. Tokyo, 1993) mineralogy. *Crystal_growth float_zone_growth superconductor oxides crystal_chemistry*
Tel. 81(3)35365707x225 Fax 81(3)35365705

Kabuto, Dr Chizuko (1945). Research assistant. Inst. Anal. Chem. Fac. of Science, Tohoku Univ., Aoba, Aoba-ku, Sendai 980, Japan. DSc (U. Tohoku, 1975) organic chemistry. *Molecular_structure highly_strained_molecular_structures*
E-mail kabuto@archi.is.tohoku.ac.jp Tel. 81(22)2221800x3577 Fax 81(22)2639207

Kageyama, Dr Hiroyuki (1955). Chief researcher. Osaka National Research Institute (AIST), MITI, Midorigaoka 1-8-31, Ikeda 563, Japan. PhD (Osaka U., 1984) applied chemistry. *EXAFS catalysis ionic_conductivity lithium_compounds*
Tel. 81(727)519618 Fax 81(727)519622

Kai, Prof. Yasushi (1943). Professor. Department of Applied Chemistry, Faculty of Engineering, Suita, Osaka 565, Japan. DEng (Osaka U., 1973) physical chemistry. *Structure_determination metalloproteins_crystals chemistry catalysts organometallic*
E-mail a63486a@center.osaka-u.ac.jp Tel. 81(6)8797408 Fax 81(6)8797409

Kaito, Prof. Chihiro (1943). Professor. Dept. of Physics, Ritsumeikan Univ., Tojiin, Kita-ku, Kyoto 603, Japan. DSc (Kyoto U., 1980) physics. *Microcrystal astrophysics gas_evaporation*
Tel. 81(75)4651111 Fax 81(75)4658241

Kakimoto, Dr Koichi (1955). Manager research. Fundamental Research Laboratories, NEC Corp., 34 Miyukigaoka, Tsukuba 305, Japan. DEng (U. Tokyo, 1985) electrical engineering. *Crystal_growth*
E-mail kakimoto@sci.cl.nec.co.jp Tel. 81(298)501148 Fax 81(298)566137

Kakitani, Prof. Dr Satoru (1924). Professor, Chief of graduate course. Applied Science, Faculty of Science., Okayama Univ. of Science, 1-1 Ridai-cho, Okayama 700, Japan. DSc (Hiroshima Univ. of Literature and Science, 1961) mineralogy. *Crystallization ceramics silicate_phosphor scanning_tunneling_microscopy (STM) atomic_force_microscopy (AFM)*
Tel. 81(86)2523161x4225 Fax 81(86)2564855

Kakudo, Em. prof. Masao (1918). [18-18 Ohara, Ashiya, Hyogo 659, Japan.] DSc (Osaka U., 1953) physical chemistry. *Structure_determination proteins crystal scattering biopolymers small-angle*
Tel. 81(797)227137

Kamigaki, Prof. Dr Nobuo (1944). Professor. Faculty of Education, Ehime Univ., 3 Bunkyo-cho, Matsuyama, Ehime 790, Japan. DSc (Hiroshima U., 1983) physics. *Point_defect*
Tel. 81(899)247111 Fax 81(899)225719

Kamimura, Dr Midori (1955). Research Scientist. Teijin Institute for Biomedical Research, 4-3-2 Asahigaoka, Hino, Tokyo 191, Japan. DSc (Tokyo Institute of Technology, 1983) chemistry. *Biomedical biochemical*
E-mail midori@alice.ik.teijin.co.jp Tel. 81(425)868218 Fax 81(425)832109

Kanda, Dr Hisao (1947). Senior researcher. National Institute for Research in Inorganic Materials, 1-1 Namiki, Tsukuba, Ibaraki 305, Japan. PhD (Tohoku U., 1985) science. *High_pressure diamond*
Tel. 81(298)513351 Fax 81(298)527449

Kanehisa, Dr Nobuko (1943). Assistant. Department of Applied Chemistry, Faculty of Engineering, Osaka University, Suita, Osaka 565, Japan. DEng (Osaka U., 1992) physical chemistry. *Structure_determination small_crystal chemistry catalysts organometallic*
E-mail a63486a@center.osaka-u.ac.jp Tel. 81(6)8775111x4322 Fax 81(6)8776994

Kaneko, Prof. Tsutomu (1941). Professor. Dept. of Applied Physics, Science Univ. of Tokyo, 1-3 Kagurazaka, Shinjuku-ku, Tokyo 162, Japan. DSc (Science Univ. of Tokyo, 1979) applied physics. *CVD kinetics whisker whisker_composite silicon_carbide iron_whisker*
Tel. 81(3)32604271 Fax 81(3)32604772

Kanke, Dr Yasushi (1961). Senior researcher. National Institute for Research in Inorganic Materials, 1-1 Namiki, Tsukuba, Ibaraki 305, Japan. [(until July 29, 1995), Geological and Geophysical Sciences, Princeton University, Guyot Hall, Princeton, NJ 08544-1003, USA.] DSc (Tohoku U., 1992) chemistry. *Mixed-valence_compounds phase_diagram structure_determination magnetism Rietveld_method phase_transition single_crystal_characterization*
E-mail kanke@phoenix.princeton.edu (until July 29, 1995) kanke@nirim.go.jp (from July 30, 1995) Tel. 81(298)513351x375 Fax 81(298)527449

Kasatani, Dr Hirofumi (1960). Associate. Department of Material Science, Shizuoka Institute of Science and Technology, 2200-2 Toyosawa, Fukuroi-shi 437, Japan. DSc (Hiroshima U., 1989) materials science. *X-ray_diffraction EXAFS*
E-mail kasatani@ms.sist.ac.jp Tel. 81(538)450167 Fax 81(538)450110

Kashida, Shoji (1944). Assistant professor. Dept. Phys., Niigata University, Ikarashi, Niigata 950-21, Japan. DSc (U. Tokyo, 1981) physics. *Modulated_structures structural_change maximum-entropy_method cooperative_phenomena solid_state_ physics instrumentation Jahn_Teller_effect semiconductor*
E-mail KASHIDA@cc.niigata-u.ac.jp Tel. 81(25)2626131 Fax 81(25)2633961

Kashino, Prof. Setsuo (1937). Professor. Dept. of Chemistry, Fac. of Sci., Okayama U., Tsushima, Okayama 700, Japan. DSc (Osaka U., 1973) chemistry. *Organic_crystal_structure solid_phase_organic_reaction*
E-mail DTFCL124@jpnoucc.bitnet Tel. 81(86)2517849 Fax 81(86)2526601

Kashiwagi, Mr Tatsuki (1964). Research Scientist. First Department, Protein Engineering Research Institute, 6-2-3 Furuedai, Suita, Osaka 565, Japan. MPharm (Tokyo U., 1991) chemistry. *Proteins_crystallography amino_acids*
E-mail dept1@pes4.peri.co.jp Tel. 81(6)8728201 Fax 81(6)8728210

Kashiwase, Prof. Yasuji (1932). Professor. Department of Applied Physics, Nagoya Univ., Chikusa-ku, Nagoya, 464-01, Japan. DSc (Nagoya U., 1965) physics. *Dynamical_diffraction X-ray_spectroscopy X-ray optics grazing_incidence anti_reflection thin_film nuclear_resonance_scattering*
Tel. 81(52)7815111x4845 Fax 81(52)7822129

Kasuga, Prof. Dr Masanobu (1941). Professor. Fac. of Engineering, Yamanashi Univ., 4 Takeda, Kofu 400, Japan. PhD (Nagoya U., 1971) engineering. *Epitaxy cadmium_compounds CVD inverse_problem*
E-mail kasuga@willow.esd.yamanashi.ac.jp Tel. 81(552)208476 Fax 81(552)208476

Kasukabe, Dr Susumu (1949). Researcher. Electronics Packaging Technology Center, Production Engineering Research Laboratory, Hitachi, Ltd, 292 Yoshida-cho, Totsuka-ku, Yokohama 244, Japan. DEng (Nagoya U., 1990) crystal growth. *Crystal_growth electron_microscopy single_crystal ultrafine_particles*
Tel. 81(45)8811241 Fax 81(45)8601624

Katano, Mr Kizuku (1959). Opto-Electronics Laboratory, Mitsubishi Kasei Corp., 1000 Higashi-mamiana, Ushiku-city, Ibaraki 300-12, Japan. MSc (Gakushuin U., 1985) *Crystal_growth III-V_compounds*
E-mail katano@rc.m-kasei.co.jp Tel. 81(298)418220 Fax 81(298)433796

Kataoka, Mr Masayuki (1962). Semiconductor Research Laboratory, Mitsubishi Electric Corp., 8-1-1 Tsukaguchi-Honmachi, Amagasaki, Hyogo 661, Japan. MSc (Waseda U., 1987) *Crystal_growth*
Tel. 81(6)4977086 Fax 81(6)4977288

Kataoka, Dr Mikio (1949). Associate Professor. Dept Earth and Space Science, Fac. Science, Osaka University, Machikaneyama, Toyonaka, Osaka 560, Japan. PhD (Osaka U., 1981) biophysics. *Biophysics folding SAXS WAXS neutron_scattering*
E-mail g62068a@center.osaka-u.ac.jp Tel. 81(6)8441151x4316 Fax 81(6)8457966

Katayanagi, Dr Katsuo (1961). Research Scientist. First Department, Protein Engineering Research Institute, 6-2-3 Furuedai, Suita, Osaka 565, Japan. DPharm (Osaka U., 1991) chemistry. *Proteins_crystallography protein_nucleic_acids_interaction*
E-mail dept1@pes4.peri.co.jp Tel. 81(6)8728201 Fax 81(6)8728210

Kato, Dr Akio (1934). Professor. Dept. of Chemical Science and Technology, Faculty of Engineering, Kyushu Univ., 6-10-1 Hakozaki, Higashi-ku, Fukuoka 812, Japan. DEng (Kyushu U., 1961) applied chemistry. *Precipitation oxides calcium_compounds*
Tel. 81(92)6411101 Fax 81(92)6515606

Kato, Prof. Hiroshi (1947). Professor. Department of Mechanical Engineering, Faculty of Engineering, Saitama Univ., 255 Shimo-okubo, Urawa, Saitama 338, Japan. DEng (U. Tokyo, 1976) solidification. *Bonding characterization stress diffusion_bonding experimental_stress_analysis*
Tel. 81(48)8522111 Fax 81(48)8562577

Kato, Prof. Norio (1923). Professor. Dept. of Physics, Faculty of Science and Technology, Meijo Univ., Tenpaku-cho, Tenpaku-ku, Nagoya 468, Japan. DSc (Nagoya U., 1953) physics. *Dynamical_diffraction crystal_growth phase_transition*
Tel. 81(52)8321151 Fax 81(52)8321170

Kato, Dr Takamasa (1947). Associate professor. Dept. of Electronic Engineering and Computer Science, Yamanashi Univ., 4-3-11 Takeda, Kofu, Yamanashi 400, Japan. DEng (U. Tokyo, 1985) semiconductor. *Epitaxy*
E-mail kato@willow.esd.yamanashi.ac.jp Tel. 81(552)521111 Fax 81(552)540376

Katsube, Dr Yukiteru (1930). Professor. Division of Protein crystallography, Institute for Protein Research, Osaka University, Suita, Osaka 565, Japan. DSc (Osaka U., 1963) polymer science. *Biocrystallography proteins structure–activity_relationship phase_determination*
E-mail katsube@pxews1.protein.osaka-u.ac.jp Tel. 81(6)8775111x3836 Fax 81(6)8762533

Katsui, Dr Akinori (1942). Professor. School of High Technology for Human Welfare, Tokai Univ., 317 Nishino, Numazu, Shizuoka 410-03, Japan. DEng (Hokkaido U., 1978) *Crystal_growth solar_cell chalcopyrites superconductors*
Tel. 81(559)681211x4406 Fax 81(559)681155

Katsumata, Dr Tooru (1956). Associate professor. Dept. of Appl. Phys., Faculty of Eng., Toyo Univ., 2100 Kujirai, Nakanodai, Kawagoe, Saitama 350, Japan. DSc (Tohoku U., 1985) crystal growth. *Crystal_growth oxides*
Tel. 81(492)311132 Fax 81(492)311031

Katsuya, Dr Yoshio (1962). Researcher. Research and Development Department, Hyogo Prefectural Institute of Industrial Research, 3-1-12 Yukihira-cho, Suma-ku, Kobe 654, Japan. DSc (Osaka U., 1989) macromolecular science. *Structure_determination proteins biological_macromolecules amylases molecular_evolution*
E-mail katsuya@kougyo1.hyogo-kg.go.jp Tel. 81(78)7314481 Fax 81(78)7357845

Kawabata, Mr Naoki (1960). Associate Engineer. LSI Research & Development Center, Ricoh Company Limited, 30-1 Saho, Yashiro-cho, Kato, Hyogo 673-14, Japan. (Science Univ. of Tokyo, 1985) *Crystal_growth*
Tel. 81(795)424729 Fax 81(795)426312

Kawado, Dr Seiji (1940). Chief research scientist. Res. Center, Sony Corp., 174 Fujitsuka-cho, Hodogaya-ku, Yokohama 240, Japan. DEng (U. Tokyo, 1982) applied physics. *X-ray_topography synchrotron_radiation crystal_imperfection semiconductors total-reflection_X-ray_fluorescence time-resolved_X-ray_diffraction*
E-mail skawado@src.sony.co.jp Tel. 81(45)3536864 Fax 81(45)3536909

Kawahara, Prof. Akira (1932). Professor. Department of Earth Sciences, Faculty of Science, Okayama Univ., Tsushima-Naka 3-1-1, Okayama 700, Japan. DSc (U. Tokyo, 1962) mineralogy. *Structure_analysis instrumentation*
E-mail ofez1011@cc.okayama-u.ac.jp Tel. 81(86)2517880 Fax 81(86)2524057

Kawakami, Mr Shoji (1954). Ceramics Science Department, National Industrial Research Institute of Nagoya, 1 Hiratemachi, Kita-ku, Nagoya 462, Japan. MSc (Kyoto U., 1978) *Float_zone_growth ceramics oxides*
Tel. 81(52)9111111

Kawakita, Dr Tetsuya (1941). Executive Engineer; Visiting Professor, Saga University. Technology and Engineering Laboratories, Ajinomoto Co., Ltd, 1-1 Suzukicho, Kawasaki 210, Japan. [31-23 3-chome, Mutsuura, Kanazawa, Yokohama 236, Japan.] PhD (U. Tokyo, 1990) separation technology. *Amino_acids separation ion_exchange membrane*
Tel. 81(44)2447167 Fax 81(44)2449463

Kawaminami, Prof. Masaru (1941). Professor. Department of Physics, College of Liberal Arts, Kagoshima University, 21-30 Kohrimoto 1-chome, Kagoshima 890, Japan. DSc (Kyushu U., 1970) physics. *Phase_transition lattice_parameter*
E-mail kawamina@cla.kagoshima-u.ac.jp Tel. 81(992)858963 Fax 81(992)858975

Kawamura, Mr Kazuhiro (1955). Manager. Waste Technology Development Division, Power Reactor and Nuclear Fuel Development Corp., 4-33 Muramatsu, Tokai-mura, Naka-gun, Ibaraki 319-11, Japan. MEng (Nagoya U., 1981) nuclear eng. *Devitrification phase_separation glasses waste nuclear_materials*
Tel. 81(292)821111 Fax 81(292)829398

Kawamura, Prof. Tsutomu (1931). Professor. Dept. of Electrical Engineering, Faculty of Engineering, Kokushikan Univ., 4-28-1 Setagaya, Setagaya-ku, Tokyo 154, Japan. DSc (U. Tokyo, 1966) mineralogy. *Electronics physical_property scanning_tunnel_microscopy semiconductor growth non-linear_property electronic_devices surface characterization laser_beam_heating thin_film thermal_property electronic_materials*
Tel. 81(3)54813263 Fax 81(3)54813253

Kawanaka, Dr Ryusuke (1935). Semiconductor Group, Sony Corp., 4-14-1 Asahi-cho, Atsugi, Kanagawa 243, Japan. (Osaka Pref. U., 1983) crystal growth. *Crystal_growth*
Tel. 81(462)305401 Fax 81(462)306233

Kawanishi, Mr So-roh (1965). Researcher. Dynamics Analysis Group, Kimura Metamelt Project, ERATO, JRDC, Satellite-2, Tsukuba Research Consortium, 5-9-9 Tokodai, Tsukuba 300-26, Japan. MS (U. Kentucky, 1992) nano crystalline materials. *Melting single_crystal DLCZ*
Tel. 81(298)475191 Fax 81(298)475089

Kawasaki, Mr Masayuki (1959). Nihon Dempa Kogyo Co., Ltd, 1275-2 Kamihirose, Sayama, Saitama 350-13, Japan. MSc (Tohoku U., 1986) *Morphology dislocation synthetic_quartz*
Tel. 81(429)527211 Fax 81(429)546269

Kifune, Dr Kouichi (1958). Lecturer. Department of Physics, Osaka Women's Univ., Daisen-cho, Sakai, Osaka 590, Japan. DSc (Hiroshima U., 1988) materials science. *Aperiodic_material boundaries crystal_growth electron_microscopy phase_transition simulation epitaxy*
Tel. 81(722)224811 Fax 81(722)385539

Kihara, Dr Kuniaki (1943). Professor. Dept. of Earth Sciences, Kanazawa University, Kanazawa 920-11, Japan. DSc (Tokyo University of Education, 1972) mineralogy. *Lattice_dynamics phase_transition structural_change anharmonicity structure_refinement structure_modelling computer_graphics*
E-mail kuniaki@icews1.ipc.kanazawa-u.ac.jp Tel. 81(762)645720 Fax 81(762)645746

Kikuchi, Prof. Makoto (1935). Professor; Division of crystal technology. Dept. Metall. Eng., Tokyo Institute of Technology, Oh-okayama, Meguro, Tokyo 152, Japan. DEng (Tokyo Inst. Tech., 1966) metallurgy. *Alloys phase_transition titanium_compounds nitrides microbeam_analysis phase_diagram*
Tel. 81(3)57343137 Fax 81(3)57342874 Telex 2466360 TITECH J

Kikuma, Dr Isao (1941). Associate professor. Faculty of Engineering, Ibaraki Univ., 4-12-1 Nakanarusawa, Hitachi, Ibaraki 316, Japan. DEng (Tohoku U., 1986) engineering. *Bridgman_Stockbarger_technique zinc_compounds bulk II–VI_compounds*
Tel. 81(294)356101 Fax 81(294)380881

Kimata, Dr Mitsuyoshi (1948). Assistant Professor. Institute of Geoscience, The University of Tsukuba, 1-1 Tennoudai, Tsukuba, Ibaraki, Japan. DSc (Tokyo University of Education, 1977) mineralogy. *X-ray_optics instrumentation X-ray_reflectivity single_crystal_characterization synchrotron_radiation*
Tel. 81(298)534209 Fax 81(298)519764

Kimoto, Dr Hiroshi (1963). Research associate. Dept. of Electrical Engineering, Kyoto Univ., Yoshida-honmachi, Sakyo, Kyoto 606-01, Japan. MEng (Kyoto U., 1988) deposition mechanism. *Semiconductor epitaxy characterization*
Tel. 81(75)7535341 Fax 81(75)7511576

Kimura, Dr Hideo (1958). Senior researcher. Physical Properties Division, National Research Institute for Metals, 1-2-1 Sengen, Tsukuba, Ibaraki 305, Japan. DEng (Tohoku U., 1991) materials science. *Microgravity phase_formation oxides*
E-mail kdeo@momokusa.nrim.go.jp Tel. 81(298)531046 Fax 81(298)531091

Kimura, Masao (1962). Senior researcher. Adv. Mater. Tech. Res. Labs., Nippon Steel Co., 1618 Ida, Nakahara-ku, Kawasaki 211, Japan. DEng (Kyoto U., 1993) physical chemistry. *X-ray_diffraction phase_transition surface powder_analysis application_of_science_to_technology*
E-mail kaiseki.lab1.nsc.co.jp Tel. 81(44)7774111 Fax 81(44)7526341

Kimura, Mr Shigeru (1963). Microelectronics Res. Labs. NEC Corp., 34 Miyukigaoka, Tsukuba, Ibaraki 305, Japan. Dr (Nagoya U., 1994) applied physics. *X-ray_topography multiple_crystal_diffractometry defect*
E-mail kimuras@lbr.cl.nec.co.jp Tel. 81(298)501189 Fax 81(298)566138

Kinoshita, Dr Kyoichi (1948). Senior scientist, Supervisor. Basic Research Laboratories, NTT Corp., 3-1 Morinosato Wakamiya, Atsugi, Kanagawa 243-01, Japan. DSc (Waseda U., 1985) crystal growth. *Bridgman_Stockbarger_technique physical_property dislocation epitaxy*
E-mail k.kino@will.ntt.jp Tel. 81(462)403354 Fax 81(462)404675

Kinoshita, Takayoshi (1966). Analytical Research Labs., Fujisawa Pharmaceutical Co., Ltd, 2-1-6 Kashima, Yodogawa-ku, Osaka 532, Japan. MSc (Osaka U., 1991) crystallography. *Biocrystallography drug_design molecular_recognition metalloproteins receptor computing*
Tel. 81(6)3901171 Fax 81(6)3071377

Kishida, Dr Satoru (1952). Associate professor. Dept. Electrical and Electronic Engineering, Tottori Univ., 4-101 Koyama-cho Minami, Tottori 680, Japan. PhD (Kobe U., 1989) *Crystal_growth*
Tel. 81(857)280321 Fax 81(857)310880

Kita, Ms Akiko (1968). Graduate student. Miki-Laboratory, Res. Lab. of Resources Utilization, Tokyo Institute of Technology, 4259 Nagatsuta, Midori-ku, Yokohama 227, Japan. MEng (Osaka U., 1992) chemistry. *Crystallography crystallization proteins biochemistry bioluminescence flavoenzymes*
E-mail akita@ce.titech.ac.jp Tel. 81(45)9221111x2232 Fax 81(45)9225179 Telex 3823553 TITNAG J

Kitadokoro, Mr Kengo (1965). Researcher. Discovery Research Laboratories I, Shionogi & Co., Ltd, 12-4 Sagisu 5 Cho-me, Fukushima-ku, Osaka 553, Japan. MS (Osaka U., 1990) applied fine chemistry. *Biochemistry biocrystallography drug proteins*
E-mail kengo@fl.lab.shionogi.co.jp Tel. 81(6)4585861x272 Fax 81(6)4580987

Kitagawa, Dr Ryuji (1949). Associate professor. Earth Planetary Systems Science, Faculty of Science, Hiroshima Univ., 1-3 Kagamiyama, Higashi-hiroshima 724, Japan. DSc (Hiroshima U., 1986) mineralogy. *Electron_microscopy micromorphology clays*
Tel. 81(824)247466 Fax 81(824)240735

Kitamura, Dr Mitsutaka (1946). Assistant professor. Dept. Chemical Engineering, Hiroshima Univ., 1-4-1 Kagamiyama, Higashi-hiroshima 724, Japan. DEng (Kyoto U., 1978) crystallization. *Crystallization nucleation polymorphism crystal_growth*
Tel. 81(824)247715 Fax 81(824)227191

Kitamura, Tatsuo (1968). Graduated Student (Master course). Department of Chemistry, Faculty of Science, Tokyo Institute of Technology, Ookayama, Meguro-ku, Tokyo 152, Japan. BSc (Tokyo Institute of Technology, 1992) chemistry. *Molecular_aggregates crystalline_state_reactions time-resolved_structure_analysis cationic_surfactants aromatic_compounds*
E-mail sanzo@chem.titech.ac.jp Tel. 81(3)37261111x2608 Fax 81(3)37206206

Kiyotani, Ms Tamiko (1965). Assistant. Lab. of Phytochemistry, Showa Coll. of Pharmaceutical Sciences, 3-3165 Higashitamagawagakuen, Machida-shi, Tokyo 194, Japan. *Plants terpenes molecular_mechanics_computing solving_structures_of_novel_triterpenoids_from_ferns*
Tel. 81(427)211575 Fax 81(427)211588

Kobayashi, Dr Akiko (1943). Associate professor. Department of Chemistry, Faculty of Science, The University of Tokyo, Hongo, Bunkyo-ku, Tokyo 113, Japan. DSc (U. Tokyo, 1971) chemistry. *Material_sciences*
Tel. 81(3)38122111 Fax 81(3)38142627

Kobayashi, Dr Hayao (1942). Professor. Department of Chemistry, Faculty of Science, Toho University, Funabashishi, Chiba 274, Japan. DSc (U. Tokyo, 1969) chemistry. *Material*
Tel. 81(474)724402 Fax 81(474)751855

Kobayashi, Prof. Jinzo (1925). Professor. Dept. Applied Physics, Waseda Univ., 3-4-1 Ohkubo, Shinjuku-ku, Tokyo, Japan. DSc (Waseda U., 1960) solid state physics. *Ferroelectric_materials crystal_optics*
Tel. 81(474)313828 Fax 81(474)372375

Kobayashi, Prof. Nobuyuki (1942). Professor. Department of Electronic and Information Engineering, Faculty of Engineering, Toyama Univ., 3190 Gofuku, Toyama 930, Japan. DSc (Nagoya U., 1972) crystal growth. *Crystal_growth melts heat_transfer mass_transfer radiation*
Tel. 81(764)411271x2837 Fax 81(764)418432

Kobayashi, Prof. Tadashi (1944). Professor. Hokuriku Univ., 3-Ho, Kanagawa-machi, Kanazawa 920-11, Japan. DSc (Kyushu U., 1972) physics. *Electron_spin_resonance phase_transition ferroelectricity scanning_tunnel_microscopy discommensuration ferroelasticity atomic_force_microscopy*
Tel. 81(762)291165x227 Fax 81(762)292781

Koda, Dr Shigetaka (1946). Associate Director. Analytical Research Labs., Fujisawa Pharmaceutical Co., Ltd, 2-1-6 Kashima, Yodogawa-ku, Osaka 532, Japan. DSc (Hiroshima U., 1987) coordination chemistry. *Polymorphism drug_design powder_diffraction drug_interaction molecular_crystal*
Tel. 81(6)3901172 Fax 81(6)3071377

Koga, Dr Kenji (1963). Researcher. National Inst. for Advanced Interdisciplinary Res., AIST MITI, Higashi, Tsukuba, Ibaraki 305, Japan. PhD (U. Tsukuba, 1991) applied physics. *Clusters liquids SAXS nucleation flash_X-ray_diffraction fractal diffuse_scattering*
E-mail kenji@nair.go.jp Tel. 81(298)542541 Fax 81(298)542549

Koide, Prof. Tsutomu (1930). Professor. Laboratory of Physical Chemistry, Osaka Kyoiku Univ., Asahigaoka, Kashiwara 582, Japan. MS (Osaka U., 1966) physical chemistry. *Solid phase_transition*
Tel. 81(729)3211x4406 Fax 81(729)3273

Koizumi, Prof. Dr Mitsue (1923). Professor. Faculty of Science & Technology, Ryukoku Univ., Seta, Otsu, Shiga 520-21, Japan. DSc (U. Tokyo, 1958) mineralogy. *Crystal_growth*
Tel. 81(775)437460 Fax 81(775)437480

Kojima, Prof. Dr Kenichi (1942). Professor; Dean, school of science. Dept. of Physics, Yokohama City Univ., 22-2 Seto, Kanazawa-ku, Yokohama 236, Japan. DEng (Tohoku U., 1971) crystal defect. *Dislocation characterization defect molecular_crystal mechanical_properties optical_properties*
E-mail kojima@yokohama-cu.ac.jp Tel. 81(45)7872171 Fax 81(45)7872172

Kojima, Yuko (1967). Research Scientist. Analytical Sciences Lab., Mitsubishi Chemical Corp., 1000 Kamoshida, Aobaku, Yokohama 227, Japan. MSc (Ochanomizu U., 1991) chemistry. *Structure materials optoelectrical_property molecular_crystal*
E-mail ykojima@atlas.rc.m-kasei.co.jp Tel. 81(45)9633156 Fax 81(45)9634206

Komatsu, Prof. Hiroshi (1935). Professor. Councillor, Crystallographic Soc. of Japan. Inst. for Materials Research, Tohoku Univ., Katahira, Aoba-ku, Sendai 980, Japan. PhD (Tokyo Kyoiku U., 1964) mineralogy. *Biomaterial crystal_growth interferometry microscopy interface biomolecules topography*
E-mail komatsu@jpnimrtu.bitnet Tel. 81(22)2152010 Fax 81(22)2152011

Komatsu, Mr Yoshinobu (1957). Assistant research associate. Research Laboratories, Mizusawa Research Division, Mizusawa Industrial Chemicals Ltd, 21 Aza Tonoda, Ohaza Nishime, Tsuruoka, Yamagata 999-75, Japan. (Hirosaki U., 1981) *Crystal_growth*
Tel. 81(235)353331 Fax 81(235)353574

Komura, Prof. Dr Shigehiro (1933). Professor. Faculty of Integrated Arts and Sciences, Hiroshima Univ., 1-7-1 Kagamiyama, Higashi-hiroshima, Hiroshima 724, Japan. DSc (U. Tokyo, 1967) physics. *Neutron diffraction phase_transition critical_phenomena small-angle_scattering*
E-mail komura@minerva.ias.hiroshima-u.ac.jp Tel. 81(824)246538 Fax 81(824)240757

Komura, Prof. Yukitomo (1924). Professor. Dept.Mechanical Engineering, Hiroshima Institute of Technology, Saeki-ku Hiroshima 731-51, Japan. [1-6-33 Ajinadai, Hatsukaichi-city, Hiroshima 738, Japan.] DSc (Osaka U., 1961) physics. *Stacking_fault electron_microscopy alloys intermetallic*
E-mail komura@cael.me.it-hiroshima.ac.jp Tel. 81(829)213121x541 Fax 81(829)396020

Kondo, Mr Hidemasa (1967). Student. Division of Biological Sciences, Graduate School of Science, Hokkaido Univ., Sapporo 060, Japan. MSc (Hokkaido U., 1993) chemistry. *Proteins crystallization purification*
E-mail kondou@polymer.hokudai.ac.jp Tel. 81(11)7162111x3809 Fax 81(11)7465232

Koto, Prof. Kichiro (1936). Professor. Fac. Integ. Arts and Sci., Tokushima Univ., Minami-Josanjima, Tokushima 770, Japan. DSc (U. Tokyo, 1969) mineralogy. *Amorphization anharmonicity antiphases condensed_matter local_structure*
E-mail koto@ias.tokushima-u.ac.jp Tel. 81(886)567239 Fax 81(886)567298 Telex 586225 TSKL J

Kouta, Mr Hikaru (1963). Material Development Center, NEC Corp., 4-1-1 Miyazaki, Miyamae-ku, Kawasaki, Kanagawa 216, Japan. MSc (Gakushuin U., 1988) *Crystal_growth*
E-mail kouta@mdc.cl.nec.co.jp Tel. 81(44)8562167 Fax 81(44)8562244

Koyama, Dr (1952). Associate professor. Dept. of Mater. & Eng., Waseda University, Ohkubo, Shinjuku-ku, Tokyo 169, Japan. DEng (Tokyo Institute of Technology, 1981) physical metallurgy. *Phase_transformation electron_microscopy metals incommensurate_structure charge_density_wave superconductivity*
Tel. 81(3)32034141x733355

Koyama, Mr Tadashi (1955). Senior researcher. Tsukuba Research Laboratory, Nippon Sheet Glass Co., Ltd, 5-4 Tokodai, Tsukuba, Ibaraki 300-26, Japan. MSc (Hokkaido U., 1981) *Glasses II–VI_compounds*
Tel. 81(298)478681 Fax 81(298)478693

Koyano, Mr Kazuo (1932). Senior Researcher. Biomedical Research Institute, Teijin Ltd, Asahigaoka 4-3-2, Hino-shi, Tokyo 191, Japan. MS (1958) chemistry. *Computer-assisted_design fibres quantum_chemistry organometallic_catalysts molecular_dynamics polymers synchrotron_radiation*
E-mail koyano@alice.ik.teijin.co.jp Tel. 81(425)868212 Fax 81(425)837541

Kozaki, Dr Shigeru (1934). Professor. The Polytechnic Univ., 4-1-1 Hashimotodai, Sagamihara, Kanagawa 229, Japan. DSc (U. Tokyo, 1974) X-ray diffraction. *Topography stress*
Tel. 81(427)639105 Fax 81(427)639114

Kuboi, Prof. Dr Ryoichi (1946). Associate professor. Dept. of Chem. Eng., Osaka Univ., Toyonaka, Osaka 560, Japan. PhD (Osaka U., 1974) chemical engineering. *Biomolecule image_processing precipitation proteins bioseparation dynamic_light_scattering crystal_size_distribution*
Tel. 81(6)8441151 Fax 81(6)8573952

Kubota, Prof. Dr Noriaki (1940). Professor. Dept. of Applied Chemistry and Molecular Science, Iwate Univ., 4-3-5 Ueda, Morioka 020, Japan. DEng (Tohoku U., 1980) industrial crystallization. *Crystallization precipitation purification control batch_crystallization optical_isomer secondary_nucleation*
Tel. 81(196)235171x2378 Fax 81(196)527144

Kubota, Yoshiki (1967). Research Assistant. Department of Natural Science, Osaka Women's University, Daisen-chō Sakai-city, Osaka 590, Japan. MEng (Nagoya U., 1991) physics. *X-ray_diffraction maximum-entropy_method electron_density_distribution metals*
E-mail kubotay@osaka-wu.ac.jp Tel. 81(722)224811x343 Fax 81(722)385539

Kuchitsu, Prof. Kozo (1927). Professor. Member of Book Series Committee. Department of Chemistry, Josai University, Sakado, Saitama 350 02, Japan. DSc (U. Tokyo, 1958) chemistry. *Anharmonic_vibration high_energy_electron_diffraction molecular_clusters molecular_structure intermolecular_interactions molecular_spectroscopy*
E-mail kuchitsu@tansei.cc.u-tokyo.ac.jp Tel. 81(492)717987 Fax 81(492)717985

Kudoh, Dr Yasuhiro (1947). Associate Professor. Institute of Mineralogy, Petrology and Economic Geology, Faculty of Science, Tohoku University, Sendai 980, Japan. DSc (U. Tokyo, 1975) mineralogy and crystallography. *Diffractometry elasticity high_pressure mineralogy diamond_anvil_high_pressure_apparatus bulk_modulus bond_compressibility*
E-mail kudoh@jpntuvm0.bitnet Tel. 81(22)2221800x3434 Fax 81(22)2626609

Kuge, Dr Ken-ichi (1952). Associate professor. Dept. of Imaging Science, Faculty of Engineering, Chiba Univ., 1-33 Yayoi-cho, Inage-ku, Chiba 263, Japan. DEng (Kyoto U., 1984) photochemistry. *Crystal_growth*
Tel. 81(43)2903454 Fax 81(43)2903490

Kukimoto, Prof. Hiroshi (1937). Professor. Imaging Science and Engineering Laboratory, Tokyo Institute of Technology, 4259 Nagatsuta, Midori-ku, Yokohama 227, Japan. DEng (U. Tokyo, 1966) applied physics. *Semiconductors optoelectronics*
Tel. 81(45)9221111 Fax 81(45)9211492

Kuma, Dr Shoji (1939). Chief researcher. Advanced Research Center, Hitachi Cable, Ltd, 5-1-1 Hitaka-cho, Hitachi, Ibaraki 319-14, Japan. DEng (Waseda U., 1983) polymeric insulation. *III–V_compounds*
Tel. 81(294)423151 Fax 81(294)432404

Kumagawa, Prof. Dr Masashi (1938). Professor. Research Institute of Electronics, Shizuoka Univ., 3-5-1 Johoku, Hamamatsu, Shizuoka 432, Japan. DEng (Tohoku U., 1967) crystal growth. *Semiconductor property vapor_growth*
E-mail m-kumagawa@rie.shizuoka.ac.jp Tel. 81(53)4711171 Fax 81(53)4740630

Kumao, Dr Akihiro (1941). Professor. Dept. of Electronics and Information Science, Kyoto Institute of Technology, Matsugasaki, Sakyo-ku, Kyoto 606, Japan. DSci (Hiroshima U., 1983) physics. *Electron_microscopy crystal_growth alloys thin_films minerals ceramics*
E-mail kumao@dj.kit.ac.jp Tel. 81(75)7247444 Fax 81(75)7247400

Kumazawa, Shintaro (1962). Research Assistant. Department of Physics, Science University of Tokyo, Yamazaki, Noda, Chiba 278, Japan. MT (Nagoya U., 1992) applied physics. *Powder profile_analysis Rietveld_method structure maximum-entropy_method electron_density_distribution nuclear_density_distribution anharmonicity phase_transition*
E-mail skuma@c1.noda.sut.ac.jp Tel. 81(471)241501x3218 Fax 81(471)239361

Kurahashi, Dr Masayasu (1943). Section Chief. Department of Analytical Chemistry, National Institute for Materials and Chemical Research, Tsukuba, Ibaraki 305, Japan. DSc (Osaka City U., 1975) chemistry. *Powder surface single_crystal interface X-ray_reflectivity*
E-mail ck227@nimc.go.jp Tel. 81(298)544626 Fax 81(298)551397

Kuroda, Dr Ekyo (1943). Professor. Nagoya College of Music, 7-1 Inabaji-cho, Nakamura-ku, Nagoya, Japan. DEng (Nagoya U., 1982) crystal growth. *Czochralski_technique solar_cell epitaxy*
Tel. 81(52)4111111 Fax 81(52)4132300

Kuroiwa, Prof. Dr Koichi (1943). Professor. Division of Electronic and Information Engineering, Faculty of Technology, Tokyo Univ. of Agriculture and Technology, 2-24-16 Nakamachi, Koganei, Tokyo 184, Japan. DEng (U. Tokyo, 1972) crystal growth. *Metallorganic_CVD thin_film dielectrics semiconductor*
E-mail kuroiwa@cc.tuat.ac.jp Tel. 81(423)887118 Fax 81(423)855395 Telex 2832663 TUATT J

Kusunoki, Dr Katsuyuki (1947). Senior researcher. Materials Physics Division, National Research Institute for Metals, 2-3-12 Nakameguro, Meguro-ku, Tokyo 153, Japan. DEng (U. Tokyo, 1978) diffusion. *Metallurgy nucleation phase_transition*
Tel. 81(3)37192271 Fax 81(3)37923337

Kusunoki, Dr Masami (1953). Research Associate. Institute for Protein Research, Osaka University, Suita, Osaka 565, Japan. DSc (Osaka U., 1980) macromolecular science. *Protein_crystallography structure–activity_relationship computer_modelling*
E-mail kusunoki@protein.osaka-u.ac.jp Tel. 81(6)8775111x3912 Fax 81(6)8762533

Kuwano, Dr Yasuhiko (1940). Senior research specialist. Material Development Center, NEC Corp., 4-1-1 Miyazaki, Miyamae-ku, Kawasaki 216, Japan. DEng (Tokyo Institute of Technology, 1991) material science. *Czochralski_technique oxides optics*
E-mail kuwano@mdc.cl.nec.co.jp Tel. 81(44)8562167 Fax 81(44)8562244

Machida, Mr Hiroshi (1953). Deputy senior researcher. Fine Ceramics Research & Development Div., Chichibu Cement Co., Ltd, 5310 Mikajiri, Kumagaya, Saitama 360, Japan. (Tokyo Institute of Technology, 1979) *Crystal_growth Czochralski_technique EFG_technique rutile*
Tel. 81(485)337461 Fax 81(485)337467

Machida, Dr Katsumi (1947). Leader. Electronics Materials Laboratory, Sumitomo Metal Mining Co., Ltd, 1-6-1 Suehiro-cho, Ohme, Tokyo 198, Japan. PhD (Tohoku U., 1990) micro-optics. *Epitaxy crystal_growth*
Tel. 81(428)311195 Fax 81(428)320252

Machida, Dr Mitsuo (1956). Associate professor. Department of Physics, Kyushu University, Fukuoka 812, Japan. DSc (Kyushu U., 1987) physics. *Phase_transition spin_resonance incommensurate*
Tel. 81(92)6411101x4186 Fax 81(92)6334525

Maiwa, Dr Koji (1960). Chemical Processing Division, National Research Institute for Metals, 2-3-12 Nakameguro, Meguro-ku, Tokyo 153, Japan. PhD (Tohoku U., 1988) mineralogy. *Aqueous_solution_growth*
Tel. 81(3)37192271 Fax 81(3)37923337

Makikawa, Mr Shinji (1960). Advanced Functional Materials Research Center, Shin-Etsu Chemical Co., Ltd, 2-13-1 Isobe, Annaka, Gunma 379-01, Japan. BSc (Osaka U., 1982) *Crystal_growth oxides*
Tel. 81(273)852301 Fax 81(273)852756

Marukawa, Prof. Kenzaburo (1937). Professor. Dept. Applied Physics, Hokkaido Univ., Sapporo 060, Japan. DSc (Kyoto U., 1967) physics. *Electron_microscopy phase_transition alloys*
Tel. 81(11)7066643 Fax 81(11)7166175

Marumo, Prof. Fumiyuki (1931). Professor. Dept. of Earth Sciences, Nihon University, Sakurajosui 3-25-40, Setagaya-ku, Tokyo 156, Japan. DSc (U. Tokyo, 1960) mineralogy. *Charge_density mineralogy phase_transition inorganic_structure*
Tel. 81(3)33291151x5205 Fax 81(3)33039899 Telex NICHIDAU29496

Maruyama, Mr Hiroyuki (1966). Student. Department of Electrical and Electronic Engineering, Toyohashi University of Technology, 1-1 Hibarigaoka, Tempaku-cho, Toyohashi 441, Japan. MEng (Toyohashi U., 1991) semiconductor. *Crystal_growth*
Tel. 81(532)470111x586 Fax 81(532)483422

Maruyama, Dr Minoru (1951). Lecturer. Faculty of Science, Osaka City Univ., 3-3-138 Sugimoto, Sumiyoshi, Osaka 558, Japan. DSc (Osaka City U., 1989) crystal growth. *Crystal_growth surface_roughening surface_melting*
E-mail f51462@jpnkudpc.bitnet Tel. 81(6)6052529 Fax 81(6)6052522

Mashiyama, Prof. Hiroyuki (1947). Professor. Dept. of Phys., Fac. of Sci., Yamaguchi Univ., Yoshida 1677-1, Yamaguchi 753, Japan. DSc (Kyoto U., 1976) condensed matter physics. *Phase_transition incommensurate ferroelectric_physics crystallography structure_transformation X-ray/neutron_scattering free_energy*
Tel. 81(839)226111x370 Fax 81(839)322041

Masuda, Mr Yukihiro (1960). R&D 3-2, Nichia Chemical Company Ltd, 491-100 Kaminaka-cho Oka, Anan-city, Tokushima 774, Japan. MEng (U. Osaka Prefecture, 1987) biological high polymer. *Crystal_growth*
Tel. 81(884)222311 Fax 81(884)231802

Masuko, Dr Akiyoshi (1935). Senior Res. Physicist. Central Res. Lab., Nippon Oil Co., Ltd, 8 Chidori-cho, Nakaku, Yokohama 231, Japan. DEng (Waseda U., 1989) engineering. *Powder_diffraction small-angle_scattering surface_structure colloid*
Tel. 81(45)6257261 Fax 81(45)6257271 Telex J27237

Masuko, Prof. Toru (1941). Professor. Dept. Materials Science & Engineering, Faculty of Engineering, Yamagata Univ., 4 Jonan, Yonezawa, Yamagata 992, Japan. DSc (Tokyo Institute of Technology, 1980) polymer science. *Crystallography molecular_crystal polymers_crystallization phase_transformation structure_determination electron_diffraction polyorganophosphazenes*
Tel. 81(238)225181x272 Fax 81(238)247241

Matsuda, Prof. (1931). Professor. Kyushu Institute of Technology, Sensui-cho, Tabata-ku, Kitakyushu 804, Japan. DEng (Kyushu U., 1972) engineering. *Property application*
Tel. 81(93)8711931x363 Fax 81(93)8817186

Matsui, Junji (1938). Fellow. Research and Development Group, NEC Corporation, 34 Miyukigaoka, Tsukuba 305, Japan. DEng (Tohoku U., 1977) engineering. *Crystal_growth characterization synchrotron_radiation crystal_defects*
E-mail matuij@tgn.cl.nec.co.jp Tel. 81(298)501116 Fax 81(298)566138

Matsui, Dr Masanori (1949). Associate Professor. Department of Earth and Planetary Sciences, Faculty of Science, Kyushu University, Hakozaki, Fukuoka 812, Japan. DSc (Kwansei Gakuin U., 1982) physical chemistry. *Computer_modelling high_pressure high_temperature mineralogy*
Tel. 81(92)6411101x4312 Fax 81(92)6322736

Matsui, Dr Yoshio (1949). Senior researcher. Editor of "Journal of Crystallographic Society of Japan". National Institute for Research in Inorganic Materials, 1-1 Namiki Tsukuba, Ibaraki 305, Japan. PhD (U. Tokyo, 1984) inorganic chemistry. *High_resolution_electron_microscopy superconductors microstructure direct_observations_of_atoms*
E-mail matsui@nirim.go.jp Tel. 81(298)513351x283 Fax 81(298)527449

Matsumoto, Mr Fumio (1962). Electronic Materials Technical Dept., Chichibu Works, Showa Denko K.K., 1505 Shimokagemori, Chichibu, Saitama 369-18, Japan. MEng (Yamagata U., 1988) *Crystal_growth semiconductor*
Tel. 81(494)236112 Fax 81(494)237787

Matsumoto, Dr Hitoshi (1953). Lecturer. Dept. of Materials Science and Engineering, National Defense Academy, 1-10-20 Hashirimizu, Yokosuka 239, Japan. DSc (Tokyo Institute of Technology, 1992) materials science. *Transformation oxidation shock_wave deformation*
Tel. 81(468)413810 Fax 81(468)445910

Matsumoto, Dr Osamu (1960). Research Associate. Faculty of Pharmaceutical Sciences, Kyoto University, Sakyo-ku, Kyoto 606-06, Japan. DSc (Kyoto U., 1990) pharmacology. *Biology structure nucleic_acids proteins*
E-mail osamu@kamo.pharm.kyoto-u.ac.jp Tel. 81(75)7534543 Fax 81(75)7534544 Telex 3823553TTTNAGJ

Matsumoto, Prof. Dr Takashi (1946). Professor. Department of Electronic Engineering, Yamanashi University, 4-3-11 Takeda, Kofu 400, Japan. DEng (U. Tokyo, 1973) crystal growth. *Crystal_growth*
Tel. 81(552)521111x5242 Fax 81(552)540376

Matsumoto, Prof. Takeo (1932). Professor. Dept. Earth Sci., Fac. of Sci., Kanazawa University, Kakumamachi Kanazawa, Ishikawa 920-11, Japan. DSc (U. Tokyo, 1962) mineralogy. *Mathematical_crystallography symmetry close_packing crystal_structure crystallographic_orbits*
Tel. 81(762)645725 Fax 81(762)645746

Matsumura, Dr Sadao (1943). Manager. Electro-Functional Material Engineering Dept., Toshiba Corp., 72 Horikawa-cho, Saiwai-ku, Kawasaki 210, Japan. Dr (Tohoku U., 1980) applied physics. *Crystal_growth piezoelectricity ferroelectricity*
Tel. 81(44)5493443 Fax 81(44)5413688

Matsumura, Dr Syo (1956). Associate Professor. Department of Nuclear Engineering, Kyushu University, 6 Hakozaki, Higashi-ku, Fukuoka 812, Japan. PhD (Kyushu U., 1992) materials science. *Dynamical_diffraction ordering phase_separation transmission_electron_microscopy structure_factor_determination ALCHEMI phase_transition_kinetics*
Tel. 81(92)6411101x5823 Fax 81(92)6417098

Matsuno, Dr Seiichi (1934). Faculty of Science and Engineering, Chuo Univ., 1-13-27 Kasuga, Bunkyo-ku, Tokyo 112, Japan. (Tokyo Institute of Technology, 1987) chemical engineering. *Apatite*
Tel. 81(3)38171910 Fax 81(3)38171895

Matsuoka, Prof. Dr Masakuni (1945). Professor. Dept. of Chemical Engineering, Tokyo Univ. of Agriculture and Technology, 2-24-16 Naka-cho, Koganei, Tokyo 184, Japan. DEng (Tokyo Institute of Technology, 1973) fractional crystallization. *Crystallization morphology purification crystal_growth organic_crystals solid_liquid_equilibrium*
E-mail mmatsuok@cc.tuat.ac.jp Tel. 81(423)887059 Fax 81(423)877944 Telex 2832 663 TUATT. J

Matsuura, Dr Yoshiki (1943). Associate Prof. Institute for Protein Research, Osaka University, Suita, Osaka 565, Japan. DSc (Osaka U., 1976) polymer science. *Biocrystallography proteins structure–activity_relationship phase_determination crystal_growth*
E-mail matsuura@pxews1.protein.osaka-u.ac.jp Tel. 81(6)8775111x3837 Fax 81(6)8762533

Matsuzaki, Takao (1945). Research fellow. Analytical Sciences Lab., Mitsubishi Chemical Corp., 1000 Kamoshida, Aobaku, Yokohama 227, Japan. DPharm (U Tokyo, 1988) chemistry. *Macromolecules proteins drug_design*
E-mail tak@rc.m-kasei.co.jp Tel. 81(45)9633156 Fax 81(45)9634206

Menda, Mr Kazunori (1961). Applied Research Dept., Olympus Optical Co., Ltd, 2-3 Kuboyama-cho, Hachioji-shi, Tokyo 192, Japan. MEng (Toyohashi Univ. of Technology, 1985) *Crystal_growth*
Tel. 81(426)917111 Fax 81(426)915709

Mihama, Prof. Dr Kazuhiro (1927). Professor. Dept. of Applied Electronics, Daido Institute of Technology, 2-21 Daido-cho, Minami-ku, Nagoya 457, Japan. DSc (U. Paris, 1961) physics. *Crystal_growth high_resolution_electron_microscopy*
Tel. 81(52)6110513 Fax 81(52)6125653

Mikida, Dr Rokuro (1938). Professor. Science Univ. of Tokyo, Suwa College, 5000-1 Toyohira, Chino, Nagano 391-02, Japan. DSc (U. Tokyo, 1985) mineralogy. *Transmission_electron_microscopy incommensurate modulated_structures binary_alloys solid_state_ionics oxides*
Tel. 81(266)731201x306 Fax 81(266)731230

Miki, Dr Kunio (1952). Professor. Department of Chemistry, Faculty of Science, Kyoto University, Sakyo-ku, Kyoto 606-01, Japan. DEng (Osaka U., 1981) chemistry. *Crystallography crystallization proteins biochemistry photosynthesis-related_proteins flavoenzymes protein–DNA_interaction biological_electron_transfer_system*
E-mail 152696@sakura.kudpc.kyoto-u.ac.jp Tel. 81(75)7534029 Fax 81(75)7534032

Minemoto, Mr Hisashi (1959). Researcher. Networks Development Center, Matsushita Electric Ind.Co., Ltd, 3-1-1 Yagumo-nakamachi, Moriguchi, Osaka 570, Japan. MS (Tsukuba U., 1985) material science. *Oxides garnet non_linear_optics magneto_optics*
Tel. 81(6)9062402 Fax 81(6)9063395

Minoda, Mr Hiroki (1964). Research-Associate. Physics Department, Tokyo Institute of Technology, Oh-okayama, Meguro-ku, Tokyo 152, Japan. MSc (Tokyo Institute of Technology, 1991) physics. *Surface crystal_growth electron_microscopy physics photoemission_electron_microscopy phase_transition STM*
E-mail hminoda@cc.titech.ac.jp Tel. 81(3)57342481 Fax 81(3)57342742

Misaki, Dr Shintaro (1962). Research Associate. Dept Life Science, Fac. Science, Himeji Inst of Techn., 1479-1 Kanaji, Kamigori, Akou, Hyogo 678-12, Japan. PhD (Okayama U., 1989) material science. *Structure_determination proteins enzymes computing DNA AIDS cancer*
E-mail misaki@sci.himeji-tech.ac.jp Tel. 81(7915)80177 Fax 81(7915)0177

Mitsuishi, Dr Tomokuni (1917). [Akamidai, Konosu, Saitama 365, Japan.] DSc (U. Tokyo, 1967) applied physics. *Phase_transformation*
Tel. 81(485)962788

Miura, Miss Keiko (1956). Researcher. New Drug Discovery Research Laboratories, Tsukuba Research Institute, Banyu Pharmaceutical Co, Ltd, Tsukuba Techno-Park, Oho, Okubo 3, Tsukuba 300-33, Japan. BPharm (U. Tokyo, 1980) pharmaceutical sciences. *Macromolecules enzyme_inhibitors crystallization drug pharmacology synchrotron_radiation*
Tel. 81(298)772000x2306 Fax 81(298)772029

Miyamae, Dr Hiroshi (1950). Associate professor. Department of Chemistry, Josai Univ., Keyakidai, 1-1 Sakado-shi, Saitama 350-02, Japan. PhD (U. Tokyo, 1978) inorganic chemistry. *Crystallography coordination_chemistry lead_compounds intermolecular_interaction*
E-mail miya@euclides.josai.ac.jp Tel. 81(492)852233x526 Fax 81(492)717985

Miyamoto, Prof. Masamichi (1949). Professor. Mineralogical Institute, Graduate School of Science, Univ. of Tokyo, Hongo, Bunkyo-ku, Tokyo 113, Japan. DSc (U. Tokyo, 1978) science. *Astronomy computer CVD diffusion planetary_science diamond meteorite*
Tel. 81(3)38122111x4548 Fax 81(3)38165714

Miyashita, Dr Satoru (1958). Research associate. Materials Properties Division (Crystal Physics), Institute for Materials Research, Tohoku Univ., Japan, 2-1-1 Katahira, Aoba-ku, Sendai 980, Japan. DSc (Kyoto U., 1989) physics. *Macromolecules crystal_growth superconductor kinetics interferometry proteins polymers*
E-mail satoru@jpnimrtu.bitnet Tel. 81(22)2276200x2329 Fax 81(22)2152011 Telex 852238(kinkenj)

Miyata, Dr Takeshi (1949). Professor. Yamanashi Inst., Gemology & Jewelry Arts, Tokojicho 1955-1, Kofu, Yamanashi 400, Japan. PhD (Tohoku U., 1980) mineralogy. *Crystal_growth crystal_form morphology*
E-mail mgh02711@niftyserve.or.jp Tel. 81(552)326671x33 Fax 81(552)336357

Miyawaki, Dr Ritsuro (1959). Senior researcher. National Industrial Research Institute of Nagoya, 1-1 Hirate, Kita, Nagoya 462, Japan. DSc (Tsukuba U., 1987) crystal chemistry. *Crystallography chemistry mineralogy structure lanthanides clay_minerals kaolinite rare earth*
Tel. 81(52)9112111x2313 Fax 81(52)9166993

Miyazawa, Dr Shintaro (1942). Research fellow. Miyazawa Research Laboratory, NTT LSI Laboratories, 3-1 Morinosato, Wakamiya, Atsugi, Kanagawa 243-01, Japan. DEng (Tohoku U., 1978) electronics. *Crystal_growth epitaxy oxides semiconductors*
Tel. 81(462)402720 Fax 81(462)404351

Miyazawa, Dr Yasuto (1939). Senior scientist. 14th Research Group, National Institute for Research in Inorganic Materials, 1-1 Namiki, Tsukuba, Ibaraki 305, Japan. PhD (Stanford U., 1972) nucleation. *Czochralski_technique crystal_growth*
Tel. 81(298)513351 Fax 81(298)527449

Mizugaki, Mr Tsutomu (1957). Nikon Inc., 1-10-1 Asamizodai, Sagamihara, Kanagawa 228, Japan. MEng (Sophia U., 1981) *Crystal_growth*
Tel. 81(427)406305 Fax 81(427)406330

Mizuno, Dr Hiroshi (1943). Group leader. Dept. Molecular Biology, National Institute of Agrobiological Resources, Kannondai 2-1-2, Tsukuba 305, Japan. DPharm (Osaka U., 1972) chemistry. *Macromolecules enzymes viruses molecular_biology*
E-mail mizuno@abr.affrc.go.jp Tel. 81(298)387014 Fax 81(298)387408

Mizuno, Dr Kaoru (1954). Associate professor. Dept. of Physics, Faculty of Science, Shimane Univ., 1060 Nishikawatsu, Matsue 690, Japan. DSc (Hiroshima U., 1986) lattice defects. *X-ray_topography dislocation vacancy synchrotron_radiation aluminium*
Tel. 81(852)326100 Fax 81(852)326409

Mizuno, Dr Yukiko (1941). Lecturer. Applied Physics Section, Institute of Low Temperature Science, Hokkaido Univ., North 19, West 8, Kita-ku, Sapporo 060, Japan. DSc (Hokkaido U., 1983) geophysics (physics of snow and ice). *Creep recrystallization interface*
Tel. 81(11)7162111 Fax 81(11)7165698

Mochizuki, Prof. Dr Katsumi (1945). Professor. Dept. of Electronic Materials, Faculty of Science and Engineering, Ishinomaki Senshu Univ., 1 Aza-Shinmito, Minamisakai, Ishinomaki 986, Japan. DEng (Tohoku U., 1975) crystal growth. *Crystal_growth luminescence III-VI_compounds I-III-VI2_compounds*
Tel. 81(225)227716 Fax 81(225)227746

Momoi, Prof. Hitoshi (1930). Professor. Dept. of Earth Sciences, Faculty of Science, Ehime Univ., 2-5 Bunkyo-cho, Matsuyama 790, Japan. DSc (Kyushu U., 1963) manganese mineralogy. *Crystal_growth*
Tel. 81(899)247111 Fax 81(899)232545

Mori, Mr Atsushi (1967). Dept. of Optical Science and Technology, Faculty of Engineering, University of Tokushima, 2-1 Minamijosanjima, Tokushima 770, Japan. MEng (Nagoya U., 1991) applied physics. *Statistical_mechanics phase_transition computer_simulation*
E-mail mori@opt.tokushima-u.ac.jp Tel. 81(886)567539 Fax 81(886)556549

Mori, Mr Shigeo (1965). Research Assistant. Department of Physics, Tokyo Institute of Technology, 12-1 2-chome, Oh-okayama, Meguro-ku, Tokyo 152, Japan. MSc (Waseda U., 1992) material science. *Phase_transition solid_state physics electron_microscopy physical_property*
Tel. 81(3)37261111x2079 Fax 81(3)37290042

Morikawa, Kosuke (1942). Director of 1st Department. First Department, Protein Engineering Research Institute, 6-2-3 Furuedai, Suita, Osaka 565, Japan. DPharm (U. Tokyo, 1972) chemistry. *Proteins_crystallography enzymes protein_nucleic_acids_interaction*
E-mail morikawa@pes4.peri.co.jp Tel. 81(6)8728211 Fax 81(6)8728210

Morimoto, Prof. Jun (1950). Professor. Dept. of Materials Science and Engineering, National Defense Academy, 1-10-20 Hashirimizu, Yokosuka, Kanagawa 239, Japan. PhD (Tokyo Institute of Technology, 1983) information processing. *Crystal_growth semiconductor AES SIMS PAS DLTS*
Tel. 81(468)413810 Fax 81(468)445910

Morimoto, Yukio (1958). Associate Professor. Faculty of Engineering, The University of Tokushima, 2-1 Minami-Josanjima, Tokushima 770, Japan. DSc (Osaka U., 1987) chemistry. *Biocrystallography proteins viruses ribosomes computing structure_biology*
E-mail morimoto@ee.tokushima-u.ac.jp Tel. 81(886)232311x4916 Fax 81(886)553160

Morinaga, Prof. Masahiko (1946). Professor. Toyohashi Univ. of Technology, Toyohashi, Japan. PhD (Northwestern U., 1978) materials science. *Diffuse_scattering lattice_distortion alloys X-ray_diffraction defect_structure*
Tel. 81(532)487854 Fax 81(532)472688

Morita, Mr Shoji (1960). Advanced Technology Research Center, Mitsubishi Heavy Industries, Ltd, 1-8-1 Sachiura, Kanazawa-ku, Yokohama, Japan. MEng (1984) *Crystal_growth*
Tel. 81(45)7711223 Fax 81(45)7711505

Moriyama, Mr Hideaki (1959). Assistant professor. Tokyo Institute of Technology, 4259 Nagatsuta, Midori-ku, Yokohama 227, Japan. DEng (Osaka U., 1987) crystallography. *Proteins*
E-mail hmoriyam@nc.titech.ac.jp Tel. 81(45)9221111x2387 Fax 81(45)9222432

Motegi, Dr Nawoto (1944). Researcher. Research and Development Center, Materials and Devices Research Laboratories, Research Laboratory I, Toshiba Corporation, 1 Komukai Toshiba-cho, Saiwai-ku, Kawasaki 210, Japan. PhD (U. Tokyo, 1947) solid state physics. *Crystal_growth*
Tel. 81(44)5492223 Fax 81(44)5201286

Mouri, Mr Mikio (1949). Research and Development Division, NEC Kansai, Ltd, 2-9-1 Seiran, Otsu, Siga 520, Japan. MEng (Kyoto U., 1974) *Crystal_growth CVD epitaxy solar_cell*
Tel. 81(775)372100x3340 Fax 81(775)376902

Muranaka, Mr Ken-ichiro (1958). Research scientist. Information Systems Dept., Nippon Shinyaku Co., Ltd, Nishioji Hachijo, Minami-ku, Kyoto 601, Japan. MS (U. Illinois, 1985) biophysics. *Crystallography crystallization proteins data_processing simulation data_analysis quantitative_structure-activity_relationship_(QSAR) deconvolution_analysis*
Tel. 81(75)3211111 Fax 81(75)3143269

Nagakura, Prof. Sigemaro (1926). Professor. Dept. of Materials Sci. and Tech., Waseda University, 3-4-1 Okubo, Shinjuku, Tokyo 169, Japan. [3-56-29 Kamariya-Nishi, Kanazawa, Yokohama 236, Japan.] DSc (Kyoto U., 1952) physics. *Electron_microscopy metallography carbides nitrides crystal_growth characterization steel*
Tel. 81(45)7883454 Fax 81(45)7883454

Nagashima, Dr Nobuya (1935). Senior scientist. Central Research Laboratories, Ajinomoto Co., Inc., 1-1 Suzuki-cho, Kawasaki-ku, Kawasaki 210, Japan. DSc (U. Tokyo, 1985) hydration of biological materials. *Crystallization crystal_structure*
Tel. 81(44)2447142 Fax 81(44)2117609

Nagashima, Dr Seiichi (1949). Associate professor. College of Engineering, Nihon Univ., 1 Nakagawara, Tokusada, Tamura-machi, Koriyama, Fukushima 963, Japan. DEng (Nihon U., 1977) thin film. *Epitaxy thin_film*
Tel. 81(249)441300 Fax 81(249)437736

Nakagawa, Dr Atsushi (1961). Research Associate. Photon Factory, National Laboratory for High Energy Physics, 1-1 Oho, Tsukuba, Ibaraki 305, Japan. PhD (Osaka U., 1989) chemistry. *Synchrotron_radiation structure_determination anomalous_dispersion biological_macromolecules multiple_anomalous_diffraction_phasing time-resolved_structure_determination data_collection*
E-mail nakagawa@kekvax.kek.jp Tel. 81(298)645646 Fax 81(298)642801 Telex 3652-534

Nakahigashi, Prof. Kiyotaka (1941). Professor. College of Integrated Arts and Sciences, Univ. of Osaka Prefecture, Gakuencho, Sakai, Osaka 593, Japan. DEng (U. Osaka Prefecture, 1976) crystal structure. *Powder maximum-entropy_method electron_density_distribution phase_transition ultrafine_particles*
E-mail kiyotaka@expy.cias.osakafu-u.ac.jp Tel. 81(722)521161x2711 Fax 81(722)552981

Nakai, Dr Izumi (1953). Associate Professor. Dept. Applied Chemistry, Fac. Sci., Science University of Tokyo, Kagurazaka, Shinjuku 162, Japan. PhD (U. Tsukuba, 1980) chemistry. *Synchrotron_radiation EXAFS X-ray_fluorescence_spectroscopy mineralogy structure_determination X-ray_powder_diffraction Rietveld_analysis XANES*
E-mail inakai@ch.kagu.sut.ac.jp Tel. 81(3)32603662 Fax 81(3)32352214

Nakamura, Prof. Akira (1930). Professor. Dept. of Applied Physics, Faculty of Science, Science Univ. of Tokyo, 1-3 Kagurazaka, Shinjuku, Tokyo 162, Japan. PhD (Osaka U., 1968) physics. *Zinc_compounds photoluminescence*
Tel. 81(3)32604271 Fax 81(3)32604772

Nakamura, Mrs Hikaru (1944). Research Scientist. Institute of Microbial Chemistry, 3-14-23 Kamiosaki, Shinagawa-ku, Tokyo 141, Japan. BSc (Science University of Tokyo, 1967) mathematics. *Antibiotics*
Tel. 81(3)34414173 Fax 81(3)34417589

Nakamura, Prof. Kazuo (1945). Professor. School of Pharmaceutical Sciences, Showa University, 1-5-8 Hatanodai, Shinagawa-ku, Tokyo 142, Japan. PhD (U. of Tokyo, 1974) pharmaceutical sciences. *Proteins_crystallography computing computer_graphics*
E-mail g30588@tansei.cc.u-tokyo.ac.jp Tel. 81(3)37848199 Fax 81(3)37825635

Nakamura, Mr Masaki (1963). Researcher. Research Center for Creating New Materials, National Institute for Research in Inorganic Materials, 1-1 Namiki, Tsukuba, Ibaraki 305, Japan. (Science Univ. of Tokyo, 1986) *Solid_state phase_equilibrium crystal_growth*
Tel. 81(298)513351 Fax 81(298)527449

Nakamura, Dr Yoshio (1958). Associate Professor. Dept. Metall. Eng., Tokyo Institute of Technology, Oh-okayama, Meguro, Tokyo 152, Japan. DEng (Tokyo Inst. Tech., 1986) metallurgy. *Transmission_electron_microscopy high-resolution_electron_microscopy substructure mixed-valence_compounds anti_phase*
Tel. 81(3)37261111x3145 Fax 81(3)37263419

Nakano, Dr Kikuo (1937). Senior researcher. Material Processing Division, National Industrial Research Institute of Nagoya, 1-1 Hirate-cho; Kita-ku, Nagoya 462, Japan. DEng (Nagoya U., 1977) metallurgy. *Crystal_growth*
Tel. 81(52)9112111 Fax 81(52)9162802

Nakata, Mr Kazuaki (1946). Associate professor. Division of Natural Science, Osaka Kyoiku Univ., 4-698-1 Asahigaoka, Kashihara 582, Japan. MEd (Osaka Kyoiku U., 1971) physics. *Powder disorder computing*
E-mail nakata@cc.osaka-kyoiku.ac.jp Tel. 81(729)763211 Fax 81(729)763273

Nakatsu, Prof. Kazumi (1928). Professor. School of Science, Kwansei Gakuin Univ., Uegahara, Nishinomiya 662, Japan. DSc (Osaka Univ, 1961) X-ray crystallography. *Anomalous_dispersion photochromism proteins low-dimensional imaging_science photographic_science functional_dyes organic_conducting_materials organometallic_compounds spectral_sensitization vector_search imaging_plate*
E-mail nakatsu@kgupyr.kwansei.ac.jp Tel. 81(798)546401 Fax 81(798)510914

Nakayama, Mr Hitoshi (1967). Faculty of Science and Engineering, Saga University, 1 Honjyo, Saga-city, Saga 840, Japan. MEng (Saga U., 1991) *Crystal_growth*
E-mail nakayamah@saga.cc.saga-u.ac.jp Tel. 81(952)245191 Fax 81(952)294441

Nakazumi, Dr Yoshihide (1919). President. Nakazumi Crystal Laboratory, 3-1 Sugaharacho, Ikeda, Osaka 563, Japan. DSc (Osaka U., 1961) chemistry. *Titanates oxides gemology flux_method flame_fusion_method*
Tel. 81(727)518832 Fax 81(727)522003

Nanba, Mr Hirokuni (1944). Manager. Basic High Technology Lab., Sumitomo Electric Ind., Ltd, 1-1-3 Shimaya, Konohana-ku, Osaka, Japan. MEng (Kyoto U., 1968) *II-VI_compounds*
Tel. 81(6)4665594 Fax 81(6)4665733

Nara, Dr Shigetoshi (1947). Associate professor. Dept. of Electrical & Electronic Engineering, Okayama Univ., 3-1-1 Tsushima-naka, Okayama 700, Japan. DSc (Kyoto U., 1977) physics. *Crystal_growth*
E-mail nara@mat.elec.okayama-u.ac.jp Tel. 81(86)2521111x8142 Fax 81(86)2561746

Nemoto, Takashi (1969). Student. Department of Chemistry, Faculty of Science, Tokyo Institute of Technology, Ookayama, Meguro-ku, Tokyo 152, Japan. MSc (Tokyo Institute of Technology, 1994) chemistry. *Chirality reactivity time-resolved_effect cobalt_compounds crystalline_state_reactions time-resolved_structure_analysis*
E-mail tnemoto@chem.titech.ac.jp Tel. 81(3)57342608 Fax 81(3)37206206

Niimura, Prof. Nobuo (1942). Group Head. Advanced Science Research Center, Japan Atomic Energy Research Institute, Tohkai-mura, Naka-gun, Ibaraki-ken, Japan. DSc (U. Tokyo, 1970) chemistry. *Neutron_diffraction small-angle_neutron_scattering proteins_structure_analysis neutron_detector*
Tel. 81(292)825906 Fax 81(292)825927

Niizeki, Dr Nobukazu (1925). PhD (Massachusetts Institute of Technology, 1957) crystallography. *Crystal_growth optoelectronics*

Ninomiya, Dr Hiroshi (1952). Research assistant. Dept. of Applied Physics, Fukuoka Univ., 8-19-1 Nanakuma, Jo-nan-ku, Fukuoka 814-01, Japan. DSc (Fukuoka U., 1991) applied physics. *Form phase_transition science_of_form computational_physics*
Tel. 81(92)8716631 Fax 81(92)8656030

Nishi, Prof. Fumito (1949). Professor. Saitama Institute of Technology, Fusaiji 1690, Okabe, 369-02 Saitama, Japan. DSc (Tokyo U., 1978) mineralogy. *Structure_determination germanates thermal_structure_change*
Tel. 81(485)852521x2413 Fax 81(485)852523

Nishida, Prof. Dr Takashi (1938). Professor. Department of Earth Sciences, Faculty of Science, Chiba University, 1-33 Yayoicho, Inage-ku, Chiba 263, Japan. DSc (U. Tokyo, 1970) mineralogy. *Mineralogy imperfection twinning polytypism crystallite_imperfection*
E-mail nishida@cucas.c.chiba-u.ac.jp Tel. 81(43)2903718 Fax 81(43)2903715

Nishihata, Dr Yasuo (1961). Research Associate. Faculty of Science, Okayama University, Tsushimanaka, Okayama 700, Japan. DSc (Kwansei Gakuin U., 1989) physics. *X-ray_diffraction EXAFS ferroelectric*
E-mail yasuon@cc.okayama-u.ac.jp Tel. 81(86)2521111x7812 Fax 81(86)2527595

Nishikawa, Dr Keiko (1948). Associate Professor. Faculty of Education, Yokohama National University, Tokiwadai, Hodogaya-ku, Yokohama 240, Japan. DSc (U. Tokyo, 1982) chemistry. *Liquid_state diffuse_scattering SAXS instrumentation fluctuation phase_transition liquid_crystal*
Tel. 81(45)3351451x2220 Fax 81(45)3331536

Nishinaga, Prof. Tatau (1939). Professor. Elected Member of Commission on Crystal Growth, IUCr; Vice President of Int. Org. Cryst. Growth (IOCG). Dept. of Electric Engineering, Univ. Tokyo, 7-3-1 Hongo, Bunkyo-ku, Tokyo 113, Japan. DEng (Nagoya Univ, 1967) electronics. *Crystal_growth semiconductor crystal_growth_under_microgravity molecular_beam_epitaxy*
Tel. 81(3)38122111x6673 Fax 81(3)56843974

Nishino, Dr Yoichi (1955). Associate professor. Dept. of Materials Science and Engineering, Nagoya Institute of Technology, Gokiso, Showa, Nagoya 466, Japan. PhD (Nagoya U., 1983) materials science. *Plasticity dislocation internal_friction EXAFS strength lattice stability thin_film*
E-mail nishino@mse.nitech.ac.jp Tel. 81(52)7322111x2525 Fax 81(52)7350487

Nishioka, Prof. Dr Kazumi (1940). Professor. Dept. of Optical Science and Technology, Univ. of Tokushima, 2-1 Minamijosanjima, Tokushima 770, Japan. PhD (Stanford U., 1970) materials science. *Nucleation interface thermodynamics dislocation elasticity fracture surface*
E-mail nishioka@opt.tokushima-u.ac.jp Tel. 81(886)567538 Fax 81(886)556549

Nishizaki, Mr Katsumi (1957). Deputy Manager. Quality Assurance Section LSI Division, Utsunomiya Works, Kawasaki Steel Corp., 166 Haga-dai, Haga-machi, Haga-gun, Tochigi 321-33, Japan. MS (Osaka U., 1981) precision engineering. *Crystal_growth*
Tel. 81(286)775312 Fax 81(286)775329

Nishizawa, Dr Shin-ichi (1965). Research Associate. Dept. of Chemical Engineering, Waseda Univ., 3-4-1 Ohkubo, Shinjuku, Tokyo 169, Japan. Dr. Eng. (Waseda U., 1994) chemical engineering. *Czochralski_technique float_zone_growth heat_transfer microgravity hydrodynamics convective_heat thermodynamics*
E-mail nisizawa@cfi.waseda.ac.jp or CXB01205@niftyserve.or.jp
Tel. 81(3)32034141x733324 Fax 81(3)32327083 Telex 232-5115 WARIKO J

Nittono, Prof. Osamu (1941). Professor. Co-editor of J. Appl. Cryst. Dept. Metall. Eng., Tokyo Inst. Tech., Oh-okayama, Meguro, Tokyo 152, Japan. DEng (Tokyo Inst. Tech., 1970) metallurgy. *Metallurgy semiconductor structural_change dynamical_diffraction diffraction_technique thin_film phase_transformation multilayered_films nanostructure X-ray_goniometry X-ray_topography*
Tel. 81(3)37261111x3145 Fax 81(3)37263419

Noda, Prof. Yasutoshi (1942). Associate professor. Dept. of Materials science, Fac. of Eng., Tohoku Univ., Aoba Aramaki, Aoba, Sendai 980, Japan. DEng (Tohoku U., 1970) materials science. *Crystal_growth structure_determination characterization chalcopyrite thermoelectric_materials compounds*
E-mail nodayasu@material.tohoku.ac.jp Tel. 81(22)2221800x4464 Fax 81(22)2682949 Telex 852246 THUCOMJ

Nonaka, Dr Takamasa (1963). Lecturer. Department of BioEngineering, Nagaoka University of Technology, Kamitomioka, Nagaoka, Niigata 940-21, Japan. DSc (Nagaoka U. Tech., 1992) protein crystallography. *Structure–activity_relationship Laue_diffraction time-resolved_effect macromolecules ribonuclease*
E-mail nonaka@voscc.nagaokaut.ac.jp Tel. 81(258)466000x4536 Fax 81(258)468163

Obinata, Mr Shiro (1944). General manager. Semiconductor Materials Laboratory, Dowa Mining Co., Ltd, 1 Sunada, Iijima, Akita 011, Japan. MEng (Tohoku U., 1969) *Crystal_growth*
Tel. 81(188)468000 Fax 81(188)469478

Ogawa, Dr Kazuhiko (1958). Research associate. Graduate School of Science and Technology, Kobe Univ., Rokkodai, Nada, Kobe, Japan. DEng (Kobe U., 1986) mechanical engineering. *Aerodynamics heat_transfer*
Tel. 81(78)8811212 Fax 81(78)8810036

Ogawa, Prof. Keiichiro (1952). Associate Professor. National Editor WDC9. Department of Chemistry, The College of Arts and Sciences, The University of Tokyo, Komaba, Meguro, Tokyo 153, Japan. DSc (U. Tokyo, 1983) chemistry. *Molecular_motion organic_chemistry molecular_crystal phase_transition*
E-mail ogawa@ramie.c.u-tokyo.ac.jp Tel. 81(3)34671171x593 Fax 81(3)34852904

Ogawa, Dr Keizo (1951). Senior Researcher. Ashigara Research Lab., Fuji Photo Film Co. Ltd, 210 Nakanuma, Minami-Ashigara, Kanagawa 250-01, Japan. PhD (Osaka U., 1980) pharmaceutical sciences. *Biomolecule crystallography drug X-ray_microscopy tennis*
Tel. 81(465)737080 Fax 81(465)737923

Ogawa, Prof. Tomoya (1930). Professor. Dept. of Physics, Gakushuin Univ., 1-5-1 Mejiro, Toshima-ku, Tokyo 171, Japan. PhD (U. Tokyo, 1966) electronics. *Crystal_growth defect light_scattering*
Tel. 81(3)39860221 Fax 81(3)35902602

Ogura, Prof. Iwao (1922). Professor Emeritus. College of Engineering, Nihon University, Koriyama, Japan. DSc (Hiroshima U., 1961) physics. *Thin_film crystal_growth surface electron_microscopy semiempirical_calculation diffusion statistical_thermodynamics*
Tel. 81(249)441300x2451 Fax 81(249)432913

Ohachi, Prof. Tadashi (1941). Professor. Dept. of Electrical Engineering, Doshisha University, Tanabe, Kyoto 610-03, Japan. DEng (Doshisha U., 1975) electrical engineering. *Crystal_growth crystal_form ionic_conductivity microwave roughening_transition*
E-mail tohachi@doshisha.ac.jp Tel. 81(7746)56329 Fax 81(7746)56811

Ohama, Prof. Nobuhiko (1940). Professor. Division of Childhood Education, Department of Literature, Seinan Gakuin University, Nishijin 6-chome 2-92, Sawara-ku, Fukuoka 814, Japan. DSc (Kyushu U., 1971) physics. *X-ray_diffraction diffractometry phase_transition crystal_physics mathematics computing*
E-mail ohama@seinan-gu.ac.jp Tel. 81(92)8411311x3415 Fax 81(92)8232506

Ohashi, Prof. Yuji (1941). Professor. Member, Commission on small molecules. Department of Chemistry, Faculty of Science, Tokyo Institute of Technology, Ookayama, Meguro-ku, Tokyo 152, Japan. DSc (U. Tokyo, 1974) chemistry. *Chirality reactivity time-resolved_effect cobalt_compounds crystalline_state_reactions time-resolved_structure_analysis*
E-mail yohashi@chem.titech.ac.jp Tel. 81(3)57342223 Fax 81(3)37206206

Ohba, Mr Hiroyuki (1958). Corporate Planning Division, Tokin Corp., Sumitomo-Seimei-Sendai-Chuou Building, 4-6-1 Chuou, Aoba-ku, Sendai 980, Japan. (Tohoku U., 1981) *Crystal_growth*
Tel. 81(22)2115414 Fax 81(22)2116593

Ohba, Prof. Shigeru (1953). Associate professor. Department of Chemistry, Faculty of Science and Technology, Keio University, Hiyoshi 3, Kohoku-ku, Yokohama 223, Japan. PhD (U. Tokyo, 1980) chemistry. *Charge_density magnetochemistry solid_state_reaction*
E-mail k12330@jpnkeio.bitnet Tel. 81(45)5631141x3912 Fax 81(45)5635967

Ohba, Dr Takuya (1953). Associate Professor. Department of Materials Science and Engineering, School of Science and Engineering, Teikyo University, Toyosatodai, Utsunomiya 320, Japan. DSc (Hiroshima U., 1984) materials science. *Diffraction structure metals phase_transition martensite composite_crystal*
E-mail ohba@koala.mse.teikyo-u.ac.jp Tel. 81(286)277166 Fax 81(286)277185

Ohgaki, Dr Masataka (1959). Research associate. Institute for Medical and Dental Engineering, Tokyo Medical and Dental University, 2-3-10 Surugadai, Kanda, Chiyoda-ku, Tokyo 101, Japan. DSc (Tokyo Inst. of Techn., 1990) crystal chemistry. *Biomaterial ceramics minerals crystal_structural_chemistry electron_density_distribution cytotoxicity_for_biomaterials*
E-mail ohgaki@tansei.cc.u-tokyo.ac.jp Tel. 81(3)52808019 Fax 81(3)52808005

Ohishi, Dr Hirofumi (1957). Assistant. Information Center, Osaka Univ. of Pharmaceutical Sciences, 2-10-65 Kawai, Matsubara, Osaka 580, Japan. PhD (U. Osaka, 1992) physical chemistry. *Crystallography_proteins crystallography nucleic_acids structure-activity_relationship carcinogens computational_study_nucleic_acids crystallization computational_study_proteins*
E-mail ohishi@pxewsl.protein.osaka-u.ac.jp Tel. 81(723)1015x286 Fax 81(723)9929

Ohkawa, Prof. Dr Tokio (1934). Professor. Polytechnic Univ., 4-1-1 Hashimotodai, Sagamihara, Kanagawa 229, Japan. DSc (Osaka U., 1984) X-ray physics. *X-ray_microscopy method*
Tel. 81(427)639005x306 Fax 81(427)639168

Ohkobchi, Mr Masato (1943). Lecturer. Dept. of Physics, Meijo Univ., 1-501 Shiogamaguchi, Tenpaku-ku, Nagoya 468, Japan. (Meijo U., 1966) *Crystal_growth*
Tel. 81(52)8321151 Fax 81(52)8321170

Ohmasa, Prof. Masaaki (1935). Professor. Member, Commission on Journals. Department of Life Science, Faculty of Science, Himeji Institute of Technology, Kanaji 1479-1, Kamigori, Akogun, Hyogo 678-12, Japan. DSc (U. of Tokyo, 1964) mineralogy. *X-ray_diffraction inorganic_materials modulated_structures phase_transition topotactic_phase_transformation diffraction_synchrotron_radiation_microcrystals symmetry*
E-mail ohmasa@sci.himeji-tech.ac.jp Tel. 81(7915)80213 Fax 81(7915)80216

Ohnaka, Prof. Dr Itsuo (1940). Professor. Dept. of Materials Science & Processing, Faculty of Engineering, Osaka Univ., 2-1 Yamadaoka, Suita, Osaka 565, Japan. DEng (U. Tokyo, 1968) mechanical engineering. *Solidification computer_simulation casting*
E-mail ohnaka@msp.eng.osaka-u.ac.jp Tel. 81(6)8797473 Fax 81(6)8797474

Ohno, Prof. Hideo (1954). Professor. Research Institute of Electrical Communication, Tohoku University, Katahira, Aobaku 980-77, Japan. PhD (U. Tokyo, 1982) electronic engineering. *Semiconductor molecular_beam_epitaxy atomic_layer_epitaxy*
E-mail ohno@riec.tohoku.ac.jp Tel. 81(22)2674748 Fax 81(22)2674748

Ohno, Dr Takehisa (1949). Associate professor. Dept. of Mechanical Engineering (Fundamental Science), Gifu National College of Technology, Shinsei-cho, Motosu-gun, Gifu 501-04, Japan. DEng (Nagoya U., 1989) crystal growth. *Diffraction_technique phase_separation*
Tel. 81(583)241101 Fax 81(583)232709

Ohsato, Associate Prof. Hitoshi (1944). Associate Professor. Nagoya Institute of Technology, Dept. of Materials Science and Engineering, Gokiso-cho, Showa-ku, Nagoya 466, Japan. DSc (U. Tokyo, 1984) mineralogy, crystal structure analysis. *Dielectric_ceramics semiconductor_thin_films epitaxy structure_analysis inorganic_materials_sciences one_dimensional_tunnel_structure*
E-mail ohsato@mse.nitech.ac.jp Tel. 81(52)7322111x2520 Fax 81(52)7322925

Ohsumi, Prof. Kazumasa (1943). Professor. CSJ executive committee member. Photon Factory (PF), National Laboratory for High Energy Physics (KEK), 1-1 Oho, Tsukuba, Ibaraki 305, Japan. DSc (U. Tokyo, 1971) mineralogy. *Synchrotron_radiation micro-crystallography symmetry minerals*
Tel. 81(298)645651 Fax 81(298)642801

Ohtsuka, Prof. Yasukuni (1935). Professor. Tohoku Institute of Technology, 35-1 Kasumi-cho, Yagiyama, Taihaku-ku, Sendai 982, Japan. MSc (Tohoku U., 1961) physics. *Epitaxy nucleation epitaxial_temperature*
Tel. 81(22)2291151 Fax 81(22)2477861

Oishi, Dr Shuji (1949). Associate professor. Dept. of Chemistry and Material Engineering, Faculty of Engineering, Shinshu Univ., 500 Wakasato, Nagano 380, Japan. DEng (Nagoya U., 1985) crystal growth. *Crystal_growth*
Tel. 81(262)264101 Fax 81(262)284295

Oka, Dr Kunihiko (1946). Senior researcher. Exotic Matter Physics, Electrotechnical Laboratory, 1-1-4 Umezono, Tsukuba, Ibaraki 305, Japan. DEng (Tsukuba U., 1992) crystal growth. *Solution growth*
E-mail koka@am.etl.go.jp Tel. 81(298)585441 Fax 81(298)545085

Okabe, Dr Nobuo (1942). Associate professor. Faculty of Pharmaceutical Sciences, Kinki Univ., 3-4-1 Kowakae, Higashiosaka, Osaka 577, Japan. DSc (Osaka U., 1970) physical chemistry. *Crystallography circular_dichroism fluorescence blood_protein-drug_interaction structure_of_chelate_compounds_of_biomolecule structure_and_function_of_proteins*
Tel. 81(6)7212332x3816 Fax 81(6)7301394

Okabe, Prof. Toshio (1942). Professor. Dept. of Physics, Fac. of Science, Toyama Univ., 3190 Gofuku, Toyama 930, Japan. DSc (Kyoto U., 1974) physics. *Whisker thin_film electron_microscopy thin_film*
E-mail okabe@jpntyavm.bitnet Tel. 81(764)411271x2314 Fax 81(764)412972

Okada, Kenji (1942). Researcher. Research and Development Center, Ricoh Company, Ltd, 16-1 Shinei-cho, Tsuzuki-ku, Yokohama 224, Japan. Dr (Science U. of Tokyo, 1965) applied chemistry. *Computing_method computer_automation computer_sciences computer_graphics dyes_compound steroids_compound organic_compound*
E-mail kokada@srd.rdc.ricoh.co.jp Tel. 81(45)5933411 Fax 81(45)5933493

Okada, Dr Yasumasa (1940). Senior research fellow. Electrotechnical Lab., Umezono, Tsukuba-shi, Ibaraki 305, Japan. PhD (Tohoku U., 1990) physics. *Semiconductors III–V_compounds dynamical diffraction vacancy X-ray diffraction single_crystal_characterization defect_structure*
E-mail yaokada@qm.etl.go.jp Tel. 81(298)585392 Fax 81(298)545085 Telex 3652570 AISTJ

Okano, Dr Yasunori (1960). Research associate. Institute for Materials Research, Tohoku Univ., 2-1-1 Katahira, Aoba-ku, Sendai 980, Japan. PhD (Waseda U., 1989) chemical engineering. *Crystal_growth convective_heat Czochralski_technique mass_transfer*
Tel. 81(22)2152102 Fax 81(22)2152101

Okazaki, Prof. Atsushi (1931). Professor. Department of Physics, Kyushu University, Fukuoka 812, Japan. DSc (Kyushu U., 1961) physics. *X-ray_high-resolution_diffractometry phase_transition Crystal_characterization effect_of_stress_on_diffraction*
Tel. 81(92)6411101x4177 Fax 81(92)6334525

Okazaki, Prof. Dr Megumi (1942). Professor. Department of Biology, Faculty of Education, Tokyo Gakugei Univ., 4-1-1 Nukuikita-machi, Koganei-city, Tokyo 184, Japan. PhD (Tokyo Univ. of Education, 1973) plant physiology & biochemistry. *Biology calcification photosynthesis carbonates marine_crystal_evolution*
Tel. 81(423)252111x2667 Fax 81(423)249832

Okino, Mr Fujio (1954). Lecturer. Shinshu Univ., 3-15-1 Tokida, Ueda 386, Japan. PhD (U. California, Berkeley, 1984) chemistry. *Fluorine_compounds graphites intercalation order-disorder battery electrochemistry fullereness*
Tel. 81(268)221215 Fax 81(268)236214

Okubo, Prof. Dr Tsuneo (1941). Associate professor. Dept. of Polymer Chemistry, Kyoto Univ., Yoshida-honmachi, Sakyo-ku, Kyoto 606-01, Japan. DEng (Kyoto U., 1970) polymer chemistry. *Crystal_growth single_crystal light_scattering diffraction poly-electrolytes*
Tel. 81(75)7535611 Fax 81(75)7535609

Okuda, Prof. Dr Takashi (1939). Professor. Dept. of Materials Science and Engineering, Nagoya Institute of Technology, Gokiso-cho, Showa-ku, Nagoya 466, Japan. DSc (Osaka U., 1969) magnetism. *Ferrites thin_film sputtering ion_beam magneto_optics electroceramics garnets*
Tel. 81(52)7322111x2514 Fax 81(52)7322925

Okuno, Dr Masayuki (1955). Research associate. Lab. Mineralogy, Dept. Earth Sciences, Kanazawa University, Kakuma-machi, Kanazawa, Ishikawa 920-11, Japan. DSc (Tokyo Institute of Technology, 1983) mineralogy. *Glasses melts_structure EXAFS minerals deformation_electron_microscopy silicates_structure_determination*
Tel. 81(762)645728 Fax 81(762)645746

Okuno, Dr Yasuo (1945). Project leader. Okuno Project, Kanagawa Academy of Science and Technology, KSP (west 614), 3-2-1 Sakado, Takatsu-ku, Kawasaki-shi, Kanagawa 213, Japan. DEng (Tohoku U., 1973) electronics. *Crystal_growth*
Tel. 81(44)8192034 Fax 81(44)8192026

Okutsu, Dr Tetsuo (1961). Researcher. Ashigara Research Laboratory, Fuji Photo Film Co., Ltd, 210 Nakanuma, Minamiashigara, Kanagawa 250-01, Japan. DSc (Tokyo Institute of Technology, 1991) physical chemistry. *Magnetic_resonance photochemistry*
Tel. 81(465)737115 Fax 81(465)737914

Okuyama, Prof. Kenji (1947). Professor. Faculty of Technology, Tokyo University of Agriculture and Technology, Koganei, Tokyo 184, Japan. DSc (Osaka U., 1977) chemistry. *Biocrystallography biomolecule molecular_crystal membranes antiferroelectricity biopolymers celluloses gels macromolecules oligopeptides oligosaccharides peptides proteins polysaccharides*
E-mail okuyamak@cc.tuat.ac.jp Tel. 81(423)887028 Fax 81(423)887209 Telex 2832663TUATTJ

Omino, Mr Akira (1947). Manager. Central Research Laboratories, Mitsui Mining Co., Ltd, 1 Kou-machi, Tochigi-shi, Tochigi 328, Japan. MEng (Tohoku U., 1974) mining and resource. *Bridgman_Stockbarger_technique crystal_growth II–VI_compounds*
Tel. 81(282)276841 Fax 81(282)277027

Ono, Prof. Dr Masatoshi (1944). Professor. Department of General Education, Health Sciences University of Hokkaido, Ishikari-Tobetsu, Hokkaido 061-02, Japan. DSc (Hokkaido U., 1973) solid state physics. *Crystal_growth*
E-mail ono@osf.hokudai.ac.jp Tel. 81(1332)31211 Fax 81(1332)30524

Ono, Mrs Yoko (1964). Research associate. Mineralogical Institute, Faculty of Science, University of Tokyo, Hongo 7-3-1, Bunkyoku, Tokyo 113, Japan. MSc (Tohoku U., 1990) crystal chemistry. *Crystal_structure*
Tel. 81(3)38122111x4547 Fax 81(3)38165714

Onoda, Dr Mitsuko (1944). Senior Researcher. National Institute for Research in Inorganic Materials, 1-1 Namiki, Tsukuba, Ibaraki 305, Japan. PhD (U. Tokyo, 1971) chemistry. *Rietveld_method diffuse_scattering sulfides incommensurate composite_crystal*
Tel. 81(298)513351

Onodera, Dr Akira (1950). Associate professor. Dept. Phys. Fac. Sci., Hokkaido Univ., Sapporo 060, Japan. DSc (Hokkaido U., 1974) physics. *Phase_transition ferroelectricity specific_heat ordering*
E-mail onodera@S1.hines.hokudai.ac.jp Tel. 81(11)7062680 Fax 81(11)7465444 Telex 932510HOKUSCJ

Onozuka, Dr Takashi (1942). Associate professor. Inst. for Materials Res., Tohoku Univ., 2-1-1 Katahira, Aoba-ku, Sendai 980, Japan. PhD (Tohoku U., 1982) material engineering. *Structural_change high-resolution_electron_microscopy superlattice diffractometry phase_transition technique modulated_structure*
Tel. 81(22)2276200x2907 Fax 81(22)2152126 Telex 852238 KINKENJ

Oonishi, Isao (1942). Professor. Department of Biomolecular Science, Faculty of Science, Toho University, 2-2-1 Miyama, Funabashi 274, Japan. DSc (Nagoya U., 1970) chemical crystallography. *Strain deformation fused_rings_organic_compounds*
E-mail ioonishi@tansei.cc.u-tokyo.ac.jp Tel. 81(474)727601 Fax 81(474)727630

Ooshima, Dr Hiroshi (1950). Associate professor. Dept. Bioapplied Chemistry, Osaka City Univ., 3-3-138 Sugimoto, Sumiyoshi-ku, Osaka 558, Japan. DEng (Osaka City U., 1986) biochemical engineering. *Crystallization proteins precipitation mechanism bioseparation*
Tel. 81(6)6052700 Fax 81(6)6052769

Osaka, Prof. Toshiaki (1941). Professor. Dept. of Materials Science and Engineering, School of Science and Engineering, Waseda Univ., Nishiwaseda, Shinjuku-ku, Tokyo 169, Japan. DEng (Waseda U., 1971) thin film. *Surface electron_diffraction thin_film electron_microscopy molecular_dynamics_simulation DV-X@_molecular_orbital_method*
Tel. 81(3)32034141x742120 Fax 81(3)32051353

Osaki, Prof. Kenji (1920). [Nampei-dai 5-9-6, Takatsuki, Osaka 569, Japan.] DSc (Osaka U., 1958) chemistry. *Molecular_crystal database*
Tel. 81(0726)953116

Osano, Yasuko T. (1960). Research Scientist. Analytical Sciences Lab., Mitsubishi Chemical Corp., 1000 Kamoshida, Aobaku, Yokohama 227, Japan. DSc (Tokyo Institute of Technology, 1992) chemistry. *Structure_property_relationship materials optoelectrical_property molecular_crystal*
E-mail osano@atlas.rc.m-kasei.co.jp Tel. 81(45)9633156 Fax 81(45)9634206

Otani, Dr Shigeki (1952). Senior researcher. The 12th Research Group, National Institute for Research in Inorganic Materials, 1-1 Namiki, Tsukuba, Ibaraki 305, Japan. PhD (Osaka U., 1984) crystal growth. *Float_zone_growth*
Tel. 81(298)513351 Fax 81(298)527449

Otobe, Mr Masanori (1967). Researcher. Oda Laboratory, Physical Electronics, Tokyo Institute of Technology, 2-12-1 O-okayama, Meguro, Tokyo 152, Japan. MEng (Tokyo Institute of Technology, 1993) electronics. *Microcrystal crystal_growth nucleation CVD*
E-mail motobe@pe.titch.ac.jp Tel. 81(3)37261111x2542 Fax 81(3)37291399

Otsuka, Prof. Kazuhiro (1937). Professor. Institute of Materials Science, Univ. of Tsukuba, Tsukuba, Ibaraki 305, Japan. DEng (U. Tokyo, 1972) metallurgy. *Phase_transition crystallography shape_memory electron_microscopy martensitic_transformation*
Tel. 81(298)535294 Fax 81(298)557440 Telex 3652580 UNTUKUJ

Oyama, Dr Yasunao (1965). Assistant. Department of Physics, Gakushuin University, 1-5-1 Mejiro, Toshima-ku, Tokyo 171, Japan. DSc (U. Tokyo, 1993) crystal chemistry. *Crystal_growth*
E-mail 940135@gakushuin.ac.jp Tel. 81(3)34786811 Fax 81(3)34015169

Ozawa, Dr Shoichi (1956). Senior research engineer. Guided Wave Optics, Opto-Technology Laboratory, The Furukawa Electric Co., Ltd, 6 Yawata-kaigandori, Ichihara, Chiba 290, Japan. PhD (U. Tokyo, 1989) crystal growth. *Crystal_growth Czochralski_technique epitaxy guided_wave_optics*
Tel. 81(436)421774 Fax 81(436)429340

Ozawa, Dr Yoshiki (1960). Research associate. Department of Material Science, Himeji Institute of Technology, Harima Science Park-City, Kamigori-cho, Hyogo 678-12, Japan. DSc (The University of Tokyo, 1988) chemistry. *Crystallography heteropoly_acids organometallic_oxides organometallic_sulfides*
E-mail ozawa@sci.himeji-tech.ac.jp Tel. 81(7915)80153 Fax 81(7915)80154

Ozeki, Dr Tomoji (1960). Research Associate. Research Laboratory of Resources Utilization, Tokyo Institute of Technology, 4259 Nagatsuta, Midori-ku, Yokohama 227, Japan. DSc (Tokyo U., 1988) chemistry. *Heteropoly_acids structural_bioinorganic_chemistry computing synchrotron_radiation*
E-mail tozeki@nc.titech.ac.jp Tel. 81(45)9245260 Fax 81(45)9245276

Paehler, Arno (1949). Senior research scientist. Analytical Sciences Lab., Mitsubishi Chemical Corp., 1000 Kamoshida, Aobaku, Yokohama 227, Japan. Dr rer.nat. (U. Goettingen, 1983) physics. *Biocrystallography computing drug_design proteins*
E-mail paehler@atlas.rc.m-kasei.co.jp Tel. 81(45)9633156 Fax 81(45)9634206

Ringer, Simon Peter (1964). Associate. Institute for Materials Research, Tohoku University, Sendai, 980-77, Japan. PhD (U. New South Wales, 1991) materials science/hsla steels/weldability. *Aluminium_alloys steels phase_transformation interface TEM atom_probe_field_ion_microscopy*
E-mail ringer@apfim.imr.tohoku.ac.jp Tel. 81(22)2152023 Fax 81(22)2152020

Sadanaga, Prof. Ryoichi (1920). Member. Japan Academy, 4-1-4 Suimeidai, Kawanishishi, Hyogo, Japan. DSc (U. of Tokyo, 1953) mineralogy. *Mathematical_crystallography*
Tel. 81(727)925100 Fax 81(727)925200

Saito, Dr Yoshio (1944). Research associate. Dept. of Electronics and Information Science, Kyoto Institute of Technology, Matsugasaki, Sakyo-ku, Kyoto 606, Japan. DSc (Kyoto U., 1985) physics. *Amorphous_phase microcrystal epitaxy energetics gas_evaporation astrophysics*
Tel. 81(75)7247442 Fax 81(75)7247442

Saito, Dr Yukio (1948). Associate professor. Dept. of Physics, Keio Univ., 3-14-1 Hiyoshi, Kohoku-ku, Yokohama 223, Japan. PhD (U. Tokyo, 1976) statistical physics. *Morphology instability Monte_Carlo theory*
E-mail yukio@rk.phys.keio.ac.jp Tel. 81(45)5631141x3973 Fax 81(45)5631761

Sakabe, Prof. Noriyoshi (1934). Professor. University of Tsukuba, Inst. of Applied Biochemistry, 1-1-1 Tennoudai, Tsukuba-shi, Ibaraki 305, Japan. DSc (Nagoya U., 1966) biophysics. *Proteins camera structure_determination actin data_collection_system synchrotron_radiation insulin*
E-mail sakabe@kekvax.kek.jp Tel. 81(298)536426 Fax 81(298)534605

Sakagami, Prof. Dr Noboru (1938). Professor. Dept. of Electrical Engineering, Akita National College of Technology, 1-1 Bunkyo-cho, Iijima, Akita 011, Japan. DEng (Tohoku U., 1977) electronics. *Crystal_growth high_purity hydrothermal_method*
Tel. 81(188)452151 Fax 81(188)573191

Sakamoto, Prof. Dr Hidekazu (1947). Associate professor. School of Science and Engineering, Teikyo Univ., 1-1 Toyosatodai, Utsunomiya, Tochigi 320, Japan. DEng (Osaka U., 1985) martensitic transformation. *Metallurgy metallography shape_memory transformation*
Tel. 81(286)277168 Fax 81(286)277185

Sakata, Dr Makoto (1944). Associate Professor. Department of Applied Physics, Nagoya University, Chikusa-ku, Nagoya, Japan. PhD (Tokyo U. Education, 1974) chemistry. *Charge_density crystallography high-precision_structure maximum_entropy_method accurate_structure_analysis anharmonic_thermal_vibration phase_transition*
E-mail a40366a@nucc.cc.nagoya-u.ac.jp Tel. 81(52)7815111x4453 Fax 81(52)7822129

Sakata, Dr Osami (1962). Research Associate. Tokyo Institute of Technology, Research Laboratory of Engineering Materials, 4259 Nagatsuta, Midoriku, Yokohama 226, Japan. Dr Eng. (Tokyo Institute of Technology, 1994) material science. *Dynamical_diffraction standing_wave surface GIXS grazing_angle_X-ray_standing_wave surface_diffraction molecular_beam*
Tel. 81(45)9245350 Fax 81(45)9225169

Sakurai, Prof. Tosio (1926). [1-4-5 Minamiotsuka, Toshima-ku, Tokyo 170, Japan.] DSc (U. Tokyo, 1962) physics. *Computing direct_method heterocycles quasicrystal*
Tel. 81(3)39466045

Sano, Mr Chiaki (1951). General manager. Refinery Technology Dept., Technology & Engineering Laboratories, Ajinomoto Co, Inc., 1-1 Suzuki-cho, Kawasaki-ku, Kawasaki 210, Japan. MS (U. Tokyo, 1977) agricultural chemistry. *Crystallization crystallography crystal_growth amino_acids*
Tel. 81(44)2447162 Fax 81(44)2117832

Sano, Prof. Masatoshi (1947). Professor. Dept. of Electrical Engineering, Faculty of Engineering, Science Univ. of Tokyo, 1-3 Kagurazaka, Shinjuku-ku, Tokyo 162, Japan. DEng (U. Tokyo, 1976) electronic engineering. *Compound_semiconductor crystal_growth luminescence electrical_conductivity*
Tel. 81(3)32604271 Fax 81(3)35261480S

Sasada, Prof. Yoshio (1926). Professor emeritus. [3-36-5 Higashi-Hongo, Midori, Yokohama 226, Japan.] DSc (Osaka U., 1958) chemistry. *Structural_chemistry reactivity proteins*
Tel. 81(45)4730674

Sasaki, Prof. Dr Akio (1932). Professor. Department of Electrical Engineering, Kyoto Univ., Yoshida Honmachi, Sakyo, Kyoto 606-01, Japan. (Kyoto U., 1976) electrical eng. *Semiconductors epitaxy optoelectronic_properties quantum_effect_properties*
Tel. 81(45)9245308 Fax 81(75)7511576

Sasaki, Prof. Kyoyu (1940). Prof. C. of Medical Techn., Nagoya University, 1-1-20 Daikominami, Higashi-ku, Nagoya 461, Japan. DSc (Nagoya U., 1973) chemistry. *Computing macromolecular_crystallography molecular_biology*
E-mail sasaki@met.nagoya-u.ac.jp Tel. 81(52)7231111x241 Fax 81(52)7230290

Sasaki, Satoshi (1951). Associate Professor. Res. Lab. Engineering Materials, Tokyo Inst. of Technology, Nagatsuta 4259, Midori-ku, Yokohama 227, Japan. DSc (U. Tokyo, 1979) crystallography. *X-ray_diffraction synchrotron_radiation crystallography mineralogy solid_state_physics*
E-mail sasaki@nc.titech.ac.jp Tel. 81(45)9221111x2308 Fax 81(45)9211015

Sasaki, Prof. Takatomo (1943). Professor. Dept. of Electrical Engineering, Osaka Univ., 2-1 Yamada-oka, Suita, Osaka 565, Japan. DEng (Osaka U., 1976) solid-state lasers. *Crystal_growth Czochralski_technique flux laser_crystal optical_nonlinear_crystal*
E-mail sasaki@pwr.eng.osaka-u.ac.jp Tel. 81(6)8797706 Fax 81(6)8797708

Sato, Dr Fumio (1949). Senior researcher. Advanced Materials & Devices Research Division, Japan Broadcasting Corp., 1-10-11 Kinuta, Setagaya-ku, Tokyo 157, Japan. DEng (Tohoku U., 1977) crystal defects. *Crystal_growth*
Tel. 81(3)54942391 Fax 81(3)54942399

Sato, Prof. Katsuaki (1942). Professor. Faculty of Technology, Tokyo Univ. of Agriculture & Technology, 2-24-16 Nakacho, Koganei, Tokyo 184, Japan. DEng (Kyoto U., 1978) magnetic semiconductor. *EPR optical_property chalcopyrites magnetooptical_materials magnetic_superlattices*
E-mail satokats@cc.tuat.ac.jp Tel. 81(423)887120 Fax 81(423)855395

Sato, Prof. Dr Kiyotaka (1946). Professor. Faculty of Applied Biological Science, Hiroshima Univ., 1-4-4 Kagamiyama, Higashi-hiroshima 724, Japan. DEng (Nagoya U., 1983) crystal growth. *Organic_crystal polymorphism proteins_crystallization*
Tel. 81(849)247935 Fax 81(849)227062

Sato, Dr Mamoru (1956). Research Associate. Division of Protein Crystallography, Institute for Protein Research, Osaka University, 3-2 Yamada-oka, Suita, Osaka 565, Japan. PhD (Osaka U., 1983) structural biology. *Protein_crystallography small-angle_X-ray_scattering computing macromolecular_assemblies time-resolved_structural_studies*
E-mail msato@pxews1.protein.osaka-u.ac.jp Tel. 81(6)8775111 Fax 81(6)8762533

Sato, Prof. Mitsuo (1932). Group leader; Director, Information Processing Center, Gunma University. Lab. Molecular Design, Department of Chemistry, Gunma University, Tenjin-cho, Kiryu, Gunma 376, Japan. DSc (Tokyo Kyouiku U., 1962) mineralogy. *Computer_modelling graph_theory structure_determination zeolites*
E-mail sato@cc.gunma-u.ac.jp Tel. 81(277)301260 Fax 81(277)301300

Sato, Dr Shoichi (1930). Research Manager. X-ray Research Institute, Rigaku Corporation, Akishima, Tokyo 196, Japan. DSc (U. Tokyo, 1980) chemistry. *Structure_determination charge_density absolute_structure superconductors phase_transition ferroelectricity coordination_compounds modulation_structure hydrogen_bonding*
Tel. 81(425)458101 Fax 81(425)467090

Sato, Takao (1959). Res. Assist. Faculty of Engineering, The University of Tokushima, 2-1 Minami-Josanjima, Tokushima 770, Japan. DSc (Tokyo Inst. Tech., 1992) chemistry. *Biocrystallography proteins structure_biology*
Tel. 81(886)232311x4912 Fax 81(886)553160

Sato, Dr Tatsuo (1949). Associate Professor. Dept. Metall. Eng., Tokyo Institute of Technology, Oh-okayama, Meguro, Tokyo 152, Japan. DEng (Tokyo Inst. Tech., 1979) metallurgy. *Decomposition electron_microscopy metallurgy phase_transition aluminium_compounds*
Tel. 81(3)37261111x3145 Fax 81(3)37263419

Satow, Prof. Yoshinori (1949). Professor. Faculty of Pharmaceutical Sciences, University of Tokyo, Hongo 7-3-1, Bunkyo-ku, Tokyo 113, Japan. PhD (U. Tokyo, 1977) pharmaceutical sciences. *Proteins biological_macromolecules methodology synchrotron_radiation immunoglobulins drug_design computing crystallization data_collection isomorphous_replacement anomalous_dispersion molecular_replacement phase_determination*
Tel. 81(3)38122111x4804 Fax 81(3)38135099

Sazaki, Mr Gen (1965). Research Assistant. Komatsu Lab., Institute for Materials Research, Tohoku University, 2-1-1 Katahira, Aoba-ku, Sendai 980-77, Japan. Dr. Eng. (Osaka City U., 1994) protein crystal growth. *Crystallization_mechanism protein dynamic_light_scattering AFM*
E-mail sazaki@komatsu.imr.tohoku.ac.jp Tel. 81(22)2152012 Fax 81(22)2152011

Sekine, Akiko (1964). Research Associate. Department of Chemistry, Faculty of Science, Tokyo Institute of Technology, Ookayama, Meguro-ku, Tokyo 152, Japan. DSc (Tokyo Institute of Technology, 1993) chemistry. *Chirality reactivity time-resolved_effect cobalt_compounds crystalline_state_reactions time-resolved_structure_analysis*
E-mail asekine@chem.titech.ac.jp Tel. 81(3)37261111x2608 Fax 81(3)37206206

Sekiwa, Mr Hideyuki (1962). Development Division, Nippon Mektron, Ltd, 831-2 Kamisohda, Isohara, Kitaibaraki, Ibaraki 319-15, Japan. MEng (Saitama U., 1986) *Czochralski_technique*
Tel. 81(293)422161 Fax 81(293)430326

Setoguchi, Dr Masahiro (1939). Senior researcher. Dept. for Energy Conversion, Osaka National Research Institute, AIST, 1-8-31 Midorigaoka, Ikeda, Osaka 563, Japan. DSc (Hiroshima U., 1987) mineralogy. *Crystal_growth phosphorus_compounds*
Tel. 81(727)519614 Fax 81(727)519622

Shibahara, Hiroyasu (1951). Associate Professor. Dept. of Chemistry, Kyoto University of Education, Fujinomori 1, Fushimi-ku Fukakusa, Kyoto 612, Japan. DEng (Osaka U., 1984) applied physics. *Electron_microscopy ceramics perovskites superconductor*
Tel. 81(75)6419281 Fax 81(75)6451734

Shibata, Mr Masatomo (1960). Research leader. Advanced Research Center, Hitachi Cable, Ltd, 5-1-1 Hitaka-cho, Hitachi-shi, Ibaraki 319-14, Japan. (Tohoku U., 1984) *Crystal_growth III–V_compounds*
Tel. 81(294)423151 Fax 81(294)432404

Shibata, Dr Noriyoshi (1951). Senior engineer. Japan Fine Ceramics Center, Japan. DEng (Nagoya U., 1982) applied physics. *CVD MBE*
Tel. 81(52)8713500 Fax 81(52)8713599

Shibata, Prof. Shuzo (1924). Professor. Shizuoka Institute of Science and Technology, Toyosawa, Fukuroi, Shizuoka 437, Japan. [490 Mikabi, Inasa, Shizuoka 431-14, Japan.] DSc (Nagoya, 1958) chemistry. *Molecular_structure electron_diffraction charge_density inorganic_crystal_structure*
Tel. 81(538)450111 Fax 81(538)450110 81(53)5241601

Shigematsu, Dr Koji (1953). Associate professor. Faculty of Education, Iwate Univ., 3-18-33 Ueda, Morioka 020, Japan. DSc (Tohoku U., 1983) physics. *Ionic_crystal_growth aqueous_solution liquid_state oxides*
Tel. 81(196)235171x2285 Fax 81(196)544214

Shimamura, Mr Kiyoshi (1966). Student. Fukuda Lab., Institute for Materials Research, Tohoku Univ., 2-1-1 Katahira, Aoba-ku, Sendai 980, Japan. MSc (Tohoku U., 1992) chemistry. *Crystal_growth oxides laser*
E-mail richard@lexus.imr.tohoku.ac.jp Tel. 81(22)2152103 Fax 81(22)2152104

Shimazu, Dr Masaji (1930). Lecturer. Institute of Earth Science, School of Education, Waseda Univ., 1-14-4 Gakuen-cho, Higashikurume-shi, Tokyo 203, Japan. DSc (Tokyo Education U., 1965) mineralogy. *Material_sciences powders optoelectronics electronic_materials*
Tel. 81(424)229733

Shimizu, Dr Kenji (1949). Associate professor. Dept. of Applied Chemistry and Molecular Science, Iwate Univ., 4-3-5 Ueda, Morioka 020, Japan. DEng (Tohoku U., 1989) chemical engineering. *Crystallization secondary_nucleation crystal_growth biocrystallization micro_gravity*
Tel. 81(0196)235171x2379 Fax 81(0196)527144

Shimizu, Mr Toshiyuki (1964). Research Scientist. Kirin Brewery Co. Ltd, Central Laboratories for Key Technology, 1-13-5 Fukuura, Kanazawa-ku, Yokohama-shi, Kanagawa 236, Japan. MPharm (Tokyo U., 1989) pharmaceutical science. *Protein_crystallography drug_design*
Tel. 81(45)7887200 Fax 81(45)7884041

Shimoi, Prof. Mamoru (1946). Associate professor. Dept. of Chemistry, Coll. of Arts and Sciences, The University of Tokyo, Komaba, Meguro, Tokyo 153, Japan. DSc (U. Tokyo, 1980) chemistry. *Boron_compounds clusters transition_metal_complexes*
E-mail shimoi@tansei.cc.u-tokyo.ac.jp Tel. 81(3)34671171x415 Fax 81(3)34852904

Shimomura, Dr Osamu (1943). Associate professor. Faculty of Engineering, Yamanashi Univ., 4-3-11 Takeda, Kofu 400, Japan. DEng (U. Tokyo, 1980) crystal growth. *Crystal_growth*
E-mail shimo@apricot.ese.yamanashi.ac.jp Tel. 81(552)521111 Fax 81(552)541708

Shimotomai, Dr Michio (1941). General manager. Technical Research Division, Kawasaki Steel Corp., Kawasaki-cho, Chiba 260, Japan. DEng (U. Tokyo, 1970) materials science. *Magnetism magnets crystal_field_calculation compound*
Tel. 81(43)2622463 Fax 81(43)2624153

Shimura, Dr Takayoshi (1964). Research Associate. Dept. of Precision Engineering, Osaka University, 2-1 Yamadaoka, Suita, Osaka 565, Japan. DEng (Nagoya U., 1993) engineering. *Surface thin_film oxides crystal_truncation_rod_scattering silicon_oxides*
E-mail shimura@prec.eng.osaka-u.ac.jp Tel. 81(6)8797282 Fax 81(6)8783819

Shindo, Prof. Dr Hitoshi (1950). Full Professor. Dept. of Applied Chemistry, Faculty of Science & Engineering, Chuo Univ., 1-13-27 Kasuga, Bunkyo-ku, Tokyo 112, Japan. DSc (U. Tokyo, 1979) chemistry. *Electrochemistry Raman scanning_tunnel_microscopy surface atomic_force_microscopy*
Tel. 81(3)38171918 Fax 81(3)38171895

Shiono, Dr Masaaki (1960). Research Associate. Department of Physics, Kyushu University, Fukuoka 812, Japan. DPhil (U. of York, 1989) physics. *Direct_method structure_determination molecular_complexes random_phasing_method Fourier_transform phase_refinement proteins*
Tel. 81(92)6411101x4187 Fax 81(92)6334525

Shiozaki, Prof. Yoichi (1936). Professor. Department of Physics, Hokkaido University, Kita 10 Nishi 8, Sapporo 060, Japan. DSc (Hokkaido U., 1971) physics. *Ferroelectricity X-ray_diffraction dielectric_property thermal_property analysis_of_Debye_Scherrer_method structural_change_associated_with_phase_transition*
E-mail shiozaki@phys.hokudai.ac.jp Tel. 81(11)7162111x3558 Fax 81(11)7465444 Telex 932510 HOKUSCJ

Shiro, Dr Motoo (1931). Manager. X-ray Res. Lab., Rigaku Corp., 3-9-12 Matsubara-cho, Akishima-shi, Tokyo 196, Japan. DSc (Osaka City U., 1976) chemistry. *Structure_determination*
Tel. 81(425)458101 Fax 81(425)467090

Soejima, Dr Yuji (1957). Associate professor. Department of Physics, Kyushu University, Fukuoka 812, Japan. DSc (Kyushu U., 1986) physics. *X-ray_high-resolution_diffractometry crystal_characterization X-ray_anomalous_dispersion phase_determination structure_determination*
Tel. 81(92)6411101x4178 Fax 81(92)6334525

Soga, Dr Tetsuo (1959). Associate professor. Instrument and Analysis Center, Nagoya Institute of Technology, Gokiso-cho, Showa-ku, Nagoya 466, Japan. DEng (Nagoya U., 1987) semiconductor eng. *Crystal_growth epitaxy heterostructure characterization*
E-mail soga@godzilla.elcom.nitech.ac.jp Tel. 81(52)7322111 Fax 81(52)7329241

Sogabe, Mr Satoshi (1967). Graduate student. Miki-Laboratory, Res. Lab. of Resources Utilization, Tokyo Institute of Technology, 4259 Nagatsuta, Midori-ku, Yokohama 227, Japan. MEng (Osaka U., 1992) chemistry. *Crystallography crystallization proteins biochemistry photosynthesis-related_proteins biological_electron_transfer_system DNA-protein_interaction*
E-mail ssogabe@nc.titech.ac.jp Tel. 81(45)9221111x2232 Fax 81(45)9225179 Telex 3823553 TITNAG J

Somiya, Prof. Dean Dr (1928). Professor, Dean. The Nishi Tokyo Univ., 3-7-19 Seijo, Setagaya, Tokyo 157, Japan. DSc (Tokyo Institute of Technology, 1962) ceramics. *X-ray_optics powder*
Fax 81(3)34156619

Sudo, Dr Toshio (1911). [14-27 Tsunashimadai, Kohoku-ku, Yokohama 223, Japan.] DSc (Imperial Univ. Tokyo, 1944) mineralogy. *Mineralogy clays*
Tel. 81(45)5449762

Suehiro, Prof. Kazuaki (1939). Professor. Dept. of Applied Chemistry, Fac. of Science and Engineering, Saga University, Honjo-machi, Saga 840, Japan. DE (Kyusyu U., 1984) polymer science. *Molecular_complexes conducting_polymers liquid_crystals clathrates isomorphism conformational_energy polymer_crystals*
Tel. 81(952)245191x2678 Fax 81(952)254628

Sueno, Dr Shigeho (1937). Professor. Inst. of Geoscience, Univ. of Tsukuba, Tsukuba 305, Japan. PhD (U. Tokyo, 1966) mineralogy. *Mineralogy structure_determination transformation SIMS superconductors diffusion_mechanism meteorites*
E-mail sueno@arsia.geo.tsukuba.ac.jp Tel. 81(298)534427 Fax 81(298)519764

Sugano, Prof. Dr Takuo (1931). Professor; President, Toyo U.; Group Director, Frontier Research Program, Inst. Phys. Chem. Res. Department of Electrical and Electronic Engineering. Toyo Univ., 2100 Kujirai, Kawagoe, Saitama 350, Japan. [Toyo University, 5-28-20 Hakusan, Bunkyo-hu, Tokyo 112, Japan.] PhD (U. Tokyo, 1959) electrical engineering. *Electronics interface microelectronics semiconductor nano-electronics quantum_transport Si-SiO2_systems*
E-mail sugano@krc.tokyo.ac.jp Tel. 81(492)311131x5463 or 81(3)39457201 Fax 81(492)331855 or 81(3)39457536

Sugawara, Dr Akira (1965). Research Associate. Dept. Metall. Eng., Tokyo Institute of Technology, Oh-okayama, Meguro, Tokyo 152, Japan. DEng (Tokyo Inst. Tech., 1994) metallurgy. *AES electron_microscopy fractal crystal_growth*
Tel. 81(3)37261111x3145 Fax 81(3)37263419

Sugawara, Dr Shigeo (1947). Associate professor. Mining College, Department of Materials Engineering and Applied Chemistry, Akita Univ., 1-1 Tegata, Akita 010, Japan. DEng (Nagoya U., 1988) crystal defect. *Crystal electron_microscopy dislocation defect*
Tel. 81(188)335261 Fax 81(188)370403

Sugawara, Mr Tamotsu (1966). Central Research Institute, Mitsubishi Materials Corp., 1-297 Kitabukuro-cho, Omiya, Saitama 330, Japan. MEng (Tokyo Institute of Technology, 1991) *Crystal growth*
Tel. 81(48)6420511 Fax 81(48)6420545

Sugawara, Dr Yoko (1952). Sr. Scient. X-ray Crystallography Lab., The Institute of Physical and Chemical Research (RIKEN), Wako, Saitama 351-01, Japan. DPharm (U. Tokyo, 1980) physical chemistry. *Biomolecule structural_change*
Tel. 81(48)4621111x3465 Fax 81(48)4624645

Sugihara, Dr Akio (1943). Executive researcher. Osaka Municipal Technical Research Institute, 6-50 1-chome, Morinomiya, Joto-ku, Osaka 536, Japan. PhD (Osaka U., 1970) polymer science. *Hydrolase enzymes lipase*
Tel. 81(6)969103x509 Fax 81(6)9632414

Sugio, Dr Shigetoshi (1958). Group leader. The Central Research Labs., The Green Cross Corporation, 2-25-1 Shodai-Ohtani, Hirakata, Osaka 573, Japan. DPhil (Osaka U., 1985) biochemistry. *Biochemistry biocrystallography computer-assisted_design biology*
E-mail sugio@greencross.co.jp Tel. 81(720)569319 Fax 81(720)575020

Sugiyama, Assoc Prof. Junji (1959). Associate Professor. Wood Research Institute, Kyoto University, Uji, Kyoto 611, Japan. PhD (U. Tokyo, 1989) agriculture. *Celluloses X-ray_diffraction electron_diffraction high_resolution_microscopy polymer_single_crystal*
E-mail junjis@wood1.kuwri.kyoto-u.ac.jp Tel. 81(774)323111x2563 Fax 81(774)333049 Telex 5453638UCLKUJ

Sugiyama, Dr Kazumasa (1959). Research associate. Institute for Advanced Materials Processing, Tohoku Univ., Katahira, Aoba-ku, Sendai 980, Japan. DSc (U. Tokyo, 1987) mineralogy. *Single_crystal characterization synchrotron_radiation powders anomalous_X-ray_scattering minerals amorphous*
Tel. 81(22)2276200x3166 Fax 81(22)2610938 Telex 0852-233 SENKEN J

Sugiyama, Mr Masaaki (1964). Research Associate. Department of Physics, Kyushu University, Fukuoka 812, Japan. MSc (Kyoto U., 1990) physics. *SAXS polymers_crystallization biophysics*
Tel. 81(92)6411101x4182 Fax 81(92)6334525

Suito, Prof. Eiji (1912). Prof. Emeritus. Kyoto Univ., National Technical College of Maizuru, 30 Kamiikeda, Kitashirakawa, Sakyo, Kyoto 606, Japan. DSc (Kyoto U., 1979) chemistry. *Electron_microscopy crystal_growth colloid crystal_morphology*
Tel. 81(75)7812737 Fax 81(75)7812737

Sukegawa, Dr Tokuzo (1935). Professor. Research Institute of Electronics, Shizuoka Univ., 3-5-1 Johoku, Hamamatsu 432, Japan. DEng (Tohoku U., 1964) electrical engineering. *Mixed_crystal electronics III–V_compounds semiconductor_devices photonics gravity_effect_on_liquid_phase_epitaxy*
Tel. 81(3)4711171 Fax 81(3)4740630

Sunagawa, Prof. Dr Ichiro (1924). Principal. Yamanashi Institute of Gemology and Jewelry Arts, 1955-1 Tokoji-machi, Kofu 400, Japan. [Kashiwa-cho 3-54-2, Tachikawa, Tokyo 190, Japan.] DSc (Hokkaido U., 1957) mineralogy. *Crystal_growth dissolution twinning morphology* mineralogy gemology
Tel. 81(552)326671 81(425)362964 Fax 81(552)336357 81(425)353637

Suwa, Dr Yoshiko (1930). Research manager. 2nd Section, STK Ceramics Laboratory Corp., 1-11 Tsukisan-cho, Minato-ku, Nagoya 455, Japan. DEng (Nagoya U., 1975) crystal chemistry. *Biomaterial calcium_phosphates*
Tel. 81(52)6523341 Fax 81(52)6523340

Suzuki, Dr Akira (1948). Senior research scientist. Corporate Research and Development Group, Central Research Laboratories, Sharp Corp., 2613-1 Ichinomoto, Tenri, Nara 632, Japan. DEng (Kyoto U., 1977) electronics. *Semiconductor CVD thin_film silicon_compounds*
Tel. 81(7436)52368 Fax 81(7436)52369

Suzuki, Dr Hideo (1956). Senior researcher. The 2nd Research Group, Central Research Laboratory, Hamamatsu Photonics k.k., 5000 Hirakuchi, Hamakita-city, Shizuoka 434, Japan. DEng (Nagoya U., 1983) applied physics. *Optics non-linear_property laser crystal_growth*
Tel. 81(53)5867111 Fax 81(53)5866180

Suzuki, Prof. Ikuo (1941). Professor. Dept. Electrical and Computer Engineering, Nagoya Institute of Technology, Gokiso, Showa, Nagoya 466, Japan. PhD (Nagoya U., 1968) applied physics. *Ferroelectricity computer_graphics electron_spin_resonance liquid_crystals*
E-mail suzuki@bama.elcom.nitech.ac.jp Tel. 81(52)7322111x2570 Fax 81(52)7336589

Suzuki, Dr Takaya (1942). Chief researcher. Hitachi Research Laboratory, Hitachi, Ltd, 7-1-1 Omika-cho, Hitachi, Ibaraki 319-12, Japan. DEng (Tohoku U., 1989) semiconductor. *Crystal_growth semiconductor epitaxy silicon_compounds*
Tel. 81(294)527537 Fax 81(294)527622

Suzuki, Prof. Toshimasa (1948). Professor. Nippon Inst. of Tech., 4-1 Gakuendai, Miyashiro, Minami-Saitama 345, Japan. DEng (Tokyo Inst. of Tech., 1979) applied physics. *Crystal_growth semiconductor superconductor scanning_tunnel_microscopy molecular_beam_epitaxy surface*
Tel. 81(480)337720 Fax 81(480)342941

Suzuki, Mr Yoshihisa (1945). Crystal Physics (Komatsu Lab.), Institute for Materials Research, 2-1-1 Katahira, Aoba-ku, Sendai 980, Japan. (Tohoku U., 1991) *Biomaterial nucleation pressure solution*
Tel. 81(22)2276200x2415 Fax 81(22)2152011

Tabata, Dr Hideyo (1941). Director, Ceramic Science Department. National Industrial Research Institute of Nagoya, Hirate-cho, Kita-ku, Nagoya 462, Japan. DSc (Osaka U., 1976) chemistry. *Crystal_growth morphology defect beryllium_compounds*
Tel. 81(52)9112111x500 Fax 81(52)9166992

Tabira, Yasunori (1963). Researcher. Research and Development Center, Ricoh Company, Ltd, 16-1 Shinei-cho, Tsuzuki-ku, Yokohama 224, Japan. MSc (Tokyo Institute of Technology, 1988) materials science. *Amorphous_phase anomalous_dispersion EELS EXAFS* chalcopyrites dielectrics glasses silicates *synchrotron_radiation incommensurate inelastic_scattering channelling transmission_electron_microscopy silicon_compounds*
E-mail ytabira@srd.rdc.ricoh.co.jp Tel. 81(45)5933411 Fax 81(45)5933493

Tachibana, Mr Masaru (1963). Research associate. Dept. of Physics, Yokohama City Univ., 22-2 Seto, Kanazawa-ku, Yokohama 236, Japan. MEng (Waseda U., 1989) molecular orbital theory. *Dislocation Czochralski_technique fluorescence topography*
Tel. 81(45)7872311 Fax 81(45)7872316

Tachikawa, Mr Shigeki (1963). Nishi Kawaguchi Bunshitsu, Marumoto Kogyo Company, 2-2-10 Nishikawaguchi, Kawaguchi, Saitama, Japan. MSc (Nihon U., 1990) petrology. *Bridgman_Stockbarger_technique single_crystal zinc_selenide*
Tel. 81(48)2577391

Tada, Dr Toshiji (1949). Manager. Analytical Research Labs., Fujisawa Pharmaceutical Co., Ltd, 2-1-6 Kashima, Yodogawa-ku, Osaka 532, Japan. DSc (Hiroshima U., 1982) coordination chemistry. *Biocrystallography drug_design molecular_recognition metalloproteins receptor*
Tel. 81(6)3901171 Fax 81(6)3071377

Tadatomo, Mr Kazuyuki (1956). Senior researcher. Fundamental Research Dept., Central Research Laboratory, Mitsubishi Cable Industries, Ltd, 4-3 Ikejiri, Itami, Hyogo 664, Japan. MEng (Osaka U., 1980) *Semiconductor epitaxy optoelectronics*
Tel. 81(727)818822 Fax 81(727)721361

Taga, Prof. Tooru (1938). Professor. Fac. of Pharm. Sci., Kyoto Univ., Sakyo-ku, Shimodachi-cho, Kyoto 606, Japan. DSc (Osaka U., 1969) X-ray crystallography. *Computer simulation oligosaccharides database X-ray_phase_determination membrane_structure maximum_entropy_method drug_design*
E-mail a50950@sakura.kudpc.kyoto-u.ac.jp Tel. 81(75)7534523 Fax 81(75)7534544

Takagi, Dr Kazumasa (1947). Chief researcher. Central Research Laboratory, Hitachi Ltd, 1-280 Higashi-Koigakubo, Kokubunji, Tokyo 185, Japan. DEng (Osaka U., 1982) crystal growth. *Crystal_growth superconductivity*
E-mail k-takagi@crl.hitachi.co.jp Tel. 81(423)231111 Fax 81(423)277722

Takagi, Dr Mieko (1919). PhD (Osaka U., 1956) X-ray crystallography. *Ferroelectric phase_transition*

Takahashi, Ms Masako (1952). Corporate R&D Center, Mitsui Mining & Smelting Co., Ltd, 1333-2 Haraichi, Ageo, Saitama 362, Japan. (Science Univ. of Tokyo, 1975) *Crystal_growth*
Tel. 81(48)7753211 Fax 81(48)7753217

Takahashi, Dr Toshio (1950). Associate Professor. Institute for Solid State Physics, University of Tokyo, Roppongi Minato-ku, Tokyo 106, Japan. PhD (U. Tokyo, 1979) applied physics. *Dynamical_diffraction optics standing_wave surface*
Tel. 81(3)34786811 Fax 81(3)34015169

Takahashi, Prof. Yasuhiro (1941). Associate Professor. Department of Macromolecular Science, Faculty of Science, Osaka University, Toyonaka, Osaka 560, Japan. DSc (Osaka U., 1973) polymer science. *Crystallography phase_transition biopolymers synthetic_polymers polymer_structure polymer_physics polymer_chemistry structural_disorder rubber_elasticity*
Tel. 81(6)8441151x4252 Fax 81(6)8558139

Takaki, Prof. Yoshito (1931). Professor. Department of Physics, Osaka Kyoiku Univ., 4-698-1 Asahigaoka, Kashihara 582, Japan. DSc (Osaka U., 1963) physics. *Indexing powder disorder*
Tel. 81(729)763211 Fax 81(729)763273

Takakura, Mr Hiroyuki (1967). Doctoral course student. Department of Physics, Kyushu University, Fukuoka 812, Japan. MSc (Kyushu U., 1992) physics. *Order-disorder orientation phase_transition maximum-entropy_method glasses*
Tel. 81(92)6411101x4180 Fax 81(92)6334525

Takama, Dr Toshihiko (1941). Associate professor. Dept. of Applied Phys., Faculty of Engineering, Hokkaido Univ, Kita-ku, Sapporo 060, Japan. DEng (Hokkaido Univ, 1984) applied physics. *Energy-dispersive_analysis dynamical_diffraction extinction charge_density crystal_characterization diffuse_scattering superconductors*
E-mail takama@sun2.hokudai.ac.jp Tel. 81(11)7162111x6644 Fax 81(11)7264336

Takano, Dr Kaoru (1946). Associate Professor. Institute of Materials Science, Univ. of Tsukuba, Tsukuba 305, Japan. DSc (Tokyo Univ. of Education, 1976) physics. *Crystal_growth high_pressure*
Tel. 81(298)535366 Fax 81(298)557440

Takata, Masaki (1959). Assistant professor. Dept. of Applied Physics, Nagoya University, Nagoya 464-01, Japan. PhD (Hiroshima U., 1988) physics. *Maximum-entropy_method charge_density metals powder_diffraction nuclear_density intermetallic_compounds*
E-mail a41024a@nucc.cc.nagoya-u.ac.jp Tel. 81(52)7815111x4455 Fax 81(52)7822129
Telex 4477355 ENUNAG-J

Takeda, Prof. Hiroshi (1934). Professor. Mineralogical Institute, Faculty of Science, Univ. of Tokyo, 7-3-1 Hongo, Tokyo 113, Japan. DSc (U. Tokyo, 1962) crystallography. *Polymorphism polytypism database mineralogy micas meteorites planetology*
Tel. 81(3)38122111x4543 Fax 81(3)38165714 Telex UTYOSCIJ33659

Takeda, Prof. Takayoshi (1944). Associate professor. Fac. of Integrated Arts and Sci., Hiroshima Univ., 1-7-1 Kagamiyama, Higashi-Hiroshima 724, Japan. DSc (Tohoku U., 1971) physics. *Neutron_scattering condensed_matter small-angle_scattering spin polarized_neutron slow_dynamics phase_transition complex_fluids biomembranes superconductivity fluxoid_lattice magnetism perovskite_oxides alloy diffraction neutron_spin_echo*
Tel. 81(824)246539 Fax 81(824)240757

Takeda, Prof. Yoshikazu (1948). Professor. Dept. of Materials Science and Engineering, Nagoya Univ., Furo-cho, Chikusa-ku, Nagoya 464-01, Japan. DEng (Kyoto U., 1980) electrical engineering. *Semiconductors epitaxy optoelectronics electronics EXAFS*
Tel. 81(52)7893363 Fax 81(52)7893239

Takei, Prof. Fumihiko (1936). Professor. Institute for Solid State Physics, The University of Tokyo, Roppongi, Minato-ku, Tokyo 106, Japan. DSc (U. Tokyo, 1969) crystal growth. *Melts oxides superconductor crystal_growth*
Tel. 81(3)34786811 Fax 81(3)34015169

Takemoto, Mr Jiro (1968). Student. Institute for Materials Research, 2-1-1 Katahira, Aoba-ku, Sendai 980, Japan. MSc (Tohoku U., 1993) crystal growth. *Solution superconductor epitaxy*
Tel. 81(22)2276200 Fax 81(22)2152011

Takenaka, Dr Takao (1947). R&D manager. SEH Isobe R&D Center, Shin-Etsu Handotai Co., Ltd, 2-13-1 Isobe, Annaka-shi, Gunma 379-01, Japan. DEng (Sophia U., 1977) epitaxial growth and application. *Semiconductor III–V_compounds*
Tel. 81(273)852511 Fax 81(273)852774

Takenaka, Dr Yasuyuki (1964). Assistant Professor. Hakodate Branch, Faculty of Education, Hokkaido Educational University, Japan. DSc (Tokyo Institute of Technology, 1993) chemistry. *Chirality reactivity time-resolved_effect cobalt_compounds crystalline_state_reactions time-resolved_structure_analysis*
E-mail takenaka@hak.hokkyodai.ac.jp Tel. 81(138)411121 Fax 81(138)423982

Takeuchi, Dr Yasuo (1960). Researcher. Instrumental Analysis Dept., Drug Analysis, Pharmaceutical Research Center, Meiji Seika Kaisha, Ltd, 760 Morooka-cho, Kohoku-ku, Yokohama 222, Japan. PhD (Tokyo U., 1992) biochemistry. *Biocrystallography computer_modelling structure-activity_relationship*
Tel. 81(45)5435125 Fax 81(45)5439771

Takeuchi, Prof. Yoshio (1924). Professor. Dept. of Earth Science, Nihon University, Sakurajosui, Setagaya-ku, Tokyo 156, Japan. DSc (U. Tokyo, 1953) mineralogy. *Mineralogy inorganic_crystal_chemistry cell_twinning*
Tel. 81(3)33291151x5210 Fax 81(3)33039899 Telex NICHIDAIJ29496

Taki, Dr Sadao (1922). DSc (Hiroshima U., 1961) crystal growth. *Crystal_growth*

Tamada, Prof. Osamu (1944). Professor. Graduate School of Human and Environmental Studies, Kyoto University, Yoshida Nihonmatsu-cho, Sakyo-ku, Kyoto 606, Japan. DSc (Kyoto U., 1980) mineralogy. *Charge_density electrostatic_potential mineralogy structure_determination*
E-mail A50608@JPNKUDPC.bitnet Tel. 81(75)7536869 Fax 81(75)7536872

Tamura, Dr Hatsue (1940). Assistant. Department of Applied Chemistry, Faculty of Engineering, Osaka Univ., Machikaneyama 1-16, Toyonaka, Osaka 560, Japan. DSc (Osaka U., 1990) physical chemistry. *Clathrates structural_disorder copper_compounds manganese_compounds sulfur_compounds conducting_compounds molybdenum_compounds*

E-mail tamura@ch.wani.osaka-u.ac.jp Tel. 81(6)8505788 Fax 81(6)8505785

Tamura, Dr Masao (1939). Director. Optoelectronics Technology Research Laboratory, Tsukuba, 5-5 Tohkohdai, Tsukuba, Ibaraki 300-26, Japan. PhD (Tokyo Institute of Technology, 1978) Si process technology. *Ion_implantation dislocation heterostructure GaAs_on_Si*

Tel. 81(298)474331 Fax 81(298)474180

Tanaka, Mr Akikazu (1952). Senior researcher. Electronics Materials Laboratory, Sumitomo Metal Mining Co., Ltd, 1-6-1 Suehiro-cho, Oume-shi, Tokyo 198, Japan. MEng (Tohoku U., 1978) *Crystal_growth semiconductors tellurides*

Tel. 81(428)311195 Fax 81(428)320252

Tanaka, Dr Akira (1946). Associate professor. Research Institute of Electronics, Shizuoka Univ., 3-5-1 Johoku, Hamamatsu 432, Japan. DEng (Tohoku U., 1983) electronic devices. *Crystal_growth alloys III–V_compounds*

Tel. 81(53)4711171 Fax 81(53)4740630

Tanaka, Prof. Isao (1948). Professor. Division of Biological Sciences, Graduate School of Science, Hokkaido Univ., Sapporo 060, Japan. DSc (Osaka U., 1979) chemistry. *Proteins DNA crystallization*

E-mail tanaks@polymer.hokudai.ac.jp. Tel. 81(11)7162111x3221 Fax 81(11)7465232

Tanaka, Prof. Isao (1958). Associate professor. Faculty of Engineering, Yamanashi Univ., 7 Miyamae, Kofu, Yamanashi 400, Japan. DEng (Tokyo Institute of Technology, 1991) inorganic materials science. *Float_zone_growth superconductor oxides nonlinear_optics melt_growth*

E-mail itanaka@yu-gate.yamanashi.ac.jp Tel. 81(552)208625 Fax 81(552)543035

Tanaka, Prof. Kiyoaki (1946). Professor. Chemistry Department, Nagoya Institute of Technology, Gokiso-cho, Showa-ku, Nagoya 466, Japan. [Hachi-mae, 1-209-406, Meitoku, Nagoya 465, Japan.] DSc (U. Tokyo, 1975) chemistry. *Accurate_measurement electron_density_distribution anharmonic_thermal_vibration multiple_diffraction chemical bonding X-ray_atomic_orbital_analysis X-ray_molecular_orbital_analysis coordination_compounds rare-earth_compounds organic_non_linear_optical_devices organic_fluorescent_compounds organic_phosphorescent_compounds synchrotron_radiation crystallographic_computing*

E-mail kiyo@tana1.kyy.nitech.ac.jp Tel. 81(52)7322111x2339 Fax 81(52)7414773

Tanaka, Dr Masahiko (1963). Research associate. Photon Factory, National Laboratory for High Energy Physics, Oho 1-1, Tsukuba, Ibaraki 305, Japan. DSc (U. Tokyo, 1992) science. *Mineralogy microtexture anomalous_dispersion synchrotron_radiation*

E-mail masahiko@kekvax.kek.jp Tel. 81(298)645643 Fax 81(298)642801

Tanaka, Michiyoshi (1938). Professor. Committee Member, Commission on Electron Diffraction, IUCr. Research Institute for Scientific Measurements, Tohoku University, Katahira 2-1-1 Aoba-ku, Sendai 980, Japan. PhD (Tokyo Institute of Technology, 1960) physics. *Convergent-beam_diffraction EELS structure_determination quasicrystal symmetry phase_transition defects four-dimensional_crystallography ferroelectricity convergent-beam_electron_diffraction*

E-mail g22065@cctu.cc.tohoku.ac.jp Tel. 81(22)2276200x2672 Fax 81(22)2276613

Tanaka, Prof. Nobuo (1949). Associate professor. Department of Applied Physics, School of Engineering, Nagoya University, 1 Furo-cho, Chikusa-ku, Nagoya 464-01, Japan. DEng (Nagoya U., 1978) physics. *High-resolution_electron_microscopy high-energy_electron_diffraction thin_film surface crystal_growth nanoanalysis*

E-mail a41263a@nucc.cc.nagoya-u.ac.jp Tel. 81(52)7894457 Fax 81(52)7893724 Telex 4477355 ENUNAG-J

Tanaka, Prof. Nobuo (1941). Professor. Faculty of Bioscience and Biotechnology, Tokyo Institute of Technology, Nagatsuta, Midori-ku, Yokohama 227, Japan. DSc (Osaka U., 1971) physical chemistry. *Crystalline_proteins molecular_replacement evolution synchrotron_radiation thermostability*

E-mail nobuo@nc.titec.ac.jp Tel. 81(45)9221111x2386 Fax 81(45)9222432

Tanaka, Dr Shigeyasu (1960). Associate professor. Dept. of Electronics, Nagoya Univ., Furo-cho, Chikusa-ku, Nagoya 464-01, Japan. DEng (Nagoya U., 1990) crystal growth. *High-resolution_electron_microscopy EELS semiconductors*

E-mail tanaka@tofu.nuee.nagoya-u.ac.jp Tel. 81(52)7815111 Fax 81(52)7826592

Tanaka, Mr Taisuke (1964). Researcher. Nissin Flour Milling Co., Ltd, Tsurugaoka, Oimachi, Iruma, Saitama 356, Japan. MSc (Tokyo Institute of Technology, 1990) materials science. *Single_crystal EXAFS ceramics crystallography*

Tel. 81(492)646211

Tanaka, Dr Toshiro (1955). Associate professor. Facul. Gen. Edu., Ehime Univ., 3 Bunkyo-cho, Matsuyama 790, Japan. DEng (Tohoku U., 1984) magnetic materials. *Magnetism superconductivity phase_transition chalcogens metals*

Tel. 81(899)247111x3882 Fax 81(899)226215

Tanaka, Mr Yoshiyuki (1967). Doctoral Student. Dept. of Pharmaceutical Science, Osaka University, 1-6 Yamadaoka, Suita, Osaka 565, Japan. MS (U. Osaka, 1993) pharmaceutical science. *Synchrotron_radiation nucleic_acids helical crystal structure_calculation*

E-mail tanaka@protein.osaka-u.ac.jp Tel. 81(6)8775111x6213

Tani, Katsuhiko (1944). Manager. Research and Development Center, Ricoh Company, Ltd, 16-1 Shinei-cho, Kohoku-ku, Yokohama 223, Japan. DSc (U. of Tokyo, 1976) mineralogy. *Symmetry group_theory quasicrystal structure semiconductors XANES EELS SR multilayer reflectivity*

E-mail tani@srd.rdc.ricoh.co.jp Tel. 81(45)5933411 Fax 81(45)5933493

Taniguchi, Prof. Tomohiko (1936). Professor. Department of Arts and Sciences, Osaka Kyoiku Univ., 4-698-1 Asahigaoka, Kashihara 582, Japan. BLE (Osaka U. Liberal Arts and Education, 1962) physics. *Computing powder space_group*

E-mail taniguti@cc.osaka-kyoiku.ac.jp Tel. 81(729)763211 Fax 81(729)763273

Taniguchi, Dr Yoshiteru (1952). Senior materials scientist. Compound Semiconductor Materials Dept., Japan Energy Corp., 187-4 Usuba, Hanakawa-cho, Kitaibaraki, Ibaraki 319-15, Japan. DEng (U. Tokyo, 1981) materials science. *Crystal_growth II–VI_compounds III–V_compounds*

Tel. 81(293)420300 Fax 81(293)435140

Tanishiro, Mr Yasumasa (1955). Research Associate. Physics Department, Tokyo Institute of Technology, Oh-okayama, Meguro-ku, Tokyo 152, Japan. MSc (Tokyo Institute of Technology, 1980) physics. *Surface phase_transition crystal_growth electron_microscopy REM–RHEED transmission_electron_microscopy_and_diffraction photoemission_electron_microscopy adsorption conductivity structure_determination STM*

E-mail ytanishi@cc.titech.ac.jp Tel. 81(3)57342481 Fax 81(3)57342742 Telex 2466360TITECHJ

Taniyama, Prof. Jou (1938). Professor. Faculty of Education, Kagawa Univ., 1-1 Saiwai-cho, Takamatsu 760, Japan. MSc (Tohoku U., 1965) mineralogy. *Education mineralogy crystal_form*

Tel. 81(878)361667 Fax 81(878)361652

Tatsuoka, Dr Hirokazu (1960). Research associate. Department of Electrical and Electronic Engineering, Faculty of Engineering, Shizuoka University, 3-5-1 Johoku, Hamamatsu 432, Japan. DEng (Shizuoka U., 1992) crystal growth. *Dislocation thin_film heterostructure epitaxy*

E-mail hiro-tatsuoka@jce.shizuoka.ac.jp Tel. 81(53)4711171 Fax 81(53)4720251

Terabe, Dr Kazuya (1962). 13th Research Group, National Institute for Research in Inorganic Materials, 1-1 Namiki, Tsukuba, Ibaraki 305, Japan. DEng (Nagoya Institute of Technology, 1992) *Film crystal_growth sol-gel*

Tel. 81(298)513351 Fax 81(298)527449

Terasaki, Dr Osamu (1943). Associate professor. Department of Physics, Faculty of Science, Tohoku Univ., Aoba-ku, Aramaki Aoba, Sendai 980, Japan. DSc (Tohoku U., 1982) solid state physics. *Zeolites clathrates alloys incommensurate electron_microscopy high-energy_electron_microscopy quasicrystal*

E-mail b21473@cctu.cc.tohoku.ac.jp Tel. 81(22)2221800x3298 Fax 81(22)2251891

Terauchi, Dr Hikaru (1942). Professor. Kwansei-Gakuin Univ., Dept. of Physics, Uegahara, Nishinomiya 662, Japan. DSc (Kwansei-Gakuin U., 1972) physics. *X-ray diffuse_scattering superlattice*

Tel. 81(798)536111x5280 Fax 81(798)510914

Terauchi, Masami (1960). Research Assistant. Research Institute for Scientific Measurements, Tohoku University, Katahira 2-1-1 Aoba-ku, Sendai 980, Japan. PhD (Tohoku U., 1988) physics. *EELS convergent-beam_diffraction four-dimensional_crystallography symmetry defects superconductors phase_transition convergent-beam_electron_diffraction*

E-mail g22065@cctu.cc.tohoku.ac.jp Tel. 81(22)2276200x2674 Fax 81(22)2276613

Togawa, Mr Shinji (1967). Dynamic Analysis Group, Kimura Metamelt Project, ERATO, JRDC, Satellite-2, Tsukuba Research Consortium, 5-9-9 Tokodai, Tsukuba 300-26, Japan. B (Kumamoto U., 1989) chemistry. *Crystal_growth*

E-mail hcc02337@niftyserve.or.jp Tel. 81(298)475191 Fax 81(298)475089

Tokizaki, Mr Eiji (1957). ERATO Kimura Metamelt Project, JRDC, Tsukuba Research Consortium, 2-Satellite, Tokodai 5-9-9, Tsukuba 300-26, Japan. MSc (U. Tsukuba, 1984) science. *Crystal_growth*

Tel. 81(298)475191 Fax 81(298)475089

Tokonami, Prof. Masayasu (1933). Professor. Mineralogical Institute, Faculty of Science, University of Tokyo, Hongo 7-3-1, Bunkyo-ku, Tokyo 113, Japan. DSc (U. Tokyo, 1966) mineralogy. *Crystal_structure structural_stability synchrotron_radiation radioactive_waste*

Tel. 81(3)38122111x4541 Fax 81(3)38165714

Tomioka, Dr Nobuo (1961). Instructor. Faculty of Pharmaceutical Sciences, University of Tokyo, Hongo Bunkyo-ku Tokyo, Japan. PhD (U. Tokyo, 1988) pharmaceutical science. *Computer-assisted_design structure-activity_relationship drug enzyme_inhibitors*

E-mail tomi@f.u-tokyo.ac.jp Tel. 81(3)38122111x4758 Fax 81(3)56890464

Tomita, Dr Ken-ichi (1928). Emeritus Professor. [1-82 Hagiwaradai_higashi, Kawanishi 666, Japan.] DSc (Osaka U., 1959) physical chemistry. *Biomolecule structure–activity_relationship proteins nucleic_acids biomolecular_complexes protein_nucleic_acid_interactions antigen_antibody_interactions*

Tel. 81(727)574875 Fax 81(727)574875

Tomoo, Koji (1964). Assistant. Lab. Physical Chemistry, Osaka University of Pharmaceutical Sciences, 2-10-65 Kawai, Matsubara, Osaka 580, Japan. MS (Osaka University of Pharmaceutical Sciences, 1988) pharmacy. *Proteins DNA enzymes crystallization structure_determination proteins_crystallization_structure_determination*

E-mail a61020g@center.osaka-u.ac.jp Tel. 81(723)321015 Fax 81(723)329929

Tomura, Dr Shinji (1952). Chief of raw materials group. Ceramic Technology Division, National Industrial Research Institute of Nagoya, 1-1 Hirate-cho, Kita-ku, Nagoya 462, Japan. DSc (Tohoku U., 1987) mineralogy. *Clays feldspars synthesis electron_microscopy hydrothermal*

Tel. 81(52)9112111x610 Fax 81(52)9166993

Tonomura, Dr Akira (1942). Senior chief research scientist. Advanced Research Laboratory, Hitachi Ltd, Hatoyama, Saitama 350-03, Japan. DEng, PhD (Nagoya Univ., Gakushuin U., 1975, 1993) electron holography. *Holography electron_beam field_emission*

Tel. 81(492)966111x216 Fax 81(492)966006

Toraya, Dr Hideo (1949). Associate Professor. Member, Commission on Powder Diffraction. Ceramics Research Laboratory, Nagoya Institute of Technology, Asahigaoka, Tajimi 507, Japan. DSc (Tokyo Inst. Tech., 1980) materials science. *Powder_diffraction Rietveld_method profile_analysis thin_film structure_determination instrumentation computer_automation*
E-mail E43517@JPNAC.bitnet Tel. 81(572)276811 Fax 81(572)276812

Toriumi, Dr Koshiro (1949). Associate professor. Department of Material Science, Himeji Institute of Technology, Harima Science Park City, Kamigori-cho, Hyogo 678-12, Japan. DSc (U. Tokyo, 1976) chemistry. *Accurate_structure_determination superstructure coordination_compound low-dimensional_materials mixed-valence_compounds synchrotron_radiation electrocrystallization organometallic_compound*
E-mail toriumi@sci.himeji-tech.ac.jp Tel. 81(7915)80153 Fax 81(7915)80154

Toyoda, Dr Kazuhiro (1959). Research Associate. Mineralogical Institute, Faculty of Science, University of Tokyo, Hongo 7-3-1, Bunkyo-ku, Tokyo 113, Japan. DSc (U. Tokyo, 1987) geochemistry. *Environment geochemistry trace_analysis mantle_plume climatic_change sediment*
Tel. 81(3)38122111x4547 Fax 81(3)38165714

Toyoda, Prof. Koichi (1933). Professor; Deputy director. Center for Joint Research, Shizuoka University, 3-5-4 Shin-Miyakoda, 1-chome, Hamamatsu 431-21, Japan. DSc (Kyoto U., 1986) physics. *Ferroelectricity information_storage*
Tel. 81(53)4285067 Fax 81(53)4285010

Toyoda, Prof. Taro (1946). Professor. Dept. of Applied Physics and Chemistry, The Univ. of Electro-Communications, 1-5-1 Chofugaoka, Chofu, Tokyo 182, Japan. DSc (Tokyo Metropolitan U., 1975) applied physics. *Crystal_growth*
Tel. 81(424)832161 Fax 81(424)847013

Tsuchiya, Mr Tomonobu (1963). Central Research Laboratory, Hitachi, Ltd, 1-280 Higashi-koigakubo, Kokubunji, Tokyo 185, Japan. MSc (Tsukuba U., 1988) *Crystal_growth*
Tel. 81(423)231111x3017 Fax 81(423)277786

Tsuda, Kenji (1964). Research Assistant. ·Research Institute for Scientific Measurements, Tohoku University, Katahira 2-1-1 Aoba-ku, Sendai 980, Japan. PhD (Tohoku U., 1991) physics. *Convergent-beam_diffraction structure_determination phase_transition quasicrystal symmetry dynamical_diffraction convergent-beam_electron_diffraction*
E-mail g22065@cctu.cc.tohoku.ac.jp Tel. 81(22)2276200x2673 Fax 81(22)2276613

Tsukihara, Prof. Tomitake (1944). Professor. Faculty of Engineering, The University of Tokushima, 2-1 Minami-Josanjima, Tokushima 770, Japan. DSc (Osaka U., 1974) chemistry. *Biocrystallography proteins viruses ribosomes structure_biology*
Tel. 81(886)232311x4910 Fax 81(886)553160

Tsunekawa, Dr Shin (1943). Research associate. Institute for Materials Research, Tohoku Univ., 2-1-1 Katahira, Sendai 980-77, Japan. DSc (Tohoku U., 1980) crystal growth and physical properties. *Ferroelasticity ferroelectricity crystallography domain_structure melt_growth*
Tel. 81(22)2152103 Fax 81(22)2152104

Tsuneyuki, Prof. Shinji (1961). Associate professor. Institute for Solid State Physics, University of Tokyo, Roppongi, Minato-ku, Tokyo 106, Japan. DSc (U. Tokyo, 1990) physics. *Computer_modelling silicates surface electron_correlation dynamical_processes_at_surfaces*
E-mail stsune@issp.u-tokyo.ac.jp Tel. 81(3)34786811x5841 Fax 81(3)34028174

Tsushima, Prof. Dr Katsutoshi (1943). Professor. Dept. of Earth Sciences, Faculty of Science, Toyama Univ., 3190 Gofuku, Toyama 930, Japan. DSc (Hokkaido U., 1980) glaciology. *Cloud_physics tribology texture regelation_of_ice*
Tel. 81(764)411271 Fax 81(764)412972

Uchida, Akira (1957). Lecturer. Department of Biomolecular Science, Faculty of Science, Toho University, 2-2-1 Miyama, Funabashi 274, Japan. DSc (Tokyo Institute of Technology, 1985) chemical crystallography. *Organometallic_compound organic_compound solid_state reaction*
E-mail auchida@tansei.cc.u-tokyo.ac.jp Tel. 81(474)727601 Fax 81(474)727630

Uchida, Dr Tsutomu (1964). Materials Division, Hokkaido National Industrial Research Institute, 2-17 Tsukisamu-higashi, Toyohira-ku, Sapporo, Hokkaido 062, Japan. DEng (Hokkaido U., 1993) applied physics. *Clathrates ice*
E-mail uchida@hniri.go.jp Tel. 81(11)8578965 Fax 81(11)8578989

Uda, Dr Satoshi (1955). Assistant manager. Advanced Products Division, Mitsubishi Materials Co., Ltd, 1-5-1 Marunouchi, Chiyoda-ku, Tokyo 100, Japan. PhD (Stanford U., 1992) materials science. *Kinetics melts solute_partitioning interface_instability congruent_melt*
E-mail 101217.257@compuserve.com Tel. 81(3)52525402 Fax 81(3)52525440

Uekusa, Dr Hidehiro (1964). Research Assistant. Department of Chemistry, Faculty of Science, Tokyo Institute of Technology, Ookayama, Meguro-ku, Tokyo 152, Japan. DSc (Keio U., 1992) chemistry. *Structural_change chirality structure_determination time-resolved_effect crystalline_state_reactions time-resolved_structure_analysis*
E-mail uekusa@chem.titech.ac.jp Tel. 81(3)37261111x2608 Fax 81(3)37206206

Ueno, Dr Satoru (1961). Assistant professor. Faculty of Applied Biological Science, Hiroshima Univ., 1-4-4 Kagamiyama, Higashi-hiroshima 724, Japan. PhD (Hiroshima U., 1992) biophysics. *Polymorphism lipids synchrotron_radiation SAXS fatty_acid biological_membrane*
Tel. 81(824)247934 Fax 81(824)227062

Umakoshi, Dr Keisuke (1960). Research associate. Department of Chemistry, Faculty of Science, Hokkaido University, Kita-ku, Sapporo 060, Japan. DSc (Osaka City U., 1988) chemistry. *Synthesis electrochemistry mixed-valence_compounds crystallography nuclear_magnetic_resonance*
E-mail umakoshi@cubane.ims.ac.jp Tel. 81(11)7063447 Fax 81(11)7063447

Umeda, Prof. Dr Takateru (1940). Dept. of Metallurgy, Faculty of Engineering, Univ. of Tokyo, 7-3-1 Hongo, Bunkyo-ku, Tokyo 113, Japan. DEng (U. Tokyo, 1968) *Solidification morphology*
Tel. 81(3)58022912 Fax 81(3)58022912

Umeno, Prof. Masataka (1939). Professor. Faculty of Engineering, Osaka Univ., 2-1 Yamadaoka, Suita 565, Japan. DEng (Osaka U., 1967) material science. *Semiconductors X-ray_diffraction field_ion_microscopy crystal_defect silicon oxidation*
Tel. 81(6)8775111 Fax 81(6)8783819

Uno, Prof. Ryosei (1924). Professor. College of Humanities and Sciences, Nihon University, 3-25-40 Sakurajosui, Setagaya-ku, Tokyo 156, Japan. DSc (Tokyo Technical Institute, 1969) solid state physics. *Powder_diffraction instrumentation solid_state_physics semiconductor chemical_bonding*
Tel. 81(3)33291151 Fax 81(3)33039899

Uragami, Prof. Takuyuki (1938). Professor. Councillor of Asian Crystallographic Association (AsCA). Okayama University of Science, Ridai-cho 1-1, Okayama 700, Japan. DSc (U. Tokyo, 1971) physics. *Dynamical_diffraction X-ray_topography*
E-mail uragami@ousic.ous.ac.jp Tel. 81(86)2523161x3118/3119 Fax 81(86)2553847

Uwaha, Dr Makio (1951). Associate professor. Dept. of Physics, Nagoya Univ., Furo-cho, Chikusa-ku, Nagoya 464-01, Japan. DSc (Nagoya U., 1980) *Physics theory statistical_thermodynamics*
E-mail g44333a@nucc.cc.nagoya-u.ac.jp Tel. 81(52)7892874 Fax 81(52)7892928

Ueyda, Dr Hiroshi (1949). Associate professor. Faculty of Science, Hokkaido Univ., North-10, West-8, Kita-ku, Sapporo 060, Japan. DSc (Hokkaido U., 1979) geophysics. *Crystal_growth*
E-mail uyeda@geophys.hokudai.ac.jp Tel. 81(11)7162111x2761 Fax 81(11)7462715

Uyeda, Emer. Prof. Natsu (1924). Emeritus professor. Kyoto Univ., Japan. [c/o Institute for Chemical Research, Kyoto University, Uji, Kyoto-Fu 611, Japan.] DSc (Kyoto U., 1958) crystal physics. *Electron_microscopy high-resolution_electron_microscopy epitaxy*
Tel. 81(774)323111 Fax 81(774)331247 Telex 5453-638UCLKU-J

Vassylyev, Dr Dmitry Glebovich (1959). Postdoctoral Fellow. First Department, Protein Engineering Research Institute, 6-2-3 Furuedai, Suita, Osaka 565, Japan. DSc. (Inst. Mol. Biol. Acad. Sci. USSR, 1989) molecular biology. *Proteins_crystallography protein_nucleic_acids_interaction*
E-mail dmitry@pes4.peri.co.jp Tel. 81(6)8728201 Fax 81(6)8728210

Wada, Prof. Koh (1941). Professor. Department of Physics, Faculty of Science, Hokkaido Univ., Sapporo 060, Japan. DSc (Hokkaido U., 1973) physics. *Phase_transition crystal_growth*
E-mail kwada@phys.hokudai.ac.jp Tel. 81(11)7162111x3517 Fax 81(11)7465444

Wada, Mr Norio (1948). High School Attached to Tokyo Metropolitan Univ., 1-1-2 Yakumo, Meguro, Tokyo 152, Japan. (Tokyo Gakugei U., 1972) *Impurity growth*
Tel. 81(3)37239966 Fax 81(3)37247041

Waizumi, Mr Kenji (1962). Lecturer. Faculty of Education, Yamaguchi University, 1677-1 Yoshida, Yamaguchi 753, Japan. MS (Tohoku U., 1988) *Crystal_growth*
Tel. 81(839)226111x687 Fax 81(839)238612

Wakahara, Dr Akihiro (1962). Research associate. Department of Electrical Engineering, Kyoto Univ., Yoshida-honmachi, Sakyo-ku, Kyoto 606-01, Japan. DEng (Toyohashi U. Technology, 1989) OMVPE growth of InN. *Crystal_growth*
E-mail wakahara@kuee.kyoto-u.ac.jp Tel. 81(75)7535298 Fax 81(75)7511576

Wakatsuki, Prof. Dr Masao (1934). Professor. Institute of Materials Science, Univ. of Tsukuba, 1-1-1 Tennohdai, Tsukuba, Ibaraki 305, Japan. DEng (U. Tokyo, 1978) material syntheses at high pressure. *Diamond growth flux high_pressure X-ray_fluorescence_analyses growth_sectors*
Tel. 81(298)535289 Fax 81(298)557440

Wakino, Dr Kikuo (1925). Corporate adviser. Murata Manufacturing Co., Ltd, 2-26-10 Nagaokakyo-shi, Kyoto 617, Japan. DEng (Osaka U., 1980) material engineering. *Compound crystal_growth crystal_characterization powder*
E-mail wakino@murata.co.jp Tel. 81(75)9519111 Fax 81(75)9511916

Wakoh, Prof. Shinya (1938). Professor. Univ. of Library and Info. Sci., Tsukuba, Ibaraki 305, Japan. DSc (U. Tokyo, 1966) physics. *Band_calculation Compton_scattering positron_annihilation metals*
E-mail wakoh@ulis.ac.jp Tel. 81(298)520511x312 Fax 81(298)524326

Waseda, Prof. Yoshio (1945). Professor & Director. Inst. for Advanced Materials Processing, Tohoku University, Katahira 2-1-1, Aoba-ku, Sendai 980, Japan. PhD (Tohoku U., 1973) materials science. *Absorption_edge anomalous_dispersion amorphous_phase GIXS anomalous_scattering synchrotron_radiation electron_distribution*
Tel. 81(22)2276200x2801 Fax 81(22)2610938 Telex 0852233SENKENJ

Watanabe, Prof. Denjiro (1926). Professor. Dept. of Materials Science, Iwaki Meisei Univ., Chuodai, Iwaki 970, Japan. DSc (Tohoku U., 1960) physics. *High-energy_electron_diffraction electron_microscopy superstructure structure_disorder lattices*
Tel. 81(246)295111x576 Fax 81(246)285415

Watanabe, Dr Jiro (1926). Emeritus Professor at Akita University. DSc (Hiroshima U., 1963) crystal physics. *Dislocation solidification dissolution*

Watanabe, Dr Koichi (1939). Associate professor. Dept. of Materials Chemistry, Gunma Univ., 1-5-1 Tenjin-cho, Kiryu, Gunma 376, Japan. DSc (Tohoku U., 1979) crystal growth. *Flux superconductor*
Tel. 81(277)301362 Fax 81(277)444599

Watanabe, Dr Nobuhisa (1962). Research Associate. Photon Factory, National Lab. for High Energy Physics, 1-1 Oho, Tsukuba, Ibaraki 305, Japan. DSc (Tsukuba U., 1989) physics. *Synchrotron_radiation Laue_diffraction proteins time-resolved_effect optics aminotransferase*

E-mail NOBUHISA@KEKVAX.KEK.JP Tel. 81(298)645645 Fax 81(298)642801

Watanabe, Mr Yoshio (1962). Toshiba Corp., 3583-5 Kawashiri, Yoshida-cho, Haibara-gun, Shizuoka 421-03, Japan. MS (Tohoku U., 1987) *Crystal_growth*

Tel. 81(548)340031 Fax 81(548)340038

Yagi, Prof. Katsumichi (1939). Professor. Commission Member (Electron Diffraction). Physics Department, Tokyo Institute of Technology, Oh-okayama Meguro-ku, Tokyo 152, Japan. DSc (Tokyo Institute of Technology, 1967) physics. *Surface solid_state physics thin_film electron_microscopy physical_property STM*

Tel. 81(3)37261111x2078 Fax 81(3)37290042

Yagi, Prof. Toshirou (1942). Professor. Electronic Materials (Phase Transition), Research Institute of Electronic Science, North 12, West 6, Kita-ku, Sapporo 060, Japan. DSc (Hokkaido U., 1972) phase transition. *Light_scattering phase_transition ferroelectrics ferroelastics critical_fluctuation*

E-mail yagi@ae.hines.hokudai.ac.jp Tel. 81(11)7062882 Fax 81(11)7064965

Yamada, Mr Hiroshi (1969). Master course student. Department of Physics, Kyushu University, Fukuoka 812, Japan. BSc (Kyushu U., 1992) physics. *X-ray_high-resolution_diffractometry crystal_characterization X-ray_anomalous_dispersion phase_determination structure_determination*

Tel. 81(92)6411101x4184 Fax 81(92)6334525

Yamada, Dr Nobusuke (1957). Tukuba Research Laboratory, Tosoh Corporation, 43 Miyukigaoka, Tukuba-shi, Ibaraki 305, Japan. DSc (U. Tsukuba, 1987) crystallography. *Crystallization electron_microscopy glasses silicon_compounds manganese_compounds symmetry_breaking*

Tel. 81(298)501042 Fax 81(298)501044

Yamagishi, Dr Hirotoshi (1951). Manager; semiconductor material department. Isobe R&D Center, Shin-Etsu Handotai Co, Ltd, Isobe, Annaka, Gunma 379-01, Japan. DEng (Tokyo Institute of Technology, 1974) engineering. *Floating_zone_growth Czochralski_technique silicon*

E-mail 101125.3521@compuserve.com Tel. 81(273)852548 Fax 81(273)852774

Yamaguchi, Dr Hiroshi (1963). Research associate. Division of Protein Crystallography, Institute for Protein Research, Osaka University, Suita, Osaka 565, Japan. DSc (Osaka U., 1991) polymer science. *Proteins structure–activity_relationship biocrystallography biochemistry biophysics*

E-mail hiroshi@pxews1.protein.osaka-u.ac.jp Tel. 81(6)8775111x3837 Fax 81(6)8762533

Yamaguchi, Dr Ko-ichi (1961). Department of Electronic Engineering, University of Electro-Communications, 1-5-1 Chofugaoka, Chofu-city, Tokyo 182, Japan. DEng (U. Tokyo, 1993) selective epitaxial growth. *Crystal_growth surface semiconductor*

E-mail kyama@electra.ee.uec.ac.jp Tel. 81(424)832161 Fax 81(424)896943

Yamaguchi, Mr Yasuhide (1960). R&D Center, Mitsui Mining & Smelting Co., Ltd, 1333-2 Haraichi, Ageo, Saitama 362, Japan. MS (Utsunomiya U., 1985) *Oxides spectroscopy*

Tel. 81(48)7753211 Fax 81(48)7756373

Yamamoto, Dr Akiji (1945). Senior Researcher. Member of Commission on Aperiodic Crystals. National Institute for Research in Inorganic Materials, Namiki 1, Tsukuba, Ibaraki 305, Japan. DSc (Kyoto U., 1981) crystallography. *Quasicrystal modulated_structures maximum-entropy_method*

Tel. 81(298)513351 Fax 81(298)527449

Yamamoto, Prof. Dr Akio (1946). Professor. Department of Electrical and Electronics Engineering, Fukui Univ., 3-9-1 Bunkyo, Fukui 910, Japan. DEng (Osaka U., 1983) anodization of compound semiconductors. *Crystal_growth III–V_compounds solar_cell MOCVD*

Tel. 81(776)278566 Fax 81(776)278749

Yamamoto, Prof. Dr Katsumi (1933). Professor. Phys. Lab., Natural Science, Osaka Women's Univ., 2-1 Daisen-cho, Sakai, Osaka 590, Japan. DSc (Kyoto U., 1980) *Crystal_structure thin_film epitaxy*

Tel. 81(722)224811x335 Fax 81(722)385539

Yamamoto, Prof. Naoki (1950). Associate Professor. Physics Department, Tokyo Institute of Technology, Oh-okayama, Meguro-ku, Tokyo 152, Japan. DSc (Tokyo Institute of Technology, 1979) physics. *Phase_transition solid_state_physics luminescence electron_microscopy physical_property radiation*

Tel. 81(3)57342481 Fax 81(3)57342742

Yamamoto, Dr Nobuyuki (1941). Associate professor. College of Engineering, Univ. of Osaka Prefecture, 1-1 Gakuen-cho, Sakai, Osaka 593, Japan. DEng (U. Osaka Prefecture, 1976) ternary chalcopyrite crystals. *Crystal_growth*

Tel. 81(722)521161x2275 Fax 81(722)593340

Yamamoto, Dr Seishi (1945). Professor. Department of Physics, Yamaguchi Univ., Faculty of Science, Yamaguchi Univ., Yamaguchi 753, Japan. DSc (Kyoto U., 1982) physics. *Diffuse_scattering molecular_crystal order–disorder phase_transition*

Tel. 81(839)226111x374 Fax 81(839)322041

Yamana, Dr Kazuo (1951). Chief of Ceramic Section. Ceramic Section, Ind. Res. Institute of Ishikawa, Kariyasu, Tsubata, Ishikawa 929-04, Japan. DSc (U. Osaka, 1980) ceramics. *Ceramics*

Tel. 81(762)88x1157

Yamanaka, Prof. Takamitsu (1942). Professor; Department Head/Member CPD. Dept. of Earth and Space Science, Faculty of Science, Osaka Univ., 1-16 Machikaneyama, Toyonaka, Osaka 560, Japan. DSc (U. of Tokyo, 1971) mineralogy, crystallography. *High_pressure_crystallography phase_transition oxides_minerals time_resolved electron_density*

E-mail b61400a@center.osaka-u.ac.jp Tel. 81(6)8505793 Fax 81(6)8505817 Telex 528607 OSAKU J

Yamane, Dr Takashi (1946). Associate professor. Department of Biotechnology, School of Engineering, Nagoya University, Furo-Cho, Chikusa-Ku, Nagoya 464-01, Japan. DSc (Osaka U., 1975) inorganic and physical chemistry. *X-ray_crystallography isomorphous_replacement molecular_replacement enzyme_inhibitors proteases proteins small_organic_molecules*

E-mail a40439a@nucc.cc.nagoya-u.ac.jp Tel. 81(52)7893342 Fax 81(52)7893223

Yamazaki, Dr Atsushi (1958). Associate professor. Department of Mineral Resources Engineering, Waseda Univ., 3-4-1, Ohkubo, Shinjuku-ku, Tokyo 169, Japan. DEng (Waseda U., 1987) science of mineral resources. *Method property application silicates minerals*

Tel. 81(3)32034141x733215 Fax 81(3)32002567

Yamazaki, Prof. Yohtaro (1945). Associate professor. Dept. Electronic Chem., Tokyo Inst. of Tech., Nagatsuta, Midori, Yokohama 227, Japan. DEng (Tokyo Institute of Technology, 1974) applied chemistry. *Computer_graphics computer_modelling magnetic_domain magnetism*

E-mail youtarou@nc.titech.ac.jp Tel. 81(45)9245411 Fax 81(45)9245433

Yanagi, Dr Kazunori (1957). Research Associate. Takarazuka Research Center, Sumitomo Chemical Co. Ltd, 2-1 4-chome Takatsukasa, Takarazuka, Hyogo 665, Japan. DSc (Tokyo Institute of Technology, 1988) chemistry. *Crystallography pesticides proteins DNA asymmetric_synthesis computing*

E-mail yanagi@ohsun01.sumitomo-chem.co.jp Tel. 81(797)742140 Fax 81(797)742133

Yao, Mr Min (1956). Student. Division of Biological Sciences, Graduate School of Science, Hokkaido Univ., Sapporo 060, Japan. MSc (Hokkaido U., 1992) chemistry. *Proteins crystallography computer_technology*

E-mail yao@polymer.hokudai.ac.jp Tel. 81(11)7162111x3221 Fax 81(11)7465232

Yase, Dr Kiyoshi (1954). Senior researcher. Dept. of Polymer Physics, National Institute of Materials and Chemical Research, 1-1 Higashi, Tsukuba, Ibaraki 305, Japan. DSc (Kyoto U., 1985) science. *Molecular_crystal deposition epitaxy long_chain_compound phthalo_cyanine fullerene*

Tel. 81(298)546259 Fax 81(298)546232

Yasuami, Dr Shigeru (1945). Assistant to Senior Manager. Research and Development Ctr., Toshiba Corporation, 1 Komukai Toshiba-cho, Saiwai-ku, Kawasaki 210, Japan. PhD (U. Nagoya, 1981) applied physics. *X-ray_diffraction thin_film X-ray_diffraction_topography X-ray_optics*

E-mail yasuami@ull.rdc.toshiba.co.jp Tel. 81(44)5492317 Fax 81(44)5492267

Yasuda, Prof. Yukio (1940). Professor. Dept. Cryst. Materials Science, School of Engineering, Nagoya Univ., Furo-cho, Chikusa-ku, Nagoya 464-01, Japan. DSc (Nagoya U., 1973) applied physics. *Thin_film semiconductors electron_micrography low-dimensional*

E-mail yasuda@alice.xtal.nagoya-u.ac.jp Tel. 81(52)7815111x6481 Fax 81(52)7826948

Yasui, Masanori (1960). Assistant. Dept. of Applied Physics and Chemistry, The University of Electro-Communications, Chofu-shi, Tokyo 182, Japan. PhD (Osaka U., 1992) biochemistry. *Molecular_crystal charge_density proteins rapid-X-ray_measurement_system_for_unstable_crystal interactions_of_protein_and_dye*

E-mail yasui@struct.pc.uec.ac.jp Tel. 81(424)832161x3822 Fax 81(424)844518

Yasuoka, Prof. Noritake (1936). Professor. Dept Life Science, Fac. Science, Himeji Inst of Techn., Kanaji 1479-1, Kamigori, Akou-gun, Hyogo 678-12, Japan. PhD (Osaka U., 1968) macromolecular science. *Structure_determination proteins enzymes computing*

E-mail yasuoka@sci.himeji-tech.ac.jp Tel. 81(7915)80179 Fax 81(7915)0177

Yokomori, Yoshinobu (1950). Associate Professor. Department of Chemistry, National Defense Academy, Hashirimizu, Yokosuka 239, Japan. DEng (U. Tokyo, 1978) applied chemistry. *Peptides folding secondary_structure zeolites aluminophosphates catalysts organic_compounds*

E-mail yokomori@jpnda Tel. 81(468)413810 Fax 81(468)445901

Yokota, Dr Masaaki (1965). Assistant professor. Department of Applied Chemistry and Molecular Science, Iwate Univ., 4-3-5 Ueda, Morioka-city, Iwate 020, Japan. DEng (Waseda U., 1993) industrial crystallization. *Crystal_growth*

Tel. 81(196)235171 Fax 81(196)527144

Yonekura, Dr Masami (1951). Associate professor. School of Agriculture, Ibaraki Univ., 3-21-1 Ami-machi, Ibaraki 300-03, Japan. PhD (Kyushu U., 1982) agriculture. *Biochemistry enzyme_inhibitors proteases*

Tel. 81(298)871261x8683 Fax 81(298)888525

Yoshimoto, Mr Noriyuki (1961). Lecturer. Faculty of Engineering, Iwate Univ., 4-3-5 Ueda, Morioka 020, Japan. PhD (Hiroshima U., 1991) crystal growth. *Crystal_growth*

Tel. 81(196)235171 Fax 81(196)528486

Yoshimura, Prof. Jun-ichi (1943). Associate professor. Inst. Inorganic Synthesis, Faculty of Engineering, Yamanashi University, 4-3-11 Takeda, Kofu 400, Japan. DSc (U. of Tokyo, 1975) applied physics. *Dynamical_diffraction X-ray_optics X-ray_topography X-ray_characterization multiple-crystal_diffractometry X-ray_interferometry synchrotron radiation crystal_growth defect silicon_compounds silicates*

Tel. 81(552)521111x5427 Fax 81(552)543035

Yoshimura, Prof. Masahiro (1942). Professor. Research Laboratory of Engineering Materials, Tokyo Institute of Technology, 4259 Nagatsuta, Midori, Yokohama 227, Japan. DSc in Eng. (Tokyo Institute of Technology, 1970) inorganic materials division. *Oxides hydrothermal*
E-mail myashima@titncj.nc.titech.ac.jp Tel. 81(45)9221111x2323 Fax 81(45)9211015

Yoshimura, Dr Yukio (1941). Associate Professor. Dept. of Physics, Ritsumeikan Univ., Tojiin, Kita-ku, Kyoto 603, Japan. DEng (Nagoya U., 1991) crystallography. *Crystal_physics phase_transition dielectric_materials*
Tel. 81(75)4651111x3693 Fax 81(75)4658241

Yuasa, Dr Hiroshi (1953). Assistant professor. Pharmaceutical Technology Lab., Tokyo Pharmaceutical University, 1432-1 Horinouti, Hachioji, Tokyo 192-03, Japan. PhD (Tokyo Pharmaceutical U., 1981) pharmaceutics. *Crystal_growth*
Tel. 81(426)765111 Fax 81(426)752605

Yumoto, Dr Hisami (1950). Associate Professor. Dept. of Materials Science and Technology, Science Univ. of Tokyo, 2641 Yamazaki, Noda, Chiba 278, Japan. PhD (Science U. of Tokyo, 1985) crystal growth. *Thin_film electroplating*
Tel. 81(471)241501 Fax 81(471)239362

Yuyama, Mr Isamu (1935). Technical Information Dept., Fuji Photo Film Co., Ltd, 210 Nakanuma, Minamiashigara, Kanagawa 250-01, Japan. BS (Science U. of Tokyo, 1966) *Crystal_growth*
Tel. 81(465)737180 Fax 81(465)737933

KAZAKHSTAN

Sub-Editor: V. S. Antoschenko

Amandosov, Dr Azamat Taumanovich (1956). Head of Laboratory. Research Institute of Experimental and Theoretical Physics, Kazakh State University, 96A Tole Bi St., Almaty 480012, Kazakhstan. Dr (Kazakh State U., 1978) physics. *Kinetics_of_growth monocrystalline aqueous_solution morphology_of_crystal_border interface_processes mass_crystallization*
E-mail /PN=KAZAKGSTAN.K.UNIV.INST/O=KAZPACK/ADMD=SOVMAIL/ C=SU@spring.com Tel. 7(3272)627686 Fax 7(3272)620111

Antoschenko, Dr Vladimir Stepanovich (1950). Chief specialist. Sub-Editor, World Directory of Crystallographers 9. Research Institute of Experimental and Theoretical Physics, Kazakh State University, 96A Tole Bi St., Almaty 480012, Kazakhstan. Dr (Kazakh State U., 1974) solid state physics. *Liquid_phase_epitaxy interfaces solid-liquid_interaction thin_film_devices processes_of_single_crystal_dissolution*
E-mail /PN=KAZAKGSTAN.K.UNIV.INST/O=KAZPACK/ADMD=SOVMAIL/ C=SU@spring.com Tel. 7(3272)677085 Fax 7(3272)677085

Bekenova, Galiya Kabeshovna (1958). Senior Research Associate; Chief, Electron Microscopy Group. Institute of Geological Sciences, National Academy of Sciences of the Republic of Kazakhstan, 69A Kabanbay Batyra St., Almaty 480100, Kazakhstan. Cand. Sci. (Inst. Geol. Sci., NAS RK, 1991) geology and mineralogy. *Electron_diffraction structure_of_minerals vanadium_bearing_minerals*
E-mail adm@geol.academ.alma-ata.su Tel. 7(3272)615066 Fax 7(3272)615314 Telex 251-258 NAUKA SU

Buranbaev, Mels Zhakanovich (1937). Head of Laboratory. Research Institute of Experimental and Theoretical Physics, Kazakh State University, 96A Tole Bi St., Almaty 480012, Kazakhstan. Magister (Kazakh State U., 1961) solid state physics. *Crystallography structure film_structure_of_organic_compounds*
E-mail /PN=KAZAKGSTAN.K.UNIV.INST/O=KAZPACK/ADMD=SOVMAIL/ C=SU@spring.com Tel. 7(3272)628129 Fax 7(3272)620111

Drobyshev, Dr Andrey Stepanovich (1950). Chief scientist. Research Institute of Experimental and Theoretical Physics, Kazakh State University, 96A Tole Bi St., Almaty 480012, Kazakhstan. Dr (Kazakh State U., 1973) physics. *Cryocrystals structure phase_transfer simulation*
Tel. 7(3272)620111 Fax 7(3272)620111

Garipogly, Dmitrij N. (1959). Scientist. Research Institute of Experimental and Theoretical Physics, Kazakh State University, 96A Tole Bi St., Almaty 480012, Kazakhstan. Magister (Kazakh State U., 1983) physics. *Cryocrystals structure phase_transfer simulation*
Tel. 7(3272)620111 Fax 7(3272)620111

Khamitov, Dr Zhenis Kh. (1956). Head of Laboratory. Kazakh State University, 71 Al-Faraby St., Almaty 480012, Kazakhstan. Dr (Kazakh State U., 1980) physics. *Metal_physics transmission_and_scanning_electron_microscopy monocrystals theory_of_dislocation microstructure radiation_damage_in_crystals*
E-mail /PN=KAZAKGSTAN.K.UNIV.INST/O=KAZPACK/ADMD=SOVMAIL/ C=SU@spring.com Tel. 7(3272)677948 Fax 7(3272)472609

Kirichenko, Dr Inessa Vladimirovna (1941). Chief scientist. Kazakh Institute of Mineral Raw Materials, 115 Bogenbay Batyra St., Almaty 480091, Kazakhstan. Dr (Kazakh State U., 1964) physics. *X-ray minerals ores clays X-ray_microanalysis*
Tel. 7(3272)618554

Kotelnikov, Petr Evgenievich (1953). Research Associate. Institute of Geological Sciences, National Academy of Sciences of the Republic of Kazakhstan, 69A Kabanbay Batyra St., Almaty 480100, Kazakhstan. Magister (Kazakh State U., 1985) physics. *X-ray_microanalysis electron_microscopy*
E-mail adm@geol.academ.alma-ata.su Tel. 7(3272)615853 Fax 7(3272)615314 Telex 251-258 NAUKA SU

Levin, Vladimir Leonidovich (1947). Senior Research Associate; Chief, X-ray Microanalysis Group. Institute of Geological Sciences, National Academy of Sciences of the Republic of Kazakhstan, 69A Kabanbay Batyra St., Almaty 480100, Kazakhstan. Cand. Sci. (Inst. Geol. Ore Deposits, Petrography, Mineralogy and Geochemistry, Moscow, 1987) geology and mineralogy. *Electron_microscopy X-ray_microanalysis_of_minerals sulfides native_metals*
E-mail adm@geol.academ.alma-ata.su Tel. 7(3272)615853 Fax 7(3272)615314 Telex 251-258 NAUKA SU

Maximov, Sergey L. (1959). Scientist. Research Institute of Experimental and Theoretical Physics, Kazakh State University, 96A Tole Bi St., Almaty 480012, Kazakhstan. Magister (Kazakh State U., 1982) physics. *Cryocrystals structure phase_transfer simulation*
Tel. 7(3272)620111 Fax 7(3272)620111

Omarov, Marat Ahmetovich (1953). Scientist. Research Institute of Experimental and Theoretical Physics, Kazakh State University, 96A Tole Bi St., Almaty 480012, Kazakhstan. Magister (Kazakh State U., 1983) solid state physics. *Diamond_films carbon thin_films liquid_phase_epitaxy_of_A3B5_compounds*
E-mail /PN=KAZAKGSTAN.K.UNIV.INST/O=KAZPACK/ADMD=SOVMAIL/ C=SU@spring.com Tel. 7(3272)677085 Fax 7(3272)677085

Petrova, Dr Nina Nikolaevna (1958). Chief of Department. Kazakh State Institute of Scientific and Technical Information, 221 Bogenbay Batyra St., Almaty 480096, Kazakhstan. Dr (Leningrad State U., 1981) physics. *X-ray_diffraction structure crystallography*
E-mail root@nti.alma-ata.su Tel. 7(3272)422181 Fax 7(3272)428059

Shalamov, Anatoliy Egorovich (1936). Scientist. Institute of Chemical Sciences, Kazakh Academy of Sciences, 106 Valihanova, Almaty 480100, Kazakhstan. Magister (Kazakh State U., 1959) physics, optics. *X-ray conformational_analysis*
Tel. 7(3272)615770

Shorin, Dr Viltor Fedorovich (1947). Chief specialist. Research Institute of Experimental and Theoretical Physics, Kazakh State University, 96A Tole Bi St., Almaty 480012, Kazakhstan. Dr (Kazakh State U., 1971) solid state physics. *Epitaxial_growth heterostructures semiconductive_A3B5_compounds solar_cells photoreceivers*
E-mail /PN=KAZAKGSTAN.K.UNIV.INST/O=KAZPACK/ADMD=SOVMAIL/ C=SU@spring.com Tel. 7(3272)677085 Fax 7(3272)677085

Skimevskaya, Elena Vasilevna (1958). Scientist. Research Institute of Experimental and Theoretical Physics, Kazakh State University, 96A Tole Bi St., Almaty 480012, Kazakhstan. Magister (Tomsk State U., 1981) physics of semiconductors and semiconductive devices. *Liquid_phase_epitaxy thin_films solution-melt thin_film_devices*
E-mail /PN=KAZAKGSTAN.K.UNIV.INST/O=KAZPACK/ADMD=SOVMAIL/ C=SU@spring.com Tel. 7(3272)677085 Fax 7(3272)677085

Slyusarev, Anatoly Petrovich (1939). Senior Research Associate; Chief of the X-ray Diffraction Group. Institute of Geological Sciences, National Academy of Sciences of the Republic of Kazakhstan, 69A Kabanbay Batyra St., Almaty 480100, Kazakhstan. Cand. Sci. (Learned Council of Physics Dept. National Acad. Sci. Republic Kazakhstan, 1973) physics and mathematics. *X-ray_diffraction structure diffractometer clay_minerals*
E-mail adm@geol.academ.alma-ata.su Tel. 7(3272)615066 Fax 7(3272)615314 Telex 251-258 NAUKA SU

Tasov, Boris Makievich (1951). Research Associate. Institute of Geological Sciences, National Academy of Sciences of the Republic of Kazakhstan, 69A Kabanbay Batyra St., Almaty 480100, Kazakhstan. Magister (Kazakh State U., 1973) physics. *X-ray_diffraction structure_of_minerals crystallography*
E-mail adm@geol.academ.alma-ata.su Tel. 7(3272)615066 Fax 7(3272)615314 Telex 251-258 NAUKA SU

Taurbaev, Dr Toktar Iskataevich (1938). Head of Laboratory. Research Institute of Experimental and Theoretical Physics, Kazakh State University, 96A Tole Bi St., Almaty 480012, Kazakhstan. Dr (Tomsk Polytechnic Institute, 1962) semiconductors and dielectrics. *Epitaxial_growth heterostructures diffusion semiconductive_A3B5_compounds thin_films solar_cells photoreceivers photodetectors*
E-mail /PN=KAZAKGSTAN.K.UNIV.INST/O=KAZPACK/ADMD=SOVMAIL/ C=SU@spring.com Tel. 7(3272)677085 Fax 7(3272)677085

KOREA

Sub-Editor: S. W. Suh

Bae, Mr In Kook (1958). Senior Researcher. Mineralogical Research Group, Korea Institute of Geology, Mining and Materials, Gajungdong 30, Yoosung-gu, Daejeon, Korea. MS (In-Ha U., 1987) ceramic engineering. *Crystal_growth_optical oxide Czochralski_technique crystal_growth_apparatus_design*
Tel. 82(42)8683123 Fax 82(42)8619720 Telex 82KIERSK K45509

Bay, Prof. J. Young (1963). Associate Professor. Department of Industrial Chemistry, Dong Yang Technical College, Seoul 152-714, Korea. MS (Seoul National U., 1987) chemical engineering. *Crystallography_environment zeolites noble_metals NOx_emission_control catalysis*
Tel. 82(2)6101818 Fax 82(2)6885494

Chae, Mr Soo Chun (1958). Senior Researcher. Mineralogical Research Group, Korea Institute of Geology, Mining and Materials, Gajungdong 30, Yoosung-gu, Daejeon, Korea. MS (Yon-Sei, 1985) geology. *Crystal_growth_optical oxide Czochralski_technique crystal_growth_apparatus_design*

Tel. 82(42)8683123 Fax 82(42)8619720 Telex 82KIERSK K45509

Choe, Prof. Hui-Woog (1948). Professor. Department of Chemistry, College of Natural Sciences, Chonbuk National University, Chonju 560-756, Korea. PhD (Freie U., Berlin, Germany, 1988) chemistry. *Crystallography_biological proteins enzymes protein_engineering*
Tel. 82(654)703418 Fax 82(654)703408

Choi, Prof. Deok (1953). Professor. Department of Physics, Myong Ji University, Yongin, Kyunggi-do, 449-728, Korea. PhD (Korea U., 1988) physics. *EPR_physical EPR_crystalline growth_crystal*
Tel. 82(335)306168 Fax 82(335)359533

Choi, Prof. Duck Kyun (1956). Assistant Professor. Department of Inorganic Materials Engineering, Hanyang University, Seoul 133-791, Korea. PhD (Stanford U., USA, 1988) material sci. & eng. *Thin_film CVD crystallization computer_simulation atomic_layer_epitaxy*
Tel. 82(2)2900506 Fax 82(2)2997148

Chung, Prof. Yong Je (1957). Assistant Professor. Dept. of Biochemistry, College of Natural Sciences, Chungbuk National University, Cheongju 360-763, Korea. PhD (U. Pittsburgh, USA, 1989) crystallography. *Crystallography_biological biochemistry proteins protein_engineering*
E-mail chungyj@cbucc.chungbuk.ac.kr Tel. 82(431)612311 Fax 82(431)674232

Hong, Prof. Sung Kwon (1956). Professor. Dept. of Polymer Science & Engineering, College of Engineering, Chungnam National University, Taejon 305-764, Korea. PhD (Case Western Reserve U., 1989) polymer physics. *Crystallization_structure polymers crystallography_macromolecular*
Tel. 82(42)8216667 Fax 82(42)8232931

Jang, Dr Youngnam (1952). Head of Lab. Mineralogical Research Group, Korea Institute of Geology, Mining and Materials, Gajungdong 30, Yoosung-gu Daejeon, Korea. PhD (Heidelberg, Germany, 1986) crystallography. *Crystal_growth_optical oxide Czochralski_technique crystal_growth_apparatus_design*
Tel. 82(42)8683120 Fax 82(42)8619720 Telex KIERSK K45509

Jeong, Dr Tae Soo (1959). Department of Physics, College of Natural Sciences, Sunchon National University, Seoul 540-742, Korea. PhD (Jeonbuk National U., 1993) physics. *Crystal_growth_physical optoelectrical_property_physical II–VI_compounds_physical*
Tel. 82(661)503640 Fax 82(661)503117

Kim, Prof. Cheol Jin (1955). Assistant Professor. Department of Inorganic Materials Engineering, College of Engineering, Gyeongsang National University, Chinju 660-701, Korea. PhD (Case Western Reserve U., USA, 1991) materials engineering. *AEM float_zone_growth superconductor EDS_quantitative crystal_growth*
Tel. 82(591)7513331 Fax 82(591)7599688

Kim, Prof. Duk Soo (1955). Assistant Professor. Department of Chemistry, College of Natural Sciences, Cheju National University, Cheju 690-756, Korea. PhD (Pusan National U., 1989) chemistry. *Crystallography_exchange zeolites_structural superconductors_synthetic catalysis adsorption*
Tel. 82(64)543543 Fax 82(64)563506

Kim, Prof. Ill won (1952). Associate Professor. Department of Physics, College of Natural Sciences, University of Ulsan, Ulsan 680-749, Korea. PhD (Busan National U., 1988) physics. *Crystal_growth_ferroelectric Czochralski_technique ferroelectricity nonlinear_properties Raman*
E-mail kimiw@munsu.ulsan.ac.kr Tel. 82(522)782323 Fax 82(522)773523

Kim, Prof. Kimoon (1954). Associate Professor. Department of Chemistry, Pohang University of Science & Technology, Pohang 790-784, Korea. PhD (Stanford U., 1986) chemistry. *Crystallography metalloprotein_model metalloprotein EXAFS fuel_cell*
E-mail kkim@vision.postech.ac.kr Tel. 82(562)2792113 Fax 82(562)2793399 Telex 82K54312

Kim, Dr Ki Soo (1955). Senior Engineer. Elec. & Magn. Materials Dept., Research Center, Ssang Yong Cement Ind. Co., 5-1 Shinsungdong Yuseong, Taejon, Korea. PhD (Stanford U., USA, 1993) materials sci. & eng. *Crystal_physics sensor electrochemistry optical_property elasticity*
Tel. 82(42)8651925 Fax 82(42)8611534

Kim, Dr Sangsoo (1958). Senior Scientist. Lucky R&D Center, Science Town, PO Box 10, Dae-Jeon 305-343, Korea. PhD (Iowa State U., USA, 1986) chemistry. *Biocrystallography molecular_recognition proteases_proteinases*
E-mail BT49873@twins.lucky.co.kr Tel. 82(42)8608651 Fax 82(42)8612056

Kim, Prof. Won Sa (1952). Professor. Department of Geology, Chungnam National University, Taejon 305-764, Korea. PhD (Carleton U., 1984) mineralogy. *Mineralogy_synthesis platinum_compounds equilibrium*
Tel. 82(42)8216428 Fax 82(42)8229690

Kim, Prof. Yang Bae (1940). Professor. Department of Manufacturing Pharmacy, College of Pharmacy, Seoul National University, Seoul 151-742, Korea. PhD (Seoul National U., 1974) pharmacy. *Crystallography_structure drug_design drug_QSAR receptor_design anti-inflammatory_compounds*
E-mail ybkim@alliant.snu.ac.kr Tel. 82(2)8807866 Fax 82(2)8880649

Ko, Taegyung (1955). Assistant Professor. Department of Ceramic Eng., College of Engineering, Inha University, Inchon 402-751, Korea. PhD (SUSB, 1988) earth sciences. *Crystallography_inorganic electroceramics ferroelectric high_pressure crystal_growth*
E-mail tako@dragon.inha.ac.kr Tel. 82(32)8607526 Fax 82(32)8743382

Ko, Prof. Dr Young Shin (1941). Professor. Department of Science Education, Seoul National University of Education, Seoul 137-072, Korea. PhD (Muenchen U., Germany, 1984) chemistry. *Graphite intercalation fullerene glassy_carbon adsorption*
Tel. 82(2)5805457, 82(2)5866658, 82(2)5876658, 82(2)5805463 Fax 82(2)5256658

Lee, Prof. Chongmu (1950). Professor. Department of Metallurgical Engineering, Inha University, Inchon 402-751, Korea. PhD (Stanford U., USA, 1984) materials science & engineering. *Thin_film_property surface thin_film_process AES XPS*
Tel. 82(32)8607536 Fax 82(32)8625546

Lee, Prof. Hong Lim (1946). Professor. Department of Ceramic Engineering, College of Engineering, Yonsei University, Seoul 120-749, Korea. PhD (Tokyo Institute of Technology, 1978) high temperature materials (ceramics). *Chemistry_crystallography carbides_nitrides oxides_thermal engineering_environment*
Tel. 82(2)3612849 Fax 82(2)3127735

Lee, Prof. Jeong Yong (1951). Associate Professor. Department of Electronic Mat., Eng., KAIST, Taejon 305-701, Korea. PhD (U. Cal., Berkeley, USA, 1986) material science. *Transmission_electron_microscopy semiconductors oxides high-resolution_TEM*
Tel. 82(42)8694216 Fax 82(42)8694210

Lee, Prof. Jung Hoo (1948). Associate Professor. Department of Geology, College of Natural Science, Chonbuk National University, Chonju 560-756, Korea. PhD (U. Michigan, 1984) geology. *Crystallography_geochemical phyllosilicates_structural high-resolution_analytical_electron_microscopy clays_crystal_chemistry mineralogy_metamorphic*
Tel. 82(652)703394 Fax 82(652)703568

Lee, Mr Jung Kyu (1968). Lucky R & D Center, Daejeon, Korea. MS (Seoul National U., 1993) chemistry. *Crystallography_biological protein receptor_recognition neurochemistry*
E-mail jklee@twins.lucky.co.kr Tel. 82(42)8608541 Fax 82(42)8612056

Lee, Prof. Soon Won (1956). Assistant Professor. Department of Chemistry, Sung Kyun Kwan University, Suwon, Kyung-ki 440-746, Korea. PhD (UC San Diego, 1989) chemistry. *Crystallography_coordination nitrogen_compounds coordination_organometallic*
Tel. 82(331)2905347 Fax 82(331)2905375

Lee, Prof. Tae Keun (1956). Assistant Professor. Dept. of Materials Engineering & Science, Seoul National Polytech. Univ., Seoul 139-743, Korea. PhD (SNU, 1991) materials engineering & science. *Czochralski_technique bismuth_compounds electrooptics new_glass SAW_device*
Tel. 82(2)9706319 Fax 82(2)9492407

Moon, Dr Sang-Jin (1956). Senior Researcher. Division of Chem. Eng., Korea Research Institute of Chem. Technology, Taejon 305-606, Korea. PhD (KAIST, 1988) chemical engineering. *Crystal growth_titanates CVD_solar_cell catalysis*
Tel. 82(42)8607517 Fax 82(42)8617022

Namgung, Prof. Hae (1942). Professor. Dept. of Chem., College of Natural Sciences, Kookmin Univ., Seoul 136-702, Korea. PhD (U. Bonn., Germany, 1988) inorganic chemistry. *Crystallography_structure synthesis conductor*
Tel. 82(2)9104763

Park, Prof. Byung Kyu (1956). Assistant Professor. Dept. of Materials Engineering & Science, Seoul National Polytech. Univ., Seoul 139-743, Korea. PhD (SNU, 1987) materials engineering & science. *Phase_transition lithium_compounds silicates vitreous_silica translucent_ceramics*
Tel. 82(2)9706631 Fax 82(2)9492407

Park, Dr Il Yeong (1959). Post-Doctor. CNS Research Team, Division of Medicinal Chemistry, Korea Research Institute of Chemical Technology, Daejon 305-606, Korea. PhD (SNU, 1992) pharmacy. *Crystallography_medicinal computer-assisted_design_medicinal drug*
Tel. 82(42)8607132 Fax 82(42)8611291

Park, Prof. Young-Han (1942). Professor. Department of Physics, College of Natural Science, Dankook University, Cheonan 330-714, Korea. PhD (Korea U., 1979) physics. *Dynamical_diffraction rocking_curves topography thermal_expansion*
Tel. 82(417)5503422 Fax 82(417)5519229

Park, Prof. Young-Ja (1942). Professor. Department of Chemistry, Sook Myung Womens University, Seoul 140-742, Korea. PhD (U. Pittsburgh, USA, 1970) crystallography. *Conformation carbohydrate calixarene*
E-mail YJPARK@SOBACK.HANA.NM.KR. Tel. 82(2)7109409 Fax 82(2)7109413

Seok, Prof. Won Kyung (1955). Professor. Department of Chemistry, College of Natural Science, Dongguk University, Seoul 100-715, Korea. PhD (U. North Carolina at Chapel Hill, USA, 1988) chemistry. *Synthesis_inorganic ruthenium_compounds_inorganic mechanism_inorganic oxidation_catalytic hemes_bioinorganic*
E-mail wonkseok@dgu4680.dongguk.ac.kr Tel. 82(2)2603216 Fax 82(2)2688204

Shin, Prof. Hyun So (1937). Professor. Department of Chemical Engineering, College of Engineering, Dongguk University, Seoul 100-715, Korea. PhD (Seoul National U., 1980) chemistry. *Crystallography_chemical organic_structure drug polymer_crystallography structure-activity_relationship*
Tel. 82(2)2655768 Fax 82(2)2756013

Song, Mr Hyun (1964). Assistant. Department of Chemical Engineering, College of Engineering, Dongguk University, Seoul 100-715, Korea. MS (Dongguk U., 1989) chemical engineering. *Crystallography_chemical organic_structure drug X-ray_fluorescence_spectroscopy high-temperature_diffractometry*
Tel. 82(2)2603363 Fax 82(2)5880374

Suh, Prof. Se Won (1951). Professor. National Sub-Editor for World Directory. Department of Chemistry, College of Natural Sciences, Seoul National University, Seoul 151-742, Korea. PhD (UCLA, 1980) chemistry. *Crystallography_biological proteins enzymes protein_engineering drug_design*
E-mail sewonsuh@alliant.snu.ac.kr Tel. 82(2)8806653 Fax 82(2)8891568

Sung, Dr Gun Yong (1960). Senior Researcher. Research Department, Electronics and Telecommunications Research Institute, Taejeon, 305-350, Korea. [PO Box 106, Yusong, Taejeon, Korea.] PhD (KAIST, 1987) materials sci. & eng.. *PVD_material superconductor_applied*
E-mail gysung@ard.etri.re.kr Tel. 82(42)8605698 Fax 82(42)8605033

Yeon, Prof. Younghee (1949). Professor. Department of Chemistry, Faculty Board, Korea Air Force Academy, Chungbuk, 363-849, Korea. PhD (U. Pittsburgh, USA, 1988) crystallography. *Crystallography_biological carbohydrates biomedicals molecular_dynamics drug_design*
E-mail afa@hanbit.kaist.ac.kr Tel. 82(431)539157 Fax 82(431)2216661

Yoon, Prof. Choon Sup (1950). Associate Professor. Department of Physics., KAIST, Daeduck, Science Town, Taejon, Korea. PhD (U. Strathclyde, UK, 1987) solid state physics. *Topography_X-ray crystal_growth_organic polar_compounds_organic thin_film_X-ray optics_non-linear*
E-mail csyoon@convex.kaist.ac.kr Tel. 82(42)8692532 Fax 82(42)8692510 Telex KOREA I. T, K45528

Yun, Prof. Hoseop (1961). Assistant Professor. Department of Chemistry, Ajou University, Suwon 441-749, Korea. PhD (Northwestern U., USA, 1988) chemistry. *Crystallography magnetochemistry chalcogenides*
Tel. 82(331)2192605 Fax 82(331)2127822

Yun, Ms Mi Kyung (1963). Research Scientist. Lucky R&D Center, Science Town PO Box10, Dae-Jeon 305-343, Korea. MS (Sookmyung Womens U., 1988) chemistry. *Crystallography_biological molecular_recognition proteases_proteinases*
E-mail BT38821@twins.lucky.co.kr Tel. 82(42)8608541 Fax 82(42)8612056

LATVIA

Sub-Editor: A. Kemme

Apinitis, Dr Smuidris (1933). Assistant professor. Technical University of Riga, Department of Chemical Technology, Azenes 14, LV 1048, Riga, Latvia. [Bikernieku 77-15, LV 1059, Riga, Latvia.] Dr Chem. (Technical U. Riga, 1970) chemistry. *Structure phosphates inorganic_chemistry*
Tel. 371(2)626298 Fax 371(8)820094

Belakov, Dr Sergejs (1961). Researcher. Latvian Institute of Organic Synthesis, Physical Organic Chemistry Lab., Aizkraukles 21, LV 1006, Riga, Latvia. Dr Phys. (Inst. of Applied Physics, Academy of Sciences of Moldova, 1987) crystallography and crystallophysics. *X-ray_structure_determination heterocycles*
E-mail inta@osi.lza.lv Tel. 371(2)559373 Fax 371(2)553493 Telex 161174 DUGA SU

Berzina, Dr Inese (1948). Researcher. Institute of Inorganic Chemistry, Lab. of Coordination Compounds, Salaspils Miera 34, LV 2169, Latvia. Dr Chem. (Inst. of Inorganic Chemistry Latvian Academy of Sciences, 1988) inorganic chemistry. *X-ray_structure_determination software*
Tel. 371(2)944783

Bondars, Dr Bruno (1951). Researcher. Institute of Inorganic Chemistry, Lab. of Dispersed Systems, Salaspils Miera 34, LV 2169, Latvia. Dr Chem. (Inst. of Inorganic Chemistry Latvian Academy of Sciences, 1981) inorganic chemistry. *Powder X-ray_structure_determination phosphorus_compounds*
E-mail bruno@mhd.iph.riga.lv Tel. 371(2)944790 Fax 371(2)944755

Borodajenko, Ms Natalya (1957). Researcher. University of Latvia, Faculty of Physics and Mathematics, X-ray Analysis Lab., Boul. Raina 19, LV 1585, Riga, Latvia. *Powder electron_diffraction structure_determination phase_transition ferroelectricity*
Tel. 371(2)227103 Fax 371(2)225039

Brante, MSc Inta (1942). Head of Laboratory. University of Latvia, Faculty of Physics and Mathematics, X-ray Analysis Lab., Boul. Raina 19, LV 1585, Riga, Latvia. MSc (U. Latvia, 1993) physics. *Powder ferroelectricity*
Tel. 371(2)227103 Fax 371(2)225039

Bremere, Ms Ingrida (1957). Researcher. Institute of Inorganic Chemistry, Lab. of Boron Compounds, Salaspils Miera 34, LV 2169, Latvia. *Powder X-ray_structure_determination software*
Tel. 371(2)944782

Brezgunov, Dr Mark (1946). Researcher. Institute of Physical Energetics, Lab. of Application of Nuclear Physics Methods for the Condensed Matter Investigation, 21 Aizkraukles Str., Riga, LV-1006, Latvia. Dr Phys. (Inst. of Physics, Latvian Academy of Sciences, 1988) solid state physics. *Neutron_diffraction Mossbauer*
E-mail phen@sun.lza.lv or fei@sun.lza.lv Tel. 371(2)558800 Fax 371()8820339 371(2)551394

Gavrilov, Dr Victor (1947). Researcher. Institute of Physical Energetics, Lab. of Application of Nuclear Physics Methods for the Condensed Matter Investigation, 21 Aizkraukles Str., Riga, LV-1006, Latvia. Dr Phys. (Inst. of Physics, Latvian Academy of Sciences, 1988) solid state physics. *Phase_transition Mossbauer*
E-mail phen@sun.lza.lv or fei@sun.lza.lv Tel. 371(2)558800 Fax 371()8820339 371(2)551394

Iolin, Prof. Eugene (1937). Professor; Head of the Laboratory. Institute of Physical Energetics, Lab. of Application of Nuclear Physics Methods for Condensed Matter Investigation, 21 Aizkraukles Str., Riga, LV-1006, Latvia. Dr Phys. habil. (Inst. of Physics, Estonian Academy of Sciences, 1978) solid state physics. *Phase_transition Mossbauer dynamic_scattering liquids*
E-mail phen@sun.lza.lv or fei@sun.lza.lv Tel. 371(2)558800 Fax 371()8820339 371(2)551394

Kapostins, MSc Peteris (1951). Lecturer. University of Latvia, Faculty of Physics and Mathematics, X-ray Analysis Lab., Boul. Raina 19, LV 1585, Riga, Latvia. MSc (U. Latvia, 1993) physics. *Powder X-ray_structure_determination*
Tel. 371(2)227103 Fax 371(2)225039

Kemme, Dr Andrejs (1941). Researcher. Sub-Editor for Latvia, WDC9. Latvian Institute of Organic Syntheses, Physical Organic Chemistry Lab., Aizkraukles 21, LV 1006, Riga, Latvia. Dr Chem. (Latvian Institute of Organic Syntheses, 1977) physical chemistry. *X-ray_structure_determination software cage_molecules*
E-mail inta@osi.lza.lv Tel. 371(2)559373 Fax 371(2)553493 Telex 161174 DUGA SU

Kozlovskis, Dr Leonids (1952). Assistant professor. Daugavpils Pedagogical University, Faculty of Physics and Mathematics, Parades Iela 1, Daugavpils, LV-5407, Latvia. Dr Phys. (Moscow State U., Russia, 1982) physics of thin solid films. *Thin_solid_film structure texture multilayers magnetic_properties*
Tel. 371(54)22922 Fax 371(54)22890

Krumina, Ms Aija (1952). Researcher. Institute of Inorganic Chemistry, Lab. of Dispersed Systems, Salaspils Miera 34, LV 2169, Latvia. *Powder phase_separation chromium_compounds manganese_compounds*
Tel. 371(2)944783 Fax 371(2)944755

Kuvaldin, Dr Boris (1939). Researcher. Institute of Physics, Lab. of Application of Nuclear Physics, Methods for the Condensed Matter Investigations, Salaspils Miera 34, LV 2169, Latvia. Dr Phys. (Inst. of Physics Latvian Academy of Sciences, 1988) physics of solid states. *Neutron_diffraction Mossbauer*
E-mail eiol@mhd.iph.riga.lv Tel. 371(2)946603

Lindina, Dr Lauma (1935). Researcher. Technical University of Riga, Department of Chemical Technology, Azenes 14, LV 1048, Riga, Latvia. Dr Chem. (Technical U. Riga, 1972) technology of silicate. *X-ray_phase_determination apparatus ceramics*
Tel. 371(2)626216 Fax 371(8)820094

Mironova, Prof. Nina (1941). Professor. Center of Nuclear Research, Salaspils Miera 30, LV 2169, Latvia. Dr Phys. habil. (U. Latvia, 1989) optics and spectroscopy. *Solid_states structure_determination structural_change_by_radiation*
Tel. 371(2)944670

Mishnev, Dr Anatoly (1952). Leading Researcher. Executive Secretary of the National Committee for Crystallography. Latvian Institute of Organic Syntheses, Physical Organic Chemistry Lab., Aizkraukles 21, LV 1006, Riga, Latvia. Dr Phys. (Inst. of Applied Physics, Academy of Sciences of Moldova, 1982) crystallography and crystallophysics. *Structure_determination direct_method software*
E-mail inta@osi.lza.lv and mishnevs@osi.lanet.lv Tel. 371(2)559373 Fax 371(2)553493 Telex 161174 DUGA SU

Ozolins, Dr Gerhards (1934). Researcher. Institute of Inorganic Chemistry, Lab. of Dispersed Systems, Salaspils Miera 34, LV 2169, Latvia. Dr Phys. (U. of Latvia, 1969) solid state physics. *Powder X-ray_precise_measurement apparatus precise_temperature_measurements_and_control*
Tel. 371(2)944783 Fax 371(2)944755

Pech, Dr Lucija (1919). Researcher. Institute of Inorganic Chemistry, Lab.of Coordination Compounds, Salaspils Miera 34, LV 2169, Latvia. Dr Chem. (Inst. of Inorganic Chemistry Latvian Academy of Sciences, 1977) inorganic chemistry. *X-ray_structure*
Tel. 371(2)944783

Purans, Dr Juris (1952). Group leader. Institute of Solid State Physics of Latvian University, Dept. of Semiconductor Technology, Exact Methods Spectroscopy Group, Kengaraga 8, LV 1063, Riga, Latvia. Dr Phys. habil. (U. Latvia, 1993) solid state physics. *X-ray_absorption_spectroscopy*
Tel. 371(2)264187

Raitman, Dr Ernst (1939). Senior Researcher. Institute of Physical Energetics, Lab. of Application of Nuclear Physics Methods for the Condensed Matter Investigation, 21 Aizkraukles Str., Riga, LV-1006, Latvia. Dr Phys. (Inst. of Physics, Latvian Academy of Sciences, 1975) solid state physics. *Dynamic_scattering Mossbauer domain_structure*
E-mail phen@sun.lza.lv or fei@sun.lza.lv Tel. 371(2)558800 Fax 371()8820339 371(2)551394

Ronis, Dr Janis (1955). Researcher. Institute of Inorganic Chemistry, Lab. of Dispersed Systems, Salaspils Miera 34, LV 2169, Latvia. Dr Chem. (Inst. of Inorganic Chemistry Latvian Academy of Sciences, 1990) inorganic chemistry. *Powder X-ray_structure_determination phosphorus_compounds*
Tel. 371(2)944773

Sedmalis, Prof. Uldis (1933). Head of Chair; Director of Institute of Silicate Materials. Technical University of Riga, Department of Chemical Technology, Azenes 14, LV 1048, Riga, Latvia. Dr Chem. habil. (Byelorussian Polytechnical Inst., 1970) technology of silicates. *Structure phosphates silicates mineralogy_of_glasses_and_ceramics*
Tel. 371(2)610066 Fax 371(8)820094

Shebanov, Dr Leonids (1949). Head of Laboratory. Institute of Solid State Physics of Latvian University, Ferroelectrics Physics Department Complex Research Lab., Kengaraga 8, LV 1063, Riga, Latvia. Dr Phys. (Inst. of Solid State Physics of Latvian U., 1980) solid state physics. *Solid_solution phase_transition structure_determination software ferroelectricity*
E-mail sebanovs@secfi.lu.lv Tel. 371(2)260896 or 371(2)261304 Fax 371(2)260543 Telex 161172 TEMA SU

Silina, Dr Elga (1941). Researcher. Institute of Inorganic Chemistry, Lab. of Coordination Compounds, Salaspils Miera 34, LV 2169, Latvia. Dr Chem. (Inst. Inorganic Chemistry Latvian Academy of Sciences, 1977) inorganic chemistry. *X-ray_structure_determination boron_compounds*
Tel. 371(2)944783 Fax 371(2)944755

Zviedre, Dr Irena (1938). Researcher. Institute of Inorganic Chemistry, Lab.of Boron Compounds, Salaspils Miera 34, LV 2169, Latvia. Dr Chem. (Inst. of Inorganic Chemistry Latvian Academy of Sciences, 1974) inorganic chemistry. *X-ray_structure_determination crystal_growth boron_compounds*
Tel. 371(2)944782

FORMER YUGOSLAVIAN REPUBLIC OF MACEDONIA

Sub-Editor: G. Jovanovski

Grupče, Dr Orhideja (1951). Scientific associate. Institute of Chemistry, Faculty of Natural Science and Mathematics, St. Kiril and Metodij University, Arhimedova 5, 91000 Skopje, Macedonia. PhD (Skopje U., 1994) chemistry. *Structural_chemistry spectra-structure_correlations*

Tel. 389(91)221033 Fax 389(91)228141

Jordanovska, Prof. Dr Vera (1938). Professor. Institute of Chemistry, Faculty of Natural Science and Mathematics, St. Kiril and Metodij University, Arhimedova 5, 91000 Skopje, Macedonia. PhD (Ljubljana U., 1982) chemistry. *Coordination_chemistry rare-earth_compounds*

Tel. 389(91)221033 Fax 389(91)228141

Jovanovski, Prof. Dr Gligor (1945). Professor. Sub-Editor for Macedonia of Ninth Edition of WDC. Institute of Chemistry, Faculty of Natural Science and Mathematics, St. Kiril and Metodij University, Arhimedova 5, 91000 Skopje, Macedonia. PhD (Zagreb U., 1981) chemistry. *Structural_chemistry metalloorganic_compound spectra-structure_correlations*

Tel. 389(91)221033 Fax 389(91)228141

Mirčeva, Dr Aneta (1946). Scientific associate. Institute of Chemistry, Faculty of Natural Science and Mathematics, St. Kiril and Metodij University, Arhimedova 5, 91000 Skopje, Macedonia. PhD (Ljubljana U., 1985) chemistry. *Crystal_structure polymers*

Tel. 389(91)221033 Fax 389(91)228141

Pocev, Prof. Dr Stefan (1940). Professor. Faculty of Technology and Metallurgy, St. Kiril and Metodij University, Ruđer Bošković 16, 91000 Skopje, Macedonia. [Bul. Partizanski Odredi 4/II-17, 91000 Skopje, Macedonia.] PhD (Zagreb U., 1978) chemistry. *Powder_diffraction crystal_structure*

Tel. 389(91)364588 Fax 389(91)257712 Telex 51471

Šoptrajanov, Prof. Dr Bojan (1937). Professor. Institute of Chemistry, Faculty of Natural Science and Mathematics, St. Kiril and Metodij University, Arhimedova 5, 91000 Skopje, Macedonia. PhD (Skopje U., 1973) chemistry. *Structural_chemistry infrared_spectroscopy crystallohydrates*

Tel. 389(91)221033 Fax 389(91)228141

Šoptrajanova, Prof. Dr Gorica (1929). Professor. Faculty of Mining, 92000 Štip, Macedonia. PhD (Beograd U., 1967) geology. *Structure minerals*

Tel. 389(92)21379

MALAYSIA

Sub-Editor: A. Hamid Othman

Abdul Aziz, Dr Yang Farina (1962). Lecturer. Department of Chemistry, Universiti Kebangsaan Malaysia, 43600 Bangi, Selangor, Malaysia. PhD (U. Reading UK, 1989) inorganic chemistry. *Structure_determination tin_compounds*

Tel. 60(3)8293901 Fax 60(3)8256086

Almashoor, Dr Syed Sheikh (1944). Associate professor. Dept of Geology, Uni. Kebangsaan Malaysia, 43600 Bangi, Selangor, Malaysia. PhD (Penn. State U., USA, 1983) igneous petrology. *Minerals X-ray_diffraction petrology geochemistry*

Tel. 60(3)8292390 Fax 60(3)8256086

Baba, Dr Ibrahim (1951). Associate professor. Dept of Chemistry, Universiti Kebangsaan Malaysia, 43600 Bangi, Malaysia. PhD (U. Reading, UK, 1977) inorganic chemistry. *Coordination_compound bioinorganic analytical_chemistry*

Tel. 60(3)8292442 Fax 60(3)8256086

Cheang, Dr Kok Keong (1949). Senior Lecturer. Schools of Materials Engineering, Uni. Sains Malaysia, Jalan Bandaraya, 30000 Ipoh, Malaysia. PhD (U. Georgia, USA, 1982) geology. *Electron_microscopy X-ray_crystallography petrology exploration_geochemistry optical_mineralogy*

Tel. 60(5)503131x2330

Chen, Dr Wei (1948). Associate professor. Dept of Chemistry, U. of Malaya, 59100 Kuala Lumpur, Malaysia. PhD (U. New South Wales, Australia, 1976) X-ray crystallography. *Single_crystal_diffractometry small_molecules_structure structural_chemistry symmetry*

Tel. 60(3)7555466x510

Fun, Dr Hoong-Kun (1946). Associate professor; in charge of X-ray Crystallography Laboratory. School of Physics, Universiti Sains Malaysia, 11800 USM Penang, Malaysia. PhD (Purdue U., Indiana, USA, 1974) experimental solid state physics. *Structure_determination Rietveld_method high_temperature_superconductor ESR NMR ENDOR laser*

E-mail hkfun@cs.usm.my Tel. 60(4)6577888 Fax 60(4)6579150 Telex MA40254

Hassan, Dr Wan Fuad (1948). Associate professor. Department of Geology, Universiti Kebangsaan Malaysia, 43600 Bangi, Selangor, Malaysia. PhD (Leeds U., UK, 1982) mineralogy and geochemistry. *Microscopy minerals X-ray_diffraction inorganic_crystals*

Tel. 60(3)8292657 Fax 60(3)8256086

Idid, Ms Sheriffah Noor (1960). Assistant director. Science and Technology Div, Ministry of Science, Technology and Environment, 14 Flr Wisma Sime Dharby, Jln Raja Laut, 50662 Kuala Lumpur, Malaysia. MSc (Imperial Coll. London, UK, 1984) physics. *X-ray_diffraction silicon_compounds gallium_arsenide*

Tel. 60(3)2938955x221 Fax 60(3)2936006

Lee, Dr Chnoog Kheng (1948). Associate professor. Dept of Chemistry, Universiti Pertanian Malaysia, 43400 Serdang, Selangor, Malaysia. PhD (Aberdeen U., UK, 1972) inorganic chemistry. *Minerals phase_determination powder_diffraction disorder structural chemistry*

Tel. 60(3)9486101

Mohamad, Dr Hamzah (1951). Associate professor. Dept of Geology, Universiti Kebangsaan Malaysia, 43600 Bangi, Selangor, Malaysia. PhD (Strathclyde U., UK, 1980) geochemistry and petrology. *Minerals powder_diffraction X-ray_diffraction*

Tel. 60(3)8292664 Fax 60(3)8256086

Ng, Dr Seik Weng (1951). Associate professor. Inst. Advanced Studies, Uni. Malaya, 59100 Kuala Lumpur, Malaysia. PhD (Oklahoma U., USA, 1983) organic chemistry. *Structure_determination tin_compounds small_molecules organometallic*

E-mail h1nswen@cc.um.my Tel. 60(3)7557000x500 Fax 60(3)7568940

Othman, Dr A. Hamid (1948). Associate professor. Sub-Editor, WDC9. Department of Chemistry, Universiti Kebangsaan Malaysia, 43600 Bangi, Selangor, Malaysia. PhD (U. Reading, UK, 1977) X-ray crystallography. *X-ray_crystallography coordination_compound macrocycles tin_compounds silver_compounds*

E-mail hamie001@vm1.cc.ukm.my Tel. 60(3)8292439 Fax 60(3)8256086 60(3)8252115

Othman, Dr Radzali (1954). Associate professor; Chairman of postgraduate studies. School of Materials Engineering, Universiti Sains Malaysia, Sri Iskandar, 31750 Tronoh, Perak, Malaysia. PhD (Sheffield U., UK, 1982) ceramics. *Electron_microscopy materials X-ray_diffraction*

E-mail mrro@kcp.usm.my Tel. 60(5)53676901x5511 Fax 60(5)3677444

Silong, Dr Sidik bin (1953). Lecturer. Dept of Chemistry, Universiti Pertania Malaysia, 43400 Serdang, Selangor, Malaysia. PhD (U. Reading, UK, 1982) structural inorganic chemistry. *Structural_chemistry catalysts GCMS*

Tel. 60(3)9486101x3621 Fax 60(3)9486646

Teh, Dr Ser Kok (1947). Lecturer. Dept of Mechanical Engineering, University Malaya, 59100 Kuala Lumpur, Malaysia. PhD (Queen Mary Coll., London, UK, 1976) material science. *Defect_structure electron_diffraction electron_microscopy ceramics alumina metal_creep*

Tel. 60(3)7553466x265 Fax 60(3)7573661

Teoh, Mr Lay Hock (1950). Principal geologist. Geological Survey of Malaysia, 20th Flr Tabung Haji Building, Jalan Tun Razak, 50736 Kuala Lumpur, Malaysia. BSc(Hons), MSc, DIC (U. Victoria, Wellington, NZ (BSc), Imperial College London, UK (MSc, DIC), 1972) geology and geochemistry; mineral exploration. *Minerals X-ray_diffraction*

Tel. 60(3)2611033 Fax 60(3)2611036

Yahya, Dr Muhamad (1947). Deputy Dean, postgraduate studies. Department of Physics, Universiti Kebangsaan Malaysia, 43600 Bangi, Malaysia. PhD (Monash U., Australia, 1979) physics. *Thin_film energy computational physics*

E-mail tdipps.cc.ukm.my Tel. 60(3)8292900 Fax 60(3)8256086 Telex UNIKEB MA 31496

MEXICO

Sub-Editor: M. Soriano García

Aguilera Herrera, Prof. Nicolás (1920). Investigador. Lab. de Investigación de Edafología, Facultad de Ciencias, UNAM, Circuito Exterior, Ciudad Universitaria, Deleg. Coyoacán, México DF 04510, México 03. PhD (U. Wisconsin, USA, 1953) soil science. *Clays soils mineralogy edaphology*
Tel. 52(5)6224922 Fax 52(5)6224828

Altuzar Coello, Mrs Patricia Eugenia (1955). Técnico Académico. Instituto de Geología, UNAM, Circuito Exterior, Ciudad Universitaria, Deleg. Coyoacán, México 04510 DF, México 03. MSc (Facultad de Ciencias, UNAM, 1990) ciencias de materiales. *X-ray_diffraction X-ray_fluorescence*
Tel. 52(5)6224283 Fax 52(5)2804864

Arreguín Espinosa, Dr Roberto Alejandro (1959). Investigador. Instituto de Química, UNAM, Circuito Exterior, Ciudad Universitaria, Deleg. Coyoacán, México D. F. 04510, México 03. PhD (Facultad de Ciencias, UNAM, 1991) biology. *Protein_structure protein_crystallization*
Tel. 52(5)6224403 Fax 52(5)6162203

Avalos Borja, Prof. Miguel (1950). Investigador. Instituto de Física, UNAM, Laboratorio de Ensenada, AP 2-681, Km. 107 Carretera Tijuana Ensenada, Ensenada B.C. 22800, México 03. PhD (U. Stanford, California, USA, 1983) electron microscopy. *High_resolution_electron_microscopy microanalysis_of_materials scanning_electron_microscopy*
E-mail leonel@ifuname.ifisicaen.unam.mx Tel. 52(617)44602 Fax 52(617)44603

Balera Perez, Mr Miguel Angel (1956). Investigador/edafología. Departamento de Investigación en Ciences Agricolas, Instituto de Ciencias de las U. Aut. de Puebla, AP 1292, Puebla 72000 Puebla, México 03. BSc (U. Autónoma de Puebla, México, 1981) industrial chemistry. *Soil_chemistry mineral_chemistry volcanic_soils*
Tel. 52(22)331179 Fax 52(22)326067

Barreto Rentería, Mr Jorge (1956). Técnico Académico. Instituto de Física, UNAM, Dept. Estado Sólido, Circuito Exterior, Ciudad Universitaria, Deleg. Coyoacán, México DF 04510, México 03. MSc (Facultad de Ciencias, UNAM, 1993) physics. *Crystal_growth superconductors*
Tel. 52(5)6225086 Fax 52(5)6162010

Bosch Giral, Prof. Pedro (1948). Investigador. Dept. of Química, UAM-Iztapalapa, Area de Catálisis, Ave. Michoacan y La Purisima S/N, Iztapalapa 09340 DF, México 03. PhD (U. Claude Bernard, France, 1976) crystal chemistry and catalysis. *Catalysis X-ray_diffraction ceramics*
Tel. 52(5)7244668 Fax 52(5)7244666

Bucio Galindo, Mr Lauro (1961). Técnico Académico. Instituto de Física, UNAM, Circuito Exterior, Ciudad Universitaria, AP 20-364, Deleg. Coyoacán, México DF 04510, México 03. BSc (U. Iberoamericana, 1990) physics. *Powder_diffraction*
E-mail galindo@ifunam.ifisicacu.unam.mx Tel. 52(5)6225039 Fax 52(5)6161535

Carrera García, Ms Luz Maria (1949). Investigador. Gerencia de Investigación Básica, Departamento de Química, Sierra Mojada 447 piso 2, Lomas de Barrilaco, México 11010 DF, México 03. MSc (Facultad de Ciencias, UNAM, 1992) material sciences. *Zeolites clays X-ray_diffraction*
Tel. 52(5)7244669 Fax 52(5)5213798

Castellanos Guzman, Dr A. Guillermo (1939). Encargado de Projectos CETAV. Investigador. Lab. de Investigación en Materiales DVTT, U. de Guadalajara, AP 2-638, Guadalajara 44281 Jalisco, México 03. PhD (Queen Mary College, London U., UK, 1981) physics. *Crystal_growth polar_crystals X-ray_structure_determination dielectric_and_optical_properties*
E-mail castel@csh.udg.mx Tel. 52(3)6561917 Fax 52(3)6565141

Castellanos Román, Mrs Maria Asunción (1943). Jefe Lab. Rayos X. Fac. de Química, UNAM, Div. Estudios de Postgrado, Ciudad Universitaria, Deleg. Coyoacán, México 04510 DF, México 03. MSc (U. Aberdeen, UK, 1979) chemistry. *Crystal_chemistry oxide_complexes*
Tel. 52(5)6223721 Fax 52(5)6162210

Cordero Borboa, Dr Adolfo (1953). Sec. Acad. UNIVERSUM. Instituto de Física, UNAM, Circuito Exterior, Cd. Univ., Deleg. Coyoacán, México 04510 DF, México 03. PhD (Fac. de Ciencias, UNAM, 1983) physics. *Crystallography_of_minerals manmade_materials_and_proteins*
Tel. 52(5)6227267 Fax 52(5)6161535 Telex 1774523-UNAMME

Cota Araiza, Prof. Leonel S. (1944). Investigador y Jefe de la Subdependencia. Laboratorio de Ensenada, Instituto de Física, UNAM, AP 2-681, Km. 107 Carretera Tijuana Ensenada, Ensenada 22800 BC, México 03. PhD (Warwick U., UK, 1974) solid state physics. *Low-energy_electron_diffraction Auger_electron_spectroscopy photoemission*
E-mail leonel@ifuname.ifisicaen.unam.mx Tel. 52(617)44602 Fax 52(617)44603

De Ita De la Torre, Prof. Antonio (1946). Investigador. Area de Ciencia de Materiales, UAM-Azcapotzalco, Av. San Pablo 180, Azcapotzalco 02200 DF, México 03. Dr Ing. (Technische Hochschule, Aachen, Germany, 1983) methodology and textures in stainless steel. *Computer_programming_in_X-ray_diffraction thermal_analysis archeometry*
Tel. 52(5)7244286 Fax 52(5)3824998

De Pablo Galan, Dr Liberto (1934). Jefe del Depto. de Geoquímica. Instituto de Geología, UNAM, Circuito Exterior, Ciudad Universitaria, AP 70-296, Deleg. Coyoacán, México 04510 DF, México 03. PhD (Ohio State U., USA, 1958) geology. *Mineralogy crystallography solid_state_chemistry*
Tel. 52(5)6224283 Fax 52(5)2804864

Domínguez Esquivel, Dr José Manuel (1948). Investigador. Instituto Mexicano del Petróleo, AP 14-805, Ave. 100 Metros #152, México 14 DF, México 03. DSc (U. Claude Bernard, France, 1977) physics. *Catalysis surface_science*
Tel. 52(5)5676600x2377

Echavarri Hernández, Dr Ariel (1939). Director. Dirección de Mineria, Geología y Energéticos, Gobierno del Estado de Sonora, Paseo de la Arboleda 30, Hermosillo 83000 Sonora, México 03. PhD (U. Paris, France, 1967) petrology. *Mineralogy petrology*
Tel. 52(62)131968

Espinosa Pérez, Ms Georgina (1959). Técnico Académico. Instituto de Química, UNAM, Circuito Exterior, Cuidad Universitaria, Deleg. Coyoacán, México 04510 DF, México 03. MSc (Facultad de Química, UNAM, 1993) organic chemistry. *Crystal_chemistry*
E-mail georgina@redvax1.dgsca.unam.mx Tel. 52(5)6224415 Fax 52(5)6162203

Gleason Villagran, Mr Roberto (1957). Técnico Académico. Facultad de Ciencias, UNAM, Circuito Exterior, Ciudad Universitaria, Deleg. Coyoacán, México 04510 DF, México 03. MSc (Facultad de Ciencias, UNAM, 1989) physics. *Crystal_growth superconductors*
Tel. 52(5)6225086 Fax 52(5)6162010

Gómez Ramírez, Dr Ricardo (1944). Investigador. Instituto de Física, UNAM, Depto. Estado Solido, AP 20364, Circuito Exterior, Ciudad Universitaria, Deleg. Coyoacán, México 04510 DF, México 03. PhD (Stanford U., USA, 1971) materials science. *Creep nucleation_in_solid-solid_deformation plastic_deformation*
Tel. 52(5)6225086 Fax 52(5)6162010

Gómez Rodríguez, Dr Alfredo (1952). Investigador. Depto. Materia Condensada, UNAM, Instituto de Física, Circuito Exterior, Cd. Universitaria, A. P. 20364, Deleg. Coyoacán, México 01000 DF, México 03. PhD (U. Warwick, UK, 1989) physics. *Crystal_diffraction quasi-crystals*
Tel. 52(5)6225084 Fax 52(5)6222010

Hernández Arana, Dr Andrés (1951). Investigador. Depto. de Química, UAM-Iztapalapa, Ave. Michoacan y La Purisima S/N, AP 55-534, Iztapalapa 09340 DF, México 03. PhD (Depto. de Química, UAM-Iztapalapa México, 1988) chemistry. *Structure_and_stability_of_proteins*
Tel. 52(5)7244674 Fax 52(5)7244666 Telex 1764296UAMME

Hernández Juárez, Mr Edilberto (1957). Técnico Académico. Instituto de Física, UNAM, Circuito Exterior, Ciudad Universitaria, AP 20-364, Deleg. Coyoacán, México 04510 DF, México 03. BSc (Facultad de Ciencias, UNAM, 1993) physics. *Crystallography*
E-mail juarez@ifunam.ifisicacu.unam.mx Tel. 52(5)6225098 Fax 52(5)6161535

Hernández Ortega, Mr Simón (1956). Técnico Académico. Instituto de Química, UNAM, Circuito Exterior, Ciudad Universitaria, Deleg. Coyoacán, México 04510 DF, México 03. BSc (U. Veracruzana, Ver. México, 1989) chemistry. *Crystal_chemistry*
Tel. 52(5)6224415 Fax 52(5)6162203

Herrera Becerra, Dr Raul (1952). Investigador. Inst. de Física, UNAM, Depto. de Materia Condensada, Circuito Exterior, Cd. Univ., AP20-364, Deleg. Coyoacán, México 01000 DF, México 03. DrC (CECESE, México, 1989) physics. *Electron_diffraction*
Tel. 52(5)6225104 Fax 52(5)6161535

Huanosta Tera, Prof. Alfonso (1944). Investigador. Inst. de Investigaciones en Materiales, UNAM, Circuito Exterior, Ciudad Universitaria, Deleg. Coyoacán, México 04510 DF, México 03. PhD (CICESE en Ensenada B. C. México, 1992) material sciences. *Ceramic_materials electron_microscopy ionic_conductivity*
Tel. 52(5)6224651 Fax 52(5)6224571

José Yacaman, Dr Miguel (1946). Director Adjunto de Investigación Científica. CONACYT, Ave. Constituyentes 1046 1er piso, Col. Lomas Altas, México 11950 DF, México 03. PhD (Facultad de Ciencias, UNAM, 1973) ciencias de materiales y caracterizacion de materiales por microscopía. *Small_particles diffraction_theory quasicrystals*
E-mail yacaman@daic.main.conacyt.mx Tel. 52(5)3277580 Fax 52(5)5708503

Lara Magaña, Ms María Eugenia (1950). Investigadora. Ciencia e Ingenieria de Materiales al Servicio de la Industria, Ojito no. 34, Deleg. Coyoacán, México 04000 DF, México 03. BSc (U. Texas, USA, 1975) pharmacy. *Crystal_chemistry organic_materials_obtained_from_plants*
Tel. 52(5)5545945

Lee Moreno, Dr José Luis (1939). Investigador. Consejo de Recursos Minerales, Niños Heroes 139, México ZP 7 DF, México 03. PhD (U. Arizona, USA, 1972) geological engineering. *Geochemistry minerals_exploration fluid_occlusions computer_applications*
Tel. 52(5)5785942 Fax 52(5)5723952

Muñoz Picone, Dr Eduardo (1937). Investigador. Instituto de Física, UNAM, Depto. Materia Condensada, Circuito Exterior, Ciudad Universitaria, AP 20-364, Deleg. Coyoacán, México 04510 DF, México 03. PhD (Facultad de Ciencias, UNAM, 1970) physics. *Crystal_growth*
E-mail picone@redvax1.dgsca.unam.mx Tel. 52(5)6225109 Fax 52(5)6161535

Murrieta Sánchez, Dr Héctor (1945). Investigador. Instituto de Física, UNAM, Depto. Estado Sólido, Circuito Exterior, Ciudad Universitaria, AP 20-364, Deleg. Coyoacán, México 01000 DF, México 03. PhD (UNAM, México, 1978) physics. *Optical_properties_in_solids*
Tel. 52(5)6225118 Fax 52(5)6162010

Oliver Salvador, Ms Maria del Carmen (1946). Investigadora. Centro de Desarrollo de Productos Bioticos del IPN, Km. 8 carretera Yautepec, Morelos, México 03. [Tonga 51, Colonia Euzkadi, México 02660 DF, México 03.] BSc (Esc. Nal. de Ciencias Biológicas, IPN, México, 1969) ingeniería bioquímica. *Enzymology protein_chemistry plant_biotechnology*
Tel. 52(5)3412095 Fax 52(739)40020

Quintana Owen, Ms Patricia (1951). Investigadora. Facultad de Química, UNAM, División Estudios de Postgrado, Ciudad Universitaria, Deleg. Coyoacán, México 04510 DF, México 03. MSc (UNAM, México, 1977) chemistry. *Phase_diagram oxide_systems Li-Zr-Si*
Tel. 52(5)6223721 Fax 52(5)6162210

Quintanar, Mr Carlos (1951). Investigador. Instituto de Física, UNAM, Depto. del Estado Sólido, Circuito Exterior, Ciudad Universitaria, Deleg. Coyoacán, México 04510 DF, México 03. BSc (Facultad de Ciencias, UNAM, 1978) physics. *Crystal_growth*
Tel. 52(5)6225086 Fax 52(5)6161535

Ramos Bernal, Dr Sergio (1945). Investigador. Instituto de Ciencias Nucleares, UNAM, Depto. de Física y Matemáticas Aplicadas, Circuito Exterior, Ciudad Universitaria, Deleg. Coyoacán, México 04510 DF, México 03. PhD (U. Manchester, UK, 1974) physics. *Radiation_damage magnetic_materials optic_properties_in_crystals*
Tel. 52(5)6224685 Fax 52(5)6162233

Reyes Chumacero, Mr Antonio (1940). Investigador. Facultad de Química, UNAM, Div. Estudios de Postgrado, Ciudad Universitaria, Deleg. Coyoacán, México 04510 DF, México 03. BSc (Facultad de Química, UNAM, 1964) engineering. *Thermodynamics condensed_phases*
Tel. 52(5)6223721 Fax 52(5)6162210

Ríos Jara, Dr David (1950). Investigador. Instituto de Investigaciones en Materiales, UNAM, Circuito Exterior, Ciudad Universitaria, AP 70-360, Deleg. Coyoacán, México 04510 DF, México 03. PhD (Inst. Nat. des Sciences Appl. de Lyon, France, 1984) sciences de materiales. *Electron_microscopy alloys Phase_transformation*
Tel. 52(5)6224641 Fax 52(5)6224571

Rivera Moras, Mr Vicente (1948). Investigador. Instituto de Investigaciones en Materiales, UNAM, Circuito Exterior, Ciudad Universitaria, AP 70-360, Deleg. Coyoacán, México 04510 DF, México 03. MSc (Facultad de Ciencias, UNAM, 1979) ciencias de los materiales. *Electron_microscopy (Lorentz) magnetic_materials*
Tel. 52(5)6224652 Fax 52(5)6224571

Riveros Rotgé, Prof. Héctor Gerardo (1940). Investigador. Instituto de Física, UNAM, Departamento de Materia Condensada, Circuito Exterior, Ciudad Universitaria, AP 20-364, Deleg. Coyoacán, México 04510 DF, México 03. PhD (Facultad de Ciencias, UNAM, 1974) solid state physics. *Crystal_growth*
Tel. 52(5)6225091 Fax 52(5)6161535

Rodríguez Romero, Dr Adela (1951). Investigadora. Instituto de Química, UNAM, Circuito Exterior, Ciudad Universitaria, Deleg. Coyoacán, México 04510 DF, México 03. PhD (Depto. de Química, UAM-Iztapalapa, 1991) chemistry. *Protein_crystallization protein_structure data_collection*
E-mail adela@redvax1.dgsca.unam.mx Tel. 52(5)6224403 Fax 52(5)6162203

Romero Romo, Dr Mario (1947). Investigador. Depto. Ciencia de Materiales UAM-Azcapotzalco, Ave. San Pablo 180, Col. Reynosa Tamaulipas, México 02200 DF, México 03. PhD (U. Liverpool, UK, 1978) metallurgy. *Chemical_corrosion metals ceramic_materials*
Tel. 52(5)7244288 Fax 52(5)5164444

Romero Sánchez, Dr Miguel (1925). General Director of NUTEK S. A. de C. V. NUTEK, S. A. de C. V., Calle 7 NORTE 356, Tehuacan 75700 Puebla, México 03. PhD (Harvard U., Cambridge, MA, USA, 1955) organic chemistry. *Technological_research_and_quality_control mineralogy crystallography geology chemistry_and_nutrition*

Tel. 52(238)30007 Fax 52(238)30214

Ruiz Mejia, Dr Carlos (1939). Investigador. Inst. de Física, UNAM, Depto. Estado Solido, Circuito Exterior, Ciudad Universitaria, Deleg. Coyoacán, AP 20-364, México 04510 DF, México 03. PhD (Facultad de Ciencias, UNAM, 1964) crystallography. *Optics formation_energies_and_defects_in_crystals*
Tel. 52(5)6225112 Fax 52(5)6162010

Salas Plza, Dr Guillermo Armando (1942). Jefe de Departamento de Geología. Depto. de Geología, Universidad de Sonora, Boulevard Transversal y Rosales, Hermosillo 83000 Sonora, México 03. PhD (Stanford U., USA, 1971) geology. *Geology mineralogy geochemistry ore_deposits*
Tel. 52(62)173181x11

Solorio Munguía, Prof. José Gregorio (1922). Investigador. Instituto de Geología, UNAM, Circuito Exterior, Ciudad Universitaria, AP70-296, Deleg. Coyoacán, México 04510 DF, México 03. PhD (U. Nac. Aut. de México, 1958) metallurgy. *Mineralogy minerals_separation crystallography geochemistry*
Tel. 52(5)6223038 Fax 52(5)2804864

Soriano Garcia, Prof. Manuel (1947). Jefe del Depto. de Bioestructura/Investigador. Sub-Editor of Ninth Edition of World Directory of Crystallographers. Instituto de Química, UNAM, Circuito Exterior, Ciudad Universitaria, Deleg. Coyoacán, México 04510 DF, México 03. PhD (SUNY, Buffalo, NY, USA, 1976) biophysics. *Crystal_structure_determination protein_crystallography crystal_chemistry teaching biologically_interesting_compounds*
E-mail soriano@redvax1.dgsca.unam.mx Tel. 52(5)6224403 Fax 52(5)6162203

Téllez Ortiz, Ms Minerva Estela (1943). Técnico Académico. Facultad de Química, UNAM, Div. Estudios de Postgrado, Ciudad Universitaria, Deleg. Coyoacán, México 04510 DF, México 03. MSc (Fac. de Química, UNAM, 1977) chemistry. *X-ray_spectroscopy X-ray_diffraction*
Tel. 52(5)6223721

Torres Villaseñor, Dr Gabriel (1944). Project Chief. Inst. de Investigaciones en Materiales, UNAM, Ciudad Universitaria, AP 70-360, Deleg. Coyoacán, México 04510 DF, México 03. PhD (Case Western U., USA, 1972) material sciences. *Alloys_of_Cu domains_in_Cu_gamma_phases*
Tel. 52(5)6224650 Fax 52(5)6224571

Toscano, Mr Ruben Alfredo (1953). Técnico Académico. Instituto de Química, UNAM, Circuito Exterior, Ciudad Universitaria, Deleg. Coyoacán, México 04510 DF, México 03. MSc (Facultad de Química, UNAM, 1993) chemistry. *Single_crystal powder dust diffraction_data*
E-mail toscano@redvax1.dgsca.unam.mx Tel. 52(5)6224415 Fax 52(5)6162203

Valenzuela Monjarás, Dr Raúl Alejandro (1946). Investigador. Inst. de Investigaciones en Materiales, UNAM, Ciudad Universitaria, AP 70-360, Deleg. Coyoacán, México 04510 DF, México 03. DSc (Fac. des Sciences, Paris, France, 1974) material sciences. *Ceramic_materials magnetic_properties*
Tel. 52(5)6224653

Vera Calderón, Ms Gloria (1938). Técnico Académico. Fac. de Química, UNAM, Div. Estudios de Postgrado, Ciudad Universitaria, Deleg. Coyoacán, México 04510 DF, México 03. BSc (U. Guanajuato, México, 1962) química industrial. *X-ray_diffraction organics*
Tel. 52(5)6223721

Villafuerte Castrejón, Ms María Elena (1948). Investigadora. Inst. de Investigaciones en Materiales, UNAM, Ciudad Universitaria, AP 70-360, Deleg. Coyoacán, México 04510 DF, México 03. MSc (Fac. de Química, UNAM, 1979) inorganic chemistry. *Crystal_chemistry ceramic_materials*
Tel. 52(5)6224646 Fax 52(5)6224571

Villena Iribe, Mr René (1954). Profesor Asociado "D". Depto. de Biotecnología, UAM-Iztapalapa, Ave. Michoacan y La Purisima S/N, Iztapalapa 09340, DF, México 03. BSc (Facultad de Química, UNAM, México, 1981) chemistry. *Protein_isolation_and_purification crystal_chemistry data_collection*
Tel. 52(5)7244717

MOLDOVA

Sub-Editor: T. I. Malinowski

Antosiac, Dr Boris Yakovlevich (1946). Scientific collaborator. X-ray Analysis Laboratory, Inst. of Applied Physics, Acad. of Sciences of Moldova, Akademia str. 5, Kishinev 277028, Moldova. PhD (Institute of Applied Physics Ac. of Sci. of Moldova, 1993) crystallography and crystal physics. *X-ray_crystal_analysis_methods coordination_compounds*
E-mail acadmold@roeam.bitnet ifa@compc.moldova.su Tel. 373(2)738154 Fax 373(2)738149

Bersuker, Prof. Isaak Borukhovich (1928). Chief of Laboratory. Lab. of Quantum Chemistry, Inst. of Chemistry, Acad. of Sciences of Moldova, Akademia str. 3, Kishinev 277028, Moldova. DSc (Leningrad State U., 1964) physics and mathematics. *Crystal_chemistry ferroelectricity structural_phase_transition*
E-mail quant@compc.moldova.su Tel. 373(2)739675 Fax 373(2)729761

Biyushkin, Prof. Victor Nikolayevich (1935). Group leader. X-ray Analysis Laboratory, Inst. of Applied Physics, Acad. of Sciences of Moldova, Akademia str. 5, Kishinev 277028, Moldova. DSc (Institute of Crystallography, Moscow, Russia, 1992) physics and mathematics. *X-ray_crystal_analysis_methods organic_compounds*
E-mail acadmold@roeam.bitnet ifa@compc.moldova.su Tel. 373(2)738154 Fax 373(2)738149

Botnariuk, Dr Vasilii Mihai (1945). Senior scientific researcher. Dept. of Semiconductors Physics, State University of Moldova, Mateevich str. 60, Kishinev 277009, Moldova. [Alexandru celbun str. 9 ap.3, Ialoveni 278272, Moldova.] PhD (State U. Moldova, 1985) semiconductor physics. *Crystal_growth_from_the_gaseous_phase epitaxial_layers_of_A3B5_compounds heterojunctions solar_cells solar_energy_conversion*

E-mail gorea@university.moldova.su Tel. 373(2)240666 Fax 373(2)240655

Bourosh, Dr Pavlina Nikolai (1959). Scientific collaborator. X-ray Analysis Laboratory, Inst. of Applied Physics, Acad. of Sciences of Moldova, Akademia 5, Kishinev 277028, Moldova. PhD (Institute of Crystallography, Moscow, 1988) crystallography and physics of the crystals. *X-ray_structure_analysis crystal_chemistry coordination_compounds*
E-mail acadmold@roeam.bitnet ifa@compc.moldova.su Tel. 373(2)738154 Fax 373(2)738149

Boyarskaya, Prof. Yuliya Stanislavovna (1928). Chief of Laboratory; principal scientific collaborator. Mechanical Properties of Crystals Laboratory, Inst. of Applied Physics, Acad. of Sciences of Moldova, Akademia 5, Kishinev 277028, Moldova. DSc (Institute of Physics of Metals, Kiev, Ukraine, 1977) solid state physics. *Dislocations point_defects microindentation electrical_phenomena_by_the_deformation acoustic_emission*
E-mail moskal@lises.moldova.su Tel. 373(2)738084 Fax 373(2)738149

Bruk, Dr Leonid Izmail (1952). Scientific researcher. Dept. of Semiconductors Physics, State University of Moldova, Mateevich str. 60, Kishinev 277009, Moldova. [Dimo str. 1/2 ap. 24, Kishinev 277030, Moldova.] PhD (State U. Moldova, 1992) semiconductor physics. *Defects impurity deep_centers II-VI_compounds heterojunctions growth_of_II-VI_compounds_layers*
E-mail gorea@university.moldova.su Tel. 373(2)240666 Fax 373(2)240655

Chumakov, Dr Yury Mikhailovich (1952). Senior scientific collaborator. X-ray Analysis Laboratory, Inst. of Applied Physics, Acad. of Sciences of Moldova, Akademia str. 5, Kishinev 277028, Moldova. PhD (Inst. of Applied Physics Acad. Sci Moldavian SSR, 1982) physics and mathematics. *X-ray_structure_analysis absolute_configuration_determination*
E-mail acadmold@roeam.bitnet ifa@compc.moldova.su Tel. 373(2)738154 Fax 373(2)738149

Daniliuk, Dr Sergei (1935). Associate professor. Dept. of Physics Semiconductors, State University of Moldova, Mateevich str. 60, Kishinev 277009, Moldova. PhD (State U. Moldova, 1970) physics and mathematics. *Growth_of_epitaxial_layers photoelectrical_properties space_time_modulation_of_light optical_bistability*
Tel. 373(2)240666

Diacon, Dr Ion Andrei (1934). Group leader. X-ray Analysis Laboratory, Inst. of Applied Physics, Acad. of Sciences of Moldova, Akademia 5, Kishinev 277028, Moldova. DSc (Institute of Crystallography, Moscow, Russia, 1989) physics and mathematics. *X-ray_structure_analysis instrumentation electron_diffraction alfa-aminoacids*
E-mail acadmold@roeam.bitnet ifa@compc.moldova.su Tel. 373(2)738148 Fax 373(2)738149

Diakonu, Dr Ion Ion (1943). Senior scientific researcher. Dept. of Semiconductors Physics, State University of Moldova, Mateevich str. 60, Kishinev 277009, Moldova. [Traian str. 8/1 ap. 57, Kishinev 277043, Moldova.] PhD (State U. Moldova, 1985) semiconductor physics. *Epitaxial_layer_growth_from_gaseous_phase_of_A3B5_compounds electrical_properties_of_p-n_junctions*
E-mail gorea@university.moldova.su Tel. 373(2)240666 Fax 373(2)240655

Donica, Dr Feodor Gavril (1937). Director. Specialized Technical-Engineering Enterprise of Solid State Electronics at the Institute of Applied Physics, Acad. of Sciences of Moldova, Academia str. 5, Kishinev 277028, Moldova. DSc (Inst. of Applied Physics, Acad. Sci. of Moldova, 1990) crystallography and crystal physics. *Crystal_physics semiconductor_devices*
Tel. 373(2)727088 Fax 373(2)739068

Donu, Dr Sofiya Vasile (1948). Scientific collaborator. X-ray Analysis Laboratory, Inst. of Applied Physics, Acad. of Sciences of Moldova, Akademia 5, Kishinev 277028, Moldova. PhD (Inst. of Applied Physics, Acad. Sci of Moldova, 1985) physics and mathematics. *X-ray_structure_analysis absolute_configuration_determination complexes_containing_alfa-aminoacids electron_diffraction*
E-mail acadmold@roeam.bitnet ifa@compc.moldova.su Tel. 373(2)738166 Fax 373(2)738149

Dvorkin, Dr Alexander Arkadyevich (1947). Senior scientific collaborator. X-ray Analysis Laboratory, Inst. of Applied Physics, Acad. of Sciences of Moldova, Akademia 5, Kishinev 277028, Moldova. PhD (Inst. of Applied Physics, Acad. Sci. of Moldova, 1975) physics and mathematics. *X-ray_structure_analysis crystal_chemistry_of_macrocycles methods_of_X-ray_technique*
E-mail acadmold@roeam.bitnet ifa@compc.moldova.su Tel. 373(2)738154 Fax 373(2)738149

Fonari, Dr Marina Semionovna (1959). Scientific collaborator. X-ray Analysis Laboratory, Inst. of Applied Physics, Acad. of Sciences of Moldova, Akademia str. 5, Kishinev 277028, Moldova. PhD (Institute of Crystallography, Moscow, Russia, 1992) crystallography and crystal physics. *X-ray_structure_analysis macrocycles crystal_chemistry*
E-mail acadmold@roeam.bitnet ifa@compc.moldova.su Tel. 373(2)738154 Fax 373(2)738149

Gashin, Prof. Petr (1943). Professor, group leader. Dept. of Physics Semiconductors, State University of Moldova, Mateevich str. 60, Kishinev 277009, Moldova. DSc (Institute of Applied Physics, Kishinev, 1990) physics and mathematics. *Growth_of_thin_layers_of_II-VI_compounds heterojunctions heterojunctions solar_cells solar_energy_conversion*
Tel. 373(2)240666

Gerbeleu, Prof. Nikolai Vasile (1931). Chief of Laboratory. Chemistry of Coordination Compounds Laboratory, Inst. of Chemistry, Acad. of Sci. of Moldova, Akademia str. 3, Kishinev 277028, Moldova. DSc (Moscow State U., Russia, 1973) *X-ray_structure_analysis crystal_chemistry_coordination_compounds phase_transitions*

E-mail acadmold@roeam.bitnet Tel. 373(2)739790 Fax 373(2)729761

Gorya, Dr Oleg Stephan (1943). Professor, group leader. Dept. of Semiconductors Physics, State University of Moldova, Mateevich str. 60, Kishinev 277009, Moldova. [Cuza-Voda str. 16/1 ap. 51, Kishinev 277060, Moldova.] PhD (State U. Moldova, 1986) physics and mathematics; semiconductor physics. *Growth_of_thin_layers_of_II-VI_compounds electrical_and_optical_properties heterojunctions solar_cells solar_energy_conversion*
E-mail gorea@university.moldova.su Tel. 373(2)240666 Fax 373(2)240655

Grabko, Dr Daria Zakharovna (1941). Leading scientific collaborator. Mechanical Properties of Crystals Laboratory, Institute of Applied Physics, Academy of Sciences of Moldova, Academia str. 5, Kishinev 277028, Moldova. DSc (Institute of Applied Physics, Ac. of Sci. of Moldova, 1992) solid state physics of semiconductors and dielectrics. *Mechanical_properties_of_crystals microindentation dislocation point_defects cathodoluminescence_by_deformation_transmission_and_scanning_electron_microscopy_methods*
E-mail acadmold@roeam.bitnet ifa@compc.moldova.su Tel. 373(2)738109 Fax 373(2)738149

Jitaru, Dr Raisa Panteleevna (1935). Leading scientific collaborator. Mechanical Properties of Crystals Laboratory, Institute of Applied Physics, Academy of Sciences of Moldova, Academia str. 5, Kishinev 277028, Moldova. PhD (Kishinev State U., 1972) crystallography and crystal physics. *Mechanical_properties_of_crystals microindentation dislocation point_defects cathodoluminescence_by_deformation_transmission_and_scanning_electron_microscopy_methods*
E-mail acadmold@roeam.bitnet ifa@compc.moldova.su Tel. 373(2)738109 Fax 373(2)738149

Kiosse, Dr Georgy Alexandrovich (1932). Group leader. X-ray Analysis Laboratory, Inst. of Applied Physics, Acad. of Sciences of Moldova, Akademia str. 5, Kishinev 277028, Moldova. PhD (Inst. of Applied Physics, Acad. Sci. of Moldova, 1975) physics and mathematics. *X-ray_structure_analysis crystal_chemistry inorganic_compounds phase_transitions*
E-mail acadmold@roeam.bitnet ifa@compc.moldova.su Tel. 373(2)739584 Fax 373(2)738149

Kon, Dr Aviv Yuliseyevich (1929). Docent. Dept. of Physics, Moldova Technical University, Av. Stephan chel Mare 168, Kishinev 277004, Moldova. PhD (Gorky U., Nizhnii Novgorod, Russia, 1967) physics and mathematics. *X-ray_structure_analysis_methods crystal_chemistry coordination_compounds*
Tel. 373(2)497020

Korotkov, Dr Vitalii Alexandr (1941). Associate professor; research group leader. Dept. of Semiconductor Physics, State University of Moldova, Mateevich str. 60, Kishinev 277009, Moldova. [Independentii str. 2/2 ap.4, Kishinev 277043, Moldova.] PhD (State U. Moldova, 1970) semiconductor physics. *Defects impurity deep_centers II-VI_compounds heterojunctions growth_of_II-VI_compounds_layers*
E-mail gorea@university.moldova.su Tel. 373(2)240666 Fax 373(2)240655

Kravtsov, Dr Victor Christophorovich (1952). Senior scientific collaborator. X-ray Analysis Laboratory, Inst. of Applied Physics, Acad. of Sciences of Moldova, Akademia str. 5, Kishinev 277028, Moldova. PhD (Inst. of Applied Physics, Acad. Sci of Moldova, 1982) physics and mathematics. *X-ray_structure_analysis absolute_configuration_determination*
E-mail acadmold@roeam.bitnet ifa@compc.moldova.su Tel. 373(2)738154 Fax 373(2)738149

Malinowski, Dr Stanislav Tadeushevich (1949). Senior scientific collaborator. Organic Synthesis Laboratory, Inst. of Chemistry, Acad. of Sciences of Moldova, Akademia str. 3, Kishinev 277028, Moldova. PhD (Institute of Crystallography, Moscow, 1978) physics and mathematics. *Structural_chemistry organic_and_inorganic_coordination_compounds*
E-mail acadmold@roeam.bitnet Tel. 373(2)739758 Fax 373(2)729761

Malinowski, Prof. Tadeush Iosifovich (1921). Full member of the Ac. of Sci. of Moldova, Chief of Laboratory. Sub-Editor of Ninth Edition of World Directory of Crystallographers. X-ray Analysis Laboratory, Inst. of Applied Physics, Acad. of Sciences of Moldova, Akademia str. 5, Kishinev 277028, Moldova. DSc (Institute of Crystallography, Moscow, Russia, 1967) physics and mathematics. *Crystallography crystal_chemistry X-ray_crystal_structure_analysis*
E-mail acadmold@roeam.bitnet ifa@compc.moldova.su Tel. 373(2)725887 Fax 373(2)738149

Mazus, Dr Mark Davidovich (1937). Group leader. X-ray Analysis Laboratory, Inst. of Applied Physics, Acad. of Sciences of Moldova, Akademia str.5, Kishinev 277028, Moldova. PhD (Inst. of Applied Physics, Acad. Sci of Moldova, 1974) physics and mathematics. *X-ray_structure_analysis coordination_and_organic_compounds*
E-mail acadmold@roeam.bitnet ifa@compc.moldova.su Tel. 373(2)738154 Fax 373(2)738149

Palistrant, Prof. Alexander Fillipovich (1933). Professor. Dept. of Geometry, State University of Moldova, Mateevich str. 60, 277003 Kishinev, Moldova. DSc (Institute of Crystallography, Moscow, Russia, 1983) physics and mathematics. *Symmetry_theory_(generalization_and_applications) colour_symmetry antisymmetry*
Tel. 373(2)251220

Palistrant, Dr Natalia Alexandrovna (1960). Scientific collaborator. Mechanical Properties of Crystals Laboratory, Institute of Applied Physics, Academy of Sciences of Moldova, Academia str. 5, Kishinev 277028, Moldova. PhD (Institute of Applied Physics, Academy of Sci. of Moldova, 1992) crystallography and crystal physics. *Mechanical_properties_of_crystals microindentation dislocation point_defects cathodoluminescence_by_deformation_transmission_and_scanning_electron_microscopy_methods*

E-mail acadmold@roearn.bitnet ifa@compc.moldova.su Tel. 373(2)738109 Fax 373(2)738149

Radautsan, Prof. Sergiu Ion (1926). Chief of Laboratory, Full member of the Ac. Sci. of Moldova. Semiconductor Compounds Laboratory, Institute of Applied Physics, Acad. of Sciences of Moldova, Akademia str. 5, Kishinev 277028, Moldova. DSc (Leningrad Polytechnical Institute, Russia, 1966) technics. *Crystal_growth binary_ternary_and_multinary_semiconductor_materials*
E-mail acadmold@roearn.bitnet ifa@compc.moldova.su Tel. 373(2)738170 Fax 373(2)738149

Raevskii, Dr Simion Dumitru (1941). Senior researcher. Dept. Semiconductors Physics, University of Moldova, Mateevich str. 60, Kishinev 277009, Moldova. [Dacia str. 38, ap. 359, Kishinev 277060, Moldova.] PhD (NPO Kvant, Moscow, 1976) physics and chemistry of solids, semiconductor physics. *Crystal_growth solid_solutions epitaxial_layers thermoelectric_materials*
E-mail gorea@university.moldova.su Tel. 373(2)240666 Fax 373(2)240655

Rusanowskii, Dr Mikhail Evstafievich (1941). Senior lecturer. Dept. of Physics, Moldova Technical University, Av. Stephan chel Mare 168, Kishinev 277004, Moldova. PhD (Institute of Applied Physics Ac. of Sci of Moldova, 1985) crystallography and crystal physics. *X-ray_structure_analysis_methods crystal_chemistry coordination_compounds*
Tel. 373(2)497020

Samus', Prof. Ivan Dmitriyevich (1926). Professor. Dept. of Physics, Moldova Technical University, Avr. Stephan chel Mare 168, Kishinev 277004, Moldova. PhD (Institute of Crystallography, Moscow, 1966) physics and mathematics. *X-ray_structure_analysis_methods crystal_chemistry coordination_compounds*
Tel. 373(2)497020

Shcurpello, Dr Anatolii Ivanovich (1948). Senior scientific collaborator. Corrosion Control Laboratory, Institute of Applied Physics of the Ac. of Sci. of Moldova, Academia str. 5, Kishinev 277028, Moldova. PhD (Institute of Applied Physics of the Ac. of Sci. of Moldova, 1980) crystallography and crystal physics. *Thin_solid_films crystal_structure defects_in_crystals*
E-mail acadmold@roearn.bitnet ifa@compc.moldova.su Tel. 373(2)738043 Fax 373(2)738149

Sherban, Dr Dormidont Arhip (1939). Chief of Laboratory. Dept. of Semiconductors Physics, State University of Moldova, Mateevich str. 60, Kishinev 277009, Moldova. [Dimo str. 29/3 ap.34, Kishinev 277030, Moldova.] PhD (State U. Moldova, 1978) semiconductor physics. *Solar_energy_conversion widezone_oxide_semiconductors solar_cells growth_of_II-VI_compounds_layers*
E-mail gorea@university.moldova.su Tel. 373(2)240666 Fax 373(2)240655

Shova, Dr Sergiu (1958). Associate Professor. Dept. of Inorganic Chemistry, State University of Moldova, Mateevich str. 60, Kishinev 277009, Moldova. PhD (Bogatsky Physico-Chemical Institute, Odessa, 1985) inorganic chemistry. *X-ray_structure_analysis coordination_chemistry_compounds*
Tel. 373(2)251299

Simashkevich, Prof. Alexei Vasilii (1929). Head, Semicond. Phys. Dept., SUM; Full member, Acad. Sci. Moldova. Dept. of Semiconductors Physics, State University of Moldova, Mateevich str. 60, Kishinev 277009, Moldova. [Sciusev. str. 55 ap. 42, Kishinev 277012, Moldova.] DSc (Leningrad Polytechnical Institute, Russia, 1979) physics and mathematics; semiconductor physics. *Epitaxial_layers_II-*

VI_and_III-V_compounds heterojunctions electrical_and_optical_properties defects solar_cells solar_energy_conversion
E-mail simash@university.moldova.su Tel. 373(2)240666 Fax 373(2)240655

Simonov, Dr Yury Alexandrovich (1937). Group leader. X-ray Analysis Laboratory, Inst. of Applied Physics, Acad. of Sciences of Moldova, Akademia str. 5, Kishinev 277028, Moldova. PhD (Gorky U., Nizhnii Novgorod, 1967) physics and mathematics. *X-ray_structure_analysis coordination_compounds organic_compounds*
E-mail acadmold@roearn.bitnet ifa@compc.moldova.su Tel. 373(2)738154 Fax 373(2)738149

Sushkevich, Dr Konstantin Dmitrievich (1943). Senior scientific collaborator. Dept. of Physics Semiconductors, State University of Moldova, Mateevich str. 60, Kishinev 277009, Moldova. PhD (State U. Moldova, 1991) physics and mathematics. *Crystal_growth solid_solutions epitaxial_layer_growth thermal_crystal_treatment p-n_junctions solar_energy_conversion*
Tel. 373(2)240666

Tsurkan, Dr Vladimir (1951). Senior scientific collaborator. Semiconductor Compounds Laboratory, Institute of Applied Physics of Moldova Academy of Sciences, Academia str. 5, Kishinev 277028, Moldova. PhD (Lebedev Physical Institute, Moscow, Russia, 1979) solid state physics. *Structural_magnetic_and_transport_properties_of_ternary_and_multinary_compounds high_Tc_superconductors crystal_growth*
E-mail acadmold@roearn.bitnet ifa@compc.moldova.su Tel. 373(2)725895 Fax 373(2)738149

Val'kovskaya, Dr Margarita Ivanovna (1938). Senior scientific collaborator. Mechanical Properties of Crystals Laboratory, Inst. of Applied Physics, Acad. of Sciences of Moldova, Akademia str. 5, Kishinev 277028, Moldova. PhD (Kishinev State U., 1966) physics and mathematics. *Physical_crystallography mechanical_crystal_properties*
E-mail acadmold@roearn.bitnet ifa@compc.moldova.su Tel. 373(2)738109 Fax 373(2)738149

Volodina, Dr Galina Fedorovna (1935). Senior scientific collaborator. X-ray Analysis Laboratory, Inst. of Applied Physics, Acad. of Sciences of Moldova, Akademia str. 5 Kishinev 277028, Moldova. PhD (Institute of Crystallography, Moscow, 1964) physics and mathematics. *X-ray_structure_analysis_methods coordination_compounds inorganic_compounds*
E-mail acadmold@roearn.bitnet ifa@compc.moldova.su Tel. 373(2)739584 Fax 373(2)738149

Zamorzayev, Prof. Alexander Mikhailovich (1927). Professor; Corresponding member of the Ac. of Sci of Moldova. Dept. of Geometry, State University of Moldova, Mateevich str. 60, Kishinev 277003, Moldova. DSc (Institute of Crystallography, Moscow, Russia, 1971) physics and mathematics. *Symmetry_theory_(generalization_and_applications)*
Tel. 373(2)251220

Zhdanova, Dr Ludmila Ivanovna (1959). Senior scientific collaborator. Corrosion Control Laboratory, Institute of Applied Physics of the Academy of Sciences of Moldova, Academia str. 5, Kishinev 277028, Moldova. PhD (Institute of Applied Physics of the Ac. of Sci. of Moldova, 1990) crystallography and crystal physics. *Crystal_structure_defects phase_analysis*
E-mail acadmold@roearn.bitnet ifa@compc.moldova.su Tel. 373(2)738043 Fax 373(2)738149

MYANMAR

Sub-Editor: S. Htoon

Hla, Miss Tin Tin (1960). Demonstrator in Physics. Dept of Physics, University of Yangon, PO Box 11041, Yangon, Myanmar. MSc (Yangon U., 1990) physics. *X-ray_diffraction crystal_growth X-ray_fluorescence*
Tel. 95(1)32772

Htoon, Prof. Dr Sein (1941). Director of Research and Head of Department. National Sub-Editor of WDC9. Nuclear Physics Lab., University of Yangon, PO Box 11041, Yangon, Myanmar. [Nuclear Physics Lab., University of Yangon, PO Box 11041, Yangon, Myanmar.] FInstP(DSc) (Inst. of Physics, London, UK, 1982) physics. *Computer_technology mathematical_physics X-ray_fluorescence crystallography nuclear_physics*
Tel. 95(1)32772

Inn, Mr Aung Paik (1939). Associate Professor. Dept of Physics, Bago Degree College, Bago, Myanmar. [Dept of Physics, Bago Degree College, Bago, Myanmar.] MSc (Yangon U., 1979) physics. *X-ray_crystallography theory physics_teaching*

Lwin, Ms Hla Myat (1956). Research Assistant. University Research Centre, Yangon University, PO Box 11041, Yangon, Myanmar. MSc (Yangon U., 1983) physics. *X-ray_diffraction scanning_electron_microscopy*

Naing, Mr Saw (1962). Demonstrator in Physics. University Research Centre, Yangon

University, PO Box 11041, Yangon, Myanmar. MSc (Yangon U., 1993) physics. *X-ray_diffraction crystal_growth X-ray_fluorescence*
Tel. 95(1)31486

Thin, Miss Kathy (1964). Demonstrator in Physics. Dept of Physics, University of Yangon, PO Box 11041, Yangon, Myanmar. MSc (Yangon U., 1992) physics. *X-ray_diffraction crystal_growth X-ray_fluorescence*
Tel. 95(1)32772

Wai, Miss Myo Sandar (1965). Demonstrator in Physics. Dept of Physics, University of Yangon, PO Box 11041, Yangon, Myanmar. MSc (Yangon U., 1991) physics. *X-ray_diffraction crystal_growth X-ray_fluorescence*
Tel. 95(1)32772

Win, Miss Ohnmar (1956). Assistant Lecturer in Physics. X-ray Fluorescence Lab., University Research Centre, PO Box 11041, Yangon, Myanmar. MSc (Yangon U., 1984) physics. *X-ray_diffraction crystal_growth X-ray_fluorescence*
Tel. 95(1)32772

Yin, Prof. Dr Soe (1941). Professor and Head of Department. Dept of Physics, Pathein Degree College, Pathein, Myanmar. PhD (U. Trieste, 1980) physics. *Line_broadening condensed_matter_theory*
Tel. 95(42)21135xPhysics

NETHERLANDS

Sub-Editor: R. Olthof-Hazekamp

Notes

1. Degrees conferred by the Netherlands universities are *doctor* (Dr) (approximately equivalent to PhD at British universities), *doctorandus* (Drs) and *ingenieur* (Ir) (these latter two between MSc and PhD).

Admiraal, Dr Gerrit (1953). Res. scient. Lab. voor Kristallografie, U. Nijmegen, Toernooiveld, 6525 ED Nijmegen, The Netherlands. Dr (U. Groningen, 1981) chemistry. *Peptides DNA_structure Patterson_method computing*
Tel. 31(80)652648

Aerts, Dr Jozef (1957). Res. scient.; section leader molecular modelling group. Akzo Nobel Central Research, Applied Mathematics Dept., Velperweg 76, PO Box 9300, 6800 SB Arnhem, The Netherlands. PhD (U. Leuven, Belgium, 1983) chemistry. *Molecular_modelling structure_and_physical_property_of_polymers*
E-mail Jos.J.Aerts@Akzo.nl Tel. 31(85)663794 Fax 31(85)665466

Altona, Prof. Dr Cornelis (1931). Prof. Dept. of Organic Chemistry, Gorlaeus Lab., U. Leiden, Einsteinweg 55, PO Box 9502, 2300 RA Leiden, The Netherlands. Dr (U. Leiden, 1964) chemistry. *Molecular_mechanics molecular_dynamics conformational_analysis biomolecule nucleotides*
Tel. 31(71)274329 or 274505 Fax 31(71)274488 or 274537

Baak, Ing. Leonardus C. (1944). Software product specialist. Nonius B.V., Röntgenweg 1, PO Box 811, 2600 AL Delft, The Netherlands. Ing. *Computing diffractometry X-ray_diffraction*
Tel. 31(15)698300 Fax 31(15)627401

Bennema, Prof. Dr Pieter (1932). Prof. Science Dept., U. Nijmegen, Toernooiveld, 6525 ED Nijmegen, The Netherlands. Dr (Delft U. Techn., 1965) crystal growth. *Crystal_growth morphology*
Tel. 31(80)653070 Fax 31(80)553450 Telex 48226 wina nl

van den Berg, Ir Adrianus J. (1940). Univ. docent. Fac. der Techn. Natuurkunde, Delft U. of Techn., Lorentzweg 1, 2628 CJ Delft, The Netherlands. Ir (Delft U. Techn, 1966) physical chemistry. *Surface_physics optical_diffraction crystal_physics powder_diffraction*
Tel. 31(15)782481 Fax 31(15)783251

Beurskens, Dr Gezina (1936). Lab. voor Kristallografie, U. Nijmegen, Toernooiveld, 6525 ED Nijmegen, The Netherlands. Dr (U. Utrecht, 1961) chemistry. *Crystal_structure_determination direct_method*
E-mail u625004@vm.uci.kun.nl Tel. 31(80)652842 or 652875 Fax 31(80)553450 Telex 48228 wina nl

Beurskens, Prof. Dr Paul T. (1934). Prof. Lab. voor Kristallografie, U. Nijmegen, Toernooiveld, 6525 ED Nijmegen, The Netherlands. Dr (U. Utrecht, 1965) chemistry. *Crystal_structure_determination direct_method Patterson_method computing automation*
E-mail u625002@vm.uci.kun.nl Tel. 31(80)652188 or 652875 Fax 31(80)553450 Telex 48228 wina nl

Beyer, Dr Ir Jenö (1942). Univ. docent; associate professor. Dept. of Mechanical Engineering, U. Twente, PO Box 217, 7500 AE Enschede, The Netherlands. Dr (U. Twente, 1982) mat. sci. *Kinetics crystallography martensitic_transformation other_phase_transformations transmission_electron_microscopy_(TEM)*
Tel. 31(53)892484 Fax 31(53)356490

de Boer, Dr Dirk K. G. (1954). Res. scient. Philips Res. Lab., Prof. Holstlaan 4, 5656 AA Eindhoven, The Netherlands. PhD (U. Groningen, 1983) physical chemistry. *Diffuse_scattering reflectometry X-ray_standing_waves X-ray_fluorescence*
E-mail deboerd@prl.philips.nl Tel. 31(40)742859 Fax 31(40)743075

de Boer, Dr Jan L. (1936). Univ. hoofddocent. Dept. of Chemical Physics, U. Groningen, Nijenborgh 4, 9747 AG Groningen, The Netherlands. Dr (U. Groningen, 1970) chemistry. *Modulated_crystal_structure diffuse_scattering instrumentation*
E-mail jldeboer@chem.rug.nl Tel. 31(50)634424 or 634565 Fax 31(50)634441

van Bolhuis, Mr Fré (1934). Crystallographer. Nijenborgh 4, 9747 AG Groningen, The Netherlands. *X-ray_diffraction instrumentation small_molecules*
Tel. 31(50)634370 Fax 31(50)634200

Bosman, Drs Wilhelmus P. J. H. (1937). Res. scient. Lab. voor Kristallografie, U. Nijmegen, Toernooiveld, 6525 ED Nijmegen, The Netherlands. Drs (U. Nijmegen, 1969) chemistry. *Computing inorganic_crystal_structure direct_method*
E-mail u625008@vm.uci.kun.nl Tel. 31(80)652591 Fax 31(80)553450 Telex 48228 wina nl

Braam, Dr Adrianus W. M. (1952). Sen. scient. Dept. of Physical Chemistry, CRO-DSM, PO Box 16, 6160 MD Geleen, The Netherlands. Dr (U. Groningen, 1981) chemistry and physics. *Semicrystalline_compounds polycrystal X-ray_crystallography inter-_and_intra-molecular_interactions morphology*
Tel. 31(46)767386 Fax 31(46)767244

Bronsveld, Dr Paul M. (1936). Sen. scient. Dept. of Applied Physics, U. Groningen, Nijenborgh 4, 9747 AG Groningen, The Netherlands. Dr (U. Toronto, Canada, 1971) physics. *Composite interface HREM*
E-mail bronsvel@phys.rug.nl Tel. 31(50)634907 Fax 31(50)634881

Bruins Slot, Dr Hilbert J. (1956). Res. scient. CAOS/CAMM Center, U. Nijmegen, Toernooiveld, PO Box 9010, 6500 GL Nijmegen, The Netherlands. Dr (U. Nijmegen, 1986) chemistry. *Computing structural_chemistry phase_determination molecular_modelling 3D_database*
E-mail bslot@caos.kun.nl Tel. 31(0)80652137 Fax 31(80)652977 Telex 48228 wina nl

Daams, Mr Jo L. C. (1942). Res. scient. Structure Analysis Department, Philips Research, Ned. Philips Bedrijven B.V., Prof. Holstlaan 4, 5656 AA Eindhoven, The Netherlands. HBO *Intermetallic_compound crystal_structure structure_property_relationship powder_diffraction_software computer_modelling transmission_electron_microscopy_(TEM) preparation_methods_for_TEM*
E-mail daams@prl.philips.nl Tel. 31(40)742664 Fax 31(40)743478

Deblieck, Dr Rudy A. C. (1956). Res. manager. PCM-MP, DSM Research, PO Box 18, 6160 MD Geleen, The Netherlands. PhD (VU Brussels, Belgium, 1986) physical chemistry. *Mechanical_properties_of_polymers*
Tel. 31(46)761661 Fax 31(46)767569 Telex 36777 dsm nl

Delhez, Dr Ir Robert (1940). Univ. docent. Fac. der Scheik. Techn. en der Materiaalk., Delft U. of Techn., Rotterdamseweg 137, 2628 AL Delft, The Netherlands. Dr (Delft U. Techn., 1978) chemistry. *Powder_diffraction instrumentation line_profile_analysis materials_science*
E-mail rob.delhez@stm.tudelft.nl Tel. 31(15)782261 Fax 31(15)786730

Dijkstra, Prof. Dr Bauke W. (1948). Prof. Dept. of Biophysical Chemistry, U. Groningen, Nijenborgh 4, 9747 AG Groningen, The Netherlands. Dr (U. Groningen, 1980) chemistry. *Proteins_crystallography proteins_engineering enzymes_mechanism*
E-mail bauke@chem.rug.nl Tel. 31(50)634381 or 634378 Fax 31(50)634800

Drenth, Prof. Dr Jan (1925). Prof. Em. Lab. voor Biofysische Chemie, U. Groningen, Nijenborgh 4, 9747 AG Groningen, The Netherlands. Dr (U. Groningen, 1957) chemistry. *Proteins X-ray structure-activity_relationship biological_macromolecules*
E-mail drenth@rugch2.chem.rug.nl Tel. 31(50)634382 Fax 31(50)634800

Driessen, Drs René A. J. (1959). Res. scient. Lab. voor Kristallografie, U. Amsterdam, Nieuwe Achtergracht 166, 1018 WV Amsterdam, The Netherlands. Drs (U. Leiden, 1982) chemistry. *Computer_modelling computer_graphics computing crystallography direct_method*
E-mail rdr@crys.chem.uva.nl Tel. 31(20)5257036 Fax 31(20)5255698

van Eijk, Ir Marcel C. P. (1966). PhD student. Dept. of Physical and Colloid Chemistry, Wageningen Agricultural U., PO Box 8038, 6703 HB Wageningen, The Netherlands. Ir (Eindhoven U. Techn., 1991) polymer techn..
E-mail mars@fenk.wau.nl Tel. 31(8370)83710 Fax 31(8370)83777 Telex 45015 nl

van Enckevort, Dr Wilhelmus J. P. (1952). Univ. hoofddocent. RIM Lab. of Solid State Chemistry, U. Nijmegen, Toernooiveld, 6525 ED Nijmegen, The Netherlands. Dr (U. Nijmegen, 1982) chemistry. *Diamond crystal_growth defect_structure optoelectrical_property*
Tel. 31(80)653433 Fax 31(80)553450

Feil, Prof. Dr Dirk (1933). Prof. -Commission on Charge, Spin & Momentum Density. Chemical Physics Lab., U. Twente, PO Box 217, 7500 AE Enschede, The Netherlands. Dr (U. Utrecht, 1961) chemistry. *Charge_density orbital_calculation scattering_factor computer_modelling hydrogen_bonding refinement maximum_entropy_method quantum_chemistry non-linear_optics*
E-mail feil@utwente.nl Tel. 31(53)892949 Fax 31(53)325710

Fleischmann, Dr Klaus D. (1937). Director. Nonius B.V., Röntgenweg 1, PO Box 811, 2600 AV Delft, The Netherlands. Dr (Munich U. Techn., Germany, 1970) physical chemistry. *Instrumentation organic_structure*
Tel. 31(15)698525 Fax 31(15)627401

Frikkee, Prof. Dr Evert (1934). Prof. Netherlands Energy Res. Foundation, ECN, PO Box 1, 1755 ZG Petten NH, The Netherlands. Dr (U. Leiden, 1973) physics. *Magnetism heavy_fermion semiconductor phase_transition neutron_diffraction*
Tel. 31(2246)4527 Fax 31(2246)4480

van der Gaast, Sjierk J. (1943). Clay mineralogist. Netherlands Inst. of Sea Research, NIOZ, PO Box 59, 1790 AB Den Burg, Texel, The Netherlands. *Clay_mineralogy structure X-ray_powder_diffraction clay/water_relationship*
E-mail gaast@nioz.nl Tel. 31(2220)69398 Fax 31(2220)19674

van Geerestein, Dr Vincent J. (1959). Res. scient.; Head of Department. CMC Dept., N.V. Organon, PO Box 20, 5340 BH Oss, The Netherlands. Dr (U. Utrecht, 1988) chemistry. *Biomolecule structure-activity_relationship*
E-mail v.geerestein@organon.ahzonobel.nl Tel. 31(4120)61882 Fax 31(4120)62539

de Gelder, René (1965). Post-doc. Lab. voor Kristallografie, U. Nijmegen, Toernooiveld 1, 6525 ED Nijmegen, The Netherlands. Dr (U. Leiden, 1992) chemistry. *Direct_method modulated_structures incommensurate computing*
E-mail u625018@vm.uci.kun.nl Tel. 31(80)652596 Fax 31(80)553450

Goedkoop, Prof. Dr Jacob A. (1921). Prof. Em. [de Rougemont - Nes 1, 1862 AB Bergen (NH), The Netherlands.] Dr (U. Amsterdam, 1952) chemistry. *Neutron_diffraction*
Tel. 31(2208)13450

Gorter, Drs Ing. Sybout (1940). Univ. docent. Vakgroep ASKAM, Gorlaeus Lab., U. Leiden, Einsteinweg 55, PO Box 9502, 2300 RA Leiden, The Netherlands. Drs (U. Leiden, 1974) chemistry. *Computing intensity_measurement inorganic_organic_structure apparatus powder-diffraction-program_exchange_bank*
E-mail askasg@rulmvs.leidenuniv.nl Tel. 31(71)274415 Fax 31(71)274537

Goubitz, Drs Kees (1953). Res. scient. Lab. voor Kristallografie, U. Amsterdam, Nieuwe Achtergracht 166, 1018 WV Amsterdam, The Netherlands. Drs (U. Amsterdam, 1981) chemistry. *Crystal_structure_determination structure–activity_relationship*
E-mail fz@crys.chem.uva.nl Tel. 31(20)5257038 Fax 31(20)5255698

de Graaff, Dr Rudolf A. G. (1941). Univ. hoofddocent. Vakgroep ASKAM, Gorlaeus Lab., U. Leiden, Einsteinweg 55, PO Box 9502, 2300 RA Leiden, The Netherlands. Dr (U. Leiden, 1974) crystallography. *Direct_method accurate_phase_determination computing organic_crystal_structure*
E-mail graaff_r@rulgca.leidenuniv.nl Tel. 31(71)274211 Fax 31(71)274537

van de Grampel, Prof. Dr Johan C. (1934). Associate Prof. Dept. of Polymer Chemistry, U. Groningen, Nijenborgh 4, 9747 AG Groningen, The Netherlands. Dr (U. Groningen, 1967) chemistry. *Molecular_structure small_molecules inorganic_polymers*
E-mail J.C.VAN.DE.GRAMPEL@CHEM.RUG.NL Tel. 31(50)634442 Fax 31(50)634400

Gros, Dr Piet (1962). Senior scientist. Dept. of Crystal and Structural Chemistry, U. Utrecht, Padualaan 8, 3584 CH Utrecht, The Netherlands. Dr (U. Groningen, 1990) chemistry. *Proteins*
E-mail gros@chem.ruu.nl Tel. 31(30)533127 Fax 31(30)533940

Haije, Dr Willem G. (1959). Res. scient. Netherlands Energy Res. Foundation, ECN, PO Box 1, 1755 ZG Petten, The Netherlands. Dr (U. Leiden, 1988) theoretical inorganic chemistry. *Electron_density inorganic_crystal group_theory magnetism neutron_diffraction phase_transition symmetry thermal_vibration*
E-mail haije@ecn.nl Tel. 31(2246)4161 Fax 31(2246)3615

Harkema, Dr Sybolt (1940). Univ. hoofddocent. Chemical Physics Lab., U. Twente, PO Box 217, 7500 AE Enschede, The Netherlands. Dr (U. Twente, 1971) chemistry. *Electron_density crown_ethers*
E-mail harkema@utwente.nl Tel. 31(53)893080 or 892950 Fax 31(53)325710 Telex 44200

Hartman, Prof. Dr Piet (1922). Prof. Em. Inst. voor Aardwetenschappen, U. Utrecht, Budapestlaan 4, PO Box 80021, 3508 TA Utrecht, The Netherlands. Dr (U. Groningen, 1953) crystallography. *Crystal_growth inorganic_structure minerals*
Tel. 31(30)535092 Fax 31(30)535030

Helmholdt, Dr Robert B. (1943). Res. scient. Secretary Dutch Cryst. Assoc. (NVK). Energy Engineering Dept., Netherlands Energy Res. Foundation, ECN, PO Box 1, 1755 ZG Petten NH, The Netherlands. Dr (U. Groningen, 1975) chemistry. *Solar_energy_conversion neutron_diffraction powder_and_single_crystal_diffraction texture_and_stress_analysis computing quality_assurance*
E-mail helmholdt@ecn.nl Tel. 31(2246)4529 Fax 31(2246)1407 Telex 57211 reacp nl

Hottenhuis, Dr Ir M. H. J. (1959). Res. scient. AKZO Res. Lab., Dept. CRT, Velperweg 76, PO Box 9300, 6800 SB Arnhem, The Netherlands. Dr (U. Nijmegen, 1988) physics. *Microscopy TEM SEM STM AFM crystal_growth*
Tel. 31(85)664333 Fax 31(85)665272

van Hummel, Gerrit J. (1945). Sen. res. asst. Chemical Physics Lab., U. Twente, PO Box 217, 7500 AE Enschede, The Netherlands. Ing. (I. H. B. O. Eindhoven, 1972) physics. *Diffractometry charge_density small_molecules organic_structures computing instrumentation*
E-mail hummel@utwente.nl Tel. 31(53)893082 Fax 31(53)325710 Telex 44200

Israël, Drs O. R. (1968). PhD student. Lab. voor Kristallografie, U. Nijmegen, Toernooiveld, 6525 ED Nijmegen, The Netherlands. Drs (U. Utrecht, 1992) chemistry. *Modulated_structure*
E-mail u625014@vm.uci.kun.nl Tel. 31(80)652648 Fax 31(80)553450

Janner, Prof. Dr Aloysio (1928). Prof. Em. Inst. voor Theoretische Fysica I, U. Nijmegen, Merellaan 15, 6581 CH Malden, The Netherlands. PhD (U. Zürich, Switzerland, 1962) philosophy. *Symmetry solid_state_physics group_theory incommensurate_structure*
E-mail alo@sci.kun.nl Tel. 31(80)653408 or 581331 Fax 31(80)652120

Jansen, Dr Jacob (1959). Post-doc. Lab. voor Kristallografie, U. Amsterdam, Nieuwe Achtergracht 166, 1018 WV Amsterdam, The Netherlands. Dr (U. Amsterdam, 1991) chemistry. *Direct_method powder_diffraction electron_diffraction*
E-mail joukj@crys.chem.uva.nl Tel. 31(20)5257041 or 31(15)782267

Janssen, Prof. Dr Ted W. J. M. (1936). Prof. Member Dutch Cryst. Assoc. (NVK). Inst. Theoretical Physics, U. Nijmegen, Toernooiveld, 6525 ED Nijmegen, The Netherlands. Dr (U. Nijmegen, 1968) theoretical physics. *Incommensurate_crystal quasicrystal lattice_dynamics electrons_and_phonons_in_quasiperiodic_systems nonlinear_effects_in_solids*
E-mail ted@sci.kun.nl Tel. 31(80)652995 Fax 31(80)652120

Kalk, Mr Kornelis H. (1944). Crystallographer. Lab. voor Biofysische Chemie, U. Groningen, Nijenborgh 4, 9747 AG Groningen, The Netherlands. *Biomolecule instrumentation*
E-mail kalk@chem.rug.nl Tel. 31(50)634388 Fax 31(50)634800

Kamphuis, Dr Ireneus G. (1952). Res. scient. Group Real Time Systems, Netherlands Energy Res. Foundation/Techn., ECN, PO Box 1, 1755 ZG Petten (NH), The Netherlands. Dr (U. Groningen, 1983) protein crystallography. *Molecular_biology instrumentation protein_crystallography*
E-mail kamphuis@ecn.nl Tel. 31(2246)4544 Fax 31(2246)1864

Kanters, Dr Jan (1928). Retired. Lab. voor Kristal- en Structuurchemie, U. Utrecht, Padualaan 8, 3584 CH Utrecht, The Netherlands. Dr (U. Utrecht, 1958) chemistry. *Structure–activity_relationship biomolecule hydrogen-bond_pattern conformation saccharides*
E-mail kanters@chem.ruu.nl Tel. 31(30)533410 Fax 31(30)533940

de Keijser, Dr Ir Thomas H. (1937). Univ. hoofddocent. Fac. der Scheik. Techn. en der Materiaalk., Delft U. of Techn., Rotterdamseweg 137, 2628 AL Delft, The Netherlands. Dr (Delft U. Techn., 1977) chemistry. *Crystallography diffraction phase_transformation thin_layer*
Tel. 31(15)784105

Keulen, Dr Evert (1932). Res. scient. / consultant. X-ray Diffraction Consultancy, Philips Analytical X-ray B.V., Koersendijk 24, 7443 PR Nijverdal, The Netherlands. Dr (U. Groningen, 1969) chemistry. *Instrumentation high_precision_diffractometry powder_diffraction*
Tel. 31(5486)16949 Fax 31(5486)18775 Telex 35000 phtc nl xlbalaa

Kiers, Dr Conradus Th. (1947). Software product specialist. Secretary Dutch Cryst. Assoc. (NVK). Dept. Crystal & Structural Chemistry, University of Utrecht, Padualaan 8, 3584 CH Utrecht, The Netherlands. [Wulverhorst 20, 3461 GJ Linschoten, The Netherlands.] Dr (U. Groningen, 1976) chemistry. *Computing data_collection direct_method area_detectors*
E-mail kiers@chem.ruu.nl Tel. 31(3480)17371 Fax 31(30)533940

Kinneging, Dr Albertus J. (1959). Applic. specialist. Member Dutch Cryst. Assoc. (NVK). Philips Analytical X-RAY, Lelyweg 1, 7602 EA Almelo, The Netherlands. Dr (U. Leiden, 1986) chemistry. *Diffraction_physics*
Tel. 31(546)839454 Fax 31(546)839598

Klop, Dr Enno A. (1960). Res. scient. AKZO Research, Dept. CRU, Velperweg 76, PO Box 9300, 6800 SB Arnhem, The Netherlands. Dr (U. Utrecht, 1989) chemistry. *Powder_diffraction small-angle_X-ray_scattering fibres anomalous_scattering X-ray_crystal_structure_determination*
E-mail klop@arnhem.arla.akzo.400net.nl Tel. 31(85)665490 Fax 31(85)665464

Koch, Dr Beatrix (1926). Retired. [Joh. Verhulststraat 8, 1071 NC Amsterdam, The Netherlands.] Dr (U. Amsterdam, 1975) crystallography. *Mineralogy powder_diffraction X-ray_spectroscopy carotenoid_structures*
Tel. 31(20)6791001

van Koningsveld, Dr Hendrikus (1942). Univ. hoofddocent. Fac. der Technische Natuurkunde, Delft U. of Techn., Lorentzweg 1, 2628 CJ Delft, The Netherlands. Dr (U. Utrecht, 1970) physical chemistry. *Zeolites crystal_structures*
E-mail havank@cad4sun.tn.tudelft.nl Tel. 31(15)782605 Fax 31(15)783251

Kooijman, Dr Huub (1964). Univ. docent. Lab. voor Kristal- en Structuurchemie, U. Utrecht, Padualaan 8, 3584 CH Utrecht, The Netherlands. Dr (U. Utrecht, 1992) chemistry. *Structural_chemistry service_crystallography*
E-mail huub@chem.ruu.nl Tel. 31(30)532533 Fax 31(30)533940

Koopmans, Prof. Dr Kasper (1927). Prof. Em. [Churchillaan 70, 2625 GW Delft, The Netherlands.] Dr (Eindhoven U. Techn., 1971) techn. sciences. *Powder_diffraction ores minerals line_profile-analysis*
Tel. 31(15)563276

Kouwijzer, Drs Milou (1967). PhD student. Lab. voor Kristal- en Structuurchemie, U. Utrecht, Padualaan 8, 3584 CH Utrecht, The Netherlands. Drs (U. Utrecht, 1991) chemistry. *Computer_modelling force_field molecular_dynamics*
E-mail milou@chem.ruu.nl Tel. 31(30)532866 Fax 31(30)533940

Krabbendam, Drs Hendrik (1934). Univ. hoofddocent. Lab. voor Kristal- en Structuurchemie, U. Utrecht, Padualaan 8, 3584 CH Utrecht, The Netherlands. Drs (U. Utrecht, 1959) chemistry. *Structure_determination_method protein_crystallography*
E-mail krabbendam@chem.ruu.nl Tel. 31(30)533414 Fax 31(30)533940

Krever, Dr Maarten (1955). Res. scient. Lab. voor Kristallografie, U. Amsterdam, Nieuwe Achtergracht 166, 1018 WV Amsterdam, The Netherlands. Dr (U. Leiden, 1989) chemistry. *Direct_method crystal_structure_determination*
Tel. 31(20)5257039

Kronenburg, Dr Martinus J. (1964). Res. asst. Lab. voor Kristallografie, U. Amsterdam, Nieuwe Achtergracht 166, 1018 WV Amsterdam, The Netherlands. Dr (U. Amsterdam, 1992) crystallography. *Direct_method*
E-mail kronenburg@fys.ruu.nl Tel. 31(20)5257031

Kroon, Prof. Dr Jan (1937). Head of laboratory. Lab. voor Kristal- en Structuurchemie, U. Utrecht, Padualaan 8, 3584 CH Utrecht, The Netherlands. Dr (U. Utrecht, 1964) chemistry. *X-ray_diffraction structure_determination_method molecular_conformation hydrogen_bonding computer_simulation biological_structure–activity_relationship biomacromolecular_structure*
E-mail kroon@chem.ruu.nl Tel. 31(30)532383 Fax 31(30)533940

Kroon-Batenburg, Dr Louise M. J. (1956). Res. scient. Lab. voor Kristal- en Structuurchemie, U. Utrecht, Padualaan 8, 3584 CH Utrecht, The Netherlands. Dr (U. Utrecht, 1985) chemistry. *Carbohydrates celluloses hydrogen_bonding molecular_mechanics nuclear_magnetic_resonance molecular_dynamics*
E-mail bate@chem.ruu.nl Tel. 31(30)532533 or 532865 Fax 31(30)533940

Kyriakidis, Dr Christos E. (1967). Researcher. Lab. of Crystallography, U. Amsterdam, Nieuwe Achtergracht 166, 1018 WV Amsterdam, The Netherlands. Dr (U. Amsterdam, 1993) crystallography. *Direct_method proteins_crystallography computing*
E-mail chrisk@crys.chem.uva.nl Tel. 31(20)5257029 Fax 31(20)5255698

van der Lee, Dr Arie (1964). Post-doc. Institut des Materiaux de Nantes, 2, Rue de la Houssiniere, 44072 Nantes Cedex 03, France. Dr (U. Groningen, 1992) inorganic chemistry. *Four-dimensional_crystallography five-dimensional_crystallography*
E-mail vdlee@cnrs-imn.fr Tel. 33(40)373907 Fax 33(40)373995

Le Loux, Drs Richard (1967). Res. asst. Lab. voor Kristallografie, U. van Amsterdam, Nieuwe Achtergracht 166, 1018 WV Amsterdam, The Netherlands. Drs (U. Leiden, 1990) chemistry. *Phase_transition bonding*
E-mail richardx@crys.chem.uva.nl Tel. 31(20)5257039 Fax 31(20)5255698

Loopstra, Prof. Dr Bert O. (1928). Prof. Em. Lab. voor Kristallografie, U. Amsterdam, Nieuwe Achtergracht 166, 1018 WV Amsterdam, The Netherlands. Dr (U. Amsterdam, 1958) crystallography. *Neutron_diffraction crystal_structure_determination*
Tel. 31(20)5257034

Maaskant, Prof. Dr Willem J. A. (1932). Prof. Lab. voor Anorganische Chemie, Gorlaeus Lab., U. Leiden, Einsteinweg 55, PO Box 9502, 2300 RA Leiden, The Netherlands. Dr (U. Leiden, 1963) theoretical chemistry. *Theoretical_physics solid_state_physics solid_state_chemistry physical_and_theoretical_chemistry organic_and_inorganic_chemistry crystallography applied_mathematics*
E-mail maaskant@rulgla.leidenuniv.nl Tel. 31(71)274214 or 274450 Fax 31(71)274537

Mahy, Dr Jan W. G. (1959). Res. scient. AKZO Nobel Central Research, Dept. RDT, Velperweg 76, PO Box 9300, 6800 SB Arnhem, The Netherlands. Dr Sc (U. Antwerpen, Belgium, 1987) physics. *Surface_spectroscopy_SIMS_XPS electron_microscopy electronic_materials inorganic_organic_interface*
Tel. 31(85)663924 Fax 31(85)665432 Telex 45204

van Maissen, Dr Kees F. (1964). Res. asst. Lab. voor Kristallografie, U. Amsterdam, Nieuwe Achtergracht 166, 1018 WV Amsterdam, The Netherlands. Dr (U. Amsterdam, 1994) chemistry. *Crystal_growth powder_diffraction*
E-mail keesm@crys.chem.uva.nl Tel. 31(20)5257029

Meetsma, Drs Auke (1946). Res. asst. Chemical Physics Dept., U. Groningen, Nijenborgh 4, 9747 AG Groningen, The Netherlands. Drs (U. Groningen, 1980) physical chemistry. *Crystal_structure_determination*
E-mail meetsma@chem.rug.nl Tel. 31(50)634368 Fax 31(50)634200

van Meurs, Dr Ir Frank (1946). Head of div. Nonius B.V., Röntgenweg 1, PO Box 811, 2600 AV Delft, The Netherlands. Dr (Delft U. Techn., 1978) chemistry. *Computing diffractometry instrumentation X-ray_diffraction structural_chemistry*
E-mail VanMeurs@nonius.nl Tel. 31(15)698507 Fax 31(15)627401

Mijlhoff, Dr Frans C. (1932). Retired. [M. fan Loanstrjitte 24, 9123 JP Metslawier, The Netherlands.] Dr (U. Amsterdam, 1964) chemistry. *Electron_microscopy EXAFS gas_phase_molecular_structure*
Tel. 31(5192)41711

Mittemeijer, Prof. Dr Ir Eric J. (1950). Prof. Lab. of Metallurgy, Delft U. of Techn., Rotterdamseweg 137, 2628 AL Delft, The Netherlands. Dr (Delft U. Techn., 1978) physical chemistry. *X-ray_diffraction line_profile_analysis electron_diffraction electron_microscopy diffusion thin_film phase_transformation surface_coating stress_analysis*
E-mail mittemeijer@stm.tudelft.nl Tel. 31(15)782207 Fax 31(15)786730

Moers, Dr Frans (1938). Univ. docent. Lab. of Crystallography, U. Nijmegen, Toernooiveld, 6525 ED Nijmegen, The Netherlands. Dr (U. Aken, Germany, 1964) science. *Chemical_crystallography*
Tel. 31(80)616161 or 652466

Moleman, Albertus C. (1949). Head of diffraction dept. v/d Waals Zeeman Lab., University of Amsterdam, Valckenierstr. 65, 1018 XE Amsterdam, The Netherlands. *Powder_diffraction data_processing computing graphics computer_simulation_of_structure instrumentation*
E-mail moleman@phys.uva.nl Tel. 31(20)5255760 or 5255621 Fax 31(20)5255788

Moonen, Dr Joost A. H. M. (1956). Group leader X-ray diffr. group. DSM Research B.V., PO Box 18, 6160 MD Geleen, The Netherlands. Dr (U. Utrecht, 1987) chemistry. *Polymers SAXS catalysts Rietveld_method*
Tel. 31(46)761676 Fax 31(46)767244

Noordik, Dr Jan H. (1944). Director CAOS/CAMM Center. Member Commission on Crystallographic Data. CAOS/CAMM Center, U. Nijmegen, Toernooiveld, 6525 ED Nijmegen, The Netherlands. Dr (U. Nijmegen, 1971) chemistry. *Computer_modelling organic_chemistry crystal_structure_determination bio_informatics automation_in_chemistry*
E-mail noordik@caos.kun.nl Tel. 31(80)653386 Fax 31(80)652977

Northolt, Dr Ir Maurits G. (1939). Res. scient. Corporate Res. Dept., AKZO, Velperweg 76, PO Box 9300, 6800 SB Arnhem, The Netherlands. Dr (U. Amsterdam, 1968) crystallography. *Polymers diffraction structure_mechanical_property_relationship*
Tel. 31(85)664056 Fax 31(85)665464 Telex 45204

Numan, Drs Milco (1969). Res. asst. Lab. voor Kristallografie, U. Amsterdam, Nieuwe Achtergracht 166, 1018 WV Amsterdam, The Netherlands. Drs (U. Amsterdam, 1993) chemistry. *Structure–activity_relationship QSAR crystal_structure_determination odour_compounds*
E-mail milcon@crys.chem.uva.nl Tel. 31(20)5257041 Fax 31(20)5255698

Olthof-Hazekamp, Drs Roeli (1937). Sen. scient. Treasurer Dutch Cryst. Assoc. (NVK); Sub-Editor WDC9. Lab. voor Kristal- en Structuurchemie, U. Utrecht, Padualaan 8, 3584 CH Utrecht, The Netherlands. Drs (U. Groningen, 1963) chemistry. *Computing*
E-mail olthof@chem.ruu.nl Tel. 31(30)532865 or 532869 Fax 31(30)533940

Peerdeman, Prof. Dr Antonius F. (1921). Prof. Em., U. Utrecht. [Prof. L. Fuchslaan 41, 3571 HE Utrecht, The Netherlands.] Dr (U. Utrecht, 1955) X-ray crystallography. *X-ray_crystallography apparatus direct_method anomalous_scattering molecular_conformation intermolecular_interaction thermodynamics*
Tel. 31(30)712460

Peerdeman, Dr Frans A. J. (1955). Sen. appl. and product specialist. Philips Analytical X-RAY, Lelyweg 1, 7602 EA Almelo, The Netherlands. MSc (Twente U., 1985) physics. *Powder_diffraction texture stress size_strain_analysis*
E-mail aplpeerdeman@amlie3.decnet.philips.nl Tel. 31(546)839432 Fax 31(546)839598 Telex 35000 phtc nl xlbalaa

Peschar, Dr René (1956). Univ. lecturer. Lab. voor Kristallografie, U. Amsterdam, Nieuwe Achtergracht 166, 1018 WV Amsterdam, The Netherlands. Dr (U. Amsterdam, 1987) chemistry. *Direct_method computing_method powder_diffraction*
E-mail rene@crys.chem.uva.nl Tel. 31(20)5257040 Fax 31(20)5255698 Telex 16460 facwn nl

Peterse, Ir Wilhelmus J. A. M. (1934). Univ. docent. Fac. der Technische Natuurkunde, Delft U. of Techn., Lorentzweg 1, 2628 CJ Delft, The Netherlands. Ir (Delft U. of Techn., 1963) physics. *X-ray_diffraction phase_transition instrumentation computing inorganic_compounds*
E-mail secvsfk@duttncb.tn.tudelft.nl Tel. 31(15)782405 or 781432 Fax 31(15)783251

van der Plas, Drs Jaco L. (1970). PhD student. Gorlaeus Lab., U. Leiden, Einsteinweg 55, PO Box 9502, 2300 RA Leiden, The Netherlands. Drs (U. Leiden, 1993) chemistry. *Direct_method Karle–Hauptman_matrices*
E-mail athplasj@rulgl.leidenuniv.nl Tel. 31(71)274414 Fax 31(71)274537

van der Plas, Prof. Dr Leendert (1928). Prof. Em. Dept. of Soil Science and Geology, Wageningen U. of Agriculture, Telefoonweg 3, 6712 GA Ede, The Netherlands. Dr (U. Leiden, 1959) petrography and mineralogy. *Clays_mineralogy feldspars_property clays_geochemistry*
Tel. 31(8380)12467 Fax 31(8370)82419

Popma, Prof. Dr Theo J. A. (1941). Rector magnificus. Dept. of Applied Physics and Electrical Engineering, U. Twente, PO Box 217, 7500 AE Enschede, The Netherlands. Dr (U. Groningen, 1970) solid state chemistry. *Materials_sciences*
E-mail th.j.a.popma@cvb.utwente.nl Tel. 31(53)892009 Fax 31(53)357956 Telex 44200

Ravelli, Drs Raimond B. G. (1968). PhD student. Lab. voor Kristal- en Structuurchemie, U. Utrecht, Padualaan 8, 3584 CH Utrecht, The Netherlands. Drs (U. Utrecht, 1992) chemistry. *Direct_methods Laue_diffraction synchrotron_radiation white-beam_radiation statistical_method diffraction_theory detector*
E-mail ravelli@chem.ruu.nl Tel. 31(30)533414 Fax 31(30)533940

Reefman, Derk (1967). Researcher. Philips Research Laboratories, Prof. Holstlaan 4, 5656 AA Eindhoven, The Netherlands. Dr (U. Leiden, 1993) physics. *Microstructure_analysis thin_film magnetic_materials ferroelectric_materials*
E-mail reefman@prl.philips.nl Tel. 31(40)742015 Fax 31(40)743478

Reiss, Drs Céleste A. (1952). Application/product specialist. Philips Analytical X-RAY, Lelyweg 1, 7602 EA Almelo, The Netherlands. Drs (U. Amsterdam, 1979) chemistry. *Direct_method crystal_structure_determination Rietveld_method*
E-mail aplreiss@amlie3.decnet.philips.nl Tel. 31(546)839454 Fax 31(546)839598

Ren, Yang (1964). PhD student. Chemical Physics Materials Science Center, U. Groningen, Nijenborgh 4, 9747 AG Groningen, The Netherlands. Master degree (Inst. of Physics, Beijing, P. R. China, 1988) solid state physics. *Modulated_crystal_structure magnetical_properties_of high_Tc_superconductors*
E-mail yren@rugch4.chem.rug.nl Tel. 31(50)634414 Fax 31(50)634441

Reus, Ir Henk J. (1942). Head R&D group. Nonius B.V., Röntgenweg 1, PO Box 811, 2600 AV Delft, The Netherlands. Ir physics. *Instrumentation*
Tel. 31(15)698432 Fax 31(15)627401

Rieck, Prof. Dr Gerard D. (1911). Prof. Em., Eindhoven U. Techn. [Mecklenburglaan 5, 5583 AG Waalre, The Netherlands.] Dr (U. Utrecht, 1945) chemistry. *Reaction recrystallization texture metals grain_growth oxidec_solids*
Tel. 31(4904)13768

Rietveld, Dr Hugo M. (1932). Retired. [Kalmanstraat 49, 1817 HW Alkmaar, The Netherlands.] PhD (U. Western Aust., Australia, 1964) physics. *Powder_diffraction computing*
Tel. 31(72)111694

Romers, Prof. Dr Cornelis (1919). Prof. Em., U. Leiden. [Nachtegaallaan 17, 2172 JP Sassenheim, The Netherlands.] Dr (U. Amsterdam, 1948) chemistry. *Methodology_and_philosophy_of_science philosophy*
Tel. 31(2522)12781

Romijn, Ir Antonius M. (1964). PhD student. Delft U. of Techn., Julianalaan 136, 2628 BL Delft, The Netherlands. Ir (Delft U. of Techn., 1990) chemical engineering. *Polymers crystallization ordering_in_amorphous_systems*
E-mail romijn@tudsv1.tudelft.nl Tel. 31(15)784319 Fax 31(15)781828

Roosenbrand, Dr Albert G. (1960). Res. scient. Koninklijke/Shell Lab. Amsterdam, Badhuisweg 3, 1031 CM Amsterdam, The Netherlands. Dr (Eindhoven U. Techn., 1990) surface science. *Modelling surface_sciences*
E-mail roosenb1@ksla.nl Tel. 31(20)6302957 Fax 31(20)6308025 Telex 11224 ksla nl

Rutten-Keulemans, Drs Elisabeth W. M. (1932). Sen. scient. Vakgroep ASKA, Gorlaeus Lab., U. Leiden, Einsteinweg 55, PO Box 9502, 2300 RA Leiden, The Netherlands. Drs (U. Leiden, 1959) chemistry. *Computing*
Tel. 31(71)123923 or 274211

Schagen, Dr Jan Dirk (1951). Consultant. jds software advice & consultancy, Westplantsoen 76, 2613 GN Delft, The Netherlands. Dr (U. Amsterdam, 1986) crystallography. *Computing instrumentation expert_system diffractometry graphics modelling*
E-mail schagen@crys.chem.uva.nl Tel. 31(15)126430 Fax 31(15)126430

Schapink, Dr Frederik W. (1931). Univ. hoofddocent. Lab. of Metallurgy, Delft U. of Techn., Rotterdamseweg 137, 2628 AL Delft, The Netherlands. Dr (Delft U. Techn., 1969) physics. *Electron_microscopy X-ray_diffraction physical_metallurgy*
Tel. 31(15)782272 Fax 31(15)786730

Schenk, Prof. Dr Hendrik (1939). Prof. Member Exec. Comm. IUCr; President Dutch Cryst. Assoc. (NVK). Lab. voor Kristallografie, U. Amsterdam, Nieuwe Achtergracht 166, 1018 WV Amsterdam, The Netherlands. Dr (U. Amsterdam, 1969) chemistry. *Direct_method crystal_structure_determination*
E-mail schenk@sara.nl Tel. 31(20)5257035 Fax 31(20)5255698

Scherrenberg, Dr Rolf (1965). Res. scient. DSM Research, PO Box 18, 6160 MD Geleen, The Netherlands. Dr (KU Leuven, Belgium, 1992) polymer science. *Polymers catalysts powder_diffraction small_angle_X-ray_scattering*
Tel. 31(46)761266 Fax 31(46)761200

Schierbeek, Dr Abraham J. (1955). Appl. scient. Nonius B.V., Röntgenweg 1, PO Box 811, 2600 AV Delft, The Netherlands. Dr (U. Groningen, 1988) protein crystallography. *Proteins_crystallography area_detector structural_biology*
E-mail Schierbeek@nonius.nl Tel. 31(15)698678 Fax 31(15)627401

Schoone, Prof. Dr Jean C. (1919). Prof. Em., U. Utrecht. [Groenlinglaan 80, 3722 VB Bilthoven, The Netherlands.] Dr (U. Utrecht, 1950) chemistry. *Computing*
Tel. 31(30)293799

van Schooneveld, Ing. Marinus (1943). Product specialist. Nonius B.V., Röntgenweg 1, PO Box 811, 2600 AV Delft, The Netherlands. Ing. (Polytechnic Institute, Rotterdam, 1965) electrical engineering. *Instrumentation*
Tel. 31(15)698503 Fax 31(15)627401

Schouten, Mr Arie (1956). Techn. asst. Lab. voor Kristal- en Structuurchemie, U. Utrecht, Padualaan 8, 3584 CH Utrecht, The Netherlands. *X-ray_diffraction powder_diffraction structural_chemistry instrumentation computing database symmetry crystal_growth phase_transition proteins*
E-mail schouten@chem.ruu.nl Tel. 31(30)533122 Fax 31(30)533940

Schreurs, Drs Antonius M. M. (1955). Res. scient. Lab. voor Kristal- en Structuurchemie, U. Utrecht, Padualaan 8, 3584 CH Utrecht, The Netherlands. Drs (U. Utrecht, 1981) chemistry. *Computing computer_graphics crystal_structure_statistics*
E-mail schreurs@chem.ruu.nl Tel. 31(30)533501 or 532869 Fax 31(30)533940

Seal, Dr Michael (1930). Consultant. Sigillum B.V., PO Box 7129, 1007 JC Amsterdam, The Netherlands. PhD (Cambridge U., UK, 1957) physics. *Crystal_growth electron_microscopy high_pressure instrumentation materials_sciences microscopy surface_structure science_and_industrial_applications_of_diamond*
Tel. 31(20)6719390 Fax 31(20)6719390

van der Sluis, Dr Paul (1962). Res. scient. Philips Res. Lab., Prof. Holstlaan 4, 5656 AA Eindhoven, The Netherlands. Dr (U. Utrecht, 1991) chemistry. *High_resolution_diffractometry instrumentation semiconductors crystal_growth*
E-mail vdsluis@prl.philips.nl Tel. 31(40)742021 or 742623 Fax 31(40)743478

Smeets, Drs Wilberthus J. J. (1958). Res. scient. Lab. voor Kristal- en Structuurchemie, U. Utrecht, Padualaan 8, 3584 CH Utrecht, The Netherlands. Drs (U. Utrecht, 1985) chemistry. *X-ray_structure_determination*
E-mail smeets@chem.ruu.nl Tel. 31(30)532533 or 532869 Fax 31(30)533940

Smit, Dr Paul H. (1949). Corporate planner. Philips Corporate Strategy & Planning, Bldg VO, PO Box 218, 5600 MD Eindhoven, The Netherlands. Dr (U. Utrecht, 1978) chemistry. *Digital_signal_processing*
Tel. 31(40)732602 Fax 31(40)786740

Smits, Mr Johannes M. M. (1948). Techn. asst. Lab. voor Kristallografie, U. of Nijmegen, Toernooiveld, 6525 ED Nijmegen, The Netherlands. *Data_collection computing powder_diffraction inorganic_and_organic_crystal_structure direct_method incommensurate_structures*
E-mail u625021@vm.uci.kun.nl Tel. 31(80)652596 Fax 31(80)553450 Telex 48228 wina nl

Smout, Ing. Emile M. (1944). Sen. project manager. Nonius B.V., Röntgenweg 1, PO Box 811, 2600 AV Delft, The Netherlands. Ing. *Instrumentation X-ray_diffraction powder_diffraction*
Tel. 31(15)698397 Fax 31(15)627401

Sonneveld, Mr Eduard J. (1945). Res. asst. TNO Plastics and Rubber Res. Inst., PO Box 6031, 2600 JA Delft, The Netherlands. *Powder_diffraction scanning_tunneling_microscopy atomic_force_microscopy*
Tel. 31(15)692052 Fax 31(15)566308

Spek, Dr Anthony L. (1944). Univ. docent. Lab. voor Kristal- en Structuurchemie, U. Utrecht, Padualaan 8, 3584 CH Utrecht, The Netherlands. Dr (U. Utrecht, 1975) chemistry. *Direct_method automation computing computer_graphics structure_determination diffractometer*
E-mail spea@chem.ruu.nl Tel. 31(30)532538 Fax 31(30)533940

Spijkerman, Ir Albert (1964). PhD student. Dept. of Chemical Physics, U. Groningen, Nijenborgh 4, 9747 AG Groningen, The Netherlands. Ir (U. Twente, 1991) physics. *Incommensurate_structure diffuse_scattering*
E-mail albert@chem.rug.nl Tel. 31(50)634415 Fax 31(50)634441

Stam, Dr Casper H. (1925). Retired. Lab. voor Kristallografie, U. Amsterdam, Chopinstraat 9, 1901 VE Castricum, The Netherlands. Dr (U. Amsterdam, 1963) chemistry. *Crystal_structure_determination crystal_optics*
Tel. 31(2518)53585

Straver, Drs Leonardus H. (1954). Appl. scient. Nonius B.V., Röntgenweg 1, PO Box 811, 2600 AV Delft, The Netherlands. Drs (U. Utrecht, 1980) chemistry. *Computing instrumentation organometallic conformation_analysis*
E-mail Straver@nonius.nl Tel. 31(15)698504 Fax 31(15)627401

Struikmans, Drs Rink (1941). Univ. docent & instructeur. Fac. der Techn. Natuurkunde, Delft U. of Techn., Lorentzweg 1, 2628 CJ Delft, The Netherlands. Drs (Free U. Amsterdam, 1970) exp. physics. *Phase_transition optical_crystallography crystal_physics instrumentation teaching condensed_matter*
E-mail secvsfk@duttncb.tn.tudelft.nl Tel. 31(15)784098 Fax 31(15)783251 Telex 38151 butud nl

Thijsse, Dr Barend J. (1950). Univ. hoofddocent. Fac. der Scheik. Techn. en der Materiaalkunde, Delft U. of Techn., Rotterdamseweg 137, 2628 AL Delft, The Netherlands. Dr (U. Leiden, 1978) physics. *X-ray_diffraction neutron_diffraction non-crystalline_solids metallic_glasses structural_relaxation ion-beam/solid-state_interactions computer_simulation_of_non-crystalline_solids*
E-mail mkmfthy@dutrex.tudelft.nl Tel. 31(15)782221 Fax 31(15)786730

Tuinstra, Prof. Dr Ir Fokke (1934). Prof. Fac. der Techn. Natuurkunde, Delft U. of Techn., Lorentzweg 1, 2628 CJ Delft, The Netherlands. Dr (Delft U. Techn., 1967) physics. *Surface_crystallography diffraction_and_scanning_tunnel_microscopy structure-property_relationship crystal_physics*
E-mail secvsfk@duttncb.tn.tudelft.nl Tel. 31(15)786112 or 781432 Fax 31(15)783251 Telex 38151 butud nl

Van Mechelen, Ing. Jan B. (1949). Assoc. res. phys. Koninklijke/Shell Laboratorium, Amsterdam, PO Box 38000, 1030 BN Amsterdam, The Netherlands. Ing. (Techn. Highschool, 1973) phys. techn. *Powder_diffraction*
E-mail mechele1@ksla.nl Tel. 31(20)6302196 Fax 31(20)6304037

van der Veen, Dr Adriaan H. (1925). Added res. Teaching economic geology. Inst. voor Aardwetenschappen, U. Utrecht, Budapestlaan 4, PO Box 80021, 3508 TA Utrecht, The Netherlands. Dr (Delft U. of Techn., 1963) mineralogy, petrology. *Quantitative_X-ray_diffraction composition high_temperature_X-ray_diffraction differential_thermal_analysis thermogravimetry quantitative_microscopy petrology mineralogy geochemistry economic_geology*
Tel. 31(30)535090 Fax 31(30)535030 Telex 40704

van de Velde, Dr George M. H. (1939). Univ. docent. Lab. of Inorganic Materials Science and Catalysis, Fac. Chemical Techn., U. Twente, PO Box 217, 7500 AE Enschede, The Netherlands. Dr (U. Twente, 1976) inorganic chemistry. *Powder_diffraction inorganic_materials_sciences*
E-mail g.m.h.vandevelde@ct.utwente.nl Tel. 31(53)892997 Fax 31(53)356024

Veldman, Drs Nora (1964). Res. asst. Vakgroep Kristal- en Structuurchemie, Bijvoet Center for Biomolecular Research, Universiteit Utrecht, Padualaan 8, 3584 CH Utrecht, The Netherlands. Drs (U. Utrecht, 1990) chemistry. *Service_structure_determination*
E-mail nora@chem.ruu.nl Tel. 31(30)533902 or 31(20)6236554 Fax 31(30)533940

Vente, Ir Jaap (1964). PhD student. Gorlaeus Lab., Solid State Chemistry, U. Leiden, Einsteinweg 55, PO Box 9502, 2300 RA Leiden, The Netherlands. Ir (Delft U. Techn., 1989) material science and engineering. *Powder_diffraction iridium_oxide_compounds*
E-mail avsvente@rulgl.leidenuniv.nl Tel. 31(71)274414 Fax 31(71)274537

Vermin, Dr Willem J. (1949). Sen. consultant. [Lindelaan 11, 2351 NV Leiderdorp, The Netherlands.] Dr (U. Leiden, 1981) crystallography. *Computing direct_method numerical_methods*
E-mail willem_vermin@sara.nl Tel. 31(71)413778

Verwer, Dr Paul (1965). Post-doc. Organon Int. B.V., Computational Medicinal Chemistry Group, AKZO Pharma Div., PO Box 20, 5340 BH Oss, The Netherlands. Dr (U. Utrecht, 1993) chemistry. *Computational_chemistry medicinal_chemistry direct_method protein_crystallography*
E-mail verwer@organon.akzo.400net.nl Tel. 31(4120)61468 Fax 31(4120)62539

Visser, Drs Jan Willem (1925). Retired. [Henry Dunantlaan 81, 2614 GL Delft, The Netherlands.] Drs (U. Amsterdam, 1955) chemistry. *Powder_diffraction automation indexing database mineralogy*
Tel. 31(15)123593 Fax 31(15)123593

Visser, Dr Rudolph J. J. (1953). Nonius B.V., PO Box 811, 2600 AV Delft, The Netherlands. Dr (U. Groningen, 1984) chemistry. *Powder_and_single_crystal_instrumentation software*
Tel. 31(15)570811 Fax 31(15)627401

Vonk, Dr Christ G. (1925). Retired. [Holsberg 17, 6129 KM Urmond, The Netherlands.] Dr (U. Groningen, 1957) physical chemistry. *Small-angle_X-ray_scattering polymers*
Tel. 31(46)330119

Vos-Looijenga, Dr Aafje (1928). [Roland Holstlaan 908, 2624 JK Delft, The Netherlands.] Dr (U. Groningen, 1952) structural chemistry. *Teaching accurate_structure_determination oligopeptides TCNQ_compounds*
Tel. 31(15)566590

de Vries, Drs Johan L. (1920). Retired. [Humperdincklaan 51, 5654 PB Eindhoven, The Netherlands.] Drs (U. Amsterdam, 1950) chemistry. *Instrumentation powder_diffraction_in_industry*
Tel. 31(40)522102

de Vries, Ir Roelof Y. (1967). PhD student. Chemical Physics Lab., U. Twente, CT-1320, PO Box 217, 7500 AE Enschede, The Netherlands. Ir (U. Twente, 1991) physics. *Charge_density scattering_factor computer computer_architecture computer_grahics computer_management computer_technology computer_modelling hydrogen_bonding refinement_method maximum-entropy_method massively_parallel_computing optimization supercomputer*
E-mail r.y.devries@ct.utwente.nl Tel. 31(53)893083 Fax 31(53)325710

van Vucht, Dr Johannes H. N. (1924). Retired. [Isodorusweg 23, 5624 KD Eindhoven, The Netherlands.] Dr (Eindhoven U. of Techn., 1963) technical sciences. *X-ray_powder_diffraction_software intermetallic_compound inorganic_compound structure-physical_property_relationship*
Tel. 31(40)446196

van de Waal, Ir Benjamin W. (1936). Sen. scient. Dept. of Physics, U. Twente, PO Box 217, 7500 AE Enschede, The Netherlands. Ir (Delft U. Techn., 1966) physics. *Molecular_packing intermolecular_force atomic_and_molecular_clusters crystal_surface crystal_growth defects non-crystalline_solids*
E-mail vhummel@utcvx.civ.utwente.nl Tel. 31(53)892954 Fax 31(53)325710

van der Wal, Dr Robert J. (1955). Sen. system progr. SARA, Ondersteuning, U. Amsterdam, Kruislaan 415, 1098 SJ Amsterdam, The Netherlands. Dr (U. Groningen, 1982) chemistry. *X-ray_crystallography accurate_electron_density supercomputer database*
E-mail rob_van_der_wal@sara.nl Tel. 31(20)5923000

Wang, Ms Yuan-Fang (1963). Res. asst. Lab. voor Kristallografie, U. Amsterdam, Nieuwe Achtergracht 166, 1018 WV Amsterdam, The Netherlands. BSc (Fudan U., China, 1985) physical chemistry. *Direct_method crystal_structure_determination*
E-mail yfw@crys.chem.uva.nl Tel. 31(20)5257029 Fax 31(20)5255698

Wiebenga, Prof. Dr Eelco H. (1913). Prof. Em., U. Groningen. [5 Allee des Cimes (la Pinede), 83420 La Croix-Valmer, France.] Dr (U. Utrecht, 1940) chemistry. *Quantum_chemistry*
Tel. 33(94)796853

Wiegers, Prof. Dr Gerrit A. (1930). Prof. Chemical Physics Dept., U. Groningen, Nijenborgh 4, 9747 AG Groningen, The Netherlands. Dr (U. Groningen, 1963) chemistry. *Inorganic_crystal_structure solid_state_chemistry_phase_transition powder_diffraction*
Tel. 31(50)634433 Fax 31(50)634441

de Wit, Drs Martin (1963). Res. asst. Lab. voor Kristallografie, U. Amsterdam, Nieuwe Achtergracht 166, 1018 WV Amsterdam, The Netherlands. Drs (U. Amsterdam, 1987) chemistry. *Powder_diffraction computing phase_transition*
Tel. 31(20)5257036

de With, Prof. Dr Gijsbertus (1950). Prof. Philips Res. Lab., Ned. Philips Bedrijven B.V., Prof. Holstlaan 4, 5656 AA Eindhoven, The Netherlands. Dr (U. Twente, 1978) chemical physics. *Ceramics glasses oxides nitrides carbides mechanical_and_chemical_properties*
E-mail with@prle.philips.nl Tel. 31(40)742132 Fax 31(40)744282

Woensdregt, Dr Cornelis F. (1937). Univ. hoofddocent. Inst. voor Aardwetenschappen, U. Utrecht, Budapestlaan 4, PO Box 80021, 3508 TA Utrecht, The Netherlands. Dr (U. Utrecht, 1990) geology. *Crystal_growth crystal_morphology surface_structure electron_microscopy X-ray_diffraction*
E-mail woens@earth.ruu.nl Tel. 31(30)535070 Fax 31(30)535030

Woning, Ing. Leo (1958). Res. asst. Hoogovens IJmuiden Refractories & Ceramics Lab., 3J-22, PO Box 10 000, 1970 CA IJmuiden, The Netherlands. Ing. (Bakhuis–Rooseboom Inst., 1988) analytical chemistry. *Powder_diffraction texture X-ray_stress scanning_electron_microscopy*
Tel. 31(2514)99572 Fax 31(2514)70489 Telex 35211 hovs nl

Zandbergen, Dr Henny W. (1950). Head group high resolution electron microscopy. National Centre HREM, Delft U. of Techn., Rotterdamseweg 137, 2628 AL Delft, The Netherlands. Dr (U. Leiden, 1981) chemistry. *Electron_microscopy electron_diffraction structure_property_relationship*
E-mail zandberg@tudsv1.tudelft.nl Tel. 31(15)782266 Fax 31(15)786730

Zoutberg, Drs Martinus C. (1960). Res. asst. Lab. voor Kristallografie, U. Amsterdam, Nieuwe Achtergracht 166, 1018 WV Amsterdam, The Netherlands. Drs (U. Amsterdam, 1986) chemistry. *Direct_method crown_ethers*
E-mail ds@crys.chem.uva.nl Tel. 31(20)5257041

NEW ZEALAND

Sub-Editor: C. E. F. Rickard

Anderson, Dr Bryan F. (1937). Senior Research Officer. Massey University, Department of Chemistry and Biochemistry, Private Bag, Palmerston North, New Zealand. PhD (U. Auckland, 1967) chemistry. *Protein_structure*

E-mail b.anderson@massey.ac.nz Tel. 64(6)3569099 Fax 64(6)3505613

Baker, Prof. Edward N. (1942). Group Leader. Member, Commission on Biological Molecules. Massey University, Department of Chemistry and Biochemistry, Palmerston North, New Zealand. PhD (U. Auckland, 1968) chemistry. *Protein_crystallography hydrogen_bonding water_structure protein_structural_relationships*

E-mail T.Baker@massey.ac.nz Tel. 64(6)3505367 Fax 64(6)3505682

Childs, Dr Cyril Walter (1941). Honorary Fellow. Chemistry Dept., Victoria University, Wellington, New Zealand. PhD (Otago U., 1967) physical chemistry. *Soils iron_compounds clays opal_phytoliths*

E-mail childs@matai.vuw.ac.nz Tel. 64(4)4721000x8183 Fax 64(4)4955241

Clark, Dr George R. (1942). Associate Professor. University of Auckland, Chemistry Department, Private Bag 92019, Auckland, New Zealand. PhD (U. Auckland, 1986) small molecule crystallography. *Metal-nucleic_acid_interaction oligonucleotide_drug_interaction organometallic_complex*

E-mail clark@ccu1.auckland.ac.nz Tel. 64(9)3737999x8294 Fax 64(9)3737422

Coombs, Prof. Douglas S. (1924). Professor Emeritus. University of Otago, Geology Department, Dunedin, New Zealand. DSc(Hon) (Geneva, 1974) mineralogy and petrology. *Mineralogy zeolites petrology geology*

E-mail dscoombs@rivendell.otago.ac.nz Tel. 64(4)4797505 Fax 64(3)4797527

Cooper, Dr Alan (1945). Associate Professor. University of Otago, Geology Department, PO Box 56, Dunedin, New Zealand. PhD (U. Otago, 1970) geology. *Mineralogy geochemistry petrology metamorphic*

E-mail ougeology@otago.ac.nz Tel. 64(3)4797515 Fax 64(3)4797527

Cutfield, Dr John F. (1945). Senior Lecturer. University of Otago, Biochemistry Department, PO Box 56, Dunedin, New Zealand. PhD (U. Auckland, 1970) chemistry. *Proteins_structure_and_function*

E-mail biocjfc@otago.ac.nz Tel. 64(3)4797836 Fax 64(3)4797866

Cutfield, Dr Susan M. (1947). Research Officer. University of Otago, Biochemistry Department, PO Box 56, Dunedin, New Zealand. DPhil (Oxford U., 1975) molecular biophysics. *Protein_crystallization proteins_structure*

E-mail biocjfc@otago.ac.nz Tel. 64(3)4797849 Fax 64(3)4797866

Gainsford, Dr Graeme J. (1945). Team Manager. NZ Institute for Industrial Research and Development (IRL), Gracefield Road, Lower Hutt, New Zealand. [NZ Institute for Industrial Research and Development (IRL), PO Box 31-310, Lower Hutt, New Zealand.] PhD (U. Canterbury, 1969) chemistry. *Computing small_molecule_chemistry Rietveld_method materials_inorganic music_barbershop*

E-mail g.gainsford@irl.cri.nz Tel. 64(4)5690000x4624 Fax 64(4)5690142

Kawachi, Dr Yosuke (1932). Senior Research Fellow. University of Otago, Geology Department, PO Box 56, Dunedin, New Zealand. PhD (Otago U., 1970) petrology. *Manganese_silicates metamorphic_minerals electron_microprobe_analysis*

E-mail ougeology@otago.ac.nz Tel. 64(3)4797507 Fax 64(3)4797527

March, Dr Frank C. (1944). Director of Information Technology Services. Victoria University of Wellington, PO Box 600, Wellington, New Zealand. PhD (Canterbury U., 1970) crystallography. *Computing information_system inorganic_structure organometallic_structure*

E-mail Frank.March@vuw.ac.nz Tel. 64(4)4715326 Fax 64(4)4715386

Nicholson, Dr Brian K. (1947). Professor. University of Waikato, Chemistry Department, Private Bag 3105, Hamilton, New Zealand. PhD (U. Otago, 1973) chemistry. *Clusters organometallic_compound*

E-mail B.Nicholson@waikato.ac.nz Tel. 64(7)8562889 Fax 64(7)8384219

Nieuwenhuyzen, Mr Mark (1963). Student. University of Canterbury, Chemistry Department, Private Bag, Christchurch, New Zealand. MSc (U. Canterbury, 1992) chemistry. *Diffractometry small_molecules_crystallography*

E-mail chem132@csc.christchurch.ac.nz Tel. 64(3)3642820 Fax 64(3)3642110

Norris, Dr Gillian E. (1948). Research Officer. Massey University, Department of Chemistry and Biochemistry, Private Bag, Palmerston North, New Zealand. PhD (Massey U., 1982) protein crystallography. *Protein_structure_and_function crystallization glycoproteins*

E-mail g.norris@massey.ac.nz Tel. 64(6)3569099x7955 Fax 64(6)3505613

Rickard, Dr Clifton E. F. (1941). Associate Professor. New Zealand Sub-Editor World Directory. University of Auckland, Chemistry Department, Private Bag 92019, Auckland, New Zealand. PhD (U. Auckland, 1967) chemistry. *Small_molecules_crystallography coordination_geometry*

E-mail rickard@ccu1.auckland.ac.nz Tel. 64(9)3737999x8289 Fax 64(9)3737422

Robinson, Dr Ward T. (1937). Reader. Member Commission on Crystallographic Teaching. University of Canterbury, Chemistry Department, Private Bag, Christchurch, New Zealand. PhD (U. Canterbury, 1963) chemistry. *Small_molecules_crystallography diffractometry teaching_crystallography*

E-mail chem132@csc.canterbury.ac.nz Tel. 64(3)3642465 Fax 64(3)3642110

Rodgers, Dr K. A. (1942). Associate Professor. University of Auckland, Geology Department, Private Bag 92019, Auckland, New Zealand. PhD (U. Auckland, 1973) geology. *Mineralogy hydroxides phosphates*

E-mail ka.rodgers@auckland.ac.nz Tel. 64(9)3737599x7414 Fax 64(9)3737435

Shelley, Dr David (1940). Associate Professor. University of Canterbury, Geology Department, Private Bag 4800, Christchurch, New Zealand. PhD (Bristol U., 1964) geology. *Mineralogy mineral_preferred_orientations*

E-mail D.Shelley@GEOL.canterbury.ac.nz Tel. 64(3)3667001x7723 Fax 64(3)3642769

Smale, Mr David (1939). Scientist. Institute of Geological and Nuclear Sciences, PO Box 30 368, Lower Hutt, New Zealand. MSc (U. Auckland, 1962) geology. *Petrology*

E-mail SRLNDXS@lhn.gns.cri.nz Tel. 64(4)5699059 Fax 64(4)5695016

Soong, Mr Raymond (1944). Clay Mineralogist. Institute of Geological and Nuclear Sciences, PO Box 30368, Lower Hutt, New Zealand. NZCS (AAVA Wellington, 1970) geology. *Coal_mineralogy geochemistry*

Tel. 64(4)5704840 Fax 64(4)5695016

Waters, Dr Joyce M. (1931). Senior Research Fellow. Convenor, NZ Committee for Crystallography. Massey University, Chemistry Department, Palmerston North, New Zealand. PhD (U. New Zealand, 1960) chemistry. *Clusters organometallic structure inorganic_structure organic_structure*

E-mail jwaters@massey.ac.nz Tel. 64(6)3569099 Fax 64(6)35505613

NIGER

Billiet, Prof. Yves (1936). Professor. Member Committee on the Nomenclature of Symmetry. Département de Chimie, Faculté des Sciences, Université de Niamey, Niger. [BP 825, Niamey, Niger.] DSc (U. Paris Sud, France, 1969) chemistry. *Symmetry group_theory phase_transition colour_symmetry teaching*
Tel. 227( )734706 Fax 227( )733997

NORWAY

Sub-Editor: K. Maartman-Moe

Notes

1. Degrees conferred by the Norwegian universities are the *Doctor philosophiae* (Dr philos.) and *Doctor technicae* (Dr techn.) (both approximately equivalent to the English DSc), *Doctor scientiarium* (Dr scient.) and *Doctor ingenieur* (Dr ing.) (both equivalent to PhD), *Canditatus realium* (Cand. real.) and *Magister scientiarium* (Mag. scient.) (both range between PhD and MSc), *Candidatus scientiarium* (Cand.scient.) and *Sivilingeniör* (siv.ing.) (approximately equivalent to MSc).

Berglund, Gunnar (1965). PhD student. Institute of Mathematical and Physical Sciences, University of Tromsø, N-9037 Tromsø, Norway. MSc (University of Tromsø, 1990) physical chemistry. *Crystallography proteins*
E-mail gunnar@chem.uit.no Tel. 47( )77644057 Fax 47( )77644756 Telex 64124 AUROB N

Bjørgo, Ingrid (1968). MSc student. Institute of Mathematical and Physical Sciences, University of Tromsø, N-9037 Tromsø, Norway. BSc (University of Tromsø, 1991) physical chemistry. *Crystallography*
Tel. 47( )77644063 Fax 47( )77644756 Telex 64124 AUROB N

Bye, Dr Erik (1945). Senor Scientist. National Institute of Occupational Health, Dept. of Occupational Hygiene, PO Box 8149 Dep., N-0033 Oslo, Norway. Dr philos. (University of Oslo, 1976) chemistry. *Minerals_dust X-ray_powder_analysis computing chemometrics structure–activity_relationships*
Tel. 47( )22466850 Fax 47( )22603276

Dahl, Prof. Tor (1938). Professor. Member of the Norwegian National Committee. Institute of Mathematical and Physical Sciences, University of Tromsø, N-9037 Tromsø, Norway. Dr philos. (University of Tromsø, 1976) chemistry. *Organic_crystal_structure intermolecular_interaction*
E-mail tord@chem.uit.no Tel. 47( )77644075 Fax 47( )77644765 Telex 64 124 AUROB N

Ehnebom, Lisbeth (1966). Research assistant. Institute of Pharmacy, University of Oslo, PO Box 1068 Blindern, N-0316 Oslo, Norway. Cand pharm (University of Oslo, 1990) pharmacy. *Pharmaceutical_organic_molecules*
E-mail lehnebom@ulrik.uio.no Tel. 47( )22855412 Fax 47( )22855947

Eliassen, Bjørn Eirik (1967). MSc student. Institute of Mathematical and Physical Sciences, University of Tromsø, N-9037 Tromsø, Norway. BSc (University of Tromsø, 1990) chemistry. *Proteins crystallography*
E-mail bjorne@mack.uit.no Tel. 47( )77614061 Fax 47( )77614765 Telex 64 124 AUROB N

Erlandsen, Heidi (1970). MSc student. Institute of Mathematical and Physical Sciences, University of Tromsø, N-9037 Tromsø, Norway. BSc (University of Tromsø, 1992) chemistry. *Crystallography proteins hemoglobins*
E-mail kjbe@mack.uit.no Tel. 47( )77644647 Fax 47( )77644765 Telex 64 124 AUROB N

Fell, Hans Jörg (1966). PhD student. Faculty of Physics and Mathematics, University of Trondheim NTH, N-7034 Trondheim, Norway. Dipl Phys (RWTH - Aachen, 1991) physics. *Conducting_polymers X-ray_scattering neutron_scattering spectroscopy*
E-mail hjfell@phys.unit.no Tel. 47( )73593589 Fax 47( )73593628 Telex 55637 NTHAD N

Fjaertoft-Pedersen, Prof. Berit (1933). Professor. Secretary Norwegian National Committee. Department of Pharmacy, University of Oslo, PO Box 1068 Oslo, N-0316 Oslo, Norway. Dr philos. (University of Oslo, 1969) physical chemistry. *Structure_determination X-ray_crystallography small_molecules transition_elements structure–activity_relationships drugs imaging_compounds antirheumatic_compound molecular_modelling*
E-mail berit.fjartoft@kjemi.uio.no Tel. 47( )22855694 Fax 47( )22855441 Telex 72705 ASTRO N

Fjellvåg, Helmer (1954). Professor. Department of Chemistry, University of Oslo, PO Box 1033 Blindern, N-0315 Oslo, Norway. MSc (University of Oslo, 1978) chemistry. *Powder_diffraction magnetic_materials defect_oxides*
E-mail helmer.fjellvag@kjemi.uio.no Tel. 47( )22855564 Fax 47( )22855565

Foss, Prof. Olav (1918). Professor emeritus. Department of Chemistry, University of Bergen, Allé gt 41, N-5007 Bergen, Norway. Dr techn (Norges tekniske høgskole, 1947) chemistry. *Inorganic_crystal_structure*
Tel. 47( )55213563 Fax 47( )55329058

Fossli, Dag (1966). MSc student. Institute of Mathematical and Physical Sciences, University of Tromsø, N-9037 Tromsø, Norway. BSc (Univ of Tromsø, 1991) chemistry. *Molecular_structure*
Tel. 47( )77614063 Fax 47( )77614765 Telex 64 124 AUROB N

Furuseth, Prof. Sigrid B. (1939). Professor. Member of the Norwegian National Committee. Department of Chemistry, University of Oslo, PO Box 1033 Blindern, N-0315 Oslo, Norway. MSc (University of Oslo, 1964) inorganic chemistry. *Structure_determination transition_elements*
E-mail sigrid.furuseth@kjemi.uio.no Tel. 47( )22855561 Fax 47( )22855565

Gjønnes, Prof. Jon Kjell (1931). Professor. Center for Materials Research, University of Oslo, Gaustadtalleen 21, N-0371 Oslo, Norway. [Department of Physics, University of Oslo, PO Box 1048 Blindern, N-0316 Oslo, Norway.] Dr philos. (University of Oslo, 1967) physics. *Electron_diffraction electron_microscopy materials_research diffraction_theory*
E-mail jon.gjonnes@fys.uio.no Tel. 47( )22958738 Fax 47( )22958747

Gjønnes, Kjersti (1961). Stipendiat. Department of Physics, University of Oslo, PO Box 1048 Blindern, N-0316 Oslo, Norway. MSc (University of Oslo, 1984) physics.
Tel. 47( )22958735 Fax 47( )22958749

Görbitz, Carl Henrik (1961). Assistant Professor. National Affiliated Centre for the Cambridge Structural Database System. Department of Chemistry, University of Oslo, PO Box 1033 Blindern, N-0315 Oslo, Norway. Dr Sc (University of Oslo, 1990) chemistry. *Database crystal_growth hydrogen_bonding peptides*
E-mail c.h.gorbitz@kjemi.uio.no Tel. 47( )22855460 Fax 47( )22855441

Groth, Per (1934). Sr Research Scientist. Department of Chemistry, University of Oslo, PO Box 1033 Blindern, N-0315 Oslo, Norway. MSc (University of Oslo, 1960) chemistry. *Computing direct_method*
E-mail per.groth@kjemi.uio.no Tel. 47( )22855492 Fax 47( )22855441

Haaland, Prof. Arne (1936). Professor. Department of Chemistry, University of Oslo, PO Box 1033 Blindern, N-0315 Oslo, Norway. Dr philos. (University of Oslo, 1970) chemistry. *Gas_electron_diffraction molecular_structure organometallic_molecules*
E-mail arne.haaland@kjemi.uio.no Tel. 47( )22855407 Fax 47( )22855441

Hagen, Prof. Kolbjørn (1943). Professor. Department of Chemistry, University of Trondheim - AVH, N-7055 Trondheim, Norway. Dr philos. (University of Trondheim, 1979) physical chemistry. *Electron_diffraction conformational_analysis*
E-mail kolbjorn.hagen@avh.unit.no Tel. 47( )73596223 Fax 47( )73595473

Hansen, Prof. Lars Kristian (1944). Professor. Institute of Mathematical and Physical Sciences, University of Tromsø, N-9037 Tromsø, Norway. MSc (University of Bergen, 1971) physical chemistry. *Proteins_crystallography X-ray_diffraction computing*
E-mail larsk@chem.uit.no Tel. 47( )77644079 Fax 47( )77644765 Telex 64 124 AUROB N

Hansen, Sissel (1963). PhD student. Institute of Mathematical and Physical Sciences, University of Tromsø, N-9037 Tromsø, Norway. MSc (University of Tromsø, 1993) physical chemistry. *Crystallography proteins phosphatases enzyme_inhibitors*
E-mail sissel@mack.uit.no Tel. 47( )77644074 Fax 47( )77644765 Telex 64 124 AUROB N

Haubeck, Dr Bjørn Christian (1957). Principal research scientist. Institutt for energiteknikk, Dept. of Physics, PO Box 40, N-2007 Kjeller, Norway. Dr ing. (University of Trondheim - NTH, 1988) physics. *X-ray_diffraction neutron_diffraction magnetic_structures*
E-mail bjorn@ife.no Tel. 47( )63806078 Fax 47( )63810920

Hauge, Dr Sverre (1932). Sr lecturer. Department of Chemistry, University of Bergen, Allé gt. 41, N-5007 Bergen, Norway. Dr philos. (University of Bergen, 1978) chemistry. *Inorganic_chemistry*
Tel. 47( )55213566 Fax 47( )55329050

Helmstad, Eldbjørg S. (1961). PhD student. Institute of Mathematical and Physical Sciences, University of Tromsø, N-9037 Tromsø, Norway. MSc (University of Tromsø, 1986) crystallography. *Molecular_dynamics proteins*
E-mail eldbjorg@trypsin.chem.uit.no Tel. 47( )77645706 Fax 47( )77644765

Helland, Ronny (1966). Student. Institute of Mathematical and Physical Sciences, University of Tromsø, N-9037 Tromsø, Norway. MSc (University of Tromsø, 1994) chemistry. *Proteins proteinases cold_adapted_enzymes*
E-mail ronny@mack.uit.no Tel. 47( )77644063 Fax 47( )77644765 Telex 64 124 AUROB N

Høier, Prof. Ragnvald (1938). Professor. Faculty of Physics and Mathematics, University of Trondheim - NTH, N-7034 Trondheim, Norway. Dr philos. (University of Oslo, 1973) physics. *Electron_diffraction electron_microscopy inorganic_materials* E-mail hoier@phys.unit.no Tel. 47()73593588 Fax 47()73593628 Telex 55 637 NTHAD N

Hordvik, Prof. Asbjørn Ivar (1928). Professor. IUCr General Secretary and Treasurer. Institute of Mathematical and Physical Sciences, University of Tromsø, N-9037 Tromsø, Norway. Dr philos. (University of Bergen, 1968) physical chemistry. *Proteins_crystallography molecular_structure sulfur_organic_compounds* E-mail asbjoern@chem.uit.no Tel. 47()77644072 Fax 47()77644765 Telex 64 124 AU-ROB N

Hough, Prof. Edward (1941). Professor. Institute of Mathematical and Physical Sciences, University of Tromsø, N-9037 Tromsø, Norway. PhD (Imperial College University of London, 1975) chemical crystallography. *Proteins_crystallography metalloenzymes phosphatases* E-mail edward@mack.uit.no Tel. 47()77644073 Fax 47()77644765 Telex 64 124 AU-ROB N

Husebye, Prof. Steinar (1933). Professor. Member, Editorial Board of 'Phosphorus, Sulfur and Silicon'. Department of Chemistry, University of Bergen, Allégt. 41, N-5007 Bergen, Norway. PhD / Dr philos. (Tulane University(USA) / University of Bergen, 1963/1970) inorganic chemistry. *X-ray_crystallography structure bonding tellurium_compounds selenium_compounds sandwich_compounds copper_compounds nickel_compounds phosphorus_compounds* E-mail steinar.husebye@kj.uib.no Tel. 47()55213551 Fax 47()55329058

Jynge, Knut (1933). Project engineer. Institute of Mathematical and Physical Sciences, University of Tromsø, N-9037 Tromsø, Norway. MSc (University at Tromsø, 1976) physical chemistry. *Proteins crystallography computing* E-mail knut@mack.uit.no Tel. 47()77644071 Fax 47()77644765 Telex 64 124 AUROB N

Karlsen, Solveig (1966). PhD student. Institute of Mathematical and Physical Sciences, University of Tromsø, N-9037 Tromsø, Norway. MSc (University of Tromsø, 1991) physical chemistry. *Proteins crystallography* E-mail solveig@chem.uit.no Tel. 47()77644065 Fax 47()77644765 Telex 64 124 AU-ROB N

Kjekshus, Prof. Arne (1932). Professor. Department of Chemistry, University of Oslo, PO Box 1033 Blindern, N-0315 Oslo, Norway. Dr philos. (University of Oslo, 1971) chemistry. *Metals oxides ceramics chalcogens transition_elements solid_solution magnetism* Tel. 47()22855560 Fax 47()22855565 Telex 72 705 ASTRO N

Klewe, Bernt (1933). Sr Lecturer. Department of Chemistry, University of Oslo, PO Box 1033 Blindern, N-0315 Oslo, Norway. MSc (University of Oslo, 1961) chemistry. *Structure small_organic_acids small_organic_bases small_organic_salts* E-mail bernt.klewe@kjemi.uio.no Tel. 47()22855463 Fax 47()22855441

Larsen, Rolf Leon (1964). PhD student. Institute of Mathematical and Physical Sciences, University of Tromsø, N-9037 Tromsø, Norway. MSc (University of Tromsø, 1988) physical chemistry. *Proteins crystallography hemoglobins* E-mail kjrl@mack.uit.no Tel. 47()77644064 Fax 47()77644765 Telex 64 124 AUROB N

Maartmann-Moe, Knut (1928). Sr Lecturer. Sub-Editor for Norway. Department of Chemistry, University of Bergen, Allé gt 41, N-5007 Bergen, Norway. MSc (University of Bergen, 1961) chemistry. *Crystal_structure_determination computing* E-mail knut.maartmann-moe@kj.uib.no Tel. 47()55213446 Fax 47()55329058

Mårdalen, Dr Jostein (1962). Post Doc. ESRF, PO Box 220, F-38043 Grenoble Cedex, France. Dr ing. (University of Trondheim - NTH, 1991) physics. *Synchrotron_radiation diffuse_scattering conducting_polymers spectroscopy* E-mail mardalen@vega.ill.fr Tel. (33)76882000 Fax (33)76882020

Marøy, Dr Kjartan (1930). Sr Lecturer. Department of Chemistry, University of Bergen, Allé gt 41, N-5007 Bergen, Norway. Dr philos. (University of Bergen, 1976) chemistry. *Inorganic_chemistry* E-mail maroy@kj.uib.no Tel. 47()55213565 Fax 47()55329058

Marthinsen, Dr Knut (1956). Research Scientist. SINTEF Applied Physics, Sem Saelands vei 7, N-7034 Trondheim, Norway. Dr ing. (University of Trondheim - NTH, 1986) physics. *Computer_modelling diffraction_theory diffraction EELS image_processing* E-mail marthinsen@hufsa.imf.unit.no Tel. 47()73593473 Fax 47()73593420

Mathiesen, Ragnvald Hermann (1965). Dr student. Faculty of Physics and Mathematics, University of Trondheim - NTH, 1991) physics. *Direct_method phase_measurement synchrotron_radiation crystallographic_computing* Tel. 47()73593584 Fax 47()73593628

Mo, Prof. Frode (1937). Professor. Faculty of Physics and Mathematics, University of Trondheim - NTH, N-7034 Trondheim, Norway. Dr techn. (University of Trondheim - NTH, 1980) crystallography. *Multiple_scattering physical_phase_determination high-precision_structure organic_chalcogenides macromolecular_crystallography applied_synchrotron_radiation* E-mail fmo@phys.unit.no Tel. 47()73593585 Fax 47()73593628

Mostad, Arvid (1929). Professor; Head of X-ray group. Department of Chemistry, University of Oslo, PO Box 1033 Blindern, N-0315 Oslo, Norway. [Ovenbakken 14 B, 1345 Osteraas, Norway.] MSc (University of Oslo, 1959) chemistry. *Structural_chemistry biological_activity protein_structure macromolecular_activity* E-mail arvid.mostad@kjemi.uio.no Tel. 47()22855415 Fax 47()22855441

Nicholson, Prof. David Graham (1944). Professor. Department of Chemistry, University of Trondheim - AVH, N-7055 Trondheim, Norway. PhD (University of London, 1969) inorganic chemistry. *Structural_inorganic_chemistry EXAFS molecular_modelling synchrotron_radiation* E-mail d.nicholson@avh.unit.no Tel. 47()73596204 Fax 47()73596255

Olsen, Dr Arne (1944). Group leader. Department of Physics, University of Oslo, PO Box 1048, N-0316 Blindern, Norway. Dr philos. (University of Oslo, 1978) physics. *Electron_microscopy diffraction microanalysis materials_science* Tel. 47()22958740 Fax 47()22958749

Raade, Gunnar (1944). Curator of Minerals. Mineralogisk-Geologisk Museum, University of Oslo, Sars gt. 1, N-0562 Oslo, Norway. MSc (University of Oslo, 1973) geology. *Mineralogy* Tel. 47()22851647 Fax 47()22851800

Riise, Bjørn (1962). Project engineer. Institute of Mathematical and Physical Sciences, University of Tromsø, N-9037 Tromsø, Norway. MSc (University of Tromsø, 1988) physical chemistry. *Proteins crystallography crystallization* E-mail bjornr@mack.uit.no Tel. 47()77644071 Fax 47()77644765 Telex 64 124 AU-ROB N

Rømming, Prof. Christian (1928). Professor. Department of Chemistry, University of Oslo, PO Box 1033 Blindern, N-0315 Oslo, Norway. Dr philos. (University of Oslo, 1968) chemistry. *Structural_chemistry macromolecular_activity* E-mail christian.romming@kjemi.uio.no Tel. 47()22855403 Fax 47()22855441

Rosenqvist, Prof. Ivan Thoralf Koss (1916). Professor emeritus. Department of Geology, University of Oslo, P O Box 1047 Blindern, N-0316 Oslo, Norway. Dr philos. (University of Oslo, 1945) geochemistry. *Clays minerals ionic_exchange biogeochemistry (weathering)* Tel. 47()22856656 Fax 47()22854215 Telex 79367 ESCON N

Saethre, Prof. Leif Jarle (1945). Professor. Department of Chemistry, University of Bergen, Allé gt 41, N-5007 Bergen, Norway. MSc (University of Bergen, 1971) physical chemistry. *Molecular_structure photoelectron_spectroscopy* E-mail Leif.Sathre@kj.uib.no Tel. 47()55213561 Fax 47()55329058

Sletten, Prof. Einar (1939). Professor. Department of Chemistry, University of Bergen, Allé gt 41, N-5007 Bergen, Norway. Dr philos. (University of Bergen, 1979) physical chemistry. *Coordination bioinorganic_material nucleic_acids nuclear_magnetic_resonance_spectroscopy* E-mail einar.sletten@kj.uib.no Tel. 47()55213352 Fax 47()55329058

Sletten, Prof. Jorunn (1941). Professor. Department of Chemistry, University of Bergen, Allé gt 41, N-5007 Bergen, Norway. Dr philos. (University of Bergen, 1976) chemistry. *Transition_elements coordination_geometry* E-mail jorunn.sletten@kj.uib.no Tel. 47()55213562 Fax 47()55329058

Smaås, Dr Arne O. (1960). Post Doc. Institute of Mathematical and Physical Sciences, University of Tromsø, N-9037 Tromsø, Norway. PhD (University of Tromsø, 1990) chemistry. *Proteins_crystallography* E-mail arnes@mach.uit.no Tel. 47()77644070 Fax 47()77644765 Telex 64 124 AU-ROB N

Stølevik, Prof. Reidar (1938). Professor. Department of Chemistry, University of Trondheim - AVH, N-7055 Trondheim, Norway. Dr philos. (University of Oslo, 1974) physical chemistry. *Conformational_structure mathematics physics* Tel. 47()73596235

Strand, Dr Tor Gogstad (1934). Sr Lecturer. Department of Chemistry, University of Oslo, PO Box 1033 Blindern, N-0315 Oslo, Norway. Dr philos. (University of Oslo, 1968) chemistry. *Electron_diffraction materials_volatile* E-mail tostrand@ulrik.uio.no Tel. 47()22855411 Fax 47()22855441

Svinning, Dr Torgeir (1948). Research scientist. SINTEF Materials Technology, N-7034 Trondheim, Norway. Dr ing. (Norwegian Institute of Technology, 1978) physics. *Materials production_technology* Tel. 47()73572037 Fax 47()73597043

Tafto, Prof. Johan (1943). Professor. Department of Physics, University of Oslo, PO Box 1048 Blindern, N-0316 Oslo, Norway. MSc (University of Oslo, 1972) physics. *Electron_microscopy superconductors intermetallics* Tel. 47()22856113 Fax 47()22856422

Thorkildsen, Dr Gunnar (1953). Sr Lecturer. Rogaland University Center, PO Box 2557 Ullaland, N-4004 Stavanger, Norway. Dr ing. (Norwegian Institute of Technology, 1983) physics. *Diffraction_theory extinction multiple_scattering MR_tomography* E-mail g-th@hsr.no Tel. 47()51874257 Fax 47()51874236

Traetteberg, Prof. Marit (1930). Professor. Department of Chemistry, University of Trondheim - AVH, N-7055 Trondheim, Norway. Dr philos. (Univerity of Trondheim, 1970) chemistry. *Electron_diffraction NMR theoretical_chemistry hydrocarbons conjugate_compounds* E-mail marit.tretteberg@avh.unit.no Tel. 47()73576225 Fax 47()73595473

PAKISTAN

Sub-Editor: M. M. Qurashi

Ahmad, Mr Ashfaq (1963). Lecturer. Centre for Solid State Physics, Punjab University, New Campus, Lahore 20, Pakistan. MSc (Punjab U., 1989) physics. *Small_crystal quasicrystal activation_energy phase nanocrystals DTXD*
Tel. 92(42)5864185

Ahmad, Mr Dabir (1938). Principal scientific officer. G-4, Gulberg Square, Block 7, Gulshan-el-Iqbal, Karachi, Pakistan. MSc (Karachi U., 1961) physics. *X-ray_crystallography instrumentation*
Tel. 92(21)468684

Ahmad, Dr Zulfiquar (1945). Director. Centre of Excellence in Mineralogy, University of Balochistan, Sariab Road, Quetta, Pakistan. PhD (U. London, 1982) geology. *Minerals X-ray_diffraction structural_chemistry electron_microscopy*

Akhtar, Mr Mohammad W. (1944). Deputy director. Geological Survey of Pakistan, 22 Ali Block, New Garden Town, Lahore 16, Pakistan. MSc (Sind U., Jamshoro, 1965) geology. *Minerals*
Tel. 92(42)855922

Akhtar, Dr Muhammad (1946). Assistant prof. 5/44-2 Model Colony, Karachi, 75100, Pakistan. PhD (U. Wales, UK, 1992) physics. *Superconductor X-ray_crystallography*
Tel. 92(21)400079

Akhter, Mr Javed (1955). Assistant Geophysicist. Geological Survey of Pakistan, 22 Ali Block, New Garden Town, Lahore 16, Pakistan. MSc (Punjab U., 1978) geophysics. *Geophysics X-ray_diffraction*

Akhter, Dr Parvez (1949). Chief research officer. National Institute of Silicon Technology, No. 25, H9 Islamabad, Pakistan. PhD (U. Sussex, UK, 1980) surface physics. *Silicon crystal_growth characterization surface solar_cell_fabrication Si-metal_interfaces*
Tel. 92(51)253504 Fax 92(51)855942 Telex 5538-NIEPK

Ali, Dr Syed Wajahat (1937). Director. Solar Energy Research Centre, Pakistan Council of Scientific & Industrial Research, Latifabad unit 2, Autobahn, Hyderabad, Pakistan. PhD (Karachi U., 1981) physical chemistry. *Structural_chemistry*
Tel. 92(221)83040

Anis, Dr Muhammad Khalid (1947). Assistant prof. Dept of Physics, University of Karachi, Karachi 75350, Pakistan. PhD (U. Brighton, UK, 1979) solid state physics. *Dielectrics HTS structure*

Anwar, Mr Muhammad (1953). Assistant director. Geological Survey of Pakistan, 16-G Model Town, Lahore, Pakistan. MSc (Punjab U., 1976) geology. *Minerals geology*
Tel. 92(42)855922

Arif, Ms Iffat (1958). Student. 233-A, G.O.R. V, Faisal Town Lahore, Pakistan. MSc (Punjab U., 1980) physics. *Biomedical_compound*

Baber, Dr Nasim (1952). Associate Professor. Department of Physics, Quaid-e-Azam University, Islamabad, Pakistan. PhD (Warsaw U. Technology, Poland, 1981) solid state physics. *Semiconductors high temperature superconductors liquid_crystals*
Tel. 92(51)829537

Baqri, Dr Syed Rafiq-ul-Hasan (1945). Director. Earth Science Division, Pakistan Museum of National History, Al Markaz F-7/2, Islamabad, Pakistan. PhD (Southampton U., UK, 1977) X-ray diffraction. *X-ray_diffraction*
Tel. 92(51)82439

Bhatti, Ms Farzana Akhtar (1965). Research fellow. H.E.J. Research Inst. of Chemistry, University of Karachi, Karachi, Pakistan. MSc (Karachi U., 1994) chemistry. *Structural_chemistry natural_products*
Tel. 92(21)4963373

Bhatti, Mr Muhammad Akram (1947). Deputy director. Geological Survey of Pakistan, Ministry of Petroleum & Natural Resources, 83-D Model Town, Lahore, Pakistan. MSc (Punjab U., 1969) geology. *Minerals geology*
Tel. 92(42)855232

Butt, Dr Khurshid Alam (1947). Group leader. Atomic Energy Minerals Centre, Ferozepur Road, Lahore, Pakistan. PhD (New Brunswick U., Canada, 1976) petrology. *Minerals X-ray_diffraction*
Tel. 92(42)870237

Butt, Mr Muhammad Hafeez (1945). Geophysicist. Geological Survey of Pakistan, Quetta, Pakistan. MSc (Punjab U., 1968) geology. *Microscopy geophysics*
Tel. 92(81)855232

Butt, Mr Nasir Parvaiz (1962). Student. Semiconductor Labs, Quaid-e-Azam University, Islamabad, Pakistan. MSc (Punjab U., 1985) solid state physics. *Superconductivity semiconductors*
Tel. 92(51)829537

Butt, Dr Noor Mohammad (1936). Director. Vice-president Pakistan Crystallographic Society. PINSTECH, Nilore, Islamabad, Pakistan. DSc, PhD (Birmingham U., UK, 1993, 1965) solid state physics. *Materials neutron_diffraction Mossbauer_spectroscopy photography house_plants*
E-mail ctc@shell.portal.com Tel. 92(51)452350 Fax 92(51)429533 Telex 5725 AT-COM PK

Chaudhary, Dr Abdul Majid (1945). Manager. Nuclear Material Division, PINSTECH Nilore, Islamabad, Pakistan. PhD (New England U., Australia, 1978) amorphous solids. *X-ray_diffraction diffuse_scattering*
Tel. 92(51)840103x240

Chaudhary, Mr G. Sarwar Alam (1944). Deputy director. Geological Survey of Pakistan, Punjab Division, 68-D Model Town Lahore 11, Pakistan. MSc (Punjab U., 1967) mineralogy. *Minerals*
Tel. 92(42)881192

Chaudhry, Mr Mohammad Anwar (1944). Assistant director. Geological Survey of Pakistan, 68-D Model Town, Lahore 11, Pakistan. MSc (Peshawar U., 1972) geology. *Minerals*
Tel. 92(42)881192

Chaudhry, Dr Muhammad Nawaz (1942). Prof. Institute of Geology, Punjab University, Quaid-e-Azam Campus, Lahore, Pakistan. PhD (U. London UK, 1967) mineralogy. *Mineralogy crystallography applied_geology*
Tel. 92(42)5866809

Chauhan, Mr Ehsanul Haq (1931). Manager. Chemistry Division, Geological Survey of Pakistan, Sariab Road, Quetta, Pakistan. MSc (Idaho U., USA, 1968) geology. *Structural_chemistry microscopy minerals*
Tel. 92(81)72617

Choudhry, Dr Muhammad Iqbal (1959). Associate Professor. Editor, Al-Chemy; Member, Executive Council of Chemical Society of Pakistan. H.E.J. Research Institute of Chemistry, University of Karachi, Karachi 75270, Pakistan. PhD (Karachi U., 1987) organic chemistry. *Natural_products structure_determination marine_invertebrates chemical_defence_system_in_lower_animals*
E-mail iqbal%hejfinst@uunet.uu.net Tel. 92(21)472780 Fax 92(21)4963373

Elahi, Mr Manzoor (1937). Senior research officer. Metallurgy group, DESTO, Chaklala, Rawalpindi, Pakistan. MSc (Brunel U., UK, 1980) metallurgy. *X-ray_diffraction*
Tel. 92(51)590859x52

Farooque, Dr Muhammad (1953). Senior research officer. Metallurgy Division, Khan Research Laboratories, PO Box 502, Rawalpindi, Pakistan. PhD (Oxford U., UK, 1992) material science. *Carbides phase_transformation microstructure*
Tel. 92(51)843397

Fatmi, Prof. Ali Nasir (1930). Deputy Director General. Geological Survey of Pakistan, Ministry of Petroleum & Natural Resources, Sariab Road, Quetta, Pakistan. PhD (Wales U., UK, 1968) geology. *Minerals*
Tel. 92(81)73564

Gilani, Mr Jamshed Ali (1956). Assistant director. Geological Survey of Pakistan, Sher Shah Block, New Garden Town, Lahore, Pakistan. MSc (Karachi U., 1977) geochemistry. *Minerals*
Tel. 92(42)78519

Habib, Mr Syed Abbas (1936). Deputy Director. Geological Survey of Pakistan, 68-D Model Town, Lahore 11, Pakistan. MSc (Victoria U., Australia, 1971) sedimentology. *Minerals*
Tel. 92(42)855923

Habiby, Dr Fakhruddin (1953). Principal research officer. Metallurgy Division, Khan Research Laboratories, PO Box 502, Rawalpindi, Pakistan. PhD (Imperial College London UK, 1991) materials engineering. *Texture single_crystal composite_compound microstructure*
Tel. 92(51)843397

Haq, Dr Anwarul (1947). Deputy director. Councillor Asian Cryst. Ass., Member Ex. Comm. Pak. Cryst. Ass.. Metallurgy Division, Khan Research Laboratories, PO Box 502, Rawalpindi, Pakistan. DSc (Karlsruhe U., Germany, 1982) metallurgy. *Texture phase thin_film*
Tel. 92(51)843397 Fax 92(51)841987

Hasan, Dr Faizul (1948). Prof. Metallurgical Engineering Dept., University of Engineering & Technology, GT Road, Lahore, Pakistan. PhD (Manchester U., UK, 1984) metallurgy. *X-ray_diffraction electron_diffraction structural_chemistry*
Tel. 92(42)339207

Hussain, Mr Hilal (1961). Scientific officer. Metallurgy Division, Khan Research Laboratories, PO Box 502, Rawalpindi, Pakistan. MSc (Punjab U., 1988) applied mathematics. *Texture_analysis*
Tel. 92(51)843397 Fax 92(51)841987

Hussain, Dr Khadim (1947). Associate professor. Centre for Solid State Physics, Punjab University, New Campus, Lahore 54590, Pakistan. PhD (Victoria U. Manchester, UK, 1981) physics. *Minerals ceramics superconductors air_pollution X-ray_diffraction*
Tel. 92(42)5864185

Ikram, Dr Nazma (1949). Professor. Centre for Solid State Physics, Punjab University, Lahore, Pakistan. PhD (Cambridge U., UK, 1976) solid state physics. *Lattice_dynamics solar_cell electron_scattering*
Tel. 92(42)586185

Iqbal, Mr Mir Waseluddin Ahmad (1926). Director. Geological Survey of Pakistan, 68-D Model Town, Lahore 11, Pakistan. MSc (U. California, USA, 1964) geology. *Minerals*
Tel. 92(42)852547

Jafry, Mr Syed Quamar Abbas (1953). Assistant director. Geological Survey of Pakistan, 16-G Model Town, Lahore, Pakistan. MSc (Punjab U., 1977) geology. *Minerals geology computing*
Tel. 92(42)852826

Jamil, Dr Ahmed (1937). Prof. Department of Chemistry, Quaid-e-Azam University, Islamabad, Pakistan. PhD (Freie U. Berlin, Germany, 1967) physical chemistry. *Crystal_structure organic_compound coordination_compound*
Tel. 92(51)829167

Kaifi, Mr F. M. Zaffar (1946). Senior research officer. Pakistan Council of Scientific & Industrial Research, Peshawar, Pakistan. MSc (New South Wales U., Australia, 1987) inorganic chemistry. *Metallic_complex minerals rock X-ray_diffraction*
Tel. 92(521)41192x5

Khalid, Mr Mohammad (1940). Senior research officer. National Physical & Standards Centre, Pakistan Council of Scientific & Industrial Research, off University Road, Karachi 75280, Pakistan. MSc (Karachi U., 1963) physics. *Structural_chemistry X-ray_crystallography X-ray_spectroscopy thermal_conductivity*
Tel. 92(21)4967603x76

Khan, Dr Abdul Quadeer (1936). Project director; chief executive. Material Sciences, Metallurgy Division, Khan Research Laboratories, PO Box 502, Rawalpindi, Pakistan. PhD, DSc (Leuven U., Belgium, 1972) physical metallurgy. *Texture_property_relationship phase_transition structural_chemistry material_sciences nuclear_technology*
E-mail 100074.172@Compuserve.Com Tel. 92(51)592806 Fax 92(51)452487 Telex 5584 PUFCO PK

Khan, Dr Abdur Rahman (1960). Senior scientific officer. Pakistan Council for Scientific & Industrial Research, Jamrud Road, Peshawar, Pakistan. PhD (U. Birmingham, UK, 1991) civil engineering (public health engineering). *Environment pollution waste_water_treatment pollution_control*
Tel. 92(521)41192x95 Fax 92(521)41476 Telex PK 52473

Khan, Dr Lajber (1951). Senior research officer. Pakistan Council for Scientific & Industrial Research, Peshawar, Pakistan. PhD (U. Wales, UK, 1988) pharmaceutical sciences. *Medicinal_plants dermatoxicology*
Tel. 92(521)41192x95

Khan, Mr Mohammad Afaq (1951). Assistant director. Geological Survey of Pakistan, Ministry of Petroleum & Natural Resources, 16-G Model Town, Lahore, Pakistan. MSc (Punjab U., 1976) geology. *Minerals geophysics*

Khan, Mr Nawazish Ali (1964). Student. Department of Physics, Semiconductors Lab., Quaid-e-Azam University, Islamabad, Pakistan. MSc (Quaid-e-Azam U., 1989) physics. *Superconductors infrared_property structure_analysis X-ray_diffraction*
Tel. 92(51)829537

Khan, Dr Shah Alam (1956). Senior research officer. Pakistan Council for Scientific & Industrial Research, Jamrud Road, Peshawar, Pakistan. PhD (U. Strathclyde, UK, 1989) chemistry. *Boron_compounds environment industrial_chemicals neutron_capture_therapy*
Tel. 92(521)41192x95

Khawaja, Dr Farid Akhtar (1949). Professor. General Secretary Pakistan Physical Society. Physics Department, Quaid-e-Azam University, Islamabad, Pakistan. PhD (Moscow State U., Russia, 1976) solid state physics. *Disorder X-ray_diffraction neutron_diffraction materials_sciences*
Tel. 92(51)219472

Khawaja, Mr Mahmood-ul-Hassan (1944). Deputy director. Geological Survey of Pakistan, Azad Kashmir, Pakistan. MSc (Punjab U., 1966) geology. *Minerals*

Mahmood, Dr Khursheed (1952). Associate professor. B1, Al-Kiran Appt, CS41, Block 7, Federal B-Area, Karachi, Pakistan. PhD (Cranfield Inst. of Tech. UK, 1989) metallurgy. *Metallic_coating corrosion metals alloys design_of_machine_components*
Tel. 92(21)6328650

Malik, Mr Zubair A. (1964). Researcher. Metallurgy Division, Khan Research Laboratories, PO Box 502, Rawalpindi, Pakistan. MSc (Baha-uddin Zakariya I. Multan, 1990) physics. *Minerals texture monocrystal_orientation*
Tel. 92(51)843397 Fax 92(51)841987

Maqsood, Dr Asghari (1947). Professor. Department of Physics, Quaid-e-Azam University, Islamabad, Pakistan. PhD (Goteborg U., Sweden, 1982) material science. *X-ray_diffraction metals*
Tel. 92(51)829472

Mian, Dr Mohammad Ashraf (1938). Professor. Institute of Applied Geology, Azad Jammu and Kashmir University, Muzaffarabad, Pakistan. PhD (Punjab U., 1976) geochemistry. *Minerals X-ray_diffraction*
Tel. 92(52)3119

Mian, Mr Muhammad Asghar (1951). Geophysicist. Geological Survey of Pakistan, 22-Ali Block, 68D, Model Town, Lahore 11, Pakistan. MSc (Punjab U., 1976) geology. *Geophysics*
Tel. 92(42)855816

Mian, Mr Mushtaq Ahmed (1939). Senior research officer. G & C Research Centre, Pakistan Council for Scientific & Industrial Research, Lahore 54600, Pakistan. MSc (Punjab U., 1964) chemistry. *Ceramics*
Tel. 92(42)878358

Mir, Mr Jan Muhammad (1937). Assistant professor. C-47, Staff Town, University of Karachi, Karachi, Pakistan. MSc (Manitoba U., Canada, 1972) mineralogy. *Sedimentation crystallography mineralogy minerals*
Tel. 92(21)473190

Naqvi, Dr Syed Ali Anwar (1947). Senior registrar. Department of Urology, Dow Medical College, Karachi, Pakistan. MSc (Karachi U., 1984) urology. *X-ray_diffraction*
Tel. 92(21)219551x247

Nasreen, Miss Shagufta (1955). Research officer. Mineral Research Division, Pakistan Council of Scientific & Industrial Research, Jamrud Road, Peshawar, Pakistan. MSc (Peshawar U., 1979) physical chemistry. *X-ray_diffraction*
Tel. 92(521)8817

Nizami, Mr Muhammad Sharif (1949). Senior research officer. Glass & Ceramics Research Centre, Pakistan Council of Scientific & Industrial Research, Lahore, Pakistan. MSc (Punjab U., 1973) chemistry. *Glasses ceramics structure_determination*
Tel. 92(42)870324x27

Pervaiz, Mr Rashed (1952). Geophysicist. Geological Survey of Pakistan, 68-D, Model Town, Lahore 11, Pakistan. MSc (Quaid-e-Azam U., Islamabad, 1975) geophysics. *Geophysics*
Tel. 92(42)855816

Qaiser, Mr Mohammad Ali (1940). Senior research officer. Mineral Research Division, Pakistan Council of Scientific & Industrial Research, Jamrud Road, Peshawar, Pakistan. MSc (Bihar U., India, 1962) physics. *Minerals structural_chemistry*
Tel. 92(521)41191x19

Qidwai, Dr Ansar Ahmad (1948). Associate professor. Department of Physics, University of Karachi, Karachi, Pakistan. PhD (Durham U., UK, 1982) physics. *Superconductors photovoltaic_compound*
Tel. 92(21)4963427

Qurashi, Dr Mazhar Mahmood (1925). President Pakistan Crystallographic Society; Sub-Editor, WDC9. Pakistan Academy of Sciences, Constitution Avenue G-5, Islamabad, Pakistan. PhD (U. Manchester, UK, 1962) physics. *Structure_determination liquids scientometrics*
Tel. 92(51)824843

Qureshi, Mr Khalid Mahmood (1953). Senior research officer. Mineralogy Division, PO Box 658, Lahore, Pakistan. MSc (Punjab U., 1983) physics. *Rock minerals natural_products X-ray_diffraction*
Tel. 92(42)870276x78

Qureshi, Dr M. Hanif (1937). Chief scientific officer. Pakistan Council for Scientific & Industrial Research, 50-A, Model Town, Lahore, Pakistan. PhD (Sheffield U., UK, 1967) materials science. *Research development*
Tel. 92(42)850055

Qureshi, Mr Mohammad Kaleem Akhter (1945). Deputy director. Geological Survey of Pakistan, Ministry of Petroleum & Natural Resources, 83-D Model Town, Lahore 54700, Pakistan. MSc (Punjab U., 1968) geology. *Geology minerals*
Tel. 92(42)852826

Rana, Mr Riaz Ahmad (1950). Assistant director. Geological Survey of Pakistan, 68-D, Model Town Lahore 11, Pakistan. MSc (Punjab U., 1977) geology. *Minerals*
Tel. 92(42)881192

Rizvi, Prof. Adibul Hassan (1938). Professor. Department of Urology, Dow Medical College, Karachi, Pakistan. FRCS (Royal College of Surgeons, UK, 1967) surgery. *X-ray_diffraction*
Tel. 92(21)219551x247

Rizvi, Dr Syed Sadrul Hasan (1933). General director. General Secretary Pakistan Crystallographic Society. National Physics & Standards Laboratory, Pakistan Council for Scientific and Industrial Research, off University Road, Karachi, Pakistan. PhD (Manchester U., UK, 1962) physics. *Structure_determination X-ray_spectroscopy instrumentation*
Tel. 92(21)4967609

Russel, Mr Nazirullah (1948). Geophysicist. Geological Survey of Pakistan, 68-D Model Town, Lahore 11, Pakistan. MSc (Punjab U., 1970) geophysics. *Geophysics*
Tel. 92(42)855816

Saeed, Mr Syed Mohammad (1943). Senior research officer. C-8 Gulberg Square, FL-1/7, Gulshan-e-Iqbal, Karachi, Pakistan. MSc (Karachi U., 1971) physics. *Instrumentation X-ray_diffraction computing*
Tel. 92(21)466577

Saghir, Mr Ahmad (1947). Principal research officer. Applied Physics, Computer & Instruments Centre, Pakistan Council for Scientific & Industrial Research, off University Road, Karachi 39, Pakistan. MSc (Karachi U., 1968) physics. *Structural_chemistry powder_diffraction computing*
Tel. 92(21)4967603x76

Salah-ud-Din, Dr (1942). Associate professor. Center for Solid State Physics, Punjab University, New Campus, Lahore 54590, Pakistan. PhD (Royal Holloway College, Egham UK, 1977) solid state physics. *X-ray_diffraction zirconium_compounds high_strength_materials transmission_electron_microscopy*

Shah, Dr Wajid Ali (1953). Senior research officer. G & C Research Centre, Pakistan Council for Scientific & Industrial Research Complex, Lahore, Pakistan. PhD (U. Sheffield, UK, 1989) materials science. *Characterization glasses ceramics*
Tel. 92(42)870324

Shahi, Mr Gulam Nabi (1953). Assistant professor; Chairman, Geology Dept. Geology Department, University of Balochistan, Sariab Road, Quetta, Pakistan. MPhil (East Anglia U., Norwich, England, 1991) geology. *Minerals sedimentology clay_mineralogy*
Tel. 92(81)43484 Fax 92(81)440323

Shaikh, Mr Mohammad Iqbal (1951). Assistant director. Geological Survey of Pakistan, 16-G Model Town, Lahore 16, Pakistan. MSc (Punjab U., 1971) geology. *Minerals*
Tel. 92(42)855922

Shaikh, Mr Mohammad Sualehin (1939). Principal research officer. Pakistan Council for Scientific & Industrial Research, D-4, University Road, Karachi 39, Pakistan. MSc (Manchester U., UK, 1966) mathematics. *Thin_film dielectrics X-ray_diffraction*
Tel. 92(21)4967603

Shaikh, Mr Qameruddin (1947). Technician. APC&I Centre, Pakistan Council for Scientific & Industrial Research Complex, Karachi, Pakistan. MSc (Karachi U., 1974) physics. *Software X-ray_diffraction X-ray_diffractometry*
Tel. 92(21)4967603x7

Shaukat, Dr Ali (1947). Associate professor. Department of Physics, University of Punjab, Lahore 54590, Pakistan. PhD (Erlangen U., Germany, 1981) crystallography. *Structural_property electronic_property semiconductors*
Tel. 92(42)5863926

Shelkh, Mr Izhar-ul-Haq (1955). Student. Physics Department, Quaid-e-Azam University, Islamabad, Pakistan. MPhil (Quaid-e-Azam U., 1986) physics. *Characterization high_temperature_superconductors material_sciences*
Tel. 92(51)253331

Shuja, Mr Tauqir Ahmad (1943). Deputy director. Geological Survey of Pakistan, Natural Resources Division, Sector H-8, Islamabad, Pakistan. MSc (Punjab U., 1967) geology. *Geology*
Tel. 92(51)825779

Siddiqi, Dr Saadat Anwar (1950). Associate professor. Centre for Solid State Physics, Punjab University, Lahore 54590, Pakistan. PhD (U. Newcastle, UK, 1984) engineering materials. *High_temperature_superconductors engineering_materials crystallography*
Tel. 92(42)5864185 Fax 92(42)5864534

Siddiqui, Mr Jawed Ahmad (1950). Lecturer. Centre of Excellence in Mineralogy, University of Balochistan, Sariab Road, Quetta, Pakistan. MSc (Karachi U., 1977) geology. *Geology X-ray_diffraction*

Siddiqui, Mr Tariq Nasser (1956). Senior engineer. Metallurgy Division, Khan Research Laboratories, PO Box 502, Rawalpindi, Pakistan. B. E. (NED Engineering U., 1980) metallurgy. *X-ray_diffraction*
Tel. 92(51)843397 Fax 92(51)841987

Sohail, Mr Mohammad (1945). Senior research officer. Pakistan Council for Scientific & Industrial Research Complex, Lahore, Pakistan. MSc (Punjab U., 1971) chemistry.

Characterization
Tel. 92(42)870326x28

Syed, Dr Kaab Akhter (1950). Assistant professor. Department of Physics, University of Karachi, Karachi, Pakistan. PhD (National U. Singapore, 1989) solid state physics. *Characterization powder fabrication_of_novel_materials rheology_of_suspension*
Tel. 92(21)471558

Syed, Mr Muhammad Tariq (1951). Manager. Product Development Cell (R&D), Bin Qasim Karachi, Pakistan. MSc (Karachi U., 1975) physics. *Refractory_compounds texture phase_analysis cold_rolled_sheets*
Tel. 92(21)7574951

Tauqir, Dr Anjum (1953). Principal engineer. Metallurgy Division, Khan Research Laboratories, PO Box 502, Rawalpindi, Pakistan. PhD (Connecticut U., USA, 1986) metallurgy. *Material_sciences metals*
Tel. 92(51)843397 Fax 92(51)841987

Tiwana, Mrs Kausar Shaheen (1966). Student. Department of Physics, Semiconductor Physics Lab., Quaid-e-Azam University, Islamabad, Pakistan. MSc (U. Punjab, 1991) physics. *High_temperature_superconductors infrared_property structure_analysis X-ray_diffraction*
Tel. 92(51)829537

Yousufzai, Mr Inayatullah Khan (1938). Director. Pakistan Institute of Cotton Research & Technology, Moulvi Tameezuddin Khan Road, Karachi, Pakistan. MSc (Strathclyde U., UK, 1966) physics. *Microscopy X-ray_diffraction*
Tel. 92(21)5682228

Yusaf, Dr Mohammad (1940). Retired. 243/A I Township, Lahore, Pakistan. PhD (Charles U., Praha, Czechoslovakia, 1973) analytical chemistry. *Materials*
Tel. 92(42)841445

POLAND

Sub-Editor: A. Pietraszko

Notes

1. The Polish degree of 'magister' (Mgr) is lower than PhD, 'doktor' (Dr) is higher than PhD and 'doktor habilitowany' (Dr hab.) is next higher than Dr. The position 'adiunkt' is equivalent to Reader.

Adamiak, Dr Dorota Anna (1948). Adiunkt. Institute of Bioorganic Chemistry, Polish Academy of Sciences, ul. Noskowskiego 12/14, 61-704 Poznań, Poland. Dr (A. Mickiewicz U. Poznań, 1975) chemistry. *Crystallization X-ray_structure_determination nucleic_acids proteins*
E-mail dorotaa@ibch.pozman.edu.pl Tel. 48(61)528503x153 Fax 48(61)520532

Anulewicz, Dr Romana Joanna (1937). Adiunkt. Faculty of Chemistry, University of Warsaw, ul. Pasteura 1, 02-093 Warszawa, Poland. Dr (U. Warszawa, 1979) chemistry. *Crystal_structure structure_determination structural_change*
E-mail ranul@chem.uw.edu.pl Tel. 48(22)222892 Fax 48(22)222892 Telex 815439 uw pl

Auleytner, Prof. Julian Jan (1922). Professor. Vice-president of Polish National Committee of Crystallography. Institute of Physics, Polish Academy of Sciences, al. Lotników 32/46, 02-668 Warszawa, Poland. Full Prof. (Council of State (proposal of Inst. Physics, Polish Acad. Sciences), 1974) solid-state_physics. *Real_structure solid_state_physics implantation X-ray_spectroscopy spectrometry electron_microscopy standing_wave_method*
E-mail auley@planif61.bitnet Tel. 48(22)436034 Fax 48(22)430926

Bąk-Misiuk, Dr Jadwiga (1942). Adiunkt. Institute of Physics, Polish Academy of Sciences, al. Lotników 32/46, 02-668 Warszawa, Poland. Dr (Inst. of Physics Polish Academy of Sciences Warszawa, 1974) physics. *Bond_method high_resolution_diffractometry thermal_expansion semiconductors thin_layer point_defect precipitation II–VI_compounds III–V_compounds*
E-mail bakmi@planif61.bitnet Tel. 48(22)4366034 Fax 48(22)430926

Barszcz, Prof. Edward (1936). Institute of Ferrous Metallurgy, 44-100 Gliwice, Poland.

Bartczak, Prof. Tadeusz Jan (1935). Associate professor. Department of X-ray and Chemical Crystallography, Institute of General and Ecological Chemistry, Technical University of Łódź, ul. Żwirki 36, 90-924 Łódź, Poland. [ul. Harcerska 5/27, 91-710 Łódź, Poland.] Dr hab. (Tech. U. Łódź, 1986) chemistry. *Metalloporphyrins cytochromes*
E-mail tjbartcz@lodz1.p.lodz.pl Tel. 48(42)313137 Fax 48(42)313103 Telex 886136 polit pl

Bojarski, Prof. Zbigniew (1921). Professor emeritus. Institute of Physics and Chemistry of Metals, University of Silesia, ul. Bankowa 12, 40-007 Katowice, Poland. [ul. Drozdów 15B, 40-530 Katowice, Poland.] Prof. (U. Silesia, Katowice, 1971) material sciences. *X-ray_crystallography structure_of_metals shape_memory_alloys*
E-mail dana@usctoux1.cto.us.edu.pl Tel. 48(32)596929 Fax 48(32)599605 Telex 315584 usk pl

Bołd, Prof. Tadeusz (1934). Institute of Ferrous Metallurgy, 44-100 Gliwice, Poland.

Borowiak, Prof. Teresa (1939). Professor. Department of Crystallography, Faculty of Chemistry, A. Mickiewicz University, Poznań, Poland. [Faculty of Chemistry, A. Mickiewicz University, ul. Grunwaldzka 6, 60-780 Poznań, Poland.] Dr hab. (A. Mickiewicz U. Poznań, 1975) chemistry. *Organic_compounds organic_crystal_chemistry biologically_active_compounds*
E-mail borowiak@plpuam11.amu.edu.pl Tel. 48(61)699181x374 Fax 48(61)658008 Telex 0413260 uam pl

Borzecka-Prokop, Dr Barbara (1953). Assistant. Faculty of Chemistry, Jagiellonian University, ul. R. Ingardena 3, 30-060 Kraków, Poland. Dr (Jagiellonian U. Kraków, 1985) chemistry. *X-ray_powder_diffraction phase_transition crystal_chemistry*
E-mail borzecka@trurl.ch.uj.edu.pl Tel. 48(12)336377x268 Fax 48(12)340515

Bronowska, Dr Wiesława (1944). Adiunkt. Institute of Physics, Technical University of Wrocław, Wybrzeże Wyspiańskiego 27, 50-370 Wrocław, Poland. Dr (Inst. Low Temp. Str. res. PASci. Wrocław, 1971) physics. *Phase_transitions ferroelectrics crystal_structure*
E-mail pawlowska@ichm.ch-pwr.wroc.udu.pl Tel. 48(71)229696 Fax 48(71)229696

Brożek-Mucha, Dr Zuzanna (1964). Assistant. Faculty of Chemistry, Jagiellonian University, ul. R. Ingardena 3, 30-060 Kraków, Poland. PhD (Jagiellonian U. Kraków, 1994) chemistry. *Crystal_structure_determination structure-physical_property_relationship phase_transitions absolute_structure absolute_chirality absolute_polarity*
E-mail brozek@trurl.ch.uj.edu.pl Tel. 48(12)336377x270 Fax 48(12)340515

Bukowska-Strzyżewska, Prof. Maria (1929). Full professor. President, Crystallography Committee, Learned Society of Łódź. Department of X-ray and Chemical Crystallography, Institute of General and Ecological Chemistry, Technical University of Łódź, ul. Żwirki 36, 90-924 Łódź, Poland. Prof. (Tech. U. Łódź, 1983) chemistry. *X-ray_crystallography crystallochemistry_of_copper(II)_bismuth(III)_lanthanides(III) metalloorganic_compounds*
Tel. 48(42)313117 Fax 48(42)313103 Telex 886136 polit pl

Ciechanowicz-Rutkowska, Dr Maria (1941). Regional Laboratory of Physicochemical Analysis and Structural Research, Jagiellonian University, Kraków, Poland. [ŚLAFiBS, Jagiellonian University, ul. R. Ingardena 3, 30-060 Kraków, Poland.] PhD (Imperial C. London UK, 1971) chemistry. *Crystal_structure_determination drug_conformation structure-activity_relationship antimalarial_and_antimuscarinic_compounds*
E-mail rutkowsk@trurl.ch.uj.edu.pl Tel. 48(12)336377x267 Fax 48(12)343859

Ciszak, Dr Ewa (1960). Adiunkt. Institute of Chemistry, Pedagogical University, ul. Oleska 48, 45-052 Opole, Poland. Dr (A. Mickiewicz U. Poznań, 1988) chemistry. *X-ray_crystallography organic_crystal_structure protein_crystallography*
Tel. 48(77)35841x356

Ciunik, Dr Zbigniew (1949). Adiunkt. Institute of Chemistry, University of Wrocław, ul. F.Joliot-Curie 14, 50-383 Wrocław, Poland. Dr (A. Mickiewicz U. Poznań, 1980) chemistry. *Carbohydrates conformation_analysis molecular_mechanics semi-empirical_methods*
E-mail ciunik@chem.uni.wroc.pl Tel. 48(71)204338 Fax 48(71)222348

Dokurno, Paweł (1967). Adiunkt. Department of Chemistry, University of Gdańsk, Sobieskiego 18, 80-952 Gdańsk, Poland. PhD (U. Gdańsk, 1994) physical chemistry and crystallography. *Crystallography crystal_lattice_energetics thermochemistry*
E-mail   chepd@halina.univ.gda.pl   pawel@ewa.chem.univ.gda.pl   Tel.
48(58)415271x224 Fax 48(58)410357 Telex 0512024 rek ug pl

Dynowska, Dr Elżbieta G. (1944). Adiunkt. Institute of Physics, Polish Academy of Sciences, al. Lotników 32/46, 02-668 Warszawa, Poland. Dr (Inst. of Physics Polish Academy of Sciences Warszawa, 1983) physics. *Powder_diffraction lattice_parameters semiconductors thin_layer superlattice solid_solution*
E-mail dynow@ifpan.edu.pl Tel. 48(22)436034 Fax 48(22)430926

Figielski, Prof. Tadeusz Institute of Physics, Polish Academy of Sciences, al. Lotników 32/46, 02-668 Warszawa, Poland.

Fruziński, Mgr inż. Andrzej (1955). Assistant. Department of X-ray and Chemical Crystallography, Institute of General and Ecological Chemistry, Technical University of Łódź, ul. Żwirki 36, 90-924 Łódź, Poland. Mgr inż. (Tech. U. Łódź, 1979) physics. *Structure–activity_relationship X-ray_crystallography*
E-mail fruziola@plearn.bitnet Tel. 48(42)3313121 Fax 48(42)313103 Telex 886136 polit pl

Gałązka, Prof. Robert Institute of Physics, Polish Academy of Sciences, al. Lotników 32/46, 02-668 Warszawa, Poland.

Gałdecka, Dr Ewa Renata (1947). Adiunkt. Institute of Low Temperature and Structure Research, Polish Academy of Sciences, ul. Okólna 2, PO Box 937, 50-950 Wrocław, Poland. Dr (Inst. of Low Temperature and Structure Research Wrocław, 1981) physics. *Computing data_collection accurate_lattice_parameter_measurements profile_analysis maximum_entropy_methods intensity_distribution_functions history_of_crystallography*
E-mail crystal@plwrtu11.bitnet Tel. 48(71)35021x145 Fax 48(71)441029 Telex 712777 int pl

Gałdecki, Prof. dr inż. Zdzisław (1924). Full professor. Chairman, Structural Analysis Subcommittee, Polish Academy of Science; Director, National Centre Affiliated to Cambridge Crystallographic Data Centre. Department of X-ray and Chemical Crystallography, Institute of General and Ecological Chemistry, Technical University of Łódź, ul. Żwirki 36, 90-924 Łódź, Poland. [ul. Curie-Skłodowskiej 51 m. 7, 50-369 Wrocław, Poland.] Prof. (Tech. U. Łódź, 1986) chemistry. *Biocrystallography structure–activity_relationship drug_modelling computing protein_crystallography chemical_crystallography*
E-mail galdecki@lodz1.p.lodz.pl Tel. 48(71)228963 Fax 48(42)313103 Telex 886136 polit pl

Gawron, Dr Marian (1950). Adiunkt. Faculty of Chemistry, A. Mickiewicz University, ul. Grunwaldzka 6, 60-780 Poznań, Poland. Dr (A. Mickiewicz U. Poznań, 1982) chemistry. *Organic_crystal_chemistry organometallic_compounds biologically_active_compounds*
E-mail mgawron@plpuam.amu.edu.pl Tel. 48(61)699181x242 Fax 48(61)658008 Telex 0413260 uam pl

Gdaniec, Dr Maria (1951). Associate professor. Faculty of Chemistry, A. Mickiewicz University, ul. Grunwaldzka 6, 60-780 Poznań, Poland. Dr hab. (A. Mickiewicz U. Poznań, 1992) chemistry. *Clathrates bioactive_compounds hydrogen_bonding X-ray_crystallography*
E-mail magdan@plpuam11.amu.edu.pl Tel. 48(61)699181x489 Fax 48(61)658008 Telex 0413260 uam pl

Gilski, Mgr Mirosław (1963). Assistant. Faculty of Chemistry, A. Mickiewicz University, ul. Grunwaldzka 6, 60-780 Poznań, Poland. Mgr (A. Mickiewicz U. Poznań, 1987) physics. *Crystallographic_computing stereochemistry hydrogen_bonding computer_simulation computer_graphics*
E-mail mirek@plpuam11.amu.edu.pl Tel. 48(61)699181x489 Fax 48(61)658008 Telex 0413260 uam pl

Głowiak, Prof. Tadeusz (1935). Professor. Institute of Chemistry, Wrocław University, ul. F.Joliot-Curie 14, 50-383 Wrocław, Poland. Prof. (U. Wrocław, 1990) crystallochemistry. *Coordination_compounds small_organic_molecules*
Tel. 48(71)204350 Fax 48(71)222348

Główka, Prof. Marek L. (1948). Associate professor; Head of Department. Department of X-ray and Chemical Crystallography, Institute of General and Ecological Chemistry, Technical University of Łódź, ul. Żwirki 36, 90-924 Łódź, Poland. Prof. (Tech. U. Łódź, 1992) chemistry. *Structure–activity_relationship drug_modelling protein_crystallography*
E-mail glowka@lodz1.p.lodz.pl Tel. 48(42)313121 Fax 48(42)313103 Telex 886136 polit pl

Godwod, Dr Krzysztof Jan Institute of Physics, Polish Academy of Sciences, al Lotników 32/46, 02-668 Warszawa, Poland.

Goliński, Dr Bohdan (1931). Adiunkt. Department of X-ray and Chemical Crystallography, Institute of General and Ecological Chemistry, Technical University of Łódź, ul. Żwirki 36, 90-924 Łódź, Poland. Dr (Tech. U. Łódź, 1965) chemistry. *Structure_determination computing structure_of_dyes_and_pigments*
Tel. 48(42)313122 Fax 48(42)313103 Telex 886136 polit pl

Górkiewicz, Mgr inż. Zbigniew (1936). Assistant. Department of X-ray and Chemical Crystallography, Institute of General and Ecological Chemistry, Technical University of Łódź, ul. Żwirki 36, 90-924 Łódź, Poland. Mgr inż. (Tech. U. Łódź, 1968) textile technology. *Dyes_crystallography organic_crystal_structure*
Tel. 48(42)313121 Fax 48(42)313103 Telex 886136 polit pl

Górski, Dr Ludwik (1936). Adiunkt. Solid State Physics Department, Institute of Atomic Energy, 05-400 Otwock-Świerk, Poland. Dr (Inst. Nuclear Research Świerk, 1978) physics. *Phase_transitions small-angle_X-ray_scattering*
Tel. 48(2)7798648 Fax 48(22)105960 Telex 812225 iea pl

Grabowski, Prof. Mieczysław Jerzy (1928). Professor. Institute of Chemistry, University of Łódź, ul. Pomorska 149/153, 90-236 Łódź, Poland. Prof. (U. Łódź, 1990) *Crystal_structure_and_properties X-ray analysis music*
E-mail mjg@plunlo51.bitnet Tel. 48(42)790447 Fax 48(42)790447

Grochowski, Dr Jacek M. (1943). Head of X-ray Division. Member, Polish Synchrotron Radiation Society - SR Crystallography. Regional Laboratory of Physicochemical Analysis and Structural Research, Jagiellonian University, Kraków, Poland. {SLAFiBS, Jagiellonian University, ul. R. Ingardena 3, 30-060 Kraków, Poland.] Dr (Jagiellonian U. Kraków, 1975) chemistry. *Chirality absolute_structure anomalous_dispersion synchrotron_radiation chiral_drugs*
E-mail grochow@trurl.ch.uj.edu.pl Tel. 48(12)336377x267 Fax 48(12)343859

Grochulski, Dr Paweł (1943). Adiunkt. Institute of Physics, Technical University of Łódź, ul. Wólczańska 219, 93-005 Łódź, Poland. Dr (Tech. U. Łódź, 1988) physics. *Proteins structure–activity_relationship drug_modelling*
Tel. 48(42)313121 Fax 48(42)313103 Telex 886136 polit pl

Gronkowski, Dr hab. Jerzy (1949). Adiunkt. Institute of Experimental Physics, Faculty of Physics, Warsaw University, ul. Hoża 69, 00-681 Warszawa, Poland. Dr hab. (U. Warszawa, 1992) physics. *Diffraction_theory defects X-ray_topography semiconductors real_crystals lattice_distortion*
E-mail gronko@fuw.edu.pl Tel. 48(22)294229 Telex 825548 uwphy pl

Hodorowicz, Prof. Stanisław A. (1941). Full professor; Head of the Jagiellonian University. Faculty of Chemistry, Jagiellonian University, ul. R. Ingardena 3, 30-060 Kraków, Poland. Prof. (Jagiellonian U. Kraków, 1987) physicochemistry. *Crystal_chemistry phase_transitions high-temperature_superconductors molybdates biologically_active_compounds crystal_structure*
E-mail hodorowi@trurl.ch.uj.edu.pl Tel. 48(12)336377x267 Fax 48(12)340515

Janczak, Dr Jan (1961). Adiunkt. Institute of Low Temperature and Structure Research, Polish Academy of Sciences, ul. Okólna 2, PO Box 937, 50-950 Wrocław, Poland. Dr (Inst. Low Temp. Str. Research PASci Wrocław, 1992) chemistry. *Coordination_chemistry intermetallic_compounds powder_diffraction Rietveld_method phase_transitions*
E-mail crystal@plwrtu11.bitnet Tel. 48(71)35021x232 Fax 48(71)441029 Telex 712777 int pl

Janecki, Dr Hieronim Piotr (1951). Surface science. Faculty of Material Sciences, Technical University, ul. Chrobrego 27, 26-600 Radom, Poland. Dr inż. (Inst. Chemistry and Nuclear Engineering Warszawa, 1989) chemistry.
E-mail hpjaneck@plearn.edu.pl Tel. 48(48)40031 Fax 48(48)23969

Jaskólski, Dr Mariusz (1952). Associate professor. Faculty of Chemistry, A. Mickiewicz University, ul. Grunwaldzka 6, 60-780 Poznań, Poland. Dr hab. (A. Mickiewicz U. Poznań, 1985) physical chemistry. *Proteins_crystallography hydrogen_bonding nucleosides_stereochemistry crystallographic_computing retroviral_proteases bacterial_asparaginases*
E-mail mariuszj@plpuam11.amu.edu.pl Tel. 48(61)699181x489 Fax 48(61)658008 Telex 0413260 uam pl

Jerzykiewicz, Mgr inż. Lucjan B. Postgraduate student. Institute of Chemistry, University of Wrocław, ul. F.Joliot-Curie 14, 50-383 Wrocław, Poland. Mgr inż. (Tech. U. Szczecin, 1991) chemistry. *Organic_synthesis*
E-mail jerzyk@plwruw11.bitnet Tel. 48(71)204350 Fax 48(71)222348

Karniewicz, Prof. Jan (1928). Professor. Institute of Physics, Technical University of Łódź, ul. Wólczańska 219, 93-005 Łódź, Poland. Prof. (Tech. U. Łódź, 1988) physics. *Crystal_growth*
E-mail kwojcie@lodz1.p.lod.edu.pl Tel. 48(42)313649 Fax 48(42)313639

Karolak-Wojciechowska, Prof. Janina (1942). Associate professor. Department of X-ray and Chemical Crystallography, Institute of General and Ecological Chemistry, Technical University of Łódź, ul. Żwirki 36, 90-924 Łódź, Poland. Prof. (Tech. U. Łódź, 1994) chemistry. *Biologically-active_small_molecules crystallography QSAR*
E-mail jkarolak@lodz1.p.lodz.pl Tel. 48(42)313122 Fax 48(42)313103 Telex 886136 polit pl

Kaszkur, Dr Zbigniew (1954). Adiunkt. Institute of Physical Chemistry, Polish Academy of Sciences, ul. Kasprzaka 44/52, 01-224 Warszawa, Poland. Dr physics.
E-mail zbig@ichf.edu.pl Tel. 48(22)323221 Fax 48(22)325276

Katrusiak, Dr hab. Andrzej (1955). Adiunkt. Faculty of Chemistry, A. Mickiewicz University, ul. Grunwaldzka 6, 60-780 Poznań, Poland. Dr hab. (A. Mickiewicz U. Poznań, 1990) *Phase_transitions crystal_chemistry high_pressure_crystallography hydrogen_bonds statistics_in_crystallography*
E-mail katran@plpuam11.bitnet Tel. 48(61)699181x443 Fax 48(61)658008 Telex 0413260 uam pl

Kępa, Dr Henryk (1944). Assistant. Institute of Experimental Physics, Warsaw University, ul. Hoża 69, 00-681 Warszawa, Poland. Dr (Warsaw U., 1981) physics. *Neutron_diffraction magnetic_semiconductors thin_films superlattices*
E-mail henkepa@fuw.edu.pl Tel. 48(22)294229 Telex 825548 uwphy pl

Kociński, Prof. Jerzy Institute of Physics, Technical University of Warsaw, ul. Koszykowa 75, 00-628 Warszawa, Poland. *Phase_transitions*

Kosturkiewicz, Prof. Zofia (1928). Professor. Faculty of Chemistry, A. Mickiewicz University, ul. Grunwaldzka 6, 60-780 Poznań, Poland. Prof. (A. Mickiewicz U. Poznań, 1978) chemistry. *X-ray_crystallography organic_crystal_chemistry*
E-mail zkostur@plpuam11.amu.edu.pl Tel. 48(61)699181x488 Fax 48(61)658008 Telex 0413260 uam pl

Kowal, Mgr Ewa (1964). Post-graduate student. Faculty of Chemistry, Jagiellonian University, ul. R. Ingardena 3, 30-060 Kraków, Poland. Mgr (Jagiellonian U. Kraków, 1989) chemistry. *X-ray_powder_diffraction phase_transitions*
E-mail dokryst@Trurl.ch.uj.edu.pl Tel. 48(12)336377x268 Fax 48(12)340515

Kozioł, Dr hab. Anna E. (1951). Adiunkt. Faculty of Chemistry, M.Curie-Skłodowska University, pl. M.Curie-Skłodowskiej 3, 20-031 Lublin, Poland. Dr hab. (Jagiellonian U. Kraków, 1993) chemistry. *Small molecules natural products chiral compounds absolute configuration crystal structure determination*
Tel. 48(81)375662 Fax 48(81)375102 Telex 0643223 umcs pl

Krajewski, Dr Janusz (1929). Institute of Organic Chemistry, Polish Academy of Sciences, ul. Kasprzaka 44/52, 01-224 Warszawa, Poland.

Krygowski, Prof. Tadeusz Marek (1937). Professor. Faculty of Chemistry, Warsaw University, ul. Pasteura 1, 02-093 Warszawa, Poland. Prof. (Warsaw U., 1983) physical chemistry. *Structural changes geometry analyses structure-activity relationship folk and classic music*
E-mail tmkryg@chem.uw.edu.pl Fax 48(22)222892 Telex 815439 uw pl

Kublak, Dr Maria (1945). Adiunkt. Institute of Chemistry, University of Wrocław, ul. F.Joliot-Curie 14, 50-383 Wrocław, Poland. Dr chemistry. *Metalloorganic complexes*
Tel. 48(71)204350 Fax 48(71)222348

Kublak, Doc. Ryszard (1944). Associate professor. Institute of Low Temperature and Structure Research, Polish Academy of Sciences, ul. Okólna 2, PO Box 937, 50-950 Wrocław, Poland. Dr hab. (Inst. Low Temp. Str. Res. PASci Wrocław, 1984) chemistry. *Intermetallic compounds organometallic compounds*
E-mail crystal@plwrtu11.bitnet Tel. 48(71)35021 Fax 48(71)441029 Telex 712777 int pl

Kubicki, Dr Maciej (1963). Adiunkt. Faculty of Chemistry, A. Mickiewicz University, ul. Grunwaldzka 6, 60-780 Poznań, Poland. Dr (A. Mickiewicz U. Poznań, 1991) chemistry. *Organic metal chemistry hydrogen bonds biologically-active compounds*
E-mail mkubicki@plpuam1l.amu.edu.pl Tel. 48(61)699181x242 Fax 48(61)658008 Telex 0413546 uam pl

Kucharczyk, Dr Damian (1951). Adiunkt. Institute of Low Temperature and Structure Research, Polish Academy of Sciences, ul. Okólna 2, PO Box 937, 50-950 Wrocław, Poland. Dr (Inst. Low Temp. Str. Res. PASci. Wrocław, 1977) physics. *Modulated structures phase transitions structure analysis diffractometers business*
E-mail crystal@plwrtu11.bitnet Tel. 48(71)35021x145 Fax 48(71)441029 Telex 712777 int pl

Kusz, Dr Joachim (1955). Assistant. Institute of Physics, University of Silesia, ul. Uniwersytecka 4, 40-007 Katowice, Poland. Dr (Academy of Mining and Metallurgy Kraków, 1990) physics. *Phase transitions modulated structures crystallographic computing electronics*
E-mail kusz@usctoux1.cto.us.edu.pl Tel. 48(32)588211x1527 Fax 48(32)588431 Telex 0315584 usk pl

Kwiatkowski, Dr Witold (1961). Assistant. Department of X-ray and Chemical Crystallography, Institute of General and Ecological Chemistry, Technical University of Łódź, ul. Żwirki 36, 90-924 Łódź, Poland. Dr (Tech. U. Łódź, 1992) chemistry. *Drug action structure-activity relationship molecular mechanics proteins crystallography*
Tel. 48(42)313121 Fax 48(32)313103 Telex 886136 polit pl

Leciejewicz, Prof. Janusz (1928). Professor. Institute of Nuclear Chemistry and Technology, ul. Dorodna 16, 03-195 Warszawa, Poland. Dr hab. (Inst. Nuclear Research Świerk, 1975) experimental physics. *Neutron diffraction material science crystal chemistry of coordination compounds*
Tel. 48(22)111313 Fax 48(22)111532 Telex 813029 ichtj pl

Lefeld-Sosnowska, Prof. dr hab. Maria (1934). Professor. Institute of Experimental Physics, Faculty of Physics, Warsaw University, ul. Hoża 69, 00-681 Warszawa, Poland. Dr hab. (Warsaw U., 1979) physics. *Diffraction theory X-ray topography real crystals semiconductors lattice distortion*
E-mail lefeld@fuw.edu.pl Tel. 48(22)294229 Telex 825548 uwphy pl

Lewiński, Dr Krzysztof (1954). Adiunkt. Faculty of Chemistry, Jagiellonian University, ul. R. Ingardena 3, 30-060 Kraków, Poland. Dr (Jagiellonian U. Kraków, 1983) chemistry. *Biocrystallography structure determination computing computer-aided education*
E-mail lewinski@Trurl.ch.uj.edu.pl Tel. 48(12)336377x270 Fax 48(12)340515

Lipkowska, Dr Zofia (1946). Adiunkt. Institute of Organic Chemistry, Polish Academy of Sciences, ul. Kasprzaka 44/52, 01-224 Warszawa, Poland. Dr (Inst. Organic Chemistry PASci, 1975) chemistry. *Small molecule crystallography structure-activity relationship organic chemistry*
E-mail ocryst@ichf.edu.pl Tel. 48(22)323221x142 Fax 48(22)325276 Telex 817097 ichf pl

Lipkowski, Prof. Janusz (1943). Professor. Institute of Physical Chemistry, Polish Academy of Sciences, ul. Kasprzaka 44/52, 01-224 Warszawa, Poland. Prof. (Inst. Physical Chemistry PASci. Warszawa, 1990) physical chemistry. *Inclusion compounds*
E-mail klatrat@alfa.ichf.edu.pl Tel. 48(22)322159 Fax 48(22)325276

Lis, Prof. Tadeusz (1947). Professor. Institute of Chemistry, University of Wrocław, ul. F.Joliot-Curie 14, 50-383 Wrocław, Poland. Dr hab. (U. Wrocław, 1979) chemistry. *Metalloorganic complexes*
Tel. 48(71)204350 Fax 48(71)222348

Luboradzki, Mgr Roman (1963). Senior assistant. Institute of Physical Chemistry, Polish Academy of Sciences, ul. Kasprzaka 44/52, 01-224 Warszawa, Poland. Magister (U. Warszawa, 1988) chemistry. *Crystal chemistry*
E-mail romek@alfa.ichf.edu.pl Tel. 48(22)323221x225 Fax 48(22)325276

Luciak, Dr Bernard (1955). Adiunkt. Institute of Physics, Technical University of Łódź, ul. Wólczańska 219, 93-005 Łódź, Poland. Dr (Tech. U. Łódź, 1988) chemistry. *X-ray crystallography computers crystal physics*
Tel. 48(42)313121 Fax 48(42)313103 Telex 886136 polit pl

Lągiewka, Prof. Eugeniusz (1939). Associate professor; Dean of Faculty. Institute of Physics and Chemistry of Metals, University of Silesia, ul. Bankowa 12, 40-007 Katowice, Poland. [40-136 Katowice, ul. Stoneczna 74 m. 17, Poland.] Dr hab. (Inst. Low Temperature and Structural Research, Polish Acad. Sci., Wrocław, 1983) physics. *X-ray diffraction Auger spectroscopy amorphous materials electrocrystallization of metals and alloys*
E-mail niewiara@usctoux1.cto.us.edu.pl Tel. 48(32)596929 Fax 48(32)599605 Telex 315572 uswz pl

Łasocha, Dr Wiesław (1957). Adiunkt. Faculty of Chemistry, Jagiellonian University, ul. R. Ingardena 3, 30-060 Kraków, Poland. Dr (Jagiellonian U. Kraków, 1986) chemistry. *Crystal chemistry superconductors*
E-mail lasocha@Trurl.ch.uj.edu.pl Tel. 48(12)336377x268 Fax 48(12)340515

Łukaszewicz, Prof. Kazimierz (1927). Professor. President of Polish National Committee on Crystallography. Institute of Low Temperature and Structure Research, Polish Academy of Sciences, ul. Okólna 2, PO Box 937, 50-950 Wrocław, Poland. Prof. (Inst. Low Temp. Str. Res. PASci. Wrocław, 1974) physics. *Phase transitions structural disorder ferroelectric materials thermal expansion precise crystal structure analysis*
E-mail crystal@plwrtu11.bitnet Tel. 48(71)35021x240 Fax 48(71)441029 Telex 712777 int pl

Matuszewski, Dr Janusz (1954). Adiunkt. Department of Inorganic Chemistry, Academy of Economics, ul. Komandorska 118, 53-345 Wrocław, Poland. Dr (U. Wrocław, 1981) chemistry. *X-ray diffraction structural chemistry inorganic phosphates powder diffraction phase transitions X-ray databases material analysis yachting*
E-mail matusz@plwrae51.bitnet Tel. 48(71)672853 Fax 48(71)672778 Telex 48(71)2427 ae pl

Maurin, Dr Jan K. (1953). Adiunkt. Solid State Physics Department, Institute of Atomic Energy, 05-400 Otwock-Świerk, Poland. Dr (U. Warszawa, 1987) chemistry. *Organic crystal chemistry hydrogen bond*
E-mail e08jm@cx1.cyf.gov.pl Tel. 48(2)7798655 Fax 48(22)105960 Telex 812225 iea pl

Michalski, Dr Edward (1951). Adiunkt. Institute of Technical Physics, Military Academy of Technology, ul. Kaliskiego 2, 01-489 Warszawa, Poland. Dr (Military Academy of Technology Warszawa, 1981) physics. *X-ray diffraction aperiodic crystals polytypes liquid crystal structure fundamental crystallography defects in crystals teaching of physics*
Tel. 48(2)6867113 Fax 48(22)362254 Telex 812535 wat pl

Miśta, Mgr Włodzimierz (1956). Post-graduate student. Institute of Low Temperature and Structure Research, Polish Academy of Sciences, ul. Okólna 2, PO Box 937, 50-950 Wrocław, Poland. Mgr (U. Wrocław, 1980) chemistry. *Powder diffraction Rietveld method heterogeneous catalysis ceramics phase analysis*
Tel. 48(71)35021x294 Fax 48(71)441029 Telex 712777 int pl

Morawiec, Prof. Henryk (1933). Professor; director of institute. Institute of Physics and Chemistry of Metals, University of Silesia, ul. Bankowa 12, 40-007 Katowice, Poland. Prof. (U. Silesia, Katowice, 1987) physics. *X-ray methods shape memory alloys structure of alloys martensitic transformation*
E-mail dana@usctoux1.cto.us.edu.pl Tel. 48(32)596929 Fax 48(32)599605 Telex 0315584 usk pl

Mucha, Mgr Dariusz (1964). Post-graduate student. Faculty of Chemistry, Jagiellonian University, ul. R. Ingardena 3, 30-060 Kraków, Poland. Mgr (Jagiellonian U. Kraków, 1990) chemistry. *Crystal structure determination Rietveld method photoconductivity physical properties*
E-mail mucha@trurl.ch.uj.edu.pl Tel. 48(12)336377x270 Fax 48(12)340515

Olczak, Mgr inż. Andrzej (1961). Assistant. Department of X-ray and Chemical Crystallography, Institute of General and Ecological Chemistry, Technical University of Łódź, ul. Żwirki 36, 90-924 Łódź, Poland. Mgr inż. (Tech. U. Łódź, 1987) physics. *Structure-activity relationship crystallographic computing X-ray crystallography teaching*
E-mail anolczak@plearn.bitnet Tel. 48(42)313121 Fax 48(42)313103 Telex 886136 polit pl

Olech, Dr Andrzej Z. (1956). Assistant. Faculty of Chemistry, Jagiellonian University, ul. R. Ingardena 3, 30-060 Kraków, Poland. Dr (Jagiellonian U. Kraków, 1992) chemistry. *Crystallization calorimetry computer modelling EPR powders texture crystal growth kinetics*
E-mail olech@Trurl.ch.uj.edu.pl Tel. 48(12)336377x270 Fax 48(12)340515

Oleksyn, Prof. Barbara J. (1940). Associate professor. Faculty of Chemistry, Jagiellonian University, ul. R. Ingardena 3, 30-060 Kraków, Poland. Dr hab. (Jagiellonian U. Kraków, 1988) chemistry. *Crystal structure determination drug conformation structure-activity relationship crystallography biocrystallography antimalarial and antimuscarinic compounds macrocycles*
E-mail oleksyn@Trurl.ch.uj.edu.pl Tel. 48(12)336377x267 Fax 48(12)340515

Oleś, Prof. Andrzej Władysław (1923). Professor; Chief of Department. Faculty of Physics and Nuclear Techniques, Academy of Mining and Metallurgy, al. Mickiewicza 30, 30-059 Kraków, Poland. [Bytomska 16/16, 30-075 Kraków, Poland.] Prof. (Jagiellonian U. Kraków, 1974) solid state physics. *Magnetic structures general physics*
E-mail oles@novell.ftj.agh.edu.pl Tel. 48(12)333740 Fax 48(12)340010 Telex 0322203 agh pl

Paciorek, Dr Włodzimierz (1953). Adiunkt. Institute of Low Temperature and Structure Research, Polish Academy of Sciences, ul. Okólna 2, PO Box 937, 50-950 Wrocław, Poland. Dr (Inst. Low Temp. Str. Res. PASci. Wrocław, 1987) physics. *Modulated structures computing symmetry*
E-mail crystal@plwrtu11.bitnet

Pajączkowska, Anna (1934). Professor. Institute of Electronics Materials Technology, ul. Wólczyńska 133, 01-919 Warszawa, Poland. Professor Dr hab. (Polish Central Commission, 1994) chemistry. *Crystal_growth real_structure X-ray_diffraction electron_microscopy optical_microscopy magnetic_properties EMP*
E-mail itme3@frodo.nask.org.pl Tel. 48(22)349949 Fax 48(22)349003

Pająk, Prof. Lucjan (1947). Adiunkt. Institute of Physics and Chemistry of Metals, University of Silesia, ul. Bankowa 12, 40-007 Katowice, Poland. Dr (U. Silesia, Katowice, 1980) chemistry. *Small_angle_scattering materials_science porosity_of_materials*
E-mail niewiara@usctoux1.cto.us.edu.pl Tel. 48(32)596929 Fax 48(32)599605 Telex 315572 uswz pl

Palosz, Dr Bogdan Fryderyk (1947). Associate Professor; Vice-Director. High Pressure Research Center, Polish Academy of Sciences, Poland. [High Pressure Research Center, Polish Academy of Sciences, Sokolowska 29, PL-01142 Warsaw, Poland.] Dr habil. (TU Warsaw, 1983) physics/crystallography. *Rietveld_method ceramics polytypism non-crystalline_computer_modelling non-crystalline_phase_refinement grain_boundary surface_phase*
E-mail palosz@iris.unipress.waw.pl Tel. 48()39123276 or 48(22)328497 Fax 48()39120331 or 48(22)324218 Telex 817618zwcpl

Paszkowicz, Dr Wojciech (1951). Adiunkt. Institute of Physics, Polish Academy of Sciences, al. Lotników 32/46, 02-668 Warszawa, Poland. Dr (Inst. of Physics PASci. Warszawa, 1992) physics. *Powder_diffraction II–VI_compounds III–V_compounds thin_films close_packing profile_analysis*
E-mail paszk@planif61.bitnet Tel. 48(22)436034 Fax 48(22)430926

Piątek, Mgr Malgorzata Katarzyna (1966). Post-graduate student. Faculty of Chemistry, Jagiellonian University, ul. R. Ingardena 3, 30-060 Kraków, Poland. Mgr (Jagiellonian U. Kraków, 1990) chemistry. *Crystallography single-crystal_structure_analysis selenium_organic_compounds structure–activity_relationship biocrystallography anti-inflammatory_and_anticancer_compounds immunochemistry computer_modelling*
E-mail dokryst@Trurl.ch.uj.edu.pl Tel. 48(12)336377x268 Fax 48(12)340515

Piecek, Mgr Wiktor (1967). Assistant. Institute of Technical Physics, Military Academy of Technology, ul. Kaliskiego 2, 01-489 Warszawa, Poland. Mgr (Military Academy of Technology Warszawa, 1991) physics. *X-ray_diffraction liquid_crystal_structure physics electrooptics applications_of_LC dispersed_LC*
Tel. 48(2)6869579 Fax 48(22)362254 Telex 812535 wat pl

Pielaszek, Dr Jerzy (1941). Head of research group; deputy director of the institute. Institute of Physical Chemistry, ul. Kasprzaka 44/52, 01-224 Warszawa, Poland. PhD (Polish Academy of Sciences, 1972) solid state physics. *X-ray_diffraction polycrystals catalysts*
E-mail jp@ichf.edu.pl Tel. 48(22)323221 Fax 48()39120238

Pietraszko, Doc. Adam (1943). Associate professor. Sub-Editor, WDC9. Institute of Low Temperature and Structure Research, Polish Academy of Sciences, ul. Okólna 2, PO Box 937, 50-950 Wroclaw, Poland. Dr hab. (Inst. Low Temp. Str. Res. PASci. Wroclaw, 1992) physics. *Phase_transitions crystal_structure_determination ferroelectrics*
E-mail crystal@plwrtu11.bitnet Tel. 48(71)35021x144 Fax 48(71)441029 Telex 712777 int pl

Ratajczak-Sitarz, Dr Malgorzata (1958). Assistant. Department of Crystallography, Faculty of Chemistry, A. Mickiewicz University, ul. Grunwaldzka 6, 60-780 Poznań, Poland. Dr (A. Mickiewicz U. Poznań, 1982) chemistry. *Hydrogen_bond structural_transformations conformational_analysis*
E-mail katran@plpuam11.bitnet Tel. 48(61)699181x443 Fax 48(61)658008 Telex 0413260 uam pl

Ratusna, Dr Alicja (1947). Adiunkt. Institute of Physics, University of Silesia, ul. Uniwersytetka 4, 40-007 Katowice, Poland. Dr hab. (Inst. Low Temp. Str. Res. PASci. Wroclaw, 1994) physics. *Crystallography phase_transitions solid_state_physics literature modern_history*
E-mail ratuszna@usctoux1.cto.us.edu.pl Tel. 48(32)588211x1501 Fax 48(32)588431 Telex 0315584 usk pl

Rychlewska, Prof. Urszula (1948). Associate professor. Faculty of Chemistry, A. Mickiewicz University, ul. Grunwaldzka 6, 60-780 Poznań, Poland. Dr hab. (A. Mickiewicz U. Poznań, 1987) chemistry. *X-ray_crystallography crystal_packing_stereochemistry MO_calculations sesquiterpenes natural_products hydrogen_bonding*
E-mail urychle@plpuam11.amu.edu.pl Tel. 48(61)699181x489 Fax 48(61)658008 Telex 0413260 uam pl

Sawka-Dobrowolska, Dr Wanda (1945). Adiunkt. Institute of Chemistry, University of Wroclaw, ul. F.Joliot-Curie 14, 50-383 Wroclaw, Poland. Dr (U. Wroclaw, 1980) chemistry. *X-ray_crystallography aminophosphonic_and_phosphinic_acids*
E-mail wanda@plwruw11.bitnet Tel. 48(71)204350 Fax 48(71)222348

Serda, Mgr Pawel (1955). Regional Laboratory of Physicochemical Analysis and, Structural Research, Jagiellonian University, ul. R. Ingardena 3, 30-060 Kraków, Poland. Mgr (Jagiellonian U. Kraków, 1979) physics. *Absolute_structure anomalous_dispersion computing synchrotron_radiation*
E-mail serda@Trurl.ch.uj.edu.pl Tel. 48(12)336377x267 Fax 48(12)343859

Skowerenda, Dr Jolanta (1945). Adiunkt. Department of X-ray and Chemical Crystallography, Institute of General and Ecological Chemistry, Technical University of Łódź, ul. Żwirki 36, 90-924 Łódź, Poland. Dr (Tech. U. Łódź, 1976) chemistry. *Organic_crystal_structure*
Tel. 48(42)313117 Fax 48(42)313103 Telex 886136 polit pl

Sobczak, Dr hab. Ewa (1948). Adiunkt. Institute of Physics, Polish Academy of Sciences, al. Lotników 32/46, 02-668 Warszawa, Poland. Dr hab. (Inst. of Physics Warszawa, 1993) physics. *EXAFS XPS white-beam_radiation X-ray_spectroscopy*
E-mail sobcz@ifpan.edu.pl Tel. 48(22)436034 Fax 48(22)430926

Sosnowska, Prof. dr hab. Izabela (1939). Professor of physics; head of structure and lattice dynamic laboratory. Institute of Experimental Physics, Warsaw University, ul. Hoża 69, 00-681 Warszawa, Poland. Dr.hab. (Warsaw U., 1973) physics. *Lattice_dynamics magnetic_materials crystal_field neutron_scattering*
E-mail izabela@fuw.edu.pl Tel. 48(2)6287252

Stadnicka, Dr Katarzyna (1943). Adiunkt. Faculty of Chemistry, Jagiellonian University, ul. R. Ingardena 3, 30-060 Kraków, Poland. Dr (Jagiellonian U. Kraków, 1973) chemistry. *Crystal_structure_determination absolute_structure phase_transitions optical_activity structure-property_relationship chirality-polarity*
E-mail stadnick@Trurl.ch.uj.edu.pl Tel. 48(12)336377x270 Fax 48(12)340515

Stępień-Damm, Dr Julia (1945). Adiunkt. Institute of Low Temperature and Structure Research, Polish Academy of Sciences, ul. Okólna 2, PO Box 937, 50-950 Wroclaw, Poland. Dr (Inst. Low Temp. Str. Res. PASci. Wroclaw, 1980) physics. *Phase_transitions crystal_structure_analysis ferroelectrics intermetallic_compounds*
E-mail crystal@plwrtu11.bitnet Tel. 48(71)35021x144 Fax 48(71)441029 Telex 712777 int pl

Stróż, Dr Danuta (1951). Adiunkt. Institute of Physics and Chemistry of Metals, University of Silesia, ul. Bankowa 12, 40-007 Katowice, Poland. Dr (Inst. Metallurgy PASci., 1984) material science. *Electron_microscopy phase_transitions shape_memory_alloys*
E-mail dana@usctoux1.cto.us.edu.pl Tel. 48(32)596929 Fax 48(32)599605 Telex 0315584 usk pl

Surowiec, Dr Marian Ryszard (1948). Adiunkt. Institute of Physics and Chemistry of Metals, University of Silesia, ul. Bankowa 12, 40-007 Katowice, Poland. Dr (U. Silesia, 1977) physics. *X-ray_topography defects_in_crystals dislocation_structure III–V_semiconducting_compounds quasicrystals*
E-mail dana@usctoux1.cto.us.edu.pl Tel. 48(32)596929 Fax 48(32)599605 Telex 0315584 usk pl

Suwińska, Dr Kinga (1954). Adiunkt. Institute of Physical Chemistry, Polish Academy of Sciences, ul. Kasprzaka 44/52, 01-224 Warszawa, Poland. Dr (Inst. Physical Chemistry PASci. Warszawa, 1984) chemistry. *Organic_crystal_structures inclusion_compounds clathrates molecular_mechanics*
E-mail kinga@alfa.ichf.edu.pl Tel. 48(22)323221x399 Fax 48(22)325276

Szczepańska, Mgr Beata Irena (1965). Technical assistant. Faculty of Chemistry, A. Mickiewicz University, ul. Grunwaldzka 6, 60-780 Poznań, Poland. Mgr (A. Mickiewicz U. Poznań, 1990) chemistry. *Small_organic_molecules molecular_packing single_crystal_X-ray_diffraction stereochemistry crystal_structure_determination hydrogen_bonding*
E-mail beatas@plpuam11.amu.edu.pl Tel. 48(61)699181x489 Fax 48(61)658008 Telex 0413260 uam pl

Szurgot, Dr Marian (1946). Adiunkt. Institute of Physics, Technical University of Łódź, ul. Wólczańska 219, 93-005 Łódź, Poland. Dr (Tech. U. Łódź, 1987) chemical engineering. *Crystal_growth etching surface_morphology X-ray_topography defects macromorphology*
Tel. 48(42)313661 Fax 48(42)313639 Telex 886136 polit pl

Śliwiński, Mgr Jan (1952). Assistant. Faculty of Chemistry, Jagiellonian University, ul. R. Ingardena 3, 30-060 Kraków, Poland. Mgr (Jagiellonian U. Kraków, 1976) chemistry. *Crystal_structure_determination drug_conformation structure–activity_relationship antimalarial_compounds macrocycles biocrystallography*
E-mail sliwinsk@Trurl.ch.uj.edu.pl Tel. 48(12)336377x268 Fax 48(12)340515

Tomaszewski, Dr Pawel E. (1952). Adiunkt. Institute of Low Temperature and Structure Research, Polish Academy of Sciences, ul. Okólna 2, PO Box 937, 50-950 Wroclaw, Poland. Dr (Inst. Low Temp. Str. Res. PASci. Wroclaw, 1985) physics. *Phase_transitions high_pressure ferroelectrics data_bases amorphization_under_pressure X-ray_diffraction precise_lattice_parameters_measurements philately*
E-mail crystal@plwrtu11.bitnet Tel. 48(71)35021x145 Fax 48(71)441029 Telex 712777 int pl

Tosik, Dr Anita (1947). Adiunkt. Department of X-ray and Chemical Crystallography, Institute of General and Ecological Chemistry, Technical University of Łódź, ul. Żwirki 36, 90-924 Łódź, Poland. Dr (Tech. U. Łódź, 1977) chemistry. *Complex_compounds organic_crystal_structures*
Tel. 48(42)313117 Fax 48(42)313103 Telex 886136 polit pl

Trzaska-Durski, Dr Zygmunt Maria (1931). Assistant professor. Faculty of Chemistry, Warsaw University of Technology, ul. Noakowskiego 3, 00-664 Warszawa, Poland. Dr (Warsaw U. of Technology, 1970) crystal structure analysis. *Crystal_symmetry X-ray_crystallography X-ray_powder_diffractometry phase_analysis*
Tel. 48(22)621007x830 Fax 48(2)6282741

Urbańczyk, Prof. Grzegorz (1928). Professor; Institute Director. Institute of Fiber Physics and Textile Finishing, Technical University of Łódź, ul. Żwirki 36, 90-924 Łódź, Poland. DSc (Tech. U. Łódź, 1970) physics. *Fibres_structure polymers fibres_physics*
Tel. 48(42)362762 Fax 48(42)362762 Telex 88-61-36

Uszyński, Dr Ignacy (1950). Physicist. Institute of Low Temperature and Structure Research, Polish Academy of Sciences, ul. Okólna 2, PO Box 937, 50-950 Wroclaw, Poland. Dr (Tech. U. Wroclaw, 1978) physics. *Structure_analysis phase_transitions powder_diffraction*
E-mail crystal@plwrtu11.bitnet Tel. 48(71)35021x145 Fax 48(71)441029 Telex 712777 int pl

Warczewski, Prof. Jerzy Zdzisław (1939). Professor; Head of the Department of Physics of Magnetic Materials. Institute of Physics, University of Silesia, ul. Uniwersytecka 4, 40-007 Katowice, Poland. Prof. (awarded by The President of the Republic of Poland, 1992) physics. *Modulated_crystals magnetic_structures magnetic_materials teaching_of_physics_and_crystallography*
E-mail warcz@usctoux1.cto.us.edu.pl Tel. 48(32)588211x1782 Fax 48(32)588431 Telex 0315584 USK PL

Waśkowska, Dr Alicja (1942). Adiunkt. Institute of Low Temperature and Structure, Research, Polish Academy of Sciences, ul. Okólna 2, PO Box 937, 50-950 Wrocław, Poland. Dr (Inst. Immunology and Exper. Therapy PASci. Wrocław, 1975) biophysics. *X-ray_crystallography phase_transitions high_pressure structure-properties_relationship*
E-mail crystal@plwrtu11.bitnet Tel. 48(71)35021x297 Fax 48(71)441029 Telex 712777 int pl

Wawrzak, Dr Zdzisław (1955). Adiunkt. Institute of Physics, Technical University of Łódź, ul. Wólczańska 219, 93-005 Łódź, Poland. Dr (Tech. U. Łódź, 1988) physical chemistry. *Protein_crystallography drug_modelling structure-activity_relationship*
Tel. 48(42)313121 Fax 48(42)313103 Telex 886136 polit pl

Wieczorek, Dr Wanda (1945). Adiunkt. Department of X-ray and Chemical Crystallography, Institute of General and Ecological Chemistry, Technical University of Łódź, ul. Żwirki 36, 90-924 Łódź, Poland. Dr (Tech. U. Łódź, 1970) chemistry. *Organic_crystallography*
Tel. 48(42)313117 Fax 48(42)313103 Telex 886136 polit pl

Wlewióra, Prof. Andrzej (1933). Professor; Head of the Department of Mineralogy and Petrology. Institute of Geological Sciences, Polish Academy of Sciences, al. Żwirki i Wigury 93, 02-089 Warszawa, Poland. Professor (State Council of Poland, 1981) natural sciences. *Crystal_chemistry_and_structure*
E-mail awiew@plearn.edu.pl Tel. 48(22)223051x142 Fax 48(22)221065

Wokulska, Dr Krystyna (1945). Adiunkt. Institute of Physics and Chemistry of Metals, University of Silesia, ul. Bankowa 12, 40-007 Katowice, Poland. Dr (U. Silesia, Katowice, 1987) physics. *X-ray_characterization_of_single_crystals*
E-mail dana@usctoux1.cto.us.edu.pl Tel. 48(32)596929 Fax 48(32)599605 Telex 0315584 usk pl

Wokulski, Prof. Zygmunt (1940). Associate professor. Institute of Physics and Chemistry of Metals, University of Silesia, ul. Bankowa 12, 40-007 Katowice, Poland. Dr hab. (Inst. Low Temp. Str. Res. PASci. Wrocław, 1992) physics. *Crystal_growth whiskers CVD nitrides carbides CVT crystal_morphology*

Wolf, Dr Wojciech (1957). Adiunkt. Department of X-ray and Chemical Crystallography, Institute of General and Ecological Chemistry, Technical University of Łódź, ul. Żwirki 36, 90-924 Łódź, Poland. Dr (Tech. U. Łódź, 1988) chemistry. *Protein_crystallography chemical_crystallography*
Tel. 48(42)313122 Fax 48(42)313103 Telex 886136 polit pl

Wolska, Dr Irena (1950). Adiunkt. Faculty of Chemistry, A. Mickiewicz University, ul. Grunwaldzka 6, 60-780 Poznań, Poland. Dr (A. Mickiewicz U. Poznań, 1985) chemistry. *Organic_compounds biologically_active_compounds organic_crystal_chemistry*
E-mail iwolska@plpuam11.amu.edu.pl Tel. 48(61)699181x374 Fax 48(61)658008 Telex 0413260 uam pl

Wołcyrz, Dr Marek (1952). Adiunkt. Institute of Low Temperature and Structure Research, Polish Academy of Sciences, ul. Okólna 2, PO Box 937, 50-950 Wrocław, Poland. Dr (Inst. Low Temp. Str. Res. PASci. Wrocław, 1982) physics. *Powder_diffraction electron_diffraction structure_determination phase_transition high-temperature_superconductors*
E-mail crystal@plwrtu11.bitnet Tel. 48(71)35021x146 Fax 48(71)441029 Telex 712777 int pl

Woźniak, Dr Krzysztof (1961). Adiunkt. Faculty of Chemistry, Warsaw University, ul. Pasteura 1, 02-093 Warszawa, Poland. Dr (U. Warszawa, 1992) chemistry. *Geometry_analysis structure-activity_relationship structure_disorder magnetic_resonance weak_interaction symmetry_breaking*
E-mail kwozniak@chem.uw.edu.pl Tel. 48(22)222892 Fax 48(22)222892 Telex 815439 uw pl

Zielińska-Rohozińska, Dr hab. Elżbieta (1938). Adiunkt. Institute of Experimental Physics, University of Warsaw, ul. Hoża 69, 00-681 Warszawa, Poland. Dr hab. (U. Warszawa, 1983) physics. *Diffraction_theory X-ray_topography real_crystals semiconductors lattice_distortion high-resolution_diffraction defects*
E-mail roh@fuw.edu.pl Tel. 48(22)294229 Telex 825548 uwphy pl

Żmija, Prof. Józef (1932). Professor. Institute of Technical Physics, Military Academy of Technology, ul. Kaliskiego 2, 01-489 Warszawa, Poland. [ul. Mendelejewa 25, 01-489 Warszawa 49, Poland.] Prof. (Military Academy of Technology Warszawa, 1975) materials science. *Liquid_crystals_technology_and_theory solid_crystals_technology liquid_crystals*
Tel. 48(2)6869731 Fax 48(22)362254 Telex 812535 wat pl

PORTUGAL

Sub-Editor: M.-M. R. R. Costa

Aires Barros, Prof. Luis António (1930). Full Professor - Head of Department. Laboratorio de Mineralogia e Petrologia, Instituto Superior Técnico, Av. Rovisco Pais, 1096 Lisboa Codex, Portugal. Agregation (U. Técnica, Lisboa, 1964) mining engineering. *Alteration environment geosciences geochemistry mineralogy petrology carbonates minerals salts silicates*
Tel. 351(1)800111 Fax 351(1)800806 Telex 63423 ISTUTL

de Almeida, Prof. Maria-José (1946). Full Professor. Departamento de Fisica, Universidade de Coimbra, 3000 Coimbra, Portugal. Agregation (U. Coimbra, 1988) physics. *Charge_density magnetization_density powder single_crystal_structure_determination binary_alloys rare-earth_compounds lanthanides fluorine_compounds SAXS*
Tel. 351(39)23675 Fax 351(39)29158 Telex 52601 DEFIUC P

Alte da Veiga, Prof. Luiz (1932). Group Leader. Departamento de Fisica, Universidade de Coimbra, 3000 Coimbra, Portugal. PhD (U. Cambridge, 1963) physics. *Charge_density magnetization_density powder single_crystal_structure_determination antitumour_compounds binary_alloys oxides*
Tel. 351(39)23675 Fax 351(39)29158 Telex 52601 DEFIUC P

Andrade, Dr Lourdes (1954). Auxiliary Professor. Departamento de Fisica, Universidade de Coimbra, 3000 Coimbra, Portugal. PhD (U. Coimbra, 1986) physics. *Charge_density magnetization_density powder single_crystal_structure_determination binary_alloys oxides zeolites uranium_compounds perovskites SAXS Rietveld_method*
E-mail lourdes@ciuc2.uc.pt Tel. 351(39)23675 Fax 351(39)29158 Telex 52601 DEFIUC P

Archer, Margarida (1968). Research Student. Instituto de Tecnologia Química e Biologica, Apartado 127, 2780 Oeiras, Portugal. Graduation (U. Nova de Lisboa, 1992) chemistry. *Proteins_crystallography*
E-mail archer@ctqb01.ctqb.pt Tel. 351(1)4426146 Fax 351(1)4418766

Barros Marques, Dr Maria Isabel (1936). Researcher. Centro de Física da Matéria Condensada, Av. Prof. Gama Pinto, 2, 1699 Lisbo Codex, Portugal. PhD physics of condensed matter. *Structure_of_ionic_liquids X-ray_diffraction EXAFS gels glasses*
E-mail marques@alf4.cc.fc.ul.pt Tel. 351(1)7950790 Fax 351(1)7954288

Basto, Maria João Cabral de Oliveira (1945). Auxiliary Researcher. Laboratorio de Mineralogia e Petrologia, Instituto Superior Técnico, Av. Rovisco Pais, 1096 Lisboa Codex, Portugal. Graduation (U. Técnica, Lisboa, 1968) industrial chemical engineering. *Diffractometry infrared mineralogy Rietveld_method synchrotron_radiation trace_analysis X-ray_fluorescence minerals sulfides XAS*
Tel. 351(1)800111 Fax 351(1)800806 Telex 63423 ISTUTL

Borges, Prof. Frederico P. S. Sodré (1942). Head of Department. Faculdade de Ciências do Porto, Praça Gomes Teixeira, 4000 Porto, Portugal. PhD (Imperial College of Science and Technology, London, UK, 1978) geology. *Deformation electron_microscopy electron_probe_micro-analysis mineralogy*
Tel. 351(2)310290 Fax 351(2)316456

Carrondo, Prof. Maria Armenia (1948). Associate Professor, Group Leader. Instituto de Tecnologia Química e Biologica, Rua da Quinta Grande, 6, 2780 Oeiras, Portugal. Agregation (U. Técnica, Lisboa, 1989) chemistry. *Structure_determination crystallography metalloproteins synchrotron_radiation*
E-mail carron@ctqb01.ctqb.pt Tel. 351(1)4426146 Fax 351(1)4418766

Castanhola Batista, Antonio Adriano (1962). Assistant Lecturer. Departamento de Fisica, Universidade de Coimbra, 3000 Coimbra, Portugal. Graduation (U. Coimbra, 1985) physics. *Residual_stress mechanics fracture diffraction_data*
Tel. 351(39)23675 Fax 351(39)29158 Telex 52601 DEFIUC P

de Castro, Prof. Baltazar (1951). Associate Professor. Faculdade de Ciências, Universidade do Porto, Praça Gomes Teixeira, 4000 Porto, Portugal. PhD (J. Hopkins U., USA, 1980) chemistry. *Catalysts zeolites organometallic membranes*
E-mail bcastro@fc.pt.up Tel. 351(2)2002476 Fax 351(2)2008628

Chaves, Prof. Maria Renata (1933). Full Professor; Member of the Directive Committee of the Centro de Fisica da Universidade do Porto; Member of the Directive Committee of IFIMUP(IMAT); Group Leader. Centro de Fisica, Universidade do Porto, Praça Gomes Teixeira, 4000 Porto, Portugal. Agregation (U. Porto, 1975) physics. *Cooperative_phenomena modulated_structures order-disorder ferroelectricity dielectrics defects ordering phosphates titanates perovskites aperiodic_material betaine_compounds*
E-mail rachaves@fc.up.pt Tel. 351(2)2001653 Fax 351(2)319267 Telex 28109 FCUP P

Coelho, Ana Maria Varela (1964). Research student. Instituto de Tecnologia Química e Biologica, Apartado 127, 2780 Oeiras, Portugal. Graduation (U. Lisboa, 1987) biochemistry. *Biochemistry time-resolved_effect proteins metalloproteins*
E-mail varela@ctqb01.ctqb.pt Tel. 351(1)4426246 Fax 351(1)4418766

Correia dos Santos, Dr António José Rebelo (1948). Auxiliary Professor. Departamento de Química, Faculdade de Ciências, Universidade de Lisboa, Campo Grande, 1700 Lisboa, Portugal. PhD (U. Lisboa, 1987) solid state chemistry. *Chemistry condensed_matter EXAFS solid_solution structure_determination XANES perovskites superconductors*
Tel. 351(1)7573141 Fax 351(1)7599404

Costa, Prof. Maria-Margarida R. R. (1945). Full Professor. Sub-Editor, World Directory of Crystallography 9. Departamento de Fisica, Universidade de Coimbra, 3000 Coimbra, Portugal. Agregation (U. Coimbra, 1981) physics. *Charge_density magnetization_density single_crystal structure_determination SAXS transition_elements binary_alloys rare-earth_compounds lanthanides fluorine_compounds chromium_compounds X-ray_magnetic_scattering*
E-mail guida@ciuc2.uc.pt Tel. 351(39)23675 Fax 351(39)29158 Telex 52601 DEFIUC P

Damas, Prof. Ana Margarida (1953). Associate Professor. Instituto de Ciências Biomédicas Abel Salazar, Largo do Prof. Abel Salazar n.2, 4000 Porto, Portugal. PhD (U. Lisboa, 1992) crystallography. *Crystallography biophysics*
E-mail andamas@ncc.up.pt Tel. 351(2)310359 Fax 351(2)2001918

Duarte, Dr Maria Teresa Leal (1958). Auxiliary Professor, Group Leader. Instituto Superior Técnico, Av. Rovisco Pais, 1000 Lisboa, Portugal. PhD (U. Técnica, Lisboa, 1989) chemistry. *Crystallography structure_determination transition_element_complexes zeolites*
E-mail d1992@beta.ist.utl.pt Tel. 351(1)8419000 Fax 351(1)3524372

Figueiredo, Prof. Maria Ondina (1938). Group Leader. Departamento de Química, Facultate de Ciências, Universidade de Lisboa, Campo Grande, 1700 Lisboa, Portugal. Agregation (U. Técnica, Lisboa, 1991) applied geology. *Crystallography diffractometry mineralogy phase_transitions EXAFS XANES systematics superconductors ceramics minerals*
Tel. 351(1)8476596 Fax 351(1)2957810

Fortes, Prof. Manuel Amaral (1938). Group Leader. Instituto Superior Técnico, Av. Rovisco Pais, 1096 Lisboa Codex, Portugal. Agregation (U. Técnica, Lisboa, 1975) physical metallurgy. *Coarsening interface polycrystal topology mosaicity graph_theory biocrystallography cellular_materials*
Tel. 351(1)8418118 Fax 351(1)8418132 Telex 63423 ISTUTL

Frazao, Dr Carlos (1959). Assistant Researcher. Instituto de Tecnologia Química e Biologica, Apartado 127, 2780 Oeiras, Portugal. PhD (TU Munich, Germany, 1988) chemistry. *Proteins_crystallography*
E-mail frazao@itqb.unl.pt Tel. 351(1)4426146 Fax 351(1)4418766

Freitas, Miguel (1957). Assistant Lecturer. Departamento de Fisica, Universidade de Coimbra, 3000 Coimbra, Portugal. Graduation (U. Coimbra, 1981) physics. *Charge_density characterization Czochralski_technique Bridgman_Stockbarger_technique low_temperature*
Tel. 351(39)23675 Fax 351(39)29158 Telex 52601 DEFIUC P

Gomes, Prof. Celso (1937). Full Professor , Group Leader. Departamento de Geociências, Universidade de Aveiro, 3800 Aveiro, Portugal. PhD (U. Leeds, UK, 1979) materials science. *Mineralogy geochemistry crystal_form crystal_growth crystallite crystallinity defects structure_determination structure_disorder polymorphism polytypism solid_solution diffractometry FTIR DTA X-ray_fluorescence high_resolution electron_microscopy scanning_electron_microscopy size_distribution ceramics paper pigments sediments silicates phyllosilicates zeolites*
Tel. 351(34)370200 Fax 351(34)381260; 351(34)28600 Telex 37373

Guimarães, Prof. Dirce Andrade (1934). Associate Professor. Departamento de Física, Universidade de Aveiro, 3800 Aveiro, Portugal. PhD (Cambridge U., UK, 1979) solid state physics. *Dielectrics devitrification domain_structure ferroelasticity ferroelectricity Fourier_transform free_energy group_theory local_order low_temperature Raman Rietveld_method phosphates*
Tel. 351(34)370200 Fax 351(34)24965

Lima-de-Faria, Prof. José (1925). Director - Head, Department of Earth Sciences. Centro de Cristalografia e Mineralogia, Instituto de Investigação Tropical, Al. D. Afonso Henriques, 41-4Esq, 1000 Lisboa, Portugal. PhD (U. Cambridge, UK, 1962) crystallography and mineralogy. *Structural_classification_of_minerals symmetry_of_structures packing_and_symmetrical_analogues condensed_models*
Tel. 351(1)8476596 Fax 351(1)3631460 Telex IICT 66932

Margarido, Dr Fernanda (1957). Auxiliary Professor. Instituto Superior Técnico, Av. Rovisco Pais, 1096 Lisboa Codex, Portugal. PhD (U. Técnica, Lisboa, 1990) metallurgical and materials engineering. *Crystal_chemistry_of_alloys metallurgy X-ray_diffraction texture microstructure kinetics melting silicon_compounds*
Tel. 351(1)8418107 Fax 351(1)8418132 Telex 63423 ISTUTL

Matias, Prof. Maria José (1946). Associate Professor. Laboratorio de Mineralogia e Petrologia, Instituto Superior Técnico, Av. Rovisco Pais, 1096 Lisboa Codex, Portugal. PhD (U. Técnica, Lisboa, 1984) applied geology (geochemistry). *Alteration geochemistry geosciences microscopy mineralogy petrography petrology carbonates minerals silicates volcanic_rocks*
Tel. 351(1)800111 Fax 351(1)800806 Telex 63423 ISTUTL

Matias, Dr Pedro M. (1959). Research Assistant. Instituto de Tecnologia Química e Biologica, Apartado 127, 2780 Oeiras, Portugal. PhD (U. Pittsburgh, USA, 1986) crystallography. *Proteins_crystallography small_molecules_crystallography crystallographic_computing*
E-mail matias@itqb.unl.pt Tel. 351(1)4426146 Fax 351(1)4418766

Matos Beja, Dr Ana (1949). Auxiliary Researcher. Departamento de Fisica, Universidade de Coimbra, 3000 Coimbra, Portugal. PhD (U. Coimbra, 1988) physics. *Charge_density magnetization_density powder single_crystal_structure_determination antitumour_compound binary_alloys oxides*
E-mail ana@ciuc2.uc.pt Tel. 351(39)23675 Fax 351(39)29158 Telex 52601 DEFIUC P

de Matos Gomes, Dr Etelvina (1957). Auxiliary Professor. Departamento de Fisica, Universidade do Minho, Largo do Paço, 4719 Braga Codex, Portugal. PhD (U. Oxford, UK, 1991) physics. *Diffraction dielectrics absolute_structure optical_activity nonlinear_property phase_transition micelles*
E-mail emg@ci.uminho.pt Tel. 351(53)604320/604334 Fax 351(53)604339

Melo Jorge, Maria da Estrela Borges (1960). Assistant Lecturer. Departamento de Química, Faculdade de Ciências, Universidade de Lisboa, Campo Grande, 1700 Lisboa, Portugal. Graduation (U. Lisboa, 1984) chemistry. *Solid_solution structure_determination perovskites*
Tel. 351(1)7573141 Fax 351(1)7599404

Mirão, José António (1967). Assistant Lecturer. Departamento de Geociências, Universidade de Évora, Apartado 94, 7001 Évora, Portugal. Graduation (U. Lisboa, 1991) geology. *Crystallography geology mineralogy sulfides*
Tel. 351(66)25572/3/4 Fax 351(66)20775 Telex 18771 UNIEVR P

Morao Dias, Dr Antonio Angelo (1952). Auxiliary Professor. Departamento de Fisica, Universidade de Coimbra, 3000 Coimbra, Portugal. PhD (U. Coimbra, 1985) mechanical engineering. *Residual_stress mechanics fracture diffraction_data coating plasticity reliability*
Tel. 351(39)23675 Fax 351(39)29158 Telex 52601 DEFIUC P

Paixão, Dr José António (1965). Assistant Lecturer. Departamento de Fisica, Universidade de Coimbra, 3000 Coimbra, Portugal. PhD (U. Coimbra, 1994) physics. *Charge_density magnetization_density actinides uranium_compounds form_factor magnetism neutron_scattering*
E-mail paixao@ciuc2.uc.pt Tel. 351(39)23675 Fax 351(39)29158 Telex 52601 DEFIUC P

Pereira, Dr Manuel Francisco da Costa (1963). Assistant Lecturer. Laboratorio de Mineralogia e Petrologia, Instituto Superior Técnico, Av. Rovisco Pais, 1096 Lisboa Codex, Portugal. Graduation (U. Técnica, Lisboa, 1990) mining engineering. *Diffractometry mineralogy phosphates*
Tel. 35N(1)800111 Fax 351(1)800806 Telex 63423 ISTUTL

Prata Pina, Jose Carlos (1949). Assistant Lecturer. Departamento de Fisica, Universidade de Coimbra, 3000 Coimbra, Portugal. PhD (U. Coimbra, 1977) electrotechnical engineering. *Residual_stress coating texture diffraction_data*
Tel. 351(39)23675 Fax 351(39)29158 Telex 52601 DEFIUC P

Rebelo, Joana Cristina Fanha (1969). Research Student. Departamento de Fisica, Universidade de Coimbra, 3000 Coimbra, Portugal. Graduation (U. Coimbra, 1992) physics. *SAXS powder Rietveld_method residual_stress uranium_compounds*
Tel. 351(39)23675 Fax 351(39)29158 Telex 52601 DEFIUC P

Reis, M. Lourdes Rodrigues Pinto De Castro (1930). Auxiliary Researcher. Instituto Geológico e Mineiro, Rua da Amieira, 4465 S.Mamede de Infesta, Portugal. MSc (Instituto Superior Técnico, Lisboa, 1961) metallurgy - crystallography. *Crystallinity diffraction identification isomorphous_replacement layer line_broadening line_profile aluminosilicates carbonates chlorides clays mineralization mineralogy pollution size_effect solid_state_structural_changes gold_compounds graphites micas minerals mixed_layers phyllosilicates silicates sulfides titanium_compounds metamorphism*
Tel. 351(2)9511915 Fax 351(2)9514040

Romao, Dr Maria Joao (1955). Assistant Lecturer. Instituto de Tecnologia Química e Biologica, Apartado 127, 2780 Oeiras, Portugal. PhD (U. Técnica, Lisboa, 1989) chemistry. *Protein_crystallography*
E-mail romao@itqb.unl.pt Tel. 351(1)4426146 Fax 351(1)4428766

Silva, Manuela Ramos Marques (1971). Research Student. Departamento de Fisica, Universidade de Coimbra, 3000 Coimbra, Portugal. Graduation (U. Coimbra, 1993) physics. *Charge_density magnetization_density powder single_crystal Rietveld_method SAXS transition_elements*
E-mail manuela@ciuc2.uc.pt Tel. 351(39)23675 Fax 351(39)29158 Telex 52601 DEFIUC P

Silva, Teresa Pereira (1963). Research Assistant. Centro de Cristalografia e Mineralogia, Instituto de Investigação Tropical, Al. D. Afonso Henriques, 41-4Esq, 1000 Lisboa, Portugal. Graduation (U. Lisboa, 1990) technological chemistry. *Diffractometry phase_transition EXAFS X-ray_fluorescence ores sulfides*
Tel. 351(1)8476596 Fax 351(1)3631460 Telex IICT 66932

Veiga, João Pedro (1968). Research Student. Centro de Cristalografia e Mineralogia, Instituto de Investigação Tropical, e Cenimat/UNL, Al. D. Afonso Henriques, 41-4Esq, 1000 Lisboa, Portugal. Graduation (U. Nova de Lisboa, 1992) physical engineering. *Diffractometry phase_transitions XANES superconductors oxides*
Tel. 351(1)8476596 Fax 351(1)2957810

ROMANIA

Sub-Editor: O.-R. Strusievicz

Ardelean, Dr Ioan (1942). Professor; Director of Research Lab., Faculty of Physics. Facultatea de Fizică, Universitatea "Babeş-Bolyai", str. Kogălniceanu nr. 1, RO-3400 Cluj-Napoca, Romania. DSc (Cluj-Napoca U., 1980) physics of the vitreous state. *Education amorphous_phase oxides boron_compounds mixed_valence_compounds polycrystals paramagnetics*
E-mail ubbmail@hercule.utcluj.ro Tel. 40(64)116101x205 Fax 40(64)111905

Balaban, Dr Aneta (1944). Research assistant. Dept of Mineralogy, Institute of Geology & Geophysics, Caransebeş str. 1, RO-78344 Bucureşti 32, Romania. DSc (Bucureşti U., 1989) mineralogy-petrography. *Deformation non-destructive_analysis crystallography mineralogy metallogenesis*
Tel. 40(1)6657530x246 Fax 40(1)3128444

Barbur, Prof. Ioan (1939). Professor. Facultatea de Fizică, Universitatea "Babeş-Bolyai", str. Kogălniceanu nr. 1, RO-3400 Cluj-Napoca, Romania. DSc (Cluj-Napoca U., 1972) magnetic resonance. *Electron_spin_resonance dielectrics ferroelectricity*
E-mail uni_cluj@pi-bucuresti.th-darmstadt.de Tel. 40(64)116101x119 Fax 40(64)111905

Burzo, Prof. Emil (1935). Professor; Dean, Faculty of Physics. Facultatea de Fizică, Universitatea "Babeş-Bolyai", str. Kogălniceanu nr. 1, RO-3400 Cluj-Napoca, Romania. DSc (Timişoara U., 1969) physics of the solid state. *Magnetic_resonance electron_spin_resonance binary_alloys superconductivity copper_compounds cobalt_compounds nickel_compounds iron_compounds uranium_compounds*
E-mail ubbmail@hercule.utcluj.ro Tel. 40(64)116101x205 Fax 40(64)111905

Clulavu, Magda Cristina (1968). Research assistant. Dept of Mineralogy, Institute of Geology & Geophysics, Caransebeş str. 1, RO-78344 Bucureşti 32, Romania. BSc (Bucureşti U., 1991) geology. *Electron_microscopy microdiffraction clays phyllosilicates*
Tel. 40(1)6657530x247 Fax 40(1)3128444

Costea, Constantin (1951). Research assistant. Dept of Mineralogy, Institute of Geology & Geophysics, Caransebeş str. 1, RO-78344 Bucureşti 32, Romania. BSc (Bucureşti U., 1979) sedimentology. *Sedimentation electron_microscopy electron_probe_microanalysis*
Tel. 40(1)6657530x281 Fax 40(1)3128444

Cristea, Corina (1959). Research assistant. Dept of Mineralogy, Institute of Geology & Geophysics, Caransebeş str. 1, RO-78344 Bucureşti 32, Romania. BSc (Bucureşti U., 1984) geology. *X-ray_diffractometry mineralogy solid_solution isomorphous_replacement*
Tel. 40(1)6657530x133 Fax 40(1)3128444

Ghergari, Prof. Lucreţia (1932). Professor. Catedra de Mineralogie-Petrometalogenie, Universitatea "Babeş-Bolyai", str. Kogălniceanu nr. 1, RO-3400 Cluj-Napoca, Romania. DSc (Cluj-Napoca U., 1973) chemistry. *Crystallography mineralogy minerals clays mineralization metallogenesis*
E-mail uni_cluj@pi-bucuresti.th-darmstadt.de Tel. 40(64)116101x161 Fax 40(64)111905

Hârtopanu, Paulina (1940). Research assistant. Dept of Mineralogy, Institute of Geology & Geophysics, Caransebeş str. 1, RO-78344 Bucureşti 32, Romania. BSc (Bucureşti U., 1967) geology. *Metallogenesis paragenesis manganese_compounds iron_compounds petrography silicates aluminosilicates apatite barium_compounds titanium_compounds*
Tel. 40(1)6657530x198 Fax 40(1)3128444

Ilincea, Gheorghe (1959). Research assistant. Dept of Mineralogy, Institute of Geology & Geophysics, Caransebeş str. 1, RO-78344 Bucureşti 32, Romania. BSc (Bucureşti U., 1984) geology. *Sulfides bismuth_compounds microdiffraction electron_microscopy reflected_light_microscopy reflectivity crystallography computer_database*
Tel. 40(1)6657530x247 Fax 40(1)3128444

Macaleţ, Conf. Viorel (1947). Assistant professor. Lab. Cristalografie & Mineralogie, Universitatea Politehnică Bucureşti, Splaiul Indipendenţei nr. 313, RO-77206 Bucureşti, Romania. DSc (Iaşi U., 1985) mineralogy. *Mineralogy petrology metallogenesis*
Tel. 40(1)6314010x595

Marincea, Ştefan (1960). Research scientist. Dept of Mineralogy, Geological Institute of Romania, Caransebeş str. 1, RO-78344 Bucureşti 32, Romania. BSc (Bucureşti U., 1985) geology. *Mineralogy paragenesis diffraction_data boron_compounds diagnostic infrared differential_thermal_analysis chemistry carbonates fluorine_compounds silicates apatite*

E-mail girbhr@roearnici.ro Tel. 40(1)6657530x198 Fax 40(1)3128444 Telex 12 286 IGRR

Mureşan, Prof. Ioan (1934). Professor. Catedra de Mineralogie-Petrometalogenie, Universitatea "Babeş-Bolyai", str. Kogălniceanu nr. 1, RO-3400 Cluj-Napoca, Romania. PhD (Bucureşti U., 1971) mineralogy–petrology. *Mineralogy microscopy crystal_growth diffractometry electron_probe_microanalysis amorphous_phase gemology*
E-mail uni_cluj@pi-bucuresti.th-darmstadt.de Tel. 40(64)116101x282 Fax 40(64)111905

Petrescu, Mircea-Ionuţ (1962). Assistant. Lab. Cristalografie & Mineralogie, Universitatea Politehnică Bucureşti, Splaiul Independenţei nr. 313, RO-77206 Bucureşti, Romania. BSc (Bucureşti U., 1987) mineralogy. *Mineralogy petrology metallogenesis*
Tel. 40(1)6314010x595

Pintea, Ioan (1955). Research assistant. Dept of Mineralogy, Institute of Geology & Geophysics, Cluj-Napoca Branch, str. Clinicilor 5-7, RO-3400 Cluj-Napoca, Romania. [Institute of Geology & Geophysics, PO Box 181, RO-3400 Cluj-Napoca 1, Romania.] BSc (Cluj-Napoca U., 1982) geology. *Crystallogeny inclusion fluids geochemistry mineralogy*
Tel. 40(64)118923 Fax 40(1)3128444 (Bucureşti)

Plugaru, Neculai (1954). Research assistant. Institutul de Fizică şi Tehnologia Materialelor, Laboratorul 12, Bucureşti - Măgurele, Romania. [PO Box MG-06, Bucureşti - Măgurele, sector 5, Romania.] DSc (Cluj-Napoca U., 1993) physics of the condensed state. *Electron_spin_resonance rare-earth_compounds float_zone_growth Laue_diffraction crystallinity magnetism superstructure structural_change thermal_expansion*
Tel. 40(1)7807040x1828 Fax 40(1)3122247

Pop, Dr Viorel (1956). Assistant professor. Facultatea de Fizică, Universitatea "Babeş-Bolyai", str. Kogălniceanu nr. 1, RO-3400 Cluj-Napoca, Romania. DSc (Cluj-Napoca U., 1993) physics of the solid state. *Magnetism paramagnetics antiferromagnetism superconductivity materials boron_compounds oxides mixed_valence_compounds*
E-mail ubbmail@hercule.utcluj.ro Tel. 40(64)116101x205 Fax 40(64)111905

Robu, Lucia (1949). Research assistant. Dept of Mineralogy, Institute of Geology & Geophysics, Caransebeş str. 1, RO-78344 Bucureşti 32, Romania. BSc (Bucureşti U., 1972) geology. *Typomorphism morphology zirconium_compounds granites crystallography isomorphism polymorphism apatite titanium_compounds computer_modelling*
Tel. 40(1)6657530x198 Fax 40(1)3128444

Simon, Prof. Simion (1949). Professor. Facultatea de Fizică, Universitatea "Babeş-Bolyai", str. Kogălniceanu nr. 1, RO-3400 Cluj-Napoca, Romania. DSc (Cluj-Napoca U., 1986) magnetic resonance. *Nuclear_magnetic_resonance layered_compounds electron_spin_resonance glasses superconductivity devitrification perovskites*
E-mail uni_cluj@pi-bucuresti.th-darmstadt.de Tel. 40(64)116101x119 Fax 40(64)111905

Stelea, Gabriela (1955). Research assistant. Dept of Mineralogy, Institute of Geology & Geophysics, Caransebeş str. 1, RO-78344 Bucureşti 32, Romania. BSc (Bucureşti U., 1979) geology. *Infrared_spectrography differential_thermal_analysis isomorphism minerals*
Tel. 40(1)6657530x247 Fax 40(1)3128444

Strusievicz, Octavian-Robert (1955). Lecturer. Sub-Editor for Romania of WDC9. Catedra de Mineralogie-Petrometalogenie, Universitatea "Babeş-Bolyai", str. Kogălniceanu nr. 1, RO-3400 Cluj-Napoca, Romania. DSc (Cluj-Napoca U., 1980) geology. *Mineralogy crystal_form geochemistry metamorphic_minerals chrysotile carbonates hydroxides chromite*
E-mail uni_cluj@pi-bucuresti.th-darmstadt.de Tel. 40(64)116101x161 Fax 40(64)111905

Udubaşa, Dr Gheorghe (1938). Chief of department. Dept of Mineralogy, Institute of Geology & Geophysics, Caransebeş str. 1, RO-78344 Bucureşti 32, Romania. DSc (Heidelberg U., 1972) mineralogy. *Mineralization paragenesis crystal_growth reflected_light_microscopy mineralogy gold_compounds graphite oxides sulfides tellurides titanium_compounds dissolution hardening nonstoichiometry*
Tel. 40(1)6657530x297 Fax 40(1)3128444

Vanghelie, Iulian (1946). Research assistant. Dept of Mineralogy, Institute of Geology & Geophysics, Caransebeş str. 1, RO-78344 Bucureşti 32, Romania. BSc (Bucureşti U., 1969) chemistry. *Mineralogy structural_crystallography X-ray_diffraction*
Tel. 40(1)6657530x133 Fax 40(1)3128444

RUSSIA

Sub-Editor: A. G. Vigdorchik

Abdullaev, Dr Abdulhamid A. (1936). Senior Researcher. Lab. of Acoustooptics of Crystals, Institute of Crystallography, Russian Academy of Sciences, Leninsky pr. 59, 117333 Moscow, Russia. PhD (Inst. of Crystallography Moscow, 1972) physics. *Crystal_growth optical_property*

Tel. 7(095)3306883 Fax 7(095)3301956

Abramov, Dr Yuri Alekseevich (1960). Lecturer. Physics Department, Mendeleev University of Chemical Technology of Russia, 9 Miusskaja Square, Moscow 125190, Russia. PhD (Karpov Inst. of Physical Chemistry, 1993) physical chemistry. *X-ray_crystallography charge_density chemical_bonding crystal_chemistry physical_properties_of_crystals*

E-mail abramov@mhti.msk.su Tel. 7(095)2585930 Fax 7(095)2004204 Telex 411744 argon

Abrosimova, Dr Galina E. Senior Researcher. Institute of Solid State Physics, Russian Academy of Sciences, Chernogolovka, Moscow Region 142432, Russia. [Institute of Solid State Physics, Chernogolovka, Moscow distr., 142432, Russia.] PhD (Moscow Inst. Steel and Alloys, 1982) solid state physics. *X-ray_crystallography phase amorphous_crystalline_structure amorphous_metallic_alloys_quasicrystals*

E-mail gabros@issp.ac.ru Tel. 7(095)5245063 Fax 7(096)5171949 Telex 412654 SERNA SU

Afanas'eva, Dr Irina N. (1968). Junior Researcher. Lab. of X-ray Diffraction Studies, Department of Solid State Physics and Chemistry, Lomonosov Academy of Fine Chemical Technology, 86 pr. Vernadskogo, Moscow 117571, Russia. MSc (Institute of Fine Chemical Technology, 1992) physical chemistry. *X-ray_diffractometry rare-earth_compounds powder_phase_formation inorganic_crystal_chemistry*

E-mail alekz@cs.msk.su alekz@red.com.ru Tel. 7(095)2480762 Fax 7(095)4307983

Afonina, Dr Galina G. (1939). Senior Researcher. Group of X-ray Structure Analysis, Institute of Geochemistry, Siberian Branch, Russian Academy of Sciences, 1a Favorsky Str., Irkutsk 664033, Russia. PhD (Irkutsk State U., 1972) physics. *Powder_diffraction silicon_compounds isomorphism unit_cell_structural_changes temperature_typomorphism*

E-mail dor@crust.irkutsk.su Tel. 7(3952)464953 Fax 7(3952)462952

Akchurin, Dr Marat Sh. (1947). Senior Researcher. Lab. of Mechanical Properties of Crystals, Institute of Crystallography, Russian Academy of Sciences, 59 Leninskii pr., 117333 Moscow, Russia. PhD (Inst. of Crystallography Moscow, 1983) solid state physics. *Scanning_electron_microscopy real_crystal mechanical_property dislocation X-ray_fluorescence_spectroscopy ESCA vacuum*

E-mail public@mechan.incr.msk.su Tel. 7(095)3308274 Fax 7(095)3308274

Akhmetov, Dr Spartak F. (1937). Chief Researcher. Lab. of Physical and Technical Research, Russian Research Institute for Synthesis of Minerals and Raw Materials, Institutskaya St. 1, Alexandrov, Vladimir Region 601600, Russia. PhD (Kazakh Polytechnic Inst., 1966) physical chemistry. *X-ray_diffractometry crystal_optics mineralogy*

Tel. 7(095)5845834 Fax 7(095)5845828

Akhmetova, Dr Galina L. (1937). Chief Researcher. Lab. of Physical and Technical Research, Russian Research Institute for Synthesis of Minerals and Raw Materials, Institutskaya St. 1, Alexandrov, Vladimir Region 601600, Russia. PhD (Kazakh Inst. of Metallurgy, 1966) physical chemistry. *X-ray_diffraction_technique mineralogy DTA*

Tel. 7(095)5845834 Fax 7(095)5845828

Alaudinov, Mr Bagomed M. (1960). Researcher. Lab. of High Temperature Crystallization, Institute of Crystallography, Russian Academy of Sciences, 59 Leninsky prosp., Moscow 117333, Russia. MSc (Daghestan State U., 1987) physics. *Crystal_growth high_temperature_crystallization X-ray_diffraction*

Tel. 7(095)1351220 Fax 7(095)1356350 or 7(095)1351011

Aldoshin, Prof. Sergei M. (1953). Head of Laboratory; Deputy Director of the Institute of Chemical Physics. Lab. of Structural Chemistry, Institute of Chemical Physics, Russian Academy of Sciences, Chernogolovka, Moscow Region 142432, Russia. DSc (Inst. Chemical Physics RAN, 1985) physical chemistry. *X-ray_crystallography organic_compound organic_crystal_chemistry molecular_structure*

E-mail sma@ich.sherna.msk.su Tel. 7(095)5245076 Fax 7(096)5153588

Aleksandrov, Prof. Kirill S. (1931). Director of Institute. Laboratory of Crystal Physics, Institute of Physics, Siberian Dept., Russian Academy of Sciences, Krasnoyarsk 660036, Russia. DSc (Inst. of Crystallography Moscow, 1967) crystallography and crystal physics. *Structural_phase-transition inorganic_compound elastic_property crystal_chemistry ferroelectricity*

E-mail aleks@iph.krasnoyarsk.su Tel. 7(391)2432635 Fax 7(391)2438923 Telex 288144 kif su

Aleksandrov, Dr Valentin V. (1959). Senior Researcher. Chair of Polymers and Crystal Physics, Physics Department, Moscow State University, Moscow 117234, Russia. PhD (Moscow State U., 1986) solid state physics. *Brillouin_spectroscopy surface_acoustics superconductor_films superlattice semiconductor superconductors physical_properties*

E-mail aleks@cryst0.phys.msu.su Tel. 7(095)9391430 Fax 7(095)9392988 Telex 411483 MGU SU

Aleksandrova, Dr Inga P. (1934). Head of Laboratory. Lab. for Radiospectroscopy of Dielectrics, Institute of Physics, Siberian Dept., Russian Academy of Sciences, Krasnoyarsk 660036, Russia. DSc (Inst. of Crystallography Moscow, 1987) solid state physics. *Structural_phase_transitions inorganic_compound_structure incommensurate_phase ferroelectricity nuclear_magnetic_resonance neutron_X-ray_diffraction*

E-mail aleks@iph.krasnoyarsk.su Tel. 7(391)2432635 Fax 7(391)2438923

Aleshko-Ozhevsky, Dr Oleg P. (1934). Chief Researcher. Lab. of X-ray Optics and Synchrotron Radiation, Institute of Crystallography, Russian Academy of Sciences, 59 Leninsky pr., Moscow 117333, Russia. DSc (Inst. of Crystallography Moscow, 1993) physics. *X-ray_topography X-ray_optics neutron_scattering*

E-mail users@crystal.msk.su Tel. 7(095)3342913 Fax 7(095)3301956

Alexeev, Mr Ilia V. (1966). Junior Researcher. Lab. of Elementary Processes of Crystal Growth, Institute of Crystallography, Russian Academy of Sciences, Leninsky pr. 59, Moscow 117333, Russia. MSc (Moscow State U., 1989) solid state physics. *Crystal_growth solid_state_physics computer_modelling*

E-mail alex@elprgr.crystal.msk.su Tel. 7(095)1354240 Fax 7(095)1351011

Aliev, Dr Zainutdin G. (1939). Senior Researcher. Branch Institute of Chemical Physics, Russian Academy of Sciences in Chernogolovka, Chernogolovka, Moscow Region 142432, Russia. PhD (Inst. Applied Physics Moldova, 1975) physics. *X-ray_crystallography inorganic_organic_crystal structure-activity_relationship*

E-mail atov@icph.sherna.msk.su Tel. 7(096)5171168 Fax 7(096)5153588

Alshits, Prof. Vladimir I. (1941). Head of Laboratory. Lab. of Mechanical Properties of Crystals, Institute of Crystallography, Russian Academy of Sciences, 59 Leninsky pr, 117333 Moscow, Russia. DSc (Inst. of Crystallography Moscow, 1977) solid state physics. *Theoretical_acoustics dislocation_theory crystal_defect solid_state_mechanics*

E-mail alshits@mechan.incr.msk.su Tel. 7(095)3308274 Fax 7(095)3308274

Anderson, Dr Alexandra A. (1947). Researcher. Lab. of X-ray Crystallography, Institute of Silicate Chemistry, Russian Academy of Sciences, 24-2 Odoevskogo str., St Petersburg 199155, Russia. PhD (Inst. Silicate Chemistry St Petersburg, 1989) physical chemistry. *X-ray_crystal_structure_determination silicates zeolites sorption phase_transition*

E-mail anderson@ihs2.spb.su Tel. 7(812)2185102 Fax 7(812)3510813 Telex 121447 VITRO SU

Andreeva, Dr Marina A. Senior Researcher. Solid State Chair, Physics Department, Moscow State University, Vorobjovy Gory, Moscow 117234, Russia. DSc (Belorussy State U., 1992) theoretical physics. *X-ray_diffraction Mossbauer_diffraction multilayer_structure surface_structure*

E-mail asi@phys.msu.su. Tel. 7(095)9391226 Fax 7(095)9390247 Telex 411483 MGU SU

Andrianova, Dr Maria E. (1937). Senior Researcher. Laboratory of X-ray Diffractometry, Institute of Crystallography, Russian Academy of Sciences, 59 Leninsky prosp., Moscow 117333, Russia. PhD (Inst. of Crystallography Moscow, 1986) crystallography and crystal physics. *X-ray_diffractometry area_detector*

E-mail CRYS@CRYS.MSK.SU Tel. 7(095)1354210 Fax 7(095)1351011

Anisimova, Dr Vera N. (1947). Researcher. Laboratory of Electromechanical Investigations of Crystals, Institute of Crystallography, Russian Academy of Sciences, Leninsky pr. 59, Moscow 117333, Russia. PhD (Inst. of Crystallography Moscow, 1986) physics. *Ferroelectricity phase_transition dielectrics specific_heat thermal_property defect*

E-mail postmaster@el-mech.incr.msk.su Tel. 7(095)3307856 Fax 7(095)1351011

Antipin, Prof. Mikhail Yu. (1951). Senior Researcher. X-ray Structural Center of the Russian Academy of Sciences, Institute of Organoelement Compounds, Russian Academy of Sciences, 28 Vavilov St., Moscow 117813, Russia. DSc (Inst. Organoelement Compounds, 1989) physical chemistry. *Charge_density thermal_motion low_temperature molecular_crystal*

E-mail mishan@xray.ineos.ac.ru Tel. 7(095)1359215 Fax 7(095)1355085

Antonov, Dr Evgenii V. (1941). Researcher. Lab. of High Temperature Crystallization, Institute of Crystallography, Russian Academy of Sciences, 59 Leninsky prosp., Moscow 117333, Russia. PhD (Inst. of General Physics Moscow, 1992) physics. *Crystal_growth high_temperature_crystallization optical_property Bridgman_Stockbarger_technique real_structure*

Tel. 7(095)1351400 Fax 7(095)1356350 or 7(095)1351011

Antsishkina, Dr Alla A. (1926). Senior Researcher. Lab. of Coordination Compounds Crystal Chemistry, Kurnakov Institute of General and Inorganic Chemistry, Russian Academy of Sciences, 31 Leninsky prosp., Moscow 117907, Russia. PhD (Inst. General and Inorganic Chemistry, 1959) inorganic chemistry. *X-ray_crystallography coordination_compound carboxylates complexonates*

Tel. 7(095)9521803 Fax 7(095)9541279

Arakcheeva, Dr Alla V. (1954). Senior Researcher. Baikov Institute of Metallurgy, Russian Academy of Sciences, Laboratory of Crystallochemistry, 49 Leninskyi pr., Moscow 117334, Russia. PhD (Institute of Crystallography Russian Academy of Sciences Moscow, 1988) chemistry. *X-ray_crystallography structure_determination complex_inorganic_compound superconductors phase_determination phase_equilibrium modulated_structures polysomatic_structures intergrowth_structures minerals ferrites*

Tel. 7(095)1359609 Fax 7(095)1358680

Aronin, Dr Alexander S. (1952). Senior Researcher. Institute of Solid State Physics, Russian Academy of Sciences, Chernogolovka, Moscow Region 142432, Russia. [Institute of Solid State Physics, Chernogolovka, Moscow distr. 142432, Russia.] PhD (Inst. Steel and Alloys Moscow, 1982) solid state physics. *Electron_microscopy amorphous_metastable_phase_determination* disorder_order_transformation
E-mail aronin@issp.ac.ru Tel. 7(096)5171637 Fax 7(096)5171949 Telex 412654 SERNA SU

Asadchikov, Dr Victor E. (1948). Senior Researcher. Laboratory of Nuclear Filters, Institute of Crystallography, Russian Academy of Sciences, 59 Leninsky pr., Moscow 117333, Russia. PhD (Inst. of Crystallography Moscow, 1983) physics. *SAXS reflectivity X-ray_diffractometer surface_study*
Tel. 7(095)1359917 Fax 7(095)1351011

Askhabov, Dr Askhab M. (1948). Laboratory Head. Lab. of Experimental Mineralogy, Institute of Geology, Komi Scientific Centre, Ural Department, Russian Academy of Sciences, 54 Pervomaiskaya Str., Syktyvkar 167000, Russia. [178 Apt. 28, Oktyabrsky Prosp., Syktyvkar 167005, Russia.] DSc (St Petersburg State U., 1989) geology and mineralogy. *Crystallography minerals crystal_growth*
Tel. 7(82122)25167 Fax 7(82122)25346

Aslanov, Prof. Leonid Aleksandrovich (1938). Head of laboratory. Member of Commission on Crystallographic Teaching. Lab. Structural Chemistry, General Chemistry Faculty, Chemistry Dept., Moscow State University, Moscow 119899, Russia. DSc (Kurnakov Institute of General and Inorganic Chemistry, 1973) physical chemistry. *Crystallography inorganic_crystal_chemistry instrumentation solid_state_chemistry* teaching philosophy history
E-mail Aslanov@struct.chem.msu.su Tel. 7(095)9395089 Fax 7(095)9390898 Telex 411483 MGU SU

Astaf'ev, Dr Serge B. (1961). Researcher. Laboratory of Crystallophysics, Institute of Crystallography, Russian Academy of Sciences, 59 Leninskii pr., Moscow 117333, Russia. PhD (Inst. of Crystallography Moscow, 1989) solid state physics. *Photorefraction bulk_photovoltaic_kinetics photoconductivity* computer_modelling computer_automation computer_graphics
E-mail root@theory.incr.msk.su Tel. 7(095)1350251 Fax 7(095)1351011

Atovmyan, Prof. Lev O. (1928). Head of Laboratory. Branch Institute of Chemical Physics, Russian Academy of Sciences in Chernogolovka, 142432 Chernogolovka, Moscow Region 142432, Russia. DSc (Inst. of General and Inorganic Chemistry Moscow, 1972) physical chemistry. *X-ray_crystallography molecular_crystal organometallic_crystal* fast_chemical_reaction solid_electrolytes energetic_compounds magnetic_compounds
E-mail atov@icph.sherna.msk.su Tel. 7(096)5171168 Fax 7(096)5153588

Avilov, Dr Anatoly S. (1943). Senior Researcher. Laboratory of Electron Diffraction, Institute of Crystallography, Russian Academy of Sciences, 59 Leninsky prosp., Moscow 117333, Russia. PhD (Inst. of Crystallography RAS Moscow, 1973) physics. *High-energy_electron_diffraction diffractometry dynamical_diffraction structure_determination* amorphous_phase
Tel. 7(095)1354010

Bagdasarov, Dr Khatchik S. (1929). Head of Laboratory. Lab. of High Temperature Crystallization, Institute of Crystallography, Russian Academy of Sciences, 59 Leninsky prosp., Moscow 117333, Russia. DSc (Inst. of Crystallography Moscow, 1972) physics. *Crystal_growth high_temperature_crystallization laser_radiation* dielectrics_microanalysis real_structure
Tel. 7(095)1356350 Fax 7(095)1356350 or 7(095)1351011

Bakakin, Dr Vladimir V. (1933). Chief Researcher. Lab. of Crystal Chemistry, Institute of Inorganic Chemistry Siberian Branch, Russian Academy of Sciences, 3 Lavrentyev prosp., Novosibirsk 630090, Russia. PhD (Moscow State U., 1963) structural mineralogy. *Inorganic_crystal_chemistry classification* structural_database structural_mineralogy solid_state_chemistry
E-mail borisov@che.nsk.su Tel. 7(3832)354366 Fax 7(3832)355960

Balagurov, Dr Anatoly M. (1945). Department Leader. Frank Lab. of Neutron Physics, Joint Institute for Nuclear Research, 141980 Dubna, 141980 Moscow region, Russia. DSc (Joint Inst. for Nuclear Research Dubna, 1989) physics. *Neutron_powder_diffraction single_crystal*
E-mail bala@nfsun7.jinr.dubna.su Tel. 7(09621)65803 Fax 7(09621)65085 Telex 911621 DUBNA SU

Balakirev, Dr Vladimir G. (1937). Chief Researcher. Lab. of Physical and Technical Research, Russian Research Institute for Synthesis of Minerals and Raw Materials, Institutskaya St. 1, Alexandrov, Vladimir Region 601600, Russia. PhD (Inst. of Geology Mineralogy and Petrography of Ore Minerals Moscow, 1977) crystallography. *Electron_microscopy solid_state_physics twinning*
Tel. 7(095)5845834 Fax 7(095)5845828

Balyunis, Dr Lyubov' E. (1955). Senior Researcher. Phase Transition Lab., Institute of Physics, Rostov State University, Pr. Stachki 194, Rostov-on-Don 344104, Russia. [2-nd Krasnodarskaya Str. 163/3, Apt 51, Rostov-on-Don 344102, Russia.] PhD (Rostov State U., 1984) physics and mathematics. *Crystal_structure phase_transition single_crystal domain_structure ferroelectrics_and_related_materials*
E-mail Balyunis@riphys.rnd.su Tel. 7(8632)285066 Fax 7(8632)285044

Bannova, Ms Irina I. (1948). Senior Researcher. Department of Crystallography and Crystallography Computing, Joint-Stock Company "Ajax", 55 Galernaja Str., St Petersburg 190000, Russia. MSc (Leningrad State U., 1972) crystallography. *X-ray_structure_determination crystal_chemistry inorganic_organic_compound*
Tel. 7(812)3125208 Fax 7(812)3122479

Barabanenkov, Dr Yury A. (1958). Researcher. Lab. of Mechanical Properties of Crystals, Institute of Crystallography, Russian Academy of Sciences, 59 Leninsky prosp., Moscow 117333, Russia. PhD (Inst. of Crystallography Moscow, 1993) solid state physics. *Crystal_defect electron_microscopy structural_chemistry complex_oxides*
E-mail public@mechan.incr.msk.su. Tel. 7(095)3307883 Fax 7(095)1351011

Baranov, Prof. Anatolii I. (1947). Chief Researcher. Laboratory of Crystallophysics, Institute of Crystallography, Russian Academy of Sciences, 59 Leninski pr., Moscow 117333, Russia. DSc (Inst. of Crystallography Moscow, 1993) solid state physics. *Phase_transition hydrogen_bonding superionic_compound ferroelectricity glasses polymers*
E-mail root@theory.incr.msk.su Tel. 7(095)1354020 Fax 7(095)1351011

Baranskii, Prof. Konstantin N. (1921). Professor. Chair of Polymer and Crystal Physics, Physics Department, Moscow State University, Moscow 119899, Russia. DSc (Physics Dept. Moscow State U., 1982) crystallography and crystal physics. *Piezoelectric_acoustic_physics electrodynamics_of_elastic_waves_in_piezoelectrics*
E-mail aleks@cryst0.phys.msu.su Tel. 7(095)9391013 Fax 7(095)9390247 Telex 411483 MGU SU

Barsukova, Dr Marina L. (1943). Researcher. Laboratory of High-Temperature Crystallization, Inst. of Crystallography, Russian Academy of Sciences, 59 Leninsky prosp., Moscow 117333, Russia. PhD (Inst. of Crystallography Moscow, 1980) crystallography and crystal physics. *Crystal_growth aqueous_solution phosphates real_structure potassium_dihydrogen_phosphate impurity_additives laser_induced_damage_threshold*
Tel. 7(095)3301792 Fax 7(095)1351011

Batsanov, Dr Andrei S. (1955). Senior Researcher. X-ray Structural Center of the Russian Academy of Sciences, Institute of Organoelement Compounds, Russian Academy of Sciences, 28 Vavilov St., Moscow 117813, Russia. PhD (Inst. Organoelement Compounds, 1983) physical chemistry. *X-ray_crystallography small_molecules organic_organometallic_crystal_chemistry*
E-mail andrei@xray.ineos.ac.ru Tel. 7(095)1359214 Fax 7(095)1355085

Bekrenev, Dr Anatoly V. (1946). Professor. Samara Polytechnic Institute, Pervomaiskaya St. 18, Samara 443002, Russia. DSc (Kharkov U., 1971) physics and mathematics. *Crystal_structure crystal_defect*

Belikova, Dr Galina S. Senior Researcher. Laboratory of Crystal Growth from Solutions, Institute of Crystallography, Russian Academy of Sciences, 59 Leninsky pr., Moscow 117333, Russia. PhD (Inst. of Crystallography Moscow, 1968) chemistry. *Crystal_growth organic_compound*
Tel. 7(095)3307883 Fax 7(095)1351011

Belimenko, Dr Lyudmila D. (1947). Chief Researcher. Lab. of Physical and Technical Research, Russian Research Institute for Synthesis of Minerals and Raw Materials, Institutskaya St. 1, Alexandrov, Vladimir Region 601600, Russia. PhD (Moscow State U., 1983) crystallography. *X-ray_topography diffractometry crystal_dislocation*
Tel. 7(095)5845834 Fax 7(095)5845828

Belokoneva, Dr Elena L. (1945). Senior Researcher. Department of Crystallography and Crystal Chemistry, Faculty of Geology, Moscow State University, Moscow 119899, Russia. PhD (Moscow State U., 1975) crystallography and crystal physics. *X-ray_crystallography minerals materials charge_density isomorphism polytypism phase_transitions structure_property_relationship*
E-mail mill%plm.phys.MSU.SU@neotext.ca Tel. 7(095)9393850 Fax 7(095)9390126 Telex 411483 MGU SU

Belov, Dr Alexander Yu. (1959). Researcher. Lab. of Crystallophysics, Institute of Crystallography, Russian Academy of Sciences, 59 Leninsky pr. 59, Moscow 117333, Russia. PhD (Inst. of Crystallography Moscow, 1987) solid state physics. *Crystal_defect dislocation lattice_distortion* plasticity fracture nanocrystal_structure_defect
E-mail root@theory.incr.msk.su Tel. 7(095)1356240 Fax 7(095)1351011

Belova, Dr Elizaveta Nikolayevna. Senior Researcher. Institute of Crystallography, Russian Academy of Sciences, 59 Leninsky prosp., Moscow 117333, Russia. PhD (Inst. of Crystallography Moscow, 1949) physics and mathematics. *Structural_analysis*
Tel. 7(095)1356140 Fax 7(095)1351011

Belugina, Dr Natalija B. (1941). Researcher. Laboratory of Electromechanical Investigations of Crystals, Institute of Crystallography, Russian Academy of Sciences, Leninsky pr. 59, Moscow 117333, Russia. PhD (Inst. of Crystallography Moscow, 1978) physics. *Ferroelectricity micromorphology defect domain_structure*
Tel. 7(095)3307856 Fax 7(095)1351011

Bendeliani, Dr Nikolai A. (1935). Group leader. Lab. of Monocrystals, Vereschagin Institute of High Pressure Physics, Troitsk, Moscow Reg. 142092, Russia. DSc (Moscow State U., 1982) inorganic chemistry. *X-ray_crystallography phase_transition crystal_growth chemistry*
E-mail scorpion@adonis.iasnet.com Tel. 7(095)3340733 Fax 7(095)3340012

Berezhkova, Dr Galina V. (1933). Chief Researcher. Lab. of Mechanical Properties of Crystals, Institute of Crystallography, Russian Academy of Sciences, 59 Leninskii pr., 117333 Moscow, Russia. PhD (Inst. of Crystallography Moscow, 1964) solid state physics. *Strength_plasticity_physics plastic_deformation superplasticity*
E-mail public@mechan.incr.msk.su Tel. 7(095)3342483 Fax 7(095)3308274

Beskrovnyi, Dr Anatoly I. (1945). Senior Researcher. Frank Lab. of Neutron Physics, Joint Institute for Nuclear Research, 141980 Dubna, 141980 Moscow region, Russia. PhD (Joint Inst. for Nuclear Research Dubna, 1989) solid state physics. *Atomic_structure superconductor*
E-mail beskr@nfsun1.jinr.dubna.su Tel. 7(095)9243914 Fax 7(09621)65085 Telex 911621 DUBNA SU

Betsofen, Dr Sergei Ya. (1946). Senior Researcher. Baikov Institute of Metallurgy, Russian Academy of Sciences, Laboratory of Crystallochemistry, 49 Leninskii pr., Moscow 117334, Russia. PhD (Institute of Aircraft Technology Moscow, 1979) physical metallurgy. *X-ray_crystallography thin_film texture strain_hardening intermetallic_compound titanium_alloys nickel_compounds*
Tel. 7(095)1359617 Fax 7(095)1358680

Beznosikov, Dr Boris V. (1930). Senior Researcher. Lab. of Crystal Physics, Institute of Physics, Siberian Dept., Russian Academy of Sciences, Krasnoyarsk 660036, Russia. PhD (Inst. of Physics Krasnoyarsk, 1978) solid state physics. *Structural_phase_transitions inorganic_compound crystal_chemistry*
E-mail aleks@iph.krasnoyarsk.su Tel. 7(391)2432635 Fax 7(391)2438923

Blinov, Dr Lev M. (1939). Laboratory Head. Lab. of Liquid Crystals, Institute of Crystallography, Russia Academy of Sciences, 59 Leninsky Pr., Moscow 117333, Russia. DSc (Inst. Crystallography Moscow, 1977) physics. *Liquid_crystal structure_property*
Tel. 7(095)3307847 Fax 7(095)1351011

Blistanov, Prof. Alexander A. (1939). Head of Department. Department of Crystallography, Institute of Steel and Alloys, 4 Leninsky prosp., Moscow 117936, Russia. DSc (Inst. of Steel and Alloys Moscow, 1972) solid state physics. *Crystal_growth defect physical_property crystallography*
Tel. 7(095)2366500 Fax 7(095)2378007

Bochkarev, Dr Alexey V. (1958). Researcher. Diffractometry Group, Engelhardt Institute of Molecular Biology, Russian Academy of Sciences, 32 Vavilov Str., Moscow 117984, Russia. PhD (Engelhardt Inst. Molecular Biology, 1992) molecular biology. *X-ray_crystallography proteins AIDS_inhibitors computer_networking*
E-mail boch@imb.mb.free.msk.su Tel. 7(095)1359944 Fax 7(095)9382187 Telex MOLBI SU

Bokiy, Prof. Georgii B. (1909). Counsellor of Director. Laboratory of X-ray Structure Analysis, Institute of Geology of Ore Deposits, Petrography, Mineralogy and Geochemistry, Russian Academy of Sciences, Staromonetny per. 35, Moscow 109017, Russia. DSc (Inst. of General and Inorganic Chemistry RAS, 1942) physical chemistry. *Systematics classification X-ray_crystallography minerals inorganic_crystal_chemistry*
E-mail ves@igem.msk.su Tel. 7(095)2308255 Fax 7(095)2302179

Bolotina, Ms Nadezhda B. (1949). Researcher. Laboratory of X-ray Structure Analysis, Institute of Crystallography, Russian Academy of Sciences, 59 Leninsky pr., Moscow 117333, Russia. MSc (Moscow State Pedagogical U., 1972) mathematics. *X-ray_analysis X-ray_diffraction_technique modulated_structure*
E-mail bolotina@rsa.crystal.msk.su Tel. 7(095)1353400 Fax 7(095)1351011

Bondarenko, Dr Vladimir I. (1956). Researcher. Institute of Crystallography, Russian Academy of Sciences, Leninsky pr. 59, Moscow 117333, Russia. PhD (Moscow Physical Engineering Institute, 1987) nuclear physics. *High_resolution_electron_microscopy software*
Tel. 7(095)1355020 Fax 7(095)1351011

Bondareva, Dr Olga S. (1948). Researcher. Lab. of Physical and Chemical Analysis, Institute of Crystallography, Russian Academy of Sciences, 59 Leninsky pr., Moscow 117333, Russia. PhD (Moscow State U., 1979) crystal chemistry. *X-ray_powder_diffraction fluorine_compounds hydrothermal_synthesis silicates_of_rare_elements*
Tel. 7(095)3307874 Fax 7(095)1351011

Borisov, Prof. Stanislav V. (1930). Head of Laboratory of Crystal Chemistry. Lab. of Crystal Chemistry, Institute of Inorganic Chemistry Siberian Branch, Russian Academy of Sciences, 3 Lavrentyev prosp., Novosibirsk 630090, Russia. DSc (Inst. Crystallography Russian Academy of Sciences, 1974) crystallography and crystal physics. *X-ray_crystallography oxides fluorides classification database structural_computing inorganic_crystal_chemistry*
E-mail borisov@che.nsk.su Tel. 7(3832)354366 Fax 7(3832)355960

Borodin, Dr Vadim L. (1955). Chief Researcher. Lab. for Synthesis of Materials by Hydrothermal Method, Russian Research Institute for Synthesis of Minerals and Raw Materials, Institutskaya St. 1, Alexandrov, Vladimir Region 601600, Russia. PhD (Moscow State U., 1980) crystallography. *Crystal_growth hydrothermal_synthesis crystal_chemistry*
Tel. 7(095)5845834 Fax 7(095)5845828

Bruskov, Dr Valery A. (1953). Technical Director. Department of Crystallography and Crystallography Computing, Joint-Stock Company "Ajax", 55 Galernaja Str., St Petersburg 190000, Russia. PhD (Lvov State U., 1985) chemistry. *X-ray_structure_determination crystal_chemistry inorganic_intermetallic_compound*
Tel. 7(812)3125208 Fax 7(812)3122479

Bubnova, Dr Rimma S. (1951). Researcher. Laboratory of Phase Equilibria, Institute of Silicate Chemistry, Russian Academy of Sciences, 2 Makarov Nab., St Petersburg 199034, Russia. PhD (Inst. of Silicate Chemistry RAS, 1987) physical chemistry. *Crystal_chemistry phase_equilibrium structure_determination X-ray_high-temperature_powder_diffraction*
Tel. 7(812)2188579

Bukin, Dr Alexander S. (1947). Senior Researcher. Lab. for Investigation of Rock-Forming Minerals by Physical Methods, Geological Institute, Russian Academy of Sciences, Pyzhevsky per. 7, Moscow 109017, Russia. PhD (Moscow State U., 1975) physics and mathematics. *Polytypism isomorphism layered_silicates*
Tel. 7(095)2308124 Fax 7(095)2318106

Bukvetsky, Dr Boris V. (1944). Senior Researcher. Institute of Chemistry, Far East Scientific Center, Russian Academy of Sciences, Pr. of the 10th Anniv. of Vladivostok 159, 690022 Vladivostok, Russia. PhD (Institute of Crystallography Moscow, 1977) physics and mathematics. *Crystal_chemistry phase_transition inorganic_compound hydrogen_bonding*
Tel. 7(432)2296590

Bulka, Dr Genrikh R. (1944). Associate Professor; Head, Lab. for Physics of Minerals and Analogs, Sci. Res. Dept. Kazan U.. Dept. of Geology, Kazan State University, 18 Lenin Str., Kazan 420008, Russia. PhD (Kazan State U., 1980) physics. *Structural_crystallography crystal_defect magnetic_resonance sedimentology mineralogy crystal_growth*
E-mail root@scikgu.kazan.su Tel. 7(8432)327062 Fax 7(8432)380994 Telex 224641 CHAIR SU

Bunina, Dr Olga A. (1954). Senior Researcher. Lab. of Crystal Physics, Institute of Physics, Rostov State University, Pr. Stachky 194, Rostov-on-Don 344104, Russia. PhD (Rostov State U., 1982) physics and mathematics. *Crystal_structure ferroelectrics phase_transitions high-Tc_superconductivity*
E-mail irina-z@riphys.rnd.su Tel. 7(8632)285066 Fax 7(8632)285044 Telex 123321, 123610 FIZIK RU

Burnasheva, Dr Veniana V. (1940). Researcher. Institute of New Chemical Problems, Russian Academy of Sciences, Chernogolovka, Moscow Region 142432, Russia. PhD (Lvov State U., 1973) chemistry. *Crystal_structure intermetallic_compound hydrides*

Bushuev, Prof. Vladimir A. (1947). Professor. Solid State Chair, Physics Department, Moscow State University, Vorobjovy Gory, Moscow 117234, Russia. DSc (Moscow State U., 1990) solid state physics. *X-ray_dynamical_diffraction Compton_scattering superstructure X-ray_diffuse_scattering computer_modelling*
E-mail asi@phys.msu.su. Tel. 7(095)9391226 Fax 7(095)9390247 Telex 411483 MGU SU

Butashin, Dr Andrey V. (1955). Senior Researcher. Lab. of Laser Crystals Physics, Institute of Crystallography, Russian Academy of Sciences, 59 Leninsky prosp., Moscow 117333, Russia. PhD (Moscow State U., 1986) crystallography and physics of crystals. *Czochralski_technique oxides crystal_growth optical_laser_crystals search growth*
E-mail boutic@plm.phys.msu.su Tel. 7(095)1356310

Butikova, Dr Irina K. (1939). Researcher. Lab. of X-ray Crystallography, Institute of Silicate Chemistry, Russian Academy of Sciences, 24-2 Odoevskogo str., St Petersburg 199155, Russia. PhD (Inst. Silicate Chemistry St Petersburg, 1987) physical chemistry. *X-ray_crystal_structure_determination silicates zeolites sorption phase_transition*
E-mail butikova@ihs2.spb.su Tel. 7(812)2185102 Fax 7(812)3510813 Telex 121447 VITRO SU

Chashchinov, Dr Yury M. (1939). Associate Professor. Chair of Mineralogy, Geological Department, St Petersburg Mining Institute, 2 21st Liniya, St Petersburg 199026, Russia. PhD (St Petersburg Mining Inst., 1972) crystal growth. *Crystal_growth crystal_defect crystal_chemistry powder_diffraction*
Tel. 7(812)2188266 Fax 7(812)2132613 Telex 121494 LGIPSU

Chernov, Prof. Alexander A. (1931). Laboratory Head. Lab. of Elementary Processes of Crystal Growth, Institute of Crystallography, Russian Academy of Sciences, Leninsky pr. 59, Moscow 117333, Russia. DSc (Inst. of Crystallography Moscow, 1970) physics. *Crystal_growth surface_research solid_state_physics*
E-mail chern@elprgr.crystal.msk.su Tel. 7(095)1354240 Fax 7(095)1351011

Chernov, Dr Alexander N. (1944). Associate Professor. Dzerzhinsk Branch, Moscow Inst. of Qualification Improvement, Gagarin St. 3, Dzerzhinsk, Nizhny Novgorod Region 606000, Russia. PhD (Inst. of Crystallography Moscow, 1970) physics and mathematics. *Crystal_chemistry organic_compound metalloorganic_compound*
Tel. 7(8313)52630

Chernyshev, Dr Vladimir Vasilievich (1955). Senior researcher. Lab. Structural Chemistry, General Chemistry Faculty, Chemistry Dept., Moscow State University, Moscow 119899, Russia. PhD (Moscow State U., 1988) physics and mathematics. *Crystallography charge_density time-of-flight_diffraction Rietveld_method philosophy*
E-mail Chern@crysma.rc.ac.ru Tel. 7(095)9395089 Fax 7(095)9390898 Telex 411483 MGU SU

Chernysheva, Dr Marina A. (1911). Researcher. Lab. of Mechanical Properties of Crystals, Institute of Crystallography, Russian Academy of Sciences, 59 Leninsky pr., 117333 Moscow, Russia. PhD (Inst. of Crystallography Moscow, 1955) crystallography and crystal physics. *Domain_structure ferroelectric_crystal real_crystal polarization_optical_technique*
E-mail public@mechan.incr.msk.su Tel. 7(095)3342483 Fax 7(095)3308274

Chevichelov, Mr Victor A. (1950). Leading Engineer. Lab. for Structural Studies, Obninsk Branch, Karpov Institute of Physical Chemistry, Obninsk, Kaluga Region 249020, Russia. MSc (Moscow Inst. of Engineering and Physics, 1979) physics. *Inorganic_compound neutron_powder_diffraction SAXS_polymers*
Tel. 7(08439)26531

Chirgadze, Prof. Yuri Nickolaevich (1935). Head of Laboratory. Member of Russian National Committee of Crystallography. Lab. of Protein Structure Analysis, Institute of Protein Research, Russian Academy of Sciences, Pushchino, Moscow Region 142292, Russia. DSc (1977, Moscow State U.) molecular biology. *Structural_biology proteins_crystallography_structure*
E-mail protres@sovam.com Tel. 7(095)9240493 Fax 7(095)9240493

Chuev, Igor L. (1964). Researcher. Institute of Chemical Physics, Russian Academy of Sciences in Chernogolovka, Chernogolovka, Moscow Region 142432, Russia. MSc (Moscow Institute of Physics and Technology, 1987). *X-ray_crystallography organic_compound organic_crystal_chemistry computing programming*
E-mail sma@icph.sherna.msk.su Tel. 7(095)5245059 Fax 7(096)5153588

Chukhovskii, Prof. Felix N. (1940). Group leader. Lab. of X-ray Optics and Synchrotron Radiation, Institute of Crystallography, Russian Academy of Sciences, 59 Leninsky prosp., Moscow 117333, Russia. DSc (Inst. of Crystallography Moscow, 1985) physics. *X-ray_diffraction electron_diffraction X-ray_topography X-ray_monochromator X-ray_diffuse_scattering*
E-mail chukhov@crystal.msk.su Tel. 7(095)3307992 Fax 7(095)3301956

Chuprunov, Prof. Evgeny V. (1951). Professor. Lab. of X-ray Diffraction Research, Research Institute of Physics and Technology, 23 Gagarin prosp., Nizhny Novgorod 603600, Russia. DSc (Inst. of Crystallography Moscow, 1991) physics. *X-ray_crystallography crystal_physics X-ray_diffractometry*
Tel. 7(8312)656365

Chvalun, Dr Sergej N. (1955). Senior Researcher. Laboratory of Polymer Structure, Karpov Institute of Physical Chemistry, 10 Obukha St., Moscow 103064, Russia. PhD (Moscow Physical and Technical Institute, 1981) physics. *SAXS_polymers WAXS_polymers structural_computer_modelling_polymers X-ray_scattering_computer_modelling_polyesters*
E-mail rebrov@nifhi.uucp.free.msk.su Tel. 7(095)2270014x2519 Fax 7(095)9752450

Darinskaya, Dr Elena V. (1942). Senior Researcher. Lab. of Mechanical Properties of Crystals, Institute of Crystallography, Russian Academy of Sciences, 59 Leninskii pr., 117333 Moscow, Russia. PhD (Inst. of Crystallography Moscow, 1984) solid state physics. *Plasticity dislocation magnetism mechanical_properties*
E-mail public@mechan.incr.msk.su Tel. 7(095)3308274 Fax 7(095)3308274

Darinskii, Dr Alexander N. (1962). Researcher. Lab. of Mechanical Properties of Crystals, Institute of Crystallography, Russian Academy of Sciences, 59 Leninskii pr., 117333 Moscow, Russia. PhD (Inst. of Crystallography Moscow, 1990) solid state physics. *Crystal_acoustics_theory electromagnetic_spin_wave_theory dislocation_theory*
E-mail adar@mechan.incr.msk.su Tel. 7(095)3308274 Fax 7(095)3308274

Dayniak, Dr Lydia G. (1947). Researcher. Lab. for Investigation of Rock-Forming Minerals by Physical Methods, Geological Institute, Russian Academy of Sciences, Pyzhevsky per. 7, Moscow 109017, Russia. PhD (Kazan State U., 1981) geology and mineralogy. *Crystal_chemistry layered_silicates spectroscopy*
Tel. 7(095)2308301 Fax 7(095)2318106

Degtyareva, Dr Valentina F. Senior Researcher. Lab. of X-ray structure Analysis, Institute of Solid State Physics, Chernogolovka, Moscow Region 142432, Russia. PhD (Inst. steel and alloys Moscow, 1975) solid state physics. *X-ray_crystallography high_pressure_structure_determination binary_alloys phase_diagrams phase_transformation*
E-mail degtyar@issp.sherna.msk.su Tel. 7(095)5245063 Fax 7(096)5171949 Telex 412654 Serna Russia

Dekaprilevich, Dr Marina O. (1941). Researcher. Institute of Organic Chemistry, Russian Academy of Sciences, 47 Leninsky prosp., Moscow 117913, Russia. PhD (Inst. Organic Chemistry, 1988) organic chemistry. *X-ray_crystallography small_molecules organic_crystal_chemistry*
E-mail xray@xray.ineos.ac.ru Tel. 7(095)1378818 Fax 7(095)1355085

Demianets, Dr Ludmila N. (1939). Laboratory Head. Lab. for Hydrothermal Synthesis, Institute of Crystallography, Russian Academy of Sciences, 59 Leninsky pr., Moscow 117333, Russia. PhD (Inst. of Crystallography Moscow, 1966) chemistry. *Crystal_growth hydrothermal_synthesis inorganic_compound crystal_chemistry high_temperature_superconductivity*
Tel. 7(095)3308156 Fax 7(095)1351011

Demishev, Dr Gennadii B. (1959). Senior Researcher. Lab. of Monocrystals, Vereschagin Institute of High Pressure Physics, Russian Academy of Sciences, Troitsk, Moscow Reg. 142092, Russia. PhD (Moscow Institute of Physics and Technology, 1988) physics and mathematics. *X-ray_crystallography quasicrystal aperiodic_material*
E-mail scorpion@adonis.iasnet.com Tel. 7(095)3340733 Fax 7(095)3340012

Denisenko, Dr Georgy A. (1945). Scientific Secretary. Institute of Crystallography, Russian Academy of Sciences, 59 Leninsky prosp., Moscow 117333, Russia. PhD (Kazan State U., 1975) physics. *Solid_state_physics*
Tel. 7(095)1356420 Fax 7(095)1351011

Dimitrova, Dr Olga V. (1948). Senior Researcher. Department of Crystallography and Crystal Chemistry, Faculty of Geology, Moscow State University, Moscow 119899, Russia. PhD (Moscow State U., 1977) crystallography and crystal physics. *Crystal_growth rare-earth_compounds hydrothermal_method*
Tel. 7(095)2032845 Fax 7(095)9390126 Telex 411483 MGU SU

Dmitrienko, Dr Vladimir E. (1949). Senior Researcher. Laboratory of Crystallophysics, Institute of Crystallography, Russian Academy of Sciences, 59 Leninsky pr., Moscow 117333, Russia. DSc (Inst. of Crystallography Moscow, 1993) solid state physics. *X-ray_diffraction_theory anisotropic_anomalous_dispersion quasicrystal synchrotron_radiation X-ray_polarization X-ray_tensor_property forbidden_reflections non-crystallographic_symmetry aperiodic_material extinction liquid_crystals*
E-mail dmitrien@theory.incr.msk.su Tel. 7(095)1356240 Fax 7(095)1351011

Dmitrieva, Dr Margarita T. Senior Researcher. Laboratory of X-ray Structure Analysis, Institute of Geology of Ore Deposits, Petrography, Mineralogy and Geochemistry, Russian Academy of Sciences, Staromonetny per. 35, Moscow 109017, Russia. PhD (Inst. of Ore Deposits, Petrography, Mineralogy and Geochemistry, 1977) crystal chemistry and mineralogy. *X-ray_crystallography minerals clusters superstructure modulated_structures titanium_compounds*
E-mail mar@igem.msk.su Tel. 7(095)2308296 Fax 7(095)2302179

Dmitrieva, Dr Tatiana V. (1933). Senior Researcher. Lab. of Resonance methods, Institute of Crystallography, Russian Academy of Science, 59 Leninsky prosp., Moscow 117321, Russia. PhD (Institute of Crystallography Moscow, 1975) physics. *Mossbauer_spectroscopy magnetic_materials phase_transition superconductor*
E-mail root@magnet.crystal.msk.su Tel. 7(095)3308329 Fax 7(095)3308329

Dolgushin, Mr Fyodor M. (1967). Junior Researcher. X-ray Structural Center of the Russian Academy of Sciences, Institute of Organoelement Compounds, Russian Academy of Sciences, 28 Vavilov St., Moscow 117813, Russia. MSc (Moscow State U., 1992) solid state physics. *Organometallic_structure metallic_clusters phase_transition*
E-mail xray@xray.ineos.ac.ru Tel. 7(095)1359214 Fax 7(095)1355085

Dorogovin, Dr Boris A. (1938). Deputy Director. Russian Research Institute for Synthesis of Minerals and Raw Materials, Institutskaya St. 1, Alexandrov, Vladimir Region 601600, Russia. PhD (Moscow State U., 1973) geology and mineralogy. *Crystal_growth gemology technology*
Tel. 7(095)5845808 Fax 7(095)5845828

Dorokhova, Dr Galina I. (1952). Chief lecturer. Department of Crystallography and Crystal Chemistry, Faculty of Geology, Moscow State University, Moscow 119899, Russia. PhD (Moscow State U., 1984) crystallography and crystal physics. *X-ray_crystallography minerals materials crystal_form crystallography gemology goniometry morphology point_group single_crystal space_group symmetry_group twins*
Tel. 7(095)9392330 Fax 7(095)9390126 Telex 411483 MGU SU

Doroshinsky, Dr Alexander L. (1933). Researcher. Institute of New Chemical Problems, Russian Academy of Sciences, Chernogolovka, Moscow Region 142432, Russia. PhD (Karpov Inst. of Physical Chemistry Moscow, 1973) chemistry. *Crystal_chemistry coordination_compound*

Doynikova, Dr Olga A. (1948). Senior Researcher. Laboratory of Electron Microscopy and Electronography, Institute of Ore Deposits, Petrology, Mineralogy and Geochemistry, Russian Academy of Sciences (IGEM RAS), Staromonetny 35, 109017 Moscow, Russia. PhD (Inst. of Ore Deposits, Petrology, Mineralogy and Geochemistry (IGEM RAS), 1982) mineralogy. *Analytical_electron_microscopy minerals selected_area_electron_diffraction*
E-mail avm@igem.msk.su Tel. 7(095)2308210 Fax 7(095)2302179

Drits, Prof. Victor A. (1932). Laboratory Head. Lab. for Investigation of Rock-Forming Minerals by Physical Methods, Geological Institute, Russian Academy of Sciences, Pyzhevsky per. 7, Moscow 109017, Russia. DSc (Inst. of Geology Mineralogy and Petrography of Ore Minerals Moscow, 1974) geology and mineralogy. *Crystal_chemistry diffractometry_spectroscopy_minerals*
Tel. 7(095)2308124 Fax 7(095)2318106

Drozdov, Dr Yuri N. (1947). Senior researcher. Laboratory of Thin Film Technology, Institute for Physics of Microstructures, Russian Academy of Sciences, 46 Ulyanov Str., Nizhny Novgorod 603600 GSP-105, Russia. PhD (Gorky State U., 1974) physics. *X-ray_diffractometry heterostructure superconductor*
E-mail svg@appl.nnov.su Tel. 7(8312)366483 Fax 7(8312)361972

Dudka, Mr Alexander P. (1959). Researcher. Laboratory of X-ray Structure Analysis, Institute of Crystallography, Russian Academy of Sciences, 59 Leninsky pr., Moscow 117333, Russia. MSc (Moscow State U., 1982) physics and mathematics. *Diffraction data_collection data_processing software*
Tel. 7(095)1351020 Fax 7(095)1351011

Dudkevich, Prof. Vladimir P. (1955). Head of Chair. Chair of Crystal Physics and Structural Analysis, Department of Physics, Rostov State University, Pr. Stachki 192, Rostov-on-Don 344090, Russia. DSc (Rostov State U., 1980) physics and mathematics. *Structural_crystallography defect_structure*
Tel. 7(8632)223413

Dyachenko, Dr Oleg A. (1939). Chief Researcher. Lab. of Crystal Chemistry, Branch Institute of Chemical Physics, Russian Academy of Sciences in Chernogolovka, Chernogolovka, Moscow Region 142432, Russia. DSc (Moscow State U., 1984) physical chemistry. *X-ray_crystallography organic_conductor_superconductor charge_transfer incommensurate physical_property*
E-mail doa@icph.sherna.msk.su Tel. 7(096)5171967 Fax 7(096)5153588

Dyachenko, Dr Vladimir V. (1936). Chief Researcher. Lab. of High Temperature Crystallization, Institute of Crystallography, Russian Academy of Sciences, 59 Leninsky prosp., Moscow 117333, Russia. PhD (Moscow Inst. of Physics and Technology, 1967) physics. *Laser_radiation monocrystal_laser_property*
Tel. 7(095)1351220 Fax 7(095)1356350 or 7(095)1351011

Dyuzheva, Dr Tatyana D. Researcher. Lab. of Monocrystals, Vereschagin Institute for High Pressure Physics, Russian Academy of Sciences, Troitsk, Moscow Region 142092, Russia. PhD (Moscow State U., 1979) crystal chemistry. *High_pressure_phase_transition crystal_growth X-ray_high_pressure_technique*
E-mail scorpion@adonis.iasnet.com Tel. 7(095)3340793 Fax 7(095)3340793

Dzyabchenko, Dr Aleksandr V. (1948). Senior Researcher. Karpov Institute of Physical Chemistry, 10 Obukha St., Moscow 103064, Russia. PhD (Karpov Inst. of Physical Chemistry, 1981) physical chemistry. *Computer-assisted_design computer_modelling semi-empirical_calculation symmetry structure_similarity database computer_network*
E-mail sdz@nifhi.rc.ac.ru Tel. 7(095)2273412 Fax 7(095)9752450

Efremov, Dr Valery A. (1950). Group Leader. COVENS Research Group, COVENS Ltd., 3-63 Maliy Kupavinsky Projezd, Moscow 105568, Russia. DSc (Moscow State U., 1993) inorganic chemistry. *Structure_determination computing powder bond_length inorganic_crystal_chemistry occupancy rare-earth_compounds oxides molybdates tungstates phase_transition*
Tel. 7(095)3087493

Efremova, Dr Elena P. (1949). Researcher. Laboratory of High-Temperature Crystallization, Inst. of Crystallography, Russian Academy of Sciences, 59 Leninsky prosp., Moscow 117333, Russia. PhD (Inst. of Crystallography Moscow, 1982) crystallography and crystal physics. *Crystal_growth aqueous_solution phosphates real_structure potassium_dihydrogen_phosphate proton_conductivity laser_induced_damage_threshold*

Tel. 7(095)3301792 Fax 7(095)1351011

Egorov-Tismenko, Dr Yury K. (1938). Associate Professor. Department of Crystallography and Crystal Chemistry, Faculty of Geology, Moscow State University, Moscow 119899, Russia. PhD (Moscow State U., 1973) crystallography and crystal physics. *X-ray_crystallography minerals materials symmetry symmetry_group geometric_crystallography colour_symmetry microcrystallography systematics*

Tel. 7(095)9394974 Fax 7(095)9390126 Telex 411483 MGU SU

Eremin, Mr Nickolai N. (1968). Researcher. Vernadsky Institute of Geochemistry and Analytical Chemistry, Russian Academy of Sciences, 19 Kosygin St., Moscow 117975, Russia. MSc (Moscow State U., 1992) geochemistry. *Crystal_structure charge_density Mossbauer_spectroscopy tin_compounds titanium_compounds computer_modelling*

Tel. 7(095)9391916 Fax 7(095)9382054

Eremkin, Dr Vladimir V. (1962). Senior Researcher. Institute of Physics, Rostov State University, Pr. Stachki 194, Rostov-on-Don 344104, Russia. [2-nd Krasnodarskaya Str. 16313, Apt 51, Rostov-on-Don 344102, Russia.] PhD (Rostov State U., 1987) physics and mathematics. *X-ray_analysis crystal_growth inorganic_compound ferroelectrics solid_solutions*

E-mail Eremkin@riphys.rnd.su Tel. 7(8632)285066 Fax 7(8632)285044

Evlanova, Dr Nina F. Researcher. Physical Department, Moscow State University, Moscow 119899, Russia. PhD (Moscow State U., 1978) physics. *Crystal_growth crystal_defect*

E-mail klimova@cryst.phys.msu.su Tel. 7(095)9392883 Fax 7(095)9390247

Fedorov, Pavel P. (1950). Chief Researcher. Laboratory of Physical and Chemical Analysis, Institute of Crystallography, Russian Academy of Sciences, 59 Leninsky pr., Moscow 117333, Russia. DSc (Moscow Inst. of Fine Chemical Technology, 1991) inorganic chemistry. *Crystal_growth X-ray_powder_diffraction fluoride differential_thermal_analysis phase_diagram*

Tel. 7(095)3307874 Fax 7(095)1351011

Fedorov, Dr Vladimir A. (1948). Researcher. Lab. of Acoustooptics of Crystals, Institute of Crystallography, Russian Academy of Sciences, Leninsky pr. 59, 117333 Moscow, Russia. PhD (Inst. of Crystallography Moscow, 1989) solid state physics. *Crystal_optics crystal_spectroscopy*

Tel. 7(095)3306883 Fax 7(095)3301956

Feigin, Prof. Lev A. (1928). Head of Laboratory. Small Angle Scattering Lab., Institute of Crystallography, Russian Academy of Sciences, Leninsky pr. 59, Moscow 117333, Russia. DSc (Institute of Crystallography Moscow, 1975) small angle scattering. *SAXS X-ray_neutron_reflectivity Langmuir-Blodgett thin_film*

E-mail feigin@saxslab.incr.msk.su Tel. 7(095)1356010 Fax 7(095)1351011

Fesenko, Prof. Evgeny G. (1918). Laboratory Head. Crystal Structure Lab., Institute of Physics, Rostov State University, Pr. Stachki 194, Rostov-on-Don 344104, Russia. DSc (Rostov State U., 1973) physics and mathematics. *Crystal_chemistry phase_transition domain_structure physical_property*

Tel. 7(8632)285100 Fax 7(8632)244311 Telex 123360 pxezo

Fesenko, Dr Oleg E. (1950). Head of Laboratory. Phase Transition Lab., Institute of Physics, Rostov State University, Pr. Stachki 194, Rostov-on-Don 344104, Russia. DSc (Azerbaidjan State U., 1989) physics and mathematics. *Phase_transition crystal_physics perovskites*

Tel. 7(8632)285066

Fetisov, Dr Gennady Vladimirovich (1947). Chief researcher. Lab. Structural Chemistry, General Chemistry Faculty, Chemistry Dept., Moscow State University, Moscow 119899, Russia. DSc (Moscow State U., 1991) physical chemistry. *X-ray_diffraction X-ray_diffractometry methodology foreign_languages*

E-mail fetisov@cry5ma.uucp.free.msk.su Tel. 7(095)9395089 Fax 7(095)9390898 Telex 411483 MGU SU

Filatov, Prof. Stanislav K. (1940). Head of department. Department of Crystallography, St Petersburg University, University Emb. 7/9, St Petersburg 199034, Russia. DSc (St Petersburg U., 1988) geology and mineralogy. *Crystal_chemistry phase_transformation structure_determination X-ray_high_temperature_powder_diffraction*

E-mail flt@dean.geol.lgu.spb.su Tel. 7(812)2189647 Fax 7(812)2181346 Telex 121481 LSU SU

Filipenko, Olga S. (1940). Senior Researcher. Institute of Chemical Physics, Russian Academy of Sciences in Chernogolovka, Chernogolovka, Moscow Region 142432, Russia. PhD (Moscow State U., 1972) physical chemistry. *X-ray_crystallography organic_compound organic_crystal_chemistry molecular_structure*

E-mail sma@icph.sherna.msk.su Tel. 7(095)5245059 Fax 7(096)5153588

Flerov, Dr Igor N. (1942). Senior Researcher. Lab. of Crystal Physics, Institute of Physics, Siberian Dept., Russian Academy of Sciences, Krasnoyarsk 660036, Russia. PhD (Inst. of Physics Krasnoyarsk, 1978) solid state physics. *Structural_phase_transition inorganic_compound thermal_property*

E-mail aleks@iph.krasnoyarsk.su Tel. 7(391)2432635 Fax 7(391)2438923

Frank-Kamenetskaya, Dr Olga V. (1945). Senior Researcher. Department of Crystallography, St Petersburg University, University Emb. 7/9, St Petersburg 199034, Russia. PhD (Leningrad State U., 1973) geology and mineralogy. *X-ray_structure_determination crystal_chemistry_inorganic_compound structural_mineralogy superconductivity*

E-mail flt@dean.geol.lgu.spb.su Tel. 7(812)2189647 Fax 7(812)2181346 Telex 121481 LSU SU

Franke, Dr Valeria D. (1945). Senior Researcher. Crystal Genesis Lab., St Petersburg University, University Emb. 7/9, St Petersburg 199034, Russia. PhD (St Petersburg U., 1982) crystallography and crystal physics. *Crystallogeny crystal_growth crystal_morphology impurity_adsorption isomorphous_replacement flotation epitaxy*

E-mail anna@dean.geol.lgu.spb.su Tel. 7(812)2189650 Fax 7(812)2181346 Telex 121481 LSU SU

Frolov, Mr Kirill V. (1970). Junior Researcher. Lab. of Resonance Methods, Institute of Crystallography, Russian Academy of Science, 59 Leninsky prosp., Moscow 117333, Russia. MSc (Physical Engineering Inst. Moscow, 1993) solid state physics. *Superconductivity superionic_materials Mossbauer_spectroscopy*

E-mail root@magnet.crystal.msk.su green@magnet.crystal.msk.su Tel. 7(095)3308329 Fax 7(095)3308329

Fundamensky, Mr Vladimir S. (1946). Group Leader. Laboratory of X-ray Structure Analysis, NPP Burevestnik, 68 Maloohtinsky Prospect, St Petersburg 195272, Russia. MSc (Leningrad State U., 1969) crystallography. *X-ray_structure_determination organic_complex_compound drug direct_method direct_method crystallography_computing X-ray_single_crystal_diffractometry*

Tel. 7(812)5280434 Fax 7(812)3122479

Furmanova, Dr Nina G. (1939). Chief Researcher. Laboratory of X-ray Structure Analysis, Institute of Crystallography, Russian Academy of Sciences, 59 Leninsky pr., Moscow 117333, Russia. DSc (Inst. of Crystallography Moscow, 1989) crystallography. *X-ray_analysis crystal_chemistry organic_compound*

E-mail simonov@rsa.crystal.msk.su Tel. 7(095)1353110 Fax 7(095)1351011

Fykin, Dr Leonid E. (1936). Senior Researcher. Lab. for Structural Studies, Karpov Institute of Physical Chemistry, Obninsk Branch, Obninsk, Kaluga Region 249020, Russia. PhD (Moscow Inst. of Engineering and Physics, 1972) solid state physics. *Neutron_diffraction_monocrystal structural_analysis superionic_conductor crystal_structure_nonstoichiometry_fluorine_compounds atomic_structure_dielectrics*

Tel. 7(08439)26575

Gabrielyan, Dr Vyacheslav T. (1940). Senior Researcher. Crystal Genesis Laboratory, St Petersburg University, University Emb. 7/9, St Petersburg U., 1978) crystallography and crystal physics. *Crystallogeny crystal_growth crystal_physics crystal_chemistry crystal_morphology isomorphism phase_diagram acoustooptic_ferroelectric_scintillating_materials_characterisation*

E-mail anna@dean.geol.lgu.spb.su Tel. 7(812)2189650 Fax 7(812)2181346 Telex 121481 LSU SU

Gagarina, Dr Elena S. (1949). Senior Researcher. Lab. of Crystal Physics, Institute of Physics, Rostov State University, Pr. Stachki 194, Rostov-on-Don 344104, Russia. PhD (Rostov State U., 1988) physics and mathematics. *Crystal_structure phase_transition inorganic_compound twinning*

Tel. 7(8632)285066

Galiulin, Dr Ravil V. (1944). Chief Researcher. Institute of Crystallography, Russian Academy of Sciences, 59 Leninsky prosp., Moscow 117333, Russia. DSc (Inst. of Crystallography Moscow, 1978) physics and mathematics. *Mathematical_crystallography*

E-mail CRYS@CRYS.MSK.SU Tel. 7(095)1353510 Fax 7(095)1351011

Galstyan, Dr Victor G. (1940). Group Leader. Lab. of Crystallization from Vapor Phase, Institute of Crystallography, Russian Academy of Sciences, 59 Leninsky pr., Moscow 117333, Russia. PhD (Inst. of Crystallography Moscow, 1972) crystal physics. *Crystal_defect electron_microscopy_of_crystals*

Tel. 7(095)3349469 Fax 7(095)1351011

Gavrilova, Dr Nadezhda D. (1937). Senior Researcher. Chair of Polymer and Crystal Physics, Physics Department, Moscow State University, Moscow 119899, Russia. DSc (Physics Dept. Moscow State U., 1990) crystallography and crystal physics. *Pyroelectricity Hydrogen_bonding Langmuir_Blodgett_film*

E-mail aleks@cryst0.phys.msu.su Tel. 7(095)9391013 Fax 7(095)9390247 Telex 411483 MGU SU

Gavrilyachenko, Dr Victor G. (1935). Associate Professor. Physics Department, Rostov State University, Pr. Stachki 192, Rostov-on-Don 344090, Russia. PhD (Rostov State U., 1971) physics and mathematics. *Domain_structure phase_transition perovskites physical_property*

Tel. 7(8632)223413

Geguzina, Dr Galina A. (1945). Senior Researcher. Physics Department, Rostov State University, Pr. Stachki 192, Rostov-on-Don 344090, Russia. PhD (Rostov State U., 1975) physics and mathematics. *Crystal_chemistry complex_oxides*

Tel. 7(8632)223756

Gel'man, Dr Yury A. (1938). Group Leader. Lab. of Elementary Processes of Crystal Growth, Institute of Crystallography, Russian Academy of Sciences, Leninsky pr. 59, Moscow 117333, Russia. PhD (Inst. of Crystallography Moscow, 1978) physics. *MBE adsorption ultra_high_vacuum II-VI_compounds*

E-mail chem@elprgr.crystal.msk.su Tel. 7(095)1354200 Fax 7(095)1351011

Genkina, Dr Elena A. (1945). Researcher. Laboratory of X-ray Structure Analysis, Institute of Crystallography, Russian Academy of Sciences, 59 Leninsky pr., Moscow 117333, Russia. PhD (Moscow State U., 1987) crystallography. *X-ray_crystallography crystal_chemistry inorganic_compound*

E-mail simonov@rsa.crystal.msk.su Tel. 7(095)1350330 Fax 7(095)1351011

Gerasimov, Dr Victor I. (1943). Associate Professor. Mordovian State U., 68a Bolshe-vitskaya st., Saransk 430000, Russia. PhD (Inst. of Applied Physics Kishinev, 1979) physics and mathematics. *X-ray_analysis*
Tel. 7(83422)99792

Geraskin, Dr Valery V. (1944). Associate Professor. Department of Crystallography, Institute of Steel and Alloys, 4 Leninsky prosp., Moscow 117936, Russia. PhD (Inst. of Steel and Alloys Moscow, 1969) solid state physics. *Crystal physical_properties defect crystallography*
Tel. 7(095)2366500 Fax 7(095)2378007

Givargizov, Dr Eugene I. (1934). Head of Laboratory. Lab. of Crystallization from Vapor Phase, Institute of Crystallography, Russian Academy of Sciences, 59 Leninsky pr., Moscow 117333, Russia. DSc (Inst. of Crystallography Moscow, 1976) crystal physics. *Crystal_growth film_whisker_growth nanostructures*
Tel. 7(095)3308265 Fax 7(095)1351011

Gladkii, Dr Vsevolod V. (1934). Group Leader. Laboratory of Electromechanical Investigations of Crystals, Institute of Crystallography, Russian Academy of Sciences, Leninsky pr. 59, Moscow 117333, Russia. DSc (Inst. of Crystallography Moscow, 1985) physics. *Thermodynamics phase_transition ferroelectricity*
E-mail postmaster@el-mech.incr.msk.su Tel. 7(095)3307856 Fax 7(095)1351011

Glazov, Dr Alexei Iv. (1942). Associate Professor. Chair of Mineralogy Crystallography and Petrography, Geological Research Faculty, St Petersburg Mining Institute, 2 21st Liniya, St Petersburg 199026, Russia. PhD (St Petersburg Mining Inst., 1976) crystallography and crystal physics. *Crystal_morphology X-ray_diffraction history_of_science*
Tel. 7(812)2188266 Fax 7(812)2132613 Telex 121494 LGIPSU

Glikin, Dr Arkady E. (1943). Laboratory Leader. Crystal Genesis Laboratory, St Petersburg University, University Emb. 7/9, St Petersburg 199034, Russia. PhD (St Petersburg U., 1978) crystallography and crystal physics. *Crystallogeny experimental_mineralogy crystal_growth crystal_chemistry crystal_morphology isomorphous_replacement phase_diagram epitaxy*
E-mail anna@dean.geol.lgu.spb.su Tel. 7(812)2189650 Fax 7(812)2181346 Telex 121481 LSU SU

Gollo, Dr Eduard Al'bertovich (1941). Head of Laboratory. Lab. of X-ray Analysis, Institute of Earth Crust, St Petersburg University, University Emb. 7/9, St Petersburg 199034, Russia. PhD (St Petersburg U., 1972) geology and mineralogy. *X-ray_powder_analysis typomorphism_phyllosilicates transformation_layer_silicates simulation_X-ray_diffraction structural_disorder*
E-mail flt@dean.geol.lgu.spb.su Tel. 7(812)2189711 Fax 7(812)2181346

Golovina, Dr Nina I. (1934). Chief Researcher. Lab. of Crystal Chemistry, Branch Institute of Chemical Physics, Russian Academy of Sciences in Chernogolovka, Chernogolovka, Moscow Region 142432, Russia. DSc (Inst. of Chemical Physics in Chernogolovka, 1987) physical chemistry. *X-ray_crystallography nitrogen_compounds structure–activity_relationship energetic_compounds*
E-mail atov@icph.sherna.msk.su Tel. 7(096)5171168 Fax 7(096)5153588

Golovko, Dr Yurii I. (1947). Senior Researcher. Lab. of Crystal Physics, Institute of Physics, Rostov State University, Pr. Stachky 194, Rostov-on-Don 344104, Russia. PhD (Rostov State U., 1980) physics and mathematics. *Crystal_structure thin_film perovskites high-Tc_superconductivity*
E-mail irina-2@riphys.rnd.su Tel. 7(8632)285066 7(8632)346919 Fax 7(8632)285044 Telex 123321, 123610 FIZIK RU

Golubev, Dr Alexandr M. (1948). Associate Professor. Department of Chemistry, Moscow N. E. Bauman State Technical University, 5 St. 2-nd Baumanskaya, Moscow 107005, Russia. PhD (Moscow State U., 1975) inorganic chemistry. *Inorganic_crystal_chemistry superstructure lattice_energy quasicrystal ionic_conductivity*
E-mail glazunov@lmm.phyche.msk.su Tel. 7(095)2636103

Golyshev, Dr Vladimir M. (1943). Associate Professor. Mordovian State U., 68a Bolshe-vitskaya st., Saransk 430000, Russia. PhD (Inst. of Crystallography Moscow, 1971) physics and mathematics. *X-ray_analysis*
Tel. 7(83422)99792

Goncharov, Dr Alexander V. (1949). Associate Professor. Department of Physics, Vladimir State Pedagogical University, Department of Physics, 11 Stroiteley Prospect, Vladimir 600024, Russia. PhD (Inst. Crystallography Moscow, 1975) crystallography. *Geometric_crystallography molecular_crystal conformational_methodology organic_crystal_chemistry*
Tel. 7(09222)72604

Gorchakova, Dr Olga E. (1942). Senior Researcher. All-Russian Institute of Scientific and Technical Information (VINITI), 20a Usievich St., Moscow 125219, Russia. PhD (Inst. of Crystallography Moscow, 1969) mineralogy. *Crystal_chemistry organic_inorganic_compound superconductivity solid_state_chemistry complex_compound*
Tel. 7(095)1554205 Fax 7(095)9430060 Telex 411249 VINITI SU

Gordienko, Mr Leonid A. (1930). Group leader. Lab. for Synthesis of Materials by Hydrothermal Method, Russian Research Institute for Synthesis of Minerals and Raw Materials, Institutskaya St.1, Alexandrov, Vladimir Region 601600, Russia. MSc (Odessa State U., 1954) mineralogy. *Crystal_growth hydrothermal_synthesis real_structure*
Tel. 7(095)5845834 Fax 7(095)5845828

Gorshkov, Prof. Anatoly Ivanovich (1929). Laboratory Head. Laboratory of Electron Microscopy, Institute of Geology of Ore Deposits, Petrography, Mineralogy and Geochemistry, Russian Academy of Sciences (IGEM RAS), Staromonetny per. 35, Moscow 109017, Russia. DSc (IGEM RAS, 1971) geology and mineralogy. *Minerals_structure morphology phase_transformation*
Tel. 7(95)2308210 Fax 7(95)2302179

Gorskaya, Dr Marina G. (1950). Senior Researcher. Department of Crystallography, St Petersburg University, University Emb. 7/9, St Petersburg 199034, Russia. PhD (St Petersburg U., 1985) crystallography. *X-ray_structure_determination crystal_chemistry_inorganic_compound structural_mineralogy X-ray_high-temperature_powder_diffraction*
E-mail flt@dean.geol.lgu.spb.su Tel. 7(812)2189647 Fax 7(812)2181346 Telex 121481 LSU SU

Gritsenko, Mr Victor Victorovich (1964). Junior Researcher. Lab. of Conductors and Catalytic Active Compounds, Branch Institute of Chemical Physics, Russian Academy of Sciences in Chernogolovka, Moscow Region Chernogolovka 142432, Russia. MSc (Rostov State U., 1989) physics. *X-ray_crystallography organic_conductor superconductor organic_crystal_chemistry*
E-mail doo@icph.sherna.msk.su Tel. 7(096)5171967 Fax 7(096)5153588

Gromilov, Dr Sergey A. (1959). Researcher. Laboratory of Crystal Chemistry, Institute of Inorganic Chemistry Siberian Branch of RAS, 3 Lavrentyev prosp., Novosibirsk 630090, Russia. PhD (Inst. Applied Physics Moldovian Academy of Sciences, 1989) crystallography and crystal physics. *X-ray_diffractometry_polycrystal_compound preparation accuracy volatility organometallic_compound*
E-mail borisov@che.nsk.su Tel. 7(3832)354366 Fax 7(3832)355960

Grunsky, Dr Oleg S. (1962). Senior Researcher. Crystal Genesis Laboratory, Geology Dept., St Petersburg University, University Emb. 7/9, St Petersburg 199034, Russia. PhD (St Petersburg U., 1988) crystallography and crystal physics. *Crystallogeny crystal_growth kinetics computer_automation Czochralski_technique*
E-mail anna@dean.geol.lgu.spb.su Tel. 7(812)2189650 Fax 7(812)2181346 Telex 121481 LSU SU

Gubkin, Mr Mikhail G. (1962). Junior Researcher. Laboratory of Electromechanical Investigations of Crystals, Institute of Crystallography, Russian Academy of Sciences, Leninsky pr. 59, Moscow 117333, Russia. MSc (Moscow State U., 1985) physics. *Magnetism conductivity manganese_compounds*
E-mail postmaster@el-mech.incr.msk.su Tel. 7(095)3307856 Fax 7(095)1351011

Gurskaya, Dr Galina Gurskaya Senior Researcher. Laboratory of X-ray Structure Analysis, Engelhardt Institute of Molecular Biology, Russian Academy of Sciences, 32 Vavilova Str., Moscow 117984, Russia. PhD (Inst. of Crystallography Russian Academy of Sciences Moscow, 1964) crystal physics of crystals. *X-ray_analysis drug nucleosides nucleotides structure–activity_relationship AIDS crystallization X-ray_crystallography*
E-mail gurg@imb.msk.su Tel. 7(095)1350237 Fax 7(095)1351405 Telex 411755 molbi su

Harutyunyan, Prof. Emil H. (1935). Chief researcher. Laboratory of Biocrystal Structure, Institute of Crystallography, Russian Academy of Sciences, 59 Leninsky prosp., Moscow 117333, Russia. DSc (Inst. of Crystallography Moscow, 1983) protein crystallography. *X-ray_structure_determination proteins_crystallography crystal_chemistry*
E-mail emharut@crys.msk.su Tel. 7(095)1352300 Fax 7(095)1351011

Ignatovich, Dr Vladimir K. (1937). Senior Researcher. Lab. of Neutron Physics, Joint Institute for Nuclear Research, Dubna, Moscow region 141980, Russia. PhD (Joint Inst. Nuclear Research, 1976) physics. *Theoretical_dynamical_diffraction Laue_diffraction high_precision_diffractometry perfect_crystal quantum_mechanics crystallography*
E-mail ignatov@nfsun1.jinr.dubna.su vignat@lnp01.jinr.dubna.su Tel. 7(09621)63377 Fax 7(09621)65085 Telex 911-621 Dubna SU

Igonin, Dr Vladimir A. (1958). Associate Professor. Department of Physics, Vladimir State Pedagogical University, Prospect Stroiteley 11, Vladimir 600024, Russia. [Uliza Gorkogo 85-b, 32, Vladimir 600005, Russia.] PhD (Moscow State Pedagogical U., 1992) solid state physics. *X-ray_analysis complex_compound silicon_compounds macrocycles X-ray_crystallography molecular_crystal*
E-mail igonin@wgpu.cityadm.vladimir.su Tel. 7(09222)72628

Ilyukhin, Dr Andrey B. (1962). Senior Researcher. Laboratory of Coordination Compounds Crystal Chemistry, Kurnakov Institute of General and Inorganic Chemistry, Russian Academy of Sciences, 31 Leninsky prosp., Moscow 117907, Russia. PhD (Inst. of Chem. Reagents and Ultrapure Chem. Substances, 1989) physical chemistry. *Packing X-ray_crystallography indium_compounds*
Tel. 7(095)9521803 Fax 7(095)9541279

Ilyushin, Prof. Alexandr S. (1943). Professor. Solid State Chair, Physics Department, Moscow State University, Vorobjovy Gory, Moscow 117234, Russia. DSc (Moscow State U., 1989) solid state physics. *X-ray_analysis structure_determination phase_transition intermetallic_compound*
E-mail asi@phys.msu.su. Tel. 7(095)9391684 Fax 7(095)9390247 Telex 411483 MGU SU

Ilyushin, Dr Grigory D. (1952). Senior Researcher. Lab. for Hydrothermal Synthesis, Institute of Crystallography, Russian Academy of Sciences, 59 Leninsky pr., Moscow 117333, Russia. PhD (Inst. of Physical Chemistry Moscow, 1982) physics. *Crystal_chemistry classification silicates germanates high_temperature_superconductor_structure*
Tel. 7(095)3308156 Fax 7(095)1351011

Imamov, Prof. Rafik M. (1938). Laboratory Leader. Lab. of Crystal Layers Diffractometry, Institute of Crystallography, Russian Academy of Sciences, 59 Leninsky prosp., Moscow 117333, Russia. DSc (Inst. of Crystallography Moscow, 1978) crystallography and crystal physics. *X-ray_diffraction standing_wave secondary_electron_emission multiple-crystal_diffractometry thin_film_structure heterostructure localization_of_impurity_atom_position*
Tel. 7(095)3306856 Fax 7(095)1351011

Indenbom, Prof. Vladimir L. (1924). Chief Researcher. Laboratory of Crystallophysics, Institute of Crystallography, Russian Academy of Sciences, 59 Leninsky prospekt, Moscow 117333, Russia. DSc (Inst. of Crystallography Moscow, 1964) solid state physics. *Diffraction_theory dynamical_diffraction defect electron_microscopy Borrmann_absorption strain_determination creep crack polarized_neutrons liquid_crystals*
E-mail root@theory.incr.msk.su Tel. 7(095)1356240 Fax 7(095)1351011

Isakov, Dr Igor V. (1935). Senior Researcher. Lab. for Structural Studies, Obninsk Branch, Karpov Institute of Physical Chemistry, Obninsk, Kaluga Region 249020, Russia. MSc (Moscow State U., 1959) physics. *Structure_high_temperature_superconductor X-ray_diffraction_defect_structures*
Tel. 7(08439)26531

Isupov, Mr Michael N. (1963). Junior Researcher. Laboratory of Biocrystal Structure, Institute of Crystallography, Russian Academy of Sciences, 59 Leninsky prosp., Moscow 117333, Russia. MSc (Moscow Physical Technical Inst., 1986) chemical physics. *Proteins_crystallography computer_modelling*
Tel. 7(095)1355420 Fax 7(095)1351011

Ivakin, Dr Gleb I. (1965). Researcher. Laboratory of Electron Diffraction, Institute of Crystallography, Russian Academy of Sciences, 59 Leninsky prosp., Moscow 117333, Russia. PhD (Inst. of Crystallography RAS Moscow, 1993) physics. *High-energy_electron_diffraction*
Tel. 7(095)1353500

Ivanov, Dr Sergey A. (1951). Senior Researcher. X-ray Laboratory, Karpov Institute of Physical Chemistry, 10 Obukha St., 103064 Moscow K-64, Russia. PhD (Karpov Inst. of Physical Chemistry, 1981) physical chemistry. *Rietveld_method phase_transition line_profile_analysis thermal_expansion ferroelectricity Debye_Waller_factor real_structure inorganic_compound X-ray_powder_diffraction*
E-mail ivan@nifhi.uucp.free.msk.su Tel. 7(095)2270014x2729 Fax 7(095)9752450

Ivanova, Ms Elena S. (1962). Junior Researcher. Laboratory of Electromechanical Investigations of Crystals, Institute of Crystallography, Russian Academy of Sciences, Leninsky pr. 59, Moscow 117333, Russia. MSc (Inst. of Crystallography Moscow, 1985) physics. *Phase_transition incommensurate ferroelectricity*
E-mail postmaster@el-mech.incr.msk.su Tel. 7(095)3307856 Fax 7(095)1351011

Kabalkina, Dr Sarra S. (1918). Senior Researcher. Lab. of Monocrystals, Vereschagin Institute of High Pressure Physics, Russian Academy of Sciences, Troitsk, Moscow Reg. 142092, Russia. DSc (Institute of Crystallography of Moscow, 1975) physics and mathematics. *X-ray_crystallography phase_transition polymorphism*
E-mail scorpion@adonis.iasnet.com Tel. 7(095)3340733 Fax 7(095)3340012

Kabalov, Dr Yurii K. (1936). Senior Researcher. Department of Crystallography and Crystal Chemistry, Faculty of Geology, Moscow State University, Moscow 119899, Russia. PhD (Moscow State U., 1974) crystallography and crystal physics. *X-ray_crystallography minerals materials Rietveld_method diffractometry*
Tel. 7(095)9394923 Fax 7(095)9390126 Telex 411483 MGU SU

Kachalov, Dr Oleg V. (1942). Chief Researcher. Lab. of Crystallophysics, Institute of Crystallography, Russian Academy of Sciences, 59 Leninsky Pr., Moscow 117333, Russia. PhD (Inst. Crystallography Moscow, 1978) solid state physics. *Light_scattering dielectric_compound Brillouin_spectroscopy Raman isomorphism*
E-mail feigin@saxslab.incr.msk.su Tel. 7(095)1355120 Fax 7(095)1351011

Kaganer, Dr Vladimir M. (1956). Senior Researcher. Lab. of Crystallophysics, Institute of Crystallography, Russian Academy of Sciences, Leninsky pr. 59, Moscow 117333, Russia. PhD (Inst. of Crystallography Moscow, 1984) solid state physics. *X-ray_topography Langmuir_monolayer diffraction_theory X-ray_diffuse_scattering liquid_crystals*
E-mail kaganer@theory.incr.msk.su Tel. 7(095)1356240 Fax 7(095)1351011

Kalinin, Dr Victor B. (1947). Senior Researcher. Laboratory of Spectroscopy, Institute of Physical Chemistry, Russian Academy of Sciences, 31 Leninsky pr., Moscow 117915, Russia. PhD (Moscow State U., 1975) inorganic chemistry. *Crystal_chemistry phase_transition superionic_conductor ferroelectricity cold_nuclear_fusion*
E-mail glazunov@lmm.phyche.msk.su Tel. 7(095)9554664 Fax 7(095)9527514

Kamentsev, Prof. Igor E. (1933). Professor. Department of Crystallography, St Petersburg University, University Emb. 7/9, St Petersburg, Russia. DSc (St Petersburg U., 1988) geology and mineralogy. *Crystallography isomorphism mineralogy order_disorder feldspar_mineralogy*
E-mail anna@dean.geol.lgu.spb.su Tel. 7(812)2189647 Fax 7(812)2181346 Telex 121481 LSU SU

Kaminsky, Dr Alexander A. (1934). Head of Laboratory. Lab. of Laser Crystals Physics, Institute of Crystallography, Russian Academy of Sciences, 59 Leninsky prosp., Moscow 117333, Russia. DSc (Inst. of Crystallography Moscow, 1974) physics and mathematics. *Crystal_physics laser_crystal*
Tel. 7(095)1352210 Fax 7(095)1351011

Kanevsky, Dr Vladimir M. (1948). Scientific Secretary. Institute of Crystallography, Russian Academy of Sciences, Acad. Sci. of Russia, 59 Leninsky prosp., Moscow 117333, Russia. PhD (Inst. Crystallography Moscow, 1987) physics and mathematics. *Real_crystal_physics*
Tel. 7(095)1356420 Fax 7(095)1351011

Kaplunnik, Dr Lidiya N. (1947). Chief lecturer. Department of Crystallography and Crystal Chemistry, Faculty of Geology, Moscow State University, Moscow 119899, Russia. PhD (Moscow State U., 1978) crystallography and crystal physics. *X-ray_crystallography minerals materials sulfides crystal_chemistry*
Tel. 7(095)9392330 Fax 7(095)9390126 Telex 411483 MGU SU

Karyakina, Dr Tatyana A. (1939). Senior Researcher. Chair of Mineralogy Crystallography and Petrography, St Petersburg Mining Institute, 2 21st Liniya, Saint Petersburg 199026, Russia. PhD (Leningrad Mining Inst., 1968) crystallography and crystal physics. *Crystal_growth morphology mineralogical_crystallography*
Tel. 7(812)2188412 Fax 7(812)2132613 Telex 121494 LGIPSU

Kashaev, Dr Anvar A. (1932). Professor. Teachers' Training Institute, Department of Technical Methods of Training, 6 Nizhnyaya Naberezhnaya St., Irkutsk 664653, Russia. DSc (St Petersburg State U., 1989) crystallography and mineralogy. *X-ray_structure_determination minerals inorganic_compounds*

Kasjanenko, Dr Evgeni V. (1950). Associate Professor. Department of Physics, Saint-Petersburg Mining Institute, 2 21st Liniya, Saint Petersburg 199026, Russia. PhD (St Petersburg State U., 1977) physics. *Solid_state_physics minerals_physics optical_spectroscopy_methods*
Tel. 7(812)2188683 Fax 7(812)2132613 Telex 121494 LGIPSU

Kataeva, Dr Olga N. (1957). Researcher. Institute of Organic and Physical Chemistry, Kazan Scientific Center, Russian Academy of Sciences, Arbuzov str. 8, Kazan 420083, Russia. PhD (Kazan Inst. of Chemical Technology, 1984) chemistry. *Structural_chemistry organic_organometallic_compound gas_electron_diffraction stereoelectronic_effects*
E-mail kataev@ksc.tat.iasnet.com Tel. 7(8432)767424 Fax 7(8432)752253

Katser, Dr Sergey B. (1962). Researcher. Laboratory of Coordination Compounds Crystal Chemistry, Kurnakov Institute of General and Inorganic Chemistry, Russian Academy of Sciences, 31 Leninsky prosp., Moscow 117907, Russia. PhD (Inst. General and Inorganic Chemistry, 1993) inorganic chemistry. *Boron_compounds clusters MO_calculation*
Tel. 7(095)2188603 Fax 7(095)9541279

Katsnelson, Prof. Albert A. (1930). Professor. Solid State Chair, Physics Department, Moscow State University, Vorobjovy Gory, Moscow 117234, Russia. DSc (Moscow State U., 1969) physics. *X-ray_diffraction condensed_matter_physics X-ray_dynamical_diffraction solid_state_physics X-ray_diffuse_scattering clustering quasicrystal computer_modelling short-range_order*
E-mail katsnelson@poly.phys.msu.su. Tel. 7(095)9394610 Fax 7(095)9390247 Telex 411483 MGU SU

Kayushina, Dr Renata L. (1926). Senior Researcher. Small Angle Scattering Lab., Institute of Crystallography, Russian Academy of Sciences, Leninsky pr. 59, Moscow 117333, Russia. PhD (Inst. of Crystallography Moscow, 1965) physics and mathematics. *X-ray_crystallography SAXS biological_macromolecules Langmuir–Blodgett_film structure_immunoglobulins_macromolecules*
E-mail feigin@saxslab.incr.msk.su Tel. 7(095)1356010 Fax 7(095)1351011

Kazimirov, Dr Alexander Yu. (1952). Senior Researcher. Lab. of X-ray Optics and Synchrotron Radiation, Institute of Crystallography, Russian Academy of Sciences, 59 Leninsky prosp., Moscow 117333, Russia. PhD (Inst. of Crystallography Moscow, 1988) physics. *X-ray_standing_wave X-ray_dynamical_diffraction X-ray_multiple_diffraction X-ray_and_synchrotron_radiation_instrumentation X-ray_monochromator*
E-mail kazim@crystal.msk.su Tel. 7(095)3307992 Fax 7(095)3301956

Khadzhi, Dr Irina P. (1937). Chief Researcher. Lab. of Physical and Technical Research, Russian Research Institute for Synthesis of Minerals and Raw Materials, Institutskaya St. 1, Alexandrov, Vladimir Region 601600, Russia. PhD (Moscow State U., 1982) physics. *Electron_microscopy mineralogy asbestos*
Tel. 7(095)5845834 Fax 7(095)5845828

Khadzhi, Dr Vadim L. (1932). Laboratory Head. Lab. for Synthesis of Materials by Hydrothermal Method, Russian Research Institute for Synthesis of Minerals and Raw Materials, Institutskaya St. 1, Alexandrov, Vladimir Region 601600, Russia. PhD (Inst. of Crystallography Moscow, 1968) crystallography. *Crystal_growth hydrothermal_synthesis crystal_chemistry*
Tel. 7(095)5845844 Fax 7(095)5845828

Khasanov, Dr Salavat S. (1956). Researcher. X-ray Laboratory, Institute of Solid State Physics, Chernogolovka, Moscow Region 142432, Russia. PhD (Institute of Solid State Physics, 1988) solid state physics. *X-ray_crystallography modulated_structures incommensurate_phase phase_transition ferroelectricity high_temperature_superconductors*
E-mail skhasan@issp.sherna.msu.su Tel. 7(095)5255063 Fax 7(095)5171949 Telex 412654 Serna su

Khatanova, Dr Nina A. Chief Lecturer. Solid State Chair, Physics Department, Moscow State University, Vorobjovy Gory, Moscow 117234, Russia. PhD (Moscow State U., 1968) solid state physics. *Intermetallic_compound phase_transition X-ray_diffraction electron_diffraction*
E-mail asi@phys.msu.su. Tel. 7(095)9392387 Fax 7(095)9390247 Telex 411483 MGU SU

Kheiker, Prof. Daniel M. (1930). Head of Laboratory. Member, Crystallographic Committee of Russia; Member, Commission on Synchrotron Radiation, IUCr. Laboratory of X-ray Diffractometry, Institute of Crystallography, Russian Academy of Sciences, 59 Leninsky prosp., Moscow 117333, Russia. DSc (Inst. of Crystallography Moscow, 1972) crystallography and crystal physics. *X-ray_diffractometry area_detector structure_biomolecule powder_diffraction*
E-mail CRYS@CRYS.MSK.SU Tel. 7(095)1356230 Fax 7(095)1351011

Khimich, Dr Tamara A. (1933). Researcher. Lab. of Resonance Methods, Institute of Crystallography, Russian Academy of Sciences, Leninsky pr. 59, Moscow 117333, Russia. PhD (Voronezh State U., 1972) physics. *Nuclear_magnetic_resonance magnetic_domain antiferromagnetism ferrites*
E-mail root@magnet.crystal.msk.su Tel. 7(095)3308329 Fax 7(095)3308329

Khisina, Dr Natasha R. (1945). Chief Researcher. Institute of Geochemistry and Analytical Chemistry, Russian Academy of Sciences, 19 Kosygin St., Moscow 117975, Russia. DSc (Inst. Geochemistry and Analyt. Chem., 1991) mineralogy. *X-ray_crystallography transmission_electron_microscopy minerals phase_transition isomorphous_replacement*
E-mail urusov@glas.apc.org Tel. 7(095)9391916 Fax 7(095)9382054 Telex 411633 TERRA SU

Khitrova, Dr Valentina I. Senior Researcher. Laboratory of Electron Diffraction, Institute of Crystallography, Russian Academy of Sciences, 59 Leninsky prosp., Moscow 117333, Russia. PhD (Inst. Crystallography RAS Moscow, 1963) physics. *High-energy_electron_diffraction chemical_reaction_mechanism structure_determination diffraction_data amorphous_phase*
Tel. 7(095)1354010

Khokhlov, Prof. Alexei R. (1954). Head of Chair. Chair of Polymer and Crystal Physics, Department of Physics, Moscow State University, Moscow 119899, Russia. DSc (Moscow State U., 1983) polymer physics. *Polymers_structure_conformation statistical_physics*
E-mail khokhlov@poly.phys.msu.su Tel. 7(095)9391013 Fax 7(095)9392988 Telex 411483 MGUSU

Khomenko, Mr Igor O. (1966). Junior researcher. Laboratory of X-ray Structure Analysis, Institute of Structural Macrokinetics, Russian Academy of Sciences, Chernogolovka, Moscow region 142432, Russia. MSc (Inst. of Physics and Technology Moscow, 1988) physical chemistry. *Time-resolved_effect combustion time-resolved_X-ray_diffraction*
E-mail postmaster@el-mech.msk.su Tel. 7(095)5245047 Fax 7(095)2017357 VNESHTECHNIKA FOR ISMAN

Khundzhua, Dr Andrey G. (1949). Professor. Solid State Chair, Physics Department, Moscow State University, Vorobjovy Gory, Moscow 117234, Russia. PhD (Moscow State U., 1980) solid state physics. *X-ray_diffraction electron_diffraction*
Tel. 7(095)9392387 Fax 7(095)9390247 Telex 411483 MGU SU

Kirikov, Dr Vladimir K. (1941). Senior Researcher. Laboratory of Electromechanical Investigations of Crystals, Institute of Crystallography, Russian Academy of Sciences, Leninsky pr. 59, Moscow 117333, Russia. PhD (Inst. of Crystallography Moscow, 1975) physics. *Phase_transition incommensurate ferroelectricity*
E-mail postmaster@el-mech.msk.su Tel. 7(095)3307856 Fax 7(095)1351011

Kirillova, Dr Nataliya I. (1945). Researcher. All-Russian Institute of Scientific and Technical Information (VINITI), 20a Usievich St., Moscow 125219, Russia. PhD (Inst. of Organo-element Compounds Moscow, 1975) chemistry. *Crystal_chemistry organic_inorganic_compound solid_state_chemistry complex_compound*
Tel. 7(095)1554205 Fax 7(095)9430060 Telex 411249 VINITI SU

Kiryanova, Dr Elena E. (1956). Senior Researcher. Crystal Genesis Lab., St Petersburg University, University Emb. 7/9, St Petersburg 199034, Russia. PhD (St Petersburg U., 1986) crystallography and crystal physics. *Crystallogeny crystal_growth crystal_morphology impurity_adsorption phase_diagram organic_impurity experimental_mineralogy*
E-mail anna@dean.geol.lgu.spb.su Tel. 7(812)2189650 Fax 7(812)2181346 Telex 121481 LSU SU

Kiselev, Prof. Nikolay A. (1928). Head of Laboratory. Institute of Crystallography, Russian Academy of Sciences, Leninsky pr. 59, Moscow 117333, Russia. DSc (Institute of Crystallography Moscow, 1964) biology. *High_resolution_electron_microscopy epitaxy_growth nanoanalysis structure_property_relationship*
Tel. 7(095)1351520 Fax 7(095)1351011

Klechkovskaya, Dr Vera V. (1938). Senior Researcher. Lab. of Electron Diffraction, Institute of Crystallography, Russian Academy of Sciences, 59 Leninsky prosp., Moscow 117333, Russia. PhD (Inst. of Crystallography RAS Moscow, 1974) physics. *High-energy_electron_diffraction structure_determination Langmuir-Blodgett_film liquid_crystals*
E-mail feigin@saxslab.incr.msk.su Tel. 7(095)1353500

Klevtsov, Dr Petr V. (1930). Chief Researcher. Laboratory of Crystal Chemistry, Institute of Inorganic Chemistry, Siberian Branch of RAS, 3 Lavrentyev prosp., Novosibirsk 630090, Russia. PhD (Inst. Crystallography Russian Academy of Sciences, 1955) crystallography and crystal physics. *Crystal_growth crystallography oxides physical_property inorganic_crystal_chemistry phase_transition*
Tel. 7(3832)354366 Fax 7(3832)355960

Klevtsova, Dr Rimma F. (1928). Chief Researcher. Lab. of Crystal Chemistry, Institute of Inorganic Chemistry, Siberian Branch of RAS, 3 Lavrentyev prosp., Novosibirsk 630090, Russia. PhD (Inst. Crystallography Russian Academy of Sciences, 1954) crystallography and crystal physics. *X-ray_crystallography molybdates tungstates classification database structural_computing inorganic_crystal_chemistry*
E-mail borisov@che.nsk.su Tel. 7(3832)354366 Fax 7(3832)355960

Klimova, Dr Anna Yu. (1943). Senior Researcher. Lab. of Crystallophysics, Shubnikov Institute of Crystallography, Russian Academy of Sciences, 59 Leninsky Pr., Moscow 117333, Russia. PhD (Inst. Crystallography Moscow, 1976) crystallography. *Optical_activity optical_property crystal_growth circular_dichroism inorganic_crystals*
E-mail root@theory.incr.msk.su Tel. 7(095)1354510 Fax 7(095)1351011

Klimova, Dr Irina P. (1955). Junior Researcher. Chair of Polymer and Crystal Physics, Physics Department, Moscow State University, Moscow 119899, Russia. PhD (Moscow State U., 1993) physics. *Crystal_growth crystal_structure physical_property oxides ferroelectrics superionic_conductors*
E-mail klimova@cryst.phys.msu.su Tel. 7(095)9392883 Fax 7(095)9392988 Telex 411483 MGU SU

Kobzareva, Dr Svetlana A. Senior Researcher. Lab. of High Temperature Crystallization, Institute of Crystallography, Russian Academy of Sciences, 59 Leninsky prosp., Moscow 117333, Russia. PhD (Inst. of Crystallography Moscow, 1966) chemistry. *Crystal_growth surface_real_structure electron_microscopy dielectrics_microanalysis*
Tel. 7(095)1350068 Fax 7(095)1356350 or 7(095)1351011

Kolesova, Dr Rimma V. (1945). Associate professor. Physics Department, Rostov State University, Pr. Stachki 192, Rostov-on-Don 344090, Russia. PhD (Rostov State U., 1967) physics and mathematics. *Crystal_structure*
Tel. 7(8632)243413

Kolin, Dr Nikolai G. (1948). Head of Department. Dept. of Radiative Physics, Obninsk Branch, Karpov Institute of Physical Chemistry, Obninsk, Kaluga Region 249020, Russia. PhD (Moscow Inst. of Steel and Alloys, 1986) physics. *III–V_compounds semiconductors nuclear_doped irradiated_semiconductors_physical_property*
Tel. 7(08439)26231

Kolobyanina, Dr Tatyana N. (1938). Senior Researcher. Laboratory of Monocrystals, Vereschagin Institute of High Pressure Physics, Russian Academy of Sciences, Troitsk, Moscow Reg. 142092, Russia. PhD (Moscow State U., 1974) physics and mathematics. *X-ray_crystallography phase_transition high_pressure*
E-mail scorpion@adonis.iasnet.com Tel. 7(095)3340733 Fax 7(095)3340012

Kolodieva, Dr Svetlana V. (1936). Chief Researcher. Lab. of Physical and Technical Research, Russian Research Institute for Synthesis of Minerals and Raw Materials, Institutskaya St. 1, Alexandrov, Vladimir Region 601600, Russia. PhD (Moscow State U., 1979) physics. *Crystal_physics piezoelectricity*
Tel. 7(095)5845834 Fax 7(095)5845828

Kompan, Dr Olga Ye. (1955). Senior Researcher. Institute of Physical and Organic Chemistry, Rostov State University, Pr. Stachki 194/3, Rostov-on-Don 344104, Russia. PhD (Rostov State U., 1981) chemistry. *X-ray_analysis single_crystal organic_compound structures_of_tautomeric_and_other_nonrigid_compounds*
E-mail root@pichko.rnd.su Tel. 7(8632)285700 Fax 7(8632)285667 Telex 123322 FOBOS SU

Kondratenko, Dr Ludmila K. (1946). Senior Researcher. Baikov Institute of Metallurgy, Russian Academy of Sciences, Laboratory of Crystallochemistry, 49 Leninskyi pr., Moscow 117334, Russia. PhD (Institute of Metallurgy Russian Academy of Sciences Moscow, 1975) physical metallurgy. *X-ray_crystallography lattice_distortion X-ray_powder_diffraction*
Tel. 7(095)1359619 Fax 7(095)1358680

Konovalikhin, Dr Sergei V. (1957). Senior Researcher. Branch Institute of Chemical Physics, Russian Academy of Sciences in Chernogolovka, Chernogolovka, Moscow Region 142432, Russia. PhD (Inst. Chemical Physics RAS, 1989) chemical physics. *X-ray_crystallography molecular_crystal structure_properties_relationship orbital_calculations*
Tel. 7(095)5245059 Fax 7(096)5153588

Konovalov, Mr Oleg V. (1961). Researcher. Small Angle Scattering Lab., Institute of Crystallography, Russian Academy of Sciences, Leninsky pr. 59, Moscow 117333, Russia. MSc (Chelyabinsk State U., 1983) solid state physics. *Reflectivity structure_determination thin_film Langmuir–Blodgett_film computing_programming direct_methods diffuse_scattering*
E-mail feigin@saxslab.incr.msk.su Tel. 7(095)1356010 Fax 7(095)1351011

Konstantinova, Dr Alisa F. (1936). Chief Researcher. Lab. of Crystallophysics, Shubnikov Institute of Crystallography, Russian Academy of Sciences, 59 Leninsky Pr., Moscow 117333, Russia. DSc (Inst. Crystallography Moscow, 1987) crystallography. *Crystal_optics anisotropic_optical_property light_absorption_spectroscopy refractive_index reflectivity rotatory_dispersion chirality circular_dichroism*
E-mail root@theory.incr.msk.su Tel. 7(095)1356150 Fax 7(095)1351011

Koptsik, Prof. Vladimir A. (1924). Professor. Member of Subcommission on N-dimensional crystallography. Chair of Polymer and Crystal Physics, Physics Department, Moscow State University, Moscow 119899, Russia. [Povarskaya St. 26, Apt. 19, Moscow 121069, Russia.] DSc (Moscow State U., 1963) crystallography and crystal physics. *Real_crystal_colour_symmetry quasicrystals_crystallography incommensurate_phases_crystallography tensor_crystal_physics_theory structure_phase_transition*
E-mail koptsik@cryst0.phys.msu.su or koptsik@plm.phys.msu.su Tel. 7(095)9391013 Fax 7(095)9392988 or 7(095)9391013 Telex 411483 MGU SU

Kornev, Dr Alexey N. (1944). Senior Researcher. Lab. of Nerve Cell Biophysics, Institute of Cell Biophysics, Russian Academy of Sciences, Pushchino, Moscow Region 142292, Russia. PhD (Inst. of Crystallography Moscow, 1973) crystallography. *Biocrystallography X-ray_crystallography biological_system_symmetry*
E-mail kornev@mars.ibioc.serpukhov.su Tel. 7(095)9255984 Telex (64) 412615 SU

Korolev, Dr Sergey V. (1962). Researcher. Laboratory of Structural and Computer Analysis, Engelhardt Institute of Molecular Biology, Russian Academy of Sciences, 32 Vavilov St., Moscow 117984, Russia. PhD (Engelhardt Inst. Molecular Biology, 1993) molecular biology. *X-ray_crystallography DNA_proteins_interaction computer_modelling computer_graphics*
E-mail korol@imb.mb.free.msk.su Tel. 7(095)1356000 Fax 7(095)9382187 Telex 411755 MOLBI SU

Kosova, Dr Tatyana B. (1940). Researcher. Lab. for Hydrothermal Synthesis, Institute of Crystallography, Russian Academy of Sciences, 59 Leninsky pr., Moscow 117333, Russia. PhD (Moscow State U., 1973) crystallography. *Crystal_growth hydrothermal_synthesis inorganic_compound silicates*
Tel. 7(095)3308156 Fax 7(095)1351011

Kotelnikova, Dr Elena N. (1945). Senior Researcher. Department of Crystallography, St Petersburg University, University Emb. 7/9, St Petersburg 199034, Russia. PhD (St Petersburg U., 1982) geology and mineralogy. *Crystal_chemistry X-ray_high-temperature_powder_diffraction phase_transformation structural_typomorphism_minerals*
E-mail ftt@dean.geol.lgu.spb.su Tel. 7(812)2189647 Fax 7(812)2181346 Telex 121481 LSU SU

Kovalchuk, Prof. Michail V. (1946). Laboratory Leader. Lab. of X-ray Optics and Synchrotron Radiation, Institute of Crystallography, Russian Academy of Sciences, 59 Leninsky prosp., Moscow 117333, Russia. DSc (Inst. of Crystallography Moscow, 1988) physics. *X-ray_standing_wave X-ray_dynamical_diffraction X-ray_and_synchrotron_radiation_instrumentation*
E-mail koval@crystal.msk.su Tel. 7(095)3307992 Fax 7(095)3301956

Kovyev, Dr Ernest K. (1960). Chief Researcher. Lab. of High Temperature Crystallization, Institute of Crystallography, Russian Academy of Sciences, 59 Leninsky prosp., Moscow 117333, Russia. PhD (Moscow State U., 1974) physics. *X-ray_diffraction diffuse_scattering real_structure crystal_defect*
Tel. 7(095)1351220 Fax 7(095)1356350 or 7(095)1351011

Krivandina, Dr Elena A. (1938). Senior Researcher. Laboratory of Physical and Chemical Analysis, Institute of Crystallography, Russian Academy of Sciences, 59 Leninsky prosp., Moscow 117333, Russia. PhD (Inst. of Crystallography Moscow, 1980) crystal growth. *Bridgman_Stockbarger_technique fluorine_compounds crystal_growth*
Tel. 7(095)3307874 Fax 7(095)1351011

Krivenko, Dr Vladimir G. (1939). Senior Researcher. Lab. of Resonance methods, Institute of Crystallography, Russian Academy of Sciences, Leninsky pr. 59, Moscow 117333, Russia. PhD (Inst. of Biophysics RAS, 1968) biophysics. *Nuclear_magnetic_resonance magnetic_domain antiferromagnetism ferrites*
E-mail root@magnet.crystal.msk.su Tel. 7(095)3308329 Fax 7(095)3308329

Kryshtop, Dr Victor M. (1945). Associate Professor. Physics Department, Rostov State University, 5 Zorge Str., Rostov-on-Don 344104, Russia. PhD (Rostov State U., 1980) physics and mathematics. *Crystal_structure perovskites*
Tel. 7(8632)223413

Kukina, Dr Galina A. Researcher. Laboratory of Coordination Compounds Crystal Chemistry, Kurnakov Institute of General and Inorganic Chemistry, Russian Academy of Sciences, 31 Leninsky prosp., Moscow, Russia. PhD (Inst. General and Inorganic Chemistry, 1962) physical chemistry. *X-ray_crystallography coordination_compound organometallic_compounds inorganic_compounds secondary_bonding*
Tel. 7(095)9521803 Fax 7(095)9541279

Kukuy, Dr Anatoly L. (1939). Senior Researcher. Department of Prospecting Exploration and Economical Estimation of Mineral Deposit, s, St Petersburg Mining Institute, 2 21st Liniya, Saint Petersburg 199026, Russia. PhD (Mining Inst. St Petersburg, 1993) crystallography mineralogy and petrography. *Morphology crystal_growth real_structure computing_programming*
Tel. 7(812)2188262 Fax 7(812)2132613 Telex 121494 LGIPSU

Kupriyanov, Dr Mikhail F. (1937). Associate Professor. Physics Department, Rostov State University, Pr. Stachki 192, Rostov-on-Don 344090, Russia. DSc (Latvian State U., 1992) physics and mathematics. *Crystal_chemistry phase_transition complex_oxides*
Tel. 7(8632)223413

Kuranova, Dr Inna P. (1933). Chief Researcher. Laboratory of Biocrystal Structure, Institute of Crystallography, Russian Academy of Sciences, 59 Leninsky prosp., Moscow 117333, Russia. DSc (Moscow State U., 1991) chemistry. *Biocrystallography proteins_crystal_growth metalloenzyme structure-activity_relationship hemoglobin protein_computer_graphics enzymes_structure_determination*
Tel. 7(095)1356220 Fax 7(095)1351011

Kurdyumov, Prof. Georgy Vyacheslavovich (1902). Full member Acad. Sci. of Russia. Russian Academy of Sciences, Leninsky pr. 14, Moscow 117901, Russia. DSc (1937) physics and mathematics. *Metals_physics real_structure*
Tel. 7(095)9381695

Kurkutova, Prof. Evdokiya N. Professor. Department of Physics, Vladimir State Pedagogical University, Department of Physics, 11 Stroiteley Prospect, Vladimir 600024, Russia. DSc (Inst. Crystallography Moscow, 1979) crystallography. *X-ray_crystallography molecular_crystal*
Tel. 7(09222)72628

Kuz'micheva, Prof. Galina M. (1947). Chief Researcher; Leader of X-ray Laboratory. Lab. of X-ray Diffraction Studies, Department of Solid State Physics and Chemistry, Lomonosov Institute of Fine Chemical Technology, 86 pr. Vernadskogo, Moscow 117571, Russia. DSc (Inst. Fine Chemical Technology Moscow, 1992) physical chemistry. *X-ray_diffractometry rare-earth_compounds structure_determination semiempirical_calculation inorganic_crystal_chemistry*
E-mail alekz@cs.msk.su alekz@red.com.ru vigdor@ecoinsys.msk.su Tel. 7(095)2480762 Fax 7(095)4307983

Kuzmin, Dr Ivan I. (1932). Senior Researcher. Dept. of Radiative Physics, Obninsk Branch, Karpov Institute of Physical Chemistry, Obninsk, Kaluga Region 249020, Russia. PhD (Inst. of Crystallography Moscow, 1970) physics. *Ferroelectric_property semiconductors thermodynamics irradiated_materials*
Tel. 7(08439)26389

Kuzmin, Prof. Runar N. (1932). Professor. Solid State Chair, Physics Department, Moscow State University, Vorobjovy Gory, Moscow 117234, Russia. DSc (Moscow State U., 1970) physics. *X-ray_diffraction Mossbauer_spectroscopy condensed_matter cooperative_phenomena bioenergetics*
E-mail asi@phys.msu.su. Tel. 7(095)9391226 Fax 7(095)9390247 Telex 411483 MGU SU

Kuzmin, Dr Vladimir S. (1946). Chief Researcher. Lab. of Phase Transformations, Kurnakov Institute of General and Inorganic Chemistry, Russian Academy of Sciences, 31 Leninsky pr., Moscow 117907, Russia. DSc (Kurnakov Institute of General and Inorganic Chemistry RAS, 1991) physical chemistry. *X-ray_crystallography small_molecules molecular_mechanics active_site_recognition adrenergic_compounds*
Tel. 7(095)9521803 Fax 7(095)9541279

Kuzmina, Dr Olga V. Researcher. Laboratory of X-ray Structure Analysis, Institute of Geology Ore Deposits, Petrology, Mineralogy and Geochemistry, Russian Academy of Sciences, Staromonetny per. 35, Moscow 109017, Russia. PhD (Inst. of Ore Deposits, Petrography, Mineralogy and Geochemistry, 1989) crystal chemistry and mineralogy. *X-ray_crystallography polytypism powder_diffraction clay_minerals*
Tel. 7(95)2308296 Fax 7(95)2302179

Kuzmina, Dr Ludmila G. (1945). Chief Researcher. Laboratory of Crystal Chemistry of Coordinational Compounds, Kurnakov Institute of General and Inorganic Chemistry, Russian Academy of Sciences, 31 Leninsky prosp., Moscow, Russia. DSc (Inst. General and Inorganic Chemistry, 1990) physical chemistry. *X-ray_crystallography coordination_compound organometallic_compounds organic_compounds secondary_bonds*
E-mail vip@solen.msk.su Tel. 7(095)9521803 Fax 7(095)9541279

Kuz'mina, Dr Mariya A. (1962). Researcher. Lab. of Modelling of Nature Crystallization, Earth Crust Research Institute, St Petersburg University, University Emb 7/9, St Petersburg 199034, Russia. PhD (St Petersburg U., 1987) crystallography and crystal physics. *Crystal_growth crystal_defect_genesis real_crystal high_temperature_superconductors*
E-mail anna@dean.geol.lgu.spb.su Tel. 7(812)2189644 Fax 7(812)2181346 Telex 121481 LSU SU

Kuznetsov, Dr Victor A. (1938). Chief Researcher. Laboratory of High-Temperature Crystallization, Inst. of Crystallography, Russian Academy of Sciences, 59 Leninsky prosp., Moscow 117333, Russia. DSc (Inst. of Crystallography, 1986) chemistry. *Crystal_growth adsorption*
Tel. 7(095)3301792 Fax 7(095)1351011

Kuznetsova, Mrs Tatyana P. Researcher. Institute of Geochemistry and Analytical Chemistry, Russian Academy of Sciences, 19 Kosygin St., Moscow 117975, Russia. MSc (Gorky State U., 1972) solid state physics. *X-ray_crystallography X-ray_diffractometry minerals silicates diffraction_technique order_disorder_structure*
Tel. 7(095)9391916 Fax 7(095)9382054 Telex 411633 TERRA SU

Lapshin, Mr Andrey E. (1957). Researcher. Lab. of X-ray Crystallography, Institute of Silicate Chemistry, Russian Academy of Sciences, 24/2 Odoevskogo st., St Petersburg 199155, Russia. PhD (Inst. of Silicate Chemistry St Petersburg, 1993) physical chemistry. *X-ray_crystal_structure_determination complex_vanadium_compounds*
E-mail lapshin@ihs2.spb.su Tel. 7(812)2185102 Fax 7(812)3510813 Telex 121447 VITRO SU

Laptev, Dr Vladimir A. (1943). Chief Researcher. Lab. for High Pressure Synthesis of Materials, Russian Research Institute for Synthesis of Minerals and Raw Materials, Institutskaya St. 1, Alexandrov, Vladimir Region 601600, Russia. PhD (Inst. of Solid State Physics Minsk, 1978) physical chemistry. *Crystal_growth high_pressure_synthesis semiconductors*
Tel. 7(095)5845834 Fax 7(095)5845828

Leitus, Dr Grigorii M. (1955). Senior Researcher. Baikov Institute of Metallurgy, Russian Academy of Sciences, Laboratory of Crystallochemistry, 49 Leninskyi pr., Moscow 117334, Russia. PhD (Institute of Steel and Alloys Moscow, 1989) physics. *X-ray_crystallography structure_determination Rietveld_method superconductors phase_determination X-ray_powder_diffraction texture modulated_structures phase_equilibrium*
Tel. 7(095)1359618 Fax 7(095)1358680

Leonyuk, Dr Lydia I. (1950). Senior Researcher. Moscow State University, Geology Faculty, Department of Crystallography and Crystal Chemistry, 119899 Moscow, Russia. PhD (Moscow State U., 1978) crystallography and crystal physics. *Crystal_growth structure_determination oxygen_compounds superconductivity*
E-mail lydia@geocr.phys.msu.su. Tel. 7(095)9392881 Fax 7(095)9390126 Telex 411483 MGU SU

Leonyuk, Dr Nikolai I. (1941). Professor. Moscow State University, Geology Faculty, Department of Crystallography and Crystal Chemistry, 119899 Moscow, Russia. DSc (Moscow State U., 1985) chemistry. *Crystal_growth morphology boron_compounds oxygen_compounds Czochralskii_technique crystallization crystal_form kinetics borosilicates silicates tantalum_compounds niobium_compounds*
E-mail lydia@geocr.phys.msu.su. Tel. 7(095)9392980 Fax 7(095)9390126 Telex 411483 MGU SU

Levin, Mr Alexandr A. (1963). Researcher. Lab. of X-ray Crystallography, Institute of Silicate Chemistry, Russian Academy of Sciences, 24-2 Odoevskogo str., St Petersburg 199155, Russia. MSc (St Petersburg State U., 1992) physics. *X-ray_crystal_structure_determination high_temperature_superconductor silicates diffraction_technique phase_transition modulated_structures*
E-mail levin@ihs2.spb.su Tel. 7(812)2185102 Fax 7(812)3510813 Telex 121447 VITRO SU

Levina, Dr Olga I. Associate Professor. Department of Physics, Vladimir State Pedagogical University, Department of Physics, 11 Stroiteley Prospect, Vladimir 600024, Russia. PhD (Inst. Crystallography Moscow, 1992) crystallography. *Conformational_methodology organic_crystal_chemistry ring_molecules*
Tel. 7(09222)72604

Li, Dr Ludmila E. (1944). Researcher. Lab. of Resonance methods, Institute of Crystallography, Russian Academy of Science, 59 Leninsky prosp., Moscow 117321, Russia. PhD (Institute of Crystallography Moscow, 1975) physics. *Crystal_field_theory method_optimization* Mossbauer_spectral_analysis computing_programming
E-mail root@magnet.crystal.msk.su Tel. 7(095)3308329 Fax 7(095)3308329

Likhushina, Dr Ekaterina V. (1956). Researcher. Solid State Chair, Physics Department, Moscow State University, Vorobjovy Gory, Moscow 117234, Russia. PhD (Moscow State U., 1985) solid state physics. *X-ray_analysis amorphous_compound*
E-mail asi@phys.msu.su. Tel. 7(095)9392387 Fax 7(095)9390247 Telex 411483 MGU SU

Lindeman, Dr Sergey V. (1958). Senior Researcher. X-ray Structural Center of the Russian Academy of Sciences, Institute of Organoelement Compounds, Russian Academy of Sciences, 28 Vavilov St., Moscow 117813, Russia. PhD (Inst. Organoelement Compounds, 1988) physical chemistry. *Organic_structure bioactive_molecules polymers_model*
E-mail serg@xray.ineos.ac.ru Tel. 7(095)1359343 Fax 7(095)1355085

Litvinov, Dr Igor A. (1953). Senior Researcher. Institute of Organic and Physical Chemistry, Kazan Scientific Center, Russian Academy of Sciences, Arbuzov str. 8, Kazan 420083, Russia. DSc (Kazan State U., 1993) chemistry. *Structural_chemistry_organic_organometallic_compound phosphorus_chemistry stereoelectronic_effects*
Tel. 7(8432)767424 Fax 7(8432)752253

Lityagina, Dr Ludmila M. (1942). Researcher. Lab. of Monocrystals, Vereschagin Institute of High Pressure Physics, Russian Academy of Sciences, Troitsk, Moscow Reg. 142092, Russia. PhD (Moscow State U., 1975) geology and mineralogy. *X-ray_crystallography structural_change high_pressure*
E-mail scorpion@adonis.iasnet.com Tel. 7(095)3340733 Fax 7(095)3340012

Loginov, Dr Evgenii B. (1951). Researcher. Laboratory of Crystallophysics, Institute of Crystallography, Russian Academy of Sciences, Leninsky prospekt 59, Moscow 117333, Russia. PhD (Institute of Crystallography Moscow, 1985) solid state physics. *Liquid_crystal_theory phase_transition phase_diagram layered_compounds_theory*
E-mail root@xray.incr.msk.su Tel. 7(095)1356240 Fax 7(095)1351011

Lomonov, Dr Vladimir A. (1948). Senior Researcher. Lab. of Acoustooptics of Crystals, Institute of Crystallography, Russian Academy of Sciences, Leninsky pr. 59, 117333 Moscow, Russia. PhD (Chemical-Technological Inst. Moscow, 1982) chemistry. *Crystal_growth Czochralski_technique*
Tel. 7(095)3306883 Fax 7(095)3301956

Loshmanov, Dr Arkady A. (1929). Senior Researcher. Laboratory of X-ray Structure Analysis, Institute of Crystallography, Russian Academy of Sciences, 59 Leninsky pr., Moscow 117333, Russia. PhD (Inst. of Metallurgy of Ferrous Metals, 1967) physics and mathematics. *Neutron_diffraction crystalline_non-crystalline_state*
Tel. 7(095)1351020 Fax 7(095)1351011

Lube, Dr Emil'.L. (1936). Senior Researcher. Lab. of High Temperature Crystallization, Institute of Crystallography, Russian Academy of Sciences, 59 Leninsky prosp., Moscow 117333, Russia. PhD (Inst. of Crystallography Moscow, 1971) physics. *Crystal_growth crystal_defect computer_modelling*
Tel. 7(095)1356500 Fax 7(095)1351011

Lunin, Dr Vladimir Yu. (1951). Head of Laboratory. Lab. of Macromolecular Crystallography, Institute of Mathematical Problems of Biology, Russian Academy of Sciences, Pushchino, Moscow Region 142292, Russia. DSc (Inst. of Crystallography Moscow, 1992) crystallography. *Biocrystallography imaging phase_determination phase_refinement*
E-mail com@impb.serpukhov.su Tel. 7(095)9233558 Telex (64) 412604 CODON SU

Lvov, Dr Yuri M. (1952). Chief Researcher. Small Angle Scattering Lab., Institute of Crystallography, Russian Academy of Sciences, Leninsky pr. 59, Moscow 117333, Russia. DSc (Institute of Crystallography Moscow, 1991) crystallography. *SAXS reflectivity thin_film heterostructure superlattice molecular_architecture_self-assembly thin_organized_films Langmuir–Blodgett_films lipids_mesophases*
E-mail feigin@saxslab.incr.msk.su Tel. 7(095)1356010 Fax 7(095)1351011

Lyubalin, Prof. Mark Dm. (1937). Group leader. Russian-American Joint Venture, "Incorporation 4T", 4/6 Ispolkomskaya str., St Petersburg 190000, Russia. DSc (St Petersburg U., 1989) crystallography and crystal physics. *Crystallography crystal_growth crystal_symmetry crystal_materials oxides semiconductor*
Tel. 7(812)2770831 Fax 7(812)2770831 Telex 121347 KRUIZ

Lyubimov, Dr Vasily N. (1936). Leading Researcher. X-ray Laboratory, Karpov Institute of Physical Chemistry, Obukha St., Moscow 103064, Russia. DSc (Inst. of Crystallography Moscow, 1988) solid state physics. *Optics acoustics piezoelectricity symmetry*
E-mail lyubimov@nifhi.uucp.free.msk.su Tel. 7(095)2270014x2346 Fax 7(095)9752450

Lyubutin, Prof. Igor S. (1938). Group leader. Lab. of Resonance Methods, Institute of Crystallography, Russian Academy of Sciences, Leninsky pr. 59, Moscow 117333, Russia. DSc (Inst. of Crystallography Moscow, 1975) physics. *Mossbauer_spectroscopy magnetic_materials phase_transition superconductivity*
E-mail root@magnet.crystal.msk.su Tel. 7(095)3308329 Fax 7(095)3308329

Lyutin, Dr Vladimir I. (1948). Chief Researcher. Lab. for Synthesis of Materials by Hydrothermal Method, Russian Research Institute for Synthesis of Minerals and Raw Materials, Institutskaya St. 1, Alexandrov, Vladimir Region 601600, Russia. PhD (Inst. of Crystallography Moscow, 1974) crystallography and crystal physics. *X-ray_diffractometry crystal_growth crystal_chemistry*
Tel. 7(095)5845834 Fax 7(095)5845828

Magaril, Dr Svetlana A. (1946). Senior Researcher. Laboratory of Crystal Chemistry, Institute of Inorganic Chemistry Siberian Branch, Russian Academy of Sciences, 3 Ak. Lavrentiev prosp., Novosibirsk 630090, Russia. PhD (Inst. Inorg. Chem. Russian Academy of Sciences, 1974) inorganic chemistry. *Crystal_structure_database structural_computing* superconducting_oxides inorganic_crystal_chemistry
E-mail borisov@che.nsk.su Tel. 7(3832)354366 Fax 7(3832)355960

Magomedova, Dr Nina S. (1945). Researcher. X-ray Laboratory, Karpov Institute of Physical Chemistry, 10 Obukha St., Moscow 103064, Russia. PhD (Karpov Institute of Physical Chemistry, 1984) physical chemistry. *X-ray_crystallography organic_phosphorus_compound small_molecules organic_organometallic_crystal_chemistry*
E-mail belsky@nifhi.uucp.free.msk.su Tel. 7(095)2270014x2748 Fax 7(095)9752450

Makarova, Dr Irina P. (1958). Researcher. Laboratory of X-ray Structure Analysis, Institute of Crystallography, Russian Academy of Sciences, 59 Leninsky pr., Moscow 117333, Russia. PhD (Inst. of Crystallography Moscow, 1990) physics and mathematics. *Structural_analysis phase_transition inorganic_compound superconductivity*
E-mail makarova@rsa.crystal.msk.su Tel. 7(095)1350400 Fax 7(095)1351011

Makhina, Dr Irina B. (1946). Laboratory Head. Lab. of Experimental Mineralogy and Treatment Technology, Russian Research Institute for Synthesis of Minerals and Raw Materials, Institutskaya St. 1, Alexandrov, Vladimir Region 601600, Russia. PhD (Moscow State U., 1980) mineralogy. *Crystal_growth hydrothermal_synthesis mineralogy*
Tel. 7(095)5845834 Fax 7(095)5845828

Malakhova, Dr Ludmila F. (1941). Senior Researcher. Member of IUCr Commission on Crystallographic Apparatus. Laboratory of X-ray Structure Analysis, Institute of Crystallography, Russian Academy of Sciences, 59 Leninsky pr., Moscow 117333, Russia. PhD (Inst. of Crystallography Moscow, 1976) crystallography. *X-ray_analysis technique apparatus*
E-mail malakh@rsa.crystal.msk.su Tel. 7(095)1353400 Fax 7(095)1351011

Maleev, Dr Andrey V. (1962). Associate Professor. Department of Physics, Vladimir State Pedagogical University, Department of Physics, 11 Stroiteley Prospect, Vladimir 600024, Russia. PhD (Inst. Applied Physics, 1990) crystallography. *Geometric_crystallography packing molecular_crystal conformational_methodology X-ray_crystallography organic_crystal_chemistry ring_molecules*
Tel. 7(09222)72604

Malenkov, Dr George G. (1938). Head of the laboratory. Lab. of Computer Simulation of Physical Chemical Processes, Institute of Physical Chemistry, Russian Academy of Sciences, 31 Leninsky pr., Moscow 117915, Russia. DSc (Ins. of Physical Chemistry RAS, 1990) physical chemistry. *Crystalline_hydrates_structure hydrogen_bonding non-crystalline_phases_structures computer_simulation science_history*
E-mail malenkov@lmm.phyche.msk.su Tel. 7(095)9554477 Fax 7(095)9527514 Telex 411029 pesum su

Malinina, Dr Lucy V. (1948). Group leader. Laboratory of X-ray Analysis, Engelhardt Institute of Molecular Biology, Russian Academy of Sciences, 32 Vavilova Str., Moscow 117984, Russia. PhD (Moscow Physics and Technology Institute, 1978) molecular biophysics. *X-ray_analysis DNA_proteins_crystallization X-ray_crystallography biomacromolecules oligonucleotides*
E-mail lucy@imb.msu.su Tel. 7(095)1359772 Fax 7(095)1351405 Telex 411755 molbi su

Malinovsky, Dr Yuri A. (1947). Senior Researcher. Laboratory of X-ray Structure Analysis, Institute of Crystallography, Russian Academy of Sciences, 59 Leninsky pr., Moscow 117333, Russia. PhD (Moscow State U., 1976) crystallography. *X-ray_crystallography crystal_chemistry inorganic_compound twinning inorganic_compounds_classification silicates rare-earth_compounds*
E-mail YMalinov@rsa.crystal.msk.su Tel. 7(095)1350330 Fax 7(095)1351011

Mal'tsev, Dr Yurii F. (1945). Associate Professor. Physics Department, Rostov State University, 5 Zorge Str., Rostov-on-Don 344104, Russia. PhD (Rostov State U., 1974) physics and mathematics. *X-ray_diffraction dynamical_theory powder_diffractometry*
Tel. 7(8632)220835

Malyushitskaya, Dr Zinaida V. Researcher. Department of Superhard Materials and Applied Investigations, Institute of High Pressure Physics, Russian Academy of Sciences, 142092 Troitsk, Moscow region 142092, Russia. PhD (Moscow State U., 1975) physical chemistry. *High_pressure_polymorphism phase_diagram_semiconductor solids_amorphization_mechanism*
E-mail scorpion@adonis.iasnet.com Tel. 7(095)3340593 Fax 7(095)3340012

Man, Dr Lucia I. Researcher. Editor, English translation of Kristallografiya (Crystallographic Reports). Lab. of Electron Diffraction, Institute of Crystallography, Russian Academy of Sciences, 59 Leninsky prosp., Moscow 117333, Russia. PhD (Inst. of Crystallography RAS Moscow, 1970) physics. *High-energy_electron_diffraction electron_microscopy mechanical_property diffraction_theory scientific_translation*
Tel. 7(095)1354010

Marin, Dr Anatoly A. (1946). Senior Researcher. Lab. of Experimental Mineralogy and Treatment Technology, Russian Research Institute for Synthesis of Minerals and Raw Materials, Institutskaya St. 1, Alexandrov, Vladimir Region 601600, Russia. PhD (Moscow State U., 1978) mineralogy. *Crystal_chemistry crystal_growth mineralogy quartz limestones*
Tel. 7(095)5845834 Fax 7(095)5845828

Marsille, Dr Irina M. Researcher. Laboratory of X-ray Structure Analysis, Institute of Geology of Ore Deposits, Petrology, Mineralogy and Geochemistry, Russian Academy of Sciences (IGEM RAS), Staromonetny per. 35, Moscow 109017, Russia. PhD (IGEM RAS, 1988) crystal chemistry and mineralogy. *X-ray_crystallography minerals order-disorder_structure*
Tel. 7(95)2308296 Fax 7(95)2302179

Maslov, Dr Andrey V. (1952). Senior Researcher. Lab. of Crystal Layers Diffractometry, Institute of Crystallography, Russian Academy of Sciences, 59 Leninsky prosp., Moscow 117333, Russia. PhD (Inst. of Crystallography Moscow, 1988) solid state physics. *X-ray_diffraction standing_wave computing thin_film_structure heterostructure localization_of_impurity_atom_position*
Tel. 7(095)3306856 Fax 7(095)1351011

Masunov, Mr Artem Eduardovich (1966). Researcher. Physical Chemistry Faculty, Chemical Dept., Moscow State University, Moscow 119899, Russia. MSc (Moscow State U., 1988) chemistry. *Crystal_packing secondary_bonding polymorphism charge_density*
E-mail masunov@indep.mepi.msk.su Tel. 7(095)2492652 Fax 7(095)3242111 Telex 411483 MCUSU

Matveeva, Dr Olga P. (1951). Senior Researcher. Lab. of Mineral Raw Material Quality Evaluation, Saint-Petersburg Mining Institute, 2 21st Liniya, St Petersburg 199026, Russia. PhD (St Petersburg State U., 1985) physics. *Solid_state_physics minerals optical_spectroscopy_methods*
Tel. 7(812)2188683 Fax 7(812)2132613 Telex 121494 LGIPSU

Maximov, Dr Boris A. (1941). Chief Researcher. Laboratory of X-ray Structure Analysis, Institute of Crystallography, Russian Academy of Sciences, 59 Leninsky pr., Moscow 117333, Russia. PhD (Inst. of Crystallography Moscow, 1969) crystallography physics of crystals. *X-ray_crystallography crystal_chemistry inorganic_structure phase_transition crystals_twinning defect_structure*
E-mail simonov@rsa.crystal.msk.su Tel. 7(095)1350330 Fax 7(095)1351011

Mchedlishvili, Dr Boris V. (1944). Head of Laboratory. Laboratory of Nuclear Filters, Institute of Crystallography, Russian Academy of Sciences, 59 Leninsky pr., Moscow 117333, Russia. DSc (Inst. of Physical Chemistry Moscow, 1992) physical chemistry. *Purification biopolymers viruses chromatography membranes membrane_filtration*
E-mail track@imb.mb.freenet Tel. 7(095)1350201 Fax 7(095)1351011

Meleshina, Dr Valentina A. Senior Researcher. Laboratory of Physical and Chemical Analysis, Institute of Crystallography, Russian Academy of Sciences, 59 Leninsky pr., Moscow 117333, Russia. PhD (Inst. of Crystallography Moscow, 1966) physical chemistry. *Crystal_growth_theory superconductor laser_crystals*
Tel. 7(095)3307874 Fax 7(095)1351011

Melik-Adamyan, Dr William R. (1937). Chief Researcher. Member of IUCr Commission on Biological Macromolecules. Laboratory of Biocrystal Structure, Institute of Crystallography, Russian Academy of Sciences, 59 Leninsky prosp., Moscow 117333, Russia. DSc (Inst. of Crystallography Moscow, 1990) protein crystallography. *X-ray_crystallography proteins_crystallization*
Tel. 7(095)1353098 Fax 7(095)1351011

Melnikov, Dr Oleg K. (1940). Senior Researcher. Laboratory of Optical Crystals Physics, Institute of Crystallography, Russian Academy of Sciences, 59 Leninskii prosp., Moscow 117333, Russia. PhD (Moscow State U., 1968) crystallography. *Crystal_growth flux crystal_chemistry crystal_growth*
Tel. 7(095)3308247 or 7(095)3308047 Fax 7(095)1351011

Melnikova, Dr Svetlana V. (1946). Researcher. Laboratory of Crystal Physics, Institute of Physics, Siberian Dept., Russian Academy of Sciences, Krasnoyarsk 660036, Russia. PhD (Inst. of Physics Krasnoyarsk, 1984) solid state physics. *Structural_phase_transition inorganic_compound optical_property*
E-mail aleks@iph.krasnoyarsk.su Tel. 7(391)2432635 Fax 7(391)2438923

Meshalkin, Dr Sergey S. (1945). Researcher. Vernadsky Institute of Geochemistry and Analytical Chemistry, Russian Academy of Sciences, 19 Kosygin St., Moscow 117975, Russia. PhD (Moscow State U., 1993) mineralogy. *X-ray_diffractometry X-ray_mineralogy crystal_structure Rietveld_method bond_length environment_protection zeolites real_structure high_precision_diffractometry software_minerals high-temperature_nonstoichiometric_phase_transition*
E-mail abasilevsky@glas.apc.org Tel. 7(095)9391916 Fax 7(095)9382054

Mikhailov, Dr Albert M. (1939). Chief researcher. Laboratory of Biocrystal Structure, Institute of Crystallography, Russian Academy of Sciences, 59 Leninsky prosp., Moscow 117333, Russia. PhD (Inst. of Crystallography Moscow, 1971) protein and virus crystallography. *X-ray_crystallography virus_structure_determination molecular_replacement Laue_diffraction three-dimensional_reconstruction macromolecules computer_modelling*
Tel. 7(095)1355420 Fax 7(095)1351011

Mill', Dr Boris V. (1936). Senior Researcher. Physics Department, Moscow State University, Moscow 119899, Russia. PhD (Inst. of Crystallography Moscow, 1966) chemistry. *Crystal_growth oxides Czochralski_technique flux_growth isomorphism gallium_compounds germanates silicates*
E-mail mill@plm.phys.msu.su Tel. 7(095)9393918

Minacheva, Dr Lidiya Kh. (1938). Senior Researcher. Laboratory of Coordination Compounds Crystal Chemistry, Kurnakov Institute of General and Inorganic Chemistry, Russian Academy of Sciences, 31 Leninsky prosp., Moscow 117907, Russia. PhD (Inst. General and Inorganic Chemistry, 1971) inorganic chemistry. *X-ray_crystallography coordination_compound podands_complexes_structural_investigation*
Tel. 7(095)9521803 Fax 7(095)9541279

Mints, Prof. Rafail I. (1944). Head of Department. Ural Polytechnic Institute, Ekaterinburg 620002, Russia. DSc (Ural Polytechnic Inst., 1965) technology. *Structural_phase_transition metastable_phase condensed_matter*

Minyukov, Dr Sergey M. (1951). Senior Researcher. Laboratory of Electromechanical Investigations of Crystals, Institute of Crystallography, Russian Academy of Sciences, Leninsky pr. 59, Moscow 117333, Russia. PhD (Inst. of Crystallography Moscow, 1984) physics. *Phase_transition_theory thermodynamics kinetics*
E-mail minyukov@el-mech.incr.msk.su Tel. 7(095)3307856 Fax 7(095)1351011

Mistryukov, Mr Alexander E. (1965). Researcher. Laboratory of Coordination Compounds Crystal Chemistry, Kurnakov Institute of General and Inorganic Chemistry, Russian Academy of Sciences, 31 Leninsky prosp., Moscow 117907, Russia. MSc (Moscow State U., 1987) crystal chemistry. *Organometallic_compound*
Tel. 7(095)9521803 Fax 7(095)9541279

Mokhov, Dr Andrey V. (1953). Senior Researcher. Laboratory of Electron Microscopy and Electronography, Institute Geology of Ore Deposits, Petrology, Mineralogy and Geochemistry, Russian Academy of Sciences (IGEM RAS), Staromonetny 35, 109017 Moscow, Russia. PhD (IGEM RAS, 1987) mineralogy. *Transmission_electron_microscopy minerals energy_dispersive_spectroscopy*
E-mail avm@igem.msk.su Tel. 7(095)2308411 Fax 7(095)2302179

Molchanov, Dr Vladimir N. (1952). Senior Researcher. Laboratory of X-ray Structure Analysis, Institute of Crystallography, Russian Academy of Sciences, 59 Leninsky pr., Moscow 117333, Russia. PhD (Inst. of Crystallography Moscow, 1982) crystallography. *X-ray_analysis X-ray_diffraction_technique inorganic_compound*
E-mail vladimol@rsa.crystal.msk.su Tel. 7(095)1353400 Fax 7(095)1351011

Morgunova, Dr Ekaterina Yu. (1960). Senior researcher. Laboratory of Biocrystal Structure, Institute of Crystallography, Russian Academy of Sciences, 59 Leninsky prosp., Moscow 117333, Russia. PhD (Inst. of Crystallography Moscow, 1991) protein crystallography. *X-ray_crystallography crystallization proteins viruses*
Tel. 7(095)1355420 Fax 7(095)1351011

Moshkin, Dr Sergey V. (1952). Senior Researcher. Lab. of Modelling of Nature Crystallization, Earth Crust Research Institute, St Petersburg University, University Emb 7/9, St Petersburg 199034, Russia. PhD (St Petersburg U., 1982) crystallography and crystal physics. *Crystal_growth crystal_defect_genesis computer_modelling high_temperature_superconductors*
E-mail anna@dean.geol.lgu.spb.su Tel. 7(812)2189644 Fax 7(812)2181346 Telex 121481 LSU SU

Mukasyan, Dr Alexander S. (1956). Head of Laboratory. Laboratory of SHS-processes mechanism, Institute of Structural Macrokinetics, Russian Academy of Sciences, Chernogolovka, Moscow Region 142432, Russia. PhD (Inst. Chemical Physics RAS, 1986) physical chemistry. *Electron_probe_microanalysis ceramics phase_formation nitrogen compounds ceramics_electron-dispersive_analysis ceramics_synthesis*
E-mail root@ism2.sherna.msk.su Tel. 7(095)5245047 Fax 7(095)2017357 (Vneshteknika for ISMAN) Telex 911652KLEN SU

Mukhamedzhanov, Dr Enver Kh. (1955). Senior Researcher. Lab. of Crystal Layers Diffractometry, Institute of Crystallography, Russian Academy of Sciences, 59 Leninsky prosp., Moscow 117333, Russia. PhD (Inst. of Crystallography Moscow, 1986) crystallography and crystal physics. *X-ray_diffraction standing_wave secondary_electron_emission multiple-crystal_diffractometry thin_film_structure heterostructure localization_of_impurity_atom_position*
Tel. 7(095)3342449 Fax 7(095)1351011

Mukhin, Dr Boris V. (1959). Researcher. Lab. of X-ray Diffraction Studies, Department of Solid State Physics and Chemistry, Lomonosov Academy of Fine Chemical Technology, 86 pr. Vernadskogo, Moscow 117571, Russia. PhD (Inst. Fine Chemical Technology, Moscow, 1993) physical chemistry. *X-ray_diffractometry rare-earth_compounds real_structure structural_disorder_compound structure_determination inorganic_crystal_chemistry*
E-mail alekz@cs.msk.su alekz@red.com.ru Tel. 7(095)2480762 Fax 7(095)4307983

Munchaev, Mr Anzor I. (1938). Researcher. Lab. of High Temperature Crystallization, Institute of Crystallography, Russian Academy of Sciences, 59 Leninsky prosp., Moscow 117333, Russia. MSc (Daghestan State U., 1962) physics. *Dielectrics_microanalysis annealing colour_center*
Tel. 7(095)1351220 Fax 7(095)1356350 or 7(095)1351011

Murashova, Dr Elena V. Researcher. Lab. of Rare Elements Chemistry and Inorganic Polymers, Kurnakov Institute of General and Inorganic Chemistry, Russian Academy of Sciences, 31 Leninsky prosp., Moscow, Russia. PhD (Inst. of Chem. Reagents and Ultrapure Chem. Substances, 1990) inorganic chemistry. *Structure_determination synthesis phosphates vanadates*
Tel. 7(095)9522487 Fax 7(095)9541279

Nabatov, Mr Victor N. (1932). Senior Researcher. Laboratory of Optical Crystals Physics, Institute of Crystallography, Russian Academy of Sciences, 59 Leninskii prosp., Moscow 117333, Russia. MSc (Moscow State U., 1958) physics. *Optical_property crystal_laser*
Tel. 7(095)3308247 or 7(095)3308047 Fax 7(095)1351011

Nadezhina, Dr Tamara N. (1946). Senior Researcher. Department of Crystallography and Crystal Chemistry, Faculty of Geology, Moscow State University, Moscow 119899, Russia. PhD (Moscow State U:, 1975) crystallography and crystal physics. *X-ray_crystallography minerals materials mineralogy crystallography physical_property crystal_growth*
Tel. 7(095)9394923 Fax 7(095)9390126 Telex 411483 MGU SU

Naumova, Dr Inessa I. Researcher. Physical Department, Moscow State University, Moscow 119899, Russia. PhD (Moscow State U., 1968) geology and mineralogy. *Crystal_growth domain_structure phase_transition niobium_compounds optical_property ferroelectricity*
E-mail ILC@compnet.msu.su Tel. 7(095)9391630 Fax 7(095)9390247

Nazimova, Dr Nina V. (1946). Researcher. Laboratory of Biocrystal Structure, Institute of Crystallography, Russian Academy of Sciences, 59 Leninsky prosp., Moscow 117333, Russia. PhD (Inst. of Crystallography Moscow, 1989) physical chemistry. *X-ray_crystallography structure–activity_relationship prostaglandins small_molecules*
Tel. 7(095)1352300 Fax 7(095)1351011

Nekrasov, Dr Yuri V. (1937). Senior researcher. Laboratory of Biocrystal Structure, Institute of Crystallography, Russian Academy of Sciences, 59 Leninsky prosp., Moscow 117333, Russia. PhD (Inst. of Crystallography Moscow, 1990) physics. *X-ray_diffractometry software data_processing_optimization*
Tel. 7(095)1353098 Fax 7(095)1351011

Nesterova, Dr Yaroslava Mikhajlovna. Researcher. Lab.Crystal Chemistry, Physical Chemistry Faculty, Chemical Dept., Moscow State University, Moscow 119899, Russia. PhD (Moscow State U., 1973) physical chemistry. *Order–disorder_structure domain_structure organic_molecular_crystal_structure aperiodic_crystal*
E-mail physch@mch.chem.msu.su Tel. 7(095)9392258 Fax 7(095)9328846 Telex 411483 MCUSU

Nevskaya, Dr Natalia A. (1944). Researcher. Group of Structure Investigations of Ribosomal Proteins, Institute of Protein Research, Russian Academy of Sciences, Pushchino, Moscow Region 142292, Russia. PhD (Ins. of Protein Research Russian Acad. Sci., 1977) molecular biology. *Structural_crystallography ribosomes RNA_proteins computer_graphics*
E-mail nevskaja@vax.ipr.serpukhov.su OR protres@sovam.com Tel. 7(095)9240493 Fax 7(095)1359984 Telex (64) 412680 BELOK SU or (871) 412680 BELOK SU

Nifontov, Mr Valentin P. (1949). Researcher. Lab. of Crystal Chemistry, Branch Institute of Chemical Physics, Russian Academy of Sciences in Chernogolovka, Chernogolovka, Moscow Region 142432, Russia. MSc (Moscow State U., 1971) physics. *High_temperature_superconductivity superionic_conductors*
E-mail atov@icph.sherna.msk.su Tel. 7(095)5245059 Fax 7(096)2655714

Nikanorova, Dr Irina A. Senior Researcher. Solid State Chair, Physics Department, Moscow State University, Vorobjovy Gory, Moscow 117234, Russia. PhD (Moscow State U., 1985) solid state physics. *X-ray_analysis phase_structure phase_transition intermetallic_compound*
E-mail asi@phys.msu.su. Tel. 7(095)9393029 Fax 7(095)9390247 Telex 411483 MGU SU

Nikiforov, Prof. Igor Ya. (1930). Head of Chair. Rostov Inst. of Agricultural Machinery, Pl. Gagarina 1, Rostov-on-Don 344010, Russia. DSc (Rostov State U., 1985) physics and mathematics. *X-ray_spectrometry perfection*
Tel. 7(8632)223413

Nikishova, Dr Lidya V. (1938). Researcher. Institute of Geology, Yakutsk Branch, Siberian Department, Russian Academy of Sciences, Lenin pr. 39, Yakutsk 677007, Russia. PhD (Irkutsk State U., 1976) physics and mathematics. *Electron_diffraction microdiffraction isomorphism polymorphism*

Nikitenko, Prof. Valerian I. (1937). Group leader. Laboratory of Real Crystal Structure, Institute of Solid State Physics, Russian Academy of Sciences, 142432 Chernogolovka, Moscow Region, Russia. DSc (Ioffe Inst. of Physics and Technology Leningrad, 1971) solid state physics. *Dynamics_defect ferromagnetic_semiconductor_superconductor magnetism deformation superconductivity*
E-mail nikiten@issp.sherna.msk.su Tel. 7(095)5245063 Fax 7(096)5171949 Telex 412654 SERNA SU

Nikolaenko, Dr Anatoli M. (1948). Senior Researcher. Lab. of X-ray Optics and Synchrotron Radiation, Institute of Crystallography, Russian Academy of Sciences, 59 Leninsky prosp., Moscow 117333, Russia. PhD (Inst. of Crystallography Moscow, 1980) physics. *X-ray_diffraction surface_research X-ray_optics X-ray_diffractometry*
E-mail vart@crystal.msk.su Tel. 7(095)3307992 Fax 7(095)3301956

Nikolskaia, Dr Natalia Kimovna (1952). Senior Researcher. Lab. of X-ray Analysis, Institute of Earth Crust, St Petersburg University, University Emb. 7/9, St Petersburg 199034, Russia. PhD (St Petersburg U., 1988) geology and mineralogy. *X-ray_analysis_powder typomorphism_phyllosilicates transformation_layer-silicates structural_disorder simulation_X-ray_diffraction*
E-mail fit@dean.geol.lgu.spb.su Tel. 7(812)2189711 Fax 7(812)2181346

Nikolskaya, Dr Larisa V. (1949). Senior Researcher. Lab. of Mineral Raw Material Quality Evaluation, St Petersburg Mining Institute, 2 21st Liniya, Saint Petersburg 199026, Russia. PhD (Moscow State U., 1977) crystallography and crystal physics. *Crystal_growth minerals_physics spectroscopy_methods*
Tel. 7(812)2188422 Fax 7(812)2132613 Telex 121494 LGIPSU

Nikonov, Dr Stanislav V. (1938). Group Leader. Group of Structure Investigations of Ribosomal Proteins, Institute of Protein Research, Russian Academy of Sciences, Pushchino, Moscow Region 142292, Russia. PhD (Institute of Crystallography Moscow, 1985) crystallography and crystal physics. *Structural_crystallography ribosomes RNA_proteins computer_graphics*
E-mail nikonov@vax.ipr.serpukhov.su OR protres@sovam.com Tel. 7(095)9240493 Fax 7(095)1359984 Telex (64) 412680 BELOK SU or (871) 412680 BELOK SU

Nizamutdinov, Dr Nazim M. (1939). Associate Professor. Dept. of Geology, Kazan State University, 18 Lenin Str., Kazan 420008, Russia. PhD (Kazan State U., 1977) physics and mathematics. *Structural_crystallography crystal_defect magnetic_resonance sedimentology mineralogy crystal_symmetry*
E-mail root@scikgu.kazan.su Tel. 7(8432)327062 Fax 7(8432)380994 Telex 224641 CHAIR SU

Noifekh, Mr Alexsander I. (1951). Senior Researcher. Lab. for Structural Studies, Obninsk Branch, Karpov Institute of Physical Chemistry, Obninsk, Kaluga Region 249020, Russia. MSc (Moscow Inst. of Engineering and Physics, 1974) *III–V_compound semiconductors oxide_superconductors nuclear_doped*
Tel. 7(08439)26272

Nosik, Dr Valery L. (1966). Engineer. Lab. of X-ray Optics and Synchrotron Radiation, Institute of Crystallography, Russian Academy of Sciences, 59 Leninsky prosp., Moscow 117333, Russia. PhD (Inst. of Crystallography Moscow, 1993) physics. *X-ray_dynamical_diffraction acoustic_vibration superlattice*
E-mail chukhov@crystal.msk.su Tel. 7(095)3307992 Fax 7(095)3301956

Novakova, Dr Alla A. (1942). Chief Lecturer. Solid State Chair, Physics Department, Moscow State University, Vorobjovy Gory, Moscow 117234, Russia. PhD (Moscow State U., 1975) solid state physics. *X-ray_diffraction condensed_matter_physics Mossbauer_spectroscopy photosynthesis amorphous_crystalline_transition*
E-mail asi@phys.msu.su. Tel. 7(095)9391226 Fax 7(095)9390247 Telex 411483 MGU SU

Novikova, Dr Natalia N. (1963). Researcher. Lab. of X-ray Optics and Synchrotron Radiation, Institute of Crystallography, Russian Academy of Sciences, 59 Leninsky prosp., Moscow 117333, Russia. PhD (Moscow State U., 1990) physics. *X-ray_standing_wave X-ray_dynamical_diffraction thin_film X-ray_and_synchrotron_radiation_instrumentation*
E-mail zhel@crystal.msk.su Tel. 7(095)3307992 Fax 7(095)3301956

Novomlinsky, Dr Leonid A. (1956). Researcher. Member of Board of Russian Association on Powder Crystallography. Laboratory of Structure Analysis, Institute of Solid State Physics, Russian Academy of Sciences, Chernogolovka, Moscow Region 142432, Russia. PhD (Moscow State U., 1988) solid state physics. *X-ray_structure_determination Rietveld_method computing disordered_incommensurate_modulated_structures high-precision_diffractometry high-resolution_diffractometry low_temperature high_temperature software_for_crystallography computer_modelling database non-ideal_structure defect stacking_fault superconductors phosphates synthetic_metals*
E-mail novom@issp.sherna.msk.su Tel. 7(095)5255063 Fax 7(096)5171949 or 7(096)5153588 Telex 412654 Serna SU

Okhrimenko, Dr Tatyana M. (1944). Senior Researcher. Laboratory of High-Temperature Crystallization, Inst. of Crystallography, Russian Academy of Sciences, 59 Leninsky prosp., Moscow 117333, Russia. PhD (Inst. of Crystallography Moscow, 1987) crystallography and crystal chemistry. *Crystal_growth aqueous_solution real_structure*
Tel. 7(095)3301792 Fax 7(095)1351011

Okinshevich, Mr Vladimir V. (1938). Researcher. Lab. of High Temperature Crystallization, Institute of Crystallography, Russian Academy of Sciences, 59 Leninsky prosp., Moscow 117333, Russia. MSc (Physical Engineering Inst. Moscow, 1962) physics of metals. *Float_zone_growth nucleation cooperative_phenomena*
Tel. 7(095)1351220 Fax 7(095)1356350 or 7(095)1351011

Oleinikov, Dr Vladimir A. (1949). Senior Researcher. Laboratory of Nuclear Filters, Institute of Crystallography, Russian Academy of Sciences, 59 Leninsky pr., Moscow 117333, Russia. PhD (Physical Engineering Inst. Moscow, 1980) physics. *Nonlinear_optics surface_research mass-spectrometry ion_electron_emission*
Tel. 7(095)1359971 Fax 7(095)1351011

Onishchina, Dr Ninel Mitrofanovna (1935). Associate Professor. Department of Mineralogy Crystallography and Petrography, Geological Research Faculty, Saint Petersburg Mining Institute, 2 21st Liniya, Saint Petersburg 199026, Russia. PhD (St Petersburg Mining Inst., 1975) crystallography and crystal physics. *Crystal_morphology crystal_growth history_of_science*
Tel. 7(812)2188266 Fax 7(812)2132613 Telex 121494 LGIPSU

Orekhov, Dr Sergey V. (1958). Researcher. Lab. Electron Diffraction, Institute of Crystallography, Russian Academy of Sciences, 59 Leninsky prosp., Moscow 117333, Russia. PhD (Inst. of Crystallography Moscow, 1992) physics. *High-energy_electron_diffraction high-resolution_diffractometry computer_automation amorphous_phase*
Tel. 7(095)1354520

Organova, Dr Natalija I. Chief Researcher. Laboratory of X-ray Structure Analysis, Institute of Geology of Ore Deposits, Petrology, Mineralogy and Geochemistry, Russian Academy of Sciences (IGEM RAS), Staromonetny per. 35, Moscow 109017, Russia. DSc (IGEM RAS, 1988) crystal chemistry and mineralogy. *X-ray_crystallography minerals modulated_structures exsolution high-resolution_electron_microscopy electron_diffraction*
E-mail natalja@igem.msk.su Tel. 7(095)2308296 Fax 7(095)2302179

Otroshchenko, Dr Ludmila P. (1940). Researcher. Laboratory of Physical and Chemical Analysis, Institute of Crystallography, Russian Academy of Sciences, 59 Leninsky pr., Moscow 117333, Russia. PhD (Moscow State U., 1981) crystallography. *Structure_determination crystal_chemistry*
Tel. 7(095)3307874 Fax 7(095)1351011

Ovchinnikov, Dr Yurii E. (1951). Senior Lecturer. Physics Department, Novosibirsk State Pedagogical Institute, 28 Vilyuiskaya St., Novosibirsk 630126, Russia. PhD (Moscow State Pedagogical Institute, 1986) solid state physics. *X-ray_crystallography small_molecules organic_organometallic_crystal_chemistry*
E-mail yuo@xray.ineos.ac.ru Tel. 7(3832)681716 Fax 7(095)1355085

Ovchinnikova, Dr Elena N. (1951). Senior Researcher. Solid State Chair, Physics Department, Moscow State University, Vorobjovy Gory, Moscow 117234, Russia. PhD (Moscow State U., 1978) solid state physics. *X-ray_diffraction Mossbauer_diffraction group_theory quasicrystal incommensurate_phase colour_symmetry computer_modelling*
E-mail asi@phys.msu.su. Tel. 7(095)9391226 Fax 7(095)9390247 Telex 411483 MGU SU

Ovsetsina, Ms Tatyana I. (1968). Researcher. Laboratory of X-ray Diffraction Research, Physics and Technology Research Institute, 23 Gagarin Pr., Nizhny Novgorod 603600, Russia. MSc (Gorky State U., 1990) physics. *X-ray_crystallography small_molecules biological_activity*
Tel. 7(8312)656365

Ozerin, Dr Alexander N. (1952). Deputy Director. Laboratory of Polymer Physics, Institute of Synthetic Polymeric Materials, Russian Academy of Science, 70 Profsoyuznaya str., Moscow 117393, Russia. DSc (Karpov Inst. of Physical Chemistry Moscow, 1992) chemistry. *SAXS_polymers computer_modelling_polymers X-ray_scattering computer_modelling silicon_compounds*
E-mail ozerin@nifhi.uucp.free.msk.su Tel. 7(095)3300597 Fax 7(095)4202229

Ozerov, Prof. Ruslan Pavlovich (1926). Head of physics chair. Mendeleev University of Chemical Technology of Russia, 9 Miusskaja Square, Moscow 125190, Russia. DSc (Inst. of Crystallography, Moscow, 1969) solid state physics. *X-ray_crystallography charge_density spin_density crystal_chemistry physical_properties_of_crystals chemical_bonding*
E-mail ozerov@mhti.msk.su Tel. 7(095)2587569 Fax 7(095)2004204 Telex 411744 argon

Panich, Dr Anatoly E. (1944). Director of NKTB. NKTB PIEZOPRIBOR, Rostov State University, 10 Melchakova Str., Rostov-on-Don 344104, Russia. PhD (Rostov State U., 1980) physics and mathematics. *Physical_property ferroelectric_piezoelectric_crystal*
Tel. 7(8632)223401 Fax 7(8632)226923 Telex 123360 pxezo

Panov, Mr Vladimir N. (1958). Chief Lecturer. Department of Physics, Vladimir State Pedagogical University, Department of Physics, 11 Stroiteley Prospect, Vladimir 600024, Russia. MSc (Rostov State U., 1981) physics. *Dirichlet_domain packing molecular_mechanics molecular_crystal X-ray_crystallography*
Tel. 7(09222)72604

Paseshnichenko, Dr Ksenia Andreevna (1959). Researcher. Lab. Structural Chemistry, General Chemistry Faculty, Chemistry Dept., Moscow State University, Moscow 119899, Russia. PhD (Moscow State U., 1985) physical chemistry. *Crystal_chemistry structure_determination hydrogen_bonding history*
E-mail Ksenia@crysma.rc.ac.ru Tel. 7(095)9395089 Fax 7(095)9390898 Telex 411483 MGU SU

Pavlov, Dr Sergey V. (1956). Researcher. Physical Department, Moscow State University, Moscow 119899, Russia. PhD (Moscow State U., 1984) physics of crystals and crystallography. *Phase_transition ferroelectricity catastrophe_theory*
Tel. 7(095)9391013 Fax 7(095)9390247

Perekalina, Dr Tatyana M. (1922). Senior Researcher. Laboratory Electromechanical Investigation of Crystals, Institute of Crystallography, Russian Academy of Sciences, Leninsky pr. 59, Moscow 117333, Russia. DSc (Inst. of Crystallography Moscow, 1973) physics. *Magnetism conductivity manganese_compounds superconductor*
E-mail postmaster@el-mech.msk.su Tel. 7(095)3307856 Fax 7(095)1351011

Perekalina, Dr Zoja B. Senior Researcher. Lab. of Crystallophysics, Shubnikov Institute of Crystallography, Russian Academy of Sciences, 59 Leninsky Pr., Moscow 117333, Russia. PhD (Inst. Crystallography Moscow, 1969) crystallography. *Circular_dichroism electrical_circular_dichroism crystal rotatory_dispersion_crystals circular_dichroism_measurement_methods*
E-mail feigin@saxslab.incr.msk.su Tel. 7(095)1354510 Fax 7(095)1351011

Perelomova, Prof. Natali V. Professor. Department of Crystallography, Institute of Steel and Alloys, 4 Leninsky prosp., Moscow 117936, Russia. PhD (Inst. of Steel and Alloys Moscow, 1972) solid state physics. *Solid_state_physics crystallography*
Tel. 7(095)2366500 Fax 7(095)2378007

Pervukhina, Dr Natalie V. (1955). Senior Researcher. Laboratory of Crystal Chemistry, Institute of Inorganic Chemistry Siberian Branch, Russian Academy of Sciences, 3 Ak. Lavrentiev prosp., Novosibirsk 630090, Russia. PhD (Inst. Inorg. Chem. Siberian Branch of Russian Academy of Sciences, 1990) inorganic chemistry. *X-ray_crystallography complex_radicals magnets inclusion_compound crystal_structural_database inorganic_crystal_chemistry*
E-mail borisov@che.nsk.su Tel. 7(3832)354366 Fax 7(3832)355960

Petropavlov, Dr Nikolai N. (1938). Senior Researcher. Institute of Cell Biophysics, Russian Academy of Sciences, Pushchino, Moscow Region 142292, Russia. PhD (Moscow State Pedagogical Inst., 1971) crystallography. *Biocrystallography crystal_structure_property phase_transition*
Tel. 7(095)9255984 Telex (64) 412615 SU

Petrovsky, Dr Vitaly A. (1946). Chief Researcher. Department of Mineralogy, Institute of Geology, Komi Scientific center of Russian Academy of Sciences, 54 Pervomayskaya Str., Syktyvkar 167610, Russia. PhD (Moscow Inst. of Geological Survey, 1981) mineralogy and crystallography. *Minerals diamond ontogeny_of_minerals crystallogenetic_modelling genesis_of_diamond_crystals*
Tel. 7(82122)71214 7(82122)25167 Fax 7(82122)25346

Petrzhik, Dr Ekaterina A. (1962). Researcher. Laboratory of Mechanical Properties of Crystals, Institute of Crystallography, Russian Academy of Sciences, 59 Leninskii pr., 117333 Moscow, Russia. PhD (Inst. of Crystallography Moscow, 1992) solid state physics. *Plasticity dislocation magnetism mechanical_properties*
E-mail public@mechan.incr.msk.su Tel. 7(095)3308274 Fax 7(095)3308274

Petukhov, Prof. Boris V. (1941). Chief researcher. Lab. of Crystallophysics, Institute of Crystallography, Russian Academy of Sciences, Leninsky pr. 59, Moscow 117333, Russia. DSc (Inst. of Crystallography Moscow, 1987) solid state physics. *Crystal_defect lattice_distortion dislocation plasticity statistical_physics nonlinear_physics soliton metastable_state_decay*
E-mail root@theory.incr.msk.su Tel. 7(095)1356240 Fax 7(095)1351011

Petushkova, Mrs Ludmila V. (1941). Researcher. Vernadsky Institute of Geochemistry and Analytical Chemistry, Russian Academy of Sciences, 19 Kosygin St., Moscow 117975, Russia. MSc (Moscow State U., 1966) geochemistry. *X-ray_crystallography X-ray_diffractometry minerals silicates diffraction_technique sulfides sulfates sulfur_compounds*
Tel. 7(095)9391916 Fax 7(095)9382054 Telex 411633 TERRA SU

Pikin, Prof. Sergey A. (1941). Deputy Director. Laboratory of Crystallophysics, Institute of Crystallography, Russian Academy of Sciences, 59 Leninsky pr., Moscow 117333, Russia. DSc (Instit. of Crystallography Moscow, 1978) solid state physics. *Liquid_crystal phase_transition ferroelectricity critical_phenomena smectic_crystal Langmuir_Blodgett_film antiferroelectricity*
E-mail root@theory.incr.msk.su Tel. 7(095)1356030 Fax 7(095)1351011

Pisarevski, Dr Yuri V. (1940). Head of Laboratory. Lab. of Acoustooptics of Crystals, Institute of Crystallography, Russian Academy of Sciences, Leninsky pr. 59, 117333 Moscow, Russia. PhD (Inst. of Crystallography Moscow, 1974) solid state physics. *Acoustooptics piezoelectricity*
Tel. 7(095)3306883 Fax 7(095)3301956

Pisarevsky, Alexander P. (1965). Researcher. X-ray Structural Center of the Russian Academy of Sciences, Institute of Organoelement Compounds, Russian Academy of Sciences, 28 Vavilov St., Moscow 117813, Russia. PhD (Moscow State U., 1992) inorganic chemistry. *X-ray_crystallography volatile_polymeric_chelates chemical_deposition_of_oxides*
E-mail alex@xray.ineos.ac.ru Tel. 7(095)1359355 Fax 7(095)1355085

Pietnev, Dr Vladimir Z. (1944). Head of Laboratory. Lab. of X-ray Structure Analysis, Institute of Bioorganic Chemistry, Russian Academy of Sciences, 16/10 Miklukho-Maklay St., Moscow 117871, Russia. DSc (Inst. of Crystallography Moscow, 1988) chemistry. *Crystal_structure peptides proteins*
E-mail pletnev@tek.siobc.msk.su Tel. 7(095)3307510 Fax 7(095)3300300

Piyasova, Prof. Ludmila M. Laboratory Head. Boreskov Institute of Catalysis, Laboratory of Structure Methods, 5 Lavrentieva Pr., Novosibirsk 630090, Russia. DSc (Institute of Catalysis Novosibirsk, 1993) crystallography and powder diffraction. *X-ray_powder_diffraction catalysts real_structure high-temperature_diffraction*
E-mail pls@catalysis.nsk.su Tel. 7(383)2354300 Fax 7(383)2355756

Podberezskaya, Dr Nina V. (1938). Chief Researcher. Laboratory of Crystal Chemistry, Institute of Inorganic Chemistry, Siberian Branch of RAS, 3 Lavrentyev prosp., Novosibirsk 630090, Russia. PhD (Gorky State U., 1971) crystallography and crystal physics. *X-ray_crystallography oxides coordination_compound database structural_computing inorganic_crystal_chemistry*
E-mail borisov@che.nsk.su Tel. 7(3832)354366 Fax 7(3832)355960

Politova, Dr Ekaterina D. (1947). Senior Researcher. X-ray Laboratory, Karpov Institute of Physical Chemistry, 10 Obukha St., Moscow 103064, Russia. PhD (Karpov Institute of Physical Chemistry, 1974) physical chemistry. *Oxides_superconductor solid_solution structural_change phase_separation oxides_ferroelectrics perovskites ceramics phase_transitions*
E-mail politova@nifhi.uucp.free.msk.su Tel. 7(095)2270014x2329 Fax 7(095)9752450

Polyakov, Dr Konstantin M. (1951). Senior researcher. Laboratory of Biocrystal Structure, Institute of Crystallography, Russian Academy of Sciences, 59 Leninsky prosp., Moscow 117333, Russia. PhD (Inst. of Crystallography Moscow, 1985) physics. *X-ray_structure_determination proteins*
E-mail kostya@imb.mb.free.net Tel. 7(095)1356020 Fax 7(095)1351011

Polyanskaya, Dr Tamara M. Senior Researcher. Laboratory of Crystal Chemistry, Institute of Inorganic Chemistry, Siberian Branch of Russian Academy of Sciences, 3 Lavrentiev prosp., Novosibirsk 630090, Russia. PhD (Gorky State U., 1972) crystallography and crystal physics. *X-ray_crystallography inclusion_compound inorganic_crystal_chemistry*
E-mail borisov@che.nsk.su Tel. 7(3832)354366 Fax 7(3832)355960

Polyansky, Dr Evgeny V. (1944). Deputy Director. Russian Research Institute for Synthesis of Minerals and Raw Materials, Institutskaya St. 1, Alexandrov, Vladimir Region 601600, Russia. PhD (Moscow State U., 1970) mineralogy. *Crystal_growth mineralogy gemology*
Tel. 7(095)5845834 Fax 7(095)5845828

Polynova, Prof. Tamara Nikitichna (1930). Associate Professor. Lab. Crystal Chemistry, Physical Chemistry Faculty, Chemical Dept., Moscow State University, Moscow 119899, Russia. PhD (Kurnakov Institute of General and Inorganic Chemistry, 1963) inorganic chemistry. *Coordination_chemistry organic_crystal_chemistry complex_compounds X-ray_research crystal_packing*
E-mail physch@mch.chem.msu.su Tel. 7(095)9392258 Fax 7(095)9328846 Telex 411483 MCUSU

Ponomarev, Dr Vasiliy I. (1940). Group leader. Laboratory of X-ray Structural Analysis, Institute of Structural Macrokinetics, Russian Academy of Sciences, Chernogolovka, Moscow Region 142432, Russia. PhD (Inst. Crystallography RAS Moscow, 1971) physical chemistry. *Lattice_dynamics*
E-mail root@ism.sherna.msk.su Tel. 7(095)5245047 Fax 7(095)2017357

Popov, Dr Alexander N. (1954). Senior Researcher. Laboratory of X-ray Diffractometry, Institute of Crystallography, Russian Academy of Sciences, 59 Leninsky prosp., Moscow 117333, Russia. PhD (Inst.of Crystallography Moscow, 1984) crystallography and crystal physics. *X-ray_diffractometry area_detector synchrotron_radiation method_Laue structure_biomolecule structure_powder structure_polymeric*
E-mail CRYS@CRYS.MSK.SU Tel. 7(095)1356230 Fax 7(095)1351011

Popova, Dr Svetlana V. Group leader. Vereschagin Institute of High Pressure Physics, Lab. of Elastic and Plastic Properties, Russian Academy of Sciences, Troitsk, Moscow Reg. 142092, Russia. DSc (Inst. of Crystallography of Moscow, 1983) physics and mathematics. *X-ray_crystallography phase_transition amorphous_phase aperiodic_material*
E-mail scorpion@adonis.iasnet.com Tel. (095)3340597 Fax 7(095)3340012

Porai-Koshits, Prof. Michail A. (1918). Principal Researcher. Laboratory of Coordination Compounds Crystal Chemistry, Kurnakov Institute of General and Inorganic Chemistry, Russian Academy of Sciences, 31 Leninsky prosp., Moscow, Russia. DSc (Inst. General and Inorganic Chemistry, 1960) physical chemistry. *X-ray_crystallography coordination_compound organometallic_compounds inorganic_compounds secondary_bonding complexonates*
E-mail vip@solen.msk.su Tel. 7(095)9521803 Fax 7(095)9541279

Portnov, Dr Vadim N. (1934). Associate Professor. Lab. of X-ray Diffraction Research, Research Institute of Physics and Technology, 23 Gagarin prosp., Nizhny Novgorod 603600, Russia. PhD (Gorky State U., 1966) physics. *Crystal_growth*
Tel. 7(8312)656365

Potekhin, Dr Konstantin A. (1954). Assistant professor. Vladimir State Pedagogical Institute, 11 Stroiteley Prosp., Vladimir 600024, Russia. PhD (Inst. of Crystallography Russian Academy of Sciences, 1980) physical chemistry. *X-ray_crystallography small_molecules organic_crystal_chemistry*
E-mail xray@yandex.su Tel. 7(092)2272604 Fax 7(095)1355085

Prokhorov, Mr Andrey I. (1964). Junior researcher. Lab. of X-ray Structural Analysis, Institute of Structural Macrokinetics, Russian Academy of Sciences, Chernogolovka, Moscow region 142432, Russia. MSc (Branch Inst. of Chemical Physics, Chernogolovka, 1990) chemical physics. *X-ray_diffraction monocrystal combustion_mechanism computer_automation_X-ray_experiments*
E-mail root@ism.sherna.msk.su Tel. 7(095)5245047 Fax 7(095)2017357 VNESH-TECHNIKA FOR ISMAN

Prokhorov, Dr Igor A. (1945). Senior Researcher. Laboratory of X-ray Structure Analysis, Kaluga Branch of Inst. of Crystallography, Russian Academy of Sciences, 2 Akademicheskaya str., Kaluga 248640, Russia. PhD (Inst. of Steel and Alloys Moscow, 1984) physics. *Single_crystal real_structure X-ray_topography X-ray_diffraction*
Tel. 7(08422)48393 Fax 7(08422)48614

Prudnikov, Dr Ilya R. (1966). Researcher. Solid State Chair, Physics Department, Moscow State University, Vorobjovy Gory, Moscow 117234, Russia. PhD (Moscow State U., 1992) solid state physics. *X-ray_dynamical_diffraction*
E-mail asi@phys.msu.su. Tel. 7(095)9392387 Fax 7(095)9390247 Telex 411483 MGU SU

Punin, Dr Yurii O. (1941). Chief Researcher. Crystal Genesis Laboratory, St Petersburg University, University Emb. 7/9, St Petersburg 199034, Russia. PhD (St Petersburg U., 1970) crystallography and crystal physics. *Crystallogeny crystal_growth crystal_defect experimental_mineralogy isomorphous_replacement crystal_morphology*
E-mail anna@dean.geol.lgu.spb.su Tel. 7(812)2189650 Fax 7(812)2181346 Telex 121481 LSU SU

Pushcharovsky, Prof. Dmitry Yu. (1944). Professor. Moscow State University, Geology Faculty, Department of Crystallography and Crystal Chemistry, 119899 Moscow, Russia. DSc (Inst. of Mineralogy, Geochemistry and Petrology of Ore Deposits of Russia, 1984) mineralogy. *X-ray_crystallography materials minerals silicates*
Tel. 7(095)9391222 Fax 7(095)9390195 Telex 411483 MGU SU

Rakin, Dr Vladimir I. (1956). Senior Researcher. Lab. of Experimental Mineralogy, Institute of Geology, Komi Scientific Centre, Ural Department, Russian Academy of Sciences, 54 Pervomaiskaya Str., Syktyvkar 167610, Russia. [31 Apt. 11, Magistralnaya Str., Syktyvkar 167022, Russia.] PhD (Moscow State U., 1985) geology and mineralogy. *Crystal_growth crystallography optical_measurements*
Tel. 7(82122)25167 Fax 7(82122)25346

Rakova, Dr Elena V. (1941). Senior Researcher. Laboratory of Electron Diffraction, Institute of Crystallography, Russian Academy of Sciences, 59 Leninsky prosp., Moscow 117333, Russia. PhD (Inst. of Crystallography RAS Moscow, 1978) physics. *RHEED surface_phase_transition lithium_compounds epitaxy*
E-mail feigin@saxslab.incr.msk.su Tel. 7(095)1354010

Rashkovich, Prof. Leonid N. (1931). Chief Researcher. Chair of Crystal Physics, Physics Department, Moscow State University, Moscow 119899, Russia. DSc (Moscow Inst. Chemical Technology, 1981) physics. *Crystal_growth_kinetics_mechanism phase_equilibrium solution_structure*
E-mail rashk@crystO.phys.msu.su Tel. 7(095)9392981 Fax 7(095)9328820 Telex 411483 MGU SU

Rastsvetaeva, Dr Ramiza K. (1936). Senior Researcher. Laboratory of X-ray Structure Analysis, Institute of Crystallography, Russian Academy of Sciences, 59 Leninsky pr., Moscow 117333, Russia. DSc (St Petersburg State U., 1991) geology and mineralogy. *X-ray_crystallography crystal_chemistry inorganic_compound single_crystal minerals*
E-mail simonov@rsa.crystal.msk.su Tel. 7(095)1350330 Fax 7(095)1351011

Rau, Dr Tamara F. (1940). Associate Professor. Department of Physics, Vladimir State Pedagogical University, Department of Physics, 11 Stroiteley Prospect, Vladimir 600024, Russia. PhD (Gorky State U., 1975) crystallography. *X-ray_crystallography molecular_crystal urea_compounds*
Tel. 7(09222)72628

Rau, Prof. Valery G. (1940). Professor. Department of Physics, Vladimir State Pedagogical University, Department of Physics, 11 Stroiteley Prospect, Vladimir 600024, Russia. DSc (Inst. Crystallography Moscow, 1985) crystallography. *X-ray_crystallography molecular_crystal theoretical_physics*
Tel. 7(09222)72628

Rebrov, Dr Alexander V. (1948). Senior Researcher. Laboratory of Polymer Structure, Karpov Institute of Physical Chemistry, 10 Obukha St., Moscow 103064, Russia. PhD (Karpov Institute of Physical Chemistry Moscow, 1984) chemistry. *SAXS_polymers computer_modelling_polymers X-ray_scattering computer_modelling silicon_compounds*
E-mail rebrov@nifhi.uucp.free.msk.su 7(095)2270014x2519 Fax 7(095)9752450

Regel, Prof. Vadim R. (1917). Chief Researcher. Lab. of Mechanical Properties of Crystals, Institute of Crystallography, Russian Academy of Sciences, 59 Leninskii pr., 117333 Moscow, Russia. DSc (A. F. Ioffe Physical-Technical Inst. Moscow, 1965) solid state physics. *Mechanical_property dislocation_theory real_crystal polymers strength_plasticity_physics*
E-mail public@mechan.incr.msk.su Tel. 7(095)3342483 Fax 7(095)3308274

Revkevich, Dr Galina P. Senior Researcher. Solid State Chair, Physics Department, Moscow State University, Vorobjovy Gory, Moscow 117234, Russia. PhD (Moscow State U., 1967) solid state physics. *X-ray_diffraction phase_formation*
E-mail asi@phys.msu.su. Tel. 7(095)9394610 Fax 7(095)9390247 Telex 411483 MGU SU

Rider, Dr Evgeny E. (1937). Senior Researcher. Lab. for Structural Studies, Obninsk Branch, Karpov Institute of Physical Chemistry, Obninsk, Kaluga Region 249020, Russia. MSc (Moscow Inst. of Engineering and Physics, 1965) physics. *Neutron_diffraction high_temperature_superconductor_structure X-ray_diffraction_defect_structures*
Tel. 7(08439)26575

Roginskaya, Dr Yuliana E. (1937). Chief Researcher. Karpov Institute of Physical Chemistry, 10 Obukha St., Moscow 103064, Russia. PhD (Karpov Institute of Physical Chemistry, 1965) physical chemistry. *Noble_metals_oxides hydroxides real_structure morphology iridium_oxide_compounds ruthenium_oxide_compounds titanium_oxide_compounds tin_oxide_compounds tantalum_oxide_compounds sol-gel_method surface_segregation XPS metastable_phases*
E-mail rogin@nifhi.uucp.free.msk.su Tel. 7(095)2270014x2336 Fax 7(095)9752450

Romanenko, Dr Galina V. (1954). Senior Researcher. Laboratory of Crystal Chemistry, Institute of Inorganic Chemistry Siberian Branch, Russian Academy of Sciences, 3 Ak. Lavrentiev prosp., Novosibirsk 630090, Russia. PhD (Inst. Inorg. Chem. Siberian Branch of Russian Academy of Sciences, 1990) inorganic chemistry. *X-ray_crystallography complex_radicals magnets crystal_structure_database structural_computing inorganic_crystal_chemistry*
E-mail borisov@che.nsk.su Tel. 7(3832)354366 Fax 7(3832)355960

Rozhansky, Prof. Vladimir N. (1923). Chief Researcher. Lab. of Mechanical Properties of Crystals, Institute of Crystallography, Russian Academy of Sciences, 59 Leninsky prosp., Moscow 117333, Russia. DSc (Inst. of Crystallography Moscow, 1963) solid state physics. *Crystal_defect electron_microscopy mechanical_properties_of_crystals*
E-mail public@mechan.incr.msk.su Tel. 7(095)3307883 Fax 7(095)1351011

Rozhdestvenskaya, Dr Ira V. (1938). Senior Researcher. Lab. of X-ray structure analysis, NPP "Burevestnik", 68 Maloohtinsky Prospect, St Petersburg 195272, Russia. PhD (Leningrad State U., 1975) physics. *X-ray_structure_determination crystal_chemistry_inorganic_compound structural_mineralogy crystallography_computing*
Tel. 7(812)5280434

Rozin, Dr Konstantin M. (1929). Associate Professor. Department of Crystallography, Institute of Steel and Alloys, 4 Leninsky prosp., Moscow 117936, Russia. PhD (Inst. of Steel and Alloys Moscow, 1963) crystal growth. *Non-homogeneous_monocrystal crystal_chemistry*
Tel. 7(095)2366500 Fax 7(095)2378007

Rubina, Dr Elena B. (1948). Senior Researcher. Baikov Institute of Metallurgy, Russian Academy of Sciences, Laboratory of Crystallochemistry, 49 Leninskyi pr., Moscow 117334, Russia. PhD (Institute of Aircraft Technology Moscow, 1984) physical metallurgy. *X-ray_crystallography texture deformation_mechanism intermetallic_compounds titanium_alloys non-crystalline_solid*
Tel. 7(095)1354757 Fax 7(095)1358680

Rudenko, Mr Sergey S. (1950). Researcher. Lab. of Mineral Raw Material Quality Evaluation, St Petersburg Mining Institute, 2 21st Liniya, St Petersburg 199026, Russia. MSc (Mining Inst. St Petersburg, 1972) crystallography and crystal physics. *Minerals_physics spectroscopy_method typomorphism_of_minerals optical_spectroscopy_methods*
Tel. 7(812)2188683 Fax 7(812383)2132613 Telex 121494 LGIPSU

Rumanova, Dr Iskra Mikhailovna Chief Researcher. Institute of History of Science and Technics, Russian Academy of Sciences, Staropansky 1/5, Moscow 103012, Russia. DSc (Inst. of Crystallography Moscow, 1971) physics and mathematics. *Structural_crystallography sciences_history*

Russo, Dr Galina V. (1951). Senior Researcher. Lab. of Modelling of Nature Crystalliz-ation, Earthcrust Research Institute, St Petersburg University, University Emb 7/9, St Petersburg 199034, Russia. PhD (St Petersburg U., 1986) crystallography and crystal physics. *Crystal_growth experimental_modelling crystallography_modelling*
E-mail anna@dean.geol.lgu.spb.su Tel. 7(812)2189644 Fax 7(812)2181346 Telex 121481 LSU SU

Ryadnov, Dr Sergei N. (1957). Senior Researcher. Lab. of High Temperature Crystalliz-ation, Institute of Crystallography, Russian Academy of Sciences, 59 Leninsky prosp., Moscow 117333, Russia. PhD (Inst. of General Physics Moscow, 1987) physical chem-istry. *Crystal_growth high_temperature_crystallization impurity_microanalysis*
Tel. 7(095)1351400 Fax 7(095)1356350 or 7(095)1351011

Rybakov, Dr Victor Borisovich (1950). Researcher. Lab. Structural Chemistry, General Chemistry Faculty, Chemistry Dept., Moscow State University, Moscow 119899, Rus-sis. PhD (Moscow State U., 1982) physical chemistry. *Structure_determination co-ordination_chemistry inorganic_complex*
E-mail Rybakov@crysma.uucp.free.msk.su Tel. 7(095)9395089 Fax 7(095)9390898 Telex 411483 MGU SU

Sadikov, Dr George G. (1932). Senior Researcher. Laboratory of Coordination Com-pounds Crystal Chemistry, Kurnakov Institute of General and Inorganic Chem-istry, Russian Academy of Sciences, 31 Leninsky prosp., Moscow 117907, Rus-sia. PhD (Inst. General and Inorganic Chemistry, 1971) inorganic chemistry. *X-ray_crystallography coordination_compound carboxylates complexonates*
Tel. 7(095)9521803 Fax 7(095)9541279

Safonova, Dr Tatiana N. (1955). Researcher. Laboratory of Biocrystal Structure, In-stitute of Crystallography, Russian Academy of Sciences, 59 Leninsky prosp., Moscow 117333, Russia. PhD (Moscow State U., 1983) physical chemistry. *X-ray_proteins_crystallography proteins_crystallization structure_determination*
Tel. 7(095)1351400 Fax 7(095)1351011

Safronov, Mr Victor V. (1966). Researcher. Small Angle Scattering Lab., Institute of Crys-tallography, Russian Academy of Sciences, Leninsky pr. 59, Moscow 117333, Russia. MSc (Moscow Institute of Physics and Technology, 1989) physics. *X-ray_reflectivity polymers Langmuir–Blodgett_film monolayer*
E-mail feigin@saxslab.incr.msk.su Tel. 7(095)1356010 Fax 7(095)1351011

Saf'yanov, Prof. Yuri N. (1949). Senior Researcher. Lab. of X-ray Diffraction Research, Research Institute of Physics and Technology, 23 Gagarin prosp., Nizhny Novgorod 603600, Russia. PhD (Gorky State U., 1976) physics. *X-ray_diffractometry het-erostructure_semiconductor rocking_curves strain_determination superlattice*
Tel. 7(8312)656365

Sakharov, Dr Boris A. (1944). Senior Researcher. Lab. for Investigation of Rock-Forming Minerals by Physical Methods, Geological Institute, Russian Academy of Sciences, Pyzhevsky per. 7, Moscow 109017, Russia. PhD (Geological Inst. Moscow, 1974) geology and mineralogy. *Diffraction_technique crystal_defect minerals*
Tel. 7(095)2308124 Fax 7(095)2318106

Samoilovich, Dr Lidiya A. (1934). Chief Researcher. Lab. of Experimental Mineral-ogy and Treatment Technology, Russian Research Institute for Synthesis of Minerals and Raw Materials, Institutskaya St. 1, Alexandrov, Vladimir Region 601600, Rus-sia. PhD (Moscow State U., 1968) physical chemistry. *Crystal_chemistry hydrother-mal_synthesis*
Tel. 7(095)5845834 Fax 7(095)5845828

Samoilovich, Prof. Mikhail I. (1937). Laboratory Head. Lab. of Physical Research, Russian Research Institute for Synthesis of Minerals and Raw Materials, Institutskaya St. 1, Alexandrov, Vladimir Region 601600, Russia. DSc (Kazan State U., 1971) physics. *Solid_state_physics EPR spectroscopy*
Tel. 7(095)5845819 Fax 7(095)5845828

Samotin, Dr Nikolay D. (1935). Senior Researcher. Lab. of Electron Microscopy, In-stitute of Ore Deposits, Petrology, Mineralogy and Geochemistry, Russian Academy of Sciences (IGEM RAS), Staromonetny 35, Moscow 109017, Russia. PhD (Inst. of Ore Deposits, Petrology, Mineralogy and Geochemistry (IGEM RAS), 1974) miner-alogy. *Analytical_electron_microscopy polytypism crystal_symmetry real_structure phase_transition crystal_growth*
Tel. 7(095)2308411 Fax 7(095)2302179

Samusina, Mrs Svetlana Nic. (1931). Chief Lecturer. Chair of Mineralogy Crystallog-raphy and Petrography, St Petersburg Mining Institute, 2 21st Liniya, St Petersburg 199026, Russia. MSc (St Petersburg Mining Inst., 1954) geology and mineralogy. *Crystal_morphology crystal_growth*
Tel. 7(812)2188266 Fax 7(812)2132613 Telex 121494 LGIPSU

Sannikov, Dr Daniil G. (1931). Chief Researcher. Laboratory of Electromechanical In-vestigations of Crystals, Institute of Crystallography, Russian Academy of Moscow, Leninsky pr. 59, Moscow 117333, Russia. DSc (Inst. of Crystallography Moscow, 1982) physics. *Phase_transition_theory solid_state group_theory ferroelectricity in-commensurate magnetism*
E-mail postmasten@el-mech.incr.msk.su Tel. 7(095)3307856 Fax 7(095)1351011

Sanzharlinsky, Dr Nikolai G. (1944). Laboratory Head. Lab. for High Pressure Syn-thesis of Materials, Russian Research Institute for Synthesis of Minerals and Raw Materials, Institutskaya St. 1, Alexandrov, Vladimir Region 601600, Russia. PhD (Inst.of Crystallography Moscow, 1981) physical chemistry. *Crystal_growth high_pressure_synthesis*
Tel. 7(095)5845834 Fax 7(095)5845828

Sapozhnikov, Dr Anatoly N. (1946). Senior Researcher. Institute of Geochem-istry, Siberian Branch, Russian Academy of Sciences, 1a Favorsky St., Irkutsk 664033, Russia. PhD (Moscow State U., 1980) geology and mineralogy. *X-ray_structure_determination inorganic_compound crystal_chemistry_minerals*
Tel. 7(9352)465953

Sarin, Dr Victor A. (1947). Senior Researcher. Frank Lab. of Neutron Physics, Joint Insti-tute for Nuclear Research, 141980 Dubna, 141980 Moscow region, Russia. PhD (Inst. of Applied Physics Kishinev, 1978) physics. *Neutron_structure_analysis*
E-mail sarin@nfsun1.jinr.dubna.su Tel. 7(09621)62132 Fax 7(09621)65085 Telex 911621 DUBNA SU

Savenko, Dr Boris N. (1946). Group Leader. Frank Lab. of Neutron Physics, Joint In-stitute for Nuclear Research, 141980 Dubna, 141980 Moscow region, Russia. PhD (Joint Inst. for Nuclear Research Dubna, 1989) physics. *Neutron_diffraction sin-gle_crystal_real_structure ferroelectric_phase_transition domain_structure neu-tron_diffuse_scattering*
E-mail savenko@nfsun1.jinr.dubna.su Tel. 7(09621)62498 Fax 7(095)9243914 or 7(09621)65085 Telex 911621 DUBNA SU

Sedov, Mr Boris B. (1941). Chief Lecturer. Department of Physics, Vladimir State Peda-gogical University, Department of Physics, 11 Stroiteley Prospect, Vladimir 600024, Russia. MSc (Vladimir State Pedagogical U., 1964) physics. *Patterson_method molecular_crystal X-ray_crystallography*
Tel. 7(09222)72604

Semenchev, Dr Aleksandr F. (1947). Associate Professor. Physics Department, Rostov State University, 5 Zorge Str., Rostov-on-Don 344104, Russia. PhD (Rostov State U., 1982) physics and mathematics. *Crystal_structure phase_transition twinning*
Tel. 7(8632)223413

Semenova, Dr Tatiyana F. (1951). Associate Professor. Department of Crystallogra-phy, St Petersburg University, University Emb. 7/9, St Petersburg 199034, Russia. PhD (Leningrad State U., 1980) crystallography. *X-ray_structure_determination crystal_chemistry_inorganic_compound structural_mineralogy X-ray high-temper-ature_powder_diffraction*
E-mail flt@dean.geol.lgu.spb.su Tel. 7(812)2189647 Fax 7(812)2181346 Telex 121481 LSU SU

Serebryanaya, Dr Nadezda R. (1937). Senior Researcher. Laboratory of Physics of Phase Transitions, Institute of Spectroscopy, Russian Academy of Sciences, Troitsk, Moscow region 142092, Russia. PhD (Inst. of Crystallography Moscow, 1970) crystallography and crystal chemistry. *Powder_diffraction high_pressure_phase_transition crys-tal_structure X-ray_high_pressure_technique*
E-mail blank@isan.msk.su Tel. 7(095)3340014 Fax 7(095)3340014

Serezhkin, Prof. Victor N. (1946). Head of Chair. Inorganic Chemistry Chair, Chem-istry Department, Samara State University, 1 Ac. Pavlov St., Samara 443011, Russia. DSc (Inst. of Inorganic Chemistry RAS Novosibirsk, 1985) inorganic chemistry. *Crys-tal_chemistry systematics coordination_compound uranium_compounds*
E-mail serezkin@univer.samara.su Tel. 7(846)2345445 Fax 7(846)2345417

Sergienko, Dr Vladimir S. (1941). Head of Laboratory. Laboratory of Coordination Compounds Crystal Chemistry, Kurnakov Institute of General and Inorganic Chem-istry, Russian Academy of Sciences, 31 Leninsky prosp., Moscow 117907 GSP-1, Russia. DSc (Inst. General and Inorganic Chemistry, 1992) physical chemistry. *X-ray_crystallography coordination_compound*
Tel. 7(095)9521803 Fax 7(095)9541279

Sevastyanov, Dr Boris K. (1930). Head of Laboratory Deputy Director of Inst. Laboratory of Optical Crystals Physics, Institute of Crystallography, Russian Academy of Sci-ences, 59 Leninskii prosp., Moscow 117333, Russia. PhD (Moscow Inst. of Physics and Technology, 1962) physics. *Optical_property crystal_laser crystal_growth*
Tel. 7(095)3308247 or 7(095)3308047 Fax 7(095)1351011

Shafranovsky, Prof. Ilarion Ilarionovich (1907). Professor. Chair of Mineralogy Crys-tallography and Petrography, Geological Research Faculty, St Petersburg Mining In-stitute, 2 21st Liniya, St Petersburg 199026, Russia. DSc (Leningrad Mining Inst., 1946) crystallography and crystal physics. *Crystal_morphology mineralogy his-tory_of_science*
Tel. 7(812)2188266 Fax 7(812)2132613 Telex 121494 LGIPSU

Shamray, Dr Vladimir F. (1937). Chief of Laboratory. Baikov Institute of Metallurgy, Russian Academy of Sciences, Laboratory of Crystallochemistry, Russian Academy of Sciences, 49 Leninskyi pr., Moscow 117334, Russia. DSc (Inst. of Physics of Metals Ministry of Metallurgy Moscow, 1980) physics. *X-ray_crystallography texture super-conductor Rietveld_method phase_determination X-ray_powder_diffraction modu-lated_structures phase_equilibrium*
Tel. 7(095)1356572 Fax 7(095)1358680

Shashkin, Dr Dmitry P. (1936). Senior Scientist. Institute of Chemical Physics, Russian Academy of Sciences, 2-b Vorobyevskoye Chaussee, Moscow 117871, Russia. PhD (Moscow State U., 1970) geology and mineralogy. *Crystal_structure_analysis*
Tel. 7(095)9397557 Fax 7(095)9382156

Shekhtman, Prof. Veniamin Sh. (1929). Vice director. X-ray Laboratory Institute of Solid State Physics, Russian Academy of Sciences, Chernogolovka, Moscow Re-gion 142432, Russia. DSc (Institute of Steel and Alloys Moscow, 1977) solid state physics. *X-ray_crystallography modulated_structures quasicrystal real_structure phase_transition ferroelectricity high_temperature_superconductors*
E-mail shekht@issp.sherna.msk.su Tel. 7(095)5255063 Fax 7(095)5171949 Telex 412654 Serna su

Shepelev, Dr Yury F. (1939). Senior Researcher. Lab. of X-ray Crystallography, Institute of Silicate Chemistry, Russian Academy of Sciences, 24-2 Odoevskogo str., St Pe-tersburg 199155, Russia. DSc (Inst. Silicate Chemistry St Petersburg, 1991) physical chemistry. *X-ray_crystal_structure_determination silicates zeolites superconductor diffraction_technique sorption phase_transition modulated_structures*
E-mail shepelev@ihs2.spb.su Tel. 7(812)2185102 Fax 7(812)3510813 Telex 121447 VITRO SU

Shibaeva, Dr Rimma P. Chief Researcher. Laboratory of X-ray Structure Analysis, Institute of Solid State Physics, Russian Academy of Sciences, Chernogolovka, Moscow Region 142432. DSc (Inst.of Crystallography Moscow, 1977) physics. *X-ray_crystallography organic_conductor structure_physical_properties_relationship conductors super-conductors*
E-mail shibaeva@issp.sherna.msk.su Tel. 7(095)5245063 Fax 7(095)5171949

Shil'nikov, Dr Valery I. (1942). Researcher. X-ray Structural Center of the Russian Academy of Sciences, Institute of Organoelement Compounds, Russian Academy of Sciences, 28 Vavilov St., Moscow 117813, Russia. MSc (Baku Azerbaijan U., 1963) mathematics. *Computing computer_modelling organic_crystal_chemistry non-bonded_interaction_potential_energy*
E-mail xray@xray.ineos.ac.ru Tel. 7(095)1359332 Fax 7(095)1355085

Shilov, Dr Gennadi V. (1956). Senior researcher. Lab. Crystal Chemistry, Branch Institute of Chem. Physics, Russian Acad. of Sciences, Chernogolovka, Moscow Region 142432, Russia. PhD (Inst. Chemical Physics Moscow, 1992) chemical physics. *X-ray_crystallography organic_inorganic_compounds*
E-mail atov@icph.sherna.msk.su Tel. 7(095)5245059 Fax 7(095)2655714

Shishkin, Dr Oleg V. (1966). Researcher. X-ray Structural Center of the Russian Academy of Sciences, Institute of Organoelement Compounds, Russian Academy of Sciences, 28 Vavilov St., Moscow 117813, Russia. PhD (Kharkov State U., 1994) organic chemistry. *X-ray_crystallography small_molecules molecular_mechanics nitrogen_compounds organic_crystal_chemistry bioactive_molecules*
E-mail xray@xray.ineos.ac.ru Tel. 7(095)1359343 Fax 7(095)1355085

Shishova, Dr Tatyana G. (1949). Associate Professor. Institute of Agriculture, 97 Gagarin Prosp., Nizhny Novgorod 603078, Russia. PhD (Inst. of Crystallography Moscow, 1977) physics. *X-ray_crystallography macromolecules organic_crystal_chemistry*
Tel. 7(8312)660694

Shmyt'ko, Dr Ivan M. (1946). Senior Researcher. Laboratory of Structure Analysis, Institute of Solid State Physics, Russian Academy of Sciences, Chernogolovka, Moscow Region 142432, Russia. PhD (Institute of Physics and Technology Moscow, 1976) solid state physics. *X-ray_crystallography phase_transition phase_kinetics incommensurate_phase ferroelasticity X-ray_analysis twinning*
E-mail shim@issp.sherna.msk.su Fax 7(096)5171949 Telex 412654 SERNA SU

Shulakov, Dr Eugene V. (1949). Senior Researcher. Laboratory of X-ray Optics and Electron Microscopy, Institute of Solid State Physics, Russian Academy of Sciences, Chernogolovka, Moscow Region 142432, Russia. PhD (Inst. Solid State Physics RAS, 1978) solid state physics. *X-ray_diffraction_theory X-ray_optics X-ray_topography X-ray_Bragg-Fresnel_diffraction white-beam_radiation_dynamical_diffraction X-ray_divergent-beam_method Laue_method_indexing_software*
E-mail shulakov@issp.sherna.msk.su Tel. 7(095)5245051 Fax 7(096)5171949 Telex 412654 SERNA SU

Shuvalov, Dr Alexander L. (1955). Senior Researcher. Lab. of Mechanical Properties of Crystals, Institute of Crystallography, Russian Academy of Sciences, 59 Leninskii pr., 117333 Moscow, Russia. PhD (Inst. of Crystallography Moscow, 1985) solid state physics. *Theoretical_acoustics solid_state_mechanics*
E-mail ashuv@mechan.incr.msk.su Tel. 7(095)3308274 Fax 7(095)3308274

Shuvalov, Prof. Lev A. (1923). Chief Researcher. Laboratory of Crystallophysics, Institute of Crystallography, Russian Academy of Sciences, 59 Leninski pr., Moscow 117333, Russia. DSc (Inst. of Crystallography Moscow, 1972) crystal physics. *Ferroelectricity ferroelasticity phase_transition symmetry superionic photorefraction crystallography*
E-mail root@theory.incr.msk.su Tel. 7(095)1350251 Fax 7(095)1351011

Silonov, Dr Valentin M. (1942). Group Leader. Solid State Chair, Physics Department, Moscow State University, Vorobjovy Gory, Moscow 117234, Russia. DSc (Moscow State U., 1990) solid state physics. *X-ray_diffraction condensed_matter_physics solid_state_electronic_theory*
E-mail katsnelson@poly.phys.msu.su. Tel. 7(095)9394308 Fax 7(095)9390247 Telex 411483 MGU SU

Silvestrova, Dr Olga Y. (1949). Researcher. Lab. of Acoustooptics of Crystals, Institute of Crystallography, Russian Academy of Sciences, Leninsky pr. 59, 117333 Moscow, Russia. MSc (Moscow Inst. of Steel and Alloys, 1971) physics. *Acoustic_elastic_property monocrystal*
Tel. 7(095)3306883 Fax 7(095)3301956

Simonov, Prof. Valentin I. (1930). Laboratory Head, Deputy Director. Co-editor of Acta Crystallographica. Laboratory of X-ray Structure Analysis, Institute of Crystallography, Russian Academy of Sciences, 59 Leninsky pr., Moscow 117333, Russia. DSc (Inst. of Crystallography Moscow, 1972) physics and mathematics. *Structural_crystallography inorganic_materials phase_transition*
E-mail simonov@rsa.crystal.msk.su Tel. 7(095)1356571 Fax 7(095)1351011

Sinay, Dr Marina Yu. (1962). Senior Researcher. Crystal Genesis Lab., St Petersburg University, University Emb. 7/9, St Petersburg 199034, Russia. PhD (St Petersburg U., 1991) mineralogy and crystallography. *Crystallogeny crystal_growth crystal_morphology phase_diagram ontogeny_minerals organic_impurity experimental_mineralogy*
E-mail anna@dean.geol.lgu.spb.su Tel. 7(812)2189650 Fax 7(812)2181346 Telex 121481 LSU SU

Sirota, Dr Mikhail I. (1945). Senior researcher. Laboratory of X-ray Structure Analysis, Institute of Crystallography, Russian Academy of Sciences, 59 Leninsky pr., Moscow 117333, Russia. PhD (Inst. of Crystallography Moscow, 1975) physics and mathematics. *Computer_software structural_analysis phase_transition*
E-mail sirota@rsa.crystal.msk.su Tel. 7(095)1352400 Fax 7(095)1351011

Sizova, Dr Natalia L. (1937). Senior Researcher. Lab. of Acoustooptics of Crystals, Institute of Crystallography, Russian Academy of Sciences, Leninsky pr. 59, 117333 Moscow, Russia. PhD (Inst. of Crystallography Moscow, 1974) solid state physics. *Mechanical_property crystal_hardness deformation dislocation*
Tel. 7(095)3306883 Fax 7(095)3301956

Skvortsova, Dr Natalya P. (1949). Researcher. Lab. of Mechanical Properties of Crystals, Institute of Crystallography, Russian Academy of Sciences, 59 Leninskii pr., 117333 Moscow, Russia. PhD (Inst. of Metals Physics Kiev, 1991) solid state physics. *Strength_plasticity_physics real_crystals plastic_deformation*
E-mail public@mechan.incr.msk.su Tel. 7(095)3342483 Fax 7(095)3308274

Smetannikova, Dr Olga G. Senior Researcher. Department of Crystallography, St Petersburg University, University Emb. 7/9, St Petersburg 199034, Russia. PhD (St Petersburg U., 1974) crystallography and crystal physics. *Crystal_chemistry isomorphous_replacement optical_property diffraction_data mineralogy crystal_morphology*
E-mail anna@dean.geol.lgu.spb.su Tel. 7(812)2189647 Fax 7(812)2181346 Telex 121481 LSU SU

Smirnov, Dr Alexey E. (1946). Researcher. Lab. of Mechanical Properties of Crystals, Institute of Crystallography, Russian Academy of Sciences, 59 Leninsky pr., 117333 Moscow, Russia. PhD (Inst. of Crystallography Moscow, 1981) solid state physics. *Thermal_chemical_dissolution etching plasticity strength track_membranes electromechanical_effect*
E-mail public@mechan.incr.msk.su Tel. 7(095)3308274 Fax 7(095)3308274

Smirnov, Mr Lev (1936). Senior Researcher. Frank Lab. of Neutron Physics, Joint Institute for Nuclear Research, 141980 Dubna, 141980 Moscow region, Russia. MSc (Moscow Inst. of Physics and Technology, 1959) physics. *Solid_state_physics phase_transition lattice_dynamics*
E-mail lsmirnov@nfsun1.jinr.dubna.su Tel. 7(095)9243914 Fax 7(09621)65085 Telex 911621 DUBNA SU

Smirnov, Prof. Yuri M. (1932). Head of chair. Applied Physics Chair, Department of Physics, Tver State University, Zhelyabova str. 33, Tver 170000, Russia. DSc (Institute of Electronic Technology Moscow, 1986) solid state physics. *Crystal_growth germanium_compounds silicon_compounds iodine_compounds crystal_physics*
Tel. 7(08222)60636 Telex 412556 GELU SU

Smirnova, Dr Irina A. (1960). Junior Researcher. Laboratory of X-ray Optics and Electron Microscopy, Institute of Solid State Physics, Russian Academy of Sciences, Chernogolovka, Moscow Region 142432, Russia. PhD (Inst. of Solid State Physics RAS, 1994) solid state physics. *X-ray_optics X-ray_acoustics*
E-mail @issp.sherna.msk.su Tel. 7(095)5245063 Fax 7(096)5171949 Telex 412654 SERNA SU

Smirnova, Dr Nina L. (1926). Senior Researcher. Crystallography and Crystal Chemistry Chair, Department of Geology, Moscow State University, Moscow 119899, Russia. PhD (Inst. of Crystallography Moscow, 1961) crystal chemistry. *Theoretical_crystal_chemistry classification inorganic_compound*
Tel. 7(095)9393875

Smirnova, Dr Sofia A. (1936). Head of Laboratory. Lab. of Crystallization from Melt, Russian Research Institute for Synthesis of Minerals and Raw Materials, Institutskaya St. 1, Alexandrov, Vladimir Region 601600, Russia. PhD (Moscow State U., 1983) physical chemistry. *Crystal_growth mineralogy*
Tel. 7(095)5845834 Fax 7(095)5845828

Smoliar-Zviagina, Dr Bela B. (1958). Researcher. Lab. for Investigation of Rock-Forming Minerals by Physical Methods, Geological Institute, Russian Academy of Sciences, Pyzhevsky per. 7, Moscow 109017, Russia. PhD (Inst. of Geology Mineralogy and Petrography of Ore Minerals Moscow, 1986) geology and mineralogy. *Crystal_chemistry structure_modelling layered_silicates*
Tel. 7(095)2308185 Fax 7(095)2318195

Smolin, Prof. Yury I. (1930). Laboratory Head. Lab. of X-ray crystallography, Institute of Silicate Chemistry, Russian Academy of Sciences, 24/2 Odoevskogo St., St Petersburg 199155, Russia. DSc (Inst. of Crystallography Moscow, 1974) physics and mathematics. *X-ray_crystallography structure_determination silicates zeolites high_temperature_superconductors complex_vanadium_compounds*
E-mail smolin@ihs2.spb.su Tel. 7(812)2185102 Fax 7(812)3510813 Telex 121447 VITRO SU

Smolsky, Dr Igor L. (1945). Laboratory Leader. Laboratory of Crystal Growth from Solutions, Institute of Crystallography, Russian Academy of Sciences, 59 Leninsky pr., Moscow 117333, Russia. PhD (Physico-Technical Inst. Ukrainian Academy of Sciences, 1977) solid state physics. *X-ray_topography real_structure crystal_growth condensed_matter*
Tel. 7(095)3307883 Fax 7(095)1351011

Smotrakov, Dr Valery G. (1944). Senior Researcher. Lab. of Crystal Physics, Institute of Physics, Rostov State University, Pr. Stachki 194, Rostov-on-Don 344104, Russia. [Zorge Str. 48/1, Apt 117, Rostov-on-Don 344090, Russia.] PhD (Rostov State U., 1971) chemistry. *Crystal_growth perovskites piezo-_and_ferroelectric_materials*
E-mail Smotrak@riphys.rnd.su Tel. 7(8632)285066 Fax 7(8632)285044

Sobolev, Dr Alexander N. (1950). Senior Researcher. Karpov Institute of Physical Chemistry, Laboratory of Organic Crystallochemistry and Structure Analysis, 10 Obukha St., Moscow 103064, Russia. PhD (Karpov Inst. of Physical Chemistry Moscow, 1988) physical chemistry. *X-ray_crystallography coordination_compound organometallic_compound organic_compound computing programming*
Tel. 7(095)2270014x2328 Fax 7(095)9752450

Sobolev, Dr Boris P. (1936). Head of laboratory. Member of Commission on Crystal Growth and Characterization. Lab. of Physical and Chemical Analysis, Institute of Crystallography, Russian Academy of Sciences, 59 Leninsky prosp., Moscow 117333, Russia. DSc (Inst. of Crystallography Moscow, 1979) chemistry. *Nonstoichiometry inorganic_fluorine_compounds structural_disorder*
Tel. 7(095)3307874 Fax 7(095)1351011

Sokolova, Dr Elena V. (1953). Chief Lecturer. Department of Crystallography and Crystal Chemistry, Faculty of Geology, Moscow State University, Moscow 119899, Russia. PhD (Moscow State U., 1980) crystallography and crystal physics. *X-ray_crystallography minerals materials single_crystal_structure_determination Rietveld_method*
Tel. 7(095)9393850 Fax 7(095)9390126 Telex 411483 MGU SU

Sokolova, Dr Nataliya G. (1939). Senior Researcher. Lab. of Mineral Raw Material Quality Evaluation, St Petersburg Mining Institute, 2 21st Liniya, Saint Petersburg 199026, Russia. PhD (St Petersburg Mining Inst., 1968) crystallography and crystal physics. *X-ray_crystallography morphology mineralogical_crystallography*
Tel. 7(812)2188455 Fax 7(812)2132613 Telex 121494 LGIPSU

Soidatov, Dr Eugeni A. (1949). Associate Professor. Department of Applied Physics and Microelectronics, Nizhny Novgorod State University, Gagarina Ave. 23, Nizhny Novgorod 603600, Russia. PhD (Inst. of Crystallography Moscow, 1979) crystallography. *X-ray_crystallography small_molecules mathematical_crystallography*
Tel. 7(8312)656475

Solodovnikov, Dr Sergei F. (1954). Senior Researcher. Laboratory of Crystal Chemistry, Institute of Inorganic Chemistry, Siberian Branch of RAS, 3 Lavrentyev prosp., Novosibirsk 630090, Russia. PhD (Inst. Inorganic Chemistry Novosibirsk, 1989) inorganic chemistry. *X-ray_crystallography inorganic_oxides superconductor_oxides structural_design crystal_structure_database structural_computing inorganic_crystal_chemistry solid_state_chemistry inorganic_synthesis*
E-mail solod@che.nsk.su Tel. 7(3832)354366 Fax 7(3832)355960

Solovyeva, Prof. Lidiya P. Chief Researcher. Boreskov Institute of Catalysis, Laboratory of Structure Methods, 5 Lavrentieva Pr., Novosibirsk 630090, Russia. DSc (Institute of Geology and Geophysics Novosibirsk, 1991) mineralogy and crystallography. *X-ray_powder_diffraction Rietveld_method real_structure computing_programming*
E-mail root@reverse.nsk.su Tel. 7(383)2354300 Fax 7(383)2355756

Sorokin, Dr Nikolai I. (1958). Senior Researcher. Lab. of Physical and Chemical Analysis, Institute of Crystallography, Russian Academy of Sciences, 59 Leninsky prosp., Moscow 117333, Russia. PhD (Inst. of Crystallography Moscow, 1989) crystal physics. *Superionic_conductivity solid_electrolytes inorganic_fluorine_compounds crystal_anion_disorder impedance_spectroscopy structure_property_relationship*
Tel. 7(095)3307874 Fax 7(095)1351011

Sorokina, Dr Kira L. (1959). Senior Researcher. Lab. of Electron Diffraction, Institute of Crystallography, Russian Academy of Sciences, 59 Leninsky prosp., Moscow 117333, Russia. PhD (Ioffe Physico-Technical Inst. St Petersburg, 1989) physics. *High-resolution_electron_microscopy dynamical_diffraction inverse_problem computer_modelling programming amorphous_phase*
Tel. 7(095)1354520

Sorokina, Dr Natalia I. (1952). Researcher. Secretary of Crystallographic Committee of Russia. Laboratory of X-ray Structure Analysis, Institute of Crystallography, Russian Academy of Sciences, 59 Leninsky pr., Moscow 117333, Russia. PhD (Inst. of Crystallography Moscow, 1984) crystallography. *X-ray_analysis crystal_chemistry inorganic_compound non-linear_property superconductivity*
E-mail simonov@rsa.crystal.msk.su Tel. 7(095)1353110 Fax 7(095)1351011

Sosfenov, Dr Nikita I. (1932). Senior researcher. Laboratory of Biocrystal Structure, Institute of Crystallography, Russian Academy of Sciences, 59 Leninsky prosp., Moscow 117333, Russia. PhD (Institute of Crystallography Moscow, 1972) crystallography and crystal physics. *X-ray_measurement_apparatus proteins_structure_determination*
Tel. 7(095)1356120 Fax 7(095)1351011

Sotman, Dr Sima S. (1946). Senior Researcher. Department of Crystallography and Crystallography Computing, Joint-Stock Company "Ajax", 55 Galernaja Str., St Petersburg 190000, Russia. PhD (Moscow State U., 1981) crystal chemistry. *Crystal_chemistry organic_complex_compound*
Tel. 7(812)3125208 Fax 7(812)3122479

Soumbatov, Mr Alexander (1961). Researcher. Karpov Institute of Physical Chemistry, 10 Obukha St., Moscow 103064, Russia. MSc (Moscow State U., 1984) physics. *Semiconductors thin_layer laser*
E-mail soumbat@nifhi.uucp.free.msk.su Tel. 7(095)2270014x2356 Fax 7(095)9752450

Sozontov, Dr Evgeny A. (1952). Senior Researcher. Laboratory of X-ray Structure Analysis, Kaluga Branch of Inst. of Crystallography, Russian Academy of Sciences, 2 Akademicheskaya str., Kaluga 248640, Russia. PhD (Inst. of Steel and Alloys Moscow, 1984) physics. *Standing_wave single_crystal_real_structure X-ray_diffraction*
Tel. 7(08422)48393 Fax 7(095)1351011

Starostina, Dr Ludmila S. (1933). Senior Researcher. Laboratory of Optical Crystals Physics, Institute of Crystallography, Russian Academy of Sciences, 59 Leninskii prosp., Moscow 117333, Russia. [Leninskii prosp. 99-20, Moscow 117421, Russia.] PhD (Moscow Engineering Physical Inst., 1965) physics. *Crystal_growth solid state_physics*
E-mail Star@nvser.msk.su Tel. 7(095)3308247 or 7(095)3308047; (home) 7(095)4342477 Fax 7(095)1351011

Stash, Mr Adam I. (1956). Researcher. X-ray laboratory, Karpov Institute of Physical Chemistry, 10 Obukha St., Moscow 103064, Russia. MSc (Moscow State U., 1983) physical chemistry. *X-ray_crystallography diffuse_scattering profile_analysis organic_organometallic_crystal_chemistry charge_density small_molecules*
E-mail adam@nifhi.uucp.free.msk.su Tel. 7(095)2270014x2328 Fax 7(095)9752450

Stefanovich, Dr Sergey Yu. (1945). Group leader. Karpov Institute of Physical Chemistry, 10 Obukha, Moscow 103064, Russia. PhD (Karpov Institute of Physical Chemistry, 1975) chemical physics. *Non_centrosymmetry_oxides ferroelectricity non-linear_optical_materials ionic_conductivity_phosphates*
E-mail stefan@nifhi.uucp.free.msk.su Tel. 7(095)2270014x2339 Fax 7(095)9752450

Stepanova, Dr Alla N. (1934). Group Leader. Lab. of Crystallization from Vapor Phase, Institute of Crystallography, Russian Academy of Sciences, 59 Leninsky pr., Moscow 117333, Russia. PhD (Inst. of Crystallography Moscow, 1974) crystal physics. *Crystal_growth whisker_growth*
Tel. 7(095)3308265 Fax 7(095)1351011

Stepantsov, Dr Evguenii A. (1951). Senior Researcher. Lab. of Mechanical Properties of Crystals, Institute of Crystallography, Russian Academy of Sciences, 59 Leninskii pr., 117333 Moscow, Russia. PhD (Inst. Crystallography Moscow, 1982) solid state physics. *Crystalline_composite grains_boundaries sintering high_temperature_superconductor_thin_films*
E-mail stepan@mechan.incr.msk.su Tel. 7(095)3308274 Fax 7(095)3308274

Stiopina, Dr Nina D. (1950). Researcher. Small Angle Scattering Lab., Institute of Crystallography, Russian Academy of Sciences, Leninsky pr. 59, Moscow 117333, Russia. PhD (Moscow State U., 1986) chemistry. *Organic_thin_film polymers_film lipids_film Langmuir–Blodgett_films*
E-mail feigin@saxslab.incr.msk.su Tel. 7(095)1356010 Fax 7(095)1351011

Stishov, Prof. Sergei M. (1937). Director of Institute. Vereschagin Institute of High Pressure Physics, Troitsk, Moscow Reg. 142092, Russia. DSc (Inst. of Crystallography Moscow, 1974) physics. *High_pressure_physics high_pressure_crystallography*
E-mail sergei@adonis.iasnet.com Tel. 7(095)3340010 Fax 7(095)8833039

Struchkov, Prof. Yuri T. (1926). Director of the X-ray Structural Centre. Vice-President of the International Union of Crystallography. X-ray Structural Center of the Russian Academy of Sciences, Institute of Organoelement Compounds, Russian Academy of Sciences, 28 Vavilov St., Moscow 117813, Russia. DSc (Inst. Organoelement Compounds, 1977) physical chemistry. *X-ray_crystallography small_molecules organic_organometallic_crystal_chemistry*
E-mail yuts@xray.ineos.ac.ru Tel. 7(095)1359271 Fax 7(095)1355085

Suvorov, Prof. Ernest V. (1937). Group leader. Member, Bureau of Russian Council of Electron Microscopy. Laboratory of X-ray Optics and Electron Microscopy, Institute of Solid State Physics, Russian Academy of Sciences, Chernogolovka, Moscow Region 142432, Russia. [Institution prospect 2-48, Chernogolovka, Moscow Region 142432, Russia.] DSc (Solid State Physics Inst. RAS, 1982) solid state physics. *X-ray_electron_dynamic_diffraction real_structure_crystal electron_microscopy metals semiconductors superconductors*
E-mail suvorov@issp.ac.ru Tel. 7(095)5849725 (directorate); 7(096)5171637 (laboratory) Fax 7(096)5171949 Telex 412654 SERNA SU

Sveshnikov, Dr Sergey V. (1939). Assistant Professor. Solid State Chair, Physics Department, Moscow State University, Vorobjovy Gory, Moscow 117234, Russia. PhD (Moscow State U., 1987) solid state physics. *X-ray_analysis amorphous_compound*
E-mail asi@phys.msu.su. Tel. 7(095)9392387 Fax 7(095)9390247 Telex 411483 MGU SU

Tafeenko, Dr Victor Aleksandrovich (1952). Senior researcher. Lab. Structural Chemistry, General Chemistry Faculty, Chemistry Dept., Moscow State University, Moscow 119899, Russia. PhD (Moscow State U., 1981) physical and organic chemistry. *Crystallography organic_crystal_chemistry structure_determination expert_systems*
E-mail Tafeenko@crysma.uucp.free.msk.su Tel. 7(095)9395089 Fax 7(095)9390898 Telex 411483 MGU SU

Tagieva, Dr Marianna M. Associate Professor. Department of Crystallography, Institute of Steel and Alloys, 4 Leninsky prosp., Moscow 117936, Russia. PhD (Inst. of Steel and Alloys Moscow, 1974) solid state physics. *Solid_state_physics crystallography*
Tel. 7(095)2366500 Fax 7(095)2378007

Talis, Dr Alexander L. (1957). Chief Researcher. Lab. of Physical and Technical Research, Russian Research Institute for Synthesis of Minerals and Raw Materials, Institutskaya St. 1, Alexandrov, Vladimir Region 601600, Russia. PhD (Moscow State U., 1989) crystallography. *Group_theory crystal_symmetry quasicrystal*
Tel. 7(095)5845834 Fax 7(095)5845828

Telegina, Dr Inna V. Chief Lecturer. Solid State Chair, Physics Department, Moscow State University, Vorobjovy Gory, Moscow 117234, Russia. PhD (Moscow State U., 1968) physics. *Phase_formation phase_determination structural_change*
E-mail asi@phys.msu.su. Tel. 7(095)9392387 Fax 7(095)9390247 Telex 411483 MGU SU

Tikhonova, Dr Anna A. Senior Researcher. Lab. of Electron Diffraction, Institute of Crystallography, Russian Academy of Sciences, 59 Leninsky prosp., Moscow 117333, Russia. PhD (Inst. of Crystallography RAS Moscow, 1973) physics. *Epitaxy defect_surface*
Tel. 7(095)1354010

Timofeeva, Dr Tatjana V. (1947). Senior Researcher. X-ray Structural Center of the Russian Academy of Sciences, Institute of Organoelement Compounds, Russian Academy of Sciences, 28 Vavilov St., Moscow 117813, Russia. PhD (Inst. Organoelement Compounds, 1982) physical chemistry. *Molecular_mechanics crystal_packing liquid_crystals*
E-mail timof@xray.ineos.ac.ru Tel. 7(095)1359343 Fax 7(095)1355085

Timofeeva, Dr Valentina A. (1923). Senior Researcher. Laboratory of Optical Crystals Physics, Institute of Crystallography, Russian Academy of Sciences, 59 Leninskii prosp., Moscow 117333, Russia. PhD (Institute of Chemistry Alma-Ata, 1949) chemistry. *Crystal_growth*
Tel. 7(095)3308247 or 7(095)3308047 Fax 7(095)1351011

Tishchenko, Prof. Galina N. Chief researcher. Laboratory of Biocrystal Structure, Institute of Crystallography, Russian Academy of Sciences, 59 Leninsky prosp., Moscow 117333, Russia. DSc (Inst. of Crystallography Moscow, 1984) crystal chemistry. *Structure_determination proteins cyclic_peptides*
Tel. 7(095)1352300 Fax 7(095)1351011

Tkachev, Dr Valeri V. (1943). Senior Researcher. Branch Institute of Chemical Physics, Russian Academy of Sciences in Chernógolovka, 142432 Chernogolovka, Moscow Region, Russia. PhD (Inst. of Chemical Physics in Chernogolovka, 1973) physical chemistry. *X-ray_crystallography inorganic_organic_compound organic_organometallic_biomolecule_crown_compounds*
E-mail vatka@cph.sherna.msk.su Tel. 7(096)5245059 Fax 7(096)5153588

Tomashpolsky, Prof. Yuri Ya. (1937). Vice-Director. Karpov Institute of Physical Chemistry, 10 Obukha St., Moscow 103064, Russia. DSc (Karpov Institute of Physical Chemistry, 1975) chemical physics. *Analytical_electron_microscopy surface secondary_electron_emission thin_film ferroelectricity superconductivity*
E-mail tomash@nifhi.uucp.free.msk.su Tel. 7(095)2971727 Fax 7(095)9752450

Tovbis, Dr Alexander B. (1940). Senior Researcher. Laboratory of X-ray Diffractometry, Institute of Crystallography, Russian Academy of Sciences, 59 Leninsky prosp., Moscow 117333, Russia. PhD (Inst.of Crystallography Moscow, 1971) mathematics. *Powder_diffraction computer_software structural_analysis*
E-mail CRYS@CRYS.MSK.SU Tel. 7(095)1351100 Fax 7(095)1351011

Treivus, Dr Eugene Borisovich (1934). Chief researcher. Crystal Genesis Laboratory, St Petersburg University, University Emb. 7/9, St Petersburg 199034, Russia. DSc (St Petersburg U., 1965) crystallography and crystal physics. *Crystal_morphology crystal_growth crystal_growth_computer_modelling*
E-mail anna@dean.geol.lgu.spb.su Tel. 7(812)2189674 Fax 7(812)2181346 Telex 121481 LSU SU

Tretyakov, Dr Vjacheslav N. (1933). Head of Laboratory. Lab. of Mineral Raw Material Quality Evaluation, St Petersburg Mining Institute, 2 21st Liniya, St Petersburg 199026, Russia. PhD (St Petersburg Mining Inst., 1978) crystallography and crystal physics. *Crystal_growth goniometry_method segregation_of_impurities crystallographically_oriented_crystals_preparation*
Tel. 7(812)2188683 Fax 7(812)2132613 Telex 121494 LGIPSU

Trofimov, Mr Victor B. (1945). Senior Researcher. Department of Crystallography, St Petersburg University, University Emb. 7/9, St Petersburg 199034, Russia. MSc (Leningrad State U., 1972) crystal chemistry and crystallography. *Crystal_chemistry X-ray_high-temperature_powder_diffraction phase_transition*
E-mail flt@dean.geol.lgu.spb.su Tel. 7(812)2189647 Fax 7(812)2181346 Telex 121481 LSU SU

Trofimova, Mrs Rimma F. (1955). Junior Researcher. Lab. of Crystal Chemistry, Branch Institute of Chemical Physics, Russian Academy of Sciences in Chernogolovka, Chernogolovka, Moscow Region 142432, Russia. MSc (Inst. of Fine Chemical Technology, 1979) chemistry. *X-ray_crystallography nitrogen_compounds structure-activity_relationship energetic_compounds*
E-mail atov@icph.sherna.msk.su Tel. 7(096)5171168 Fax 7(096)5153588

Trubkin, Dr Nikolay V. (1949). Senior Researcher. Laboratory of Electron Microscopy, Institute of Geology of Ore Deposits, Petrology, Mineralogy and Geochemistry, Russian Academy of Sciences (IGEM RAS), Staromonetny 35, 109017 Moscow, Russia. PhD (IGEM RAS, 1985) mineralogy. *Analytical_electron_microscopy materials experimental_high_resolution_electron_microscopy*
Tel. 7(095)2308411 Fax 7(095)2302179

Tsikhotsky, Dr Evgeny S. (1946). Senior Researcher. Physics Department, Rostov State University, Pr. Stachki 192, Rostov-on-Don 344090, Russia. PhD (Rostov State U., 1975) physics and mathematics. *Crystal_research perfection*
Tel. 7(8632)292174

Tsinober, Dr Leonid I. (1924). Chief Researcher. Lab. of Physical and Technical Research, Russian Research Institute for Synthesis of Minerals and Raw Materials, Institutskaya St. 1, Alexandrov, Vladimir Region 601600, Russia. PhD (Inst. of Crystallography Moscow, 1962) crystallography. *X-ray_crystallography real_structure twinning symmetry*
Tel. 7(095)5845834 Fax 7(095)5845828

Tsirelson, Prof. Vladimir Grigorievich (1948). Professor. Consultant, IUCr Commission on Charge, Spin and Momentum Density. Mendeleev University of Chemical Technology of Russia, 9 Miusskaja Square, Moscow 125190, Russia. DSc (Karpov Inst. of Physical Chemistry, 1990) solid state physics. *X-ray_crystallography charge_density spin_density crystal_chemistry physical_properties_of_crystals chemical_bonding*
E-mail tsirel@mhti.msk.su Tel. 7(095)2585930 Fax 7(095)2004204 Telex 411744 argon

Tsybulya, Dr Sergey V. (1959). Senior Researcher. Boreskov Institute of Catalysis, Laboratory of Structure Methods, 5 Lavrentieva Pr., Novosibirsk 630090, Russia. PhD (Institute of Applied Physics Kishineu, 1989) crystallography and physics of crystals. *X-ray_powder_diffraction Rietveld_method real_structure computing_programming*
E-mail root@reverse.nsk.su Tel. 7(383)2354300 Fax 7(383)2355756

Tzalenchuk, Dr Alexander Ya. (1962). Researcher. Lab. of Mechanical Properties of Crystals, Institute of Crystallography, Russian Academy of Sciences, 59 Leninsky pr., 117333 Moscow, Russia. PhD (Inst. of Crystallography Moscow, 1988) solid state physics. *High_temperature_superconductivity tunnelling thin_film grain_boundaries*
E-mail sasha@mechan.incr.msk.su Tel. 7(095)3308274 Fax 7(095)3308274

Udalova, Dr Valentina V. (1932). Senior Researcher. Institute of Crystallography, Russian Academy of Sciences, 59 Leninsky prosp., Moscow 117333, Russia. PhD (Inst. of Crystallography Moscow, 1974) physics and mathematics. *Structure_analysis electron_diffraction*
Tel. 7(095)1353510 Fax 7(095)1351011

Urusov, Prof. Vadim S. (1936). Head of department. Consultant, IUCr Commission on Charge, Spin and Momentum Densities; Member, National Committee of Russian Crystallographers. Moscow State University, Geology Faculty, Department of Crystallography and Crystal Chemistry, Moscow 119899, Russia. DSc (Institute of Geochemistry and Anal. Chem. of Russia Acad.of Sciences, 1975) geochemistry crystal chemistry. *Density_distribution crystal_energetics isomorphous_replacement structural_modelling interatomic_interaction bond_strength atomic_size solid_solubility structural_stability*
E-mail urusov@glas.apc.org Tel. 7(095)9395575 Fax 7(095)9390126 Telex 411483 MGU SU

Urusovskaya, Prof. Aida A. (1929). Chief Researcher. Lab. of Mechanical Properties of Crystals, Institute of Crystallography, Russian Academy of Sciences, 59 Leninskii pr., 117333 Moscow, Russia. DSc (Inst. of Crystallography Moscow, 1981) solid state physics. *Mechanical_property point_defect dislocation real_structure ion_implantation polarization irradiation*
E-mail public@mechan.incr.msk.su Tel. 7(095)3308274 Fax 7(095)3308274

Urzhumtsev, Dr Alexander G. (1951). Senior Researcher. Lab. of Macromolecular Crystallography, Institute of Mathematical Problems of Biology, Russian Academy of Sciences, Pushchino, Moscow Region 142292, Russia. PhD (Inst. of Crystallography Moscow, 1985) crystallography. *Biocrystallography imaging phase_determination phase_refinement*
E-mail root@impb.serpukhov.su Tel. 7(095)9233558 Telex (64) 412604 CODON SU

Vainshtein, Prof. Boris K. (1921). Director of Institute. Co-editor of Acta Crystallographica D. Laboratory of Biocrystal Structure, Institute of Crystallography, Russian Academy of Sciences, 59 Leninsky prosp., Moscow 117333, Russia. DSc (Inst. of Crystallography Moscow, 1955) structural crystallography. *X-ray_crystallography diffraction_theory biological_macromolecules crystal_structure_analysis_theory electron_microscopy*
Tel. 7(095)1356541 Fax 7(095)1351011

Vartanyants, Dr Ivan A. (1956). Senior Researcher. Lab. of X-ray Optics and Synchrotron Radiation, Institute of Crystallography, Russian Academy of Sciences, 59 Leninsky prosp., Moscow 117333, Russia. PhD (Moscow Physical Engineering Inst., 1984) physics. *X-ray_dynamical_diffraction X-ray_standing_wave X-ray_diffraction_theory X-ray_optics X-ray_diffractometry X-ray_diffuse_scattering*
E-mail vart@crystal.msk.su Tel. 7(095)3307992 Fax 7(095)3301956

Vasil'ev, Dr Evgeny K. (1922). Senior Researcher. Institute of the Earth Crust, Siberian Branch, Russian Academy of Sciences, Inst. of the Earth Crust, 128 Lermontov St., Irkutsk 664033, Russia. PhD (Irkutsk State U., 1966) physics. *X-ray_powder_diffraction inorganic_compound database_minerals qualitative_quantitative_phase_analysis_rocks data_collection_processing_inorganic_compounds crystal_chemistry_minerals*
E-mail dor@crust.irkutsk.su OR san@cora.irkutsk.su Tel. 7(3952)464456 Telex 231283 zemlya

Vasiliev, Dr Alexander D. (1947). Group leader. Laboratory of Crystal Physics, Institute of Physics, Siberian Dept., Russian Academy of Sciences, Krasnoyarsk 660036, Russia. PhD (Inst. of Crystallography Moscow, 1986) crystallography and crystal physics. *X-ray_structure_determination small_molecules inorganic_compound phase_transition high-temperature_superconductors*
E-mail aleks@iph.krasnoyarsk.su Tel. 7(391)2432635 Fax 7(391)2438923

Vasiliev, Dr Alexandr L. (1956). Senior Researcher. Institute of Crystallography, Russian Academy of Sciences, Leninsky pr. 59, Moscow 117333, Russia. PhD (Institute of Crystallography Moscow, 1992) solid state physics. *High_resolution_electron_microscopy epitaxy boundaries high_temperature_superconductivity*
Tel. 7(095)1350010 Fax 7(095)1351011

Velichkina, Dr Tat'yana S. (1925). Associate Professor. Chair of Polymer and Crystal Physics, Physics Department, Moscow State University, Moscow 119899, Russia. PhD (Physics Dept. Moscow State U., 1954) crystallography and crystal physics. *Brillouin_spectroscopy phase_transitions*
E-mail aleks@cryst0.phys.msu.su Tel. 7(095)9391013 Fax 7(095)9390247 Telex 411483 MGU SU

Venevtsev, Prof. Yurii N. (1926). Head of laboratory. Karpov Institute of Physical Chemistry, 103064 Moscow, Obukha 10, Russia. DSc (Lebedev Physical Institute, 1970) solid state physics. *Superconductivity magnetism ferroelectricity oxides phase_transition*
E-mail venev@nifhi.uucp.msk.su Tel. 7(095)2270014x2327 Fax 7(095)9752450

Veremeichik, Dr Tamara F. (1945). Senior Researcher. Lab. of Crystallophysics, Shubnikov Institute of Crystallography, Russian Academy of Sciences, 59 Leninsky Pr., Moscow 117333, Russia. PhD (Inst. Crystallography Moscow, 1977) physics. *Crystal_field_theory metastable_impurity_electronic_spectrum Jahn-Teller-effect circular_dichroism*
E-mail root@theory.incr.msk.su Tel. 7(095)1351300 Fax 7(095)1351011

Vergasov, Dr Vladimir L. (1955). Senior Researcher. Lab. of Electron Diffraction, Institute of Crystallography, Russian Academy of Sciences, 59 Leninsky prosp., Moscow 117333, Russia. PhD (Inst. of Crystallography RAS Moscow, 1985) physics. *High-energy_electron_diffraction high-resolution_electron_microscopy*

dynamical_diffraction electron_diffraction_theory electron_holography transmission_electron_microscopy secondary_electron_emission convergent-beam_electron_diffraction
Tel. 7(095)1354520

Vigdorchik, Dr Asya G. (1949). Researcher. Sub-Editor, Russia, Ninth Edition of World Directory of Crystallographers. Laboratory of X-ray Structure Analysis, Institute of Crystallography, Russian Academy of Sciences, 59 Leninsky pr., Moscow 117333, Russia. PhD (Institute of Crystallography Moscow, 1993) chemistry. *X-ray_structure_determination coordination_compound rare-earth_compounds nitrates*
E-mail asyavig@rsa.crystal.msk.su Tel. 7(095)1350330 Fax 7(095)1351011

Vilkov, Prof. Lev Vasilyevich (1931). Professor. Chemistry Department, Moscow State University, Moscow 119899, Russia. DSc (Moscow State U., 1969) chemistry. *Structural_analysis organic_compound electron_diffraction_gases physical_methods_in_chemistry*
E-mail vilkov@eldiff.chem.msu.su Tel. 7(095)9392637 Fax 7(095)9328846

Vinogradov, Dr Alexander V. (1964). Junior Researcher. Lab. of Acoustooptics of Crystals, Institute of Crystallography, Russian Academy of Sciences, Leninsky pr. 59, 117333 Moscow, Russia. MSc (Moscow State U., 1987) physics. *Acoustooptics computer_automation_modelling*
Tel. 7(095)3306883 Fax 7(095)3301956

Vinokurov, Prof. Vladimir M. (1921). Professor of Chair. Dept. of Geology, Kazan State University, 18 Lenin Str., Kazan 420008, Russia. DSc (Kazan State U., 1966) geology and mineralogy. *Crystallography spectroscopy physical_property magnetic_resonance mineralogy*
E-mail root@scikgu.kazan.su Tel. 7(8432)327062 Fax 7(8432)380994 Telex 224641 CHAIR SU

Virovets, Dr Alexander V. (1966). Researcher. Lab.of Crystal Chemistry, Institute of Inorganic Chemistry Siberian Branch, Russian Academy of Sciences, 3 Lavrentyev prosp., Novosibirsk 630090, Russia. PhD (Inst. of Inorg. Chem. Siberian Branch of Russian Academy of Sciences, 1993) inorganic chemistry. *X-ray_crystallography small_molecules clusters computing computer-aided_education database structural_computing inorganic_crystal_chemistry solid_state_chemistry*
E-mail borisov@che.nsk.su OR educ@xray.nsk.su Tel. 7(3832)354366 Fax 7(3832)355960

Vlasov, Dr Michael Yu. (1959). Researcher. Lab. of Modelling of Nature Crystallization, Earthcrust Research Institute, St Petersburg University, University Emb 7/9, St Petersburg 199034, Russia. PhD (St Petersburg Technology Inst., 1986) chemistry. *Solid_state_chemistry solid_electrolytes diffusion structural_change*
E-mail anna@dean.geol.lgu.spb.su Tel. 7(812)2189644 Fax 7(812)2181346 Telex 121481 LSU SU

Vlasov, Dr Vasily Platonovich (1941). Senior Researcher. Lab. of Elementary Processes of Crystal Growth, Institute of Crystallography, Russian Academy of Sciences, Leninsky pr. 59, Moscow 117333, Russia. PhD (Inst. of Crystallography Moscow, 1978) physics. *Epitaxy nucleation electron*
E-mail chern@elprgr.crystal.msk.su Tel. 7(095)1354160 Fax 7(095)1351011

Volkov, Dr Vladimir V. (1953). Senior Researcher. Lab. of Crystallophysics, Institute of Crystallography, Russian Academy of Sciences, 59 Leninsky Pr., Moscow 117333, Russia. PhD (Inst. Crystallography Moscow, 1990) crystallography. *Small-angle_scattering structure_determination ribosomes proteins chemometrics spectral_analysis spectral_decomposition computing linear_algebra nonlinear_optimization electron_spin_resonance titanates*
E-mail feigin@saxslab.incr.msk.su Tel. 7(095)1350229 Fax 7(095)1351011

Voloshin, Dr Alexey E. (1960). Researcher. Lab. of Crystal Growth from Solutions, Institute of Crystallography, Russian Academy of Sciences, 59 Leninsky pr., Moscow 117333, Russia. PhD (Inst. of Fine Chemical Technology Moscow, 1986) physical chemistry. *X-ray_topography real_structure image_processing diffraction_technique elasticity_theory*
Tel. 7(095)3307883 Fax 7(095)1351011

Voronkova, Dr Valentina I. (1936). Senior Researcher. Chair of Polymer and Crystal Physics, Physics Department, Moscow State University, Moscow 119899, Russia. PhD (Moscow State U., 1969) physics. *Crystal_growth crystal_structure physical_property oxides ferroelectrics superionic_conductors superconductors*
E-mail voronk@cryst.phys.msu.su Tel. 7(095)9392883 Fax 7(095)9392988 Telex 411483 MGU SU

Vrublevskaja, Dr Zoja V. Researcher. Laboratory of X-ray Structure Analysis, Institute of Ore Deposits, Petrography, Mineralogy and Geochemistry, Russian Academy of Sciences (IGEM RAS), Staromonetny per.35, Moscow 109017, Russia. PhD (IGEM RAS, 1974) crystal chemistry and mineralogy. *X-ray_crystallography minerals systematics polymorphism electron_diffraction polytypism carbon_compounds clays*
E-mail Zoja@igem.msk.su Tel. 7(095)2308296 Fax 7(095)2302179

Yakovlev, Prof. Ivan A. (1912). Professor. Chair of Polymer and Crystal Physics, Physics Department, Moscow State University, Moscow 119899, Russia. DSc (Moscow State U., 1958) crystallography and crystal physics. *Brillouin_spectroscopy phase_transitions*
E-mail aleks@cryst0.phys.msu.su Tel. 7(095)9391013 Fax 7(095)9392988 Telex 411483 MGU SU

Yakubovich, Dr Olga V. (1950). Senior Researcher. Department of Crystallography, Geological Faculty, Moscow State University, 119899 Moscow, Russia. PhD (Moscow State University, 1978) crystallography and crystal physics. *X-ray_crystallography minerals materials phosphate_crystal_chemistry electron_density_distribution Rietveld_method*
Tel. 7(095)9393850 Fax 7(095)9390195 Telex 411483 MGU SU

Yakunin, Mr Andrew N. (1961). Researcher. Laboratory of Polymer Structure, Karpov Institute of Physical Chemistry, 10 Obukha, Moscow 103064, Russia. MSc (Moscow Physical and Technical Institute, 1983) chemistry. *Phase_transition_theory critical_phenomena polymers amorphous_phase_heterostructure X-ray_small-angle_scattering polymers field_theory symmetry_breaking*
E-mail yakunin@nifhi.uucp.free.msk.su Tel. 7(095)2270014x2519 Fax 7(095)9752450

Yakushkin, Dr Eugene D. (1952). Researcher. Laboratory of Electromechanical Investigations of Crystals, Institute of Crystallography, Russian Academy of Sciences, Leninsky pr. 59, Moscow 117333, Russia. PhD (Inst. of Crystallography Moscow, 1988) physics. *Ferroelectricity thermal_property ultrasonics real_structure phase_transition ionic_conductivity*
E-mail postmaster@el-mech.incr.msk.su Tel. 7(095)3307856 Fax 7(095)1351011

Yamnova, Dr Natalya A. (1950). Senior Researcher. Department of Crystallography and Crystal Chemistry, Faculty of Geology, Moscow State University, Moscow 119899, Russia. PhD (Moscow State U., 1976) crystallography and crystal physics. *X-ray_crystallography materials minerals rare-earth_compounds silicates carbonates*
Tel. 7(095)9392330 Fax 7(095)9390126 Telex 411483 MGU SU

Yanovskii, Dr Vladimir K. (1931). Senior Researcher. Chair of Polymer and Crystal Physics, Physics Department, Moscow State University, Moscow 119899, Russia. PhD (Moscow Inst. Chemical Technology, 1963) physical chemistry. *Crystal_growth crystal_structure physical_property oxides ferroelectrics superionic_conductors superconductors*
E-mail yanov@cryst.phys.msu.su Tel. 7(095)9392883 Fax 7(095)9392988 Telex 411483 MGU SU

Yanovsky, Dr Alexander I. (1957). Senior Researcher. X-ray Structural Center of the Russian Academy of Sciences, Institute of Organoelement Compounds, Russian Academy of Sciences, 28 Vavilov St., Moscow 117813, Russia. PhD (Inst. Organoelement Compounds, 1983) physical chemistry. *Organometallic_structure carboranes metallic_clusters computing*
E-mail yan@xray.ineos.ac.ru Tel. 7(095)1359214 Fax 7(095)1355085

Yanulova, Dr Ludmila A. (1942). Researcher. Institute of Geology, Komi Department, Russian Academy of Sciences, 54 Pervomayskaya Str., Syktyvkar 167610, Russia. PhD (Inst. of Geology and Geochemistry Sverdlovsk, 1972) geology and mineralogy. *Crystal_chemistry sulphides*
Tel. 7(82122)25160 Fax 7(82122)25346

Yanusova, Dr Ludmila G. (1949). Researcher. Small Angle Scattering Lab., Institute of Crystallography, Russian Academy of Sciences, Leninsky pr. 59, Moscow 117333, Russia. PhD (Inst. of Steel and Alloys Moscow, 1980) physics and mathematics. *SAXS reflectivity thin_film superlattice optical_elastic_properties_of_crystals*
E-mail feigin@saxslab.incr.msk.su Tel. 7(095)1356010 Fax 7(095)1351011

Yatsenko, Dr Alexandr Vasilievich (1960). Senior researcher. Lab. Structural Chemistry, General Chemistry Faculty, Chemistry Dept., Moscow State University, Moscow 119899, Russia. PhD (Moscow State U., 1988) coordination crystal chemistry. *Organic_crystal_chemistry hydrogen_bonding organic_synthesis*
E-mail Yatsenko@crysma.uucp.free.msk.su Tel. 7(095)9395089 Fax 7(095)9390898 Telex 411483 MGU SU

Yufit, Dr Dmitrii S. (1957). Researcher. X-ray Structural Center of the Russian Academy of Sciences, Institute of Organoelement Compounds, Russian Academy of Sciences, 28 Vavilov St., Moscow 117813, Russia. PhD (Moscow State U., 1984) physical chemistry. *Organic_structure small_molecules sulfur_compounds polycyclic_molecules*
E-mail dima@xray.ineos.ac.ru Tel. 7(095)1359355 Fax 7(095)1355085

Yushkin, Prof. Nikolay P. (1936). Director of Institute; Academician of Russian Acad. Sci. Institute of Geology, Komi Department, Russian Academy of Sciences, 54 Pervomayskaya Str., Syktyvkar 167610, Russia. [46, apt. 84, Kirova Str., Syktyvkar 167000, Russia.] DSc (St Petersburg Mining Inst., 1968) geology and mineralogy. *Crystallography crystallogeny minerals history_of_crystallography_and_mineralogy*
Tel. 7(82122)20037 Fax 7(82122)25346

Zadorozhnaya, Dr Ludmila A. (1944). Researcher. Lab. of Crystallization from Gas Phase, Institute of Crystallography, Russian Academy of Sciences, 59 Leninsky prosp., Moscow 117333, Russia. PhD (Moscow State U., 1977) crystal growth. *CVD crystal_growth semiconductor_film*
Tel. 7(095)3308265 Fax 7(095)1351011

Zaharova, Dr Maria I. Professor. Solid State Chair, Physics Department, Moscow State University, Vorobjovy Gory, Moscow 117234, Russia. DSc (Moscow State U., 1948) solid state physics. *Phase_transition X-ray_electron_diffraction*
E-mail asi@phys.msu.su Tel. 7(095)9392387 Fax 7(095)9390247 Telex 411483 MGU SU

Zaitsev, Dr Sergey M. (1951). Associate Professor. Physics Department, Rostov State University, Pr. Stachki 192, Rostov-on-Don 344090, Russia. PhD (Rostov State U., 1979) physics and mathematics. *X-ray_structure_analysis*
Tel. 7(8632)220867

Zakharchenko, Dr Irina N. (1946). Senior Researcher. Lab. of Crystal Physics, Institute of Physics, Rostov State University, Pr. Stachky 194, Rostov-on-Don 344104, Russia. PhD (Rostov State U., 1978) physics and mathematics. *X-ray_structure_analysis phase_transition crystal_defect thin_ferroelectrics_films*
E-mail irina-z@riphys.rnd.su Tel. 7(8632)285066 Fax 7(8632)285044 Telex 123321, 123610 FIZIK RU

Zakharov, Dr Boris G. (1937). Head of Laboratory; Director. Laboratory of X-ray Structure Analysis, Kaluga Branch of Institute of Crystallography, Russian Academy of Sciences, 2 Akademicheskaya str., Kaluga 248640, Russia. [53-8 Komarov str., Kaluga 248600, Russia.] DSc (Inst. of Steel and Alloys Moscow, 1984) technology of materials of electronic technique. *Crystal_growth real_structure X-ray_diffraction microgravity*
E-mail zakharov@academ.kaluga.su Tel. 7(08422)48614 (work) 7(08422)94869 (home) Fax 7(08422)48614

Zakharov, Prof. Lev N. (1950). Senior Researcher. Lab. of Technology of Organometallic Compounds, Inst. of Organometallic Chemistry, Russian Academy of Sciences, 49 Tropinina St., Nizhny Novgorod 603600 GSP-445, Russia. PhD (Inst. of Applied Physics Kishinev, 1981) physics. *X-ray_crystallography small_molecules organic_organometallic_crystal_chemistry*
E-mail imoc@infotel.msk.su Tel. 7(8312)664370 Fax 7(8312)661497

Zalesskii, Prof. Andrei V. (1930). Chief Researcher. Institute of Crystallography, Russian Academy of Sciences, Laboratory of Resonance methods, Leninsky pr. 59, Moscow 117333, Russia. DSc (Inst. of Crystallography Moscow, 1985) physics. *Nuclear_magnetic_resonance magnetic_domain antiferromagnetism ferrites*
E-mail root@magnet.crystal.msk.su Tel. 7(095)3308329 Fax 7(095)3308329

Zasimov, Dr Victor S. (1940). Senior Researcher. Solid State Chair, Physics Department, Moscow State University, Vorobjovy Gory, Moscow 117234, Russia. PhD (Moscow State U., 1973) solid state physics. *X-ray_diffraction Mossbauer_spectroscopy crystal_field*
E-mail asi@phys.msu.su. Tel. 7(095)9392387 Fax 7(095)9390247 Telex 411483 MGU SU

Zassourskaya, Mrs Larissa Alexandrovna (1952). Junior researcher. Lab. Crystal Chemistry, Physical Chemistry Faculty, Chemical Dept., Moscow State University, Moscow 119899, Russia. MSc (Moscow State U., 1979) physical chemistry. *Organic_crystal_chemistry coordination_chemistry complex_compounds X-ray_research crystal_packing hydrogen_bonding*
E-mail physch@mch.chem.msu.su Tel. 7(095)9392258 Fax 7(095)9328846 Telex 411483 MCUSU

Zavodnik, Dr Valery E. (1941). Chief Researcher. X-ray Laboratory, Karpov Institute of Physical Chemistry, 10 Obukha St., 103064 Moscow, Russia. PhD (Inst. of Applied Physics, 1990) crystallography and crystal physics. *X-ray_crystallography charge_density accurate_intensity profile_analysis diffuse_scattering phase_transition data_processing*
E-mail zaval@nifhi.uucp.free.msk.su Tel. 7(095)2270014x2328 Fax 7(095)9752450

Zayakina, Dr Nadezhda V. (1943). Senior Researcher. Institute of Geology, Yakutsk Branch, Siberian Department, Russian Academy of Sciences, Lenin pr. 39, Yakutsk 677007, Russia. PhD (Moscow State U., 1976) geology and mineralogy. *Crystal_structure_determination crystal_chemistry silicates minerals*

Zheludeva, Dr Svetlana I. (1948). Senior Researcher. Lab. of X-ray Optics and Synchrotron Radiation, Institute of Crystallography, Russian Academy of Sciences, 59 Leninsky prosp., Moscow 117333, Russia. PhD (Moscow State U., 1976) physics. *X-ray_standing_wave X-ray_dynamical_diffraction thin_film X-ray_and_synchrotron_radiation_instrumentation*
E-mail zhel@crystal.msk.su Tel. 7(095)3307992 Fax 7(095)3301956

Zhirnov, Dr Victor V. (1966). Junior researcher. Lab. of Crystallization from Vapor Phase, Institute of Crystallography, Russian Academy of Sciences, 59 Leninsky pr, Moscow 117333, Russia. PhD (Institute of Physics and Technology Moscow, 1992) solid state electronics. *Growth_whisker diamond_deposition molecular_beam_epitaxy silicides*
E-mail zhirnov@cvdlab.incr.msk.su Tel. 7(095)3308265 Fax 7(095)1351011

Zhmurova, Dr Zinaida I. (1930). Chief Researcher. Laboratory of Physical and Chemical Analysis, Institute of Crystallography, Russian Academy of Sciences, 59 Leninsky pr., Moscow 117333, Russia. PhD (Inst. of Crystallography Moscow, 1970) physical chemistry. *Crystal_growth Bridgman_Stockbarger_technique*
Tel. 7(095)3307874 Fax 7(095)1351011

Zhukhlistov, Dr Anatolyi P. (1938). Senior Researcher. Laboratory of Electronography, Institute of Ore Deposits, Petrology, Mineralogy and Geochemistry, Russian Academy of Sciences (IGEM RAS), Staromonetny per. 35, Moscow 109017, Russia. PhD (IGEM RAS, 1963) mineralogy. *Structural_crystallography structural_mineralogy electron_diffraction polytypism*
Tel. 7(95)2308296 Fax 7(95)2302179

Zhukhlistova, Dr Nadezhda E. (1941). Senior researcher. Laboratory of Biocrystal Structure, Institute of Crystallography, Russian Academy of Sciences, 59 Leninsky prosp., Moscow 117333, Russia. PhD (Institute of Cryst. Moscow, 1983) crystallography and crystal physics. *X-ray_crystallography small_molecules peptides structure-activity_relationship steroids*
Tel. 7(095)1352300 Fax 7(095)1351011

Zhukov, Dr Sergey Gennadievich (1962). Researcher. Lab. Structural Chemistry, General Chemistry Faculty, Chemistry Dept., Moscow State University, Moscow 119899, Russia. PhD (Moscow State U., 1990) solid state physics. *Solid_state_physics charge_density phase_transition computing*
E-mail Zhukov@crysma.uucp.free.msk.su Tel. 7(095)9395089 Fax 7(095)9390898 Telex 411483 MGU SU

Zhurov, Mr Vladimir V. (1963). Researcher. Karpov Institute of Physical Chemistry, X-ray Laboratory, 10 Obukha St., Moscow 103064, Russia. MSc (Moscow Institute of Physics and Technology, 1986) chemical physics. *Rietveld_method structural_phase_transition thermal_expansion ferroelectricity Debye_Waller_factor real_structure inorganic_compound X-ray_powder_diffraction_software electron_density_distribution line_profile_analysis X-ray_powder_diffraction*
E-mail zhurov@nifhi.uucp.free.msk.su Tel. 7(095)2270014x2729 Fax 7(095)9752450

Zhurova, Dr Elizabeth A. (1964). Researcher. Laboratory of X-ray Structure Analysis, Institute of Crystallography, Russian Academy of Sciences, 59 Leninsky pr., Moscow 117333, Russia. PhD (Inst. of Physical Chemistry Moscow, 1993) physical chemistry. *Anharmonicity density_distribution perovskites solid_solution*
E-mail simonov@rsa.crystal.msk.su Tel. 7(095)1350330 Fax 7(095)1351011

Zibrov, Dr Igor P. (1955). Senior Researcher. Lab. of Mechanical properties of Crystals, Institute of Crystallography, Russian Academy of Sciences, 59 Leninsky prosp, Moscow 117333, Russia. PhD (Moscow Inst. of Steel and Alloys, 1983) physical chemistry. *X-ray_diffraction high_pressure_chemistry oxides tungsten_oxides_crystal_chemistry*
E-mail public@mechan.incr.msk.su. Tel. 7(095)3307883 Fax 7(095)1351011

Zinenko, Dr Victor I. (1942). Chief Researcher. Lab. of Crystal Physics, Institute of Physics, Siberian Dept., Russian Academy of Sciences, Krasnoyarsk 660036, Russia. DSc (Inst. of Physics Krasnoyarsk, 1984) solid state physics. *Structural_phase_transition inorganic_compound*
E-mail aleks@iph.krasnoyarsk.su Tel. 7(391)2432635 Fax 7(391)2438923

Zlatkin, Dr Alex T. (1954). Researcher. Lab. of Liquid Crystals, Institute of Crystallography, Russian Academy of Sciences, 59 Leninsky Pr., Moscow 117333, Russia. PhD (Inst. of Crystallography Moscow, 1991) physics. *Scanning_tunnel_microscopy Langmuir_Blodgett_film*
E-mail lev@glas.apc.org Tel. 7(095)3307847 Fax 7(095)1351011

Zlokazov, Dr Victor B. (1941). Senior Researcher. Frank Lab. of Neutron Physics, Joint Institute for Nuclear Research, Dubna, Moscow Region 141980, Russia. PhD (Joint Inst. for Nuclear Research Dubna, 1977) physics and mathematics. *Neutron_diffraction data_processing*
E-mail zlokazov@main1.jinr.dubna.su Tel. 7(09621)65476 Fax 7(095)9752381 Telex 911621 Dubna SU

Zorkaya, Dr Olga N. (1953). Researcher. Lab. Crystal Chemistry, Physical Chemistry Faculty, Chemical Dept., Moscow State University, Moscow 119899, Russia. PhD (Inst. of Applied Physics Moldavian Academy of Sciences, 1986) chemical crystallography. *X-ray_crystallography small_molecules organic_crystal_chemistry*
E-mail physch@mch.msu.su Tel. 7(095)9395434 Fax 7(095)9328846 Telex 411483 MGU SU

Zorkii, Prof. Petr M. (1933). Professor. Lab. Crystal Chemistry, Physical Chemistry Faculty, Chemical Dept., Moscow State University, Moscow 119899, Russia. DSc (Moscow State U., 1973) physical chemistry. *Crystallography crystal_chemistry hydrogen_bonding symmetry philosophy_of_science history_of_science graph_theory*
E-mail physch@mch.chem.msu.su Tel. 7(095)9395434 Fax 7(095)9328846 Telex 411483 MGU SU

Zubenko, Dr Vasily V. (1930). Associate Professor. Solid State Chair, Physics Department, Moscow State University V, Vorobjovy Gory, Moscow 117234, Russia. PhD (Moscow State U., 1968) physics. *Phase_formation phase_determination structural_change*
E-mail asi@phys.msu.su. Tel. 7(095)9392387 Fax 7(095)9390247 Telex 411483 MGU SU

Zvezdinskaya, Dr Larissa V. (1948). Senior Researcher. Laboratory of Uranium Geology, Institute of Ore Deposits, Petrology, Mineralogy and Geochemistry, Russian Academy of Sciences, Staromonetny per. 35, Moscow 109017, Russia. PhD (Moscow State U., 1979) crystal chemistry and mineralogy. *X-ray_crystallography minerals polymorphism*
Tel. 7(95)2308296 Fax 7(95)2302179

Zvyagin, Prof. Boris Borisovich (1921). Laboratory Head. Laboratory of Electronography, Institute of Ore Deposits, Petrology, Mineralogy and Geochemistry, Russian Academy of Sciences, Staromonetny per. 35, Moscow 109017, Russia. DSc (Inst. of Crystallography Moscow, 1963) physics and mathematics. *Structural_crystallography structural_mineralogy electron_diffraction polytypism*
Tel. 7(95)1356514 Fax 7(95)2302179

SERBIA

Sub-Editor: D. Poleti

Andjelković, Dr Katarina (1957). Assistant Professor. Chemistry Department, Faculty of Sciences, Studentski trg 16, Belgrade, Serbia. [Chemistry Department, Faculty of Sciences, PO Box 550, 11001 Belgrade, Serbia.] PhD (U. Belgrade, 1992) chemistry. Structure_determination_of_transition_elements coordination_compound conformation_analysis optical_activity trans-effect
Tel. 381(11)3282111x737 Fax 381(11)638785

Bogdanović, Mr Goran (1965). Research Associate. Laboratory of Theoretical Physics and Physics of Condensed Matter, Institute of Nuclear Sciences - Vinča, Vinča, Belgrade, Serbia. [Laboratory of Theoretical Physics and Physics of Condensed Matter, Institute of Nuclear Sciences - Vinča, PO Box 522, 11001 Belgrade, Serbia.] BSc (U. Belgrade, 1991) chemistry. Inorganic_compound coordination_compound
Tel. 381(11)4440871x766 Fax 381(11)8363020

Čelijević, Dr Valerija (1948). Assistant Professor. Laboratory for Inorganic and Coordination Chemistry, Institute for Chemistry, Faculty of Sciences, Trg Dositeja Obradovića 3, 21000 Novi Sad, Serbia. PhD (U. Novi Sad, 1989) chemistry. Coordination_compound synthesis_and_characterization_of_coordination_compounds
Tel. 381(21)350122x411 Fax 381(21)55662

Dimitrijević, Dr Radovan (1947). Associate Professor. Crystallographic Laboratory, Faculty of Mining and Geology, Djušina 7, 11000 Belgrade, Serbia. PhD (U. Belgrade, 1985) crystallography. Zeolites minerals
Tel. 381(11)635217 Fax 381(11)635217

Divjaković, Dr Vladimir (1946). Professor. Faculty of Sciences, University of Novi Sad, Trg Dositeja Obradovića 4, 21000 Novi Sad, Serbia. PhD (U. Bern, 1976) mineralogy-crystallography. Polymers physics
Tel. 381(21)350122x411 Fax 381(21)55662

Djurić, Dr Stevan (1931). Associate Professor. Crystallographic Laboratory, Faculty of Mining and Geology, Studentski trg 16, Belgrade, Serbia. [Crystallographic Laboratory, Faculty of Mining and Geology, Djušina 7, 11000 Belgrade, Serbia.] PhD (U. Belgrade, 1980) geology. Instrumentation clays materials_for_electronics
Tel. 381(11)635217 Fax 381(11)635217

Erić, Mrs Suzana (1963). Research Associate. Geoinstitut, Rovinjska 12, 11000 Beograd, Serbia. BSc (U. Belgrade, 1986) mineralogy. Minerals
Tel. 381(11)4880506x139

Horvat-Gabriel, Miss Margareta (1968). Research Associate. Institute of Physics, Faculty of Sciences, University of Novi Sad, Trg Dositeja Obradovića 4, 21000 Novi Sad, Serbia. BSc (U. Novi Sad, 1993) physics - crystallography. Biocrystallography conformation_analysis hydrogen_bonding molecular_packing
Tel. 381(21)55318 Fax 381(21)55662

Ivanović, Miss Ivana (1971). Research Assistant. Chemistry Department, Faculty of Sciences, Studetski trg 16, Belgrade, Serbia. [Chemistry Department, Faculty of Sciences, PO Box 550, 11001 Belgrade, Serbia.] BSc (U. Belgrade, 1993) inorganic chemistry. Inorganic_compound coordination_compound
Tel. 381(11)635425 Fax 381(11)638785

Ivegeš, Miss Erika (1966). Research Assistant. Laboratory for Inorganic and Coordination Chemistry, Institute for Chemistry, Faculty of Sciences, Trg Dositeja Obradovića 3, 21000 Novi Sad, Serbia. BSc (U. Novi Sad, 1989) coordination chemistry. Coordination_compound synthesis_and_characterization_of_coordination_compounds
Tel. 381(21)350122x411 Fax 381(21)55662

Janjić, Dr Svetislav (1931). Professor. Institute for Copper, Zeleni Bulevar 35, 19210 Bor, Serbia. PhD (U. Belgrade, 1979) mineralogy - crystallography. Minerals inorganic_compound
Tel. 381(30)35216 Fax 381(30)35247 Telex 19137

Kapor, Dr Agneš (1950). Professor. Institute of Physics, Faculty of Sciences, University of Novi Sad, Trg Dositeja Obradovića 4, Serbia. PhD (U. Novi Sad, 1981) solid state physics - crystallography. Biocrystallography organic_compound molecular_packing conformation_analysis hydrogen_bonding
Tel. 381(21)55318 Fax 381(21)55662

Karanović, Dr Ljiljana (1950). Associate Professor. Crystallographic Laboratory, Faculty of Mining and Geology, Studentski Trg 16, Belgrade, Serbia. [Crystallographic Laboratory, Faculty of Mining and Geology, Djušina 7, 11000 Belgrade, Serbia.] PhD (U. Belgrade, 1985) direct methods in crystallography. Structure_determination minerals powder_method solid_state materials_science
Tel. 381(11)635217 Fax 381(11)635217

Krstanović, Dr Ilija (1927). Professor. Crystallographic Laboratory, Faculty of Mining and Geology, Studentski Trg 16, Belgrade, Serbia. [Crystallographic Laboratory, Faculty of Mining and Geology, Djušina 7, 11000 Belgrade, Serbia.] PhD (U. Belgrade, 1961) mineralogy. Minerals inorganic_compound
Tel. 381(11)635217 Fax 381(11)635217

Lazar, Dr Dušan (1951). Research Assistant. Institute of Physics, Faculty of Sciences, Trg Dositeja Obradovića 4, 21000 Novi Sad, Serbia. PhD (U. Novi Sad, 1993) crystallography. Biocrystallography estrogens biologically_active_compounds
Tel. 381(21)55318 Fax 381(21)55662

Leković, Mrs Nada (1941). Research Associate. Laboratory for Special Materials, "IRI-TEL" - Beograd, Batajnički Put 23, 11000 Belgrade, Serbia. BSc (U. Belgrade, 1966) inorganic chemistry. Vitreous_state electronics microelectronics thin_film
Tel. 381(11)109477 Fax 381(11)108801 Telex 12237YUEiZEP

Leovac, Dr Vukadin (1943). Professor. Laboratory for Inorganic and Coordination Chemistry, Institute of Chemistry, Faculty of Sciences, Trg Dositeja Obradovića 3, 21000 Novi Sad, Serbia. PhD (U. Novi Sad, 1978) coordination chemistry. Coordination_compound synthesis_and_characterization_of_coordination_compounds
Tel. 381(21)350122x411 Fax 381(21)55662

Malinar, Dr Mijat (1932). Associate Professor. Chemistry Department, Faculty of Sciences, Studentski Trg 16, Belgrade, Serbia. [Chemistry Department, Faculty of Sciences, PO Box 550, 11001 Belgrade, Serbia.] PhD (U. Belgrade, 1976) coordination chemistry. Coordination_compound amino_acids chelates conformation optical_activity absolute_configuration
Tel. 381(11)3282111x742 Fax 381(11)638785

Niketić, Dr Svetozar R. (1944). Professor. Chemistry Department, University of Belgrade, Studentski Trg 16, Belgrade, Serbia. [Chemistry Department, University of Belgrade, PO Box 550, 11001 Belgrade, Serbia.] PhD (The Technical University of Denmark, Lyngby, 1975) coordination chemistry - molecular mechanics. Computing crystal_field database intermolecular_interaction chirooptical_properties stereochemistry
E-mail xpmfh01%yubgss21%fon@moumee.calstatela.edu Tel. 381(11)3282111x747 Fax 381(11)638785

Nikolić, Dr Slobodanka (1951). Scientific Associate. IHTM, Karnegijeva 4/I, 11000 Belgrade, Serbia. PhD (U. Belgrade, 1985) mechanism of crystal growth. Crystal_growth imperfection kinetic_and_mechanism_of_crystal_growth materials_science
Tel. 381(11)3222281 Fax 381(11)182995

Pejović, Mrs Verica (1940). Research Associate. Laboratory for Special Materials, "IRI-TEL" - Beograd, Batajnički Put 23, 11000 Belgrade, Serbia. BSc (U. Belgrade, 1965) organic chemistry. Ceramics glasses materials_for_electronics
Tel. 381(11)105042x356 Fax 381(11)108801 Telex 12237YUEiZEP

Poleti, Dr Dejan (1952). Associate Professor. National Sub-Editor for WDC9. Department of General and Inorganic Chemistry, Faculty of Technology and Metallurgy, Karnegijeva 4, Belgrade, Serbia. [Department of General and Inorganic Chemistry, Faculty of Technology and Metallurgy, PO Box 494, 11001 Belgrade, Serbia.] PhD (U. Belgrade, 1988) technical sciences. Inorganic_compound coordination_compound powder_method solid_state materials_science
Tel. 381(11)3228839 Fax 381(11)3220847

Prelesnik, Dr Bogdan (1938). Scientific Adviser. Laboratory of Theoretical Physics and Physics of Condensed Matter, Institute of Nuclear Sciences - Vinča, Vinča, Belgrade, Serbia. [Laboratory of Theoretical Physics and Physics of Condensed Matter, Institute of Nuclear Sciences - Vinča, PO Box 522, 11001 Belgrade, Serbia.] PhD (U. Bern, 1975) crystallography. Inorganic_compound coordination_compound organic_compound heterogeneous_ice_nucleation
Tel. 381(11)4440871x716 Fax 381(11)8363010

Radaković, Mrs Aleksandra (1956). Research Assistant. Crystallographic Laboratory, Faculty of Mining and Geology, Studentski trg 16, Belgrade, Serbia. [Crystallographic Laboratory, Faculty of Mining and Geology, Djušina 7, 11000 Belgrade, Serbia.] BCs (U. Belgrade, 1980) mineralogy. Cements minerals powder_method
Tel. 381(11)635217 Fax 381(11)635217

Radmilović, Dr Velimir (1948). Associate Professor. Faculty of Technology and Metallurgy, Karnegijeva 4, Belgrade, Serbia. [Faculty of Technology and Metallurgy, PO Box 494, 11001 Belgrade, Serbia.] PhD (U. Belgrade, 1985) metallurgy. Phase_transition electron_microscopy structure_of_ordered_phases
Tel. 381(11)3228671 Fax 381(11)3220847

Rakić, Mr Srdjan (1965). Research Assistant. Institute of Physics, Faculty of Sciences, Trg Dositeja Obradovića 4, 21000 Novi Sad, Serbia. BSc (U. Novi Sad, 1991) physics. Organic_compound biocrystallography hydrogen_bonding conformation_analysis
Tel. 381(21)55318 Fax 381(21)55662

Ribár, Dr Béla (1930). Professor. Institute of Physics, Faculty of Sciences, Trg Dositeja Obradovića 4, 21000 Novi Sad, Serbia. PhD (U. Bern, 1969) crystallography. Natural_products biomolecule coordination_compound
E-mail ribar%unsim%etfbg%fon.uucp@moumee.calstatela.edu Tel. 381(21)334024 Fax 381(21)55662

Rodić, Dr Dubravko (1955). Scientific Associate. Laboratory of Theoretical Physics and Physics of Condensed Matter, Institute of Nuclear Sciences - Vinča, Vinča, Belgrade, Serbia. [Laboratory of Theoretical Physics and Physics of Condensed Matter, Institute og Nuclear Sciences - Vinča, PO Box 522, 11001 Belgrade, Serbia.] PhD (U. Belgrade, 1992) solid state physics. Powder_method magnetism magnetic_domain magnetic_semiconductors solid_state_physics intermetallics
Tel. 381(11)8363020 Fax 381(11)8363010

Stanković, Dr Slobodanka (1941). Professor. Institute of Physics, Faculty of Sciences, Trg Dositeja Obradovića 4, 21000 Novi Sad, Serbia. PhD (U. Novi Sad, 1980) crystallography. Organic_compound biocrystallography steroids estrogens structure-activity_relationships
Tel. 381(21)350122x308 Fax 381(21)55622

Škrbić, Dr Željko (1951). Research Assistant. Institute of Physics, Faculty of Sciences, Trg Dositeja Obradovića 4, 21000 Novi Sad, Serbia. PhD (U. Novi Sad, 1993) polymer physics. Polymers polymer_physics
Tel. 381(21)55318 Fax 381(21)55662

Tančić, Mr Pavle (1965). Research Associate. Geoinstitut, Rovinjska 12, 11000 Belgrade, Serbia. BSc (U. Belgrade, 1990) silicates. *Minerals inorganic_compound*
Tel. 381(11)4880506x117

Tomić, Mr Zoran (1963). Research Associate. Laboratory of Theoretical Physics and Physics of Condensed Matter, Institute of Nuclear Sciences - Vinča, Vinča, Belgrade, Serbia. [Laboratory of Theoretical Physics and Physics of Condensed Matter, Institute of Nuclear Sciences - Vinča, PO Box 522, 11000 Belgrade, Serbia.] BSc (U. Belgrade, 1989) inorganic chemistry. *Inorganic_compound coordination_compound*
Tel. 381(11)8363020 Fax 381(11)8363010

Valčić, Dr Andreja (1933). Professor. Faculty of Technology and Metallurgy, Karnegijeva 4, Belgrade, Serbia. [Faculty of Technology and Metallurgy, PO Box 494, 11001 Belgrade, Serbia.] PhD (U. Belgrade, 1962) materials science. *Crystal_growth imperfection materials_science*
Tel. 381(11)3222281 Fax 381(11)3220847

Vučković, Dr Gordana (1953). Assistant Professor. Chemistry Department, Faculty of Sciences, Studentski Trg 16, Belgrade, Serbia. [Chemistry Department, Faculty of Sciences, PO Box 550, 11001 Belgrade, Serbia.] PhD (U. Belgrade, 1987) coordination chemistry of transition elements. *Macrocycles transition_elements_coordination_compound amino_acids_coordination_compound cyclam_complexes*
Tel. 381(11)3282111x685 Fax 381(11)638785

SINGAPORE

Sub-Editor: L. L. Koh

Ang, Dr Siau-Gek (1958). Senior Lecturer. Dept. of Chemistry, Nat. U. of Singapore, Kent Ridge, Singapore 0511, Singapore. PhD (U. Cambridge, UK, 1987) organic chemistry. *Transition_metal_clusters metalloprotein_chemistry*
E-mail chmangsg@leonis.nus.sg Tel. 65()7722840 Fax 65()7791691 Telex UNISPO RS 33943

Chowdari, Prof. B. V. R. (1943). Assoc. Prof. Dept. of Physics, Nat. U. of Singapore, Kent Ridge, Singapore 0511, Singapore. PhD (I. I. T., Kanpur, India, 1968) physics. *Glasses solid_state_battery thin_films materials_science*
E-mail phychow@leonis.nus.sg Tel. 65()7722956 Fax 65()7776126 Telex UNISPO RS 33943

Chung, Dr Mui-Fatt (1936). Sr. Lect. Dept. of Physics, Nat. U. of Singapore, Kent Ridge, Singapore 0511, Singapore. PhD (U. New South Wales, Australia, 1966) surface physics. *Surface electron_diffraction Auger_spectroscopy*
E-mail phycmf@leonis.nus.sg Tel. 65()7722618 Fax 65()7776126 Telex UNISPO RS 33943

Koh, Dr Lip Lin (1935). Assoc. Prof. Sub-Editor of World Directory of Crystallographers. Dept. of Chemistry, Nat. U. of Singapore, Kent Ridge, Singapore 0511, Singapore. PhD (Boston U. USA, 1964) physical chemistry. *X-ray_crystallography*
E-mail chmkohll@leonis.nus.sg Tel. 65()7722847 Fax 65()7791691 Telex UNISPO RS 33943

Kuok, Dr Meng Hau (1951). Sr. Lect. Dept. of Physics, Nat. U. of Singapore, Kent Ridge, Singapore 0511, Singapore. PhD (U. Canterbury, New Zealand, 1978) solid state physics. *Raman_spectroscopy infrared_spectroscopy fluorescence_spectroscopy*
E-mail phykmh@leonis.nus.sg Tel. 65()7722609 Fax 65()7776126 Telex UNISPO RS 33943

Mok, Dr Kum-fun Assoc. Prof. Dept. of Chemistry, Nat. U. of Singapore, Kent Ridge, Singapore 0511, Singapore. PhD (Victoria U. New Zealand, 1965) inorganic chemistry. *Coordination_chemistry structural_chemistry*
E-mail chmmokkf@leonis.nus.sg Tel. 65()7722669 Fax 65()7791691 Telex UNISPO RS 33943

Ng, Prof. Ser Choon (1937). Prof. Dept. of Physics, Nat. U. of Singapore, Kent Ridge, Singapore 0511, Singapore. PhD (McMaster U., Canada, 1967) solid state physics. *X-ray_diffraction neutron_diffraction light_scattering*
E-mail phyngsc@leonis.nus.sg Tel. 65()7722610 Fax 65()7776126 Telex UNISPO RS 33943

Tan, Dr Hock Siew (1950). Sr. Lect. Dept. of Physics, Nat. U. of Singapore, Kent Ridge, Singapore 0511, Singapore. PhD (U. Rochester, USA, 1980) solid state physics. *Solid_structure*
E-mail phytanhs@leonis.nus.sg Tel. 65()7722985 Fax 65()7776126 Telex UNISPO RS 33943

Teh, Dr Hung Chuan (1941). Sr. Lect. Information Systems and Computer Sci. Dept., Nat. U. of Singapore, Kent Ridge, Singapore 0511, Singapore. PhD (McMaster U., Canada, 1972) solid state physics. *Neutron_diffraction computer_graphics*
E-mail isctehhc@leonis.nus.sg Tel. 65()7722912 Fax 65()7794580 Telex UNISPO RS 33943

Xu, Dr Yan (1963). Research Fellow. Dept. of Chemistry, Nat. U. of Singapore, Kent Ridge, Singapore 0511, Singapore. PhD (Royal Inst. and Imperial C., U. of London, 1991) solid state chemistry. *Zeolites structural_chemistry*
E-mail chmxuy@leonis.nus.sg Tel. 65()7726582 Fax 65()7791691 Telex UNISPO RS 33943

SLOVAKIA

Sub-Editor: M. Čerňanský

Notes

1. The first standard degree awarded by universities and technical universities (after 5 years study) corresponds to MSc in the British system of degrees. The contraction 'Dr' is used for the title 'Doctor' which is conferred by universities. The abbreviation 'Ing.' indicates the title 'Ingenieur' of the persons graduated from technical universities. The higher degree PhD is indicated inside the country by CSc (candidatus scientiarum) - at least 3 years post-graduate study. The highest degree DSc, locally indicated by DrSc (doctor scientiarum) - for outstanding contributions to the development of the scientific field. The abbreviation 'Prof.' is indicated in addition to a title for the highest current position of a university teacher - (full) professor. The abbreviation 'Doc.' is used for the 'Docent' - associate professor.

2. Additional abbreviations used:
Acad. Sci. - Academy of Sciences
Tech. U. - Technical University
P. A. - private address

Baran, Ing. Peter (1962). Senior lecturer. Dept. of Inorganic Chemistry, Faculty of Chemical Technology, Slovak Tech. U., Radlinského 9, 812 37 Bratislava, Slovak Republic. PhD (Slovak Tech. U., Bratislava, 1992) chemistry. *Coordination preparation magnetism molecular_structure crystal_structure complexes*
E-mail baran@cvt.stuba.sk Tel. 42(7)56021x618 Fax 42(7)493198

Čaplovič, Ing. Ľubomír (1955). Senior lecturer. Dept. of Material Engineering and Thermal Treatment, Faculty of Material Technology, Slovak Tech. U., Gen. Svobodu 52, 917 24 Trnava, Slovak Republic. MSc (Slovak Tech. U., Bratislava, 1978) material sciences. *Electron_microscopy X-ray_diffraction metallic_alloys electron_diffraction*
Tel. 42(805)32919x64

Čaplovičová, Ing. Mária (1957). Senior lecturer. Dept. of Material Engineering and Thermal Treatment, Faculty of Material Technology, Slovak Tech. U., Gen. Svobodu 52, 917 24 Trnava, Slovak Republic. MSc (Slovak Tech. U., Bratislava, 1982) material sciences. *Electron_microscopy X-ray_diffraction electron_diffraction ferromagnetics*
Tel. 42(805)32919x61

Černák, Dr Juraj (1956). Senior lecturer. Dept. of Inorganic Chemistry, Faculty of Sciences, Šafárik U., Moyzesova 11, 041 54 Košice, Slovak Republic. PhD (Slovak Tech. U., Bratislava, 1986) chemistry. *Bimetallic_compounds cyanide thermal_property coordination structure clathrates education inclusion copper_compounds nickel_compounds silver_compounds zinc_compounds*
E-mail cernakju@kosice.upjs.sk Tel. 42(95)6228114x18 Fax 42(95)6222124 Telex

77562 UPJSC

Červeň, Dr Ivan (1933). Associate professor. Dept. of Physics, Faculty of Electrotechnical Eng., Slovak Tech. U., Ilkovičova 3, 812 19 Bratislava, Slovak Republic. PhD (Slovak Tech. U., Bratislava, 1980) physics. *Thin_film texture Langmuir_Blodgett_film*
E-mail icerven@elf.stuba.sk Tel. 42(7)351846 Fax 42(2)727427

Dobročka, Dr Edmund (1955). Senior lecturer. Dept. of Solid State Physics, Faculty of Mathematics and Physics, Comenius U., Mlynská dolina F2, 842 15 Bratislava, Slovak Republic. PhD (Comenius U., Bratislava, 1988) physics. *Real_structure X-ray_topography semiconductor electron_microscopy education stress strain lattice_defects powder_diffraction*
E-mail dobrocka@mff.uniba.sk Tel. 42(7)720003x252 Fax 42(7)725882

Dunaj-Jurčo, Doc. Ing. Michal (1936). Associate professor. Chairman, Regional Committee of Czech and Slovak Crystallographers, IUCr. Dept. of Inorganic Chemistry, Faculty of Chemical Technology, Slovak Tech. U., Radlinského 9, 812 37 Bratislava, Slovak Republic. PhD (Slovak Tech. U., Bratislava, 1967) chemistry. *Mixed_valence transition_metal preparation structural_chemistry education coordination_chemistry diffractometry*
E-mail dunaj@cvt.stuba.sk Tel. 42(7)326021x617 Fax 42(7)493198

Ďurovič, Ing. Slavomil (1929). Senior scientist. Dept of Theoretical Chemistry, Inst. of Inorganic Chemistry, Slovak Acad. Sci., Dúbravská cesta, 842 36 Bratislava, Slovak Republic. [Pusta 1., 841 04 Bratislava, Slovak Republic.] PhD (Comenius U., Bratislava, 1962) geology and mineralogy. *Order-disorder polytypism layered_silicates structural_disorder*
E-mail uachduro@savba.sk Tel. 42(7)3782305 Fax 42(7)373541 Telex 71734

Faktor, Ing. Dušan (1965). Senior lecturer. Dept. of Microelectronics, Faculty of Electrotechnical Eng., Slovak Tech. U., Ilkovičova 3, 812 19 Bratislava, Slovak Republic. MSc (Slovak Tech. U., Bratislava, 1988) technical physics. *Crystal_growth semiconductors surface_structure*
E-mail vesely@elf.stuba.sk Tel. 42(7)351329 Fax 42(7)723480 Telex 92521

Fejdi, Doc. Dr Pavel (1949). Associate professor. Dept. of Mineralogy and Petrology, Faculty of Sciences, Comenius U., Mlynská dolina, 842 15 Bratislava, Slovak Republic. PhD (Comenius U., Bratislava, 1980) mineralogy. *Mineralogy crystal_structure minerals silicates X-ray_diffraction optical_microscopy disordered_structures*
Tel. 42(7)720003x352 Fax 42(7)729064

Gyepesová, Dr Dalma (1936). Scientist. Department of Theoretical Chemistry, Inst. of Inorganic Chemistry, Slovak Acad. Sci, Dúbravská cesta, 842 36 Bratislava, Slovak Republic. PhD (Comenius U., Bratislava, 1972) chemistry. *Crystal_structure layered_silicates intercalates vanadium_compounds*
E-mail uachgyep@savba.sk Tel. 42(7)3782175 Fax 42(7)373541 Telex 71734

Harmatha, Ing. Ladislav. Scientist. Dept. of Microelectronics, Faculty of Electrotechnical Eng., Slovak Tech. U., Ilkovičova 3, 812 19 Bratislava, Slovak Republic. PhD (Slovak Tech. U., Bratislava, 1984) technical physics. *Microelectronics semiconductor heterostructure defect*
E-mail harmatha@elf.stuba.sk Tel. 42(7)351127 Fax 42(7)723480 Telex 92521

Havlík, Doc. Ing. Tomáš (1953). Associate professor. Dept. of Non-ferous Metallurgy, Faculty of Metallurgy, Tech. U., Letná 9/A, 043 85 Košice, Slovak Republic. PhD (Tech. U., Košice, 1983) metallurgy. *Database kinetics diffraction_technique minerals sulfides superconductors waste real_structure high-temperature_X-ray_diffraction qualitative_and_quantitative_phase_determination*
E-mail havlik@ccsun.tuke.sk Tel. 42(95)399063x499 Fax 42(95)37048

Hollý, Ing. Alois (1941). Department head. Dept. of Metallography, Research and Testing Institute, VSŽ-Steel Works, Ltd, 044 54 Košice, Slovak Republic. PhD (Tech. U., Košice, 1984) metallurgy. *Texture phase_determination metals steels texture_of_oriented_steel_sheets_for_cars*
Tel. 42(95)733348 Fax 42(95)733927

Jackuliak, Doc. Dr Quido (1937). Associate professor. Dept. of Technical Physics, Faculty of Electrotechnical Eng., University of Transport and Communication, Veľký Diel, 010 26 Žilina, Slovak Republik. PhD (Slovak Tech. U., Bratislava, 1978) technical physics. *Structural_change thin_layer_real_structure qualitative_phase_determination metals powder_diffraction*
E-mail qjackk@fpedas.utc.sk Tel. 42(89)54927 Fax 42(89)54927

Janovec, Ing. Jozef (1956). Scientist. Inst. of Material Research, Slovak Acad. Sci., Solovjevova 47, 043 53 Košice, Slovak Republic (Tech. U., Košice, 1986) metallurgy. *Carbides phase_determination phase_transition cohesion segregation grain_boundary microanalysis*
Tel. 42(95)38115

Jergel, Ing. Matej (1954). Scientist. Dept. of Metal Physics, Inst. of Physics, Slovak Acad. Sci., Dúbravská cesta 9, 842 28 Bratislava, Slovak Republic. PhD (Slovak Acad. Sci., Bratislava, 1985) physics. *GIXS multilayer synchrotron_radiation thin_film*
E-mail fyzijerm@savba.sk Tel. 42(7)3782481 Fax 42(7)376085 Telex 93373

Koman, Doc. Ing. Marian (1953). Associate professor. Secretary, Regional Committee of Czech and Slovak Crystallographers, IUCr. Dept. of Inorganic Chemistry, Faculty of Chemical Technology, Slovak Tech. U., Radlinského 9, 812 37 Bratislava, Slovak Republic. PhD (Slovak Tech. U., Bratislava, 1981) chemistry. *Structure_determination chemistry complexation coordination X-ray_diffraction crystal_structure*

E-mail koman@cvt.stuba.sk Tel. 42(7)326021x623 Fax 42(7)493198

Koreň, Dr Branislav (1954). Senior lecturer. Dept. of Chemical Physics, Faculty of Chemical Technology, Slovak Tech. U., Radlinského 9, 812 37 Bratislava, Slovak Republic. PhD (Slovak Tech. U., Bratislava, 1985) physics. *Conductivity scattering_factor barium_compounds titanium_compounds layered_compounds*
Tel. 42(7)326021x758 Fax 42(7)493198

Kožíšek, Ing. Jozef (1952). Senior lecturer. Dept. of Inorganic Chemistry, Faculty of Chemical Technology, Slovak Tech. U., Radlinského 9, 812 37 Bratislava, Slovak Republic. PhD (Slovak Tech. U., Bratislava, 1982) chemistry. *Structural_chemistry chemical_bonding coordination X-ray_diffraction molecular_structure database*
E-mail kozisek@cvt.stuba.sk Tel. 42(7)56021x612 Fax 42(7)493198

Machajdík, Ing. Daniel (1944). Scientist. Dept. of Superconductivity, Electrotechnical Institute, Slovak Acad. Sci, Dúbravská cesta, 809 32 Bratislava, Slovak Republic. PhD (Slovak Acad. Sci., Bratislava, 1979) technical physics. *Superconductors texture crystal_structure X-ray_diffraction powder_diffraction phase_determination thin_film*
E-mail machajdik@savba.sk Tel. 42(7)3782311 Fax 42(7)375816

Majková, Dr Eva (1950). Scientist. Dept. of Metal Physics, Inst. of Physics, Slovak Acad. Sci., Dúbravská cesta 9, 842 28 Bratislava, Slovak Republic. PhD (Slovak Acad. Sci., Bratislava, 1980) physics. *Multilayers metals amorphous_phase electrical_conductivity phase_determination thin_film semiconductors disordered_structure*
E-mail majkova@savba.sk Tel. 42(7)372479 Fax 42(7)376085 Telex 93373

Morháčová, Dr Eva (1956). Scientist. Inst. of Inorganic Chemistry, Slovak Acad. Sci, Dúbravská cesta 9, 842 36 Bratislava, Slovak Republic. PhD (Slovak Acad. Sci., Bratislava, 1992) physics. *Material_sciences software X-ray_diffraction crystal_structure X-ray_topography microanalysis crystal_growth symmetry inorganic_crystals*
E-mail uachmorh@savba.sk Tel. 42(7)3782175 Fax 42(7)373541 Telex 71734

Mrafko, Dr Peter (1940). Department head. Member of the Regional Committee of Czech and Slovak Crystallographers, IUCr. Dept. of Metal Physics, Inst. of Physics, Slovak Acad. Sci., Dúbravská cesta 9, 842 28 Bratislava, Slovak Republic. PhD (Slovak Acad. Sci., Bratislava, 1975) physics. *Amorphous_solids quasicrystal X-ray_diffraction simulation glassy_metals*
E-mail fyzimraf@savba.sk Tel. 42(7)3782243 Fax 42(7)376085 Telex 93373

Pavelčík, Doc. Ing. František (1945). Associate professor. Dept. of Inorganic Chemistry, Faculty of Sciences, Comenius U., Mlynská dolina, 842 15 Bratislava, Slovak Republic. PhD (Slovak Tech. U., Bratislava, 1977) chemistry. *Patterson_method computer_automation coordination_compounds conformation_analysis*
E-mail pavelcik@fns.uniba.sk Tel. 42(7)720003x745 Fax 42(7)729064

Radzo, Prof. Dr Vendelín (1929). Professor. Dept. of Geology and Mineralogy, Faculty of Mining, Tech. U., Park Komenského 15, 043 84 Košice, Slovak Republic. PhD (Tech. U., Košice, 1962) geology and mineralogy. *X-ray_diffraction silicates thermal_analysis infrared_spectroscopy phase_determination*
Tel. 42(95)32721 Telex 366 18

Sivý, Dr Július (1956). Research worker, teacher; assistant professor. Dept. of Analytical Chemistry, Faculty of Pharmacy, Comenius U., Odbojárov 10, 832 32 Bratislava, Slovak Republic. [Leskova 21, 811 04 Bratislava, Slovak Republic.] PhD (Comenius U., Bratislava, 1993) physics. *Software X-ray_diffraction small_molecules modelling_molecular_mechanics quantum_mechanics*
E-mail sivy@.magist.fpharm.uniba.sk Tel. 42(7)60451x151 Fax 42(7)60388

Ševčík, Ing. Jozef (1938). Group leader. Member of IUCr Commission on Biological Macromolecules. Inst. of Molecular Biology, Slovak Acad. Sci., Dúbravská cesta, 842 51 Bratislava, Slovak Republic. DSc (Comenius U., Bratislava, 1990) molecular biology. *Protein_crystallization protein_crystallography structure-activity_relationship_of_enzymes*
E-mail umbisevc@savba.savba.sk Tel. 42(7)3782072 Fax 42(7)372316

Šutta, Dr Pavol (1944). Scientist. Dept. of Physics, Faculty of Logistic, Military Academy, 031 19 Liptovský Mikuláš, Slovak Republic. PhD (Comenius U., Bratislava, 1987) physics. *Ceramics semiconductors thin_film X-ray_diffraction phase_determination education real_structure texture*
Tel. 42(849)25241x2475 Fax 42(849)22237

Valach, Doc. Ing. Fedor (1946). Associate professor. Dept. of Chemical Physics, Faculty of Chemical Technology, Slovak Tech. U., Radlinského 9, 812 37 Bratislava, Slovak Republic. PhD (Slovak Tech. U., Bratislava, 1972) chemistry. *Bonding superconductivity statistics X-ray_diffraction*
E-mail valach@cvt.stuba.sk Tel. 42(7)326021x756 Fax 42(7)493198

Valigura, Doc. Ing. Dušan (1948). Associate professor. Dept. of Inorganic Chemistry, Faculty of Chemical Technology, Slovak Tech. U., Radlinského 9, 812 37 Bratislava, Slovak Republic. PhD (Slovak Tech. U., Bratislava, 1977) chemistry. *Coordination_compounds chemical_bonding infrared_spectroscopy molecular_structure preparation_chemistry complexes*
E-mail valigura@cvt.stuba.sk Tel. 42(7)495257 Fax 42(7)493198

SLOVENIA

Sub-Editor: I. Leban

Arhar, Mr Andrej (1942). Researcher. DONIT-Tesnit, 61240 Medvode, C. komandanta Staneta 35, Slovenia. MSc (U. Ljubljana, 1975) inorganic chemistry. *Inorganic_chemistry structure_chemistry*
Tel. 386(61)613331 Fax 386(61)612030

Barbo, Mr Martin (1964). Researcher. Krka Pharmaceuticals, Dept. of Chemistry, 68000 Novo mesto, Cesta herojev 45, Slovenia. Dipl. Eng. Chem. (U. Ljubljana, 1990) chemistry. *Organic_synthesis organic_structure*
Tel. 386(68)22711 Fax 386(68)21537

Blinc, Prof. Dr Robert (1933). Professor of Physics; Head, Solid State Physics Dept., J. Stefan Inst. Vice-President, Slovenian Academy of Sciences and Arts. Inst. J. Stefan, 61000 Ljubljana, Jamova 39, Slovenia. PhD (U. Ljubljana, 1959) physics. *Experimental_condensed_matter_physics solid_state_spectroscopy physics_of_disordered_systems NMR_of_phase_transitions*
E-mail robert.blinc@ijs.si Tel. 386(61)1259199 Fax 386(61)219385 Telex 31296 JOSTIN SI

Bole, Mrs Meta (1964). Research Assistant. Dept. of Geology, University of Ljubljana, 61000 Ljubljana, Aškerčeva 12, Slovenia. [61000 Ljubljana, Peričeva 7, Slovenia.] Dipl. Eng. Geol. (U. Ljubljana, 1989) mineralogy, technical mineralogy. *Mineralogy applied_mineralogy powder_diffraction*
Tel. 386(61)1254121

Brenčič, Prof. Dr Jurij V. (1940). Professor of Inorganic Chemistry. Dept. of Chem. and Chem. Techn., University of Ljubljana, 61001 Ljubljana, POB 537, Slovenia. PhD (U. Ljubljana, 1969) inorganic chemistry. *Coordination_chemistry chromium_compounds molybdenum_compounds tungsten_compounds*
E-mail jurij.brencic@uni-lj.si Tel. 386(61)1264344 Fax 386(61)1258220

Bukovec, Prof. Dr Nataša (1946). Assoc. Professor. Dept. of Chem. and Chem. Techn., University of Ljubljana, 61001 Ljubljana, POB 537, Slovenia. PhD (U. Ljubljana, 1978) inorganic chemistry. *Thermal_analysis synthetic_chemistry*
Tel. 386(61)1264344 Fax 386(61)1258220

Bukovec, Prof. Dr Peter (1946). Professor of Inorganic Chemistry. Dept. of Chem. and Chem. Techn., University of Ljubljana, 61001 Ljubljana, POB 537, Slovenia. PhD (U. Ljubljana, 1972) inorganic chemistry. *Inorganic_compound superconductors synthetic_chemistry*
Tel. 386(61)1264344 Fax 386(61)1258220

Cvjetović, Mr Srdjan (1967). Assistant Researcher. Inst. J. Stefan, 61000 Ljubljana, Jamova 39, Slovenia. [Krimeja 5, 51000 Rijeka, Croatia.] MSc (U. Ljubljana, 1992) inorganic chemistry. *Synthetic_chemistry crystallography*
E-mail srdjan.cvjetovic@ijs.si

Čeh, Dr Boris (1954). Docent. Dept. of Chem. and Chem. Techn., University of Ljubljana, 61001 Ljubljana, POB 537, Slovenia. PhD (U. Ljubljana, 1985) inorganic chemistry. *Tungsten_compounds synthetic_chemistry liposomes*
Tel. 386(61)1264344 Fax 386(61)1258220

Čeh, Dr Miran (1958). Researcher. Inst. J. Stefan, 61000 Ljubljana, Jamova 39, Slovenia. PhD (U. Ljubljana, 1991) ceramics. *Inorganic_material perovskites microanalysis*
E-mail miran.ceh@ijs.si Tel. 386(61)1259199 Fax 386(61)1261029

Demšar, Dr Alojz (1953). Docent. Dept. of Chem. and Chem. Techn., University of Ljubljana, 61001 Ljubljana, POB 537, Slovenia. PhD (U. Ljubljana, 1987) inorganic chemistry. *Synthetic_chemistry inorganic_compound thermal_analysis*
E-mail alojz.demsar@uni-lj.si Tel. 386(61)1264344 Fax 386(61)1258220

Dražič, Dr Goran (1957). Research Assistant. Inst. J. Stefan, 61000 Ljubljana, Jamova 39, Slovenia. PhD (U. Ljubljana, 1990) solid state chemistry. *Electron_microscopy microanalysis ceramics*
E-mail goran.drazic@ijs.si Tel. 386(61)1259199 Fax 386(61)273677

Gabršček, Dr Sergij (1952). Director. National Examination Centre, 61000 Ljubljana, Šmartinska 134 a, Slovenia. [61000 Ljubljana, Bratovševa pl. 21, Slovenia.] PhD (U. Ljubljana, 1991) materials science. *Materials_science*
Tel. 386(61)1401112 Fax 386(61)445969

Golič, Prof. Dr Ljubo (1932). Professor of Structural Chemistry. President of the National Committee. Dept. of Chem. and Chem. Techn., University of Ljubljana, 61001 Ljubljana, POB 537, Slovenia. PhD (U. Ljubljana, 1965) chemistry. *Hydrogen_bonding inorganic_structure structure_determination* zeolites aluminophosphates heterocyclic_aminoacids
E-mail ljubo.golic@uni-lj.si Tel. 386(61)1264344 Fax 386(61)1258220

Jesih, Dr Adolf (1953). Scientist. Inst. J. Stefan, 61000 Ljubljana, Jamova 39, Slovenia. PhD (U. Ljubljana, 1989) inorganic chemistry. *Synthetic_chemistry crystallography*
E-mail adolf.jesih@ijs.si Tel. 386(61)1259199 Fax 386(61)1261029

Kaučič, Prof. Dr Venčeslav (1950). Head of Laboratory for Inorganic Chemistry. Member of National Committee. National Institute of Chemistry, Hajdrihova 19, POB 30, 61115 Ljubljana, Slovenia. DPhil (U. Ljubljana, 1977) crystallography. *Structure_chemistry zeolites aluminophosphates*
E-mail slavko@phoenix.ki.si Tel. 386(61)1232061 Fax 386(61)1259244

Kolar, Prof. Dr Drago (1932). Professor. Dept. of Chem. and Chem. Techn., University of Ljubljana, 61001 Ljubljana, POB 537, Slovenia. PhD (U. Ljubljana, 1964) chemistry. *Ceramics solid_state*
Tel. 386(61)1264344 Fax 386(61)1258220

Leban, Prof. Dr Ivan (1947). Professor of Inorganic Chemistry. Secretary of National Committee; Sub-Editor, WDC9. Dept. of Chem. and Chem. Techn., University of Ljubljana, 61001 Ljubljana, POB 537, Slovenia. DPhil (U. York, England, 1974) crystallography. *Crystal_structure_determination structural_chemistry inorganic_chemistry mineralogy*
E-mail ivan.leban@uni-lj.si Tel. +386(61)1264344 Fax +386(61)1258220

Lutar, Dr Karel (1947). Senior Scientist. Inst. J. Stefan, 61000 Ljubljana, Jamova 39, Slovenia. PhD (U. Ljubljana, 1980) inorganic chemistry. *Synthetic_chemistry crystallography*
E-mail karel.lutar@ijs.si Tel. 386(61)1259199 Fax 386(61)219385

Majcen Le Marechal, Dr Alenka (1947). Associate Professor; Head of the Laboratory for Chemistry and Organic Dyes. Technical Faculty, University of Maribor, 62000 Maribor, Smetanova 17, Slovenia. PhD (U. Ljubljana, 1988) organic chemistry. *Organic_synthesis chemistry*
Tel. 386(62)25461 Fax 386(62)225013 Telex 33334 si tfmb

Makovec, Mr Darko (1963). Assistant. Inst. J. Stefan, 61000 Ljubljana, Jamova 39, Slovenia. MSc (U. Ljubljana, 1992) ceramics. *Ceramics dielectrics transmission_electron_microscopy HREM X-ray_diffraction*
Tel. 386(61)1259199 Fax 386(61)1261029

Marinković, Prof. Dr Velibor (1929). Professor of Physical Metallurgy. Member of National Committee. Department of Geology, Mining and Metallurgy, University of Ljubljana, 61000 Ljubljana, POB 331, Slovenia. DPhil (U. Ljubljana, 1965) chemistry. *Microstructure electron_microscopy tunnelling_microscopy surfaces_and_interfaces*
E-mail velibor.marinkovic@ijs.si Tel. 386(61)1254121 Fax 386(61)224312

Meden, Dr Anton (1963). Assistant Lecturer. Dept. of Chem. and Chem. Techn., University of Ljubljana, 61001 Ljubljana, POB 537, Slovenia. Dr (U. Ljubljana, 1994) inorganic chemistry. *Structure_chemistry data_collection*
E-mail tone.meden@uni-lj.si Tel. 386(61)1264344 Fax 386(61)1258220

Milićev, Prof. Dr Svetozar (1934). Research Adviser. Inst. J. Stefan, 61000 Ljubljana, Jamova 39, Slovenia. PhD (U. Ljubljana, 1972) spectroscopy. *Fluorine_compounds spectroscopy*
Tel. 386(61)1259199 Fax 386(61)1261029

Mirtič, Prof. Dr Breda (1950). Assoc. Professor. Dept. of Geol., Miner. and Metallurgy, University of Ljubljana, 61000 Ljubljana, Aškerčeva 12, Slovenia. PhD (U. Ljubljana, 1991) mineralogy and techn. mineralogy. *Mineralogy applied_mineralogy ceramics refractories*
Tel. 386(61)1254121

Mišić, Mr Miha (1949). Researcher. IGGG-Inst. for Geology, Geophysics and Geotechnics, 61000 Ljubljana, Dimičeva 14, Slovenia. [61000 Ljubljana, Rimska 20, Slovenia.] MSc (U. Ljubljana, 1992) clay minerals. *Clay_minerals computing data_collection structures_of_phyllosilicates*
Tel. 386(61)1682461 Fax 386(61)1682557

Modec, Miss Barbara (1967). Assistant. Dept. of Chem. and Chem. Techn., University of Ljubljana, 61001 Ljubljana, POB 537, Slovenia. MSc (U. Ljubljana, 1991) chemistry. *Coordination_chemistry molybdenum_compounds spectroscopy*
Tel. 386(61)1264344 Fax 386(61)1258220

Pejovnik, Prof. Dr Stane (1946). Director. National Institute of Chemistry, 61000 Ljubljana, Hajdrihova 19, Slovenia. PhD (U. Ljubljana, 1977) chemistry. *Ceramics superconductors inorganic_chemistry*
Tel. 386(61)1232061 Fax 386(61)1259244

Petrič, Mrs Desanka (1961). Head of Quality Control. Cementarna Trbovlje, 61420 Trbovlje, Kolodvorska c. 5, Slovenia. Dipl. Eng. Chem. (U. Ljubljana, 1985) chemistry. *Mineralogy clinker analysis_silicates*
Tel. 386(601)22544 Fax 386(601)24167

Petrič, Dr Marko (1962). Assistant. Dept. of Wood Science and Technology, Biotechnical Faculty, University of Ljubljana, 61000 Ljubljana, Rožna dolina C.VIII/34, Slovenia. PhD (U. Ljubljana, 1994) chemistry. *Inorganic_compound biological_activity fungicides insecticides*
Tel. 386(61)1231161 Fax 386(61)272297

Petriček, Miss Saša (1962). Assistant. Dept. of Chem. and Chem. Techn., University of Ljubljana, 61001 Ljubljana, POB 537, Slovenia. MSc (U. Ljubljana, 1990) inorganic chemistry. *Ternary_sulfides ternary_oxides quaternary_oxides superconductors*
E-mail sasa.petricek@uni-lj.si Tel. 386(61)1264344 Fax 386(61)1258220

Pirc, Mrs Milojka (1947). Researcher. Salonit-Anhovo, 65210 Anhovo, Slovenia. MSc (U. Ljubljana, 1984) inorganic chemistry. *Cements powder_diffraction*
Tel. 386(65)51030 Fax 386(65)51226

Polanc, Prof. Dr Slovenko (1948). Professor of Organic Chemistry. Dept. of Chem. and Chem. Techn., University of Ljubljana, 61001 Ljubljana, POB 537, Slovenia. PhD (U. Ljubljana, 1975) organic chemistry. *Heterocyclic_compound organic_synthesis sonochemistry pharmaceuticals*
E-mail slovenko.polanc@uni-lj.si Tel. 386(61)1264344 Fax 386(61)1258220

Pompe, Mrs Dobruša (1962). Researcher. TDR, Institute, 62342 Ruše, Slovenia. Dipl. Eng. Chem. (U. Ljubljana, 1988) inorganic chemistry. *Inorganic_chemistry*
Tel. 386(62)661108 Fax 386(62)511504

Prelovšek, Prof. Dr Peter (1947). Professor of Physics. Inst. J. Stefan, 61000 Ljubljana, Jamova 39, Slovenia. PhD (U. Ljubljana, 1975) physics. *Solid_state physics*
E-mail peter.prelovsek@ijs.si Tel. 386(61)1259199 Fax 386(61)1261029

Prodan, Dr Albert (1944). Senior Scientific Officer. Member of National Committee. Inst. J. Stefan, Ljubljana, Slovenia. [University of Ljubljana, "Jožef Stefan" Institute, 61111 Ljubljana, PO Box 100, Jamova 39, Slovenia.] PhD (U. Zagreb, Croatia, 1974) physics. *Electron_microscopy tunnelling_microscopy modulated_structures*
E-mail albert.prodan@ijs.si Tel. 386(61)1259199 Fax 386(61)219385 or 386(61)273677 Telex 31296 JOSTIN SI

Rečnik, Mr Aleksander (1962). Research Assistant. Inst. J. Stefan, 61000 Ljubljana, Jamova 39, Slovenia. MSc (U. Ljubljana, 1990) chemical process engineering. *Ceramics transmission_electron_microscopy twinning solid_state_chemistry*
E-mail aleksander.recnik@ijs.si Tel. 386(61)1259199 Fax 386(61)1261029

Remškar, Mrs Maja (1960). Research Assistant. Inst. J. Stefan, 61000 Ljubljana, Jamova 39, Slovenia. MSc (U. Ljubljana, 1990) solid state physics. *Electron_microscopy conductor_semiconductor charge_density*
Tel. 386(61)1259199 Fax 386(61)1261029

Sinur, Miss Amalija (1969). Research Assistant. Dept. of Chem. and Chem. Techn., University of Ljubljana, 61001 Ljubljana, POB 537, Slovenia. Dipl. Eng. Chem. (U. Ljubljana, 1992) chemistry. *Structure_chemistry computing data_collection*
E-mail amalija.sinur@uni-lj.si Tel. 386(61)1264344 Fax 386(61)1258220

Šegedin, Prof. Dr Primož (1948). Assoc. Professor of Inorg. Chem. Dept. of Chem. and Chem. Techn., University of Ljubljana, 61001 Ljubljana, POB 537, Slovenia. PhD (U. Ljubljana, 1987) coordination chemistry. *Coordination_chemistry copper_compounds molybdenum_compounds chemical_education*
Tel. 386(61)1264344 Fax 386(61)1258220

Tišler, Prof. Dr Miha (1926). Professor of Organic Chemistry. Dept. of Chem. and Chem. Techn., University of Ljubljana, 61001 Ljubljana, POB 537, Slovenia. PhD (U. Ljubljana, 1954) organic chemistry. *Organic_chemistry heterocyclic_chemistry*
E-mail miha.tisler@uni-lj.si Tel. 386(61)1264344 Fax 386(61)1258220

Turel, Mr Iztok (1964). Assistant. Dept. of Chem. and Chem.Techn., University of Ljubljana, 61001 Ljubljana, POB 537, Slovenia. MSc (U. Ljubljana, 1991) bioinorganic chemistry. *Bioinorganic_chemistry*
E-mail iztok.turel@uni-lj.si Tel. 386(61)1264344 Fax 386(61)1258220

Turk, Dr Dušan (1959). Researcher. Inst. J. Stefan, 61000 Ljubljana, Jamova 39, Slovenia. PhD (Technische U. Muenchen, Germany, 1992) protein crystallography. *Proteins_crystallography*
E-mail dusan.turk@ijs.si Tel. 386(61)1259199 Fax 386(61)273594

Turk, Prof. Dr Vito (1937). Professor of Biochemistry. Inst. J. Stefan, 61000 Ljubljana, Jamova 39, Slovenia. PhD (U. Ljubljana, 1968) chemistry. *Biochemistry proteins*
E-mail vito.turk@ijs.si Tel. 386(61)1259199 Fax 386(61)1261029

Udovic, Mr Boris (1951). Postgraduate Student. Dept. of Chem. and Chem. Techn., University of Ljubljana, 61000 Ljubljana, POB 537, Slovenia. Dipl. Eng. Chem. (U. Ljubljana, 1990) inorganic chemistry. *Synthesis structure_chemistry crystallography superconductors photochemistry*
Tel. 386(61)1264344 Fax 386(61)1258220

Zabukovec, Miss Nataša (1968). Research Assistant. National Inst. of Chemistry, 61115 Ljubljana, Hajdrihova 19, Slovenia. Dipl. Eng. Chem. (U. Ljubljana, 1992) inorganic chemistry. *Synthesis_zeolites zeolites_catalysis*
E-mail natasa.zabukovec@kipc.kibk.si Tel. 386(61)1259199 Fax 386(61)1258220

Žorž, Mr Mirjan (1955). Director of Quality Control. Lek Pharmaceuticals, 61000 Ljubljana, Verovškova 57, Slovenia. [61290 Grosuplje, Prešernova 53, Slovenia.] MSc (U. Ljubljana, 1986) chemistry. *Analytical_chemistry crystallography_of_minerals*
Tel. 386(61)1682161 Fax 386(61)1684581

SOUTH AFRICA

Sub-Editor: G. J. Kruger

Alberts, Prof. Hermanus Lambertus (1941). Professor. Department of Physics, Rand Afrikaans University, PO Box 524, Johannesburg 2000, South Africa. PhD (Rand Afrikaans U., 1970) physics. *Magnetism magnetic_properties elastic_properties*
Tel. 27(11)4892330 Fax 27(11)7267723

Ball, Prof. Anthony (1939). Departmental head. Department of Materials Engineering, University of Cape Town, Private Bag, Rondebosch 7700, Cape Town, South Africa. DEng (U. Birmingham, UK, 1987) physical metallurgy. *Materials deformation tribology microstructure*
Tel. 27(21)6503173 Fax 27(21)6503726

Basson, Dr Stephen Smuts (1942). Professor. Department of Chemistry, University of the Orange Free State, PO Box 339, Bloemfontein 9300, South Africa. DSc (U. Orange Free State, 1969) chemistry. *Rhodium_compounds iridium_compounds complex_cyanides*
Tel. 27(51)4012348 Fax 27(51)8279

Beukes, Prof. Gerhardus Johannes (1943). Associate professor. Department of Geology, University of the Orange Free State, Bloemfontein, South Africa. [Department of Geology, U. Orange Free State, PO Box 339, Bloemfontein 9300, South Africa.] DSc (U. Orange Free State, 1973) geology. *X-ray_diffractometry X-ray_fluorescence_spectrometry applied_mineralogy photography natural_history*
E-mail geogb@rs.uovs.ac.za Tel. 27(51)4012393 Fax 27(51)478501 Telex 276666ZA

Billing, Mr David Gordon (1966). Lecturer. Department of Chemistry, Technicon Witwatersrand, Johannesburg, South Africa. [Department of Chemistry, Technicon Witwatersrand, PO Box 17011, Doornfontein 2028, South Africa.] MSc (U. Witwatersrand, 1991) chemistry. *Low_temperature powder_diffraction X-ray_fluorescence charge_density_studies*
E-mail 009dgbp@witsvma.wits.ac.za Tel. 27(11)4062327 Fax 27(11)4020475

Boeyens, Prof. Jan Christoffel Antonie (1934). Professor. Department of Chemistry, University of the Witwatersrand, PO Wits, Johannesburg 2050, South Africa. DSc (U. Pretoria, 1964) physical and theoretical chemistry. *Disorder structural_theory molecular_mechanics phase_transition*
E-mail jboeyens@aurum.chem.ac.za Tel. 27(11)7164097 Fax 27(11)3397967

Bourne, Dr Susan Ann (1965). Research Officer. Department of Chemistry, University of Cape Town, Rondebosch 7700, South Africa. PhD (U. Cape Town, 1991) chemistry. *Organic_clathrates*
E-mail xraysue@psipsy.uct.ac.za Tel. 27(21)6502570 Fax 27(21)6503788

Brown, Prof. Michael Ewart (1938). Professor. Department of Chemistry, Rhodes University, PO Box 94, Grahamstown 6140, South Africa. PhD (Rhodes U., 1966) physical chemistry. *Solid_phase_reaction kinetics mechanisms*
E-mail chmb@hippo.ru.ac.za Tel. 27(461)318254 Fax 27(461)25109

Bühmann, Dr Dieter (1941). Senior Specialist Scientist. Geological Survey of South Africa, 280 Pretoria Road, Pretoria-Silverton, South Africa. [Geological Survey of South Africa, Private Bag X112, Pretoria 0001, South Africa.] Dr.rer.nat (U. Göttingen, 1974) mineralogy. *Clays mineralogy clay_mineralogy*

Tel. 27(12)8411293 Fax 27(12)8411221 Telex 350286=SAGEO

Cairn, Prof. Mino Rodolfo (1949). Associate professor. Department of Chemistry, University of Cape Town, Private Bag, Rondebosch 7700, South Africa. PhD (U. Cape Town, 1975) chemistry. *Clathrates drug_polymorphism molecular_complexation drug_structure_reactivity_relationships*
E-mail xraymino@psipsy.uct.ac.za Tel. 27(21)6503071 Fax 27(21)6503788

Caveney, Dr Robert John (1941). Director of Research. De Beers Industrial Diamond Division, PO Box 916, Johannesburg 2000, South Africa. PhD (U. Witwatersrand, 1970) physics. *Crystal_growth diamond_structure high_pressure composite_materials defects*
Tel. 27(11)4906126 Fax 27(11)8352337

Coetzee, Ms Anita (1969). Teaching Assistant. Department of Chemistry, University of Cape Town, Private Bag, Rondebosch 7700, South Africa. MSc (Rhodes U., 1992) chemistry. *Inclusion_complex structure_properties thermal_properties_of_inclusion_compounds*
E-mail xrayani@psipsy.uct.ac.za Tel. 27(21)6502562 Fax 27(21)6503788

Comins, Dr Neville Raymond (1945). Programme Manager. Speciality Metals, Div. of Mat. Sci. and Tech., CSIR, PO Box 395, Pretoria 0001, South Africa. PhD (Cambridge U., UK, 1971) physics. *Electron_microscopy physical_metallurgy*
E-mail ncomins@mattek.csir.co.za Tel. 27(12)8413420 Fax 27(12)8414395 Telex 3-21312SA

Copperthwaite, Dr Richard George (1945). Group Leader, Catalysis. Research Department, AECI Ltd, Modderfontein 1645, South Africa. PhD (U. London, UK, 1971) chemistry. *Catalysts catalytic_processes*
Tel. 27(11)6052391 Fax 27(11)6083200

Coville, Prof. Neil John (1945). Professor. Chemistry Dept, University of the Witwatersrand, Johannesburg, South Africa. [Chemistry Dept, University of the Witwatersrand, PO Wits, Johannesburg 2050, South Africa.] PhD (McGill U., Canada, 1973) chemistry. *Organometallic_chemistry catalysis*
E-mail ncoville@aurum.chem.wits.ac.za Tel. 27(11)7162311 Fax 27(11)3397967 Telex 4-27155A

Crawford, Mr John Lawrence (1937). Senior lecturer. Department of Physics, University of the Witwatersrand, PO Wits, Johannesburg 2050, South Africa. BSc(Hons.) (U. Witwatersrand, 1959) physics. *Electron_microscopy defect_structure*
Tel. 27(11)7162287 Fax 27(11)4031926

Davies, Dr Geoffrey John (1948). Assistant Research Manager. De Beers Diamond Research Laboratory, PO Box 916, Johannesburg 2000, South Africa. PhD (U. Reading, UK, 1972) physics. *High_pressure diamond_physics*
Tel. 27(11)8353232 Fax 27(11)8352337

Davies, Dr Gladstone (1952). Chief Research Officer. Div. Building Tech., CSIR, PO Box 395, Pretoria 0001, South Africa. PhD (U. Witwatersrand, 1983) geology. *Powder_diffraction materials_science*
Tel. 27(12)8412507 Fax 27(12)8414680

De Villiers, Dr Johan Pietèr Roos (1942). Director. Mineralogy Division, MINTEK, Private Bag X3015, Randburg 2125, South Africa. PhD (U. Illinois, USA, 1969) mineralogy. *Minerals inorganic_phase_determination*
Tel. 27(11)7094745 Fax 27(11)7094564

Dillen, Prof. Jan L. M. (1955). Associate professor. Department of Chemistry, University of Pretoria, Pretoria 0002, South Africa. PhD (U. Antwerp, Belgium, 1981) physical chemistry. *Computing molecular_mechanics molecular_modelling*
E-mail jdillen@scinet.up.ac.za Tel. 27(12)4202527 Fax 27(12)432864

du Plessis, Prof. Paul de Villiers (1940). Professor. Department of Physics, University of the Witwatersrand, PO Wits, Johannesburg 2050, South Africa. DSc (U. Orange Free State, 1966) physics. *Neutron_scattering magnetism heavy_fermion_physics 4f_and_5f_materials phonon_properties elastic_properties*
Tel. 27(11)7164420 Fax 27(11)3398262

du Toit, Mr Johan (1966). Junior lecturer. Department of Chemistry, University of the Witwatersrand, PO Wits, Johannesburg 2050, South Africa. MSc (U. Witwatersrand, 1991) chemistry. *Oxides structure_determination microwave_synthesis*
E-mail 009jdtp@witsvma.wits.ac.za Tel. 27(11)7163826 Fax 27(11)3397967

Eales, Prof. Hugh Victor (1929). Professor. Department of Geology, Rhodes University, PO Box 94, Grahamstown 6140, South Africa. PhD (Rhodes U., 1961) geology. *Spinel_minerals petrology geochemistry*
Tel. 27(461)22023x310 Fax 27(461)25049

Engel, Prof. Dennis Walter (1939). Professor. Department of Physics, University of Durban-Westville, Private Bag X54001, Durban 4000, South Africa. Dr. rer. nat. (Tech. U. München, Germany, 1971) physics. *Anomalous_scattering powder_diffraction*
Tel. 27(31)8202226 Fax 27(31)8202383

Field, Prof. John (1946). Professor. Department of Chemistry, University of Natal, P O Box 375, Pietermaritzburg 3201, South Africa. PhD (U. Cambridge, UK, 1973) inorganic chemistry. *Coordination_chemistry*
E-mail field@chem.unp.ac.za Tel. 27(331)2605239 Fax 27(331)2605009

Fourie, Dr Jacobus Theodor (1930). Chief specialist scientist. Div. Mat. Sci. and Tech., CSIR, PO Box 395, Pretoria 0001, South Africa. DSc (U. Pretoria, 1956) physics. *Transmission_electron_microscopy plastic_deformation metals*
Tel. 27(12)8413386 Fax 27(12)8414395

Glasser, Prof. Leslie (1935). Professor. Department of Chemistry, University of the Witwatersrand, Johannesburg, South Africa. [Department of Chemistry, University of the Witwatersrand, PO Wits 2050, South Africa.] D. I. C. (Imperial Coll., U. London, 1960) chemical engineering. *Electrical_property materials hydrogen_bonding computer_modelling protein protein_folding*
E-mail glasser@aurum.chem.wits.ac.za Tel. 27(11)7162070 Fax 27(11)3397967

Griffith, Ms Vivienne Jean (1966). Teaching assistant. Department of Chemistry, University of Cape Town, Private Bag, Rondebosch 7700, South Africa. BSc(Hons) (U. Cape Town, 1988) biochemistry. *Drug_interaction cyclodextrins cyclodextrins drug_interaction*
E-mail vivienne@psipsy.uct.ac.za Tel. 27(21)6502562 Fax 27(21)6503788

Heckroodt, Prof. Renier Oelof (1935). Professor. Department of Civil Engineering, University of Cape Town, Private Bag, Rondebosch 7700, Cape Town, South Africa. DSc (U. Pretoria, 1968) geology. *Clays mineralogy clay_mineralogy*
Tel. 27(21)6503176 Fax 27(21)6503726

Heyns, Prof. Anton Michal (1939). Professor; Head of Department of Chemistry, Vice Dean of Science Faculty. Department of Chemistry, University of Hatfield, Pretoria 0002, South Africa. PhD (U. South Africa, 1968) chemistry. *Infrared_spectroscopy Raman_spectroscopy powder_diffraction ionic_solids*
E-mail AHEYNS@SCINET.UP.AC.ZA Tel. 27(12)4202512 or 27(12)4202588 Fax 27(12)432863 Telex 322723 SA

Horsfield, Mr Edgar Charles (1942). Senior lecturer. Department of Physics, University of Durban-Westville, Private Bag X54001, Durban 4000, South Africa. MSc (U. Natal, 1969) crystallography. *Organometallic_complex*
Tel. 27(31)8202662 Fax 27(31)8202780 Telex ech@pixie.udw.ac.za

Hutton, Dr Alan Thomas (1953). Senior lecturer. Editor of S. Afr. J. Chem. Department of Chemistry, University of Cape Town, Private Bag, Rondebosch 7700, Cape Town, South Africa. PhD (U. Cape Town, 1980) chemistry. *Chemical_crystallography organometallic_chemistry coordination_chemistry*
E-mail athutton@psipsy.uct.ac.za Tel. 27(21)6502550 Fax 27(21)6503788

Irving, Dr Anne (1940). Senior lecturer. Department of Chemistry, University of Cape Town, Private Bag, Rondebosch 7700, Cape Town, South Africa. PhD (U. Leeds, UK, 1969) X-ray crystallography. *Organic_structure inorganic_structure*
E-mail xrayanne@psipsy.uct.ac.za Tel. 27(21)6502564 Fax 27(21)6503788

Kessler, Dr Susanne (1962). Senior Research Officer. De Beers Industrial Diamond Division, Diamond Research Laboratory, Johannesburg, South Africa. [De Beers Diamond Research Laboratory, P O Box 916, Johannesburg 2000, South Africa.] PhD (Max-Planck-Inst. fur Metallforschung Stuttgart, 1993) materials science; crystallization in ceramics. *Diamond ceramics crystal_growth high_pressure high_temperature SEM TEM crystal_defects phase_diagrams single_crystal_techniques*
Tel. 27(11)4906447 Fax 27(11)8352337

Kruger, Prof. Gert Jacobus (1943). Professor. South African Sub-Editor of Ninth Edition of World Directory of Crystallographers; Member, IUCr Commission on Crystallographic Computing. Department of Chemistry and Biochemistry, Rand Afrikaans University, Johannesburg, South Africa. [Department of Chemistry and Biochemistry, RAU, PO Box 524, Auckland Park 2006, South Africa.] DSc (Potchefstroom U., 1970) chemistry. *Computing direct_method powder_diffraction gold_complexes*
E-mail kruger@chemie.rau.ac.za gjk@rau3.rau.ac.za Tel. 27(11)4892368 Fax 27(11)4892363

Laing, Dr Mary Elizabeth (1935). Part-time lecturer. 61 Baines Road, Durban 4001, South Africa. PhD (U. California, Los Angeles, USA, 1964) chemistry. *Organic_structure inorganic_structure*
Tel. 27(31)251951

Laing, Prof. Michael John (1937). Professor. Department of Chemistry, University of Natal, King George V Ave, Durban 4001, South Africa. PhD (U. California, Los Angeles, USA, 1965) inorganic chemistry. *Coordination_complex strained_organic_compound polymorphism aromatic_organic_compounds*
Tel. 27(31)8163103 Fax 27(31)8163091

Leipoldt, Prof. Johann Gotlieb (1940). Professor. Department of Chemistry, University of the Orange Free State, PO Box 339, Bloemfontein 9300, South Africa. DSc (U. Orange Free State, 1969) inorganic chemistry. *Transition_elements_compound transition_metal_complexes*
Tel. 27(51)4012497 Fax 27(51)8279

Le Roux, Dr Stephanus David (1947). Consulting scientist. Atomic Energy Corporation, Pelindaba, Pretoria, South Africa. [Atomic Energy Corporation, PO Box 582, Pretoria 0001, South Africa.] PhD (Purdue U., USA, 1975) solid state physics. *Powder_diffraction texture_analysis neutron_scattering materials_science amorphous_silicon_photovoltaics*
Tel. 27(12)3165575 Fax 27(12)3165135

Levendis, Dr Demetrius (1957). Lecturer. Chemistry Dept, University of the Witwatersrand, PO Wits, Johannesburg 2050, South Africa. PhD (U. Witwatersrand, 1984) crystallography. *Disorder phase_transition electron_density conglomerate_molecular_crystals*
E-mail demi@aurum.chem.wits.ac.za Tel. 27(11)7162348 Fax 27(11)3397967

Liles, Mr David Charles (1950). Chief research officer. Division of Materials Science Technology, CSIR, Pretoria, South Africa. [MATTEK, CSIR, PO Box 395, Pretoria 0001, South Africa.] BSc(Hons) (Loughborough U., 1973) chemistry. *Inorganic_materials transition_elements powder_diffraction transition_metal_complexes*
E-mail dliles@mattek.csir.co.za Tel. 27(12)8413571 Fax 27(12)8414395 Telex 3-21312SA

Maske, Prof. Siegfried (1928). Professor. Department of Geology, University of the Witwatersrand, PO Wits, Johannesburg 2050, South Africa. DSc (Stellenbosch U., 1964) geology. *Mineralogy geochemistry ore_genesis sulfide_minerals*
Tel. 27(11)7162799 Fax 27(11)4031926

Nabarro, Prof. Frank Reginald Nunes (1916). Professor emeritus. University of the Witwatersrand, PO Wits, Johannesburg 2050, South Africa. DSc FRS (U. Birmingham, UK, 1953) metallurgy (dislocation theory). *Crystal_defect*
Tel. 27(11)7162175 Fax 27(11)3398262

Nassimbeni, Prof. Luigi Renzo (1939). Professor. Department of Chemistry, University of Cape Town, Private Bag, Rondebosch 7700, Cape Town, South Africa. PhD (U. Cape Town, 1969) physical chemistry. *Inclusion_complex structure_and_thermal_properties_of_inclusion_compounds*
E-mail xrayluig@psipsy.uct.ac.za Tel. 27(21)6502569 Fax 27(21)6503788

O'Neill, Dr Françoise Marcelle (1965). Junior lecturer. Department of Chemistry, University of the Witwatersrand, PO Wits, Johannesburg 2050, South Africa. PhD (U. Witwatersrand, 1993) chemistry. *Bimetallic_compounds bond_order_of_dimetal_centres*
Tel. 27(11)7163826 Fax 27(11)4031926

Paige-Green, Mr Philip (1952). Engineering geologist. Div. Road Transport Tech., CSIR, PO Box 395, Pretoria 0001, South Africa. MSc (U. Natal, 1975) geology. *Minerals clays road_materials*
Tel. 27(12)8412924 Fax 27(12)8413232

Pipkin, Dr Noel John (1942). Assistant Research Manager. Department of Physics, De Beers Diamond Research Laboratory, PO Box 916, Johannesburg 2000, South Africa. PhD (U. Newcastle-upon-Tyne, UK, 1967) metallurgy. *Diamond characterization graphite cubic_boron_nitride*
Tel. 27(11)8353232x240 Fax 27(11)8352337

Pretorius, Dr Jan Andries (1949). Research fellow. Chemical Computation Group, Research & Development Department, AECI Ltd, PO Modderfontein, 1645, South Africa. PhD (U. South Africa, 1978) physical chemistry. *Molecular_modelling X-ray_powder_diffraction*
Tel. 27(11)6052763 Fax 27(11)6083200 27(11)6052540

Retief, Dr Johannes Jacobus (1941). Principal scientist. R&D Division, SASTECH, PO Box 1, Sasolburg 9570, South Africa. PhD (U. Orange Free State, 1978) physics. *Powder_diffraction catalysts coal_minerals*
Tel. 27(16)7082940 Fax 27(16)7082826

Reynhardt, Prof. Eduard Christiaan (1944). Professor; Head of Department. Department of Physics, University of South Africa, PO Box 392, Pretoria 0001, South Africa. PhD (U. South Africa, 1971) physics. *Phase_transition phase_separation molecular_reorientation_in_solids phase_transitions_in_solids*
E-mail reynhec@alpha.unisa.ac.za Tel. 27(12)4298062 Fax 27(12)4293434

Richter, Dr Paul Wilhelm (1946). Chief Research Officer. Div. Mat. Sci. Tech., CSIR, PO Box 395, Pretoria 0001, South Africa. PhD (U. South Africa, 1971) physical chemistry. *High_pressure thermal_analysis ceramics crystal_growth materials research*
Tel. 27(12)8412434 Fax 27(12)8414395

Roodt, Prof. Andreas (1956). Associate professor. Department of Chemistry, University of the Orange Free State, Bloemfontein 9300, South Africa. PhD (U. Orange Free State, 1987) inorganic chemistry. *Kinetics coordination_chemistry*
E-mail char@uovsvm1.uovs.ac.za Tel. 27(51)4012923 Fax 27(51)474152

Roos, Dr Hester Marita (1963). Senior Lecturer. Department of Chemistry, University of Pretoria, Hatfield, Pretoria, South Africa. [Department of Chemistry, University of Pretoria, Pretoria 0002, South Africa.] PhD (U. Pretoria, 1989) chemistry. *Organic_structure organometallic_structure structure-activity_relationship molecular_modelling*
E-mail hroos@scinet.up.ac.za Tel. 27(12)4203089 Fax 27(12)432863

Rutherford, Prof. John Stewart (1938). Professor. Department of Chemistry, University of Transkei, Private Bag X1, UNITRA, Umtata, Transkei, South Africa. PhD (McMaster U., Canada, 1967) physical chemistry. *Small_molecules bonding discrete_mathematics_applications bonding_in_solids*
E-mail chjr@unitrix.utr.ac.za Tel. 27(471)26811 Telex 734TT

Schoch, Dr Aylva Ernest (1933). Research professor. Vice-President, Geological Society of South Africa. Department of Geology, University of the Orange Free State, Bloemfontein 9301, South Africa. [Kapt. Goodman str. 9, Bloemfontein 9301, South Africa] DSc (Stellenbosch U., 1972) igneous petrology. *Order-disorder non-linear_variations_in_rock-forming_minerals*
E-mail geoaes@rs.uovs.ac.za Tel. 27(51)4012593 Fax 27(51)478501

Schöning, Prof. Friedrich Richard Ludwig (1923). Professor emeritus. Department of Chemistry, University of the Witwatersrand, PO Wits, Johannesburg 2050, South Africa. PhD (U. Witwatersrand, 1959) physics. *Diffraction_physics crystal_defect non-crystalline_materials*
Tel. 27(11)7163086 Fax 27(11)3397967

Schutte, Prof. Casper Jan Hendrik (1934). Professor; responsible for educational technology. Chief Executive Director, Science, Technology and Informatics, University of South Africa, PO Box 392, Pretoria 0001, South Africa. Dr (U. Amsterdam, Netherlands, 1960) physical chemistry. *Infrared_spectroscopy Raman_spectroscopy molecular_vibrations_of_solids*
E-mail schutcjh@alpha.unisa.ac.za Tel. 27(12)4296022 Fax 27(12)4293405

Scott, Ms Janet Lesley (1964). Lecturer. Department of Chemistry, University of Cape Town, Private Bag, Rondebosch 7700, South Africa. BSc(Hons) (U. Cape Town, 1989) chemistry. *Organic_clathrates*
E-mail xrayjan@psipsy.uct.ac.za Tel. 27(21)6502563 Fax 27(21)6503788

Sewell, Dr Bryan Trevor (1953). Associate professor. Electron Microscope Unit, University of Cape Town, Rondebosch 7700, Cape Town, South Africa. PhD (London, UK, 1981) protein crystallography. *Computer_graphics macromolecular_structure chromatin electron_microscope_tomography*
E-mail sewell@uctvax.uct.ac.za Tel. 27(21)6502817 Fax 27(21)6891528

Shabalala, Mr Innocent Philani (1966). Teaching assistant. Department of Chemistry, University of Cape Town, Private Bag, Rondebosch 7700, Cape Town, South Africa. BSc(Hons) (U. Cape Town, 1992) chemistry. *Drug_polymorphism drug_solvation*
E-mail xrayinno@psipsy.uct.ac.za Tel. 27(21)6502562 Fax 27(21)6503788

Sommerville, Mrs Polly Baker Melville (1924). Research assistant. Department of Chemistry, University of Natal, King George V Ave., Durban 4001, South Africa. MSc (U. Natal, 1970) chemistry. *Organic_structure*
Tel. 27(31)8163090

Spalding, Dr Dennis Raymond (1942). Senior lecturer. Department of Physics, University of Natal, King George V Ave., Durban 4001, South Africa. PhD (Cambridge U., UK, 1969) physics. *High_temperature_superconductors*
Tel. 27(31)8162775 Fax 27(31)8162214

Subramony, Mr Loganathan (1946). Lecturer. School of Health Sciences, M L Sultan Technikon, PO Box 1334, Durban 4000, South Africa. MSc (U. Durban-Westville, 1984) physics. *Organometallic_complex*
Tel. 27(31)316681x2203

Van Rooyen, Prof. Petrus Hendrik (1949). Professor. Department of Chemistry, University of Pretoria, Hatfield, Pretoria, South Africa. [Department of Chemistry, University of Pretoria, Pretoria 0002, South Africa.] PhD (Rand Afrikaans U., 1979) chemistry. *Molecular_design organic_structure organometallic_structure molecular_modelling*
E-mail pvrooyen@scinet.up.ac.za Tel. 27(12)4202519 Fax 27(12)432863

Van Schalkwyk, Prof. Theunis Gabriel Dirkse (1920). Professor. Department of Chemistry, University of Cape Town, Private Bag, Rondebosch 7700, South Africa. MSc (Stellenbosch U., 1943) physics. *Organic_structure inorganic_structure*
E-mail xrayvan@psipsy.uct.ac.za Tel. 27(21)6502568 Fax 27(21)6503788

Winder, Ms Nicola (1962). Senior Scientific Officer. Department of Chemistry, University of Cape Town, Private Bag, Rondebosch 7700, Cape Town, South Africa. BSc(Hons) (U. Cape Town, 1984) chemistry. *Inclusion_complex thermal_property structure_and_thermal_properties_of_inclusion_compounds*
E-mail nicola@psipsy.uct.ac.za Tel. 27(21)6502538 Fax 27(21)6503788

SPAIN

Sub-Editor: **X. Solans**

Notes

1. Degrees conferred by Spanish Universities are *Doctor* (DSc), *Graduado* (Grad) and *Licenciado* (MSc).

2. The occupational titles in Universities are in decreasing order: *Catedratico, Prof. Titular, Prof. Asociado, Prof. Ayudante, Laboral,* and *Becario.*

3. The occupational titles in CSIC are in decreasing order: *Prof. Investigacion, Investigador, Colaborador, Tecnico Superior,* and *Becario.*

Abad-Ortega, Mr Maria del Mar (1964). Becario. Dep. Mineralogía y Petrologia, Universidad de Granada, Fuentenueva s/n, 18002-Granada, Spain. Grad (U. Granada, 1988) geology. *X-ray_diffraction minerals*
Tel. 34(58)243368 Fax 34(58)243368

Aguiló, Dr Magdalena (1953). Prof. Titular. Dept. de Química, Universidad Rovira i Virgili, Pza. Imperial Tarraco 1, 43005-Tarragona, Spain. DSc (U. Barcelona, 1983) physics. *Crystal_growth morphology_structure_relationship*
Tel. 34(77)225254x2205 Fax 34(77)243319

Alamo-Serrano, Dr Jaime (1946). Prof. Titular. Dept. de Química Inorgánica, Universidad de Valencia, Avda. Dr Moliner 50, 46100-Burjassot, Spain. DSc (U. Valencia, 1971) chemistry. *Powder_diffraction properties-structure_relationships thermal_expansion ion_transport*
Tel. 34(6)3864856 Fax 34(6)3864322

Albert de la Cruz, Mr Armando (1967). Becario. Inst. Rocasolano, CSIC, Serrano 119, 28006-Madrid, Spain. DSc (U. Autónoma de Madrid, 1990) chemistry. *Crystal_packing structure_determination fractal geometry proteins_crystallography*
E-mail xalbert@roca.csic.es Tel. 34(1)5619400 Fax 34(1)5642431 Telex 42182

Alcobé, Mr Xavier (1962). Laboral 1. Servicio Científico-Técnico, Universidad de Barcelona, Lluís Solé i Sabaris s/n, 08028-Barcelona, Spain. Grad (U. Autónoma de Barcelona, 1985) geology. *Powder_diffraction phase_transitions diffraction_techniques molecular_alloys*
Tel. 34(3)3398949 Fax 34(3)3393798

Alonso, Dr José Antonio (1958). Colaborador Cientifico, CSIC. Instituto de Ciencia de Materiales de Madrid, CSIC, Serrano 113, 28006-Madrid, Spain. DSc (U. Complutense Madrid, 1984) chemistry. *Structural_inorganic_chemistry solid_state_chemistry high_pressure superconducting_oxides transition_metal_oxides perovskite_oxides neutron_diffraction powder_diffraction magnetic_structures*
E-mail immja25@cc.csic.es Tel. 34(1)5854907 Fax 34(1)5618471

Alvarez, Dr Aurelio (1935). Prof. Titular. Dept. Geología, Univ. Autónoma de Barcelona, 08193-Bellaterra, Spain. DSc (U. Barcelona, 1974) geology. *Minerals*
E-mail igmn0@ebccuab1.bitnet.es Tel. 34(3)5811611 Fax 34(3)5811263 Telex 52040 EDUCIE

Alvarez, Dr M. Angeles (1951). Prof. Titular. Dept. Geología, Univ. Sevilla, Apdo. 553, 41071-Sevilla, Spain. [Dept. Crystallography - F. Chemistry, c/. Prof. García González s/n, Sevilla, Spain.] DSc (U. Autónoma de Madrid, 1976) chemistry. *Minerals X-ray_diffraction TEM ATEM HRTEM clay_minerals crystal_growth_mechanism*
Tel. 34(5)4625060 Fax 34(5)4557141

Alvarez-Larena, Dr Angel (1959). Investigador. Dept. Geología, Univ. Autónoma de Barcelona, 08193-Bellaterra, Spain. DSc (Inst. Químico de Sarriá, 1989) chemistry. *Structural_disorder charge_density crystal_structure_determination*
E-mail igcr0@cc.uab.es Tel. 34(3)5812621 Fax 34(3)5811263 Telex 52040 EDUCIE

Amigó, Prof. José-Maria (1940). Catedrático. Dept. Geología, Univ. Valencia, Avda. Dr Moliner 50, 46100-Burjassot, Spain. DSc (U. Barcelona, 1966) geology. *Minerals crystal_structure_determination powder_diffraction*
E-mail Jose.M.Amigo@uv.es Tel. 34(6)3864603 Fax 34(6)3864372

Apreda, Dr M. Carmen (1943). Comentarado. Inst. Rocasolano, CSIC, Serrano 119, 28006-Madrid, Spain. DSc (U. Nacional de la Plata, Argentina, 1976) physics. *Crystal_structure_determination*
Tel. 34(1)2619400

Aragón de la Cruz, Prof. Francisco (1933). Prof. Investigación. Inst. Química-Inorgánica, CSIC, Serrano 113, 28006-Madrid, Spain. DSc (U. Complutense Madrid, 1960) chemistry. *Inorganic_crystals*
Tel. 34(1)4111772

Aramburu, Mr Ibon (1966). Becario. Dept. Física, Univ. del País Vasco, Appdo. 644, 48080-Bilbao, Spain. Grad (U. País Vasco, 1989) physics. *X-ray_diffraction*
E-mail wmbarlei@lg.ehu.es Tel. 34(4)4647700x2488 Fax 34(4)4648500 Telex 33259 ehupv e

Arana, Prof. Rafael (1942). Catedrático. Dept. Química Agrícola, Geología y Edafología, Universidad de Murcia, Campus Univ. de Espinardo, 30100-Murcia, Spain. DSc (U. Granada, 1972) geology. *Optical_crystallography mineralogy X-ray_diffraction computing mathematical_crystallography*
Tel. 34(68)833000x2208 Fax 34(68)833902

Aranzabe, Mrs Ana (1965). Prof. Asociado. Dept. Química Inorgánica, Univ. del País Vasco, Appdo. 644, 48080-Bilbao, Spain. PhD (U. País Vasco, 1993) chemistry. *Structural_chemistry crystal_structure_determination coordination_chemistry bioinorganic_chemistry*
E-mail qibargaa@lg.ehu.es Tel. 34(4)4647700x2450 Fax 34(4)4648500 Telex 33259 ehupv

Arrieta, Prof. Juan Manuel (1952). Catedrático. Dept. Química Inorganica, Univ. del País Vasco-E.H.U., Appdo. 644, 48080-Bilbao, Spain. DSc (U. País Vasco, 1980) chemistry. *Crystal_structure_determination polyoxometalate_chemistry*
E-mail qiparugj@lg.ehu.es Tel. 34(4)4647700x2981 Fax 34(4)4648500

Arriortua, Prof. Maribel (1950). Catedrático. Member of Spanish National Committee. Dept. Mineralogía y Petrología, Univ. del País Vasco, Appdo. 644, 48080-Bilbao, Spain. DSc (U. País Vasco, 1981) chemistry. *Crystal_structure_determination structural_chemistry*
E-mail npparmai@lg.ehu.es Tel. 34(4)4647700x2555 Fax 34(4)4648500

Balcázar, Dr José Luis (1929). Prof. Titular. Univ. Nacional de Educación a Distancia, Senda del Rey s/n, 28040-Madrid, Spain. DSc (U. Valladolid, 1970) chemistry. *Crystal_structure_determination minerals powder_diffraction*
Tel. 34(1)3987361 Fax 34(1)3986697

Bassas, Mr Josep (1963). Laboral 2. Servicio Científico-Técnico, Universidad de Barcelona, Lluís Solé i Sabarís s/n, 08028-Barcelona, Spain. Grad (U. Barcelona, 1987) physics. *Powder_diffraction phase_transitions diffraction_techniques crystal_structure_determination*
Tel. 34(3)3398949 Fax 34(3)3393798

Bastida, Dr Joaquín (1955). Prof. Titular. Dept. Geología, Univ. Valencia, Avda. Dr Moliner 50, 46100-Burjassot, Spain. DSc (U. Autónoma de Barcelona, 1980) geology. *Teaching X-ray_diffraction minerals*
Tel. 34(6)3864393 Fax 34(6)3864372

Bayón, Dr J. Carlos (1954). Professor. Dept. Química, Univ. Autónoma de Barcelona, 08193-Bellaterra, Spain. DSc (U. Autónoma de Barcelona, 1981) chemistry. *Structural_chemistry organometallic_chemistry homogenous_catalysis asymmetric_catalysis*
E-mail iqin3@cc.uab.es Tel. 34(3)5812889 Fax 34(3)5812477 Telex 52040 EDUCI E

Bermúdez-Polonio, Dr Joaquín (1930). Investigador Científico. Inst. Química Inorgánica Elhuyar, CSIC, Serrano 113, 28006-Madrid, Spain. DSc (U. Complutense de Madrid, 1964) chemistry. *X-ray_spectroscopy properties_and_structure_relationships X-ray_diffraction inorganic_crystals*
Tel. 34(1)4111772

Bernalte, Prof. Antoni (1927). Professor of Materials Science; Head, Dept. Physics of Materials, UNED. Univ. Nac. de Educación a Distancia, Apdo. 60141, E-28080 Madrid, Spain. [UNED-Facultad de Ciencias, Apdo. 60141, E-28080 Madrid, Spain.] PhD (U. California, Berkeley, USA, 1968) physics. *Defect_structures mathematical_crystallography dislocations mechanical_thermal_and_electrical_properties_ (metals_and_ceramics)*
E-mail Antonio.Bernalte@uned.es Tel. 34(1)3987172 34(1)3987185 Fax 34(1)3986697 Telex 4526 - 47844

Briansó, Prof. José Luis (1944). Catedrático. Dept. Geología, Univ. Autónoma de Barcelona, 08193-Bellaterra, Spain. DSc (U. Barcelona, 1972) geology. *Crystal_structure*
E-mail igcr0@cc.uab.es Tel. 34(3)5812034 Fax 34(3)5811263 Telex 52040 EDUCIE

Bruque, Dr Sebastián (1947). Prof. Titular. Fac. Ciencias, Univ. de Málaga, Apdo. 59, 29071-Málaga, Spain. DSc (U. Málaga, 1979) chemistry. *Powder_diffraction inorganic_crystal_structure_determination*
E-mail bruque@ccuma.uma.es Tel. 34(5)2131878 Fax 34(5)2132000

Caballero, Prof. M. Antonio (1942). Catedrático. Dep. Estructura y Propiedades de los Materiales, Univ. de Cádiz, Appdo. 40, 11510-Pto. Real, Spain. DSc (U. Complutense de Madrid, 1972) geology. *Crystallography defects diffraction topography*
E-mail nono@c2v1.uca.es Tel. 34(56)830210 Fax 34(56)834924

Cabeza, Dr Javier A. (1958). Prof. Titular. Dept. Química Org. e Inorg., Univ. Oviedo, 33071-Oviedo, Spain. PhD (U. Zaragoza, 1983) chemistry. *Inorganic_chemistry organometallic_chemistry structural_chemistry*
E-mail jac@dwarf1.quimica.uniovi.es Tel. 34(8)5103501 Fax 34(8)5103446 Telex UNIOV 84322

Cabré, Mr Roger (1966). Prof. Titular. Dept. de Química, Univ. Rovira i Virgili, Pza. Imperial Tarraco 1, 43005-Tarragona, Spain. Grad (U. Barcelona, 1986) physics. *Crystal_growth computing Superconductors*
Tel. 34(77)225254x2205 Fax 34(77)243319

Calvet, Dr Teresa (1960). Prof. Titular. Dep. Cristalografía, Mineralogía y Depósitos Minerales, Univ. Barcelona, Martí y Franquès s/n, 08028-Barcelona, Spain. DSc (U. Barcelona, 1990) geology. *X-ray_diffraction syncrystallization*
Tel. 34(3)4021350 Fax 34(3)4021340

Cardellach, Dr Esteve (1949). Prof. Titular. Dept. Geología, Univ. Autónoma de Barcelona, 08193-Bellaterra, Spain. DSc geology. *Minerals*
Tel. 34(3)5811611

Carriedo, Dr Gabino-Alejandro (1952). Catedrático. Dept. Química Organometálica, Univ. Oviedo, Julián Clavería s/n, 33071-Oviedo, Spain. DSc (U. Valladolid, 1981) chemistry. *Organometallic_structural_chemistry*
Tel. 34(85)103449 Fax 34(85)103446

Casabo, Prof. Jaume (1941). Catedrático. Dept. Química, Univ. Autónoma de Barcelona, 08193-Bellaterra, Spain. DSc (U. Barcelona, 1972) chemistry. *Structural_chemistry*
Tel. 34(3)5811369

Cascales, Dr Concepcion (1959). Colaborador científico CSIC. Inst. de Ciencia de Materiales de Madrid, CSIC, Serrano 113, E-28006 Madrid, Spain. DSc (U. Complutense de Madrid, 1986) chemistry. *Structural_inorganic_chemistry solid_state_chemistry high_Tc_superconducting_oxides rare_earth_luminescent_materials neutron_diffraction X-ray_powder_diffraction X-ray_single_crystal_diffraction*
E-mail immcc53@cc.csic.es Tel. 34(1)5854907 Fax 34(1)5618471

Castiñeiras, Prof. Alfonso (1942). Catedrático. Dept. Química Inorgánica, Univ. Santiago de Compostela, Campus Universitario, 15706-Santiago, Spain. DSc (U. Santiago, 1974) chemistry. *Structural_chemistry crystal_structure_determination*
E-mail sdzzs001@seins.usc.es Tel. 34(81)594636

Castro, Dr Alicia (1958). Research staff; Manager of research projects. Instituto de Ciencia de Materiales, CSIC, Serrano 113, 28006-Madrid, Spain. DSc (U. Complutense de Madrid, 1984) chemistry. *Structural_chemistry solid_state_chemistry*
E-mail imma26@cc.csic.es Tel. 34(1)5854902 34(1)5854913 Fax 34(1)5618471

Cendon, Mr Dionisio (1968). Prof. Asociado. Dept. Geología, Univ. Autónoma de Barcelona, 08193-Bellaterra, Spain. Grad (U. Autónoma de Barcelona, 1992) geology. *Structural_disorder solid_solution diffuse_scattering EXAFS*
E-mail igcr0@cc.uab.es Tel. 34(3)5812034 Fax 34(3)5811263 Telex 52040 EDUCIE

Claramunt, Prof. Rosa María (1948). Catedrático. Univ. Nacional de Educación a Distancia, Ciudad Universitaria, 28040-Madrid, Spain. DSc (U. Barcelona, 1973) chemistry. *Structural_chemistry*
E-mail rosa.claramunt@human.uned.es Tel. 34(1)3987322 Fax 34(1)3986697

Colacio, Dr Enrique (1957). Prof. Titular. Dep. Química Inorgánica, Univ. Granada, Fuentenueva s/n, 18002-Granada, Spain. DSc (U. Granada, 1989) chemistry. *Structural_chemistry*
E-mail ecolacio@ugr.es Tel. 34(58)243236 Fax 34(58)243322

Coll, Prof. Miquel (1955). Investigador. Centro de Investigación y Desarrollo, CSIC, Jordi Girona 18, 08034-Barcelona, Spain. DSc (U. Barcelona, 1986) biology. *Protein_crystallography*
E-mail mcoll@eq.upc.es Tel. 34(3)4016687 Fax 34(3)4016600

Conde, Prof. Alejandro (1947). Catedrático. Dep. Física de la Materia Condensada, Univ. Sevilla, Appdo. 1065, 41080-Sevilla, Spain. DSc (U. Sevilla, 1972) physics. *Non-crystalline magnetism devitrification microstructure diffraction electron_microscopy DSC TGA electrical_resistivity crystal_structure lattice_energy X-ray_diffraction*
E-mail conde@Cica.es Tel. 34(5)4616615 Fax 34(5)4612097

Conde, Dr Clara Francisca (1952). Prof. Titular. Dep. Física de la Materia Condensada, Univ. Sevilla, Appdo. 1065, 41080-Sevilla, Spain. DSc (U. Sevilla, 1981) physics. *Crystal_structure_determination amorphous_materials*
Tel. 34(5)4616615

Cortés, Dr Roberto (1959). Prof. Titular. Dept. Química Inorgánica, Univ. del País Vasco, Appdo. 644, 48080-Bilbao, Spain. DSc (U. País Vasco, 1990) chemistry. *Structural_chemistry*
Tel. 34(4)5131666

Coy-Yll, Prof. Ramon (1940). Catedrático; Research adviser. Institut d'Estudis Catalans, Carmen 47, 08001-Barcelona, Spain. DSc (U. Barcelona, 1964) geology. *Minerals dynamical_properties optical_spectroscopy*
E-mail rcy@as400.iec.es Tel. 34(3)3185516 Fax 34(3)4122994

Cuevas-Diarte, Prof. Miquel Angel (1948). Catedratico. Dep. Cristalografía, Mineralogía y Depósitos Minerales, Univ. Barcelona, Martí y Franquès s/n, 08028-Barcelona, Spain. DSc (U. Barcelona, 1979) geology. *Isomorphism syncrystallization phase_diagram molecular_binary_alloys thermoanalysis diffraction_data*
E-mail mangel@natura.geo.ub.es Tel. 34(3)4021350 Fax 34(3)4021340

Cumbrera, Dr Francisco Luis (1954). Catedrático. Dep. Física de la Materia Condensada, Univ. Extremadura, Carretera de Portugal, 06071-Badajoz, Spain. DSc (U. Sevilla, 1982) physics. *X-ray_diffraction amorphous_materials*
Tel. 34(24)274800 Fax 34(24)271304 Telex 28638

Diánez-Millán, Dr M. Jesus (1950). Colaborador. Dep. Física de la Materia Condensada, Univ. Sevilla, Appdo. 1065, 41080-Sevilla, Spain. DSc (U. Sevilla, 1985) physics. *X-ray_diffraction organic_structure lattice_energy*
E-mail dianez@cica.es Tel. 34(5)4616615 Fax 34(5)4612097

Diaz, Prof. Francesc (1953). Catedrático. Dept. de Química, Univ. Rovira i Virgili, Pza. Imperial Tarraco 1, 43005-Tarragona, Spain. DSc (U. Barcelona, 1982) physics. *Crystal_growth*
Tel. 34(77)225254x2220 Fax 34(77)243319

Diéguez, Dr Ernesto (1948). Prof. Titular. Dep. Física Aplicada, Univ. Autónoma de Madrid, 28049-Madrid, Spain. DSc (U. Autónoma de Madrid, 1983) physics. *Crystal_growth*
Tel. 34(1)3978579

Domenech, Dr M. Victoria (1940). Colaborador. Dep. Cristalografía, Mineralogía y Depósitos Minerales, Univ. Barcelona, Martí y Franquès s/n, 08028-Barcelona, Spain. DSc (U. Oviedo, 1981) chemistry. *Crystal_structure_determination*
Tel. 34(3)4021352 Fax 34(3)4021340

Domínguez, Prof. Esther (1947). Catedrático. Dept. Química Orgánica, Univ. del País Vasco, Appdo. 644, 48080-Bilbao, Spain. DSc (U. País Vasco, 1975) chemistry. *Crystal_structure_determination*
Tel. 34(4)4647700x2622 Fax 34(4)4648500

Espinet, Prof. Pablo (1949). Catedrático. Dept. Química Inorgánica, Univ. Valladolid, Prado de la Magdalena s/n, 47005-Valladolid, Spain. DSc (U. Zaragoza, 1975) chemistry. *Structural_chemistry*
Tel. 34(83)423231 Fax 34(83)423013

Esteve-Cano, Mr Vicente J. (1960). Becario. Dept. C. Experimentales, Univ. Jaume I, Appdo. 224, 12080-Castellón, Spain. MSc (U. Valencia, 1984) chemistry. *Computing Rietveld_method condensed_matter synchrotron_radiation*
E-mail estevev@exp.uji.es Tel. 34(64)345700 Fax 34(64)254585

Estop, Dr Eugenia (1950). Prof. Titular. Dept. Geología, Univ. Autónoma de Barcelona, 08193-Bellaterra, Spain. DSc (U. Barcelona, 1980) geology. *Molecular_crystals structural_disorder mixed_crystals X-ray_diffraction neutron_diffuse_scattering EXAFS solid-state_NMR*
E-mail igcr0@ccuab1.uab.es Tel. 34(3)5811611 Fax 34(3)5811263

Estrada, Dr M. Dolores (1953). Prof. Titular. Dep. Física de la Materia Condensada, Univ. Sevilla, Appdo. 1065, 41080-Sevilla, Spain. DSc (U. Sevilla, 1984) physics. *Crystal_structure_determination lattice_energy X-ray_diffractometry*
Tel. 34(5)4616615 Fax 34(5)4612097

Ezpeleta, Mr José María (1965). Prof. Asociado. Dept. Física, Univ. del País Vasco, Appdo. 644, 48080-Bilbao, Spain. Grad (U. País Vasco, 1988) physics. *X-ray_diffraction*
E-mail wmpezzarj@lg.ehu.es Tel. 34(4)4647700x2488 Fax 34(4)4648500 Telex 33259 ehupv e

Farran, Mr Joan (1967). Becario. Dept. Geología, Univ. Autónoma de Barcelona, 08193-Bellaterra, Spain. Grad (U. Autónoma de Barcelona, 1990) geology. *Structure_determination tellurium_compounds selenium_compounds*
E-mail igcr0@cc.uab.es Tel. 34(3)5812034 Fax 34(3)5811263 Telex 52040 EDUCIE

Faus-Paya, Prof. Juan (1946). Catedrático. Dept. Química Inorgánica, Univ. Valencia, Avda. Dr Moliner 50, 46100-Burjassot, Spain. DSc (U. Valencia, 1971) chemistry. *Structural_chemistry*
Tel. 34(6)3864856 Fax 34(6)3864322

Fayos, Prof. José (1940). Prof. Investigación. Inst. Rocasolano, CSIC, Serrano 119, 28006-Madrid, Spain. DSc (U. Complutense de Madrid, 1967) physics. *Close_packing clustering cohesive_energy crystal_field design dynamics inelastic_scattering packing phase_transition*
Tel. 34(1)5619400 Fax 34(1)5642431

Fenoll Hach-Alí, Prof. Purificacion (1935). Catedrática; Head, Research Group of Mineralogy, Petrology and Ore Deposits. Dep. Mineralogía y Petrología, Univ. Granada, Fuentenueva s/n, 18002-Granada, Spain. DSc (U. Granada, 1966) chemistry–geology. *X-ray_diffraction minerals ore_microscopy*
E-mail pfenoll@ugr.es Tel. 34(58)243338 Fax 34(58)243368 Telex Univ. GR78435 EDUCIE

Fernández, Dr Carlos José (1951). Prof. Titular. Dep. Geología, Univ. Oviedo, Arias de Velasco s/n, 33005-Oviedo, Spain. DSc (U. Oviedo, 1982) geology. *Minerals*
Tel. 34(85)103096 Fax 34(85)233911

Fita, Dr Ignacio (1953). Prof. Titular. Dep. Ingeniería Química, Univ. Politécnica de Catalunya, Diagonal 647, 08028-Barcelona, Spain. DSc (U. Autónoma de Barcelona, 1981) biology. *Structural_biology*
E-mail fita@eq.upc.es Tel. 34(3)4016685 Fax 34(3)4016600

Florencio, Dr Feliciana (1924). Investigador. Inst. Rocasolano, CSIC, Serrano 119, 28006-Madrid, Spain. DSc (U. Complutense de Madrid, 1954) chemistry. *Crystal_structure_determination*
Tel. 34(1)2619400

Foces-Foces, Prof. Concepcion (1946). Prof. Investigación. Inst. Rocasolano, CSIC, Serrano 119, 28006-Madrid, Spain. DSc (U. Complutense de Madrid, 1974) physics. *Crystal_structure_determination computing hydrogen_bonding inclusion*
E-mail xconcha@roca.csic.es Tel. 34(1)5619400 Fax 34(1)5642431

Fonseca, Dr Isabel (1933). Titulado Superior. Inst. Rocasolano, CSIC, Serrano 119, 28006-Madrid, Spain. DSc (U. Complutense de Madrid, 1980) chemistry. *Crystal_structure_determination drugs_packing*
Tel. 34(1)5619400 Fax 34(1)5642431

Font-Altaba, Prof. Manuel (1923). Prof. Emeritus. Dep. Cristalografía, Mineralogía y Depósitos Minerales, Univ. Barcelona, Martí y Franqués s/n, 08028-Barcelona, Spain. DSc (U. Barcelona, 1954) chemistry. *Crystal_structure_determination crystal_growth minerals*
E-mail ubacmd10@ebcesca1.bitnet.es Tel. 34(3)4021343 Fax 34(3)4021340

Font-Bardia, Dr Merce (1956). Laboral 1. Dep. Cristalografía, Mineralogía y Depósitos Minerales, Univ. Barcelona, Martí y Franqués s/n, 08028-Barcelona, Spain. DSc (U. Barcelona, 1990) geology. *Crystal_structure_determination*
E-mail ubacmd10@ebcesca1.bitnet.es Tel. 34(3)4021343 Fax 34(3)4021340

Forteza, Mr Matilde (1954). Prof. Ayudante. Dep. Geología, Univ. Sevilla, Appdo. 553, 41071-Sevilla, Spain, 41071-Sevilla, Spain. Grad (U. Sevilla, 1978) pharmacy. *Minerals*
Tel. 34(5)4625060

Fuente-Cullell, Dr Carlos de la (1941). Prof. Titular. Dep. Cristalografía, Mineralogía y Depósitos Minerales, Univ. Barcelona, Martí y Franqués s/n, 08028-Barcelona, Spain. DSc (U. Barcelona, 1972) geology. *Minerals clay_minerals ceramic_chemistry ceramic_physics_properties*
Tel. 34(3)4021351 Fax 34(3)4021340

Fuertes, Dr Amparo (1959). Colaborador. Inst. de Ciencia de Materiales de Barcelona, CSIC, Campus de la UAB, 08193-Bellaterra, Spain. DSc (U. Valencia, 1986) chemistry. *Structural_chemistry crystal_structure_determination*
Tel. 34(3)3302716

Fuertes-Martínez, Dr José Félix (1955). Prof. Titular. Dep. Física, Univ. Oviedo, Calvo Sotelo s/n, 33007-Oviedo, Spain. DSc (U. Oviedo, 1988) chemistry. *Crystal_growth*
Tel. 34(85)103319 Fax 34(85)103324

Gaete, Dr Walter (1934). Prof. Titular. Dept. Química, Univ. Autónoma de Barcelona, 08193-Bellaterra, Spain. DSc (U. Sta. Maria, Chile, 1960) chemistry. *Structural_chemistry*
Tel. 34(3)5811010

Galan, Prof. Emilio (1942). Catedrático; Head of research group on Applied Mineralogy. Dept. Cristalografía y Mineralogía, Univ. Sevilla, Apdo. 553, 41071-Sevilla, Spain. DSc (U. Madrid, 1972) geology. *Clay_minerals non-metallic_minerals*
Tel. 34(5)4625060 Fax 34(5)4557141

Gali, Prof. Salvador (1949). Catedrático. Dep. Cristalografía, Mineralogía y Depósitos Minerales, Univ. Barcelona, Martí y Franqués s/n, 08028-Barcelona, Spain. DSc (U. Barcelona, 1976) geology. *Crystal_structure_determination*
E-mail ubacmd11@ebcesca1.bitnet.es Tel. 34(3)4021347 Fax 34(3)4021340

Galindo, Dr Agustin (1960). Prof. Titular. Dept. Química Inorgánica, Univ. Sevilla, Appdo. 553, 41071-Sevilla, Spain. DSc (U. Sevilla, 1986) chemistry. *Structural_chemistry*
E-mail galindo@cica.es Tel. 34(5)4557157 Fax 34(5)4557134

Gamasa, Dr Pilar (1951). Prof. Titular. Dept. Química Organometálica, Univ. Oviedo, Julián Clavería s/n, 33071-Oviedo, Spain. DSc (U. Zaragoza, 1977) chemistry. *Organometallic_chemistry*
Tel. 34(85)103460 Fax 34(85)103446 Telex UNIOV 84322

García-Aranda, Dr Miguel Angel (1966). Prof. Asociado. Dept. Química Inorgánica, Univ. Málaga, Appdo. 59, 29071-Málaga, Spain. DSc (U. Málaga, 1992) chemistry. *Crystallography solid_state_chemistry*
E-mail g_aranda@ccuma.uma.es Tel. 34(52)132022 Fax 34(52)132000

García-Blanco, Prof. Severino (1922). Prof. Emeritus. Inst. Rocasolano, CSIC, Serrano 119, 28006-Madrid, Spain. DSc (U. Complutense de Madrid, 1946) chemistry. *Crystal_structure_determination*
Tel. 34(1)5619400

García-Granda, Dr Santiago (1955). Prof. Titular. Dept. Química-Física y Analítica, Univ. Oviedo, Julián Clavería s/n, 33006-Oviedo, Spain. DSc (U. Oviedo, 1984) chemistry. *Chemistry computer_automation QSAR*
E-mail sgg@dwarf1.quimica.uniovi.es Tel. 34(85)103477 Fax 34(85)103480 Telex 84322 EDUCI-E

García-Rodriguez, Dr Antonio (1945). Catedrático. Dep. Química Inorgánica, Univ. Granada, Fuentenueva s/n, 18002-Granada, Spain. DSc (U. Granada, 1972) chemistry. *Structural_chemistry*
E-mail frey@ugr.es Tel. 34(58)243324 Fax 34(58)243322

García-Ruiz, Dr Joaquín (1951). Prof. Titular. Dep. Física de la Materia Condensada, Univer. Zaragoza, Pza. S. Francisco, 50009-Zaragoza, Spain. DSc (U. Zaragoza, 1981) physics. *X-ray_absorption_spectroscopy*
Tel. 34(76)353557

García-Ruiz, Dr Juan Manuel (1953). Prof. Investigación. Inst. Andaluz de Geol. Mediterranea, Fuentenueva s/n, 18002-Granada, Spain. DSc (U. Madrid, 1980) geology. *Crystal_growth growth_textures biological_macromolecules biomineralization*
E-mail jmgarcia@ugr.es Tel. 34(58)243360 Fax 34(58)243384

Garrido, Mr Carlos (1968). Becario. Inst. Andaluz de Geol. Mediterranea, Fuentenueva s/n, 18002-Granada, Spain. Grad (U. Granada, 1990) geology. *Minerals*
E-mail jdmartin@ugr.es Tel. 34(58)243368 Fax 34(58)271873

Gervilla, Dr Fernando (1961). Prof. Asociado. Dep. Mineralogía y Petrología, Univ. Granada, Fuentenueva s/n, 18002-Granada, Spain. DSc (U. Granada, 1985) geology. *X-ray_diffraction minerals*
Tel. 34(58)243368 Fax 34(58)243368

Gimeno, Prof. José (1947). Catedrático. Dept. Química Organometálica, Univ. Oviedo, Julián Clavería s/n, 33071-Oviedo, Spain. DSc (U. Zaragoza, 1972) chemistry. *Organometallic_structural_chemistry*
Tel. 34(85)103461 Fax 34(85)103446

Gómez-Sal, Prof. José Carlos (1948). Catedrático. Facultad de Ciencias, Univ. Cantabria, Avda. de los Castros s/n, 39005-Santander, Spain. DSc (U. Complutense de Madrid, 1976) physics. *Magnetic_structures intermetallic_compounds*
Tel. 34(42)201506 Fax 34(42)201402 Telex 35861 EDUCIE

Gómez-Sal, Dr M. Pilar (1951). Prof. Titular. Dept. Química Inorgánica, Univ. Alcalá de Henares, Campus Universitario, 28871-Alcalá de Henares, Spain. DSc (U. Complutense de Madrid, 1978) chemistry. *Crystal_structure_determination molecular_complexes organometallic*
E-mail qinorgomez@alcala.es Tel. 34(1)8854656 Fax 34(1)8854660

Gomis-Rüth, Dr Franz Xavier (1964). Research scientist. Institut de Biología Fonamental, Univ. Autónoma de Barcelona, 08193-Bellaterra, Spain. PhD (Ludwig-Maximilian U. (München), 1992) chemistry. *Protein_crystallography*
E-mail gomis@luz.uab.es Tel. 34(3)5812807 Fax 34(3)5812011

González-Calbet, Dr José M. (1952). Full Professor; Head of Department. Dept. Química Inorgánica, Univ. Complutense de Madrid, Campus Universitario, 28040-Madrid, Spain. DSc (U. Complutense de Madrid, 1979) chemistry. *Solid_state_chemistry inorganic_chemistry*
Tel. 34(1)3944342 Fax 34(1)3944352

González-Duarte, Prof. Pilar (1945). Catedrático. Dept. Química, Univ. Autónoma de Barcelona, 08193-Bellaterra, Spain. DSc (U. Barcelona, 1975) chemistry. *Structural_chemistry*
E-mail iqin1@ebccuab1.bitnet.es Tel. 34(3)5811363 Fax 34(3)5812477 Telex 52040 EDUCI E

González-Mañas, Dr Marina (1955). Prof. Titular. Dep. Estructura y Propiedades de los Materiales, Univ. de Cádiz, Appdo. 40, 11510-Pto. Real, Spain. DSc (U. Cádiz, 1990) chemistry. *Crystallography defects diffraction topography domain_structure twinning crystal_growth*
E-mail nono@c2v1.uca.es Tel. 34(56)830210 Fax 34(56)834924

González-Platas, Mr Javier (1966). Prof. Asociado. Dep. Física Fundamental y Experimental, Univ. La Laguna, 38204 La Laguna, Spain. Grad (U. La Laguna, 1990) physics. *Structural_chemistry crystal_structure_determination*
E-mail amg@iac.es Tel. 34(22)603274 Fax 34(22)603684

González-Silgo, Cristina (1966). Prof. Asociado. Dep. Física Fundamental y Experimental, Univ. La Laguna, 38204 La Laguna, Spain. Grad (U. La Laguna, 1990) physics. *Structural_chemistry crystal_structure_determination*
E-mail amg@iac.es Tel. 34(22)603274 Fax 34(22)603684

Gregorkiewitz, Prof. Miguel (1946). Investigador. Instituto de Ciencia de Materiales, CSIC, Serrano 115 bis, 28006-Madrid, Spain. DSc (Tech. Hochschule, Darmstadt, Germany, 1980) natural sciences. *Crystal_structure_determination X-ray_diffraction synchrotron pseudo-symmetry ionic_conductivity frame-work_silicates micas*
E-mail immmg18@cc.csic.es Tel. 34(1)5854883 Fax 34(1)5624526

Gutierrez-Puebla, Dr Enrique (1952). Investigador. Inst. de Ciencia de Materiales, CSIC, Serrano 113, 28006-Madrid, Spain. DSc (U. Complutense de Madrid, 1978) chemistry. *Crystal_structure_determination*
E-mail gutierrez@quim.ucm.es Tel. 34(1)3944285 Fax 34(1)3944284

Gutierrez-Zorrilla, Dr Juan Manuel (1957). Prof. Titular. Dept. Química Inorgánica, Univ. del País Vasco, Appdo. 644, 48080-Bilbao, Spain. DSc (U. País Vasco, 1984) chemistry. *Crystal_structure_determination structural_chemistry bioinorganic_chemistry coordination_chemistry*
E-mail qipguloj@lg.ehu.es Tel. 34(4)4647700x2450 Fax 34(4)4648500 Telex 32098 educi. e

Guzman, Carmen (1962). Becario. Univ. del País Vasco, Appdo. 644, 48080-Bilbao, Spain. Grad (U. País Vasco, 1986) chemistry. *Crystal_structure_determination structural_chemistry*
E-mail qibgumic@lg.ehu.es Tel. 34(4)4647700x2450 Fax 34(4)4648500

Hermoso-Domínguez, Dr Juan Antonio (1964). Becario. Inst. Rocasolano, CSIC, Serrano 119, 28006-Madrid, Spain. DSc (U. Complutense de Madrid, 1992) physics. *Crystal_structure_determination proteins_crystallography*
E-mail xjuan@roca.csic.es Tel. 34(1)5619400 Fax 34(1)5642431 Telex 42182

Hernández-Cano, Prof. Félix (1941). Prof. Investigación. Inst. Rocasolano, CSIC, Serrano 119, 28006-Madrid, Spain. DSc (U. Complutense de Madrid, 1969) physics. *Crystal_structure_determination computing teaching molecular_crystal_channel intermolecular_packing*
E-mail xcano@roca.csic.es Tel. 34(1)5619400 Fax 34(1)5642431

Hidalgo-Laguna, Mr Miguel Angel (1964). Laboral. Dep. Mineralogía y Petrología, Univ. Granada, Fuentenueva s/n, 18002-Granada, Spain. Grad (U. Granada, 1986) geology. *Crystal_structure_determination electron_microprobe*
E-mail jmartin@ugr.es Tel. 34(58)243339 Fax 34(58)243368

Iglesias, Prof. Juan Eugenio (1942). Profesor de Investigacion; Director de la Sede C del ICMM (CSIC). Instituto de Ciencia de Materiales de Madrid, CSIC, Serrano 115 bis, 28006-Madrid, Spain. PhD (U. Texas, USA, 1971) chemical eng.. *Structural_chemistry X-ray_diffraction infrared_spectroscopy ionic_conductors*
E-mail jeiglesias@cc.csic.es Tel. 34(1)5622838 Fax 34(1)5624526

Insausti, Mr María Teresa (1965). Prof. Asociado. Dept. Química Inorgánica, Univ. del País Vasco, Appdo. 644, 48080-Bilbao, Spain. Grad (U. País Vasco, 1988) chemistry. *Structural_chemistry*
E-mail npbinpem@lg.ehu.es Tel. 34(4)4647700x2450 Fax 34(4)4648500

Jiménez-Garay, Prof. Rafael (1946). Catedrático; Head of Department. Dep. Estructura y Propiedades de los Materiales, Univ. de Cádiz, Appdo. 40, 11510-Pto. Real, Spain. DSc (U. Sevilla, 1973) physics. *Amorphous_phase diffraction SAXS glasses chalcogenides metallic_glasses*
E-mail jimenez@czv1.uca.es Tel. 34(56)830210 Fax 34(56)834924

Jiménez-Morales, Dr Francisco (1959). Prof. Titular. Dept. Física de la Materia condensada, Univ. Sevilla, Appdo. 1065, 41080-Sevilla, Spain. DSc (U. Sevilla, 1987) physics. *Crystal_growth ferroelectricity calorimetry*
E-mail jimenez@Cica.es Tel. 34(5)4616615 Fax 34(5)4612097

Julve-Olcina, Prof. Miguel (1953). Catedrático. Dept. Química Inorgánica, Univ. Valencia, Avda. Dr Moliner 50, 46100-Burjassot, Spain. DSc (U. Valencia, 1981) chemistry. *Structural_chemistry magnetic_properties coordination_compounds coordination_chemistry*
E-mail julve@mac.uv.es Tel. 34(6)3864300x4856 Fax 34(6)3864322 Telex 64298 EDUCI E

Labrador, Dr Manuel (1952). Prof. Titular. Dpto. Cristal.lografia, Mineralogia i Dipósits minerals, Univ. Barcelona, Martí i Franqués s/n, 08028-Barcelona, Spain. DSc (U. Barcelona, 1990) geology. *Crystallography X-ray_and_neutron_diffraction DSC molecular_crystals molecular_alloys structure_refinement Rietveld_method energy rigid_body_analysis energetic_compounds*
E-mail manuell@natura.geo.ub.es Tel. 34(3)4021354 Fax 34(3)4021340

Lahoz, Dr Fernando J. (1958). Investigador. Member of the Executive Committee of the Spanish Crystallographic Association (GEC). Inst. Ciencia de Materiales de Aragón, CSIC, 50009-Zaragoza, Spain. [ICMA, Facultad de Ciencias, Univ. de Zaragoza, 50009-Zaragoza, Spain.] PhD (Zaragoza U., 1984) chemistry. *Crystal_structure_determination structural_chemistry organometallic_complexes metallic_hydrides transition_elements*
E-mail lahoz@cc.unizar.es or lahoz@icma0.unizar.es Tel. 34(76)554559 Fax 34(76)567920 Telex 58198 EDUCI E

Lezama, Dr Luis (1964). Prof. Titular. Dept. Química Inorgánica, Univ. del País Vasco, Appdo. 644, 48080-Bilbao, Spain. DSc (U. País Vasco, 1991) chemistry. *Structural_chemistry*
Tel. 34(4)4647700x2449 Fax 34(4)4648500

Liso-Rubio, Prof. M. Jesus (1944). Catedrático. Universidad de Extremadura, Carretera de Portugal, 06071-Badajoz, Spain. DSc (U. Complutense de Madrid, 1969) pharmacy. *Minerals*

Llu-González, Mr Malva (1959). Doctorando. Dept. Geología, Univ. Valencia, Avda. Dr Moliner 50, 46100-Burjassot, Spain. Grad (U. Valencia, 1990) physics. *X-ray_diffraction*
E-mail evalrx@espec.uv.es Tel. 34(6)3864576 Fax 34(6)3864322

Llamas-Saiz, Dr Antonio Luis (1966). Becario. Inst. Rocasolano, CSIC, Serrano 119, 28006-Madrid, Spain. DSc (U. Complutense de Madrid, 1993) physics. *Crystal_structure_determination computing hydrogen_bonding inclusion*
E-mail xantonio@roca.csic.es Tel. 34(1)5619400 Fax 34(1)5642431 Telex 42182

Lloveras, Dr Joaquim (1946). Prof. Titular. Dep. Chemical Engineering, Univ. Politécnica de Catalunya, Diagonal 647, 08028-Barcelona, Spain. DSc (U. Politecnica de Catalunya, 1974) engineering. *Fiber_diffraction small-angle_scattering structural_biology*
Tel. 34(3)2495800x236

López-Acevedo, Dr M. Victoria (1953). Prof. Titular. Dep. Cristalografía y Mineralogía, Univ. Complutense de Madrid, 28040-Madrid, Spain. DSc (U. Complutense de Madrid, 1983) geology. *Crystallography crystal_growth*
Tel. 34(1)3944880 Fax 34(1)3944872

López-Aguayo, Prof. Francisco (1945). Catedrático. Dep. Estructura y Propiedades de los Materiales, Univ. de Cádiz, Appdo. 40, 11510-Pto. Real, Spain. DSc (U. Complutense de Madrid, 1972) geology. *Geology minerals*
E-mail lopez-aguayo@czv1.uca.es Tel. 34(56)836712 Fax 34(56)834924

López-Castro, Prof. Amparo (1928). Prof. Investigación. Instituto de Ciencia de Materiales de Sevilla, CSIC, Reina Mercedes s/n, 41080-Sevilla, Spain. DSc (U. Complutense de Madrid, 1954) physics. *Crystal_structure_determination X-ray_diffraction*
Tel. 34(5)4616615 Fax 34(5)4612097

Lopez-Galindo, Dr Alberto (1960). Investigador. Dep. Mineralogía y Petrología, Univ. Granada, Fuentenueva s/n, 18002-Granada, Spain. DSc (U. Granada, 1986) geology. *X-ray_diffraction minerals*
E-mail alberto@ugr.es Tel. 34(58)243342 Fax 34(58)243368

Lopez-Gonzalez, Prof. Juan de D. (1924). Catedrático. Dept. Química Inorgánica, Univ. Nacional Educaión a Distancia, Campus Universitario, 28040-Madrid, Spain. DSc (U. Complutense de Madrid, 1949) chemistry. *Structural_chemistry*
Tel. 34(1)2439431

Lopez-Soler, Dr Angel (1940). Profesor de Investigación. Instituto Jaime Almera, CSIC, Martí y Franqués s/n, 08028-Barcelona, Spain. DSc (U. Barcelona, 1968) geology. *Microscopy electron_microscopy gemology*
E-mail alopez@u.ija.csic.es Tel. 34(3)3302716 Fax 34(3)4110012

Luque, Dr Antonio (1962). Prof. Titular. Dept. Química Inorgánica, Univ. del País Vasco, Appdo. 644, 48080-Bilbao, Spain. DSc (U. País Vasco, 1990) chemistry. *Structural_chemistry crystal_structure_determination*
E-mail qipluam@lg.ehu.es Tel. 34(4)4647700x2450 Fax 34(4)4648500

Madariaga, Dr Gotzon (1959). Prof. Titular. Dept. Física Materia Condensada, Univ. del País Vasco, Appdo. 644, 48080-Bilbao, Spain. DSc (U. País Vasco, 1985) physics. *Physics condensed_matter phase_transition incommensurate_phase modulated_structures*
E-mail wmpmameg@lg.ehu.es Tel. 34(4)4647700 Fax 34(4)4648500 Telex 33259 ehupv-e

Marcos, Dr Celia (1954). Prof. Titular. Dep. Geología, Univ. Oviedo, Arias de Velasco s/n, 33005-Oviedo, Spain. DSc (U. Oviedo, 1985) geology. *Crystal_growth diffraction_technique electron_microprobe-analysis gemology reflectance structure_determination*
Tel. 34(85)103100 Fax 34(85)103103

Marin-Elena, Dr José Manuel (1955). Prof. Titular. Dept. Química Inorgánica, Univ. Sevilla, 41012-Sevilla, Spain. DSc (U. Sevilla, 1981) chemistry. *Structural_chemistry*
Tel. 34(5)4629061

Márquez-Delgado, Prof. Rafael (1929). Catedrático. Dep. Física de la Materia Condensada, Univ. Sevilla, Appdo. 1065, 41080-Sevilla, Spain. DSc (U. Complutense de Madrid, 1957) physics. *Ceramics_compounds mechanics electron_microscopy DSC organic_compounds non-crystalline_compounds diffractometry structural_analysis*
Tel. 34(5)4616615 Fax 34(5)4612097

Martí-Artoy, Mr Xavier (1963). Becario. Dept. Geología, Univ. Autónoma de Barcelona, 08193-Bellaterra, Spain. Grad (U. Autónoma de Barcelona, 1986) geology. *Structural_chemistry*
Tel. 34(3)5811611

Martín-Izard, Dr Agustín (1955). Prof. Titular. Dep. Geología, Univ. Oviedo, Arias de Velasco s/n, 33005-Oviedo, Spain. DSc (U. Salamanca, 1985) geology. *Minerals*
Tel. 34(85)103095 Fax 34(85)233911

Martín-Ramos, Dr José Daniel (1949). Prof. Titular. Dep. Mineralogía y Petrología, Univ. Granada, Fuentenueva s/n, 18002-Granada, Spain. DSc (U. Granada, 1977) geology. *Crystal_structure_determination powder_methods minerals*
E-mail jdmartin@ugr.es Tel. 34(58)243339 Fax 34(58)243368

Martín-Vivaldi, Dr Juan Luis (1950). Prof. Titular. Dep. Cristalografía y Mineralogía, Univ. Complutense de Madrid, 28040-Madrid, Spain. DSc (U. Complutense de Madrid, 1983) geology. *Crystallography crystal_growth*
Tel. 34(1)3944876 Fax 34(1)3944872

Martínez, Mr Benjamin (1960). Colaborador. Inst. de Ciencia de Materiales de Barcelona, CSIC, Campus UAB, 08193-Bellaterra, Spain. Grad (U. Barcelona, 1984) physics. *Crystal_structure_determination*
Tel. 34(3)5801853

Martínez, Dr Francisco (1945). Prof. Titular. Dpto. Química Inorgánica, Universidad Zaragoza, 50009-Zaragoza, Spain. DSc (U. Zaragoza, 1975) chemistry. *Organometal-lic_chemistry mixed-valence_structural_chemistry clusters*
Tel. 34(76)452347

Martínez-Carrera, Prof. Sagrario (1925). Prof. Investigación. Inst. Rocasolano, CSIC, Serrano 119, 28006-Madrid, Spain. DSc (U. Complutense de Madrid, 1949) chemistry. *Crystal_structure_determination*
Tel. 34(1)5619400

Martínez-Ripoll, Prof. Martín (1946). Prof. Investigación. Secretary of Spanish Crystallographic Committee. Inst. Rocasolano, CSIC, Serrano 119, 28006-Madrid, Spain. DSc (U. Complutense de Madrid, 1970) chemistry. *Crystallography diffractometry computing proteins crystal_structure_determination*
E-mail xmartin@roca.csic.es Tel. 34(1)5619400 Fax 34(1)5642431 Telex 42182

Martínez-Sarrión, Dr María Luisa (1940). Prof. Titular. Dep. Química Inorgánica, Univ. Barcelona, Diagonal 647, 08028-Barcelona, Spain. DSc (U. Murcia, 1968) chemistry. *Phase_transitions ferroelectricity solid_chemistry*
E-mail est.solid@ub.es Tel. 34(3)4021225 Fax 34(3)4111492

Mesa, Dr José Luis (1960). Prof. Titular. Dept. Química Inorgánica, Univ. del País Vasco, Appdo. 644, 48080-Bilbao, Spain. DSc (U. País Vasco, 1987) chemistry. *Structural_chemistry*
Tel. 34(4)4647700x2449 Fax 34(4)4648500

Millán-Muñoz, Dr Maria (1950). Prof. Titular. Dep. Física de la Materia Condensada, Univ. Sevilla, Appdo. 1065, 41080-Sevilla, Spain. DSc (U. Sevilla, 1980) physics. *Non-crystalline magnetism microstructure electron_microscopy DSC TGA electrical_resistivity X-ray_diffractometry*
Tel. 34(5)4616615 Fax 34(5)4612097

Miravitlles, Prof. Carlos (1942). Prof. Investigación. Chairman of Spanish Crystallographic Committee. Inst. de Ciencia de Materiales de Barcelona, CSIC, Campus UAB, 08193-Bellaterra, Spain. DSc (U. Barcelona, 1971) pharmacy. *Crystal_structure_determination materials_science*
Tel. 34(3)5801853 Fax 34(3)5805729

Molina, Mr José (1966). Becario. Dep. Mineralogia y Petrología, Univ. Granada, Fuentenueva s/n, 18002-Granada, Spain. Grad (U. Granada, 1990) geology. *Minerals*
E-mail jdmartin@ugr.es Tel. 34(58)243348 Fax 34(58)243368

Molina-Molina, Dr José (1951). Prof. Titular. Dep. Química Orgánica, Univ. Granada, Fuentenueva s/n, 18002-Granada, Spain. DSc (U. Granada, 1980) chemistry. *Organic_computational_chemistry*
E-mail jmolina@ugr.es Tel. 34(58)243186 Fax 34(58)243322

Molins, Dr Elies (1957). Investigador. Inst. de Ciencia de Materiales de Barcelona, CSIC, Campus UAB, 08193-Bellaterra, Spain. DSc (U. Barcelona, 1985) physics. *Electron_density_distribution materials_science*
E-mail elies.molins@icmab.es Tel. 34(3)5801853 Fax 34(3)5805729

Monfort, Dr Montse (1949). Prof. Titular. Dep. Química Inorgánica, Univ. Barcelona, Diagonal 647, 08028-Barcelona, Spain. DSc (U. Barcelona, 1983) chemistry. *Structural_chemistry magnetism*
Tel. 34(3)4021264 Fax 34(3)4111492

Monge, Dr M. Angeles (1951). Investigador. Inst. de Ciencia de Materiales, CSIC, Serrano 113, 28006-Madrid, Spain. DSc (U. Complutense de Madrid, 1978) chemistry. *Crystal_structure_determination*
E-mail monge@quim.ucm.es Tel. 34(1)3944285 Fax 34(1)3944284

Moreiras, Prof. Damaso (1947). Catedrático. Dep. Geología, Univ. Oviedo, Arias de Velasco s/n, 33005-Oviedo, Spain. DSc (U. Oviedo, 1980) geology. *Minerals structure_determination diffraction gemology crystallography*
Tel. 34(85)103094 Fax 34(85)233911

Moreno-Carcamo, Mr Abel (1966). Candidate for the degree of doctor. Instituto Andaluz de Geologia Mediterranea, CSIC, University of Granada, Fuentenueva s/n, 18002 Granada, Spain. Graduate (U. Autónoma de Puebla (México), 1989) chemistry. *Crystal_growth protein_crystallization protein_crystallography*
E-mail amoreno@ugr.es Tel. 34(58)243360 Fax 34(58)243384

Moreno-Carretero, Dr Miguel (1958). Prof. Titular. Dep. Química Inorgánica, Col. Univ. Santo Reino, 23071-Jaen, Spain. DSc (U. Granada, 1983) chemistry. *Structural_chemistry*
Tel. 34(53)212150

Moreno-Sánchez, Dr José María (1962). Prof. Asociado. Dep. Química Inorgánica, Univ. Granada, Fuentenueva s/n, 18002-Granada, Spain. DSc (UNED, 1992) chemistry. *Structural_chemistry*
E-mail jmoreno@ugr.es Tel. 34(58)243236 Fax 34(58)243322

Morón, Dr M. Carmen (1960). Colaborador. Instituto Ciencia de Materiales de Aragón, CSIC, Ciudad Universitaria, 50009-Zaragoza, Spain. DSc (U. Zaragoza, 1988) physics. *X-ray_diffraction neutron_diffraction structural_correlations solid_state_chemistry magnetism*
E-mail nina@cc.unizar.es Tel. 34(76)563296 Fax 34(76)567920

Muñoz-Roca, Dr M. Carmen (1958). Prof. Titular. Dept. Física Aplicada (EUITI), Univ. Politécnica de Valencia, Camino de Vera s/n, 46071-Valencia, Spain. DSc (U. Valencia, 1988) physics. *Structure_determination magnetism conductivity*
E-mail munoz@evalun11.bitnet.es Tel. 34(6)3877525 Fax 34(6)3877189

Mzayek, Mr Elias (1957). Becario. Inst. de Química Inorgánica, CSIC, Serrano 113, 28006-Madid, Spain. Grad (U. Alepo, Syria, 1981) chemistry. *Structural_chemistry*
Tel. 34(1)4111772

Navarro, Dr Carmen (1947). Prof. Titular. Dept. Química Inorgánica, Univ. Autónoma de Madrid, 28049-Madrid, Spain. DSc (U. Autónoma de Madrid, 1976) chemistry. *Structural_chemistry*
Tel. 34(1)3974356

Nieto, Mr José (1968). Doctorando. Dep. Mineralogía y Petrología, Univ. Granada, Fuentenueva s/n, 18002-Granada, Spain. Grad (U. Granada, 1991) geology. *Minerals X-ray_diffraction*
E-mail jdmartin@ugr.es Tel. 34(58)243348 Fax 34(58)243368

Nieto-García, Dr Fernando (1955). Prof. Titular. Dep. Mineralogía y Petrología, Univ. Granada, Fuentenueva s/n, 18002-Granada, Spain. DSc (U. Granada, 1982) geology. *Minerals*
E-mail fnieto@goliat.ugr.es Tel. 34(58)243342 Fax 34(58)243368

Nogués-Carulla, Dr Joaquim (1946). Prof. Titular; Vice-Dean, Faculty of Geology. Dep. Cristalografía, Mineralogía y Depósitos Minerales, Univ. Barcelona, Martí y Franqués s/n, 08028-Barcelona, Spain. Doctor in geology (U. Barcelona, 1976) opaque minerals (reflectivity). *Mineralogy gemology*
Tel. 34(3)4021349 Fax 34(3)4021340

Núñez, Dr Pedro (1956). Prof. Titular. Dep. Química Inorgánica, Univ. La Laguna, E-38200 La Laguna, Tenerife, Canary Islands, Spain. DSc (U. La Laguna, 1978) chemistry. *Structural_chemistry*
E-mail pnunez@ull.es Tel. 34(22)603770 Fax 34(22)603634

Ochando, Dr Luis E. (1964). Prof. Ayudante. Dept. Geología, Univ. Valencia, Avda. Dr Moliner 50, 46100-Burjassot, Spain. DSc (U. Valencia, 1988) chemistry. *Structure_determination minerals powder_diffraction*
E-mail Luis.E.Ochando@uv.es Tel. 34(6)3864603 Fax 34(6)3864372

Ortega-Huertas, Dr Miguel (1949). Prof. Titular. Dep. Mineralogía y Petrología, Univ. Granada, Fuentenueva s/n, 18002-Granada, Spain. DSc (U. Granada, 1978) geology. *Minerals X-ray_diffraction*
Tel. 34(58)243342 Fax 34(58)243368

Otalora, Mr Fermín (1965). Becario. Inst. Andaluz de Geol. Mediterranea, Fuentenueva s/n, 18002-Granada, Spain. Grad (U. Granada, 1988) geology. *Crystal_growth*
E-mail otalora@ugr.es Tel. 34(58)243360 Fax 34(58)271873

Otero-Diaz, Prof. L. Carlos (1951). Catedrático. Dept. Química Inorgánica, Univ. Baleares, Palma de Mallorca, Spain. DSc (U. Complutense de Madrid, 1979) chemistry. *Solid_state_chemistry*

Palacio, Prof. Fernando (1944). Prof. Investigación. Instituto Ciencia de Materiales de Aragón, CSIC, Ciudad Universitaria, 50009-Zaragoza, Spain. DSc (U. Zaragoza, 1974) chemistry. *Structural_and_magnetic_phase_transition neutron_powder_diffraction, synchrotron_powder_diffraction Rietveld_method pseudosymmetry magnetism crystal_growth high_pressure_diffraction*
E-mail palacio@cc.unizar.es Tel. 34(76)563296 Fax 34(76)567920

Palomo-Delgado, Dr M. Inmaculada (1957). Prof. Titular. Dep. Mineralogía y Petrología, Univ. Granada, Fuentenueva s/n, 18002-Granada, Spain. DSc (U. Granada, 1987) geology. *Minerals X-ray_diffraction*
Tel. 34(58)243340 Fax 34(58)243368

Paniagua, Dr Andrés (1959). Prof. Asociado. Dep. Geología, Univ. Oviedo, Arias de Velasco s/n, 33005-Oviedo, Spain. Grad (U. Oviedo, 1984) geology. *Minerals crystallography*
Tel. 34(85)103094 Fax 34(85)233911

Perales, Dr Aurea (1929). Investigador. Inst. Rocasolano, CSIC, Serrano 119, 28006-Madrid, Spain. DSc (U. Valencia, 1952) chemistry. *Absolute_configuration biocrystallography chirality conformation crystallography natural_products proteins*
Tel. 34(1)5619400

Pérez-Garrido, Dr Simeon (1943). Prof. Titular. Dep. Física de la Materia Condensada, Univ. Sevilla, Appdo. 1065, 41080-Sevilla, Spain. DSc (U. Sevilla, 1971) physics. *Crystal_structure_determination X-ray_diffraction organic_structure*
Tel. 34(5)4616615 Fax 34(5)4612097

Pérez-Mato, Prof. Juan Manuel (1952). Catedrático. Chairman, IUCr Ad-Interim Commission on Aperiodic Crystals. Dept. Física, Univ. del País Vasco, Appdo. 644, 48080-Bilbao, Spain. DSc (U. País Vasco, 1980) physics. *X-ray_diffraction phase_transitions modulated_structures quasi-crystals*
E-mail wmppemam@lg.ehu.es Tel. 34(4)4647700x2492 Fax 34(4)4648500 Telex 33259 ehupv

Pertierra, Mrs Pilar (1966). Prof. Asociado. Dept. Química-Física y Analítica, Univ. Oviedo, Julián Clavería s/n, 33006-Oviedo, Spain. Grad (U. Oviedo, 1991) chemistry. *Powder_diffraction structure_resolution*
E-mail ppc@dwarf1.quimica.uniovi.es Tel. 34(85)103477 Fax 34(85)103480 Telex 84322 EDUCI-E

Piniella, Dr Juan Francisco (1952). Prof. Titular. Dept. Geología, Univ. Autónoma de Barcelona, 08193-Bellaterra, Spain. DSc (U. Autónoma de Barcelona, 1985) chemistry. *Structural_chemistry charge_density*
E-mail igcr0@cc.uab.es Tel. 34(3)5811163 Fax 34(3)5811263 Telex 52040 EDUCIE

Pizarro, Dr José Luis (1961). Prof. Titular. Dept. Mineralogía y Petrología, Univ. del País Vasco, Appdo. 644, 48080-Bilbao, Spain. DSc (U. País Vasco, 1991) geology. *Crystal_structure_determination minerals*
E-mail npbpisaj@lg.ehu.es Tel. 34(4)4647700x2555 Fax 34(4)4648500

Plana-Llevat, Dr Feliciano (1946). Investigador. Instituto Jaime Almera, CSIC, Martí y Franqués s/n, 08028-Barcelona, Spain. DSc (U. Barcelona, 1974) geology. *X-ray diffractometry minerals microanalysis clays_minerals*
E-mail fplana@u.ija.csic.es Tel. 34(3)3302716 Fax 34(3)4110012

Polvorinos, Dr Angel Jesus (1952). Prof. Titular. Dept. Geología, Univ. Sevilla, Apdo. 553, 41071-Sevilla, Spain. DSc (U. Sevilla, 1981) geology. *Minerals*
Tel. 34(5)4625060

Portal-Olea, Dr M. Dolores (1949). Prof. Titular. Dep. Química Orgánica, Univ. Granada, Fuentenueva s/n, 18002-Granada, Spain. Grad (U. Granada, 1981) chemistry. *Organic_computational_chemistry*
E-mail portal@ugr.es Tel. 34(58)243186 Fax 34(58)243320

Prieto, Prof. Manuel (1950). Catedrático. Dep. Geología, Univ. Oviedo, Arias de Velasco s/n, 33005-Oviedo, Spain. DSc (U. Complutense de Madrid, 1982) geology. *Minerals crystal_growth crystallization mineralogy*
Tel. 34(85)103088 Fax 34(85)233911

Puig-Molina, Mr Anna (1968). Becario. Dept. Geología, Univ. Autónoma de Barcelona, 08193-Bellaterra, Spain. Grad (U. Autónoma de Barcelona, 1991) geology. *Structural_chemistry charge_density*
E-mail igcr0@cc.uab.es Tel. 34(3)5812034 Fax 34(3)5811263 Telex 52040 EDUCIE

Puigjaner, Prof. Luis (1935). Catedrático; Director, Environment Center (LCMA). Dep. Enginyeria Química, Univ. Politècnica de Catalunya, Diagonal 647, 08028-Barcelona, Spain. DSc (U. Politécnica de Madrid, 1986) chemistry engineering. *Structural_biology computing X-ray_diffraction*
E-mail lpc@eq.upc.es Tel. 34(3)4016678 Fax 34(3)4017150 Telex 52821UPC

Queralt-Mitjans, Dr Ignasi (1953). Colaborador. Inst. Jaime Almera, CSIC, Martí y Franqués s/n, 08028-Barcelona, Spain. DSc (U. Barcelona, 1988) geology. *Diffraction clays_minerals ceramic_processing basaltic_rocks_minerals*
E-mail iqueralt@u.ija.csic.es Tel. 34(3)3302716 Fax 34(3)4110012

Quirós, Dr Miguel (1962). Prof. Asociado. Dep. Química Inorgánica, Univ. Granada, Fuentenueva s/n, 18071-Granada, Spain. DSc (U. Granada, 1989) chemistry. *Structural_chemistry*
E-mail mquiros@ugr.es Tel. 34(58)243325 Fax 34(58)243322

Rasines, Prof. Isidoro (1927). Prof. Investigación. Member, European Community Training Council. Instituto de Ciencia de Materiales, CSIC, Serrano 113, 28006 Madrid, Spain. DSc (U. Complutense de Madrid, 1970) chemistry. *Crystal_structure_determination superconductors physical_properties_of_novel_compounds*
E-mail immir09@cc.csic.es Tel. 34(1)5854907 Fax 34(1)5618471 Telex csice e 42182

Rausell-Colom, Prof. José Antonio (1932). Prof. Investigación. Instituto de Ciencia de Materiales, CSIC, Serrano 115 bis, 28006-Madrid, Spain. DSc (U. Complutense de Madrid, 1962) chemistry. *Structural_chemistry minerals small-angle_scattering*
Tel. 34(1)4111772

Reventós, Dr María-Mercedes (1943). Prof. Titular. Dept. Geología, Univ. Valencia, Avda. Dr Moliner 50, 46100-Burjassot, Spain. DSc (U. País Vasco, 1987) geology. *Structure_determination teaching powder_diffraction*
Tel. 34(6)3864603 Fax 34(6)3864372

Ribas, Prof. Joan (1943). Catedrático. Dep. Química Inorgánica, Univ. Barcelona, Diagonal 647, 08028-Barcelona, Spain. DSc (U. Barcelona, 1974) chemistry. *Structural_chemistry magnetochemistry*
E-mail int.magnet@ub.es Tel. 34(3)4021264 Fax 34(3)4111492

Riera, Prof. Victor (1936). Catedrático. Dept. Química Organometálica, Univ. Oviedo, Julián Clavería s/n, 33071-Oviedo, Spain. DSc (U. Oviedo, 1959) chemistry. *Structural_chemistry*
Tel. 34(85)103463 Fax 34(85)103446 Telex UNIOV 84322

Rius, Prof. Jordi (1954). Investigador. Inst. de Ciencia de Materiales de Barcelona, CSIC, Campus UAB, 08193-Bellaterra, Spain. DSc (U. Marburg, Germany, 1980) natural sciences. *Structure_determination direct_methods Patterson_methods powder_methods computing mineralogy ionic conduction*
E-mail jordi.rius@icmab.es Tel. 34(3)5801853 Fax 34(3)5805729

Rodriguez-Carvajal, Dr Juan (1953). Colaborador. Inst. de Ciencia de Materiales de Barcelona, CSIC, Campus UAB, 08193-Bellaterra, Spain. DSc (U. Barcelona, 1984) physics. *X-ray_diffraction neutron_diffraction powder_diffraction*
Tel. 34(3)3302716

Rodriguez-Clemente, Prof. Rafael (1948). Profesor Investigación. Crystal_growth committee. Inst. de Ciencia de Materiales de Barcelona, CSIC, Campus UAB, 08193-Bellaterra, Spain. DSc (U. Barcelona, 1974) geology. *Crystal_growth morphology nucleation CVD precipitation*
Tel. 34(3)5801853 Fax 34(3)5805729

Rodriguez-Gallego, Prof. Manuel (1935). Catedrático. Dep. Mineralogía y Petrología, Univ. Granada, Fuentenueva s/n, 18002-Granada, Spain. DSc (U. Granada, 1960) pharmacy. *Scattering defect_structures*
E-mail jdmartin@ugr.es Tel. 34(58)243339 Fax 34(58)243368

Rodriguez-Gordillo, Dr José (1946). Prof. Titular. Dep. Mineralogía y Petrología, Univ. Granada, Fuentenueva s/n, 18002-Granada, Spain. DSc (U. Granada, 1976) chemistry. *Minerals*
Tel. 34(58)243343 Fax 34(58)243368

Rodriguez-Navarro, Mr Alejandro (1969). Doctorando. Instituto Andaluz de Geología Mediterranea-CSIC, Avda. de Fuentenueva s/n, 18002-Granada, Spain. Grad (U. Granada, 1992) physics. *Textures thin-film diamond ceramic eggshell biomineral biomimetic competitive_crystal_growth preferred_orientation_of_material history*
E-mail rodriguez@ugr.es Tel. 34(58)243360 Fax 34(58)271873

Rodriguez-Roldan, Dr Ana (1951). Becario. Dept. Química Inorgánica, Univ. Complutense de Madrid, Campus Universitario, 28040-Madrid, Spain. DSc (U. Complutense de Madrid, 1983) chemistry. *Crystal_structure_determination*
Tel. 34(1)4491850

Rodriguez-Romero, Mr F. Victor (1968). Becario. Dep. Física Fundamental y Experimental, Univ. La Laguna, 38204 La Laguna, Spain. Grad (U. La Laguna, 1987) chemistry. *Structural_chemistry crystal_structure_determination*
E-mail amg@iac.es Tel. 34(22)603274 Fax 34(22)603684

Rojo, Prof. Teofilo (1951). Catedrático. Dept. Química Inorgánica, Univ. del País Vasco, Appdo. 644, 48080-Bilbao, Spain. DSc (U. País Vasco, 1981) chemistry. *Structural_chemistry*
Tel. 34(4)4647700x2448 Fax 34(4)4648500

Román, Prof. Pascual (1947). Catedrático. Dept. Química Inorgánica, Univ. del País Vasco, Appdo. 644, 48080-Bilbao, Spain. DSc (U. País Vasco, 1976) chemistry. *Structural_chemistry crystal_structure_determination*
E-mail qipropop@lg.ehu.es Tel. 34(4)4647700x2620 Fax 34(4)4648500

Romero, Dr Antonio (1958). Becario. Inst. de Química Inorgánica, CSIC, Serrano 113, 28006-Madrid, Spain. DSc (U. Complutense de Madrid, 1988) chemistry. *Crystal_structure_determination*
Tel. 34(1)4111772

Romero-Garzón, Mr José (1960). Laboral. Dep. Mineralogía y Petrología, Univ. Granada, Fuentenueva s/n, 18002-Granada, Spain. Grad (U. Granada, 1982) geology. *Crystal_structure_determination textures*
E-mail jdmartin@ugr.es Tel. 34(58)243339 Fax 34(58)243368

Romero-Molina, Dr M. Angustias (1950). Prof. Titular. Dep. Química Inorgánica, Univ. Granada, Fuentenueva s/n, 18071-Granada, Spain. DSc (U. Granada, 1980) chemistry. *Structural_chemistry*
E-mail mquiros@ugr.es Tel. 34(58)243236 Fax 34(58)243322

Ros, Prof. Josep (1953). Professor. Dept. Química, Univ. Autónoma de Barcelona, 08193-Bellaterra, Spain. DSc (U. Autónoma de Barcelona, 1981) chemistry. *Structural_chemistry organometallic_and_inorganic_chemistry*
E-mail iqikaos@cc.uab.es Tel. 34(3)5811889 Fax 34(3)5812477 Telex 52040 EDUCI E

Rosa de la, Dr Nicolás (1953). Prof. Titular. Dep. Estructura y Propiedades de los Materiales, Univ. de Cádiz, Appdo. 40, 11510-Pto. Real, Spain. DSc (U. Sevilla, 1985) physics. *Amorphous_phase Anomalous_dispersion SAXS ceramic*
E-mail rosa@czv1.uca.es Tel. 34(56)830966 Fax 34(56)837205

Ruiz, Dr Xavier (1953). Catedrático EU. Dept. de Química, Univ. Rovira i Virgili, Pza. Imperial Tarraco 1, 43005-Tarragona, Spain. DSc (U. Barcelona, 1991) physics. *Crystal_growth computing*
Tel. 34(77)225254x2205

Ruiz-Pérez, Dr Catalina (1957). Prof. Titular. Dep. Física Fundamental y Experimental, Univ. La Laguna, 38204 La Laguna, Spain. DSc (U. Valencia, 1987) physics. *Structural_chemistry crystal_structure_determination*
E-mail amg@iac.es Tel. 34(22)603272 Fax 34(22)603684

Ruiz-Valero, Dr Caridad (1957). Becario. Inst. de Química Inorgánica, CSIC, Serrano 113, 28006-Madrid, Spain. DSc (U. Complutense de Madrid, 1982) chemistry. *Crystal_structure_determination*
Tel. 34(1)4111772

Rull, Prof. Fernando (1948). Catedrático. Universidad de Valladolid, Prado de la Magdalena s/n, 47005-Valladolid, Spain. DSc (U. Valladolid, 1976) chemistry. *Spectroscopy crystal_growth*
Tel. 34(83)423195 Fax 34(83)423013

Salas-Peregrín, Dr Juan Manuel (1952). Prof. Titular. Dep. Química Inorgánica, Univ. Granada, Fuentenueva s/n, 18071-Granada, Spain. DSc (U. Granada, 1979) chemistry. *Structural_chemistry*
E-mail mquiros@ugr.es Tel. 34(58)243325 Fax 34(58)243322

Salvado, Mr Miguel Angel (1966). Prof. Asociado. Dept. Química-Física y Analítica, Univ. Oviedo, Julián Clavería s/n, 33006-Oviedo, Spain. Grad (U. Oviedo, 1990) chemistry. *Powder_X-ray_diffraction Rietveld_method*
E-mail mass@dwarf1.quimica.uniovi.es Tel. 34(85)103477 Fax 34(85)103480 Telex 84322 EDUCI-E

San José, Mr Ana M. (1966). Becario. Dept. Química Inorgánica, Univ. del País Vasco, Appdo. 644, 48080-Bilbao, Spain. Grad (U. País Vasco, 1989) chemistry. *Structural_chemistry crystal_structure_determination*
E-mail qibsawea@lg.ehu.es Tel. 34(4)4647700x2450 Fax 34(4)4648500

Sánchez-Navas, Dr Antonio (1961). Prof. Asociado. Dep. Mineralogía y Petrología, Univ. Granada, Fuentenueva s/n, 18002-Granada, Spain. DSc (U. Granada, 1985) geology. *Powder_X-ray_diffraction HRTEM minerals*
Tel. 34(58)243355 Fax 34(58)243368

Sánchez-Sánchez, Dr Purificación (1958). Prof. Titular. Dep. Química Inorgánica, Univ. Granada, Fuentenueva s/n, 18071-Granada, Spain. DSc (U. Granada, 1984) chemistry. *Structural_chemistry*
E-mail mquiros@ugr.es Tel. 34(58)243325 Fax 34(58)243322

Santos-Sánchez, Dr Alberto (1955). Prof. Titular. Dep. Estructura y Propiedades Materiales, Universidad de Cádiz, Apdo. 40, 11510-Pto. Real, Spain. DSc (U. Complutense de Madrid, 1988) geology. *Crystal_growth minerals phase_transition*
Tel. 34(56)830210 Fax 34(56)834924

Sanz-Aparicio, Dr Juliana (1959). Colaborador. Inst. Rocasolano, CSIC, Serrano 119, 28006-Madrid, Spain. DSc (U. Complutense de Madrid, 1987) chemistry. *Crystal_structure_determination proteins_crystallography drugs_packing*
E-mail xjulia@roca.csic.es Tel. 34(1)5619400 Fax 34(1)5642431 Telex 42182

Sanz-Ruiz, Prof. Francisco (1941). Catedrático. Dept. Termología, Univ. Valencia, Avda. Dr Moliner 50, 46100-Burjassot, Spain. DSc (U. Complutense de Madrid, 1969) physics. *Crystal_structure_determination amorphous_diffraction*
Tel. 34(6)3864300x3280

Sebastian, Dr Eduardo Manuel (1949). Prof. Titular. Dep. Mineralogía y Petrologia, Univ. Granada, Fuentenueva s/n, 18002-Granada, Spain. DSc (U. Granada, 1979) geology. *X-ray_diffraction minerals*
Tel. 34(58)243340 Fax 34(58)243368

Sola, Dr Joan (1952). Prof. Titular. Dept. Química, Univ. Autónoma de Barcelona, 08193-Bellaterra, Spain. DSc (U. Autónoma de Barcelona, 1982) chemistry. *Structural_chemistry*
E-mail iqin2@cc.uab.es Tel. 34(3)5811372 Fax 34(3)5812477 Telex 52040 EDUCIE

Solans, Prof. Joaquim (1940). Catedrático. Dep. Cristalografía, Mineralogía y Depósitos Minerales, Univ. Barcelona, Martí y Franqués s/n, 08028-Barcelona, Spain. DSc (U. Barcelona, 1966) geology. *Structural_chemistry physical_property*
Tel. 34(3)4021352 Fax 34(3)4021340

Solans, Prof. Xavier (1949). Catedrático. Sub-Editor of Ninth Edition of World Directory of Crystallographers; Member of Spanish Crystallographic Committee; Spanish delegate in IUCr and ECC. Dep. Cristalografía, Mineralogía y Depósitos Minerales, Univ. Barcelona, Martí y Franqués s/n, 08028-Barcelona, Spain. DSc (U. Barcelona, 1977) physics. *Crystal_structure_determination phase_transitions diffraction_data structural_chemistry physics_property diffraction_technique*
E-mail ubacmd10@ebcesca1.bitnet.es Tel. 34(3)4021343 Fax 34(3)4021340

Solé, Mr Rosa (1966). Prof. Ayudante. Dept. de Química, Univ. Rovira i Virgili, Pza. Imperial Tarraco 1, 43005-Tarragona, Spain. Grad (U. Barcelona, 1986) physics. *Crystal_growth computing hexaferrites*
Tel. 34(77)225254x2205

Suades, Dr Joan (1954). Prof. Titular. Dept. Química, Univ. Autónoma de Barcelona, 08193-Bellaterra, Spain. DSc (U. Autónoma de Barcelona, 1983) chemistry. *Structural_chemistry*
E-mail iqin2@cc.uab.es Tel. 34(3)5811010 Fax 34(3)5812477 Telex 52040 EDUCI E

Subirana, Prof. Juan A. (1936). Catedrático. Dept. Enginyeria Química, Univ. Politécnica de Catalunya, Diagonal 647, 08028-Barcelona, Spain. DSc (U. Complutense de Madrid, 1960) chemistry. *Fibres_diffraction structural_biology*
E-mail subirana@eq.upc.es Tel. 34(3)4016688 Fax 34(3)4016600 Telex 52821UPC

Tauler, Dr Esperanza (1953). Prof. Titular. Dep. Cristalografía, Mineralogía y Depósitos Minerales, Univ. Barcelona, Martí y Franqués s/n, 08028-Barcelona, Spain. DSc (U. Barcelona, 1983) geology. *X-ray_diffraction molecular_crystal isomorphism DTA*
Tel. 34(3)4021350 Fax 34(3)4021340

Tena, Dr M. Angeles (1963). Prof. Titular. Dep. Química, Univ. Jaume I, 12080-Castellón, Spain. DSc (U. Jaume I, 1992) chemistry. *Structural_chemistry X-ray_diffraction crystallography*
E-mail tena@mestral.uji.es Tel. 34(64)345700x8202 Fax 34(64)345654

Tomás, Dr Milagros (1954). Investigador. Dpto. Química Inorgánica, Universidad Zaragoza, 50009-Zaragoza, Spain. DSc (U. Zaragoza, 1979) chemistry. *Structural_chemistry organometallic_compounds mixed-valence_compounds clusters*
E-mail milagros@cc.unizar.es Tel. 34(76))552347 Fax 34(76))567920

Torres-Ruiz, Dr José (1952). Prof. Titular. Dep. Mineralogía y Petrologia, Univ. Granada, Fuentenueva s/n, 18002-Granada, Spain. DSc (U. Granada, 1980) geology. *Minerals*
Tel. 34(58)243354 Fax 34(58)243368

Traveria-Cros, Dr Adolfo (1928). Investigador. Instituto Jaime Almera, CSIC, Martí y Franqués s/n, 08028-Barcelona, Spain. DSc (U. Barcelona, 1964) geology. *X-ray_diffraction X-ray_fluorescence EDX*
E-mail atraveria@u.ija.csic.es Tel. 34(3)3302716 Fax 34(3)4110012

Urtiaga, Dr Karmele (1964). Prof. Titular. Dept. Mineralogía y Petrologia, Univ. del País Vasco, Appdo. 644, 48080-Bilbao, Spain. DSc (U. País Vasco, 1983) geology. *Crystal_structure_determination*
E-mail nppurgrm@lg.ehu.es Tel. 34(4)4647700x2474 Fax 34(4)4648500

Valin, Dr Mariluz (1956). Prof. Titular. Dep. Geología, Univ. Oviedo, Arias de Velasco s/n, 33005-Oviedo, Spain. DSc (U. Oviedo, 1986) geology. *Minerals crystal_structure_determination*
Tel. 34(85)103100 Fax 34(85)103153

Vallet-Regi, Dr Maria (1946). Prof. Titular. Dept. Química Inorgánica, Univ. Complutense de Madrid, Campus Universitario, 28040-Madrid, Spain. DSc (U. Complutense de Madrid, 1975) chemistry. *Solid_state_chemistry*
Tel. 34(1)4491850

Van der Maelen, Dr Juan Francisco Javier (1959). Prof. Asociado. Dept. Química-Física y Analítica, Univ. Oviedo, Julián Clavería s/n, 33006-Oviedo, Spain. DSc (U. Oviedo, 1991) chemistry. *Algorithm-resolution_and_refinement computer_automation organometallic_characterization group_theory MO_calculation linear_algebra*
E-mail fvu@dwarf2.quimica.uniovi.es Tel. 34(85)103480 Fax 34(85)103480 Telex 84322 EDUCI-E

Vegas, Prof. Angel (1947). Investigador. Inst. Rocasolano, CSIC, Serrano 119, 28006-Madrid, Spain. DSc (U. Complutense de Madrid, 1975) chemistry. *Inorganic_crystal_structure_determination crystal_chemistry clusters charge_density*
E-mail xangel@roca.csic.es Tel. 34(1)5619400 Fax 34(1)5642431 Telex 42182

Veintemillas, Dr Sabino (1957). Colaborador. Inst. de Ciencia de Materiales de Barcelona, CSIC, Campus UAB, 08193-Bellaterra, Spain. DSc (U. Complutense de Madrid, 1986) chemistry. *Crystal_growth*
E-mail sabino@icmab.es Tel. 34(3)5801853

Velilla, Dr Nicolas (1952). Prof. Titular. Dep. Mineralogía y Petrologia, Univ. Granada, Fuentenueva s/n, 18002-Granada, Spain. DSc (U. Granada, 1983) geology. *Minerals*
Tel. 34(58)243343 Fax 34(58)243368

Vendrell, Dr Marius (1949). Prof. Titular. Dep. Cristalografía, Mineralogía y Depósitos Minerales, Univ. Barcelona, Martí y Franqués s/n, 08028-Barcelona, Spain. DSc (U. Barcelona, 1978) geology. *Minerals*
Tel. 34(3)4021348 Fax 34(3)4021340

Vicente, Prof. José (1943). Catedrático. Dep. Química Inorgánica, Fac. Química, Universidad Murcia, Aptdo. 4021, E-30001 Murcia, Spain. DSc (U. Zaragoza, 1973) chemistry. *Structural_chemistry inorganic organic organometallic gold palladium platinum mercury*
E-mail jvs@fcu.um.es Tel. 34(68)364143 Fax 34(68)364143

Vila, Dr Eladio (1953). Titulado Superior. Inst. de Química Inorgánica, CSIC, Serrano 113, 28006-Madrid, Spain. DSc (U. Complutense de Madrid, 1980) chemistry. *X-ray_spectroscopy*
Tel. 34(1)4111772

Zuñiga, Dr Fco. Javier (1953). Prof. Titular. Dept. Física, Univ. del País Vasco, Appdo. 644, 48080-Bilbao, Spain. DSc (U. País Vasco, 1980) physics. *Phase_transitions crystal_structure_determination modulated_structures*
Tel. 34(4)4647700x2492 Fax 34(4)4648500

SWEDEN

Sub-Editor: Y. Andersson

Ahlzén, Dr Per-Johan (1962). Research engineer. SECO TOOLS AB, Fagersta. [SECO TOOLS AB, S-737 82 Fagersta, Sweden.] PhD (Uppsala U., 1990) solid state chemistry. *Composite CVD PVD*
Tel. 46(223)40272 Fax 46(223)11860

Åhman, Johan I. (1968). PhD student. Dept. of Inorganic chemistry, Chalmers University of Technology. [Dept. of Inorganic Chemistry, Chalmers University of Technology, S-412 96 Göteborg, Sweden.] MSc (1992) *Crystallography growth flux dielectrics*
E-mail johan@ctkem.phc.chalmers.se Tel. 46(31)7722871 Fax 46(31)7722846

Albertsson, Prof. Jörgen (1939). Professor. Commission on Journals; Co-editor of Acta Crystallographica. Dept. of Inorganic Chemistry, Chalmers University of Technology. [Dept. of Inorganic Chemistry, Chalmers University of Technology, S-412 96 Göteborg, Sweden.] PhD (Lund U., 1972) inorganic chemistry. *Structural_chemistry solid_state_chemistry instrumentation structure-property_relationships*
E-mail jalb@ctkem.phc.chalmers.se Tel. 46(31)7722851 Fax 46(31)7722846

Al-Karadaghi, Dr Salam (1953). Postdoc. Dept. of Molecular Biophysics, Lund University. [Dept. of Molecular Biophysics, Box 124, S-221 00 Lund, Sweden.] PhD (Moscow U. Stockholm U., 1982 1993) crystallography. *Crystallography proteins*
E-mail salam@robin.mbfys.lth.se Tel. 46(46)104566 Fax 46(46)104543

Andersson, Dr Inger A. (1949). Associate professor. Dept. of Molecular Biology, Swedish University of Agricultural Sciences. [Dept. of Molecular Biology, Biomedical Center, Box 590, S-751 24 Uppsala, Sweden.] PhD (U. des Saarlandes, 1980) biochemistry. *Macromolecular_crystallography enzymatic_catalysis photosynthesis*
E-mail inger@xray.bmc.uu.se Tel. 46(18)174523 Fax 46(18)536971 Telex 76132 BIOMED

Andersson, Dr Margaretha (1963). Research scientist. Inst. of Chemistry, University of Uppsala. [Dept. of Inorganic Chemistry, Box 531, S-751 21 Uppsala, Sweden.] PhD (Uppsala U., 1992) inorganic chemistry. *Materials_sciences structural_chemistry phase_determination*
E-mail margaretha.andersson@kemi.uu.se Tel. 46(18)183726 Fax 46(18)508542

Andersson, Prof. Sten (1931). Professor. Inorganic Chemistry 2, Lund University. [Inorganic Chemistry 2, Chemical Center, Box 124, S-221 00 Lund, Sweden.] DSc (Stockholm U., 1967) chemistry. *Inorganic_chemistry*
Tel. 46(31)108227 Fax 46(31)104012

Andersson, Dr Yvonne (1947). Senior lecturer. Sub-Editor for Sweden of Ninth Edition of World Directory of Crystallographers. Inst. of Chemistry, University of Uppsala. [Dept. of Inorganic Chemistry, Box 531, S-751 21 Uppsala, Sweden.] PhD (Uppsala U., 1984) inorganic chemistry. *Synthesis hydrides magnetic_structure neutron_diffraction*
E-mail yvonne.andersson@kemi.uu.se Tel. 46(18)183780 Fax 46(18)508542

Åsbrink, Dr Stig (1929). Senior lecturer. Dept. of Inorganic Chemistry, Stockholm University. [Dept. of Inorganic Chemistry, Arrhenius Laboratory, Stockholm University,

S-106 91 Stockholm, Sweden.] PhD (Stockholm U., 1973) chemical crystallography. *Inorganic_compounds high_pressure phase_transition synchrotron_radiation*
E-mail stig@struc.su.se Tel. 46(8)162387 Fax 46(8)152187

Bemm, Ulf (1968). PhD student. Dept. of Structural Chemistry, Stockholm University. [Dept. of Structural Chemistry, Arrhenius Laboratory, Stockholm University, S-106 91 Stockholm, Sweden.] BSc (Sundsvall U., 1991) *Alkoxides ceramics gels hydrolysis*
E-mail ulfb@struc.su.se Tel. 46(8)162382 Fax 46(8)152187 Telex 8105199 univers

Berger, Dr Rolf A. (1946). Lecturer. Inst. of Chemistry, University of Uppsala. [Dept. of Inorganic Chemistry, Box 531, S-751 21 Uppsala, Sweden.] PhD (Uppsala U., 1978) chemical crystallography. *Phase_transition condensed_matter magnetochemistry synthesis chalcogenides*
Tel. 46(18)183704 Fax 46(18)508542

Bolt, Rindert (1959). Postdoc. Dept. of Inorganic Chemistry, Chalmers University of Technology. [Dept. of Inorganic Chemistry, Chalmers University of Technology, S-412 96 Göteborg, Sweden.] (1992) solid state chemistry. *Crystal_growth dielectric nonlinear_optics*
E-mail bolt@ctkem.phc.chalmers.se Tel. 46(31)7722882 Fax 46(31)7722846

Boström, Dr Nils Dan (1954). Senior lecturer. Dept. of Inorganic Chemistry, Umeå University. [Dept. of Inorganic Chemistry, Umeå University, S-901 87 Umeå, Sweden.] PhD (Umeå U., 1988) chemical crystallography. *Mineralogy order-disorder molybdates aluminium_compounds*
E-mail dan@oorgserv.chem.umu.se Tel. 46(90)165467 Fax 46(90)136310

Bovin, Jan-Olov (1943). Professor. National Center for HREM, Inorganic Chemistry 2, Lund University. [Chemical Center, PO Box 124, Lund, Sweden.] PhD (Lund U., 1975) inorganic chemistry. *Electron_microscopy photography*
Tel. 46(46)104769 Fax 46(46)104012

Brisander, Mr Magnus (1970). PhD student. Dept. of Structural Chemistry, Stockholm University. [Dept. of Structural Chemistry, Arrhenius Laboratory, Stockholm University, S-106 91 Stockholm, Sweden.] BSc (Linköping U., 1993) organic chemistry. *Structure_determination organic organometallic organic_synthesis dynamics activity*
E-mail magnus@struc.su.se Tel. 46(8)161440 Fax 46(8)152187

Bryntse, Dr Ingrid (1956). Research associate. Dept. of Inorganic Chemistry, Stockholm University. [Dept. of Inorganic Chemistry, Arrhenius Laboratory, Stockholm University, S-106 91 Stockholm, Sweden.] PhD (Stockholm U., 1993) inorganic chemistry. *Solid_state superconductors electron_microscopy profile_analysis*
E-mail ingridb@inorg.su.se Tel. 46(8)162434 Fax 46(8)152187

Carlson, Stefan (1965). PhD student. Dept. of Structural Chemistry, Stockholm University. [Dept. of Structural Chemistry, Arrhenius Laboratory, Stockholm University, S-106 91 Stockholm, Sweden.] BSc (Stockholm U., 1993) *Structure_determination transition_metals high_pressure*
E-mail sgkl@struc.su.se Tel. 46(8)161440 Fax 46(8)152187

Carlsson, Anna B. V. (1968). PhD student. National Center for HREM, Inorganic Chemistry 2, Lund University. [Chemical Center, PO Box 124, S-221 00 Lund, Sweden.] MSc (Lund U., 1991) chemical engineering. *High-resolution_electron_microscopy image_processing three-dimensional_reconstruction inorganic_chemistry*
E-mail anna.carlsson@oorg2.lth.se Tel. 46(46)108231 Fax 46(46)104012

Csöregh, Dr Ingeborg (1942). Associate professor. Dept. of Structural Chemistry, Stockholm University. [Dept. of Structural Chemistry, Arrhenius Laboratory, Stockholm University, S-106 91 Stockholm, Sweden.] PhD (Stockholm U., 1983) structural chemistry. *Structure_determination organic_compounds organometallic_compounds structure-activity_relationship supramolecular_chemistry inclusion_compounds*
E-mail ics@struc.su.se Tel. 46(8)162381 Fax 46(8)152187

Delaplane, Dr Robert G. (1942). Researcher. Inst. of Chemistry, University of Uppsala. [Dept. of Inorganic Chemistry, Box 531, S-751 21 Uppsala, Sweden.] PhD (Northwestern U., 1969) neutron diffraction. *Amorphous_phase crystallography neutron_diffraction disorder*
E-mail robert.delaplane@kem.uu.se Tel. 46(18)183773 Fax 46(18)508542 Telex 76088 TSLISV S

Edström, Dr Kristina (1958). Lecturer. Inst. of Chemistry, University of Uppsala. [Dept. of Inorganic Chemistry, Box 531, S-751 21 Uppsala, Sweden.] PhD (Uppsala U., 1990) chemistry. *Crystal_structures ionic_conductors structure_property_relationship*
E-mail kristina.edstrom@kemi.uu.se Tel. 46(18)183712 Fax 46(18)508542 Telex 76088 TSLISV S

Eklund, Prof. Hans (1940). Professor. Dept. of Molecular Biology, Swedish University of Agricultural Sciences. [Dept. of Molecular Biology, Biomedical Center, PO Box 590, S-751 24 Uppsala, Sweden.] PhD (Swedish U. Agri. Sci., 1976) chemistry. *Proteins reductases*
E-mail hasse@xray.bmc.uu.se Tel. 46(18)174559 Fax 46(18)536971

Engström, Dr Ingvar O. J. (1934). Senior lecturer. Inst. of Chemistry, University of Uppsala. [Dept. of Inorganic Chemistry, Box 531, S-751 21 Uppsala, Sweden.] PhD (Uppsala U., 1970) chemistry. *Structural_chemistry*
Tel. 46(18)183740 Fax 46(18)508542

Eriksson, Dr Anders (1945). Lecturer. Inst. of Chemistry, University of Uppsala. [Dept. of Inorganic Chemistry, Box 531, S-751 21 Uppsala, Sweden.] DSc (Uppsala U., 1981) inorganic chemistry. *Vibrational_spectroscopy coordination_compounds*
E-mail anders.eriksson@kemi.uu.se Tel. 46(18)183770 Fax 46(18)508542

Eriksson, Dr Lars (1960). Lecturer. Dept. of Structural Chemistry, Stockholm University. [Dept. of Structural Chemistry, Arrhenius Laboratory, Stockholm University, S-106 91 Stockholm, Sweden.] PhD (Stockholm U., 1990) crystallography. *Computer structure_determination symmetry*
E-mail lerik@eros.misu.su.se Tel. 46(8)162394 Fax 46(8)152187

Eriksson, Dr Sten Gunnar (1958). Research scientist. Dept. of Inorganic Chemistry, University of Göteborg. [Dept. of Inorganic Chemistry, University of Göteborg, S-412 96 Göteborg, Sweden.] PhD (Göteborg U., 1990) chemical crystallography. *Rietveld_method synthesis superconductivity superstructure neutron_scattering structure_property_relationship*
Tel. 46(31)7722857 Fax 46(31)167194

Ersson, Nils Olov (1942). Research engineer. Inst. of Chemistry, University of Uppsala. [Dept. of Inorganic Chemistry, Box 531, S-751 21 Uppsala, Sweden.] PhD (Uppsala U., 1985) inorganic chemistry. *Camera powder_diffraction diffractometry computing*
Tel. 46(18)183739 Fax 46(18)508542

Ertan, Ms Anne (1957). PhD student. Dept. of Structural Chemistry, Stockholm University. [Dept. of Structural Chemistry, Arrhenius Laboratory, Stockholm University, S-106 91 Stockholm, Sweden.] BSc (Stockholm U., 1984) chemistry. *Structure_determination macrocyclic_ligands coordination*
E-mail anne@struc.su.se Tel. 46(8)162381 Fax 46(8)152187

Forslund, Dr Bertil S. (1943). Research associate. Dept. of Inorganic Chemistry, Stockholm University. [Dept. of Inorganic Chemistry, Arrhenius Laboratory, Stockholm University, S-106 91 Stockholm, Sweden.] PhD (Stockholm U., 1984) inorganic chemistry. *Crystal_growth high_pressure inorganic_materials*
Tel. 46(8)162353 Fax 46(8)152187

Gallardo, Olga (1955). PhD student. Dept. of Structural Chemistry, Stockholm University. [Dept. of Structural Chemistry, Arrhenius Laboratory, Stockholm University, S-106 91 Stockholm, Sweden.] BSc (Moscow State U., 1978) chemistry. *Heteromolecular_host/guest complexes hydrogen_bonds small_active_molecules dicarboxylic_acid_hosts*
E-mail olga@struc.su.se Tel. 46(8)162381 Fax 46(8)152187 Telex unives s 8105199

Glaser, Julius (1948). Senior lecturer. Inorganic Chemistry, The Royal Institute of Technology. [Inorganic Chemistry, The Royal Institute of Technology, S-100 44 Stockholm, Sweden.] DSc (Royal Inst. Techn., 1987) inorganic chemistry. *Coordination_chemistry NMR thallium_uranium_chemistry*
E-mail julius@inorg.kth.se Tel. 46(8)7908151 Fax 46(8)212626

Grins, Dr Jekabs (1952). Research associate. Dept. of Inorganic Chemistry, Stockholm University. [Dept. of Inorganic Chemistry, Arrhenius Laboratory, Stockholm University, S-106 91 Stockholm, Sweden.] PhD (Stockholm U., 1980) chemistry. *Solid_state_chemistry material_sciences*
Tel. 46(8)162365 Fax 46(8)152187

Gullman, Jan O. (1943). Senior research officer. Central Board of National Antiquities, Conservation Institute. [Central Board of National Antiquities, Box 5405, S-114 84 Stockholm, Sweden.] PhD (Uppsala U., 1987) chemical crystallography. *Characterisation damage inorganic_materials degradation of materials*
Tel. 46(8)7839340 Fax 46(8)6614277

Gustafsson, Torbjörn (1949). Research engineer. Inst. of Chemistry, University of Uppsala. [Dept. of Inorganic Chemistry, Box 531, S-751 21 Uppsala, Sweden.] PhD (Uppsala U., 1987) chemistry. *Crystal_structure low_temperature instrumentation phase_transitions*
E-mail torbjorn.gustafsson@kemi.uu.se Tel. 46(18)183766 Fax 46(18)508542 Telex 76088 TSLISV S

Hansen, Doc. Staffan (1954). Lecturer, docent. National Center for HREM, Inorganic Chemistry 2, Lund University. [Chemical Center, PO Box 124, S-221 00 Lund, Sweden.] PhD (Lund U., 1985) inorganic chemistry. *Crystal_chemistry electron_microscopy catalytic_oxides cement minerals*
E-mail staffan.hansen@oorg2.lth.se Tel. 46(46)108233 Fax 46(46)104012

Hårsta, Dr Anders (1952). Associate professor. Thin Film and Surface Chemistry Group, Inst. of Chemistry, University of Uppsala. [Dept. of Inorganic Chemistry, Box 531, S-751 21 Uppsala, Sweden.] PhD (Uppsala U., 1985) inorganic chemistry. *CVD epitaxy rocking_curves*
Tel. 46(18)183723 Fax 46(18)503056

Hassler, Eivind (1939). Inst. of Chemistry, University of Uppsala. [Inst. of Chemistry, Box 531, S-751 21 Uppsala, Sweden.] PhD (Uppsala U., 1970) inorganic chemistry. *Boron_compounds phosphorus_compounds*
Tel. 46(18)183725

Hebert, Dr Hans (1951). Associate professor. Centrum för Strukturbiokemi, Karolinska Institutet. [Centrum för Strukturbiokemi, Karolinska Institutet, Novum, S-141 57 Huddinge, Sweden.] PhD (1979) chemical crystallography. *Biomolecule membrane electron_microscopy image_processing*
E-mail hans@ondine.csb.ki.se Tel. 46(8)6089219 Fax 46(8)6089290

Hegedüs, Zsolt (1948). Departmental engineer. Dept. of Inorganic Chemistry, Stockholm University. [Dept. of Inorganic Chemistry, Arrhenius Laboratory, Stockholm University, S-106 91 Stockholm, Sweden.] PhD (U. Cluj-Napoca, 1983) solid state chemistry. *Solid_state ceramics superconductors luminescent_compounds*
Tel. 46(8)162434 Fax 46(8)152187

Hermansson, Dr Kersti (1951). Senior lecturer. Member of the Commission on Charge, Spin and Momentum Densities. Inst. of Chemistry, University of Uppsala. [Dept. of Inorganic Chemistry, Box 531, S-751 21 Uppsala, Sweden.] PhD (Uppsala U., 1984) inorganic chemistry. *Computer_modelling quantum_mechanics surface water*
E-mail kersti@kemi.uu.se Tel. 46(18)183767 Fax 46(18)508542

Hong, Sam-Hyo (1940). Ericsson Components AB, Microelectronic Systems Technology. [Ericsson Components AB, Isafjordsgatan 16, Kista, S-164 81 Stockholm, Sweden.] PhD (Stockholm U., 1982) chemical crystallography. *Electronics diffusion electron_microscopy system_integration*
E-mail eka.ekashx@memo.ericsson.se Tel. 46(8)7574690 Fax 46(8)7574161

Hovmöller, Sven (1947). Lecturer. Dept. of Structural Chemistry, Stockholm University. [Dept. of Structural Chemistry, Arrhenius Laboratory, Stockholm University, S-106 91 Stockholm, Sweden.] PhD (Stockholm U., 1980) electron crystallography. *Electron_crystallography image_processing structure_determination high-resolution_electron_microscopy*
Tel. 46(8)162380 Fax 46(8)152187

Ivarsson, Dr Gun J. M. (1943). Senior lecturer. Dept. of Inorganic Chemistry, Umeå University. [Dept. of Inorganic Chemistry, Umeå University, S-901 87 Umeå, Sweden.] PhD (Umeå U., 1983) chemical crystallography. *Inorganic_crystal_structures biochemical_systems*
E-mail oorg@se um dc 51 Tel. 46(90)165467 Fax 46(90)136310

Jagner, Susan (1940). Senior lecturer. Dept. of Inorganic Chemistry, Chalmers University of Technology. [Dept. of Inorganic Chemistry, Chalmers University of Technology, S-412 96 Göteborg, Sweden.] PhD (Göteborg U., 1970) inorganic chemistry. *Copper_compounds silver_compounds single_crystal order-disorder organometallic*
E-mail susan@ctkem.phc.chalmers.se 46(31)7722852 Fax 46(31)7722846

Jahnberg, Dr Lena (1937). Senior lecturer. Dept. of Inorganic Chemistry, Stockholm University. [Dept. of Inorganic Chemistry, Arrhenius Laboratory, Stockholm University, S-106 91 Stockholm, Sweden.] DSc (Stockholm U., 1972) chemistry. *High-resolution_electron_microscopy X-ray_diffraction oxides*
Tel. 46(8)162368 Fax 46(8)152187

Jansson, Dr Kjell (1959). Researcher. Dept. of Inorganic Chemistry, Stockholm University. [Dept. of Inorganic Chemistry, Arrhenius Laboratory, Stockholm University, S-106 91 Stockholm, Sweden.] PhD (Stockholm U., 1988) chemistry. *Structure amorphous_metals electron_microscopy*
Tel. 46(8)162372

Joelson, Thorleif (1950). Research scientist. Dept. of Molecular Biology, University of Uppsala. [Dept of Molecular Biology, Biomedical Center, Box 590, S-751 24 Uppsala, Sweden.] PhD (Swedish U. Agri. Sci., 1988) molecular biology. *AIDS viruses biomolecule immunobiology*
Tel. 46(18)174988 Fax 46(18)536971 Telex 76132 BIOMED S4

Jones, T. Alwyn (1947). Research professor. Co-editor Acta Crystallographica D. Dept. of Molecular Biology, University of Uppsala. [Dept. of Molecular Biology, Biomedical Center, Box 590, S-751 24 Uppsala, Sweden.] PhD (London U., 1973) biophysics. *Macromolecules crystallography computing*
E-mail alwyn@xray.bmc.uu.se Tel. 46(18)174982 Fax 46(18)536971

Karlsson, Magnus E. J. (1969). PhD student. Dept. of Inorganic Chemistry, Umeå University. [Dept. of Inorganic Chemistry, Umeå University, S-901 87 Umeå, Sweden.] MSc (1992) chemistry. *Inorganic_aluminium_compounds organic_aluminium_compounds*
E-mail magnus@oorgserv.chem.umu.se Tel. 46(90)165445 Fax 46(90)136310

Klerkegaard, Prof. Peder (1928). Professor. National Committee of Crystallography; Nobel Committee for Chemistry; Royal Academy of Sciences. Dept. of Structural Chemistry, Stockholm University. [Dept. of Structural Chemistry, Arrhenius Laboratory, Stockholm University, S-106 91 Stockholm, Sweden.] DSc (Stockholm U., 1962) chemical crystallography. *Structure_determination organic_inorganic_compounds instrumentation*
Tel. 46(8)162385 Fax 46(8)152187

Kihlborg, Prof. Lars H. E. (1930). Professor. Member of Commission on Electron Diffraction. Dept. of Inorganic Chemistry, Stockholm University. [Dept. of Inorganic Chemistry, Arrhenius Laboratory, Stockholm University, S-106 91 Stockholm, Sweden.] DSc (Uppsala U., 1963) inorganic chemistry. *High-resolution_electron_microscopy diffraction inorganic_materials oxides solid_state_chemistry*
E-mail larsk@inorg.su.se Tel. 46(8)162370 Fax 46(8)152187 Telex univers s 8105199

Kritikos, Mikael (1961). Research associate. Dept. of Structural Chemistry, Stockholm University. [Dept. of Structural Chemistry, Arrhenius Laboratory, Stockholm University, S-106 91 Stockholm, Sweden.] PhD (Stockholm U., 1991) structural chemistry. *Magnetochemistry solid_state transition_elements X-ray_diffraction symmetry*
E-mail mkr@struc.su.se Tel. 46(8)162382 Fax 46(8)152187 Telex Univers s 8105199

Ladenstein, Prof. Dr Rudolf (1943). Professor. Karolinska Institute, Center of Structural Biochemistry, Sweden. [Karolinska Institute, Center of Structural Biochemistry, S-14157 Huddinge, Sweden.] Prof. (Karolinska Institute/Sweden, 1990) structural biochemistry. *Biocrystallography crystallization catalysis proteins protein_folding protein_stability thermostable_proteins*
E-mail rla@barra.csb.ki.se Tel. 46(8)6089222 Fax 46(8)6089290

Lampe-Önnerud, Christina (1967). PhD student. Inorganic Chemistry, University of Uppsala. [Dept. of Inorganic Chemistry, Box 531, S-751 21 Uppsala, Sweden.] *Battery vanadium_compounds Rietveld_method thin_film*
E-mail kia.lampe-onnerud@kemi.uu.se Tel. 46(18)183769 Fax 46(18)508542

Langer, Dr Vratislav (1949). Associate professor. Dept. of Inorganic Chemistry, University of Göteborg. [Dept. of Inorganic Chemistry, University of Göteborg and Chalmers Univ. of Technology, S-412 96 Göteborg, Sweden.] PhD (Charles U. Praha, 1978) physics. *Structure_determination proteins powders diffractometry*
E-mail langer@amos.inoc.chalmers.se Tel. 46(31)7722877 Fax 46(31)167194

Larsson, Ann-Kristin (1963). PhD student. Inorganic Chemistry 2, Lund University. [Inorganic Chemistry 2, Chemical Center, PO Box 124, S-221 00 Lund, Sweden.] MSc (Lund U., 1988) chemical engineering. *Crystallography metals electron_microscopy coordination superstructures*
Tel. 46(46)108112 Fax 46(46)104012

Larsson, Tommy (1966). PhD student. Inst. of Chemistry, University of Uppsala. [Dept. of Inorganic Chemistry, Box 531, S-751 21 Uppsala, Sweden.] *Neutron_diffraction hydrides band_calculation*
E-mail tommy.larsson@kemi.uu.se Tel. 46(18)183726 Fax 46(18)508542

Lashgari, Kianosh (1968). PhD student. Dept. of Structural Chemistry, Stockholm University. [Dept. of Structural Chemistry, Arrhenius Laboratory, Stockholm University, S-106 91 Stockholm, Sweden.] BSc (Stockholm U., 1993) chemical crystallography.
E-mail kia@struc.su.se Tel. 46(8)162381 Fax 46(8)152187

Lidin, Dr Sven (1961). Research assistent. Inorganic Chemistry 2, Lund University. [Chemical Center, PO Box 124, S-221 00 Lund, Sweden.] PhD (Lund U., 1990) solid state chemistry. *Intermetallics superstructure modelling minimal_surfaces*
Tel. 46(46)108232 Fax 46(46)104012

Liljas, Dr Lars (1947). Lecturer. Dept. of Molecular Biology, University of Uppsala. [Dept. of Molecular Biology, Biomedical Center, Box 590, S-751 24 Uppsala, Sweden.] PhD (Uppsala U., 1977) biochemistry. *Viruses*
E-mail lars@xray.bmc.uu.se Tel. 46(18)174204 Fax 46(18)536971

Lindahl, Tommie (1937). R & D. CEA AB. [CEA AB, Box 174, S-645 23 Strängnäs, Sweden.] MSc (Stockholm U., 1969) chemistry. *X-ray_film*
Tel. 46(152)12930 Fax 46(152)15665

Lindqvist, Ylva (1947). Principal investigator. Dept. of Molecular Biology, Swedish University of Agricultural Sciences. [Dept. of Molecular Biology, Biomedical Center, Box 590, S-751 24 Uppsala, Sweden.] PhD (1977) protein crystallography. *Protein_crystallography enzymes_structure-activity relationship structure_determination*
E-mail ylva@xray.bmc.uu.se Tel. 46(18)174489 Fax 46(18)536971

Lövqvist, Karin (1967). PhD student. Inorganic Chemistry 1, Lund University. [Inorganic Chemistry 1, Chemical Center, PO Box 124, S-221 00 Lund, Sweden.] BSc (Lund U., 1991) chemical crystallography. *Crystallography platinum_compounds coordination_compounds*
E-mail karin.lovqvist@inorgk1.lu.se Tel. 46(46)108108 Fax 46(46)104439

Lundberg, Doc. Bruno K. S. (1939). Senior lecturer. Dept. of Inorganic Chemistry, Umeå University. [Dept. of Inorganic Chemistry, Umeå University, S-901 87 Umeå, Sweden.] PhD (Stockholm U., 1972) chemical crystallography. *Computer_modelling inorganic_crystal_structures biochemical_systems*
E-mail oorgk@biovax.umdc.umu.se Tel. 46(90)165155 Fax 46(90)136310

Lundberg, Dr Monica (1938). Associate professor. Dept. of Inorganic Chemistry, Stockholm University. [Dept. of Inorganic Chemistry, Arrhenius Laboratory, Stockholm University, S-106 91 Stockholm, Sweden.] PhD (Stockholm U., 1971) crystallography. *High-resolution_electron_microscopy crystallography powder_diffraction transition_elements X-ray materials*
Tel. 46(8)162368 Fax 46(8)152187

Lundgren, E. A. Lennart (1950). Senior researcher. National Institute of Occupational Health, Technical Division. [National Institute of Occupational Health, ITA, S-171 84 Solna, Sweden.] BSc (Uppsala U., 1973) chemistry. *Air biological_tissues fibrous_aerosols microscopy*
E-mail lennartl@nioh.se Tel. 46(8)7309742 Fax 46(8)828678 Telex 15816 ARBSKY S

Lundström, Prof. Torsten (1929). Professor. Inst. of Chemistry, University of Uppsala. [Dept. of Inorganic Chemistry, Box 531, S-751 21 Uppsala, Sweden.] DSc (Uppsala U., 1969) inorganic chemistry. *Refractory_compounds superconductors phase_diagram crystal_growth*
E-mail torsten@akka.kemi.uu.se Tel. 46(18)183722 Fax 46(18)508542 Telex 76088 TSLISV S

Lyxell, Dan-Göran (1945). Lecturer. Dept. of Inorganic Chemistry, Umeå University. [Dept. of Inorganic Chemistry, Umeå University, S-901 87 Umeå, Sweden.] MSc (Umeå U., 1969) physics. *Structure_determination large_angle_scattering molybdates LAXS*
E-mail danke@oorgserv.umu.se Tel. 46(90)166328 Fax 46(90)136310

Magnéli, Prof. Arne (1914). Emeritus professor. Commission for Crystallography Teaching: Member 1954–60 Chairman 1957–60; Executive Committee: Member 1972–81 President 1975–78; Swedish National Committee for Crystallography: Member 1953–, Chairman 1974–82. Dept. of Inorganic Chemistry, Stockholm University. [Dept. of Inorganic Chemistry, Arrhenius Laboratory, Stockholm University, S-106 91 Stockholm, Sweden.] DSc (Uppsala U., 1950) chemistry. *Structural_inorganic_chemistry non-stoichiometry*
Tel. 46(8)161265 and 46(8)6739500 Fax 46(8)152187 and 46(8)155670

Marinder, Bengt-Olov (1927). Dept. of Inorganic Chemistry, Stockholm University. [Dept. of Inorganic Chemistry, Arrhenius Laboratory, Stockholm University, S-106 91 Stockholm, Sweden.] DSc (Stockholm U., 1986) chemical crystallography. *Diffraction high-resolution_electron_microscopy transition_elements oxides twinning_on_unit_cell_level*
E-mail bom@inorg.su.se Tel. 46(8)161255 Fax 46(8)152187 Telex 8105199

Nikkola, Matti Johannes (1961). Research scientist. Dept. of Molecular Biology, Swedish University of Agricultural Sciences. [Dept. of Molecular Biology, Biomedical Center, Box 590, S-751 24 Uppsala, Sweden.] PhD (Swedish U. Agri. Sci., 1991) protein engineering. *Protein biology mutagenesis protein_engineering*
E-mail matti@xray.bmc.uu.se Tel. 46(18)174000 Fax 46(18)536971

Nord, Anders Gunnar (1942). Senior scientist. Conservation Institute of National Antiquities. [Conservation Institute of National Antiquities, Box 5405, S-114 84 Stockholm, Sweden.] PhD (Stockholm U., 1974) chemical crystallography. *Mineral_structures powder_diffraction Rietveld_method*
Tel. 46(8)7839339 Fax 46(8)6614277

Norén, Bertil Nils (1931). Lecturer. Inorganic Chemistry 1, Lund University. [Inorganic Chemistry 1, Chemical Center, PO Box 124, S-221 00 Lund, Sweden.] PhD (Lund U., 1970) coordination chemistry. *Coordination organometallic complex compound*
E-mail bertil.noren@inorgk1.lu.se Tel. 46(46)108109 Fax 46(46)146030 Telex 33553

Noréus, Dag (1951). Lecturer. Dept. of Structural Chemistry, Stockholm University. [Dept. of Structural Chemistry, Arrhenius Laboratory, Stockholm University, S-106 91 Stockholm, Sweden.] DSc (Royal Inst. Techn. Stockholm, 1983) neutron physics. *X-ray_neutron_diffraction inelastic_neutron_scattering hydrides archeometallurgy*
E-mail dag@struc.su.se Tel. 46(8)161253 Fax 46(8)152187 Telex 8105199 UNIVERS

Norrestam, Rolf (1937). Professor. Member of Commission on Computing; Secretary of the Swedish National Committee for Crystallography. Dept. of Structural Chemistry, Stockholm University. [Dept. of Structural Chemistry, Arrhenius Laboratory, Stockholm University, S-106 91 Stockholm, Sweden.] PhD (1972) structural chemistry. *Computing structure_determination sol-gel*
E-mail rolf@struc.su.se Tel. 46(8)162025 Fax 46(8)152187

Nygren, Prof. Mats (1938). Professor. Dept. of Inorganic Chemistry, Stockholm University. [Dept. of Inorganic Chemistry, Arrhenius Laboratory, Stockholm University, S-106 91 Stockholm, Sweden.] PhD (Stockholm U., 1972) chemistry. *Solid_state_chemistry preparation sol-gel*
Tel. 46(8)162366 Fax 46(8)152187

Olander, Kerstin (1966). PhD student. Dept. of Inorganic Chemistry, Chalmers University of Technology. [Dept. of Inorganic Chemistry, Chalmers University of Technology, S-412 96 Göteborg, Sweden.] MSc (1991) chemical engineering. *Crystallography crystal_growth*
E-mail kerstin@ctkem.phc.chalmers.se Tel. 46(31)7722871 Fax 46(31)7722846

Olovsson, Dr Gunnar (1953). Inst. of Chemistry, University of Uppsala. [Dept. of Inorganic Chemistry, Box 531, S-751 21 Uppsala, Sweden.] PhD (Uppsala U., 1991) chemistry. *Charge_density hydrogen_bonding radical_salts organic_metals*
Tel. 46(18)183720 Fax 46(18)508542 Telex 76088 TSLISV S

Olovsson, Prof. Ivar (1928). Professor. Chairman of the Swedish National Committee of Crystallography. Inst. of Chemistry, University of Uppsala. [Dept. of Inorganic Chemistry, Box 531, S-751 21 Uppsala, Sweden.] DSc (Uppsala U., 1961) chemistry. *Charge_density hydrogen_bonding hydrates carboxylates reaction_pathways neutron_diffraction*
Tel. 46(18)183721 Fax 46(18)508542 Telex 76088 TSLISV S

Olson, Mrs Solveig (1944). Research assistent. Dept. of Inorganic Chemistry, Chalmers University of Technology and University of Göteborg. [Dept. of Inorganic Chemistry, University of Göteborg, S-412 96 Göteborg, Sweden.] *Coordination_compounds structural_compounds*
E-mail solveig@ctkem.phc.chalmers.se Tel. 46(31)7722881 Fax 46(31)7722846

Oskarsson, Dr Åke (1942). Senior lecturer. Inorganic Chemistry 1, Lund University. [Inorganic Chemistry 1, Chemical Center, PO Box 124, S-221 00 Lund, Sweden.] PhD (Lund U., 1974) chemical crystallography. *Crystallography small_molecules coordination_compounds*
E-mail ake.oskarsson@inorgk1.lu.se Tel. 46(46)108102 Fax 46(46)104439

Österberg, Prof. Ragnar (1932). Professor. Dept. of Chemistry, Swedish University of Agricultural Sciences. [Dept. of Chemistry, Box 7015, S-750 07 Uppsala, Sweden.] PhD (Göteborg U., 1966) structures in solution. *SAXS humic_compounds proteins fractal SANS*
Tel. 46(18)301485 Fax 46(18)673476

Rundqvist, Prof. Stig O. (1929). Professor. Inst. of Chemistry, University of Uppsala. [Dept. of Inorganic Chemistry, Box 531, S-751 21 Uppsala, Sweden.] DSc (Uppsala U., 1963) inorganic chemistry. *Solid_state_chemistry alloys hydrides*
Tel. 46(18)183718 Fax 46(18)508542 Telex 76088 TSLISV S

Sandström, Magnus K. E. (1945). Associate professor. Dept. of Chemistry, Royal Institute of Technology. [Dept. of Chemistry, Royal Institute of Technology, S-100 44 Stockholm, Sweden.] DSc (1978) chemical crystallography. *Coordination EXAFS solution structure mercury_compounds Jahn–Teller_compounds*
Tel. 46(8)7908156 Fax 46(8)212626

Schneider, Gunter (1953). Principal investigator. Dept. of Molecular Biology, Swedish University of Agricultural Sciences. [Dept. of Molecular Biology, Biomedical Center, Box 590, S-751 24 Uppsala, Sweden.] PhD (Saarbrücken, 1983) biochemistry. *Crystallography proteins enzymes mechanism protein_engineering*
E-mail gunter@xray.bmc.uu.se Tel. 46(18)174006 Fax 46(18)536971

Sjöberg, Bo (1941). Professor. Dept. of Medical Biochemistry, University of Göteborg. [Dept. of Medical Biochemistry, Medicinaregatan 9, 413 90 Göteborg, Sweden.] PhD (1974) physical biochemistry. *Biomaterial structure small_angle_scattering computer_modelling*
E-mail Bo.Sjoberg@medkem.gu.se Tel. 46(31)7733458 Fax 46(31)416108

Sjölin, H. Lennart G. (1949). Associate professor. Dept of Inorganic Chemistry, Chalmers University of Technology and University of Göteborg. [Dept. of Inorganic Chemistry, Chalmers University of Technology, S-412 96 Göteborg, Sweden.] PhD (Göteborg U., 1979) chemistry. *Biocrystallography maximum-entropy_method neutron_diffraction*
E-mail lennart@ctkem.phc.chalmers.se Tel. 46(31)7722876 Fax 46(31)167194

Ståhl, Dr Kenny (1953). Senior lecturer. Inorganic Chemistry 2, Lund University. [Inorganic Chemistry 2, PO Box 124, S-221 00 Lund, Sweden.] PhD (Lund U., 1983) chemical crystallography. *X-ray_synchrotron_radiation neutron_diffraction zeolites non-linear_optics*
E-mail kenny.stahl@oorg2.lth.se Tel. 46(46)108117 Fax 46(46)104012

Stålhandske, Claes Ivar (1941). Senior lecturer. Inorganic Chemistry 2, Lund University. [Inorganic Chemistry 2, Chemical Center, PO Box 124, S-221 00 Lund, Sweden.] PhD (Lund U., 1980) inorganic chemistry. *Inorganic_crystal_structures*
E-mail claes-ivar.stalhandske@oorg2.lth.se Tel. 46(46)108234 Fax 46(46)104012

Stomberg, Rolf S. O. (1933). Senior Lecturer; Director of Studies; Deputy Head of the Department. Dept. of Inorganic Chemistry, Chalmers University of Technology and University of Göteborg. [Dept. of Inorganic Chemistry, Chalmers University of Technology and University of Göteborg, S-412 96 Göteborg, Sweden.] PhD (Chalmers U. Techn., 1965) chemical crystallography. *X-ray_structure_determination lignin_model_compounds peroxo_compounds*
E-mail rolf@ctkem.phc.chalmers.se Tel. 46(31)7722874 Fax 46(31)7722846

Strandberg, Prof. Bror (1930). Professor; Deputy Chairman. Dept. of Molecular Biology, University of Uppsala. [Dept. of Molecular Biology, Biomedical Center, Box 590, S-751 24 Uppsala, Sweden.] PhD (1967) protein crystallography. *Structure function inhibition drug_design proteins viruses protein-nucleic_acid_interaction*
E-mail bror@xray.bmc.uu.se Tel. 46(18)513453 Fax 46(18)536971

Strandberg, Dr Rolf A. G. (1938). Senior lecturer. Dept. of Inorganic Chemistry, Umeå University. [Dept. of Inorganic Chemistry, Umeå University, S-901 87 Umeå, Sweden.] PhD (Umeå U., 1974) chemical crystallography. *Inorganic_crystal_structures*
E-mail oorg@se um dc 51 Tel. 46(90)165467 Fax 46(90)136310

Strid, Prof. Karl-Gustav A. (1940). Professor. Biomaterials Group, Inst. for Surgical Science, University of Göteborg. [Inst. for Surgical Science, medicinaregatan 8, S-413 90 Göteborg, Sweden.] DSc (Chalmers U. Techn., 1976) physics. *Biomaterials image_processing instrumentation spectroscopy*
Tel. 46(31)818805 Fax 46(31)163152 Telex 5428097

Ström, Carin (1961). Research scientist. Dept. of Inorganic Chemistry, University of Göteborg. [Dept. of Inorganic Chemistry, University of Göteborg, S-412 96 Göteborg, Sweden.] PhD (1993) chemical crystallography. *Rietveld_method superconductivity synchrotron_radiation high_pressure neutron_scattering structure_property_relationships*
Tel. 46(31)7722857 Fax 46(31)167194

Sundberg, Dr Margareta (1944). Associate professor. Dept. of Inorganic Chemistry, Stockholm University. [Dept. of Inorganic Chemistry, Arrhenius Laboratory, Stockholm University, S-106 91 Stockholm, Sweden.] PhD (Stockholm U., 1981) inorganic chemistry. *High-resolution_electron_microscopy structure_determination order-disorder oxides*
E-mail marsu@inorg.su.se Tel. 46(8)162434 Fax 46(8)152187

Svensson, Christer (1945). Associate professor. Inorganic Chemistry 2, Lund University. [Inorganic Chemistry 2, PO Box 124, S-221 00 Lund, Sweden.] PhD (Lund U., 1978) inorganic chemistry. *Chemistry computing diffractometry dielectrics*
E-mail christer.svensson@oorg2.lth.se Tel. 46(46)104295 Fax 46(46)104012

Svensson, Göran (1955). Research scientist. Dept. of Inorganic Chemistry, Chalmers University of Technology. [Dept. of Inorganic Chemistry, Chalmers University of Technology, S-412 96 Göteborg, Sweden.] DSc (1987) chemical crystallography. *Crystal_growth coordination computing crystallography*
E-mail hsg@ctkem.phc.chalmers.se Tel. 46(31)7722882 Fax 46(31)7722846

Svensson, Gunnar (1960). Research associate. Dept. of Inorganic Chemistry, Stockholm University. [Dept. of Inorganic Chemistry, Arrhenius Laboratory, Stockholm University, S-106 91 Stockholm, Sweden.] PhD (Stockholm U., 1989) inorganic chemistry. *Electron_microscopy inorganic_chemistry chemical_crystallography*
E-mail gunnar@inorg.su.se Tel. 46(8)162365 Fax 46(8)152187

Svensson, L. Anders (1959). Research associate. Dept. of Molecular Biophysics, Lund University. [Dept. of Molecular Biophysics, Chemical Center, PO Box 124, S-221 00 Lund, Sweden.] PhD (1989) protein crystallography. *Proteins crystallography*
E-mail anders.svensson@mbfys.lu.se Tel. 46(46)104566 Fax 46(46)104543

Tellgren, I. G. Roland (1930). Head of the Neutron Diffraction Group. Member of Commission on Powder Diffraction. Inst. of Chemistry, Uppsala University. [Dept. of Inorganic Chemistry, Box 531, S-751 21 Uppsala, Sweden.] PhD (Uppsala U., 1975) chemistry. *Neutron_diffraction metal_hydrides magnetic_structures powder_diffraction*
E-mail roland.tellgren@kemi.uu.se Tel. 46(18)183776 Fax 46(18)508542

Thomas, Prof. John O. ("Josh") (1944). Professor. Inst. of Chemistry, University of Uppsala. [Dept. of Inorganic Chemistry, Box 531, S-751 21 Uppsala, Sweden.] PhD (London U., 1969) chemical crystallography. *Electrochemistry MD_simulation oxides polymers in_situ_diffraction*
E-mail josh.thomas@kemi.uu.se Tel. 46(18)183763 Fax 46(18)555604

Thomasson, Ronnie (1959). Inorganic Chemistry 2, Lund University. [Inorganic Chemistry 2, PO Box 124, S-221 00 Lund, Sweden.] PhD (Lund U., 1991) inorganic chemistry. *Zeolites adsorption mineralogy*
E-mail ronnie.thomasson@oorg2.lth.se Tel. 46(46)108223 Fax 46(46)104012 Telex 33533 LUNIVER S

Törnroos, Karl Wilhelm (1956). Research associate. Dept. of Structural Chemistry, Stockholm University. [Dept. of Structural Chemistry, Arrhenius Laboratory, Stockholm University, S-106 91 Stockholm, Sweden.] PhD (Stockholm U., 1989) structural chemistry. *Vibration_analysis geometry_analysis charge_density silicon_compounds computer_graphics*
E-mail kwt@struc.su.se Tel. 46(8)162363 Fax 46(8)152187 Telex 8105199 UNIVERS

Unge, Torsten (1945). Associate professor. Dept. of Molecular Biology, University of Uppsala. [Dept. of Molecular Biology, Biomedical Center, Box 590, S-751 24 Uppsala, Sweden.] PhD (Uppsala U., 1979) molecular biology. *Viruses AIDS inhibition*
E-mail torsten@xray.bmc.uu.se Tel. 46(18)174985 Fax 46(18)536971

Uppenberg, Jonas (1963). PhD student. Dept. of Molecular Biology, University of Uppsala. [Dept. of Molecular Biology, Biomedical Center, Box 590, S-751 24 Uppsala, Sweden.] Cand (1989) chemistry. *Macromolecules biochemistry lipases*
E-mail jonas@xray.bmc.uu.se Tel. 46(18)174543 Fax 46(18)536971

Valegård, Karin (1958). Dept. of Molecular Biology, University of Uppsala. [Dept. of Molecular Biology, Biomedical Center, Box 590, S-751 24 Uppsala, Sweden.] PhD (Uppsala U., 1990) virus crystallography. *Crystallization computer_graphics virus_crystallography*
E-mail karin@xray.bmc.uu.se Tel. 46(18)174204 Fax 46(18)536971 Telex 76132 BIOMED S4

Vannerberg, Nils-Gösta (1930). Director of research and development. Eka Nobel AB. [Eka Nobel AB, S-445 80 Bohus, Sweden.] PhD (Chalmers U. Techn., 1959) inorganic chemistry. *Inorganic_chemistry catalysts vitreous_state*
Tel. 46(31)587000 Fax 46(31)587400 Telex 2435 ekagbg

Wahlberg, Dr Anders (1945). Lecturer. Inst. of Chemistry, University of Uppsala. [Inst. of Chemistry, Box 531, S-751 21 Uppsala, Sweden.] PhD (Uppsala U., 1985) chemistry. *Classification combinational_theory amphiphilic_salts structure_relationship*
Tel. 46(18)183669 Fax 46(18)508542 Telex 76088 TSLISV S

Wallenberg, L. Reine (1957). Associate professor. Inorganic Chemistry 2, Lund University. [Inorganic Chemistry 2, Chemical Center, Box 124, S-221 00 Lund, Sweden.] DSc (Lund U., 1987) inorganic chemistry. *Electron_microscopy thin_film catalysis*

superlattices
Tel. 46(46)108233 Fax 46(46)104012

Werner, Prof. Per-Erik (1931). Professor. Dept. of Structural Chemistry, Stockholm University. [Dept. of Structural Chemistry, Arrhenius Laboratory, Stockholm University, S-106 91 Stockholm, Sweden.] PhD (Stockholm U., 1971) chemical crystallography. *Powder_diffraction inorganic_crystal_structures computing*
E-mail pew@struc.su.se Tel. 46(8)162393 Fax 46(8)152187

Westdahl, Marianne (1951). PhD student. Dept. of Structural Chemistry, Stockholm University. [Dept. of Structural Chemistry, Arrhenius Laboratory, Stockholm University, S-106 91 Stockholm, Sweden.] Cand (1990) chemistry. *Computing structure_determination powder_diffraction*
E-mail mw@struc.su.se Tel. 46(8)161254 Fax 46(8)152187

Zou, Xiaodong (1964). PhD student. Dept. of Structural Chemistry, Stockholm University. [Dept. of Structural Chemistry, Arrhenius Laboratory, Stockholm University, S-106 91 Stockholm, Sweden.] MSc (1986) crystallography. *Electron_crystallography electron_diffraction high-resolution_electron_microscopy image_processing*
E-mail zou@hybris.sunet.se Tel. 46(8)162380 Fax 46(8)152187

SWITZERLAND

Sub-Editor: W. B. Schweizer

Armbruster, Dr Thomas Michael Ludwig (1950). Lecturer. [Lab. chem. miner. Crystallogr., University of Bern, Freiestr. 3, CH-3012 Bern, Switzerland.] Prof. (U. Bern, 1994) mineralogical crystallography. *Crystallography optics crystal_chemistry zeolites*
E-mail ARMBRUSTER@KRIST.UNIBE.CH Tel. 41(31)6314266 or 6318496 Fax 41(31)6314499 or 6313996

Banner, Dr David William (1946). Senior scientist. Pharmaceutical Research-New Technologies, B65/312, F. Hoffmann-La Roche Ltd, CH-4002 Basel, Switzerland. DPhil (U. Oxford, 1972) protein crystallography. *Biocrystallography macromolecular_structure recognition_molecular computer-assisted_design_drug*
E-mail David.Banner@Roche.Com Tel. 41(61)6887587 Fax 41(61)6887408 Telex 962292 hlr ch

Bärlocher, Dr Christian (1944). Senior research scientist. [Laboratorium für Kristallographie, ETH-Zentrum, CH-8092 Zürich, Switzerland.] PhD (Imperial Coll. London, 1973) physical chemistry. *Powder_diffraction zeolites synchrotron_radiation computing*
E-mail Ch.Baerlocher@kristall.erdw.ethz.ch Tel. 41(1)6323749 Fax 41(1)6321133

Baumann, Dr Jürgen R. (1957). Project Leader. Landis & Gyr Business Support AG, Postbox, CH-6301 Zug, Switzerland. [Landis & Gyr Business Support AG, Felsenrainstr. 17, CH-8052 Zürich, Switzerland.] Dr rer.nat. (U. Konstanz, 1990) solid state physics. *REM metallography solid_phase*
E-mail juergen.r.baumann@lgbs.lg-ch.ch Tel. 41(42)242770 Fax 41(42)244331

Bürgi, Prof. Hans-Beat (1942). Prof. Laboratorium für Chemische und Mineralogische Kristallographie, Universität Bern. [Laboratorium für Chemische und Mineralogische Kristallographie, Universität Bern, Freiestrasse 3, CH-3012 Bern, Switzerland.] Dr (ETH Zürich, 1969) chemistry. *Structure_correlation motion_in_crystal packing_disorder chemical_reactivity_and_structure*
E-mail hbuergi@krist.unibe.ch Tel. 41(31)6314282 Fax 41(31)6313996

Baumgarte, Dr Andreas (1960). Postdoctoral Research Assistant. [Laboratory for Cristallography, ETH-Zentrum, Sonneggstr. 5, CH-8092 Zürich, Switzerland.] Dr rer. nat. (U. Osnabrück, Germany, 1992) inorganic chemistry. *Quasicrystal structure_determination crystal_chemistry*
E-mail Andreas.Baumgarte@kristall.erdw.ethz.ch Tel. 41(1)6326403 Fax 41(1)6321133

Bernardinelli, Dr Gérald (1945). [Laboratoire de Cristallographie, 24, quai Ernest Ansermet, Université de Genève, CH-1211 Genève 4, Switzerland.] DSc (U. Genève, 1978) chemistry/crystallography. *Organic_compound structure_reactivity clathrates computer_graphics*
E-mail bernard@sc2a.unige.ch Tel. 41(22)7026202 Fax 41(22)7812192 Telex ch-342 11 59 siad

Blanc, Dr Eric (1963). Assistant. [Institut de Cristallographie, Université de Lausanne, CH-1015 Lausanne, Switzerland.] DSc (U. Lausanne, 1993) physics. *Computing absorption_correction graph_theory fullerenes*
E-mail eblanc@ulys.unil.ch Tel. 41(21)6922360 Fax 41(21)6922307

Bonin, Dr Michel G. (1956). Post Doc. [Institut de Cristallographie, Université de Lausanne, BSP Dorigny, Rm 517, CH-1015 Lausanne, Switzerland.] PhD (U. Lille I, 1993) physics. *Ferroelectrics incommensurate_structure*
E-mail Michel.Bonin@Ulys.Unil.CH Tel. 41(21)6923779 Fax 41(21)6923605

Brinkmann, Prof. Detlef (1931). Prof. [Physik Institut, U. Zürich, Physik-Institut, Winterthurerstr. 190, CH-8057 Zürich, Switzerland.] Dr (U. Zurich, 1961) exp. physics. *Crystallography solid_state_physics minerals NMR*
E-mail brinkman@physik.unizh.ch Tel. 41(1)2575751 or 41(1)2575721 Fax 41(1)2575704

Burkhard, Dr Andreas (1947). Res. scient. [CIBA-GEIGY, K-127.6, CH-4002 Basel, Switzerland.] Dr phil II (U. Basel, 1977) mineralogy. *Powder_diffraction microscopy minerals*
Tel. 41(61)6964014 Fax 41(61)6963649

Bussien Gaillard, Mr Valérie (1964). PhD Student. [Institut de Cristallographie, Université de Lausanne, BSP Dorigny, Rm 517, CH-1015 Lausanne, Switzerland.] Chimiste diplômée (U. de Lausanne, 1989) chemical physics. *Incommensurate_structures*
E-mail Valerie.BussienGaillard@Ulys.Unil.CH Tel. 41(21)6923781 Fax 41(21)6923605

Cenzual, Dr Karin (1954). Postdoctoral Research Assistant. [Département de Chimie Minérale, Analytique et Appliquée, Université de Genève, 30, quai Ernest-Ansermet, CH-1211 Genève 4, Switzerland.] DSc (U. Geneva, 1988) crystallography. *Crystal_chemistry inorganic_database*
E-mail cenzual@cgeuge52 Tel. 41(22)7026034 Fax 41(22)3296102

Chapuis, Prof. Gervais (1944). Prof. Chairman of the Commission on aperiodic crystals. [Université de Lausanne, Institut de Cristallographie, BSP, 1015 Lausanne, Switzerland.] PhD (ETH Zürich, 1971) crystallography. *Aperiodic_material incommensurate_crystal phase_transitions synchrotron_radiation superspace_symmetry (computer_aided)_crystallographic_teaching*
E-mail gervais.chapuis@ic.unil.ch Tel. 41(21)6923771 Fax 41(21)6923775

Cowan-Jacob, Dr Sandra (1962). Post-Doc. [Dept. of Biotechnology, Pharmaceuticals Div., K-681.5.02, Ciba-Geigy Ltd, CH-4002 Basel, Switzerland.] PhD (Melbourne U., Australia, 1988) biochemistry. *Protein_crystallography*
E-mail JACOBSA1@FMI.CH Tel. 41(61)6965004 Fax 41(61)6964069

Currao, Mr Antonio (1966). Doctorand. [Laboratorium für Anorganische Chemie, ETH-Zentrum, Universitätsstr. 6, CH-8092 Zürich, Switzerland.] Dipl. Chem. organometallic chemistry. *Silicon_compound solid_state_chemistry silicides*
E-mail toni@fest.ac.chem.ethz.ch Tel. 41(1)6322863 Fax 41(1)2620718

Daly, Dr John J. (1931). Scientific expert. [Pharmaceutical Research, F. Hoffmann-LaRoche, CH-4002 Basel, Switzerland.] PhD (U. Leeds, UK, 1959) inorganic and structural chemistry. *Structures_of_molecules small_molecules computing large_molecules*
E-mail john.daly@roche.com Tel. 41(61)6886046 Fax 41(61)6887408

Dobler, Prof. Max (1937). Prof. [Lab. für organische Chemie, ETH Zentrum, CH-8092 Zürich, Switzerland.] Dr sc techn. (ETH Zürich, 1963) chemical crystallography. *Biological_molecules molecular_modelling*
E-mail dobler@czheth5a.bitnet Tel. 41(1)6324509 Fax 41(1)6321072

Dubler, Prof. Erich (1939). Prof. [Anorganisch-Chemisches Institut, Univ. Zürich, Winterthurerstrasse 190, CH-8057 Zürich, Switzerland.] Dr (U. of Zürich, 1970) chemistry. *Bioinorganic_chemistry metal_based_drugs metal_complexes_of_purine_derivatives*
E-mail k55611@czhrzu1a Tel. 41(1)2574621 or 41(1)2574610 Fax 41(1)3638611

Dunitz, Prof. Jack David (1923). Professor em. [Lab. für Organische Chemie, ETH Zentrum, Universitätstr. 16, CH-8092 Zürich, Switzerland.] PhD (Glasgow U., 1947) chemistry. *Phase_transitions thermal_motion crystal_and_molecular_structure polymorphism supramolecular_chemistry*
E-mail jdd@ezrz1.vmsmail.ethz.ch Tel. 41(1)6322892 Fax 41(1)6321109

Egli, Dr Martin (1961). Lecturer. [Organic Chemistry Laboratory, Swiss Federal Institute of Technology, ETH-Zentrum, Universitätstrasse 16, CH-8092 Zürich, Switzerland.] PhD (ETH Zürich, 1988) chemistry. *Macromolecular_structure nucleic_acids molecular_modelling molecular_evolution*
E-mail egli@aeolus.vmsmail.ethz.ch Tel. 41(1)6324506 Fax 41(1)6321072

Engel, Dr Nora (1953). Chargée de recherche. [Dépt. de Minéralogie, Muséum d'Histoire naturelle, Case postale 6434, 1211 Genève 6, Switzerland.] DSc (U. Genève, 1986) crystallography. *Mineral_structures crystal_chemistry classification computing mineralogy gemology museology*
Tel. 41(22)7359130 Fax 41(22)7353445

Estermann, Dr Michael Alexander (1961). Postdoctoral Research Assistant. [Laboratorium für Kristallographie, Eidgenössische Technische Hochschule (ETH) Zürich, ETH Zentrum, CH-8092 Zürich, Switzerland.] Dr sc nat. (ETH Zürich, 1991) physics. *Powder_diffraction direct_method patterson_method Zeolites structure_determination quasicrystal*
E-mail michael.estermann@kristall.erdw.ethz.ch Tel. 41(1)6326404 Fax 41(1)6321133

Fässler, Dr Thomas F. (1959). Research assistant. [Laboratorium für Anorganische Chemie, ETH-Zentrum, Universitätsstr. 6, CH-8092 Zürich, Switzerland.] Dr (U. Heidelberg, 1988) organometallic chemistry. *Organometallic_synthesis solid_state_chemistry crystallography tight_binding_methods electron_localisation surface_reconstruction extended_Hückel_calculations*
E-mail faessler@inorg.chem.ethz.ch Tel. 41(1)6322285 Fax 41(1)6321149

Fehlmann, Dr Mel (1940). Oberassistent. [Laboratorium für Kristallographie, Eidgenössische Technische Hochschule (ETH) Zürich, ETH Zentrum, CH-8092 Zürich, Switzerland.] PhD (Zurich U., 1965) crystallography. *Synchrotron_radiation X-ray_topography electron_microscopy*
E-mail M.Fehlmann@kristall.erdw.ethz.ch Tel. 41(1)6323747 Fax 41(1)6321133

Fischer, Dr Berthold (1959). Postdoctoral Research Assistant. Anorganisch Chemisches Institut der Universität Zürich, Switzerland. [Anorganisch Chemisches Institut der Universität Zürich, Winterthurerstr. 190, CH-8057 Zürich, Switzerland.] Dr rer.nat. (U. Tübingen, 1991) chemistry. *Biocoordination biocrystallography bioinorganic molybdenum_compounds oxidoreductases*
E-mail k556115@czhrzu1a. Tel. 41(1)2574612 Fax 41(1)3638611 Telex 817251unizi

Fischer, Dr Peter (1937). Leader of LNS neutron diffraction group. [Laboratory for Neutron Scattering ETHZ & Paul Scherrer Institut, CH-5232 Villigen PSI, Switzerland.] Dr (ETH Zurich, 1966) solid state physics. *Solid_state_physics neutron_scattering crystallography magnetism*
E-mail PETER.FISCHER@CVAX.PSI.CH Tel. 41(56)992904 Fax 41(56)992939 Telex 827417 psi ch

Flack, Howard D. (1943). Member, Sub-committee of Statistical Descriptors; Consultant, Commission on Computing; Consultant, Committee for the Maintenance of the CIF Standard (COMCIFS); Member, Committee on Electronic Publishing, Dissemination and Storage of Information. [Laboratoire de Cristallographie, Université de Genève, 24 quai Ernest-Ansermet, CH-1211 Genève 4, Switzerland.] PhD (U. London, 1968) crystallography. *Crystallography*
E-mail flack@scsun.unige.ch Tel. 41(22)7026249 Fax 41(22)7812192 Telex ch-421159 siad

Galetti, Prof. Giulio (1937). Prof. [Institut de Minéralogie, U. de Fribourg, Pérolles, CH-1700 Fribourg, Switzerland.] Dr (U. Padova, Italy, 1971) geochemistry. *X-ray_diffraction inorganic_crystal*
Tel. 41(37)826268 Fax 41(37)826584

Gingl, Franz (1959). Assistant. Laboratoire de Cristallographie, Université Genève 14, quai Ernest Ansermet, CH-1211 Genève 4, Switzerland. [Laboratoire de Cristallographie, Université Genève, 14, quai Ernest Ansermet, CH-1211 Genève 4, Switzerland.] Dr rer.nat. (U. Tübingen, 1990) chemistry. *Hydrides X-ray_and_neutron_powder_diffraction solid_state chemistry*
E-mail gingl@sc2a.unige.ch Tel. 41(22)7026372 Fax 41(22)7812192

Giovanoli, Prof. Rudolf (1936). Prof. [Anorg. Chem. Inst., Universität Bern, Freiestr. 3, Postfach 906, CH-3000 Bern 9, Switzerland.] Dr (U. Bern, 1965) chemistry. *Topotaxy structural_texture metallic_oxides finely_divided_solids oxidehydroxides structure-texture_relationships interconversion_reactions*
Tel. 41(31)6314317 Fax 41(31)6314499

Gladyshevskii, Dr Roman (1958). Postdoctoral Research Assistant. [Département Physique de la Matière Condensée, Université de Genève, 24, quai Ernest-Ansermet, 1211 Genève 4, Switzerland.] DSc (Moscow State U., 1987) inorganic chemistry. *Crystal_chemistry alloys superconductors*
E-mail gladych@sc2a.unige.ch Tel. 41(22)7026264 Fax 41(22)7812192 Telex ch-421159 siad

Grimmer, PD Hans (1941). Research scientist. [Lab. for Neutron Scattering, Paul Scherrer Institut, CH-5232 Villigen PSI, Switzerland.] PhD (Edinburgh U., 1969) mathematical physics. *Powder_diffraction layered_structures symmetry tensor_properties*
E-mail hans.grimmer@cvax.psi.ch Tel. 41(56)992421 Fax 41(56)992199 Telex 41(56)827417

Grosse Kunstleve, Ralf W. (1964). PhD student. [Laboratory of Crystallography, ETH-Zentrum, CH-8092 Zürich, Switzerland.] Diploma (Ruhr-U. Bochum, 1992) crystallography. *Powder_diffraction synchrotron_radiation structure_solution zeolites "Al"_techniques*
E-mail ralf@kristall.erdw.ethz.ch or ralf@czheth5a.bitnet Tel. 41(1)6323727 Fax 41(1)6321133

Grütter, Dr Markus Gerhard (1947). Group leader. [Dept. of Biotechnology, Pharmaceuticals Div., K-681.5.45, Ciba-Geigy Ltd, CH-4002 Basel, Switzerland.] PhD (U. Basel, Germany, 1976) biophysics. *Protein_crystallography protein_chemistry protein_engineering*
E-mail GRUETTER@FMI.CH Tel. 41(61)6966328 Fax 41(61)6964069

Gramlich, Dr Volker (1941). Wiss. Adj. [Inst. für Kristallographie und Petrographie, ETH Zürich, ETH-Zentrum, CH-8092 Zürich, Switzerland.] Dr sc. nat. (ETH Zürich 1971) crystallography. *Crystal_structure_analysis*
E-mail u5570@czheth5a Tel. 41(1)6323771 Fax 41(1)2620075

Günter, Prof. John Ralph (1943). Prof. [Institute for Inorganic Chemistry, University of Zürich, Winterthurerstr. 190, CH-8057 Zürich, Switzerland.] PhD (U. Zürich, 1981) inorganic chemistry. *Inorganic_solids crystal_chemistry topotactic_reactions electron_microscopy*
Tel. 41(1)2574646 Fax 41(1)3638611

Haibach, Mr Torsten (1966). PhD Student. [Laboratorium für Kristallographie, ETH-Zentrum, CH-8032 Zürich, Switzerland.] Diploma (Ludwig-Maximilians-U. Munich, 1992) crystallography. *Quasicrystal synchrotron_radiation maximum_entropy*
E-mail torsten@kristall.erdw.ethz.ch Tel. 41(1)6323774 Fax 41(1)6321133

Hauser, Dr Jürg (1954). [Universität Bern, Lab. für chemische und mineralogische Kristallographie, Freiestr. 3, CH-3012 Bern, Switzerland.] DSc (U. Bern, 1985) chemistry. *Computing computer_graphics thermal_motion scientific_visualization*
E-mail hauser@krist.unibe.ch Tel. 41(31)6314390 Fax 41(31)6314499

Hennig, Dr Michael (1963). Assistant. Biozentrum der Universität Basel, Switzerland. [Biozentrum der Universität Basel, Klingelbergstr. 70, CH-4056 Basel, Switzerland.] Dr (Humboldt-U. Berlin, 1992) biochemistry. *Biocrystallography proteins crystallization structure_determination*
E-mail hennig@urz.unibas.ch Tel. 41(61)2672092

Hepp, Dr Alfred (1939). [Chesa Arcada, CH-7482 Bergün, Switzerland.] Dr phil. (U. Zürich, 1981) crystallography. *Silicates_structure graph_theory symmetry*

Hollenstein, Mr Sandro (1965). PhD student. [Lab. für organische Chemie, ETH Zentrum, Universitätsstr. 16, CH-8092 Zürich, Switzerland.] Dipl. Chem. ETH (ETH Zürich, 1989) organic chemistry. *Chemistry crystal_structure geometry*
E-mail sholle@czheth5a Tel. 41(1)6324501 Fax 41(1)2514633

Hulliger, Dr Jürg (1953). Prof. [Inst für Anorganische, analytische und physikalische Chemie, U. Bern, Freiestrasse 3, CH-3012 Bern, Switzerland.] Dr phil. (U. Zürich, 1984) chemistry. *Crystal_growth solid_state_properties_of_inorganic_and_organic_materials non-linear_optics lasers atomic_force_microscopy crystal_defects optical_microscopy*
Tel. 41(31)6314241 Fax 41(31)6313993

Hummel, Dr Wolfgang (1958). Group leader. [Paul Scherrer Institut, Waste Management Lab., CH - 5232 Villigen PSI, Switzerland.] PhD (U. Bern, 1990) crystallography. *Chemical_thermodynamics metalloorganic_complexation computer_modelling toxic_waste structure-property_relations thermodynamic_databases*
E-mail hummel@cageir5a.bitnet Tel. 41(56)992994 Fax 41(56)992821

Jansonius, Prof. Johan N. (1932). Prof. [Biozentrum der Universitaet Basel, Abteilung Strukturbiologie, Klingelbergstr. 70, CH-4056 Basel, Switzerland.] Dr phil. (U. Groningen, The Netherlands, 1967) chemistry. *Protein_crystallography enzyme_mechanisms*
Tel. 41(61)2672080 Fax 41(61)2672109

Kallen, Dr Joerg A. (1957). Team leader. SANDOZ Pharma AG, PKF/DAT/DDG/-PROTEIN-XRAY, 503/1208, CH-4002 Basel, Switzerland. DSc (Biozentrum U. Basel, 1985) biophysics. *Protein_ligands drug_design immunosuppressants linear_dichroism thermodynamics*
E-mail JOERG.KALLEN@PKFDDB.PHARMA.SANDOZ.CH Tel. 41(61)3245579 Fax 41(61)3242686

Ketterer, Dr Jürgen (1949). Research and Development. ILFORD AG, Switzerland. [ILFORD AG, Industriestr. 15, CH-1701 Freiburg, Switzerland.] Dr rer.nat. (U. Freiburg/Br., 1985) crystallography. *X-ray diffractometry Rietveld_method crystallography micro_crystal_engineering silver_halide_technology*
Tel. 41(37)214849 Fax 41(37)215213 Telex 942429

Kostorz, Dr Gernot (1941). Prof. of Physics; Head, Institut für Angewandte Physik. ETH Zürich, CH-8092 Zürich, Switzerland. [Institut für Angewandte Physik, ETH Zürich, CH-8093 Zürich, Switzerland.] Dr rer nat (U. Göttingen, Germany, 1968) physics. *Alloys defect plasticity phase_transformation microstructure order-disorder short-range_order neutron_scattering non-computerized_communication*
E-mail kostorz@iap.ethz.ch Tel. 41(1)6333399 Fax 41(1)6331105 Telex 823474 ehpz ch

Kostrewa, Dr Dirk (1961). Research Scientist. Pharmaceutical Research-New Technologies, B65/312, F. Hoffmann-La Roche Ltd, CH-4002 Basel, Switzerland. Dr rer.nat. (FU. Berlin, 1991) chemistry. *Biocrystallography macromolecular_structure recognition_molecular computer-assisted_design_drug*
E-mail Dirk.Kostrewa@Roche.Com Tel. 41(61)6887750 Fax 41(61)6887408 Telex 962292 hlr ch

Kühnle, Mr Florian N. M. (1968). PhD Student. [Lab. für Organische Chemie, ETH Zürich, Universitätsstrasse16, CH-8092 Zürich, Switzerland.] Dipl. Chem. (ETH Zurich, 1992) chemistry. *Organic_compound guest_host_complexes metal_organic_compound catalysis*
E-mail florian@czheth5a.bitnet Tel. 41(1)6324781 Fax 41(1)2620529

Laube, Dr Thomas (1952). Associate prof. [Lab. für Organische Chemie, ETH Zürich, Universitätsstrasse 16, CH-8092 Zürich, Switzerland.] Dr sc nat. (ETH Zürich, 1984) organic chemistry. *Chemistry mathematics crystallography carbocation_structures*
E-mail CPLUS@CZHETH5A.bitnet Tel. 41(1)6322935 Fax 41(1)2514633

Linden, Dr Anthony (1957). Departmental crystallographer. [Organisch-chemisches Institut, Universität Zürich - Irchel, Winterthurerstrasse 190, CH-8057 Zürich, Switzerland.] PhD (U. Melbourne, 1986) inorganic chemistry. *Computing hydrogen_bonding molecular_conformation service organic_compound organometallic_compound crystal_packing education*
E-mail alinden@oci.unizh.ch Tel. 41(1)2574228 Fax 41(1)3619895

Lohse, Mr Christian (1967). PhD student. [Lab. für organische Chemie, ETH Zentrum, Universitätsstrasse 16, CH-8092 Zürich, Switzerland.] Dipl. Nat. ETH (ETH Zürich, 1992) organic chemistry. *Synthesis crystallization structure_determination*
E-mail clohse@czheth5a Tel. 41(1)6324501 Fax 41(1)2514633

Lubini, Mr Paolo (1965). PhD Student. [Laboratorium für Organische Chemie, ETH Zürich, ETH Zentrum, Universitätsstrasse 16, CH-8092 Zürich, Switzerland.] Dipl. Chem (ETH Zürich, 1990) physical chemistry. *Computing DNA_X-ray_crystallography photography*
E-mail lubini@ezrz1.vmsmail.ethz.ch Tel. 41(1)6322909 Fax 41(1)2514633

Luján, Mr Marcos (1965). Assistant. [Dép. de Chimie Minérale Analytique et Appliqué, Université de Genève, 30 quai Ernest-Ansermet, 1211 Genève, Switzerland.] MSc (U. Genève, 1991) chemistry. *Crystal_structure_determination solid_state_chemistry magnetism optical_crystallography*
E-mail lujan@sc2a.unige.ch Tel. 41(21)7026033

McCusker, Dr Lynne B. (1951). Research associate. Member of the Commission on Powder Diffraction (CPD). [Laboratorium für Kristallographie, ETH-Zentrum, CH-8092 Zurich, Switzerland.] PhD (U. of Hawaii, 1980) chemistry. *Zeolites powder_diffraction synchrotron_radiation*
E-mail Lynne.McCusker@kristall.erdw.ethz.ch Tel. 41(1)6323721 Fax 41(1)6321133

Meier, Prof. Walter M. (1926). Em. Prof. [Inst. für Kristallographie, ETH Zürich, ETH-Zentrum, 8092 Zürich, Switzerland.] DSc (London U., 1983) physical chemistry. *Zeolites applied_chemistry*
Tel. 41(1)6323752 Fax 41(1)2620075

Meyer zu Altenschildesche, Mr Holger (1963). Doctorand. [Laboratorium für Anorganische Chemie, ETH-Zentrum, Universitätsstr. 6, CH-8092 Zürich, Switzerland.] Dipl. Chem. (Ruhr-U. Bochum, 1988) chemistry. *Zeolites solid_state_nuclear_magnetic_resonance*
E-mail meyer@inorg.chem.ethz.ch Tel. 41(1)6325984 Fax 41(1)6321149

Mez, Dr Hans-Christian (1935). Head, science staff of the executive committee. Zentralsekretariat, Ciba-Geigy AG, Basel Switzerland. [ZS 1, K-141.1.11, Ciba-Geigy AG, Postfach, CH-4002 Basel, Switzerland.] Dr Sc (ETH Zürich, 1961) chemistry. *Computing social*
E-mail wmez@chbs.ciba.com Tel. 41(61)6962104 Fax 41(61)6966428

Moor, Dr Robert (1948). Leader of environmental protection. [Swiss Federal Propellant Plant, 3752 Wimmis, Switzerland.] Dr sc nat. (ETH Zürich, 1983) natural science. *Theoretical_crystallography*
E-mail PFMR@A1@ACLAB Tel. 41(33)552296 Fax 41(33)552273

Nesper, Prof. Reinhard (1949). Prof. of inorganic chemistry. [Laboratorium für Anorganische Chemie, ETH Zürich, Universitätsstr. 6, CH-8092 Zürich, Switzerland.] Dr habil. (U. Stuttgart, Germany, 1978) inorganic chemistry. *Solid_state_chemistry electronic_structure_of_solids*
E-mail nesper@fest.ac.chem.ethz.ch Tel. 41(1)6323069 Fax 41(1)6321149

Oberhänsli, Dr Willi. E. (1930). [Talmattstr. 36, CH-4125 Riehen, Switzerland.] PhD (U. of Wisconsin, 1964) physical chemistry. *Direct_methods organic_molecules*
Tel. 41(61)6015033

de Oñate Martinez, Mr Javier Enrique (1962). Research Associate. Laboratorium für Kristallographie, ETH-Zentrum, CH-8092 Zürich, Switzerland. [Laboratorium für Kristallographie, ETH-Zentrum, Sonneggstr. 5, CH-8092 Zürich, Switzerland.] Dipl. Phys. (TU Dresden, 1988) solid state physics. *Powder_diffraction synchrotron_radiation zeolites*
E-mail J.deOnate@kristall.erdw.ethz.ch Tel. 41(1)6323727 Fax 41(1)6321133

Parthé, Prof. Erwin (1928). Professor em. [Département de Chimie Minérale, Analytique et Appliqué, Université de Genève, 30 quai Ernest-Ansermet, CH-1211 Genève 4, Switzerland.] PhD (U. Wien, 1954) physical chemistry. *Inorganic_structural_chemistry*
E-mail parthe@sc2a.unige.ch Tel. 41(22)7026034 Fax 41(22)3296102 Telex ch-421159siad

Petcher, Dr Trevor J. (1943). Coordinator, Preclinical Research. [Sandoz Pharma Ltd, Preclinical Research, Bldg. 386/1025, CH-4002 Basel, Switzerland.] PhD (U. Sheffield, UK, 1967) chemical crystallography. *Drug_design peptide_conformational_analysis molecular_modelling molecular_graphics research_management peptidomimetics*
E-mail petcher@pkfltg.pharma.sandoz.ch Tel. 41(61)3244851 Fax 41(61)3242141

Petter, Prof. Walter (1926). Retired. [Sternenstr. 24, CH-8002 Zürich, Switzerland.] Dr sc nat. (ETH Zürich, 1969) physics. *Crystallography*
Tel. 41(1)2024296

Pfeffer-Hennig, Dr Sabine (1966). Head of Laboratory. SANDOZ Pharma AG, Switzerland. [SANDOZ Pharma AG, Bau 360/1003, CH-4002 Basel, Switzerland.] Dr (Humboldt-U. Berlin, 1992) crystallography. *Polymorphism biocrystallography differential_thermal_analysis X-ray_powder_diffraction Rietveld_method*
E-mail sabine.pfeffer@trdafe.pharma.sandoz.ch Tel. 41(61)3247412 Fax 41(61)3249275

Piontek, Dr Klaus (1952). Research associate. [Laboratory of Biochemistry, Swiss Federal Institute of Technology (ETH), Universitätstr. 16, CH-8092 Zürich, Switzerland.] Dr rer. nat. (U. Freiburg(Germany), 1984) chemistry. *Biological_macromolecular_crystallography structure_function phase_determination glycoproteins hemoproteins*
E-mail u4609@czheth5a.bitnet Tel. 41(1)6323141 Fax 41(1)6321121

Priestle, Dr John Peter (1954). Research scientist; group leader. Core Drug Discovery Technologies, Pharmaceuticals Research, Ciba-Geigy Ltd, CH-4002 Basel, Switzerland. [K-681.5.43, Ciba-Geigy Ltd, CH-4002 Basel, Switzerland.] PhD (U. Texas at Austin, 1982) biochemistry. *Protein_crystallography refinement crystallographic_computing*
E-mail PRIESTLE@FMI.CH Tel. 41(61)6961965 Fax 41(61)6969301

Rahuel, Mr Joseph (1950). Research scientist. Dept. of Biotechnology, Pharmaceuticals Div., Ciba-Geigy Ltd, CH-4002 Basel, Switzerland. [K-681.5.03, Ciba-Geigy Ltd, CH-4002 Basel, Switzerland.] Dipl. U. Tech. (Inst. U. Tech., Rennes, France, 1971) chemistry. *Protein_crystallography data_collection*
E-mail RAHUEL@FMI.CH Tel. 41(61)6965004 Fax 41(61)6964069

Rheinwald, Dr Gerd (1963). Assistant. [Institut de Chimie, Université de Neuchâtel, Avenue de Bellevaux 51, CH-2000 Neuchâtel, Switzerland.] PhD (U. Neuchâtel, 1994) inorganic chemistry. *Synthesis clusters small_molecules ruthenium_compound computer_graphics molecular_mechanics*
E-mail gerd.rheinwald@ich.unine.ch Tel. 41(38)252815 Fax 41(38)214081

Richmond, Prof. Timothy J. (1948). Prof. [Institut für Molekularbiologie und Biophysik, ETH-Zürich, ETH-Hönggerberg, CH-8093 Zürich, Switzerland.] PhD (Yale U., 1975) molecular biophys. and biochem.. *Proteins nucleic_acids structure_determination biophysics chromatin nucleosome*
E-mail richmond@xray.vmsmail.ethz.ch Tel. 41(1)6332470 Fax 41(1)6331073

Rieger, Dr Wolfhart H. (1939). General manager. [TKT Metoxit AG, CH-8420 Thayngen, Switzerland.] PhD (U. Wien, Austria, 1965) physical chemistry. *High_temperature_compound ceramics intermetallic_phases*
Tel. 41(53)391050 Fax 41(53)393965

Rihs, Ms Greti (1943). [Ciba-Geigy AG, K127.626, Postfach, CH-4002, Switzerland.] Dipl. Chem. (ETH Zürich, 1966) chemistry. *Charge_transfer_complex structure_activity_relationships*
Tel. 41(61)6964004 Fax 41(61)6963649

Scheel, Mr Hans J. (1937). Group leader. [Crystal Growth Group, Inst. of Micro- and Optoelectronics, Dept. of Physics, Swiss Federal Institute of Technology (EPFL), Cristallogenese-IMO, Ch. de Bellerive 34, 1007 Lausanne, Switzerland.] *Crystal_growth epitaxy high_Tc_superconductors semiconductors Czochralski_growth flux_growth liquid_phase_epitaxy crystal_growth_technology characterization history_of_crystal_growth_and_crystallography growth_mechanisms hydrodynamics*
E-mail Hans.Scheel@dpqm.epfl.ch Tel. 41(21)6934452 Fax 41(21)6934750

Schicht, Dr Rudolf (1934). Industrie Consultant, Switzerland. [Industrie Consultant, Hohrain, CH-8874 Mühlehorn, Switzerland.] Dr rer.nat. (RWTH Aachen, 1970) mineralogy. *Crystallization chemisorption infrared_emission bonding dehydration gels*
Tel. 41(58)321622 Fax 41(58)3219544

Schirmer, Dr Tilman (1954). Assistant professor. [Biozentrum der Universitaet Basel, Abteilung Strukturbiologie, Klingelbergstr. 70, CH-4056 Basel, Switzerland.] Dr rer nat. (Technical U. Munich, Germany, 1985) structural biology. *Protein_crystallography membrane_proteins*
E-mail schirmer@urz.unibas.ch Tel. 41(61)2672089 Fax 41(61)2672109

Schmalle, Dr Helmut Willi (1941). Research assoc. Anorganisch-Chemisches Institut, Univ. Zürich, Winterthurerstrasse 190, CH-8057 Zürich, Switzerland. Dr (Hamburg U. Germany, 1977) crystallography. *Crystal_structure_determination crystal_chemistry allergenic_compounds*
E-mail schmalle@aci.unizh.ch Tel. 41(1)2574650 Fax 41(1)3638611

Schobinger-Papamantellos, Dr Penelope (1937). Res. assoc. [Laboratorium für Kristallographie, Eidgenössische Technische Hochschule (ETH) Zürich, ETH Zentrum, CH-8092 Zürich, Switzerland.] PhD (U. Wien, Austria, 1962) physical chemistry, crystallography. *Magnetic_structures neutron_diffraction X-ray_diffraction*
E-mail Nelly@kristall.erdw.ethz.ch Tel. 41(1)6323773 Fax 41(1)6321133

Schönholzer, Mr Peter (1937). [F. Hoffmann - La Roche, Dept. PRTP-X, 4002 Basel, Switzerland.] *Structure_determination_of_small_molecules powder_diffraction quantitative_analysis*
E-mail SCHOENHP@ROCBI.DNET.ROCHE.COM Tel. 41(61)6882902 Fax 41(61)6887408

Schwarzenbach, Prof. Dieter (1936). Prof. Co-editor Acta Crystallographica. [Institut de Cristallographie, Université de Lausanne BSP Dorigny, CH-1015 Lausanne, Switzerland.] PhD (ETH Zürich, 1965) crystallography. *Accurate_structure charge_density refinement_method teaching_crystallography*
E-mail dieter.schwarzenbach@ic.unil.ch Tel. 41(21)6923772 Fax 41(21)6923605

Schweizer, Dr W. Bernd (1947). Res. assoc. Member of the Data Commission; national editor of WDC9. [Lab. für Organische Chemie, ETH Zürich, Universitätsstr. 16, CH-8092 Zürich, Switzerland.] Dr sc nat. (ETH Zürich, 1977) chemical crystallography. *Crystallography conformation computing database reaction_pathways*
E-mail schweizer@org.chem.ethz.ch Tel. 41(1)6324507 Fax 41(1)6321072

Schwer, Dr Hansjörg (1960). Postdoctoral Research Assistant. Laboratorium für Festkörperphysik, ETH-Hönggerberg, CH-8093 Zürich, Switzerland. Dr (U. Freiburg/Br., 1990) mineralogy. *X-ray single_crystal diffraction structural_change superconductors*
E-mail schwer@ezinfo.vmsmail.ethz.ch Tel. 41(1)6332256 Fax 41(1)6331072

Seiler, Mr Paul (1945). Res. assoc. [Lab. für Organische Chemie, ETH Zentrum, CH-8092-Zürich, Switzerland.] *Crystallography*
E-mail seiler@org.chem.ethz.ch Tel. 41(1)6324508 Fax 41(1)6321072

Shklover, Dr Valery (1946). Scientific coworker. [Institut für Kristallographie und Petrographie, ETH Zentrum, CH-8092 Zürich, Switzerland.] DSc (Moscow INEOS, 1978) physical chemistry. *Inorganic_structures solid_state_chemistry zeolites thin_film energy_harvesting_and_storage intercalation_compounds polymers nanostructures*
E-mail shklover@ikp.ethz.ch Tel. 41(1)6323775 Fax 41(1)2620075

Smit, Dr Jan Derk Geert (1945). Visiting Research Scientist, F. Hoffmann-LaRoche Ltd. [F. Hoffmann-LaRoche, Pharmaceutical Div., Schachenstr. 74, CH-8906 Bonstetten, Switzerland.] Dr phil (U. Groningen NL, 1973) protein crystallography. *Proteins enzymes haems drug_design glycolysis diseases*
Tel. 41(1)7002328 Fax 41(61)6887408

Steurer, Prof. Walter (1950). Prof. Member of the Commission on Aperiodic Crystals. [Inst. für Kristallographie, ETH-Zentrum, CH-8092 Zürich, Switzerland.] Dr.rer.nat.habil. (Munich U., 1987) crystallography and mineralogy. *Quasicrystal aperiodic_material maximum-entropy_method structural_disorder higher_dimensional_structure_analysis*
E-mail w.steurer@kristall.erdw.ethz.ch Tel. 41(1)6326650 Fax 41(1)6321133

Stoeckli-Evans, Prof. Helen (1944). Prof. assoc. in chemical crystallography. [Institut de Chimie, Université de Neuchâtel, Avenue de Bellevaux 51, CH-2000 Neuchâtel, Switzerland.] PhD (U. of Salford UK, 1969) physical chemistry. *Structure_analysis service_crystallography coordination_polymers molecular_mechanics inorganic_synthesis*
E-mail stoeckli-evans@ICH.UNINE.CH Tel. 41(38)252815 Fax 41(38)214081

Toldo, Dr I. G. Luca (1965). PhD student. [ETH Zürich, Institute of Molecular Biology and Biophysics, HPM F 11 ETH-Hönggerberg, CH-8093 Zürich, Switzerland.] Dr (U. Udine (Italy), 1989) animal production sciences. *Protein_crystallography drug_design SAXS direct_method neural_networks macintosh_programming ROBOT low-gravity crystallization oncogenes immune_system histone_octamer nucleosome proteins low-temperature molecular_graphics DNA-protein_interactions*
E-mail Luca@aeolus.ethz.ch Tel. 41(1)6332460 Fax 41(1)3714873

Vedani, Dr Angelo (1952). Project leader. [Biographics Laboratory, Swiss Institute for Alternatives to Animal Testing (SIAT), Aeschstrasse 14, CH-4107 Ettingen, Switzerland.] PhD (U. Zürich, 1981) inorganic chemistry. *Computer-aided_drug_design software_development structural_databases force_field_development replacement_of_animal_models_in_medical-research space_group_Fm3m mountain_climbing music*
E-mail vedani@czheth5a Tel. 41(61)7218958 Fax 41(61)7218958

Weber, Dr Hans Peter (1936). Preclinical Res. Bldg. 503/560, Sandoz AG, CH-4002 Basel, Switzerland. Dr sc nat. (ETH Zürich, 1964) X-ray analysis. *X-ray_analysis biological_molecules molecular_modelling drug_design molecular_mechanics quantum_chemistry*
Tel. 41(61)3244343 Fax 41(61)3248001

Weber, Prof. Hans-Peter (1941). Project leader; Director, Swiss–Norwegian Beam Line at ESRF. [Institut de Cristallographie, BSP, Université de Lausanne, CH-1015 Lausanne, Switzerland.] PhD (U. Chicago, 1971) geophysics. *Synchrotron_radiation high_pressure modelling*
E-mail hans-peter.weber@ic.unil.ch Tel. 41(21)6923773 Fax 41(21)6923605 Telex 455 110 UNIL

Werk-Albers, Dr Margit L. (1956). [Eschenstr. 8, CH-9000 St. Gallen, Switzerland.] Dr rer.nat. (U. Hamburg, 1984) crystallography. *Inorganic_crystal_structure phase_transition structural_change*
Tel. 41(71)238505

Winkler, Dr Fritz K. (1944). Section head. [Pharmaceutical Research-New Technologies, B65/312, F. Hoffmann-La Roche Ltd, CH-4002 Basel, Switzerland.] Dr sc techn. (ETH Zürich, 1973) chemistry. *Biocrystallography macromolecular_structure recognition_molecular computer-assisted_drug_design*
E-mail Fritz.Winkler@Roche.Com Tel. 41(61)6885187 Fax 41(61)6887408 Telex 962292 hlr ch

Wörle, Mr Michael D. (1962). PhD student. [Laboratorium für Anorganische Chemie, ETH-Zentrum, Universitätsstr. 6, CH-8092 Zürich, Switzerland.] Dipl. Chem. (U. Stuttgart, 1990) chemistry. *Solid_state_chemistry crystallography*
E-mail woerle@inorg.chem.ethz.ch Tel. 41(1)6326743 Fax 41(1)6321149

Yvon, Prof. Klaus (1943). Prof. [Laboratoire de Cristallographie, U. Genève, 24 Quai E. Ansermet, CH-1211 Genève 4, Switzerland.] Dr phil. (U. Wien, 1967) physics. *Crystallography bonding_hydrides alloys hydrogen_storage powder_diffraction*
E-mail yvonk@sc2a.unige.ch Tel. 41(22)7026231 Fax 41(22)7812192

Zehnder, Margareta (1942). Akad. Adjunkt. Delegate. Inst. für anorganische Chemie, Universität Basel, Spitalstrasse 51, CH-4056 Basel, Switzerland. Prof. (1973, Universität Basel) Dr, chemistry. *Metal_complexes crystal_and-molecular_structure refinement*
E-mail zehnder@urz.unibas.ch Tel. 41(61)3225668 Fax 41(61)3225668

SYRIA

Ali, Prof. Shamsuddin (1939). Dean, faculty of sciences. Physics Dept, Albaath University, PO Box 77, Homs, Syria. PhD (Hull U., UK, 1969) solid state physics. *Condensed_matter*
Tel. 963(31)411722 Fax 963(31)431847

TADJIKISTAN

Sub-Editor: D. Patchadjanov

Baratova, Zebo (1950). Head of Laboratory. Institute of Chemistry, Tadjik Academy of Sciences, Aini 299/2, 734063 Dushanbe, Republic of Tadjikistan. Candidate of sciences (Institute of Chemistry, Tadjik Academy of Sciences, 1981) coordination chemistry. *Coordination_chemistry ecology structure_of_inorganic_compounds*
Tel. 7(3772)274420

Faisiev, Dr Abulkhak (1950). Professor and Dean, Geological Faculty. Geological Faculty, Tadjik State University, Rudaki str. 17, 734016 Dushanbe, Republic of Tadjikistan. Dr, Geomineralogical Sciences (Geological Faculty, Tadjik State U., 1981) mineralogy. *Mineralogy crystallography geochemistry*
Tel. 7(3772)223571

Patchadjanov, Prof. Daler (1937). Vice-President, Academy of Sciences; Head of Laboratory, Institute of Chemistry. Sub-Editor, World Directory of Crystallographers 9. Presidium of the Academy of Sciences of the Republic of Tadjikistan, Rudaki str. 33, 734025 Dushanbe, Republic of Tadjikistan. [Institute of Chemistry, Aini 299, 734063 Dushanbe, Republic of Tadjikistan.] DSc, Professor, Academician (Presidium of the Tadjik Academy of Sciences, Institute of Chemistry, 1976) geochem-

istry, analytical chemistry. *Geochemistry analytical_chemistry ecology composition_and_structure_of_materials_(minerals_and_alloys)*
Tel. 7(3772)226991 Fax 7(3772)234917

Salakhutdinov, Dr Mals (1938). Deputy Director, Head of Laboratory. "Umarov" Physical and Technical Institute, Tadjik Academy of Sciences, Aini 299/1, 734063 Dushanbe, Republic of Tadjikistan. Dr, Physico-mathematical Sciences (Umarov Physical-Technical Institute, Tadjik Academy of Sciences, 1993) physics of solids. *Structure_and_phase_transitions_of_liquid_crystals thermodynamic_and_kinetic_properties_of_crystals physics_of_solids*
Tel. 7(3772)251726

Suyarov, Kurbon (1958). Assistant Professor. Chemical Faculty, Tadjik State University, Rudaki str. 17, 734016 Dushanbe, Republic of Tadjikistan. Candidate of chemical sciences (Chemical Faculty, Tadjik State U., 1989) structure of complex compound. *Chemistry_of_complex_compounds chemistry_of_water_solutions*
Tel. 7(3772)227276

TAIWAN

Sub-Editor: Y. Wang

Chang, Hou-Cheng (1946). Senior Scientist. Member. Chemical Systems Research Division, Chung Shun Institute of Science & Technology, Lung-Tan, Taiwan, Republic of China. [PO Box 90008-17-11, 2 Chung Shun Road, Chia An Village, Lung-Tan, Taiwan, Republic of China.] PhD (Weizmann Inst. of Science, Israel, 1981) chemistry. *Rietveld_method structure_determination crystallite powders*
Tel. 886(2)3145384x354039 Fax 886(3)4711057

Chang, Prof. Shih-Lin (1946). Professor. Department of Physics, National Tsing Hua University, Hsinchu, Taiwan, Republic of China. [Department of Physics, National Tsing Hua University, Hsinchu, Taiwan, Republic of China.] PhD (Polytechnic Inst. of Brooklyn, 1975) physics. *Dynamical_diffraction multibeam phase_determination*
E-mail slchang@srrcx1.srrc.gov.tw Tel. 886(35)715131x3216 Fax 886(35)723052

Chen, Prof. Ruey-Hong (1947). Professor. Department of Physics, National Taiwan Normal University, Taipei, Taiwan, Republic of China. [Department of Physics, National Taiwan Normal University, Taipei, Taiwan, Republic of China.] PhD (U. Pittsburgh, 1977) crystallography. *Phase_transition sulfate ferroelastic*
E-mail RHCHEN%PHY01.DNET@NTNUNEWS.SCC.NTNU.EDU.TW Tel. 886(2)9346620x152 or 128 Fax 886(2)9326408

Cheng, Mr Hei-Ying (1964). Post-Doctor. Department of Chemistry, National Taiwan University, Taipei, Taiwan, Republic of China. [Department of Chemistry, National Taiwan University, Taipei, Taiwan, Republic of China.] PhD (National Taiwan U., 1992) chemistry. *Data_collection transition_elements bonding*
Tel. 886(2)3630231x2111 Fax 886(2)3636359

Cheng, Ms Ming-Chu (1954). Technician. Department of Chemistry, National Taiwan University, Taipei, Taiwan, Republic of China. [Department of Chemistry, National Taiwan University, Taipei, Taiwan, Republic of China.] B (Taiwan U., 1978) chemistry. *X-ray crystallography*
E-mail mccheng@chem38.ch.ntu.edu.tw Tel. 886(2)3630231x2325 Fax 886(2)3636359

Chiang, Professor Michael Yen-Nan (1954). Associate Professor. Department of Chemistry, National Sun Yat-Sen University, Kaohsiung, Taiwan, Republic of China. [Department of Chemistry, National Sun Yat-Sen University, Kaohsiung, Taiwan, Republic of China.] PhD (U. of Southern California, 1984) chemistry. *Structure_determination transition_elements bond disorder aggregates bioinorganic*
E-mail bc35a29@student.nsysu.edu.tw Tel. 886(7)5316171x3529 Fax 886(7)5615598

Chiu, Kuan-Cheng Chiu (1954). Associate Professor. Department of Physics, Chung-Yuan Christian University, Chung-Li, Taiwan, Republic of China. [Department of Physics, Chung-Yuan Christian University, Chung-Li, Taiwan 32023, Republic of China.] PhD (U. of Utah, 1986) physics. *Crystal_growth morphology simulation II–VI_compound C60 HgI2*
E-mail phys1@phys720.cycu.etu.tw Tel. 886(3)4563177x3213 Fax 886(3)4563177x3299

Hseu, Prof. Tzong-Hsiung (1941). Professor. Institute of Life Science, National Tsing Hua University, Hsinchu, Taiwan, Republic of China. [Institute of Life Science, National Tsing Hua University, Hsinchu, Taiwan 30043, Republic of China.] PhD (U. of Washington, 1972) chemistry. *Biomolecule enzymes molecular*
E-mail lshth@life.nthu.edu.tw Tel. 886(35)715131x3451 Fax 886(35)715934

Huang, Dr Eugene (1954). Associate Research Fellow. Institute of Earth Sciences, Academia Sinica, Nankang, Taipei, Taiwan, Republic of China. [Institute of Earth Sciences, Academia Sinica, PO Box 1-55, Nankang, Taipei, Taiwan, Republic of China.] PhD (Cornell U., 1987) geology. *X-ray diffraction oxides phase_transition spectroscopy silicates lattice_vibration*
E-mail eugene@earth.sinic.edu.tw Tel. 886(2)7839910x501 Fax 886(2)7839871

Hung, Dr H. H. (1959). Associate Research Scientist. Users Division, Synchrotron Radiation Research Center, Hsinchu, Taiwan, Republic of China. [Synchrotron Radiation Research Center, Hsinchu Science-Based Industrial Park, Hsinchu, Taiwan, Republic of China.] PhD (National Tsing-Hua U., 1992) physics. *Phase_transition binary_alloys surface dynamical_diffraction_theory GIXS*
E-mail hhhung@twnsrrc1 Tel. 886(35)780281x7323 Fax 886(35)781881

Jang, Peter (1951). Assistant Research Scientist. Synchrotron Radiation Research Center, No. 1, R & D Road VI, Hsinchu Science-Based Industrial Park, Hsinchu, Taiwan, Republic of China. [Synchrotron Radiation Research Center, No. 1, R & D Road VI, Hsinchu Science-Based Industrial Park, Hsinchu 30077, Taiwan, Republic of China.] MS (National Cheng Kung U., 1981) physics. *Multiple_scattering molecules phase absorption_spectroscopy superconductor quantitative*
E-mail LYJANG@SRRCX1.SRRC.GOV.TW Tel. 886(35)780281x7119 Fax 886(35)781881

Jean, Dr Yuch-Cheng (1957). Associate Research Scientist. Synchrotron Radiation Research Center, Hsinchu Science-Based Industrial Park, Hsinchu, Taiwan, Republic of China. [Synchrotron Radiation Research Center, No. 1 R&D Road VI, Hsinchu Science-Based Industrial Park, Hsinchu, Taiwan, Republic of China.] PhD (Tamkang U., 1988) chemistry. *Biocrystallography conformational_change biochemistry carcinogenesis DNA Anticancer_compounds anomalous_dispersion phase_determination biomolecule*
E-mail YCJEAN@SRRCX1.SRRC.GOV.TW Tel. 886(35)780281x7321 Fax 886(35)781881

Lee, Dr Chih-Hao (1955). Associate Research Scientist. Synchrotron Radiation Research Center, Hsinchu, Taiwan, Republic of China. [Synchrotron Radiation Research Center, Hsinchu, Taiwan, Republic of China.] PhD (National Tsing Hua University, 1987) nuclear engineering. *Diffractometer semiconductors X-ray surface metals interfacial*
E-mail chlee@srrcx1.srrc.gov.tw Tel. 886(35)780281x7117 Fax 886(35)781881

Lee, Hsin-Yi (1954). Assistant Research Scientist. Synchrotron Radiation Research Center, Hsinchu Science-Based Industrial Park, Hsinchu, Taiwan, Republic of China. [Synchrotron Radiation Research Center, No. 1 R&D Road VI, Hsinchu Science-Based Industrial Park, Hsinchu, Taiwan, Republic of China.] MS (National Sun Yat-Sen University, 1984) material. *SAXS diffraction_technique scattering Rietveld_method anomalous_dispersion phase_transition time_resolved_effect*
E-mail HYLEE@SRRCX1.SRRC.GOV.TW Tel. 886(35)780281x7321 Fax 886(35)781881

Liao, Fen-ling (1963). Department of Chemistry, National Tsing Hua University, Hsinchu, Taiwan 30043, Republic of China. [Department of Chemistry, National Tsing Hua University, Hsinchu, Taiwan 30043, Republic of China.] BS (National Kaohsiung Normal College, 1986) chemistry. *Crystal_structure_analysis small_molecules*
E-mail flliao@chem.nthu.edu.tw Tel. 886(35)716640 Fax 886(35)711082

Liaw, Shwu-Huey (1965). Associate Professor. Institute of Molecular Medicine, School of Medicine, National Taiwan University, Taipei, Taiwan. [Institute of Molecular Medicine, School of Medicine, National Taiwan University, Taipei, Taiwan.] PhD (U. of California, Los Angeles, 1992) biochemistry. *Structure proteins macromolecular regulation_and_reaction_mechanisms_of_enzymes*
E-mail shwu@ccms.ntu.edu.tw Tel. 886(2)3970800x2702 Fax 886(2)3210977

Liaw, Dr Yen-Chywan (1960). Assistant Research Fellow. Institute of Molecular Biology, Academia Sinica, Taipei, Taiwan 11529, Republic of China. [Institute of Molecular Biology, Academia Sinica, Taipei, Taiwan 11529, Republic of China.] PhD (Taiwan U., 1986) chemistry. *Structure proteins macromolecular toxins drugs*
E-mail mbycliaw@twnas886.bitnet Tel. 886(2)7899199 Fax 886(2)7826085

Lii, Kwang-Hwa (1954). Research Fellow. Institute of Chemistry, Academia Sinica, Nankang, Taipei, Taiwan, Republic of China. [Institute of Chemistry, Academia Sinica, Nankang, Taipei, Taiwan, Republic of China.] PhD (Iowa State U., 1985) chemistry. *Solid_state phosphates structural magnetochemistry*
E-mail LII@CHEM.SINICA.EDU.TW Tel. 886(2)7824045 Fax 886(2)7831237

Lin, Dr Hsi-Che (1936). Research Scientist. Materials Research Laboratories, Industrial Technology Research Institute, Bldg.77, 195 Chung-Hsin Road, Sec. 4, Chutung, Hsinchu, Taiwan 31015, R. O. C. [Materials Research Laboratories, Industrial Technology Research Institute, Bldg.77, 195 Chung-Hsin Road, Sec. 4, Chutung, Hsinchu, Taiwan 31015, R. O. C..] PhD (Ohio State U., 1967) mineralogy. *Phase_equilibrium oxides X-ray powder diffraction phase transformation*
E-mail CA013@MRLD.MRL.ITRI.ORG.TW Tel. 886(35)916848/916922 Fax 886(35)820262/820288 Telex 34684 MRL

Lin, Dr Kusu-Jiuh (1964). Postdoctoral Fellow. Institute of Chemistry, Academia Sinica, Taipei, Taiwan, Republic of China. [Institute of Chemistry, Academia Sinica, Taipei, Taiwan 11529, Republic of China.] PhD (National Taiwan U., 1993) chemistry. *Deformation_density metal_carbene hydrogen_bonding*
Tel. 886(2)7821889

Lin, Prof. Szu-Bin (1938). Professor. Department of Geology, National Taiwan University, Taipei, Taiwan, R. O. C. [Department of Geology, National Taiwan University, 245 Choushan Rd, Taipei, Taiwan, Republic of China.] PhD (McMaster U., Canada, 1971) mineralogy and geology. *Diffraction minerals X-ray inorganic_materials*
Tel. 886(2)3630231x2343 Fax 886(2)3636095

Liu, Prof. Ling-Kang (1950). Research Fellow. Institute of Chemistry, Academia Sinica, Nankang, Taipei 11529, Taiwan ROC. [128, Yen-Chiu-Yuan Road, Sec. 2, Nankang, Taipei 11529, Taiwan ROC.] PhD (U.of Texas at Austin, 1978) chemistry. *Structure_determination iron_compounds organometallic chiral nuclear_magnetic_resonance phosphorus_compounds*
E-mail liuu@chem.sinica.edu.tw Tel. 886(2)7899785x242 Fax 886(2)7831237

Lu, Prof. Tian-Huey (1939). Professor. Department of Physics, National Tsing Hua University, Hsinchu, Taiwan, Republic of China. [Department of Physics, National Tsing Hua University, Hsinchu, Taiwan 300, Republic of China.] Doctor of Engineering (Nagoya U., Japan, 1989) applied chemistry. *Structure_determination transition_elements solid biophysics protein crystal*
E-mail THLU@PHYS.NTHU.EDU.TW Tel. 886(35)715131x3218 Fax 886(35)723052

Peng, Prof. Shie-Ming (1949). Professor. Department of Chemistry, National Taiwan University, Taipei, Taiwan, Republic of China. [Department of Chemistry, National Taiwan University, Taipei, Taiwan, Republic of China.] PhD (U. Chicago, 1975) chemistry. *Coordination_chemistry*
Tel. 886(2)3630231x2111 Fax 886(2)3636359

Pong, Prof. Pong (1955). Associate Professor. Department of Physics, Tamkang University, Tamsui, Taiwan, Republic of China. [Department of Physics, Tamkang University, Tamsui, Taiwan, Republic of China.] PhD (U. Notre Dame, 1990) physics. *EXAFS semiconductor structural EELS*
E-mail tkut030@twnmoe10 Tel. 886(2)6252333 Fax 886(2)6252316

Sheu, Mr Hwo-Shuenn (1959). Assistant Research Scientist. Synchrotron Radiation Research Center, Hsinchu, Taiwan, Republic of China. [Synchrotron Radiation Research Center, Hsinchu, Taiwan, Republic of China.] M. S. (U. Tamkang, 1984) chemistry. *Synchrotron_radiation resonant_scattering Rietveld_method non-linear* E-mail HSHEU@SRRCX1.SRRC.GOV.TW Tel. 886(35)780281x7121

Shieh, Dr Ming-huey (1959). Associate Professor. Department of Chemistry, National Taiwan Normal University, Taipei, Taiwan, Republic of China. [Department of Chemistry, National Taiwan Normal University, Taipei, Taiwan, Republic of China.] PhD (Rice U., 1989) chemistry. *Synthesis metal_cluster main_group_transition_metal layer compound* Tel. 886(2)9350749x308 Fax 886(2)9324249

Shiu, Prof. Kom-Bei (1951). Professor. Department of Chemistry, National Cheng Kung University, Tainan, Taiwan, Republic of China. [Department of Chemistry, National Cheng Kung University, Tainan, Taiwan, Republic of China.] PhD (U. Michigan, 1984) chemistry. *Structure_property_correlation transition_elements bioinorganic_compounds inorganic_material* E-mail kbshiu@mail.ncku.edu.tw Tel. 886(6)2757575x65351 Fax 886(6)2740552

Tang, Dr Chia-Pin (1946). Senior Scientist. Department of Chemistry, Chung-Shan Institute of Science and Technology, Lung-Tan, Taiwan, Republic of China. [PO Box 90008-17, Lung-Tan, Taiwan, Republic of China.] PhD (Weizmann Inst. of Science, 1979) chemistry. *Molecular_packing organic_acids comformation hydrogen_bond* Tel. 886(3)4712201x358156

Tsai, Ji-Ching (1965). Assistant Scientist. Member. Chemical Systems Research Division, Chung Shun Institute of Science & Technology, Lung-Tan, Taiwan, Republic of China. [PO Box 90008-17-11, 2 Chung Shun Road, Chia An Village, Lung-Tan, Taiwan, Republic of China.] MS (Ching-Hau U., 1990) chemistry. *Quantitative_analysis powders* Tel. 886(2)3145384x354039 Fax 886(3)4711057

Ueng, Prof. Chuen-Her (1948). Professor. Department of Chemistry, National Taiwan Normal University, Taipei, Taiwan, Republic of China. [Department of Chemistry, National Taiwan Normal University, Taipei, Taiwan, Republic of China.] PhD (National Taiwan University, 1987) chemistry. *Synthesis_molybdenum_compounds_structural synthesis_chromium_tungsten_compounds_structural* Tel. 886(2)9350749-307 Fax 886(2)9324249

Wang, Dr Jinn-Lung (1952). Senior Specialist. Chemical Systems Research Division, Chung-Shan Institute of Science and Technology, Lung-Tan, Taiwan, Republic of China. [PO Box 90008-17, Lung-Tan, Taiwan, Republic of China.] PhD (Weizmann Inst. of Science, 1993) materials and interfaces. *Crystal_growth polar_compounds materials grazing_incidence ice* Tel. 886(3)4712201x358249 Fax 886(3)4711057

Wang, Prof. Ju-Chun (1958). Associate Professor. Department of Chemistry, Soochow University, Taipei, Taiwan, Republic of China. [Department of Chemistry, Soochow University, Taipei, Taiwan 11102, Republic of China.] PhD (Louisiana State U., 1987) chemistry. *Bonding_theory low_valent_transition_elements* E-mail BIT005@TWNSCU10 Tel. 886(2)8819471x6805 Fax 886(2)8837060

Wang, Sue-Lein (1953). Professor. Department of Chemistry, National Tsing Hua University, Hsinchu, Taiwan 30043, Republic of China. [Department of Chemistry, National Tsing Hua University, Hsinchu, Taiwan 30043, Republic of China.] PhD (Iowa State U., 1985) chemistry. *Powder single_crystal X-ray diffraction transition_elements arsenate EXAFS* E-mail slwang@chem.nthu.edu.tw Tel. 886(35)716640 Fax 886(35)711082

Wang, Dr Wen-Ching (1961). Instructor. Department of Life Science, National Tsing Hua University, Hsinchu, Taiwan, Republic of China. [Department of Life Science, National Tsing Hua University, Hsinchu, Taiwan 30043, Republic of China.] PhD (California Inst. of Technology, 1992) chemistry. *Structure_determination proteins molecular immunology* E-mail lswwc@life.nthu.edu.tw Tel. 886(35)715131x3458 Fax 886(35)717237

Wang, Prof. Yu (1943). Professor. Sub-Editor, WDC9. Department of Chemistry, National Taiwan University, Taipei, Taiwan, Republic of China. [Department of Chemistry, National Taiwan University, Taipei, Taiwan, Republic of China.] PhD (U. of Illinois, 1973) chemistry. *Charge_density transition_elements accurate hydrogen_compound* E-mail ac7b0001@twnmoe10 Tel. 886(2)3630231x2325 Fax 886(2)3636359

Wu, Dr Jong-Chang (1957). Associate professor. Foo Yinjunior College of Nursing & Medical Technology, Taliao, Kaoshiung Hsein, Taiwan, R.O.C.. [Foo Yinjunior College of Nursing & Medical Technology, Taliao, Kaoshiung Hsein, Taiwan, R.O.C..] PhD (U. New Orleans, 1990) chemistry. Tel. 886(7)7811151x217

Yeh, Dr Taun-ran (1944). Associate Head. Physics Division, Institute of Nuclear Energy Research, Lungtan, Taiwan, Republic of China. [PO Box 3-4, Lungtan, Taiwan, Republic of China.] PhD (City U. New York, 1979) physics. *Neutron_scattering instrumentation X-ray_microscope* Tel. 886(34711400x7304 Fax 886(3)4711408

Yu, Prof. Shu-Cheng (1942). Professor. Department of Earth Science, National Cheng Kung University, Tainan, Taiwan, Republic of China. [Department of Earth Science, National Cheng Kung University, Tainan, Taiwan, Republic of China.] PhD (Pennsylvania State U., 1976) mineralogy. *High_pressure_minerals_behaviour CVD diamond* Tel. 886(6)2757575x65420 Fax 886(6)2752532

Yuan, Dr Hanna S. (1961). Associate Fellow. Institute of Molecular Biology, Academia Sinica, Taipei, Taiwan, Republic of China. [Institute of Molecular Biology, Academia Sinica, Taipei, Taiwan 11529, Republic of China.] PhD (U. Southern California, 1988) chemistry. *Structure proteins macromolecular protein–DNA* E-mail mbyuan@ccvax.sinica.edu.tw Tel. 886(2)7899197 Fax 886(2)7826085

THAILAND

Sub-Editor: P. Phavanantha

Anantachai, Mrs Suda Yasarawana (1940). Lecturer. Physics Dept, Chiangmai Univ, Huey Keow Road, Chiangmai 50002, Thailand. MSc (London U., 1973) crystallography. *Diffraction_technique damage_in_semiconductors* Tel. 66(53)221934x51 Fax 66(53)222268

Anugul, Mrs Surang (1935). Associate professor. Chemistry Dept., Chulalongkorn U., Phya Thai Road, Bangkok 10330, Thailand. MS (Oregon State U., USA, 1961) inorganic chemistry. *Inorganic_structure* Tel. 66(2)2527019 Fax 66(2)2541309

Busaracome, Mr Suwin (1948). Lecturer. Geology Dept., Khon Kaen U., Khon Kaen 40002, Thailand. MSc (Victoria U. Wellington New Zealand, 1978) geochemistry. *Crystal_growth mineralogy petrology gemology X-ray_crystallography* Tel. 66(43)236199x1320

Chaichit, Dr Narongsak (1947). Lecturer. Physics Dept., Silpakorn U., Nakorn Pathom 73000, Thailand. PhD (Monash U. Australia, 1982) X-ray crystallography. *Superconductors natural_products organometallic_structure* Tel. 66(34)255093 Fax 66(34)255820

Chaikum, Dr Nitirampai Latavalya (1947). Assistant professor. Chemistry Dept., Mahidol U., Rama 6 Road, Bangkok 10400, Thailand. PhD (Flinders U., South Australia, 1976) X-ray crystallography. *Structure_determination organic_and_inorganic_structures* E-mail scnck@mucc.mahidol.ac.th Tel. 66(2)2461358-74x1208 Fax 66(2)2477050

Chaikum, Dr Nopadol (1949). Associate professor. Chemistry Dept., Mahidol U., Rama 6 Road, Bangkok 10400, Thailand. PhD (Otago U. New Zealand, 1976) geochemistry. *Minerals clays* Tel. 66(2)2461360x124 Fax 66(2)2477050

Choosang, Mrs Pilai (1937). Assistant professor. Chemistry Dept., Chulalongkorn U., Phya Thai Road, Bangkok 10330, Thailand. BSc (Chulalongkorn U., 1959) chemistry. *Alloys structure_determination* Tel. 66(2)2527019 Fax 66(2)2541309 Telex 20217UNICHULTH

Haller, Kenneth (1951). Director, Instrumentation Center. The Petroleum and Petrochemical College, Chulalongkorn University, Phyathai Road, Chula 501 12, Bangkok 10330, Thailand. PhD (U. Arizona, 1978) chemistry. *Small_molecule_structure incommensurate_and_modulated_structures instrumentation_and_software* E-mail haller@chulkn.chula.ac.th Tel. 66(2)2184104 Fax 66(2)2154459

Hoonnivathana, Mr Ekachai (1956). Assistant professor. Physics Dept., Kasetsart U., Phaholyothin Road, Bangkok 10900, Thailand. MSc (Chulalongkorn U., 1984) physics. *Physical_metallurgy inorganic_and_organometallic_structures* Tel. 66(2)5795529 Fax 66(2)5790514

Jinawath, Dr Supatra (1945). Associate professor. Materials Sci. Dept., Chulalongkorn U., Phya Thai Road, Bangkok 10330, Thailand. PhD (Leeds U. UK, 1974) mineral science. *Ceramics mineralogy cement_chemistry* Tel. 66(2)2511954 Fax 66(2)2155523 Telex 20217UNICHULTH

Jirajesda, Mr Jate (1955). Scientist. Geological Survey Div., Dept. of Mineral Resources, Rama 6 Road, Bangkok 10400, Thailand. BSc (Khon Kaen U., 1980) chemistry. *Powder_diffraction fluorescence* Tel. 66(2)2821164

Kaenkeo, Miss Watcharaporn (1958). Geology Dept., Khon Kaen U., Khon Kaen 40002, Thailand. BSc (Chiangmai U., 1980) geology. *Minerals crystallography* Tel. 66(43)236199x1320

Kamolchote, Mr Poonsak (1953). Lecturer. Chemistry Dept., Silpakorn U., Nakorn Pathom 73000, Thailand. MSc (Chiangmai U., 1977) physical chemistry. *Polymer_structures structure_determination* Tel. 66(34)255797 Fax 66(34)255820

Keow-kam-nerd, Dr Kanchana (1937). Associate professor. Chemistry Dept., Chiangmai U., Huey Keow Road, Chiangmai 50002, Thailand. PhD (U. de Besancon France, 1970) chemical technology - inorganics. *Ceramics silicate_technology* Tel. 66(53)221934x21 Fax 66(53)222268

Khantaprab, Dr Chaiyudh (1942). Associate professor. Geology Dept, Chulalongkorn U., Bangkok 10330, Thailand. PhD (Imperial Coll. London, 1972) sedimentology (geology). *Sedimentation crystallography* sedimentology environmental_geology
Tel. 66(2)2525931 Fax 66(2)2527847 Telex 20217UNICHULTH

Kritayakirana, Mrs Rungsri (1940). Assistant professor. Physics Dept., Chulalongkorn U., Bangkok 10330, Thailand. MS (Northeastern U., 1969) physics. *Powder_diffraction*
Tel. 66(2)2529987 Fax 66(2)2531150

Nimgirawath, Mrs Kloy (1944). Assistant professor. Physics Dept., Silpakorn U., Nakorn Pathom 73000, Thailand. MSc (U. New South Wales, Australia, 1975) X-ray crystallography. *Organic_structure* inorganic_structure
Tel. 66(34)255093 Fax 66(34)255820

Padmasuta, Mrs Soontari (1939). Scientist. Geological Survey Division, Dept. of Mineral Resources, Rama 6 Road, Bangkok 10400, Thailand. BSc (Chulalongkorn U., 1963) physics. *Minerals identification powder_diffraction fluorescence*
Tel. 66(2)2821164

Pakawatchai, Dr Chaveng (1941). Lecturer. Chemistry Dept., Prince of Songkla U., Haad Yai, Songkla 90112, Thailand. PhD (U. Western Australia, 1984) crystallography. *Organic_inorganic_structure*
E-mail chaveng@ratree.psu.ac.th Tel. 66(74)211030x2660 Fax 66(74)212918

Phaovibul, Dr Orapin (1941). Professor. Chemistry Dept., Mahidol U., Rama 6 Road, Bangkok 10400, Thailand. PhD (The Free U. Berlin, 1971) physics. *Physical_property polymers liquid_crystal* structure–properties_relation
E-mail scopv@mucc.mahidol.ac.th Tel. 66(2)2461360x1609 Fax 66(2)2477050

Phavanantha, Dr Phathana (1942). Associate professor. Consultant, Commission on Crystallographic Teaching; Sub-Editor, WDC9. Physics Dept., Crystallography Lab., Science Faculty, Chulalongkorn U., Phya Thai Road, Bangkok 10330, Thailand. PhD (Imperial Coll. London, 1970) crystallography. *Natural_products structure-activity_relationship gemology crystal_growth* education computer_sciences industry
E-mail fscippv@chulkn.chula.ac.th Tel. 66(2)2529987 Fax 66(2)2531150

Pisutha-Arnond, Dr Visut (1951). Assistant professor. Geology Dept., Chulalongkorn U., Phythai Road, Bangkok 10330, Thailand. PhD (Penn. State U. USA, 1982) inorganic geochemistry. *Ores mineralogy* ore_deposit_research
Tel. 66(2)2525931 Fax 66(2)2527847

Pongsapich, Dr Wasant (1942). Associate professor. Geology Dept., Chulalongkorn U., Phya Thai Road, Bangkok 10330, Thailand. PhD (U. Washington USA, 1974) geology. *Chemical analysis mineralogical_determination*
Tel. 66(2)2527019 Fax 66(2)2527847

Puttajakr, Mrs Taswal (1955). Assistant professor. Physics Dept., King Mongkut's Inst. of Techn. Thonburi, Suksawad Road, Bangkok 10140, Thailand. MSc (Chulalongkorn U., 1982) high energy physics. *X-ray diffraction gemology* crystallization_of_materials
Tel. 66(2)4270039x6203 Fax 66(2)4278050

Ratanasthien, Dr Benjavun (1946). Associate professor. Geological Sci. Dept., Chiangmai U., Huey Keow Road, Chiangmai 50002, Thailand. PhD (Aston U., Birmingham, UK, 1975) geochemistry. *Clays minerals diffraction_technique* energy
Tel. 66(53)221699x129 Fax 66(53)222268

Rukvichai, Mr Surapol (1950). Assistant professor. Applied Physics Dept., King Mongkut's Inst. of Techn., Bangkok 10520, Thailand. MSc (Chulalongkorn U., 1976) physics. *Inorganic_structure alloys*
Tel. 66(2)3266052x395 Fax 66(2)3269981

Sangariyavanich, Mrs Archara (1948). Nuclear physicist. Physics Dept., Office of Atomic Energy for Peace, Vibhavadee Road, Bangkok 10900, Thailand. MSc (Chulalongkorn U., 1973) solid state physics. *Materials_sciences*
Tel. 66(2)5790138x332

Satittada, Miss Gannaga (1955). Associate professor. Physics Dept., King Mongkut's Inst. of Techn. Thonburi, Suksawad Road, Bangkok 10140, Thailand. MSc (Chulalongkorn U., 1977) physics. *Diffraction_technique X-ray_crystallography powder_diffraction*
Tel. 66(2)4270039x6212 Fax 66(2)4278050

Silskulsuk, Mr Buncha (1955). Assistant professor. Physics Dept., Srinakharinwirot U., Sukhumvit 23, Bangkok 10110, Thailand. MSc (Chulalongkorn U., 1984) physics. *Diffraction_technique alloys*
Tel. 66(2)2583989

Siripisarnpipat, Dr Sutatip (1947). Assistant professor. Chemistry Dept., Kasetsart U., Phaholyothin Road, Bangkok 10900, Thailand. PhD (U. of Missouri-Columbia USA, 1981) inorganic chemistry. *Organic_structure* coordination_compounds
Tel. 66(2)5790658 Fax 66(2)5790514

Siripitayananon, Dr Jintana (1954). Lecturer. Chemistry Dept., Chiangmai U., Chiangmai 50200, Thailand. PhD (Australian Nat. U., 1986) crystallography. *Diffraction_technique polymers superconductors diffuse_scattering* molecular_modelling
E-mail jin-sc@chiangmai.ac.th Tel. 66(53)221699x3331 Fax 66(53)222268 Telex 43553 UNICHIM TH

Siriratwatanakul, Mr Narin (1952). Assistant professor. Physics Dept., Chiangmai U., Huey Keow Road, Chiangmai 50002, Thailand. MSc (Chiangmai U., 1981) physics. *Superconductors damage_in_semiconductors*
Tel. 66(53)221934x51 Fax 66(53)222268

Suddhiprakarn, Dr Anohsloe (1948). Assistant professor. Dept. of Soils, Kasetsart U., Bangkok 10903, Thailand. PhD (U. of Western Australia, 1978) mineralogy. *Mineralogy*
Tel. 66(2)5792028 Fax 66(2)5799538

Sukapaddhi, Mr Narong (1941). Section Head. Eng. Techn. Service, Thai Oil Refinery Co. Ltd, Km 124 1/2 Sukhumvit Road, Au Udom, Sriracha, Chonburi 20210, Thailand. MSc (London U., 1970) X-ray crystallography. *Metallography electron_diffraction dislocation* polycrystalline_texture
Tel. 66(38)351555x1500 Fax 66(38)351554

Tanasugarn, Dr Lerson (1955). Lecturer. Biochemistry Dept., Faculty of Science, Chulalongkorn University, Bangkok 10330, Thailand. PhD (Harvard U., 1985) biology. *Biochemistry macromolecules biocrystallography computing* biotechnology industry
E-mail fsciltn@chulkn.chula.ac.th Tel. 66(2)2511953 Fax 66(2)2792419

Thanomkul, Dr Srinuan Chaiwasie (1936). Associate professor. Physics Dept., Chulalongkorn U., Phya Thai Road, Bangkok 10330, Thailand. PhD (U. of Trondheim Norway, 1974) X-ray crystallography. *Diffraction_technique* organic_and_inorganic_structures
Tel. 66(2)2529987 Fax 66(2)2531150

Thinapong, Dr Pongchan Chananont (1948). Assistant professor. Chemistry Dept., Mahidol U., Rama 6 Road, Bangkok 10400, Thailand. PhD (Birmingham U., 1981) chemistry. *Structure–activity_relationship* biologically_active_compounds
Tel. 66(2)2451364x1209 Fax 66(2)2477050

Tontrakoon, Mr Jeerapong (1950). Assistant professor. Physics Dept., Chiangmai U., Huey Keow Road, Chiangmai 50002, Thailand. MSc (Chiangmai U., 1978) physics. *Superconductors damage_in_semiconductors*
Tel. 66(53)221934x51 Fax 66(53)222268

Tooptakong, Dr Uncharee Methong (1954). Lecturer. Chemistry Dept., Silpakorn U., Nakorn Pathom 73000, Thailand. PhD (Australian Nat. U., 1985) X-ray crystallography. *Structure_determination* organic_and_inorganic_structures
Tel. 66(34)255797 Fax 66(34)255820

Treechairusme, Mr Kamchai (1955). Lecturer. Physics Dept., Silpakorn U., Nakorn Pathom 73000, Thailand. MSc (Chulalongkorn U., 1985) physics. *Inorganic_structure superconductors*
Tel. 66(34)255093 Fax 66(34)255820

Tunkasiri, Dr Tawee (1943). Associate professor. Physics Dept., Chiangmai U., Huey Keow Road, Chiangmai 50002, Thailand. PhD (U. Surrey, UK, 1975) physics. *Superconductors damage_in_semiconductors*
Tel. 66(53)221934x51 Fax 66(53)222268

Uttamasil, Dr Lek (1943). Director. Res. Inst. of Metal and Material Sci., Chulalongkorn U., Phya Thai Road, Bangkok 10330, Thailand. PhD (Ohio State U. USA, 1971) ceramic engineering. *Clays minerals high-temperature_materials*
Tel. 66(2)2184106 Fax 66(2)2528960

Wongshaiboon, Dr Sajee (1946). Assistant professor. Physics Dept., Chulalongkorn U., Phya Thai Road, Bangkok 10330, Thailand. PhD (Uppsala U., Sweden, 1981) chemistry. *Physical_property crystal_structure* general crystallography
Tel. 66(2)2529987 Fax 66(2)2531150

TURKEY

Sub-Editor: **D. Ülkü**

Akkurt, Dr Mehmet (1958). Assistant professor. Department of Physics, Faculty of Arts and Sciences, Erciyes University, 38039 Kayseri, Turkey. PhD (Erciyes U., 1987) crystallography. *Structure_determination diffraction*
E-mail akkurt@trerun.bitnet Tel. 90(352)4374938

Armağan, Dr Nizamettin (1942). Professor. Fizik Bölümü, Fen Fakültesi, Ege Üniversitesi, Bornova, 35100 Izmir, Turkey. PhD (University College,, Cardiff UK, 1970) physics. *Structure*
E-mail efefiz01@tream.bitnet Tel. 90(232)3881892

Aydın, Metin (1964). Research Assistant. Department of Engineering Physics, Faculty of Sciences, University of Ankara, 06100 Ankara, Turkey. MSc (Ankara U., 1993) crystallography. *Structure MO_calculation metalloproteins*
Tel. 90(312)2126720 Fax 90(312)2232396

Aydınuraz, Dr Arsın (1941). Professor. Fizik Bölümü, Fen Fakültesi, Ankara Üniversitesi, Tandoğan, 06100 Ankara, Turkey. PhD (U. Ankara, 1969) physics. *Crystallography structure_determination*
Tel. 90(312)2126720x1234 Fax 90(312)2232395

Aytaş, Dr S. Işık (1942). Associate Professor. Marmara Research Institute, Department of Physics, Gebze, Izmit, Turkey. PhD (Surrey U., UK, 1971) physics. *Lattice_distortion*
Tel. 90(262)6412300 Fax 90(262)6412309

Büyükgüngör, Dr Orhan (1954). Professor. Department of Science, Ondokuz Mayıs University, 55139 Kurupelit, Samsun, Turkey. PhD (Hacettepe U., 1983) X-ray crystallography. *Pyridine_complexes*
E-mail orhanb@tromuni.bitnet Tel. 90(362)4576000

Buyum, Muharrem (1967). Research Assistant. Department of Physics, Faculty of Sciences, Ankara University, Beşevler, 06100 Ankara, Turkey. BSc (Ankara U., 1993) solid state physics. *Structure disorder defects phase_transition surface_physics_of_metals transmission_electron_microscopy_(TEM) scanning_electron_microscopy_(SEM) scanning_tunnelling_microscopy_(STM)*
Tel. 90(312)2126720x1266 Fax 90(312)2232395

Coşkun, Gül (1972). Research Assistant. Department of Engineering Physics, Faculty of Sciences, University of Ankara, 06100 Ankara, Turkey. BSc (Ankara U., 1993) crystallography. *Structure MO_calculation metalloproteins*
Tel. 90(312)2126720 Fax 90(312)2232396

Durlu, Dr T. Nuri (1945). Professor. Department of Physics, Faculty of Sciences, Ankara University, Beşevler, 06100 Ankara, Turkey. DPhil (Oxford U., 1974) material science. *Phase_transition metals lattice_distortion*
Tel. 90(312)2126720

Elerman, Dr Yalçın (1951). Professor. Department of Engineering Physics, Faculty of Sciences, University of Ankara, 06100 Ankara, Turkey. PhD (Ankara U., 1978) crystallography. *Structure MO_calculation metalloproteins*
E-mail elerman@vitruvius.arch.metu.edu.tr Tel. 90(312)2126720 Fax 90(312)2232395

Elmalı, Dr Ayhan (1965). Associate Professor. Department of Engineering Physics, Faculty of Sciences, University of Ankara, 06100 Ankara, Turkey. PhD (Ankara U., 1993) crystallography. *Structure MO_calculation metalloproteins*
E-mail elerman@vitruvius.arch.metu.edu.tr Tel. 90(312)2126720 Fax 90(312)2232395

Ercan, Dr Filiz (1961). Research Associate. Department of Engineering Physics, Hacettepe University, 06532 Beytepe, Ankara, Turkey. PhD (Hacettepe U., 1994) crystallography. *Crystallography diffraction structure_determination powder*
E-mail f_ercan@trhun.bitnet Tel. 90(312)2352551 Fax 90(312)2352550

Erdönmez, Dr Ahmet (1950). Professor. Department of Science, Ondokuz Mayıs University, 55139 Kurupelit, Samsun, Turkey. PhD (Hacettepe U., 1980) X-ray crystallography. *Transition_series_complexes*
E-mail ahmete@tromuni.bitnet Tel. 90(362)4576000

Ergin, Dr Ömer (1949). Associate Professor. Balkesir Üniversitesi Necatibey Eğitim Fakültesi, Fizik Eğitimi Anabilim Dalı, 10100 Balıkesir, Turkey. PhD (Atatürk Üniversitesi, 1980) crystallography. *Structure_analysis_of_organic_compound metal_organic_complexes*
Tel. 90(266)2412762 Fax 90(266)2412909

Gözel, Dr Güller (1965). Assistant Professor. Department of Chemistry, Middle East Technical University, 06531 Ankara, Turkey. PhD (Middle East Technical U., 1993) inorganic chemistry. *Structure phosphates sulfides borophosphates bioceramics*
E-mail a09984@trmetu.bitnet Tel. 90(312)2101000x3208 Telex 42761 odtk tr

Gündoğdu, M. Niyazi (1951). Associate Professor. Department of Geological Engineering, Hacettepe University, 06532 Beytepe, Ankara, Turkey. PhD (Hacettepe U., 1982) geology. *X-ray_fluorescence_spectroscopy electron_probe_microanalysis rocks*
Tel. 90(312)2352542 Fax 90(312)2352862

Güneş, Bilal (1966). Research assistant. Department of Science Education, Physics branch, Gazi University, 06300 Beşevler, Ankara, Turkey. MSc (Gazi U., 1993) solid state physics. *Structure_determination*
Tel. 90(312)3970099

Hökelek, Dr Tuncer (1957). Associate Professor. Department of Engineering Physics, Hacettepe University, 06532 Beytepe, Ankara, Turkey. PhD (Hacettepe U., 1986) crystallography. *Crystallography diffraction structure_determination thin_films*
E-mail f_tuncer@trhun.bitnet Tel. 90(312)2352551 Fax 90(312)2352550

Ide, Dr Semra (1964). Research Associate. Department of Engineering Physics, Hacettepe University, 06532 Beytepe, Ankara, Turkey. PhD (Hacettepe U., 1994) crystallography. *Structure_determination powder*
E-mail Fide@eti.cc.hun.edu.tr Tel. 90(312)2352551 Fax 90(312)2352550

Kızılyallı, Dr Meral (1935). Professor. Department of Chemistry, Middle East Technical University, 06531 Ankara, Turkey. [Mebusevleri Ergin sok. 37/4, 06531 Ankara, Turkey.] PhD (London U., 1973) inorganic chemistry. *Structure rare_earth_compounds phosphates sulfides borophosphates bioceramic superconductivity*
E-mail a09984@trmetu.bitnet Tel. 90(312)2101000x3208 Fax 90(312)2101280 Telex 42761 odtk tr

Kabak, Mehmet (1962). Research Assistant. Department of Engineering Physics, Faculty of Sciences, Ankara University, 06100 Ankara, Turkey. MSc (Ankara U., 1992) crystallography. *Structure MO_calculation metalloproteins*
Tel. 90(312)2126720

Kandil, Şebnem (1971). Research Assistant. Department of Science Education, Physics branch, Gazi University, 06500 Beşevler, Ankara, Turkey. MSc (Gazi U., 1994) solid state physics. *Structure_determination*
Tel. 90(312)2502118

Kendi, Dr Engin (1945). Professor. Department of Engineering Physics, Hacettepe University, 06532 Beytepe, Ankara, Turkey. PhD (Hacettepe U., 1974) crystallography. *Structure_determination powder*
E-mail Fkendi@eti.cc.hun.edu.tr Tel. 90(312)2352551 Fax 90(312)2352550

Navruz, Nurten (1967). Research Assistant. Department of Physics, Science Faculty, Ankara University, Beşevler, 06100 Ankara, Turkey. MSc (Ankara U., 1992) solid state physics. *Structure order-disorder defects phase_transition*
Tel. 90(312)2126720

Özbey, Dr Süheyla (1957). Associate Professor. Department of Engineering Physics, Hacettepe University, 06532 Beytepe, Ankara, Turkey. PhD (Hacettepe U., 1988) crystallography. *Crystallography diffraction structure_determination powder*
E-mail Fozbey@eti.cc.hun.edu.tr Tel. 90(312)2352551 Fax 90(312)2352550

Özenbaş, Dr Macit (1951). Professor. Department of Metallurgical Engineering, Middle East Technical University, 06531 Ankara, Turkey. PhD (Middle East Technical U., 1981) material sciences. *Superconductivity thin_film thick_film crystallization ferroelectricity EDX scanning_electron_microscopy*
E-mail macitk@vm.cc.metu.edu.tr Tel. 90(312)2101000x2532 Fax 90(312)2101267

Özkan, Dr Hüsnü (1945). Professor. Department of Physics, Middle East Technical University, 06531 Ankara, Turkey. PhD (Marquette U., 1975) material science. *Phase_transition structure radiation superconductivity*
E-mail hozkan@trmetu.bitnet Tel. 90(312)2101000x3270 Telex 42761 odtk tr

Öztürk, Sema (1967). Research Assistant. Department of Physics, Faculty of Arts and Sciences, Erciyes University, 38039 Kayseri, Turkey. MSc (Erciyes U., 1991) crystallography. *Structure_determination diffraction*
E-mail akkurt@trerun.bitnet Tel. 90(352)4374938x1824

Peder, Murat (1969). Research Assistant. Department of Engineering Physics, Faculty of Sciences, University of Ankara, 06100 Ankara, Turkey. BSc (Ankara U., 1991) crystallography. *Structure MO_calculation metalloproteins*
Tel. 90(312)2126720 Fax 90(312)3367822

Sarı, Musa (1968). Research Assistant. Institute of Science and Technology, Gazi University, 06500 Beşevler, Ankara, Turkey. MSc (Gazi U., 1993) solid states physics. *Structure_determination*
Tel. 90(312)3455793

Soylu, Dr Hüseyin (1933). Professor. Department of Science Education, Gazi University, 06500 Beşevler, Ankara, Turkey. PhD (Hacettepe U., 1973) solid state physics. *Structure_determination organometallics*
Tel. 90(312)2351383

Tarımcı, Dr Çelik (1945). Associate Professor. Department of Engineering Physics, Faculty of Sciences, University of Ankara, 06100 Ankara, Turkey. PhD (U. Pittsburgh, 1975) crystallography–nuclear quadrupole resonance. *Small_molecules structures*
Tel. 90(312)2126720

Taş, Dr Cüneyt A. (1961). Assistant Professor. Department of Metallurgical Engineering, Middle East Technical University, 06531 Ankara, Turkey. PhD (Iowa State U., 1993) ceramic engineering/material science. *Rietveld_method ceramics structure_determination rare_earths crystallization phase_determination aluminosilicates*
E-mail tas@rorqual.cc.metu.edu.tr Tel. 90(312)2101000x5910 Fax 90(312)2101267

Temel, Abidin (1959). Research Assistant. Department of Geological Engineering, Hacettepe University, 06532 Beytepe, Ankara, Turkey. PhD (Hacettepe U., 1992) geological engineering. *X-ray_fluorescence_spectroscopy diffraction_techniques electron_probe_microanalysis rocks*
Tel. 90(312)2352542 Fax 90(312)2352862

Timunçin, Dr Muharrem (1941). Professor. Department of Metallurgical Engineering, Middle East Technical University, 06531 Ankara, Turkey. PhD (U. Missouri-Rolla, 1969) metallurgical engineering. *Synthesis decomposition dielectrics ferroelectricity optoelectronics alumina_barium_components bioceramics ferrites magnets porous_materials transducer thick_film semiconductor*
Tel. 90(312)2101000x2517 Fax 90(312)2101267

Ülkü, Dr Dinçer (1940). Professor. Sub-Editor for 9th Edition of World Directory of Crystallographers. Department of Engineering Physics, Hacettepe University, 06532 Beytepe, Ankara, Turkey. Dr.rer.nat. (U. München, 1965) crystallography. *Diffraction structure_determination small_molecules superconductors*
E-mail Fulku@eti.cc.hun.edu.tr Tel. 90(312)2352551 Fax 90(312)2352550

Usanmaz, Dr Ali (1945). Professor. Department of Chemistry, Middle East Technical University, 06531 Ankara, Turkey. PhD (Polytechnic Institute of Brooklyn, 1974) polymer chemistry (X-ray minor). *Polymerization structure_determination* radiochemistry

E-mail a09984@trmetu.bitnet Tel. 90(312)2101000x3225 Telex 42761 odtk tr

Uztetik Amour, Dr Ayşe (1959). Assistant Professor. Department of Chemistry, Middle East Technical University, 06531 Ankara, Turkey. PhD (Middle East Technical U., 1992) inorganic chemistry. *Structure phosphates sulfides bioceramics*
E-mail a09984@trmetu.bitnet Tel. 90(312)2101000x3208 Telex 42761 odtk tr

Yıldız, Çiğdem (1967). Research Assistant. Department of Physics, Faculty of Sciences, University of Ankara, Beşevler, 06100 Ankara, Turkey. BSc (Ankara U., 1990) solid state physics. *Structure OD defects phase_transitions*
Tel. 90(312)2126720

Yağbasan, Dr Rahmi (1949). Associate Professor. Department of Science Education, Gazi University, 06500 Beşevler, Ankara, Turkey. PhD (Hacettepe U., 1980) solid state physics. *Structure_determination*
Tel. 90(312)4901275

TURKMENISTAN

Sub-Editor: A. N. Karryev

Alecksanyn, Mrs Svetlana N. (1949). Researcher. Physico-Technical Institute of the Academy of Sciences of Turkmenistan, 15 Gogol St., Ashgabat, 744000, Turkmenistan. BSc (Turkmenistan State U., Ashgabat, Turkmenistan, 1973) physics of solid state. *X-ray_diffractometry semiconductor structure_composition high_temperature_superconductor*
Tel. 7(3632)298697

Annaorazov, Dr Murad P. (1950). Assistant Professor. Turkmenistan State University, 31 Saparmurat Turkmenbashi Shayoli, Ashgabat, 744014, Turkmenistan. DSc (M. V. Lomonosov Moscow State U., Moscow, Russia, 1993) physics of magnetic phenomena. *Physical_property magnetic_phase_transition compounds_alloys energy_conversion*
Tel. 7(3632)254214

Berkeliev, Prof. Amandurdy B. (1936). Director of the Physico-Technical Institute of Academy of Sciences. Physico-Technical Institute of the Academy of Sciences of Turkmenistan, 15 Gogol St., Ashgabat, 744000, Turkmenistan. DSc (Institute of Physics of the Academy of Sciences of Azerbaijan, 1987) physics of semiconduc-

tors. *Technology_physical_properties_3-5_compounds_heterostructure semiconductor_solar_cell*
Tel. 7(3632)254285 7(3632)290104 Fax 7(3632)295403

Karryev, Mr Alemi N. (1949). Manager of Laboratory. Sub-Editor, World Directory of Crystallographers 9. Physico-Technical Institute of the Academy of Sciences of Turkmenistan, 15 Gogol St., Ashgabat, 744000, Turkmenistan. PhD (P. N. Lebedev Physical Institute of Russian Academy of Sciences, 1987) physics of semiconductors. *Physical_property amorphous_semiconductor electron_probe_microanalysis scanning_electron_microscopy*
Tel. 7(3632)298697

Meretliev, Mr Shamyrat (1958). Manager of Scientific-Production Centre. Physico-Technical Institute of the Academy of Sciences of Turkmenistan, 15 Gogol St., Ashgabat, 744000, Turkmenistan. PhD (Electrotechnical Institute, St Petersburg, Russia, 1987) physics of semiconductors. *High_temperature_superconductivity energy_conversion*
Tel. 7(3632)291588

UKRAINE

Sub-Editor: B. Y. Kotur

Akselrud, Dr Lev Grygorovych (1948). Leading scientist. Dept. of Inorganic Chemistry, Lviv State University. [Dept. of Inorganic Chemistry, Lviv State University, Kyryla i Mefodiya Str. 6, 290005 Lviv, Ukraine.] PhD (Lviv U., 1980) chemistry. *X-ray_structure_analysis computing oxides superconductivity intermetallic_compounds*
E-mail pchem@chem.franko.lviv.ua Tel. 7(0322)794506

Alaverdova, Dr Olga Georgivna (1938). Associate professor; Deputy Head of Department. Kharkiv State Polytechnical University, Frunze Str.21, 310002 Kharkiv, Ukraine. [apts. 58, Derevyanko Str. 14, 310103 Kharkiv, Ukraine.] PhD (Kharkiv Polytechnical Inst., 1975) technics. *X-ray_structure_analysis crystal_structure fault phase_analysis*
E-mail fedorenko@polyt.kharkov.ua Tel. 7(0572)400831 Fax 7(0572)400801 Telex 311034 VZOR

Arinkin, Dr Oleksandr Victorovych (1948). Lecturer. Dept. of Physics of Metals and Semiconductors, Kharkiv Polytechnical Inst.. [Dept. of Physics of Metals and Semiconductors, Kharkiv Polytechnical Inst., Frunze Str. 21, 310002 Kharkiv, Ukraine.] PhD (Kharkiv Polytechnical Inst., 1978) physics and mathematics. *X-ray_structure_analysis fault diffuse_scattering*
Tel. 7(0572)400861

Bagmut, Dr Oleksandr Grygorovych (1949). Postdoctoral research assistant. Kharkiv Polytechnical Inst.. [Kharkiv Polytechnical Inst., Frunze Str. 21, 310002 Kharkiv, Ukraine.] PhD (Kharkiv Polytechnical Inst., 1979) physics and mathematics. *Electron_microscopy thin_film_properties thin_film_structure*
Tel. 7(0572)400562 Fax 7(0572)400801 Telex 311034 VZOR SU

Bartoshynsky, Dr Sergij Mykolayovych (1953). Senior scientist. Dept. of Mineralogy, Lviv State University. [Dept. of Mineralogy, Lviv State University, Hrushevskogo Str. 4, 290005 Lviv, Ukraine.] PhD (Lviv U., 1987) geology and mineralogy. *Minerals crystallography physics*
Tel. 7(0322)743343

Bartoshynsky, Dr Volodymyr Zbignevych (1957). Junior scientist. Inst. of Geochemistry Mineralogy and Ore Formation, Acad. Sci. Ukraine. [Inst. of Geochemistry Mineralogy

and Ore Formation, Acad. Sci. Ukraine, Palladina Prosp. 34, 252680 Kyiv, Ukraine.] PhD (Inst. of Geochemistry and Physics of Minerals, Acad. Sci. Ukraine, Kyiv, 1990) geology and mineralogy. *Crystallography minerals X-ray_structure_analysis*
E-mail pan@igpm.kiev.ua (Relcom) Tel. 7(044)4440242 Fax 7(044)4441270

Bartoshynsky, Prof. Zbignev Vladyslavovych (1929). Professor. Member of Ukrainian Crystallographic Committee. Dept. of Mineralogy, Lviv State University. [Dept. of Mineralogy, Lviv State University, Hrushevskogo Str. 4, 290005 Lviv, Ukraine.] DSc (Inst. of Mineral Raw Materials, Moscow, 1983) geology and mineralogy. *Crystallography minerals crystal_morphology minerals_physics diamond_mineralogy*
Tel. 7(0322)743343

Belan, Dr Bogdana Dmytrivna (1953). Senior scientist. Dept. of Inorganic Chemistry, Lviv State University. [Dept. of Inorganic Chemistry, Lviv State University, Kyryla i Mefodiya Str. 6, 290005 Lviv, Ukraine.] PhD (Lviv U., 1988) chemistry. *Crystal_chemistry intermetallic_compound_synthesis*
E-mail bodak@chem.franko.lviv.ua Tel. 7(0322)742388

Belotskii, Prof. Dmytro Petrovych (1918). Lab. head. Inst. of Thermoelectricity, Acad. Sci. and Ministry of Education of Ukraine, 274000 Chernivtsi, Ukraine. [Gogola Str. 3, apt. 3, 274000 Chernivtsi, Ukraine.] DSc (Inst. for Problems of Materials Science, Acad. Sci. Ukraine, Kyiv, 1972) chemistry. *Thermoelectric_materials technology crystal_chemistry*
Tel. 7(03722)24415

Belyavina, Dr Nadiya Mykolayivna (1950). Senior scientist. Dept. of Physics, Faculty of Physics, Kyiv University. [Dept. of Physics, Faculty of Physics, Kyiv University, Akad. Glushkova Prosp. 6, 252000 Kyiv, Ukraine.] PhD (Kyiv U., 1983) physics and mathematics. *X-ray_structure_analysis intermetallic_compound_crystal_structure gallium_compounds*
Tel. 7(044)2660513

Bilonizhka, Dr Petro Mykhajlovych (1935). Assistant professor. Dept. of Mineralogy, Lviv State University. [Dept. of Mineralogy, Lviv State University, Hrushevskogo

Str. 4, 290005 Lviv, Ukraine.] PhD (Lviv U., 1972) geology and mineralogy. *Crystal_chemistry geochemistry*
Tel. 7(0322)743343

Bodak, Prof. Oksana Ivanivna (1942). Dept. head. Member of Ukrainian Crystallographic Committee. Dept. of Inorganic Chemistry, Lviv State University. [Dept. of Inorganic Chemistry, Lviv State University, Kyryla i Mefodiya Str. 6, 290005 Lviv, Ukraine.] DSc (Inst. for Problems of Materials Science, Akad. Sci. Ukraine, Kyiv, 1984) chemistry. *Intermetallic_compound_crystal_chemistry*
E-mail bodak@chem.franko.lviv.ua Tel. 7(0322)742388

Bondar, Dr Anatolij Adolfovych (1956). Senior scientist. Inst. for Problems of Materials Science, Acad. Sci. Ukraine. [Inst. for Problems of Materials Science, Acad. Sci. Ukraine, Krzhyzhanivskogo Str. 3, 252180 Kyiv, Ukraine.] PhD (Inst. for Problems of Materials Science, Acad. Sci. Ukraine, Kyiv, 1987) chemistry. *Intermetallic_compound_crystal_structure phase_diagram carbon_compounds*
Tel. 7(044)4443090 Fax 7(044)4440492 Telex 131257 STAN SU

Borisova, Dr Svitlana Serafymivna (1953). Scientist. Dept. of Physics of Metals and Semiconductors, Kharkiv Polytechnical Inst.. [Dept. of Physics of Metals and Semiconductors, Kharkiv Polytechnical Inst., Frunze Str. 21, 310002 Kharkiv, Ukraine.] PhD (Kharkiv Polytechnical Inst., 1988) physics and mathematics. *X-ray_structure_analysis layered_compounds crystal_structure*
Tel. 7(0572)400831

Bugaichuk, Dr Mykola Trokhymovych (1957). Senior scientist. Inst. for Problems of Materials Science, Acad. Sci. Ukraine. [Inst. for Problems of Materials Science, Acad. Sci. Ukraine, Krzhyzhanivskogo Str. 3, 252180 Kyiv, Ukraine.] PhD (Inst. for Problems of Materials Science, Acad. Sci. Ukraine, Kyiv, 1988) physics and mathematics. *X-ray_structure_analysis intermetallic_compound_crystal_structure computing*
Tel. 7(044)4443228 Fax 7(044)4442078 Telex 131257 STAN SU

Bulanova, Dr Marina Vadymivna (1956). Senior scientist. Inst. for Problems of Materials Science, Acad. Sci. Ukraine. [Inst. for Problems of Materials Science, Acad. Sci. Ukraine, Krzhyzhanivskogo Str. 3, 252180 Kyiv, Ukraine.] PhD (Inst. for Problems of Materials Science, Acad. Sci. Ukraine, Kyiv, 1989) chemistry. *Phase_diagram intermetallic_compound_synthesis titanium_compounds rare-earth_compounds*
Tel. 7(044)4443090 Fax 7(044)4442078 Telex 131257 STAN SU

Buyanov, Dr Yurij Ivanovych (1937). Senior scientist. Inst. for Problems of Materials Science, Acad. Sci. Ukraine. [Inst. for Problems of Materials Science, Acad. Sci. Ukraine, Krzhyzhanivskogo Str. 3, 252180 Kyiv, Ukraine.] PhD (Inst. for Problems of Materials Science, Acad. Sci. Ukraine, Kyiv, 1970) chemistry. *Crystal_chemistry intermetallic_compound_synthesis*
Tel. 7(044)4443090 Telex 131257 STAN SU

Chaban, Dr Nadiya Fedorivna (1942). Assistant professor. Dept. of Analytical Chemistry, Lviv State University. [Dept. of Analytical Chemistry, Lviv State University, Kyryla i Mefodiya Str. 6, 290005 Lviv, Ukraine.] PhD (Lviv U., 1973) chemistry. *Intermetallic_compound_crystal_structure boron_compounds*
Tel. 7(0322)725740

Cheremskij, Dr Petro Grygorovych (1942). Leading scientist. Dept. of Physics of Metals and Semiconductors, Kharkiv Polytechnical Inst.. [Dept. of Physics of Metals and Semiconductors, Kharkiv Polytechnical Inst., Frunze Str. 21, 310002 Kharkiv, Ukraine.] PhD (Kharkiv Polytechnical Inst., 1972) technics. *X-ray_structure_analysis crystal_structure fault thin_film_physics*
Tel. 7(0572)400861 Fax 7(0572)400801 Telex 311034 VZOR SU

Chernega, Dr Oleksandr Mykolayovych (1955). Lab. head. Member of Ukrainian Crystallographic Committee. Inst. of Organic Chemistry, Acad. Sci. Ukraine. [Inst. of Organic Chemistry, Acad. Sci. Ukraine, Murmanska Str. 5, 253660 Kyiv, Ukraine.] PhD (Inst. of Element Organic Chemistry, Acad. Sci. USSR, Moscow, 1985) chemistry. *X-ray_structure_analysis low_temperature_phase organic_phosphorus_compounds*
Tel. 7(044)5510646 Fax 7(044)5528308 Telex 131494 BUTON SU MARKOVSKY

Chykhrij, Dr Stepan Ivanovych (1963). Teaching assistant. Dept. of Analytical Chemistry, Lviv State University. [Dept. of Analytical Chemistry, Lviv State University, Kyryla i Mefodiya Str. 6, 290005 Lviv, Ukraine.] PhD (Lviv U., 1990) chemistry. *Intermetallic_compound_crystal_chemistry phosphorus_compounds*
Tel. 7(0322)725740

Dvorina, Dr Ljudmyla Andrijivna (1934). Dept. head. Inst. for Problems of Materials Science, Acad. Sci. Ukraine. [Inst. for Problems of Materials Science, Acad. Sci. Ukraine, Krzhyzhanivskogo Str. 3, 252180 Kyiv, Ukraine.] DSc (Inst. for Problems of Materials Science, Acad. Sci. Ukraine, Kyiv, 1990) technics. *X-ray_structure_analysis alloys intermetallic_compound_synthesis*
Tel. 7(044)4441501 Fax 7(044)4440492

Fedorchuk, Dr Anatolij Oleksandrovych (1958). Scientist. Dept. of Inorganic Chemistry, Lviv State University. [Dept. of Inorganic Chemistry, Lviv State University, Kyryla i Mefodiya Str. 6, 290005 Lviv, Ukraine.] PhD (Lviv U., 1991) chemistry. *Phase_diagram intermetallic_compound_crystal_structure intermetallic_compound_physical_property*
E-mail bodak@chem.franko.lviv.ua Tel. 7(0322)742388

Fedorenko, Prof. Anatolij Ivanovych (1937). Dept. head. Dept. of Physics of Metals and Semiconductors, Kharkiv Polytechnical Inst.. [Dept. of Physics of Metals and Semiconductors, Kharkiv Polytechnical Inst., Frunze Str. 21, 310002 Kharkiv, Ukraine.] DSc (Kharkiv Polytechnical Inst., 1980) physics and mathematics. *Film physics superlattice structure epitaxy_layer_crystallography superconductor_superlattice*
Tel. 7(0572)400096 Fax 7(0572)400801 Telex 311034 VZOR SU

Fedyna, Dr Mykhajlo Fedorovych (1960). Senior scientist. Dept. of Inorganic Chemistry, Lviv State University. [Dept. of Inorganic Chemistry, Lviv State University, Kyryla i Mefodiya Str. 6, 290005 Lviv, Ukraine.] PhD (Lviv U., 1988) chemistry. *X-ray_structure_analysis intermetallic_compound_crystal_structure*
E-mail bodak@chem.franko.lviv.ua Tel. 7(0322)794378

Finkel', Dr Vitalij Oleksandrovych (1934). Lab. head. Kharkiv Inst. of Physics and Technology. [Kharkiv Inst. of Physics and Technology, Akademichna Str. 1, 310108 Kharkiv, Ukraine.] DSc (Kharkiv Polytechnical Inst., 1976) physics and mathematics. *Superconductor crystal_structure magnetism physics*
E-mail kfti%kfti.kharkov.ua@relay.ussr.eu.net Tel. 7(0572)356021 Fax 7(0572)351738 Telex 115175 DEKAN SU

Fomina, Dr Larisa Petrivna (1954). Senior scientist. Dept. of Physics of Metals and Semiconductors, Kharkiv Polytechnical Inst.. [Dept. of Physics of Metals and Semiconductors, Kharkiv Polytechnical Inst., Frunze Str. 21, 310002 Kharkiv, Ukraine.] PhD (Kharkiv Polytechnical Inst., 1984) physics and mathematics. *X-ray_structure_analysis film*
Tel. 7(0572)400861

Gamarnik, Dr Moisei Yankelevych (1936). Leading scientist. Inst. of Geochemistry Mineralogy and Ore Formation, Acad. Sci. Ukraine. [Inst. of Geochemistry Mineralogy and Ore Formation, Acad. Sci. Ukraine, Palladina Prosp. 34, 252680 Kyiv, Ukraine.] DSc (Kharkiv U., 1992) physics and mathematics. *Microcrystal_particles_structure*
E-mail pan@igpm.kiev.ua (Relcom) Tel. 7(044)4440540 Fax 7(044)4441270

Gladkikh, Dr Liliya Ivanivna (1934). Assistant professor. Kharkiv Polytechnical Inst.. [Kharkiv Polytechnical Inst., Frunze Str. 21, 310002 Kharkiv, Ukraine.] PhD (Kharkiv Polytechnical Inst., 1966) technics. *X-ray_diffractometry internal_strain superalloys crystal_structure high_pressure_phase_transition*
Tel. 7(0572)400486 Fax 7(0572)400801 Telex 311034 VZOR SU

Gladyshevskii, Prof. Evgen Ivanovych (1924). Professor. Chairman of Ukrainian Crystallographic Committee. Dept. of Inorganic Chemistry, Lviv State University. [Dept. of Inorganic Chemistry, Lviv State University, Kyryla i Mefodiya Str. 6, 290005 Lviv, Ukraine.] DSc (Moscow U., 1968) chemistry. *Intermetallic_compound_crystal_chemistry X-ray_structure_analysis*
E-mail evgig@chem.franko.lviv.ua Tel. 7(0322)742388

Gorogotska, Dr Ljudmyla Ivanivna (1935). Senior scientist. Inst. of Geochemistry Mineralogy and Ore Formation, Acad. Sci. Ukraine. [Inst. of Geochemistry Mineralogy and Ore Formation, Acad. Sci. Ukraine, Palladina Prosp. 34, 252680 Kyiv, Ukraine.] PhD (Inst. of Geology, Mineralogy and Petrography, Acad. Sci. USSR, Moscow, 1967) geology and mineralogy. *Mineralogy crystal_chemistry X-ray_structure_analysis*
E-mail pan@igpm.kiev.ua (Relcom) Tel. 7(044)4441266 Fax 7(044)4441270

Grigorov, Dr Sergej Nikolajevich (1944). Associate professor. Dept. of Theoretical and Experimental Physics, Kharkiv Polytechnical University.. [Dept. of Theoretical and Experimental Physics, Kharkiv Polytechnical University., Frunze Str. 21, 310002 Kharkiv, Ukraine.] PhD (Kharkiv Polytechnical U., 1971) technics. *Thin-film_growth electron_microscopy grain_boundaries_structure solar_cells*
E-mail grigorovs@polyt.kharkov.ua Tel. 7(0572)400092 Fax 7(0572)400601 Telex 311034 VZOR SU

Grin', Dr Yurij Mykolayovych (1955). Leading scientist. Dept. of Inorganic Chemistry, Lviv State University. [Dept. of Inorganic Chemistry, Lviv State University, Kyryla i Mefodiya Str. 6, 290005 Lviv, Ukraine.] PhD (Lviv U., 1980) chemistry. *Intermetallic_compound_crystal_chemistry gallium_compounds magnetic_susceptibility electrical_resistivity*
E-mail bodak@chem.franko.lviv.ua Tel. 7(0322)742388

Hordijtchuk, Dr Oleg Volodymyrovych (1956). Senior scientist. Inst. for Problems of Materials Science, Acad. Sci. Ukraine. [Inst. for Problems of Materials Science, Acad. Sci. Ukraine, Krzhyzhanivskogo Str. 3, 252180 Kyiv, Ukraine.] PhD (Inst. for Problems of Materials Science, Acad. Sci. Ukraine, Kyiv, 1988) chemistry. *Phase_diagram synthesis rare-earth_compounds carbon_compounds transition_elements*
Tel. 7(044)4443090 Fax 7(044)4442078 Telex 131257 STAN SU

Kalinichenko, Dr Anatolij Mykhajlovych (1941). Senior scientist. Inst. of Geochemistry Mineralogy and Ore Formation, Acad. Sci. Ukraine. [Inst. of Geochemistry Mineralogy and Ore Formation, Acad. Sci. Ukraine, Palladina Prosp. 34, 252680 Kyiv, Ukraine.] PhD (Inst. of Geochemistry and Physics of Minerals, Acad. Sci. Ukraine, Kyiv, 1975) geology and mineralogy. *Nuclear_magnetic_resonance EPR solid_solution crystal_defect*
E-mail pan@igpm.kiev.ua (Relcom) Tel. 7(044)4443160 Fax 7(044)4441270

Kalychak, Dr Yaroslav Mykhajlovych (1947). Assistant professor., Dept. of Inorganic Chemistry, Lviv State University. [Dept. of Inorganic Chemistry, Lviv State University, Kyryla i Mefodiya Str. 6, 290005 Lviv, Ukraine.] PhD (Lviv U., 1977) chemistry. *X-ray_structure_analysis intermetallic_compound_synthesis indium_compounds*
E-mail bodak@chem.franko.lviv.ua Tel. 7(0322)742388

Karpec, Dr Myroslav Vasyliovych (1958). Senior scientist. Inst. for Problems of Materials Science, Acad. Sci. Ukraine. [Inst. for Problems of Materials Science, Acad. Sci. Ukraine, Krzhyzhanivskogo Str. 3, 252180 Kyiv, Ukraine.] DSc (Inst. for Problems of Materials Science, Acad. Sci. Ukraine, Kyiv, 1986) physics and mathematics. *X-ray_structure_analysis pseudosymmetry computing*
Tel. 7(044)4443228 Fax 7(044)4442078 Telex 131257 STAN SU

Khaenko, Prof. Borys Volodymyrovych (1935). Lab. head. Member of Ukrainian Crystallographic Committee. Inst. for Problems of Materials Science, Acad. Sci. Ukraine. [Inst. for Problems of Materials Science, Acad. Sci. Ukraine, Krzhyzhanivskogo Str. 3, 252180 Kyiv, Ukraine.] DSc (Central Scientific Inst. of Black Metallurgy, Moscow, 1981) physics and mathematics. *X-ray_structure_analysis alloys ordering*
Tel. 7(044)4443228 Fax 7(044)4442078 Telex 131257 STAN SU

Kharchenko, Dr Olga Ivanivna (1946). Senior scientist. Dept. of Inorganic Chemistry, Lviv State University. [Dept. of Inorganic Chemistry, Lviv State University, Kyryla i Mefodiya Str. 6, 290005 Lviv, Ukraine.] PhD (Lviv U., 1977) chemistry. *X-ray_structure_analysis phase_diagram intermetallic_compound_synthesis*
E-mail bodak@chem.franko.lviv.ua Tel. 7(0322)794378

Kinzhibalo, Dr Volodymyr Vasyliovych (1947). Assistant professor. Dept. of Inorganic Chemistry, Lviv State University. [Dept. of Inorganic Chemistry, Lviv State University, Kyryla i Mefodiya Str. 6, 290005 Lviv, Ukraine.] PhD (Lviv U., 1978) chemistry. *Phase_diagram intermetallic_compound_crystal_structure*
E-mail bodak@chem.franko.lviv.ua Tel. 7(0322)742388

Klymiv, Dr Ivan Mykhajlovych (1940). Lab. head. Inst. of Physical Optics. [Inst. of Physical Optics, Dragomanova Str. 23, 290005 Lviv, Ukraine.] PhD (Lviv U., 1985) physics and mathematics. *Crystal_physics crystal_optics non-linear_optics optoelectronics*
Tel. 7(0322)729567 Fax 7(0322)742089

Kolupajeva, Dr Zoya Ivanivna (1949). Senior scientist. Dept. of Physics of Metals and Semiconductors, Kharkiv Polytechnical Inst.. [Dept. of Physics of Metals and Semiconductors, Kharkiv Polytechnical Inst., Frunze Str. 21, 310002 Kharkiv, Ukraine.] PhD (Kharkiv Polytechnical Inst., 1982) physics and mathematics. *X-ray_structure_analysis*
Tel. 7(0572)400831

Konyk, Dr Maria Bogdanivna (1957). Scientist. Dept. of Inorganic Chemistry, Lviv State University. [Dept. of Inorganic Chemistry, Lviv State University, Kyryla i Mefodiya Str. 6, 290005 Lviv, Ukraine.] PhD (Lviv U., 1989) chemistry. *X-ray_structure_analysis metallography intermetallic_compound_synthesis*
E-mail evgig@chem.franko.lviv.ua Tel. 7(0322)794378

Kosevich, Prof. Vadim Markovych (1931). Dept. head. Dept. of Theoretical and Experimental Physics, Kharkiv Polytechnical Inst.. [Dept. of Theoretical and Experimental Physics, Kharkiv Polytechnical Inst., Frunze Str. 21, 310002 Kharkiv, Ukraine.] DSc (Physics Technical Inst. of Low Temperatures, Acad. Sci. Ukraine, Kharkiv, 1970) physics and mathematics. *Film fault physics electron_microscopy*
Tel. 7(0572)400564 Fax 7(0572)400801 Telex 311034 VZOR SU

Kosmachev, Dr Sergej Mykhajlovych (1946). Associate professor. Dept. of Theoretical and Experimental Physics, Kharkiv Polytechnical University. [Dept. of Theoretical and Experimental Physics, Kharkiv Polytechnical University, Frunze Str. 21, 310002 Kharkiv, Ukraine.] PhD (Kharkiv Polytechnical U., 1974) technics. *Thin-film_growth electron_microscopy grain_boundaries_structure solar_cells*
E-mail kosmachev@polyt.kharkov.ua Tel. 7(0572)400092 Fax 7(0572)400601 Telex 311034 VZOR SU

Kotur, Dr Bogdan Yaroslavovych (1952). Assistant professor. Secretary of Ukrainian Crystallographic Committee; Sub-Editor of ninth edition of World Directory of Crystallographers. Dept. of Inorganic Chemistry, Lviv State University. [Dept. of Inorganic Chemistry, Lviv State University, Kyryla i Mefodiya Str. 6, 290005 Lviv, Ukraine.] PhD (Lviv U., 1978) chemistry. *X-ray_structure_analysis intermetallic_compound_crystal_chemistry intermetallic_compound_synthesis phase_diagram rare-earth_compounds*
E-mail kotur@chem.franko.lviv.ua Tel. 7(0322)742388

Koz'ma, Dr Oleksandr Oleksiyovych (1939). Assistant professor. Dept. of Physics of Metals and Semiconductors, Kharkiv Polytechnical Inst.. [Dept. of Physics of Metals and Semiconductors, Kharkiv Polytechnical Inst., Frunze Str. 21, 310002 Kharkiv, Ukraine.] PhD (Kharkiv Polytechnical Inst., 1969) technics. *X-ray_structure_analysis film radiation fault*
Tel. 7(0572)400861

Krochuk, Dr Vasyl' Maksymovych (1954). Postdoctoral research assistant. Inst. of Geochemistry Mineralogy and Ore Formation, Acad. Sci. Ukraine. [Inst. of Geochemistry Mineralogy and Ore Formation, Acad. Sci. Ukraine, Palladina Prosp. 34, 252680 Kyiv, Ukraine.] PhD (Inst. of Geochemistry and Physics of Minerals, Acad. Sci. Ukraine, Kyiv, 1982) geology and mineralogy. *Crystallography minerals crystal_morphology crystal_growth twinning*
E-mail pan@igpm.kiev.ua (Relcom) Tel. 7(044)4440570 Fax 7(044)4441270

Kubili, Dr Vasyl' Zenonovych (1954). Senior scientist. Inst. for Problems of Materials Science, Acad. Sci. Ukraine. [Inst. for Problems of Materials Science, Acad. Sci. Ukraine, Krzhyzhanivskogo Str. 3, 252180 Kyiv, Ukraine.] PhD (Inst. for Problems of Materials Science, Acad. Sci. Ukraine, Kyiv, 1986) chemistry. *Phase_diagram intermetallic_compound_crystal_structure carbon_compounds*
Tel. 7(044)4443090 Fax 7(044)4440492 Telex 131257 STAN SU

Kukol', Dr Victor Valerijevych (1956). Scientist. Inst. for Problems of Materials Science, Acad. Sci. Ukraine. [Inst. for Problems of Materials Science, Acad. Sci. Ukraine, Krzhyzhanivskogo Str. 3, 252180 Kyiv, Ukraine.] PhD (Inst. for Problems of Materials Science, Acad. Sci. Ukraine, Kyiv, 1991) physics and mathematics. *X-ray_structure_analysis intermetallic_compound_crystal_structure pseudosymmetry computing*
Tel. 7(044)4443228 Fax 7(044)4442078 Telex 131257 STAN SU

Kurdyumov, Prof. Oleksandr Vyacheslavovych (1938). Dept. head. Member of Ukrainian Crystallographic Committee. Inst. for Problems of Materials Science, Acad. Sci. Ukraine. [Inst. for Problems of Materials Science, Acad. Sci. Ukraine, Krzhyzhanivskogo Str. 3, 252180 Kyiv, Ukraine.] DSc (Inst. for Problems of Materials Science, Acad. Sci. Ukraine, Kyiv, 1977) physics and mathematics. *X-ray_structure_analysis electron_microscopy high_pressure_phase*
Tel. 7(044)4443401 Fax 7(044)4442078

Kuz'ma, Prof. Yurij Bogdanovych (1934). Dept. head. Dept. of Analytical Chemistry, Lviv State University. [Dept. of Analytical Chemistry, Lviv State University, Kyryla i Mefodiya Str. 6, 290005 Lviv, Ukraine.] DSc (Lviv U., 1974) chemistry. *X-ray_structure_analysis phase_diagram intermetallic_compound_crystal_chemistry boron_compounds phosphorus_compounds*
Tel. 7(0322)725740

Kuznetsov, Dr Gennadij Vasyliovych (1937). Senior scientist. Inst. of Geochemistry Mineralogy and Ore Formation, Acad. Sci. Ukraine. [Inst. of Geochemistry Mineralogy and Ore Formation, Acad. Sci. Ukraine, Palladina Prosp. 34, 252680 Kyiv, Ukraine.] PhD (Inst. of Geochemistry and Physics of Minerals, Acad. Sci. Ukraine, Kyiv, 1976) geology and mineralogy. *Crystal_chemistry luminescence minerals*
E-mail pan@igpm.kiev.ua (Relcom) Tel. 7(044)4440305 Fax 7(044)4441270

Kvasnitsa, Dr Victor Mykolayovych (1942). Lab. head. Inst. of Geochemistry Mineralogy and Ore Formation, Acad. Sci. Ukraine. [Inst. of Geochemistry Mineralogy and Ore Formation, Acad. Sci. Ukraine, Palladina Prosp. 34, 252680 Kyiv, Ukraine.] DSc (Lviv U., 1993) geology and mineralogy. *Crystallography minerals diamond graphites*
E-mail pan@igpm.kiev.ua (Relcom) Tel. 7(044)4440242 Fax 7(044)4441270

Lebedeva, Dr Marina Volodymyrivna (1941). Assistant professor. Dept. of General and Experimental Physics, Kharkiv Polytechnical Inst.. [Dept. of General and Experimental Physics, Kharkiv Polytechnical Inst., Frunze Str. 21, 310002 Kharkiv, Ukraine.] PhD (Kharkiv Polytechnical Inst., 1973) technics. *Physics structure condensed_layer*
Tel. 7(0572)400341 Fax 7(0572)400801 Telex 311034 VZOR SU

Legkova, Dr Galyna Victorivna (1949). Senior scientist. Inst. of Geochemistry Mineralogy and Ore Formation, Acad. Sci. Ukraine. [Inst. of Geochemistry Mineralogy and Ore Formation, Acad. Sci. Ukraine, Palladina Prosp. 34, 252680 Kyiv, Ukraine.] PhD (Inst. of Geochemistry and Physics of Minerals, Acad. Sci. Ukraine, Kyiv, 1985) geology and mineralogy. *Crystallography minerals electron_probe_microanalysis*
E-mail pan@igpm.kiev.ua (Relcom) Tel. 7(044)4440242 Fax 7(044)4441270

Lisnyak, Dr Semen Stepanovych (1929). Assistant professor. Inst. of Oil and Gas, Ivano-Frankivsk. [Inst. of Oil and Gas, Karpatska Str. 15, 284000 Ivano-Frankivsk, Ukraine.] PhD (Ural Polytechnical Inst., 1956) technics. *Crystal_chemistry spinel*
Tel. 7(03422)99394

Listovnichii, Dr Victor Evgenovych (1934). Leading scientist. Inst. for Problems of Materials Science, Acad. Sci. Ukraine. [Inst. for Problems of Materials Science, Acad. Sci. Ukraine, Krzhyzhanivskogo Str. 3, 252180 Kyiv, Ukraine.] DSc (Inst. for Problems of Materials Science, Acad. Sci. Ukraine, Kyiv, 1992) chemistry. *Phase_diagram chemical_bonding intermetallic_compound_synthesis*
Tel. 7(044)4443090 Fax 7(044)4442078 Telex 131257 STAN SU

Litovchenko, Dr Anatolij Stepanovych (1941). Dept. head. Inst. of Geochemistry Mineralogy and Ore Formation, Acad. Sci. Ukraine. [Inst. of Geochemistry Mineralogy and Ore Formation, Acad. Sci. Ukraine, Palladina Prosp. 34, 252680 Kyiv, Ukraine.] DSc (Inst. for Problems of Materials Science, Acad. Sci. Ukraine, Kyiv, 1989) physics and mathematics. *Crystal_magnetism crystal_electrical_conductivity crystal_lattice_distortion fault*
E-mail pan@igpm.kiev.ua (Relcom) Tel. 7(044)4443160 Fax 7(044)4441270

Litvin, Dr Oleksandr Lukych (1927). Leading scientist. Inst. of Geochemistry Mineralogy and Ore Formation, Acad. Sci. Ukraine. [Inst. of Geochemistry Mineralogy and Ore Formation, Acad. Sci. Ukraine, Palladina Prosp. 34, 252680 Kyiv, Ukraine.] DSc (Inst. of Geochemistry and Physics of Minerals, Acad. Sci. Ukraine, Kyiv, 1977) geology and mineralogy. *Crystallography minerals X-ray_structure_analysis rock_formation_minerals*
E-mail pan@igpm.kiev.ua (Relcom) Tel. 7(044)4440242 Fax 7(044)4441270

Lomnytska, Dr Yaroslava Fedorivna (1952). Assistant professor. Dept. of Analytical Chemistry, Lviv State University. [Dept. of Analytical Chemistry, Lviv State University, Kyryla i Mefodiya Str. 6, 290005 Lviv, Ukraine.] PhD (Lviv U., 1983) chemistry. *X-ray_structure_analysis phase_diagram intermetallic_compound_crystal_structure phosphorus_compounds*
Tel. 7(0322)725740

Malykhin, Dr Sergij Volodymyrovych (1956). Senior scientist. Dept. of Physics of Metals and Semiconductors, Kharkiv Polytechnical Inst.. [Dept. of Physics of Metals and Semiconductors, Kharkiv Polytechnical Inst., Frunze Str. 21, 310002 Kharkiv, Ukraine.] PhD (Kharkiv Polytechnical Inst., 1990) physics and mathematics. *X-ray_structure_analysis tensometry radiation fault*
Tel. 7(0572)400861

Manyako, Dr Mykola Bogdanovych (1956). Senior scientist. Dept. of Inorganic Chemistry, Lviv State University. [Dept. of Inorganic Chemistry, Lviv State University, Kyryla i Mefodiya Str. 6, 290005 Lviv, Ukraine.] PhD (Lviv U., 1989) chemistry. *Intermetallic_compound_crystal_structure phase_diagram aluminium_compounds*
E-mail bodak@chem.franko.lviv.ua Tel. 007(0322)742388

Markiv, Dr Vasyl' Yakovych (1937). Assistant professor. Dept. of Physics, Faculty of Physics, Kyiv University. [Dept. of Physics, Faculty of Physics, Kyiv University, Akad. Glushkova Prosp. 6, 252000 Kyiv, Ukraine.] PhD (Lviv U., 1966) chemistry. *X-ray_structure_analysis intermetallic_compound_synthesis intermetallic_compound_crystal_structure gallium_compounds*
Tel. 7(044)2660513

Matjash, Prof. Ivan Vasyliovych (1930). Dept. head. Inst. of Geochemistry Mineralogy and Ore Formation, Acad. Sci. Ukraine. [Inst. of Geochemistry Mineralogy and Ore Formation, Acad. Sci. Ukraine, Palladina Prosp. 34, 252680 Kyiv, Ukraine.] DSc (Kyiv U., 1972) physics and mathematics. *Crystallography crystal_physics crystal_magnetism fault*
E-mail pan@igpm.kiev.ua (Relcom) Tel. 7(044)4443160 Fax 7(044)4441270

Matjushenko, Dr Mykola Mykolayovych (1930). Senior scientist. Kharkiv Physical Technical Inst., Ukrainian Scientific Centre. [Kharkiv Physical Technical Inst., Ukrainian Scientific Centre, Akademichna Str. 1, 310108 Kharkiv, Ukraine.] PhD (Kharkiv U., 1967) physics and mathematics. *X-ray_structure_analysis crystal_chemistry space_group symmetry*
Tel. 7(0572)356714

Matkovsky, Prof. Orest Ilyarovych (1929). Dept. head. Dept. of Mineralogy, Lviv State University. [Dept. of Mineralogy, Lviv State University, Hrushevskogo Str. 4, 290005 Lviv, Ukraine.] DSc (Inst. of Geochemistry and Physics of Minerals, Acad. Sci. Ukraine, Kyiv, 1975) geology and mineralogy. *Minerals crystallography mineralogy*
Tel. 7(0322)743343

Mel'nik, Dr Yurij Mykhajlovych (1927). Senior scientist. Faculty of Geology, Lviv State University. [Faculty of Geology, Lviv State University, Hrushevskogo Str. 4, 290005 Lviv, Ukraine.] PhD (Lviv U., 1963) geology and mineralogy. *Minerals crystallography crystal_morphology*
Tel. 7(0322)728056

Mel'nikov, Dr Volodymyr Stepanovych (1938). Lab. head. Inst. of Geochemistry Mineralogy and Ore Formation, Acad. Sci. Ukraine. [Inst. of Geochemistry Mineralogy and Ore Formation, Acad. Sci. Ukraine, Palladina Prosp. 34, 252680 Kyiv, Ukraine.] PhD (Inst. of Geochemistry and Physics of Minerals, Acad. Sci. Ukraine, Kyiv, 1974) geology and mineralogy. *X-ray_structure_analysis minerals superconductor*
E-mail pan@igpm.kiev.ua (Relcom) Tel. 7(044)4440570 Fax 7(044)4441270

Mikhailov, Dr Igor Fedorovych (1949). Leading scientist. Dept. of Physics of Metals and Semiconductors, Kharkiv Polytechnical Inst.. [Dept. of Physics of Metals and Semiconductors, Kharkiv Polytechnical Inst., Frunze Str. 21, 310002 Kharkiv, Ukraine.] DSc (Kharkiv Polytechnical Inst., 1989) physics and mathematics. *X-ray_structure_analysis film*
Tel. 7(0572)400861

Mokra, Dr Ivanna Romanivna (1950). Assistant professor. Dept. of Inorganic Chemistry, Lviv State University. [Dept. of Inorganic Chemistry, Lviv State University, Kyryla i Mefodiya Str. 6, 290005 Lviv, Ukraine.] PhD (Lviv U., 1979) chemistry. *X-ray_structure_analysis intermetallic_compound_crystal_structure*
E-mail bodak@chem.franko.lviv.ua Tel. 7(0322)794378

Movchan, Dr Mykola Prokopovych (1938). Dept. head. Dept. of Radiogeochemistry of Environment, Inst. of Geochemistry Mineralogy and Ore Formation, Acad. Sci. Ukraine. [Dept. of Radiogeochemistry of Environment, Inst. of Geochemistry Mineralogy and Ore Formation, Acad. Sci. Ukraine, Palladina Prosp. 34, 252680 Kyiv, Ukraine.] PhD (Inst. of Geochemistry and Physics of Minerals, Acad. Sci. Ukraine, Kyiv, 1971) geology and mineralogy. *Clays minerals crystallography sorption radioactivity*
E-mail pan@igpm.kiev.ua (Relcom) Tel. 7(044)4440043 Fax 7(044)4440060

Mrooz, Dr Oksana Yaroslavivna (1957). Senior scientist. Sci. Research Inst. of Materials, Concern Electron, Lviv. [Sci. Research Inst. of Materials, Concern Electron, Stryjska Str. 202, 290031 Lviv, Ukraine.] PhD (Lviv U., 1988) chemistry. *X-ray_structure_analysis phase_diagram intermetallic_compound_synthesis*
Tel. 7(0322)638303 Fax 7(0322)632228

Mykhalenko, Dr Svitlana Ivanivna (1946). Senior scientist. Dept. of Analytical Chemistry, Lviv State University. [Dept. of Analytical Chemistry, Lviv State University, Kyryla i Mefodiya Str. 6, 290005 Lviv, Ukraine.] PhD (Lviv U., 1976) chemistry. *Intermetallic_compound_crystal_structure intermetallic_compound_synthesis boron_compounds*
Tel. 7(0322)725740

Mykhalichko, Dr Borys Myronovych (1958). Scientist. Dept. of Inorganic Chemistry, Lviv State University. [Dept. of Inorganic Chemistry, Lviv State University, Kyryla i Mefodiya Str. 6, 290005 Lviv, Ukraine.] PhD (Lviv U., 1991) chemistry. *Complex_compound_crystal_structure catalysis*
E-mail margm@chem.franko.lviv.ua Tel. 7(0322)794378

Mykolajchuk, Prof. Oleksij Gordijovych (1931). Dept. head. Dept. of X-ray Metal Physics, Lviv State University. [Dept. of X-ray Metal Physics, Lviv State University, Kyryla i Mefodiya Str. 8, 290005 Lviv, Ukraine.] PhD (Lviv U., 1966) physics and mathematics. *Electron_microscopy epitaxy structure amorphous_materials zone_structure*
Tel. 7(0322)794116

Mys'kiv, Dr Maryan Grygorovych (1947). Professor. Dept. of Inorganic Chemistry, Lviv State University. [Dept. of Inorganic Chemistry, Lviv State University, Kyryla i Mefodiya Str. 6, 290005 Lviv, Ukraine.] DSc (Inst. of General and Inorganic Chemistry, Acad. Sci. USSR, Moscow, 1991) chemistry. *Complex_compound_synthesis complex_compound_crystal_chemistry*
E-mail margm@chem.franko.lviv.ua Tel. 7(0322)794378

Nesterov, Dr Volodymyr Mykolayovych (1959). Lecturer. Dept. of Chemistry, Lugansk State Pedagogical Inst.. [Dept. of Chemistry, Lugansk State Pedagogical Inst., Oboronna Str. 2, 348011 Lugansk, Ukraine.] PhD (Inst. of Organic Chemistry, Acad. Sci. USSR, Moscow, 1988) chemistry. *Organic_compound_crystal_chemistry stereochemistry*
Tel. 7(0642)535396 Fax 7(0642)533127

Nikolaychuk, Dr Grygorij Pavlovych (1958). Assistant professor. Dept. of Theoretical and Experimental Physics, Kharkiv Polytechnical Inst.. [Dept. of Theoretical and Experimental Physics, Kharkiv Polytechnical Inst., Frunze Str. 21, 310002 Kharkiv, Ukraine.] PhD (Kharkiv Polytechnical Inst., 1990) physics and mathematics. *Film growth structure electron_microscopy laser_plasmas*
Tel. 7(0572)400564 Fax 7(0572)400801 Telex 311034 VZOR SU

Nosenko, Prof. Anatolij Erofijovych (1940). Professor. Faculty of Physics, Lviv State University. [Faculty of Physics, Lviv State University, Dragomanova Str. 50, 290005 Lviv, Ukraine.] DSc (Kyiv U., 1989) physics and mathematics. *Crystal_physics fault EPR complex_oxides*
Tel. 7(0322)794326 Fax 7(0322)727981

Oleksyn, Dr Oksana Yaroslavivna (1966). Scientist. Dept. of Inorganic Chemistry, Lviv State University. [Dept. of Inorganic Chemistry, Lviv State University, Kyryla i Mefodiya Str. 6, 290005 Lviv, Ukraine.] PhD (Lviv U., 1990) chemistry. *X-ray_structure_analysis intermetallic_compound_crystal_chemistry*
E-mail bodak@chem.franko.lviv.ua Tel. 7(0322)794506

Olijnyk, Dr Volodymyr Volodymyrovych (1959). Postdoctoral research assistant. Dept. of Inorganic Chemistry, Lviv State University. [Dept. of Inorganic Chemistry, Lviv State University, Kyryla i Mefodiya Str. 6, 290005 Lviv, Ukraine.] PhD (Lviv U., 1986) chemistry. *Complex_compound_crystal_chemistry complex_compound_synthesis*
E-mail margm@chem.franko.lviv.ua Tel. 7(0322)794378

Oryshchyn, Dr Stepan Vasyliovych (1955). Senior scientist. Dept. of Analytical Chemistry, Lviv State University. [Dept. of Analytical Chemistry, Lviv State University, Kyryla i Mefodiya Str. 6, 290005 Lviv, Ukraine.] PhD (Lviv U., 1984) chemistry. *Intermetallic_compound_crystal_structure phosphorus_compounds*
Tel. 7(0322)725740

Panchekha, Dr Petro Oleksiyovych (1938). Associate professor; Deputy Head of Department of Science. Kharkiv State Polytechnical University, Frunze Str.21, 310002 Kharkiv, Ukraine. [apts. 58, Derevyanko Str. 14, 310103 Kharkiv, Ukraine.] PhD (Kharkiv Polytechnical Inst., 1968) technics. *Fault crystal_structure phases_interaction*
E-mail boiko@polyt.kharkov.ua Tel. 7(0572)224432 Fax 7(0572)400801 Telex 311034 VZOR

Pavlyshyn, Prof. Volodymyr Ivanovych (1940). Dept. head. Inst. of Geochemistry Mineralogy and Ore Formation, Acad. Sci. Ukraine. [Inst. of Geochemistry Mineralogy and Ore Formation, Acad. Sci. Ukraine, Palladina Prosp. 34, 252680 Kyiv, Ukraine.] DSc (St.-Petersburg U., 1981) geology and mineralogy. *Minerals crystallography*
E-mail pan@igpm.kiev.ua (Relcom) Tel. 7(044)4440242 Fax 7(044)4441270

Pavlyuk, Dr Volodymyr Vasyliovych (1958). Assistant professor. Dept. of Inorganic Chemistry, Lviv State University. [Dept. of Inorganic Chemistry, Lviv State University, Kyryla i Mefodiya Str. 6, 290005 Lviv, Ukraine.] DSc (Lviv U., 1993) chemistry. *Intermetallic_compound_crystal_chemistry phase_diagram lithium_compounds*
E-mail pavl@chem.franko.lviv.ua Tel. 7(0322)742388

Pecharsky, Dr Vitaly Kostiantynovych (1954). Assistant professor. Dept. of Inorganic Chemistry, Lviv State University. [Dept. of Inorganic Chemistry, Lviv State University, Kyryla i Mefodiya Str. 6, 290005 Lviv, Ukraine.] PhD (Lviv U., 1979) chemistry. *X-ray_structure_analysis computing intermetallic_compound_crystal_chemistry*
E-mail vitkp@chem.franko.lviv.ua Tel. 7(0322)794506

Pinegin, Dr Volodymyr Ivanovych (1949). Senior scientist. Dept. of Physics of Metals and Semiconductors, Kharkiv Polytechnical Inst.. [Dept. of Physics of Metals and Semiconductors, Kharkiv Polytechnical Inst., Frunze Str. 21, 310002 Kharkiv, Ukraine.] PhD (Kharkiv Polytechnical Inst., 1987) physics and mathematics. *X-ray_structure_analysis film radiation scattering*
Tel. 7(0572)400861

Pugachov, Prof. Anatolij Tarasovych (1940). Professor; Head of Electron Diffraction Laboratory. Dept. of Physics of Metals and Semiconductors, Kharkov State Polytechnical University. [Dept. of Physics of Metals and Semiconductors, Kharkov State Polytechnical University, Frunze Str. 21, 310002 Kharkov, Ukraine.] DSc (Kharkov State U., 1982) physics and mathematics. *Film physics crystal_structure fault superconductivity nanocrystals electron_diffraction*
E-mail pugachov@polyt.kharkov.ua Tel. 7(0572)400831 Fax 7(0572)400601 Telex 311034 VZOR SU

Romaka, Dr Ljuba Petrivna (1958). Senior scientist. Dept. of Inorganic Chemistry, Lviv State University. [Dept. of Inorganic Chemistry, Lviv State University, Kyryla i Mefodiya Str. 6, 290005 Lviv, Ukraine.] PhD (Lviv U., 1985) chemistry. *X-ray_structure_analysis intermetallic_compound_synthesis physical_properties_intermetallic_compounds*
E-mail bodak@chem.franko.lviv.ua Tel. 7(0322)794503

Salamakha, Dr Petro Stepanovych (1957). Teaching assistant. Dept. of Inorganic Chemistry, Lviv State University. [Dept. of Inorganic Chemistry, Lviv State University, Kyryla i Mefodiya Str. 6, 290005 Lviv, Ukraine.] PhD (Lviv U., 1989) chemistry. *Phase_diagram intermetallic_compound_crystal_structure*
E-mail bodak@chem.franko.lviv.ua Tel. 7(0322)742388

Semenova, Dr Olena Leonidivna (1945). Senior scientist. Inst. for Problems of Materials Science, Acad. Sci. Ukraine. [Inst. for Problems of Materials Science, Acad. Sci. Ukraine, Krzhyzhanivskogo Str. 3, 252180 Kyiv, Ukraine.] PhD (Inst. for Problems of Materials Science, Acad. Sci. Ukraine, Kyiv, 1981) chemistry. *Phase_diagram martensitic_transformation transition_elements rare-earth_compounds*
Tel. 7(044)4443090 Fax 7(044)4442078 Telex 131257 STAN SU

Sereda, Dr Sergij Volodymyrovych (1959). Scientist. Inst. of Organic Chemistry, Acad. Sci. Ukraine. [Inst. of Organic Chemistry, Acad. Sci. Ukraine, Murmanska Str. 5, 253660 Kyiv, Ukraine.] PhD (Inst. of Element Organic Chemistry, Acad. Sci. USSR, Moscow, 1987) chemistry. *X-ray_structure_analysis organic_compound_crystal_chemistry*
Tel. 7(044)5510646

Shevchenko, Dr Ljudmyla Larionivna (1932). Assistant professor. Dept. of Analytical Chemistry, Kyiv University. [Dept. of Analytical Chemistry, Kyiv University, Volodymyrska Str. 64, 252017 Kyiv, Ukraine.] PhD (Inst. of General and Inorganic Chemistry, Acad. Sci. Ukraine, Kyiv, 1960) chemistry. *Complex_compound_crystal_chemistry*
Tel. 7(044)5553705

Shopa, Dr Yaroslav Ivanovych (1956). Assistant professor. Faculty of Physics, Lviv State University. [Faculty of Physics, Lviv State University, Kyryla i Mefodiya Str. 8, 290005 Lviv, Ukraine.] PhD (1983) physics and mathematics. *Crystal_optics opto-electronics phase_transition anisotropy*
Tel. 7(0322)794780

Shpyrka, Dr Zinoviya Mykhajlivna (1957). Scientist. Dept. of Inorganic Chemistry, Lviv State University. [Dept. of Inorganic Chemistry, Lviv State University, Kyryla i Mefodiya Str. 6, 290005 Lviv, Ukraine.] PhD (Lviv U., 1990) chemistry. *X-ray_structure_analysis intermetallic_compound_synthesis*
E-mail evgig@chem.franko.lviv.ua Tel. 7(0322)794378

Sichevich, Dr Olga Mykhajlivna (1956). Senior scientist. Dept. of Inorganic Chemistry, Lviv State University. [Dept. of Inorganic Chemistry, Lviv State University, Kyryla i Mefodiya Str. 6, 290005 Lviv, Ukraine.] PhD (Lviv U., 1986) chemistry. *Phase_diagram intermetallic_compound_crystal_structure physical_properties_intermetallic_compounds*
E-mail bodak@chem.franko.lviv.ua Tel. 7(0322)742388

Skolozdra, Prof. Roman Volodymyrovych (1941). Professor. Dept. of Inorganic Chemistry, Lviv State University. [Dept. of Inorganic Chemistry, Lviv State University, Kyryla i Mefodiya Str. 6, 290005 Lviv, Ukraine.] DSc (Inst. for Problems of Materials Science, Acad. Sci. Ukraine, Kyiv, 1990) chemistry. *X-ray_structure_analysis phase_diagram intermetallic_compound_crystal_chemistry tin_compounds physical_properties_intermetallic_compounds*
E-mail bodak@chem.franko.lviv.ua Tel. 7(0322)794503

Sobol', Dr Oleg Valentynovych (1963). Senior scientist. Dept. of Physics of Metals and Semiconductors, Kharkiv Polytechnical Inst.. [Dept. of Physics of Metals and Semiconductors, Kharkiv Polytechnical Inst., Frunze Str. 21, 310002 Kharkiv, Ukraine.] PhD (Kharkiv Polytechnical Inst., 1993) physics and mathematics. *X-ray_structure_analysis radiation fault tensometry*
Tel. 7(0572)400344 Fax 7(0572)400801 Telex 311034 VZOR SU

Sokil, Dr Anatolij Opanasovych (1937). Assistant professor. Kharkiv Polytechnical Inst.. [Kharkiv Polytechnical Inst., Frunze Str. 21, 310002 Kharkiv, Ukraine.] PhD (Kharkiv Polytechnical Inst., 1970) technics. *Fault crystal_structure electron_microscopy*
Tel. 7(0572)400564 Fax 7(0572)400801 Telex 311034 VZOR SU

Starodoob, Dr Pavlo Korniyovych (1942). Assistant professor. Dept. of Inorganic Chemistry, Lviv State University. [Dept. of Inorganic Chemistry, Lviv State University, Kyryla i Mefodiya Str. 6, 290005 Lviv, Ukraine.] PhD (Lviv U., 1988) chemistry. *Phase_diagram intermetallic_compound_crystal_structure*
E-mail bodak@chem.franko.lviv.ua Tel. 7(0322)742388

Synlushko, Dr Vasyl' Grygorovych (1948). Assistant professor. Dept. of X-ray Metal Physics, Lviv State University. [Dept. of X-ray Metal Physics, Lviv State University, Kyryla i Mefodiya Str. 8, 290005 Lviv, Ukraine.] PhD (Lviv U., 1974) physics and mathematics. *X-ray_emission_spectroscopy structure metals alloys*
Tel. 7(0322)727064

Tarashchenko, Dr Arkadij Mykolayovych (1936). Dept. head. Inst. of Geochemistry Mineralogy and Ore Formation, Acad. Sci. Ukraine. [Inst. of Geochemistry Mineralogy and Ore Formation, Acad. Sci. Ukraine, Palladina Prosp. 34, 252680 Kyiv, Ukraine.] DSc (Inst. of Geology Mineralogy and Petrography, Acad. Sci. USSR, Moscow, 1974) geology and mineralogy. *Minerals physics crystal_chemistry*
E-mail pan@igpm.kiev.ua (Relcom) Tel. 7(044)4440305 Fax 7(044)4441270

Telovich, Dr Roman Volodymyrovych (1945). Lab. head. Inst. of Metal Physics, Acad. Sci. Ukraine. [Inst. of Metal Physics, Acad. Sci. Ukraine, Vernadskogo Str. 36, 252680 Kyiv, Ukraine.] PhD (Inst. of Metal Physics, Acad. Sci. Ukraine, Kyiv, 1980) physics and mathematics. *X-ray_structure_analysis crystallography iron_compounds computing electron_microscopy*
Tel. 7(044)4449582 Fax 7(044)4442561

Tretyachenko, Dr Ljudmyla Oleksandrivna (1926). Senior scientist. Inst. for Problems of Materials Science, Acad. Sci. Ukraine. [Inst. for Problems of Materials Science, Acad. Sci. Ukraine, Krzhyzhanivskogo Str. 3, 252180 Kyiv, Ukraine.] PhD (Inst. for Problems of Materials Science, Acad. Sci. Ukraine, Kyiv, 1966) chemistry. *Intermetallic_compound_crystal_chemistry*
Tel. 7(044)4443090 Fax 7(044)4440492 Telex 131257 STAN SU

Tyvanchuk, Dr Anna Teodorivna (1947). Senior scientist. Dept. of Inorganic Chemistry, Lviv State University. [Dept. of Inorganic Chemistry, Lviv State University, Kyryla i Mefodiya Str. 6, 290005 Lviv, Ukraine.] PhD (Lviv U., 1980) chemistry. *X-ray_structure_analysis phase_diagram intermetallic_compound_synthesis*
E-mail bodak@chem.franko.lviv.ua Tel. 7(0322)794378

Val'ter, Dr Anton Antonovych (1933). Dept. head. Inst. of Geochemistry Mineralogy and Ore Formation, Acad. Sci. Ukraine. [Inst. of Geochemistry Mineralogy and Ore Formation, Acad. Sci. Ukraine, Palladina Prosp. 34, 252680 Kyiv, Ukraine.] DSc (Inst. of Mineral Raw Materials, Moscow, 1980) geology and mineralogy. *Minerals crystallography mineralogy*
E-mail pan@igpm.kiev.ua (Relcom) Tel. 7(044)4441041 Fax 7(044)4441270

Velikanova, Prof. Tamara Yakivna (1933). Dept. head. Inst. for Problems of Materials Science, Acad. Sci. Ukraine. [Inst. for Problems of Materials Science, Acad. Sci. Ukraine, Krzhyzhanivskogo Str. 3, 252180 Kyiv, Ukraine.] DSc (Inst. for Problems of Materials Science, Acad. Sci. Ukraine, Kyiv, 1991) chemistry. *X-ray_structure_analysis intermetallic_compound_crystal_chemistry inclusion_compound*
Tel. 7(044)4443090 Fax 7(044)4440492 Telex 131257 STAN SU

Vlokh, Prof. Orest Grygorovych (1934). Inst. director. Inst. of Physical Optics. [Inst. of Physical Optics, Dragomanova Str. 23, 290005 Lviv, Ukraine.] DSc (Inst. of Crystallography, Acad. Sci. USSR, Moscow, 1979) physics and mathematics. *Crystal_physics crystal_optics non-linear_optics phase_transition*
Tel. 7(0322)742089 Fax 7(0322)742089

Voroshilov, Prof. Yurij Vitaliyovych (1938). Dept. head. Member of Ukrainian Crystallographic Committee. Dept. of Chemistry of Solids and Semiconductors, Uzhgorod State University. [Dept. of Chemistry of Solids and Semiconductors, Uzhgorod State University, Gorkogo Str. 46, 294000 Uzhgorod, Ukraine.] DSc (Moscow U., 1985) chemistry. *X-ray_structure_analysis crystal_chemistry crystal_properties*
Tel. 7(03122)35091

Voznyak, Dr Dmytro Kostiantynovych (1938). Lab. head. Inst. of Geochemistry Mineralogy and Ore Formation, Acad. Sci. Ukraine. [Inst. of Geochemistry Mineralogy and Ore Formation, Acad. Sci. Ukraine, Palladina Prosp. 34, 252680 Kyiv, Ukraine.] PhD (Inst. of Geological Sciences, Acad. Sci. Ukraine, Kyiv, 1971) geology and mineralogy. *Phase_transition diamond mineralogy*
E-mail pan@igpm.kiev.ua (Relcom) Tel. 7(044)4440570 Fax 7(044)4441270

Yanchuk, Dr Edward Oleksandrovych (1937). Assistant professor. Dept. of Mineralogy, Lviv State University. [Dept. of Mineralogy, Lviv State University, Hrushevskogo Str. 4, 290005 Lviv, Ukraine.] PhD (Lviv U., 1973) geology and mineralogy. *Minerals crystallography mineralogy geochemistry X-ray_analysis*
Tel. 7(0322)743343

Yanson, Dr Tamara Ivanivna (1947). Lab. head. Dept. of Inorganic Chemistry, Lviv State University. [Dept. of Inorganic Chemistry, Lviv State University, Kyryla i Mefodiya Str. 6, 290005 Lviv, Ukraine.] PhD (Lviv U., 1975) chemistry. *Crystal_structure phase_diagram intermetallic_compound_crystal_chemistry aluminium_compounds*
E-mail tamya@chem.franko.lviv.ua Tel. 7(0322)742388

Yasinska, Dr Angelina Andrijivna (1922). Assistant professor. Dept. of Mineralogy, Lviv State University. [Dept. of Mineralogy, Lviv State University, Hrushevskogo Str. 4, 290005 Lviv, Ukraine.] PhD (Lviv U., 1951) geology and mineralogy. *Minerals crystallography mineralogy cosmic_mineralogy*
Tel. 7(0322)743343

Zarechniuk, Prof. Oleg Safoniyovych (1923). Leading scientist. Dept. of Inorganic Chemistry, Lviv State University. [Dept. of Inorganic Chemistry, Lviv State University, Kyryla i Mefodiya Str. 6, 290005 Lviv, Ukraine.] DSc (Moscow U., 1983) chemistry. *Intermetallic_compound_crystal_chemistry aluminium_compounds*
E-mail bodak@chem.franko.lviv.ua Tel. 7(0322)742388

Zaremba, Dr Vasyl' Ivanovych (1962). Teaching assistant. Dept. of Inorganic Chemistry, Lviv State University. [Dept. of Inorganic Chemistry, Lviv State University, Kyryla i Mefodiya Str. 6, 290005 Lviv, Ukraine.] PhD (Lviv U., 1990) chemistry. *X-ray_structure_analysis intermetallic_compound_crystal_structure indium_compounds*
E-mail bodak@chem.franko.lviv.ua Tel. 7(0322)794378

Zavalii, Dr Petro Yuliyanovych (1957). Postdoctoral Research Assistant. Dept. of Inorganic Chemistry, Lviv State University. [Dept. of Inorganic Chemistry, Lviv State University, Kyryla i Mefodiya Str. 6, 290005 Lviv, Ukraine.] PhD (Lviv U., 1983) chemistry. *X-ray_structure_analysis computing intermetallic_compounds*
E-mail vitkp@chem.franko.lviv.ua Tel. 7(0322)794506

Zubarev, Dr Evgeny Nikolayevich (1958). Teaching assistant. Metal and Semiconductor Physics Dept., Kharkiv State Polytechnic University. [Metal and Semiconductor Physics Dept., Kharkiv State Polytechnic University, Frunze Str. 21, Kharkiv 310002, Ukraine.] PhD (Kharkiv Polytechnical Inst., 1987) physics and mathematics. *Film physics superlattice structure electron_microscopy*
E-mail fedorenko@polyt.kharkov.ua Tel. 7(0572)400932 Fax 7(0572)400601 Telex 311034 VZOR SU

Zubenko, Dr Oleksandr Ivanovych (1953). Assistant professor. Dept. of Analytical Chemistry, Kyiv University. [Dept. of Analytical Chemistry, Kyiv University, Volodymyrska Str. 64, 252017 Kyiv, Ukraine.] PhD (Kyiv U., 1982) chemistry. *Inorganic_compound_crystal_chemistry*
Tel. 7(044)2210264

UNITED KINGDOM

Sub-Editor: F. H. Allen

Notes

1. Telephone exchange codes shown are effective from 1 April 1995. Dialling difficulties prior to that date may be resolved by omitting the initial '1' from the exchange code, except for some special cases indicated in individual entries.

Abell, Dr Stuart (1944). Senior Lecturer. School of Metallurgy and Materials, University of Birmingham, Birmingham, B15 2TT, England. PhD (Surrey U., 1970) solid state physics. *Rare_earth_compounds magnetic_materials oxide_superconductors crystal_growth_characterisation*
Tel. 44(121)4145168 Fax 44(121)4145232

Abrahams, Dr Jan Pieter (1961). Postdoctoral Fellow. MRC Laboratory of Molecular Biology, Hills Road, Cambridge, CB2 2QH, England. PhD (Leiden U., 1991) biochemistry. *Biocrystallography bioenergetics biochemistry biosynthesis*
E-mail jpa@mrc-lmb.cam.ac.uk Tel. 44(1223)402212 Fax 44(1223)213556

Acharya, Dr K. Ravi (1955). Reader. School of Biology and Biochemistry, University of Bath, Claverton Down, Bath, BA2 7AY, England. PhD (Bangalore U., 1982) X-ray crystallography. *Crystallography structure_determination antigens molecular_modelling lactose_synthesis angiogenesis superantigens*
E-mail bsskra@midge.bath.ac.uk Tel. 44(1225)826238 Fax 44(1225)826449

Adams, Dr Margaret Joan (1939). Fellow and Tutor at Somerville College. Co-Editor Acta Crystallographica. Laboratory of Molecular Biophysics, Rex Richards Building, South Parks Road, Oxford, OX1 3QU, England. DPhil (Oxford U., 1968) chemistry: protein crystallography. *Proteins dehydrogenases NAD(P)*
E-mail margaret@biop.ox.ac.uk Tel. 44(1865)275391 Fax 44(1865)510454

Alcock, Dr N. W. (1939). Reader. Former Member, Commission on Journals. Department of Chemistry, University of Warwick, Coventry, CV4 7AL, England. PhD (Cambridge U., 1963) chemistry. *Absorption_correction bonding_intermolecular main_group_compounds macrocycles actinides*
E-mail msrbb@csv.warwick.ac.uk Tel. 44(1203)523228 Fax 44(1203)524112

Alexeev, Dr Dimitriy (1956). Postdoctoral Research Assistant. Department of Biochemistry, The University of Edinburgh, Hugh Robson Building, George Square, Edinburgh, EH8 9XD, Scotland. PhD (Moscow Institute of Crystallography, 1988) DNA fibre diffraction. *Macromolecular_crystallography DNA proteins*
E-mail dima@castle.edinburgh.ac.uk Tel. 44(131)6503721 Fax 44(131)6503711

Allan, Dr David R. (1967). Postdoctoral Research Assistant. Department of Physics, The University of Edinburgh, James Clerk Maxwell Buildings, The King's Building, Mayfield Road, Edinburgh, EH9 3JZ, Scotland. [(from January 1995), Bayerisches Forschungsinstitut für Experimentelle Geochemie und Geophysik, Universität Bayreuth, D-95440 Bayreuth, Germany.] PhD (U. Edinburgh, 1993) physics: high-pressure crystallography. *High_pressure_crystallography X-ray_diffraction neutron_diffraction single-crystal_diffraction powder_diffraction inorganic_structures electrooptic_materials mineral_structures instrumentation*
E-mail egnr44@castle.ed.ac.uk; from January 1995 david.allan@uni-bayreuth.de Tel. 44(131)6505293; from January 1995 49(921)553700 Fax 44(131)6624712; from January 1995 49(921)553769 Telex 727442 UNIVED G

Allen, Dr Frank Harmsworth (1944). Deputy Director CCDC. Editor, Acta Crystallographica Section B; Chair, BCA Chemical Crystallography Group; Sub-Editor, WDC9. Cambridge Crystallographic Data Centre, 12 Union Road, Cambridge, CB2 1EZ, England. PhD (Imperial Coll. London, 1968) physical chemistry. *Structural_database molecular_systematics statistical_method information_science*
E-mail fha1@chemcrys.cam.ac.uk Tel. 44(1223)336425 Fax 44(1223)336033

Allen, Dr Simon C. (1967). Postdoctoral Research Assistant. Department of Biochemistry, University of Bath, Claverton Down, Bath, BA2 7AY, England. PhD (Bristol U., 1992) biochemistry. *Biochemistry structure_determination molecular_replacement multiple_isomorphous_replacement protein_structure_and_function angiogenesis*
E-mail s.c.allen@midge.bath.ac.uk Tel. 44(1225)826x3091 Fax 44(1225)826449

Andrews, Dr Steven John (1959). Senior Research Scientist. ICI Chemicals and Polymers Ltd, PO Box 8, The Heath, Runcorn, Cheshire, WA7 4QD, England. PhD (U. Edinburgh, 1984) chemistry: crystallography. *Synchrotron_radiation Rietveld_method energy_dispersive_analysis inorganic_structure_determination Laue_diffraction structure_solution_from_microcrystals*
E-mail asj@dl.ac.uk Tel. 44(1928)513831 Fax 44(1928)581178

Anwar, Dr Jamshed (1957). Lecturer. BCA Industrial Group Committee Member. Department of Pharmacy, King's College London, Manresa Road, London, SW3 6LX, England. PhD (Birkbeck Coll. London, 1990) crystallography. *Phase_transition computer_modelling powder_X-ray_diffraction polymorphism*
E-mail udkj060@elm.cc.kcl.ac.uk Tel. 44(171)3334782 Fax 44(171)3515307

Arndt, Dr U. W. (1924). Project Grant Holder. MRC Laboratory of Molecular Biology, Hills Road, Cambridge, CB2 2QH, England. PhD (Cambridge U., 1949) physics. *Instrumentation proteins macromolecular_structure*
Tel. 44(1223)248011 Fax 44(1223)213556 Telex 81532

Artymiuk, Dr Peter J. (1952). Research Fellow. Krebs Institute, Department of Molecular Biology and Biophysics, University of Sheffield, Sheffield, S10 2TN, England. DPhil (Oxford U., 1979) molecular biophysics. *Protein_crystallography biomolecule_structure_comparison protein_refinement_methods protein_bonding_interactions*

Arzt, Mrs Steffi (1964). Research Associate. Clarendon Laboratory, University of Oxford, Department of Physics, Parks Road, Oxford OX1 3PU, England. Dipl. Min. (U. Köln, 1991) mineralogy/crystallography. *Optical activity polarization crystallography*
Tel. 44(865)272334 Fax 44(865)272400 Telex 83295mucloxg

Aslamin, Stephen (1963). Research Student. Department of Materials Science, University of Cambridge, Pembroke Street, Cambridge, CB2 3QZ, England. BSc (1989) physics. *Transmission_electron_microscopy phase_transformations thin_films interface_characterization geophysics data_analysis electromagnetic_wave_theory superconductivity*
E-mail sa124@phx.cam.ac.uk Tel. 44(1233)334341 Fax 44(1233)334567

Aspden, Dr Richard M. (1955). MRC Senior Fellow. Department of Orthopaedics, University of Aberdeen, Polwarth Building, Foresterhill, Aberdeen, AB9 2ZD, Scotland. PhD (Manchester U., 1981) biophysics: protein organisation. *Biomaterials biomechanics tissues computer_modelling*
E-mail r.aspden@abdn.ac.uk Tel. 44(1224)681818[52767] Fax 44(1224)685373

Attfield, Dr J. Paul (1962). Lecturer. Department of Chemistry, University of Cambridge, Lensfield Road, Cambridge, CB2 1EW, England. DPhil (Oxford U., 1987) chemistry. *Rietveld_method anomalous_dispersion_disordered_material superconductors ordered_structure_magnetic metal_oxides_synthesis oxo-salts transition_metal_compounds*
E-mail jpa14@phx.cam.ac.uk Tel. 44(1223)336332 Fax 44(1223)336362

Bacon, Prof. George E. (1917). Emeritus Professor. [Windrush Way, Guiting Power, Cheltenham, GL 54 5US, England.] ScD (Cambridge U., 1964) physics. *Neutron_diffraction hydrogen_bonding magnetism*
Tel. 44(1451)850631

Badcock, Tracey Dianne (1968). Research Scientist. Coates Lorilleux International, Cray Avenue, St Mary Cray, Orpington, Kent, England. BSc (Napier U., Edinburgh, Scotland, 1991) applied chemistry. *Structure_property_relationship structure_colour_relationship colour_physics pigments dispersion*
Tel. 44(1680)527080 Fax 44(1680)839613 Telex 896365 inkdev g

Bagley, Arthur George (1947). Director. Hiltonbrookes, Yew Tree Cottage, Knutsford Road, Cranage, Holmes Chapel, Cheshire, CW4 8EP, England. *Manufacture_and_sale_of_analytical_X-ray_equipment*
Tel. 44(1477)534140 Fax 44(1477)534140

Bahar, Susanna (1969). Research Student. Department of Chemistry, University of Durham, South Road, Durham, DH1 3LE, England. BA (Oxford U., 1992) chemistry. *Chemical_crystallography crystallisation packing peptides*
E-mail susanna.bahar@durham.ac.uk Tel. 44(191)3744703 Fax 44(191)3743745

Bailey, Dr Susan (1958). Senior Scientific Officer. Daresbury Laboratory, Daresbury, Warrington, Cheshire, WA4 4AD, England. PhD (Birkbeck Coll. London, 1987) protein crystallography. *Proteins metalloproteins reductases enzymes*
E-mail s.bailey@dl.ac.uk Tel. 44(1925)603530 Fax 44(1925)603174

Baker, Dr Patrick J. (1961). Research Fellow. The Krebs Institute, Department of Molecular Biology, University of Sheffield, PO Box 594, Sheffield, S10 2UH, England. PhD (Sheffield U., 1988) macromolecular crystallography. *Dehydrogenases domain_motion optimized_anomalous_scattering generators_and_detectors*
E-mail p.baker@sheffield.ac.uk Tel. 44(114)2824325 [44(742)824325 before 1 April 1995] Fax 44(114)2728697 [44(742)728697 before 1 April 1995]

Balchin, Dr Anthony Arthur (1932). Chartered Physicist. [23 Wilmington Close, Hassocks, West Sussex, BN6 8QB, England.] PhD (U. London, 1959) crystallography. *Layer_compounds structural_chelation polytypic_dichalcogenides crystal_growth industrial_applications superconductors quasicrystals optical_analogues*
Tel. 44(1273)843624

Barnes, Dr Hazel A. (1954). [5 Scotswood Crescent, Wormit, Newport-on-Tay, Fife, DD6 8PU, Scotland.] PhD (Dundee U., 1991) electron-donor acceptor complexes. *Structure_determination molecular_complexes nuclear_magnetic_resonance QSAR*
E-mail jbarnes@dundee.ac.uk Tel. 44(1382)541627

Barnes, Dr John Conquest (1935). Senior Lecturer. Chemistry Department, University of Dundee, Dundee, DD1 4HN, Scotland. DSc (Dundee U., 1986) inorganic and structural chemistry. *Structure_determination thermoanalysis carboxylates rare_earth_compounds*
E-mail jbarnes@dundee.ac.uk Tel. 44(1382)223181x4705 44(1382)344705 Fax 44(1382)202830

Barnes, Prof. Paul (1942). Professor in Applied Crystallography. BCA Council Member. Industrial Materials Group, Department of Crystallography, Birkbeck College, Malet Street, London, WC1E 7HX, England. DSc physics. *Applied-crystallography_materials diffraction_technique computer_modelling phase_kinetics general_microscopy time-resolved_diffraction edge-topography_of_polytypes*
E-mail P.Barnes@cryst.bbk.ac.uk Tel. 44(171)6316817 Fax 44(171)6316803

Barrett, Mrs Stella B. (1946). Technical Editor. Cambridge Crystallographic Data Centre, 12 Union Road, Cambridge, CB2 1EZ, England. BSc (Birmingham U., 1966) chemistry. *Organic_database metalloorganic_database information_system*
E-mail sbb1@chemcrys.cam.ac.uk Tel. 44(1223)336032 Fax 44(1223)336033

Barrow, Dr M. J. (1946). Senior Lecturer. Department of Applied Chemical and Physical Sciences, Napier University, Edinburgh, EH10 5DT, Scotland. PhD (Victoria U. Manchester, 1972) chemical crystallography. *Chemical_crystallography organic_pigments acetylenes molecular_graphics*
Tel. 44(131)4552621 Fax 44(131)4477989

Bax, Dr Benjamin (1962). Postdoctoral Research Fellow. Department of Crystallography, Birkbeck College, Malet Street, London, WC1E 7HX, England. PhD (Birkbeck Coll. London, 1989) protein crystallography. *Proteins quaternary_structure crystallization computer_modelling*
E-mail b.bax@cryst.bbk.ac.uk Tel. 44(171)6316824 Fax 44(171)6316803

Baxter, Ian (1970). Research Student. School of Chemistry and Applied Chemistry, University of Wales Cardiff, PO Box 912, Cardiff, CF1 3TB, Wales. BSc (1992) chemistry. *Structure_determination organometallics computer_modelling*
E-mail baxter@cardiff.ac.uk Tel. 44(1222)874950

Beagley, Dr Brian (1936). Reader. Department of Chemistry, UMIST, PO Box 88, Manchester, M60 1QD, England. DSc (U. Birmingham, 1981) structural chemistry. *Structural_chemistry least-squares_refinement software zeolites EXAFS_of_solution aluminophosphates wave_theory*
E-mail mcdssbb@mh1.mcc.ac.uk Tel. 44(161)2004522 Fax 44(161)2367677 Telex 666094

Beanland, Dr Richard (1965). Research Associate. Department of Materials Science and Engineering, The University of Liverpool, PO Box 147, Liverpool, L69 3BX, England. PhD (Liverpool U., 1991) materials science. *Interface epitaxy dislocation semiconductor electron_microscope diffraction multilayer*
E-mail beanland@liv.ac.uk Tel. 44(151)7945381 Fax 44(151)7944675 Telex 627095 UNILPL G

Beddell, Dr Christopher Raymond. Wellcome Research Laboratories, Langley Court, Beckenham, Kent, BR3 3BS, England. DPhil (Oxford U., 1971) molecular biophysics. *Macromolecular_structure macromolecular_function biophysics_analysis*
Tel. 44(181)6582211 Fax 44(181)6633788

Beddoes, Roy L. (1938). Senior Experimental Officer. Department of Chemistry, University of Manchester, Oxford Road, Manchester, M13 9PL, England. MA (U. Cambridge, 1965) natural sciences. *Molecular_structure_determination*
E-mail bed@v5.ch.man.ac.uk Tel. 44(161)2754705 Fax 44(161)2754598

Beevers, Dr Cecil Arnold (1908). Retired. Chemistry Department, The University of Edinburgh, West Mains Road, Edinburgh, EH9 3JJ, Scotland. DSc (Liverpool U., 1943) X-ray crystallography. *X-ray_crystallography*
Tel. 44(131)6501000 Fax 44(131)6504743

Bell, Anthony Martin Thomas (1963). Junior Research Assistant. DRAL Daresbury Laboratory, Warrington, Cheshire, WA4 4AD, England. MSc (Keele U., 1992) chemistry. *Powder_diffraction Rietveld_method synchrotron_radiation structure_determination silicates zeolites minerals*
E-mail a.m.t.bell@dl.ac.uk Tel. 44(1925)603123 Fax 44(1925)603124

Bennett, Dr Pauline M. (1944). Scientific Staff. MRC Muscle and Cell Motility Unit, The Randall Institute, King's College London, 26-29 Drury Lane, London, WC2B 5RL, England. PhD (King's Coll. London, 1977) biophysics. *Electron_microscopy image_processing muscles biophysics*
E-mail udbp147@bay.cc.kcl.ac.uk Tel. 44(171)8368851 Fax 44(171)4979078

Berry, Dr Amanda S. (1965). Assistant Technical Editor. International Union of Crystallography, 5 Abbey Square, Chester, CH1 2HU, England. PhD (U. Liverpool, 1990) marine geochemistry. *Electronic_publishing database_preparation editing checking*
E-mail ab@iucr.ac.uk Tel. 44(1244)342878 Fax 44(1244)314888 Telex 669755 OFFICE G attn UNICRYSTAL

Beveridge, Dr David (1951). Research Chemist. Ilford Ltd, Rajar Works, Town Lane, Mobberley, Knutsford, Cheshire, WA16 7JL, England. DPhil (Oxford U., 1978) chemistry. *Silver_compounds analytical_phase_refinement powder halides X-ray_fluorescence_spectroscopy mineralogy polytypism*
Tel. 44(1565)650000 Fax 44(1565)872734

Bewley, Maria Christine (1967). Research Student. Astbury Building, Department of Biochemistry and Molecular Biology, University of Leeds, Leeds, LS2 9JT, England. BSc (Leeds U., 1990) biophysics. *X-ray_crystallography*
E-mail maria@leeds.ac.uk Tel. 44(113)2332592 [44(532)332592 before 1 April 1995] Fax 44(113)2333167 [44(532)333167 before 1 April 1995]

Bird, Craig M. (1971). Research Student. Cambridge Crystallographic Data Centre, 12 Union Road, Cambridge, CB2 1EZ, England. BSc (U. Salford, 1992) chemistry. *Hydrogen_bonding chemistry database*
E-mail cmb10@chemcrys.cam.ac.uk Tel. 44(1223)336012 Fax 44(1223)336033

Blake, Dr Alexander John (1954). Research Fellow/Crystallographer. BCA Chemical Crystallography Group Committee Member. Department of Chemistry, The University of Edinburgh, King's Buildings, West Mains Road, Edinburgh, EH9 3JJ, Scotland. PhD (Aberdeen U., 1980) structural chemistry. *Data_collection low_temperature low_melting_compounds macrocycles handling_thermolabile_samples oil_film/ flash_cooling_methods*
E-mail ajb01@ed.ac.uk Tel. 44(131)6504737 Fax 44(131)6504743

Bloomer, Dr Anne C. (1945). MRC Staff Scientist. BCA Treasurer. MRC Laboratory of Molecular Biology, Hills Road, Cambridge, CB2 2QH, England. DPhil (U. Oxford, 1972) molecular biophysics. *Macromolecular_structure_determination biological_structure-activity_relationship molecular_immunology analytical_crystallography therapeutic_antibodies*
E-mail acb1@mrc-lmb.cam.ac.uk Tel. 44(1223)248011 Fax 44(1223)213556 Telex 81532

Blow, Emeritus Prof. David Mervyn (1931). Senior Research Fellow. Blackett Laboratory, Imperial College, London, SW7 2BZ, England. [1 Meeting Street, Appledore, Bideford, N. Devon, EX39 1RH, England.] PhD (Cambridge U., 1957) physics. *Protein_structure enzyme_mechanism protein_crystallisation*
E-mail d.blow@ic.ac.uk Tel. 44(171)5947683 44(1237)421106 Fax 44(171)5890191

Blundell, Prof. Tom Leon (1942). Honorary Director and Professor; from 1 October 1996 Sir William Dunn Professor of Biochemistry, U. Cambridge. Chief Executive: UK Biotechnology and Biological Sciences Research Council; Director of School of Crystallography, Erice. ICRF Unit of Structural Molecular Biology, Department of Crystallography, Birkbeck College, Malet Street, London, WC1E 7HX, England. DPhil (Oxford U., 1967) crystallography. *Structural_molecular_biology structure_based_design protein_crystallography protein_structure_prediction*
Tel. 44(171)6316856 Fax 44(171)6316805

Bond, Charles Simon (1970). Research Student. Structural Chemistry Group, Department of Chemistry, University of Manchester, Oxford Road, Manchester, M13 9PL, England. BSc (Manchester U., 1992) chemistry with industrial experience. *Reductases_inhibitors_modelling structure-activity_relationship biomedical_crystallization synchrotron_radiation organic_synthesis popular_science*
E-mail charlie@v7.chemistry.manchester.ac.uk Tel. 44(161)2754689

Borkakoti, Dr Nivedita Neera (1949). Research Scientist. Roche Products Ltd, PO Box 8, Welwyn Garden City, Hertfordshire, AL7 3AY, England. PhD (U. London, 1978) crystallography. *Proteins computer_assisted_design computer_modelling biocrystallography*
E-mail borkakon@rocbi.dnet.roche.com Tel. 44(1707)366375 Fax 44(1707)373540

Bowen, Prof. Alun W. (1942). Section Leader. Materials and Structures Department, Defence Research Agency, Farnborough, Hampshire, GU14 6TD, England. PhD physical metallurgy. *Microstructure texture residual_stress AEM line_broadening transformation precipitation fracture*
Tel. 44(1252)392208 Fax 44(1252)376370

Bowen, Prof. D. Keith (1940). Professor, Director of Centre. Centre for Nanotechnology and Microengineering, Department of Engineering, University of Warwick, Coventry, CV4 7AL, England. DPhil (Oxford U., 1967) metallurgy. *X-ray_topography X-ray_reflection diffraction material_characterization nanotechnology precision_engineering*
E-mail d.k.bowen@warwick.ac.uk Tel. 44(1203)523133 Fax 44(1203)418922 Telex 311904 UNIVWK G

Boys, Dr Bill (1954). Postdoctoral Research Assistant. Department of Biochemistry, Hugh Robson Building, George Square, Edinburgh, E11 1EE, Scotland. DPhil (Oxford U., 1984) biochemistry. *Crystallisation*
E-mail bill@biovax.ed.ac.uk Tel. 44(131)6503721 Fax 44(131)6503711

Bracke, Dr Ben R. F. (1965). Postdoctoral Research Associate. School of Chemistry and Applied Chemistry, University of Wales Cardiff, PO Box 912, Cardiff, CF1 3TB, Wales. PhD (1993) chemistry. *Crystallography MO-calculations modelling charge_density crystal_field_modelling solid_state_calculations multipole_refinements*
E-mail Bracke@cardiff.ac.uk Tel. 44(1222)874950 Fax 44(1222)874029

Brady, Dr Robert Leo (1961). Lecturer. Department of Biochemistry, University of Bristol, Bristol, BS8 1TD, England. DPhil (U. York, 1987) chemical crystallography. *Crystallization macromolecular_crystallography antibodies glycoproteins*
E-mail brady@bsa.bristol.ac.uk Tel. 44(117)9287436 Fax 44(117)9288274

Bright, Dr Alan (1948). Physical Chemist. Schering Agrochemicals Ltd, Chesterford Park, Saffron Walden, Essex, CB10 1XL, England. PhD (1974) chemistry. *Interactions molecular_mechanics charge_transfer polymorphism NMR_spectroscopy computer_automation*
Tel. 44(1799)530123x3493 Fax 44(1799)30021 Telex 817300

Britton, Dr K. Linda (1965). Research Associate. The Krebs Institute, Department of Molecular Biology, University of Sheffield, PO Box 594, Sheffield, S10 2UH, England. PhD (Sheffield U., 1991) crystallography. *Biocrystallography crystallization dehydrogenases isomorphous_replacement homology chirality conformational_change least-squares_refinement*
E-mail k.britton@sheffield.ac.uk Tel. 44(114)2824325 [44(742)824325 before 1 April 1995] Fax 44(114)2728697 [44(742)728697 before 1 April 1995]

Brown, Dr David Summers (1937). Senior Lecturer. Department of Chemistry, Loughborough University of Technology, Loughborough, Leicestershire, LE11 3TU, England. PhD (U. Nottingham, 1962) crystallography. *Structural_chemistry small_angle_scattering*
E-mail d.s.brown@lut.ac.uk Tel. 44(1509)222558 Fax 44(1509)233163 Telex 34319

Bruno, Dr Ian J. (1969). Scientific Officer. Cambridge Crystallographic Data Centre, 12 Union Road, Cambridge, CB2 1EZ, England. PhD (U. Sheffield, 1994) chemical information. *Chemical_database information_science computing software*
E-mail ian_b@chemcrys.cam.ac.uk Tel. 44(1223)336022 Fax 44(1223)336033

Bryant, Dr Patrick Kevin (1953). Protein Crystallographer. Department of Physical Sciences, The Wellcome Research Laboratories, Langley Court, Beckenham, Kent, BR3 3BS, England. PhD (Aston U., 1988) chemical crystallography. *Macromolecular_crystallography data_collection drug_design education*
E-mail pb5@ib.rl.ac.uk Tel. 44(181)6395285 Fax 44(181)6633788

Bullock, Dr James F. (1962). Senior Research Chemist. Zeneca Specialties, Specialties Research Centre, PO Box 42, Hexagon House, Manchester, M9 8ZS, England. DPhil (Oxford U., 1987) materials science. *Organic_dyes organic_pigments X-ray_crystallography crystallization*
Tel. 44(161)7212169 Fax 44(161)7956005

Bunning, Dr John David (1949). Senior Lecturer. Division of Applied Physics, Sheffield Hallam University, Pond Street, Sheffield, S1 1WB, England. PhD (Leeds U., 1980) liquid crystals. *Structure physical_property liquid_crystals*
E-mail j.d.bunning@shu.ac.uk Tel. 44(114)2533355 [44(742)533355 before 1 April 1995] Fax 44(114)2533066 [44(742)533066 before 1 April 1995]

Bushnell-Wye, Dr Graham (1950). Senior Scientific Officer. Daresbury Laboratory, Daresbury, Warrington, Cheshire, WA4 4AD, England. PhD (London U., 1983) crystallography. *Applied_crystallography resonant_scattering synchrotron_radiation diffraction disordered_systems*
E-mail g.bushnell-wye@daresbury.ac.uk Tel. 44(1925)603623 Fax 44(1925)603174

Cahn, Prof. Robert W. (1924). Senior Associate. Editor of various book series and of journal Intermetallics. Department of Materials Science and Metallurgy, University of Cambridge, Pembroke Street, Cambridge, CB2 3QZ, England. ScD (Cambridge U., 1963) metallurgy. *Physical_metallurgy metallurgical_transformations recrystallization order-disorder scientific_editing scientific_popularisation*
Tel. 44(1223)334381 Fax 44(1223)334748 Telex 81240 CAMSPL G

Campbell, Dr John W. (1944). Software Support Scientist. DRAL Daresbury Laboratory, Daresbury, Warrington, WA4 4AD, England. PhD (Edinburgh U., 1969) chemistry. *Laue_diffraction data_processing synchrotron_radiation software*
E-mail j.campbell@daresbury.ac.uk Tel. 44(1925)603528

Carlile, Dr Colin (1946). Head of Spectroscopy and Support Division, ISIS. ILL Scientific Council. ISIS Pulsed Source, Rutherford Appleton Laboratory, Chilton, Didcot, Oxon, OX11 OQX, England. PhD (U. Birmingham, 1973) physics: neutron-scattering. *Slow_neutron_spectroscopy pulsed_neutron_scattering high-resolution_diffractometry molecular_crystals micas graphites molecular_tunnelling_spectroscopy*
E-mail cjc@isise.rl.ac.uk Tel. 44(1235)445684 Fax 44(1235)445383

Cartwright, Dr Michael (1940). Lecturer. Chemical Systems Group, Cranfield University (RMCS), Shrivenham, Swindon, Wiltshire, SN6 8LA, England. PhD (London U., 1974) chemistry. *Phase_transitions solid_solutions explosives polymers_interaction Rietveld_method stacking_faults nuclear_magnetic_resonance*
Tel. 44(1793)785357 Fax 44(1793)783192

Castleden, Dr Ian R. (1961). Scientific Officer. Cambridge Crystallographic Data Centre, 12 Union Road, Cambridge, CB2 1EZ, England. PhD (U. Western Australia, 1990) direct methods. *Database artificial_intelligence expert_system structure_determination maximum_entropy_method*
E-mail irc10@chemcrys.cam.ac.uk Tel. 44(1223)336034 Fax 44(1223)336033

Catlow, Prof. C. Richard A. (1947). Professor. Royal Institution, 21 Albemarle Street, London, W1X 4BS, England. DPhil (Oxford U., 1974) solid state chemistry. *Computer_modelling synchrotron_radiation catalysis zeolites powder_diffraction ceramics*
E-mail richard@ricx.ri.ac.uk Tel. 44(171)4092992 Fax 44(171)6293569

Cernik, Dr Robert Joseph (1954). Head of X-ray Diffraction in Materials Science. Member, IUCr Commission for Powder Diffraction. Daresbury Laboratory, Warrington, WA4 4AD, England. PhD (U. Wales, 1984) X-ray crystallography. *Synchrotron_radiation powder_diffraction structure_solution instrumentation school_governor*
E-mail r.j.cernik@dl.ac.uk Tel. 44(1925)603238 Fax 44(1925)603124

Champion, Dr John A. (1930). Retired. [191 London Road, Twickenham, Middlesex, TW1 1EJ, England.] PhD (U. London, 1961) physics. *Materials_metrology metallurgy crystal_growth micrography geology archaeology*
Tel. 44(181)8927898

Champness, Dr Pamela (1942). Reader. Department of Geology, University of Manchester, M13 9PL, England. PhD (U. Cambridge, 1969) mineralogy. *Analytical_electron_microscopy mineralogy metamorphic_reactions radiolysis*
Tel. 44(161)2753808 Fax 44(161)2753947

Chayen, Dr Naomi E. (1955). Postdoctoral Research Fellow; supervision of student projects. Biophysics Section, Physics Department, Imperial College, London, SW7 2BZ, England. PhD (Brunel U., 1984) biochemistry. *Protein_crystallization solubility automation microgravity histochemistry pharmacology*
Tel. 44(171)5947631 Fax 44(171)5890191

Cheetham, Dr Graham Mark Thomas (1967). Research Fellow. MRC Laboratory of Molecular Biology, Hills Road, Cambridge, CB2 2QH, England. PhD (Liverpool U., 1993) chemical crystallography. *Biological_diffraction macromolecular_structure microcrystallography computing*
E-mail gmtc@mrc-lmb.cam.ac.uk Tel. 44(1223)248011 Fax 44(1223)213556

Chen, Mr Yu Wai (1961). Research Student. Centre for Protein Engineering, Medical Research Council Centre, Hills Road, Cambridge, CB2 2QH, England. BSc (Imperial Coll. London, 1990) chemistry and biochemistry. *X-ray_crystallography proteins nuclear_magnetic_resonance mutagenesis computer_graphics protein_folding microcomputer_interface*
E-mail ywc@mrc-lmb.cam.ac.uk Tel. 44(1223)402152 Fax 44(1223)402140

Chippindale, Dr Ann Mary (1961). Research Fellow. Chemical Crystallography Laboratory, 9 Parks Road, Oxford, OX1 3PD, England. DPhil (Oxford U., 1987) chemistry. *Solid_state_chemistry X-ray_diffraction neutron_diffraction intercalates microporous_materials*
E-mail kryst4@vax.ox.ac.uk Tel. 44(1865)272600 Fax 44(1865)272690

Chisholm, Dr James Edwin (1945). Senior Scientific Officer. Health and Safety Executive, Broad Lane, Sheffield, S3 7HQ, England. PhD (1973) mineralogy. *Dust silicates qualitative_analysis asbestos quantitative_analysis*
Tel. 44(114)2892646 [44(742)892646 before 1 April 1995] Fax 44(114)2892500 [44(742)892500 before 1 April 1995] Telex 54556 A/B HSERLS G

Chrosch, Dr Jutta (1961). Postdoctoral Research Assistant. Department of Earth Sciences and Interdisciplinary Research Centre for Superconductivity, University of Cambridge, Downing Street, Cambridge CB2 3EQ, England. Dr rer.nat. (U. Köln, 1993) crystallography. *X-ray_diffuse_scattering superconductivity computing electrooptics nonlinear_optics*

Clackson, Dr Stephen Gregory (1961). Postdoctoral Fellow. Department of Computer Science, University College of Wales, Aberystwyth, SY23 3DB, Wales. PhD (Royal Holloway Coll. London, 1989) physics: X-ray crystallography. *X-ray_topography diffuse_scattering apparatus synchrotron_radiation computing diamond gallium_arsenide*
E-mail sgc93@aber.ac.uk Tel. 44(1823)698625 Fax 44(1970)622455

Clark, Dr Simon Martin (1958). Senior Scientific Officer. DRAL, Daresbury Laboratory, Warrington, WA4 4AD, England. PhD (Birkbeck Coll. London, 1993) crystallography. *High_pressure high_temperature phase_transitions powder_diffraction time_resolved_powder_diffraction*
E-mail smc@dl.ac.uk Tel. 44(1925)603123 Fax 44(1925)603174

Clark-Jones, Mrs Louise Elizabeth (1965). Editorial Assistant. International Union of Crystallography, 5 Abbey Square, Chester, CH1 2HU, England. BSc (U. Manchester, 1988) biochemistry. *Electronic_publishing information_science biocrystallography editing biological_sciences medical_sciences*
E-mail lj@iucr.ac.uk Tel. 44(1244)342878 Fax 44(1244)314888 Telex 669755 OFFICE G attn UNICRYSTAL

Cleasby, Dr Anne (1959). Research Scientist. Glaxo Research and Development Ltd, Biomolecular Structure Department, Greenford Road, Greenford, Middlesex, UB6 0HE, England. PhD (Birkbeck Coll. U. London, 1984) protein crystallography. *Protein_structure drug_design crystallization enzyme_mechanism*
E-mail ac12789@ggr.co.uk Tel. 44(181)9664188 Fax 44(181)9662156

Clegg, Prof. William (1949). Professor. Co-editor, Acta Crystallographica. Department of Chemistry, University of Newcastle, Newcastle upon Tyne, NE1 7RU, England. ScD (Cambridge U., 1989) chemistry. *Structure_determination data_collection structural_chemistry computer_aided_education synchrotron_radiation*
E-mail w.clegg@newcastle.ac.uk Tel. 44(191)2226649 Fax 44(191)2226929

Clifton, Dr Ian Jeffrey (1959). Postdoctoral Research Assistant. Laboratory of Molecular Biophysics, Rex Richards Building, South Parks Road, Oxford, OX1 3QU, England. DPhil (Oxford U., 1993) molecular biophysics. *Biocrystallography Laue_diffraction software rapid_data_collection*
E-mail ian.clifton@linacre.oxford.ac.uk Tel. 44(1865)275387 Fax 44(1865)510454

Cole, Mr Jason C. (1970). Research Student. Department of Chemistry, University of Durham, South Road, Durham, DH1 3LE, England. BSc (Bristol U., 1991) chemistry. *Database symmetry computing small_molecules*
E-mail j.c.cole@durham.ac.uk Tel. 44(191)3744703 Fax 44(191)3743745

Coles, Simon J. (1971). Research Assistant. School of Chemistry and Applied Chemistry, University of Wales Cardiff, PO Box 912, Cardiff, CF1 3TB, Wales. BSc (1992) chemistry. *Structure_determination computer_modelling organometallics catalysis quantum_mechanics*
E-mail coles.s.j@cardiff.ac.uk Tel. 44(1222)874950

Collison, Dr David (1952). Senior Lecturer in Chemistry. Chemistry Department, The University, Manchester, M13 9PL, England. PhD (Manchester U., 1980) chemistry. *Metalloproteins metalloenzymes EPR coordination_chemistry clusters_in_coordination_complexes ESEEM_ENDOR single_crystal_spectroscopy*
E-mail David.Collison@man.ac.uk Tel. 44(161)2754660 Fax 44(161)2754598

Colston, Sally (1963). Research Student. Birkbeck College, Department of Crystallography, Malet Steet, London, WC1E 7HX, England. MSc (Birkbeck College, U. London, 1985) crystallography. *X-ray_powder_diffraction cement_hydration cement_microstructure industrial_materials*
E-mail s.colston@cr.bbk.ac.uk Tel. 44(171)6316861 Fax 44(171)6316803

Conway, Mr Sean (1966). Editorial Assistant. International Union of Crystallography, 5 Abbey Square, Chester, CH1 2HU, England. MPhil (Manchester Polytechnic, 1990) chemistry. *Electronic_publishing editing chemical_nomenclature*
E-mail sc@iucr.ac.uk Tel. 44(1244)342878 Fax 44(1244)314888 Telex 669755 OFFICE G attn UNICRYSTAL

Cooper, Dr Jonathan (1963). Lecturer. Department of Crystallography, Birkbeck College, Malet Street, London, WC1E 7HX, England. PhD (London U., 1990) crystallography. *Proteins structure_determination*
E-mail j.cooper@uk.ac.bbk.cryst.mv3b Tel. 44(171)6316832 Fax 44(171)6316803

Cooper, Prof. Malcolm John (1944). Professor; Chairman, Faculty of Science. President, European Synchrotron Radiation Society. Department of Physics, University of Warwick, Coventry, CV4 7AL, England. PhD (Cambridge U., 1967) X-ray scattering. *Compton_scattering synchrotron_radiation magnetism non_destructive_testing*
E-mail csmc@spec.warwick.sc.uk Tel. 44(1203)523379 (direct) 44(1203)523379 (Department) Fax 44(1203)692016

Copley, Dr Royston C. B. (1966). Postdoctoral Research Associate. Department of Chemistry, University of Durham, South Road, Durham, DH1 3LE, England. DPhil (Oxford U., 1993) inorganic chemistry. *Inorganic_chemical_crystallography cluster_compounds charge_density*
E-mail r.c.b.copley@durham.ac.uk Tel. 44(191)3744704 Fax 44(191)3743745

Cousins, Christopher Stanley George (1934). Senior Lecturer. Physics Department, Exeter University, Stocker Road, Exeter, Devon, EX4 4QL, England. MA (Oxford U., 1958) physics. *Synchrotron_radiation resonant_scattering energy-dispersive_analysis differential_X-ray_spectroscopy*
E-mail c.cousins@cen.exeter.ac.uk Tel. 44(1392)264116 Fax 44(1392)264111 Telex 42894 EXUNIV G

Cox, Dr Ernest Gordon (1906). Retired. [117 Hampstead Way, London, NW11 7JN, England.] DSc (Bristol U., 1936) crystallography. *Carbohydrates coordination history_of_crystallography agricultural_research horticultural_research*
Tel. 44(181)4552618

Cox, Dr Philip John (1947). Lecturer. The Robert Gordon University, School of Pharmacy, School Hill, Aberdeen, AB9 1FR, Scotland. PhD (Glasgow U., 1972) chemical crystallography. *Natural_products carbohydrate organometallic pharmaceutical_chemistry*
E-mail p.j.cox@rgu.ac.uk Tel. 44(1224)262535 Fax 44(1224)626559

Crennell, Kate Mary (1933). Consultant. Formerly BCA Council Member (1989–1993). Rutherford Appleton Laboratory, Chilton, Didcot, Oxon, OX11 0QX, England. ['Greytops', The Lane, Chilton, Didcot, Oxon, OX11 0SE, England.] MS (U. Pittsburgh, 1960) physics. *Computer_graphics_molecular computer_aided_education computer_modelling scattering_neutron visualisation*
E-mail kmc@isise.rl.ac.uk Tel. 44(1235)834357 Fax 44(1235)445720

Crennell, Dr Susan J. (1966). Postdoctoral Research Assistant. Department of Biology and Biochemistry, University of Bath, Claverton Down, Bath, BA2 7AY, England. DPhil (Oxford U., 1991) crystallography. *Enzymatic_proteins time-of-flight_powder_diffraction*
E-mail bsssjc@bath.ac.uk Tel. 44(1225)826826x4302 Fax 44(1225)826449

Cruickshank, Prof. Durward William John (1924). Emeritus Professor. Chemistry Department, UMIST, Manchester, M60 1QD, England. [105 Moss Lane, Alderley Edge, Cheshire, SK9 7HW, England.] ScD (Cambridge U., 1961) crystallography. *Synchrotron_radiation macromolecules Laue_diffraction structural_chemistry*
Tel. 44(1625)582656 Fax 44(161)2367677

Dacombe, Mr Michael H. (1950). Executive Secretary IUCr. International Union of Crystallography, 2 Abbey Square, Chester, CH1 2HU, England. BSc (Leeds U., 1972) chemistry and earth sciences. *Information_science publishing*
E-mail execsec@iucr.ac.uk Tel. 44(1244)345431 Fax 44(1244)344843 Telex 669755 OFFICE G attn UNICRYSTAL

Darlington, Dr Charles Nicholas Wright (1945). Lecturer. BCA Physical Crystallography Committee Member. School of Physics, University of Birmingham, Birmingham, B15 2TT, England. PhD (U. Cambridge, 1971) physical crystallography. *Phase_transition structural_change perovskites superconductivity*
Tel. 44(121)4144729 Fax 44(121)4144719

David, Prof. William I. F. (1955). Professor. Vice-Chairman of BCA Physical Crystallography Group. ISIS Science Division, Rutherford Appleton Laboratory, Chilton, Didcot, Oxon, OX11 0QX, England. DPhil (Oxford U., 1981) physics: crystallography. *Powder_diffraction phase_transitions structure_determination pulsed_neutrons high_resolution_diffractometry molecular_crystals computer_graphics maximum_entropy_method*
E-mail wifd@isise.rl.ac.uk Tel. 44(1235)445179 Fax 44(1235)445720 Telex 83159 RUTHLB G

Davies, Dr Gideon J. (1964). Postdoctoral Research Assistant. Department of Chemistry, University of York, Heslington, York, YO1 5DD, England. PhD (Bristol U., 1990) biochemistry. *Biocrystallography enzymatic_catalysis enzymes*
E-mail davies@yorvic.york.ac.uk Tel. 44(1904)432596 Fax 44(1904)410519

Davies, Dr John Edward (1947). Senior Scientist. Cambridge Crystallographic Data Centre, 12 Union Road, Cambridge, CB2 1EZ, England. PhD (Monash U., 1974) crystallography. *Database*
E-mail jed2@chemcrys.cam.ac.uk Tel. 44(1223)336027 Fax 44(1223)336033

Dent Glasser, Dr Lesley Scott (1932). Director SATRO North Scotland. SATRO North Scotland, Marischal College, University of Aberdeen, AB9 1AS, Scotland. DSc (Aberdeen U., 1972) crystallography. *Education communicating_science enhancing_science_education hands-on_science_centres*
Tel. 44(1224)273161 Fax 44(1224)273160

Derwent, Frank William (1912). Retired. [43 Halls Farm Close, Winchester, Hampshire, SO22 6RE, England.] MSc (London U., 1948) physics: microscopy. *Biocrystallography surface lattice_distortion epitaxy*
Tel. 44(1962)880478

Diamond, Dr Robert (1929). Scientific Staff. IUCr Executive Committee 1984–93; Convenor, IUCr Finance Committee 1984–93; IUCr Computing Commission 1975–1984 (Chairman 1978–1981); British Crystallographic Association, President 1990–1992. MRC Laboratory of Molecular Biology, Hills Road, Cambridge, CB2 2QH, England. PhD (Cambridge U., 1956) physics. *Optimization thermal_motion macromolecular_geometric_fitting computer_graphics International_Tables*
E-mail rd10@mrc-lmb.cam.ac.uk Tel. 44(1223)248011[2210] Fax 44(1223)213556 Telex 81532

Dodson, Eleanor (1936). Research Associate. IUCr Electronic Publishing Committee. Department of Chemistry, University of York, Heslington, York, YO1 5DD, England. BA(Hons) (U. Melbourne, 1955) maths/english. *Crystallographic_computing_techniques*
E-mail ccp4@yorvic.york.ac.uk Tel. 44(1904)432520 Fax 44(1904)410519

Dodson, Prof. George Guy (1937). Professor of Biochemistry. Department of Chemistry, University of York, York, YO1 5DD, England. [NIMR, The Ridgeway, London NW7 1AA, England.] PhD (U. New Zealand, 1962) crystallography. *Crystallography proteins biochemistry synchrotron_radiation*
E-mail ggd@yorvic.york.ac.uk Tel. 44(1904)432520 (York) 44(181)9593666x2510 (London) Fax 44(1904)410519 (York) 44(181)9064477 (London)

Drew, Dr Michael G. B. (1941). Senior Lecturer. Department of Chemistry, The University, Whiteknights, Reading, RG6 2AD, England. PhD (Imperial Coll. London, 1966) chemical crystallography. *Computer_modelling molecular_mechanics quantum_mechanics crystal_packing host-guest_complexes*
E-mail scsdrew@reading.ac.uk Tel. 44(1734)875123x7427 Fax 44(1734)311610

Duggleby, Dr Helen (1965). Postdoctoral Research Assistant. Department of Chemistry, University of York, Heslington, York, YO1 5DD, England. DPhil (York U., 1990) protein crystallography. *Macromolecular_crystallography proteins structure_determination carbohydrate_degradation antibiotic_biosynthesis*
E-mail swift@yorvic.york.ac.uk Tel. 44(1904)432588 Fax 44(1904)410519

Duhlev, Dr Rumen (1956). Scientific Editor. World Scientific Publishing Co., 57 Shelton Street, London, WC2 H9HE, England. PhD (Bulgarian Acad. Sci., 1985) chemistry: inorganic crystallography. *Systematics structural_prediction composition_relationship inorganic_compounds*
E-mail wspc@wspc.demon.co.uk Tel. 44(171)8360888 Fax 44(171)8362020

Dunn, Dr Cameron R. (1962). Research Officer. University of Bristol, Department of Biochemistry, School of Medical Sciences, University Walk, Bristol, BS8 1TD, England. PhD (Bristol U., 1989) biochemistry and crystallography. *Computer_management macromolecular_crystallography catalytic_conformational_change computer_networking*
E-mail c.dunn@bristol.ac.uk Tel. 44(117)9303030x4359 [44(272)303030x4359 before 1 April 1995] Fax 44(117)9303497 [44(272)303497 before 1 April 1995] Telex 445938 BSUNIV G

Dyson, David John (1939). Head, X-ray and Electron Optics. Chairman BCA Industrial Group. British Steel Technical, Swinden Laboratories, Moorgate, Rotherham, South Yorkshire, S60 3AR, England. BSc (U. Durham, 1960) physics. *Computing steels_ceramics_dusts quantitative_diffraction electron_diffraction EDX WDS Texture*
Tel. 44(1709)825287 Fax 44(1709)825337 Telex 547279

Edgington, Dr P. R. (1963). VMS Systems Manager. Member COMCIFS. Cambridge Crystallographic Data Centre, 12 Union Road, Cambridge, CB2 1EZ, England. PhD (Glasgow U., 1993) computational chemistry. *Chirality database CIF_file_processing*
E-mail pre10@chemcrys.cam.ac.uk Tel. 44(1223)336026 Fax 44(1223)336033

Edmondson, Mr Michael (1940). Group Leader of Physical Sciences Group. Zeneca Specialties, Research Centre, Blackley, Manchester, M9 8ZS, England. MInstP (Institute of Physics, 1969) physics. *Polymorphism powder diffractometry organic_pigments differential_thermal_analysis EDX image_processing*
Tel. 44(161)7212431 Fax 44(161)7956005

Edwards, Dr Alison Jeanine (1962). Postdoctoral Research Assistant. Chemical Crystallography Laboratory, 9 Parks Road, Oxford, OX1 3PD, England. PhD (U. Melbourne, 1990) inorganic chemistry. *Chemical_crystallography absolute_configuration polymorphism phase_transition solid_state_nuclear_magnetic_resonance_spectroscopy*
E-mail aje@vax.ox.ac.uk Tel. 44(1865)270832 Fax 44(1865)270820

Elliott, Prof. James Cornelis (1937). Professor of Biophysics. Department of Child Dental Health, The London Hospital Medical College, Turner Street, London, E1 2AD, England. PhD (London Hosp. Med. Coll., 1964) crystallography. *Rietveld_method calcium_compounds X-ray_microscopy phosphates*
E-mail rdbc003@lhmc.ac.uk Tel. 44(171)3777639 Fax 44(171)3777677

Elsegood, Dr Mark Robert James (1965). University Demonstrator. Department of Chemistry, Bedson Building, University of Newcastle, Newcastle-upon-Tyne, NE1 7RU, England. PhD (Univ. Coll. London, 1991) chemistry: organometallic. *Chemical_crystallography low_temperature education chemistry_synthetic ruthenium_compounds_organometallic*
E-mail mark.elsegood@newcastle.ac.uk Tel. 44(191)2226000[7110] Fax 44(191)2611182

Errington, Dr William (1940). Lecturer. Department of Chemistry, University of Warwick, Coventry, CV4 7AL, England. PhD (Univ. Coll. London, 1965) chemistry. *Transition_elements gallium_compounds natural_products absolute_configuration*
E-mail msrpq@csv.warwick.ac.uk Tel. 44(1203)524803 Fax 44(1203)524112

Etheridge, Dr Joanne Postdoctoral Research Associate. Department of Materials Science and Metallurgy, University of Cambridge, Pembroke Street, Cambridge, CB2 3QZ, England. PhD (Royal Melbourne Inst. Tech., 1994) physics. *Electron_microscopy electron_diffraction superconductors*
E-mail je10006@cus.cam.ac.uk Tel. 44(1223)334469 Fax 44(1223)334437 Telex 81240 CAMSPL G

Evans, Dr Philip Richard (1946). Scientific Staff. MRC Laboratory of Molecular Biology, Hills Road, Cambridge, CB2 2QH, England. DPhil (Oxford U., 1973) protein crystallography. *Macromolecules RNA computing proteins*
E-mail pre@mrc-lmb.cam.ac.uk Tel. 44(1223)402211 Fax 44(1223)213556

Fawcett, Dr John (1947). Senior Experimental Officer. Department of Chemistry, Leicester University, University Road, Leicester, LE1 7RH, England. PhD (Leicester U., 1980) chemistry. *Organometallic_crystallography metalloorganic_fluorine_compounds anticancer_compounds*
E-mail jxf@leicester.ac.uk Tel. 44(116)2523616 [44(533)523616 before 1 April 1995] Fax 44(116)2523789 [44(533)523789 before 1 April 1995]

Fenn, Dr Ruth H. (1938). Principal Lecturer. School of Chemistry, Physics and Radiography, University of Portsmouth, Park Building, King Henry I Street, Portsmouth, Hampshire, PO1 2DZ, England. PhD (London U., 1964) crystallography. *High_temperature_superconductors crystal_structure_determination neutron_diffraction*
E-mail fennr@cv.port.ac.uk Tel. 44(1705)842151 Fax 44(1705)842157

Ferguson, Dr Ian Forster (1931). Retired. [1 Ingle Head, Fulwood, Preston, PR2 3NR, England.] PhD (London U., 1961) inorganic chemistry. *Powder_diffraction computing Auger_spectroscopy line_broadening atomic_energy ceramics corrosion*
Tel. 44(1772)717935 Fax 44(1772)717935

Fewster, Dr Paul Frederick (1950). Senior Principal Scientist. Philips Research Laboratories, Cross Oak Lane, Redhill, RH1 5HA, England. DSc (London U., 1994) crystallography. *Multiple_crystal_diffractometry topography theoretical_dynamical_diffraction polycrystal_materials*
E-mail fewster@prl.philips.co.uk Tel. 44(1293)815714 Fax 44(1293)815500

Field, Mrs Jennifer (1948). Technical Editor. Cambridge Crystallographic Data Centre, 12 Union Road, Cambridge, CB2 1EZ, England. BSc (U. London, 1970) chemistry. *Organic_database metalloorganic_database information_system*
E-mail jclf10@chemcrys.cam.ac.uk Tel. 44(1223)336031 Fax 44(1223)336033

Finney, Prof. John Leslie (1943). Quain Professor of Physics. Department of Physics and Astronomy, University College London, Gower Street, London, WC1E 6BT, England. PhD (Birkbeck Coll. London, 1968) crystallography. *Non-crystalline_condensed_matter biomolecule_structure water_structure hydrogen_bonding aqueous_solution ice_structures non-aqueous_solution*
E-mail jlf@v1.ph.ucl.ac.uk Tel. 44(171)3807850 Fax 44(171)3807145 Telex 28722 UCPHYS G

Flanders, Dr David John (1948). Software Support. Digital Equipment Co., Customer Support Centre, Jays Close, Basingstoke, Hampshire, RG22 4DE, England. PhD (1985) structural chemistry. *Computing computer_graphics computer_management*
E-mail flanders@uvo.dec.com Tel. 44(1256)488552 Fax 44(1256)488401 Telex 858489

Fletcher, Dr Steven Reginald (1946). Analytical Group Manager. Chairman BCA Industrial Group 1992–1994. ICI Chemicals and Polymers Ltd, PO Box 8, The Heath, Runcorn, Cheshire, WA7 4QD, England. PhD (Imperial Coll. London, 1972) chemical crystallography. *Powder line_profile_analysis*
Tel. 44(1928)513601 Fax 44(1928)581178

Foreman, Mrs Kathleen (1947). Technical Editor. Cambridge Crystallographic Data Centre, 12 Union Road, Cambridge, CB2 1EZ, England. BSc (Edinburgh U., 1967) chemistry. *Organic_database metalloorganic_database information_system*
E-mail klf10@chemcrys.cam.ac.uk Tel. 44(1223)336031 Fax 44(1223)336033

Forsyth, Prof. J. Bruce (1932). Senior Scientist. Secretary Commission on Charge, Spin and Momentum Densities; Commission on Neutron Scattering; Vice-President BCA. ISIS Diffraction Division, Rutherford Appleton Laboratory, Chilton, Oxon, OX11 0QX, England. PhD (Cambridge U., 1959) physics. *Magnetization_density diffractometry synchrotron_radiation structure_magnetic computing apparatus cryogenics instrumentation*
E-mail jbf@isise.rl.ac.uk Tel. 44(1235)446116 Fax 44(1235)445720

Fotinou, Dr Constantina (1963). Postdoctoral Research Assistant. Department of Chemistry, University of Glasgow, Glasgow, G12 8QQ, Scotland. PhD (Crete U., 1993) biochemistry. *Crystallization monoclonal_Fab_fragments hormones membranes*
E-mail dina@chem1.gla.ac.uk Tel. 44(141)3398855[6180] Fax 44(141)3304888 Telex 777070 UNIGLA

Frampton, Dr Christopher S. (1959). Team leader. Roche Products Ltd, PO Box 8, Welwyn Garden City, Hertfordshire, AL7 3AY, England. PhD (Essex U., 1985) inorganic chemistry. *Charge_density high_resolution_diffractometry chemical_crystallography pharmaceutical_activity total_density_analysis topological_properties_of_charge_distribution*
E-mail framptoc@rwep03.dnet.roche.com Tel. 44(1707)366031 Fax 44(1707)373504

Francis, Mr John Godfrey (1941). Senior Scientific Officer. Natural History Museum, Department of Mineralogy, Cromwell Road, London, SW7 5BD, England. BSc (London U., 1966) geology. *Minerals_identification minerals_characterization*
E-mail j.francis@nhm.ac.uk Tel. 44(171)9389274 Fax 44(171)9389268

Franks, Prof. Albert. Emeritus NPL Fellow. National Physical Laboratory, Teddington, Middlesex, TW11 0LW, England. DSc (London U., 1976) X-ray optics. *Grazing_incidence interferometry mirrors*
E-mail af@newton.npl.co.uk Tel. 44(181)9436515 Fax 44(181)9432945 Telex 262344 NPL G

Freemont, Dr Paul Simon (1959). ICRF Head of Laboratory. Protein Structure Laboratory, Imperial Cancer Research Fund, 44 Lincoln's Inn Fields, London, WC2A 3PX, England. PhD (Aberdeen U., 1984) biochemistry: protein sequencing. *Protein_crystallography structural_biology nucleic_acid_crystallography macromolecular_crystallisation protein_biochemistry protein_structure_and_function*
E-mail p_freemont@icrf.ac.uk Tel. 44(171)2693291 Fax 44(171)2693093

Freer, Dr Andrew (1946). Research Assistant. Department of Protein Crystallography, Chemistry Department, University of Glasgow, Glasgow, G12 8QQ, Scotland. PhD (Glasgow U., 1980) crystallography. *Protein_data_collection membrane_proteins pharmaceutical_compounds direct_methods crystallizing_membrane bound_proteins*
E-mail andy@chem1.gla.ac.uk Tel. 44(141)3398855 Fax 44(141)3304888

Fuentes-Mateos, Mr Angel Manuel (1966). Research Student. Institute of Food Research, Protein Engineering Department, Earley Gate, Whiteknights Road, Reading, RG6 2EF, England. MSc (U. Oviedo, 1992) food biotechnology. *Proteins biochemistry yeast_expression_systems crystallization*
E-mail fuentes@afrc.ac.uk Tel. 44(1734)357000x254 Fax 44(1734)267917

Fuller, Prof. Watson (1935). Head of Department. Department of Physics, Keele University, Keele, Staffordshire, ST5 5BG, England. PhD (1961) biophysics. *Biophysics DNA polymeric_material synchrotron_radiation SAXS WAXS Neutron_diffraction*
Tel. 44(1782)583326 Fax 44(1782)711093 Telex 36113 UNKLIB G

Fülöp, Dr Vilmos (1961). Postdoctoral Research Assistant. Central Research Institute for Chemistry, 1025 Budapest, Pusztaszeri ut 59-67, Hungary. [Laboratory of Molecular Biophysics, University of Oxford, South Parks Road, Oxford, OX1 3QU, England.] PhD (Eötvös U. Budapest, 1989) chemistry: chemical crystallography. *Biocrystallography crystallization Laue_diffraction metalloproteins time-resolved_diffraction*
E-mail vilmos@biop.ox.ac.uk Tel. 44(1865)275378 Fax 44(1865)510454

Furley, Jonathon (1967). [52 Herschell Road, Leigh-on-Sea, Essex, SS9 2NH, England.] PhD (Brunel U., 1994) materials/microtexture. *Superconductors electron-microscopy microtexture grain_boundary_engineering*
E-mail jon.furley@brunel.ac.uk Tel. 44(1702)78777

Garman, Dr Elspeth F. (1954). Research Assistant. Laboratory of Molecular Biophysics, Rex Richards Building, University of Oxford, South Parks Road, Oxford, OX1 3QU, England. DPhil (Oxford U., 1980) nuclear structure physics. *Macromolecular_crystallography data_collection structure_determination*
E-mail elspeth@biop.ox.ac.uk Tel. 44(1865)275398 Fax 44(1865)510454

Geddes, Dr Alexander J. (1941). Senior Lecturer. Department of Biochemistry and Molecular Biology, University of Leeds, Leeds, LS2 9JT, England. PhD (1966) biophysics: fibre diffraction. *Macromolecular_crystallography drug_computer-assisted_design proteins_sequence_database binding_enzyme_inhibitors*
E-mail bph6ajg@biovax.leeds.ac.uk Tel. 44(113)2333041 [44(532)333041 before 1 April 1995] Fax 44(113)2333167 [44(532)333167 before 1 April 1995]

Giles, Mr Raymond Richard (1944). Senior Physicist. Alcan Chemicals Ltd, Chalfont Park, Gerrards Cross, Bucks, SL9 0QB, England. *Crystallography ceramics scanning_electron_microscopy EDS microscopy characterization size_distribution*
Tel. 44(1753)887373 Fax 44(1735)881556

Gilmore, Dr Chris (1946). Reader in Chemistry. Department of Chemistry, University of Glasgow, Glasgow, G12 8QQ, Scotland. PhD (Bristol U., 1968) crystallography. *Direct_method powder electron_microscope structure_determination*
E-mail chris@tcrystal.glasgow.ac.uk Tel. 44(141)3304419 Fax 44(141)3304419 Telex 777070 UNIGLA

Glasser, Prof. Fredrik Paul (1929). Professor. Department of Chemistry, University of Aberdeen, Meston Walk, Aberdeen, AB9 2UE, Scotland. DSc (U. Aberdeen, 1969) chemistry. *Phase_equilibria chemical_crystallography oxide_chemistry*
Tel. 44(1224)272906 Fax 44(1224)272921 Telex 73458 UNIABN G

Glazer, Dr Anthony Michael (1943). Lecturer. Editor of Journal of Applied Crystallography. Physics Department, Clarendon Laboratory, University of Oxford, Parks Road, Oxford, OX1 3PU, England. PhD (Univ. Coll. London, 1968) crystallography. *Absolute_configuration optical_property phase_transition physical_property symmetry perovskite_structures disorder*
E-mail glazer@physics.ox.ac.uk Tel. 44(1865)272290 Fax 44(1865)272290 Telex 83154 CLAROX

Glazier, Edward James (1948). Senior Scientific Officer. Metropolitan Police, Forensic Science Laboratory, 109 Lambeth Road, London, SE1 7LP, England. MSc (Thames Poly, London, 1977) chemical analysis. *Debye-Scherrer powder forensic_microanalysis drug*
Tel. 44(171)2306392 Fax 44(171)2306253 Telex 892733

Gildewell, Christopher (1944). Reader. School of Chemistry, University of St. Andrews, St. Andrews, Fife, KY16 9ST, Scotland. ScD (Cambridge U., 1990) chemistry. *Conducting_polymers hydrogen_bonding organometallic_clusters sandwich_compounds*
E-mail cg@st-andrews.ac.uk Tel. 44(1334)463839 Fax 44(1334)463808

Glykos, Mr Nicholas M. (1968). Research Student. Astbury Building, University of Leeds, Leeds, LS2 9JT, England. BSc (Athens U., 1990) biology. *Structure_determination protein-nucleic_acid_complex*
E-mail bmb5nimg@biovax.leeds.ac.uk Tel. 44(113)2333027 [44(532)333027 before 1 April 1995] Fax 44(113)2336017 [44(532)336017 before 1 April 1995]

Gould, Dr Robert Ozburn (1938). Senior Lecturer. Department of Chemistry, The University of Edinburgh, Edinburgh, EH9 3JJ, Scotland. PhD (U. St. Andrews, 1963) chemistry. *Symmetry intermolecular_interaction disorder direct_method*
E-mail r.o.gould@ed.ac.uk Tel. 44(131)6504806 Fax 44(131)6504743

Gould, Dr Sheila E. B. (1940). Director BMMU. Beevers Miniature Models Unit, Department of Chemistry, The University of Edinburgh, West Mains Road, Edinburgh, EH9 3JJ, Scotland. PhD (Edinburgh U., 1965) chemistry. *Structural_modelling biological_chemistry*
E-mail bmmu@castle.ed.ac.uk Tel. 44(131)6504824 Fax 44(131)6504743 Telex 727442 UNIVED G

Gourley, David Gilmour (1970). Research Student. The Robertson Laboratories, Department of Chemistry, Glasgow University, Glasgow, G12 8QQ, Scotland. BSc (Glasgow U., 1992) chemistry with medicinal chemistry. *Protein_crystallography*
E-mail davidg@chem.gla.ac.uk Tel. 44(141)3398855 Fax 44(141)3304888

Gover, Dr Sheila (1948). Postdoctoral Research Assistant. Laboratory of Molecular Biophysics, Rex Richards Building, South Parks Road, Oxford, OX1 3QU, England. PhD (Cambridge U., 1975) physical crystallography. *Computing macromolecular_structure_determination dehydrogenases diffuse_scattering*
E-mail sheila@biop.ox.ac.uk Tel. 44(1865)275392 Fax 44(1865)510454

Greasley, Samantha E. (1970). Research Student. Department of Biochemistry and Molecular Biology, Astbury Building, University of Leeds, Leeds, LS2 9JT, England. BSc (Leeds U., 1991) molecular biophysics. *X-ray_crystallography*
E-mail sam@biovax.leeds.ac.uk Tel. 44(113)2332592 [44(532)332592 before 1 April 1995] Fax 44(113)2333167 [44(532)333167 before 1 April 1995]

Greenhough, Dr Trevor J. (1948). Senior Lecturer. Department of Physics, Keele University, Keele, Staffs, ST5 5BG, England. PhD (U. Surrey, 1976) crystallography. *Protein_structure synchrotron_radiation data_processing Laue_diffraction C_reactive_proteins software*
E-mail green@dlpx1.dl.ac.uk Tel. 44(1782)583405 Fax 44(1782)711093

Gutteridge, Mr Walter Alfred (1931). Manager - Industrial Affairs. British Cement Association, Century House, Telford Avenue, Crowthorne, Berks., RG11 6YS, England. MSc (London U., 1967) crystallography. *Powder_diffraction cement minerals*
Tel. 44(1344)725712 Fax 44(1344)727205

Habash, Dr Jarjis (1943). Scientist. Structural Chemistry Group, Chemistry Department, Manchester University, Manchester, M13 9PL, England. PhD (Sheffield U., 1980) crystallography. *Crystallography Laue_diffraction proteins synchrotron_radiation*
E-mail jh@v5.chemistry.manchester.ac.uk Tel. 44(161)2754688 Fax 44(161)2754734

Hajdu, Dr Janos (1948). Senior MRC Staff Scientist. Laboratory of Molecular Biophysics, Oxford University, South Parks Road, Oxford, OX1 3QU, England. DSc (Hung. Acad. Sci., 1973) chemistry and biochemistry. *Biocrystallography crystallisation Laue_diffraction fast_diffraction_methods time-resolved_crystallography*
E-mail janos@biop.ox.ac.uk Tel. 44(1865)275763 Fax 44(1865)510454

Halfpenny, Dr Joan Christine (1954). Principal Lecturer. Department of Chemistry and Physics, Nottingham Trent University, Clifton Lane, Nottingham, NG11 8NS, England. PhD (U. Lancaster, 1978) crystallography. *X-ray_diffraction structural_chemistry mercury_compounds*
E-mail chp3halfpjc@cluster.ntu.ac.uk Tel. 44(115)9418418x3312 [44(602)418418x3312 before 1 April 1995] Fax 44(115)9486636 [44(602)486636 before 1 April 1995]

Halliwell, Mary A. G. (1942). Scientific Consultant. Philips Analytical X-ray, Lelyweg 1, 7602EA Almelo, The Netherlands. [172 High Road, Trimley St Mary, Ipswich, IP10 0SS, England.] MSc (Chelsea Coll. London, 1967) solid state physics. *X-ray_rocking_curves epitaxy_semiconductor perfection X-ray_topography diffraction_space_mapping*
Tel. 44(1394)279753 Fax 44(1394)279753

Hamor, Dr Thomas Senior Lecturer. School of Chemistry, University of Birmingham, Birmingham B15 2TT, England. DSc (U. Birmingham, 1974) chemistry. *Coordination structure-activity_relationship structure_determination bonding*
E-mail hamorta@ibm3090.bham.ac.uk Tel. 44(121)4144360 Fax 44(121)4144403

Hao, Dr Quan (1963). Lecturer. Department of Applied Physics, De Montfort University, Leicester, LE1 9BH, England. PhD (Academia Sinica, 1988) crystallography. *Direct_method Laue_diffraction anomalous_dispersion computing*
E-mail qhao@dmu.ac.uk Tel. 44(116)2577125 [44(533)577125 before 1 April 1995] Fax 44(116)2577135 [44(533)577135 before 1 April 1995]

Harding, Dr Marjorie M. (1934). Reader in Chemistry. Co-Editor Acta Crystallographica. Chemistry Department, Liverpool University, PO Box 147, Liverpool, L69 3BX, England. DPhil (Oxford U., 1961) crystallography. *Synchrotron_radiation Laue_diffraction microcrystal structure_determination*
E-mail mmh@liv.ac.uk Tel. 44(151)7943535 Fax 44(151)7943588

Harlos, Dr Karl (1948). Postdoctoral Research Assistant. Laboratory of Molecular Biophysics, The Rex Richards Building, South Parks Road, Oxford, OX1 3QU, England. PhD (Tech. U. Braunschweig, 1980) biochemistry. *Macromolecular_crystallography X-ray_instrumentation crystallization*
E-mail karl@biop.ox.ac.uk Tel. 44(1865)275398 Fax 44(1865)275182

Harris, Dr Gillian W. (1960). Senior Researcher. Protein Crystallography, Department of Protein Engineering, IFR Reading Laboratory, Earley Gate, Whiteknights Road, Reading, RG6 2EF, England. PhD (U. Witwatersrand, 1986) chemistry. *Macromolecular_structure_determination least-squares_refinement phase_determination diffuse_scattering TLS_refinement*
E-mail gillian.harris@bbsrc.ac.uk Tel. 44(1734)357142 Fax 44(1734)267917

Harris, Dr Kenneth David Maclean (1963). Lecturer. Department of Chemistry, University College London, 20 Gordon Street, London, WC1H 0AJ, England. PhD (Cambridge U., 1988) solid state chemistry. *Inclusion_compounds hydrogen_bonding incommensurate_phases diffraction solid_state_NMR_spectroscopy EXAFS_spectroscopy dynamic_properties_of_solids*
E-mail kenneth.harris@ucl.ac.uk Tel. 44(171)3877050x4683 Fax 44(171)3807463

Harris, Dr Stephanie Ellen (1967). Research Scientist. BCA Executive Secretary. 7 Staplow Road, Worcester, WR5 2LZ, England. PhD (Bristol U., 1991) crystallography. *Structural_database structural_systematics*
E-mail s.e.harris@aston.ac.uk Tel. 44(1905)358550 Fax 44(1905)358550

Hart, Prof. Michael (1938). Consultant. IUCr Executive Committee. Department of Physics, Schuster Laboratory, The University, Manchester, M13 9PL, England. DSc (U. Bristol, 1971) physics. *Synchrotron_radiation dynamical_diffraction polarization perfect_crystal X-ray_optics*
E-mail 100331.2750@CompuServe.COM Tel. 44(1565)632893 Fax 44(1565)632893

Harvey, Dr Ian (1964). Lecturer in synchrotron radiation. School of Applied Sciences, De Montfort University, Leicester, LE1 9BH, England. [Daresbury Laboratory, Warrington, WA4 4AD, England.] PhD (U. Warwick, 1990) chemistry. *Metalloproteins metalloenzymes bioinorganic_structure_determination EXAFS X-ray_fluorescence_spectroscopy*
E-mail iha@dl.ac.uk Tel. 44(1925)603658 Fax 44(1925)603124

Hasnain, Prof. Samar (1952). Professor. Editor, Journal of Synchrotron Radiation. Molecular Biophysics Group, Daresbury Laboratory, Warrington, Cheshire, WA4 4AD, England. PhD (Manchester U., 1976) physics. *Biophysics synchrotron_radiation EXAFS scattering*
E-mail sj@dlva.daresbury.ac.uk Tel. 44(1925)603273 Fax 44(1925)603174

Hatt, Ben Albert (1931). Retired. [23 Tancred Road, High Wycombe, Bucks., HP13 5EQ, England.] MSc (London U., 1957) crystallography. *Industry_X-ray_diffraction intermetallic_phase_equilibrium intermetallic_phase_transition CVD XPS metallography*
Tel. 44(1491)528055

Haworth, Dr Colin W. (1932). Honorary Lecturer (Retired). Department of Engineering Materials, Sheffield University, Sir Robert Hadfield Building, PO Box 600, S1 4DU, England. DPhil (Oxford U., 1958) metallurgy. *Metallurgy electron_probe_microanalysis diffusion diffractometry*
Tel. 44(114)2301486 [44(742)301486 before 1 April 1995] Fax 44(114)2754325 [44(742)754325 before 1 April 1995]

Heale, Gillian (1949). Technical Editor. Cambridge Crystallographic Data Centre, 12 Union Road, Cambridge, CB2 1EZ, England. BSc (U. London, 1971) biology and chemistry. *Organic_database metalloorganic_database information_system*
E-mail gh106@chemcrys.cam.ac.uk Tel. 44(1223)336031 Fax 44(1223)336033

Helliwell, Prof. John Richard (1953). Professor of Structural Chemistry. Co-Editor, Acta Crystallographica; Editor, Journal of Synchrotron Radiation; Chairman Commission on Synchrotron Radiation (1989-1993). Section of Structural Chemistry, Department of Chemistry, University of Manchester, Manchester, M13 9PL, England. DPhil (U. Oxford, 1978) molecular biophysics. *Macromolecular_crystallography synchrotron_radiation structural_chemistry anomalous_dispersion_methods Laue_method lectins enzymes*
E-mail hell@man.ac.uk Tel. 44(161)2754686 Fax 44(161)2754734 Telex 666517 UNIMAN

Hempstead, Dr P. D. (1966). Postdoctoral Research Assistant. Molecular Biology and Biotechnology, University of Sheffield, Western Bank, Sheffield, S10 2TN, England. PhD (Sheffield U., 1992) chemical crystallography. *Protein_crystallography biological_mineralization*
E-mail p.hempstead@sheffield.ac.uk Tel. 44(114)2824190 [44(742)824190 before 1 April 1995] Fax 44(114)2728697 [44(742)728697 before 1 April 1995]

Henderson, Prof. C. M. B. (1938). Professor of Petrology. Department of Geology, University of Manchester, Manchester, M13 9PL, England. DSc (Durham U., 1992) mineralogy and petrology. *Petrology mineralogy aluminosilicate_phase-transitions EXAFS*
E-mail chenderson@fs2.ge.man.ac.uk Tel. 44(161)2753812 Fax 44(161)2753947

Henderson, Dr Richard (1945). Research Scientist. MRC Laboratory of Molecular Biology, Hills Road, Cambridge, CB2 2QH, England. PhD (Cambridge U., 1970) molecular biology. *Protein_structure electron_microscopy protein_crystallography*
E-mail rh15@mrc-lmb.cam.ac.uk Tel. 44(1223)402215 Fax 44(1223)213556 Telex 81532

Hendriksen, Dr Barry A. (1943). Senior Chemist. Pharmaceutical Research, Eli Lilly and Co., Lilly Research Centre, Erl Wood Manor, Windlesham, Surrey, GU20 6PH, England. PhD (U. Nottingham, 1968) physical chemistry. *Characterization polymorphism crystallization morphology*
Tel. 44(1276)853260 Fax 44(1276)853392

Hibbs, David E. (1970). Research Assistant. School of Chemistry and Applied Chemistry, University of Wales Cardiff, PO Box 912, Cardiff, CF1 3TB, Wales. BSc (1992) chemistry. *Biochemistry biomolecule structure_determination chirality computer_modelling amino_acids organic_synthesis sequencing drug*
E-mail hibbs.de.@cardiff.ac.uk Tel. 44(1222)874950

Holden, David (1969). Research Student. Department of Physics, Keele University, Keele, Staffs, ST5 5BG, England. BSc (John Moores U., 1991) applied physics. *Biophysics crystallography structure_determination Laue_method time_resolved_crystallography*
E-mail phd85@seql.keele.ac.uk Tel. 44(1782)621111 Fax 44(1782)711093 Telex 36113 UNKLIB G

Holmes, Dr Gillian F. (1965). Senior Editorial Assistant. International Union of Crystallography, 5 Abbey Square, Chester, CH1 2HU, England. PhD (U. Manchester, 1991) radio astronomy. *Electronic_publishing database_preparation editing checking*
E-mail gh@iucr.ac.uk Tel. 44(1244)342878 Fax 44(1244)314888 Telex 669755 OFFICE G attn UNICRYSTAL

Howard, Prof. Judith Ann Kathleen (1945). Professor of Structural Chemistry. President of BCA; UK IRC Representative. Department of Chemistry, University of Durham, South Road, Durham, DH1 3LE, England. DSc (Bristol, 1986) chemical crystallography. *Charge_density low-temperature_crystallography instrumentation neutron_diffraction chemical_crystallography crystallographic_systematics*
E-mail j.a.k.howard@durham.ac.uk Tel. 44(191)3744647 Fax 44(191)3743745

Howard, Dr Sean (1964). SERC Advanced Research Fellow. Project team of Charge, Spin and Momentum Densities Commission. School of Chemistry and Applied Chemistry, University of Wales Cardiff, PO Box 912, Cardiff, CF1 3TB, Wales. PhD (U. Birmingham, 1990) theoretical chemistry. *Charge_density X-ray_diffraction quantum_chemistry*
E-mail howardst@cardiff.ac.uk Tel. 44(1222)874950 Fax 44(1222)874029

Howie, Dr R. Alan (1940). Senior Research Officer. Department of Chemistry, University of Aberdeen, Meston Walk, Aberdeen, AB9 2UE, Scotland. PhD (U. Aberdeen, 1973) chemistry. *Structure_determination inorganic_compounds organic_compounds organometallic_compounds*
E-mail r.a.howie@aberdeen.ac.uk Tel. 44(1224)272907 Fax 44(1224)272921 Telex 73458 uniabn g

Howlin, Dr Brendan (1959). Lecturer. Department of Chemistry, University of Surrey, Guildford, Surrey, GU2 5XH, England. PhD (Essex U., 1984) chemistry. *Macromolecular_crystallography computer_modelling computer_aided_education Mossbauer proteins_polymers_inorganics*
E-mail chs1bh@surrey.ac.uk Tel. 44(1483)300800x2592 Fax 44(1483)300803

Hoy, Vanessa Jane (1971). Research Student. Chemistry Department, University of Durham, South Road, Durham, DH1 3LE, England. BSc (Durham U., 1993) chemistry. *Chemical_crystallography neutron_diffraction database*
E-mail v.j.hoy@durham.ac.uk Tel. 44(191)3744703 Fax 44(191)3743745

Hoyland, Dr Michael A. (1965). Research and Development Assistant. International Union of Crystallography, 5 Abbey Square, Chester, CH1 2HU, England. PhD (U. Birmingham, 1989) solid state physics. *Electronic_publishing information_storage*
E-mail mh@iucr.ac.uk Tel. 44(1244)342878 Fax 44(1244)314888 Telex 669755 OFFICE G attn UNICRYSTAL

Hughes, David (1955). Research Technician. Department of Chemistry, University of Wales Cardiff, PO Box 912, Cardiff, CF1 3TB, Wales. MPhil (U. Cardiff, 1994) pharmacy and crystallography. *Drug_design cancer aminoglutethimide*
E-mail sacdsh@cardiff.ac.uk Tel. 44(1222)874950 Fax 44(122)874029

Hughes, Dr David Lewis (1941). Principal Scientific Officer. Nitrogen Fixation Laboratory, University of Sussex, Brighton, BN1 9RQ, England. PhD (U. British Columbia, 1971) chemical crystallography. *Single_crystal_analysis structure_determination transition_element_complexes metalloenzymes nitrogen_fixation cofactor_cluster_complexes*
E-mail hughes@afrc.ac.uk Tel. 44(1273)678198 Fax 44(1273)678133

Hukins, Prof. David (1947). MacRobert Professor of Physics. Department of Bio-Medical Physics and Bioengineering, University of Aberdeen, Foresterhill, Aberdeen, AB9 2ZD, Scotland. DSc (U. Manchester, 1991) biophysics. *Biomaterial biomechanics biomedical_calcification medical_image_processing applications_in_orthopaedics*
E-mail d.hukins@abdn.ac.uk Tel. 44(1224)681818x53495 Fax 44(1224)685645 Telex 73458 UNIABN G

Hull, Dr Stephen (1959). Senior Scientific Officer. ISIS Science Division, Rutherford Appleton Laboratory, Chilton, Didcot, Oxon OX11 0QX, England. PhD (1985) physics. *Defect_clusters neutron_diffraction superionic_conductors high_pressure I–VII_compounds*
E-mail sh@isise.rl.ac.uk Tel. 44(1235)821900x6628 Fax 44(1235)445720

Humphreys, Prof. Colin J. (1941). Professor and Head of Department. Department of Materials Science and, Metallurgy, University of Cambridge, Pembroke Street, Cambridge, CB2 3QZ, England. PhD (U. Cambridge, 1967) physics. *Electron_microscopy electron_diffraction accurate_structure_factor microlithography materials_science*
Tel. 44(1223)334457 Fax 44(1223)334437 Telex 81240 CAMSPL G

Hunter, Dr William N. (1958). Senior Lecturer. Department of Chemistry, University of Manchester, Oxford Road, Manchester, M13 9PL, England. PhD (Glasgow U., 1982) chemistry. *Crystallisation biocrystallography nucleic_acids proteins*
Tel. 44(161)2754712 Fax 44(161)2754598

Hurley, Mr Patrick W. (1937). Self Employed Consultant. [Endian, Gore Tree Road, Hemingford Grey, Huntingdon, Cambs., PE18 9BP, England.] *Powder X-ray_fluorescence_spectroscopy radiation_protection diffractometry technical_writing training*
Tel. 44(1480)469458 Fax 44(1480)469458

Hursthouse, Prof. Michael B. (1941). Professor. Department of Chemistry, University of Wales Cardiff, PO Box 912, Cardiff, CF1 3TB, Wales. PhD (U. London, 1965) structural inorganic chemistry. *Systematics_metal_complexes structure_determination charge_density computer_modelling solid_state_reactions phase_transitions*
E-mail hursthouse@cardiff.ac.uk Tel. 44(1222)874068 Fax 44(1222)874029

Hussain-Bates, Dr Bilquis (1961). Postdoctoral Research Assistant. Department of Crystallography, Birkbeck College, Malet Street, London WC1E 7HX, England. PhD (Queen Mary and Westfield Coll. U. London, 1993) chemical crystallography. *Peptides crystallization purification circular_dichroism indium_compounds*
E-mail ubcg25e@mv3b.cryst.bbk.ac.uk Tel. 44(171)6316455 Fax 44(171)4368918

Hutchings, Dr Michael Thomas (1937). Manager : NDT Materials Characterisation. NDT Department, Technical Services Division, AEA Technology, 521.1 Harwell, Didcot, Oxon, OX11 0RA, England. DPhil (Oxford U., 1963) physics. *Crystalline_disorder magnetism residual_stress neutron_scattering positron_annihilation*
E-mail mike.hutchings@aea.orgn.uk Tel. 44(1235)435232 Fax 44(1235)432274

Ibberson, Dr Richard M. (1965). Instrument scientist responsible for the High Resolution Powder Diffractometer (HRPD). ISIS Science Division, Rutherford Appleton Laboratory, Chilton, Didcot, Oxon, OX11 0QX, England. PhD (Reading U., 1993) physics. *Neutron_powder_diffraction high_resolution_diffractometry small_molecules_organic phase_transitions*
E-mail rmi@isise.rutherford.ac.uk Tel. 44(1235)445871 Fax 44(1235)445720 Telex 83159 RUTHLAB G

Isaacs, Prof. Neil William (1945). Professor. Department of Chemistry, University of Glasgow, Glasgow, G12 8QQ, Scotland. PhD (U. Queensland, 1970) chemistry. *Proteins glycoproteins membranes enzymes*
E-mail neil@chem.gla.ac.uk Tel. 44(141)3398855x5954 Fax 44(141)3304888 Telex 777070 UNIGLA

Jackson, Dr Robert A. (1957). Lecturer. Department of Chemistry, Keele University, Keele, Staffordshire, ST5 5BG, England. PhD (Univ. Coll. London, 1984) computational chemistry. *Computer_modelling_solids*
E-mail cha41@cc.keele.ac.uk Tel. 44(1782)583042 Fax 44(1782)712378

Jakubovics, Dr John Paul (1938). Lecturer. Department of Materials, Parks Road, Oxford, OX1 3PH, England. PhD (Cambridge U., 1965) physics. *Electron_microscopy magnetic_domain magnetism multilayer*
E-mail john.jakubovics@materials.oxford.ac.uk Tel. 44(1865)273722 Fax 44(1865)273789

Janes, Dr Robert William (1956). Lecturer in Crystallography. Department of Crystallography, Birkbeck College, University of London, Malet Street, London, WC1E 7HX, England. PhD (Birkbeck Coll. London, 1992) crystallography. *Proteins_peptides_drug macromolecular_crystallography magnetic_resonance biochemistry*
E-mail ubcg06w@ccs.bbk.ac.uk Tel. 44(171)6316857 Fax 44(171)6316803

Jeffreys, Dr J. A. D. (1927). Honorary Senior Lecturer; Retired. Department of Pure and Applied Chemistry, University of Strathclyde, Glasgow, G1 1XL, Scotland. DPhil (Oxford U., 1952) chemistry. *Camera_method organic_compounds spectroscopy_and_molecular_structure synthesis_of_organic_compounds*
E-mail cbas14@vaxa.strath.ac.uk Tel. 44(141)5524400x2285 Fax 44(141)5525664

Jeyaratnam, Mailoo (1929). Senior Scientific Officer. Health and Safety Executive, OMH Laboratory, Broad Lane, Sheffield, S3 7HQ, England. MSc (Strathclyde U., 1968) forensic science. *Diffraction diffraction_technique diffraction_theory crystallinity microscopy X-ray_spectrometry food_science law*
Tel. 44(114)2892645 [44(742)892645 before 1 April 1995] Fax 44(114)2892850 [44(742)892850 before 1 April 1995]

Jhoti, Dr Harren (1962). Research Scientist. Glaxo Research and Development Ltd, Biomolecular Structure Department, Greenford Road, Greenford, Middlesex, UB6 0HE, England. PhD (Birkbeck Coll. U. London, 1989) protein crystallography. *Protein_structure drug_design molecular_modelling*
E-mail hj17786@ggr.co.uk Tel. 44(181)9662182 Fax 44(181)9662156

Johnson, Prof. Louise Napier (1940). Professor of Molecular Biophysics. Laboratory of Molecular Biophysics, University of Oxford, Rex Richards Building, South Parks Road, Oxford, OX1 3QU, England. PhD (London U., 1965) biophysics. *Biocrystallography Laue_diffraction structure-activity_relationship enzymes_phosphorylases*
E-mail louise@biop.ox.ac.uk Tel. 44(1865)275365 Fax 44(1865)510454

Johnson, Dr Michael William (1944). Division Head. Rutherford Appleton Laboratory, Chilton, Didcot, Oxon, OX11 0QX, England. PhD (U. London, 1971) crystallography. *Instrumentation inverse_problem engineering computer_graphics*
E-mail mwj@isise.rl.ac.uk Tel. 44(1235)445418 Fax 44(1235)445383

Johnson, Dr Owen (1958). Scientific Officer. Cambridge Crystallographic Data Centre, 12 Union Road, Cambridge, CB2 1EZ, England. PhD (U. Bradford, 1985) chemical crystallography. *Computing database crystallography*
E-mail oj100@chemcrys.cam.ac.uk Tel. 44(1223)336023 Fax 44(1223)336033

Jones, Prof. Derry Wynn (1928). Honorary Professorial Fellow. Chemistry and Chemical Technology, University of Bradford, Bradford, West Yorkshire, BD7 1DP, England. [11 Meadow Close, Harden, Bingley, West Yorkshire, BD16 1JB, England.] DSc (Bradford U., 1980) structural chemistry. *Polycyclic_carcinogens neutron_diffraction nuclear_magnetic_resonance carbides chemical_education fossil_fuels organic_crystal_chemistry*
Tel. 44(1535)273963 Fax 44(1274)385350 Telex 51309 UNIBFD G

Jones, Dr Edith Yvonne (1960). Royal Society University Research Fellow. Laboratory of Molecular Biophysics, The Rex Richards Building, South Parks Road, Oxford, OX1 3QU, England. DPhil (Oxford U., 1985) molecular biophysics. *Macromolecular_crystallography immunology*
E-mail yvon@biop.oxford.ac.uk Tel. 44(1865)275181 Fax 44(1865)275182

Jones, Dr Richard Hywel (1957). Lecturer. Department of Chemistry, University of Keele, Keele, Staffordshire, ST5 5BG, England. PhD (U. Birmingham, 1984) chemistry. *Aluminosilicates heterogeneous_catalysis synchrotron_radiation Rietveld_method microporous_solids time-resolved_studies neutron_diffraction*
E-mail cha35@seq1.kl.ac.uk Tel. 44(1782)583039 Fax 44(1782)712378 Telex 36113

Jones, Dr William (1949). University Lecturer. Chemistry Department, Cambridge University, Lensfield Road, Cambridge, CB2 1EW, England. PhD (U. Wales, 1974) chemistry. *Solid_state_reactivity hydrogen_bonding structure_determination electron_microscopy crystal_engineering intercalation_compounds pillared_layered_solids*
E-mail wj10@phx.cam.ac.uk Tel. 44(1223)336468 Fax 44(1223)336362 Telex 81240camsplg

Kariuki, Dr Benson M. (1962). Postdoctoral Research Associate. Chemistry Department, Liverpool University, PO Box 147, Liverpool, L69 3BX, England. PhD (Cambridge U., 1991) chemistry. *X-ray_structure_determination structure–activity_relationship synchrotron_radiation*
E-mail bmk@xss2.ch.liv.ac.uk Tel. 44(151)7943515 Fax 44(151)7943588

Keller, Mr Peter A. (1963). Research Assistant. Department of Biochemistry, University of Bath, Bath, BA2 7AY, England. MSc (Birkbeck Coll. London, 1992) crystallography. *Proteins antibodies*
E-mail p.a.keller@bath.ac.uk Tel. 44(1225)826826x4302 Fax 44(1225)826449

Kelly, Mr Eric (1939). Unemployed. [44 Ffordd Penymynydd, Penyffordd, Chester, CH4 0LD, England.] BA (Open U., 1976) general science. *Powder diffraction_analysis*
Tel. 44(1244)549536

Kennard, Dr Olga (1924). Director CCDC. Cambridge Crystallographic Data Centre, 12 Union Road, Cambridge, CB2 1EZ, England. ScD (Cambridge U., 1971) crystallography. *Oligonucleotides structural_biology structural_chemistry structural_database information_systems*
E-mail ok10@chemcrys.cam.ac.uk Tel. 44(1223)336408 Fax 44(1223)336033

Kepert, Mr Cameron John (1970). Research Student. Royal Institution of Great Britain, 21 Albemarle Street, London, W1X 4BS, England. BSc (U. Western Australia, 1991) chemistry. *BEDT–TTF superconductivity magnetism charge_density lanthanide*
E-mail cameron@ricx.ri.ac.uk Tel. 44(171)4092992 Fax 44(171)6293569

Kidd, Dr Patricia (1961). Postdoctoral Research Assistant. Department of Materials Science and Engineering, University of Surrey, Guildford, Surrey, GU2 5XH, England. DPhil (Oxford U., 1983) materials science. *Multiple-crystal_diffractometry strain_determination multilayers semiconductors*
Tel. 44(1483)300800x2421 Fax 44(1483)31040

King, Mrs Susan E. (1950). Technical Editor. International Union of Crystallography, 5 Abbey Square, Chester, CH1 2HU, England. BSc (U. Leeds, 1973) biophysics. *Electronic_publishing editing journal_publication*
E-mail sk@iucr.ac.uk Tel. 44(1244)342878 Fax 44(1244)314888 Telex 669755 OFFICE G attn UNICRYSTAL

Kitson, Dr David H. (1958). Support Scientist. Biosym Technologies, Ltd, Unit 17, Intec 2, Wade Road, Basingstoke, RG24 0NE, England. PhD (U. Glasgow, 1983) chemistry. *Homology_proteins molecular_mechanics potential_energy structure-activity_relationship_drug*
E-mail david@biosym.demon.co.uk Tel. 44(1256)817577 Fax 44(1256)817600

Klug, Dr Aaron (1926). Director. MRC Laboratory of Molecular Biology, Hills Road, Cambridge, CB2 2QH, England. PhD (Cambridge U., 1953) physics. *Crystallography_proteins_nucleic_acids image_reconstruction electron_microscopy DNA_interactions gene_expression*
Tel. 44(1223)248011 Fax 44(1223)412231

Knight, Kevin Steven (1956). Instrument Scientist. ISIS Science Division, Rutherford Appleton Laboratory, Chilton, Didcot, Oxon, OX11 0QX, England. MSc (Birkbeck Coll. London, 1982) crystallography. *Neutron_powder_diffraction X-ray_powder_diffraction mineralogy ionic_conductors*
E-mail ksk@isise.rl.ac.uk Tel. 44(1235)445220 Fax 44(1235)445720 Telex 83159 RUTHLB G

Körber, Dr Fritjof Carl Friedrich (1952). Senior Lecturer. School of Biomolecular Sciences, Liverpool John Moores University, Byrom Street, Liverpool, L3 3AF, England. PhD (U. of Leeds, 1984) biophysics: macromolecular crystallography. *Macromolecular_crystallography porphyrins small_angle_scattering low-temperature_crystallography*
E-mail bmsfkorb@livjm.ac.uk Tel. 44(151)2312255 Fax 44(151)2981946

Kumaraswamy, Vidya Sagar (1966). Research Student. Department of Physics, University of Keele, Keele, Staffs, ST5 5BG, England. BSc (John Moores U., 1991) applied physics. *Biophysics crystallography diffuse_scattering structure_determination cryocooled_crystallography*
E-mail phd87@seql.keele.ac.uk Tel. 44(1782)621111x7941 Fax 44(1782)711093 Telex 36113 UNKLIB G

Ladd, Dr Marcus F. C. (1926). Department of Chemistry, University of Surrey, Guildford, Surrey, England. DSc (London U., 1975) crystallography. *Structure_analysis symmetry_computing*
Tel. 44(1438)300800

Lake, Mr Philip G. (1956). Development Analyst. The Wellcome Foundation Ltd, Temple Hill, Dartford, Kent, DA1 5AH, England. BSc (Bristol U., 1978) chemistry. *Pharmaceuticals polymorphism organics quantification*
Tel. 44(1332)223488x1971 Fax 44(1332)289285

Lang, Prof. Andrew Richard (1924). Emeritus Professor. H. H. Wills Physics Laboratory, University of Bristol, Tyndall Avenue, Bristol, BS8 1TL, England. PhD (Cambridge U., 1953) crystallography. *Crystal_growth electron_microscopy X-ray_topography diamond*
Tel. 44(117)9288954 [44(272)288954 before 1 April 1995] Fax 44(117)9255624 [44(272)255624 before 1 April 1995]

Langford, Dr J. Ian (1935). Reader. Co-Editor J. Applied Crystallography; Co-opted member, Commission on Powder Diffraction. School of Physics and Space Research, University of Birmingham, Birmingham, B15 2TT, England. DSc (Birmingham U., 1988) powder diffraction. *Powder_diffraction line_profile_analysis microstructure search_and_match*
E-mail xt-1@i.ph.bham.ac.uk Tel. 44(121)4144662 Fax 44(121)4144719

Langridge, Sarah-Jane (1971). Research Student. The Krebs Institute, Department of Molecular Biology, University of Sheffield, PO Box 594, Sheffield, S10 2UH, England. BSc (Sheffield U., 1993) biochemistry and microbiology. *Protein_crystallography macromolecular_structure enzymes molecular_crystals metabolism molecular_replacement structure_determination*
E-mail s.j.langridge1@sheffield.ac.uk Tel. 44(114)2824242 [44(742)824242 before 1 April 1995] Fax 44(114)2728697 [44(742)728697 before 1 April 1995]

Lapthorn, Dr Adrian J. (1966). Postdoctoral Research Assistant. Robertson Laboratories, Department of Chemistry, University of Glasgow, Glasgow, G12 8QQ, Scotland. PhD (Liverpool Polytechnic, 1992) crystallography. *Protein_crystallography biochemistry glycoproteins structure*
E-mail adrian@chem.gla.ac.uk Tel. 44(141)3398855x6180 Fax 44(141)3304888

Leake, Dr John Anthony (1939). Lecturer. Department of Materials Science and Metallurgy, University of Cambridge, Pembroke Street, Cambridge, CB2 3QZ, England. PhD (Cambridge U., 1965) physics. *Identification multilayer size_distribution texture characterisation_of_materials*
Tel. 44(1223)334331 Fax 44(1223)334373

Lehmann, Dr Christian W. (1964). Research Scientist. Department of Chemistry, University of Durham, South Road, Durham, DH1 3LE, England. Dr.rer.nat. (Freie U. Berlin, 1990) crystallography. *Charge_density neutron_powder_diffraction synchrotron_radiation*
E-mail c.w.lehmann@durham.ac.uk Tel. 44(191)3744702 Fax 44(191)3844737

Leonard, Dr Gordon A. (1959). Postdoctoral Research Fellow. Department of Chemistry, University of Manchester, Oxford Road, Manchester, M13 9PL, England. PhD (Surrey U., 1984) chemical crystallography. *Biocrystallography DNA RNA synchrotron_radiation*
E-mail gordon@v5.chemistry.manchester.ac.uk Tel. 44(161)2754689 Fax 44(161)2754598

Leonidas, Dr Demetrios D. (1964). Postdoctoral Research Officer. Biochemistry Department, University of Bath, Bath, BA2 7AY, England. PhD (Athens U., 1992) biochemistry. *Structure–activity_relationship crystallization_methods proteins Charcot_Leyden_crystal_protein*
E-mail bssddl@midge.bath.ac.uk Tel. 44(1225)826826x3091

Leslie, Dr Andrew G. W. (1949). Staff Scientist. MRC Laboratory of Molecular Biology, Hills Road, Cambridge, CB2 2QH, England. PhD (Manchester U., 1970) crystallography. *Macromolecular_crystallography X-ray_data_processing*
E-mail andrew@mrc-lmb.cam.ac.uk Tel. 44(1223)248011 Fax 44(1223)213556 Telex 81532

Lieberman, Harvey (1969). Research Student. School of Chemistry and Applied Chemistry, University of Wales Cardiff, PO Box 912, Cardiff, CF1 3TB, Wales. PhD (U. Wales, 1990) chemistry. *X-ray_crystallography computer_modelling charge_density drug_design*
E-mail lieberman@cardiff.ac.uk Tel. 44(1222)874950 Fax 44(1222)874029

Lightfoot, Dr Philip (1961). Senior Scientific Officer. Department of Chemistry, University of St. Andrews, St. Andrews, Fife, KY16 9ST, Scotland. DPhil (Oxford U., 1987) chemical crystallography. *Powder_diffraction oxides superconductors zeolites neutron_diffraction solid_electrolytes polymer_electrolytes*
E-mail pl@st-and.ac.uk Tel. 44(1334)463841 Fax 44(1334)463808

Lindley, Prof. Peter F. (1942). Head of Structural Biology Division. Co-Editor of Acta Crystallographica. DRAL Daresbury Laboratory, Daresbury, Warrington, Cheshire, WA4 4AD, England. PhD (Bristol U., 1966) chemistry. *Macromolecular_crystallography synchrotron_radiation plasma_proteins eye_lens_structure*
E-mail p.f.lindley@dl.ac.uk Tel. 44(1925)603324 Fax 44(1925)603124

Lipscomb, Ms Karen J. (1964). Technical Editor. Cambridge Crystallographic Data Centre, 12 Union Road, Cambridge, CB2 1EZ, England. MSc (Sheffield U., 1988) chemistry. *Organic_database metalloorganic_database information_system*
E-mail kjl12@chemcrys.cam.ac.uk Tel. 44(1223)336031 Fax 44(1223)336033

Lorimer, Prof. Gordon Winston (1941). Professor; Head of Department. Manchester Materials Science Centre, University of Manchester/UMIST, Grosvenor Street, Manchester, M1 7HS, England. PhD (U. Cambridge, 1968) metallurgy. *Analytical_electron_microscopy aluminium_compounds magnesium_compounds alloys X-ray_microanalysis_of_thin_specimens*
E-mail g.lorimer@fs1.mt.umist.ac.uk Tel. 44(161)2003567 Fax 44(161)2003636 Telex 666094

Loveday, Dr John S. (1963). Research Fellow. Department of Physics, The University of Edinburgh, Mayfield Road, Edinburgh, EH9 3JZ, Scotland. [ISIS Facility R3, Rutherford Appleton Laboratory, Chilton, Didcot, Oxon, OX11 0QX, England.] PhD (Bristol U., 1989) physics. *Physical_crystallography high_pressure pulsed_neutron_diffraction_techniques structured_phase_transitions*
E-mail jsl01@isise.nd.rl.ac.uk Tel. 44(1235)446873 Fax 44(1235)445720

Lowe, Dr Philip R. (1948). Departmental Superintendent. Department of Pharmaceutical and Biological Sciences, Aston University, Birmingham, B4 7ET, England. PhD (Aston U., 1984) chemical crystallography. *Chemical_crystallography*
Tel. 44(121)3593611x4190 Fax 44(121)3590733

Ludlam-Brown, Dr Ian R. (1959). Section Head. Inhaled Technology Unit, Glaxo Manufacturing Services, Priory Street, Ware, Hertfordshire, SG12 0DJ, England. PhD (1992) pharmaceutical technology. *Powder crystallisation_X-ray diffraction_thermal characterisation laser_diffractometry process_control*
Tel. 44(1920)463993 Fax 44(1920)863128

Macdonald, Dr John Emyr (1959). Lecturer. Department of Physics and Astronomy, University of Wales Cardiff, PO Box 913, Cardiff, CF2 3YB, Wales. DPhil (Oxford U., 1985) physics. *Grazing_incidence surface thin_film strain*
E-mail macdonald@cf.ac.uk Tel. 44(1222)874458 Fax 44(1222)874056

Mackay, Prof. Alan Lindsay (1926). Professor Emeritus. Department of Crystallography, Birkbeck College, University of London, Malet Street, London, WC1E 7HX, England. [22 Lanchester Road, London N6 4TA, England.] DSc (London U., 1986) crystallography. *Computer_graphics systematics structure flexi-crystallography generalised_crystallography*
E-mail ubcg04m@ccs.bbk.ac.uk Tel. 44(181)8834810 Fax 44(171)6316803

Maclean, John K. F. (1971). Research Student. Department of Chemistry, Glasgow University, Glasgow, G12 8QQ, Scotland. BSc (Glasgow U., 1993) chemistry. *Protein_crystallography*
E-mail johnm@chem1.glasgow.ac.uk Tel. 44(141)3398855x6180 Fax 44(141)3304374

Macrae, Clare F. (1964). Computer Officer. Cambridge Crystallographic Data Centre, 12 Union Road, Cambridge, CB2 1EZ, England. BA (Oxford U., 1987) chemistry. *Computing mol:cular_modelling crystallography_databases*
E-mail cfm10@chemcrys.cam.ac.uk Tel. 44(1223)336024 Fax 44(1223)336033

Maginn, Dr Stephen James (1965). Senior Applications Support Scientist. Secretary, BCA Cher ical Crystallography Group. Molecular Simulations, 240/250 The Quorum, Barnwell Road, Cambridge, CB5 8RE, England. PhD (Liverpool U., 1990) chemistry: crystallography. *Molecular_modelling polymorphism crystal_growth_morphology phase_transitions*
E-mail steve_maginn@msicam.co.uk Tel. 44(1223)413300 Fax 44(1223)413301

Main, Dr Peter (1939). Reader in Physics. Department of Physics, University of York, York, YO1 5DD, England. PhD (Manchester U., 1963) physics: X-ray crystallography. *Direct_method phase_determination image_processing double_Patterson_method*
E-mail pm1@vaxa.york.ac.uk Tel. 44(1904)432265 Fax 44(1904)432214

Mallinson, Dr Paul (1943). Programmer. Secretary, IUCr Commission on Charge, Spin and Momentum Densities Project 'XD'. Chemistry Department, University of Glasgow, Glasgow, G12 8QQ, Scotland. PhD (Essex U., 1970) chemical crystallography. *Charge_density quantum_mechanics computing molecular_inclusion organometallics*
E-mail paul@chem.gla.ac.uk Tel. 44(141)3398855x4409 Fax 44(141)3304888

Malone, Dr John Francis (1944). Senior Lecturer. School of Chemistry, Queen's University, Belfast, BT9 5AG, Northern Ireland. PhD (Leeds U., 1969) chemistry. *Stereochemistry absolute_configuration structure–activity_relationship computer_modelling*
E-mail j.malone@v2.qub.ac.uk Tel. 44(1232)245133x4423 Fax 44(1232)382117

Manojlović-Muir, Dr Ljubica (1931). Reader. Member of Editorial Board, J. Fluorine Chemistry. Chemistry Department, University of Glasgow, Glasgow, G12 8QQ, Scotland. PhD (U. Belgrade, 1963) chemical crystallography. *X-ray_neutron_diffraction chemical_crystallography organometallic_complexes metal_clusters hydrogen_bonding*
E-mail ljubica@chem1.gla.ac.uk Tel. 44(141)3398855x4506 Fax 44(141)3304888 Telex 777070 UNIGLA

Marvin, Dr D. A. (1934). Senior Research Associate. Department of Biochemistry, University of Cambridge, Cambridge, CB2 1QW, England. PhD (King's Coll. London, 1960) physics. *Macromolecular_structure–activity_relationship biological_fibres structural_phase_transition helical_macromolecules*
E-mail marvin@mrc-lmb.cam.ac.uk Tel. 44(1223)353090

Matthew, Prof. James Andrew Davidson (1938). Professor and Head of Department. Department of Physics, University of York, Heslington, York, YO1 5DD, England. PhD (Aberdeen U., 1964) physics. *Surface_crystallography electron_spectroscopy atomic_scattering_factors history_of_physics*
Tel. 44(1904)432200 Fax 44(1904)432214

McCullough, Dr Kevin J. (1952). Senior Lecturer. Department of Chemistry, Heriot-Watt University, Riccarton, Edinburgh, EH14 4AS, Scotland. PhD (Strathclyde U., 1979) chemistry. *Structure_determination_organic structure_determination_organometallic conformation_analysis organic_peroxides*
E-mail chekjm@clust.hw.ac.uk Tel. 44(131)4495111x4120 Fax 44(131)4513180

McKee, Dr Vickie (1955). Reader. Chemistry Department, Queen's University, Belfast, BT9 5AG, Northern Ireland. PhD (Queen's U. Belfast, 1979) inorganic chemistry. *Macrocycles transition_metals bioinorganic_chemistry*
E-mail v.mckee@qub.ac.uk Tel. 44(1232)245133 Fax 44(1232)382117

McKie, Dr Christine H. (1931). Lecturer. Department of Earth Sciences, University of Cambridge, CB2 3EQ, England. PhD (U. Cambridge, 1958) crystallography. *Mineralogy polytypism_silicates structural_change computer_aided_education*
E-mail chk1@esc.cam.ac.uk Tel. 44(1223)333400x33479 Fax 44(1223)333450

McKie, Dr Duncan (1930). University Lecturer. Department of Earth Sciences, Downing Street, Cambridge, CB2 3EQ, England. PhD (Cambridge U., 1962) mineralogy. *Structural_mineralogy carbonates anomalous_scattering chemical_crystallography history_of_mineralogy*
Tel. 44(1223)333404 Fax 44(1223)333450

McLachlan, Dr Andrew D. (1935). Scientific Staff. MRC Laboratory of Molecular Biology, Hills Road, Cambridge, CB2 2QH, England. ScD (Cambridge U., 1988) biology: protein structure. *Proteins homology maximum-entropy_method Fourier_transform*
E-mail admcl@mrc-lmb.cam.ac.uk Tel. 44(1223)248011 Fax 44(1223)213556

McMahon, Mr Brian (1956). Research and Development Officer. Coordinating Secretary, Committee for the Maintenance of the CIF standard (COMCIFS). International Union of Crystallography, 5 Abbey Square, Chester, CH1 2HU, England. MA (U. Oxford, 1978) physics. *Electronic_publishing database computing CIF structure_checking computer_networks*
E-mail bm@iucr.ac.uk Tel. 44(1244)342878x27 Fax 44(1244)314888 Telex 669755 OFFICE G attn UNICRYSTAL

McMahon, Dr Malcolm Iain (1965). Postdoctoral Research Associate. Department of Physics, The University of Edinburgh, Mayfield Road, Edinburgh, EH9 3JZ, Scotland. [DRAL Daresbury Laboratory, Daresbury, Warrington, Cheshire, WA4 4AD, England.] PhD (U. Edinburgh, 1991) physics. *Physical_crystallography structural_phase_transition high_pressure diffraction_technique*
Tel. 44(1925)603639 Fax 44(1925)603174

Meek, Dr Keith Michael Andrew (1951). Chairman of Biophysics Research Group. BCA Biological Structures Group Committee Member. The Open University, Oxford Research Unit, Foxcombe Hall, Boars Hill, Oxford, OX5 1HR, England. PhD (Manchester U., 1976) biophysics. *Fibrous_diffraction fibrous_proteins biological_macromolecules synchrotron_radiation collagen_structure corneal_transparency corneal_pathologies*
Tel. 44(1865)327001 Fax 44(1865)326322

Milledge, Dr H. Judith (1927). Emeritus Reader. Crystallography and Mineral Physics Unit, Department of Geological Sciences, University College London, Gower Street, London, WC1E 6BT, England. [46 Gibson Square, Islington, London N1 0RA, England.] DSc (U. London, 1963) crystallography. *Diamond_analysis isotope_solid_solutions X-ray_real_time_imaging restrained_least_squares X-ray_techniques FTIR/Raman_defect_spectroscopy growth/Dissolution Morphology*
Tel. 44(171)3889415 Fax 44(171)3887614

Mills, Dr Alan (1945). Postdoctoral Research Assistant. Crystallography Department, Birkbeck College, University of London, Malet Street, London, WC1E 7HX, England. PhD (Leeds U., 1973) physics. *Proteins proteinases Laue_diffraction time_resolved_effect aspartic_proteases lectins ribonuclease*
E-mail a.mills@cryst.bbk.ac.uk Tel. 44(171)6316551 Fax 44(171)4368918

Mills, Mr Owen S. Reader. Department of Chemistry, University of Manchester, Manchester, M13 9PL, England. BSc (Liverpool U., 1945) chemistry. *Organic_chemistry organometallic_chemistry superconductors*
E-mail osm@man.ac.uk Tel. 44(161)2754680 Fax 44(161)2754598

Mitchell, Dr Geoffrey Robert (1949). Reader in Polymer Physics. University of Reading, Department of Physics, Whiteknights, Reading, RG6 2AF, England. PhD (CNAA, 1983) materials science and polymer physics. *Polymers computer_modelling fibres liquid_crystals*
E-mail g.r.mitchell@reading.ac.uk Tel. 44(1734)318573 Fax 44(1734)750203 Telex 847813 RULIB G

Monahan, Miss Ailsa M. K. (1970). Editorial Assistant. International Union of Crystallography, 5 Abbey Square, Chester, CH1 2HU, England. BSc (U. Glasgow, 1992) chemistry with computer applications. *Electronic_publishing chemical_nomenclature editing*
E-mail am@iucr.ac.uk Tel. 44(1244)342878 Fax 44(1244)314888 Telex 669755 OFFICE G attn UNICRYSTAL

Moody, Dr Peter C. E. (1956). Postdoctoral Research Fellow. Protein Structure Laboratory, Chemistry Department, University of York, York, YO1 5DD, England. PhD (Imperial Coll. London, 1984) biophysics. *Macromolecular_crystallography carcinogenesis antibiotics-binding enzyme DNA_repair DNA-binding_proteins*
E-mail moody@yorvic.york.ac.uk Tel. 44(1904)432590 Fax 44(1904)410519

Moore, Dr A. Moreton (1943). Reader. Editor of Crystallography Reviews. Department of Physics, Royal Holloway, University of London, Egham, Surrey, TW20 0EX, England. PhD (U. Bristol, 1973) physics. *X-ray_topography multiple-crystal_diffractometry diamond_diffuse_scattering synchrotron_radiation crystal_growth morphology*
E-mail m.moore@vax.rhbnc.ac.uk Tel. 44(1784)443441 Fax 44(1784)472794 Telex 935504

Moore, Caroline (1972). Research Student. Department of Physics, University of Durham, South Road, Durham, DH1 3LE, England. BSc (1993) physics. *Bragg_intensity X-ray_topography diffractometry reflectivity*
E-mail c.d.moore@durham.ac.uk Tel. 44(191)3742114 Fax 44(191)3743749

Moore, Miss Katie A. (1968). Editorial Assistant. International Union of Crystallography, 5 Abbey Square, Chester, CH1 2HU, England. BSc (U. Edinburgh, 1991) astrophysics. *Electronic_publishing information_science editing*
E-mail cm@iucr.ac.uk Tel. 44(1244)342878 Fax 44(1244)314888 Telex 669755 OFFICE G attn UNICRYSTAL

Moore, Dr Madeleine H. (1960). Lecturer. Chemistry Department, York University, Heslington, York, YO1 5DD, England. PhD (U. Cape Town, 1987) chemistry. *Oligonucleotide_structure protein_structure inclusion_compounds DNA_interactions teaching_crystallography data_collection DNA_protein_drug_interactions*
E-mail moore@yorvic.york.ac.uk Tel. 44(1904)432584 Fax 44(1904)410519

Moseley, Dr Patrick Timothy (1943). Technical Director. European Editor, Journal of Power Sources. Capteur Sensors and Analysers Ltd, 66 Milton Park, Abingdon, Oxon, OX14 4RY, England. [Ivy Cottage, South Row, Chilton, Didcot, Oxon, OX11 0RT, England.] DSc (Durham U., 1994) chemical crystallography and materials chemistry. *Microstructure oxides conductivity catalysis sensors*
Tel. 44(1235)821323 Fax 44(1235)820632

Moss, Dr David S. (1941). Reader; Chairman of Department. Department of Crystallography, Birkbeck College, Malet Street, London WC1E 7HX, England. PhD (U. London, 1967) chemical crystallography. *Proteins computer_modelling diffuse_scattering_proteins computer_aided_education eye_lens_proteins MHC_proteins*
E-mail d.moss@cryst.bbk.ac.uk Tel. 44(171)6316802 Fax 44(171)6316803

Motherwell, Dr William David Samuel (1941). Principal Scientist. Cambridge Crystallographic Data Centre, 12 Union Road, Cambridge, CB2 1EZ, England. PhD (St. Andrews U., 1967) chemical crystallography. *Packing computer_graphics computing database*
E-mail wdsm@chemcrys.cam.ac.uk Tel. 44(1223)336021 Fax 44(1223)336033

Muir, Dr Kenneth Walter (1941). Reader. Co-editor, Acta Crystallographica. Chemistry Department, Glasgow University, Glasgow, G12 8QQ, Scotland. PhD (U. Glasgow, 1967) chemical crystallography. *Charge_density crystallographic_computing metal–loorganic_structures cluster_compounds_transition_elements inclusion_compounds*

E-mail ken@chem1.gla.ac.uk Tel. 44(141)3398855x5354 Fax 44(141)3304888 Telex 777070 UNIGLA

Muirhead, Dr Hilary Reader in Biochemistry. Department of Biochemistry, University of Bristol, Bristol, BS8 1TD, England. PhD (Cambridge U., 1964) physics. *Proteins allostery computer_modelling*

E-mail h.muirhead@bristol.ac.uk Tel. 44(117)9288593 [44(272)288593 before 1 April 1995] Fax 44(117)9288274 [44(272)288274 before 1 April 1995]

Murray-Rust, Dr Judith (1946). Scientific Officer. Crystallography Department, Birkbeck College, Malet Street, London, WC1E 7HX, England. PhD (Stirling U., 1971) chemistry: crystallography. *Proteins receptors growth_factors*

E-mail ubcg09j@cr.bbk.ac.uk j_murray@icrf.icnet.uk Tel. 44(171)6316214 Fax 44(171)6316590

Murray-Rust, Dr Peter (1941). Research Associate. Consultant, IUCr COMCIFS. Glaxo Research and Development Ltd, Biomolecular Structure Department, Greenford Road, Greenford, Middlesex, UB6 0HE, England. DPhil (Oxford U., 1968) crystallography. *Protein_structure computing information_storage distance_learning collaborative_computing World_Wide_Web bioinformatics*

E-mail pmr1716@ggr.co.uk Tel. 44(181)9663075 Fax 44(181)9662156

Myles, Dr Dean A. A. (1963). Postdoctoral Research Fellow. Department of Physics, Keele University, Keele, Staffordshire, ST5 5BJ, England. PhD (Keele U., 1992) protein crystallography. *Proteins_crystallography proteins_crystallization synchrotron_radiation*

Tel. 44(1782)621111x7941 Fax 44(1782)711093 Telex 36113 UNKLIB G

Neidle, Prof. Stephen (1946). Professor of Biophysics; Chairman, Section of Structural Biology; Head of Laboratories, The Institute of Cancer Research. Cancer Research Campaign, Biomolecular Structure Unit, The Institute of Cancer Research, Cotswold Road, Sutton, Surrey, SM2 5NG, England. PhD (Imperial Coll. London, 1970) chemical crystallography. *DNA anticancer_compounds computer-assisted_design X-ray_structure_determination drug-DNA_interactions*

E-mail steve@bms-unit.icr.ac.uk Tel. 44(181)6438901 Fax 44(181)6431675

Nelmes, Prof. Richard J. (1943). Professor. Chairman of High Pressure Group of Commission on Crystallographic Apparatus. Department of Physics and Astronomy, The University of Edinburgh, Mayfield Road, Edinburgh, EH9 3JZ, Scotland. [ISIS Facility R3, Rutherford Appleton Laboratory, Chilton, Didcot, Oxon, OX11 0QX, England.] ScD (Cambridge U., 1982) physics. *Physical_crystallography structural_phase_transition high_pressure diffraction_technique*

E-mail rjn01@isise.rl.ac.uk Tel. 44(1235)445285 Fax 44(1235)445720

Nguti, Nde Donatus (1968). Research Student. Department of Physics, Keele University, Keele, Staffs, ST5 5BG, England. BSc (Keele U., 1991) physics and mathematics. *X-ray_anomalous_scattering*

E-mail phd04@seq1.cc.keele.ac.uk Tel. 44(1782)621111

Noreland, Dr Jakob (1964). Visiting Research Fellow. Rutherford Appleton Laboratory, Chilton, Didcot, Oxon, OX11 0QX, England. PhD (U. Uppsala, 1992) inorganic chemistry. *Neutron_Compton_scattering hydrogen_bonding structure_determination cation_radical_salts*

E-mail jakobn@kemi.uu.se Tel. 44(1235)445882 Fax 44(1235)445720 Telex 83159 RUTHLB G

North, Prof. Anthony Charles Thomas (1931). Astbury Professor of Biophysics. Secretary-General, International Union for Pure and Applied Biophysics. Department of Biochemistry and Molecular Biology, University of Leeds, Leeds, LS2 9JT, England. PhD (King's Coll. London, 1955) biophysics. *Proteins_structure computer_modelling proteins_database*

E-mail actm@biovax.leeds.ac.uk Tel. 44(113)2333023 [44(532)333023 before 1 April 1995] Fax 44(113)2333167 [44(532)333167 before 1 April 1995]

Nunn, Dr Christine Margaret (1961). Macromolecular Crystallographer. Institute of Cancer Research, 15 Cotswold Road, Sutton, Surrey, SM2 5NG, England. PhD (Bristol U., 1987) chemistry: chemical crystallography. *X-ray_crystallography DNA protein computing drug_DNA_complexes*

E-mail c.nunn@icr.ac.uk Tel. 44(181)6438901[4536] Fax 44(181)7707893

Nyburg, Prof. Stanley C. (1924). Hon. Senior Research Fellow. Department of Chemistry, King's College, University of London, Strand, London, WC2R 2LS, England. DSc (London U., 1973) crystallography and thermodynamics. *Intermolecular_interaction alkanes thermodynamics long_chain_aliphatic_compounds*

E-mail udca097@bay.cc.kcl.ac.uk Tel. 44(171)8732161 Fax 44(171)8732810

O'Hara, Dr Bernard P. (1957). Research Fellow. Biomolecular Structure Group, Department of Biochemistry and Molecular Biology, University College, Gower Street, London, WC1E 6BT, England. PhD (Birkbeck Coll. London, 1992) crystallography. *Computer_graphics protein_structure_determination protein_crystallization*

E-mail o'hara@bsm.bioc.ucl.ac.uk Tel. 44(171)3877050x2209

O'Neil, Paul Andrew. Postdoctoral Research Assistant. Chemistry Department, Imperial College of Science and Technology, London, SW7 2AY, England. *Low_temperature_data_collection direct_methods organometallic_lithium_compounds supramolecular_chemistry*

Orpen, Prof. A. Guy (1955). Professor; Head, Inorganic Chemistry. School of Chemistry, University of Bristol, Bristol, BS8 1TS, England. PhD (U. Cambridge, 1979) chemical crystallography. *Chemical_crystallography database modelling EXAFS*

E-mail csd@siva.bris.ac.uk Tel. 44(117)9287648 Fax 44(117)9290509

Owston, Dr Philip G. (1921). Retired. [Afton Cottage, Berks Hill, Chorleywood, Herts., WD3 5AJ, England.] DSc (London U., 1976) chemical crystallography. *Structure_determination structure-activity_relationship metalloorganic_catalysts*

Tel. 44(1923)283708

Page, Dr James E. (1915). Retired. [127 Northumberland Road, Harrow, HA2 7RB, England.] DSc (London U., 1957) physical methods. *Antibiotics polymorphism*

Tel. 44(181)8668871

Palmer, Dr Rex Alfred (1936). Reader in structural crystallography. ARSCMB Lecturer 1993. Department of Crystallography, Birkbeck College, Malet Street, London, WC1E 7HX, England. PhD (Birkbeck Coll. London, 1962) crystallography. *Drug_mechanism protein nucleic_acid carbohydrate structure-activity_relationship molecular_recognition lectins ribosome_inactivating_proteins toxins immunotoxin_design*

E-mail ubcg06p@cu.bbk.ac.uk Tel. 44(171)6316830 Fax 44(171)6316803

Pape, Ian (1972). Research Student. Department of Physics, University of Durham, South Road, Durham, DH1 3LE, England. BSc (1993) physics. *Bragg_intensity X-ray_topography diffractometry reflectivity*

E-mail ian.pape@durham.ac.uk Tel. 44(191)3742114 Fax 44(191)3743749

Parsons, Prof. Ian (1939). Professor of Mineralogy. Department of Geology and Geophysics, The University of Edinburgh, West Mains Road, Edinburgh, EH9 3JW, Scotland. PhD (Durham U., 1963) geology. *Phase_equilibrium phase_kinetics scanning_electron_microscopy transmission_electron_microscopy mineralogy diffraction porosity polymorphism strain*

E-mail iparsons@srv0.glg.ed.ac.uk Tel. 44(131)6504839 Fax 44(131)6683184

Parsons, Dr Mark R. (1962). Postdoctoral Research Fellow. Department of Biochemistry and Molecular Biology, University of Leeds, Leeds, LS2 9JT, England. PhD (U. Leeds, 1989) protein crystallography. *Macromolecular_crystallography molecular_recognition nucleic_acid_binding_proteins proteins*

E-mail mark@biovax.leeds.ac.uk Tel. 44(113)2333027 [44(532)333027 before 1 April 1995] Fax 44(113)2333167 [44(532)333167 before 1 April 1995]

Parsons, Dr Simon (1965). Postdoctoral Fellow. Department of Chemistry, The University of Edinburgh, Kings Buildings, West Mains Road, Edinburgh, EH9 3JJ, Scotland. PhD(1991) chemistry. *Inorganic_structures main-group_compounds*

E-mail ajb03@festival.ed.ac.uk Tel. 44(131)6504804 Fax 44(131)6504743

Partridge, Mr Ben L. (1971). Research Student. Cambridge Crystallographic Data Centre, 12 Union Road, Cambridge, CB2 1EZ, England. BSc (Bristol U., 1992) chemistry. *Crystallography nucleic_acids structure_determination refinement_method DNA RNA*

E-mail blp1000@cus.cam.ac.uk Tel. 44(1223)336012 Fax 44(1223)336033

Passalacqua, Edward F. (1967). Research Student. Department of Biochemistry, University of Bath, Claverton Down, Bath, BA2 7AY, England. BSc (U. Wales, 1991) biochemistry: protein crystallography. *Biocrystallography bacterial_toxin immunobiology perfection structure-function_relationships staphylococcal_enterotoxins toxic_shock_syndrome*

E-mail bspefp@midge.bath.ac.uk Tel. 44(1225)826826x3091 Fax 44(1225)826449

Paton, Mr John D. (1946). Research Assistant. Chemistry Department, University of Dundee, Dundee, DD1 4HN, Scotland. MSc (Dundee U., 1979) chemical crystallography. *Structure_determination X-ray_powder_diffraction X-ray_fluorescence*

E-mail j.d.paton@dundee.ac.uk Tel. 44(1382)223181x4705

Pauptit, Dr Richard A. (1954). Group Leader, Protein Crystallography. Zeneca Pharmaceuticals, Mereside, Alderley Park, Macclesfield, Cheshire, SK10 4TG, England. PhD (U. British Columbia, 1981) chemical crystallography. *Proteins*

E-mail pauptit@zeneca-ph.co.uk Tel. 44(1625)516133 Fax 44(1625)583074

Pearl, Dr Laurence Harris (1956). Lecturer. Department of Biochemistry and Molecular Biology, University College London, Gower Street, London, WC1E 6BT, England. PhD (Birkbeck Coll. London, 1983) protein crystallography. *Biocrystallography crystallisation enzymes proteins*

E-mail pearl@bsm.bioc.ucl.ac.uk Tel. 44(171)3807372 Fax 44(171)3807193

Perutz, Dr Max F. (1914). Member of Staff of Medical Research Council. Laboratory of Molecular Biology, Hills Road, Cambridge, CB2 2QH, England. PhD (U. Cambridge, 1940) X-ray crystallography. *Protein_crystallography molecular_biology haemoglobin evolution literary_work*

Tel. 44(1223)248011 Fax 44(1223)213556 Telex 82532

Pettifer, Dr Robert F. (1948). Senior Lecturer. Department of Physics, University of Warwick, Coventry, CV4 7AL, England. PhD (Warwick U., 1978) physics. *Absorption_spectroscopy_experimental absorption_spectroscopy_theoretical synchrotron_radiation_experimental wavelength_absolute NMR MAD optics_X-ray*

E-mail pettifer@warwick.ac.uk Tel. 44(1203)523919 Fax 44(1203)692016

Phillips, Prof. Simon E. V. (1950). Professor. Department of Biochemistry and Molecular Biology, University of Leeds, Leeds, LS2 9JT, England. PhD (Univ. Coll. London, 1974) chemistry. *Crystallography proteins nucleic_acids*

E-mail sevp@biovax.leeds.ac.uk Tel. 44(113)2333027 [44(532)333027 before 1 April 1995] Fax 44(113)2333167 [44(532)333167 before 1 April 1995] Telex 556473 UNILDS G

Pickersgill, Dr Richard W. (1959). Protein Crystallographer; Head of Protein Crystallography Section. Institute of Food Research, Reading Laboratory, Protein Crystallography, Earley Gate, Whiteknights Road, Reading, RG6 2EF, England. DPhil (U. Oxford, 1984) protein crystallography. *Macromolecular_crystallography enzymes proteins macromolecules protein_architecture protein_engineering*

E-mail richard.pickersgill@bbsrc.ac.uk Tel. 44(1734)357143 Fax 44(1734)267917

Pike, Ashley C. W. (1969). Research Student. Department of Biochemistry, University of Bath, Claverton Down, Bath, BA2 7AY, England. BSc (Bath U., 1992) biochemistry. *Biocrystallography biochemistry structure_determination molecular_replacement structure_function_relationships lactose_synthesis glycosyltransferases*

E-mail a.c.w.pike@midge.bath.ac.uk Tel. 44(1225)826826x3091 Fax 44(1225)826449

Pirie, Dr John D. (1939). Honorary. Physics Unit, Fraser Noble Building, University of Aberdeen, Aberdeen, AB9 2UE, Scotland. PhD (Aberdeen U., 1965) physics. *Thermal_diffuse_scattering lattice_dynamics*
E-mail nph012@aberdeen.ac.uk Tel. 44(1224)272499

Pitchford, Nigel Aaron (1969). Research Student. Chemistry Department, University of Durham, South Road, Durham, DH1 3LE, England. BA (U. Oxford, 1991) chemistry. *Database conformational_change chemometrics coordination*
E-mail n.a.pitchford@durham.ac.uk Tel. 44(191)3744703 Fax 44(191)3743745

Plant, Dr John Stewart (1945). Lecturer. Computer Centre, Keele University, Keele, Staffordshire, ST5 5BG, England. PhD (Sheffield U., 1970) neutron diffraction. *Antiferromagnetism computing neutron_crystallography vitamins neutron_polarisation_analysis garnets*
E-mail cca08@keele.ac.uk Tel. 44(1782)583064

Platts, James (1972). Research Assistant. Department of Chemistry, University of Wales Cardiff, PO Box 912, Cardiff, CF1 3TB, Wales. BSc (1993) chemistry. *Charge_density MO_calculation quantum_mechanics bonding*
E-mail platts@cardiff.ac.uk Tel. 44(1222)874950

Pond, Prof. Robert Charles (1946). Professor. Department of Materials Science and, Engineering, University of Liverpool, PO Box 147, Liverpool, L69 3BX, England. PhD (Bristol U., 1973) physics. *Interfacial_structure interface_property topology crystal_defect*
Tel. 44(151)7944660 Fax 44(151)7944675 Telex 627095 UNILPL G

Powell, Dr Annie (1959). Lecturer. School of Chemical Sciences, University of East Anglia, Norwich, NR4 7TJ, England. PhD (1985) chemistry. *Bioinorganic_model_compound molecular_magnetic transition_element_clusters macromolecular_structure_determination macromolecular_crystallization metalloproteins liquid_crystals*
E-mail c004@cpcma.uea.ac.uk Tel. 44(1603)56161x3140 Fax 44(1603)259396

Powell, Dr Harold Roger (1958). Postdoctoral Research Assistant. Cambridge Crystallographic Data Centre, 12 Union Road, Cambridge, CB2 1EZ, England. PhD (Queen Mary Coll. London, 1986) metalloorganic chemistry. *Data_collection data_processing cluster_chemistry biomolecules*
E-mail hrp1000@cus.cam.ac.uk Tel. 44(1223)336015 Fax 44(1223)336033

Price, Dr Sarah Lois (1956). Lecturer. University College London, Department of Chemistry, 20 Gordon Sreet, London, WC1H OA3, England. PhD (U. Cambridge, 1980) theoretical chemistry. *Computer_modelling electrostatics polymorphism*
E-mail sally@ri.ricx.ac.uk Tel. 44(171)3877050x4622 Fax 44(171)3807463

Prince, Dr Stephen M. (1966). Postdoctoral Research Assistant. Robertson Laboratories, Department of Chemistry, University of Glasgow, Glasgow, G12 8QQ, Scotland. PhD (Liverpool John Moores U., 1993) crystallography. *Protein_crystallography*
E-mail steve@chem.gla.ac.uk Tel. 44(141)3398855x6180 Fax 44(141)3304888

Raithby, Dr Paul Robert (1951). Assistant Director of Research. Member of Chemical Crystallography Committee of BCA. Department of Chemistry, University of Cambridge, Lensfield Road, Cambridge, CB2 1EW, England. PhD (Queen Mary Coll. London, 1976) inorganic chemistry. *X-ray_transition_elements molecular_clusters X-ray_data_collection osmium_compounds*
E-mail prr1@phx.cam.ac.uk Tel. 44(1223)336323 Fax 44(1223)336362

Ralph, Dr Brian (1939). Dean of Technology. Department of Materials Technology, Brunel University, Uxbridge, Middlesex, UB8 3PH, England. ScD (Cambridge U., 1980) materials science. *Metallurgy metallography electron_microscopy phase_transitions materials_science microdiffraction textures defects*
Tel. 44(1895)203319 Fax 44(1895)812636

Redfern, Dr Simon A. T. (1962). Lecturer. Department of Earth Sciences, University of Cambridge, Downing Street, Cambridge, CB2 3EQ, England. PhD (Cambridge U., 1990) mineralogy. *Ferroelasticity phase_transition mineralogy aluminosilicates*
E-mail satr@esc.cam.ac.uk Tel. 44(1223)333475 Fax 44(1223)333450

Reid, Mrs Jean S. (1941). Technical Editor. Cambridge Crystallographic Data Centre, 12 Union Road, Cambridge, CB2 1EZ, England. BSc (Glasgow U., 1962) chemistry. *Organic_database metalloorganic_database information_system*
E-mail jsr11@chemcrys.cam.ac.uk Tel. 44(1223)336032 Fax 44(1223)336033

Reid, Dr John Sinclair (1942). Lecturer; Physics Unit. Cruickshank Building, Fraser Noble Building, Aberdeen University, Aberdeen, AB9 2UE, Scotland. PhD (Aberdeen U., 1970) physics. *Diffuse_scattering energy-dispersive_analysis absorption_correction lattice_dynamics history_of_instruments teaching meteorological_optics*
E-mail j.s.reid@uk.ac.abdn Tel. 44(1224)272507 Fax 44(1224)272497 Telex 73458 UNIABN G

Rendle, Dr David Forbes (1946). Principal Scientific Officer. R & D Section, Metropolitan Police, Forensic Science Laboratory, 109 Lambeth Road, London, SE1 7LP, England. PhD (Guelph U., 1972) chemistry. *Powder_diffraction drugs_dyes computing instrumentation dye_HPLC*
Tel. 44(171230)6170 Fax 44(171230)6253

Rice, Prof. David W. (1952). Professor; Director, Krebs Institute. The Krebs Institute, Department of Molecular Biology, University of Sheffield, PO Box 594, Sheffield, S10 2UH, England. DPhil (Oxford U., 1979) protein crystallography. *Protein_crystallography protein_crystallization enzymes structure-activity_relationship*
E-mail d.rice@sheffield.ac.uk Tel. 44(114)2824242 [44(742)824242 before 1 April 1995] Fax 44(114)2728697 [44(742)728697 before 1 April 1995] Telex 547216 UGSHEF G

Ristic, Dr Radoljub I. (1949). Postdoctoral Research Fellow. University of Strathclyde, Pure and Applied Chemistry, 295 Cathedral Street, Glasgow, G1 1XL, Scotland. PhD (U. Belgrade, 1985) solid state physics. *Crystal_growth crystallization X-ray_topography X-ray_optics computing*
E-mail cbes24@vaxa.strath.ac.uk. Tel. 44(141)5524400x2261 Fax 44(141)5525664

Roberts, Dr Kevin John (1950). Reader in Physical Chemistry. Department of Pure and Applied Chemistry, University of Strathclyde, Glasgow, G1 1XL, Scotland. PhD (Portsmouth Polytechnic, 1978) crystal characterisation. *Crystallisation X-ray_spectroscopy computer_modelling diffraction synchrotron_radiation_techniques_for_examining_real_chemical_interface*
Tel. 44(141)5534174 Fax 44(141)5523770 Telex 77472 UNSLIB G

Robertson, Dr John Harry (1923). Retired. Chairman, IUCr–OUP Book Series Committee. School of Chemistry, University of Leeds, Leeds, LS2 9JT, England. PhD (Edinburgh U., 1949) chemical crystallography. *Organic_structures inorganic_structures biological_structures physical_crystallography crystallographic_books philosophy_of_scientific_method*
Tel. 44(113)2336406 [44(532)336406 before 1 April 1995] Fax 44(113)2336565 [44(532)336565 before 1 April 1995]

Roe, Dr Stephen Mark (1965). Higher Scientific Officer. AFRC IPSR Nitrogen Fixation Laboratory, University of Sussex, Brighton, BN1 9RQ, England. PhD (U. Warwick, 1989) chemical crystallography. *X-ray_biocrystallography nitrogenases metalloenzymes*
E-mail roesm@afrc.ac.uk Tel. 44(1273)678198 Fax 44(1273)678133

Roebuck, Dr Peter H. A. (1952). Works Manager. Kanthal Limited, Inveralmond Industrial Estate, Perth, PH1 3ED, Scotland. PhD (U. Newcastle-upon-Tyne, 1978) engineering ceramics. *High_temperature_ceramics electrical_ceramics refractory_compounds high_temperature_furnaces*
Tel. 44(1738)20931 Fax 44(1738)20936 Telex 76460 KANTHAL G

Rogers, Prof. Donald (1921). Retired. [11 Salvington Crescent, Bexhill-on-Sea, East Sussex, TN39 3NP, England.] PhD (U. London, 1944) physics.
Tel. 44(1424)222157 Fax 44(1424)211242

Rogers, Dr Keith (1959). Lecturer. Cranfield University (RMCS), Shrivenham, Swindon, Wiltshire, SN6 8LA, England. PhD (U. Wales, 1984) physical crystallography. *Thin_film phase_transition ion_beam_analysis computer_aided_education radiography*
E-mail rogersk@rmcs.cranfield.ac.uk Tel. 44(1793)785399 Fax 44(1793)783878

Rohl, Dr Andrew L. (1966). Postdoctoral Research Assistant. The Royal Institution of Great Britain, 21 Albemarle Street, London, W1X 4BS, England. DPhil (Oxford, 1993) inorganic chemistry. *Computer_modelling crystal_growth sulfates modelling_growth_inhibitors_on_inorganic_surfaces*
E-mail andrew@ricx.ri.ac.uk Tel. 44(171)4092992 Fax 44(171)6293569

Ross, Prof. Donald Keir (1939). Professor. Department of Physics, University of Salford, Salford, Manchester, M5 4WT, England. DSc (U. Birmingham, 1972) physics.
Tel. 44(161)7455881

Rout, Joanne (1961). Technical Manager. ICI Chemicals and Polymers Ltd, Saffil Business, Tanhouse Lane, Widnes, Cheshire, WA8 0RY, England. BSc (U. Sussex, 1982) materials science. *Crystallisation polymorphism phase_transition*
Tel. 44(1928)517233 Fax 44(1928)517714

Rowland, Dr R. Scott (1962). Senior Scientist. Cambridge Crystallographic Data Centre, 12 Union Road, Cambridge, CB2 1EZ, England. PhD (U. Alabama at Birmingham, 1990) biochemistry. *Non-bonded_interaction database computer_modelling molecular_structure*
E-mail Scott.Rowland@chemcrys.cam.ac.uk Tel. 44(1223)336026 Fax 44(1223)336033

Russell, Dr David R. (1939). Senior Lecturer. Department of Chemistry, University of Leicester, University Road, Leicester, LE1 7RH, England. PhD (U. Glasgow, 1963) chemistry: inorganic fluorine chemistry. *Chemical_crystallography structure_determination*
E-mail drr@leicester.ac.uk Tel. 44(116)2522090 [44(533)522090 before 1 April 1995] Fax 44(116)2523789 [44(533)523789 before 1 April 1995] Telex 347250 LEICUN G

Sanderson, Ian (1966). Research Student. Biomolecular Structure Unit, Department of Biochemistry and Molecular Biology, University College London, Gower Street, London, WC1E 6BT, England. BSc (King's Coll. London, 1989) chemistry. *Bio_molecule*
E-mail sanders@bsm.bioc.ucl.ac.uk Tel. 44(171)3877050x2209 Fax 44(171)3877193

Sawyer, Dr Lindsay (1944). Senior Lecturer. Department of Biochemistry, The University of Edinburgh, George Square, Edinburgh, EH8 9XD, Scotland. PhD (U. Edinburgh, 1971) physics: protein crystallography. *Protein_crystallography computer_modelling small_molecule-macromolecule_interactions*
E-mail l.sawyer@edinburgh.ac.uk Tel. 44(131)6503729 Fax 44(131)6503711

Scarrott, Keith (1957). Senior Research Engineer. BNR EUROPE Ltd, London Road, Harlow, Essex, CM17 9NA, England. CChem MRSC (Royal Society of Chemistry, 1991) chemistry. *High_resolution_diffractometry III-V_compounds_characterization computer_modelling*
Tel. 44(1279)403340 Fax 44(1279)402826 Telex 81151 BNR HW G

Schwalbe, Dr Carl Hellmuth Walter (1942). Senior Lecturer. BCA Chemical Crystallography Group Committee Member. Department of Pharmaceutical and Biological Sciences, Aston University, Aston Triangle, Birmingham, B4 7ET, England. PhD (Harvard U., 1970) chemistry. *Structure_determination drugs MO-calculations structure-activity_relationships graphics*
E-mail c.h.schwalbe@aston.ac.uk Tel. 44(121)3593611x4201 Fax 44(121)3590733

Shafirstein, Dr Gal (1961). Postdoctoral Consultant. Division of Materials Metrology, National Physical Laboratory, Queens Road, Teddington, Middlesex, TW11 0LW, England. DSc (Tech. Inst. of Techn. (Israel), 1993) materials science. *X-ray_boron_compounds PVD_nitrides strain orientation PVD_nitrides_adhesion*
E-mail gals@newton.npl.co.uk Tel. 44(181)9437114 Fax 44(181)9432989

Sharff, Dr Andrew J. (1965). Research Associate. CRC Biomolecular Structure Unit, Institute of Cancer Research, 15 Cotswold Road, Sutton, Surrey, SM2 5NG, England. DPhil (Oxford U., 1991) molecular biophysics. *Proteins structure_determination computer_modelling drug_design binding_proteins protein-ligand_interactions hinge_bending cancer_research*
E-mail andrew@bms-unit.icr.ac.uk Tel. 44(181)6438901 Fax 44(181)7707893

Shekunov, Dr Boris Yu (1962). Postdoctoral Research Fellow. Department of Pure and Applied Chemistry, University of Strathclyde, 295 Cathedral Street, Glasgow, G1 1XL, Scotland. PhD (Moscow State U., 1990) crystallophysics. *Crystal_growth crystallization non-destructive_analysis computer_modelling optics mathematics_non-linear-phenomena*
E-mail cbas131@vaxa.strath.ac.uk Tel. 44(141)5524400x2261 Fax 44(141)5525664

Sherwood, Prof. John N. (1933). Professor of Physical Chemistry; Vice Principal Elect 1994-1996. Chairman, British Association for Crystal Growth 1994-1996. Department of Pure and Applied Chemistry, University of Strathclyde, 295 Cathedral Street, Glasgow, G1 1XL, Scotland. DSc (Durham U., 1976) physical chemistry. *X-ray_topography crystal_growth crystallization*
E-mail J.N.Sherwood@uk.ac.strath Tel. 44(141)5524400x2288 Fax 44(141)5525664 Telex 77472 UNSLIB G

Shrive, Dr Annette K. (1966). Research Fellow. Department of Physics, Keele University, Keele, Staffs, ST5 5BG, England. PhD (Keele U., 1991) physics. *Diffraction_theory synchrotron_radiation Laue_diffraction physics*
E-mail aks@cxa.dl.ac.uk Tel. 44(1782)621111x7941 Fax 44(1782)711093

Sinn, Prof. Ekkehard (1945). Professor of Inorganic Chemistry. School of Chemistry, University of Hull, Hull, HU6 7SJ, England. PhD (U. New South Wales, 1969) inorganic. *Bio_coordination crystallography magnetism*
Tel. 44(1482)466353 Fax 44(1482)466410

Skakle, Jan (1968). Postdoctoral Research Fellow. Department of Chemistry, University of Aberdeen, Meston Walk, Aberdeen, AB9 2UE, Scotland. PhD (U. Aberdeen, 1994) chemistry: novel complex oxides. *Rietveld_method structure_determination inorganic_compounds powder_diffraction oxide_superconductors*
E-mail j.skakle@aberdeen.ac.uk Tel. 44(1224)272916 Fax 44(1224)272921

Skarżyński, Dr Tadeusz (1953). Principal Research Analyst. Department of Biomolecular Structure, Glaxo Research and Development Ltd, Greenford Road, Greenford, Middlesex, UB6 0HE, England. PhD (Łódź U., 1981) chemistry. *Crystallography proteins drug_design*
E-mail ts14913@ggr.co.uk Tel. 44(181)9664188 Fax 44(181)9662156

Slawin, Alexandra Martha Zoya (1961). Senior experimental officer. Molecular Structure Laboratory, Dept. Chemistry, Loughborough University, Leics, LE11 0GP, England. [Dept. Chemistry, Loughborough University, Loughborough, Leics, LE11 0GP, England.] BSc chemistry. *General_chemistry inorganic_and_organic_chemistry living*
E-mail A.M.Slawin@lut.ac.uk Tel. 44(1509)222586 Fax 44(1509)233163

Small, Colin John (1960). Principal Metallurgist X-ray Diffraction Laboratory. Rolls-Royce plc., Po Box 31, Derby, DE24 8BJ, England. BMet (Sheffield U., 1983) metallurgy. *Powder_diffraction metallurgical_superalloys single_crystal_orientation phase_equilibrium*
Tel. 44(1332)240210 Fax 44(1332)240327 Telex 37645

Small, Dr Ronald W. H. (Sam) (1921). Reader Emeritus. Chemistry Department, The University, Lancaster, LA1 4YA, England. DSc (U. Birmingham, 1982) crystallography. *Polymorphism phase_transition_organic*
E-mail r.small@cent1.lancs.ac.uk Tel. 44(1524)65201x3339

Smart, Dr Lesley E. (1947). Lecturer. Open University, Department of Chemistry, Walton Hall, Milton Keynes, MK7 6AA, England. PhD (Southampton U., 1971) raman spectroscopy. *Chemical_crystallography solid_state_chemistry*
E-mail l.e.smart@open.ac.uk Tel. 44(1908)653191 Fax 44(1908)653744

Smith, Dr Arnold John (1931). Lecturer in Inorganic Chemistry. BCA Secretary. Department of Chemistry, University of Sheffield, Sheffield, S3 7HF, England. PhD (London U., 1957) inorganic chemistry. *Lanthanides actinides coordination geometry_analysis high_coordination_numbers*
E-mail a.j.smith@sheffield.ac.uk Tel. 44(114)2824476 [44(742)824476 before 1 April 1995] Fax 44(114)2738673 [44(742)738673 before 1 April 1995]

Smith, Gallienus William (1924). Research Officer. Chemistry Department, University of Surrey, Guildford, Surrey, GU2 5XH, England. [1 West Farm Avenue, Ashtead, Surrey, KT21 2LD, England.] MSc (London U., 1953) crystallography. *Inorganic_structure_determination crystal_structure_determination powder_method organic_structure_determination*
Tel. 44(1483)3008009582

Smith, Mr Garry T. (1971). Research Student. Chemistry Department, University of Durham, South Road, Durham, DH1 3LE, England. BSc (Glasgow U., 1993) chemistry. *Charge_density quantum_mechanics inverse_problem maximum_entropy_method*
E-mail g.t.smith@durham.ac.uk Tel. 44(191)3744704 Fax 44(191)3743745

Smith, Paul Raymond (1965). Senior Research Assistant. Foods Research Department, Unilever Research, Colworth Laboratory, Sharnbrook, Bedford, MK44 1LQ, England. BA (U. Oxford, 1988) materials science. *Crystal_growth lipids X-ray_diffraction optical_microscopy habit_modification polymorphism fats*
E-mail Paul.R.Smith@URCGB.SPRINT.COM Tel. 44(1234)222554 Fax 44(1234)222401

Smith, Dr Ronald I. (1965). Instrument Scientist. The ISIS Facility, Rutherford Appleton Laboratory, Chilton, Oxfordshire, OX11 0QX, England. PhD (U. Aberdeen, 1992) chemistry. *Solid_state_chemistry time-of-flight_diffraction incommensurate_structure_factors*
E-mail ris01@isise.rl.ac.uk Tel. 44(1235)445683 Fax 44(1235)445720

Snell, Edward H. (1969). Research Student. Chemistry Department, University of Manchester, Oxford Road, Manchester, M13 9PL, England. BSc (Liverpool Polytechnic, 1992) applied physics. *Laue_diffraction crystallography synchrotron_radiation computing proteins macromolecules*
E-mail eddie@v7.ch.man.ac.uk Tel. 44(161)2754689 Fax 44(161)2754598

Soar, Martin (1959). Chemist. BP Research and Engineering Centre, Chertsey Road, Sunbury-on-Thames, Middlesex, TW16 7LN, England. MSc (Birkbeck Coll. London, 1991) crystallography. *High_temperature_diffraction high_pressure_diffraction catalysts Rietveld_method in-situ_structure_determination size_effect neutron_diffraction*
Tel. 44(1932)763724 Fax 44(1932)762177

Spratt, Stephen Brian Douglas (1949). Electron Microscopist. Johnson Matthey Technology Centre, Blounts Court, Sonning Common, Reading, RG4 9NH, England. *Energy_dispersive_analysis diffraction image_processing platinum_compounds safety diffractometry X-ray_fluorescence_spectroscopy Debye_Scherrer*
Tel. 44(1734)722811x2143 Fax 44(1734)723030

Squire, Dr Gavin Daniel (1966). Principal Scientist. British Gas PLC., Gas Research Centre, Ashby Road, Loughborough, Leicestershire, LE11 3QU, England. PhD (U. Reading, 1991) catalysis. *Catalysis*
Tel. 44(1509)282336 Fax 44(1509)283115

Squire, Dr John Michael (1945). Reader in Biophysics. Member BCA Council; Editor, BCA Crystallography News; Chairman, BCA Biological Structure Group 1991-1994; Chairman CCP13 (fibre diffraction). Biophysics Section, Blackett Laboratory, Imperial College, London, SW7 2BZ, England. PhD (King's Coll. London, 1969) biophysics. *Biophysics X-ray_diffraction electron_microscopy image_processing muscle_time_resolved_X-ray_diffraction low-angle_diffraction fibrous_proteins*
E-mail j.squire@ic.ac.uk Tel. 44(171)5947691 Fax 44(171)5890191

Stanbury, Howard J. (1953). Senior Editor. IUCr Book Series Committee. SMJ Division, Oxford University Press, Walton Street, Oxford, OX2 6DP, England. BSc (U. East Anglia, 1974) biological sciences. *Publishing software*
E-mail stanburh@oup.co.uk Tel. 44(1865)267669 Fax 44(1865)267680 Telex 837330 OXPRES G

Stedman, Dr Nicola J. (1966). Editorial Assistant. International Union of Crystallography, 5 Abbey Square, Chester, CH1 2HU, England. DPhil (Oxford U., 1994) solid state chemistry. *Electronic_publishing*
E-mail ns@iucr.ac.uk Tel. 44(1244)342878 Fax 44(1244)314888 Telex 669755 OFFICE G attn Unicrystal

Steward, Prof. Edward George (1923). Emeritus Professor. City University, Northampton Square, London EC1V 0HB, England. DSc (U. London, 1974) molecular medicine. *Structure-property_relationships Fourier_optics*
Tel. 44(171)4778121 Fax 44(171)4778121

Stillman, Dr Timothy J. (1961). Research Associate. The Krebs Institute, Department of Molecular Biology, University of Sheffield, PO Box 594, Sheffield, S10 2UH, England. PhD (Lancaster U., 1988) crystallography. *Protein_crystallography protein_crystallization enzymes structure-activity_relationship*
E-mail t.stillman@sheffield.ac.uk Tel. 44(114)2824242 [44(742)824242 before 1 April 1995] Fax 44(114)2728697 [44(742)728697 before 1 April 1995]

Stoddard, Miss Jillian Kaye (1968). Senior Editorial Assistant. International Union of Crystallography, 5 Abbey Square, Chester, CH1 2HU, England. BSc (U. Sheffield, 1989) pharmacology and chemistry. *Electronic_publishing editing*
E-mail jk@iucr.ac.uk Tel. 44(1244)342878 Fax 44(1244)314888 Telex 669755 OFFICE G attn UNICRYSTAL

Stoddart, Christopher Paul (1964). Postdoctoral Research Assistant. Department of Pharmaceutical Technology, University of Bradford, West Yorkshire, BD7 1DP, England. PhD (Staffordshire U., 1994) physical and computational crystallography. *Lattice_dynamics Rietveld_method organic_indexing carbonate_formation*
E-mail c.p.stoddart@bradford.ac.uk Tel. 44(1274)733466x4754 Fax 44(1274)384769

Strickland, Mr Peter R. (1956). Managing Editor. International Union of Crystallography, 5 Abbey Square, Chester, CH1 2HU, England. BSc (U. Sheffield, 1977) chemistry. *Electronic_publishing information_science chemical_nomenclature editing*
E-mail ps@iucr.ac.uk Tel. 44(1244)342878 Fax 44(1244)314888 Telex 669755 OFFICE G attn UNICRYSTAL

Sutton, Dr Brian J. (1954). Lecturer. The Randall Institute, King's College London, 26 Drury Lane, London, WC2B 5RL, England. DPhil (Oxford U., 1980) biophysics. *Immunoglobulins enzymes antibiotics antiallergenics beta_lactamases receptors*
E-mail brian@helios.rai.kcl.ac.uk Tel. 44(171)8368851x223 Fax 44(171)4979078

Sweet, Mrs Tracy K. N. (1969). Research Student. School of Chemistry and Applied Chemistry, University of Wales Cardiff, PO Box 912, Cardiff, CF1 3TB, Wales. BSc (U. Wales Cardiff, 1992) chemistry. *Structure_determination inorganic_complexes computer_modelling*
E-mail sweet@cardiff.ac.uk Tel. 44(1222)874950 Fax 44(1222)874029

Tanner, Prof. Brian Keith (1947). Professor of Physics. Department of Physics, University of Durham, South Road, Durham, DH1 3LE, England. DPhil (Oxford U., 1972) materials science. *High-resolution_diffractometry topography synchrotron_radiation GIXS*
E-mail b.k.tanner@durham.ac.uk Tel. 44(191)3742137 Fax 44(191)3743749

Taylor, Prof. Charles A. (1922). Emeritus Professor. [9 Hill Deverill, Warminster, Wilts., BA12 7EF, England.] DSc (Manchester U., 1959) physics. *Optical_transformation education diffraction disorder*
Tel. 44(1985)840574

Taylor, Dr Robin (1951). Cambridge Crystallographic Data Centre, 12 Union Road, Cambridge, CB2 1EZ, England. PhD (Cambridge U., 1976) chemical crystallography. *Computer_modelling agrochemical_computer-assisted_design database hydrogen_bonding*
Tel. 44(1223)336408 Fax 44(1223)336033

Thomas, Dr Noel W. (1958). Lecturer. School of Materials, University of Leeds, Leeds, LS2 9JT, England. PhD (U. Cambridge, 1984) chemistry: organic solid state. *Crystal_chemistry electroceramics structural_computer_modelling ferroelectric_oxides crystal_structure-physical-property_relationships*
E-mail n.w.thomas@leeds.ac.uk Tel. 44(113)2332538 [44(532)332538 before 1 April 1995] Fax 44(113)2422531 [44(532)422531 before 1 April 1995]

Thomas, Dr Pamela Anne (1962). Lecturer in Physics. Member BCA Physical Crystallography Group Committee. Department of Physics, University of Warwick, Coventry, CV4 7AL, England. PhD (U. Oxford, 1987) physics: crystallography. *Structure-activity_relationship optical_property high-resolution_diffractometry domain_structure teaching anomalous_dispersion topography*
E-mail phrve@csv.warwick.ac.uk Tel. 44(1203)523354 Fax 44(1203)692016

Thornton, Prof. Janet M. (1949). Professor; Director of Structural Biology. Biochemistry and Molecular Biology Department, University College, Gower Street, London, WC1E 6BT, England. PhD (King's Coll. London, 1973) biophysics. *Proteins structure computer_modelling prediction nuclear_magnetic_resonance drugs design molecular_recognition*
E-mail thornton@bsm.bioc.ucl.ac.uk Tel. 44(171)3807048 Fax 44(171)3807198

Thornton-Pett, Dr Mark (1957). Senior Experimental Officer. School of Chemistry, University of Leeds, Leeds, LS2 9JT, England. PhD (Queen Mary Coll. London, 1982) chemistry. *Boron_compounds carboranes metallacarboranes organometallics metallaheteroboranes polyphosphines heteroboranes*
E-mail chm6mtp@cif.leeds.ac.uk Tel. 44(113)2336423 [44(532)336423 before 1 April 1995] Fax 44(113)2336565 [44(532)336565 before 1 April 1995] Telex 556473 UNILDS G

Tocher, Dr Derek A. (1957). Lecturer. Christopher Ingold Laboratories, Department of Chemistry, University College, 20 Gordon Street, London, WC1H 0AJ, England. PhD (U. Edinburgh, 1983) chemistry. *Crystallography ruthenium_compounds rhodium_compounds anticancer_compounds*
E-mail ucca51l@pyr.ucl.ac.uk Tel. 44(171)3877050x4709 Fax 44(171)3807463

Tolworthy, Benjamin J. (1962). Post-Graduate Research Assistant. Nottingham Trent University, Department of Chemistry and Physics, Clifton Lane, Nottingham, NG11 8NS, England. BSc (Nottingham Trent U., 1987) applied chemistry. *X-ray_diffraction liquid_crystals powder_diffraction structural_chemistry amides*
E-mail hc154bjt01@uk.ac.ntu.cluster Tel. 44(115)9418418x3318 [44(602)418418x3318 before 1 April 1995] Fax 44(115)9486636 [44(602)486636 before 1 April 1995]

Torabi, Ali Asgar (1951). Visiting Lecturer. Office A4-11C, Chemistry Department, Glasgow University, Glasgow, G12 8QQ, Scotland. MSc (U. Tehran, 1977) physical chemistry. *Metalloorganic_structures inclusion_compounds*
E-mail torabi@chem.gla.ac.uk Tel. 44(141)3398855x4380 Fax 44(141)3304888 Telex 777070 UNIGLA

Truter, Prof. Mary R. (1925). Visiting Professor. Chemistry Department, University College London, 20 Gordon Street, London, WC1H 0AJ, England. DSc (U. London, 1965) chemistry. *Structure_determination macrocycles main-group_compounds transition_elements*
Tel. 44(171)3877050x4657 Fax 44(171)3807463

Tucker, Ian Malcolm (1964). Research Scientist. Unilever Research, Port Sunlight Laboratory, Quarry Road East, Bebington, Wirral, L63 3JW, England. BSc (Bristol U., 1985) physics. *Identification_quantitative_qualitative SAXS structure phase_determination clays small_angle_neutron_scattering reflectivity zeolites*
Tel. 44(151)4713103 Fax 44(151)4711800

Turrillas, Dr Xavier (1954). Postdoctoral Research Fellow. Industrial Materials Group, Crystallography Department, Birkbeck College, University of London, Malet Street, London, WC1E 7HX, England. PhD (U. Grenoble, 1984) materials science. *Energy_dispersive_analysis neutron_powder_diffractometry synchrotron_radiation time_resolved_effect zirconia_zeolites_cements Rietveld_method nuclear_magnetic_resonance solid_state_modelling*
E-mail ubcg91v@ccs.bbk.ac.uk Tel. 44(171)6316853 Fax 44(171)4368918

Vickers, Mary Elizabeth (1953). Senior Research Assistant. Department of Materials Science and Metallurgy, University of Cambridge, Pembroke Street, Cambridge, CB2 3QZ, England. BSc (U. Bristol, 1975) chemical physics. *X-ray_diffraction SAXS Polymers*
E-mail mev20@cam.ac.uk Tel. 44(1223)334369 44(1223)334300 Fax 44(1223)334567

Vincent, Dr Jan E. (1945). Technical Editor. Cambridge Crystallographic Data Centre, 12 Union Road, Cambridge, CB2 1EZ, England. PhD (Cambridge U., 1972) virology. *Organic_database metalloorganic_database information_system*
E-mail jev10@chemcrys.cam.ac.uk Tel. 44(1223)336031

Walker, Miss Elaine M. (1970). Research Student. Department of Pure and Applied Chemistry, University of Strathclyde, 295 Cathedral Street, Glasgow, G1 1XL, Scotland. BSc (Strathclyde U., 1992) pure chemistry. *Computer_modelling crystalline_morphology surface_analysis_theoretical polymorphic_solvents*
E-mail cbar101@vaxa.strath.ac.uk Tel. 44(141)5524400x3850 Fax 44(141)5523770

Wallace, Dr Bonnie Ann (1951). Reader in Crystallography. Department of Crystallography, Birkbeck College, University of London, Malet Street, London, WC1E 7HX, England. PhD (Yale U., 1977) molecular biophysics and biochemistry. *Macromolecular_crystallography structure_biological_membranes circular_dichroism biophysics*
E-mail ubcg91c@ccs.bbk.ac.uk Tel. 44(171)6316857 Fax 44(171)6316803

Wallis, Dr John D. (1954). Lecturer. Chemical Laboratory, University of Kent, Canterbury, CT2 7NH, England. DPhil (Oxford U., 1979) organic synthesis and chemical crystallography. *Organic_materials intramolecular_interaction intermolecular_interaction sulfur_compounds stereoelectronic_effects reaction_coordinates*
E-mail jdw@ukc.ac.uk Tel. 44(1227)764000x3547 Fax 44(1227)475475

Wallwork, Dr Stephen Collier (1925). Retired. Department of Chemistry, University of Nottingham, University Park, Nottingham, NG7 2RD, England. [15 Elm Avenue, Beeston, Nottingham, NG9 1BU.] DPhil (U. of Oxford, 1950) chemical crystallography. *Structure_determination charge_transfer hydrogen_bonding organic_complexes*
Tel. 44(115)9257048 [44(602)257048 before 1 April 1995]

Watkin, Dr David John (1942). Research Assistant. Member, IUCr Computing Commission. Chemical Crystallography Laboratory, 9 Parks Road, Oxford, OX1 3PD, England. PhD (Birmingham U., 1968) chemical crystallography. *Computer refinement software design*
E-mail watkin@vax.ox.ac.uk david.watkin@chemcryst.oxford.ac.uk Tel. 44(1865)270826 Fax 44(1865)272690

Watson, Dr David Gilfillan (1934). Principal Scientist. Cambridge Crystallographic Data Centre, 12 Union Road, Cambridge, CB2 1EZ, England. PhD (U. Glasgow, 1960) chemical crystallography. *Organic_database metalloorganic_database information_system*
E-mail dgw@chemcrys.cam.ac.uk Tel. 44(1223)336394 Fax 44(1223)336033

Watson, Kimberly A. (1963). Research Student. Laboratory of Molecular Biophysics, The Rex Richards Building, South Parks Road, Oxford, OX1 3QU, England. MSc (Queen's U. Kingston, 1990) chemistry. *Drug_design crystallization_macromolecular synchrotron_radiation enzyme_inhibitors computer_assisted_design recognition data_processing QSAR*
E-mail kim@biop.ox.ac.uk Tel. 44(1865)275379 Fax 44(1865)510454

Webster, Dr Michael (1938). Lecturer. Department of Chemistry, University of Southampton, Southampton, SO17 1BJ, England. PhD (U. London, 1962) inorganic chemistry. *Chemical_crystallography crystallization computing structural_database*
E-mail m.webster@soton.ac.uk Tel. 44(1703)593513 Fax 44(1703)593781

Weight, Tony (1968). Editorial assistant. International Union of Crystallography, 5 Abbey Square, Chester CH1 2HU, England. PhD (Keele U., 1993) astrophysics.
E-mail tw@iucr.ac.uk Tel. 44(1244)342878 Fax 44(1244)314888

Welch, Prof. Alan J. (1949). Professor. Department of Chemistry, Heriot-Watt University, Edinburgh, EH14 4AS, Scotland. PhD (U. London, 1974) chemistry. *Chemistry boron_compounds MO_calculations organometallic_compounds boron-neutron_capture_therapy copper_compounds silver_compounds gold_compounds*
Tel. 44(131)4513217 Fax 44(131)4513180

West, Prof. Anthony Roy (1947). Professor of Chemistry. Chemistry Department, Aberdeen University, Meston Walk, Aberdeen, AB9 2UE, Scotland. DSc (Aberdeen U., 1984) solid state chemistry. *Solid_state_materials oxide_phase_diagrams electrical_ceramics oxide_solid_solutions*
Tel. 44(1224)272918 Fax 44(1224)272938

Weston, Dr Simon Alan (1965). Postdoctoral Scientist. Zeneca Pharmaceuticals, Mereside, Alderley Park, Macclesfield, SK10 4TG, England. PhD (U. Portsmouth, 1992) crystallography. *Biomolecule proteins_DNA_drug macromolecular_interaction catalytic_enzymes transcription computing*
E-mail weston@zeneca-ph.co.uk Tel. 44(1625)516137 Fax 44(1625)583074 Telex 669095 ICIPHA G

Westrip, Mr Simon P. (1966). Senior Editorial Assistant. International Union of Crystallography, 5 Abbey Square, Chester, CH1 2HU, England. BSc (U. Exeter, 1988) chemistry. *Electronic_publication information_science editing chemical_nomenclature*
E-mail sw@iucr.ac.uk Tel. 44(1244)342878 Fax 44(1244)314888 Telex 689755 OFFICE G attn UNICRYSTAL

Wheatley, Dr P. J. (1921). Retired Lecturer. University Chemical Laboratory, Lensfield Road, Cambridge, CB2 1EW, England. DPhil (U. Oxford, 1950) physical chemistry. *Molecular_structure*
Tel. 44(1223)336032

Whitaker, Dr Alan (1932). Senior Lecturer (part-time). Department of Physics, Brunel University, Uxbridge, Middlesex, UB8 3PH, England. PhD (Birkbeck Coll. London, 1966) crystallography. *X-ray_crystallography structure_determination organic_pigments ceramic_phase_diagrams organic_dyes*
E-mail alan.whitaker@brunel.ac.uk Tel. 44(1895)27400x2406 Fax 44(1895)272391

Whittaker, Dr E. J. W. (1921). Retired. [60 Exeter Road, Kidlington, Oxon, OX5 2DZ, England.] PhD (London U., 1956) X-ray crystallography. *Amphibole_minerals chrysotile four-dimensional_crystallography*
Tel. 44(1865)373592

Wilkinson, Kay (1968). Research Assistant. The Krebs Institute, Department of Molecular Biology, University of Sheffield, PO Box 594, Sheffield, S10 2UH, England. BSc (Sheffield U., 1988) chemistry with biochemistry. *Biocrystallography protein_crystallization dehydrogenases isomorphous_replacement*
E-mail K.W.Grabham@Sheffield.ac.uk Tel. 44(114)2824325 [44(742)824325 before 1 April 1995] Fax 44(114)2728697 [44(742)728697 before 1 April 1995]

Williams, Dr David Arfon (1962). Researcher. Hitachi Cambridge Laboratory, Cavendish Laboratory, Madingley Road, Cambridge, CB3 0HE, England. PhD (Cambridge U., 1987) physics: microelectronics. *Semiconductor phonon_resonance charge_density_wave low_temperature non-equilibrium_phonons strongly_corr- elated_systems*
E-mail daw10@phx.cam.ac.uk Tel. 44(1223)467944 Fax 44(1223)467942

Williams, Dr David J. (1940). Reader. Department of Chemistry, Imperial College, Lon- don, SW7 2AY, England. PhD (Imperial Coll. London, 1979) crystallography. *Molec- ular_structure molecular_interactions macrocycles chemical_crystallography*
Tel. 44(171)2258344 Fax 44(171)5893869

Willis, Prof. B. T. M. (1927). Retired. Chemical Crystallography Laboratory, Univer- sity of Oxford, 9 Parks Road, Oxford, OX1 3PD, England. DSc (U. London, 1968) physics. *International_Tables_for_Crystallography time-of-flight_diffraction Isat- tice_vibration neutron_scattering*
E-mail cjc@uk.ac.rl.isise Tel. 44(1865)270832 Fax 44(1865)272690

Willoughby, Anne Margaret Elizabeth (1971). Research Student. Department of Materi- als Science and, Metallurgy, Pembroke Street, Cambridge, CB2 3QZ, England. BA (U. Cambridge, 1992) natural sciences. *X-ray_diffraction structure_determination computer_modelling polyolefins*
E-mail amew1@cam.ac.uk Tel. 44(1223)334335 Fax 44(1223)334567

Wilson, Prof. Arthur J. C. (1914). Retired. Editor Volume C of International Tables for Crystallography; Member Commission on Crystallographic Nomenclature. St John's College, Cambridge CB2 1TP, England. [25 Kings Road Cambridge CB3 9DY Eng- land] PhD (U. Cambridge, 1942) physics. *Statistics space_groups symmetry*
E-mail ajcw@cam.ac.uk

Wilson, Dr Chick (1961). Group Leader (Crystallography). ISIS Facility, Ruther- ford Appleton Laboratory, Chilton, Didcot, Oxon, 0X11 0QX, England. PhD (U. Dundee, 1985) physics: X-ray crystallography. *Laue_time-of-flight_diffraction Pat- terson_methods DNA organic_phase_transitions*
E-mail ccw@isise.rl.ac.uk Tel. 44(1235)445137 Fax 44(1235)445720

Wilson, Claire (1969). Research Student. Department of Chemistry, University of Durham, South Road, Durham, DH1 3LE, England. BSc (Bristol U., 1990) chemistry. *Chemi- cal_crystallography charge_density neutron_diffraction*
E-mail claire.wilson@durham.ac.uk Tel. 44(191)3744703 Fax 44(191)3743745

Wilson, Prof. Herbert Rees (1929). Professor Emeritus. Department of Biological and Molecular Sciences, University of Stirling, Stirling, FK9 4LA, Scotland. [Lower Bryanston, St Margaret's Drive, Dunblane, FK15 0DP, Scotland.] PhD (U. Wales, 1952) physics. *DNA_structure virus_structure nucleosides image_processing liter- ature theatre art*
Tel. 44(1786)473171x7761 Fax 44(1786)464994

Wilson, Sarah (1971). Research Student. Cambridge Crystallographic Data Centre, 12 Union Road, Cambridge, CB2 1EZ, England. BSc (Manchester U., 1993) chemistry. *Nucleic_acids synthesis crystal_growth crystallography*
Tel. 44(1223)336012 Fax 44(1223)336033

Windsor, Dr Colin G. (1938). Senior Scientist. AEA Technology, B 521, Harwell Laboratory, Oxon, OX11 0RA, England. DPhil (Oxford U., 1960) physics. *Neu- tron_diffraction small_angle_scattering residual_stress alloy_clustering*

Tel. 44(1235)435101 Fax 44(1235)432726

Winter, Dr Marcus John (1956). Product Manager. Siemens plc., Sir William Siemens House, Princess Road, Manchester, M20 8UR, England. PhD (U. Southampton, 1980) chemical physics. *X-ray_diffraction X-ray_fluorescence X-ray_scattering Tech- niques*
Tel. 44(161)4465267 Fax 44(161)4465289 Telex 8951091 SIELON G

Wonacott, Dr Alan John (1941). Research Leader. Glaxo Research and Development Ltd, Biomolecular Structure Department, Greenford Road, Greenford, Middlesex, UB6 0HE, England. PhD (King's Coll. U. London, 1966) biophysics. *Biocrystallogra- phy drug_design data_collection crystallization enzyme_inhibitors proteases antivi- ral_compounds crystal_spectrophotometry*
E-mail aw7896@ggr.co.uk Tel. 44(181)9662181 Fax 44(181)9662156

Wood, Dr Ian G. (1952). Lecturer. Department of Geological Sciences, University College London, Gower Street, London, WC1E 6BT, England. PhD (U. London, 1977) crystal- lography. *Phase_transition powder_diffraction optical_birefringence mineralogy*
Tel. 44(171)3877050 Fax 44(171)3887614

Woolfson, Prof. Michael Mark (1927). Professor Emeritus. Physics Department, Univer- sity of York, York, YO1 5DD, England. DSc (Manchester U., 1960) physics. *Di- rect_methods anomalous_scattering biological_structures*
Tel. 44(1904)432230 Fax 44(1904)432214 Telex 57933

Wright, Dr Helen (1957). Researcher. School of Computer Studies, University of Leeds, Leeds, LS2 9JT, England. DPhil (U. York, 1983) theoretical crystallography. *Comput- ing computer_graphics direct_method visualization multimedia*
E-mail helenw@scs.leeds.ac.uk Tel. 44(113)2336793 [44(532)336793 before 1 April 1995] Fax 44(113)2335468 [44(532)335468 before 1 April 1995]

Yip, Kitty (1971). Research Student. The Krebs Institute, Department of Molecular Biol- ogy, University of Sheffield, PO Box 594, Sheffield, S10 2UH, England. BSc (Glasgow U., 1993) chemistry and medicinal chemistry. *Protein_crystallography macromolec- ular_structure*
E-mail k.s.yip@sheffield.ac.uk Tel. 44(114)2824242 [44(742)824242 before 1 April 1995] Fax 44(114)2728697 [44(742)728697 before 1 April 1995]

Young, Robert James (1965). Research Assistant. The Randall Institute, Division of Biomedical Sciences, King's College London, 26-29 Drury Lane, London, WC2B 5RL, England. PhD (King's Coll. London, 1994) biochemistry. *Biophysics computer_modelling protein_crystallography immunochemistry mutagenesis pro- tein_engineering*
E-mail robert@helios.rai.kcl.ac.uk Tel. 44(171)8368851 Fax 44(171)4979078

Zeng, Zheng (1968). Research Student. Department of Applied Chemical and, Phys- ical Sciences, Napier University, 10 Colinton Road, Edinburgh, EH10 5DT, Scot- land. BSc (Lancashire Polytechnic, 1991) chemistry. *X-ray_crystallography crys- tal_growth computer_modelling organic_synthesis single_crystal polymerization solid_state non-linear_property*
E-mail j.zeng@central.napier.ac.uk Tel. 44(131)4552630 Fax 44(131)4557989

Zussman, Prof. Jack (1924). Retired. Department of Geology, University of Manchester, Manchester, M13 9PL, England. PhD (Cambridge U., 1952) crystallography. *Miner- alogy geochemistry crystallography asbestos*
Tel. 44(161)2753937 Fax 44(161)2753947

USA

Sub-Editor: M. M. Teeter

Notes

1. In the addresses, the following two-letter abbreviations are used for states and territories:

AL Alabama	IA Iowa	MT Montana	RI Rhode Island
AK Alaska	ID Idaho	NC North Carolina	SC South Carolina
AR Arkansas	IL Illinois	ND North Dakota	SD South Dakota
AZ Arizona	IN Indiana	NE Nebraska	TN Tennessee
CA California	KS Kansas	NH New Hampshire	TX Texas
CO Colorado	KY Kentucky	NJ New Jersey	UT Utah
CT Connecticut	LA Louisiana	NM New Mexico	VA Virginia
CZ Canal Zone	MA Massachusetts	NV Nevada	VI Virgin Islands
DC District of Columbia	MD Maryland	NY New York	VT Vermont
DE Delaware	ME Maine	OH Ohio	WA Washington
FL Florida	MI Michigan	OK Oklahoma	WI Wisconsin
GA Georgia	MN Minnesota	OR Oregon	WV West Virginia
GU Guam	MO Missouri	PA Pennsylvania	WY Wyoming
HI Hawaii	MS Mississippi	PR Puerto Rico	

Aarif, Dr Atta M. (1953). Staff Crystallographer. Chemistry Dept., University of Utah, Salt Lake City, UT 84112, USA. PhD (U. London, Queen Mary College, 1983) inorganic chemistry. *Structure_determination crystallographic_computing*
E-mail xray@chemistry.chem.utah.edu xray@utahcca.bitnet Tel. 1(801)5815320 Fax 1(801)5818433

Abad-Zapatero, Dr Celerino (Cele) (1947). Group Leader. Member. Protein Crystallography Lab., Department D-46Y, AP-9A, 100 Abbott Park Rd., Abbott Park, IL 60064, USA. PhD (U. Texas, Austin, 1978) biophysics. *Structure_based_drug_design structure_based_protein_engineering aspartic_proteinases direct_methods_for_proteins computer_graphics*
E-mail abad@randb.abbott.com Tel. 1(708)9370294 Fax 1(708)9372625

Abboud, Dr Khalil A. (1952). Assistant Research Scientist. Department of Chemistry, University of Florida, Gainesville, FL 32611, USA. PhD (Louisiana State University, 1985) inorganic chemistry. *Single_crystals small_molecule computer_structure-activity_relationship drug_design quasicrystals non-linear_property biocrystallography*
E-mail kaabboud@ufpine.bitnet khalil@xrdvax.chem.ufl.edu Tel. 1(904)3925948 Fax 1(904)3928758

Abdel-Meguid, Dr Sherin S. (1946). Associate Director. Department of Macromolecular Sciences, SmithKline Beecham, 709 Swedeland Road, King of Prussia, PA 19406, USA. PhD (U. Nebraska-Lincoln, 1977) chemistry. *Single_crystal computer-assisted_design enzyme_inhibitors proteins structure-based_drug_design viruses protein_crystallization homology*
E-mail abdelmeguiss%phvax.dnet@smithkline.com Tel. 1(215)2707942 Fax 1(215)2704091

Abergel, Chantal (1961). Visiting Fellow. National Institutes of Health, Bethesda, MD 20892, USA. PhD (St Jerome University France, 1990) biophysics. *Biological_macromolecules immunology*
E-mail chantal@bandol.niddk.nih.gov Tel. 1(301)4021781 Fax 1(301)4960201

Abraham, Dr Donald J. (1936). Professor and Chairman, Department of Medicinal Chemistry. Medical College of Virginia/Virginia Commonwealth University, Department of Medicinal Chemistry, 410 N. 12th St., Box 540, R.B. Smith Bldg., Rm 548, Richmond, VA 23298-0540, USA. [Department of Medicinal Chemistry, MCV Box 980540, Richmond, VA 23298-0540, USA.] PhD (Purdue U., 1963) organic chemistry. *Drug_binding drug_design macromolecular_interactions allostery sickle_cell_anemia allosteric_effectors proteins_involved_in_cancer_pathways Alzheimers_proteins molecular_modelling hydrophobic_fields*
E-mail dabraham@gems.vcu.edu Tel. 1(804)7868483 Fax 1(804)3717625

Abrahams, Dr Sidney Cyril (1924). Adjunct Professor of Physics. Chairman, Commission on Crystallographic Nomenclature; Representative to IUPAC Interdivisional Committee on Nomenclature and Symbols. Physics Department, Southern Oregon State College, Ashland, OR 97520, USA. DSc (U. Glasgow, UK, 1957) crystallography. *Physical_property_prediction_and_measurement*
E-mail sca@cauchy.sosc.osshe.edu Tel. 1(503)5526496, 1(503)4827942 Fax 1(503)5526415

Adams, Dr W. Wade (1946). Senior Scientist. Materials Directorate, Wright Laboratory, Wright-Patterson AFB, OH 45433-7702, USA. [WL/MLPJ Bldg 651, 3005 P St, Suite 1, WPAFB, OH 45433-7702, USA.] PhD (U. Massachusetts, 1984) polymer science and engineering. *Morphology computer_modelling microscopy aerospace computer-aided_education polymeric_liquid_crystals structure_determination synchrotron_radiation*
E-mail adamsww@ml.wpafb.af.mil Tel. 1(513)2556652x3171 Fax 1(513)2551128

Adman, Dr Elinor T. (1941). Research Professor. Member, Commission on Journals. Department of Biological Structure SM-20, University of Washington, Seattle, WA 98195, USA. PhD (Brandeis U., 1967) physical chemistry. *Proteins metalloproteins refinement copper_proteins iron_sulfur_proteins solvent_structure*
E-mail adman@u.washington.edu Tel. 1(206)5436589 Fax 1(206)5431524 Telex 4740096 UW UI

Afshar, Carol (1956). Scientific assistant. The Institute for Cancer Research, Fox Chase Cancer Center, 7701 Burholme Ave., Philadelphia, PA 19111, USA. BS (Spring Garden College of Technology, 1980) biochemistry. *Protein_crystallography proteins reactions*
E-mail ce_afshar@fccc.edu Tel. 1(215)7282220 Fax 1(215)7282863

Agard, Dr David A. (1953). Professor. Dept of Biochemistry and Biophysics, UCSF, Box 0448, San Francisco, CA 94143-0448, USA. PhD (California Institute of Technology, 1980) biological chemistry. *Structure_function protein_folding biological_macromolecules computer_modelling drug_design chromosome_structure proteases*
E-mail agard@msg.ucsf.edu Tel. 1(415)4762521 Fax 1(415)4761902

Aldas, H. Oswaldo (1953). Graduate student. Department of Chemistry, University of Texas at Austin, Austin, TX 78712-1167, USA. MA (U. Texas at Austin, 1985) physical chemistry. *Solid state_chemistry computer_modelling materials science_teaching science_philosophy science_history*
E-mail oaldas@utxvms.cc.utexas.edu Tel. 1(512)4714042 Fax 1(512)4718696

Aleshin, Dr Alexander E. (1961). Visiting Scientist. Biochemistry and Biophysics Department, Molecular Biology Building, Iowa State University, Ames, IA 50011, USA. PhD (Moscow Institute of Crystallography, 1992) protein crystallography. *Structure-activity_relationship proteins structure_determination computer_modelling low_temperature_crystallography affinity carbohydrate_hydrolysis conformational_change*
E-mail aleshin@iastate.edu Tel. 1(515)2940567 Fax 1(515)2940453

Alexander, Prof. Leroy E. (1910). Retired. Carnegie-Mellon University. [68401 Hill St., Sturgis, MI 49091, USA.] PhD (U. Minnesota, 1943) physical chemistry. *Polycrystalline_X-ray_diffraction organic_structures polymer_structure*
Tel. 1(616)6512850

Alexander, Dr Richard S. (1965). Research investigator. DuPont Merck, Experimental Station E228-340A, PO Box 80228, Wilmington, DE 19880-0228, USA. PhD (U. Pennsylvania, 1992) chemistry. *Proteins computer_modelling rational_drug_design*
E-mail alexanrs@lldmpc.dnet.dupont.com Tel. 1(302)6953216 Fax 1(302)6958667

Alkire, Dr Randall W. (1953). Technical staff. Argonne National Lab., 9700 S. Cass Ave, Bldg. 202, Argonne, IL 60439, USA. [NSLS Bldg. 725, X8, Upton, NY 11973, USA.] PhD (U. Missouri-Rolla, 1982) physical chemistry. *Single_crystal_diffraction instrumentation synchrotron_radiation neutron absolute_structure_factors*
E-mail alkire@bnl.gov Tel. 1(516)2825025 Fax 1(516)2827956

Allen, James P. (1955). Assistant Professor. Department of Chemistry and Biochemistry, Arizona State University, Tempe, AZ 85287-1604, USA. PhD (U. Illinois-Urbana, 1982) physics. *Membrane_proteins electron_transfer redox_proteins photosynthesis*
E-mail allenj@asuchm.la.asu.edu allen@prtnx1.la.asu.edu Tel. 1(602)9658241 Fax 1(602)9652747

Allen, Joseph H. (1924). Retired. 3629 Swallow Lane, Irvine, TX 75062, USA. BS (Texas Tech., 1949) chemistry.

Allen, Dr Karen N. (1962). Assistant Professor. Department of Physiology, Boston University School of Medicine, 80 East Concord Street, Boston, MA 02118-2394, USA. PhD (Brandeis U., 1989) biochemistry. *Enzymatic_catalysis conformational_change enzyme_inhibitors enzymatic_mechanism kinetics*
E-mail allen@med-xtal.bu.edu Tel. 1(617)6384398 Fax 1(617)6384273

Allersma, Ties (1936). Senior Research Associate. PPG Industries, Glass Technology Center, Box 11472, Pittsburgh, PA 15238, USA. MS (Delft U., 1965) physics. *X-ray_diffraction*
Tel. 1(412)8208793 Fax 1(412)8208580

Ammon, Dr Herman L. (1936). Professor. Department of Chemistry and Biochemistry, University of Maryland, College Park, MD 20742, USA. PhD (U. Washington, 1963) organic chemistry. *Energetic_materials density_prediction structure_prediction*
E-mail ha3@umail.umd.edu ammon@hlammon.umd.edu Tel. 1(301)4051824 Fax 1(301)3149121

Amzel, Dr L. Mario (1942). Professor. Department of Biophysics and Biophysical Chemistry, Johns Hopkins School of Medicine, 725 N. Wolfe St., Baltimore, MD 21205, USA. PhD (U. Buenos Aires, 1968) physical chemistry. *Binding_proteins recognition computer_modelling homology_prediction oxidations/phosphorylations*
E-mail mario@jhuigf.med.jhu.edu marioi@jhuigf.bitnet Tel. 1(410)9553955 Fax 1(410)9550637

Anderson, Amy C. (1969). Graduate student. Department of Biophysics, Harvard University, Dana Farber Cancer Institute, 44 Binney St. D1040, Boston, MA 02115, USA. BS (MIT, 1991) biology. *RNA RNA_chemistry RNA_structure nucleic_acids hairpins pseudoknots*
E-mail amyac@green.dfci.harvard.edu Tel. 1(617)6324754 Fax 1(617)6324393

Anderson, Dr Daniel H. (1956). Research Faculty. Molecular Biology Institute, University of California, Los Angeles, Los Angeles, CA 90024-1570, USA. PhD (UC San Diego, 1986) biochemistry. *Peptides proteins data_collection area_detectors*
E-mail dha@uclaue.mbi.ucla.edu Tel. 1(310)2063642 Fax 1(310)2063914

Anderson, Dr Gary D. (1943). Professor. Department of Chemistry, Marshall University, Huntington, WV 25755, USA. PhD (Florida State U., 1972) organic chemistry. *Computer_modelling*
E-mail m043005@marshall.wvnet.edu m043005@marshall.bitnet Tel. 1(304)6966594

Anderson, John E. (1952). Senior Staff Investigator. W. M. Keck Structural Biology Laboratory, Cold Spring Harbor Laboratory, 1 Bungtown Road, Cold Spring Harbor, NY 11724-2220, USA. PhD (Harvard University, 1985) biophysics. *Macromolecular_crystallography molecular_complexes conformational_change nervous_system neurotransmitter_receptors ion_channels protein–DNA_interactions theoretical_neurobiology*
E-mail anderson@cshl.org Tel. 1(516)3678822 Fax 1(516)3678873

Anderson, Oren P. (1942). Professor; Department Chair. Department of Chemistry, Colorado State University, Fort Collins, CO 80523, USA. PhD (Northwestern U., 1968) chemistry. *Coordination_chemistry polydentate_chelates porphyrins mixed-valence_compounds protein_crystallography*
E-mail opa@lamar.colostate.edu Tel. 1(303)4916339 Fax 1(303)4911801

Anderson, Timothy A. (1967). Graduate Student. Department of Biochemistry, University of Minnesota, 4-225 Millard Hall, 435 Delaware St. S.E., Minneapolis, MN 55455, USA. MS (U. Wisconsin-Madison, 1991) biochemistry. *Biocrystallography refinement_method lipoproteins structure–activity_relationship*
E-mail anderson@dccc.med.umn.edu Tel. 1(612)6252115

Anderson, Dr Wayne F. (1948). Professor of Molecular Pharmacology and Biological Chemistry, Northwestern University Medical School, 303 E. Chicago, IL 60611, USA. PhD (Yale U., 1975) molecular biophysics and biochemistry. *Macromolecular_crystallography computer_modelling proteins nucleic_acids protein_DNA_recognition*
E-mail wf-anderson@nwu.edu Tel. 1(312)5031697 Fax 1(312)5030796

Andrews, Larry (1941). Editor. International Centre for Diffraction Data (ICDD), Newtown Square Corporate Campus, 12 Campus Blvd., Newtown Sq., PA 19073-3273, USA. PhD (U. Washington, Seattle, 1975) chemistry. *Lattices computer_modelling reduced_cells inclusion_compounds protein_molecule_interactions*
E-mail andrews@icdd.com Tel. 1(610)3259814 Fax 1(610)3259823

Antonio, Dr Mark (1954). Scientist. Argonne National Laboratory, Chemistry Division, 9700 S. Cass Ave, Argonne, IL 60439-4831, USA. PhD (Michigan State U., 1983) inorganic chemistry. *Synchrotron_radiation XANES EXAFS Mossbauer heteropoly_acids polyoxoanions lanthanides catalysis*
E-mail mrantonio@anlchm.chm.anl.gov Tel. 1(708)2529267 Fax 1(708)2524225

Arai, G. J. Cons. Eng. Zenith Electronics Corp, 2407 North Ave, Melrose Park, IL 60160-1120, USA. PhD (U. Leiden, The Netherlands, 1960) inorganic chemistry. *Analytical_chemistry X-R-F SEM*
Tel. 1(708)4508380 Fax 1(708)4508398

Araki, Takaharu (1929). Senior Associate Editor; retired December 31, 1994. Chemical Abstracts Service, 2540 Olentangy River Road, Columbus, OH 43210, USA. [15306 NE 17th Street, Apt. H 162, Bellevue, WA 98007-4301, USA.] DSc (Kyoto U., Japan, 1961) mineralogy. *Minerals ceramics structure_determination computing*
Tel. 1(206)6039369

Archer, Ronald D. (1932). Professor of Chemistry. Chemistry Dept., University of Massachusetts, Amherst, MA 01003-4510, USA. PhD (U. Illinois, Urbana, 1959) inorganic chemistry. *Synthesis structure metallic_chelates metallic_polymers transition_metal lanthanide_coordination*
E-mail archer@chemistry.mass.edu Tel. 1(413)5451521 Fax 1(413)5454490

Arents, Dr Gina (1944). Research Associate. Department of Biology, Johns Hopkins University, Baltimore, MD 21218, USA. PhD (Johns Hopkins U., 1987) biophysics. *Chromatin macromolecules proteins computer_modelling refinement_method homology computer_graphics*
E-mail bio._zga@jhuvms.hcf.jhu.edu Tel. 1(410)5168590

Arevalo, Jairo H. (1960). Scientific Associate. Department of Molecular Biology, The Scripps Research Institute, 10666 North Torrey Pines Rd., La Jolla, CA 92037, USA. PhD (The Scripps Research Institute, 1993) macromolecular and cellular structure and chemistry. *Drug_design proteins steroids antibodies receptors crystallization molecular_recognition*
E-mail arevalo@scripps.edu Tel. 1(619)5543473 Fax 1(617)5546105

Arjunan, Dr P. Arjunan (1956). Research Associate. Department of Crystallography, University of Pittsburgh, Pittsburgh, PA-15260, USA. PhD (Indian Institute of Science, India, 1985) organic chemistry. *Structure_DNA structure_proteins computer_modelling dynamics_proteins drug_computer-assisted_design synthesis_oligonucleotides nuclear_magnetic_resonance photochemistry_molecular_complexes*

E-mail arjun@pittvms Tel. 1(412)6249300 Fax 1(412)6241882

Armstrong, Prof. Richard N. (1948). Professor. Department of Chemistry and Biochemistry, University of Maryland, College Park, MD 20910, USA. PhD (Marquette U., 1975) organic chemistry. *Enzyme_mechanisms protein_structure stereochemistry membranes protein_evolution xenobiotic_metabolism*
E-mail ra19@umail.umd.edu Tel. 1(301)4051812 Fax 1(301)4057956

Armstrong, Ronald W. (1934). Professor. 1112 Brice Dr., Edgewater, MD 21037, USA. PhD (Carnegie-Mellon U., 1958) metallurgical engineering. *Dislocation deformation fracture topography energetic_crystals*
Tel. 1(301)4055291 Fax 1(301)3149477

Armstrong, Shelly R. (1966). Graduate student. Department of Biochemistry, Cornell University, Ithaca, NY 14853, USA. MS (U. Alabama, Birmingham, 1991) biochemistry. *Structure_determination proteins enzymatic_mechanism nucleosides synchrotron_radiation mutagenesis crystallization membranes_transport*
E-mail shelly@vgx.tn.cornell.edu Tel. 1(607)2552174 Fax 1(607)2552428

Armstrong, Dr William H. (1954). Associate Professor. Department of Chemistry, Boston College, Chestnut Hill, MA 02167, USA. PhD (Stanford U., 1982) inorganic chemistry. *Small_molecules bioinorganic_chemistry*
E-mail armstrong@hermes.bc.edu Tel. 1(617)5528077 Fax 1(617)5522705

Arnold, Dr Edward (1957). Associate Professor. Center for Advanced Biotechnology and Medicine, Dept. of Chemistry/Rutgers University, 679 Hoes Lane, Piscataway, NJ 08854, USA. PhD (Cornell U., 1982) organic chemistry. *Viruses polymerases macromolecular_structure_determination antiviral_compounds AIDS common_cold_viruses reverse_transcriptases molecular_recognition/evolution drug_design protein_engineering synchrotron_radiation refinement conformational_changes*
E-mail arnold@rhino.cabm.rutgers.edu Tel. 1(908)2355323 Fax 1(908)2354850

Arnone, Dr Arthur (1942). Professor. Department of Biochemistry, 4-470 Bowen Science Building, The University of Iowa, Iowa City, IA 52242, USA. PhD (Massachusetts Institute of Technology, 1970) physical chemistry. *Macromolecular_crystallography proteins hemoglobins allostery*
E-mail amone@eve.biochem.uiowa.edu Tel. 1(319)3357882 Fax 1(319)3359570

Aronson, Hans-Erik G. (1963). Graduate Student. Department of Biochemistry and Molecular Biophysics, Columbia University, 630 West 168th Street, New York, NY 10032, USA. MPhil biochemistry and molecular biophysics. *Synchrotron_radiation protein_structure immunology catalytic_antibodies anomalous_dispersion cell_cell_adhesion signal_transduction computer_modelling*
E-mail aronson@cuhhca.hhmi.columbia.edu Tel. 1(212)3051951 Fax 1(212)3057379

Arora, Dr Satish K. (1942). Research Professor. Department of Crystallography, University of Pittsburgh, Pittsburgh, PA 15260, USA. PhD (U. Poona, 1970) physical chemistry. *Structural_drugs DNA proteins NMR computer_modelling cancer*
E-mail sarora@pitt.vms.edu Tel. 1(412)6249300 1(412)6247282 Fax 1(412)6241882

Arval, Andrew S. (1963). Senior Research Programmer. The Scripps Research Institute, Mail Drop MB4, 10666 North Torrey Pines Rd., La Jolla, CA 92037, USA. MS (U. California at San Diego, 1989) bioengineering. *Proteins crystallography computer_programming*
E-mail arvai@scripps.edu Tel. 1(619)5544485 Fax 1(619)5546880

Arvanitis, Dr Georgia Associate Professor. Trenton State College, Hillwood Lakes, CN4700, Trenton, NJ 08650, USA. PhD (Princeton U., 1987) organometallic chemistry. *Inorganic platinum_compounds oligonucleotides DNA*
E-mail arvanit@tscvm.trenton.edu Tel. 1(609)7712917 Fax 1(609)7713167

Athappilly, Dr Francis K. (1956). Associate Research Scientist. Department of Biochemistry and Molecular Biophysics, College of Physicians and Surgeons of Columbia University, 630 West 168 Street, New York, NY 10032, USA. PhD (Indian Institute of Science, Bangalore, India, 1984) molecular biophysics. *Crystallography proteins genetic_engineering drug_design*
E-mail francis@cuhhca.hhmi.columbia.edu Tel. 1(212)3052219 Fax 1(212)3057379

Atkinson, Dr David (1944). Professor of Biophysics, Research Professor of Biochemistry. Department of Biophysics, Boston University School of Medicine, 80 East Concord Street, Boston, MA 02118, USA. PhD (Council for National Academic Awards, England, 1975) biophysics - X-ray diffraction. *Lipoproteins lipids proteins macromolecules crystallography small_angle_scattering electron_microscopy computer_modelling*
E-mail atkinson@med-biophi.bu.edu Tel. 1(617)6384015 Fax 1(617)6384041

Atoji, Masao Associate Editor. 702 86th Place, Downers Grove, IL 60516-4951, USA. PhD (Osaka U., Japan, 1956) physical chemistry/crystallography. *Physics chemistry crystallography semiconductor*
Tel. 1(708)9851248

Attard, Alfred E. Sr. Physicist. DOD-R53, 9800 Savage Road, Ft. Meade, MD 20755, USA. [5434 Phelps Luck Drive, Columbia, MD 21045, USA.] PhD (Illinois Institute of Technology, 1962) physics. *Structure_determination computer_modelling diffraction_physics nonlinear_optics materials_science*

Atwood, Dr David (1965). Assistant Professor. Department of Chemistry, North Dakota State University, Fargo, North Dakota 58105, USA. PhD (U. Texas at Austin, 1992) chemistry. *Organometallics*
E-mail datwood@plains.nodak.edu Tel. 1(701)2378747 Fax 1(701)2378831

Augustin, Director, Rolf M. (1932). International Business Development. Polaroid Corporation, 575 Technology Square, Cambridge, MA 2139, USA. BA (Princeton U., 1954) public and international affairs.
Tel. 1(617)5774477 Fax 1(617)5775756

Axtell, III, Enos A. (1967). Graduate student. Department of Chemistry, Michigan State University, East Lansing, MI 48824-1322, USA. BS (U. Missouri, 1989) chemistry. *Novel_chalcogenides molten_salt_synthesis bulk_single_crystal_growth*
E-mail axtell@msucem.bitnet Tel. 1(517)3559715x177

Babu, Dr Y. S. (1952). Vice President. Biocryst Pharmaceuticals Ltd, 2190 Parkway Lake Dr, Birmingham, AL 35244, USA. PhD (Indian Institute of Science, 1979) crystallography. *Enzyme_structure proteins drug_design modelling receptor_structure*
E-mail babu@orion.cmc.uab.edu Tel. 1(205)4444606 Fax 1(205)4444640

Badger, John (1962). Senior Research Associate. Rosenstiel Basic Medical Sciences Research Center, Brandeis University, Waltham, MA 02254, USA. DPhil (York U., 1986) physics. *Structural_change macromolecules hydration_structure computer_graphics macromolecular_diffuse_scattering protein_electrostatics insulin*
E-mail badger@brandeis Tel. 1(617)7364910 Fax 1(617)7362405

Baeyens, Katrien J. (1968). Res. Associate. Department of Chemistry, Calvin Laboratory, University of California, Berkeley CA 94720, USA. MS (K. U. Leuven, Belgium, 1991) pharmaceutical sciences. *Biocrystallography macromolecular_crystallization DNA RNA*
E-mail hldebondt@lbl.bitnet Tel. 1(510)4865449 Fax 1(510)4866059

Bailey, Larry K. (1952). Senior staff scientist. Conoco Inc., PO Box 1267, Ponca City, OK 74602-1267, USA. MS (Eastern New Mexico U., 1977) chemistry. *Minerals carbon_materials catalysts petroleum environmental*
Tel. 1(405)7675012

Bailey, Prof. Sturges (1919). Professor. Dept. Geology and Geophysics, University of Wisconsin-Madison, 1215 W. Dayton St., Madison, WI 53706, USA. PhD (U. Cambridge, 1955) physics (crystallography). *Mineralogy phyllosilicates polytypism clays*
E-mail bailey@geology.wisc.edu Tel. 1(608)2621806 Fax 1(608)2620693 Telex 265452

Baird, Dr Herbert Wallace (1936). Professor (Retired), Wake Forest U. [PO Box 849, Walkertown, NC 27051-0849, USA.] PhD (U. Wisconsin-Madison, 1963) physical chemistry.
E-mail hbaird@delphi.com Tel. 1(910)5954480 Fax 1(910)5959370

Bajorath, Dr Jurgen (1960). Senior Scientist. Immunology, Bristol-Myers Squibb, 3005 First Avenue, Seattle, WA 98121, USA. PhD (FU Berlin, 1988) biochemistry. *Protein_structure structural_similarity protein_modelling protein_ligand_complexes*
E-mail bajorath@protos.bms.com Tel. 1(206)7273612 Fax 1(206)7273602

Baker, Tim S. (1949). Professor of Biological Sciences. Dept. Biological Sciences, Purdue University, West Lafayette, IN 47907, USA. PhD (U. California, Los Angeles, 1976) biochemistry. *Electron_microscopy image_analysis virus_structure image_reconstruction macromolecule_structure*
E-mail tsb@bragg.bio.purdue.edu Tel. 1(317)4945645 Fax 1(317)4961189

Baldwin, Dr Eric T. (1960). Scientist. The Upjohn Company, 301 Henrietta Street, Kalamazoo, MI 49001-0199, USA. [Analytical Chemistry, The Upjohn Company, MS 7255-209-102, Kalamazoo, MI 49001-0199, USA.] PhD (U. North Carolina, 1990) biochemistry. *Macromolecular_crystallography chemotherapy crystallization data_collection*
E-mail etbaldwi@upj.com Tel. 1(616)3856815 Fax 1(616)3857522

Baldwin, Kenneth J. (1948). Research Associate. Department of Earth and Space Sciences, State University of New York, Stony Brook, NY 11794-2100, USA. MS (SUNY Stony Brook, 1973) geochemistry. *Computer_automation instrumentation diffraction synchrotron_radiation*
E-mail baldwin@sbmp04.ess.sunysb.edu kjbaldwin@ccmail.sunysb.edu Tel. 1(516)6328206 Fax 1(516)6328140

Ball, Richard G. (1950). Senior Research Fellow. Merck Research Laboratories, MS R50-105, PO Box 2000, Rahway, NJ 07065, USA. PhD (The University of Western Ontario, 1978) chemical crystallography. *Bioorganic_crystallography drug_structure charge_density structural_accuracy computing*
E-mail ball@merck.com Tel. 1(908)5945341 Fax 1(908)5946100

Ban, Nenad (1966). Graduate Student. Department of Biochemistry, University of California at Riverside, Riverside, CA 92521, USA. BS (U. Zagreb, 1989) molecular biology. *Proteins immunoglobulins isomorphous_replacement molecular_replacement drug_design molecular_modelling crystal_growth tRNA_synthetases*
E-mail nenadban@ucrac1.ucr.edu Tel. 1(909)7873397 Fax 1(909)7873790

Banaszak, Leonard (1933). Dietrich Professor of Biochemistry. Dept. of Biochemistry, 4-225 Millard Hall, University of Minnesota, 435 Delaware St. S.E., Minneapolis, MN 55455, USA. PhD (Loyola U. Chicago, Illinois, 1961) biochemistry. *Protein_crystallography protein_structure/function protein/lipid_interactions enzyme_mechanism*
E-mail len_b@dccc.med.umn.edu Tel. 1(612)6266597 1(612)6252170 Fax 1(612)6252163

Bancroft, Daniel P. (1959). Research Assistant Professor. Biochemistry Dept., School of Medicine, University of Utah, 50 North Medical Drive, Salt Lake City, UT 84132, USA. PhD (Texas A&M University, 1986) inorganic chemistry. *Purple_acid phosphatases metalloenzymes metalloproteins protein_structure platinum anticancer_drugs*
E-mail bancroft@uxray.med.utah.edu Tel. 1(801)5853921 Fax 1(801)5817959

Barber, Patrick G. (1942). Professor of Chemistry. Route 2 Box 29B, Keysville, VA 23947-9514, USA. PhD (Cornell U., 1969) physical chemistry/crystallography. *Liquid_crystals crystal_growth crystal_polymers*
E-mail pbarber@lwcvm1.lwc.edu Tel. 1(804)3952573 Fax 1(804)3952652

Barbour, Dr Leonard James (1965). Postdoctoral Fellow. Department of Chemistry, University of Missouri, Columbia, MO 65211, USA. PhD (U. Cape Town, 1994) chemistry. *Inclusion_complex structure_and_thermal_properties_of_inclusion_compounds calixarenes*
E-mail chemlb@mizzou1.missouri.edu Tel. 1(314)8821811 Fax 1(314)8822754

Barkigia, Dr Kathleen M. (1951). Chemist. Brookhaven National Laboratory, Department of Applied Science, Bldg. 815, Upton, NY 11973, USA. PhD (Georgetown U., 1978) inorganic chemistry. *X-ray_and_neutron_structures porphyrin_structure hydrogen_bonding conformational_change_to_property_relations*
E-mail barkigia@ckb.chm.bnl.gov Tel. 1(516)2827661 Fax 1(516)2825815

Barnes, Dr Charles L. (1949). Crystallographer/Research Investigator. Department of Chemistry, University of Missouri, Columbia, MO 65211, USA. PhD (U. Tennessee, 1980) biochemistry. *Structural_chemistry natural_products peptides*
E-mail chemclb@mizzou1.missouri.edu chemclb@mizzou1.bitnet Tel. 1(314)8822962 Fax 1(314)8822754

Barnes, Dr John D. (1939). Physicist. Mechanical Performance and Structure Group, NIST Polymers Division, Bld 224, Rm A209, Gaithersburg, MD 20899, USA. PhD (Catholic Univ of America, 1972) physics. *SAXS small-angle_scattering Pole_figure Polymers microstructure texture crystallinity*
E-mail johnbarnes@enh.nist.gov johnbar@polysaxs.nist.gov Tel. 1(301)9756786 Fax 1(301)9772018

Barnhart, Dr David M. (1933). Staff crystallographer. Department of Chemistry-BG-10, University of Washington, Seattle, WA 98195, USA. PhD (Oregon State U., 1964) physical chemistry. *Structure_determination*
E-mail barnhart@uwchem.chem.washington.edu Tel. 1(206)5430210 Fax 1(206)6858665

Bartell, Prof. Lawrence S. (1923). Professor. Department of Chemistry, University of Michigan, Ann Arbor, MI 48109, USA. PhD (U. Michigan, 1951) physical chemistry. *Molecular_clusters nucleation_dynamics electron_diffraction MD_simulations intermolecular_forces electron_holography*
E-mail l.s.bartell@um.cc.umich.edu Tel. 1(313)7647375 Fax 1(313)7474865

Bartelmehs, Dr Kurt L. (1962). Research Associate. Dept. of Geological Sciences, Virginia Polytechnic Institute and State University, Blacksburg, Virginia 24061, USA. PhD (Virginia Polytechnic Institute and State U., 1993) geology. *Computer_graphics modelling statistics thermal_motion*
E-mail aufinger@vtvm1.cc.vt.edu Tel. 1(703)2316521 Fax 1(703)2313386

Barton, Dr Randolph (1941). Research Associate. DuPont Co., Experimental Station-PO Box 80302, Wilmington, DE 19880-0302, USA. PhD (The Johns Hopkins U., 1968) physical chemistry. *Fibrous_polymers WAXS SAXS diffraction_technique*
E-mail bartonr@csoc.dnet.dupont.com Tel. 1(302)6952578 Fax 1(302)6959811

Basu, Dr Sankar P. Physicist. Ctr. for Devices and Rad. Health, 1390 Piccard Dr., Rockville, MD 20850, USA. [9508 Lumber Jack Row, Columbia, MD 21046, USA.] PhD (U. Oklahoma, 1977) biophysics. *X-ray_diffraction_macromolecules laser_biophysics UV_radiation_effect*
E-mail spb@fdadr.cdrh.fda.gov Tel. 1(301)5941307 Fax 1(301)5942358

Batalia, Michael A. Department of Chemistry and Biochemistry, University of Texas at Austin, Mail Code 15300, Austin, TX 78712, USA. BA (U. Chicago, 1989) chemistry. *Proteins computer_modelling protein_toxins*
E-mail Batalia@utbc01.cm.utexas.edu Tel. 1(512)4713625

Batterman, Prof. Boris W. (1930). Walter S. Carpenter Professor of Engineering. Cornell U., Clark Hall, Ithaca, NY 14853, USA. PhD (Massachusetts Institute of Technology, 1956) physics. *Dynamical_diffraction optics crystallography*
E-mail bwb1@cornell.edu Tel. 1(607)2550917 Fax 1(607)2559001

Bau, Dr Robert (1944). Professor. Department of Chemistry, University of Southern California, Los Angeles, CA 90089, USA. PhD (U. California at Los Angeles, 1968) inorganic chemistry. *Neutron_diffraction metal_hydrides organometallic_chemistry metal_clusters*
E-mail bau@chem1.usc.edu Tel. 1(213)7402692 Fax 1(213)7400930

Bauer, Cary B. (1968). Graduate Student. Department of Chemistry, Northern Illinois University, DeKalb, IL 60115, USA. anticipated PhD 12/94 (Northern Illinois U., 1994) inorganic chemistry. *Chemistry coordination heavy_atom_radiochemistry separation_science inclusion synthesis*
E-mail cary@rdr.chem.niu.edu Tel. 1(815)7536886 Fax 1(815)7534802

Baughman, Russell G. (1946). Professor. Div. Science, Northeast Missouri State University, Kirksville, MO 63501-0828, USA. PhD (Iowa State U., 1977) physical chemistry. *Pesticides charge_transfer porphyrins anti-viral small_molecules*
E-mail sc05@nemomus.edu Tel. 1(816)7854627 Fax 1(816)7854045

Baures, Paul W (1963). Graduate student. Department of Medicinal Chemistry, HS Unit F, 308 Harvard St. SE, Minneapolis, MN 55455-0343, USA. MS (U. Minnesota, 1985) chemistry.
E-mail baur0001@gold.tc.umn.edu Tel. 1(612)6243885

Bayya, Shyam S. (1964). Postdoctoral research associate. NYS College of Ceramics, Alfred University, Alfred, NY 14802-1296, USA. PhD (Alfred U., 1993) ceramics. *Superconductivity thick_films_synthesis diffraction EXAFS aerosol crystal_growth thin_films*
E-mail bayya@xray.alfred.edu Tel. 1(607)8712392

Bazan, Dr J. Fernando (1960). Staff Scientist. Protein Machine Group, Dept. of Molecular Biology, DNAX Research Institute of Molecular and Cellular Biology, 901 California Ave., Palo Alto, CA 94043-1104, USA. PhD (U. California, Berkeley, 1989) biophysics. *Proteins computer_modelling homology_prediction evolution proteins_cytokines proteins_receptors proteins_enzymes*
E-mail bazan@cgl.ucsf.edu dnax@cellbio.stanford.edu Tel. 1(415)4961115 Fax 1(415)4961200

Beamer, Dr Lesa J. (1963). Postdoctoral Fellow. Molecular Biology Institute, UCLA, 405 Hilgard Avenue, Los Angeles, CA 90024, USA. PhD (Johns Hopkins University School of Medicine, 1991) biophysics. *Biology crystal_growth proteins chaperones*
E-mail lesa@uclaue.mbi.ucla.edu Tel. 1(310)8258901 Fax 1(310)2067286

Bear, Richard S. (1908). Professor. 1515 E. Franklin St., Unit 43, Chapel Hill, NC 27514, USA. PhD (U. California, Berkeley, 1933) chemistry. *Structure_of_collagen American_leading_arts+science+humanist_society*

Becker, Dr Joseph W. (1943). Senior Research Fellow. Merck Research Laboratories, PO Box 2000 (R50-105), Rahway, NJ 07065-0900, USA. PhD (Stanford U., 1970) physical chemistry. *Proteins macromolecules*
E-mail joseph_becker@merck.com Tel. 1(908)5943418 Fax 1(908)5946100

Bedarkar, Sudhir (1951). Visiting research scientist. Dept. of Pharmacology, Rm.I-352, Yale University School of Medicine, 333 Cedar Street, New Haven, CT 06510, USA. PhD (London U., UK, 1982) biophysics, macromolecular crystallography.
E-mail sudhir@pharm.med.yale.edu Tel. 1(203)7856232 Fax 1(203)7856232

Bednowitz, Dr Allan L. (1939). Assoc. Dir. & Sr. Consultant. InnoVal Systems Solutions, Inc., 604 Mamaroneck Avenue, Harrison, NY 10528, USA. PhD (Polytechnic Institute of Brooklyn, 1966) chemical physics. *Molecular_graphics computing database multimedia desk_top_publishing*
E-mail bednowitz@aol.com Tel. 1(914)8353838 Fax 1(914)8353857

Bedzyk, Prof. Michael J. (1951). Associate Professor. Materials Science Department, Northwestern University, 2225 Sheridan Ave., Evanston, IL 60208, USA. [Argonne National Laboratory, Materials Science Division Bld. 223, Argonne, IL 60439, USA.] PhD (SUNY Albany, 1982) physics. *Synchrotron_radiation dynamical_diffraction surface interferometry X-ray_standing_waves*
E-mail bedzyk@anlsrs.msd.anl.gov Tel. 1(708)4913570 1(708)2527763 Fax 1(708)4917820

Beers, Dr William W. (1955). Senior Research Chemist. Technology Division, General Electric - Lighting, NELA Park, 1975 Noble Rd, Cleveland, OH 44112, USA. PhD (Iowa State U., 1983) inorganic chemistry. *Inorganic_luminescence Rietveld_method microstrain crystallinity ceramics oxides powders symmetry_breaking*
E-mail beers@liso.dnet.ge.com Tel. 1(216)2663149 Fax 1(216)2662987

Beese, Lorena S. Asst. professor. Dept. of Biochem., Box 3711, Duke University Medical Center, Durham, NC 27710, USA. PhD (Brandeis U., 1984) biophysics. *Macromolecules_DNA polymerases protein-nucleic_acid_interactions signal_transduction_proteins*
E-mail beese@brass.biochem.duke.edu Tel. 1(919)6815267 Fax 1(919)6848885

Beiter, Dr Thomas A. (1947). Postdoctoral researcher. Chemistry Department, Miami University, Oxford, OH 45056, USA. [PO Box 3532, Mansfield, OH 44907, USA.] PhD (Miami U. (Ohio), 1992) physical chemistry. *Mathematical_methods*
Tel. 1(419)8642501

Bell, Dr Jeffrey A. (1954). Assistant Professor. Department of Chemistry, Rensselaer Polytechnic Inst., Troy, NY 12180, USA. PhD (Cornell U., 1985) biochemistry. *Biomaterial proteins crystal_growth hydration protein_engineering*
E-mail bell@xray.chem.rpi.edu Tel. 1(518)2764075 Fax 1(518)2764045

Bellamy, Henry D. (1948). Research Associate. Stanford Synchrotron Radiation Laboratory, Mail Stop 69, Box 4349, Stanford, CA 94301, USA. PhD (Washington U., 1985) molecular biology. *Proteins MAD-phasing Laue_method cryocrystallography*
E-mail bellamy@ssrl01.slac.stanford.edu bellamy@ssrl750.bitnet Tel. 1(415)9263107 Fax 1(415)9264100

Belt, Roger F. (1929). Airtron Division-Litton, 200 E. Hanover Ave., Morris Plains, NJ 07960, USA. PhD (State U. Iowa, 1956) physical chemistry. *Crystal_growth*
Tel. 1(201)5395500 Fax 1(201)5392210

Bennett, Dr Dennis W. (1946). Associate Professor. Department of Chemistry, University of Wisconsin-Milwaukee, Milwaukee, WI 53201, USA. PhD (U. Utah, 1978) inorganic chemistry. *Chemical_crystallography bonding_theory macromolecular_crystallography crystallography_education theoretical_crystallography*
E-mail dwbent@csd4.csd.uwm.edu Tel. 1(414)2295276 Fax 1(414)2295530

Bennett, J. Michael (1939). Research Associate. Mobil Research and Development Corporation, Paulsboro Research Laboratory, Post Office Box 480, Paulsboro, NJ 08066-480, USA. PhD (Aberdeen U., 1966) chemistry. *Zeolite_structure_chemistry microporous_materials*
E-mail jmbennett%vsjmb.dnet@dal.mobil.com Tel. 1(609)2242533 Fax 1(609)2243608

Bennett, Melanie J. (1967). Graduate Student. Molecular Biology Institute, University of California-Los Angeles, 405 Hilgard Avenue, Los Angeles, CA 90024, USA. BA (Reed College, 1989) physics. *Proteins toxins*
E-mail bennett@uclaue.mbi.ucla.edu Tel. 1(310)2063642

Benning, Dr Matthew M. (1961). Assistant Scientist. University of Wisconsin, Institute for Enzyme Research, 1710 University Ave., Madison, WI 53705-4098, USA. PhD (Northern Illinois U., 1988) inorganic chemistry. *Proteins electron_transfer lipid_transport*
E-mail Benning@enzyme.wisc.edu Tel. 1(608)2620529 Fax 1(608)2652904

Beno, Dr Mark A. (1951). Staff Scientist. Materials Science Division, Argonne National Laboratory, 9700 South Cass Ave., Argonne, IL 60439, USA. PhD (Ohio State U., 1979) physical chemistry. *Synchrotron_radiation crystallography instrumentation neutron_diffraction organic_conductors high_Tc_structure new_powder_diffraction_techniques detectors*
E-mail Mark_Beno@qmgate.anl.gov beno@anlsrs.msd.anl.gov Tel. 1(708)2523507 Fax 1(708)2527777

Benson, James E. (1933). Senior Chemist. Materials Char. Laboratory, General Electric Co., Cleveland, OH 44117, USA. MS (Iowa State U., 1963) physical chemistry. *Powder_diffraction automation Rietveld_analysis*
E-mail benson@cle.dnet.ge.com Tel. 1(216)2668370 Fax 1(216)2668380

Benson, Timothy E. (1969). Graduate Student. Department BCMP, Harvard Medical School, 240 Longwood Ave, Boston, MA 02115, USA. BA/MAT (Johns Hopkins U., 1990) chemistry/teaching. *Enzymes mechanism peptidoglycan_biosynthesis structure/function_relationships*
E-mail benson@bcmp.med.harvard.edu Tel. 1(617)4320930

Berendzen, Joel (1961). Staff Member. Biophysics Group (P-6), M715, Los Alamos National Laboratory, Los Alamos, NM 87545, USA. PhD (U. Illinois at Urbana-Champaign, 1990) biophysics. *Dynamics proteins Laue*
E-mail joelb@lanl.gov Tel. 1(505)6652552 Fax 1(505)6652343

Beres, Dr John J. (1947). Manager. Gen Corp Research, 2990 Gilehrist Road, Akron, Ohio 44305, USA. [6014 Echodell NW, N. Canton, Ohio 44720, USA.] PhD (Carnegie-Mellon U., 1975) physical chemistry. *Polymer SAXS WAXS*
Tel. 1(216)7946315 Fax 1(216)7946375

Berghuis, Albert M. (1960). Postdoctoral Fellow. Howard Hughes Medical Institute, University of Texas, Southwestern Medical Center, 5323 Harry Hines Boulevard, Dallas, TX 75235-9050, USA. PhD (U. British Columbia, Vancouver, Canada, 1993) biochemistry. *Proteins heme_proteins signal_transduction error_analysis crystallographic_computing*
E-mail albert@howie.swmed.edu Tel. 1(214)6485015 Fax 1(214)6486336

Berliner, Prof. Lawrence (1941). Professor of Chemistry. Dept of Chemistry, The Ohio State Univ, 120 W 18th Ave, Columbus, OH 43210, USA. PhD (Stanford Univ, 1967) physical chemistry. *Proteins paramagnetic_resonance electron_spin_resonance fluorescence spin_labeling*
E-mail berliner.2@osu.edu Tel. 1(614)2920134 Fax 1(614)2921532

Berliner, Ronald R. (1943). Sr. Research Scientist. Res. Reactor, Univ of Missouri, Res. Park, Columbia, MO 65211, USA. PhD (U. Illinois-Urbana, 1972) physics. *Phase_transitions cement instrumentation*
E-mail berliner@reactor.murr.missouri.edu Tel. 1(314)8825235 Fax 1(314)8823443

Berman, Dr Helen M. (1943). Professor. Department of Chemistry, Rutgers University, Wright Reiman Lab., Taylor Rd., PO Box 939, Piscataway, NJ 08855, USA. PhD (U. Pittsburgh, 1967) crystallography. *Nucleic_acids proteins databases hydration*
E-mail berman@dnarna.rutgers.edu Tel. 1(908)4454667 Fax 1(908)4455958

Berman, Dr Lonny E. (1960). Physicist. National Synchrotron Light Source, Brookhaven National Laboratory, Bldg. 725D, Upton, New York 11973, USA. PhD (Cornell U., 1988) applied physics. *X-ray_standing_waves X-ray_scattering X-ray_optics*
E-mail berman@bnl.gov berman@bnl.bitnet Tel. 1(516)2825333 Fax 1(516)2823238

Bernal, Ivan (1931). Professor. University of Houston, Cullen Blvd., Houston, TX 77204-5641, USA. PhD (Columbia U., 1963) chemical physics. *Stereochemistry optical_activity spontaneous_resolution_of_chiral_species*
E-mail chem4m@jetson.uh.edu Tel. 1(713)7432718 Fax 1(713)7432709

Bernstein, Ms Frances C. (1942). Computer Analyst. Protein Data Bank, Dept. of Chemistry, Brookhaven National Laboratory, Upton, NY 11973, USA. MS (New York U., 1965) mathematics. *Macromolecular_database macromolecular_structure computing*
E-mail bernstei@bnl.gov Tel. 1(516)2824382 Fax 1(516)2825751 Telex 6852516 BNL DOE

Bernstein, Herbert J. (1944). Consultant. Bernstein + Sons, 5 Brewster Lane, PO Box 177, Bellport, NY 11713, USA. PhD (New York U., 1968) mathematics. *Theoretical_crystallography lattice_identification cell_reduction group_theory*
E-mail yaya@aip.org Tel. 1(516)2861999 1(516)2822261 Fax 1(516)2861999 1(516)2825815

Berry, Chester R. (1919). Retired. 37 Heritage Dr., S. Orleans, MA 02662, USA. [PO Box 10, S. Orleans, MA 02662, USA.] PhD (Cornell U., 1946) physics. *Crystal_growth crystal_imperfections*
Tel. 1(508)2556206

Betts, Laurie (1956). Postdoctoral Scientist. PO Box 603, Chapel Hill, NC 27514, USA. PhD (U. North Carolina Chapel Hill, 1991) biochemistry. *Structure_based_drug_design nucleic_acid_structure protein_folding cytokines*
E-mail lb14133@usav01.glaxo.com Tel. 1(919)9413209 1(919)9673038 Fax 1(919)9413704 Telex 802813

Bhat, T. N. Senior scientist. Blds 322, PRI, NCIFCRDS, PO Box B, Frederick, MD 21702, USA. PhD (Indian Institution of Science, 1976) physics. *Structure_proteins modelling phase_protein computer_programming*
E-mail bhat@ncifcrf.gov Tel. 1(301)8461987 Fax 1(301)8466066

Bianchet, Dr Mario A. (1958). Research Associate. Department of Biophysics and Biophysical Chemistry, Johns Hopkins Medical School, 725 N Wolfe st., MD 21205, USA. PhD (National University of La Plata, 1988) physics. *Protein_structure crystallography_methods computer_modelling bioenergetics homology_prediction drug_design*
E-mail bianchet@cthulu.med.jhu.edu bianchet@jhuigf.bitnet Tel. 1(410)9558715 Fax 1(410)9550637

Bienenstock, Arthur (1935). Professor and Stanford Synchrotron Radiation Laboratory Director. Department of Materials Science and Engineering, Stanford University, Stanford, CA 94305-2205, USA. [Stanford Synchrotron Radiation Laboratory, Stanford Linear Accelerator Center, MS 69, PO Box 4349, Stanford, CA 94309-0210, USA.] PhD (Harvard U., 1962) applied physics. *Amorphous_materials synchrotron_radiation anomalous_scattering small_angle_scattering phase_separation*
E-mail a@ssrl01.slac.stanford.edu a@ssrl750.bitnet Tel. 1(415)9263153 Fax 1(415)9264100

Bigelow, Dr Wilbur C. (1923). Professor Emeritus. Department of Materials Engineering, University of Michigan, Ann Arbor, MI 48109-2136, USA. [Department of Materials Engineering, University of Michigan, Ann Arbor, MI 48109-2136, USA.] PhD (U. Michigan, 1952) physical chemistry. *Electron_diffraction metal_phases powder_diffraction*
E-mail wil_bigelow@mse.engin.umich.edu Tel. 1(313)7643321 Fax 1(313)7634788

Bilderback, Dr Donald H. (1947). Associate Director. Cornell High Energy Synchrotron Source, 281 Wilson Laboratory, Cornell University, Ithaca, New York 14853, USA. PhD (Purdue U., 1975) solid state physics. *X-ray_optics microdiffraction grazing_incidence_mirrors diffraction_technique synchrotron_radiation capillary_optics high_heatload_monochromators Laue_diffraction*
E-mail dhb2@cornell.edu Tel. 1(607)2550916 Fax 1(607)2559001

Bingman, Dr Craig A. (1962). Postdoctoral Fellow. Department of Biochemistry and Mol. Biophysics, Columbia University, 650 W. 168th Street, New York, NY 10032, USA. PhD (U. Wisconsin–Madison, 1991) biochemistry. *Nucleic_acid_structure protein_nucleic_acid_complexes*
E-mail bingman@cuhhca.hhmi.columbia.edu Tel. 1(212)3051951 Fax 1(212)3057379

Birdsall, David L. Researcher. Dept of Biochemistry and Biophysics, Univ. of Calif. at San Francisco, San Francisco, CA 94143-0448, USA. PhD (Purdue U., 1985) biochemistry. *Computer_modelling NA/protein_interactions X-ray_crystallography*
E-mail birdsall@msg.ucsf.edu Tel. 1(415)4960953 Fax 1(415)4960961

Birdsall, William J. (1944). Research Specialist. University of Pennsylvania, School of Veterinary Medicine, New Bolton Center, 382 West Street Road, Kennett Square, PA 19348-1692, USA. PhD (Penn. State U., 1971) inorganic chemistry. *Purines pyrroles pyrimidines analytical_chemistry toxicology*
E-mail Birdsall@upenn.nbc.edu Tel. 1(610)4445800 Fax 1(610)4440892

Birktoft, Dr Jens J. (1942). Senior Scientist. Roche Research Center, Hoffmann LaRoche Inc., Nutley, NJ 07110, USA. Cand. Scient. (U. Copenhagen, Denmark, 1967) biochemistry. *Proteins enzyme_mechanism proteolytic_enzymes protein:protein_interactions homology_prediction evolution blood_coagulation*
E-mail birktofj@rnch01.dnetroche.com Tel. 1(201)2358154 Fax 1(201)2352682

Birnbaum, Dr George I. (1931). Retired. [10264 Wild Apple Circle, Gaithersburg, MD 20879, USA.] PhD (Columbia U., USA, 1961) chemistry. *Proteins macromolecular_crystallization microgravity biological_structure*
E-mail celia@iris7.carb.nist.gov Tel. 1(301)9901050

Bish, Dr David L. (1952). Technical Staff Member. Geology and Geochemistry, Mail Stop D469, Los Alamos National Laboratory, Los Alamos, NM 87545, USA. PhD (Pennsylvania State U., 1977) mineralogy and petrology. *Rietveld_refinement quantitative_x_ray_diffraction_analysis natural_zeolites clay_mineralogy silicate_mineralogy thermal_analysis*
E-mail bish@essxrd.lanl.gov Tel. 1(505)6671165 Fax 1(505)6723337

Bjorkman, Dr Pamela J. (1956). Assistant Professor. Division of Biology 156-29, Howard Hughes Medical Institute, California Institute of Technology, Pasadena, CA 91125, USA. PhD (Harvard U., 1984) biochemistry and molecular biology. *Proteins immunology tertiary_structure thermoanalysis peptides lymphocytes*
E-mail bjorkman@citray.caltech.edu Tel. 1(818)3958350 1(818)3958351 Fax 1(818)7923683

Blaber, Dr Michael (1958). Assistant Professor. Institute of Molecular Biophysics, Florida State University, Tallahassee, Florida 32306, USA. PhD (U. California at Irvine, 1990) biological chemistry. *Proteins structure_function de _novo_design proteolysis*
E-mail blaber@sb.fsu.edu Tel. 1(904)6445870 Fax 1(904)5611406

Blanton, Thomas N. (1959). Analytical Chemist. Eastman Kodak Co., Kodak Park B49, Rochester, NY 14652-3712, USA. MS (Rochester Institute of Technology, 1986) materials science and engineering. *Powder_diffraction polymer_diffraction in-situ_temperature_diffraction*
E-mail blanton@Kodak.COM Tel. 1(716)7223323 Fax 1(716)4773029

Blessing, Robert H. (1941). Senior research scientist. Hauptman-Woodward Medical Research Institute (formerly the Medical Foundation of Buffalo), 73 High Street, Buffalo, NY 14203, USA. PhD (Ohio U., 1970) phys. and inorg. chem.. *Crystallography high_resolution_diffractometry high_precision_structures electron_density_distribution*
E-mail blessing@hwi.buffalo.edu Tel. 1(716)8569600 Fax 1(716)8524846

Blevins, Dr Richard A. (1956). Research Fellow. Department of Bioinformatics, Merck & Co., Inc., 126 E. Lincoln Ave., Rahway, NJ 07065-0900, USA. PhD (Michigan State U., 1985) physical chemistry. *Biocrystallography computer_graphics algorithm software*
E-mail blevins@merck.com Tel. 1(908)5947329 Fax 1(908)5942929

Bolin, Prof. Jeffrey T. (1952). Associate Professor. Department of Biological Sciences/LILY, Purdue University, W. Lafayette, IN 47907-1392, USA. PhD (U. of California, San Diego, 1982) chemistry. *Macromolecular_crystallography anomalous_diffraction enzymes metalloproteins*
E-mail b4b@mace.cc.purdue.edu Tel. 1(317)4944922 Fax 1(317)4961189

Bond, Andrew H. (1968). Graduate Student. Department of Chemistry, Northern Illinois University, DeKalb, IL 60115, USA. anticipated PhD 12/94 (Northern Illinois U., 1994) inorganic chemistry. *Chemistry coordination heavy_atom_radiochemistry separation_science inclusion synthesis*
E-mail andyb@rdr.chem.niu.edu Tel. 1(815)7536886 Fax 1(815)7534802

Bond, Elizabeth Rivette (1958). Powder Diffraction Specialist. F. L. Hartley Research Center, Unocal Corp., 376 S. Valencia Ave., USA. [2639 N. Grand Ave., Santa Ana, CA 92701, USA.] BS geological sciences. *X-ray_powder_diffraction inorganic_phase_determination analytical_service minerals clays corrosion catalysts_measurement agriculture_chemical cements*

E-mail stanerb@an.unocal.com 73441.624@compuserve.com Tel. 1(714)5771684 1(714)8354765 Fax 1(714)5771610

Bonham, Russell A (1931). Professor. Chemistry Department, Indiana University, Bloomington, IN 47405, USA. PhD (Iowa State U., 1958) physical chemistry. *Charge_density Compton_scattering scattering_synchrotron_radiation plasmas electron_scattering X-ray_scattering gas_phase*
E-mail bonham@iubacs.bitnet Tel. 1(812)8554843 Fax 1(812)8558300

Borgstahl, Dr Gloria E. O. (1962). Research Associate. Department of Molecular Biology, The Scripps Research Institute, La Jolla, CA 92037, USA. PhD (U. Iowa, 1992) biochemistry. *Allostery proteins structural_disorder computer_modelling hemoglobin superoxide_dismutase*
E-mail gloria@scripps.edu Tel. 1(619)5544484 Fax 1(619)5542880

Borhani, Dr David W. (1960). Research Chemist. Southern Research Institute, 2000 Ninth Avenue South, Birmingham, AL 35205, USA. PhD (Massachusetts Institute of Technology, 1986) organic chemistry. *Proteins enzyme_mechanism viruses synchrotron_radiation*
E-mail dborhani@orion.cmc.uab.edu Tel. 1(205)5812555 Fax 1(205)5812870

Boskey, Adele L. (1943). Director Research Operations. 4 Winding Way, North Caldwell, NJ 07006, USA. [The Hospital for Special Surgery, 535 E. 70 St., New York, NY 10021, USA.] PhD (Boston U., 1970) physical chemistry. *Calcification_bone mineralization*
E-mail aboskey@aol.com Tel. 1(201)2267463 Fax 1(212)4725331

Bossart Whitaker, Dr Patricia J. (1954). Postdoctoral Fellow. National Institutes of Health, NIAMS, Building 6, Room 425, 6 Center Dr. MSC 2755, Bethesda, MD 20892-2755, USA. PhD (U. Alabama, Birmingham, 1992) microbiology. *Macromolecular_crystallography rational_drug_design enzymes time-resolved_crystallography complex_protein_interactions*
E-mail pat@kayak.niams.nih.gov Tel. 1(301)4023223 Fax 1(301)4023417

Bourne, Dr Philip E. (1953). Senior Associate. US Representative IUCr Computing Commission; Member Macromolecular CIF Committee. Howard Hughes Medical Institute, Department of Biochemistry and Molecular Biophysics, Columbia University, 630 W. 168th Street, New York NY 10032, USA. PhD (Flinders University of South Australia, 1980) physical chemistry. *Computational_crystallography data_representation user_interfaces macromolecular_structure*
E-mail system@cuhhca.hhmi.columbia.edu Tel. 1(212)3053657 Fax 1(212)3057379

Bowie, Dr James U. (1959). Assistant Professor. Molecular Biology Institute, UCLA, Los Angeles, CA 90024, USA. PhD (Massachusetts Institute of Technology, 1989) biochemistry. *Protein_structure protein_folding structure_prediction membrane_proteins*
E-mail bowie@uclaue.mbi.ucla.edu Tel. 1(310)2064747 Fax 1(310)2064749

Bowman, Allen L. (1931). Retired. 3861 S. Via Dl Tejedor, Gree Valley, AZ 85614-5412, USA. PhD (Iowa State U., 1958) fused salts.
Tel. 1(602)6258607

Boyington, Jeffrey C. (1965). Postdoctoral Fellow. Johns Hopkins Univ. School of Medicine, Dept of Biophysics, 725 N. Wolfe St., Baltimore, MD 21205, USA. PhD (Johns Hopkins U. Sch. Med., 1994) biophysics. *Protein_crystallography pi-helices lipoxygenases*
E-mail jeff@cthulu.med.jhu.edu Tel. 1(410)9558715 Fax 1(410)9550637

Braden, Dr Bradford C. (1951). Scientist. Center for Advanced Research in Biotechnology, University of Maryland Biotechnology Institute, and National Institute of Standards and Technology, 9600 Guldelsky Drive, Rockville, MD 20850, USA. PhD (Indiana U., 1978) biophysics. *Proteins hemoglobin immune_system amyloidosis*
E-mail bcb@iris8.carb.nist.gov Tel. 1(410)7386123 Fax 1(410)7386225

Brady, John (1952). Associate Professor. Dept. of Food Sci., Cornell University, Stocking Hall, Ithaca, NY 14853, USA. PhD (SUNY-Stony Brook, 1980) chemistry.
E-mail jwb7@cornell.edu.usa Tel. 1(607)2552897 Fax 1(607)2544868

Bragg, Dr Robert H. (1919). Professor, Emeritus. Department of Materials Science and Mineral Engineering, University of California, Berkeley, CA 94720, USA. [2 Admiral Dr. #373, Emeryville, CA 94608, USA.] PhD (Illinois Institute of Technology, 1960) physics. *Structure_pregraphite_carbon_materials small_angle_scattering structural_transformation_in_carbon*
Tel. 1(510)6556283 Fax 1(510)6435792

Brammer, Lee (1963). Assistant Professor. Department of Chemistry, University of Missouri-St. Louis, 8001 Natural Bridge Rd., St. Louis, MO 63121-4499, USA. PhD (U. Bristol, 1987) inorganic chemistry. *Charge_density neutron_diffraction organometallic_hydrides hydrogen_bonding organometallic_chemistry organometallic_clusters clathrates*
E-mail slbramm@slvaxa.umsl.edu slbramm@umslvaxa.bitnet Tel. 1(314)5535345 Fax 1(314)5535342

Brandhuber, Barbara (1959). Group leader. Synergen, 1885 33rd St., Boulder, CO 80301, USA. PhD (U. Colorado-Boulder, 1988) chemistry.
Tel. 1(303)5411445 Fax 1(303)4415535

Breiter, Deborah R. (1965). Postdoctoral Associate. Department of Biochemistry, University of Minnesota, 4-225 Millard Hall, 435 Delaware St. S.E., Minneapolis, MN 55455, USA. PhD (University of Wisconsin-Madison, 1991) physical chemistry. *Structure_function directed-mutagenesis*
E-mail breiter@dccc.med.umn.edu Tel. 1(612)6252115 Fax 1(612)6252163

Brennan, Dr Richard G. (1955). Assistant Professor. Department of Biochemistry and Molecular Biology, Oregon Health Sciences University, 3181 S.W. Sam Jackson Park Road, Portland, OR 97201-3098, USA. PhD (U. Wisconsin-Madison, 1984) biochemistry. *Protein_DNA receptors kinases purine_metabolism homology_prediction modelling heme_proteins*
E-mail brennanr@ohsu.edu Tel. 1(503)4944427 Fax 1(503)4948393

Brennan, Dr Sean M. (1955). Physicist. Stanford Synchrotron Radiation Laboratory, Stanford Linear Accelerator Center, Box 4349, MS 69, Stanford, CA 94309, USA. PhD (Stanford U., 1982) materials science and engineering. *Grazing_incidence surface_structure X-ray_optics reflectivity*
E-mail bren@ssrl01.slac.stanford.edu Tel. 1(415)9263173 Fax 1(415)9264100

Brenner, Dr Charles (1961). Postdoctoral fellow. Rosenstiel Center, Rm 652, Brandeis University, Waltham, MA 02254-9110, USA. PhD (Stanford U., 1993) cancer biology. *Enzymes structure–activity_relationship proteases carcinogenesis yeast_molecular_biology cell_cycle*
E-mail brenner@auriga.rose.brandeis.edu Tel. 1(617)7364908 Fax 1(617)7362405

Brenner, Mr Stephen A. (1937). Res. Chemist. Naval Res. Lab., Code 6030, Washington, DC 20375, USA. [Naval Res. Lab., Code 6030, Washington, DC 20375, USA.] MA (Boston U., 1962) physical chemistry. *Mathematics chemistry physics computer science*
E-mail brenner@nrl.navy.mil Tel. 1(202)7673496 Fax 1(202)7676874

Brese, Dr Nathaniel E. (1965). Senior Project Engineer, Specialty Phosphors. OSRAM Sylvania, Inc., Hawes St., Towanda, PA 18848, USA. PhD (Arizona State U., 1991) solid state chemistry. *Bonding nitrides neutron_diffraction crystal_chemistry*
E-mail nathanb@chemres.tn.cornell.edu Tel. 1(717)2685461 Fax 1(717)2685350

Brinen, Dr Linda S. (1966). Postdoctoral Associate. Department of Biochemistry and Biophysics, School of Medicine, University of California, San Francisco, San Francisco, California 94143-0448, USA. PhD (Cornell U., 1993) chemistry/biophysics. *Phosphorylation dephosphorylation signal_transduction protein_purification protein_expression molecular_mechanics phasing_methodology*
E-mail brinen@msg.ucsf.edu Tel. 1(415)4765051 Fax 1(415)4761902

Britton, Doyle (1930). Professor. Dept. of Chemistry, University of Minnesota, Minneapolis, MN 55455, USA. PhD (Calif. Inst. Tech., 1955) chemistry. *Packing intermolecular_interactions*
E-mail britton@chemsun.chem.umn.edu Tel. 1(612)6259535 Fax 1(612)6267541

Broach, Dr Robert W. (1949). Senior Research Associate. UOP Research Center, 50 E. Algonquin Rd., Des Plaines, IL 60017, USA. PhD (U. Wisconsin, Madison, 1977) chemistry. *Crystallography synchrotron_radiation computer_modelling catalysts zeolites*
E-mail rwbroach@uop.com Tel. 1(708)3913313 Fax 1(708)3913719

Brock, Prof. Carolyn Pratt (1946). Professor, Co-Editor, Acta Crystallographica. Department of Chemistry, University of Kentucky, Lexington, KY 40506-0055, USA. PhD (Northwestern U., 1972) physical chemistry. *Crystal_packing thermal_motion*
E-mail cpbrock@ukcc.uky.edu Tel. 1(606)2571959 Fax 1(606)3231069

Brooks Jr, Dr Frederick P. (1931). Kenan Professor. Department of Computer Science, University of North Carolina, Chapel Hill, NC 27599-3175, USA. [413 Granville Road, Chapel Hill, NC 27514, USA.] PhD (Harvard U., 1956) applied mathematics. *Computer_graphics molecular_graphics computer_architecture*
E-mail brooks@cs.unc.edu Tel. 1(919)9621931 Fax 1(919)9422529 1(919)9621799

Brown Jr, Prof. Gordon E. (1943). Kirby Professor of Earth Sciences and Professor, Stanford Synchrotron Radiation Laboratory. Dept. of Geological and Environmental Sciences, Stanford University, Stanford, CA 94305-2115, USA. PhD (Virginia Polytechnic Institute and State U., 1970) mineralogy and crystallography. *Synchrotron_radiation mineral_chemistry geochemistry oxide_surfaces*
E-mail gordon@pangea.stanford.edu Tel. 1(415)7239168 Fax 1(415)7252199

Brown, Dr Jerry H. (1962). Post Doc. Department of Biochemistry and Molecular Biology, 7 Divinity Avenue, Harvard University, Cambridge, MA 02138, USA. PhD (Harvard U., 1991) chemistry. *Proteins modelling immunology HLA*
E-mail brown@xtal0.harvard.edu Tel. 1(617)4955043 Fax 1(617)4959613

Brown, Leo D. (1948). Research Associate. Synthetic Fuels Research, Exxon Research and Development Labs, PO Box 2226, Baton Rouge, LA 70821-2226, USA. PhD (U. California, Berkeley, 1974) inorganic chemistry. *Inorganic_chemistry silicate_minerals optical_microscopy powder_diffraction*
Tel. 1(504)3594519

Brown, Dr Raymond S. (1946). Instructor. Department of Microbiology and Molecular Genetics, Harvard Medical School, 200 Longwood Avenue, Boston, MA 02115, USA. [HHMI Lab. of Molecular Medicine, Children's Hospital, 300 Longwood Avenue, Boston, MA 02115, USA.] PhD (London U., 1990) structural biology. *Transcription zinc_fingers protein–DNA_complexes gene_regulation RNA_structure*
E-mail raybrown@xtal0.harvard.edu raybrown@huxtal.bitnet Tel. 1(617)7356240 Fax 1(617)7300506

Browner, Dr Michelle F. (1957). Staff Researcher II. Molecular Structure Department, Syntex, Palo Alto, CA 94303, USA. PhD (Baylor College of Medicine, 1986) cell biology. *Protein_structure enzymatic_mechanisms structure-based_drug_design allosteric_enzymes bound_ligand_interactions conformational_changes*
E-mail browner@arvax.syntex.com Tel. 1(415)3542238 Fax 1(415)3547363

Bruck, Dr Michael A. (1953). Assistant Staff Scientist. University of Arizona, Department of Chemistry, Tucson, AZ 85721, USA. PhD (U. Southern California, 1985) inorganic chemistry. *Service crystallography*
E-mail mbruck@xray0.chem.arizona.edu Tel. 1(602)6214168 Fax 1(602)6218407

Brumberger, Harry (1926). Professor. Dept. of Chemistry, Syracuse University, Syracuse, NY 13244-1200, USA. PhD (Polytechnic Institute of Brooklyn, 1955) physical chemistry. *Small_angle_scattering sintering catalysis ceramics*
E-mail hbrumber@mailbox.syr.edu Tel. 1(315)4435923 Fax 1(315)4434070

Brunger, Dr Axel T. (1956). Professor. Howard Hughes Medical Institute and Department of Molecular Biophysics and Biochemistry, Yale University, New Haven, CT 06520, USA. PhD (Technische U. München, 1982) biophysics. *Structural_molecular_biology computational_chemistry NMR_spectroscopy X-ray_crystallography transport_in_biological_systems*
E-mail brunger@laplace.csb.yale.edu Tel. 1(203)4325067 Fax 1(203)4326946

Bruns, Christopher M. (1964). Graduate student. Section of Biochemistry, Molecular and Cell Biology, Cornell University, Ithaca, NY 14850, USA. BA (Reed College, 1986) biology.
E-mail bruns@penelope.bio.cornell.edu Tel. 1(607)2558432 Fax 1(607)2552428

Bryan, Clinton D. (1960). Assistant Professor. Department of Physical Sciences, Cameron University, Lawton, OK 73505, USA. MS (U. Kansas, 1988) physical organic chemistry. *Small_molecule_structure_determination*
Tel. 1(405)5812246

Bryan, Robert F. (1933). Prof. of Chemistry. Book Review Editor, Member Commission on Journals. University of Virginia, Dept. of Chemistry, University of Virginia, Charlottesville, VA 22901, USA. PhD (U. of Glasgow, 1957) chemical crystallography. *Structural_chemistry liquid_crystals*
E-mail rfb6w@virginia.edu Tel. 1(804)9243619 Fax 1(804)9243710

Bryant, Garold L. (1958). Chemist. Physical and Analytical Chemistry, 301 Henrietta Street, The Upjohn Company, Kalamazoo, MI 49001, USA. MS (Iowa State U., 1986) geology. *Macromolecular_structure powder high_temperature small_molecules*
E-mail glbryant@upj.com Tel. 1(616)3857998 Fax 1(616)3857522

Bryant, Dr Stephen H. (1954). Senior Investigator. National Center for Biotechnology Information, National Library of Medicine, National Institutes of Health, Building 38A, Room 8N805, 8600 Rockville Pike, Bethesda, MD 20894, USA. PhD (Johns Hopkins Medical School, 1981) protein crystallography. *Protein_database molecular_modelling protein_threading contact_potential*
E-mail bryant@ncbi.nlm.nih.gov Tel. 1(301)4962475 Fax 1(301)4809241

Bryden, John H. (1920). Professor of chemistry. California State University, Fullerton, 800 State College Blvd., Fullerton, CA 92634, USA. PhD (UCLA, 1951) physical chemistry.
Tel. 1(714)7733833 Fax 1(714)8793997

Bu, Dr Xianhui (1964). Staff Crystallographer. Department of Chemistry, University of California, Santa Barbara, CA 93106, USA. PhD (State U. New York at Buffalo, 1991) solid state chemistry. *Zeolites_structure computer_modelling diffractometer_hardware Rietveld powder_diffraction crystallographic_software*
E-mail xianhui@sbxray.ucsb.edu xianhui@voodoo.bitnet Tel. 1(805)8932399 Fax 1(805)8934120

Buchanan, Dr David R. (1934). Professor and Associate Dean. Department of Textile Engineering, Chemistry, and Science, College of Textiles, North Carolina State University, Raleigh, NC 27695-8301, USA. PhD (The Ohio State U., 1962) physical chemistry. *Amorphous_phase SAXS fibres polymers*
E-mail david_buchanan@ncsu.edu Tel. 1(919)5156649 Fax 1(919)5153057

Budai, John D. Research Staff. Bldg. 3025, MS 6030, Oak Ridge National Laboratory, Oak Ridge, TN 37831, USA. PhD (Cornell U., 1982) physics. *Synchrotron_radiation epitaxy ion_implantation*
E-mail xry@ornl.gov Fax 1(615)5744143

Bugg, Dr Charles E. (1941). Chairman and Chief Executive Officer. Editor-in-Chief, Acta Crystallographica. BioCryst Pharmaceuticals, Inc., 2190 Parkway Lake Dr, Birmingham, AL 35244, USA. PhD (Rice U., 1965) physical chemistry. *Protein_crystallography drug_design protein_crystal_growth*
E-mail bugg@orion.cmc.uab.edu Tel. 1(205)4444600 Fax 1(205)4444640

Bullock, Timothy L. Institute of Molecular Biology, University of Oregon, Eugene, OR 97401, USA.
E-mail tim@uoxray.uoregon.edu Tel. 1(510)3465192

Bunick, Dr Gerard J. (1947). Senior Staff Member Coordinator, ORNL Structural Biology Programs. Biology Division, PO Box 2009 MS 8077, Oak Ridge National Laboratory, Oak Ridge, TN 37831-8077, USA. PhD (U. Pennsylvania, 1975) biological chemistry. *Nucleoproteins DNA macromolecular_crystallography phosphatases SAXS small-angle_neutron_scattering neutron_crystallography protein_crystallization*
E-mail bunick@biovx1.bio.ornl.gov bunick@biou03.bio.ornl.gov bunickgj@ornl.gov Tel. 1(615)5762685 Fax 1(615)5741274

Bunning, Dr Timothy J. (1966). Materials Engineer. Materials Directorate, WL/MLPJ, BLDG. 651, 3005 P. Street, Ste.1, Wright-Patterson Air Force Base, Ohio 45433-7702, USA. PhD (U. Connecticut, 1992) chemical engineering. *Morphology liquid_crystals nlo_materials structure/property_rel synchrotron_radiation electron_microscopy diffraction synthesis*
E-mail bunnintj@mlgate.ml.wpafb.af.mil Tel. 1(513)2553808 Fax 1(513)2551128

Burbank, Dr Robinson D. (1921). Bell Telephone Laboratories, Retired, 45 Woodland Ave., Summit, NJ 07901, USA. PhD (Mass. Inst. of Technology, 1950) inorganic chemistry. *Structure_interhalogen_compounds structure_noble_gas_compounds structure_inorganic_compounds structure_thin_films phase_transformations*
Tel. 1(908)2737967

Burkhart, Brian M. (1968). Graduate Student. Department of Chemistry, The Ohio State University, Columbus, OH 43210, USA. PhD (The Ohio State U., 1994) biological chemistry. *Proteins nucleic_acids*
E-mail brian?obiot@mps.ohio-state.edu Tel. 1(614)2922926 Fax 1(614)2922524

Burley, Stephen K. (1957). Head of laboratory. Rockefeller University/HHMI, 1230 York Avenue, New York, NY 10021, USA. MD (Harvard Medical School, 1987) medicine. *Protein–nucleic_acid_interactions molecular_recognition transcriptional_regulation*
E-mail burley@rockvax.rockefeller.edu Tel. 1(212)3278336 Fax 1(212)3278337

Burmeister, Dr Wilhelm Pascal (1964). Postdoc. Department of Biology 1526-29, Californian Institute of Technology, Pasadena, CA 91125, USA. PhD (U. Joseph Fourier, Grenoble, France, 1992) physics. *Proteins biocrystallography biochemistry immunology glycoproteins synchrotron_radiation computer_modelling detector*
E-mail burmeister@citray.caltech.edu Tel. 1(818)3568351 Fax 1(818)7923683

Burnett, Dr Roger M. (1941). Professor, Chairman of the Structural Biology Program. The Wistar Institute, 3601 Spruce Street, Philadelphia, PA 19107, USA. PhD (Purdue U., 1970) protein crystallography. *Macromolecules viruses antibodies molecular_recognition molecular_interactions antiviral_compounds*
E-mail burnett@wistb.wistar.upenn.edu Tel. 1(215)8982201 Fax 1(215)8983868

Burnham, Prof. Charles (1933). Prof. of Mineralogy. Dept. of Earth and Planetary Sciences, Harvard University, 20 Oxford St., Cambridge, MA 02138, USA. PhD (MIT, 1961) crystallography and mineralogy. *Silicate_mineralogy computer_modelling phase_transitions order-disorder*
E-mail cwb@eps.harvard.edu Tel. 1(617)4952484 Fax 1(617)4958839

Burns, Dr John H (1930). Sr. Res. Staff. Oak Ridge National Lab., PO BOX 2008, Oak Ridge, TN 37831-6119, USA. PhD (Rice U., 1955) chemistry. *Small_molecule_structures*
Tel. 1(615)5745018

Burshtein, Dr Izya F. (1942). Teaching Assistant. Computer Sciences Department, University of Wisconsin - Madison, 1210 W. Dayton St., Madison, WI 53706, USA. [5401 Meadowood Dr., Madison, WI 53711, USA.] PhD (Institute of Applied Physics Kishinev, USSR, 1977) crystallography. *Patterson_function automatic_solution multifunctional_ligands*
E-mail izya@boursin.cs.wisc.edu Tel. 1(608)2730571 Fax 1(608)2629777

Busing, Dr William R. (1923). Retired. Chemistry Division, Oak Ridge National Laboratory, PO Box 2008, Oak Ridge, TN 37831-6197, USA. [317 Louisiana Avenue, Oak Ridge, TN 37830, USA.] PhD (Princeton U., 1949) phys. chem.. *Least-squares_refinement computer_modelling fibrous_polymers*
E-mail wrb@ornlstc wrb@ornl.gov wrb%ornl.gov@umcgate.stc10 Tel. 1(615)4835700

Butcher, Ray J. (1945). Associate Professor. Chemistry Dept, Howard University, Washington, DC 20059, USA. PhD (U. Canterbury, New Zealand, 1974) inorganic chemistry/X-ray crystallography. *Antiferromagnetism biocoordination magnetochemistry*
E-mail raybutcher@aol.com Tel. 1(202)8066886 Fax 1(703)5035546

Buzatu, Dan A. (1968). Research Assistant. Department of Chemistry, University of New Orleans, New Orleans, LA 70148, USA. BS (Ohio State U., 1991) psychology. *Antiestrogen_compounds drug_receptor_interaction structure_determination charge_density biocrystallography carcinogenesis*
E-mail edscm3@uno.edu Tel. 1(504)2867217 Fax 1(504)2866860

Byram, Susan K. (1945). Product Manager, Crystallographic Systems. Siemens Industrial Automation, Inc., Analytical Instrumentation Division, 6300 Enterprise Lane, Madison, WI 53719-1173, USA. MSc (U. Toronto, 1970) crystallography. *Real-time_control instrumentation area_detectors diffractometry*
Tel. 1(608)2763041 Fax 1(608)2763006

Byrn, Dr Stephen R. (1944). Charles B. Jordan Professor, Head; Director, Center for AIDS Research. Department of Industrial and Physical Pharmacy, 1333 R. Heine Pharmacy Building, Purdue University, West Lafayette, IN 47907-1333, USA. PhD (U. Illinois, Champaign-Urbana, 1970) organic chemistry/physical chemistry. *Structure_disorder molecular_mobility computer_modelling solid_state_NMR solid_state_reactivity*
E-mail bym@vm.cc.purdue.edu Tel. 1(317)4941460 Fax 1(317)4946790

Cachau, Dr Raul E. (1959). Associate Scientist. National Cancer Institute, Frederick Research and Development Center, Frederick, MD 21702, USA. PhD (U. La Plata, 1986) physical chemistry. *Proteins computer_modelling water_structure*
E-mail cachau@ncifcrf.gov cachau@oliver.bitnet Tel. 1(301)8466062 Fax 1(301)8466066

Caffrey, Assoc. Prof. Martin (1950). Associate Professor. Dept. Chemistry, 120 W 18th Ave., The Ohio State University, Columbus, OH 43210, USA. PhD (Cornell U., 1982) biochemistry. *Membrane_structure membrane_proteins lipids phase_transitions phase_transition kinetics signal_transduction*
E-mail caffrey+@osu.edu Tel. 1(614)2928437 Fax 1(614)2921532

Cahn, Dr John W. (1928). Senior Fellow. Materials Science and Engineering Laboratory, National Institute of Standards and Technology, Gaithersburg, MD 20899, USA. PhD (U. Cal. at Berkeley, 1953) physical chemistry. *Quasicrystals phase_transitions metals microstructure_evolution crystal_growth*
E-mail cahn@enh.nist.gov cahn@nbsenh.bitnet Tel. 1(301)9755664 Fax 1(301)9267975

Cai, Li (1968). Department of Biology, Massachusetts Institute of Technology, Cambridge, MA 02139, USA. BS (Mass. Inst. Tech., 1991)
E-mail lcai@mit.edu Tel. 1(617)2534506

Calandra, Peter M. (1939). President. Innovative Tech Inc., 2 New Pasture Rd., Newburyport, MA 01950, USA. BS (Canisius College, 1962) chemistry. *Small_angle_scattering synchrotron*
Tel. 1(508)4624415 Fax 1(508)4623338

Camerman, Dr Arthur (1939). Research Professor. Neurology Department, RG-27, University of Washington, Seattle, WA 98195, USA. PhD (U. British Columbia, 1964) chemistry. *Biological_molecules structure-activity-relationships drug_design*
E-mail neurol@max.u.washington.edu Tel. 1(206)5432340 Fax 1(206)6858100

Cameron Jr, Robert P. (1958). Assistant Professor. Dept. of Chemistry, Samford University, Birmingham, AL 35229, USA. PhD (U. New Orleans, 1990) physical chemistry. *Electron_density deformation_density metal_hydrides proteins*
E-mail rpcamero@samford Tel. 1(205)8702055 Fax 1(205)8702329

Campana, Dr Charles F. (1947). Senior Applications Scientist. Siemens Industrial Automation, Inc., Analytical Instrumentation Division, 6300 Enterprise Lane, Madison, WI 53719-1173, USA. PhD (U. Wisconsin - Madison, 1975) inorganic chemistry. *Chemical_crystallography instrumentation structure_determination accurate_intensity_data_collection*
E-mail campana@bert.chem.wisc.edu Tel. 1(608)2763042 Fax 1(608)2763006

Campobasso, Nino (1966). Graduate Student. Department of Biological Sciences, Purdue University, West Lafayette, IN 47097, USA. BS (U. Chicago, 1988) chemistry. *Proteins metalloenzymes MAD_phasing computing data_collection*
E-mail ncamp@snow.bio.purdue.edu Tel. 1(317)4940108 Fax 1(317)4961189

Cantrell, Joseph S. (1932). Professor. Chemistry Department, Miami University, Oxford, OH 45056, USA. PhD (Kansas State U., 1961) physical chemistry. *Structural_disorder_alloys organic computer_modelling triclinic_indexing*
E-mail jcantrel@miamiu.acs.muohio.edu Tel. 1(513)5292834 Fax 1(513)5297284

Carducci, Michael D. (1965). Graduate Student. Department of Chemistry, University of Arizona, Tucson, AZ 85721, USA. BS (U. California, Irvine, 1987) chemistry. *Inorganic charge_density_distributions crystallographic_computing computer_graphics magnetism*
E-mail mcarducci@xray.chem.arizona.edu carducci@arizvms.bitnet Tel. 1(602)6216335 Fax 1(602)6218407

Cargill III, G. Slade (1943). Professor of Materials Science and Metallurgy. Dept. Chem. Engr., Mat. Sci. and Mining Engr., Columbia University, 1144 S. W. Mudd Building, 500 E 120th St., New York, NY 10027, USA. PhD (Harvard U., 1969) applied physics. *Microdiffraction microstrain EXAFS scanning_electron_microscopy amorphous_materials*
E-mail gsc15@columbia.edu Tel. 1(212)8546167 Fax 1(212)8546278

Carlson, Ernest H. (1933). Associate Professor. Geology Dept, Kent State University, Kent, OH 44242-0001, USA. PhD (McGill U., Montreal, 1966) geology. *Geochemistry mineralogy X-ray_crystallography*
Tel. 1(216)6723778 Fax 1(216)6727949

Carpenter, Gene B. (1922). Professor Emeritus. Department of Chemistry, Brown University, Providence, RI 02912, USA. PhD (Harvard U., 1947) physical chemistry. *Structure_determination small_molecules*
E-mail gbc@brownvm.brown.edu Tel. 1(401)8633389 Fax 1(401)8632594

Carpenter, John M. (1935). Senior Physicist; Technical Director, IPNS. Intense Pulsed Neutron Source, Building 360, Argonne National Laboratory, Argonne, IL 60439, USA. PhD (U. Michigan, 1963) nuclear engineering. *Amorphous_phase neutron_diffraction neutron_spectroscopy pulsed_neutrons neutron_source instrument_development*
E-mail jmcarpenter@anl.gov Tel. 1(708)2525519 Fax 1(708)2524163

Carperos, Dr Vasili (1958). Lab Manager. Department of Chemistry, University of Colorado, Boulder, CO 80309, USA. PhD (Emory U., 1985) physical chemistry. *Macromolecular_crystallography area_detector data_processing refinement cryo_crystallography diffractometry small_molecules computer_management*
E-mail carperos@mmol.colorado.edu Tel. 1(303)4920970 Fax 1(303)4925894

Carr, Martin J. (1949). Manager. Department 1822, MS 0342, Sandia Nat. Labs, Albuquerque, NM 87185, USA. PhD (RPI, 1976) materials engineering. *Electron back_reflection_electron_Kikuchi_pattern search_and_match JCPDS ICDD PDF crystal_data TEM SEM*
E-mail mjcarr@sandia.gov Tel. 1(505)8440993 Fax 1(505)8447910

Carrell, Dr H. L. (1940). Sr. Research Associate. The Institute for Cancer Research, The Fox Chase Cancer Center, 7701 Burholme Ave, Philadelphia, PA 19111, USA. PhD (U. Southern California, 1966) physical chemistry. *Proteins enzymatic_mechanism metalloenzymes small_molecules computer_graphics*
E-mail carrell@fccc.edu Tel. 1(215)7282220 Fax 1(215)7283574

Carroll, Dr Patrick J. (1948). Research Specialist. Department of Chemistry, University of Pennsylvania, 3301 Spruce St., Philadelphia, PA 19104-6323, USA. PhD (Temple U., 1977) inorganic chemistry. *Service small_molecules*
E-mail carroll@doe.chem.upenn.edu Tel. 1(215)8983505 Fax 1(215)5732112

Carson, Dr William Michael (1951). Adjunct Assistant Professor. Center for Macromolecular Crystallography, UAB, Birmingham, AL 35294, USA. PhD (U. Texas, 1980) chemistry. *Proteins graphics computer_modelling structure_based_drug_design*
E-mail carson@luna.cmc.uab.edu carson@uabcmc.bitnet Tel. 1(205)9341983 Fax 1(205)9340480

Carter, Charles W. Jr. (1945). Professor. Department of Biochemistry and Biophysics, CB 7260, UNC-CH, Chapel Hill, NC 27599-7260, USA. PhD (U. California, San Diego, 1972) biology. *Experimental_design enzyme_mechanisms phase_determination crystal_growth aminoacyl_tRNA_synthetases quaternary_structure*
E-mail carter@med.unc.edu Tel. 1(919)9663263 Fax 1(919)9662852

Cartz, Dr Louis (1926). Professor. Department of Mechanical and Industrial Engineering, Marquette University, 1515 W. Winsconsin Avenue, Milwaukee, WI 53233, USA. PhD (U. London, UK, 1954) crystallography. *Thermal_expansion nondestructive_testing cements ceramics radiography polarized_light_microscopy*
E-mail 6055cartzl@vmsd.csd.mu.edu Tel. 1(414)2883517 1(414)2883510 Fax 1(414)2887082

Celikel, Reha (1949). Research Associate. The Scripps Research Institute, 10666 N. Torrey Pines Rd., SBR-8, La Jolla, CA 92037, USA. PhD (Leeds U., UK, 1979) molecular biophysics. *Protein_purification protein_crystallization protein_structure blood proteins protein_crystallography*
E-mail reha@scripps.edu Tel. 1(619)5542974 Fax 1(619)5546779

Cervantes-Lee, Dr Francisco J. (1950). Research Scientist. Chemistry Department, University of Texas at El Paso, El Paso TX. 79968-0513, USA. PhD (U. Aberdeen, 1981) solid state chemistry. *Organometallic_chemistry mineralogy_materials X-ray_spectroscopy selected_area_electron_diffraction*
E-mail ex02@utep.bitnet Tel. 1(915)7477553 Fax 1(915)7475748

Chacko, Dr Susan (1962). Visiting Fellow. Lab of Molecular Biology, NIDDK, NIH, Bldg 5, Rm 334, Bethesda, MD 20892, USA. PhD (U. Illinois at Urbana-Champaign, 1991) biophysics. *Proteins computer_modelling immunoglobulins*
E-mail susanc@helix.nih.gov susan@nihklmb.bitnet Tel. 1(301)4024496 Fax 1(301)4960201

Chadha, Raj (1950). Director, X-ray facility. The Scripps Research Institute, 10666 N. Torrey Pines Road, La Jolla, CA 92037, USA. PhD (Panjab U., India, 1979) inorganic chemistry. *Small_molecule_crystallography disordered_structures painting*
Tel. 1(619)5547073 Fax 1(619)5546414

Chakoumakos, Bryan C. (1955). Research Staff Member. Solid State Division, Oak Ridge National Lab., PO Box 2008, Oak Ridge, TN 37831-6393, USA. PhD (Virginia Tech, 1984) geological sciences. *Structure_systematics superconductors neutron_diffraction X-ray_crystallography silicates phosphates minerals*
E-mail kou@ornlstc.bitnet kou@solid.ssd.ornl.gov Tel. 1(615)5745235 Fax 1(615)5746268

Chandra, Dhanesh (1944). Associate Professor. Chemistry and Metallurgy Dept., Mackay School of Mines, University of Nevada at Reno, Reno, NV 89557-0047, USA. PhD (U. Denver, 1976) metallurgy and material science. *Phase_transformation hydrides plastic_crystals*
Tel. 1(702)7844960 Fax 1(702)7844316

Chandra, Naveen (1954). Postdoctoral fellow. Department of Biochemistry, Temple University, Philadelphia, PA 19140, USA. [56] AHB Fels Institute for Cancer Research and Molecular Biology, 3307 N Broad St., Philadelphia, PA 19140, USA.] PhD (IIT Bombay, India, 1985) chemistry. *Structure_determination_macromolecules DNA-binding_proteins disordered_structures refinement_problematic_structures*
E-mail naveen@sgi1.fels.temple.edu Tel. 1(215)7078290 or 1(215)7074161 Fax 1(215)7071454

Chandrasegaran, Srinivasan (1954). Associate Professor. Dept. Environmental Health Sci., Johns Hopkins University School of Hygiene & Public Health, 615 N. Wolfe St., Baltimore, MD 21205, USA. PhD (Georgetown U., 1982) chemistry. *Restriction endonucleases modification methylases protein_engineering*
Tel. 1(410)9550023 Fax 1(410)9550617

Chandrasekaran, Dr Rengaswami (1939). Professor. Whistler Center for Carbohydrate Research, 1160 Smith Hall, Purdue University, West Lafayette, IN 47907-1160, USA. PhD (Madras U., 1966) X-ray crystallography. *Fiber_diffraction computer_modelling nucleic_acids polysaccharides conformational_analysis_of_macromolecules*
E-mail eok@mace.cc.purdue.edu chandrar@foodsci.purdue.edu Tel. 1(317)4944923 Fax 1(317)4947953

Chandrasekhar, Dr K. Senior Research Associate. Department of Biological Sciences, 319 Clapp Hall, University of Pittsburgh, Pittsburgh, PA 15260, USA. PhD (U. Madras, India, 1979) crystallography and biophysics. *Macromolecular_structure reaction_paths symmetry_of_non-rigid_molecules structure_function structural_homology hydrogen_bonding databases*
E-mail sekhar1@vms.cis.pitt.edu chandra@jmr2.xtal.pitt.edu Tel. 1(412)6244638 Fax 1(412)6244759

Chandross, Dr Ronald J. (1935). Owner. RJ Computing, 68 Oak Forest Circle, Charlottesville, VA 22901, USA. PhD (Mass Inst. of Tech., 1961) physical chemistry. *Proteins small_angle_scattering computing data_processing collagen_structure area_detectors*
E-mail rjc0w@darwin.clas.virginia.edu rjc0w@virginia.bitnet Tel. 1(804)9784871

Chaney, Dr Michael O. (1943). Senior Research Scientist. Technology Core, MC625, Lilly Research Laboratories, Lilly Corporate Center, Indianapolis, IN 46285, USA. PhD (Indiana U., 1969) biochemistry. *Computer_modelling protein_folding drug_design protein_structure*
E-mail chaney_michael_o@lilly.com Tel. 1(317)2764135 Fax 1(317)2771125

Chang, ChiehYing Y. (1951). Research Associate. Bristol-Myers Squibb Pharmaceutical Research Institute, PO Box 4000, Room H.3107, Princeton, NJ 08543, USA. MS (Southern Illinois U. at Edwardsville, 1983) biochemistry. *Proteins protein_crystallization*
Tel. 1(609)2525536 Fax 1(609)2526030

Chang, Chong-Hwan (1950). Senior Research Scientist. DuPont Merck Pharmaceutical Company, DuPont Experimental Station E228/316F, Wilmington, DE 19880-0228, USA. PhD (Univ of Pittsburgh, 1981) crystallography. *Protein_crystallography structure_based_drug_design*
E-mail changch@lldmpc.dnet.dupont.com Tel. 1(302)6951787 Fax 1(302)6958667

Chang, Ning-Leh (1964). Postdoctoral Fellow. Department of Chemistry, University of Texas at Austin, Austin, TX 78712, USA. PhD (U. Texas at Austin, 1994) analytical chemistry. *Drug_design proteins computer_modelling solid_state_chemistry hydrogen_bonding*
E-mail cmak153@utxvms.cc.utexas.edu Tel. 1(512)4718605 Fax 1(512)4718696

Chantal, Abergel (1961). Visiting Fellow. National Institutes of Health, Bld 5 Room 303, Bethesda, MD 20892, USA. PhD (St Jerome University France, 1990) biophysics. *Biological_molecules immunology*
E-mail chantal@bandol.niddk.nih.gov Tel. 1(301)4021781 Fax 1(301)4960201

Chapman, Dr Michael S. (1961). Assistant Professor. Institute of Molecular Biophysics, Florida State University, Tallahassee, FL 32306, USA. PhD (UCLA, 1987) chemistry/biochem. *Viruses proteins drug_design methods*
E-mail chapman@sb.fsu.edu Tel. 1(904)6448354 Fax 1(904)6443257

Chasen, Edith (1947). Asst. Prof. Adj. Physics Department, St. John's University, Grand Central and Utopia Pkwys., Jamaica, New York 11439, USA. [86-02 Park Lane South, Woodhaven, New York 11421, USA.] A. M. (Boston U., 1970) geology. *Mineralogy crystallography computer_sciences telecommunications geophysics*
E-mail chasene@sjuvm.stjohns.edu Tel. 1(718)9906161

Chastain Jr, Roser V. (1938). 1962 Dowden Cr., Poolesville, MD 20837, USA. PhD (U. Washington, Seattle, 1965) physical chemistry. *Structure_analysis software*
Tel. 1(301)9727487

Chattopadhyay, Dr Debasish (1957). Postdoctoral Research Scientist. Physical and Analytical Chemistry, 301 Henrietta Street, The Upjohn Company, Kalamazoo, MI 49001, USA. PhD (Jadavpur U., 1989) chemistry. *Drug_design macromolecular_structure*
E-mail dchattop@upj.com Tel. 1(616)3857501 Fax 1(616)3857522

Cheer, Dr Clair J. (1937). Professor. Department of Chemistry, University of Rhode Island, Kingston, RI 02881, USA. PhD (Wayne State U., 1964) organic chemistry. *Small_molecules solid_state_photochemistry computer_modelling molecular_recognition strained_organics chiral_space_groups asymmetric_synthesis*
E-mail ccheer@chm.uri.edu Tel. 1(401)7922103 Fax 1(401)7925072

Cheetham, Prof. Anthony K. (1946). Professor of Materials, Director of MRL. Materials Research Laboratory, University of California, Santa Barbara, CA 93106, USA. DPhil (Oxford (UK), 1971) solid state chemistry. *Inorganic_materials synchrotron_X-rays neutron_diffraction zeolites*
E-mail akc@iristew.ucsb.edu Tel. 1(805)8938767 Fax 1(805)8938797

Chen, Dr Celia (1949). Research Associate. Center for Advanced Research in Biotechnology, University of Maryland Biological Institute, University of Maryland, USA. [CARB, 9600 Gudelsky Drive, Rockville, MD 20850, USA.] PhD (Columbia U., 1967) biochemistry. *Solubility crystallization biocrystallography*
E-mail celia@indigo8.carb.nist.gov Tel. 1(301)7386272 Fax 1(301)7386255

Chen, Haydn H. (1948). Professor of Materials Science & Engineering and Director, UNI-CAT. Dept. of Materials Science & Engineering, University of Illinois, 1304 W. Green Street, Urbana, IL 61801, USA. PhD (Northwestern U., 1977) materials science and engineering. *Diffuse_scattering applied_crystallography synchrotron_radiation phase_transformations time-resolved_scattering_studies semiconductor_and_ceramic_thin_films synchrotron_X-ray_instrumentation*
E-mail chen@uimrl7.mrl.uiuc.edu Tel. 1(217)2444666 1(217)3337636 Fax 1(217)3332736

Chen, Lirong (1954). Graduate student. Dept. of Crystallography, Univ. of Pittsburgh, 304 Thaw Hall, Pittsburgh, PA 15260-3630, USA. PhD (U. Pittsburgh, 1994) crystallography. *DNA proteins charge_density drug_design*
E-mail lchst1@vms.cis.pitt.edu Tel. 1(412)6249305 Fax 1(412)6241882

Chen, Dr Longyin (1947). Research Associate. Procter & Gamble Pharmaceutics, Miami Valley Laboratories, PO Box 538707, Cincinnati, OH 45253-8707, USA. PhD (Indiana U., 1988) biochemistry. *Proteins drug-design structure-activity modelling computer-programming natural-product*
E-mail chenl3@pg.com Tel. 1(513)6270806 Fax 1(513)6271196

Chen, Dr Shun-Le (1943). Research associate. Department of Molecular Biophysics and Biochemistry, Yale University, 260 Whitney Ave., New Haven, CT 06511, USA. PhD (City University of New York, 1993) physical chemistry. *Enzyme structure DNA recombination virus_structure medicine_natural_products*
E-mail chen@hhvms8.csb.yale.edu Tel. 1(203)4325623 Fax 1(203)4325175

Chen, Xin (1963). PhD candidate. Department of Chemistry, Washington State University, Pullman, WA 99164-4630, USA. MS (Washington State U., 1993) physical chemistry. *Phase_transition structural_disorder commensurate_and_incommensurate_phases*
E-mail 60753903@wsuvm1.csc.wsu.edu Tel. 1(509)3325677 Fax 1(509)3358867

Chen, Zhiwei (1950). Research Associate. Dept. of Biochemistry and Molecular Biophysics, Washington University School of Medicine, 660 South Euclid Ave., St. Louis, MO 63110, USA. BS (Fuzhow U., China, 1978) chemistry. *Protein_crystallography*
E-mail chen-z@cryst2.wustl.edu Tel. 1(314)3621079 Fax 1(314)3627183

Chen, Dr Zhongguo (1943). Research Fellow. WP44-B122, Merck Research Laboratory, West Point, PA 19486, USA. PhD (Technical U. Munich, Germany, 1982) biochemistry. *Proteins drug_design thrombosis synchrotron virus*
E-mail zhongguo_chen@msdrl.com Tel. 1(215)6524199 Fax 1(215)6527310

Cheng, Mr Beisong (1966). Graduate student. Department of Chemistry and Biochemistry, University of Notre Dame, Notre Dame, IN 46556, USA. BS (U. Sci. Tech. China, 1989) inorganic chemistry. *Metalloporphyrins magnetism electronic_spectrum unusual_bonding area_detectors_for_small_molecules*
Tel. 1(219)6316816 Fax 1(219)6316652

Cheng, Dr R. Holland Research Associate. Department of Biological Sciences, Purdue University, West Lafayette, IN 47907, USA. PhD (Purdue U., 1992) structural biology. *Virus_structural protein_nucleic_acid_interaction computer_modelling drug_design*
E-mail rhc@bragg.bio.purdue.edu o2s@mace.cc.purdue.edu Tel. 1(317)4945643 Fax 1(317)7434315

Cheng, Dr Xiaodong (1962). Senior Staff Investigator. Cold Spring Harbor Laboratory, PO Box 100, Cold Spring Harbor, New York 11724, USA. PhD (SUNY st Stony Brook, 1989) protein crystallography and physics. *Methyltransferases endonucleases protein_kinases zinc_amidase*
E-mail cheng@cshl.org Tel. 1(516)3678894 Fax 1(516)3678873

Cherbavaz, Dr Diana B. (1959). Postdoctoral Fellow. Dept. of Biochem. & Biophys. S-1070, Univ. of California at San Francisco, 513 Parnassus Ave. Box 0448, San Francisco, CA 94143-0448, USA. PhD (Brandeis U., 1991) biophysics. *Crystallography crystallization biophysics structure–activity_relationship membrane_proteins ion_channel_proteins membrane_associated_receptor_molecules kinetics*
E-mail cherbavaz@msg.ucsf.edu Tel. 1(415)4763937 Fax 1(415)4761902

Chi, Young-in (1960). Graduate Student (PhD candidate). Biological Sciences Department, Purdue University, West Lafayette, IN 47907, USA. [3031 Courthouse Dr #2A, West Lafayette, IN 47906, USA.] BS (Central Methodist College, 1988) biology. *Proteins macromolecule_structure computer_modelling protein-toxins biophysics immunology*
E-mail chiyi@scylla.bio.purdue.edu Tel. 1(317)4947249 Fax 1(317)4961189

Chidester, Connie Senior Research Scientist. Physical and Analytical Chemistry, 7255-209-1, The Upjohn Company, Kalamazoo, MI 49007-4940, USA. MA (Western Michigan U., 1968) mathematics. *Drug_receptor_modelling structure–activity_relationships biologically_active_small_molecules area_detectors*
E-mail cgchides@upj.com Tel. 1(616)3857624 Fax 1(616)3857522

Chirgadze, Dr Nickolay Y. (1965). Senior Scientist. Macromolecular X-ray Crystallography, Lilly Research Labs, Eli Lilly and Company, Indianapolis, IN 46285, USA. PhD (Institute of Crystallography Academy of the Sciences of the USSR, 1991) physics and mathematics. *Data_collection structure_determination computer_modelling drug_design blood_coagulation_factors phosphatase synchrotron-radiation low_temperature*
E-mail nyc@lilly.com Tel. 1(317)2769703 Fax 1(317)2769722

Chiu, Professor Wah (1947). Professor. Department of Biochemistry, Baylor College of Medicine, One Baylor Plaza, Houston, TX 77030, USA. PhD (U. California, Berkeley, 1975) biophysics. *Electron_crystallography_macromolecules virus_structure 2-dimensional_crystallization_proteins*
E-mail wah@bcm.tmc.edu Tel. 1(713)7986985 Fax 1(713)7969438

Choe, Dr Senyon Assistant Professor. Structural Biology Laboratory, The Salk Institute, La Jolla, CA 92037, USA. PhD (UC Berkeley, 1987) biophysics. *Membrane_macromolecules_structure molecular_recognition*
E-mail choe@crick.salk.edu Tel. 1(619)4534100x166 Fax 1(619)5468526

Choi, Dr Chang Sun (1926). Research Physicist. Energetics and Warhead Div., ARDEC, Picatinny Arsenal, NJ 07806, USA. [Reactor Radiation Div., NIST, Gaithersburg, MD 20899, USA.] PhD. (Kyungpook National U., Korea, 1968) physics. *Neutron_diffraction texture micro_structure explosive_structure*
E-mail choi@enh.nist.gov Tel. 1(301)9756225 Fax 1(301)9219847

Choi, Jungwon (1963). Postdoctoral associate. Department of Chemistry, Cornell University, Ithaca, NY 14853-1301, USA. PhD (U. Pittsburgh, 1994) crystallography. *Immunosuppressants proteins proteins–DNA_interaction computer-assisted_design pharmacology molecular_machines*
E-mail choi@chemres.tn.cornell.edu Tel. 1(607)2556145 Fax 1(607)2551253

Chook, Yuh Min Chook (1965). Grad. student. Box 36, Dept. of Chemistry, Harvard University, 12 Oxford St., Cambridge, MA 02138, USA. AB (Bryn Mawr College, 1988) biology and chemistry. *Protein_crystallography enzymes signal_transduction cytoskeletal_proteins*
E-mail chook@huray.harvard.edu Tel. 1(617)4954767 Fax 1(617)4953330

Chow, Paul (1957). Visiting Prof. University of Houston, Dept of Physics, 617SR1, 4800 Calitoon, Houston, TX 77204-5506, USA. PhD (Illinois, 1988) physics. *Structure X-ray_diffraction*
E-mail chow@xray.phys.uh.edu Tel. 1(713)7433541 Fax 1(713)7433589

Christensen, Dr Arild (1939). Senior Member of the Technical Staff. Scintag, Inc., 707 Kifer Road, Sunnyvale, CA 94086, USA. PhD (U. Oslo, 1966) physical chemistry. *Diffraction_technique crystallography instrumentation residual_stress pole_figure thin_film*
E-mail 72203.420@compuserve.com Tel. 1(408)7377200x123 Fax 1(408)7379841

Christianson, David W. (1961). Associate Professor. Dept. of Chemistry, University of Pennsylvania, Philadelphia, PA 19104-6323, USA. PhD (Harvard, 1987) chemistry. *Metalloenzyme inhibitor serpin protein_engineering*
E-mail chris@xtal.chem.upenn.edu Tel. 1(215)8985714 Fax 1(215)5732201

Chu, Prof. Benjamin (1932). Distinguished Professor. Department of Chemistry, State University of New York, Stony Brook, NY 11794-3400, USA. PhD (Cornell U., 1959) chemistry. *SAXS polymer_blends polymer_colloids supramolecules chain_conformation_and_dynamics*
Tel. 1(516)6327928 1(516)6327892 Fax 1(516)6327960

Chu, Shirley Shan-C. (1929). Professor Emeritus. 12 Duncannon Court, Dallas, TX 75225-1809, USA. PhD (U. Pittsburgh, 1961) physical chemistry. *Structure_small_molecule photovoltaic_semiconductors photovoltaics solar_energy_conversion*
Tel. 1(214)3682046 Fax 1(214)3682536

Church, William Bret (1960). Applications Scientist. Biosym Technologies, Inc., 1190 Saratoga Avenue, Suite #210, San Jose, CA 95129, USA. PhD (U. Sydney, 1991) inorganic chemistry. *Structure_modelling proteins interactions*
E-mail wbc@biosym.com Tel. 1(408)2363252 Fax 1(408)2490640

Churchill, Prof. Melvyn R. (1940). Professor of Chemistry. Department of Chemistry, FNSM Complex, North Campus, State University of New York at Buffalo, Buffalo, NY 14260, USA. PhD (Imperial College, U. London, 1964) inorganic chemistry. *Small_molecules inorganic organometallic disorder*
E-mail cbexray@ubvms.cc.buffalo.edu Tel. 1(716)6456800x2155 Fax 1(716)6456963

Clancy, Ms Laura Lee (1950). Research Chemist. Physical and Analytical Chemistry, 301 Henrietta Street, The Upjohn Company, Kalamazoo, MI 49001, USA. MS (U. Pittsburgh, 1986) crystallography. *Protein_structure protein_crystal_growth microgravity_experiments synchrotron_radiation*
E-mail llclancy@upj.com Tel. 1(616)3849794 Fax 1(616)3857522

Clarage, Dr James B. (1963). Research Associate. Department of Biochemistry, Rice University, PO Box 1892, Houston, TX 77251, USA. PhD (Brandeis U., 1990) physics. *Diffuse_scattering macromolecules molecular_dynamics disorder water_structure fractional_calculus*
E-mail clarage@rice.edu Tel. 1(713)5278101x3346

Clardy, Jon (1943). Horace White Professor. Department of Chemistry - Baker Laboratory, Cornell University, Ithaca, NY 14853-1301, USA. PhD (Harvard U., 1969) organic chemistry. *Chemotherapy immune_regulation natural_products cell_signalling drug_protein_interactions*
E-mail jcc12@cornell.edu Tel. 1(607)2557583 Fax 1(607)2551253 Telex WUI 6713054

Clark, Dr Edward S. (1930). Professor. Dept. of Materials Science and Eng., University of Tennessee, Knoxville, TN 37996, USA. PhD (U. California (Berkeley), 1956) physical chemistry. *Polymers_crystallinity polymers_crystallite polymers_processing*
E-mail esclark@utkux1.utk.edu Tel. 1(615)9745340 Fax 1(615)9745340

Clark, Joan R. (1920). Retired. 56 Citation Drive, Los Altos, CA 94024-7136, USA. PhD (The Johns Hopkins University, 1958) crystallography. *Crystallography mineralogy*
Tel. 1(415)9600628

Clarke, Dr Michael J. (1946). Professor. Department of Chemistry, Boston College, Chestnut Hill, MA 02167, USA. PhD (Stanford U., 1974) inorganic chemistry. *Transition_metal_complexes*
E-mail Clarke@hermes.bc.edu Clarke@bcvms.bitnet Tel. 1(617)5523624 Fax 1(617)5522705

Clarke, Neil D. (1958). Assistant Professor. 708 WBSB, Dept. Biophysics and Biophysical Chemistry, Johns Hopkins, Baltimore, MD 21205, USA. PhD (MIT, 1987) biology. *Biochemistry computing proteins oligonucleotides DNA repair protein_folding structure_prediction*
E-mail neil.clarke@qmail.bs.jhu.edu Tel. 1(410)6140338 Fax 1(410)5506910

Clarke, Prof. Roy (1947). Director of Applied Physics. Harrison M. Randall Laboratory of Physics, University of Michigan, Ann Arbor, MI 48109, USA. PhD (London U., Queen Mary College, 1973) physics. *Synchrotron_radiation molecular_beam_epitaxy RHEED phase_transition deputy_director MHATT–CAT*
E-mail royc@umich.edu Tel. 1(313)7644466 Fax 1(313)7642193

Claus, Albert C. (1931). Associate Professor. 15928 W. Woodbine Circle, Mundelein, IL 60060, USA. PhD (California Institute of Technology, 1956) physical chemistry. *Twinning metals*
Tel. 1(312)5083534

Clawson, David K. (1963). Physical Chemist. Physical Chemistry and Supercomputing Applications Research, Eli Lilly & Co., Indianapolis, IN 46285, USA. [Department of Biochemistry and Molecular Biology, Indiana University, Indianapolis, IN 46202, USA.] MS (Purdue U., 1989) organic chemistry. *Structure_modelling*
E-mail dkc@lilly.com Tel. 1(317)2761010 Fax 1(317)2769722

Clayton, William R. (1938). Miller Brewing Co, 3939 West Highland Blvd, Milwaukee, WI 53201, USA. PhD (Texas A&M University, 1974) physical chemistry.
Tel. 1(414)9313789 Fax 1(414)9313789

Clearfield, Prof. Abraham (1927). Professor; Director, Materials Science and Engineering Program. Department of Chemistry, Texas A&M University, College Station, TX 77843, USA. PhD (Rutgers U., 1954) physical chemistry. *Diffraction_powder_neutron_X-ray inorganic_ion_exchange catalysis material_chemistry*
E-mail clearf@acxrd.tamu.edu Tel. 1(409)8452936 Fax 1(409)8454719

Cody, Dr Vivian (1943). Senior Research Scientist II. Hauptman-Woodward Medical Research Institute (formerly the Medical Foundation of Buffalo), 73 High St, Buffalo, New York 14203-1196, USA. PhD (U. Cincinnati, 1969) chemistry. *Proteins computer_modelling drug_design structure_function anticancer hormone receptor protein_toxins*
E-mail Cody@hwi.buffalo.edu Tel. 1(716)8569600 Fax 1(716)8524846

Cohen, Dr Carolyn (1929). Professor. Rosenstiel Center, Brandeis University, Waltham, MA 02254, USA. PhD (MIT, 1954) biophysics. *Protein_structure_and_dynamics muscle_regulation coiled_coils electron_microscopy fibrinogen_fibrin*
E-mail ccohen@brandeis.bitnet ccohen@binah.cc.brandeis.edu Tel. 1(617)7362446 Fax 1(617)7362405

Cohen, Dr Gerson H. (1939). Research Chemist. National Institutes of Health, Bldg 5, Rm 335, 9000 Rockville Pike, Bethesda, MD 20892, USA. PhD (Cornell U., 1965) physical chemistry. *Proteins macromolecular_refinement computers automated_data_collection computer_networks computer_graphics macromolecular_structure_comparison*
E-mail ghc@vger.niddk.nih.gov ghc@nihklmb.bitnet Tel. 1(301)4024495 Fax 1(301)4960201

Cohen, Dr Jerome B. (1932). Dean. Northwestern University, McCormick School of Engineering, 2145 Sheridan Road, Evanston, IL 60208, USA. ScD (M. I. T., 1957) metallurgy. *Residual_stress thermodynamics ordering clustering local_order phase_transitions defects_in_oxides polymers catalysis*
E-mail j-cohen3@nwu.edu Tel. 1(708)4915220 1(708)4913243 Fax 1(708)4918539

Cole, Dr Brent (1960). Research Scientist II. X-ray Crystallography, BioCryst Pharmaceuticals, Inc., 2190 Parkway Lake Drive, Birmingham, AL 35244-2812, USA. PhD (Oklahoma State U., 1986) organic chemistry. *Proteins drug_design complement structure_based_drug_design*
E-mail cole@orion.cmc.uab.edu Tel. 1(205)4444624 Fax 1(205)4444640

Cole, Henderson (1924). Retired, IBM. 122 W. King St., Danbury, CT 06811, USA. PhD (MIT, 1952) physics. *Dynamical_effects automotion*
Tel. 1(203)7970527

Colella, Prof. Roberto (1935). Professor. Purdue University, Physics Department, West Lafayette, Indiana 47907, USA. Dr (U. Milan (Italy), 1958) physics. *Dynamical_theory multiple_diffraction phase_problem quasicrystals interferometry (with X-rays)*
E-mail colella@physics.purdue.edu Tel. 1(317)4943029 Fax 1(317)4940706

Collins, Dr Douglas M. (1939). Senior Scientist. Laboratory for the Structure of Matter, 6030 Naval Research Laboratory, Washington, DC 20375-5341, USA. PhD (Rutgers U., 1966) physical chemistry. *Maximum-entropy_method charge_density proteins protein_interaction anomalous_dispersion structural_chemistry quantum_mechanics computing*
E-mail collins1@lsm.nrl.navy.mil Tel. 1(202)7674134 1(202)7673496 Fax 1(202)7676874

Collins, Dr Edward J. (1961). Postdoctoral Fellow. Department of Molecular and Cellular Biology, Harvard University, 7 Divinity Ave, Cambridge, MA 02138, USA. PhD (U. Texas at Austin, 1990) chemistry. *Biophysics affinity folding immunology*
E-mail collins@xtal220.harvard.edu Tel. 1(617)4955043 Fax 1(617)4959613

Conant, John W. (1924). Visiting Scientist. Los Alamos National Laboratory, PO Box 159, Tesuque, NM 87574, USA. PhD (U. Iowa, 1955) chemistry.
Tel. 1(505)6652185

Concha, Dr Nestor O. (1957). Research Associate. CARB, University of Maryland, 9600 Gudelsky Dr, Rockville, MD 20851, USA. PhD (Boston U., 1993) physiology–protein crystallography. *Molecular_recognition proteins catalysis protein_dynamics calcium_binding membrane_proteins water_structure*
E-mail concha@elan1.carb.nist.gov Tel. 1(301)7386211 Fax 1(301)7386255

Connolly, Dr Michael L. (1951). 1259 El Camino Real, #184, Menlo Park, CA 94025, USA. PhD (U. California, 1981) biophysics. *Proteins computer_modelling computational_chemistry computer_graphics*
E-mail connolly@netcom.com Tel. 1(415)7800321 Fax 1(415)3264203

Cook Jr, William R. (1927). Corporate Secretary. 684 Quilliams Rd, Cleveland Heights, OH 44121-1955, USA. PhD (Case Western Reserve U., 1971) geology. *Minerals crystal_growth*
Tel. 1(216)3819003 Fax 1(216)4866103

Cook, Dr William J. (1949). Professor. Center for Macromolecular Crystallography, University of Alabama at Birmingham, UAB Station, THT-79, Birmingham, AL 35294, USA. MD PhD (U. Alabama at Birmingham, 1974 (MD) 1976 (PhD)) biochemistry. *Proteins structure_determination drug_design ubiquitin_mediated_proteolysis*
E-mail cook@orion.cmc.uab.edu Tel. 1(205)9348585 Fax 1(205)9340480

Copeland, Richard F. (1938). Professor. 1131 Lakewood Pard Dr, Daytona Beach, FL 32117-3940, USA. PhD (Texas A&M University, 1965) chemistry. *Crystallographic_computing information_retrieval*
E-mail copeland@cookman.edu Tel. 1(904)2538020

Cople, Valerie (1961). NIH IRTA Fellow. National Institute of Dental Research, National Institutes of Health, Bldg. 30, Rm.226, Bethesda, MD 20892, USA. PhD (Massachusetts Institute of Technology, 1990) physical chemistry. *Protein_nucleic_acid_structure protein_nucleic_acid_function protein_molecular_dynamics*
Tel. 1(301)4966307 Fax 1(301)4020824

Copley, Dr John R. D. (1944). Physicist. Materials Science and Engineering Laboratory, National Institute of Standards and Technology, Gaithersburg, MD 20899, USA. PhD (McMaster U., 1970) physics. *Disorder neutron_scattering inelastic_scattering neutron_spectroscopy neutron_optics rotational_disorder fullerenes*
E-mail john@rdstrad.nist.gov copley@enh.nist.gov Tel. 1(301)9756220 1(301)9755133 Fax 1(301)9219847

Coppens, Prof. Philip (1930). Distinguished Professor. Executive Committee, President 1993–1996. Department of Chemistry, 732 Natural Sciences and Mathematics Complex, State University of New York, University at Buffalo, Buffalo, New York 14260-3000, USA. PhD (U. Amsterdam, 1960) physical chemistry. *Charge_densities accuracy synchrotron_radiation incommensurately_modulated_structures anomalous_dispersion superconductors low-dimensional_conductors composite_structures low- temperature_techniques area_detectors*
E-mail che9990@ubvms or che9990@ubvms.cc.buffalo.edu Tel. 1(716)6456800x2217 Fax 1(716)6456948

Cordes, A. Wallace (1934). Professor. Department of Chemistry and Biochemistry, University of Arkansas, Fayetteville, AR, USA. PhD (U. Illinois, 1960) inorganic chemistry. *Small_molecule_structure_determination*
E-mail wcordes@uafsysb.uark.edu Tel. 1(501)5754601 Fax 1(501)5754049

Corfield, Peter W. R. (1937). Professor and Chair of Math. and Science. Department of Math and Science, The King's College, Briarcliff Manor, NY 10510-9985, USA. PhD (Durham University, UK, 1963) inorganic chemistry. *Protein_toxins water_structure small_molecules metal_complexes*
Tel. 1(914)9445527 Fax 1(914)9445636

Corliss, Lester M. (1919). Senior Chemist–Retired. 74 Forest at Duke Drive, Durham, NC 27705-5610, USA. PhD (Harvard U., 1950) chemical physics. *Neutron_physics magnetism critical_phenomena*

Coulter, Dr Charles L. (1933). Program Director. National Institutes of Health, National Center for Research Resources, Bethesda, MD 20892, USA. PhD (UCLA, 1960) physical chemistry. *Instrumentation structural_biology*
E-mail cco@nihcu.bitnet cco@cu.nih.gov Tel. 1(301)5947934 Fax 1(301)5949187

Cowan, Dr Paul L. (1950). Physicist. Physics Division, Argonne National Laboratory, Argonne, IL 10439, USA. PhD (Pennsylvania State U., 1977) physics. *X-ray_back_reflection X-ray_inelastic_scattering X-ray_resonant_scattering synchrotron_radiation field_ion_microscopy Auger_spectroscopy ion_channelling X-ray_monochromator*
E-mail cowan@anl.gov cowan@anlphy.bitnet Tel. 1(708)2524055 Fax 1(708)2526210 Telex 687-1701 DOE-ANL

Cowburn, David (1945). Associate Professor, Head of Laboratory. The Rockefeller University, 1230· York Avenue, New York, NY 10021-6399, USA. DSc (London U., 1981) biophysics. *NMR src_homology signal_transduction protein_modelling protein_nucleic_acids peptides*
E-mail cowburn@rockvax.rockefeller.edu Tel. 1(212)3278270 Fax 1(212)3277566

Cowley, Dr John M. (1923). Regents Professor. Dept. of Physics and Astronomy, Box 871504, Arizona State University, Tempe, AZ 85287-1504, USA. DSc (U. Adelaide, 1957) physics. *Electron_diffraction electron_microscopy disorder surface_structure diffraction_theory electron_holography*
E-mail cowleyj@physt.la.asu.edu Tel. 1(602)9656459 Fax 1(602)9657954

Cox, Dr David E. (1934). Senior Physicist. Member, IUCr Commission on Powder Diffraction. Department of Physics, Brookhaven National Laboratory, Building 510B, Upton, NY 11973, USA. PhD (London U., UK, 1958) inorganic chemistry. *Rietveld_method zeolite_phase_transition maximum_entropy_method magnetic_structure_determination powder_diffraction_techniques linear_PSD*
E-mail cox@bnlx7a.nsls.bnl.gov Tel. 1(516)2823818 Fax 1(516)2822739 Telex 16852516 bnl doe

Craven, Bryan M. (1932). Professor of Crystallography. Co-editor Acta Cryst. Department of Crystallography, University of Pittsburgh, Pittsburgh, PA 15260, USA. PhD (U. New Zealand, 1957) chemistry. *Charge_density thermal_vibration neutron_diffraction lipids nucleic_acids*
E-mail craven@vms.cis.pitt.edu Tel. 1(412)6249300 Fax 1(412)6241882

Crippen, Dr Gordon M. (1945). Professor. College of Pharmacy, University of Michigan, Ann Arbor, Michigan 48109, USA. PhD (Cornell U., 1971) chemistry. *Computational_chemistry protein_folding_theory distance_geometry computer-aided_drug_design*
E-mail crip@phar.umich.edu Tel. 1(313)7639722 Fax 1(313)7632022

Crist, Buckley (1941). Professor. Dept. of Materials Sci. & Eng., Northwestern Univ., Evanston, IL 60208-3108, USA. PhD (Duke U., 1966) physical chemistry. *Polymers defects deformation thermodynamics*
E-mail crist@ccmatsci.ms.nwu.edu Tel. 1(708)4913279 Fax 1(708)4917820

Croft, Dr William J (1926). Research chemist. Army Research Laboratory, Ceramics Division, Watertown, MA 02172, USA. PhD (Columbia U., 1954) crystallography (mineralogy). *X-ray_powder_diffraction electron_diffraction ceramics optical_microscopy borides*
E-mail wcroft@watertown-emh1.army.mil Tel. 1(617)9235358 Fax 1(617)9235385

Cromer, Don T. Staff Member. 11052 Academy Ridge Rd. NE, Albuquerque, NM 87111-6871, USA. [INC-4, MS C346, Los Alamos Nat. Lab., PO Box 1663, Los Alamos, NM 87545, USA.] PhD (U. Wisconsin, 1953) chemistry. *Intermetallic_structures charge_density computer_programming scattering_factors anomalous_dispersion*
E-mail hdjp25a@prodigy.com Tel. 1(505)6672424

Crouse, Richard E. (1950). Research Specialist. 10 East Main St., Oxford, MA 01540-1724, USA. *X-ray_instrumentation cryogenic_systems*
Tel. 1(508)9872232 Fax 1(508)9872527

Crowther, Robert L. (1949). Scientist. Hoffmann-LaRoche, 340 Kingsland Street, Nutley, NJ 07110, USA. AB (Manchester College, 1971) chemistry. *Proteins structure_determination synchrotron_radiation direct_methods*
E-mail crowther@mch01.dnet.roche.com Tel. 1(201)2352922 Fax 1(201)2352682

Cudney, Robert (1959). Research Associate. Department of Biochemistry, University of California, Riverside, CA 92521, USA. [Hampton Research, 5225 Canyon Crest #71336, Riverside, CA 92507, USA.] MS/MBA (U. California/Pepperdine U., 1989) biochemistry/business. *Proteins biochemistry crystallization peptides antibodies nucleic_acids crystal_growth*
E-mail cudney@ucrac1.ucr.edu cudney@ucrvms.bitnet xtalrox@aol.com Tel. 1(909)7898932 Fax 1(909)7898265

Cunane, Louise M. (1955). Postdoctoral Research Associate. Washington University School of Medicine, Department of Biochemistry and Molecular Biophysics, Box 8231, 660 S. Euclid Ave, St. Louis, MO 63110, USA. PhD (The Flinders University of South Australia, 1991) chemistry. *Biocrystallography enzymes structure_determination charge_density*
E-mail cunane_l@crystl.wustl.edu Tel. 1(314)3624646 Fax 1(314)3627183

Czerwinski, Dr Edmund W. (1940). Associate Professor. Department of Human Biological Chemistry and Genetics F-63, The University of Texas Medical Branch, Galveston, Texas 77555-0653, USA. PhD (Indiana U., 1971) biochemistry. *Structure_function proteins nucleic_acids small_molecules computer_modelling*
E-mail czerwins@beach.utmb.edu Tel. 1(409)7723287

Dahl, Dr Lawrence F. (1929). Professor. Department of Chemistry, 1101 University Ave., University of Wisconsin, Madison, WI 53706-1322, USA. PhD (Iowa State U., 1956) physical-inorganic chemistry. *Clusters bonding structural_change symmetry structural_disorder inorganic_organometallic_platinum*
E-mail dahl@chem.wisc.edu Tel. 1(608)2625859 Fax 1(608)2620381

Dai, Yongshan (1957). Research Mineralogist. Garber Research Center, Harbison-Walker Refractories, 1001 Pittsburgh-Mckeesport Blvd., West Mifflin, PA 15122, USA. PhD (Miami U., OH, 1990) mineralogy and crystallography. *Powder_single_crystal quantitative_analysis_with_Rietveld_techniques*
Tel. 1(412)4693889

Daniels, Dr Lee M. (1956). Research Scientist. Department of Chemistry, Texas A&M University, College Station, TX 77843-3255, USA. PhD (Texas A&M U., 1984) inorganic chemistry. *Disorder polymorphism computer_management*
E-mail daniels@tamu.edu Tel. 1(409)8453726 Fax 1(409)8459351

Dann, Jeffrey N. (1946). Senior project engineer. OSRAM SYLVANIA INC., Hawes Street, Towanda, PA 18848-0504, USA. MS (Polytechnic Institute of Brooklyn, 1970) physics. *Powder_diffraction*
Tel. 1(717)2685343 Fax 1(717)2685330 Telex 834610

David, Dr Peter Rensis (1959). Cell Biology Department, Stanford University Medical Center, Stanford, CA 94305-5400, USA. PhD (Harvard U., 1990) physical chemistry. *Computer_modelling membrane_receptors protein_crystallography electron_microscopy computer_modelling computer_drug_design pharmacology synchrotron_radiation transcriptases metallo_enzymes cell_signalling*
E-mail david@lama.stanford.edu Tel. 1(415)7250754 1(415)7238393 Fax 1(415)7238464

Davies, Dr Christopher (1965). Research associate. Dept. of Microbiology, Duke University, Durham, NC 27710, USA. PhD (Bristol U., UK, 1991) protein crystallography.
E-mail davies@iris.biochem.duke.edu chris@whitesws.mc.duke.edu Tel. 1(919)6843864 Fax 1(919)6848735

Davies, David R. (1927). Chief, Section on Molecular Structure. NIDDK, National Institutes of Health, Bldg 5, Room 338, 9000 Rockville Pike, Bethesda, MD 20892, USA. DPhil (Oxford U., 1952) chemical crystallography. *Proteins nucleic_acids antibodies enzymes cytokines receptors polynucleotides*
E-mail drd@vger.niddk.nih.gov Tel. 1(301)4944295 1(301)4944295 Fax 1(301)4960201

Davies II, Dr Jay F. (1955). Research Scientist. Crystallography Group, Agouron Pharmaceuticals, 3565 General Atomics Court, San Diego, CA 92121, USA. PhD (U. California, San Diego, 1990) chemistry. *Protein_structures drug_design*
E-mail davies@agouron.com Tel. 1(619)6223027

Davis, Prof. Raymond E. (1938). Professor. Department of Chemistry and Biochemistry, University of Texas at Austin, Austin, TX 78712-1167, USA. PhD (Yale U., 1965) physical chemistry. *Organic_solid_state_chemistry crystal_packing hydrogen_bonding pseudosymmetry chiral_recognition chiral_resolution organic_crystallography chemical_education*
E-mail redavis@utxvms.cc.utexas.edu Tel. 1(512)4714440 Fax 1(512)4718696

Davison, Daniel B. (1955). Assistant Professor. University of Houston, Department of Biochemical and Biophysical Sciences, BCHS-5934, 4800 Calhoun, Houston, TX 77203-5934, USA. PhD (SUNY at Stony Brook, 1985) biology.
E-mail davison@uh.edu Tel. 1(713)7438366 Fax 1(713)7438366

Day, Cynthia S. (1952). President. Crystalytics Company, 1701 Pleasant Hill Road, Lincoln, Nebraska 68523, USA. PhD (U. Nebraska-Lincoln, 1978) crystallography.
Tel. 1(402)4212797 Fax 1(402)4212797

Day, Delbert E. (1936). Curators Professor. Materials Research Center, School of Mines and Metallurgy, University of Missouri-Rolla, Rolla, MO 65401-0249, USA. PhD (Penn State U., 1961) ceramic engineering. *Ceramics glass biomaterials*
E-mail mrc@umrvmb.umr.edu Tel. 1(314)3414354 Fax 1(314)3412071

Day, John S. (1947). Senior Research Assistant. University of California, Riverside, Department of Biochemistry, University of California, Riverside, Riverside, California 92521, USA. MS (U. California, Riverside, 1983) biochemistry.
E-mail jday@ucrac1.ucr.edu Tel. 1(909)7873397 Fax 1(909)7873790

Dealwis, Dr Chris G. (1964). Research Associate. Department of Biochemistry and Molecular Biology, Cummings Building, Moffat Lab., University of Chicago, East 58th Street, Chicago, IL 60637, USA. PhD (U. London, 1993) protein crystallography. *Computer_modelling protein-RNA_interactions electrostatic_interactions protein_toxins molecular_recognition*
E-mail dealwis@circe.uchicago.edu Tel. 1(312)7021801 Fax 1(312)7020439

DeBoer, Barry G. (1942). Engineering Specialist. OSRAM Sylvania Inc., Hawes St., Towanda, PA 18848, USA. PhD (U. of California, Berkeley, 1968) chemistry. *Luminescence inorganic_structure inorganic_simulation inorganic_synthesis phosphors lighting computing*
Tel. 1(717)2685430 Fax 1(717)2685350

De Bondt, Dr Hendrik L. (1964). Postdoc. Department of Chemistry, Calvin Laboratory, University of California, Berkeley CA 94720, USA. PhD (K. U. Leuven, Belgium, 1991) pharmaceutical sciences. *Biocrystallography macromolecules computer_modelling electrostatic_potential*
E-mail hldebondt@lbl.bitnet Tel. 1(510)4865449 Fax 1(510)4866059

De Camp, Dr Wilson H. (1936). Supervisory Chemist. Division of Topical Drug Products (HFD-540), Food and Drug Administration, 5600 Fishers Lane, Rockville, MD 20857, USA. PhD (U. Maryland, 1970) physical chemistry. *Stereochemistry polymorphism*
E-mail decamp@fdacd.bitnet Tel. 1(301)5946386 Fax 1(301)5946589

DeHaven, Patrick W. (1949). Advisory Engineer. 203 Cherry Hill Drive, Poughkeepsie, NY 12603, USA. [IBM Corporation, Z/40E, 1580 Route 52, Hopewell Junction, NY 12533, USA.] PhD (Iowa State U., 1976) physical chemistry. *Microdiffraction topography powder diffractometry high_temperature XRD residual_stress_measurement*
E-mail dehaven@fshvmcc.vnet.ibm.com Tel. 1(914)8946859 1(914)8949652

Deisenhofer, Johann (1943). Professor. HHMI and Department of Biochemistry, UT Southwestern Medical Center, 5323 Harry Hines Boulevard Y4.206, Dallas, Texas 75235-9050, USA. Dr rer. nat. (Technical University Munich, 1974) experimental physics. *Macromolecular_crystallography protein_structure protein_function biophysics membrane_protein_structure*
E-mail jd@howie.swmed.edu Tel. 1(214)6485089 Fax 1(214)6485095

Dejus, Roger J. (1954). Assistant Physicist. Argonne National Laboratory, The Advanced Photon Source, 9700 South Cass Ave., Argonne, IL 60439, USA. PhD (Lith, Sweden, 1986) material science. *Computer_modelling diffraction image_processing synchrotron_radiation*
E-mail dejus@anlaps.aps.anl.gov Tel. 1(708)2525557 Fax 1(708)2523222

Delaney, Matthew S. (1927). Professor of Mathematics. Department of Physical Science and Mathematics, Mount St. Mary's College, 12001 Chalon Rd., Los Angeles, CA 90049, USA. [PO Box 2007, Seal Beach, CA 90740, USA.] PhD (Ohio State U., 1971) mathematics. *Mathematical_crystallography*
Tel. 1(310)4762237 Fax 1(914)8926256

DeLucas, Dr Lawrence J. (1950). Director. Center for Macromolecular Crystallography, University of Alabama at Birmingham, 79-THT, BHSB 262, 1918 University Boulevard, Birmingham, Alabama 35294-0005, USA. PhD (U. Alabama at Birmingham, 1982) biochemistry. *Protein_crystal_growth protein_structure X-ray_diffraction crystallography microgravity*
Tel. 1(205)9345329 Fax 1(205)9340480

De Rosier, David J. (1939). Professor. Biology Department, Brandeis University, 215 South St., Waltham, MA 02254, USA. PhD (University of Chicago, 1965) biophysics. *Electron_crystallography actin flagellum*
E-mail der@auriga.rose.brandeis.edu Tel. 1(617)7362426 Fax 1(617)7362419

Deschamps, Dr Jeffrey R. (1957). Research Chemist. Laboratory for the Structure of Matter, Code 6030, Naval Research Laboratory, Washington, DC 20375, USA. PhD (U. Delaware, 1984) marine studies, biochemistry. *Hyperpurification proteins enkephalins computer_modelling metalloenzymes metal-binding*
E-mail deschamps@lsm.nrl.navy.mil Tel. 1(202)7670656 Fax 1(202)7676874

Desper, Dr Richard (1937). Research Chemist. Polymer Research Branch, Army Research Laboratory, Watertown, MA 02172, USA. PhD (U. Massachusetts, 1967) physical chemistry. *Synthetic_polymers fibrous_polymers molecular_dynamics synchrotron_radiation liquid_crystal_polymers low_momentum_transfer_scattering*
E-mail rdesper@watertown-emh1.army.mil rdesper%watertown-emh1.army.mil@mitvma.mit.edu Tel. 1(617)9235216 Fax 1(627)9235385

DeTitta, George T. (1947). Senior Research Scientist. Member: Commission on Crystallographic Apparatus. Hauptman-Woodward Medical Research Institute (formerly the Medical Foundation of Buffalo), 73 High Street, Buffalo, NY 14203, USA. PhD (U. Pittsburgh, 1973) crystallography/biochemistry. *Macromolecular_crystal_growth phase_problem charge_density_studies cryo-crystallography biotin_biochemistry*
E-mail detitta@hwi.buffalo.edu chegdt@ubvms.cc.buffalo.edu Tel. 1(716)8569600 Fax 1(716)8524846

De Vos, Abraham M. (1955). Senior scientist. Dept. Protein Engineering, Genentech, Inc., 460 Point San Bruno Blvd, So. San Francisco, CA 94080, USA. PhD (U. Utrecht, 1985) chemistry. *Protein_structure molecular_recognition*
E-mail devos@gene.com Tel. 1(415)2252523 Fax 1(415)2253734

Dexter, Dr David D. (1940). Director of Campus Information Systems. Computer Center, Mount Union College, Alliance, OH 44601, USA. PhD (Georgetown U., 1968) physical chemistry. *Computing*
E-mail dexterdd@muc.edu Tel. 1(216)8232854 Fax 1(216)8234687

Dickerson, Richard E. (1931). Professor of Biochemistry and Geophysics. Molecular Biology Institute, 611 Circle Drive East, University of California, Los Angeles, Los Angeles, CA 90024, USA. PhD (U. Minnesota, 1957) physical chemistry. *DNA X-ray_crystallography protein molecular_evolution*
E-mail red@uclaue.mbi.ucla.edu Tel. 1(310)8255864 Fax 1(310)8250982

DiGabriele, Anna D. (1963). Postdoctoral Fellow. Department of Biochemistry and Molecular Biophysics, Columbia University, 630 West 168th Street BB-2-208, New York, NY 10032, USA. PhD (Yale U., 1991) biophysical chemistry. *Proteins structure_determination DNA crystal_disorder signal_transduction tyrosine_kinase_receptors cell_surface*
E-mail digabri@cuhhca.hhmi.columbia.edu Tel. 1(212)3058236 Fax 1(212)3057379

DiMarco, John D (1963). Associate Research Scientist. Bristol Myers Squibb, Pharmaceutical Research Institute, PO Box 4000, Princeton, NJ 08543-4000, USA. MS (Polytechnic University (Brooklyn), 1987) physics. *Structure_determination solid_state_transformation structural_disorder TGA DSC Circular_dichroism*
E-mail dimarco@bms.com Tel. 1(609)2525150 Fax 1(609)2526012

Ding, Dr Jianping (1961). Assistant Research Professor. CABM and Rutgers University, 679 Hoes Lane, Piscataway, NJ 08854, USA. PhD (Fudan U., Shanghai, 1987) chemistry. *Macromolecular_crystallography protein structure_and_function drug_design HIV-1_reverse_transcriptase polymerase hydrogen_bonding synchrotron*
E-mail ding@biovax.rutgers.edu Tel. 1(908)2354498 Fax 1(908)2354850

Dixon, Dr Melinda M. (1961). Research Fellow. Biophysics Research, IST, 2200 Bonisteel Blvd., Ann Arbor, MI 48109-2099, USA. PhD (U. Oregon, 1991) physics. *Proteins direct_methods*
E-mail lindy@norway.biop.umich.edu Tel. 1(313)7632199

Doedens, Robert J. (1937). Professor. Dept. of Chemistry, University of California, Irvine, CA 92717, USA. PhD (U. Wisconsin, 1965) chemistry.
E-mail rdoeden@uci.edu Tel. 1(714)8246605 Fax 1(714)8248571

Doolittle, Russell (1931). Professor. Center Molecular Genetics, University of California, San Diego, La Jolla, CA 92093-0634, USA. PhD (Harvard U., 1962) biochemistry. *Polymerization protein_structure evolution fibrinogen*
E-mail rdoolittle@ucsd.edu Tel. 1(619)5344417 Fax 1(619)5344985

Dorset, Dr Douglas L. (1942). Principal Research Scientist. Member, Commission on Electron Diffraction, IUCr. Electron Diffraction Department, Hauptman-Woodward Medical Research Institute (formerly the Medical Foundation of Buffalo), 73 High Street, Buffalo, NY 14203, USA. PhD (U. Maryland, 1971) biophysics. *Electron_crystallography direct_phasing structure_analysis binary_organic_solids polymer_structure lipid_structure paraffin_crystals*
E-mail eddd1%mfb@ubvms.bitnet Tel. 1(716)8569600 Fax 1(716)8524846

Double, Dr Sylvie (1965). Post-doc. Department of Biochemistry and Biophysics, University of North Carolina at Chapel Hill, Chapel Hill, NC 27599-7260, USA. PhD (U. North Carolina, 1993) biochemistry. *Proteins tRNA_binding_proteins*
E-mail sdouble@med.unc.edu Tel. 1(919)9663263 Fax 1(919)9662852

Doukov, Tzanko I. (1963). Graduate Student - PhD in Crystallography. Department of Chemistry, University of Nebraska-Lincoln, 729 Hamilton Hall, Lincoln, NE 68588-0304, USA. MS (Sofia U., Sofia, Bulgaria, 1986) analytical and inorganic chemistry. *Crystallization proteins nucleic_acids computer_modelling sequence_analysis*
E-mail tzanko@prophet1.unl.edu Tel. 1(402)4728894 Fax 1(402)4729402

Downing, Dr Kenneth H. (1945). Senior Scientist. Life Science Division, Lawrence Berkeley Laboratory, Berkeley, CA 94720, USA. PhD (Cornell U., 1974) applied physics. *Electron_crystallography proteins instrumentation electron_scattering*
E-mail khdowning@lbl.gov Tel. 1(510)4865941 Fax 1(510)4866488

Downs, James W. (1952). Associate Professor. Department of Geological Sciences, Ohio State University, Columbus, OH 43210, USA. PhD (Virginia Tech, 1983) geological sciences. *Electron_density crystal_chemistry mineralogy*
E-mail jdowns@geo1s.mps.ohio-state.edu Tel. 1(614)2926290 Fax 1(614)2927688

Downs, Dr Robert T. (1955). Research Associate. Carnegie Institution of Washington, 5251 Broad Branch Rd. NW, Washington, DC 20015, USA. PhD (Virginia Polytechnic Institute and State U., 1992) geology. *Mineralogy thermal_motion MO_calculation computer_graphics*
E-mail downs@gl.ciw.edu Tel. 1(202)6862410x2469 Fax 1(202)6862419

Doyne, Prof. Thomas H. (1927). Professor. Department of Chemistry, Villanova University, Villanova, PA 19085-1699, USA. PhD (Penn State U., 1957) biochemistry. *Structure_bioactive peptides lipids*
E-mail doyne@ucis.vill.edu Tel. 1(610)6454874 Fax 1(610)6457167

Drennan, Catherine Luschinsky (1963). Graduate Student. Department of Biological Chemistry, University of Michigan, Ann Arbor, MI 48109, USA. MS (U. Michigan, 1990) biological chemistry. *Proteins*
E-mail cathy@norway.biop.umich.edu Tel. 1(313)7632199 Fax 1(313)7643323

Duax, William L. (1939). Executive Director. Editor IUCr Newsletter; Program Chair XVII Congress and General Assembly. Hauptman-Woodward Medical Research Institute (formerly the Medical Foundation of Buffalo), 73 High St., Buffalo, NY 14203, USA. PhD (U. of Iowa, 1967) physical chemistry. *X-ray_crystallography_of_steroids ionophores macromolecular_steroid_conformation structure-activity_relationships endocrinology*
E-mail duax@hwi.buffalo.edu Tel. 1(716)8569600 Fax 1(716)8524846

Duchamp, Dr David James (1939). Director, Physical and Analytical Chemistry. Physical and Analytical Chemistry, 301 Henrietta Street, The Upjohn Company, Kalamazoo, MI 49001, USA. PhD (California Institute of Technology, 1965) chemistry. *Biological_molecules molecular_mechanics computing potential_energy*
E-mail djducham@upj.com Tel. 1(616)3857766 Fax 1(616)3857522

Duesler, Eileen N. (1943). Research Scientist. Department of Chemistry, University of New Mexico, Albuquerque, New Mexico 87131, USA. [2920 Charleston NE, Albuquerque, NM 87110-2704, USA.] PhD (U. California, Berkeley, 1973) inorganic chemistry. *Small_molecule structural_chemistry*
E-mail   duesler@bootes.unm.edu   Tel.   1(505)2776649   1(505)2770505   Fax 1(505)2772609

Durley, Rosemary C. E. (1948). Research Associate. Department of Biochemistry and Molecular Biophysics, Washington University School of Medicine, Campus Box 8231, 660 S. Euclid Avenue, St. Louis, MO 63011, USA. PhD (U. London, UK, 1972) chemical crystallography. *Structure_determination proteins metalloenzymes electron_transfer copper_proteins*
E-mail durley_r@cryst2.wustl.edu Tel. 1(314)3624646 Fax 1(314)3627183

Dwiggins, Claudius W. (1933). Retired. 1211 S Keeler, Bartleville, OK 74003, USA. PhD (U. Arkansas, 1958) physical chemistry. *Small_angle_scattering colloids micelles*
Tel. 1(918)3368546

Dyar, M. Darby (1958). Assistant Professor. Dept. of Geology and Astronomy, West Chester University, West Chester, PA 19383, USA. PhD (MIT, 1985) geochemistry. *Mineralogy mica amphibole olivine*
E-mail ddyar@wcupa.edu Tel. 1(610)4362213 Fax 1(610)4362213

Dyda, Dr Fred (1962). IRTA fellow. Lab of Molecular Biology, NIDDK, NIH, Bethesda, MD 20892, USA. PhD (U. Pittsburgh, 1992) protein crystallography. *Viral_proteins protein_toxins*
E-mail   dyda@vger.niddk.nih.gov   dyda@nihklmb.bitnet   Tel.   1(301)4024496   Fax 1(301)4960201

Dyda, Dr Fred P. (1962). Postdoctoral Fellow. Laboratory of Molecular Biology, National Institute of Diabetes and Digestive and Kidney Diseases, 9000 Rockville Pike Bethesda, MD 20892, USA. PhD (U. Pittsburgh, 1992) crystallography. *Viral proteins structure determination toxins*
E-mail dyda@ulti.niddk.nih.gov Tel. 1(301)4024496 Fax 1(301)4960201

Eades, John Alwyn (1939). Director of the Center for Microanalysis of Materials. Center for Microanalysis of Materials, Materials Research Laboratory MC-230, University of Illinois, 104 S. Goodwin, Urbana, Illinois 61801-2985, USA. PhD (Cambridge U., UK, 1967) physics. *RHEED electron_microscopy convergent-beam_diffraction symmetry EBSP*
E-mail eades@uimrl7.mrl.uiuc.edu Tel. 1(217)3338396 Fax 1(217)2442278

Ealick, Dr Steven E. (1951). Professor. Section of Biochemistry, Molecular and Cell Biology, Cornell University, 207 Biotechnology Building, Ithaca, New York 14853, USA. PhD (U. Oklahoma, 1976) physical chemistry. *Protein_structure enzymes drug_design synchrotron_radiation MAD_phasing purine_pyrimidine metabolism*
E-mail see3@cornell.edu Tel. 1(607)2557961 Fax 1(607)2552428

Earnest, Dr Thomas N. (1950). Staff Scientist. Lawrence Berkeley Laboratory, University of California, Berkeley, CA 94720, USA. PhD (Boston U., 1987) physics. *Proteins synchrotron_radiation electron_microscopy membrane_proteins*
E-mail earnest@rhoda.lbl.gov Tel. 1(510)4864603 Fax 1(510)4866488

Eckert, Dr Juergen (1947). Staff Member. LANSCE, MS H805, Los Alamos National Laboratory, Los Alamos, NM 87545, USA. PhD (Princeton U., 1975) materials science. *Hydrides molecular_vibration molecular_crystal hydrogen_bonding dihydrogen_complexes spectroscopy battery_materials zeolites*
E-mail juergen@lanl.gov Tel. 1(505)6652374 Fax 1(505)6652676

Edmundson, Allen B. (1932). Chair of Macromolecular Crystallography. Harrington Cancer Center, 1500 Wallace Blvd., Amarillo, TX 79106, USA. PhD (Rockefeller U., 1961) biochemistry. *Antibodies peptides opiates macromolecules complexation mechanisms_binding nucleic_acids peptides opiates*
E-mail allen@marvin.ama.ttuhsc.edu Tel. 1(806)3594673 Fax 1(806)3545887

Edwards, Dr Steven L. (1950). Visiting assistant Professor. University of California, Davis, Section of Cellular and Molecular Biology, Room 149 Briggs Hall, Davis, CA 95616, USA. PhD (U. California, San Diego, 1971) physical chemistry. *Redox_proteins enzyme_intermediates Laue_diffraction*
Tel. 1(916)7523611

Egami, Takeshi (1945). Professor. Dept. of Materials Science & Engineering, University of Pennsylvania, LRSM, 3231 Walnut Street, Philadelphia, PA 19104-6272, USA. PhD (U. Pennsylvania, 1971) materials science. *Superconductivity neutron_scattering synchrotron_radiation magnetism metallic_glasses electronic_structure correlated_electron_systems condensed_matter_physics*
E-mail   egami@pdfvax.lrsm.upenn.edu   Tel.   1(215)8985138   Fax 1(215)5732128 1(650)4713861

Eggleston, Dr Drake S. (1954). Assistant Director. Member, US National Committee for Crystallography (through 1997). Department of Physical and Structural Chemistry, SmithKline Beecham Pharmaceuticals, 709 Swedeland Road, UW-2950, King of Prussia, PA 19406, USA. PhD (U. North Carolina, Chapel Hill, 1983) inorganic chemistry. *Peptide_structure_function molecular_design molecular_recognition chiral_discrimination small_macromolecular_crystallography structure_computation_correlations protein_structure*
E-mail   eggleston@phvax.dnet@smithkline.com   Tel.   1(610)2706690   Fax 1(610)2706608

Eick, Harry A. (1929). Professor. Dept. of Chemistry, Michigan State University, E. Lansing, MI 48824-1322, USA. PhD (U. Iowa, 1956) physical chemistry. *Synthesis rare-earth_compounds superconductivity Rietveld_method*
E-mail   eick@msucem.cem.msu.edu   eick@msucem.bitnet   Tel.   1(517)3559715x168 1(517)3559715x171 Fax 1(517)3531793

Eigenbrot, Dr Charles W. (1954). Scientist. Genentech, Inc., Protein Engineering, 460 Pt. San Bruno Blvd., South San Francisco, CA 94080, USA. PhD (U. California, Berkeley, 1981) inorganic chemistry. *Structure_proteins*
E-mail charlie@gene.com Tel. 1(415)2252106

Einspahr, Dr Howard Martin (1943). Senior Scientist. Physical and Analytical Chemistry, 301 Henrietta Street, The Upjohn Company, Kalamazoo MI 49007 USA. PhD (U. Pennsylvania, 1970) chemistry. *Proteins macromolecular_structure macromolecular_structure_determination macromolecular_crystallization protein_structure_and_function structure_based_drug_design*
E-mail hmeinspa@upj.com Tel. 1(616)3855492 Fax 1(616)3857522

Eisenberg, Dr David S. (1939). Professor. Department of Chemistry and Biochemistry, UCLA, 405 Hilgard Ave., Los Angeles, CA 90024-1569, USA. DPhil (Oxford U., 1964) theoretical chemistry. *Protein_structure protein_folding water*
E-mail   david@uclaue.mbi.ucla.edu   david@uclaue.bitnet   Tel.   1(310)8253754 1(310)8251402 Fax 1(310)2063914

Eisenberg, Richard (1943). Professor of Chemistry and Chair. Department of Chemistry, University of Rochester, Rochester, NY 14627, USA. PhD (Columbia U., 1967) chemistry. *Structure-reactivity_relationships metal_hydrides homogeneous_catalysis metal_dithiolates transition_metal photochemistry*
E-mail rse7@chem.chem.rochester.edu Tel. 1(716)2755573 Fax 1(716)4736889

Elango, Dr Elango N. (1957). Research Associate. Department of Biochemistry, University of Minnesota, 4-225 Millard Hall, 435 Delaware St SE, Minneapolis, MN 55455, USA. PhD (U. Madras, 1989) physics. *Proteins computer_modelling crystallographic_programming homology_prediction*
E-mail elango@dc-crystal.med.umn.edu Tel. 1(612)6252115 Fax 1(612)6252163

Elder, Richard C. (1939). Professor and Director Biomedical Chemistry Research Center. Department of Chemistry, University of Cincinnati, Cincinnati, OH 45221-0172, USA. PhD (Massachusetts Institute of Technology, 1964) inorganic chemistry. *EXAFS computer_modelling metallodrugs bioinorganic_chemistry synchrotron_X-rays imaging_agents wide_angle_scattering*
E-mail dick.elder@uc.edu elder@ucbeh.bitnet Tel. 1(513)5569224 Fax 1(513)5569239

El-Kabbani, Dr Ossama (1959). Research Assistant Professor. Center for Macromolecular Crystallography, University of Alabama at Birmingham, Birmingham, AL 35294, USA. PhD (U. Saskatchewan, 1987) crystallography of biological molecules. *Protein_structures data_collection computer_modelling drug_design NADPH_oxidoreductases TIM_barrels reaction_centers membrane_proteins*
E-mail elkabbani@orion.cmc.uab.edu elkabbani@uabcmc.bitnet Tel. 1(205)9346003 Fax 1(205)9340480

Elkins, Patricia A. (1956). Graduate student. Dept. of Biological Sciences, Purdue University, West Lafayette, IN 47907, USA. BA (Rice U., 1978) behavioral science.
E-mail elkins@charybdis.bio.purdue.edu Tel. 1(317)4949247 Fax 1(317)4961189

Ely, Kathryn R. (1944). Senior Staff Scientist. La Jolla Research Found., 10901 N. Torrey Pines Rd., La Jolla, CA 92037-1062, USA. PhD (U. Utah, 1981) protein crystallography. *Protein_crystallography nucleic_acid_crystallography matrix_proteins receptors graphics_modelling*
E-mail ely@ljcrf.edu Tel. 1(619)4556480 Fax 1(619)4550181

Emerson, Dr Kenneth (1931). Professor - Retired. Department of Chemistry and Biochemistry, Montana State University, Bozeman, MT 59717-0342, USA. PhD (U. Minnesota, 1961) physical chemistry. *Incommensurate_structures transition_metal_halides magnetochemistry*
E-mail uchke@earth.oscs.mont.edu uchke@mtsunix1.bitnet Tel. 1(406)9945393 1(406)9945423 Fax 1(406)9945407

Enemark, Prof. John H. (1940). Prof. of Chemistry. Department of Chemistry, University of Arizona, Tucson, Arizona 85721, USA. PhD (Harvard U., 1966) chemistry. *Bioinorganic_chemistry molybdenum enzymes deformation_density_studies intramolecular_electron_transfer*
E-mail jenemark@ccit.arizona.edu Tel. 1(602)6212245 Fax 1(602)6218407

Enwall, Eric L. (1940). Director, Analytical Services Center. Chemistry Department, University of Oklahoma, 620 Parrington Oval, Norman, OK 73019, USA. PhD (Montana State U., 1969) chemistry. *Computing instrumentation*
E-mail eric-enwall@uokor.edu Tel. 1(405)3252843 Fax 1(405)3256111

Eppelsheimer, Dr Daniel (1941). Library Assist. MIT, 145-100, 77 Mass. Ave., Cambridge, MA 02139, USA. Dr.rer.nat. (U. Heidelberg, 1981) mineralogy. *Crystal physics Rietveld_analysis*
E-mail dseppels@mit.edu Tel. 1(617)6741167

Erman, Mary J. (1936). Sr Research Associate. Molecular Biophysics, Hauptman-Woodward Medical Research Institute (formerly the Medical Foundation of Buffalo), 73 High St, Buffalo, NY 14203, USA. BS (D'youville College, 1958) biology. *Crystal_growth proteins membrane protein_sequence amino_acid_mutations*
Tel. 1(716)8569600 Fax 1(716)8524846

Ernst, Stephen Richard (1939). Engineering-Science Research Associate. University of Texas, Department of Chemistry and Biochemistry, Austin, Texas 78712, USA. PhD (U. Utah, 1972) physical chemistry.
E-mail ernst@utbc01.cm.utexas.edu Tel. 1(512)4711105 Fax 1(512)4718696

Esser, Dr Lothar (1964). Research Associate. Howard Hughes Medical Institute, Room Y4.210, 5323 Harry Hines Blvd, Dallas, Texas 75235-9050, USA. PhD (Technische U. Berlin, 1992) inorganic chemistry. *Chemistry computer_graphics proteins direct_method organometallic rare-earth_compounds macromolecules*
E-mail esser@howie.swmed.edu Tel. 1(214)6485093 Fax 1(214)6485095

Evans Jr, Dr Howard T. (1919). Research Chemist. U.S.Geological Survey, National Center MS959, Reston, Virginia 22092, USA. [6107 Roseland Drive, Rockville, Maryland 20852, Usa.] PhD (Massachusetts Institute of Technology, 1948) inorganic chemistry. *Crystal_chemistry crystal_structure mineralogy geochemistry silicates sulfides vanadium_compounds heteropolyacids*
E-mail hevans@xray.er.usgs.gov Tel. 1(703)6486762 Fax 1(703)6486789

Faber, Catherine (1963). Graduate Student. Department of Biology, University of Utah, Salt Lake City, UT 84102, USA. [Crystallography Lab., 1500 Wallace Blvd., Amarillo, TX 79106, USA.] BS (U. Oregon, 1986) biochemistry. *Computer_modelling antibodies DNA_binding_proteins tetanus_toxin IgM_structure*
E-mail cat@trillian.ama.ttuhsc.edu Tel. 1(806)3545875x278

Fackler Jr, Prof. John P. (1944). Distinguished Professor. Department of Chemistry, Texas A&M University, College Station, TX 77843-3255, USA. PhD (Massachusetts Institute of Technology, 1960) inorganic chemistry. *Gold heavy_metals luminescence group-11 bio-inorganic Zeigler-catalysts simpler*
E-mail fackler@chemvx.tamu.edu Tel. 1(409)8450648 Fax 1(409)8459351

Faerman, Dr Carlos H. (1954). Research Associate. Biochemistry, Molecular and Cell Biology, Cornell University, Ithaca, NY 14853, USA. PhD (U. Toronto, 1987) intermolecular forces. *Computer_modelling electrostatic_properties enzymes*
E-mail carlos@penelope.bio.cornell.edu Tel. 1(607)2558432

Fait, Dr James F. (1957). Software Engineer. Siemens Industrial Automation, Analytical Instruments Division, 6300 Enterprise Lane, Madison, WI 52719-1173, USA. PhD (Montana State U., 1987) inorganic chemistry. *Diffractometry*
E-mail fait@bert.chem.wisc.edu Tel. 1(608)2763040 Fax 1(608)2763015

Fanwick, Dr Phillip E. (1947). Crystallographer. Department of Chemistry, Purdue University, West Lafayette, IN 47907, USA. PhD (Iowa State U., 1977) inorganic chemistry. *Service structure_determination thermal_motion*
E-mail fanwick@xray.chem.purdue.edu Tel. 1(317)4944572 Fax 1(317)4940239

Farber, Gregory K. (1962). Assistant Professor. Department of Chemistry, The Pennsylvania State University, 152 Davey Laboratory, University Park, PA 16802, USA. PhD (Massachusetts Institute of Technology, 1988) physical chemistry. *Proteins time-resolved_crystallography enzyme_substrate_complexes computational_crystallography*
E-mail farber@retina.chem.psu.edu Tel. 1(814)8651554 Fax 1(814)8653314

Farmer, Dr Barry L. (1947). Professor. Department of Materials Science and Engineering, Thornton Hall, University of Virginia, Charlottesville, VA 22903-2442, USA. PhD (Case Western Reserve U., 1974) macromolecular science. *Polymers computer_modelling simulations structural_transitions*
E-mail farmer@virginia.edu Tel. 1(804)9240605 Fax 1(804)9825660

Farrar, William W. (1940). Professor. Dept. Biol. Sciences, Eastern Kentucky U., Richmond, KY 40475, USA. PhD (Virginia Polytechnic Inst. and State Univ, 1970) biochemistry. *Crystallography folding kinetics viruses*
E-mail farr@mace.cc.purdue.edu biofarra@acs.eku.edu Tel. 1(606)6221531 Fax 1(317)4941662

Faulk, John W. (1940). Administrative Manager. 2473 Mary Drive, Sulphur, LA 70663, USA. BS (Menceese State, 1964) physics. *Diffraction*
Tel. 1(318)5837554 Fax 1(318)5832872

Fauman, Dr Eric B (1965). Post-doc. Biophysics Research Group, 3250 Chemistry, 930 N. University, Ann Arbor, MI 48109-1055, USA. PhD (U. California, San Francisco, 1993) biochemistry and biophysics. *Proteins computer_modelling anisotropy Debye_Waller_factor*
E-mail fauman@umich.edu Tel. 1(313)7633384 Fax 1(313)7643323

Fay, Robert C. (1936). Professor. Dept. of Chem., Baker Lab., Cornell University, Ithaca, NY 14853-1301, USA. PhD (U. Illinois at Urbana, 1962) inorganic chemistry. *Stereochemistry metal_complexes chelates early_transition_metal_chemistry*
Tel. 1(607)2553636 Fax 1(607)2554137

Feher, George (1924). Professor. Department of Physics, 0319, University of California at San Diego, La Jolla, CA 92903-0319, USA. [9500 Gilman Drive, La Jolla, CA 92903, USA.] PhD (U. California, Berkeley, 1954) physics. *Photosynthesis electron_transfer reaction_center rhodobacter_sphaeroides biophysics energy_conversion*
Tel. 1(619)5344389 Fax 1(619)5340173

Feldman, Robert E. (1939). Assistant Professor of Physics. 425 Lincoln Blvd Apt 2U, Hauppauge, NY 11788-2913, USA. PhD (Polytechnic Institute, 1969) borrmann effect. *Diffraction_theory*
Tel. 1(516)3483372

Ferguson-Miller, Shelagh (1942). Professor and Associate Chair. Department of Biochemistry, Biochemistry Building, Michigan State University, East Lansing, MI 48824-1319, USA. PhD (U. Wisconsin-Madison, 1971) biochemistry. *Cytochrome_oxidase detergents membrane_proteins proton_transfer electron_transfer*
E-mail fergus20@pilot.msu.edu Tel. 1(517)3550199 Fax 1(517)3539334

Fernando, Dr Quintus (1926). Professor. Department of Chemistry, University of Arizona, Tucson, AZ 85721, USA. PhD (U. Louisville, 1953) analytical chemistry. *Microbeam_analysis imaging trace_analysis toxicology metals environment ESCA EPR*
E-mail fernandq@ccit.arizona.edu Tel. 1(602)6212105 Fax 1(602)6218407

Ferrara, Dr Joseph David (1961). Vice President of X-ray Diffraction. Molecular Structure Corporation, 3200 Research Forest Drive, The Woodlands, Texas 77381-4238, USA. PhD (Case Western Reserve U., 1988) physical organometallic chemistry. *Data_collection diffractometry image_processing oscillation_camera*
E-mail jdf@msc.com Tel. 1(713)3631033 Fax 1(713)3643628

Ferre-D'Amare, Adrian R. (1966). Graduate Fellow. Laboratory of Molecular Biophysics, Rockefeller University, 1230 York Avenue, New York, NY 10021, USA. BSc (Instituto Tecnologico de Monterrey, 1990) chemistry. *DNA_proteins_recognition thermodynamics*
E-mail ferreda@rockvax.rockefeller.edu Tel. 1(212)3278339 Fax 1(212)3278337

Fettinger, Dr James C. (1959). Director, X-ray Crystallographic Facility. Department of Chemistry and Biochemistry, University of Maryland at College Park, College Park, MD 20742, USA. PhD (State U. New York at Buffalo, 1987) inorganic chemistry. *Structural_disorder*
E-mail jf96@umail.umd.edu Tel. 1(301)4051861 Fax 1(301)3149121

Finer-Moore, Dr Janet S. (1951). Associate Research Biochemist. Department of Biochemistry and Biophysics, University of California, S964, San Francisco, CA 94143-0448, USA. PhD (Iowa State U., 1978) physical chemistry. *Proteins X-ray_crystallography enzymatic_mechanism enzyme_inhibitors*
E-mail finer@msg.ucsf.edu Tel. 1(415)4763937 Fax 1(415)4761902

Finger, Dr Larry W. (1940). Crystallographer. Geophysical Laboratory, Carnegie Institution of Washington, 5251 Broad Branch Road, NW, Washington DC 20015-1305, USA. PhD (U. Minnesota, 1967) mineralogy/crystallography. *Mineralogy structure_determination pressure temperature crystal_chemistry crystallographic_software*
E-mail finger@gl.ciw.edu Tel. 1(202)6862410x2464 Fax 1(202)6862419

Finkenstadt, Victoria L. (1967). Graduate Research Assistant. Whistler Center for Carbohydrate Research, Purdue University, 1160 Smith Hall, West Lafayette, IN 47907-1160, USA. BS (McPherson College (McPherson, KS), 1989) chemistry. *Computer_modelling diffuse_scattering structure_determination structure_activity_relationship biomedical*
E-mail vicki@kiwi.foodsci.purdue.edu eog@mace.cc.purdue.edu vicki@foodsci.purdue.edu Tel. 1(317)4944914 Fax 1(317)4947953

Finzel, Dr Barry Craig (1956). Senior Scientist. Physical and Analytical Chemistry, 301 Henrietta Street, The Upjohn Company, Kalamazoo, MI 49001, USA. PhD (U. California, San Diego, 1983) chemistry. *Macromolecular_structure computer_modelling computing*
E-mail bcfinzel@upj.com Tel. 1(616)3849744 Fax 1(616)3857522

Fisher, Dr Andrew J. (1965). Postdoc. Institute for Enzyme Research, University of Wisconsin, 1710 University Ave., Madison, WI 53705, USA. PhD (Purdue U., 1992) biophysics. *Proteins viruses*
E-mail andy@enzyme.wisc.edu Tel. 1(608)2620529

Fitzgerald, Dr Paula M. D. (1949). Senior Research Fellow. Member Biological Macromolecules Commission; Member Committee for the Maintenance of the CIF Standard. Merck Research Laboratories, PO Box 2000, Ry50-105, Rahway, New Jersey 07065, USA. PhD (The Johns Hopkins University, 1977) biophysics. *Macromolecular_crystallography molecular_replacement drug_design structure-based_drug_design*
E-mail paula_fitzgerald@merck.com Tel. 1(908)5945510 Fax 1(908)5945510

Fletterick, Robert (1943). Professor. Department of Biochemistry, University of California, San Francisco CA 94143-0448, USA. PhD (Cornell U., 1970) chemistry. *Protein_structure*
E-mail flett@msg.ucsf.edu Tel. 1(415)4765080 Fax 1(415)4761902

Flick, Karen E. (1969). Graduate Student. Department of Molecular and Cell Biology, University of California, Berkeley, CA 94720, USA. AB (Harvard College, 1991) biochemical sciences. *Proteins protein–DNA_interactions*
E-mail flick@ucxray4.berkeley.edu Tel. 1(510)6436357 Fax 1(510)6439290

Flippen-Anderson, Judith L. (1941). Chemist. Member of organizing committee for 1996 IUCr meeting; secretary of the Small Molecule Commission. Code 6030, Naval Research Laboratory, Washington, DC 20375-5341, USA. MS (Arizona State U., 1965) physical chemistry. *Neuropeptides opiates explosives small_molecules crystal_growth hydrogen_bonding structure–activity_relationship*
E-mail flippen@roentgen.nrl.navy.mil Tel. 1(202)7673463 Fax 1(202)7676874

Florio, John V. (1925). Volunteer. The Metropolitan Museum of Art, Objects Conservation Department, 5th Avenue at 83nd St., New York, NY 10028, USA. [31 Marion Drive, North Haven, CT 06473-2018, USA.] PhD (Iowa State U., 1952) physical chemistry. *Conservation paintings ceramics metals*
Tel. 1(212)5703858

Folting-Streib, Kirsten. (1932). Staff Crystallographer. Indiana Univ. Molecular Structure Center, Chemistry Department, Indiana University, Bloomington, IN 47405-4001, USA. PhD (Royal Danish School of Pharmacy, 1964) organic chemistry. *Organic_crystal_structures organometallic_compounds*
E-mail kstreib@indiana.edu Tel. 1(812)8556821

Foord, Eugene E. (1946). Geologist-Mineralogist. US Geological Survey, MS905, Denver Federal Center, Lakewood, CO 80225, USA. PhD (Stanford U., 1976) geology-mineralogy. *Mineralogy silicates Ta-compounds Nb-compounds pegmatites Nb-Ta_minerals*
E-mail efoord@usgs.gov Tel. 1(303)2364755 Fax 1(303)2365603

Forest, Katrina (1966). Postdoctoral Fellow. The Scripps Research Institute MB-4, 10666 N. Torrey Pines Rd., San Diego, CA 92037, USA. PhD (Princeton U., 1993) molecular biology. *Proteins crystallography electron_microscopy infectious_disease toxins*
E-mail forest@scripps.edu Tel. 1(619)5549262 Fax 1(619)5546880

Foris, C. M. E.I. du Pont de Nemours Co., 228 Experimental Station, PO Box 80228, Wilmington, DE 19880-0228, USA. *Powder_diffraction Guinier_techniques inorganic_crystal_chemistry*
E-mail foris@esvax.dnet.dupont.com Tel. 1(302)6953687 Fax 1(302)6951351 Telex 835420

Fornoff, Mario (1932). Sales/marketing Manager. PO Box 4262, Wilmington, DE 19807-4262, USA. ChE (U. Cincinnati, 1957) chemical engineering. *Powder_diffraction*
E-mail fornoff@icdd.com Tel. 1(610)3259810 Fax 1(610)3259823

Foster, Dr Mark D. (1959). Assistant Professor. Department of Polymer Science, The University of Akron, Akron, OH 44325-3909, USA. PhD (U. Minnesota, 1987) chemical engineering. *Polymers small_angle_scattering neutron_scattering thin_films surfaces interfaces diblock_copolymers morphology interdiffusion*
E-mail foster@frank.polymer.uakron.edu Tel. 1(216)9725323 Fax 1(216)9725461

Foundling, Dr Stephen I. (1958). Assistant Member. Protein Studies Program, Oklahoma Medical Research Foundation, Lab. of Crystallography, 825 N.E. 13th Street, Oklahoma City, OK 73104-5097, USA. PhD protein crystallography. *Proteins enzymes enzyme_ligand_complex ligand_design proteases protease_inhibitors molecular_modelling*
E-mail Stephen-Foundling@omrf.uokhsc.edu Tel. 1(405)2717543 1(405)2717588 Fax 1(405)2713980

Fox, Kristin M. (1966). Graduate Student. Department of Biochemistry, Molecular and Cell Biology, 221 Biotechnology Building, Cornell University, Ithaca, NY 14853, USA. BS (Lafayette College, 1988) chemistry. *Macromolecular_crystallography enzymes crystal_growth flavoproteins*
E-mail fox@penelope.bio.cornell.edu Tel. 1(607)2558432 Fax 1(607)2552428

Foxman, Dr Bruce M. (1942). Professor. Department of Chemistry, Brandeis University, PO Box 9110, Waltham, MA 02254-9110, USA. PhD (M. I. T., 1968) inorganic chemistry. *Solid_state_reactions coordination_polymers*
E-mail foxman1@binah.cc.brandeis.edu foxman1@brandeis.bitnet Tel. 1(617)7362532 Fax 1(617)7362516 Telex 7607715 BRAND UC

Franzen, Hugo F. (1934). Professor. Chemistry Dept., Iowa State University, Ames, IA 50011, USA. PhD (U. Kansas, 1962) physical chemistry. *High_temperature_diffraction transitions refractory_solids*
E-mail franzen@ameslab.gov Tel. 1(515)2945773 Fax 1(515)2945718

Fratini, Albert V. (1939). Professor and Chair. Department of Chemistry, University of Dayton, Dayton, OH 45469-2357, USA. PhD (Yale U., 1965) physical chemistry. *Structure_and_morphology ordered_polymers organic_compounds nonlinear_optical_materials modelling_of_macromolecular_structure*
E-mail fratini@udavxb.oca.udayton.edu Tel. 1(513)2292849 Fax 1(513)2292635

Frederick, Christin (1952). Assist. Prof. DANA Farber Cancer Inst., 44 Binney St., Boston, MA 02115, USA. PhD (U. PGH, 1980) biochemistry. *RNA_structure DNA-protein_interactions protein–protein_interactions*
E-mail caf@amber.dfci.harvard.edu Tel. 1(617)6323984 Fax 1(617)6324393

Freed, Robert Lowell (1938). Professor. Geology Dept, Trinity University, 715 Stadium Drive, San Antonio, TX 78212, USA. PhD (University of Michigan, 1966) mineralogy.
E-mail bfreed@geology.trinity.edu Tel. 1(210)7367609 Fax 1(210)7368264

Freeman, Dr Clive M. (1962). Scientist. BIOSYM Technologies, Inc., 9685 Scranton Road, San Diego, CA 92121, USA. PhD (University College London, 1986) physical chemistry. *Inorganic molecular quantum intermolecular powder_diffraction structure_prediction*
E-mail clive@biosym.com Tel. 1(619)5465592 Fax 1(619)4580136

Freer, Dr Stephan T. (1933). Principal Scientist. Agouron Pharmaceuticals, Inc., 3565 General Atomics Court, San Diego, CA 92122, USA. PhD (U. Washington, 1964) biochemistry. *Proteins drug computer_modelling simulation artificial_intelligence computer_science*
E-mail freer@agouron.com Tel. 1(619)6223048 Fax 1(619)6223299

Fremont, Dr D. H. (1964). Research Associate. Dept. of Molecular Biology, The Scripps Research Institute, La Jolla CA 92037, USA. PhD (U. California, San Diego, 1993) chemistry.
E-mail fremont@scripps.edu Tel. 1(619)5543473 Fax 1(619)5546105

French, Alfred D. (1943). Chemist. Southern Regional Research Center, US Dept. of Agriculture, PO Box 19687, New Orleans, LA 70179-0687, USA. PhD (Arizona State U., 1971) physical chemistry. *Carbohydrates modelling fiber_diffraction*
E-mail afrench@nola.srrc.usda.gov Tel. 1(504)2864250 Fax 1(504)2864419

Frenz, Bert (1945). President. B. A. Frenz & Associates, 209 University Dr East, College Station, TX 77840, USA. PhD (Northwestern U., 1972) inorganic chemistry. *Software computers*
Tel. 1(409)8469042 Fax 1(409)2601382

Freund, Prof. Dr Friedemann (1933). Professor. Department of Physics, San Jose State University, San Jose, CA 95192-0106, USA. [Code SSX MS 239-4, NASA Ames Research Center, CA 94035 Moffet Field, USA.] Dr (U. Marburg, 1959) mineralogy. *Solid_state physics catalysis defect proton_conductivity dielectrics*
Tel. 101(415)6045183 Fax 101(415)6041088

Frevel, Ludo K. (1910). ICDD representative. 1205 W. Park Drive, Midland, MI 48640, USA. PhD (The Johns Hopkins University, 1934) X-ray diffraction. *X-ray_diffraction electron_diffraction catalysis applied_mathematics*
Tel. 1(517)8328983

Freymann, Douglas (1957). Postdoctoral Fellow. Department of Biochemistry and Biophysics, University of California, San Francisco, CA 94143-0448, USA. PhD (Harvard U., 1989) biochemistry. *Proteins RNA*
E-mail freymann@msg.ucsf.edu Tel. 1(415)4761123 Fax 1(415)4765233

Friedman, Dr Jonathan M. (1959). Assistant Professor. Department of Chemistry, University of Houston, Houston, TX 77204-5641, USA. PhD (Harvard U., 1986) organic chemistry. *Macromolecular_structure_determination synthetic_enzyme_inhibitors computer-assisted_design_software experimental_intermolecular_potential energy_parameter_measurement protein_ligand_interactions mechanism_based_inhibitors structure_based_inhibitors nucleic_acid_base-pairing*
E-mail friedman@kitten.chem.uh.edu Tel. 1(713)7432747 Fax 1(713)7432709

Fronczek, Frank R. (1948). Research Associate. Department of Chemistry, Louisiana State University, Baton Rouge, LA 70803, USA. PhD (California Institute of Technology, 1975) chemistry. *Service_crystallography natural_products crown_compounds hydrogen_bonding coordination_chemistry inclusion_compounds macrocycles heterocycles terpenes*
E-mail fronz@chxray.dnet.lsu.edu Tel. 1(504)3888270 Fax 1(504)3883458

Frueh, Prof. Alfred J. (1919). Professor Emeritus. Dept. of Geology and Geophysics, University of Connecticut U-45, Storrs, CT 06268, USA. [23 Bundy Lane, Storrs, CT 06268, USA.] PhD (MIT, 1949) mineralogy. *Mineral structure disorder*
Tel. 1(203)4861385 Fax 1(203)4861383

Fujii, Gary (1958). Associate Director. Vestar Inc., 650 Cliffside Dr., San Dimas, CA 91773, USA. PhD (UCLA, 1991) physical chemistry. *Membrane_proteins liposomes protein_membrane_interactions protein_engineering gene_therapy viral_structure_and_function*
E-mail fujii@uclave.mbi.ucla.edu Tel. 1(909)3944000 Fax 1(909)5928530

Fullenwider, Malcolm A. (1940). President. R.K. Labs, 2970 Mac Arthur Rd., PO Box 2, Whitehall, PA 18052, USA. PhD (U. Pennsylvania, 1969) chemistry. *Crystal_electrochemistry tetramethyl_adamantane_crystallography*
Tel. 1(215)4351452

Fuoss, Paul Henry (1953). Distinguished Member of Technical Staff. 7C-201, AT&T Bell Laboratories, 600 Mountain Avenue, Murray Hill, NJ 07974-0636, USA. PhD (Stanford U., 1980) materials science. *Surface_physics crystal_growth amorphous_materials X-ray_physics*
E-mail fuoss@physics.att.com Tel. 1(908)5824951 Fax 1(908)5824941

Furey, Dr William (1952). Research Chemist, Adj. Associate Professor. Biocrystallography Laboratory, VA Medical Center, PO Box 12055, University Drive C, Pittsburgh, PA 15240, USA. [Dept. of Crystallography, 304 Thaw Hall, University of Pittsburgh, Pittsburgh, PA 15260, USA.] PhD (Rutgers U., 1977) physical chemistry/crystallography. *Structure_determination proteins bacterial_toxins enzymes computing crystallization area_detector phase_determination*
E-mail 300531@vms.cis.pitt.edu 300531@pittvms.bitnet Tel. 1(412)6923517

Furnas, Thomas C. (1922). President. Molecular Data Corporation, 2869 Scarborough Road, Cleveland, OH 44118, USA. PhD (MIT, 1952) physics. *Instrumentation small_angle_scattering intensity_measurements X-ray_fluorescence*
Tel. 1(216)3816328 Fax 1(216)9324718

Fuzek, John F. (1921). Retired-Eastman Kodak Chem. Div. 4603 Mitchell Rd, Kingsport, TN 37994-2125, USA. PhD (U. Tennessee, 1947) physical chemistry. *Fiber_polymer_structure*
Tel. 1(615)2885362

Gaier, James R. (1952). Research Scientist. Electro Physics Branch, NASA Lewis Research Center, 21000 Brookpark Rd., Cleveland, OH 44135, USA. PhD (Michigan State U., 1983) chemistry. *Proteins intercalation_compounds*
E-mail segaier@lims02.lerc.nasa.gov Tel. 1(216)4336686 Fax 1(216)4332221

Galitsky, Dr Nikolai M. (1950). Research Scientist. Hauptman-Woodward Medical Research Institute (formerly the Medical Foundation of Buffalo), 73 High Street, Buffalo, New York 14203-1196, USA. PhD (Institute of Bioorganic Chemistry, Moscow, Russia, 1978) bioorganic chemistry, crystallography. *Peptides proteins structure_determination function pore_formation_membrane*
E-mail galitsky@nexus.hwi.buffalo.edu Tel. 1(716)8569600 Fax 1(716)8524846

Gallagher, Dr D. Travis (1957). Research Fellow. C.A.R.B., 9600 Gudelsky Dr, Rockville, MD 20850, USA. PhD (U of Texas at Austin, 1990) chemistry. *Proteins crystallization DNA_binding folding lattice_formation water_structure education*
E-mail travis@ibm5.carb.nist.gov Tel. 1(301)7386202 Fax 1(301)7386255

Gallucci, Dr Judith C. (1953). Department crystallographer. Department of Chemistry, 120 West 18th Avenue, The Ohio State University, Columbus, OH 43210, USA. PhD (U. Massachusetts, 1979) inorganic chemistry. *Small_molecules structure_determination*
E-mail gallucci.1@osu.edu Tel. 1(614)2924039 Fax 1(614)2921685

Gamble, Theresa R. (1966). Graduate Student. Graduate Group in Biophysics, University of California, San Francisco, San Francisco, CA 94143-0448, USA. [M.S. #27, 460 Point San Bruno Blvd, South San Francisco, CA 94080, USA.] BS (Massachusetts Institute of Technology, 1988) electrical engineering. *Proteins neutron_crystallography structure_determination hormones_estrogen cancer_breast womens_health*
E-mail gamble@genie.com Tel. 1(415)2251172 Fax 1(415)2253734

Ganesh, Dr Venkatapathy (1961). Research Associate. Department of Chemistry, Michigan State University, East Lansing, MI 48824, USA. PhD (Indian Inst. Tech., Madras, India, 1989) physics. *Macromolecular_crystallography computer_modelling drug_design knowledge_based_design homology_prediction*
E-mail ganesh@cemvax.cem.msu.edu ganesh@msucem.bitnet Tel. 1(517)3559715x251 Fax 1(517)3531793

Gangloff, Andy (1960). Asst. Vice President, Service Manager. Enraf-Nonius, 390 Central Ave., Bohemia, NY 11716, USA.
E-mail sangloff@delftny.com Tel. 1(516)5892885 X13 Fax 1(516)5892068

Ganguli, Subrata (1963). Graduate Student. Dept. of Micro/Immuno, University of Illinois at Chicago, E-703 MSB, M/C 790, 901 South Wolcott Av., Chicago, IL 60612-7344, USA. MTech (Indian Institute of Technology, 1987) biochem. eng. *Protein structure function*
E-mail u25128@uicvm.uic.edu Tel. 1(312)9962314

Gantzel, Dr Peter Kellogg (1934). Volunteer, UCSD. 8308 Paseo del Ocaso, La Jolla, CA 92037, USA. PhD (UCLA, 1962) physical chemistry. *Disorder twinning phase_transition small_molecules thermal_parameters*
E-mail pgantzel@ucsd.edu Tel. 1(619)4596440 Fax 1(619)5344671

Gao, Dr Yan (1954). Chemist. GE CRD, PO Box 8, K-1, 2C35, Schenectady, NY 12301, USA. PhD (SUNY Buffalo, 1990) physical chemistry. *Diffraction_technique single_crystal powder synchrotron_radiation applied_crystallography anomalous_scattering modulated_structures*
E-mail gao@crd.ge.com Tel. 1(518)3876249 Fax 1(518)3876972

Garavito, Dr R. Michael (1952). Assistant Professor. Department of Biochemistry and Molecular Biology, The University of Chicago, 920 East 58th Street, Chicago, IL 60637, USA. PhD (Purdue U., 1978) biochemistry/biophysics. *Macromolecular_crystallography proteins catalysis membranes carcinogenesis free_radicals computer_modelling electron_microscopy*
E-mail garavito@biovax.uchicago.edu Tel. 1(312)7029481 Fax 1(312)7020439

Garbauskas, Dr Mary F. (1953). Research staff. GE Corporate Research and Development, PO Box 8 K-1 2C35, Schenectady, NY 12301, USA. PhD (U. California Los Angles, 1979) physical chemistry. *Plastics X-ray_characterization grazing_incidence SAXS WAXS X-ray_fluorescence X-ray_powder_diffraction polymers superconductors service_crystallography*
E-mail garbauskas@crd.ge.com Tel. 1(518)3875797 Fax 1(518)3876945

Gardner, Dr Kenn Corwin H. (1947). Research Associate. DuPont CR&D, Experimental Station, Box 80356, Wilmington, DE 19880-0356, USA. PhD (Case Western Reserve U., 1974) macromolecular science. *Fiber_diffraction polymers biopolymers*
E-mail gardner@esvax.dnet.dupont.com Tel. 1(302)6952408 Fax 1(302)6958207

Garman, Scott C. (1964). Graduate Student. Program on Higher Degrees in Biophysics, Harvard University, Cambridge, MA 02139, USA. [Department of Biochemistry and Molecular Biology, Box 142, 7 Divinity Avenue, Cambridge, MA 02139, USA.] AB (Princeton U., 1987) chemistry. *Proteins twinning low_temperature synchrotron_radiation parasites glycoproteins structure_disorder*
E-mail garman@xtal0.harvard.edu garman@huxtal.bitnet Tel. 1(617)4954091 Fax 1(617)4959613

Garvey, Roy G. (1941). Associate Professor. North Dakota State University, Department of Chemistry, Fargo, ND 58105-5516, USA. PhD (U. Utah, 1966) inorganic chemistry. *Transition_metal coordination_complex computer_models ligation oxovanadium*
E-mail nu25304@ndsuvm1 Tel. 1(701)2318697

Gattl, Domenico L. (1958). Postdoctoral Fellow. Biophysics Research Division, University of Michigan, Ann Arbor, MI 48109, USA. PhD (U. Bari, Italy, 1987) biochemistry.
E-mail mimo@norway.biop.umich.edu Tel. 1(313)7632199

Gelb, Dr Steven J. (1959). Research Assistant Professor. Department of Chemistry, University of Pittsburgh, Pittsburgh, PA 15260, USA. PhD (U. Delaware, 1989) analytical chemistry. *Small_molecule_structure_determination organometallic_organic_hydrogen_bonding molecular_recognition*
E-mail geib@vms.cis.pitt.edu Tel. 1(412)6241131

Geiger, Dr James H. (1962). Postdoctoral fellow. Department of MB&B, Yale University, New Haven, CT 06510, USA. PhD (Princeton U., 1989) physical organic chemistry. *Proteins transcription_initiation stereochemistry steric_hindrance unnatural_amino_acids*
E-mail geiger@csbarg.med.yale.edu Tel. 1(203)7374433 Fax 1(203)7826940

Geil, Phillip H. (1930). Professor of Polymer Science and Engineering. Polymer Div. Materials Sci. and Eng., University of Illinois, 1304 W. Green St., Urbana, IL 61801, USA. PhD (U. Wisconsin, 1957) physics. *Electron_diffraction polymers morphology liquid_crystal_polymers fiber_structure processing_morphology_property_relationships*
Tel. 1(217)3330149 Fax 1(217)3332736

Geiser, Dr Urs (1956). Chemist. Chemistry and Materials Science Divisions, Argonne National Laboratory, Argonne, IL 60439, USA. PhD (Washington State U., 1985) chemistry. *Superconductors superconductivity spin_resonance structure_determination organic_superconductors molecular_metals fullerenes*
E-mail geiser@anchx4.chm.anl.gov geiser@anlchm.bitnet Tel. 1(708)2523509 Fax 1(708)2524470

George, Dr Clifford F. (1941). Research Physicist. Code 6030, Naval Research Laboratory, Washington, DC 20375, USA. PhD (Catholic U., 1978) physics. *Energetic_materials neuropeptides density_prediction*
E-mail george@lsmnic.nrl.navy.mil Tel. 1(202)7673463 Fax 1(202)7676874

Georgiadis, Millie M. (1962). Assistant Professor. Waksman Institute, Department of Chemistry, Rutgers University, Hoes Lane, Piscataway, NJ 08855, USA. PhD (U. California, Los Angeles, 1990) biochemistry.
E-mail georgiadis@mbcl.rutgers.edu Tel. 1(908)4454643 1(908)4454456 Fax 1(908)4455753

Getzoff, Dr Elizabeth D. (1954). Associate Member. Molecular Biology Department MB4, The Scripps Research Institute, 10666 N. Torrey Pines Road, La Jolla, CA 92037, USA. PhD (Duke U., 1982) biochemistry. *Conformational_change computer_modelling structure_determination time-resolved_effect protein_crystallography protein_design protein_structure_analysis metalloproteins*
E-mail edg@scripps.edu Tel. 1(619)5542878 Fax 1(619)5546880

Ghose, Subrata (1932). Professor of Mineral Physics. Dept. Geological Sciences, University of Washington, Seattle, WA 98195, USA. PhD (U. Chicago, 1959) mineralogy-crystallography. *Mineralogy_crystallography lattice-dynamics phase_transitions high-temperature high-pressure_silicates*
E-mail ghose@ghose.geology.washington.edu Tel. 1(206)5437378 Fax 1(206)5433836

Ghosh, Dr Debashis (1952). Senior Research Scientist. Molecular Biophysics, Hauptman-Woodward Medical Research Institute (formerly the Medical Foundation of Buffalo), 73 High Street, Buffalo, NY 14203-1153, USA. PhD (U. Pittsburgh, 1981) crystallography. *Proteins enzymes crystallography structure-activity_relationship short-chain_dehydrogenases esterases kinases protein_folding methods*
E-mail ghosh@hwi.buffalo.edu Tel. 1(716)8569600 Fax 1(716)8524846

Ghosh, Partho (1962). Postdoctoral Fellow. Department of Biochemistry, Harvard University, 7 Divinity Avenue, Cambridge, MA 02138, USA. PhD (UCSF, 1992) biochemistry and biophysics. *Immunology conformational_change histocompatibility low_temperature signal_transduction transmembrane ion_channels*
E-mail ghosh@xtal220.harvard.edu Tel. 1(617)4954091 Fax 1(617)4959631

Ghosh, Dr Sutapa (1959). Postdoctoral Researcher. The Wistar Institute, 3601 Spruce Street, Philadelphia, PA 19104, USA. PhD (Calcutta U., 1991) physics. *Macromolecular_crystallography*
E-mail ghosh@wistb.wistar.upenn.edu Tel. 1(215)8982202 Fax 1(215)8983868

Gibbs, Doon Physicist. Physics Department, Brookhaven Nat'l Lab., Upton, NY 11973, USA. PhD (U. Illinois-CU, 1983) physics. *Surfaces magnetism phase_transition*
E-mail doon@solids.bnl.gov Tel. 1(516)2824608 Fax 1(516)2822739

Giese Jr, Rossman F. (1936). Professor. 2960 Bowen Road, Elma, NY 14059-9441, USA. PhD (Columbia U., 1962) geology. *Mineralogy surface_properties_minerals*
E-mail glgclay@ubvms.cc.buffalo.edu Fax 1(716)6453999

Gilardi, Dr Richard D. (1940). Research Chemist. Laboratory for the Structure of Matter, Naval Research Lab., Washington, DC 20375-5341, USA. PhD (Univ Maryland, 1966) physical chemistry. *Energetic_compounds electrostatics packing_analysis density_prediction cubanes heterocycles nitrogen_compounds*
E-mail gilardi@lsmnic.nrl.navy.mil Tel. 1(202)7673463 Fax 1(202)7676874

Gilbert, William A. (1953). Research Associate Professor. Department of Biochemistry and Molecular Biology, Spaulding Life Sciences Building, University of New Hampshire, Durham, New Hampshire 03824-3544, USA. PhD (U. Florida, 1978) biochemistry. *Software network Macintosh*
E-mail gilbert@unh.edu Tel. 1(603)8622958 Fax 1(603)4271327

Gilfrich, John V. (1927). Consultant. 8710 Lowell Street, Bethesda, MD 20817-3218, USA. BA (American International College, 1949) chemistry. *X-ray_spectrometry crystal_properties synchrotron_radiation*
Tel. 1(301)3655070 Fax 1(301)3655070

Gilje, John W. (1939). Professor. Chemistry Dept., Univ. of Hawaii, 2545 The Mall, Honolulu, HI 96822-2275, USA. PhD (U. Michigan, 1965) inorganic chemistry.
E-mail gilje@hawaii.edu Tel. 1(808)9565733 Fax 1(808)9565908

Gilliland, Dr Gary L. (1948). Associate Director. CARB (NIST/UMBI), 9600 Gudelsky Dr, Rockville, MD 20850, USA. PhD (Rice U., 1979) biochemistry. *Protein crystallization database enzymes hemoglobin allostery water proteases*
E-mail gary@ibm3.carb.nist.gov Tel. 1(301)7386262 Fax 1(301)7386255

Ginell, Dr Stephan L. (1949). Scientist. Structural Biology Center, Biological and Medical Research Division, Argonne National Laboratory, 9700 Cass Ave, Argonne, IL 60439-4833, USA. [Argonne National Laboratory, c/o Biology Dept., Bldg 463, Brookhaven National Laboratory, Upton, NY 11973, USA.] PhD (Roswell Park Memorial Inst., SUNY at Buffalo, 1980) biophysics. *DNA RNA proteins computer_modelling water_structure synchrotron_radiation low_temperature_crystallography radiation_damage*
E-mail ginell@bnlsbc.nsls.bnl.gov Tel. 1(516)2823018 Fax 1(516)2827956

Giranda, Dr Vincent L. (1960). Senior Research Scientist. Sterling Winthrop Inc., 1250 S. Collegeville Rd., Collegeville, PA 19426-0900, USA. MD/PhD (Temple U., 1987) biochemistry.
E-mail giranda@kodak.com Tel. 1(610)9835869 Fax 1(610)9835559

Glaeser, Dr Robert M. (1937). Professor. Department of Molecular and Cell Biology, Stanley/Donner ASU, University of California, Berkeley, CA 94720, USA. PhD (U. California, Berkeley, 1964) biophysics. *Dynamical_diffraction electron_microscopy high-energy_electron_diffraction high-resolution_electron_microscopy*
E-mail rmglaeser@lbl.bitnet Tel. 1(510)6422905 Fax 1(510)4866488

Glass, Howard L. (1942). Production manager, photonics. Johnson Mattmey Electronics, East 15128 Euclid Avenue, Spokane, Washington 99216, USA. PhD (Rutgers U., 1969) physics. *Crystal growth epitaxy*
Tel. 1(509)9242200 Fax 1(509)9248617

Glinka, Charles Joseph (1947). Physicist. Reactor Div., National Bureau of Standards, Gaithersburg, MD 20899, USA. PhD (U. Maryland, 1975) solid state physics. *Neutron_small-angle_scattering slow_neutron porosity morphology*
E-mail glinka@enh.nist.gov Tel. 1(301)9756242 Fax 1(301)9219847

Glucksman, Dr Marc J. (1956). Assistant Professor. BOX 1065- Fishberg Research Center in Neurobiology, Mount Sinai Medical Center, 1 Gustave Levy Place, New York, NY 10029, USA. PhD (Columbia U., 1989) biochemistry and molecular biophysics. *Enzyme_substrate proteins computer_modelling drug_design neurobiology_receptors homology_prediction spectroscopy*
E-mail glux@msvax.mssm.edu glux@msvax.mssm.bitnet image@cuccfa.columbia.edu Tel. 1(212)2419233 Fax 1(212)9969785

Glusker, Dr Jenny P. (1931). Senior Member. Editor, Acta Crystallographica D; IUCr–OUP Book Series Committee. The Institute for Cancer Research, The Fox Chase Cancer Center, 7701 Burholme Avenue, Philadelphia, PA 19111, USA. DPhil, DSc (Oxford U., and College of Wooster, Ohio, 1957, and 1985) chemistry. *Protein_structure carcinogen_antitumour agents carcinogen-nucleic_acid_interactions intermolecular_interactions crystallography vitamin B12*
E-mail jp_glusker@fccc.edu Tel. 1(215)7282220 Fax 1(215)7282863

Godden, Dr Jeff W. (1957). Senior Fellow. Biological Structure SM-20, University of Washington, Seattle, WA 98195, USA. PhD (U. Washington, 1992) biochemistry. *Metalloenzymes proteins computer_modelling electron_transfer numerical_methods copper_enzymes bee_keeping*
E-mail godden@u.washington.edu Tel. 1(206)5438865 Fax 1(206)5431524

Goldberg, Dr Stephen Z. (1947). Professor. Department of Chemistry, Adelphi Univ., Box 701, Garden City, NY 11530, USA. PhD (U. California (Berkeley), 1973) inorganic chemistry. *Archaeology structure_determination bioinorganic_structure computer-aided_education art_conservation chemistry computing MO_calculations philosophy_of_science*
E-mail goldberg@sable.adelphi.edu goldberg@adlibv.adelphi.edu Tel. 1(516)8774147 Fax 1(516)8774191

Golden, Barbara L. (1967). Graduate Student. Department of Microbiology, Duke University Medical Center, Box 3020, Durham, North Carolina 27710, USA. BS (U. Chicago, 1989) chemistry.
E-mail barb@whitesws.mc.duke.edu Tel. 1(919)6843864 Fax 1(919)6848735

Goldish, Elihu (1928). Adjunct Prof. Calif. State University Long Beach, Department of Geology, Long Beach, CA 90840, USA. PhD (Calif. Inst. Tech., 1956) chemistry. *X-ray_powder_diffraction physical_chemistry*

Goldsmith, Elizabeth J. (1945). Associate Professor. Department of Biochemistry, The University of Texas Southwestern Medical Center at Dallas, 5323 Harry Hines Blvd., Dallas, TX 75235-9050, USA. PhD (UCLA, 1972) physical chemistry. *Proteins regulation protein kinases protease_inhibitors conformational_regulation_of_protein_molecules*
E-mail betsy@howie.swmed.edu Tel. 1(214)6485009 Fax 1(214)6486336

Goldsmith, Julian R. (1918). Chas E. Merriam Distinguished Service Professor Emeritus. Geophysical Sci. Dept., University of Chicago, 5734 Ellis Ave., Chicago, IL 60637, USA. [5631 Blackstone, Chicago, IL 60637, USA.] PhD (U. Chicago, 1947) geochemistry.

Goldstein, Dr Barry M. (1952). Associate Professor. Department of Biophysics, University of Rochester Medical Center, 601 Elmwood Avenue, Rochester, NY 14642, USA. PhD MD (U. Rochester, 1982) biophysics. *Dehydrogenases enzyme_inhibitors antitumour_compounds non-bonded_interaction MO_calculation computer-assisted_design nuclear_magnetic_resonance conformation*
E-mail bmg@bphvax.biophysics.rochester.edu Tel. 1(716)2755095 Fax 1(716)2756007

Goldstone, Joyce A. (1949). Deputy Center Director. LANSCE MS H805, Los Alamos National Laboratory, Los Alamos, NM 87545, USA. PhD (State U. of New York at Stony Brook, 1978) materials science. *Actinide_structures*
E-mail jag@lanl.gov Tel. 1(505)6673629 Fax 1(505)6652676

Gong, Dr Minfang (1965). Postdoctor. Dana Farber Cancer Institute, Rm.1040, 44 Binney Street, Boston, MA 02115, USA. (1992) physics. *Structure protein-structure DNA-structure complex-structure anti-cancer_drug_structural_study*
E-mail gong@amber.dfci.harvard.edu Tel. 1(617)6324754 Fax 1(617)6324393

Goodsell, Dr David S. (1961). Assistant Researcher. Molecular Biology Institute, University of California, Los Angeles, CA 90024, USA. PhD (U. California at Los Angeles, 1987) biochemistry. *Macromolecular_interaction nucleic_acids computer_graphics computer_assisted_design drug_design biochemical_education*
E-mail goodsell@uclaue.mbi.ucla.edu Tel. 1(310)2068278 Fax 1(310)8250982

Goonesekere, Nalin C. W. Graduate Student. Department of Chemistry, Princeton University, Princeton, NJ 08544, USA. MSc (Princeton U., 1990) biophysics. *Proteins signal_transduction cell_motility drug_design computer_modelling molecular_dynamics international_affairs education*
E-mail nalin_goones@chemvax.princeton.edu Tel. 1(609)2582826 Fax 1(609)2586746

Gouaux, J. Eric (1961). Assistant Professor. 920 East 58th St., University of Chicago, Chicago, IL 60637, USA. PhD (Harvard U., 1989) chemistry. *Membrane_proteins ion_channels signal_transduction protein_engineering novel_detergents crystallization_strategies gene_synthesis*
E-mail gouaux@biovax.uchicago.edu Tel. 1(312)7026490 Fax 1(312)7020439

Gougoutas, Dr Jack Zanos (1939). Sr. Research Fellow. Bristol Myers Squibb, Pharmaceutical Research Institute, PO Box 4000, Princeton, NJ 08543-4000, USA. PhD (Harvard U., 1963) chemistry. *Solid_state_reaction topotaxy organic_compounds structure_properties_relationships isoteres*
Tel. 1(609)2524562 Fax 1(609)2526012

Grant, Robert (1955). Postdoctoral Fellow. Dept. Biological Chemistry and Molecular Pharmacology, Harvard Medical School, 240 Longwood Ave., Boston, MA 02115, USA. PhD (U. Arizona, 1987) biochemistry and molecular biology. *X-ray_crystallography electron_microscopy virus_structure proteins drug_design*
E-mail grant@ganesh.harvard.med.edu Tel. 1(617)4323919

Graves, Bradford J. (1954). Research Investigator. Department of Physical Chemistry, Hoffmann-LaRoche Inc., Nutley, NJ 07110, USA. PhD (U. North Carolina, 1980) chemistry. *Proteins computer_modelling anomalous_dispersion homology_modelling protein_drug_interaction*
E-mail graves@rmch01.dnet.roche.com Tel. 1(201)2355815 Fax 1(201)2352682

Gray, Dr Terry M. (1958). Associate Professor. Department of Chemistry, Calvin College, 3201 Burton Street SE, Grand Rapids, MI 49546, USA. PhD (U. Oregon, 1985) molecular biology. *Proteins T4_lysozyme computer_modelling protein_stability molecular_dynamics_simulations protein_folding mutational_analysis*
E-mail grayt@calvin.edu Tel. 1(616)9577187 Fax 1(616)9576501

Greenberg, Dr Berton L. (1940). Senior Member of Tech. Staff. Philips Laboratories, 345 Scarborough Road, Briarcliff Manor, NY 10510, USA. PhD (Stevens Inst. of Tech., 1979) materials science. *Powder_diffraction epitaxial_structures quantitative_phase_analysis*
E-mail blg@philabs.philips.com Tel. 1(914)9456074 Fax 1(914)9456375

Gress, Mary E. (1946). Program Manager, Photochemical and Radiation Sciences. Chemical Sciences Division, ER-141 GTN, Department of Energy, Washington, DC 20585, USA. PhD (Iowa State U., 1973) physical chemistry. *X-ray neutron small_molecule_crystallography photochemistry radiation_chemistry*
Tel. 1(301)9035820 Fax 1(301)9034110

Gribskov, Dr Michael (1958). Principal Scientist, UCSD Adjunct Asst. Professor of Biology. San Diego Supercomputer Center, PO Box 85608, San Diego, CA 92186-9784, USA. [(Courier packages only), San Diego Supercomputer Center, 10100 Hopkins Dr, La Jolla, CA 92093-0505, USA.] PhD (U. Wisconsin - Madison, 1985) molecular biology. *Protein_structure protein_evolution macromolecular_sequence_analysis homology_modelling database*
E-mail gribskov@sdsc.edu Tel. 1(619)5348312 Fax 1(619)5345117

Griffen, Dr Dana T. (1943). Professor; Chair, Department of Geology, Brigham Young University. Department of Geology, Brigham Young University, Provo, UT 84602, USA. PhD (Virginia Tech, 1975) mineralogy. *Single_crystal_diffraction mineralogy silicates electron_microscopy Mossbauer_spectroscopy phase_transition EDS_analysis*
E-mail dtg@geology.byu.edu Tel. 1(801)3782305 Fax 1(801)3788143

Griffin, Dr Jane F. (1933). Head, Molecular Biophysics Department and Associate Research Director. Molecular Biophysics Department, Hauptman-Woodward Medical Research Institute (formerly the Medical Foundation of Buffalo), 73 High Street, Buffalo, New York 14203-1196, USA. PhD (State U. New York at Buffalo, 1974) physical chemistry. *Biomolecule_structure–activity_relationship_steroids_hormones structure–activity_relationship_opiates_peptides_glycosides absolute_configuration charge_density*
E-mail griffin@hwi.buffalo Tel. 1(716)8569600 Fax 1(716)8524846

Griffith, Dr Diana L. (1957). Research Scientist. Structural Biology, Brandeis University, Rosenstiel Center, Waltham, MA 02254, USA. PhD (Brandeis, 1990) biophysics. *Protein_crystallization macromolecular_crystallography protein_structure_function*
E-mail diana@auriga.rose.brandeis.edu Tel. 1(617)7364937 Fax 1(617)7362405

Griffith, Elizabeth H. (1935). Assistant Department Chairman/Associate Professor Research. Department of Chemistry and Biochemistry, University of South Carolina, Columbia, SC 29208, USA. PhD (U. South Carolina, 1970) physical chemistry. *Structural_chemistry non-linear_optical_materials bioinorganic_model_systems undergraduate_instruction*
E-mail griffith@psc.psc.scarolina.edu Tel. 1(803)7772687 Fax 1(803)7779521

Gromek, Dr Jack M. (1953). Research Associate. Institute of Materials Science, University of Connecticut, Storrs, CT 06269, USA. PhD (U. Pennsylvania, 1983) physical inorganic chemistry. *Computer_modelling SAXS polymer_structure programming*
E-mail gromek@uconnvm.uconn.edu gromek@uconnvm.bitnet Tel. 1(203)4864622 Fax 1(203)4864622

Grosse, David A. (1954). Associate Professor. Department of Chemistry, Wright State University, Dayton OH 45435, USA. PhD (Texas Christian U., 1982) chemistry. *Macrocyclic_silver_compounds non-linear_optical properties computer_modelling computing_methods*
E-mail dgrossie@desire.wright.edu Tel. 1(513)8732210 Fax 1(513)8733301

Groy, Dr Thomas L. (1954). Associate Research Specialist. Department of Chemistry, Arizona State University, Tempe, AZ 85287-1604, USA. PhD (Arizona State U., 1982) physical chemistry. *Computer_modelling diffraction_theory phase_determination*
E-mail groy@asuchm.la.asu.edu Tel. 1(602)9651511 Fax 1(602)9652747

Gruner, Dr Sol M. (1950). Professor. Department of Physics, Princeton University, Princeton, NJ 08544, USA. PhD (Princeton U., 1977) physics. *X-ray_detectors liquid_crystals polymers lipids protein_motions materials_science physics_of_paracrystalline_materials*
E-mail gruner@pupgg.princeton.edu Tel. 1(609)2584334 Fax 1(609)4993512

Gschneidner Jr, Prof. Karl A. (1930). Distinguished Professor. Ames Laboratory, Iowa State University, Ames, IA 50011-3020, USA. PhD (Iowa State U., 1957) physical chemistry. *Rare-earth_materials magnetic_behavior alloy_theory prep_metals purification*
E-mail cagey@ameslab.gov Tel. 1(515)2947931 Fax 1(515)2943709 Telex 283359

Guan, Yue (1960). Postdoctoral Research Associate. Molecular Biology-MB4, The Scripps Research Institute, 10666 N. Torrey Pines Rd., La Jolla, CA 92037, USA. PhD (U. Illinois, 1994) biophysics. *Proteins protein_DNA_complex drug_design computer_modelling*
E-mail guan@scripps.edu Tel. 1(619)5542880 Fax 1(619)5546880

Guddat, Dr Luke M. Research Associate. Harrington Cancer Center, 1500 Wallace Blvd., Amarillo, TX 79106, USA. PhD (Melbourne U., 1989) protein crystallography. *Immune_system_proteins ligand_protein_interactions molecular_signalling*
E-mail luke@arthur.ama.ttuhsc.edu Tel. 1(806)3594673 1(806)3545875x277 Fax 1(806)3545887

Guggenheim, Stephen (1948). Professor of Geology. Dept. Geological Sciences-186, 845 West Taylor St., Chicago, IL 60607-7059, USA. PhD (U. Wisconsin-Madison, 1976) geology. *Mineralogy crystallography silicate_mineralogy X-ray_diffraction electron_diffraction phyllosilicates*
E-mail u10368@uicvm.uic.edu Tel. 1(312)9963263 Fax 1(312)4132279

Gulbis, Jacqueline M. (1961). Postdoc. Box 3, Laboratory of Molecular Biophysics, The Rockefeller University, 1230 York Avenue, New York, NY 10021, USA. PhD (La Trobe U., Australia, 1990) crystallog.of small biol. active molecules. *Proteins protein–DNA_interactions*
E-mail gulbisj@jkdec.rockefeller.edu Tel. 1(212)3277915

Guo, Dr Hwai-Chen (1960). CRI postdoctoral fellow. Department of Biochemistry and Molecular Biology, Harvard University, 7 Divinity Avenue, Cambridge, MA 02138, USA. PhD (Cornell U., 1990) biochemistry and molecular biology. *Macromolecular_complex receptor immunology computer_modelling proteins/nucleic_acids_interaction transcription_factors glycoproteins*
E-mail guo@xtal0.harvard.edu guo@huxtal.bitnet Tel. 1(617)4954091 Fax 1(617)4959613

Gurel, Ogan (1964). MD/PhD Trainee. Dept. of Biochem & Mol. Biophysics, Columbia University of Physicians and Surgeons, 630 West 168th St., New York, NY 10032, USA. MPhil (Columbia U., 1991) biochemistry & molecular biophysics. *Proteins growth_factors molecular_dynamics membrane_proteins*
E-mail gurel@cuhhca.hhmi.columbia.edu Tel. 1(212)3051951 Fax 1(212)3057379

Gursky, Dr Olga (1960). Postdoc. 104 Auburndale Ave., West Newton, MA 02165, USA. PhD (Brandeis U., 1991) physics. *Structural_disorder insulin allostery*
E-mail gursky@auriga.rose.brandeis.edu Tel. 1(617)7364940 Fax 1(617)7362405

Guthrie, Dr George D. (1962). Staff Scientist. Geology and Geochemistry, Los Alamos National Laboratory, Los Alamos, NM 87545, USA. PhD (Johns Hopkins U., 1989) mineralogy/crystallography. *Mineral_structures computer_modelling TEM XRD biological_effects catalysis surface_structure surface_chemistry*
E-mail GGUTHRIE@LANL.GOV Tel. 1(505)6656340 Fax 1(505)6653285

Hackert, Prof. Marvin L. (1944). Shive Prof. of Biochemistry. Dept. Chemistry and Biochemistry, University of Texas at Austin, Austin, TX 78712, USA. PhD (Iowa State U., 1970) physical chemistry. *Protein_crystallography PLP,Prv_enzymes hemoglobins multienzyme_complexes*
E-mail hackert@utbc01.cm.utexas.edu Tel. 1(512)4711105 Fax 1(512)4718696

Hadfield, Dr Andrea T. (1967). Postdoctoral Research Assistant. Structural Studies, Department of Biological Sciences, Lilly Hall of Life Sciences, Purdue University, West Lafayette, IN 47907, USA. DPhil (U. Oxford, 1992) protein crystallography. *Viruses proteins*
E-mail ah@mace.cc.purdue.edu Tel. 1(317)4944507 Fax 1(317)4961189

Haeffner, Dean R. (1960). Assistant Physicist. Advanced Photon Source, Argonne National Laboratory, Argonne, IL 60439, USA. PhD (Northwestern U., 1991) materials science. *materials* E-mail dean_haeffner@qmgate.anl.gov haeffner@anlaps.aps.anl.gov Tel. 1(708)2524312 Fax 1(708)2523222

Haendler, Dr Helmut M. (1913). Professor Emeritus. Department of Chemistry, University of New Hampshire, Durham, NH 03824-3598, USA. [309 Lee Hook Road, Lee, NH 03824-6417, USA.] PhD (The University of Washington, 1940) inorganic chemistry. *Metal-organic_complexes*
E-mail h_haendler@unhh.unh.edu Tel. 1(603)6593942

Haggerty, Brian S. (1966). Graduate Student (PhD)/Staff Crystallographer at Univ. of Delaware. 142 E. St. Marks Place, Valley Stream, NY 11580, USA. BS (Washington and Lee U., 1988) chemistry.
E-mail bhaggerty@wotan.duch.udel.edu Tel. 1(516)5611245

Hagler, A. T. (1941). Chief Scientific Officer/Founder. Biosym Technologies, Inc., 9685 Scranton Road, San Diego, CA 92121, USA. PhD (Cornell U., 1970) biophysical chemistry. *Molecular_modelling drug_design*
Tel. 1(619)5465514 Fax 1(619)4505041

Haile, Prof. Sossina (1966). Assistant Professor. Dept. of Materials Science and Engineering, FB-10, University of Washington, Seattle, WA 98195, USA. PhD (MIT, 1992) MSE. *Rietveld*
E-mail smhaile@u.washington.edu Tel. 1(206)6857869 Fax 1(206)5433100

Hall, Dr Michael D. (1963). Postdoctoral Fellow. Department of Biochemistry, University of Minnesota, 4-225 Millard Hall, 435 Delaware St. SE, Minneapolis, MN 55455, USA. PhD (Washington U., 1993) biochemistry. *Proteins*
E-mail hall@dccc.med.umn.edu Tel. 1(612)6252115 Fax 1(612)6252163

Haltiwanger, R. Curtis (1947). Investigator. SmithKline Beecham Pharmaceuticals, 709 Swedeland Road, King of Prussia, PA 19406, USA. [SmithKline Beecham Pharmaceuticals, Dept UW 2950, PO Box 1539, King of Prussia, PA 19406, USA.] MS (U. Virginia, 1972) chemistry. *Absolute_configuration area_detectors crystal_growth hydrogen_bonding*
E-mail haltiwangerc@sb.com Tel. 1(610)2704097 Fax 1(610)2706996

Hamill, Dr Gregory P. (1949). Manager, technical applications XRD. Rigaku USA, Inc., 199 Rosewood Drive, Suite 200, Danvers, MA 01923, USA. PhD (Calif. Inst. of Tech., 1977) applied physics/materials science. *Materials_science X-ray_diffraction_physics instrumentation analytical_sciences technical_software_development*
Tel. 1(508)7772446 Fax 1(508)7773594

Hamilton, Jean A. (1938). Professor of Biochemistry and Molecular Biology. Biochemistry and Molecular Biology Dept., Indiana U. Sch. of Medicine, 635 Barnhill Dr., Indianapolis, IN 46223, USA. PhD (Glasgow U., 1962) chemistry. *Crystallography proteins biological_complexes structure/function_of_proteins*
E-mail lks@biochem1.iupui.edu Tel. 1(317)2747403 Fax 1(317)2744686

Hamlin, Ronald C. (1946). President. Area Detector Systems Corp., 12550 Stowe Drive, Poway, CA 92064-6804, USA. PhD (U. California, 1975) physics. *Area_detectors instrumentation macromolecular small_molecule*
E-mail adsc@netcom.com Tel. 1(619)4860444 Fax 1(619)4860722

Hammack, Professor William S. (1961). Associate Professor. Department of Chemical Engineering, Carnegie Mellon University, Pittsburgh, PA 15213-3890, USA. PhD (U. Illinois-Urbana, 1988) chemical engineering. *High_pressure_crystallography pressure-induced_amorphization_disordering*
E-mail Hammack@cmu.edu Tel. 1(412)2682227 Fax 1(412)2687139

Han, Dr Fusen (1946). Senior Research Scientist. Physical and Analytical Chemistry, 301 Henrietta Street, The Upjohn Company, Kalamazoo, MI 49001, USA. PhD (Chinese Academy of Sciences, 1983) physical chemistry. *Structure_small_molecule structure_methods computer_program computer_program*
Tel. 1(616)3855271 Fax 1(616)3857522

Han, Dr Gye Won (1959). Postdoc. Department of Chemistry, University of California, San Diego, La Jolla, CA 92093-0317, USA. PhD (U. Pittsburgh, 1993) crystallography. *Crystallization structure_determination protein_structure hemoproteins lipid_structure thermal_vibration*
E-mail gwhan@chem.ucsd.edu Tel. 1(619)5342011 Fax 1(619)5346128

Han, Shaoxu (1955). PhD student. Dept of Geophysical Sciences, University of Chicago, 5734 S. Ellis, Chicago, IL 60637, USA. MS (China University of Geosciences, 1982) crystallography. *Crystallography zeolites mineral_and_inorganic_structures X-ray_diffraction*
E-mail sh14@midway.uchicago.edu Tel. 1(312)7028109 Fax 1(312)7029505

Hannick, Linda I. (1947). Scientist-contractor. Laboratory for the Structure of Matter, Code 6030, Naval Research Laboratory, Washington, DC 20375, USA. PhD (U. New Orleans, 1986) physical chemistry. *Biocrystallography accurate_area_detector_data luminescence mutagenesis synchrotron_radiation low-temperature structure-activity_relationships chelation*
E-mail hannick@lsm.nrl.navy.mil Tel. 1(202)7670657 1(202)7670656 Fax 1(202)7676874

Hanson, Jonathan C. (1941). Senior Associate Chemist Local contact NSLS Beam line X7B. Chemistry Department, Bldg 555, PO Box 5000, Upton, NY 11973-5000, USA. PhD (U. of Michigan, 1969) chemistry. *Synchrotron zeolites biominerals*
E-mail Hanson1@bnl.gov Tel. 1(516)2824378 1(516)2825707 Fax 1(516)2825815

Hanson, Leif (1953). Graduate Student–PhD candidate. Harrington Cancer Center, 1500 Wallace Blvd, Amarillo, TX 79106, USA. MS (U. Utah, 1988) biology. *Biology kinetics phylogeny immunology entomology biological_systematics*
E-mail leif@arthur.ama.ttuhsc.edu Tel. 1(806)3545875x278 Fax 1(806)3545887

Hardgrove, Dr George L. Jr. (1933). Professor of Chemistry. Department of Chemistry, St. Olaf College, Northfield, MN 55057, USA. PhD (U. California Berkeley, 1959) physical chemistry. *Small-molecule_crystallography NMR_spectroscopy*
E-mail hardgrov@stolaf.edu Tel. 1(507)6463659 Fax 1(507)6463968

Hardy, Dr Larry W. (1954). Associate Professor. Department of Pharmacology, University of Massachusetts Medical Center, 55 Lake Avenue North, Worcester, MA 01655, USA. [Program in Molecular Medicine, Biotech 2, 373 Plantation Street, Worcester, MA 01605, USA.] PhD (U. California, Berkeley, 1983) biochemistry. *Enzymatic_kinetics enzymatic_active_site antifolates enzyme_inhibitors phosphoprotein_phosphatases DNA_repair protein_folding kinetic_isotope_effects*

E-mail lwhardy@umassmed.ummed.edu Tel. 1(508)8564900 or 1(508)8566744 Fax 1(508)8564289

Hargrave, Karl D. Associate director. Dept of Medical Chemistry, Boehringer Ingelheim Pharmaceuticals, Inc, 900 Ridgebury Road, Ridgefield, CT 06877, USA. PhD (Duke U., 1977) chemistry. *Chemistry immune_regulation computer-assisted_design pharmaceuticals*
Tel. 1(203)7985136 Fax 1(203)7916072

Harlow, Richard L. (1942). Senior Research Associate. Central Research and Development, E228/316d, E. I. DuPont de Nemours & Co., Inc., Wilmington, DE 19880-0228, USA. [7 Shull Dr., Newark, DE 19711, USA.] PhD (Syracuse U., 1971) chemistry. *Synchrotron_radiation diffraction catalysts condensed_matter R_Harlow_Foundation_for_Disabused_Crystallographers*
E-mail harlow@esvax.dnet.dupont.com Tel. 1(302)6952097 Fax 1(302)6951351

Harp, Joel M. (1950). Research Associate. Univ. of Tennessee, Biology Div., Oak Ridge National Lab., Oak Ridge, TN 37831-8077, USA. *Biocrystallography biophysics proteins nucleic_acids*
E-mail harp@biovx1.bio.ornl.gov Tel. 1(615)5741210 Fax 1(615)5741274

Harris, Lisa J. (1968). PhD Graduate Student. Department of Biochemistry, University of California, Riverside, Riverside, CA 92521, USA. BS (Baylor U., 1990) biology. *Antibodies proteins molecular_replacement conformational_change antibody_conformations complement_system*
E-mail lisajo@ucrac1.ucr.edu Tel. 1(909)7873397 Fax 1(909)7873790

Harrison, Robert W. (1957). Assistant professor. Department of Pharmacology, Thomas Jefferson University, 233 South 10th Street, Philadelphia, PA 19107, USA. PhD (Yale U., 1985) molecular biophysics and biochemistry. *Drug_design computer_modelling structure-activity_relationships direct_methods algorithm_design protein_folding statistical_mechanics parallel_computing*
E-mail harrison@asterix.jci.tju.edu Tel. 1(215)9554592 Fax 1(215)9232117

Harrison, Prof. Stephen C. (1943). Investigator, HHMI; Professor of Biochemistry, Harvard. Howard Hughes Medical Institute, Harvard University, Fairchild Biochemistry Bldg., Cambridge, MA 02138, USA. PhD (Harvard U., 1968) biophysics. *Viruses protein/DNA_interactions macromolecular_assemblies large-unit-cell_methods*
Tel. 1(617)4954090 Fax 1(617)4959613

Harrison, Dr William T. A. (1960). Assistant research professor. Department of Chemistry, University of Houston, 4800 Calhoun Street, Houston, TX 77204-5641, USA. PhD (Oxford U., England, 1986) inorganic chemistry. *Powder_diffraction neutron_diffraction computing zeolites computer_typesetting_graphics*
E-mail harrison@madmax.chem.uh.edu harrison@sbxray.ucsb.edu Tel. 1(713)7432789 Fax 1(713)7432787

Hartsuck, Dr Jean A. (1939). Associate Member. Oklahoma Medical Research Foundation, 825 N.E. 13th Street, Oklahoma City, OK 73104, USA. PhD (Harvard U., 1964) physical chemistry. *Biocrystallography enzymes kinetics zymogen aspartic_proteinases*
E-mail Hartsuck@omrf.omrf.uokhsc.edu Tel. 1(405)2717293 Fax 1(405)2717249

Hasemann, Dr Charles A. (1961). Associate. Howard Hughes Medical Institute, University of Texas Southwestern Medical Center, 5323 Harry Hines Blvd., Dallas, TX 75235-9050, USA. PhD (U. Texas Southwestern Medical Center, 1990) immunology. *Structure-determination proteins immunology structure-activity_relationship hemoproteins synchrotron_radiation*
E-mail hasemann@howie.swmed.edu Tel. 1(214)6485058 Fax 1(214)6485095

Hassell, Anne M. (1946). Senior Scientist. Glaxo, Inc., Dept. of Bioanalytical & Structural Chemistry, 5 Moore Drive, Research Triangle Park, NC 27709, USA. MS (U. Georgia, 1972) microbiology. *Protein_crystallography protein_crystallization*
E-mail amhl2478@usav01.glaxo.com Tel. 1(919)9413228 Fax 1(919)9413411

Hastings, Dr Jerome. Biller (1948). Senior Scientist. Nat. Synchrotron Light Source, Brookhaven National Lab., Bldg. 725D, Upton, NY 11973, USA. PhD (Cornell U., 1975) applied physics. *Synchrotron radiation diffraction applications X-ray_physics*
E-mail hastings@bnlc11.bitnet Tel. 1(516)2823930 Fax 1(516)2823238

Hatada, Dr Marcos H. (1955). Principal Research Scientist. ARIAD Pharmaceuticals, 26 Landsdowne St., Cambridge, MA 02139, USA. PhD (Michigan State U., 1982) physical chemistry. *Macromolecular_crystallography*
E-mail marcos@ariad.com Tel. 1(617)4940400x247

Hau, Dr Herbert (1941). Associate Clinical Professor. Restorative Dept., Tufts School of Dental Medicine, 1 Kneeland St., Boston, MA 02111, USA. [228 Harrison Ave., G/Floor, Tai-Tung Village, Boston, MA 02111, USA.] D. M. D./PhD (Harvard Univ of Dental Medicine/Boston U., 1977/1970) dental medicine.
Tel. 1(617)4235055

Hauptman, Dr Herbert A. (1917). President. Hauptman-Woodward Medical Research Institute (formerly the Medical Foundation of Buffalo), 73 High Street, Buffalo, NY 14203-1196, USA. PhD (U. Maryland, 1955) mathematics. *Direct_methods organic_crystal_structures*
E-mail hauptman@hwi.buffalo.edu Tel. 1(716)8569600 Fax 1(716)8524846

Head, Dr James F. (1947). Professor. Department of Physiology, Boston University School of Medicine, 80 E. Concord St., Boston, MA 02118, USA. PhD (Birmingham U., England, 1974) biochemistry. *Conformational_change proteins ligand_binding regulation calcium_compounds biocrystallography structure-activity_relationship SAXS*
E-mail jfh@medxtal.bu.edu Tel. 1(617)6384396 Fax 1(617)6384273

Hedman, Dr Britt G. M. (1949). Senior Research Associate. Member Commission on Synchrotron Radiation. Stanford Synchrotron Radiation Laboratory, Stanford University, SLAC, MS 69, PO Box 4349, Stanford, CA 94309, USA. PhD (Umeå U., Sweden, 1978) chemistry. *X-ray_absorption_spectroscopy bio-inorganic_chemistry polarized_XAS synchrotron_radiation active_sites ligand_spectroscopy electronic_structure protein_crystallography*

E-mail hedman@ssrl750.bitnet hedman@ssrl01.slac.stanford.edu Tel. 1(415)9263052 Fax 1(415)9264100 Telex STANFRD STNU 348402

Heeg, Mary Jane (1952). Staff Crystallographer. Department of Chemistry, Wayne State University, Detroit, Michigan 48202, USA. PhD (U. Cincinnati, 1978) inorganic chemistry. *Structural_inorganic_chemistry EXAFS Tc_Re_complexes*
E-mail mheeg@waynest1.bitnet Tel. 1(313)5772587 Fax 1(313)5771377

Hegde, Rashmi S. Assistant Professor. Skirball Institute for Biomolecular Medicine, New York Univ. Medical Center, 550 First Avenue, New York, NY 10016, USA. PhD (U. Pittsburgh, 1989) medicinal chemistry. *Transcriptional_regulation protein-DNA_interactions*
E-mail rashmi.hegde@mcska.med.nyu.edu Tel. 1(212)2637751 Fax 1(212)2638951

Heine, Dr Andreas (1963). Research Associate. The Scripps Research Institute, Department of Molecular Biology, 10666 North Torrey Pines Road, La Jolla, CA 92037, USA. PhD (Georg-August-U. Göttingen, Germany, 1993) inorganic chemistry. *Crystallization proteins low_temperature_data_collection*
E-mail aheine@scripps.edu Tel. 1(619)5549866 Fax 1(619)5546105

Henderson, Stephen (1956). Staff. PO Box 5862, Oak Ridge, TN 37831, USA. PhD (Victoria U. Wellington, 1986) physical chemistry. *Small_angle_scattering detector instrumentation*
E-mail hendersonj@bioax1.bio.ornl.gov Tel. 1(615)5740963 Fax 1(615)5741274

Hendricks, Dr Robert W. (1937). Professor of Materials Science and Engineering; Director, Residual Stress Laboratory; Director, X-ray diffraction Laboratory. Materials Science and Engineering Department, 127 Holden Hall, Virginia Polytechnic Institute and State University, Blacksburg, VA 24061-0237, USA. PhD (Cornell U., 1964) materials science and engineering. *X-ray_diffraction small-angle_scattering physical_metallurgy solid_state_physics residual_stress X-ray_detectors instrumentation diffuse_scattering*
E-mail robert.hendricks@vt.edu Tel. 1(703)2316917 Fax 1(703)2318919

Hendrixson, Dr Thomas L. (1960). Postdoctoral Research Associate. Whistler Center for Carbohydrate Research, 1160 Smith Hall, Purdue University, West Lafayette, IN 47907-1160, USA. PhD (Iowa State U., 1989) physical chemistry. *Patterson_method pattern_recognition computer_modelling computer_algorithms computer_graphics image_processing_theory antimony_compounds halides polysaccharides polymers*
E-mail thom@kiwi.foodsci.purdue.edu eoh@mace.cc.purdue.edu Tel. 1(317)4945248 Fax 1(317)4947953

Henley, Christopher Lee (1955). Associate Professor. Department of Physics, Cornell University, Clark Hall, Ithaca, NY 14853, USA. PhD (Harvard U., 1983) physics. *Quasicrystals*
E-mail cph@lassp.cornell.edu Tel. 1(607)2555056 Fax 1(607)2556428

Henling, Larry (1955). Chemist. Beckman Institute, California Institute of Technology, Pasadena, CA 91125-0001, USA. MS (Caltech, 1981) inorganic chemistry.
E-mail lmh@xray.caltech.edu Tel. 1(818)3952735 Fax 1(818)4494159

Henry, Rodger (1964). D-418, Building AP9, Abbott Laboratories, Abbott Park, IL 60064, USA. MS (Northern Illinois U., 1990) chemistry. *Crystallization hydrogen_bonding pharmacology polymorphism*
E-mail henryr@nmrb.abbott.com Tel. 1(607)9384246 Fax 1(708)9386470

Henslee, Dr Walter (1946). Technical Manager. Process Analyzer Resource Center, B-2020, Dow USA - Texas Operations, 2301 N. Brazosport Blvd., Freeport, TX 77541-3257, USA. PhD (U. Texas, 1974) organometallic single crystals. *Solid_solutions catalysts ceramics*
Tel. 1(409)2384531

Herbette, Leo G. (1953). Associate Professor; Director, Biomolecular Structure Analysis Center. Biomolecular Structure Analysis, University of Connecticut Health Center, Farmington, CT 06030, USA. PhD (U. Pennsylvania, 1980) biophysics. *Membranes drug_design computer_modelling ion_channels G-protein_coupled_receptors*
E-mail herbette@bsac.uchc.edu Tel. 1(203)6792951 Fax 1(203)6791989

Hermans, Jan (1933). Professor. Department of Biochemistry and Biophysics, University of North Carolina, Chapel Hill, NC 27599-7260, USA. PhD (U. Leiden, 1958) physical chemistry. *Peptide_conformation protein_conformation conformational_stability computer_simulation*
E-mail hermans@mcnc.org Tel. 1(919)9664644 Fax 1(919)9662852

Herron, Dr James N. (1954). Associate Professor. Department of Pharmaceutics, 421 Wakara Way, Suite 316, Salt Lake City, UT 84108, USA. PhD (U. Illinois at Urbana-Champaign, 1981) microbiology. *Proteins immunoglobulin_structure antigen-antibody_complexes molecular_graphics immunochemistry biosensors molecular_immunology*
E-mail herron@bioiris.med.utah.edu herron@uucc.cc.utah.edu Tel. 1(801)5817303 Fax 1(801)5817848

Herzberg, Dr Osnat (1949). Professor. Center for Advanced Research in Biotechnology, University of Maryland, 9600 Gudelsky Drive, Rockville, MD 20850, USA. PhD (Weizmann Institute of Science, 1982) chemistry. *Protein_structure_and_function protein_crystallography*
E-mail osnat@elan1.carb.nist.gov Tel. 1(301)7386245 Fax 1(301)7386255

Hester, Dr Gerko (1963). Postdoctoral research fellow. Department of Medicinal Chemistry, Medical College of Virginia/Virginia Commonwealth University, Box 540, Richmond, VA 23298, USA. PhD (Swiss Federal Institute of Technology Zurich Switzerland, 1993) X-ray crystallography. *Proteins lectins computer_modelling aldolases*
E-mail ghester@gems.vcu.edu ghester@vcuvax.bitnet Tel. 1(804)7861208

Higgins, Dr John B. (1947). Research Associate. Mobil Central Research, PO Box 1025, Princeton, NJ 08540, USA. PhD (Virginia Polytechnic Institute, 1978) geological science. *Zeolite_crystallography powder_diffraction microporous_materials tetrahedral_frameworks*
E-mail jbhiggins@crl.mobil.com Tel. 1(609)7374215 Fax 1(609)7375217

Hill, Dr Christopher P. (1958). Assistant Professor. Biochemistry Department, University of Utah, Salt Lake City, UT 84132, USA. DPhil (U. York, 1987) chemistry. *Protein_structure HIV ubiquitin_system*
E-mail chris@uxray.med.utah.edu Tel. 1(801)5855536 Fax 1(801)5817959

Himmel, Daniel (1961). Graduate Student. Biology Dept. (Biophysics Program), Bassine 235, Brandeis University, 415 South Street, Waltham, MA 02254-9110, USA. MS (U. Pittsburgh, 1986) biological sciences, biophysics. *Muscle_proteins X-ray_crystallography molecular_dynamics calcium_compounds crystallization*
E-mail himmel@auriga.rose.brandeis.edu himmel@binah.cc.brandeis.edu Tel. 1(617)7362475 Fax 1(617)7362419

Hingerty, Dr Brian E. (1948). Research Staff. Health Sciences Research Division, Oak Ridge National Laboratory, PO Box 2009, MS-8077, Oak Ridge, TN 37831-8077, USA. PhD (Princeton U., 1974) biophysics. *DNA computer_modelling neutron_diffraction supercomputers structure_prediction water_structure hydrogen_bonds*
E-mail beh@ornl.gov u16127@nersc.gov Tel. 1(615)5740844 Fax 1(615)5741274

Hiremath, Chaitanya (1962). Postdoctoral Fellow. Department of Biological Chemistry and Molecular Pharmacology, Harvard Medical School, 240 Longwood Avenue, Boston, MA 02115, USA. PhD (Indian Institute of Science, Bangalore, 1991) virus crystallography. *Virus structure crystallography drug_design protein_structure*
E-mail chetan@ganesh.med.harvard.edu Tel. 1(617)4323919 Fax 1(617)4324360

Hirth, John Price (1930). Professor. Mechanics and Material Engineering Dept., Washington State University, Pullman, WA 99164-2920, USA. PhD (Carnegie Institute of Technology, 1957) metallurgy. *Dislocations phase_transformations interfaces*
E-mail hirth@mme.wsu.edu Tel. 1(509)3358654 Fax 1(509)3354662

Ho, Dr Douglas M. (1951). Staff Crystallographer. Princeton University, Department of Chemistry, Princeton, NJ 08544-1009, USA. PhD (U. Southern California, 1981) inorganic chemistry. *X-ray_diffraction service_crystallography small_molecules pharmaceutical natural_products electronic_materials*
E-mail ho@chemvax.princeton.edu Tel. 1(609)2583160 Fax 1(609)2586746

Ho, Dr Joseph Xiaomin (also known as Xiao-min He) (1944). Senior research scientist. ES 76, Marshall Space Flight Center, Huntsville, AL 35812, USA. PhD (U. Pittsburgh, 1984) crystallography. *Crystallographic_computing molecular_thermal_vibrations protein_structure_determination serum_albumin antibody*
Tel. 1(205)5445531 Fax 1(205)5449305

Hobbs, Linn W. (1944). John F. Elliott Professor. MIT, Dept. Materials Science & Engineering, Room 13-4062, Cambridge, MA 02139, USA. DPhil (Oxford U., 1972) science of materials. *Electron_diffraction high_resolution_transmission_electron_microscopy small-angle_scattering neutron_diffraction double-crystal_X-ray_diffraction point_and_extended_defects amorphization medium_range_order_in_glasses*
E-mail hobbs@mit.edu Tel. 1(617)2536835 Fax 1(617)2521020 Telex 921473MIT-CAM

Hodgson, Keith Owen (1947). Professor. Department of Chemistry, Stanford University, Stanford, CA 94305, USA. PhD (UC Berkeley, 1972) inorganic chemistry. *Bio_inorganic_chemistry anomalous_dispersion synchrotron_radiation X-ray_absorption_spectroscopy*
E-mail Hodgson@ssrl750.bitnet Hodgson@ssrl01.slac.stanford.edu Tel. 1(415)7231328 Fax 1(415)7234817 Telex STANFRD STNU 348402

Hodsdon, Dr John M. (1938). Owner and operator. Longridge Farm, 85 Daniel Webster Highway, Meredith, NH 03253-5611, USA. PhD (U. California-Berkeley, 1970) biochemistry. *Enzyme_structure protein_refinement accuracy protein_disorder*
Tel. 1(603)2796126

Hogle, Dr James M. (1951). Professor. Department of Biological Chemistry and Molecular Pharmacology, Harvard Medical School, 240 Longwood Avenue, Boston, MA 02115, USA. PhD (U. Wisconsin-Madison, 1978) biochemistry. *Virus_structure viruses viral_proteins drug_design antivirals data_collection_methods receptors*
E-mail hogle@hogles.med.harvard.edu Tel. 1(617)4323918 Fax 1(617)4324360

Hol, Dr Wim (1945). Professor of Biological Structure, and Head of Biomolecular Structure Center. University of Washington, Department of Biological Structures SM20, Seattle, WA 98195, USA. PhD (U. Groningen, 1971) protein crystallography. *Protein_crystallography drug_design vaccine_development infectious_diseases*
E-mail hol@gouda.bchem.washington.edu Tel. 1(206)6857044 Fax 1(206)5431524

Holbrook, Dr Stephen R. (1948). Staff Scientist. Structural Biology Division, Lawrence Berkeley Laboratory, Melvin Calvin Building, University of California, Berkeley, CA 94720, USA. PhD (U. Oklahoma, 1974) physical chemistry. *RNA_structure protein_structure_prediction computer_modelling crystallographic_refinement protein_nucleic_acid_interactions DNA_structure*
E-mail srholbrook@lbl.gov srholbrook@lbl.bitnet Tel. 1(510)4864304 Fax 1(510)4866059

Holden, Professor Hazel M. (1955). Institute for Enzyme Research, University of Wisconsin, 1710 University Ave, Madison, WI 53705, USA. PhD (Washington U., St Louis MO, 1982) biochemistry. *Proteins electron_transport enzyme_mechanism*
E-mail Holden@enzyme.wisc.edu Tel. 1(608)2624988 Fax 1(608)2652904

Holden, James Richard (1928). Retired from NSWC, White Oak, MD. HCR 31, Box 347, Bath, ME 04530-9609, USA. PhD (State U. Iowa, 1955) physical chemistry. *Crystal_structure prediction organic_small_molecules*
E-mail jholden@bruin.bowdoin.edu Tel. 1(207)4439628

Holland, Dr Debra Reid (1962). Postdoctoral fellow. Institute of Molecular Biology, University of Oregon, Eugene, OR 97403, USA. PhD (U. Oregon, 1993) chemistry. *Proteins growth_factors cell_regulation metal_binding*
E-mail holland@uoxray.uoregon.edu Tel. 1(503)3465874 Fax 1(503)3465870

Hollander, Dr Frederick J. (1946). Staff Crystallographer. Department of Chemistry, University of California, Berkeley, CA 94720, USA. PhD (U. C. Berkeley, 1972) physical chemistry. *Problem_structures small_molecules hydrogen_bonding instructional_software packing_forces*
E-mail flieg@garnet.berkeley.edu Tel. 1(510)6428444

Holmes, Margaret (1953). Associate professor. Fred Hutchinson Cancer Research Center, 1124 Columbia St., Seattle, WA 98104, USA. [Fred Hutchinson Cancer Research Center, 1100 Fairview Ave. N., C3-168, Seattle, WA 98109, USA.] PhD (U. Oregon, 1980) chemistry. *Antibodies proteins crystallography*
E-mail mholmes@fred.fhcrc.org Tel. 1(206)6676837 Fax 1(206)6676524

Holt, Dr Elizabeth M. (1939). Professor. Department of Chemistry, Oklahoma State University, Stillwater, OK 74078, USA. PhD (Brown U., 1966) chemistry. *Structure_Cu(I) structure_Ca structure_phosphates*
E-mail chememh@osucc.bitnet Tel. 1(405)7445949 Fax 1(405)7446007

Hom, Tommy (1949). Philips Electronic Instruments Company, 85 McKee Drive, Mahwah, NJ 07430, USA. PhD (Polytechnic Institute of New York, 1979) physics. *Powder_diffraction X-ray_optics instrumentation_computers*
Tel. 1(201)5296188 Fax 1(201)5295084

Honzatko, Richard B. (1954). Associate Professor. Dept of Biochemistry and Biophysics, Molecular Biology Bldg. Rm 4206, Iowa State University, Ames, IA 50011, USA. PhD (Harvard U., 1982) physical chemistry.
E-mail honzatko@iastate.edu Tel. 1(515)2946116 Fax 1(515)2940453

Hoover, David M. (1967). Student. Department of Biological Chemistry, The University of Michigan, Ann Arbor, Michigan 48109, USA. [930 North University Avenue, Ann Arbor, Michigan 48109-1055, USA.] BS (The University of Illinois, 1990) biochemistry. *Macromolecular_structure_determination molecular_replacement free_radical binding_macromolecules flavoproteins protein_engineering*
E-mail david@norway.biop.umich.edu Tel. 1(313)7632199

Hope, Prof. Håkon (1930). Professor Emeritus. Department of Chemistry, University of California, Davis, Davis, CA 95616, USA. Cand. Real. (U. Oslo, 1958) *Apparatus biocrystallography service thin film crystallography_in_chemistry*
E-mail hhope@ucdavis.edu Tel. 1(916)7520957 Fax 1(916)7528995

Horton, Dr John R. (1962). Postdoctoral Scientist. Cold Spring Harbor Laboratory, W.M. Keck Structural Biology Laboratory, PO Box 100, Cold Spring Harbor, NY 11724, USA. PhD (Columbia U., 1992) biochemistry/molecular biophysics. *Structure-activity_relationship structure_determination anomalous_dispersion macromolecular_crystallography cell_cycle/development neural_development selenomethionyl_proteins MAD_phasing*
E-mail horton@cshl.org Tel. 1(516)3678816 1(516)3678371 Fax 1(516)3678873

Horton, Nancy C. (1964). Graduate student. Student member. Department of Chemistry, University of Pennsylvania, 231 S. 34th Street, Philadelphia, PA 19104-6323, USA. BS (Southern Illinois U., 1986) chemistry. *Proteins structure computer_assisted_design biocrystallography X-ray crystallography allostery*
E-mail horton@xtal04.med.upenn.edu Tel. 1(215)8980711 Fax 1(215)8984217

Hossain, Dr M. Bilayet (1937). Research Scientist. Department of Chemistry and Biochemistry, University of Oklahoma, 620 Parrington Oval, Norman, OK 73019-0370, USA. PhD (London U., 1965) crystallography. *Structure_determination biomolecules marine_natural_products anticancer_compounds molecular_mechanics graphics*
E-mail aa0024@uokmvsa.bitnet Tel. 1(405)3255831 Fax 1(405)3256111

Houdusse, Dr Anne M. (1966). Postdoctoral fellow. Rosenstiel Center, Brandeis University, 415 South St., Waltham, MA 02254, USA. PhD (Pasteur Institute, Paris, France, 1992) protein crystallography. *Muscle_proteins cytoskeleton_proteins regulation_by_calcium crystallization immunology antibodies surface_recognition dehydrogenases enzymes EF-hands*
E-mail anne@auriga.rose.brandeis.edu houdusse@brandeis.bitnet 1(617)7362468 Fax 1(617)7362405

Howard, Dr Andrew J. (1954). Senior Scientist. Molecular Simulations Inc., c/o Center for Advanced Research in Biotechnology, 9600 Gudelsky Drive, Rockville, MD 20850, USA. PhD (U Calif. San Diego, 1982) physics. *Crystallography methodology software data_processing profile_analysis mathematical_crystallography*
E-mail ahoward@harry.carb.nist.gov ahoward@msi.com Tel. 1(301)7386122 Fax 1(301)7386255

Hriljac, Dr Joseph A. (1960). Chemist. Department of Applied Science, Brookhaven National Laboratory, Building 510B, Upton, NY 11973, USA. PhD (Northwestern U., 1987) inorganic chemistry. *Microcrystalline_compounds synchrotron_radiation neutron_scattering powder porous_solids solid_oxides structure_determination phase_transition*
E-mail hriljac@bnlc16.bnl.gov Tel. 1(516)2827762 Fax 1(516)2823137

Hruby, Dr Victor J. (1938). Regents Professor. Department of Chemistry, University of Arizona, Tucson, AZ 85721, USA. PhD (Cornell U., 1965) chemistry. *Peptides_conformation conformation_bioactivity synthesis peptidomimetics_receptors intercellular_communication molecular_recognition*
E-mail hruby@cgf.chem.Arizona.edu hruby@ccit.Arizona.edu Tel. 1(602)6216332 Fax 1(602)6218407

Hsiou, Yu Hsiou (1964). Graduate Student. Department of Chemistry ad Biochemistry, University of California at Los Angeles, Los Angeles, CA 90024-1569, USA. PhD (U. California, 1993) inorganic chemistry. *Structural_disorder computer_modelling*
E-mail hsiou@uclac1.chem.ucla.edu hsiou@uclach.bitnet Tel. 1(310)8250656 Fax 1(310)2064038

Hsu, Barbara T. (1957). *Staff.* Department of Chemistry, 147-75 CH, Caltech, Pasadena, CA 91125, USA. BS (Caltech, 1979) chemistry. *Proteins DNA instrumentation computer electron_transfer proteases carboxypeptidases electrostatic_interaction*
E-mail hsu@citray.caltech.edu Tel. 1(818)395-8404 1(818)395-8392 Fax 1(818)568-9430

Hsu, I-Nan Professor. Department of Chemistry, California State University, 18111 Nordhoff, Northridge, CA 91330, USA. PhD (U. Oklahoma, 1971) physical chemistry. *Proteins drug_interactions salicylates nicotinates molecular_complexes metal_complexes*
Tel. 1(818)8853366 Fax 1(818)8852912

Huang, Dr De-Bin (1951). Postdoctoral appointee. Center for Mechanistic Biology and Biotechnology, Argonne National Laboratory, Argonne, IL 60439, USA. PhD (University of Pittsburgh, 1991) crystallography. *Crystallography hydrogen_bonding immunoglobulins proteins carbohydrates clusters*
E-mail huang@anlbem.bim.anl.gov Tel. 1(708)2523887 Fax 1(708)2525517

Huang, Dr Rui H. (1940). Specialist. Dept. of Chemistry, Michigan State University, East Lansing, MI 48824, USA. PhD (Michigan State U., 1987) physical chemistry. *Structure_alkalides_electrides X-ray_powder*
E-mail huang@msucem.bitnet Tel. 1(517)3559715X292 Fax 1(517)3531793

Huang, Dr Ting (1942). Research Staff Member. Co-editor of JAC. IBM Research Division, Almaden Research Center, 650 Harry Road, San Jose, CA 95120-6099, USA. PhD (Polytechnic U., Brooklyn, 1972) physics (X-ray diffraction). *Powder_diffraction thin-film_characterization crystallographic_computer_programming*
E-mail huang@icdd.com Tel. 1(408)9272375 Fax 1(408)9272100

Hubbard, Camden R. (1944). Leader of the Diffraction and Thermophysical Properties User Centers, Oak Ridge National Laboratory, High Temperature Materials Laboratory, PO Box 2008, Bldg. 4515, MS 6064, Oak Ridge, TN 37831-6064, USA. PhD (Iowa State U., 1971) physical chemistry/crystallography. *Power_diffraction residual_stress neutron_diffraction thermophysical_properties*
E-mail hubbardcr@ornl.gov Tel. 1(615)5744472 Fax 1(615)5744913

Hubbard, Dr Stevan R. (1957). Associate Research Scientist. Department of Biochemistry and Molecular Biophysics, Columbia University, 650 West 168th Street, New York, NY 10032, USA. PhD (Stanford U., 1988) applied physics. *Protein_kinases signal_transduction*
E-mail hubbard@cuhhca.hhmi.columbia.edu Tel. 1(212)3058236 Fax 1(212)3057379

Hubbell, John H. (1925). Physicist/consultant. Secretary, IUCr X-ray Attenuation Project; Editor-in-Chief, Radiation Physics and Chemistry; President, International Radiation Physics Society, 1994-1997. Room C-312 Radiation Physics Bldg, National Institute of Standards and Technology, Gaithersburg, MD 20899, USA. MS (U. Michigan, 1950) physics. *(X-ray)attenuation_coefficient Compton_scattering incoherent_scattering X-ray_fluorescence distributed-source_radiation_field_calculations*
Tel. 1(301)9755550 Fax 1(301)8697682 Telex 197674 NIST UT

Huber, Andrew H. (1967). Graduate Student. Division of Biology 156-29, California Institute of Technology, Pasadena, CA 91125, USA. BS (Cornell U., 1989) biochemistry. *Structure_proteins*
E-mail ahh@citray.caltech.edu Tel. 1(818)3958351 Fax 1(818)7923683

Huber, Susan R. (1967). Research support specialist. Department of Chemistry, Yale University, New Haven, CT 06511, USA. MA (U Arizona, 1992) inorganic chemistry. *Service_crystallography*
E-mail huber@bragg.yale.edu Tel. 1(203)4323930 Fax 1(203)4326144

Huddle, Benjamin P. (1941). Professor & Chairperson. Department of Chemistry, Roanoke College, Salem, VA 24153, USA. PhD (U. North Carolina, 1968) physical chemistry.
E-mail huddle@acc.roanoke.edu Tel. 1(703)3752440 Fax 1(703)3894236

Huffman, Dr Charles C. (1941). Senior Scientist. Department of Chemistry, Molecular Structure Center, Indiana University, Bloomington, IN 47405, USA. PhD (Indiana U., 1974) physical chemistry. *Low_temperature single_crystal structure_determination computer_automation software powder_diffraction service_crystallography*
E-mail huffman@indiana.edu Tel. 1(812)8556742 Fax 1(812)8559433

Hughes, John M. (1952). Chair and Professor. Geology Dept., Miami Univ., Oxford, OH 45056, USA. PhD (Dartmouth College, 1981) earth sciences. *Mineralogy rare_earth_elements vanadium crystal_chemistry*
E-mail hughes@miavx1.muohio.edu Tel. 1(513)5293216 Fax 1(513)5291542

Hughes, Dr Robert E. (1924). President. Associated Universities, Inc., 1400 16th Street, NW Ste 730, Washington, DC 20036-2217, USA. PhD (Cornell U., 1952) physical chemistry. *Structural_chemistry inorganic macromolecular_polymer*
E-mail rehughes@bnl.gov Tel. 1(202)4621676 Fax 1(202)2327161

Humblet, Dr Christine (1953). Senior Director, Biomolecular Structures & Drug Design. Department of Chemistry, Parke-Davis Pharmaceutical Res., Warner-Lambert Cie, 2800 Plymouth Road, Ann Arbor, MI 48106-1047, USA. PhD (Facultés Universitaires de Namur, Belgium, 1978) chemistry. *Computer_modelling crystal_structures protein_structures medicinal_chemistry structure-activity structure-function*
E-mail humblet@aa.wl.com Tel. 1(313)9967034 1(313)9967389 Fax 1(313)9982782

Hunt, Richard E. (1931). Owner. 7609 Range Rd., Alexandria, VA 22306-2425, USA. BS (Wagner College, 1956) chemistry.
Tel. 1(703)7680836 Fax 1(703)7680836

Huo, Wen (1963). Postdoctoral. Department of Chemistry, University of Nebraska-Lincoln, Lincoln, NE 68588-0304, USA. PhD (U. Canterbury, 1992) chemistry. *Computer-assisted_design proteins structure-activity_relationship vacancy crystallization computer_modelling*
E-mail wen@prophet1.unl.edu whuo@unlinfo.unl.edu Tel. 1(402)4728894

Hurley, Dr James H. (1963). Senior Staff Fellow. Laboratory of Molecular Biology, Bldg. 5 Rm. 435, NIDDK, National Institutes of Health, Bethesda, MD 20892-0580, USA. PhD (UC San Francisco, 1990) biophysics. *Protein_structure protein_phosphorylation signal_transduction enzymes*
E-mail hurley@tove.niddk.nih.gov Tel. 1(301)4024703 Fax 1(301)4960201

Hurley, Dr Thomas D. (1961). Assistant Professor. Department of Biochemistry and Molecular Biology, Indiana University School of Medicine, 635 Barnhill Drive MS454, Indianapolis, IN 46202, USA. PhD (Indiana U., 1990) biochemistry. *Structure catalysis energetics computer-assisted design enzyme_mechanisms_kinetics*
E-mail hurley@biochem6.iupui.edu Tel. 1(317)2782008 Fax 1(317)2744686

Huxley, Hugh E. (1924). Professor. Rosenstiel Center, Brandeis University, Waltham, MA 02254-9110, USA. PhD (Cambridge U., UK, 1952_1964) structural biology. *Muscle contraction X-ray synchrotron electromicroscopy*
E-mail heh@auriga.rose.brandeis.edu Tel. 1(617)7362401 Fax 1(617)7362405

Hyde, Dr C. Craig (1956). Special Expert and Head, X-ray crystallography group. Laboratory of Structural Biology Research/NIAMS/NIH, Bldg. 6, Rm 425, 6 Center Dr. MSC 2755, Bethesda, MD 20892-2755, USA. PhD (U. Iowa, 1985) biochemistry. *Biocrystallography enzymes proteins structural_biology*
E-mail cch@discus.niams.nih.gov Tel. 1(301)4024574 Fax 1(301)4026030

Hynes, Dr Thomas R. (1961). Research Scientist. Central Research, Pfizer Inc., Eastern Point Road, Groton, CT 06340, USA. PhD (Stanford U., 1989) cell biology. *Proteins conformational_change enzyme_inhibitors mutagenesis loop_conformation protein_engineering molecular_recognition*
E-mail hynestr@pfizer.com Tel. 1(203)4416357 Fax 1(203)4413783

Ibers, James A. (1930). Morrison Professor. Department of Chemistry, Northwestern University, 2145 Sheridan Rd., Evanston, IL 60208-3113, USA. PhD (California Institute of Technology, 1954) chemistry. *Synthesis_structure_chalcogenides synthesis_structure_porphyrins synthesis_structure_coordination_compounds*
E-mail ibers@chem.nwu.edu Tel. 1(708)4915449 Fax 1(708)4917713 Telex 446116 NUCHEM

Ice, Dr Gene Emery (1950). Senior Staff Scientist. Metals and Ceramics Division, Oak Ridge National Laboratory, Rm B260 4500S, PO Box 2008, Oak Ridge, TN 37831-6118, USA. PhD (U. Oregon, 1979) physics. *Anomalous_dispersion inelastic_scattering diffuse_scattering order_disorder X-ray_optics synchrotron_radiation microdiffraction*
E-mail gei@ornl.gov Tel. 1(615)5742744 Fax 1(615)5747659

Ilag, Dr Leodevico L. (1964). Postdoctoral Res. Assoc. Department of Biological Sciences, Purdue University, West Lafayette, IN 47909, USA. PhD (U. Tennessee-Memphis, 1991) molecular biology. *Virus_assembly chaperonins molecular_genetics virus_receptors structure_prediction DNA_binding_proteins membrane_proteins*
E-mail b35@mace.cc.purdue.edu Tel. 1(317)4941662

Inouye, Dr Hideyo (1951). Assistant Professor. Neurology Research, Children's Hospital, Harvard Medical School, 300 Longwood Avenue, Boston, MA 02115, USA. PhD (Kyoto U., 1979) inorganic chemistry in pharmaceutical sciences. *Diffuse_scattering fiber_diffraction myelin amyloid sequence_analysis macromolecular_assembly*
E-mail inouye@amy.tch.harvard.edu Tel. 1(617)7356102 Fax 1(617)7300636

Irving, Thomas C (Tom) (1955). Staff Scientist. Div. Biology, Illinois Institute of Technology, Chicago, IL 60616, USA. PhD (U. Guelph, 1989) biophysics. *Small-angle_diffraction muscle synchrotron_radiation service_crystallography*
E-mail irving@biocat1.iit.edu Tel. 1(312)5673489 Fax 1(312)5673576

Jabri, Evelyn (1966). Graduate student. Department of Biochemistry, Molecular and Cell Biology, Cornell University, Ithaca, NY 14850, USA. BA (U. Colorado, 1989) chemistry and molecular, cellular, and developmental biology. *Proteins metallo_enzymes MAD_phasing*
E-mail jabri@penelope.bio.cornell.edu Tel. 1(607)2558432 Fax 1(607)2552428

Jacobsen, Chris (1960). Assistant Professor. Dept. Physics, SUNY Stony Brook, Stony Brook, NY 11794-3800, USA. PhD (SUNY Stony Brook, 1989) physics. *Microscopy soft_X-ray imaging optics X-ray_microscopy*
E-mail jacobsen@xray1.physics.sunysb.edu Tel. 1(516)6328093 Fax 1(516)6328101

Jacobson, Dr Bruce L. (1962). Assistant Scientist. Institute for Enzyme Research, University of Wisconsin, 1710 University Avenue, Madison WI 53705, USA. PhD (Rice U., 1990) biochemistry. *Amino_acids anisotropy mutagenesis refinement*
E-mail genejock@enzyme.wisc.edu Tel. 1(608)2620529 Fax 1(608)2652904

Jacobson, David H. (1965). Doctoral Graduate Student. Dept. of Biological Chemistry and Molecular Pharmacology, Harvard University Medical School, 240 Longwood Avenue, Boston, MA 02115, USA. MS (U. California, San Diego, 1990) biochemistry. *Virus_structure proteins drug_design refinement_methods biotechnolgy loop_modelling crystallization_methods*
E-mail jacobson@ganesh.med.harvard.edu Tel. 1(617)4323919 Fax 1(617)7380516

Jacobson, Dr Raymond H. (1963). Postdoctoral research associate. Institute of Molecular Biology, University of Oregon, Eugene, OR 97403, USA. PhD (U. Oregon, 1993) physics. *Glycosidases protein_structure synchrotron_radiation*
E-mail ray@uoxray.uoregon.edu Tel. 1(503)3464080 Fax 1(503)3465870

Jacobson, Dr Robert A. (1932). Professor. Chemistry Dept., Iowa State University, Ames, Iowa 50011, USA. PhD (U. Minn., 1959) physical chemistry. *Patterson_method phase_determination computer_modelling Fourier_transform*
E-mail raj@vaxld.ameslab.gov Tel. 1(515)2941144 Fax 1(515)2945233

Jain, Sanjeev (1962). Asst. Professor. Regular member. Dept. of Biochem., St. Louis University School of Medicine, 1402 S. Grand Blvd., St. Louis MO 63104-1028, USA. PhD M. D. (U. Wisconsin-Madison, 1990) biochemistry, medicine. *Proteins nucleic_acids computer-assisted_design phase_determination lysosomal_enzymes*
E-mail jains@sluava.slu.edu Tel. 1(314)5778140

Janakiraman, Dr Musiri N. (1955). Postdoctoral Research Scientist. Physical and Analytical Chemistry Research, The Upjohn Company, Kalamazoo, MI 49001-0199, USA. PhD (Iowa State U., 1988) physical chemistry. *Proteins crystallization heavy_atoms data_processing molecular_replacement structure_determination refinement_method*
E-mail raman@upj.com Tel. 1(616)3857501 Fax 1(616)3857522

Jap, Bing K. Group Leader, Sr. Staff Scientist. Cell & Molecular Biology Division, Lawrence Berkeley Lab., Donner Lab., Rm 212, Berkeley, CA 94720, USA. PhD (U. California, Berkeley, 1975) biophysics. *Dynamical_diffraction high_resolution_electron_microscopy membranes proteins electron_crystallography channel_proteins receptors*
E-mail jap@rhoda.lbl.gov Tel. 1(510)4867104 Fax 1(510)4866488

Jardetzky, Ted Department of Biochemistry, Molecular Biology and Cell Biology, Northwestern University, 2153 Sheridan Road, Evanston, IL 60208-3500, USA. PhD
E-mail jardetz@tochtli.biochem.nwu.edu Tel. 1(708)4674048 Fax 1(708)4671380

Jasinski, Dr Jerry P. (1940). Professor. Department of Chemistry, Keene State College, 229 Main Street, Keene, NH 03431, USA. [12 Orchard Lane, Springfield, VT 05156, USA.] PhD (U. Wyoming, 1974) inorganic chemistry. *Coordination_chemistry luminescence_spectroscopy computer_modelling small_molecule_crystallography undergraduate_research polymer_education laser_dye_compounds*
E-mail jjasinski@keene.edu Tel. 1(603)3582563 Fax 1(603)3582257

Jaszczak, Prof. John A. (1961). Assistant Professor and Adjunct Curator. Department of Physics, Michigan Technological University, Houghton, MI 49931, USA. [1400 Townsend Dr., Houghton, MI 49931-1295, USA.] PhD (Ohio State U., 1989) physics. *Graphite diamond quasicrystals crystal_shapes molybdenite twinning epitaxy mineral_collecting*
E-mail jaszczak@phy.mtu.edu Tel. 1(906)4872255 Fax 1(906)4872933

Jeanloz, Raymond Professor. 301 McCone Hall, Dept. Geology and Geophysics, University of California, Berkeley, CA 94720-4767, USA. PhD (California Institute of Technology, 1979) geology and geophysics.
E-mail jeanloz@uclink.berkeley.edu Tel. 1(510)6422639 1(510)6423993 Fax 1(510)6439980

Jedrzejas, Dr Marek J. (1961). Postdoctoral fellow. Center for Macromolecular Crystallography, University of Alabama at Birmingham, THT-79, Birmingham, AL 35294-0005, USA. PhD (Cleveland State U., 1993) physical chemistry. *Proteins drug_design viruses computer_modelling*
E-mail jedrzeja@orion.cmc.uab.edu jedrzejas@uabcmc.bitnet Tel. 1(205)9757627 Fax 1(205)9340480

Jeffrey, Prof. George Alan (1915). University Professor Emeritus. Department of Crystallography, University of Pittsburgh, Pittsburgh, PA 15260, USA. DSc (U. Birmingham, UK, 1953) chemistry. *Small_molecules carbohydrates hydrogen_bonding liquid_crystals*
Tel. 1(412)6249300 Fax 1(412)6241882 Telex 812466

Jensen, Dr Lyle H. (1915). Professor Emeritus. Dept. of Biological Structure, SM-20, University of Washington, Seattle, WA 98195, USA. PhD (U. Washington, 1943) chemistry. *Proteins structure_activity_relationship accuracy refinement_method structure_solution refinement precision water*
Tel. 1(206)5431983 Fax 1(206)5431524

Jessen, Dr Sven Michael (1962). Postdoctoral fellow. Waksman Institute for Microbiology, Rutgers University, PO Box 759, Piscataway, NJ 08855-0759, USA. Dr rer. nat. (U. Kiel, FRG, 1991) crystallography. *AIDS biocrystallography molecular_modelling intramolecular_hydrogen_bonding*
E-mail jessen@mbcl.rutgers.edu jessen@biovax.bitnet Tel. 1(908)4454456 Fax 1(908)4455735

Jesser, William A. (1939). Thomas G. Digges Professor and Chairperson. Dept. of Materials Science and Engineering, University of Virginia, Thornton Hall, Charlottesville, VA 22901, USA. PhD (U. Virginia, 1966) physics. *Epitaxy interfaces misfit_dislocations electron_microscopy chemical_vapor_deposition*
E-mail waj@virginia.edu Tel. 1(804)9825654 1(804)9825643 Fax 1(804)9825660

Ji, Dr Xinhua (1948). Research Associate Professor. Center for Advanced Research in Biotechnology, 9600 Gudelsky Dr., Rockville, MD 20850, USA. PhD (U. Oklahoma, 1990) physical chemistry. *Enzymes catalysis proteins structure_activity_relationship natural_products*
E-mail ji@indigo10.carb.nist.gov Tel. 1(301)7386246 Fax 1(301)7386255

Jin-an Feng, Jan Feng (1960). Postdoctoral. Molecular Biology Institute, University of California, Los Angeles, CA 90024, USA. PhD (U. Southern California, 1990) inorganic chemistry. *Structure proteins protein_DNA protein_drug computer_modelling*
E-mail feng@uclaue.mbi.ucla.edu Tel. 1(310)2068270 Fax 1(310)8250982

Jircitano, Dr Alan J. (1955). Assistant Professor. Division of Science, Penn State University - Behrend College, Station Road, Erie, PA 16563, USA. PhD (U. Kansas, 1982) inorganic chemistry. *Macrocycles coordination_chemistry homogeneous_catalysis*
E-mail a0j@psuvm.psu.edu a0j@psuvm.bitnet Tel. 1(814)8986400 Fax 1(814)8986213

John, Kevin (1970). Graduate student. University of Pittsburgh, Dept. of Chemistry, 219 Parkman Ave., Pittsburgh, PA 15260, USA. BS (U. Pittsburgh, 1992) chemistry. *Non-linear_property chemistry molybdenum_compounds tungsten_compounds metal-metal_multiple_bonds*
E-mail kdjst@vms.cis.pitt.edu Tel. 1(412)6241131 Fax 1(412)6248552

Johnson, Dr Carroll K. (1929). Senior Research Staff Member. Chemistry Division, Oak Ridge National Laboratory, Building 4500N, MS-6197, Oak Ridge, TN 37831-6197, USA. PhD (Massachusetts Institute of Technology, 1959) biophysics. *Computer_modelling neutron_diffraction biomolecules space_group_orbifolds critical_point_topology artificial_intelligence*
E-mail ckj@ornl.gov Tel. 1(615)5744975 Fax 1(615)5765235

Johnson Jr, Dr Gerald G. (1939). Associate Professor Computer Science. Assistant Director of IMRL at PSU; Chairman of Board of JCPDS–ICDD. 123 ROB/MRL/PSU, University Park, PA 16802-4800, USA. PhD (The Pennsylvania State U., 1965) solid state science. *Powder_diffraction computer_analysis searching_of_crystallographic_databases pattern_recognition*
E-mail johnson@vax1.mrl.psu.edu Tel. 1(814)8651637 Fax 1(814)8637845

Johnson, Dr John E. (1945). Professor. Department of Biological Sciences, Purdue University, West Lafayette, IN 47907, USA. PhD (Iowa State U., 1972) physical chemistry. *Viruses protein_microscopy structure_refinement_radiation*
E-mail b4h@mace.cc.purdue.edu Tel. 1(317)4945911 Fax 1(317)4961189

Johnson, Dr Paul L. (1941). Chemist. Analytical Chemistry Laboratory, Chemical Technology Division, Argonne National Laboratory, 9700 S. Cass Avenue, Argonne, IL 60439, USA. PhD (Washington State U., 1968) physical chemistry. *Computing superconductor synchrotron_radiation neutron_diffraction*
E-mail pj@anl.gov Tel. 1(708)2529398

Johnson, Quintin President. Materials Data, Inc., PO Box 791, Livermore, CA 94551-0791, USA. PhD (UC Berkeley, 1961) physical chemistry. *Computing automation powder_pattern data_bases*
E-mail sgbw64a@prodigy.com Tel. 1(510)4491084 Fax 1(510)3731659

Johnston, Steven C. (1964). Graduate Student. Biochemistry Department, University of Utah, Salt Lake City, UT 84132, USA. BA (Colorado College, 1987) biology. *Protein_structure ubiquitin_system*
E-mail steve@uxray.med.utah.edu Tel. 1(801)5853919 Fax 1(801)5817959

Jones, Dr Daniel S. (1943). Associate Professor. Department of Chemistry, The University of North Carolina at Charlotte, Charlotte, NC 28223, USA. PhD (Harvard U., 1971) physical chemistry.
E-mail fch00dsj@unccvm.uncc.edu fch00dsj@unccvm.bitnet Tel. 1(704)5474438 Fax 1(704)5473151

Jones, Dr Noel D. (1937). Vice-President of Drug Design. Molecular Structure Corporation, 3200 Research Forest Drive, The Woodlands, TX 77381, USA. PhD (California Institute of Technology, 1964) chemistry. *Structure_proteins automation_crystallization synchrotron_radiation structure-based_drug_design*
E-mail ndj@msc.com Tel. 1(713)3631033 Fax 1(713)3643628

Jordan, Steven R. (1952). Crystallography Section Head. Glaxo, 5 Moore Dr, Research Triangle Park, NC 27709, USA. PhD (U. Arizona, 1983) chemistry. *Macromolecular_crystallography drug_design*
E-mail srj22978@glaxo.com Tel. 1(919)9413211 Fax 1(919)9413411

Jorgensen, Dr James D. (1948). Senior Physicist and Group Leader. Member, US National Committee for Crystallography. Materials Science Division, Bldg. 223, Argonne National Laboratory, Argonne, IL 60439, USA. PhD (Brigham Young U., 1975) physics. *Neutron_diffraction superconducting_materials fast-ion_conductors*
E-mail jim_jorgensen@qmgate.anl.gov Tel. 1(708)2525513 Fax 1(708)2527777

Joseph-McCarthy, Dr Diane (1964). Postdoctoral Fellow. Dept. Biol. Chem. and Molec. Pharm., Harvard Medical School, C-2, 240 Longwood Ave., Boston, MA 02115, USA. PhD (MIT, 1992) physical chemistry. *Proteins viruses computer_modelling structure_refinement drug_design molecular_dynamics structure_prediction*
E-mail joseph@vp2.med.harvard.edu joseph@tammy.harvard.edu Tel. 1(617)4323919

Joshua-Tor, Leemor (1961). Postdoctoral Research Fellow. Division of Chemistry 147-75CH, California Institute of Technology, Pasadena, CA 91125, USA. PhD (The Weizmann Institute of Science, 1990) chemistry. *Biological_crystallography proteins nucleic_acids structure molecular_recognition regulation signal_transduction*
E-mail leemor@citray.caltech.edu Tel. 1(818)3958392 Fax 1(818)5689430

Julian, Maureen M. Adjunct Professor. Dept Geological Science, Virginia Polytech & State University, Blacksburg, VA 24061, USA. [3863 Red Fox Drive, Roanoke, VA 24017, USA.] PhD (Cornell U., 1966) physical chemistry. *Silicon_nitride ab_initio_calculations diamond lattice_dynamics history_crystallography women_in_crystallography*
E-mail ireland@vtvm1.cc.vt.edu Tel. 1(703)3456212 Fax 1(703)2313386

Juo, Zong Sean (1965). Postdoctoral Fellow (as of 03/95). Molecular Biology Institute, University of California, Los Angeles, 405 Hilgard Ave, Los Angeles, CA 90024, USA. PhD (as of 03/95) (U. California, Los Angeles, 1994) biochemistry. *Protein–DNA_interaction transcription_factors*
E-mail juo@uclaue.mbi.ucla.edu Tel. 1(310)2068270 Fax 1(310)8250982

Kaduk, Dr James A. (1952). Associate Research Scientist. Amoco Corporation, Amoco Research Center, 150 W. Warrenville Road, PO Box 3011, Naperville, IL 60566, USA. PhD (Northwestern U., 1977) inorganic chemistry. *Solid_state_chemistry catalysts zeolites materials_science intermolecular_interactions computational_chemistry*
E-mail kaduk@amoco.com Tel. 1(708)4204547 Fax 1(708)4205252

Kahr, Bart E (1961). Assistant professor. Chem. Dept., Purdue University, Brown Laboratories, West Lafayette, IN 47907-1393, USA. PhD (Princeton U., 1988) stereochemistry. *Crystal_growth solid_solutions molecular_crystals history_of_crystallography physical_organic_chemistry*
E-mail bartkahr@purccvm Tel. 1(317)4945257 Fax 1(317)4940239

Kaler, Eric W. (1956). Professor. Dept. of Chemical Engineering, University of Delaware, Newark, DE 19716, USA. PhD (U. Minnesota, 1982) chemical engineering. *Small-angle_scattering complex_fluids surfactants*
E-mail kaler@che.udel.edu Fax 1(302)8314466

Kamer, Greg (1959). Outside Consultant. 47 Knoll Crest Ct., W. Lafayette, IN 47906-4511, USA. BS (Purdue U., 1981) biology.
E-mail o2g@mace.cc.purdue.edu kamer@rhino.cabm.rutgers.edu Tel. 1(317)7436451

Kamitori, Shigehiro (1961). Postdoctoral fellow. Dept. of Chem., Univ. of Kansas, Lawrence, KS 66045 0046, USA. PhD (Osaka City U., Japan, 1989) biophysical chemistry. *Proteins DNA computer_modelling crystallization time-resolved_effect RNA cyclodextrins*
E-mail kamitori@kuxry3.chem.ukans.edu xray2@kuhub.cc.ukans.edu Tel. 1(913)8644347 Fax 1(913)8645396

Kampermann, Sabine P. (1964). Graduate student. Department of Crystallography, University of Pittsburgh, 3943 O'Hara St., Pittsburgh, PA 15232, USA. Dipl.-Ing. (FH) (Fachhochschule Isny, 1986) physics. *Charge_density electrostatic_potentials thermal_vibration*
E-mail sabine@vms.cis.pitt.edu Tel. 1(412)6249300 Fax 1(412)6241882

Kanatzidis, Dr Mercouri G. (1957). Professor. Michigan State University, Department of Chemistry, East Lansing, MI 48824, USA. PhD (U. Iowa, 1984) inorganic chemistry. *Solid_state_chemistry intercalation_compounds inorganic_synthesis*
E-mail kanatzidis@cemvax.cem.msu.edu Tel. 1(517)3559715 Fax 1(517)3531793

Kantardjieff, Dr Katherine A. (1957). Assistant Professor. Department of Chemistry and Biochemistry, California State University, Fullerton, CA 92634, USA. PhD (U. California, Los Angeles, 1988) physical chemistry. *Protein_structure computer_modelling protein_toxins liposomes ceruloplasmin*
E-mail kant@csu.fullerton.edu kathyk@uclaue.mbi.ucla.edu Tel. 1(714)7733752 Fax 1(714)4495316

Kapulsky, Alexander (1954). Postdoctoral fellow. Rosenstiel Basic Medical Sciences, Brandeis University, Waltham, MA 02254, USA. PhD (Institute of Crystallography, Moscow, 1986) structure of hormonal steroids. *Proteins solution_structure cation_binding_analysis*
E-mail alek@auriga.rose.brandeis.edu. Tel. 1(617)7364911 Fax 1(617)7362405

Karen, Dr Vicky Lynn. Research Chemist. National Institute of Standards and Technology, A215 Materials Building, Reactor Radiation Division, Materials Science and Engineering Laboratory, National Institute of Standards and Technology, Gaithersburg, Maryland 20899, USA. PhD physical chemistry. *Crystal_database lattice_theory converse_transformation symmetry neutron_diffraction*
E-mail karen@tiber.nist.gov Tel. 1(301)9756255 Fax 1(301)9752128

Karipides, Anastas (1937). Professor. Department of Chemistry, Miami University, Oxford, Ohio 45056, USA. PhD (U. Illinois(Champaign-Urbana), 1964) inorganic chemistry. *Fluoroorganics hydrogen_bonding computer_modelling crystal_packing inorganic_carboxylates*
E-mail ak85chmf@miamiu.muohio.edu Tel. 1(513)5292813 Fax 1(513)5297284

Karle, Dr Isabella L. (1921). Head, X-ray crystallographer. Naval Research Laboratory, Code 6030, Washington, DC 20375-5341, USA. PhD (U. Michigan, 1944) physical chemistry. *Natural_products peptides ion_transport structure_solution structure_analysis_methods biologically_interesting_molecules polypeptides ionophores photo_rearrangement_products*
E-mail karle2@lsm.nrl.navy.mil Tel. 1(202)7672624 Fax 1(202)7676874

Karle, Dr Jean Marianne (1950). Res. chemist. Pharmacology Dept., Walter Reed Army Inst. of Research, Washington, DC 20307-5100, USA. [Pharmacology Dept., Walter Reed Army Inst. of Research, Washington, DC 20307-5100, USA.] PhD (Duke U., 1976) chemistry. *Conformational_analysis antimalarials organic_molecules*
Tel. 1(301)4275177 Fax 1(301)4276569

Karle, Prof. Jerome (1918). Chief scientist. Laboratory for the Structure of Matter, Code 6030, Naval Research Laboratory, Washington, DC 20375-5341, USA. PhD (U. Michigan, 1944) physical chemistry. *Structure_analysis_methods diffraction_applications X-ray electron neutron_diffraction*
E-mail williams6@lsm.nrl.navy.mil Tel. 1(202)7672665 Fax 1(202)7676874 Telex TWX7108220147

Karplus, Dr P. Andrew (1957). Associate Professor. Section of Biochemistry Molecular and Cell Biology, Cornell University, Ithaca, NY 14853, USA. PhD (U. Washington, 1984) biochemistry. *Protein_structure enzyme_mechanism computer_modelling drug_design homology_prediction X-ray_crystallography water_structure*
E-mail andy@penelope.bio.cornell.edu Tel. 1(607)2555701 Fax 1(607)2552428

Kastner, Dr Margaret E. (1950). Associate Professor. Department of Chemistry, Bucknell University, Lewisburg, PA 17837, USA. PhD (U of Notre Dame, 1979) inorganic chemistry. *Transition_metal_complexes instructional_materials*
E-mail kastner@bucknell.edu Tel. 1(717)5243258 Fax 1(717)5241739

Katti, Suresh K. (1954). Staff Scientist. Institute for Chemistry, Miles Research Center, 400 Morgan Lane, West Haven, CT 06516, USA. PhD (Indian Institute of Science, 1981) crystallography. *Proteins crystallography computer_aided_design*
E-mail katti@mrc.com Tel. 1(203)9372857 1(203)9372853 Fax 1(203)9376923

Katz, Darin S. (1968). Graduate Student. Department of Chemistry, University of Pennsylvania, 231 S. 34th Street, Philadelphia, PA 19104-6323, USA. BS (Pennsylvania State U., 1990) biological chemistry. *Serpins homology_modelling peptide_synthesis rational_drug_design*
E-mail darin@xtal.chem.upenn.edu katz@a.chem.upenn.edu Tel. 1(215)8982227 Fax 1(215)5732112

Katz, Prof. J. Lawrence (1927). Professor of Biomedical Engineering. Department of Biomedical Engineering, Case Western Reserve University, Cleveland, OH 44106, USA. PhD (Polytechnic Institute of Brooklyn, 1957) physics. *Bone_biomechanics bone_microstructure bone_biomaterials ultrasonics anisotropic_elasticity composite_mechanics*
E-mail jlk9@po.cwru.edu Tel. 1(216)3684050 Fax 1(216)3684069

Katz, Lewis (1923). Professor Emeritus. Dept. Chemistry, University of Connecticut, Storrs, CT 06268, USA. PhD (U. Minnesota, 1951) chemistry.

Kaufman, Hershall W. (1940). Professor. Dept. Oral Biology, SUNY at Stony Brook, School of Dental Medicine, Stony Brook, NY 11794-8702, USA. PhD (U. Manitoba, 1967) oral biology. *Calcification mineralization statistics calcium_compounds laser saliva dental_plaque*
E-mail hkaufman@ccmail.sunysb.edu hkaufman@sbccmail.bitnet Tel. 1(516)6328925 Fax 1(516)6329707

Kavanaugh, Jeffrey S. (1964). Postdoctoral Fellow. Department of Biochemistry, University of Iowa, Iowa City, IA 52242, USA. PhD (U. Iowa, 1992) biochemistry. *Macromolecular_crystallography proteins hemoglobins allostery*
E-mail jeffk@eve.biochem.uiowa.edu Tel. 1(319)3357883 Fax 1(319)3359570

Ke, Dr Hengming (1948). Associate Professor. Department of Biochemistry and Biophysics, The University of North Carolina, Chapel Hill, NC 27599, USA. PhD (Harvard U., 1989) biophysics. *Structure_determination proteins direct_methods immunophilins*
E-mail hke@med.unc.edu Tel. 1(919)9662244 Fax 1(919)9662852

Keder, Nancy L. (1955). Applications Chemist. Assay Technology, Inc., 1070 E. Meadow Circle, Palo Alto, CA 94303, USA. [1x172 Ashcroft Way, Sunnyvale, CA 94087, USA.] PhD (U. California, Los Angeles, 1984) chemistry. *Small_molecule powder_diffraction software*
E-mail nlkeder@aol.com Tel. 1(408)7362387 Fax 1(415)4240336

Keefe, Dr Lisa J. (1961). Alexander Hollaender Postdoctoral Fellow. Biological and Medical Research Division, Argonne National Laboratory, 9700 Cass Avenue, Argonne, IL 60439-4833, USA. [Biology Department, Brookhaven National Laboratory, Upton, NY 11973, USA.] PhD (Johns Hopkins University School of Medicine, 1992) biophysics and biophysical chemistry. *Proteins DNA cell_regulation synchrotron_radiation signal_transduction proteases protein_kinases DNA-binding_proteins*
E-mail keefe@bnlsbc.nsls.bnl.gov Tel. 1(516)2827742 Fax 1(516)2823407

Keller, Dr Ludwig (1978). CAMET Research, Inc., 6409-F Camino Vista, Goleta, CA 93117, USA. Dr.rer.nat. mineralogy.
Tel. 1(805)6851665 Fax 1(805)6859082

Kelly, Dr Judith A. (1944). Associate Professor Department Head. Department of Molecular and Cell Biology, University of Connecticut, 75 North Eagleville Road, U Box 125, Storrs, CT 06269-3125, USA. PhD (U. Connecticut, 1977) biophysics. *Enzyme_function proteins computer_modelling penicillin_binding_proteins tobacco_mosaic_virus_coat_protein_mutants*
E-mail kelly@uconnvm.bitnet kelly@uconnvm.uconn.edu Tel. 1(203)4864353 1(203)4864622 Fax 1(203)4864331

Keramidas, Vassilis G. (1938). Executive Director. Photonic & Electro Division, Bellcore Rm NVC 3Z-375, 331 Newman Springs Rd, Red Bank, NJ 07701, USA. PhD (Penn State U., 1973) solid state science. *Photonic_electronic_materials information_storage energy_storage*
Tel. 1(908)7583353 Fax 1(908)7584372

Kerfeld, Dr Cheryl A. (1961). Postdoctoral Fellow. Molecular Biology Institute, UCLA, 405 Hilgard Avenue, Los Angeles, CA 90024-1570, USA. PhD (UCLA, 1993) biology. *Cytochromes pigment_proteins membrane_proteins energy_transduction*
E-mail kerfeld@uclaue.mbi.ucla.edu Tel. 1(310)8258901 Fax 1(310)2063914

Keszler, Douglas A. (1957). Associate Professor. Dept. of Chemistry, Oregon State University, Corvallis, OR 97331, USA. PhD (Northwestern U., 1985) inorganic chemistry. *Chemistry solid_state optical_property synthesis*
Tel. 1(503)7376736 Fax 1(503)7372062

Khan, Dr Masood A. (1947). Staff Crystallographer Assistant Professor. Department of Chemistry and Biochemistry, University of Oklahoma, 620 Parrington Oval, Norman, OK 73019-0370, USA. PhD (U. Victoria, British Columbia, Canada, 1976) inorganic chemistry. *Structure_determination chemistry_inorganic chemistry_organometallic*
E-mail ab2026@uokmvsa.bitnet mkhan@chemdept.chem.uoknor.edu Tel. 1(405)3254542 Fax 1(405)3256111

Khan, Dr Saeed I. (1956). Staff Crystallographer. Department of Chemistry and Biochemistry, University of California Los Angeles, 405 Hilgard Ave, Los Angeles, CA 90024, USA. PhD (U. Southern California, 1985) inorganic chemistry.
E-mail khan@uclac1.chem.ucla.edu khan@uclach.bitnet Tel. 1(310)8255940 Fax 1(310)2064038

Kim, Dr Jung-Ja P. (1941). Associate Professor. Department of Biochemistry, Medical College of Wisconsin, 8701 Watertown Plank Road, Milwaukee WI 53226, USA. PhD (Cornell U., 1969) physical chemistry. *Protein_structure nucleic_acid_structure enzymes structure_function enzyme_mechanism electron_transfer*
E-mail jjkim@post.its.mcw.edu Tel. 1(414)2578479 Fax 1(414)2572008

Kim, Prof. Sung-Hou Professor. Department of Chemistry and Lawrence Berkeley Laboratory, University of California, Berkeley, CA 94720, USA. [Melvin Calvin Laboratory, University of California, Berkeley, CA 94720, USA.] PhD (U. Pittsburgh, 1966) X-ray crystallography. *Protein_structure DNA/RNA_structure protein_folding drug_discovery signal_transduction cell_cycle conformational_changes*
Tel. 1(510)4864333 Fax 1(510)4865272

King, Dr Hubert E. (1949). Staff Physicist. Corporate Research Science Laboratories, Exxon Research and Engineering Co., Route 22 East, Annandale, NJ 08801, USA. PhD (SUNY Stony Brook, 1979) mineral physics. *Anvil_cell sulfides silicates high_pressure thermal_expansion synchrotron_radiation alkanes hydrates*
E-mail heking@erenj.com Tel. 1(908)7302888 Fax 1(908)7303042

King, M. V. (1922). Research Scientist. Ultrastructure Analysis Section, NY State Dept. of Health, Wadworth Ctr-Labs & Res., Box 509, Albany, NY 12201, USA. PhD (U. Minnesota, 1949) chemistry.
Tel. 1(518)4864971 Fax 1(518)4748590

Kingma, Dr Kathleen J. (1962). Postdoctoral Fellow. Geophysical Laboratory, Carnegie Inst. of Washington, 5251 Broad Branch Rd., N.W., Washington, DC 20015-1305, USA. PhD (The Johns Hopkins U., 1994) mineral physics. *Mineralogy high_pressure microscopy spectroscopy*
E-mail kingma@gl.ciw.edu Tel. 1(202)6862410x2495 Fax 1(202)6862419

Kipper, Dr Marianne Byrn (1939). Staff Research Associate. Department of Chemistry, University of California Los Angeles, Los Angeles, California 90024, USA. PhD (U. California Berkeley, 1966) bio-organic chemistry. *Porphyrins clathrates systematics structural_disorder*
E-mail kipper@uclac3.chem.ucla.edu kipper@uclach.bitnet Tel. 1(310)8250656 Fax 1(310)2064038

Kirchner, Dr Richard M. (1941). Professor of Chemistry. Department of Chemistry, Manhattan College, Bronx, NY 10471, USA. PhD (U. Washington (Seattle), 1971) inorganic chemistry. *Zeolite-like_materials structures_from_powder_data Rietveld_refinement molecular_sieve_structures*
E-mail spf6@manvax.cc.mancol.edu spf6@manvax.bitnet Tel. 1(718)9200293 Fax 1(718)5484910

Kirschner, Daniel A. (1944). Associate Professor. Neurology Research - Enders 2, Children's Hospital, 300 Longwood Avenue, Boston MA 02115, USA. PhD (Harvard U., 1972) biophysics. *Membrane_structure structural_neurochemistry neuropathology X-ray_diffraction neutron_diffraction fiber_diffraction macromolecular_assemblies electron_microscopy*
E-mail dkirsch@amy.tch.harvard.edu Tel. 1(617)7356103 Fax 1(617)7300636

Kirz, Janos (1937). Professor. Physics Department, SUNY at Stony Brook, Stony Brook, NY 11794, USA. PhD (UC Berkeley, 1963) physics. *X-ray_microscopy*
E-mail kirz@sbhep.physics.sunysb.edu Tel. 1(516)6328106 Fax 1(516)6328101

Kissel, Lynn (1949). Physicist. Physics and Space Technology, Lawrence Livermore National Laboratory, PO Box 808, Livermore, CA 94550, USA. PhD (U. Pittsburgh, 1977) physics. *Anomalous_dispersion scattering_factor form_factor theory*
E-mail lkissel@llnl.gov Tel. 1(510)4237228 Fax 1(510)4237228

Kissinger, Dr Charles R. (1956). Research Scientist. Agouron Pharmaceuticals, Inc., 3565 General Atomics Court, San Diego, CA 92121, USA. PhD (U. Washington, 1989) biological structure. *Macromolecular_crystallography drug_design computer_modelling protein-nucleic_acid_interactions*
E-mail crk@agouron.com Tel. 1(619)6227930 Fax 1(619)6227999

Kistenmacher, Dr Thomas J. (1943). Principal Scientist. Applied Physics, Johns Hopkins University, Laruel, MD 20723-6099, USA. PhD (U. Illinois, 1969) inorganic chemistry. *Thin-film_epitaxy crystal_growth_sputtering space_group_symmetry precession_diffractometry*
E-mail tjk@aplcomm.jhuapl.edu Tel. 1(301)9536215 Fax 1(301)9536904

Klei, Herbert E. (1960). Graduate Student. Department of Molecular and Cell Biology, University of Connecticut, Storrs, CT 06269-3125, USA. MS (U. Washington, 1984) chemical engineering. *Proteins computer_modelling homology_prediction structure_prediction*
E-mail klei@vxray.ims.uconn.edu klei@uconnvm.bitnet Tel. 1(203)4862099 1(203)4864331

Klein, Dr Cheryl L. (1956). Professor. Dept. of Chemistry, Xavier University, 7325 Palmetto Street, New Orleans, LA 70125, USA. PhD (U. New Orleans, 1982) physical chemistry. *Electron_densities medicinal_chemistry pharmaceuticals_neuroleptics biocrystallography carcinogenesis drug_receptor_interaction narcotics*
E-mail clkcm@uno.edu Tel. 1(504)4837377 Fax 1(504)4821561

Klemens, P. G. (1950). Professor Emeritus. Dept. of Physics, University of Connecticut, Storrs, CT 06268, USA. PhD (U. Oxford, 1950) theoretical physics. *Thermal_conductivity_of_solids*
Tel. 1(203)4863134 Fax 1(203)4863346

Klimasauskas, Dr Saulius (1958). Visiting Scientist. W.M.Keck Structural Biology Laboratory, Cold Spring Harbor Laboratory, 1 Bungtown Rd, Cold Spring Harbor, NY 11724-2220, USA. PhD (Institute of Organic Synthesis, Riga, Latvia, 1987) bioorganic chemistry. *Macromolecular_recognition protein-DNA_interactions DNA_methyltransferases*
E-mail klimasau@cshl.org Tel. 1(516)3678816 Fax 1(516)3678873

Knighton, Dr Daniel R. (1962). Postdoctoral Fellow. Agouron Pharmaceuticals, Inc., 3565 General Atomics Court, San Diego, CA 92121-1122, USA. PhD (U. California, San Diego, 1991) biochemistry. *Protein_kinases dihydrofolate_reductases AIDS proteins phosphoproteins*
E-mail knighton@agouron.com Tel. 1(619)6223006 Fax 1(619)6223299

Knobler, Carolyn B. (1934). Research Chemist. Department of Chemistry and Biochemistry, University of California, Los Angeles, CA 90024, USA. PhD (Pennsylvania State U., 1959) inorganic chemistry. *Structure_carboranes molecular_recognition structure_boron_compounds*
E-mail knoblert@uclach.chem.ucla.edu Tel. 1(310)2066626 Fax 1(310)2064038

Knox, Dr James R. (1941). Professor. Department of Molecular and Cell Biology, University of Connecticut, Storrs, CT 06269-3125, USA. PhD (Boston U., 1967) physical chemistry. *Enzyme_structure_and_function computer_modelling drug_binding water_structure low_angle_solution_scattering*
E-mail knox@uconnvm.bitnet Tel. 1(203)4863133 Fax 1(203)4864745

Kobe, Bostjan (1965). Research Fellow. Department of Biochemistry, University of Texas Southwestern Medical Center, 5323 Harry Hines Blvd., Dallas, TX 75235-9050, USA. PhD (U. Texas Southwestern, 1994) molecular biophysics. *Proteins crystallization drug protein_structure protein_folding drug_design*
E-mail kobe@howie.swmed.edu Tel. 1(214)6485036 Fax 1(214)6485095

Koch, Stephen A. (1948). Professor. Department of Chemistry, State University of New York, Stony Brook, NY 11794-3400, USA. PhD (MIT, 1975) inorganic chemistry. *Bioinorganic_sulfides transition_metal organometallic metalloproteins*
E-mail skoch@ccmail.sunysb.edu skoch@sbccmail.bitnet Tel. 1(516)6327944 1(516)6327931 Fax 1(516)6327960

Koelsch, Gerald (1961). Graduate Student. Oklahoma Medical Research Foundation, 825 N.E. 13th Street, Oklahoma City, OK 73104, USA. BS (Southern Nazarene U., 1988) chemistry. *Biocrystallography enzymes modelling zymogen aspartic_proteinases*
E-mail koelsch@omrf.omrf.uokhsc.edu Tel. 1(405)2717251 Fax 1(405)2717249

Koeppe II, Roger E. (1949). Professor. Department of Chemistry and Biochemistry, University of Arkansas, Fayetteville, Arkansas 72701, USA. PhD (California Institute of Technology, 1976) chemistry and biochemistry. *Membrane_channels*
E-mail rk27487@uafsysb.uark.ued Tel. 1(501)5754601 Fax 1(501)5754049

Koetzle, Dr Thomas F. (1943). Senior Chemist. Chemistry Department, PO Box 5000, Brookhaven National Laboratory, Upton, NY 11973-5000, USA. PhD (Harvard U., 1970) chemistry. *Organometallic_chemistry metal_hydrides neutron_scattering synchrotron_radiation macromolecules structural_databases*
E-mail koetzle@chm.chm.bnl.gov Tel. 1(516)2824384 Fax 1(516)2825815 Telex 6852516 BNL DOE

Kohn, Jack A. (1925). Retired/consultant. 65 Wigwam Rd., Locust, NJ 07760, USA. PhD (U. Michigan, 1950) mineralogical crystallography. *Electronic_materials twinning polytypism research_and_development_management*
Tel. 1(908)8722295

Koknat, Friedrich W. (1938). Professor. Youngstown State University, Chemistry Dept., Youngstown, OH 44555-0001, USA. PhD (U. Giessen, Germany, 1965) inorganic chemistry. *Transition_metal_cluster_compounds*
Tel. 1(216)7423668 Fax 1(216)7421579

Kokotailo, George T. (1919). Retired. [98 N. American St., Woodbury, NJ 08096, USA.] PhD (Temple U., 1955) solid state physics. *Zeolite-synthesis structure powder-diffraction catalysis model_building elemental_distribution*
Tel. 1(609)8456508 Fax 1(215)5732093

Kolatkar, Anand R. (1965). Graduate student. Rice University, Dept. Biochemistry and Cell Biology, PO Box 1892, Houston, TX 77251-1892, USA. [6100 S. Main St., 406 Biology Building, Dept. Biochemistry and Cell Biology, Houston, TX 77005, USA.] BA (Augustana College, 1987) chemistry/biology. *Diffuse_scattering structural_disorder computer_modelling RNA computer_graphics synchrotron_radiation*
E-mail anandk@bioc.rice.edu Tel. 1(713)5278101x3346 Fax 1(713)5285154

Kolatkar, Dr Prasanna R. (1963). Post-doc. Department of Biological Sciences, Purdue University, West Lafayette, IN 47906, USA. PhD (U. Texas at Austin, 1991) biochemistry. *Proteins virus glycoproteins receptors virus_receptor_interactions*
E-mail bvs@mace.cc.purdue.edu Tel. 1(317)4944910 Fax 1(317)4961189

Komiya, Dr Hiromi (1956). Assistant Scientist. Division of Chemistry and Chemical Engineering 147-75CH, California Institute of Technology, Pasadena, CA 91107, USA. PhD (UC Berkeley, 1987) biochemistry. *Allostery_function protein_structure secondary_structure_prediction electron_transfer photosynthesis*
E-mail komiya@citray.caltech.edu Tel. 1(818)3958392 Fax 1(818)5689430

Kong, Dr Xiangpeng (1955). Assistant Professor. Skirball Institute, 550 First Ave., New York, NY 10016, USA. PhD (SUNY at Stony Brook, 1987) statistical mechanics. *Structure proteins dynamics computer_modelling*
E-mail kong@mcbi-34.med.nyu.edu Tel. 1(212)2638950 Fax 1(212)2638951

Konnert, Dr John (L941). Research chemist. Code 6030, Naval Research Lab., Washington, D.C. 20375-5000, USA. PhD (U. Minnesota, l967) physical chemistry. *Restrained_least_squares electron_microscopy scanning_microscopy*
E-mail konnert@lsm.nrl.navy.mil Tel. 1(202)7672735 Fax 1(202)7676874

Konnert, Judith A. (1941). Chemist. US Geological Survey, National Center MS 959, Reston, VA 22092, USA. BA (College of Wooster, 1963) chemistry. *Electron_microprobe_analysis WDS EDS crystallography computing scanning_electron_microscopy*
E-mail jkonnert@lithos.er.usgs.gov Tel. 1(703)6486763

Kopelman, Raoul (1933). Kasimir Fajans Professor of Chemistry, Physics and Applied Physics. The University of Michigan, Department of Chemistry, Ann Arbor, MI 48109-1055, USA. PhD (Columbia U., 1960) chemistry. *Nanocrystals molecular_crystals super_resolution_microscopy spectroscopy chemical_kinetics*
E-mail Raoul.kopelman@umich.edu Tel. 1(313)7647541 Fax 1(313)7474865

Kopka, Mary L. (1938). Research Faculty. 259A MBI, Los Angeles, CA 90024, USA. BS (DePaul U., 1960) chemistry. *Macromolecules DNA drugs recognition*
E-mail kopka@uclaue.mbi.ucla.edu Tel. 1(310)2068278 Fax 1(310)8250982

Korp, James D. (1950). Assistant Professor. Department of Chemistry, University of Houston, Houston, TX 77204-5641, USA. PhD (U. Texas, 1975) analytical chemistry. *Service_crystallography single_crystal small_molecule*
Tel. 1(713)7432801

Kossiakoff, Anthony A (1945). Director of Protein Engineering. Genentech, Inc., 460 Point San Bruno Blvd, South San Francisco, CA 94080, USA. PhD (U. Delaware, 1972) physical chemistry. *Neutron_diffraction structure/function protein_interaction dynamics water_structure*
E-mail koss@gene.com Tel. 1(415)2251332 Fax 1(415)2253734

Kostiner, Dr Edward S. (1940). Professor of Chemistry. Department of Chemistry, U-60, University of Connecticut, 215 Glenbrook Road., Storrs, CT 06269-3060, USA. PhD (Polytechnic Institute of Brooklyn, 1966) chemistry. *Solid_state_inorganic_chemistry crystal_chemistry crystal_growth*
E-mail chemadm3@uconnvm.uconn.edu chemadm3@uconnvm.bitnet   Tel. 1(203)4863220 Fax 1(203)4862981

Koszelak, Stan (1953). Research Biochemist. Department of Biochemistry, University of California, Riverside, CA 92521, USA. PhD (U. Oklahoma Health Sciences Center, 1984) biochemistry. *Crystal_growth microgravity biochemistry crystallization*
E-mail koszelak@ucrac1.ucr.edu koszelak@ucrvms.bitnet Tel. 1(909)7877396 Fax 1(909)7873790

Kourinov, Dr Igor Kourinov (1959). PostDoctoral researcher. Jefferson Cancer Institute, Thomas Jefferson University, BLSB room 826, 233 S. 10th Street, Philadelphia PA 19107, USA. PhD (Institute of Chemical Physics, Moscow, 1988) physics, biophysics. *Protein biomolecular_dynamics X-ray_crystallographic_refinement computer_modelling Mossbauer_spectroscopy water_structure*
E-mail kurinov@calvin.jci.tju.edu kurinov@asterix.jci.tju.edu Tel. 1(215)9554594 Fax 1(215)9232117

Kovari, Dr Ladislau Z. (1956). Postdoctoral Research Associate. Department of Biological Sciences, Purdue University, 1392 Lilly Hall of Life Sciences, Room B102C, IN 47907, USA. PhD (U. Tennessee, Memphis, 1992) microbiology and immunology. *Proteins viruses HIV cancer homology_prediction Fab_protein_complexes protein_expression protein_purification*
E-mail laci@mace.cc.purdue.edu Tel. 1(317)4944908 Fax 1(317)4961189

Kraatz, Paul (1940). Principal Engineer. Northrop B-2 Division, 8900 E. Washington Blvd., M/S T 244/GK, Pico Rivera, CA 90660, USA. PhD (U. Minnesota, Minneapolis, 1972) geology & geophysics. *Thin_film_structure_properties epitaxy*
Tel. 1(310)9425816

Krause, Dr Kurt L. (1956). Assistant Professor. Department of Biochemistry, University of Houston, Houston, TX 77204-5934, USA. MD, PhD (Baylor College of Medicine, Harvard U., 1980, 1986) physical chemistry. *Proteins enzymes allosterism catalysis folding_stability*
E-mail kkrause@uh.edu Tel. 1(713)7438370 Fax 1(713)7438373

Krause-Bauer, Dr Jeanette A. (1960). Crystallographer and systems manager. University of Cincinnati, Department of Chemistry, Mail Location 172, Cincinnati, OH 45221, USA. PhD (Ohio State U., 1989) inorganic chemistry. *Inorganic_organic_structures small_molecule_crystallography organometallic_coordination_chemistry structure_function_relationships crystallization*
E-mail krause@uchem.oa.uc.edu Tel. 1(513)5569226 Fax 1(513)5569239

Kraut, Joseph (1926). Professor of Chemistry and Biochemistry. Department of Chemistry, University of California/San Diego, La Jolla, CA 92093-0317, USA. PhD (Caltech, 1954) physical chemistry. *Protein_structure_function_and_evolution enzyme_mechanisms*
E-mail jkraut@ucsd.edu Tel. 1(619)5343366 Fax 1(619)5346128

Krawiec, Dr Mariusz (1963). Postdoctoral fellow. Department of Chemistry, Texas Christian University, Fort Worth, TX 76129, USA. [3495 South Hills Ave., Apt. 2035, Fort Worth, TX 76109, USA.] PhD (Texas Christian U., 1994) chemistry.
E-mail krawiec@gamma.is.tcu.edu Tel. 1(817)9217195 Fax 1(817)9217330

Krawitz, Aaron (1943). Professor of Mech and Aero Engineering and Sr. Res. Scientist, MURR. Dept. Mech & Aero Engrg and MURR, University of Missouri, Columbia, MO 65211, USA. PhD (Northwestern U., 1972) material science. *Neutron_scattering residual_stress*
E-mail krawitz@murrvax Tel. 1(314)8827671 Fax 1(314)8823443

Kretsinger, Robert H. (1937). Professor. Department of Biology, University of Virginia, Charlottesville, VA 22901, USA. PhD (Mass. Inst. Tech., 1964) biophysics. *EF-Hand_proteins annexins area_detectors glutathione_transferase protein_evolution protein_flexibility tandem_repeats cell_signalling*
E-mail rhk5@virginia.edu Tel. 1(804)9825764 Fax 1(804)9825626

Krieger, Monty Dept. of Biology, Rm 68-483, MIT, 77 Mass. Ave, Cambridge, MA 02139, USA.

Kroeger, Kenneth Scott (1966). Graduate Student. Department of Chemistry and Biochemistry, University of Colorado at Boulder, Boulder, CO 80309, USA. BS (U. Dayton, 1989) chemistry.
E-mail kroeger@cuchem.colorado.edu Tel. 1(303)4924503

Krueger, Dr Susan (1957). Research Physicist. Reactor Radiation Division, NIST, Bldg. 235/Room E151, Gaithersburg, MD 20899, USA. PhD (U. Maryland, 1987) physics. *Small-angle_scattering reflectivity macromolecular_structure conformational_changes biopolymers lipids ligand_binding computer_modelling DNA/protein_complexes*
E-mail kruegers@enh.nist.gov Tel. 1(301)9756734 Fax 1(301)9219847

Kryger, Gitay (1964). Graduate Student. Rosenstiel Center, Brandeis University, Waltham, MA 02254, USA. BSc (Tel-Aviv U., 1989) chemistry. *Proteins computer_modelling cryogenics halobacteria multiple-wavelength_anomalous_scattering*
E-mail kryger@auriga.rose.brandeis.edu Tel. 1(617)7364907 Fax 1(617)7362405

Kuhn, Dr Leslie A. (1959). Assistant Professor. Michigan State University, Department of Biochemistry, East Lansing, MI 48824-1319, USA. PhD (U. Pennsylvania, 1989) biophysics. *Proteins computer_modelling structural_motifs bound_water inhibitor_and_drug_design recognition_motifs*
E-mail kuhn@agua.bch.msu.edu Tel. 1(517)3538745 Fax 1(517)3539334

Kuilnig, Rudolph K. (1918). Retired. Box 424, Mcclellan Rd., RD 1, Nassau, NY 12123-9743, USA. PhD (U. Ottawa, Ont., 1958) chemistry.
Tel. 1(518)7663827

Kumar, Satyendra (1954). Associate Professor. Dept. of Physics, Kent State University, Kent, OH 44242, USA. PhD (U. Illinois, 1981) physics. *Liquid_crystals solid_state_physics liquid_crystal_displays*
E-mail satyen@xray.kent.edu Tel. 1(216)6722566 Fax 1(216)6722796

Kumar, Vinod D. (1956). Assistant Professor. Department of Pharmacology, Jefferson Cancer Institute, Thomas Jefferson University, 233 South 10th Street, Philadelphia, PA 19107, USA. PhD (Wayne State U., 1987) biochemistry. *Structure_proteins nucleic_acids computer_modelling calcium_binding_proteins*
E-mail kumar@calvin.jci.tju.edu Tel. 1(215)9554569

Kundrot, Prof. Craig E. (1960). Assistant Professor. Department of Chemistry and Biochemistry, University of Colorado, Boulder, CO 80309-0215, USA. PhD (Yale U., 1987) molecular biophysics. *Proteins rna water allostery folding energetics*
E-mail kundrot@cechem.colorado.edu Tel. 1(303)4920855 Fax 1(303)4925894

Kunz, Martin (1963). Postdoc. Center for High Pressure Research, Earth & Space Sciences, SUNY, Stony Brook, NY 11794-2100, USA. PhD (U. Bern, Switzerland, 1991) mineralogy. *Mineralogy silicates powder_diffraction high_pressure computer_modelling lattice_distortion*
E-mail kunz@sbmp01.ess.sunysb.edu Tel. 1(516)6328058 Fax 1(516)6328140

Kuriyan, John (1960). Professor. Laboratories of Molecular Biophysics, Howard Hughes Medical Institute, The Rockefeller University, 1230 York Avenue, New York, NY 10021, USA. PhD (M. I. T., 1986) physical chemistry. *Protein_structure DNA_replication oncogenes redox_proteins*
E-mail kuriyan@rocky2.rockefeller.edu Tel. 1(212)3278342 Fax 1(212)3278618

Kuser, Dr Paula R. (1966). Post-Doc. The Wistar Institute, 3601 Spruce Street, Philadelphia, PA 19102, USA. PhD (Birkbeck College, U. London, 1993) crystallography. *Proteins crystallography virus computer_modelling synchrotron adenovirus prediction*
E-mail pkuser@wistb.wistar.upenn.edu Tel. 1(215)8982202 Fax 1(215)8983868

Kuyper, Lee (1949). Section Head, Division of Organic Chemistry. Burroughs Wellcome Co., 3030 Cornwallis Road, Research Triangle Park, NC 27709, USA. PhD (U. Arkansas, 1977) organic chemistry. *Structure-based_drug_design*
E-mail lkuyper@bwco.com Tel. 1(919)3154343 Fax 1(919)3150430

Lachgar, Dr Abdessadek (1958). Assistant Professor. Department of Chemistry, Wake Forest University, Winston-Salem, NC 27109, USA. PhD (U. Nantes, France, 1987) solid state chemistry. *Structure_synthesis_oxides_nitrides_conductivity_magnetism synthesis_new_materials_structure-bonding-properties_relationship*
E-mail lachgar@hbar.phys.wfu.edu Tel. 1(919)7594676

Lager, George (1948). Professor. Dept. of Geography and Geoscience, University of Louisville, Louisville, KY 40292-0001, USA. PhD (U. British Columbia, 1976) geological science.
E-mail galage01@ulkyvx.louisville.edu Tel. 1(502)8526821 Fax 1(502)8520884

Lando, Jerome B. (1932). Professor; Technical Director Edison Polymer Innovation Corp. Macromolecular Science, Case Western Reserve University, University Circle, Cleveland, OH 44106, USA. PhD (Polytechnic Institute of Brooklyn, 1963) chemistry. *Polymers structure thin_films reactions solid state reactions*
E-mail hyw@po.cwru.edu Tel. 1(216)3686366 Fax 1(216)3684028

Langs, Dr David A. (1941). Senior Research Scientist. Department of Molecular Biophysics, Hauptman-Woodward Medical Research Institute (formerly the Medical Foundation of Buffalo), 73 High St., Buffalo, NY 14203, USA. PhD (State University of New York at Buffalo, 1968) inorganic chemistry. *Biomolecular_ion_channels cardiovascular_agents drug_receptor_interactions computational_methods molecular_replacement direct_methods*
E-mail langs@hwi.buffalo.edu Tel. 1(716)8569600 Fax 1(716)8524846

Larsen, Dr Teresa A. (1958). Research Associate. The Scripps Research Institute, MB5, 10666 N. Torrey Pines Road, La Jolla, CA 92037, USA. [Post Office Box 1433, La Jolla, CA 92038-1433, USA.] PhD (U. California at Los Angeles, 1990) biochemistry. *Molecular_imaging molecular_modelling DNA_replication computer-assisted_design molecular_computer_animation computer-aided_education educational_video*
E-mail larsen@scripps.edu Tel. 1(619)5817278 1(619)5544392 Fax 1(619)5546860

Larson, Allen C. (1928). Self-employed consultant. 14 Cerrado Loop, Santa Fe, NM 87505-8248, USA. PhD (Washington U., St. Louis, MO, 1956) chemistry. *Crystallography data_processing least-squares_refinement symmetry computing texture color_symmetry single_crystal Rietveld_method*
E-mail alarson@lanl.gov larson@odin.lansce.lanl.gov   Tel.   1(505)4664792 1(505)6672942

Larson, Bennett C. (1941). Section Head. Bldg 3025, Oak Ridge National Lab., PO Box 2008, Oak Ridge, TN 37831, USA. PhD (U. Missouri, 1970) physics. *Time-resolved_diffraction defects diffuse_scattering*
E-mail bcl@ornl.gov Tel. 1(615)5745506 Fax 1(615)5744143

Larson, Elizabeth M. (1951). Assistant Professor. Grand Canyon University, College of Science and Allied Health, 3300 W. Camelback Rd., Phoenix, AZ 85061-1097, USA. PhD (Arizona State U., 1985) inorganic chemistry. *Synchrotron_radiation EXAFS XANES time-resolved_effects amorphous_and_electronic_materials*
E-mail elizabeth@hera.lanl.gov Tel. 1(602)5892714 Fax 1(602)5892716

Larson, Dr Jay Michael (1943). Manager of Engineering. Engine Components Operation, Eaton Corp., 19218 B. Drive South, Marshall, MI 49068, USA. PhD (U. Washington, 1970) materials engineering; international marketing. *Internal_combustion_engine_valve_train_systems antique_automobiles blacksmithing home_restorations*
E-mail larson@ecmecd.dnet.etn.com Tel. 1(616)7810372 Fax 1(616)7810307

Larson, Dr Steven Bland (1949). Assistant Research Biochemist. Dept. of Biochemistry, University of California, Riverside, CA 92521, USA. PhD (Brigham Young U., 1980) analytical chemistry. *Small_molecules protein_structure virus_structure*
E-mail larson@ucrac1.ucr.edu Tel. 1(909)7873397 Fax 1(909)7873790

Lattman, Dr Eaton D. (1940). Professor; graduate program director. Editor-in-Chief 'Proteins: Structure, Function and Genetics'. Department of Biophysics and Biophys. Chem., Johns Hopkins University School of Medicine, 725 N. Wolfe St., Baltimore, Maryland 21205-2185, USA. PhD (Johns Hopkins U., 1969) biophysics. *Macromolecules stability_and_functional_mutants_of_staph._nuclease protein_folding profilin actin-binding_proteins methods_in_protein_crystallography complexes_between_ribosomal_proteins_and_RNA*
E-mail lattman@email.bs.jhu.edu Tel. 1(410)9551210 Fax 1(410)9550637

Laudise, Dr Robert (1930). ADJ Chemical Director. ATT Bell Labs, Room 1A-264, Murray Hill, NJ 07974-0636, USA. PhD (MIT, 1956) chemistry. *Crystal_growth*
Tel. 1(908)5826220 Fax 1(908)5822521

Lavie, Arnon (1965). Graduate student. Brandeis University, Rosenstiel 6th fl, Waltham, MA 02254, USA. BSc (Tel-Aviv U., 1989) chemistry. *Crystallography structure structure–activity_relationship synchrotron_radiation data_collection Laue_diffraction*
E-mail lavie@binah.cc.brandeis.edu Tel. 1(617)7364906 Fax 1(617)7362405

Lawson, A. C. (1946). Staff, Los Alamos National Laboratory. 300 Aragon Avenue, Los Alamos, NM 87544, USA. PhD (U. California, San Diego, 1972) physics. *Anomalous_dispersion Debye_Waller_factor pulsed_neutron Rietveld_method disorder ferroelasticity ferroelectricity lattice_stability magnetism small-angle_scattering*
E-mail lawson@lanl.gov Tel. 1(505)6678844 Fax 1(505)6652676

Lawson, Dr Catherine L. (1960). Associate Biophysicist. Biology Department, Brookhaven National Laboratory, Upton, NY 11973, USA. PhD (U. Chicago, 1987) biophysics and theoretical biology. *Biocrystallography recognition structure–activity relationships molecular_replacement crystallization protein_DNA_interactions Lyme_disease*
E-mail lawson@bnlcl1.bnl.gov lawson@bnlcl1.bitnet Tel. 1(516)2827667 Fax 1(516)2823407

Lawton, Stephen L. (1939). Sr. Research Associate. Mobil Res.& Dev. Corp., Research Dept., PO Box 480, Paulsboro, NJ 08066-0480, USA. MS (Iowa State U., 1966) inorganic chemistry. *Crystallography aluminosilicate_structures*
Tel. 1(609)224-2167 Fax 1(609)224-3608

Lebioda, Dr Lukasz (1943). Associate Professor. Department of Chemistry and Biochemistry, University of South Carolina, Columbia, SC 29208, USA. PhD (Jagiellonian U, Poland, 1972) physical chemistry. *Structure-function_relationship inhibitor_design enzymes homology_prediction*
E-mail lebioda@spike.psc.scarolina.edu d130011@univscvm.bitnet Tel. 1(803)7772140 Fax 1(803)7779521

Le Du, Dr Marie H. (1962). Postdoctoral Associate. Section of Biochemistry, Molecular and Cell Biology, Cornell University, 209, Biotechnology Building, Ithaca, NY 14853, USA. PhD (Université d'Aix-Marseille II, France, 1992) protein crystallography. *Biocrystallography crystallization data_collection phase_determination protein_toxins glycoproteins protein_complexes*
E-mail ledu@afterdec.tn.cornell.edu Tel. 1(607)2552174 Fax 1(607)2552428

Lee, Byungkook (BK) (1941). Section Chief, Molecular Modeling. Bldg. 37, Room 4B15, National Institutes of Health, Bethesda, MD 20892, USA. PhD (Cornell U., 1967) physical chemistry. *Hydrophobicity computer_modelling protein_stability protein_folding molecular_graphics immunotoxin molecular_design*
E-mail bkl@helix.nih.gov Tel. 1(301)4966580 1(301)4020436 Fax 1(301)4021344

Lee, Dr Peter L. (1958). Post Doc. MSD, 223, Argonne National Laboratory, 9700 South Cass Ave., Argonne, IL 60439, USA. [Bldg. 510E, Brookhaven National Laboratory, Upton, NY 11973, USA.] PhD (SUNY at Buffalo, 1991) physical chemistry. *Synchrotron_radiation anomalous_dispersion time-resolved_effect modulated_structure DAFS thin_film_crystallography*
E-mail lee@anlsrs.msd.anl.gov Tel. 1(516)2822210 Fax 1(516)2825239

Lee, Sukyeong (1966). Graduate student. Department of Biological Sciences, Purdue University, West lafayette IN 47907, USA. MS (Seoul National U., 1991) chemistry.
E-mail sukyeong@mace.cc.purdue.edu Tel. 1(317)4044908

Lee, Xavier (1937). Senior research scientist, Head of the protein crystallography laboratory. Department of Cancer Biology, Research Institute of the, Cleveland Clinic Foundation, 9500 Euclid Av., Cleveland, Ohio 44195, USA. PhD (U. Grenoble, Grenoble, France, 1981) molecular biology. *Interaction_recognition_macromolecules neutron_scattering_for_the_conformation_of_proteins_in_solution*
E-mail lee@xtal.ri.ccf.org Tel. 1(216)4457270 Fax 1(216)4449329

LeGeros, Racquel Z. (1935). Professor. Dental Materials Science, New York University, 345 East 24th Street, New York, NY 10010, USA. PhD (New York U., 1967) biochemistry. *Calcium_phosphates apatite bone_substitutes coatings implants*
Tel. 1(212)9989580 Fax 1(212)9954244

Lenhert, P. Galen (1933). Professor of Physics, Emeritus. Physics Dept., Vanderbilt University, BOX 1807, Station B, Nashville, TN 37235, USA. PhD (Johns Hopkins U., 1960) biophysics. *Crystal_structure inorganic_biological computer_programming_for_data_collection*
E-mail lenherpg@ctrvax.vanderbilt.edu Tel. 1(615)3436045 Fax 1(615)3437263

Leonowicz, Michael E. (1949). 201 Migazee Trail, Medford Lakes, NJ 08055, USA. PhD (Cornell U., 1976) physical chemistry. *Catalysis electron_microscopy framework_structure powder_structure_determination synchrotron_radiation catalysts layered_compounds oxides zeolites*
Tel. 1(609)9539506

Lesburg, Charles A. (1969). Graduate Fellow. Department of Chemistry, University of Pennsylvania, 231 South 34th Street, Philadelphia, PA 19104, USA. AB (Harvard U., 1991) chemistry. *Protein-metal_interactions hydrogen_bonds design_of_metal_binding_sites modelling crystallization_methods graphical_representation_of_macromolecules*
E-mail lesburg@a.chem.upenn.edu @xtal.chem.upenn.edu @anchor.chem.upenn.edu Tel. 1(215)8982227 Fax 1(215)5732112

Lessinger, Leslie (1943). Professor of Chemistry. Dept. of Chem., Barnard College, Columbia University, New York, NY 10027-6598, USA. PhD (Harvard U., 1972) chemistry. *Small_biological_molecules basic_salts*
Tel. 1(212)8548461

Le Trong, Isolde (1949). Research Technologist. Department of Biostructure, University of Washington, Seattle, WA 98195, USA. MSc (U. Hohenheim, 1984) genetics. *Crystallization data_collection area_detector data_processing*
E-mail isolde@u.washington.edu Tel. 1(206)5434496 Fax 1(206)5431524

Levinthal, Peter Steven (1966). Visualization Support Specialist. Convex Computer Corporation, 3000 Waterview Parkway, Richardson, TX 833851, USA. BS (Purdue U., 1990) computer science. *Computer_graphics structure–activity_relationship information_system supercomputer proteins viruses database structure_prediction semiotics hypermedia*
E-mail petelev@convex.com Tel. 1(214)4974107 Fax 1(214)4974500

Levy, Dr Henri A. (1913). Retired. [116 Meadow Road, Oak Ridge, TN 37830, USA.] PhD (California Institute of Technology, 1938) chemistry. *Computer_modelling electron microscope tomography*
E-mail levy@biovx1.bio.ornl.gov levy@biovx1.ornl.gov@mitvma.bitnet Tel. 1(615)5740820 1(615)4839567 Fax 1(574)1274

Li, Chi-Tang (1934). Senior Analytical Specialist. Dow Corning Corp., Mail CO42C1, Midland, MI 48686-0994, USA. PhD (Montana State U., 1964) physical chemistry (crystallography). *Powder_diffraction silicone*
Tel. 1(517)4966058 Fax 1(517)4966824

Li, Hui-Ying Dept Chemistry, University of South Carolina, Columbia, SC 29208, USA.

Li, Thomas Tien-Hsiung (1967). Research Associate. Biophysics Department, Johns Hopkins Univ., School of Medicine, WBSB #604, 725 N. Wolfe St., Baltimore, MD 21205, USA. PhD (U. South Florida, 1990) inorganic and analytical chemistry. *Proteins crystallization crystallography data_collection data_processing structure DNA-binding_proteins*
E-mail tli@hubeta.med.jhu.edu Tel. 1(410)9553967 Fax 1(410)9550637

Li, Ying (1943). Research Biochemist. WP44-B122, Merck Research Laboratory, West Point, PA 19486, USA. BS (Beijing Polytechnic U., China, 1965) chemistry. *Proteins drug_design thrombosis crystallization*
E-mail ying_li@msdrl.com Tel. 1(215)6525084 Fax 1(215)6527310

Li, Dr Yong J. (1933). Associate Professor. Department of Chemistry, University of Puerto Rico, Rio Piedras, PR 00931-23346, USA. PhD (SUNY at Buffalo, 1985) inorganic chemistry. *Structure–activity_relationship charge_density hydrogen_bonding drug_receptor_interaction biomolecule organometallic computer_modelling method_development instrumentation absolute_configuration*
E-mail y_ll@upr1.upr.clu.edu Tel. 1(809)7640000 Fax 1(809)7642890

Li, Dr Youli (1960). Research Scientist. Rosenstiel Center, Brandeis University, Waltham, MA 02254-9110, USA. PhD (Brandeis U., 1991) physics. *Structural_disorder proteins computer_modelling water_structure diffuse_scattering*
E-mail youli@sad.rose.brandeis.edu youli@glad.rose.brandeis.edu Tel. 1(617)7362425 Fax 1(617)7262405

Liang, Dr Jiin-Yun (1958). Research Associate. Department of Chemistry, Harvard University, 12 Oxford Street, Cambridge, MA 02138, USA. PhD (Harvard U., 1987) biophysics. *Structure_and_function enzyme_catalysis fructose-1,6-bisphosphatase theoretical_simulation_modelling*
E-mail liang@hubeta.harvard.edu Tel. 1(617)4954767 Fax 1(617)4953330

Liang, Dr Li (1943). Senior Research Associate. Department of Chemistry, Xavier University of Louisiana, 7325 Palmetto Street, New Orleans, LA 70125, USA. PhD (Academia Sinica, 1981) biophysics. *Charge_density proteins photosynthesis drug_receptor_interaction biocrystallography photoreaction_center structure_determination*
E-mail edscrm4@uno.edu Tel. 1(504)4867411x375 Fax 1(504)2866860

Liao, Dr Der-Ing Liao (1961). Postdoctoral research associate. Center for Advanced Research in, Biotechnology, 9600 Gudelsky Drive, Rockville MD 20850, USA. PhD (U. Oregon, Eugene, 1990) physics–protein crystallography. *Proteases glycoproteins proteins_evolution enzymes lectins cytoskeleton*
E-mail liao@iris7.carb.nist.gov Tel. 1(301)7386247 Fax 1(301)7386255

Liebman, Michael N. (1947). Program Manager, Bioinformatics. Amoco Techn. Co., Bioinfo Gro, 150 Warrenville Rd., Naperville, IL 60563-8460, USA. PhD (Michigan State U., 1977) physical chemistry. *Protein_structure_function pathway_modelling neural_networks distributed_databases*
E-mail mliebman@amoco.com Tel. 1(708)9617850 Fax 1(708)4203845

Lim, Dr Louis W. (1950). Assistant Professor. Department of Medical Biochemistry, Southern Illinois University School of Medicine, Carbondale, IL 62901-6503, USA. PhD (Washington U., 1979) molecular biology. *Proteins mechanism molecular_interaction crystallization nucleic_acids structural_stability drug_design*
E-mail lim@qc-cmols.siu.edu lim@qm-c-som.siu.edu Tel. 1(618)4535002 Fax 1(618)4536440

Lind, M. David (1934). Retired. 1690 Stoddard Avenue, Thousand Oaks, CA 91360, USA. PhD (Cornell U., 1962) physical chemistry. *X-ray_crystallography crystal_growth*
Tel. 1(805)4958936

Lingafelter, Dr Edward C. (1914). Professor Emeritus. Department of Chemistry, BG-10, University of Washington, Seattle, WA 98195, USA. PhD (U. California, Berkeley, 1939) chemistry. *Structure_chelates*
E-mail ling@uwchem.chem.washington.edu Tel. 1(206)5431686

Lippard, Stephen J. (1940). Arthur Amos Noyes Professor. Department of Chemistry, 18-290 MIT, Cambridge, MA 02139, USA. PhD (MIT, 1965) inorganic chemistry. *Inorganic_chemistry bioinorganic_chemistry anticancer_drugs methane_monooxygenase organometallic_chemistry DNA_chemistry*
E-mail lippard@lippard.mit.edu Tel. 1(617)2531892 Fax 1(617)2588150

Lipscomb, Leigh Ann (1967). Postdoctoral Research Associate. Member. School of Chemistry and Biochemistry, Georgia Institute of Technology, Atlanta, GA 30332-0400, USA. [500 Northside Circle Apt. S5, Atlanta, GA 30309, USA.] PhD (Georgia Institute of Technology, 1992) analytical chemistry. *DNA_drug_complexes proteins intercalation metalloporphyrins water_structure*
E-mail cmldwll@thor.gatech.edu Tel. 1(404)8948320 Fax 1(404)8947452

Lipscomb, Prof. William Nunn (1919). Abbott & James Lawrence Professor of Chemistry, Emeritus. Department of Chemistry, Harvard University, 12 Oxford Street, Cambridge, MA 02138, USA. PhD (California Institute of Technology, 1946) chemistry. *Structures enzymes proteins organic_compounds inorganic_compounds low_temperatures*
E-mail lipscomb@chemistry.harvard.edu Tel. 1(617)4954098 Fax 1(617)4953330

Litvin, Daniel Bernard (1940). Professor of Physics. Co-Editor Volume E of the International Tables for Crystallography. Department of Physics, Penn State Berks Campus, The Pennsylvania State University, PO Box 7009, Reading, PA 19610-6009, USA. DSc (Technion - Israel Institute of Technology, 1971) physics. *Solid_state_physics crystallography*
E-mail u3c@psuvm.psu.edu Tel. 1(610)3204856 Fax 1(610)3204857

Liu, Dr Hongying (Hattie) (1962). Crystallographer. Molecular Structure Corporation, 3200 Research Forest Drive, The Woodlands, TX 77381, USA. PhD (Georgia Institute of Technology, 1989) chemistry. *Small_molecules*
E-mail hhl@msc.com Tel. 1(713)3631033 Fax 1(713)3643628

Lloyd, Michael A. (1967). Graduate student. Department of Chemistry, University of Kentucky, Lexington KY 40506-00553, USA. BS (Jacksonville U., 1993) chemistry.
E-mail malloy00@ukcc.uky.edu

Lobkovsky, Dr Emil B. (1941). Research Associate. Baker Laboratory, Cornell University, Ithaca, NY 14853, USA. PhD (Moscow State U., 1974) inorganic chemistry. *Natural_products aluminium_compounds data_collection Rietveld_method packing*
E-mail emil@chemres.tn.cornell.edu Tel. 1(607)2556145 Fax 1(607)2551253

Loeb, Arthur L. (1923). Senior lecturer and honorary associate; Teaching Faculty. Dept. of Visual and Environmental Studies, Carpenter Center for the Visual Arts, Harvard University, Cambridge, MA 02138, USA. PhD (Harvard, 1949) chemical physics. *Systematics symmetry graphics solid_state polyhedra*
Tel. 1(617)4951950 Fax 1(617)8649424

Loehlin, James H. (1934). Professor of Chemistry. Department of Chemistry, Wellesley College, Wellesley, MA 02181, USA. PhD (Massachusetts Institute of Technology, 1960) physical chemistry. *Hydrogen_bonding intermolecular_chains crystallography_education intermolecular_patterns*
E-mail jloehlin@lucy.wellesley.edu Tel. 1(617)2833043 Fax 1(617)2833642

Lolis, Elias (1962). Assistant Professor. Department of Pharmacology, Yale University School of Medicine, 333 Cedar Street, New Haven, CT 06510, USA. PhD (Massachusetts Institute of Technology, 1989) chemistry. *Proteins cytokines receptors*
E-mail lolis@eliris.med.yale.edu Tel. 1(203)7856233 Fax 1(203)7857670

Loll, Dr Patrick J. (1958). Research Associate. Department of Biochemistry and Molecular Biology, University of Chicago, 920 E. 58th St., Chicago, IL 60637, USA. PhD (Johns Hopkins U School of Medicine, 1989) biophysics. *Membrane_proteins drug_design protein_crystallization data_collection*
E-mail loll@biovax.uchicago.edu Tel. 1(312)7020286 Fax 1(312)7020439

Long, Dr Gabrielle Gibbs Supervisory Physicist. National Institute of Standards and Technology, Building 223, Room A163, Gaithersburg, MD 20899, USA. PhD (Polytechnic Institute of Brooklyn, 1972) physics. *Microstructure synchrotron_radiation small_angle_scattering ceramics materials_processing disordered_materials*
E-mail gabrielle@enh.nist.edu gabrielle@nbsenh.bitnet Tel. 1(301)9755975 Fax 1(301)9752128 Telex 45897

Long, Dr Marianna M. (1944). Associate Director of Center for Macromolecular Crystallography and Associate Professor, School of Medicine. Center for Macromolecular Crystallography, University of Alabama at Birmingham, UAB Station, Box THT-79, Birmingham, AL 35294-0005, USA. PhD (U. Alabama at Birmingham, 1972) biochemistry. *Technology_transfer microgravity_research protein_crystal_growth flight_hardware_design*
E-mail long@orion.cmc.uab.edu Tel. 1(205)9348991 Fax 1(205)9340480

Love, Warner Edwards (1922). Professor. Thomas C. Jenkins Department of Biophysics, Johns Hopkins University, 3400 N. Charles St., Baltimore, MD 21218, USA. PhD (U. Pennsylvania, 1951) physiology. *Hemoglobins*
Tel. 1(410)5167250 Fax 1(410)5164118

Lovejoy, Brett (1965). Research Investigator I. Glaxo Inc., Five Moore Drive, Research Triangle Park, NC 27709, USA. PhD (UCLA, 1992) molecular biology. *Structure-based_drug_design*
E-mail bal30785@ussglv.glaxo.com Tel. 1(919)9413046 Fax 1(919)8413411 Telex 802813

Low, Prof. Barbara W. (1920). Professor Emeritus. Department of Biochemistry and Molecular Biophysics, Columbia University, 630 West 168th Street, New York, NY 10032, USA. DPhil (Oxford U., 1948) chemistry. *Structure_function neurotoxins membrane_receptor_proteins protein_protein water_protein interactions homology prediction_non-helical_protein_structure inclusion_compounds*
Tel. 1(212)3053896 Fax 1(212)3057932

Lowe-Ma, Dr Charlotte K. (1951). Research Scientist. Member of Board of Directors, International Centre for Diffraction Data; Secretary, American Crystallographic Association. Research Department, Code 474230D, Naval Air Warfare Center - Weapons Div., China Lake, CA 93555, USA. PhD (California Institute of Technology, 1980) chemistry. *Diffraction_technique structural_chemistry inorganic_crystallography sulfides superconductors explosives*
E-mail Charlotte_LoweMa@CL_63SMTP_gw.ChinaLake.Navy.Mil   Tel. 1(619)9391607 Fax 1(619)9391617

Lowrey, Alfred H. (1940). Research Chemist. Laboratory for the Structure of Matter, Naval Research Laboratory, Washington, DC 20375, USA. [Code 6030, Naval Research Laboratory, Washington, DC 20375, USA.] PhD (Yale U., 1966) theoretical chemistry. *Theoretical_chemistry computer_modelling spectroscopy thermal_motion electron_diffraction_gas_phase environmental_tobacco_smoke*
E-mail lowrey@lsm.nrl.navy.mil Tel. 1(202)7679456 1(301)2200614 Fax 1(202)7676874

Ludwig, Martha L. (1931). Professor. Member. US Natl. Comm.. Biophysics Research Division, University of Michigan, 930 N. University Ave., Ann Arbor, MI, USA. PhD (Cornell U. Medical College, 1956) biochemistry. *Electron_transfer flavoproteins metalloproteins site-mutants*
E-mail ludwig@norway.biop.umich.edu Tel. 1(313)7472736 1(313)7632199 Fax 1(313)7643323

Luecke, Hartmut "Hudel" (1962). Research Associate. Stanford Synchrotron Radiation Laboratory, POB 4349, MS 69, Stanford, CA 94309, USA. PhD (Rice U., Houston, 1990) biochemistry. *Protein_crystallography large_molecular_assemblies very_high_resolution proteasome annexins binding_proteins synchrotron_radiation*
E-mail hudel@slac.stanford.edu Tel. 1(415)9264944 Fax 1(415)9264100

Lukas, Dr Thomas J. (1951). Research Associate Professor. Department of Pharmacology, 432 MRB Vanderbilt University, Nashville, TN 37232-6600, USA. PhD (Rutgers U., 1979) organic chemistry. *Computer_modelling calcium_binding_proteins amphiphilic_peptides X-ray_crystallography calcium_modulated_proteins motif_structure*
E-mail watterdm@ctrvax.vanderbilt.edu Tel. 1(615)3224403 Fax 1(617)3227192

Lynch, Dr Vincent M. (1952). Service Crystallographer. Dept. of Chemistry and Biochemistry, University of Texas at Austin, Austin, TX 78712, USA. PhD (U. Florida, 1973) inorganic chemistry. *Macrocycles natural_products gallium_compounds lanthanides*
E-mail cmgc055@utxvms.cc.utexas.edu Tel. 1(512)4714042 Fax 1(512)4718696

Ma, Yiqun (1956). Research Associate. 264 Materials Res Lab., University of Illinois at Urbana-Champaign, 104 S. Goodwin Ave., Urbana, IL 61801, USA. PhD (Northwestern U., 1990) materials science and engineering. *Surface_metallurgy diffraction semiconductors RHEED MBE*
E-mail y_ma@uimr17.mrl.uiuc.edu Tel. 1(217)3330191 Fax 1(217)2442278

Mack, Joseph P. G. (1974). Staff Scientist. Crystallography, Bldg 6/B2-12, NIAMS, 9000 Rockville Pike, Bethesda, MD 20892, USA. PhD (Sydney U., 1974) biochemistry. *Protein recombination kinetics cooperativity*
E-mail mack@ncifcrf.gov mack@guppy.niams.nih.gov Tel. 1(301)4023223 Fax 1(301)4020009

Mackie, Dr Paul E. (1942). Senior Research Scientist. STL/GTRI, Georgia Institute of Technology, Atlanta, GA 30332, USA. [836 Summerchase Trail, Woodstock, GA 30188, USA.] PhD (Georgia Institute of Technology, 1972) physics. *Computer_modelling infrared_signatures solar_energy*
E-mail paul.mackie@gtri.gatech.edu Tel. 1(404)8539146 Fax 1(404)8948515

Madden, John J. (1943). Associate Professor. Depts of Psychiatry & Biochemistry, Emory University School of Medicine, Box AF, Atlanta, GA 30322, USA. [Human Genetics, GMHI, 1256 Briarcliff Rd., Atlanta, GA 30306, USA.] PhD (Emory U., 1968) biochemistry. *Beta-endorphin receptors DNA_repair UV_effects mu_receptors opiates addiction*
Tel. 1(404)8948644 Fax 1(404)8948502

Madrid, Marcela (1955). Scientific Specialist. Pittsburgh Supercomputing Center, Pittsburgh, PA 15213, USA. PhD (Instituto Balseiro, 1985) physics. *Proteins computer_modelling nuclear_magnetic_resonance structure_determination crystallography molecular_mechanics*
E-mail mmadrid@psc.edu Tel. 1(412)2685135 Fax 1(412)2685832

Magill, J. H. (1928). Professor Emeritus. Department of Material Science & Engineering, University of Pittsburgh, Pittsburgh, PA 15261, USA. [1719 Theodan Dr., Pittsburgh, PA 15216, USA.] PhD/DSc (Queen's U. Belfast, N. Ireland, 1956/1990) physical-organic chemistry/materials science. *Polymer_properties crystal_morphology diffusion kinetics*
E-mail magill@civengl.civ.pitt.edu Tel. 1(412)6249727 Fax 1(412)6241108

Makinen, Marvin W. (1939). Professor. Dept. Biochemistry & Molecular Biology, University of Chicago, 920 E. 58th St., Chicago, IL 60637, USA. MD/DPhil (U. Penn/Oxford U., 1968/1976) medicine/molecular biophysics. *Enzyme_structure magnetic_resonance metalloenzymes molecular_dynamics protein_structure enzyme_kinetics*
E-mail makinen@biovax.uchicago.edu Tel. 1(312)7021080 Fax 1(312)7020439

Makowski, Lee (1949). Director. Institute of Molecular Biophysics, Florida State University, Tallahassee, FL 32306, USA. PhD (MIT, 1976) electrical engineering. *Virus_structure membrane_structure macromolecular_assemblies partially_ordered_materials*
E-mail makowski@sb.fsu.edu Tel. 1(904)6440451 Fax 1(904)5611406

Malkin, Dr Alexander Joseph Research Biochemist. Department of Biochemistry, University of California, Riverside, CA 92521, USA. PhD (Institute of Crystallography Soviet Academy of Sciences, 1989) biological sciences. *Biochemistry crystallization nucleation light_scattering biocrystallography biomolecule viruses proteins*
E-mail alsov@ucrac1.ucr.edu alsov@ucrvms.bitnet Tel. 1(909)7873397 Fax 1(909)7873790

Malley, Mary F. (1953). Research Investigator. Bristol Myers Squibb, Pharmaceutical Research Institute, PO Box 4000, Princeton, NJ 08543-4000, USA. BA (Rutgers U., 1975) chemistry. *X-ray_crystallography drug_design*
E-mail malley@bms.com Tel. 1(609)2524986 Fax 1(609)2526012

Maloney, Peter C. (1941). Professor. Dept. of Physiology, Johns Hopkins Univ., Sch. of Med., 725 North Wolfe St., Baltimore, MD 21205, USA. PhD (Brown U., 1972) biological sciences. *Membrane_proteins*
E-mail pmalone@wpo.bs.jhu.edu Tel. 1(410)9558325 Fax 1(410)9550461

Mandal, Sanjay K. (1965). Postdoctoral Research Associate. Department of Chemistry, Boston College, Chestnut Hill, MA 02167, USA. PhD (Texas A&M U., 1992) inorganic chemistry. *Synthetic_modelling_metalloenzymes structure-function_relationship catalysis_structure_of_intermediates coordination_chemistry_transition_metals*
E-mail mandal@bcvms.bc.edu Tel. 1(617)5528076 Fax 1(617)5522705

Mandel, Gretchen Associate Professor. Research Service #151, VA Medical Center, Medical College of WI, Milwaukee, WI 53295, USA. PhD (U. Pennsylvania, 1972) crystallography. *X-ray_crystallography silicosis kidney_stone_diseases urinary_tract_diseases urinary_calculi pneumoconioses*
Tel. 1(414)3842000 Fax 1(414)3825320

Mandel, Neil S. Professor; Associate Chief of Staff for Research & Development. Research Service #151, VA Medical Center/Medical Col. of Wisc., Milwaukee, WI 53295, USA. PhD (U. Pennsylvania, 1971) chemistry. *X-ray_crystallography biophysics crystal-related_diseases kidney_stone urinary_tract_diseases infrared_spectroscopy urinary_calculi*
Tel. 1(414)3842000 Fax 1(414)3825320

Mao, Chen (1961). Graduate Student. Sect. Biochem. Molecular & Cell Biology, Cornell Univ., Ithaca, NY 14853, USA. MS (Fudan U., 1983) laser chemistry. *Protein crystallography drug_design phosphorylases*
E-mail mao@vgx.tn.cornell.edu Tel. 1(607)2552174

Mao, Ho-Kwang (1941). Geophysicist. Geophysical Laboratory, Carnegie Institution of Washington, 5251 Broad Branch Road, N.W., Washington, DC 20015-1305, USA. PhD (U. Rochester, 1968) geological sciences. *High-pressure diamond_cell geophysics geochemistry material_sciences*
E-mail mao@gl.ciw.edu Tel. 1(202)6862410x2467 Fax 1(202)6862419

Marcotte, Edward M. (1967). Graduate student. Dept. of Chem. & Biochem., Univ. of Texas at Austin, WEL 5.252, MC A5300, Austin, TX 78712, USA. BS (U. Texas, 1991) microbiology. *Proteins computer_modelling molecular_complexes maximum_entropy_method protein_engineering nanotechnology*
E-mail marcotte@utbc01.cm.utexas.edu Tel. 1(512)4713625 Fax 1(512)4718696

Markman, M. Sc Ofer (1964). PhD Student. Member. Department of Chemistry, Boston College, Chestnut Hill, MA 02167, USA. MSc (The Weizmann Institute of Science, 1990) life sciences. *Toxins phospholipid_protein_interactions intermediate_filaments NMR antibodies_insulin autoimmunity_diabetes*
E-mail markman@bcchem.bc.edu Tel. 1(617)5523615 Fax 1(617)5522705

Marko, Eric (1960). Director of Marketing. Spellman High Voltage, 7 Fairchild Ave., Plainview, NY 11803, USA. MBA (Hofstra U., 1993)
Tel. 1(516)3498686 Fax 1(516)3498699

Marks, Laurence D. (1954). Professor. Northwestern University, Department of Materials Science And Engineering, Evanston, IL 60208-3108, USA. PhD (Cambridge U., 1980)
Tel. 1(708)4913996 Fax 1(708)4917820

Markwell, Mary Ann (1948). Health Scientist Administrator. National Center for Research Resources, Westwood Bldg., Rm. 8A15, 5333 Westbard Ave., Bethesda, MD 20892, USA. PhD (Michigan State U., 1975) biochemistry. *Membrane_proteins disease-related_structures*
Tel. 1(301)5947934 Fax 1(301)5949187

Marmorstein, Ronen (1962). Assistant professor. The Wistar Institute, 3601 Spruce Street, Philadelphia, PA 19104, USA. PhD (U. Chicago, 1989) chemistry. *Proteins nucleic_acids cyclin-dependent_kinases*
E-mail marmor@wista.wistar.upenn.edu Tel. 1(215)8985006 Fax 1(215)8983868

Marsh, Richard E. (1922). Senior Research Associate, Emeritus. Co-editor, Acta Crystallographica C. Beckman Institute, 139-74, California Institute of Technology, Pasadena, CA 91125, USA. PhD (U. California, Los Angeles, 1950) chemistry. *Crystal_structures*
E-mail rem@xray.caltech.edu Tel. 1(818)3952738 Fax 1(818)4494159

Martin, Kenneth L. (1947). Research Assistant. Department of Chemistry, University of New Orleans, New Orleans, LA 70148, USA. BS (U. New Orleans, 1992) chemistry. *Charge_density chemical_bonding exciton_structure structure_determination explosives photochemistry laser_dyes*
E-mail edscm3@uno.edu Tel. 1(504)2867217 Fax 1(504)2866860

Martin, Yvonne Connolly (1936). Sr. Project Leader. D-47E, AP9A, Abbott Laboratories, 1 Abbott Park Rd., Abbott Park, IL 60064, USA. PhD (Northwestern U., 1964) chemistry. *Drug-design 3D-databases*
E-mail martiny@abbott.com Tel. 1(708)9375372 1(708)9374981 Fax 1(708)9372625

Martinez, Sergio E. (1964). Graduate research assistant. Department of Biological Sciences, Purdue University, West Lafayette, IN 47907, USA. BS (Cornell U., 1986) biophysics. *Proteins*
E-mail xta@mace.cc.purdue.edu sergio@pandora.bio.purdue.edu Tel. 1(317)4949247 Fax 1(317)4961189

Massa, Louis (1940). Professor. Hunter College, City University of New York, 695 Park Ave, New York, NY 10021, USA. PhD (Georgetown U., 1966) physics. *Quantum_mechanics crystallography infrared_signatures*
E-mail massa@hvaxgr.hunter.cuny.edu Tel. 1(212)7725330 Fax 1(212)7725332

Mastropaolo, Donald (1945). Research Associate Professor. Neurology Department, RG-27, University of Washington, Seattle, WA 98195, USA. PhD (Rutgers U., 1974) physical chemistry. *Biocrystallography structure-activity-relationships computer_modelling direct_methods crystal_growth drug_design*
E-mail neurol@max.u.washington.edu Tel. 1(206)5432340 Fax 1(206)6858100

Mathews, Dr F. Scott Department of Biochemistry, Washington University School of Medicine, 660 S. Euclid, Campus Box 8231, St. Louis, MO 63110, USA.
E-mail mathews_s@crystl.wustl.edu mathews_s@crystl.wustl.bitnet Tel. 1(314)3621080 Fax 1(314)3627183

Matthews, Dr Brian W. (1938). Professor of Physics, Member, Institute of Molecular Biology, Investigator, Howard Hughes Medical Institute. Institute of Molecular Biology, Howard Hughes Medical Institute, University of Oregon, Eugene, OR 97403, USA. PhD (U. Adelaide, 1964) protein crystallography. *Protein_crystallography macromolecular_structures*
E-mail brian@uoxray.uoregon.edu Tel. 1(503)3462572 Fax 1(503)3465870

Matthews, Dr David A. (1943). Director of Crystallography/Senior Research Fellow. Agouron Pharmaceuticals, 3565 General Atomics Court, San Diego, CA 92121, USA. PhD (U. Illinois, Urbana, 1971) physical chemistry. *Structure determination_proteins mechanisms_enzymes computer-assisted_design_drug inhibitor_binding*
E-mail davem@agouron.com Tel. 1(619)6223016 Fax 1(619)6223299

Matyi, Richard J. (1953). Associate Professor. 1500 Johnson Drive, Madison, WI 53706, USA. PhD (Northwestern U., 1983) materials science and engineering. *High_resolution_X-ray_diffraction diffuse_X-ray_scattering semiconductors molecular_beam_epitaxy*
E-mail matyi@engr.wisc.edu Tel. 1(608)2631716 Fax 1(608)2628353

Maverick, Dr Emily F. (1929). Lecturer (Professor Emeritus, Los Angeles City College). Department of Chemistry and Biochemistry, University of California, Los Angeles, Los Angeles, CA 90024-1569, USA. PhD (U. California, Los Angeles, 1972) analytical chemistry. *Structural_disorder thermal_motion molecular_complexes packing*
E-mail maverick@.uclac1.chem.ucla.edu maverick@uclach.bitnet Tel. 1(310)8251259 Fax 1(310)2064038

Mazany, Anthony (1954). Senior R and D Scientist. The BFGoodrich Research and Development Center, 9921 Brecksville Road, Brecksville, OH 44141-3289, USA. PhD (Case Western Reserve U., 1984) inorganic and organometallic chemistry. *Coordination_compounds gold_chemistry intercalated_materials polymers*
Tel. 1(216)4475559 Fax 1(216)4475249 Telex ITT 423294

McCammon, Dr J. Andrew (1947). J. E. Mayer Chair of Theoretical Chemistry. Department of Chemistry and Biochemistry, University of California at San Diego, La Jolla, CA 92093-0365, USA. PhD (Harvard, 1976) chemical physics. *Proteins computer_modelling*
E-mail jmccammon@ucsd.edu Tel. 1(619)5343575 Fax 1(619)5346255

McCarthy, Dr/Prof. Gregory J. (1943). Professor and Chair. Department of Chemistry, North Dakota State University, Fargo, ND 58105-5516, USA. PhD (Penn State Univ, 1969) solid state science. *Crystal_chemistry powder_X-ray_diffraction mineral_structures*
E-mail gmccarth@vm1.nodak.edu Tel. 1(701)2317193 Fax 1(701)2318831

McCauley, James W. (1940). Dean-NYS State College of Ceramics. New York State College of Ceramics At Alfred University, Alfred, NY 14802, USA. PhD (Penn State U., 1968) solid state science. *Applied_crystallography ceramics materials_science crystal_growth*
E-mail mccauley@ceramics.bitnet Tel. 1(607)8712411 Fax 1(607)8712344

McCrone, Lucy B. (1928). Research Microscopist. McCrone Research Institution, 2820 South Michigan Ave, Chicago, IL 60616_3292, USA. BA (Wellesley College, 1945) chemistry. *Optical_crystallography*
Tel. 1(312)8427100 Fax 1(312)8421078

McCrone, W. C. (1916). Chairman of the Board. McCrone Research Institution, 2820 South Michigan Ave, Chicago, IL 60616-3292, USA. PhD (Cornell U., 1942) microscopy. *Optical_crystallography*
Tel. 1(312)8427100 Fax 1(312)8421078

McDonald, Dr Neil Q. (1963). Postdoctoral Fellow. Department of Biochemistry, College of Physicians and Surgeons, Columbia University, Room 208, 630 W 168th St, New York, NY 10032, USA. PhD (Birkbeck College, London, 1991) protein crystallography. *Polypeptide_growth_factors receptors signal_transduction mad*
E-mail mcdonald@cuhhca.hhmi.columbia.edu Tel. 1(212)3052219 Fax 1(212)3057379

McGrath, Mary E. (1958). Scientist. Khepri Pharmaceuticals, Inc., 260 Littlefield Ave., South San Francisco, CA 94080, USA. PhD (U. California, Riverside, 1986) biochemistry. *Structure-function_proteases protease_inhibitors*
E-mail mcgrath@khepri.com Tel. 1(415)7943518 Fax 1(415)7943599

McGuire, Dr Nancy K. (1956). Project Leader. Central R&D, Materials Science and Development, 1702 Building, The Dow Chemical Company, Midland MI 48674-1702, USA. PhD (Arizona State U., 1985) solid state chemistry. *WAXS small_angle_scattering light_scattering polymers computer_modelling synchrotron_radiation semicrystalline_compounds diffraction*
E-mail nmcguire@dow.com Tel. 1(517)6367633 Fax 1(517)6389623

McLean, Dr John (1937). Lab director. National Environmental Services, 4320 N. Campbell Ave, Suite 230, Tucson, Arizona 85728-5819, USA. [National Environmental Services, PO Box 65810, Tucson, Arizona 85728-5819, USA.] PhD (U. Pittsburgh, 1968) earth and planetary sciences.
Tel. 1(602)7903391 Fax 1(602)2994910

McMillan, Dr Martin (1959). Research Scientist. Analytical Technology Division, Eastman Kodak Company, Building 49 Kodak Park, Rochester, NY 14652-3712, USA. [83 South St., Pittsford, NY 14534-2042, USA.] PhD (Yale U., 1987) chemical engineering. *Direct_methods molecular_packing anomalous_scattering powder_diffraction electron_crystallography synchrotron_radiation X-ray_optics*
E-mail mcmillan@kodak.com Tel. 1(716)5887059 Fax 1(716)4773029

McMullan, Dr Richard K. (1929). Chemist. Chemistry Department, Brookhaven National Laboratory, Upton, NY 11973, USA. PhD (Iowa State U., 1956) inorganic chemistry. *Neutron_diffraction structure hydrogen_bonding clathrate_hydrates inclusion_phenomena thermal_motion*
E-mail mcmullan@bnlchm.bitnet Tel. 1(516)2824380 Fax 1(516)2825815

McMurdie, H. F. (1905). Consultant. National Institute of Standards & Technology, Gaithersburg, MD 20899, USA. BS (Northwestern U., 1928) chemistry. *Phase_diagram diffraction_data polymorphism preparation*
Tel. 1(301)9755792 Fax 1(301)9752128

McPherson, Dr Alexander (1944). Professor. Department of Biochemistry, University of California, Riverside, CA 92521, USA. PhD (Purdue U., 1970) biological sciences. *Proteins biochemistry crystallization viruses biocrystallography biomolecule antibodies Fab_fragments*
E-mail mcpherson@ucrac1.ucr.edu mcpherson@ucrvms.bitnet Tel. 1(909)7875391 Fax 1(909)7873790

McRee, Dr Duncan E. (1957). Assistant Member. Molecular Biology Department MB4, The Scripps Research Institute, 10666 N. Torrey Pines Road, La Jolla, CA 92037, USA. PhD (Duke U., 1984) biochemistry. *Protein_structure function crystallography computing metalloproteins*
E-mail dem@scripps.edu Tel. 1(619)5549235 Fax 1(619)5546880

McTigue, Michele M. (1964). Postdoctoral Fellow. Department of Molecular Biology, The Scripps Research Institute, La Jolla, CA 92037, USA. PhD (U. California, San Diego, 1992) chemistry. *Protein_structure drug_design cell_cycle parasites*
E-mail mctigue@scripps.edu Tel. 1(619)5549093 Fax 1(617)5546880

McWhan, Dr Denis B. (1935). Department Chairman, NSLS. Co-editor Journal of Synchrotron Radiation. National Synchrotron Light Source, Bldg 725B, Brookhaven Nat. Labs., Upton, NY 11973, USA. PhD (U. California at Berkeley, 1961) physical chemistry. *Synchrotron_radiation magnetism phase_transitions*
E-mail mcwhan@bnl.gov Tel. 1(516)2823927 Fax 1(516)2825842 Telex 6852516 BNL DOE

Medrud, Ronald C. (1934). Staff Scientist. Analytical Sciences Unit, Chevron Research & Technology Co., 100 Chevron Way, 50-1254, Richmond, CA 94802-0627, USA. PhD (Univ. of Iowa, 1963) physical chemistry. *Powder_diffraction zeolites computer_modelling Rietveld_method material_characterization high-resolution diffractometry*
E-mail trcme@rrc.chevron.com Tel. 1(510)2424090 Fax 1(510)2425320

Mendelson, Dr Robert A. (1941). Professor. Dept. of Biochem. and Biophysics and C.V.R.I., University of California, San Francisco, CA 94143, USA. [University of California, Box 0524, San Francisco, CA 94143, USA.] PhD (U. Iowa, 1967) physics. *Proteins contractility proteins_muscle proteins_motors*
E-mail mendel@musl.ucsf.edu Tel. 1(415)4761827 Fax 1(415)4768173

Mercola, Dan (1940). Associate; P. I., Antisense R&D. San Diego Regional Cancer Center, 3099 Science Park Road, San Diego, CA 92121, USA. PhD (UCLA, 1969) biophysics. *Growth_factor_structure gene_therapy transcription_factors antisense pathology*
Tel. 1(619)4505990x234 Fax 1(619)4503251

Merritt, Ethan A. (1952). Research Asst Professor. Dept Biological Structure SM-20, University of Washington, Seattle, WA 19895, USA. PhD (U. Wisconsin-Madison, 1980) molecular biology. *Synchrotron_radiation anomalous_scattering computer_graphics proteins enterotoxins*
E-mail merritt@xray.bchem.washington.edu Tel. 1(206)5431421 Fax 1(206)5431524

Meshii, Mike (1939). John Evans Professor. Northwestern University, Department Matls Science & Engineering, The Technological Institute, Evanston, IL 60208, USA. PhD (Northwestern U., 1959) material science. *Phase_identification microstructure mechanical_properties*
E-mail messhi@ccmatsci.ms.nwu.edu Tel. 1(708)4913213 Fax 1(708)4676573

Messick, Julian (1933). Corp. Sec. & Gen. Mgr. Newtown Sq Corp Camp #12, Newtown Sq., PA 19073, USA. BS (Widener, 1955) chemistry. *Management data_base_mgt*
Fax 1(610)3259829

Metzger, Prof. Robert M. (1940). Prof. of chemistry. Dept. of Chemistry, University of Alabama, Tuscaloosa, AL 35487-0336, USA. PhD (Cal Tech, 1969) chemistry. *Organic_metals unimolecular_devices Langmuir–Blodgett_films cohesive_energies*
E-mail rmetzger@ua1vm.uz.edu Tel. 1(205)3485952 Fax 1(205)3489104

Meyer, Dr Edgar F. (1935). Professor. Department of Biochemistry and Biophysics, Texas A&M University, College Station, TX 77843-2128, USA. PhD (U. Texas at Austin, 1963) chemistry. *Protein_structure-function_relationships computer_modelling inhibitor/drug design water_structure*
E-mail meyer@monoc.tamu.edu meyer@bigraf.bitnet Tel. 1(409)8451744 Fax 1(409)8459274

Meyer, Frank H. (1915). Physics Professor Emeritus, U. Wisc. Editor, Reciprocity (journal of ISUS Inc.). ISUS Inc, 1580 E. atkins Ave., Salt Lake City, UT 84106, USA. [1103 15th Avenue SE, Minneapolis, MN 55414-2407, USA.] MS/MA (Polytech University/University of Minnesota, 1951/1968) physics/philosophy of science. *Revalued_unified_physics forces_of_solid_cohesion*
Tel. 1(612)3316086

Meyers, Edward A. (1930). Professor. Dept. of Chemistry, Texas A&M University, College Station, TX 77840-3255, USA. PhD (U. Minnesota, 1955) chemistry. *Crystal_structures organic inorganic*
Tel. 1(409)8452544 Fax 1(409)8454719

Mighell, Dr Alan D. Research Chemist; Director, Crystal and Electron Diffraction Data Center. National Institute of Standards and Technology, A215 Materials Building, Reactor Radiation Division, Materials Science and Engineering Laboratory, National Institute of Standards and Technology, Gaithersburg, Maryland 20899, USA. PhD (Princeton U., 1963) physical chemistry. *Crystallographic_database lattice_theory neutron_diffraction structure_determination symmetry*
E-mail mighela@tiber.nist.gov Tel. 1(301)9756254 Fax 1(301)9752128

Mikkola, Donald E. (1938). Professor. Dept. Metallurgical and Materials Engineering, Michigan Tech University, Houghton, MI 49931-1295, USA. PhD (Northwestern U., 1964) materials science. *Structure_property_relations intermetallics diffraction shape_memory_materials*
E-mail demikkol@mtu.edu Tel. 1(906)4872636 Fax 1(906)4872934

Milberg, Morton E. (1926). Research Professor. 5448 E. Placita Apan, Tucson, AZ 85718-6318, USA. PhD (Cornell U., 1949) physical chemistry. *Structure glass ceramics polymers*
Tel. 1(602)5290249 Fax 1(602)3222993

Milburn, Michael V. (1964). Visiting Scientist. Laboratory of Molecular Medicine, Children's Hospital, 300 Longwood Avenue, Boston, MA 02115, USA. PhD (U. California, Berkeley, 1991) biophysical chemistry. *Drug_design proteins homology_prediction ligand_binding*
E-mail milburn@xtal0.harvard.edu mvm25452@glaxo.com Tel. 1(617)7356240

Milillo, Prof. Frank (1943). Professor of Mechanical Engineering. Dept. Mechanical Engineering, Union College, Schenectady, NY 12308, USA. PhD (Polytechnic Institute of Brooklyn, 1975) physical metallurgy. *Phase_transformations interstitial_alloys*
Tel. 1(518)3886264 Fax 1(518)3886789

Millane, Dr Rick P. (1954). Associate Professor. Whistler Center for Carbohydrate Research, Purdue University, West Lafayette, IN 47907-1160, USA. PhD (U. Canterbury, New Zealand, 1981) electrical engineering. *Diffraction_fibrous polysaccharides disorder phase_refinement diffraction_theory quasicrystal inverse_problem modelling*
E-mail ojg@mace.cc.purdue.edu Tel. 1(317)4949272 Fax 1(317)4947953

Miller, Prof. Donald F. (1927). Retired. DPM Consulting & Computational Services, PO Box 423, Waveland, MS 39576, USA. PhD (Polytechnic Institute of Brooklyn, 1962) physics. *Macromolecules molecular-mechanics*
E-mail miller@jupiter.cast.msstate.edu xray1@biophy.phys.clemson.edu Tel. 1(601)4675522

Miller, Dr Lance L. (1961). Associate Physicist. Ames Laboratory, Iowa State University, Ames, IA 50011, USA. PhD (Iowa State U., 1988) physical chemistry. *Magnetism transport_properties electronic_structure crystal_chemistry*
E-mail miller@ameslab.gov Tel. 1(515)2946816 Fax 1(515)2940689

Miller, Mitchell D. (1969). Graduate Student. Department of Biochemical and Biophysical Sciences, University of Houston, Houston, TX 77204-5934, USA. BS (U. Houston, 1991) biochemical and biophysical sciences. *Biochemistry proteins macromolecules macromolecular_crystallization*
E-mail mitch@bragg1.bchs.uh.edu Tel. 1(713)7438371 Fax 1(713)7438373

Miller, Robert L. (1929). Emeritus. 2209 Westbury Dr., Midland, MI 48642-3254, USA. PhD (Brown U., 1954) chemical physics. *Polymer_structure*
Tel. 1(517)6315358

Miller, Prof. Russ (1958). Associate Professor. Department of Computer Science, 224 Bell Hall, State University of New York at Buffalo, Buffalo, NY 14260, USA. PhD (SUNY-Binghamton, 1985) mathematics. *Direct_methods parallel_algorithms computational_geometry image_analysis*
E-mail miller@cs.buffalo.edu Tel. 1(716)6453180x113 Fax 1(716)6453464

Miller, Stephen T. (1970). Graduate Student. Department of Biophysics, Harvard University, Cambridge, MA 02138, USA. [Box 89, 7 Divinity Ave., Cambridge, MA 02138, USA.] BA (Princeton U., 1992) chemistry. *Entropy proteins viruses computer_modelling virology immunology graphics_systems*
E-mail smiller@husc8.harvard.edu Tel. 1(617)4955043 Fax 1(617)4959613

Milnes, A. G. (1922). Professor Emeritus. Department of Electrical & Computer Engineering, Carnegie-Mellon University, Pittsburgh, PA 15213, USA. DSc (U. Bristol, 1956) electrical engineering. *Gallium_antimonide I–VIII–V_semiconductors*
E-mail amilnes+@andrew.cmu.edu Tel. 1(412)2682463 Fax 1(412)2682860

MingLuo, Ming Luo, Associate Prof. (1959). Associate Professor. Room 284, Basic Health Sciences Buidling, Center for Macromolecular Crystallography, University of Alabama at Birmingham, Birmingham, AL 35294, USA. PhD (Purdue U., 1987) virus structure, biology. *Virus_structure drug_design protein_crystallography*
E-mail ming@orion.cmc.uab.edu ming@uabcmc.bitnet Tel. 1(205)9344259 Fax 1(205)9340480

Minor, Dr Wladek (1946). Research Assistant Scientist. Department of Biological Sciences, Purdue University, West Lafayette, IN 47907, USA. PhD (Warsaw U., 1978) physics. *Crystallography proteins computer_graphics synchrotron_radiation computational_crystallography*
E-mail xte@mace.cc.purdue.edu Tel. 1(317)4940879 Fax 1(317)4961189

Mirsky, Dr Kira (1935). Adj. Professor. Physics Dept., Southern Oregon State College, Ashland, OR 97520, USA. PhD (Academy of Sciences, USSR, 1967) physics, mathematics. *Interatomic_interactions intermolecular_interactions ferroelectricity*
Tel. 1(503)4825858 Fax 1(503)5526415

Mitchell, David (1967). Graduate Student. Department of Chemistry & Biochemistry, University of Texas at Austin, 24th & Speedway, Welch Hall, Austin, TX 78712, USA. BS (Beloit College, 1989) biochemistry. *Proteins protein_folding hemoglobins protein evolution*
Tel. 1(512)4718605 Fax 1(512)4718696

Mitchell, Donald J. (1938). Professor of Chemistry; Director, Science Outreach. Dept. of Chemistry, Juniata College, 1900 Moore St., Huntingdon, PA 16652, USA. [1700 Moore St., Huntingdon, PA 16652, USA.] PhD (Vanderbilt U., 1965) physical chemistry. *Polymers structure orientation crystallinity science_outreach_to_basic_education*
E-mail mitchell@juncol.juniata.edu Tel. 1(814)6434310x566 Fax 1(814)6433620

Moews, Paul C. (1933). Research Associate. Institute of Materials Science, Box U-136, University of Connecticut, Storrs, CT 06269, USA. PhD (Cornell U., 1960) inorganic chemistry. *Beta_lactamases molecular_replacement protein_structure_refinement computer_graphics large_scale_computation*
E-mail moews@uconnvm.bitnet moews@uconnvm.uconn.edu Tel. 1(203)4864622 Fax 1(203)4864745

Moini, Ahmad (1963). Senior Research Chemist. Mobil Research & Development, Central Research Lab., PO Box 1025, Princeton, NJ 08543, USA. PhD (Texas A&M U., 1986) chemistry. *Zeolites synthesis geochemistry*
E-mail amoini@crl.mobil.com Tel. 1(609)7374835 Fax 1(609)7375217

Molin-Case, JoAnn (1939). Research Associate. Department of Biochemistry, Medical College of Wisconsin, 8701 W. Watertown Plank Road, Milwaukee, WI 53226, USA. PhD (U. Wisconsin-Madison, 1967) physical chemistry. *Protein computer_modelling biochemistry crystal_growth direct_methods refinement_methods flavoproteins*
E-mail molincas@post.its.mcw.edu Tel. 1(414)4564305 Fax 1(414)2668497

Momany, Cory (1961). Postdoctoral fellow. Department of Biological Sciences, Lily B-146, Purdue University, West Lafayette, IN 47907, USA. PhD (U. Texas at Austin, 1991) chemistry.
E-mail momany@mace.cc.purdue.edu Tel. 1(317)4944925 Fax 1(317)4961189

Mondragon, Alfonso (1958). Assistant Professor. Department of Biochemistry, Molecular Biology and Cell Biology, Northwestern University, 2153 Sheridan Road, Evanston, IL 60208-3500, USA. PhD (Cambridge U., 1985) biophysics.
E-mail a-mondragon@nwu.edu mondrago@tochtli.biochem.nwu.edu Tel. 1(708)4917726 Fax 1(708)4671380

Montague, Professor Daniel G. (1937). Professor. Department of Physics, Willamette University, 900 State Street - D184, Salem, OR 97301-3922, USA. PhD (U. Southern California, 1966) nuclear physics. *Neutron_diffraction X-ray_diffraction liquids computer_modelling*
E-mail montague@willamette.edu Tel. 1(503)3706422 Fax 1(503)3706148

Montfort, Dr William R. (1958). Department of Biochemistry, University of Arizona, Tucson, AZ 85721, USA. PhD (U. Texas, Austin, 1985) biochemistry. *Macromolecular_crystallography*
E-mail montforth%biotec@arizona.edu Tel. 1(602)6211884 Fax 1(602)6219288

Monzingo, Arthur F. (1952). Research Associate. Department of Chemistry and Biochemistry, University of Texas at Austin, Austin, TX 78712, USA. PhD (U. Texas at Austin, 1979) chemistry. *Proteins plant_toxins ribosome_inactivating_proteins*
E-mail monzingo@utbc01.cm.utexas.edu Tel. 1(512)4713625 Fax 1(512)4718696

Moore, Donald L. (1928). President. Instruments & Technology, Inc., 21 W. Van Buren Ave., Naperville, IL 60540, USA. BS (U. Cincinnati, 1952) metallurgical engineering.
Tel. 1(708)3557748 Fax 1(708)3557754

Moore, Keren M. (1944). Senior Research Associate. Center for Macromolecular Crystallography, University of Alabama at Birmingham, University Station, THT 79, Birmingham, AL 35294-0005, USA. PhD (U. Alabama at Birmingham, 1982) bio-inorganic chemistry. *Crystal_growth reductases metalloenzymes antibodies active_site_structure substrate_binding drug_design*
Tel. 1(205)9340117 Fax 1(205)9340480

Moore, Peter B. (1939). Professor. Department of Chemistry, Yale University, 225 Prospect St., New Haven, CT 06511, USA. [Department of Chemistry, Yale University, PO Box 208107, New Haven, CT 06520-8107, USA.] PhD (Harvard U., 1966) biophysics. *RNA_structure ribonucleoproteins NMR crystallography*
E-mail moore@neutron.chem.yale.edu Tel. 1(203)4323995 Fax 1(203)4326144

Moreland, James (1946). Director of Quality, Wacker Siltronic Corp, PO Box 83180, Portland, OR 97283, USA. PhD (U. of California at Irvine, 1974) chemistry. *Defects electronic_materials*
Tel. 1(503)2417512 Fax 1(503)2417598

Moriarty, John L. (1932). Metallurgist. 1917 Perry St., Davenport, IA 52803-2920, USA. PhD (U. Iowa, 1960) physical chemistry. *X-ray_materials metals weer_surfaces tooling sensors*
Tel. 1(319)3234686

Moring, Dr Jill (1943). Assistant Professor. Department of Psychiatry MC-1410, University of Connecticut Health Center, 263 Farmington Avenue, Farmington, CT 06030, USA. PhD (U. Connecticut, 1986) solid state inorganic chemistry. *Crystallography proteins small-angle_scattering membranes biophysics G-proteins ethanol*
E-mail moring%bsac.dnet@mbcg.uchc.edu Tel. 1(203)6793990 Fax 1(203)6791296

Morosin, Bruno (1934). Distinguished Member Technical Staff. Sandia National Labs, Organ. 1131, Mailstop 0345, Albuquerque, NM 87185-0345, USA. PhD (U. Washington, 1959) physical chemistry. *Crystal_physics structures materials properties crystal_growth shock_induced solid_state_chemistry*
Tel. 1(505)8448169 Fax 1(505)8445459

Moss, Simon C. (1934). Professor. Physics Department, University of Houston, Houston, TX 77204-5506, USA. Sc. D. (MIT, 1962) materials science. *Disorder diffuse_scattering phase_transitions artificial_structures fullerenes quasicrystals superconductors glasses*
Tel. 1(713)7433539 Fax 1(713)7433589

Mottonen, Dr James M. (1958). Research Associate. Department of Biochemistry, University of Texas Southwestern Medical Center at Dallas, Dallas, TX 75235, USA. PhD (MIT, 1989) chemistry. *Proteins computer-aided_education proteases chemotaxis serpins computerized_conferencing*
E-mail mottonen@howie.swmed.edu Tel. 1(214)6895056 Fax 1(214)6466336

Muckelbauer, Jodi (1959). Post-doc. 1392 Lilly Hall of Life Sciences, Dept of Biological Sciences, Purdue University, West Lafayette, IN 47907, USA. PhD (Purdue U., 1993) crystallography. *Crystallography virology*
E-mail eie@mace.cc.purdue.edu Tel. 1(317)4944911 Fax 1(317)4961189

Mueller, Dr Christoph (1960). Postdoctoral. Harvard University, 7 Divinity Avenue, Cambridge, MA 02138, USA. PhD (U. Freiburg, Germany, 1991) crystallography.
E-mail mueller@xtal0.harvard.edu Tel. 1(617)4958909

Mueller, Melvin H. (1918). Sr. Scientist, Retired. Argonne Natl Lab., IPNS Div Bldg 360, Argonne, IL 60439, USA. PhD (U. Illinois, 1949) chemistry. *Rietveld pulsed_neutron powder lattice_parameter silicon_compounds uranium_compounds*
E-mail mueller@anlpns.pns.anl.gov Tel. 1(708)2523554 Fax 1(708)2524163

Mueser, Dr Timothy C. (1961). IRTA Fellow. NIH/NIAMS, Bldg 6 Rm B2-34, 9000 Rockville Pike, Bethesda, MD 20892, USA. PhD (U. Nebraska - Lincoln, 1989) biophysics. *Proteins cooperativity allosterism DNA_replication*
E-mail tcm@killi.niams.nih.gov Tel. 1(301)4023223 Fax 1(301)4023417

Mui, Suet C. (1961). Postdoctoral Fellow. Boehringer Ingelheim Pharmceuticals, Inc., 900 Ridgebury Rd., PO Box 368, Ridgefield, CT 06877, USA. PhD (U. California, Riverside, 1993) biochemistry. *Protein_crystallography*
Tel. 1(203)7916431 Fax 1(203)7916072

Muir, Dr James A. (1938). Dr. Southwest Texas State University, 601 University Drive, San Marcos, TX 78666, USA. PhD (Northwestern U., 1966) solid state physics. *Inorganic_chemistry transition_metal_complexes bioinorganic_chemistry*
E-mail jm20@admin.swt.edu Tel. 1(512)3925195 Fax 1(512)2458095

Muir, Dr Mariel M. (1939). Professor. School of Science, Southwest Texas State University, 601 University Drive, San Marcos, TX 78666, USA. PhD (Northwestern U., 1965) inorganic chemistry. *Inorganic_chemisry transition_metal_complexes bioinorganic_chemistry*
E-mail mm11@admin.swt.edu Tel. 1(512)2452119 Fax 1(512)2458095

Muldawer, Leonard (1920). Professor Emeritus; Chairman, Executive Committee, Chautauqua Program. Physics Dept., Temple University, Philadelphia, PA 19122, USA. PhD (MIT, 1948) physics. *Order-disorder_metals martensitic_transformations experimental_solid_state_physics science_education*
Tel. 1(215)2047668 Fax 1(215)2045652

Munshi, Dr Sanjeev (1962). Research Associate. Department of Biological Sciences, Purdue University, West Lafayette, IN 47907, USA. PhD (Indian Institute of Science, 1990) biophysics. *Homology_prediction molecular_evolution virus_structure protein_structure computer_modelling*
E-mail eid@mace.cc.purdue.edu sanjeev@bragg.bio.purdue.edu Tel. 1(317)4940831 Fax 1(317)4961189

Murall, Ramachandran Murali (1958). Postdoctoral Scientist. The Wistar Institute, University of Pennsylvania, 3601 Spruce Street, Philadelphia, PA 19104, USA. PhD (Madras U., India, 1987) physics. *Macromolecules computer_modelling software small_molecules molecular_recognition viruses enzymes Fab_fragments*
E-mail murali@wistb.wistar.upenn.edu Tel. 1(215)8982202 Fax 1(215)8983868

Murmann, R. Kent (1927). Professor Emeritus. Chemistry Department, University of Missouri, Columbia, MO 65211, USA. PhD (Northwestern U., 1953) inorganic chemistry. *Coordination_compounds copper nickel rhenium*
E-mail chem0676@mizzou1 Tel. 1(314)8822826 Fax 1(314)8822754

Murphy, Michael E. P. (1967). Senior Fellow. Department of Biological Structure SM-20, School of Medicine, University of Washington, Seattle, WA 98195, USA. PhD (U. British Columbia, 1993) biochemistry. *Proteins oxygenases copper_compounds cytochrome computer_graphics metalloproteins*
E-mail murphy@gouda.bchem.washington.edu Tel. 1(206)5434496 Fax 1(206)5431524

Murray, Dr Henry H. (1954). Senior Chemist. Exxon Research and Engineering, Corporate Research, Route 22 East, Annandale, NJ 08804, USA. PhD (Yale U., 1981) organometallic chem. *Transition_metal_sulfides computer_modelling electrochemistry pyrazole_complexes bimetallic_complexes*
E-mail murray@erenj.com Tel. 1(908)7302714 Fax 1(908)7303042

Murthy, Dr Krishna H. M. (1952). Assistant Professor. Fels Institute for Cancer Research, Temple University Medical School, 557 Allied Health Building, 3307 North Broad Street, Philadelphia PA 19140, USA. PhD (Indian Institute of Science, 1981) biophysics. *Proteins DNA complexes drug_design anomalous_diffraction phasing_methods synchrotron_radiation transcription_factor*
E-mail murthy@sgi1.fels.temple.edu Tel. 1(215)7072481 Fax 1(215)7071454

Murthy, Dr Sanjeeva N. (1949). Research Scientist. Alliedsignal Inc., Research and Technology, 101 Columbia Road, PO BOX 1021, Morristown, NJ 07962, USA. PhD (U. Connecticut, 1976) materials science. *Small_angle_scattering short_range_order morphology wide_angle_scattering polymer_structure*
E-mail murthy@research.allied.com Tel. 1(201)4553764 Fax 1(201)4555295

Mylvaganam, Dr Shankari E. (1961). Postdoctoral Fellow. The Scripps Research Institute, 10666 N. Torrey Pines Rd, La Jolla, CA 92037, USA. PhD (Birkbeck College, U. London, UK, 1987) crystallography. *Proteins structure_function conformational_change crystallization novel_structures hemoglobins antibodies*
E-mail shankari@scripps.edu Tel. 1(619)5549261 Fax 1(619)5546880

Nachman, Dr Joseph (1953). Senior Research Specialist. Member. Department of Microbiology and Immunology, University of Illinois - Chicago, MC790, Box 6998, Chicago, IL 60680, USA. [Department of Biology, Brookhaven National Laboratory, Bldg. 463, Upton, NY 11973, USA.] PhD (U. Cambridge, UK, 1986) chemistry. *Proteins enzymatic_reactions Laue_diffraction synchrotron_radiation*
E-mail nachman@bnlvgx.bio.bnl.gov Tel. 1(516)2824454 Fax 1(516)2823407

Naismith, Jim (1968). Fellow. Howard Hughes Med Inst, University of Texas Southwestern, 5323 Harry Hines Blvd, Dallas TX75214-9060, USA. PhD (Manchester, UK, 1992) protein crystallography. *Proteins regulatory_proteins metalloproteins*
E-mail naismith@howie.swmed.edu Tel. 1(214)6485015 Fax 1(214)6485066

Narayan, Jagdish (1945). Distinguished University Professor. Director, Div. Materials Research, NSF (1990–92), Washington, DC 20550. Dept. of Materials Sciences and Engineering, North Carolina State University, Box 7916, Raleigh, NC 27695, USA. PhD (U. California, 1971) materials science and engineering. *Advanced_materials_and_processing diamond_thin_films high-temp_superconductor atomic-scale_characterization*
E-mail narayan@mat.mte.ncsu.edu Tel. 1(919)5157874 Fax 1(919)5157642

Narayana, Dr Sthanam V. L. (1952). Assistant Professor. Center for Macromolecular Crystallography, University of Alabama at Birmingham, Birmingham, AL 35294, USA. PhD (Indian Institute of Technology-Bombay, 1980) X-ray crystallography. *Proteins computer_modelling drug_design enzyme_inhibitors homology_prediction crystallization*
E-mail narayana@uabcmc.bitnet Tel. 1(205)9340119 Fax 1(205)9340121

Narendra, Dr Narayana N (1959). Post graduate research Chemist. Department of Chemistry, 0317, UCSD, La Jolla, CA 92093, USA. PhD (Indian Institute of Science, 1986) X-ray crystallography. *Structure biomolecules nucleic-acid proteins*
E-mail narayana@chem.ucsd.edu Tel. 1(619)5347227

Navia, Manuel A. (1946). Vice President, Senior Scientist. Vertex Pharmaceuticals Incorporated, 40 Allston Street, Cambridge, MA 02139, USA. PhD (U. Chicago, 1974) biophysics. *Proteins structure_based_drug_design*
E-mail navia@vpharm.com navia@altus.com Tel. 1(617)5763111 Fax 1(617)5762109

Navrotsky, Prof. Alexandra (1943). Professor. Dept of Geolog. and Geophys. Sci. and Princeton Materials Institute, Princeton University, Princeton, NJ 08544, USA. PhD (U. Chicago, 1967) chemistry. *Minerals oxides ceramics high_pressure order_disorder thermodynamics*
E-mail alex@weasel.princeton.edu Tel. 1(609)2584674 Fax 1(609)2581274

Neidhart, Dr David J. (1962). Medical Student. Department of Medicine, Washington University School of Medicine, 660 South Euclid Ave., St. Louis, MO 63110, USA. [813 Forest Trace Dr, Chesterfield, MO 63017, USA.] PhD (MIT, 1989) biological chemistry. *Proteins computer_modelling medicine*
E-mail neidhart_d@wums.wustl.edu Tel. 1(314)5372880

Nelson, Dr Hillary C. M. Assistant Professor. University of California, Dept. of Molecular and Cell Biology, 229 Stanley Hall, Berkeley, CA 94720, USA. PhD (M. I. T., 1985) biology/biochemistry. *Protein-DNA_interactions protein-protein_interactions protein_structure DNA_structure*
E-mail nelson@ucxray4.berkeley.edu Tel. 1(510)6422100 Fax 1(510)6439290

Newcomer, Marcia E. (1954). Associate Professor. Biochemistry Department, Vanderbilt University School of Medicine, Nashville, TN 37232-0146, USA. PhD (Rice U., 1979) biochemistry. *Protein_structure*
E-mail newcommer@lhmrba.hh.vanderbilt.edu Tel. 1(615)3437333 Fax 1(615)3224349

Newman, Dr Robert A. (1952). Research Leader. Analytical Sciences, Inorganic Materials Science and Characterization (IMS&C), 1897G Building, Dow Chemical, Midland, MI 48667, USA. PhD (Iowa State U., 1981) inorganic chemistry. *Ceramics thin_films H.T._XRD new_XRD_technology 2-dimensional_XRD_instrumentation preferred_orientation_polymers/fibers*
E-mail ra_newman@na1.dow.com Tel. 1(517)6364001 Fax 1(517)6365453

Newnham, Robert E. (1929). Alcoa Professor of Solid State Science; Associate Director, Materials Research Lab.. Materials Research Laboratory, Pennsylvania State University, University Park, PA 16802, USA. PhD (Cambridge U., 1960) crystallography. *Crystal_physics crystal_chemistry electroceramics composite_materials*
Tel. 1(814)8651612 Fax 1(814)8657593

Newsam, Professor John M. Senior Director, Solid State. Biosym Technologies Inc., 9685 Scranton Road, San Diego, CA 92121, USA. DPhil (Oxford U., 1980) chemistry. *Computer_modelling neutron_diffraction powder_crystallography catalysis structure_solution zeolites education planar_faulting*
E-mail jmnewsa@biosym.com Tel. 1(619)5465391 Fax 1(619)4580136

Newton, Dr M. Gary (1939). Professor. Department of Chemistry, University of Georgia, Athens, GA 30602-2556, USA. PhD (Georgia Institute of Technology, 1965) organic chemistry. *Bioorganic_structure organic_structure organometallic_structure phosphoorganic_structure computer_modelling*
E-mail newton@sunchem.chem.uga.edu Tel. 1(706)5421969 Fax 1(706)5429454

Ng, Joseph David (1962). Postdoctoral Fellow. Department of Biochemistry, University of California, Riverside, CA 92521, USA. PhD (U. California, Riverside, 1992) biochemistry. *Proteins biochemistry crystallization mutagenesis biocrystallography biomolecule protein engineering*
E-mail joseph@ucrac1.ucr.edu joseph@ucrvms.bitnet Tel. 1(909)7873397 Fax 1(909)7873790

Nichols, Monte C. (1938). Sr. Member of Tech. Staff. MC 9404, Sandia National Laboratories, Livermore, CA 94551-0969, USA. MS (U. Arizona, 1960) physical chemistry. *Descriptive_mineralogy X-ray_micro_tomography X-ray_micro_fluorescence crystallographic_databases mineralogical_databases X-ray_powder_diffraction*
E-mail monte@cms1.llnl.gov Tel. 1(510)2942906 Fax 1(510)2943410

Nicklitenko, Dr Alex V. (1960). Research Associate. Department of Biochemistry, Howard Hughes Medical Institute, Bailor College of Medicine, 1200 Moursund Ave., Houston, TX 77030, USA. PhD (Institute of Crystallography Russian Acad. of Sci., 1992) protein crystallography. *Proteins neurotoxins computing protein_folding computer_modelling crystallization structure_prediction*
E-mail alex@bragg2.bchs.uh.edu Tel. 1(713)7986564 Fax 1(713)7976718

Nicolosi, Joseph (1950). Technical Manager. Philips Electronic Instruments, 85 Mckee Drive, Mahway, NJ 07430, USA. PhD (Polytechnic Institute of New York, 1982) physics (crystallography). *Spectroscopy X-ray_analysis*
Tel. 1(914)5296217 Fax 1(914)5294760

Niederhut, Monica M. (1963). Graduate Student. Department of Crystallography, University of Pittsburgh, Pittsburgh, PA 15260, USA. BS (U. Nevada, Las Vegas, 1986) applied physics. *Small_molecules charge_density electrostatic_potential ferroelectricity*
E-mail mmnst5@vms.cis.pitt.edu Tel. 1(412)6249300 Fax 1(412)6241882

Nienaber, Dr Vicki L. (1964). Post-doc. Department of Crystallography and Biophysical Chemistry, The DuPont Merck Pharmaceutical Company, Experimental Station, Post Office Box 80228, Wilmington, DE 19880-0228, USA. PhD (Ohio State U., 1990) chemistry. *Proteins*
E-mail nienabvl@lldmpc.dnet.dupont.com Tel. 1(302)6954697 Fax 1(302)6958667

Nikolov, Dimitar B. (1966). Graduate Student. Laboratories of Molecular Biophysics, The Rockefeller University, New York, NY 10021, USA. MSc (Sofia U., Bulgaria, 1991) physics/biology. *Structural_biology molecular_recognition transcription_factors homology_prediction*
E-mail nikolod@rockvax.rockefeller.edu nikolod@rockvax.bitnet Tel. 1(212)3278339 Fax 1(212)3278337

Nogales, Dr Evangelina (1965). Postdoctoral fellow. Life Science Division, Lawrence Berkeley Laboratory, Berkeley, CA 94720, USA. PhD (U. Keele, England, 1993) physics. *Electron_crystallography proteins X-ray_solution_scattering synchrotron_radiation tubulin*
E-mail enm@rhoda.lbl.gov Tel. 1(510)4866437 Fax 1(510)4866488

Noll, Bruce C. (1962). Graduate student. Department of Chemistry, University of California, Davis, CA 95616, USA. BS (Metropolitan State College of Denver, 1988) chemistry. *Bioinorganic_chemistry computer_modelling structural_disorder*
E-mail bcnoll@ucdavis.edu bnoll@xray1.ucdavis.edu Tel. 1(916)7520192 Fax 1(916)7528995

Noonan, Joseph F. (1959). Systems Maganer. Molecular Structure Corp., 3200 Research Forest Drive, The Woodlands, TX 77381, USA. BS (Auburn U.-Montgomery, 1983) chemistry. *Computing computer_management computer_graphics charge_density*
E-mail jfn@msc.com Tel. 1(713)3631033 Fax 1(713)3643628

Nordman, Prof. Christer E. (1925). Professor. Department of Chemistry, University of Michigan, Ann Arbor, MI 48109, USA. PhD (U. Minnesota, 1953) physical chemistry. *Biological_structure patterson_method computing*
E-mail chris.nordman@um.cc.umich.edu Tel. 1(313)7647326 Fax 1(313)7474865

Norton, Dr Michael L. (1955). Associate Professor. Department of Chemistry, Marshall University, Huntington, WV 25755, USA. PhD (Arizona State U., 1982) solid state chemistry. *Superconductors DNA computer_modelling synthetic_metals materials_by_design STM*
Tel. 1(304)6966627 Fax 1(304)6962600

Nunes, Dr Anthony C. (1942). Professor. Department of Physics, University of Rhode Island, Kingston, RI 02881, USA. PhD (Massachusetts Institute of Technology, 1969) solid state physics. *Magnetic_particles surface_magnetism magnetic_colloids neutrons ordered_colloids small_angle_scattering photonic_crystals instrumentation*
E-mail acnunes@uriacc.uri.edu Tel. 1(401)7922048 Fax 1(401)7922380

Oeder Manning, Nancy (1952). Scientific Associate. Protein Data Bank, Department of Chemistry, Bldg.555, Brookhaven National Laboratory, PO Box 5000, Upton, NY 11973-5000, USA. MS (Rensselaer Polytechnic Institute, 1990) chemistry. *Protein_structure membrane_proteins*
E-mail oeder@bnl.gov Tel. 1(516)2825744

Oh, Dr Byung-Ha (1961). Associate Senior Investigator. Department of Macromolecular Sciences, SmithKline Beecham, 709 Swedeland Road, King of Prussia, PA 19406, USA. [261 Murray Dr., #B, King of Prussia, PA 19406, USA.] PhD (U. Wisconsin-Madison, 1989) biochemistry. *Structural_disorder proteins computer_modelling amphiphilic_toxins enzyme_mechanism enzyme_regulation homology_prediction inhibitor_design allosteric_regulation*
E-mail ohbh%phvax.dnet@smithkline.com Tel. 1(610)2707017 Fax 1(610)2704091

Ohashi, Yoshikazu (1941). Principal Software Engineer. Cognex, One Vision Dr., Natick, MA 01760-2059, USA. PhD (Harvard U., 1973) geology. *Silicates minerals computer_modelling machine_vision image_analysis*
E-mail yoshi@cognex.com Tel. 1(508)6503214 Fax 1(508)6503336

Ohlendorf, Dr Douglas H. (1950). Associate Professor. 4-225 Millard Hall, University of Minnesota, 435 Delaware Street S.E., Minneapolis, MN 55455-0347, USA. PhD (Washington U., 1978) biochemistry. *Biochemistry computer_modelling macromolecules structure–activity_relationship protein_engineering metalloproteins superantigens catalysis*
E-mail ohlen@dccc.med.umn.edu Tel. 1(612)6248436 1(612)6255175 Fax 1(612)6252163

Ohrt, Jean M. (1923). Retired. 33 Tamarack St, Buffalo, NY 14220-1730, USA. BA (U. Buffalo, 1949) biology.
Tel. 1(716)8255113

Ojala, William H. (1954). Research Associate. Biomedical Engineering Center, Box 107 UMHC, University of Minnesota, Minneapolis, MN 55455, USA. PhD (U. Minnesota, 1986) chemistry.
E-mail bojala@crystal.med.umn.edu Tel. 1(612)6261900 Fax 1(612)6251121

Oliveira, Marcos de Oliveira (1958). Postdoctoral research associate. Department of Chemistry and Biochemistry, University of Texas at Austin, Austin, TX 78712, USA. PhD (Purdue U., 1993) structural biology. *Crystallography_macromolecular computer_modelling molecular_replacement isomorphous_replacement homology_prediction maximum_entropy_method*
E-mail oliveira@utbc01.cm.utexas.edu Tel. 1(512)4718605

Oliver, Joel D. (1945). Section Head. 3073 Buell Rd., Cincinnati, OH 45251, USA. PhD (U. Texas, 1971) physical chemistry. *Proteins computer_modelling enzymes molecular_design homology_modelling water_structure structure_refinement data_collection*
E-mail oliver@pg.com Tel. 1(513)6271437 Fax 1(513)6271233

Olkowski, Ms Jennifer L. (1968). Physical Chemist; small-molecule crystallography laboratory. Small Molecule Crystallography, Eli Lilly and Company, Lilly Corporate Center, Indianapolis IN 46285-1513, USA. BS (Miami U., 1990) chemistry. *Crystallization data_collection structure_based_drug_design*
E-mail olkowski_jennifer_l@lilly.com Tel. 1(317)2761480 Fax 1(317)2765431

Olmstead, Dr Marilyn M. (1943). Staff Research Associate. Department of Chemistry, University of California, Davis. PhD (U. Wisconsin, 1969) inorganic chemistry. *Main_group bioinorganic fullerenes charge_transfer*
E-mail olmstead@chem.ucdavis.edu Tel. 1(916)7526668 1(916)7523534 Fax 1(916)7528995

Olsen, Dr Kenneth W. (1944). Professor and Chair. Department of Chemistry, Loyola University Chicago, 6525 N. Sheridan Road, Chicago, IL 60626, USA. PhD (Duke U., 1972) biochemistry. *Proteins computer_modelling hemoglobin homology_prediction molecular_dynamics evolution*
E-mail #114kwo@luccpua.it.luc.edu Tel. 1(312)5083121

Olson, Dr Arthur J. (1946). Member/Professor. Commission on Crystallographic Computing. Department of Molecular Biology, The Scripps Research Institute, La Jolla, CA 92037, USA. PhD (U. California, Berkeley, 1975) physical chemistry. *Molecular_computer_graphics protein_interaction drug_design viral_structure docking_computations antibody_antigen_interactions visualization_technology*
E-mail olson@scripps.edu Tel. 1(619)5549702 Fax 1(619)5546860

Olson, Dr David H. (1937). Senior Res. Consultant. Mobil R & D Corp, PO BOX 1025, Princeton, NJ 08543-1025, USA. PhD (Iowa State U., 1963) physical chemistry. *Zeolite_crystal_chemistry structure_property_relationships zeolite_catalysis*
E-mail dholson@crl.mobil.com Tel. 1(609)7374253 Fax 1(609)7374722

Onan, Kay (1949). Associate Dean. Offie of the Dean, 400 Meserve, Northeastern University, Boston, MA 02115, USA. PhD (Duke U., 1975) chemistry. *Transition_element_complexes heterocycles conformation*
E-mail konan@casdn.neu.edu Tel. 1(617)3735172 Fax 1(617)3732942

Ordway, Fred (1922). President. Instart Corp., 5205 Elsmere Ave., Bethesda, MD 20814-5732, USA. PhD (1949) physical chemistry. *Structural_disorder computer_modelling instrumentation water_structure*
E-mail fred.ordway@f412.n109.z1.fidonet.org Tel. 1(301)5302280

Oren, Deena A. (1961). Graduate Student. Chemistry Dept., Rutgers University and C.A.B.M., 679 Hoes Ln, Piscataway, NJ 08854-5638, USA. MSc (Hebrew U., Israel, 1986) chemistry. *Proteins nucleotides enzymatic_mechanisms viruses protein_ligand_interactions structure_function*
E-mail oren@biovax.rutgers.edu Tel. 1(908)2354482 Fax 1(908)2354850

Orme-Johnson III, William H. (1938). Professor of Chemistry and Housemaster (Bexley Hall). Department of Chemistry, Bldg 18-023, Mass.Inst. of Technology, Cambridge, MA 02138, USA. PhD (U. Texas, 1964) chemistry (biochemistry). *Enzymology X-ray_scattering metalloproteins nitrogen_fixation steroidogenesis renal_lithogenesis*
E-mail whoj@mit.edu Tel. 1(617)2531862 Fax 1(617)2531998

Osslund, Dr Timothy D. (1957). Research Scientist. Molecular Structure Group, Amgen, Thousand Oaks, CA 91320, USA. PhD (UCLA, 1992) protein X-ray crystallography. *Peptide_mimetics proteins computer_modelling homology_prediction protein_toxins*
E-mail osslund@amgen.com Tel. 1(805)4992403 Fax 1(805)4995788

Ostrofsky, Bernard (1922). Retired Senior Research Associate. 3041 Falstaff Rd., Baltimore, MD 21209, USA. BS (City College of New York, 1945) physical chemistry. *Crystallography*
Tel. 1(410)7645667

Otwinowski, Zbyszek (1956). Southwestern Medical Center, University of Texas, 5323 Harry Hines Blvd, Dallas, TX 75235-9038, USA. PhD (U. Chicago, 1989) biochemistry and molecular biology. *Detectors data_collection phasing software*
E-mail zbyszek@mix.swmed.edu Tel. 1(214)6485098 Fax 1(214)6485095

Ou, Chia-Chih (1945). Principal Scientist. Construction Products Div., WR Grace & Co., 62 Whittmore Avenue, Cambridge, MA 02140, USA. PhD (Rutgers U., 1977) physical chemistry. *Clay_minerals liquid_crystals*
Tel. 1(617)4984805 Fax 1(617)8766896

Pabo, Dr Carl O. (1952). Professor. Howard Hughes Medical Institute, Department of Biology, Massachusetts Institute of Technology, Cambridge, MA 02139, USA. PhD (Harvard U., 1980) biochemistry and molecular biology. *Protein–DNA_interactions protein_design*
E-mail pabo@pabo1.mit.edu Tel. 1(617)2538865 Fax 1(617)2538728

Pacalo, Rosemary E. Gerald (1959). Postdoctoral Research Associate. Department of Chemistry, Arizona State University, Tempe, AZ 85287-1604, USA. PhD (SUNY at Stony Brook, 1993) geophysics. *High_pressure synthesis crystal_structure_determination elasticity_systematics phase_transition Brillouin_spectroscopy synchrotron_radiation nitrides*
E-mail aarep@acvax.inre.asu.edu Tel. 1(602)9656634 1(602)9658861 Fax 1(602)9650474

Padlan, Eduardo (1940). Visiting Scientist. National Institutes of Health, Bethesda, MD 20892, USA. PhD (Johns Hopkins U., 1968) biophysics. *Biological_macromolecules immunology*
E-mail eap@vger.niddk.nih.gov Tel. 1(301)4021780 Fax 1(301)4960201

Padmanabhan, Dr Kaillathe (1958). Research Associate. Department of Chemistry, Michigan State University, East Lansing, MI 48824-1322, USA. PhD (Indian Institute of Science Bangalore India, 1986) organic chemistry. *Blood_proteins crystallization structure_solution computer_modelling*
E-mail pappan@cemvax.cem.msu.edu Tel. 1(517)3559715x251 Fax 1(517)3531793

Pagoaga, Dr M. Katherine (1952). Computer Specialist. National Institute of Standards and Technology, 325 Broadway, Mailstop 883.04, Boulder, CO 80303-3328, USA. PhD (U. Maryland, 1983) geochemistry. *Crystallographic_programming uranium_minerals minerals small_molecules*
E-mail pagoaga@bldr.nist.gov pagoaga@nist.gov Tel. 1(303)4975104 Fax 1(303)4973012

Palenik, Dr Gus J. (1933). Professor. Department of Chemistry, University of Florida, Gainesville, FL 32611-2046, USA. PhD (U. Southern California, 1960) inorganic chemistry. *Lanthanides aluminum_compounds bismuth_compounds transition_elements*
E-mail crystal@nervm.nerdc.ufl.edu crystal@nervm.bitnet Tel. 1(904)3926734 Fax 1(904)3923255

Palmer, Kenneth J. (1910). Retired. 1324 Peartree Lane, Medford, OR 97504, USA. PhD (California Institute of Technology, 1938) electron diffraction.
Tel. 1(503)8576424

Pandi, Dr Veerapandian (1951). Head of Scientific Computing Center. La Jolla Cancer Research Foundation, 10901 North Torrey Pines Road, La Jolla, CA-92037, USA. PhD (Indian Institute of Science, 1984) protein crystallography. *Protein_crystallography modelling drug_design interleukin-1 sequence_alignment teaching_aids*
E-mail pandi@ljcrf.edu Tel. 1(619)4556480 Fax 1(619)4532242

Pandit, Dr Jay (1957). Senior Research Scientist. Central Research Division, Pfizer Inc., Eastern Point Road, Groton, CT 06340, USA. PhD (Indian Institute of Science, Bangalore, 1985) physics. *Proteins computer_modelling homology_prediction enzyme_inhibitors renins phosphorylases*
E-mail pandit@pfizer.com Tel. 1(203)4413738 Fax 1(203)4413754

Pangborn, Robert N. (1951). Professor and Chair, Intercollege Graduate Program in Materials. Department of Engineering Science & Mechanics, Penn State University, 227 Hammond Building, University Park, PA 16802, USA. [322 E. Irvin Avenue, State College, PA 1601, USA.] PhD (Rutgers U., 1979) mechanics & materials science. *Failure_analysis fatigue residual_stress fracture coatings composite_materials nondestructive_evaluation*
E-mail mpesm@engr.psu.edu Tel. 1(814)8630721

Pani, Dr Kodandapani Ramadurgam. (1961). Res. associate. La Jolla Cancer Research Foundation, 10901 N. Torrey Pines Road, La Jolla CA 92037, USA. PhD (IISc Bangalore India, 1990) molecular biophysics. *Solvent_structure proteins water_mediated_transformations bacterial_LDH Fab_opg2 ms2 drug_binding*
E-mail kpani@ljcrf.edu Tel. 1(619)4556480 Fax 1(619)4502049

Parge, Dr Hans E. (1955). Senior Scientist. Agouron Pharmaceuticals, Inc., Research Laboratories, General Atomics Court, San Diego, CA 92121, USA. PhD (Trinity College Dublin, 1982) crystallography. *Macromolecules crystallisation recognition crystallography*
E-mail parge@agouron.com Tel. 1(619)6223000 Fax 1(619)6223299

Parise, John B. (1953). Associate Professor. Department of Earth and Space Sciences, State University of New York, Stony Brook, NY 11794-2100, USA. PhD (James Cook University of North Queensland, 1981) geochemistry. *Solid_state sulfur_compounds diffraction minerals zeolites synchrotron_radiation neutron_diffraction*
E-mail parise@sbmp01.ess.sunysb.edu jparise@ccmail.sunysb.edu Tel. 1(516)6328196 Fax 1(516)6328140

Park, Chang H. Park (1947). Group Leader. Abbott Laboratories, D46Y, AP9A, Abbott Park, IL 60064, USA. PhD (U. Kansas, 1982) physical chemistry. *Proteins*
E-mail parkc@randb.abbott.com Tel. 1(708)9370488 Fax 1(708)9372625

Park, Hee-Won (1963). Graduate Student. c/o Diesenhofer Lab., 5323 Harry Hines Blvd, Dallas, TX 75235-9050, USA. prePhD (U. Texas Southwestern Medical Center, 1995) biochemistry and biophysics. *Crystallography proteins DNA_binding electron_transfer*
E-mail park@howie.swmed.utexas.edu Tel. 1(214)6485091 Fax 1(214)6485095

Parkin, Dr Sean R. (1966). Post Doc. Department of Chemistry, University of California, Davis, CA 95616, USA. PhD (U. C. Davis, 1993) crystallography. *Cryocrystallography biological_macromolecules methods_development computer_programming structural_chemistry surface_structure*
E-mail srparkin@ucdavis.edu Tel. 1(916)7526668

Parris, Dr Kevin D. (1960). Staff Fellow. Laboratory of Molecular Biology/NIDDK, National Institutes of Health, Bethesda, MD 20892, USA. PhD (U. Pittsburgh, 1988) organic chemistry. *Protein_crystallography computer_modelling mechanism allostery chemistry diffraction_techniques immunology*
E-mail parris@vger.niddk.nih.gov Tel. 1(301)4024495 Fax 1(301)4960201

Partin, Daniel E. (1965). Post-Doc. Member. Department of Chemistry and Biochemistry, Arizona State University, Tempe, AZ 85287, USA. PhD (Arizona State U., 1993) solid state chemistry. *Powder_diffraction high_pressure pulsed_neutron structure_determination hydroxides spectroscopy pulsed_neutron_diffraction*
E-mail agdep@acvax.inre.asu.edu Tel. 1(602)9656570 Fax 1(602)9650474

Paruchuri, Mohan (1959). Research associate. Department of Materials Science and Engineering, Carnegie Mellon University, Pittsburgh, PA 15213, USA. PhD (Osmania U., India, 1989) materials science/physics. *Lattice_parameter strain_determination thermal_expansion ultra_high_pressure intermetallics*
E-mail mpiz@andrew.cmu.edu Tel. 1(412)2683625 Fax 1(412)2687596

Pathak, Dr Dushyant (1962). Postdoctoral Associate. Department of Molecular Biophysics and Biochemistry, Yale University, Box 6666, 260 Whitney Avenue, New Haven, CT 06511, USA. PhD (Northwestern U., 1989) biochemistry, molecular biology and cell biology. *Protein_structure protein_folding protein_nucleic_acid_interactions enzymes molecular_dynamics*
E-mail dushyant@csbarg.med.yale.edu dushyant@laplace.csb.yale.edu Tel. 1(203)4325066

Pattanayek, Rekha (1953). Research Instructor. Department of Biochemistry, Vanderbilt University Medical Center, 21st Av. South, Nashville, TN 37232-0146, USA. PhD (Saha Inst. of Nuclear Physics, Calcutta U., India, 1986) physics(X-ray crystallography). *Crystal_structure_proteins retinoic_acid_binding_protein fiber_diffraction macromolecular_assembly_studies*
E-mail menlab@lhmrba.hh.vanderbilt.edu Tel. 1(615)3437334 Fax 1(615)3224349

Patton, Mr Alan T. (1955). Instructor. Department of Physical Science, Mesa Community College, 1733 W. Southern Ave., Mesa, AZ 85257, USA. MS (U. California, Los Angeles, 1991) organic chemistry.
E-mail 4cap000@mc.maricopa.edu 4cap000@mc.bitnet Tel. 1(602)9450575

Pattridge, Kitty (1948). Research Associate. Biophysics Research Division, 3050 Chemistry Building, 930 N. University, The University of Michigan, Ann Arbor, Michigan 48109, USA. MS (The University of Michigan, 1975) geological sciences. *Proteins structure_refinement structure_comparison computer_modelling computer_graphics computer_management data_collection data_processing*
E-mail pattridg@umich.edu Tel. 1(313)7634698 Fax 1(313)7643323

Paul, Prof. Iain (1938). Professor. Department of Chemistry, University of Illinois, Urbana, IL 61801, USA. PhD (U. Flasgow, 1962) chemistry. *Organic_solid_state symmetry phase_transitions*
Tel. 1(217)3333007 Fax 1(217)2448068

Pavalow, Melvin (1923). Retired (Emeritus Professor). 4 Dale Ave., Syoset, NY 11791, USA. PhD (Adelphi U., 1974) X-ray crystallography.
Tel. 1(516)9354761

Pavkovic, Stephen F. (1932). Professor. Dept. of Chem., Loyola Univ. Chicago, 6525 N. Sheridan Rd, Chicago, IL 60626-5385, USA. PhD (Ohio State U., 1964) inorganic chemistry. *Small_molecules*
E-mail spavkov@luccpua.it.luc.edu Tel. 1(312)5083100 Fax 1(312)5083086

Pavlovsky, Dr Alexander G. (1947). Scientist Associate. ABL-Basic Research Program, NCI-FCRFDC, PO Box B, Bldg 539, Frederick, MD 21702-1201, USA. PhD (Institute of Molecular Biology Academy of Sciences of the USSR, 1979) protein crystallography / molecular biology. *Protein_structure lymphokines growth_hormones receptor_binding structural_homology computer_modelling*
E-mail pavlovsk@ncifcrf.gov Tel. 1(301)8465031 Fax 1(301)8465991

Peascoe, Roberta A. (1951). Postdoctoral Fellow. Department of Medicinal Chemistry, Medical College of Virginia, Virginia Commonwealth University, Box 540, Richmond, VA 23298-0540, USA. PhD (Texas A&M U., 1990) inorganic chemistry.
E-mail rpeascoe@ruby.vcu.edu rpeascoe@vcuvax Tel. 1(804)3717406

Peat, Tom (1963). Grad student. Department of Biochemistry, Hendrickson Lab., Columbia University, 650 W. 168 St., New York, NY 10032, USA. MS biochemistry. *Protein_crystallography*
E-mail peat@cuhhca.hhmi.columbia.edu Tel. 1(212)3058236 Fax 1(212)3057379

Pecharsky, Dr Vitalij K. (1954). Visiting Scientist. Ames Laboratory, Iowa State University, Ames, IA 50011-3020, USA. PhD (Lviv State U., Ukraine, 1979) inorganic chemistry. *Intermetallides crystallographic_computing powder_data_processing low_temperature magnetic_behaviors heat_capacity*
E-mail vitkp@ameslab.gov viktp@alisuvax.bitnet Tel. 1(515)2942728 Fax 1(515)2943709

Pecoraro, Vincent L. (1956). Professor. University of Michigan, Ann Arbor, MI 48109-1055, USA. PhD (UC Berkeley, 1981) inorganic chemistry. *Manganese vanadium metallacrowns metallapeptides*
E-mail vincent.pecoraro@chem.lsa.umich.edu Tel. 1(313)7631519 Fax 1(313)9367628

Peek, Mary E. (1969). Graduate Student. School of Chemistry and Biochemistry, Georgia Institute of Technology, Atlanta, GA 30332, USA. BA (Hollins College, 1990) chemistry. *DNA_distortion DNA_structure DNA_flexibility DNA_protein_interactions*
E-mail peek@max.chemistry.gatech.edu Tel. 1(404)8948338 Fax 1(404)8947452

Peiser, H. Steffen (1917). [638 Blossom Drive, Rockville, MD 20850, USA.] MA (Cambridge/UK, 1943) chemistry. *Atomic_weights avogadro_constant* Fax 1(301)7621173

Pence, Laura E. (1965). Postdoctoral Fellow. Department of Chemistry, Massachusetts Institute of Technology, Cambridge, MA 02139, USA. PhD (Michigan State U., 1992) inorganic chemistry. *Small_molecule model_compounds heavy_atoms education iron_compounds manganese_compounds rhodium_compounds*
E-mail pence@lippard.mit.edu Tel. 1(617)2531823 Fax 1(617)2588150

Penner-Hahn, Dr James E. (1957). Associate Professor. Department of Chemistry, University of Michigan, 930 N. University Avenue, Ann Arbor, MI 48109-1055, USA. PhD (Stanford U., 1984) chemistry. *EXAFS magnetic_resonance synchrotron_radiation WAXS oxygen_evolution photosynthesis*
E-mail james.penner-hahn@umich.edu Tel. 1(313)7647324 Fax 1(313)7474865

Pennington, Dr William T. (1955). Director, Molecular Structure Center. Department of Chemistry, Clemson University, Clemson, SC 29634-1905, USA. PhD (U. Arkansas, 1983) inorganic chemistry. *Charge_transfer iodine_compounds polyiodides small_molecules*
E-mail xraylab@clemson.clemson.edu xraylab@clemson.bitnet Tel. 1(803)6564200 Fax 1(803)6566613

Perona, Dr John J. (1961). Assistant Professor. Department of Chemistry, and Interdepartmental Program in Biochemistry and Molecular Biology, University of California at Santa Barbara, Santa Barbara, CA 93106, USA. PhD (Yale U., 1990) molecular biophysics and biochemistry. *RNA_structure protein-nucleic_acid_interactions enzyme_specificity*
E-mail jperona@sbmm1.ucsb.edu Tel. 1(805)8937389 Fax 1(805)8934120

Perozzo, Mary Ann (1961). Research Chemist. Laboratory for the Structure of Matter, Code 6030, Naval Research Laboratory, 4555 Overlook Avenue, S.W., Washington, DC 20375-5000, USA. MS (The Catholic U of America, 1992) biophysical chemistry. *Macromolecular_crystallization proteins structure_determination metalloproteins bioluminescent_proteins organophosphorus_acid_anhdrase phosphotriesterase*
E-mail perozzo@lsm.nrl.navy.mil Tel. 1(202)7670657 Fax 1(202)7676874

Pershan, Peter (1934). Professor. 205C Pierce Hall, Harvard University, Cambridge, MA 02138, USA. PhD (Harvard U., 1960) physics. *Liquid_surfaces order-disorder synchrotron scattering*
E-mail pershan@das.harvard.edu Tel. 1(617)4953214 Fax 1(617)4959837

Petersen, Donald R. (1929). Principal investigator. Greenleaf Associates, Post Office Box 1785, Midland, MI 48641-1785, USA. PhD (Caltech, 1955) physical chemistry. *Powder_diffraction crystallographic_computing statistical_analysis_experimental_data*
Tel. 1(517)8358356 Fax 1(517)8352441

Petersen, Jeffrey L. (1947). Professor. Department of Chemistry, West Virginia University, Morgantown, WV 26506-6045, USA. PhD (U. Wisconsin-Madison, 1974) physical chemistry. *Organometallic_chemistry X-ray_diffraction catalysis synthesis early_transition_metal_compounds reactivity*
E-mail u0523@wvnvm Tel. 1(304)2933435 Fax 1(304)2934904

Peters-Libeu, Clare (1963). Graduate Student. Department of Biological Structure, SM20, University of Washington, Seattle, WA 98195, USA. MS (San Francisco State U., 1990) physics. *High_precision_structures metalloproteins protein_kinases electrostatics computer_modelling protein_phosphorylation bioenergetics electrostatic_potential phosphorylases redox_proteins electron_transfer cell_signaling*
E-mail libeu@u.washington.edu Tel. 1(206)5431089

Petsko, Dr Gregory A. (1948). Lucille P. Markey Professor of Biochemistry and Chemistry. Rosenstiel Basic Medical Sciences Research Center, Brandeis University, Waltham, MA 02254-9110, USA. DPhil (Oxford U., 1973) molecular biophysics. *Enzyme_catalysis protein_dynamics transporters genetic_selection enzyme_evolution isomerases racemases*
E-mail petsko@binah.cc.brandeis.edu petsko@brandeis.bitnet Tel. 1(617)7362474 Fax 1(617)7362405

Pett, Dr Virginia B. (1941). Associate Professor. Secretary, Small Molecules Special Interest Group, Am. Crystallographic Assn. Department of Chemistry, The College of Wooster, Wooster, OH 44691, USA. PhD (Wayne State U., 1979) inorganic chemistry. *Computer_modelling proteins organo-cobalt_complexes*
E-mail pett@acs.wooster.edu pett@wooster.bitnet Tel. 1(216)2632114 Fax 1(216)2632386

Petz, John I. (1936). Retired Consultant. Alumina Dept, Reynolds Metals, 103 Hawthorne, Portland, TX 78374, USA. MS (U. Arkansas, 1959) physics. *Liquid_state powders*
Tel. 1(512)6437871 Fax 1(512)6437871

Pflaum, Wolfgang (1944). [20 Cresthaven, Irvine, CA 92714-3315, USA.] BS (Harvey Mudd College, 1966) physical chemistry. *Solid_state_materials*
Tel. 1(714)5595082

Pfluger, Prof. Clarence E. (1930). Faculty Associate. Keene State College, Department of Chemistry, 229 Main Street, Keene, NH 03431-4183, USA. PhD (U. of Texas at Austin, 1958) chemistry. *Solid_state_photochemistry photochromism hydrogen_bonding packing interactions*
E-mail pfluger@newpisgah.keene.edu Tel. 1(603)3582563 Fax 1(603)3582257

Pflugrath, James W. (1957). Sr Staff Investigator. WM Keck Structural Biology Laboratory, Cold Spring Harbor Lab., PO Box 100, Cold Spring Harbor, NY 11724, USA. PhD (Rice U., 1984) biochemistry. *Macromolecules protein_kinases data_collection synchrotron_radiation signal_transduction protein/protein_interactions protein/DNA_interactions*
E-mail pflugrath@cshl.org Tel. 1(516)3678821 Fax 1(516)3678873

Phillips Jr, Prof. George N. (1952). Professor. Dept. of Biochemistry and Cell Biology, Rice University, Houston, TX 77251-1892, USA. PhD (Rice U., 1977) biochemistry. *Proteins structure_disorder diffuse_scattering refinement_method computational_crystallography muscle_proteins myoglobin tropomyosin*
E-mail georgep@rice.edu Tel. 1(713)5274910 Fax 1(713)2855154

Phillips, Dr James Christopher (1952). Senior Applications Scientist. Siemens Analytical X-ray, 6300 Enterprise Lane, Madison, WI 53719-1173, USA. PhD (Stanford U., 1979) applied physics. *Instrumentation methods structure_determination methods_sources instrumentation*
Tel. 1(608)2763032 Fax 1(608)2763006

Phillips, T. J. (1919). Professor Emeritus. 1750 S. Flannery Rd., Baton Rouge, LA 79816, USA. PhD (Ohio State U., 1958) physical chemistry.

Phillips, Walter C. (1936). Senior Scientist. Rosenstiel Basic Medical Sciences Research Center, Brandeis University, Waltham, MA 02254-9110, USA. PhD (MIT, 1964) condensed matter physics. *X-ray_detectors X-ray_instrumentation macromolecular_assemblies*
E-mail walter@dark.rose.brandeis.edu Tel. 1(617)7362542 Fax 1(617)7362405

Phizackerley, Prof. Richard Paul (1945). Professor (Research). Stanford Synchrotron Radiation Laboratory, Stanford Linear Accelerator Center, PO Box 4349, Bin 69, Stanford University, CA 94309-0210, USA. PhD (U. Cambridge, UK, 1971) physics. *Protein_crystallography synchrotron_radiation anomalous_scattering structure_and_function_of_macromolecules instrumentation Laue_diffraction*
E-mail phiz@ssrl01.slac.stanford.edu phiz@ssrl750.bitnet Tel. 1(415)9263431 Fax 1(415)9264100

Pichla, Susan P. (1966). Graduate Student. The Wistar Institute, 3601 Spruce Street, Philadelphia, PA 19107, USA. BA (Gettysburg College, 1988) chemistry. *Macromolecules antibodies molecular_recognition Fab_fragments antigens*
E-mail susan@wistb.wistar.upenn.edu Tel. 1(215)8982202 Fax 1(215)8983868

Picot, Daniel (1955). Research associate. Department of Biochemistry and Molecular Biology, The University of Chicago, 920 E 58th Street, Chicago, IL 60615, USA. PhD (U. Basel Switzerland, 1986) biophysics. *Proteins crystallography structure_determination structure-activity_relationship membrane_proteins detergent crystallization*
E-mail picot@biovax.uchicago.edu picot@ibpc.fr Tel. 1(312)7020286 Fax 1(617)7020439

Pieper, Dr Ursula (1963). Research Associate. Center of Advanced Research in Biotechnology, 9600 Gudelsky Dr, Rockville MD 20850, USA. Dr.rer.nat. (U. Göttingen, 1993) inorganic chemistry and crystallography. *Structural_disorder proteins computing crystallization*
E-mail ursula@elan1.carb.nist.gov Tel. 1(301)7386210 Fax 1(301)7386255

Pierpont, Cortlandt G. (1942). Professor of Chemistry. Department of Chemistry, University of Colorado, Boulder, CO 80309, USA. PhD (Brown U., 1971) chemistry. *Coordination_complexes quinone_ligands*
E-mail cortlandt.pierpont@colorado.edu Tel. 1(303)4928420 Fax 1(303)4925894

Pignataro, Mrs Edith H. (1925). Retired. [230 Jay St. APT. 1-G, Brooklyn, NY 11201-1948, USA.] MS (Polytechnic Inst. of Brooklyn, 1954) physics (X-ray crystallography).

Pinkerton, Dr A. Alan (1943). Professor. Department of Chemistry, University of Toledo, Toledo, OH 43606, USA. PhD (U. Alberta, Canada, 1971) inorganic chemistry. *Small_molecule-crystallography experimental_charge_densities NMR lanthanide_and_actinide_chemistry phosphorus_and_sulfur_chemistry*
E-mail apinker@uoft02.utoledo.edu Tel. 1(419)5374568 Fax 1(419)5374033

Pique, Michael E. (1951). Programmer. The Scripps Research Institute MB-5, 10666 N. Torrey Pines Road, La Jolla, CA 92037, USA. MS (U. North Carolina, Chapel Hill, 1980) computer science. *Three-dimensional_computer_graphics scientific_visualization*
E-mail mp@scripps.edu Tel. 1(619)5549775

Pjura, Dr Philip E. (1956). Postdoctoral Research Associate. University of Delaware, Department of Chemical Engineering, Newark, DE 19716, USA. PhD (California Institute of Technology, 1987) chemical biology. *Nucleic_acids proteins anticancer_compounds protein_stability thermodynamics crystallization computer_graphics*
E-mail pjura@che.udel.edu Tel. 1(302)8314132 Fax 1(302)8311048

Pley, Heinz W. (1963). Graduate student. Dept. Cell Biology, Stanford University Medical Center, Stanford CA 94305-5400, USA. MS (U. Regensburg, 1990) biology. *Catalytic_RNA low_temperature_crystallography oligonucleotides nucleic_acids*
E-mail pley@cellbio.stanford.edu Tel. 1(415)7236622 Fax 1(415)7238464

Pluth, Joseph J. (1943). Senior Research Associate. Department of Geophysical Sciences, CARS, MRL, The University of Chicago, 5734 S. Ellis Ave., Chicago, IL 60637, USA. PhD (U. of Washington) physical chemistry. *Zeolites microcrystallography powder_diffraction synchrotron_radiation structure_solution_from_microcrystal_and_powder_data*
E-mail pluth@geo1.uchicago.edu Tel. 1(312)7028109 Fax 1(312)7029505

Ponader, Carl W. (1958). Senior Research Scientist. SP-FR-O1-8, Corning, NY 14831, USA. PhD (Stanford U., 1988) geochemistry. *Electron_X-ray_absorption_fine_structure electron_spin_resonance Raman glasses*
E-mail ponader_cw@corning.com Tel. 1(607)9743364 Fax 1(607)9743675

Ponder, Dr Jay W. (1957). Assistant Professor. Biochemistry and Molecular Biophysics, Box 8231, Washington University School of Medicine, 660 South Euclid Avenue, St. Louis, MO 63110, USA. PhD (Harvard U., 1984) organic chemistry. *Proteins computer_modelling molecular_dynamics protein_engineering protein_folding*
E-mail ponder@comet.wustl.edu Tel. 1(314)3624195 Fax 1(314)3627183

Poojary, Damodara (1953). Research Associate. Department of Chemistry, Texas A & M University, College Station, TX 77843, USA. PhD (I. I. Sc., Bangalore, India, 1984) crystallography. *Material_science powder_diffraction electron_diffraction structure_solution_from_powder_data*
E-mail mdpo711@tamxrd.tamu.edu Tel. 1(409)8453812 Fax 1(409)8454719

Porter, Leigh C. (1955). Assistant Professor. Department of Chemistry, University of Texas at El Paso, El Paso, TX 79968-0513, USA. PhD (U. California, Irvine, 1984) inorganic chemistry. *Small_molecule_crystallography*
Tel. 1(915)7477579 Fax 1(915)7475748

Post, Jeffrey E. (1954). Chairman/Curator. Smithsonian Institution, NHB 119, Washington, DC 20560, USA. PhD (Arizona State U., 1981) crystallography/geochemistry. *Manganese_minerals oxides Rietveld powder_X-ray_diffraction*
E-mail mnhms001@sivm.si.edu Tel. 1(202)3574009 Fax 1(202)3572476

Potenza, Joseph A. (1941). Professor of Chemistry, Provost, Dean of Graduate School. Chemistry Department, Rutgers University, New Brunswick, NJ 08903, USA. PhD (Harvard U., 1967) chemistry. *Copper_compounds chelates twinning electron_spin_resonance*
Tel. 1(908)9327461 Fax 1(908)9328184

Poulos, Thomas L. (1947). Professor of Biochemistry and Biophysics, Dept. of Molecular Biol. and Biochem., Univ. Calif, Irvine, Irvine, CA 92717-3900, USA. PhD (U. Calif. San Diego, 1972) biology. *Protein_crystallography engineering heme_enzyme_structure_and_function*
E-mail poulos@uci.edu Tel. 1(714)8567020 Fax 1(714)8568540

Powell, Dr Douglas R. (1953). Senion Scientist. Department of Chemistry, 1101 University Ave., University of Wisconsin, Madison, WI 53706-1322, USA. PhD (Iowa State U., 1980) physical chemistry. *Computing service diffractometry experi_system*
E-mail powell@chem.wisc.edu Tel. 1(608)2634694 Fax 1(608)2620381

Pratt, David W. (1937). Professor. Dept of Chemistry, University of Pittsburgh, Pittsburgh, PA 15260, USA. PhD (UC Berkeley, 1967) physical chemistry. *Molecular_structure_determination_by_eigenstate_resolved_spectroscopy*
E-mail pratt@vms.cis.pitt.edu Tel. 1(412)6248660 Fax 1(412)6248552

Preusch, Peter (1953). Program Administrator. National Institute of General Medical Science, 45 Center Drive, MSC 6200, Bethesda, MD 20892-6200, USA. PhD (Cornell U., 1979) biochemistry. *Biological_membranes proteins crystallization research_administration*
E-mail pc6@cu.nih.gov Tel. 1(301)5941832 Fax 1(301)4802802

Prewitt, Dr Charles T. (1933). Director. Geophysical Laboratory, Carnegie Institution of Washington, 5251 Broad Branch Road, N.W., Washington, DC 20015-1305, USA. PhD (Massachusetts Institute of Technology, 1962) geology and geophysics. *Geochemistry geophysics mineralogy solid_state_chemistry high_pressure high_temperature chalcogenides silicates*
E-mail prewitt@gl.ciw.edu Tel. 1(202)6862410x2450 Fax 1(202)6862419

Prince, Dr Edward (1928). Research Physicist. Member, Commission on Neutron Diffraction. Reactor Radiation Division, National Institute of Standards and Technology, Gaithersburg, MD 20899, USA. PhD (U. Cambridge, UK, 1952) physics. *Neutron_diffraction instrumentation refinement_method Rietveld_method powder_diffraction maximum_entropy fast_Fourier_transform*
E-mail prince@enh.nist.gov prince@nbsenh.bitnet Tel. 1(301)9756230 Fax 1(301)9219847 Telex 197674NISTSTUT

Prive, Dr Gilbert G. (1960). Postdoctoral Fellow. Molecular Biology Institute, UCLA, Los Angeles, CA 90024-1570, USA. PhD (U. California, Los Angeles, 1988) biochemistry. *Proteins DNA membranes computer_modelling active_transport sequence_homology hydrophobic_effect transmembrane_signalling*
E-mail prive@uclaue.mbi.ucla.edu Tel. 1(310)2063642 1(310)2065055 Fax 1(310)2063914

Proctor, Peter d'urphee (1953). Graduate student. Case Western Reserve University, Department of Biochemistry, 10900 Euclid Avenue, Cleveland, Ohio 44106-4935, USA. BS (U. Washington, 1986) physical oceanography. *Blue_copper_proteins electron_transport_proteins electron_transport oceanography*
E-mail proctor@cwbio.bioc.cwru.edu Tel. 1(216)3688743 Fax 1(216)3684544

Profeta Jr, Salvatore (1951). Director, NCSC Research Institute. North Carolina Supercomputing Center, Research Institute, 3029 Cornwallis Road, Research Triangle Park, NC 27709-2889, USA. PhD (U. Georgia, 1978) physical and organic chemistry. *Structural_biology small_molecule_structure conformational_analysis conformational_polymorphism computer_assisted_drug_discovery computational_chemistry*
E-mail sal@mcnc.org sal@ncsc.org Tel. 1(919)2481100 Fax 1(919)2481101

Pulliam, Curtis R. (1957). Associate Professor. Department of Chemistry, Utica College of Syracuse University, 1600 Burrstone Road, Utica, NY 13502, USA. PhD (U. Wisconsin-Madison, 1986) inorganic chemistry. *Iron-sulfur_clusters electrochemistry transition_metal_clusters*
E-mail crp@uc1.ucsu.edu Tel. 1(315)7923140 Fax 1(315)7923292

Pulsinelli, Phillip D. (1943). Associate Professor. Department of Pharmaceutical Sciences/School of Pharmacy, University of Pittsburgh, 528 Salk Hall, Pittsburgh, PA 15261, USA. PhD (U. Pittsburgh, 1971) crystallography. *Protein_structure receptors drug_structure*
E-mail pulsinelli@druginfonet.pharm.epid.pitt.edu Tel. 1(412)6488529 Fax 1(412)6488490 Telex 199126

Purdy, Samuel (1926). Senior Research Associate. TRC-D, National Steel Corp., Trenton, MI 48183, USA. BS (Lehigh U., 1948) metallurgical engineering. *Metallography structure_metals*

Tel. 1(313)6762682 Fax 1(313)6762030

Qiu, Xiayang (1963). Senior Fellow. Dept. of Biological Structures, SM-20, Univ. of Washington, School of Medicine, Seattle, WA 98195, USA. PhD (Michigan State U., 1993) physical chemistry. *Crystal_structures proteins DNA drug_design protein-DNA_interaction blood_coagulation*

E-mail xiayang@gouda.bchem.washington.edu Tel. 1(206)5438865 Fax 1(206)5431524

Quicksall, Carl O. (1941). Vice President of Science & Technology. Mallinckrodt Chemical, Inc., PO Box 5439, St Louis, MO 63147, USA. [405 Redwood Forest Dr., Manchester, MO 63021, USA.] PhD (Princeton U., 1971) physical chemistry. *Industrial_crystallography*

E-mail 73317.23@compuserve.com Tel. 1(314)5391225 1(314)3940188 Fax 1(314)5395866

Quillin, Michael L. (1967). Graduate Student. Department of Biochemistry and Cell Biology, Rice University, 6100 S. Main, Houston, TX 77005, USA. [Department of Biochemistry and Cell Biology, Rice University, PO Box 1892, Houston, TX 77251, USA.] BS (St. Mary's University, 1989) biology and chemistry. *Structure_determination hemoproteins mutagenesis structure–activity_relationship*

E-mail mike@keckiris.rice.edu Tel. 1(713)5278101x3346 Fax 1(713)2855154

Quintana, Dr John P. Senior Research Scientist. DND-CAT Synchrotron Research Center, Northwestern University, Evanston, IL 60208, USA. [DND-CAT Synchrotron Research Center, APS/ANL Sector-5, Bldg. 400, 9700 South Cass Avenue, Argonne, IL 60439, USA.] PhD (Northwestern U., 1991) materials science and engineering. *Synchrotron_instrumentation computer_modelling X-ray_optics crystallography diffuse_scattering residual_stress materials_science*

E-mail jpq@nwu.edu Tel. 1(708)2520223 Fax 1(708)2520226

Quiocho, Dr Florante A. Professor and Investigator, HHMI. Department of Biochemistry, Baylor College of Medicine, Houston, TX 02167, USA. PhD (Yale U., 1966) biochemistry. *Structural_biology protein_crystallography molecular_recognition*

E-mail faq@dino.qci.bioch.bcm.tmc.edu Tel. 1(713)7986565 Fax 1(713)7976718

Quirk, Dr Stephen (1963). Postdoctoral Fellow. Dept. of Biophysics and Biophysical Chemistry, Johns Hopkins Medical School, Room 607 WBSB, 725 N. Wolfe St., Baltimore, MD 21205, USA. PhD (Johns Hopkins U., 1991) biology. *Enzymology substrate_design protein_complexes evolution*

E-mail quirk@jhuigf.med.jhu.edu quirk@jhuigf.bitnet Tel. 1(410)9558388 Fax 1(410)9550637

Rabalais, J. W. (1944). Cullen Professor of Chemistry and Physics. Dept. of Chemistry, University of Houston, Houston, TX 77004, USA. PhD (Louisiana State U., 1970) electronic spectroscopy. *Chemisorption surfaces structure reconstruction ion_beams film_deposition*

E-mail chem1w@jetson.uh.edu Tel. 1(713)7433282 Fax 1(713)7432709

Rabenberg, Llewellyn (1956). Associate Professor. Department of Mechanical Engineering, Center for Materials Science and Engineering, University of Texas at Austin, Austin, TX 78712-1063, USA. PhD (U. California at Berkeley, 1983) materials science. *Electron_microscopy electron_diffraction materials_science*

E-mail lew@mcl.cc.utexas.edu Tel. 1(512)4713178 Fax 1(512)4717681

Rabinowitz, Israel (1935). President. Cerechem Corp., 4187 Carpinteria Ave. Unit 6, Carpinteria, CA 93013, USA. PhD (Rutgers, 1966) biochemistry.

Tel. 1(805)5663410 Fax 1(805)5663416

Rader, Stephen D. (1964). Graduate Student. Graduate Group in Biophysics, Box 0448, University of California, San Francisco, San Francisco, CA 94117, USA. BA (Swarthmore College, 1987) biochemistry. *Protein_folding enzymology molecular_dynamics chaperonins RNA_splicing*

E-mail rader@cgl.ucsf.edu Tel. 1(415)4765143 Fax 1(415)4761902

Radha, Akella (1960). Research Associate. Whistler Center for Carbohydrate Research, 1160 Smith Hall, Purdue University, West Lafayette, IN 47907-1160, USA. PhD (Indian Institute of Technology, Madras, India, 1988) X-ray crystallography. *Macromolecular_crystallography molecular_modelling polysaccharides DNA structure_function_relationships*

E-mail jpj@mace.cc.purdue.edu Tel. 1(317)4944924 Fax 1(317)4947953

Rafalko, Patrice White (1951). Assist. Scientist. The BOC Group Technical Center, 100 Mountain Ave., Murray Hill, NJ 07974-2064, USA. MA (U of Texas, Austin, 1977) chemistry, crystallography. *Small_molecule single_crystal powder X-ray_fluorescence*

Tel. 1(908)7716380 Fax 1(908)7716488

Raines, Ronald T. (1958). Assistant Professor. Biochemistry Department, University of Wisconsin, 420 Henry Mall, Madison, WI 53706-1569, USA. PhD (Harvard U., 1986) chemistry. *Protein_folding protein_engineering enzymology biomolecular_recognition*

E-mail raines@biochem.wisc.edu Tel. 1(608)2628588 Fax 1(608)2623453

Rao, Dr Jejjala Krishna Mohana (1943). Scientist Associate. Natl Cancer Inst - Frederick Cancer R&D Center, ABL - Basic Research Program, PO Box B, Frederick, MD 21702, USA. PhD (Indian Inst. of Science, 1971) X-ray crystallography. *Proteins viruses computer_modelling homology_prediction molecular_evolution software_development*

E-mail rao@ncifcrf.gov Tel. 1(301)8465031 Fax 1(301)8465991

Rao, Narasinga (1938). Director of Faculty Research. Treasurer, CGA-17, IUCr Congress in 1996; Treasurer, ACA, 1989–1995. University of Central Oklahoma, 100 N. University Drive, AD218, Edmond, OK 73034, USA. PhD (SUNY, Buffalo, 1973) bio-physics crystallography. *Crystal_structure conformation protein_structure protein_function*

E-mail snrao@aix1.ucok.edu Tel. 1(405)3412980x2524 Fax 1(405)3303830

Rao, Dr S. T. (1937). Research Scientist. Biotechnology Center, The Ohio State University, 1060 Carmack Road, Columbus, OH 43210, USA. PhD (U. Madras, 1967) physics. *Proteins nucleic_acids computer_modelling computer_programming structural_homology protein-DNA_interactions hydration*

E-mail raost%biot@mps.ohio-state.edu Tel. 1(614)2922926 Fax 1(614)2922524

Rao, Dr Sudhakara S. P. (1958). Research Associate. Rosenstiel Center, Brandeis University, Waltham, MA 02254, USA. [Rigaku/USA, Inc., 199 Rosewood Drive, Danvers, MA 01923, USA.] PhD (Indian Institute of Science, Bangalore, India, 1986) inorganic chemistry. *Protein_crystallography image_analysis proteins coiled_coil_proteins modelling simulation*

E-mail rao@brandeis rao@auriga.rose.brandeis.edu Tel. 1(508)7772446 Fax 1(508)7773594

Rao, Dr Usha (1962). Postdoctoral Fellow. Department of Chemistry, Boston College, Chestnut Hill, MA 02167, USA. PhD (Boston College, 1992) chemistry. *Protein_crystallography amphiphilic_toxins rational_drug_design homology_prediction computer_modelling molecular_mechanics_dynamics*

E-mail rao@bcchem.bc.edu rao@bcchem.bitnet Tel. 1(617)5523905 Fax 1(617)5522705

Rardin, Dr R. Lynn (1963). American Cancer Society Postdoctoral Fellow. Rosenstiel Basic Medical Sciences Research Center, Brandeis University, 415 South Street, Waltham, MA 02254-9110, USA. [10 Riverside St. Apt. #3-4, Watertown, MA 02172-2661, USA.] PhD (Massachusetts Institute of Technology, 1991) inorganic chemistry. *Bioinorganic_chemistry molecular_complexes metalloproteins active_site_metals small_molecules iron_compounds dismutases area_detector computer_management*

E-mail rardin@auriga.rose.brandeis.edu rardin@brandeis.bitnet Tel. 1(617)7364941 Fax 1(617)7362405

Rath, Dr Nigam P. (1958). Research Assistant Professor. Department of Chemistry, University of Missouri - St. Louis, 8001 Natural Bridge Road, St. Louis, MO 63121, USA. PhD (Oklahoma State U., 1985) organometallic chemistry. *Crystallography structure_determination structural_disorder computing*

E-mail c1906@slvaxa.umsl.edu c1906@umslvaxa.bitnet Tel. 1(314)5535333 Fax 1(314)5535342

Rath, Virginia L. Postdoctoral Fellow. Department of Molecular Biophysics and Biochemistry, Yale University, New Haven, CT 06511, USA. [Department of Molecular Biophysics and Biochemistry, Yale University, 260 Whitney Avenue, PO Box 208114, Bass 413, New Haven, CT 06520-8114, USA.] PhD (U. California at San Francisco, 1991) biochemistry.

E-mail rath@csb.yale.edu Tel. 1(203)4325795 Fax 1(203)4323282

Ravichandran, Kurumbail G. (1960). Research Associate. Howard Hughes Medical Institute, Room Y4.210, 5323 Harry Hines Blvd, Dallas, Texas 75235-9050, USA. PhD (Michigan State U., 1989) physical chemistry. *Coagulation oxygenases macromolecular_structure crystallography metalloenzymes protein_kinases reductases receptor*

E-mail ravi@howie.swmed.edu Tel. 1(214)6485011 Fax 1(214)6485095

Ray, Dr Sanjoy (1964). Postdoc. Department of Biochemistry and Molecular Biology, Harvard University, Cambridge, MA 02138, USA. PhD (Imperial College London, 1990) biological chemistry. *Proteins computer_modelling*

E-mail ray@xtal220.harvard.edu ray@huxtal.bitnet Tel. 1(617)4954091 Fax 1(617)4959613

Rayment, Professor Ivan (1951). Professor. Institute for Enzyme Research, University of Wisconsin, 1710 University Ave, Madison, WI 53705, USA. PhD (U. Durham, England, 1975) chemistry. *Proteins muscle macromolecular_assemblies*

E-mail ivan@enzyme.wisc.edu Tel. 1(608)2620437 Fax 1(608)2652904

Reddy, Dr Jaya G. (1964). Research Associate. Department of Biochemistry, Michigan State University, East Lansing, MI 48824, USA. PhD (Madras U., India, 1992) biophysics. *Protein_crystallography molecular_biology transcription_factors protein-DNA_interactions nuclear_magnetic_resonance protein_biochemistry*

E-mail triezenbergs@clvax1.cl.msu.edu Tel. 1(517)3363668 Fax 1(517)3539334

Reddy, Dr Swaminatha B. (1946). Department Manager. Hughes STX Corp., 7701 Greenbelt Road, Suite 400, Greenbelt, MD 20770, USA. PhD (Indian Institute of Science, 1974) physics. *Computing nucleic_acids modelling macromolecular*

E-mail sreddy@ccmail.stx.com Tel. 1(301)4414045 Fax 1(301)4411853

Reddy, Vijay S. (1962). Postdoctoral Research Associate. Department of Biological Sciences, Lily Hall, #B142, Purdue University, West Lafayette IN 47907, USA. PhD (Indian Institute of Science, 1991) biophysics. *Crystallography virus_structure computer_modelling protein_folding*

E-mail reddyv@flash.cc.purdue.edu Tel. 1(317)4940831 Fax 1(317)4961189

Redford, Susan M. (1964). Graduate Student. Molecular Biology Dept., MB4, The Scripps Research Institute, 10666 N. Torrey Pines Rd., La Jolla, CA 92037-1112, USA. BA (U. California, San Diego, 1987) biochemistry and cell biology. *Stability protein_engineering catalysis allostery design ageing_process conformational_change*

E-mail smr@riscsm.Scripps.edu smr@Scripps.edu Tel. 1(619)5542880 Fax 1(619)5542880 1(619)5546880

Redinbo, Matthew R. (1966). Graduate student. Department of Chemistry and Biochemistry, and Molecular Biology Institute, University of California, Los Angeles, 405 Hilgard Avenue, Los Angeles, CA 90024, USA. BS (U. California, Davis, 1990) biochemistry. *Proteins photosynthesis twinning macromolecules computer_modelling methylation*

E-mail redinbo@uclaue.mbi.ucla.edu Tel. 1(310)8258901 Fax 1(310)2063914

Reeber, Dr Robert R. (1937). Materials Engineer Adjunct Professor. Department of Geology, CB#3315, University of North Carolina-CH, Chapel Hill, NC 27599-3315, USA. [Army Research Office, PO Box 12211, RTP, NC 27709, USA.] PhD (The Ohio State U., 1968) industrial mineralogy. *Thermal_expansion mineral_physics phase_transitions_crystal_growth*

E-mail rrrcryst@email.unc.edu reeber@aro-emh1.army.mil Tel. 1(919)5494318 Fax 1(919)5494310 or 1(919)9664519

Reeder, Dr Richard J. (1953). Professor. Department of Earth and Space Sciences, State University of New York at Stony Brook, Stony Brook, NY 11794-2100, USA. PhD (U. California, Berkeley, 1980) geochemistry. *Order-disorder phase_transition carbonates crystal_growth electron_microscopy geochemistry minerals high_temperature*

E-mail reeder@sbmp04.ess.sunysb.edu Tel. 1(516)6328208 Fax 1(516)6328240

Reeke Jr, Dr George N. (1943). Associate Professor. Laboratory of Biological Modelling, The Rockefeller University, 1230 York Avenue, New York, NY 10021, USA. PhD (Harvard U., 1969) physical chemistry. *Computing nervous_system proteins recognition neural_modelling simulation_software*

E-mail reeke@lobimo.rockefeller.edu Tel. 1(212)3277627 Fax 1(212)3277469

Rees, Douglas C. (1952). Professor. Division of Chemistry and Chemical Engineering, 147-75CH, California Institute of Technology, Pasadena, CA 91125, USA. PhD (Harvard, 1980) biophysics. *Biocrystallography biochemistry macromolecules enzymes*

E-mail rees@citray.caltech.edu Tel. 1(818)3958393 Fax 1(818)5689430

Regan, Michael J. (1966). Graduate Student. Department of Applied Physics, Stanford University, Stanford, CA 94305, USA. PhD (Stanford U., 1993) applied physics. *Phase_separation small-angle_scattering amorphous_solids X-ray_scattering*

E-mail regan@ssrl01.slac.stanford.edu regan@ssrl750.bitnet Tel. 1(415)9263953 Fax 1(415)9264100

Reibensplies, Dr Joseph H. (1958). Staff crystallographer. Department of Chemistry, Texas A & M University, College Station, Texas 77843, USA. PhD (Colorado State U., 1987) inorganic chemistry. *Data_processing inorganic_chemistry small_molecule macromolecule molecular_modelling*

E-mail reibensplies@chemvx.tamu.edu Tel. 1(409)8459125 Fax 1(409)8454719

Reid Jr, Austin (1957). Senior Research Supervisor. Iler Research Center, DuPont, 1 Quality Lane, New Johnsonville, TN 37134, USA. PhD (Auburn U., 1982) inorganic chemistry. *Inorganic_structural_chemistry small_molecule_structures amorphous_materials*

Tel. 1(615)5357439 Fax 1(615)5357606

Reinisch, Karin M. (1966). Graduate student. Chemistry Department, Harvard University, 12 Oxford Street, Cambridge, MA 02138, USA. BA (Harvard Radcliffe, 1989) chemistry. *Proteins protein_DNA_complexes*

E-mail reinisch@huray.harvard.edu reinisch@hubeta.harvard.edu Tel. 1(617)4954767 Fax 1(617)4953350

Reis Jr, Arthur H. (1946). Associate Provost. Office of the Provost, Brandeis University, Irving 104, Waltham, MA 02254-9110, USA. PhD (Harvard U., 1972) chemistry. *Small_molecules neutron_scattering electron_absorption_X-ray_fine_structure neural_networks*

E-mail reis@binah.cc.brandeis.edu Tel. 1(617)7362106 Fax 1(617)7363457

Remington, S. James (1950). Associate Professor. Institute of Molecular Biology, University of Oregon, Eugene, OR 97404, USA. PhD (U. Oregon, 1977) physics. *Proteins*

E-mail jim@uoxray.uoregon.edu Tel. 1(503)3465190 Fax 1(503)3465192 1(503)3465870

Ren, Zhong (1964). Research Associate. Department of Biochemistry and Molecular Biology, The University of Chicago, 920 East 58th Street, Chicago, IL 60637, USA. PhD (Academia Sinica, 1990) protein crystallography. *Laue_method time-resolved_crystallography synchrotron_radiation structural_biology anomalous_scattering MAD protein_structure*

E-mail renz@circe.uchicago.edu Tel. 1(312)7023603 Fax 1(312)7020439

Rey, Dr Felix (1957). Research Associate. Harvard University, Dept. of Biochemistry, 7 Divinity Ave., Cambridge, MA 02138, USA. PhD (U. Paris, XI, 1988) biochemistry. *Structural_virology receptors-complexes glycoproteins DNA_binding_proteins*

E-mail rey@xtal0.harvard.edu Tel. 1(617)4954091 Fax 1(617)4959613

Reynolds, Dr Ross (1951). Associate Professor. Physics Department, 234 Loutit Hall, Grand Valley State University, Allendale, MI 49401, USA. PhD (U. Oregon, 1983) physics. *Macromolecular_structure instrumentation inorganic_molecules synchrotron_radiation*

E-mail reynold_R@gvsu.edu Tel. 1(616)8952278 Fax 1(616)8958506

Rhee, Sangkee (1961). Graduate Student. Department of Biochemistry, 4-411 BSB, The University of Iowa, Iowa City, IA 52241, USA. PhD (The University of Iowa, 1993) X-ray crystallography. *Macromolecules_structure proteins enzymes_mechanism proteins_dynamics*

E-mail rhee@eve.biochem.uiowa.edu Tel. 1(319)3357883

Rheingold, Arnold L. (1940). Professor. Department of Chemistry, University of Delaware, Newark, DE 19716, USA. PhD (U. Maryland, 1969) inorg. chem.. *Organometallic small_molecule*

E-mail arnrhein@strauss.udel.edu Tel. 1(302)8318720 Fax 1(302)8316335

Rhodes, Dr Gale (1943). Professor. Department of Chemistry, University of Southern Maine, Portland, ME 04103, USA. PhD (North Carolina, 1971) organic chemistry. *Proteins computer_modelling crystallography_education science_and_literature chemical_education*

E-mail rhodes@portland.caps.maine.edu Tel. 1(207)7804734 Fax 1(207)7804933

Ribbe, Paul H. (1935). Professor of Mineralogy. Department of Geological Science, Virginia Polytech. Inst., Blacksburg, VA 24061-0420, USA. PhD (Cambridge U., 1963) crystallography.

E-mail ribbe@vt.edu Tel. 1(703)2316880 Fax 1(703)2313386

Ricci, Dr John S. (1940). Professor. Department of Chemistry, University of Southern Maine, Portland, ME 04103, USA. PhD (State University of New York at Stony Brook, 1969) inorganic chemistry. *Neutron_diffraction organometallics hydrogen_bonding*

E-mail ricci@portland.caps.maine.edu Tel. 1(207)7804736

Rice, Marybeth (1956). Graduate Student. Stanford Synchrotron Radiation Laboratory, PO Box 4349, MS 69, Stanford, CA 94309-0210, USA. MS (Stanford U., 1986) electrical engineering. *Amorphous_materials anomalous_small-angle_X-ray_scattering phase_separation synchrotron_radiation metal–germanium_alloys*

E-mail rice@ssrl01.slac.stanford.edu rice@ssrl750.bitnet Tel. 1(415)9263041 Fax 1(415)9264100

Richards, Frederic M. (1925). Professor Emeritus. Department of Molecular Biophysics and Biochemistry, Yale University, 260 Whitney Ave, New Haven, CT 06520-8114, USA. PhD (Harvard U., 1952) biochemistry. *Proteins_(all_aspects)*

E-mail richards@hhvms8.decnet.yale.edu Tel. 1(203)4325620 Fax 1(203)4325175

Richardson, David C. (1940). Professor. Dept. of Biochemistry, Duke University, Durham, NC 27710, USA. PhD (MIT, 1967) inorganic chemistry. *Protein_structure protein_design computer_graphics computer_modelling enzyme_function molecular_biophysics*

E-mail dcr@suna.biochem.duke.edu Tel. 1(919)6846010 Fax 1(919)6848885

Richardson, Jane S. (1941). James B. Duke Professor of Biochemistry. 211 Nanaline Duke Bldg., Dept. of Biochemistry, Duke University, Durham, NC 27710, USA. MA (Harvard U., 1966) philosophy. *Proteins protein_design macromolecular_graphics protein_folding*

E-mail jsr@suna.biochem.duke.edu Tel. 1(919)6846010 Fax 1(919)6848885

Richardson, Dr John F. (1954). Associate Professor. Department of Chemistry, University of Louisville, Louisville, KY 40292, USA. PhD (U. Western Ontario, 1981) inorganic chemistry. *Main_group computer_modelling hydrogen_bonding catalysis long-range_contacts*

E-mail jfrich01@ulkyvm.louisville.edu Tel. 1(502)8527069 Fax 1(502)8528149

Richman, Marc H. (1936). Professor of Engineering. Box D, Division of Engineering, Brown University, Providence, RI 02912, USA. ScD (MIT, 1963) physical metallurgy. *Structure_properties microstructure forensic*

E-mail mrichman@brownvm.brown.edu Tel. 1(401)8632317 Fax 1(401)8631137

Ringe, Dr Dagmar (1942). Lucille P. Markey Associate Professor of Biochemistry and Chemistry. Rosenstiel Basic Medical Sciences Research Center, Brandeis University, Waltham, MA 02254-9110, USA. PhD (Boston U., 1968) organic chemistry. *Enzyme_catalysis protein_structure aminotransferases proteases G-proteins*

E-mail ringe@binah.cc.brandeis.edu ringe@brandeis.bitnet Tel. 1(617)7362473 Fax 1(617)7362405

Ringrose, Sharon (1967). Graduate student/research assistant. Department of Chemistry, Iowa State University, Ames, Iowa 50011, USA. [Ames Laboratory, 44 Spedding Hall, Ames, IA 50011, USA.] BS (Oklahoma State U., 1989) chemistry. *Structure_determination small_molecules Patterson_methods*

E-mail ringrose@instate.edu Tel. 1(515)2948477 Fax 1(515)2145233

Rizvi, Safia Khalil (1963). Graduate Research Assistant. Department of Chemistry and Biochemistry, University of Oklahoma, 620 Parrington Oval, Norman, OK 73019-0370, USA. MSc (U. Oklahoma, 1990) chemistry. *Crystallography database purification protein biochemistry data_collection peptide refinement_method toxicology*

E-mail aa4024@uokmvsa.bitnet Tel. 1(405)3257612 Fax 1(405)3256111

Robbins, Arthur H. (1946). Staff Scientist. Institute for Crystallography, Miles Research Center, 400 Morgan Lane, West Haven, CT 06516, USA. PhD (U. Pittsburgh, 1975) crystallography. *Proteins crystallography metalloproteins computer_aided_design*

E-mail art@mrc.com Tel. 1(203)9372814 1(203)9372853 Fax 1(203)9376923

Roberts, David L. (1965). Postdoctoral Fellow. Department of Biochemistry, Medical College of Wisconsin, 8701 W. Watertown Plank Road, Milwaukee, WI 53226, USA. PhD (U. Wisconsin-Milwaukee, 1993) chemistry. *Proteins crystallography computer_modelling biochemistry isomorphous_replacement homology_prediction flavoproteins*

E-mail droberts@post.its.mcw.edu Tel. 1(414)4564305 Fax 1(414)2668497

Roberts, Dr Sue A. (1950). Crystallographic Specialist. Department of Biochemistry, Biol. Sci. West 526, University of Arizona, Tucson, AZ 85721, USA. PhD (Washington State U., 1978) physical chemistry. *Protein_crystallography computer_modelling*

E-mail roberts@biosci.arizona.edu Tel. 1(602)6218171 Fax 1(602)6219288

Robertson, Dr J. Lee (1961). Postdoc. Solid State Division, Bldg. 7962, MS 6393, Oak Ridge National Laboratory, Oak Ridge, TN 37831-6393, USA. PhD (U. Houston, 1991) physics. *Diffuse_scattering amorphous_phase metallurgy quasicrystal neutron_scattering*

E-mail robertsonjl@ornl.gov Tel. 1(615)5745243 Fax 1(615)5746268

Robertus, Dr Jon D. (1945). Professor. U. Texas. Department of Chem. and Biochem., University of Texas, Austin, TX 78712, USA. PhD (U. California-San Diego, 1972) cell biology. *Protein_engineering X-ray_crystallography plant_toxins chitinases drug_design enzyme_mechanisms*

E-mail robertus@utbc01.cm.utexas.edu Tel. 1(512)4713175 Fax 1(512)4718696

Robie, Stephen B. (1955). Senior Applications Scientist. Scintag, Inc., 707 Kifer Road, Sunnyvale, CA 94086, USA. PhD (Rensselaer Polytechnic Institute, 1982) analytical chemistry. *Diffraction_technique instrumentation powder high_temperature low_temperature reflectivity thin_film*
E-mail 72203.420@compuserve.com Tel. 1(408)7377200x120 Fax 1(408)7379841

Robinson, Ian K. (1955). Professor. Physics Department, University of Illinois at Urbana-Champaign, 1110 West Green St, Urbana IL 61801, USA. PhD (Harvard U., 1981) biophysics. *Surfaces interfaces phase_transitions proteins*
E-mail robinson@uimr17.mrl.uiuc.edu Tel. 1(217)2442949 Fax 1(217)3336126

Robinson, Paul D. (1939). Senior Scientist. Department of Geology, Southern Illinois University, Carbondale, IL 62901, USA. MS (Southern Illinois U., 1963) geology/mineralogy. *Structure_analysis crystal_chemistry minerals electron_optics*
E-mail robinson@qm.c-geo.siu.edu Tel. 1(618)4537373 Fax 1(618)4537393

Robinson, Victoria Lee (1966). Graduate Student (PhD). Department of Biochemistry, University of Iowa, Iowa 52242, USA. MS (Villanova U., 1992) chemistry. *Macromolecular_crystallography proteins hemoglobin allostery*
E-mail vikki@eve.biochem.uiowa.edu Tel. 1(319)3357883 Fax 1(319)3359570

Roderick, Dr Steven L. (1956). Assistant Professor. Department of Biochemistry, Albert Einstein College of Medicine, 1300 Morris Park Avenue, Bronx, NY 10461-1602, USA. PhD (Washington University School of Medicine, 1986) molecular biology. *Biochemistry proteins macromolecules*
E-mail roderick@aecom.yu.edu Tel. 1(718)4302784 Fax 1(718)8920703

Rodgers, Dr David W. (1958). Postdoc. Department of Biochemistry and Molecular Biology, Harvard University, Cambridge, MA 02138, USA. PhD (Cornell U., 1987) biochemistry. *Macromolecular_crystallography proteins nucleic_acid transcription data_collection low_temperature*
E-mail rodgers@xtal0.harvard.edu rodgers@huxtal.bitnet Tel. 1(617)4954091 Fax 1(617)4959613

Rogers, Robin D. (1957). Professor. Associate Editor 'Journal of Chemical Crystallography'. Department of Chemistry, Northern Illinois University, DeKalb, IL 60115, USA. PhD (U. Alabama, 1982) inorganic chemistry. *Complexation coordination cyclic_polyethers macrocyles molecular_recognition acyclic_polyethers*
E-mail t40rdr1@mvs.cso.niu.edu Tel. 1(815)7536886 Fax 1(815)7534802

Rognlie, David (1934). President. Blake Industries, Inc., 660 Jerusalem Road, Scotch Plains, NJ 07076-2099, USA. BS (U. North Dakota, 1956) electrical engineering. *Diffraction instrumentation*
Tel. 1(908)2337240 Fax 1(908)2331354

Rohrer, Dr Douglas C. (1942). Sr. Research Scientist. Computer Aided Drug Discovery, MS 7247-267-131, Upjohn Laboratories, 301 Henrietta St., Kalamazoo, MI 49007-4940, USA. PhD (Case Western Reserve U., 1970) chemistry. *Computer_graphics computer_modelling molecular_mechanics structure–activity_relationships computational_drug_design molecular_similarity*
E-mail dcrohrer@upj.com Tel. 1(616)3857729 Fax 1(616)3858488

Roof, Dr Raymond B. (1929). Retired. 2700 Vista Grande N.W., Unit #102, Albuquerque, New Mexico 87120, USA. PhD (U. Michigan, 1955) mineralogy/crystallography.
Tel. 1(505)8394817

Rosauer, Ruth (1953). Lab. manager. Harrington Cancer Center, 1500 Wallace Blvd., Amarillo, TX 79106, USA. MA (U. of Colorado, 1990) economics. *Crystal_growth crystallization immunology purification statistics*
E-mail ruth@marvin.ama.ttuhsc.edu Tel. 1(806)3594673x278 Fax 1(806)3545887

Rose, Dr John P. (1953). Assistant Professor. Department of Crystallography, University of Pittsburgh, Pittsburgh, PA 15260, USA. PhD (Rutgers U., 1980) physical chemistry. *Proteins protein_interactions instrumentation structure_validation*
E-mail rose@bcl5.xtal.pitt.edu Tel. 1(412)6249300 Fax 1(412)6243219 1(412)6241882

Rosemond, M. Jane Cox (1959). Research Scientist III. Molecular Genetics and Microbiology, Burroughs Wellcome Company, 3030 Corwallis Rd., Research Triangle Park, NC 27709, USA. MS (Emory U., 1986) biochemistry. *Macromolecular_crystallography structure_based_drug_design*
E-mail mcb28@bwco.com Tel. 1(919)3153620 Fax 1(919)3150201

Rosenstein, Dr Robert Daniel (1922). 6350 Genesee Ave., Apt.119A, San Diego, CA 92122, USA. Fil. Lic. (U. Uppsala, Sweden, 1961) inorganic chemistry.
E-mail rdr@scripps.edu Tel. 1(619)4575894 Fax 1(619)5546860

Rosenzweig, Amy C. (1967). Postdoctoral fellow. Department BCMP, Harvard Medical School, Dana Farber Cancer Institute, XRC Lab D1040, 44 Binney St., Boston, MA 02115, USA. PhD (Massachusetts Institute of Technology, 1993) inorganic chemistry. *Metalloenzymes diiron_proteins homology_prediction phasing_methods*
E-mail amy@amber.dfci.harvard.edu Tel. 1(617)6324399 Fax 1(617)6324393

Ross, Dr Fred K. (1942). Senior Research Scientist. Research Reactor Center, University of Missouri-Columbia, Columbia, MO 65211, USA. PhD (U. Illinois-Urbana, 1969) inorganic chemistry. *X-ray_neutron charge_density_distribution structure_determination diffractometry gamma-ray_diffraction MO_calculations instrumentation*
E-mail rossho@hailwood.murr.missouri.edu ross@murrvax.bitnet Tel. 1(314)8825237 Fax 1(314)8823443

Ross II, Dr Charles R. (1957). Struc. Chem. Res. Spec. Department of Chemistry, 718 HaH, University of Nebraska, Lincoln, NE 68588-0304, USA. PhD (UCLA, 1986) geology. *Structural_disorder small_molecules minerals computer_modelling orientational_disorder incommensurate_ordering*
E-mail ross@prophet1.unl.edu Tel. 1(402)4728894

Rossi, Dr Miriam (1952). Associate Professor, Chair. Vassar College, Department of Chemistry, Box 484, Poughkeepsie, NY 12601, USA. PhD (The Johns Hopkins U., 1979) chemistry. *Small_biological_molecules computer_modelling teaching_crystallography*
E-mail rossi@vassar.edu rossi@vassar.bitnet Tel. 1(914)4375746

Rossmann, Michael G. (1930). Hanley Professor of Biological Sciences. ACA member. Department of Biological Sciences, Purdue University, West Lafayette, Indiana 47907, USA. PhD (U. Glasgow, Scotland, 1956) chemical crystallography. *Virus_structure protein_evolution viral_receptors phase_problem data_processing synchrotron_radiation nucleotide_binding_structures*
E-mail b4p@mace.cc.purdue.edu Tel. 1(317)4944911 Fax 1(317)4961189

Rotella, Dr Frank J. (1949). Chemist. IPNS Division, Bldg. 360, Argonne National Laboratory, 9700 S. Cass Ave., Argonne, IL 60439, USA. PhD (SUNY at Buffalo, 1979) chemistry. *Rietveld_method neutron_powder_diffraction_data metal_hydrides superconductors single_crystal structure_determination organometallic_compounds*
E-mail rotella@anlpns.pns.anl.gov Tel. 1(708)2525785 Fax 1(708)2524163

Roth, Robert S. (1926). Retired/Consultant. B214 Materials Bldg., National Institute of Standards and Technology, Washington, DC 20234, USA. PhD (U. Illinois, 1951) geology/mineralogy. *Phase_diagram oxides phase_equilibrium titanates niobates tantalates dielectrics electronic_ceramics*
E-mail bob@credit.nist.gov Tel. 1(301)9756116 Fax 1(301)9908729

Rotonda, Jennifer (1963). Senior Research Chemist. Merck Research Laboratories, PO Box 2000 (R50-105), Rahway, NJ 07065-0900, USA. PhD (Rutgers U., 1993) chemistry. *Proteins macromolecules*
E-mail rotonda@merck.com Tel. 1(908)5943416 Fax 1(908)5946100

Rould, Mark (1962). Biology Dept., MIT/HHMI, Cambridge, MA 02139, USA. PhD (Yale, 1991) molecular biophysics and biochemistry. *Proteins nucleic_acids*
E-mail rould@pabo1.mit.edu Tel. 1(617)2585881 Fax 1(617)2538728

Roy, Rustum (1924). Prof. of the Solid State. Founder Materials Res. Lab. (Penn State) Materials Res. Society. Materials Research Lab., Pennsylvania State University, University Park, PA 16802, USA. PhD (Penn State, 1948) ceramics. *Crystal_chemistry materials_synthesis*
E-mail rxr3@psuvm.psu.edu Tel. 1(814)8653421 Fax 1(814)8637040

Royer, Dr William E. (1954). Assistant Professor. Program in Molecular Medicine, University of Massachusetts Medical Center, 373 Plantation Street, Worcester, MA 01605, USA. PhD (Johns Hopkins U., 1984) biophysics. *Proteins macromolecules hemoglobins allostery*
E-mail royer@darwin.ummed.edu Tel. 1(508)8566912 Fax 1(508)8564289

Rozwarski, Denise A. (1964). Postdoc. Section of Biochemistry, Molecular and Cell Biology, Cornell University, Ithaca, NY 14853, USA. PhD (U. Texas at Austin, 1992) biochemistry.
E-mail denise@penelope.bio.cornell.edu Tel. 1(607)2558432

Rubin, J. Ronald (1947). Research Fellow. Chemistry Department, Parke-Davis Pharmaceuticals, Ann Arbor, MI 48105, USA. PhD (U. Wisconsin, Madison, 1974) molecular biology. *Protein_structure protein–nucleic_acid_interactions protein–drug_interactions drug_structures macromolecular_structure*
E-mail rubinj@aa.wl.com Tel. 1(313)9983298 Fax 1(313)9982782

Ruble, Dr John R. (1946). Principal research asst. Department of Crystallography, University of Pittsburgh, 304 Thaw Hall, Pittsburgh, PA 15260, USA. PhD (U. Pittsburgh, 1975) crystallography. *Small_molecular_charge_density*
E-mail ruble@pittvms.bitnet Tel. 1(412)6249300 Fax 1(412)6241882

Rudenko, Gabrielle (1967). Graduate student. Dept. of Biological Structure, Health Sciences Building SM-20, School of Medicine, University of Washington, Seattle, WA 98195, USA. (U. Amsterdam, 1990) biochemistry. *Protein_crystallization protein_structure_determination*
E-mail rudenko@gouda.bchem.washington.edu Tel. 1(206)5431089

Ruderman, Pres. Warren (1920). President. INRAD, Inc, 181 Legrand Ave., Northvale, NJ 07647, USA. PhD (Columbia U., 1949) chemical physics. *Crystal_structure crystal_perfection*
Tel. 1(201)7671910 Fax 1(201)7679644

Rudman, Prof. Reuben (1937). Professor of Chemistry. Department of Chemistry, Adelphi University, Garden City, NY 11530, USA. PhD (Polytechnic Inst. Brooklyn, 1967) inorganic chemistry. *Low_temperature order-disorder phase_transition orientational_disorder*
E-mail rudman@auvax1.adelphi.edu Tel. 1(516)8774133

Rudnick, Suzanne E. (1951). Associate Professor. Chemistry Department, Manhatten College, Manhatten College Parkway, Reverdale, NY 10471, USA. PhD (Boston U., 1979) chemistry. *Proteins*
Tel. 1(718)9200340 Fax 1(718)5484910

Rudolf, Dr Philip R. (1955). Analytical Sciences, 1897G Building, The Dow Chemical Company, Midland, MI 48667, USA. PhD (Texas A&M U., 1983) inorganic chemistry. *Synchrotron_radiation image_processing powder single_crystal scattering software diffractometry diffraction_technique electron_microscopy*
E-mail prudolf@dow.com Tel. 1(517)6365183 Fax 1(517)6365453

Ruggles, Dr James A. (1964). Damon Runyon–Walter Winchell Postdoctoral Fellow. Department of Chemistry and Biochemistry, University of Colorado, Boulder, Campus Box 215, Boulder, CO 80309, USA. PhD (U. Colorado, Boulder, 1991) biochemistry. *Protein_nucleic-acid crystal_structure telomeres telomerase antifreeze_proteins ice_nucleation_proteins structured_RNA*
E-mail ruggles@cuchem.colorado.edu Tel. 1(303)4925591 Fax 1(303)4925894

Ruhlandt-Senge, Dr Karin (1961). Assistant Professor. Department of Chemistry, 1-014 Center for Science and Technology, Syracuse University, Syracuse, NY 13244, USA. Dr (Philipps U., Germany, 1991) inorganic chemistry/crystallography. *Inorganic_chemistry crystallography organometallic_chemistry bioinorganic_chemistry*
E-mail kruhland@mailbox.syr.edu Tel. 1(315)4431306 Fax 1(315)4434070

Rupp, Mag. Dr Bernhard (1956). Group Leader, Crystallography. Biology and Biotechnology Program, L-452, University of California, Lawrence Livermore National Laboratory, PO Box 808, Livermore, CA 94551, USA. PhD (U. Vienna, 1984) physical chemistry. *Crystallographic_computing macromolecular_crystallography X-ray_instrumentation magnetic_structure valence_fluctuations*
E-mail rupp1@llnl.gov Tel. 1(510)4233273 Fax 1(510)4222282

Russell, Thomas Paul (1952). Research Staff Member. IBM Almaden Research Center, K13/802, 650 Harry Road, San Jose, CA 95120, USA. PhD (U. Massachusetts, 1979) polymer science and engineering. *Polymers surfaces interfaces morphology copolymers mixtures structure*
E-mail trussell@almaden.ibm.com Tel. 1(408)9271638 Fax 1(408)9273310

Ruud, Clayton O. (1934). Professor. Department of Industrial and Manufacturing Engineering, Pennsylvania State University, University Park, PA 16802, USA. [207 Hammond, State College, PA 16801, USA.] PhD (U. Denver, 1980) material science. *Strain powder polycrystal stress texture composition*
E-mail corie@engr.psu.edu Tel. 1(814)8630974 Fax 1(814)8634745

Rux, Dr John J. (1963). Postdoctoral Research Fellow. The Wistar Institute, 3601 Spruce Street, Philadelphia, PA 19104, USA. PhD (U. South Carolina, 1991) biological chemistry. *Macromolecules viruses antibodies glycoproteins molecular_recognition molecular_interactions antiviral_compounds*
E-mail rux@wistb.wistar.upenn.edu Tel. 1(215)8982202 Fax 1(215)8983868

Ryba, Earle (1934). Associate Professor. The Pennsylvania State University, 304 Steidle, University Park PA 16802, USA. PhD (Iowa State U., 1960) physical metallurgy. *Quasicrystal_structure alloys polymeric_thin_film X-ray_reflectivity intermetallic_compound_structures GIXD synchrotron_radiation*
E-mail ryba@ems.psu.edu Tel. 1(814)8653760 Fax 1(814)8653760

Rydel, Dr Timothy J. (1959). Scientist. The Procter & Gamble Company, Corporate Research Division, Miami Valley Laboratories, PO Box 398707, Cincinnati, OH 45239-8707, USA. PhD (Michigan State U., 1991) physical chemistry. *Macromolecular_crystallography proteases enzyme_inhibitors lipases*
E-mail proctor!mv1.pg.com!rydel_tj@ms.uky.edu Tel. 1(513)6271604 Fax 1(513)6271233

Sabat, Dr Michal (1947). Director, Molecular Structure Laboratory. Department of Chemistry, University of Virginia, Charlottesville, VA 22901, USA. PhD (U. Wroclaw, Poland, 1976) inorganic chemistry. *Bioinorganic_chemistry nucleic_acids molecular_modelling quantum_chemistry drug_design*
E-mail sabat@molmod.chem.virginia.edu Tel. 1(804)9247862 Fax 1(804)9243710

Sack, John S. (1953). Senior Research Investigator. Department of Macromolecular Crystallography, Bristol-Myers Squibb Pharmaceutical Research Institute, PO Box 4000, Princeton, NJ 08543-4000, USA. PhD (Johns Hopkins U., 1981) biophysics. *Crystallography proteins computer_graphics*
E-mail sack@bms.com Tel. 1(609)2525559 Fax 1(609)2526030

Sadowski, Lucian M. (1954). Mechanical engineering. US Army Armament Research Development and Engineering Center, Close Combat Armament Center/SMCAR CCL E, Picatinny Arsenal, NJ 07806-50000, USA. MS (Florida Inst. of Tech., 1981) engineering management. *Ballistics physics astronomy explosive photography*
Tel. 1(201)7247072 Fax 1(201)7243793

Safo, Dr Martin K. (1958). Post doctor. Dept. of Medicinal Chemistry (VCU), Medical College of Virginia, PO Box 541, Richmond, VA 23298, USA. PhD (U. Notre Dame, 1991) inorganic chemistry. *Macromolecule_crystallography drug_design molecular_modelling*
E-mail msafo@gems.vcu.edu Tel. 1(804)3717406

Sakon, Dr Joshua (1966). Postdoc. Department of Biochemistry, Cornell University, Ithaca, NY 14853, USA. PhD (U. Wisconsin-Madison, 1993) biophysics. *Macromolecule_structure*
E-mail joshua@penelope.bio.cornell.edu Tel. 1(607)2558432

Salemme, F. Raymond (1945). President and CEO. 107 Marshall Bridge Road, Kennett Sq., PA 19348, USA. PhD (U. California, 1972) chemistry. *Structure_based_drug_design*
Tel. 1(215)2228950 Fax 1(215)2228960

Samudzi, Cleopas T. (1958). Assistant Professor. University of Missouri-Columbia, Biochemistry Department, Columbia, MO 65211, USA. PhD (U. Pittsburgh, 1990) macromolecular crystallography/biological sciences. *X-ray_crystallography structure-function_relationships protein–nucleic_acid_interactions ligand–receptor_interactions*
E-mail cleo1@mercury.biochem.missouri.edu bccleos@muccmail.missouri.edu Tel. 1(314)8828339 Fax 1(314)8844812

Sands, Dr Donald E. (1929). Professor. Department of Chemistry, University of Kentucky, Lexington, KY 40506-0055, USA. PhD (Cornell U., 1955) physical chemistry. *Properties_of_crystals tensors statistics thermodynamics*
E-mail sands@ukcc.uky.edu Tel. 1(606)2577082 Fax 1(606)3231069

Sanishvili, Dr R. (1961). Research Associate. Department of Biological Sciences, The University of Illinois at Chicago, Chicago IL 60680, USA. [LMB, Bio. Sci. UIC, 845 W. Taylor St., Chicago, IL 60607-7060, USA.] PhD (Institute of Crystallography, Moscow, 1989) protein crystallography. *Antigen_antibody cytochromes_c protein_structure computer_modelling regulatory_proteins DNA–protein electrostatic_interactions_proteins*
E-mail sanishvili@anbiw4.sbc.anl.gov Tel. 1(312)9965190 Fax 1(312)4132691

Santarsiero, Dr Bernard D. (1952). Manager/Scientist. Macromolecular Crystallography, Molecular Structure Corporation, 3200 Research Forest Drive, The Woodlands, TX 77381-4238, USA. PhD (U. Washington, 1980) physical chemistry. *Protein_structure crystallographic_computing rational_drug_design host–guest_complexes*
E-mail bds@msc.com Tel. 1(713)3631033 Fax 1(713)3643628

Saper, Dr Mark A. (1954). Assistant Professor. Biophysics Research Division, University of Michigan, 930 N. University Avenue, Ann Arbor, MI 48109-1055, USA. PhD (Rice U., 1983) biochemistry. *Proteins signal_transduction tyrosine phosphatases software*
E-mail saper@umich.edu Tel. 1(313)7643353 Fax 1(313)7643323

Sappenfield, Dr Eric L. (1949). Postdoctoral Fellow. Department of Chemistry, Baylor University, PO Box 97348, Waco, Texas 76798-7348, USA. PhD (Baylor U., 1987) physical chemistry. *Small_molecule crystallographic_computing lanthanides teaching powder_diffractometry zeolytic_compounds*
E-mail sappenfilde@baylor.edu vaxsdp@baylor.edu Tel. 1(817)7553311x4823 1(817)7553311x6853 Fax 1(817)7552403

Sarko, Anatole (1930). Professor and Chairman. Department of Chemistry, SUNY College of ESF, Syracuse, NY 13210, USA. PhD (State U. New York, 1966) polymer chemistry. *Polysaccharides carbohydrates cellulose diffraction_analysis computer_modelling*
E-mail asarko@mailbox.syr.edu Tel. 1(315)4706855 Fax 1(315)4706856

Sarma, Dr Raghu (1937). Associate Professor. Dept. of Biochemistry and Cell Biology, State Univ. of New York, Stony Brook, NY 11794-5215, USA. PhD (U. Madras, India, 1963) biophysics. *Protein_structures lipase_structures enzyme_mechanisms protein_crystallization*
E-mail rsarma@sbccmail.sunysb.edu Tel. 1(516)6328558 Fax 1(516)6328575

Sauter, Dr Nicholas K. (1961). Postdoctoral Fellow. Department of Biochemistry and Biophysics, University of California, San Francisco, San Francisco, CA 94143-0448, USA. PhD (Harvard U., 1991) biophysics. *Glucocorticoid_receptor hsp90*
E-mail sauter@msg.ucsf.edu Tel. 1(415)4764480 Fax 1(415)4766129

Sawaya, Michael R. (1967). Graduate Student. Department of Chemistry, University of California at San Diego, La Jolla, CA 92093-0317, USA. MS (U. California at San Diego, 1992) biochemistry. *Enzyme_catalysis conformational_change dehydrogenases polymerases*
E-mail msawaya@chem.ucsd.edu mrs@chem.ucsd.edu Tel. 1(619)5342011 Fax 1(619)5342618

Sax, Martin. Associate Chief Staff Research and Development. Biocrystallography Lab., Oakland VA Hospital, Pittsburgh, PA 15240, USA. *Proteins toxins enzymes structure_function*
Tel. 1(412)6833000 Fax 1(412)6834917

Sayre, Dr David (1924). Guest Scientist. Department of Physics, State University of New York, Stony Brook, NY 11794, USA. [3 Harbor Road, St. James, NY 11780, USA.] PhD (Oxford, 1951) chemical crystallography. *X-ray_microscopy image_reconstruction non-crystallographic_structure_phase_determination_technique single_cell_diffraction*
E-mail sayre@bnl.gov Tel. 1(516)5846349 Fax 1(516)6328101

Scapin, Giovanna (1961). Research Assistant. Department of Biochemistry, A. Einstein College of Medicine, 1300 Morris Park Avenue, Bronx, NY 10461, USA. PhD (Padova U., Italy, 1989) protein crystallography. *Proteins phosphoribosyltransferases lipid_binding_proteins neutron_diffraction*
E-mail scapin@zugbug.bioc.aecom.yu.edu Tel. 1(718)4302743 Fax 1(718)5975692

Scaringe, Raymond P. (1950). Senior Research Scientist. 1ML/CNTD/IRAD, Mail Drop 02102, Research Laboratories, Eastman Kodak Co., Rochester, NY 14650, USA. PhD (U. North Carolina, 1976) inorganic chemistry. *Packing morphology polymorphism crystallization prediction_of_packing_structure growth_modifier_design*
E-mail scaringe@kodak.com Tel. 1(716)4777052 Fax 1(716)5887611

Schaber, Peter M. (1953). Professor. Department of Chemistry, Canisius College, Buffalo, NY 14208, USA. PhD (State U. New York at Buffalo, 1980) inorganic chemistry. *Small_molecules*
E-mail schaber@canisius.bitnet Tel. 1(716)8882351 Fax 1(716)8883112

Schaefer, Dale W. (1941). Manager; Technical Advisor, DOE Office of Economic Competitiveness. Sandia National Laboratories, Dept. 5200, Albuquerque, NM 87185-1393, USA. [DP-14, US Dept. of Energy, 1000 Independence Ave. SW, Washington, DC 20585, USA.] PhD (Mass. Inst. Technology, 1967) physical chemistry. *Small-angle_X-ray_scattering small-angle_neutron_scattering inelastic_neutron_scattering polymers ceramics porous_materials amorphous_materials*
E-mail dwschae@sandia.gov Tel. 1(202)5861014 Fax 1(202)2875410

Schaefer, Dr William P. (1931). Senior Research Associate, Emeritus. Div. of Chem. and Chem. Eng., Mail Code 139-74, California Institute of Technology, Pasadena, CA 91125, USA. PhD (UCLA, 1960) analytical chemistry. *Structural_chemistry transition_metal_compounds oxygen_compounds*
E-mail wps@xray.caltech.edu Tel. 1(818)3952739 Fax 1(818)4494159

Scheidt, Dr W. Robert (1942). Professor. Department of Chemistry and Biochemistry, University of Notre Dame, Notre Dame, IN 46556, USA. PhD (U. Michigan, 1968) inorganic chemistry. *Metalloporphyrins magnetism electronic_spectrum unusual_bonding radical_salts electron_paramagnetic_resonance area_detectors_for_small_molecules*
E-mail scheidt.1@nd.edu Tel. 1(219)6315939 1(219)6316816 Fax 1(219)6316652

Schevitz, Dr Richard W. (1943). Research Scientist. Lilly Research Laboratories, Eli Lilly and Company, Indianapolis, IN 46285, USA. PhD (Purdue U., 1970) protein crystallography. *Proteins drug_design*
E-mail schevitz@lilly.com Tel. 1(317)2762128 Fax 1(317)2769722

Schewe-Miller, Dr Irmgard M. (1960). Postdoctoral Research Associate. Ames Laboratory DOE, Iowa State University, USA. [Ames Laboratory DOE, Iowa State University, 339 Spedding Hall, Ames, IA 50011, USA.] Dr rer.nat. (U. Stuttgart, 1990) chemistry. *Chalcogenides solids structure_determination bonding*
E-mail schewe@iastate.edu Tel. 101(515)2943513 Fax 101(515)2945718

Schiffer, Dr Marianne (1935). Senior Scientist. Center for Mechanistic Biology and Biotechnology, Argonne National Laboratory, Argonne, IL 60439, USA. PhD (Columbia U., 1965) biochemistry. *Proteins immunoglobulins photoreaction_centre conformational_change computer_modelling membrane_proteins structure_prediction*
E-mail schiffer@anlbem.bim.anl.gov schiffer@anlbem.bitnet Tel. 1(708)2523883 Fax 1(708)2525517

Schildbach, Joel F. (1964). Postdoctoral Fellow. Dept. of Biology, 68-559, MIT, 77 Massachusetts Avenue, Cambridge, MA 02139, USA. PhD (Harvard U., 1992) immunology. *Affinity biochemistry immunochemistry recognition DNA–binding_proteins*
E-mail joel@rosa.mit.edu Tel. 1(617)2533149 Fax 1(617)2538699

Schioler, Dr Liselotte. Director, Diamond film and ceramic technology. BDM Technologies, Inc., 7915 Jones Branch Drive, McLean, VA 22102-3396, USA. ScD (MIT, 1989) ceramic science.
Tel. 1(703)8487137 Fax 1(703)8485144

Schirber, James E. (1931). Research Scientist/Manager. Department 1002 Saudia Labs, Albuquerque, NM 87185-0335, USA. PhD (Iowa State U., 1960) physics. *High_pressure superconductivity fullerenes*
Tel. 1(505)8448134 Fax 1(505)8444045

Schlemper, Dr Elmer O. (1939). Professor. Department of Chemistry, University of Missouri, Columbia, MO 65211, USA. PhD (U. Minnesota, 1965) chemistry. *Neutron_diffraction electron_density transition_metal_complexes radiopharmaceuticals*
E-mail chemeos@mizzou1.missouri.edu chemeos@mizzou1.bitnet Tel. 1(314)8828374 Fax 1(314)8822754

Schlenker, John L. Mobil Research and Development Corporation, Central Research Laboratory, PO Box 1025, Princeton, NJ 08540, USA. PhD (Virginia Polytechnic Institute and State U., 1976) mineralogy/geology. *Powder_diffraction*
Tel. 1(609)7374001

Schmidt, Paul W. (1926). Professor. 503 South Garth Avenue, Columbia, MO 65203, USA. [Department of Physics and Astronomy, University of Missouri, Columbia, MO 65211, USA.] PhD (U. Wisconsin, 1953) physics. *SAXS SANS porous_solids fractals collimator_correction disordered_solids data_processing*
E-mail phys1111@mizzoul.missouri.edu (somewhat unreliable) Tel. 1(314)8828241 Fax 1(314)8824195

Schneider, Dr Dieter K. (1947). Biophysicist. Biology Department, Brookhaven National Laboratory, Upton, NY 11973, USA. PhD (U. Basel, 1977) biophysics. *Neutron_small-angle_scattering neutron_crystallography biocrystallography neutron_instrumentation fibre_diffraction liquids*
E-mail schneider@bnlh9b.bio.bnl.gov Tel. 1(516)2823423 Fax 1(516)2823407 Telex 6852516 BNL DOE

Schoenborn, Prof. Benno P. (1936). Senior Fellow. Life Science Division, University of California, Los Alamos National Laboratory, Los Alamos, NM 87545, USA. PhD; DSc (U. NSW Australia, 1962) solid state physics. *Neutron_diffraction protein_crystallography protein_dynamics instrumentation water_structure hydrogen_bonding super_Laue diffraction*
E-mail schoenbom@flovax.lanl.gov Tel. 1(505)6652033 Fax 1(505)6653024

Schomaker, Verner (1914). (UW:)Prof. Emeritus (Caltech:)Faculty Associate. (Summer) Department of Chemistry BG-10, University of Washington, Seattle, WA 98195, USA, (Winter) Division of Chemistry and Chemical Engineering, Beckman Institute 139-74, Caltech, Pasadena, CA 91106, USA. PhD (Caltech, 1938) physical chemistry. *Rigidbody_analysis structure_determination least-squares_refinement zeolites thermodynamics coordination*
E-mail (UW:) scho@uwachem.bitnet (Caltech:) vs%xray@hamlet.bitnet Tel. (UW:) 1(206)5431686 (Caltech:) 1(818)3972737

Schreiner, Dr Walter N. (1941). President. IC Laboratories, PO Box 721, Amawalk, NY 10501, USA. PhD (Virginia Polytechnic Institute and State U., 1973) physics. *Powder_diffraction phase_identification crystallite_size_analysis Rietveld_quantitative_analysis*
Tel. 1(914)9622477 Fax 1(914)9625564

Schroeder, Dr LeRoy W. (1943). Chief, Chemistry Group. Food & Drug Administration, HFZ-150, 5600 Fishers Lane, Rockville, MD 20857, USA. PhD (Northwestern U., 1969) physical chemistry. *Molecular_modelling diffusion surfaces computational_chemistry_estimation_of_properties*
E-mail lys@fdadr.cdrh.fda.gov Tel. 1(301)4437003

Schubert, Heidi L. (1968). PhD candidate. Department of Biological Chemistry, University of Michigan, Ann Arbor, MI 48109-0606, USA. BS (Miami U., OH, 1990) chemistry. *Phosphatases proteins conformational_change crystallography tyrosine_phosphatases enzyme_substrate_interactions*
E-mail heidi@oscar.biop.umich.edu Tel. 1(313)7633384 Fax 1(313)7643323

Schuller, David J. (1961). Postdoctoral Assistant. Department of Molecular Biology, University of California, Irvine, CA 92717-3900, USA. PhD (Washington U. at St Louis, 1993) molecular biology. *Allostery noncrystallographic_symmetry protein_flexibility proteins modelling phase_refinement electron_transport*
E-mail schuller@indigo2.biomol.uci.edu Tel. 1(714)8244322

Schuller, Ivan (1946). Professor. Physics Department 0319, University of California, San Diego, La Jolla, CA 92093-0319, USA. PhD (Northwestern U., 1976) physics.
Tel. 1(619)5342540 Fax 1(619)5347161

Schultz, Dr Arthur J. (1947). Chemist; Instrument Scientist for IPNS Single-Crystal Diffractometer. IPNS, Bldg. 360, Argonne National Laboratory, Argonne, IL 60439-4814, USA. PhD (Brown U., 1973) inorganic chemistry. *Time-of-flight_diffraction pulsed_neutron single_crystal superconductors phase_transitions organic_superconductors high_pressure high_Tc*
E-mail ajschultz@anl.gov Tel. 1(708)2523465 Fax 1(708)2524163

Schultz, L. Wayne (1968). Graduate Student. Department of Chemistry, Cornell University, Baker Laboratory, Ithaca, NY 14853-1301, USA. MS (Cornell U., 1992) analytical chemistry. *Proteins*
E-mail schultz@chemres.tn.cornell.edu Tel. 1(607)2556145 Fax 1(607)2551253

Schulze-Gahmen, Dr Ursula (1960). Postdoc. Dept. Chemistry, UC Berkeley, Berkeley, CA 94720, USA. PhD (U. Heidelberg, Germany, 1988) biology. *Protein_structure structure_function_correlation immunology signal_transduction drug_design*
E-mail uschulze-gahmen@lbl.gov Tel. 1(510)4864338 Fax 1(510)4866059

Schwartz, Dr Kenneth B. (1954). Senior Staff Scientist. Raychem Corporation, Corporate Technology, MS 122/6404, 300 Constitution Drive, Menlo Park, CA 94025, USA. PhD (State U. New York Stony Brook, 1982) earth and space sciences. *Polymer_crystallography powder_diffraction neutron_diffraction Rietveld_method*
E-mail kschwart@rd.raychem.com Tel. 1(415)3613957 Fax 1(415)3612723

Schwartz, Lyle H. (1936). Director, Materials Science and Engineering Laboratory, Director, Materials Science and Engineering Laboratory, National Institute of Standards and Technology, Bldg 223, Rm B309, Gaithersburg, MD 20899, USA. PhD (Northwestern U., 1964) material science.
E-mail schwartz@micf.nist.gov Tel. 1(301)9755658 Fax 1(301)9268349

Schwartz, Mr Steven A. (1954). Doctoral Candidate. Electrical Engineering Department, Washington University, Box 1161, 1 Brookings Drive, University City, MO 63130-4899, USA. MS (U. Missouri, Columbia, 1984) applied mathematics. *Protein_crystallography structure_determination algorithm refinement*
E-mail s.a.schwartz@ieee.org sxs@saturn.wustl.edu Tel. 1(314)9357547 Fax 1(314)9354842

Schweiker, Viloya L. (1956). Research Associate. Department of Chemistry and Biochemistry, Campus Box 215, University of Colorado, Boulder, CO 80309, USA. PhD (CU Boulder, 1982) physical inorganic chemistry.
E-mail vi%cechem@vaxf.colorado.edu Tel. 1(303)4920970 Fax 1(303)4925894

Sclar, Charles B. (1925). Prof. Emeritus. Committee on Crystallography at High Pressure and Temperature 1968–1972. Department of Earth & Environmental Sciences, Lehigh University, Bethlehem, PA 18015-3188, USA. PhD (Yale U., 1951) geology. *Mineralogy*
E-mail cbs1@lu.edu Tel. 1(610)7583660 Fax 1(610)7583677

Seaton, Dr Barbara A. (1952). Associate Professor. Department of Physiology, Boston University School of Medicine, 80 E. Concord St., Boston, MA 02118, USA. PhD (MIT, 1983) biochemistry. *Conformational_change proteins ligand_binding regulation calcium_compounds biocrystallography biological_membranes*
E-mail seaton@medxtal.bu.edu Tel. 1(617)6385061 Fax 1(617)6384273

See, Dr Ronald F. (1960). Post-doc. Dept. of Chem., Harvard University, 12 Oxford St, Cambridge, MA 02138, USA (State U. New York, Buffalo, 1993) inorganic chemistry. *Hydrogen_bonding theoretical_bonding allostery philosophy_of_science*
E-mail see@hubeta.harvard.edu Tel. 1(617)4954097 Fax 1(617)4953330

Seely, Dr Oliver (1939). Professor. Department of Chemistry, California State University, Dominguez Hills, Carson, CA 90747, USA. PhD (U. Illinois, 1966) physical chemistry. *Peptides proteins*
E-mail oliver@dhvx20.csudh.edu Tel. 1(310)5163778 Fax 1(310)5164268

Seeman, Prof. Nadrian Charles (1945). Professor. Department of Chemistry, New York University, New York, NY 10003, USA. PhD (U. Pittsburgh, 1970) crystallography/biochemistry. *Macromolecular_design DNA_recombination nucleic_acid_topology DNA_branched_junctions nanotechnology*
E-mail seeman@ucfcluster.nyu.edu Tel. 1(212)9988395 Fax 1(212)2607905

Seff, Prof. Karl (1938). Professor. Department of Chemistry, University of Hawaii, 2545 The Mall, Honolulu, HI 96822-2275, USA. PhD (Mass. Inst. of Tech., 1964) physical chemistry. *Zeolite_chemistry intrazeolitic_structure zeolites crystallography*
E-mail kseff@helium.chem.hawaii.edu Tel. 1(808)956-7665 Fax 1(808)956-5908 Telex 7238861 HIGCY HR

Sehnke, Dr Paul C. (1958). Asst. In. Interdisciplinary Center for Biotechnology Research/ Dept. of Hort. Sci., University of Florida, Gainesville, FL 32611, USA. PhD (Purdue U., 1990) biophysics. *Transcription_factors viral_proteins plant_toxins viruses zinc_fingers calcium-binding_transcription_factors protein_expression*
E-mail pcs@nervm.nerdc.ufl.edu Tel. 1(904)3921928x329 Fax 1(904)3926479

Senadhi, Vijay-Kumar (1953). Assistant Professor. Department of Biochemistry, Temple University, Philadelphia, PA 19140, USA. [561, AHB, 3307 N Broad St., Philadelphia, PA 19140, USA.] PhD (IIT-Bombay, 1982) biophysics. *Structure_determination_macromolecules DNA-binding_proteins structure_based_drug_design crystallization_macromolecules disordered_structures refinement_problematic_structures*
E-mail vijay@sgi1.fels.temple.edu Tel. 1(215)2214161 Fax 1(215)2218290

Servos, Kurt (1928). Professor of Geology, Emeritus. 1281 Mills St #9, Menlo Park, CA 94025-32077, USA. MS (Yale, 1954) geology. *Crystal_symmetry morphology geometrical_crystallography mineral_crystal_structures*
Tel. 1(415)3221245

Shackelford, James F. (1944). Professor and Associate Dean. Division of Materials Science & Engineering, Department of Chemical Engineering & Materials Sciences, University of California, Davis, CA 95616, USA. PhD (U. California at Berkeley, 1971) materials sciences and engineering. *Amorphous_solids glasses biomaterials*
E-mail jfshckelford@ucdavis.edu Tel. 1(916)7520556 Fax 1(916)7528058

Shaffner, Thomas J. (1941). Research Manager. Texas Instruments, Inc., MS 147, Dallas, TX 75265, USA. PhD (Vanderbilt U., 1969) physics. *Semiconductors characterization microelectronics X-ray_diffraction topography data_processing*
E-mail shaffner@resbld.csc.ti.com Tel. 1(214)9956764 Fax 1(214)9957785 Telex 6829291

Shan, Lin (1956). Research Associate. Department of Protein Crystallography, Harrington Cancer Center, 1500 Wallace Blvd., Amarillo, TX 79106, USA. PhD (U. Utah, 1989) protein chemistry and protein crystal growth. *Biocrystallography proteins computer_modelling chromatography immunology cyanobacteria gene_regulation*
E-mail lin@ford.ama.ttu.edu Tel. 1(806)3545875x278 Fax 1(806)3545887

Shang, Maoyu (1945). Assistant Faculty Fellow. Department of Chemistry & Biochemistry, University of Notre Dame, Notre Dame, IN 46556, USA. PhD (Fujian Inst. of Res. on Structure of Matter, 1985) structural chemistry. *Small-molecule protein diffractometer_service*
E-mail mshang@vyasa.helios.nd.edu Tel. 1(219)6316220 Fax 1(219)6316652

Sharma, Amit (1968). Graduate student. Department of Biochemistry, Northwestern University, 2153 Sheridan Road, Evanston, IL 60208, USA. MS (Northwestern U., 1992) biochemistry. *Crystallization_proteins DNA-binding_proteins structure_elucidation crystallographic_computing protein_overexpression protein_purification domain_structure viral_proteins evolution_domains*
E-mail sharma@tochtli.biochem.nwu.edu Tel. 1(708)4915430 Fax 1(708)4671380

Sharma, Dr Brahama D. (1931). Professor. Departments of Chemistry, Cal State University L.A. and L.A. Pierce College, 1520 Rose Villa Street, Pasadena, CA 91106-3525, USA. PhD (U. Southern California, 1961) physical chemistry. *Molecular_structure inorganic organic biological minerals proteins DNA hydrogen_bonding crystallography diffraction computers lattice molecular dynamics*
Tel. 1(818)7938724

Sharp, Paul R. (1952). Professor of Chemistry. 123 Chemistry, University of Missouri, Columbia, MO 65211, USA. PhD (MIT, 1980) inorganic chemistry. *Single_crystal chemistry clusters transition_elements non-linear_optical organometallic synthesis*
E-mail chemprs@showme.missouri.edu Tel. 1(314)8827715 Fax 1(314)8822754

Sharrah, Paul C. (1914). Emeritus Professor of Physics. University of Arkansas - Physics 202, Fayetteville, Arkansas 72701, USA. PhD (U. Missouri, 1942) physics. *X-ray_and_neutron_diffraction atomic_distribution_in_liquids*
E-mail psharrah@comp.uark.edu Tel. 1(501)5752506 or 1(501)5757932 Fax 1(501)5754580

Shechtman, Dan (1941). Scientist. B 309/223, NIST, Gaithersburg, MD 20899-0001, USA. PhD (Technion, Haifa, Israel, 1972) materials science. *Quasicrystals structural_defects*
E-mail danny@enh.nist.gov Tel. 1(301)9755766 Fax 1(301)9268349

Sheldon, Robert I. (1945). Research Staff. Nuclear Materials Technology Division, Mail Stop E502, Los Alamos National Laboratory, Los Alamos, NM 87545, USA. PhD physical chemistry. *Single_crystal powder Rietveld_method high_temperature*
E-mail rsheldon@lanl.gov Tel. 1(505)6650144 Fax 1(505)6654775

Shen, Qun (1959). Staff Scientist/Adj. Associate Professor. Cornell High Energy Synchrotron Source, Cornell University, Ithaca, NY 14853, USA. PhD (Purdue U., 1987) physics. *Synchrotron_radiation diffraction_physics phase_problem X-ray_polarimetry charge_density_studies magnetic_structures n-beam_diffraction*
E-mail qs11@cornell.edu Tel. 1(607)2550923 Fax 1(607)2559001

Sheriff, Dr Steven (1951). Research Fellow. Bristol-Myers Squibb Pharmaceutical Research Institute, PO Box 4000, Princeton, NJ 08543, USA. PhD (U. Washington, 1979) biochemistry. *Macromolecular_crystallography proteins antibody_antigen_complexes structure_function*
E-mail sheriff@bms.com Tel. 1(609)2525934 Fax 1(609)2526030

Sheu, Eric (1953). Sr. Research Physicist. Texaco Research Center, PO Box 509, Beacon, NY 12508, USA. PhD (MIT, 1987) applied radiation physics.
E-mail eysheu@texaco.com Tel. 1(914)8387120

Shieh, Dr Huey-Sheng (1946). Principal Crystallographer. Monsanto, BB4K, 700 Chesterfield Parkway N., Chesterfield, MO 63198-0001, USA. PhD (U. Pennsylvania, 1975) chemistry. *Protein_structure drug_design structure-activity_relationship protein_crystallization computer_graphics*
E-mail hsshie@bb1t.monsanto.com Tel. 1(314)5376025 Fax 1(314)5376480

Shimoni, Liat (1965). Graduate Student. Fox Chase Cancer Center, 7701 Burholme Ave., Philadelphia, PA 19111, USA. MSc (Ben-Gurion U., 1992) chemistry. *Crystallography proteins molecular_interaction molecular_structure*
E-mail liat@lindo.fccc.edu Tel. 1(215)7282220 Fax 1(215)7283574

Shindyalov, Ilya (1959). Associate Research Scientist. Department of Biochemistry & Molecular Biophysics, Columbia University, 630 W. 168th Street, New York, NY 10032, USA. PhD (Institute of Cytology and Genetics, Novosibirsk, Russia, 1988) biology. *Database proteins prediction homology molecular_evolution protein_modelling*
E-mail shindyal@cuhhca.hhmi.columbia.edu Tel. 1(212)3054270 Fax 1(212)3057379

Shiono, Dr Ryonosuke (1923). Associate Professor of Crystallography (retired). Department of Crystallography, University of Pittsburgh, Pittsburgh, PA 15260, USA. DSc (Osaka (Imperial) U., 1960) physics. *Molecular_crystal computer_graphics*
E-mail shiono1@vms.cis.pitt.edu Tel. 1(412)6249302 Fax 1(412)6241882

Shipley, Prof. G. Graham (1937). Professor of Biophysics. Biophysics Department, Boston University School of Medicine, 80 East Concord St., Boston, MA 02118, USA. PhD (U. Nottingham, 1963) X-ray crystallography. *Macromolecular_structure membrane_diffraction receptor_ligand_interactions bacterial_toxins*
E-mail shipley@med-biopha.bu.edu Tel. 1(617)6384009 Fax 1(617)6384041

Shoemaker, Dr Clara B. (1921). Professor of Chemistry, Emeritus. Dept. of Chemistry, Oregon State University, Corvallis, OR 97331-4003, USA. [3453 NW Hayes Ave, Corvallis, OR 97330, USA.] PhD (Leiden U. (Netherlands), 1950) inorganic chemistry. *X-ray_crystallography inorganic_chemistry alloys crown_compounds quasicrystal related_alloys*
E-mail shoemake@ccmail.orst.edu Tel. 1(503)7376730 Fax 1(503)7372062

Shoemaker, Dr David P. (1920). Professor of Chemistry, Emeritus. Dept. of Chemistry, Oregon State University, Corvallis, OR 97331-4003, USA. PhD (Calif. Inst. of Tech., 1947) chemistry. *Physical_chemistry alloys hydrides zeolites tetrahedrally-close-packed_metal_phases*
E-mail shoemakd@ccmail.orst.edu Tel. 1(503)7376730 Fax 1(503)7372062

Shoham, Dr Menachem (1944). Associate Professor. Department of Biochemistry, School of Medicine, Case Western Reserve University, 10900 Euclid Ave, Cleveland, OH 44106-4935, USA. PhD (Weizmann Inst of Science, 1979) chemistry. *Peptide_antibodies redox_proteins protein_toxins computer_modelling protein_stability thermophilic_proteins halophilic_proteins acidophiles*
E-mail shoham@cwbio.bioc.cwru.edu Tel. 1(216)3684665 Fax 1(216)3684544

Shoja, Massud (1948). Director. Chemistry Department, Fordham University, Bronx, NY 10458, USA. PhD (Fordham, 1979) physical chemistry. *Flavonoids*
Tel. 1(718)8174452

Short, Dr Michael A. (1930). Senior engineer; R&D Project Manager. Hughes Missile Systems Company, Rancho Cucamonga, CA, USA. [24832 Weyburn Drive, Laguna Hills, CA 92653, USA.] PhD (Penn State U., 1961) X-ray diffraction. *X-ray_physics X-ray_instrumentation X-ray_fluorescence X-ray_diffraction microwaves*
E-mail 00k7803@ccmail.emis.hac.com Tel. 1(714)9518273 Fax 1(909)4834061

Shu, Ms Fang (1964). Graduate Student. Biology Department, Brookhaven National Laboratory, Upton, NY 11973, USA. MA (The State University of New York, 1991) physics. *Water_structure refinement_method single_crystal diffraction proteins structure_determination*
E-mail fshu@bnlc11.bnl.gov Tel. 1(516)2823421

Shui, Xiuqi (1951). Postdoctor. SmithKline Beecham Pharmaceuticals, UW2950, PO Box 1539, King of Prussia, PA 19406, USA. PhD (Wayne State U., 1992) inorganic chemistry. *Small_molecule peptide*
E-mail shuix1%phv074.dnet@sb.com Tel. 1(610)2706499 Fax 1(610)2706996

Shulmeister, Dr Vladimir M. (1948). PostDoc. Lawrence Berkeley Laboratory, m/s 30230, 1 Cyclotron Rd., Berkeley, CA 94720, USA. PhD (Shubnikov Crystal. Institute, Moscow, 1981) crystallography, crystal physics. *Protein_crystallography membrane_proteins computer_modelling instrumentation*
E-mail vmshulmeister@lbl.gov vladimir@lcbvax.cchem.berkeley.edu Tel. 1(510)4864338

Siegel, Lester A. (1925). 44 Strawberry Hill Ave., Apt 10E, Stamford, CT 06902-2632, USA. PhD (Massachusetts Institute of Technology, 1948) physics. *X-ray_diffraction*
Tel. 1(203)3592030

Siegrist, Dr Theo (1955). Member of Technical Staff. AT&T Bell Laboratories, Room 1D-348, 600 Mountain Ave, Murray Hill, NJ 07974, USA. PhD (Swiss Federal Institute of Technology, 1982) solid state physics. *Inorganic_structures oxides inter-metallic_phases thin_films powder_diffraction Rietveld*
E-mail tsi@allwise.mh.att.com Tel. 1(908)5825253 Fax 1(908)5822521

Sieker, Dr Larry C. (1933). Senior Research Associate. Dept. of Biological Structure, SM-20, University of Washington, Seattle, WA 98195, USA. PhD (U. Washington, 1981) biological structure. *Proteins structure-activity_relationship carcinogenesis crystallization metalloproteins refinement precision water*
E-mail sieker@gouda.bchem.washington.edu sieker@ibs.fr Tel. 1(206)5436541 Fax 1(206)5431524

Sigler, Paul B. (1934). Professor. BCMM 154, Yale University/Howard Hughes Medical Institute, 295 Congress Ave, New Haven, CT 06510, USA. MD/PhD (MRC Lab/U. Cambridge, 1965) biochemistry.
E-mail sigler@csb.yale.edu Tel. 1(203)7374441 Fax 1(203)7763550

Silcox, John (1935). Professor of Applied Physics. School of Applied and Engineering Physics, Cornell University, 235 Clark Hall, Ithaca, NY 14853, USA. PhD (Cambridge U., 1961) physics. *Electron_diffraction microscopy scattering*
E-mail jsilcox@msc.cornell.edu Tel. 1(607)2553332 Fax 1(607)2557658

Silva, Dr Abelardo M. (1948). Senior Scientist. National Cancer Institute, Building 322, Frederick, Maryland 21702-1201, USA. PhD (U. Nacional de La Plata, Argentina, 1979) physics. *Proteins aspartic_proteases icosahedral_viruses methods_in_crystallography drug_design ion_channel_structures proteolitic_mechanisms molecular_modelling*
E-mail abelardo@ncifcrf.gov Tel. 1(301)8461977 Fax 1(301)8466066

Silverton, Enid W. Research Chemist. NIH, Building 5, Room 334, Bethesda, MD 20892, USA. PhD chemistry. *Protein_crystallography*
E-mail enid@vger.niddk.nih.gov Tel. 1(301)4024496 Fax 1(301)4960201

Silverton, Dr James V. (1934). Scientist. National Institutes of Health, Bethesda, MD 20892, USA. [NIH, NHLBI-LBC, 10/7N-307, 10 Center Dr, MSC-1676, Bethesda, MD 20892-1676, USA.] PhD (U. Glasgow, 1963) chemistry. *Drug_conformation direct_methods absolute_configuration atomic_scale_microscopy quantum_mechanics molecular_mechanics computer_methods*
E-mail jvs@helix.nih.gov jvs@nihcu.bitnet Tel. 1(301)4961515 Fax 1(301)4023404

Sime, Rodney J. (1931). Professor Emeritus. Department of Chemistry, California State University, Sacramento, CA 95819-6057, USA. [609 Shangri Lane, Sacramento, CA 95825-5504, USA.] PhD (U. Washington, 1959) physical chemistry. *Small_moiety education*

E-mail rodsime@csus.edu Tel. 1(916)2786659 Fax 1(916)2786664

Simmons, Charles (1948). Associate Professor of Chemistry. Brigham Young University - Hawaii, MSc Division, Laie, HI 96762, USA. PhD (U. Hawaii, 1980) inorganic. *Structural_characterization_of_copper Jahn-Teller_complexes cobalt-dioxygen_complexes*

Tel. 1(808)2933813 Fax 1(808)2933825

Simonsen, Prof. Stanley H. (1918). Professor Emeritus. Department of Chemistry and Biochemistry, The University of Texas at Austin, Austin, TX 78712-1167, USA. PhD (U. Illinois (Urbana), 1949) analytical chemistry. *Heterocyclic_compounds cyclic_conformation structure_property_relationships enantioselective_catalysts packing transition_element_coordination_compounds*

E-mail cmal761@utxvms.cc.utexas.edu Tel. 1(512)4715755 Fax 1(512)4710985

Simpson, H. D. (1937). Staff Consultant. 3772 Hamilton Street, Irvine, CA 92714, USA. PhD (Texas U., 1969) chemistry. *Catalysis hydrotreating thermochemistry_structure bonding*

E-mail strrhds@ptl.unocal.com Tel. 1(714)5771584 Fax 1(714)5773176

Singu, Phirtu (1933). Director X-ray Crystallization Lab. Dept. of Chem., North Carolina State University, Box 8204, Raleigh, NC 27695-8204, USA. PhD (U. Colorado, 1965) physical chemistry.

Tel. 1(919)5157362 Fax 1(919)5155079

Sinha, Dr Sunil K. (1939). Senior Research Associate. Corporate Research, Exxon Research and Engineering Co., Annandale, NJ 08801, USA. PhD (Cambridge U., 1964) physics. *Surface/interface_structure thin_films multilayers diffuse_scattering small_angle_scattering reflectivity polymers magnetism*

E-mail sksinha@erenj.com Tel. 1(908)7302875 Fax 1(908)7303042

Skarstad, Paul (1942). Director of Research and Development. MEDTRONIC, 6700 Shingle Creek Parkway, Brooklyn Center, MN 55430, USA. PhD (Cornell U., 1971) physical chemistry. *Solid_state_ionics structure-property_relations*

Tel. 1(612)5691217 Fax 1(612)5691284

Skelton, Dr Earl F. (1940). Section Head. Condensed Matter and Radiation Sciences Division, Naval Research Laboratory, Washington, DC 20375-5345, USA. PhD (Rensselaer Polytechnic Institute, 1967) physics. *High-pressure_synchrotron_radiation microdiffraction_superconductors*

E-mail skelton@anvil.nrl.navy.mil Tel. 1(202)7673014 Fax 1(202)7674868

Skinner, Dr Matthew M. (1962). Postdoctoral Fellow. Life Science Division, Los Alamos National Laboratory, Los Alamos, NM 87545, USA. PhD (U. California, San Diego, 1991) protein crystallography. *Protein_crystallography protein_folding proteins protein-nucleic_acid protein_NMR*

E-mail mms@prov2.lanl.gov Tel. 1(505)6652501 Fax 1(505)6653024

Skita, Dr Victor (1957). Assistant Professor. Department of Biochemistry, c/o Biomolecular Structure Analysis Center/MC-2017, University of Connecticut Health Center, 263 Farmington Avenue, Farmington, CT 06030-2017, USA. PhD (U. Pennsylvania, 1985) biophysics. *Small_angle_scattering diffraction_theory membranes surfactants thin_film pulmonary_medicine synchrotron_radiation*

E-mail skita@bsac.uchc.edu Tel. 1(203)6794421 Fax 1(203)6791912

Skrzypczak-Jankun, Ewa (1948). Chemistry Instrumentation Supervisor. Chemistry Department, University of Toledo, Bowman Oddy Laboratories, 2801 W. Bancroft St., Toledo, OH 43606-3390, USA. PhD (A. Mickiewicz U., 1976) chemistry. *Service_crystallography small_molecules proteins powders_and_thin_films molecular_modelling*

E-mail eskrzyp@uoft02.utoledo.edu Tel. 1(419)5377861 1(419)5377897 Fax 1(419)5374033

Slaughter, Maynard (1934). Professor. Department of Chemistry & Geochemistry, Colorado School of Mines, Golden, CO 80401, USA. PhD solid state chemistry. *Silicate_chemistry/structure clay_geochemistry*

Tel. 1(303)2733648 Fax 1(303)2733629

Sleight, Arthur W. (1939). Professor. Department of Chemistry, Orgon State University, Corvallis, OR 97331-4003, USA. PhD (U. Connecticut, 1963) inorganic chemistry. *Powder Rietveld_method defect order-disorder*

E-mail sleighta@ccmail.orst.edu Tel. 1(503)7376749 Fax 1(503)7372062

Smith, Albert E. (1908). 72 San Mateo Rd., Berkeley, CA 94707, USA. MS (U. California at Berkeley, 1935) chemical physics. *Structure_metals organics*

Tel. 1(510)5243697

Smith, Dr Craig D. (1954). Research Associate. THT 79, BHS 233a, Center for Macromolecular Crystallography, University of Alabama at Birmingham, Birmingham, AL 35294-0005, USA. PhD (U. Alabama at Birmingham, 1986) biophysical sciences. *Proteins area_detector low_temperature crystal_growth biocrystallography data_collection data_processing drug*

E-mail smith@orion.cmc.uab.edu Tel. 1(205)9347233 Fax 1(205)9340480

Smith, David J. (1948). Professor. Center for Solid State Science, Arizona State University, Tempe, AZ 85287, USA. PL.D. & DSc (U. Melbourne, Australia, 1978 & 1988) physics. *High_resolution_electron_microscopy*

E-mail smithd@csss.la.asu.edu Tel. 1(602)9654540 Fax 1(602)9659004

Smith Jr, Deane Kingsley (1930). Professor of Mineralogy. Member, Commission on Powder Diffraction; Editor, World Directory of Powder Diffraction Programs. Department of Geosciences, The Pennsylvania State University, 239 Deike Building, University Park, PA 16802, USA. [1652 Princeton Drive, State College, PA 16803, USA.] PhD (U. Minnesota, 1956) geology. *Silicate_mineralogy sulfide_mineralogy mineral_structures oxide_ceramics powder_diffraction geometrical_crystallography uranium_mineralogy cement*

E-mail smith@psumrl1.psu.edu Tel. 1(814)8655782 1(814)8637845 Fax 1(814)8637845

Smith, Deborah R. (1949). Graduate Student. Department of Chemistry, PO Box 939, Rutgers University, Piscataway, NJ 08854-0759, USA. BS (Saint Mary's College, 1971) medical technology. *Water_structure protein_toxins proteins homology_prediction*

E-mail dsmith@dnarna.rutgers.edu Tel. 1(908)9322200 Fax 1(908)9325958

Smith, Douglas L. (1937). Retired. [758 Hickory Lane, Berwyn, PA 19312-1439, USA.] PhD (U. Wisconsin, 1962) physical chemistry. *Proteins drugs*

Tel. 1(610)6447804

Smith, Dr Francine R. (1958). Assistant Professor. Program in Molecular Medicine, University of Massachusetts Medical Center, 373 Plantation Street, Worcester, MA 01605, USA. PhD (The Johns Hopkins U., 1985) biology. *Proteins macromolecules hemoglobins allostery*

E-mail fran@darwin.ummed.edu Tel. 1(508)8566903 Fax 1(508)8564289

Smith, Dr G. David (1941). Senior Research Scientist. Hauptman-Woodward Medical Research Institute (formerly the Medical Foundation of Buffalo), 73 High Street, Buffalo, NY 14217, USA. PhD (Ohio U., 1968) physical chemistry. *Proteins direct_method conformational_change insulin oligopeptides peptaibols computer_modelling polymorphism*

E-mail smith@odin.hwi.buffalo.edu Tel. 1(716)8569600 Fax 1(716)8524846

Smith, Graham M. (1947). Senior Investigator. Merck Research Laboratories, WP42-3, West Point, PA 19486, USA. PhD (State U. New York at Buffalo, 1974) organic chemistry. *Proteins computer_modelling solvent_effects inhibitor_binding*

E-mail graham_smith@merck.com Tel. 1(215)6527620 Fax 1(215)6526913

Smith, Dr Janet L. (1951). Associate Professor. Department of Biological Sciences, Purdue University, West Lafayette, IN 47907-1392, USA. PhD (U. Wisconsin-Madison, 1978) biochemistry. *Protein_crystallography protein_structure anomalous_diffraction*

E-mail smithj@ewald.bio.purdue.edu smithj@purccvm.bitnet Tel. 1(317)4949246 Fax 1(317)4961189

Smith, Dr Joseph V. (1928). Professor. Department of Geophysical Sciences, University of Chicago, Chicago, IL 60637, USA. PhD (Cambridge U., 1951) physics. *Mineralogy*

E-mail smith@geo1.uchicago.edu Tel. 1(312)7028110 Fax 1(312)7020157

Smith, Dr Phillip Ross (1945). Associate Professor of Cell Biology; Director, Research Computing Resource. Department of Cell Biology, NYU Medical Center, 550 First Avenue, New York, NY 10016, USA. PhD (U. Cambridge, 1971) high energy physics. *Electron_microscopy image_processing image_reconstruction software*

E-mail smithp01@mcrcr.med.nyu.edu Tel. 1(212)2635356 Fax 1(212)2638139

Smith, Robert W. (1956). Assistant Professor. Department of Physics, University of Nebraska at Omaha, Omaha, NB 68182-0266, USA. PhD (Oregon State U., 1989) inorganic chemistry.

Tel. 1(402)5543592 Fax 1(402)5543100

Smith, Thomas James (1959). Associate Professor. Dept. Biology, Lilly Hall B135, Purdue University, West Lafayette, IN 47907, USA. PhD (U. Rochester, 1985) biochemistry. *Allostery antibodies liquid_proteins viral_proteins*

E-mail tom@bragg.bio.purdue.edu Tel. 1(317)4948038 Fax 1(317)4961189

Smith, Dr Ward W. (1949). Principal Scientist. Agouron Pharmaceuticals Inc., 3565 General Atomics Court, San Diego, CA 92121, USA. PhD (The University of Michigan, 1977) biological chemistry. *Area-detector biomolecule computer-assisted_design computing data_collection instrumentation*

E-mail wwsmith@agouron.com Tel. 1(619)6223028 Fax 1(619)6223299

Smithteale, Michael J. PhD candidate. Center for Macromolecular Crystallography, University of Alabama at Birmingham, THT 79, BHS 210, 1918 University Blvd., Birmingham, AL 35294-0005, USA. BS (Florida State U., 1986) biochemistry. *Structure_determination proteins drug_design protein_design compliment_proteins photosynthetic_proteins energy_production*

E-mail teale@orion.cmc.uab.edu teale@uabcmc.bitnet Tel. 1(205)9340124 Fax 1(205)9340480

Smolucttowski, Roman (1910). Professor Emeritus. Department of Astronomy and Physics, University of Texas, Austin, TX 78712, USA. PhD (U. Groningen, 1936) physics.

Tel. 1(512)4711305

Snow, Mark E. (1959). Section Leader. Schering-Plough Res. Inst., 2015 Galloping Hill Rd., K-15, Kenilworth, NJ 07033, USA. PhD (Johns Hopkins Medical School, 1986) biophysics. *Proteins computer_modelling homology_prediction structure_based_drug_design*

Tel. 1(908)2983575 Fax 1(908)2987545

Snyder, James A. (1962). Research Assistant. Department of Chemistry, University of New Orleans, New Orleans, LA 70148, USA. BS (U. Nevada, Las Vegas, 1989) mathematics and chemistry. *Charge_density chemical_bonding MO_calculation least-squares_refinement LCAO_method diffraction_theory electron_density_distribution*

E-mail edscm3@uno.edu Tel. 1(504)2867217 Fax 1(504)2866860

Snyder, Robert L. (1941). Professor. Co-chair size–strain round robin for Commission on Powder Diffraction. NYS College of Ceramics, Alfred University, Alfred, NY 14802, USA. PhD (Fordham U., 1968) physical chemistry. *Powder_diffraction structure_property_relationships high_Tc_superconductivity automation*
E-mail snyder@xray.alfred.edu Snyder@ceramics.bitnet Tel. 1(607)8712438 Fax 1(617)8712392

Socci, Edward Peter (1968). Graduate Student. Department of Materials Science and Engineering, University of Virginia, Charlottesville, VA 22903-2442, USA. MS (U. Virginia, 1993) materials science and engineering. *Cholesterol_esters liquid_crystals nlo_materials theory fiber_diffraction thin_films computer_modelling optical_limiting*
E-mail eps8d@Virginia.EDU Tel. 1(804)9825682 Fax 1(804)9825660

Soltis, S. Michael. Staff Scientist. Stanford Synchrotron Radiation Laboratory, PO Box 4349, MS 69, Stanford CA 94309, USA. PhD (UCLA, 1988) chemistry. *Laue enzymes proteins*
E-mail soltis@ssrl01.slac.stanford.edu Tel. 1(415)9263050 Fax 1(415)9263050

Sommerer, Dr Shaun O. (1962). Assistant Professor. Department of Chemistry, Penn State University at Erie, Erie, PA 16563-1200, USA. PhD (U. Florida, 1991) inorganic chemistry. *Computer_modelling small_molecules*
E-mail sos1@psuvm.psu.edu sos1@psuvm.bitnet Tel. 1(814)8986401 Fax 1(814)8986213

Somorjai, Gabor A. (1935). Professor of Chemistry. Department of Chemistry, University of California, Berkeley, CA 94720, USA. PhD (U. California, 1960) chemistry. *Adsorption catalysis chemisorption surface AES chemistry interface low-energy_electron_diffraction scanning_tunnelling_microscopy single_crystal carbon_compounds ice noble_metals oxides*
E-mail somorjai@garnet.berkeley.edu Tel. 1(510)6424053 Fax 1(510)6439668

Sosnick, Tobin R. (1961). Postdoctoral. Department of Biochemistry and Biophysics, Univ. of Pennsylvania, Philadelphia, PA 19104-6059, USA. PhD (Harvard U., 1989) applied physics. *Protein_folding protein_stability*
E-mail tobin@hxiris.med.upenn.edu Tel. 1(215)8986580 Fax 1(215)8982415

Spangler, Dr Brenda D. (1939). Adjunct Assistant Professor. Department of Chemistry and Department of Biological Sciences, Northern Illinois University, DeKalb, IL 60115, USA. PhD (Northern Illinois U., 1984) biochemistry. *Bacterial_protein_toxins proteins crystallization molecular_recognition*
E-mail spangler@cz.chem.niu.edu Tel. 1(815)7533106 Fax 1(815)7534802

Sparks, Cullie J. (1929). Group Leader, X-ray Diffraction. Metals & Ceramic Div. 6118, Oak Ridge National Lab., PO Box 2008, Bldg 4500S, Oak Ridge, TN 37831-6118, USA. PhD (U. Kentucky, 1957) materials science. *Short-range_order defect materials alloys synchrotron_radiation anomalous_dispersion*
Tel. 1(615)5746996 Fax 1(615)5747659

Sparks, Dr Robert Allen (1928). Consultant. [1800 Pearl St. W. Apt. 4, Tillamook, OR 97141, USA.] PhD (UCLA, 1958) physical chemistry. *Instrumentation automation structure_determination computing_methods protein_folding*
Tel. 1(503)8428237

Specht, Eliot D. (1959). Research Staff. Metals and Ceramics Division, Oak Ridge National Laboratory, PO Box 2008, MS 6118, Oak Ridge, TN 37831-6118, USA. PhD (Massachusetts Institute of Technology, 1987) physics. *Anomalous_dispersion synchrotron_radiation MBE interface*
E-mail esy@ornl.gov Tel. 1(615)5747682 Fax 1(615)5747659

Speir, Jeffrey A. (1967). Graduate Student. 1392 Lilly Hall of Life Sciences, Department of Biological Sciences, Purdue University, West Lafayette, IN 47907-1392, USA. BA (U. California, San Diego, 1989) chemistry/biochemistry. *Viruses proteins molecular_replacement electron_microscopy_image_reconstruction viral_assembly viral_polymorphism tricornaviruses purification*
E-mail b4f@mace.cc.purdue.edu Tel. 1(317)4940832 Fax 1(317)4961189

Spielberg, Nathan (1926). Professor. Department of Physics, Kent State University, Kent, OH 44242, USA. PhD (Ohio State U., 1952) physics. *Liquid_crystals X-ray_diffractometry X-ray_fluorescence*
E-mail nspielbe@kentvm.kent.edu nspielbe@kentvm.bitnet Tel. 1(216)6722881 Fax 1(216)6722959

Sprang, Dr Stephen R. (1949). Associate Professor. Howard Hughes Medical Institute, University of Texas Southwestern Medical Center, 5323 Harry Hines Blvd, Dallas, TX 75235-9050, USA. PhD (U. Wisconsin, Madison, 1978) biochemistry. *Allostery proteins catalysis conformational-change phosphorylases G-proteins cytokines receptors*
E-mail sprang@howie.swmed.edu Tel. 1(214)6485008 Fax 1(214)6486336

Springer, James P. (1950). Director. Merck Research Laboratories, Department of Biophysical Chemistry, RY50-100, Rahway, NJ 07065-0900, USA. PhD (Iowa State U., 1976) physical chemistry. *Proteins small_molecules drugs biochemistry computer_modelling nuclear_magnetic_resonance structure-activity_relationships calorimetry*
E-mail james_springer@merck.com Tel. 1(908)5945496 Fax 1(908)5946100

Sproul, Gordon Duane (1944). Professor. Chemistry Department, University of South Carolina, 80 Carteret St., Beaufort, SC 29902, USA. PhD (U. Illinois, 1971) inorganic chemistry. *Electronegativity coordination_polymers*
Tel. 1(803)5214162 Fax 1(803)5214162

Spurlino, Dr John C. (1958). Research Investigator. Biophysics and Computational Chemistry, Sterling Winthrop Inc., 1250 S. Collegeville Rd., PO Box 5000, Collegeville, PA 19426-0900, USA. PhD (Rice U., 1988) biochemistry. *Proteins computer_modelling structure_prediction graphics rational_drug_design neurological_proteins*
E-mail jcs@kodak.com Tel. 1(215)9835890 Fax 1(215)9836908

Squattrito, Philip J. (1960). Associate Professor. Department of Chemistry, Central Michigan University, Mt. Pleasant, MI 48859, USA. PhD (Northwestern U., 1987) inorganic chemistry. *Inorganic layered_materials coordination_chemistry*
E-mail 3clwp5s@cmuvm.csv.cmich.edu Tel. 1(517)7744407 Fax 1(517)7747106

Sridhar, Dr Vaidehi. IRTA fellow. NIAMS, National Institutes of Health, Bldg 6, B2-12, 9000, Rockville Pike, Bethesda, MD 20892, USA. PhD (Massachusetts Institute of Technology, 1993) biochemistry. *Structure_function_relationship protein_crystallography protein_engineering protein_expression*
E-mail vaidehi@betta.niams.nih.gov Tel. 1(301)4023223 Fax 1(301)4023417

Srikrishnan, Thamarapu (1943). Cancer Research Scientist. Department of Biophysics, Roswell Park Cancer Institute, 666 Elm Street, Buffalo, NY 14263, USA. PhD (U. Madras, Madras, India, 1969) X-ray crystallography. *Structures_of_bio_molecules structure_of_immunomodulators X-ray_diffraction_techniques direct_methods structure_carcinogens_and_chemotherapeutic_drugs*
E-mail srikrishnan@sc3103.med.buffalo.edu Tel. 1(716)8458302 Fax 1(716)8458899

Srinivasan, A. R. (1946). Research Assistant Professor. Rutgers University, Chemistry Department, New Brunswick, NJ 08903, USA. PhD (U. Madras, India, 1974) biomolecular physics. *Computer_graphics computer_modelling conformation semi-empirical_calculation DNA_structure conformation_rings conformation_wheels Holliday_junctions DNA_packing triple_and_tetra_helices*
E-mail srini@chemf.rutgers.edu Tel. 1(908)4459239 Fax 1(908)4455958

Sriram, Mahalingam (1964). Postdoctoral Associate. Member. Argonne National Laboratory, 9700 S. Cass Avenue, Bldg. 202 Rm. A121, Argonne, IL-60439, USA. PhD (U. Illinois Urbana-Champaign, 1994) biophysics. *Macromolecular_structure computer_modelling protein_structure_prediction*
E-mail mahal@anlbem.bim.anl.gov Tel. 1(708)2525254 Fax 1(708)2525517

St. Charles, Robert J. (1954). Postdoctoral Fellow. Department of Pharmacology, Jefferson Cancer Institute, Thomas Jefferson University, 233 South 10th Street, Philadelphia, PA 19107, USA. PhD (Wayne State U., 1989) biochemistry. *Protein_structure cytokines*
E-mail rstc@calvin.jci.tju.edu Tel. 1(215)9554571

Stadelmaier, Hans H. (1922). Prof. Emeritus. Dept. of Materials Science and Eng., North Carolina State Univ., Box 7907, Raleigh, NC 27695-7907, USA. Dr. rer. nat. (U. Stuttgart, Germany, 1956) physics. *Electron_microscopy intermediate_phase interstitial magnetism magnets metallurgy_metals ferromagnetic phase_diagram*
E-mail stadelmaier@mat.mte.ncsu.edu Tel. 1(919)5152349 Fax 1(919)5157724

Stalick, Dr Judith K. (1943). Research Chemist. Reactor E151, National Institute of Standards and Technology, Gaithersburg, MD 20899, USA. PhD (Northwestern U., 1969) inorganic chemistry. *Neutron_diffraction Rietveld_method quantitative_phase_determination superconductors powder_diffraction database*
E-mail stalick@enh.nist.gov Tel. 1(301)9756223 Fax 1(301)9219847

Stallings, Dr William C. (1947). Fellow. Monsanto Corporate Research/Searle BB4K, 700 Chesterfield Parkway North, St. Louis, Missouri 63198, USA. PhD (U. Pennsylvania, 1975) physical chemistry. *Crystallography proteins macromolecules small_molecules enzyme_mechanisms metalloproteins structure_based_design synchrotron_radiation*
E-mail wcstal@bb1t.monsanto.com wcstal@ccmail.monsanto.com Tel. 1(314)5377236 Fax 1(314)5377425

Stanfield, Dr Robyn L. (1958). Senior Research Associate. Department of Molecular Biology, The Scripps Research Institute, 10666 N. Torrey Pines Road, La Jolla, CA 92037, USA. PhD (U. Texas at Austin, 1986) biochemistry. *Immunoglobulins proteins*
E-mail robyn@scripps.edu Tel. 1(619)5543473 Fax 1(619)5546105

Stanko, Dr Joseph A. (1941). Associate Professor. Department of Chemistry, SCA 240, University of South Florida, Tampa, FL 33620, USA. PhD (U. Illinois Urbana-Champaign, 1966) inorganic chemistry. *Platinum_compounds palladium_compounds anticancer_compounds electrical_conductivity superconductors phosphates*
E-mail stan@chuma.cas.usf.edu Tel. 1(813)9742373 Fax 1(813)9743203

Stanton, Dr Martin (1957). Senior Research Associate. Rosenstiel Center, Brandeis University, Waltham, MA 02254, USA. PhD (Brandeis U., 1993) biophysics. *Area_detector synchrotron_radiation biophysics computer_modelling data_processing data_collection*
E-mail marty@dark.rose.brandeis.edu Tel. 1(617)7362424 Fax 1(617)7362405

Staples, Richard J. (1961). Senior Research Associate/Lecturer. Department of Chemistry, Texas A&M University, College Station, TX 77843-3255, USA. PhD (U. Toledo, 1989) chemistry. *Organometallic inorganic coordination sulfur gold*
E-mail staples@chemvx.tamu.edu Tel. 1(409)8454837 Fax 1(409)8454719

Starr, Jonathan E. (1955). Engineer. [20350 Stevens Creek Blvd, Apt. 106, Cupertino, CA 95014, USA.] MS (U. California, Santa Barbara, 1994) electrical engineering. *Silicon silicon_carbide gallium_nitride*
E-mail jstarr@cup.hp.com Tel. 1(408)2555750

States, Dr David J. (1953). Associate Professor. Institute for Biomedical Computing, Washington University, Box 8036, St. Louis, MO 63110, USA. MD, PhD (Harvard U., 1983) biophysics. *Computer_modelling proteins databases homology_prediction genome_analysis molecular_sequence molecular_evolution*
E-mail states@ibc.wustl.edu Tel. 1(314)3622134 Fax 1(314)3620234

Staudenmann, Jean-Louis (1940). Physicist. Quantum Metrology Division, National Inst. of Standards and Technology, Bldg. 221, Room A141, Gaithersburg, MD 20899, USA. PhD (U. Geneva, 1975) solid state and X-ray diffraction. *Diffraction_instrumentation*
Tel. 1(301)9754866 Fax 1(301)9753038

Stauffacher, Cynthia Vianne (1948). Associate professor. Dept. of Biological Sciences, Purdue Univ., West Lafayette, IN 47907, USA. PhD (UCLA, 1977) physical chemistry. *Crystallography biophysics enzymes membrane_associated_proteins channel_proteins cholesterol biosynthesis*
E-mail   b4u@mace.cc.purdue.edu   cyndy@charybdis.bio.purdue.edu   Tel. 1(317)4944937 Fax 1(317)4961189

Stebbins, Jonathan (1954). Associate Professor. Department of Geological Sciences, Stanford University, Stanford, CA 94305-2115, USA. PhD (U. Cal. at Berkeley, 1983) geology. *Inorganic_materials oxides silicates minerals nuclear_magnetic_resonance_spectroscopy thermodynamic_properties high_temperature chemistry glasses amorphous_materials oxide_melts*
E-mail stebbins@pangea.stanford.edu Tel. 1(415)7231140 Fax 1(415)7252199

Stec, Dr Boguslaw (1956). Research Associate. Department of Chemistry, Boston College, Chestnut Hill, MA 02167, USA. PhD (Jagiellonian U., 1986) theoretical physics. *Protein_structure structure_function_relationship metal_ions_catalysis protein_motions biological_mathematical_models protein_folding modelling*
E-mail stec@bcchem.bc.edu stec@bcchem.bitnet Tel. 1(617)5523615 Fax 1(617)5522705

Steele, Dr Ian M. (1944). Senior Research Associate. Department of Geophysical Sciences, University of Chicago, 5734 S. Ellis Ave., Chicago, IL 60637, USA. PhD (U. Illinois at Urbana, 1971) mineralogy. *Mineralogy electron_probe_microanalysis crystallography structure lead_acid_battery meteorites lunar_mineralogy*
E-mail steele@geo1.uchicago.edu Tel. 1(312)7028109 Fax 1(312)7029505

Stehle, Thilo (1962). Post-doc. Dept. of Biochem and Mol. Biology, Harvard University, 7 Divinity Ave., Cambridge, MA 02138, USA. (Freiburg, Germany, 1992) X-ray crystallography. *Virus_structure X-ray_methods receptor_binding carbohydrate_structures*
E-mail stehle@xtal0.harvard.edu Tel. 1(617)4955043 Fax 1(617)4959613

Steinfink, Dr Hugo (1924). Professor. Coeditor, Acta Crystallographica. Department of Chemical Engineering, University of Texas, Austin, TX 78712, USA. PhD (Polytechnic U. New York, 1954) physical chemistry. *Physical_properties_electrical_magnetic solid_state_mineralogy_clays synthesis_oxides_chalcogenides superconductors*
E-mail hugo@che.utexas.edu Tel. 1(512)4715233 Fax 1(512)4717060

Steinrauf, Dr Larry K. (1931). Professor. Department of Biochemistry and Molecular Biology, Indiana University School of Medicine, 635 Barnhill Drive, Indianapolis, IN 46202-5122, USA. PhD (U. Washington, Seattle, 1957) biochemistry. *Protein_crystallography antibiotics*
E-mail lks@biochem1.iupui.edu Tel. 1(317)2747544 Fax 1(317)2744868

Steitz, Prof. Thomas A (1940). Professor. Department of Molecular Biophysics and Biochemistry, Howard Hughes Medical Institute, Yale University, 266 Whitney Avenue, Bass Center-Room 418, PO Box 208114, New Haven, CT 06520-8114, USA. PhD (Harvard U., 1966) biochemistry and molecular biology. *Protein_crystallography protein–DNA_interaction protein–RNA_interaction enzyme_mechanism*
Tel. 1(203)4325617 1(203)4325619 Fax 1(203)4323282

Stenkamp, Dr Ronald E. (1948). Associate Professor. Dept. of Biological Structure, SM-20, University of Washington, Seattle, WA 98195, USA. PhD (U. Washington, 1975) physical chemistry. *Proteins computer_modelling environment_protection metalloproteins structure_solution refinement precision water*
E-mail stenkamp@u.washington.edu Tel. 1(206)6851721 Fax 1(206)5431524

Stephens, Peter (1951). Professor. Physics Department, SUNY-Stony Brook, Stony Brook, NY 11794, USA. PhD (MIT, 1978) physics. *Synchrotron_radiation powder surface fullerenes quasicrystals*
E-mail pstephens@sunysb.edu Tel. 1(516)6328156 Fax 1(516)6328874

Stern, Charlotte L. (1963). X-ray Crystallographer. Department of Chemistry, Northwestern University, Evanston, IL 60208-3113, USA. BS (U. Illinois, 1986) chemistry. *Service small_molecules*
E-mail stern@mv3600.chem.nwu.edu Tel. 1(708)4912950 Fax 1(708)4917713

Stern, Edward A. (1930). Professor. Physics Dept. FM-15, University of Washington, Seattle, WA 98195, USA. PhD (CalTech, 1955) physics. *X-ray_absorption_fine_structure phase_transitions disordered_materials high_temperature_superconductivity electron_energy_loss metalloproteins*
E-mail stern@phys.washington.edu Tel. 1(206)5432023 1(206)5430435 Fax 1(206)5439525

Stern, Lawrence (1961). Professor. Department of Chemistry, Room 16-611, Massachusetts Institute of Technology, Cambridge, MA 02139, USA. PhD (MIT, 1989) chemistry.
E-mail stern@mit.edu Tel. 1(617)2532849 Fax 1(617)2587500

Stevens, Dr Edwin D. (1947). Distinguished Professor. Secretary, Commission on Charge, Spin, and Momentum Densities. Department of Chemistry, University of New Orleans, New Orleans, LA 70148, USA. PhD (U. California, Davis, 1973) physical chemistry. *Charge_density electrostatic_potential chemical_bonding hydrogen_bonding biocrystallography carcinogenesis drug_receptor inorganic_chemistry*
E-mail edscm@uno.edu Tel. 1(504)2866856 Fax 1(504)2866860

Stevens, Prof. Raymond C. (1963). Assistant Professor. Department of Chemistry, University of California, Berkeley, CA 94720, USA. PhD (U. Southern California, 1988) chemistry. *Protein crystallography neurobiology neurotoxins neuroreceptors electron_diffraction electron_microscopy*
E-mail stevens@neuron1.berkeley.edu Tel. 1(510)6438285 Fax 1(510)6439290

Stewart, James McDonald (1931). Professor. Formerly Dept Chem/Biochem, University of Maryland, College Park, MD 20742, USA. [PO Box 472, McConnellsburg, PA 17233, USA.] PhD (U. Washington, Seattle, 1958) physical chemistry. *Crystallographic_computing*
E-mail stewart1@lsm.nrl.navy.mil Tel. 1(717)4855990

Stezowski, Dr John J. (1942). Professor. US Editor, Zeitschrift fuer Kristallographie; Co-Editor Acta Crystallographica (B, C & D); Co-editor Crystallography Reviews. Department of Chemistry, University of Nebraska Lincoln, Lincoln, NE 68588-0304, USA. PhD (Michigan State U., 1969) chemistry. *Organic_materials molecular_biophysics structure–activity_relationships computer_modelling*
E-mail jjs@unlinfo.unl.edu Tel. 1(402)4728570 Fax 1(402)4720168

Stock, Ann (1957). Assistant Professor. Center for Advanced Biotechnology and Medicine, 679 Hoes Lane, Piscataway, NJ 08854-5638, USA. PhD (U. California, Berkeley, 1986) biochemistry. *Biochemistry catalysis chemotaxis conformational_change signal_transduction*
E-mail stock@mbcl.rutgers.edu Tel. 1(908)2354844 1(908)2355164 Fax 1(908)2355318

Stock, Stuart R. (1955). Associate Professor. School of Materials Science and Engineering, Georgia Institute of Technology, Atlanta, GA 30332-0245, USA. PhD (U. Illinois at Urbana-Champaign, 1983) metallurgical engineering. *X-ray_diffraction microtomography stress/strain_measurements nondestructive_evaluation synchrotron_radiation*
E-mail stuart.stock@mse.gatech.edu Tel. 1(404)8946882 Fax 1(404)8539140

Stoddard, Barry L. (1963). Assistant Member. Division of Basic Sciences, Fred Hutchinson Cancer Research Center, 1100 Fairview Ave N., Seattle, WA 98109, USA. [Division of Basic Sciences, Fred Hutchinson Cancer Research Center, 1124 Columbia St. A3-023, Seattle, WA 98104, USA.] PhD (Massachusetts Institute of Technology, 1990) biophysical chemistry. *Laue_diffraction metalloenzymes superoxides photochemistry catalysis receptor microgravity_crystal_growth DNA_binding_proteins computer_assisted_drug_design*
E-mail bstoddard@fred.fhcrc.org Tel. 1(206)6674031 Fax 1(206)6673515

Stokes, David L. (1956). Assistant Professor. Department of Molecular Physiology and Biological Physics, University of Virginia Health Sciences Center, Charlottesville, VA 22908, USA. PhD (Brandeis U., 1986) biophysics. *High-resolution_electron_microscopy image_analysis low_temperature three-dimensional_reconstruction*
E-mail dls4n@virginia.edu Tel. 1(804)9823412 Fax 1(804)9821616

Stouch, Terry (1956). Senior Research Investigator II. Bristol-Myers Squibb Pharmaceutical Research Institute, PO Box 4000, Princeton, NJ 08543-4000, USA. PhD (Pennsylvania State U., 1985) chemistry. *Computer_simulation biomolecules membranes drug_design protein_structure water_structure*
E-mail stouch@bms.com Tel. 1(609)2757234 Fax 1(609)2756030 Telex 4754082 BRMYSQ PRIN

Stout, Dr Charles David (1947). Associate Member. Department of Molecular Biology, The Scripps Research Institute, La Jolla, CA 92037, USA. PhD (U. Wisconsin, 1976) biochemistry. *Protein_crystallography metalloproteins metalloenzymes fertilization_proteins nucleic_acid_structure iron–sulfur_proteins protein–RNA_complexes*
E-mail dave@scripps.edu Tel. 1(619)5548738 Fax 1(619)5546188

Stout, George Hubert (1932). Retired. [7037 27th Ave NE, Seattle, WA 98115, USA.] PhD (Harvard, 1956) organic chemistry. *Solution_methods erroneous_structures*
E-mail stout@xray0.bchem.washington.edu Tel. 1(206)5232039

Stout, Dr Thomas J. (1964). Postdoctoral Associate. Department of Biochemistry and Biophysics, School of Medicine, University of California, San Francisco, San Francisco, California 94143-0448, USA. PhD (Cornell U., 1992) chemistry. *Signal_transduction protein–ligand_interactions rational_drug_design phosphorylation direct_methods molecular_mechanics data_collection_methodology*
E-mail stout@msg.ucsf.edu Tel. 1(415)4763937 Fax 1(415)4761902

Stowell, Dr Joseph G. (1951). Director of Laboratories. Department of Medicinal Chemistry and Pharmacognosy, 1333 R. Heine Pharmacy Building, Purdue University, West Lafayette, IN 47907-1333, USA. PhD (U. California, Davis, 1980) organic chemistry. *Structure_disorder molecular_graphics molecular_mobility polymorphs*
E-mail stowell@vm.cc.purdue.edu Tel. 1(317)4947090

Streib, Dr William E. (1931). Crystallographer. Chemistry Department, Molecular Structure Center, Indiana University, Bloomington, IN 67605, USA. PhD (U. Minnesota, 1962) physical chemistry. *Instrumentation structures organic inorganic*
Tel. 1(812)8556821

Strickland, Corey L. (1990). Graduate Student. Department of Biochemistry, Molecular and Cell Biology, 221 Biotechnology Building, Cornell University, Ithaca, NY 14853, USA. BS (Pennsylvania State U., 1990) biochemistry. *Macromolecular_crystallography enzymes drug_design molecular_modelling*
E-mail corey@penelope.bio.cornell.edu Tel. 1(607)2558432 Fax 1(607)2552428

Strong, Roland K. (1963). Assistant Member. Fred Hutchinson Cancer Research Center, Division of Basic Sciences, Mail Stop A3-023, 1124 Columbia St., Seattle, WA 98104, USA. PhD (Harvard U., 1990) biophysics. *Proteins glycoproteins immunology recognition*
E-mail rstrong@fred.fhcrc.org Tel. 1(206)6675587 Fax 1(206)6677730

Stroud, Dr Robert M. (1942). Professor, Biophysics and Biochemistry; Professor, Pharmaceutical Chemistry, Editor, Ann. Rev. Biophysics & Biomolecular Structure. Dept. of Biochemistry and Biophysics, UCSF, Box 0448, San Francisco, CA 94143-0448, USA. PhD (U. London, Birkbeck College, 1968) biological crystallography. *Structure–function biological_macromolecules drug_design membrane_proteins phosphorylation trypsin thymidylate_synthase*
E-mail stroud@msg.ucsf.edu Tel. 1(415)4764224 Fax 1(415)4761902

Stroud, William J. (1961). Postdoctoral Research Associate. Whistler Center for Carbohydrate Research, Smith Hall, Purdue University, West Lafayette, IN 47907, USA. PhD (Purdue U., 1993) polymer crystallography. *Structural_disorder computer_modelling polymers_structure_determination phase_refinement* diffraction_theory image_processing mathematics computer_management
E-mail bill@kiwi.foodsci.purdue.edu Tel. 1(317)4944914 Fax 1(317)4947953

Strouse, Prof. Charles E. (1944). Professor. Department of Chemistry and Biochemistry, University of California, Los Angeles, Los Angeles, CA 90024-1569, USA. PhD (U. Wisconsin, 1969) physical chemistry. *Lattice_clathrates phase_transitions porphyrins instrumentation*
E-mail strouse@uclac0.chem.ucla.edu Tel. 1(310)4752800 Fax 1(310)4759291

Stubbs, Prof. Gerald James (1947). Professor. Department of Molecular Biology, Vanderbilt University, Box 1820, Station B, Nashville, TN 37235, USA. DPhil (Oxford U., 1972) molecular biophysics. *Diffraction_theory proteins fibrous_proteins viruses fiber_diffraction_theory macromolecular_assemblies*
E-mail stubbsgj@ctrvax.vanderbilt.edu gerald%mobv01@ctrvax.vanderbilt.edu Tel. 1(615)3222018 Fax 1(615)3436707

Stuckey, Dr Jeanne A. (1964). Research Fellow. Biophysics Research Division, Chemistry Building, 930 North University, Ann Arbor, MI 48109-1055, USA. PhD (Wayne State University School of Medicine, 1992) biochemistry. *Protein_structure protein_tyrosine_phosphatases signal_transduction* platelet_factor_4 heparin blood_clotting
E-mail jass@oscar.biop.umich.edu Tel. 1(313)7633384 Fax 1(313)7643323

Stura, Dr Enrico A. (1955). Assistant Professor. Department of Molecular Biology, MB-13, The Scripps Research Institute, 10666 N. Torrey Pines Road, La Jolla, CA 92037, USA. DPhil (Oxford U., 1981) molecular biophysics. *Protein_crystallization Fab_complex_crystallization malaria_vaccine* structural_immunology antigen_antibody_interaction
E-mail stura@scripps.edu Tel. 1(619)5542456 1(619)5548271 Fax 1(617)5546105

Su, Shu-Chun (1940). Lab Supervisor. Research Center, Hercules, Inc., 500 Hercules Rd., Wilmington, DE 19808-1599, USA. PhD (Virginia Polytechnic Institute, 1985) mineralogy.
E-mail su@ssnet.com Tel. 1(302)9953498 Fax 1(302)9944135 Telex 835479

Su, Ying (1964). Graduate student. Dept. of Chemistry 0317, University of California at San Diego, La Jolla, CA 92093-0317, USA. PhD (UCSD, 1994) protein crystallography. *Proteins structures functions crystallization*
E-mail ysu@ucsd.edu Tel. 1(619)5344241

Su, Dr Zhengwei (1963). Postdoctoral Research Associate. Department of Chemistry, State University of New York at Buffalo, Amherst, NY 14260-3000, USA. PhD (State U. New York at Buffalo, 1993) chemical crystallography. *Small_molecule computer_modelling charge_density electrostatic_properties computational_chemistry*
E-mail che9985@ubvms.cc.buffalo.edu; zsu@acsu.buffalo.edu Tel. 1(716)6456800x2233 Fax 1(716)6456948

Subbiah, S. (1961). Research Associate. Member of ACA. D-109 Fairchild Center, Stanford Medical School, Stanford, CA 94305, USA. PhD (Harvard, 1988) biophysics. *Phase_problem protein_folding homology_modelling conformational_change*
E-mail subbiah@cellbio.stanford.edu Tel. 1(415)7250754 Fax 1(415)7238464

Sudbeck, Elise (1966). Graduate Student. Chemistry Department, University of Minnesota, 139 Smith Hall, 207 Pleasant Street S.E., Minneapolis, MN 55455-0431, USA. BS (U. South Dakota, 1988) chemistry. *Hydrogen-bonding service_crystallography molecular_recognition*
E-mail sudbeck@chemsun.chem.umn.edu Tel. 1(612)6254385 1(612)6261717 Fax 1(612)6267541

Sullenger, Dr Don B. (1929). Retired. 135 Bethel Road, Centerville, OH 45458, USA. PhD (Cornell U., 1969) physical chemistry. *Direct_methods boron_crystal_chemistry applied_crystallography*
Tel. 1(513)4337904

Sun, Dr Daopin (1962). Visiting Associate. LMB, NIDDK, National Institutes of Health, Bethesda, MD 20892, USA. PhD (U. Oregon, 1990) protein crystallography. *Proteins protein_folding protein_structure receptors* tgf-beta_superfamily receptors_complexes
E-mail sun@vger.niddk.nih.gov Tel. 1(301)4024497 Fax 1(301)4960201

Sundaralingam, Dr Muttaiya. Professor and Ohio Eminent Scholar. Department of Chemistry and Biochemistry, The Ohio State University, Columbus, OH 43210, USA. PhD (U. Pittsburgh, 1961) chemistry/crystallography. *Proteins nucleic_acids water_structure protein_folding*
E-mail msundo%bio@mps.ohio-state.edu Tel. 1(614)2922999 Fax 1(614)2922524

Sundaramoorthy, Munirathinam (1960). Postgraduate researcher. Department of Molecular Biology and Biochemistry, University of California, Irvine, CA 92717, USA. PhD (Indian Institute of Science, 1990) molecular biophysics. *Protein_crystallography*
E-mail sundar@indigo1.biomol.uci.edu Tel. 1(714)8564322 Fax 1(714)8568540

Swain, Amy L. Senior Scientist. Hoffmann-LaRoche Inc., 340 Kingsland Street, Bldg. 76/1516, Nutley, NJ 07110-1199, USA. PhD (U. South Carolina, 1988) chemistry-crystallography. *Enzyme protein_structure-function*
E-mail swaina@rnch01.dnet.roche.com Tel. 1(201)2355585 Fax 1(201)2352682

Swaminathan, Dr Kunchit (1961). Research Associate. Department of Chemistry, University of Pennsylvania, Philadelphia, PA 19104, USA. PhD (Indian Institute of Technology, Bombay, India, 1989) X-ray crystallography. *Protein_structure*
E-mail nathan@a.chem.upenn.edu Tel. 1(215)8988613 Fax 1(215)8982037

Swaminathan, Dr S. Research Chemist. Biocrystallography Lab., VA Medical Center, PO Box 12055, University Drive, Pittsburgh, PA 15240, USA. PhD (The University of Madras, 1980) physics. *Structure_proteins hydration_proteins charge density* toxins_structure electrostatic_potential
E-mail swami@vms.cis.pitt.edu Tel. 1(412)6923517

Swanson, Rosemarie (1942). Research Scientist. Biochemistry and Biophysics, Texas A&M University, College Station, TX 77843-2128, USA. PhD (Stanford U., 1969) chemical physics. *Entropy macromolecular_cooperative_phenomena proteins computer_management* evolution sequence_similarity artificial_life
E-mail rosman@tamu.edu Tel. 1(409)8456842 Fax 1(409)8459274

Swanson, Dr Stanley M. (1938). Research scientist. Biochemistry Dept., Texas A&M U., College Station, TX 77843-2128, USA. PhD (Stanford U., 1968) theoretical physics. *Computer_graphics density_imaging density_statistics quantum_theory*
E-mail stan@monoc.tamu.edu Tel. 1(409)8456846 Fax 1(409)8459274

Sweet, Dr Robert M. (1943). Biologist. Biology Department, Brookhaven National Laboratory, Upton, New York 11973-5000, USA. PhD (U. Wisconsin, Madison, 1970) phys. chem. *Protein_structure synchrotron_radiation dynamic_crystallography Laue_diffraction protease_mechanism*
E-mail SWEET@BNL.GOV Tel. 1(516)2823401 Fax 1(516)2823407

Swepston, Paul N. (1954). Manager of Crystallography. Biosym Technologies, Inc., 9685 Scranton Road, San Diego, CA 92121-0136, USA. PhD (U. Arkansas, 1981) inorganic chemistry. *Software graphics structure_analysis diffractometry*
E-mail pns@biosym.com Tel. 1(619)5979776 Fax 1(619)4580136

Swink, Laurence N. (1934). President. 1617 E Julie Dr, Tempe, AZ 85283-3101, USA. PhD (Brown U., 1969) physical chemistry.
Tel. 1(602)8399489

Swinnea, J. Steven (1953). Research Associate/Lecturer. Department of Chemical Engineering, Center For Materials Science and Engineering, Mail Code 66201, The University of Texas at Austin, Austin, TX 78712, USA. PhD (The University of Texas at Austin, 1981) chemical engineering. *Chemical_physical_property solid_state_chemistry structure-property_relationship powder_diffraction structure_determination* software computing numerical_analysis
E-mail swinnea@che.utexas.edu Tel. 1(512)4713173 Fax 1(512)4717060

Switendick, Alfred C. (1931). Consultant. Theoretical Chemistry, Idaho National Engineering Laboratory, PO Box 1625, Idaho Falls, ID 834152208, USA. [1900 Parkwood Apt A207, Idaho Falls, ID 83401-6116, USA.] PhD (MIT, 1963) solid state physics. *Charge_density hydrides boron_compounds interstitial_compounds*
E-mail azs@inel.gov Tel. 1(208)5261414 Fax 1(208)5268541

Syed, Dr Ashfaquzzaman (1952). European Product Manager. Eastman Kodak Company, Office Imaging EAME Region, 901 Elmgrove Road, Rochester, NY 14607, USA. PhD (Gorakhpur U., 1979) physics.
Tel. 1(716)7260032 Fax 1(716)7266154

Syed, Dr Rashid (1960). Research Associate. US Editor of 9th Edition of World Directory of Crystallographers. Department of Molecular Biology MB13, The Scripps Research Institute, 10666 N. Torrey Pines Rd., La Jolla, CA 92037, USA. PhD (The University of Toledo, 1987) chemistry. *Structural_disorder proteins computer_modelling amphiphilic_toxins* homology_prediction proteins_toxins water_structure
E-mail syed@scripps.edu Tel. 1(619)5543474 Fax 1(619)5546105

Szebenyi, Doletha M. E. (1947). Research Associate. Section of Biochemistry, Molecular and Cell Biology, 209 Biotechnology Building, Cornell University, Ithaca, New York 14853, USA. PhD (U. Connecticut, 1972) physical chemistry. *Biological_macromolecules synchrotron_radiation Laue_diffraction multiple_wavelength_phase_determination* metalloproteins calcium_compounds data_processing
E-mail szebenyi@batman.tn.cornell.edu szebenyi@robin.tn.cornell.edu Tel. 1(607)2552174 Fax 1(607)2552428

Tabernero, Lydia (1961). Postdoctoral fellow. Bristol-Myers Squibb, PO Box 4000, Princeton, NJ 08543, USA. PhD (U. Politecnica de Catalunya, Barcelona, Spain, 1991) biological sciences. *Protein_structure DNA_structure macromolecules_interactions*
E-mail lydia@nkyene.bms.com Tel. 1(609)2524339 Fax 1(609)2526030

Tainer, Dr John A. (1951). Associate Member. Molecular Biology Department MB4, The Scripps Research Institute, 10666 N. Torrey Pines Road, La Jolla, CA 92037, USA. PhD (Duke U., 1982) biochemistry. *Protein_crystallography computer_graphics metalloenzymes protein_engineering* superoxide_dismutases cell_cycle_control DNA-repair_enzymes bacterial_pili
E-mail ja@scripps.edu Tel. 1(619)5548119 Fax 1(619)5546880

Takahara, Patricia M. (1968). Graduate student. Massachusetts Institute of Technology, 77 Massachusetts Ave. Room 18-128, Cambridge, MA 02139, USA. BS (U. California, Berkeley, 1990) chemistry. *DNA_structure anticancer_drugs RNA_structure virus_structure*
E-mail takahara@lippard.mit.edu Tel. 1(617)2531824

Takusagawa, Dr Fusao (1946). Director. Departments of Chemistry and Biochemistry, University of Kansas, Lawrence, KS 66045, USA. PhD (Osaka City U., 1974) physical chemistry. *DNA_RNA DNA_drug drug_design proteins*
E-mail xraymain@kuhub.cc.ukans.edu xraymain@ukanvax.bitnet Tel. 1(913)8644727 Fax 1(913)8645396

Tanner, Jack (1961). Postdoctoral. Department of Biochemical and Biophysical Sciences, University of Houston, Houston, TX 77204-5934, USA. PhD (Brown U., 1988) chemistry. *Proteins molecular_dynamics macromolecules macromolecular_crystallization*
E-mail tanner@Bragg1.bchs.uh.edu Tel. 1(713)7438371 Fax 1(713)7438372

Taylor Jr, Dr Ivan F. (1944). Senior User Services Consultant. Information Services, Texas Christian University, Fort Worth, TX 76129-0001, USA. [TCU, PO Box 32883, Fort Worth, Texas 76129-0001, USA.] PhD (U. South Carolina, 1971) physical chemistry. *Computer_modelling*
E-mail taylor@gamma.is.tcu.edu cc071co@tcuamus.bitnet Tel. 1(817)9217695 Fax 1(817)9217110

Taylor, Jean Ellen (1944). Professor. Mathematics Department, Rutgers University, New Brunswick, NJ 08903, USA. PhD (Princeton U., 1973) mathematics (geometric measure theory). *Morphology growth kinetics computer_modelling*
E-mail taylor@math.rutgers.edu Tel. 1(908)4453484 Fax 1(908)4455530

Taylor, Dr Robert Cooper (1917). Emeritus Professor of Chemistry. US Editor 6th, 7th, 8th Editions World Directory of Crystallographers. Dept. of Chemistry, University of Michigan, Ann Arbor, MI 48109-1055, USA. PhD (Brown U., 1947) physical chemistry. *Raman_spectroscopy infrared_spectroscopy force_constant_structure*
E-mail usem708@umichum.bitnet Tel. 1(313)7472120 Fax 1(313)7474865

Teeter, Dr Martha M. (1944). Associate Professor. US Sub-Editor of 9th Edition of World Directory of Crystallographers. Department of Chemistry, Boston College, Chestnut Hill, MA 02167, USA. PhD (Pennsylvania State U., 1973) inorganic chemistry. *Structural_disorder proteins computer_modelling amphiphilic_toxins homology_prediction protein_toxins water_structure*
E-mail teeter@bcchem.bc.edu teeter@bcchem.bitnet Tel. 1(617)5523615 Fax 1(617)5522705

Teller, Dr David C. (1938). Professor. Department of Biochemistry SJ-70, University of Washington, Seattle, WA 98195, USA. PhD (U. California, Berkeley, 1965) biochemistry. *Proteins coagulation evolution factor_XIII physical_biochemistry analytical_ultracentrifugation*
E-mail teller@gouda.bchem.washington.edu teller@u.washington.edu Tel. 1(206)5431756 1(206)5436047 Fax 1(206)6851792

Teller, Raymond (1946). Manager, Inorganic Analysis. BP Chemical, 4440 Warrensville Road, Cleveland, OH 44202, USA. PhD (U. Southern California, 1978) inorganic crystallography. *Solid_state_chemistry neutron_diffraction reaction_mechanism in_situ_crystallography catalysis EXAFS*
E-mail tellerrp@rcwpo2.usaclv.msnet.bp.com Tel. 1(216)5815953 Fax 1(216)5815621

Temple, Dr Brenda R. S. (1957). Postdoctoral Fellow. Department of Biochemistry and Molecular Biology, 7 Divinity Avenue, Harvard University, Cambridge, MA 02138, USA. PhD (Cornell U., 1989) applied physics. *Laue virus*
E-mail temple@xtal0.harvard.edu Tel. 1(617)4955043

Templeton, Prof. David H. (1920). Professor Emeritus. Department of Chemistry, University of California, Berkeley, CA 94720, USA. [1244 Brewster Drive, El Cerrito, CA 94530, USA.] PhD (U. California, Berkeley, 1947) physical chemistry. *Anomalous_dispersion synchrotron_radiation theory polarized_dispersion*
E-mail lilo@lbl.bitnet Tel. 1(510)4865615 Fax 1(510)4865596

Templeton, Dr Lieselotte K. (1918). Retired. Department of Chemistry, University of California, Berkeley, CA 94720, USA. [1244 Brewster Drive, El Cerrito, CA 94530, USA.] PhD (U. California, Berkeley, 1950) physical chemistry. *Anomalous_dispersion synchrotron_radiation computing polarized_dispersion*
E-mail lilo@lbl.bitnet Tel. 1(510)4865615 Fax 1(510)4865596

Tench, Alan Howard (1946). Director Research and Development. 2044 Walnut St., Boulder, CO 80302, USA. PhD (U. Colorado, 1972) physical chemistry. *Direct_methods carcinogens carcino_stats*
Tel. 1(303)4441730

Ten Eyck, Dr Lynn F. (1942). Specialist. Department of Chemistry 0654, University of California, San Diego, La Jolla, CA 92093-0654, USA. PhD (Princeton U., 1970) biochemical sciences. *Proteins computational_methods computer_modelling hemoglobin molecular_databases*
E-mail lteneyck@ucsd.edu teneyck@sdsc.edu teneyck@sdsc.bitnet Tel. 1(619)5348189 Fax 1(619)5348193

Terwilliger, Dr Thomas C. (1956). Staff Scientist. Life Sciences Division, Mail Stop M880, Los Alamos National Laboratory, Los Alamos, NM 87645, USA. PhD (UCLA, 1982) molecular biology. *Proteins*
E-mail terwill@prov2.lanl.gov Tel. 1(505)6670072 Fax 1(505)6653024

Tesh, Kris F. (1965). Service Crystallographer. Molecular Structure Corporation, 3200 Research Forest Drive, The Woodlands, TX 77381, USA. PhD (Vanderbilt U., 1992) inorganic chemistry. *Main_group_chemistry catalysts bioinorganic_chemistry chemical_education*
E-mail kft@msc.com Tel. 1(713)3631033 Fax 1(713)3643628

Tesmer, John J. G. (1967). Graduate Student. Department of Biological Sciences, Purdue University, W. Lafayette, IN 47907, USA. BA (Rice U., 1990) biochemistry and english. *Proteins computer_modelling synchrotron_radiation amidotransferases MAD*
E-mail tesmen@pandora.bio.purdue.edu Tel. 1(317)4949247 Fax 1(317)4961189

Thackeray, Dr Michael Makepeace (1949). Group Leader. Chemical Technology Division, Argonne National Laboratory, Argonne, IL 60439, USA. [Chemical Technology Division, Argonne National Laboratory, 9700 South Cass Avenue, Argonne, IL 60439, USA.] PhD (U. Cape Town, 1977) chemistry. *Solid_state_chemistry battery_technology intercalation_compounds*
E-mail thackeray@cmt.anl.gov Tel. 1(708)2529184 Fax 1(708)2524176

Tham, Fook S. (1956). Instrumental Specialist. 304 8th St., Troy, NY 12180, USA. PhD (Rensselaer Poly. Inst., 1987) chemistry. *Structure_small_molecule X-ray_fluorescence*
Tel. 1(518)2766648 Fax 1(508)2764887

Thanki, Narmada. Postdoctoral Fellow. Macromolecular Structure Laboratory, ABL-Basic Research Program, NCI-FCRDC, PO Box B, Frederick, MD 21702, USA. PhD (Birkbeck College, U. London, 1991) *Protein_structure computer_modelling proteases signal_transduction energy_calculations protein_hydration homology_prediction*
E-mail thanki@ncifcrf.gov Tel. 1(301)8465031 Fax 1(301)8465991

Thayer, Dr Maria M. (1961). Postdoctoral Fellow. The Scripps Research Institute, Dept. of Molecular Biology, MB4, 10666 North Torrey Pines Road, La Jolla, CA 92037, USA. PhD (U. Colorado, Boulder, 1989) biochemistry. *Protein_structure DNA_repair_enzymes*
E-mail mmt@scripps.edu Tel. 1(619)5549261 Fax 1(619)5546880

Thiel, Dr Daniel J. (1963). Postdoctoral Fellow. CHESS, Cornell University, Ithaca, NY 14853, USA. PhD (Cornell U., 1992) applied physics. *Biocrystallography biophysics Laue_diffraction microdiffraction EXAFS photorearrangement CCD_detectors*
E-mail thiel@msc.cornell.edu thiel@cmlches.bitnet Tel. 1(607)2557163 Fax 1(607)2559001

Thielke, Mr Harry G. (1929). Retired. 212 Westhaven Road, Greenville, NC 27834, USA. MS (U. Delaware, 1962) chemistry. *X-ray powder_diffraction*
Tel. 1(919)7568141

Thiessen, Kris (1977). Student. Student member. Athens High School, PO Box 109, Athens, AL 35611, USA. chemistry/biochemistry. *Proteins molecular_modelling microgravity_crystal_growth gel_crystallography gel_crystallography proteins*
Tel. 1(205)2326244 Fax 1(205)3501952

Thiyagarajan, Dr P. (1952). Staff Physicist. Intense Pulsed Neutron Division, Bldg 200, JA105, Argonne National Laboratory, 9700 South Cass Avenue, Argonne, IL 60439, USA. PhD (U. Madras, 1979) physics. *Technique_instrumentation neutron_X-ray_scattering membrane_proteins polymers structure_interaction micelles_microemulsions disordered_materials*
E-mail thiyaga@anlpns.pns.anl.gov thiyaga@anlpns.bitnet Tel. 1(708)2523593 Fax 1(708)2523822

Thoden, James B. (1965). Postdoctoral Fellow. Institute for Enzyme Research, University of Wisconsin, 1710 University Avenue, Madison, WI 53705, USA. PhD (U. Wisconsin-Madison, 1993) inorganic chemistry. *Enzymes metallic_clusters organometallic*
E-mail jim@enzyme.wisc.edu Tel. 1(608)2620529 Fax 1(608)2650904

Tibbitts, Dr Thomas T. (1960). Lecturer. Merkert Chemistry Center, Boston College, Chestnut Hill, MA 02167-3860, USA. PhD (Brandeis U., 1990) biophysics. *Metalloenzymes membranes fibres software mutagenesis electron_microscopy image_processing scanning_tunnel_microscopy*
E-mail tibbitts@vmax.bc.edu Tel. 1(617)5524248 Fax 1(617)5522705

Tillinger, Martin H. (1943). Tillinger Consulting Corp., 972 Sheffield RD., Teaneck, NJ 07666, USA. PhD physics. *Computing*
E-mail 70544.1372@compuserve.com Tel. 1(201)8334355 Fax 1(201)8331216

Tilton Jr, Robert F. (1957). Staff Scientist. Miles Research Center, Miles. Inc., 400 Morgan Ln., West Haven, CT 06516, USA. PhD (U. C. San Francisco, 1984) pharmaceutical chemistry. *Protein_dynamics protein_structure water_structure drug_design xenon*
E-mail tilton@mrc.com Tel. 1(203)937-2829 Fax 1(203)937-2650

Tippin, Douglas B. (1966). Student. Biotechnology Center, Ohio State University, 1060 Carmack Rd, Columbus, Ohio 43210, USA. MS (Baylor, 1990) chemistry.
E-mail black%biot@mps.ohio-state.edu Tel. 1(614)2922926 Fax 1(614)2922524

Tischler, Jonathan Z. (1953). Research Staff. Solid State Div., Oak Ridge National Laboratory, PO Box 2008, Oak Ridge, TN 37831-6030, USA. PhD (Cornell U., 1982) physics. *Diffuse_scattering resonant_scattering thin_films time-resolved_effect synchrotron_radiation*
E-mail zzt@ornl.gov Tel. 1(615)5746505 Fax 1(615)5744143

Toby, Dr Brian H. (1958). Senior Principal Research Chemist. Principal developer of the CIF dictionary for powder diffraction; member COMCIFS. Air Products and Chemicals, Inc., 7201 Hamilton Blvd., Allentown, PA 18195, USA. PhD (Caltech, 1987) physical chemistry. *Zeolites powder_diffraction structure_from_powder_diffraction synchrotron_diffraction neutron_diffraction CIF*
E-mail tobybh@ttown.apci.com Tel. 1(610)4814198 Fax 1(610)4816517

Tomchick, Dr Diana R. (1960). Postdoctoral Research Associate. Purdue University, Lilly Hall of Life Sciences, Department of Biological Sciences, West Lafayette, IN 47907, USA. PhD (U. Wisconsin-Madison, 1990) inorganic chemistry. *Structure-activity_relationship allostery multiple_wavelength_anomalous_phasing synchrotron_radiation enzyme_inhibitors metalloenzymes*
E-mail tomchick@bragg.bio.purdue.edu Tel. 1(317)4949247 Fax 1(317)4961189

Toney, Michael F. (1957). Research Staff Member. IBM Almaden Research Center, 650 Harry Road, San Jose, CA 95120-6099, USA. PhD (U Washington, 1983) physics. *Interfacial_structure_determination metallic_multilayer grazing_incidence electrochemistry water_structure reflectivity anomalous_scattering*
E-mail toney@almaden.ibm.com toney@almaden.bitnet Tel. 1(408)9271182 Fax 1(408)9272305

Tong, Dr Liang (1963). Senior Scientist. Department of Medicinal Chemistry, Boehringer Ingelheim Pharmaceuticals, Inc., 900 Ridgebury Road / PO Box 368, Ridgefield, CT 06877, USA. PhD (U. California, Berkeley, 1989) biophysical chemistry. *Protein_crystallography small_molecule_crystallography crystallographic_computing*
E-mail b4w@mace.cc.purdue.edu Tel. 1(203)7985139 Fax 1(203)7916072

Torardi, Dr Charlie C. (1951). Senior Research Chemist. DuPont Central Research, Experimental Station, Wilmington, DE 19880-0356, USA. PhD (Iowa State U., 1981) inorganic chemistry. *Solid_state_chemistry*
Tel. 1(302)6952236 Fax 1(302)6951664

Tormo, Jose (1960). Postdoctoral Fellow. Department of Molecular Biophysics & Biochemistry, Yale University, 266 Whitney Avenue, Box 208114, New Haven, CT 06520-8114, USA. PhD (U. Politecnica Catalunya (Spain), 1992) biology. *Protein_structure antibodies protein–DNA_interactions*
E-mail tormo%hhvms@venus.ycc.yale.edu Tel. 1(203)4325795 Fax 1(203)4323282

Toth, Dr Kenneth S. (1966). Student. Center for Macromolecular Crystallography, University of Alabama at Birmingham, THT 79, BHS 233, 1918 University Blvd., Birmingham, AL 35294-0005, USA. PhD (U. Alabama at Birmingham, 1993) biochemistry. *Proteins viruses computer_graphics artificial_intelligence*
E-mail ktoth@orion.cmc.uab.edu = ktoth@uabcmc.bitnet Tel. 1(205)9341611 Fax 1(205)9340480

Towns, Robert L. R. (1940). Professor, Associate Dean. Department of Chemistry, Cleveland State University, 1980 E 24th, Cleveland, OH 44115, USA. PhD (U. Texas, 1969) physical chemistry. *Structure_small_organic structure_heterocyclics XRF*
Tel. 1(216)6872468 Fax 1(216)6879298

Transue, Thomas R. (1968). Graduate Student Research Assistant. Biophysics Research Division, 930 N. University Avenue, Ann Arbor, MI 48109-1055, USA. BA (Oberlin College, 1990) biology, mathematics. *Amaranthin lectins protein T-antigen_disaccharide*
E-mail transue@norway.biop.umich.edu Tel. 1(313)7633384 Fax 1(313)7643323

Trefonas, Louis Marco (1931). Professor of Chemistry. Department of Chemistry, University of Central Florida, Orlando, FL 32186, USA. PhD (U. Minnesota, 1959) physical chemistry. *Small_molecules cancer pharmaceuticals*

Tronrud, Dr Dale E. (1957). Research Associate. Member. Institute of Molecular Biology, University of Oregon, Eugene, OR 97402, USA. PhD (U. Oregon, 1986) protein crystallography. *Fourier_transform ideal_structure least-squares_refinement_method restrained_least_squares software zinc_peptidase philosophy_of_science*
E-mail dale@uoxray.uoregon.edu Tel. 1(503)3465176 Fax 1(503)3465870

Troup, Jan Marshall (1946). President. Molecular Structure Corporation, 3200 Research Forest Drive, The Woodlands, TX 77381, USA. PhD (Texas A&M U., 1974) chemistry. *Inorganic_structures X-rays data_collection_techniques powder_structures experimental_electron_density area_detector_development*
E-mail jmt@msc.com Tel. 1(713)3631033 Fax 1(713)3643628

Trueblood, Kenneth N. (1920). Professor Emeritus. Department of Chemistry and Biochemistry, University of California, Los Angeles, CA 90024-1569, USA. PhD (California Institute of Technology, 1947) chemistry. *Thermal_motion physical_chemistry host–guest_compounds*
E-mail trueblood@uclash.chem.ucla.edu Tel. 1(310)8251259 Fax 1(310)2064038

Trus, Dr Benes L. (1946). Chief, Image Proc. Res. Section, CBEL, DCRT, Building 12A, Room 2033, National Institutes of Health, Bethesda, Maryland 20892-5624, USA. PhD (California Institute of Technology, 1972) physical chemistry. *Image_processing electron_microscopy image_reconstruction viral_structure*
E-mail trus@helix.nih.gov Tel. 1(301)4962250 Fax 1(301)4022867

Tsai, Prof. Chun-che (1937). Professor of Chemistry. Department of Chemistry, Kent State University, Kent, OH 44242-0001, USA. PhD (Indiana U., Bloomington, IN, 1968) physical chemistry. *Molecular_design drug_design nucleic_acids_DNA_RNA computer_modelling drug–nucleic_acid_interaction*
E-mail ctsai@kentvm.bitnet Tel. 1(216)6722989 Fax 1(216)6723816

Tulinsky, Dr Alexander (1928). Professor of Chemistry. Department of Chemistry, Michigan State University, East Lansing, MI 48824, USA. PhD (Princeton U., 1956) physical chemistry. *X-ray_crystallography blood_proteins*
E-mail tulinsky@cemvax.cem.msu.edu tulinsky@msucem.bitnet Tel. 1(517)3559715x250 Fax 1(517)3531793

Tuomi, Donald (1920). Retired. 626 S. Kaspar, Arlington Heights, IL 60005-2320, USA. PhD (Ohio State U., 1952) physical chemistry. *Thermoelectric_materials batteries*
Tel. 1(708)3923003

Turley, June Williams (1929). Project Manager. 1208 Wakefield Drive, Midland, MI 48640, USA. PhD (The Pennsylvania State U., 1957) agricultural & biological chemistry. *Active_site biopolymers crystal structure–activity_relationship chaos_theory biotechnology*
Tel. 1(517)8355811 Fax 1(517)8353755

Turley, Stewart (1950). Research Technician. University of Washington, Biological Structure Dept., Health Science SM20, Seattle, WA 98195, USA. MSc (Leeds U., 1972) combustion. *Crystallization data_collection area_detectors data_processing copper_proteins*
E-mail turley@u.washington.edu Tel. 1(206)5434496 Fax 1(206)5431524

Twigg, Pam (1960). Visiting Scientist. 127 Water Oak Ct, Harvest, AL 35749-9305, USA. MS (U. Alabama at Birmingham, 1986) biomedical engineering. *Proteins*
Tel. 1(205)5445494 Fax 1(205)5441777

Ulmer, Dr Kevin M. (1951). Chief Scientific Officer. SEQ, Ltd, 30 Margin Street, Cohasset, MA 02025, USA. PhD (Massachusetts Institute of Technology, 1978) biological oceanography. *Proteins homology computer_modelling DNA homology_prediction processive_exonucleases*
E-mail kmulmer@biotechnet.com Tel. 1(617)3832858 Fax 1(617)3832852

Umland, Timothy C. (1966). Research crystallographer. Biocrystallography Laboratory, VA Medical Center, PO Box 12055, Pittsburgh, PA 15240, USA. PhD (U. Pittsburgh, 1994) protein crystallography. *Proteins neurological_toxins protein–lipid_complexes*
E-mail umland@vms.cis.pitt.edu Tel. 1(412)6923511 Fax 1(412)6241882

Vajdos, Felix (1968). Graduate Student. Biochemistry Department, University of Utah, Salt Lake City, UT 84132, USA. BS (Texas A&M U., 1991) biochemistry. *Proteins structure_function*
E-mail felix@uxray.med.utah.edu Tel. 1(801)5853919 Fax 1(801)5817959

Valente, Edward J. (1949). Professor of Chemistry. Department of Chemistry, Mississippi College, Box 4065, Clinton, MS 39058-4065, USA. PhD (U. Washington, 1977) physical-organic chemistry. *Conformation organic_diastereomers structure-property_relationships hydrogen_bonding*
E-mail valente@mc.edu Tel. 1(601)9253424 Fax 1(601)9253852

van der Helm, Dr Dick (1933). George Lynn Cross Research Professor. Department of Chemistry and Biochemistry, University of Oklahoma, 620 Parrington Oval, Norman, OK 73019-0370, USA. DSc (U. Amsterdam, Netherlands, 1960) chemistry. *Molecular_structure natural_products siderophores membrane_proteins conformation peptides peptide_chelates anticancer_agents*
E-mail dvdhelm@chemdept.chem.uoknor.edu ax1024@uokmvsa.bitnet Tel. 1(405)3255831 Fax 1(405)3256111

VanDerveer, Dr Donald G. (1947). Research Scientist. School of Chemistry and Biochemistry, Georgia Institute of Technology, Atlanta, GA 30332-0400, USA. PhD (Brown U., 1974) inorganic chemistry. *Service_crystallography computer_management*
E-mail vanderveer@boss.chemistry.gatech.edu Tel. 1(404)8948517 Fax 1(404)8947452

Vanderhoff-Hanaver, Peggy A. (1959). Postdoctoral Fellow. Michigan State University, Department of Chemistry, East Lansing, MI 48824-1322, USA. PhD (Rutgers University - Newark, 1989) physical chemistry. *Blood_protein*
E-mail vanderhoff@cemvax.cem.msu.edu Tel. 1(517)3559715x253 Fax 1(517)3531793

Vander Sande, John B. (1944). Professor. Room 1-206, MIT, 77 Massachusetts Ave., Cambridge, MA 02139, USA. PhD (Northwestern U., 1970) materials science. *Electron_microscopy ceramics*
Tel. 1(617)2536933 Fax 1(617)2538549

van der Woerd, Mark J. (1962). Graduate Student. Center for Macromolecular Crystallography, University of Alabama at Birmingham, University Station, THT 79, Birmingham, AL 35294-0005, USA. MSc (Eindhoven University of Technology, 1988) chemical engineering. *Anomalous_dispersion structure–activity_relationship folding crystal_growth EPR MO_calculation*
E-mail woerd@orion.cmc.uab.edu woerd@uabcmc Tel. 1(205)9348990 Fax 1(205)9340480

Van Hove, Dr Michel A. (1947). Senior Scientist. Materials Sciences Division, Lawrence Berkeley Laboratory, Berkeley, CA 94720, USA. PhD (U. Cambridge, 1974) solid state physics. *Surface_science surface_structure low_energy_electron_diffraction photo-electron_diffraction*
E-mail vanhove@lbl.bitnet vanhove@lbl.gov Tel. 1(510)4866160 Fax 1(510)4864995 Telex 9103662037

van Meurs, Frank (1946). Marketing Manager. Enraf-Nonius, 390 Central Ave., Bohemia, NY 11716, USA. PhD (Delft University of Technology, 1978) chemistry. *Computing_diffractometry_instrumentation X-ray_diffraction structural_chemistry*
E-mail vanmeurs@delftny.com. Tel. 1(516)5892885x33 Fax 1(516)5892068

Van Nordstrand, Robert (1917). Retired from Chevron Research Co. 520 Montecillo Rd., San Rafael, CA 94903, USA. MS (U. Michigan, 1939) physical chemistry. *Zeolites catalysis alumina SAXS graphic_arts*
Tel. 1(415)4727062

Van Roey, Patrick (1952). Research scientist. Wadsworth Center for Labs. and Res., New York State Department of Health, Empire State Plaza, PO Box 509, Albany, NY 12201-0509, USA. PhD (U. Calgary, 1980) chemistry. *Proteins carbohydrates intermolecular_interaction*
E-mail vanroey@tethys.ph.albany.edu Tel. 1(518)4741444 Fax 1(518)4747992

Varughese, Dr Kottayil Iype (1946). Associate Project Scientist. Dept of Biology, University of California, San Diego, La Jolla, CA 92093-0317, USA. PhD (U. Madras, 1974) X-ray crystallography. *Protein_crystallography structure function peptide_conformation*
E-mail kvarughese@ucsd.edu kvarughese@ucsd.bitnet Tel. 1(619)5347227 Fax 1(619)5342565

Veblen, David R. (1947). Professor. Earth and Planetary Sci. Dept, Johns Hopkins University, Baltimore, MD 21218, USA (Harvard U., 1976) geological sciences. *Defects reactions transmission_electron_microscopy*
E-mail dveblen@aol.com Tel. 1(410)5168487 Fax 1(410)5167933

Venugopal, Dr Manju Grover (1963). Research Associate. Department of Biochemistry, Robert Wood Johnson Medical School, University of Medicine and Dentistry of New Jersey, 675 Hoes Lane, Piscataway, NJ 08854, USA. [808, NW 89th Ave, Plantation, FL 33324, USA.] PhD (Rensselaer Polytechnic Institute, 1992) biochemistry/biophysics. *Crystallization collagen_peptides membrane_protein biophysical_characterization computer_modelling membrane/insecticidal/anti-bacterial_proteins*
E-mail venugopa@rwja.umdnj.edu Tel. 1(305)4247039 Fax 1(908)2354783

Verlinde, Dr Christophe L. M. J. (1958). Senior Fellow. Department of Biological Structure, University of Washington SM-20, Seattle, WA 98195, USA. PhD (U. Leuven, 1988) pharmacy. *Proteins computer_modelling drug_design*
E-mail verlinde@gouda.bchem.washington.edu Tel. 1(206)5438865 Fax 1(206)5431524

Vijayalakshmi, Dr J. (1950). Research Associate. Department of Chemistry, Michigan State University, East Lansing, MI 48824-1322, USA. PhD (Madras U., India, 1987) crystallography and biophysics. *Crystallography protein_small_molecule serine_proteases inhibitor_design molecular_modelling protein_folding programming statistical_data_analysis*
E-mail viji@cem.msu.edu viji@msucem.bitnet Tel. 1(517)3559715x251 Fax 1(517)3551793

Villafranca, Dr J. Ernesto (1951). Director of Biophysics. Agouron Pharmaceuticals Inc., 3565 General Atomics Court, San Diego, CA 92121, USA. PhD (U. Texas, Austin, 1980) chemistry. *Protein_engineering protein_structure structure-based_drug_design protein–ligand_interactions*
E-mail ern@humbert@agouron.com Tel. 1(619)6223000 Fax 1(619)6223299

Vincent, Dr Beverly R. (1961). Service Crystallography Manager. Molecular Structure Corporation, 3200 Research Forest Drive, The Woodlands, Texas, USA. PhD (Dalhousie U., 1988) chemistry. *Hydrogen_bonding conformation computing thermal_vibration crystallographic_computing graphical_user_interfaces*
E-mail brv@msc.com Tel. 1(713)3631033 Fax 1(713)3643628

Vitali, Dr Jacqueline (1950). Research Associate. Department of Biochemistry, Wayne State University, 540 East Canfield, Detroit, MI 48201, USA. [#4E, 1579 Rhinelander Avenue, Bronx, NY 10461, USA.] PhD (State University of New York at Buffalo, 1986) biophysical sciences. *Macromolecular_crystallography*
E-mail vitali%xray.dnet@rocdec.roc.wayne.edu Tel. 1(313)5771506

Vlasse, Dr Marcus (1933). Senior scientist. Code ES74, NASA, Marshall Space Flight Center, MSFC, AL 35812, USA. PhD, DSc (U. Pittsburgh, U. Bordeaux, France, 1965, 1980) crystallography, materials science. *Crystal_growth crystallographic_analysis materials_processing_in_space superconductivity organic_inorg_materials preparation_and_characterization nonlinear_optical_materials*
E-mail vlasse@ssl.msfc.nasa.gov Tel. 1(205)5447781 Fax 1(205)5442102

Voet, Donald (1938). Associate Professor. Chemistry Department, University of Pennsylvania, Philadelphia, PA 19104, USA. PhD (Harvard U., 1967) chemistry. *Proteins DNA enzymes cytokines coauthor_biochemistry_textbook*
E-mail voet@dv.chem.upenn.edu Tel. 1(215)8986457 Fax 1(215)8985747

Vogelaar, Dr Nancy (1961). Research Staff. Department of Chemistry, Princeton University, Princeton, NJ 08544, USA. PhD (California Institute of Technology, 1989) biophysical chemistry. *Crystallization proteins structure_determination homology_prediction biomolecule X-ray diffraction*
E-mail vogelaar@chemvax.princeton.edu Tel. 1(609)2582827 Fax 1(609)2586746

Vojtechovsky, Dr Jaroslav (1964). Postdoctoral fellow. Department of Chemistry, Rutgers University, Piscataway, NJ 08855, USA. PhD (Charles U., Prague, 1982) physical chemistry. *DNA_crystallography protein_crystallography solvent_network*
E-mail vojtech@ndbviz.rutgers.edu Tel. 1(908)9322200 Fax 1(908)9325958

Volz, Dr Karl W. (1950). Assistant Professor. Department of Microbiology and Immunology, University of Illinois at Chicago, Chicago, IL 60612, USA. PhD (U. California, San Diego, 1981) chemistry. *Protein_structure_function_and_genetics signal_processing*
E-mail volz@uicbal Tel. 1(312)9962314

Von Dreele, Dr Robert B. (1943). Technical Staff Member. LANSCE, MS H805, Los Alamos National Laboratory, Los Alamos, NM 87545, USA. PhD (Cornell U., 1971) inorganic chemistry. *Neutron_diffraction X-ray_diffraction powder_diffraction high_pressure crystal_chemistry texture*
E-mail vondreele@lanl.gov Tel. 1(505)6673630 Fax 1(505)6652676

Vreeland Jr, Thad. Professor Emeritus. Mail Code 138-78, California Institute of Technology, 1201 East California Blvd., Pasadena, CA 91125, USA. PhD *X-ray_rocking_curve shock_consolidation*
E-mail tv@hyperfine.caltech.edu Tel. 1(818)3954431 Fax 1(818)7956132

Vyas, Meenakshi (1949). Research Associate. Dept. of Biochemistry, Baylor College of Medicine, 1 Baylor Plaza, Houston, TX 77030, USA. PhD (Indian Institute of Technology, Bombay, 1976) X-ray crystallography of organic compounds. *Macromolecular_crystallography crystallization carbohydrates structure–activity_relationship Fab_fragments enzymes*
E-mail mina@dino.qci.bioch.bcm.tmc.edu Tel. 1(713)7986563 Fax 1(713)7976718

Vyas, Dr Nand K. (1949). Research Assistant Professor. Department of Biochemistry, Baylor College of Medicine, One Baylor Plaza, Houston, TX 77030, USA. PhD (Indian Institute of Technology, Bombay, 1977) chemistry. *Proteins carbohydrates immunoglobins nucleic_acids homology_prediction protein_carbohydrates antibody_antigen protein_ligand*
E-mail nand@dino.qci.bioch.bcm.tmc.edu Tel. 1(713)7966563 1(713)7986564 Fax 1(713)7976718

Waksman, Dr Gabriel (1957). Assistant Professor. Department of Biochemistry and Molecular Biophysics, Washington University, School of Medicine, 660 South Euclid Avenue, Box 8231, Saint Louis, USA. PhD (Paris U., 1982) biochemistry. *X-ray_crystallography signal_transduction*
Tel. 1(314)3620260

Walter, Dr Mark R. (1962). Assistant Professor. Department of Pharmacology, University of Alabama at Birmingham, Birmingham, AL 35205, USA. PhD (U. Alabama at Birmingham, 1989) macromolecular structure. *Protein_structure molecular_recognition protein_engineering protein_homology structure_based_drug_design*
E-mail walter@str.cmc.uab.edu walter@uabcmc.bitnet Tel. 1(205)9349279 Fax 1(205)9340480

Walter, Norman M. (1923). Retired. PO Box 417, Round Lake, NY 12151-0417, USA. MS (Penn State U., 1954) physics.
Tel. 1(518)8992130

Walter, Richard L. (1965). Graduate student. Section of Biochemistry, Cell & Molecular Biology, Cornell University, 209 Biotechnology Bldg., Ithaca, NY 14853, USA. BA (College of Wooster, 1987) biology. *Proteins synchrotron_radiation low_temperature_crystallography anomalous_dispersion biochemistry crystallization area_detector mosaicity Laue_diffraction*
E-mail walter@vgx.tn.cornell.edu Tel. 1(607)2552174 Fax 1(607)2552480

Wang, Prof. Andrew H.-J. (1945). Prof. of Biophysics & Chemistry. Dept. of Cell & Structural Biology, 506 Morrill Hall, Univ. of Illinois at Urbana-Champaign, Urbana, IL-61801, USA. PhD (U. Illinois, 1974) chemistry. *Macromolecular_crystallography NMR structure_function_of_nucleic_acids*
E-mail wang@c.scs.uiuc.edu wang@uiucscs.bitnet Tel. 1(217)2446637 Fax 1(217)2443181

Wang, Bi-Cheng (1938). Professor. Department of Crystallography, University of Pittsburgh, Pittsburgh, PA 15260, USA. PhD (U. Arkansas, 1969) chemistry. *Protein_crystallography RNA_polymerase neurophysin glutathione_transferase*
E-mail wang@bcl1.xtal.pitt.edu Tel. 1(412)6249310 Fax 1(412)6243219 1(412)6241882

Wang, Z. L. (1960). Scientist. Metallurgy Division, NIST, Gaithersburg, MD 20899, USA. PhD (Arizona State U., 1987) physics. *Transmission_electron_microscopy EELS*
E-mail zwang@enh.nist.gov Tel. 1(301)9756182 Fax 1(301)9267975

Warburton, Dr William K. (1942). President. X-ray Instrumentation Associates, 2513 Charleston Road, STE 207, Mountain View, CA 94043-1607, USA. PhD (Harvard U., 1972) physics. *Detector_development disordered_materials synchrotron_radiation*
E-mail warburton@ssrl01.slac.stanford.edu Tel. 1(415)9039980 Fax 1(415)9039887

Ward, Dr Donald L. (1943). Academic Specialist/Crystallographer. Department of Chemistry, Michigan State University, East Lansing, MI 48824-1322, USA. PhD (Montana State U., 1972) chemistry. *Single_crystal_structure_determination education computer_software service_crystallography glass_collimator scientific_visualization*
E-mail ward@cemvax.cem.msu.edu Tel. 1(517)3559715x217 Fax 1(517)3531793

Ward, Keith B. (1943). Research Biophysicist. Laboratory for the Structure of Matter, Code 6030, Naval Research Laboratory, 4555 Overlook Ave. SW, Washington, DC 20375-5341, USA. PhD (Johns Hopkins U., 1974) biophysics. *Macromolecular_crystallization proteins structure_determination metalloproteins metal_ion_biosensors bioluminescent_proteins OPA_anhydrolases*
E-mail Ward@lsm.nrl.navy.mil Tel. 1(202)7672735 Fax 1(202)7676874

Warren, Stephen G. (1945). Associate Professor. Atmospheric Sciences AK-40, University of Washington, Seattle, WA 98195, USA. PhD (Harvard U., 1973) physical chemistry. *Ice*
E-mail sgw@cloudy.atmos.washington.edu Tel. 1(206)5437230 Fax 1(206)5430308

Waser, Jurg (1916). Retired Professor, Caltech. 6120 Terryhill Drive, La Jolla, CA 92037, USA. PhD (Caltech, 1944) X-ray crystallography.
Tel. 1(619)4545622

Watenpaugh, Dr Keith Donald (1939). Senior Scientist. Commission on Crystallographic Computing; Commission on Synchrotron Radiation. Physical and Analytical Chemistry, 301 Henrietta Street, The Upjohn Company, Kalamazoo, MI49001, USA. PhD (Montana State U., 1967) physical chemistry. *Macromolecular_structure computing methodology synchrotron_radiation*
E-mail kdwatenp@upj.com Tel. 1(616)3855481 Fax 1(616)3857522

Watkins, Steven F. (1940). Assoc. Prof., Director of Graduate Studies. Department of Chemistry, Louisiana State University, Baton Rouge, LA 70803, USA. PhD (U. Wisconsin (Madison), 1967) physical chemistry. *Superconductors natural_products computer_modelling chemical_information*
E-mail watkins@chxray.dnet.lsu.edu Tel. 1(504)3883467 1(504)3888270 Fax 1(504)3883458

Watowich, Dr Stan J. Postdoctoral Fellow. Department of Biochemistry, Harvard University, 7 Divinity Avenue, Cambridge, MA 02138, USA. PhD (U. Chicago, 1986) physical chemistry. *Protein–protein_interactions structure–function binding molecular_modelling kinases signal_transduction*
E-mail watowich@xtal0.harvard.edu watowich@xtal220.harvard.edu Tel. 1(617)4954091 Fax 1(617)4959613

Watson, Prof. William H. (1931). Professor. Department of Chemistry, Texas Christian University, Fort Worth, TX 76129, USA. [Box 32908, Fort Worth, TX 76129, USA.] PhD (Rice U., 1958) physical chemistry. *Strained_molecules natural_products organic_conductors*
E-mail watson@gamma.is.tcu.edu Tel. 1(817)9217195 Fax 1(817)9217330

Watt, Dr William (1955). Scientist. Physical and Analytical Chemistry, Upjohn Company, 301 Henrietta St., Kalamazoo, MI 49007, USA. PhD (Michigan State U., 1990) physical chemistry. *Flavoproteins electron_transport_proteins biologically_active_small_molecules area_detectors*
E-mail wwatt@upj.com Tel. 1(616)3854585 Fax 1(616)3857522

Watts, Ethel J. (1928). Retired from Aerospace Corporation. 5841 Ghent Drive, Huntington Beach, CA 92649-4640, USA. MS (Pennsylvania State U., 1952) physics. *Ultrastructure_of_inorganic_materials semiconductors contaminant_identification liquid_crystals*
Tel. 1(714)8464315

Weakley, Dr Timothy J. R. (1933). Research Associate. Department of Chemistry, University of Oregon, Eugene, OR 97403, USA. PhD (Oxford U., 1959) chemistry. *Coordination_compounds inorganic_crystals molecular_crystals structural_chemistry*
E-mail tweakley@oregon.uoregon.edu Tel. 1(503)3464620 Fax 1(503)3464643

Weaver, Arthur J. (1960). Senior Research Fellow. Mayo Clinic, Rochester, MN 55905, USA. PhD, MD (Mayo Graduate School, 1989) biochemistry. *Proteins crystallization toxins anticancer_compounds membrane_proteins cryocrystallography crystallization_techniques*
E-mail art@mayo.edu Tel. 1(507)2841947 Fax 1(507)2841086

Weber, Christian H. (1968). Graduate Student. Department of Biological Chemistry, University of Michigan, Ann Arbor, MI 48109, USA. [Biophysics Research Division, University of Michigan, Chemistry Building, 930 North University Blvd., Ann Arbor, MI 48109-1055, USA.] BA (U. Virginia, 1989) chemistry and biology. *Proteins protein_folding conformational_change fractal complex_dynamics*
E-mail chris@norway.biop.umich.edu Tel. 1(313)7634698 1(313)7632199 Fax 1(313)7643353

Weber, Irene T. (1953). Associate Professor. Department of Pharmacology, Jefferson Cancer Institute, Thomas Jefferson University, 233 South 10th Street, Philadelphia, PA 19107, USA. PhD (U. Oxford, 1978) molecular biophysics. *Structure_proteins structure_nucleic_acids computer_modelling enzymes*
E-mail weber@calvin.jci.tju.edu Tel. 1(215)9554575 Fax 1(215)9232117

Weber, Patricia C. (1952). Senior Director, Structural Chemistry. Co-editor, Journal of Applied Crystallography. Schering-Plough Research Institute, K-15-3/3855, 2015 Galloping Hill Road, Kenilworth, NJ 07033-0539, USA. PhD (U. Arizona, 1979) chemistry. *Structure-based_drug_design protein_crystallization*
Tel. 1(908)2983426 Fax 1(908)2987305

Wechsler, Dr Barry A. (1951). Senior Staff Physicist. Hughes Research Laboratories, 3011 Malibu Canyon Road, Malibu, CA 90265, USA. PhD (State U. New York at Stony Brook, 1981) earth and space sciences. *Crystal_growth dielectric_and_optical_properties laser_and_nonlinear_optical_materials*
E-mail bwechsler@msmail4.hac.com Tel. 1(310)3175639 Fax 1(310)3175483 Telex 652310 HACRESL MLBU

Wedekind, Joseph (1968). Graduate student. Enzyme Institute, 1710 University Ave., Madison, WI 53713, USA. BS (U. California-Riverside, 1990) biochemistry. *Biological_macromolecular_crystallography enzymes synchrotron_radiation metalloproteins enzyme_mechanisms protein_structure_and_function*
E-mail joe@enzyme.wisc.edu Tel. 1(608)2620529 Fax 1(608)2652904

Weeks, Dr Charles M. (1944). Senior Research Scientist. Hauptman-Woodward Medical Research Institute (formerly the Medical Foundation of Buffalo), 73 High Street, Buffalo, NY 14203-1196, USA. PhD (State University of New York at Buffalo, 1970) biophysics. *Biocrystallography computing direct_method proteins allostery structure-activity_relationship phase_determination*
E-mail weeks@hwi.buffalo.edu Tel. 1(716)8569600 Fax 1(716)8524846

Wei, Mingyi (1962). PhD candidate. Department of Chemistry, Washington State University, Pullman, WA 99164-4630, USA. MS (Washington State U., 1993) physical chemistry. *Jahn–Teller_effect phase_transition structural_disorder*
E-mail 60833903@wsuvm1.csc.wsu.edu Tel. 1(509)3325677 Fax 1(509)3358867

Weichsel, Andrzej (1961). Research Associate. Department of Biochemistry, BSW 526, University of Arizona, Tucson, AZ 85721, USA. PhD (U. Wroclaw, Poland, 1990) chemistry. *Proteins protein_crystallization computer_modelling protein_chemistry conformational_change DNA_binding_proteins protein_complexes phosphate_esters*
E-mail weichsel@biosci.arizona.edu Tel. 1(602)6218171 Fax 1(602)6219288

Weis, William I. (1959). Assistant Professor. Department of Structural Biology, Fairchild Building, Stanford University School of Medicine, Stanford, CA 94305, USA. PhD (Harvard U., 1988) biochemistry. *Proteins carbohydrates refinement_method anomalous_dispersion*
E-mail weis@cellbio.stanford.edu Tel. 1(415)7254623 Fax 1(415)7238464

Weiss, George H. (1930). Chief, Physical Sciences Lab. Div. Computer Res. and Techn., NIH, 9000 Rockville Pike, Bldg. 12A, Rm 2007, Bethesda, MD 20892, USA. PhD (U. Maryland, 1958) applied mathematics. *Statistical_physics*
E-mail ghw@cu.nih.gov Tel. 1(301)4961135 Fax 1(301)4024544

Weiss, Dr Manfred S. (1963). Postdoctoral Fellow. Molecular Biology Institute, University of California at Los Angeles, 405 Hilgard Ave., Los Angeles, CA 90024, USA. PhD (Freiburg U., 1992) biochemistry. *Protein_structure protein_crystallography molecular_recognition structure_prediction membrane_proteins protein_complexes*
E-mail msw@uclaue.mbi.ucla.edu Tel. 1(310)8251402 Fax 1(310)2063914

Weissman, Larry (1947). System Programmer. Center for Bioengineering, WD-12, University of Washington, Seattle, WA 98195, USA. PhD (UCLA, 1979) biochemistry. *Macromolecules*
E-mail larryw@bioeng.washington.edu Tel. 1(206)6852011 Fax 1(206)6853000

Weissmann, Prof. Sigmund (1917). Professor Emeritus 1987. Editor of ICDD. Mat. Sci. Deartment, College of Eng., Rutgers University, Piscataway, NJ 08855, USA. [112 Clarendon Court, Metuchen, NJ 08840-1507, USA.] PhD (PUN, 1952) physical chemistry. *Lattice_defects physical_properties_relationship X-ray_topography dynamical_theory physical_metallurgy*
Tel. 1(908)5484539

Wery, Dr Jean-Pierre (1958). Senior Scientist. Lilly Research Laboratories, Eli Lilly and Company, Indianapolis, IN 46285, USA. PhD (U. Liege, 1987) protein crystallography. *Proteins viruses drug_design*
E-mail wery_jean-pierre@lilly.com Tel. 1(317)2769701 Fax 1(317)2769722

Wesenberg, Gary (1954). Associate Scientist. Institute for Enzyme Research, University of Wisconsin, Madison, WI 53705-4098, USA. PhD (U. Wisconsin, 1983) physical chemistry. *Computing proteins*
E-mail gary@enzyme.wisc.edu wesenberg@wiscmacc.bitnet U. Tel. 1(608)2652904 Fax 1(608)2652904

Wesson, Laura (1959). Technical staff. 205 Molecular Biology Inst., University of California at Los Angeles, 405 N. Hilgard Ave., Los Angeles, CA 90025, USA. *Applications_of_mathematics_to_crystallography protein_structure_prediction*
E-mail laura@mbi.ucla.edu

Westbrook, Dr Edwin M. (1948). Scientist. Structural Biology Center, Argonne National Laboratory, Bldg. 202, 9700 South Cass Avenue, Argonne, IL 60439, USA. PhD, MD (U. Chicago, 1981) biophysics and medicine. *Structural_biology X-ray_optics detector_design synchrotrons microbial_toxins enzymes medical_imaging*
E-mail westbrook@anbiw4.bim.anl.gov 33260@anlcv1.bitnet Tel. 1(708)2523983 Fax 1(708)2526126

Wester, Dennis W. (1949). Senior Research Scientist. Pacific Northwest Laboratory, P7-25, Richland, WA 99352 USA, USA. PhD (U. Florida, 1975) inorganic chemistry. *Actinides chelates ion_exchanger macrocycles*
E-mail dw_wester@pnl.gov Tel. 1(509)3764522 Fax 1(509)3723861

Wheeler, Dr Kraig A. (1965). Graduate Professor. Department of Chemistry, Delaware State University, 1200 North Dupont Highway, Dover, DE 19901, USA. PhD (Brandeis U., 1992) organic chemistry. *Hydrogen_bonding reactivity structure_prediction symmetry*
Tel. 1(302)7394934 Fax 1(302)7394934

Whitby, Frank G. (1962). Student. Rice University, Dept. Biochemistry and Cell Biology, PO Box 1892, Houston, TX 77251, USA. BS (U. Denver, 1986) chemistry. *Macromolecular_structures tropomyosin muscle_physiology proteins*
E-mail frankw@rice.edu Tel. 1(713)5278750x3346

White, Clinton L. (1968). Graduate Fellow. Department of Biochemistry, University of Alabama-Birmingham, Birmingham, AL 35294, USA. [Center for Macromolecular Crystallography, THT Box 79, Birmingham, AL 35294, USA.] BS (Georgia Tech, 1990) chemistry. *Viral_proteins drug_design DNA_binding_proteins protein_stability computer_modelling protein_folding*
E-mail white@orion.cmc.uab.edu Tel. 1(205)9757627 Fax 1(205)9340480

White, Dr Joe L. (1921). Prof. Emeritus of Soil Mineralogy and Chemistry. Dept. of Agronomy, Purdue University, West Lafayette, IN 47907, USA. PhD (U. Wisconsin, 1947) soil mineralogy and chemistry. *Clay_minerals crystallinity aluminium_hydroxide vaccine_adjuvants line-broadening layer_silicates*
Tel. 1(317)4472697 Fax 1(317)4946508

White, Dr John G. (1922). Professor Emeritus. Chemistry Department, Fordham University, Bronx, NY 10458, USA. [323 E. Beckwith, Missoula, MI 59801, USA.] PhD (Glasgow U., 1947) chemical crystallography. *Crystal_structures_of_organic_and_inorganic_materials*
Tel. 1(406)7287725

White, Peter S. (1947). Director of the X-ray Facility. CB 3290 Venable Hall, Department of Chemistry, University of North Carolina at Chapel Hill, Chapel Hill, NC 27599-3290, USA. PhD (Dalhousie U., 1973) inorganic chemistry. *Computer_programming diffractometer_control graphics direct_methods*
E-mail pwhite@pyrite.chem.unc.edu Tel. 1(919)9621689 Fax 1(919)9622388

White, Scott A. (1964). Research Associate. Dept. of Biochemistry and Molecular Biophysics, Washington University Medical School, St Louis, MO 63110, USA. [Dept. of Biochemistry, Campus Box 8231, Washington University Medical School, 660 South Euclid Ave, St Louis, MO 63110, USA.] PhD (U. Edinburgh, 1990) bioinorganic chemistry. *Proteins biocrystallography area_detector computer_modelling flavins iron_sulphur_clusters cytochromes electron_transfer*
E-mail white_s@crystl.wustl.edu Tel. 1(314)3624646 Fax 1(314)3627183

White, Prof. Stephen H. (1940). Professor. Department of Physiology and Biophysics, University of California, Irvine, CA 92717-4560, USA. PhD (U. Washington, 1969) biophysics. *Protein_folding_in_membranes structure_of_bilayers peptide-bilayer_interactions structure_prediction protein_evolution*
E-mail shwhite@uci.edu blanco@helium.biomol.uci.edu Tel. 1(714)8247122 Fax 1(714)8248540

White, Dr Stephen W. (1952). Associate Professor. Department of Microbiology, Duke Medical Center, Box 3020, Durham, NC 27710, USA. DPhil (Oxford U., 1978) molecular biophysics. *DNA-binding_proteins RNA-binding_proteins ribosome_structure RNA_structure*
E-mail white@whitesws.mc.duke.edu Tel. 1(919)6844529 Fax 1(919)6848735

Whitlow, Marc D. (1956). Head of Biophysics. Berlex Biosciences, 15049 San Pablo Avenue, Richmond, CA 94804-0099, USA. [Berlex Biosciences, PO Box 4099, Richmond, CA 94804-0099, USA.] PhD (Boston U., 1986) biochemistry. *Drug_design protein_engineering macromolecular antibody single-chain_Fv*
Tel. 1(510)6694575 Fax 1(510)6694244

Wien, Michelle W. (1969). Graduate student. Biophysics, Hogle Lab., Harvard Medical School, 240 Longwood Ave BCMP, Boston, MA 02115, USA. BA (Wesleyan U., 1991) molecular biology and biochemistry. *Viruses viral-complexes virus-receptor_and_virus–Fab_interactions*
E-mail mich@pachyderm.med.harvard.edu Tel. 1(617)4323919 Fax 1(617)4321216 1(617)4324360

Wierda, Dr Derk A. (1962). Assistant Professor. Department of Chemistry, Saint Anselm College, Manchester, NH 03102, USA. PhD (Harvard U., 1990) inorganic chemistry. *Small_molecules inorganic conformation_analysis catalysis chiral_induction*
E-mail dwierda@anselm.edu Tel. 1(603)6417148 Fax 1(603)6417116

Wiff, Donald R. (1936). Section Head, Gen Corp Inc., Corporate Res., 2990 Gilchrist Rd, Akron, OH 44305-4489, USA. PhD (Texas A&M U., 1966) theoretical physics. *Polymers molecular_modelling nonlinear_optics nanocomposites molecular_composites*
E-mail wiff@iris.gencorp.com Tel. 1(216)791-6339 Fax 1(216)794-6375

Wiley, Don C. (1944). Professor of Biochemistry & Biophysics. Harvard University, Department of Molecular and Cellular Biology, 7 Divinity Avenue, Cambridge, MA 02138, USA. PhD (Harvard, 1971) biophysics.
E-mail wiley@xtal0.harvard.edu; adm.: garnett@xtal0.harvard.edu Tel. 1(617)4951808 Fax 1(617)4959613

Wilkinson, Dr Angus P. (1966). Assistant Professor of Chemistry. School of Chemistry and Biochemistry, Georgia Institute of Technology, Atlanta, GA 30332-0400, USA. DPhil (U. Oxford, England, 1992) chemistry. *Anomalous_dispersion ferroelectricity Rietveld_method zeolites*
E-mail angus.wilkinson@chemistry.gatech.edu Tel. 1(404)8944036 Fax 1(404)8947452

Willett, Prof. Roger D. (1936). Professor and Chairman. Department of Chemistry, Washington State University, Pullman, WA 99163, USA. PhD (Iowa State U., 1962) physical chemistry and physics. *Phase_transitions copper_compounds magnetic_susceptibility modulates crystal_chemistry magnetism EPR*
E-mail willett@wsuvm1.bitnet Tel. 1(509)3353925 Fax 1(509)3358867

Williams, Graheme J. B. (1942). President, USA office. Enraf-Nonius, 390 Central Ave., Bohemia, NY 11716-3147, USA. PhD (U. Alberta, 1972) biochemistry. *Instrumentation proteins small_molecules computer_programs*
E-mail williams@delftny.com Tel. 1(516)5892885 Fax 1(516)5892068

Williams, Dr John A. (1940). Director of Materials Technology. Circon ACMI, 300 Stillwater Ave, Stamford, CT 06902, USA. PhD (Pennsylvania State U., 1970) ceramic science. *Small-angle_X-ray_scattering liquid_structure glass_structure*
Tel. 1(203)3288663 Fax 1(203)3288789

Williams, Dr Loren Dean (1955). Assistant Professor. School of Chemistry and Biochemistry, Georgia Institute of Technology, Atlanta, GA 30332-0400, USA. PhD (Duke U., 1985) physical chemistry. *DNA intercalation hydration DNA_damage_and_serine_ proteases*
E-mail loren.williams@chemistry.gatech.edu Tel. 1(404)8949752 Fax 1(404)8948338

Williard, Prof. Paul G. (1950). Professor. Department of Chemistry, Box H, Brown University, Providence, RI 02912-9000, USA. PhD (Columbia U., 1976) chemistry. *Synthetic_organic_chemistry organometallic_chemistry lithium_compounds computer_graphics vitamin_D steroid_receptors*
E-mail pgw@brownvm.brown.edu pgw@brown.bitnet paul_williard@brown.edu Tel. 1(401)8633589 1(401)8633768 Fax 1(401)8632594 Telex BRN TLX CTR PVD

Wilson, Dr Charles (1964). Assistant professor. Sinsheimer Laboratories, Department of Biology, University of California, 1156 High Street, Santa Cruz, CA 95064, USA. PhD (U. California, San Francisco, 1991) biophysics. *Ribozymes RNA phase_refinement*
E-mail wilson@biology.ucsc.edu Tel. 1(408)4595126 Fax 1(408)4595126

Wilson, David K. (1964). Graduate Student. HHMI - Baylor College of Med., One Baylor Plaza, Houston, TX 77030, USA. BA (Rice U., 1990) biophysics. *Protein_structure enzyme_mechanism drug_design*
E-mail wookie@dino.qci.bioch.bcm.tmc.edu Tel. 1(713)7985024 Fax 1(713)7976718

Wilson, Dr Ian A. (1949). Member. Member of Commision on Biological Macromolecules. Department of Molecular Biology, The Scripps Research Institute, 10666 N. Torrey Pines Road, La Jolla, CA 92037, USA. DPhil (Oxford U., 1976) molecular biophysics. *Antibody_antigen_interactions lymphokine_receptors MHC_complexes folate_dependent_enzymes*
E-mail wilson@scripps.edu Tel. 1(619)5549706 Fax 1(619)5546105

Wing, Richard M. (1936). Professor. Dept. of Chem., Univ. of California, Riverside, CA 92521-0403, USA. PhD (State University of New York at Buffalo, 1965) chemistry. *Organometallics DNA mechanism spectroscopy non-linear_materials molecular_transduction*
E-mail dickwing@ucrvms.bitnet Tel. 1(909)7873503 Fax 1(909)7874713

Wingert, Dr Lavinia (1943). Research Assistant Professor. Department of Crystallography, University of Pittsburgh, Pittsburgh, PA 15260, USA. PhD (U. Pittsburgh, 1986) crystallography. *Proteins charge_density liquid_crystals quasicrystal diabetes*
E-mail wingert@pittvms.bitnet Tel. 1(412)6923511 Fax 1(412)6241882

Winick, Herman (1932). Deputy Associate Director, Professor. SSRL, PO Box 4349, Bin 69, Stanford, CA 94305, USA. PhD (Columbia U., 1957) physics. *Synchrotron_radiation_sources free_electron_lasers*
E-mail winick@slac.stanford.edu Tel. 1(415)9263155 Fax 1(415)9264100

Wismer, Robert (1945). Chemistry Professor. Chemistry Department, Millersville University, Millersville, PA 17551-0302, USA. PhD (Iowa State U., 1972) physical chemistry.
Tel. 1(717)8723661

Wlodawer, Dr Alexander (1946). Laboratory Director. Macromolecular Structure Laboratory, National Cancer Institute - FCRDC, Bldg. 539, PO Box B, Frederick, MD 21702, USA. PhD (UCLA, 1974) molecular biology. *Protein_structure crystallographic_techniques neutron_diffraction*
E-mail wlodawer@ncifcrf.gov Tel. 1(301)8465036 Fax 1(301)8466128

Wolberger, Dr Cynthia (1957). Assistant Professor. Dept of Biophysics and Biophysical Chemistry, Johns Hopkins University School of Medicine, 725 N. Wolfe Street, Baltimore, MD 21205-2185, USA. PhD (Harvard U., 1987) biophysics. *Proteins nucleic_acids*
E-mail cynthia@cthulu.med.jhu.edu Tel. 1(410)9550728 Fax 1(410)9550637

Wolf, Dr Sharon G. Postdoctoral Fellow, Life Sciences Division, Donner Laboratory, Lawrence Berkeley Laboratory, Berkeley, CA 94720, USA. PhD (Weizmann Institute of Science, 1991) structural chemistry. *Proteins electron_crystallography*
E-mail wolf@csa.lbl.gov wolf@rhoda.lbl.gov Tel. 1(510)4866437 Fax 1(510)4866488

Wolff, Dr Gunther A. (1918). G.A. Consultants, NPO, 3776 Northampton Road, Cleveland Heights, OH 44121-2027, USA. ScD (Techn. U. Berlin, BRD, 1948) physical chemistry. *Crystal_growth crystal_form-crystallography surface kinetics III-V_compounds materials_engineering device_applications semiconductors intermetallic_compounds*
Tel. 1(216)3817284

Wong, Ms Rosalind Y. (1940). Research Chemist. US Dept. of Agriculture, Western Regional Research Center, 800 Buchanan Street, CA 94710, USA. BS (San Jose State U., 1965) chemistry. *Structure_determination small_molecules agriculture-natural_products stereochemistry structure-activity_relationship computer_modelling*
E-mail rozwong@pw.usda.gov Tel. 1(510)5595842 Fax 1(510)5595777

Wong-Ng, Dr Winnie (1947). Research Chemist. A215 MATLS, NIST, Gaithersburg, MD 20899, USA. PhD (Louisiana State U., 1974) inorganic chemistry. *Phase_equilibrium modelling structure_determination superconductors crystal_chemistry*
E-mail wongng@tiber.nist.gov Tel. 1(301)9755791 Fax 1(301)9752128

Wood, Elizabeth A. (1912). Retired. 2217 Applewood Dr., Freehold, NJ 07728-3982, USA. PhD (Bryn Mawr, 1939) geology.
Tel. 1(908)7805216

Wood, Dr John S. (1936). Professor of Chemistry. Chemistry Dept., University of Massachusetts, Amherst, MA 01003, USA. PhD (Manchester U., UK, 1962) crystallography and inorganic chemistry. *Charge density Jahn–Teller_compounds transition_elements magnetic_exchange molecular_orbital_calculations*
E-mail a12r000@ucs.umass.edu Tel. 1(413)5456080 Fax 1(413)5454490

Worcester, Prof. David L. (1944). Responsible for a group of students. Biology Division, University of Missouri, Columbia, MO 65211, USA. PhD (Harvard U., 1971) physics. *Neutron_diffraction small-angle_scattering structural_biology macromolecules*
E-mail worcester@biosci.mbp.missouri.edu Tel. 1(314)8826864 Fax 1(314)8820123

Workman, S. Thomas (1926). Retired CEO of Siemens Anal. X-ray Instr. 700 Seaside Ave., Absecon, NJ 08201, USA. BS (St Joseph's U., 1950) physics. *Instrumentation automation software computer support_FSU_scientists*
Tel. 1(609)6459660

Worthington, C. R. Professor of Biological Sciences. Carnegie-Mellon Univ., Mellon Inst-4400-5 Ave, Pittsburgh, PA 15213, USA. PhD *Biological_structure X-ray_diffraction*
Tel. 1(402)2683198

Worthylake, David K. (1956). Graduate Student. Biochemistry Department, University of Utah, Salt Lake City, UT 84132, USA. MS (U. Oregon, 1991) physics. *Protein_structure ubiquitin_system*
E-mail david@uxray.med.utah.edu Tel. 1(801)5853919 Fax 1(801)5817959

Wright, Dr Christine S. (1940). Associate Professor. Department of Biochemistry and Mol. Biophysics, Medical College of Virginia, Richmond, VA 23298, USA. PhD (U. California, San Diego, 1969) chemistry. *Protein_structure computer_modelling glycoproteins lectins protein_carbohydrate_interactions protein_evolution*
E-mail cswright@gems.vcu.edu Tel. 1(804)3717405 Fax 1(804)3717625

Wright, Harlan Tonie (1941). Associate Professor. Dept. Biochemistry, MCV, Virginia Commonwealth Univ., Box 98-0614, Richmond, VA 23298-0614, USA. PhD (U. Calif. San Diego, 1968) chemistry. *Proteins enzymes inhibitors folding*
E-mail xrdproc@gems.vcu.edu Tel. 1(804)8286139 Fax 1(804)8281473

Wright, Dr William V. (1931). Research Professor. Department of Computer Science, University of North Carolina, Chapel Hill, NC 27599-3175, USA. [104 Campbell Lane, Chapel Hill, NC 27514, USA.] PhD (U. North Carolina, 1972) computer science. *Interactive_computer_graphics molecular_graphics*
E-mail wright@cs.unc.edu Tel. 1(919)9621838 1(919)9421144 Fax 1(919)9621799

Wu, Edward Chia-Kuei (1962). PhD student. Department of Crystallography, University of Pittsburgh, Pittsburgh, PA 15260, USA. MS (National Chin-Hua U., 1988) physics. *Proteins crystallography crystal_growth diffraction_theory molecular_mechanics*
E-mail chwst1@vms.cis.pitt.edu chwst1@unixd2.cis.pitt.edu Tel. 1(412)6243219 Fax 1(412)6241882

Wu, Dr Hao (1964). Postdoctoral Fellow. Department of Biochemistry and Molecular Biophysics, Columbia University Health Sciences, 630 W. 168th St., New York, NY 10032, USA. PhD (Purdue U., 1992) biochemistry. *Structure CD4 MHC hCG software*
E-mail wu@cuhhca.hhmi.columbia.edu Tel. 1(212)3051846 Fax 1(212)3057379

Wu, Jin (1943). X-ray manager. Jefferson Cancer Institute, X-ray Lab., 233 South 10th Street, Philadelphia, PA 19107, USA. MS (Syracuse U., 1989) chemistry. *Structure_proteins computer_modelling data_collection_technique*
E-mail jinwu@calvin.jci.tju.edu Tel. 1(215)9554588 Fax 1(215)9232117

Wu, Dr Ping (1963). Postdoctoral Research Associate. Department of Chemistry, Brookhaven National Laborary, Upton, NY 11973, USA. PhD (Brown U., 1990) inorganic chemistry. *Material_synthesis characterization single_crystal thin_film solid_state_chemistry structure_determination chalcogenides oxides*
E-mail wu@chm.chm.bnl.gov Tel. 1(516)2827857 Fax 1(516)2825815

Wu, Shan (1960). Scientist. Genencor International Inc., 180 Kimball Way, South San Fransisco, CA 94080, USA. PhD (U. Catholique Louvain, Belgium, 1988) crystallography. *Protein_crystallography enzyme_mechanism protein_engineering*
Tel. 1(415)7427221 Fax 1(415)5838269

Wuensch, Prof. Bernhardt John (1933). Professor of Ceramics. Department of Materials Science and Engineering, Massachusetts Institute of Technology, 77 Massachusetts Ave., Cambridge, MA 02139-4307, USA. PhD (Mass. Inst. of Tech., 1963) crystallography. *X-ray_and_neutron_scattering sulfides point_defects diffusion fast-ion_conductors*
Tel. 1(617)2536889 Fax 1(617)2587874

Xia, Dr Di (1961). Research Associate. Howard Hughes Medical Institute and Department of Biochemistry, University of Texas, Southwest Medical Center at Dallas, Dallas, Texas 75235-9050, USA. PhD (Purdue U., 1992) structure virology. *Crystallographic structural_virology membrane_protein_structure programming structure_comparison*
E-mail dixia@howie.swmed.edu Tel. 1(214)6485093 Fax 1(214)6485095

Xiang, Shibin (1962). Postdoctoral Research Associate. Department of Biochemistry & Biophysics, University of North Carolina, Chapel Hill, NC 27599, USA. PhD (Institute of Physics, Chinese Academy of Sciences, 1990) solid state physics. *Proteins computer_modelling direct_method XANES EXAFS maximum_entropy_method crystallography*
E-mail shibin@med.unc.edu Tel. 1(919)9663263 Fax 1(919)9662852

Xiao, Dr Gaoyi (1959). Research Associate. Center for Advanced Research in Biotechnology, 9600 Gudelsky Drive, Rockville, MD 20850, USA. PhD (U. of Oklahoma, 1993) physical chemistry. *Protein computer_modelling drug_design high_order_X-ray_data*
E-mail gaoyi@indigo2.carb.gov Tel. 1(301)7386126 Fax 1(301)7386255

Xiayang, Dr Qiu (1963). Senior Fellow. Department of Biological Structures, SM-20, School of Medicine, University of Washington, Seattle, Washington 98195, USA. PhD (Michigan State U., 1993) protein crystallography. *Protein–DNA_interaction rational_drug_design blood_coagulation*
E-mail xiayang@gouda.bchem.washington.edu Tel. 1(206)5438865 Fax 1(206)5431524

Xie, Xiaoling (1960). Student. Rosenstiel Center, Brandeis University, 415 South Street, Waltham, MA 02254, USA. MA (Brandeis U., 1988) physics. *Proteins structure-activity_relationship Laue_diffraction crystal_growth time_resolved calcium_binding_proteins phasing_methods*
E-mail xie@brandeis.bitnet xiaoling@auriga.rose.brandeis.edu Tel. 1(617)7362475 Fax 1(617)7362405

Xu, Dr Weixin (1956). Research Scientist. Wyeth-Ayerst Research, 865 Ridge Road, Monmouth Jct., New Jersey, NJ 08852, USA. PhD (U. Texas at Austin, 1988) biochemistry. *Proteins hormones DNA drug drug_discovery*
E-mail XUW@prince.mm.wyeth.com Tel. 1(908)2744471 Fax 1(908)2744292

Xu, Zhang-Bao (1944). Senior Research Scientist. American Cyanamid Company, MRD. Biomedical Research Computing, Pearl River, NY 10965, USA. [190/206D, Lederie Lab., Pearl River, NY 10965, USA.] PhD (U. Pittsburgh, 1986) protein crystallography. *Macromolecular_crystallization X-ray_data_collection protein_structure computing_method direct_methods*
Tel. 1(914)7323787 Fax 1(914)7325682

Xue, Dr Yafeng (1961). Postdoctoral fellow. Department of Chemistry, Harvard University, 12 Oxford Street, Cambridge, MA 02138, USA. PhD (U. Umeå, Sweden, 1992) biochemistry. *Protein crystallographic_computing crystallization*
E-mail xue@hubeta.harvard.edu Tel. 1(617)4954767 Fax 1(617)4953330

Xuong, Dr Nguyen-huu (1933). Professor of Physics, Chemistry and Biology. Department of Chemistry, 0359, University of California, San-Diego, 9500 Gilman Drive, La Jolla, CA 92093-0359, USA. PhD (U. California, Berkeley, 1962) physics. *Protein_structures area_detector multiple_wavelength*
E-mail xuong@chem.ucsd.edu Tel. 1(619)5342501 Fax 1(619)5346174

Yadav, Dr Prem N. (1959). Research Associate. Department of Biochemistry and Mol. Biology, UMD-New Jersey Medical School, Newark, NJ 07103, USA. PhD (Banaras Hindu U., 1988) biophysics. *Computer_simulation molecular_modelling protein_modelling DNA_polymerase homology_modelling protein–DNA_interaction drug_design*
E-mail pyadav@njmsa.umdnj.edu Tel. 1(201)9824053 Fax 1(201)9825594

Yalkovsky, Ralph (Rafael) (1917). Professor Emeritus Geology and Oceanography; science writer. National Association of Science Writers, PO Box 398, Grand Island, NY 14072, USA. PhD (U. Chicago, 1956) marine geology and sedimentation. *Oceanography sedimentology heavy_minerals international_law_of_the_sea public_policy*

Yamano, Akihito (1961). Graduate Student. Department of Chemistry, Boston College, Chestnut Hill, MA 02167, USA. MS (Tokyo Institute of Technology, 1987) biochemistry. *Structure_determination software instrumentation structural_disorder proteins computer_modelling water_structure powder X-ray_fluorescence_spectroscopy*
E-mail yamano@bcchem.bc.edu yamano@bchem.bitnet Tel. 1(617)5523905 Fax 1(617)5522705

Yan, Youwei (1946). Research Associate. Biochemistry and Molecular Biology, Harvard University, 7 Divinity Ave., Cambridge, MA 002138, USA. PhD equivalent (Academia Sinica, 1986) crystallography. *Crystallography receptors enzyme_drugs*
E-mail yan@xtal0.harvard.edu Tel. 1(617)4955043 Fax 1(617)4959613

Yang, Fan (1968). Graduate Student. Department of Biochemistry and Cell Biology, Rice University, Houston, TX 77251, USA. BS (Nanjing U., P. R. China, 1990) physics. *X-ray_crystallography hemoproteins mutagenesis kinetics myoglobin low_pH FTIR structure_determination*
E-mail fyfy@bioc.rice.edu Tel. 1(713)5278750x3346 Fax 1(713)2855154

Yang, Jian (1964). Research Associate. Department of Biochemistry and Molecular Biophysics, Washington University School of Medicine, St Louis, MO 63110, USA. [660 South Euclid Ave/Box 8231, St Louis, MO 63110, USA.] MSc (Institute of Biophysics, Academia Sinica, Beijing, 1989) protein crystallography. *X-ray_crystallography hemoglobin*
E-mail yang_j@crystl.wustl.edu Tel. 1(314)3624646 Fax 1(314)3627183

Yang, Wei (1963). Post-doc. Department of Biophysics & Biochemistry, Yale University, PO Box 1184, New Haven, CT 06520, USA. PhD (Columbia U., 1991) biochemistry & biophysics. *Protein nucleic-acid macromolecular_crystallography*
E-mail yangwei@hhvms8.csb.yale.edu Tel. 1(203)4325795 Fax 1(203)4323282

Yeates, Dr Todd O. (1961). Assistant Professor. Department of Chemistry and Biochemistry, University of California, Los Angeles, Los Angeles, CA 90024-1569, USA. PhD (U. California, Los Angeles, 1988) biochemistry. *Proteins molecular_replacement twinning isomorphous_replacement structure_verification topology direct_methods structure_prediction*
E-mail yeates@uclaue.mbi.ucla.edu Tel. 1(310)2064866 Fax 1(310)2063914

Yee, Dr Vivien C. (1964). Senior Fellow. Biological Structure SM-20, University of Washington, Seattle, WA 98195, USA. PhD (U. British Columbia, 1990) physical chemistry. *Proteins crystallography coagulation transglutaminases modelling*
E-mail yee@gouda.bchem.washington.edu Tel. 1(206)5436047 Fax 1(206)5431524

Yeh, Joanne I. Graduate student. Department of Chemistry, University of California - Berkeley, 234 MCL, Berkeley, CA 94720, USA. PhD (U. California, Berkeley, 1994) physical chemistry. *Protein_crystallography protein_design signal_transduction*
E-mail Joanne@lcbvax.cchem.berkeley.edu Tel. 1(510)4864355 Fax 1(510)4866059

Yennawar, Dr Hemant P. (1962). Postdoctoral Fellow. Department of Chemistry, 152 Davey Laboratory, The Pennsylvania State University, University Park, PA 16802, USA. PhD (Indian Institute of Science, 1992) physics, X-ray crystallography. *Protein_structure macromolecular_crystallography enzyme_chemistry computer_programming computer_graphics*
E-mail hemant@retina.chem.psu.edu Tel. 1(814)8658383 Fax 1(814)8653314

Yennawar, Dr Neela H. (1964). Postdoctoral Fellow. Department of Chemistry, 152 Davey Laboratory, The Pennsylvania State University, University Park, PA 16802, USA. PhD (Indian Institute of Science, 1992) physics, X-ray crystallography. *Protein_structure macromolecular_crystallography enzyme_chemistry computer_programming computer_graphics*
E-mail neela@retina.chem.psu.edu Tel. 1(814)8658383 Fax 1(814)8653314

You, Dr Hoydoo (1953). Staff Scientist. Materials Science Division, Argonne National Laboratory, 9700 South Cass Ave., Argonne, IL 60439, USA. PhD (U. Washington, 1985) physics. *Synchrotron_radiation crystallography instrumentation neutron_diffraction surface interface electrochemistry ultra_high_vacuum reflectivity scattering*
E-mail you@anlaps.aps.anl.gov you@anlsrs.msd.anl.gov Tel. 1(708)2523429 Fax 1(708)2527777

Young, Prof. R. A. (1921). Professor Emeritus of Physics. Chairman of Commission on Powder Diffraction. School of Physics, Georgia Institute of Technology, Atlanta, GA 30332-0430, USA. PhD (The Polytechnic Institute of Brooklyn, 1959) physics. *Structural_change diffraction_theory powder_diffraction Rietveld_method size_effect microstrain diffraction_applications crystal_physics atomic_scale_mechanisms*
E-mail ph268ry@vm1.gatech.edu Tel. 1(404)8945208 Fax 1(404)8539958

Young Jr, Dr Victor G. (1956). Staff Crystallographer. Department of Chemistry, 1711 Gilman Hall, Iowa State University, Ames, IA 50011-3111, USA. PhD (Arizona State U., 1985) inorganic chemistry. *Small_molecule neutron_diffraction Rietveld_analysis powder_diffraction*
E-mail vyoung@alisuvax.bitnet isuvax::young@inst2.chem.iastate.edu Tel. 1(515)2947956 Fax 1(515)2940105

Youngs, Prof. Wiley (1949). Professor of Chemistry. Chemistry Dept., University of Akron, Akron, OH 44325-3601, USA. PhD (State U. New York at Buffalo, 1980) inorganic chemistry. *Cyclynes alkynes crystallography liquid_crystals*
E-mail youngs@atlas.chemistry.uakron.edu Tel. 1(216)9725362 Fax 1(216)9727370

Yu, Chang-An (1937). Regents Professor. Department of Biochemistry and Molecular Biology, Oklahoma State University, Stillwater, OK 74078, USA. PhD (U. Illinois, 1969) biochemistry. *Bioenergetics membrane_protein*
E-mail cayuq@okway.okstate.edu Tel. 1(405)7446612 Fax 1(405)7447799

Yuan, Jie (1954). Graduate Student. Department of Chemistry, University of Cincinnati, ML 0172, Cincinnati, OH 45221-0172, USA. MS (U. Cincinnati, 1989) inorganic chemistry. *Inorganic small-molecule computer_modelling*
E-mail jie.yuan@uc.edu Tel. 1(513)5560195 1(513)5569230 Fax 1(513)5569239

Yuhasz, Stacieann (1958). Postdoctoral fellow. Biophysics Department, Johns Hopkins School of Medicine, 725 N. Wolfe St., Baltimore, MD 21205, USA. PhD (Johns Hopkins U., 1987) biophysics. *Antibody beta-amyloid*
E-mail yuhasz%jhuigf@cunyum.cuny.edu Tel. 1(410)9558715 Fax 1(410)9550637

Yuvaniyama, Jirundon (1968). PhD student. Biophysics Research Division, Chemistry Bldg. Room 3250, 930 N. University, Ann Arbor, MI 48109-1055, USA. BSc (Mahidol U., Thailand, 1991) chemistry. *Proteins computer_modelling*
E-mail ji@oscar.biop.umich.edu Tel. 1(313)7633384 Fax 1(313)7643323

Zabel, Dr Volker (1944). Senior Research Associate. University of Tennessee Graduate School of Biomedical Sciences, Oak Ridge National Laboratory, PO Box 2009, MS 8077, Oak Ridge, TN 37831-8077, USA. Dr rer.nat. (Freie U. Berlin, 1976) chemical crystallography. *Proteins hydrogen_bonding neutron_diffraction water_structure synchrotron_radiation anomalous_scattering*
E-mail zabel@biovx1.bio.oml.gov Tel. 1(615)5740844 Fax 1(615)5741274

Zacharias, Dr David Edward (1926). Senior research associate. Molecular Structure Department, Institute for Cancer Research, Fox Chase Cancer Center, Philadelphia, PA 19111-2421, USA. PhD (U of Pittsburgh, 1969) X-ray crystallography. *Crystal_structures small_molecules geometric_analysis*
E-mail zacharia@lindo.spot.fccc.edu Tel. 1(215)7282220 Fax 1(215)7283574

Zakrzewski, Marek. Scientist. Analytical Chemistry, Procter & Gamble Pharmaceuticals, PO Box 191, Norwich, NY 13815, USA. PhD (Dalhousie U., 1990) physical chemistry. *Structural_disorder computer_modelling powder_diffraction*
E-mail zakrzewskim@pg.com Tel. 1(607)3356853 Fax 1(607)3352100

Zalkin, Allan (1926). Retired. 81 Edgecroft Rd., Berkeley, CA 94707-1412, USA. PhD (U. California, Berkeley, 1951) chemistry. *Chemistry inorganic_computing instrumental_computing powder_diffraction*
Tel. 1(510)5244798 Fax 1(510)5246744

Zavalij, Dr Peter Yu. (1957). Research Assistant. Materials Research Center, Department of Chemistry, State University of New York at Binghamton, Binghamton, NY 13902-6000, USA. PhD (Lviv State U., Ukraine, 1983) inorganic chemistry. *Intercalates layered_compounds modulated_structures computing computer_modelling structure_determination density_distribution*
E-mail zavalij@bingvaxa.bitnet Tel. 1(607)7774623 Fax 1(607)7774623

Zawadzke, Dr Laura E. (1963). Postdoctoral Fellow. Biophysics and Biophysical Chemistry Department, Johns Hopkins University, School of Medicine, 725 N. Wolfe St., Baltimore, MD 21205, USA. PhD (Massachusetts Institute of Technology, 1990) biochemistry. *Chiral_protein_crystallography metal_binding_proteins direct_method peptide_synthesis enzymes*
E-mail lauraz@jhuigf.med.jhu.edu Tel. 1(410)9556636 Fax 1(410)5506910

Zeng, Huang (1962). Graduate Student. Department of Chemistry, Washington State University, Pullman, WA 99164-4630, USA. BS (Zhongshan U., China, 1983) physical chemistry. *Solid_state incommensurate_phases solid_state_NMR_in_phase_transitions*
E-mail 60771913@wsuvm1.bitnet Tel. 1(509)3359125

Zeng, Ke (1957). Graduate student. University of Pittsburgh, Dept. of Crystallography, University of Pittsburgh, Pittsburgh, PA 15260, USA. MS (Wuhan U. Tech., Wuhan, P. R. China, 1985) material science. *Crystallography macromolecules spectroscopy computer_graphics dynamics magnetochemistry crystal_field EXAFS*
E-mail kzest@vms.cis.pitt.edu kzest@unixd.cis.pitt.edu Tel. 1(412)6243219

Zhang, Dr Cai Xj. (1959). Postdoctor. Institute of Molecular Biology, HHMI, University of Oregon, Eugene, OR 97403, USA. PhD (Oregon U., 1992) physics. *Proteins computer_modelling protein_stability molecular_replacement water_structure alanine_substitution*
E-mail chk@uoxray.uoregon.edu Tel. 1(503)3465874 Fax 1(503)3465874

Zhang, Dr Faming (1964). Instructor. Lilly Research Laboratories, Eli Lilly and Company, Lilly Corporate Center, Indianapolis, IN 46285, USA. PhD (Institute of Biophysics, 1990) biochemistry. *Proteins structural_modelling kinases synchrotron structure-based_drug_design crystallization*
E-mail zfm@lilly.com Tel. 1(317)2478828 1(317)2776110 Fax 1(317)2769722

Zhang, Dr Gongyi (1964). Visiting Fellow. Laboratory of Molecular Biology, NIDDK, NIH, Building 5, Room 435, 9000 Rockville Pike, Bethesda, MD 20892, USA. PhD (Institute of Biophysics, Academia Sinica, P. R. China, 1993) protein crystallography. *Proteins kinase_structure*
E-mail zhang@vger.niddk.nih.gov zhang@tove.niddk.nih.gov Tel. 1(301)4024705 Fax 1(301)4960201

Zhang, Hong (1964). Graduate Student. Department of Molecular Biology, Vanderbilt University, Nashville, TN 37235, USA. BS (Peking U., 1987) genetics. *Proteins computer_modelling toxins data_processing*
E-mail hong%mobv01@ctrvax.vanderbilt.edu Tel. 1(615)3222012 Fax 1(615)3436707

Zhang, Dr Hongming (1942). Director of X-ray facility. Department of Chemistry, Southern Methodist University, Dallas, Texas 75275, USA. PhD (U. of Alabama, 1989) inorganic chemistry. *Computing structural_chemistry carborane_compounds lanthanide_complexes*
E-mail hzhang@sun.cis.smu.edu Tel. 1(214)7684141 Fax 1(214)7684089

Zhang, Dr Kam Y. J. (1963). Postdoctoral Research Molecular Biologist. Molecular Biology Institute and Department of Chemistry and Biochemistry, University of California, Los Angeles, 405 Hilgard Avenue, Los Angeles, CA 90024, USA. PhD (U. York, England, 1989) physics. *Protein_structure phase_problem protein_folding*
E-mail kam@uclaue.mbi.ucla.edu Tel. 1(310)8251402 Fax 1(310)2063914

Zhang, Xiaohua (1956). Dept. Biology, MIT, Rm 16-739, Cambridge, MA 02139-4307, USA. PhD (Rutgers U., 1989) chemistry. *Macromolecular_crystallography structural_biology membrane_receptors*
E-mail zhang@xtal.mit.edu Tel. 1(617)2537006 Fax 1(617)2536738

Zhang, Dr Yiping (1960). Staff Researcher. Syntex Discovery Research, Mail Stop R6-002, 3401 Hillview Avenue, Palo Alto, CA 94304, USA. PhD (Harvard U., 1989) inorganic chemistry. *Proteins crystallographic_computing structure-based_drug_design*
E-mail zhang@saffron.syntex.com Tel. 1(415)8134451 Fax 1(415)3547363

Zhao, Dong (1960). Research Associate. Dept. of Chemistry, Univ. of Missouri-St. Louis, 8001 Natural Bridge Rd., St. Louis, MO 63121, USA. PhD (U. Southern California, 1992) inorganic. *Structural_transition metal_hydrogen bond_hydrides main_group_materials chemistry*
E-mail sdzhao@slvaxa.umsl.edu Tel. 1(314)5536483 Fax 1(314)5535342

Zhao, Rui (1969). Department of Biological Sciences, Lilly Hall, Purdue University, West Lafayette, IN 47907, USA. BS (U. Sci. Tech. China, 1991) biology. *Virus_structure protein_structure virus_assembly antiviral_drug*
E-mail b45@mace.cc.purdue.edu Tel. 1(317)4941662 Fax 1(317)4961189

Zheng, Dr Jianhua. Senior Scientist. ImmunoPharmaceutics. Inc., 11011 Via Frontera, San Diego, CA 92127, USA. PhD (U. Calif. San Diego, 1993) biochemistry, protein crystallography. *Protein_structure protein_function rational_drug_design protein_kinase*
E-mail jhz%aditya@uunet.uu.net Tel. 1(619)4518400

Zhou, Ms Lan (1965). PhD candidate. Center for Macromolecular Crystallography, University of Alabama at Birmingham, Birmingham, AL 35294, USA. MS (Institute of Biophysics, Academia Sinica, 1990) molecular biophysics. *Virus_structure protein_crystallography drug_design crystallization homology_prediction*
E-mail lane@orion.cmc.uab.edu Tel. 1(205)9341611 Fax 1(205)9340480

Zhu, Dr Nai-jue (1942). Senior Research Associate. Department of Chemistry, Xavier University of Louisiana, 7325 Palmetto Street, New Orleans, LA 70125, USA. PhD (Academia Sinica, 1981) inorganic chemistry. *Charge_density chemical_bonding electron_density_distribution electrostatic_potential biocrystallography proteins drug_receptor_interaction antipsychotic_compounds*
E-mail mpcm@uno.edu Tel. 1(504)4867411x375 Fax 1(504)2866860

Zhu, Xiaotian (1966). Postdoc. CD8/MHC. Dept. of Biochem., 630 W. 168th St., Columbia University, New York, NY 10032, USA. PhD (California Institute of Technology, 1993) biophysical chemistry. *Signal_transduction protein_structure immune system*
E-mail zhu@cuhhca.hhmi.columbia.edu Tel. 1(212)3056996 Fax 1(212)3057379

Zolensky, Michael E. (1955). Space Scientist. SN2/NASA, Johnson Space Center, Houston, TX 77058, USA. PhD (Pennsylvania State U., 1983) mineralogy and geochemistry. *Mineralogy meteorites*
E-mail zolensky@curate.jsc.nasa.gov Tel. 1(713)4835128 Fax 1(713)4835347

Zompa, Leverett J. (1938). Professor. Department of Chemistry, University of Massachusetts, Harbor Campus, 100 Morrisey Blvd., Boston, MA 02125-3393, USA. PhD (Boston College, 1964) inorganic chemistry. *Inorganic complexation stereochemistry chelation macrocyclic_complexes*
Tel. 1(617)2876149 Fax 1(617)2657173

Zuccola, Harmon J. (1964). Postdoctoral fellow. Dept of BCMP, Harvard Med. School, Boston, MA 02115, USA. PhD (Georgia Inst. of Technology, 1992) chemistry. *Protein_structure structure_function homology_prediction viruses*
E-mail harmon@ganesh.med.harvard.edu Tel. 1(617)4321216

Zwell, Leo (1915). Bibliographer. JCPDS-ICDD, 12 Campus Blvd, Newtown Sq., PA 19703, USA. [117 S. Chester Rd., Swarthmore, PA 19081, USA.] BS (Brooklyn College, 1934) physics. *Powder_diffraction*
Tel. 1(610)3259814 Fax 1(610)3259823

URUGUAY

Sub-Editor: R. Mariezcurrena

Gonzalez, Mr Oscar (1955). Assistant. Lab. Cristalografia, Facultad de Quimica, Universidad de la Republica, General Flores 2124, Montevideo, Uruguay. [Facultad de Quimica, Universidad de la Republica, Casilla de Correos 1157, Montevideo, Uruguay.] *Crystal_structure_determination Rietveld_method bioinorganic_compound*
E-mail crystal@insqim.edu.uy Tel. 598(2)941880x46 Fax 598(2)941906

Mariezcurrena, Prof. Raul (1939). Professor. Sub-Editor of World Directory of Crystallography. Catedra de Fisica, Facultad de Quimica, Universidad de la Republica, General Flores 2124, Montevideo, Uruguay. [Catedra de Fisica, Facultad de Quimica, General Flores 2124, Casilla de Correos 1157, CP 11800 Montevideo, Uruguay.] Dr (Universidad de la Republica, 1975) pharmaceutical chemistry. *Crystal_structure_determination organic_compound proteins_crystallography*
E-mail crystal@insqim.edu.uy Tel. 598(2)941880x40 Fax 598(2)941906

Mombru, Mr Alvaro (1966). Lecturer. Lab. Cristalografia, Facultad de Quimica, Uni-versidad de la Republica, General Flores 2124, Montevideo, Uruguay. [Facultad de Quimica, Universidad de la Republica, Casilla de Correos 13118, CP 11800 Montevideo, Uruguay.] MSc (Universidad de la Republica, 1991) chemistry. *Crystal_structure_determination Rietveld_method neutron_diffraction magnetism superconductivity inorganic_compound Mossbauer*
E-mail crystal@insqim.edu.uy Tel. 598(2)941880x46 Fax 598(2)941906

Victoria, Mr Nelson (1961). Assistant. Inst. Fisica, Facultad de Ingenieria, Universidad de la Republica, Julio Herrera y Reissig 565, Montevideo, Uruguay. [Inst. Fisica, Facultad de Ingenieria, Universidad de la Republica, Casilla de Correos 30, CP 11000, Montevideo, Uruguay.] MSc (Universidad de la Republica, 1993) chemistry. *Powder_method corrosion electronics*
E-mail nelson@fising.edu.uy Tel. 598(2)710905 Fax 598(2)715446

UZBEKISTAN

Sub-Editor: B. T. Ibragimov

Beketov, Dr Kajrat M. (1964). Institute of Bioorganic Chemistry, Academy of Sciences of Republic Uzbekistan, Kh. Abdullaev Str. 83, Tashkent 143, Uzbekistan. Magistr. chemistry. *Organic_crystal_chemistry inclusion_compound biomolecule bioorganic_chemistry*
Tel. 7(3712)627071

Ibragimov, Prof. Bakhtijar T. (1949). Professor. Sub-Editor World Directory. Institute of Bioorganic Chemistry, Academy of Science Uzbekistan, Kh. Abdullaev Str. 83, Tashkent 143, Uzbekistan. PhD (State U. Moscow, 1991) physical chemistry. *Organic_crystal_chemistry inclusion_compound biomolecule bioorganic_chemistry*
E-mail root@ibc.tashkent.su Tel. 7(3712)627071 Fax 7(3712)627071

Khudojrov, Dr Achmat B. (1953). Institute of Chemistry, Academy of Science of Uzbekistan, Kh. Abdullaev Str. 77, Tashkent 170, Uzbekistan. PhD (Inst. of Chemistry, Academy of Science Uzbekistan, 1989) chemistry. *Inorganic_crystal_chemistry molecular_structure biological_activity*
Tel. 7(3712)627959

Makhmudov, Dr Mirzakadir K. (1960). Institute of Chemistry of Plant Substances, Kh. Abdullaev Str. 77, Tashkent 170, Uzbekistan. [M. R. Mirzo Str. 4, Tashkent 175, Uzbekistan.] Magistr. (Institute of Chemistry, Tashkent, 1993) physical chemistry. *Organic_crystal_chemistry molecular_structure biological_activity alkaloids terpenoids molecular_mechanics_technique quantum_chemistry*
E-mail root@icsp.tashkent.su Tel. 7(3712)627207 Fax 7(3712)891475

Sharipov, Prof. Khasan T. (1947). Chief of department. Institute of Chemistry, Academy of Science Uzbekistan, Kh. Abdullaev Str. 77, Tashkent 170, Uzbekistan. PhD (Institute of Chemistry, Academy of Science Uzbekistan, 1990) physical chemistry. *Inorganic_crystal_chemistry molecular_structure biological_activity inorganic_compound spectrochemistry coordination_compound*
Tel. 7(3712)627959

Talipov, Dr Samat A. (1952). Institute of Bioorganic Chemistry, Academy of Sciences of Republic Uzbekistan, Kh. Abdullaev Str. 83, Tashkent 143, Uzbekistan. PhD (Institute of Crystallography, Academy of Sciences USSR, 1989) crystallography and crystallophysics. *Organic_crystal_chemistry inclusion_compound biomolecule bioorganic_chemistry*
Tel. 7(3712)627071

Tashkhodzhaev, Dr Bakhodir (1948). Head of Research Group. Institute of Chemistry of Plant Substances, Kh. Abdullaev Str. 77, Tashkent 170, Uzbekistan. [Shumanai Str. 2/1, Tashkent 169, Uzbekistan.] PhD (Dept of Chemistry, Moscow U., 1977) physical chemistry. *Organic_crystal_chemistry molecular_structure biological_activity alkaloids terpenoids steroids molecular_mechanics_technique quantum_chemistry*
E-mail root@icsp.tashkent.su Tel. 7(3712)627207 Fax 7(3712)891475

Yusupova, Dr Ilihamia M. (1954). Junior Scientific officer. Institute of Chemistry of Plant Substances, Kh. Abdullaev Str. 77, Tashkent 170, Uzbekistan. [G. Petrov str. 9, Apt 86, Tashkent 700007, Uzbekistan.] Magistr. (Institute of Chemistry, Tashkent, 1993) physical chemistry. *Organic_crystal_chemistry molecular_structure biological_activity alkaloids terpenoids steroids*
E-mail root@icsp.tashkent.su Tel. 7(3712)627207 Fax 7(3712)891475

VENEZUELA

Sub-Editor: M. V. Capparelli

Notes

1. First degrees awarded by Venezuelan Universities for scientific and technical subjects are Licenciado (Lic.) and Ingeniero (Ing.) (5 years of courses). Degrees awarded for graduated studies are Magister Scientiarum (MSc) (two years of courses and reasearch) and Doctor (Dr) (three-four years of courses and reasearch). Equivalent graduated studies at IVIC lead to degrees of MSc and Philosophus Scientiarum (PhSc).

2. The following abbreviations have been used.
  (a) For institutions:
  Depto. = Departamento
  Esc. = Escuela
  Fac. = Facultad
  Inst. = Instituto
  U. = Universidad
  IUT-RC = Inst. Universitario de Tecnología - Región Capital
  IVIC = Inst. Venezolano de Investigaciones Científicas
  UCV = U. Central de Venezuela
  UDO = U. de Oriente
  ULA = U. de Los Andes
  UNEG = U. Nacional Experimental de Guayana
  USB = U. Simón Bolívar
  (b) For titles/degrees: see above
  (c) For addresses:
  Apdo. = Apartado
  Vzla. = Venezuela

Alamo, Lic. Lorenzo (1958). Professional staff. Centro de Biofísica y Bioquímica (IVIC), Km 11 Carretera Panamericana - Miranda, Venezuela. [Apdo. 21827, Caracas 1020-A, Venezuela.] Lic. (UCV, 1984) biology. *SAXS electron_microscopy image_processing muscles contractile_systems*
E-mail alamo@ivic.ivic.ve Tel. 58(2)5011231 Fax 58(2)5011093

Aquino Hernández, Ing. Rosa Cruz (1958). Professional Staff. Lab. Geológico El Chaure, Corpoven S.A., Apdo. 4326 - Puerto La Cruz, Anzoátegui, Venezuela. MSc (UCV, 1993) geological sciences. *Petrography sedimentology*
Tel. 58(81)73875 Fax 58(81)603700

Arce Sagüés, Dr Alejandro Jaime (1951). Researcher. Centro de Química - IVIC, Km 11 Carretera Panamericana, Miranda, Venezuela. [Apdo. 21827, Caracas 1020-A, Venezuela.] PhD (U. Wales - UK, 1980) chemistry. *Metals_clusters fullerenes*
E-mail aarce@quimica.ivic.ve Tel. 58(2)5011386 Fax 58(2)5011350

Bello, Prof. Alfredo (1957). Associate Professor. Depto. de Física - USB, Sartenejas - Baruta, Caracas 1080, Venezuela. [Apdo. 89000, Caracas 1081-A, Venezuela.] PhD (Case Western Reserve U. - USA, 1984) physics. *Crystal_defect polymers insulators*
E-mail abello@usb.ve Tel. 58(2)9063526 Fax 58(2)9063627

Bonyuet Lee, Prof. Dickar (1954). Associate Professor. Depto. de Física, Esc. de Ciencias - UDO Nucleo Sucre, Av. Universidad - Cerro Colorado, Cumaná - Sucre, Venezuela. [Apdo. 245, Cumaná 6101-A, Sucre, Venezuela.] MSc (Brown U. - USA, 1983) physics. *Condensed_matter metals*
E-mail nsucre@dino.conicit.ve Tel. 58(93)514256x3451

Camillo, Prof. Francesco Emilio (1955). Associate Professor. Depto. de Tecnología de Materiales, IUT-RC, Km 8 Carretera Panamericana, Miranda, Venezuela. [Apdo. 40347, Caracas 1040-A, Venezuela.] Engineer (Ecole Nat. Sup. Ceramique Ind. Sevres - France, 1978) ceramics. *Crystallinity phase_determination ceramics clays*
Tel. 58(2)6820048 Fax 58(2)6812754

Capparelli, Dr Mario Vicente (1937). Academic consultant. Sub-Editor WDC9; Chairman Natl. Comm. Cryst. Centro de Química - IVIC, Km 11 Carretera Panamericana, Miranda, Venezuela. [Apdo. 21827, Caracas 1020-A, Venezuela.] PhD (U. London - UK, 1984) crystallography. *Single_crystal X-ray_diffraction structure_determination topochemistry structural_chemistry*
E-mail mcappare@dino.conicit.ve Tel. 58(2)5011353 Fax 58(2)5011350

Casanova Bautista, Dr Rodrigo (1951). Full Professor. Depto. de Física, Fac. de Ciencias - ULA, La Hechicera, Mérida, Mérida, Venezuela. [Apdo. 40 Mérida, 5251-A Mérida, Venezuela.] PhD (U. Pennsylvania - USA, 1981) physics. *LEED photoemission SEXAFS*
E-mail rodrigo@ciens.ula.ve Tel. 58(74)401339 Fax 58(74)401286

Cisneros, Ing. Oswaldo (1952). Consultant. C. A. Voladuras y Excavaciones, Edificio Banco Construcción, 9-4 Av. Municipal, Puerto La Cruz - Anzoátegui, Venezuela. Ing. (UCV, 1985) geology. *Minerals gemstones*
Tel. 58(81)664165 Fax 58(81)675661

Coronel, Dr Domingo Alberto (1958). Professor. Depto. de Tecnología de Materiales, IUT-RC, Km 8 Carretera Panamericana, Miranda, Venezuela. [Apdo. 40347, Caracas 1040-A, Venezuela.] Dr (U. Nice - France, 1990) earth sciences. *Powder_diffraction qualitative_analysis data_processing quantitative_analysis*
Tel. 58(2)6812428 Fax 58(2)6812754

Cosenza, Prof. Erika (1956). Assistant Professor. Depto. de Química, IUT-RC, Km 8 Carretera Panamericana, Miranda, Venezuela. [Apdo. 40347, Caracas 1040-A, Venezuela.] MSc (U. Poitiers - France, 1980) chemistry. *Catalysis coke_structure zeolites molecular_sieves*
Tel. 58(2)6812467 Fax 58(2)6812754

Delgado Quiñones, Prof. José Miguel (1956). Associate Professor. Depto. de Química, Fac. de Ciencias - ULA, La Hechicera - Mérida, Mérida, Venezuela. [Apdo. 40 - Mérida, 5251-A Mérida, Venezuela.] PhD (MIT - USA, 1988) materials science. *Single_crystal X-ray_diffraction powder_diffraction inorganic_compound structure_properties_relationships metal_organic_compounds*
E-mail migueld@ciens.ula.ve Tel. 58(74)401372 Fax 58(74)401286

Díaz de Delgado, Prof. Graciela Carlota (1959). Assistant Professor. Depto. de Química, Fac. de Ciencias - ULA, La Hechicera - Mérida, Mérida, Venezuela. [Apdo. 40, Mérida-5251A, Mérida, Venezuela.] PhD (Brandeis U. - USA, 1988) chemistry. *Solid_reaction topochemistry carboxylates unsaturated_carboxylic_acids*
E-mail diaz@ciens.ula.ve Tel. 58(74)401372 Fax 58(74)401286

Donat Reyes, Prof. Maximiliano José (1958). Professor. Depto. de Tecnología de Materiales, IUT-RC, Km 8 Carretera Panamericana, Miranda, Venezuela. [Apdo. 40347, Caracas 1040-A, Venezuela.] DES (Ecole Nat. Sup. Ceramique Industr., Limoges, France, 1984) ceramics. *Crystallinity silicates thermal_and_mechanical_properties*
Tel. 58(2)6823302x227 Fax 58(2)6812754

Freites, Prof. Juan Alfredo (1964). Professor. Esc. de Física y Matemática, Fac. de Ciencias - UCV, Ciudad Universitaria - Los Chaguaramos, Caracas 1050, Venezuela. [Apdo. 47586, Caracas 1041-A, Venezuela.] MSc (USB, 1993) materials science. *Electron_diffraction structure_determination quasicrystal*
E-mail cmeuv@dino.conicit.ve Tel. 58(2)6718378 Fax 58(2)2846645

García Morales, Lic. María Inmaculada (1966). Professional staff. Depto. de Producción, INTEVEP S.A., Los Teques - Miranda, Venezuela. [Apdo. 76343, Caracas 1070-A, Venezuela.] Lic. (UCV, 1991) chemistry. *Powder_diffraction cements*
Tel. 58(2)9087929 Fax 58(2)9087818

Gómez de Andérez, Prof. Dora (1954). Associate Professor. Secretary National Committee for Crystallography. Depto. de Química, Fac. de Ciencias - ULA, La Hechicera - Mérida, Mérida, Venezuela. [Apdo. 40 - Mérida, 5251-A Mérida, Venezuela.] PhD (U. York - UK, 1990) physics. *X-ray_diffraction synchrotron_radiation semiconductors*
E-mail gomez@ciens.ula.ve Tel. 58(74)401372 Fax 58(74)401286

Gómez Rodríguez, Prof. Leonir José (1958). Associate Professor. Depto. de Física - UNEG, Villa Asia - Calle China, Puerto Ordaz - Bolívar, Venezuela. [Apdo. 302 Puerto Ordaz, 8015-A Bolívar, Venezuela.] Lic. (UDO, 1985) physics. *Powder diffraction alloys coal*
Tel. 58(86)695061 58(86)695593 Fax 58(86)220296 58(86)228814

González, Lic. Aleida del Carmen (1960). Professional staff. Centro de Física - IVIC, Km 11 Carretera Panamericana, Miranda, Apdo. 21827 Caracas 1020-A, Venezuela. MSc (IVIC, 1993) physics. *Neutron_scattering magnetism computational_physics*
E-mail ale@ivic.ivic.ve Tel. 58(2)5011312 Fax 58(2)5011148

González, Dr Gema (1951). Professor. Esc. de Física y Matemática, Fac. de Ciencias - UCV, Ciudad Universitaria - Los Chaguaramos, Caracas 1050, Venezuela. [Apdo. 47586, Caracas 1041-A, Venezuela.] PhD (U. London - UK, 1981) materials science. *Electron_diffraction structure_determination quasicrystal*
E-mail cmeucv@dino.conicit.ve Tel. 58(2)6718378 Fax 58(2)2846645

González Rodríguez, Dr Oscar Victorio (1939). Professor. Dpto. de Física, Esc. de Ciencias - UDO Nucleo Sucre, Av. Universidad - Cerro Colorado, Cumaná - Sucre, Venezuela. [Apdo. 245, Cumaná 6101-A, Sucre, Venezuela.] Dr (U. Strasbourg - France, 1977) solid state physics. *Solid_state intercalation_compound*
E-mail nsucre@dino.conicit.ve Tel. 58(93)514256 Fax 58(93)514754

Granados, Lic. Maristela (1960). Professional Staff. Centro de Biofísica y Bioquímica-IVIC, Km 11 Carretera Panamericana Miranda, Venezuela. [Apdo. 21827, Caracas 1020-A, Venezuela.] Lic. (UCV, 1986) biology. *SAXS electron_microscopy image_processing contractile_systems*
E-mail granados@ivic.ivic.ve Tel. 58(2)5011231 Fax 58(2)5011093
Greaves, Dr Eduardo D. (1942). Professor. Depto. de Física, USB - Sarenejas Baruta, Caracas 1080-A, Venezuela. [Apdo. 89000, Caracas 1081-A, Venezuela.] PhD (London U. - UK, 1979) physics. *X-ray_diffraction X-ray_fluorescence minerals gamma_lasers*
E-mail egreaves@dino.conicit.ve Tel. 58(2)9633022 Fax 58(2)9623230
Guerrero, Lic. José Reinaldo (1950). Professional staff. Centro de Biofísica e Bioquímica - IVIC, Km 11 Carretera Panamericana Miranda, Venezuela. [Apdo. 21827, Caracas 1020-A, Venezuela.] MSc (IVIC, 1990) biology. *SAXS electron_microscopy image_processing muscles contractile_systems*
E-mail guerrero@ivic.ivic.ve Tel. 58(2)5011231 Fax 58(2)5011093
Khan, Prof. Alí (1936). Full Professor. Member National Comm. for Crystallography. Depto. de Química, Esc. de Ciencias - UDO Núcleo Sucre, Av. Universidad - Cerro Colorado, Cumaná - Sucre, Venezuela. [Apdo. 245, Cumaná 6101-A, Sucre, Venezuela.] PhD (Rensselaer Polytech. Inst. - USA, 1968) chemistry. *Semiconductors synthesis characterization powder_diffraction*
E-mail nsucre@dino.conicit.ve Tel. 58(93)653612x3402 Fax 58(93)652110
Laredo, Prof. Estrella (1940). Full Professor. Depto. de Física, USB, Sartenejas, Baruta, Caracas 1080, Venezuela. [Apdo. 89000, Caracas 1081-A, Venezuela.] Dr 3e cycle (U. Paris - France, 1965) crystallography. *Polymers crystal_defects*
E-mail elaredo@usb.ve Tel. 58(2)9063526 Fax 58(2)9063627
Lira-Olivares, Prof. Dr Joaquín (1937). Full Professor. Depto. de Ciencias de los Materiales - USB, Sartenejas - Baruta, Caracas 1081-A, Venezuela. [Apdo. 89000, Caracas 1081-A, Venezuela.] PhD (U. California - USA, 1974) engineering science. *Metallurgy materials material_sciences nanocrystalline_materials metalceram*
E-mail jlira@dino.conicit.ve Tel. 58(2)9633022 Fax 58(2)9621695 58(2)713424
Mahmoudi, Dr Mounir (1959). Professional staff. Lab. Geológico I.T. Aplicada, COR-POVEN S.A., Puerto La Cruz - Anzoátegui, Venezuela. [Apdo. 4326 Puerto La Cruz, Anzoátegui, Venezuela.] Dr (U. Paris-Sud - France, 1986) sedimentology. *Mineralogy petrography seismology sedimentology*
Tel. 58(81)603801 Fax 58(81)603700
Marcano Fermín, Dr Cenis María (1949). Associate Professor. Depto. de Química, Esc. de Ciencias - UDO Nucleo Sucre, Av. Universidad - Cerro Colorado, Cumaná, Venezuela. [Apdo. 245 - Cumaná, 6101-A Sucre, Venezuela.] Dr (U. Complutense - Spain, 1986) inorganic chemistry. *Powder_diffraction III-V_compounds semiconductors 4f_series_compounds*
E-mail nsucre@dino.conicit.ve Tel. 58(93)514256 58(93)515398 Fax 58(93)514754
Marcó Parra, Lic. Lué Merú (1964). Technician. Depto. de Física - USB, Sartenejas - Baruta, Caracas 1080, Venezuela. [Apdo. 89000, Caracas 1081-A, Venezuela.] Lic. (USB, 1991) chemistry. *Powder_diffraction X-ray_fluorescence ceramics rare-earth_compounds*
E-mail rtorre@usb.ve Tel. 58(2)9063590
Mateu Suay, Dr Leonardo (1939). Researcher. Centro de Biofísica e Bioquímica - IVIC, Km 11 Carretera Panamericana Miranca, Venezuela. [Apdo. 21827, Caracas 1020-A, Venezuela.] Dr (U. Paris - France, 1974) physics. *Membranes structure function myelin_structure myelin_structure_function demyelinating_diseases kidney_stone_crystallization*
E-mail lmateu@cbb.ivic.ve Tel. 58(2)5011454 Fax 58(2)5011455
Mendialdua Lecue, Dr Juan Angel (1942). Professor. Depto. de Física, Fac. de Ciencias - ULA, La Hechicera - Mérida, Mérida, Venezuela. [Apdo. 512, Mérida 5251, Venezuela.] DSc (U. Sci. et Tech. - Lille - France, 1983) physics. *AES LEED photoemission ISS*
E-mail juanm@ciens.ula.ve Tel. 58(74)401339 Fax 58(74)401286
Michinel Machado, Prof. José Luis (1951). Professor. Esc. de Física e Matemática, Fac. de Ciencias - UCV, Ciudad Universitaria - Los Chaguaramos - Caracas 1050, Venezuela. [Apdo. 47586, Caracas 1041-A, Venezuela.] MSc (IVIC, 1987) biophysics. *SAXS X-ray_diffraction*
E-mail jmichinel@dino.conicit.ve Tel. 58(2)6718378 Fax 58(2)6718378
Montes, Dr Arturo (1951). Associate Professor. Depto. de Química, IUT-RC, Km 8 Carretera Panamericana - Miranda, Venezuela. [Apdo. 40347, Caracas 1040-A, Venezuela.] DSc (U. Poitiers - France, 1987) catalysis. *Catalysis coke_structure zeolites molecular_sieves*
Tel. 58(2)6812467 Fax 58(2)6812754
Murgich, Dr Juan (1944). Researcher. Centro de Química - IVIC, Km 11 Carretera Panamericana - Miranda, Venezuela. [Apdo. 21827, Caracas 1020-A, Venezuela.] Dr (U. Córdoba - Argentina, 1974) physics. *Incommensurate_phase phase_transition compounds*
E-mail jmurgich@quimica.ivic.ve Tel. 58(2)5011302 Fax 58(2)5011091
Ng Lee, Prof. Yolanda Mi Lien (1954). Associate Professor. Member National Comm. for Cryst. Esc. de Física y Matemática, Fac. de Ciencias - UCV, Ciudad Universitaria - Los Chaguaramos, Caracas 1050, Venezuela. [Apdo. 47586, Caracas 1041-A, Venezuela.] MSc (UCV, 1992) physics. *Powder_diffraction EPR superconductors solid_state_physics*
E-mail yolanda@dino.conicit.ve Tel. 58(2)6629785 Fax 58(2)6629734

Padrón, Dr Raúl (1950). Researcher; senior investigator; Head, Structural Biology Laboratory. Structural Biology Laboratory, IVIC, Venezuela. [IVIC-Biofísica, Apdo. 21827, Caracas 1020-A, Venezuela.] PhD (IVIC, 1980) biology. *SAXS electron_microscopy image_processing muscles contractile_systems*
E-mail padron@ivic.ivic.ve Tel. 58(2)5011098 Fax 58(2)5011093
Rao, Dr Mokka Narasinga (1916). Professor. Member National Comm. for Crystallography. Coordinación de Postgrado UNEG, Edif. General de Seguros, Av. Las Américas, Puerto Ordaz - Bolívar, Venezuela. [Apdo. 302, Puerto Ordaz 8015-A, Bolívar, Venezuela.] PhD (Wisconsin U. - USA, 1952) engineering. *Metallurgical_process aluminium_alloys alloys*
Tel. 58(86)225504 Fax 58(86)225673
Rivera Ocando, Dr Angela Valentina (1945). Full Professor. Depto. de Química, Fac. de Ciencias - ULA, La Hechicera - Mérida, Mérida, Venezuela. [Apdo. 40 - Mérida, 5251-A Mérida, Venezuela.] PhD (U. Cambridge - UK, 1979) chemistry. *Information_science*
E-mail funmr@dino.conicit.ve Tel. 58(74)401372 Fax 58(74)401286
Rodríguez, Dr Alfonso (1951). Associate Professor. Depto. de Física, Fac. de Ciencias - ULA, La Hechicera - Mérida, Mérida, Venezuela. [Apdo. 40 Mérida, 5251-A Mérida, Venezuela.] Dr (U. Sci. et Tech. - Lille - France, 1989) materials science. *AES LEED photoemission ISS*
E-mail alfonso@ciens.ula.ve Tel. 58(74)401339 Fax 58(74)401286
Rodríguez de Aguirrezabala, Dr María Asunción (1948). Associate Professor. Dpto. de Química, Esc. de Ciencias - UDO Nucleo Sucre, Av. Universidad, Cerro Colorado, Cumaná - Sucre, Venezuela. [Apdo. 245, Cumaná 6101-A, Sucre, Venezuela.] DSc (U. Complutense - Spain, 1990) chemistry. *Powder_diffraction III-V_compounds semiconductors*
E-mail nsucre@dino.conicit.ve Tel. 58(93)514256 58(93)515368 Fax 58(93)514754
Rojas, Dr Carlos Eduardo (1950). Professor. Centro de Microscopía Electrónica, Fac. de Ciencias - UCV, Ciudad Universitaria, Los Chaguaramos, Caracas 1050, Venezuela. [Apdo. 47411, Caracas 1041-A, Venezuela.] PhD (U. London - UK, 1983) physics. *Electron_diffraction electron_microscopy surface_physics*
E-mail cmeucv@dino.conicit.ve Tel. 58(2)6624752 Fax 58(2)6624752
Rueda, Prof. Fulgencio (1948). Professor. Dpto. de Física, Fac. de Ciencias - ULA, La Hechicera - Mérida, Mérida, Venezuela. [Apdo. 40, Mérida 5251-A, Mérida, Venezuela.] Lic. (ULA, 1977) physics. *Minerals*
E-mail rueda@ciens.ula.ve Tel. 58(74)401339 Fax 58(74)401286
Salazar Arcay, Prof. Esmina Noelia (1958). Assistant Professor. Depto. de Química - UNEG, Villa Asia - Calle China, Puerto Ordaz - Bolívar, Venezuela. [Apdo. 302 Puerto Ordaz, 8015-A Bolívar, Venezuela.] Ing. (U. Carabobo - Venezuela, 1986) chemical engineering. *Powder_diffraction alloys metals aluminium_alloys*
Tel. 58(86)695061 Fax 58(86)220296 58(86)228814
Schorin, Dr Hasso (1940). Researcher. Centro de Química - IVIC, Km 11 Carretera Panamericana - Miranda, Venezuela. [Apdo. 21827, Caracas 1020-A, Venezuela.] Dr (U. Tübingen - Germany, 1972) geochemistry. *Powder_diffraction bauxite_mineralogy minerals characterization laterites*
E-mail hshorin%quimica@ivic.ivic.ve Tel. 58(2)5011401 Fax 58(2)5011350
Sojo, Dr Pedro Rafael (1945). Associate Professor. Esc. de Química, Fac. de Ciencias - UCV, Ciudad Universitaria, Los Chaguaramos, Caracas 1050, Venezuela. [Apdo. 47586, Caracas 1041-A, Venezuela.] PhD (UMIST - UK, 1984) catalysis. *Perovskites catalysis solid_state_chemistry*
Tel. 58(2)6626592 58(2)6627543 Fax 58(2)6627121
Suárez, Prof. Nery (1948). Full Professor. Depto. de Física, USB, Sartenejas, Baruta, Caracas 1080, Venezuela. [Apdo. 89000, Caracas 1081-A, Venezuela.] MSc (UMIST - UK, 1977) solid state electronics. *Polymers crystal_defects*
E-mail nsuarez@usb.ve Tel. 58(2)9063526 Fax 58(2)9063627
Urbina de Navarro, Dr Caribay Teresa (1952). Professor. Centro de Microscopía Electrónica, Fac. de Ciencias - UCV, Ciudad Universitaria, Los Chaguaramos, Caracas 1050, Venezuela. [Apdo. 47411, Caracas 1041-A, Venezuela.] Dr (UCV, 1988) chemistry. *Electron_diffraction electron_microscopy catalysts*
E-mail cmeucv@dino.conicit.ve Tel. 58(2)6624752 Fax 58(2)6624752
Vargas, Dr Rodolfo (1955). Researcher. Member Natl. Comm. for Crystallography. Centro de Biofísica y Bioquímica - IVIC, Km 11 Carretera Panamericana - Miranda, Venezuela. [Apdo. 21827, Caracas 1020-A, Venezuela.] Dr (U. Rennes - France, 1981) physics. *SAXS powder_diffraction membranes_structure lipids_structure*
E-mail rvargas@cbb.ivic.ve rvargas@dino.conicit.ve Tel. 58(2)5011455 Fax 58(2)5011455
Vonasek, Lic. Eva María (1953). Professional Staff. Centro de Biofísica y Bioquímica - IVIC, Km 11 Carretera Panamericana - Miranda, Venezuela. [Apdo. 21827, Caracas 1020-A, Venezuela.] MSc (IVIC - Venezuela, 1984) biochemistry. *SAXS biological_crystal proteins myelin*
E-mail evonasek@ivic.ivic.ve Tel. 58(2)5011454 Fax 58(2)5011455
Zorrilla, Mr Oscar Enrique (1942). Technician. Lab. Geológico El Chaure, COR-POVEN S.A. - Puerto La Cruz, Anzoátegui, Venezuela. [Apdo. 4326, Puerto La Cruz, Anzoátegui, Venezuela.] Technician (Esc. Técnica Industrial - Bolívar - Vzla., 1965) geology. *Petrography powder_diffraction minerals sedimentology*
Tel. 58(81)603891 Fax 58(81)603700

VIETNAM

Sub-Editor: D. M. Nghiep

An, Prof. Nguyen (1938). Associate professor. Dept. Physics, National University of Hanoi, DHTH 90 NguyenTrai, Hanoi, Vietnam. DSc (Humboldt-U. Berlin, 1984) metal physics. *Crystal_structure physical_property*
Tel. 84(4)584069 Fax 84(4)583061

Anh, Mrs Pham Lan (1960). Research Assistant. X-ray Laboratory, Institute of Industrial Chemistry, Vien HHCN 2 PhamNguLao, Hanoi, Vietnam. BSc (Hanoi U. Techn., 1982) electronics. *Structure minerals chemical_compounds*
Tel. 84(4)267500 Fax 84(4)256562

Ba, Prof. Hoang Trong (1938). Associate professor. Dept. Machinery, Pedagogical Institute of Thuduc, DHSP Thuduc, Vietnam. PhD (Polytechnic Inst. Lvow, 1975) physical metallurgy. *Structure_analysis mechanical_property metallic_materials inorganic_compounds*
Tel. 84(8)968641

Chi, Prof. Nguyen Van (1942). Associate professor. Materials Science Center, Hanoi University of Technology, DHBK Hanoi, Vietnam. PhD (Bergakademie Freiberg, 1977) metal physics. *Diffusion microanalysis*
Tel. 84(4)691332 Fax 84(4)692006

Chu, Dr Nguyen Huu (1940). Lecturer. Dept. Technology of Metals, Danang Polytechnic Institute, DHBK Danang, Vietnam. PhD (Bergakademie Freiberg, 1973) physical metallurgy. *Metallic_alloys*
Tel. 84(51)21224

Doan, Mr Nguyen Dai (1950). Research Assistant. Lab. Metallography, Institute of Ferrous Metallurgy, Vien LKDen, Thuongtin, Vietnam. BSc (Kim Il Sung U., 1972) metal physics. *Plastic_deformation metallic_structure X-ray_diffraction*
Tel. 84(34)53255

Duc, Dr Luong Hong (1949). Lecturer. Center of New Materials, Polytechnic Institute of HoChiMinh City, DHBK TP HCM, Vietnam. PhD (Polytechn. Inst. Leningrad, 1985) physical metallurgy. *Structure_analysis metallic_alloys*
Tel. 84(8)658687 Fax 84(48)650161

Duc, Mr Nguyen Van (1965). Research Assistant. Materials Science Center, Hanoi University of Technology, DHBK Hanoi, Vietnam. BSc (Inst. of Metals and Alloys, Moscow, 1989) physical metallurgy. *Computing X-ray_diffraction*
Tel. 84(4)691332 Fax 84(4)692006

Dzuong, Prof. Lecong (1940). Professor. AsCA councillor. Materials Science Center, Hanoi University of Technology, DHBK Hanoi, Vietnam. DSc (PH Halle, 1987) metal physics. *Physical_metallurgy X-ray_diffraction rapid_solidification electron_diffraction*
Tel. 84(4)691332 Fax 84(4)692006

Ha, Dr Cao The (1952). Lecturer. Dept. Physical Chemistry, National University of Hanoi, DHTH 19 LeThanhTong, Hanoi, Vietnam. PhD (Nat. U. Hanoi, 1992) kinetics and catalytic chemistry. *Kinetics catalytic_chemistry dental_materials*
Tel. 84(4)261854 Fax 84(4)583061

Hân, Prof. Trinh (1938). Associate professor. Dept. Geology, National University of Hanoi, DHTH 90 NguyenTrai, Hanoi, Vietnam. PhD (State U. Moscow, 1969) crystallography. *Structure minerals gemology*
Tel. 84(4)581421 84(4)581420 84(4)533534 Fax 84(4)583061

Hien, Dr Nguyen Van (1957). Research Assistant. Dept. Measurement and Material Control, Research Institute of Technology, Vien CN Tuliem, Hanoi, Vietnam. PhD (Minsk Polytechn. Inst., 1984) physical metallurgy and heat treatment. *Material_control*
Tel. 84(4)343757

Hien, Prof. Than Duc (1939). Professor; Director of the Research Institute. Cryogenic Laboratory, National University of Hanoi, DHTH 90 NguyenTrai, Hanoi, Vietnam. DSc (Warsaw U., 1991) magnetic materials. *Physical_property superconductors magnetic_materials materials_science*
Tel. 84(4)585281 84(4)692518 Fax 84(4)692963

Huan, Dr Le Van (1950). Research Assistant. Inst. of materials, National Center of Natural Sciences and Technology, TT KHTNCNQG, Hanoi, Vietnam. PhD (Inst. Solid State Physics, Moscow, 1986) solid state physics. *Structure semiconductors*
Tel. 84(4)352129 Fax 84(4)352483

Huynh, Dr Tran Dai (1937). Lecturer. Dept. Physics, National University of Hanoi, DHTH 90 NguyenTrai, Hanoi, Vietnam. PhD (State U. Moscow, 1975) metal physics. *Structure metal_physics solid_state*
Tel. 84(4)584069 Fax 84(4)583061

Khang, Prof. Quan Han (1937). Associate professor. Dept. Physics, University of HoChiMinh City, DHTH TP HoChiMinh, Vietnam. PhD (Leningrad U., 1966) crystallography. *Crystallography structure minerals*
Tel. 84(8)353193

Nghi, Prof. Nguyen Hoang (1944). Associate professor. Inst. Techn. Physics, Hanoi University of Technology, DHBK Hanoi, Vietnam. PhD (Inst. of Metals and Alloys, Moscow, 1983) metal physics. *Electron_microscopy X-ray_diffraction magnetic_amorphous_alloys*
Tel. 84(8)696949 Fax 84(4)692006

Nghiep, Prof. Do Minh (1945). Associate professor. Sub-Editor, WDC9. Materials Science Center, Hanoi University of Technology, DHBK Hanoi, Vietnam. PhD (TU Dresden, 1980) metal physics. *X-ray_diffraction plastic_deformation rapid_solidification magnetic_materials dental_materials*
Tel. 84(4)691332 Fax 84(4)692006

Ngoan, Dr Dao Cong (1953). Lecturer. Dept. Inorganic Chem., National University of Hanoi, DHTH 19 LeThanhTong, Hanoi, Vietnam. PhD (Wroclaw U., 1988) crystallography. *Structure crystallography*
Tel. 84(4)253503 Fax 84(4)583061

Nhieu, Prof. Pham Van (1948). Associate professor. Dept. Physical Chemistry, National University of Hanoi, DHTH 19 LeThanhTong, Hanoi, Vietnam. PhD (Bacu U., 1984) theoretical chemistry. *Structure physical_chemistry*
Tel. 84(4)261854 Fax 84(4)583061

Nhung, Dr Nguy Tuyet (1946). Lecturer. Dept. Geology, National University of Hanoi, DHTH 90 NguyenTrai, Hanoi, Vietnam. PhD (Leningrad U., 1979) crystallography. *Crystal_structure minerals*
Tel. 84(4)581421 Fax 84(4)583061

Phac, Dr Le Van (1952). Research Assistant. Dept. Metallurgical Techn., Research Institute of Technology, Vien CN Tuliem, Hanoi, Vietnam. PhD (Technical U. Praha, 1984) metallurgical technology. *Metallic_alloys*
Tel. 84(4)333512

Phuc, Prof. Nguyen Khoa (1939). Associate professor. Research Inst. of Technology, Vien CN Tuliem, Hanoi, Vietnam. PhD (Bergakademie Freiberg, 1974) physical metallurgy. *Structure diffusion*
Tel. 84(4)343758

Phuc, Dr Phan Vinh (1938). Group leader. Inst. of Materials, National Center of Natural Sciences and Technology, TT KHTNCNQG, Hanoi, Vietnam. PhD (Acad. of Sciences, Berlin, 1976) analytical chemistry. *Structure X-ray_fluorescence*
Tel. 84(4)352129 Fax 84(4)352483

Quan, Mr Do Hong (1958). Assistant. Dept. Silicates and Civil Engineering, Hanoi University of Technology, DHBK Hanoi, Vietnam. BSc (Hanoi U. Techn., 1981) inorganic chemistry. *Crystallography mineralogy silicates*
Tel. 84(4)692517 Fax 84(4)692006

Quy, Dr Le Cong (1949). Research Assistant. Inst. of Materials, Nat. Center of Natural Sciences and Technology, TT KHTNCNQG, Hanoi, Vietnam. PhD (Inst. of Crystallography, Moscow, 1986) crystallography. *Crystallography superconductors*
Tel. 84(4)352129 Fax 84(4)352483

Quyen, Dr Nguyen Hong (1950). Group leader. Inst. of Materials, National Center of Natural Sciences and Technology, TT KHTNCNQG, Hanoi, Vietnam. PhD (ZFW Dresden, 1980) metal physics. *Magnetic_materials structure superconductors*
Tel. 84(4)352129 Fax 84(4)352483

Sinh, Dr Nguyen Huy (1950). Lecturer. Cryogenic Lab., National University of Hanoi, DHTH 90 NguyenTrai, Hanoi, Vietnam. PhD (Nat. U. Hanoi, 1990) solid state physics. *Physical_property crystal_structure superconductors*
Tel. 84(4)585281 Fax 84(4)256562

Soa, Prof. Nguyen Dinh (1939). Associate professor. Dept. Chemistry, Polytechnic Institute of HoChiMinh City, DHBK TP HoChiMinh, Vietnam. PhD (Inst. Metallurgy Moscow, 1976) structure analysis. *Structure chemical_property inorganic_materials organic_compounds metallic_alloys*
Tel. 84(8)658687 Fax 84(48)650161

Soc, Prof. Le Nguyen (1937). Associate professor. X-ray Laboratory, Institute of Industrial Chemistry, Vien HHCN 2 PhamNguLao, Hanoi, Vietnam. MSc (Nat. U. Hanoi, 1964) crystallography. *Structure chemical_compounds*
Tel. 84(4)267500 Fax 84(4)256562

Sua, Dr Nguyen Van (1948). Group leader. Research Dept., Institute of Ferrous Metallurgy, Vien LKDen, Thuongtin, Vietnam. PhD (College of Mining and Metallurgy, Ostrava, 1986) physical metallurgy. *Plastic_deformation electron_diffraction metallic_alloys*
Tel. 84(34)53255

Tai, Dr Luu Tuan (1949). Lecturer. Cryogenic Lab., National University of Hanoi, DHTH 90 NguyenTrai, Hanoi, Vietnam. PhD (Nat. U. Hanoi, 1990) solid state physics. *Magnetic_property crystal_structure rare-earth_compound intermetallic_compound*
Tel. 84(4)585281 Fax 84(4)256562

Thang, Dr Tran Quoc (1949). Lecturer. Materials Science Center, Hanoi University of Technology, DHBK Hanoi, Vietnam. PhD (INP Grenoble, 1987) metal physics. *Electron_microscopy X-ray_diffraction aluminium_alloys*
Tel. 84(4)692003 Fax 84(4)692006

That, Prof. Ta Van (1938). Associate professor. Dept. Foundry and Heat Treatment, Hanoi University of Technology, DHBK Hanoi, Vietnam. PhD (Cracow Polytechnic Inst., 1972) physical metallurgy. *Phase_transformation metallic_alloys*
Tel. 84(4)692462 Fax 84(4)692006

Thiem, Prof. Lam Ngoc (1940). Associate professor. Dept. Physical Chemistry, National University of Hanoi, DHTH 19 LeThanhTong, Hanoi, Vietnam. PhD (U. Kharcow, 1968) theoretical and structural chemistry. *Structure*
Tel. 84(4)261854 Fax 84(4)583061

Thu, Dr Nguyen Thi (1947). Lecturer. Dept. Chemistry, Hanoi Teacher Training College, DHSP1 Tuliem, Hanoi, Vietnam. PhD (Teacher Training College Hanoi, 1991) physical chemistry. *Physical_property crystal_structure catalysts*
Tel. 84(4)344842 Fax 84(4)254679

Thu, Dr Trinh Le (1944). Vice director. Center of Geological Analysis and Test, TT PTTN diachat, Hanoi, Vietnam. PhD (State U. Moscow, 1984) crystal chemistry. *Structure minerals*
Tel. 84(4)583108 Fax 84(4)254734

Thuc, Prof. Dao Dinh (1930). Associate professor. Dept. Physical Chemistry, National University of Hanoi, DHTH 19 LeThanhTong, Hanoi, Vietnam. PhD (F. Schiller U. Jena, 1970) theoretical chemistry. *Structure X-ray_diffraction*
Tel. 84(4)261854 Fax 84(4)583061

Tri, Prof. Nguyen Van (1936). Professor. Inst. Techn. Physics, Hanoi University of Technology, DHBK Hanoi, Vietnam. DSc (TU Ilmenau, 1990) applied physics. *Molecular-electronic_structure dynamics magnetic_resonance disordered_materials*
Tel. 84(8)691671 Fax 84(4)692006

Trung, Mr Nguyen Tan (1950). Research Assistant. Center of Geological Analysis and Test, TT PTTN diachat, Hanoi, Vietnam. BSc (State U. Moscow, 1973) geology. *Microanalysis minerals*
Tel. 84(4)583108 Fax 84(4)254734

Truong, Dr Chu Thien (1946). Lecturer. Materials Science Center, Hanoi University of Technology, DHBK Hanoi, Vietnam. PhD (PH Halle, 1988) physical metallurgy. *X-ray_diffraction decomposition aluminium_alloys*

Tel. 84(4)691332 Fax 84(4)692006

Tuyen, Mr Nguyen Cuong (1948). Research Assistant. Center of Geological Analysis and Test, TT PTTN diachat, Hanoi, Vietnam. BSc (Nat. U. Hanoi, 1970) geology. *X-ray_diffraction minerals*
Tel. 84(4)583108 Fax 84(4)254734

Van, Prof. Dang Ung (1945). Associate professor. Dept. Physical Chemistry, National University of Hanoi, DHTH 19 LeThanhTong, Hanoi, Vietnam. DSc (Nat. U. Hanoi, 1991) physical chemistry. *Structure chemical_thermodynamics computer_simulation Monte_Carlo_treatment*
E-mail Hanoi@coombs.anu.edu.au Tel. 84(4)581282 Fax 84(4)583061

Vinh, Dr Ngo Van (1951). Research Assistant. Dept. Chemistry, Research Institute of Technology, Vien CN Tuliem, Hanoi, Vietnam. PhD (Technical U. Brno, 1987) materials technology. *Piezoelectric_ceramic*
Tel. 84(4)345672

Vong, Dr Vo (1941). Group leader. Inst. of Materials, National Center of Natural Science and Technology, Vien KHVatlieu TT KHTNCNQG, Hanoi, Vietnam. PhD (Centre of Electron Microscopy Res., Halle, 1974) metal physics. *Electron_microscopy structure*
Tel. 84(4)352129 Fax 84(4)352483

Vuong, Dr Nguyen Van (1950). Research Assistant. Inst. of Materials, National Center of Natural Science and Techn., Vien KHVatlieu TT KHTNCNQG, Hanoi, Vietnam. PhD (Humboldt U. Berlin, 1984) metal physics. *Structure semiconductors*
Tel. 84(4)352129 Fax 84(4)352483

NAME INDEX

Anstis, Dr Geoffrey Richard *Australia*.
Antipin, Prof. Mikhail Yu. *Russia*.
Antolić, Ms Snježana *Croatia*.
Antonio, Dr Mark *USA*.
Antonopoulos, Prof. Ioannis *Greece*.
Antonov, Dr Evgenii V. *Russia*.
Antoschenko, Dr Vladimir Stepanovich *Kazakhstan*.
Antosiac, Dr Boris Yakovlevich *Moldova*.
Antsishkina, Dr Alla A. *Russia*.
Antson, Olli Kalervo *Finland*.
Anugul, Mrs Surang *Thailand*.
Anulewicz, Dr Romana Joanna *Poland*.
Anwar, Dr Jamshed *UK*.
Anwar, Mr Muhammad *Pakistan*.
Anwar, Prof. Yehia *Egypt*.
Aoki, Prof. Dr Yoshikazu *Japan*.
Aoki, Prof. Katsuyuki *Japan*.
Aoki, Dr Yoshihira *Japan*.
Apinitis, Dr Smuidris *Latvia*.
Apostolov, Prof. Andrei *Bulgaria*.
Apostolov, Dr Anton *Bulgaria*.
Apreda, Dr M. Carmen *Spain*.
Aquilano, Prof. Dino *Italy*.
Aquino Hernández, Ing. Rosa Cruz *Venezuela*.
Arafa, Prof. Salah *Egypt*.
Aragón de la Cruz, Prof. Francisco *Spain*.
Araí, G. J. *USA*.
Arakcheeva, Dr Alla V. *Russia*.
Araki, Takaharu *USA*.
Aramburu, Mr Ibon *Spain*.
Arana, Prof. Rafael *Spain*.
Aranzabe, Mrs Ana *Spain*.
Arató, Dr Péter *Hungary*.
Aravindakshan, Dr Cheetambadi *India*.
Arce Sagüés, Dr Alejandro Jaime *Venezuela*.
Archer, Margarida *Portugal*.
Archer, Ronald D. *USA*.
Ardelean, Dr Ioan *Romania*.
Arents, Dr Gina *USA*.
Arevalo, Jairo H. *USA*.
Argay, Mr Gyula *Hungary*.
Arguello, Dr Zoraide Primerano *Brazil*.
Argyrakis, Prof. Panos *Greece*.
Arhar, Mr Andrej *Slovenia*.
Arif, Ms Iffat *Pakistan*.
Arima, Dr Yoshiyasu *Japan*.
Arimitsu, Dr Yutaka *Japan*.
Arinkin, Dr Oleksandr Victorovych *Ukraine*.
Arjunan, Dr P. Arjunan *USA*.
Armağan, Dr Nizamettin *Turkey*.
Armbruster, Dr Thomas Michael Ludwig *Switzerland*.
Armstrong, Prof. Richard N. *USA*.
Armstrong, Ronald W. *USA*.
Armstrong, Shelly R. *USA*.
Armstrong, Dr William H. *USA*.
Arndt, Dr U. W. *UK*.
Arni, Dr Raghuvir Krishnaswamy *Brazil*.
Arnold, Prof Dr Heinrich Günther Alfred *Germany*.
Arnold, Dr Edward *USA*.
Arnone, Dr Arthur *USA*.
Arnoux, Dr Bernardette *France*.
Aronin, Dr Alexander S. *Russia*.
Aronson, Hans-Erik G. *USA*.
Arora, Dr Narinder Kumar *India*.
Arora, Dr Satish K. *USA*.
Aroyo, Dr Mois *Bulgaria*.
Arreguín Espinosa, Dr Roberto Alejandro *Mexico*.
Arrieta, Prof. Juan Manuel *Spain*.
Arriortua, Prof. Maribel *Spain*.
Arryanto, Dr Yateman *Indonesia*.
Artioli, Prof. Gilberto *Italy*.
Artymiuk, Dr Peter J. *UK*.
Arvai, Andrew S. *USA*.
Arvanitis, Dr Georgia *USA*.
Arzt, Mrs Steffi *UK*.
Asada, Dr Eiichi *Japan*.
Asadchikov, Dr Victor E. *Russia*.
Asadov, Prof Ysif Gazanfar oglu *Azerbaijan*.
Asami, Harumi *Japan*.
Asaoka, Mr Hidehito *Japan*.

Ashida, Mr Atsushi *Japan*.
Ashida, Dr Sakichi *Japan*.
Ashida, Prof. Tamaichi *Japan*.
Aslamah, Stephen *UK*.
Askhabov, Dr Askhab M. *Russia*.
Aslanian, Dr Selma *Bulgaria*.
Aslanov, Prof. Leonid Aleksandrovich *Russia*.
Aspden, Dr Richard M. *UK*.
Astaf'ev, Dr Serge B. *Russia*.
Atanassov, Dr Vassil *Bulgaria*.
Athappilly, Dr Francis K. *USA*.
Atkinson, Dr David *USA*.
Atoji, Masao *USA*.
Atovmyan, Prof. Lev O. *Russia*.
Atria, Dr Ana María *Chile*.
Attard, Alfred E. *USA*.
Attfield, Dr J. Paul *UK*.
Attie, Prof. Abdelkader *Egypt*.
Atwood, Dr David *USA*.
Auffermann, Dr Gudrun *Germany*.
Augsburger, Ms Marta Susana *Argentina*.
Augustin, Director, Rolf M. *USA*.
Auleytner, Prof. Julian Jan *Poland*.
Auriemma, Dr Finizia *Italy*.
Authier, Prof. André *France*.
Avalos Borja, Prof. Miguel *Mexico*.
Averbuch-Pouchot, Dr Marie-Thérèse *France*.
Avey, Hugh Philip *Australia*.
Avilov, Dr Anatoly S. *Russia*.
Avramov, Dr Isak *Bulgaria*.
Axtell, III, Enos A. *USA*.
Aydın, Metin *Turkey*.
Aydınmuraz, Dr Arsın *Turkey*.
Aytaş, Dr S. Işık *Turkey*.
Azer, Dr Nazima *Egypt*.
Ba, Prof. Hoang Trong *Vietnam*.
Baak, Ing. Leonardus C. *Netherlands*.
Baars, Dr Jan W. *Germany*.
Baba, Dr Ibrahim *Malaysia*.
Babel, Prof. Dr Dietrich *Germany*.
Baber, Dr Nasim *Pakistan*.
Babu, Dr Y. S. *USA*.
Bacchi, Dr Alessia *Italy*.
Bachechi, Dr Fiorella *Italy*.
Bachet, Mr Bernard *France*.
Backhaus, Dr Karl-Otto *Germany*.
Bacon, Prof. George E. *UK*.
Badcock, Tracey Dianne *UK*.
Bader, Dr Yehia Abdelhamid *Egypt*.
Badger, John *USA*.
Badurek, Dr Gerald *Austria*.
Bae, Mr In Kook *Korea*.
Bärlocher, Dr Christian *Switzerland*.
Bärnighausen, Prof. Dr Hartmut *Germany*.
Baeyens, Katrien J. *USA*.
Bagchi, Prof. Subodh Nath *Canada*.
Bagdasarov, Dr Khatchik S. *Russia*.
Baggio, Dr Ricardo Fortunato *Argentina*.
Baggio, Dr Sergio *Argentina*.
Bagley, Arthur George *UK*.
Bagmut, Dr Oleksandr Grygorovych *Ukraine*.
Bahadur, Mr S. Asath *India*.
Bahar, Susanna *UK*.
Bahgat, Prof. Alaa Eldien *Egypt*.
Bai, Prof. Chunli *China*.
Bailey, Larry K. *USA*.
Bailey, Prof. Sturges *USA*.
Bailey, Dr Susan *UK*.
Baird, Dr Herbert Wallace *USA*.
Bajorath, Dr Jurgen *USA*.
Bakakin, Dr Vladimir V. *Russia*.
Baker, Anthony Thomas *Australia*.
Baker, Prof. Edward N. *New Zealand*.
Baker, Dr Patrick J. *UK*.
Baker, Tim S. *USA*.
Bąk-Misiuk, Dr Jadwiga *Poland*.
Bakshi, Eduard Nieolaevitch *Australia*.
Bakshi, Mr Pradip Kumar *Canada*.
Balaban, Dr Aneta *Romania*.
Balagurov, Dr Anatoly M. *Russia*.

Balaic, David Xavier *Australia*.
Balakirev, Dr Vladimir G. *Russia*.
Balasingh, Dr C. *India*.
Balasubramanian, Mr T. *India*.
Balcárek, Ing. Jan *Czech Republic*.
Balcázar, Dr José Luis *Spain*.
Balchin, Dr Anthony Arthur *UK*.
Baldrian, Dr Josef *Czech Republic*.
Baldwin, Dr Eric T. *USA*.
Baldwin, Kenneth J. *USA*.
Balen, Mr Dražen *Croatia*.
Balen, Ms Milka *Croatia*.
Balera Perez, Mr Miguel Angel *Mexico*.
Balian, Dr Minas K. *Armenia*.
Balić Žunić, Tonči *Denmark*.
Ball, Prof. Anthony *South Africa*.
Ball, Richard G. *USA*.
Balyunis, Dr Lyubov' E. *Russia*.
Balzar, Dr Davor *Croatia*.
Balzarotti, Prof. Adalberto *Italy*.
Bamzai, Mr Krishen Kumar *India*.
Ban, Nenad *USA*.
Banaszak, Leonard *USA*.
Bancroft, Daniel P. *USA*.
Bandoli, Prof. Giuliano *Italy*.
Banerjee, Dr Srikumar *India*.
Banić, Ms Zrinka *Croatia*.
Banner, Dr David William *Switzerland*.
Bannova, Ms Irina I. *Russia*.
Bansal, Dr Manju *India*.
Baqri, Dr Syed Rafiq-ul-Hasan *Pakistan*.
Barabanenkov, Dr Yury A. *Russia*.
Barakat, Prof. Nayel *Egypt*.
Baran, Ing. Peter *Slovakia*.
Baran, Dr Zbigniew *Brazil*.
Baranov, Prof. Anatolii I. *Russia*.
Baranskii, Prof. Konstantin N. *Russia*.
Baratova, Zebo *Tadjikistan*.
Barbagelata, Mr Franco *Chile*.
Barber, Patrick G. *USA*.
Barbier, Mr Bruno *Germany*.
Barbier, Dr Jacques *Canada*.
Barbieri, Prof. Renato *Italy*.
Barbo, Mr Martin *Slovenia*.
Barbour, Dr Leonard James *USA*.
Barbur, Prof. Ioan *Romania*.
Barelli, Dr Nilso *Brazil*.
Baricco, Dr Marcello *Italy*.
Barkigia, Dr Kathleen M. *USA*.
Barnea, Zwi *Australia*.
Barnes, Dr Charles L. *USA*.
Barnes, Dr Hazel A. *UK*.
Barnes, Dr John Conquest *UK*.
Barnes, Dr John D. *USA*.
Barnes, Prof. Paul *UK*.
Barnhart, Dr David M. *USA*.
Barreto Renteria, Mr Jorge *Mexico*.
Barrett, Mrs Stella B. *UK*.
Barrington-Leigh, Dr John *Canada*.
Barros Marques, Dr Maria Isabel *Portugal*.
Barrow, Dr M. J. *UK*.
Barry, John C. *Australia*.
Barsukova, Dr Marina L. *Russia*.
Barszcz, Prof. Edward *Poland*.
Bartczak, Prof. Tadeusz Jan *Poland*.
Bartell, Prof. Lawrence S. *USA*.
Bartelmehs, Dr Kurt L. *USA*.
Bartels, Mrs Heike *Germany*.
Bartels, Dr Matthias *Germany*.
Bartl, Prof. Dr Hans *Germany*.
Barton, Dr Randolph *USA*.
Barton, Prof. Richard J. *Canada*.
Bartoshynsky, Dr Sergij Mykolayovych *Ukraine*.
Bartoshynsky, Dr Volodymyr Zbignevych *Ukraine*.
Bartoshynsky, Prof. Zbignev Vladyslavovych *Ukraine*.
Baruchel, Dr José *France*.
Basha, Dr Ahmed Fouad *Egypt*.
Bassas, Mr Josep *Spain*.
Bassignana, Dr Isabella C. *Canada*.
Bassiouny, Prof. Mohamed Khafagi *Egypt*.

Basso, Prof. Riccardo *Italy.*
Basson, Dr Stephen Smuts *South Africa.*
Bastida, Dr Joaquín *Spain.*
Basto, Maria João Cabral de Oliveira *Portugal.*
Basu, Dr Sankar P. *USA.*
Batalia, Michael A. *USA.*
Batchelor, Dr Raymond John *Canada.*
Bats, Dr Jan Willem *Germany.*
Batsanov, Dr Andrei S. *Russia.*
Battaglia, Prof. Luigi Pietro *Italy.*
Batterman, Prof. Boris W. *USA.*
Battezzati, Prof. Livio *Italy.*
Bau, Dr Robert *USA.*
Baudet-Maze, Dr Mona *France.*
Bauer, Cary B. *USA.*
Bauer, Prof. Dr Günther *Austria.*
Baughman, Russell G. *USA.*
Baum, Dr Elke *Germany.*
Baumann, Dr Jürgen R. *Switzerland.*
Baumgarte, Dr Andreas *Switzerland.*
Baumgartner, Dr Oswald *Austria.*
Baures, Paul W *USA.*
Bautsch, Prof. Dr Hans-Joachim *Germany.*
Bavoux, Dr Claude *France.*
Bax, Dr Benjamin *UK.*
Baxter, Ian *UK.*
Bay, Prof. J. Young *Korea.*
Bayliss, Peter *Australia.*
Bayón, Dr J. Carlos *Spain.*
Bayya, Shyam S. *USA.*
Bazan, Dr J. Fernando *USA.*
Beagley, Dr Brian *UK.*
Beamer, Dr Lesa J. *USA.*
Beanland, Dr Richard *UK.*
Bear, Richard S. *USA.*
Beauchamp, Prof. André *Canada.*
Becherer, Prof. Dr Gerhard *Germany.*
Beck, Prof. Dr Horst P. *Germany.*
Beck, Prof. Dr Johannes *Germany.*
Becker, Prof. Dr Gerd *Germany.*
Becker, Dr Joseph W. *USA.*
Becker, Dr Paul *France.*
Bedarida, Prof. Federico *Italy.*
Bedarkar, Sudhir *USA.*
Beddell, Dr Christopher Raymond *UK.*
Beddoes, Roy L. *UK.*
Bednowitz, Dr Allan L. *USA.*
Bedzyk, Prof. Michael J. *USA.*
Beers, Dr William W. *USA.*
Beese, Lorena S. *USA.*
Beevers, Dr Cecil Arnold *UK.*
Behlke, Prof. Dr Joachim *Germany.*
Behm, Dr Helmut *Germany.*
Behrens, Prof. Dr Ulrich *Germany.*
Behrens, Dr Heinrich *Germany.*
Behrens, Dr Peter *Germany.*
Beiter, Dr Thomas A. *USA.*
Bekenova, Galiya Kabeshovna *Kazakhstan.*
Beketov, Dr Kajrat M. *Uzbekistan.*
Bekrenev, Dr Anatoly N. *Russia.*
Belakov, Dr Sergejs *Latvia.*
Belan, Dr Bogdana Dmytrivna *Ukraine.*
Bélanger, Ms Suzanne *Canada.*
Bélanger-Gariépy, Ms Francine *Canada.*
Belikova, Dr Galina S. *Russia.*
Belimenko, Dr Lyudmila D. *Russia.*
Bell, Anthony Martin Thomas *UK.*
Bell, Dr Jeffrey A. *USA.*
Bellamy, Henry D. *USA.*
Bello, Prof. Alfredo *Venezuela.*
Bellon, Prof. Pier Luigi *Italy.*
Bellotto, Dr Maurizio *Italy.*
Belluso, Dr Elena *Italy.*
Belokoneva, Dr Elena L. *Russia.*
Belotskil, Prof. Dmytro Petrovych *Ukraine.*
Belov, Dr Alexander Yu. *Russia.*
Belova, Dr Elizaveta Nikolayevna *Russia.*
Belt, Roger F. *USA.*
Beltran Abrego, Dr José Ramón *Brazil.*
Belugina, Dr Natalija B. *Russia.*

Belyavina, Dr Nadiya Mykolayivna *Ukraine.*
Bemm, Ulf *Sweden.*
Bendeliani, Dr Nikolai A. *Russia.*
Bender, Dr Hugo *Belgium.*
Benedetti, Prof. Ettore *Italy.*
Benedict, Dr Ulrich *Germany.*
Benetollo, Dr Franco *Italy.*
Benghiat, Dr Victor *Israel.*
Bengochea, Dr Amado Leandro *Argentina.*
Benna, Dr Piera *Italy.*
Bennema, Prof. Dr Pieter *Netherlands.*
Bennett, Dr Dennis W. *USA.*
Bennett, J. Michael *USA.*
Bennett, Melanie J. *USA.*
Bennett, Dr Pauline M. *UK.*
Bennett, Dr William *Germany.*
Benning, Dr Matthew M. *USA.*
Beno, Dr Mark A. *USA.*
Bensch, Dr Wolfgang *Germany.*
Bensimon, Dr Corinne *Canada.*
Benson, James E. *USA.*
Benson, Timothy E. *USA.*
Bentley, Dr Graham *France.*
de Benyacar, Prof. María Angélica Rodriguez *Argentina.*
Benz, Prof. Dr Klaus-Werner *Germany.*
Beran, Prof. Dr Anton *Austria.*
Berar, Dr Jean-François *France.*
Berendzen, Joel *USA.*
Beres, Dr John J. *USA.*
Beretka, Julius *Australia.*
Berezhkova, Dr Galina V. *Russia.*
Berg, Dr Liselotte *Germany.*
Berg, Rolf W. *Denmark.*
Berger, Dr Hans *Germany.*
Berger, Dr Rolf A. *Sweden.*
Bergerhoff, Prof. Dr Günter *Germany.*
Berghuis, Albert M. *USA.*
Berglund, Gunnar *Norway.*
Bergmann, Dr Ernst Michael *Canada.*
Bergunde, Dr Thomas *Germany.*
Berkeliev, Prof Amandurdy B. *Turkmenistan.*
Berkovitch-Yellin, Dr Ziva *Israel.*
Berliner, Prof. Lawrence *USA.*
Berliner, Ronald R. *USA.*
Berman, Dr Helen M. *USA.*
Berman, Dr Lonny E. *USA.*
Bermanec, Dr Vladimir *Croatia.*
Bermúdez-Polonio, Dr Joaquín *Spain.*
Bernal, Ivan *USA.*
Bernalte, Prof. Antoni *Spain.*
Bernardinelli, Dr Gérald *Switzerland.*
Bernhardt, Prof. Dr Wolfgang *Germany.*
Bernstein, Ms Frances C. *USA.*
Bernstein, Herbert J. *USA.*
Bernstein, Prof. Joel *Israel.*
Berry, Dr Amanda S. *UK.*
Berry, Chester R. *USA.*
Bersuker, Prof. Isaak Borukhovich *Moldova.*
Bertaut, Dr Erwin Felix *France.*
Berthet-Colominas, Dr Carmen *France.*
Berthold, Mr Thomas *Germany.*
Berti, Dr Giovanni *Italy.*
Bertolasi, Prof. Valerio *Italy.*
Berzina, Dr Inese *Latvia.*
Beskrovnyi, Dr Anatoly I. *Russia.*
Besoain, Dr Eduardo *Chile.*
Bessiere, Dr Michel *France.*
Besson, Dr Jean-Michel *France.*
Betai, Mr Badal Kumar *India.*
Betsofen, Dr Sergei Ya. *Russia.*
Betts, Laurie *USA.*
Betzel, Dr Christian *Germany.*
Beukes, Prof. Gerhardus Johannes *South Africa.*
Beurskens, Prof. Dr Paul T. *Netherlands.*
Beurskens, Dr Gezina *Netherlands.*
Beveridge, Dr David *UK.*
Bewley, Maria Christine *UK.*
Beyer, Dr Ir Jenö *Netherlands.*
Bezirganian, Dr Siranoush E. *Armenia.*
Bezirganyan, Dr Hakob P. *Armenia.*

Bezirganyan, Prof. Petros H. *Armenia.*
Bezjak, Prof. Dr Aleksandar *Croatia.*
Beznosikov, Dr Boris V. *Russia.*
Bhagwat, Dr Vasant *India.*
Bhaktapriya, Dr S. R. Y. *India.*
Bharati Rao, Ms T. *India.*
Bhat, Ms Sushma *India.*
Bhat, T. N. *USA.*
Bhatia, Dr Subhash Chander *India.*
Bhattacharya, Ms Archana *India.*
Bhattacharya, Dr Ramendranarayan *India.*
Bhattacherjee, Dr Satyananda *India.*
Bhatti, Ms Farzana Akhtar *Pakistan.*
Bhatti, Mr Muhammad Akram *Pakistan.*
Bi, Prof. Ruchang *China.*
Biagini-Cingi, Prof. Marina *Italy.*
Bianchet, Dr Mario A. *USA.*
Bianchi Orlandini, Dr Annabella *Italy.*
Bianchi, Dr Riccardo *Italy.*
Bienenstock, Arthur *USA.*
Bieniok, Dr Anna *Germany.*
Bigelow, Dr Wilbur C. *USA.*
Bigi, Prof. Adriana *Italy.*
Bigoli, Prof. Francesco *Italy.*
Bilderback, Dr Donald H. *USA.*
Billiet, Prof. Yves *Niger.*
Billing, Mr David Gordon *South Africa.*
Bilonizhka, Dr Petro Mykhajlovych *Ukraine.*
Binas, Dr Horst *Germany.*
Bingman, Dr Craig A. *USA.*
Bino, Prof. Avi *Israel.*
Bird, Craig M. *USA.*
Bird, Prof. Peter Hans *Canada.*
Birdsall, David L. *USA.*
Birdsall, William J. *USA.*
Birktoft, Dr Jens J. *USA.*
Birnbaum, Dr George I. *USA.*
Birnbaum, Dr Karin Bjámer *Canada.*
Bish, Dr David L. *USA.*
Bissert, Dr Gertrud *Germany.*
Bist, Dr B. M. S. *India.*
Biswas, Mr Gautam *India.*
Biswas, Dr Subhash Chandra *India.*
Bittencourt, Dr Diomar da Rocha *Brazil.*
Biyushkin, Prof. Victor Nikolayevich *Moldova.*
Bjørgo, Ingrid *Norway.*
Bjorkman, Dr Pamela J. *USA.*
Blaber, Dr Michael *USA.*
Blake, Dr Alexander John *UK.*
Blanc, Dr Eric *Switzerland.*
Blanke, Mrs Frauke *Germany.*
Blanton, Thomas N. *USA.*
Blaschko, Prof. Dr Oskar *Austria.*
Blasi, Prof. Achille *Italy.*
Blaton, Prof. Norbert Maurice *Belgium.*
Blažina, Dr Želimir *Croatia.*
Blessing, Robert H. *USA.*
Blevins, Dr Richard A. *USA.*
Blinc, Prof. Dr Robert *Slovenia.*
Blinov, Dr Lev M. *Russia.*
Blistanov, Prof. Alexander A. *Russia.*
Bliznakov, Prof. Georgi *Bulgaria.*
Blödner, Mr Ralph-Uwe *Germany.*
Blom, Dr Nick S. *Canada.*
Blomberg, Merja Kristiina *Finland.*
Bloomer, Dr Anne C. *UK.*
Blow, Emeritus Prof. David Mervyn *UK.*
Blüthgen, Mr Waldemar *Germany.*
Bluhm, Dr Karsten *Germany.*
Blundell, Prof. Tom Leon *UK.*
Bocchi, Dr Claudio *Italy.*
Bocelli, Dr Gabriele *Italy.*
Bochkarev, Dr Alexey V. *Russia.*
Bodak, Prof. Oksana Ivanivna *Ukraine.*
Bode, Dr Wolfram *Germany.*
Bodot, Prof. Hubert *France.*
Böcskei, Dr Zsolt *Hungary.*
Bögge, Dr Hartmut *Germany.*
Böhm, Prof. Dr Horst *Germany.*
Boehm, Dr James M. *Australia.*

Boehnke, Dr Undine-Constanze *Germany.*
Boese, Dr Roland *Germany.*
Boesman, Prof. Etienne Roland *Belgium.*
Boeyens, Prof. Jan Christoffel Antonie *South Africa.*
Bogdanović, Mr Goran *Serbia.*
Bohatý, Prof. Dr Ladislav *Germany.*
Bohm, Prof. Dr Joachim *Germany.*
Bohnen, Mr Frank Michael *Germany.*
Bohr, Jakob *Denmark.*
Bojarski, Prof. Zbigniew *Poland.*
Bokiy, Prof. Georgii B. *Russia.*
Bokd, Prof. Tadeusz *Poland.*
Bole, Mrs Meta *Slovenia.*
Bolin, Prof. Jeffrey T. *USA.*
Boller, Prof. Dr Herbert *Austria.*
Bollmann, Dr Ulrich *Germany.*
Bolognesi, Prof. Martino *Italy.*
Bolotina, Ms Nadezhda B. *Russia.*
Bolt, Rindert *Sweden.*
Bolte, Dr Michael *Germany.*
Bombicz, Dr Petra *Hungary.*
Bombieri, Prof. Gabriella *Italy.*
Bonamartini-Corradi, Prof. Anna *Italy.*
Bonamico, Dr Mario *Italy.*
Bonazzi, Dr Paola *Italy.*
Bond, Andrew H. *USA.*
Bond, Charles Simon *UK.*
Bond, Elizabeth Rivette *USA.*
Bondar, Dr Anatolij Adolfovych *Ukraine.*
Bondarenko, Dr Vladimir I. *Russia.*
Bondareva, Dr Olga S. *Russia.*
Bondars, Dr Bruno *Latvia.*
Bonefačić, Prof. Dr Antun M. *Croatia.*
Boneto, Ms Rita *Argentina.*
Bonev, Dr Ivan *Bulgaria.*
Bonham, Russell A *USA.*
Bonin, Dr Michel G. *Switzerland.*
Bonnet, Dr Michel *France.*
Bonpunt, Dr Louis *France.*
Bonse, Prof. Dr Dr hc. Ulrich *Germany.*
Bonyuet Lee, Prof. Dickar *Venezuela.*
Booth, Dr Andrew Donald *Canada.*
Bora, Mr Mohendra Nath *India.*
Borbély, Mr Andrés *Hungary.*
Borchardt-Ott, Dr Walter *Germany.*
Bořecký, Ing. Karel *Czech Republic.*
Borges, Prof. Frederico P. S. Sodré *Portugal.*
Borgstahl, Dr Gloria E. O. *USA.*
Borhani, Dr David W. *USA.*
Borisov, Prof. Stanislav V. *Russia.*
Borisova, Dr Svitlana Serafymivna *Ukraine.*
Borkakoti, Dr Nivedita Neera *UK.*
Borodajenko, Ms Natalya *Latvia.*
Borodin, Dr Vadim L. *Russia.*
Borowiak, Prof. Teresa *Poland.*
Bortel, Mr Gábor *Hungary.*
Borzęcka-Prokop, Dr Barbara *Poland.*
Bosch Giral, Prof. Pedro *Mexico.*
Bose, Mr Shyamal Kumar *India.*
Boskey, Adele L. *USA.*
Bosman, Drs Wilhelmus P. J. H. *Netherlands.*
Bossart Whitaker, Dr Patricia J. *USA.*
Bostanov, Dr Vesselin *Bulgaria.*
Boström, Dr Nils Dan *Sweden.*
Botnariuk, Dr Vasilii Mihai *Moldova.*
Botoshansky, Dr Mark *Israel.*
Bott, Raymond Clinton *Australia.*
Bottomley, Prof. Frank *Canada.*
Bouckaert, Ms Julie *Belgium.*
Bouillot, Prof. Jacques *France.*
Bourne, Dr Philip E. *USA.*
Bourne, Dr Susan Ann *South Africa.*
Bourosh, Dr Pavlina Nikolai *Moldova.*
Bourret, Dr Alain *France.*
Bovin, Jan-Olov *Sweden.*
Bovio, Prof. Bruna *Italy.*
Bowen, Prof. Alun W. *UK.*
Bowen, Prof. D. Keith *UK.*
Bowie, Dr James U. *USA.*
Bowman, Allen L. *USA.*

Boyarskaya, Prof. Yuliya Stanislavovna *Moldova.*
Boyington, Jeffrey C. *USA.*
Boys, Dr Bill *UK.*
Boys, Dr Daphne *Chile.*
Boysen, Dr Hans H. *Germany.*
Bozopoulos, Prof. Anastasios *Greece.*
Bozukov, Latchezar Nikolov *Bulgaria.*
Braam, Dr Adrianus W. M. *Netherlands.*
Bracke, Dr Ben R. F. *UK.*
Braden, Dr Bradford C. *USA.*
Brádler, Ing. Jaroslav *Czech Republic.*
Brady, John *USA.*
Brady, Dr Robert Leo *UK.*
Braga, Prof. Dario *Italy.*
Bragg, Dr Robert H. *USA.*
Bram, Mr Andreas *Germany.*
Brammer, Lee *USA.*
Brand, Prof. Dr Paul *Germany.*
Brandhuber, Barbara *USA.*
Brandmüller, Prof. Dr Josef *Germany.*
Brandon, Dr James Kenneth *Canada.*
Brandstätter, Dr Franz *Austria.*
Brante, MSc Inta *Latvia.*
Brasseur, Prof. Henri Alphonse Lambert *Belgium.*
Brayer, Prof. Gary David *Canada.*
Brehm, Lotte *Denmark.*
Breiter, Deborah R. *USA.*
Breiter, Prof. Dr Manfred W. *Austria.*
Bremere, Ms Ingrida *Latvia.*
Brenčič, Prof. Dr Jurij V. *Slovenia.*
Brendel, Mr Uwe *Germany.*
Brennan, Dr Richard G. *USA.*
Brennan, Dr Sean M. *USA.*
Brenner, Dr Charles *USA.*
Brenner, Mr. Stephen A. *USA.*
Bresciani Pahor, Prof. Nevenka *Italy.*
Brese, Dr Nathaniel E. *USA.*
Brezgunov, Dr Mark *Latvia.*
Březina, Ing. Bohuslav *Czech Republic.*
Briansó, Prof. José Luis *Spain.*
Bridson, Dr John N. *Canada.*
Brigatti, Prof. Maria Franca *Italy.*
Bright, Dr Alan *UK.*
Brinen, Dr Linda S. *USA.*
Brinkmann, Prof. Detlef *Switzerland.*
Brisander, Mr Magnus *Sweden.*
Brisse, Dr François *Canada.*
Bristoti, Dr Anildo *Brazil.*
Brito, Dr Ivan *Chile.*
Britten, Dr James Francis *Canada.*
Britton, Doyle *USA.*
Britton, Dr K. Linda *UK.*
Broach, Dr Robert W. *USA.*
Brock, Prof. Carolyn Pratt *USA.*
Brokmeier, Dr Heinz-Günther *Germany.*
Bronger, Prof. Dr Welf *Germany.*
Bronowska, Dr Wiesława *Poland.*
Bronsveld, Dr Paul M. *Netherlands.*
Brooks Jr, Dr Frederick P. *USA.*
Brown, Dr David Summers *UK.*
Brown, Dr Ian David *Canada.*
Brown, Dr Jerry H. *USA.*
Brown Jr, Prof. Gordon E. *USA.*
Brown, Leo D. *USA.*
Brown, Prof. Michael Ewart *South Africa.*
Brown, Dr Raymond S. *USA.*
Browner, Dr Michelle F. *USA.*
Brożek-Mucha, Dr Zuzanna *Poland.*
Brückner, Prof. Sergio *Italy.*
Brüderl, Mr Georg *Germany.*
Bruck, Dr Michael A. *USA.*
Bruins Slot, Dr Hilbert J. *Netherlands.*
Bruk, Dr Leonid Izmail *Moldova.*
Brumberger, Harry *USA.*
Brunger, Dr Axel T. *USA.*
Brunie, Dr Simone *France.*
Bruno, Prof. Emiliano *Italy.*
Bruno, Prof. Giuseppe *Italy.*
Bruno, Dr Ian J. *UK.*
Bruns, Christopher M. *USA.*

Bruque, Dr Sebastián *Spain.*
Bruskov, Dr Valery A. *Russia.*
Bruvo, Mr M. *Croatia.*
Bruzzone, Prof. Giacomo *Italy.*
Bryan, Clinton D. *USA.*
Bryan, Robert F. *USA.*
Bryant, Garold L. *USA.*
Bryant, Dr Patrick Kevin *UK.*
Bryant, Dr Peter John *Australia.*
Bryant, Dr Stephen H. *USA.*
Bryden, John H. *USA.*
Bryntse, Dr Ingrid *Sweden.*
Bu, Dr Xianhui *USA.*
Bubnova, Dr Rimma S. *Russia.*
Buchal, Dr Antonín *Czech Republic.*
Buchanan, Dr David R. *USA.*
Buchwald, Vagn Fabritius *Denmark.*
Bucio Galindo, Mr Lauro *Mexico.*
Buck, Prof. Dr Peter *Germany.*
Budai, John D. *USA.*
Budevski, Prof. Evgeni *Bulgaria.*
Budurov, Prof. Stoyan *Bulgaria.*
Bühmann, Dr Dieter *South Africa.*
Bürgi, Prof. Hans-Beat *Switzerland.*
Böttner, Dr Wolfgang *Germany.*
Büyükgüngör, Dr Orhan *Turkey.*
Bugaichuk, Dr Mykola Trokhymovych *Ukraine.*
Bugg, Dr Charles E. *USA.*
Bukin, Dr Alexander S. *Russia.*
Bukovec, Prof. Dr Nataša *Slovenia.*
Bukovec, Prof. Dr Peter *Slovenia.*
Bukowska-Strzyżewska, Prof. Maria *Poland.*
Bukvetsky, Dr Boris V. *Russia.*
Bulanova, Dr Marina Vadymivna *Ukraine.*
Bulka, Dr Genrikh R. *Russia.*
Bullock, Dr James F. *UK.*
Bullock, Timothy L. *USA.*
Bunge, Prof. Dr Dr hc. Hans-Joachim *Germany.*
Bunick, Dr Gerard J. *USA.*
Bunina, Dr Olga A. *Russia.*
Bunning, Dr John David *UK.*
Bunning, Dr Timothy J. *USA.*
Buranbaev, Mels Zhakanovich *Kazakhstan.*
Burbank, Dr Robinson D. *USA.*
Burgázy, Dr Frank *Germany.*
Burggraf, Prof. Charles *France.*
Burggraf, Dr Christiane *France.*
Buri, Dr Stavri *Albania.*
Burkel, Prof. Dr Eberhard *Germany.*
Burkhard, Dr Andreas *Switzerland.*
Burkhart, Brian M. *USA.*
Burla, Dr Maria Cristina *Italy.*
Burley, Stephen K. *USA.*
Burmeister, Dr Wilhelm Pascal *USA.*
Burnasheva, Dr Veniana V. *Russia.*
Burnett, Dr Roger M. *USA.*
Burnham, Prof. Charles *USA.*
Burns, Dr John H *USA.*
Burschka, Dr Christian *Germany.*
Burshtein, Dr Izya F. *USA.*
Bursill, Leslie A. *Australia.*
Burzlaff, Dr Hans *Germany.*
Burzo, Prof. Emil *Romania.*
Busaracome, Mr Suwin *Thailand.*
Buschmann, Dr Jürgen F. *Germany.*
Busetta, Dr Bernard *France.*
Busetti, Prof. Vilma *Italy.*
Bushnell, Prof. Gordon William *Canada.*
Bushnell-Wye, Dr Graham *UK.*
Bushuev, Prof. Vladimir A. *Russia.*
Busing, Dr William R. *USA.*
Bussien Gaillard, Mr Valérie *Switzerland.*
Butashin, Dr Andrey V. *Russia.*
Butcher, Ray J. *USA.*
Butikova, Dr Irina K. *Russia.*
Butler, Dr Brent Dennis *Australia.*
Butt, Dr Khurshid Alam *Pakistan.*
Butt, Mr Muhammad Hafeez *Pakistan.*
Butt, Mr Nasir Parvaiz *Pakistan.*
Butt, Dr Noor Mohammad *Pakistan.*

Buyanov, Dr Yurij Ivanovych *Ukraine.*
Buyers, FRSC, Dr William James Leslie *Canada.*
Buyum, Muharrem *Turkey.*
Buzatu, Dan A. *USA.*
Bye, Dr Erik *Norway.*
Byram, Susan K. *USA.*
Byriel, Karl Alwyn *Australia.*
Byrn, Dr Stephen R. *USA.*
Caballero, Prof. M. Antonio *Spain.*
Cabanillas, Mr Edgardo Domingo *Argentina.*
Cabeza, Dr Javier A. *Spain.*
Cabré, Mr Roger *Spain.*
Cachau, Dr Raul E. *USA.*
Caffrey, Assoc. Prof. Martin *USA.*
Caglioti, Prof. Giuseppe *Italy.*
Cahn, Dr John W. *USA.*
Cahn, Prof. Robert W. *UK.*
Cai, Dr Guanliang *China.*
Cai, Li *USA.*
Caira, Prof. Mino Rodolfo *South Africa.*
Cairati, Dr Paolo A. *Italy.*
Calamiotou, Prof. Maria S. *Greece.*
Calandra, Peter M. *USA.*
Calestani, Prof. Gianluca *Italy.*
Callebaut, Dr Isabelle *France.*
Callegari, Dr Athos *Italy.*
Callens, Dr Freddy Johan *Belgium.*
Calleri, Prof. Mariano *Italy.*
Calligaris, Prof. Mario *Italy.*
Calvet, Dr Teresa *Spain.*
Calzadilla, Prof. Octavio *Cuba.*
Camalli, Dr Mercedes *Italy.*
Camerman, Dr Arthur *USA.*
Camerman, Dr Norman *Canada.*
Cameron Jr, Robert P. *USA.*
Cameron, Prof. Theodore Stanley *Canada.*
Camillo, Prof. Francesco Emilio *Venezuela.*
Cammenga, Dr Heiko Karl *Germany.*
Campana, Dr Charles F. *USA.*
Campanelli, Dr Anna Rita *Italy.*
Campbell, Dr John W. *UK.*
Campbell, Dr Robert Laurence *Canada.*
Campelo Farias, Prof. Carlinda *Brazil.*
Campobasso, Nino *USA.*
Campos, Dr Cícero *Brazil.*
Candeloro De Sanctis, Prof. Sofia *Italy.*
Caneiro, Dr Alberto *Argentina.*
Canepa, Dr Horacio Ricardo *Argentina.*
Canizares, Mr Hian *Cuba.*
Cannas, Prof. Mario *Italy.*
Cannillo, Dr Elio *Italy.*
Cantrell, Joseph S. *USA.*
Cao, Dr Rong *China.*
Capasso, Prof. Sante *Italy.*
Capelle, Dr Bernard *France.*
Čapková, Doc. Dr Pavla *Czech Republic.*
Čaplovič, Ing. Ľubomír *Slovakia.*
Čaplovičová, Ing. Mária *Slovakia.*
Capo, Dr Luis *Cuba.*
Capparelli, Dr Mario Vicente *Venezuela.*
Cappuccio, Dr Giorgio *Italy.*
Caranoni, Prof. Claude *France.*
Carbonin, Prof. Susanna *Italy.*
Carbonio, Dr Raul Ernesto *Argentina.*
Cardellach, Dr Esteve *Spain.*
Cardin, Dr Christine *Ireland.*
Cardoso de Lima, Dr João *Brazil.*
Carducci, Michael D. *USA.*
Cargill III, G. Slade *USA.*
Carlile, Dr Colin *UK.*
Carlson, Ernest H. *USA.*
Carlson, Stefan *Sweden.*
Carlsson, Anna B. V. *Sweden.*
Carnasciali, Dr Maria M. *Italy.*
Carpenter, Gene B. *USA.*
Carpenter, John M. *USA.*
Carperos, Dr Vasili *USA.*
Carr, Martin J. *USA.*
Carr, Paul David *Australia.*
Carrell, Dr H. L. *USA.*

Carrera García, Ms Luz Maria *Mexico.*
Carriedo, Dr Gabino-Alejandro *Spain.*
Carroll, Dr Patrick J. *USA.*
Carrondo, Prof. Maria Armenia *Portugal.*
Carson, Dr William Michael *USA.*
Carter, Charles W. Jr. *USA.*
Cartwright, Dr Michael *UK.*
Cartz, Dr Louis *USA.*
Caruso, Dr Francesco *Italy.*
Carvalho, Dr Carlos A. M. *Brazil.*
Casabo, Prof. Jaume *Spain.*
Casalone, Dr Gianluigi *Italy.*
Casanova Batista, Dr Rodrigo *Venezuela.*
Casanova, Mr Jorge Ramón *Argentina.*
Cascales, Dr Concepcion *Spain.*
Cascarano, Dr Giovanni Luca *Italy.*
Cashion, John Dixon *Australia.*
Cassedane, Dr Jeannine *Brazil.*
Castanhola Batista, Antonio Adriano *Portugal.*
Castellano, Dr Eduardo Ernesto *Brazil.*
Castellanos Guzman, Dr A. Guillermo *Mexico.*
Castellanos Román, Mrs Maria Asunción *Mexico.*
Castellari, Prof. Carlo *Italy.*
Castiñeiras, Prof. Alfonso *Spain.*
Castleden, Dr Ian R. *UK.*
Castro, Dr Alicia *Spain.*
Castro, Prof. Baltazar de *Portugal.*
Caticha Ellis, Prof. Stephenson *Brazil.*
Catlow, Prof. C. Richard A. *UK.*
Catti, Prof. Michele *Italy.*
Caucia, Dr Franca *Italy.*
Cavarelli, Dr Jean *France.*
Caveney, Dr Robert John *South Africa.*
Čeh, Dr Boris *Slovenia.*
Čeh, Dr Miran *Slovenia.*
Čejka, Ing. Jiří *Czech Republic.*
Čejková, Dr Iva *Czech Republic.*
Celikel, Reha *USA.*
Cellai, Dr Luciano *Italy.*
Celotti, Dr Giancarlo *Italy.*
Cendon, Mr Dionisio *Spain.*
Cenzual, Dr Karin *Switzerland.*
Čepera, Dr Milan *Czech Republic.*
Černák, Dr Juraj *Slovakia.*
Čerňanský, Ing. Marian *Czech Republic.*
Cernik, Dr Robert Joseph *UK.*
Černý, Dr Petr *Canada.*
Černý, Dr Radovan *Czech Republic.*
Cerpa, Mr Arisbel *Cuba.*
Cerrini, Dr Silvio *Italy.*
Cervantes-Lee, Dr Francisco J. *USA.*
Červeň, Dr Ivan *Slovakia.*
Červinka, Dr Ladislav *Czech Republic.*
Češljević, Dr Valerija *Serbia.*
Čevelík, Dr Radomír *Czech Republic.*
Chaban, Dr Nadiya Fedorivna *Ukraine.*
Chacko, Dr K. K. *India.*
Chacko, Dr Susan *USA.*
Chadha, Dr Gopal Krishan *India.*
Chadha, Raj *USA.*
Chae, Mr Soo Chun *Korea.*
Chaichit, Dr Narongsak *Thailand.*
Chaikum, Dr Nitirampai Latavalya *Thailand.*
Chaikum, Dr Nopadol *Thailand.*
Chakoumakos, Bryan C. *USA.*
Chakrabarti, Dr Chandra *India.*
Chakrabarti (Chatterjee), Dr Mrs Chandana *India.*
Chakrabarti, Dr Pinak *India.*
Chakrabarty, Mr Dipak Kumar *India.*
Chakrabarty, Mr Subhasis *India.*
Chakrabarty, Mr Sugoto *India.*
Champion, Dr John A. *UK.*
Champness, Dr Pamela *UK.*
Chandra, Dhanesh *USA.*
Chandra, Naveen *USA.*
Chandrasegaran, Srinivasan *USA.*
Chandrasekaran, Dr Rengaswami *USA.*
Chandrasekhar, Dr K. *USA.*
Chandrasekhar, Prof. Sivaramakrishna *India.*
Chandrasekharaiah, Dr M. N. *India.*

Chandross, Dr Ronald J. *USA.*
Chaney, Dr Michael O. *USA.*
Chang, ChiehYing Y. *USA.*
Chang, Chong-Hwan *USA.*
Chang, Hou-Cheng *Taiwan.*
Chang, Ning-Leh *USA.*
Chang, Prof. Shih-Lin *Taiwan.*
Chang, Prof. Wenrui *China.*
Chantal, Abergel *USA.*
Chapman, Dr Michael S. *USA.*
Chapuis, Prof. Gervais *Switzerland.*
Charland, Dr Jean-Pierre *Canada.*
Chasen, Edith *USA.*
Chashchinov, Dr Yury M. *Russia.*
Chastain Jr, Roser V. *USA.*
Chatterjee, Ajoy *Australia.*
Chatterjee, Dr Amitava *India.*
Chatterjee, Dr Sanat Kumar *India.*
Chattopadhyay, Dr Debasish *India.*
Chattopadhyay, Dr Debasish *USA.*
Chattopadhyay, Dr Tapan *France.*
Chaudhary, Dr Abdul Majid *Pakistan.*
Chaudhary, Mr G. Sarwar Alam *Pakistan.*
Chaudhry, Mr Mohammad Anwar *Pakistan.*
Chaudhry, Dr Muhammad Nawaz *Pakistan.*
Chaudhuri, Dr Ahindra Kumar *India.*
Chauhan, Mr Ehsanul Haq *Pakistan.*
Chaves, Prof. Maria Renata *Portugal.*
Chayen, Dr Naomi E. *UK.*
Cheang, Dr Kok Keong *Malaysia.*
Cheer, Dr Clair J. *USA.*
Cheetham, Prof. Anthony K. *USA.*
Cheetham, Dr Graham Mark Thomas *UK.*
Chelliah, Dr Mahadevan *India.*
Chen, Prof. Benming *China.*
Chen, Dr Celia *USA.*
Chen, Haydn H. *USA.*
Chen, Dr Jie *Canada.*
Chen, Prof. Liquan *China.*
Chen, Lirong *USA.*
Chen, Dr Longyin *USA.*
Chen, Prof. Minqin *China.*
Chen, Prof. Ruey-Hong *Taiwan.*
Chen, Prof. Shizhi *China.*
Chen, Dr Shun-Le *USA.*
Chen, Dr Wei *Malaysia.*
Chen, Xin *USA.*
Chen, Mr Yu Wai *UK.*
Chen, Zhiwei *USA.*
Chen, Dr Zhongguo *USA.*
Cheng, Mr. Beisong *USA.*
Cheng, Mr Graham Cheng-Hsun *Hong Kong.*
Cheng, Mr. Hei-Ying *Taiwan.*
Cheng, Ms Ming-Chu *Taiwan.*
Cheng, Dr Pei-Tak *Canada.*
Cheng, Dr R. Holland *USA.*
Cheng, Dr Xiaodong *USA.*
Chenite, Dr Abdellatif *Canada.*
Cherbavaz, Dr Diana B. *USA.*
Cheremskij, Dr Petro Grygorovych *Ukraine.*
Chernega, Dr Oleksandr Mykolayovych *Ukraine.*
Chernov, Prof. Alexander A. *Russia.*
Chernov, Dr Alexander N. *Russia.*
Chernyshev, Dr Vladimir Vasilievich *Russia.*
Chernysheva, Dr Marina A. *Russia.*
Chetal, Prof. Amritlal *India.*
Cheung, Dr Kung Kai *Hong Kong.*
Chevichelov, Mr Victor A. *Russia.*
Chevrier, Dr Bernard *France.*
Chi, Prof. Nguyen Van *Vietnam.*
Chi, Young-in *USA.*
Chiang, Professor Michael Yen-Nan *Taiwan.*
Chiari, Prof. Giacomo *Italy.*
Chidambaram, Dr Rajagopala *India.*
Chidester, Connie *USA.*
Chieh, Prof. Chung *Canada.*
Chiesi-Villa, Prof. Angiola *Italy.*
Chikaura, Dr Yoshinori *Japan.*
Childs, Dr Cyril Walter *New Zealand.*
Chinnakali, Mr K. *India.*

Chippindale, Dr Ann Mary *UK.*
Chiragov, Prof. Mamed Isa oglu *Azerbaijan.*
Chirgadze, Dr Nickolay Y. *USA.*
Chirgadze, Prof. Yuri Nickolaevich *Russia.*
Chisholm, Dr James Edwin *UK.*
Chiu, Kuan-Cheng Chiu *Taiwan.*
Chiu, Professor Wah *USA.*
Choe, Prof. Hui-Woog *Korea.*
Choe, Dr Senyon *USA.*
Choi, Dr Chang Sun *USA.*
Choi, Prof. Deok *Korea.*
Choi, Prof. Duck Kyun *Korea.*
Choi, Jungwon *USA.*
Chomiller, Dr Jacques *France.*
Chook, Yuh Min Chook *USA.*
Choosang, Mrs Pilai *Thailand.*
Chornik, Dr Boris *Chile.*
Choudhary, Dr Muhammad Iqbal *Pakistan.*
Chow, Paul *USA.*
Chowdarl, Prof. B. V. R. *Singapore.*
Christensen, Dr Arild *USA.*
Christensen, Dr Axel Nørlund *Denmark.*
Christensen, Finn Erland *Denmark.*
Christianson, David W. *USA.*
Christidis, Prof. Panayiotis Chrysostomos *Greece.*
Christofides, Prof. George *Greece.*
Chroach, Dr Jutta *UK.*
Chu, Prof. Benjamin *USA.*
Chu, Dr Nguyen Huu *Vietnam.*
Chu, Shirley Shan-C. *USA.*
Chuev, Igor I. *Russia.*
Chukhovskil, Prof. Felix N. *Russia.*
Chumakov, Dr Yury Mikhailovich *Moldova.*
Chung, Liping *Australia.*
Chung, Dr Mui-Fatt *Singapore.*
Chung, Prof. Yong Je *Korea.*
Chuprunov, Prof. Evgeny V. *Russia.*
Church, William Bret *USA.*
Churchill, Prof. Melvyn R. *USA.*
Chvalun, Dr Sergej N. *Russia.*
Chykhrlj, Dr Stepan Ivanovych *Ukraine.*
Clani, Prof. Gianfranco *Italy.*
Cld, Dr Hilda *Chile.*
Ciechanowicz-Rutkowska, Dr Maria *Poland.*
Çina, Dr Aleksander *Albania.*
Cini, Dr Renzo *Italy.*
Ciraflci, Prof. Salvino *Italy.*
Cirilli, Dr Maurizio *Italy.*
Císařová, Dr Ivana *Czech Republic.*
Cisneros, Ing. Oswaldo *Venezuela.*
Ciszak, Dr Ewa *Poland.*
Ciulavu, Magda Cristina *Romania.*
Ciunik, Dr Zbigniew *Poland.*
Clackson, Dr Stephen Gregory *UK.*
Clancy, Ms Laura Lee *USA.*
Clapp, Rodney Alexander *Australia.*
Clarage, Dr James B. *USA.*
Claramunt, Prof. Rosa Maria *Spain.*
Clardy, Jon *USA.*
Clark, Dr Edward S. *USA.*
Clark, Dr George R. *New Zealand.*
Clark, Joan R. *USA.*
Clark, Dr Malcolm John Roy *Canada.*
Clark, Dr Simon Martin *UK.*
Clarke, Dr Michael J. *USA.*
Clarke, Neil D. *USA.*
Clarke, Prof. Roy *USA.*
Clark-Jones, Mrs Louise Elizabeth *UK.*
Claus, Albert C. *USA.*
Claus, Mrs Cornelia *Germany.*
Claus, Mr Karl Heinz *Germany.*
Clausen, Kurt Nørgaard *Denmark.*
Clauws, Prof. Paul Cyriel *Belgium.*
Clawson, David K. *USA.*
Clayton, William R. *USA.*
Clearfield, Prof. Abraham *USA.*
Cleasby, Dr Anne *UK.*
Clegg, Prof. William *UK.*
Clemente, Prof. Dore Augusto *Italy.*
Clifton, Dr Ian Jeffrey *UK.*

Cockayne, Prof David *Australia.*
Coda, Prof. Alessandro *Italy.*
Codding, Dr Penelope W. *Canada.*
Cody, Dr Vivian *USA.*
Coelho, Ana Maria Varela *Portugal.*
Coetzee, Ms Anita *South Africa.*
Cohen, Dr Carolyn *USA.*
Cohen, Dr Gerson H. *USA.*
Cohen, Dr Jerome B. *USA.*
Cohen, Dr Shmuel *Israel.*
Cohen-Addad, Dr Claudine *France.*
Coiro, Dr Vincenza Maria *Italy.*
Cojazzi, Dr Gianna *Italy.*
Colacio, Dr Enrique *Spain.*
Colapietro, Prof. Marcello *Italy.*
Cole, Dr Brent *USA.*
Cole, Henderson *USA.*
Cole, Mr Jason C. *UK.*
Colella, Prof. Roberto *USA.*
Coles, Simon J. *UK.*
Coll, Prof. Miquel *Spain.*
Collins, Dr Douglas M. *USA.*
Collins, Dr Edward J. *USA.*
Collins, Dr M. F. *Canada.*
Collison, Dr David *UK.*
Colloc'h, Dr Nathalie *France.*
Colman, Peter Malcolm *Australia.*
Colmanet, Silvano Francesco *Australia.*
Colombo, Dr Arturo *Italy.*
Colston, Sally *UK.*
Comes, Dr Robert *France.*
Comins, Dr Neville Raymond *South Africa.*
Comodi, Dr Paola *Italy.*
Conant, John W. *USA.*
Concha, Dr Nestor O. *USA.*
Conconi, María Susana *Argentina.*
Conde, Prof. Alejandro *Spain.*
Conde, Dr Clara Francisca *Spain.*
Connolly, Dr Michael L. *USA.*
Convert, Dr Pierre *France.*
Conway, Mr Sean *UK.*
Cook Jr, William R. *USA.*
Cook, Dr William J. *USA.*
Coombs, Prof. Douglas S. *New Zealand.*
Cooper, Dr Alan *New Zealand.*
Cooper, Dr Jonathan *UK.*
Cooper, Prof. Malcolm John *UK.*
Copeland, Richard F. *USA.*
Copie, Valerie *USA.*
Copley, Dr John R. D. *USA.*
Copley, Dr Royston C. B. *UK.*
Coppens, Prof. Philip *USA.*
Copperthwaite, Dr Richard George *South Africa.*
Corbett, Madeline *Australia.*
Cordero Borboa, Dr Adolfo *Mexico.*
Cordes, A. Wallace *USA.*
Corfield, Peter W. R. *USA.*
Corliss, Lester M. *USA.*
Coronel, Dr Domingo Alberto *Venezuela.*
Corradini, Prof. Paolo *Italy.*
Correia dos Santos, Dr António José Rebelo *Portugal.*
Correia Neves, Prof. José Marques *Brazil.*
Cortelezzi, Dr Cesar Rafael *Argentina.*
Cortés, Dr Roberto *Spain.*
Cosenza, Prof. Erika *Venezuela.*
Coşkun, Gül *Turkey.*
Costa, Prof. Giorgio Andrea *Italy.*
Costa Gouveia, Prof. Albany H. *Brazil.*
Costa, Prof. Maria-Margarida R. R. *Portugal.*
Costa Viana, Prof. Carlos Sérgio da *Brazil.*
Costamagna, Dr Juan Alberto *Chile.*
Costea, Constantin *Romania.*
Cota Araiza, Prof. Leonel S. *Mexico.*
Coulombe, Mr René *Canada.*
Coulter, Dr Charles L. *USA.*
Cousins, Christopher Stanley George *UK.*
Coville, Prof. Neil John *South Africa.*
Cowan, Dr Paul L. *USA.*
Cowan-Jacob, Dr Sandra *Switzerland.*
Cowburn, David *USA.*

Cowie, Prof. Martin *Canada.*
Cowley, Dr John M. *USA.*
Cox, Dr David E. *USA.*
Cox, Dr Ernest Gordon *UK.*
Cox, Dr Philip John *UK.*
Coyle, Richard Alan *Australia.*
Coy-Yll, Prof. Ramon *Spain.*
Craievich, Dr Aldo Félix *Brazil.*
Craig, Donald Chadwick *Australia.*
Cranswick, Lachlan Michael David *Australia.*
Craven, Bryan M. *USA.*
Crawford, Mr John Lawrence *South Africa.*
Creagh, Dudley Cecil *Australia.*
Crennell, Kate Mary *UK.*
Crennell, Dr Susan J. *UK.*
Crippen, Dr Gordon M. *USA.*
Crist, Buckley *USA.*
Cristea, Corina *Romania.*
Croft, Dr William J *USA.*
Cromer, Don T. *USA.*
Crouse, Richard E. *USA.*
Crowther, Robert L. *USA.*
Cruickshank, Prof. Durward William John *UK.*
Cruz, Dr Carlos *Cuba.*
Cruz, Dr Francisco *Cuba.*
Csöregh, Dr Ingeborg *Sweden.*
Cudney, Robert *USA.*
Cuevas-Diarte, Prof. Miquel Angel *Spain.*
Cuff, Christopher *Australia.*
Cui, Prof. Wenyuan *China.*
Cui, Prof. Xiushan *China.*
Culot, Ms Christine M. S. *Belgium.*
Cumbrera, Dr Francisco Luis *Spain.*
Cunane, Louise M. *USA.*
Cunningham, Prof. Des *Ireland.*
Curbelo, Dr Ciro *Cuba.*
Curien, Prof. Hubert *France.*
Curmi, Paul *Australia.*
Currao, Mr Antonio *Switzerland.*
Curzon, Prof. Albert Edward *Canada.*
Cusatis, Dr César *Brazil.*
Cutfield, Dr John F. *New Zealand.*
Cutfield, Dr Susan M. *New Zealand.*
Cvjetović, Mr Srdjan *Slovenia.*
Cygler, Dr Mirek *Canada.*
Czank, Priv. Doz. Dr Michael *Germany.*
Czerwinski, Dr Edmund W. *USA.*
Czugler, Dr Mátyás *Hungary.*
Daams, Mr Jo L. C. *Netherlands.*
Dabkowska, Dr Hanna A. *Canada.*
Dabkowski, Dr Antoni *Canada.*
Dacombe, Mr Michael H. *UK.*
Däweritz, Dr Lutz *Germany.*
Dago, Dr Angel *Cuba.*
Dahan, Dr Françoise *France.*
Dahl, Dr Lawrence F. *USA.*
Dahl, Prof. Tor *Norway.*
Dahlems, Mr Thomas *Germany.*
Dai, Yongshan *USA.*
Dal Negro, Prof. Alberto *Italy.*
Dalei, Mrs Snehlata *India.*
Daly, Dr John J. *Switzerland.*
Damas, Prof. Ana Margarida *Portugal.*
Daniels, Dr Lee M. *USA.*
Daniels, Dr Peter *Germany.*
Danielsen, Jacob *Denmark.*
Daniliuk, Dr Sergei *Moldova.*
Dann, Jeffrey N. *USA.*
Dapporto, Prof. Paolo *Italy.*
Daran, Dr Jean-Claude *France.*
Darinskaya, Dr Elena V. *Russia.*
Darinskii, Dr Alexander N. *Russia.*
Darlington, Dr Charles Nicholas Wright *UK.*
Dartyge, Prof. Elisabeth *France.*
Das, Mr Amit Kumar *India.*
Das, Dr Birendra Nath *India.*
Das , Dr Indu Mohan *India.*
Das, Dr Kalyan *India.*
Das, Dr Sabita *India.*
Datta , Mr Manuja *India.*

Dattagupta, Prof. Jiban Karati *India*.
Dautant, Dr Alain *France*.
Dauter, Dr Zbigniew *Germany*.
David, Dr Peter Rensis *USA*.
David, Prof. William I. F. *UK*.
Davies, Dr Christopher *USA*.
Davies, David R. *USA*.
Davies, Dr Geoffrey John *South Africa*.
Davies, Dr Gideon J. *UK*.
Davies, Dr Gladstone *South Africa*.
Davies II, Dr Jay F. *USA*.
Davies, Dr John Edward *UK*.
Davis, Paul Christopher *Australia*.
Davis, R. Lindsay *Australia*.
Davis, Prof. Raymond E. *USA*.
Davis, Timothy John *Australia*.
Davison, Daniel B. *USA*.
Day, Cynthia S. *USA*.
Day, Delbert E. *USA*.
Day, John S. *USA*.
Dayal, Dr Radha Raman *India*.
Dayniak, Dr Lydia G. *Russia*.
de Almeida, Prof. Maria-José *Portugal*.
De, Dr Amitabha *India*.
De Angelis, Prof. Giuseppe *Italy*.
de Bergevin, Dr François *France*.
de Boer, Dr Dirk K. G. *Netherlands*.
de Boer, Dr Jan L. *Netherlands*.
De Bondt, Dr Hendrik L. *USA*.
De Bondt, Dr Hendrik Leon Augusta Jozef *Belgium*.
De Camp, Dr Wilson H. *USA*.
de Gelder, René *Netherlands*.
de Graaff, Dr Rudolf A. G. *Netherlands*.
De Gryse, Prof. Roger Marc *Belgium*.
De Ita De la Torre, Prof. Antonio *Mexico*.
de Keijser, Dr Ir Thomas H. *Netherlands*.
De, Dr Madhusudhan *India*.
de Matos Gomes, Dr Etelvina *Portugal*.
de Médicis, Dr Rinaldo M. *Canada*.
De Montis, Prof. Pierfranco *Italy*.
De Munno, Prof. Giovanni *Italy*.
de Oñate Martinez, Mr Javier Enrique *Switzerland*.
De Pablo Galan, Dr Liberto *Mexico*.
De Poi Biasi, Prof. Carla *Italy*.
De Ranter, Prof. Camiel Joseph *Belgium*.
De Rosa, Prof. Claudio *Italy*.
De Rosier, David J. *USA*.
De Santis, Prof. Pasquale *Italy*.
de Souza, Dr Christina Franco *Brazil*.
De Villiers, Dr Johan Pieter Roos *South Africa*.
De Vos, Abraham M. *USA*.
de Vries, Ir Roelof Y. *Netherlands*.
de Vries, Drs Johan L. *Netherlands*.
De Winter, Dr Hans Louis Jos *Belgium*.
de Wit, Drs Martin *Netherlands*.
de With, Prof. Dr Gijsbertus *Netherlands*.
Dealwis, Dr Chris G. *USA*.
Deblieck, Dr Rudy A. C. *Netherlands*.
DeBoer, Barry G. *USA*.
Decanniere, Dr Klaas *Germany*.
Declercq, Prof. Jean-Paul *Belgium*.
Degtyareva, Dr Valentina F. *Russia*.
DeHaven, Patrick W. *USA*.
Deisenhofer, Johann *USA*.
Deiseroth, Prof. Dr Hans-Jörg *Germany*.
Dejus, Roger J. *USA*.
Dekaprilevich, Dr Marina O. *Russia*.
Del Piero, Dr Gastone *Italy*.
Del Prà, Prof. Antonio *Italy*.
Delaney, Matthew S. *USA*.
Delaney, William Timothy *Australia*.
Delaplane, Dr Robert G. *Sweden*.
Delbaere, Dr Louis Theophil Joseph *Canada*.
Delettré, Dr Jean *France*.
Delgado Quiñones, Prof. José Miguel *Venezuela*.
Delhez, Dr Ir Robert *Netherlands*.
Deliens, Dr Michel *Belgium*.
Delineshev, Dr Svetoslav *Bulgaria*.
Della Giusta, Prof. Antonio *Italy*.
DeLucas, Dr Lawrence J. *USA*.

Demartin, Prof. Francesco *Italy*.
Demianets, Dr Ludmila N. *Russia*.
Demishev, Dr Gennadii B. *Russia*.
Demšar, Dr Alojz *Slovenia*.
Deng, Dr Shuiquan *China*.
Denicoló, Dr Ireno *Brazil*.
Denisenko, Dr Georgy A. *Russia*.
Denoyer, Dr Françoise *France*.
Dent Glasser, Dr Lesley Scott *UK*.
Deopura, Dr B. L. *India*.
Depero, Dr Laura Eleonora *Italy*.
Depmeier, Prof. Dr Wulf *Germany*.
Derewenda, Dr Zygmunt *Canada*.
Derwent, Frank William *UK*.
Deschamps, Dr Jeffrey R. *USA*.
Desiraju, Prof. Gautam R. *India*.
Desmedt, Dr Amelia N. M. G. *Belgium*.
Desper, Dr Richard *USA*.
Despujols, Prof. Jacques *France*.
Destro, Prof. Riccardo *Italy*.
DeTitta, George T. *USA*.
Deutsch, Prof. M. *Israel*.
Dexter, Dr David D. *USA*.
Dhanaraj, Dr V *India*.
Dhaneshwar, Dr Narayandatta Nagesh *India*.
Di Blasio, Prof. Benedetto *Italy*.
Di Vaira, Prof. Massimo *Italy*.
Diacon, Dr Ion Andrei *Moldova*.
Diakonu, Dr Ion Ion *Moldova*.
Diamond, Dr Robert *UK*.
Diánez-Millán, Dr M. Jesus *Spain*.
Dias Rodrigues, Mrs Ana Maria Gonçalves *Brazil*.
Díaz de Delgado, Prof. Graciela Carlota *Venezuela*.
Diaz, Prof. Francesc *Spain*.
Dichman, Dr Klaus *Canada*.
Dickerson, Richard E. *USA*.
Dideberg, Dr Otto *France*.
Diéguez, Dr Ernesto *Spain*.
Diehl, Dr Roland *Germany*.
Dietrich, Prof. Dr Burkhart *Germany*.
DiGabriele, Anna D. *USA*.
Dijkstra, Prof. Dr Bauke W. *Netherlands*.
Dillen, Prof. Jan L. M. *South Africa*.
DiMarco, John D *USA*.
Dimitriadis, Prof. Charalambos *Greece*.
Dimitrijević, Dr Radovan *Serbia*.
Dimitrova, Dr Olga V. *Russia*.
Ding, Dr Jianping *USA*.
Diodati, Dr Francisco Piero *Argentina*.
Dionysiou-Kouimtzi, Prof. Semiramis *Greece*.
Dirl, Prof. Dr Rainer *Austria*.
Divjaković, Dr Vladimir *Serbia*.
Dixon, Dr Melinda M. *USA*.
Djarova, Dr Maria *Bulgaria*.
Djinovic Carugo, Dr Kristina *Italy*.
Djurić, Dr Stevan *Serbia*.
Dlouhá, Ing. Maja *Czech Republic*.
Dmitrienko, Dr Vladimir E. *Russia*.
Dmitrieva, Dr Margarita T. *Russia*.
Dmitrieva, Dr Tatiana V. *Russia*.
Doan, Mr Nguyen Dai *Vietnam*.
Dobler, Prof. Max *Switzerland*.
Dobrev, Dr Dobri *Bulgaria*.
Dobročka, Dr Edmund *Slovakia*.
Dodson, Eleanor *UK*.
Dodson, Prof. George Guy *UK*.
Doedens, Robert J. *USA*.
Dörfel, Dr Ilona *Germany*.
Doi, Mitsunobu *Japan*.
Dokurno, Pawel *Poland*.
Dolgushin, Mr Fyodor M. *Russia*.
Dolling, Dr Gerald *Canada*.
Dolmella, Dr Alessandro *Italy*.
Domenech, Dr M. Victoria *Spain*.
Domeneghetti, Prof. Maria Chiara *Italy*.
Domenicano, Prof. Aldo *Italy*.
Domiano, Prof. Paolo *Italy*.
Domínguez Esquivel, Dr José Manuel *Mexico*.
Domínguez, Prof. Esther *Spain*.
Dominguez Perez, Mr Roberto *France*.

Donat Reyes, Prof. Maximiliano José *Venezuela*.
Dong, Prof. Yicheng *China*.
Doni, Prof. Efthimia G. *Greece*.
Donica, Dr Feodor Gavril *Moldova*.
Donu, Dr Sofiya Vasile *Moldova*.
Doolittle, Russell *USA*.
Dorogovin, Dr Boris A. *Russia*.
Dorokhova, Dr Galina I. *Russia*.
Doroshinsky, Dr Alexander L. *Russia*.
Dorset, Dr Douglas L. *USA*.
Dosière, Dr Marcel *Belgium*.
Double, Dr Sylvie *USA*.
Doukov, Tzanko I. *USA*.
Dovesi, Prof. Roberto *Italy*.
Downing, Dr Kenneth H. *USA*.
Downs, James W. *USA*.
Downs, Dr Robert T. *USA*.
Doyne, Prof Thomas H. *USA*.
Doynikova, Dr Olga A. *Russia*.
Dräger, Prof. Dr Martin *Germany*.
Drake, Prof. John E. *Canada*.
Drašner, Dr Antun *Croatia*.
Dražič, Dr Goran *Slovenia*.
Drennan, Catherine Luschinsky *USA*.
Drennan, Dr John *Australia*.
Drenth, Prof. Dr Jan *Netherlands*.
Dressler, Dr Ludwig *Germany*.
Drew, Dr Michael G. B. *UK*.
Driesel, Dr Wolfgang *Germany*.
Driessen, Drs René. A. J. *Netherlands*.
Dristas, Dr Jorge Anastasio *Argentina*.
Drits, Prof. Victor A. *Russia*.
Drobyshev, Dr Andrey Stepanovich *Kazakhstan*.
Drouin, Mr Marc *Canada*.
Drozdov, Dr Yuri N. *Russia*.
du Boulay, Douglas John *Australia*.
du Plessis, Prof. Paul de Villiers *South Africa*.
du Toit, Mr Johan *South Africa*.
Duarte, Dr Maria Teresa Leal *Portugal*.
Duax, William L. *USA*.
Dubler, Prof. Erich *Switzerland*.
Dubourg, Dr Antoine *France*.
Dubský, Ing. Jiří *Czech Republic*.
Duc, Dr Luong Hong *Vietnam*.
Duc, Mr Nguyen Van *Vietnam*.
Duchamp, Dr David James *USA*.
Ducruix, Dr Arnaud *France*.
Dudka, Mr Alexander P. *Russia*.
Dudkevich, Prof. Vladimir P. *Russia*.
Dünkel, Dr Lothar *Germany*.
Duesler, Eileen N. *USA*.
Duggleby, Dr Helen *UK*.
Duhlev, Dr Rumen *UK*.
Duke, Dr Norma *Canada*.
Dunaj-Jurčo, Doc. Ing. Michal *Slovakia*.
Dunitz, Prof. Jack David *Switzerland*.
Dunn, Dr Cameron R. *UK*.
Dupont, Dr Léon *Belgium*.
Duque, Mr Julio *Cuba*.
Durant, Prof. François *Belgium*.
Durchschlag, Dr Helmut *Germany*.
Durif, Dr André *France*.
Durley, Rosemary C. E. *USA*.
Durlu, Dr T. Nuri *Turkey*.
Ďurovič, Ing. Slavomil *Slovakia*.
Durruthy, Mr Obel *Cuba*.
Dutta, Dr BishnuPada *India*.
Dužević, Prof. Dr Davor *Croatia*.
Dvorina, Dr Ljudmyla Andrijivna *Ukraine*.
Dvorkin, Dr Alexander Arkadyevich *Moldova*.
Dwiggins, Claudius W. *USA*.
Dwivedi, Dr Ganpat Lal *India*.
Dyachenko, Dr Oleg A. *Russia*.
Dyachenko, Dr Vladimir V. *Russia*.
Dyar, M. Darby *USA*.
Dyda, Dr Fred P. *USA*.
Dyda, Dr Fred *USA*.
Dym, Orly *Israel*.
Dynowska, Dr Elżbieta G. *Poland*.
Dyson, David John *UK*.

Dyuzheva, Dr Tatyana D. *Russia.*
Dzuong, Prof. Lecong *Vietnam.*
Dzyabchenko, Dr Aleksandr V. *Russia.*
Eades, John Alwyn *USA.*
Eales, Prof. Hugh Victor *South Africa.*
Ealick, Dr Steven E. *USA.*
Earnest, Dr Thomas N. *USA.*
Eberhard, Prof. Dr Emil *Germany.*
Ebied, Dr Mohamed *Egypt.*
Echavarri Hernández, Dr Ariel *Mexico.*
Echeverría, Mr Gustavo Alberto *Argentina.*
Eckert, Dr Juergen *USA.*
Economou, Prof. Nicolas Alkiviadis *Greece.*
Eder, Prof. Dr Otto J. *Austria.*
Edgington, Dr P. R. *UK.*
Edmondson, Mr Michael *UK.*
Edmundson, Allen B. *USA.*
Edström, Dr Kristina *Sweden.*
Edwards, Alison J. *Australia.*
Edwards, Dr Alison Jeanine *UK.*
Edwards, Karen Jennifer *Australia.*
Edwards, Dr Steven L. *USA.*
Effenberger, Dr Herta *Austria.*
Effendy, Dr *Indonesia.*
Efremov, Dr Valery A. *Russia.*
Efremova, Dr Elena P. *Russia.*
Egami, Takeshi *USA.*
Egert, Prof. Dr Ernst *Germany.*
Eggleston, Dr Drake S. *USA.*
Eggleton, Richard A. *Australia.*
Egikian, Dr David *Armenia.*
Egli, Dr Martin *Switzerland.*
Egner, Dr Ursula *Germany.*
Egorov-Tismenko, Dr Yury K. *Russia.*
Ehnebom, Lisbeth *Norway.*
Ehrlich, Grant M. *France.*
Ehses, Dr Karl-Heinz *Germany.*
Eichhorn, Dr Gerd *Germany.*
Eichhorn, Dr Klaus D. *Germany.*
Eichler, Dr František *Czech Republic.*
Eick, Harry A. *USA.*
Eigenbrot, Dr Charles W. *USA.*
Einspahr, Dr Howard Martin *USA.*
Einstein, Prof. Frederick W. B. *Canada.*
Elramjian, Dr Ferdinand H. *Armenia.*
Elramjian, Dr Tigran H. *Armenia.*
Eiríksson, Dr Vésteinn Rúni *Iceland.*
Eisenberg, Dr David S. *USA.*
Eisenberg, Dr Henryk *Israel.*
Eisenberg, Richard *USA.*
Eisenschmidt, Dr Christian *Germany.*
Eisenstein, Dr Miriam *Israel.*
Eklund, Prof. Hans *Sweden.*
Elahi, Mr Manzoor *Pakistan.*
Elango, Dr Elango N. *USA.*
Elango, Mr N. *India.*
Elbadri, Dr Ahmed Fouad *Egypt.*
Elcombe, Dr Margaret Marion *Australia.*
Elder, Richard C. *USA.*
Eldosoky, Dr Shokry *Egypt.*
Eleftheriadis, Prof. George *Greece.*
Elerman, Dr Yalçin *Turkey.*
Elf, Dr Frank *Germany.*
Elhagry, Mr Magdy A. *Egypt.*
Elhinnaur, Prof. Essam E. *Egypt.*
Eliassen, Bjørn Eirik *Norway.*
Eliopoulos, Prof. Elias E. *Greece.*
Elisabettini, Ms Paola *Belgium.*
El-Kabbani, Dr Ossama *USA.*
Elkardy, Dr Mhamed Abdelmoneim *Egypt.*
Elkins, Patricia A. *USA.*
Elkoly, Dr Esmat M. *Egypt.*
Ellena, Mr Javier Alcides *Argentina.*
Ellern, Dr Arkady M. *Israel.*
Elliott, Prof. James Cornelis *UK.*
Ellner, Dr Martin Oliver *Germany.*
Elmaghraby, Dr E. Mohamed *Egypt.*
Elmali, Dr Ayhan *Turkey.*
Elsayed, Prof. Karimat *Egypt.*
Elsegood, Dr Mark Robert James *UK.*

Elshabiny, Prof. Aida M. *Egypt.*
Elshanshury, Dr Ismail *Egypt.*
Elsharkawi, Dr Mohamed Abdelhamid *Egypt.*
Elshazli, Prof. Elshazli Mohamed *Egypt.*
Elshora, Dr A. Ebrahim *Egypt.*
Ely, Kathryn R. *USA.*
Emerson, Dr Kenneth *USA.*
Endoh, Dr Hisamitsu *Japan.*
Endriss, Mr Axel *Germany.*
Enemark, Prof. John H. *USA.*
Engel, Prof. Dennis Walter *South Africa.*
Engel, Dr Nora *Switzerland.*
Engel, Dr Walter *Germany.*
Engelen, Dr Bernward *Germany.*
Engelhardt, Dr Günter *Germany.*
Engels, Prof. Dr Siegfried *Germany.*
Englert, Dr Ulli *Germany.*
Engström, Dr Ingvar O. J. *Sweden.*
Enright, Dr Gary *Canada.*
Enwall, Eric L. *USA.*
Epelboin, Prof. Yves *France.*
Eppelsheimer, Dr Daniel *USA.*
Ercan, Dr Filiz *Turkey.*
Ercit, Dr Timothy Scott *Canada.*
Erdönmez, Dr Ahmet *Turkey.*
Eremin, Mr Nickolai N. *Russia.*
Eremkin, Dr Vladimir V. *Russia.*
Erez, Prof. Gidon *Israel.*
Ergin, Dr Ömer *Turkey.*
Erić, Mrs Suzana *Serbia.*
Eriksson, Dr Anders *Sweden.*
Eriksson, Dr Lars *Sweden.*
Eriksson, Dr Sten Gunnar *Sweden.*
Erlandsen, Heidi *Norway.*
Erman, Mary J. *USA.*
Ermer, Prof. Dr Otto *Germany.*
Ermrich, Dr Martin *Germany.*
Ernst, Stephen Richard *USA.*
Errington, Dr William *UK.*
Ersson, Nils Olov *Sweden.*
Ertan, Ms Anne *Sweden.*
Esaka, Dr Hisao *Japan.*
Espinet, Prof. Pablo *Spain.*
Espinosa Pérez, Ms Georgina *Mexico.*
Esser, Dr Lothar *USA.*
Estermann, Dr Michael Alexander *Switzerland.*
Esteve-Cano, Mr Vicente J. *Spain.*
Estienne, Dr Jacques *France.*
Estop, Dr Eugenia *Spain.*
Estrada, Dr M. Dolores *Spain.*
Eswara Prasad, Dr Gummuluri *India.*
Etheridge, Dr Joanne *UK.*
Eto, Mr Tetsujiro *Japan.*
Ettmayer, Prof. Dr Peter *Austria.*
Euler, Dr Harald *Germany.*
Euler, Dr Robert *Germany.*
Euthymiou, Prof. Paraskevi *Greece.*
Evans Jr, Dr Howard T. *USA.*
Evans, Dr Philip Richard *UK.*
Evans, Dr Stephen *Canada.*
Everaert, Mr Dirk Herman Maria *Belgium.*
Evlanova, Dr Nina F. *Russia.*
Evrard, Miss Christine *Belgium.*
Evrard, Prof. Guy Henri *Belgium.*
Eysel, Prof. Dr Walter *Germany.*
Ezpeleta, Mr José María *Spain.*
Faber, Catherine *USA.*
Fábry, Dr Jan *Czech Republic.*
Fackler Jr, Prof. John P. *USA.*
Faerman, Dr Carlos H. *USA.*
Fässler, Dr Thomas F. *Switzerland.*
Fagherazzi, Prof. Giuliano *Italy.*
Faigel, Dr Gyula *Hungary.*
Faisiev, Dr Abulkhak *Tadjikistan.*
Fait, Dr James F. *USA.*
Fajardo, Dr Fabio *Cuba.*
Faktor, Ing. Dušan *Slovakia.*
Falk, Dr Michael *Canada.*
Fallon, Dr Gary D. *Australia.*
Fan, Prof. Chenggao *China.*

Fan, Prof. Haifu *China.*
Fan, Prof. Yuguo *China.*
Fanchon, Dr Eric *France.*
Fanfani, Prof. Luca *Italy.*
Fanter, Dr Detlef *Germany.*
Fantini, Dr Marcia C. A. *Brazil.*
Fanwick, Dr Phillip E. *USA.*
Farag, Prof. Ibrahim S. A. *Egypt.*
Farber, Gregory K. *USA.*
Fares, Dr Vincenzo *Italy.*
Farkas-Jahnke, Dr Mária *Hungary.*
Farmer, Dr Barry L. *USA.*
Farooque, Dr Muhammad *Pakistan.*
Farran, Mr Joan *Spain.*
Farrar, Dr David H. *Canada.*
Farrar, William W. *USA.*
Fatmi, Prof. Ali Nasir *Pakistan.*
Faulk, John W. *USA.*
Fauman, Dr Eric B *USA.*
Faure, Prof. René *France.*
Faus-Paya, Dr Juan *Spain.*
Fawcett, Dr John *UK.*
Fay, Robert C. *USA.*
Fayed, Prof. Leila *Egypt.*
Fayek, Prof. Mohamed Khorshed *Egypt.*
Fayos, Prof. José *Spain.*
Fedeli, Prof. Walter *Italy.*
Fedorchuk, Dr Anatolij Oleksandrovych *Ukraine.*
Fedorenko, Prof. Anatolij Ivanovych *Ukraine.*
Fedorov, Pavel P. *Russia.*
Fedorov, Dr Vladimir A. *Russia.*
Fedyna, Dr Mykhajlo Fedorovych *Ukraine.*
Feher, George *USA.*
Fehlmann, Dr Mel *Switzerland.*
Feidenhans'l, Robert Krarup *Denmark.*
Feigin, Prof. Lev A. *Russia.*
Feil, Prof. Dr Dirk *Netherlands.*
Fejdi, Doc. Dr Pavel *Slovakia.*
Feldman, Robert E. *USA.*
Fell, Hans Jörg *Norway.*
Felsteiner, Joshua *Israel.*
Feneau-Dupont, Ms Janine *Belgium.*
Fenn, Dr Ruth H. *UK.*
Fenoll Hach-Alí, Prof. Purificacion *Spain.*
Ferguson, Dr George *Canada.*
Ferguson, Dr Ian Forster *UK.*
Ferguson, Prof. Robert Bury *Canada.*
Ferguson-Miller, Shelagh *USA.*
Fernandes, Dr Nelson Gonçalves *Brazil.*
Fernández, Dr Carlos José *Spain.*
Fernández de Rapp, Ms María Emilia *Argentina.*
Fernández, Ing. Juan Carlos *Argentina.*
Fernandis, Mr Jacob Richard *India.*
Fernando, Dr Quintus *USA.*
Ferracini, Prof. Elena *Italy.*
Ferrara, Dr Joseph David *USA.*
Ferrari, Dr Claudio *Italy.*
Ferrari-Belicchi, Prof. Marisa *Italy.*
Ferraris, Prof. Giovanni *Italy.*
Ferre-D'Amare, Adrian R. *USA.*
Ferreira de Souza, Prof. Milton *Brazil.*
Ferrero, Prof. Adele *Italy.*
Ferretti, Dr Maurizio *Italy.*
Ferretti, Dr Valeria *Italy.*
Fesenko, Prof. Evgeny G. *Russia.*
Fesenko, Dr Oleg E. *Russia.*
Fethiere, Dr James *Canada.*
Fetisov, Dr Gennady Vladimirovich *Russia.*
Fettinger, Dr James C. *USA.*
Fewster, Dr Paul Frederick *UK.*
Fiala, Doc. Dr Jaroslav *Czech Republic.*
Fichera, Dr Annamaria *Italy.*
Field, Dr Donald William *Australia.*
Field, Mrs Jennifer *UK.*
Field, Prof. John *South Africa.*
Fiermans, Prof. Lucien Victor August *Belgium.*
Figgis, Brian N. *Australia.*
Figleiski, Prof. Tadeusz *Poland.*
Figueiredo, Prof. Maria Ondina *Portugal.*
Figueiredo Neto, Prof. Antônio Martins *Brazil.*

Filatov, Prof. Stanislav K. *Russia*.
Filhol, Dr Alain *France*.
Filipenko, Olga S. *Russia*.
Filippakis, Dr Sophocles *Greece*.
Filippini, Dr Giuseppe *Italy*.
Filizova, Dr Lyudmila *Bulgaria*.
Finer-Moore, Dr Janet S. *USA*.
Finger, Dr Larry W. *USA*.
Finkel', Dr Vitalij Oleksandrovych *Ukraine*.
Finkenstadt, Victoria L. *USA*.
Finlayson, Trevor Roy *Australia*.
Finney, Prof. John Leslie *UK*.
Finster, Prof. Dr Joachim *Germany*.
Finzel, Dr Barry Craig *USA*.
Fischer, Dr Berthold *Switzerland*.
Fischer, Prof. Dr Karl *Germany*.
Fischer, Prof. Dr Werner *Germany*.
Fischer, Dr Peter *Switzerland*.
Fisher, Dr Andrew J. *USA*.
Fita, Dr Ignacio *Spain*.
Fitch, Dr Andrew *France*.
Fitzgerald, Dr Paula M. D. *USA*.
Fjaertoft-Pedersen, Prof. Berit *Norway*.
Fjellvåg, Helmer *Norway*.
Flack, Howard D. *Switzerland*.
Flanders, Dr David John *UK*.
Fleet, Dr Michael Edward *Canada*.
Fleischmann, Dr Klaus D. *Netherlands*.
Flensburg, Claus *Denmark*.
Flerov, Dr Igor N. *Russia*.
Fletcher, Neville Horner *Australia*.
Fletcher, Dr Steven Reginald *UK*.
Fletterick, Robert *USA*.
Flick, Karen E. *USA*.
Flippen-Anderson, Judith L. *USA*.
Flörke, Dr Ulrich *Germany*.
Florencio, Dr Feliciana *Spain*.
Florio, John V. *USA*.
Foces-Foces, Prof. Concepcion *Spain*.
Follner, Prof. Dr Heinz *Germany*.
Folting-Streib, Kirsten *USA*.
Fomina, Dr Larisa Petrivna *Ukraine*.
Fonari, Dr Marina Semionovna *Moldova*.
Fonseca, Dr Eugenia *Chile*.
Fonseca, Dr Isabel *Spain*.
Fontaine, Dr Alain *France*.
Fontaine, Dr Frédéric *Belgium*.
Font-Altaba, Prof. Manuel *Spain*.
Font-Bardia, Dr Merce *Spain*.
Foord, Eugene E. *USA*.
Foreman, Mrs Kathleen *UK*.
Forest, Katrina *USA*.
Foresti, Prof. Elisabetta *Italy*.
Foris, C. M. *USA*.
Formoso, Dr Milton Luiz *Brazil*.
Fornasini, Prof. Maria L. *Italy*.
Fornoff, Mario *USA*.
Forslund, Dr Bertil S. *Sweden*.
Forsyth, Prof. J. Bruce *UK*.
Fortes, Prof. Manuel Amaral *Portugal*.
Forteza, Mr Matilde *Spain*.
Fortier, Dr Suzanne *Canada*.
Forwood, Christopher Thomas *Australia*.
Foss, Prof. Olav *Norway*.
Fossli, Dag *Norway*.
Foster, Dr Mark D. *USA*.
Fotinou, Dr Constantina *UK*.
Foundling, Dr Stephen I. *USA*.
Fourie, Dr Jacobus Theodor *South Africa*.
Fourme, Prof. Roger *France*.
Fox, Kristin M. *USA*.
Foxman, Dr Bruce M. *USA*.
Frahm, Dr Ronald R. *Germany*.
Frampton, Dr Christopher S. *UK*.
Franceschi, Prof. Enrico *Italy*.
Francis, Mr John Godfrey *UK*.
Francisco, Dr Regina Helena Porto *Brazil*.
Francke, Mr Rudolf *Germany*.
Frank, Prof. Dr Walter *Germany*.
Franke, Dr Valeria D. *Russia*.

Frank-Kamenetskaya, Dr Olga V. *Russia*.
Franks, Prof. Albert *UK*.
Fransolet, Dr André-Mathieu *Belgium*.
Franzen, Hugo F. *USA*.
Franzosi, Dr Paolo *Italy*.
Fraser, Dr Marie *Canada*.
Fratini, Albert V. *USA*.
Frazao, Dr Carlos *Portugal*.
Frederick, Christin *USA*.
Freed, Robert Lowell *USA*.
Freeman, Dr Clive M. *USA*.
Freeman, Hans Charles *Australia*.
Freemont, Dr Paul Simon *UK*.
Freer, Dr Andrew *UK*.
Freer, Dr Stephan T. *USA*.
Freire D'Aguiar, Dr Manoel Marcos *Brazil*.
Freitas, Miguel *Portugal*.
Freites, Prof. Juan Alfredo *Venezuela*.
Fremont, Dr D. H. *USA*.
French, Alfred D. *USA*.
Frenz, Bert *USA*.
Freund, Prof. Dr Friedemann *USA*.
Frevel, Ludo K. *USA*.
Frey, Prof. Dr Friedrich *Germany*.
Frey, Dr Michel *France*.
Freymann, Douglas *USA*.
Friedman, Dr Jonathan M. *USA*.
Friese, Mrs Karen *Germany*.
Frigeri, Dr Cesare *Italy*.
Frikkee, Prof. Dr Evert *Netherlands*.
Fritzsch, Dr Günter *Germany*.
Fröhlich, Dr Armin *Germany*.
Frolov, Mr Kirill V. *Russia*.
Frolow, Dr Felix *Israel*.
Fronczek, Frank R. *USA*.
Frueh, Prof. Alfred J. *USA*.
Fruziński, Mgr inż. Andrzej *Poland*.
Frydenvang, Karla *Denmark*.
Frye, Mr Thomas *Germany*.
Ftikos, Prof. Christos *Greece*.
Fu, Prof. Heng *China*.
Fu, Prof. Pingqiu *China*.
Fu, Prof. Zhengmin *China*.
Fu, Prof. Zhuji *China*.
Fülöp, Dr Vilmos *UK*.
Fuente-Cullell, Dr Carlos de la *Spain*.
Fuentes, Dr Juan *Cuba*.
Fuentes, Dr Luis *Cuba*.
Fuentes-Mateos, Mr Angel Manuel *UK*.
Fuertes, Dr Amparo *Spain*.
Fuertes-Martínez, Dr José Félix *Spain*.
Fuess, Prof. Dr-Ing. Hartmut *Germany*.
Fötterer, Mr Klaus *Germany*.
Fuhrmann, Mr Peter *Germany*.
Fujii, Gary *USA*.
Fujii, Mr Mitsuhiro *Japan*.
Fujii, Dr Satoshi *Japan*.
Fujii, Mr Takashi *Japan*.
Fujii, Mr Tomomi *Japan*.
Fujimoto, Mr Hiroshi *Japan*.
Fujimura, Dr Norifumi *Japan*.
Fujinaga, Dr Masao *Canada*.
Fujishita, Mr Hideshi *Japan*.
Fujita, Mr Keiichiro *Japan*.
Fujita, Prof. Shigeo *Japan*.
Fujiwara, Prof. Takaji *Japan*.
Fujiwara, Miss Yohko *Japan*.
Fukamachi, Prof. Tomoe *Japan*.
Fukami, Prof. Dr Takeshi *Japan*.
Fuke, Dr Shunro *Japan*.
Fukuda, Prof. Akeharu *Japan*.
Fukuhara, Dr Akira *Japan*.
Fukui, Prof. Takashi *Japan*.
Fukuyama, Dr Keiichi *Japan*.
Fulfaro, Dr Roberto *Brazil*.
Fullenwider, Malcolm A. *USA*.
Fuller, Prof. Watson *UK*.
Fun, Dr Hoong-Kun *Malaysia*.
Fundamensky, Mr Vladimir S. *Russia*.
Fuoss, Paul Henry *USA*.

Furey, Dr William *USA*.
Furley, Jonathon *UK*.
Furmanova, Dr Nina G. *Russia*.
Furnas, Thomas C. *USA*.
Furuhata, Mr Yoshio *Japan*.
Furukawa, Dr Yoshinori *Japan*.
Furuseth, Prof. Sigrid B. *Norway*.
Fusegawa, Mr Izumi *Japan*.
Fushino, Mr Tetsuo *Japan*.
Fuzek, John F. *USA*.
Fykin, Dr Leonid E. *Russia*.
Gabalaah, Dr G. Abdelstar *Egypt*.
Gabe, Dr Eric James *Canada*.
Gaber, Dr Abdelfattah *Egypt*.
Gable, Dr Robert William *Australia*.
Gabrielian, Dr Karen T. *Armenia*.
Gabrielyan, Dr Vyacheslav T. *Russia*.
Gabriček, Dr Sergij *Slovenia*.
Gad, Prof. Gamal Mohamed *Egypt*.
Gadgil, Mr. Ajit K. *India*.
Gaebel, Mr Rainer *Germany*.
Gaete, Dr Walter *Spain*.
Gaffar, Prof. M. A. *Egypt*.
Gagarina, Dr Elena S. *Russia*.
Gaier, James R. *USA*.
Gainsford, Dr Graeme J. *New Zealand*.
Gait, Dr Robert Irwin *Canada*.
Gajhede, Michael *Denmark*.
Galan, Prof. Emilio *Spain*.
Gałązka, Prof. Robert *Poland*.
Gałdecka, Dr Ewa Renata *Poland*.
Gałdecki, Prof. dr inż. Zdzisław *Poland*.
Galetti, Prof. Giulio *Switzerland*.
Gali, Prof. Salvador *Spain*.
Galindo, Dr Agustin *Spain*.
Galitsky, Dr Nikolai M. *USA*.
Galiulin, Dr Ravil V. *Russia*.
Gallagher, Dr D. Travis *USA*.
Gallardo, Olga *Sweden*.
Galli, Prof. Ermanno *Italy*.
Gallois, Dr Bernard *France*.
Gallucci, Dr Judith C. *USA*.
Galstyan, Dr Victor G. *Russia*.
Galy, Dr Jean *France*.
Gamarnik, Dr Moisei Yankelevych *Ukraine*.
Gamasa, Dr Pilar *Spain*.
Gambardella, Miss Maria Teresa do Prado *Brazil*.
Gamble, Theresa R. *USA*.
Gan, Miss Bee Kwan *Australia*.
Ganazzoli, Dr Fabio *Italy*.
Gandais, Dr Madeleine *France*.
Ganesan, Mr V. *India*.
Ganesh, Dr Venkatapathy *USA*.
Ganev, Ing. Nikolaj *Czech Republic*.
Gangloff, Andy *USA*.
Ganguli, Subrata *USA*.
Ganis, Prof. Paolo *Italy*.
Gantzel, Dr Peter Kellogg *USA*.
Gao, Dachao *Australia*.
Gao, Dr Yan *USA*.
Garavito, Dr R. Michael *USA*.
Garbauskas, Dr Mary F. *USA*.
García Morales, Lic. María Inmaculada *Venezuela*.
García-Aranda, Dr Miguel Angel *Spain*.
García-Blanco, Prof. Severino *Spain*.
García-Granda, Dr Santiago *Spain*.
García-Rodríguez, Dr Antonio *Spain*.
García-Ruiz, Dr Joaquín *Spain*.
García-Ruiz, Dr Juan Manuel *Spain*.
Gardner, Dr Kenn Corwin H. *USA*.
Garg, Dr Ajay Kumar *India*.
Garín, Dr Jorge Leonidas *Chile*.
Garipogly, Dmitrij N. *Kazakhstan*.
Garland, Mrs María Teresa *Chile*.
Garman, Dr Elspeth F. *UK*.
Garman, Scott C. *USA*.
Garnier, Dr Emmanuel *France*.
Garrett, Richard Federick *Australia*.
Garrett, Dr Thomas P. J. *Australia*.
Garrido, Mr Carlos *Spain*.

Garsche, Mr Markus *Germany.*
Garvey, Roy G. *USA.*
Gashin, Prof. Petr *Moldova.*
Gaspard, Prof. Jean-Pierre *Belgium.*
Gasparian, Dr Laura G. *Armenia.*
Gasparri-Fava, Prof. Giovanna *Italy.*
Gasymov, Dr Vagif Akber ogly *Azerbaijan.*
Gatehouse, Dr Bryan Michael K. C. *Australia.*
Gatti, Domenico L. *USA.*
Gautham, Mr Namasivayam *India.*
Gauthier, Prof. Jean-Pierre *France.*
Gavezzotti, Prof. Angelo *Italy.*
Gavrilov, Dr Victor *Latvia.*
Gavrilova, Dr Nadezhda D. *Russia.*
Gavrilyachenko, Dr Victor G. *Russia.*
Gavuzzo, Dr Enrico *Italy.*
Gawron, Dr Marian *Poland.*
Gay, Hebe Dina *Argentina.*
Gazzano, Dr Massimo *Italy.*
Gazzoni, Prof. Giuseppe *Italy.*
Gdaniec, Dr Maria *Poland.*
Ge, Prof. Zhongjiu *China.*
Gebert, Dr Walter *Germany.*
Geddes, Dr Alexander J. *UK.*
Geguzina, Dr Galina A. *Russia.*
Gelb, Dr Steven J. *USA.*
Geiger, Dr Charles A. *Germany.*
Geiger, Dr James H. *USA.*
Geiger, Dr Martin *Germany.*
Gell, Phillip H. *USA.*
Gelse, Prof. Herman J. V. H. *Belgium.*
Gelser, Dr Urs *USA.*
Geist, Dr Volker *Germany.*
Gel'man, Dr Yury A. *Russia.*
Genkina, Dr Elena A. *Russia.*
George, Dr Amand *France.*
George, Dr Clifford F. *USA.*
Georgiadis, Millie M. *USA.*
Gérard, Dr François *France.*
Gerasimov, Dr Victor I. *Russia.*
Geraskin, Dr Valery V. *Russia.*
Gerbeleu, Prof. Nikolai Vasile *Moldova.*
Germain, Prof. Gabriel *Belgium.*
Gernat, Mrs Christine *Germany.*
Gerson, Andrea Ruth *Australia.*
Gerstenberg, Michael Christian *Denmark.*
Gertel-Kloos, Dr Heike *Germany.*
Gervasio, Prof. Giuliana *Italy.*
Gervilla, Dr Fernando *Spain.*
Gerward, Dr Leif *Denmark.*
Gesemann, Prof. Dr-Ing. Dr Renate *Germany.*
Getzoff, Dr Elizabeth D. *USA.*
Ghefs, Dr Marianne *France.*
Ghergari, Prof. Lucreţia *Romania.*
Ghermani, Dr Nour-Eddine *France.*
Ghilardi, Dr Carlo Alfredo *Italy.*
Ghose, Subrata *USA.*
Ghosh, Dr Debashis *USA.*
Ghosh, Ms Manuja *India.*
Ghosh, Dr Mrs Minakshi *India.*
Ghosh, Partho *USA.*
Ghosh, Miss Soma *India.*
Ghosh, Mr Sutapa *India.*
Ghosh, Dr Sutapa *USA.*
Giacovazzo, Prof. Carmelo *Italy.*
Gianfagna, Dr Antonio *Italy.*
Gibbons, Dr Cyril Stephen *Canada.*
Gibbs, Doon *USA.*
Gieren, Prof. Dr Alfred *Austria.*
Gies, Prof. Dr Hermann *Germany.*
Giese Jr, Rossman F. *USA.*
Giester, Mag. Dr Gerald *Austria.*
Giglio, Prof. Edoardo *Italy.*
Gilabert, Mr Ulises *Argentina.*
Gilani, Mr Jamshed Ali *Pakistan.*
Gilardi, Dr Richard D. *USA.*
Gilbert, William A. *USA.*
Gilboa, Dr A. Joseph *Israel.*
Giles, Mr Raymond Richard *UK.*
Gilfrich, John V. *USA.*

Gilje, John W. *USA.*
Gille, Dr Peter *Germany.*
Gille, Dr Wilfried *Germany.*
Gilli, Prof. Gastone *Italy.*
Gilli, Dr Paola *Italy.*
Gilliland, Dr Gary L. *USA.*
Gilmore, Dr Chris *UK.*
Gilski, Mgr Mirosław *Poland.*
Gimeno, Prof. José *Spain.*
Ginderow, Dr Daria *France.*
Ginell, Dr Stephan L. *USA.*
Gingl, Franz *Switzerland.*
Giordano, Prof. Federico *Italy.*
Giorgi, Mr *France.*
Giovanoli, Prof. Rudolf *Switzerland.*
Giranda, Dr Vincent L. *USA.*
Giri, Mr Anil K. *India.*
Giri, Mr Siba Narayan *India.*
Girirajan, Dr K. S. *India.*
Gíslason, Dr Haflidi Pétur *Iceland.*
Giunchi, Dr Giovanni *Italy.*
Givargizov, Dr Eugene I. *Russia.*
Gjønnes, Prof. Jon Kjell *Norway.*
Gjønnes, Kjersti *Norway.*
Gjudjenov, Mrs Karin *Germany.*
Gladić, Mr Jadranko *Croatia.*
Gladkii, Dr Vsevolod V. *Russia.*
Gladkikh, Dr Liliya Ivanivna *Ukraine.*
Gladyshevskii, Prof. Evgen Ivanovych *Ukraine.*
Gladyshevskii, Dr Roman *Switzerland.*
Glaeser, Dr Robert M. *USA.*
Glaisher, Robert William *Australia.*
Glaremin, Mr Peter *Germany.*
Glaser, Julius *Sweden.*
Glass, Howard L. *USA.*
Glasser, Prof. Fredrik Paul *UK.*
Glasser, Prof. Leslie *South Africa.*
Glatter, Prof Dr Otto *Austria.*
Glazer, Dr Anthony Michael *UK.*
Glazier, Edward James *UK.*
Glazov, Dr Alexei Iv. *Russia.*
Gleason Villagran, Mr Roberto *Mexico.*
Glidewell, Christopher *UK.*
Glikin, Dr Arkady E. *Russia.*
Glinka, Charles Joseph *USA.*
Glinnemann, Dr Jürgen *Germany.*
Głowiak, Prof. Tadeusz *Poland.*
Główka, Prof. Marek L. *Poland.*
Glucksman, Dr Marc J. *USA.*
Glumoff, Dr Tuomo *Finland.*
Glusker, Dr Jenny P. *USA.*
Glykos, Mr Nicholas M. *UK.*
Godavarthi, Mr Bhagavannarayana *India.*
Goddard, Dr Richard John *Germany.*
Godden, Dr Jeff W. *USA.*
Godwod, Dr Krzysztof Jan *Poland.*
Göbel, Dr Herbert E. *Germany.*
Goedkoop, Prof. Dr Jacob A. *Netherlands.*
Görbitz, Carl Henrik *Norway.*
Göris, Dr Helmar *Germany.*
Goeta, Mr Andrés Eduardo *Argentina.*
Göttlicher, Mr Jörg *Germany.*
Götz, Dr Wolfgang *Germany.*
Götzinger, Dr Michael Alois *Austria.*
Gözel, Dr Güller *Turkey.*
Gollo, Dr Eduard Al'bertovich *Russia.*
Goldberg, Prof. Israel *Israel.*
Goldberg, Dr Stephen Z. *USA.*
Golden, Barbara L. *USA.*
Goldgur, Mr Yehuda *Israel.*
Goldish, Elihu *USA.*
Goldman, Adrian *Finland.*
Goldsmith, Elizabeth J. *USA.*
Goldsmith, Julian R. *USA.*
Goldstein, Dr Barry M. *USA.*
Goldstone, Joyce A. *USA.*
Golič, Prof. Dr Ljubo *Slovenia.*
Goliński, Dr Bohdan *Poland.*
Golovina, Dr Nina I. *Russia.*
Golovko, Dr Yurii I. *Russia.*

Golubev, Dr Alexandr M. *Russia.*
Golyshev, Dr Vladimir M. *Russia.*
Gomes, Mr Albert Cardinal *India.*
Gomes, Mr Ariel *Cuba.*
Gomes, Prof. Celso *Portugal.*
Gómez de Andérez, Prof. Dora *Venezuela.*
Gómez Ramírez, Dr Ricardo *Mexico.*
Gómez Rodríguez, Dr Alfredo *Mexico.*
Gómez Rodríguez, Prof. Leonir José *Venezuela.*
Gómez-Sal, Prof. José Carlos *Spain.*
Gómez-Sal, Dr M. Pilar *Spain.*
Gomis-Rüth, Dr Franz Xavier *Spain.*
Gomm, Dr Martin *Germany.*
Goncharov, Dr Alexader V. *Russia.*
Gonda, Prof. Dr Takehiko *Japan.*
Gong, Dr Minfang *USA.*
Gonschorek, Dr Walter *Germany.*
González, Dr Gema *Venezuela.*
González, Lic. Aleida del Carmen *Venezuela.*
Gonzalez, Mr Oscar *Uruguay.*
González Rodríguez, Dr Oscar Victorio *Venezuela.*
Gonzalez-Calbet, Dr José M. *Spain.*
González-Duarte, Prof. Pilar *Spain.*
González-Mañas, Dr Marina *Spain.*
González-Platas, Mr Javier *Spain.*
González-Silgo, Mr Cristina *Spain.*
Goodman, Peter *Australia.*
Goodsell, Dr David S. *USA.*
Goonesekere, Nalin C. W. *USA.*
Gorchakova, Dr Olga E. *Russia.*
Gordienko, Mr Leonid A. *Russia.*
Górkiewicz, Mgr inż. Zbigniew *Poland.*
Gorogotska, Dr Ljudmyla Ivanivna *Ukraine.*
Gorshkov, Prof. Anatoly Ivanovich *Russia.*
Gorskaya, Dr Marina G. *Russia.*
Górski, Dr Ludwik *Poland.*
Gorter, Drs Ing. Sybout *Netherlands.*
Gorya, Dr Oleg Stephan *Moldova.*
Gosmanová, Dr Galina *Czech Republic.*
Gospodinov, Prof. Marin *Bulgaria.*
Goswami, Prof. K. N. *India.*
Goswami, Dr S. Niranjana N. *India.*
Goto, Ms Midori *Japan.*
Goto, Dr Yoshiaki *Japan.*
Gotoh, Mr Hirohito *Japan.*
Gouaux, J. Eric *USA.*
Goubitz, Drs Kees *Netherlands.*
Gougoutas, Dr Jack Zanos *USA.*
Gould, Dr Robert Ozburn *UK.*
Gould, Dr Sheila E. B. *UK.*
Gountsidou, Vassiliki *Greece.*
Gourley, David Gilmour *UK.*
Gover, Dr Sheila *UK.*
Graafsma, Dr Heinz *France.*
Gråbæk, Lars Friis *Denmark.*
Grabko, Dr Daria Zakharovna *Moldova.*
Grabowski, Prof. Mieczysław Jerzy *Poland.*
Graf, Dr Hans Anton *Germany.*
Graham, Dr A. Ronald *Canada.*
Gramaccioli, Prof. Carlo Maria *Italy.*
Gramlich, Dr Volker *Switzerland.*
Granados, Lic. Maristela *Venezuela.*
Grant, Robert *USA.*
Granzin, Dr Joachim *Germany.*
Grattan-Bellew, Dr Patrick Edward *Canada.*
Gravereau, Prof. Pierre *France.*
Graves, Bradford J. *USA.*
Gray, Dr Terry M. *USA.*
Graziani, Prof. Giorgio *Italy.*
Grdenić, Prof. Dr Drago *Croatia.*
Greasley, Samantha E. *UK.*
Greaves, Dr Eduardo D. *Venezuela.*
Greedan, Dr John E. *Canada.*
Greenberg, Dr Berton L. *USA.*
Greenblatt, Harry (Tzvi) *Israel.*
Greene, Mr Fernando *Chile.*
Greenhough, Dr Trevor J. *UK.*
Gregoriadis, Prof. Panagiotis *Greece.*
Gregorkiewitz, Prof. Miguel *Spain.*
Greis, Dr Ortwin *Germany.*

Grenier, Ms Lucie *Canada.*
Greploni, Dr Fabrizia *Italy.*
Gress, Mary E. *USA.*
Grey, Dr Ian Edward *Australia.*
Gribakov, Dr Michael *USA.*
Grice, Dr Joel Denison *Canada.*
Griewatsch, Mr Carsten *Germany.*
Griffen, Dr Dana T. *USA.*
Griffin, Dr Jane F. *USA.*
Griffith, Dr Diana L. *USA.*
Griffith, Elizabeth H. *USA.*
Griffith, Ms Vivienne Jean *South Africa.*
Griger, Dr Ágnes *Hungary.*
Grigorian, Dr Arshak H. *Armenia.*
Grigorov, Dr Sergej Nikolajevich *Ukraine.*
Grimmer, PD Hans *Switzerland.*
Grin, Dr Yuri N. *Germany.*
Grin', Dr Yurij Mykolayovych *Ukraine.*
Grins, Dr Jekabs *Sweden.*
Gritsenko, Mr Victor Victorovich *Russia.*
Groat, Dr Lee Andrew *Canada.*
Grochowski, Dr Jacek M. *Poland.*
Grochulski, Dr Pawel *Canada.*
Grochulski, Dr Pawel *Poland.*
Gromek, Dr Jack M. *USA.*
Gromilov, Dr Sergey A. *Russia.*
Gronkowski, Dr hab. Jerzy *Poland.*
Gros, Dr Piet *Netherlands.*
Grosse, Prof. Dr Peter *Germany.*
Grosse Kunstleve, Ralf W. *Switzerland.*
Grosse, Mr Mirco *Germany.*
Grossie, David A. *USA.*
Groth, Per *Norway.*
Groy, Dr Thomas L. *USA.*
Grozdanov, Mr. Lyudmil *Bulgaria.*
Gruber, Doc. Dr Boris *Czech Republic.*
Gruber, Mr. Karl *Austria.*
Gruehn, Prof. Dr Reginald *Germany.*
Grütter, Dr Markus Gerhard *Switzerland.*
Grundvig, Sidsel *Denmark.*
Grundy, Dr Harry Douglas *Canada.*
Gruner, Dr Sol M. *USA.*
Grunsky, Dr Oleg S. *Russia.*
Grzeta, Dr Biserka *Croatia.*
Gschneidner Jr, Prof. Karl A. *USA.*
Gu, Prof. Xiaocheng *China.*
Gu, Prof. Yuanxin *China.*
Guagliardi, Dr Antonietta *Italy.*
Guan, Yue *USA.*
Guardiola, Dr Rene *Cuba.*
Gubkin, Mr Mikhail G. *Russia.*
Guda, Dr Bardhyl *Albania.*
Guddat, Dr Luke W. *USA.*
Gündoğdu, M. Niyazi *Turkey.*
Güneş, Bilal *Turkey.*
Günter, Dr Christina *Germany.*
Günter, Prof. John Ralph *Switzerland.*
Günther, Prof. Dr H. Fritz *Germany.*
Guerin, Dr Diego Marcelo *Argentina.*
Guerra, Prof. Gaetano *Italy.*
Guerrero, Lic. José Reinaldo *Venezuela.*
Guggenheim, Stephen *USA.*
Guha, Dr Romel *India.*
Guha, Dr Sankarananda *India.*
Guillin, Mr Jacques *France.*
Guimarães, Prof. Dirce Andrade *Portugal.*
Guimpel, Julio J. *Argentina.*
Guindi, Prof. Amin Riad *Egypt.*
Guinier, Prof. André *France.*
Gulbis, Jacqueline M. *USA.*
Gullman, Jan O. *Sweden.*
Guo, Prof. Changlin *China.*
Guo, Dr Hwai-Chen *USA.*
Gupta , Mr Manoj Kumar *India.*
Gupta, Dr Satish Chander *India.*
Gupta, Miss Sunita *India.*
Gupta, Dr Vijai Prakash *India.*
Gupta, Prof. Vishwambhar Dayal *India.*
Gurbán, Mr Sándor *Hungary.*
Gurel, Ogan *USA.*

Gurskaya, Dr Galina Gurskaya *Russia.*
Gursky, Dr Olga *USA.*
Gururow, Dr Tayur N. *India.*
Guseinov, Dr Gahraman Gusein oglu *Azerbaijan.*
Guss, Dr J. Mitchell *Australia.*
Gustafsson, Torbjörn *Sweden.*
Guthrie, Dr George D. *USA.*
Gutierrez-Puebla, Dr Enrique *Spain.*
Gutierrez-Zorrilla, Dr Juan Manuel *Spain.*
Gutteridge, Mr Walter Alfred *UK.*
Gutzov, Prof. Ivan *Bulgaria.*
Guzikevich-Guerstein, Gali *Israel.*
Guzman, Mr Carmen *Spain.*
Gweifel, Dr Ismail *Egypt.*
Gyepesová, Dr Dalma *Slovakia.*
Ha, Dr Cao The *Vietnam.*
Haake, Mrs Annegret *Germany.*
Haaland, Prof. Arne *Norway.*
Haapala, Ilmari J. *Finland.*
Hårsta, Dr Anders *Sweden.*
Haase, Prof. Dr Wolfgang *Germany.*
Haav, Dr Aksel *Estonia.*
Habash, Dr Jarjis *UK.*
Habib, Mr Syed Abbas *Pakistan.*
Habiby, Dr Fakhruddin *Pakistan.*
Habuka, Hitoshi *Japan.*
Hackert, Prof. Marvin L. *USA.*
Had, Ing. Jiří *Czech Republic.*
Hadan, Dr Marianne *Germany.*
Hadfield, Dr Andrea T. *USA.*
Haditsch, Prof. Dr Johann Georg *Austria.*
Hädicke, Dr Erich *Germany.*
Haeffner, Dean R. *USA.*
Hämäläinen, Keijo Johannes *Finland.*
Hämäläinen, Reijo Pertti *Finland.*
Haendler, Dr Helmut M. *USA.*
Hähnert, Dr Irmela *Germany.*
Härtwig, Dr Jürgen *France.*
Hafiz, Prof. Mohamed M. *Egypt.*
Haga, Dr Yumiko *Japan.*
Hagelstein, Dr Michael *France.*
Hagen, Prof. Kolbjørn *Norway.*
Haggerty, Brian S. *USA.*
Hagiya, Dr Kenji *Japan.*
Hagler, A. T. *USA.*
Hahn, Prof. Dr Theo *Germany.*
Hahn, Mr Michael *Germany.*
Haibach, Mr Torsten *Germany.*
Haibach, Mr Torsten *Switzerland.*
Haije, Dr Willem G. *Netherlands.*
Haile, Prof. Sossina *USA.*
Hajdu, Dr Ferenc *Hungary.*
Hajdu, Dr Janos *UK.*
Hakanen, Arvi Tapani *Finland.*
Halder, Dr Sujit Kumar *India.*
Halfpenny, Dr Joan Christine *UK.*
Hall, Prof. Eric O. *Australia.*
Hall, Dr Michael D. *USA.*
Hall, Sydney Reading *Australia.*
Haller, Kenneth *Thailand.*
Halliwell, Mary A. G. *UK.*
Haltiwanger, R. Curtis *USA.*
Halwax, Dr Erich Johann *Austria.*
Hamada, Dr Kensaku *Japan.*
Hambley, Trevor William *Australia.*
Hamdi, Prof. Hassan Mahmoud *Egypt.*
Hamidov, Prof Hamidov *Azerbaijan.*
Hamill, Dr Gregory P. *USA.*
Hamilton, Jean A. *USA.*
Hamlin, Ronald C. *USA.*
Hammack, Professor William S. *USA.*
Hammad, Prof. Elsayed Mohamed *Egypt.*
Hammond, Mr Lloyd C. *Australia.*
Hamodrakas, Dr Stavros *Greece.*
Hamor, Dr Thomas *UK.*
Han, Dr Fusen *USA.*
Han, Dr Gye Won *USA.*
Han, Shaoxu *USA.*
Hán, Prof. Trinh *Vietnam.*
Han, Prof. Yuzhen *China.*

Hanan, Ms Salima Shaukat *India.*
Hange, Mr Ferenc *Hungary.*
Hangleiter, Dr Thomas *Germany.*
Hanke, Dr Kurt *Germany.*
Hannick, Linda I. *USA.*
Hansen, Doc. Staffan *Sweden.*
Hansen, Mr Harly Andreas Smedemark *Germany.*
Hansen, Prof. Lars Kristian *Norway.*
Hansen, Dr Niels *France.*
Hansen, Sissel *Norway.*
Hanson, Jonathan C. *USA.*
Hanson, Leif *USA.*
Hao, Dr Quan *UK.*
Haq, Dr Anwarul *Pakistan.*
Harada, Prof. Jimpei *Japan.*
Harada, Dr Shigeharu *Japan.*
Harbrecht, Prof. Dr Bernd *Germany.*
Hardgrove, Dr George L. Jr. *USA.*
Harding, Dr Marjorie M. *UK.*
Hardy, Prof. Antoine *France.*
Hardy, Dr Larry W. *USA.*
Harel, Michal *Israel.*
Hargittai, Prof. István *Hungary.*
Hargittai, Prof. Magdolna *Hungary.*
Hargrave, Karl D. *USA.*
Hariharan, Ms Meena *India.*
Harkema, Dr Sybolt *Netherlands.*
Harle, Säde Pirjo Anneli *Finland.*
Harlos, Dr Karl *UK.*
Harlow, Richard L. *USA.*
Harmatha, Ing. Ladislav *Slovakia.*
Harms, Mr Joerg M. *Germany.*
Haroutyunian, Dr Levon A. *Armenia.*
Haroutyunian, Dr Valery S. *Armenia.*
Harp, Joel M. *USA.*
Harris, Dr Gillian W. *UK.*
Harris, Dr Kenneth David Maclean *UK.*
Harris, Lisa J. *USA.*
Harris, Dr Stephanie Ellen *UK.*
Harrison, Robert W. *USA.*
Harrison, Prof. Stephen C. *USA.*
Harrison, Dr William T. A. *USA.*
Hart, Prof. Michael *UK.*
Hartl, Prof. Dr Hans *Germany.*
Hartley, Richard H. *Australia.*
Hartman, Prof. Dr Piet *Netherlands.*
Hartmann, Prof. Ervin *Hungary.*
Hartmann, Mr Frank *Germany.*
Hártopanu, Paulina *Romania.*
Hartsuck, Dr Jean A. *USA.*
Hartung, Prof. Dr Helmut *Germany.*
Harutyunyan, Prof. Emil H. *Russia.*
Harvey, Dr Ian *UK.*
Hasan, Dr Faizul *Pakistan.*
Hasan, Dr Zinab Abdelkhlek *Egypt.*
Hasebe, Dr Masami *Japan.*
Hasebe, Prof. *Japan.*
Hasegawa, Dr Masashi *Japan.*
Hašek, Dr Jindřich *Czech Republic.*
Hasemann, Dr Charles A. *USA.*
Haser, Dr Richard *France.*
Hashimoto, Prof. Dr Mituru *Japan.*
Hashizume, Daisuke *Japan.*
Hashizume, Prof. Hiroo *Japan.*
Hasnain, Prof. Samar *UK.*
Hassan, Prof. Mohamed Youssef *Egypt.*
Hassan, Dr Wan Fuad *Malaysia.*
Hassell, Anne M. *USA.*
Hassler, Eivind *Sweden.*
Hastings, Dr Jerome Biller *USA.*
Hata, Dr Tadashi *Japan.*
Hata, Dr Yasuo *Japan.*
Hatada, Dr Marcos H. *USA.*
Hatt, Ben Albert *UK.*
Hatzikraniotis, Evripidis *Greece.*
Hatzisymeon, Constantinos *Greece.*
Hau, Dr Herbert *USA.*
Hauback, Dr Bjørn Christian *Norway.*
Hauge, Dr Sverre *Norway.*
Hauptman, Dr Herbert A. *USA.*

Hausen, Dr Hans-Dieter *Germany.*
Hauser, Dr Jürg *Switzerland.*
Havlík, Doc. Ing. Tomáš *Slovakia.*
Haworth, Dr Colin W. *UK.*
Hawthorne, Dr Frank Christopher *Canada.*
Hay, Dr David Gilbert *Australia.*
Hayakawa, Dr Yasuhiro *Japan.*
Hayashi, Dr Koya *Japan.*
Hayashi, Dr Shigeyuki *Japan.*
Hazell, Alan Charles *Denmark.*
Hazes, Dr Bart *Canada.*
He, Dr Xiao-min (*a.k.a.* Joseph Xiaomin Ho) *USA.*
Head, Dr James F. *USA.*
Heale, Gillian *UK.*
Heba, Dr Zein Elabdeen Kamel *Egypt.*
Hebert, Dr Hans *Sweden.*
Heckroodt, Prof. Renier Oelof *South Africa.*
Hedman, Dr Britt G. M. *USA.*
Heeg, Mary Jane *USA.*
Hegde, Rashmi S. *USA.*
Hegedűs, Zsolt *Sweden.*
Heide, Prof. Dr Klaus *Germany.*
Heikinheimo, Pirkko *Finland.*
Heime, Prof. Dr Klaus *Germany.*
Heimstad, Eldbjørg S. *Norway.*
Heine, Dr Andreas *USA.*
Heinemann, Dr Frank Wilhelm *Germany.*
Heinemann, Dr Udo *Germany.*
Helgesson, Dr Göran *Sweden.*
Helin, Sari Katariina *Finland.*
Helland, Ronny *Norway.*
Heller-Kallai, Lisa *Israel.*
Helliwell, Prof. John Richard *UK.*
Hellner, Prof. Dr Erwin E. *Germany.*
Helmholdt, Dr Robert B. *Netherlands.*
Helmreich, Dr Dieter *Germany.*
Hempel, Dr Andrew *Canada.*
Hempstead, Dr P. D. *UK.*
Henderson, Prof. C. M. B. *UK.*
Henderson, Dr Grant S. *Canada.*
Henderson, Dr Richard *UK.*
Henderson, Stephen *USA.*
Hendricks, Dr Robert W. *USA.*
Hendriksen, Dr Barry A. *UK.*
Hendrixson, Dr Thomas L. *USA.*
Henke, Dr Henning *Germany.*
Henkel, Prof. Dr Gerald *Germany.*
Henley, Christopher Lee *USA.*
Henling, Larry *USA.*
Hennig, Dr Michael *Switzerland.*
Henríquez, Mr Fernando *Chile.*
Henry, Rodger *USA.*
Henslee, Dr Walter *USA.*
Hepp, Dr Alfred *Switzerland.*
Herbette, Leo G. *USA.*
Herbstein, Prof. Frank H. *Israel.*
Herbst-Irmer, Dr Regine *Germany.*
Herceg, Dr Marija *Croatia.*
Herdade, Dr Sílvio B. *Brazil.*
Herdtweck, Dr Eberhardt F. Ch. *Germany.*
Heredia, Mr Eduardo Armando *Argentina.*
Hergold-Brundić, Dr Antonija *Croatia.*
Hermans, Jan *USA.*
Hermansson, Dr Kersti *Sweden.*
Hermida, Mr Jorge Daniel *Argentina.*
Hermoso-Domínguez, Dr Juan Antonio *Spain.*
Hernández Arana, Dr Andrés *Mexico.*
Hernández Juárez, Mr Edilberto *Mexico.*
Hernández Ortega, Mr Simón *Mexico.*
Hernández-Cano, Prof. Félix *Spain.*
Héroux, Ms Annie *Canada.*
Herrera Becerra, Dr Raul *Mexico.*
Herrera, Ms Victoria *Cuba.*
Herres, Dr Nikolaus *Germany.*
Herron, Dr James N. *USA.*
Herting-Agthe, Dr Susanne *Germany.*
Hervé, Dr Francisco *Chile.*
Herzberg, Dr Armin *Germany.*
Herzberg, Dr Osnat *USA.*
Hesse, Dr Karl-Friedrich *Germany.*

Hester, Dr Gerko *USA.*
Hester, James R. *Australia.*
Hewaidy, Prof. I. F. *Egypt.*
Hewat, Dr Alan W. *France.*
Hewat, Dr Elizabeth Ann *France.*
Heydenreich, Prof. Dr Johannes *Germany.*
Heyding, Dr Robert Donald *Canada.*
Heyns, Prof. Anton Michal *South Africa.*
Hibbs, David E. *UK.*
Hibiya, Taketoshi *Japan.*
Hidalgo-Laguna, Mr Miguel Angel *Spain.*
Hiebl, Dr Kurt *Austria.*
Hien, Dr Nguyen Van *Vietnam.*
Hien, Prof. Than Duc *Vietnam.*
Higashi, Dr Akira *Japan.*
Higatsberger, Prof. Dr Dr h.c. Michael J. *Austria.*
Higgins, Dr John B. *USA.*
Higgins, Dr Tim *Ireland.*
Higuchi, Prof. Taiichi *Japan.*
Higuchi, Yoshiki *Japan.*
Hiismäki, Pekka Eljas *Finland.*
Hikosaka, Prof. Masamichi *Japan.*
Hildebrandt, Prof. Dr Gerhard *Germany.*
Hilgenfeld, Dr Rolf *Germany.*
Hill, Dr Christopher P. *USA.*
Hill, Dr Roderick Jeffrey *Australia.*
Hiller, Prof. Dr Wolfgang Paul *Germany.*
Hilmi, Prof. Mohamed Ezzeldin *Egypt.*
Hilmy, Prof. Mohamed Ezzeldin *Egypt.*
Hiltunen, Lassi Ilmari *Finland.*
Himmel, Daniel *USA.*
Hingerty, Dr Brian E. *USA.*
Hinrichs, Dr Winfried *Germany.*
Hinrichsen, Prof. Dr Georg *Germany.*
Hinz, Dr Dietrich *Germany.*
Hippler, Mr Bernd *Germany.*
Hirabayashi, Prof. Makoto *Japan.*
Hiraga, Prof. Kenji *Japan.*
Hiramatsu, Dr Kazumasa *Japan.*
Hirano, Prof. Dr Shin-ichi *Japan.*
Hirasawa, Prof. Dr Izumi *Japan.*
Hirata, Dr Hiroshi *Japan.*
Hirayama, Prof. Noriaki *Japan.*
Hiremath, Mr Chaitanya N. *India.*
Hiremath, Chaitanya *USA.*
Hirotsu, Prof. Ken *Japan.*
Hirth, John Price *USA.*
Hjorth, Michael *Denmark.*
Hla, Miss Tin Tin *Myanmar.*
Ho, Dr Douglas M. *USA.*
Ho, Dr Joseph Xiaomin (*a.k.a.* Xiao-min He) *USA.*
Hobbs, Linn W. *USA.*
Hockless, Dr David *Australia.*
Hodeau, Dr Jean Louis *France.*
Hodgson, Keith Owen *USA.*
Hodorowicz, Prof. Stanisław A. *Poland.*
Hodsdon, Dr John M. *USA.*
Höbler, Dr Hans-Joachim *Germany.*
Höche, Prof. Dr Hans-Reiner *Germany.*
Höfer, Dr Harry *Germany.*
Hoeffken, Dr Hans Wolfgang *Germany.*
Höhling, Prof. Dr Hans Jürgen *Germany.*
Höhne, Dr Wolfgang *Germany.*
Høier, Prof. Ragnvald *Norway.*
Hökelek, Dr Tuncer *Turkey.*
Hölsä, Jorma Pertti Kalervo *Finland.*
Hölzel, Mr Alexander R. *Germany.*
Hönle, Dr Wolfgang Johannes *Germany.*
Hofeldt, Mr Jochen *Germany.*
Hoffmann, Dr Prof. Dr Wolfgang *Germany.*
Hoffmann, Dr Klaus *Germany.*
Hofmann, Mr Rolf R. *Germany.*
Hofmeister, Dr Wolfgang *Germany.*
Hogle, Dr James M. *USA.*
Hoier, Dr Helga *Germany.*
Hol, Dr Wim *USA.*
Holbrook, Dr Stephen R. *USA.*
Holden, David *UK.*
Holden, Professor Hazel M. *USA.*
Holden, James Richard *USA.*

Holinski, Dr Rüdiger *Germany.*
Holland, Dr Debra Reid *USA.*
Hollander, Dr Frederick J. *USA.*
Hollenstein, Mr Sandro *Switzerland.*
Holý, Ing. Alois *Slovakia.*
Holmes, Dr Gillian F. *UK.*
Holmes, Margaret *USA.*
Holt, Dr Elizabeth M. *USA.*
Holý, Doc. Dr Václav *Czech Republic.*
Holzapfel, Prof. Dr Wilfried B. *Germany.*
Hom, Tommy *USA.*
Homma, Prof. Teiichi *Japan.*
Hon, Dr Ping-Kay *Hong Kong.*
Honda, Dr Kazumasa *Japan.*
Hong, Prof. Maochun *China.*
Hong, Sam-Hyo *Sweden.*
Hong, Prof. Sung Kwon *Korea.*
Hongisto, Ossi Valtteri *Finland.*
Honkimäki, Veijo *Finland.*
Honma, Mr Shigeru *Japan.*
Honoré, Tage *Denmark.*
Honzatko, Richard B. *USA.*
Hooft, Dr Rob W. W. *Germany.*
Hoogewijs, Prof. Robert *Belgium.*
Hoonnivathana, Mr Ekachai *Thailand.*
Hoover, David M. *USA.*
Hope, Prof. Håkon *USA.*
Hoppe, Prof. Dr Dr h.c.mult. Rudolf *Germany.*
Hordijtchuk, Dr Oleg Volodymyrovych *Ukraine.*
Hordvik, Prof. Asbjørn Ivar *Norway.*
Hori, Dr Kayako *Japan.*
Horiuchi, Dr Hiroyuki *Japan.*
Horn, Dr Ernst *Japan.*
Horsfield, Mr Edgar Charles *South Africa.*
Horton, Dr John R. *USA.*
Horton, Nancy C. *USA.*
Horvat-Gabriel, Miss Margareta *Serbia.*
Horváth, Dr Zsolt E. *Hungary.*
Horvatić, Mr Davor *Croatia.*
Hosemann, Prof. Dr Dr h.c. Rolf *Germany.*
Hoshikawa, Prof. Dr Keigo *Japan.*
Hoshino, Prof. Sadao *Japan.*
Hoskins, Dr Bernard Foster *Australia.*
Hosomi, Dr Satoru *Japan.*
Hospital, Dr Michel *France.*
Hossain, Dr M. Bilayet *USA.*
Hosur, Dr Madhusudhan V. *India.*
Hottenhuis, Dr Ir M. H. J. *Netherlands.*
Houdusse, Dr Anne M. *USA.*
Hough, Prof. Edward *Norway.*
Hountas, Prof. Athanasios S. *Greece.*
Hovestreydt, Dr Eric Robert *Germany.*
Hovmöller, Sven *Sweden.*
Howard, Dr Andrew J. *USA.*
Howard, Dr Christopher J. *Australia.*
Howard, Prof. Judith Ann Kathleen *UK.*
Howard, Dr Sean *UK.*
Howell, Dr P. Lynne *Canada.*
Howie, Dr R. Alan *UK.*
Howlin, Dr Brendan *UK.*
Hoy, Vanessa Jane *UK.*
Hoyer, Prof. Dr Walter *Germany.*
Hoyland, Dr Michael A. *UK.*
Hozawa, Prof. Dr Mitsunori *Japan.*
Hrdý, Dr Jaromír *Czech Republic.*
Hriljac, Dr Joseph A. *USA.*
Hruby, Dr Victor J. *USA.*
Hseu, Prof. Tzong-Hsiung *Taiwan.*
Hsiou, Yu Hsiou *USA.*
Hsu, Barbara T. *USA.*
Hsu, I-Nan *USA.*
Hsu, Rebeka Min-fang *Australia.*
Htoon, Prof. Dr Sein *Myanmar.*
Hu, Prof. Shengzhi *China.*
Hu, Dr Shu-Hong *Australia.*
Hu, Dr Xiaorui *Germany.*
Hu, Dr Zhengwei *China.*
Hua, Prof. Ziqian *China.*
Huan, Dr Le Van *Vietnam.*
Huang, Dr De-Bin *USA.*

Huang, Dr Eugene *Taiwan*.
Huang, Prof. Jinling *China*.
Huang, Prof. Jinshun *China*.
Huang, Prof. Qichen *China*.
Huang, Dr Rui H. *USA*.
Huang, Prof. Taishan *China*.
Huang, Dr Ting *USA*.
Huanosta Tera, Prof. Alfonso *Mexico*.
Hubbard, Camden R. *USA*.
Hubbard, Dr Stevan R. *USA*.
Hubbell, John H. *USA*.
Huber, Andrew H. *USA*.
Huber, Dr Carol P. *Canada*.
Huber, Susan R. *USA*.
Huddle, Benjamin P. *USA*.
Hübener, Dr Rainer *Germany*.
Hümmer, Prof. Dr Kurt *Germany*.
Huffman, Dr John C. *USA*.
Hughes, Dr David Lewis *UK*.
Hughes, David *UK*.
Hughes, John M. *USA*.
Hughes, Dr Robert E. *USA*.
Hukins, Prof. David *UK*.
Hull, Dr Stephen *UK*.
Hulliger, Dr Jürg *Switzerland*.
Humblet, Dr Christine *USA*.
Huml, Dr Karel *Czech Republic*.
Hummel, Dr Hans-Ulrich *Germany*.
Hummel, Dr Wolfgang *Switzerland*.
Humphreys, Prof. Colin J. *UK*.
Hung, Dr H. H. *Taiwan*.
Hunt, Richard E. *USA*.
Hunter, Dr William N. *UK*.
Huo, Wen *USA*.
Huq, Fazlul *Australia*.
Hurley, Dr James H. *USA*.
Hurley, Mr Patrick W. *UK*.
Hurley, Dr Thomas D. *USA*.
Hursthouse, Prof. Michael B. *UK*.
Husák, Ing. Michal *Czech Republic*.
Husebye, Prof. Steinar *Norway*.
Hussain, Mr Hilal *Pakistan*.
Hussain, Dr Khadim *Pakistan*.
Hussain-Bates, Dr Bilquis *UK*.
Hutcheon, Dr Wendy Lou (Brooks) *Canada*.
Hutchings, Dr Michael Thomas *UK*.
Hutton, Dr Alan Thomas *South Africa*.
Huxley, Hugh E. *USA*.
Huynh, Dr Tran Dai *Vietnam*.
Hybler, Dr Jiří *Czech Republic*.
Hyde, Bruce *Australia*.
Hyde, Dr C. Craig *USA*.
Hynes, Dr Rosemary Catherine *Canada*.
Hynes, Dr Thomas R. *USA*.
Iandelli, Prof. Aldo *Italy*.
Ianelli, Dr Sandra *Italy*.
Iannelli, Dr Pio *Italy*.
Ibata, Dr Koichi *Japan*.
Ibberson, Dr Richard M. *UK*.
Iberl, Angelika R. *Germany*.
Ibers, James A. *USA*.
Ibragimov, Prof. Bakhtijar T. *Uzbekistan*.
Ice, Dr Gene Emery *USA*.
Ichimiya, Prof. Ayahiko *Japan*.
Ichinose, Prof. Noboru *Japan*.
Ide, Dr Semra *Turkey*.
Idid, Ms Sheriffah Noor *Malaysia*.
Ido, Prof. Toshiyuki *Japan*.
Iglesias, Prof. Juan Eugenio *Spain*.
Ignatovich, Dr Vladimir K. *Russia*.
Igonin, Dr Vladimir A. *Russia*.
Ihringer, Dr Jörg *Germany*.
Iijima, Dr Kenji *Japan*.
Iijima, Dr Kinya *Japan*.
Iijima, Dr Sumio *Japan*.
Iijima, Prof. Takao *Japan*.
Iishi, Prof. Dr Kazuaki *Japan*.
Iitaka, Prof. Yoichi *Japan*.
Ikemiya, Mr Norihito *Japan*.
Ikram, Dr Nazma *Pakistan*.

Ilag, Dr Leodevico L. *USA*.
Ilakovac, Ms Vita *Croatia*.
Ilinca, Gheorghe *Romania*.
Ilyukhin, Dr Andrey B. *Russia*.
Ilyushin, Prof. Alexandr S. *Russia*.
Ilyushin, Dr Grigory D. *Russia*.
Imakuma, Dr Kengo *Brazil*.
Imamov, Prof. Rafik M. *Russia*.
Imashimizu, Dr Yuji *Japan*.
Imhof, Dr Wolfgang *Germany*.
Immirzi, Prof. Attilio *Italy*.
Imperatori, Dr Patrizia *Italy*.
Inaoka, Dr Kimio *Japan*.
Inbanathan, Mr S. Stephen R. *India*.
Indenbom, Prof. Vladimir L. *Russia*.
Infante, Mr Guillermo *Cuba*.
Iñiguez Rodriguez, Dr Adrian Mario *Argentina*.
Inn, Mr Aung Paik *Myanmar*.
Ino, Prof. Shozo *Japan*.
Inoue, Dr Morio *Japan*.
Inoue, Dr Tsuyoshi *Japan*.
Inouye, Dr Hideyo *USA*.
Insausti, Mr María Teresa *Spain*.
Ioku, Dr Koji *Japan*.
Iolin, Prof. Eugene *Latvia*.
Ipohorski Lenkiewicz, Dr Miguel *Argentina*.
Iqbal, Mr Mir Waseluddin Ahmad *Pakistan*.
Irngartinger, Prof. Dr Hermann *Germany*.
Irving, Dr Anne *South Africa*.
Irving, Thomas C (Tom) *USA*.
Isaacs, Prof. Neil William *UK*.
Isakov, Dr Igor V. *Russia*.
Ishibashi, Dr Hiroyuki *Japan*.
Ishibashi, Mr Hisato *Japan*.
Ishibashi, Mr Tsutomu *Japan*.
Ishibashi, Prof. Yoshihiro *Japan*.
Ishida, Prof. Ishida *Japan*.
Ishihara, Mr Makoto *Japan*.
Ishii, Prof. Makoto *Japan*.
Ishii, Prof. Mitsuru *Japan*.
Ishikawa, Dr Tetsuya *Japan*.
Ishitani, Dr Akihiko *Japan*.
Ishizaki, Mr Junya *Japan*.
Ishizawa, Dr Nobuo *Japan*.
Ishizuka, Dr Kazuo *Japan*.
Isnard, Mr Olivier *France*.
Isoda, Prof. Dr Seiji *Japan*.
Isogami, Dr Mineo *Japan*.
Israël, Drs O. R. *Netherlands*.
Isupov, Mr Michael N. *Russia*.
Ito, Prof. Taichiro *Japan*.
Ito, Prof. Tetsuzo *Japan*.
Itoh, Dr Hideaki *Japan*.
Itoh, Dr Hiroyuki *Japan*.
Itoh, Dr Nobuo *Japan*.
Itri, Dr Rosangela *Brazil*.
Ivakin, Dr Gleb I. *Russia*.
Ivaldi, Prof. Gabriella *Italy*.
Ivanković, Mr Hrvoje *Croatia*.
Ivanov, Dr Sergey A. *Russia*.
Ivanova, Ms Elena S. *Russia*.
Ivanović, Miss Ivana *Serbia*.
Ivarsson, Dr Gun J. M. *Sweden*.
Ivegeš, Miss Erika *Serbia*.
Iwai, Dr Kunimoto *Japan*.
Iwaki, Prof. Dr Toshihiro *Japan*.
Iwami, Prof. Dr Motohiro *Japan*.
Iwanaga, Prof. Dr Hiroshi *Japan*.
Iwanov, Dr Dantsho *Bulgaria*.
Iwasaki, Prof. Fujiko *Japan*.
Iwasaki, Prof. Hiroshi *Japan*.
Iwasaki, Dr Hitoshi *Japan*.
Iwata, Dr Yutaka *Japan*.
Iyengar, Dr Leela *India*.
Izard, Tina *Australia*.
Izui, Prof. Kazuhiko *Japan*.
Izumi, Dr Kunihide *Japan*.
Izumi, Prof. Yoshinobu *Japan*.
Jabri, Evelyn *USA*.
Jackson, Dr Robert A. *UK*.

Jackuliak, Doc. Dr Quido *Slovakia*.
Jacobi, Dr Hans *Germany*.
Jacobs, Prof. Dr Herbert *Germany*.
Jacobsen, Chris *USA*.
Jacobsen, Dr Bruce L. *USA*.
Jacobson, David H. *USA*.
Jacobson, Dr Raymond H. *USA*.
Jacobson, Dr Robert A. *USA*.
Jääskeläinen, Sirpa *Finland*.
Järvinen, Dr Matti J. *Finland*.
Jafarov, Dr Jafarov *Azerbaijan*.
Jafry, Mr Syed Quamar Abbas *Pakistan*.
Jagadeesh Kumar, Dr N. *India*.
Jagadish, Chennupati *Australia*.
Jagannathan, Mr N. R. *India*.
Jagner, Susan *Sweden*.
Jagodzinski, Prof. Dr Dr h.c. Heinz E. *Germany*.
Jahn, Dr Irmin Rudolf *Germany*.
Jahnberg, Dr Lena *Sweden*.
Jain, Dr Arun Kumar *India*.
Jain, Sanjeev *USA*.
Jakkal, Mr Vasant Shankar *India*.
Jakubovics, Dr John Paul *UK*.
James, Dr Margaret Ann *Canada*.
James, Prof. Michael N. G. *Canada*.
James, Veronica Jean *Australia*.
James-Surcouf, Dr Evelyne *France*.
Jamil, Dr Ahmed *Pakistan*.
Janakiraman, Dr Musiri N. *USA*.
Janczak, Dr Jan *Poland*.
Janecki, Dr Hieronim Piotr *Poland*.
Janes, Dr Robert William *UK*.
Jang, Peter *Taiwan*.
Jang, Dr Wen-Jye *Japan*.
Jang, Dr Youngnam *Korea*.
Janin, Prof. Joël *France*.
Janjić, Dr Svetislav *Serbia*.
Janner, Prof. Dr Aloysio *Netherlands*.
Janot, Prof. Christian *France*.
Janovec, Prof. Dr Václav *Czech Republic*.
Janovec, Ing. Jozef *Slovakia*.
Jansen, Prof. Dr Martin *Germany*.
Jansen, Dr Jacob *Netherlands*.
Jansonius, Prof. Johan N. *Switzerland*.
Janssen, Prof. Dr Ted W. J. M. *Netherlands*.
Jansson, Dr Kjell *Sweden*.
Jap, Bing K. *USA*.
Jarchow, Prof. Dr Otto *Germany*.
Jardetzky, Ted *USA*.
Jardin, Dr Christian *France*.
Jasinski, Dr Jerry P. *USA*.
Jaskólski, Dr Mariusz *Poland*.
Jaszczak, Prof. John A. *USA*.
Jauch, Dr Wolfgang *Germany*.
Jayadevan, Dr Naduviledath Chennuvittil *India*.
Jayanty, Dr Ashok *India*.
Jayashree, Ms A. N. *India*.
Jean, Dr Yuch-Cheng *Taiwan*.
Jeanloz, Raymond *USA*.
Jedrzejas, Dr Marek J. *USA*.
Jeffrey, Prof. George Alan *USA*.
Jeffreys, Dr J. A. D. *UK*.
Jeglitsch, Prof. Dr Dr h.c. Franz *Austria*.
Jeitschko, Prof. Dr Wolfgang *Germany*.
Jennings, Dr M. C. *Canada*.
Jensen, Anette Frost *France*.
Jensen, Birthe *Denmark*.
Jensen, Dorte Juul *Denmark*.
Jensen, Dr Lyle H. *USA*.
Jensen, Stig Jorgo *Denmark*.
Jeong, Dr Tae Soo *Korea*.
Jergel, Ing. Matej *Slovakia*.
Jerzykiewicz, Mgr inż. Lucjan B. *Poland*.
Jesih, Dr Adolf *Slovenia*.
Jessen, Dr Sven Michael *USA*.
Jesser, William A. *USA*.
Jex, Prof. Dr Hartmut *Germany*.
Jeyabalan Moses, Mr Mohandas Prabhu *India*.
Jeyaratnam, Mailoo *UK*.
Jhoti, Dr Harren *UK*.

Ji, Dr Xinhua *USA.*
Jiang, Prof. Shusheng *China.*
Jiang, Prof. Xiaolong *China.*
Jiang, Dr Xiaoming *China.*
Jiang, Prof. Yandao *China.*
Jiménez-Garay, Prof. Rafael *Spain.*
Jiménez-Morales, Dr Francisco *Spain.*
Jin, Prof. Xianglin *China.*
Jin-an Feng, Jan Feng *USA.*
Jinawath, Dr Supatra *Thailand.*
Jirajesda, Mr. Jate *Thailand.*
Jircitano, Dr Alan J. *USA.*
Jitaru, Dr Raisa Panteleevna *Moldova.*
Jockel, Mr Dietmar *Germany.*
Joelson, Thorleif *Sweden.*
Jørgensen, Jens *Denmark.*
Jogl, Mr. Gerwald *Austria.*
Johanson, Bo Stefan *Finland.*
John, Kevin *USA.*
Johnsen, Ole *Denmark.*
Johnson, Dr Andrew William Syme *Australia.*
Johnson, Dr Carroll K. *USA.*
Johnson, Dr John E. *USA.*
Johnson Jr, Dr Gerald G. *USA.*
Johnson, Prof. Louise Napier *UK.*
Johnson, Dr Michael William *UK.*
Johnson, Dr Owen *UK.*
Johnson, Dr Paul L. *USA.*
Johnson, Quintin *USA.*
Johnston, Steven C. *USA.*
Jones, Dr Daniel S. *USA.*
Jones, Prof. Derry Wynn *UK.*
Jones, Dr Edith Yvonne *UK.*
Jones, Dr Noel D. *USA.*
Jones, Dr Richard Hywel *UK.*
Jones, Dr Stephen John *Canada.*
Jones, T. Alwyn *Sweden.*
Jones, Dr William *UK.*
Jordan, Steven R. *USA.*
Jordanovska, Prof. Dr Vera *(FYR) Macedonia.*
Jorgensen, Dr James D. *USA.*
José Yacaman, Dr Miguel *Mexico.*
Joseph-McCarthy, Dr Diane *USA.*
Joshi, Dr Shri Krishna *India.*
Joshua-Tor, Leemor *USA.*
Joswig, Dr Werner *Germany.*
Jovanovski, Prof. Dr Gligor *(FYR) Macedonia.*
Julian, Maureen M. *USA.*
Juo, Zong Sean *USA.*
Jurković, Prof. Dr Ivan B. *Croatia.*
Jynge, Knut *Norway.*
Kabak, Mehmet *Turkey.*
Kabaleeswaran, Mr V *India.*
Kabalkina, Dr Sarra S. *Russia.*
Kabalov, Dr Yurii K. *Russia.*
Kabelka, Dr Heinz *Austria.*
Kablah, Prof. Lotfi *Egypt.*
Kabsch, Dr Wolfgang *Germany.*
Kabuto, Dr Chizuko *Japan.*
Kacerovský, Ing. Pavel *Czech Republic.*
Kachalov, Dr Oleg V. *Russia.*
Kaduk, Dr James A. *USA.*
Kadziola, Anders *Denmark.*
Kaenkeo, Miss Watcharaporn *Thailand.*
Kaftory, Prof. Menahem *Israel.*
Kaganer, Dr Vladimir M. *Russia.*
Kagarakis, Prof. Constantine *Greece.*
Kageyama, Dr Hiroyuki *Japan.*
Kahlert, Prof. Dr Hartmut *Austria.*
Kahr, Bart E *USA.*
Kai, Prof. Yasushi *Japan.*
Kaid, Ms Mayssa Fathi *Egypt.*
Kaifi, Mr F. M. Zaffar *Pakistan.*
Kaischew, Prof. Rostislaw *Bulgaria.*
Kaiser, Dr Volker *Germany.*
Kaitner, Dr Branko *Croatia.*
Kaito, Prof. Chihiro *Japan.*
Kakimoto, Dr Koichi *Japan.*
Kakitani, Prof. Dr Satoru *Japan.*

Kakudo, Em. prof Masao *Japan.*
Kalbe, Mrs Ute *Germany.*
Kalceff, Dr Walter *Australia.*
Kaler, Eric W. *USA.*
Kalinichenko, Dr Anatolij Mykhajlovych *Ukraine.*
Kalinin, Dr Victor B. *Russia.*
Kalk, Mr Kornelis H. *Netherlands.*
Kallen, Dr Joerg A. *Switzerland.*
Kallio, Mr Pekka Yrjö Juhani *Finland.*
Kálmán, Prof. Alajos *Hungary.*
Kalsbeek, Nicoline *Denmark.*
Kalyanaraman, Dr A. R *India.*
Kalychak, Dr Yaroslav Mykhajlovych *Ukraine.*
Kambas, Prof. Costas *Greece.*
Kamel, Prof. Omar A. *Egypt.*
Kamel, Prof. Raafat Wasef *Egypt.*
Kamenar, Prof. Dr Boris *Croatia.*
Kameníček, Dr Jiří *Czech Republic.*
Kamentsev, Prof. Igor E. *Russia.*
Kamer, Greg *USA.*
Kamigaki, Prof. Dr Nobuo *Japan.*
Kamimura, Dr Midori *Japan.*
Kaminsky, Dr Alexander A. *Russia.*
Kamitori, Shigehiro *USA.*
Kammann, Mrs Elke *Germany.*
Kamolchote, Mr Poonsak *Thailand.*
Kampermann, Sabine P. *USA.*
Kamphuis, Dr Ireneus G. *Netherlands.*
Kanagaraj, Dr Sekar *India.*
Kanatzidis, Dr Mercouri G. *USA.*
Kanda, Dr Hisao *Japan.*
Kandil, Şebnem *Turkey.*
Kandler, Mr Andreas *Germany.*
Kanehisa, Dr Nobuko *Japan.*
Kaneko, Prof. Tsutomu *Japan.*
Kanellis, Prof. George *Greece.*
Kanevsky, Dr Vladimir M. *Russia.*
Kanke, Dr Yasushi *Japan.*
Kannan, Dr K. K. *India.*
Kannan, Dr S. *India.*
Kansikas, Dr Jarno Juhani *Finland.*
Kant, Dr Rajni *India.*
Kantardjieff, Dr Katherine A. *USA.*
Kanters, Dr Jan *Netherlands.*
Kaplunnik, Dr Lidiya N. *Russia.*
Kapon, Dr Moshe *Israel.*
Kapor, Dr Agneš *Serbia.*
Kapostins, MSc Peteris *Latvia.*
Kappenstein, Prof. Charles *France.*
Kapulsky, Alexander *USA.*
Kar, Dr (Mrs) Tanusree *India.*
Karakhanian, Dr Robert K. *Armenia.*
Karakostas, Prof. Theodoros *Greece.*
Karanović, Dr Ljiljana *Serbia.*
Karapetian, Dr Harutyun A. *Armenia.*
Karapetian, Dr Sveta V. *Armenia.*
Karen, Dr Vicky Lynn *USA.*
Karipides, Anastas *USA.*
Kariuki, Dr Benson M. *UK.*
Karl, Prof. Dr Norbert *Germany.*
Karle, Dr Isabella L. *USA.*
Karle, Dr Jean Marianne *USA.*
Karle, Prof. Jerome *USA.*
Karlsen, Solveig *Norway.*
Karlsson, Magnus E. J. *Sweden.*
Karmazin, Dr Lubomír *Czech Republic.*
Karniewicz, Prof. Jan *Poland.*
Karolak-Wojciechowska, Prof. Janina *Poland.*
Karpec, Dr Myroslav Vasyliovych *Ukraine.*
Karplus, Dr P. Andrew *USA.*
Karryev, Mr Alemi N. *Turkmenistan.*
Karthe, Mr P. *India.*
Kartheuser, Dr Edward Peter *Belgium.*
Karvinen, Saila Marjatta *Finland.*
Karyakina, Dr Tatyana A. *Russia.*
Kasatani, Dr Hirofumi *Japan.*
Kashaev, Dr Anvar A. *Russia.*
Kashchiev, Prof. Dimcho *Bulgaria.*
Kashida, Shoji *Japan.*
Kashino, Prof. Setsuo *Japan.*

Kashiwagi, Mr. Tatsuki *Japan.*
Kashiwase, Prof. Yasuji *Japan.*
Kashyap, Dr Ram Prasad *India.*
Kasjanenko, Dr Evgeni V. *Russia.*
Kastner, Dr Berthold *Germany.*
Kastner, Dr Margaret E. *USA.*
Kastrup, Jette Sandholm Jensen *Denmark.*
Kasuga, Prof. Dr Masanobu *Japan.*
Kasukabe, Dr Susumu *Japan.*
Kaszkur, Dr Zbigniew *Poland.*
Kataeva, Dr Olga N. *Russia.*
Katagas, Prof. Christos *Greece.*
Katano, Mr Kizuku *Japan.*
Kataoka, Mr Masayuki *Japan.*
Kataoka, Dr Mikio *Japan.*
Katayanagi, Dr Katsuo *Japan.*
Kato, Dr Akio *Japan.*
Kato, Prof. Hiroshi *Japan.*
Kato, Prof. Norio *Japan.*
Kato, Dr Takamasa *Japan.*
Katrusiak, Dr hab. Andrzej *Poland.*
Katser, Dr Sergey B. *Russia.*
Katsnelson, Prof. Albert A. *Russia.*
Katsube, Dr Yukiteru *Japan.*
Katsui, Prof. Dr Akinori *Japan.*
Katsumata, Dr Tooru *Japan.*
Katsuya, Dr Yoshio *Japan.*
Katti, Suresh K. *USA.*
Katz, Darin S. *USA.*
Katz, Prof. J. Lawrence *USA.*
Katz, Lewis *USA.*
Katzke, Mrs Hannelore *Germany.*
Kaučič, Prof. Dr Venčeslav *Slovenia.*
Kaufman, Hershall W. *USA.*
Kavanaugh, Jeffrey S. *USA.*
Kavounis, Prof. Constantin A. *Greece.*
Kawabata, Mr Naoki *Japan.*
Kawachi, Dr Yosuke *New Zealand.*
Kawado, Dr Seiji *Japan.*
Kawahara, Prof. Akira *Japan.*
Kawakami, Mr Shoji *Japan.*
Kawakita, Dr Tetsuya *Japan.*
Kawaminami, Prof. Masaru *Japan.*
Kawamura, Mr Kazuhiro *Japan.*
Kawamura, Prof. Tsutomu *Japan.*
Kawanaka, Dr Ryusuke *Japan.*
Kawanishi, Mr So-roh *Japan.*
Kawasaki, Mr Masayuki *Japan.*
Kayden, Dr Catherine Sheila *Canada.*
Kayushina, Dr Renata L. *Russia.*
Kazarian, Dr Levon *Armenia.*
Kazimirov, Dr Alexander Yu. *Russia.*
Ke, Dr Hengming *USA.*
Keder, Nancy L. *USA.*
Keefe, Dr Lisa J. *USA.*
Keitel, Mr Thomas *Germany.*
Kek, Dr Stefan *Germany.*
Keller, Prof. Dr Paul *Germany.*
Keller, Dr Egbert *Germany.*
Keller, Dr Ludwig *USA.*
Keller, Mr Peter A. *UK.*
Kelly, Mr Eric *UK.*
Kelly, Dr Judith A. *USA.*
Kelly, Dr Pat *Australia.*
Kelly, Dr Thomas *Ireland.*
Kemme, Dr Andrejs *Latvia.*
Kemmler-Sack, Prof. Dr Sibylle *Germany.*
Kendi, Dr Engin *Turkey.*
Kennard, Colin Harold Leslie *Australia.*
Kennard, Dr Olga *UK.*
Keow-kam-nerd, Dr Kanchana *Thailand.*
Kepa, Dr Henryk *Poland.*
Kepert, Mr Cameron John *UK.*
Keramidas, Constantinos *Greece.*
Keramidas, Vassilis G. *USA.*
Kerestedjian, Dr Thomas *Bulgaria.*
Kerfeld, Dr Cheryl A. *USA.*
Kern, Mr Arnt *Germany.*
Kern, Prof. Raymond *France.*
Kessler, Dr Susanne *South Africa.*

Keszler, Douglas A. *USA.*
Ketolainen, Prof. Pertti Pekka Juhani *Finland.*
Ketterer, Dr Jürgen *Switzerland.*
Keulen, Dr Evert *Netherlands.*
Khadr, Prof. Moustafa *Egypt.*
Khadzhi, Dr Irina P. *Russia.*
Khadzhi, Dr Vadim L. *Russia.*
Khaenko, Prof. Borys Volodymyrovych *Ukraine.*
Khalid, Mr Mohammad *Pakistan.*
Khalifa, Dr B. Abdelmeguid *Egypt.*
Khalifa, Prof. Berlant *Egypt.*
Khamitov , Dr Zhenis Kh. *Kazakhstan.*
Khan, Dr Abdul Quadeer *Pakistan.*
Khan, Dr Abdur Rahman *Pakistan.*
Khan, Prof. Alí *Venezuela.*
Khan, Dr Lajber *Pakistan.*
Khan, Dr Masood A. *USA.*
Khan, Mr Mohammad Afaq *Pakistan.*
Khan, Mr Nawazish Ali *Pakistan.*
Khan, Dr Saeed I. *USA.*
Khan, Dr Shah Alam *Pakistan.*
Khang, Prof. Quan Han *Vietnam.*
Khantaprab, Dr Chaiyudh *Thailand.*
Kharchenko, Dr Olga Ivanivna *Ukraine.*
Kharisun, *Australia.*
Khasanov, Dr Salavat S. *Russia.*
Khatanova, Dr Nina A. *Russia.*
Khawaja, Dr Farid Akhtar *Pakistan.*
Khawaja, Mr Mahmood-ul-Hassan *Pakistan.*
Kheiker, Prof. Daniel M. *Russia.*
Khidr, Prof. Fatma Abdelhakim *Egypt.*
Khimich, Dr Tamara A. *Russia.*
Khisina, Dr Natasha R. *Russia.*
Khitrova, Dr Valentina I. *Russia.*
Khokhlov, Prof. Alexei R. *Russia.*
Kholif, Dr Mahmoud *Egypt.*
Khomenko, Mr Igor O. *Russia.*
Khudojrov, Dr Achmat B. *Uzbekistan.*
Khundzhua, Dr Andrey G. *Russia.*
Kidd, Dr Patricia *UK.*
Kierkegaard, Prof. Peder *Sweden.*
Kiers, Dr Conradus Th. *Netherlands.*
Kies, Dr Jörg *Germany.*
Kießling, Dr Frank-Michael *Germany.*
Kifune, Dr Kouichi *Japan.*
Kihara, Dr Kuniaki *Japan.*
Kihlborg, Prof. Lars H. E. *Sweden.*
Kilranen, Dr Kalle *Estonia.*
Kikuchi, Prof. Makoto *Japan.*
Kikuma, Dr Isao *Japan.*
Killen, Peter David *Australia.*
Kim, Prof. Cheol Jin *Korea.*
Kim, Prof. Duk Soo *Korea.*
Kim, Prof. Ill won *Korea.*
Kim, Dr Jung-Ja P. *USA.*
Kim, Dr Kimoon *Korea.*
Kim, Dr Ki Soo *Korea.*
Kim, Dr Sangsoo *Korea.*
Kim, Prof. Sung-Hou *USA.*
Kim, Prof. Won Sa *Korea.*
Kim, Prof. Yang Bae *Korea.*
Kimata, Dr Mitsuyoshi *Japan.*
Kimmel, Dr Giora *Israel.*
Kimoto, Mr Tsunenobu *Japan.*
Kimura, Dr Hideo *Japan.*
Kimura, Masao *Japan.*
Kimura, Mr Shigeru *Japan.*
King, Prof. Geoffrey S. D. *Belgium.*
King, Dr Hubert E. *USA.*
King, Prof. Hubert Wylam *Canada.*
King, M. V. *USA.*
King, Mrs Susan E. *UK.*
Kingma, Dr Kathleen J. *USA.*
Kinneging, Dr Albertus J. *Netherlands.*
Kinoshita, Dr Kyoichi *Japan.*
Kinoshita, Takayoshi *Japan.*
Kinzhibalo, Dr Volodymyr Vasyliovych *Ukraine.*
Kiosse, Dr Georgy Alexandrovich *Moldova.*
Kipper, Dr Marianne Byrn *USA.*
Kiralj, Mr Rudolf *Croatia.*

Kirchhoff, Dr Andreas *Germany.*
Kirchner, Prof. Dr Elisabeth Ch. *Austria.*
Kirchner, Dr Richard M. *USA.*
Kirfel, Prof. Dr Armin Harald *Germany.*
Kirichenko, Dr Inessa Vladimirovna *Kazakhstan.*
Kirikov, Dr Vladimir K. *Russia.*
Kirillova, Dr Nataliya I. *Russia.*
Kirkova, Prof. Elena *Bulgaria.*
Kirov, Mr Georgi Kirilov *Bulgaria.*
Kirov, Prof. Georgi Nikolov *Bulgaria.*
Kirs, Dr Juho *Estonia.*
Kirschner, Daniel A. *USA.*
Kiryanova, Dr Elena E. *Russia.*
Kirz, Janos *USA.*
Kiselev, Prof. Nikolay A. *Russia.*
Kishida, Dr Satoru *Japan.*
Kishk, Dr Fawzi Mohamed *Egypt.*
Kisi, Dr Erich *Australia.*
Kissel, Lynn *USA.*
Kissinger, Dr Charles R. *USA.*
Kistenmacher, Dr Thomas J. *USA.*
Kita, Ms Akiko *Japan.*
Kitadokoro, Mr Kengo *Japan.*
Kitagawa, Dr Ryuji *Japan.*
Kitamura, Dr Mitsutaka *Japan.*
Kitamura, Tatsuo *Japan.*
Kitson, Dr David H. *UK.*
Kivekäs, Raikko Terjo Ilari *Finland.*
Kivikoski, Dr Jussi Heikki *Finland.*
Kiyotani, Ms Tamiko *Japan.*
Kızlyallı, Dr Meral *Turkey.*
Kjær, Kristian *Denmark.*
Kjekshus, Prof. Arne *Norway.*
Kjeldgaard, Morten *Denmark.*
Klapdor, Mr Martin Frank *Germany.*
Klapper, Prof. Dr Helmut *Germany.*
Klechkovskaya, Dr Vera V. *Russia.*
Klee, Prof. Dr Wilfrid Edgar *Germany.*
Klei, Herbert E. *USA.*
Klein, Dr Cheryl L. *USA.*
Klein, Mr Volker *Germany.*
Kleint, Prof. Dr Christian *Germany.*
Klemens, P. G. *USA.*
Klement, Dr Ulrich *Germany.*
Klevtsov, Dr Petr V. *Russia.*
Klevtsova, Dr Rimma F. *Russia.*
Klewe, Bernt *Norway.*
Klimakow, Dr Alexander *Germany.*
Klimasauskas, Dr Saulius *USA.*
Klimm, Dr Detlef *Germany.*
Klimova, Dr Anna Yu. *Russia.*
Klimova, Dr Irina P. *Russia.*
Klinga, Martti *Finland.*
Klinkhammer, Dr Karl Wilhelm *Germany.*
Klintschar, Mr. Gerd *Austria.*
Klop, Dr Enno A. *Netherlands.*
Klosi, Dr Fatos *Albania.*
Klug, Dr Aaron *UK.*
Klug, Dr Dennis D. *Canada.*
Klymiv, Dr Ivan Mykhajlovych *Ukraine.*
Kniep, Prof. Dr Rüdiger *Germany.*
Knight, Kevin Steven *UK.*
Knighton, Dr Daniel R. *USA.*
Knížek, Dr Karel *Czech Republic.*
Knobler, Carolyn B. *USA.*
Knoch, Dr Falk *Germany.*
Knop, Prof. Osvald *Canada.*
Knossow, Dr Marcel *France.*
Knox, Dr James R. *USA.*
Ko, Prof. Dr Young Shin *Korea.*
Ko, Taegyung *Korea.*
Kobayashi, Dr Akiko *Japan.*
Kobayashi, Dr Hayao *Japan.*
Kobayashi, Prof. Jinzo *Japan.*
Kobayashi, Prof. Nobuyuki *Japan.*
Kobayashi, Prof. Tadashi *Japan.*
Kobe, Bostjan *USA.*
Kobzareva, Dr Svetlana A. *Russia.*
Koch, Dr Beatrix *Netherlands.*
Koch, Prof. Dr Elke *Germany.*

Koch, Stephen A. *USA.*
Kociński, Prof. Jerzy *Poland.*
Kocman, Dr Vladimir *Canada.*
Koda, Dr Shigetaka *Japan.*
Kodama, Dr Hideomi *Canada.*
Kodandapani, Mr R. *India.*
Ködderitzsch, Dr Horst *Germany.*
Koellner, Dr Gertraud *Germany.*
Köhler, Dr Rolf *Germany.*
Koelsch, Gerald *USA.*
König de Perazzo, Ms Patricia Veronica *Argentina.*
Köpke, Dr Jürgen *Germany.*
Koeppe II, Roger E. *USA.*
Körber, Dr Fritjof Carl Friedrich *UK.*
Körber, Dr Wolfgang *Germany.*
Kőszegi, Mr László *Hungary.*
Koetzle, Dr Thomas F. *USA.*
Koga, Dr Kenji *Japan.*
Koh, Dr Lip Lin *Singapore.*
Kohl, Prof. Dr Helmut *Germany.*
Kohlbeck, Dr Franz *Austria.*
Kohli, Dr Vijay Kumar *India.*
Kohn, Jack A. *USA.*
Koide, Prof. Tsutomu *Japan.*
Koizumi, Prof. Dr Mitsue *Japan.*
Kojić-Prodić, Dr Biserka *Croatia.*
Kojima, Prof. Dr Kenichi *Japan.*
Kojima, Yuko *Japan.*
Kokanian, Dr Edward P. *Armenia.*
Kokkou, Prof. Sokrates *Greece.*
Koknat, Friedrich W. *USA.*
Kokotailo, George T. *USA.*
Kolar, Prof. Dr Drago *Slovenia.*
Kolatkar, Anand R. *USA.*
Kolatkar, Dr Prasanna R. *USA.*
Kolega, Ing. Michal *Czech Republic.*
Kolesova, Dr Rimma V. *Russia.*
Kolin, Dr Nikolai G. *Russia.*
Kolobyanina, Dr Tatyana N. *Russia.*
Kolodieva, Dr Svetlana V. *Russia.*
Kolupajeva, Dr Zoya Ivanivna *Ukraine.*
Koman, Doc. Ing. Marian *Slovakia.*
Komarek, Prof. Dr Kurt L. *Austria.*
Komatsu, Prof. Hiroshi *Japan.*
Komatsu, Mr Yoshinobu *Japan.*
Komiya, Dr Hiromi *USA.*
Komninou, Prof. Philomela *Greece.*
Kompan, Dr Olga Ye. *Russia.*
Komura, Prof. Dr Shigehiro *Japan.*
Komura, Prof. Yukitomo *Japan.*
Kon, Dr Aviv Yuliseyevich *Moldova.*
Kondo, Mr Hidemasa *Japan.*
Kondratenko, Dr Ludmila K. *Russia.*
Kong, Dr Xiangpeng *USA.*
Konguetsof, Dr Helen *Greece.*
Konnert, Dr John *USA.*
Konnert, Judith A. *USA.*
Konovalikhin, Dr Sergei V. *Russia.*
Konovalov, Mr Oleg V. *Russia.*
Konstantinov, Dr Ivan *Bulgaria.*
Konstantinova, Dr Alisa F. *Russia.*
Konyk, Dr Maria Bogdanivna *Ukraine.*
Kooijman, Dr Huub *Netherlands.*
Koopmans, Prof. Dr Kasper *Netherlands.*
Kopelman, Raoul *USA.*
Kopf, Dr Jürgen *Germany.*
Kopka, Mary L. *USA.*
Kopský, Dr Vojtěch *Czech Republic.*
Koptsik, Prof. Vladimir A. *Russia.*
Koreň, Dr Branislav *Slovakia.*
Koritsánszky, Dr Tibor *Hungary.*
Korkhin, Yaacov M. *Israel.*
Kornev, Dr Alexey N. *Russia.*
Korolev, Dr Sergey V. *Russia.*
Korotkov, Dr Vitalii Alexandr *Moldova.*
Korp, James D. *USA.*
Korvenranta, Jorma Artturi *Finland.*
Kosevich, Prof. Vadim Markovych *Ukraine.*
Kosmachev, Dr Sergej Mykhajlovych *Ukraine.*
Kosmopoulos, Dr John *Greece.*

Kosova, Dr Tatyana B. *Russia.*
Kossiakoff, Anthony A *USA.*
Kosten, Dr Klaus *Germany.*
Kostiner, Dr Edward S. *USA.*
Kostorz, Dr Gernot *Switzerland.*
Kostov, Prof. Ivan *Bulgaria.*
Kostov, Dr Ruslan I. *Bulgaria.*
Kostrewa, Dr Dirk *Switzerland.*
Kosturkiewicz, Prof. Zofia *Poland.*
Košutić Hulita, Mrs Nada *Croatia.*
Koszelak, Stan *USA.*
Kotelnikov, Petr Evgenievich *Kazakhstan.*
Kotelnikova, Dr Elena N. *Russia.*
Kotila, Sirpa Kristiina *Finland.*
Koto, Prof. Kichiro *Japan.*
Kotru, Dr P. N. *India.*
Kotsanidis, Dr Panagiotis *Greece.*
Kotsis, Dr Konstadinos T. *Greece.*
Kotur, Dr Bogdan Yaroslavovych *Ukraine.*
Kotzev, Dr Joseph *Bulgaria.*
Koumelis, Dr Christos *Greece.*
Kourinov, Dr Igor Kourinov *USA.*
Kouta, Mr Hikaru *Japan.*
Kouwijzer, Drs Milou *Netherlands.*
Kovachev, Dr Peter *Bulgaria.*
Kovács, Prof. Alessandro Leone *Italy.*
Kovalchuk, Prof. Michail V. *Russia.*
Kovari, Dr Ladislau Z. *USA.*
Kovyev, Dr Ernest K. *Russia.*
Kowal, Mgr Ewa *Poland.*
Koyama, Dr *Japan.*
Koyama, Mr Tadashi *Japan.*
Koyano, Mr Kazuo *Japan.*
Kozaki, Dr Shigeru *Japan.*
Kozioł, Dr hab. Anna E. *Poland.*
Kožíšek, Ing. Jozef *Slovakia.*
Kozlovskis, Dr Leonids *Latvia.*
Koz'ma, Dr Oleksandr Oleksiyovych *Ukraine.*
Kraatz, Paul *USA.*
Krabbendam, Drs Hendrik *Netherlands.*
Krämer, Prof. Dr Volker *Germany.*
Kräußlich, Dr Jürgen *Germany.*
Krajewski, Dr Adriano *Italy.*
Krajewski, Dr Janusz *Poland.*
Králová, Dr Rudolfa *Czech Republic.*
Krane, Dr Hans-Georg *Germany.*
Kranold, Prof. Dr Rainer *Germany.*
Kratky, Dr Christoph *Austria.*
Kratky, Prof. Dr Dr h.c. mult. Otto *Austria.*
Kraus, Prof. Dr Ivo *Czech Republic.*
Krauß, Dr Norbert *Germany.*
Krause, Dr Christian *Germany.*
Krauße, Dr Joachim *Germany.*
Krause, Dr Kurt L. *USA.*
Krause-Bauer, Dr Jeanette A. *USA.*
Krausová, Dr Dagmar *Czech Republic.*
Kraut, Joseph *USA.*
Kravtsov, Dr Victor Christophorovich *Moldova.*
Krawiec, Dr Mariusz *USA.*
Krawitz, Aaron *USA.*
Krebs, Prof. Dr Bernt *Germany.*
Krestev, Mr Venelin *Bulgaria.*
Kresteva, Dr Manya *Bulgaria.*
Kretschmar, Mr Martin *Germany.*
Kretsinger, Robert H. *USA.*
Kretzschmar, Dr Ulrike *Germany.*
Kreutz, Dr Ernst Wolfgang *Germany.*
Krever, Dr Maarten *Netherlands.*
Kriegel, Mr Steffen *Germany.*
Krieger, Monty *USA.*
Krischner, Prof. Dr Harald *Austria.*
Krishna, Prof. Padmanabhan *India.*
Krishnaiah, Dr Musali *India.*
Krishnamachari, Dr Ramaswamy *India.*
Krishnaswamy, Dr S. *India.*
Kritayakirana, Mrs Rungsri *Thailand.*
Kritikos, Mikael *Sweden.*
Krivandina, Dr Elena A. *Russia.*
Krivenko, Dr Vladimir G. *Russia.*
Krochuk, Dr Vasyl' Maksymovych *Ukraine.*

Kroeger, Kenneth Scott *USA.*
Kroll, Prof. Dr Herbert *Germany.*
Kronenburg, Dr Martinus J. *Netherlands.*
Kroon, Prof. Dr Jan *Netherlands.*
Kroon-Batenburg, Dr Louise M. J. *Netherlands.*
Krstanović, Dr Ilija *Serbia.*
Krüger, Prof. Dr Carl *Germany.*
Krueger, Dr Susan *USA.*
Krug, Prof. Dr Detlef *Germany.*
Kruger, Prof. Gert Jacobus *South Africa.*
Krumina, Ms Aija *Latvia.*
Kryger, Gitay *USA.*
Krygowski, Prof. Tadeusz Marek *Poland.*
Kryschi, Dr Carola *Germany.*
Kryshtop, Dr Victor M. *Russia.*
Kuang, Prof. Bao *China.*
Kuban, Dr Ralf-Jürgen *Germany.*
Kuběna, Doc. Dr Josef *Czech Republic.*
Kubiak, Doc. Ryszard *Poland.*
Kubiak, Dr Maria *Poland.*
Kubicki, Dr Maciej *Poland.*
Kubili, Dr Vasyl' Zenonovych *Ukraine.*
Kuboi, Prof. Dr Ryoichi *Japan.*
Kubota, Prof. Dr Noriaki *Japan.*
Kubota, Yoshiki *Japan.*
Kucharczyk, Dr Damian *Poland.*
Kucharski, Dr Edward Stanislaw *Australia.*
Kuchitsu, Prof. Kozo *Japan.*
Kudoh, Dr Yasuhiro *Japan.*
Kühlbrandt, Dr Werner *Germany.*
Kühn, Prof. Dr Robert *Germany.*
Kühnle, Mr Florian N. M. *Switzerland.*
Küppers, Prof. Dr Horst *Germany.*
Kuge, Dr Ken-ichi *Japan.*
Kuhn, Dr Leslie A. *USA.*
Kuhs, Prof. Dr Werner F. *Germany.*
Kukimoto, Prof. Hiroshi *Japan.*
Kukina, Dr Galina A. *Russia.*
Kukol', Dr Victor Valerijevych *Ukraine.*
Kukuy, Dr Anatoly L. *Russia.*
Kulda, Dr Jiří *France.*
Kuma, Dr Shoji *Japan.*
Kumagawa, Prof. Dr Masashi *Japan.*
Kumao, Dr Akihiro *Japan.*
Kumar, Satyendra *USA.*
Kumar, Mr V. Amarendra *India.*
Kumar, Dr Vinay *India.*
Kumar, Vinod D. *USA.*
Kumaraswamy, Vidya Sagar *UK.*
Kumazawa, Shintaro *Japan.*
Kumereson, Dr P. *India.*
Kundra, Dr Krishan Dev *India.*
Kundrot, Prof. Craig E. *USA.*
Kundu, Dr Mohanlal *India.*
Kunrath, Mr José Irineu *Brazil.*
Kunstelj, Dr Drago *Croatia.*
Kunte, Mr Vivek P *India.*
Kunz, Martin *USA.*
Kuo, Prof. Ke-hsin *China.*
Kuok, Dr Meng Hau *Singapore.*
Kupriyanov, Dr Mikhail F. *Russia.*
Kurahashi, Dr Masayasu *Japan.*
Kuranova, Dr Inna P. *Russia.*
Kurdyumov, Prof. Georgy Vyacheslavovich *Russia.*
Kurdyumov, Prof. Oleksandr Vyacheslavovych *Ukraine.*
Kuriyan, John *USA.*
Kurki-Suonio, Prof. Dr Kaarle V. J. *Finland.*
Kurkutova, Prof. Evdokiya N. *Russia.*
Kuroda, Dr Ekyo *Japan.*
Kuroiwa, Prof. Dr Koichi *Japan.*
Kuser, Dr Paula R. *USA.*
Kusunoki, Dr Katsuyuki *Japan.*
Kusunoki, Dr Masami *Japan.*
Kusz, Dr Joachim *Poland.*
Kuvaldin, Dr Boris *Latvia.*
Kuwano, Dr Yasuhiko *Japan.*
Kuyper, Lee *USA.*
Kužel, Dr Radomír Jr *Czech Republic.*
Kuz'ma, Prof. Yurij Bogdanovych *Ukraine.*

Kuzmany, Prof. Dr Hans *Austria.*
Kuz'micheva, Prof. Galina M. *Russia.*
Kuzmin, Dr Ivan I. *Russia.*
Kuzmin, Prof. Runar N. *Russia.*
Kuzmin, Dr Vladimir S. *Russia.*
Kuzmina, Dr Ludmila G. *Russia.*
Kuz'mina, Dr Mariya A. *Russia.*
Kuzmina, Dr Olga V. *Russia.*
Kuznetsov, Dr Gennadij Vasyliovych *Ukraine.*
Kuznetsov, Dr Victor A. *Russia.*
Kuznetsova, Mrs Tatyana P. *Russia.*
Kvasnitsa, Dr Victor Mykolayovych *Ukraine.*
Kvick, Prof. Åke *France.*
Kwiatkowski, Dr Witold *Poland.*
Kyriakidis, Dr Christos E. *Netherlands.*
Kyriakos, Prof. Dimitris S. *Greece.*
Kyröläinen, Antero Johannes *Finland.*
Labaki, Dr Lucila *Brazil.*
Labbé, Prof. Philippe *France.*
Labib, Dr Fawkia *Egypt.*
Labrador, Dr Manuel *Spain.*
Lachgar, Dr Abdessadek *USA.*
Lacmann, Prof. Dr-Ing. Rolf *Germany.*
Ladd, Dr Marcus F. C. *UK.*
Ladenstein, Prof. Dr Rudolf *Sweden.*
Lähdeniemi, Matti J. I. *Finland.*
Lager, George *USA.*
Laggner, Prof. Dr Peter *Austria.*
Łągiewka, Prof. Eugeniusz *Poland.*
Lagomarsino, Dr Stefano *Italy.*
Lahiri, Dr Barendra Nath *India.*
Lahoz, Dr Fernando J. *Spain.*
Lahti, Seppo I. *Finland.*
Lai, Prof. Luhua *China.*
Lai, Dr Ting Fong *Hong Kong.*
Laine, Ensio Sulo Uolevi *Finland.*
Laing, Dr Mary Elizabeth *South Africa.*
Laing, Prof. Michael John *South Africa.*
Lajzerowicz, Prof. Janine *France.*
Lake, Mr Philip G. *UK.*
Lakshminarayanan, Mr Muthuvijayan *India.*
Lal, Dr Krishan *India.*
Lamas, Mr Diego Germán *Argentina.*
Lamba, Dr Doriano *Italy.*
Lambert, Prof. Marianne *France.*
Lambert, Dr Ulrich *Germany.*
Lambert-Smith, John Ernie Warwick *Australia.*
Lamotte-Brasseur, Dr Josette *Belgium.*
Lampe-Önnerud, Christina *Sweden.*
Lando, Jerome B. *USA.*
Lanfranchi, Prof. Maurizio *Italy.*
Lang, Prof. Andrew Richard *UK.*
Lange, Dr Joachim Reinhard *Germany.*
Langer, Prof. Dr Klaus *Germany.*
Langer, Dr Vratislav *Sweden.*
Langford, Dr J. Ian *UK.*
Langridge, Sarah-Jane *UK.*
Langs, Dr David A. *USA.*
Lanzavecchia, Dr Salvatore *Italy.*
Lapshin, Mr Andrey E. *Russia.*
Laptev, Dr Vladimir A. *Russia.*
Lapthorn, Dr Adrian J. *UK.*
Lara Magaña, Ms María Eugenia *Mexico.*
Laredo, Prof. Estrella *Venezuela.*
Lariucci, Dr Carlito *Brazil.*
Larsen, Finn Krebs *Denmark.*
Larsen, Ingrid Kjoeller *Denmark.*
Larsen, Rolf Leon *Norway.*
Larsen, Sine *Denmark.*
Larsen, Dr Teresa A. *USA.*
Larson, Allen C. *USA.*
Larson, Bennett C. *USA.*
Larson, Elizabeth M. *USA.*
Larson, Dr Jay Michael *USA.*
Larson, Dr Steven Bland *USA.*
Larsson, Ann-Kristin *Sweden.*
Larsson, Tommy *Sweden.*
Lascombe, Dr Marie-Bernard *France.*
Lashgari, Kianosh *Sweden.*
Łasocha, Dr Wiesław *Poland.*

Lattman, Dr Eaton D. *USA*.
Laube, Dr Gert *Germany*.
Laube, Dr Thomas *Switzerland*.
Laudise, Dr Robert *USA*.
Lavie, Arnon *USA*.
Lawrence, Michael Colin *Australia*.
Lawson, A. C. *USA*.
Lawson, Dr Catherine L. *USA*.
Lawton, Stephen L. *USA*.
Lazar, Dr Dušan *Serbia*.
Le Du, Dr Marie H. *USA*.
Le Loux, Drs Richard *Netherlands*.
Le Page, Dr Yvon *Canada*.
Le Roux, Dr Stephanus David *South Africa*.
Le Trong, Isolde *USA*.
Lea, Prof. Sydney George *Canada*.
Leadbetter, Dr Alan *France*.
Leake, Dr John Anthony *UK*.
Leban, Prof. Dr Ivan *Slovenia*.
Lebech, Bente *Denmark*.
Lebedeva, Dr Marina Volodymyrivna *Ukraine*.
Lebioda, Dr Lukasz *USA*.
Lebuis, Dr Anne-Marie *Canada*.
Leccabue, Dr Fabrizio *Italy*.
Lechat, Dr Johannes Rudiger *Brazil*.
Leciejewicz, Prof. Janusz *Poland*.
Leckebusch, Dr Rudolf *Botswana*.
Lecomte, Prof. Claude E. P. *France*.
Lee, Byungkook (BK) *USA*.
Lee, Dr Chih-Hao *Taiwan*.
Lee, Dr Chnoog Kheng *Malaysia*.
Lee, Prof. Chongmu *Korea*.
Lee, Prof. Hong Lim *Korea*.
Lee, Hsin-Yi *Taiwan*.
Lee, Prof. Jeong Yong *Korea*.
Lee, Prof. Jung Hoo *Korea*.
Lee, Mr Jung Kyu *Korea*.
Lee Moreno, Dr José Luis *Mexico*.
Lee, Dr Peter L. *USA*.
Lee, Prof. Soon Won *Korea*.
Lee, Sukyeong *USA*.
Lee, Prof. Tae Keun *Korea*.
Lee, Xavier *USA*.
Lefebvre, Dr Simone *France*.
Lefeld-Sosnowska, Prof. dr hab. Maria *Poland*.
LeGeros, Racquel Z. *USA*.
Legkova, Dr Galyna Victorivna *Ukraine*.
Legros, Prof. Jean-Pierre *France*.
Lehmann, Mrs Alice *Germany*.
Lehmann, Dr Christian W. *UK*.
Lehmann, Dr Mogens Steen *France*.
Lehtinen, Martti Kalevi *Finland*.
Lehto, Vesa-Pekka *Finland*.
Leipner, Dr Hartmut S. *Germany*.
Leipoldt, Prof. Johann Gotlieb *South Africa*.
Leiro, Jarkko Albert *Finland*.
Leiserowitz, Prof. Leslie *Israel*.
Leite, Dr Cirano Rocha *Brazil*.
Leitus, Dr Grigorii M. *Russia*.
Leković, Mrs Nada *Serbia*.
Lengauer, Dr Christian Leopold *Austria*.
Lengauer, Dr Walter Oskar Franz *Austria*.
Lenhert, P. Galen *USA*.
Lenstra, Dr Albert T. H. *Belgium*.
Lenz, Mrs Annett *Germany*.
Leonard, Dr Gordon A. *UK*.
Leonardsen, Erik *Denmark*.
Leong, Mr Weng Kee *Canada*.
Leoni, Prof. Leonardo *Italy*.
Leonidas, Dr Demetrios D. *UK*.
Leonowicz, Michael E. *USA*.
Leonyuk, Dr Lydia I. *Russia*.
Leonyuk, Dr Nikolai I. *Russia*.
Leovac, Dr Vukadin *Serbia*.
Leporati, Prof. Enrico *Italy*.
Leppä-aho, Jaakko Antero *Finland*.
Lerf, Dr Anton *Germany*.
Lesburg, Charles A. *USA*.
Leskelä, Markku Antero *Finland*.

Leslie, Dr Andrew G. W. *UK*.
Lessinger, Leslie *USA*.
Leuschner, Dr Dieter *Germany*.
Leute, Prof. Dr Volkmar *Germany*.
Levendis, Dr Demetrius Christos *South Africa*.
Leventouri, Dr Dora *Greece*.
Leverett, Prof. Peter *Australia*.
Levin, Mr Alexandr A. *Russia*.
Levin, Vladimir Leonidovich *Kazakhstan*.
Levina, Dr Olga I. *Russia*.
Levinthal, Peter Steven *USA*.
Levonian, Dr Levon V. *Armenia*.
Levy, Dr Henri A. *USA*.
Lewiński, Dr Krzysztof *Poland*.
Lewit-Bentley, Dr Anita *France*.
Lezama, Dr Luis *Spain*.
Li, Chi-Tang *USA*.
Li, Prof. Fanghua *China*.
Li, Prof. Genpei *China*.
Li, Hui-Ying *USA*.
Li, Dr Ludmila E. *Russia*.
Li, Ms Mei *China*.
Li, Prof. Runshen *China*.
Li, Thomas Tien-Hsiung *USA*.
Li, Prof. Wanmao *China*.
Li, Ying *USA*.
Li, Dr Yong J. *USA*.
Li, Dr Youli *USA*.
Liang, Prof. Dongcai *China*.
Liang, Dr Jiin-Yun *USA*.
Liang, Prof. Jingkui *China*.
Liang, Dr Li *USA*.
Liao, Dr Der-Ing Liao *USA*.
Liao, Fen-ling *Taiwan*.
Liaw, Shwu-Huey *Taiwan*.
Liaw, Dr Yen-Chywan *Taiwan*.
Licci, Dr Francesca *Italy*.
Licheri, Prof. Giovanni *Italy*.
Lidin, Dr Sven *Sweden*.
Liebau, Prof. Dr Friedrich *Germany*.
Lieberman, Harvey *UK*.
Liebman, Michael N. *USA*.
Liehr, Dr Günter *Germany*.
Lightfoot, Dr Philip *UK*.
Lii, Kwang-Hwa *Taiwan*.
Likhushina, Dr Ekaterina V. *Russia*.
Liles, Mr David Charles *South Africa*.
Liljas, Dr Lars *Sweden*.
Lim, Dr Louis W. *USA*.
Lima-de-Faria, Prof. José *Portugal*.
Lin, Prof. Chichang *China*.
Lin, Dr Hsi-Che *Taiwan*.
Lin, Dr Kuau-Jiuh *Taiwan*.
Lin, Prof. Shaofan *China*.
Lin, Prof. Szu-Bin *Taiwan*.
Lin, Prof. Xisheng *China*.
Lin, Prof. Yonghua *China*.
Lin, Prof. Yujuan *China*.
Lin, Prof. Zhengjiong *China*.
Lincoln, Frank John *Australia*.
Lind, M. David *USA*.
Lindahl, Tommie *Sweden*.
Lindegaard-Andersen, Asger *Denmark*.
Lindeman, Dr Sergey V. *Russia*.
Lindemann, Prof. Dr Willi *Germany*.
Linden, Dr Anthony *Switzerland*.
Lindgreen, Holger *Denmark*.
Lindina, Dr Lauma *Latvia*.
Lindley, Prof. Peter F. *UK*.
Lindner, Prof. Dr Hans Jörg *Germany*.
Lindqvist, Kristian Vilhelm *Finland*.
Lindqvist, Ylva *Sweden*.
Lindroos, Veikko Kalervo *Finland*.
Lingafelter, Dr Edward C. *USA*.
Linke, Mr Dieter *Germany*.
Lipkowska, Dr Zofia *Poland*.
Lipkowski, Prof. Janusz *Poland*.
Lippard, Stephen J. *USA*.
Lippmann, Prof. Dr Friedrich *Germany*.
Lippmann, Mr Thomas *Germany*.

Lipscomb, Ms Karen J. *UK*.
Lipscomb, Leigh Ann *USA*.
Lipscomb, Prof. William Nunn *USA*.
Liquori, Prof. Alfonso Maria *Italy*.
Lira-Olivares, Prof. Dr Joaquín *Venezuela*.
Lis, Prof. Tadeusz *Poland*.
Lisnyak, Dr Semen Stepanovych *Ukraine*.
Liso-Rubio, Prof. M. Jesus *Spain*.
Listovnichii, Dr Victor Evgenovych *Ukraine*.
Litovchenko, Dr Anatolij Stepanovych *Ukraine*.
Litster, Dr Stephen *Canada*.
Litvin, Daniel Bernard *USA*.
Litvin, Dr Oleksandr Lukych *Ukraine*.
Litvinov, Dr Igor A. *Russia*.
Lityagina, Dr Ludmila M. *Russia*.
Liu, Prof. Hanqin *China*.
Liu, Dr Hongying (Hattie) *USA*.
Liu, Prof. Ling-Kang *Taiwan*.
Liu, Prof. Shixiong *China*.
Liu, Prof. Xiaolan *China*.
Liu-González, Mr Malva *Spain*.
Livnah, Oded *Israel*.
Llamas-Saiz, Dr Antonio Luis *Spain*.
Llanos, Dr Jaime *Chile*.
Lloveras, Dr Joaquim *Spain*.
Lloyd, Douglas James *Australia*.
Lloyd, Michael A. *USA*.
Lobkovsky, Dr Emil B. *USA*.
Lobsanov, Dr Yuri D. *Canada*.
Locchi, Dr Stelio *Italy*.
Lock, Prof. Colin James Lyne *Canada*.
Loeb, Arthur L. *USA*.
Löchner, Dr Ulrich *Germany*.
Loehlin, James H. *USA*.
Loeksmanto, Prof. Dr Waloejo *Indonesia*.
Löns, Dr Friedrich Jürgen *Germany*.
Lövqvist, Karin *Sweden*.
Loginov, Dr Evgenii B. *Russia*.
Lohse, Mr Christian *Switzerland*.
Lois, Mr Christos *Germany*.
Loizos, Mr Zafiris *Greece*.
Lokanatha, Mr S. *India*.
Lolis, Elias *USA*.
Loll, Dr Patrick J. *USA*.
Lombardo, Prof. Eduardo *Argentina*.
Lomnytska, Dr Yaroslava Fedorivna *Ukraine*.
Lomonov, Dr Vladimir A. *Russia*.
Long, Dr Gabrielle Gibbs *USA*.
Long, Dr Marianna M. *USA*.
Longueville, Dr Willy *France*.
Lontos, Dr Charalambos *Greece*.
Loopstra, Prof. Dr Bert O. *Netherlands*.
Lopez de Diego, Heidi *Denmark*.
Lopez, Dr Silio *Cuba*.
López-Acevedo, Dr M. Victoria *Spain*.
López-Aguayo, Prof. Francisco *Spain*.
López-Castro, Prof. Amparo *Spain*.
Lopez-Galindo, Dr Alberto *Spain*.
Lopez-Gonzalez, Prof. Juan de D. *Spain*.
Lopez-Soler, Dr Angel *Spain*.
Lorentzen, Torben *Denmark*.
Loreto, Prof. Lucio *Italy*.
Lorimer, Prof. Gordon Winston *UK*.
Loris, Mr Remy *Belgium*.
Loshmanov, Dr Arkady A. *Russia*.
Lotfy, Prof. Mohamed *Egypt*.
Lottermoser, Dr Werner *Austria*.
Loub, Prof. Dr Josef *Czech Republic*.
Louër, Dr Daniel *France*.
Lough, Dr Alan John *Canada*.
Loupias, Prof. Geneviève *France*.
Lovas, Dr György Antal *Hungary*.
Love, Warner Edward *USA*.
Loveday, Dr John S. *UK*.
Lovejoy, Brett *USA*.
Lovey, Dr Francisco Carlos *Argentina*.
Low, Prof. Barbara W. *USA*.
Lowe, Dr Philip R. *UK*.
Lowe-Ma, Dr Charlotte K. *USA*.
Lowrey, Alfred H. *USA*.

Lu, Prof. Guangying *China.*
Lu, Prof. Jiaxi *China.*
Lu, Prof. Kunquan *China.*
Lu, Prof. Shaofang *China.*
Lu, Prof. Tian-Huey *Taiwan.*
Lu, Ms Yang *China.*
Lube, Dr Emil' L. *Russia.*
Lubini, Mr Paolo *Switzerland.*
Luboradzki, Mgr Roman *Poland.*
Lucas, Dr Brian William *Australia.*
Lucchesi, Dr Sergio *Italy.*
Lucchetti, Prof. Gabriella *Italy.*
Luciak, Dr Bernard *Poland.*
Ludlam-Brown, Dr Ian R. *UK.*
Ludwig, Martha L. *USA.*
Luecke, Hartmut "Hudel" *USA.*
Lueken, Prof. Dr Heiko *Germany.*
Luger, Prof. Dr Peter *Germany.*
Lulé, Dr Marija *Croatia.*
Luján, Mr Marcos *Switzerland.*
Lukas, Dr Thomas J. *USA.*
Lukaszewicz, Prof. Kazimierz *Poland.*
Lukaszewski, George Michael *Australia.*
Lumme, Paavo Olavi *Finland.*
Lundberg, Doc. Bruno K. S. *Sweden.*
Lundberg, Dr Monica *Sweden.*
Lundgren, E. A. Lennart *Sweden.*
Lundström, Prof. Torsten *Sweden.*
Lunin, Dr Vladimir Yu. *Russia.*
Luo, Prof. Jingchu *China.*
Luo, Dr Yao Guang *Canada.*
Luque, Dr Antonio *Spain.*
Lutar, Dr Karel *Slovenia.*
Lutz, Prof. Dr Heinz Dieter *Germany.*
Lux, Prof. Dr Benno *Austria.*
Luzzati, Dr Vittorio *France.*
Lvov, Dr Yuri M. *Russia.*
Lwin, Ms Hla Myat *Myanmar.*
Lynch, Denis Francis *Australia.*
Lynch, Dr Vincent M. *USA.*
Lyubalin, Prof. Mark Dm. *Russia.*
Lyubimov, Dr Vasily N. *Russia.*
Lyubutin, Prof. Igor S. *Russia.*
Lyutin, Dr Vladimir I. *Russia.*
Lyxell, Dan-Göran *Sweden.*
Ma, Prof. Lidun *China.*
Ma, Prof. Xingqi *China.*
Ma, Yiqun *USA.*
Ma, Prof. Zhesheng *China.*
Mårdalen, Dr Jostein *Norway.*
Maartmann-Moe, Knut *Norway.*
Maas, Prof. Dr Gerhard *Germany.*
Maaskant, Prof. Dr Willem J. A. *Netherlands.*
Macalet, Conf. Viorel *Romania.*
Macdonald, Dr John Emyr *UK.*
Machajdík, Ing. Daniel *Slovakia.*
Machida, Mr Hiroshi *Japan.*
Machida, Dr Katsumi *Japan.*
Machida, Dr Mitsuo *Japan.*
Maciček, Dr Josef *Bulgaria.*
Mack, Joseph P. G. *USA.*
Mackay, Prof. Alan Lindsay *UK.*
Mackay, Dr Maureen Florence *Australia.*
Mackenzie, Dr James Kenneth *Australia.*
Mackle, Dr Paul E. *USA.*
Maclean, John K. F. *UK.*
Macrae, Clare F. *UK.*
Madariaga, Dr Gotzon *Spain.*
Madden, Dean R. *Germany.*
Madden, John J. *USA.*
Madhusudan, Mr *India.*
Madhusudan, Mr *India.*
Madhusudhana, Dr N. V. *India.*
Madrid, Marcela *USA.*
Madsen, Ian Charles *Australia.*
Madureira Filho, Prof. José Barbosa de *Brazil.*
Mady, Prof. Khairy A. *Egypt.*
Mäki, Jouko Kalervo *Finland.*
Mändar, Dr Hugo *Estonia.*
Maene, Mr Norbert *Belgium.*

Maenhout - Van Der Vorst, Dr W. M. R. *Belgium.*
Maes, Dr Dominique *Belgium.*
Maes, Mr Stefan *Belgium.*
Magarill, Dr Svetlana A. *Russia.*
Magerl, Dr Andreas *France.*
Magill, J. H. *USA.*
Maginn, Dr Stephen James *UK.*
Magnéli, Prof. Arne *Sweden.*
Magomedova, Dr Nina S. *Russia.*
Mahalingam, Mr Bhuvaneswari *India.*
Mahata, Dr Akhil *India.*
Maheswaran, Saravanamuthu *Australia.*
Mahmood, Dr Khursheed *Pakistan.*
Mahmoudi, Dr Mounir *Venezuela.*
Mahy, Dr Jan W. G. *Netherlands.*
Mai, Prof. Zhenhong *China.*
Maier, Dr Horst *Germany.*
Main, Dr Peter *UK.*
Maiti, Dr Gobinda Chandra *India.*
Maiwa, Dr Koji *Japan.*
Maixner, Dr Jaroslav *Czech Republic.*
Maiza, Dr Pedro Jose *Argentina.*
Majcen Le Marechal, Dr Alenka *Slovenia.*
Majková, Dr Eva *Slovakia.*
Majumdar, Dr Amit *Australia.*
Mak, Prof Thomas Chung Wai *Hong Kong.*
Makarova, Dr Irina P. *Russia.*
Makhina, Dr Irina B. *Russia.*
Makhmudov, Mr Mirzakadir K. *Uzbekistan.*
Makikawa, Mr Shinji *Japan.*
Makinen, Marvin W. *USA.*
Makovec, Mr Darko *Slovenia.*
Makovicky, Emil *Denmark.*
Makowski, Lee *USA.*
Maksić, Prof. Dr Zvonimir *Croatia.*
Malachevsky, Maria Teresa *Argentina.*
Malakhova, Dr Ludmilla F. *Russia.*
Malby, Ms Robyn L. *Australia.*
Malcherek, Mr Thomas *Germany.*
Maleev, Dr Andrey V. *Russia.*
Maleev, Dr Michael *Bulgaria.*
Malenkov, Dr George G. *Russia.*
Malgrange, Prof. Cécile *France.*
Malicskó, Ass. Prof. Dr László *Hungary.*
Malik, Mr Zubair A. *Pakistan.*
Malinar, Dr Mijat *Serbia.*
Malinina, Dr Lucy V. *Russia.*
Malinovsky, Dr Yuri A. *Russia.*
Malinowski, Dr Stanislav Tadeushevich *Moldova.*
Malinowski, Prof. Tadeush Iosifovich *Moldova.*
Malinowski, Prof. Yordan *Bulgaria.*
Malkin, Dr Alexander Joseph *USA.*
Malley, Mary F. *USA.*
Mallinson, Dr Paul *UK.*
Malone, Dr John Francis *UK.*
Maloney, Peter C. *USA.*
Malpezzi, Dr Luciana *Italy.*
Malta, Dr Viscardo *Italy.*
Mal'tsev, Dr Yurii F. *Russia.*
Malykhin, Dr Sergij Volodymyrovych *Ukraine.*
Malyushitskaya, Dr Zinaida V. *Russia.*
Mammi, Prof. Mario *Italy.*
Man, Dr Lucia I. *Russia.*
Manassero, Prof. Mario *Italy.*
Mancini, Prof. Annamaria *Italy.*
Mandal, Dr Pradip Kumar *India.*
Mandal, Sanjay K. *USA.*
Mande, Mr Sekhar Chintamani *India.*
Mandel, Gretchen *USA.*
Mandel, Neil S. *USA.*
Mandia, Ilir *Albania.*
Mangani, Dr Stefano *Italy.*
Manghi, Estela Margarita *Argentina.*
Mani, Mr A. *India.*
Manickkavachgam, Dr Ramanathan *India.*
Mannherz, Prof. Dr Hans Georg *Germany.*
Manninen, Seppo Olavi *Finland.*
Manojlović-Muir, Dr Ljubica *UK.*
Manolikas, Prof. Constantinos *Greece.*
Manotti-Lanfredi, Prof. Anna Maria *Italy.*

Manoukian, Dr Hasmik M. *Armenia.*
Manríquez, Dr Víctor *Chile.*
Mansikka, Kauko *Finland.*
Mantler, Dr Michael K. *Austria.*
Mantovani, Prof. Giorgio *Italy.*
Manyako, Dr Mykola Bogdanovych *Ukraine.*
Mao, Chen *USA.*
Mao, Ho-Kwang *USA.*
Mao, Mr Zhihua *China.*
Maqsood, Dr Asghari *Pakistan.*
Marbec, Miss Emma Rosa *Argentina.*
Marcano Fermín, Dr Cenis María *Venezuela.*
March, Dr Frank C. *New Zealand.*
Marchessault, Dr Robert Henry *Canada.*
Marchetti, Dr Fabio *Italy.*
Marciante, Mrs Clara *Italy.*
Marcó Parra, Lic. Lué Merú *Venezuela.*
Marcos, Dr Celia *Spain.*
Marcotte, Edward M. *USA.*
Maresch, Prof. Walter *Germany.*
Marezio, Dr Massimo *France.*
Margarido, Dr Fernanda *Portugal.*
Mariezcurrena, Prof. Raul *Uruguay.*
Marigo, Prof. Antonio *Italy.*
Marin, Dr Anatoly A. *Russia.*
Marincea, Ştefan *Romania.*
Marinder, Bengt-Olov *Sweden.*
Marin-Elena, Dr José Manuel *Spain.*
Marinković, Prof. Dr Velibor *Slovenia.*
Marinov, Dr Miko *Bulgaria.*
Mariolacos, Dr Konstantin *Germany.*
Marjanović, Ms Tihana *Croatia.*
Markiv, Dr Vasyl' Yakovych *Ukraine.*
Markman, M. Sc Ofer *USA.*
Marko, Eric *USA.*
Markov, Prof. Ivan *Bulgaria.*
Marković, Mr Berislav *Croatia.*
Marks, Laurence D. *USA.*
Markwell, Mary Ann *USA.*
Marmorstein, Ronen *USA.*
Marøy, Dr Kjartan *Norway.*
Marongiu, Prof. Giaime *Italy.*
Márquez-Delgado, Prof. Rafael *Spain.*
Marsau, Prof. Pierre *France.*
Marsh, Richard E. *USA.*
Marsille, Dr Irina M. *Russia.*
Marthi, Katalin *Denmark.*
Marthinsen, Dr Knut *Norway.*
Martí-Artoy, Mr Xavier *Spain.*
Martin, Dr Jennifer L. *Australia.*
Martin, Kenneth L. *USA.*
Martin, Yvonne Connolly *USA.*
Martinelli, Prof. Giuliano *Italy.*
Martínez, Mr Benjamin *Spain.*
Martínez, Dr Francisco *Spain.*
Martínez, Mr Luiz Gallego *Brazil.*
Martínez, Sergio E. *USA.*
Martínez, Dr Vicente de Paul *Chile.*
Martínez-Carrera, Prof. Sagrario *Spain.*
Martínez-Ripoll, Prof. Martín *Spain.*
Martínez-Sarrión, Dr María Luisa *Spain.*
Martín-Izard, Dr Agustín *Spain.*
Martín-Ramos, Dr José Daniel *Spain.*
Martín-Vivaldi, Dr Juan Luis *Spain.*
Martin-Vosshage, Dr Dagmar *Germany.*
Martirosian, Dr Aida H. *Armenia.*
Martorana, Prof. Antonio *Italy.*
Marukawa, Prof. Kenzaburo *Japan.*
Marumo, Prof. Fumiyuki *Japan.*
Maruyama, Mr Hiroyuki *Japan.*
Maruyama, Dr Minoru *Japan.*
Marvin, Dr D. A. *UK.*
Mas, Dra Graciela Raquel *Argentina.*
Mascarenhas, Dr Sérgio *Brazil.*
Mascarenhas, Prof. Yvonne Primerano *Brazil.*
Masciocchi, Dr Norberto *Italy.*
Mashiyama, Prof. Hiroyuki *Japan.*
Masi, Mr Dante *Italy.*
Maske, Prof. Siegfried *South Africa.*
Maslen, Edward Norman (Ted) *Australia.*

Maslov, Dr Andrey V. *Russia.*
Mason, Dr Sax Anton *France.*
Massa, Prof. Dr Werner *Germany.*
Massa, Louis *USA.*
Massarotti, Prof. Vincenzo *Italy.*
Massaux, Dr Michel Louis *France.*
Mastelaro, Dr Valmor Roberto *Brazil.*
Mastropaolo, Donald *USA.*
Masuda, Mr Yukihiro *Japan.*
Masuko, Dr Akiyoshi *Japan.*
Masuko, Prof. Toru *Japan.*
Masunov, Mr Artem Eduardovich *Russia.*
Masut, Prof. Remo A. *Canada.*
Matak, Ms Dijana *Croatia.*
Mateu Suay, Dr Leonardo *Venezuela.*
Mathews, Dr F. Scott *USA.*
Mathews, Mr Irimpan I. *India.*
Mathiesen, Ragnvald Hermann *Norway.*
Mathieson, Alexander McLeod *Australia.*
Mathur, Dr Balbir Kumar *India.*
Matias, Prof. Maria José *Portugal.*
Matias, Dr Pedro M. *Portugal.*
Matijašić, Dr Ivanka *Croatia.*
Matjash, Prof. Ivan Vasyliovych *Ukraine.*
Matjushenko, Dr Mykola Mykolayovych *Ukraine.*
Matković , Dr Boris *Croatia.*
Matković, Dr Prošper *Croatia.*
Matković, Dr Tanja *Croatia.*
Matković-Čalogović, Dr Dubravka *Croatia.*
Matkovsky, Prof. Orest Ilyarovych *Ukraine.*
Matos Beja, Dr Ana *Portugal.*
Matsuda, Prof. *Japan.*
Matsui, Junji *Japan.*
Matsui, Dr Masanori *Japan.*
Matsui, Dr Yoshio *Japan.*
Matsumoto, Prof. Dr Takashi *Japan.*
Matsumoto, Mr Fumio *Japan.*
Matsumoto, Dr Hitoshi *Japan.*
Matsumoto, Dr Osamu *Japan.*
Matsumoto, Prof. Takeo *Japan.*
Matsumura, Dr Sadao *Japan.*
Matsumura, Dr Syo *Japan.*
Matsuno, Dr Seiichi *Japan.*
Matsuoka, Prof. Dr Masakuni *Japan.*
Matsuura, Dr Yoshiki *Japan.*
Matsuzaki, Takao *Japan.*
Matte, Mr Allan *Canada.*
Mattes, Prof. Dr Rainer *Germany.*
Mattevi, Dr Andrea *Italy.*
Matthew, Prof. James Andrew Davidson *UK.*
Matthews, Dr Brian W. *USA.*
Matthews, Dr David A. *USA.*
Matthews, Prof. Frederick White *Canada.*
Matthys, Dr Paul Frederik André Edmond *Belgium.*
Mattia, Prof. Carlo Andrea *Italy.*
Matuszewski, Dr Janusz *Poland.*
Matveeva, Dr Olga P. *Russia.*
Matyi, Richard J. *USA.*
Maue, Mrs Erika *Germany.*
Maurin, Dr Jan K. *Poland.*
Maurus, Mr Robert *Canada.*
Mautner, Dr Franz Andreas *Austria.*
Maverick, Dr Emily F. *USA.*
Maveyraud, Mr Laurent *France.*
Mavridis, Prof. Aristides A. *Greece.*
Maximov, Dr Boris A. *Russia.*
Maximov, Sergey L. *Kazakhstan.*
Maxwell, Prof. George *Canada.*
May, Dr Roland Peter *France.*
Mayer, Mrs Bärbel *Germany.*
Mayer, Dr Helmut *Austria.*
Mayer, Dr Itzchak *Israel.*
Mayo, Sheridan Clare *Australia.*
Mayr, Dr Michael *Austria.*
Mazany, Anthony *USA.*
Mazus, Dr Mark Davidovich *Moldova.*
Mazza, Prof. Fernando *Italy.*
Mazzarella, Prof. Lelio *Italy.*
Mazzaro, Dr Irineu *Brazil.*
Mazzi, Prof. Fiorenzo *Italy.*

Mazzochi, Dr Vera *Brazil.*
Mcardle, Prof. Patrick *Ireland.*
McCammon, Dr J. Andrew *USA.*
McCarthy, Dr/Prof Gregory J. *USA.*
McCauley, James W. *USA.*
McCrone, Lucy B. *USA.*
McCrone, W. C. *USA.*
McCullough, Dr Kevin J. *UK.*
McCusker, Dr Lynne B. *Switzerland.*
McDonald, Dr Neil Q. *USA.*
McDonald, Dr Robert *Canada.*
McGrath, Mary E. *USA.*
McGuire, Dr Nancy K. *USA.*
Mchedlishvili, Dr Boris V. *Russia.*
McIntyre, Dr Garry james *France.*
McKee, Dr Vickie *UK.*
McKenzie, David Robert *Australia.*
McKie, Dr Christine H. *UK.*
McKie, Dr Duncan *UK.*
McLachlan, Dr Andrew D. *UK.*
McLaren, Alexander Clark *Australia.*
McLaughlin, George *Australia.*
McLean, Dr John *USA.*
McMahon, Mr Brian *UK.*
McMahon, Dr Malcolm Iain *UK.*
McMillan, Dr Martin *USA.*
McMorrow, Desmond Francis *Denmark.*
McMullan, Dr Richard K. *USA.*
McMurdie, H. F. *USA.*
McPhalen, Dr Catherine A. *Canada.*
McPherson, Dr Alexander *USA.*
McRee, Dr Duncan E. *USA.*
McTigue, Michele M. *USA.*
McWhan, Dr Denis B. *USA.*
Mealli, Dr Carlo *Italy.*
Medeiros Rodrigues, Dr Maria Mabel *Brazil.*
Meden, Dr Anton *Slovenia.*
Medimorec, Mr Stanislav S. *Croatia.*
Medrud, Ronald C. *USA.*
Meek, Dr Keith Michael Andrew *UK.*
Meetsma, Drs Auke *Netherlands.*
Meier, Prof. Walter M. *Switzerland.*
Meille, Dr Stefano Valdo *Italy.*
Meinnel, Prof. Jean *France.*
Meissner, Mrs Elke *Germany.*
Mejstříková, Ing. Lubomíra *Czech Republic.*
Meleshina, Dr Valentina A. *Russia.*
Melian, Ms Maria Antonia *Cuba.*
Melik-Adamyan, Dr William R. *Russia.*
Melka, Dr Karel *Czech Republic.*
Melkonian, Dr Haroutyun S. *Armenia.*
Mellini, Prof. Marcello *Italy.*
Melledge, Dr H. Judith *UK.*
Mel'nik, Dr Yurij Mykhajlovych *Ukraine.*
Melnikov, Dr Oleg K. *Russia.*
Mel'nikov, Dr Volodymyr Stepanovych *Ukraine.*
Melnikova, Dr Svetlana V. *Russia.*
Melo Jorge, Maria da Estrela Borges *Portugal.*
Melzer, Dr Rolf *Germany.*
Menchetti, Prof. Silvio *Italy.*
Menda, Mr Kazunori *Japan.*
Mendelson, Dr Robert A. *USA.*
Mendialdua Lecue, Dr Juan Angel *Venezuela.*
Mentzafos, Prof. Dimitrios E. *Greece.*
Menzinger, Prof. Filippo *Italy.*
Merati, Dr Felicita *Italy.*
Mercola, Dan *USA.*
Mereiter, Prof. Dr Kurt *Austria.*
Meretliev, Mr Shamyrat *Turkmenistan.*
Meriani, Prof. Sergio *Italy.*
Merisalo, Matti Juhani *Finland.*
Merlino, Prof. Stefano *Italy.*
Merlo, Prof. Franco *Italy.*
Mernagh, Ms Victoria *Ireland.*
Merritt, Ethan A. *USA.*
Mertin, Dr Wilhelm *Germany.*
Mesa, Dr José Luis *Spain.*
Meshalkin, Dr Sergey S. *Russia.*
Meshli, Mike *USA.*
Mesropian, Dr Mesrop H. *Armenia.*
Messerschmidt, Dr Albrecht M. W. *Germany.*

Messick, Julian *USA.*
Mestnik, Dr José *Brazil.*
Meštrović, Mr Ernest *Croatia.*
Metzger, Prof. Robert M. *USA.*
Meunier-Piret, Dr Jacqueline *Belgium.*
Meyer, Prof. Dr Gerd *Germany.*
Meyer, Dr Edgar F. *USA.*
Meyer, Frank H. *USA.*
Meyer zu Altenschildesche, Mr Holger *Switzerland.*
Meyer-Ehmsen, Prof. Dr Gerhard *Germany.*
Meyerheim, Dr Holger L. *Germany.*
Meyers, Edward A. *USA.*
Mez, Dr Hans-Christian *Switzerland.*
Mian, Dr Mohammad Ashraf *Pakistan.*
Mian, Mr Muhammad Asghar *Pakistan.*
Mian, Mr Mushtaq Ahmed *Pakistan.*
Miao, Prof. Fangming *China.*
Michaelides, Dr Adonis *Greece.*
Michailov, Dr Evgeni *Bulgaria.*
Michailov, Mr Michail *Bulgaria.*
Michalski, Dr Edward *Poland.*
Michel, Prof. André Gustave *France.*
Michel, Prof. Dr Bernd *Germany.*
Michel, Prof. Dr Hartmut *Germany.*
Michel, Prof. Karl H. *Belgium.*
Michinel Machado, Prof. José Luis *Venezuela.*
Mighell, Dr Alan D. *USA.*
Mihama, Prof. Dr Kazuhiro *Japan.*
Mihichuk, Prof. Lynn Michael *Canada.*
Miida, Dr Rokuro *Japan.*
Mijlhoff, Dr Frans C. *Netherlands.*
Mikelsaar, Dr Raik-Hiio *Estonia.*
Mikenda, Dr Werner *Austria.*
Mikhailov, Dr Albert M. *Russia.*
Mikhailov, Dr Igor Fedorovych *Ukraine.*
Mikhov, Dr Michael *Bulgaria.*
Miki, Dr Kunio *Japan.*
Mikkola, Donald E. *USA.*
Mikula, Dr Pavol *Czech Republic.*
Milat, Dr Ognjen *Croatia.*
Milberg, Morton E. *USA.*
Milburn, Michael V. *USA.*
Milchev, Dr Alexander *Bulgaria.*
Milchev, Dr Andrei *Bulgaria.*
Milićev, Dr Svetozar *Slovenia.*
Milillo, Prof. Frank *USA.*
Milinković, Mr Vjekoslav *Croatia.*
Milius, Dr Wolfgang *Germany.*
Mill', Dr Boris V. *Russia.*
Millane, Dr Rick P. *USA.*
Millán-Muñoz, Dr Maria *Spain.*
Milledge, Dr H. Judith *UK.*
Miller, Prof. Donald P. *USA.*
Miller, Dr Lance L. *USA.*
Miller, Mitchell D. *USA.*
Miller, Robert L. *USA.*
Miller, Prof. Russ *USA.*
Miller, Stephen T. *USA.*
Millini, Dr Roberto *Italy.*
Mills, Dr Alan *UK.*
Mills, Mr Owen S. *UK.*
Milnes, A. G. *USA.*
Miloshev, Prof. Georgi *Bulgaria.*
Minacheva, Dr Lidiya Kh. *Russia.*
Minčeva-Stefanova, Prof. Yordanka *Bulgaria.*
Minemoto, Mr Hisashi *Japan.*
MingLuo, Ming Luo, Associate Prof. *USA.*
Minkoff, Prof. I. *Israel.*
Minoda, Mr Hiroki *Japan.*
Minor, Dr Wladek *USA.*
Mints, Prof. Rafail I. *Russia.*
Minyukov, Dr Sergey M. *Russia.*
Mir, Mr Jan Muhammad *Pakistan.*
Mirão, José António *Portugal.*
Miravitlles, Prof. Carlos *Spain.*
Mirčeva, Dr Aneta (FYR) *Macedonia.*
Mironova, Prof. Nina *Latvia.*
Mirsky, Dr Kira *USA.*
Mirtič, Dr Dr Breda *Slovenia.*
Mirwald, Prof. Dr Peter W. *Austria.*

Misaki, Dr Shintaro *Japan.*
Mishnev, Dr Anatoly *Latvia.*
Mišič, Mr Miha *Slovenia.*
Misra, Prof. Nirmal Kumar *India.*
Misra, Prof. Somnath *India.*
Miśta, Mgr Włodzimierz *Poland.*
Mistryukov, Mr Alexander E. *Russia.*
Mitchell, Dr Crighton Maurice *Canada.*
Mitchell, David *USA.*
Mitchell, Donald J. *USA.*
Mitchell, Dr Geoffrey Robert *UK.*
Mitchell, Dr Keith A. R. *Canada.*
Mitra, Prof. Girija Bhushan *India.*
Mitsuishi, Dr Tomokuni *Japan.*
Mittemeijer, Prof. Dr Ir Eric J. *Netherlands.*
Miura, Miss Keiko *Japan.*
Miyamae, Dr Hiroshi *Japan.*
Miyamoto, Prof. Masamichi *Japan.*
Miyashita, Dr Satoru *Japan.*
Miyata, Dr Takeshi *Japan.*
Miyawaki, Dr Ritsuro *Japan.*
Miyazawa, Dr Shintaro *Japan.*
Miyazawa, Dr Yasuto *Japan.*
Mizugaki, Mr Tsutomu *Japan.*
Mizuno, Dr Hiroshi *Japan.*
Mizuno, Dr Kaoru *Japan.*
Mizuno, Dr Yukiko *Japan.*
Mo, Prof. Frode *Norway.*
Mobed, Dr A. Mohamed *Egypt.*
Mochizuki, Prof. Dr Katsumi *Japan.*
Modec, Miss Barbara *Slovenia.*
Modlich, Mr Jörn *Germany.*
Möck, Dr Peter *Germany.*
Möhling, Dr Werner *Germany.*
Moers, Dr Frans *Netherlands.*
Moews, Paul C. *USA.*
Moguš-Milanković, Dr Andrea *Croatia.*
Mohamad, Dr Hamzah *Malaysia.*
Mohamed, Mr Ahmed Aly *Egypt.*
Mohamed, Mr Almoktar Mohamed *Egypt.*
Mohamed, Dr Esmat A. *Egypt.*
Mohamed, Dr Galal A. *Egypt.*
Mohanlal, Dr Sembu Krishnaiyer *India.*
Mohanty, Mr. Arun K. *India.*
Moini, Ahmad *USA.*
Mok, Dr Kum-fun *Singapore.*
Mokhov, Dr Andrey V. *Russia.*
Mokra, Dr Ivanna Romanivna *Ukraine.*
Molchanov, Dr Vladimir N. *Russia.*
Moleman, Albertus C. *Netherlands.*
Molin, Gianmario *Italy.*
Molina, Mr José *Spain.*
Molina-Molina, Dr José *Spain.*
Molin-Case, JoAnn *USA.*
Molins, Dr Elies *Spain.*
Moliterni, Dr Anna Grazia *Italy.*
Momany, Cory *USA.*
Mombru, Mr Alvaro *Uruguay.*
Momoi, Prof. Hitoshi *Japan.*
Monaco, Prof. Hugo L. *Italy.*
Monahan, Miss Ailsa M. K. *UK.*
Monari, Dr Magda *Italy.*
Mondragon, Alfonso *USA.*
Monfort, Dr Montse *Spain.*
Monge, Dr M. Angeles *Spain.*
Mongiorgi, Prof. Romano *Italy.*
Montague, Professor Daniel G. *USA.*
Montenero, Prof. Angelo *Italy.*
Montes, Dr Arturo *Venezuela.*
Montfort, Dr William R. *USA.*
Montgomery, Prof. Henry *Canada.*
Monzingo, Arthur F. *USA.*
Moodie, Alexander Forbes *Australia.*
Moody, Dr Peter C. E. *UK.*
Moon, Dr Sang-Jin *Korea.*
Moon, Prof Tony *Australia.*
Moonen, Dr Joost A. H. M. *Netherlands.*
Moor, Dr Robert *Switzerland.*
Moore, Dr A. Moreton *UK.*
Moore, Caroline *UK.*

Moore, Donald L. *USA.*
Moore, Miss Katie A. *UK.*
Moore, Keren M. *USA.*
Moore, Dr Madeleine H. *UK.*
Moore, Peter B. *USA.*
Mootz, Prof. Dr Dietrich *Germany.*
Morandi, Prof. Noris *Italy.*
Morao Dias, Dr Antonio Angelo *Portugal.*
Moras, Dr Dino *France.*
Morawiec, Prof. Henryk *Poland.*
Moreau, Prof. Jean-Michel *France.*
Moreau, Prof. Jules *Belgium.*
Moreiras, Prof. Damaso *Spain.*
Moreland, James *USA.*
Moreno Fuquen, Rodolfo *Colombia.*
Moreno-Carcamo, Mr Abel *Spain.*
Moreno-Carretero, Dr Miguel *Spain.*
Moreno-Sánchez, Dr José María *Spain.*
Moret, Dr Massimo *Italy.*
Morgenroth, Dr Wolfgang *Germany.*
Morgunova, Dr Ekaterina Yu. *Russia.*
Morháčová, Dr Eva *Slovakia.*
Mori, Mr Atsushi *Japan.*
Mori, Mr Shigeo *Japan.*
Moriarty, John L. *USA.*
Morikawa, Kosuke *Japan.*
Morimoto, Prof. Jun *Japan.*
Morimoto, Yukio *Japan.*
Morinaga, Prof. Masahiko *Japan.*
Moring, Dr Jill *USA.*
Morita, Mr Shoji *Japan.*
Moriyama, Mr Hideaki *Japan.*
Morniroli, Prof. Jean-Paul *France.*
Mornon, Dr Jean-Paul *France.*
Morón, Dr M. Carmen *Spain.*
Morosin, Bruno *USA.*
Mortensen, Kell *Denmark.*
Morton, Dr Allan James *Australia.*
Mosca, Dr Dante Homero *Brazil.*
Moseley, Dr Patrick Timothy *UK.*
Moshkin, Dr Sergey V. *Russia.*
Moss, Dr David S. *UK.*
Moss, Simon C. *USA.*
Mosset, Prof. Alain *France.*
Mostad, Arvid *Norway.*
Mostafa, Mr Golam *India.*
Mostafa, Mr Said H. *Egypt.*
Mosyak, Ms Lidia G. *Israel.*
Motegi, Dr Nawoto *Japan.*
Motherwell, Dr William David Samuel *UK.*
Mottana, Prof. Annibale *Italy.*
Mottonen, Dr James M. *USA.*
Moureau, Dr Florence G. C. G. *Belgium.*
Mouri, Mr Mikio *Japan.*
Moustakali-Mavridis, Dr Irene E. *Greece.*
Movchan, Dr Mykola Prokopovych *Ukraine.*
Moze, Prof. Oscar *Italy.*
Mrafko, Dr Peter *Slovakia.*
Mrooz, Dr Oksana Yaroslavivna *Ukraine.*
Mrvoš-Sermek, Dr Draginja *Croatia.*
Mucha, Mgr Dariusz *Poland.*
Muckelbauer, Jodi *USA.*
Muddle, Dr Barry C *Australia.*
Mudher, Dr Khush Dev Singh *India.*
Mühlberg, Prof. Dr Manfred *Germany.*
Mueller, Dr Christoph *USA.*
Müller, Prof. Dr Eberhard *Germany.*
Müller, Prof. Dr Gerd *Germany.*
Müller, Prof. Dr Gerhard *Germany.*
Müller, Prof. Dr Horst *Germany.*
Mueller, Melvin H. *USA.*
Müller, Prof. Dr Ulrich *Germany.*
Müller, Dr Jürgen Joachim *Germany.*
Müller, Dr Paul *Germany.*
Müller, Mrs Sabine *Germany.*
Müller-Fahrnow, Dr Anke *Germany.*
Münninghoff, Dr Günter *Germany.*
Mueser, Dr Timothy C. *USA.*
Mugnoli, Prof. Angelo *Italy.*
Muhonen, Dr Heikki Juhani *Finland.*

Mui, Suet C. *USA.*
Muir, Dr James A. *USA.*
Muir, Dr Kenneth Walter *UK.*
Muir, Dr Mariel M. *USA.*
Muirhead, Dr Hilary *UK.*
Mujica, Dr Carlos *Chile.*
Mukasyan, Dr Alexander S. *Russia.*
Mukhamedzhanov, Dr Enver Kh. *Russia.*
Mukherjee, Dr Alok Kumar *India.*
Mukherjee, Dr Amal Bikash *India.*
Mukherjee, Dr Biswanath *India.*
Mukherjee (Mondal), Dr (Mrs) Monika *India.*
Mukherjee, Dr Partha Sarathi *India.*
Mukhin, Dr Boris V. *Russia.*
Mukhopadhyay, Mr Bishnu Prasad *India.*
Mukhopadhyay, Dr (Mrs) Anuradha *India.*
Mukhopadhyay, Dr Pradip *India.*
Muldawer, Leonard *USA.*
Munchaev, Mr Anzor I. *Russia.*
Mundt, Dr Otto *Germany.*
Munirathinam, Mr Nethaji *India.*
Muñoz Picone, Dr Eduardo *Mexico.*
Muñoz-Roca, Dr M. Carmen *Spain.*
Munshi, Mr Sanjeev Kumar *India.*
Munshi, Dr Sanjeev. *USA.*
Mura, Dr Pasquale *Italy.*
Murad, Dr Enver *Germany.*
Murali, Ramachandran Murali *USA.*
Muralidharan, Mr K. V. *India.*
Muranaka, Mr Ken-ichiro *Japan.*
Murashova, Dr Elena V. *Russia.*
Mureşan, Prof. Ioan *Romania.*
Murgich, Dr Juan *Venezuela.*
Murmann, R. Kent *USA.*
Murphy, Michael E. P. *USA.*
Murray, Dr Henry H. *USA.*
Murray-Rust, Dr Judith *UK.*
Murray-Rust, Dr Peter *UK.*
Murrieta Sánchez, Dr Héctor *Mexico.*
Murta, Prof. Clécio *Brazil.*
Murthy, Dr Krishna H. M. *USA.*
Murthy, Mr Mathur R N *India.*
Murthy, Dr Sanjeeva N. *USA.*
Musatti, Prof. Amos *Italy.*
Mustafayev, Dr Nariman *Azerbaijan.*
Mutikainen, Dr Ilpo Pellervo *Finland.*
Mykhaienko, Dr Svitlana Ivanivna *Ukraine.*
Mykhalichko, Dr Borys Myronovych *Ukraine.*
Mykolajchuk, Prof. Oleksij Gordijovych *Ukraine.*
Myles, Dr Dean A. A. *UK.*
Mylvaganam, Dr Shankari E. *USA.*
Mys'kiv, Dr Maryan Grygorovych *Ukraine.*
Mzayek, Mr Elias *Spain.*
Nabarro, Prof. Frank Reginald Nunes *South Africa.*
Nabatov, Mr Victor N. *Russia.*
Nachman, Dr Joseph *USA.*
Nadezhina, Dr Tamara N. *Russia.*
Nägele, Dr Erhard *Germany.*
Näsäkkälä, Matti Eerik *Finland.*
Nag, Dr Dilip Kumar *India.*
Nag, Mrs Jhumjhumi *India.*
Naga, Prof. Mohamed Abdelhamid *Egypt.*
Nagakura, Prof. Sigemaro *Japan.*
Nagashima, Dr Nobuya *Japan.*
Nagashima, Dr Seiichi *Japan.*
Nagbhushana Rao, Mr Chemboli *India.*
Nagel, Dr Wolfgang *Germany.*
Nagendra, Mr *India.*
Nagl, Dr Ante *Croatia.*
Naing, Mr Saw *Myanmar.*
Nair, Ms Bindu *India.*
Naismith, Jim *USA.*
Nakagawa, Dr Atsushi *Japan.*
Nakahigashi, Prof. Kiyotaka *Japan.*
Nakai, Dr Izumi *Japan.*
Nakamura, Prof. Akira *Japan.*
Nakamura, Mrs Hikaru *Japan.*
Nakamura, Prof. Kazuo *Japan.*
Nakamura, Mr Masaki *Japan.*
Nakamura, Dr Yoshio *Japan.*

Nakano, Dr Kikuo *Japan.*
Nakata, Mr Kazuaki *Japan.*
Nakatsu, Prof. Kazumi *Japan.*
Nakayama, Mr Hitoshi *Japan.*
Nakazumi, Dr Yoshihide *Japan.*
Nakhla, Dr Fakhry *Egypt.*
Namgung, Prof. Hae *Korea.*
Nanba, Mr Hirokuni *Japan.*
Nandi, Mr Asok Kumar *India.*
Nandi, Dr Ranjan Kumar *India.*
Nanev, Prof. Christo *Bulgaria.*
Napolitano, Prof. Roberto *Italy.*
Naqvi, Dr Syed Ali Anwar *Pakistan.*
Nar, Dr Herbert *Germany.*
Nara, Dr Shigetoshi *Japan.*
Narasimhamurthy, Mr Narasappa *India.*
Narasimhan, Dr P. *India.*
Narayan, Jagdish *USA.*
Narayana, Dr Sthanam V. L. *USA.*
Narda, Dra Griselda Edith *Argentina.*
Nardelli, Prof. Mario *Italy.*
Nardin, Prof. Giorgio *Italy.*
Narendra, Dr Narayana N *USA.*
Nasreen, Miss Shagufta *Pakistan.*
Nassimbeni, Prof. Luigi Renzo *South Africa.*
Nastopoulos, Dr Vassilios *Greece.*
Natarajan, Dr S. *India.*
Natarajan, Dr Subramanian *India.*
Natesan, Mr Elango *India.*
Nath, Mr Kashi *India.*
Naud, Dr Jean *Belgium.*
Naudon, Dr André *France.*
Nauer-Gerhardt, Carola U. *Germany.*
Naumova, Dr Inessa I. *Russia.*
Navarra, Dr Gabriele *Italy.*
Navarro, Dr Carmen *Spain.*
Navia, Manuel A. *USA.*
Navrotsky, Prof. Alexandra *USA.*
Navruz, Nurten *Turkey.*
Nazimova, Dr Nina V. *Russia.*
Neckel, Prof. Dr Adolf *Austria.*
Neder, Dr Reinhard B. *Germany.*
Neff, Prof. Dr Hans *Germany.*
Nehasil, Dr Miroslav *Czech Republic.*
Neidhart, Dr David J. *USA.*
Neidle, Prof. Stephen *UK.*
Nekrasov, Dr Yuri V. *Russia.*
Nelmes, Prof. Richard J. *UK.*
Nelson, Dr Hillary C. M. *USA.*
Nemoto, Takashi *Japan.*
Nenonen, Pertti Olavi *Finland.*
Nenow, Prof. Dimiter *Bulgaria.*
Nesper, Prof. Reinhard *Switzerland.*
Nesterov, Dr Volodymyr Mykolayovych *Ukraine.*
Nesterova, Dr Yaroslava Mikhajlovna *Russia.*
Neubert, Mr Michael *Germany.*
Neumann, Dr Wolfgang *Germany.*
Nevskaya, Dr Natalia A. *Russia.*
Newcomer, Marcia E. *USA.*
Newman, Dr Robert A. *USA.*
Newnham, Robert E. *USA.*
Newsam, Professor John M. *USA.*
Newton, Dr M. Gary *USA.*
Ng, Joseph David *USA.*
Ng Lee, Prof. Yolanda Mi Lien *Venezuela.*
Ng, Dr Seik Weng *Malaysia.*
Ng, Prof. Ser Choon *Singapore.*
Nghi, Prof. Nguyen Hoang *Vietnam.*
Nghiep, Prof. Do Minh *Vietnam.*
Ngoan, Dr Dao Cong *Vietnam.*
Nguti, Nde Donatus *UK.*
Nguyen-Ba, Dr Chanh *France.*
Nhieu, Prof. Pham Van *Vietnam.*
Nhung, Dr Nguy Tuyet *Vietnam.*
Nichols, Monte C. *USA.*
Nicholson, Dr Brian K. *New Zealand.*
Nicholson, Prof. David Graham *Norway.*
Nickel, Prof. Dr Klaus G. *Germany.*
Nickitenko, Dr Alex V. *USA.*
Nicolò, Dr Francesco *Italy.*

Nicolosi, Joseph *USA.*
Niederhut, Monica M. *USA.*
Niedermayr, Dr Gerhard *Austria.*
Nieger, Dr Martin *Germany.*
Nielsen, Anders *Denmark.*
Nielsen, Kurt *Denmark.*
Nieminen, Kari Veikko Juhani *Finland.*
Nienaber, Dr Vicki L. *USA.*
Nierlich, Dr Martine *France.*
Nieto, Mr José *Spain.*
Nieto-García, Dr Fernando *Spain.*
Nieuwenhuyzen, Mr Mark *New Zealand.*
Nifontov, Mr Valentin P. *Russia.*
Nigam, Dr Gur Dayal *India.*
Nigli, Ms Selina *India.*
Nihtianova, Dr Diana *Bulgaria.*
Niimura, Prof. Nobuo *Japan.*
Niizeki, Dr Nobukazu *Japan.*
Nikanorova, Dr Irina A. *Russia.*
Niketić, Dr Svetozar R. *Serbia.*
Nikiforov, Prof. Igor Ya. *Russia.*
Nikishova, Dr Lidya V. *Russia.*
Nikitenko, Prof. Valerian I. *Russia.*
Nikkola, Matti Johannes *Sweden.*
Nikolaenko, Dr Anatoli M. *Russia.*
Nikolaychuk, Dr Grygorij Pavlovych *Ukraine.*
Nikolić, Dr Slobodanka *Serbia.*
Nikolov, Dimitar B. *USA.*
Nikolskaia, Dr Natalia Kimovna *Russia.*
Nikolskaya, Dr Larisa V. *Russia.*
Nikonov, Dr Stanislav V. *Russia.*
Nikulin, Dr Andrei Yurievich *Australia.*
Nimgirawath, Mrs Kloy *Thailand.*
Nimmo, John Kenneth *Australia.*
Ninomiya, Dr Hiroshi *Japan.*
Nippus, Dr Michael *Germany.*
Nirmala, Ms Kusuma Anantharamiah *India.*
Nishi, Prof. Fumito *Japan.*
Nishida, Prof. Dr Takashi *Japan.*
Nishihata, Dr Yasuo *Japan.*
Nishikawa, Dr Keiko *Japan.*
Nishinaga, Prof. Tatau *Japan.*
Nishino, Dr Yoichi *Japan.*
Nishioka, Prof. Dr Kazumi *Japan.*
Nishizaki, Mr Katsumi *Japan.*
Nishizawa, Dr Shin-ichi *Japan.*
Nissen, Poul *Denmark.*
Nitsche, Mr Robert *Germany.*
Nitschmann, Dr Günter Max Alfred *Germany.*
Nittono, Prof. Osamu *Japan.*
Nizami, Mr Muhammad Sharif *Pakistan.*
Nizamutdinov, Dr Nazim M. *Russia.*
Noda, Prof. Yasutoshi *Japan.*
Nørskov-Lauritsen, Dr Leif *Denmark.*
Nogales, Dr Evangelina *USA.*
Nogués-Carulla, Dr Joaquim *Spain.*
Noll, Bruce C. *USA.*
Nonaka, Dr Takamasa *Japan.*
Noonan, Joseph F. *USA.*
Noor, Mrs Sahina Begum *India.*
Noordik, Dr Jan H. *Netherlands.*
Norberg, Ms Bernadette *Belgium.*
Nord, Anders Gunnar *Sweden.*
Nordman, Prof. Christer E. *USA.*
Noreland, Dr Jakob *UK.*
Norén, Bertil Nils *Sweden.*
Noréus, Dag *Sweden.*
Norrestam, Rolf *Sweden.*
Norris, Dr Gillian E. *New Zealand.*
North, Prof. Anthony Charles Thomas *UK.*
Northolt, Dr Ir Maurits G. *Netherlands.*
Norton, Dr Michael L. *USA.*
Nosenko, Prof. Anatolij Erofijovych *Ukraine.*
Nosik, Dr Valery L. *Russia.*
Nouet, Prof. Jean *France.*
Novakova, Dr Alla A. *Russia.*
Novikova, Dr Natalia N. *Russia.*
Novoa, Mr Hector *Cuba.*
Novomlinsky, Dr Leonid A. *Russia.*

Novosel Radović, Dr Vjera *Croatia.*
Nowack, Mrs Ellen Carla *Germany.*
Nuffield, Prof. Edward Wilfrid *Canada.*
Numan, Drs Milco *Netherlands.*
Nunes, Dr Anthony C. *USA.*
Núñez, Dr Pedro *Spain.*
Nunn, Dr Christine Margaret *UK.*
Nunzi, Prof. Antonio *Italy.*
Nyborg, Jens *Denmark.*
Nyburg, Prof. Stanley C. *UK.*
Nygren, Prof. Mats *Sweden.*
O Brien, Mr Terence P. *Ireland.*
Oberhänsli, Dr Willi. E. *Switzerland.*
Oberti, Dr Roberta *Italy.*
Obinata, Mr Shiro *Japan.*
Obretenov, Dr Willy *Bulgaria.*
Ochando, Dr Luis E. *Spain.*
O'Connor, Brian *Australia.*
Oeder Manning, Nancy *USA.*
Oesten, Dr Rüdiger *Germany.*
Österberg, Prof. Ragnar *Sweden.*
Özbey, Dr Süheyla *Turkey.*
Özenbaş, Dr Macit *Turkey.*
Özkan, Dr Hüsnü *Turkey.*
Öztürk, Sema *Turkey.*
Oganesian, Dr Levone A. *Armenia.*
Ogawa, Dr Kazuhiko *Japan.*
Ogawa, Prof. Keiichiro *Japan.*
Ogawa, Dr Keizo *Japan.*
Ogawa, Prof. Tomoya *Japan.*
Ogura, Prof. Iwao *Japan.*
Oh, Dr Byung-Ha *USA.*
Ohachi, Prof. Tadashi *Japan.*
Ohama, Prof. Nobuhiko *Japan.*
O'Hara, Dr Bernard P. *UK.*
Ohashi, Yoshikazu *USA.*
Ohashi, Prof. Yuji *Japan.*
Ohba, Mr Hiroyuki *Japan.*
Ohba, Prof. Shigeru *Japan.*
Ohba, Dr Takuya *Japan.*
Ohgaki, Dr Masataka *Japan.*
Ohishi, Dr Hirofumi *Japan.*
Ohkawa, Prof. Dr Tokio *Japan.*
Ohkohchi, Mr Masato *Japan.*
Ohmasa, Prof. Masaaki *Japan.*
Ohnaka, Prof. Dr Itsuo *Japan.*
Ohno, Prof. Hideo *Japan.*
Ohno, Dr Takehisa *Japan.*
Ohrt, Jean M. *USA.*
Ohsato, Associate Prof. Hitoshi *Japan.*
Ohsumi, Prof. Kazumasa *Japan.*
Ohtsuka, Prof. Yasukuni *Japan.*
Oishi, Dr Shuji *Japan.*
Ojala, William H. *USA.*
Oka, Dr Kunihiko *Japan.*
Okabe, Dr Nobuo *Japan.*
Okabe, Prof. Toshio *Japan.*
Okada, Kenji *Japan.*
Okada, Dr Yasumasa *Japan.*
Okano, Dr Yasunori *Japan.*
Okazaki, Prof. Atsushi *Japan.*
Okazaki, Prof. Dr Megumi *Japan.*
Okhrimenko, Dr Tatyana M. *Russia.*
Okino, Mr Fujio *Japan.*
Okinshevich, Mr Vladimir V. *Russia.*
Okrusch, Prof. Dr Martin *Germany.*
Okubo, Prof. Dr Tsuneo *Japan.*
Okuda, Prof. Dr Takashi *Japan.*
Okuno, Dr Masayuki *Japan.*
Okuno, Dr Yasuo *Japan.*
Okutsu, Dr Tetsuo *Japan.*
Okuyama, Prof. Kenji *Japan.*
Olander, Kerstin *Sweden.*
Olcese, Prof. Giorgio L. *Italy.*
Olczak, Mgr inż. Andrzej *Poland.*
Olech, Dr Andrzej Z. *Poland.*
Oleinikov, Dr Vladimir A. *Russia.*
Oleksin, Prof. Barbara J. *Poland.*
Oleksyn, Dr Oksana Yaroslavivna *Ukraine.*

Oleś, Prof. Andrzej Władysław *Poland.*
Olijnyk, Dr Volodymyr Volodymyrovych *Ukraine.*
Oliva, Dr Glaucius *Brazil.*
Oliveira Lopes, Prof. César *Brazil.*
Oliveira, Marcos de Oliveira *USA.*
Oliver, Joel D. *USA.*
Oliver Salvador, Ms Maria del Carmen *Mexico.*
Olivier-Deyris, Mrs Laurence *France.*
Olivieri, Dr Johnny Rizzieri *Brazil.*
Olkowski, Ms Jennifer L. *USA.*
Ollis, David *Australia.*
Olmstead, Dr Marilyn M. *USA.*
Olovsson, Dr Gunnar *Sweden.*
Olovsson, Prof. Ivar *Sweden.*
Olsen, Dr Arne *Norway.*
Olsen, Dr Kenneth W. *USA.*
Olson, Dr Arthur J. *USA.*
Olson, Dr David H. *USA.*
Olson, Mrs Solveig *Sweden.*
Olsson, Christina L. *Australia.*
Olthof-Hazekamp, Drs Roeli *Netherlands.*
Omarov, Marat Ahmetovich *Kazakhstan.*
Omino, Mr Akira *Japan.*
Onan, Kay *USA.*
Ondráček, Ing. Jan *Czech Republic.*
Ondruš, Ing. Petr *Czech Republic.*
O'Neil, Paul Andrew *UK.*
O'Neill, Dr Françoise Marcelle *South Africa.*
Onishchina, Dr Ninel Mitrofanovna *Russia.*
Ono, Prof. Dr Masatoshi *Japan.*
Ono, Mrs Yoko *Japan.*
Onoda, Dr Mitsuko *Japan.*
Onodera, Dr Akira *Japan.*
Onozuka, Dr Takashi *Japan.*
Oonishi, Isao *Japan.*
Ooshima, Dr Hiroshi *Japan.*
Orama, Olli Antero *Finland.*
Ordway, Fred *USA.*
Orekhov, Dr Sergey V. *Russia.*
Oren, Deena A. *USA.*
Organova, Dr Natalija I. *Russia.*
Orioli, Prof. Pierluigi *Italy.*
Orme-Johnson III, William H. *USA.*
Orpen, Prof. A. Guy *UK.*
Ortega-Huertas, Dr Miguel *Spain.*
Oryshchyn, Dr Stepan Vasyliovych *Ukraine.*
Osaka, Prof. Toshiaki *Japan.*
Osaki, Prof. Kenji *Japan.*
Osano, Yasuko T. *Japan.*
Osičková, Dr Jana *Czech Republic.*
Oskarsson, Dr Åke *Sweden.*
Osmanzadeh Shener, Dr Osmanzadeh *Azerbaijan.*
Osslund, Dr Timothy D. *USA.*
Ostrofsky, Bernard *USA.*
Oszlányi, Dr Gábor *Hungary.*
Otalora, Mr Fermín *Spain.*
Otani, Dr Shigeki *Japan.*
Otero-Díaz, Prof. L. Carlos *Spain.*
Othman, Dr A. Hamid *Malaysia.*
Othman, Dr Radzali *Malaysia.*
Otobe, Mr Masanori *Japan.*
Otroshchenko, Dr Ludmila P. *Russia.*
Otsuka, Prof. Kazuhiro *Japan.*
Otten, Mr Peter *Germany.*
Otto, Prof. Dr Hans Hermann *Germany.*
Otwinowski, Zbyszek *USA.*
Ou, Chia-Chih *USA.*
Ouladdiaf, Dr Bachir *France.*
Ovanesian, Dr Karine L. *Armenia.*
Ovchinnikov, Dr Yurii E. *Russia.*
Ovchinnikova, Dr Elena N. *Russia.*
Ovsetsina, Ms Tatyana I. *Russia.*
Owen, Mr Charles Gordon *Canada.*
Owston, Dr Philip G. *UK.*
Oyama, Dr Yasunao *Japan.*
Ozawa, Dr Shoichi *Japan.*
Ozawa, Dr Yoshiki *Japan.*
Ozeki, Dr Tomoji *Japan.*
Ozerin, Dr Alexander N. *Russia.*
Ozerov, Prof. Ruslan Pavlovich *Russia.*

Ozolins, Dr Gerhards *Latvia.*
Paakkari, Prof. Timo L. P. *Finland.*
Pabo, Dr Carl O. *USA.*
Pabst, Dr Ingeborg *Germany.*
Pacalo, Rosemary E. Gerald *USA.*
Paciorek, Dr Włodzimierz *Poland.*
Padian, Eduardo *USA.*
Padmanabhan, Dr Kaillathe *USA.*
Padmasuta, Mrs Soontari *Thailand.*
Padrón, Dr Raúl *Venezuela.*
Paehler, Arno *Japan.*
Page, Dr James E. *UK.*
Pagoaga, Dr M. Katherine *USA.*
Pai, Dr Emil F. *Canada.*
Paige-Green, Mr Philip *South Africa.*
Paiva Santos, Dr Carlos de Oliveira *Brazil.*
Paixão, Dr José António *Portugal.*
Pajączkowska, Anna *Poland.*
Pajak, Prof. Lucjan *Poland.*
Pajunen, Aarne Veikko *Finland.*
Pakawatchai, Dr Chaveng *Thailand.*
Pakkanen, Tapani Antti *Finland.*
Pakkanen, Dr Tuula T. *Finland.*
Pal, Mr Hiranmay *India.*
Palacio, Prof. Fernando *Spain.*
Palenik, DR. Gus J. *USA.*
Palenzona, Prof. Andrea *Italy.*
Palistrant, Prof. Alexander Fillipovich *Moldova.*
Palistrant, Dr Natalia Alexandrovna *Moldova.*
Paljević, Dr Matija *Croatia.*
Palmer, Kenneth J. *USA.*
Palmer, Dr Rex Alfred *UK.*
Palomo-Delgado, Dr M. Inmaculada *Spain.*
Palosz, Dr Bogdan Fryderyk *Poland.*
Pan, Prof. Kezhen *China.*
Pan, Dr Nitya Ranjan *India.*
Panagos, Prof. Athanasios *Greece.*
Panchekha, Dr Petro Oleksiyovych *Ukraine.*
Pandey, Dr Dhananjai *India.*
Pandi, Dr Veerapandian *USA.*
Pandit, Dr Jay *USA.*
Pandya, Prof. Janardhan Rameshchandra *India.*
Paneque, Mr Armando *Cuba.*
Pang, Dr Li *Canada.*
Pangborn, Robert N. *USA.*
Pani, Dr Kodandapani Ramadurgam. *USA.*
Pani, Dr Marcella *Italy.*
Paniagua, Dr Andrés *Spain.*
Panich, Dr Anatoly E. *Russia.*
Pannetier, Dr Jean *France.*
Panov, Mr Vladimir N. *Russia.*
Paorici, Prof. Carlo *Italy.*
Papadimitriou, Prof. Leonidas *Greece.*
Papadopoulos, Dr Dimitrios *Greece.*
Papaioannou, Prof. John *Greece.*
Papavinasam, Dr E *India.*
Pape, Ian *UK.*
Papunen, Prof. Heikki Tapani *Finland.*
Parak, Prof. Dr Fritz *Germany.*
Paraskevopoulos, Prof. Konstantinos *Greece.*
Parge, Dr Hans E. *USA.*
Parise, John B. *USA.*
Parisi, Francisco Eduardo Alberto *Argentina.*
Parisini, Dr Emilio *Italy.*
Parissakis, Prof. George *Greece.*
Park, Prof. Byung Kyu *Korea.*
Park, Chang H. Park *USA.*
Park, Hee-Won *USA.*
Park, Dr Il Yeong *Korea.*
Park, Prof. Young-Han *Korea.*
Park, Prof. Young-Ja *Korea.*
Párkányi, Dr László *Hungary.*
Parker, Dr Michael William *Australia.*
Parkin, Dr Sean R. *USA.*
Parris, Dr Kevin D. *USA.*
Parsons, Prof. Ian *UK.*
Parsons, Dr Mark R. *UK.*
Parsons, Dr Simon *UK.*
Parthasarathy, Dr S. *India.*
Parthasarathy, Dr Tiruvallur Eachambadi *India.*

Parthé, Prof. Erwin *Switzerland.*
Partin, Daniel E. *USA.*
Partridge, Mr Ben L. *UK.*
Paruchuri, Mohan *USA.*
Parvez, Dr Masood *Canada.*
Pascard, Dr Claudine *France.*
Pasero, Dr Marco *Italy.*
Paseshnichenko, Dr Ksenia Andreevna *Russia.*
Pashov, Prof. Nikolai *Bulgaria.*
Passaglia, Prof. Elio *Italy.*
Passalacqua, Edward F. *UK.*
Pasti, Mr Fabio *Italy.*
Paszkowicz, Dr Wojciech *Poland.*
Patchadjanov, Prof. Daler *Tadjikistan.*
Patel, Dr Prabhudas Revandas *India.*
Patel, Dr Rajan Prafulbhai *India.*
Patel, Dr Tankadhar *India.*
Pathak, Dr Dushyant *USA.*
Pathinettam, Dr Pandian *India.*
Paton, Mr John D. *UK.*
Pattabhi, Dr Mrs Vasantha *India.*
Pattanayek, Rekha *USA.*
Patton, Mr Alan T. *USA.*
Pattridge, Kitty *USA.*
Paufler, Prof. Dr Peter *Germany.*
Paul, Prof. Iain *USA.*
Paulitsch, Prof. Dr Peter *Germany.*
Paulmann, Mr Carsten *Germany.*
Paulus, Dr Erich Friedrich *Germany.*
Paunov, Dr Michael *Bulgaria.*
Pauptit, Dr Richard A. *UK.*
Pavalow, Melvin *USA.*
Pavel, Prof. Nicolae Viorel *Italy.*
Pavelčík, Doc. Ing. František *Slovakia.*
Pavese, Dr Alessandro *Italy.*
Pavie Cardoso, Dr Lisandro *Brazil.*
Pavkovic, Stephen F. *USA.*
Pavlov, Konstantin *Germany.*
Pavlov, Dr Sergey V. *Russia.*
Pavlović, Ms Gordana *Croatia.*
Pavlovsky, Dr Alexander G. *USA.*
Pavlyshyn, Prof. Volodymyr Ivanovych *Ukraine.*
Pavlyuk, Dr Volodymyr Vasyliovych *Ukraine.*
Payne, Dr Nicholas Charles *Canada.*
Pazdernik, Prof. LeRoy J. *Canada.*
Pearl, Dr Laurence Harris *UK.*
Pearson, Ms Céline *Canada.*
Pearson, Prof. William Burton *Canada.*
Peascoe, Roberta A. *USA.*
Peat, Tom *USA.*
Pech, Dr Lucija *Latvia.*
Pecharsky, Dr Vitalij K. *USA.*
Pecharsky, Dr Vitaly Kostiantynovych *Ukraine.*
Pecoraro, Vincent L. *USA.*
Peder, Murat *Turkey.*
Pedersen, Jan Skov *Denmark.*
Pedone, Prof. Carlo *Italy.*
Pedregosa, Dr Jose Carmelo *Argentina.*
Peek, Mary E. *USA.*
Peerdeman, Prof. Dr Antonius F. *Netherlands.*
Peerdeman, Dr Frans A. J. *Netherlands.*
Peeters, Ms Anik *Belgium.*
Peeters, Prof. Oswald Maurice *Belgium.*
Pel, Prof. Guangwen *China.*
Peiser, H. Steffen *USA.*
Pejović, Mrs Verica *Serbia.*
Pejovnik, Prof. Dr Stane *Slovenia.*
Pelizzi, Prof. Giancarlo *Italy.*
Pellinghelli, Prof. Maria Angela *Italy.*
Pelosi, Dr Giorgio *Italy.*
Penavić, Dr Maja *Croatia.*
Pence, Laura E. *USA.*
Peneva, Dr Stefka *Bulgaria.*
Peng, Prof. Shie-Ming *Taiwan.*
Pennartz, Dr Paul Ulrich *Germany.*
Penner-Hahn, Dr James E. *USA.*
Pennington, Dr William T. *USA.*
Pense, Prof. Dr Jürgen K. E. *Germany.*
Pepe, Dr Gérard *France.*
Perales, Dr Aurea *Spain.*

Perdikatsis, Dr Basilios *Greece.*
Perego, Dr Giovanni *Italy.*
Pereira, Dr Manuel Francisco da Costa *Portugal.*
Perekalina, Dr Tatyana M. *Russia.*
Perekalina, Dr Zoja B. *Russia.*
Perelomova, Prof. Natali V. *Russia.*
Perez, Ms Hiram *Cuba.*
Perez, Dr Serge *France.*
Pérez-Garrido, Dr Simeon *Spain.*
Pérez-Mato, Prof. Juan Manuel *Spain.*
Perona, Dr John J. *USA.*
Perozzo, Mary Ann *USA.*
Perrin, Prof. Monique *France.*
Pershan, Peter *USA.*
Pertierra, Mrs Pilar *Spain.*
Pertlik, Prof. Dr Franz *Austria.*
Perutz, Dr Max F. *UK.*
Pervaiz, Mr Rashed *Pakistan.*
Pervukhina, Dr Natalie V. *Russia.*
Peschar, Dr René *Netherlands.*
Peshev, Prof. Pavel *Bulgaria.*
Pessa, Markus *Finland.*
Petcher, Dr Trevor J. *Switzerland.*
Peterse, Ir Wilhelmus J. A. M. *Netherlands.*
Petersen, Donald R. *USA.*
Petersen, Jeffrey L. *USA.*
Petersen, Jens Frølund Winthe'r *Denmark.*
Petersen, Ole V *Denmark.*
Peters-Libeu, Clare *USA.*
Peterson, Dr Ronald Charles *Canada.*
Petkov, Dr Valeri *Bulgaria.*
Petraccone, Prof. Vittorio *Italy.*
Petrás, Mr László *Hungary.*
Petratos, Dr Kyriacos *Greece.*
Petrescu, Mircea-Ionuţ *Romania.*
Petrič, Mrs Desanka *Slovenia.*
Petrič, Dr Marko *Slovenia.*
Petriček, Miss Saša *Slovenia.*
Petříček, Dr Václav *Czech Republic.*
Petroff, Prof. Jean-François *France.*
Petropavlov, Dr Nikolai N. *Russia.*
Petrosian, Dr Ashot G. *Armenia.*
Petrov, Prof. Kostadin *Bulgaria.*
Petrov, Dr Ognyan *Bulgaria.*
Petrova, Dr Nina Nikolaevna *Kazakhstan.*
Petrovsky, Dr Vitaly A. *Russia.*
Petrzhik, Dr Ekaterina A. *Russia.*
Petsko, Dr Gregory A. *USA.*
Pett, Dr Virginia B. *USA.*
Petter, Prof. Walter *Switzerland.*
Pettifer, Dr Robert F. *UK.*
Petukhov, Prof. Boris V. *Russia.*
Petushkova, Mrs Ludmila V. *Russia.*
Petz, John I. *USA.*
Pfeffer-Hennig, Dr Sabine *Switzerland.*
Pflaum, Wolfgang *USA.*
Pfluger, Prof. Clarence E. *USA.*
Pflugrath, James W. *USA.*
Phac, Dr Le Van *Vietnam.*
Phaovibul, Dr Orapin *Thailand.*
Phavanantha, Dr Phathana *Thailand.*
Philipov, Mr Alexander *Bulgaria.*
Philipp, Dr George *Egypt.*
Phillips, Dr James Christopher *USA.*
Phillips Jr, Prof. George N. *USA.*
Phillips, Prof. Simon E. V. *UK.*
Phillips, T. J. *USA.*
Phillips, Walter C. *USA.*
Phipps, Dr Barry *Canada.*
Phizackerley, Prof. Richard Paul *USA.*
Phuc, Prof. Nguyen Khoa *Vietnam.*
Phuc, Dr Phan Vinh *Vietnam.*
Piątek, Mgr Małgorzata Katarzyna *Poland.*
Pichla, Susan P. *USA.*
Pichon-Pesme, Dr Virginie *France.*
Pickardt, Prof. Dr-Ing. Joachim *Germany.*
Pickersgill, Dr Richard W. *UK.*
Picot, Daniel *USA.*
Piecek, Mgr Wiktor *Poland.*
Pielaszek, Dr Jerzy *Poland.*

Pieper, Dr Ursula *USA.*
Pierer, Dr Gerhard *Germany.*
Pierpont, Cortlandt G. *USA.*
Pierrot, Dr Marcel *France.*
Pietraszko, Doc. Adam *Poland.*
Pietsch, Mr Hans-Henning *Germany.*
Pietsch, Dr Ullrich *Germany.*
Pietzuch, Mr Walter *Germany.*
Pifferi, Dr Augusto *Italy.*
Pignataro, Mrs Edith H. *USA.*
Pike, Ashley C. W. *UK.*
Pikin, Prof. Sergey A. *Russia.*
Pilati, Dr Tullio *Italy.*
Pilz, Mr Edgar *Germany.*
Pilz, Mrs Katrin *Germany.*
Pinegin, Dr Volodymyr Ivanovych *Ukraine.*
Piniella, Dr Juan Francisco *Spain.*
Pinkerton, Dr A. Alan *USA.*
Pintea, Ioan *Romania.*
Piontek, Dr Klaus *Switzerland.*
Pipkin, Dr Noel John *South Africa.*
Pique, Michael E. *USA.*
Pirard, Mr Bernard G. G. *Belgium.*
Pirc, Mrs Milojka *Slovenia.*
Piret, Prof. Paul *Belgium.*
Pirie, Dr John D. *UK.*
Piro, Dr Oscar Enrique *Argentina.*
Pirttimäki, Jukka Olavi *Finland.*
Pisarevski, Dr Yuri V. *Russia.*
Pisarevsky, Alexander P. *Russia.*
Pisutha-Arnond, Dr Visut *Thailand.*
Pitchford, Nigel Aaron *UK.*
Pitkänen, Ilkka Pellervo *Finland.*
Pizarro, Dr José Luis *Spain.*
Pjura, Dr Philip E. *USA.*
Plana-Llevat, Dr Feliciano *Spain.*
Plant, Dr John Stewart *UK.*
Platikanova, Dr Vesselina *Bulgaria.*
Platts, James *UK.*
Platzbecker, Mr Rolf *Germany.*
Plesken, Prof. Dr Wilhelm *Germany.*
Pleštil, Ing. Josef *Czech Republic.*
Pletinckx, Mr Jurgen *Belgium.*
Pletnev, Dr Vladimir Z. *Russia.*
Pley, Heinz W. *USA.*
Plugaru, Neculai *Romania.*
Pluth, Joseph J. *USA.*
Plyasova, Prof. Ludmila M. *Russia.*
Pocev, Prof. Dr Stefan *(FYR) Macedonia.*
Pochetti, Dr Giorgio *Italy.*
Pochettino, Dr Alberto Antonio *Argentina.*
Podberezskaya, Dr Nina V. *Russia.*
Podder, Dr Alok *India.*
Podjarny, Dr Alberto *France.*
Podlahová, Doc. Dr Jana *Czech Republic.*
Pöllmann, Dr Herbert *Germany.*
Pöttgen, Dr Rainer *Germany.*
Pohl, Prof. Dr Dieter *Germany.*
Pohl, Prof. Dr Siegfried *Germany.*
Pohl, Mr Ehmke *Germany.*
Polanc, Prof. Dr Slovenko *Slovenia.*
Polborn, Dr Kurt Volkmar *Germany.*
Polcarová, Dr Milena *Czech Republic.*
Polekhina, Galina *Denmark.*
Poleti, Dr Dejan *Serbia.*
Polidori, Dr Giampiero *Italy.*
Politova, Dr Ekaterina D. *Russia.*
Poll, Dr Wolfgang *Germany.*
Polvorinos, Dr Angel Jesus *Spain.*
Polyakov, Dr Konstantin M. *Russia.*
Polyanskaya, Dr Tamara M. *Russia.*
Polyansky, Dr Evgeny V. *Russia.*
Polychroniadis, Prof. Evstathios K. *Greece.*
Polynova, Prof. Tamara Nikitichna *Russia.*
Pomes, Dr Ramon *Cuba.*
Pompe, Mrs Dobruša *Slovenia.*
Ponader, Carl W. *USA.*
Pond, Prof. Robert Charles *UK.*
Ponder, Dr Jay W. *USA.*
Pong, Prof. Pong *Taiwan.*

Pongsapich, Dr Wasant *Thailand.*
Ponomarev, Dr Vasiliy I. *Russia.*
Poojary, Damodara *USA.*
Poojary, Dr M. Damodara *India.*
Poortmans, Dr Freddy *Belgium.*
Pop, Dr Viorel *Romania.*
Popma, Prof. Dr Theo J. A. *Netherlands.*
Popov, Dr Alexander *Bulgaria.*
Popov, Dr Alexander N. *Russia.*
Popova, Dr Svetlana V. *Russia.*
Popović, Prof. Dr Stanko *Croatia.*
Poppi, Prof. Luciano *Italy.*
Poppleton, Bruce James *Australia.*
Porai-Koshits, Prof. Michail A. *Russia.*
Porta, Prof. Piero *Italy.*
Portal-Olea, Dr M. Dolores *Spain.*
Portalone, Dr Gustavo *Italy.*
Porter, Leigh C. *USA.*
Portnov, Dr Vadim N. *Russia.*
Porzio, Dr William *Italy.*
Post, Jeffrey E. *USA.*
Post, Dr Michael Leonard *Canada.*
Postma, Dr Johannes Petrus Maria *Germany.*
Potekhin, Dr Konstantin A. *Russia.*
Potenza, Joseph A. *USA.*
Poulin, Mrs Suzie *Canada.*
Poulios, Prof. Ioannis *Greece.*
Poulos, Thomas L. *USA.*
Powell, Dr Annie *UK.*
Powell, Dr Brian Mathieson *Canada.*
Powell, Dr Douglas R. *USA.*
Powell, Dr Harold Roger *UK.*
Pradhan, Dr Dukhabandhu *India.*
Pradhan, Dr Swapan Kumar *India.*
Prado, Fernando *Argentina.*
Prager, Dr Peter Robert *Australia.*
Prakash, Mr Balaji *India.*
Prandl, Prof. Dr Wolfram *Germany.*
Prasad, Dr Lata *Canada.*
Prasad, Dr Narayan *India.*
Prasad, Dr Ravindra *India.*
Prasad, Dr Satya Murti *India.*
Prasad , Dr Y. R. Ananth *India.*
Prata Pina, Jose Carlos *Portugal.*
Pratt, David W. *USA.*
Precigoux, Dr Gilles *France.*
Preisinger, Prof. Dr Anton *Austria.*
Prekesnik, Dr Bogdan *Serbia.*
Prelovšek, Prof. Dr Peter *Slovenia.*
Pretorius, Dr Jan Andries *South Africa.*
Preuß, Dr Heinz Hans Walter *Germany.*
Preusch, Peter *USA.*
Preut, Dr Hans *Germany.*
Previde Massara, Dr Elisabetta *Italy.*
Prewitt, Dr Charles T. *USA.*
Price, Dr Sarah Lois *UK.*
Priestle, Dr John Peter *Switzerland.*
Prieto, Prof. Manuel *Spain.*
Prifis, Prof. George *Greece.*
Prince, Dr Edward *USA.*
Prince, Dr Stephen M. *UK.*
Pring, Allan *Australia.*
Prive, Dr Gilbert G. *USA.*
Proctor, Peter d'urphee *USA.*
Prodan, Dr Albert *Slovenia.*
Profeta Jr, Salvatore *USA.*
Profi, Ms. Stella *Greece.*
Prokhorov, Mr Andrey I. *Russia.*
Prokhorov, Dr Igor A. *Russia.*
Proserpio, Dr Davide M. *Italy.*
Prudnikov, Dr Ilya R. *Russia.*
Pučálka, Dr Vladimír *Czech Republic.*
Pugachov, Prof. Anatolij Tarasovych *Ukraine.*
Pugazhenthi, Ms Umarani *Canada.*
Puigjaner, Prof. Luis *Spain.*
Puig-Molina, Mr Anna *Spain.*
Pulcinelli, Dr Sandra Helena *Brazil.*
Puliti, Dr Raffaella *Italy.*
Pulliam, Curtis R. *USA.*
Pulsinelli, Phillip D. *USA.*

Pulya, Umamaheswara Sastry *India.*
Punin, Dr Yurii O. *Russia.*
Punkkinen, Matti *Finland.*
Puntarec, Mr Vitomir *Croatia.*
Punte, Prof. Graciela *Argentina.*
Puranik, Mrs Vedavati Gururaj *India.*
Purans, Dr Juris *Latvia.*
Purdy, Samuel *USA.*
Pushcharovsky, Prof. Dmitry Yu. *Russia.*
Puttajakr, Mrs Taswal *Thailand.*
Pyka, Dr Niels Michael *France.*
Pyykkö, Prof. Pekka *Finland.*
Qaiser, Mr Mohammad Ali *Pakistan.*
Qidwai, Dr Ansar Ahmad *Pakistan.*
Qiu, Xiayang *USA.*
Quagliata, Prof. Claudio *Italy.*
Quail, Dr J. Wilson *Canada.*
Quan, Mr Do Hong *Vietnam.*
Quartieri, Dr Simona *Italy.*
Queiroz do Amaral, Dr Lia *Brazil.*
Queisser, Prof. Dr Hans Joachim *Germany.*
Queralt-Mitjans, Dr Ignasi *Spain.*
Quere, Mr Luc *Belgium.*
Quicksall, Carl O. *USA.*
Quillin, Michael L. *USA.*
Quinones, Dr Jose *Cuba.*
Quintana, Dr John P. *USA.*
Quintana Owen, Ms Patricia *Mexico.*
Quintanar, Mr Carlos *Mexico.*
Quiocho, Dr Florante A. *USA.*
Quirk, Dr Stephen *USA.*
Quirós, Dr Miguel *Spain.*
Qurashi, Dr Mazhar Mahmood *Pakistan.*
Qureshi, Mr Khalid Mahmood *Pakistan.*
Qureshi, Dr M. Hanif *Pakistan.*
Qureshi, Mr Mohammad Kaleem Akhter *Pakistan.*
Quy, Dr Le Cong *Vietnam.*
Quyen, Dr Nguyen Hong *Vietnam.*
R. Oldenbourg Verlag, *Germany.*
Raabe, Dr Gerhard P. *Germany.*
Raade, Gunnar *Norway.*
Rabalais, J. W. *USA.*
Rabenberg, Llewellyn *USA.*
Rabie, Dr Farida Hamed *Egypt.*
Rabinovich, Prof. Dov *Israel.*
Rabinowitz, Israel *USA.*
Rabu, Dr Pierre *France.*
Radaković, Mrs Aleksandra *Serbia.*
Radautsan, Prof. Sergiu Ion *Moldova.*
Rader, Stephen D. *USA.*
Radha, Akella *USA.*
Radmilović, Dr Velimir *Serbia.*
Radnai, Dr Tamás *Hungary.*
Radnóczy, Dr György *Hungary.*
Radović, Ms Nikol *Croatia.*
Radwan, Prof. Mostafa Mohsen *Egypt.*
Radzo, Prof. Dr Vendelín *Slovakia.*
Rae, Professor Alan David *Australia.*
Raevski, Dr Simion Dumitru *Moldova.*
Rafalko, Patrice White *USA.*
Raftery, Tony *Australia.*
Ragab, Prof. Dr Abdelghani *Egypt.*
Raghavacharyalu, Dr Iyyunni Venkata Veera *India.*
Raghunatha, Mr Chary *India.*
Ragimov, Dr Kerim Ragimov *Azerbaijan.*
Rahuel, Mr Joseph *Switzerland.*
Raines, Ronald T. *USA.*
Rainov, Dr Nikola *Bulgaria.*
Raithby, Dr Paul Robert *UK.*
Raitman, Dr Ernst *Latvia.*
Rajagopal, Mr Hariharasubramonia Iyer *India.*
Rajala, Riitta Helena *Finland.*
Rajan, Mr R. D. *India.*
Rajan, Dr S. S. *India.*
Rajaram, Dr Ramaswamy Karunandam *India.*
Rajashekharan, Dr T. *India.*
Raju, Dr I. V. K. Bhagavan *India.*
Raju, Dr K. S. *India.*
Rakić, Mr Srdjan *Serbia.*
Rakin, Dr Vladimir I. *Russia.*

Rakova, Dr Elena V. *Russia.*
Ralph, Dr Brian *UK.*
Ram Kishore, Dr *India.*
Ram, Dr Purushottam *India.*
Ramachandran , Prof. Gopalasamudram N. *India.*
Ramadan, Prof. Ahmed Ahmed *Egypt.*
Ramakrishnan, Prof. Chandrasekharan *India.*
Ramanadham, Dr Muthyala *India.*
Ramaswamy, Dr Krishnamachari *India.*
Ramaswamy, Mr S. *India.*
Rambaud, Dr Jöelle *France.*
Ramm, Dr Matthias *Germany.*
Ramos Bernal, Dr Sergio *Mexico.*
Ramos Parente, Dr Carlos Benedicto *Brazil.*
Rana, Mr Riaz Ahmad *Pakistan.*
Randaccio, Prof. Lucio *Italy.*
Ranganath, Dr G. S. *India.*
Ranger, Dr Georges Joseph *Canada.*
Ranløv, Jens *Denmark.*
Rao, Dr Jejjala Krishna Mohana *USA.*
Rao, Dr Keshavamurthy Narayana Swamy *India.*
Rao, Dr Mokka Narasinga *Venezuela.*
Rao, Narasinga *USA.*
Rao, Dr S. T. *USA.*
Rao, Dr Sudhakara S. P. *USA.*
Rao, Dr Usha *USA.*
Raoux, Dr Denis *France.*
Rardin, Dr R. Lynn *USA.*
Rashkov, Prof. Stefan *Bulgaria.*
Rashkovich, Prof. Leonid N. *Russia.*
Rasines, Prof. Isidoro *Spain.*
Rasmussen, Hanne *Denmark.*
Rasmussen, Svend Erik *Denmark.*
Raston, Prof. Colin Llewellyn *Australia.*
Rastsvetaeva, Dr Ramiza K. *Russia.*
Ratajczak-Sitarz, Dr Malgorzata *Poland.*
Ratanasthien, Dr Benjavun *Thailand.*
Rath, Dr Nigam P. *USA.*
Rath, Virginia L. *USA.*
Ratna, Miss B. R. *India.*
Ratuszna, Dr Alicja *Poland.*
Raty, Jean-Yves *Belgium.*
Rau, Dr Tamara F. *Russia.*
Rau, Prof. Valery G. *Russia.*
Rauch, Prof. Dr Helmut *Austria.*
Raudsepp, Dr Mati *Canada.*
Rausell-Colom, Prof. José Antonio *Spain.*
Rautioaho, Risto Heikki *Finland.*
Ravelli, Drs Raimond B. G. *Netherlands.*
Raves, Drs Mia L. *Israel.*
Ravichandran, Kurumbail G. *USA.*
Ravichandran, Dr Veena *India.*
Ravishankar, Mr Ramachandran *India.*
Ray, Dr Mrs *India.*
Ray , Dr Pankaj Narayan *India.*
Ray , Dr Pradip Kumar *India.*
Ray , Dr Sanjoy *USA.*
Rayment, Professor Ivan *USA.*
Read, Dr Randy John *Canada.*
Rebelo, Joana Cristina Fanha *Portugal.*
Rebrov, Dr Alexander V. *Russia.*
Rečnik, Mr Aleksander *Slovenia.*
Reddy, Dr Jaya G. *USA.*
Reddy, Dr Swaminatha B. *USA.*
Reddy, Vijay S. *USA.*
Redfern, Dr Simon A. T. *UK.*
Redford, Susan M. *USA.*
Redinbo, Matthew R. *USA.*
Redler, Dr László *Hungary.*
Reeber, Dr Robert R. *USA.*
Reeder, Dr Richard J. *USA.*
Reefman, Derk *Netherlands.*
Reeke Jr, Dr George N. *USA.*
Rees, Dr Bernard *France.*
Rees, Douglas C. *USA.*
Regan, Michael J. *USA.*
Regel, Prof. Vadim R. *Russia.*
Regueira Teodósio, Prof. Joel *Brazil.*
Rehfeldt, Dr Angeline *Germany.*
Rehse, Dr Peter *Canada.*

Reibenspies, Dr Joseph H. *USA.*
Reid, Mrs Jean S. *UK.*
Reid, Dr John Sinclair *UK.*
Reid Jr, Austin *USA.*
Reimers, Dr Walter *Germany.*
Reinemer, Dr Peter *Germany.*
Reinen, Prof. Dr Dirk *Germany.*
Reinisch, Karin M. *USA.*
Reinke, Dr Helmut *Germany.*
Reis Jr, Arthur H. *USA.*
Reis, M. Lourdes Rodrigues Pinto De Castro *Portugal.*
Reiss, Drs Céleste A. *Netherlands.*
Reissner, Dr Michael *Austria.*
Remington, S. James *USA.*
Remškar, Mrs Maja *Slovenia.*
Ren, Yang *Netherlands.*
Ren, Zhong *USA.*
Renaud, Dr Jean-Paul *France.*
Rendle, Dr David Forbes *UK.*
Renner, Ing. Oldřich *Czech Republic.*
Rentzeperis, Prof. Panagiotis I. *Greece.*
Restivo, Dr Roderic John *Canada.*
Retief, Dr Johannes Jacobus *South Africa.*
Rettig, Dr Steven John *Canada.*
Reuber-Kürbs, Dr-Ing. Ellen *Germany.*
Reus, Ir Henk J. *Netherlands.*
Reventós, Dr María-Mercedes *Spain.*
Revkevich, Dr Galina P. *Russia.*
Rey, Dr Felix *USA.*
Reyes Chumacero, Mr Antonio *Mexico.*
Reynaers, Prof. Harry *Belgium.*
Reynhardt, Prof. Eduard Christiaan *South Africa.*
Reynolds, Philip Andrew *Australia.*
Reynolds, Dr Ross *USA.*
Rhee, Sangkee *USA.*
Rheingold, Arnold L. *USA.*
Rheinwald, Dr Gerd *Switzerland.*
Rhodes, Dr Gale *USA.*
Ribár, Dr Béla *Serbia.*
Ribas, Prof. Joan *Spain.*
Ribbe, Paul H. *USA.*
Ricci, Dr John S. *USA.*
Rice, Prof. David W. *UK.*
Rice, Marybeth *USA.*
Richards, Frederic M. *USA.*
Richardson, David C. *USA.*
Richardson, Jane S. *USA.*
Richardson, Dr John F. *USA.*
Richardson, Dr Mary Frances *Canada.*
Richman, Marc H. *USA.*
Richmond, Prof. Timothy J. *Switzerland.*
Richter, Dr Paul Wilhelm *South Africa.*
Richter, Dr Rainer *Germany.*
Richter, Dr Waltraut *Germany.*
Rickard, Dr Clifton E. F. *New Zealand.*
Rider, Dr Evgeny E. *Russia.*
Rieck, Prof. Dr Gerard D. *Netherlands.*
Riedel, Mr Gernot *Germany.*
Rieder, Prof. Dr Milan *Czech Republic.*
Rieger, Dr Wolfhart H. *Switzerland.*
Riella, Mr Humberto Gracher *Brazil.*
Riera, Prof. Victor *Spain.*
Ries, Mr Ronald *Germany.*
Rietveld, Dr Hugo M. *Netherlands.*
Rigault de la Longrais, Prof. Germain *Italy.*
Rigotti, Dr Graciela Ester *Argentina.*
Rihs, Ms Greti *Switzerland.*
Riise, Bjørn *Norway.*
Rimm, Mr Karel *Estonia.*
Rinaldi, Prof. Romano *Italy.*
Rinaudo, Prof. Caterina *Italy.*
Ringe, Dr Dagmar *USA.*
Ringer, Simon Peter *Japan.*
Ringrose, Sharon *USA.*
Rini, Dr James *Canada.*
Rios Jara, Dr David *Mexico.*
Ripamonti, Prof. Alberto *Italy.*
Rissanen, Kari *Finland.*
Ristic, Dr Radoljub I. *UK.*
Ritter, Dr Clemens *France.*

Rius, Prof. Jordi *Spain.*
Riva di Sanseverino, Prof. Lodovico *Italy.*
Riva, Dr Fernando *Italy.*
Rivera Moras, Mr Vicente *Mexico.*
Rivera Ocando, Dr Angela Valentina *Venezuela.*
Rivero, Prof. Blas Eduardo *Argentina.*
Riveros Rotgé, Prof. Héctor Gerardo *Mexico.*
Rizvi, Prof. Adibul Hassan *Pakistan.*
Rizvi, Safia Khalil *USA.*
Rizvi, Dr Syed Sadrul Hasan *Pakistan.*
Rizzi, Dr Menico *Italy.*
Rizzoli, Dr Corrado *Italy.*
Robarick, Dr Eckhard *Germany.*
Robbins, Arthur H. *USA.*
Roberts, Mr Andrew Clifford *Canada.*
Roberts, David L. *USA.*
Roberts, Dr Kevin John *UK.*
Roberts, Dr Sue A. *USA.*
Robertson, Prof. Beverly Ellis *Canada.*
Robertson, Dr J. Lee *USA.*
Robertson, Dr John Harry *UK.*
Robertson, Ms Katherine Nancy *Canada.*
Robertus, Dr Jon D. *USA.*
Robie, Stephen B. *USA.*
Robinson, Ian K. *USA.*
Robinson, Paul D. *USA.*
Robinson, Victoria Lee *USA.*
Robinson, Dr Ward T. *New Zealand.*
Robu, Lucia *Romania.*
Rochon, Dr Fernande D. *Canada.*
Roderick, Dr Steven L. *USA.*
Rodgers, Dr David W. *USA.*
Rodgers, Dr John R. *Canada.*
Rodgers, Dr K. A. *New Zealand.*
Rodić, Dr Dubravko *Serbia.*
Rodrigues, Dr Antonio Ricardo Droher *Brazil.*
Rodrigues da Silva, Dr Rilson *Brazil.*
Rodrigues, Dr Edson *Brazil.*
Rodríguez, Dr Alfonso *Venezuela.*
Rodríguez, Dr Carlos *Cuba.*
Rodríguez de Aguirrezabala, Dr María A. *Venezuela.*
Rodríguez Romero, Dr Adela *Mexico.*
Rodríguez-Carvajal, Dr Juan *France.*
Rodríguez-Carvajal, Dr Juan *Spain.*
Rodríguez-Clemente, Prof. Rafael *Spain.*
Rodríguez-Gallego, Prof. Manuel *Spain.*
Rodríguez-Gordillo, Dr José *Spain.*
Rodríguez-Navarro, Mr Alejandro *Spain.*
Rodríguez-Roldan, Dr Ana *Spain.*
Rodríguez-Romero, Mr F. Victor *Spain.*
Roe, Dr Stephen Mark *UK.*
Roebuck, Dr Peter H. A. *UK.*
Rømming, Prof. Christian *Norway.*
Roeschmann, Mr Carlos *Chile.*
Roetti, Prof. Carla *Italy.*
Rogers, Prof. Donald *UK.*
Rogers, Dr Keith *UK.*
Rogers, Robin D. *USA.*
Roginskaya, Dr Yuliana E. *Russia.*
Rogl, Dr Peter Franz *Austria.*
Rognlie, David G. *USA.*
Rohl, Dr Andrew L. *UK.*
Rohrer, Dr Douglas C. *USA.*
Rojas, Dr Carlos Eduardo *Venezuela.*
Rojo, Prof. Teofilo *Spain.*
Romaka, Dr Ljuba Petrivna *Ukraine.*
Román, Prof. Pascual *Spain.*
Romanenko, Dr Galina V. *Russia.*
Romao, Dr Maria Joao *Portugal.*
Romero, Dr Antonio *Spain.*
Romero Romo, Dr Mario *Mexico.*
Romero Sánchez, Dr Miguel *Mexico.*
Romero-Garzón, Mr José *Spain.*
Romero-Molina, Dr M. Angustias *Spain.*
Romers, Prof. Dr Cornelis *Netherlands.*
Romijn, Ir Antonius M. *Netherlands.*
Rondeau, Dr Jean-Michel *France.*
Ronis, Dr Janis *Latvia.*
Rontoyianni, Dr Aliki *Greece.*
Roodt, Prof. Andreas *South Africa.*

Roof, Dr Raymond B. *USA.*
Roos, Dr Hester Marita *South Africa.*
Roosenbrand, Dr Albert G. *Netherlands.*
Root, Dr J. H. *Canada.*
Roquet, Ms Françoise Jeanne Valentine *France.*
Ros, Prof. Josep *Spain.*
Rosa de la, Dr Nicolás *Spain.*
Rosa, Dr Rodolfo *Italy.*
Rosauer, Ruth *USA.*
Rose, Dr David Richard *Canada.*
Rose, Dr John P. *USA.*
Rosemond, M. Jane Cox *USA.*
Rosenqvist, Prof. Ivan Thoralf Koss *Norway.*
Rosenstein, Dr Robert Daniel *USA.*
Rosenzweig, Amy C. *USA.*
Ross, Prof. Donald Keith *UK.*
Ross, Dr Fred K. *USA.*
Ross II, Dr Charles R. *USA.*
Rossi, Dr Marco *Italy.*
Rossi, Dr Miriam *USA.*
Rossmanith, Prof. Dr Elisabeth *Germany.*
Rossmann, Michael G. *USA.*
Rossouw, Chris *Australia.*
Rostomian, Dr Armand *Armenia.*
Rostomian, Dr Armen M. *Armenia.*
Roszak, Dr Aleksander W. *Canada.*
Rotella, Dr Frank J. *USA.*
Roth, Dr Georg *Germany.*
Roth, Robert S. *USA.*
Rothammel, Dr Walter *Germany.*
Rothbauer, Dr Richard *Germany.*
Rotonda, Jennifer *USA.*
Rould, Mark *USA.*
Rout, Joanne *UK.*
Routsi, Prof. Christine *Greece.*
Rouvinen, Juha *Finland.*
Roveri, Prof. Norberto *Italy.*
Rowland, Dr R. Scott *UK.*
Roy, Rustum *USA.*
Roychowdhury, Dr Mrs S. *India.*
Roychowdhury, Dr Priyobroto *India.*
Royer, Dr William E. *USA.*
Rozenberg, Haim *Israel.*
Rozhansky, Prof. Vladimir N. *Russia.*
Rozhdestvenskaya, Dr Ira V. *Russia.*
Rozin, Dr Konstantin M. *Russia.*
Rozsondai, Dr Béla *Hungary.*
Rozwarski, Denise A. *USA.*
Rubbo, Prof. Marco *Italy.*
Rubin, J. Ronald *USA.*
Rubina, Dr Elena B. *Russia.*
Ruble, Dr John R. *USA.*
Ruck, Dr Michael *Germany.*
Rudenko, Gabrielle *USA.*
Rudenko, Mr Sergey S. *Russia.*
Ruderman, Pres. Warren *USA.*
Rudert, Dr Rainer *Germany.*
Rudman, Prof. Reuben *USA.*
Rudnick, Suzanne E. *USA.*
Rudolf, Dr Philip R. *USA.*
Rudolph, Prof. Dr Peter *Germany.*
Rueda, Prof. Fulgencio *Venezuela.*
Ruggles, Dr James A. *USA.*
Ruhlandt-Senge, Dr Karin *USA.*
Ruiz Mejia, Dr Carlos *Mexico.*
Ruiz, Dr Xavier *Spain.*
Ruiz-Pérez, Dr Catalina *Spain.*
Ruiz-Valero, Dr Caridad *Spain.*
Rukvichai, Mr Surapol *Thailand.*
Rull, Prof. Fernando *Spain.*
Rumanova, Dr Iskra Mikhailovna *Russia.*
Rundqvist, Prof. Stig O. *Sweden.*
Rupp, Mag. Dr Bernhard *USA.*
Rusanowskii, Dr Mikhail Evstafievich *Moldova.*
Ruslan, Dr Peter R. *Armenia.*
Russel, Mr Nazirullah *Pakistan.*
Russell, Dr David R. *UK.*
Russell, Thomas Paul *USA.*
Russev, Dr Krassimir *Bulgaria.*
Russo, Dr Galina V. *Russia.*

Rutherford, Prof. John Stewart *South Africa.*
Rutten-Keulemans, Drs Elisabeth W. M. *Netherlands.*
Ruud, Clayton O. *USA.*
Rux, Dr John J. *USA.*
Ryadnov, Dr Sergei N. *Russia.*
Ryba, Earle *USA.*
Rybakov, Dr Victor Borisovich *Russia.*
Rychlewska, Prof. Urszula *Poland.*
Rydel, Dr Timothy J. *USA.*
Rypniewski, Dr Wojciech R. *Germany.*
Saalfeld, Prof. Dr Horst *Germany.*
Sabat, Dr Michal *USA.*
Sabatino, Dr Piera *Italy.*
Sabbioni, Dr Cristina *Italy.*
Sabelli, Dr Cesare *Italy.*
Sabine, Terence Murray *Australia.*
Sabrowsky, Prof. Dr Horst *Germany.*
Sacanian, Dr Martin S. *Armenia.*
Sacerdoti, Prof. Michele *Italy.*
Sachsenröder, Mr Steffen *Germany.*
Sack, John S. *USA.*
Sadanaga, Prof. Ryoichi *Japan.*
Sadek, Prof. Gamil *Egypt.*
Sadikov, Dr George G. *Russia.*
Sadowski, Lucian M. *USA.*
Saeed, Mr Syed Mohammad *Pakistan.*
Saenger, Prof. Dr-Ing. Wolfram H. E. *Germany.*
Saethre, Prof. Leif Jarle *Norway.*
Safo, Dr Martin K. *USA.*
Safonova, Dr Tatiana N. *Russia.*
Safro, Dr Mark G. *Israel.*
Safronov, Mr Victor V. *Russia.*
Saf'yanov, Prof. Yuri N. *Russia.*
Saghir, Mr Ahmad *Pakistan.*
Saha, Mr Bishwa Nath *India.*
Sahalos, John *Greece.*
Sahaymary, Mrs J. James *India.*
Sahm, Prof. Dr-Ing. Peter R. *Germany.*
Sahu, Dr Bhola Nath *India.*
Sahu, Dr Ram Gopal *India.*
Said, Dr Fikria *Egypt.*
Saito, Dr Yoshio *Japan.*
Saito, Dr Yukio *Japan.*
Sakabe, Prof. Noriyoshi *Japan.*
Sakagami, Prof. Dr Noboru *Japan.*
Sakamoto, Dr Dr Hidekazu *Japan.*
Sakata, Dr Makoto *Japan.*
Sakata, Dr Osami *Japan.*
Sakharov, Dr Boris A. *Russia.*
Sakon, Dr Joshua *USA.*
Sakurai, Prof. Tosio *Japan.*
Salah-ud-Din, Dr *Pakistan.*
Salakhutdinov, Dr Mals *Tadjikistan.*
Salamakha, Dr Petro Stepanovych *Ukraine.*
Salas Piza, Dr Guillermo Armando *Mexico.*
Salas-Peregrin, Dr Juan Manuel *Spain.*
Salazar Arcay, Prof. Esmina Noelia *Venezuela.*
Salem, Dr Safia Mahmoud *Egypt.*
Salemme, F. Raymond *USA.*
Salminen, Tiina *Finland.*
Salo, Päivi Tuulikki *Finland.*
Salvado, Mr Miguel Angel *Spain.*
Salviati, Dr Giancarlo *Italy.*
Samanta, Ms Chitra *India.*
Samantaray, Dr Biswas Kumar *India.*
Samoilovich, Dr Lidiya A. *Russia.*
Samoilovich, Prof. Mikhail I. *Russia.*
Samotin, Dr Nikolay D. *Russia.*
Samudzi, Cleopas T. *USA.*
Samus', Prof. Ivan Dmitriyevich *Moldova.*
Samusina, Mrs Svetlana Nic. *Russia.*
San José, Mr Ana M. *Spain.*
Sánchez-Navas, Dr Antonio *Spain.*
Sánchez-Sánchez, Dr Purificación *Spain.*
Sanderson, Ian *UK.*
Sands, Dr Donald E. *USA.*
Sandström, Magnus K. E. *Sweden.*
Sangariyavanich, Mrs Archara *Thailand.*
Sanishvili, Dr R. *USA.*
Sanjeeviraja, Dr C. *India.*

287

Sankaran, Ms Hema *India*.
Sankaranarayanan, Mr Rajan *India*.
Sanker, Mr B. N. *India*.
Sannikov, Dr Daniil G. *Russia*.
Sano, Mr Chiaki *Japan*.
Sano, Prof. Masatoshi *Japan*.
Sansoni, Prof. Mirella *Italy*.
Santarsiero, Dr Bernard D. *USA*.
Santini, Dr Antonello *Italy*.
Santos, Dr Pérsio de Souza *Brazil*.
Santos, Dr Regina Helena de Almeida *Brazil*.
Santos-Sánchez, Dr Alberto *Spain*.
Sanz-Aparicio, Dr Juliana *Spain*.
Sanzharlinsky, Dr Nikolai G. *Russia*.
Sanz-Ruiz, Prof. Francisco *Spain*.
Saper, Dr Mark A. *USA*.
Sapozhnikov, Dr Anatoly N. *Russia*.
Sappenfield, Dr Eric L. *USA*.
Šare-Lahodny, Prof. Dr Olga *Croatia*.
Sari, Musa *Turkey*.
Sariel, Joseph *Israel*.
Sarin, Dr Victor A. *Russia*.
Sarkar, Ms Chitra *India*.
Sarkar, Dr Satyabrata *India*.
Sarko, Anatole *USA*.
Sarma, Dr J. A. R. P. *India*.
Sarma, Dr Raghu *USA*.
Sasada, Prof. Yoshio *Japan*.
Sasaki, Prof. Dr Akio *Japan*.
Sasaki, Prof. Kyoyu *Japan*.
Sasaki, Satoshi *Japan*.
Sasaki, Prof. Takatomo *Japan*.
Sastry, Dr G. V. S. *India*.
Sasvári, Dr Kálmán *Hungary*.
Sathyamurthy, Ms Padma *India*.
Satittada, Miss Gannaga *Thailand*.
Sato, Prof. Dr Kiyotaka *Japan*.
Sato, Dr Fumio *Japan*.
Sato, Prof. Katsuaki *Japan*.
Sato, Dr Mamoru *Japan*.
Sato, Prof. Mitsuo *Japan*.
Sato, Dr Shoichi *Japan*.
Sato, Takao *Japan*.
Sato, Dr Tatsuo *Japan*.
Satow, Prof. Yoshinori *Japan*.
Sauter, Dr Nicholas K. *USA*.
Sauvage, Dr Michèle *France*.
Savenko, Dr Boris N. *Russia*.
Savithramma, Miss K. L. *India*.
Sawaya, Michael R. *USA*.
Sawka-Dobrowolska, Dr Wanda *Poland*.
Sawyer, Dr Lindsay *UK*.
Sax, Martin *USA*.
Sayre, Dr David *USA*.
Sazaki, Mr Gen *Japan*.
Scandale, Prof. Eugenio *Italy*.
Scapin, Giovanna *USA*.
Scardi, Prof. Paolo *Italy*.
Scaringe, Raymond P. *USA*.
Scarrott, Keith *UK*.
Scattarin, Prof. Vladinmiro *Italy*.
Šćavnićar, Prof. Dr Stjepan *Croatia*.
Schaber, Peter M. *USA*.
Schaefer, Dale W. *USA*.
Schaefer, Mr Klaus *Germany*.
Schäfer, Dr Wolfgang *Germany*.
Schaefer, Dr William P. *USA*.
Schaetzle, Dr Peter *Germany*.
Schagen, Dr Jan Dirk *Netherlands*.
Schapink, Dr Frederik W. *Netherlands*.
Schattschneider, Dr Peter *Austria*.
Scheel, Mr Hans J. *Switzerland*.
Scheidt, Dr W. Robert *USA*.
Schenk, Dr Dr Hendrik *Netherlands*.
Schenk, Prof. Dr Manfred *Germany*.
Scherrenberg, Dr Rolf *Netherlands*.
Schetelich, Mr Christoph *Germany*.
Schevitz, Dr Richard W. *USA*.
Schewe-Miller, Dr Irmgard M. *USA*.
Schiavinato, Prof. Giuseppe *Italy*.

Schicht, Dr Rudolf *Switzerland*.
Schierbeek, Dr Abraham J. *Netherlands*.
Schiffer, Dr Marianne *USA*.
Schildbach, Joel F. *USA*.
Schilling, Dr Frank Rüdiger *Germany*.
Schiltz, Mr Marc *France*.
Schinzer, Mr Carsten *Germany*.
Schioler, Dr Liselotte *USA*.
Schirber, James E. *USA*.
Schirmer, Dr Tilman *Switzerland*.
Schlemper, Dr Elmer O. *USA*.
Schlenker, John L. *USA*.
Schlenker, Prof. Michel *France*.
Schloemer, Prof. Dr Hermann Johannes *Germany*.
Schlünzen, Mr Frank *Germany*.
Schmahl, Dr Wolfgang W. *Germany*.
Schmalle, Dr Helmut Willi *Switzerland*.
Schmid, Dr Siegbert *Australia*.
Schmidbauer, Dr Martin *Germany*.
Schmidt, Prof. Dr Günther *Germany*.
Schmidt Nielsen, Søren *Denmark*.
Schmidt, Paul W. *USA*.
Schmitz, Dr Werner *Germany*.
Schneider, Bohdan *Czech Republic*.
Schneider, Dr Dieter K. *USA*.
Schneider, Prof. Dr Jochen R. *Germany*.
Schneider, Gunter *Sweden*.
Schneider, Dr Julius *Germany*.
Schneider, Mr Thomas R. *Germany*.
Schneider, Dr Walter *Germany*.
Schnick, Prof. Dr Wolfgang *Germany*.
Schnieders, Mr Felix *Germany*.
Schobinger-Papamantellos, Dr Penelope *Switzerland*.
Schoch, Dr Aylva Ernest *South Africa*.
Schöllhorn, Prof. Dr Robert *Germany*.
Schoenborn, Prof. Benno P. *USA*.
Schönholzer, Mr Peter *Switzerland*.
Schöning, Prof. Friedrich Richard Ludwig *South Africa*.
Schollmeyer, Dr Dieter *Germany*.
Schomaker, Verner *USA*.
Schoone, Prof. Dr Jean C. *Netherlands*.
Schorin, Dr Hasso *Venezuela*.
Schouten, Mr Arie *Netherlands*.
Schrag, Dr Joseph D. *Canada*.
Schramm, Dr Volker *Germany*.
Schranz, Dr Wilfried *Austria*.
Schreiner, Dr Walter N. *USA*.
Schreuder, Dr Herman A. *France*.
Schreuer, Mr Jürgen *Germany*.
Schreurs, Drs Antonius M. M. *Netherlands*.
Schröder, Dr-Ing. Jens *Germany*.
Schroeder, Dr LeRoy W. *USA*.
Schröder, Dr Winfried *Germany*.
Schryvers, Dr Dominique *Belgium*.
Schubert, Prof. Dr Ulrich *Austria*.
Schubert, Prof. Dr Ulrich *Germany*.
Schubert, Heidi L. *USA*.
Schüller, Prof. Dr Karl-Heinz *Germany*.
Schürmann, Dr Kay Uwe *Germany*.
Schuerman, Ms Geertrui Simone Maria *Belgium*.
Schuller, David J. *USA*.
Schuller, Ivan *USA*.
Schultz, Dr Arthur J. *USA*.
Schultz, Dr György *Hungary*.
Schultz, L. Wayne *USA*.
Schulz, Prof. Dr Heinz Hermann *Germany*.
Schulze-Gahmen, Dr Ursula *USA*.
Schumann, Dr Bernd *Germany*.
Schuster, Prof. Dr Hans-Uwe *Germany*.
Schuster, Dr Isabelle *France*.
Schuster, Dr Julius *Austria*.
Schuster, Dr Manfred R. *Germany*.
Schutte, Prof. Casper Jan Hendrik *South Africa*.
Schwahn, Dr Dietmar *Germany*.
Schwalbe, Dr Carl Hellmuth Walter *UK*.
Schwartz, Dr Kenneth B. *USA*.
Schwartz, Lyle H. *USA*.
Schwartz, Mr Steven A. *USA*.
Schwarz, Prof. Dr Karlheinz *Austria*.
Schwarz, Dr Ulrich *Germany*.

Schwarzenbach, Prof. Dieter *Switzerland*.
Schweda, Dr Eberhard *Germany*.
Schwegle, Dr Wolfgang *France*.
Schweiker, Viloya L. *USA*.
Schweizer, Dr W. Bernd *Switzerland*.
Schwer, Dr Hansjörg *Switzerland*.
Sclar, Charles B. *USA*.
Scordari, Prof. Fernando *Italy*.
Scott, Ms Janet Lesley *South Africa*.
Scudder, Marcia Lorraine *Australia*.
Seal, Dr Alpana *India*.
Seal, Prof. Arun Kumar *India*.
Seal, Dr Michael *Netherlands*.
Seaton, Dr Barbara A. *USA*.
Sebastian, Dr Eduardo Manuel *Spain*.
Secco, Prof. Anthony Silvio *Canada*.
Sedmalis, Prof. Uldis *Latvia*.
Sedov, Mr Boris B. *Russia*.
See, Dr Ronald F. *USA*.
Seely, Dr Oliver *USA*.
Seeman, Prof. Nadrian Charles *USA*.
Seetharaman, Mr Venkataramakrishanan *India*.
Seff, Prof. Karl *USA*.
Šegedin, Prof. Dr Primož *Slovenia*.
Sehnke, Dr Paul C. *USA*.
Seidl, Dr Erwin *Austria*.
Seifert, Prof. Dr Karl-Friedrich *Germany*.
Seiler, Mr Paul *Switzerland*.
Sekine, Akiko *Japan*.
Sekiwa, Mr Hideyuki *Japan*.
Self, Dr Peter Geoffrey *Australia*.
Selladurai, Mr S. *India*.
Sellar, Jeffrey Ronald *Australia*.
Semenchev, Dr Aleksandr F. *Russia*.
Semenova, Dr Olena Leonidivna *Ukraine*.
Semenova, Dr Tatiyana F. *Russia*.
Semertzidis, Mr Michel *France*.
Sen, Dr Deb Kumar *India*.
Sen Gupta, Dr Amitava *India*.
Sen Gupta, Prof. Siba Prasad *India*.
Sen, Miss Mina *India*.
Sen, Dr Ranjit Kumar *India*.
Sen, Dr Suchitra *India*.
Sen, Mr Udayaditya *India*.
Senadhi, Vijay-Kumar *USA*.
Sengupta, Ms Suparna *India*.
Seok, Prof. Won Kyung *Korea*.
Sequeira, Dr Anisbert S. *India*.
Serafin, Dr Michael *Germany*.
Serda, Mgr Paweł *Poland*.
Serebryanaya, Dr Nadezda R. *Russia*.
Sereda, Dr Sergij Volodymyrovych *Ukraine*.
Serezhkin, Prof. Victor N. *Russia*.
Serglenko, Dr Vladimir S. *Russia*.
Serimaa, Ritva Elina *Finland*.
Serra, Mr Alberto *Cuba*.
Servidori, Dr Marco *Italy*.
Servos, Kurt *USA*.
Seshasayee, Ms Maha *India*.
Setiaji, Dr Bambang *Indonesia*.
Setoguchi, Dr Masahiro *Japan*.
Sevastyanov, Dr Boris K. *Russia*.
Ševčík, Ing. Jozef *Slovakia*.
Sewell, Dr Bryan Trevor *South Africa*.
Sgarabotto, Prof. Paolo *Italy*.
Sgualdino, Dr Giulio *Italy*.
Shaanan, Dr Boaz *Israel*.
Shabalala, Mr Innocent Philani *South Africa*.
Shach, Romem *Israel*.
Shackelford, James F. *USA*.
Shaffner, Thomas J. *USA*.
Shafirstein, Dr Gal *UK*.
Shafranovsky, Prof. Ilarion Ilarionovich *Russia*.
Shah, Dr Wajid Ali *Pakistan*.
Shahi, Mr Gulam Nabi *Pakistan*.
Shaikh, Mr Mohammad Iqbal *Pakistan*.
Shaikh, Mr Mohammad Sualehin *Pakistan*.
Shaikh, Mr Qameruddin *Pakistan*.
Shaked, Prof. Hagai *Israel*.
Shakked, Prof. Zippora *Israel*.

Shalamov, Anatoliy Egorovich *Kazakhstan*.
Shamah, Prof. Abdelmoniem Mohamed *Egypt*.
Shamray, Dr Vladimir F. *Russia*.
Shan, Lin *USA*.
Shang, Maoyu *USA*.
Shao, Prof. Meicheng *China*.
Sharff, Dr Andrew J. *UK*.
Sharipov, Prof. Khasan T. *Uzbekistan*.
Sharma, Amit *USA*.
Sharma, Dr Brahama D. *USA*.
Sharma, Mr Braj Bhushan *India*.
Sharma, Mr Girish B. *India*.
Sharma, Mr K. K. *India*.
Sharma, Dr Surinder Dutt *India*.
Sharp, Paul R. *USA*.
Sharrah, Paul C. *USA*.
Shashidhar, Dr R. *India*.
Shashkin, Dr Dmitry P. *Russia*.
Shatzky-Schwartz, Michal *Israel*.
Shaukat, Dr Ali *Pakistan*.
Shcurpello, Dr Anatolii Ivanovich *Moldova*.
Shebanovs, Dr Leonids *Latvia*.
Shechtman, Dan *USA*.
Sheikh, Mr Izhar-ul-Haq *Pakistan*.
Shekhtman, Prof. Veniamin Sh. *Russia*.
Shekunov, Dr Boris Yu *UK*.
Sheldon, Robert I. *USA*.
Sheldrick, Prof. Dr George M. *Germany*.
Sheldrick, Prof. Dr William Stephen *Germany*.
Shelley, Dr David *New Zealand*.
Shen, Prof. Fuling *China*.
Shen, Prof. Jinchuan *China*.
Shen, Qun *USA*.
Shen, Wei *Australia*.
Shepelev, Dr Yury F. *Russia*.
Sherban, Dr Dormidont Arhip *Moldova*.
Sheriff, Dr Steven *USA*.
Sherwood, Prof. John N. *UK*.
Sheu, Eric *USA*.
Sheu, Mr. Hwo-Shuenn *Taiwan*.
Shevchenko, Dr Ljudmyla Larionivna *Ukraine*.
Shi, Dashuang *Australia*.
Shi, Mr Lei *China*.
Shi, Prof. Nicheng *China*.
Shibaeva, Dr Rimma P. *Russia*.
Shibahara, Hiroyasu *Japan*.
Shibata, Mr Masatomo *Japan*.
Shibata, Dr Noriyoshi *Japan*.
Shibata, Prof. Shuzo *Japan*.
Shieh, Dr Huey-Sheng *USA*.
Shieh, Dr Ming-huey *Taiwan*.
Shigematsu, Dr Koji *Japan*.
Shil'nikov, Dr Valery I. *Russia*.
Shilov, Dr Gennadi V. *Russia*.
Shimamura, Mr Kiyoshi *Japan*.
Shimazu, Dr Masaji *Japan*.
Shimizu, Dr Kenji *Japan*.
Shimizu, Mr. Toshiyuki *Japan*.
Shimol, Prof. Mamoru *Japan*.
Shimomura, Dr Osamu *Japan*.
Shimon, Dr Linda J. W. *Israel*.
Shimoni, Liat *USA*.
Shimotomai, Dr Michio *Japan*.
Shimura, Dr Takayoshi *Japan*.
Shin, Prof. Hyun So *Korea*.
Shindo, Prof. Dr Hitoshi *Japan*.
Shindyalov, Ilya *USA*.
Shiono, Dr Masaaki *Japan*.
Shiono, Dr Ryonosuke *USA*.
Shiozaki, Prof. Yoichi *Japan*.
Shipley, Prof. G. Graham *USA*.
Shirinyan, Dr Grikor *Armenia*.
Shiro, Dr Motoo *Japan*.
Shishkin, Dr Oleg V. *Russia*.
Shishova, Dr Tatyana G. *Russia*.
Shiu, Prof. Kom-Bei *Taiwan*.
Shivaprakash, Dr N. C. *India*.
Shklover, Dr Valery *Switzerland*.
Shmueli, Prof. Uri *Israel*.
Shmyt'ko, Dr Ivan M. *Russia*.

Shnulin, Dr Anatoly Shnulin *Azerbaijan*.
Shoemaker, Dr Clara B. *USA*.
Shoemaker, Dr David P. *USA*.
Shoham, Gil *Israel*.
Shoham, Dr Menachem *USA*.
Shoja, Massud *USA*.
Shopa, Dr Yaroslav Ivanovych *Ukraine*.
Shorin, Dr Viltor Fedorovich *Kazakhstan*.
Short, Dr Michael A. *USA*.
Shoukri, Prof. Nasri *Egypt*.
Shova, Dr Sergiu *Moldova*.
Shpyrka, Dr Zinoviya Mykhajlivna *Ukraine*.
Shrive, Dr Annette K. *UK*.
Shu, Ms Fang *USA*.
Shui, Xiuqi *USA*.
Shuja, Mr Tauqir Ahmad *Pakistan*.
Shulakov, Dr Eugene V. *Russia*.
Shulmeister, Dr Vladimir M. *USA*.
Shumov, Dr Dimiter *Bulgaria*.
Shuvalov, Dr Alexander L. *Russia*.
Shuvalov, Prof. Lev A. *Russia*.
Shvelashvili, Prof. Arsen *Georgia*.
Sica, Dr Filomena *Italy*.
Sichevich, Dr Olga Mykhajlivna *Ukraine*.
Síchová, Dr Hana *Czech Republic*.
Siddiqi, Dr Saadat Anwar *Pakistan*.
Siddiqui, Mr Jawed Ahmad *Pakistan*.
Siddiqui, Mr Tariq Nasser *Pakistan*.
Sieber, Dr Norbert H. W. *Germany*.
Siegel, Lester A. *USA*.
Sieger, Dr Peter *Germany*.
Siegrist, Dr Theo *USA*.
Sieker, Dr Larry C. *USA*.
Sielecki, Dr Anita R. *Canada*.
Sieler, Prof. Dr Joachim *Germany*.
Sigler, Paul B. *USA*.
Sigvaldason, Dr Gudmundur *Iceland*.
Siivola, Prof. Jaakko Uolevi *Finland*.
Sikirica, Prof. Dr Milan *Croatia*.
Sikka, Dr Satinder Kumar *India*.
Silcox, John *USA*.
Silina, Dr Elga *Latvia*.
Silong, Dr Sidik bin *Malaysia*.
Silonov, Dr Valentin M. *Russia*.
Silskulsuk, Mr Buncha *Thailand*.
Silva, Dr Abelardo M. *USA*.
Silva Campos, Dr Helio *Brazil*.
Silva, Dr Elisa *Chile*.
Silva, Manuela Ramos Marques *Portugal*.
Silva, Teresa Pereira *Portugal*.
Silverton, Enid W. *USA*.
Silverton, Dr James V. *USA*.
Silvestre, Dr Jean-Paul *France*.
Silvestrova, Dr Olga Y. *Russia*.
Simard, Dr Michel *Canada*.
Simashkevich, Prof. Alexei Vasilii *Moldova*.
Sime, Rodney J. *USA*.
Simmons, Charles *USA*.
Simon, Dr Kálmán *Hungary*.
Simon, Prof. Simion *Romania*.
Simone, Dr Carlos Alberto de *Brazil*.
Simonov, Prof. Valentin I. *Russia*.
Simonov, Dr Yury Alexandrovich *Moldova*.
Simonsen, Ole *Denmark*.
Simonsen, Prof. Stanley H. *USA*.
Simov, Dr Stefan *Bulgaria*.
Simpson, H. D. *USA*.
Sinay, Dr Marina Yu. *Russia*.
Singh, Dr Bhanu Pratap *India*.
Singh, Dr Govind *India*.
Singh, Dr S. N. *India*.
Singh, Mr Surendra Prakash *India*.
Singu, Phirtu *USA*.
Sinh, Dr Nguyen Huy *Vietnam*.
Sinha, Dr Sunil K. *USA*.
Sinha, Dr Umesh Chandra *India*.
Sinn, Prof. Ekkehard *UK*.
Sinojmeri, Dr Agim *Albania*.
Sinur, Miss Amalija *Slovenia*.
Sirdeshmukh, Dr Dinker *India*.

Siripisarnpipat, Dr Sutatip *Thailand*.
Siripitayananon, Dr Jintana *Thailand*.
Siriratwatanakul, Mr Narin *Thailand*.
Sironi, Prof. Angelo *Italy*.
Sirota, Dr Mikhail I. *Russia*.
Sitter, Prof. Dr Helmut *Austria*.
Sivakumar, Dr K. *India*.
Sivonen, Seppo Juhani *Finland*.
Sivý, Dr Július *Slovakia*.
Sizova, Dr Natalia L. *Russia*.
Sjöberg, Bo *Sweden*.
Sjölin, H. Lennart G. *Sweden*.
Skakle, Jan *UK*.
Skála, Dr Roman *Czech Republic*.
Skalicky, Prof. Dr Peter *Austria*.
Skarstad, Paul *USA*.
Skarżyński, Dr Tadeusz *UK*.
Skelton, Brian Warwick *Australia*.
Skelton, Dr Earl F. *USA*.
Skimevskaya, Elena Vasilevna *Kazakhstan*.
Skinner, Dr Matthew M. *USA*.
Skipworth, Ms Margaret Helene *Australia*.
Skita, Dr Victor *USA*.
Skolozdra, Prof. Roman Volodymyrovych *Ukraine*.
Skoulika, Dr Stavroula *Greece*.
Skowerenda, Dr Jolanta *Poland*.
Skowron, Dr Aniceta *Canada*.
Škrbić, Dr Željko *Serbia*.
Skrzypczak-Jankun, Ewa *USA*.
Skvortsova, Dr Natalya P. *Russia*.
Slade, Phillip Garland *Australia*.
Slaughter, Maynard *USA*.
Slawin, Alexandra Martha Zoya *UK*.
Sleight, Arthur W. *USA*.
Sletten, Prof. Einar *Norway*.
Sletten, Prof. Jorunn *Norway*.
Śliwiński, Mgr Jan *Poland*.
Slovenec, Prof. Dr Dragutin *Croatia*.
Slyusarev, Anatoly Petrovich *Kazakhstan*.
Smalås, Dr Arne O. *Norway*.
Smale, Dr David *New Zealand*.
Small, Colin John *UK*.
Small, Dr Ronald W. H. (Sam) *UK*.
Smart, Dr Lesley E. *UK*.
Smeets, Drs Wilberthus J. J. *Netherlands*.
Smetannikova, Dr Olga G. *Russia*.
Smirnov, Dr Alexey E. *Russia*.
Smirnov, Mr Lev *Russia*.
Smirnov, Prof. Yuri M. *Russia*.
Smirnova, Dr Irina A. *Russia*.
Smirnova, Dr Nina L. *Russia*.
Smirnova, Dr Sofia A. *Russia*.
Šmit, Dr Ivan *Croatia*.
Smit, Dr Jan Derk Geert *Switzerland*.
Smit, Dr Paul H. *Netherlands*.
Smith, Albert E. *USA*.
Smith, Dr Arnold John *UK*.
Smith, Dr Craig D. *USA*.
Smith, David J. *USA*.
Smith Jr, Deane Kingsley *USA*.
Smith, Deborah R. *USA*.
Smith, Douglas L. *USA*.
Smith, Dr Francine R. *USA*.
Smith, Dr G. David *USA*.
Smith, Gallienus William *UK*.
Smith, Mr Garry T. *UK*.
Smith, Dr Graham *Australia*.
Smith, Graham M. *USA*.
Smith, Dr Janet L. *USA*.
Smith, Dr Joseph V. *USA*.
Smith, Katherine Leah *Australia*.
Smith, Paul Raymond *UK*.
Smith, Dr Phillip Ross *USA*.
Smith, Robert W. *USA*.
Smith, Dr Ronald I. *UK*.
Smith, Prof. T. Fred *Australia*.
Smith, Thomas James *USA*.
Smith Jr., Prof. Vedene H. *Canada*.
Smith, Dr Ward W. *USA*.
Smithteale, Michael J. *USA*.

Smits, Mr Johannes M. M. *Netherlands.*
Smolander, Dr Kimmo Juhani Nils-Eric *Finland.*
Smoilar-Zviagina, Dr Bela B. *Russia.*
Smolin, Prof. Yury I. *Russia.*
Smolsky, Dr Igor L. *Russia.*
Smoluchttowski, Roman *USA.*
Smotrakov, Dr Valery G. *Russia.*
Smout, Ing. Emile M. *Netherlands.*
Snell, Edward H. *UK.*
Snow, Mark E. *USA.*
Snyder, James A. *USA.*
Snyder, Robert L. *USA.*
Soa, Prof. Dr Ernst-Adolf *Germany.*
Soa, Prof. Nguyen Dinh *Vietnam.*
Soar, Martin *UK.*
Soares de Vasconcelos, Dr Dionicarlos *Brazil.*
Sobczak, Dr hab. Ewa *Poland.*
Sobol', Dr Oleg Valentynovych *Ukraine.*
Sobolev, Dr Alexander N. *Russia.*
Sobolev, Dr Boris P. *Russia.*
Sobry, Dr Roger *Belgium.*
Soc, Prof. Le Nguyen *Vietnam.*
Socci, Edward Peter *USA.*
Soejima, Dr Yuji *Japan.*
Sørensen, Mr Ole *Denmark.*
Søfofte, Inger *Denmark.*
Soga, Dr Tetsuo *Japan.*
Sogabe, Mr Satoshi *Japan.*
Sohail, Mr Mohammad *Pakistan.*
Sojo, Dr Pedro Rafael *Venezuela.*
Sokil, Dr Anatolij Opanasovych *Ukraine.*
Sokolova, Dr Elena V. *Russia.*
Sokolova, Dr Nataliya G. *Russia.*
Sola, Dr Joan *Spain.*
Solans, Prof. Joaquim *Spain.*
Solans, Prof. Xavier *Spain.*
Soldatos, Prof. Triantafyllos *Greece.*
Soldatov, Dr Eugeni A. *Russia.*
Solé, Mr Rosa *Spain.*
Soledade Jr, Prof. Teomar *Brazil.*
Soliman, Mr F. Abdelaal *Egypt.*
Soliman, Dr Mohamed Soliman *Egypt.*
Solodovnikov, Dr Sergei F. *Russia.*
Solorio Munguía, Prof José Gregorio *Mexico.*
Solovyeva, Prof Lidiya P. *Russia.*
Soltis, S. Michael *USA.*
Soman, Ms Jayashree *India.*
Somiya, Prof. Dean Dr *Japan.*
Sommer, Dr Olaf *Germany.*
Sommerer, Dr Shaun O. *USA.*
Sommer-Larsen, Peter *Denmark.*
Sommerville, Mrs Polly Baker Melville *South Africa.*
Somorjai, Gabor A. *USA.*
Sondermann, Dr Ulrich *Germany.*
Song, Mr Hyun *Korea.*
Song, Prof. Shiying *China.*
Sonneveld, Mr Eduard J. *Netherlands.*
Soong, Mr Raymond *New Zealand.*
Sopková, Dr Jana *Czech Republic.*
Šoptrajanov, Prof. Dr Bojan *(FYR) Macedonia.*
Šoptrajanova, Prof. Dr Gorica *(FYR) Macedonia.*
Soriano Garcia, Prof Manuel *Mexico.*
Sorokin, Dr Nikolai I. *Russia.*
Sorokina, Dr Kira L. *Russia.*
Sorokina, Dr Natalia I. *Russia.*
Sosfenov, Dr Nikita I. *Russia.*
Sosnick, Tobin R. *USA.*
Sosnowska, Prof. dr hab. Izabela *Poland.*
Sotman, Dr Sima S. *Russia.*
Soumbatov, Mr Alexander *Russia.*
Sousa, Dr José Carlos de *Brazil.*
Souza, Ms Dulce Helena Ferreira *Brazil.*
Souza, Prof. Irineu Marques *Brazil.*
Sowa, Dr Heidrun *Germany.*
Soylu, Dr Hüseyin *Turkey.*
Sozontov, Dr Evgeny A. *Russia.*
Spackman, Mark A. *Australia.*
Spadon, Prof. Paola *Italy.*
Spagna, Dr Riccardo *Italy.*

Spalding, Dr Dennis Raymond *South Africa.*
Spangler, Dr Brenda D. *USA.*
Sparks, Cullie J. *USA.*
Sparks, Dr Robert Allen *USA.*
Spassov, Dr Tony *Bulgaria.*
Specht, Eliot D. *USA.*
Speir, Jeffrey A. *USA.*
Spek, Dr Anthony L. *Netherlands.*
Spengler, Mr Roland Helmuth Michael *Germany.*
Speziali, Dr Nivaldo Lúcio *Brazil.*
Spielberg, Nathan *USA.*
Spijkerman, Ir Albert *Netherlands.*
Spindler, Dr Peter *Austria.*
Spink, John Arthur *Australia.*
Spinney, Mr Richard *Canada.*
Spiriet, Dr Marie-Rose *Belgium.*
Spodine, Dr Evgenia *Chile.*
Sprang, Dr Stephen R. *USA.*
Spratt, Stephen Brian Douglas *UK.*
Springer, James P. *USA.*
Sproul, Gordon Duane *USA.*
Spurlino, Dr John C. *USA.*
Squattrito, Philip J. *USA.*
Squire, Dr Gavin Daniel *UK.*
Squire, Dr John Michael *UK.*
Sridhar Prasad, Mr G. *India.*
Sridhar, Dr Vaidehi *USA.*
Srikrishnan, Thamarapu *USA.*
Srinivasa, Dr V. K. *India.*
Srinivasan, Dr A. R. *USA.*
Srinivasan, Prof. R. *India.*
Srinivasan, Dr Sampat *India.*
Sriram, Mahalingam *USA.*
Srivastava, Dr Ramesh Chandra *India.*
St. Charles, Robert J. *USA.*
Ståhl, Dr Kenny *Sweden.*
Stålhandske, Claes Ivar *Sweden.*
Stadelmaier, Hans H. *USA.*
Stadnicka, Dr Katarzyna *Poland.*
Stalick, Dr Judith K. *USA.*
Stalke, Dr Dietmar *Germany.*
Stallings, Dr William C. *USA.*
Stam, Dr Casper H. *Netherlands.*
Stanbury, Howard J. *UK.*
Stanfield, Dr Robyn L. *USA.*
Stanko, Dr Joseph A. *USA.*
Stanković, Dr Slobodanka *Serbia.*
Stanton, Dr Martin *USA.*
Staples, Richard J. *USA.*
Starodoob, Dr Pavlo Korniyovych *Ukraine.*
Starostina, Dr Ludmila S. *Russia.*
Starr, Jonathan E. *USA.*
Stash, Mr Adam I. *Russia.*
Stasi, Prof. Francesca *Italy.*
States, Dr David J. *USA.*
Staudenmann, Jean-Louis *USA.*
Stauffacher, Cynthia Vianne *USA.*
Staykov, Dr Georgi *Bulgaria.*
Stebbins, Jonathan *USA.*
Stec, Dr Boguslaw *USA.*
Stedman, Dr Nicola J. *UK.*
Steeb, Prof. Dr Siegfried *Germany.*
Steele, Dr Ian M. *USA.*
Stefanov, Mr Dechko *Bulgaria.*
Stefanović, Mr Aleksandar *Croatia.*
Stefanovich, Dr Sergey Yu. *Russia.*
Steffes-Tun, Mr Wolfgang *Germany.*
Stehle, Thilo *USA.*
Stein, Zafra *Israel.*
Steinberger, Prof. Itzhak *Israel.*
Steiner, Prof. Dr Michael *Germany.*
Steiner, Prof. Dr Walter *Austria.*
Steiner, Dr Thomas *Germany.*
Steinfink, Dr Hugo *USA.*
Steinhart, Dr Miloš *Czech Republic.*
Steinke, Prof. Dr Ursula *Germany.*
Steinrauf, Dr Larry K. *USA.*
Steins, Dr Manfred *Germany.*
Steinthórsson, Dr Sigurdur *Iceland.*
Steitz, Prof. Thomas A *USA.*

Stelea, Gabriela *Romania.*
Stenkamp, Dr Ronald E. *USA.*
Stepanova, Dr Alla N. *Russia.*
Stepantsov, Dr Evguenii A. *Russia.*
Stephen, Mr Suresh *India.*
Stephens, Peter *USA.*
Stępień-Damm, Dr Julia *Poland.*
Stergiou, Prof. Anagnostis *Greece.*
Stergioudis, Prof. George *Greece.*
Stern, Charlotte L. *USA.*
Stern, Edward A. *USA.*
Stern, Lawrence *USA.*
Steurer, Prof. Walter *Switzerland.*
Stevens, Dr Edwin D. *USA.*
Stevens, Prof. Raymond C. *USA.*
Stevenson, Andrew Wesley *Australia.*
Steward, Prof. Edward George *UK.*
Stewart, Glen Alan *Australia.*
Stewart, James McDonald *USA.*
Stezowski, Dr John J. *USA.*
Stillman, Dr Timothy J. *UK.*
Stiopina, Dr Nina D. *Russia.*
Stishov, Prof. Sergei M. *Russia.*
Stock, Ann *USA.*
Stock, Stuart R. *USA.*
Stoddard, Barry L. *USA.*
Stoddard, Miss Jillian Kaye *UK.*
Stoddart, Christopher Paul *UK.*
Stoeckli-Evans, Prof. Helen *Switzerland.*
Stølevik, Prof. Reidar *Norway.*
Stoemenos, Prof. Ioannis *Greece.*
Stoinova, Prof. Margarita *Bulgaria.*
Stojanoff, Dr Vivian *Brazil.*
Stojanova, Dr Liljana *Bulgaria.*
Stokes, David L. *USA.*
Stomberg, Rolf S. O. *Sweden.*
Stouch, Terry *USA.*
Stout, Dr Charles David *USA.*
Stout, George Hubert *USA.*
Stout, Dr Thomas J. *USA.*
Stowell, Dr Joseph G. *USA.*
Stoyanov, Dr Stoyan *Bulgaria.*
Stoyanova, Dr Valeriya *Bulgaria.*
Stoychev, Dr Nikola *Bulgaria.*
Strähle, Prof. Dr Joachim *Germany.*
Strand, Dr Tor Gogstad *Norway.*
Strandberg, Prof. Bror *Sweden.*
Strandberg, Dr Rolf A. G. *Sweden.*
Strasdeit, Dr Henry *Germany.*
Straver, Drs Leonardus H. *Netherlands.*
Streib, Dr William E. *USA.*
Streltsov, Dr Victor A. *Australia.*
Streltsova, Dr Natalie *Australia.*
Strickland, Corey L. *USA.*
Strickland, Mr Peter R. *UK.*
Strid, Prof. Karl-Gustav A. *Sweden.*
Ström, Carin *Sweden.*
Strong, Roland K. *USA.*
Stroud, Dr Robert M. *USA.*
Stroud, William J. *USA.*
Strouse, Prof. Charles E. *USA.*
Stróż, Dr Danuta *Poland.*
Struchkov, Prof. Yuri T. *Russia.*
Struikmans, Drs Rink *Netherlands.*
Strukan, Mr Neven *Croatia.*
Strumpel, Dr Marianna *Germany.*
Strusievicz, Octavian-Robert *Romania.*
Strynadka, Dr Natalie *Canada.*
Stuart, Dr Sue-Anne *Australia.*
Stubbs, Prof. Gerald James *USA.*
Stubičar, Dr Mirko *Croatia.*
Stuckenschmidt, Dr Elli *Germany.*
Stuckey, Dr Jeanne A. *USA.*
Studnička, Dr Václav *Czech Republic.*
Stuhrmann, Prof. Dr Heinrich B. *Germany.*
Stumpfl, Prof. Dr Eugen F. *Austria.*
Stura, Dr Enrico A. *USA.*
Su, Shu-Chun *USA.*
Su, Ying *USA.*
Su, Dr Zhengwei *USA.*

Sua, Dr Nguyen Van *Vietnam.*
Suades, Dr Joan *Spain.*
Suárez, Prof. Nery *Venezuela.*
Subbiah, S. *USA.*
Subhadra, Dr K. G. *India.*
Subirana, Prof. Juan A. *Spain.*
Subramanian, Prof. E. *India.*
Subramanian, Dr K. *India.*
Subramanya, Mr H. S. *India.*
Subramony, Mr Loganathan *South Africa.*
Šubrtová, Ing. Věra *Czech Republic.*
Suck, Dr Dietrich *Germany.*
Sudaramoorthy, Mr M. *India.*
Sudarsanakumar, Dr Chellappan Pillai *India.*
Sudbeck, Elise *USA.*
Suddhiprakarn, Dr Anohsloe *Thailand.*
Sudo, Dr Toshio *Japan.*
Suehiro, Prof. Kazuaki *Japan.*
Sueno, Dr Shigeho *Japan.*
Suffritti, Prof. Giuseppe Baldovino *Italy.*
Sugano, Prof. Dr Takuo *Japan.*
Sugawara, Dr Akira *Japan.*
Sugawara, Dr Shigeo *Japan.*
Sugawara, Mr Tamotsu *Japan.*
Sugawara, Dr Yoko *Japan.*
Sugihara, Dr Akio *Japan.*
Sugio, Dr Shigetoshi *Japan.*
Sugiyama, Assoc Prof. Junji *Japan.*
Sugiyama, Dr Kazumasa *Japan.*
Sugiyama, Mr Masaaki *Japan.*
Suguna, Dr K. *India.*
Suh, Prof. Se Won *Korea.*
Sui, Ms Xiaoling *Canada.*
Suito, Prof. Eiji *Japan.*
Sukapaddhi, Mr Narong *Thailand.*
Sukegawa, Prof. Dr Tokuzo *Japan.*
Sullenger, Dr Don B. *USA.*
Sun, Dr Daopin *USA.*
Sunagawa, Prof. Dr Ichiro *Japan.*
Sundaralingam, Dr Muttaiya *USA.*
Sundaramoorthy, Mr M. *India.*
Sundaramoorthy, Munirathinam *USA.*
Sundararajan, Dr Pudupadi R. *Canada.*
Sundberg, Dr Margareta *Sweden.*
Sunder, Dr Sham *Canada.*
Sundius, Tom Robert *Finland.*
Sung, Dr Gun Yong *Korea.*
Suoninen, Prof. Eero Juhani *Finland.*
Suortti, Prof. Pekka *Finland.*
Suresh, Dr C. G. *India.*
Suresh, Mr K. A. *India.*
Suri, Mr D. K. *India.*
Suri, Mr D. K. *India.*
Surowiec, Dr Marian Ryszard *Poland.*
Suryanarayana, Dr Challapalli *India.*
Suryanarayana, Dr Shambhuni V. *India.*
Sushkevich, Dr Konstantin Dmitrievich *Moldova.*
Susse, Prof. Dr Peter *Germany.*
Sussman, Prof. Joel Leonard *Israel.*
Suta, Ms Elizabeth *India.*
Sutherland, Dr Elizabeth Eldred *Australia.*
Šutta, Dr Pavol *Slovakia.*
Sutton, Dr Brian J. *UK.*
Sutton, Dr Mark *Canada.*
Suvorov, Prof. Ernest V. *Russia.*
Suwa, Dr Yoshiko *Japan.*
Suwalsky, Dr Mario *Chile.*
Suwińska, Dr Kinga *Poland.*
Suyarov, Kurbon *Tadjikistan.*
Suzuki, Dr Akira *Japan.*
Suzuki, Dr Carlos Kenichi *Brazil.*
Suzuki, Dr Hideo *Japan.*
Suzuki, Prof. Ikuo *Japan.*
Suzuki, Dr Takaya *Japan.*
Suzuki, Prof. Toshimasa *Japan.*
Suzuki, Mr Yoshihisa *Japan.*
Svensson, Christer *Sweden.*
Svensson, Göran *Sweden.*
Svensson, Gunnar *Sweden.*
Svensson, L. Anders *Sweden.*

Sveshnikov, Dr Sergey V. *Russia.*
Svinning, Dr Torgeir *Norway.*
Svisero, Dr Darcy Pedro *Brazil.*
Swain, Amy L. *USA.*
Swainson, Dr Ian Peter *Canada.*
Swaminathan, Dr Kunchit *USA.*
Swaminathan, Dr S. *USA.*
Swanson, Rosemarie *USA.*
Swanson, Dr Stanley M. *USA.*
Sweet, Dr Robert M. *USA.*
Sweet, Mrs Tracy K. N. *UK.*
Swepston, Paul N. *USA.*
Swink, Laurence N. *USA.*
Swinnea, J. Steven *USA.*
Switendick, Alfred C. *USA.*
Syarif, Dr Shamid *Indonesia.*
Syed, Dr Ashfaquzzaman *USA.*
Syed, Dr Kaab Akhter *Pakistan.*
Syed, Mr Muhammad Tariq *Pakistan.*
Syed, Dr Rashid *USA.*
Sygusch, Prof. Jurgen *Canada.*
Syniushko, Dr Vasyl' Grygorovych *Ukraine.*
Szczepańska, Mgr Beata Irena *Poland.*
Szebenyi, Doletha M. E. *USA.*
Szent-Királyi, Mrs Zsuzsanna *Hungary.*
Szkaradzinska, Ms Maria *Canada.*
Szurgot, Dr Marian *Poland.*
Szymański, Dr Jan Tomasz *Canada.*
Tabata, Dr Hideyo *Japan.*
Tabernero, Lydia *USA.*
Tabira, Yasunori *Japan.*
Tachibana, Mr Masaru *Japan.*
Tachikawa, Mr Shigeki *Japan.*
Tada, Dr Toshiji *Japan.*
Tadatomo, Mr Kazuyuki *Japan.*
Tadini, Prof. Carla *Italy.*
Tafeenko, Dr Victor Aleksandrovich *Russia.*
Tafto, Prof. Johan *Norway.*
Taga, Prof. Tooru *Japan.*
Tagieva, Dr Marianna M. *Russia.*
Tahoun, Prof. Salah *Egypt.*
Tai, Dr Luu Tuan *Vietnam.*
Taikina-aho, Olavi Seppo Allan *Finland.*
Tainer, Dr John A. *USA.*
Takagi, Dr Kazumasa *Japan.*
Takagi, Dr Mieko *Japan.*
Takahara, Patricia M. *USA.*
Takahashi, Ms Masako *Japan.*
Takahashi, Dr Toshio *Japan.*
Takahashi, Prof. Yasuhiro *Japan.*
Takaki, Prof. Yoshito *Japan.*
Takakura, Mr Hiroyuki *Japan.*
Takama, Dr Toshihiko *Japan.*
Takano, Dr Kaoru *Japan.*
Takata, Masaki *Japan.*
Takeda, Prof. Hiroshi *Japan.*
Takeda, Prof. Takayoshi *Japan.*
Takeda, Prof. Yoshikazu *Japan.*
Takei, Prof. Fumihiko *Japan.*
Takemoto, Mr Jiro *Japan.*
Takenaka, Dr Takao *Japan.*
Takenaka, Dr Yasuyuki *Japan.*
Takeuchi, Dr Yasuo *Japan.*
Takeuchi, Prof. Yoshio *Japan.*
Taki, Dr Sadao *Japan.*
Takla, Dr Maher Azmi *Egypt.*
Takusagawa, Dr Fusao *USA.*
Talipov, Dr Samat A. *Uzbekistan.*
Talis, Dr Alexander L. *Russia.*
Talukdar, Dr Amarendra Nath *India.*
Tamada, Prof. Osamu *Japan.*
Tamm, Prof. Jüri *Estonia.*
Tamura, Dr Hatsue *Japan.*
Tamura, Dr Masao *Japan.*
Tan, Dr Hock Siew *Singapore.*
Tanaka, Mr Akikazu *Japan.*
Tanaka, Dr Akira *Japan.*
Tanaka, Prof. Isao *Japan.*
Tanaka, Prof. Isao *Japan.*
Tanaka, Prof. Kiyoaki *Japan.*

Tanaka, Dr Masahiko *Japan.*
Tanaka, Michiyoshi *Japan.*
Tanaka, Prof. Nobuo *Japan.*
Tanaka, Prof. Nobuo *Japan.*
Tanaka, Dr Shigeyasu *Japan.*
Tanaka, Mr Taisuke *Japan.*
Tanaka, Dr Toshiro *Japan.*
Tanaka, Mr Yoshiyuki *Japan.*
Tanasugarn, Dr Lerson *Thailand.*
Tančić, Mr Pavle *Serbia.*
Tang, Dr Chia-Pin *Taiwan.*
Tang, Prof. Youqi *China.*
Tani, Katsuhiko *Japan.*
Taniguchi, Prof. Tomohiko *Japan.*
Taniguchi, Dr Yoshiteru *Japan.*
Tanishiro, Mr Yasumasa *Japan.*
Taniyama, Prof. Jou *Japan.*
Tanner, Prof. Brian Keith *UK.*
Tanner, Jack *USA.*
Tanus, Ms Mercedes *Cuba.*
Tarashchan, Dr Arkadij Mykolayovych *Ukraine.*
Tarımcı, Dr Çelik *Turkey.*
Tarján, Prof. Imre *Hungary.*
Tarna, Toivo *Finland.*
Tarricone, Dr Aldo *Italy.*
Tas, Dr Clüneyt A. *Turkey.*
Tashkhodzhaev, Dr Bakhodir *Uzbekistan.*
Tashko, Dr Artan *Albania.*
Tasov, Boris Makievich *Kazakhstan.*
Tasset, Dr Francis *France.*
Tatsuoka, Dr Hirokazu *Japan.*
Tauler, Dr Esperanza *Spain.*
Tauqir, Dr Anjum *Pakistan.*
Taurbaev, Dr Toktar Iskataevich *Kazakhstan.*
Tavale, Dr Sudam Shankar *India.*
Távora, Prof. Elysiário *Brazil.*
Taylor, Prof. Charles A. *UK.*
Taylor, Jean Ellen *USA.*
Taylor, Dr John Charles *Australia.*
Taylor Jr, Dr Ivan F. *USA.*
Taylor, Max Ronald *Australia.*
Taylor, Dr Peter *Canada.*
Taylor, Dr Robert Cooper *USA.*
Taylor, Dr Robin *UK.*
Tazzoli, Prof. Vittorio *Italy.*
Tebbe, Prof. Dr Karl-Friedrich *Germany.*
Teeter, Dr Martha M. *USA.*
Tegze, Dr Miklós *Hungary.*
Teh, Dr Hung Chuan *Singapore.*
Teh, Dr Ser Kok *Malaysia.*
Telegina, Dr Inna V. *Russia.*
Teller, Dr David C. *USA.*
Teller, Raymond *USA.*
Téllez Ortiz, Ms. Minerva Estela *Mexico.*
Tellgren, I. G. Roland *Sweden.*
Telovich, Dr Roman Volodymyrovych *Ukraine.*
Temel, Abidin *Turkey.*
Tempelhoff, Dr Klaus *Germany.*
Temple, Dr Brenda R. S. *USA.*
Templeton, Prof. David H. *USA.*
Templeton, Dr Lieselotte K. *USA.*
Ten Eyck, Dr Lynn F. *USA.*
Tena, Dr M. Angeles *Spain.*
Tench, Alan Howard *USA.*
Teoh, Mr Lay Hock *Malaysia.*
Terabe, Dr Kazuya *Japan.*
Terasaki, Dr Osamu *Japan.*
Terauchi, Dr Hikaru *Japan.*
Terauchi, Masami *Japan.*
Terwilliger, Dr Thomas C. *USA.*
Terzis, Dr Aristides *Greece.*
Terzyan, Dr Simon S. *Armenia.*
Tesh, Kris F. *USA.*
Teske, Dr Christoph Ludwig *Germany.*
Tesmer, John J. G. *USA.*
Tewari, Dr Raghavendra *India.*
Thabet, Dr Atef *Egypt.*
Thackeray, Dr Michael Makepeace *USA.*
Tham, Fook S. *USA.*
Thang, Dr Tran Quoc *Vietnam.*

Thanki, Narmada USA.
Thanomkul, Dr Srinuan Chaiwasie Thailand.
That, Prof. Ta Van Vietnam.
Thayer, Dr Maria M. USA.
Theobald, Prof. François France.
Thiel, Dr Daniel J. USA.
Thielke, Mr. Harry G. USA.
Thiem, Prof. Lam Ngoc Vietnam.
Thiessen, Kris USA.
Thijsse, Dr Barend J. Netherlands.
Thin, Miss Kathy Myanmar.
Thinapong, Dr Pongchan Chananont Thailand.
Thirup, Søren Denmark.
Thiyagarajan, Dr P. USA.
Thoden, James B. USA.
Thoft, Nina Bjørn Denmark.
Thomas, Prof. John O. ("Josh") Sweden.
Thomas, Dr Noel W. UK.
Thomas, Mr P. Muthiah India.
Thomas, Dr Pamela Anne UK.
Thomasson, Ronnie Sweden.
Thompson, Dr John Gerard Australia.
Thorkildsen, Dr Gunnar Norway.
Thornton, Prof. Janet M. UK.
Thornton-Pett, Dr Mark UK.
Thorup, Niels Denmark.
Thozet, Dr Alain France.
Thu, Dr Nguyen Thi Vietnam.
Thu, Dr Trinh Le Vietnam.
Thuc, Prof. Dao Dinh Vietnam.
Thurn, Dr Herbert Germany.
Thygesen, Mr Jesper Germany.
Tibballs, Dr John E Australia.
Tibbitts, Dr Thomas T. USA.
Tibljaš, Mr Darko Croatia.
Tieghi, Prof. Giuseppe Italy.
Tiekink, Edward Richard Tom Australia.
Tigges, Mr Hartmut Germany.
Tiitta, Antero Tapani Finland.
Tikhonova, Dr Anna A. Russia.
Tillinger, Martin H. USA.
Tillmanns, Prof. Dr Ekkehart Austria.
Tilton Jr, Robert F. USA.
Timmins, Dr Peter France.
Timofeeva, Dr Tatjana V. Russia.
Timofeeva, Dr Valentina A. Russia.
Timunçin, Dr Muharrem Turkey.
Tinant, Dr Bernard Guy André Francois Belgium.
Tintorero, Mr Oscar Cuba.
Tippelt, Mrs Birgit Germany.
Tippin, Douglas B. USA.
Tiripicchio, Prof. Antonio Italy.
Tiripicchio-Camellini, Prof. Marisa Italy.
Tischler, Jonathan Z. USA.
Tishchenko, Prof. Galina N. Russia.
Tišler, Prof. Dr Miha Slovenia.
Tiwana, Mrs Kausar Shaheen Pakistan.
Tkachev, Dr Valeri V. Russia.
Tkalčec, Dr Emilija Croatia.
Toby, Dr Brian H. USA.
Tocher, Dr Derek A. UK.
Todd, Dr Donald Douglas Australia.
Töpel-Schadt, Dr Jutta Germany.
Törnroos, Karl Wilhelm Sweden.
Togawa, Mr Shinji Japan.
Toivonen, Jukka Tapio Finland.
Tokizaki, Mr Eiji Japan.
Tokonami, Prof. Masayasu Japan.
Toldo, Dr I. G. Luca Switzerland.
Tolentino, Dr Hélio Cesar Nogueira Brazil.
Tolworthy, Benjamin J. UK.
Tomás, Dr Milagros Spain.
Tomashpolsky, Prof. Yuri Ya. Russia.
Tómasson, Cand. Real. Jens Iceland.
Tomaszewski, Dr Paweł E. Poland.
Tomchick, Dr Diana R. USA.
Tomić, Dr Sanja Croatia.
Tomić, Mr Zoran Serbia.
Tomioka, Dr Nobuo Japan.
Tomita, Dr Ken ichi Japan.

Tomita, Dr Koychi Brazil.
Tomoo, Koji Japan.
Tomov, Dr Ivan Bulgaria.
Tomura, Dr Shinji Japan.
Tonejc, Dr Andelka Croatia.
Tonejc, Prof. Dr Anton Croatia.
Toney, Michael F. USA.
Tong, Dr Hua Harry Canada.
Tong, Dr Liang USA.
Tonomura, Dr Akira Japan.
Tontrakoon, Mr Jeerapong Thailand.
Tooptakong, Dr Uncharee Methong Thailand.
Topalova-Kalitzova, Dr Maria Bulgaria.
Torabi, Ali Asgar UK.
Torardi, Dr Charlie C. USA.
Toraya, Dr Hideo Japan.
Toriumi, Dr Koshiro Japan.
Torkkeli, Hannu Mika Finland.
Tormo, Jose USA.
Torres Villaseñor, Dr Gabriel Mexico.
Torres-Ruiz, Dr José Spain.
Torriani, Dr Iris Linares Brazil.
Toscano, Mr Ruben Alfredo Mexico.
Tosi, Prof. Giorgio Italy.
Tosik, Dr Anita Poland.
Tosson, Dr Salama Egypt.
Toth, Dr Kenneth S. USA.
Tóth, Dr Lajos Hungary.
Toudic, Dr Bertrand France.
Tougard, Dr Pierre France.
Toupet, Dr Loïc France.
Toussaint, Prof. Jean Belgium.
Tovbis, Dr Alexander B. Russia.
Towns, Robert L. R. USA.
Toyoda, Dr Kazuhiro Japan.
Toyoda, Prof. Koichi Japan.
Toyoda , Prof. Taro Japan.
Traetteberg, Prof. Marit Norway.
Trah, Dr Hans-Peter Germany.
Trampert, Dr Achim Peter Germany.
Transue, Thomas R. USA.
Traub, Prof. Wolfie Israel.
Trauth, Dr Jürgen Germany.
Traveria-Cros, Dr Adolfo Spain.
Treechairusme, Mr Kamchai Thailand.
Trefonas, Louis Marco USA.
Trefry, Dr Michael Australia.
Treivus, Dr Eugene Borisovich Russia.
Tretyachenko, Dr Ljudmyla Oleksandrivna Ukraine.
Tretyakov, Dr Vjacheslav N. Russia.
Tri, Prof. Nguyen Van Vietnam.
Triantafyllou, Spyros Greece.
Tribaudino, Dr Mario Italy.
Trigubo, Ms Alicia Beatriz Argentina.
Trömel, Prof. Dr Martin Germany.
Trofimov, Mr Victor B. Russia.
Trofimova, Mrs Rimma F. Russia.
Trojko, Mr Rudolf Croatia.
Tronrud, Dr Dale E. USA.
Trontsios, Prof. George Greece.
Trosti-Ferroni, Prof. Renza Italy.
Trotter, Prof. James Canada.
Troup, Jan Marshall USA.
Trubkin, Dr Nikolay V. Russia.
Trueblood, Kenneth N. USA.
Trung, Mr Nguyen Tan Vietnam.
Truni, Dr Karapet G. Armenia.
Truni, Dr Lusine K. Armenia.
Truong, Dr Chu Thien Vietnam.
Trus, Dr Benes L. USA.
Truter, Prof. Mary R. UK.
Trzaska-Durski, Dr Zygmunt Maria Poland.
Tsai, Prof. Chun-che USA.
Tsai, Ji-Ching Taiwan.
Tschulena, Dr Guido R. Germany.
Tse, Dr John S. Canada.
Tsikhotsky, Dr Evgeny S. Russia.
Tsinober, Dr Leonid I. Russia.
Tsintsadze, Prof. Givi Georgia.
Tsirambides, Prof. Ananias Greece.

Tsirelson, Prof. Vladimir Grigorievich Russia.
Tsuchiya, Mr Tomonobu Japan.
Tsuda, Kenji Japan.
Tsukihara, Prof. Tomitake Japan.
Tsunekawa, Dr Shin Japan.
Tsuneyuki, Prof. Shinji Japan.
Tsurkan, Dr Vladimir Moldova.
Tsushima, Prof. Dr Katsutoshi Japan.
Tsybulya, Dr Sergey V. Russia.
Tucker, Ian Malcolm UK.
Tuinstra, Prof. Dr Ir Fokke Netherlands.
Tulinsky, Dr Alexander USA.
Tun, Dr Zin Canada.
Tunkasiri, Dr Tawee Thailand.
Tuomi, Donald USA.
Tuomi, Turkka Olavi Finland.
Turco, Prof. Guy France.
Turel, Mr Iztok Slovenia.
Turk, Prof. Dr Vito Slovenia.
Turk, Dr Dušan Slovenia.
Turley, June Williams USA.
Turley, Stewart USA.
Turner, Dr Mary Canada.
Turpeinen, Urho Taneli Finland.
Turrillas, Dr Xavier UK.
Turunen, Dr Markus Finland.
Tuyen, Mr Nguyen Cuong Vietnam.
Twigg, Pam USA.
Tyvanchuk, Dr Anna Teodorivna Ukraine.
Tzalenchuk, Dr Alexander Ya. Russia.
Uchida, Akira Japan.
Uchida, Dr Tsutomu Japan.
Uda, Dr Satoshi Japan.
Udalova, Dr Valentina V. Russia.
Udovic, Mr Boris Slovenia.
Udron, Dr Dominique Brazil.
Udubaşa, Dr Gheorghe Romania.
Uecker, Mrs Doris-Christiane Germany.
Uecker, Mr Reinhard Germany.
Uekusa, Dr Hidehiro Japan.
Ülkü, Dr Dinçer Turkey.
Ueng, Prof. Chuen-Her Taiwan.
Ueno, Dr Satoru Japan.
Ughetto, Dr Giovanni Italy.
Ugliengo, Prof. Piero Italy.
Ugozzoli, Dr Franco Italy.
Uhlig, Dr Stefan Helmut Germany.
Ulbricht, Prof. Dr Heinz Germany.
Ullrich, Prof. Dr Hans-Jürgen Germany.
Ulmer, Dr Kevin M. USA.
Uma Devi, Ms B India.
Umakoshi, Dr Keisuke Japan.
Umeda, Prof. Dr Takateru Japan.
Umeno, Prof. Masataka Japan.
Umland, Timothy C. USA.
Ungár, Prof. Tamás Hungary.
Ungaretti, Prof. Luciano Italy.
Unge, Torsten Sweden.
Uno, Prof. Ryosei Japan.
Untersteller, Mr Eugen Germany.
Uppenberg, Jonas Sweden.
Uragami, Prof. Takuyuki Japan.
Urbańczyk, Prof. Grzegorz Poland.
Urbina de Navarro, Dr Caribay Teresa Venezuela.
Urland, Prof. Dr Werner Germany.
Urtiaga, Dr Karmele Spain.
Urusov, Prof. Vadim S. Russia.
Urusovskaya, Prof. Aida A. Russia.
Urzhumtsev, Dr Alexander G. Russia.
Usanmaz, Dr Ali Turkey.
Usha, Ms R. India.
Usher, Dr Brian Francis Australia.
Uszyński, Dr Ignacy Poland.
Uttamasil, Dr Lek Thailand.
Uustare, Dr Teet Estonia.
Uwaha, Dr Makio Japan.
Uyeda, Emer. Prof. Natsu Japan.
Uyeda, Dr Hiroshi Japan.
Uztetik Amour, Dr Ayşe Turkey.
Vaccari, Prof. Giuseppe Italy.

Vagg, Robert Sylvester *Australia.*
Vahvaselkä, Dr Aino Margit *Finland.*
Vahvaselkä, Dr Sakari *Finland.*
Vainshtein, Prof. Boris K. *Russia.*
Vajdos, Felix *USA.*
Valach, Doc. Ing. Fedor *Slovakia.*
Valčić, Dr Andreja *Serbia.*
Valdrè, Prof. Ugo *Italy.*
Valegård, Karin *Sweden.*
Valente, Edward J. *USA.*
Valenzuela Monjarás, Dr Raúl Alejandro *Mexico.*
Valero, Mr Ricardo *Chile.*
Valigura, Doc. Ing. Dušan *Slovakia.*
Valin, Dr Mariluz *Spain.*
Valkonen, Jussi Uolevi *Finland.*
Val'kovskaya, Dr Margarita Ivanovna *Moldova.*
Valle, Dr Giovanni Carlo *Italy.*
Vallet-Regí, Dr Maria *Spain.*
Val'ter, Dr Anton Antonovych *Ukraine.*
Valvoda, Prof. Dr Václav *Czech Republic.*
Van Acker, Mr Karel *Belgium.*
Van Alsenoy, Dr Christian *Belgium.*
van Bolhuis, Mr Fré *Netherlands.*
Van, Prof. Dang Ung *Vietnam.*
van de Grampel, Prof. Dr Johan C. *Netherlands.*
van de Velde, Dr George M. H. *Netherlands.*
van de Waal, Ir Benjamin W. *Netherlands.*
van den Berg, Ir Adrianus J. *Netherlands.*
Van den Bossche, Dr Guy Ghislain Remy *Belgium.*
van der Gaast, Sjierk J. *Netherlands.*
Van der Heijden, Dr Simon Petrus Nicolaas *Canada.*
van der Helm, Dr Dick *USA.*
van der Lee, Dr Arie *Netherlands.*
Van der Maelen, Dr Juan Francisco Javier *Spain.*
van der Plas, Prof. Dr Leendert *Netherlands.*
van der Plas, Drs Jaco L. *Netherlands.*
van der Sluis, Dr Paul *Netherlands.*
van der Veen, Dr Adriaan H. *Netherlands.*
van der Wal, Dr Robert J. *Netherlands.*
van der Woerd, Mark J. *USA.*
Van Dyck, Dr Dirk Ernest Maria *Belgium.*
van Eijk, Ir Marcel C. P. *Netherlands.*
van Enckevort, Dr Wilhelmus J. P. *Netherlands.*
van Geeresteln, Dr Vincent J. *Netherlands.*
Van Houtte, Prof. Paul *Belgium.*
Van Hove, Dr Michel A. *USA.*
van Hummel, Gerrit J. *Netherlands.*
van Koningsveld, Dr Hendrikus *Netherlands.*
Van Landuyt, Prof. Jef Florent *Belgium.*
van Maissen, Dr Kees F. *Netherlands.*
Van Mechelen, Ing. Jan B. *Netherlands.*
Van Meervelt, Dr Luc *Belgium.*
van Meurs, Frank *USA.*
van Meurs, Dr Ir Frank *Netherlands.*
Van Nordstrand, Robert *USA.*
Van Roey, Patrick *USA.*
Van Rooyen, Prof. Petrus Hendrik *South Africa.*
Van Schalkwyk, Prof. Theunis G. D. *South Africa.*
van Schooneveld, Ing. Marinus *Netherlands.*
Van Smaalen, Dr Sander *Germany.*
Van Tendeloo, Prof. Gustaaf *Belgium.*
van Vucht, Dr Johannes H. N. *Netherlands.*
Vana, Prof. Dr Norbert *Austria.*
Vander Sande, John B. *USA.*
Vanderhoff-Hanaver, Peggy A. *USA.*
VanDerveer, Dr Donald G. *USA.*
Vandonselaar, Mrs Margaret *Canada.*
Vaney, Dr Marie-Christine *France.*
Vanghelie, Iulian *Romania.*
Vanhellemont, Dr Jan H. *Belgium.*
Vannerberg, Nils-Gösta *Sweden.*
Varadarajan, Dr Raghavan *India.*
Vardanian, Dr David M. *Armenia.*
Varela, Dr José Arana *Brazil.*
Vargas, Dr Rodolfo *Venezuela.*
Varghese, Dr Joseph Noozhumurry *Australia.*
Varschavsky, Ari *Chile.*
Vartanyants, Dr Ivan A. *Russia.*
Varughese, Dr Kottayil Iype *USA.*
Vasanth, Dr K. L. *India.*

Vasil'ev, Dr Evgeny K. *Russia.*
Vasilev, Dr Alexander D. *Russia.*
Vasiliev, Dr Alexandr L. *Russia.*
Vassilev, Dr Ivan *Bulgaria.*
Vassylyev, Dr Dmitry Glebovich *Japan.*
Vavelidis, Prof. Michael *Greece.*
Veblen, David R. *USA.*
Vedani, Dr Angelo *Switzerland.*
Vega, Mr Daniel Roberto *Argentina.*
Vegas, Prof. Angel *Spain.*
Veiga, João Pedro *Portugal.*
Veintemillas, Dr Sabino *Spain.*
Veith, Prof. Dr Michael *Germany.*
Velankar, Mr Sameer Sudhakar *India.*
Veldman, Drs Nora *Netherlands.*
Velichkina, Dr Tat'yana S. *Russia.*
Velikanova, Prof. Tamara Yakivna *Ukraine.*
Velilla, Dr Nicolas *Spain.*
Vellieux, Dr Fred *France.*
Vencato, Dr Ivo *Brazil.*
Vendrell, Dr Marius *Spain.*
Venevtsev, Prof. Yurii N. *Russia.*
Venkataraman, Dr Ravichandran *India.*
Venkatesan, Dr Kailash *India.*
Vennik, Prof. Joost *Belgium.*
Vente, Ir Jaap *Netherlands.*
Venudhar, Dr Y. C. *India.*
Venugopal, Dr Manju Grover *USA.*
Venugopalan, Mr P. *India.*
Venzo, Dr Alfonso *Italy.*
Vera Calderón, Ms Gloria *Mexico.*
Verbist, Prof. Jacques Jozef *Belgium.*
Verboven, Ms Christel Clara Remigius *Belgium.*
Veremeichik, Dr Tamara F. *Russia.*
Vergasov, Dr Vladimir L. *Russia.*
Verhulst, Mr Koen *Belgium.*
Verlinde, Dr Christophe L. M. J. *USA.*
Verma, Dr Ajit Ram *India.*
Vermin, Dr Willem J. *Netherlands.*
Vermishian, Dr Garnik A. *Armenia.*
Versaci, Dr Raúl Antonio *Argentina.*
Verwer, Dr Paul *Netherlands.*
Vesselinov, Mr Iliya *Bulgaria.*
Vettier, Dr Christian *France.*
Veysseyre, Dr Renée *France.*
Vezzalini, Prof. Maria Giovanna *Italy.*
Vlani, Prof. Robert *France.*
Vicat, Prof. Jean *France.*
Vicente, Prof. José *Spain.*
Vickers, Mary Elizabeth *UK.*
Vicković, Dr Ivan *Croatia.*
Victoria, Mr Nelson *Uruguay.*
Viczián, Dr István *Hungary.*
Vidal, Prof. Geneviève *France.*
Vidal, Miss Haydeé Marta *Argentina.*
Vidal, Dr Jean-Pierre *France.*
Vielhaber, Dr Edmund Antonius *Germany.*
Vigdorchik, Dr Asya G. *Russia.*
Vijayalakshmi, Dr J. *USA.*
Vijayan, Prof. Mamannamana *India.*
Vijayan, Dr Mrs Kalyani *India.*
Vila, Dr Eladio *Spain.*
Vilkov, Prof. Lev Vasilyevich *Russia.*
Villadsen, Jørgen *Denmark.*
Villafranca, Dr J. Ernesto *USA.*
Villafuerte Castrejón, Ms María Elena *Mexico.*
Villaroel, Prof. Hugo S. *Brazil.*
Villena Iribe, Mr René *Mexico.*
Villeret, Mr Vincent *France.*
Vimla, Ms T. M. *India.*
Viña, Mr Raúl Oscar *Argentina.*
Vincent, Dr Beverly R. *USA.*
Vincent, Prof. Henri *France.*
Vincent, Dr Jan E. *UK.*
Vinh, Dr Ngo Van *Vietnam.*
Vinhas, Dr Laércio A. *Brazil.*
Vinković, Dr Mladen *Croatia.*
Vinogradov, Dr Alexander V. *Russia.*
Vinokurov, Prof. Vladimir M. *Russia.*
Virovets, Dr Alexander V. *Russia.*

Visser, Drs Jan Willem *Netherlands.*
Visser, Dr Rudolph J. J. *Netherlands.*
Viswamitra, Prof. Mysore Ananthamurthy *India.*
Vitali, Prof. Gianfranco *Italy.*
Vitali, Dr Jacqueline *USA.*
Vitanov, Prof. Todor *Bulgaria.*
Viterbo, Prof. Davide *Italy.*
Vittal, Dr Jagadese *Canada.*
Vlasov, Dr Michael Yu. *Russia.*
Vlasov, Dr Vasily Platonovich *Russia.*
Vlasse, Dr Marcus *USA.*
Vlokh, Prof. Orest Grygorovych *Ukraine.*
Völlenkle, Dr Horst *Austria.*
Voet, Donald *USA.*
Vogelaar, Dr Nancy *USA.*
Vogl, Prof. Dr Gero *Austria.*
Vojtěchovský, Dr Jaroslav *Czech Republic.*
Vojtechovsky, Dr Jaroslav *USA.*
Volkmann, Mr Niels Heinz Paul *Germany.*
Volkov, Dr Vladimir V. *Russia.*
Volodina, Dr Galina Fedorovna *Moldova.*
Voloshin, Dr Alexey E. *Russia.*
Volz, Dr Karl W. *USA.*
von Böhlen, Dr Klaus J. *Germany.*
Von Dreele, Dr Robert B. *USA.*
von Philipsborn, Prof. Dr Henning *Germany.*
von Schnering, Prof. Dr h.c. Hans Georg *Germany.*
Vonasek, Lic. Eva María *Venezuela.*
Vong, Dr Vo *Vietnam.*
Vonk, Dr Christ G. *Netherlands.*
Vora, Dr Rasiklal Amulakhbhai *India.*
Vorbach, Dr Angelika *Germany.*
Vorderwisch, Dr Peter *Germany.*
Voronkova, Dr Valentina I. *Russia.*
Voroshilov, Prof. Yurij Vitaliyovych *Ukraine.*
Vos-Looljenga, Dr Aafje *Netherlands.*
Voutsas, Prof. George *Greece.*
Voznyak, Dr Dmytro Kostiantynovych *Ukraine.*
Vratislav, Ass. Prof. Ing. Stanislav *Czech Republic.*
Vreeland Jr, Thad *USA.*
Vrielink, Dr Alice *Canada.*
Vrublevskaja, Dr Zoja V. *Russia.*
Vučić, Dr Zlatko *Croatia.*
Vučković, Dr Gordana *Serbia.*
Vuilhorgne, Dr Marc *France.*
Vuong, Dr Nguyen Van *Vietnam.*
Vyas, Mr K *India.*
Vyas, Meenakshi *USA.*
Vyas, Dr Nand K. *USA.*
Wacker, Dr Klaus *Germany.*
Wada, Prof. Koh *Japan.*
Wada, Mr Norio *Japan.*
Wadhawan, Dr Vinod Kumar *India.*
Wagenfeld, Heinrich Karsten *Australia.*
Wagner, Dr Gerald *Germany.*
Wagner, Mrs Trixie *Germany.*
Wagner, Dr Ulrike Gabriella *Austria.*
Wahlberg, Dr Anders *Sweden.*
Wai, Miss Myo Sandar *Myanmar.*
Waizumi, Mr Kenji *Japan.*
Wakahara, Dr Akihiro *Japan.*
Wakatsuki, Prof. Dr Masao *Japan.*
Wakatsuki, Dr Soichi *France.*
Wakino, Dr Kikuo *Japan.*
Wakoh, Prof. Shinya *Japan.*
Waksman, Dr Gabriel *USA.*
Walitzi, Prof. Dr Eva Maria *Austria.*
Walker, Miss Elaine M. *UK.*
Walker, Dr Nigel P. C. *Germany.*
Wallace, Dr Bonnie Ann *UK.*
Wallenberg, L. Reine *Sweden.*
Wallis, Dr John D. *UK.*
Wallrafen, Dr Franz *Germany.*
Wallwork, Dr Stephen Collier *UK.*
Walsh, Dr Martin *Ireland.*
Walsöe de Reca, Dr Noemí Elisabeth *Argentina.*
Walter, Dr Franz *Austria.*
Walter, Dr Mark R. *USA.*
Walter, Norman M. *USA.*
Walter, Richard L. *USA.*

Wan, Ms Jiayi *China*.
Wang, Prof. Andrew H.-J. *USA*.
Wang, Bi-Cheng *USA*.
Wang, Prof. Dacheng *China*.
Wang, Prof. Jinling *China*.
Wang , Dr Jinn-Lung *Taiwan*.
Wang, Prof. Ju-Chun *Taiwan*.
Wang, Dr Naiding *Germany*.
Wang, Prof. Qiguang *China*.
Wang, Prof. Ruji *China*.
Wang, Sue-Lein *Taiwan*.
Wang, Mr Weibin *Canada*.
Wang, Dr Wen-Ching *Taiwan*.
Wang, Dr Xiqu *Germany*.
Wang, Prof. Yu *Taiwan*.
Wang, Ms Yuan-Fang *Netherlands*.
Wang, Prof. Yuming *China*.
Wang, Z. L. *USA*.
Wang, Prof. Zutao *China*.
Warburton, Dr William K. *USA*.
Warczewski, Prof. Jerzy Zdzisław *Poland*.
Ward, Dr Donald L. *USA*.
Ward, Keith B. *USA*.
Warhanek, Prof. Dr Hans *Austria*.
Warmiński, Tadeusz, Piotr *Australia*.
Warner, Dr Joanne Kathleen *Australia*.
Warren, Stephen G. *USA*.
Wartchow, Dr Rudolf *Germany*.
Waseda, Prof. Yoshio *Japan*.
Waser, Jurg *USA*.
Waśkowska, Dr Alicja *Poland*.
Watanabe, Prof. Denjiro *Japan*.
Watanabe, Dr Jiro *Japan*.
Watanabe, Dr Koichi *Japan*.
Watanabe, Dr Nobuhisa *Japan*.
Watanabe, Mr Yoshio *Japan*.
Watenpaugh, Dr Keith Donald *USA*.
Waters, Dr Joyce M. *New Zealand*.
Watkin, Dr David John *UK*.
Watkins, Steven F. *USA*.
Watowich, Dr Stan J. *USA*.
Watson, Dr David Gilfillan *UK*.
Watson, Kimberly A. *UK*.
Watson, Prof. William H. *USA*.
Watt, Dr William *USA*.
Watts, Ethel J. *USA*.
Watts, John Andrew *Australia*.
Wawrzak, Dr Zdzisław *Poland*.
Weakley, Dr Timothy J. R. *USA*.
Weaver, Arthur J. *USA*.
Weber, Christian H. *USA*.
Weber, Dr Hans Peter *Switzerland*.
Weber, Prof. Hans-Peter *Switzerland*.
Weber, Irene T. *USA*.
Weber, Dr Jürgen *Germany*.
Weber, Patricia *USA*.
Webster, Dr Michael *UK*.
Wechsler, Dr Barry A. *USA*.
Weckert, Dr Edgar F. *Germany*.
Wedekind, Joseph *USA*.
Weeks, Dr Charles M. *USA*.
Wegener, Dr Wolter *Belgium*.
Wei, Mingyi *USA*.
Wei, Prof. Xincheng *China*.
Weichsel, Andrzej *USA*.
Weight, Tony *UK*.
Weis, William I. *USA*.
Weise, Mrs Sylvia *Germany*.
Weiss, George H. *USA*.
Weiss, Dr Manfred S. *USA*.
Weiss, Dr Zdeněk *Czech Republic*.
Weissman, Larry *USA*.
Weissmann, Prof. Sigmund *USA*.
Weitzel, Dr Hans *Germany*.
Welberry, (Thomas) Richard *Australia*.
Welch, Prof. Alan J. *UK*.
Wenda, Prof. Dr Richard *Germany*.
Wendschuh-Josties, Dr Michael Thomas *Germany*.
Wenig, Prof. Dr Werner *Germany*.
Werk-Albers, Dr Margit L. *Germany*.

Werner, Prof. Per-Erik *Sweden*.
Werthmann, Mrs Ulrike *Germany*.
Wery, Dr Jean-Pierre *USA*.
Wesenberg, Gary *USA*.
Wesson, Laura *USA*.
West, Prof. Anthony Roy *UK*.
Westbrook, Dr Edwin M. *USA*.
Westdahl, Marianne *Sweden*.
Wester, Dennis W. *USA*.
Westhof, Prof. Eric *France*.
Weston, Dr Simon Alan *UK*.
Westrip, Mr Simon P. *UK*.
Wetzel, Mrs Andrea *Germany*.
Wetzig, Prof. Dr Klaus *Germany*.
Weyrich, Prof. Dr-Ing. Dr phil.h.c. Wolf *Germany*.
Wheatley, Dr P. J. *UK*.
Wheeler, Dr Kraig A. *USA*.
Whitaker, Dr Alan *UK*.
Whitby, Frank G. *USA*.
White, Allan Henry *Australia*.
White, Mr Andre *Canada*.
White, Clinton L. *USA*.
White, Dr Joe L. *USA*.
White, Dr John G. *USA*.
White, John William *Australia*.
White, Peter S. *USA*.
White, Scott A. *USA*.
White, Prof. Stephen H. *USA*.
White, Dr Stephen W. *USA*.
White, Timothy John *Australia*.
Whitfield, Harold John *Australia*.
Whitlow, Marc D. *USA*.
Whitlow, Dr Simon Hugh *Canada*.
Whittaker, Dr E. J. W. *UK*.
Whöler, Dr Annick *France*.
Wicks, Dr Frederick John *Canada*.
Wiebenga, Prof. Dr Eelco H. *Netherlands*.
Wieczorek, Dr Wanda *USA*.
Wiedemann, Mr Peter *Germany*.
Wieder, Dr Thomas *Germany*.
Wiegers, Prof. Dr Gerrit A. *Netherlands*.
Wiehl, Dr Leonore *Germany*.
Wielunski, Leszek *Australia*.
Wien, Michelle W. *USA*.
Wierda, Dr Derk A. *USA*.
Wieser, Prof. Dr Egbert *Germany*.
Wiewióra, Prof. Andrzej *Poland*.
Wiff, Donald R. *USA*.
Wildner, Mag. Dr Manfred *Austria*.
Wiley, Don C. *USA*.
Wilke, Prof. Dr Wolfgang *Germany*.
Wilken, Dr Gerdt *Germany*.
Wilkins, Dr Stephen William *Australia*.
Wilkinson, Dr Angus P. *USA*.
Wilkinson, Dr Clive *France*.
Wilkinson, Kay *UK*.
Will, Prof. Dr Georg *Germany*.
Willaime, Prof. Christian *France*.
Willett, Prof. Roger D. *USA*.
Williams, Dr David Arfon *UK*.
Williams, Dr David J. *UK*.
Williams, Mr Donald Allan *Australia*.
Williams, Geoffrey Allan *Australia*.
Williams, Graheme J. B. *USA*.
Williams, Dr Ian Duncan *Hong Kong*.
Williams, Dr John A. *USA*.
Williams, Dr Loren Dean *USA*.
Willard, Prof. Paul G. *USA*.
Willis, Dr Anthony C. *Australia*.
Willis, Prof. B. T. M. *UK*.
Willoughby, Anne Margaret Elizabeth *UK*.
Wilmanns, Dr Matthias *Germany*.
Wilson, Alan Richard *Australia*.
Wilson, Prof. Arthur J. C. *UK*.
Wilson, Dr Charles *USA*.
Wilson, Dr Chick *UK*.
Wilson, Claire *UK*.
Wilson, David K. *USA*.
Wilson, Prof. Herbert Rees *UK*.
Wilson, Dr Ian A. *USA*.

Wilson, Sarah *UK*.
Win, Miss Ohnmar *Myanmar*.
Winder, Ms Nicola *South Africa*.
Windsor, Dr Colin G. *UK*.
Wing, Richard M. *USA*.
Wingert, Dr Lavinia *USA*.
Winick, Herman *USA*.
Winkler, Dr Björn *Germany*.
Winkler, Dr Fritz K. *Switzerland*.
Winnacker, Prof. Dr Albrecht *Germany*.
Wintenberger, Dr Micheline *France*.
Winter, Prof. Dr Hannspeter *Austria*.
Winter, Prof. Dr Werner *Germany*.
Winter, Dr Gabriele *Germany*.
Winter, Dr Marcus John *UK*.
Wiskemann, Dr René *Germany*.
Wismer, Robert *USA*.
Withers, Dr Ray *Australia*.
Witte, Prof. Dr Helmut *Germany*.
Wittke, Prof. Oscar *Chile*.
Wittmüss, Mrs Dörte *Germany*.
Witz, Dr Jean *France*.
Wlodawer, Dr Alexander *USA*.
Wobrauschek, Dr Peter *Austria*.
Wögerbauer, Dr Rupert *Germany*.
Wölfel, Prof. Dr Erich R. *Germany*.
Woensdregt, Dr Cornelis F. *Netherlands*.
Wörle, Mr Michael D. *Switzerland*.
Woitok, Dr Joachim *Germany*.
Wokulska, Dr Krystyna *Poland*.
Wokulski, Prof. Zygmunt *Poland*.
Wolberger, Dr Cynthia *USA*.
Wokyrz, Dr Marek *Poland*.
Wolf, Dr Eberhard Günter *Germany*.
Wolf, Dr Sharon G. *USA*.
Wolf, Dr Wojciech *Poland*.
Wolff, Dr Gunther A. *USA*.
Wolmershäuser, Dr Gotthelf *Germany*.
Wolska, Dr Irena *Poland*.
Wonacott, Dr Alan John *UK*.
Wondratschek, Prof. Dr Hans *Germany*.
Wong, Ms Rosalind Y. *USA*.
Wong, Dr Wing Tak *Hong Kong*.
Wong, Dr Yau-Shing *Hong Kong*.
Wong-Ng, Dr Winnie *USA*.
Wongshaiboon, Dr Sajee *Thailand*.
Woning, Ing Leo *Netherlands*.
Wood, Elizabeth A. *USA*.
Wood, Dr Gordon H. *Canada*.
Wood, Dr Ian G. *UK*.
Wood, Dr John S. *USA*.
Woolfson, Prof. Michael Mark *UK*.
Worcester, Prof. David L. *USA*.
Workman, S. Thomas *USA*.
Worthington, C. R. *USA*.
Worthylake, David K. *USA*.
Worzala, Prof. Dr Horst *Germany*.
Wouters, Mr Johan F. A. M. *Belgium*.
Woźniak, Dr Krzysztof *Poland*.
Wright, Dr Christine S. *USA*.
Wright, Harlan Tonie *USA*.
Wright, Dr Helen *UK*.
Wright, Dr William V. *USA*.
Wu, Prof. Bomu *China*.
Wu, Edward Chia-Kuei *USA*.
Wu, Dr Hao *USA*.
Wu, Jin *USA*.
Wu, Dr Jong-Chang *Taiwan*.
Wu, Dr Ping *USA*.
Wu, Shan *USA*.
Wu, Mr Wanguo *China*.
Wu, Prof. Xintao *China*.
Wuensch, Prof. Bernhardt John *USA*.
Wulff, Dr Harm *Germany*.
Wunderlich, Dr Hartmut *Germany*.
Wutzler, Mrs Ronny *Germany*.
Wyns, Prof. Lode *Belgium*.
Xia, Dr Di *USA*.
Xia, Mr Qi *Canada*.
Xia, Prof. Zongxiang *China*.

Xian, Prof. Dingchang *China*.
Xiang, Shibin *USA*.
Xiao, Dr Gaoyi *USA*.
Xiayang, Dr Qiu *USA*.
Xie, Xiaoling *USA*.
Xu, Prof. Jingyang *China*.
Xu, Dr Weixin *USA*.
Xu, Prof. Xiaojie *China*.
Xu, Dr Yan *Singapore*.
Xu, Zhang-Bao *USA*.
Xue, Dr Yafeng *USA*.
Xuong, Dr Nguyen-huu *USA*.
Yadav, Dr Asheshwar *India*.
Yadav, Dr Prem N. *USA*.
Yadav, Dr Tapaswi *India*.
Yadava, Dr Bishwanath *India*.
Yadava, Dr Vijay Singh *India*.
Yağbasan, Dr Rahmi *Turkey*.
Yagi, Prof. Katsumichi *Japan*.
Yagi, Prof. Toshirou *Japan*.
Yahya, Dr Muhamad *Malaysia*.
Yakinthos, Prof. Ioannis *Greece*.
Yakovlev, Prof. Ivan A. *Russia*.
Yakubovich, Dr Olga V. *Russia*.
Yakunin, Mr Andrew N. *Russia*.
Yakushkin, Dr Eugene D. *Russia*.
Yalkovsky, Ralph (Rafael) *USA*.
Yamada, Mr Hiroshi *Japan*.
Yamada, Dr Nobusuke *Japan*.
Yamagishi, Dr Hirotoshi *Japan*.
Yamaguchi, Dr Hiroshi *Japan*.
Yamaguchi, Dr Ko-ichi *Japan*.
Yamaguchi, Mr Yasuhide *Japan*.
Yamamoto, Dr Akiji *Japan*.
Yamamoto, Prof. Dr Akio *Japan*.
Yamamoto, Prof. Dr Katsumi *Japan*.
Yamamoto, Prof. Naoki *Japan*.
Yamamoto, Dr Nobuyuki *Japan*.
Yamamoto, Dr Seishi *Japan*.
Yamana, Dr Kazuo *Japan*.
Yamane, Dr Takashi *Japan*.
Yamano, Akihito *USA*.
Yamazaki, Dr Atsushi *Japan*.
Yamazaki, Prof. Yohtaro *Japan*.
Yamnova, Dr Natalya A. *Russia*.
Yan, Dr Xiaoqian *Canada*.
Yan, Youwei *USA*.
Yanagi, Dr Kazunori *Japan*.
Yanchuk, Dr Edward Oleksandrovych *Ukraine*.
Yaneva, Dr Svetlana *Bulgaria*.
Yang, Prof. Daniel Shun-Chung *Canada*.
Yang, Fan *USA*.
Yang, Prof. Guangdi *China*.
Yang, Mr Jian *Canada*.
Yang, Jian *USA*.
Yang, Prof. Qingchuan *China*.
Yang, Wei *USA*.
Yanovskii, Dr Vladimir K. *Russia*.
Yanovsky, Dr Alexander I. *Russia*.
Yanson, Dr Tamara Ivanivna *Ukraine*.
Yanulova, Dr Ludmila A. *Russia*.
Yanusova, Dr Ludmila G. *Russia*.
Yao, Mr Min *Japan*.
Yao, Prof. Xinkan *China*.
Yase, Dr Kiyoshi *Japan*.
Yasinska, Dr Angelina Andrijivna *Ukraine*.
Yasuami, Dr Shigeru *Japan*.
Yasuda, Prof. Yukio *Japan*.
Yasui, Masanori *Japan*.
Yasuoka, Prof. Noritake *Japan*.
Yatsenko, Dr Alexandr Vasilievich *Russia*.
Ye, Prof. Chuntang *China*.
Yeates, Dr Todd O. *USA*.
Yee, Dr Vivien C. *USA*.
Yeh, Joanne I. *USA*.
Yeh, Dr Taun-ran *Taiwan*.
Yennawar, Dr Hemant P. *USA*.
Yennawar, Dr Neela H. *USA*.
Yeon, Prof. Younghee *Korea*.

Yeritsian, Dr Grant *Armenia*.
Yıldız, Çiğdem *Turkey*.
Yin, Prof Dr Soe *Myanmar*.
Yip, Kitty *UK*.
Ylinen, Dr Eero E. *Finland*.
Yokomori, Yoshinobu *Japan*.
Yokota, Dr Masaaki *Japan*.
Yonath, Prof. Ada *Israel*.
Yonath, Prof. Dr Ada *Germany*.
Yonekura, Dr Masami *Japan*.
Yoon, Prof. Choon Sup *Korea*.
Yoshimoto, Mr Noriyuki *Japan*.
Yoshimura, Prof. Jun-ichi *Japan*.
Yoshimura, Prof. Masahiro *Japan*.
Yoshimura, Dr Yukio *Japan*.
You, Dr Hoydoo *USA*.
Young Jr, Dr Victor G. *USA*.
Young, Prof. R. A. *USA*.
Young, Robert James *UK*.
Youngs, Prof. Wiley *USA*.
Youssef, Prof. I. Mourad *Egypt*.
Yousufzai, Mr Inayatullah Khan *Pakistan*.
Ysker, Dr Jan Stinus *Germany*.
Yu, Chang-An *USA*.
Yu, Prof. Ruihuang *China*.
Yu, Prof. Shu-Cheng *Taiwan*.
Yu, Prof. Xiufen *China*.
Yuan, Dr Hanna S. *Taiwan*.
Yuan, Jie *USA*.
Yuasa, Dr Hiroshi *Japan*.
Yufit, Dr Dmitrii S. *Russia*.
Yuhasz, Stacieann *USA*.
Yumoto, Dr Hisami *Japan*.
Yun, Prof. Hoseop *Korea*.
Yun, Ms Mi Kyung *Korea*.
Yusaf, Dr Mohammad *Pakistan*.
Yushkin, Prof. Nikolay P. *Russia*.
Yusupova, Dr Ilihamia M. *Uzbekistan*.
Yuvaniyama, Jirundon *USA*.
Yuyama, Mr Isamu *Japan*.
Yvon, Prof. Klaus *Switzerland*.
Zabel, Dr Volker *USA*.
Zabukovec, Miss Nataša *Slovenia*.
Zaccai, Dr Joseph *France*.
Zacharias, Dr David Edward *USA*.
Zachau-Christiansen, Birgit *Denmark*.
Zadorozhnaya, Dr Ludmila A. *Russia*.
Zagari, Prof. Adriana *Italy*.
Zaghete, Dr Maria Aparecida *Brazil*.
Zaghloul, Prof. Mohamed Zaki *Egypt*.
Zaghloul, Prof. Dr Zaki M. *Egypt*.
Zaharova, Dr Maria I. *Russia*.
Zaitsev, Dr Sergey M. *Russia*.
Žák, Doc. Dr Zdirad *Czech Republic*.
Zakharchenko, Dr Irina N. *Russia*.
Zakharov, Dr Boris G. *Russia*.
Zakharov, Prof. Lev N. *Russia*.
Zakrzewski, Marek *USA*.
Zalba, Dr Patricia Eugenia *Argentina*.
Zalesskii, Prof. Andrei V. *Russia*.
Zalkin, Allan *USA*.
Zamir, Sharona *Israel*.
Zamorzayev, Prof. Alexander Mikhailovich *Moldova*.
Zanazzi, Prof. Pierfrancesco *Italy*.
Zandbergen, Dr Henny W. *Netherlands*.
Zannetti, Prof. Roberto *Italy*.
Zanotti, Prof. Giuseppe *Italy*.
Zanotti, Dr Lucio *Italy*.
Zarechniuk, Prof. Oleg Safoniyovych *Ukraine*.
Zaremba, Dr Vasyl' Ivanovych *Ukraine*.
Zarka, Dr Albert *France*.
Zasimov, Dr Victor S. *Russia*.
Zassourskaya, Mrs Larissa Alexandrovna *Russia*.
Zaumseil, Dr Peter *Germany*.
Zavalii, Dr Petro Yuliyanovych *Ukraine*.
Zavalij, Dr Peter Yu. *USA*.
Zavodnik, Dr Valery E. *Russia*.
Zawadzke, Dr Laura E. *USA*.
Zaworotko, Dr Michael John *Canada*.
Zayakina, Dr Nadezhda V. *Russia*.

Zaytzev-Bashan, Anat *Israel*.
Zedler, Dr Achim *Germany*.
Zefiro, Dr Livio *Italy*.
Zegers, Ms Ingrid *Belgium*.
Zehnder, Margareta *Switzerland*.
Zellinger, Prof. Dr Anton *Austria*.
Zelada, Mr Gabriel *Chile*.
Zelwer, Dr Charles *France*.
Zeman, Ing. Jiří *Czech Republic*.
Zemann, Emer. Prof. Dr Josef *Austria*.
Zeng, Huang *USA*.
Zeng, Ke *USA*.
Zeng, Zheng *UK*.
Zerbi, Prof. Giuseppe *Italy*.
Zeyen, Dr Claude Mathias Emile *France*.
Zhang, Dr Cai Xj. *USA*.
Zhang, Dr Faming *USA*.
Zhang, Dr Gongyi *USA*.
Zhang, Prof. Hanqing *China*.
Zhang, Hong *USA*.
Zhang, Dr Hongming *USA*.
Zhang, Dr Kam Y. J. *USA*.
Zhang, Prof. Ruilin *China*.
Zhang, Prof. Shaohui *China*.
Zhang, Prof. Shiwei *China*.
Zhang, Xiaohua *USA*.
Zhang, Dr Yiping *USA*.
Zhang, Prof. Ze *China*.
Zhang, Prof. Zhishun *China*.
Zhao, Dong *USA*.
Zhao, Rui *USA*.
Zhdanova, Dr Ludmila Ivanovna *Moldova*.
Zheludeva, Dr Svetlana I. *Russia*.
Zheng, Dr Jianhua *USA*.
Zheng, Prof. Peiju *China*.
Zheng, Prof. Qi-Tai *China*.
Zheng, Prof. Weitao *China*.
Zheng, Dr Yueqing *Germany*.
Zhirnov, Dr Victor V. *Russia*.
Zhmurova, Dr Zinaida I. *Russia*.
Zhou, Prof. Kangjing *China*.
Zhou, Ms Lan *USA*.
Zhou, Prof. Zhongyuan *China*.
Zhou, Prof. Zonghua *China*.
Zhu, Dr Nai-jue *USA*.
Zhu, Dr Wenjie *China*.
Zhu, Xiaotian *USA*.
Zhu, Prof. Ying *China*.
Zhukhlistov, Dr Anatolyi P. *Russia*.
Zhukhlistova, Dr Nadezhda E. *Russia*.
Zhukov, Dr Sergey Gennadievich *Russia*.
Zhurov, Mr Vladimir V. *Russia*.
Zhurova, Dr Elizabeth A. *Russia*.
Zibrov, Dr Igor P. *Russia*.
Zidarova, Dr Bogdana *Bulgaria*.
Ziel, Mr Rainer *Germany*.
Zielińska-Rohozińska, Dr hab. Elżbieta *Poland*.
Ziemer, Dr Burkhard *Germany*.
Zikmund, Dr Zdeněk *Czech Republic*.
Zimmerman, Miss Rosa *Argentina*.
Zimmermann, Dr Hans Dieter *Denmark*.
Zimmermann, Dr Helmuth W. *Germany*.
Zinenko, Dr Victor I. *Russia*.
Zink, Mr Ullrich N. *Germany*.
Zipper, Prof. Dr Peter *Austria*.
Zlatkin, Dr Alex T. *Russia*.
Zlokazov, Dr Victor B. *Russia*.
Zlosilo, Mr Mario *Chile*.
Žmija, Prof. Józef *Poland*.
Zobel, Dr Dieter *Germany*.
Zobetz, Dr Erich *Austria*.
Zocchi, Prof. Marcello *Italy*.
Zolensky, Michael E. *USA*.
Zompa, Leverett J. *USA*.
Zorkaya, Dr Olga N. *Russia*.
Zorkii, Prof. Petr M. *Russia*.
Zorn, Dr Gerhard M. *Germany*.
Zorrilla, Mr Oscar Enrique *Venezuela*.
Žorž, Mr Mirjan *Slovenia*.
Zosi, Prof. Gianfranco Luigi *Italy*.

Zotov, Dr Nikolay *Bulgaria.*
Zotov, Dr Nikolay Stamenov *Germany.*
Zou, Dr Jin *Australia.*
Zou, Xiaodong *Sweden.*
Zoutberg, Drs Martinus C. *Netherlands.*
Zachach, Dr Siegfried *Germany.*
Zsoldos, Mrs Éva *Hungary.*

Zsoldos, Dr Lehel *Hungary.*
Zubak, Vilma Maria *Australia.*
Zubarev, Dr Evgeny Nikolayevich *Ukraine.*
Zubenko, Dr Oleksandr Ivanovych *Ukraine.*
Zubenko, Dr Vasily V. *Russia.*
Zuccola, Harmon J. *USA.*
Zukerman Schpector, Dr Julio *Brazil.*

Zuñiga, Dr Fco. Javier *Spain.*
Zussman, Prof. Jack *UK.*
Zvezdinskaya, Dr Larissa V. *Russia.*
Zviedre, Dr Irena *Latvia.*
Zvyagin, Prof. Boris Borisovich *Russia.*
Zwell, Leo *USA.*